Revised Edition

A TEXTBOOK OF Applied Electronics

[For the Students of B.E./ B.Tech., B. Sc. (Engg.) Sec. A and B of A.M.I.E., Grade I.E.T.E Diploma and other Competitive Examinations]

Dr. R.S. SEDHA

B.E. (Hons), M.Tech. (IIT Delhi)
Ph.D. (U.K.), FIETE, SMIEEE
SINGAPORE

S. CHAND PUBLISHING

S Chand And Company Limited

(ISO 9001 Certified Company)

S.CHAND
PUBLISHING

S Chand And Company Limited

(ISO 9001 Certified Company)

Head Office: D-92, Sector–2, Noida – 201301, U.P. (India), Ph. 91-120-4682700

Registered Office: A-27, 2nd Floor, Mohan Co-operative Industrial Estate, New Delhi – 110 044
Phone: 011-49731800

www.**schandpublishing.com**; e-mail: **info@schandpublishing.com**

Marketing Offices:

Chennai	:	Ph: 23632120; chennai@schandpublishing.com
Guwahati	:	Ph: 2738811, 2735640; guwahati@schandpublishing.com
Hyderabad	:	Ph: 40186018; hyderabad@schandpublishing.com
Jalandhar	:	Ph: 4645630; jalandhar@schandpublishing.com
Kolkata	:	Ph: 23357458, 23353914; kolkata@schandpublishing.com
Lucknow	:	Ph: 4003633; lucknow@schandpublishing.com
Mumbai	:	Ph: 25000297; mumbai@schandpublishing.com
Patna	:	Ph: 4011400; patna@schandpublishing.com

First Edition 1990
Subsequent Editions and Reprints 1996, 97, 98, 99, 2000, 2001, 2002, 2003, 2004, 2005, 2006
Multicolour Revised Edition 2006, Reprint 2008, 2008 (Twice), 2010, 2012
Revised Edition 2013

Low Priced Students' Paperback Edition (LPSPE) 2021, Reprints 2022, 2023 (Twice), 2024, 2025 (Twice)

Reprint 2026

ISBN : 978-93-5501-068-1 **Product Code:** H3APE53ETRN10ENAA21L

PRINTED IN INDIA

By Shri Shyamlal Printing Press Private Limited, Plot 20/4, Site-IV, Industrial Area Sahibabad, Ghaziabad – 201 010 and Published by S. Chand And Company Limited, A-27, 2nd Floor, Mohan Co-operative Industrial Estate, New Delhi – 110 044.

Preface to the Fourth Edition

The field of electronics engineering has been advancing at a great pace. Since the first edition of this book released in 1990, there has been a tremendous progress in integrated circuit design, manufacturing processes, technology and applications. Products with touch screens have become common these days. Mobile broadband and high-bandwidth Internet connection networks have made the Digital media so fast and reliable. The recent deployment of Optical fibre networks in advanced countries has removed further the bottlenecks that plagued previous systems. High definition display systems are also a commonplace. Video capture and broadcast has become so easy that even the mobile phones are capable of producing high definition video clips and sending over the wireless network and uploading to public sites such as YouTube or Facebook in very little time allowing millions to access to media quickly and freely.

There has been a tremendous progress in Electronic Design Automation (EDA or ECAD) Software Tools for design, simulation, analysis and verification and manufacturing preparation of electronic devices and printed circuit boards.

With each edition of this book, I am amazed at the new information that scientists, engineers and technologists have gathered about the electronic materials, their characteristic and applications. Although there are great achievements in the last 20 years, researchers face even great challenges in the future.

In this edition, I have included new material in many chapters particularly from 11 to 36. This will help the readers to enhance their conceptual framework about electronic devices and circuits. The chapter 16 on Integrated circuits has been expanded to include steps involved in integrated circuit design. In chapter, 26, topics on Darlington connection of BJT, parallel Darlington transistor, dc bias operation of Darlington circuit, ac analysis of Darlington pair, feedback pair, dc bias operation and ac analysis of feedback pair have been included. In Chapter 17, topics on series and parallel combination of SCR, protection circuit, SCR crowbar circuit and Snubber circuit have been included. In Chapter 27 on Power Amplifiers, topics on Class AB amplifier, Class D amplifier and Class E amplifier have been included. In Chapter 31 on Non Sinusoidal oscillators, topics on Ring oscillators and Franklin oscillator have been included. In Chapter 35 on Operation Amplifiers, topics on averaging amplifier, linear amplifier and subtractor have been included. In Chapter 36 on Operational Amplifier Applications, topics on Instrumentation amplifier, precision rectifier, precision half-wave and precision full-wave rectifier, log and antilog amplifiers, voltage-to-current and current-to-voltage converters have been included. More photographs of the electronic devices have been included in this edition to help the readers to develop better understanding.

To help students for preparing and doing well in the examination, I have included "Examination Questions" from several Indian Universities until Dec 2011, as solved examples. I have expanded the section on "Descriptive Questions" also to include the theory type examination questions. The section on "Multiple Choice Questions" has also been expanded in several chapters.

The mistakes and misprints which had crept in the previous edition, have been eliminated in this edition.

I wish to express my sincere thanks to my friends and colleagues in various technical universities and engineering colleges for recommending this book to their students. I also express my sincere thanks to the students for recommending this book to their friends and relatives. I hope they will continue to patronize this book in the future also.

Any errors, omissions and suggestions for the improvement of this book brought to my notice will be thankfully acknowledged and incorporated in the next edition.

Dr. R.S. Sedha
Singapore
Email: rssedha@gmail.com

Preface to the First Edition

I take an opportunity to present this standard treatise entitled **A Textbook of Applied Electronics** to the students of Degree, Diploma, A.M.I.E. (I) and I.E.T.E. in particular as well as B.Sc. of all the Indian Universities in general. The object of this book is to present the subject matter in a most concise, compact, to the point and lucid manner.

While writing this treatise, I have constantly kept in mind the requirements of all the students, regarding the latest as well as changing trend of their examinations. To make it really useful for the students, latest examination questions of various Indian and foreign universities as well as other important examining bodies have been included. The book has been written in an easy style, with full details and illustrations. All along the approach to the subject matter, every care has been taken to arrange the matter from simpler to harder, known to unknown, in a self-study style. In short, it is expected that the book will meet the crying need of the students, for whom it has been produced.

Although every care has been taken to check mistakes and misprints, yet it is very difficult to claim perfection. Any errors, omissions and suggestions, for the improvement of this book, brought to my notice will be thankfully acknowledged and incorporated in the next edition.

R.S. SEDHA

Acknowledgements

The author is thankful to the following Indian as well as foreign universities and examining bodies whose examination papers have been included in the text of the subject by way of illustration. Moreover their syllabi have been kept in view while writing this treatise.

1. Dr. B.R.A. University
2. Alagappa University
3. Aligarh University
4. Allahabad University
5. Andhra University
6. Anna University
7. Annamalai University
8. Bharthiar University
9. Banaras University
10. Bangalore University
11. Bhopal University
12. Bihar University
13. Bombay University
14. Calcutta University
15. Calicut University
16. Cochin University
17. Chatrapati Shahu Ji Maharaj University
18. Delhi University
19. Gauhati University
20. Gorakpur University
21. Gujarat University
22. Haryana Technical University
23. Indore University
24. Jabalpur University
25. Jadavpur University
26. Jawaharlal Nehru Technological University (JNTU)
27. Jiwaji University
28. Kanpur University
29. Karnataka University
30. Kerala University
31. Kurukshetra University
32. Madras University
33. Mahatma Gandhi University
34. Madurai University
35. Mysore University
36. Nagpur University
37. Osmania University
38. Punjab University
39. Punjab Technical University
40. Patna University
41. Pune University
42. Punjabi University
43. Purvanchal University
44. Rajasthan University
45. Ranchi University
46. Roorkee University
47. Shivaji University
48. Gautam Buddha Technical University
49. Utkal University
50. Vikram University
51. Visveswaraiah Technological University (VTU)
52. West Bengal University of Technology
53. Birmingham University
54. Cambridge University
55. London University
56. Oxford University
57. Institutution of Electronics and Telecommunication Engineers (IETE)
58. Institution of Engineers (India)
59. Union Public Service Commision (UPSC)- Civil Services and Engineering Services

The author wants to acknowledge a number of people who have been immensely helpful in making this book possible. My family has been more than usual "patient and understanding". They are a constant reminder of the rich life beyond the electronics world. The author is deeply indebted to the people at S.Chand for providing continuous support while preparing this text.

Contents

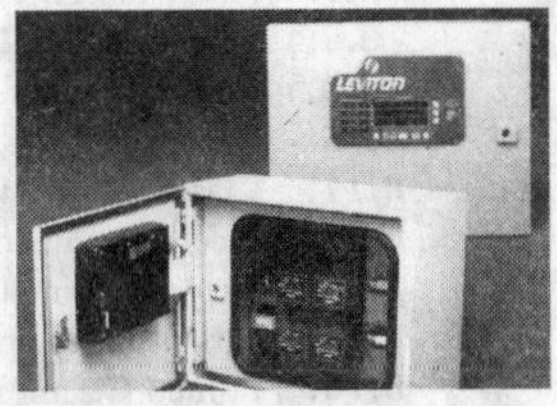
LEVITON

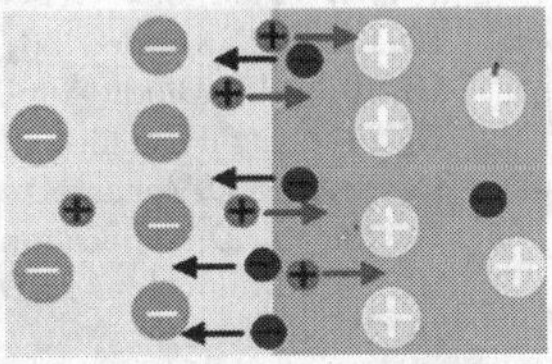

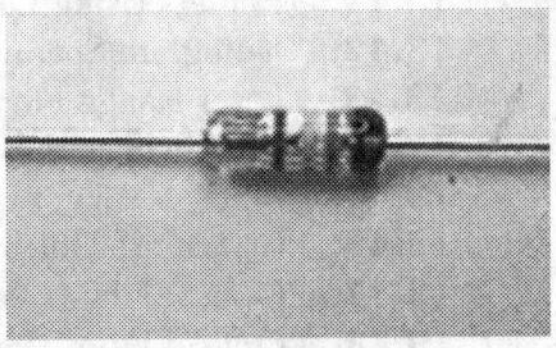

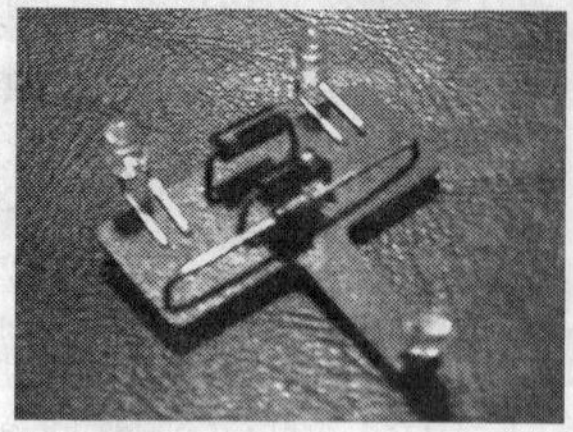

C1815
Y 51

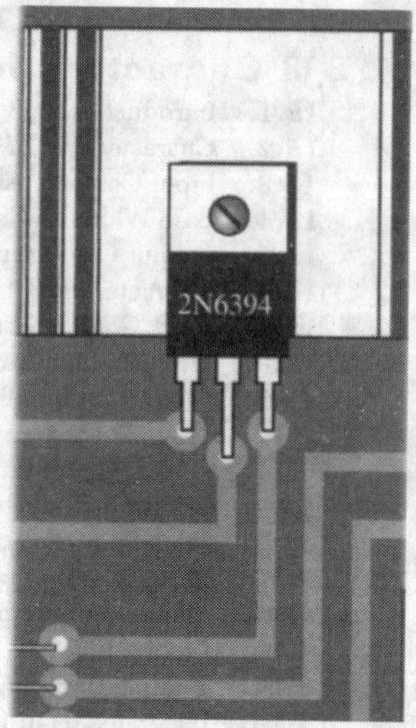
2N6394

Typical Reading
CONDUCTION TIME

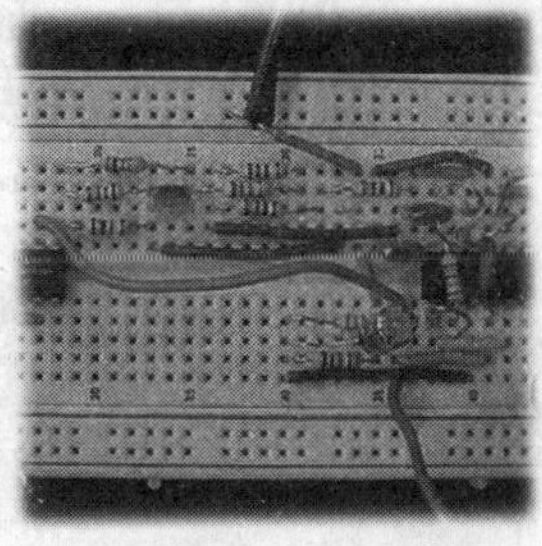

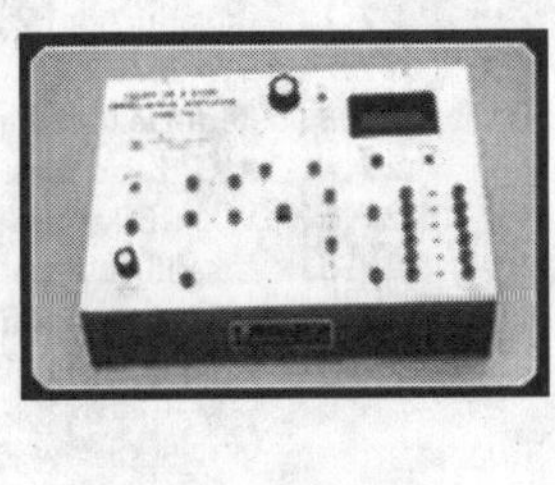

ZETEX

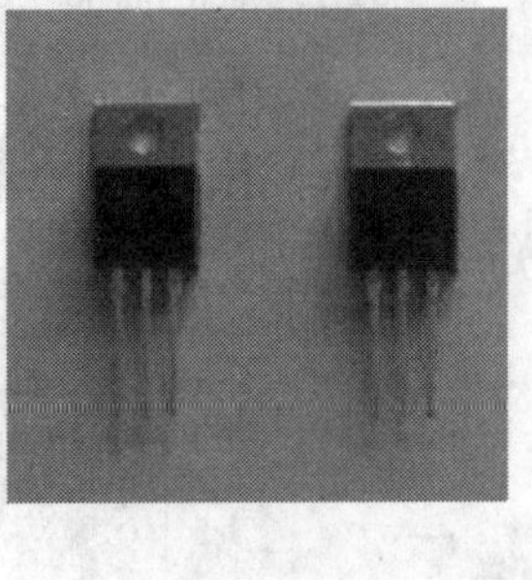

Photek-T
[MP3 Av

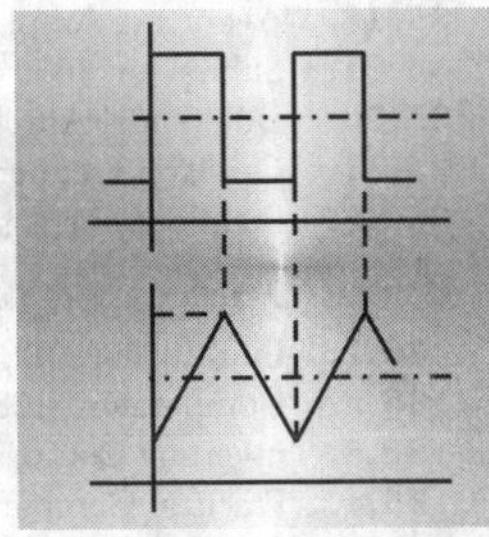

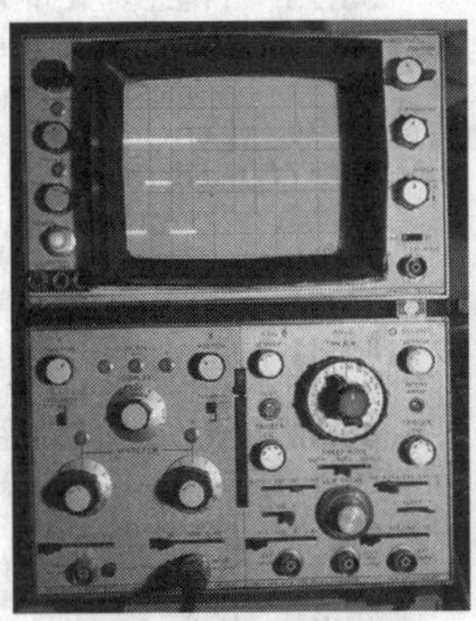

National Semiconductor
The Sight & Sound of Information
LMV422

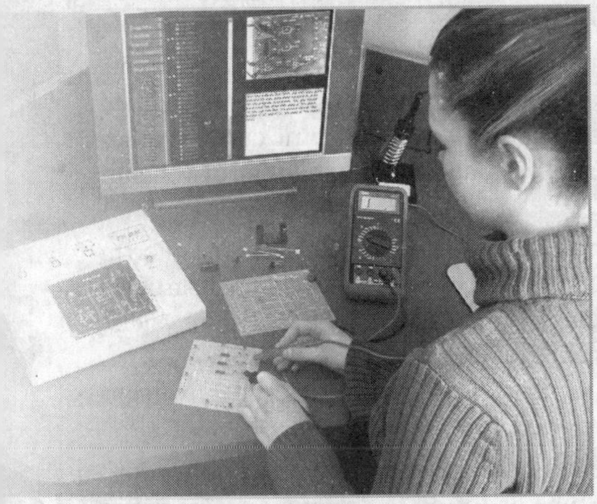

1

Chapter

Introduction

Outline

1. Definition.
2. Beginning and Development of Electronics.
3. Overview of Electronic Industry.
4. Current Scenario of Electronics Industry in India.
5. Major Players in Electronic Industry.
6. Modern Trends in Electronics.
7. Applications of Electronics.
8. Communication and Entertainment Applications.
9. Instrumentations and Control Applications.
10. Defence Applications.
11. Applications in Medicine.
12. Computer Aided Design of Electronic Circuits.
13. Careers in the field of Electronics.

1.1. Definition

The word 'electronics' has been derived from a Greek word 'elektron', which means the study of behaviour of an electron under different conditions of externally applied fields. The Institution of Radio Engineers (IRE) has given a standard definition of electronics in the proceedings of I.R.E. Vol. 38 (1950) as that field of science and engineering, which deals with the study, design and use of devices, which depends on the conduction of electricity through a vacuum, gas or semiconductor. Until recently, it was considered as an integral part of Electrical Engineering. But due to its tremendous advancement, during the last few decades, it has achieved its own place in the field of science and technology.

These days, the field of electronics has become the most important branch of engineering in our society. As a matter of fact, it is a field, in which rapid developments are taking place every day. The electronic devices and gadgets are being used in almost all industries for quality control and automation. Because of the growing applications of electronics, in almost all fields, the students of all engineering disciplines have to be taught electronics at the undergraduate level.

1.2. Beginning and Development of Electronics

The era of electronics is believed to have dawned with the experiments involving electric currents in glass vacuum tubes. Such experiments were performed around 1850 by a German scientist named Geissler. He found that if the air is removed from a glass tube, it glows when an electric current passes through it. Around 1878. British scientist, Sir William Crookes found that the current in vacuum tubes seemed to consist of particles.

As a matter of fact, the electron was discovered in the 1890's. A French physicist, Perrin demonstrated that current in a vacuum tube consists of the movement of negatively charged particles in a given direction. Some of the properties of these particles were measured by a British physicist Thomson. These negatively charged particles were later on, named as electrons. An American physicist, Milikan, in 1909 measured the charge on an electron. As a result of these discoveries, the movement of electrons could be controlled, and thus the electron era started.

Fleming, in 1904, invented a vacuum tube that allowed electrical current only in one direction. This tube was named as Fleming valve or diode vacuum tube and used to detect electromagnetic waves. In 1907, Lee deForest, an American scientist invented a tube, which could amplify weak electrical a.c. signals. This tube was named as triode vacuum tube.

Around 1915, few great improvements were made in the triode vacuum tube. It was used in oscillator circuit. It was also used in telephone system, which made possible the communication between the different continents. The tetrode was invented in 1916 by a German engineer and a pentode was invented in 1926 by a Dutch engineer. The tetrode, alongwith the pentode, greatly improved the triode. The first television picture tube, called the kinescope, was developed in 1920's by an American researcher.

Several types of microwave tubes were developed during the World War II. Such tubes made possible the modern RADAR and other communication systems. The magnetron was invented in Britain in 1939. In the same year, klystron microwave tube was developed by two Americans. The travelling wave tube (TWT) was invented in 1943 by an American scientist.

The era of solid state electronics began with the invention of transistor in 1947 at Bell laboratory. The inventors were Walter Britain, John Bardeen and William Shockley. The transistors were produced commercially in 1951. The companies like RCA, Raythesn, General Electric, Westinghouse and Western Electric were the first to fabricate transistors.

In 1958, it was thought that germanium and silicon can be used to build an entire circuit, called monolithic circuit. Resistors were formed with the bulk semiconductor or by diffusing one semiconductor into another. The capacitors were formed by using a metallic layer and the semiconductor for the plates and an oxide layer for the dielectric. In 1959, Noyce gave an idea for making multiple devices on a single piece of silicon, in order to make an interconnection between devices as part of manufacturing process. It could reduce the size, weight and cost per device. The monolithic-circuit was named an integrated circuit (IC). The IC's were produced, commercially in 1961 by Fairchild and Texas instruments.

1.3. Overview of Electronic Industry in India

The Electronic Industry in India took off around 1965 with an orientation towards space and defence technologies. This was rigidly controlled and initiated by the government. This was followed by developments in consumer electronics mainly with transistor radios, Black & White TV, Calculators and other audio products. Colour Televisions soon followed. In 1982-a significant year in the history of television in India - the government allowed thousands of colour TV sets to imported into the country to coincide with the broadcast of Asian Games in New Delhi. 1985 saw the advent of Computers and Telephone exchanges, which were succeeded by Digital Exchanges in 1988. The period between 1984 and 1990 was the golden period for electronics during which the industry witnessed continuous and rapid growth.

From 1991 onwards, there was first an economic crises triggered by the Gulf War which was followed by political and economic uncertainties within the country. Pressure on the electronics industry remained though growth and developments have continued with digitalisation in all sectors, and more recently the trend towards convergence of technologies. After the software boom in mid 1990s India's focus shifted to software. On the other hand, the hardware sector was treated with indifference by successive governments. Moreover the steep fall in custom tariffs made the hardware sector suddenly vulnerable to international competition. In 1997, the International Trade Agreement (ITA) was signed at the WTO (World Trade Organisation) where India committed itself to total elimination of all customs duties on IT hardware by 2005. In the subsequent years. a number of companies turned sick and had to be closed down. At the same time companies like Moser Bear, Samtel Colour, Celetronix etc. have made a mark globally.

1.4. Current Scenario of Electronics Industry in India

The electronic industry in India constitutes less than 1% of the global electronic industry. However, the demand in Indian market is growing rapidly. The output of Electronic hardware industry in India is worth US $13.0 Billion at present (2011). India is also an exporter of a vast range of electronic components and products for the following segments:

1. Display Technology
2. Entertainment electronics
3. Optical storage devices
4. Passive components
5. Electro-mechanical components
6. Telecom equipments
7. Transmission and signalling equipments
8. Semiconductor designing
9. Electronic manufacturing services

India is well known for its ability in developing software. But on the hardware front, the progress in rather slow. Already 50 electronics manufacturing services (EMS)/Original Design Manufactures (ODMS) providers are operating in India, ranging from global players including. Flextronics and solectron to indigenous firms including Dettron, TVS Electronics, and Saharra.

1.5. Major Players in Electronic Industry

1. ***Solectron.*** Solectron Centum Electronics Limited is the leading Indian company offering state of art solutions for Frequency Control Products (FCP), Electronic Manufacturing Service (EMS) and Hybrid Micro Circuits (HMC). Solectron has a manufacturing unit and design cente in Bangalore and a post manufacturing centre in Mumbai. The EMS operation focuses primarily on the domestic market.

2. ***Flextronics.*** Flextronics entered India in 2001 when it purchased a Motorola facility. Flextronics maintains a Bangalore facility with 18,000 sq ft and over 300 employees. The products manufactured are engine management Ccrd, TV tuners set top box, energy meters, cellular phone, networking cards and WLL wall sets.

3. ***Jabil Circuits.*** Jabil Circuit operates a 51,000 sq. ft. plant in Pimpri, Pune (Maharashtra), which the provider took from Philips in 2002. The Pimpri plant manufactures TV analog monitor cards and certain audio products for Philips. All production today is for the Indian market.

4. *Samtel Group.* The Samtel group is the largest Indian integrated manufacturer of a wide range of display devices like Colour and B&W TV picture tubes, tubes for avionics, medical and industrial applications, CRT glass, electron guns, heaters and cathodes, deflection yokes and engineering services. Samtel has acquired a facility in Germany to manufacture high tech, high resolution CRTs for demanding applications such as aircraft avionics, medical monitors and a variety of industrial applications through a continuous focus on R&D.

Another major player and exporter in this segment is Hotline Group which manufactures B/W and Colour Picture Tubes.

5. *Moser Bear India Ltd. (MBIL).* In the Optical Storage Devices segment, Moser Bear India Ltd., is today the world's third-largest optical CD maker in an industry dominated by Japan and Taiwan. MBIL supplies to a number of branded players like Sony, Veratim, TDK, Maxwell, Imation, Samsung etc., and has collaborative R&D programs as well as reciprocal training programs with these world class companies.

6. *Hical Magnetics.* India has several world class companies manufacturing electromechanical and wound components. These include MNCs as well as Indian companies such as Hical Magnetics, a QS9000 company, which designs and manufactures professional grade transformers. Hical also offers world class design services for the state-of-the art power converters and qualification services for both power converters and magnetics. Hical has also established a subsidiary in USA with focus on giving Design, Engineering and Sales support to their North American customers.

7. *Tyco Electronics.* Tyco Electronics, an MNC, with a modern facility in Bangalore specialises in high grade connectros such as fibre optic interconnection systems and smart card connectors.

8. *Midas Communications.* Midas communications has championed the telecommunication revolution throughout rual and urban India, since its inception in 1994. CorDECT (Midas's Flagship product) connect over 3 million people around the world to the global economy. The corDECT technology is based on the Digital Enhanced Cordless Telecommunications standard specified by the European Telecommunications Standards Institute (ETSI).

9. *Elin Electronics Ltd.* A major player and pioneer in the Entertainment electronics segment is Elin Electronics Ltd, manufacturing Tape Deck and CD Mechanisms. The company also manufactures AC & DC Micro Motors for OEM's and provides Contract Manufacturing Services.

10. *Cadence Design.* Electronic Chip Design is another growth area for the electronics industry, which can provide support for growth of manufacturing as well as greater value addition. World leaders such as Cadence have set up shop in India and more and more companies are relocating design centres due to technical skill available here.

1.6. Modern Trends in Electronics

The development of an integrated circuit (IC) in 1961 was a great achievement in the field of electronics. But, as there is no end to the man's desire, the growth accelerated further every year. Rapid developments have been made in the integrated circuit technology. Thus starting from the *small-scale integration (SSI) in 1961, then **medium-scale integration (MSI) in 1966, and ***large-scale integration (LSI) in 1970. Now in 1980's we have a very large-scale integration (VLSI). It means that now we have a technology, which can fabricate about 1 million or more components on a single chip.

It will be interesting to know that the use of vacuum tube has become almost obsolete in 1960's. It was replaced by the solid state devices (bipolar transistors and field-effect transistors). Similarly

* It indicates that the number of components fabricated on a single-chip was less than 100.

** It indicates that the number of components fabricated on a single-chip was 100 to 1000.

*** It indicates that number of components fabricated on a single-chip is between 1000 to 10,000.

the developments in integrated circuit technology, has made the use of individual transistors as unnecessary. These days, we see all electronic systems made of IC's. The developments which, have taken place in last 35 years, may be understood from the fact that today's microcomputer is 1000 times faster, has a large memory, thousands times more reliable, consumes negligible power, occupies 1/30,000 the volume and costs 1/10 000 as compared to the first large electronic computer.

The microelectronics (term used for the fabrication of VLSI-circuits) has made a great impact on communication control and computer industries. The applications, which were difficult to realize earlier, have now become possible. In last decade, silicon based technology has dominated the sphere of electronics and will continue to play the role for the next decade as well. However, after that gallium arsenide (Gas) might play a significant role in electronics.

Overview of Integrated Circuit (IC) Design Houses

There are over seventy IC design centers in India. The cities like Bangalore and Noida (near Delhi) are the places which houses the leading companies like Texas Instruments, IBM, Infineon, ST Microelectronics, Motorola, Philips, Intel India, Cadence, Wipro New Logic, Cypress Semiconductors, Quantumthink Technologies and Spike Infotech. Refer to Fig. below for the distribution of design centres in different parts of India.

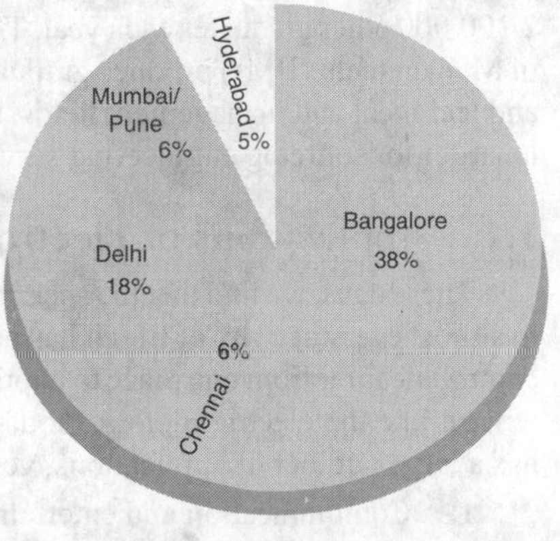

IC Design centers in India.

Most of the design engineering companies are in Bangalore and Delhi/Noida.

All these design houses need trained people with the required hands on skills in the field of CAD/EDA tools, back end tools (DRC, LVS, RCX), QA test structures for PDK components, experience with Python, PERL, TCL, C/C++ programming. CMOS device and Circuit design.

The major growth drivers and trends in the electronics industry are:

1. **Consumer Products**– Flat panel TVs, High Definition TVs, iPods, iPads, iPhones, digital cameras and set– top boxes.
2. **Communication products**– 3G handsets, TV reception on handsets, mobile services.
3. Industrial electronics such as Radio Frequency Identification (RFID), green electronics, optical recognition.
4. Electronics automotive products such as global positioning systems (GPS), hybrid car and electronics for safety purpoes.
5. Electronics use for casino gaming purposes will also be trendy.

Market will be driven by automotive electronics (which is rising rapidly); computers and communications (which are still big), consumer and industrial (which is expected to rebound) and electronics products focused on high technology, multi-function, networking base, with high quality audio visual effects products and complex and high precision electronics.

Challenges and Opportunities of Electronics Industry

Challenges. The major challenges facing the Indian electronic manufacturing market are an infrastructure that needs to be improved at the earliest possibility, casing of foreign investment procedures, which is underway, and a restructured government tariff that now makes domestically manufactured goods more expensive that imported goods with zero tariff.

There are also other problems, which are hampering the growth of the Indian electronics industry. Some of them are:

(*i*) Lack of World-class infrastructure.

(*ii*) Lack of clear-cut government policy for the industry.

(*iii*) Very little expenditure in Research and Development area.

(*iv*) Power of Marketing not harnessed to the maximum

Opportunities. While the Electronics industry in India is currently small, there are several advantage that India offers that can be effectively leveraged to achieve higher growth. These can be categorised under the following three heads:

(*i*) Manpower

(*ii*) Market Demand

(*iii*) Policy and Regulatory Support

India produces over 500 PhDs, 200,000 engineers, 300,000 non-engineering postgraduates and 2,100,000 other graduates each year. The Indian Institute of Technology (IITs) and the Indian Institute of Management (IIMs) produce graduates and post graduates with best-inclass skills and capabilities and technical and management fields. India's capabilities in IT and engineering make it an attractive location for sourcing engineering services such as Research & Development (R&D) and design.

1.7. Applications of Electronics

These days, we find that the sphere of electronics has spread so vast, that it plays an important role in almost every activity of life. It has penetrated into our homes, our places of work and our means of communication from one place to another. It is believed that by the end of this century, the electronics will be like the electric motor of today (*i.e.*, its presence will go unnoticed). Although, electronics has a large number of applications, yet the following are important from the subject point of view:

1. Communication and entertainment applications.
2. Instrumentation and control applications.
3. Defence applications.
4. Applications in medicine.
5. Computer-aided design of electronic circuits.

Now we shall discuss all these applications, in brief, in the following pages.

1.8. Communication and Entertainment Applications

It will be interesting to know that in the beginning of 20th century, the electronics was used only in the field of telephony and telegraphy. It required a pair of conducting wires. But, these days, we have cordless telephones and mobile phones which send or receive information from one place to another, without the use of any wire. Such a communication is known as wireless communication. Todays, we have a push button telephones with memory. Today we have mobile phone handsets which has a email touch panel for interactive gaming and music, 8.1M Pixel camera, high-definition video recording can open multiple applications at once and so on. We have several voice over internet protocol devices like skype phone, Google phone etc. Such telephones can also record the messages in the absence of its owner. It is possible to have video conference from anywhere on the earth, or infact millions of kilometers in space. It is surprising that someone sitting at a space center in NASA/ISRO can instruct a space vehicle, million kilometers away, to turn on its TV camera and send pictures back to the earth.

These days, the radio and television provide a means of communication, education and entertainment. The electronics gadgets like stereos, tape recorders, DVD players, MP-3 players and

video games etc., are widely used for entertainment. The use of integrated circuits ICs have improved both the reliability and performance of radios and television receivers. It has also helped to bring down their cost, because of the simple procedures adopted during the assembly stages.

We have digital cameras and digital camcorders where the still pictures/video can be stored on a memory stick (flash memory). The pictures can be downloaded to a computer to keep a record in albums for sharing with the relatives and friends through the Internet. The MP-3 player can be used to download the music from the Internet and listen at your convenience.

1.9. Instrumentation and Control Applications

The instrumentation plays an important role in our industry. It helps in precise measurement and testing of a number of quantities of interest. The electronics is also playing an increasingly important role in instrumentation. The cathode ray oscilloscopes (CRO's) frequency counters, pulse and signal generators, digital multimeters (DMM's) power supplies etc., are some of the instruments, which are found in every industry and research organisations.

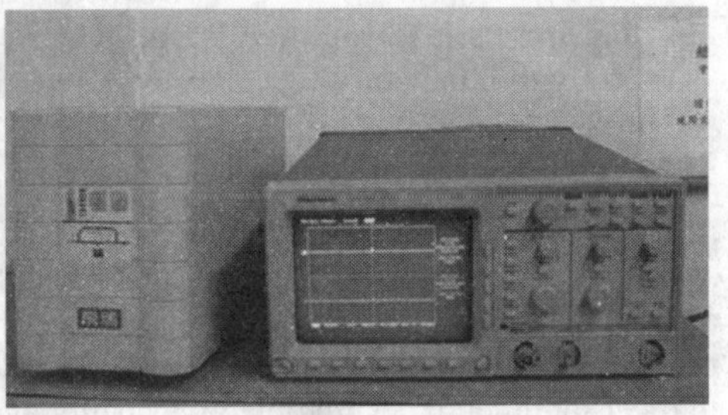

The control industry has been drastically effected by the introduction of semiconductor electronics. In some traditional areas, such as motor speed control and power rectifiers and inverters, the silicon controlled rectifier (SCR) has replaced the gas filled tube (thyratron). Most of the control systems, which are available today, are possible only because of the recent dramatic advances in electronics. The control system designs, which were not practical earlier, because of the cost (only five years ago) are already becoming obsolete today. The introduction of microprocessors, microcomputers and digital integrated circuits has led to "smart" instruments and continuously increasing variety of digital control systems. With microelectronics, the computers have become integral component of our control systems.

The control, instrumentation and industry has state of the art reliable SCADA, PLC/Data acquistion systems. These are being applied across various sections of the process industry. Latest AC drive systems from smaller to very high power levels are available for large engineering industries like steel plants and metal industries. World class UPS (un-interruptable power supply) systems are available for the emerging digital economy.

1.10. Defence Applications

The role of electronics in defence started with the introduction of RADAR during the World War II. The RADAR is an abbreviation used for Radio Detection and Ranging. A radar is capable of detecting and finding exact location of enemy aircraft. These days, we have aeroplanes, ships, anti-aircraft guns and guided missiles, underwater robots completely controlled by electronics.

Electronic systems used commonly on aircraft, artificial satellites and spacecraft are called Avionic systems. Such systems are used for aircraft communication, navigation, glass cockpit, flight control systems, collision avoidance systems, weather systems and management systems.

1.11. Applications in Medicine

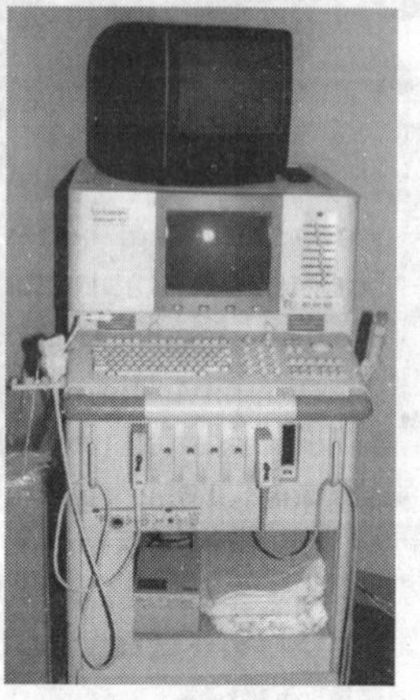

There are variety of electronics device instruments and technology being used in health care industry. Some examples of electronic devices and instruments used in health care are pacemakers, infusion pumps, heart-lung machine, dialysis. Besides this .... (remove these days).

These days, big hospitals and drug research organisations are utilising an increasing role of electronic instrumentation. In many developed countries, the computers are used to hold patients records and assist in arriving at correct diagnosis. A large number of electronic instruments are also used in the diagnosis and treatment of various diseases. Some of the important ones are given below :

1. ***Electrocardiograph (ECG).*** It is used to find the condition of the heart of a patient.
2. ***X-ray equipment.*** It is used for taking pictures of internal bone structures. It is also used for the treatment of some diseases.
3. ***Ultrasound scanner.*** It is used to take pictures and examine the functions of a brain, kidney etc. It is also used for the treatment of some diseases.
4. ***Electrical enceplograph (EEG).*** It is used for neurological investigations.
5. ***Cathode ray oscilloscope.*** It is used for studying muscle actions within the body.
6. ***Magnetic Resonance Imaging (MRI).*** It is used for investigating turmours or other defects within the body.

1.12. Computer-Aided Design of Electronic Circuits

The circuit designers, of previous generations, used to build a circuit on printed boards, and then study its performance. These days, it is possible to use a computer not only as a means of simulating a particular system design, but also to model the effects of statistical variations in the manufacturing process. The computer is also used to optimise parameters in a given system design. It is also used to meet some set of design objectives better and to explore various alternative designs. It is generally done to determine the relative trade-off between the various competing objectives. These days, the VLSI (very large-scale integrated) circuits are designed by means of a computer. Even for a product design, the engineer can design the product in the computer, carryout the studies on various parameters by simulation on the computer. Rapid prototyping is also done by using computers.

1.13. Careers in the Field of Electronics

The field of electronics is very diverse and career opportunities are available in many areas. Since electronics is currently found in so many different applications ranging from toys to spacecrafts and new technology is being developed at a very fast rate, its future appears limitless. There is hardly any area of our lives that is not enhanced to some degree by electronics technology. Those who acquire a sound, basic knowledge of electrical and electronic principles and are willing to continue learning will always be in demand.

The importance of obtaining a through understanding of the basic principles contained in this text cannot be overemphasized. Most employers prefer to hire people who have both a through grounding in the basics and the ability and eagerness to grasp new concepts and techniques. If you have a good training in basics, an employer will train you in the specifics of the job to which you are assigned.

There are many types of job classification for which a person with training in electronics technology may qualify. A few of the most common job functions are discussed briefly in the following paragraphs.

Scientists and Engineers. Personnel in this category are involved in designing, prototyping and testing electronic systems in research and development laboratories during the development phase of the product.

Industrial Manufacturing Engineers. Manufacturing personnel are involved in testing of electronic products at the assembly-line level or in the maintenance and trouble shooting of electronic and electromechanical system used in testing and manufacturing of products. Virtually every type of manufacturing plant, regardless of its product, uses automated equipment that is electronically controlled.

Technical Sales and Marketing. Technically trained people are in demand as sales and marketing representatives for high technology products. The ability both to understand technical concepts and to communicate the technical aspects of the product to a potential customer is very valuable. In this area, as in technical writing, competency in expressing yourself orally, and in writing is essential. Actually, being able to communicate well is very important in any technical job category because you must be able to record data clearly and explain procedures, conclusions, and actions taken so that others can readily understand what you are doing.

Technical Writers. These personal compile technical information and then use the information to write and produce manuals and audiovisual materials. A broad knowledge of a particular system and the ability to clearly explain its principles and operation are essential.

Besides this, the diploma engineering people can find job as a customer support officers, laboratory technician, field service technician, service-shop technician, engineering assistant and many more.

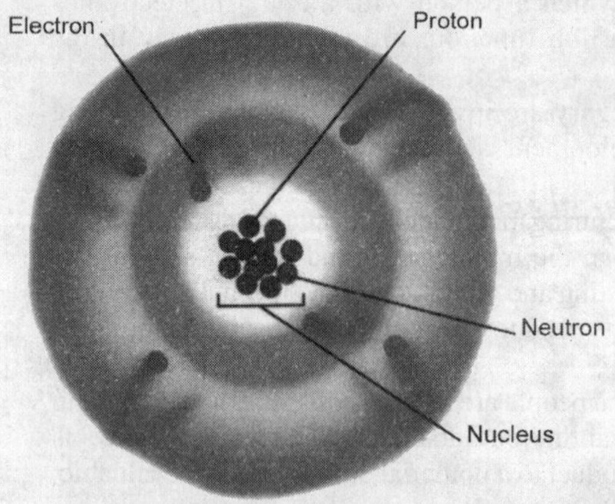

2 Chapter

Structure of Solids

Outline

1. Introduction.
2. Structure of Solids.
3. Types of Crystalline Structures.
4. Structure of an Atom.
5. Atomic Number.
6. Atomic Mass.
7. Electron Orbits.
8. Distribution of Electrons in Atoms.
9. Electron Sub-orbits or Sub-shells.
10. Valence Electrons.
11. Ionized Atoms.
12. Application of Energy to an Atom.
13. Energy Possessed by an Electron in the Orbit.
14. Energy Level Diagram.
15. Bonding in Solids.
16. Ionic Bond.
17. Covalent Bond.
18. Metallic Bond.
19. Energy Bands.
20. Simplified Representation of Energy Level Diagram of a Solid.
21. Insulators, Conductors and Semiconductors.

2.1. Introduction

It has been established since long that the operation of electronic devices depends upon the *motion of charged particles* within them. Therefore, we must have an understanding of the forces, which control the motion of these particles. But as a matter of fact, it is the physical structure of a solid, which exerts this control. Therefore, we shall study the structure and arrangement of atoms in solids. Though this topic is very large, yet we shall only study the structure of solids with a brief introduction to crystal structure.

2.2. Structure of Solids

A detailed study of X-ray scattering and many others have revealed that most of the solids are crystalline in structure. It means that a solid consists of *atoms* or *molecules* (technically speaking, ions) which are arranged in a periodic fashion.

There is always some basic *arrangement* of atoms, which is repeated throughout the entire solid material. Such an arrangement of atoms, within a solid, is called *crystal lattice*. Such solids are called *crystalline solids*. However, there are some other solid materials, which do not have crystalline structure. Such solid materials are called *non-crystalline* or *amorphous solids*. All metals and elements like silicon and germanium are crystalline materials while wood, plastic, paper, glass etc. are amorphous materials.

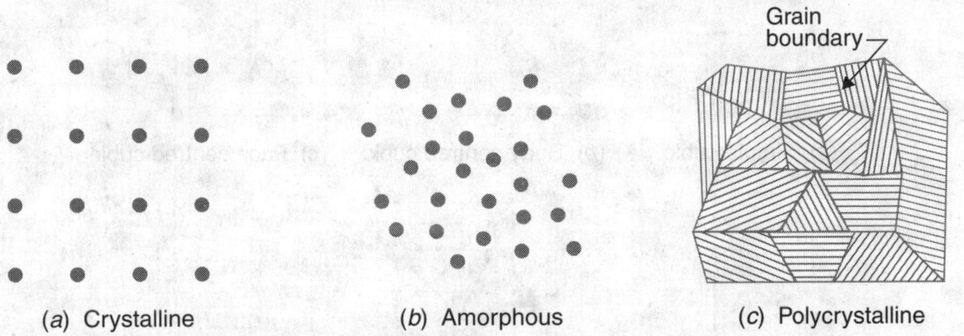

Fig. 2.1. Structure of solids.

A crystalline material may be in the form of single crystal or an aggregate of many crystals known as *polycrystalline*. A polycrystalline material is always separated by well defined boundaries called grain boundaries. Fig. 2.1 (*a*), (*b*) and (*c*) distinguishes among crystalline, amorphous and polycrystalline structures.

In a crystalline structure, since there are many different ways of placing atoms in a volume, the distances and orientation between atoms can take many forms. However, in every case, the *lattice* contains a volume called a *unit cell*. The unit cell is a representative of the entire lattice, and is regularly repeated throughout the crystal. Fig. 2.2 shows a two-dimensional arrangement of atoms within a unit cell *OABC*. This cell has an atom at each corner, which is shared with the adjacent cells. The importance of the unit cell is that we can analyse the crystal by determining its volume. For example, from the unit cell, we can find out the distance between one atom and the next nearest atom. This can be used to calculate the forces holding the crystal lattice together. However, the most important point is that the properties of crystal lattice determine the allowed energies of the electrons, which participate in the conduction process.

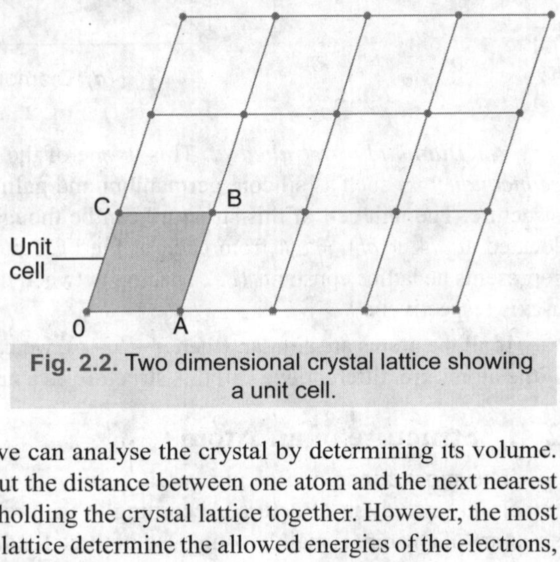

Fig. 2.2. Two dimensional crystal lattice showing a unit cell.

2.3. Types of Crystalline Structures

Although there are many types of crystalline structures, yet the following four are important from the subject point of view:

1. ***Simple cubic (SC)***. The unit cell of this structure contains *atoms at all the corners of a cube* as shown in Fig. 2.3 (*a*). The compound cesium chloride (CsCl) possess this type of structure.

2. ***Body centred cubic (BCC)***. The unit cell of this structure is a *simple cube, with an additional atom at the body centre of the cube* as shown in Fig. 2.3 (*b*). The crystals such as tungsten, vanadium, chromium and alkali metals (*i.e.*, sodium, potassium etc.) have a body centred cubic structure.

3. ***Face centred cubic (FCC)***. The unit cell of this structure is a *simple cube with an additional atom in the centre of each six faces of the cube* as shown in Fig. 2.3 (*c*). The crystals such as gold, silver, copper, nickel, lead etc. have a face centred cubic structure.

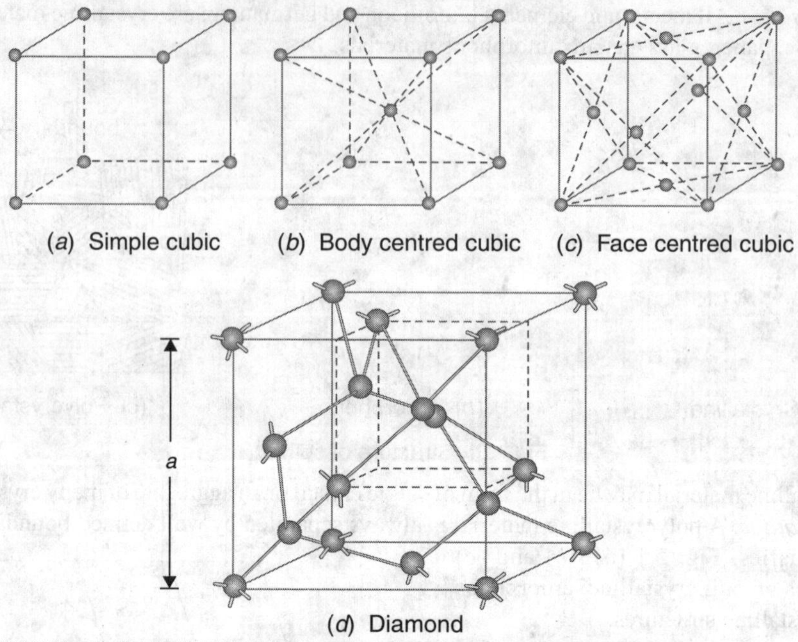

Fig. 2.3.

4. ***Diamond or tetrahedral.*** This is one of the most important crystal structure, because the *semiconductors* such as silicon, germanium and gallium arsenide (GaAs) have this type of crystal structure. The unit cell of this structure can be thought of as an FCC structure with an extra atom located at $a/4$, + $b/4$, + $c/4$ from each of the FCC atoms as shown in Fig. 2.1 (*d*). The a, b and c represents he lattice constants (*i.e.*, spacing between the adjacent atoms, along the x-axis, y-axis and z-axis respectively.

If all the atoms are similar, (then $a = b = c$), we call this structure as diamond lattice. However, if the atoms are different, we call this structure as a zinc blends lattice.

2.4. Structure of an Atom

The most fundamental unit of a solid is an atom, and it is important to study its structure. According to the present concept, the structure of an atom is mainly composed of three fundamental particles *i.e.*, electrons, protons and neutrons. An electron has a negative charge equal to 1.6×10^{-19} coulomb A neutron has no charge at all. The protons and neutrons are located in a small nucleus at the centre of an atom. The electrons revolve around the nucleus in fixed circular as well as elliptical orbits called energy levels or main shells as shown in Fig. 2.4.

Neils Bohr
(1885–1962)

The electrons moving in the various energy levels, occupy a vast empty space around the nucleus. The aggregate of all the electrons forms an electron cloud around the nucleus, by the electromagnetic force of attraction between the nucleus and the electrons. In a normal atom, the number of protons is equal to the number of electrons. Thus the entire atom is electrically neutral. The mass of an electron is 9.1×10^{-31} kg. However, the mass of a proton or a neutron is 1840 times that of an electron. Thus the mass of an atom is mainly due to protons and neutrons only. Atoms can be observed individually using special instruments such as the scanning tunneling microscope.

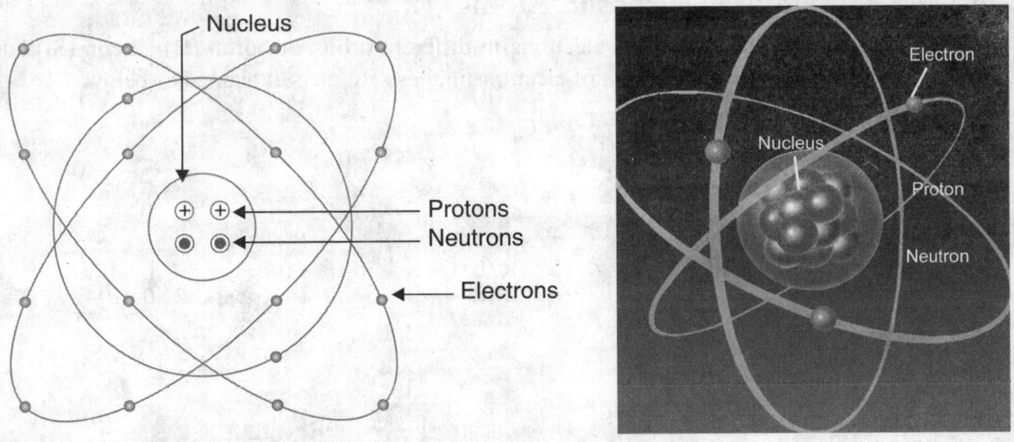

Fig. 2.4. Structure of an atom.

2.5. Atomic Number

The atomic number (Z) of an atom is equal to the number of protons (or electrons) contained in an atom. For example, hydrogen (H) atom has one proton in its nucleus (or one electron around its nucleus). Therefore atomic number of hydrogen atom is 1. The atomic numbers of some important atoms are: $Z = 5$ for boron, $Z = 6$ for carbon, $Z = 13$ for aluminum, $Z = 14$ for silicon, Z = 15 for phosphorus and $Z = 32$ for germanium.

2.6. Atomic Mass

The atomic mass (also called atomic weight) of an atom is equal to the total number of protons and neutrons contained in its nucleus. For example, a silicon atom has 14 protons and 14 neutrons in its nucleus. Therefore, its atomic mass is 28. Similarly, a germanium atom has 32 protons and 41 neutrons. Therefore, its atomic mass is 73.

2.7. Electron Orbits

We have already discussed in Article 2.4 that an electron revolves around the nucleus in fixed orbits, called energy levels or shells. These orbits are represented by the letters K, L, M, N, etc., counted from the nucleus to outwards. Thus the K-shell is closest to the nucleus. The next is L-shell and so on. Sometimes K, L, M, N etc. are also designated as 1, 2, 3, 4 etc.

The maximum number of electrons, in any shell, is given by the relation $2n^2$ where n is an integer and is called principal quantum number. Thus for K-shell, $n = 1$. Therefore, it can have a maximum number of $2(1)^2 = 2$ electrons. Similarly, L-shell, with $n = 2$ can have $2\ (2)^2 = 8$ electrons. The M-shell with the $n = 3$ can accommodate a maximum of $2(3)^2 = 18$ electrons. The N-shell with $n = 4$ can have a maximum of $2(4)^2 = 32$ electrons. However, the distribution of electrons in an atom is subject to the following two rules:

1. The outer most shell of an atom cannot accommodate more than 8 electrons, even if it has a capacity to accommodate more electrons.
2. The shell of an atom just prior to the outer most shell, cannot accommodate more than 8 electrons.

The above mentioned points help us in understanding the distribution of elections in the atom as discussed below:

2.8. Distribution of Electrons in Atoms

Fig. 2.5 shows the distribution of electrons in different orbits of boron (B) silicon (Si) and germanium (Ge) atoms. The distribution of electrons in these atoms is explained as below.

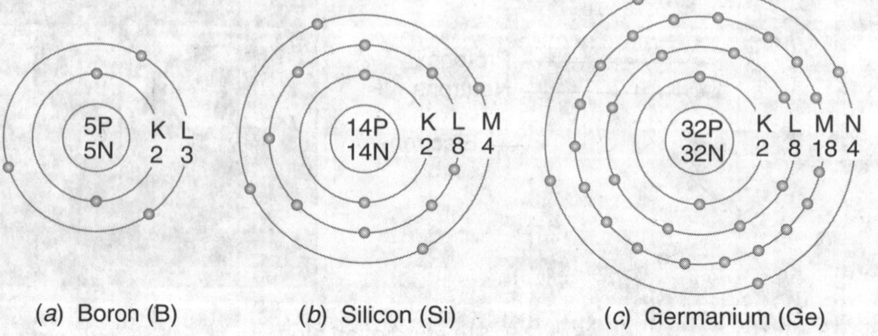

Fig. 2.5. Distribution of electrons in different atoms.

We know that the atomic number of born is 5. It means that it has 5 protons and 5 electrons. Two electrons occupy the *K*-shell, which is then said to be completely filled. The other 3 electrons occupy the *L*-shell as shown in Fig. 2.5 (*a*). Similarly, the atomic number of silicon (Si) is 14. It has 14 protons and 14 electrons. The 14 electrons are distributed as 2 in the *K*-shell, 8 in *L*-shell and 4 in *M*-shell as shown in Fig. 2.5 (*b*). The atomic number of germanium (Ge) is 32. It has 32 protons and 32 electrons. These electrons are distributed as 2 in the *K*-shell, 8 in *L*-shell, 18 in *M*-shell and 4 in *N*-shell as shown Fig. 2.5 (*c*).

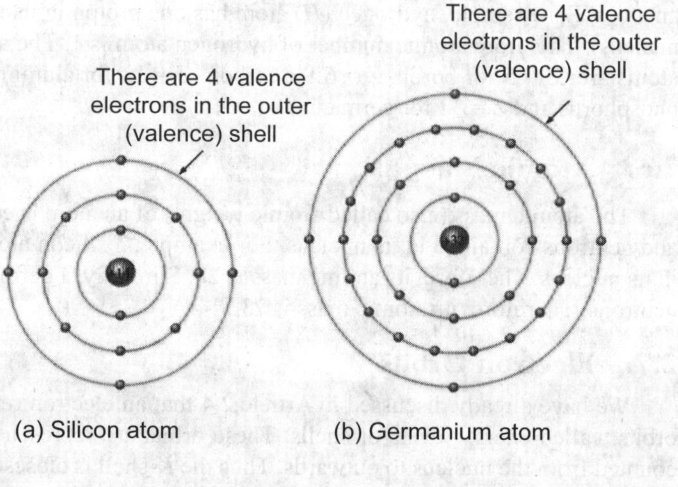

Example 2.1.

Phosphorus atom has Z = 15. Give its electronic distribution, indicating which orbits are full and which are not.

Solution.

The phosphorus atom has 15 electrons. These are distributed in the various shells as given below:

K-shell	2 electrons	Full
L-shell	8 electrons	Full
M-shell	5 electrons	Incomplete

2.9. Electron Sub-orbits or Sub-shells

It has been found that each electron orbit (or main-shell) in an atom consists of number of sub-orbits or sub-shells. The number of sub-orbits is equal to the number of its principal quantum number

(n). For example, for K-shell ($n = 1$), there is only one orbit, which is its own sub-orbit. However, for L-shell ($n = 2$), there are two sub-orbits. For M-shell ($n = 3$), there are three sub-orbits and for N-shell ($n = 4$), there are four sub-orbits.

These sub-orbits are designed by the letters *s*, *p*, *d*, *f*, etc. The total number of electrons (which equals $2n^2$) are distributed in the various sub-orbits as shown in Table 2.1.

Table 2.1: Distribution of electrons in sub-orbits

Orbit	*Total No. of electrons*	*No. of the sub-orbits*	*Name of the sub-orbits*	*Distribution of electrons in sub-orbits*
K ($n = 1$)	2	1	1*s*	2
L ($n = 2$)	8	2	2*s*, 2*p*	2, 6
M ($n = 2$)	18	3	3*s*, 3*p*, 3*d*	2, 6, 10
N ($n = 4$)	32	4	4*s*, 4*p*, 4*d*, 4*f*	2, 6, 10, 14

It is evident from the above table that K-orbit has one sub-orbit, which is also the main orbit. This sub-orbit is called ls sub-orbit and contains only two electrons. The L-orbit has two sub-orbits called 2*s* and 2*p* sub-orbits. The 2*s* sub-orbit contains 2 electrons and 2*p* sub-orbit contains 6 electrons, thus making up a total of 8 electrons. The M-orbit has three sub-orbits called 3*s*, 3*p* and 3*d*, which contains 2, 6 and 10 electrons respectively, thus making a total of 18 electrons. The N-orbit has four sub-orbits called 4*s*, 4*p*, 4*d* and 4*f* which contains 2, 6, 10 and 14 electrons respectively. Thus making a total of 32 electrons. The arrangement of these sub-orbits, for first three orbits is as known in Fig. 2.6 (*a*).

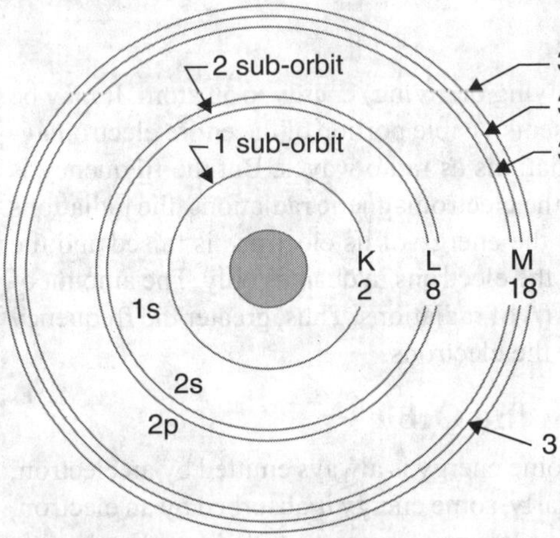

(*a*) Sub-orbits for first three orbits.

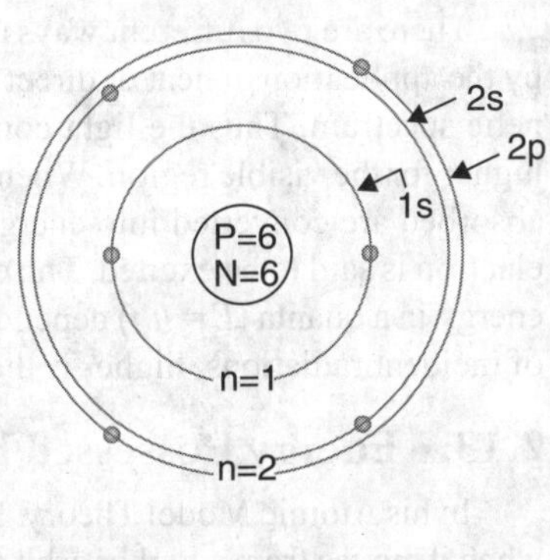

(*b*) Sub-orbits for carbon atom.

Fig. 2.6.

Fig. 2.6 (*b*) shows a carbon atom having a total of 6 electrons. Two of these electrons fill up the K-orbit ($n = 1$) which is also its own sub-orbit. The other 4 electrons fill up the L-orbit in such a way that 2 electrons are in the 2*s* sub-orbit and other 2 in the 2*p* sub-orbit. It may be noted that 2*p* sub-orbit can contain four more electrons. Therefore the L-orbit of the carbon atom is said to be partially filled.

2.10. Valence Electrons

The electrons, which exists in the *outermost shell* of an atom, are called valence electrons. For example, Fig. 2.5 shows that boron has 3 valence electrons, while silicon and germanium have 4 valence electrons. It will be interesting to know that it is the number of valence electrons, which determine the action of an another element to form compounds. If the outermost shell of an electron is completely filled, the element is said to be stable. It means that this atom is not capable of combining with the atoms of other element to form a compound. However, if the outermost shell of an atom is not complete, the atom is susceptible of forming compound. The general nature of these formations is that an atom will accept valence electrons, from another atom, so that the combined electrons fill up the outer shell completely. In turn, an atom may contribute its valence electrons to those of another atom to form a compound. The general rule for the acceptance or contribution of valence electrons is that if an atom contains more than 4 valence electrons, it will usually receive electrons from another atom. However, if an atom contains less than 4 valence electrons, it will usually contribute (*i.e.*, give away) its electrons to those of the other atom.

2.11. Ionized Atoms

The atoms have an important property that electric charge on the nucleus is always equal to the negative charge as determined by the number of electrons. Thus for hydrogen, the nucleus has a positive charge of one unit as its atom contains only one electron. Similarly, silicon atom has a positive charge of 14 units and germanium has a charge of 32 units. It means that every atom, itself, is electrically neutral. However, if an atom loses or gains an electron, it will be in a charged condition. The atom in a charged condition is said to be an ionized atom or an ion. If an atom loses its electron (or electrons), it becomes a positive ion. On the other hand, if an atom gains an electron (or electrons), it becomes a negative ion.

2.12. Application of Energy to an Atom

There are two different ways of applying (*i.e.*, supplying or giving) energy to an atom. It may be by the application of heat or direct light. The light represents visible portion of the entire electromagnetic spectrum. Thus the light consists of the same radiations as radio waves. But the frequency is higher for the visible region. Whenever a body receives the electromagnetic radiations, the radiations absorbed are converted into energy. As a result of this, the energy of its electrons is raised and the electron is said to be excited. The energy is absorbed by the electrons in quantas only. The amount of energy in a quanta ($E = h.f$) depends upon the frequency (f) of radiations. Thus, greater the frequency of incident radiations, higher is the energy absorbed by the electrons.

2.13. Energy Possessed by an Electron in the Orbit

In his Atomic Model Theory, Bohr suggested that some energy is always emitted by an electron, when it moves from a higher orbit to a lower orbit. Similarly, some energy is absorbed by an electron, when it moves from a lower orbit to a higher orbit. The amount of energy emitted (or absorbed) is given by the relation:

$$W_2 - W_1 = h.f \text{ (in joules)}$$

where W_1 = Energy of the initial orbit,

W_2 = Energy of the final orbit,

h = Planck's constant whose value is 6.626×10^{-34} in joules sec, and

f = Frequency of radiation.

It has been found that the total energy (both kinetic and potential) possessed by an electron, when it revolves in the nth orbit of an atom, with atomic number Z, is given by the expression:

$$W_n = -21.76 \times 10^{-19} \frac{Z^2}{n^2} \text{ (in joules)}$$

A joule is a unit of energy in S.I. system. But in the study of electronics, the energies involved are much smaller. Therefore we use a much smaller unit of energy called electron volt (eV) which is defined as :

$$1 \text{ eV} = 1.602 \times 10^{-19}$$

The energy of an electron (in electron volts) is given by the expression,

$$E_n = -\frac{21.76 \times 10^{-19}}{1.6 \times 10^{-19}} \times \frac{Z^2}{n^2} = -13.6 \times \frac{Z^2}{n^2} \text{ (in electron volts)}$$

The above equation is a general expression and can be used to determine the energy possessed by the electron in any orbit. It may be noted that energy in joules is designated by W_n and energy in electron volt (eV) is designed by E_n.

2.14. Energy Level Diagram

In the last article, we have obtained an expression for the total energy of an electron, when it is revolving in some orbit around its nucleus. Now let us obtain the energies E_1, E_2 E_3*....... etc., by using the relation :

$$E = -13.6 \frac{Z^2}{n^2}$$

If we plot the energies corresponding to *K*, *L*, *M*,...... etc. shells along the positive side of the vertical axis by horizontal lines, we shall obtain a diagram a shown in Fig. 2.7. This diagram is popularly known as energy level diagram.

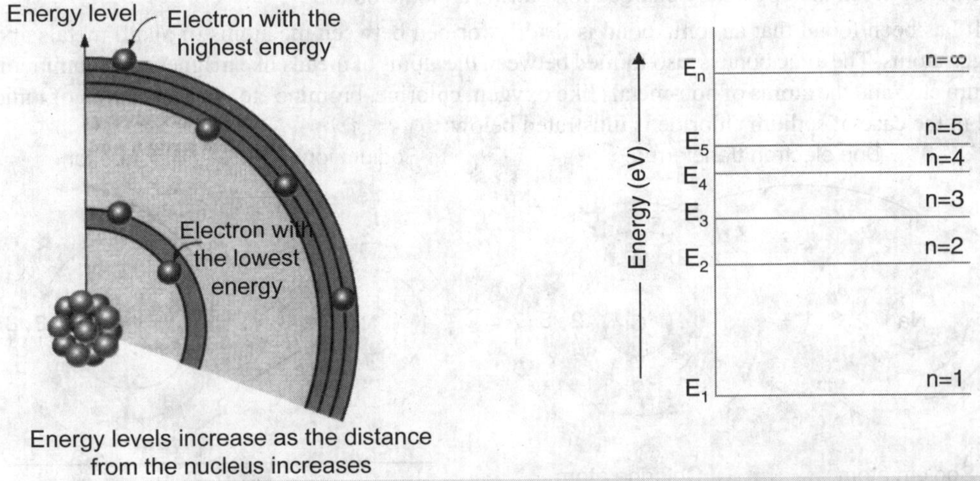

Fig. 2.7. Energy level diagram.

It will be interesting to know that when the electron is present in the first orbit ($n = 1$), it is said to be in its normal or ground state. And if it is present in the higher orbits, it is said to be in the excited state.

* Actually, the values of energy levels are negative and hence should be plotted on the negative side of the vertical axis by horizontal lines. But in actual practice, it is more convenient to indicate energy levels on the positive side of the vertical axis.

2.15. Bonding in Solids

The rigidity of a solid arises from the fact that the atoms in the solid are held together by interatomic forces, called interatomic bonds. The spatial arrangement of atoms, in a solid, is strongly influenced by the nature of these bonds, which in turn is influenced by the electronic structure of the atoms. The interatomic bonds, in solids, have the following two important characteristics:

1. ***Bond energy.*** The energy corresponding to the position of the atoms, at which the bonds are formed, is called bond energy and is designated by W_0. It is expressed in kilojoules per mole of bonds (kJ/mole). The bonds, which possess higher bond energy, are called stronger bonds and those which possess smaller bond energy are called weaker bonds.
2. ***Bond length.*** It may be defined as centre-to-centre distance of the bonding atoms. The stronger bonds have smaller bond length, while weaker bonds have larger bond length.

As mentioned above, the bond energy is used to classify the various types of bonds in solids in two broad categories *i.e.*, stronger or primary bonds and weak or secondary bonds. Now we shall discuss the formation of only primary bonds in solids. There are three types of primary bonds *i.e.*, ionic, covalent and metallic. Since the formation of each such bond is different, therefore, we shall discuss the formation of each of the above mentioned primary bonds.

2.16. Ionic Bond

The bond formed between two oppositely charged ions, which are produced by the transfer of electrons from one atom to another, is called ionic bond or electrovalent bond. In the formation of an ionic bond, the atoms acquire electronic arrangement or configuration of *noble gases (or inert gases). We know that an atom, which loses the electron, acquires a positive charge and becomes a positive ion. On the other hand, an atom which gains the electron, acquires a negative charge and becomes a negative ion. In this way, both the atoms become oppositely charged ions. The electrostatic force of attraction between the oppositely charged ions forms an ionic bond.

It has been found that an ionic bond is readily formed between the atoms of alkali metals and halogen atoms. The ionic bond is also formed between the atoms of metals like magnesium, aluminum, calcium etc., and the atoms of non-metals like oxygen, chlorine, bromine etc. The formation of ionic bond in the case of sodium chloride is illustrated below:

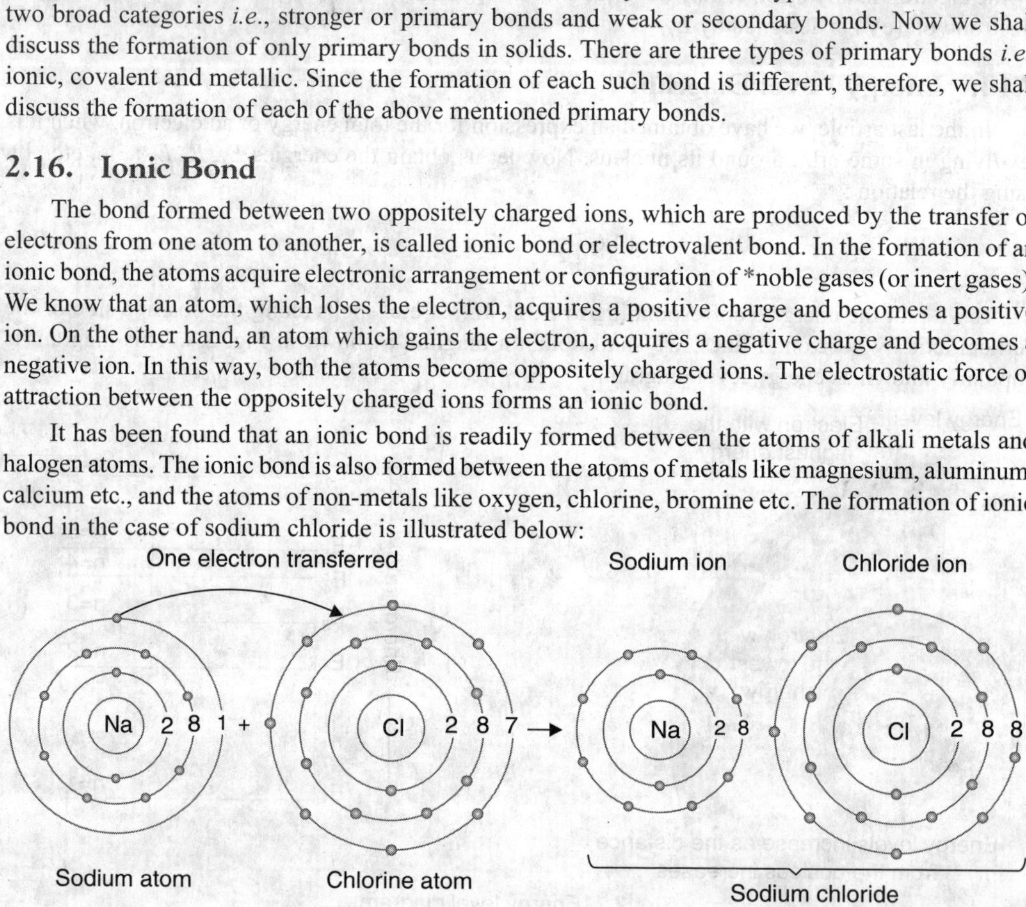

Fig. 2.8. Formation of sodium chloride.

We know that atomic number of sodium is 11, and its electronic configuration is 2, 8, 1. Similarly,

* The elements like, helium, neon, argon, krypton etc. are known as noble gases or inert gases. Such elements do not form bonds. The arrangement of electrons in these elements is such that their outermost orbits are completely filled and hence are said to be in stable arrangement.

atomic number of chlorine is 17 and its electronic configuration is 2, 8, 7. The atomic structure of both these atoms is as shown in Fig. 2.8 (*a*) and (*b*). The sodium atom has 1 electron in its outermost energy level, whereas the chlorine atom has 7 electrons. We know that sodium atom has a tendency to lose its electron in the outermost energy level to have a stable configuration of neon inert gas. After losing its outermost electron, sodium atom becomes a sodium ion (with configuration 2, 8). Moreover, chlorine atom has a tendency to gain or accept an electron (lost by the sodium atom) in its outer most energy level to have a stable configuration of argon inert gas (with configuration 2, 8, 8). After gaining an electron in its outermost energy level, the chlorine atom becomes chlorine ion. The binding forces between sodium ion and chloride ion gives rise to the formation of sodium chloride as shown in Fig 2.8 (*c*).

The compounds containing ionic bonds are called *ionic solids*. The ionic solids consist of oppositely charged ions. As a result of this, the ions are tightly bound to one another by the electrostatic force of attraction. Therefore a large amount of energy is required to break away an ionic bond. The ionic solids have some interesting properties which are mentioned as below:

1. These are rigid, unidirectional and crystalline in nature.
2. These are bad conductors of electricity in their solid state but good conductors of electricity in their molten state.
3. They have high melting and boiling temperatures.

2.17. Covalent Bond

The bond, which is formed by the sharing of electrons between two atoms, is called covalent bond. The sharing of electrons takes place in such a way that each atom, in the resulting molecule, gets the stable electron arrangement of an inert or a noble gas. A covalent bond is formed when both the reacting atoms need electrons to achieve the inert gas electron configuration. This happens in non-metals *i.e.*, the elements which have 5, 6 and 7 electrons in the outermost energy levels of their atoms. In addition to this, covalent bonds are also formed in elements containing 4 electrons in their outermost energy level. The example of such elements are carbon, silicon and germanium etc. The formation of a covalent bond in silicon is as illustrated below:

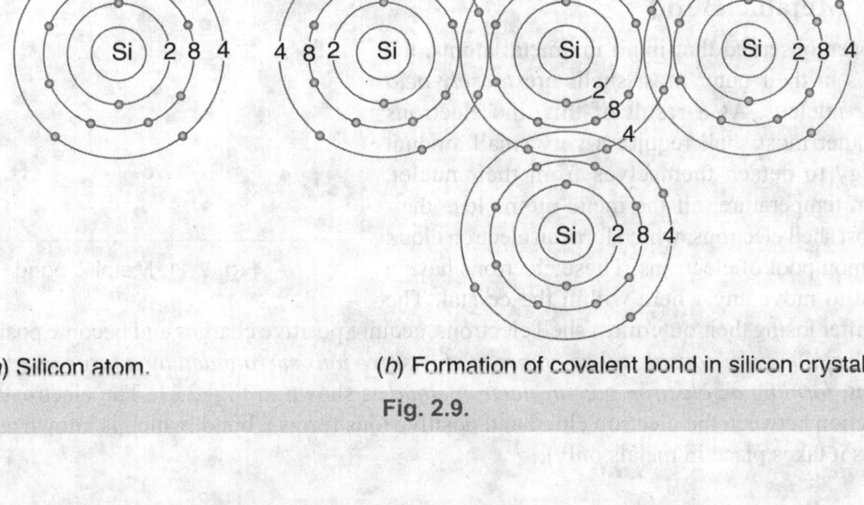

(*a*) Silicon atom. (*b*) Formation of covalent bond in silicon crystal.

Fig. 2.9.

We know that atomic number of silicon atom is 14 and its electronic configuration is 2, 8, 4. The arrangement of these 14 electrons is as shown in Fig. 2.9 (*a*). The silicon atom has 4 electrons in its outermost shell and, it needs 4 more electrons to complete its outermost shell to become stable. It gets these electrons by sharing four electrons, one each from its neighbouring silicon atoms, as shown in Fig. 2.9 (*b*). Similarly, each neighbouring silicon atom shares one electron each with the other neighbouring silicon atoms. It means that a pair of electrons is shared between any two silicon atoms.

It is convenient to represent the covalent bond in silicon structure as shown in Fig. 2.10. In this figure, + 4 indicates, a nucleus and the inner (10) electrons of silicon atom. A part of closed (or broken) lines represent a covalent bond between two silicon atoms, while two dark circles within dashed lines represent a pair of shared electrons.

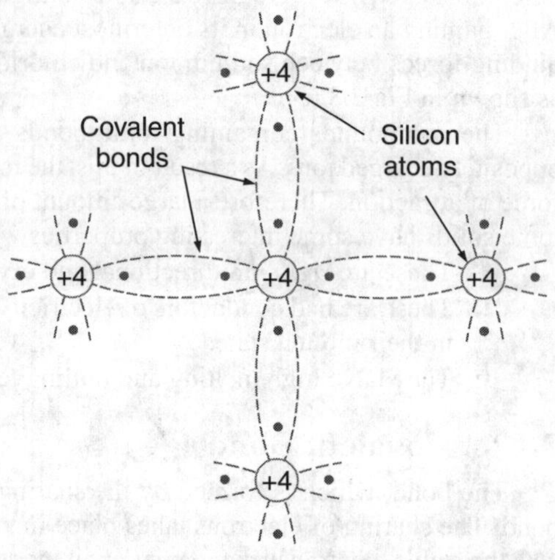

Fig. 2.10. Covalent bond in silicon structure.

A covalent bond formed by the sharing of one pair of electrons between the atoms (as in silicon) is called a *single covalent bond.* There are some molecules or compounds, which are formed by the sharing two or three pairs of electrons. A covalent bond, formed by the sharing of two pair of electrons between the atoms, is called a *double covalent bond.* Similarly, a covalent bond formed by the sharing of three pairs of electrons between the atom is known as a *tripple covalent bond.*

The compounds, containing covalents bonds, are called *covalent compounds.* They are known as compounds, because they can exist as solids, liquids and gases under normal conditions of temperature and pressure. The covalent compounds are not as strong as the ionic solids. Some of the interesting properties of covalent solids are as follows:

1. They are rigid, directional and crystalline in nature.
2. Usually, they are bad conductors of electricity. However, some of the solids such as silicon and germanium can be made to conduct electricity by adding certain impurities in the controlled amount.
3. They have low melting and boiling temperatures as compared to ionic solids.

2.18. Metallic Bond

It has been observed that in all the metal atoms, the electrons in their outer most shells are *loosely held* by their nucleus. As a result of this, the electrons in the outer most shell require a very small amount of energy to detach themselves from their nuclei. At room temperature, all the metal atoms lose their outermost shell electrons, which form an electron cloud or common pool of electrons. These electrons have a freedom to move any where within the crystal. The atoms, after losing their outermost shell electrons, acquire positive charges and become positive ions. Thus *a metal may be considered as an array of positive ions surrounded by a large number of free electrons, forming an electron gas or electron cloud* as shown in Fig. 2.11. The electrostatic force of attraction between the electron cloud and positive ions forms a bond, which is known as metallic bond (as it takes place in metals only).

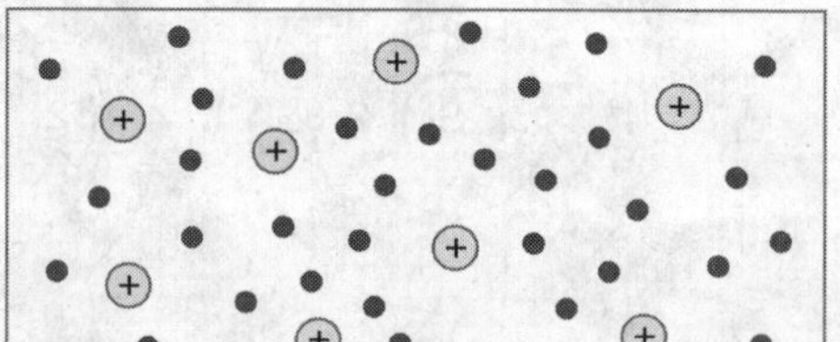
Fig. 2.11. Metallic bond.

It may be noted that in metallic bonding, the electrons are not bound to any atom or between two atoms as in the case of covalent bond. This freedom makes metallic bonds *non-directional.* The solids containing metallic bonds, are called metallic solids. Some of the interesting properties of metallic solids are as follows:

1. They are rigid, non-directional and crystalline in nature.
2. They are good conductors of heat as well as electricity.
3. They have low melting and boiling temperatures as compared to the ionic and covalent solids.

2.19. Energy Bands

It has been found that when atoms come together to form a molecule, they exert an influence on one another. As a result of this, the allowed energy levels 'split' into closely spaced levels. A similar situation exists, when many atoms come together to form a solid. The splitted energy levels form continuous bands of allowed energy, which the electrons may occupy. Since the motion of electrons in a solid is controlled by the energy levels, which they occupy, therefore the nature of these bands is of primary interest in determining the electrical properties of the solid. This has been explained in the following example.

Consider an isolated atom of a carbon. We know that each carbon atom ($Z = 6$) has 6 electrons, two of which occupy the *K*-shell whereas the other four are in the *L*-shell. The energy level diagram of a carbon atom, alongwith its state of electron occupancy, is shown in Fig. 2.12 (*a*).

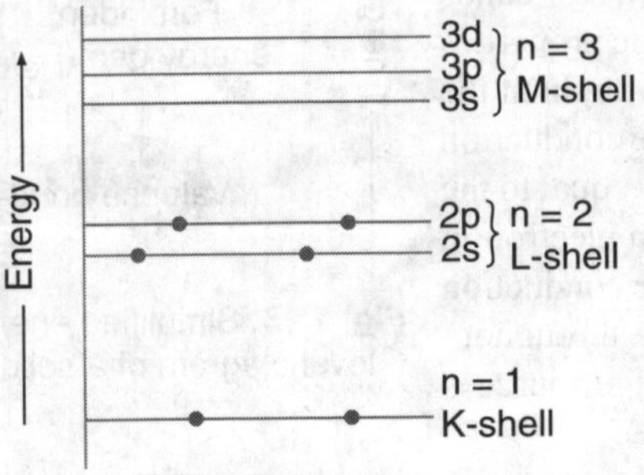

(*a*) Energy level diagram of an isolated carbon atom.

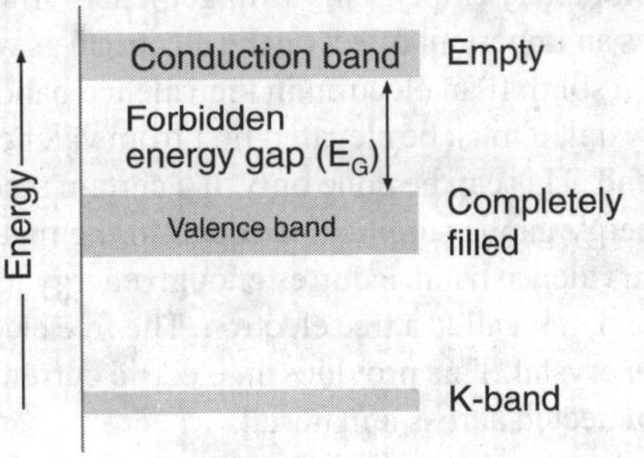

(*b*) Energy level diagram of a diamond crystal.

Fig. 2.12.

Now consider imaginary formation of a diamond crystal from isolated carbon atoms. When the carbon atoms are brought closer to each other, then every atom exerts some force on its neighbouring atoms. Because of this, a coupling will exist between the electrons of one atom and those of its neighbours. This coupling is strongest for the electrons in the higher lying levels.

These levels correspond to the outermost shells of the atoms. As a result of this coupling, the splitting of the energy levels take place. The splitting is greatest for the outermost shells and negligible for the innermost shells. After splitting, the separation between the energy levels is small. But since the number of atoms in a crystal is very large ($\approx 10^{23}$ atoms/cm^3), the total spread between the maximum and minimum energy may be several electronvolts (eV). This large number of discrete, but closely spaced energy levels, are called energy bands. The *K*-shell of an atom forms one energy band, called *K*-band (or l*s*-band). Similarly, the bands are formed by 2*s*, 2*p* ; 3*s*, 3*p*, 3*d* and high lying sub-orbits of the atoms. For the diamond crystal, the bands formed by 2*s* and 2*p*, after overlapping splits into two bands separated by a forbidden energy gap (E_G) as shown in the Fig. 2.12 (*b*). The lower band contains all the 4 valence electrons of the *L*-shell of the isolated atom and is called valence band. This band is completely filled with the electrons. The upper band is an empty band *i.e.*, it contains no electrons. The empty band corresponds to the first excited state, where the valence electrons will move under the influence of an external field. The electrons are free to move in this band. This empty band, which lies just above the valence band, is called conduction band. The conduction band and valence band are separated by a forbidden energy gap (E_G) as shown in the diagram.

Notes: **1.** The number of energy bands in a solid is equal to the number of energy levels in any isolated atom of the same substance.

2. Each band consists of a large number of very closely spaced discrete energy levels. The number of such levels in each band is the same as that of number of atoms in the solids.

3. In considering the electrical properties, we are only interested in the two upper bands. These two bands are the conduction band and the valence band. The electrons, having energies in the conduction band, are those which take part in the conduction of current.

2.20. Simplified Representation of Energy Level Diagram of a Solid

Fig. 2.13 shows a simplified representation of an energy level diagram of a solid. In this diagram, only the conduction and the valence bands are shown. The inner (or lower) bands, which are completely filled, are not shown in the diagram. This has been done due to the fact that the inner bands are not important in determining the electrical as well as electronic properties of a solid. It may be noted that at absolute zero (0 K) temperature, the electrons will occupy the lowest energy levels available to them. Thus at 0 K, the valence band is filled, while the conduction band is completely empty. This arrangement of filled and empty energy bands has an important effect on the electrical as well as electronic properties of a solid. If an electron in the valence band is to move throughout the crystal, it must be elevated first from valence band into the conduction band. This can be done only, if a certain amount of energy equal to the energy gap is supplied or added to the material. When an electron in the valence band acquires enough energy to move into the conduction band, it is called a free electron. The free electron can move throughout the crystal. This provides an electric current, when a voltage source is connected across a material.

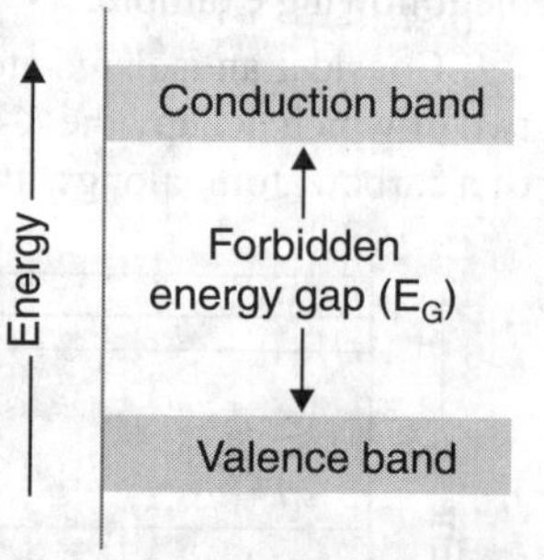

Fig. 2.13. Simplified energy level diagram of a solid.

2.21. Insulators, Conductors and Semiconductors

As a matter of fact, every solid has its own characteristic energy band structure. This variation in band structure is responsible for the wide range of electrical properties observed in various materials. Now we shall discuss the electrical properties of insulators, conductors and semiconductors.

Insulators.

1. ***Insulators.*** The materials, in which the conduction and valence bands are separated by a wide energy gap (≈ 15 eV) as shown in Fig. 2.14 (*a*) are called insulators (also referred to as dielectric materials). A wide energy gap means that a large amount of energy is required, to free the electrons, by moving them from the valence band into the conduction band. Since at room temperature, the valence electrons of an insulator do not have enough energy to jump into the conduction band, therefore insulators do not have an ability to conduct current. Thus insulators have very high resistivity (or extremely low conductivity) at room temperatures. However, if the temperature is raised, some of the valence electrons may acquire energy and jump into the conduction band. It causes the resistivity of insulators to decrease. Therefore the insulators have negative-temperature coefficient of resistance.

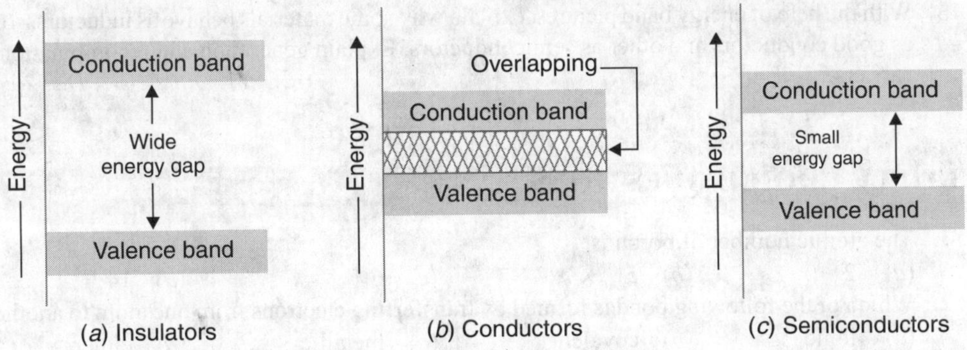

Fig. 2.14.

2. ***Conductors.*** The materials, in which conduction and valence bands overlap as shown in Fig. 2.14 (*b*) are called conductors. The overlapping indicates a large number of electrons available for conduction. Hence the application of a small amount of voltage results in a large amount of current.

3. ***Semiconductors.*** The materials, in which the conduction and valence bands are separated by a small energy gap ($\approx$ 1 eV) as shown in Fig. 2.14 (*c*) are called semiconductors. Silicon and germanium are the commonly used semiconductors. A small energy gap means that a small amount of energy is required to free the electrons by moving them from the valence band into the conduction band. The semiconductors behave like insulators at 0 K, because no electrons are available in the conduction band. However, at room temperature, a significant number of electrons are available in the conduction band. If the temperature is further increased (above room temperature), more valence electrons will acquire energy to jump into the conduction band. Thus like insulators, semiconductors also have negative temperature coefficient of resistance. It means that conductivity of semiconductors increases with the increase in temperature.

DESCRIPTIVE QUESTIONS

1. What are the various types of crystal structures?
2. Define the terms atomic number and atomic mass.
3. What are valence electrons? Do they help in bonding the solids?
4. Explain the procedure for bonding in a solid.
5. Differentiate between an ionic bond and a covalent bond.
6. Describe the insulators, conductors and semiconductors using energy band diagram.
7. Derive expression for energy of electron. *(Anna University., April/May 2008)*
8. Explain about semiconductor, Insulator & Conductor with neat sketch. *(JNTU., June 2009)*
9. Differentiate among conductor, insulator and semiconductor using energy band concept. *(GBTU/MTU., 2005-06, 2008-09)*
10. Explain the energy band diagram for conductors, insulators and Semiconductors. *(GBTU/MTU., 2008-09, 2004-05, Anna University, April/May 2008)*
11. Distinguish conductors, Insulators and semiconductors according to energy band diagram. *(GBTU/MTU., 2008-09, 2005-06)*
12. Show the energy level diagram in insulators, semiconductors and conductor materials. *(GBTU/MTU., 2005, 2007)*

13. With the help of energy band pictures, explain why some materials behave as inductors, some as good conductors and other as semiconductors. Explain generation and recombination of carriers. *(R.G.P.V., June 2011, Dec 2008)*

MULTIPLE CHOICE QUESTIONS

1. The atomic number of boron is
   (*a*) 2 (*b*) 5 (*c*) 6 (*d*) 14
2. Which of the following bond is formed by transferring electrons from one atom to another?
   (*a*) ionic (*b*) covalent (*c*) metallic (*d*) either '*a*' or '*b*'
3. Which of the following solids have least electrical conductivity?
   (*a*) ionic (*b*) covalent (*c*) metallic (*d*) all of them
4. In a semiconductor material, the energy gap is
   (*a*) large (*b*) small (*c*) zero (*d*) infinity

ANSWERS

1. (*b*) **2.** (*a*) **3.** (*a*) **4.** (*b*)

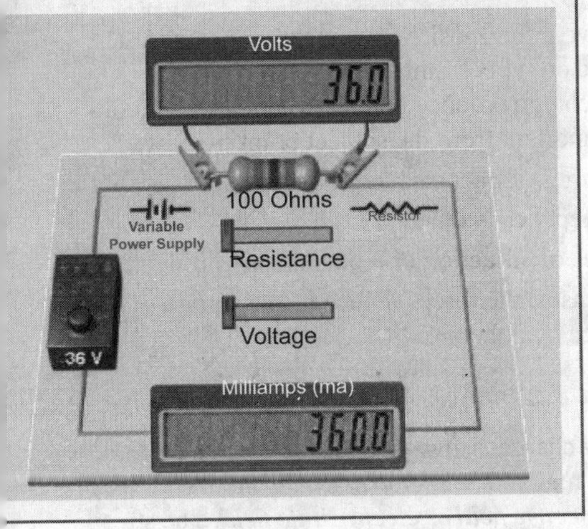

3

Chapter

Electricity and Ohm's Law

Outline

3.1. Introduction

We have already discussed in the last chapter that an atom is composed of three fundamental particles namely electrons, protons and neutrons. An electron has a negative charge equal to 1.6×10^{-19} coulomb. A proton has a positive charge equal to the electron charge (*i.e.*, 1.6×10^{-19} C). It will be interesting to know that if we rub a hard rubber, pen or comb on a sheet of paper, the rubber will attract a corner of the paper, if it is free and easy to move. The paper and rubber give the evidence of a static electric charge. The work of rubbing results in separating electrons and protons to produce a charge of excess electrons on the surface of the rubber and a charge of excess protons on the paper. This is an example of static electricity, because electrons or protons are not in motion.

In actual practice, a charge of many billions of electrons or protons is necessary for common applications of electricity. Therefore, it is convenient to define a practical unit called *coulomb* (*C*). It is equal to a charge of 6.25×10^{18} electrons or protons (*i.e.*, reciprocal of 1.6×10^{-19} coulomb) stored in an insulator or dielectric material (*i.e.*, a material which does not conduct electricity). The charge is designated by the symbol Q or q.

It may be noted that a Coulomb of charge can have either positive or negative polarity. The negative polarity is assigned to the static charge produced on rubber, amber and plastic materials. On the other hand, positive polarity is assigned to the static charge produced on glass and other vitreous materials. Following are some of the laws, which are important from the subject point of view:

1. Charges of opposite polarity (called unlike charges) attract each other.
2. Charges of same polarity (called like charges) repel each other.
3. If the charges are closer to each other, the forces of attraction or repulsion are stronger.

However, as the distance between the charges increases, the forces reduce as the square of the distance, *e.g.*, if the distance is doubled, the forces reduce by a factor of 4.

3.2. Electric Field

It has been found that the region around a stationery charge is always under stress. And a force acts on a positive and negative charge when placed in that region. *The region in which the stress exists (i.e., where the forces acts), is called an electric field, dielectric field or electrostatic field.* The stress is represented by imaginary lines of force. All the lines of force form the electric field. The lines of force are called imaginary because they cannot be seen. The direction of lines of force at any point, is the direction along, which a unit positive charge placed at that point would move if free to do so.

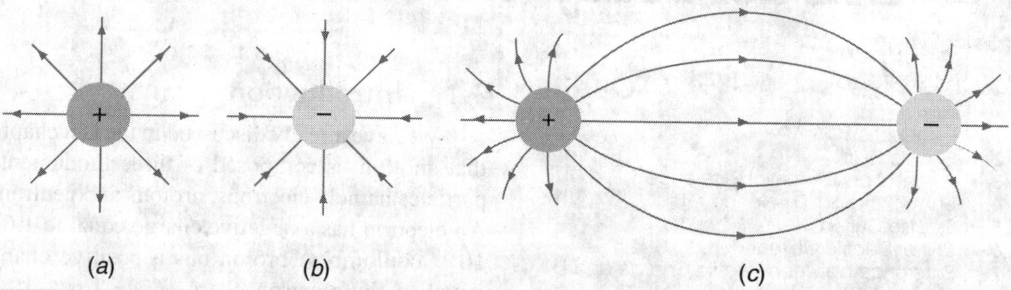

Fig. 3.1. Electric lines of force.

It will be interesting to know that the lines of force emanate from a positive charge as shown in Fig. 3.1 (*a*) and end on a negative charge as shown in Fig. 3.1 (*b*). If two charges of opposite priority are placed closer to each other, the lines of force will be as shown in Fig. 3.1 (*c*).

3.3. Potential and Potential Difference

The term *potential* (or electric potential) may be defined as the energy capable of doing some work. It will be interesting to know that an electric charge has a potential to do some work by moving another charge either by attraction or repulsion. When we consider two charges, they have a difference of potential.

A charge is a result of work done in separating electrons and protons. Because of the separation, there is an accumulation of electrons and protons at different places and there is stress and strain associated with these opposite charges. The stress and strain is due to the fact that, normally, the charges have a tendency to balance each other, which produces a neutral condition. The accumulated electrons are drawn tight and are straining themselves to be attracted towards protons, in order to return to the neutral condition. Similarly, the work of producing the charge causes a condition of stress in the protons, which are trying to attract electrons and return to the neutral condition. Because of these forces, the charge of electrons and protons has potential, as it is ready to give back the work put in for producing the charge.

It may be noted that if one charge is different from the other there is a difference of potential between them. For example, consider a positive charge of 3 coulomb and a negative charge of

1 coulomb separated by a certain space containing electrons as shown in Fig. 3.2 (*a*). These charges have a certain amount of potential corresponding to the amount of work which these charges can do. The work to be done is moving the electrons from one place to another.

Now suppose a charge of 1 C can move 3 electrons. Then a charge of + 3 C can attract 9 electrons towards right. However, the charge of + 1 C can attract 3 electrons towards left. Now the net result is that 6 electrons can be moved towards right, *i.e.*, towards the more positive charge. Similarly, consider two charges of 0 C and 2 C as shown in Fig. 3.2 (*b*). The charge of 0 C has no effect on the electrons. But + 2 C attracts 6 electrons to its side.

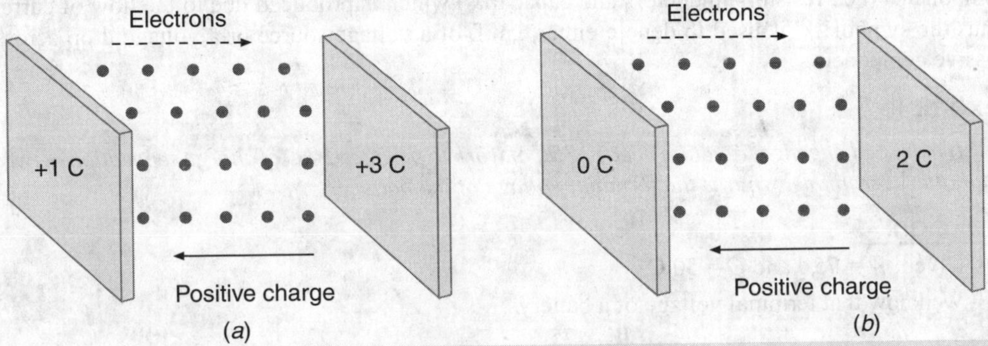

Fig. 3.2. Potential difference.

It is evident from the above discussion that the net number of electrons moved in the direction of more positive charge depends on the difference of potential, between the two charges. This difference corresponds to 2 C for both the cases shown in Fig. 3.2 (*a*) and (*b*). It will be interesting to know that if two charges are equal, in magnitude and are of the some polarity (*i.e.*, either positive or negative), there is no potential difference between them. Hence no work can be done in moving electrons between the two identical charges.

3.4. Unit of Potential Difference

The unit of potential difference is called *volt*. The volt (V) is a measure of the amount of work required to move an electric charge. Mathematically,

$$\text{Volt} = \frac{\text{Work done (in joules)}}{\text{Charge (in coulombs)}}$$

or

$$V = \frac{W}{Q}$$

where V = The potential difference or voltage in volts,
W = Work done in joules, and
Q = Charge in coulombs.

When one joule (J) of work is required to move one coulomb (C) of charge between two points, the potential difference is equal to one volt. Mathematically,

$$1 \text{ Volt} = \frac{1 \text{ joule}}{1 \text{ coulomb}}$$

We know that one coulomb of charge is equal to the charge of 6.25×10^{18} electrons. Therefore one volt may also be defined as one joule of work required to move 6.25×10^{18} electrons between two points *i.e.*,

$$1 \text{ Volt} = \frac{6.25 \times 10^{18} \text{ electrons}}{1 \text{ coulomb}}$$

The potential difference is designated by the symbol V or *v* (where V stands for voltage). As a matter of fact, the volt unit is so common that potential difference is generally called as voltage. Thus *voltage is equal to the potential difference between the two points*. If the potential difference between the two points is large it is called a high voltage. Similarly, if the potential difference is small, then it is called to as a low voltage.

The voltage has two senses or plays two roles in electronic circuits. In one sense, it is a measure of potential difference across the standard voltage sources such as batteries. This potential difference is called *electromotive force* (*e.m.f.*). In the other sense, it is used to denote voltage drop across passive components (*i.e.*, resistors, inductors and capacitors) which is produced due to the flow of current. Thus the symbol '*V*' is used to denote either e.m.f. of a voltage source or a voltage drop across a passive component.

Example 3.1.

If a dry cell chemical reaction needed 75 J of work to produce 50 C of charge separation between the battery terminals, what is the terminal voltage of the battery?

Solution.

Given: $W = 75$ J and $Q = 50$ C.

We know that terminal voltage of a battery,

$$V = \frac{W}{Q} = \frac{75}{50} = 1.5 \text{ V} \textbf{ Ans.}$$

Example 3.2.

Find the amount of charge that can be separated by a 1.5 V battery, if the energy produced by the chemical reaction is 7.5 J.

Solution.

Given: Terminal voltage of a battery, $V = 1.5$ V and energy produced *i.e.*, work done, $(W) = 7.5$ J

Let Q = Amount of charge that can be separated by 1.5 V battery.

We know that the potential difference (V),

$$1.5 = \frac{W}{Q} = \frac{75}{Q} \quad \text{or} \quad Q = \frac{7.5}{1.5} = 5 \text{ C} \textbf{ Ans.}$$

3.5. Electric Current

The movement of electric charge is called an *electric current*. The electric current is always produced due to the potential difference between the two charges. The potential difference forces the charge to move. The charges that can be made to move are called charge carriers or mobile charges.

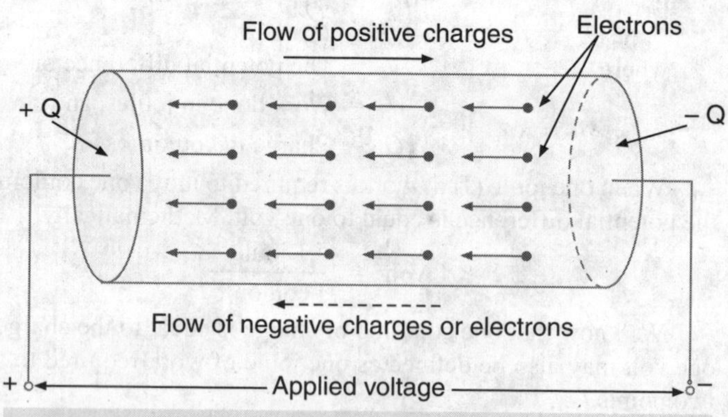

Fig. 3.3. Electric current in copper wire.

We know that valence electrons, in metals, are loosely bound to their nuclei. These electrons may be considered as if they are dissociated from their parent atoms. Therefore, they are known as free electrons. As a result of this, these electrons can be made to move quite easily by a small potential difference, applied across the metal ends. For example, if a potential difference is connected across two ends of a copper wire, as shown in Fig. 3.3, the applied voltage forces the free electrons to move, which constitute an electric current. This current is a drift of electrons from the point of negative charge at one end, moving through the wire and returning to the positive charge at the other end. As the electrons move in one direction, we can think of a positive and equal charge moving in the opposite directions. This positive charge is moving through the wire from one end to the other end.

It will be interesting to know that the number of free electrons, which can be made to move through the wire, to produce electric current, depends upon the amount of the applied voltage. The higher the applied voltage, greater is the value of current produced and *vice versa*. It may be noted that no current is produced, if the applied voltage is zero.

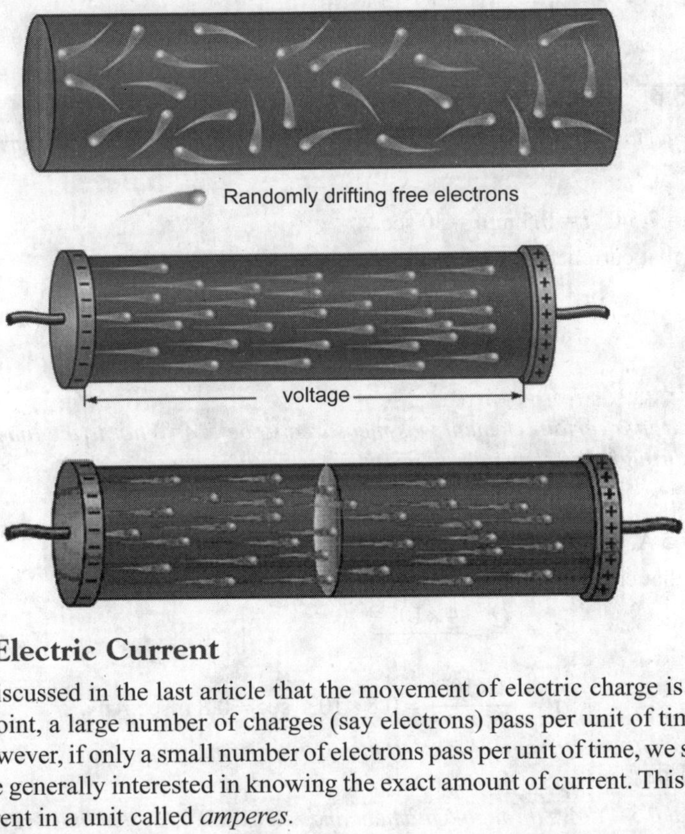

3.6. Unit of Electric Current

We have already discussed in the last article that the movement of electric charge is called electric current. If at any point, a large number of charges (say electrons) pass per unit of time, we say that current is large. However, if only a small number of electrons pass per unit of time, we say that current is small. But we are generally interested in knowing the exact amount of current. This is achieved by expressing the current in a unit called *amperes*.

Thus the practical unit of current is ampere (abbreviated as A). It is a measure of the amount of charges moving past a given point per unit time. *One ampere* (A) *is the amount of current, which causes one coulomb of charge* (equal to 6.25×10^{18} electrons) *to pass a given point in one second.* Mathematically,

$$1 \text{ Ampere} = \frac{1 \text{ coulomb}}{1 \text{ second}} = \frac{6.25 \times 10^{18} \text{ electrons}}{1 \text{ second}}$$

The electric current is designated by the symbol I or i (where I stands for intensity). The amount of I gives an idea as to how intense or concentrated the electrons flow. for example, a current of five amperes is of higher intensity than one ampere.

Sometimes, the current is also known as amperage. It will be interesting to know that in electronic circuits, usually the current is expressed in smaller units called *milliamperes* (mA) and *microamperes* (mA). Such that

$$1 \text{ mA} = 10^{-3} \text{ A and } 1 \text{ mA} = 10^{-6} \text{ A}$$

3.7. Difference between Electric Charge and Current

We know that an electric charge is the quantity of electricity accumulated in an insulator. In other words, charge is a static electricity, at rest, without any motion. When the charges move in a conductor, the current (I) indicates the intensity of motion of the charges. This characteristic gives a fundamental definition of current. Mathematically, current

$$I = \frac{Q}{t} \text{ in amperes}$$

where Q = charge in coulombs, and

t = Time in seconds.

Example 3.3.

If the charge flowing in a certain element is 7.5 C for 0.5 min, what is the current in this element?

Solution.

Given: $Q = 7.5$ C; $t = 0.5$ min = 30 sec.

We know that current in the element,

$$I = \frac{Q}{t} = \frac{7.5}{30} = 0.25 \text{ A } \textbf{Ans.}$$

Example 3.4.

The current in a certain element was measured to be 5 A. What is the time in which 4 mC of charge flows through this element?

Solution.

Given: $I = 5$ A, $Q = 4$ mC $= 4 \times 10^{-3}$ C

We know that current in the element, (I),

$$5 = \frac{Q}{t} = \frac{4 \times 10^{-3}}{t}$$

$$t = \frac{4 \times 10^{-3}}{5} = 0.8 \times 10^{-3} \text{ sec} = 0.8 \text{ msec } \textbf{Ans.}$$

Example 3.5.

A current of 0.3 A flows through a filament and releases 9.45 J of heat energy in 5 seconds. Find the voltage across filament.

Solution.

Given: $I = 0.3$ A, $W = 9.45$ J and $t = 5$ sec

Let V = Voltage across the filament.

We know that the current through a filament, (I),

$$0.3 = \frac{Q}{t} = \frac{Q}{5}$$

$$\therefore \quad Q = 0.3 \times 5 = 1.5 \text{ C}$$

and voltage across the filament,

$$V = \frac{W}{Q} = \frac{9.45}{1.5} = 6.3 \text{ V } \textbf{Ans.}$$

3.8. General Nature of Electric Current

We have already discussed in Art. 3.5 that the flow of an electric current, in a metal conductor, is due to the movement of charges. If the metal conductor is copper, then the flow of current is due to the movement of free charges, which are free electrons of copper atoms. In this case, the moving charges have a negative polarity.

The flow of current in some materials may also be due to the movement of positive charges. A common example of this occurs in *P-type solid semiconductor materials*. In such materials, the current flows due to *holes*, which have a positive charge. This will be discussed in detail in chapter on semiconductors.

3.9. Electron Current and Conventional Current Flow

We know that in a piece of copper wire, the only charges which flow are the free electrons. These free electrons move under the influence of an applied voltage out of the negative terminal of the battery, through the wire of the positive terminal as shown in Fig. 3.4 (*a*). This direction of charge flow is known as *electron current flow*.

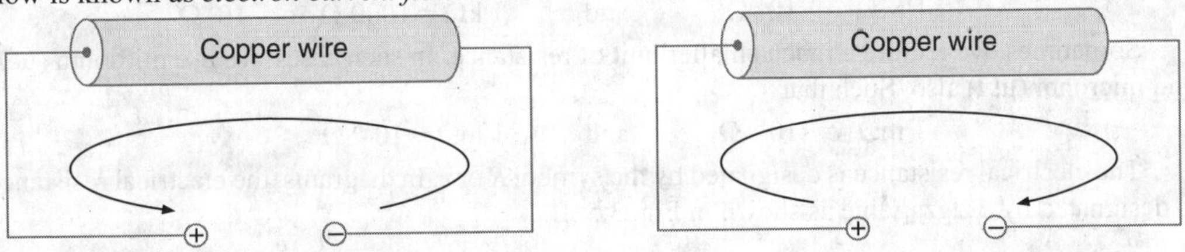

(*a*) Electron current flow. (*b*) Conventional current flow.

Fig. 3.4.

As the electrons move in one direction, it is possible to think of a positive and equal charge moving in opposite direction as shown in Fig. 3.4 (*b*). In this case, positive charge flows out of the positive terminal of the battery through the wire to the negative terminal. This direction of positive charge flow is called *conventional current flow.*

It is evident from the above discussion that electron flow is in a direction opposite to that of a conventional flow. It is well established today that an *electric current in a conductor actually results due to movement of negative charges from negative terminal of the battery to the positive terminal.* But even now, many people prefer to regard current as the movement of positive charges from the positive terminal of the battery to the negative terminal. This view has been able to sustain, because once we think above the atomic level, it makes no difference whether the charge flows from positive to negative or from negative to positive. Mathematically, we get the same answers either way. Thus even though the electron flow is the truth, but still the conventional flow is used in actual practice.

3.10. Electrical Resistance

The term '*electrical resistance*' may be defined as *a property of a substance due to which it opposes the flow of current through it.* It results due to the collision, which takes place between electrons and atoms (or molecules) of a substance due to the flow of current. The collision produces heat, which results in a loss of electrical energy. The resistance reduces magnitude of the current, which can be produced by the applied voltage.

It will be interesting to know that the conductors have very little resistance, while insulators have a large amount of resistance. For example, the atoms of a copper wire has large number of free electrons, which can be moved easily by the applied voltage. Therefore copper wire has little opposition (or resistance) to the flow of electrons, when voltage is applied. This corresponds to a low value of resistance for copper. On the other hand, carbon has fewer number of free electrons than that of copper. As a result, when the same amount of voltage is applied to carbon, less current will flow. Thus we say that carbon has a higher value of resistance than copper.

3.11. Units of Electrical Resistance

The practical unit of electrical resistance is ohm. It is designated by a Greek capital letter omega,(Ω). A conductor has a resistance of one ohm, if it allows one ampere current to flow through it, when a voltage of one volt is applied across its terminals. A good conductor like copper wire can have a resistance of 0.03 Ω for a one metre length. The resistance—wire heating element in a 750 W toaster has a resistance of 70.5 Ω and the tungsten filament in a 100 W 230 V light bulb has a resistance of 529 Ω.

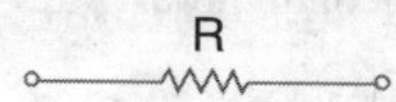

Fig. 3.5. Symbol for resistance.

The insulators have a high value of electrical resistance. Therefore, in order to express the resistance of these materials, much bigger units are used. Such as megaohm (MΩ) and kilohm (kΩ) *i.e.*,

$$1\ \text{M}\Omega = 10^6\ \Omega \qquad \text{and} \qquad 1\ \text{k}\Omega = 1000\ \Omega \ \text{ or } \ 10^3\ \Omega$$

Sometimes, we require a much smaller unit of resistance. In such cases, we use milliohm (μΩ) and microhm (μΩ) also. Such that

$$1\ \text{m}\Omega = 10^{-3}\ \Omega \qquad \text{and} \qquad 1\ \text{m}\Omega = 10^{-6}\ \Omega$$

The electrical resistance is designated by the symbol R or r. In diagrams, the electrical resistance is designated by a zigzag line as shown in Fig. 3.5.

3.12. Factors Affecting Electrical Resistance

The amount of electrical resistance (R) offered by a conductor depends upon the following factors:

1. ***Nature of the material.*** The various materials have different number of free electrons available for producing current. For example, good conductors like copper and silver have large number of free electrons. However, carbon and nichrome have a smaller number of free electrons than those of copper or silver. Higher the number of free electrons, lesser is the electrical resistance.
2. ***Length of the material.*** The resistance of any conductor is directly proportional to its length. Thus any increase in the length of a conductor, increases its resistance and vice versa.
3. ***Cross-sectional area of the material.*** The resistance of any conductor is inversely proportional to its cross-sectional area. Thus any increase in the cross-sectional area of a conductor, decreases its resistance and vice versa.
4. ***Temperature of the material.*** The resistance of any conductor also depends upon the temperature. It increases with the increase in temperature and vice versa.

Neglecting the effect of temperature for the time being, the electrical resistance of a conductor is given by the relation :

$$R = \rho \times \frac{l}{A}$$

where ρ = A constant called specific resistance or resistivity of a material. The resistivity is always expressed by a Greek lower case letter ρ (called *rho*).

l = Length of the conductor, and

A = Area of cross-section of the conductor.

Example 3.6.

Fig. 3.6 shows a metal frame (called chassis) used to give support to a printed circuit board (PCB).

Calculate the resistance between the left and right ends of the conductor. Given the resistivity of metal, $\rho = 2.83 \times 10^{-8}$ Ω-m.

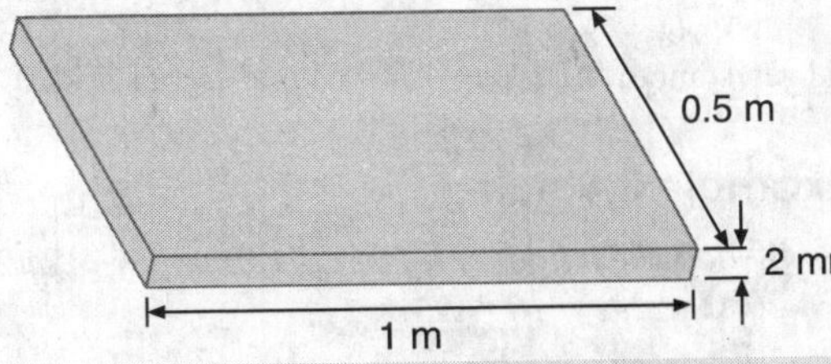

Fig. 3.6.

Solution.

Given: $\rho = 2.83 \times 10^{-8}$ Ω-m; Length (l) = 1 m;

Width = 0.5 m and thickness = 2 mm = 0.002 m.

We know that area of cross-section of the conductor,

$$A = 0.5 \times 0.002 = 0.001 \text{ m}^2$$

and resistance of a conductor,

$$R = \rho \times \frac{l}{A} = 2.83 \times 10^{-8} \times \frac{1}{0.001} = 2.803 \times 10^{-8}\ \Omega$$

$$= 28.3\ \mu\Omega \textbf{ Ans.} \qquad ...(\because 1\ \mu\Omega = 10^{-6}\ \Omega)$$

Example 3.7.

A rectangular carbon block has dimensions 1.0 cm × 1.0 cm × 50 cm. What is the resistance measured (a) between the two square ends? (b) between two opposite rectangular faces? Resistivity of carbon at 20°C is 3.5×10^{-5} Ω-m.

Solution.

Given: Dimension = 1.0 cm × 1.0 cm × 50 cm and $\rho = 3.5 \times 10^{-5}$ Ω-m.

(*a*) *Resistance between two square ends*

We know that length of the block in this case,

$$l = 50 \text{ cm} = 0.5 \text{ m and } A = 1 \times 1 \text{ cm}^2 = 1 \times 10^{-4} \text{ m}^2$$

$$\therefore \text{ Resistance, } R = \rho \times \frac{l}{A} = (3.5 \times 10^{-5}) \times \frac{0.5}{1 \times 10^{-4}} = 0.175\ \Omega \textbf{ Ans.}$$

(*b*) *Resistance between two rectangular ends*

We know that length of the block in this case,

$$l = 1 \text{ cm} = 0.01 \text{ m and } A = 1 \times 50 = 50 \text{ cm}^2 = 50 \times 10^{-4} \text{ m}^2$$

$$\therefore \quad R = \rho \times \frac{l}{A} = 3.5 \times 10^{-5} \times \frac{0.01}{50 \times 10^{-4}} = 7 \times 10^{-5}\ \Omega \textbf{ Ans.}$$

Example 3.8.

Find the resistance of a 120 m copper wire which has a circular cross-section with a diameter of 0.25 cm. Assume that this conductor is at room temperature. Take resistivity of the copper as 1.7×10^{-8} Ω-m.

Solution.

Given: l = 120 m, d = 0.25 cm = 2.5×10^{-3} m and $\rho = 1.7 \times 10^{-8}$ Ω-m.

We know that area of cross section of the wire,

$$A = \frac{\pi d^2}{4} = \frac{\pi \times (2.5 \times 10^{-3})^2}{4} = 4.908 \times 10^{-6} \text{ m}^2$$

and resistance,
$$R = \rho \times \frac{l}{A} = (1.7 \times 10^{-8}) \times \frac{120}{4.908 \times 10^{-6}} = 0.416\ \Omega \textbf{ Ans.}$$

Example 3.9.

What is the length of a 10 Ω aluminum wire at 20°C, if its diameter is known to be 0.15 cm. Take ρ at 20°C = 2.8 × 10^{-8} Ω.-m.

Solution.

Given: $R = 10\ \Omega$; $d = 0.15$ cm $= 1.5 \times 10^{-3}$ m; $\rho = 2.8 \times 10^{-8}$ Ω.m.

Let l = length of the aluminium wire,

We know that area of cross-section of the wire,

$$A = \frac{\pi d^2}{4} = \frac{\pi \times (1.5 \times 10^{-3})^2}{4} = 1.767 \times 10^{-6} \text{ m}^2$$

and resistance of the wire, (R),

$$10 = \rho \times \frac{l}{A} = (2.8 \times 10^{-8}) \times \frac{l}{1.767 \times 10^{-6}} = (1.585 \times 10^{-2})\, l$$

$$\therefore \quad l = \frac{10}{1.585 \times 10^{-2}} = 631 \text{ m } \textbf{Ans.}$$

Example 3.10.

Two copper wires are used to connect the dc supply to a motor, which is 150 m away. If the resistivity of the copper is 1.7 × 10^{-8} Ω-m, what is the approximate diameter of the wire, if the total resistance of the wire used is about 0.722 Ω.

Solution.

Given: Total length of the copper wire (l) = 2 × 150 = 300 m; $\rho = 1.7 \times 10^{-8}$ Ω-m and $R = 0.722\ \Omega$.

Let d = diameter of the copper wire, and

We know that resistance of a copper wire (R),

$$0.722 = \rho \times \frac{l}{A} = (1.7 \times 10^{-8}) \times \frac{300}{A} = \frac{510 \times 10^{-8}}{A}$$

$$\therefore \quad A = \frac{(510 \times 10^{8})}{0.722} = 7.064 \times 10^{-6} \text{ m}^2$$

and area of cross-section, (A).

$$7.064 \times 10^{-6} = \frac{\pi d^2}{4} \quad \text{or} \quad d^2 = \frac{(7.064 \times 10^{-6}) \times 4}{\pi} = 8.994 \times 10^{-6}$$

or
$$d = \sqrt{8.994} = 3 \times 10^{-3} \text{ m} = 3 \text{ mm } \textbf{Ans.}$$

Example 3.11.

A 200 m length of copper wire of a certain diameter has a resistance of 1.5 Ω. What will be the resistance of a silver wire of 10 m length and of the same diameter as that of a copper wire. Take resistivity of copper and silver as 1.7 × 10^{-8} Ω-m and 1.6 × 10^{-8} Ω-m respectively.

Solution.

Given: $l_C = 200$ m; $R_C = 1.5\ \Omega$; $l_S = 10$ m; $\rho_C = 1.7 \times 10^{-8}\ \Omega$-m and $\rho_S = 1.6 \times 10^{-8}\ \Omega$-m.

Let A = Area of the copper wire

$$A_C = \frac{\pi d^2}{4}$$

and resistance of a copper wire, (R_C)

$$1.5 = \rho \cdot \frac{l_C}{A_C} = (1.7 \times 10^{-8}) \times \frac{200}{A_C} = \frac{340 \times 10^{-8}}{A_C}$$

$$\therefore \quad A_C = \frac{340 \times 10^{-8}}{1.5} = 226.7 \times 10^{-8}\ \text{m}^2$$

Since the diameter of the silver wire is the same as that of a copper wire, therefore $A_S = A_C$ = 226.7 × 10^{-8} m^2. Therefore, resistance of a silver wire,

$$R_S = \rho_S \times \frac{l_S}{A_S} = 1.6 \times 10^{-8} \times \frac{10}{226.7} = 0.07\ \Omega \textbf{ Ans.}$$

3.13. Effect of Temperature on Electrical Resistance

It has been found that all the atoms, in a material, are in a continuous state of motion called nucleous. In a solid material, the motion consists of vibrations around a point in the structure. As the temperature increases, the vibrations also increase. The reason for the same is that a greater amount of thermal energy is converted into mechanical energy. Since the number of collisions between electrons and atoms are affected by the vibrations of the atoms, therefore resistance of a material is affected by the temperature. The effect of temperature is different for conductors (pure metals and alloys), semiconductors and insulators as discussed below:

1. ***Conductors.*** The resistance of pure metals increases with the increase in temperature. This increase is large and fairly regular for normal range of temperature. The metals have a positive temperature-coefficient of resistance. The resistance of alloys also increases with the increase in temperature. But this increase is relatively small and irregular. It will be interesting to know that for some alloys like constantan (or Eureka) and manganin, the increase in temperature is negligible over a wide range of temperature.
2. ***Semiconductors and insulators.*** The resistance of semiconductors (*i.e.*, silicon and germanium) and insulators (*i.e.*, mica, glass, paper and rubber) decreases with the increase in temperature. Therefore, such materials have negative temperature-coefficient of resistance. The electrolytes such as sulphuric acid and water have also a negative value of temperature-coefficient of resistance.

3.14. Temperature Coefficient of Electrical Resistance

Consider a metallic conductor having a resistance R at a temperature t (say 20°C). Let us heat this conductor to a temperature t_x, where its resistance increases to R_x. Then considering normal range of temperature, it is found that the increase in resistance, $\Delta R = R_x - R$ depends upon the following factors:

1. Directly on its initial resistance (R),
2. Directly on the rise in temperature ($t_x - t$), and
3. On the nature of the material.

Combining the above three factors, the increase in resistance is given by the relation,

$$\Delta R = R_x - R = \alpha \cdot R.\,(t_x - t) \qquad \text{...(i)}$$

where α is a constant known as temperature-coefficient of resistance of a material. Its value is different for different materials. The equation (*i*) may also be written as:

$$\alpha = \frac{\Delta R}{R\,(t_x - t)} = \frac{R_x - R}{R\,(t_x - t)}$$

Substituting $R = 1$ W and $t_x - t = 1°C$ in the above equation,

$$\alpha = \Delta R$$

Thus temperature-coefficient of resistance (α) may be defined as the increase in resistance per °C rise in temperature per original resistance. The unit of α is Ω/°C/Ω. And its value is, usually, expressed at 20°C.

The equation (*i*) may also be written as:

$$R_x = R\,[1 + \alpha \cdot (t_x - t)] \qquad \text{...(ii)}$$

Equation (*ii*) may be used to determine resistance (R_x) at any temperature (t_x) provided the value of R is given at temperature t. The equation (*ii*) may also be written in the resistivity and conductivity forms. We know that resistance,

$$R = \rho \cdot \frac{l}{A}$$

$$\therefore \quad \rho = \frac{R \cdot A}{l}$$

and resistivity at temperature t_x,

$$\rho_x = \frac{R_x \cdot A}{l} = R\,[1 + \alpha\,(t_x - t)] \cdot \frac{A}{l}$$

$$= \rho\,[1 + \alpha\,(t_x - t)] \text{ in ohm-m}$$

We also know that conductivity is a reciprocal of resistivity, therefore conductivity at any temperature t_x,

$$\sigma_x = \frac{1}{\rho_x} = \frac{1}{\rho[1 + \alpha\,(t_x - t)} = \frac{\sigma}{1 + \alpha\,(t_x - t)}$$

where σ = Conductivity at temperature t (usually 20°C) in siemens (s) or mhos/metre,
α = Temperature-coefficient of resistance, and
$t_x - t$ = Rise in temperature.

The temperature-coefficient (*a*) of some metals and alloys at 20°C is as shown in Table 3.1.

Table 3.1: Temperature-coefficient of some metals and alloys at 20°C.

S. No.	*Material*	*Temperature-coefficient (α) in Ω/°C /Ω at 20°C*
1.	Aluminum	3.9×10^{-3}
2.	Constantan	4×10^{-6}
3.	Copper	3.93×10^{-3}
4.	Gold	3.4×10^{-3}
5.	Nichrome	4×10^{-3}
6.	Silver	3.8×10^{-3}
7.	Tungsten	4.46×10^{-3}

Example 3.12.

Determine the resistance of a tungsten filament at both 800°C at 2250°C. Given the resistance, $R = 3.49\ \Omega$ at 20°C, and α at 20°C = $4.5 \times 10^{-3}\ \Omega/°C/\Omega$.

Solution.

Given: $R = 3.49\ \Omega$; $t = 20°C$ and $\alpha = 4.5 \times 10^{-3}\ \Omega/°C/\Omega$.

We know that resistance of the tungsten filament at 800°C,

$$R_{800} = R\,[1 + \alpha\,(t_x - t)]$$
$$= 3.49\,[1 + (4.5 \times 10^{-3}) \times (800 - 20)] = 15.7\ \Omega\ \textbf{Ans.}$$

and resistance of the tungsten filament at 2250°C,

$$R_{2250} = R\,[1 + \alpha\,(t_x - t)]$$
$$= 3.49\,[1 + (4.5 \times 10^{-3}) \times (2250 - 20)] = 38.5\ \Omega\ \textbf{Ans.}$$

Example 3.13.

A copper wire wound resistor has resistance of 10 kΩ at 20°C. What is its resistance at 80°C and –25°C ? Assume a *for copper = 0.0039 Ω/°C/Ω at 20°C.*

Solution.

Given: $R = 10\ k\Omega$; $t = 20°C$ and $\alpha = 0.0039\ \Omega/°C/\Omega$.

We know that resistance of copper wire at 80°C,

$$R_{80} = R\,[1 + \alpha\,(t_x - t)] = 10\,[1 + 0.0039\,(80 - 20)]\ k\Omega$$
$$= 12.3\ k\Omega\ \textbf{Ans.}$$

and resistance of copper wire at –25°C,

$$R_{-25} = R\,[1 + (t_x - t)] = 10\,[1 + 0.00039\,(-25 - 20)]\ k\Omega$$
$$= 8.2\ k\Omega\ \textbf{Ans.}$$

Example 3.14.

Determine the resistance of a gold alloy wire used in the manufacture of electronic components. The gold alloy's resistivity is $14 \times 10^{-8}\ \Omega$-m and its temperature coefficient is $5.8 \times 10^{-4}\ \Omega/°C/\Omega$, both of which are specified at 20°C. The wire used has a length of 3 meters and 13 micrometer (μm) diameter and is operating at 200°C.

Solution.

Given: $\rho = 14 \times 10^{-8}\ \Omega$-m; $\alpha = 5.8 \times 10^{-4}\ \Omega/°C/\Omega$; $t = 20°C$; $l = 3$ m and $d = 13\ \mu m = 13 \times 10^{-6}$ m.

We know that area of cross-section of gold wire,

$$A = \frac{\pi \times d^2}{4} = \frac{\pi \times (13 \times 10^{-6})^2}{4} = 1.327 \times 10^{-10}\ m^2$$

and resistance at 20°C, $$R = \rho \times \frac{l}{A} = 14 \times 10^{-8} \times \frac{3}{1.327 \times 10^{-10}} = 3164\ \Omega$$

∴ Resistance at 200°C,

$$R_{200} = R\,[1 + \alpha\,(t_x - t)] = 3164\,[1 + (5.8 \times 10^{-4})\,(200 - 20)]\ \Omega$$
$$= 3494\ \Omega\ \textbf{Ans.}$$

3.15. Units of Electrical Resistivity

We have already derived a relation in Art. 3.12 for the electrical resistance of a conductor,

$$R = \rho \times \frac{l}{A} \qquad ...(i)$$

or $$\rho = \frac{R \times A}{l} \qquad ...(ii)$$

Now, if we substitute $l = 1$ m and $A = 1$ m^2, in the above equation, then $\rho = R$. Thus the specific resistance or resistivity of a material may be defined as the resistance between the opposite faces of a cube of 1 m side.

From equation (*ii*), we find that if the resistance (R) is in ohms, area (A) in m^2 and length (l) in m, then the unit of resistivity will be ohm-metre (briefly written as Ω-m). However, if area is in cm^2 and length is cm, then the units of resistivity will be in ohm-cm (*i.e.*, Ω-cm). As a matter of fact, the resistivity of a material helps us in comparing the resistances of different materials according to their nature. A higher value of resistivity means more resistance. The resistivities of some important conducting materials at 20°C (*i.e.*, normal room temperature) are shown in Table 3.2.

Table 3.2: Resistivities of conducting materials at 20°C.

S. No.	*Material*	*Resistivity (ρ) at 20°C in Ω-m*
1.	Aluminium	2.83×10^{-8}
2.	Constantan (60% Cu – 40% Ni)	49×10^{-8}
3.	Copper	1.72×10^{-8}
4.	Gold	2.44×10^{-8}
5.	Iron	9.55×10^{-8}
6.	Manganin (84% Cu – 12% Mn – 4% Ni)	$44 - 48 \times 10^{-8}$
7.	Nichrome (60% Cu – 25% Fe – 15% Cr)	100×10^{-8}
8.	Nickel	9.0×10^{-8}
9.	Silver	1.64×10^{-8}
10.	Tungsten	5.5×10^{-8}

It may be noted from the Table 3.2 that silver, copper and aluminum have lower values of resistivity. As a matter of fact, silver has the lowest value of resistivity and thus is the best conductor. However, copper is as good and much cheaper than silver. Thus almost all good conducting wires are made of copper. Aluminum is also cheap and lightweight. But it is not as good a conductor as copper. A conductor made of aluminum with the same dimensions as a conductor made of copper, would be 1.65 times larger in resistance.

Tungsten and iron have a higher values of resistivity. The metallic alloys have a still higher values of resistivity. Because of this property, these alloys are used for applications in heating elements such as a toaster, incandescent bulb or a heater.

3.16. Resistivities of Semiconductors and Insulators

The resistivities of semiconductors and insulators are much higher than that of conductors. The resistivities of some important semiconductors and insulators at 20°C are shown in Table 3.3.

Table 3.3: Resistivities of semiconductors and insulators at 20°C.

S. No.	*Material*	*Resistivity (ρ) at 20°C in Ω-m*
1.	Carbon	4×10^{-5}
2.	Germanium	0.45
3.	Silicon	2500
4.	Paper	10^{10}
5.	Mica	5×10^{11}
6.	Glass	10^{12}

Insulators provide very high resistances and thus allow an insignificant amount of current to flow. Such materials are used to form the insulating coating of conductors (copper wires) so that they can be grouped together without current flow between them. They are also used in making capacitors as discussed later in Chapter 7. Insulators are used to support power transmission lines and telephone lines.

Semiconductors (*i.e.*, carbon, silicon and germanium) are neither good conductors nor good insulators. Silicon and germanium are used primarily in the production of diodes, transistors and all types of integrated circuits. The resistivity of these materials can be carefully controlled by adding the required amount of other elements usually aluminum, gallium, phosphorus, zinc or arsenic. Carbon is not a useful semiconductor. It is used primarily to produce commercial resistor elements.

3.17. Conductance and Conductivity

The term '*conductance*' *may be defined as *reciprocal of resistance* (*R*). As a matter of fact the resistance of a material measures the opposition, which it offers to the flow of current, while the conductance of a material is a measure of the inducement, which it offers to the flow of current. Mathematically, conductance,

$$G = \frac{1}{R} = \frac{1}{\rho} \times \frac{A}{l} = \sigma \times \frac{A}{l}$$

The unit of conductance is *siemens* (S) or *mhos* (ohm spelled backwards). *The reciprocal of resistivity* (ρ) *is called conductivity* (σ). The unit of conductivity is *siemens/metre* (S/m) or *mhos/metre* (Ω/m). However, it may be noted that if the resistivity of a material is in ohm-cm, then conductivity is in mhos/cm (Ω/cm). Sometimes, it is convenient to represent the units of conductivity in any one of the following ways:

$$\text{mhos/metre} = \Omega/\text{metre} \text{ or } \Omega^{-1}\text{- m}^{-1} \text{ or } (\Omega\text{-m})^{-1}$$
$$\text{mhos/cm} = \Omega/\text{cm} \text{ or } \Omega^{-1}\text{-cm}^{-1} \text{ or } (\Omega\text{-cm})^{-1}$$

Example 3.15.

Determine the conductance of a gold conductor with a resistance of 10 mΩ.

Solution.

Given: $R = 10 \text{ m}\Omega = 10 \times 10^{-3}\,\Omega$

We know that the conductance,

$$G = \frac{1}{R} = \frac{1}{10 \times 10^{3}} = 0.1 \times 103 = 100 \text{ siemens } \textbf{Ans.}$$

Example 3.16.

Calculate the conductance of a 10 kΩ resistor.

Solution.

Given: $R = 10 \text{ k}\Omega = 10 \times 10^{3}\,\Omega$

We have that conductance,

$$G = \frac{1}{R} = \frac{1}{10 \times 10^{3}} = 1 \times 10^{-4} \text{ S} = 100\ \mu\text{S } \textbf{Ans.}$$

Example 3.17.

A circuit has 50 µS of conductance. Calculate the value of circuit resistance.

* This definition is true for d.c. circuits. In a.c. circuits, conductance has a different meaning.

Solution.

Given: Conductance, $G = 50\ \mu S = 50 \times 10^{-6}$ S

We know that conductance, (G),

$$50 \times 10^{-6} = \frac{1}{R}$$

or $$R = \frac{1}{50 \times 10^{-6}\,\Omega} = 20{,}000\ \Omega = 20\ k\Omega\ \textbf{Ans.}$$

Example 3.18.

A silicon conductor has a voltage of 18 V and a current of 60 μA. Determine the conductance of the semiconductor.

Solution.

Given: $V = 18$ Volts and $I = 60\ \mu A = 60 \times 10^{-6}$ A.

We know that the *resistance,

$$R = \frac{V}{I} = \frac{18}{60 \times 10^{-6}} = 0.3 \times 10^{6}\ \Omega$$

and conductance, $$G = \frac{1}{R} = \frac{1}{0.3 \times 10^{6}} = 3.33 \times 10^{-6} = 3.33\ \mu\ \text{Siemens.}\ \textbf{Ans.}$$

3.18. Electric Circuit or Network

The electric circuit or a network may be defined as a combination of electrical devices and in such a way that when they are connected together to form a conducting path, fulfill a desired function such as to light bulb, heat a toasters etc. Consider a simple circuit for lighting an incandescent bulb as shown in Fig. 3.7 (*a*). The bulb emits light when the tungsten-filament wire inside the bulb is white hot, which produces an incandescent glow.

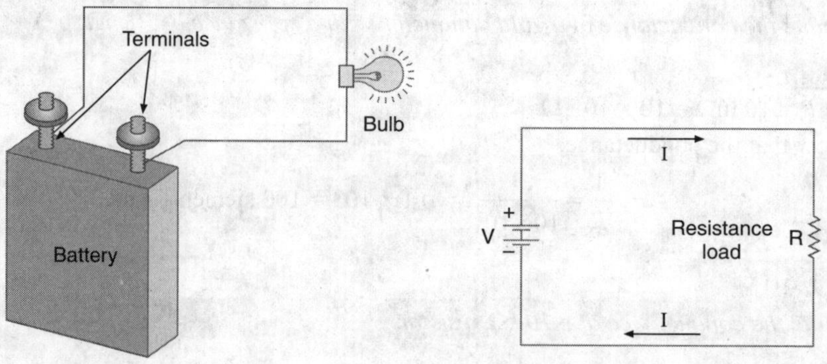

(*a*) Wiring diagram (*b*) Schematic diagram

Fig. 3.7.

It may be noted that the tungsten-filament cannot produce current without a voltage source or battery. This battery provides a voltage across its two output terminals. This voltage is connected across the filament of the bulb by means of two wires, which produce current through the filament. Fig. 3.7 (*b*) shows the schematic diagram of the circuit. The battery and the tungsten-filament of the bulb are replaced by their symbols. The connecting wires are shown simply as straight lines because their resistance is negligible (≈ 0.01 Ω) as compared to that of the filament (≈ 300 Ω). An electric circuit has the following three important characteristics :

* For details, please refer to Art.3.23.

1. There must be a source of voltage. Without the applied voltage, the current cannot flow.
2. There must be a complete path for current flow from one side of the applied voltage source, through the external circuit and returning to the other side of the voltage source.
3. Usually, the current path has some value of resistance. The purpose of resistance, in the circuit, is either to generate the current or to the limit the amount of current.

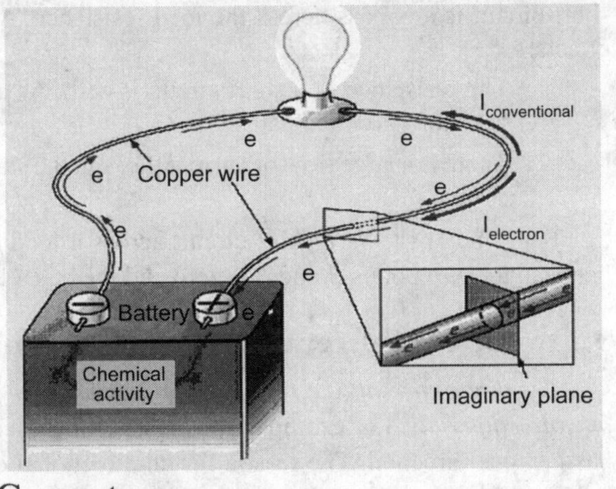

3.19. Load Resistance and Load Current

The part of a circuit, which is connected to the voltage source, is called *load resistance.* It determines how much energy the source will supply. For example, in the circuit of Fig. 3.7 (*b*) the bulb's filament is the load resistance for the battery. The resistance of the filament determines as to how much current the battery will produce.

The current, which flows through the load resistance, is called the *load current.* It will be interesting to know that a lower value of load resistance means a higher value of load current and vice versa. Generally, the term 'load' is used instead of load current. Thus a heavy load means a higher value of load current. And a light load means a lower value of load current.

3.20. Open and Short Circuits

A circuit is said to be '*open*' when any part of the path is open or broken, and there is no continuity in the conducting path. The open-circuit may be in the connecting wires or in the load resistance. An open-circuit gives rise to the following two effects:

1. The resistance of an open circuit is infinitely high.
2. There is no current in an open circuit.

Fig. 3.8 (*a*) shown an open circuit in a load-resistance. As a result of this, the resistance between the two points *A* and *B* is infinity and there is no current through the load resistance.

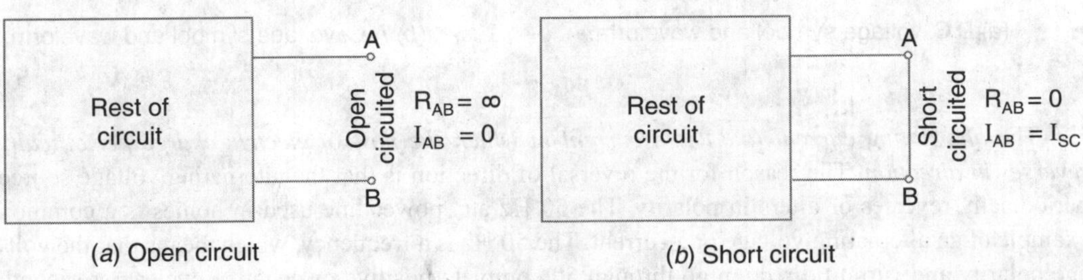

Fig. 3.8.

A circuit is said to be '*short*' when the voltage source has a closed path across its terminals. And the resistance is, practically, zero. This results in too much current in the circuit. Usually, the short-circuit is a bypass across the load resistance. A short-circuit gives rise to the following two effects:

1. The resistance of a short circuit is zero. As a result of this there is no voltage drop across the short-circuited paths.
2. There is a large flow of current through the short-circuit. And it is referred to as short-circuit current.

Fig. 3.8 (*b*) shows a short-circuit across a load resistance. As a result of this, the resistance between the two points *A* and *B* is zero and there is a large flow of current through the short-circuit.

3.21. Direct Current (D.C.) and Alternating Current (A.C.)

The direct current (d.c.) is a current in which the flow of electric charges is unidirection i.e., in one direction only. For example, the current illustrated in Fig. 3.7 (*b*) is a direct current, because it has just one direction. The reason for unidirectional current is that the battery maintains the same polarity of output voltage. It may be noted that d.c. voltage source can change the amount of its output voltage. But if the same polarity is maintained, direct current will flow in just one direction. A battery is a constant or steady d.c. voltage source, because it has fixed polarity and its output voltage has a steady value. Fig. 3.9 (*a*) shows the symbol of a d.c. voltage source and waveform of a d.c. voltage. In the waveform, the amount of d.c. voltage is plotted against time.

As seen, the waveform is a straight horizontal line. In actual practice, we can have a d.c. voltage which varies with respect to time. But so long it does not reverses its polarity, we define it as a d.c. voltage.

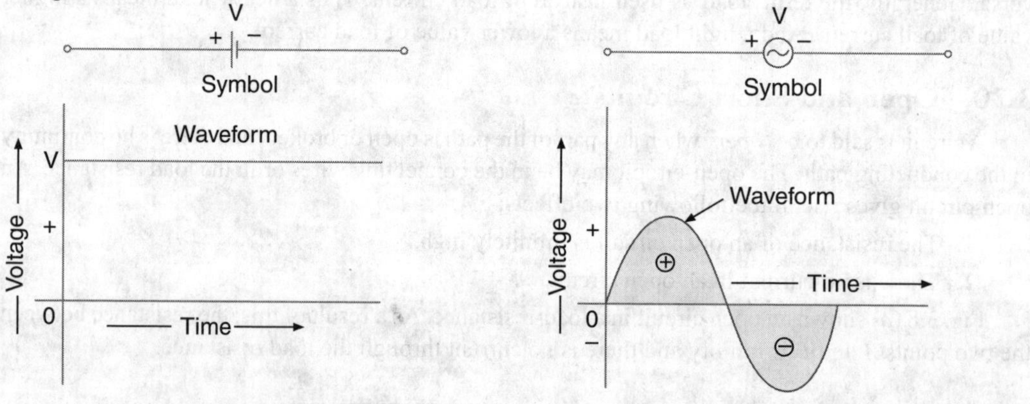

(*a*) D.C. voltage symbol and waveform. (*b*) A.C. voltage symbol and waveform.

Fig. 3.9.

The *alternating current (a.c.) is a current in which the flow of electric charge periodically reverses in direction*. The reason for the reversal of direction is that the alternating voltage source periodically reverses or alters in polarity. The 50 Hz a.c. power line used in homes is a common example of an alternating voltage or a current. The 50 Hz is a frequency, which means that the voltage polarity and current direction go through 50 complete positive or negative cycles per second. Fig. 3.9 (*b*) shows the symbol of an a.c. voltage source and waveform of an alternating voltage. In the waveform, the amount of a.c. voltage is plotted against time. It may be noted that in the a.c. voltage waveform, the magnitude of voltage varies during the positive and negative cycles (between the reversals in polarity).

3.22. Comparison of D.C. Voltage and A.C. Voltage

Table 3.4 shows some important points of d.c. voltage and a.c. voltage.

Table 3.4 : Comparison of d.c. voltage and a.c. voltage.

S. No.	*D.C. voltage*	*A.C. voltage*
1.	It has a fixed polarity.	It reverses in polarity.
2.	It can be steady or vary in magnitude.	It varies between reversals in polarity.
3.	Its steady value cannot be stepped up or down by a transformer.	It can be stepped up or down by a transformer for electric power distribution.
4.	It is easy to measure by a meter called d.c. voltmeter.	It is difficult to measure because it varies with time. It is measured by a meter called a.c. voltmeter. This meter reads the root-mean square value of a.c. voltage.

3.23. Ohm's Law

It gives the relationship between the voltage, current and resistance for good conductors. The Ohm's Law states that *the electric current passing through a conductor is directly proportional to the potential difference applied across the conductor*. Mathematically, the current (in amperes)

$$I \propto V$$

$$= \frac{V}{R} \qquad \text{...}(i)$$

German physicist
George Ohm (1789-1854)

where R = A constant of proportionality, known as resistance of the conductor in ohms.

The equation (*i*) may also be expressed in the following two different forms:

$$V = I \cdot R \qquad \text{...}(ii)$$

and

$$R = V/I \qquad \text{...}(iii)$$

The meaning of equation (*ii*) is that whenever there is a current through a conductor, the potential difference or voltage drop across its two ends is equal to the product of current (I) passing through it and the resistor (R) of the conductor. Similarly, the equation (*iii*) says that the ratio of potential-difference (V) across the two ends of a conductor to the current (I) passing through the conductor is constant and is equal to the resistance (R) of the conductor.

Example 3.19.

A certain soldering iron has a resistance of 600 Ω when operated from a 230 volts power line. How much current does it take from the power line?

Solution.

Given: $R = 600\ \Omega$ and $V = 230$ volts.

We know that current taken by the soldering iron,

$$I = \frac{V}{R} = \frac{230}{600} = 0.38 \text{ A } \textbf{Ans.}$$

Example 3.20.

Fig. 3.10 shows a moving coil loudspeaker. The coil has a resistance of 8 Ω. The maximum current it can carry safely is 2.5 A. If it carries more than 2.5 A, the wire in the coil may overheat or the vibrations of the speaker cone may get so violent that the cone would tear off.

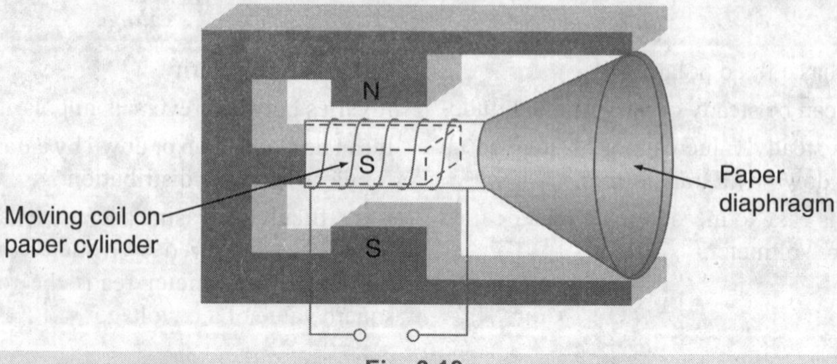

Fig. 3.10.

Find the maximum safe voltage, which can be applied to the speaker coil.

Solution.

Given: $R = 8\ \Omega$ and $I = 2.5$ A.

We know that the safe voltage that can be applied to the speaker coil,

$$V = I \cdot R = 2.5 \times 8 = 20 \text{ volts } \textbf{Ans.}$$

Example 3.21.

A certain relay coil has a resistance of 1.5 kΩ. The coil requires a current of 16 mA in order to energize (i.e., close its contacts). How much voltage must be applied to the relay coil to energize it?

Solution.

$R = 1.5\ \text{k}\Omega = 1.5 \times 10^3\ \Omega$ and $I = 16\ \text{mA} = 16 \times 10^{-3}$ A.

We know that the voltage that must be applied to the relay coil to energize it,

$$V = I \cdot R = (16 \times 10^{-3}) \times (1.5 \times 10^3) = 24 \text{ volts } \textbf{Ans.}$$

Example 3.22.

Typically, a seven segment readout device of a digital clock requires a current of 20 mA/segment for adequate visibility. If the readout device is driven by a source of 5 volts, how much resistance should be inserted into the circuit of each segment ? Assume that a segment has no resistance of its own.

Solution.

Given: Current/segment = 20 mA = 20×10^{-3} A and $V = 5$ volts.

We know that the resistance that should be inserted into the circuit of each segment,

$$R = \frac{V}{I} = \frac{5}{20 \times 10^{-3}} = 250\ \Omega \ \textbf{Ans.}$$

Example 3.23.

Fig. 3.11 shows the graph of a curve tracer for a certain resistance. Each horizontal mark on the graph represent 2 V and each vertical mark represents 5 mA.

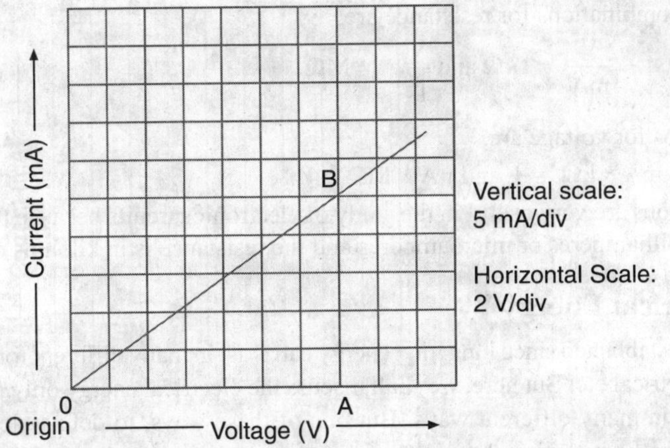

Fig. 3.11.

What is the value of resistance?

Solution.

Given: Each horizontal mark on the graph = 2 V and each vertical mark = 5 mA.

In order to determine the value of *R*, it is necessary to read the voltage and current at any single point along the line. Let us choose the point *B* on the given curve. This point is 7 divisions to the right of the origin horizontally and 5 divisions up vertically. The voltage corresponding to point *B*,

$$V = 7 \text{ div} \times (2 \text{ V/div}) = 14 \text{ volts}$$

and current,

$$I = 5 \text{ div} \times (5 \text{ mA/div}) = 25 \text{ mA} = 0.025 \text{ A}$$

∴ Value of resistance,

$$R = \frac{V}{I} = \frac{14}{0.025} = 560\ \Omega \ \textbf{Ans.}$$

3.24. Typical Values of Voltage (V) and Current (I)

The electronic devices *i.e., diodes, transistors* and *integrated circuits* generally operate with a d.c. supply of 3, 5, 6, 9, 12, 20 or 28 volts. The current is, usually, in the range of few microamperes (or milliamperes) to about 5 amperes. However, some electronic equipments use high-voltages also, which can be as high as 25000 V (25 kV). In that case, the current flowing in the circuit is usually kept very small, otherwise tremendous amount of power is necessary. The current can have a low value, when there is a large amount of resistance in the circuit.

3.25. Short-Cuts in Ohm's Law Problems

Sometimes, it is very convenient to take few arithmetic short-cuts in Ohm's Law problems. These are made by using engineering units of current, voltage and resistance. The methods for taking short-cuts for current, resistance and voltage are discussed as below.

When the voltage is in volts and resistance in kilohms, then the current is in milliamperes *i.e.*,

$$\frac{\text{V}}{k\Omega} = \frac{\text{V}}{10^3\Omega} = 10^{-3} \times \frac{\text{V}}{\Omega} = 10^{-3} \text{ A} = \text{mA}$$

and when the voltage is in volts and resistance in megohms, then the current is in microamperes. *i.e.*,

$$\frac{\text{V}}{M\Omega} = \frac{\text{V}}{10^6\Omega} = 10^{-6} \times \frac{\text{V}}{\Omega} = 10^{-6} \text{ A} = \mu\text{A}$$

Similarly, combinations for resistance are,

$$\frac{\text{V}}{\text{mA}} = k\Omega \text{ and } \frac{\text{V}}{\mu}\text{A} = \text{M}\Omega$$

and combinations for voltage are,

$$\text{mA} \times \text{k}\Omega = V \text{ and } \mu\text{A} \times \text{M}\Omega = V$$

These relations are very useful in the study of electronic circuits because the current is, usually, in the units of milliamperes or microamperes and the resistance is in kilohms or megohms.

3.26. Electrical Energy

It has been established since long, that energy can exist in many different forms such as chemical, mechanical, electrical etc. But here, we shall discuss the electrical energy only. The electrical energy may be defined in many different ways. But one of these ways, to define electrical energy, is *the ability to produce heat by causing charge to move from one location to another*. For example, a battery contains electrical energy, because it can cause some charge to move from one terminal of a resistance to the other. The electrical energy, contained in the battery, is equal to the battery voltage (V) multiplied by the total charge, which is present on the battery plates.

The symbol of energy is *E* and its *unit is joule* (J). One joule is the amount of energy converted to heat when one coulomb of charge moves through a resistance with a voltage of one volt existing across it. Mathematically,

$$1 \text{ Joule} = 1 \text{ volt} \times 1 \text{ coulomb}$$

3.27. Electrical Power

It may be defined as *the rate of producing or using electrical energy as well as the rate of doing work*. Mathematically, power,

$$P = \frac{W}{t}$$

where W = Work done or energy in joules, and

t = Time in seconds.

The unit of electrical power is *watt* (W). One watt of power is equal to the work done in one second by a voltage of one volt in moving one coulomb of charge. Mathematically,

$$1 \text{ watt} = \frac{1 \text{ joule}}{1 \text{ second}} = \frac{1 \text{ volt} \times 1 \text{ coulomb}}{1 \text{ second}} = 1 \text{ volts} \times 1 \text{ ampere}$$

$\therefore$ Power, $P = V \times I$ in watts. ...(*i*)

Thus power in watts is equal to the product of voltage and current. The relation between *P*, *V* and *I* given by equation (*i*) is called power formula. It can also be expressed as:

$$I = \frac{P}{V} \quad \text{or} \quad V = \frac{P}{I}$$

Note. We know that in large number of applications, electrical power is converted into mechanical power and *vice versa*. Sometimes, the mechanical power is expressed in horse power (h.p). The relationship between watt and horse power is given by,

$$1 \text{ h.p.} = 746 \text{ W}$$

The above relation is easy to remember as 1 h.p. is approximately equal to 3/4 kilowatt (kW) where 1 kilowatt = 1000 watts.

Example 3.24.

Determine the rate at which an electric water heater can convert the electrical energy into heat energy, if 64 kJ of heat is produced in 40 seconds.

Solution.

Given: $W = 64$ kJ $= 64 \times 10^3$ J and $t = 40$ s

We know that the rate at which electrical energy is converted into heat energy,

$$P = \frac{W}{t} = \frac{64 \times 10^3}{40} = 1600 \text{ watt} = 1.6 \text{ kW } \textbf{Ans.}$$

Example 3.25.

A toaster takes 5 A from the 230 watt power line. How much power is consumed by the toaster?

Solution.

Given: $I = 5$ A and $V = 230$ volts

We know that power consumed by the toaster,

$$P = V \times I = 230 \times 5 = 1150 \text{ W} = 1.15 \text{ kW } \textbf{Ans.}$$

Example 3.26.

How much current flows through the filament of a 36 watt soldering iron connected to a 230 volts power line.

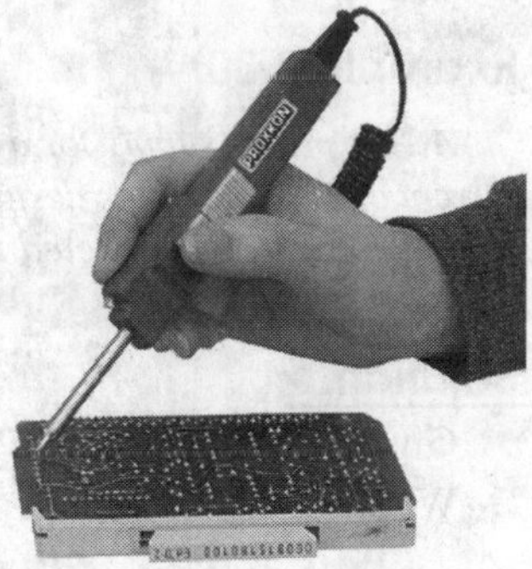

Solution.

Given: $P = 36$ W and $V = 230$ volts.

We know that current through the filament,

$$I = \frac{P}{V} = \frac{36}{230} = 0.16 \text{ A } \textbf{Ans.}$$

Example 3.27.

Calculate the energy used (in kWh) to run twelve 150 W light bulbs for ten hours.

Solution.

Given: Power rating of each bulb, $P = 150$ W ; Number of bulbs = 12, Time, $t = 10$ hours

Let W = Energy used in kWh

We know that total power consumed by 12 bulbs,

$$P = 150 \times 10 = 1500 \text{ W} = 1.5 \text{ kW}$$

and the total power (P),

$$1.5 = \frac{W}{t} = \frac{W}{10} \text{ kW/h}$$

$$\therefore \quad W = 1.5 \times 10 = 15 \text{ kWh } \textbf{Ans.}$$

Example 3.28.

Calculate the energy used (in kWh) to run a 500 W stereo system and a 2400 W air conditioner for three hours.

Solution.

Given: $P_{stereo} = 500$ W ; $P_{air\ con} = 2400$ W

Let E = Energy used in kWh

We know that total power consumed,

$$P = P_{stereo} + P_{air\ con} = 500 + 2400 + 2900 \text{ W}$$
$$= 2.9 \text{ kW}$$

and the power, (P),

$$2.9 = \frac{W}{t} = \frac{W}{3} \text{ kW/h}$$

$\therefore$ $$W = 2.9 \times 3 = 8.7 \text{ kWh } \textbf{Ans.}$$

Example 3.29.

A television has a 230 V input and an average power rating of 180 W. What is the television's average input current.

Solution.

Given: $V = 230$ volts ; $P = 180$ W

We know that input current,

$$I = \frac{P}{V} = \frac{180}{230} = 0.783 \text{ A } \textbf{Ans.}$$

Example 3.30.

A 24 V power supply for a deepawali decoration lighting system can provide a maximum output current of 2 A. The lights designed to be used with the system contain 0.5 W bulbs. How many lights (or bulbs) can be connected to the power supply without exceeding 80% of its maximum power rating.

Solution.

Given: $V = 24$ volts ; $I = 2$ A ; Power rating of each light bulb = 0.5 W

We know that maximum power,

$$P = V \times I = 24 \times 2 = 48 \text{ W.}$$

and 80% of this power rating,

$$P_{80\%} = 48 \times \frac{80}{100} = 38.4 \text{ W}$$

Number of lights (or bulbs) required,

$$= \frac{P_{80\%}}{\text{Power rating of each bulb}} = \frac{38.4}{0.5}$$
$$= 76.8$$

Since the actual number of lights have to be an integer whole number, therefore the number of lights = 76 **Ans.**

3.28. Power Dissipation in the Resistance

We have already discussed in Art. 3.10 that when a current flows in a resistance, some heat is produced due to the collision between the moving electrons and atoms (or molecules). The heat is an evidence of the fact that some power is used in producing the current. The power dissipated by the resistance in the form of heat must be supplied by the voltage source, otherwise it cannot maintain the voltage required to produce current. Since the power is dissipated in the resistance of a circuit, therefore, it is convenient to express power in term of the resistance (R). We know that the power,

$$P = V \times I = (I \cdot R) \times I = I^2 \times R \qquad ...(\because V = I \cdot R)$$

$$= V \times \frac{V}{R} = \frac{V^2}{R} \qquad ...\left(\because I = \frac{V}{R}\right)$$

It may be noted that any one of these three power formulae can be used to calculate the power dissipated in a resistance. It will be interesting to know that in some applications, power dissipation is desirable, because the component must produce heat to do its job. For example, a 100 W bulb must dissipate this power to make the filament white so that it can produce light. Similarly, a 750 W toaster must dissipate this amount of power to produce the necessary amount of heat. However, in other applications, heat may be undesirable because the purpose of the resistance in the circuit is to provide current through it. But in any case, whenever there is a flow of current, it dissipates power equal to $I^2 \cdot R$.

The components, which utilize the power dissipated in their resistance, such as light bulbs and toaster are generally rated in terms of power. The power rating is at normal voltage, which is 230 V of the power line. For example, a 750 W, 230 V toaster has this rating because it dissipates 750 W in the resistance of the heating element, when connected across 230 V power line. It may be noted from the power formula, that if the source voltage is low, it requires a high current for the same power and vice versa.

3.29. Power Formulae

We have already discussed in the last article, the three basic forms of power formulae. But each such form of power formula can be further expressed in three different forms as shown below:

$$P = V \times I \quad \text{or} \quad P = I^2 \times R \quad \text{or} \quad P = \frac{V^2}{R}$$

$$\text{or} \quad I = \frac{P}{V} \quad \text{or} \quad I = \sqrt{\frac{P}{R}} \quad \text{or} \quad R = \frac{P}{I^2} \quad \text{or} \quad R = \frac{V^2}{P}$$

$$\text{or} \quad V = \frac{P}{I} \quad \text{or} \quad V = \sqrt{P \times R}$$

Example 3.31.

A certain relay coil has a resistance of 750 Ω and requires a current of 32 mA in order to energize (operate its contacts). How much power, does the relay coil consume, while it is energized?

Solution.

Given: $R = 750\ \Omega$ and $I = 32\text{ mA} = 32 \times 10^{-3}$ A.

We know that power consumed by relay coil,

$$P = I^2 \times R = (32 \times 10^{-3})^{-2} \times 750 \text{ W}$$

$$= 0.768 \text{ W} = 768 \text{ mW} \textbf{ Ans.}$$

Example 3.32.

Determine the power rating of a microwave oven, if it has a resistance of 36 Ω. The oven is attached to a 230 V power line.

Solution.

Given: $R = 36\ \Omega$ and $V = 230$ volt.

We know that power rating of the oven,

$$P = \frac{V^2}{R} = \frac{(230)^2}{36} = 1470 \text{ W} = 1.47 \text{ kW } \textbf{Ans.}$$

Example 3.33.

A certain soldering iron takes 36 W from a 230 V source. What is the resistance of its heating element?

Solution.

Given: $P = 36$ W and $V = 230$ volt.

We know that resistance of the heating element,

$$R = \frac{V^2}{P} = \frac{(230)^2}{36} = 1469\ \Omega \textbf{ Ans.}$$

Example 3.34.

Fig. 3.12 shows the component part of a music system. A low voltage music source, such as a phonograph cartridge, delivers a small (millivolt) signal to an audio amplifier. The amplifier boosts the signal to a higher level and delivers electrical power to speaker. The speaker converts electrical power into two forms (1) accoustic vibrations (sound), (2) heating of the air in the speaker enclosure. The resistance of most high fidelity speakers is about 8 Ω.

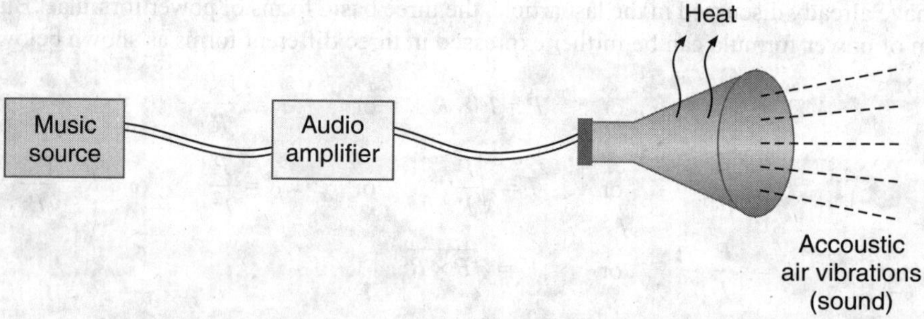

Fig. 3.12.

(*a*) *If the speaker has a maximum continuous power handling capacity of 60 W, what is the maximum continuous current that it can carry ?*

(*b*) *If the original speaker is replaced by a more rugged speaker having a maximum continuous power rating of 120 W, what is the new maximum continuous current carrying capacity?*

Solution.

Given: Resistance of the speaker (R) = 8 Ω and powery (P) = 60 W.

(*a*) *Maximum continuous current*

We know that the current,

$$I = \frac{\sqrt{P}}{R} = \frac{\sqrt{60}}{8} = 2.74 \text{ A } \textbf{Ans.}$$

(*b*) *Maximum new current*

In this case resistance of the speaker (R) = 8 Ω and power handling capacity of the speaker (P) = 120 W

We know that the new current

$$I = \frac{\sqrt{P}}{R} = \frac{\sqrt{120}}{8} = 3.87 \text{ A } \textbf{Ans.}$$

Example 3.35.

If a current through the fixed resistor is increased by 50%, then find the percentage increase in the power dissipation of the resistor.

Fixed resistor 30 k ohm 5%.

Solution.

Given: Increase in current = 50%

Let I = Value of the previous current, and

I' = New value of current (*i.e.*, after the previous current is increased by 50%) equal to 1.5 I

We know that power dissipation of the resistor with previous value of current,

$$P = I^2 \cdot R \quad ...(i)$$

and power dissipation with the increased value of current,

$$P' = I'^2 \cdot R = (1.5\, I)^2 = 2.25\, (I^2R) \quad ...(ii)$$

Percentage increase in the power dissipation

$$= \frac{P' - P}{P} \times 100 = \frac{2.25(I^2 R) - I^2 R}{I^2 R} \times 100$$

$$= \frac{2.25 - 1}{1} \times 100 = 125\% \textbf{ Ans.}$$

Example 3.36.

A 6 V battery with a capacity of 2 Ah is used to energise a torchlight rated at 1.2 W. If the torchlight is turned ON, for how long will it last?

Solution.

Given: Battery Voltage, V = 6 Volts, Capacity of the battery = 2 Ah and power rating of torchlight = 1.2 W.

Let R = Resistance of the torchlight and

t = time in hours which it will last.

We know that power rating of a torch light (P)

$$1.2 = \frac{V^2}{R} = \frac{(6)^2}{R} = \frac{36}{R} \quad \text{or} \quad R = \frac{36}{1.2} = 30\ \Omega$$

∴ Current drawn by the bulb of the torchlight from the battery.

$$I = \frac{V}{R} = \frac{6}{30} = 0.2 \text{ A}$$

We also know that capacity of a battery is basically a charge stored on the battery, which is actually a product of current (A) and time (in hours). Therefore, capacity of the battery,

$$2\ Ah = I \times t = 0.2 \times t$$

$$t = 2/0.2 = 10 \text{ hours } \textbf{Ans.}$$

3.30. Electric Shock

An electric shock is a potential source of injury to any one working in electrical and electronics industry. It has been experienced, when you touch a live conductor, while the power is on. The shock is a sudden involuntary contraction of the muscles with a feeling of pain, caused by the current through the body. If a shock is large enough, it can be fatal also. Therefore safety must always be the rule.

It has been found that the greatest hazard is from high voltage circuits that can supply appreciable amount of power. The resistance of a human body is also an important factor. If you hold a conducting wire in each hand, the resistance of the body across the conductors is 10 to 50,000 Ω. Holding the conductors tighter, lowers the body resistance. If we hold only one conductor, our body resistance is much higher. As a matter of fact, higher the resistance, smaller is the current that can flow through our body.

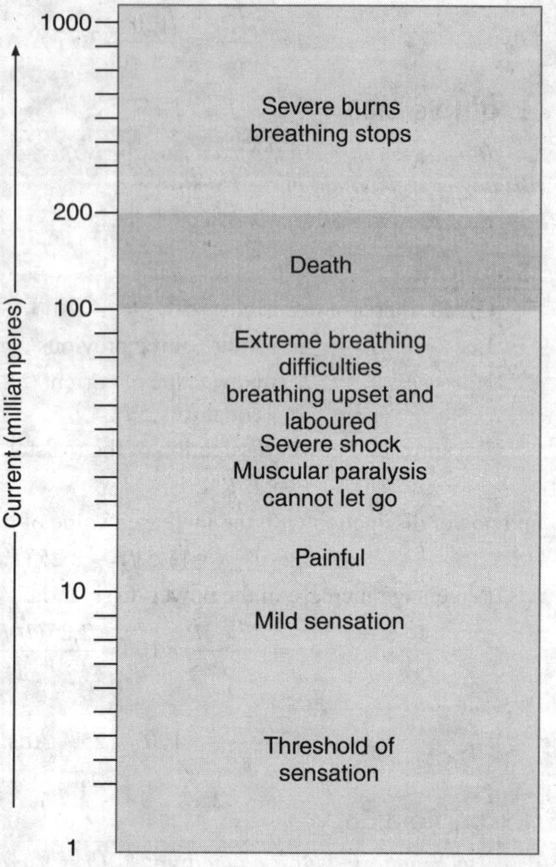

Fig. 3.13. Physiological effects of current.

The physiological effects of electric current can generally be predicted by a chart as shown in Fig. 3.13. It may be noted that it is the current that does the damage. As seen from the chart, the currents above 100 milliamperes (or only one-tenth of an ampere) are fatal. A workman who has contacted currents above 200 milliamperes may live to see another day if given rapid treatment. Currents below 100 milliamperes can be serious and painful.

Thus we must exercise extreme caution, while working on live circuits. Even circuits, operating with low voltage, should be carefully handled. Some of the basic safety procedures for avoiding electric shock and safe working are as follows:

1. Wherever possible, shut the circuit OFF by unplugging the power source before working on an equipment.
2. Before working on a circuit, study the circuit diagram, read the operation manual and pay more attention to the references on the safety hazards.
3. Never assume that a circuit is OFF, even when a switch is open or OFF connected to the earth.
4. Keep your self insulated from the earth, when working on power line circuits, because one side of the line is, usually, connected to the earth. The metal chasis of radio and television receivers is often connected to the power ground.
5. Some resistors in electronic circuits get very hot especially those which carry high currents. Watch those 5 and 10 W resistors. They will burn the skin of your fingers. Stay away from them until they cool off.

6. Be careful for all capacitors, which may still retain some charge. Not only can you get a dangerous and sometimes fatal shock, you may also get a burn from electric discharge. If the rated voltage of electrolytic capacitors is exceeded or their polarities reversed, they may get very hot and may actually burst.
7. Watch that hot soldering iron or gun. Don't place it on the bench where your arm might accidently hit it. Never keep it awaym, while still hot. Some innocent unsuspecting person may pick it up.
8. Hot solder can be particularly uncomfortable in contact with your skin. Wait for soldered joints to cool. When desoldering joints, don't shake hot solder off, so that you or your neighbour might get hit in the eyes or on his clothes or body.

NUMERICAL PROBLEMS

1. What is the amount of energy gained by 10 C of charge when passing through a 9 V battery.? **(Ans. 90 J)**
2. If a battery provides 80 J of work to 16 C of charge, find the terminal voltage of the battery. **(Ans. 5 V)**
3. Find the amount of charge that can be separated by 1.5 V battery, if the energy produced by the chemical reaction is 7.5 J. **(Ans. 5 C)**
4. A piece of silver wire has a resistance of 1 Ω. What will be the resistance of a manganin wire of one-third length and one-third diameter, if resistivity of manganin is 30 times than that of silver? **(Ans. 90 Ω)**
5. The resistivity of an alloy is 40×10^{-8} Ω-m. A sheet of this material is 10 cm long, 5 cm wide and 0.14 mm thick. Determine resistance between the) opposite ends. **(Ans. 80×10^{-12} Ω)**
6. A piece of aluminium wire 100 m long has a square cross-section of 5 mm on a side. Calculate its resistance. Take ρ as 2.83×10^{-8} Ω-m at 20°C. **(Ans. 0.113 Ω)**
7. Find the resistance at 20°C of (*a*) a copper wire 1 m long with a circular cross-section diameter 1 mm. (*b*) a constantan wire 2.5 m long with a circular cross-section diameter of 1 mm. Take resistivity for copper = 1.72×10^{-8} Ω-m and for constantan, = 49×10^{-8} Ω-m. **(Ans. 0.547 Ω, 1.56 Ω)**
8. Determine the resistance from end to end of a cylinder of intrinsic silicon at 20°C, if it is 2 cm long and 4 cm in diameter. Take, resistivity of intrinsic silicon at 20°C as 640 Ω-m. **(Ans. 10.2 kΩ)**
9. The copper foil conductor of a printed circuit board is 20.3 cm in length, 0.76 mm in width and 5.08×10^{-2} mm in thickness. Assuming ρ = 1.72×10^{-8} Ω-m.at 20°, determine the resistance at copper foil conductor. **(Ans. 90.5 mΩ)**
10. Two resistances of field coils with copper conductors of a dynamo is 120 Ω at 25°C. After working for 6 hours on full load, the resistance of the coil increases to 140 Ω. Calculate the mean temperature rise of the field coil. Take temperature coefficient of the copper material as 0.0039 at 20°C. **(Ans. 43.8°)**
11. Calculate the voltage drop across a 560 Ω resistance when the current flowing in it is 20 mA. **(Ans. 1.120 V)**
12. What is the resistance of a lamp, if a 6 V battery results in a 100 mA current flow in it? **(Ans. 60 Ω)**
13. If the voltage drop across a 2 kΩ resistor is 8 V, what is the current flowing in the resistor? **(Ans. 4 mA)**

14. A certain radio receiver presents an effective resistance of 300 Ω to the input signal on the antena downlead. The signal voltage is 700 μV. How much signal current will flow in the antena downlead? **(Ans. 2.3 μA)**

15. A certain speaker has a resistance of 4 Ω. To produce barely audible sound, the speaker must carry 150 mA current. Find the minimum voltage to drive the speaker. **(Ans. 600 mV)**

16. Determine the resistance of a carbon rod that has 20 mA of current in the rod with a voltage of 100 V across the rod. **(Ans. 5 kΩ)**

17. Determine the power delivered to a 15 ampere air conditioner, if it has a resistance of 15Ω. **(Ans. 3.4 kΩ)**

DESCRIPTIVE QUESTIONS

1. Briefly define each of the following, giving its unit and symbol:Charge, Potential, Potential difference, Current resistance and Conductance.
2. What is the essential difference between direct current and alternating current?
3. State and explain the three versions of Ohm's Law relating voltage, current and resistance.
4. Derive the three formulae for electric power.
5. State two safety precautions to follow when working on electric circuits.

MULTIPLE CHOICE QUESTIONS

1. In a certain wire, 4 C of charge moves past a point in 0.5 second. The current in amperes is
(*a*) 8 A (*b*) 4 A (*c*) 2 A (*d*) 1 A

2. The resistance of a conductor is
(*a*) directly proportional to its area of cross-section,
(*b*) inversely proportional to its length,
(*c*) dependent upon the nature of the material
(*d*) all of them.

3. The temperature coefficient of resistance is
(*a*) positive for semiconductors and insulators
(*b*) negative for conductors and semiconductors
(*c*) negative for semiconductors and insulators
(*d*) none.

4. Which of the following statement is not made by Ohm's Law?
(*a*) current is directly proportional to voltage,
(*b*) current is inversely proportional to resistance,
(*c*) current is directly proportional to the resistance
(*d*) none of these.

5. The unit for power is
(*a*) joules (*b*) joules-sec (*c*) joules/sec (*d*) joules/min

ANSWERS

1. (*a*) **2.** (*c*) **3.** (*c*) **4.** (*c*) **5.** (*c*)

4

Chapter

D.C. Resistive Circuits

Outline

4.1. Introduction

We have already discussed in the last chapter that an electrical (or an electronic) circuit is a combination of electrical (or an electronic) devices, components and conductors. The main function of such a circuit is to fulfil a desired requirement such as amplification, filtering or oscillation. The various types of electrical (or electronic) circuits may be divided under the following two heads:

1. ***D.C. circuits.*** These are the circuits in which the current moves along one direction only at all times.
2. ***A.C. circuits.*** These are the circuits in which the current reverses its direction periodically with the time.

In this book, we shall discuss only the d.c. circuits. The d.c. circuits consisting of resistances only are called d.c. resistive circuits. These circuits are of the following three types:

1. Series circuits, **2.** Parallel circuits, and **3.** Series-parallel circuits.

It is very essential to understand the following three important factors of a d.c. resistive circuit:

1. Current direction and voltage polarity of circuit components,

2. Voltage notation, and
3. Voltage reference point.

4.2. Current Direction and Voltage Polarity of Circuit Components

The current direction and voltage polarity of electric components such as resistances, inductance coils and capacitors is important to study the behaviour of electric circuits. We know that the conventional direction of a current flow is that it flows out from a positive terminal of battery, passes through the circuit components and finally returns to the negative terminal as shown in Fig. 4.1 (*a*).

We also know that if a current flows through a resistance, it produces a voltage drop (called I.R. drop) across it. This voltage drop means that there exists a potential difference across two ends of a resistance. The end, through which the conventional current enters the resistance, is taken as positive and the other end as negative as shown in Fig. 4.1 (*b*).

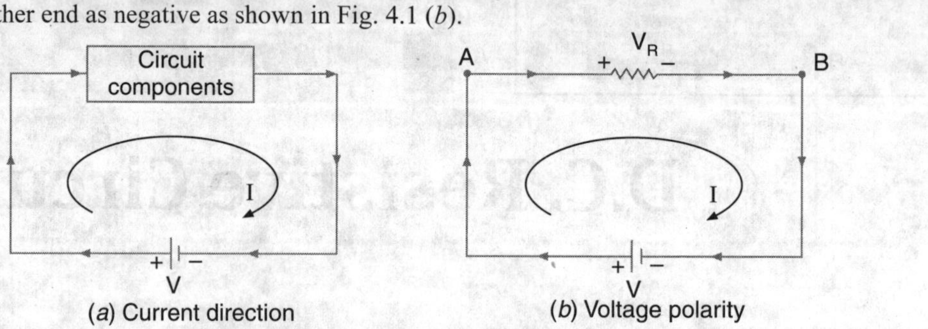

Fig. 4.1.

4.3. Voltage Notation

There is a system of notation for indicating both the magnitude and polarity of the voltages in an electric circuit. Consider a simple circuit consisting of three resistances and a voltage source connected as shown in Fig. 4.2. The voltage drops across three resistances are shown in the figure with plus (+) and minus (–) signs. The symbols *A*, *B*, *C* and *D* represent the nodes. These nodes are used to form a symbolic representation or a notation for the voltages. The notation is indicated by using a double letter subscript with a letter *V*, *e.g.*, V_{AB}, V_{CD} and V_{DA}. The notation V_{AB} is read as a voltage from node *A* to node *B*. Similarly, the notation V_{BC} is read as a voltage from node *B* to node *C* and and so on. In this system of notation, the second subscript letter always stands for a reference point. Thus for a voltage V_{AB}, the node *B* stands for a reference point.

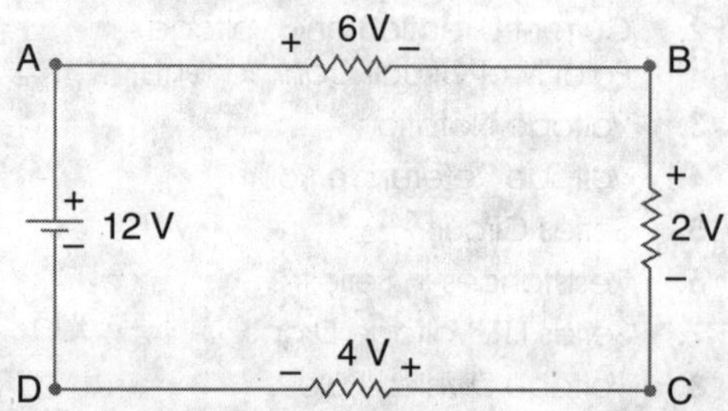

Fig. 4.2.

The polarity of a designated voltage is assigned a positive sign, if the first node is positive with respect to the second node. For example, the voltage V_{AB} is designated as + 6 V, because the node *A* is positive with respect to node *B*. The polarity of a designated voltage is assigned a negative sign, if the first node is negative with respect to the second node. For example, the voltage V_{BA} is designated as – 6 V, because the node *B* is negative with respect to the node *A*.

4.4. Voltage Reference Point

The voltages in electric circuits are measured from various points with respect to a common reference point. This reference point may be metal chassis or a particular conductor as in printed wiring circuits. In many circuits, the reference point may be connected to an earth ground through wiring of the power supply cord.

There are several schematic symbols, which are used to indicate the reference point for voltage measurements. But the three symbols as shown in Fig. 4.3 (*a*), (*b*), and (*c*) are more common. The symbol shown in Fig. 4.3(*a*) indicates a physical connection of the common connection point (*i.e.,* the reference point) in the circuit to the metal chassis. The symbol shown in Fig. 4.3 (*b*) indicates a physical connection of the common connection point to the earth. This symbol indicates that the reference point is grounded. The symbol shown in Fig. 4.3 (*c*) indicates only a common connection point, which is the reference point for voltage measurements. This symbol does not indicate a chassis or a ground connection.

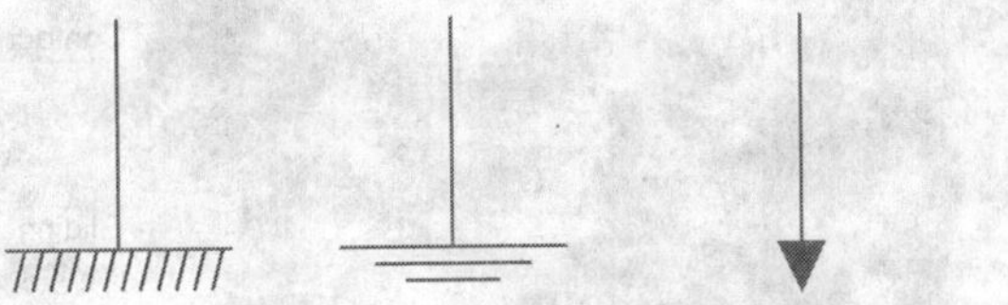
(*a*) Metal chassis (*b*) Earth (*c*) Reference point

Fig. 4.3.

4.5. Series Circuit

A circuit, in which various resistances are connected across a source in such a way that there is only one path for the current to flow, is called a series circuit. Fig. 4.4 (*a*) shows a simple circuit in which three resistances R_1, R_2 and R_3 are connected in series across a battery. It may be noted that the series circuit has only two wires at each node. One of the wires carries the current entering the node and the other carries the current leaving that node.

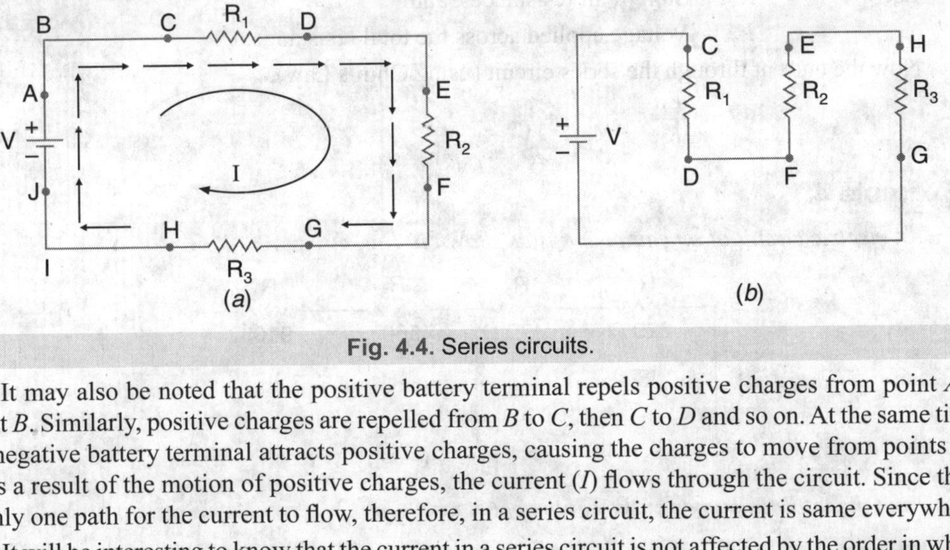

(*a*) (*b*)

Fig. 4.4. Series circuits.

It may also be noted that the positive battery terminal repels positive charges from point *A* to point *B*. Similarly, positive charges are repelled from *B* to *C*, then *C* to *D* and so on. At the same time, the negative battery terminal attracts positive charges, causing the charges to move from points *I* to *J*. As a result of the motion of positive charges, the current (*I*) flows through the circuit. Since there is only one path for the current to flow, therefore, in a series circuit, the current is same everywhere.

It will be interesting to know that the current in a series circuit is not affected by the order in which the components are connected in series. For example, R_2 is connected in a reverse order compared with R_1 and R_3 as shown in Fig. 4.4 (*b*). But they are in series. The current through each resistance is the same, because there is only one path.

Fig. 4.5 (*a*) shows a torch light. If you carefully observe the internal arrangement of a sliding switch, bulb and batteries, you will find that they all are connected as a simple series circuit as shown in Fig. 4.5 (*b*).

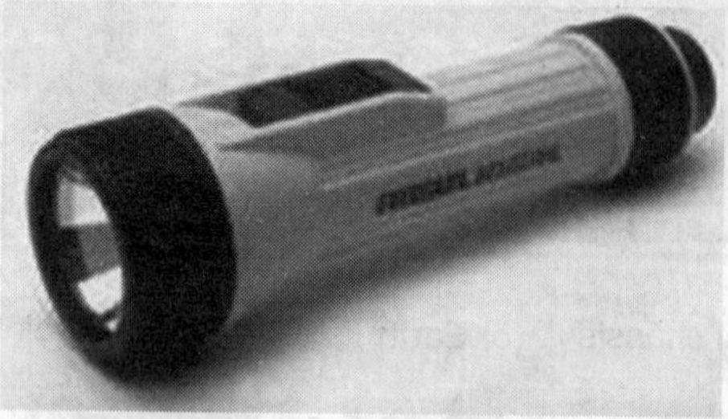

(a)

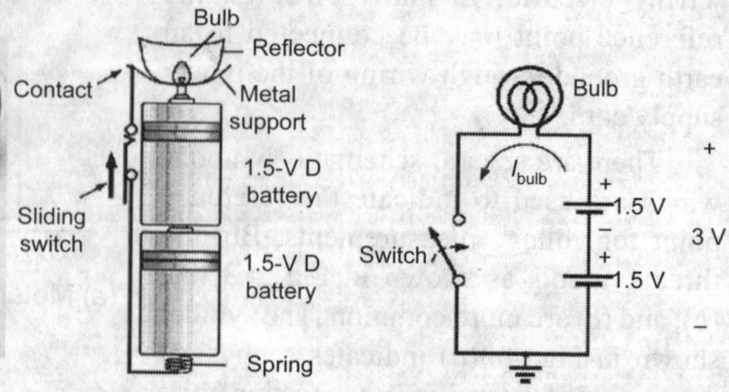

(b)

Fig. 4.5.

4.6. Resistances in Series

Consider a series circuit connected across a voltage source as shown in Fig. 4.4 (*a*). We know that in such a circuit, the current is same through all the resistances. Therefore, the total resistance (also called equivalent resistance) across the voltage source is the sum of individual resistances. Mathematically total resistance,

$$R_T = R_1 + R_2 + R_3 + \ldots\ldots$$

The above relation is called series resistance formula. It is applicable to any number of resistances, whether equal or not, as long as they are in the same series circuit.

Let R_T = Sum of all resistances, and

V_T = Voltage applied across the total resistance.

Now the current through the series circuit (using Ohm's Law),

$$I = \frac{V_T}{R_T}$$

Example 4.1.

An electric circuit of an equipment is as shown in Fig. 4.6.

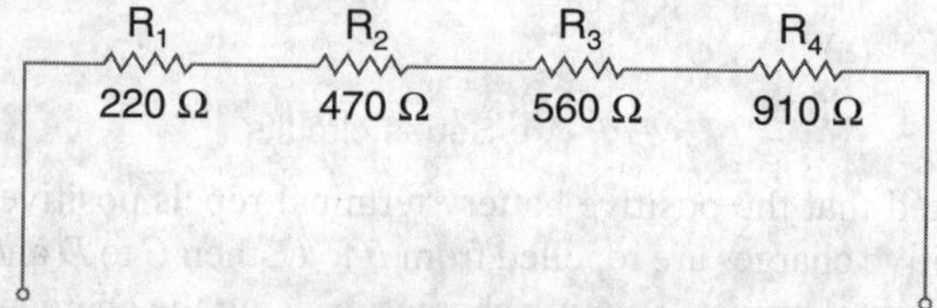

Fig. 4.6.

Find the total resistance of the equipment.

Solution.

Given: $R_1 = 220\ \Omega$; $R_2 = 470\ \Omega$; $R_3 = 560\ \Omega$ and $R_4 = 910\ \Omega$.

We know that total resistance of the equipment (or circuit),

$$R_T = R_1 + R_2 + R_3 + R_4 = 220 + 470 + 560 + 910\ \Omega = 2160\ \Omega \textbf{ Ans.}$$

Example 4.2.

Fig. 4.7 shows a circuit in which three resistances with different values of R_1, R_2 and R_3.

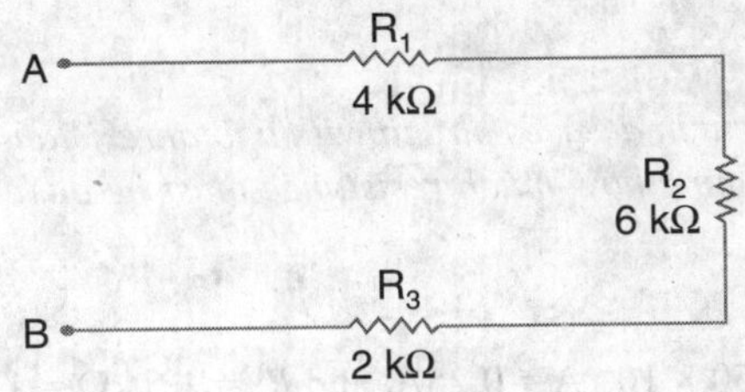

Fig. 4.7.

Find the equivalent resistance between the terminals A and B of this circuit.

Solution.

Given: $R_1 = 4$ kΩ; $R_2 = 6$ kΩ and $R_3 = 2$ kΩ.

We know that equivalent resistance between terminal *A* and *B*,

$$R_T = R_1 + R_2 + R_3 = 4 + 6 + 2 = 12 \text{ k}\Omega \textbf{ Ans.}$$

Example 4.3.

The series circuit of Fig. 4.8 contains three resistances, one of which is unknown.

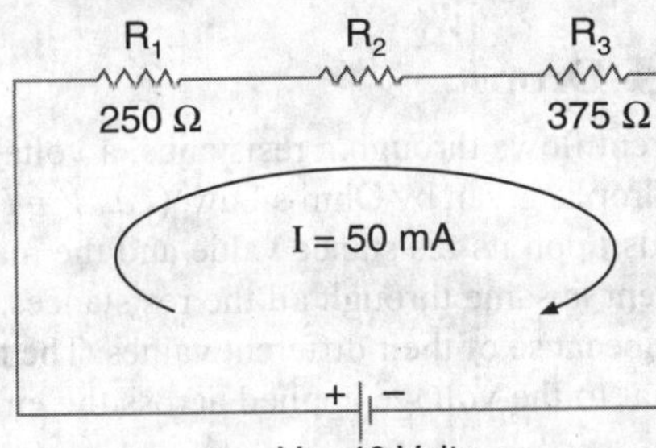

Fig. 4.8.

From the other information provided, find the total resistance in the circuit and the value of unknown resistance R_2.

Solution.

Given: $V = 40$ Volts, $I = 50$ mA $= 50 \times 10^{-3}$A; $R_1 = 250\ \Omega$ and $R_3 = 375\ \Omega$.

Total resistance in the circuit

We know that current in a series circuit (*I*),

$$50 \times 10^{-3} = \frac{V}{R_T} = \frac{40}{R_T}$$

$$\therefore \quad R_T = \frac{40}{50 \times 10^{-3}} = 800\ \Omega \textbf{ Ans.}$$

Unknown resistance R_2

We also know that total resistance (R_T)

$$800 = R_1 + R_2 + R_3 = 250 + R_2 + 375 = R_2 + 625$$

$$\therefore \quad R_2 = 800 - 625 = 175\ \Omega \textbf{ Ans.}$$

Example 4.4.

There is a 250 mA current in a series circuit with a total resistance of 1.5 kΩ. The current must be reduced by 25%. Determine how much resistance must be added in order to accomplish this reduction in current.

Solution.

Given: $I = 250\text{ mA} = 250 \times 10^{-3}\text{A} = 0.25\text{A}$, and $R_T = 1.5\text{ k}\Omega = 1.5 \times 10^3\ \Omega = 1500\ \Omega$

Let V_S = Applied voltage across the circuit,

We know that the current through the circuit, (*I*),

$$0.25 = \frac{V_S}{R_T} = \frac{V_S}{1500} \quad \text{or} \quad V_S = 0.25 \times 1500 = 375\text{ V}$$

Since the current is to be reduced by 2.5%, therefore the new value of current in the circuit,

$$I' = 0.75 \times 0.25 = 0.1875\text{ A}$$

Total resistance in the circuit,

$$R' = \frac{V_S}{I'} = \frac{375}{0.1875} = 2000\ \Omega.$$

and value of the resistance in the circuit which must be added in order to reduce the current,

$$R' = R' - R_T = 2000 - 1500 = 500\ \Omega \textbf{ Ans.}$$

4.7. Series I.R. Voltage Drops

We know that when a current flows through a resistance, a voltage drop is produced across it. The magnitude of this voltage drop is given by Ohm's Law, (*i.e.*, $V = I \cdot R$). This implies that voltage drop across a resistance depends upon its resistance value and the magnitude of current. We know that in a series circuit, the current is same through all the resistances. But the voltage drops across the resistances will be different because of their different values. The total voltage drop (*i.e.*, sum of individual voltage drops is equal to the voltage applied across the circuit. Consider a series circuit as shown in Fig. 4.4 (*a*).

Let V_1 = Voltage drop across R_1 $(I \cdot R_1)$

V_2 = Voltage drop across R_2 $(I \cdot R_2)$, and

V_3 = Voltage drop across R_3 $(I \cdot R_3)$

Then total voltage drop,

$$= V_1 + V_2 + V_3$$

Thus, in general for a series circuit, the applied voltage is equal to the total or sum of the individual *I. R.* drops. Mathematically the applied voltage,

$$V_T = V_1 + V_2 + V_3 + \ldots\ldots$$

Example 4.5.

Three resistances are connected in series in a circuit as shown in Fig. 4.9. If the voltage drop across R_2 is 6 V, what are the voltage drops across other resistances?

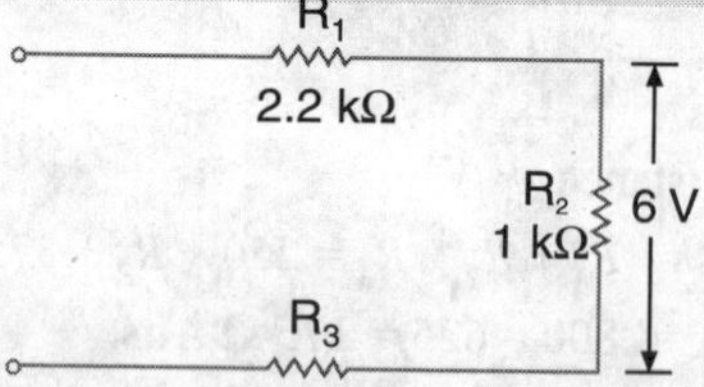

Fig. 4.9.

Solution.

Given: $R_1 = 2.2\ \text{k}\Omega$; $R_2 = 1\ \text{k}\Omega$; $R_3 = 3.3\ \text{k}\Omega$ and voltage drop across R_2 $(V_2) = 6$ V.

Let, V_1 = Voltage drop across resistance R_1, and

V_2 = Voltage drop across resistance R_2.

We know that current through resistance R_2,

$$I = \frac{V_2}{R_2} = \frac{6}{1 \times 10^3} = 6 \times 10^{-3}\ \text{A} = 6\ \text{mA}$$

Since current in the series circuit the same everywhere, therefore current through R_1 and R_3 is the same as that of R_2.

∴ Voltage drop across R_3.

$$V_3 = I \cdot R_3 = 6 \times 10^{-3} \times 2.2 \times 10^3 = 13.2\ \text{V}\ \textbf{Ans.}$$

and voltage drop across $R_2V_2 = I \cdot R_2 = 6 \times 10^{-3} \times 3.3 \times 10^3 = 19.8\ \text{V}$ **Ans.**

4.8. Voltage Division Rule

The voltage division rule states that voltage drop across any resistance (or a combination of resistances) in a series circuit is equal to the ratio of that resistance to the total resistance multiplied by the source voltage. Mathematically the voltage drop

$$V_x = \frac{R_x}{R_T} \times V$$

where R_x = Any resistance,

R_T = Total resistance of a circuit, and

V = Source voltage.

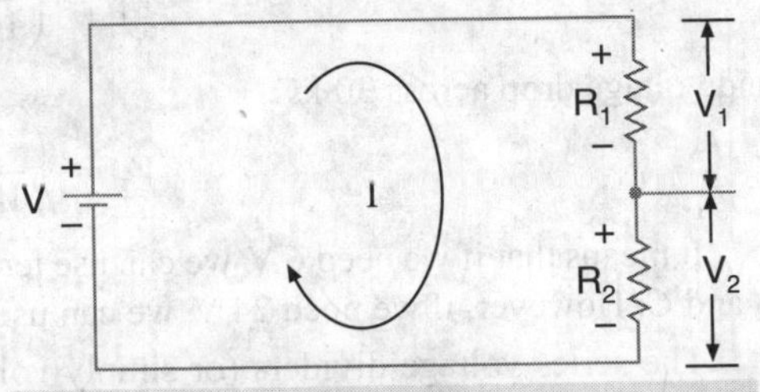

Fig. 4.10.

Consider a simple series circuit consisting of two resistances R_1 and R_2 as shown in Fig. 4.10. We know that total resistance of such a circuit is $R_T + R_1 + R_2$. Then according to voltage division rule, the voltage drop across R_1,

$$V_1 = \left(\frac{R_1}{R_1 + R_2}\right) \times V$$

and voltage drop across R_2,

$$V_2 = \left(\frac{R_2}{R_1 + R_2}\right) \times V$$

4.9. Series Voltage Divider

A series voltage divider consists of two or more resistances R_1, R_2... etc. connected in series across a voltage source or battery as shown in Fig. 4.11 (*a*). The current flowing through these resistances gives rise to the voltage drops V_1, V_2 ... etc., which are proportional to their resistances values. If we need voltage V_1 only we can use terminals *A* and *B*. However, if we need voltage V_2, we can use terminals *B* and *C*. On the other hand, if we require the full source voltage, we can use terminal *A* and *C*.

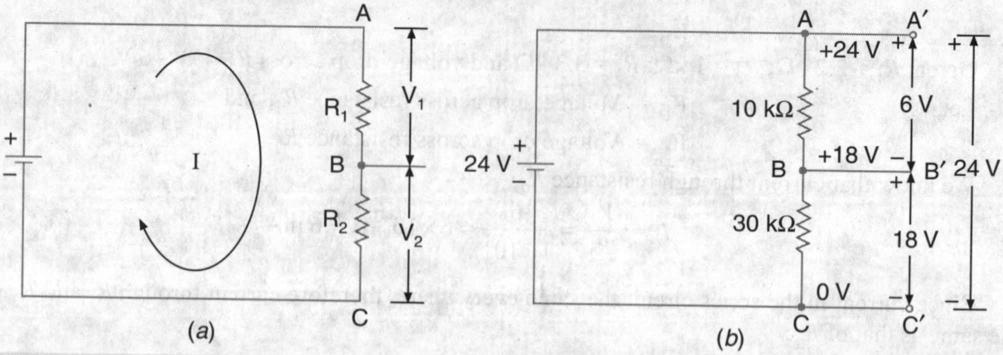

Fig. 4.11. Series voltage.

Consider a series voltage divider circuit as shown in Fig. 4.11 (*b*). In this circuit, two resistances of 10 kΩ and 30 kΩ are connected in series across a 24 V source. Now according to voltage division rule, voltage drop across 10 kΩ,

$$V_1 = \left(\frac{10}{10+30}\right) 24 = 6 \text{ V}$$

and voltage drop across 30 kΩ,

$$V_2 = \left(\frac{30}{10+30}\right) 24 = 18 \text{ V}$$

It means that if we need 6 V, we can use terminal *A* and *B*. If we need 18 V we can use terminals *B* and *C*. However, if we need 24 V, we can use terminals *A* and *C*.

The series voltage dividers (or simply voltage dividers) are widely used in amplifiers, where many different voltage levels are required for many solid state devices. The voltage divider allows to provide different voltage levels from a single power supply. This ensures proper circuit operation at a minimal cost.

Example 4.6.

Fig. 4.12 shows a voltage divider circuit with resistance R_1, R_2 and R_3 connected across a 30 V battery.

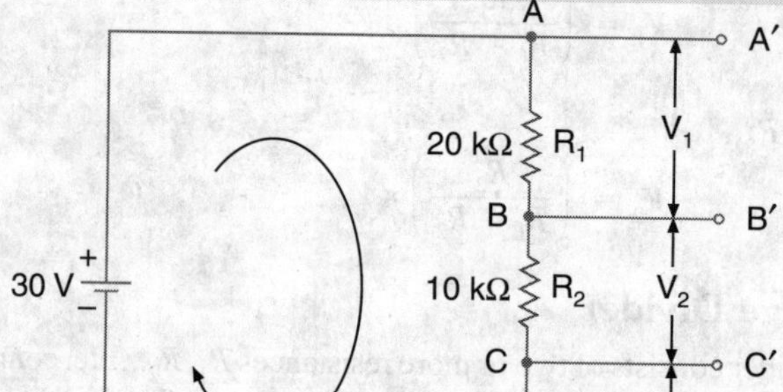

Fig. 4.12.

Determine (a) the voltage drops across R_1, R_2 and R_3 and (b) the voltage at each node.

Solution.

Given: $V = 30$ Volts; $R_1 = 20\ k\Omega$; $R_2 = 10\ k\Omega$ and $R_3 = 70\ k\Omega$.

We know that total resistance of the circuit,

$$R_T = R_1 + R_2 + R_3 = 20 + 10 + 70 = 100\ k\Omega$$

(*a*) *Voltage drop across R_1, R_2 and R_3*

We know that voltage drop across R_1

$$V_1 = \left(\frac{R_1}{R_T}\right)V = \left(\frac{20}{100}\right) \times 30 = 6 \text{ V } \textbf{Ans.}$$

Similarly, voltage drop across R_2,

$$V_2 = \left(\frac{R_2}{R_T}\right)V = \left(\frac{10}{100}\right) \times 30 = 3 \text{ V } \textbf{Ans.}$$

and voltage drop across R_3,

$$V_3 = \left(\frac{R_3}{R_T}\right)V = \left(\frac{70}{100}\right) \times 30 = 21 \text{ V } \textbf{Ans.}$$

(*b*) *Voltage at each node*

We also know that voltage at node *C*,

$$V_C = V_3 + V_D = 21 + 0 = 21 \text{ V } \textbf{Ans.}$$

Similarly, voltages at node *B*

$$V_B = V_2 + V_C = 3 + 21 = 24 \text{ V } \textbf{Ans.}$$

and voltage at node *A*.

$$V_A = V_1 + V_B = 6 + 24 = 30 \text{ V } \textbf{Ans.}$$

4.10. Total Power in a Series Circuit

The power needed to produce current in each series resistance is used up in the form of heat. Therefore, the total power used is the sum of the individual values of power dissipated in each component of the circuit. Mathematically, total power,

$$P_T = P_1 + P_2 + P_3 + ...$$

where $P_1, P_2, P_3......$ = Individual values of power dissipated in each component of the circuit.

The total power dissipated in the circuit is always equal to the power generated by the source.

Example 4.7.

Two resistance R_1 and R_2 are connected in series. How much resistance R_1 must be added in series with a 100 Ω R_2 to limit the current to 0.3 A with a 120 V applied ? Show the schematic diagram. How much power is dissipated in each resistance.

Solution.

Given: $R_2 = 100\ \Omega$ $I = 0.3$ A and $V_T = 120$ V.

Value of the each resistance

Let R_T = Total series resistance.

We know that current (I),

$$0.3 = \frac{V_T}{R_T} = \frac{120}{R_T}$$

$$\therefore \quad R_T = 120/0.3 = 400\ \Omega$$

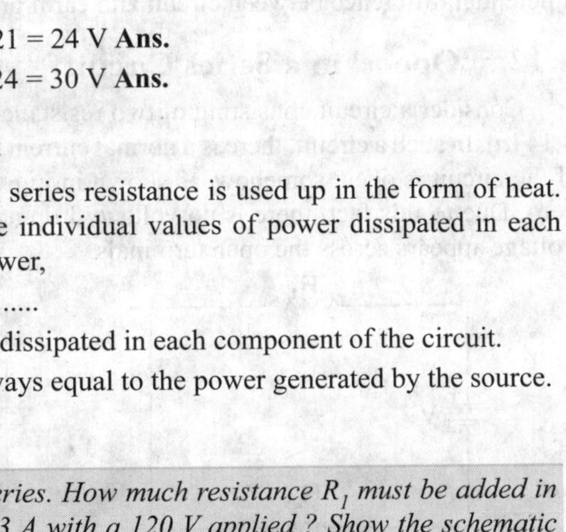

Fig. 4.13.

and total series resistance (R_T)

$$400 = R_1 + R_2 = R_1 + 100$$

$$R_1 = 400 - 100 = 300\ \Omega \textbf{ Ans.}$$

Power dissipated in each resitance.

We also know that the power dissipated by resistance R_1,

$$P_1 = I^2 \times R_1 = (0.3)^2 \times 300 = 27 \text{ W} \textbf{ Ans.}$$

and the power dissipated by resistance R_2,

$$P_2 = I^2 \times R_2 = (0.3)^2 \times 100 = 9 \text{ W} \textbf{ Ans.}$$

Figure 4.13 shows the schematic diagram illustrating R_1 and R_2 connected in series.

4.11. Circuit Ground

The term *ground* derives from the method used in ac power supply lines, in which one side of the line is neutralized by connecting it to a water pipe or a metal rod driven into the ground. This method of grounding is called **earth ground.** However, in most electronic equipments, a large conductive area on a printed circuit board or the metal chassis that houses the assembly is used as the common or reference point, called the circuit ground or chassis ground. This ground provides a convenient way of connecting all common points within the circuit back to one side of the battery or other energy source.

It may be noted that the circuit or chassis ground does not necessarily have to be connected to earth ground. However, in many cases it is earth grounded in order to prevent a shock hazard due to a potential difference between circuit and earth ground.

4.12. 'Opens' in a Series Circuit

Consider a circuit consisting of two resistances R_1 and R_2 connected in series as shown in Fig. 4.14 (*a*). In such a circuit, there is a normal current flow and voltage drops across both the resistances. If the circuit is open, somehow, as shown in Fig. 4.14 (*b*), the current through the circuit becomes zero. Due to this fact, there is no voltage drop across the resistances R_1 and R_2. All of the source voltage appears across the open terminals.

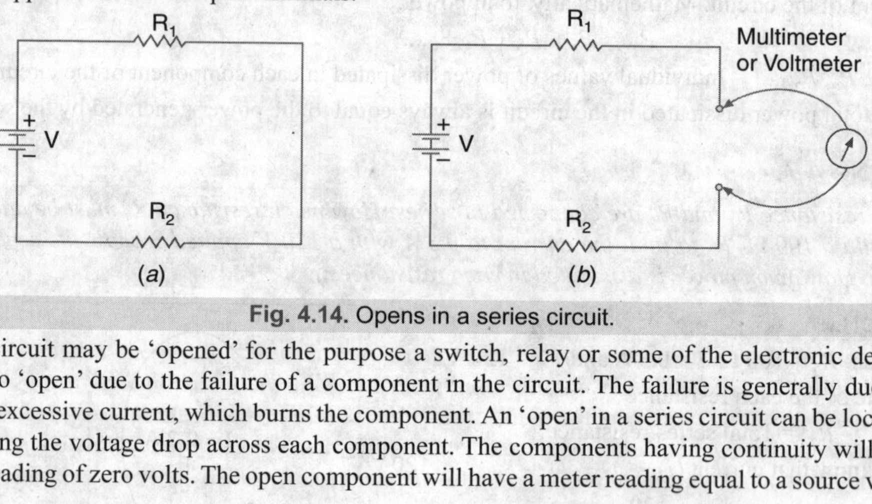

Fig. 4.14. Opens in a series circuit.

A circuit may be 'opened' for the purpose a switch, relay or some of the electronic device. It may also 'open' due to the failure of a component in the circuit. The failure is generally due to the flow of excessive current, which burns the component. An 'open' in a series circuit can be located by measuring the voltage drop across each component. The components having continuity will have a meter reading of zero volts. The open component will have a meter reading equal to a source voltage.

4.13. 'Shorts' in a Series Circuit

We have already discussed that a short in a series circuit represents zero resistance between the terminals, which are shortened. A short may also occur due to a component failure or wiring error in a circuit. It may cause some problem due to excessively high current or due to zero resistance, in the circuit. This, in turn, causes power dissipation to increase many times. It may. even, result in burning out the circuit components.

Example 4.8.

A circuit connected across 10 V battery has three resistances of 1Ω, 2Ω and 3Ω respectively. Determine the power dissipated in (a) the entire circuit; (b) the circuit when 3 Ω resistance is shortened, and (c) the circuit when both the 3Ω and 2Ω resistances are shortened.

Solution.

Given: $V = 10$ Volts; $R_1 = 1\ \Omega$; $R_2 = 2\ \Omega$; $R_3 = 3\ \Omega$.

(a) Power dissipated in the entire circuit

First of all, consider the entire circuit as shown in Fig. 4.15 (*a*). We know that total resistance,

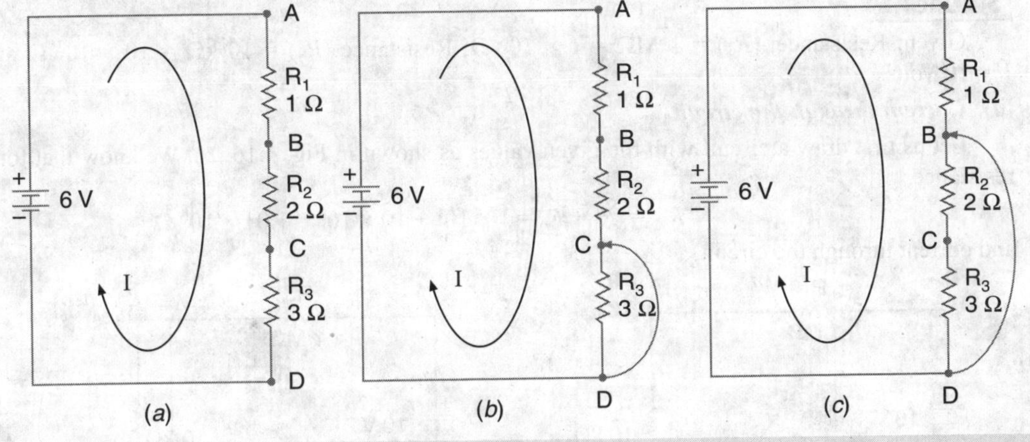

Fig. 4.15.

$$R_T = R_1 + R_2 + R_3 = 1 + 2 + 3 = 6\ \Omega$$

and the current,
$$I = \frac{V}{R_T} = \frac{6}{6} = 1\ \text{A}$$

∴ Power dissipated in the entire circuit,

$$P = I^2 \cdot R_T = (1)^2 \times 6 = 6\ \text{W}\ \textbf{Ans.}$$

(b) Power dissipated in the circuit when 3 Ω resistance is shortened

Now consider the circuit when 3 Ω resistance is shortened as shown in Fig. 4.15 (*b*). We know that the total resistance,

$$R_T = R_1 + R_2 = 1 + 2 = 3\ \Omega$$

and current,
$$I = \frac{V}{R} = \frac{6}{3} = 2\ \text{A}$$

∴ Power dissipated in the circuit,

$$P = I^2 \times R = (2)^2 \times 3 = 12\ \text{W}\ \textbf{Ans.}$$

(c) Power dissipated in the circuit when both the 3 Ω and 2 Ω resistances are shortened

Now consider the circuit when both the 3 Ω and 2 Ω resistances are shortened as shown in Fig. 4.15 (c). We know that the resistance,

$$R = R_1 = 1\ \Omega$$

and the current, $I = \frac{V}{R} = \frac{6}{1} = 6 \text{ A}$

$\therefore$ Power dissipated in the circuit,

$$P = I^2 \times R = (6)^2 \times 1 = 36 \text{ W } \textbf{Ans.}$$

It may be noted from the above answered that the current in a circuit, increases excessively because of a short circuit. As a result of this, some components get heated up and eventually burn out.

Example 4.9.

A 1 MΩ resistance is connected in series with a 10 kΩ resistance. The combination is connected across a 10 V battery. Determine (a) current through the circuit, (b)current, when 10 kΩ resistance is shortened in the circuit.

Solution.

Given: Resistance, (R_1) = 1 MΩ = 1 × 10^6 Ω; Resistance (R_2) = 10 kΩ = 10 × 10^3 Ω and V = 10 Volts.

(a) Current through the circuit

Let us first draw a circuit with the given values as shown in Fig. 4.16 (*a*). We know that total resistance,

$$R_T = R_1 + R_2 = 1 \times 10^6 + 10 \times 10^3 = 1.01 \times 10^6 \ \Omega$$

and current through the circuit,

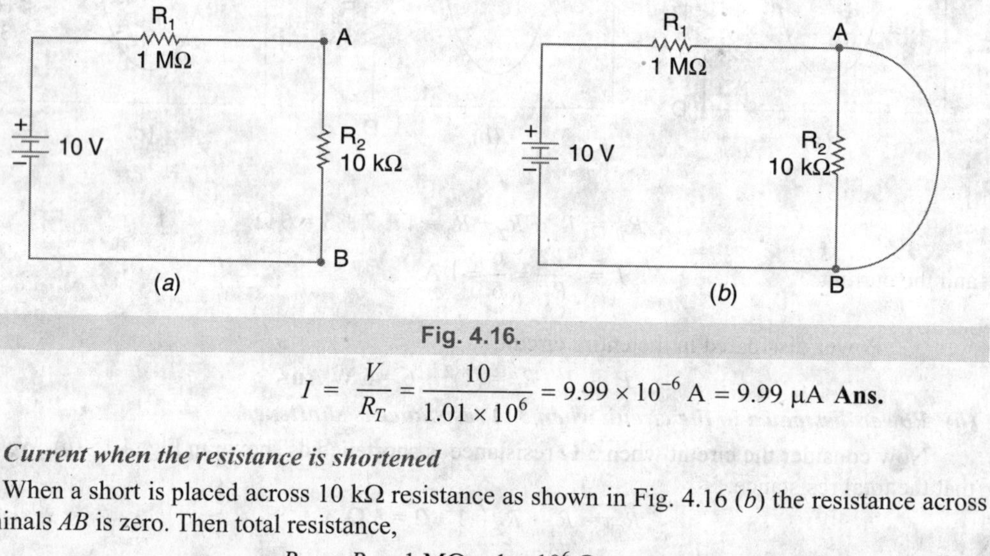

Fig. 4.16.

$$I = \frac{V}{R_T} = \frac{10}{1.01 \times 10^6} = 9.99 \times 10^{-6} \text{ A} = 9.99 \ \mu\text{A } \textbf{Ans.}$$

(b) Current when the resistance is shortened

When a short is placed across 10 kΩ resistance as shown in Fig. 4.16 (*b*) the resistance across terminals *AB* is zero. Then total resistance,

$$R_T = R_1 = 1 \text{ M}\Omega = 1 \times 10^6 \ \Omega$$

$\therefore$ Current in the circuit, $I = \frac{V}{R_T} = \frac{10}{1 \times 10^6} = 10 \times 10^{-6} \text{ A} = 10 \mu\text{A } \textbf{Ans.}$

Note: We see that short circuiting of 10 kΩ resistance keeps the circuit current almost the same as above. Therefore, we say that even if 10 kΩ resistance remains connected in the circuit, it does not play any useful role. In other words, a 10 kΩ resistance may be thought of as a short circuit in comparison to 1 MΩ resistance.

In general, for two series resistance, if the large resistance is 100 times the smaller resistance, the total resistance is virtually equal to the larger resistance.

4.14. Parallel Circuit

The circuit in which the various resistances are connected across the source in such a way that there is more than one path for the current to flow, is called a parallel circuit. Fig. 4.17 shows a simple circuit in which three resistances, R_1, R_2 and R_3 are connected in parallel across a battery.

It may be noted that the points, *A*, *B*, *C*, *E* and *G* are really equivalent to a direct connection to the positive terminal of a battery because the connecting wires have practically no resistance. Similarly, the points, *H*, *F*, *D*, *I* and *J* are the same as a direct connection at the negative terminal of the battery. Since the resistances R_1, R_2 and R_3 are directly connected across the two terminals of a battery, therefore all these resistances must have the same voltage as that of the battery.

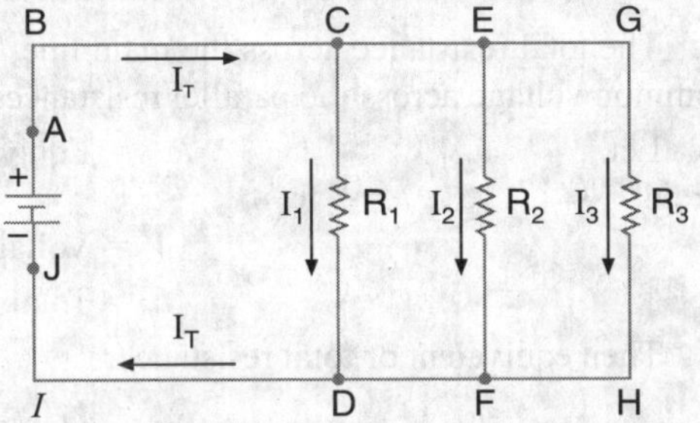

Fig. 4.17. Parallel circuit.

It is evident from the above discussion, that the voltage across all the components connected in parallel is *the same.* A common application of parallel circuits is house wiring, which is connected to the 230 V power line. The house wiring may be connected to many lights and appliances such as fans, toasters, geysers etc.

The current flowing in circuit of Fig. 4.17 has three separate paths. Each path is called a branch. Thus the circuit has three branches, namely *CD*, *EF* and *GH*. All the parallel branches are connected to a voltage source by a pair of the connecting leads, namely, *ABC* and *DIJ*. The pair of leads, connecting all the branches to the terminals of the voltage source, is called the *main line.*

According to Ohm's Law, the current is equal to the voltage applied across the circuit divided by the resistance between the two points where the voltage is applied.

Thus current through R_1,

$$I_1 = \frac{V}{R_1}$$

Similarly, current through R_2, $I_1 = \dfrac{V}{R_2}$

and current through R_3, $I_3 = \dfrac{V}{R_3}$

The total current in the main line, is equal to the sum of the individual branch currents. Mathematically, the total current,

$$I_T = I_1 + I_3 + I_1 + ...$$

The above formula applies for any number of parallel branches, whether the resistances are equal or not.

Example 4.10.

Three branches R_1, R_2 and R_3 are connected across a certain voltage supply and draw a total line current of 750 mA. The R_1 branch takes 200 mA, the R_3 branch takes 150 mA. What is the current drawn by R_2 branch?

Solution.

Given: $I_T = 750$ mA; $I_1 = 200$ mA and $I_3 - 150$ mA.

Let I_2 = Current drawn by R_2 branch.

We know that total line current (I_T),

$$750 = I_1 + I_2 + I_3 = 200 + I_2 + 150 = I_2 + 350$$

$$I_2 = 750 - 350 = 400 \text{ mA } \textbf{Ans.}$$

4.15. Resistances in a Parallel Circuit

The total resistance across the main line, in a parallel circuit, may be obtained by dividing the common voltage across the parallel resistances by the total current of all the branches.

Let R_{eq} = Equivalent resistance (*i.e.*, total resistance) of the parallel circuit,

V = Voltage across the branches, and

I_T = Total main line current.

Then equivalent or total resistance,

$$R_{eq} = V/I_T$$

It may be noted from the above relation that for a constant applied voltage, higher the value of I_T, lower is the value of R_{eq} and *vice versa.* If the value of main line current is not known, then equivalent resistance is obtained from a reciprocal resistance formula which is derived as below:

We know that total main line current I_T is the sum of all the individual branch currents. *i.e.*

$$\therefore \quad I_T = I_1 + I_2 + I_3 + ...$$

or

$$\frac{V}{R_{eq}} = \frac{V}{R_1} + \frac{V}{R_2} + \frac{V}{R_3} + ...$$

Dividing both sides by V,

$$\frac{1}{R_{eq}} = \frac{1}{R_1} + \frac{2}{R_2} + \frac{3}{R_3} + ...$$

or

$$R_{eq} = \frac{1}{\frac{1}{R_1} + \frac{I}{R_2} + \frac{1}{R_3} + ...}$$

The above relation is called reciprocal resistance formula, and is used for any number of parallel resistances.

Example 4.11.

A circuit consists of three resistances R_1, R_2 and R_3 connected in parallel as shown in Fig. 4.18. Determine equivalent resistance, and branch currents in the circuit.

Solution.

Given: $V = 12$ Volts; $R_1 = 4\ \Omega$; $R_2 = 6\ \Omega$ and $R_3 = 12\ \Omega$

Equivalent resistance

We know that equivalent resistance in the circuit,

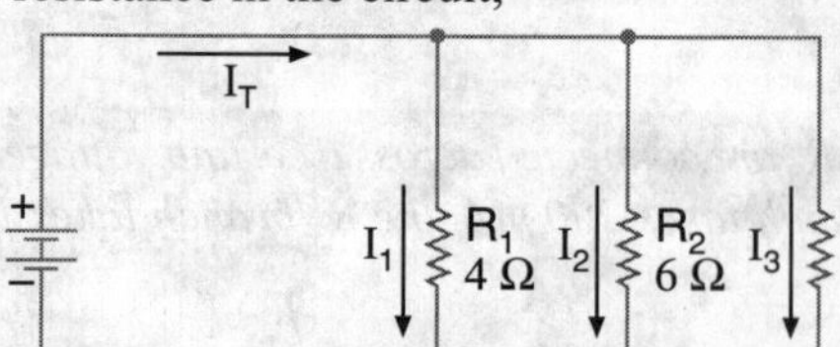

Fig. 4.18.

$$R_{eq} = \frac{1}{\frac{1}{R_1}+\frac{1}{R_2}+\frac{1}{R_3}} = \frac{1}{\frac{1}{4}+\frac{1}{6}+\frac{1}{12}} = 2\ \Omega\ \textbf{Ans.}$$

Branch currents

We also know that branch current through resistance R_1,

$$I_1 = \frac{V}{R_1} = \frac{12}{4} = 3\ \text{A}\ \textbf{Ans.}$$

Similarly, branch current through resistance R_2,

$$I_2 = \frac{V}{R_2} = \frac{12}{6} = 2\ \text{A}\ \textbf{Ans.}$$

and branch current through resistance R_3,

$$I_3 = \frac{V}{R_3} = \frac{12}{12} = 1\ \text{A}\ \textbf{Ans.}$$

4.16. Equivalent Resistance of a two Branched Circuit

It will be interesting to know that in the field of electronics, two branched parallel circuit is one of the most commonly used circuits. Such circuits are used in tuned amplifiers (used in television receivers and radios) and oscillators. Consider a simple two branched parallel circuit consisting of two resistances R_1 and R_2 as shown in Fig. 4.19. We know that equivalent resistance in this case,

$$R = \frac{1}{\frac{1}{R_1}+\frac{1}{R_2}} = \frac{R_1 \cdot R_2}{R_1 + R_2}$$

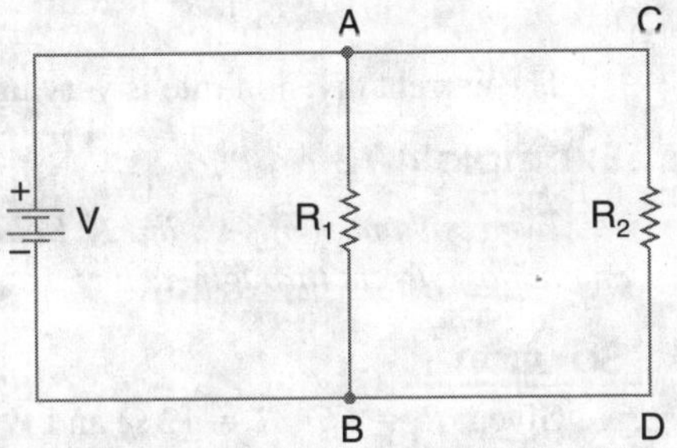

Fig. 4.19. Equivalent resistance of a two branched parallel circuit.

The above relation shows that equivalent resistance of a parallel circuit consisting of two resistances may be determined by the ratio of product of the resistances $(R_1 \cdot R_2)$ to the sum of two resistances $(R_1 + R_2)$. This relation is called *product-over-sum* formula. Sometimes, it is very convenient to represent the product-over-sum formula as:

$$R_{eq} = R_1 \parallel R_2$$

where || means a parallel combination resistance of R_1 and R_2.

Example 4.12.

Find the equivalent resistance of a two branched circuit, if $R_1 = R_2 = 10\ k\Omega$.

Solution.

Given: $R_1 = R_2 = 10\ \text{k}\Omega$.

We know that equivalent resistance,

$$R_{eq} = \frac{R_1 \cdot R_2}{R_1 + R_2} = \frac{10 \times 10}{10 + 10} = 5\ \text{k}\Omega\ \textbf{Ans.}$$

Notes: 1. If two resistances with equal values are connected in parallel, the equivalent resistance is reduced to half of the resistance value.

2. If a parallel circuit consists of many branches with equal resistances the value is divided by the number of branches. For example if there are three branches and every branch has 60 Ω resistance, then equivalent resistance is 60/3 = 20 Ω .

3. If a circuit consists of the parallel resistances with one of the resistance at least 100 times larger than the other, the total resistance or equivalent resistance is virtually equal to the smaller resistance.

4.17. Current Division Rule

A parallel circuit acts as a current divider. The current division rule states that total current (I_T) divides, itself, among the parallel branches in a manner inversely proportional to their resistance values. Now consider a circuit consisting of two resistances connected in parallel as shown in Fig. 4.20.

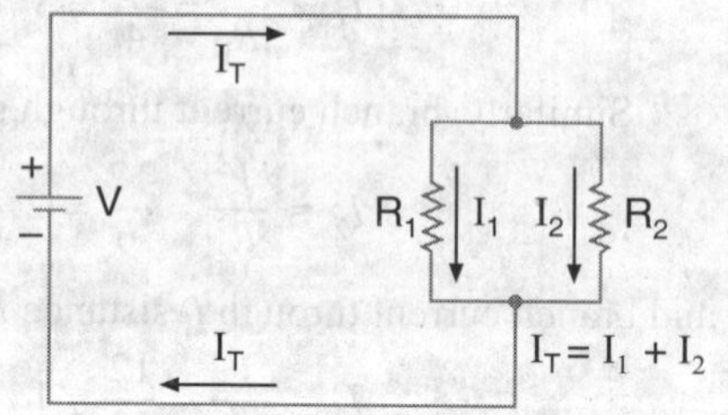

Fig. 4.20. Current division rule.

Let I_1 = Current through resistance R_1, and

I_2 = Current through resistance R_2.

We know that total current,

$$I_T = I_1 + I_2$$

$$\therefore \quad I_1 = \left(\frac{R_2}{R_1 + R_2}\right) I_T \text{ and } I_2 = \left(\frac{R_1}{R_1 + R_2}\right) I_T$$

The current division rule is very useful in analysing electric circuits by Norton's Theorem.

Example 4.13.

Two resistances of 4 Ω and 12 Ω are connected in parallel across a 6 V battery. Determine the amount of current through each resistance.

Solution.

Given: $R_1 = 4\ \Omega$; $R_2 = 12\ \Omega$ and $V = 6$ Volts.

We know that equivalent resistance:

$$R_{eq} = R_1 \parallel R_2 = \frac{R_1 \cdot R_2}{R_1 + R_3} = \frac{4 \times 12}{4 + 12} = 3\ \Omega$$

$$\therefore \quad \text{Total current,} \quad I_T = \frac{V}{R_{eq}} = \frac{6}{3} = 2 \text{ A}$$

We also know that current through R_1,

$$I_1 = \left(\frac{R_2}{R_1 + R_2}\right) I_T = \left(\frac{12}{4 + 12}\right) \times 2 = 1.5 \text{ A} \textbf{ Ans.}$$

and current through R_2,

$$I_2 = \left(\frac{R_1}{R_1 + R_2}\right) I_T = \left(\frac{4}{4 + 12}\right) \times 2 = 0.5 \text{ A} \textbf{ Ans.}$$

4.18. Total Power in a Parallel Circuit

The total power, in a parallel circuit, is the sum of the individual values of power dissipated in the circuit. It is also equal to the total power generated by the source. Mathematically, total power,

$$P_T = P_1 + P_2 + P_3 + ...$$

where, P_1, P_2, P_3...... are the individual values of power dissipated in the branches.

It may be noted that total power formula is the same for a series and a parallel-circuit. It is due to the fact that series or parallel connections can change the distribution of voltage and current. But power is the rate at which energy is supplied by the source to the circuit and cannot change with a circuit arrangement.

Example 4.14.

A certain series circuit is made up of a 1/8 W resistor, a 1/4 W resistor and a 1/2 W resistor. The total resistance is 2400 Ω. If each of the resistors is operating at its maximum power level, determine, (a) current (b) applied voltage and (c) value of each resistor.

Solution.

Given: $P_{R_1} = 1/8$ W; $P_{R_2} = 1/4$ W; $P_{R_3} = 1/2$ W and $R_T = 2400\ \Omega$.

(a) *Current through the series circuit*

Let $I =$ Total current in the circuit and

$V_S =$ Applied voltage to the circuit,

We know that total power dissipated by the circuit,

$$P_T = P_{R_1} + P_{R_2} + P_{R_3} + \frac{1}{8} + \frac{1}{4} + \frac{1}{2} = \frac{7}{8}\text{ W}$$

and also the total power dissipated, (P_T),

$$\frac{7}{8} = I^2 \cdot R_T = I^2 \cdot 2400;$$

$$I^2 = \frac{7}{8 \times 2400} = 3.646 \times 10^{-4} \text{ or } I = 0.0191 \text{ A} = 19.1 \text{ mA. } \textbf{Ans.}$$

(b) *Applied Voltage across the circuit*

We know that the applied voltage across the circuit,

$$V_S = I \cdot R_T = 19.1 \times 10^{-3} \times 2400 = 45.84 \text{ V } \textbf{Ans.}$$

(c) *The value of each resistor*

We know that power dissipated by the resistor R_1, (P_{R_1}),

$$\frac{1}{8} = I^2 \cdot R_1 = (0.0191)^2 \times R_1 = 3.648 \times 10^{-4} R_1$$

$$\therefore \quad R_1 = \frac{7}{8 \times 2400} = 3.646 \times 10^{-4} \text{ or } I = 0.0191 \text{ A} = 19.1 \text{ mA. } \textbf{Ans.}$$

Similarly power dissipated by the resistor R_2, (P_{R_2}),

$$\frac{1}{4} = I^2 \cdot R_2 = (0.0191)^2 \times R_2 = 3.648 \times 10^{-4} R_2$$

$$\therefore \quad R_2 = \frac{1}{4 \times (3.648 \times 10^{-4})} = 685.2\ \Omega \textbf{ Ans.}$$

and power dissipated by the resistor R_3, (P_{R_3}),

$$\frac{1}{2} = I^2 \cdot R_3 = (0.0191)^2 \times R_3 = 3.648 \times 10^{-4} R_3$$

$$R_3 = \frac{1}{2 \times (3.648 \times 10^{-4})} = 1370.4\ \Omega\ \textbf{Ans.}$$

4.19. Opens in a Parallel Circuit

We have already mentioned that an 'open' between any two points of a circuit represents infinite resistance. An open in a parallel circuit may be of two types, namely an open in the main line and an open in the branch line. In order to illustrate both these types, consider a circuit consisting of two resistances R_1 and R_2 as shown in Fig. 4.21 (*a*). Here, the current flows in both the branches depending upon their resistance values. Let us consider the following cases:

1. ***Open in main line.*** The conducting wires like *AC*, *CE*, *BD*, and *DF* of Fig. 4.21 (*a*) are called main lines. If a circuit opens at any point on these lines as shown in Fig. 4.21 (*b*), it is called an open in the main line. It may be noted that an open in the line *AC* or *BD* prevents the flow of current in all the branches, while an open in line *CE* or *DF* prevents the flow of current in branch *EF* only. To locate an 'open' in the main line, we measure the voltages across all the main lines. In the normal circuit, voltage across these terminals is zero. If any one of such lines is 'open', the voltage across that open line is equal to the applied source voltage.

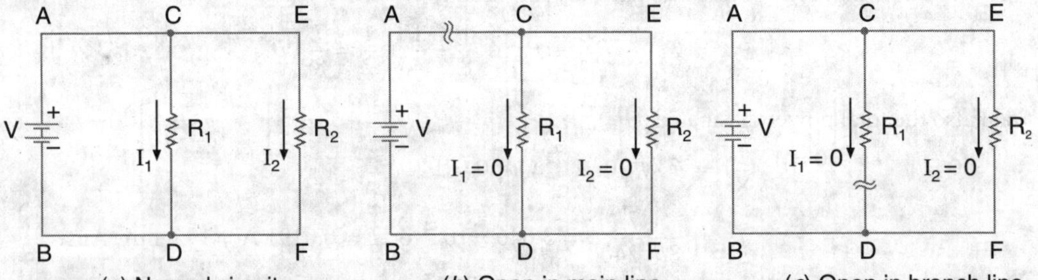

(*a*) Normal circuit. (*b*) Open in main line. (*c*) Open in branch line.

Fig. 4.21.

2. ***Open in branch line.*** The parallel paths like *CD* and *EF* of Fig. 4.21 (*c*) are called branch lines. If a circuit opens at any point in the branch line as shown in Fig. 4.21 (*c*), it is called an open in the branch line. It may be noted that an open in any branch line prevents the flow of current in that branch only. To locate an open in the branch line, we measure the currents in each branch of the circuit. In a normal circuit, all the branches have a certain current value. However, if any branch is open, the current in that branch is equal to zero.

4.20. Shorts in a Parallel Circuit

We have already mentioned that a 'short' between any two points of a circuit represents zero resistance. A 'short' in a parallel circuit may occur, if both the terminals of any branch are short-circuited. Due to this, a short becomes directly connected across the applied voltage source and draws almost an infinite value of current, because the resistance of the connecting wires is negligible. As a result of this, the wires may get heated and even burn out unless the circuit is not protected by a fuse.

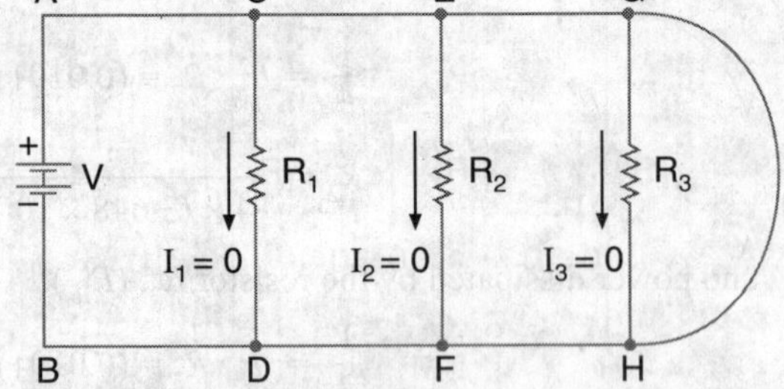

Fig. 4.22. Short in a parallel circuit.

Fig. 4.22 shows a parallel circuit consisting of three resistances R_1, R_2, R_3 and a short across resistance R_3. It will be interesting to know that a short across R_3, also short-circuits all the other branches (or resistances). As a result of this, current in all the resistances (R_1, R_2 and R_3) is zero. Hence the components in the short-circuited branches are not damaged. It is possible to protect any branch from short-circuited branch by separating them by a resistance. For example, the branches *CD* and *EF* may be protected from the short-circuited branch *GH*, by placing a resistance between the branches *EF* and *GH* (*i.e.,* a resistance between the terminals *E* and *G* or between *F* and *H*).

4.21. Series Parallel Circuits

We have already discussed, series and parallel circuits. But in practice, we have a combination of series and parallel circuits. Such circuits are used to provide various amounts of current and voltages with single power supply. Usually, the electronic circuits are of this type, because they generally use only one voltage source. In such circuits, some of the components are connected in series and other in parallel.

Example 4.15.

Fig. 4.23 shows a series-parallel circuit.

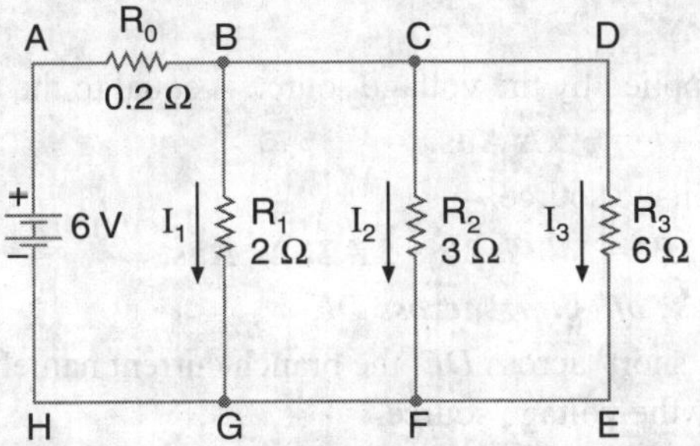

Fig. 4.23.

Find (a) total resistance (or equivalent resistance) of the circuit. (b) value of branch currents, (c)current and power supplied by the voltage source. (d) current and power supplied by the voltage source, if an accidental short occurs across the branch DE.

Solution.

Given: $V = 6$ Volt, $R_0 = 0.2\ \Omega$; $R_1 = 2\ \Omega$, $R_2 = 3\ \Omega$ and $R_3 = 6\ \Omega$

(a) Total resistance

We know that branches *BG*, *CF* and *DE* form a parallel circuit. Therefore its equivalent resistance,

$$R_{eq} = \frac{1}{\frac{1}{R_1}+\frac{1}{R_2}+\frac{1}{R_3}} = \frac{1}{\frac{1}{2}+\frac{1}{3}+\frac{1}{6}} = 1\ \Omega$$

Now, let us connect the equivalent resistance of the parallel circuit back to the original circuit as shown in Fig. 4.24. It is a simple series circuit and its total resistance,

$$R = R_o + R_{eq} = 0.2 + 1\ \Omega$$
$$= 1.2\ \Omega \textbf{ Ans.}$$

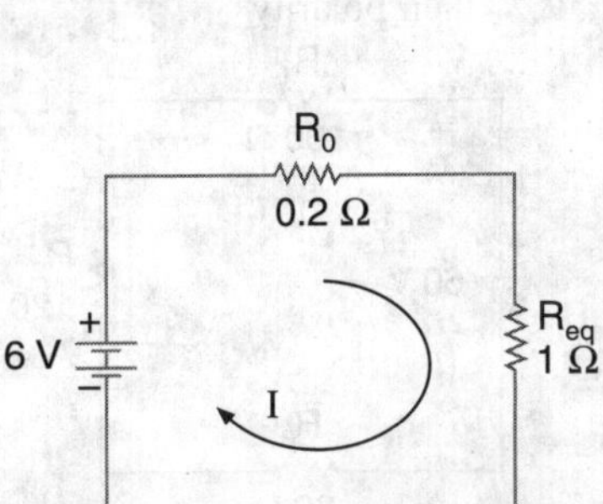

Fig. 4.24.

(b) Branch currents

We know that magnitude of total current,

$$I = \frac{V}{R} = \frac{6}{1.2} = 5\ \text{A}$$

∴ Voltage drop across R_o.

$$V_0 = I \cdot R_0 = 5 \times 0.2 = 1 \text{ V}$$

and voltage drop across R_{eq},

$$V_{eq} = V - V_0 = 6 - 1 = 5 \text{ V}$$

We know that voltage drop across all the three branches namely *BG*, *CF* and *DE* is same and is equal to 5 V. Therefore, branch current through resistance R_1,

$$I_1 = \frac{V}{R_1} = \frac{5}{2} = 2.5 \text{ A } \textbf{Ans.}$$

Similarly, branch current through resistance R_2,

$$I_2 = \frac{V}{R_2} = \frac{5}{3} = 1.67 \text{ A } \textbf{Ans.}$$

and branch current through resistance R_3,

$$I_3 = \frac{V}{R_3} = \frac{5}{6} = 0.83 \text{ A } \textbf{Ans.}$$

(c) *Current and power*

We know that current supplied by the voltage source is equal to the total current. *i.e.*,

$$I = 5 \text{ A } \textbf{Ans.}$$

and power supplied by the voltage source,

$$P = V \cdot I = 6 \times 5 = 30 \text{ W } \textbf{Ans.}$$

(d) *Current and power in case of 'short' across DE*

We know that in case of 'short' across *DE*, the branch current namely, I_1, I_2 and I_3 will be zero. Therefore current supplied by the voltage source,

$$I = \frac{V}{R_0} = \frac{6}{0.2} = 30 \text{ A } \textbf{Ans.}$$

and power supplied by the voltage source,

$$P = V \cdot I = 6 \times 30 = 180 \text{ watt } \textbf{Ans.}$$

NUMERICAL PROBLEMS

1. For the series circuit shown in Fig. 4.25 find (*a*) the total resistance of the circuit, and (*b*) the current which flows in the circuit. **(Ans. R_T = 300 Ω and I = 0.2 A)**
2. For the series circuit of Fig. 4.25, find the voltage drop across each resistor and indicate their polarity. **(Ans. V_1 = 20 V, V_2 = 24 V, V_3 = 16 V).**

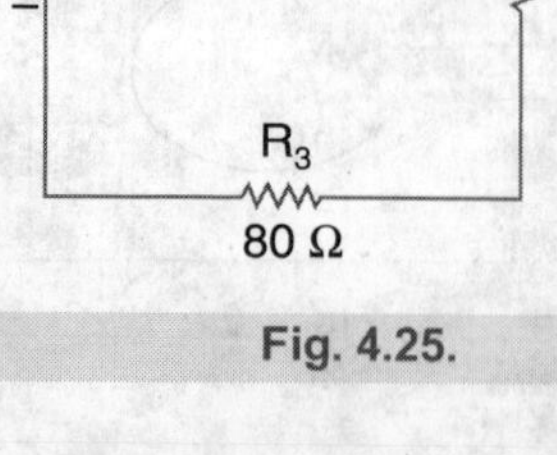

Fig. 4.25.

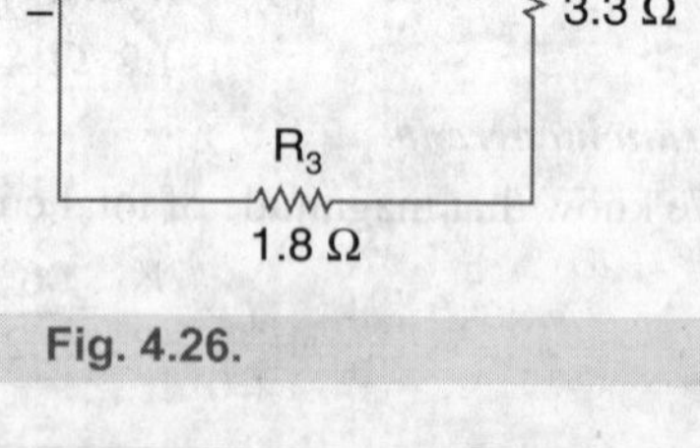

Fig. 4.26.

3. Fig. 4.26 shows a series circuit. For this circuit, find (*a*) the total resistance, (*b*) the current in the circuit, and (*c*) the individual voltage drops across the resistors.
(Ans. R_T = 6.3 kΩ; I = 7.9 mA; V_1 = 9.5 V, V_2 = 26.2 V and V_3 = 14.3 V)
4. Three 50 Ω resistances are in series across on applied voltage of 180 V. How much is the voltage drop across each resistance? **(Ans. 60 V)**
5. Two resistances of R_1 = 10 Ω and R_2 = 20 Ω are in series across an applied voltage of 60 V. What is the power dissipated by each resistance? **(Ans. P_1 = 40 Ω and P_2 = 80 Ω)**
6. Using the equivalent resistance, calculate the value of I_T in Fig. 4.27, Also find the values of I_1 and I_2. **(Ans. I_T = 4 mA, I_1 = 3 mA and I_2 = 1 mA)**

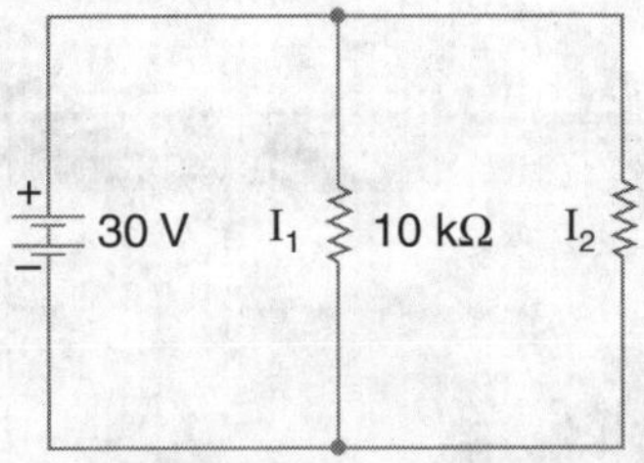

Fig. 4.27.

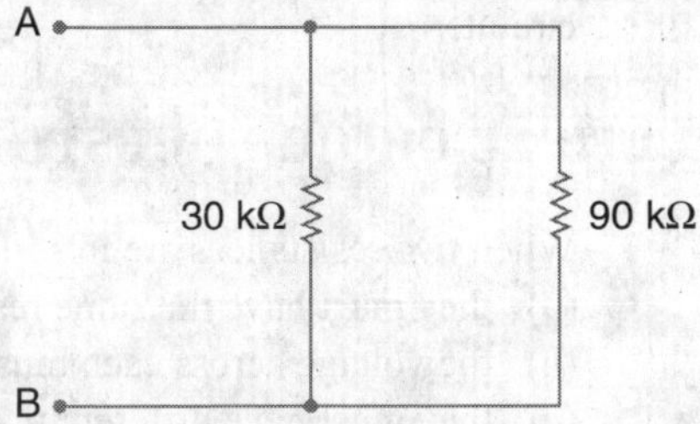

Fig. 4.28.

7. What is the equivalent resistance between the terminals *AB* of the circuit shown in Fig. 4.28. **(Ans. 22.5 kΩ)**
8. Two 10 MΩ resistances are in parallel. What is their equivalent resistance? **(Ans. 5 MΩ)**
9. The equivalent resistance of a three branch parallel circuit is 10 Ω. If R_1 = 47 Ω and R_2 = 22 Ω, determine the value of R_2. **(Ans. 30 Ω)**
10. Determine the value of resistance that must be placed in parallel with 24 kΩ to yield an equivalent resistance of 1.5 kΩ. **(Ans. 1.6 kΩ)**
11. Two resistances R_1 = 820 Ω and B_2 = 680 Ω are connected in parallel with a 22 volt source. Determine the value of I_1 and I_2. **(Ans. I_1 = 26.8 mA and I_2 = 32.4 mA)**
12. Using the product-over-sum formula, find the equivalent resistance of the two resistor combination shown in Fig. 4.29. Verify by applying the reciprocal formula. **(Ans. 353 Ω)**

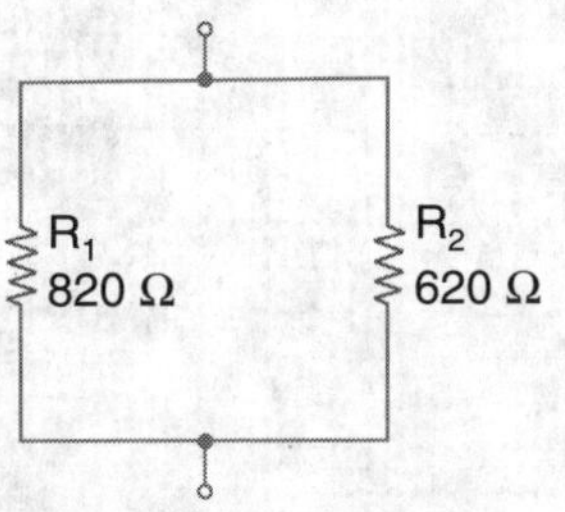

Fig. 4.29.

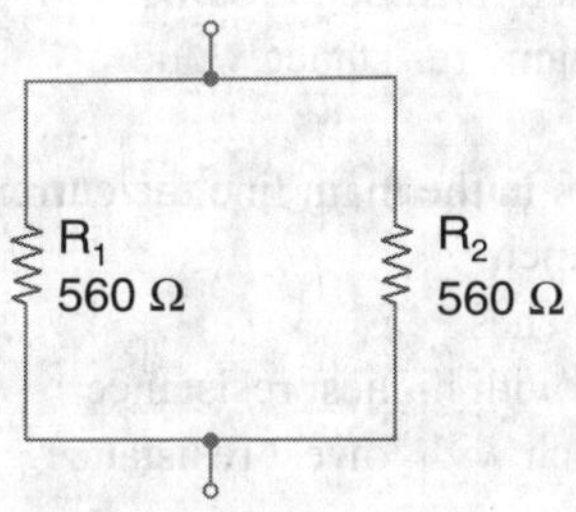

Fig. 4.30.

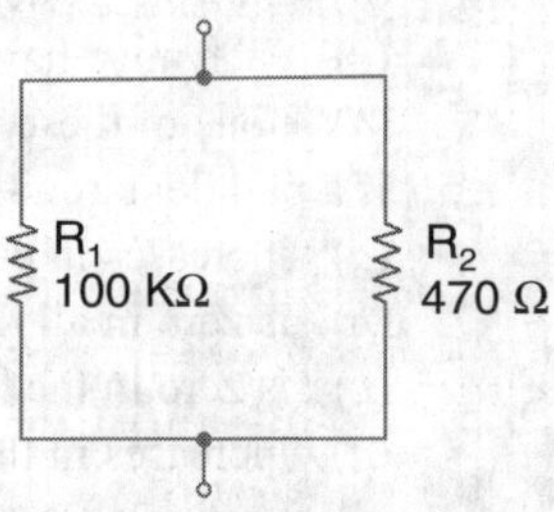

Fig. 4.31.

13. What is the equivalent resistance of a two resistance parallel combination of Fig. 4.30. Verify by applying the reciprocal formula. **(Ans. 280 Ω)**
14. For the two parallel resistances shown in Fig. 4.31, what is the approximate value of equivalent resistance? Also find the exact value of resistance by applying product-over-sum formula. **(Ans. 470 Ω, 467.8 Ω)**

DESCRIPTIVE QUESTIONS

1. Explain clearly the terms voltage notation and voltage reference point.
2. What do you mean by a voltage division rule? Explain it with a suitable example.
3. Give some of the applications where voltage dividers are used.
4. Explain the following terms: (*a*) 'open' in a series circuit and (*b*) 'short' in a series circuit.
5. Derive an expression for the equivalent resistance of a two branched circuit.
6. What do you understand by a current division rule? Explain it with suitable example.
7. Explain the following terms: (*a*) 'Opens' in a parallel circuit and (*b*) 'Shorts' in a parallel circuit.

MULTIPLE CHOICE QUESTIONS

1. When two resistances are in series
   (*a*) they must have the same resistance value
   (*b*) the voltage across each must be the same
   (*c*) they must have different resistance values
   (*d*) there is only one path for current through both resistances
2. The sum of series $I \cdot R$ voltage drops is
   (*a*) less than the smallest voltage drop
   (*b*) equal to the average value of all the voltage drops
   (*c*) equal to the applied voltage
   (*d*) usually more than the applied voltage
3. When one resistance in a series circuit is open
   (*a*) current is maximum in the normal resistances
   (*b*) current is zero in all the resistances
   (*c*) voltage is zero across the open resistance
   (*d*) current increases in the voltage source
4. With two resistances connected in parallel
   (*a*) the current through each must be the same
   (*b*) the voltage across each must be the same
   (*c*) each must have the same resistance value
   (*d*) none of these
5. If a parallel circuit is open in the main line the current
   (*a*) increases in each branch
   (*b*) is zero in all the branches
   (*c*) is zero in the branch with highest resistance
   (*d*) increases in the branch with lowest resistance

ANSWERS

1. (*d*) **2.** (*c*) **3.** (*b*) **4.** (*b*) **5.** (*b*)

5 Chapter

Kirchhoff's Laws and Network Theorems

Outline

1. Introduction
2. Kirchhoff's Laws
3. Kirchhoff's Current Law
4. Kirchhoff's Voltage Law
5. Thevenin's Theorem
6. Procedure for Applying the Thevenin's Theorem
7. Norton's Theorem
8. Procedure for Applying Norton's Theorem
9. Superposition Theorem
10. Maximum Power Transfer Theorem
11. Applications of Maximum Power Transfer Theorem.

5.1. Introduction

The Kirchhoff's Laws and network theorems are important tools in the analysis of electrical and electronic circuits. In this chapter, we shall discuss, Kirchhoff's Current Law, Kirchhoff's Voltage Law and the network theorems such as Thevenin's, Norton's, superposition and maximum power transfer theorems.

5.2. Kirchhoff's Laws

These laws are essential for the analysis of electronic devices and circuits, covered in this book. These laws are applicable to both the d.c. and a.c. circuits. Following are two types of Kirchhoff's Laws:

1. Kirchhoff's Current Law (KCL).
2. Kirchhoff's Voltage Law (KVL)

5.3. Kirchhoff's Current Law

This law states that *in any network, the algebraic sum of currents, meeting at a point (or a junction), is zero.* In other words, the sum of the current entering a junction is always equal to the sum of currents leaving it.

Fig. 5.1 shows a circuit with one d.c. voltage source and three resistive branches. Now let us adopt a sign convention that the current entering the junction is positive and that leaving the junction is negative. We see that the current entering the junction *A*

$$= I_T$$

and current leaving the junction *A*

$$= -(I_1 + I_2 + I_3)$$

∴ According to the Kirchhoff's Current Law,

$$I_T - (I_1 + I_2 + I_3) = 0$$

$$I_T = I_1 + I_2 + I_3$$

or

$$I_{in} = I_{out}$$

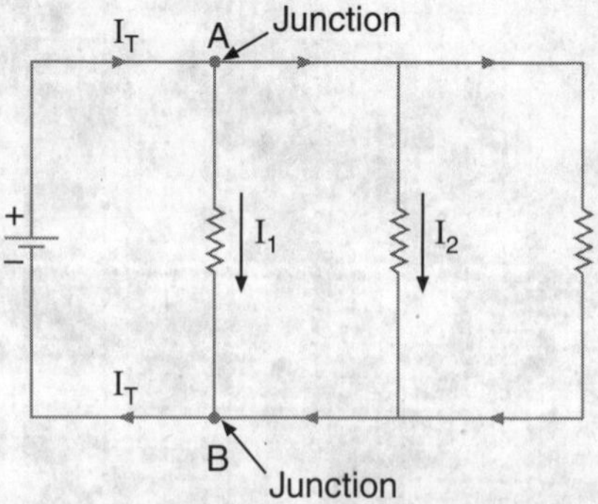

Fig. 5.1. Kirchhoff's Current Law.

Example 5.1.

Three resistors on the PCB board are connected as shown in Fig. 5.2.

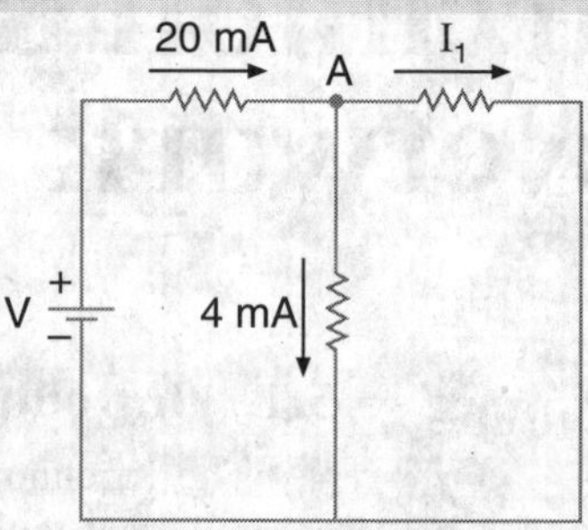

Fig. 5.2.

Find the value of the current I_1.

Solution.

Given: Current entering the junction *A* $I_T = 20$ mA and the current leaving the junction *A*

$$= I_1 + 4 \text{ mA}$$

According to Kirchhoff's Law, the current entering the *A* (I_T),

$$20 = I_1 + 4 \text{ mA} \quad \text{or} \quad I_1 = 20 - 4 = 16 \text{ mA } \textbf{Ans.}$$

5.4. Kirchhoff's Voltage Law

The law states that *in any network, the sum of all the voltages around a closed path (or a loop) is always zero.* In other words, if we start at any point in the circuit and go around a path, which returns to the starting point, we shall find that if all the voltages are added the result is equal to zero. Mathematically, Kirchhoff's Voltage law may be expressed as:

$$\Sigma I \cdot R + \Sigma \text{ e.m.f.'s} = 0$$

where $\Sigma I \cdot R$ = Sum of voltage drops across resistors, and

Σ e.m.f.'s = Sum of battery voltages.

Gustav R. Kirchhoff (1824-1887)

Following procedure is adopted for applying the Kirchhoff's Voltage Law to any circuit:

1. First of all, select any point in the circuit as a starting point (say point *A*) as shown in Fig. 5.3. Now let us move along the circuit in a clockwise direction (*i.e.*, in a direction as marked

by arrow in Fig. 5.3) and return to the starting point.

2. Go on adding algebraically, the values (*i.e.*, the magnitudes) of each voltage source and voltage drop across resistance. Care should always be taken to write down the first sign we see (+ or –) and the magnitude of the voltage source (or voltage drop across resistance).
3. After returning back to the starting point, equate the sum of all the voltages in the circuit to zero.

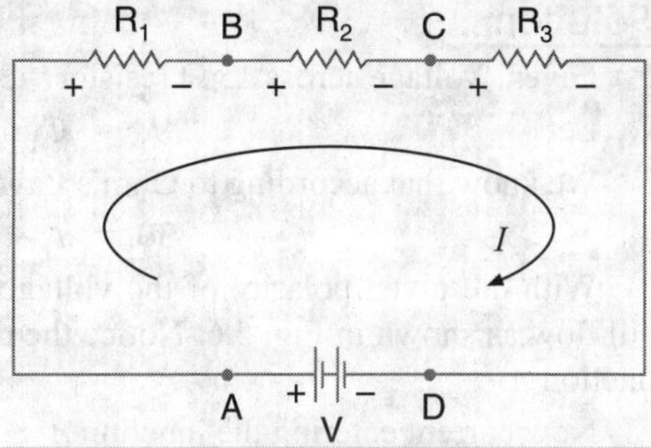

Fig. 5.3. Illustrating Kirchhoff's Voltage Law.

Example 5.2.

A simple circuit of an electrical gadget is shown in Fig. 5.4. Find the value of the unknown resistance R, by applying Kirchhoff's Voltage Law.

Take the value of series circuit current as 1 A.

Solution.

Given: $I = 1$ A

Applying Kirchhoff's Voltage Law,

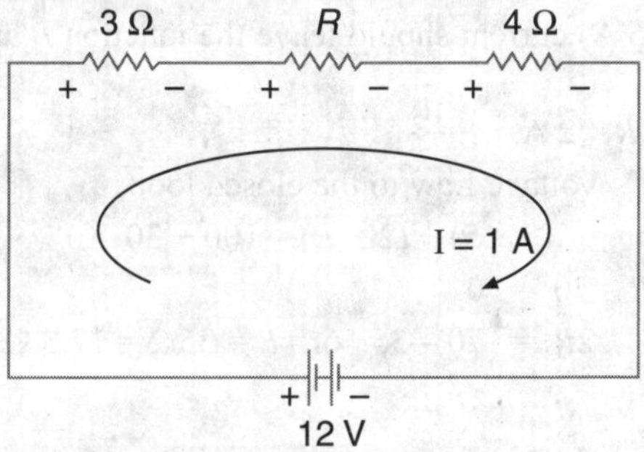

Fig. 5.4.

$$(1 \times 3) + (1 \times R) + (1 \times 4) - 12 = 0$$

$$3 + R + 4 - 12 = 0 \quad \text{or} \quad R = 5\ \Omega \ \textbf{Ans.}$$

Example 5.3.

Using Kirchoff's and Ohm's Laws, find the value of resistor, R in the circuit shown in Fig. 5.5.

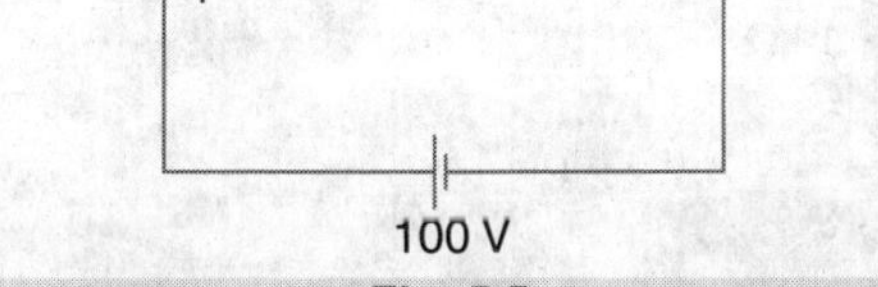

Fig. 5.5.

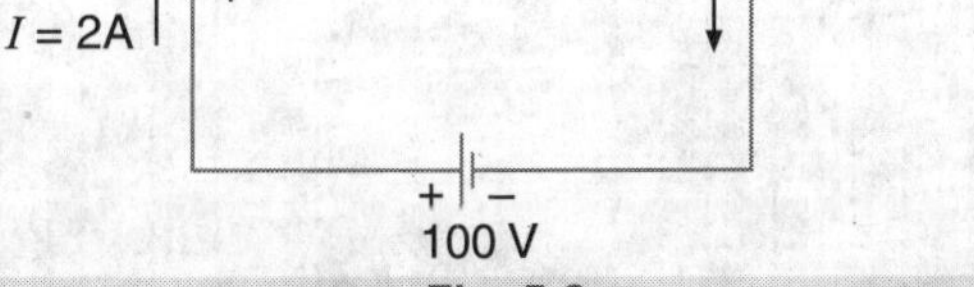

Fig. 5.6.

Solution.

Given: Voltage across 15 Ω resistor = 30 V.

Let I_1 = Current through 15 Ω resistor.

We know that according to Ohm's Law, voltage across 15 Ω resistor,

$$30 = I_1 \times 15 \text{ or } I_1 = 30/15 = 2 \text{ A}$$

With the given polarity of the voltage across 15 W resistor, the current through the resistor will flow as shown in Fig. 5.6. Notice the current direction indicated by an arrow marked towards junction *A*.

Now current entering the junction *A*

$$= 5 + 2 = 7 \text{ A}$$

Now as per the Kirchhoff's Current Law, the current leaving the junction '*A*' should also be equal to 7 A. This current flows from junction 'A' to 'B' as shown in Fig. 5.6.

Thus current through 5 W resistor,

$$I_2 = 7 \text{ A}$$

Now current entering the junction *B*.

$$= 7 \text{ A}$$

According to Kirchhoff's current law, the current leaving the junction *B* should also be equal 7 A. Since the current of 2 A and 3 A (total 5 A) is already shown to be leaving the junction *B*.

Another 2 A (*i.e.*, 7 A – 5 A) current should leave the junction *B* and flow through *R* as shown in Fig. 5.6.

Current through resistor *R* = 2 A.

Now applying Kirchhoff's Voltage Law to the closed loop,

$$(7 \times 5) + (2 \times R) - 100 + 30 = 0$$

$$35 + 2R - 70 = 0$$

or $$2R = 70 - 35 \quad \text{or} \quad R = 35/2 = 17.5\ \Omega$$ **Ans.**

Example 5.4.

Fig. 5.7 shows a d.c. equivalent circuit of an electronic component.

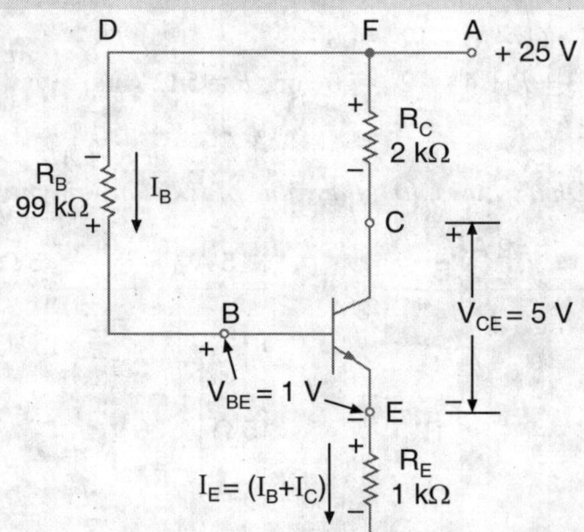

Fig. 5.7.

Determine the values of the currents I_B and I_C.

Solution.

In order to determine the current I_B and I_C, we will have to apply the Kirchhoff's Law to the closed paths *ADBEGA,* and *AFCEGA* respectively. It may be noted that in the given figure, the negative and of the battery is grounded. Therefore the points *A* and *G* are connected through a battery of an e.m.f. equal to 25 volts.

Closed path (or base-loop) ADBEGA

Let us start from point *A*, and move in an anticlockwise direction towards points *D, B, E, G* and return back to *A*. As we move down, we get the following voltages.

1. *Voltage drop across R_B.* Its value is $(I_B \cdot R_B)$. And the sign is positive, because while moving from *D* to *B*, the first sign is positive.
2. *Voltage drop V_{BE}.* Its value is equal to 1 V. And the sign is positive, because while moving from *B* to *E*, the first sign is positive.
3. *Voltage drop across R_E.* Its value is $(I_E \cdot R_E)$. And the sign is positive, because while moving from *E* to *G*, the first sign is positive.
4. *Voltage V_{CC}* . Its value is equal to 25 V. And the sign is negative, because while moving from *G* to *A*, the first sign is negative.

Now summing up all the voltages (*i.e.*, voltage drops and e.m.f's.) with their proper signs around the base-loop,

$$I_B \cdot R_B + V_{BE} + I_E \cdot R_E - V_{CC} = 0$$

or

$$I_B \cdot R_B + V_{BE} + (I_B + I_C) R_E - V_{CC} = 0 \qquad ...(\because I_E = I_B + I_C)$$

Substituting the given values in the above equation,

$$99 I_B + 1 + 1 \times (I_B + I_C - 25) = 0 \quad \text{or} \quad 100 I_B + I_C = 24 \qquad ... (i)$$

Closed path (or collector loop) ACEGA

Summing up all the voltages (*i.e.*, voltage drops and e.m.f's) with their proper sign around the collector loop, and using a similar approach as discussed for base-loop.

$$I_C \cdot R_C + V_{CE} + I_E \cdot R_E - V_{CC} = 0$$

Substituting the given values in the above equation,

$$2I_C + 5 + 1 \times (I_B + I_C) - 25 = 0 \quad \text{or} \quad I_B + 3I_C = 20 \qquad ... (ii)$$

Solving the equation (*i*) and (*ii*),

$$I_C = 1976/299 = 6.61 \text{ mA} \ \textbf{Ans.}$$

and

$$I_B = 20 - (3 \times 6.61) = 0.17 \text{ mA} \ \textbf{Ans.}$$

5.5. Thevenin's Theorem

It states that *any complex network can be replaced by a series circuit consisting of an ideal voltage source and a resistance.* Consider a box containing a circuit connected to its terminals *A* and *B* as shown in Fig. 5.8 (*a*). Then according to Thevenin's Theorem, the entire circuit connected to *A* and *B* can be replaced by a single voltage source, (V_{TH}) in series with a single resistance (R_{TH}) across the same terminals as shown in Fig. 5.8 (*b*).

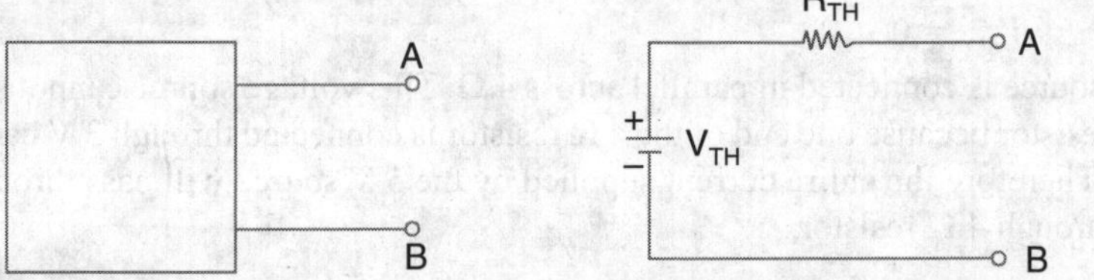

(*a*) Complex network. (*b*) Thevenin's equivalent circuit.

Fig. 5.8.

The term V_{TH} is called the Thevenin's voltage. It is, actually, the open circuit voltage existing between the terminals *A* and *B*. Similarly, the term R_{TH} is called the Thevenin's resistance. It is the resistance of the circuit as seen back from the terminals A and B with all the sources removed by their internal resistance, if any.

5.6. Procedure for Applying the Thevenin's Theorem

The following procedure is, generally, used for Thevenin's Theorem to the electrical circuits:

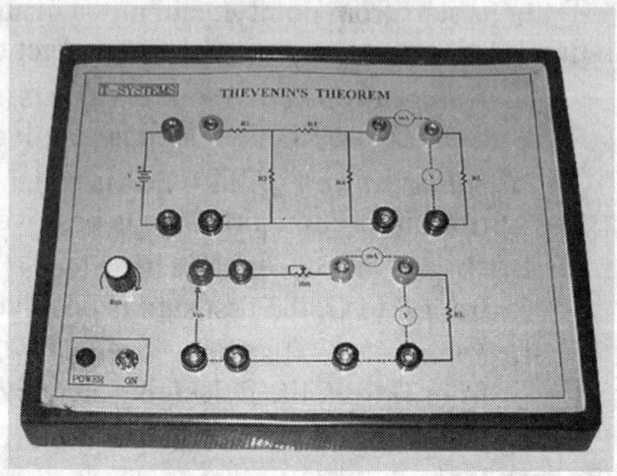

1. First of all, disconnect the load resistor (if any) from the terminals *A* and *B* (*i.e.*, the load terminals).
2. With the load terminals open, determine the open-circuit voltage between them by any known method.
3. Remove the sources by their internal resistances, if any. *i.e.*, an ideal voltage source is replaced by a short-circuit and a current source by an open-circuit. Now find the equivalent resistance (R_{TH}) of the circuit, when looking back from the terminals *A* and *B*.

Example 5.5.

For the circuit shown in Fig. 5.9, determine the Thevenin's equivalent circuit across the A-B terminals.

Solution.

To determine the Thevenin's equivalent circuit, let us apply the Thevenin's theorem to the given circuit as discussed below:

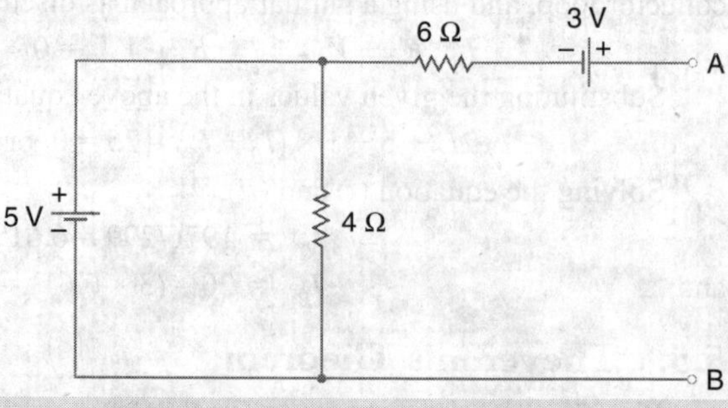

Fig. 5.9.

1. Since, there is no load resistance in the circuit. therefore, let us jump to step 2.
2. Moreover as the positive terminal of the battery is connected to terminal 'A' which is open, therefore the 3 V voltage source cannot send any current through 6 Ω or 4 Ω resistor.

   As a result of this, there is no current through 6 Ω resistor. Therefore the voltage drop across 6 Ω resistor,

$$V_{6\Omega} = 0$$

The 5 V source is connected in parallel across 4 Ω. This voltage source cannot send any current through 4 Ω resistor because one end of the 4 Ω resistor is connected through 3 V battery to the open terminal '*A*'. Therefore the entire current supplied by the 5 V source will pass through 4 Ω resistor. The current through 4 Ω resistor,

$$I = \frac{5}{4} = 1.25 \text{ A}$$

and voltage drop across 4 Ω resistor,

$$V = I \cdot R = 1.25 \times 4 = 5 \text{ V}$$

Note. This point can checked by inspection that 5 V source and the 4 Ω resistors are in parallel.Therefore entire 5 V is dropped across 4 Ω resistor.

We know that the total voltage drop across the terminals '*AB*' (called open-circuit voltage or Thevenin's voltage), is equal to the sum of voltage across 3 V source, voltage drop across 6 Ω resistor and voltage drop across 4 Ω resistor *i.e.*,

$$V_{OC} = 3 + V_{6\,\Omega} + V_{4\,\Omega} = 3 + 0 + 5 = 8 \text{ V}$$

3. Now replace the voltage source of 5 V and 3 V by a short circuit as shown in Fig. 5.10 (*a*). Looking from the terminals *A* and *B*, in the direction of arrow, we find 4 Ω resistor appears in parallel with a short circuit. The equivalent resistance of such a combination is the short circuit itself. Thus the equivalent resistance between the terminals *A* and *B* (R_{TH}) is 6 Ω as shown in Fig. 5.10 (*b*). Combining the 'open-circuit voltage' and the equivalent resistance (R_{TH}), the Thevenin's equivalent circuit is as shown in Fig. 5.10 (*c*).

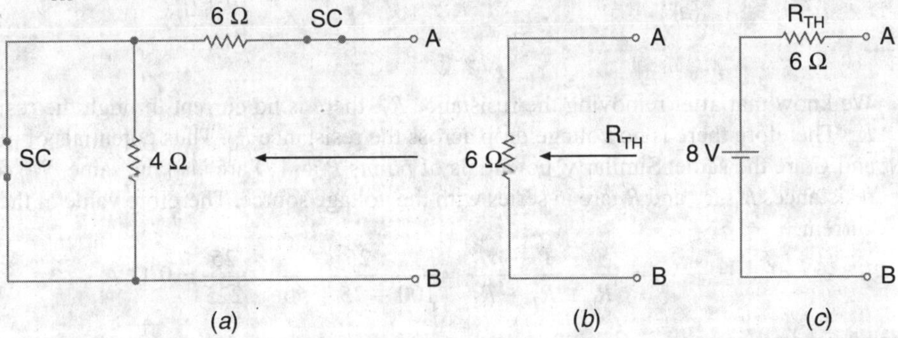

Fig. 5.10.

Example 5.6.

A circuit for an equipment is given in Fig. 5.11.

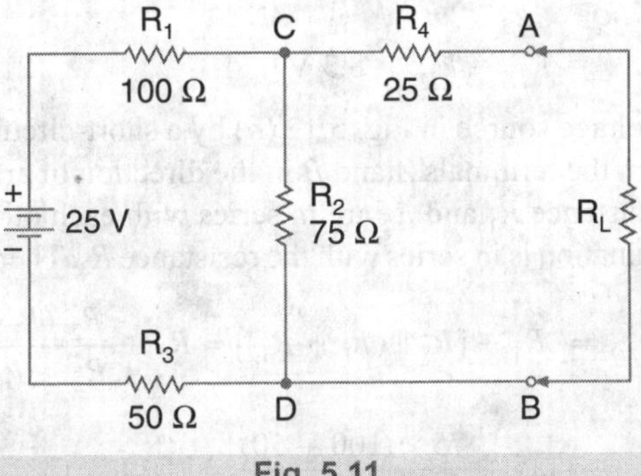

Fig. 5.11.

Determine (a) Thevenin's equivalent circuit and (b) current through the resistance R_L , if its value is 250 Ω.

Solution.

Given: $V = 25$ volts; $R_1 = 100\ \Omega$; $R_2 = 75\ \Omega$; $R_3 = 50\ \Omega$ and $R_4 = 25\ \Omega$.

(a) Thevenin's equivalent circuit

First of all, let us apply the Thevenin's Theorem to the given circuit as discussed below:

1. Disconnect the resistance R_L from the terminals A and B of the given circuit. The resulting circuit is as shown in Fig. 5.12 (*a*).

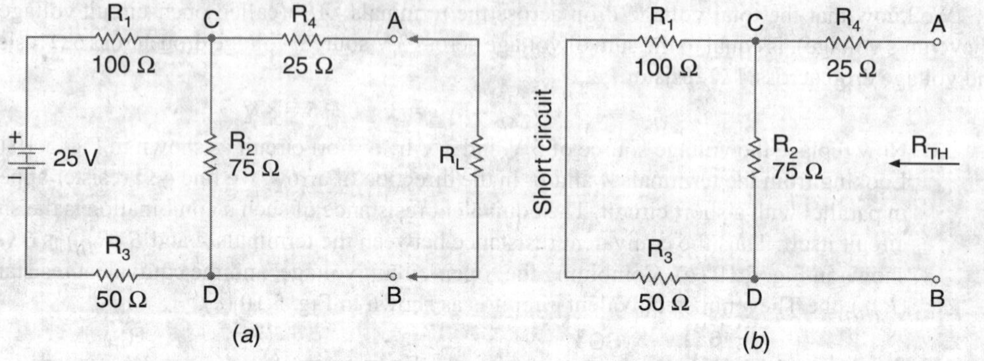

Fig. 5.12.

2. We know that after removing the resistance R_L, there is no current through the resistance R_4. Therefore there is no voltage drop across the resistance R_4. Thus potentials of points A and C are the same. Similarly, potentials of points B and D are also the same. We see that resistances R_1, R_2 and R_3 are in series with the voltage source. Therefore value of the series current,

$$I = \frac{V}{R_1 + R_2 + R_3} = \frac{25}{100 + 75 + 50} = \frac{25}{225} = 0.11 \text{ A}$$

and the voltage drop across resistance R_2,

$$V_{R_2} = I \cdot R_2 = 0.11 \times 75 = 8.3 \text{ V}$$

We know that voltage across terminals A and B (called open circuit voltage) is equal to the voltage drop across resistance R_2.

$$V_{DC} = V_{R_2} = 8.3 \text{ V}$$

and Thevenin's voltage $= V_{TH} = V_{DC} = 8.3$ V

3. Now replace the voltage source in Fig. 5.12 (*a*) by a short-circuit as shown in Fig. 5.12 (*b*). When we look from the terminals A and B in the direction of arrow shown in Fig. 5.12 (*b*), we find that the resistance R_1 and R_3 are in series with each other and from a parallel path with R_2. This combination is in series with the resistance R_4. Therefore equivalent resistance,

$$R_{TH} = R_4 + [R_2 \parallel (R_1 + R_3)] = R_4 + \frac{R_2\ (R_1 + R_3)}{R_2 + R_1 + R_3}$$

$$= 25 + \frac{75 \times (100 + 50)}{75 + 100 + 50} = 75\ \Omega$$

Thus the given circuit may be replaced by a Thevenin's equivalent circuit as shown in Fig. 5.13 (*a*). It consists of a Thevenin's voltage source (V_{TH}) of 8.3 V with Thevenin's equivalent resistance (R_{TH} of 75 Ω) **Ans.**

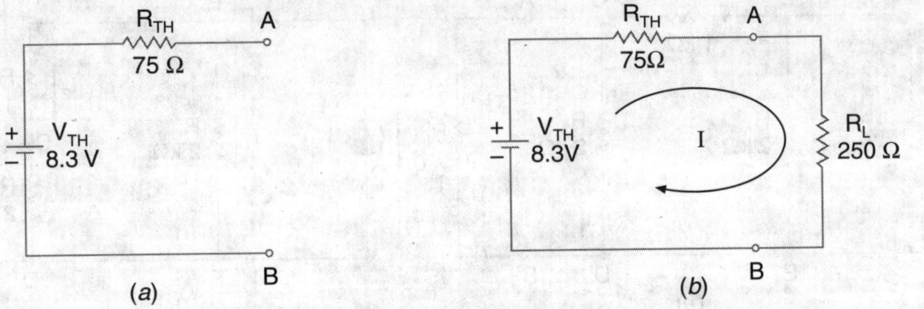

Fig. 5.13.

(b) Current through the resistance R_L

Now connect the resistance R_L across the terminals *A* and *B* of the Thavenin's equivalent circuit as shown in Fig. 5.13 (*b*). We know that current through the resistance R_L.

$$I = \frac{V_{TH}}{R_{TH} + R_L} = \frac{8.3}{75 + 250} = 0.026 \text{ A } \textbf{Ans.}$$

Example 5.7.

The circuit shown in Fig. 5.14 is a part of a circuit used in a wireless communication system.

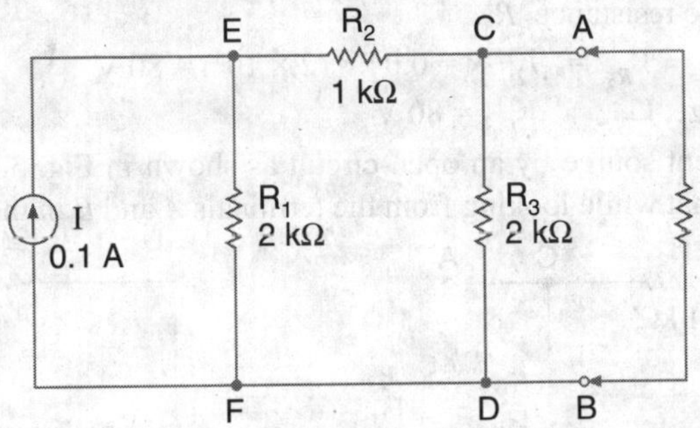

Fig. 5.14.

Determine the Thevenin's Equivalent Circuit.

Solution.

Given: $R_1 = 2$ kΩ; $R_2 = 1$ kΩ and $R_3 = 2$ kΩ.

Let us apply Thevenin's Theorem to the given circuit as discussed below:

1. Disconnect the resistance R_L from the terminals *A* and *B* of the given circuit. The resulting circuit is as shown in Fig. 5.15 (*a*).
2. We know that after removing the resistance R_L, there is no current through the branches *CA* and *DB*. Moreover, the potential of points *C* and *A* is the same. Similarly, the points *D* and *B* are also at the same potential. It means that voltage across the terminals *C* and *D* is the same as that across the open circuited terminals *A* and *B*.

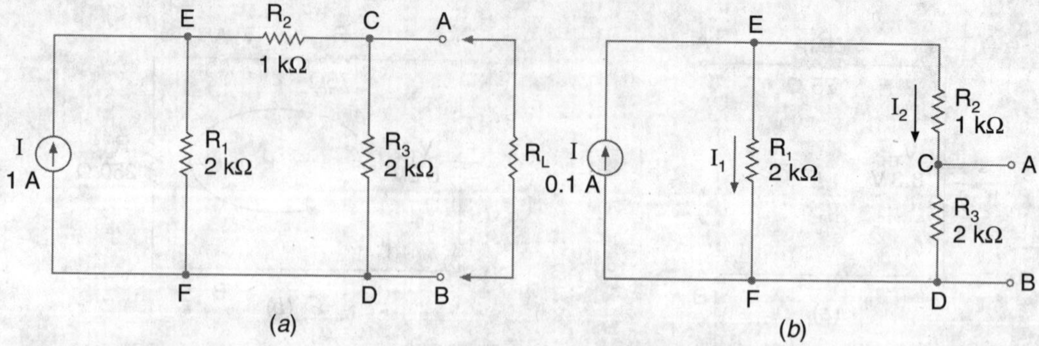

Fig. 5.15.

To find this voltage, we redraw the circuit as shown in Fig. 5.15 (*b*). It may be noted from this circuit that R_1 forms a parallel path with the series combination of R_2 and R_3, *i.e.*, $R_3 \parallel (R_2 + R_3)$. Now according to current division rule, the total current I is divided between two parallel branches in a manner inversely proportional to their resistive values. Therefore current through the branch *ECD*,

$$I_2 = I \times \frac{R_1}{R_1 + (R_2 + R_3)}$$

$$= 0.1 \times \frac{2 \times 10^3}{(2 \times 10^3) + (1 \times 10^3) + (2 \times 10^3)} = 0.04 \text{ A}$$

and voltage drop across the resistance, R_3,

$$V_{R_3} = I_2 \cdot R = 0.04 \times (2 \times 10^3) = 80 \text{ V}$$

$\therefore$ Thevenin's voltage, $V_{TH} = V_{R_3} = 80$ V

3. Replace the current source by an open-circuit as shown in Fig. 5.16 (*a*). It may be noted from this figure that while looking from the terminals *A* and *B* in the direction of the arrow,

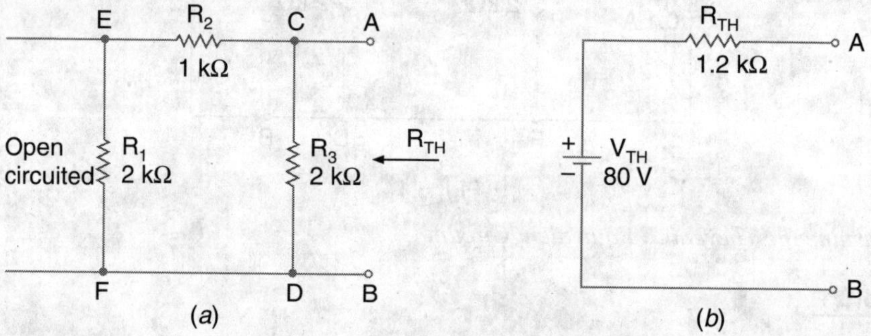

Fig. 5.16.

we find that the resistance R_3 forms a parallel combination with series combination of resistances R_1 and R_2. Therefore Thevenin's equivalent resistance,

$$R_{TH} = R_3 \parallel (R_1 + R_2) = \frac{R_3\ (R_1 + R_2)}{(R_3 + R_1 + R_2)} \text{ k}\Omega$$

$$= \frac{2\,(2+1)}{2+1+2} = 1.2 \text{ k}\Omega$$

Thus Thevenin's equivalent circuit for the given circuit, consists of a Thevenin's voltage source (V_{TH}) of 80 volts in series with Thevenin's equivalent resistance (R_{TH}) of 1.2 k Ω as shown in Fig. 5.16 (*b*). **Ans.**

Example 5.8.

Find the Thevenin's equivalent resistance between the points A and B the circuit shown in Fig. 5.17.

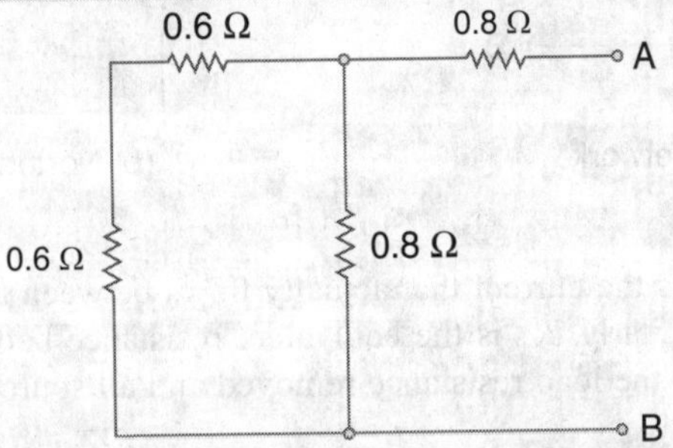

Fig. 5.17.

Solution.

Given the circuit shown in Fig. 5.17, and redrawn as shown in Fig. 5.18.

Let R_{TH} = Thevenin's equivalent resistance between the points A and B.

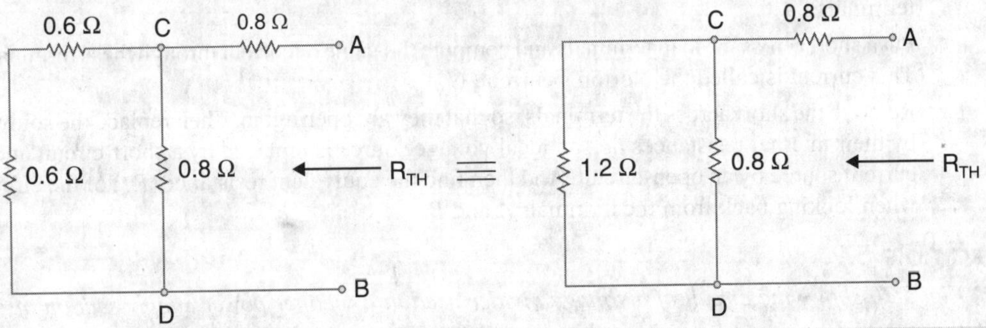

Fig. 5.18.

Notice that when looking from points A and B, we have 0.8 Ω resistance connected in series with the parallel combination of 0.8 Ω and 1.2 Ω (a series combination of two 0.6 Ω resistances).

$$\therefore \quad R_H = 0.8 + 0.8 \parallel 1.2$$

$$= 0.8 + \frac{0.8 \times 1.2}{0.8 + 1.2} = 0.8 + 0.48$$

$$= 1.28\ \Omega \ \textbf{Ans.}$$

5.7. Norton's Theorem

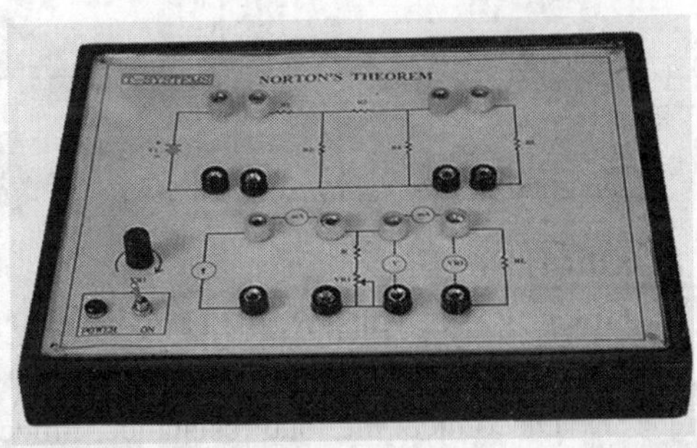

It states that *any completed network can be replaced by a parallel circuit consisting of an ideal current source and a resistance.* Consider a box containing a circuit connected to its terminals A and B as shown in Fig. 5.19 (a). According to the Norton's Theorem, the entire circuit, connected between A and B, can be replaced by a single current source (I_N) in parallel with a single resistance (R_N) across the terminals as shown in Fig. 5.19 (b).

(a) Complex network. (b) Norton's equivalent circuit.

Fig. 5.19.

As a matter of fact, I_N is the current that actually flows between the terminals A and B, when they are short-circuited. Similarly, R_N is the equivalent resistance between the terminals A and B of the original network with the load resistance removed and all sources replaced by their internal resistances.

5.8. Procedure for Applying the Norton's Theorem

The following procedure is, generally, adopted for the application of Norton's Theorem to the electrical circuits:

1. First of all, disconnect the load resistor (if any) from the terminals A and B (*i.e.*, load terminals).
2. Put a short across the load terminals and compute the value of current through these terminals. This current is called the Norton's current, (I_N).
3. Remove the short across the terminals, so that they are open again. Then replace the sources by their internal resistances *i.e.*, an ideal voltage source is replaced by a short-circuit and a current source by an open-circuit. And then find the equivalent resistance (R_N) of the circuit when looking back from the terminals A and B.

Example 5.9.

Fig. 5.20 shows the circuit of a voltage divider used in a rectifier power supply section of an oscilloscope.

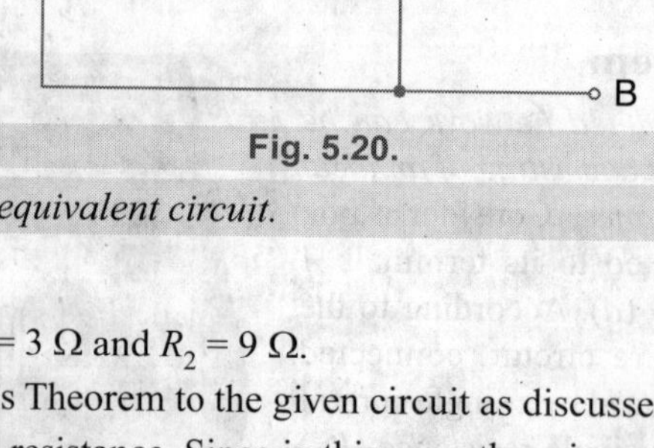

Fig. 5.20.

Determine the Norton's equivalent circuit.

Solution.

Given: $V = 12$ volts; $R_1 = 3\ \Omega$ and $R_2 = 9\ \Omega$.

Let us apply the Norton's Theorem to the given circuit as discussed below:

1. Disconnect the load resistance. Since in this case, there is no load resistance, therefore go to the next step.

2. Put a short-circuit across the terminals A and B of the given circuit as shown Fig. 5.21 (*a*). Now we know that a short-circuit, in parallel with any resistance, is equivalent to a short-circuit. Therefore, the circuit of Fig. 5.21 (*a*) is reduced to a simple one as shown in Fig. 5.21 (*b*). From this circuit, we find that current through the shorted circuit,

$$I_{SC} = V_{R_1} = \frac{12}{3} = 4 \text{ A}$$

$\therefore$ Norton's current, $\quad I_N = I_{SC} = 4 \text{ A}$

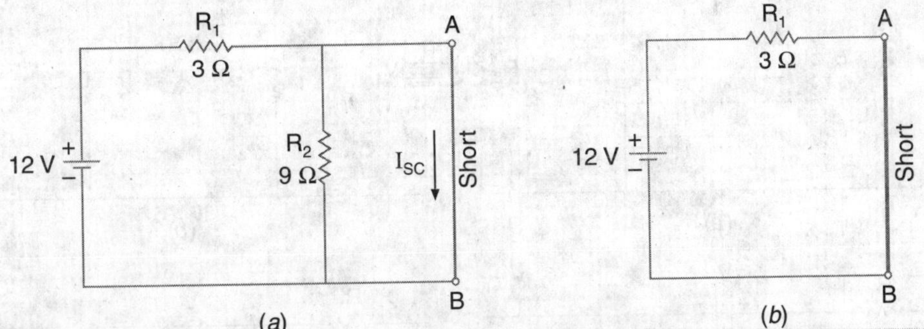

Fig. 5.21.

3. Remove the short from the terminals A and B, so that they are again open. Replace the voltage source by its internal resistance (*i.e.*, a short-circuit a shown in Fig. 5.22 (*a*). From this circuit, when looking from the terminals A and B, in the direction of the arrow, we find that R_1 and R_2 are in parallel to each other. Therefore their equivalent resistance.

$$R_N = R_1 \parallel R_2 = \frac{R_1 \cdot R_2}{R_1 + R_2} = \frac{3 \times 9}{3 + 9} = 2.25 \ \Omega$$

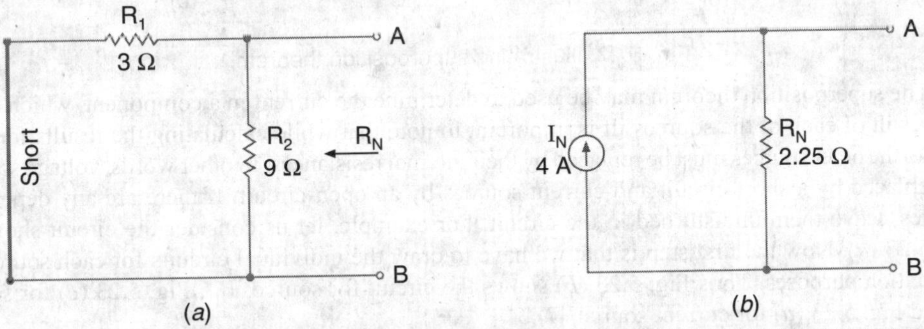

Fig. 5.22.

Thus the Norton's equivalent circuit for the given resistive network consists of current source of 4 A with internal equivalent resistance of 2.25 Ω as shown in Fig. 5.22 (*b*).

Notes: 1. The Norton's equivalent circuit can be used to determine Thevenin's equivalent circuit. We know that Thevenin's voltage,

$$V_{TH} = I_N \cdot R_N$$

and the Thevenin's equivalent resistance is the same as that of Norton's equivalent resistance. *i.e.*,

$$R_{TH} = R_N$$

2. The Thevenin's equivalent resistance may also be used to determine Norton's equivalent circuit. We know that Norton's current.

$$I_N = V_{TH}/R_{TH} \quad \text{and} \quad R_N \text{ is equal to } R_{TH}$$

5.9. Superposition Theorem

This theorem is very useful in analysing circuits, where more than one voltage or current source is present. The superposition theorem states that the *voltage or current, present in a component, is equal to the sum of voltages or currents, which exist independently.*

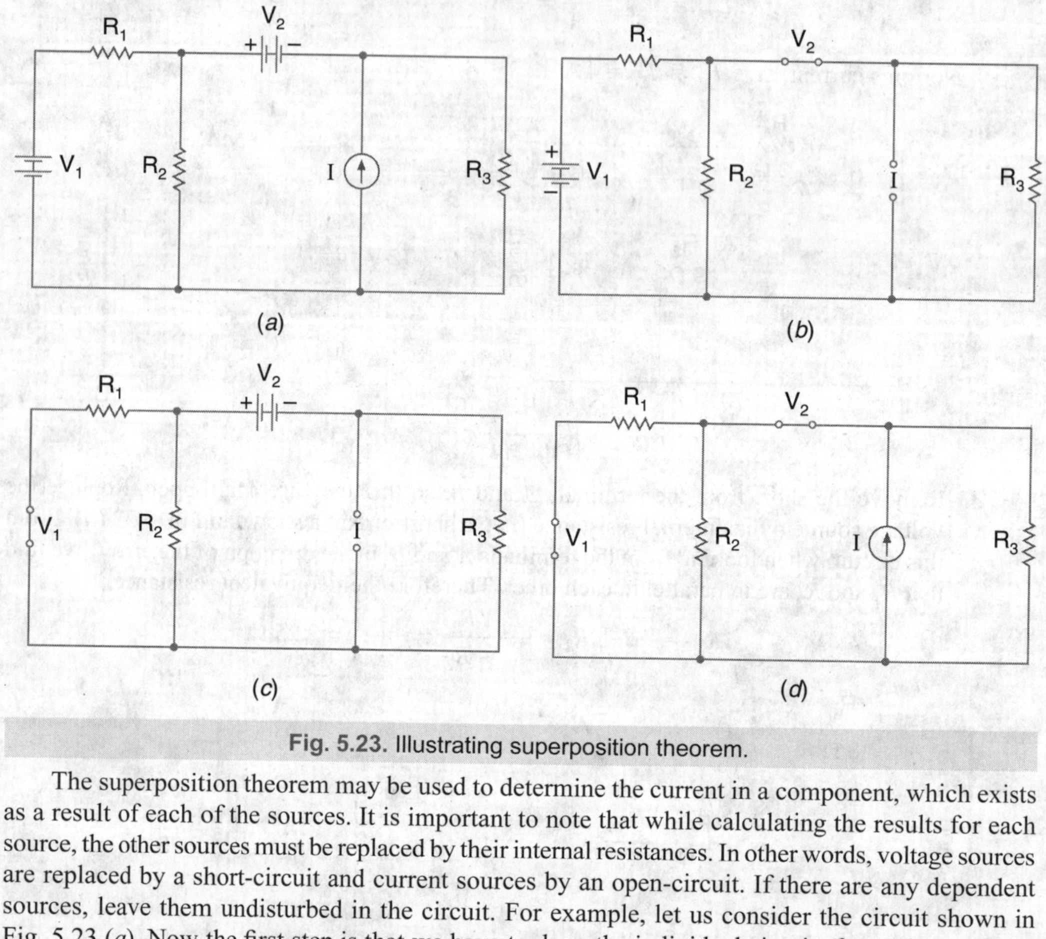

Fig. 5.23. Illustrating superposition theorem.

The superposition theorem may be used to determine the current in a component, which exists as a result of each of the sources. It is important to note that while calculating the results for each source, the other sources must be replaced by their internal resistances. In other words, voltage sources are replaced by a short-circuit and current sources by an open-circuit. If there are any dependent sources, leave them undisturbed in the circuit. For example, let us consider the circuit shown in Fig. 5.23 (*a*). Now the first step is that we have to draw the individual circuits for each source for calculation purposes. Thus Fig. 5.23 (*b*) shows the circuit for source, V_1 , Fig. 5.23 (*c*) for source V_2 and Fig. 5.23 (*d*) for current source, (*I*).

It will be interesting to know that the superposition theorem is applicable to the circuits containing resistors, inductors and capacitors. It is not applicable to the circuits containing active devices such as diodes, transistors etc. It is because of the fact, that resistors, inductors and capacitors are considered as linear and bilateral components. A linear component is one, whose current is directly proportional to the voltage drop across it. While a bilateral component is one in which the current can flow in either direction. The active devices (*i.e.*, diode, transistors etc.) are neither linear nor bilateral devices.

Example 5.10.

Using the superposition theorem, determine the voltage drop and current across the resistor R_2 *as shown in Fig. 5.24.*

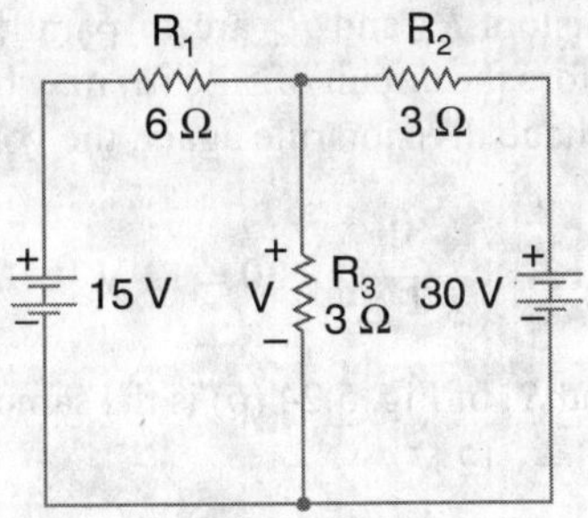

Fig. 5.24.

Solution.

Given: $V = 15$ Volts; $R_1 = 6\ \Omega$; $R_2 = 3\ \Omega$ and $R_2 = 3\ \Omega$.

1. First of all let, us draw the equivalent circuits for each source operating independently. Fig. 5.25 (*a*) shows the equivalent circuit for 15 V source acting independently. Similarly, Fig. 5.25 (*a*) also shows the equivalent circuit for 30 V source acting independently.

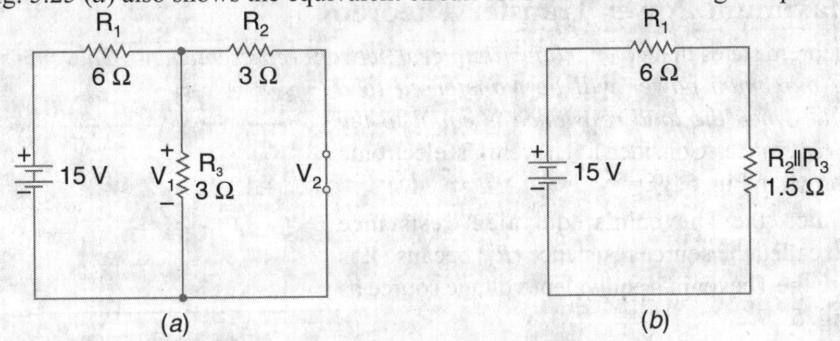

Fig. 5.25.

2. Let us now calculate the voltage drop across R_3 by each source. Fig. 5.25 (*a*) shows that R_2 and R_3 are in parallel and their equivalent resistance $R_2 \parallel R_3$ (3 ‖ 3) is equal to 1.5 Ω. Therefore the circuit of Fig. 5.25 (*a*) may be redrawn in a simple way as shown in Fig. 5.25 (*b*). Now using voltage division rule, the voltage drop across 1.5 Ω resistor (due to 15 V source),

$$V_1 = \left(\frac{1.5}{6 + 1.5}\right) \times 15 = 3 \text{ V}$$

Since the voltage drop across R_2 and R_3 of Fig. 5.25 (*b*) is the same, therefore voltage drop across R_3 , due to 15 V source

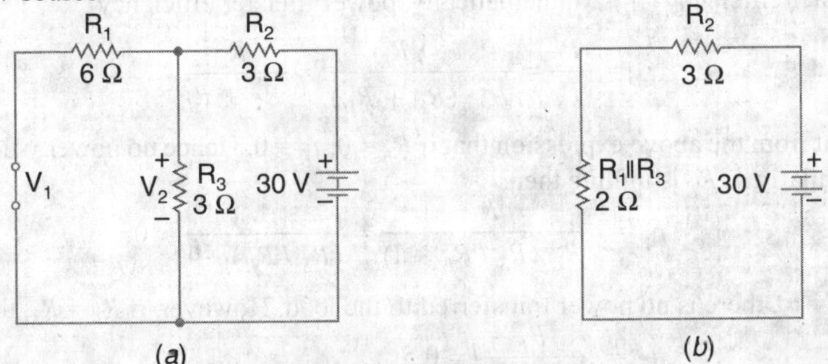

Fig. 5.26.

Fig. 5.26 (*a*) shows that resistors R_1 and R_3 are in parallel and their equivalent resistance $R_1 \| R_3$ (6 || 3) equal to 2 Ω. Therefore the circuit of 5.26 (*a*) may be redrawn in simple way as shown in Fig. 5.26 (*b*). Now using the voltage division rule again, the voltage drop across 2 Ω resistor (due to 3 V source),

$$V_2 = \left(\frac{2}{3+2}\right) \times 30 = 12 \text{ V}$$

Since the voltage across R_1 and R_3 of Fig. 5.24 (*b*) is the same, therefore, voltage drop across R_3

$$V_2 = 12 \text{ V}$$

3. Now add the individual voltage drops V_1 and V_2 across R_3 algebraically. Since, V_1 and V_2 have same polarities, therefore the total voltage drop across R_3,

$$V = V_1 + V_2 = 3 + 12 = 15 \text{ V } \textbf{Ans.}$$

and current through resistor R_3,

$$I = \frac{V}{R_3} = \frac{15}{3} = 5 \text{ A } \textbf{Ans.}$$

5.10. Maximum Power Transfer Theorem

This theorem states that *given a linear bilateral network represented by a Thevenin's equivalent circuit, the maximum power will be transferred to a resistive load, when the load resistance is equal to the Thevenin's resistance.* Consider a Thevenin's electronic circuit as shown in Fig. 5.29.

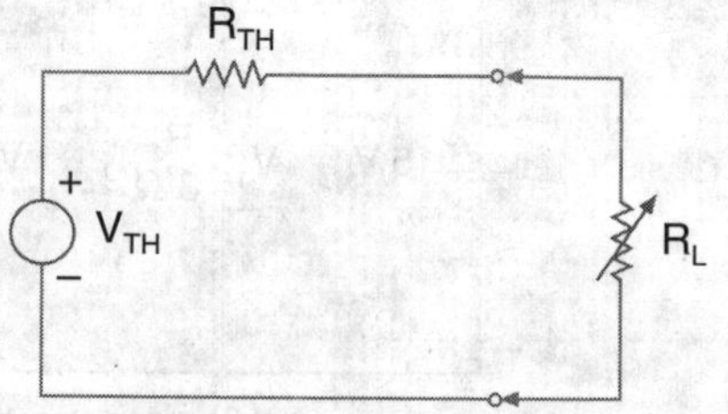

Fig. 5.27. Illustrating maximum power transfer theorem.

Sometimes, the Thevenin's equivalent resistance (R_{TH}) is also called the source resistance (R_S) because it is in series with the Thevenin's equivalent voltage source as shown in Fig. 5.27.

We know that the current through the load resistanc.

$$I = \frac{V_{TH}}{R_S + R_L}$$

and power developed in the load resistance,

$$P_L = I^2 \,.\, R_L = \left(\frac{V_{TH}}{R_S + R_L}\right)^2 \times R_L$$

Moreover power supplied by the source,

$$P_S = I^2 \,(R_S + R_L)$$

The ratio of the power developed in load resistance to the power supplied by the source is called the power transfer efficiency (η_t). Mathematically, power transfer efficiency

$$\eta_t = \frac{P_L}{P_S} = \frac{I^2 \cdot R_L}{I^2 \cdot (R_S + R_L)} = \frac{R_L}{R_S + R_L}$$

It is evident from the above expression that if $R_L = 0$, $\eta_t = 0$. Hence no power will be transferred to the load. Similarly, if R_L is infinite then,

$$\eta_t = \frac{R_L}{R_L\,(R_S / R_L + 1)} = \frac{1}{(R_S / R_L) + 1} \approx 0$$

Thus if $R_L = \infty$, there is no power transferred to the load. However, if $R_L = R_S$, then

$$\eta_t = \frac{R_L}{R_S + R_L} = 0.5$$

It means that if $R_L = R_S$ the power transfer efficiency is 50% and is maximum. The efficiency higher than this value cannot be obtained.

5.11. Applications of the Maximum Power Transfer Theorem

The maximum power transfer theorem has a great importance in electronic circuits, especially those related to communication systems. A common example is a television (TV) receiver matched to the television antenna. The equivalent circuit of a TV antenna may be represented by a Thevenin's voltage (V_{TH}) and a Thevenin resistance (R_{TH}) as shown in Fig. 5.28.

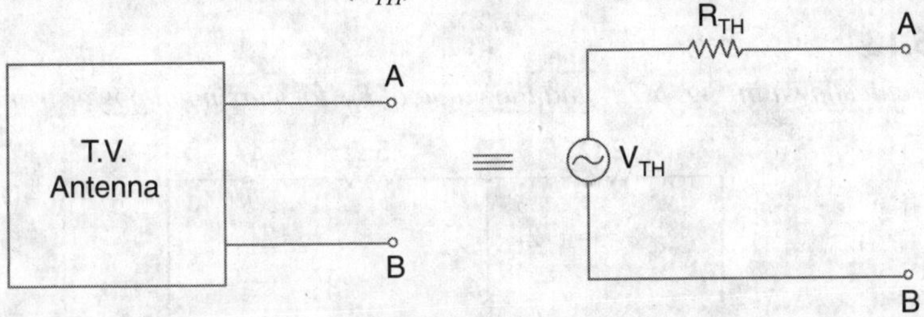

Fig. 5.28. TV antenna with its equivalent circuit.

The quality of a TV picture depends on the power supplied to the TV receiver. The greater the power supplied to the TV receivers, better will be the picture quality. This is possible only if the load resistance (R_L) of the TV receiver is equal to Thevenin resistance of the TV antenna. The Thevenin resistance of TV antenna is 300 ohms. Therefore, TV receivers are built with a resistance of 300 ohms. The telephone and microwave systems also utilize the principle of maximum power transfer. The matching resistance in telephone systems is 600 ohms, while in microwave systems, the matching resistance is 50 ohms.

Example 5.11.

A radar antenna has a Thevenin voltage of 100 mV and a Thevenin resistance of 50 Ω. What should be the value of load resistance to obtain the maximum load power? What is the maximum load power?

Solution.

Given: $V_{TH} = 100\ \mu V = 100 \times 10^{-6}$ V and $R_{TH} = R_S = 50\ \Omega$.

Value of load resistance

We know that for maximum load power, the load resistance must be equal to the Thevenin's resistance, *i.e.*,

$$R_L = R_{TH} = 50\ \Omega \textbf{ Ans.}$$

Maximum load power

We also know that maximum load power,

$$P_L = \left(\frac{V_{TH}}{R_{TH} + R_L}\right)^2 \times R_L = \left(\frac{100 \times 10^{-6}}{50 + 50}\right)^2 \times 50$$

$$= 50 \times 10^{-12}\ \text{W} = 50\ \text{pW} \textbf{ Ans.}$$

Example 5.12.

A TV antenna has a Thevenin voltage of 20 mV and a Thevenin resistance of 300 Ω. When the receiver is matched to the antenna, what is the power transferred to the receiver?

Solution.

Given: $V_{TH} = 20\text{ mV} = 20 \times 10^{-3}$ V; $R_{TH} = 300\ \Omega$ and $R_L = 300\ \Omega$.

We know that the power transferred to the receiver,

$$P_L = \left(\frac{V_{TH}}{R_{TH} + R_L}\right)^2 \times R_L = \left(\frac{20 \times 10^{-3}}{300 + 300}\right)^2 \times 300 \text{ W}$$

$$= 0.33 \times 10^{-6} \text{ W} = 0.33\ \mu\text{W } \textbf{Ans.}$$

Example 5.13.

For the circuit shown in Fig. 5.29, find the value of R_L for maximum power transfer.

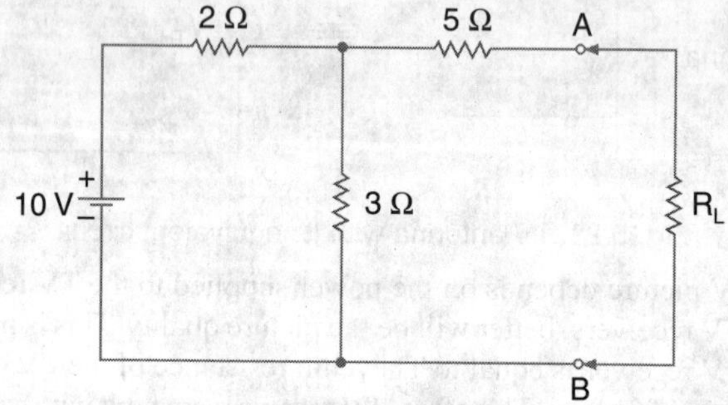

Fig. 5.29.

Solution.

We know that for maximum power transfer, the value of load resistance (R_L) must be equal to the Thevenin's equivalent resistance across the terminals *A* and *B*.

To find the Thevenin's equivalent resistance, replace the voltage source of 10 V by a short circuit as shown in Fig. 5.30 (*a*).

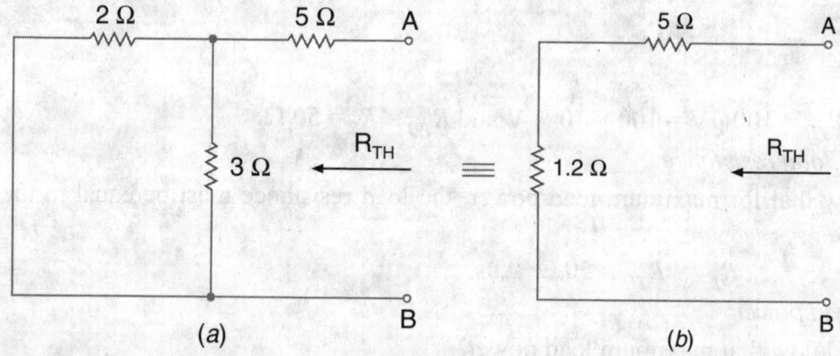

Fig. 5.30.

Looking in the direction of arrow, we find that 2 Ω and 3 Ω are in parallel and their equivalent resistance,

$$R_{eq} = \frac{2 \times 3}{2 \times 3} = \frac{6}{3} = 1.2\ \Omega$$

This equivalent resistance of 1.2 Ω appears in series with 5 Ω resistance as shown in Fig. 5.30(*b*).

∴ Thevenin's equivalent resistance across the terminals *A* and *B* $R_{TH} = 5 + 1.2 = 6.2\ \Omega$

and load resistance, $R_L = 6.2\ \Omega$ **Ans.**

NUMERICAL PROBLEMS

1. Thevenise the circuit shown in Fig. 5.31. Also determine the current through a load resistance R_L, if its value is 4 kΩ. **(Ans. V_{TH} = 5 V; R_{TH} = 4 kΩ and I = 0.25 A)**

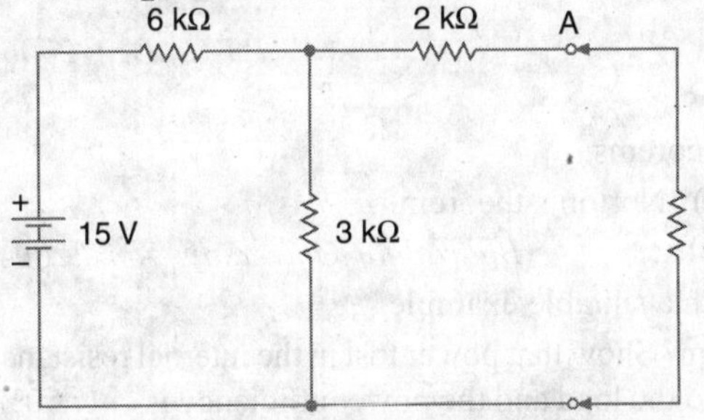

Fig. 5.31

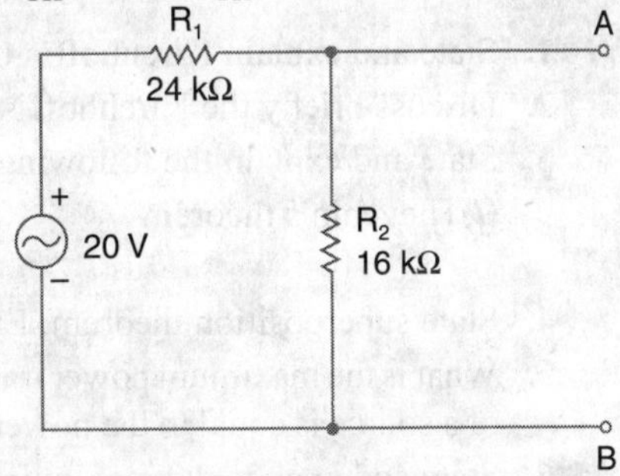

Fig. 5.32

2. Obtain the Thevenin's equivalent circuit for the network of Fig. 5.32. **(Ans. V_{TH} = 8 V and R_{TH} = 9.6 kΩ)**

3. Find the Thevenin equivalent circuit of the network shown in Fig. 5.33. Also determine the current through load resistance. **(Ans. V_{TH} = 6.67 V, I = 0.436 A and R_{TH} = 3.3 Ω)**

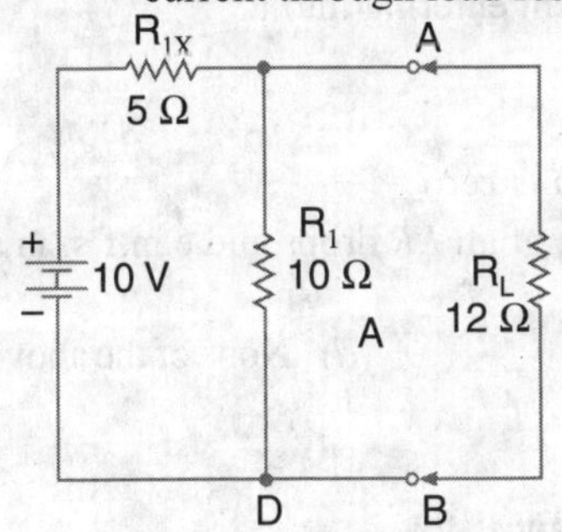

Fig. 5.33.

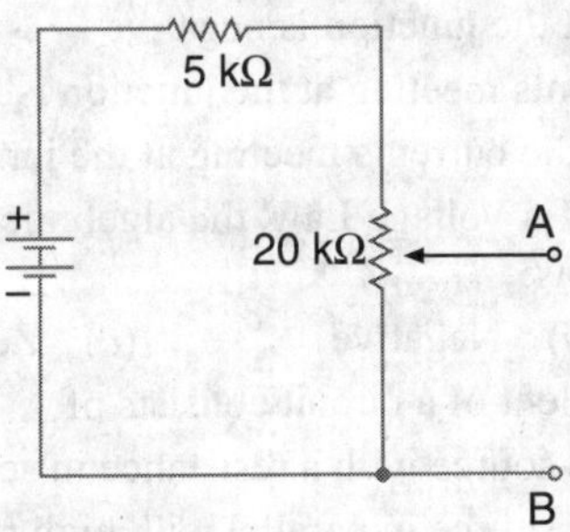

Fig. 5.34.

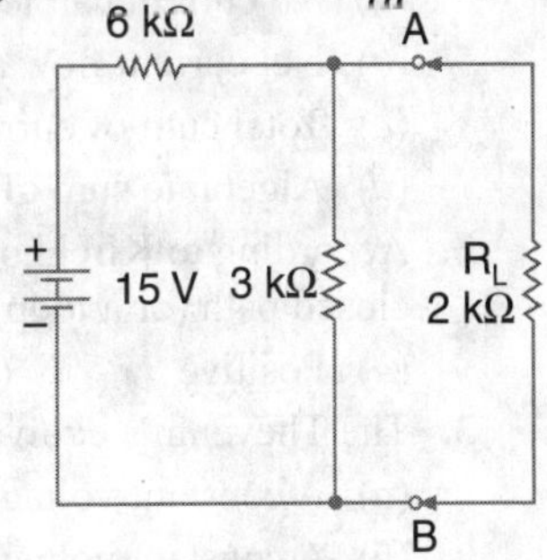

Fig. 5.35.

4. The network shown in Fig. 5.34 is connected to an electronic circuit whose resistance is 10 Ω. What is the voltage between the terminals A and B when the wiper is (a) at the middle and (b) at the top. **(Ans. 6.67 V; 10 V)**

5. Nortonize the circuit shown in Fig. 5.35. Also determine the current through R_L. **(Ans. I_N = 2.5 A; R_N = 2 kΩ and I_L = 1.25 A)**

6. Find the Norton's equivalent circuit for the network shown in Fig. 5.36. **(Ans. I_N = 4 mA and R_N = 2 kΩ)**

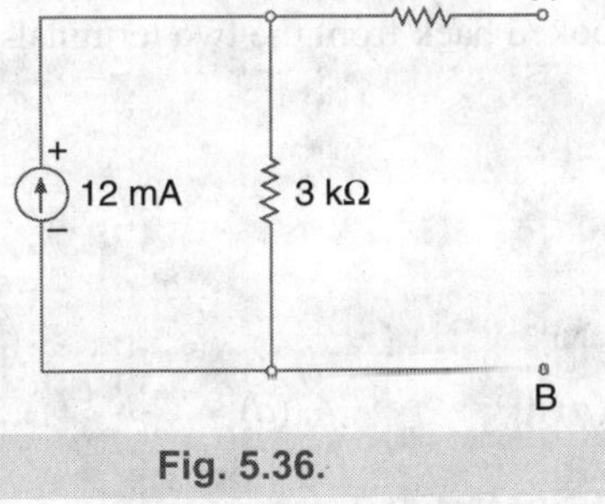

Fig. 5.36.

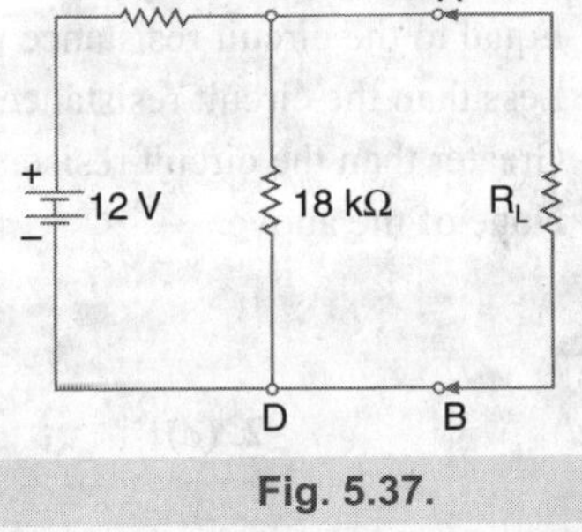

Fig. 5.37.

7. What is the value of R_B in Fig. 5.37, which can supply maximum power to the load? What is the value of load power? **(Ans. 6 kΩ and 2.67 mΩ)**

DESCRIPTIVE QUESTIONS

1. State and explain Kirchhoff's Current Law. *(P.T.U., Dec. 2006)*
2. Discuss briefly the Kirchhoff's Voltage Law.
3. State and explain the following network theorems:
   (*i*) Thevenin's theorem (*ii*) Norton's theorem.
   (Bharathiar University., April 2004)
4. State superposition theorem. Explain it with a suitable example.
5. What is the maximum power transfer theorem? Show that power lost in the internal resistance of a source is equal to the power delivered to the load and the power efficiency is only 50%.
6. State and explain the maximum power-transfer theorem and discuss its application.
   (Bharathiar University., April 2004)

MULTIPLE CHOICE QUESTIONS

1. According to Kirchhoff's Current Law:
   (*a*) No current can leave the junction without some current entering into it
   (*b*) Net current flow at the junction is negative
   (*c*) Total cum of currents meeting at the junction is zero
   (*d*) Algebraic sum of the currents meeting at the junction is zero
2. According to Kirchhoff's Voltage Law, the algebraic sum of all *I.R* drops and e.m.f's. in a closed path (or a loop) is:
   (*a*) Positive (*b*) Negative (*c*) Zero (*d*) None of the above
3. The Thevenin's equivalent of a circuit consists of
   (*a*) A constant voltage source with a resistance in series
   (*b*) A constant voltage source in parallel with high resistance
   (*c*) A constant current source with a resistance in parallel
   (*d*) None of the above
4. The Thevenin's equivalent resistance (R_{TH}) is found by
   (*a*) Short circuiting the given two terminals
   (*b*) Removing voltages sources by their internal resistances
   (*c*) Removing current sources by their internal resistance
   (*d*) Both (*b*) and (*c*)
5. In order to obtain maximum power from the load terminals of a circuit, the resistance across the load terminals should be
   (*a*) Equal to the circuit resistance when looked back from the two terminals
   (*b*) Less than the circuit resistance
   (*c*) Greater than the circuit resistance
   (*d*) None of the above

ANSWERS

1. (*d*) **2.** (*c*) **3.** (*a*) **4.** (*d*) **5.** (*a*)

Chapter

6

A.C. Fundamentals

Outline

6.1. Introduction

We have already discussed in the chapter on Electricity and Ohm's Law that an ac (*i.e.*, alternating current) is an electric current, which changes in both magnitude and direction with the time. The alternating current is produced by a voltage source whose terminal voltage polarity keeps alternating or reversing with the time. Such a alternating voltage is generated by alternator and electronic oscillators (or signal generators) etc. In this chapter, we shall discuss some important terms related to alternating current, which are important from the subject point of view.

6.2. Types of Alternating Current or Voltage

The alternating current (or voltage) may be broadly classified into the following two types :

1. ***Sinusoidal.*** The alternating current (or a voltage), whose value varies in a manner, which is similar to a *sine* or *cosine function*, is known as sinusoidal alternating current (or a voltage). The waveform of a sinusoidal alternating current is shown in Fig. 6.1 (*a*) and (*b*).

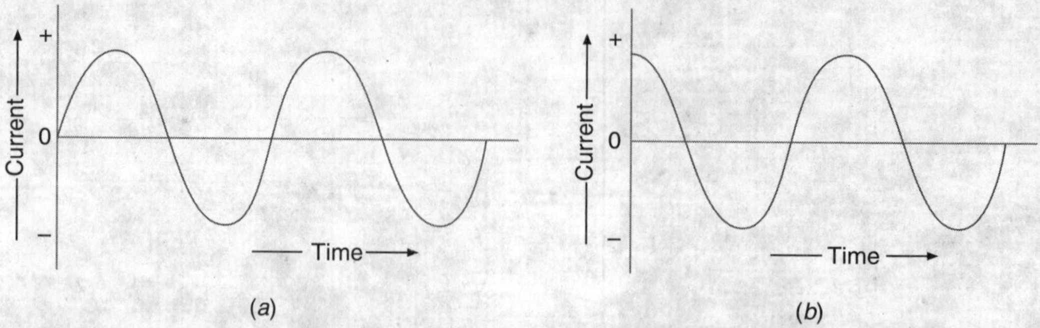

Fig. 6.1. Sinusoidal alternating current waveforms.

The sinusoidal alternating current waveform is one of the most useful waveforms in the field of electronics. Such waveforms are radiating from radio, television and telephone aerials throughout the world everyday. The power supplied to the industrial, commercial or even residential purposes (*i.e.*, 50 Hz mains supply) has also a sinusoidal waveform.

2. ***Non-sinusoidal.*** In addition to the sinusoidal alternating current, the other alternating currents in use are square, triangular and sawtooth. The waveforms of such alternating currents are as shown in Fig. 6.2 (*a*), (*b*) and (*c*) respectively.

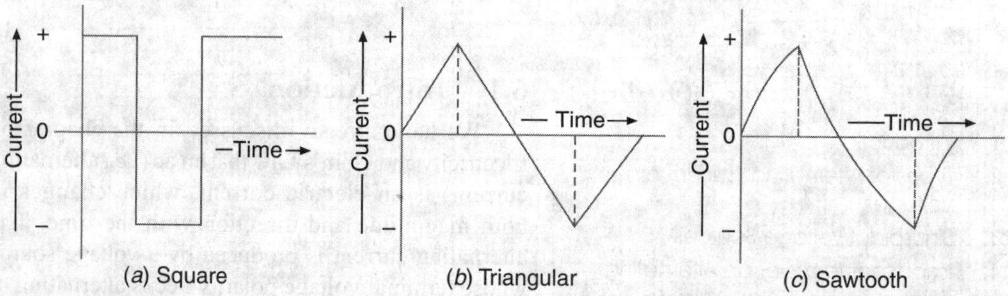

Fig. 6.2.

The square waveform is used quite frequently in computer circuits, triangular waveform in industrial control systems and the sawtooth waveform in oscilloscopes and television receivers.

The alternating current waveforms, as discussed above, are known as *symmetrical waveforms* because their positive and negative half cycles are equal in magnitude as well as time. However, in certain applications, it is desirable to have waveforms with unequal positive and negative half cycles. Such waveforms are called *unsymmetrical waveforms*.

6.3. Sinusoidal Voltage Sources

There are two basic methods of generating sinusoidal voltages namely, electromagnetic and electronic. The sinusoidal waveforms are produced electromagnetically by ac generators and electronically by ac generators.

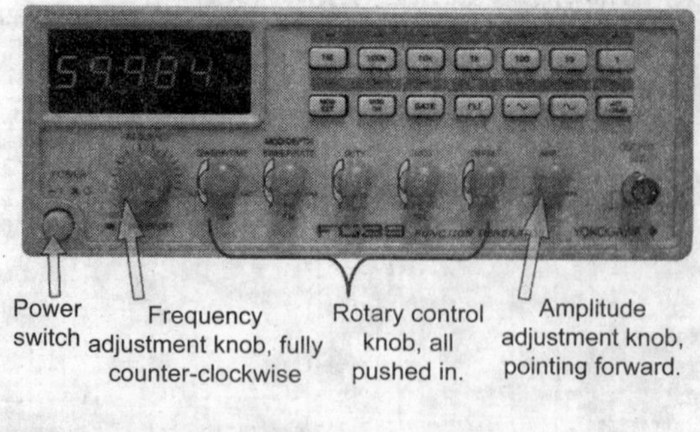

An ac generator consists of a coil placed in a magnetic field. As this coil is made to rotate in the magnetic field, a voltage is produced across two ends of the coil. After one complete revolution of the coil, one full cycle of the sine wave voltage is produced. As the coil continues to rotate, the repetitive cycles of sine wave are produced.

The signal generator is an instrument that electronically produces sine waves for use in testing or controlling electronic circuits and systems. There are a large variety of signal generators. Some of them are special purpose instruments that produce only one type of waveform in a limited frequency range. While others are programmable instruments that produce a wide range of frequencies and waveforms. All signal generators basically consist of an oscillator, which is an electronic circuit that produces sine wave voltages. The amplitude and frequency of such waves can be adjusted.

6.4. Important Terms of an Alternating Current

We have already mentioned in the Art. 6.2 that a sinusoidal alternating current waveform is the most commonly used waveform in the field of electronics. Therefore we shall use this waveform to define some important terms, which are used quite frequently in the electronic circuits. It may be noted that the definition of any important term is the same whether it is meant for current or voltage.

Following are some of the terms related with the sinusoidal alternating current, which are important from the subject point of view :

1. Cycle
2. Time period
3. Frequency
4. Amplitude
5. Peak-to-peak value
6. Instantaneous value
7. Root-mean square value
8. Average value
9. Phase
10. Phase difference.

Now we shall discuss all these important terms in the following pages.

6.5. Cycle

One complete set of positive and negative values of an alternating current as shown in Fig. 6.3 is called a cycle.

6.6. Time Period

The time taken by an alternating current to complete one cycle as shown in Fig. 6.3 is called its time period and is designated by the letter (T). Mathematically, time period is the reciprocal of number of cycles. For example, if an alternating current makes 50 cycles in one second, then its time period

$$\frac{1}{50} = 0.02 \text{ second}$$

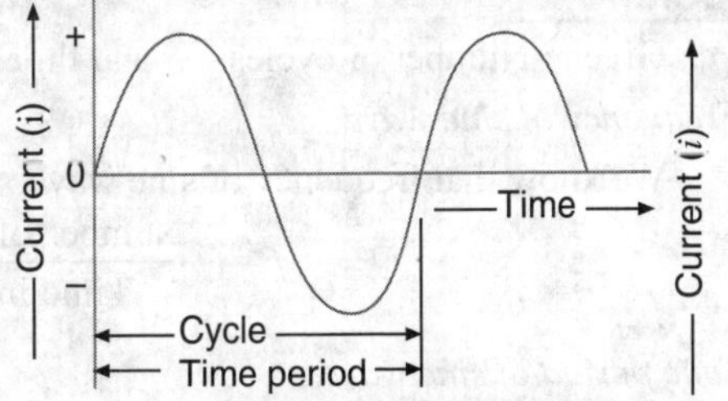

Fig. 6.3. Sinusoidal alternating current waveform illustrating cycle, time period and frequency.

Example 6.1.

An oscilloscope shows 10 cycles of a sinusoidal wave occurring in 1 millisecond. What is the time period ?

Solution.

Given : No. of cycles = 10 and time period = 1 millisecond.

We know that time period *i.e.*, time taken by one cycle

$$= \frac{\text{Time}}{\text{Number of cycles}} = \frac{1}{10} = 0.1 \text{ millisecond } \textbf{Ans.}$$

Example 6.2.

The output of a half-wave rectifier gives only the positive half-cycles which occurs in 0.01 second. What is the time period of the rectifier input? Assume the rectifier input as a sinusoidal alternating current.

Solution.

Given : Time period of positive half-cycle = 0.01 sec.

We know that sinusoidal alternating current waveform has equal positive and negative half-cycles. Therefore time period of the complete cycle.

$$T = 0.01 + 0.01 = 0.02 \text{ sec } \textbf{Ans.}$$

6.7. Frequency

The number of cycles per second, made by an alternating current in called frequency (f). Mathematically, the frequency,

$$f = \frac{\text{Number of cycles}}{\text{Times in sec}}$$

It is also defined as the reciprocal of the time period (T) of the alternating current. *i.e.*,

$$f = 1/T \text{ Hz} \qquad \text{or} \qquad T = 1/f \text{ sec}$$

The unit of frequency is hertz (Hz). But hertz is a much smaller unit for use in the electronics field. Therefore, we use much larger units known as kilohertz (kHz), megahertz (MHz) and gigahertz (GH). Such that

$$1 \text{ kHz} = 10^3 \text{ Hz}; 1 \text{ MHz} = 10^6 \text{ Hz} \qquad \text{and} \qquad 1 \text{ GHz} = 10^9 \text{ Hz}$$

Example 6.3.

An oscilloscope shows 5 cycles of a sine wave occurring in 10 msec. What is the frequency and the time period of the sine wave?

Solution.

Given : Number of cycles = 5 and time period 10 m sec = 10×10^{-6} sec

Frequency of sine wave

We know that frequency of sine wave,

$$f = \frac{\text{Number of cycles}}{\text{Time in sec}} = \frac{5}{10 \times 10^{-6}} = 0.5 \text{ MHz } \textbf{Ans.}$$

Time period of sine wave

We also know that time period,

$$T = \frac{1}{f} = \frac{1}{0.5 \times 10^6} = 2 \text{ } \mu \text{ sec. } \textbf{Ans.}$$

Example 6.4.

Channel 4 of a television receiver operates at a frequency of 69 MHz. What is its time period ?

Solution.

Given : $f = 69 \text{ MHz} = 69 \times 10^6$ Hz

We know that the time period,

$$T = \frac{1}{f} = \frac{1}{69 \times 10^6} = 0.0145 \times 10^{-6} \text{ sec.}$$

$$= 14.5 \text{ } n\text{sec } \textbf{Ans.} \qquad \text{... } (\because 1 \text{ } n\text{sec} = 10^{-9} \text{ sec})$$

6.8. Amplitude (or Peak Value)

The maximum value of positive or negative half-cycle of on alternating current is called its amplitude or peak value as shown in Fig. 6.4. It is designated by the symbol I_m, I_{max}, I_p or I_{peak}.

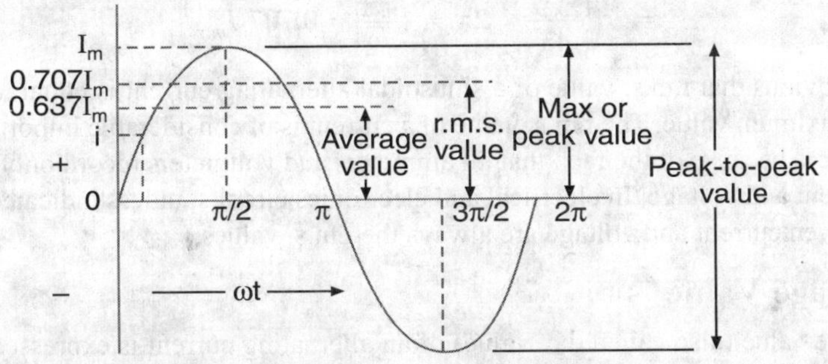

Fig. 6.4. Sinusoidal alternating current waveform illustrating amplitude, instantaneous value, r.m.s. value and average value.

6.9. Peak-to-peak Value

The sum of positive and negative peak values is called a peak-to-peak value. It is designated by the symbol I_{p-p} or $I_{peak-to-peak}$ and is as shown in Fig. 6.4. The peak to peak value of a sinusoidal alternating current is equal to 2 times the peak value.

6.10. Instantaneous Value

The value of an alternating current, at any particular moment, is called its instantaneous value. It is designated by the letter '*i*' as shown in Fig. 6.4. Mathematically the instantaneous value of a sinusoidal alternating current,

$$i = I_{max.} \sin \omega t = I_{max.} \sin \theta$$

where $I_{max.}$ = Maximum or peak value,

ω = Angular frequency (whose value is equal to $2\pi f$ in radians per second), and

t = Time in seconds.

6.11. Root Mean Square Value

The root mean square value (abbreviated as r.m.s. value) is expressed by that *steady (d.c.) current, which while flowing through a given circuit for a given time, produces the same heat as produced by an alternating current when flowing through the same circuit for the same time*. It is also known as *effective value* or *virtual value* of an alternating current. However, in actual practice, the terms r.m.s. and effective value are used more extensively. The r.m.s. value is designated by the symbol I_{rms} or *I*. But in actual practice, the symbol *I* is more commonly used. The r.m.s. value may be determined by taking the mean of the squares of the instantaneous values of current over one complete cycle. Mathematically, the r.m.s. value,

$$I = \sqrt{\frac{1}{2\pi}\int_0^{2\pi} i^2 . d\theta}$$

Substituting $i = I_{max} \cdot \sin\theta$ in above expression,

$$I = \sqrt{\frac{1}{2\pi}\int_0^{2\pi} I_{max^2} . \sin^2\theta \, d\theta} = \sqrt{\frac{I_{max^2}}{2\pi}\int_0^{2\pi} \sin^2\theta \, d\theta}$$

$$= \sqrt{\frac{I_{max}^2}{2\pi}\int_0^{2\pi}\left(\frac{1-\cos^2\theta}{2}\right)d\theta} = \sqrt{\frac{I_{max}^2}{4\pi}\left|\theta - \frac{\sin^2\theta}{2}\right|_0^{2\pi}}$$

$$= \sqrt{\frac{I_{max}^2}{4\pi}\times 2\pi} = \frac{I_{max}}{\sqrt{2}} = 0.707\ I_{max}$$

It is thus obvious that r.m.s. value of a sinusoidal alternating current is equal to the product of 0.707 and its maximum value. The r.m.s. value of a current is of considerable importance in the field of electronics. It is because of the fact, that all ammeters and voltmeters record only r.m.s. values of alternating current and voltage. In electrical and electronics circuits, unless indicated otherwise, the values of the given current and voltage are always the r.m.s. values.

6.12. Average Value

The average value (also called d.c. value) of an alternating current is expressed by that steady (d.c.) current, which transfers across any circuit, the same charge as transferred by that of an alternating current during the same time. It is designated by the symbol I_{av} or I_{dc}. In the case of a symmetrical alternating current waveform (*i.e.*, a current whose positive and negative half cycles are equal), the average value over a complete cycle is zero. Hence in that case, the average value is obtained by adding or integrating the instantaneous values of current over one half-cycle only. Thus average value of alternating current,

$$I_{dc} = \frac{1}{\pi}\int_0^{\pi} i\cdot d\theta = \frac{1}{\pi}\int_0^{\pi} I_{max}\cdot\sin\theta\cdot d\theta$$

$$= \frac{I_{max}}{\pi}\int_0^{\pi}\sin\theta\cdot d\theta = \frac{I_{max}}{\pi}\left|-\cos\theta\right|_0^{\pi}$$

$$= \frac{I_{max}}{\pi}\left|-(-1)-(-1)\right| = \frac{2\,I_m}{\pi}$$

$$= 0.637\,I_{max}$$

It is thus obvious that average value of a symmetrical sinusoidal alternating current is equal to the product of 0.637 and its maximum value.

6.13. Relations for Sinusoidal Alternating Voltage

The relations for sinusoidal alternating voltage are similar to those of an alternating current. Thus an instantaneous value of an alternating voltage,

$$v = V_{max}\sin\omega t = V_{max}\cdot\sin\theta$$

where V_{max} is the maximum value of an alternating voltage. The r.m.s. or effective value of a sinusoidal alternating voltage,

$$V = 0.707\ V_{max}$$

and its average value, $V_{dc} = 0.637\ V_{max}$

Example 6.5.

A sinusoidal signal has a maximum value of 20 mV. What is the r.m.s. and average values of the signal ?

Solution.

Given : $V_{max} = 20$ mV

We know that r.m.s. value of the sinusoidal signal,

$$V = 0.707\ V_{max} = 0.707\times 20 = 14.14\ \text{mV}$$ **Ans.**

and its average value, $V_{dc} = 0.637\ V_{max} = 0.637\times 20 = 12.74$ mV **Ans.**

6.14. Phase

The fraction of a cycle or time-period, which has elapsed since an alternating current (or voltage) last passed a given reference point, is called its phase. Usually, the reference point is taken as the starting point itself. The phase may be expressed in any one of the following three different ways :

1. Angle expressed in radians or degrees,
2. Fraction of a time-period, and
3. Time measured in seconds.

Consider a sinusoidal alternating current waveform as shown in Fig. 6.5 (*a*). The phase at point *A* may be expressed as follows :

1. Angle expressed in radians or degrees = $\pi/4$ or 45°
2. Fraction of a time-period = $\dfrac{\pi/4}{2\pi} = \dfrac{1}{8}$th of a cycle.
3. Time measured in seconds = $\dfrac{T}{8} = \dfrac{1}{50 \times 8} = \dfrac{1}{400}$ sec

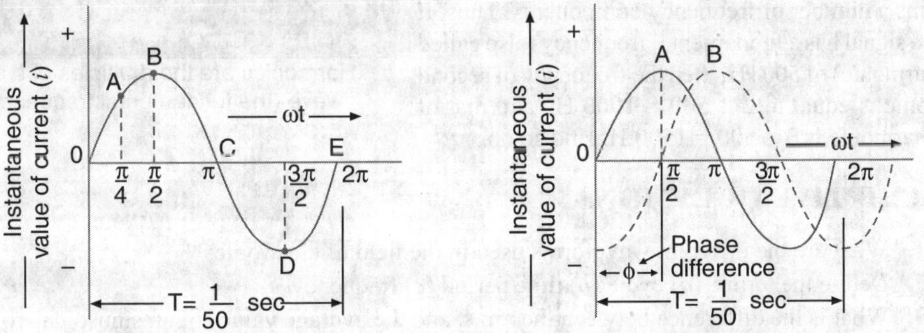

(*a*) Phase (*b*) Phase difference

Fig. 6.5.

6.15. Phase Difference

The difference in phases between two alternating currents (or voltages), of the same frequency, is called phase difference. It is designated by letter 'ϕ' and may be expressed in any one of the following three ways :

1. Angle in radians or degrees,
2. Fraction of a time-period, and
3. Time measured in seconds.

Consider two alternating currents as shown in Fig. 6.5 (*b*). The phase difference between these two waveforms may be expressed as follows :

Angle in radians or degrees = $\pi/2$ or 90°

$$\text{Fraction of a time-period} = \frac{\pi/2}{2\pi} = \frac{1}{4}\text{ of a cycle.}$$

$$\text{Time measured in seconds} = \frac{T}{4} = \frac{1}{50 \times 4} = \frac{1}{200}\text{ second.}$$

The above definition of phase difference does not give any indication regarding which current, reaches its maximum value first. This problem may be solved by using the term '*lead*' or '*lag*' with the phase difference. A leading current is that which reaches its maximum or zero value earlier as compared to the other. Similarly, a lagging current is that which reaches its maximum or zero value later than the other. For example, in Fig. 6.5 (*b*), the phase difference between currents *A* and *B* may now be expressed as *A* leading *B* by 90°. Similarly, the phase difference may also be expressed as *B* lagging *A* by 90°.

If we take *A* as the reference current, the equations for the instantaneous values of *A* and *B* are,

$$i_A = i_{max} \sin \omega t$$

and
$$i_B = I_{max} \cdot \sin\left(\omega t - \frac{\pi}{4}\right) = I_{max} \cdot \sin(\omega t - 90°)$$

6.16. Harmonics

The term 'harmonics' means the frequencies, which are an integral multiples of the lowest or fundamental frequency. It is useful in the signals with a complex waveform. *i.e.*, a waveform, which contains a number of frequency components. For example, the voice or a music signal, picked up by a microphone, contains a number of frequency components. Thus, if a voice signal has a fundamental frequency (also called first harmonic) of 500 Hz, then the frequency of second harmonic is equal to 2 × 500 = 1000 Hz and that of third harmonic is 3 × 500 = 1500 Hz and so on.

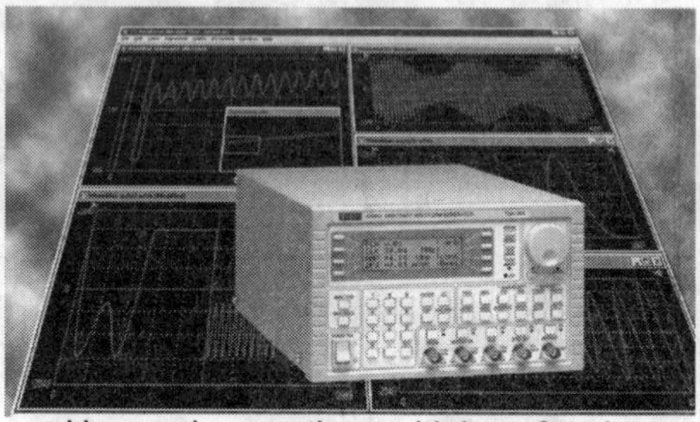

Harmonics are the multiples of a sine wave (the fundamental frequency).

DESCRIPTIVE QUESTIONS

1. What are the different waveforms used in the field of electronics ?
2. Define the terms, (*a*) cycle (*b*) time period (*c*) frequency.
3. What is the difference between the r.m.s. and the average values of an sinusoidal voltage ?
4. Explain briefly, what do you understand by the terms phase and phase difference.
5. What do you understand by average current repetitive peak current, non-repetitive current, peak-inverse voltage and reverse saturation current? (*GBTU/MTU., 2005-06, 2006-07*)
6. What do you understand by r.m.s value and average value as applied an A.C circuit. (*PTU., May 2007*)

MULTIPLE CHOICE QUESTIONS

1. The direction of an a.c. current
   (*a*) is in one direction only (*b*) keeps on changing with the time
   (*c*) keeps on reversing with the time (*d*) none of the above
2. The equation of an alternating current is given by the relation, *V* = 50 sin (q + 45°). Its r.m.s. value is
   (*a*) 50 V (*b*) 35.4 V (*c*) 70.7 V (*d*) 100 V
3. A sine wave has a period of 0.2 millisecond. Its frequency is :
   (*a*) 5 (*b*) 50 (*c*) 500 (*d*) 5000
4. The mains supply used for residential purposes has a frequency of 50 Hz, its period is :
   (*a*) 0.02 sec (*b*) 0.2 sec (*c*) 2 sec (*d*) 20 sec
5. A sinusoidal current has an r.m.s. value of 10 V. Its peak-to-peak value is
   (*a*) 7.07 V (*b*) 14.14 V (*c*) 6.37 V (*d*) 0.628 V

ANSWERS

1. (*c*) **2.** (*b*) **3.** (*d*) **4.** (*a*) **5.** (*b*)

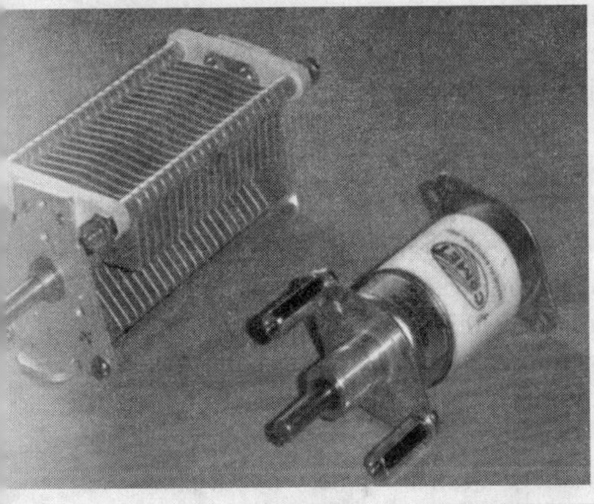

7

Chapter

Passive Circuits Elements

Outline

7.1. Introduction

A passive circuit element refers to any device or a component; which does not introduce gain or does not have directional function. In other words, such components conduct current in either direction and therefore are known as bilateral devices. In actual practice, the most commonly used passive circuit elements are of the following three types:

1. Resistors 2. Inductors 3. Capacitors

7.2. Resistors

A resistor is an electrical component, which has been manufactured with a specified amount of resistance. The resistors can conduct current in both the

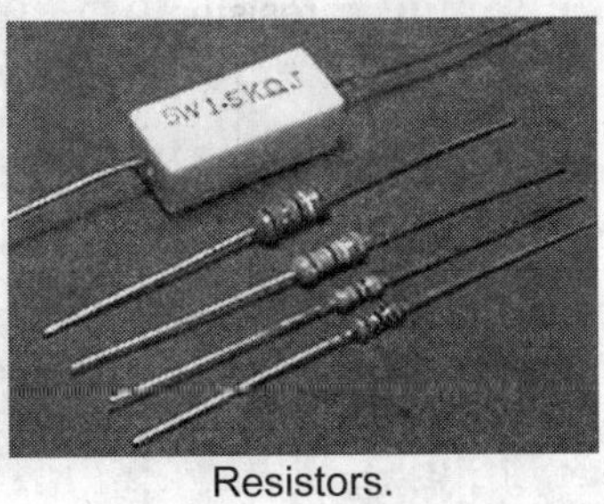

Resistors.

directions. Thus the resistors may be connected in an electric circuit without concern for lead polarisation. The resistors are used mainly for two purposes, namely controlling the flow of electric current and providing desired amounts of voltage in electric or electronic circuits.

7.3. Resistor Specifications

The resistors are specified in terms of their resistance values, tolerance, power rating and thermal stability. By tolerance, we mean the allowed variation permitted in the normal or marked value of the resistor. It means that the actual value of the resistor may be either greater or smaller than that of the indicated value, by a factor given by the specified tolerance. Thus resistors are manufactured with a specified tolerance. For example, a 5000 Ω resistor with a tolerance of ± 10% will have an actual resistance value anywhere between 4500 Ω and 5500 Ω or in other words 500 Ω greater or smaller than the rated value.

The power rating of a resistor is given by the maximum wattage it can dissipate, without excessive heating. Since the power rating is proportional to the square of a current, therefore current must not be higher than its safe value. If the current exceeds the safe value, the resistance will burn out. Usually, carbon composition resistors will fail, if operated at or near the rated power values. In this case, the resistor will not burn out. But the failure is gradual, which takes many months. It changes gradually to a much lower amount of resistance. This causes an improper operation of an electronic circuit particularly in amplifier. Thus in order to increase the life of resistors, these must be used at much less power than their rated values. A standard practice is to use a power dissipation of about half of the rating of the resistor.

The thermal stability of a resistor is indicated by the temperature coefficient specification, which is usually expressed in parts per million per degree centigrade (± ppm/°C). The smaller value of temperature coefficient will have less variation in the resistance value. Therefore, smaller value of temperature coefficient means a higher thermal stability of a resistor.

7.4. Classification of Resistors

Fig. 7.1 shows the classification of resistors in the form of a family tree. The resistors are basically of two types, namely linear resistors and non-linear resistors. Each type is further subdivided into many types as shown in the figure.

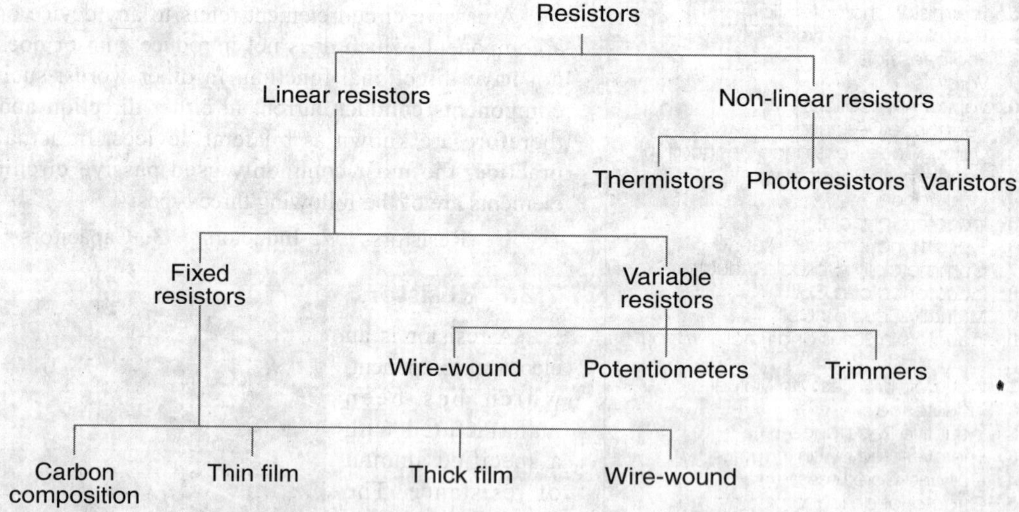

Fig. 7.1. Resistor family tree.

1. ***Linear resistors.*** The resistors, through which the current is directly proportional to the applied voltage, are called linear resistors. Such resistors have a property that their resistance values do not change with the variation in applied voltage, temperature or light intensity. The linear resistors are of two types namely fixed resistors and variable resistors.
2. ***Non-linear resistors.*** The resistors, through which the current is not directly proportional to the applied voltage, are called non-linear resistors. Such resistors have a property that their resistance values change with the variation in applied voltage, temperature or light intensity. The non-linear resistors are of three types namely thermistor, photo-resistor and varistor.

7.5. Fixed Resistors

The fixed resistors are those, whose value do not change with the variation in applied voltage, temperature and light intensity. Such resistors are available in various shapes and sizes, with both axial and radial leads as shown in Fig. 7.2.

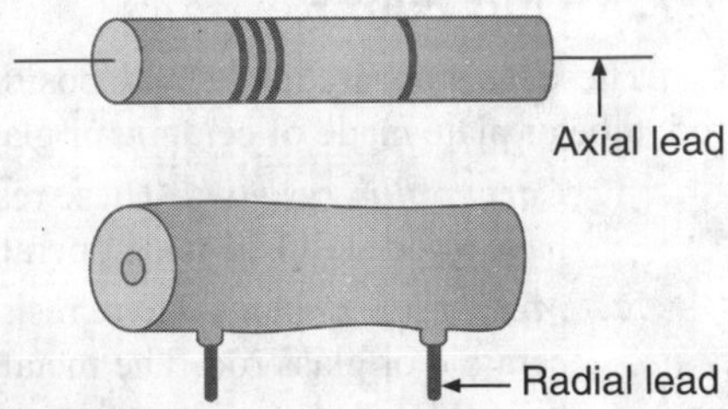

Fig. 7.2. Fixed resistors.

In addition to this, the fixed resistors are available with lugs for installation by soldering or mounting with screws and rivets. The fixed resistors are of the following types:

1. Carbon composition resistors.
2. Thin film resistors.
3. Thick film resistors.
4. Wire-wound resistors.

7.6. Carbon Composition Resistors

These resistors are made by mixing carbon powder and insulating binders to produce the desired value of resistance. The resulting resistance values are within ± 10% of the desired value. However, the resistors with ± 5% tolerance are also obtained through special techniques. Usually, the resistance element is a simple rod of carbon powder, which is enclosed in a plastic case for insulation and mechanical strength as shown in Fig. 7.3 (*a*). The two ends of the carbon resistance element are joined to metal caps with leads of tinned wire. The leads are provided for soldering the resistor into a circuit.

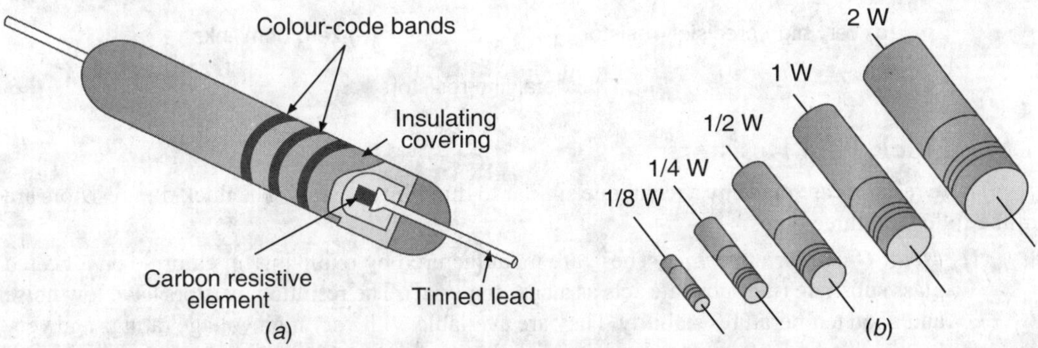

Fig. 7.3. Carbon composition resistors.

The carbon composition resistors are available in resistance values ranging 1 Ω to 22 MΩ (Refer to Appendix A) and power ratings of $\frac{1}{8}, \frac{1}{4}, \frac{1}{3}, \frac{1}{2}, \frac{2}{3}$, 1 and 2 watts. The size of these resistors varies with the power ratings as shown in Fig. 7.3 (*b*). Fig. 7.4 (*a*) through (*d*) shows the few commonly available carbon composition resistors. These resistors are widely used in electronic equipments.

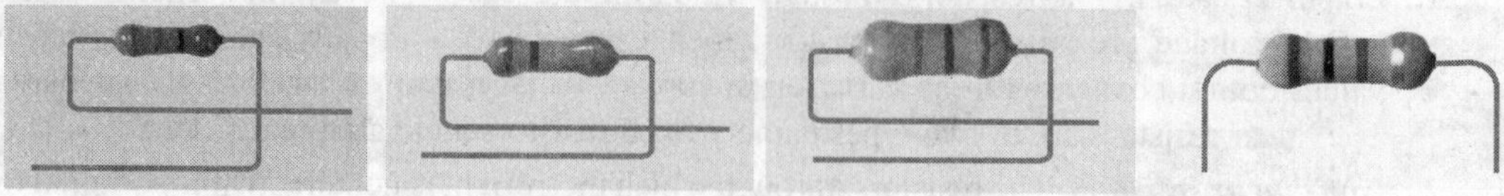

(*a*) 0.25 W resistor (*b*) 0.33 W resistor (*c*) 0.66 W resistor (*d*) 1 W resistor

Fig. 7.4. Carbon film resistors.

7.7. Thin Film Resistors

These resistors are made by depositing a very thin layer of conducting material on an insulated rod, tube or plate made of ceramic or glass. Thin film resistors are of the following two types:

1. ***Carbon film resistors.*** These resistors are cheaper than carbon composition resistors. They have good stability, wide operating frequency range and low noise.
2. ***Metal film resistors.*** There resistors are made by depositing a very thin layer of metal on a ceramic or glass rod. The metal film is spiral cut to the desired resistance. These resistors have tolerances ranging from ± 0.025% to 2%, of the desired value. Fig. 7.5 (*a*) through (*e*) shows some metal film resistors. Notice the package used for very high precision resistor in Fig. 7.5 (*d*).

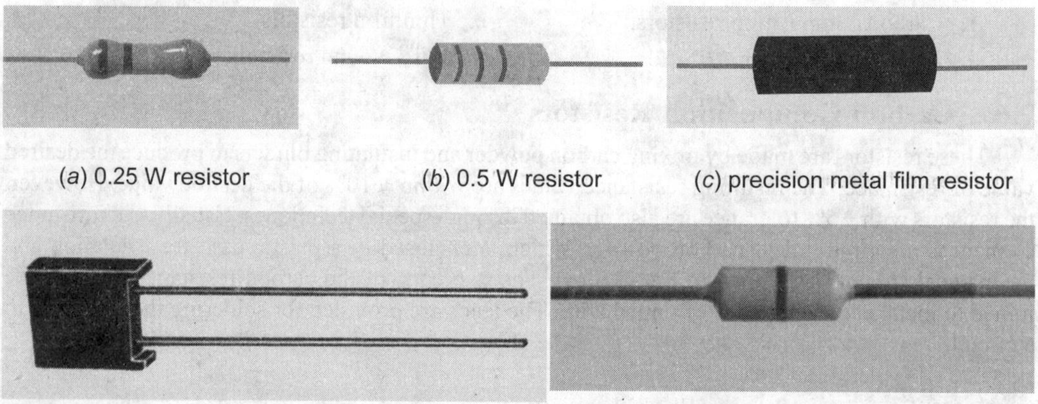

(*a*) 0.25 W resistor (*b*) 0.5 W resistor (*c*) precision metal film resistor

(*d*) very high precision resistor (*e*) zero ohm links

Fig. 7.5. Metal film resistors.

7.8. Thick Film Resistors

These resistors are made by a technique similar to thin film resistors. The thick film resistors are of the following three types:

1. ***Metal Oxide film resistors.*** These are manufactured by oxidizing tin chloride on a heated glass substrate (the substrate acts as a base material). The resulting resistors have low noise and good temperatures stability. They are available with very high voltage ratings and very high resistance values. These resistors are available up to 10^6 MΩ on special order.
2. ***Bulk property film resistors.*** These are made of metal film, which is photoetched to provide close resistance tolerances ranging from ± 0.1% to 1%. These resistors have extremely low noise and are virtually non-inductive. They have a very low temperature coefficient and can work at high frequencies.

3. ***Cermet film resistors.*** These are made by placing a coating of metal alloy alongwith insulating materials on a ceramic substrate. The combination is then fired into a ceramic metal called cermet. The cermet film resistors are available as axial lead resistors and resistor networks as shown in Fig. 7.6 (*a*) and (*b*).

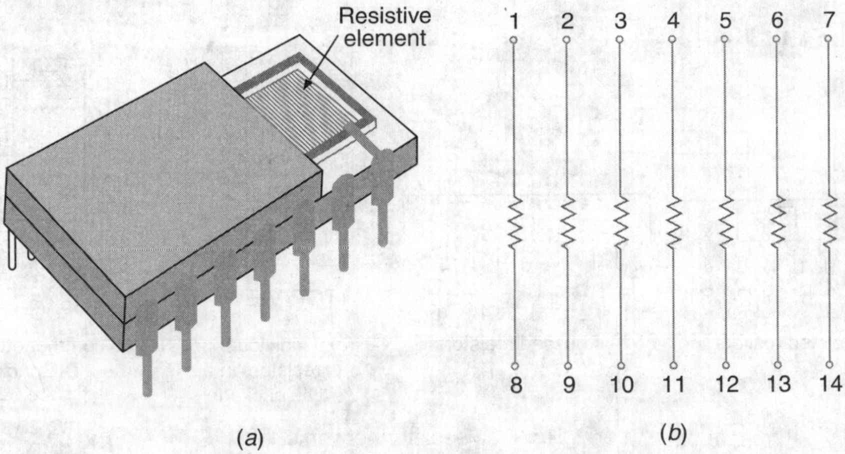

Fig. 7.6. Cermet film resistors.

The cermet material provides the highest resistance than any other resistive material. The cermet resistors are available with values up to 500 MΩ and tolerances ranging from ± 0.5% to 2%.

Fig. 7.7 (*a*) through (*d*) shows some cermet film (also called chip resistors). Notice the resistors used in Fig. 7.7 (*a*), (*b*) and (*c*). Such resistor are used as surface mount devices.

(*a*) Chip resistor-0.063 W (*b*) Chip resistor-2 W (*c*) Chip resistor arrays (*d*) High voltage resistor

Fig. 7.7. Thick film chip resistors.

Fig. 7.8 shows resistor networks in a variety of configurations. As seen, the resistor network shown in Fig. 7.8 (*a*) contains 4 isolated resistors. On the other hand, the resistor network shown in Fig. 7.8 (*b*) has also 4 resistors but one end is connected together. Such a combination is called bussed resistors. Fig. 7.8 (*c*) shows 8 individual resistors while Fig. 7.8 (*d*) shows 15 resistors with one end connected together. Notice the resistor networks indicated in Fig. 7.8 (*a*) and (*b*) are available in single-in-line (SIL) package while those shown in Fig. 7.8 (*c*) and (*d*) are available in dual-in-line (DIL) package. You will learn about the different types of packages in chapter on Integrated circuits.

The resistor networks packages are available that contain three, four or five individual resistors or commoned configurations (*i.e.* with one end connected together) of four, five, seven or nine resistors. They are ideal for high density circuit design and typical applications include pull up/down, D-to-A conversions etc.

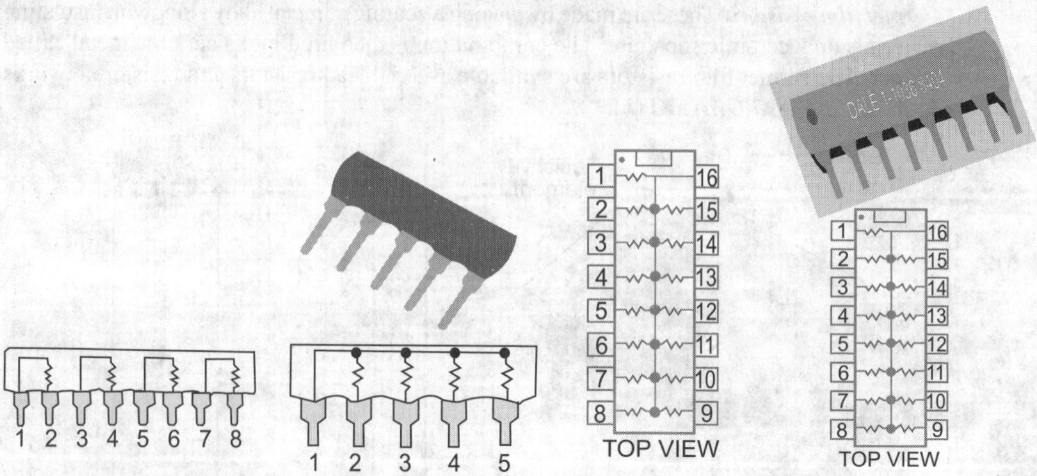

(a) 4 isolated resitors (b) 4 bussed resistors (c) 8 individual resistors in a DIL package (d) 15 commoned resistors in a DIL package

Fig. 7.8. Resistor networks.

7.9. Wire-Wound Resistors

These resistors are made by winding resistive wire such as nichrome (a nicklet-chromium alloy) on a ceramic film. The wire is then coated with an insulative material, which is either a vitreous enamel or silicon ceramic material. Refer to Fig. 7.9. The wire-wound resistors are costly as compared to other types of resistors. But these resistors have excellent electrical properties such as low noise, good time-stability and good overload (high current) characteristics. The wire-wound resistors are suitable for use in d.c. and audio-frequency applications. They cannot be used in high frequency applications due to the inductance and capacitance present in the resistor.

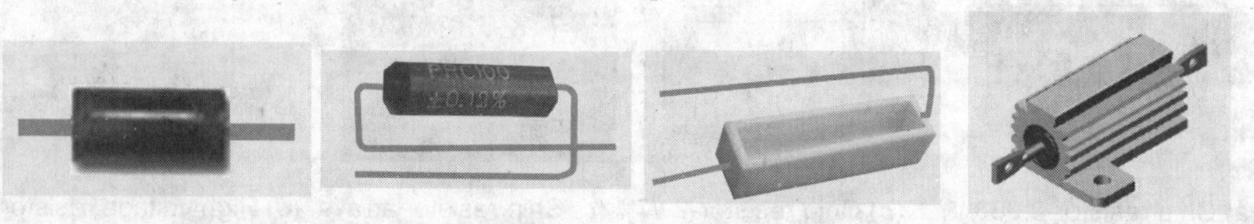

(a) 0.5 W precision (b) 6 W vitreous enamel (c) 4 W ceramic (d) 10 W aluminum housed

Fig. 7.9. Wire-wound fixed resistor.

Fig. 7.9 (*a*) through (*e*) shows wire-wound fixed resistors in various shapes and sizes. Notice the material used to cover the wire-wound resistors. The resistor shown in Fig. 7.9 (*a*) is a 0.5 W wire-wound resistor. Since the power rating is low, it does not require any heat sink. However the wire-wound resistors shown in Fig. 7.9 (*c*) and (*d*) requires metal heat sinks to dissipate heat to the surroundings.

7.10. Variable Linear Resistors

These resistors, like fixed resistors, are used to control current flow and provide desired amounts of voltage in electric circuits. But unlike fixed resistors, the resistance values of variable linear resistors can be varied from 0 to a specified value. Fig. 7.10 (*a*) shows a simple application of a variable resistor used to limit current in a lamp. By adjusting the position of the variable resistor, the intensity of lamp can be changed from full brightness to total darkness.

Fig. 7.10 (*b*) shows, another application of a variable resistor, where it is used to vary the gain of amplifier.

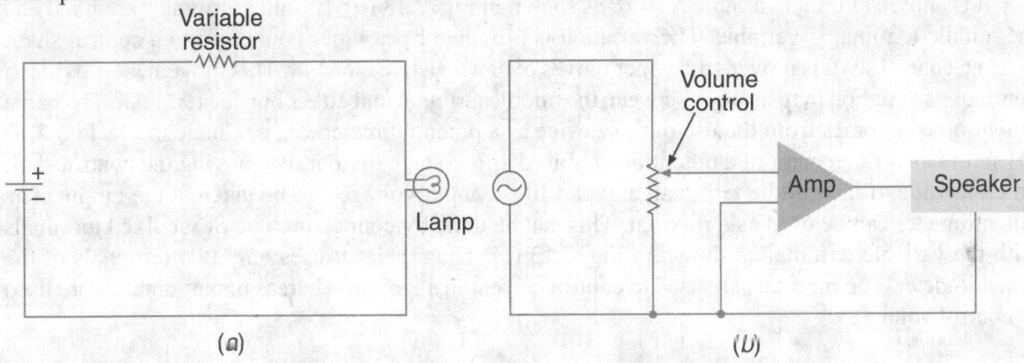

Fig. 7.10. Applications of variable resistors.

When a variable resistor is used in this manner, it is called a volume control. The variable resistors are of the following three types:

1. Variable wire-wound **2.** Potentiometer and **3.** Trimmer

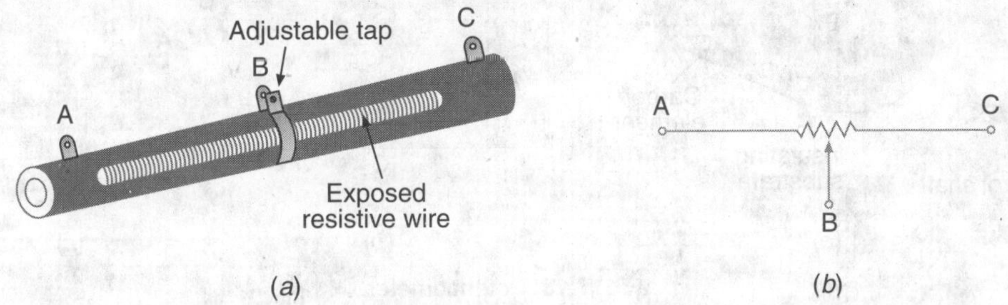

Fig. 7.11. Variable wire wound resistors.

7.11. Variable Wire-Wound Resistors

These resistors are made of nichrome wire wound on a ceramic core and covered with an insulative coating. A window is left in the insulating cover, which exposes the resistive wire as shown in Fig. 7.11 (*a*). An adjustable tap B rides along the exposed wire, which makes electrical contact with the wire. Fig. 7.11 (*b*) shows the schematic symbol of a variable wire-wound resistor.

The points *A* and *C* represent the end points, and *B* indicates the adjustable point. The variable wire-wound resistors are used in power supplies and low-frequency a.c. circuits. Because of their inductive and capacitive properties, the wire-wound resistors are not suitable for high frequency applications. It may be noted that the wire-wound resistors are not used for continuously variable control. But they are 'preset' at a given value and held at that value for extended period of operation. The wire-wound resistors are available with resistance values ranging from 1 Ω to 150 kΩ with ± 5% to ± 10% tolerances and power rating from 3 to 200 W.

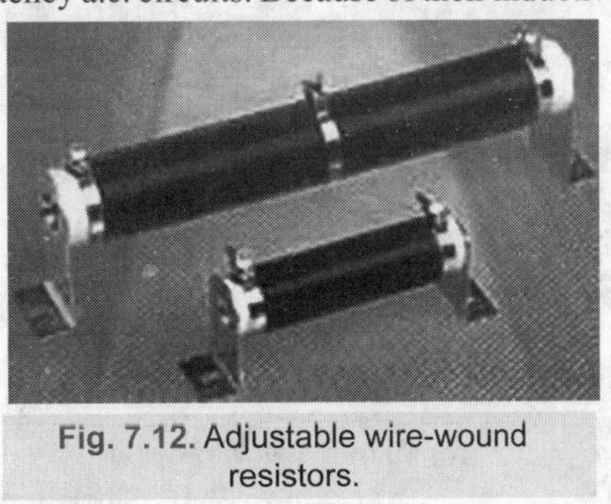

Fig. 7.12. Adjustable wire-wound resistors.

Fig. 7.12 shows two commercially available variable (or adjustable) wire-wound resistors.

7.12. Potentiometer

It is a three terminal variable resistor as shown in Fig. 7.13 (*a*). Its outer terminals are fixed and the middle terminal is variable. The variation is provided by a wiper connected to a control shaft. When a control shaft is moved, the wiper moves over a resistive element. This movement provides a continuous variation in resistance between the middle terminal and either outside terminal. The name potentiometer comes from the use of this device as a potential meter *i.e.*, a voltage meter. Fig. 7.13 (*b*) shows an arrangement of a potentiometer used as a voltage divider. By moving the control shaft on either side of the middle terminal, any part of the input voltage can be taken at the output. The potentiometer can be used as a rheostat. This can be done by connecting one of the fixed terminals with the variable terminal as shown in Fig. 7.13 (*c*). Thus rheostat uses only two terminals of the potentiometer. The rheostats are used to control current in a circuit, whereas potentiometers are used to control voltage.

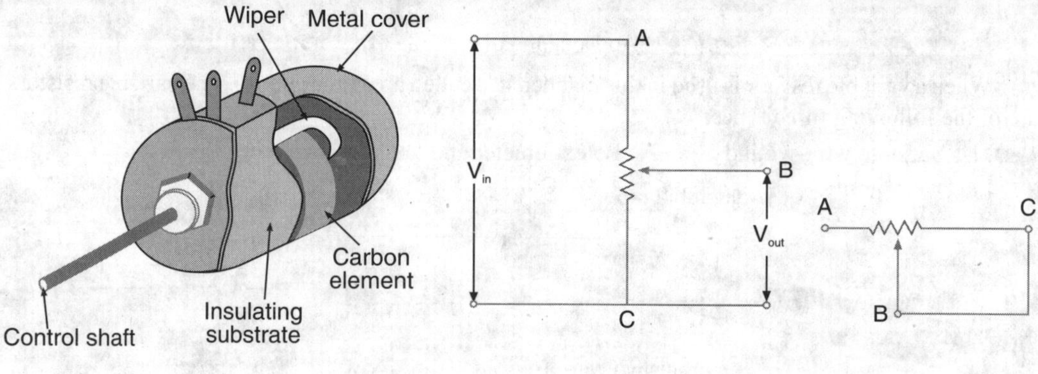

(*a*) (*b*) (*c*)

Fig. 7.13. Potentiometer.

The resistive element of a potentiometer is made from material such as carbon composition, carbon film, cermet and wire. These days, the carbon-composition potentiometers are the most widely used potentiometers. They have a low cost and long life, low noise and good reliability. Carbon composition potentiometers may be used as volume controls, tone controls, linearity and brightness controls in radio and television receivers.

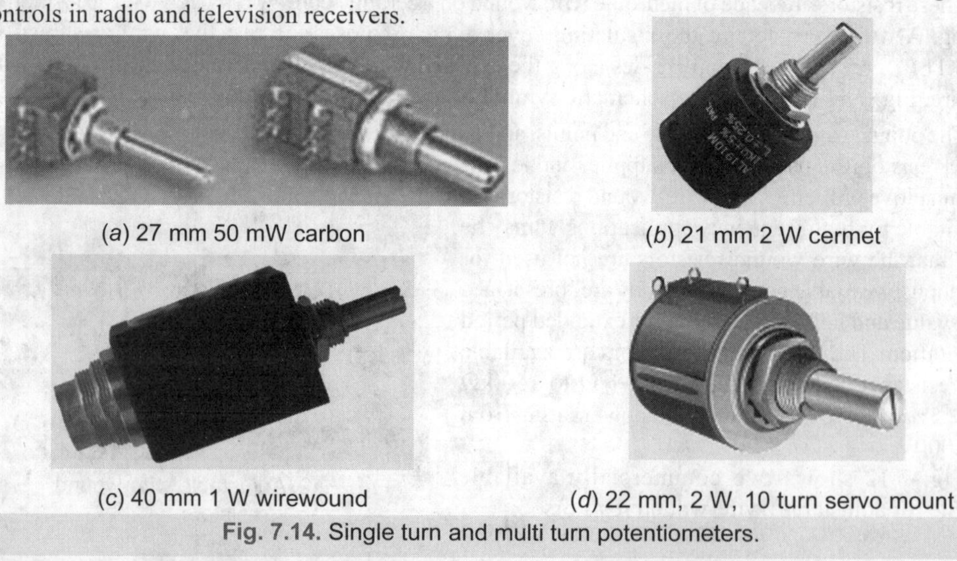

Fig. 7.14. Single turn and multi turn potentiometers.

Fig. 7.14 shows some of the commercially available potentiometers. Notice the dimensions of these devices. Also notice that the devices shown in Fig. 7.14 (*a*), (*b*) and (*c*) are single turn potentiometers while the potentiometer shown in Fig. 7.14 (*d*) is a 10 turn potentiometer. This potentiometer is suitable for connecting to the small dc or ac motors for position control.

7.13. Trimmers

These are used in electronic circuits to trim the circuit to the required operating conditions by inserting a small screw driver into a slot and turning one or more times. The trimmers are available both in single-turn and multi-turns as shown in Fig. 7.15 (*a*) and (*b*).

The materials used in the construction of a trimmer are carbon composition, carbon film cermet and wire. The trimmers are available for resistance values ranging from 50 Ω to 5 MΩ, with a power rating from 1/4 to 3/4 watt.

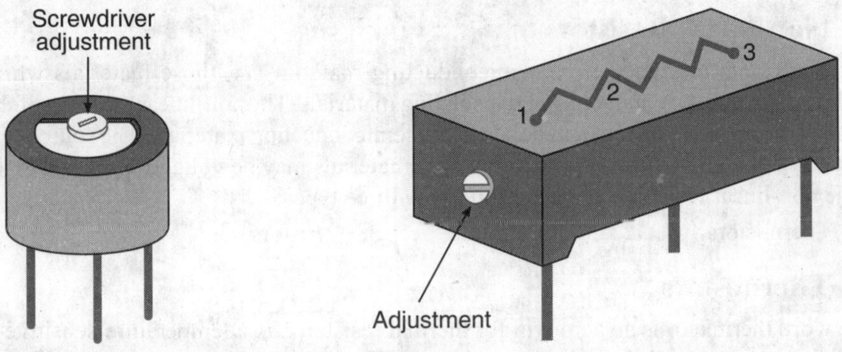

(*a*) (*b*)

Fig. 7.15.

Fig. 7.16 (*a*) through (*d*) shows some of the commercially available single turn trimmers. Notice the different shapes and sizes. Also notice that trimmers shown in Fig. 7.16 (*a*) and (*b*) can be adjusted by using a finger. While those shown in Fig. 7.16 (*c*) and (*d*) are adjusted by using a screw driver. The trimmer in Fig. 7.16 (*c*) is adjustable from top while the one in Fig. 7.16 (*d*) is adjustable on the side. So the actual selection depends upon the application. Fig 7.17 (*a*), (*b*) and (*c*) shows some of the commercially available multi turn trimmers. Notice the size, the shape and the position of the adjustable point.

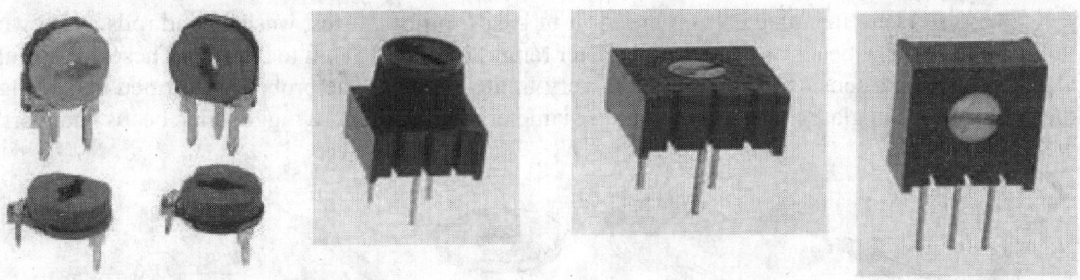

(*a*) Carbon trimmers, 10 mm 150 mW (*b*) Cermet trimmer – finger adjust (*c*) Cermet trimmer – top adjust (*d*) Cermet trimmer – side adjust

Fig. 7.16. Single turn trimmers.

(a) 6 mm, 0.25 W top adjust (b) 6 mm, 0.25 W side adjust (c) 19 mm 0.5 W rectangular

Fig. 7.17. Multi turn trimmers.

7.14. Non-linear Resistors

These resistors are made from semiconducting material *i.e.*, those materials which possess conductivity between the conducting and insulating materials. The non-linear property of the resistors arises due to the breaking of covalent bonds in the semiconducting materials. The source of energy for the creation of charge carriers in semiconducting materials may be voltage, temperature or incident light. The non-linear resistors are of the following three types:

1. Thermistors. 2. Photoresistors. 3. Varistors

7.15. Thermistors

The word thermistor is an acronym for thermal resistor, *i.e.*, a temperature-sensitive-resistor. It is used to detect very small changes in temperature. The variation in temperature is reflected through an appreciable variation of the resistance of the device. Thermistors with both negative-temperature-coefficient (NTC) and positive temperature-crefficient (PTC) are available, But NTC thermistors are more common. The negative-temperature coefficient means that the resistance decreases with the increases in temperature. And the positive-temperature-coefficient means that the resistance increases with the increase in temperature.

The NTC thermistors are manufactured by *sintering semiconductor ceramic materials prepared from mixtures of metallic oxides of cobalt, nickel, manganese etc. These materials have high negative temperature-coefficient. The PTC thermistors are made from doped barium titanate semiconducting material. This material has a very large change in resistance for a small change in temperature.

Thermistors are manufactured in the form of beads, probes, discs, washers and rods as shown in Fig. 7.18 (*a*). The beads are made in diameter ranging from 0.15 mm to 2.5 mm. These are useful where temperature sensing must be done in very limited spaces. The probes are formed by sealing thermistor beads in glass rods upto 25 mm in diameter. These are more rugged than beads and work well in liquids.

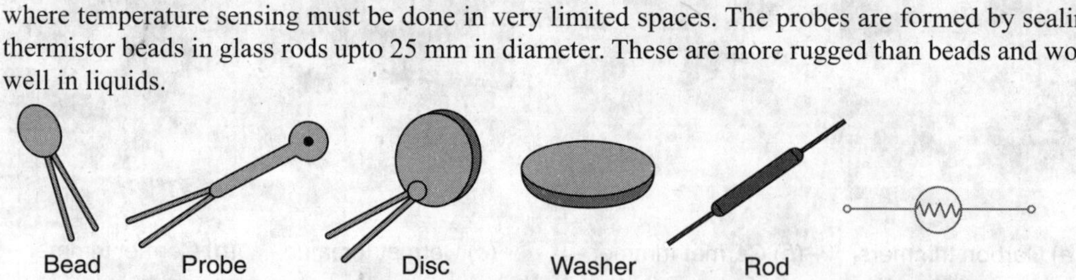

(*a*) Various types of thermistors. (*b*) Schematic symbol.

Fig. 7.18.

* Sintering is a process in which powdered materials are fused together by the application of heat.

The discs and washers are made to meet certain industrial requirements. The discs are used in moderate power applications in conjunction with time response applications. The rods and washers are used for high power applications. Figure 7.18 (*b*) shows the schematic symbol of a thermistor.

Fig. 7.19 shows one of the commercially available thermistors. Notice that the thermistor is package in a surface-mount technology (SMT) chip.

Fig. 7.19. SMT chip thermistor.

7.16. Thermistor Characteristic

Fig. 7.20 shows the Thermistor characteristic (temperature-resistance characteristic) for the NTC and PTC thermistors. It may be noted that NTC thermistor has a resistance of about 10 kΩ at –50°C and 100 Ω at 150°C respectively. It means that a temperature change of 200°C has resulted in a 100 : 1 change in resistance.

In case of PTC thermistor, the curve rises rapidly. The temperature where the resistance increases rapidly is called switching point as shown in the figure. The switching point temperature depends upon the particular type of PTC thermistor. The typical values of the switching point temperature may range from 30° to 160°C.

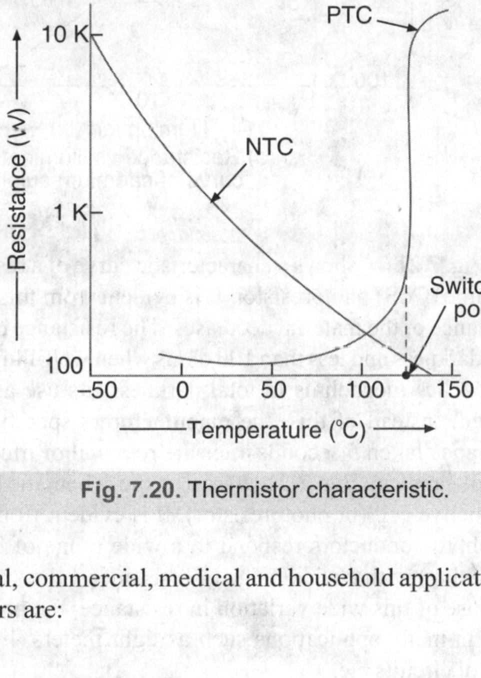

Fig. 7.20. Thermistor characteristic.

7.17. Thermistor Applications

Thermistors are extensively used in industrial, commercial, medical and household applications. Some of the important applications of thermistors are:

1. Temperature measurement and control.
2. Liquid level measurement.
3. Temperature compensation in electronic circuits.
4. Flow rate measurement.
5. Radio-frequency (R.F.) power measurement.
6. Time-delay circuits.

7.18. Photoresistors

These are also called photo-conductive cells. The photoresistors are made from semiconductor materials, whose resistance changes when illuminated with light energy. Such materials (known as photo-conductors) are cadmium sulphide (CdS), cadmium selenide (CdSe) and lead sulphide (PbS). When these materials are exposed to light, the covalent bonds are broken. This produces charge carriers (*i.e.*, electrons in the conduction band and holes in the valence band). The amount of illumination on the surface of the material determines the number of electron-hole pairs generated in the material. This in turn determines the resistance of the photo-conductive cells. The greater the amount of light falling on the surface (called surface illumination), greater will be the number of electron-hole pairs generated and therefore lower will be the value of resistance of the material. The lower the amount of light falling on the surface, higher will be the value of resistance of the material. Thus the resistance of the semiconductor varies inversely with the amount of light.

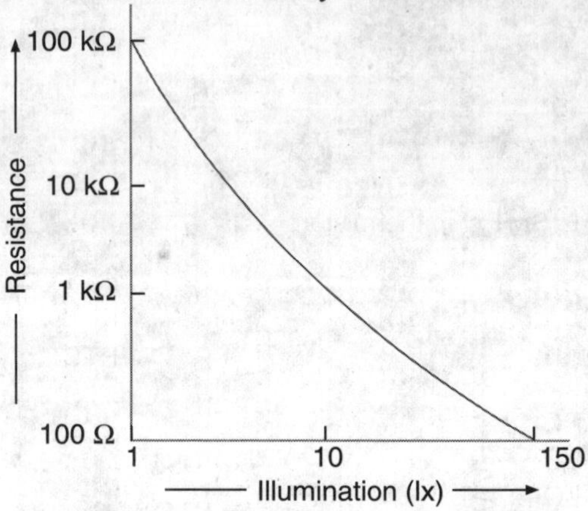

(*a*) Resistance vs illumination curve of cadmium sulphide.

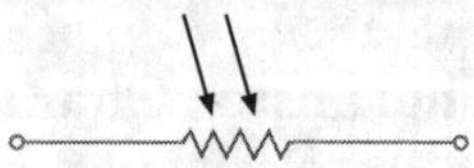

(*b*) Symbol.

Fig. 7.21.

Fig. 7.21 (*a*) shows a characteristic curve of a resistance versus illumination for a typical cadmium sulphide (CdS) photoresistor. It is evident from this figure that with the increase in illumination, the resistance of the material decreases. The resistance of photo-conductors may be several megaohms in total darkness and less than 100 ohms when well-illuminated. Since the resistance of photo-conductors approaches megaohms in total darkness, its use as a specification is limited. Instead of this, the manufacturers specify a minimum dark resistance taken 5 seconds after the removal of illumination of 20 lux (2 foot candles). Fig. 7.21 (*b*) shows the schematic symbol of a photo-conductive cell (or photoresistor). It is evident from Fig. 7.21 (*a*) that the photo-conductors respond to a wide range of illumination (1 lux to 100 lux) with a wide range of resistance values (100 Ω to 100 kΩ). Because of this wide variation in resistance, the photo-conductors are used in many applications such as light meters, light activated relay control circuits etc.

Fig. 7.22 shows the commercially available photoresistor.

Fig. 7.22. Commercially available photoresistor. (Photo condine cells)

7.19. Varistors

The word varistor is an acronym from variable resistor. The varistors are voltage dependent resistors (VDR's). These are used to protect circuitry from high energy voltage transients (also called surges) by rapidly changing from high stand by resistance to low conducting resistance. This action of a varistor clamps the voltage to a safe level. A high-energy voltage transient is an abnormal short living disturbance in the circuit, which is produced by switching operation, a sudden fault in electrical equipment or lightening stroke.

A typical volt-ampere (V-I) characteristic of a varistor is shown in Fig. 7.23 (*a*). In this figure, the solid curve represents region for continuous operation, while the dashed curve represents surge response. It may be noted from the characteristic curve that if a transient changes a voltage in a circuit from 120 to 170 volts, the current suddenly jumps from 100 mA to 400 mA. It means a 300% increase of current in the circuit. This sudden increase in the current may damage the circuit. The varistor protects the circuit from destructive energy by dissipating energy in its body. Fig. 7.23 (*b*) shows a schematic symbol of a varistor.

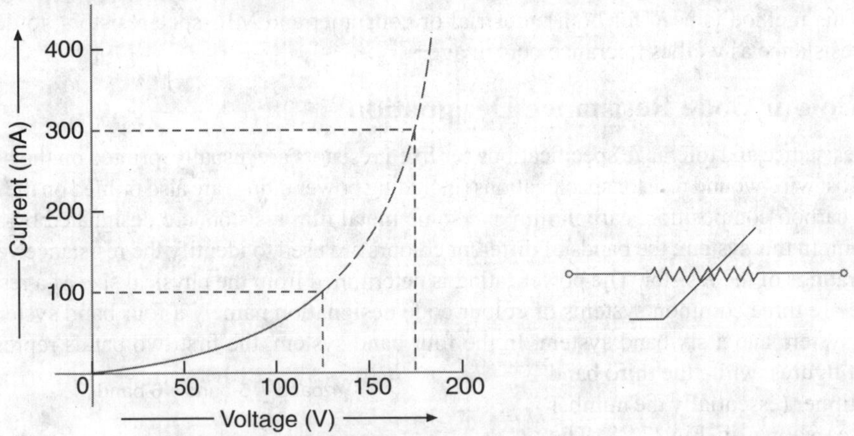

(*a*) V.I. characteristic (*b*) Symbol

Fig. 7.23. Varistors.

The varistors are available in a variety of packages. These are capable of handling instantaneous currents up to 2000 A with ac operating voltage ranging from 12 V to 660 V and at temperatures from –40°C to + 85°C.

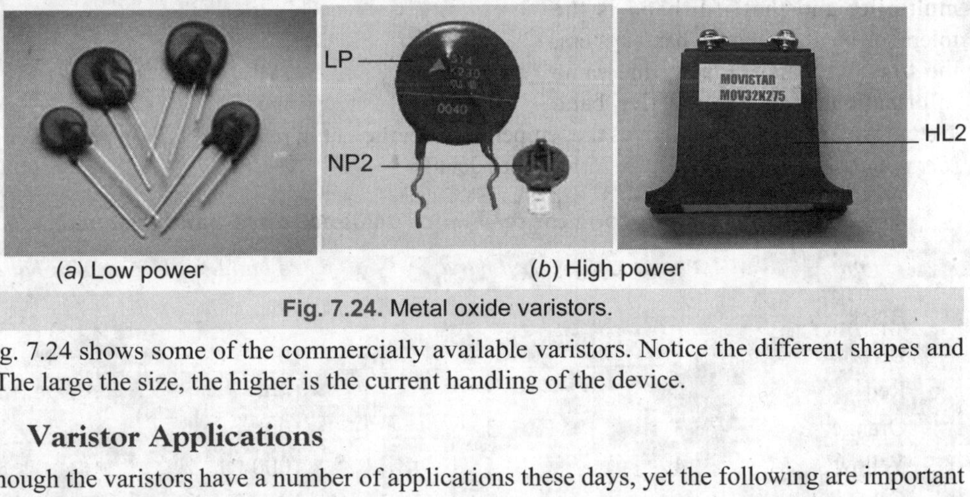

(*a*) Low power (*b*) High power

Fig. 7.24. Metal oxide varistors.

Fig. 7.24 shows some of the commercially available varistors. Notice the different shapes and sizes. The large the size, the higher is the current handling of the device.

7.20. Varistor Applications

Though the varistors have a number of applications these days, yet the following are important from the subject point of view:

1. Transient suppression in inductive and transformer switching circuits.
2. Switch contact are suppression.
3. Protection of circuits.

7.21. Resistance Designation

A number of methods are available to mark the resistance values on the body of a resistor. Some of the important methods are as discussed below:

1. The resistance values are printed on the body of the resistor. This method is widely used for marking industrial or consumer resistors.
2. The resistance values are printed on the body of the resistor as a series of numbers or number and the letter *R*. This method is used for marking MIL-spec (Military specification resistors).
3. The resistance values are printed on the body of a resistor as a code using coloured bands. This method is used for both industrial or consumer and MIL-spec resistors, to designate resistance as well as tolerance percentage.

7.22. Colour Code Resistance Designation

The resistance and tolerance specifications for fixed resistors are, usually, printed on the body of a resistor. Most wire-wound resistor specifications (including power rating) are also printed on the resistor. However, carbon-composition, carbon film and some metal film resistors are designated by a colour code system. In this system, the bands of different colours are used to identify the resistance value and tolerance ratings of the resistor. The power rating is determined from the physical size of a resistor.

There are three common systems of colour code designation namely a four band system and a five band system and a six band system. In the four band system, the first two bands represent the significant figures, while the third band is the multiplier (essentially the number of zeros) as shown in Fig. 7.25. The fourth band represents the tolerance band.

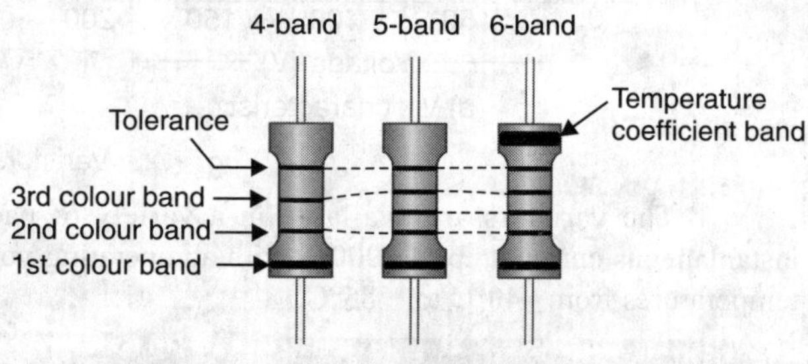

Fig. 7.25.

In the five band system, the first three bands represent the significant figures while the fourth band is the multiplier and the fifth band is the tolerance band. In the six band system, the first five bands convey the same information as that in the five band system while the sixth band gives the temperature coefficient in parts per million per degree celsius (*i.e.* ppm/°C). Refer to the Fig. 7.25 for more details.

Table 7.1: Colour code for resistance designation (4-band systems).

Colour	*1st band*	*2nd band*	*3rd band*	*4th band*
Black	0	0	× 1	–
Brown	1	1	× 10	± 1%
Red	2	2	× 100	± 2%
Orange	3	3	× 100	–
Yellow	4	4	× 1000	–
Green	5	5	× 10 000	± 0.5%
Blue	6	6	× 100 000	± 0.25%
Violet	7	7	× 1000 000	± 0.1%
Grey	8	8	–	± 0.05%
White	9	9	–	± 1%
Gold	–	–	× 0.1	± 5%
Silver	–	–	× 0.01	± 10%

Table 7.2: Colour code for resistance designation (5-band systems).

Colour	*1st band*	*2nd band*	*3rd band*	*4th band*	*5th band*
Black	0	0	0	× 1	–
Brown	1	1	1	× 10	± 1%
Red	2	2	2	× 100	± 2%
Orange	3	3	3	× 100	–
Yellow	4	4	4	× 1000	–
Green	5	5	5	× 10 000	± 0.5%
Blue	6	6	6	× 100 000	± 0.25%
Violet	7	7	7	× 1000 000	± 0.1%
Grey	8	8	8	–	± 0.05%
White	9	9	9	–	± 1%
Gold	–	–	–	× 0.1	± 5%
Silver	–	–	–	× 0.01	± 10%

Table 7.3: Colour code for resistance designation (6-band systems).

Colour	*1st band*	*2nd band*	*3rd band*	*4th band*	*5th band*	*6th band*
Black	0	0	0	× 1	–	–
Brown	1	1	1	× 10	± 1%	100
Red	2	2	2	× 100	± 2%	50
Orange	3	3	3	× 100	–	15
Yellow	4	4	4	× 1000	–	25
Green	5	5	5	× 10 000	± 0.5%	–
Blue	6	6	6	× 100 000	± 0.25%	10
Violet	7	7	7	× 1000 000	± 0.1%	5
Grey	8	8	8	–	± 0.05%	–
White	9	9	9	–	± 1%	1
Gold	–	–	–	× 0.1	± 5%	–
Silver	–	–	–	× 0.01	± 10%	–

Table 7.1 shows the colour code for 4-band resistors. Similarly table 7.2 shows the colour code for 5-band resistors and table 7.3, for 6-band resistors. Notice the colour code for all the three systems (*i.e.* 4-band, 5-band or 6-band) is the same. That is the black represents the minimum value (0) and white the maximum value (9). The brown (as a tolerance band) represents a tolerance of ± 1%, red ± 2%, gold ± 5% and a plain (or no colour) represents a tolerance of ± 20%. The following example illustrates the use of 4-band colour code system.

For example, suppose we have to specify the 4-band colour code for 10 Ω ± 10% resistor. Look at Table 7.1 and check that the first significant figure (or first digit) of the specified value (10 Ω) is 1. Therefore the first band will be of a brown colour. The second significant figure of the specified value (10 Ω) is 0. Therefore the second band will be of a black colour. The third band will again be of a black colour, because we need a multiplier of 1 (× 1) for 10 Ω resistance. Finally ± 10% is represented by a silver colour band. Therefore 10 Ω ± 10% will have a colour code designation as shown in Fig. 7.26 (*a*).

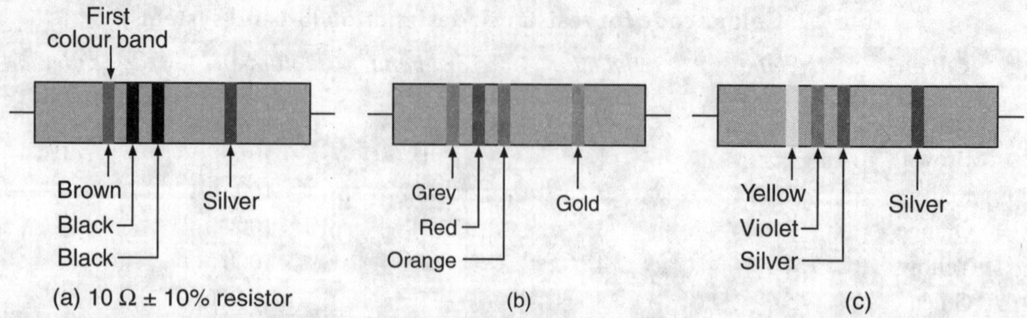

Fig. 7.26.

Now consider a 4-band colour code resistor as shown in Fig. 7.26 (*b*). The grey corresponds to first band its value is 8. Similarly, red corresponds to second band its value is 2. Therefore the first two significant figures of the resistance values is 82. The third band is of orange colour. Therefore the multiplier (number of zeros) is 1000. Hence the resistance value is 82 000 Ω. The tolerance band is of gold colour which corresponds to a tolerance of ± 5%. Therefore the resistance value can lie anywhere between 82 000 ± 5% *i.e.*, 73 800 Ω and 90 200 Ω.

Sometimes, the resistance values less than 10 Ω are also used. For such resistors, the third band is either of gold or silver. It serves as a fractional multiplier. If the third band is gold, multiply the first two digits of the resistance value by 0.1. If it is silver, then multiply by 0.01. However, the fourth band represents the tolerance value as discussed earlier. Thus we can check that the resistance value of the resistor shown in Fig. 7.26 (*c*) is 0.47 ± 10% Ω.

Example 7.1.

Write down a four band colour code for the following resistors:

(a) 150 Ω ± 5% *(b) 2.7 k Ω ± 10%*

Solution.

Given: Resistors = 150 Ω ± 5% and 2.7 kΩ ± 10%

(a) Colour code for 150 Ω ± 5%

We know that in this case, the first significant figure is 1, second is 5 and the multiplying factor is 10. Therefore,

1	5	× 10	± 5%
↓	↓	↓	↓
Brown	Green	Brown	Gold

Thus the four band colour code for 150 W ± 5% resistor is brown, green, brown and gold. **Ans.**

(b) Colour code for 2.7 kΩ ± 10%

We know that 2.7 kΩ = 2700 Ω. In this case, the first significant figure is 2, second is 7 and the multiplier is 100. Therefore

2	7	× 100	± 10%
↓	↓	↓	↓
Red	Violet	Red	Silver

Thus the four band colour code for 2.7 kΩ ± 10% resistor is red, violet, red and silver. **Ans.**

Example 7.2.

Specify a four band colour code for the following resistors:

(a) 56 kΩ ± 20% *(b) 10 MΩ ± 10%*

Solution.

Given: Resistors - 56 kΩ ± 20% and 10 MΩ ± 10%

(a) ***Colour code for 56 kΩ ± 20%***

We know that 56 kΩ = 56 000 Ω. In this case, the first significant figure is 5, the second is 6 and the multiplier is 1000. Therefore

5	6	× 1000	±20%
↓	↓	↓	↓
Green	Blue	Orange	Plain

Thus the four band colour code for 56 kΩ ± 20% resistor is green, blue and orange. There is no colour band for the tolerance value (because it is plain). **Ans.**

(b) ***Colour code for 10 MΩ ± 10%***

We know that 10 MΩ = 10 000 000 Ω. In this case, the first significant figure is 1, the second is zero and the multiplier is 1 000 000. Therefore

1	0	× 1 000 000	± 10%
↓	↓	↓	↓
Brown	Black	Blue	Silver

Thus the four band colour code for 10 MΩ ± 10% resistor is brown, black, blue and silver. **Ans.**

Example 7.3.

Determine the resistance and tolerance ratings for the following four band colour codes:

(*a*) *Brown, Black, Yellow —*

(*b*) *Yellow, Violet, Black and Gold,*

(*c*) *Green, Blue, Brown and Silver.*

Solution.

(a) ***Resistance for brown, black, yellow, —***

We know that in this case,

Brown	Black	Yellow	—
↓	↓	↓	↓
1	0	× 10 000	± 20%

∴ Resistance value is 10 × 10 000 = 100 kΩ ± 20%. **Ans.**

(b) ***Resistance for yellow, violet, black and gold***

We know that in this case,

Yellow	Violet	Black	Gold
↓	↓	↓	↓
4	7	× 1	± 5%

∴ Resistance value is 47 × 1 = 47 Ω ± 5%. **Ans.**

(c) ***Resistance for green, blue, brown and silver***

We know that in this case,

Green	Blue	Brown	Silver
↓	↓	↓	↓
5	6	× 10	+ 10%

∴ Resistance value is 56 × 10 = 560 Ω ± 10%. **Ans.**

7.23. Troubles with Resistors

The most common trouble with the resistors is an 'open' (*i.e.*, its value becomes infinity). This happens, when an excessive current flows through the resistance and it gets overheated. Usually, the colour of the resistor changes after it gets overheated. This indicates that the resistance is lost and it has opened. Such resistors should be discarded straight away. However, it is usually better to check its resistance value with an ohmmeter on an appropriate scale. It may be noted that the resistor must be disconnected from the remaining circuit before checking it.

7.24. Inductance

It is the property of a coil, *which opposes changes in current by means of energy storage in the form of magnetic field.* It has been observed that a current carrying conductor or a coil has a magnetic field (or flux) associated with it. The strength of this magnetic field is directly proportional to the amount of current in the coil. As long as the current in the coil is constant, the flux is also constant. Now, if the current through the coil changes (*i.e.*, increases or decreases), the flux linked with the coil also changes. This change in flux induces a voltage or e.m.f. in the coil. The polarity of induced voltage is such that it opposes the change in current through the coil. The magnitude of this voltage is directly proportional to the rate of change of current *i.e.*,

$$V \alpha \frac{dI}{dt}$$

$$= L \cdot \frac{dI}{dt}$$

where V = Magnitude of the induced voltage,

dI/dt = Rate of change of current, and

L = Constant of proportionality known as inductance of the coil.

The inductance is a measure of the energy stored in the coil in the form of magnetic field. The unit of inductance is henry (H). In actual practice, 'henry' is an extremely large unit. Therefore we use much smaller, units like millihenry (mH) or microhenry (μH), Such that

$$1 \text{ mH} = 10^{-3} \text{ H} \quad \text{and} \quad 1 \text{ μH} = 10^{-6} \text{ H}$$

Sometimes, the inductance of a coil is also referred to as a self-inductance, in order to distinguish it from mutual inductance. It is important when we have a number of coils in a circuit and we have a self-inductance as well as the mutual inductance in the circuit.

7.25. Inductors (or Coils)

An inductor (or a coil) is an electrical component, which is manufactured with a specified amount of inductance. The inductors are used in tuning and filter circuits. (These concepts will be discussed in later chapters). They are used in radio receivers as a built-in antenna coil to pick up radio signals. They are also used in transformers and coupled circuits to transfer (or couple) energy from one circuit to another.

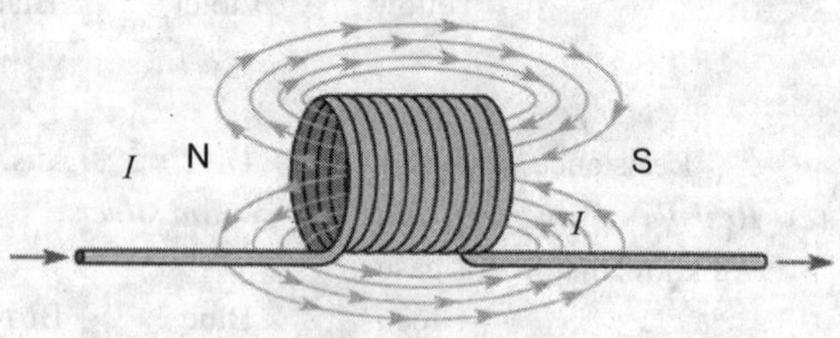

In many applications, the inductors are used to minimize alternating currents, while permitting flow of direct current. In such applications, the inductor is called a choke. Thus we have chokes for audio-frequency range and radio-frequency range. These chokes are called audio-frequency chokes (AFC) and radio-frequency chokes (RFC). The inductors are generally specified with inductance value and current capacity.

7.26. Types of Inductors

The inductors are of two types namely fixed inductors and variable inductors depending upon the type of core, whether it is fixed or variable. Now we shall discuss both the types of inductors.

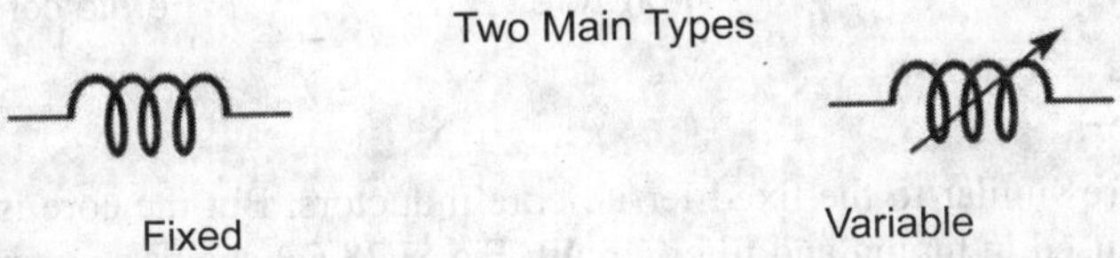

1. Fixed inductors

The fixed inductors are of three types namely air inductors, iron-core inductors and ferrite core inductors.

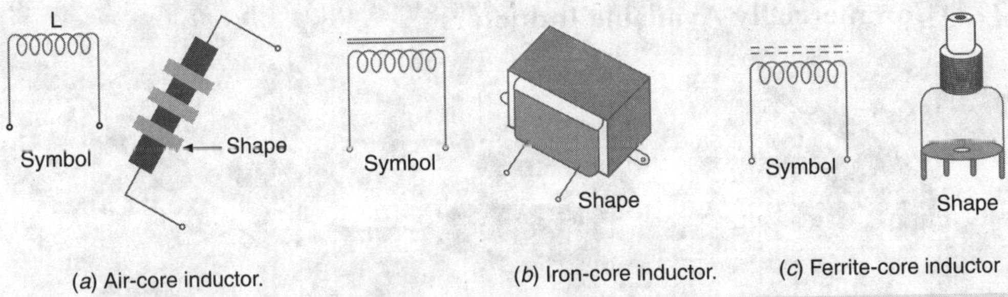

Fig. 7.27. Fixed inductors.

(*a*) ***Air-core inductor.*** This inductor is made of coils of wire wound on a former made of simple cardboard. Figure 7.27 (*a*) shows a symbol and shape of an air-core inductor. The air-core inductor has a very low value of inductance. They are suitable for radio-frequency (RF) applications.

(*b*) ***Iron-core inductor.*** This inductor is made of coils of wire wound on solid iron-core. The iron core is laminated to avoid eddy current loss. A laminated core consists of thin iron laminations pressed together, but insulated from each other, Fig. 7.27 (*b*) shows a symbol and shape of an iron-core inductor. The iron-core inductors are very suitable for audio-frequency (AF) applications.

(*c*) ***Ferrite-core inductor.*** This inductor is made of coils of wire wound on a ferrite core. A ferrite is a magnetic material consisting of fine particles of iron, cobalt or nickel embedded in an insulator binder. The ferrite core has a very low eddy current loss. Therefore ferrite-core inductors are used at higher frequencies.

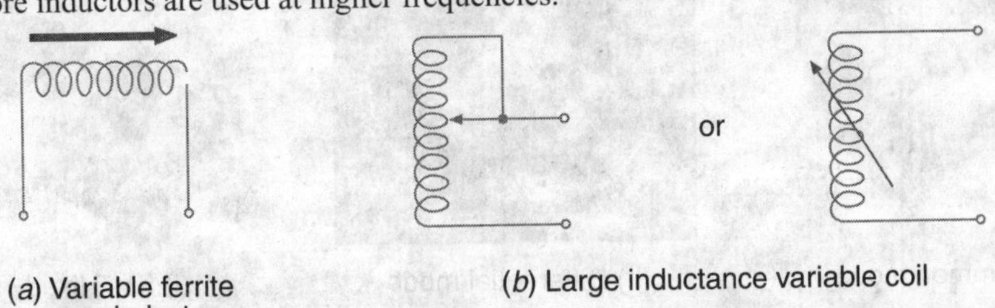

Fig. 7.28. Shows a symbol and shape of a ferrite-core inductor.

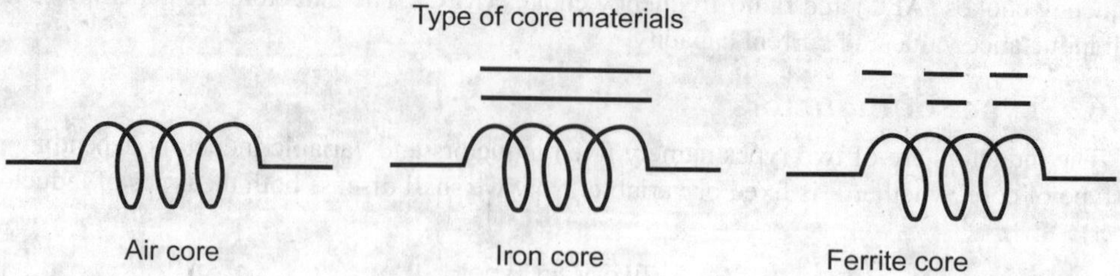

2. Variable inductors

These inductors are similar to the fixed ferrite-core inductors. But the core is adjustable. The variable inductors are used in tuning and filter circuit. Fig. 7.28 (*a*) shows a symbol for a variable ferrite core inductor. In case of large inductance coils, a variable inductor is made of a long coil wound on a ferrite core provided with a slider contact. The slider contact can be used to vary the inductance of a coil. Fig. 7.28 (*b*) shows a symbol for large inductance variable coils.

7.27. Commercially Available Inductors

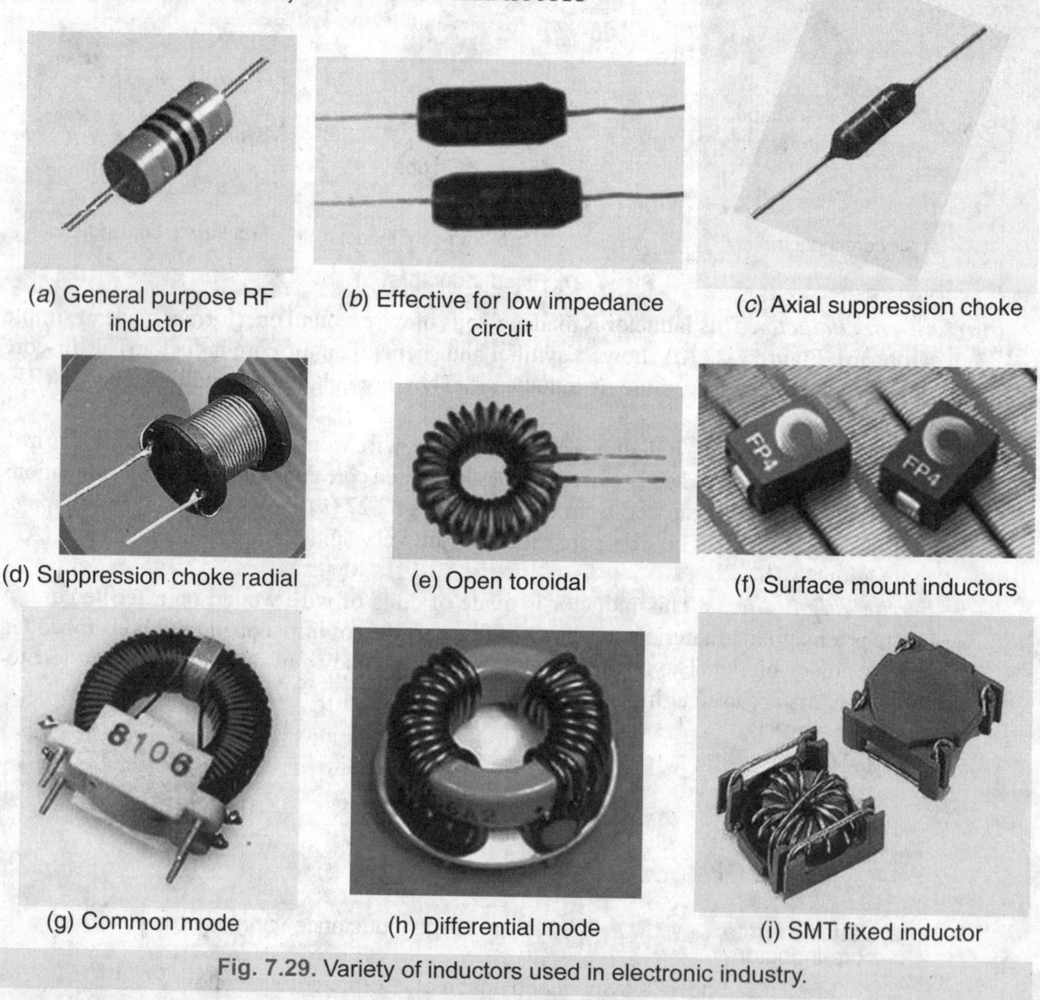

(*a*) General purpose RF inductor
(*b*) Effective for low impedance circuit
(*c*) Axial suppression choke
(d) Suppression choke radial
(e) Open toroidal
(f) Surface mount inductors
(g) Common mode
(h) Differential mode
(i) SMT fixed inductor

Fig. 7.29. Variety of inductors used in electronic industry.

Fig. 7.29 (*a*) is a resin coated general purpose RF inductor. This type of inductor has a small size and is useful for dc supply line filtering and low power switching regulator applications. The inductors are available in this size from 1 μH to 10 mH. Fig. 7.29 (*b*) shows inductors effective for low impedance circuit. It is suitable for preventing overshooting of digital signals such as clock signal or for limitation of high frequency noises.

Fig. 7.29 (*c*) shows an axial suppression choke. The inductance is indicated by colour bands. It comprises of a ferrite drum core with copper wire winding. It is suitable for RF blocking and filtering, decoupling interference suppression, antenna systems, energy saving lamps and automotive electronics. The inductors of this type are available with the inductance in the range of 1 μH to 39 μH. Fig. 7.29 (*d*) shows general purpose inductors (RF chokes) which uses a ferrite core with a copper wire winding. These are suitable for power decoupling in logic circuits, I.F. tuned circuit applications and filters etc. The low Q (Refer to Art. 7.35) high current inductors are especially suited to low-medium power dc-to-dc converters. The inductors of this type are available with inductance in the range of 100 μH ($Q = 60$) to 1000 μH ($Q = 20$).

Fig. 7.29 (*e*) suppression choke open toroidal is designed for use with phase angle control applications operating at 240 Vac. The inductance of this type of component is about 4 mH and the power rating is 100 W. Fig. 7.29 (*f*) shows general purpose surface mount inductors. These are ideal for circuits requiring a sharp frequency response due to the high Q (quality) factor. The Q is usually around 25. The applications include communications equipment, vehical security systems, and portable equipment.

Fig. 7.29 (*g*) and (*h*) shows inductors which have a PCB mounting frame. It is ideal for computers, vending machines and instrumentation. The voltage rating is 250 Vac. Fig. 7.29 (*i*) shows an SMT fixed inductor. The inductors of this type can handle high current, wide inductance range and are suitable for noise filtering, dc-to-dc cónverters.

7.28. Inductance of a Coil

The inductance of a coil can be determined from its physical parameters. The physical parameters are the length, area of cross-section and number of turns of a coil as shown in Fig. 7.30.

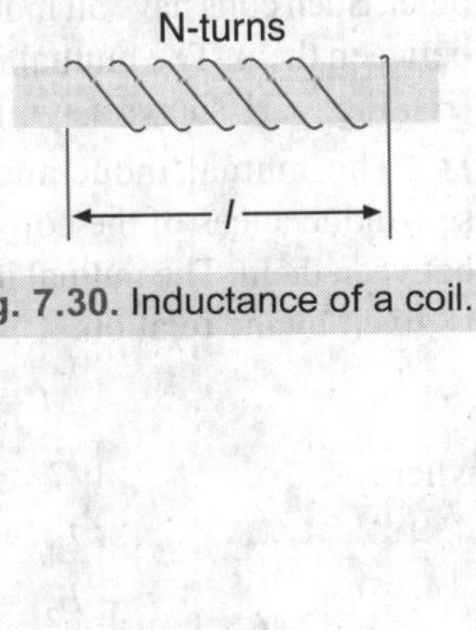

Fig. 7.30. Inductance of a coil.

Let l = Length of core on which the coil is wound in metres (m)

A = Area of cross-section of a core in square metres (m^2)

N = Number of turns of coil

μ_0 = Permeability of free space ($= 4\pi \times 10^{-7}$ H/m), and

μ_r = Relative permeability of the core material

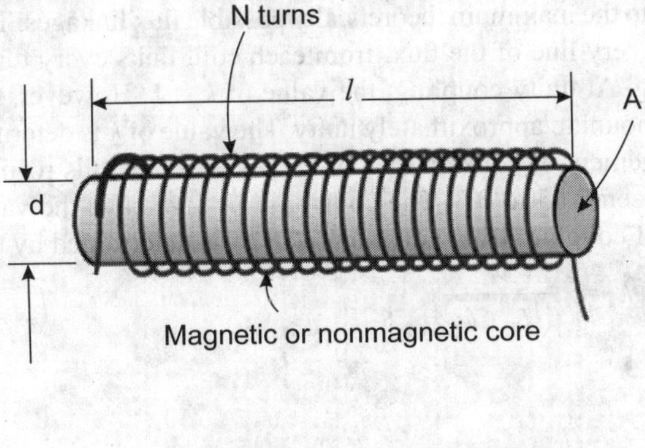

Now the inductance of a coil is given by the relation,

$$L = \frac{\mu_0 \mu_r \; A \, N^2}{l}$$

It is evident from the above relation that the inductance of coil depends directly on the permeability, area of cross-section of the core and square of the number of turns of a coil. It also depends inversely on the length of the core on which the coil is wound.

Example 7.4.

A choke coil consists of 150 turns wound upon a high permeability core. If $\mu_r = 3540$, $\mu_0 = 4\pi \times 10^{-7}$ H/m, coil length is 5 cm and cross-sectional area is 5×10^{-4} m^2, find the value of coil inductance.

Solution.

Given: $N = 150$; $\mu_r = 3540$; $\mu_0 = 4\pi \times 10^{-7}$ H/m; l = 5 cm = 0.05 m and $A = 5 \times 10^{-4}$ m^2.

We know that the coil inductance,

$$L = \frac{\mu_0 \mu_r \; A \, N^2}{l} = \frac{(4\pi \times 10^{-7}) \times 3540 \times (5 \times 10^{4}) \times (150)^2}{0.05} \text{ H}$$

$$= 1 \text{ H } \textbf{Ans.}$$

7.29. Mutually Coupled Coils

Sometimes, two coils are placed so close to each other that the changing magnetic flux of one coil, linking with the other, produces some voltage in the other coil. In such a case, the two coils are said to be mutually coupled with each other. Such coils have an inductance called mutual inductance between them. The mutual inductance is expressed in henry (H). Fig. 7.31 shows a symbol of mutually coupled coils.

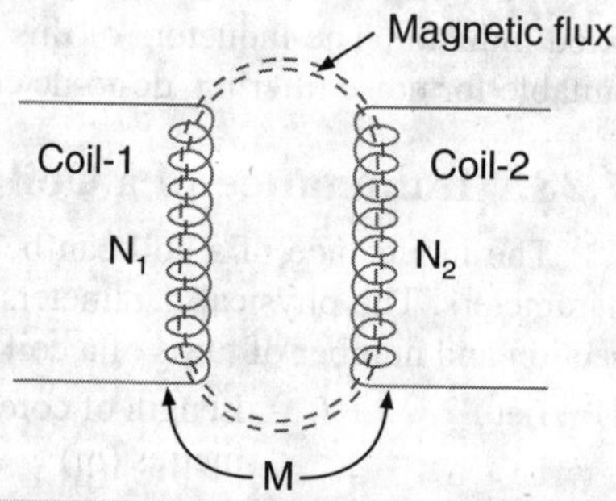

Fig. 7.31. Mutually coupled coils.

The mutual inductance of a coil depends upon the self-inductances of the coils and the coefficient of coupling between them. The mutual inductance between the two coils is given by the relation,

$$M = k\sqrt{L_1 \cdot L_2} \qquad ...(i)$$

where
k = Coefficient of coupling,
L_1 = Self-inductance of coil 1, and
L_2 = Self-inductance of coil 2.

The coefficient of coupling (k) may be defined as the ratio of actual flux linkages between the mutually coupled coils to the maximum theoretically possible flux linkages. The maximum theoretical linkage occurs, when every line of the flux, from each coil, links every turn of the other coil. This is called unity coupling. At unity coupling, the value of $k = 1$. However, the practical values of k range from very low amount to approximately unity. The value of k is determined by the type of core material and the arrangement of coils on the core. Thus air-core coils wound on the same material have a value of k between 0.05 to 0.3. The coils on iron-core have the value of k almost equal to unity. The coefficient of coupling from equation (i) may be determined by the relation,

$$k = \frac{M}{\sqrt{L_1 \cdot L_2}}$$

The mutual inductance is the basic of operation of transformers, which are extremely useful devices in the field of electronics. Transformers allow a.c. voltage and current levels to change efficiently. The transformers used at power (50 Hz) and audio (20 Hz - 20 kHz) frequencies have $k \approx 1$, while the transformers used at radio frequencies (frequencies above 20 kHz) have a very low value of k ranging from 0.002 to about 0.2.

Example 7.5.

What is the coefficient of coupling between the two coupled coils, if $L_1 = 40\ \mu H$, $L_2 = 80\ \mu H$ and $M = 11.3\ \mu H$?

Solution.

Given: $L_1 = 40\ \mu\text{H} = 40 \times 10^{-6}$ H; $L_2 = 80\ \mu\text{H} = 80 \times 10^{-6}$ H and $M = 11.3\ \mu\text{H} = 11.3 \times 10^{-6}$ H.

We know that coefficient of coupling.

$$k = \frac{M}{\sqrt{L_1 \cdot L_2}} = \frac{11.3 \times 10^{-6}}{\sqrt{(40 \times 10^{-6}) \times (80 \times 10^{-6})}}$$

$= 0.2$ **Ans.**

7.30. Inductors in Series and Parallel without Mutual Inductance

Consider two inductors connected in series and parallel as shown in Fig. 7.32 (*a*) and (*b*).

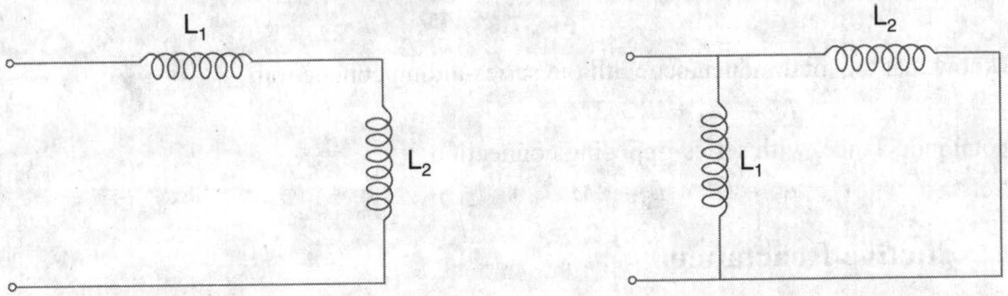

(*a*) Series connection (*b*) Parallel connection

Fig. 7.32.

Let L_1 = Self-inductance of inductor 1, and

L_2 = Self-inductance of inductor 2.

Though there are many parameters for such inductors, yet the total inductance is important from the subject point of view. Now we shall discuss the total inductance for both types of inductors one by one.

1. *Series connection.* We know that in this case, the total inductance is the sum of all inductances present in the circuit. Therefore, total inductance,

$$L = L_1 + L_2$$

In general, for any number of inductors, connected in series, the total inductance,

$$L = L_1 + L_2 + ... + L_n$$

2. *Parallel connection.* We know that in this case the total inductance,

$$L = \frac{1}{\frac{1}{L_1} + \frac{1}{L_2}} = \frac{L_1 \cdot L_2}{L_1 + L_2}$$

In general, for any number of inductor connected in parallel, the total inductance,

$$L = \frac{1}{\frac{1}{L_1} + \frac{1}{L_2} + \ldots\ldots + \frac{1}{L_n}}$$

7.31. Inductors in Series with Mutual Inductance

Consider two inductors connected in series with mutual inductance (*M*) between them as shown in Fig. 7.33. If the magnetic field polarity of each coil is the same, so that they aid each other, the inductors are said to have series-aiding connection as shown in Fig. 7.33 (*a*). On the other hand, if the magnate field polarity of each coil is opposite to each other, then the inductors are said to have series-opposing connection as shown in Fig. 7.33 (*b*).

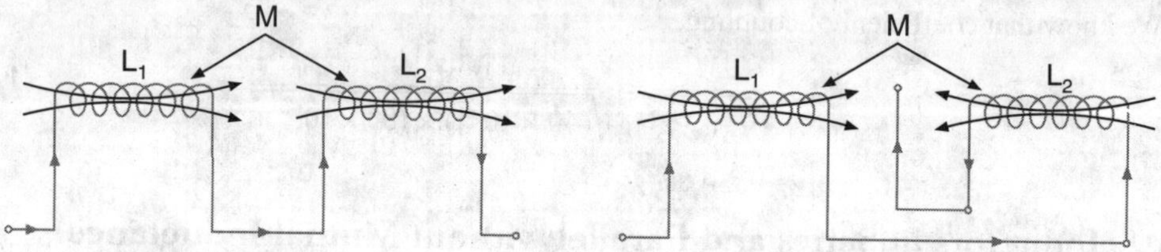

(*a*) Series-aiding connection (*b*) Series-opposing connection

Fig. 7.33.

We know that the total inductance with the series-aiding connection,

$$L = L_1 + L_2 + 2M$$

and the total inductance, with series-opposing connection,

$$L = L_1 + L_2 - 2M$$

7.32. Inductive Reactance

Whenever an a.c. current passes through an inductor, it reacts to the current by producing an induced voltage, which opposes the flow of current in the inductor. This opposition is called inductive reactance. It depends upon the frequency of a.c. current and the size of the inductor. The inductive reactance (in ohms) is given by the relation,

$$X_L = \omega \cdot L = 2\pi \cdot f \cdot L$$

where ω = Angular frequency in radians/sec whose value is equal to $2\pi f$, and

L = Self-inductance of an inductor.

It may be noted from the above relation that if the frequency (f) is equal to zero, then $X_L = 0$. It implies that inductor offers no reactance to the passage of a.c. current with zero frequency (*i.e.,* direct current). However, if the frequency is increased, the inductive reactance also increases.

A practical inductor has always some value of resistance (though small) called d.c. resistance (R). Such an inductor offers opposition in the from of impedance (Z). Mathematically, the value of impedance,

$$Z = \sqrt{R^2 + X_L^2} = \sqrt{R^2 + \omega^2 . L^2}$$

7.33. Energy Stored by an Inductor

We have already discussed in Art. 7.24 that the strength of magnetic flux is directly proportional to the amount of the current passing through an inductor. Thus, if current passing through an inductor increases, the magnetic flux also increases. This increase in flux represents energy taken from the circuit or the applied voltage source. This energy is stored in the magnetic field generated by the current passing through the inductor. When the current passing through the inductor decreases, the magnetic flux also decreases. The decreasing flux returns energy stored in the magnetic field to the circuit. The amount of energy stored by an inductor is given by the relation,

$$= \frac{1}{2} L \cdot I^2 \text{ (in joules)}$$

where L = Self-inductance of an inductor, and

I = Current passing through an inductor.

It may be noted that energy storage in the inductor takes place only when current passing through an inductor changes its value. Moreover, the inductors do not dissipate power like resistors.

7.34. Coil and Core Losses

It has been found that in a practical inductor a small amount of stored energy is lost as heat and is not returned to the circuit. The source of this lost energy is in the coil and core of the inductor.

An inductor has resistance inherent in the copper wire, which is used in making the coil. This resistance is known as d.c. resistance of the coil or inductor. The energy dissipated by this coil resistance represents coil losses. Thus a practical inductor is represented by a coil of inductance (L) in series with a small resistance (R_O) as shown in Fig. 7.34. The value of R_O depends on the length and thickness of the wire used in making a coil. Some energy is also lost, when some flux fails to link the turns. This flux is called flux leakage. It uses same energy to establish it. But it does no work in return.

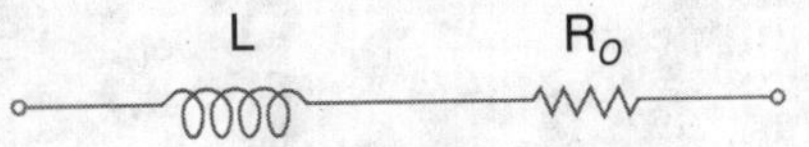

Fig. 7.34. Practical inductor.

The core losses include hysteresis and eddy current losses. The hysteresis loss is caused by retentivity of the core material. While the eddy current loss is caused by the currents induced in the magnetic core, the core losses waste energy in the form of heat. These losses are minimized by using a laminated core made up of thin silicon steel coated with an insulating material.

7.35. Q-factor of an Inductor

The ability of a reactive device to store energy, as compared to the dissipation of energy within the device, is called Q-factor or simply Q. It is known as figure of merit. A high Q- factor means little energy dissipation with respect to energy storage, while a low Q-factor means that energy dissipation is as large as energy storage. Mathematically, the Q-factor is given by the relation,

$$= \frac{\text{Energy Stored}}{\text{Energy dissipated}}$$

We know that energy stored in an inductor is proportional to the inductive reactance (X_L), while energy dissipated is proportional to d.c. resistance (R_O) of a coil. Thus if I is the current passing through inductor, then

$$\text{Energy stored} = I^2 \cdot X_L$$

and

$$\text{energy dissipated} = I^2 \cdot R_O$$

$\therefore$ Q-factor of an inductor,

$$Q = \frac{I^2 . X_L}{I^2 . R_O} = \frac{X_L}{R_O} \quad ...(i)$$

$$= \frac{\omega \cdot L}{R_O} \qquad ...(\because X_L = \omega \cdot L)$$

$$= \frac{2\,\pi\,f \cdot L}{R_O} \qquad ...(\because \omega = 2\pi f)$$

The value of Q-factor for coils may be as low as 5 to as high as 100. It is evident from equation (*i*) that smaller the value of resistance (R_O), higher is the value of Q-factor. A high Q is desirable in tuning circuits, because it makes the circuit more selective and sensitive.

Example 7.6.

The Q-factor of a 15 μH coil, when operating at 10 MHz is 90. Find the d.c. resistance of the coil.

Solution.

Given: $Q = 90$; $L = 15\ \mu H = 15 \times 10^{-6}$ H and $f = 10$ MHz $= 10 \times 10^6$ Hz.

We known that Q-factor of coil,

$$90 = \frac{2\,\pi\,f \cdot L}{R_O} = \frac{2\,\pi \times (10 \times 10^6) \times (15 \times 10^{-6})}{R_O} = \frac{942.5}{R_O}$$

$$R_O = \frac{942.5}{90} = 10.5\ \Omega \quad \textbf{Ans.}$$

7.36. Capacitance

It is the property of a capacitor, *which opposes the changes in voltage by means of energy storage in the form of electrostatic energy.* We know that a simple capacitor consists of two parallel conducting plates, separated by an insulating medium called dielectric. Consider such a capacitor connected across a d.c. voltage source, through a switch 'S' as shown in Fig. 7.35. If the switch 'S' is closed, the capacitor plates start charging. As long as the voltage (V) across the plates is constant, the charge on the plates is also constant. However, if the voltage increases, the charge on the plates also increases and vice versa. In other words, charge on the plates (Q) is directly proportional to the voltage (V) across the plates, *i.e.*,

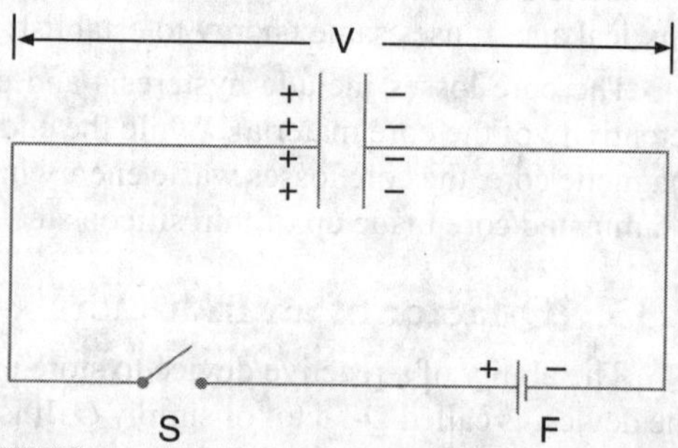

Fig. 7.35.

$$Q \propto V$$
$$= C.V$$

where C is a constant of proportionality and is called capacitance. It is a measure of the energy stored by the capacitor in the form of electrostatic energy. The unit of capacitance is farad (written as F). A farad is an extremely large unit of capacitance. Therefore we use much smaller units like microfarad (μF) nonofarad (nF) and picofarad (pF). Such that

$$1\mu F = 10^{-6}\text{ F},\ 1\text{ nF} = 10^{-9}\text{ F and } 1\text{ pF} = 10^{-12}\text{ F}$$

In the field of electronics, it is quite frequent to convert the values of capacitors from one unit to another to represent the values in a convenient manner. Table 7.4 gives the conversions from farads, microfarads and picofarads.

Table 7.4: Conversions for farads, microfarads and picofarads

To convert from	*To*	*Move the decimal point*
Farads	Microfarads	6 places to the right ($\times 10^6$)
Farads	Picofarads	12 places to the right ($\times 10^{12}$)
Microfarads	Farads	6 places to the left ($\times 10^{-6}$
Microfarads	Picofarads	6 places to the right ($\times 10^6$)
Picofarads	Farads	12 places to the left ($\times 10^{-12}$)
Picofarads	Microfarads	6 places to the left ($\times 10^{-6}$)

7.37. Factors Affecting Capacitance

The capacitance of a capacitor depends on the following three factors:

1. ***Plate area.*** It affects the capacitance directly *i.e.*, capacitance increases with the increase in plate area (A).
2. ***Plate separation.*** It affects the capacitance inversely *i.e.*, capacitance decrease with the increase in plate separation (d).
3. ***Type of dielectric.*** It also affects the capacitance directly. It is measured with the parameter called relative permittivity, or dielectric constant (ε_r). Higher the value of relative permittity, larger will be the value of capacitance.

Combining the above three factors, the capacitance,

$$C \propto \frac{\varepsilon_r \cdot A}{d}$$

$$= \frac{\varepsilon_o \cdot \varepsilon_r \cdot A}{d}$$

where ε_r is a constant of proportionality is known as absolute permittivity. Its value is 8.854×10^{-12} farads/meter (F/m). The relative permittivities (ε_r) of some important dielectric materials are shown in Table. 7.5

Table 7.5 : Relative permittivities (ε_r) of some important dielectric materials.

S. No.	*Material*	ε_r	*S. No.*	*Material*	ε_r
1.	Air or Vacuum	1	**6.**	Mica	3 – 6
2.	Aluminum oxide	10	**7.**	Paper (waxed)	3 – 5
3.	Bakelite	4.5 – 5.5	**8.**	Polystyrene	2.6
4.	Ceramic	6 – 7500	**9.**	Tantalum oxide	11
5.	Glass	7 – 8			

Example 7.7.

Determine the capacitance of a parallel plate capacitor having a plate area of 0.04 m² and a plate separation of 0.02 m. The dielectric is mica, which has a dielectric constant of 5.0.

Solution.

Given: $A = 0.04\ \text{m}^2$, $d = 0.02$ m and $\varepsilon_r = 5.0$

We know that capacitance of the parallel plate capacitor,

$$C = \frac{\varepsilon_o \cdot \varepsilon_r \cdot A}{d} = \frac{(8.85 \times 10^{-12}) \times 5.0 \times 0.04}{0.02} = 1.77 \times 10^{-11}$$

$$= 17.7 \text{ pF } \textbf{Ans.}$$

Example 7.8.

A ceramic capacitor has a plate area of 0.2 m^2. Determine the thickness of the dielectric, if the capacitance value is 0.428 μF. The dielectric constant of ceramic is 1200.

Solution.

Given: $A = 0.2$ m^2, $C = 0.428$ μF $= 0.428 \times 10^{-6}$ F and $\varepsilon_r = 1200$.

Let d = Thickness of the dielectric.

We know that capacitance (C),

$$0.428 \times 10^{-6} = \frac{\varepsilon_o \cdot \varepsilon_r \cdot A}{d} = \frac{(8.85 \times 10^{-12}) \times 1200 \times 0.2}{d} = \frac{2.124 \times 10^{-9}}{d}$$

$$d = \frac{2.124 \times 10^{-9}}{0.428 \times 10^{-6}} = 4.96 \times 10^{-3} \text{ m} = 4.96 \text{ mm } \textbf{Ans.}$$

7.38. Capacitors

A capacitor is an electronic component manufactured with a specified amount of capacitance. This component has an ability to charge or store energy, which neither a resistance nor an inductor can do. It opposes any change of voltage in the circuit in which it is connected. A capacitor is used in a number of applications, which are highly specialized. Some of the important applications of a capacitor are in starting motors, blocking d.c. current, passing a.c. current, filtering unwanted signals, tuning circuit to a specific frequency, coupling the electronic circuits, bypassing signals etc.

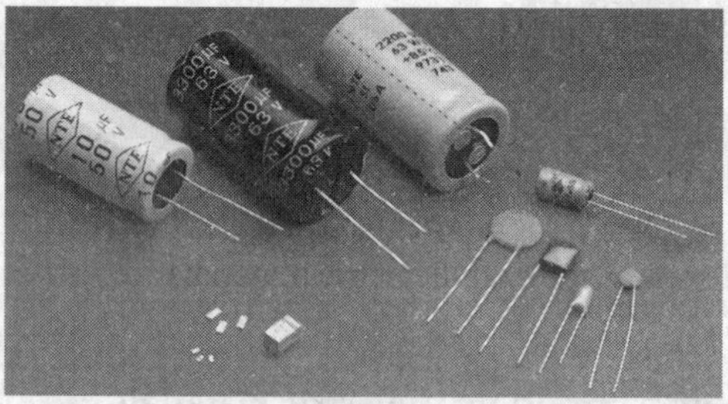

A capacitor consists of two conducting plates separated by a dielectric. The dielectric is an insulating material in which an electric field can be established with little or no leakage current. The dielectrics commonly used in actual practice are air, mica, ceramic, paper, plastic, aluminum, tantalum etc.

7.39. Capacitor Specifications

The capacitors are specified with the type of a dielectric, capacitance value, working voltage, capacitance tolerance and its physical size. Depending upon the type of dielectric used, the capacitors are specified as ceramic capacitors, tantalum capacitor etc. The capacitance of a capacitor is specified in either microfarads (μF) or picofarads (pF). The capacitors are manufactured with capacitance values that range from few picofarads to several thousands of microfarads.

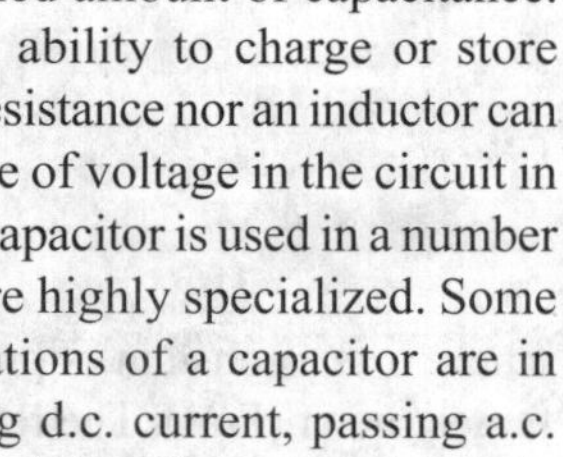

The working voltage (also called voltage rating) of a capacitor is the maximum voltage at which the capacitor may be operated continuously at a specified temperature. The working voltage of capacitors ranges from few volts to several thousands of volts. The capacitors, used more commonly in electronic circuits, have a working voltage of 5 volts. On the other hand, the capacitors used in colour television receivers may have a voltage rating as high as 50 kV.

The capacitor tolerance is expressed in terms of a percentage of specified capacitance value. It may also be given as a minimum rated value. The capacitance value and the working voltage of a capacitor is, usually, printed on its body. It will be interesting to know that sometimes the physical size of a capacitor becomes an important consideration. For instance, while mounting on printed circuit boards (PCB's), it is necessary that capacitors to be used should be as small as possible in size.

7.40. Classification of Capacitors

The capacitors are commonly classified on the basis of dielectric material used for their manufacturing. It is because of the fact that characteristics of a capacitor are mainly due to the properties of a dielectric. Fig. 7.36 shows different types of capacitors manufactured these days. The capacitors may be divided into two classes, namely fixed capacitors and variable capacitors. Each type is further subdivided in to many types as shown in the figure.

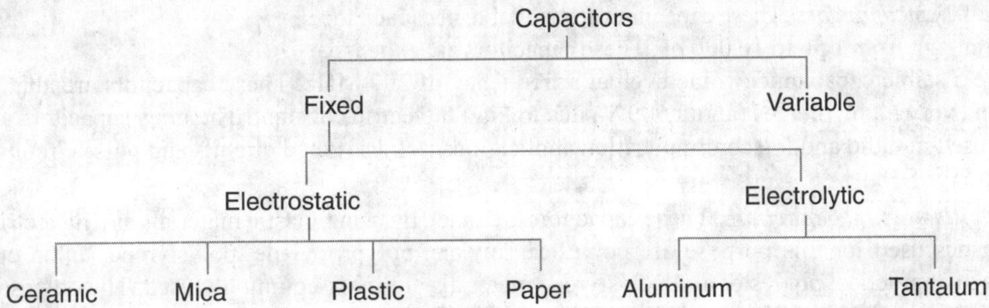

Fig. 7.36. Capacitor family tree.

7.41. Fixed Capacitors

The capacitors, in which the capacitance value cannot be varied by any means (*i.e.*, either by changing plate separation or area) are called fixed capacitors. These capacitors are broadly of two types namely electrostatic (or non-electrolytic) and electrolytic capacitors. Both these capacitor types are discussed in the following pages.

7.42. Electrostatic Capacitors

These capacitors are made up of two metal conductors (called plates) separated by a dielectric. They are characterized by very low leakage current and high leakage resistance. Following are the various types of electrostatic capacitors:

1. ***Ceramic capacitors.*** These capacitors are made by using various ceramic materials as dielectrics. Usually, a powdered mixture of barium-strontium-titanate is used as a ceramic material. The ceramic capacitors have a large capacitance in a small package and can have values from less than 10 pF to more than 1μF. These capacitors have a very high dielectric constant and therefore are usually smaller in size than paper or mica capacitors for the same capacitance value. However, ceramic capacitors have lower breakdown voltage than that of mica or paper type capacitors. The leakage resistance in ceramic capacitors is very high. But it is not as high as that of paper or mica type. These capacitors are available in both disc and tubular types and are used as bypassing and coupling capacitors in electronic circuits. The ceramic capacitors are cheaper than paper or mica type capacitors and have excellent performance upto 200 MHz.

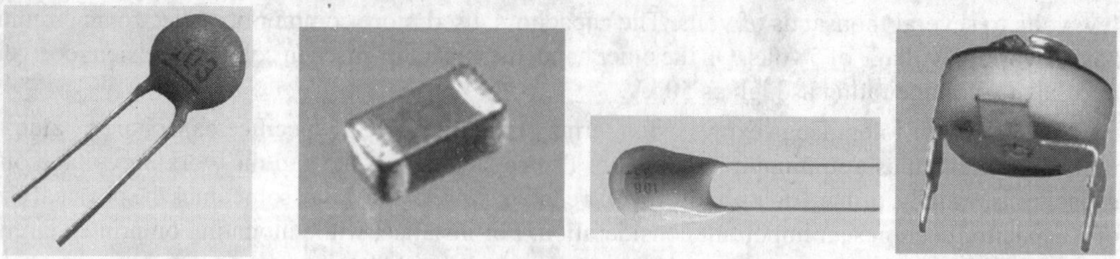

(*a*) Disc (*b*) SMT Ceramic multilayer (*c*) Radial (*d*) Trimmer

Fig. 7.37.

Fig. 7.37 (*a*), (*b*), (*c*) and (*d*) shows four different types of ceramic capacitors. Fig. 7.37 (*a*) is a disc type capacitor while Fig. 7.37 (*b*) is a surface mount (SMT) type capacitor. Similarly, Fig. 7.37 (*c*) is a radial type capacitor and Fig. 7.37 (*d*) is a capacitor which can be directly mounted on the PCB (printed circuit board). All these ceramic capacitors can be used for bypass, decoupling, fittering, sample and hold and in timing circuits.

Fig. 7.38. Mica capacitor

2. ***Mica capacitors.*** These capacitors are available in capacitance values ranging from 1 pF to 10 000 pF. These capacitors are expensive but have a stable capacitance value even at a frequency of 200 MHz. These capacitors are able to withstand very high voltages (about 500 V) due to high dielectric constant. The mica capacitors are widely used in radio and telecommunication applications (*i.e.* RF tuned circuits and pulse circuits). Refer to Fig. 7.38.

3. ***Plastic film capacitors.*** These capacitors are made by using plastic materials as a dielectric. The plastics used for this purpose are polyester, (mylar), polypropylene, polystyrene, teflon etc. But in actual practice, polyester and polystyrene materials are more commonly used. The polyester capacitors are available with capacitance values ranging from 0.001 μF to 10 μF, while polystyrene capacitors, from 5 pF to 0.05μF. The teflon capacitors may be operated up to 250°C. The polystyrene capacitors have very low leakage and good high frequency properties. The plastic film capacitors provide a wide choice in cost versus performance.

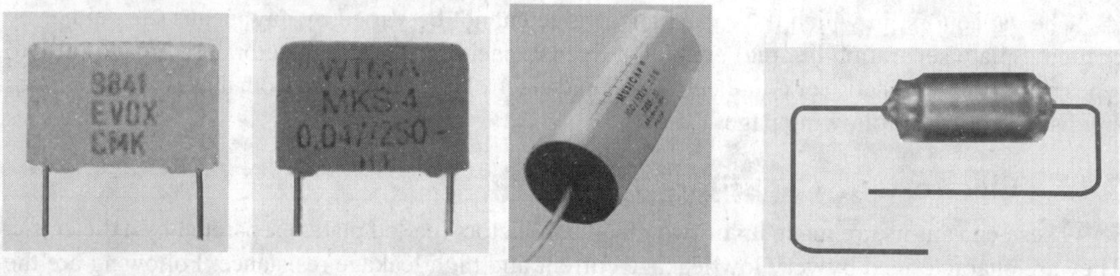

(*a*) Polypropylene/ polycarbonate (*b*) Metallised polyester film (*c*) Film foil polypropylene (*d*) Polystyrene

Fig. 7.39. Plastic film capacitors.

Fig. 7.39 (*a*) is a polypropylene polycarbonate capacitor. It ideal for automotive applications, filters, oscillators and timing circuits. It is available with 63 Vdc (40 Vac) to 400 Vdc (200 Vac). Fig. 7.39 (*b*) suitable for many industrial applications. Fig. 7.39 (*c*) has a very low loss suitable for continuous use at high ac voltages. It will withstand fast rise time pulses and has a excellent high frequency performance. Working voltage 1000 Vdc (350 Vac) and tolerance ± 20%. Fig. 7.39 (*d*) plain foil tubular with a polystyrene dielectric. These are physically smaller than mica.

4. ***Paper capacitors.*** These capacitors are made by using strips of aluminum foil with treated paper as a dielectric. The foil and the paper is rolled in a ribbon form. The leads are connected to the aluminum foil and taken out at each end. The capacitor may be sealed in a wax paper or plastic. The paper capacitors are available with capacitance values ranging from 0.001 μF to 1 μF. These capacitors have an excellent high voltage characteristics. Their working voltage may be as high as 2000 volts.

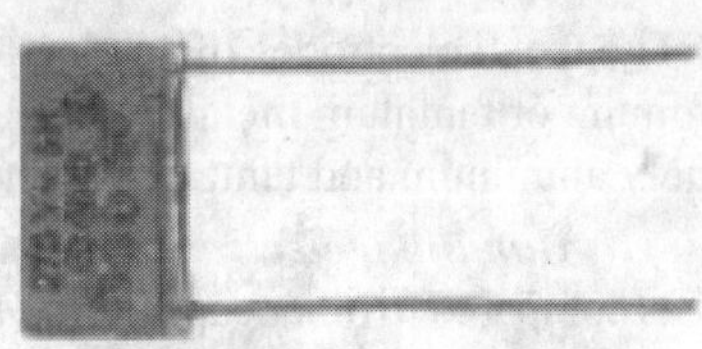

Fig. 7.40. Paper capacitor

Fig. 7.40 shows a picture of a metallised paper capacitor. The paper capacitors are designed to suppress radio frequency interference specifically from industrial and domestic applications operating on 240 Vac. These capacitors have excellent stability and reliability.

7.43. Electrolytic Capacitors

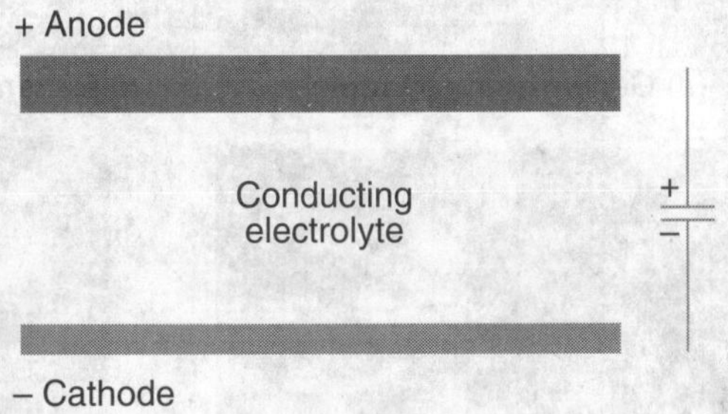

Fig. 7.41. Electrolytic capacitor.

These capacitors are made up of metal plates, which have a definite polarity separated by a thin metal oxide dielectric as shown in Fig. 7.41. The metal oxide is a conductive compound having dielectric constant between 8 and 25. Usually, it is in a paste form, though it can be in a liquid form also. The metal oxide film is deposited on one of the metal plates. This plate acts as a positive electrode or anode. The capacitor is formed by using either a conducting electrolyte as a second electrode or a semiconductor such as manganese dioxide. The electrolyte used is either in a liquid form or in the form of a paste, which saturates a paper or a gauge. The capacitor is packed in metal cylinder. The cathode is connected to the cylinder. The cylinder is usually, enclosed in a paper tube or cardboard tube, in order to insulate it form outside.

When a voltage of correct polarity is applied to a capacitor, a very thin insulating layer of oxygen atoms forms between the anode and the oxide layer. A reversal of polarity removes the insulating layer, thereby allowing very high currents. Thus electrolytic capacitors are known as polarised capacitors. They must be connected in a circuit according to the 'plus' (+) and 'minus' (–) markings on the body of a capacitor. If the capacitor is connected with a reverse polarity, it will act as a short circuit and get overheated, due to excessive leakage current, and it can explode also.

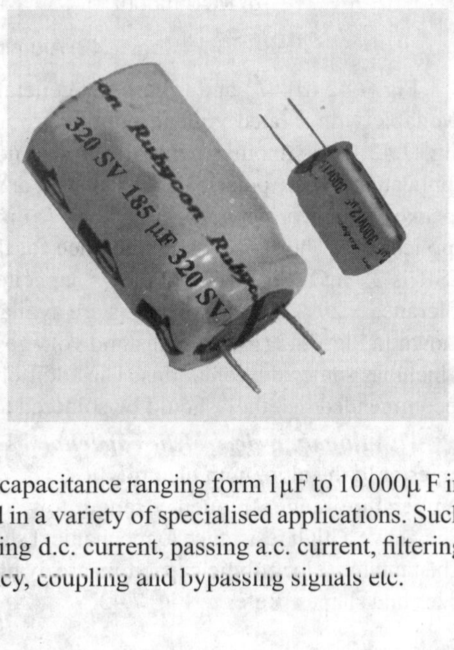

Some electrolytic capacitors are made with two capacitors in one cylinder. These capacitors are internally connected in series in an inverse arrangement, so that one of them is always working as a capacitor. This prevents the high leakage current, when operated with reverse polarity. Such capacitors are known as non-polarized electrolytic capacitors.

The electrolytic capacitors possess a large value of capacitance ranging form 1μF to 10 000μ F in very compact sizes. The electrolytic capacitors are used in a variety of specialised applications. Such application include their uses in starting motors, blocking d.c. current, passing a.c. current, filtering unwanted signals, tuning currents to a specific frequency, coupling and bypassing signals etc.

7.44. Types of Electrolytic Capacitors

The metal electrodes of an electrolytic capacitor, as discussed in last article, can be either of an aluminum or tantalum metal. Accordingly, the electrolytic capacitors can be divided into two types namely aluminum and tantalum electrolytic capacitors.

1. ***Aluminum electrolytic capacitors.*** These capacitors have higher working voltage and low cost as compared to other capacitors of similar capacity. But they are susceptible to certain compounds like freon, trichloroethylene and carbon tetrachloride etc.

(*a*) General purpose radial

(*b*) General purpose axial

(*c*) General purpose

(*d*) Non-polarised

(*e*) High current

(*f*) Long file high grade

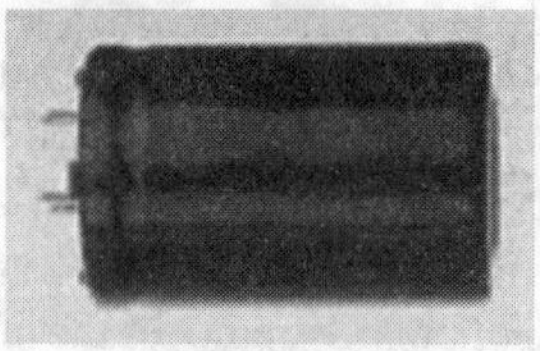

(*g*) PCB mount

(*h*) PCB mount with additional pins

Fig. 7.42. Aluminum electrolytic capacitors

Fig. 7.42 (*a*), (*b*) and (*c*) are the general purpose capacitors. The capacitors of these types are available with a rated voltage ranging from 16 V to 100 V. The capacitance tolerance is ± 20%. Fig. 7.42 (*d*) is another general purpose non-polarized capacitors. These are ideally suited for applications where polarity is not constant or reverse voltage may be applied, *e.g.* interstage coupling, speaker crossover networks. Fig. 7.42 (*e*) is designed for high current professional power supply applications. The capacitance tolerance for this capacitor is 10% + 30%, voltage rating is 40 V and ESR is 35 mΩ. Fig. 7.42 (*f*) shows capacitors that have long life, high performance, capacitance tolerance ± 20%. These capacitors are available with voltage rating is 63 V to 500 V. The capacitor shown in Fig. 7.42 (*g*) can withstand voltage up to 1000 Vdc. Fig. 7.42 (*h*) is a PCB mount capacitor which has some additional pins. The additional pins are meant for mechanical connection only. It is recommended that they should be soldered to the printed circuit board on isolated pads.

2. ***Tantalum electrolytic capacitors.*** These capacitors are superior to the aluminum electrolytics. They have a longer life, temperature range, frequency range, low leakage current etc. They have a high resistance to mechanical shock and vibrations. The tantalum electrolytic capacitors are available in a variety of sizes and shapes. Refer to Fig. 7.43.

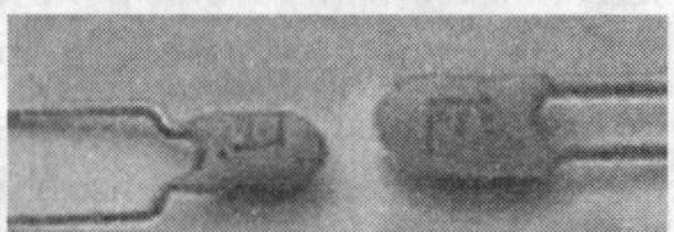

Fig. 7.43. Tantalum dipped capacitors.

7.45. Disadvantages of Electrolytic Capacitors

We have already discussed that electrolytic capacitors have a large capacitances values ranging form 1 μF to 10 000 μF. They have a large number of applications both in electrical and electronic circuits. But they have certain disadvantages, which can not be ignored. Following are some of important ones:

1. They have lower working voltages than other capacitors of similar capacity.
2. They have much higher leakage current and leakage resistance.
3. They have poor storage life, as their dielectric loses its properties over long periods of service.

7.46. Variable Capacitors

The capacitors, in which the capacitance value may be changed by some means, are called variable capacitors. The capacitance value is, usually, changed either by varying the area between the plates or by adjusting the spacing between them. The variable capacitors have capacitance in picofarad (pF) range. These capacitors are made by using air, mica, ceramic or plastic as a dielectric. The variable capacitors, using air as a dielectric, are used as ganged capacitors in radio receivers. The variable capacitors, using other dielectrics, are called trimmers or padders.

7.47. Variable Capacitor with Air as a Dielectric

This capacitor consists of two sets of metal plates, separated from each other, by air as shown in Fig. 7.44 (*a*). One set of the plates is fixed while the other set is connected to a shaft and can be rotated. The fixed set of plates is insulated from the body of a capacitor on which it is mounted. The set of moving plates can be moved in or out of a fixed set of plates with the help of a suitable knob connected to a shaft. As the plates are moved in and out of fixed plates, the capacitance value varies. The capacitance is minimum, when the moving plates are completely out and maximum when these are completely in.

Sometimes, two or more such capacitors are operated by a single shaft. Such a capacitor is known as ganged capacitor. Figure 7.44 (*b*) shows a 2-gang capacitor. The ganged capacitors are used in radio receivers for tuning to different radio stations.

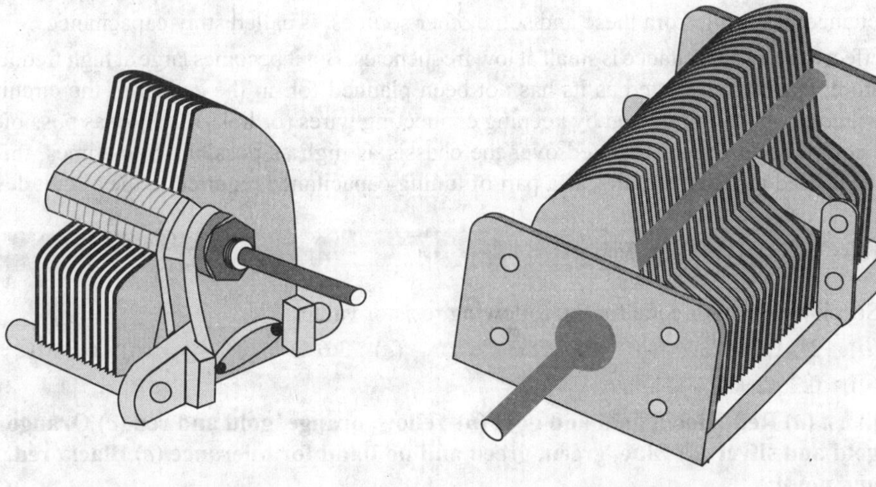

(*a*) Variable capacitor (*b*) Ganged capacitor

Fig. 7.44.

7.48. Trimmers and Padders

These capacitors are variable capacitors and are used exclusively for making fine adjustments on the total capacitance of a device. A trimmer consists of two small flexible metal plates separated by a dielectric (ceramic, mica or plastic). The spacing between the plates can be changed by means of a screw adjustment. As the screw is rotated inwards, the plates are compressed and its capacitance is increased. The capacitance value of trimmers can be adjusted over a range from 5 pF to 30 pF.

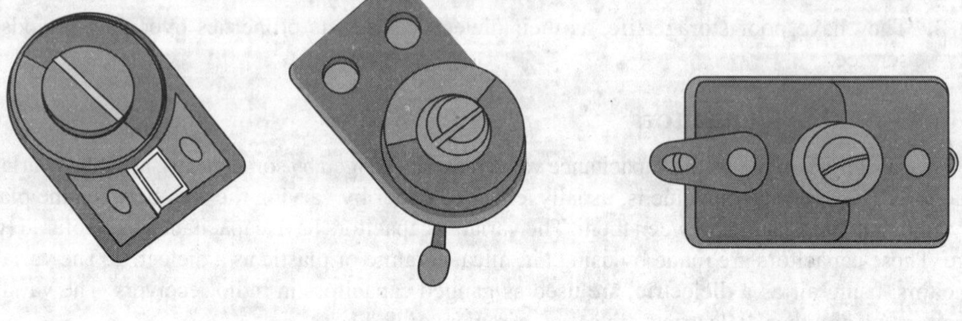

(*a*) Ceramic type (*b*) Mica type

Fig. 7.45. Some commonly used trimmers and padders.

The padders are similar to trimmers in appearance. But padders are larger in size or may have more plates. The capacitance value of padders can be adjusted over a range from 10 pF to 500 pF. Trimmers and padders are available in a number of shapes. Some of the common types available are as shown in Fig. 7.45 (*a*) and (*b*).

7.49. Stray Capacitance

It is a well known fact that any pair of conductors, separated by air or other dielectric, forms a capacitor. Therefore, small amounts of capacitance (about 1 to 10 pF) is present between conductors within a wire cable. In an electronic circuit, such capacitance is present between circuit wiring and metal chassis. It is also present between adjacent or opposite conductors on a printed circuit board. The capacitance, resulting from these and some other sources, is called stray capacitance.

The effect of stray capacitance is small at low frequencies. But it becomes large at high frequencies. It is therefore, usually, unwanted as its has not been planned for, in the design of the circuit. The stray capacitance can be minimized by keeping connecting wires (or links) as short as possible. The leads and components should be fixed over the chassis as high as possible. Sometimes, the stray capacitance is used in tuned circuits, as a part of tuning capacitance required in the circuit design.

NUMERICAL PROBLEMS

1. Specify the colour code for the following resistor values:

(*a*) 2.7 Ω ± 5% (*b*) 4.3 kΩ ± 2% (*c*) 3.6 Ω ± 10% (*d*) 6.8 MΩ ± 20%
(*e*) 0.2 Ω ± 5%

[Ans. (*a*) Red, violet, gold and gold (*b*) Yellow, orange, gold and red (*c*) Orange, blue, gold and silver (*d*) Blue, green, green and no band for tolerance (*e*) Black, red, silver and gold]

2. Record both the values and tolerances of the resistors with the following colour codes:
   (*a*) Brown, gray, yellow, silver (*b*) Red, red, green, gold
   (*c*) Red, black, green, silver (*d*) Yellow, violet, orange
   (*e*) Orange, black, gold
   (Ans. (*a*) 180 kΩ ± 10% (*b*) 2.2 MΩ ± 5% (*c*) 2 MΩ 10% (*d*) 47 kΩ ± 20% (*e*) 3 Ω ± 20%)
3. Two unconnected coils are mutually coupled. Given that L_1 = 350 mH, L_2 = 200 mH and k = 0.6. What is the value of mutual inductance? **(Ans. 158.7 mH)**
4. Two coils are mutually coupled with k = 0.4. Given that L_1 = 150 mH and L_2 = 200 mH. What is the value of mutual inductance? **(Ans. 69.2 mH)**
5. The resistance of a 50 mH coil is 30 ohms. What is Q-factor of the coil at 5 kHz? **(Ans. 52.4)**
6. The Q-factor of a 100 mH inductor is 80, when operated in 400 kHz range. What is the d.c. resistance (R_0) of the inductor? **(Ans. 31.4 Ω)**
7. Determine the capacitance of a parallel plate capacitor having a plate area of 0.01 m^2 and a plate separation of 0.02 m. The dielectric is mica, which has a dielectric constant of 5. **(Ans. 22.1 μF)**

DESCRIPTIVE QUESTIONS

1. What do you understand by the term 'passive circuit elements'?
2. Describe the important specifications of a resistor.
3. Explain briefly the following:
   (*a*) Linear resistor and (*b*) Non-linear resistors.
4. What are the different types of fixed resistors? Give the range of wattage for each resistor.
5. Give the constructional features of the following resistors:
   (*a*) Carbon-composition resistor and (*b*) Thin film resistors
6. Explain briefly, the basic construction of a variable wire-wound resistor and potentiometer.
7. Write a short note on any one of the following:
   (*a*) Thermistors (*b*) Photoresistors and (*c*) Varistors
8. What are the different types of inductors? Explain them with their constructional features. Give some important applications of inductors.
9. Write a short note on coil and core losses.
10. What is a capacitor? Give its three applications.
11. Describe the different types of capacitors.
12. Explain briefly about the basic construction of an electrolytic capacitor.
13. What are the different types of electrolytic capacitors? Explain them briefly.
14. Write a short note on variable capacitors.
15. Explain with examples colour coding and tolerance of resistor. *(Calicut University., Dec. 2007)*
16. Explain with figures the construction of different types of resistors. Give applications of each type of resistor also. *(M.G University., Aug.2007)*
17. Describe the carbon composition resistor, its color coding and tolerance? *(M.G University.,May/June. 2006)*

18. What are the advantages of color coding? *(M.G University., Dec. 2005)*
19. What are the different types of capacitors? Explain their characteristics *(Calicut University., Dec. 2006)*
20. Briefly discuss the function of resistors and capacitor as basic circuit component. *(M.G University., April/May 2003)*
21. Explain with figures the construction of different types of capacitors. Discuss the specific application of each type. *(M.G University., Dec. 2005)*
22. With neat sketches, describe the constructional details of Electrolytic capacitor? What are its specialties compared to other types? Where it is used? *(M.G University., Dec. 2007)*
23. Describe how electrolytic capacitors are made. *(M.G University., April/May 2004)*
24. Describe the different type of ceramic capacitor their characteristics and applications. *(M.G University., May/June. 2006)*
25. Write an essay on the following:
    (*a*) Capacitive filter (*b*) Inductive filters. *(Bhartiya University., April 2004)*
26. Define insulators with examples. *(M.G University.,April/May 2003)*
27. What are the different types of variable inductors? *(M. G University., Aug.2007)*
28. What are the characteristics of inductor? *(Calicut University., Dec. 2006)*
29. Discuss the function of inductor in basic circuit components. *(M.G University.,April/May 2003)*
30. Explain the meaning of coefficient of coupling. *(M.G University.,April/May 2003)*

MULTIPLE CHOICE QUESTIONS

1. A resistor has colour-code bands brown, black green and gold. Its nominal value is:
   (*a*) 1 kΩ ± 10% (*b*) 10 kΩ ± 5% (*c*) 1000 kΩ ± 10% (*d*) 1 MΩ ± 10%
2. A resistor used in colour television has the following colour bands: yellow, violet, orange and silver. Its nominal value is:
   (*a*) 4.7 kΩ ± 10% (*b*) 47 kΩ ± 5% (*c*) 47 kΩ ± 10% (*d*) none of these
3. A resistor with a nominal value of 2.7 kΩ ± 10% is used in a cassette recorder. Its colour code is:
   (*a*) Red, violet, red and silver (*b*) Red, violet, orange and silver
   (*c*) Red, violet, yellow and gold (*d*) Red, violet, red and gold
4. Ferrite cores are useful for
   (*a*) low frequencies (*b*) high frequencies
   (*c*) very high frequencies (*d*) none of these
5. Which of the following capacitors can have the highest capacitance value?
   (*a*) Mica (*b*) Paper (*c*) Ceramic (*d*) Electrolytic

ANSWERS

1. (*c*) **2.** (*c*) **3.** (*a*) **4.** (*b*) **5.** (*d*)

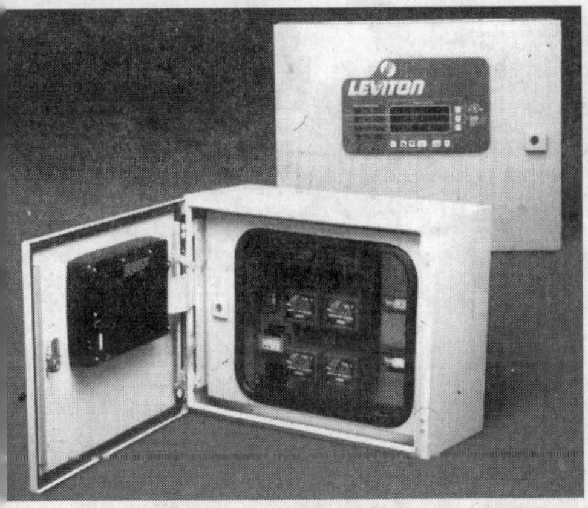

8

Chapter

Circuit Control and Protective Devices

Outline

8.1. Introduction

As a matter of fact, the control of electric current and the protection of electronic circuits and devices, form excessive value of current, is a very important consideration in the modern design offices. If this provision is not provided, the equipment may get damaged and people injured. The switches, fuses, circuit breakers and relays are the important devices developed for this purpose. We shall discuss all these devices which are discussed in the following pages.

8.2. Switch

A switch is a device, which is used in series with the load and source to switch the power 'ON' and 'OFF' to the load. When the switch is in its closed position (or 'ON' position) the current flows through the load, because resistance of the switch contacts is very small (usually few milliohms). However, when the switch is in its open position (*i.e.*, 'OFF' position), the current does not flow through the load because of the high resistance between the switch contacts. Since the entire source voltage is across the switch contacts, therefore, the air gap between the switch contacts must have a breakdown voltage rating higher than the voltage of the source.

At the time of selection of a switch for its 'ON-OFF' applications, both the load current and source voltage must be taken into account. The life of a switch is shortened, if it is used to control current higher than that of its rated value. The standard practice is to a operate a switch below its maximum current rating. For a given switch, the voltage and current ratings may be different for a.c. operation. The d.c. switch ratings are, usually, lower for a given switch.

The deterioration of switch contacts also affects the life of a switch. The deterioration of contacts may be due to oxidation, pitting or weld formation. Thus end-of-life of a switch is not a mechanical failure. But it results when contact resistance becomes very high. If 'R' is the value of contact resistance and 'I' the amount of current passing through the switch (when it is in 'ON' position), then power dissipated by the switch,

Switch

$$P = I^2 \cdot R$$

8.3. Switching Actions

A switch has, generally, two kinds of switching actions namely momentary-contact action and maintained-contact action. Both these actions are explained as below :

1. ***Momentary-contact action.*** Fig. 8.1 shows this type of action in a manual switch. Figure 8.1 (*a*) shows a condition, when the switch is not actuated. But when the switch is actuated (by pressing the push button), the moveable contact is transferred from the normally-closed (NC) contact to normally-open (NO) contact as shown in Fig. 8.1 (*b*). When the push button is released, the moveable contact returns to NC contact by the spring action as shown in Fig. 8.1 (*c*).

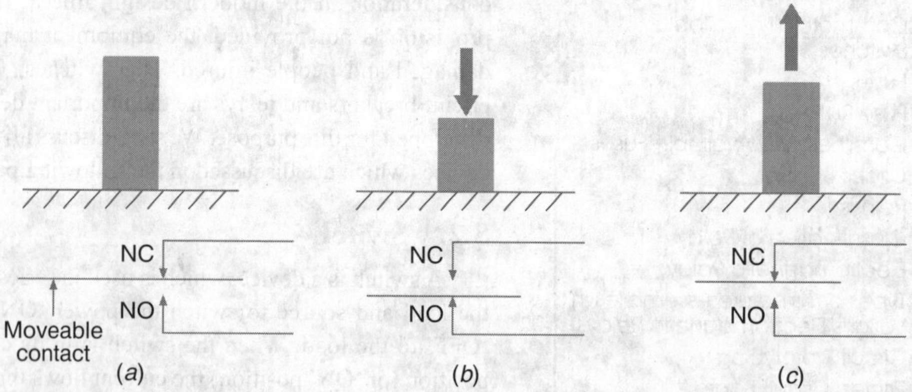

Fig. 8.1. Action of a momentary-contact manual switch.

2. ***Maintained-contact action.*** Fig. 8.2 shows this type of action in a manual switch: Fig. 8.2 (*a*) shows a condition, when the switch is not actuated. But when the switch is actuated, the moveable contact is transferred from contact 1 to contact 2 as shown in Fig. 8.2 (*b*). A latch holds these contacts together, so that when the actuator is released, they remain closed as shown in Fig. 8.2 (*c*). A second application of the force to the switch releases the latch. This causes the movable contact to return to the contact 1 as shown in Fig. 8.2 (*d*).

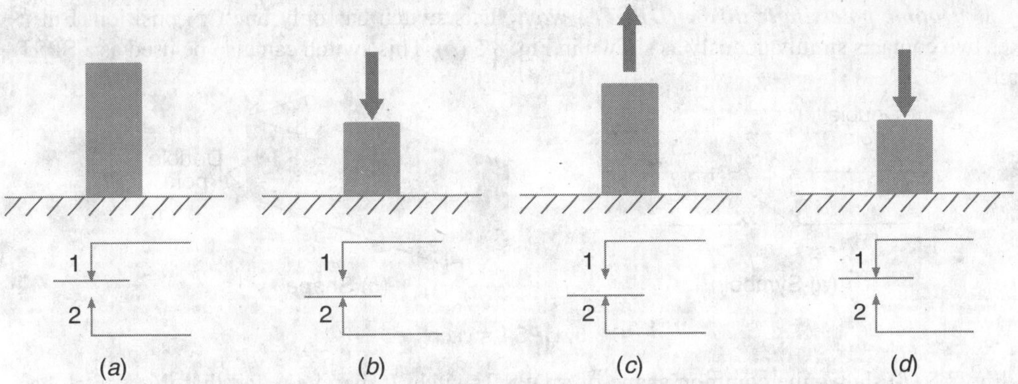

Fig. 8.2. Action of a maintained-contact manual switch.

8.4. Types of Switches

There are number of types and forms of switches used in ON-OFF control of electric circuits. The switches may be classified according to the mode of operation, construction or function such as manual switches, limit switches, precision switches etc. The choice depends upon the application and the value of maximum current in the circuit. Here we shall discuss the various types of manual switches only.

1. ***Single pole single throw (SPST) switch.*** This switch has only one lever (or arm) for making contact (known as a pole) as shown in Fig. 8.3 (*a*). It can connect or disconnect one side of a line, or a single wire circuit.

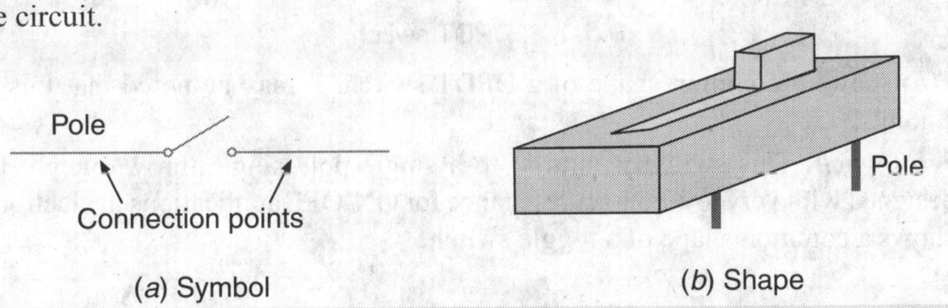

Fig. 8.3. SPST switch.

Figure 8.3 (*b*) shows a common shape of SPST switch. It may be noted that this switch has only two connection points.

2. ***Single pole double throw (SPDT) switch.*** This switch has two ON positions (1 and 2) and one OFF position as shown in Fig. 8.4 (*a*). It has one lever that can be moved into either of the two ON positions. This switch can connect either of the two circuits.

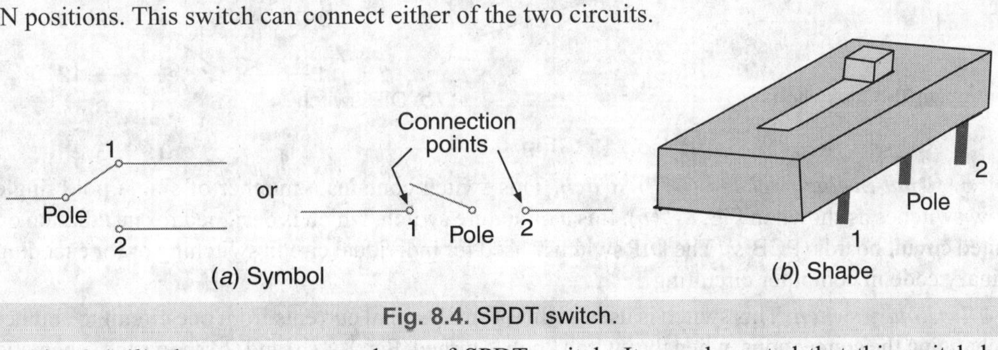

Fig. 8.4. SPDT switch.

Fig. 8.4 (*b*) shows a common shape of SPDT switch. It may be noted that this switch has 3 connection points.

3. ***Double pole single throw (DPST) switch.*** This switch has only one ON position. But it makes two contacts simultaneously as shown in Fig. 8.5 (*a*). This switch can also be used as a SPST switch.

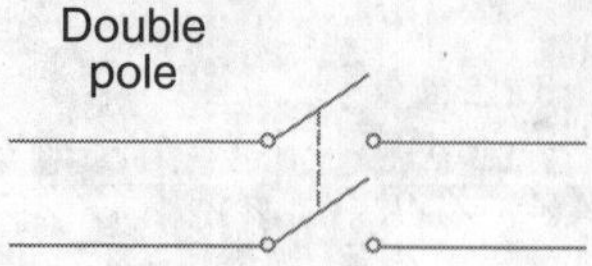

(*a*) Symbol

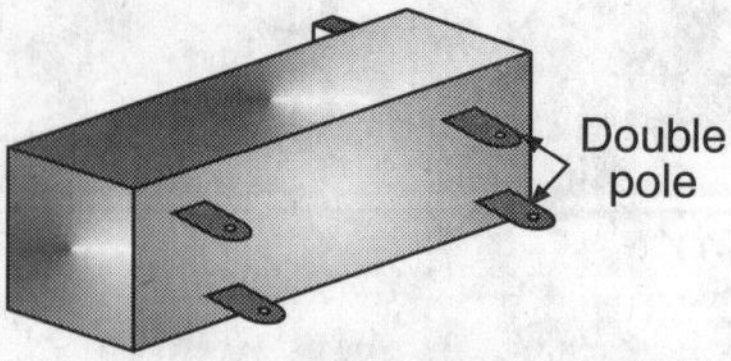

(*b*) Shape

Fig. 8.5. DPST switch.

Fig. 8.5 (*b*) shows the common shape of a DPST switch. It may be noted that this switch has 4 connection points.

4. ***Double pole double throw (DPDT) switch.*** The switch has two ON positions and makes two contacts simultaneously as shown in Fig. 8.6 (*a*). This switch can also be used as two DPST switches together.

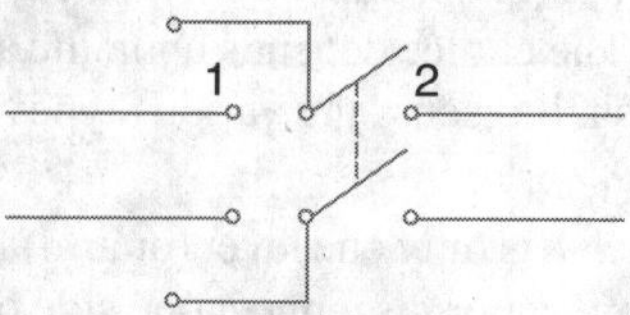

(*a*)

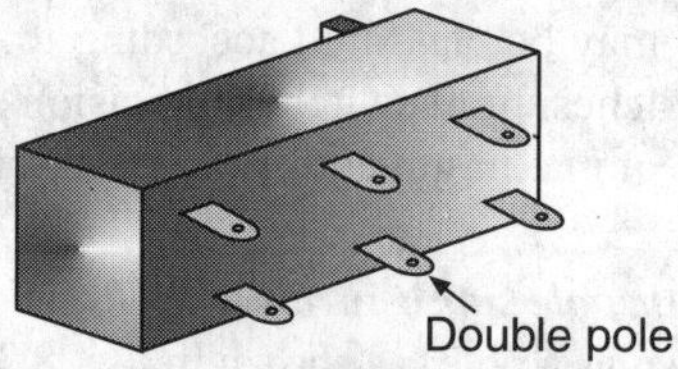

(*b*)

Fig. 8.6. DPDT switch.

Fig. 8.6 (*b*) shows a common shape of a DPDT switch. It may be noted that this switch has 6 connection points.

5. ***Toggle switch.*** This switch is similar to a single pole single throw switch. It has been specifically designed with very low contact resistance for ON-OFF applications in electronic circuits. Fig. 8.7 (*a*) shows a common shape of a toggle switch.

(*a*) Toggle switch

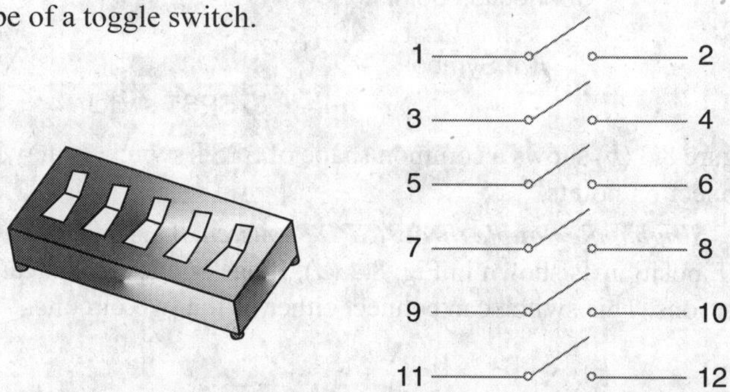

(*b*) DIP switch

Fig. 8.7.

6. ***Dual-in-line package (DIP) switch.*** This switch contains a number of single pole single throw switches as shown in Fig. 8.7 (*b*). It is a miniature switch, which is designed for installation on printed circuit boards (PCB's). The DIP switch is used for individual circuit switching or for encoding a binary code in computer circuiting.

7. ***Rotary switch.*** This switch is used to direct small signal currents from one circuit to another by changing the connections, when the switch knob is turned. Fig. 8.8 (*a*) and (*b*) shows the schematic symbol and shape of a common rotary switch.

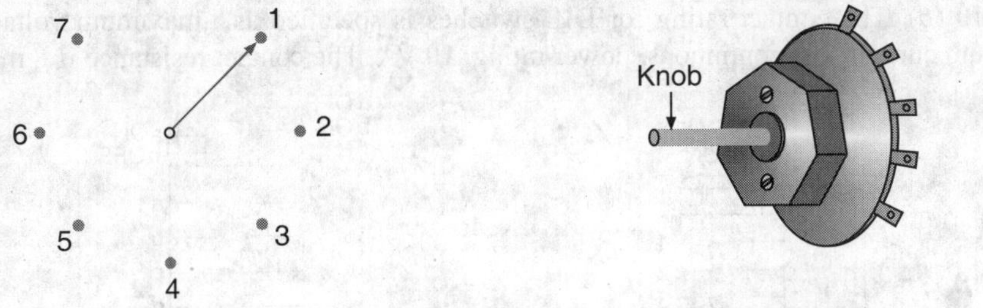

(a) Symbol (b) Shape

Fig. 8.8. Rotary switch.

A rotary switch is used in television receivers for channel selection, in radio receivers for band selection and in multimeters for range selection of voltage, current or resistance measurement.

8.5. Commercially Available Switches

There is a large variety of switches used in electronic industry. The selection of a switch depends upon many factors like type of action (momentary or maintained) required, size, shape an application. Fig. 8.9 (*a*) through (*j*) shows many different types of the switches used in electronics industry.

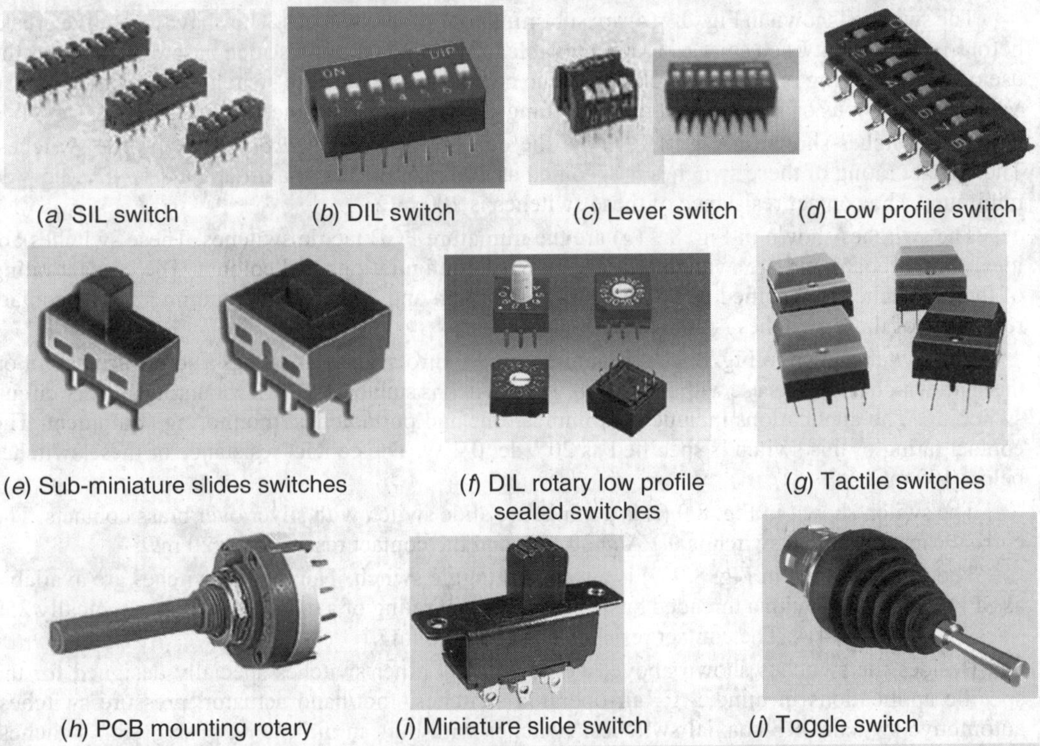

(*a*) SIL switch (*b*) DIL switch (*c*) Lever switch (*d*) Low profile switch

(*e*) Sub-miniature slides switches (*f*) DIL rotary low profile sealed switches (*g*) Tactile switches

(*h*) PCB mounting rotary (*i*) Miniature slide switch (*j*) Toggle switch

Fig. 8.9.

The switches shown in Fig. 8.9 (*a*) are a single-in-line (SIL) sub-miniature switches. These switches are *end stackable*. The contact rating of these switches is 0.4 VA, 20 V dc maximum. The contact resistance is 20 mΩ. The switches shown in Fig. 8.9 (b) are dual-in-line sub-miniature switches. These switches can be arranged in a choice of configurations and actions. For example, a 4-pole SPST switch showin in Fig. 8.10 (*a*) can be configured as a 2-pole changeover switch as shown in

Fig. 8.10 (*b*). The contact rating for DIL switches is specified as : maximum voltage, 100 Vdc, maximum current, 1 A continuous, power rating, 10 VA. The contact resistance is 8 mΩ.

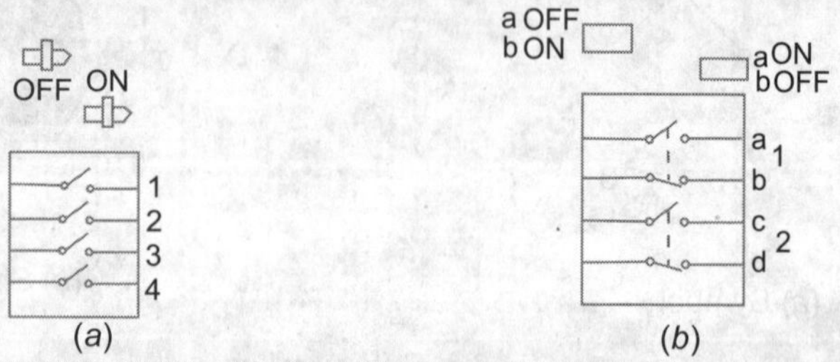

Fig. 8.10. A 4-pole DIL sub-miniature switch configured as 2-pole changeover.

The switch shown in Fig. 8.9 (*c*) is an in-line *lever* switch. The switches of this type are suitable for edge-of-board locations for timing, counting and coding circuits. The current rating of these switches is specified as 50 Vdc, 100 mA continuous and 25 Vdc, 25 mA. The contact resistance is 50 mΩ at 100 Vdc.

The switches shown in Fig. 8.9 (*d*) is a low profile in-line switches. These switches are suitable for immersion cleaning and automated assembly. The contact rating of the switches is specified as 100 mA at 50 Vdc and 10 μA at 3.5 Vdc. The contact resistance is 100 mΩ.

The switches shown in Fig. 8.9 (*e*) are sub-miniature slides switches. They have a positive break-before-make-action. Each switch has a removable top sealing tape and sealed base which allows the use of *flow soldering* production technique. The contact rating of these switches is specified as 250 Vac, 2A continuous or 28 Vdc, 250 mA switching. The contact resistance is 18 mW.

The switches shown in Fig. 8.9 (*f*) are the dual-in-line rotary *encoded* low profile switches. The contact rating of these switches is specified as 100 mA, 30 Vdc maximum and 10 μA, 3.5 Vdc minimum. The contact resistance of these switches is 200 mΩ.

The switches shown in Fig. 8.9 (*g*) are the miniature PCB tactile switches. These switches are ideal for keyboard and are available in a variety of configurations and colours. The contact rating of these switches is specified as 50 V, 50 mA maximum and 20 mV, 1 mA minimum. The contact resistance of these switches is 50 mΩ.

The switch shown in Fig. 8.9 (*h*) is a miniature PCB rotary switch. It has a screw driver slot on the shaft which allows its use without a knob. This switch is suitable for PCB mounting and dry circuit switching. The applications include communications and portable test/monitoring equipment. The contact rating of this switch is specified as 20 Vdc, 0.4 W. The contact resistance of these switches below 100 mΩ.

The switch shown in Fig. 8.9 (*i*) is a miniature slide switch with silver over brass contacts. The current capacity of this switch is 0.1 A@ 30 Vdc and the contact resistance is 20 mΩ.

The switch shown in Fig. 8.9 (*j*) is a standard toggle switch. The toggle switches are available as SPDT and DPDT with a threaded bushes. The contact rating of a toggle is switch is typically 250 Vac, 2 A ; 30 Vdc, 4 A. The contact resistance is about 20 mΩ.

Besides the switches shown above, we have many other switches specially designed for the specific applications in mind, *e.g.*, air-operated switches, foot/hand actuator, pressure switches, automotive switches RF coaxial switches, switches with built in indicators, push button switches, microswitches, heavy duty switches etc. The discussion on all these switches is beyond the scope of this book. The reader is advised to refer to the manufacturer's catalogue.

8.6. Fuses

The fuses are used to protect the electric circuit against excessive current. The excessive current may flow due to overload or short-circuit. If the fuse is not provided, the circuit components may get damaged. The fuse is connected in series with the electric circuit as shown in Fig. 8.11 (*a*).

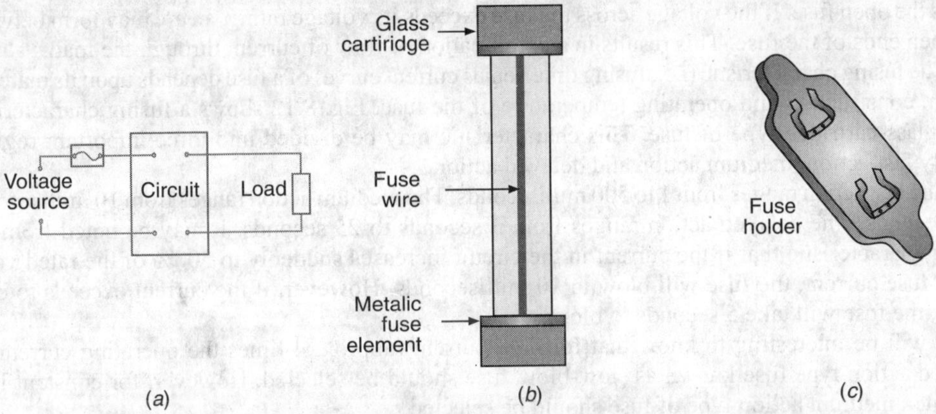

Fig. 8.11. Fuse wire.

A fuse consists of fine wire fixed in between the metallic fuse elements and sealed in a glass cartridge as shown in Fig. 8.11 (*b*). The glass cartridge is fixed in fuse-holders as shown in Fig. 8.11 (*c*). The fuse wire is generally made of tin coated copper (or nickel) and aluminum. This type of fuse is called glass-cartridge fuse and is used in all electronic instruments.

Under normal load conditions, the current flowing through the fuse wire, does not produce excessive heat. Therefore the fuse does not blow. However, if an excessive current (more than the rated value of a fuse) flows through the circuit, the fuse wire gets heated and the fuse blows.

8.7. Fuse Ratings

The fuses are rated by current capacity, voltage and fusion characteristics. The selection of a fuse, for a particular application, requires the knowledge of each of these ratings. The current rating of a fuse varies from 2 mA to many amperes of current. When a fuse blows, it should be replaced with another fuse with the same current, voltage and fusing characteristic. But the replacement should be done only after the reason for the failure of the fuse is determined. As a general rule, the fuse current rating is selected as 25% more than the load current. For example, if the load current is 1 A, then the fuse should have a current rating of 1.25 A.

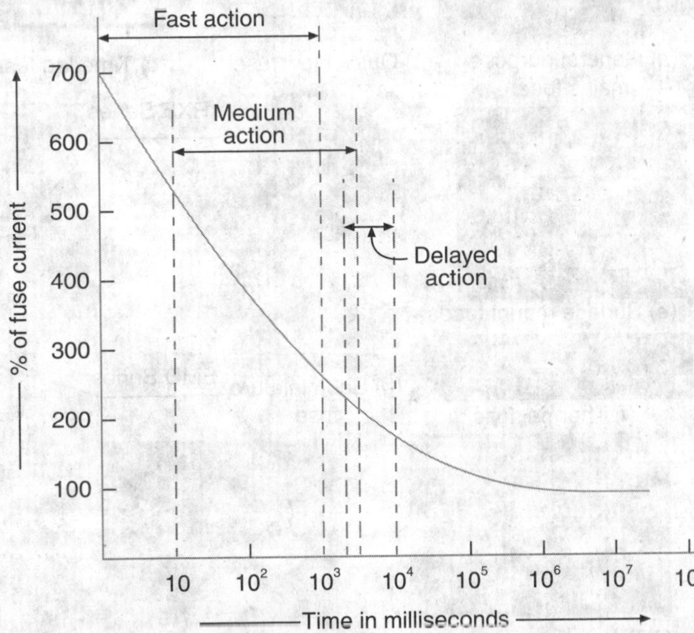

Fig. 8.12. Fusing characteristic.

The voltage rating of a fuse becomes important only when the fuse opens. At this instant, the entire source voltage is dropped across the open fuse. If the voltage across the fuse exceeds its voltage rating, an arc may form between the open ends of the fuse. This results in a continuation of flow of current through the load.

The fusing characteristic (*i.e.*, fusing time versus current curve) of a fuse depends upon its material, design, construction and operating temperature of the fuse. Fig. 8.12 shows a fusing characteristic for a glass-cartridge type of fuse. This characteristic may be divided into three important regions namely fast action, medium action and delayed action.

The fast action ranges from 1 to 500 milliseconds. The medium action ranges from 10 milliseconds to 5 seconds. The delayed action ranges from 3 seconds to 25 seconds. It may be noted from the fusing characteristic that if the current in the circuit increases suddenly to 500% of the rated value of the fuse current, the fuse will blow in 10 milliseconds. However, if the current exceeds to only 200%, the fuse will take 5 seconds to blow.

It will be interesting to know that for high current (approx. 4 times the operating current), a delayed action type fuse known as slow-blow-fuse should be selected. However, for constant load current, a medium action type of fuse should be selected.

8.8. Commercially Available Fuses

As a matter of fact, fuses provide an intentionally weakened part of an electrical circuit, acting as a safety link in the event of current loads. The excessive current is interrupted by the melting of the fuse wire through which the current flows.

Fuses are produced in several physical dimensions, shapes and internal constructions. However fuses with two dimensions namely 20 × 5 mm and 32 × 6.3 mm are more common in the electronics equipment. The fuses with 20 × 5 mm dimension are used in Europe and many other parts of the world whereas fuses with 32 × 6.3 mm are more common in U.S.A. and Canada. Fig. 8.13 (*a*) through (*i*) shows fuses for a variety of applications. Fuse shown in (*a*) is a general type fuse with a rated voltage of 240 Vac and are available with current rating ranging from 1 A to 10 A, (*b*) shows a *quick blow* type fuse. This type of fuse has a glass body, rated voltage of 250 Vac and a current rating ranging from 250 mA to 15 A.

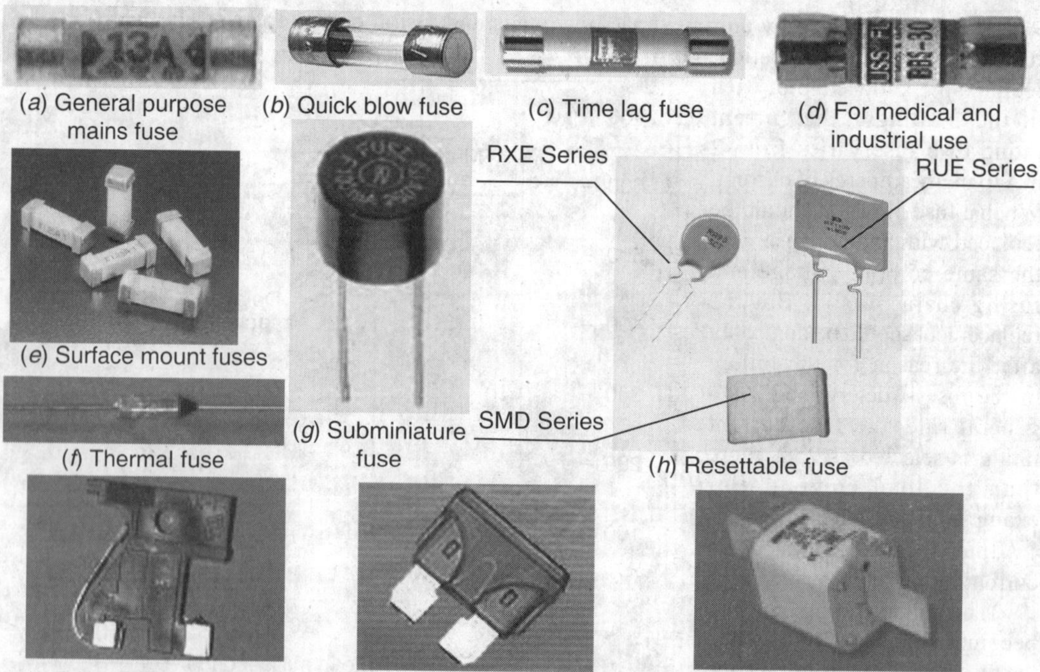

(*a*) General purpose mains fuse (*b*) Quick blow fuse (*c*) Time lag fuse (*d*) For medical and industrial use

(*e*) Surface mount fuses (*f*) Thermal fuse (*g*) Subminiature fuse (*h*) Resettable fuse

(*i*) Telecommunication fuse (*j*) Fuse for automotive applications (*k*) Used in switchgear and

Fig. 8.13. Commercially available fuses.

Fig. 8.13 (*c*) shows a anti-surge time lag fuse. It has a ceramic body. The breaking capacity of this type of fuse is 1500 A at 250 V. Fig. 8.13 (*d*) shows a quick acting fuse for medical and industrial applications. The breaking capacity for this type of fuse is 10 kA at the rated voltage. The fuses shown in Fig. 8.13 (*e*) are extremely small thin film surface mount type fuses. These fuses are available with a current rating ranging from 200 mA to 4 A. The fuse shown in Fig. 8.13 (*f*) has a axial lead and it protects equipment against fire risk or thermal damage due to overload. The operating current range for this type of a fuse range from 5 mA to 15 A, the rated current is 10 A. The fuse shown in Fig. 8.13 (*g*) is a high performance space saving fuse. It can be mounted on a PCB using a fuse holder. The fuses shown in Fig. 8.13 (*h*) are of automatically resettable fuses. The maximum interrupting current for these type of fuses is 40 A. The main applications of these fuses are in security systems, power supplies and transformers, power switching devices, instrumentation and control, computers and peripherals, audio visual and telecommunication.

Fig 8.13 (*i*) is a special fuse for telecommunication. This fuse is for multiple panel mounting. This fuse comes in different colours which in reality is for indicating different current rating. For example red colour fuse has a current rating of 500 mA, orange, 2 A and red-white, 10 A. All the fuses of this type have a ac voltage rating of 125 V and a dc voltage rating of 60 V. The fuse shown in Fig. 8.13 (*j*) is a fuse that is designed specifically for automotive applications. It has a rated voltage of 32 V and the current rating ranges from 1 A to 40 A. The fuse shown in Fig. 8.13 (*k*) is an ideal for protection of switchgear and motor loads. The fuses of this type are available with current rating ranging from 63 A to 160 A. The breaking capacity of this type of fuse is 80 kA.

8.9. Circuit Breaker

The circuit breakers are the control devices, which are widely used in industrial, commercial and household applications. The function of a circuit breaker is to carry a rated current, until an overload occurs. At overload, the breaker opens the circuit automatically. Some of the breakers are designed to respond quickly, whereas others have a time delay mechanism. Short time delays (a few seconds) are used to protect electronic equipments. The circuit breakers, with longer delays (upto 30 seconds), are used with small motors and some heating equipment.

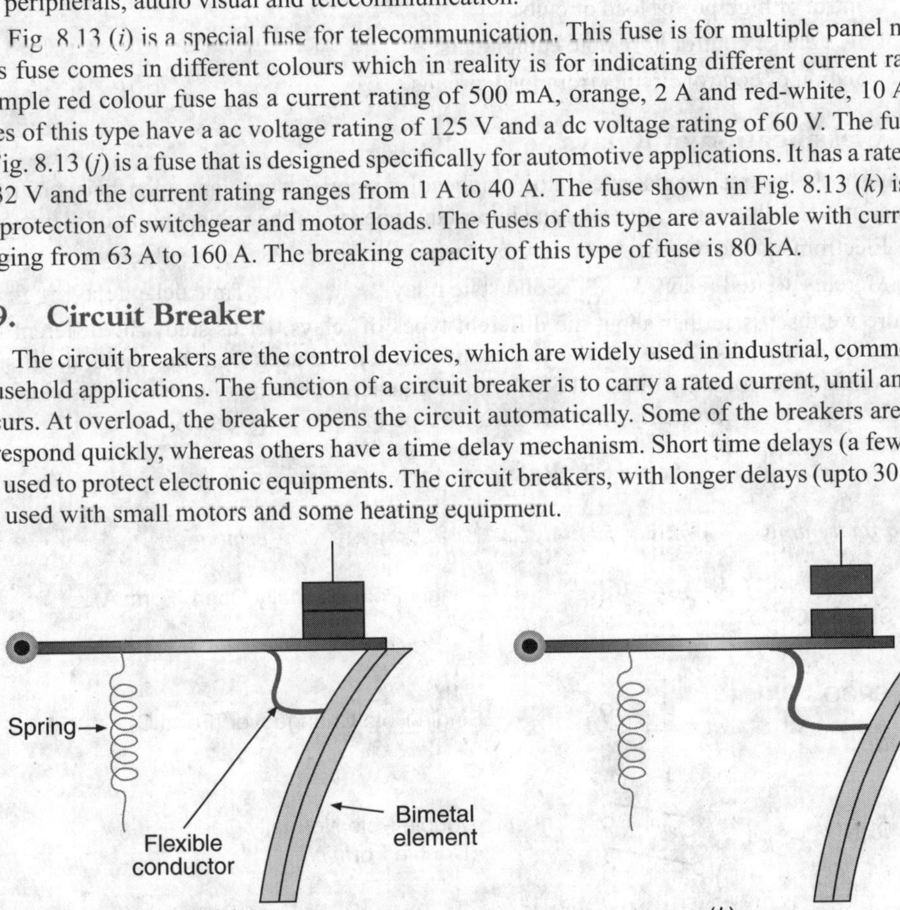

Fig. 8.14. Circuit breaker.

The circuit breakers may be broadly classified into three categories namely thermal, thermal-magnetic and hydraulic-magnetic. Fig. 8.14 (*a*) shows a thermal circuit breaker. It consists of a thermal bimetal element, a flexible conductor and a spring. The operation of a circuit breaker is as follows :

In the normal operation, the circuit current flows through the switch contacts and the bimetal element. As long as magnitude of the current remains below the rated value of the breaker, the thermal element does not react sufficiently to trip off the switch. However, if the current increases beyond the rated value (usually 25% greater than the rated value), the bimetal element expands due to the increased temperature. This causes the element to bend and open the switch contacts as shown in Fig. 8.14 (*b*). Higher overload currents create more expansion and therefore, the bimetal elements bends more quickly. This reduces the time delay of opening the switch contacts of the circuit breaker.

8.10. Relays

A relay is a device, which functions as an electrically operated switch. In response to an electrical signal, known as control signal, the relay opens or closes its switch contacts in some prearranged and fixed combination. The contacts may be in the same circuit as the operating signal or in another circuit (or a combination of circuits). The relays are widely used in industry as control devices. Following are some of the important applications, where the relays are used :

1. Control of high power load circuits.
2. Low voltage control of remote equipments.
3. Isolation of control circuits from load circuits.

8.11. Classification of Relays

Though the relays may be classified into a number of categories, depending upon their construction or application yet following are some of the important types from the subject point of view :

1. Electromagnetic relay
2. Crystal can relay
3. Dry reed relay
4. Mercury wetted relay
5. Solid state relay
6. Time delay relay

Before we discuss further about the different types of relays, let us study the different types of contacts used in relays. Table 8.1 shows this information. As seen from this table, the first row indicates single-pole normally open (SPNO) type of contact. This type of contact is also known as single-pole single-throw normally open (SPST-NO) type of contact or form A.

Table 8.1 : Relay contact information

Contact name	*Contact symbol*	*Description*
SPNO SPST-NO		Single Pole Normally Open (Form A)
SPCO SPDT	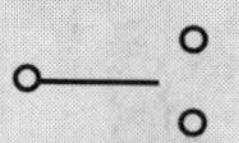	Single Pole Changeover (Form C)
DPNO	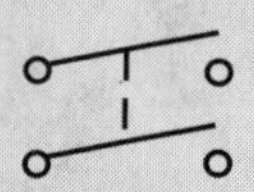	Double Pole Normally Open (Double Form A)
DPCO DPCT	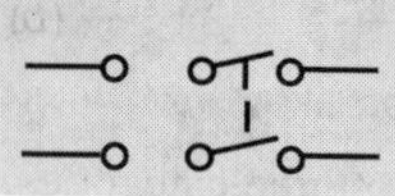	Double Pole Changeover (Double Form C)

The second row indicates single-pole changeover (SPCO) type of contact. It is also referred to as single-pole double-throw (SPDT) or form C type of contact. The third row indicates double-pole normally open (DPNO) type of contact. It is also called double form A type of contact. Finally the last row indicates double-pole changeover (DPCO) type of contact. It is also known as double-pole double-throw (DPDT) or double form C type of contact.

There are several relay manufacturers in the world but the list below shows a list of some leading manufacturers :

1. OMRON
2. Idec
3. FUJITSU
4. MEDER Electronic
5. Finder
6. CRYDOM

Fig. 8.15 (*a*) through (*i*) shows several different types of relays used in the electronic industry. The relays shown in Fig. 8.15 (*a*) are called reed relays. These relays have no coil inside and are operated by using an external magnet. The reed relays have high operating speed, low bounce times, low contact resistance, and high mechanical life. They are useful in a variety of applications from logic level to high voltage switching-ideal for relays, process control sensors, and proximity switches.

The relay shown in Fig. 8.15 (*b*) is a single-in-line (SIL) SPNO (single pole normally open) PCB reed relay. This device can handle a maximum current of 0.5 A, maximum voltage of 200 Vdc. Its contact resistance is ar\ound 200 mΩ.

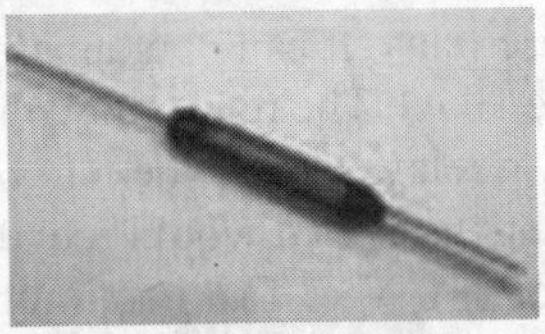

(*a*) Form A and C reed switch

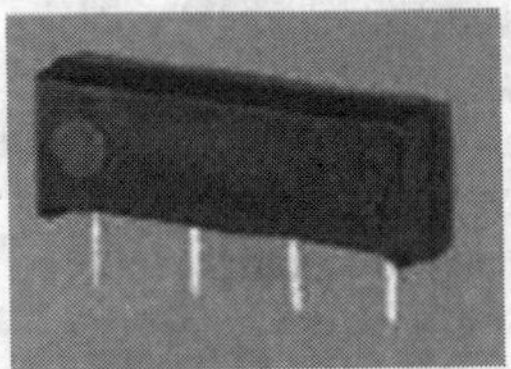

(*b*) SIL PCB reed relay SPNO (Form A)

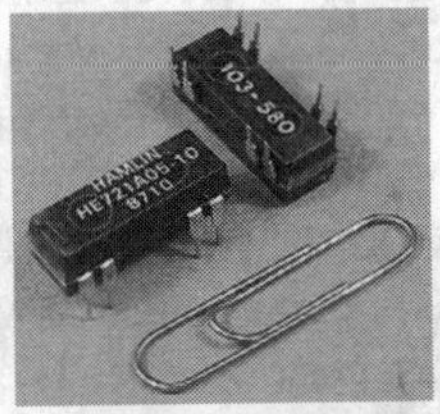

(*c*) Encapsulated PCB reed ralay SPNO

(*d*) Subminiature PCB relay-low power DPDT

(*e*) Metal can relay

(*f*) PCB power relay

(*g*) Relay timer

(*h*) Plug-in interface relay 5 A DPCO, 18 A SPCO

(*i*) Hermetically Sealed relay - 3 A 4PCO

Fig. 8.15. Different types of relays used in electronic industry.

The relay shown in Fig. 8.15 (*c*) is a reed relay encapsulated in a nickel-plated steel case for magnetic screening. The typical contact ratings of this type of relay are : maximum switched current of 0.75 A, maximum switched voltage of 200 Vdc and contact resistance of 150 mΩ. The relay shown in Fig. 8.15 (*d*) is a low power DPDT (double-pole double-throw) subminiature PCB relay. This relay has a gold clad, silver-palladium bifurcated crossbar contacts for reliable low power switching. Its contact rating is 1 A @ 30 Vdc, 0.3 A @ 125 Vdc. It can switch a maximum current 2 A dc or ac, maximum voltage of 220 Vdc or 250 Vac. The contact resistance of this device is 50 mΩ and the operate time does not exceed 5 ms.

The relay shown in Fig. 8.15 (*e*) is DPCO (double-pole changeover) metal can relay packaged in TO5 package. This type of relay is suitable for control and signal switching from dc to 1 GHz. The capacitance between the open contacts is 0.4 pF. The contact rating for this device is 1 A, 28 Vdc, the contact resistance is 100 mΩ and the operation time and release time are 2 ms each.

The relay shown in Fig. 8.15 (*f*) is a power relay suitable for PCB mounting. Its maximum switching voltage is around 250 Vac to 400 Vac and the switching power is 2500 VA. The nominal voltage for a typical relay of this type is 12 Vdc, nominal current, 40 mA and coil resistance, 300 Ω. The relay finds its applications in air conditioning, ceramic industry machines, thermal control and wood processing machines.

The relay shown in Fig. 8.15 (*g*) is called SPDT timer relay. This device has a timer which operates in 4 modes *namely on delay*, *interval*, *flicker off-start* and *flicker on-start*. The timer range of a typical device of this type is from 0.1 s to 10 ms, the rated supply voltage, 12 Vdc, output current rating, 3 A @ 250 Vac resistive load.

The relay shown in Fig. 8.15 (*h*) is a new generation general purpose relay. It can operate as a 5 A DPCO or 10 A SPCO device. It had a high switching capacity (1 pole 10 A). The relay shown in Fig. 8.15 (*i*) is a 3 A, 4PCO hermetically sealed relay. Hermetic sealing in relays allows them to be used in environments which are very dusty and in atmospheres containing gases that would reduce relay performance. The contact rating for this device is 3 A, 24 Vdc or 110 Vac. The maximum switched current is 3 A, the maximum switched voltage is 125 Vdc/125 Vac.

8.12. Electromagnetic Relay

An electromagnetic relay consists of an iron core and its surroundings coil of wire, iron yoke (or frame) armature and a set of contacts. They are arranged in a manner as shown in Fig. 8.16.

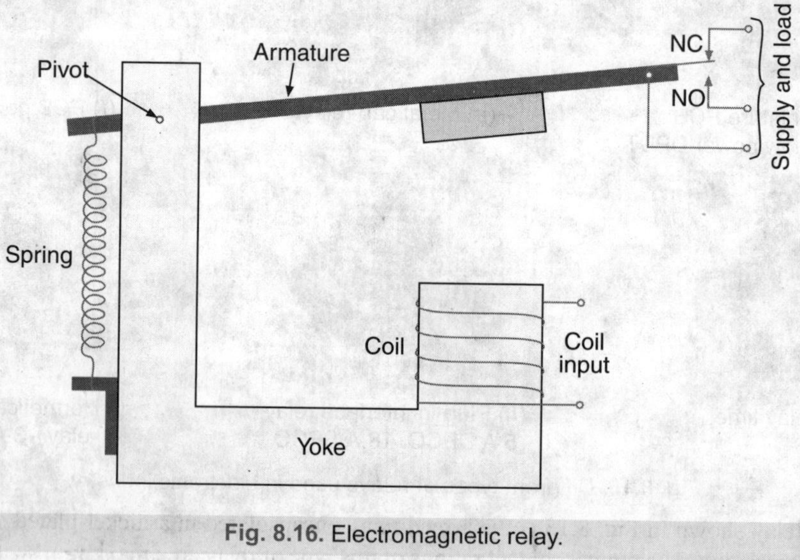

Fig. 8.16. Electromagnetic relay.

The iron-yoke provides a low reluctance path for the magnetic flux. It is shaped in such a way, so that a magnetic circuit can be closed by the armature and a set of contacts. The armature is a moveable piece of iron. It is hinged to the yoke, and held by a spring in such a way that there is an air gap in the magnetic circuit. There are three contacts in an electromagnetic relay namely normally-closed (NC) or break contact, normally-open (NO) or make contact and a moveable contact. The NC and NO contacts are placed on a mounting bracket, while the moveable contact is placed on an armature.

When a voltage is applied at the coil input of the relay, it magnetizes the yoke. This causes the armature to move towards the yoke. As a result of this, the armature opens the NC contact and closes the NO contact. When the voltage is removed from the coil input, a spring returns the armature to its original position. This operation takes place in few milliseconds.

The electromagnetic relays are available separately for d.c. and a.c. operations. The d.c. relays are available with standard operating voltages of 6 V, 12 V, 24 V, 48 V and 110 V. On the other hand, a.c. relays are available with standard operating voltages of 6 V, 12 V, 24 V, 48 V, 120 V and 240 V. The d.c. electromagnetic relays are faster than a.c. relays.

8.13. Types of Armature Assemblies in Various Electromagnetic Relays

Fig. 8.17 shows different types of armature assemblies found on various types of electromagnetic relays. Fig. 8.17 (*a*) indicates the clapper-type armature. As seen from the figure, it has a long arm connected at one end to a pivot point. When the coil is magnetized, the middle of the arm is pulled to the coil. This will cause the movable contacts to move to the left until they touch the stationery contacts. When power is removed from the coil, a spring will pull the contacts back to their open position.

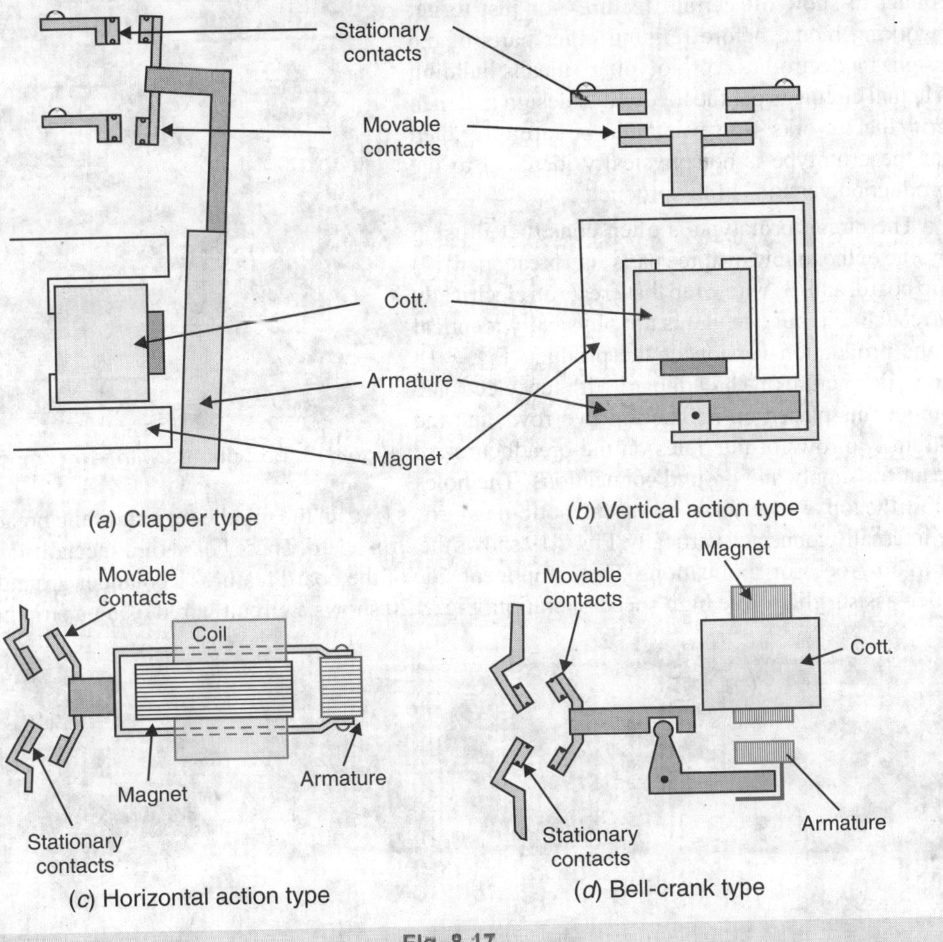

Fig. 8.17.

The vertical action type armature shown in Fig. 8.17 (*b*) is shaped like a letter *C* but in a backward direction. The contacts are connected to the top part of the armature and the bottom part is pulled to the coil when it becomes magnetized. When the armature is pulled up, the movable contacts will touch the stationery contacts. When power is turned off to the coil, the weight of the armature will cause it to move down to its original position where the contacts are open.

The horizontal action type armature shown in Fig. 8.17 (*c*) has a stationery contacts indicated on the far left, and a movable armature. The armature in this device moves to the left when the coil is energised. This causes the contacts to close. When the current is turned off and the coil is no longer magnetized, a spring will move the armature to the right.

The bell-crank type armature has an *L*-shaped arm connected to a pivot point. The armature in this type of relay pulls one end of the arm up when the coil is energised. This action shifts the movable contacts to the left so that they press against the stationery contacts. When current is turned off to the coil, the weight of the armature will cause the end of the arm to drop. This inturn will pull the movable contacts back to the right.

8.14. Circuit Prototyping

"Prototyping" is a process of creating a prototype product to show off certain features, or just to get a working model before refining other parts of the design. In electronics, "prototyping" means building an actual circuit as per the theoretical design and then verify that it works. However it may be carefully noted that the prototype is not physically identical to the "production version" of the product.

Fig. 8.18. Breadboard with few components.

The circuit prototype is often constructed using any one of the following three parts : (1) breadboard (2) strip board, and (3) wire wrap that create an electrically correct circuit, but one that is not physically identical to the production version of the product. Fig. 8.18 shows the picture of a breadboard with few electronic components placed on it. The top two rows and the bottom two rows of the holes on the breadboard are meant for supply and ground connections. The holes below the top two rows and above the bottom two rows (*i.e.*, in the middle region) of the breadboard are internally connected vertically. Fig. 8.19 shows the strip board. These boards are specially designed for microprocessor applications. The component side of the board features a colander ground plane which assists glitch-free high speed operation. Fig. 8.20 shows a circuit wired using a strip board.

(*a*) (*b*)

Fig. 8.19. Strip board : (*a*) top-wiring side and (*b*) bottom component side.

Fig. 8.20. Circuit wired on a strip board.

Fig. 8.21 shows the picture of a board used for wire wrapping along with the necessary parts and tools. The socket is used to connect the electronic components and the wire. It has 0.43 mm (0.017 inch) square posts, 25 mm (1 inch) long, and spaced at 2.5 mm (0.1 inch) intervals. Each socket's post can hold up to 3 wire connections. The wire used to wrap the components is of 30 AWG or 0.25 mm diameter. It is made of silver-plated soft copper insulated with fluorocarbon.

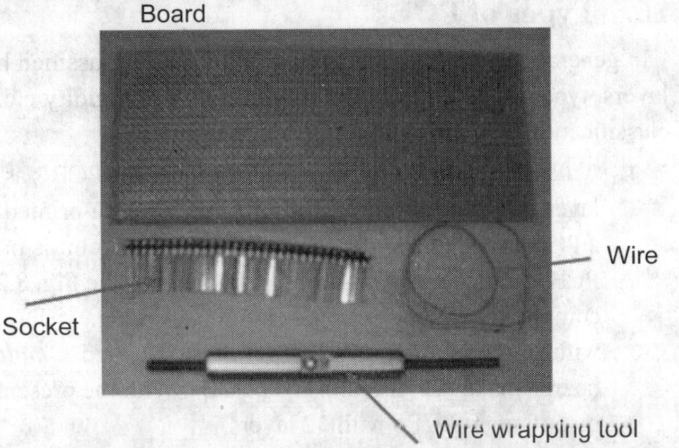

Fig. 8.21. Wire wrapping board with necessary parts and tools.

The manual wire wrap tool resembles a small pen. It has two holes at the ends. To wrap, approximately 6 mm of insulated wire is placed in a hole near the edge of the tool. The hole in the centre of the tool is placed over the post and the tool is rapidly twisted in the clockwise direction. The result is that 1.5 to 2 turns of insulated wire are wrapped around the post, and a top of that, 7 to 9 turns of bare wire are wrapped around the post. Fig. 8.22 shows a circuit using a wire wrapping technique.

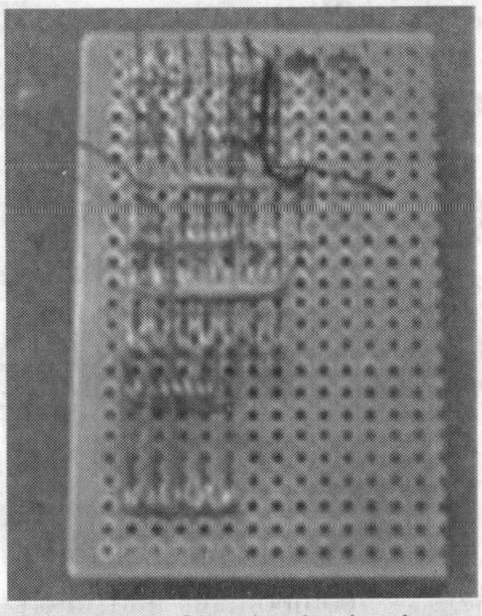

Fig. 8.22. Circuit wired using wire wrapping technique.

8.15. Printed Circuit Board

A printed circuit board (or PCB) is an electronic circuit in which the connections between the components (like integrated circuits (ICs), resistors, capacitors, inductors, relays etc.) are formed by etching a metallic coating (usually copper) on one or both sides of the insulating board. Fig. 8.23 shows a typical printed circuit board (or a PCB) with several IC's, resistors, capacitors and connections.

Fig. 8.23. A typical printed circuit board used in a computer (Courtesy : notova inc.)

8.16. Types of PCBs

In general printed circuit boards (PCBs) can be classified by different methods such as number of layers, types of base materials, conductor types, rigidity etc. A brief description of each method of classification is given below :

1. ***Number of layers*** : This method classifies the PCB's as a single-layer, double-layer or multi-layer PCBs. A simple PCB may have a circuit printed only on one side of the board. Such a PCB is called a *single-layer* PCB. If the circuit is printed on both sides of the board, the PCB is called a *double-layer* PCB. Refer to the Fig. 8.24. But these days with the increased complexity of the electronic circuit packaging, we have the circuit printed on several layers with in the board itself. Such a PCB is called a *multi-layer* PCB. The computer mother boards make use of multi-layer PCBs. With the present state of technology, it is possible to manufacture PCBs with 32 layers.
2. ***Type of Base materials*** : This method classifies the PCBs on the basis of the nature of the base material used for the board construction. The materials used for the fabrication of the PCBs are Fibre-glass (FR 4), Polyamide, Cyanate ester, Getek, PTFE, Reinforced teflon, Goreply, Teflon Composite and Glassteel and Polyester Reinforcement.
3. ***Conductor types*** : This method classifies the PCBs on the basis of the thickness of the copper foil and width of the conductor paths used on the board. The copper thickness could vary over the range from 0.0014" to 0.0042". Whereas the conductor width could be any one of the following values : 0.005", 0.010", 0.020" or 0.030". The two dimensions *i.e.*, thickness of the copper foil and the width of the conductive path on the board, defines the cross-sectional area of the conductor. This inturn determines the maximum electric current capability of the conductor.
4. ***Rigidity*** : This method classifies the PCB's as rigid or flexible. The rigid PCBs are used in most of the equipments for providing support for circuit components and interconnect them electrically. To provide sufficient rigidity, the boards are made with thickness of 1.6 mm (0.63"). The flexible PCBs are used to transport signal from one board to the another or from one system to another. The base materials used for the rigid and flexible PCBs are different.

(*a*) Single-layer PCB.

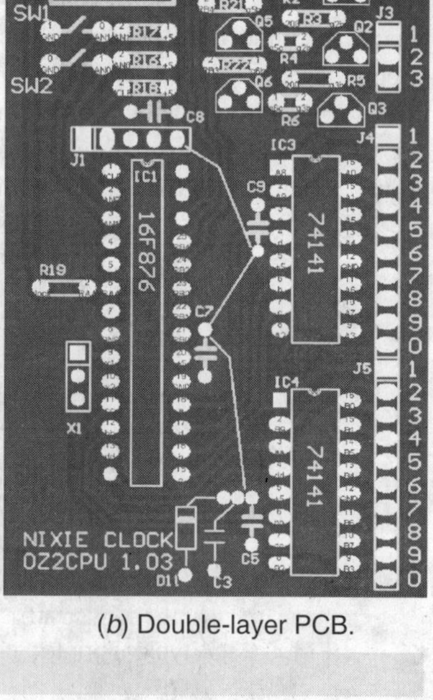

(*b*) Double-layer PCB.

Fig. 8.24.

8.17. Board Construction

We have already discussed in the last article that circuit boards are classified as either flexible or rigid. This refers to the base material. Flexible boards are made of sandwich of copper foil laminated between two pieces of soft plastic (usually a polyamide or polyester film). The plastic permits bending and shaping of the board.

Copper foil is laminated or glued into the surface of the base material. Boards are produced with copper foil on either or both sides of the substrate to make single or double sided boards as shown in Fig. 8.25 (*a*) and (*b*) respectively. The base material consists of some insulating material combined with a plastic (usually fibre-glass) under heat and pressure. The copper foil used in printed circuits is usually 0.04 to 0.1 mm thick. Its support called the substrate or base is usually a fibre glass material.

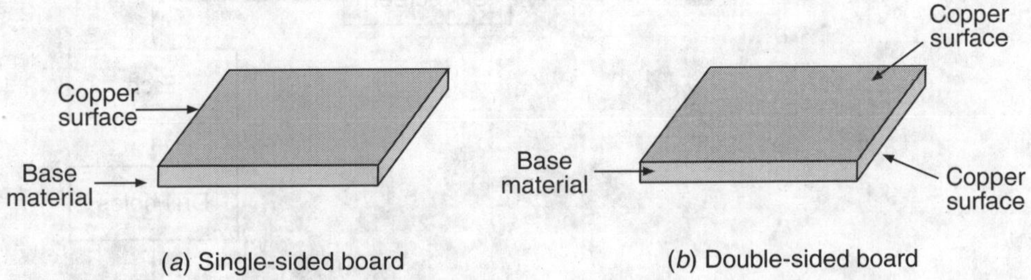

Fig. 8.25. A circuit board consisting of a thin layer of copper foil and a supporting base.

Design of a PCB

1. Schematic capture or Schematic entry. It is done through EDA tool. There are many EDA tools for creating schematic some of the EDA tools are ORCAD by Cadence, Proteus, Mentor Graphics.

2. Card dimensions and template. These are decided based on required circuiting and case of the PCB. Determine the fixed components and heat sinks if required.

3. Deciding stark layers of the PCB. 4 to 12 layers or more depending on desison complexity. Ground plane and power plane are decided. Signal planes where signal planes are routed in top layer well as internal layers.

4. Line impedance determination. It is done by using dielectric layer thickness, routing copper thickness and trace width. Trace separation is also taken into account in case of different signals. Microstrip, stripline or dual stripline can be used to route signals.

5. Placement of components. While placing components, thermal considerations and geometry are taken into account. Vias and land are marked.

6. Routing the signal trace. For optimal EMI performance, high frequency signals are rounted in internal layers between power ground planes as power planes behaves as ground for AC.

7. Gerber file. It is generated for manufacturing a PCB.

8.18. Steps Involved in the Development of a PCB

Fig. 8.26 shows the basic steps involved in the development of a circuit board. We start with an electronic schematic diagram of the circuit. This diagram is the actual wiring layout of the circuit in a shorthand form using standard electronic symbols. The schematic diagram shows how the circuit is wired in a graphic format.

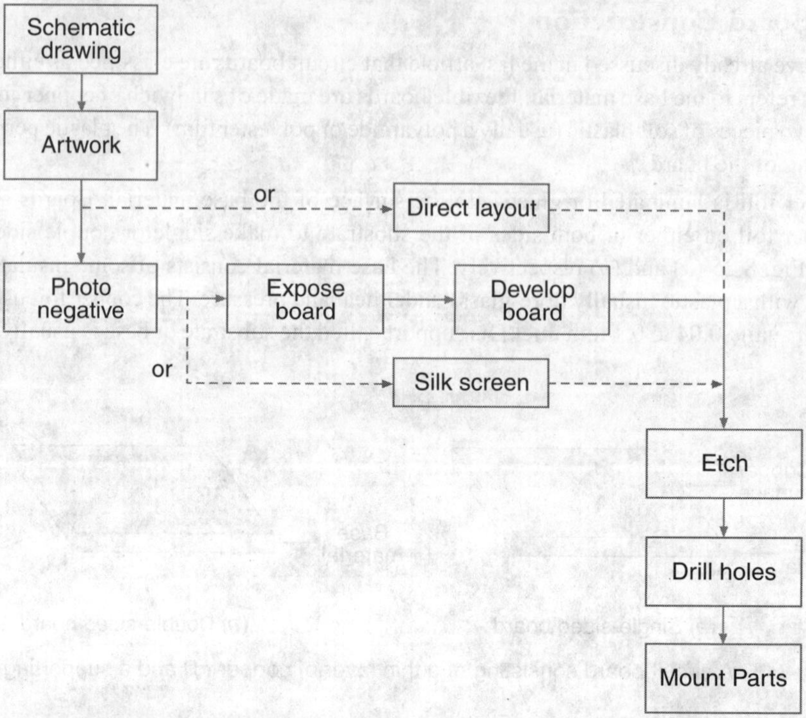

Fig. 8.26. The steps involved in the development of a printed circuit board.

The next step is to translate the schematic diagram into the artwork used to make printed circuit board. This is probably the most challenging part of the whole process. Artwork may seem difficult at first, but once some simple rules are understood, it becomes both interesting and satisfying.

There are two methods available for the third step : (1) layout the artwork directly on the printed circuit board material or (2) layout the artwork on a piece of photographic film. This decision depends upon the person making the board and factors such as the complexity of the circuit, number of identical boards to be made, materials available and whether someone else has already laid out the board.

Once this decision has been made, the artwork can be reduced on the circuit board material. One method requires artwork to be transferred to a piece of photographic film. The film is developed and placed on top of a specially prepared circuit board. Exposing the photographic negative and sensitized board to light transfers the artwork to the circuit board. In the other method, the artwork is redrawn automatically on the copper board material with an ink or paint resist. This second method is excellent for the reproduction of one-of-a kind boards. The board is then exposed and developed.

Before *etching*, a resist material is put on top of the copper foil. The resist protects the copper foil so that it will not be removed when the circuit board is etched. All the foil, except where the resist is placed, is removed from the board by etching. The etching solution, attacks the foil and removes it from the board. The end result is the electronic circuit attached to the substrate material. The board is removed from the etching solution, cleaned and inspected.

After inspection, holes are drilled in the board to be used for mounting parts, like resistors, transistors, capacitors and ICs. The parts are mounted and soldered to the board. The printed circuit board is ready for use now.

8.19. Advantages of PCBs

The printed circuit board offers the following advantages :

1. efficient conductor routing
2. greater circuit reproduction
3. greater flexibility in circuit packaging
4. reduced wiring errors
5. easier component mounting
6. simpler troubleshooting of problems
7. reduced size and weight, *i.e.*, more compact.

DESCRIPTIVE QUESTIONS

1. What is a switch? Explain briefly about its action.
2. Describe the different types of switches. What kind of switch is used in television and radio receivers ?
3. Explain briefly the fuse current *vs* time characteristic for a glass-cartridge type of fuse.
4. What is a circuit breaker ? Explain its operation.
5. Describe the operation of an electromagnetic relay.
6. What is a PCB ? Explain it briefly.
7. What are the various types of PCBs ? Describe them in detail.
8. Explain the principle of working different types of relays. (*Calicut University., June 2007*)

MULTIPLE CHOICE QUESTIONS

1. Which of the following switches is used for channel selection in television receivers
   (*a*) Toggle switch (*b*) DPDT switch
   (*c*) Rotary switch (*d*) None of the above
2. The main purpose of using a fuse in the circuit is to
   (*a*) Prevent the flow of excessive current (*b*) Protect the battery
   (*c*) Protect the circuit components (*d*) Open the circuit.
3. The fuse wire is made of
   (*a*) Tungsten (*b*) Tin coated zine (*c*) Tin coated nickel (*d*) all of these
4. The circuit breaker is a control device, which disconnects the circuit from the mains supply, if the load current flows
   (*a*) More than the rated value of current (*b*) Less than the rated value of current
   (*c*) Equal to the rated value of current (*d*) either 'b' or 'c'
5. Which of the following device can be used to isolate the control circuit from the load
   (*a*) Toggle switch (*b*) Circuit breaker
   (*c*) Relay (*d*) None of the above

ANSWERS

1. (*c*) **2.** (*a*) **3.** (*c*) **4.** (*a*) **5.** (*c*)

9

Chapter

Voltage and Current Sources

Outline

9.1. Introduction

We know that every electronic circuit requires some source of energy for its operation. The energy may be supplied either in the form of a voltage or current. The source, which supplies energy in the form of a voltage, is called voltage source. And the source, which supplies energy in the form of a current, is called current source. All voltage and current sources may be broadly classified into two categories depending upon whether a source supplies d.c. (direct current) or a.c. (alternating current) to the load. Some of the commonly used d.c. sources are batteries, d.c. generators and regulated d.c. power supplies. And the commonly used a.c. sources are alternators, oscillators or signal generators. In the field of applied electronics, the batteries, regulated d.c. power supplies and oscillators are the main sources of energy.

9.2. Batteries

The term 'battery' is derived from the expression 'battery of cells'. Thus a battery consists of two or more number of similar cells. A cell is a fundamental source of energy. It consists of a combination of materials, which produce direct current (d.c.) electrical energy from its internal chemical reactions.

The cells can be divided into two classes namely primary cells and secondary cells. Primary cells are those, which cannot be recharged after they are discharged; whereas the secondary cells are those which can be recharged. The carbon-zinc cells, manganese-alkaline cells, mercury cells, silver-oxide cells etc., are the primary cells; whereas the lead acid cell is an example of a secondary cell. The carbon-zinc cells are the most widely used commercial dry cells. The manganese alkaline cells are used in radios, TV sets, tape recorders, cameras and cordless type home appliances. The mercury and silver-oxide cells are used in cameras, hearing aids, electronic watches and test instruments. The lead acid cells are mostly used for making automobile batteries. The pictures below shows variety of batteries used in electronic equipment.

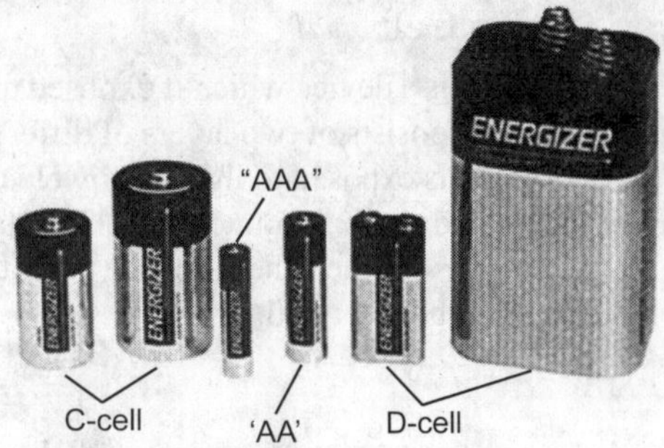

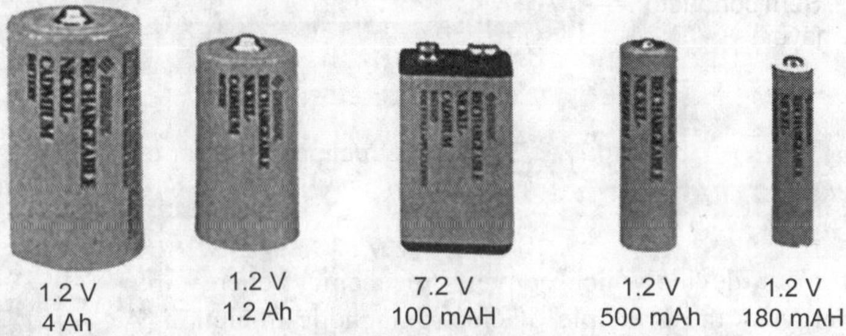

(a) LithodeTM lithium-iodine cell 2.8 V, 870 mAh Long-life power sources with printed circuit board mounting capability.

(b) Lithode-iodine pacemaker cell 2.8 V, 2.0 Ah.

9.3. Regulated D.C. Supplies

These are the most commonly used types of energy sources in an electronics laboratory. A regulated d.c. supply contains a rectifier, filter and a voltage regulator. The rectifier converts the alternating voltage (which is easily available from the main supply) into a fixed voltage. The filter smoothens the rectifier output and a voltage regular keeps the d.c. output voltage independent at any variation in the rectifier input voltage or load resistance. The battery eliminator, used with a transistor radio or calculator, is a regulated d.c. supply source.

9.4. Solar Cell

A solar cell is a device which if exposed to sunlight, produces a voltage is across its terminals. A basic solar cell consists of two layers of different semiconductors joined together to form a junction. When one layer is exposed to light, many electrons acquire enough energy to break away from their parent atoms and cross the junction. This process forms negative ions on one side of the junction and positive ions on the other. Thus a potential difference is developed. Fig. 9.1 shows the basic construction of the solar cell.

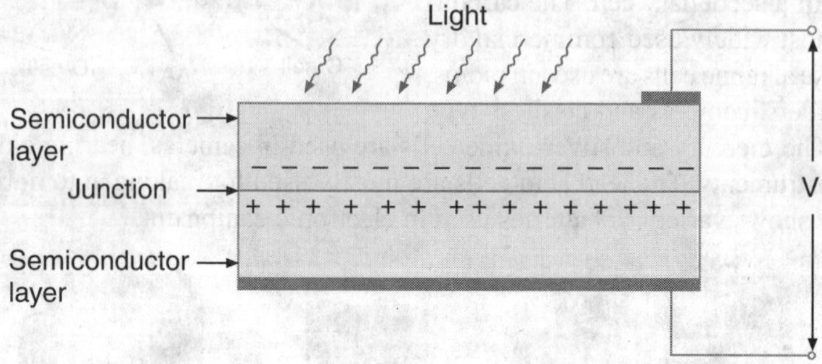

Fig. 9.1. Basic construction of a solar cell.

9.5. A Generator

Generator is a device which converts mechanical energy into electrical energy by the principle of electromagnetic induction. A conductor is rotated through a magnetic field and a voltage is produced across the conductor. The details of generator is beyond the scope of this book.

This calculator makes use of solar energy.

9.6. Oscillators or Signal Generators

An oscillator or a signal generator is an equipment, which supplies a.c. voltages. This voltage is used as a signal to test the operation (or response) of the different electronic circuits such as amplifiers filters, closed loop control systems etc. The frequency of the a.c. signal supplied by the signal generators can be varied over a wide frequency range and are capable of giving triangular or square wave outputs in addition to sinusiodal-wave output.

9.7. Internal Resistance of a Source

It has been found that all sources of electrical energy have some internal resistance. The value of this resistance is generally few ohms. It is due to this internal resistance that a source does not behave ideally. Thus when a voltage source supplies power to a load, its terminal voltage (*i.e.*, the voltage available at its terminals) drops as compared to its open-circuit voltage. For example, a cell used in a torch has a voltage of 1.5 V across its terminals (or electrodes) when it is open-circuited (*i.e.*, nothing connected to it). However, when it is connected to a bulb, its voltage reduces below 1.5 V. Such a reduction in the terminal voltage of the cell is due to the internal resistance of the torch. The internal resistance of a source may be due to any one or more of the following reasons :

1. The resistance of an electrolyte between the electrodes in the case of dry cell.
2. The resistance of an armature winding in the case of d.c. generator or an alternator.
3. The output resistance of an active device like a transistor or vacuum tube in case of oscillator, signal generator or regulated d.c. supply.

Example 9.1.

Fig. 9.1 (a) and (b) shows a dry cell of 1.5 V connected across a bulb and an equivalent circuit, where the cell is replaced by a constant voltage source of 1.5 V in series with internal resistance R_S (equal to 0.2 Ω)

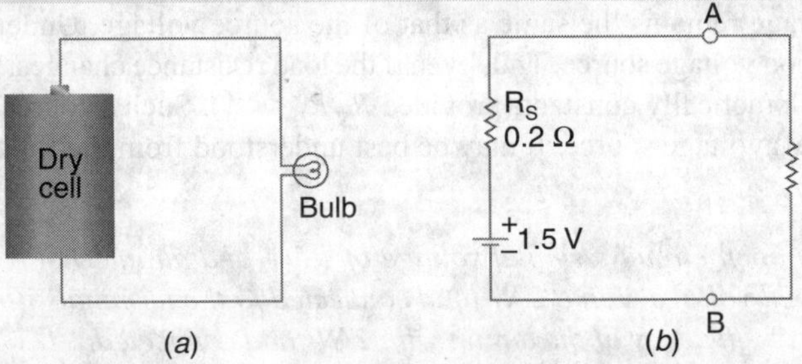

Fig. 9.2.

If the bulb is replaced by a load resistor R_L (equal to 1Ω) find the voltage drop due to the internal resistance.

Solution.

Given : $V = 1.5$ volt; $R_S = 0.2\ \Omega$ and $R_L = 1\ \Omega$.

We know that the total resistance in the circuit,

$$R_T = R_S + R_L = 0.2 + 1 = 1.2\ \Omega$$

and current in the circuit,

$$I = \frac{V}{R_T} = \frac{1.5}{1.2} = 1.25 \text{ A}$$

∴ Voltage across the terminal AB,

$$V_{AB} = I \times R_L = 1.25 \times 1 = 1.25 \text{ V}$$

and voltage drop due to the internal resistance

$$= 1.5 - 1.25 = 0.25 \text{ V } \textbf{Ans.}$$

9.8. Concept of a Voltage Source

Consider a d.c. source connected across a variable load resistance as shown in Fig. 9.3.

Let V_S = Open-circuit voltage *i.e.*, a voltage, which exists across the source terminals when nothing is connected to it.

R_S = Internal resistance of the source, and

R_L = Variable load resistance.

We know that the total resistance

$$R_T = R_S + R_L$$

and current,

$$I = \frac{V_S}{R_T} = \frac{V_S}{R_S + R_L}$$

∴ Voltage across the terminals AB (which is also equal to the voltage across the load),

$$V_{AB} = V_L = I \times R_L$$

$$= \frac{V_S}{R_S + R_L} \times R_L$$

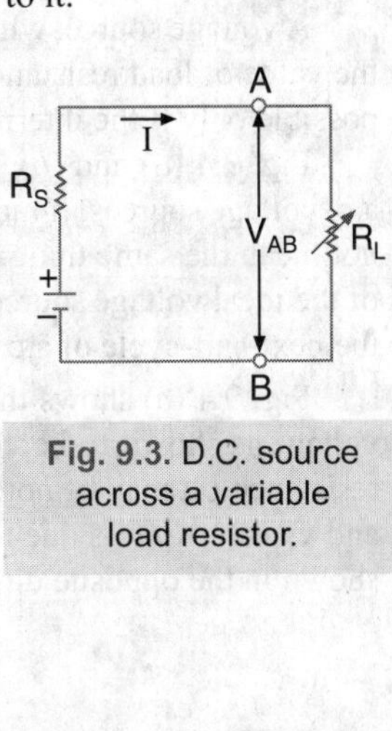

Fig. 9.3. D.C. source across a variable load resistor.

$$= \frac{V_S}{1 + \frac{R_S}{R_L}}$$

It may be noted from the above expression that if the ratio R_S/R_L is small as compared to unity, then the terminal voltage remains the same as that of the source voltage. Under this condition, the source behaves as a good voltage source. Thus even if the load resistance changes, the terminal voltage of the source remains practically constant (provided $R_S/R_L << 1$). Such a source is then said to be a good (though not ideal) voltage source. It may be best understood from the following example.

Example 9.2.

A dry cell has an open-circuit terminal voltage of 1.5 V and an internal resistance of 0.4 Ω. Determine, (a) terminal voltage with a 2 W load connected to the terminals, (b) power dissipated by the 2 Ω load and (c) efficiency of the circuit with 2 W load connected.

Solution.

Given : $V_S = 1.5$ V; $R_S = 0.4\ \Omega$ and $R_L = 2\ \Omega$.

(*a*) ***Terminal voltage***

We know that current in the circuit,

$$I = \frac{V_S}{R_S + R_L} = \frac{1.5}{0.4 + 2} = 0.625 \text{ A}$$

and terminal voltage,

$$V_T = I \times R_L = 0.625 \times 2 = 1.25 \text{ V } \textbf{Ans.}$$

(*b*) ***Power dissipated***

We know that power dissipated by the load resistance,

$$P_L = I^2 \times R_L = (0.625)^2 \times 2 = 0.78 \text{ W } \textbf{Ans.}$$

(*c*) ***Efficiency of the circuit***

We also know that power supplied by the voltage source,

$$P_S = I^2 (R_S + R_L) = (0.625)^2 \times (0.4 + 2) = 0.94 \text{ W}$$

and efficiency of the circuit,

$$\eta = \frac{P_L}{P_S} = \frac{0.78}{0.94} = 0.83 = 83\% \textbf{ Ans.}$$

9.9. Ideal Voltage Source

A voltage source, which produces a terminal voltage (or load voltage) and does not depend upon the value of load resistance, is called an ideal voltage source or perfect voltage source. This may be possible only if the internal resistance of a voltage source is zero.

Fig. 9.4 (*a*) and (*b*) shows the circuit symbol and reference direction for current of a d.c. and a.c. voltage sources. It may be noted that the marking of positive and negative on an a.c. source does not mean the same thing as that of a d.c. source. In an a.c. source, it means that the upper terminal of the ideal voltage source is positive, with respect to the lower terminal at that particular instant. In the next half-cycle of a.c., the markings will become opposite.

Fig. 9.4 (*c*) shows the V-I characteristic of an ideal voltage source. In this figure, the terminal voltage is shown to be constant (equal to V_S) for all values of load current (or all values of load resistance). It may be noted that as the value of load resistance is reduced, the load current increases and vice versa. It is due to this fact that the variation of load current (I_L) and load resistance (R_L) is shown in the opposite directions along the horizontal axis of the V-I characteristic.

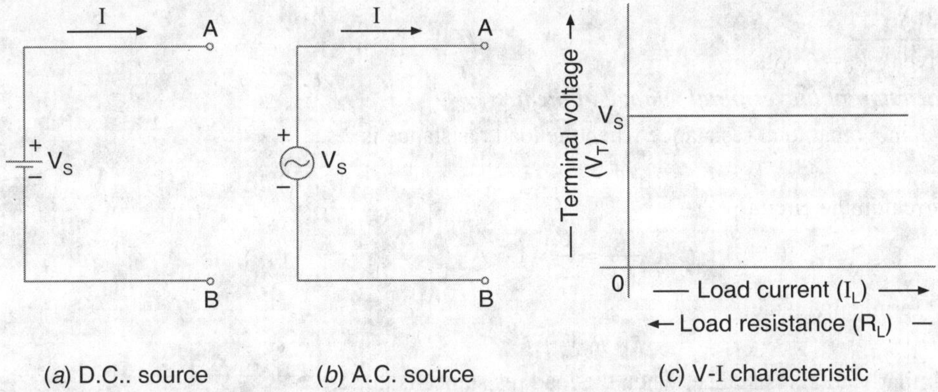

Fig. 9.4. Circuit symbols and V-I characteristic for an ideal voltage source.

9.10. Practical Voltage Source

It will be interesting to know that in actual practice, there is no voltage source, which is an ideal one. Moreover, it is also not practically possible. It can be easily understood by considering a voltage source with its terminals shorted (*i.e.*, load resistance is made zero) so that the load current should approach infinity. It means that a source can supply infinite amount of power to a short- circuit. But this is not possible, in actual practice. Therefore, an ideal voltage source does not exist in actual practice. However, the concept of an ideal voltage source is very helpful in understanding the circuits containing a practical voltage source.

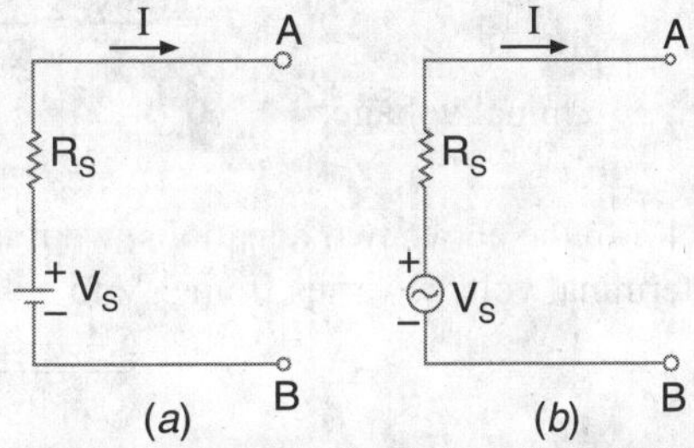

Fig. 9.5. Circuit symbols for practical d.c. and a.c. sources.

A practical voltage source (also called real voltage source) consists of an ideal voltage source in series with a resistance. This resistance is known as internal resistance of the source. Fig. 9.5 (*a*) and (*b*) shows the circuit symbols and reference direction for current of the practical d.c. and a.c. sources.

We shall now discuss that what should be the characteristic of a source, so that it could be considered as a good enough constant voltage source. We know that an ideal voltage source has zero internal resistance. But in actual practice, no source is an ideal one. Therefore, it is necessary to determine how much should be the value of internal resistance, so that a source may be considered as a practical voltage source.

Note: The mathematical concept of practical voltage source has already been discussed in Art. 9.8.

Example 9.3.

Fig. 9.6 (a) and (b) shows two practical voltage sources having variable load resistance. Determine the variation in the terminal voltage in both the cases. Also give a comparison of both the cases.

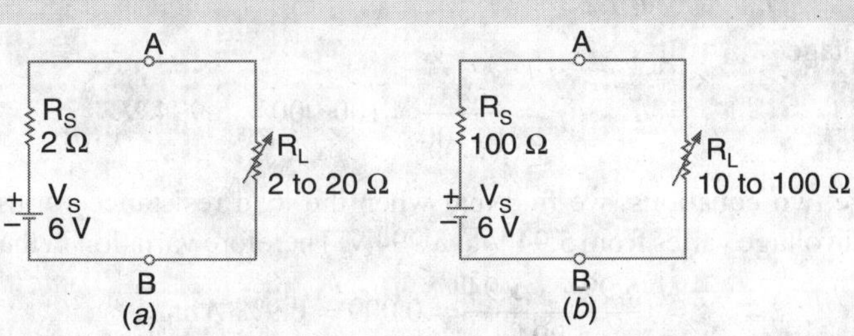

Fig. 9.6.

Solution.

Given : $V_S = 6$ volts

(a) *Variation in the terminal voltage in the first case*

We know that total resistance when the load resistance is 2 Ω.

$$R_T = 2 + 2 = 4\ \Omega$$

and current in the circuit,

$$I = \frac{V_S}{R_T} = \frac{6}{4} = 1.5\ \text{A}$$

∴ Terminal voltage,

$$V_T = I \times R_L = 1.5 \times 2 = 3\ \text{V} \quad ....(i)$$

Similarly, total resistance, when the load resistance is 20 Ω,

$$R_T = 2 + 20 = 22\ \Omega$$

and current in the circuit,

$$I = \frac{V_S}{R_T} = \frac{6}{22} = 0.273\ \text{A}$$

∴ Terminal voltage,

$$V_T = I \times R_L = 0.273 \times 20 = 5.46\ \text{V} \quad ....(ii)$$

From the above two equations, we find that when the load resistance varies from 2 Ω to 20 Ω, the terminal voltage varies from 3 V to 5.46 V. Therefore variation in the terminal voltage.

$$= \frac{5.46 - 30}{5.46} = 0.45 = 45\%\ \textbf{Ans.}$$

(b) *Variation in the terminal voltage in the second case*

We know that total resistance when the load resistance is 10 kΩ

$$R_T = 100 + (10 \times 10^3) = 10\ 100\ \Omega$$

and current in the circuit

$$I = \frac{V_S}{R_T} = \frac{6}{10\ 100}\ \text{A}$$

∴ Terminal voltage,

$$V_T = I \times R_L = \frac{6}{10\ 100} \times 10\ 000 = 5.94\ \text{V} \quad ...(iii)$$

Similarly total resistance, when the load resistance is 100 kΩ,

$$R_T = 100 + (100 \times 10^3) = 100\ 100\ \Omega$$

and current in the circuit

$$I = \frac{V_S}{R_T} = \frac{6}{100\ 100}\ \text{A}$$

∴ Terminal voltage

$$V_T = I \times R_L = \frac{6}{10\ 100} \times 100\ 000 = 5.994\ \text{V} \quad ....(iv)$$

From the above two equations, we find that when the load resistance varies from 10 kΩ to 100 kΩ, the terminal voltage varies from 5.94 V to 5.994 V. Therefore variation in the terminal voltage.

$$= \frac{5.994 - 5.94}{5.994} = 0.009 = 0.9\%\ \textbf{Ans.}$$

Comparison of both the cases

From the above variations in both the cases, we find that in the first case the variation of terminal voltage is 45%, whereas in the second case, it is 0.9% only. Thus the first case, can not be termed as a constant voltage source, whereas the second case can be termed as a constant voltage source.

Moreover, from the given data, we also find that internal resistance in the first case is 2 Ω and that in the second case is 10 kΩ. Thus we conclude that it is not the absolute value of the internal resistance, which decides whether a source is good constant voltage source or not. But it is the ratio of internal resistance to the load resistance (*i.e.*, R_S/R_L), which is important from this point of view. From the given data, we also find that this ratio in the first case varies from 1 to 0.1, whereas it varies in the second ratio from 0.06 to 0.0006. Hence, smaller the ratio of R_S/R_L, better is the source as a practical constant voltage source. **Ans.**

Note: In actual practice, a source will provide a practical constant terminal voltage, if the ratio R_S/R_L is at least 0.05. In other words, the load resistance must be at least 20 times larger than the internal resistance of the voltage source.

9.11. V-I Characteristic of a Practical Voltage Source

Fig. 9.7 (*a*) shows the V-I characteristic of a practical voltage source. It may be noted from this figure that as the load current increases (*i.e.*, R_L decreases), the terminal voltage decreases. This happens because of the existence of internal resistance within the voltage source. As the load current increases, the voltage drop across the internal resistance also increases. Hence the terminal voltage decreases below the open-circuit voltage. In general, the terminal voltage is given by the relation,

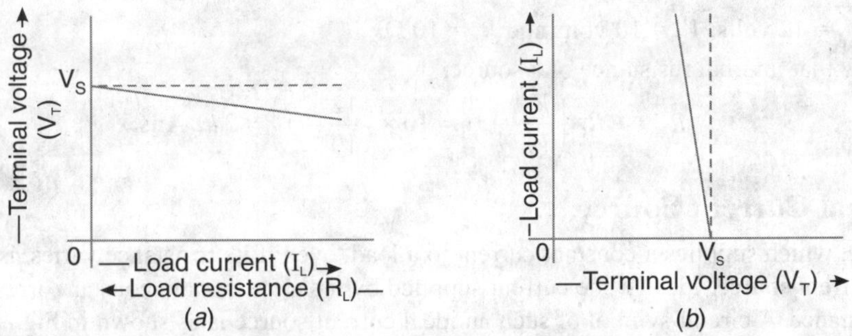

Fig. 9.7. V-I characteristic of a practical voltage source.

$$V_T = V_S - I \cdot R_S$$

where V_S = Open-circuit voltage,

I = Current in the circuit, and

R_S = Internal resistance.

Sometimes, it is preferred to take terminal voltage on the horizontal axis and load current along the vertical axis. In that case, the V-I characteristic of a practical voltage source is shown in Fig. 9.7 (*b*).

9.12. Procedure for Determining Internal Resistance

The internal resistance of a practical voltage source may be determined by using the procedure given below :

1. First of all, measure the voltage across the terminals of a source with the load removed from the circuit as shown in Fig. 9.8 (*a*). This voltage is equal to the ideal source voltage (V_S).

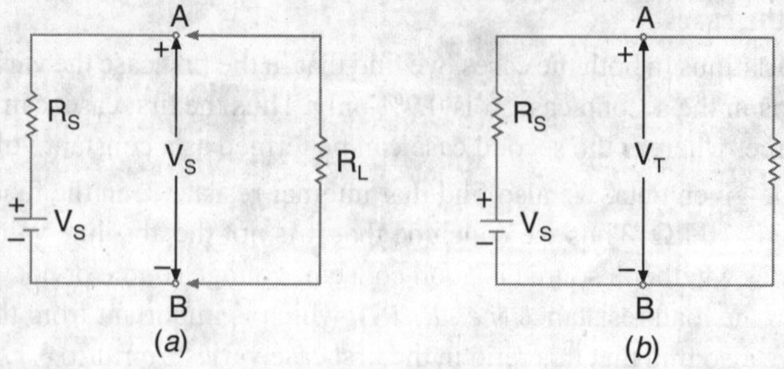

Fig. 9.8.

2. Now measure the voltage across the terminals of source with the load connected across the source as shown in Fig. 9.8 (*b*). Let this voltage be equal to V_T.

Now the internal resistance of a voltage source is given by the relation,

$$R_S = R_L\left(\frac{V_S}{V_T} - 1\right)$$

Example 9.4.

In a certain experiment, the open-circuit voltage of a source is found to be equal to 12 V and the terminal voltage of 10 V with a load resistance of 10 Ω connected across the source. What is the internal resistance of the source ?

Solution.

Given : $V_S = 12$ volts; $V_T = 10$ volts and $R_L = 10\ \Omega$.

We know that internal resistance of a source,

$$R_S = R_L\left(\frac{V_S}{V_T} - 1\right) = 10 \times \left(\frac{12}{10} - 1\right) = 2\ \Omega \textbf{ Ans.}$$

9.13. Ideal Current Source

A source, which supplies a constant current to a load, even if its resistance varies, is known as a constant current source. Ideally, the current supplied by it should remain constant, irrespective of the load resistance. A circuit symbol of such an ideal current source is as shown in Fig. 9.9(*a*). The arrow inside the circle indicates the direction in which the current flows in the circuit when a load is connected across it.

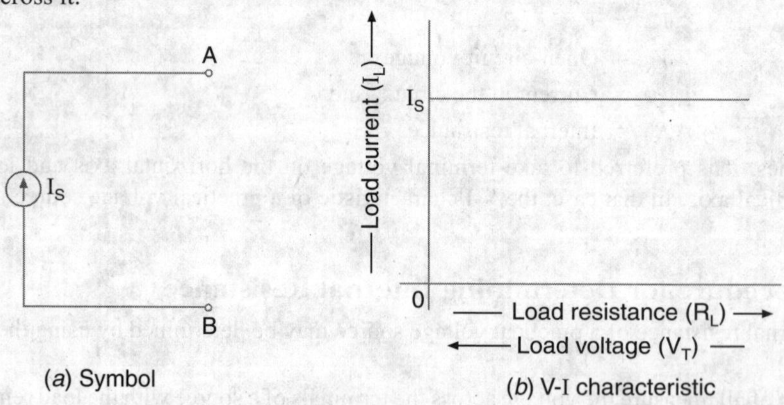

Fig. 9.9. Ideal current source.

Fig. 9.9 (*b*) shows the V-I characteristic of an ideal current source. In this figure, the load current (*i.e.*, the current supplied to the load by the source) is shown to be constant (equal to I_S) for all values of the load resistance (R_L). It may be noted that as the value of load resistance is increased, the terminal voltage also increases. It is due to the fact, that the variation of terminal voltage is shown in the same direction as that of load resistance along the horizontal axis of the V-I characteristic.

9.14. Practical Current Source

In actual practice, there is no current source, which is an ideal or a perfect one. An ideal current source is merely a theoretical idea. It can be understood from the discussion given below :

Consider a variable load resistance (R_L) connected to an ideal constant current source as shown in Fig. 9.10 (*a*). As discussed in the last article, the current supply by the source should remain constant (equal to I_S) for all values of load resistance. It means, that even if load resistance is made infinite (*i.e.*, terminal *A* and *B* are open-circuited) the current through it should remain equal to I_S. But in actual practice, there is no such source, which could satisfy this condition. It is because of the fact that if the load resistance is of infinite value, then there is no conducting path external to the source, which exists between the terminals *A* and *B*. Hence it is not possible for the current to flow between the terminals *A* and *B*. If the source could maintain a current equal to I_S, through an infinite load resistance, there would be infinite value of voltage drop across the load. Then it would have consumed infinite amount of power from the source. As a matter of fact, no practical source can ever supply infinite amount of power.

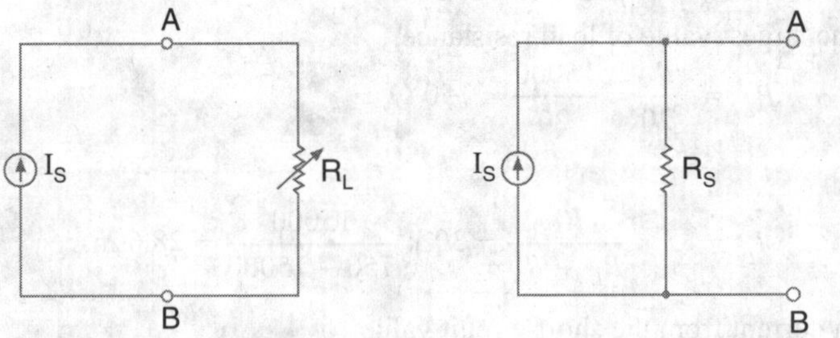

(*a*) Variable load connected across an ideal current source.

(*b*) Circuit symbol for a practical current source.

Fig. 9.10.

It is evident from the above discussion that there must exist some path within the source, itself, through which the current I_S can flow, when the load resistance is of infinite value. This internal path has a resistance (R_S) and is called internal resistance of the current source. Thus a practical current source consists of an ideal current source in parallel with the internal resistance (R_S). Fig. 9.10 (*b*) shows the circuit symbol of a practical current source. A practical current source supplies current (I_S) to a short-circuit load (*i.e.*, when $R_L = 0$). It is due to this fact that I_S is called a short-circuit current. When the load resistance is increased, the load current falls below I_S.

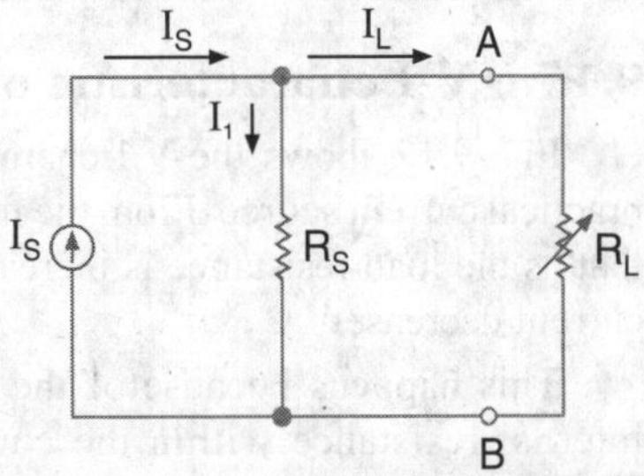

Fig. 9.11. Practical current source connected to a load.

Now let us find the conditions under which a source can work as a good current source. Consider the circuit shown in Fig. 9.11. Here, the load resistance (R_L) is connected across a current source. The current source has a short-circuit, (I_S) and an internal resistance (R_S).

As seen, the current I_S divides it into two parts namely I_L and I_1. The part I_L flows through the load and I_1 through the internal resistance, Thus,

$$I_S = I_1 + I_L$$

or

$$I_1 = I_S - I_L$$

The voltage drop across R_S, is equal to $(I_1 \cdot R_S)$and that across R_L is $(I_L \cdot R_S)$. Since the resistances R_S and R_L are in parallel, therefore, the voltage drop across them should be equal, *i.e.*,

$$I_1 \cdot R_S = I_L \cdot R_L \quad \text{or} \quad (I_S - I_L)\, R_S = I_L \cdot R_L$$

$$\therefore \quad I_L = \frac{I_S}{R_S + R_L} \times R_S = \frac{I_S}{1 + R_L / R_S}$$

The equation (*i*) indicates that the load current (I_L) will remain almost equal to I_S, provided the ratio R_L/R_S is small as compared to unity. In other words, larger the value of internal resistance (R_S), as compared to the load resistance (R_L), smaller is the ratio R_L/R_S and hence better is the practical current source.

Example 9.5.

Find the largest value of load resistance to provide constant current from a current source with $I_S = 30$ mA and $R_S = 15$ kΩ. Also find the variation of current from the short-circuit value.

Solution.

Given : $I_S = 30$ mA and $R_S = 15$ kΩ $= 15\,000$ Ω

We know that the largest value of load resistance,

$$R_L = \frac{R_S}{20} = \frac{15000}{20} = 750\ \Omega$$

and the load current,

$$I_L = I \times \frac{R_S}{R_L + R_S} = 30 \times \frac{15000}{750 + 15000} = 28.6 \text{ mA}$$

∴ Variation in current from the short-circuit value

$$= \frac{I_S - I_L}{I_S} = \frac{30 - 28.6}{30} = 0.047 \text{ or } 4.7\%$$

This result indicates that if we select $R_L = 750$ Ω (equal to $R_S/20$), the variation in current is 4.7% only, which is reasonably good for all practical purposes.

9.15. V-I characteristic of a Practical Current Source

Fig. 9.12 shows the V-I characteristic of a practical current source. From the figure, we find that as the load resistance is increased, the load current decreases.

This happens because of the existence of internal resistance within the current source. We know that as the load resistance increases, the current drawn by internal resistance also increases. Hence, the load current decreases below the short-circuit current (I_S).

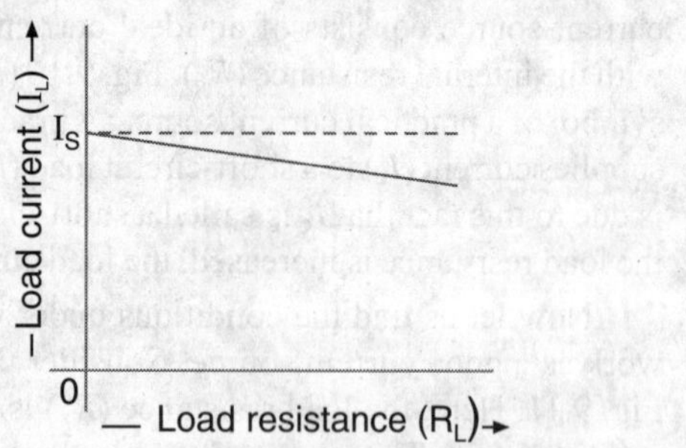

Fig. 9.12.

9.16. Equivalence between Voltage Source and Current Source

As a matter of fact, a voltage source is not different from a current source. A source may work either as a voltage source or a current source. It depends upon the working conditions. If the value of load resistance (R_L) is very large as compared to the internal resistance (R_S), the source is treated as a voltage source. On the other hand, if R_L is very small as compared to R_S, the source is treated as a current source. From the circuit point of view, it does not matter, whether the source is a voltage source or a current source. It is possible to convert a voltage source into a current source and vice versa.

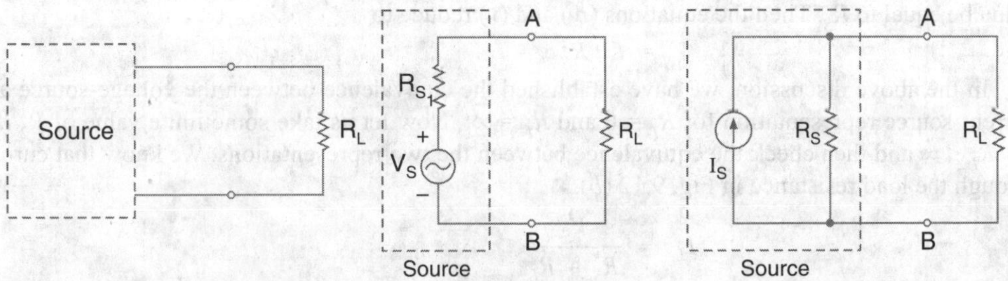

(*a*) Source connected to a load. (*b*) Voltage-source representation. (*c*) Current-source representation.

Fig. 9.13.

In order to establish equivalence between a voltage source and current source, consider an a.c. source connected to the load resistance (R_L) as shown in Fig. 9.13 (*a*). We know that this source may be treated as a voltage source or current source as shown in Fig. 9.13 (*b*) and (*c*) respectively. The voltage source representation consists of an ideal voltage source (V_S) in series with a source resistance (R_{S_1}). The current-source representation consists of an ideal current source (I_S) in parallel with source resistance (R_{S_2}). Since both these representations are of the same source, therefore, they must give the same results. This may be proved as given below :

Suppose in Fig. 9.13 (*b*), the value of load resistance (R_L) is reduced to zero, (*i.e.*, the terminals *A* and *B* are short-circuited), then current through the short-circuited load,

$$I_L = V_S / (R_{S_1}) \quad ...(i)$$

Now consider Fig. 9.13 (*c*) and let the value of load resistance (R_L) is reduced to zero. In this case, the value of current through the short-circuited load is equal to the source current I_S (the total resistance of a parallel combination found by R_{S_2} and a short-circuit is equivalent to a short-circuit). Thus we find that if the voltage source representation and the current-source representation are to be equal to I_S. *i.e.*,

$$I_L = I_S = V_S / R_{S_1} \quad ...(ii)$$

Again consider the two representation of the source and let R_L be disconnected from the source *i.e.*, the terminals *A* and *B* are open-circuited. In the voltage-source representation, the open-circuit terminal voltage is equal to the source voltage *i.e.*,

$$V_T = V_S \quad ...(iii)$$

In the current-source representations, the open-circuit terminal voltage,

$$V_T = I_S \cdot / R_{S_2} \quad ...(iv)$$

and for the two representations to be equal, the equations (*iii*) and (*iv*) must be equal *i.e.*,

$$V_T = V_S = I_S \cdot R_{S_2} \qquad ...(v)$$

From equation (*ii*) we find,

$$V_S = I_S \cdot R_{S_1} \qquad ...(vi)$$

Equating the two values V_S from equations (*v*) and (*vi*),

$$I_S \cdot R_{S_1} = I_S \cdot R_{S_2}$$

or

$$R_{S_1} = R_{S_2}$$

The above relation indicates that source resistances in both representations are the same. Let this value be equal to R_S. Then the equations (*iv*) and (*v*) reduce to

$$V_S = I_S \cdot R_S$$

In the above discussion, we have established the equivalence between the voltage-source and current-source representation for $R_L = 0$ and $R_L = \infty$. Now let us take some finite value of R_2 *i.e.*, $0 < R_L < \infty$ and then check the equivalence between the two representations. We know that current through the load resistance in Fig. 9.12 (*b*),

$$I_{L_1} = \frac{V_S}{R_S + R_L}$$

and the current through the load resistance in Fig. 9.12 (*c*),

$$I_{L_2} = I_S \times \frac{R_S}{R_S + R_L} = \frac{V_S}{R_S + R_L} \qquad ...(\because I_S \cdot R_S = V_S)$$

$$\therefore \qquad I_{L_1} = I_{L_2}$$

The above relation indicates that whatever be the value of R_L, the current drawn by the load resistance in either voltage-source representation or the current-source representation is the same. Thus the equivalence between a voltage-source and current source is completely established.

9.17. Conversion of Voltage Source into Current Source

Consider a practical voltage source (V_S) with an internal resistance (R_S) as shown in Fig. 9.14 (*a*). In order to convert it into its equivalent current-source representation, let us short-circuit the terminals *A* and *B*, and then determine the current through this short-circuit. We know that short-circuit current,

$$I_S = V_S/R_S$$

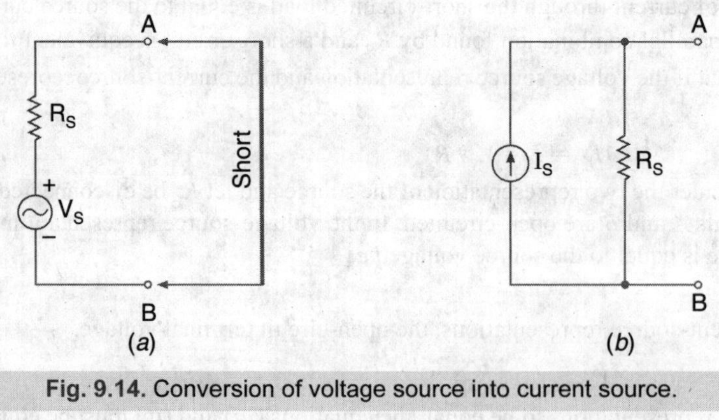

Fig. 9.14. Conversion of voltage source into current source.

We have already discussed in the last article, that the source resistance in either voltage-source representation or current-source representation is the same. Therefore, current-source has the same internal resistance as that of a voltage-source. Thus, the equivalent current-source may be represented by a current-source (I_S) in parallel with a source resistance as shown in Fig. 9.14 (*b*).

Example 9.6.

Fig. 9.15 (a) shows a voltage source having an open-circuit voltage of 12 V and an internal resistance of 3 Ω. Find its equivalent current-source representation.

Solution.

Given : $V_S = 12$ volts and $R_S = 3\ \Omega$.

We know that the short-circuit current,

$$I_S = \frac{V_S}{R_S} = \frac{12}{3} = 4\text{ A}$$

∴ The current-source representation of the given voltage-source is shown in Fig. 9.15 (*b*).

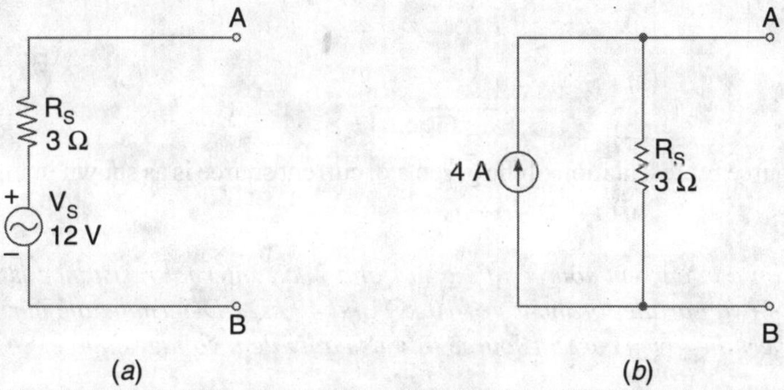

Fig. 9.15.

9.18. Conversion of Current Source into Voltage Source

Consider a practical current source (I_S) with an internal resistance (R_S) as shown in Fig. 9.16 (*a*). In order to convert it into its equivalent voltage-source representation, let us find the open-circuit terminal voltage. We know that open-circuit terminal voltage,

$$V_S = I_S \cdot R_S$$

Since the internal resistance of the voltage-source is the same as that of a current source, therefore, the voltage-source representation is as shown in Fig. 9.16 (*b*).

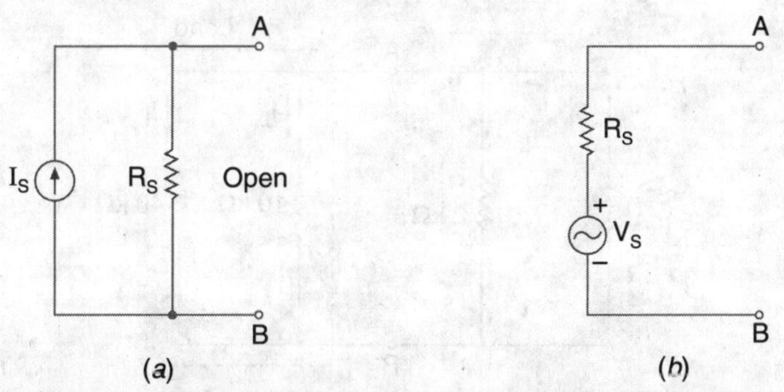

Fig. 9.16. Conversion of current source into voltage source.

Example 9.7.

Find the equivalent voltage-source of an a.c. current source shown in Fig.9.16(a).

Solution.

Given : $I_S = 5$ mA $= 5 \times 10^{-3}$ A and $R_S = 2$ kΩ $= 2 \times 10^3$ Ω.

We know that the open-circuit terminal voltage,

$$V_S = I_S \cdot R_S = (5 \times 10^{-3}) \times (2 \times 10^3)\ 10 \text{ V}$$

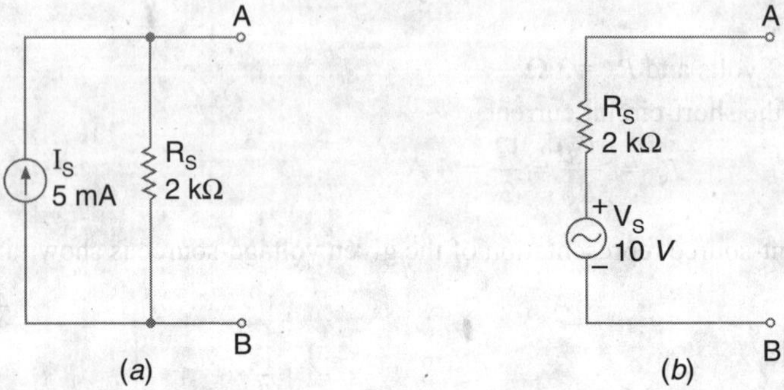

Fig. 9.17.

∴ Voltage-source representation of the given a.c. current source is as shown in Fig. 9.17 (*b*). **Ans.**

Example 9.8.

In a circuit, an a.c. current source of 1.5 mA and 2 kΩ internal resistance is connected to a load consisting of two parallel branches of 10 kΩ and 40 kΩ . Determine the current through 40 kW branch. Convert the given current source into its equivalent voltage source.

Solution.

Given : $I_S = 1.5$ mA and $R_S = 2$ kΩ.

Fig. 9.18 shows the given current source connected to a load consisting of two parallel branches of 10 kΩ and 40 kΩ.

We know that the load resistance,

$$R_L = 10 \parallel 40 = \frac{10 \times 40}{10 + 40} \text{k}\Omega = 8 \text{ k}\Omega$$

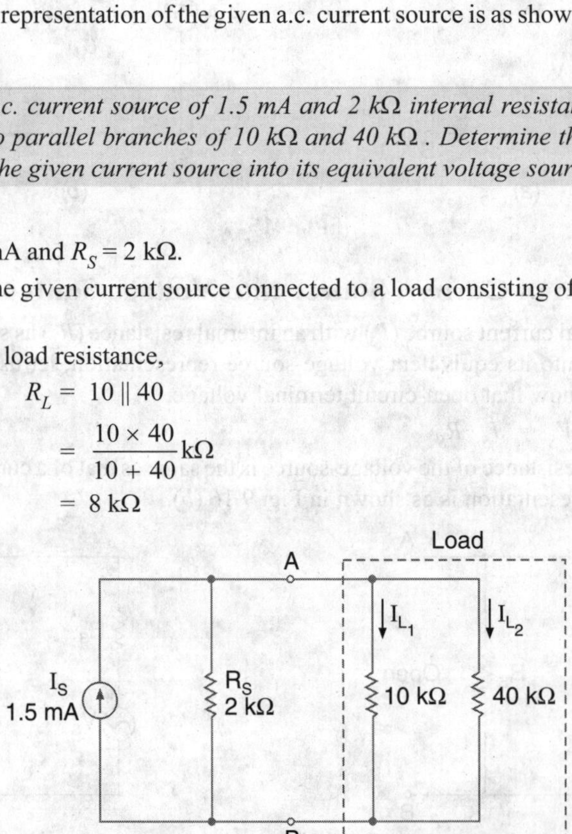

Fig. 9.18.

and the current drawn by the load resistance,

$$I_L = I_S \times \frac{R_S}{R_L + R_S}$$

$$= 1.5 \times \frac{2}{8+2} \text{ mA}$$

$$= 0.3 \text{ mA}$$

This load current is divided into two parts *i.e.*, part I_{L_1} flowing through 10 kW load and I_{L_2} through 40 kΩ load.

$$\therefore \qquad I_{L_2} = I_L \times \frac{10}{10+40} = 0.3 \times 0.2 = 0.06 \text{ mA } \textbf{Ans.}$$

We know that the open-circuit terminal voltage to the given current source,

$$V_S = I_S \cdot R_S = (1.5 \times 10^{-3})\ (2 \times 10^3) \text{ Volts}$$

$$= 3 \text{ Volts}$$

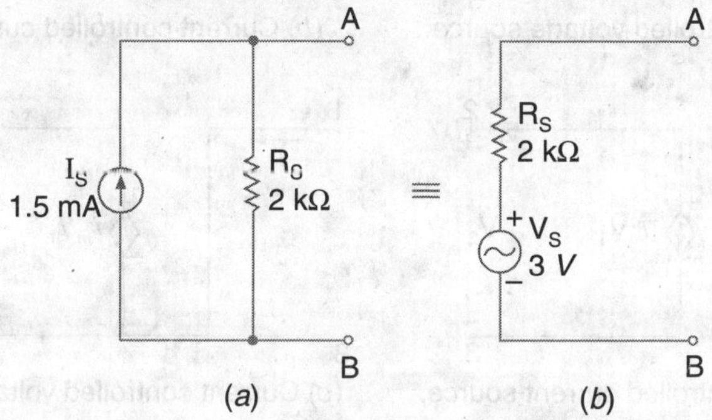

Fig. 9.19.

The internal resistance of the equivalent voltage-source is the same as that of the given current source. Therefore, equivalent voltage-source representation of the given current-source is as shown in Fig. 9.19 (*b*). **Ans.**

9.19. Controlled or Dependent Sources

So far we have discussed the sources, which deliver the specified voltage or current independent of the circuit to which they are connected. Such sources are known as independent sources. There is another class of sources, called controlled or dependent sources. These sources consist of four terminals (instead of two as in independent sources) as shown on Fig. 9.20 (*a*). There are two input and two output terminals. However, most of the sources (which are, usually electronic devices such as vacuum tubes, bipolar transistors, field-effect transistors and photo cell etc.) have one terminal, which is common to both input and output as shown in Fig. 9.20 (*b*).

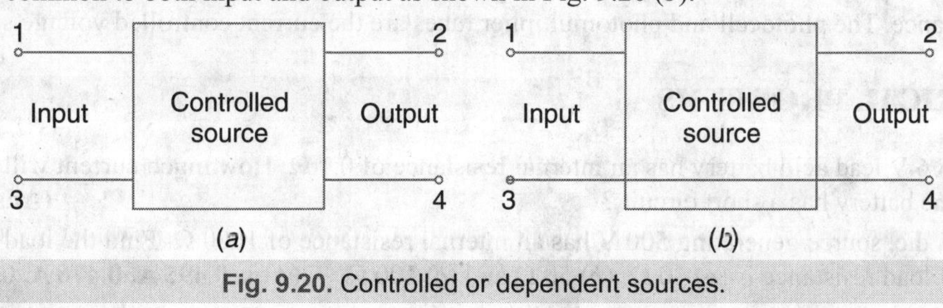

Fig. 9.20. Controlled or dependent sources.

The terminal 1 and 3 are the input or control terminals, while terminals 2 and 4 are the output terminals of the controlled source. The output of controlled sources is controlled by (or dependent upon) the signal applied at the input terminals. There are four types of controlled sources as discussed below:

1. *Voltage controlled voltage source.* In this source, the control voltage (V_1) is applied to the input terminals as shown in Fig. 9.21 (*a*). And the voltage source (of strength $A_v \cdot V_1$) acts at output terminals. The parameter (A_v) relates the strength of the source, with the control voltage, and is called voltage gain. The vacuum tubes and operational Amplifiers (OPAMP) are the voltage controlled voltage sources.

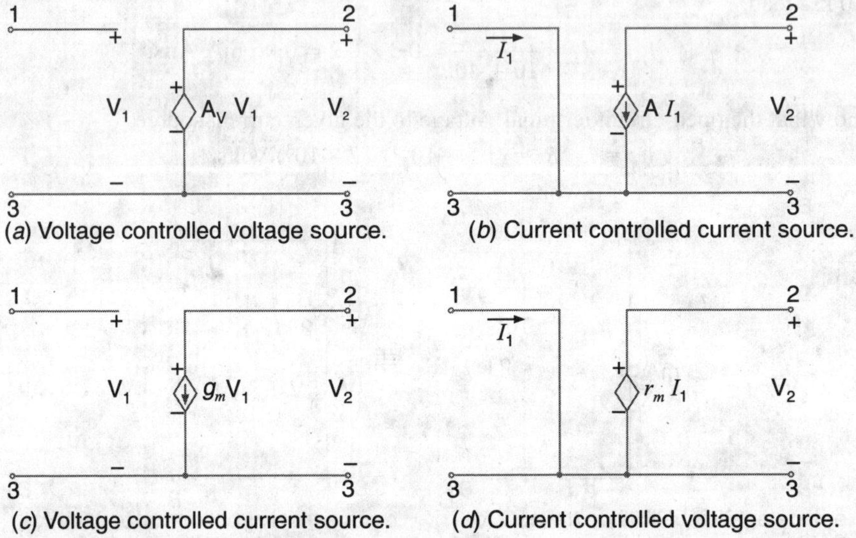

Fig. 9.21.

2. *Current controlled current source.* In this source, the control current (I_1) is applied to the input terminals as shown in Fig. 9.21 (*b*). And the current source (of strength AI_1) acts at output terminals. The parameter (A) relates the strength of the source to the control current and is called current gain. The bipolar junction transistor (BJT) is a current controlled current source.

3. *Voltage controlled current source.* In this source, the control voltage (V_1) is applied to the input terminals as shown in Fig. 9.21 (*c*). And the current source (of strength $g_m \cdot V_1$) acts at the output terminals. The parameter (g_m) relates the strength of the source to the control voltage and is known as transconductance or mutual conductance. The field-effect transistor is a voltage controlled current source.

4. *Current controlled voltage source.* In this source, the control current (I_1) is applied to the input terminals as shown in Fig. 9.21 (*d*). And the voltage source (of strength $r_m \cdot I_1$) acts at the output terminals. The parameter (r_m) relates to the strength of the source to the control current and is called transresistance. The photocell and photomultiplier tubes are the current controlled voltage sources.

NUMERICAL PROBLEMS

1. A 6 V lead acid battery has an internal resistance of 0.1 Ω. How much current will flow, if the battery has a short circuit ? **(Ans. 60 A)**
2. A d.c. source generating 500 V has an internal resistance of 1000 Ω. Find the load current if load resistance is (*a*) 10 Ω, (*b*) 50 Ω and (*c*) 100 Ω. **(Ans. 0.495 A, 0.476 A, 0.455 A)**

3. The output voltage of a battery drops from 90 V at zero load to 60 V with a load current 50 mA.
   (*a*) How much is the internal resistance of the battery ?
   (*b*) How much is R_L for this load current ?
   (*c*) How much R_L reduces the load voltage to one-half the open-circuit (or no load) voltage.

(Ans. (*a*) 600 Ω (*b*) 1200 Ω (*c*) 600Ω)

DESCRIPTIVE QUESTIONS

1. Explain briefly the constant voltage and constant current sources. What is their utility ?
2. Explain the difference between an ideal voltage source and a practical voltage source. Illustrate with a suitable example.
3. What is the difference between an ideal current source and a practical current source ?
4. How the internal resistance of a voltage source and a current source is determined ?

MULTIPLE CHOICE QUESTIONS

1. Thirty zinc-carbon dry cells, each of 1.5 V, are connected in series. The total output voltage is,
   (*a*) 1.5 V (*b*) 30 V (*c*) 45 V (*d*) 60 V
2. A constant voltage source has
   (*a*) low internal resistance (*b*) high internal resistance
   (*c*) minimum efficiency (*d*) minimum current capacity
3. A d.c. source has an output of 10 V in an open-circuit, which drops to 5 V with a load current of 50 mA and an R_L of 1000 Ω. The internal resistance of the source is equal to
   (*a*) 25 Ω (*b*) 50 Ω (*c*) 100 Ω (*d*) 1000 Ω
4. A d.c. current source has a short-circuit current of 2 mA and an open-circuit voltage of 50 V. The internal resistance is
   (*a*) 5 kΩ (*b*) 10 kΩ (*c*) 20 kΩ (*d*) 25 kΩ
5. Which of the following has a high internal resistance ?
   (*a*) A constant-voltage source (*b*) A constant-current source
   (*c*) Both of these (*d*) None of these

ANSWERS

1. (*c*) **2.** (*a*) **3.** (*c*) **4** . (*d*) **5.** (*b*)

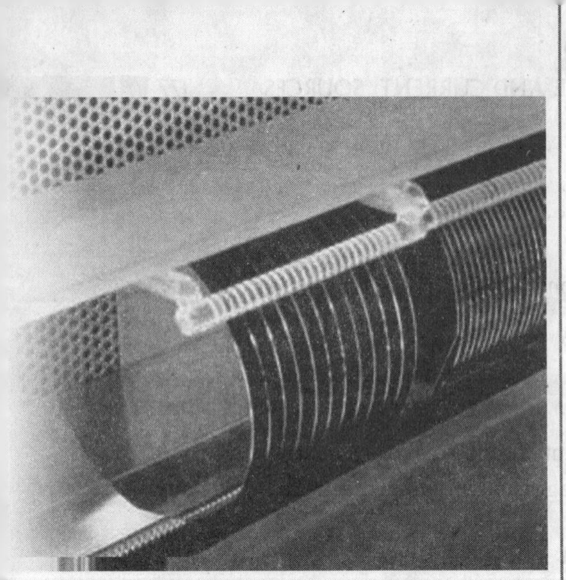

10
Chapter
Semiconductors

Outline

1. Introduction
2. Types of Semiconductors
3. Intrinsic Semiconductors
4. Generation of Electrons and Holes
5. Mechanism of Holes contributing in conductivity
6. Free Charge Carriers in Semiconductors
7. Generation and Recombination of Carriers
8. Extrinsic Semiconductors
9. N-type Semiconductors
10. P-type Semiconductors
11. Majority and Minority charge carriers
12. Mobile Charge Carriers and Immobile Ions
13. Conduction in metals
14. Electric Current in a Conductor
15. Conductivity of a Conductor
16. Conduction in Intrinsic Semiconductors
17. Conductivity of N- and P-type Semiconductors
18. Mass Action Law
19. Charge Densities in N- and P-type Semiconductors
20. Variation in Semiconductor Parameters
21. Intrinsic Concentration
22. Mobility
23. Conductivity
24. Energy Gap
25. Diffusion Current in Semiconductor
26. Total Current in a Semiconductor
27. Einstein's Relationship for a Semiconductor
28. Effect of temperature on semiconductor
29. The Hall Effect
30. Experiment Determination of Carrier Concentration and Mobility
31. Hall Effect Applications.

10.1. Introduction

The materials, whose electrical conductivity lie between those of conductors and insulators, are known are *semiconductors*. This means the conductivity is in the range of 10^{-3} to 10^{-8} S/cm. Semiconductors are the foundation of modern electronics including radio, television, computers, telephones and many more.

Common semiconductors are crystalline solids but amorphous and liquid semiconductors are also known. The examples of such materials

Silicon (Si)

are germanium (Ge), silicon (Si), gallium arsenide (GaAs), cadmium sulfide (CdS), lead telluride etc.

In terms of energy bands, the semiconductor may be defined as those materials, which at room temperature, have

1. Partially filled conduction band.
2. Partially filled valence band.

3. A very narrow energy gap, E_G ($\approx$ 1 eV) between conduction and valence bands.

We know that at absolute zero temperature (*i.e.*, 0 K) there are no electrons in the conduction band of semiconductors and the valence band is completely filled. Thus the semiconductors behave like perfect insulators at 0 K. However, as the temperature is increased, the width of energy gap reduces. Consequently, some of the electrons jump into conduction band and semiconductors show some conductivity. It is thus obvious, that the conductivity of semiconductors increases with the increase in temperature. The electrical conductivity of semiconductors is in the range of 10^{-3} to 10^{-6} per ohm per cm.

10.2. Types of Semiconductors

The semiconductors are of the following two types :

1. Intrinsic semiconductors
2. Extrinsic semiconductors

10.3. Intrinsic Semiconductors

A semiconductor, which is in its extremely pure form, is known as an *intrinsic* (or pure) semiconductor. The nature of semiconductors is such that even a small amount of certain impurities can change their electrical properties drastically. It is due to this fact, that a semiconductor would not be called truly intrinsic, unless the impurity level is very small. For germanium, the impurity level is less than 1 part in 10^8 parts. And for silicon, it is less than 1 part in 10^{12} parts. In actual practice, however, a semiconductor material, with somewhat larger impurity concentrations, than those mentioned above, is still called as an intrinsic.

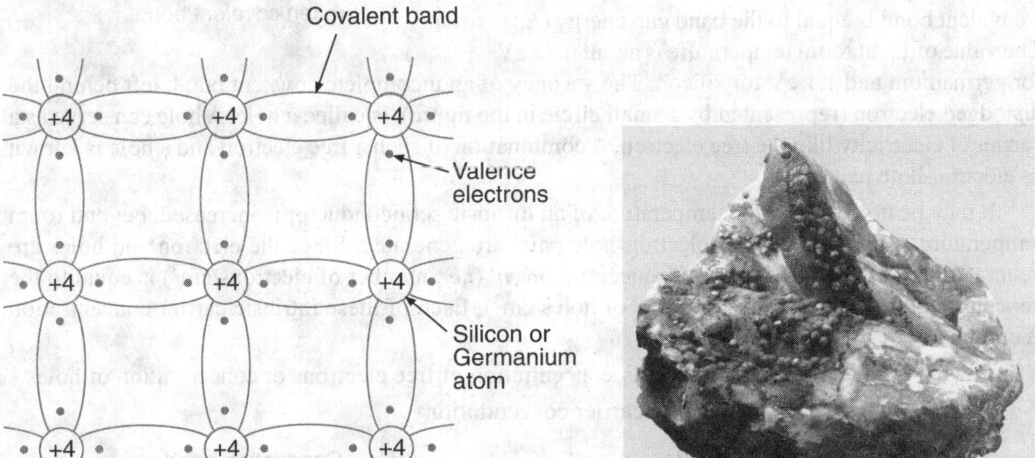

Fig. 10.1. Two dimensional representation of a silicon or a germanium crystal.

The silicon and germanium are the two most widely used intrinsic semiconductors. The crystal structure of these materials consists of a regular repetition in three dimensions of a unit cell having the form of tetrahedron, with one atom at each vertex. The two dimensional representation of such a structure is shown in Fig. 10.1. For example, the silicon has a total of 14 electrons in its atomic structure. Each atom in a silicon crystal contributes 4 valence electrons, so that the atom is tetravalent. The four valence electrons are shown by the dots in the figure. The circle, with + 4 designation, represents the central portion of the atom, including the nucleus and the electrons except those in the outermost orbit. This portion is often known as core of the atom. These atoms are bonded together by a force, which results from the sharing of the neighbouring atoms. This bonding force is called a covalent bond and is shown by a pair of curved lines in the figure.

10.4. Generation of Electrons and Holes

At very low temperatures (*i.e.*, in the range of – 273°C or 0 K), an intrinsic semiconductor will have exactly the ideal structure as shown in Fig. 10.1. In that structure, all the valence electrons are tightly held by the parent atoms and through covalent bonds by other atoms. Since the electrons are not free to move, through the crystal structure, therefore, they cannot conduct electricity. Thus an intrinsic semiconductor, at absolute zero temperature (*i.e.*, 0 K), behaves like an insulator.

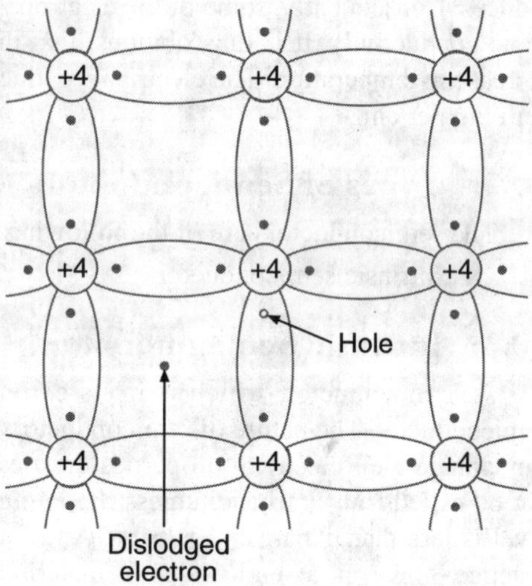

Fig. 10.2. Intrinsic semiconductor with a broken covalent bond.

Now, let us assume that the temperature of the semiconductor is increased to room temperature (*i.e.*, 27°C or 300 K). At this temperature, some of the covalent bonds are broken. This situation is as shown in Fig. 10.2. Here an electron, which was a part of covalent bond earlier, is now shown as a dislodged electron. This electron is free to move anywhere within the crystal. The energy required to break such a covalent bond is equal to the band gap energy (E_G). The value of E_G at room temperature is about 0.72 eV for germanium and 1.1 eV for silicon. The vacancy of an incomplete covalent bond, left behind the dislodged electron (represented by a small circle in the figure), is called a hole. A hole can serve as a carrier of electricity like the free electron. A combination of such a free electron and a hole is known as electron-hole pair.

It may be noted that if the temperature of an intrinsic semiconductor is increased, beyond room temperature, a large number of electron-hole pairs are generated. Since the electron and holes are generated in pairs, the free electron concentration 'n' (*i.e.*, number of electrons/cm^3) is equal to the concentration of holes 'p' (*i.e.*, number of holes/cm^3). Each of these intrinsic carrier concentration is commonly denoted by the letter n_i.

Thus for an intrinsic semiconductor, concentration of free electrons or concentration of holes,

$$n = \text{Intrinsic carrier concentration}$$
$$= p = n_i$$

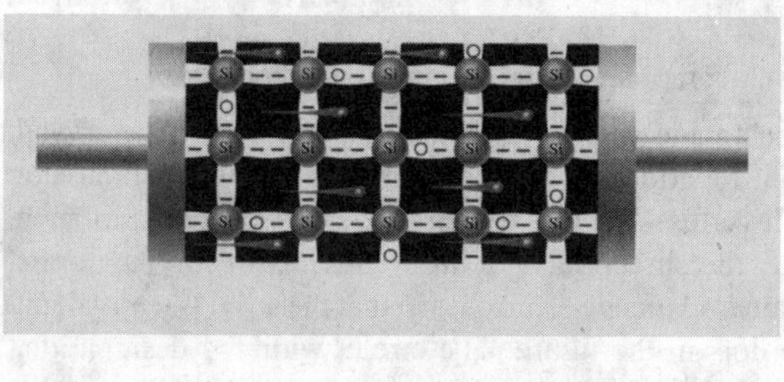

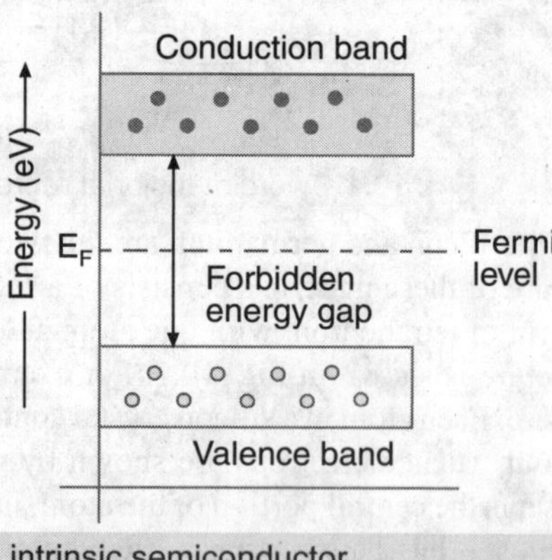

Fig. 10.3. Energy band diagram for an intrinsic semiconductor.

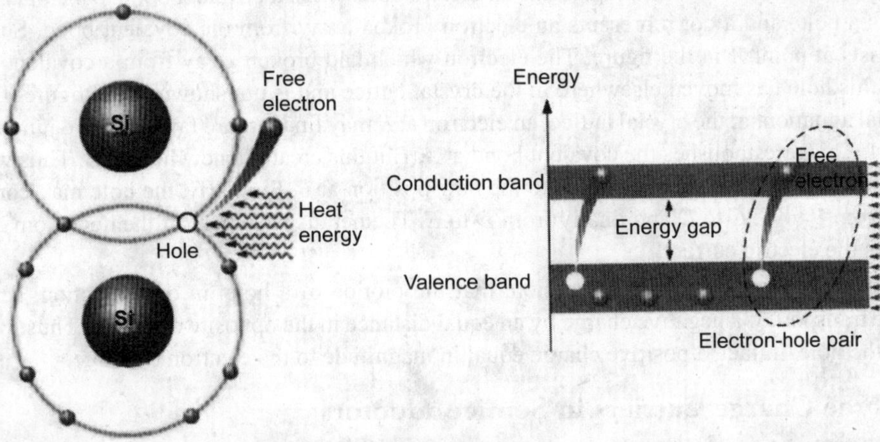

Fig. 10.3 shows the energy band diagram for an intrinsic semiconductor at room temperature. Here the horizontal line marked E_F indicates the *Fermi level. It may be noted that the Fermi level lies at the middle of the forbidden energy gap.

10.5. Mechanism of Holes Contributing in Conductivity

Consider the structure of an intrinsic semiconductor as shown in Fig. 10.4. Now, let us imagine that a hole is created by an electron breaking away from the covalent bond. At this instant, it is possible that an electron from the neighbouring atom leaves its covalent bond to fill up the hole. However, this electron generates a new hole in its initial position. When this occurs, it appears that a hole has moved in a direction opposite to that of the electron. This hole, in its new position, may now be filled up by an electron from another covalent bond. And the hole will move, correspondingly, one more step in a direction opposite to the motion of the electron. This process is illustrated in Fig. 10.4.

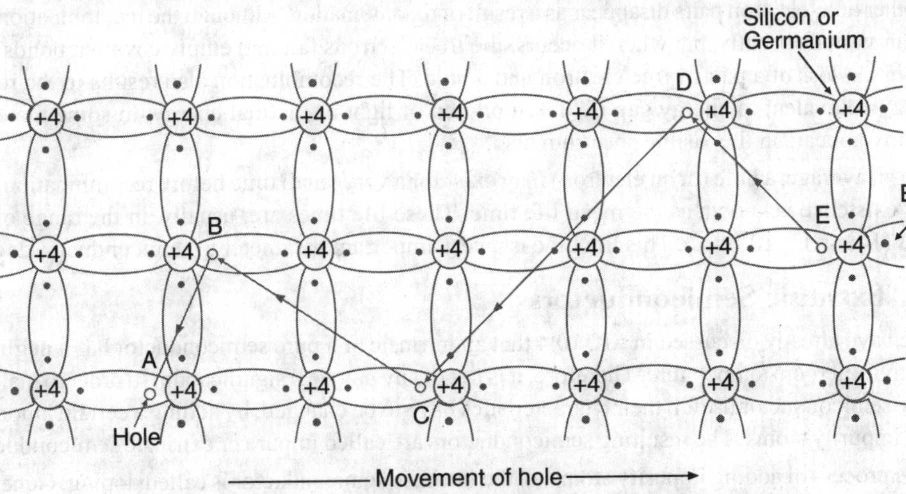

Fig. 10.4. Mechanism of holes contributing to conductivity.

* The Fermi level (E_F) is the maximum energy level, which is occupied by an electron at absolute zero (0K) temperature. At higher temperatures, some electrons are excited into higher energy states. Then Fermi level corresponds to the centre of gravity of conduction electrons and holes weighed according to their energies.

Here a circle with a dot in it, represents an electron present in a covalent bond. An empty circle represents a hole, and a dot represents an electron broken away from the covalent bond. Suppose a hole exists at point *A* in the figure. The electron which had broken away from a covalent bond, creating, this hole has moved elsewhere in the crystal lattice and is not shown in the figure. Due to the thermal agitation, in the crystal lattice, an electron at *B* may find it more favourable to jump to the position *A*. This re-establishes the covalent bond at *A*. But this creates a new hole at *B*. Thus we can say that the hole at *A* has simply moved to the new position at *B*. Similarly, the hole may continue to move from *B* to *C*, *C* to *D* and finally from *D* to *E*. The net displacement of the hole from *A* to *E* constitutes the electric current.

It is evident from the above discussion, that the motion of a hole, in one direction, actually means the transport of a negative charge by an equal distance in the opposite direction. Thus, we can consider the hole, in fact, a positive charge equal in magnitude to the electronic charge.

10.6. Free Charge Carriers in Semiconductors

We know that the conduction properties of a semiconductor depend upon the number of valence electrons. When a sufficient energy (thermal or electrical) is given to valence electrons, they break away, from their parent atoms, and take part in conduction through the crystal. Such valence electrons are called conduction band electrons. These electrons leave behind the holes in the valence band. The electrons and holes, generated in such a way, are called free charge carriers. They are free to move anywhere within the crystal. Hence, such electrons and holes (called charge carriers) are known as free charge carriers. It will be interesting to know that the number of free charge carriers is an important factor, because it determines the property of a semiconductor.

10.7. Generation and Recombination of Carriers

We have already discussed in Art. 10.4 that in a pure semiconductor, the number of holes is equal to the number of electrons. However, thermal agitation continues to produce new hole-electron pairs, while other hole-electron pairs disappear as a result of recombination. Although the recombination does not occur very frequently, but when it occurs, the free electrons fall into empty covalent bonds. This results into a loss of a pair of free electron and a hole. The recombination also results in the release of energy equivalent to energy gap (E_G) as a photon of light or thermal energy. In some materials, the energy appears in the visible spectrum also.

On an average, a hole (or an electron) may exist for a very small time before recombination. This time of existence is known as the mean life time. These life times are, usually, in the range of 1 to 1000 μs (1 $\mu s = 1 \times 10^{-6}$ sec). This life time is a very important parameter in semiconductor devices.

10.8. Extrinsic Semiconductors

We have already discussed in Art. 10.4 that an intrinsic or a pure semiconductor has a negligible conductivity at room temperature. Therefore, it is not of any practical signification. In order to make the intrinsic semiconductor useful, their characteristics have to be changed, by adding a certain amount of desired impurity atoms. The resulting semiconductors are called impure or extrinsic semiconductors.

The process of adding impurity atoms, to the intrinsic semiconductor is called doping. Generally, the impurities are added at the rate of only one atom of impurity per 10^6 to 10^{10} semi conductor atoms. As a result of this, the impurity atoms are separated in the crystal by hundreds or thousands of semiconductor atoms in all directions. These impurity atoms are so scarce, that they do not exert any influence on the manner in which the crystal grows. It may be noted that an impurity atom enters the crystal by substituting for one of germanium or silicon atoms.

The purpose of adding impurity is to increase either the number of free electrons or holes in a semiconductor. Generally, two types of impurity atoms are added to the semiconductor namely the impurity atoms containing 5 valence electrons (called pentavalent impurity atoms) and the impurity atoms containing 3 valence electrons (called trivalent impurity atoms). Depending upon the type of impurity atoms added to the semiconductor, the resulting semiconductor (*i.e.*, extrinsic semiconductor) may be of the following two types :

1. N-type semiconductors.
2. P-type semiconductors.

10.9. N-Type Semiconductors

The semiconductors, which are obtained by introducing pentavalent impurity atoms, (*i.e.*, atoms containing 5 valence electrons) are known as *N*-type semiconductors. The pentavalent impurity is an element from Group V of the Periodic Table. The elements in this group contain 5 valence electrons. The typical examples of such elements are phosphorus (P), antimony (Sb), arsenic (As) and bismuth (Bi) (refer to Table 10.1). These elements donate excess (negative) electron carriers. Therefore, such elements are known as donor or N-type impurities.

Table 10.1: Pentavalent impurities

S. No.	*Element*	*Symbol*
1.	Phosphorus	P
2.	Antimony	Sb
3.	Arsenic	As
4.	Bismuth	Bi

It has been observed that when a pentavalent impurity is added to a pure semiconductor, it displaces some of its atoms. Fig. 10.5. shows the structure of a silicon crystal lattice containing an antimony atom at the central position. It may be noted that out of 5 valence electrons, 4 electrons will form covalent bonds by sharing one electron each with the electrons of the neighbouring atoms. The 5th electron is an extra electron, and is loosely bound with the antimony atom. This extra electron, if detached from the antimony atom, will be available as a carrier of the current. The energy required to detach this electron is of the order of 0.05 eV for silicon and 0.01 eV for germanium.

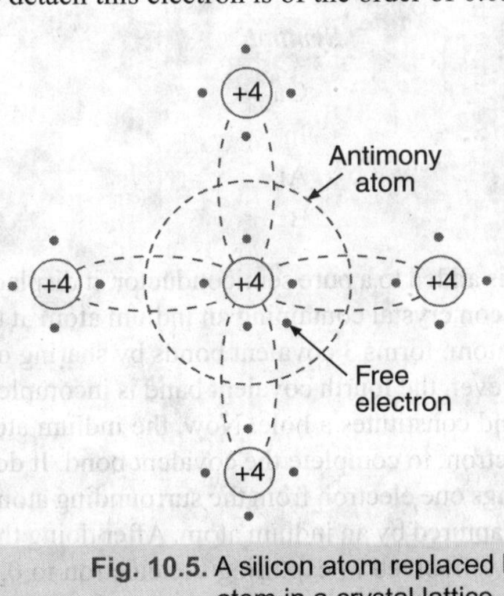

Fig. 10.5. A silicon atom replaced by stationary atom in a crystal lattice.

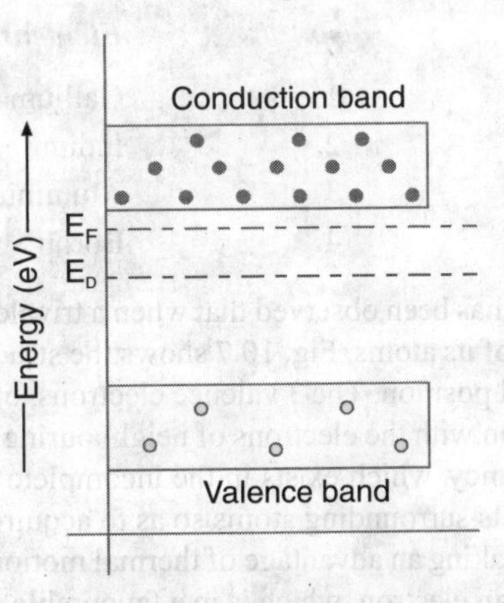

Fig. 10.6. Energy bond diagram of an N-type semiconductor.

It will be interesting to know that if a pure semiconductor is doped with N-type impurities, the number of electrons in the conduction band increases above a level that would be available in a pure semiconductor. However, the number of holes in the valence band decreases below a level, that would be available in a pure semiconductor. It is due to the fact that a large number of electrons present in the semiconductor increases the rate of recombination of electrons with holes. Because of the greater number of electrons in conduction band than that of holes in the valence band. The Fermi level shifts upward towards the bottom of the conduction band as shown in Fig. 10.6.

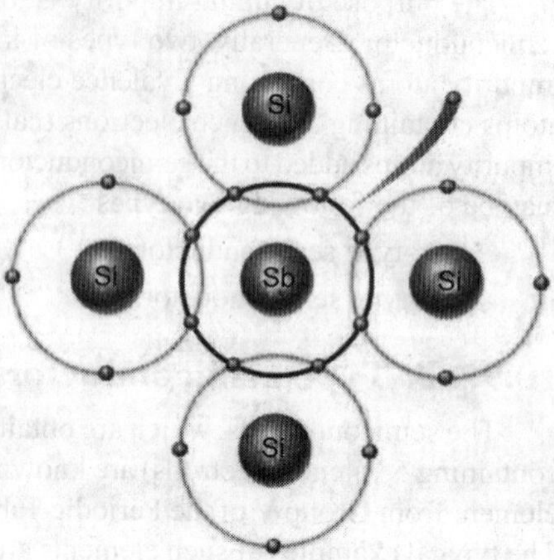

The addition of a pentavalent impurity atoms gives away or donates its 5th electron as discussed above. After the donation, the impurity atom becomes a positively charged ion and is known as a donor ion. This ion is bounded by four covalent bonds in the crystal lattice and therefore, cannot be made to move within the lattice. As a result of this, donor ions do not contribute to the conduction of current in N-type semiconductor. Thus in an *N*-type semiconductor, the current flows due to the movement of electrons and holes. But a major part of the current flows due to the movement of electrons. Therefore, the electrons in an N-type semiconductor are known as majority carriers and holes as minority carriers.

10.10. P-Type Semiconductors

The semiconductors, which are obtained by introducing a trivalent impurity atoms (*i.e.*, atoms containing 3 valence electrons) are known as *P*-type semiconductors. The trivalent impurity is an element of group III of the Periodic Table. The elements in this group contain 3 valence electrons. Typical examples of such elements are gallium (Ga), indium (In), aluminium (Al), Boron (B) etc (refer to Table 10.2). These elements make available positive charge carriers because they create holes, which can accept electrons. Therefore, such elements are known as acceptor or P-type impurities.

Table 10.2 : Trivalent imprities

No.	*Element*	*Symbol*
1.	Gallium	Ga
2.	Indium	In
3.	Aluminium	Al
4.	Boron	B

It has been observed that when a trivalent impurity is added to a pure semiconductor, it displaces some of its atoms. Fig. 10.7 shows the structure of a silicon crystal containing an indium atom at the central position. The 3 valence electrons, of an indium atom, forms 3 covalent bonds by sharing one electron with the electrons of neighbouring atoms. However, the fourth covalent band is incomplete. A vacancy, which exists in the incomplete covalent bond constitutes a hole. Now, the indium atom seeks its surrounding atoms so as to acquire the 4th electron, to complete the covalent bond. It does so by taking an advantage of thermal motion, which brings one electron from the surrounding atoms. Thus, an electron, which is in a favourable position, is captured by an indium atom. After doing this, the indium atom becomes an immobile ion. The energy, involved in capturing the electron to 0.05 eV for silicon and 0.01 eV for germanium.

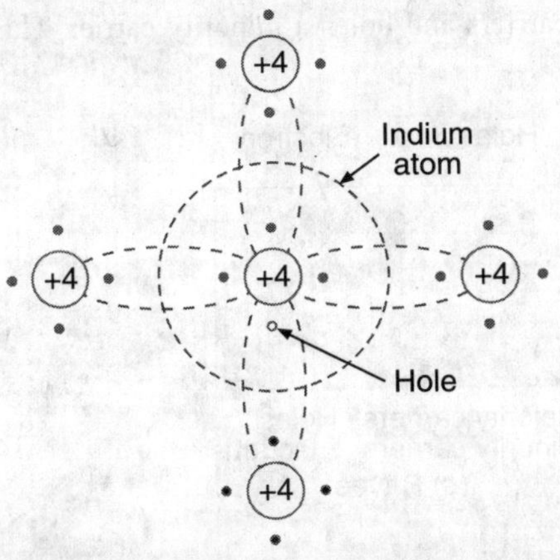

Fig. 10.7. A silicon atom replaced by indium atom in a crystal lattice.

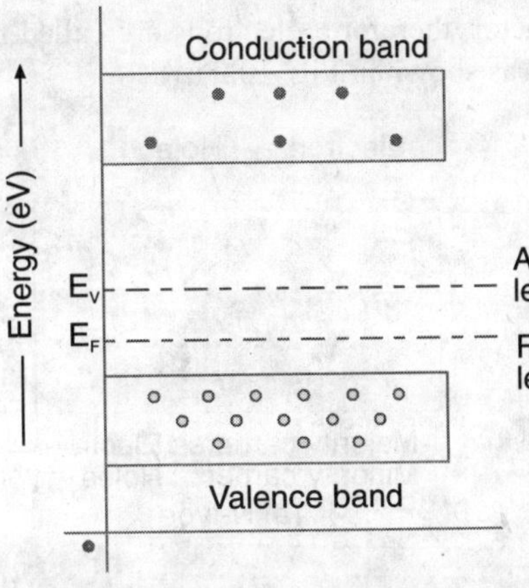

Fig. 10.8. Energy bond diagram of a P-type semiconductor.

It will be interesting to know that if a pure semiconductor is doped with P-type impurities, the number of holes in the valence band increases above a level, which would be available in a pure semiconductor. However, the number of electrons in the conduction band decreases below a level, which would be available in a pure semiconductor. It is due to the fact, that a large number of holes present in the semiconductor increases the rate of recombination of holes with electrons. Because of the greater number of holes in the valence band than that of electrons in the conduction band, the Fermi level shifts downwards towards the top of the valence band as shown in Fig. 10.8.

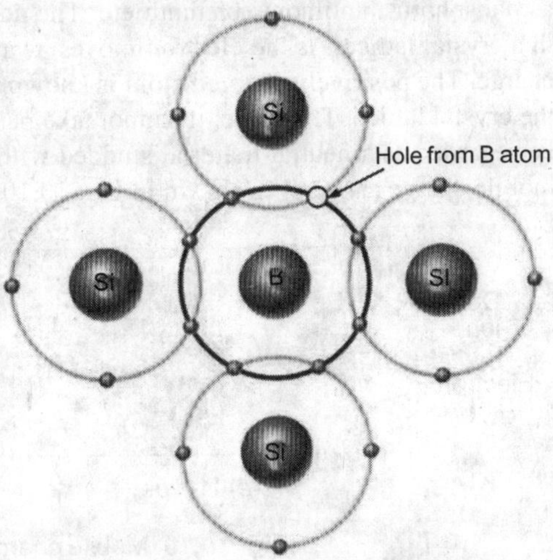

The addition of a trivalent impurity atom takes or accepts its fourth electron as discussed above. After accepting, the impurity atom becomes a negatively charged ion and is known as an acceptor ion. This ion is bounded by four covalent bonds in the crystal lattice and therefore, cannot be made to move within the lattice. As a result of this, the acceptor ions do not contribute to the conduction of current in a P-type semiconductor. Thus in a P-type semiconductor, the current flows due to the movement of holes and electrons. But a major part of the current flows due to the movement of holes. Therefore, holes in a P-type semiconductor are known as majority carriers and electrons as minority carriers.

10.11. Majority and Minority Charge Carriers

We have already discussed that a pure (or intrinsic) semiconductor has no free charge carriers at absolute zero (0 K) temperature. However, as the temperature of the semiconductor is raised to room temperature, some of the covalent bonds are broken, due to which few electron-hole pairs are created. These electron-hole pairs are called thermally generated charge carriers. The number of these charge carriers is, usually, very small. However, an extrinsic semiconductor (either N- or P-type) has a larger number of charge carriers. For instance, an N-type semiconductor has a large number of electrons. Besides this, N-type semiconductor also contains fewer number of thermally generated holes. Since the number of electrons, are much larger than the number of holes in a N-type semiconductor, therefore, electrons are called majority carriers and holes a minority carrier. This situation is as shown in Fig. 10.9 (*a*).

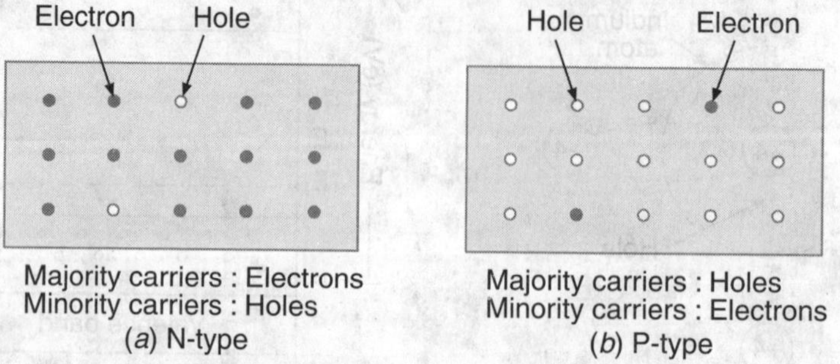

Fig. 10.9.

Similarly, in a P-type semiconductor, the number of holes are much larger than the number of electrons. Therefore, in a P-type semiconductor, holes are called majority carriers and electrons as minority carriers. This situation is as shown in Fig. 10.9 (*b*).

10.12. Mobile Charge Carriers and Immobile Ions

We know that an N-type semiconductor is formed by the addition of pentavalent atoms such as phosphorus, antimony, bismuth etc. The addition of such atom results in an additional electron in a crystal lattice. As the electron moves away from the parent atom, the atom acquires a positive charge. The positively charged atom is known as a donor ion. The position of a donor ion is fixed in the crystal lattice. Therefore, it cannot take part in conduction. Thus an N-type semiconductor may be regarded as a neutral material studded with donor ions (*i.e.*, immobile ions) and electrons (*i.e.*, mobile charge carriers) as shown in Fig. 10.10 (*a*).

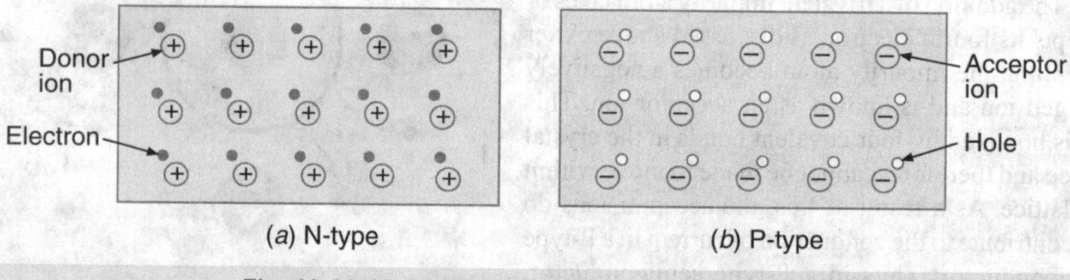

Fig. 10.10. Mobile charge carriers and immobile ions.

Similarly, a P-type semiconductor is formed by the addition of trivalent atoms such as gallium, indium, aluminium etc. The addition of each such atom results in an additional hole in a crystal lattice. As a hole moves away from the parent atom, the atom acquires a negative charge. This negatively charged atom is known as an acceptor ion. The position of an acceptor ion is fixed in the crystal lattice. In other words, it is immobile and therefore, cannot take part in conduction. Thus a P-type material may be regarded as a neutral material studded with acceptor ions (*i.e.*, immobile ions) and holes (mobile charge carriers) as shown in Fig. 10.10 (*b*).

10.13. Conduction in Metals

In order to study the characterstics of a conduction in semiconductor, it is important to discuss first, the conduction in metals. We shall derive few relations for drift velocity, current density and conductivity for metals. Later on, we shall use these relations to develop similar relations for semiconductors.

We know that valence electrons in a metal move continuously at random. During their movement, the valence electrons collide with heavy (stationery ions). After each collision, the direction of its movement is changed. The average distance between collisions is known as mean free path. Since the motion of electrons is at random, therefore, there will be as many electrons passing through a unit area in the metal in any direction as in the opposite direction in a given time. Hence the average value of the current is zero.

Now, if an electric field is applied to a metal, than due to the electrostatic force, the electrons would be accelerated and the velocity would increase indefinitely, if there would have been no collision with the ions. However, at each collision with the ion, the electron loses some energy. A steady-state condition is reached, when a finite value of drift velocity is obtained. The drift velocity is in a direction opposite to that of the applied electric field. Now let us obtain an expression for the drift velocity in a metal.

Let E = Strength of the applied electric field,

F = Force experienced by an electron due to the applied electric field,

a = Acceleration, which the electron obtains due to the force experienced by it,

t = Time between the collisions,

v = Drift velocity,

q = Charge of an electron, and

m = Mass of an electron.

We know that force experienced by the electron due to applied electric field.

$$F = q \cdot E$$

and acceleration, $$a = \frac{\text{Force}}{\text{Mass}} = \frac{F}{m} = \frac{q \cdot E}{m}$$

$\therefore$ Drift velocity, $$v = a \cdot t = \frac{q \cdot E}{m} \cdot t = \mu \cdot E$$

where μ is equal to $\frac{q \cdot t}{m}$ and is known as mobility of an electron. The mobility is expressed in cm^2/volt-sec or m^2/(volt-sec) and its value is different for different metals. The electron mobility is a measure of how quickly an electron can be move through a metal. When pulled by electric field.

Note. $1\ cm^2/V.s = 10^{-4}\ m^2/V.s$

10.14. Electric Current in a Conductor

Consider a piece of a metal or conductor in which electrons are uniformly distributed as shown in Fig. 10.11.

Let N = No. of free electrons distributed in the conductor

L = Length of the conductor

A = Cross-sectional area of the conductor

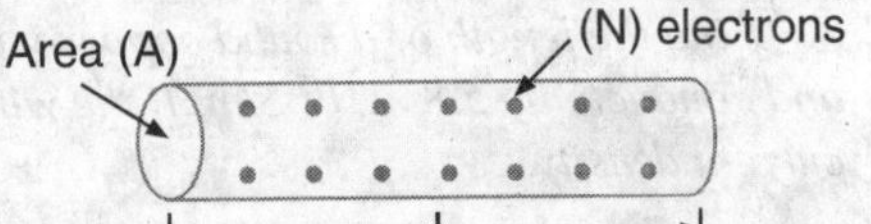

Fig. 10.11. Metal or conductor.

We know that when an electric field (E) is applied to the conductor, the electorns start moving in the conductor. Therefore, avarage velocity of the electrons

$$= \frac{L}{T}$$

where T is the time taken by the electrons to travel through a distance.

The number of electrons passing through any area per second,

$$= \frac{N}{T}$$

∴ Total charge passing through any area per second (called current),

= Charge of an electron × Number of electrons per second passing through any area

$$= q \times \frac{N}{T}$$

Multiplying and dividing the right hand side of the above equation by L,

$$= q \times \frac{N}{T} \times \frac{L}{L} = \frac{q \cdot N \cdot v}{L} \qquad \ldots\left(\because v = \frac{L}{T}\right)$$

Thus the current in a conductor,

$$I = \frac{q \cdot N \cdot v}{L}$$

and the current per unit area (called current density),

$$J = \frac{I}{A} = \frac{q \cdot N \cdot v}{L \cdot A}$$

Since the term '$L \cdot A$' is the volume of the conductor containing N electrons, therefore, concentration of electrons per unit volume,

$$n = \frac{N}{L \cdot A}$$

and $J = q \cdot n \cdot v$

10.15. Conductivity of a Conductor

We have already discussed in the last article that electric current density (*i.e.*, electric current per unit area),

$$J = q \cdot n \cdot v$$

Substituting the value of v from Art. 10.13 in the above equation,

$$J = q \cdot n \cdot (\mu \cdot E) = q \cdot n \cdot \mu \cdot E = \sigma \cdot E$$

where σ (equal to $q \cdot n \cdot \mu$) is called conductivity of a metal. The above relation indicates that electric current density (J) within a metal is directly proportional to the applied electric field (E). Now we know that resistivity of a metal is reciprocal of conductivity, *i.e.*,

$$\rho = \frac{1}{\sigma} = \frac{1}{q \cdot n \cdot \mu}$$

and $\sigma = \frac{1}{\rho} = q \cdot n \cdot \mu$

Example 10.1.

Find the length of a round copper wire of diameter 1 mm which will have a resistance of 1 kΩ and conductivity 5.8 × 10⁷ S/m. If the wire is subjected to an electric field of 10 mV/m, what is the current density. (*Pune Univ. 1994*)

Solution.

Given: Resistance of a copper wire (R) = 1 kΩ = 1000 Ω; Conductivity (σ) = 5.8×10^7 S/m; diameter (d) = 1 mm = 0.001 m.

Length of Copper wire

Let l = length of the copper wire (m).

We know that resistance of a wire (R),

$$1000 = \rho \cdot \frac{l}{a} = \frac{1}{\sigma} \cdot \frac{l}{A} = \frac{1}{\sigma} \cdot \frac{l}{\pi (d^2 / 4)}$$

$$= \frac{1}{5.8 \times 10^7} \times \frac{l}{\pi \times (0.001)^2 / 4}$$

$$= \frac{4l}{182.12}$$

$$l = \frac{1000 \times 182.12}{4} = 45.5 \times 10^3 \text{ m} = 45.5 \text{ km} \textbf{ Ans.}$$

Current desnity in copper wire

Given: $E = 10 \text{ mV/m} = 10 \times 10^{-3} \text{ V/m}$

We known that current density,

$$J = \sigma E$$

$$= (5.8 \times 10^7) \times (10 \times 10^{-3})$$

$$= 5.8 \times 10^5 \text{ A/m}^2 \textbf{ Ans.}$$

Example 10.2.

A copper wire of 2 mm diameter with conductivity of 5.8×10^7 Siemens/m and electron mobility of 0.0032 m^2/V-s is subjected to an electric field of 20 mV/m. Find (a) the charge density of free electrons. (b) the current density (c) the current flowing in the wire (d) the electron drift velocity.

(UPSC Engg Services 2002)

Solution.

Given: Diameter of copper wire (d) = 2 mm = 2×10^{-3} m; Conductivity, (σ) = 5.8×10^7 S/m; electron mobility, (μ) = 0.0032 m^2/V-s; Electric field (E) = 20 mV/m.

(a) *Charge density of free electrons*

Let n = number of free electrons per unit volume of the copper wire (*i.e.*, electron density) per m^3.

We know that conductivity (σ),

$$5.8 \times 10^7 = q \cdot n \cdot \mu = 1.6 \times 10^{-19} \times n \times 0.0032$$

$$\therefore \quad n = \frac{5.8 \times 10^7}{1.6 \times 10^{-19} \times 0.0032} = 1.132 \times 10^{29}/\text{m}^3 \textbf{ Ans.}$$

(b) *Current density*

We know that current density,

$$J = \sigma E = 5.8 \times 10^7 \times 20 \times 10^{-3} = 1.6 \times 10^6 \text{ A/m}^2 \textbf{ Ans.}$$

(c) *Current flowing in the wire*

Let I = the value of current flowing in the wire (A)

We know that are of cross-section of wire,

$$A = \frac{\pi d^2}{4} = \frac{\pi \times (2 \times 10^{-3})^2}{4} = 3.16 \times 10^{-6} \text{ m}^2$$

We know that current density (J),

$$1.16 \times 10^6 = \frac{I}{3.16 \times 10^{6}}$$

or $\quad I = 1.16 \times 3.16 = 3.67 \text{ A}$ **Ans.**

(d) Electron Drift velocity

We know that drift velocity,

$$v = \mu E = 0.0032 \times (20 \times 10^{-3}) = 64 \text{ m/s } \textbf{Ans.}$$

Example 10.3.

A conduction wire has a resistivity of 1.54 × 10⁻⁸ Ω-m at room temperature. There are 5.8 × 10²⁸ conduction electrons per m³. Calculate the mobility and relaxation time of electrons.

(UPSC Engg Services 2000)

Solution.

Given: $\rho = 1.54 \times 10^{-8}\ \Omega\text{-m}$ and $n = 5.8 \times 10^{28}$.

Let μ = mobility and

t = relaxation time (*i.e.*, time between the collisions)

Mobility

We know that resistivity (ρ),

$$1.54 \times 10^{-8} = \frac{1}{\sigma}$$

$$\therefore \quad \sigma = \frac{1}{1.54 \times 10^{-8}} = 0.649 \times 10^{8} \text{ S/m}$$

We also know that conductivity (σ)

$$0.649 \times 10^{8} = q \cdot n \cdot \mu$$

$$= 1.60 \times 10^{-19} \times 5.8 \times 10^{28} \times \mu$$

$$\therefore \quad \mu = \frac{0.649 \times 10^{8}}{9.28 \times 10^{9}} = 0.699 \text{ m}^2\text{/V-s } \textbf{Ans.}$$

Relaxation time

We know that mobility (μ),

$$0.699 = \frac{q \cdot t}{m} = \frac{1.6 \times 10^{-19} \times t}{9.1 \times 10^{-31}}$$

$$= 0.176 \times 10^{12}\, t$$

$$\therefore \quad t = \frac{0.699}{0.176 \times 10^{12}} = 3.976 \times 10^{-12} \text{ s} = 3.976 \text{ ps } \textbf{Ans.}$$

10.16. Conduction in Intrinsic Semiconductors

We have already discussed in Art. 10.4 that in an intrinsic semiconductor at room temperature (*i.e.*, 300 K), the thermal energy is sufficient to produce a large number of electron hole pairs. When an electric field is applied, the current flows through the semiconductor. The flow of current, in a semiconductor, is due to the movement of free electrons in one direction and holes in the opposite direction. We know that electric current density in a metal (which is due to the movement of free electrons) is given by the relation,

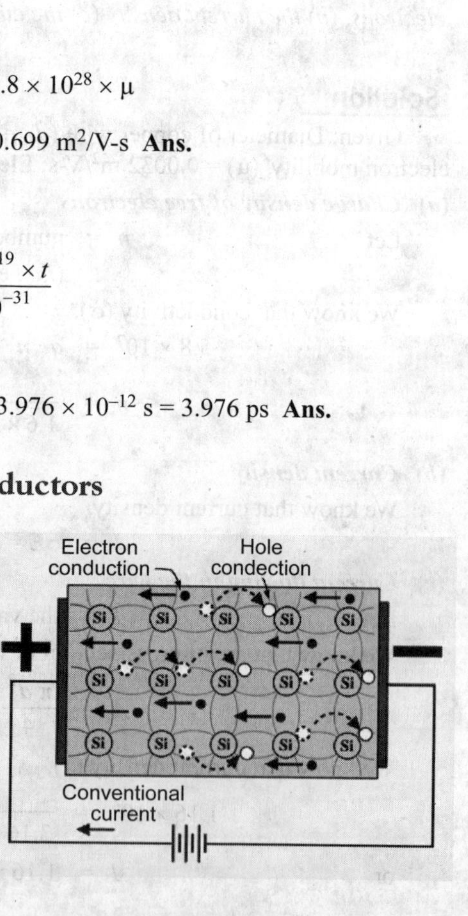

$$J = q \cdot n \cdot \mu \cdot E$$

Using the above relation, we can write an equation for the current density in an intrinsic semiconductor. Thus current density, which results due to the movement of electrons only, is given by the relation,

$$J_n = q \cdot n \cdot \mu_n \cdot E$$

where
q = Charge of an electron,
n = Number of electrons per unit volume also called (electron concentration) in an intrinsic semiconductor,
μ_n = Mobility of electrons in a semiconductor, and
E = Applied electric field strength.

A similar expression can be written for current density due to holes in an intrinsic semiconductor and is given by the relation,

$$J_p = q \cdot p \cdot \mu_p \cdot E$$

where
q = Charge of a hole and is equivalent to that of an electron,
p = Number of holes per unit volume also called the concentration in an intrinsic semiconductor,
μ_p = Mobility of holes in a semiconductor, and
E = Applied electric field strength.

The total current density with in an intrinsiç semiconductor,

$$\begin{aligned} J &= J_n + J_p \\ &= q \cdot n \cdot \mu_n \cdot E + q \cdot p \cdot \mu_p \cdot E \quad \text{...(i)} \\ &= q\,(n \cdot \mu_n + p \cdot \mu_p)\,E \quad \text{...(ii)} \\ &= \sigma \cdot E \end{aligned}$$

where σ equal to $q \cdot (n \cdot \mu_n + p \cdot \mu_p)$ is called conductivity of a semiconductor. It is evident from equation (*ii*) that current density within a semiconductor is directly proportional to applied electric field (E).

We know that for an intrinsic semiconductor, the number of holes is equal to the number of electrons *i.e.*, $p = n = n_i$ where n_i is called intrinsic carrier concentration. Substituting $n = p = n_i$ in equation (*ii*),

$$J = q \cdot n_i\,(\mu_n + \mu_p)\,E$$

and conductivity of an intrinsic semiconductor,

$$\sigma_i = q \cdot n_i\,(\mu_n + \mu_p)$$

It is evident from the above relation that conductivity of an intrinsic semiconductor depends upon its intrinsic concentration (n_i) and the mobility of electrons (m_n) and holes (m_p).

Notes. 1. The value of q (*i.e.*, charge of electron or hole) is taken as 1.6×10^{-19} coulomb.

2. Typically electron mobility for Si at room temperature (300K) is $cm^2/V.s$ and hole mobility is around 500 $cm^2/V.s$.

Example 10.4.

The mobilities of free electrons and holes in pure germanium are 0.38 and 0.18 m^2/V-s. The corresponding values for pure silicon are 0.13 and 0.05 m^2/V-s respectively. Find the values of intrinsic conductivity for both the materials. Assume $n_i = 2.5 \times 10^{19}/m^3$ for Ge and $n_i = 1.5 \times 10^{16}/m^3$ for S_i at room temperature.

Solution.

Given : For germanium, $\mu_n = 0.38\ m^2/V$-s; $\mu_p = 0.18\ m^2/V$-s; $n_i = n = p = 2.5 \times 10^{19}/m^3$ and for silicon, $\mu_n = 0.13\ m^2/V$-s; $\mu_p = 0.05\ m^2/V$-s and $n_i = n = p = 1.5 \times 10^{16}/m^3$.

(a) Intrinsic conductivity for germanium

We know that the intrinsic conductivity for germanium,

$$\sigma_i = q \cdot n_i (\mu_n + \mu_p) = (1.6 \times 10^{-19}) \times (2.5 \times 10^{19}) \times (0.38 + 0.18)\ (\Omega\text{-m})^{-1}$$
$$= 2.24\ (\Omega\text{-m})^{-1}\ \textbf{Ans.}$$

(b) Intrinsic conductivity for silicon

We also know that intrinsic conductivity for silicon,

$$\sigma_i = q \cdot n_i (\mu_n + \mu_p)$$
$$= (1.6 \times 10^{-19}) \times (1.5 \times 10^{16}) \times (0.13 + 0.05)\ (\Omega\text{-m})^{-1}$$
$$= 4.32 \times 10^{-4}\ (\Omega\text{-m})^{-1}\ \textbf{Ans.}$$

Example 10.5.

Calculate the intrinsic conductivity of silicon at room temperature if $n_i = 1.41 \times 10^{16}\ m^{-3}$, μ_n = 0.145 m2/V-s, $\mu_p = 0.05\ m^2/V\text{-}s$ and $q = 1.6 \times 10^{-19}$ C. What are the individual contributions made by electrons and holes ? (*Electronic Engg. Nagpur Univ. 1991*)

Solution.

Given: $n = 1.41 \times 10^{16}\ \text{m}_{-3}$, $\mu_n = 0.145\ \text{m}^2/\text{V-s}$, $\mu_p = 0.05\ \text{m}^2/\text{V-s}$ and $q = 1.6 \times 10^{-19}$ C

We know that conductivity of an intrinsic semiconductor,

$$\sigma_i = q \cdot n_i (\mu_n + \mu_p)$$
$$= \underbrace{q\, n_i\, \mu_n}_{\text{Contibution by electrons}} + \underbrace{q\, n_i\, \mu_p}_{\text{Contibution by holes}}\ (\Omega\text{-m})^{-1}$$
$$= (1.6 \times 10^{-19}) \times (1.41 \times 10^{16}) \times 0.145 + (1.6 \times 10^{-19}) \times (1.41 \times 10^{16}) \times 0.05$$
$$= 0.325 \times 10^{-3} + 0.112 \times 10^{-3}\ \Omega\text{-m}^{-1}$$

$\therefore$ Contribution by electrons $= 0.325 \times 10^{-3}\ (\Omega\text{-m})^{-1}$ **Ans.**

and Contribution by holes $= 0.112 \times 10^{-3}\ (\Omega\text{-m})^{-1}$ **Ans.**

Example 10.6.

A specimen of silicon is 0.2 mm long and has a cross-section of 0.2 × 0.2 mm. One volt impressed across the bar results in a current of 8 mA. Assuming that the current is due to electrons, calculate (a) concentration of free electrons and (b) the drift velocity. Given at 300 K, $\mu_n = 1300\ cm^2/V\text{-}s$ and $q = 1.60 \times 10^{19}$ C.

Solution.

Given : $l = 0.2$ mm $= 0.2 \times 10^{-3}$ m; $A = 0.2 \times 0.2 = 0.04\ \text{mm}^2 = 0.04 \times 10^{-6}\ \text{m}^2$; $V = 1$ volt; $I = 8$ mA $= 8 \times 10^{-3}$ A; $\mu_n = 1300\ \text{cm}^2/\text{V-s} = 0.13\ \text{m}^2/\text{V-s}$.

(a) Concentration of free electrons

Let n = Concentration of free electrons, and

ρ = Resistivity of the specimen.

We know that resistance of a specimen,

$$R = \frac{V}{I} = \frac{1}{8 \times 10^{-3}} = 125\ \Omega$$

and resistance of a specimen (R),

$$125 = \rho \times \frac{l}{A} = \rho \times \frac{0.2 \times 10^{-3}}{0.04 \times 10^{-6}} = 5000\ \rho$$

$\therefore \qquad \rho = 125/5000 = 0.025\ \Omega\text{-m}$

and conductivity, $\quad \sigma = \dfrac{1}{\rho} = \dfrac{1}{0.025} = 40$

We know that the conductivity of a silicon specimen (σ),

$$40 = q \cdot n \cdot \mu_n = (1.60 \times 10^{-19}) \times n \times 0.13 = 0.208 \times 10^{-19}\, n$$

$$n = \frac{40}{(0.208 \times 10^{-19})} = 1.92 \times 10^{21}/\text{m}^3 \textbf{ Ans.}$$

(b) Drift velocity

Let $\qquad v =$ Drift velocity of free electrons in silicon.

We know that current density,

$$J = \frac{I}{A} = \frac{8 \times 10^{-3}}{0.04 \times 10^{-6}} = 2 \times 10^5 \text{ A/m}^2$$

and drift velocity,

$$v = \frac{J}{n \cdot q} = \frac{2 \times 10^5}{1.92 \times 10^{21} \times 1.60 \times 10^{-19}} = 651 \text{ m/sec } \textbf{Ans.}$$

Example 10.7.

Find the intrinsic carrier concentration of germanium, if its intrinsic resistivity at 300 K is 0.47 Ω-m. It is given that the electronic charge is 1.6 × 10^{-19} *coulombs and that electron and hole mobilities at 300 K are 0.39 and 0.19 m*^{2}*/volt-sec respectively.*

Solution.

Given: $\rho = 0.47\ \Omega\text{-m}$; $q = 1.6 \times 10^{-19}$ C; $\mu_n = 0.39\ \text{m}^2/\text{V-sec}$ and $\mu_p = 0.19\ \text{m}^2/\text{V-sec}$.

Let $\qquad n_i =$ Intrinsic carrier concentration of germanium.

We know that conductivity of a semiconductor,

$$\sigma = \frac{1}{\rho} = \frac{1}{0.47} = 2.13\ (\Omega\text{-m})^{-1}$$

and conductivity in intrinsic semiconductor (σ),

$$2.13 = q \cdot n\,(\mu_n + \mu_p) = 1.6 \times 10^{-19} \times n_i\,(0.39 + 0.19)$$

$$= 0.93 \times 10^{-19}\, n_i$$

$$n_i = \frac{2.13}{(0.93 \times 10^{-19})} = 2.3 \times 10^{19}/\text{m}^3 \textbf{ Ans.}$$

10.17. Conductivity of N-and P-type Semiconductors

We know that conductivity of an intrinsic semiconductor,

$$\sigma_i = q \cdot n_i\,(\mu_n + \mu_p) = q\,(n_i \cdot \mu_n + n_i \cdot \mu_p)$$

Now substituting $n_i = n = p$ in the above relation,

$$\sigma_i = q\,(n \cdot \mu_n + p \cdot \mu_p)$$

We also know that for N-type semiconductor ($n >> p$), therefore conductivity,

$$\sigma = q \cdot n \cdot \mu_n$$

and for a P-type semiconductor ($p >> n$), therefore, conductivity,

$$\sigma = q \cdot p \cdot \mu_p$$

10.18. Mass Action Law

We have already discussed in Art. 10.9 that the addition of N-type impurities decreases the number of holes below a level, which would have been available in a pure semiconductor. Similarly, the addition of P-type impurities decreases the number of electrons below a level, which would have been available in a pure semiconductor. It has been experimentally found that under thermal equilibrium, the product of the number of holes and the number of electrons is constant and is independent of the amount of donor and acceptor impurity doping. This relation is known as mass action law and is given by the relation,

$$n \cdot p = n_i^2$$

where n = Free electron concentration (*i.e.*, number of free electrons per m^3),

p = Hole concentration (*i.e.*, number of holes per m^3), and

n_i = Intrinsic concentration.

10.19. Charge Densities in N- and P-type Semiconductors

We have already discussed in the last article the relationship between free electron concentration (n) and hole concentration (p), which is known as mass action law. These concentrations are further related by the law of electrical neutrality as discussed below:

Let N_D = Concentration of donor atoms (*i.e.*, the number of positive charges/m^3 contributed by donor ions),

p = Hole concentration (*i.e.*, number of holes/m^3),

N_A = Concentration of acceptor atoms (*i.e.*, the number of negative charges/m^3 contributed by acceptor ions), and

n = Free electron concentration (*i.e.*, number of free electrons/m^3).

∴ Total number of positive charges/m^3 (called total positive charge density) in a semiconductor

$$= N_D + p$$

and the total number of negative charges/m^3 (called total negative charge density) in a semiconductor

$$= N_A + n$$

Since the semiconductor, as a whole, is electrically neutral, therefore, magnitude of total positive charge density is equal to total negative charge density, *i.e.*,

$$N_D + p = N_A + n \qquad \text{... (i)}$$

Now consider an N-type semiconductor. We know that in N-type semiconductor, the concentration of acceptor atoms is zero (*i.e.*, $N_A = 0$), and also the number of free electrons is much larger than the number of holes, (*i.e.*, $n >> p$). Therefore, for an N-type semiconductor, the equation (*i*) may be written as :

$$N_D = n$$

The above relation shows that in an N-type semiconductor, the free electron concentration (n) is equal to the number of donor atoms/m^3. The concentration of holes in an N-type semiconductor may be obtained by using mass action law (*i.e.*, $n \,.\, p = n_i^2$) and its value is given by the relation,

$$p = \frac{n_i^2}{n} = \frac{n_i^2}{N_D} \qquad \text{... (ii)}$$

Similarly, in a P-type semiconductor, the concentration of donor atoms is zero (*i.e.*, $N_D = 0$) and the number of holes is much larger than the number of free electrons (*i.e.*, $p >> n$). Therefore, for a P-type semiconductor, the equation (*i*) may be written as :

$$p = N_A$$

The above relation shows that in a P-type semiconductor, the hole concentration (p) is equal to the number of acceptor atoms/m³. The concentration of free electrons may be obtained by using mass action law (*i.e.*, $n \cdot p = n_i^2$) and its value is given by the relation,

$$n = \frac{n_i^2}{p} = \frac{n_i^2}{N_A} \quad \text{... (iii)}$$

In order to avoid unnecessary confusion regarding the type of semiconductor under consideration, it is better to add subscript 'N' for a N-type semiconductor and subscript 'P' for a P-type semiconductor. Thus if

n_N = Free electron concentration in a N-type semiconductor,
p_N = Hole concentration in a N-type semiconductor,
n_P = Free electron concentration in a P-type semiconductor, and
p_P = Hole concentration in P-type semiconductor.

Then the equations (*ii*) and (*iii*) may be rewritten as :

$$p_N = \frac{n_i^2}{n_N} = \frac{n_i^2}{N_D}$$

and

$$n_N = \frac{n_i^2}{p_P} = \frac{n_i^2}{N_A}$$

It will be interesting to know that it is possible to add donor atoms into a P-type semiconductor. Similarly, it is also possible to add acceptor atoms into an N-type semiconductor. If an equal concentration of donor atoms and acceptor atoms is added into an intrinsic semiconductor, it remains intrinsic semiconductor. It is due to the fact that holes supplied by the acceptor atoms combine with the free electrons supplied by the donor atoms to give no additional carriers. The above statement indicates that if the concentration of donor atoms added to a P-type semiconductor exceeds the concentration of acceptor ions, then the semiconductor is changed from a P-type to N-type. This concept is used in the preparation of P-N junction, which is an essential part of all electronic devices

Example 10.8.

A sample of silicon is doped with phosphorus to a density of $10^{21}/m^3$ as well as with boron to a density of $5 \times 10^{20}/m^3$. What will be conductivity of the silicon sample. The electron mobility in silicon is $0.18\ m^2/V\text{-}s$.

Solution.

Given : $N_D = 10^{21}/\text{m}^3$ and $N_A = 5 \times 10^{20}/\text{m}^3$

Since phosphorous atoms are donor and boron atoms are acceptor, therefore, net donor density,

$$= N_D' = N_D - N_A = 10^{21} - 5 \times 10^{20} = 5 \times 10^{20}/\text{m}^3$$

Number of free electrons, $n = N_D' = 5 \times 10^{20}/\text{m}^3$

∴ We know that conductivity of silicon,

$$\sigma = q \cdot n \cdot m_n = (1.6 \times 10^{-19}) \times (5 \times 10^{20}) \times 0.18\ (\Omega\text{-m})^{-1}$$
$$= 14.4\ (\Omega\text{-m})^{-1}\ \textbf{Ans.}$$

Example 10.9.

Calculate the doner concentration in N-type germanium having resistivity of 100 Ω-m. Take $q = 1.6 \times 10^{-19}$ C and $\mu_n = 0.36\ m^2\ V^{-1}\ s^{-1}$.

Solution.

Given: Resistivity of N-type germanium (ρ) = 100 Ω-m, $q = 1.6 \times 10^{-19}$ C and $\mu_n = 0.36\ \text{m}^2\ \text{V}^{-1}\ \text{s}^{-1}$.

Let N_D = Donor concentration in N-type germanium

We know that conductivity of N-type germanium,

$$\sigma = \frac{1}{p} = \frac{1}{100} = 0.01\ (\Omega\text{-m})^{-1}$$

We also know that conductivity of N-type germanium (s),

$$0.01 = q\,n\,\mu_n$$

$$= 1.6 \times 10^{-19} \times n \times 0.36$$

$$\therefore \quad n = \frac{0.01}{(1.6 \times 10^{-19}) \times 0.36} = 1.74 \times 10^{-17} \text{ atoms/m}^3$$

and $N_D \approx n = 1.74 \times 10^{17}$ atoms/m³ **Ans.**

Example 10.10.

Determine the concentration of the free electrons and holes in a sample of Ge at 300 K which has a concentration of donor atoms equal to 2×10^{14} atoms/cm³ and a concentration of acceptor ions equal to 3×10^{14} atoms/cm³. (Nagpur Univ. Winter 2003)

Solution.

Given: $N_D = 2 \times 10^{14}$ atoms/cm³; $N_A = 3 \times 10^{14}$ atoms/cm³,

Let n = number of free electrons, and

p = number of holes.

We know that number of free electrons,

$$n = \frac{n_i^2}{N_A} \qquad \text{...}(i)$$

where n_i = intrinsic concentration = 2.3×10^{19}

Substituting the value of n_i (= 2.3×10^{19}) and N_A (= 3×10^{14}) in equation (*i*), we get

$$n = \frac{(2.3 \times 10^{19})}{3 \times 10^{14}} = 1.76 \times 10^{24} \text{ electrons/cm}^3 \quad \textbf{Ans.}$$

We know that number of holes,

$$p = \frac{n_i^2}{N_D}$$

$$= \frac{(2.3 \times 10^{19})^2}{2 \times 10^{14}} = 2.65 \times 10^{24} \text{ holes/cm}^3 \quad \textbf{Ans.}$$

Example 10.11.

A germanium PN junction at 300 K has the following parameters :

$N_D = 5 \times 10^{18}$/cm³, $N_A = 6 \times 10^{16}$/cm³, $n_i = 1.5 \times 10^{10}$/cm³. Calculate the minority electron density in the P-region and the minority hole density in the N-region. (Banglore Univ. July 2003)

Solution.

Given: $N_D = 5 \times 10^{18}$/cm³, $N_A = 6 \times 10^{16}$/cm³, $n_i = 1.5 \times 10^{10}$/cm³

Let n = number of electrons (or electron density) in the P-region and

p = number of holes (or hole density) in the N-region.

We know that number of electrons, in the P-region,

$$n = \frac{n_i^2}{N_A} = \frac{(1.5 \times 10^{10})^2}{6 \times 10^{16}} = 0.375 \times 10^4 = 3750 \textbf{ Ans.}$$

and the number of hole in the N-region,

$$p = \frac{n_i^2}{N_D} = \frac{(1.5 \times 10^{10})^2}{5 \times 10^8} = 0.45 \times 10^{12} \textbf{ Ans.}$$

Example 10.12.

A cylinderical piece of silicon having a diameter of 1 mm is diped with $10^{20}/m^3$ atoms of phosphorus which are fully ionized. What length of this silicon would be required to give a resistance of 1 kW if electron mobility in silicon is 0.1 m^2/V-s. *(Pune Univ. 1994)*

Solution.

Given: diameter of silicon piece (d) = 1 mm = 0.001 m; Number of phosphorus atom (N_D) = $10^{20}/m^3$; Resistance (R) = 1 kΩ = 1000 Ω; electron mobility (μ_n) = 0.1 m^2/V-s.

Let l = length of silicon piece (m),

We know that number of free electrons,

$$n = N_D = 10^{20}/m^3$$

We also know that conductivity,

$$\sigma = q \cdot n \cdot \mu_n$$
$$= 1.6 \times 10^{-19} \times 10^{20} \times 0.1$$
$$= 1.6 \text{ S/m}$$

and the resistance of silicon piece (R),

$$1000 = \frac{1}{\sigma} \cdot \frac{l}{A} = \frac{1}{1.6} \times \frac{l}{\pi\left(\frac{d^2}{4}\right)}$$

$$= \frac{1}{1.6} \times \frac{l}{\pi \times \frac{(0.001)^2}{4}}$$

$$= 7.96 \times 10^{+5}\, l$$

$$\therefore \quad l = \frac{1000}{7.96 \times 10^5} = 1.26 \times 10^{-3} \text{ m} = 1.26 \text{ mm} \textbf{ Ans.}$$

Example 10.13.

Find the concentration of holes and electrons in a P-type germanium at 300 K, if the conductivity is 100 per ohm per cm. Also find these values for N-type silicon, if the conductivity is 0.1 per ohm per cm. Given for germanium, $n_i = 2.5 \times 10^{13}/cm^3$; $\mu_n = 3800\ cm^2/V$-sec; $\mu_p = 1800\ cm^2/V$-sec and for silicon $n_i = 1.5 \times 10^{10}/cm^3$; $\mu_n = 1300\ cm^2/V$-sec and $m_p = 500\ cm^2/V$-sec. *(Banglore Univ. 2004)*

Solution.

Given : For germanium $\sigma = 100\ \Omega/cm^{-1}$;

For silicon, $\sigma = 0.1\ (\Omega\text{-cm})^{1}$; $n_i = 1.5 \times 10^{10}/cm^3$;

$\mu_n = 1300\ cm^2/V$-sec, and $\mu_p = 500\ cm^2/V$-sec

$n_i = 2.5 \times 10^{13}/\text{cm}^3$; $\mu_n = 3800$ cm²/V-sec; $\mu_p = 1800$ cm²/V-sec and

Let p = Concentration of holes, and

n = Concentration of electrons.

Concentration of P-type germanium

We know that conductivity of a P-type germanium (σ),

$$100 = q \cdot p \cdot \mu_p = 1.602 \times 10^{-19} \times p \times 1800 = 2.88 \times 10^{-16} \times p$$

$$\therefore \quad p = \frac{100}{2.88 \times 10^{-16}} = 3.47 \times 10^{17}/\text{cm}^3 \textbf{ Ans.}$$

and the concentration of electrons,

$$n = \frac{n_i^2}{p} = \frac{(2.5 \times 10^{13})^2}{3.47 \times 10^{17}} = 1.8 \times 10^9/\text{cm}^3 \textbf{ Ans.}$$

Concentration in N-type silicon

We also know that conductivity of an N-type silicon, (σ),

$$0.1 = q \cdot n \cdot \mu_n = 1.602 \times 10^{-19} \times n \times 1300 = 2.08 \times 10^{-16} \times n$$

$$\therefore \quad n = \frac{0.1}{2.08 \times 10^{-16}} = 4.8 \times 14^{14}/\text{cm}^3 \textbf{ Ans.}$$

and the number of holes,

$$p = \frac{n_i^2}{n} = \frac{(1.5 \times 10^{10})^2}{4.8 \times 10^{14}} = 4.7 \times 10^5/\text{cm}^3 \textbf{ Ans.}$$

Example 10.14.

In a germanium sample, a donor type impurity is added to the extent of 1 atom per 10^8 germanium atoms. Show that resistivity of the germanium sample drops to 3.7 ohm-cm. Given : $\mu_n = 3800$ cm²/V-sec; $\mu_p = 1800$ cm²/V-sec; $n_i = 2.5 \times 10^{13}/cm^3$; $N_{ge} = 4.41 \times 10^{22}/cm^3$ and $q = 1.602 \times 10^{-19}$ coulombs.

Solution.

Given : $\mu m = 3800$ cm²/V-sec; $\mu_p = 1800$ cm²/V-sec; $n_i = 2.5 \times 10^{13}/\text{cm}^3$; $N_{ge} = 4.41 \times 10^{22}/\text{cm}^3$ and $g = 1.602 \times 10^{-19}$ coulombs.

We know that the number of donor atoms,

$$N_D = \frac{N_{ge}}{10^8} = \frac{4.41 \times 10^{22}}{10^8} = 4.41 \times 10^{14}/\text{cm}^3$$

Since the number of free electrons (n) is approximately equal to the donor atoms (N_D), therefore, $n \approx 4.41 \times 10^{14}/\text{cm}^3$. We know that number of holes,

$$p = \frac{n_i^2}{N_D} = \frac{(2.5 \times 10^{13})^2}{4.41 \times 10^{14}} = 1.417 \times 10^{12}/\text{cm}^3$$

It may be noted that the number of electrons (4.41×10^{14}) is very large as compared to the number of holes (1.417×10^{12}) in a doped germanium. Therefore, conductivity of an N-type germa- nium,

$$\sigma = q \cdot n \cdot \mu_n = (1.602 \times 10^{-19}) \times (4.41 \times 10^{14}) \times 3800$$

$$= 0.268\ (\Omega\text{-cm})^{-1}$$

$\therefore$ Resistivity of a doped germanium,

$$\rho = \frac{1}{\sigma} = \frac{1}{0.268} = 3.73 \text{ ohm-cm} \textbf{ Ans.}$$

Example 10.15.

Find the resistivity of intrinsic silicon. What will this change to, when the silicon is doped with a pentavalent impurity to the extent of one impurity atom for each 50 million silicon atoms ? Following data may be used.

Number of silicon atoms = $4.96 \times 10^{22}/cm^3$
Intrinsic carrier concentration = $1.52 \times 10^{10}/cm^3$
Electron charge = 1.6×10^{-19} *coulomb*
Electron mobility = $0.135\ m^2/Volt\text{-}sec$
Hole mobility = $0.048\ m^2/Volt\text{-}sec$

Solution.

Given : $N_{si} = 4.96 \times 10^{22}$ /cm^3; $n_i = 1.52 \times 10^{10}$/cm^2; $q = 1.6 \times 10^{-19}$ C; $\mu_n = 0.135$ m^2/V-sec = 1350 cm^2/V-sec and $\mu_p = 0.048$ m^2/V-sec = 480 cm^2/V-sec.

Resistivity of intrinsic silicon

We know that conductivity of an intrinsic silicon,

$$\sigma = q \cdot n_i (\mu_n + \mu_p) = (1.6 \times 10^{-19}) \times (1.52 \times 10^{10}) \times (1350 + 480)\ (\Omega\text{-cm})^{-1}$$
$$= 4.5 \times 10^{-16}\ (\Omega\text{-cm})^{-1}$$

∴ Resistivity of intrinsic silicon,

$$\rho = \frac{1}{\sigma} = \frac{1}{4.5 \times 10^{-6}} = 2.22 \times 10^5\ \Omega\text{-cm}\ \textbf{Ans.}$$

Resistivity of doped silicon

We know that if the silicon is doped with a pentavalent impurity to the extent of one impurity atom for each 50 million (50×10^6) silicon atoms, then the number of donor atoms,

$$N_D = \frac{N_{si}}{50 \times 10^6} = \frac{4.96 \times 10^{22}}{50 \times 10^6} = 9.92 \times 10^{14}\ /\text{cm}^3$$

∴ Number of free electrons in a semiconductor,

$$n = N_D = 9.92 \times 10^{14}/\text{cm}^3$$

We also know that number of holes in a doped silicon,

$$p = \frac{n_i^2}{n} = \frac{(1.52 \times 10^{10})^2}{9.92 \times 10^{14}} = 2.33 \times 10^5\ /\text{cm}^3$$

Since the number of electrons (equal to 9.92×10^{14}) in a doped silicon is much larger than the number of holes (equal to 2.33×10^5), therefore, conductivity of a doped silicon,

$$\sigma = q \cdot n \cdot \mu_n = (1.6 \times 10^{-19}) \times (9.92 \times 10^{14}) \times 1350 = 0.21\ (\Omega\text{-cm})^{-1}$$

∴ Resistivity of a doped silicon,

$$\rho = \frac{1}{\sigma} = \frac{1}{0.21} = 4.67\ \Omega\text{-cm}\ \textbf{Ans.}$$

10.20. Variation in Semiconductor Parameters

The most useful semiconductor parameters are **1.** Intrinsic concentration (n_i), **2.** Mobility (μ), **3.** Conductivity (σ), and **4.** Energy gap (E_G).

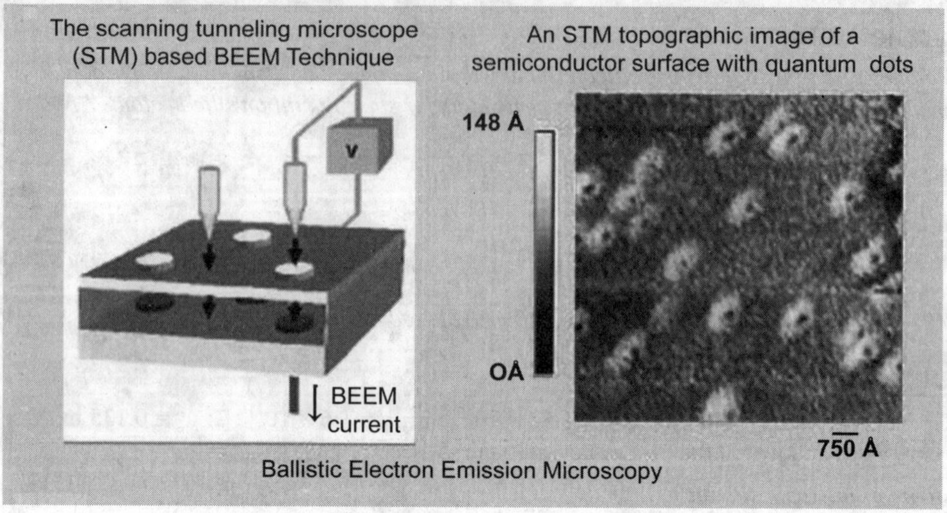

As a matter of fact most of the semiconductors operate on a wide range of temperature. Therefore, variations of these parameters, with temperature, are very important. The typical values of these parameters for silicon and germanium are as shown in Table 10.1.

Table 10.3 : Properties of intrinsic silicon and germanium

S. No.	*Property*	*Silicon*	*Germanium*
1.	Atomic Number	14	32
2.	Atomic weight	28.1	72.6
3.	Density (in gm/cm^3)	2.33	5.32
4.	Number of atoms/cm^3	5.0×10^{22}	4.4×10^{22}
5.	Energy gap (E_G) at 0 K (in eV)	1.21	0.785
6.	Energy gap (E_G) at 300 K (in eV)	1.1	0.72
7.	Resistivity at 300 K (in W-cm)	2.3×10^5	45
8.	Electron mobility (m$_n$) at 300 K (in cm^2/Volt-sec)	1300	3800
9.	Hole mobility (m$_p$) at 300 K (in cm^2/Volt-sec)	500	1800
10.	Electron diffusion constant (D_n) at 300 K (in cm^2/sec)	34	99
11.	Hole diffusion constant (D_p) at 300 K (in cm^2/sec)	13	47

10.21. Intrinsic Concentration

In an intrinsic semiconductor, the concentration (n_i) increases with the increase in temperature. The dependence of intrinsic concentration on the temperature is given by the relation,

$$n_i^2 = A_0 \cdot T^3 \cdot e^{-E_{G0} \cdot k \cdot T}$$

where A_0 = A constant independent of the temperature,

T = Temperature (in K),

E_{G_0} = Energy gap at 0 K (in eV), and

k = Boltzmann's constant (in eV/K).

The increase in n_i^2 with the temperature has an effect on the charge densities in extrinsic semiconductor. For example in *N*-type semiconductor, the number of free electrons (n) do not change appreciably with the increase in temperature, but number of holes (p) increases. Similarly, for a *P*-type semiconductor, the number of free electrons (n) increases with the increase in temperature, but the number of holes (p) remains constant.

10.22. Mobility

The mobility (μ) of an intrinsic semiconductor varies as T^{-m} over a temperature range (T) of 100 to 400 K. For silicon, the value of $m = 2.4$ for electrons and 2.2 for holes. Similarly, for germanium, the value of $m = 1.7$ for electrons and 2.3 for holes. The mobility of an intrinsic semiconductor decreases with the increase in temperature because at higher temperatures, the number of carriers are more and they are more energetic also. This causes an increased number of collisions with the atoms and thus the mobility decreases.

10.23. Conductivity

The conductivity (σ) of an intrinsic semiconductor depends upon the number of hole-electron pairs and mobility. The number of hole-electron pairs increases with the increase in temperature, while its mobility decreases. However, the increase in hole-electron pairs is greater than the decrease in their mobilities. Therefore, the conductivity of an intrinsic semiconductor increases with the increase in temperature. The conductivity at any temperature (T) is given by the relation,

$$\sigma = \sigma_0 (1 + \alpha \cdot \Delta T)$$

α = Temperature coefficient, *i.e.*, the rate of which conductivity increases per degree Kelvin, and

ΔT = Rise in temperature = $(T - T_0)$

The conductivity of extrinsic semiconductors decreases with the increase in temperature. It is because of the fact, that with the increase in temperature, the number of majority carriers is nearly constant, but mobility decreases. This causes the conductivity to decrease.

Example 10.16.

An intrinsic semiconductor (silicon) has 5×10^{28} atoms/m³ at 20°C room temperature. At this temperature, there are 1.5×10^{16} electron-hole pairs. Find the conductivity of silicon (σ) at 20°C. If the above material is doped with indium atoms at the rate of 1 atom per 10^7 silicon atoms, find the conductivity of the doped material at room temperature. If the conductivity increases at the rate of 5% per °C then find the conductivity of silicon at 34°C also.

Note: *Hole mobility,* $\mu_p = 0.048$ *m²/Volt-sec*

Electron mobility, $\mu_n = 0.135$ *m²/Volt-sec*

Charge of an electron, $q = 1.602 \times 10^{-19}$ *coulombs*

Solution.

Given : $\mu_p = 0.048$ m²/Volt-sec; $\mu_n = 0.135$ m²/Volt-sec and $q = 1.602 \times 10^{-19}$ C.

(a) ***Conductivity of intrinsic silicon***

Given : $N_{si} = 5 \times 10^{28}/\text{m}^3$ and $n_i = 1.5 \times 10^{16}$

We know that conductivity of intrinsic silicon,

$$\sigma = q \cdot n_i (\mu_n + \mu_p) = (1.602 \times 10^{-19}) \times (1.5 \times 10^{16}) \times (0.135 + 0.048)$$

$$= 0.44 \times 10^{-3}\ (\Omega\text{-m})^{-1}\ \textbf{Ans.}$$

(b) Conductivity of doped silicon

We know that if intrinsic silicon is doped with indium atoms at the rate of 1 atom in 10^7 silicon atoms, then number of indium atoms,

$$N_A = \frac{N_{si}}{10^7} = \frac{5 \times 10^{28}}{10^7} = 5 \times 10^{21} /\text{m}^3$$

Number of holes, $p = N_A = 5 \times 10^{21}/\text{m}^3$

We also know that number of free electrons,

$$n = \frac{n_i^2}{p} = \frac{(1.5 \times 10^{16})^2}{5 \times 10^{21}} = 4.5 \times 10^{10} /\text{m}^3$$

Since the number of holes (equal to 5×10^{21}) in a doped silicon is much larger than the number of free electrons (equal to 4.5×10^{10}), therefore, the doped material is a P-type semiconductor. We know that conductivity of a P-type silicon,

$$\sigma = q \cdot p \cdot \mu_p = (1.602 \times 10^{-19}) \times (5 \times 10^{21}) \times 0.048\ (\Omega\text{-m})^{-1}$$

$$= 38.45\ (\Omega\text{-m})^{-1}\ \textbf{Ans.}$$

(c) Conductivity of silicon at 34°C

Given : $\alpha = 0.05/°\text{C}$; $\Delta T = 34 - 20 = 14°\text{C}$; and $\sigma_{20} = 0.44 \times 10^{-3}\ (\Omega\text{-m})^{-1}$

We also know that conductivity at 34°C,

$$\sigma_{34} = \sigma_{20}\,(1 + \alpha \cdot \Delta T) = (0.44 \times 10^{-3}) \times (1 + 0.05 \times 14)\ (\Omega\text{-m})^{-1}$$

$$= 0.75 \times 10^{-3}\ (\Omega\text{-m})^{-1}\ \textbf{Ans.}$$

10.24. Energy Gap

The energy required to break a covalent bond in a semiconductor is known as energy gap. It is equal to the difference of energy levels between the conduction band and valence band of the semiconductor crystal structure. It has been found that energy gap decreases with the increase in temperature. The energy gap at any temperature (T) is expressed by the relation,

$$E_G = E_{G_0} - \beta \cdot T \qquad \text{... } (i)$$

where E_{G_0} = Energy gap at 0 K. Its value for silicon is equal to 1.21 eV and for germanium 0.785 eV, and

β = A constant, whose value depends upon the nature of the material. Its value for silicon is equal to 3.60×10^{-4} and for germanium, 2.23×10^{-4}.

Substituting values of E_{G_0} (equal to 1.21) and β (equal to 3.60×10^{-4}) for silicon in equation (*i*),

$$E_G = 1.21 - 3.60 \times 10^{-4}\ T$$

∴ Energy gap for silicon at room temperature (*i.e.*, 300 K),

$$E_G = 1.21 - 3.60 \times 10^{-4} \times 300 = 1.1 \text{ eV}$$

Similarly, substituting the values of E_{G_0} (equal to 0.785) and β (equal to 2.23×10^{-4}) for germanium in equation (*i*),

$$E_G = 0.785 - 2.23 \times 10^{-4}\ T$$

∴ Energy gap for germanium at room temperature (*i.e.*, 300 K),

$$E_G = 0.785 - (2.23 \times 10^{-4} \times 300) = 0.72 \text{ eV}$$

10.25. Diffusion Current in a Semiconductor

We know that if some voltage is applied to a semiconductor bar, the holes move towards the negative terminal of the battery and electrons towards the positive terminal. This combined effect of the movement of holes and electrons constitute electric current, which is known as drift current.

It is possible for an electric current to flow in a semiconductor even in the absence of an applied voltage, provided a concentration gradient exists in the material. A concentration gradient exists, if the number of either electrons or holes is greater in one region of a semiconductor as compared to the rest of the region. It has been observed that when the concentration gradient of carriers (either electrons or holes) exists in a material, the carriers tend to move from the region of higher concentration to the region of lower concentration. This process is called diffusion and the electric current, produced due to this process, is known as diffusion current.

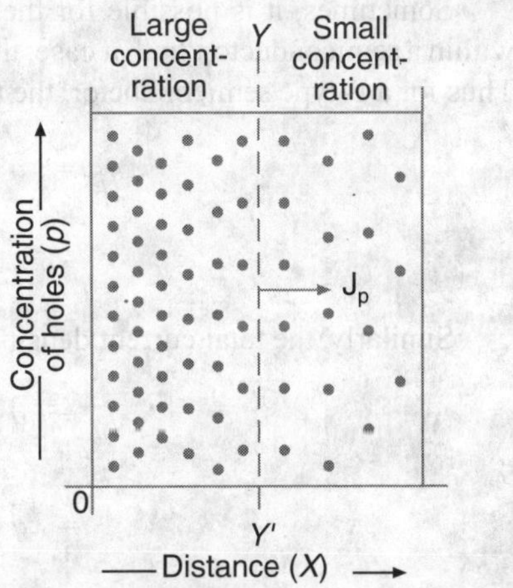

Fig. 10.12. Illustrating diffusion holes in a P-type semiconductor.

Consider a bar of P-type semiconductor having a non-uniform concentration of holes as shown in Fig. 10.12. Now consider an imaginary surface *Y-Y'* (shown by a dotted line in the figure). The density of holes on left side of *Y-Y'* is larger than that of the right side. This means that there exists a concentration gradient (*dp*/*dx*) whose value is given by the relation :

$$\frac{dp}{dx} = \frac{p_0 - p_x}{x}$$

where p_0 = Concentration of holes at left end of the semiconductor baa and may be referred to as an origin plane,

p_x = Concentration of holes at the imaginary surface *Y-Y'*, and

x = Distance of the imaginary surface from the origin plane.

The negative sign indicates that the hole concentration decreases with an increase in the value of *x*. Because of the concentration gradient (*dp*/*dx*), the holes from left of the *Y-Y'* surface diffuse towards the right side. This produces an electric current, called diffusion current. The magnitude of diffusion current per unit area (called diffusion density) is given by the relation,

$$J_p = -q D_p \cdot \frac{dp}{dx}$$

where q = Charge of a hole in coulombs,

D_p = Diffusion constant for holes in m²/sec, and

$\frac{dp}{dx}$ = Concentration gradient.

The negative sign indicates that concentration gradient is negative with the increasing value of *x*. Similarly, it can be shown that for an N-type semiconductor, the diffusion current density is given by the relation,

$$J_n = -q \cdot D_n \cdot \frac{dn}{dx}$$

where D_n is the diffusion constant for electrons in m^2/sec.

10.26. Total Current in a Semiconductor

Sometimes, it is possible for the electric field concentration gradient to exist simultaneously within a semiconductor. In that case, the total current is the sum of drift current and diffusion current. Thus for a P-type semiconductor, the total current per unit area (*i.e.*, total current density),

$$J_p = q \cdot \mu_p \cdot p \cdot E + \left(-q \cdot D_p \cdot \frac{dp}{dx}\right)$$

$$= q \cdot \mu_p \cdot p \cdot E - q \cdot D_p \cdot \frac{dp}{dx}$$

Similarly, the total current density for an N-type semiconductor,

$$J_n = q \cdot \mu_n \cdot n \cdot E + \left(-q \cdot D_n \cdot \frac{dn}{dx}\right)$$

$$= q \cdot \mu_n \cdot n \cdot E - q \cdot D_n \cdot \frac{dn}{dx}$$

10.27. Einstein's Relationship for a Semiconductor

According to the Einstein's relationship for a semiconductor, the ratio of diffusion constant (D_p or D_n) to the mobility (μ_p or μ_n) of the charge carriers is constant and is equal to the volt equivalent of the temperature (V_T). Mathematically, the Einstein relationship is expressed as :

$$\frac{D_p}{\mu_p} = \frac{D_n}{\mu_n} = V_T$$

where D_p = Diffusion constant for holes in m²/sec,

D_n = Diffusion constant for electrons in m²/sec,

μ_p = Hole mobility in m²/V-sec,

μ_n = Electron mobility in m²/V-sec, and

V_T = Volt equivalent of temperature whose value is equal to $T/11600$

where T is the temperature in K. The value of V_T is 26 mV at room temperature (*i.e.*, 300 K).

The importance of Einstein's relationship is that it can be used to determine the value of D_p (or D_n), if the mobility of holes (or electrons) is measured experimentally. The values of D_p and D_n for an intrinsic silicon and germanium are as given below :

D_p = 34 cm²/sec (= 0.0034 m²/sec)

and D_p = 13 cm²/sec (= 0.0013 m²/sec)

Similarly, for germanium the corresponding values are :

D_p = 47 cm²/sec (= 0.0047 m²/sec)

and D_n = 97 cm²/sec (= 0.0097 m³/sec)

Example 10.17.

Find the diffusion coefficients of holes and electrons for germanium at 300 K. The carrier mobilities in cm²/volt-sec at 300 K for electrons and holes are respectively 3600 and 1700. Density of carriers is 2.5 × 10²³. Boltzmann constant k = 1.38 × 10²³.

(*UPSC Engg Services 1998*)

Solution.

Given: μ_n = 3600 m²/V-sec, μ_p = 1700 m²/V-sec.

We know that the diffusion coefficient of holes,

$$D_p = \mu_p \cdot V_T$$
$$= 1700 \times \frac{T}{11600}$$
$$= 1700 \times \frac{300}{11600} = 4.36 \times 10^{-3}\ \text{m}^2/\text{s}\ \textbf{Ans.}$$

and the diffusion coefficient of electrons,

$$D_n = \mu_n V_T$$
$$= 3600 \times \frac{T}{11600}$$
$$= 3600 \times \frac{300}{11600} = 9.2 \times 10^{-3}\ \text{m}^2/\text{sec}\ \textbf{Ans.}$$

10.28. Effect of Temperature on Semiconductors

The product of the majority and minority carrier densities in an intrinsic or extrinsic semiconductor, is fixed at a constant temperature. The ratio of majority to minority carrier densities is in the rage of 10^8 to 10^{24} at 300 K. This indicates that in an extrinsic semiconductor, the contribution of minority carriers to the current flow is insignificant. However, we have also discussed in Art. 10.21, that the intrinsic carrier density (n_i) increases with an increase in temperature. Therefore, in an extrinsic semiconductor, both the majority and minority carrier densities increase with the increase in temperature. Since thermal generation produces electron-hole pairs, therefore, the ratio of majority to minority carrier densities reduces as the temperature increases. This ratio continues to decrease until a temperature called *transition-temperature* is reached. At this temperature, the density of thermally generated carriers is the same as the density of the carriers due to doping. Thus for N-type, $n_i = N_D$ and for P-type, $n_i = N_A$.

However, at and above the transition-temperature, the number of donor electrons in a N-type semiconductor is not equal to the number of donor atoms. Similarly, in a P-type semiconductor, the number of acceptor electrons is not equal to the number of acceptor atoms. This happens due to the significant proportion of the thermally generated carriers. Therefore, an extrinsic semiconductor loses its properties. It acquires properties those of a intrinsic semiconductor as the electron and hole densities are of the same order.

The generation of electron-hole pairs, at elevated temperature, is important in certain semiconductor devices, because it increases the current flow resulting from the rise in carrier densities. This causes the power dissipation of the device to increase. These in turn, increase the temperature of the semiconductor device due to which the carrier densities further increase. Beyond a certain limit, it results in the failure of the device. This effect is called *thermal run away*. It will be taken up again in the subsequent chapters.

10.29. The Hall Effect

It was discovered in 1874 by E.H. Hall, which effect is used to determine the nature of a semiconductor material (*i.e.*, whether it is P-type or N-type), mobile carrier concentration and its mobility. Hall discovered that if a specimen of semiconductor carrying a current is placed in a transverse magnetic field (B) then an electric field, is induced in a direction perpendicular to both I and B. This phenomenon is known as Hall effect. It is also observed in metals.

Consider a semiconductor (P-type or N-type) in which the current (I) and magnetic field (B) are applied in the directions as shown in Fig. 10.13. Because of this, a force (F) acts on the charge carriers pushing them towards the back as shown in the figure. The current (I) may be due to the free electrons moving in a direction opposite to that of the current or due to holes moving in the same direction as the current. Thus whether the carriers are electrons or holes, they will be pushed towards the back. This means that the force on the carriers is independent of the type of carriers. But it is dependent on the current.

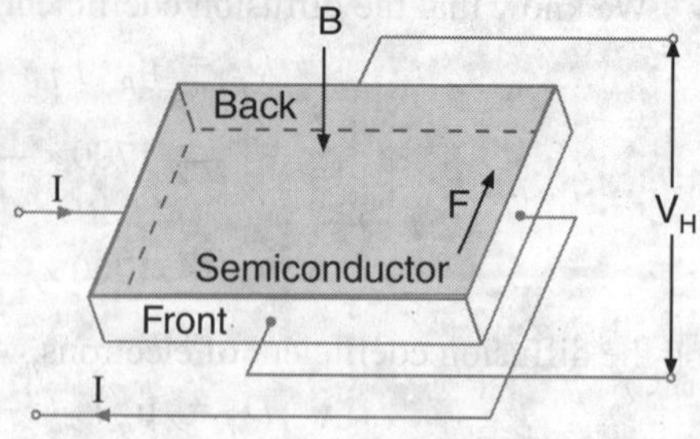

Fig. 10.13. Hall effect.

It will be interesting to know that when the mobile carriers (*i.e.*, electrons or holes) are pushed towards the back, the front becomes depleted and the semiconductor loses its neutrality. Now there is an excess of mobile carriers at the back and an excess of opposite charge due to the ionized impurity atoms at the front. If the semiconductor is N-type material, the electrons will be in excess at the back surface and this surface become negatively charged with respect to the front surface. This gives rise to a potential difference called the Hall Voltage (V_H) between the front and back surfaces as shown in the figure. On the other hand, if the semiconductor is P-type material, then holes will be in excess at the back and this surface becomes positively charged with respect to the front surface producing thereby a Hall Voltage (V_H). Thus for a P-type semiconductor, the polarity of V_H is in the reverse direction.

10.30. Experimental Determination of Carrier Concentration and Mobility

Consider an N-type semiconductor carrying a current (I) and placed in a transverse magnetic field (B) in a direction as shown in Fig. 10.14. Then due to the forces exerted on the electrons by the electric and magnetic field, a Hall Voltage (V_H) is induced across the semiconductor as shown in the figure.

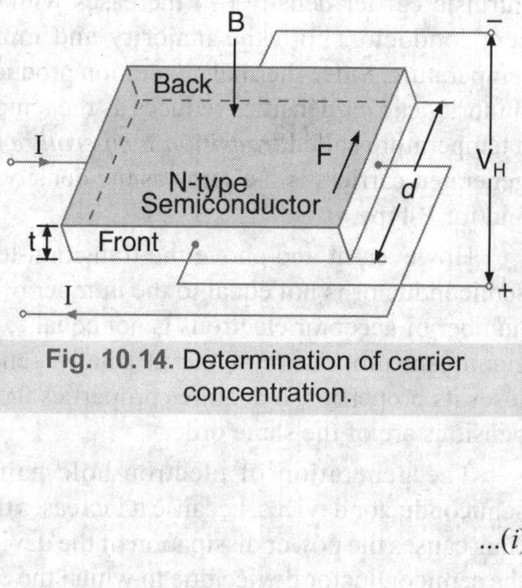

Fig. 10.14. Determination of carrier concentration.

In order to maintain a steady flow of current through the semiconductor (called equilibrium state) the force exerted on the electrons due to the electric field and magnetic field must balance each other, *i.e.*,

$$F_{(electric)} + F_{(magnetic)} = 0$$

$$n \cdot q \cdot E + B \cdot J = 0$$

$$E = -\frac{B \cdot J}{n \cdot q} \quad ...(i)$$

where J = Current density,

n = Electron concentration, and

q = Electronic charge.

It is customary to introduce a term called Hall coefficient in equation (*i*), which is defined as follows. The Hall coefficient,

$$R_H = -\frac{1}{n \cdot q}$$

Substituting the above value of R_H in equation (*i*),

$$E = B \cdot J \cdot R_H \quad ...(ii)$$

We know that the electric field,

$$E = V_H/d$$

where V_H = Hall voltage, and

$$d = \text{Distance between the front and back surface.}$$

We also know that current density is defined as the current per unit area,

$$J = \frac{I}{A} = \frac{I}{d \cdot t}$$

where t is the thickness of the semiconductor in the direction of the magnetic field. Now substituting the values of E and J in equation (*ii*),

$$\frac{V_H}{d} = B \times \frac{I}{d \cdot t} \times \frac{1}{n \cdot q}$$

$$\therefore \quad V_H = \frac{B \cdot I}{n \cdot q \cdot t}$$

or $$n = \frac{B \cdot I}{V_H \cdot q \cdot t} \quad \text{...(iii)}$$

It means that by measuring the values of B, I, V_H and t, the concentration of free electrons can be determined in an N-type semiconductor. The equation (*iii*) is also valid to determine free electrons in a metal. A similar procedure may be used to determine the concentration of holes (p) in a P-type semiconductor, which is given by the relation :

$$p = \frac{B \cdot I}{V_H \cdot q \cdot t}$$

The Hall coefficient for a P-type semiconductor is given by the relation :

$$R_H = +\frac{1}{p \cdot q}$$

The Hall coefficient for a P-type semiconductor is positive, whereas it is negative for a N-type semiconductor. It is true because the Hall Voltage in a P-type semiconductor is of opposite polarity to that in a N-type semiconductor. We know that electrical conductivity for an extrinsic semiconductor is given by the relation :

$$\sigma = n \cdot q \cdot \mu_n \quad \text{... (For N-type semiconductor)}$$

$$= p \cdot q \cdot \mu_p \quad \text{... (For P-type semiconductor)}$$

where μ_n = Electron mobility, and

μ_p = Hole mobility

Thus, if the conductivity of a semiconductor is also measured alongwith R_H, then mobility can be determined from the following relations :

$$\mu_n = \frac{\sigma}{n \cdot q} = \sigma \cdot R_H \quad \text{... (For N-type semiconductor)}$$

and $$\mu_p = \frac{\sigma}{p \cdot q} = \sigma \cdot R_H \quad \text{... (For P-type semiconductor)}$$

Example 10.18.

A sample of N-type semiconductor has a Hall coefficient of 160 cm³/coulomb. If its resistivity is 0.16 ohm-cm estimate the electron mobility in the sample.

Solution.

Given : $R_H = 160$ cm^3/coulomb; $\rho = 0.16$ ohm-cm.

We know that electron mobility,

$$\mu_n = \sigma \cdot R_H = \frac{1}{\rho} \cdot R_H = \frac{1}{0.16} \times 160 = 1000 \text{ cm}^2/\text{volt-sec} \textbf{ Ans.}$$

Example 10.19.

A current of 50 A is passed through a metal strip, which is subjected to a magnetic flux of density 1.2 Wb/m². The magnetic field is directed at right angles to the current direction and the thickness of the strip in the direction of the magnetic field is 0.5 mm. The Hall voltage is found to be 100 V. Calculate the number of conduction electrons per cubic metre in the metal.

Solution.

Given : $I = 50$ A; $B = 1.2$ Wb/m²; $t = 0.5$ mm $= 0.5 \times 10^{-3}$ m and $V_H = 100$ volts.

We know that number of conduction electrons,

$$n = \frac{B \cdot I}{V_H \cdot q \cdot t} = \frac{1.2 \times 50}{100 \times 1.6 \times 10^{-19} \times 0.5 \times 10^{-3}} = 7.51 \times 10^{21}/\text{m}^3 \textbf{ Ans.}$$

Example 10.20.

An N-type semiconductor has a resistivity of 20×10^{-2} ohm-m. The mobility of the electrons through a separate experiment was found to be 100×10^{-4} m² V⁻¹ s⁻¹. Find the number of electron carriers per m³.

Solution.

Given : $\rho = 20 \times 10^{-2}$ ohm-m; $\mu = 100 \times 10^{-4}$ m² V^{-1} s^{-1}.

Let n = Number of electron carriers per m³.

We know that mobility (μ)

$$100 \times 10^{-4} = \frac{\sigma}{n \cdot q} = \frac{1}{\rho \cdot n \cdot q} = \frac{1}{20 \times 10^{-2} \times n \times 1.6 \times 10^{-19}} = \frac{1}{32 \times 10^{-21} \times n}$$

$$n = \frac{1}{(100 \times 10^{-4}) \times (32 \times 10^{-21})} = 3.1 \times 10^{21}/\text{m}^3 \textbf{ Ans.}$$

Example 10.21.

The Hall-coefficient of a specimen of doped silicon is found to be 3.66×10^{-4} m³C; the resistivity of the specimen is 8.93×10^{-3} Ω-m. Find the mobility and density of charge carriers assuming single carriers conduction. (UPSC Engg. Services 2002)

Solution.

Given: $R_H = 3.66 \times 10^{-4}$ m³C ; $\rho = 8.93 \times 10^{-3}$ Ω-m

Let μ = mobility and

n = density of charge carriers.

We know that mobility of charge carriers,

$$\mu = \sigma R_H = \frac{1}{\rho} \cdot R_H = \frac{1}{8.93 \times 10^{-3}} \times 3.66 \times 10^{-4}$$

$$= 0.041 \text{ m}^2/\text{V-s}$$

and the Hall coefficient, (R_H),

$$3.66 \times 10^{-4} = \frac{1}{n \cdot q} = \frac{1}{n \times 1.6 \times 10^{-19}}$$

$$\therefore \quad n = \frac{1}{3.66\times10^{-4}\times1.6\times10^{-19}} = 1.7\times10^{22}/\text{m}^3 \quad \textbf{Ans.}$$

Example 10.22.

Resistivity of a sample semiconductor is 9 mW-m. Its holes have mobility of 0.03 m²/V-s. Calculate the value of Hall-coefficient of the sample. State assumptions made.

(UPSC Engg. Services 2002)

Solution.

Given: $\rho = 9\ \text{m}\Omega\text{-m} = 9\times10^{-3}\ \Omega\text{-m};\ \mu_p = 0.03\ \text{m}^2/\text{V-s}$.

We know that conductivity,

$$\sigma = \frac{1}{\rho} = \frac{1}{9\times10^{-3}} = 111.1\ \text{S/m}.$$

We also know that mobility (μp),

$$0.03 = \sigma\cdot R_H = 111.1\times R_H$$

$$\therefore \quad R_H = \frac{0.03}{111.1} = 2.7\times10^{-4}\ \text{m}^3\text{C} \quad \textbf{Ans.}$$

The assumption is that conductivity in a P-type semiconductor is due to majority carriers (holes) only. In other words the conductivity due to minority carriers (electrons) is neglected.

10.31. Hall Effect Applications

The Hall effect is used to determine whether the semiconductor is N-type or P-type. It is also used to measure semiconductor parameters like electron or hole concentration and mobility. The Hall effect is used in a magnetic field meter to sense the angular position of static magnetic fields. It is also used in an instrument called Hall effect multiplier, which gives the output proportional to the product of two input signals. Hall devices for such applications are made from N-type Indium antimonide (InSb), because it has a very high electron mobility as well as high R_H and high sensitivity.

NUMERICAL PROBLEMS

1. Determine the conductivity and resistivity of an intrinsic sample of silicon at normal room temperature (*i.e.*, 300 K). Assume electron mobility, $\mu_n = 1350\ \text{cm}^2/(\text{V-sec})$; hole mobility, $\mu_p = 480\ \text{cm}^2/(\text{V/sec})$, intrinsic conc entration at 300 K = 1.52×10^{10} and charge of an electron or hole = 1.6×10^{-19} coulomb. **(Ans. 4.45×10^{-6} ; 2.25×10^{5})**
2. Density of free electrons in pure or intrinsic germanium at 300 K is about 2.4×10^{19} electron per m³. Determine the resistance of a bar of germanium 1 mm × 2 mm and 10 mm long. Given $\mu_n = 0.39$; $\mu_p = 0.19$ and $q = 1.6\times10^{-19}$ coulomb. **(Ans. 2250 Ω)**
3. A sample of germanium is doped to the extent of 10^{14} donor atoms/cm³ and 7×10^{13} acceptor atoms/cm³. At the temperature of 300 K, the resistivity of pure germanium is 60 Ω-cm. If the applied electric field is 2 V/cm, find the total conduction current density.
4. Given that pure silicon has 4.99×10^{-22} atoms/cm³ and the intrinsic concentration $n_i = 1.5\times10^{10}/\text{cm}^3$ at 300 K. Calculate its resistivity and number of holes and free electrons. If there is a donor atom per 10^{12} atoms of silicon, calculate the number of holes and electrons and resistivity of the sample. Other constants are: $q = 1.6\times10^{-19}$ coulomb; $\mu_n = 1300\ \text{cm}^2/\text{V-sec}$ and $\mu_p = 500\ \text{cm}^2/\text{V sec}$).
(Ans. $\rho_1 = 2.3\times10^{5}\ \Omega\text{-cm}$; $n_i = n = p = 1.5\times10^{10}/\text{cm}^3$; $\rho = 9.3\times10^{4}$; $n = 4.99\times10^{10}/\text{cm}^3$ and $p = 0.45\times10^{10}/\text{cm}^3$)

5. Silicon has a conductivity of 5×10^{-4} ohm^{-1} m^{-1} when pure. How many aluminum atom/m^3 are required so that a saturation conductivity of 200 ohm^{-1} m^{-1} could be achieved in silicon using aluminum as an impurity? Given mobility of holes in silicon is 0.05 m^2 v^{-1} s^{-1} and of electrons is 0.13 m^2 v^{-1} s^{-1}. **(Ans. 2.5 × 10^{22} m^3)**
6. A current of 10 A is passed through a metal strip which is subjected to a magnetic flux density of 1 Wb/m^2. The magnetic field is directed at right angles to the current direction and the thickness of the strip in the direction of the magnetic field is 1 mm. The Hall voltage is found to be 10 V. Calculate the number of conduction electrons per cubic metre in the metal. **(Ans. 6.25 × 10^{15})**
7. The Hall effect is used to determine the mobility of holes in a P-type silicon bar. Assume the bar resistivity as 200 000 Ω-cm, the magnetic field β = 0.1 Wb/m^2 and $d = t = 3$ mm. The measured value of current and Hall voltages are 10 mA and 50 mV respectively. Find the hole mobility (μ_p). **(Ans. 7.5 m/sec)**

DESCRIPTIVE QUESTIONS

1. What is meant by intrinsic semiconductor ?
2. (*a*) Explain what a hole is ? How do they move in intrinsic semiconductor ?
   (*b*) What is the role of doping of impurities in pure silicon or germanium ?
3. Name three acceptor and three donor materials for doping of a semiconductor.
4. Draw roughly, to scale, the energy band diagrams for a metal, an N-type semiconductor and an insulator at room temperature.
5. Explain the diffusion and drift current for a semiconductor.
6. Give a simple diagram of intrinsic silicon crystal structure and explain briefly the generation and movements of electrons and holes. How does doping modify the conductivity of semiconductors? What are P-type and N-type of semiconductors ?
7. Discuss the meaning of the term covalent bond and its significance in relation to the conduction of electricity in intrinsic P-type and N-type semiconductors. Draw the energy level diagrams in each case. Develop an expression for the resistivity of an intrinsic semiconductor from the concept of free electron and mobility of charge carriers ?
8. Describe the Hall effect and explain its relation to mechanical force exerted by magnetic field on a conduction.
9. Explain N- type and P- type semiconductor. (*JNTU., June 2009*)
10. Differentiate between N-type and P-type semiconductor materials. (*GBTU/MTU., 2006-07*)
11. Define and explain P and N- type materials. What is donor impurity. (*M.G University., April/May 2003*)
12. Differentiate between Donor and Acceptor impurities. (*GBTU/MTU., 2007, 2004-05*)
13. Draw the structure of P-type material and explain the effect of acceptor impurities on structure (*GBTU/MTU., 2007*)
14. Draw the intrinsic structure of P-type material and explain the effect of accepter impurities on the energy band structure. (*GBTU/MNTU., 2004-05*)
15. Sketch the energy band picture for an intrinsic N-type and P-type semiconductor. Indicate the position of the Fermi, the donor and the accepter levels. (*GBTU/MNTU., 2002-03*)
16. Explain what you mean by donor and acceptor impurities. (*GBTU/MNTU., 2004-05*)
17. Explain the Fermi level with reference to the intrinsic and extrinsic semiconductors. (*GBTU/MTU., 2008-09*)

18. Define a hole in a semiconductor. Indicate pictorially how a hole contributes to conduction. *(GBTU/MTU., 2008-09)*
19. What is resistivity? How does resistivity of a semiconductor related with carrier mobility? Explain *(GBTU/MTU., 2008-09)*
20. How the carrier motilities related with resistivity of a semiconductor? Does a hole in a semiconductor contribute to a flow current? If yes how and if no, how? *(GBTU/MTU., 2006)*
21. Give the equation for conductivity in an extrinsic semiconductor. Discuss the effect of temperature on the energy gap and mobility of a semiconductor. *(GBTU/MTU., 2009)*
22. Write four properties of semiconductors. *(M.G University., Dec. 2007)*
23. State mass action law for extrinsic semiconductor. Give the expression for minority carrier concentrations in extrinsic semiconductors of P and N-type. *(GBTU/MTU., 2009)*
24. With neat schematic diagrams, explain the P and N-type's semiconductors. *(M.G University., Dec. 2007)*
25. Define an intrinsic material, a negative temperature coefficients and covalent bonding list three materials that have negative temperature coefficients. *(GBTU/MTU., 2005-06)*
26. How does the resistance of metals and semi conductors vary with temperature. *(PTU., May 2007)*
27. Explain thermal generation of carriers in intrinsic semiconductors. *(Anna University., April/May 2008)*
28. What is meant by N-type impurities in a semiconductor? *(Anna University., April/May 2008)*
29. Draw and explain the energy band diagrams of metal semiconductor junction. *(Anna University., April/May 2008)*
30. How do you increase the conductivity of intrinsic semiconductor? *(Anna University., Nov/Dec 2007)*
31. Define Mass Action Law. *(JNTU., June 2009)*
32. Distinguish between intrinsic and extrinsic semiconductors. What properties of a semiconductor are determined from Hall Effect experiment? Explain briefly. *(GBTU/MNTU., 2005-06)*
33. State the Einstein relationship for semiconductor. *(JNTU., June 2009)*
34. What do you mean by "diffusivity" and "mobility" of holes ? *(PBUT., 2008)*
35. Distinguish between majority and minority carriers. Explain the formation of a depletion layer in a PN – junction *(PBUT., Dec. 2004)*
36. Define mobility and conductivity of an intrinsic semiconductor *(PBUT., 2008)*
37. How current flow in a semiconductor is different from that in a metal? Explain. *(PBUT., Dec 2005)*
38. Distinguish between intrinsic and extrinsic semiconductors. What properties of a semiconductor are determined from Hall effect experiment? Explain briefly. *(GBTU/MTU., 2005-06)*
39. What is Hall effect? *(Anna University., Nov/Dec 2007)*
40. Explain the Hall effect and derive the expression for the hall coefficient. *(GBTU/MTU., 2008-09)*
41. State and explain Hall effect. Give the application of Hall Effect. *(Anna University., April/May 2008, GBTU/MTU 2003-04)*

42. Discuss different type of semiconductor materials with the help of energy band diagram. *(R.G.P.V., June 2009)*
43. Differentiate between intrinsic semiconductor and extrinsic semiconductor. *(R.G.P.V., June 2011)*
44. Differentiate between diffusion current and drift current. *(R.G.P.V., June 2011)*
45. Explain Hall effect and its applications. *(R.G.P.V., June 2011, June 2010, Feb 2010, June 2009)*
46. Explain the following terms and give their units:
    (*a*) Drift Velocity
    (*b*) Diffusion
    (*c*) Mobility
    (*d*) Conductivity *(Nagpur University, Summer 2003)*
47. Explain law of mass action. Give in brief temperature dependence of intrinsic concentration. *(Nagpur University, Summer 2003)*
48. Describe the mechanism of current flow in semiconductors. *(Nagpur University, Summer 2000)*
49. Prove that the minority carrier concentration, in a N-type semiconductor bar which was momentarily illuminated, decrease exponentially with time. *(GTU., May 2011)*
50. Describe the Hall effect and also explain how it is helpful in determining the different properties of semiconducting material. *(GTU., Dec. 2011)*
51. Explain the generation of holes and electrons in an intrinsic semiconductor. *(GTU., Dec. 2011)*
52. What is a semiconductor? Which are the commonly used semiconductors? How are N-type and P-type semiconductor obtained? *(RGTU.,Feb.2010)*

MULTIPLE CHOICE QUESTIONS

1. The bonding forces in compound semiconductors such as GaAs arise from
   (*a*) ionic bonding (*b*) metallic bonding
   (*c*) covalent bonding (*d*) combination of ionic and covalent bonding
   (UPSC Engg. Services 2002)
2. For silicon, the energy gap at 300 K is
   (*a*) 1.1 W (*b*) 1.1 J (*c*) 1.1 eV (*d*) all of these
3. The forbidden gap for germanium is,
   (*a*) 0.12 eV (*b*) 0.72 eV (*c*) 7.2 eV (*d*) none of these
4. The most commonly used semiconducting material in manufacturing of electronic devices is
   (*a*) germanium (*b*) silicon
   (*c*) gallium arsenide (*d*) indium antimonide
5. The electric charge on a hole is
   (*a*) 1.6×10^{-19} C (*b*) 1.6×10^{-8} C (*c*) 1.6×10^{8} C (*d*) 1.6 C
6. Electron-hole pairs are produced by
   (*a*) recombination (*b*) thermal energy (*c*) ionization (*d*) doping

7. Recombination takes place when
   (*a*) an electron falls into a hole
   (*b*) a positive and a negative ion bond together
   (*c*) a valence electron becomes a conduction
   (*d*) a crystal is formed
8. The process of adding impurities to a pure semiconductor is called
   (*a*) mixing (*b*) doping (*c*) diffusing (*d*) refining
9. An intrinsic semiconductor at the absolute zero temperature
   (*a*) behaves like an insulator
   (*b*) has a large number of holes
   (*c*) has few holes and same number of electrons
   (*d*) behaves like a metallic conductor
10. With a rise in temperature around 300 K, the conductivity of an intrinsic semiconductor,
   (*a*) remains constant (*b*) increases
   (*c*) decreases (*d*) becomes zero
11. The donor type impurities
   (*a*) create excess hole
   (*b*) can be added to germanium but not to silicon
   (*c*) must have only thre valence electrons
   (*d*) must have only five valence electrons
12. The pentavalent impurities like antimony and phosphorus, added to intrinsic semiconductors are called
   (*a*) acceptor or P-type impurities (*b*) donor or P-type impurities
   (*c*) acceptor or N-type impurities (*d*) donor or N-type impurities
13. Impurities like boron, gallium or indium are added to intrinsic semicondutor to form
   (*a*) N-type doped semiconductor (*b*) P-type doped semiconductor
   (*c*) A junction diode (*d*) all of these
14. In a N-type semiconductor, the position of Fermi-level
   (*a*) is lower than the centre of energy gap(*b*) is at the centre of energy gap
   (*c*)is higher than the centre of energy gap (*d*) all of these
15. Which of the following doping will produce a P-type semiconductor,
   (*a*) germanium with phosphorus (*b*) silicon with phosphorus
   (*c*) germanium with antimony (*d*) silicon with indium
16. The mobility of electrons in a material is expressed in units of:
   (*a*) V/s (*b*) m^2/V-s (*c*) m^2/s (*d*) J/K
17. If for intrinsic silicon at 27°C, the charge concentration and monbilities of free electrons and holes are 1.5×10^{16} per m^3, 0.13 m^2/(Vs) and 0.05 m^2 (Vs) respectively the conductivity will be :
   (*a*) 2.4×10^{-3} $(\Omega\text{-m})^{-1}$ (*b*) 3.15×10^{-3} $(\Omega\text{-m})^{-1}$
   (*c*) 5×10^{-4} $(\Omega\text{-m})^{-1}$ (*d*) 4.32×10^{-4} $(\Omega\text{-m})^{-1}$
18. In a metal
   (*a*) the electrical conduction is by electrons and holes
   (*b*) the conductivity decreases with the rise in temperature
   (*c*) the conduction band is empty
   (*d*) none of the above

19. The conductivity of a conductor may be increased by

(*a*) increasing its temperature (*b*) decreasing its temperature
(*c*) incrasing its vibration (*d*) changing its shape

20. The energy gap in a semiconductor

(*a*) increases with temperature (*b*) does not change with temperature
(*c*) decreases with temperature (*d*) is zero

21. The Hall-effect occurs only in

(*a*) metals (*b*) N-type semiconductors
(*c*) intrinsic semiconductor (*d*) all of the above

22. If an intrinsic semiconductor is doped with a very small amount of boron, then in the extrinsic semiconductor so formed, the number of electrons and holes will,

(*a*) decrease (*b*) increase and decrease respectively
(*c*) increase (*d*) decrease and increase respectively

23. Two initially identical samples A and B of pure germanium are doped with donors to concentrations of 1×10^{20} and 3×10^{20} respectively. If the hole concentration in A is 9×10^{12}, then the hole concentration in B at the same temperature will be

(*a*) 3×10^{12} m^{-3} (*b*) 7×10^{12} m^{-3} (*c*) 11×10^{12} m^{-3} (*d*) 27×10^{12} m^{-3}

24. Half effect is observed in a specimen when it (metal or a semiconductor) is carrying current and is placed in a magnetic field. The resultant electric field inside the specimen will be in

(*a*) a direction normal to both current and magnetic field
(*b*) the direction of current
(*c*) a direction antiparallel to the magnetic field
(*d*) an arbitrary direction depending upon the conductivity of the specimen.

25. The unit of Hall-coefficient is

(*a*) Vm^3 A^{-1} Wb^{-1} (*b*) Vm^2 A Wb^{-1} (*c*) Vm^3 A Wb^{-1} (*d*) Vm^2 A^{-2} Wb

(*Chatrapati Shahuji Maharaj Univ. 2004*)

ANSWERS

1. (*a*)	2. (*c*)	3. (*b*)	4. (*b*)	5. (*a*)
6. (*b*)	7. (*a*)	8. (*b*)	9. (*a*)	10. (*b*)
11. (*a*)	12. (*d*)	13. (*b*)	14. (*c*)	15. (*d*)
16. (*b*)	17. (*d*)	18. (*b*)	19. (*b*)	20. (*c*)
21. (*b*)	22. (*d*)	23. (*a*)	24. (*a*)	25. (*a*)

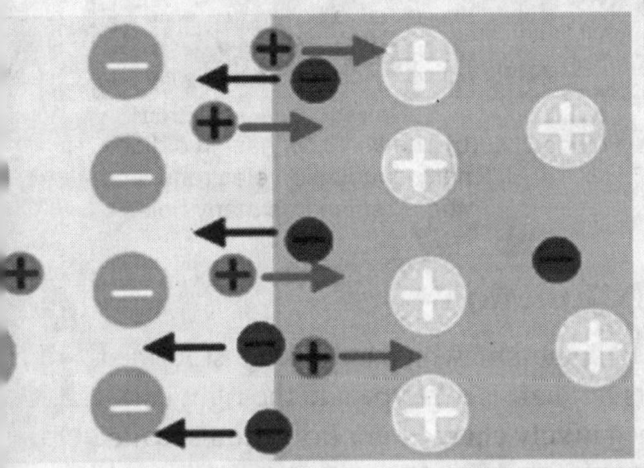

11 Chapter

PN Junction

Outline

11.1. Introduction

We have already discussed about the P-type and N-type semiconductors in the last chapter. As a matter of fact, the P-type and N-type semiconductors, taken separately, are of little use in actual practice. If we join a piece of P-type semiconductor to a piece of N-type semiconductor such that the crystal structure remains continuous at the boundary as shown in Fig. 11.1, a PN junction is formed. Such a PN junction forms a very useful device and is called a semiconductor diode, PN junction diode or simply a crystal diode.

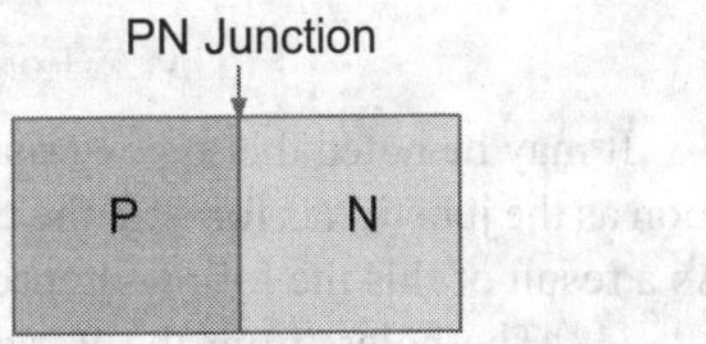

Fig. 11.1. PN junction

It will be interesting to know that a PN junction cannot be formed by simply joining or welding the two pieces together, because it would produce a discontinuous crystal structure. Special fabrication techniques are used to prepare PN junctions.

The PN junction, itself, is an important device. Besides this, all semiconductor devices contain at least one PN junction. Therefore, it is very important to understand the behaviors of a PN junction, when connected in an electric circuit.

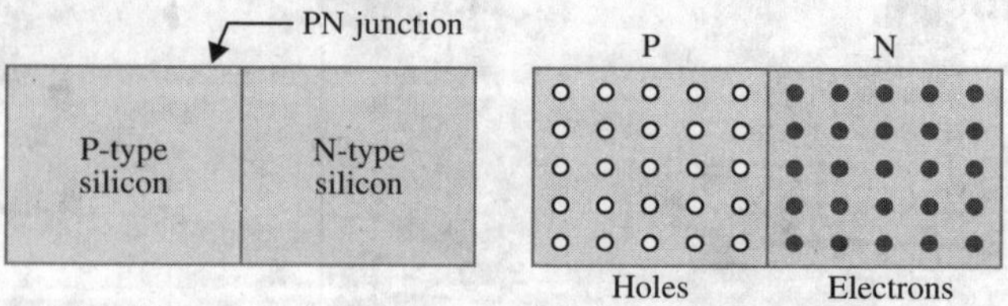

(a) P-N Junction in a diode. (b) The N region has, in a diode many conduction electrons, and the P region has many holes.

11.2. Formation of Depletion Layer in a PN Junction

Figure 11.2 (*a*) shows a semiconductor consisting of a PN junction, which has been just formed. It may be noted that this entire sample is a single crystal. Its left half is a P-type and the right half is N-type. The P-region has holes (as majority carriers) and negatively charged impurity atoms, called negative ions (or acceptor ions). The N-region has free electrons (as majority carriers) and positively charged impurity atoms, called positive ions (or donor ions). The minority carriers in the P-and N-regions are not shown just for simplicity.

We know that holes and electrons are mobile charges, and therefore, are known as moble charge carriers. On the other hand, positive and negative ions are immobile charges and therefore, do not take part in the conduction. The sample, as a whole, is electrically neutral and so are the P- and *N*-regions, considered separately. Thus in the *P*-region, the total positive charge on the holes is equal to the total negative charges on free electrons and immobile ions. Similarly, in the N-region, the total negative charge on free electrons is equal to the total positive charge on holes and immobile ions.

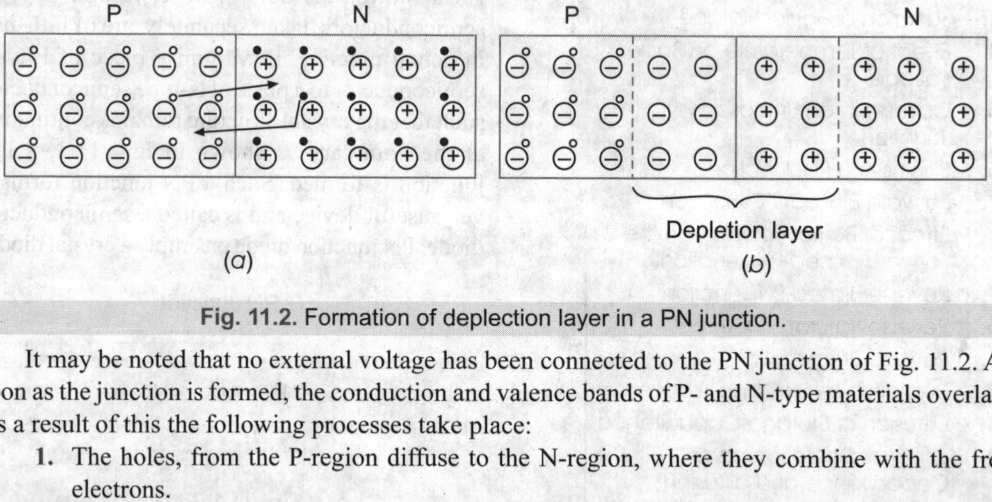

Fig. 11.2. Formation of deplection layer in a PN junction.

It may be noted that no external voltage has been connected to the PN junction of Fig. 11.2. As soon as the junction is formed, the conduction and valence bands of P- and N-type materials overlap. As a result of this the following processes take place:

1. The holes, from the P-region diffuse to the N-region, where they combine with the free electrons.
2. The free electrons, from the N-region diffuse to the P-region, where they combine with holes.
3. The diffusion of holes (from P-region to N-region) and free electrons (from N-region to P-region) takes place due to the reason that there is a difference of concentrations in the two regions *i.e.,* the P-region has more number of holes, whereas N-region has more number of free electrons. This difference in concentration creates a concentration gradient across the junction. It results in the diffusion of mobile charge carries across the junction. Moreover, the electrons and holes move at random in all directions due to thermal energy. In doing so, some carriers manage to crossover the junction.

4. The diffusion of holes and free electrons across the junction takes place for a short time. After a few recombination of holes and free electrons, in the vicinity of the junction, a restraining force is automatically set up. This force is produced due to depletion region, which exists on either side of the junction. As a result of this, further diffusion of holes and free electrons from one region to the other is stopped by this depletion layer. The formation of this depletion is explained below:

We know that as soon as the PN junction is formed, some of the holes in P-region and the free electrons in the N-region diffuse in each other and disappear due to recombination. In this process, the negative acceptor ions in the P-region and positive donor ions in the N-region are left uncovered (or uncompensated) in the vicinity of junction as shown in Fig. 11.2 (*b*). The additional holes, trying to diffuse to the N-region, are repelled by the uncovered positive charge of the donor ions. Similarly, the electrons, trying to diffuse into the P-region, are repelled by the uncovered negative charge of the acceptor ions. As a result of this, the further diffusion of free electrons and holes across the junction is stopped.

The region containing the uncovered accepter and donor ions, in the vicinity of the junction, is called depletion region. Since this region has immobile (or fixed) ions, which are electrically charged, therefore, the depletion region is also known as space-charge region. Moreover, as the uncovered charges within the depletion region exists in the form of parallel rows or plates of opposite charges, therefore, it is known as depletion layer. The depletion layer behaves like an insulator. Because of the presence of rows of fixed charges, the depletion layer possesses capacitance. It will be discussed in detail in Art. 11.12.

It will be interesting to know that the width of depletion layer depends upon the doping level of the impurity in *N*-type or *P*-type semiconductor. The higher the doping level, the thinner will be the depletion layer and vice-versa. It is due to the fact that a highly doped PN junction contains a large number of electrons and holes. Because of this, a diffusing charge carrier (either free electron or hole) has not to travel across the junction for recombination.

11.3. Junction or Barrier Voltage

We have already discussed in the last article that the depletion layer of a PN junction has no mobile charge curriers. But it contains fixed rows of oppositely charged ions on its two sides. Because of this charge separation, an electric potential (designated as V_B) is established across the junction, even when the junction is not connected to any external voltage source as shown in Fig. 11.3. This electric potential is called *junction* or *potential barrier.*

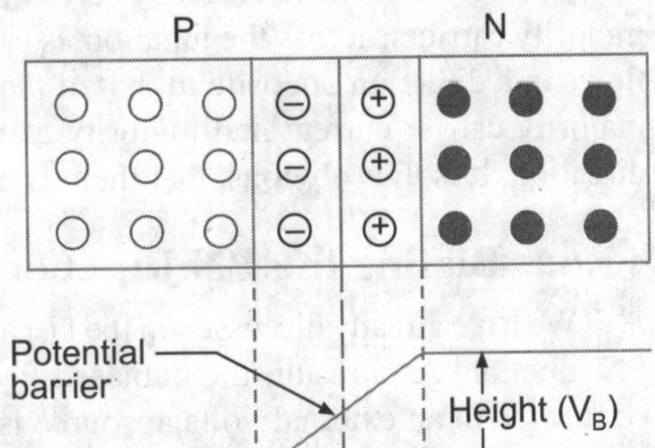

Fig. 11.3. Junction or barrier voltage.

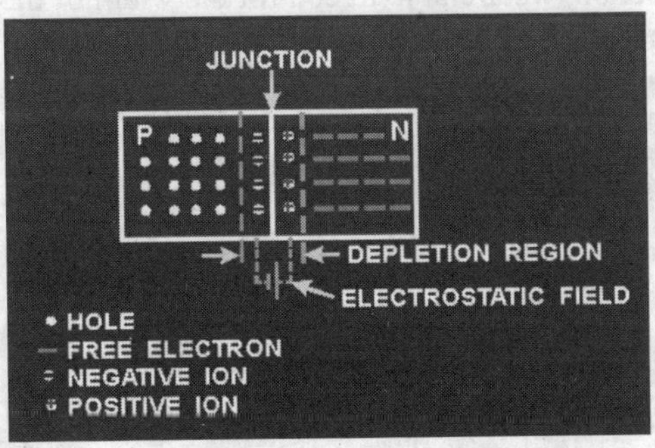

It will be interesting to know that this potential barrier exerts a repelling force on the mobile charge carriers, trying to crossover the junction. This force stops the mobile charge carriers to crossover the junction, unless the energy is supplied from an external source. At room temperature, (*i.e.*, at 300 K), the value of V_B

$$V_B = 0.6 \text{ V for silicon and } 0.2 \text{ V for germanium.}$$

11.4. Effect of Temperature on Barrier Voltage

The barrier voltage of a PN junction depends upon three factors namely density, electronic charge and temperature. For a given PN junction, the first two factors are constant, thus making the value of V_B dependent only on temperature. It has been observed that for both germanium and silicon the value of V_B decrease by 2 mV/°C. Mathematically, the decreases in barrier voltage,

$$\Delta V_B = -\ 0.002 \times \Delta t$$

where D*t* is the increase in temperature in °C.

11.5. Currents in an Unbiased PN Junction

A PN junction, across which *no external voltage* source is connected, is known as *unbiased* PN *junction.* Now, consider such an unbiased PN junction as shown in Fig. 11.3. We know that immediately after the formation of a PN junction, a depletion layer is formed in the vicinity of junction. Because of this, there exists a barrier potential (V_B), which stops the further diffusion (or crossover) of carriers across the junction.

The significance of V_B is that, a hole in the P-region requires an energy (equal to $q \cdot V_B$) in order to crossover the junction. Similarly an electron in the N-region requires the same amount of energy (equal to $q \cdot V_B$) in order to crossover the junction. If the junction is at room temperature, thermal energy is added continuously. As a result of this, few holes and electrons will acquire enough energy to get over the potential barrier, and diffuse the junction. Since the diffusion of electrons and holes is *opposite* in direction, therefore, there is a single current across the junction. This component of current is called a current due to majority carriers or simply *majority carrier current.*

There is another component of current, which flows through the junction. This current is due to the diffusion of minority carriers across the junction. We know that thermal energy causes electron-hole pairs to be generated within the semiconductor materials. Such electron-hole pairs are assisted in diffusing across the junction by barrier potential (V_B). The current produced due to the diffusion of minority carriers, across the junction, is called *minority carrier current.* The minority carrier current flows in a direction opposite to that of the majority carrier current. In an unbiased PN junction, the majority carrier current and minority carrier current are equal in magnitude and flow in opposite direction. It is thus obvious, that there is no net flow of current across the junction.

11.6. Biasing the PN Junction

We have already discussed in the last article that there is no net flow of current across an unbiased PN junction. As a result, the unbiased PN junction is of no use in actual practice. A PN junction, connected to an external voltage source is called a *biased* PN *junction.* Such a PN junction finds a variety of applications in semiconductor devices.

By applying an external voltage across a PN junction, we are able to control the width of the depletion layer. This allows us to control the resistance of the PN junction and also the amount of current that can pass through the device.

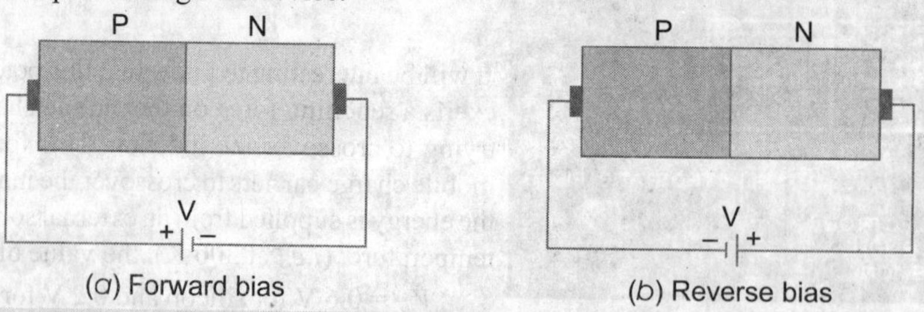

Fig. 11.4. Biasing the PN junction.

There are two ways of connecting voltage source to a PN junction as discussed below:

1. ***Forward bias.*** In this case, positive terminal of the voltage source is connected to the P-side and negative terminal to the N-side as shown in Fig. 11.4 (*a*). A large amount of current flows through the junction under this condition.
2. ***Reverse bias.*** In this case, positive terminal of the voltage source is connected to the N-side and negative terminal to the P-side as shown in Fig. 11.4 (*b*). Practically, no current flows through the junction under this condition.

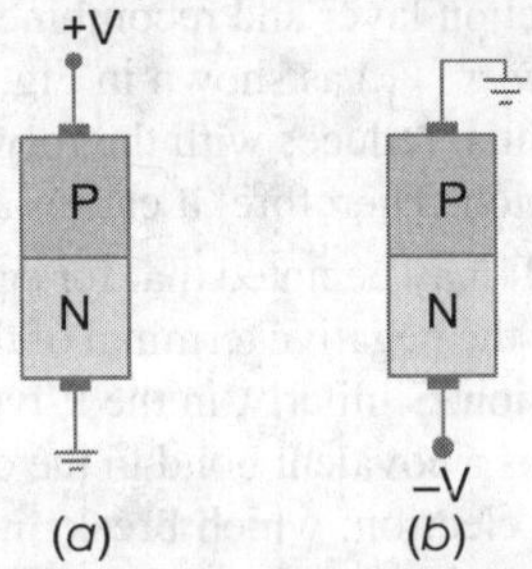

Fig. 11.5. Forward biasing a PN junction.

The application of forward bias across a PN junction shown in Fig. 11.4 (*a*) can also be shown as in Fig. 11.5 (*a*). Another way of applying the forward bias is as shown in Fig. 11.5 (*b*).

Similarly, the application of reverse bias across a PN junction shown in Fig. 11.4 (*b*) can also be shown as in Fig. 11.6 (*a*). Another way of applying the reverse bias is as shown in Fig. 11.6 (*b*).

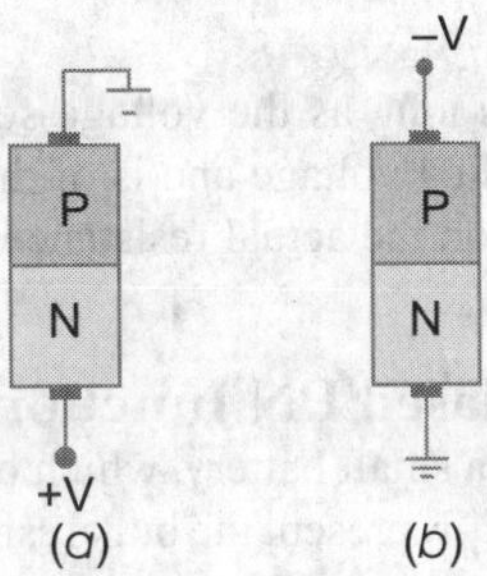

Fig. 11.6. Reverse biasing a PN junction.

Now we shall discuss the forward and reverse-biased PN junctions in greater details in the following pages.

11.7. Forward Biased PN Junction

We have already discussed in the last article that if we connect voltage source to the PN junction, such that the positive terminal is connected to the P-region and negative terminal to the N-region, the PN junction is said to be forward biased as shown in Fig. 11.7.

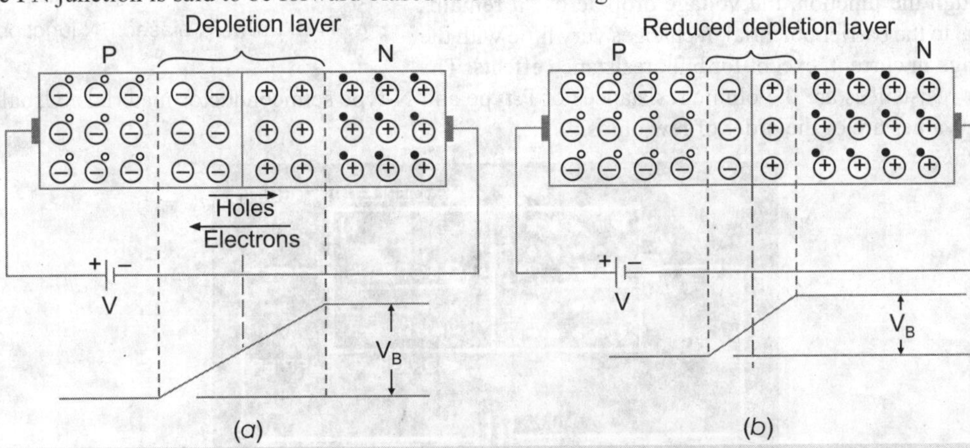

Fig. 11.7. PN junction with forward bias.

When a PN junction is forward biased as shown in Fig. 11.7 (*a*) the holes are repelled by the positive terminal of the voltage source and are forced to move towards the junction. Similarly, the electrons are repelled by the negative terminal of the voltage source and move towards the junction. Because of their acquired energy (from the voltage source), some of the holes and electrons enter the depletion layer and recombine themselves. *This reduces the width as well as height of the potential barriers* (V_B) as shown in Fig. 11.7 (*b*). In other words, the width of depletion layer and the barrier potential reduces with the forward bias. As a result of this, more majority carriers diffuse across the junction. Therefore, it causes a large current to flow through the PN junction.

It may be noted that for each recombination of free electron and hole, which occurs, an electron from the negative terminal of the voltage source enters the N-type region. Then it moves towards the junction. Similarly, in the P-type region near the positive terminal of the voltage source, an electron breaks a covalent bond in the crystal and enters the positive terminal of the voltage source. Thus for each electron, which breaks its bond, a hole is created. This hole drifts towards the junction. The current through the external circuit, is due to the movement of electrons only. On the other hand, the current within the PN junction is the sum of electron current (in the N-region) and hole current (in the P-region).

The current in the external circuit continues to flow as flow as long as the voltage source is present in the circuit. The current increases with the increase in applied voltage and is of the order of several milliamperes. The maximum value of current depends upon the actual resistance, called bulk resistance of the semiconductor material.

11.8. Effect of Barrier Potential on the forward Biased PN Junction

The barrier potential of the depletion layer can be considered as a small battery, which opposes the external d.c. voltage as shown in Fig. 11.8. The resistance *Rp* and R_n represent the bulk resistances of the P-type and N-type semiconductors.

The PN junction does not permit the current to flow, until the external bias voltage overcomes the barrier potential, (*i.e.*, $V \ngtr V_B$). For example, the *silicon* PN junction, does not conduct as long as the external applied voltage is below 0.7 V. Similarly, the germanium PN junction does not conduct as long as the external voltage is below 0.3 V.

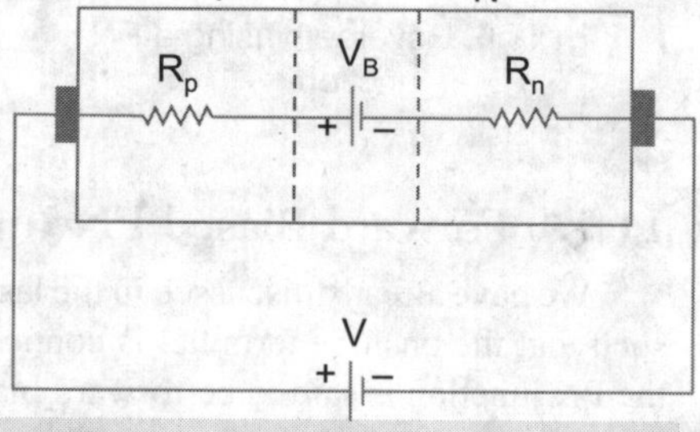

Fig. 11.8. Forward biased PN junction.

It may be noted, that once the current starts flowing through the junction, the voltage drop across it remains equal to the barrier potential. It changes very little with the change in current, except for bulk resistance effects. The bulk resistances are the ohmic resistances of P-type and N-type semiconductor materials. Usually, their values are of the order of few ohms.

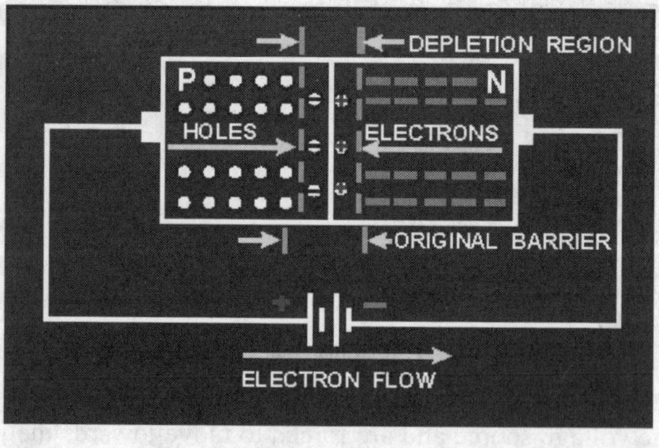

11.9. Reverse Biased *PN* Junction

We have already discussed in Art. 11.6 that if we connect a voltage source to a PN junction, such that positive terminal of the voltage source is connected to the N-region and negative to the P-region, then the PN junction is said to be reverse biased. Fig. 11.9 shows a reverse biased PN junction.

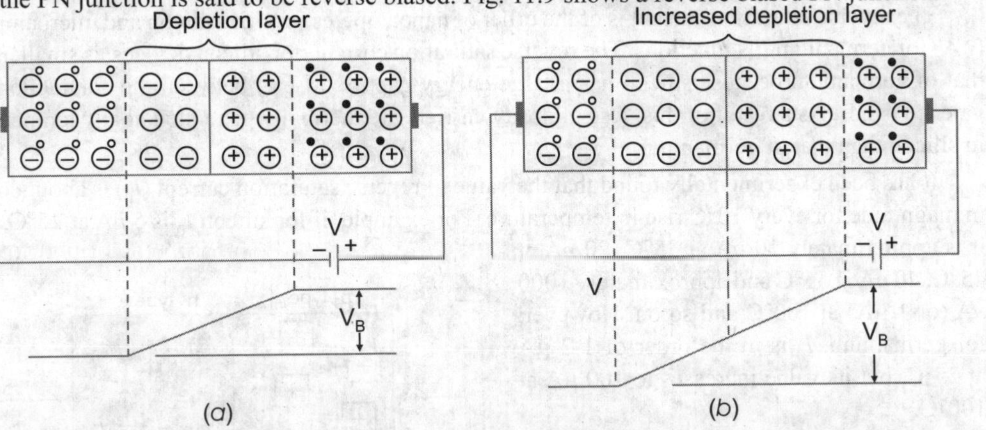

Fig. 11.9. PN junction with reverse bias.

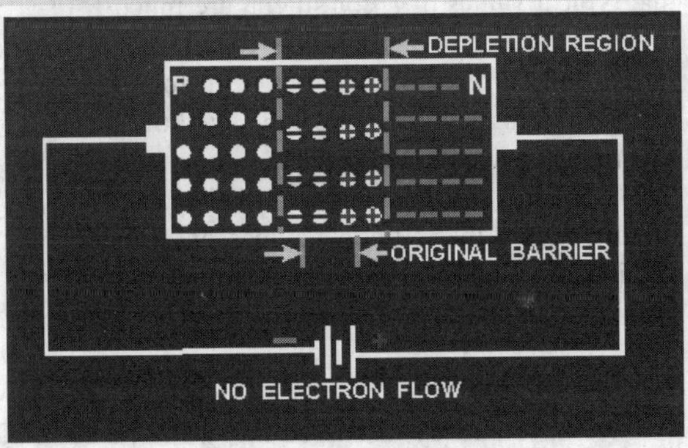

When a PN junction is reverse biased, as shown in Fig. 11.9 (*a*), the holes in the P-region are attracted towards the negative terminal of the voltage source. And the electrons in the N-region are attracted to the positive terminal of the voltage source. Thus the majority carriers are drawn away from the junction. *This widens the depletion layer and increases the barrier potential* as shown in Fig. 11.9 (*b*).

The increased barrier potential makes it very difficult for the majority carriers to diffuse across the junction. Thus there is no current due to majority carriers in a reverse biased PN junction. In other words, the junction offers very high resistance under reverse biased condition. However, the barrier potential helps the minority carriers in crossing the junction. As a matter of fact, as soon as a minority carrier is generated, it is swept (*i.e.,* drifted) across the junction because of the barrier potential. Hence a small amount of current does flow through the reverse biased PN junction. The amount of this current depends upon the number of minority carriers diffusing across the junction. This, in turn, depends upon the generation of minority carriers within the P-region and N-region. It may be noted that generation of minority carriers is dependent upon the temperature and is independent of the applied reverse voltage. Therefore, the current, due to the flow of minority carriers, remains the same whether the applied voltage is increased or decreased. Because of this reason, the current is known as reverse saturation current.

11.10. Reverse Saturation Current

We have already discussed that in a reverse biased PN junction, practically no current flows due to majority carriers. However, a small amount of current does flow due to the diffusion of minority carriers across the junction. The minority carriers are the hole-electron pairs generated throughout the semiconductor material as a result of thermal energy. The current, so produced, is known as reverse saturation current and is designated by I_0 (or I_s) The word saturation implies that the reverse current cannot be increased by increasing the reverse bias across the PN junction. But it can be increased with the increase in temperature.

The reverse saturation current is of the order of nanoamperes (nA) for silicon and microamperes (μA) for germanium PN junctions. The reverse saturation current, for silicon devices, is smaller than that of germanium because silicon has a large energy gap (≈ 1.1 eV) between the conduction and valence bands. As a result of this, less minority carriers are thermally generated at any temperature in silicon than that of germanium.

It has been experimentally found that the value of reverse saturation current (I_0) is about double in magnitude for every 10°C rise in temperature. For example, if for silicon I_0 is 5 nA at 25°C, then it is approximately 10 nA at 35°C, 20 nA at 45°C, 40 nA at 55°C and approximately 1000 nA (or 1 μA) at 100°C and so on. However, for germanium, I_0 is in the order of 1-2 μA at 25°C, but its value increases to 100 μA at 100°C.

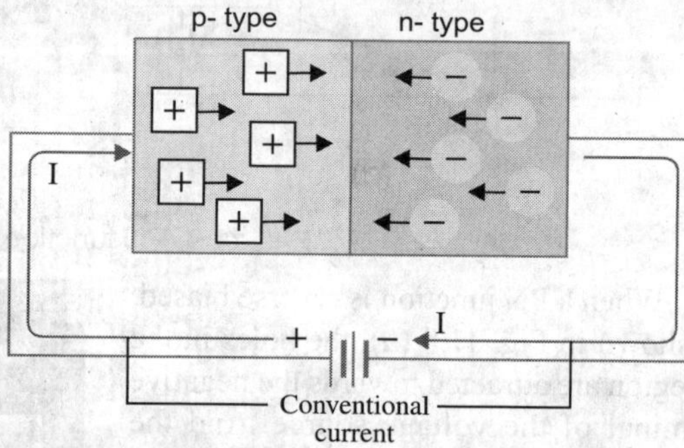

Under a reverse bias condition, there is almost no current through the diode.

It is evident from the above discussion that typical values of I_0 for silicon are much lower than that of germanium for similar power and current levels. It has been found that even at high temperatures, the level of I_0 for silicon devices do not reach the same high levels obtained for germanium. It is because of this reason that silicon devices enjoy a significantly higher level of development and utilization in design.

11.11. Reverse Breakdown

We have already discussed that a PN junction allows a very small amount of current (called reverse saturation current), when it is reverse biased. This current is due to the movement of minority carriers across the junction and is independent of the applied reverse voltage. If the reverse bias is increased to a large value, the current through the junction increase abruptly. The voltage, at which this action (*i.e.,* abrupt increase) occurs is known as breakdown voltage. At this voltage, the crystal structure breaks down. It may be noted, that the junction break down is avoided in normal applications. If the excess reverse bias voltage is removed, the crystal structure will return to normal, provided that overheating has not permanently damaged the crystal. The following two processes cause junction breakdown due to the increase in reverse voltage:

1. ***Zener breakdown.*** In this case the breakdown occurs in junctions, which are heavily doped. The heavily doped junctions have a narrow depletion layer. When the reverse voltage is increased, the electric field at the junction also increases. A strong electric field causes a covalent bond to break from the crystal structure. As a result of this, a large number of minority carriers are generated and a large current flows through the junction.
2. ***Avalanche breakdown.*** In this case, the increased reverse voltage increases the amount of energy imparted to minority carriers, as they diffuse across the junction. As the reverse voltage is increased, further, the minority carriers acquire a large amount of energy (or momentum). When these carriers collide with silicon (or germanium) atoms, within the crystal structure,

they impart sufficient energy to break a covalent bond and generate additional carriers (*i.e.,* electron-hole pairs). These additional carriers pick up energy from the applied voltage and generate still more carriers. As a result of this, the reverse current increases rapidly. This cummulative process of carrier generation (or multiplication) is known as avalanche breakdown or avalanche multiplication.

11.12. Depletion Layer Capacitance

We have already discussed that when a PN junction is formed, a layer of positive and negative impurity ions, called depletion layer, is formed on either side of the junction. It is also known as depletion-region, space-charge region or transition region. The depletion-layer acts as a dielectric (*i.e.,* non-conductive) medium between P-region and N-region. We know that the P-region and N-region on either side of the junction, has a low resistance. Therefore, these regions act as two plates of a capacitor, separated by a dielectric (*i.e.,* depletion layer) as shown in Fig. 11.10.

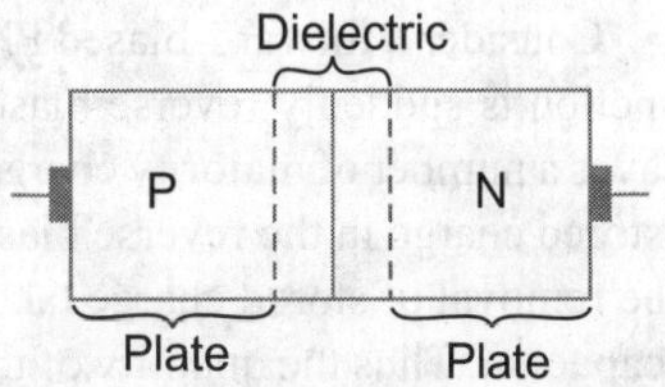

Fig. 11.10. Depletion layer capacitance.

The capacitance formed in a junction area is called depletion layer capacitance. It is also called depletion region-capacitance, space charge capacitance, transition region capacitance or simply junction capacitance. The capacitance of a parallel plate capacitor is given by the equation, $C = \varepsilon.\frac{A}{d}$

where ε is the permittivity of the dielectric (insulator) between the plates of area A separated by a distance d. Since the depletion layer width (d) increases with the increase in reverse bias voltage, the resulting depletion layer capacitance will decrease with the increased reverse bias.

The depletion layer capacitance depends upon the nature of a PN junction, semiconductor material and magnitude of the applied reverse voltage. It is given by the relation,

$$C_T = \frac{K}{(V_B - V)^n}$$

where K = A constant, depending upon the nature of semiconductor material,

V_B = Barrier voltage. It is 0.6 V for silicon and 0.2 V for germanium,

V = Applied reverses voltage, and

n = A constant depending upon the nature of the junction.

It is evident from the above relation that the value of depletion layer capacitance (C_T) can be controlled by varying the applied reverse voltage. This property of variable capacitance, possessed by a reverse biased PN junction, is used in the construction of a device called varicap or varacter.

Notes: 1. The value of constant K is given by the relation:

$$K = A \times \frac{\varepsilon \cdot q}{2}\left(\frac{N_A \cdot N_D}{N_A + N_D}\right)$$

2. The value of n is taken as 1/2 for step or abrupt junction, 1/3 for linearly graded junction.

11.13. Diffusion or Storage Capacitance

We have already discussed in the last that a PN junction behaves like a capacitor. The capacitance, which exists in a forward-biased junction, is called a *diffusion* or *storage capacitance.* It is different from the depletion layer capacitance, which exists in a reverse-biased junction. The diffusion capacitance arises due to the arrangement of minority carrier density. And its value is much larger than the depletion layer capacitance. Its value for an abrupt junction is given by the relation:

$$C_D = \frac{\tau \cdot I}{\eta \cdot V_T}$$

where t = Mean life time of the carriers,
I = Value of forward current,
η = A constant (whose value is 1 for germanium and 2 for silicon), and
V_T = Volt equivalent of temperature.

It is evident from the above relation, that diffusion capacitance is directly proportional to the forward current (I). The effect of diffusion capacitance may be understood from the following discussion:

Consider a forward-biased PN junction carrying a current of (I) amperes through it. Suppose the junction is suddenly reverse biased. It causes the forward current to reduce quickly to zero. But it leaves a number of majority charge carriers stored within the depletion region. This charge represents a stored charge in the reverse biased condition and must be removed out of the space charge region. The removal of stored charge takes a finite time. It represents an effect similar to the discharging of a capacitor. Thus the quantity of the stored charge represents the magnitude of diffusion capacitance.

The typical value of diffusion capacitance (C_D) is about 0.02 μF, which is 5000 times the depletion layer capacitance. The effect of diffusion capacitance is negligible for a reverse-biased PN junction.

11.14. Effect of Depletion Layer Capacitance and Diffusion Capacitance

Fig. 11.11 shows a depletion layer and diffusion capacitance versus applied bias for a silicon diode. From the figure, we find that the value of C_D increases with the increase in applied bias while value of C_T decreases with the increase in reverse bias.

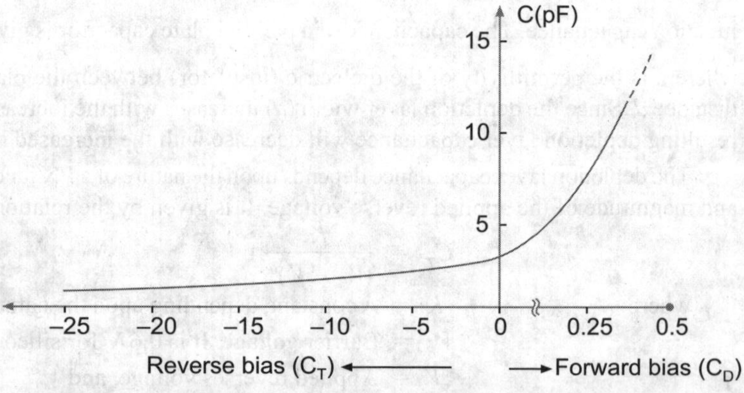

Fig. 11.11. Depletion layer and diffusion capacitance versus applied bias for a silicon diode.

It is a well established fact that the electronic devices are inherently sensitive to very high frequencies. Most shunt capacitive effects can be ignored at low frequencies, because the reactance, $X_C = 1/2\pi fC$ is very large (equivalent to open circuit). However, at very high frequencies, the value of X_C becomes sufficiently small due to the high value of f. This introduces a low reactance (or shorting) path for the current. Thus the depletion layer capacitance and diffusion capacitance do not effect much the performance of PN junction device at low frequencies. Hence, we can ignore their presence in the PN junction at low frequencies. However, their effect is significant at high frequencies.

11.15. Reverse Recovery Time

We have already discussed in the previous article that in the forward-bias there are large numbers of electrons from the N-type materials move towards P-type material and large number of holes from the P-type material move towards N-type material. It establishes a large number of minority carriers in each material.

If the applied voltage is reversed to establish a reverse-bias situation, we would ideally like to see the diode change instantaneously from the conduction state to the nonconduction state. However, because of the large number of minority carriers in each material, the diode current will simply reverse as shown in Fig. 11.12 and stay at this measurable level for the period of time t_s (storage time) required for the minority carriers to return to their majority-carrier state in the opposite material.

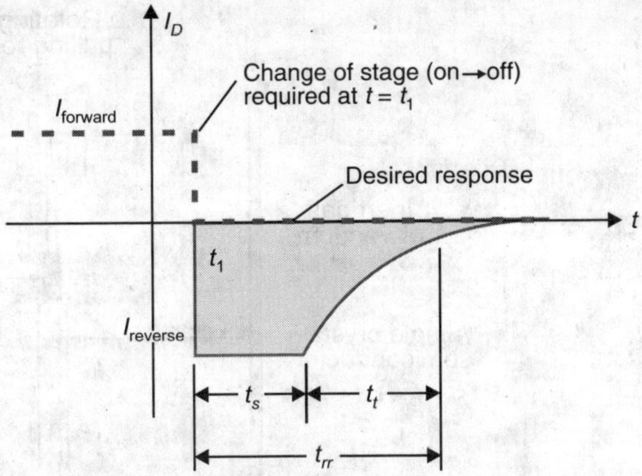

Fig. 11.12.

In essence, the diode will remain in the short-circuit state with a current determined by the network parameters. Eventually, when this storage phase has passed, the current will reduce in level to that associated with the nonconduction state. This second period of time is denoted by t_t (transition interval). The reverse recovery time is the sum of these two intervals $t_{rr} = t_s + t_t$

11.16. Fabrication of PN Junction

There are a number of techniques for the fabrication of PN junctions:

1. Grown junction
2. Alloy junction
3. Diffused junction
4. Epitaxial growth
5. Point contract junction

The first process in the manufacture of any semiconductor device is to obtain the semiconductor material in an extremely pure form. In actual practice, the accepted impurity level is less than one part of impurity in ten billion parts of the semiconductor material. To begin with, the material is heated to a temperature above its melting point in an inert gas atmosphere. This gives us a material with uncertain impurity and in polycrystalline form. The material is further purified by a process called zone refining process. This process is based on the fact that when a molten semiconductor solidifies prograssively, most of the impurities tend to remain in the liquid phase and will be swept to the end that solidifies in the last. The next process is to obtain single crystals of silicon or germanium, which are required for the fabrication of devices. This is obtained by using a process called Czochralski. These crystals are then cut into wafers. Let us now discuss the basic processes commonly used in the manufacturing of a PN junction.

11.17. Grown Junction

This junction is prepared by using either the Czochralski or floating zone technique. The apparatus used for Czochralski technique is as shown in Fig. 11.13. It consists of a graphite crucible, a quartz container, a rotating pulling rod and the induction heating coils placed around the graphite crucible. The graphite crucible contains the molten semiconductor material (*i.e.,* silicon or germanium). A single crystal semiconductor (called seed) of the desired impurity level is placed at the lower end of the rotating rod.

First of all, a single semiconductor seed is immersed in the molten semiconductor material. Then it is gradually withdrawn, while the rod holding the seed is slowly rotating. The PN junctions are grown by first adding P-type impurities to the melt and then changing it to N-type.

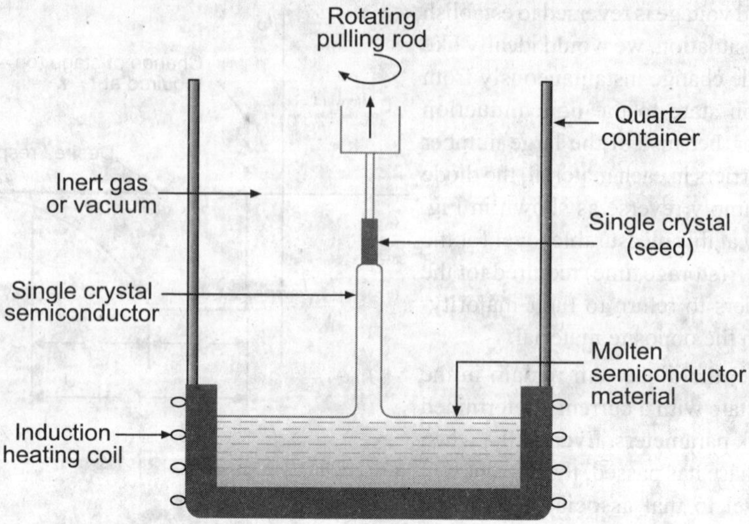

Fig. 11.13. Apparatus for Czochralski technique.

11.18. Alloy Junction

The alloy junction technique produces PN junction which have high (PIV) peak-inverse-voltage and current ratings. Such junctions have large capacitance, due to their large junction area. In the alloy junction technique, a small dot of indium is placed on the N-type silicon wafer as shown in Fig. 11.14 (*a*).

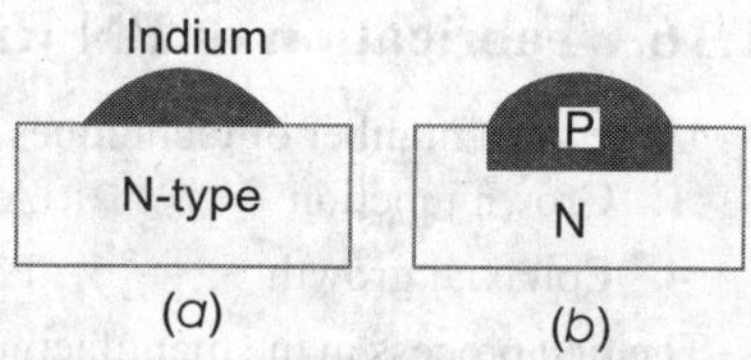

Fig. 11.14. Alloy junction.

It is then heated to a temperature of about 150°C. At this temperature, the indium melts and dissolves some of the silicon. Then its temperature is lowered and silicon refreezes to form a single crystal having a PN junction as shown in Fig. 11.14 (*b*). The junction location obtained by this method is difficult to control precisely.

11.19. Diffused Junction

This technique gives us a precise control of impurity concentration for manufacturing *PN* junctions. The diffused junctions may be obtained by the following three methods:

1. *Diffused mesa junction.* In this method, an N-type silicon wafer called substrate (or base) is exposed to a gaseous impurity of P-type as shown in Fig. 11.15 (*a*). Then the wafer is heated to a sufficiently high temperature at which the impurities diffuse slowly into the surface of the wafer. After the diffusion, portions of the surface are protected and the rest are etched out as shown in Fig. 11.15 (*b*).

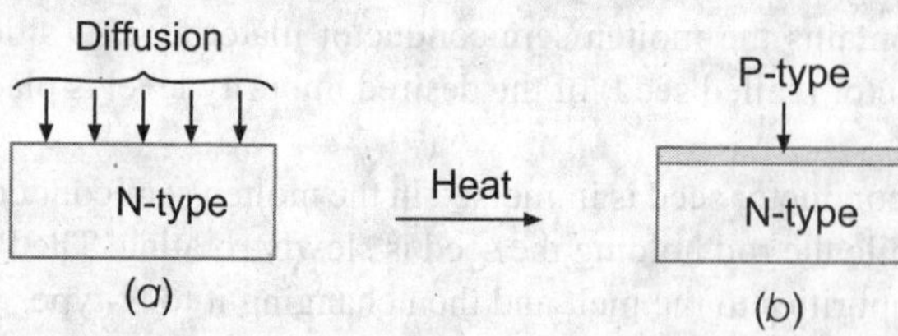

Fig. 11.15. Diffused mesa junction.

2. ***Diffused planar junction.*** It is the most commonly used method of fabricating the semiconductor devices and integrated circuits, because it gives a better degree of control over the lateral geometry of the diffused junction. This is achieved by using a silicon dioxide layer, which can prevent most donor and acceptor impurities from diffusing through it.

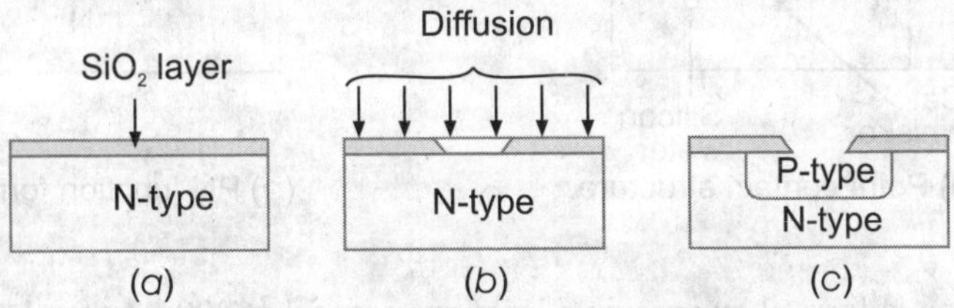

Fig. 11.16. Diffused planar junction.

A very thin layer of silicon dioxide (approximately 1 μm) is thermally grown on N-type silicon wafer as shown in Fig. 11.16 (*a*). Then with the use of lithographic techniques, portions of the oxide layer can be removed and the windows (or patterns) are cut in the oxide, as shown in Fig. 11.16 (*b*) for introducing P-type impurity. This P-type impurity will diffuse only through the exposed silicon surface, and PN junction will be formed as shown in Fig. 11.16 (*c*). This process of forming the *PN* junction is called a planar process.

3. ***Ion implantation.*** It is a process of introducing energetic charged atomic particles into a substrate for changing the electrical, metallurgical or chemical properties of the substrate. The ion implantation method gives the most precise control of an impurity concentration at present.

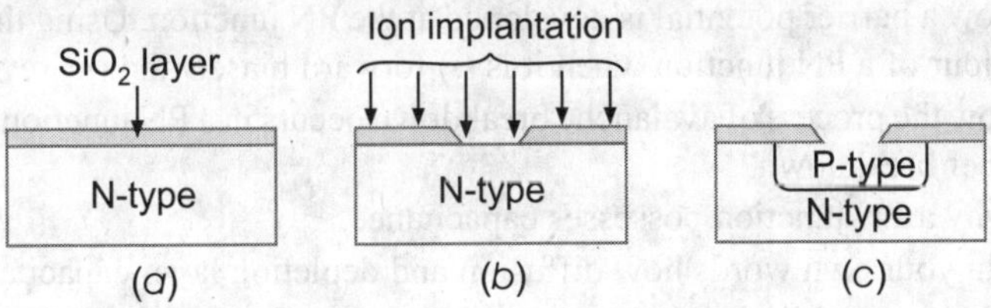

Fig. 11.17. Ion implantation.

The formation of a PN junction proceeds by growing a silicon dioxide (SiO_2) layer on the substrate as shown in Fig. 11.17 (*a*). The layer is then removed by lithographic techniques and the windows are cut in the oxide as shown in Fig. 11.17 (*b*). The ions of P-type impurity are implanted through the exposed silicon surface and the PN junction is formed as shown in Fig. 11.17 (*c*).

11.20. Epitaxial Junction

This junction differs from the diffused junction only in one aspect that the junction is fabricated not on the substrate, but on the epitaxial layer grown above the substrate. The epitaxial junctions have an advantage of low resistance.

11.21. Point Contact Junction

It consists of an N-type semiconductor (silicon or germanium) wafer, whose one face is soldered to a metallic base and the other face has a phosphor bronze (or tungsten) spring (called Cat's whisker) present against it as shown in Fig. 11.18 (*a*). The whole assembly is encapsulated in a ceramic or glass envelop to give it mechanical strength.

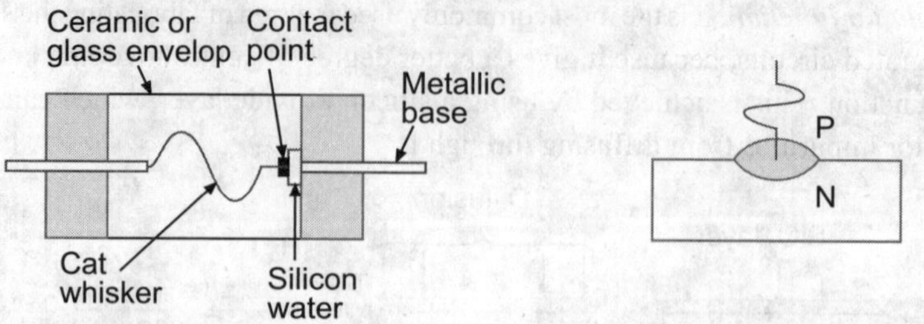

(a) Point contact structure. (b) PN-junction formed.

Fig. 11.18.

The PN junction is formed by passing a large current (about 200 μA) for 1 to 100 millisecond duration. The junction is formed at the contact point because of the metling of silicon surface and diffusion of the whisker material into the surface at that point as shown in Fig. 11.18 (*b*).

The point contact junctions have very low value of the capacitance. Because of this, such junctions are very useful for operation at the frequencies as high as 10 GHz.

DESCRIPTIVE QUESTIONS

1. What is a PN junction? Explain the formation of depletion layer in a PN junction.
2. Explain the formation of the depletion region in an open circuit PN junction.
3. Briefly explain (without derivations) the behaviour of junction in forward and reverse bias mode.
4. Discuss how a barrier potential is developed at the PN junction. Using this concept, discuss the behaviour of a PN junction when it is (*i*) forward biased and (*ii*) reverse biased.
5. Explain how the process of avalanche breakdown occurs in a PN junction. How it is different from a zener breakdown?
6. Explain why a PN junction possesses capacitance.
7. Describe in your own words how diffusion and depletion layer capacitance differ.
8. Explain PN junction diode. (*GBTU/MNTU., 2004-05*)
9. Draw and explain the equivalent circuit of a PN Junction diode. (*GBTU/MNTU., 2004-05*)
10. Explain the two breakdown mechanisms of a reverse bias diode. (*GBTU/MTU., 2009-10*)
11. Discus the different types of junction break down that can occur in a reverse biased diode. Explain the shape of the break down diode characteristics. What will be their thermal coefficient? (*GBTU/MTU., 2008-09*)
12. Discuss the capacitive effect of PN junction under reverse bias. What will happen to capacitance if the polarity of the applied voltage is changed explain? (*GBTU/MTU., 2006-07*)
13. Describes the condition established by forward and reverse bias condition on PN junction diode and how the resulting current is affected? (*GBTU/MTU., 2005-06*)
14. How does the diffusion capacitance CD vary with the dc diode current? Explain. (*GBTU/MTU., 2005-06*)
15. Explain the following terms in context with a semiconductor diode.
    (*a*) Potential Barrier (*b*) Depletion layer
    (*c*) Breakdown voltage (*d*) Peak inverse voltage
    (*e*) knee voltage (*GBTU/MTU., 2005-06*)
16. Explain the behavior of PN junction at no bias, reverse bias and forward bias. Sketch the V-I characteristic of PN junction diode. (*GBTU/MTU., 2006-07*)

17. What are the differences between Diffusion and Transition Capacitance? *(GBTU/MTU., 2006-07)*

18. What is the effect of temperature on the reverse saturation current of a diode? Explain. *(GBTU/MTU., 2008-09)*

19. What is potential barrier? Explain its significance in a PN junction. *(GBTU/MTU., 2009) (R.G.P.V., Dec. 2003)*

20. What is the transition capacitance? Drive an expression for transition capacitance. *(GBTU/MTU., 2009)*

21. What do you mean by Zener and avalanche breakdown in the barrier layer of a semiconductor? *(GBTU/MTU., 2005-06)*

22. What is meant by breakdown? Explain the various types of breakdown phenomenon. *(GBTU/MTU., 2008-09, 2002-03)*

23. Give the theory of current components of PN junction diode. *(Anna University., Nov/Dec. 2007, GBTU/MTU, 2001-02, 02-03)*

24. Derive an expression for the diffusion capacitance of a PN junction diode. *(Anna University., Nov/Dec 2007)*

25. How do the transition region width and contact potential across a PN junction vary with the applied bias voltage? *(Anna University., Nov/Dec 2007)*

26. Mention the two mechanisms of breakdown in a PN junction. *(Anna University., Nov/Dec 2007)*

27. Describe the operation of PN junction diode with neat diagram and derive for current. *(Anna University., April/May 2008)*

28. What is avalanche breakdown? *(Anna University., April/May 2008)*

29. Define Diffusion Current. *(Anna University., April/May 2008)*

30. Capacitance of diodes. *(Anna University., Nov/Dec 2007)*

31. What is meant by depletion region? *(Anna University., Nov/Dec 2007)*

32. Write a note on temperature dependence of breakdown voltages. *(Anna University., Nov/Dec 2007)*

33. Explain junction capacitance with reference to a PN diode. *(VTU., Dec. 08, Jan. 09)*

34. Explain the two breakdown mechanism of a reverse bias diode. *(PBUT., 2009-10)*

35. Distinguish between Zener breakdown and avalanche breakdown *(PBUT., 2009, GBTU/MTU., 2008-09, 2007-08, Nagpur University., 2008, PTU., Dec. 2006)*

36. Distinguish between Zener breakdown and avalanche breakdown. Which of these can be used in high voltage applications ? *(PBUT., April 2005)*

37. Explain the formation of potential barrier in a PN junction. Why is silicon preferred to germanium in the manufacturing of semiconductor device. *(GBTU/MNTU., 2003-03)*

38. What do you understand by depletion region at PN junction ? What is the effect of forward and reverse biasing of PN junction on the depletion region ? Explain with necessary diagrams. *(JNTU., May/June 2006)*

39. Explain Transition capacitance. *(Nagpur University., 2008, Nagpur University, Summer 2008, 2003, Anna University., Nov/Dec 2007)*

40. Explain the term "Transition capacitance CT" of a PN junction diode. *(JNTU., Aug/Sep2006)*

41. Explain the process of breakdown of a PN junction diode due to (*a*) Avalanche effect (*b*) Zener effect. *(JNTU., May/June 2006, GBTU/MTU 2004-05)*

42. With reference to PN junction diode, (*a*) distinguish between drift current and diffusion current (*b*) distinguish between diffusion capacitance and transition capacitance. *(JNTU., Aug/Sep 2006, GBTU/MTU 2005-06)*
43. Explain about diffusion capacitance in detail. *(JNTU., June 2009)*
44. Derive an expression for diffusion capacitance. *(JNTU, June 2009, Nagpur University, Winter 2006)*
45. What are the various application of PN junction diode ? Explain. *(JNTU., Aug/Sep 2006)*
46. What is potential barrier in a diode? How it gets established? *(PTU., Dec 2006)*
47. Discuss the behavior of PN junction diode under forward and reverse biased conditions. *(PTU., Dec 2007)*
48. Define term junction capacitance. Name different type of capacitances a diode posse. *(PTU., Dec 2007)*
49. What is the relation between the transition capacitance and reverse bias voltage in a PN diode? *(PTU., May 2007)*
50. Discuss the process of formation of depletion layer at PN junction. *(R.G.P.V., June 2011)*
51. Differentiate between transition capacitance and diffusion capacitance. *(R.G.P.V., June 2011)*
52. Discuss the capacitance effect in semiconductor diode.*(Nagpur University, Summer 2006)*
53. Discuss how does the diode voltage (at constant current) and reverse saturation current PN diode vary with temperature. *(Nagpur University, Winter 2004)*
54. It is predicted that, for germanium the reverse saturation current should increase by 0.11 per *(Nagpur University, Summer 2003)*
55. Explain the construction and working of photodiode and draw its V-I characteristics. *(Nagpur University, Summer 2003)*
56. Discuss the various current components in forward biased PN junction. *(Nagpur University, Winter 2002)*
57. What is charge control description of diode? *(Nagpur University, Winter 2002)*
58. For what voltage will the reverse saturation current in a PN junction germanium diode reach 90 % of its saturation value at room temperature? *(R.G.P.V., June 2003)*
59. What is depletion layer? How is it formed in a semiconductor diode? What is the effect of forward and reverse bias on the depletion layer? Explain. *(R.G.P.V., June 2008)*
60. Discuss transition and diffusion capacitance in PN junction. *(R.G.P.V., June 2010, Dec 2010)*
61. Explain the terms P-type, N-type materials, barrier potential and the reason for the development of this potential, depletion region? *(M.G University., May/June 2006)*
62. Discuss the temperature coefficient of breakdown voltage for breakdown diodes. *(Nagpur University, Winter 2002)*
63. Derive the expression for the "Transistor Capacitance" in case of reverse biased PN junction diode. *(Nagpur University, Summer 2003)*
64. Write short notes on :
    (*a*) Depletion Layer *(R.G.P.V., Dec. 2005, 2003)*
    (*b*) Breakdown Voltage *(R.G.P.V., Dec. 2005)*
    (*c*) Reverse Saturation Current. *(R.G.P.V., June 2008, Dec. 2003)*
    (*d*) Knee voltage *(R.G.P.V., June 2008, Dec. 2005)*
    (*e*) Barrier potential *(R.G.P.V., Dec. 2003)*

(*f*) Transition capacitances (*R.G.P.V., Dec. 2003*)

65. Explain the formation of barrier potential in open circuited PN junction diode. Also derive the expression for barrier potential. (*GTU., Dec. 2011*)

66. What is the difference between drift and diffusion. (*WBUT., Dec. 2011–12*)

MULTIPLE CHOICE QUESTIONS

1. When a PN junction is formed, diffusion current causes
(*a*) mixing of current carriers (*b*) foward bias
(*c*) reverse bias (*d*) barrier potential

2. The depletion region of a PN junction is one, that is depleted of
(*a*) Atoms (*b*) Mobiles charges
(*c*) Immobile charges (*d*) Velocity of the carriers

3. The depletion region within a PN junction is reduced when the junction has :
(*a*) Zero bias (*b*) Forward bias (*c*) Reverse bias (*d*) all of these

4. A silicon PN junction in forward conduction has a voltage drop closer to
(*a*) 0.1 V (*b*) 0.7 V (*c*) 1.7 V (*d*) 2.1 V

5. For a reverse biased PN junction, the current through the junction increases abruptly at
(*a*) breakdown voltage (*b*) 0 V
(*c*) 0.2 eV (*d*) 7.2 eV

6. The reverse saturation current of a PN junction varies with temperature (T) as
(*a*) T (*b*) $1/T$
(*c*) Independent of T (*d*) T^2

7. The transition capacitance of a reverse biased PN junction having uniform doping on both sides, varies with junction voltage (V_R) as
(*a*) $1/V_B$ (*b*) V_B (*c*) $V_B^{-1/2}$ (*d*) $V_B^{\ 2}$

8. The leakage current of a PN junction is caused by
(*a*) heat energy (*b*) chemical energy
(*c*) barrier potential (*d*) majority carriers

9. The junction capacitance of a linearly graded junction varies with the applied reverse bias, V_R as
(*a*) V_R^{-1} (*b*) $V_R^{\ -1/2}$ (*c*) $V_R^{\ -1/3}$ (*d*) $V_R^{\ 1/2}$
(*UPSC Engg. Services 2002*)

10. The diffusion capacitance of a forward biased P^+N junction diode with a steady current I depends on
(*a*) width of the depleted region (*b*) mean life-time of the holes
(*c*) mean life-time of the electrons (*d*) junction area
(A P^+N junction diode is a diode with very heavily doped *P*-region)
(*UPSC Engg. Services 2002*)

ANSWERS

1. (*a*)	**2.** (*b*)	**3.** (*b*)	**4.** (*b*)	**5.** (*a*)
6. (*a*)	**7.** (*a*)	**8.** (*b*)	**9.** (*c*)	**10.** (*a*)

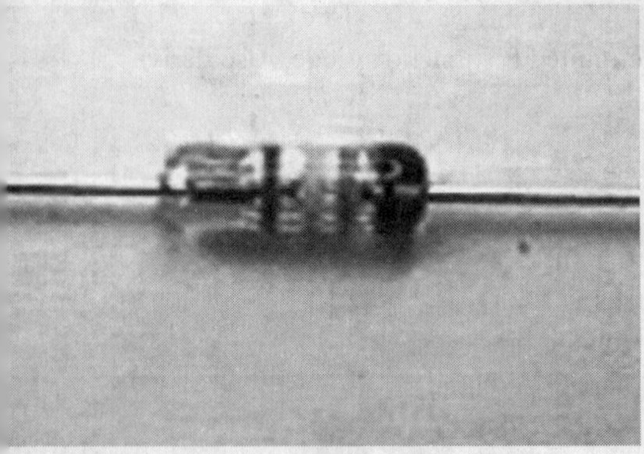

12

Chapter

PN Junction Diode

Outline

1. Introduction
2. V-I Characteristics of a PN Junction Diode
3. Forward Characteristic
4. Reverse Characteristic
5. Diode Current Equation
6. Effect of Temperature on Diode Characteristic
7. The Ideal Diode
8. Static and Dynamic Resistance of a Diode
9. The Real Diode
10. Compair on of Si and Ge Semiconductor Diodes
11. The diode as a circuit elements
12. Diode Applications
13. Power and Current Ratings of a Diode
14. Peak inverse voltage
15. Diode Specification Sheets
16. Testing of Diode with an Ohmmeter
17. Diode Circuits with DC Voltage Source
18. Parallel configuration of diode
19. Diode Circuits with D.C. and A.C. Voltage Sources.

12.1. Introduction

A PN junction diode consists of a PN junction, formed either in *germanium* or *silicon* crystal. The diode has two terminals namely *anode* and *cathode*. The anode refers to the P-type region and cathode refers to the N-type region as shown in Fig. 12.1 (*a*). Its circuit symbol is as shown in Fig. 12.1 (*b*). The arrow head, shown in the circuit symbol, points the direction of current flow, when it is forward biased. It may be noted that it is the same direction in which the movement of holes takes place.

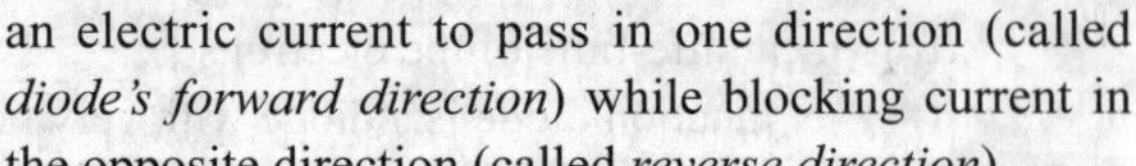

(*a*) Construction (*b*) Symbol

Fig. 12.1. PN junction diode.

The most common function of a diode is to allow an electric current to pass in one direction (called *diode's forward direction*) while blocking current in the opposite direction (called *reverse direction*).

The commercially available diodes, usually have some notations to identify the P and N terminals (or leads). The standard notation consists of type numbers preceded by IN, such as IN 240 and IN 1250. Here 240 and 1250 correspond to colour bands. In some diodes, the schematic symbol of a diode is painted or the colour dots are marked on the bodies. Diodes are usually referred to as D for diode on PCBs (Printed Circuit Boards). Sometimes the abbreviation CR for crystal diode is also used.

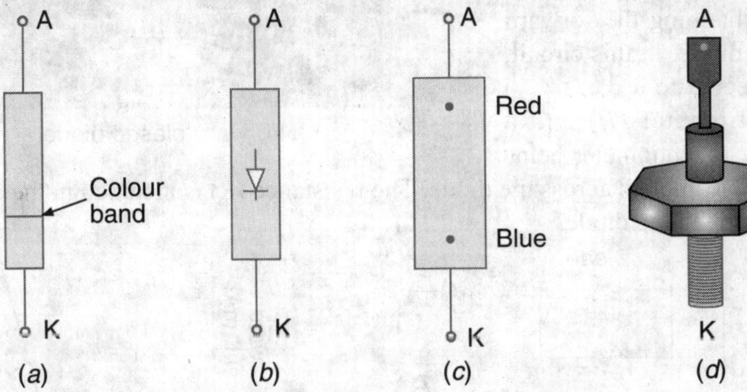

Fig. 12.2. Low, medium and high current diodes.

The diode shown in Fig. 12.2 (*a*) has colour band located near one of the ends. The end, which is near the colour band, is identified as cathode (K). And the other end, obviously, is the anode (A). The diode of Fig. 12.2 (*b*), has a schematic symbol actually painted as its cathode (K) and the other end as anode. The diodes of Fig. 12.2 (*a*) and (*b*), can pass a forward current of 100 mA and are known as low current diodes.

The diode shown in Fig. 12.2 (*c*) has colour dots marked on its body. The end lying rear the blue dot is a cathode, while the other end is anode. Sometimes, the diode of Fig. 12.2 (*c*) is shown bigger in size than that of diodes shown in Fig. 12.2 (*a*) and (*b*). The diodes of this size can pass a forward current of 500 mA and are known as medium current diodes. Fig. 12.2 (*d*) shows a diode, which can pass a forward current of several amperes. Therefore, it is known as a power diode or a high current diode.

12.2. V-I Characteristic of a PN Junction Diode

It is very important to know how a device responds (or behaves) when it is connected in an electrical circuit. This information is obtained by means of a graph known as its volt-ampere (or *V-I*) characteristics or simply characteristics.

It is a graph between the voltage applied across the terminals of a device and the current that flows through it. Figure 12.3 shows the *V-I* characteristics of a typical PN junction diode with respect to break down voltage (V_{BR}). It may be noted that the complete graph can be divided into two parts namely forward characteristic and reverse characteristic.

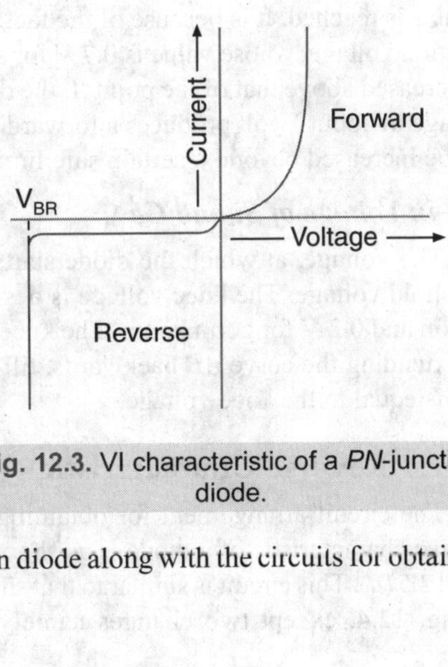

Fig. 12.3. VI characteristic of a *PN*-junction diode.

Now we shall discuss the forward characteristics and reverse characteristics of a PN junction diode along with the circuits for obtaining them in the laboratory.

12.3. Forward Characteristic

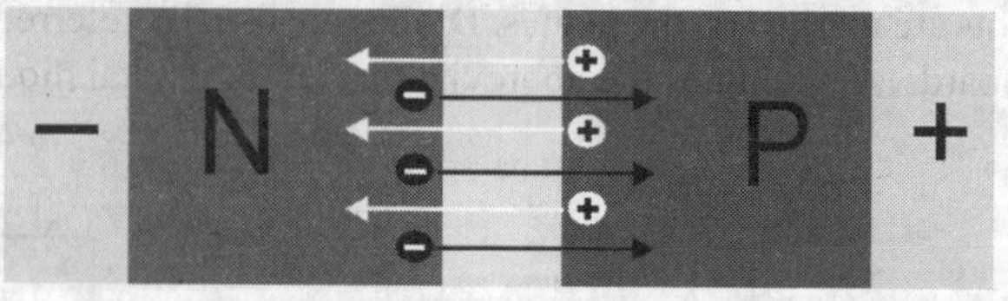

Forward biased diode.

Fig. 12.4 (*a*) shows the circuit arrangement for obtaining the forward characteristic of a diode. In this circuit, the diode is connected to a d.c. (V_{AA}) through a potentiometer (*P*) and a resistance (*R*). The potentiometer helps in varying the voltage applied across the diode. The resistance (*R*) is included in the circuit, so as to limit the current through the diode.

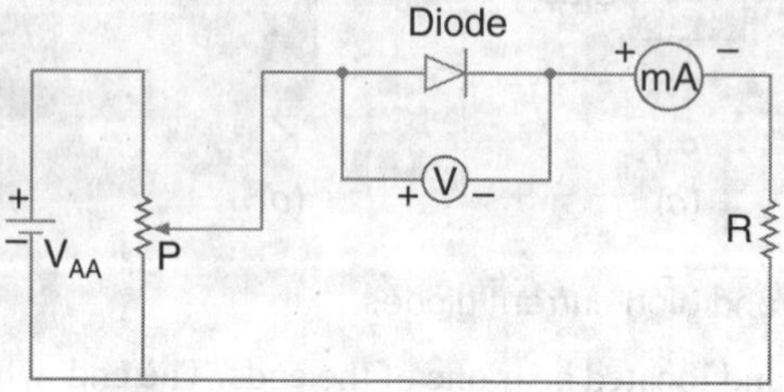

(*a*) Circuit arrangement.

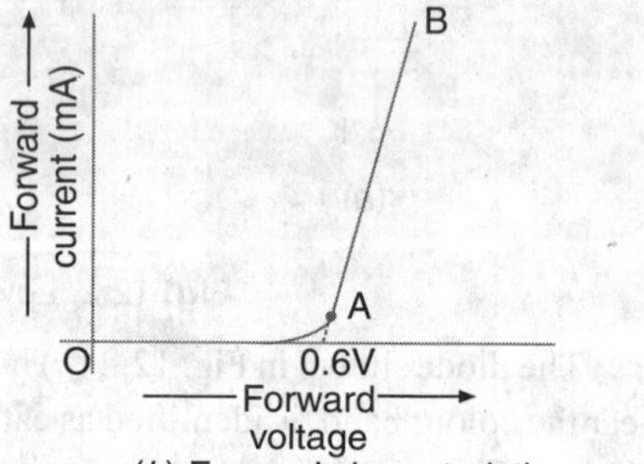

(*b*) Forward characteristic.

Fig. 12.4.

It will be interesting to know that if excessive current is permitted to flow through the diode, it may get permanently damaged. A voltmeter is connected across the diode to measure the voltage, whereas a milliammeter measures current in the circuit.

The positive terminal of the voltage source connected to the anode of a diode and negative terminal to the cathode. Hence the diode is forward-biased. Let us gradually increase the voltage in small steps of about 0.1 V and record the corresponding values of diode current. Now, if we plot a graph with voltage across the diode, along the horizontal axis and diode current, along the vertical axis, we shall obtain a curve *OAB* as shown in Fig. 12.4 (*b*). The curve *OAB* is called the forward characteristic of a silicon PN junction diode.

A careful study of the forward characteristic indicates that there is no diode current till the point *A* is reached. It is because of the fact, that the external applied voltage is being opposed by the junction voltage, whose value is 0.7 V for silicon and 0.3 V for germanium. However, as the voltage is increased above that of the point *A*, the diode current increases rapidly. It has been observed that a voltage of about 1 volt produces a forward current of about 20 to 50 mA. The applied voltage should not be increased beyond a certain safe limit, otherwise the diode is likely to burn out.

Cut-in Voltage of Si and Ge

The voltage, at which the diode starts conducting, is called a knee voltage, cut-in voltage or threshold voltage. The knee voltage is designated either by V_K or V_g. Its value is equal to 0.6 V for silicon and 0.2 V for germanium. The knee voltage may be obtained from the forward characteristic by extending the curve *AB* backwards, till it meets the horizontal axis. The value on the horizontal axis is equal to the knee voltage.

12.4. Reverse Characteristic

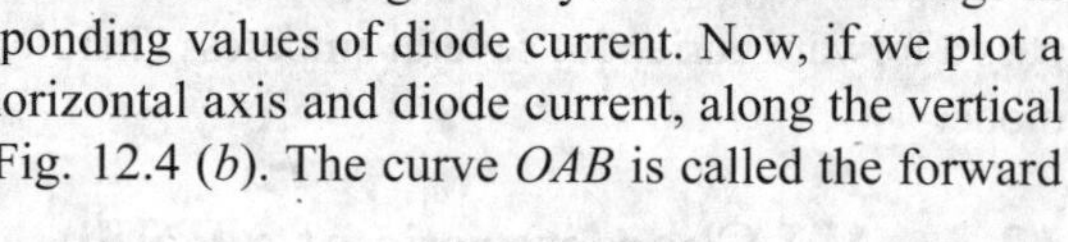

Reverse biased diode.

The circuit arrangement for obtaining the reverse characteristic of a diode is as shown in Fig. 12.5 (*a*). This circuit is similar to that shown in Fig. 12.4, except two changes namely the diode terminals are reversed and the milliammeter is replaced by a microammeter.

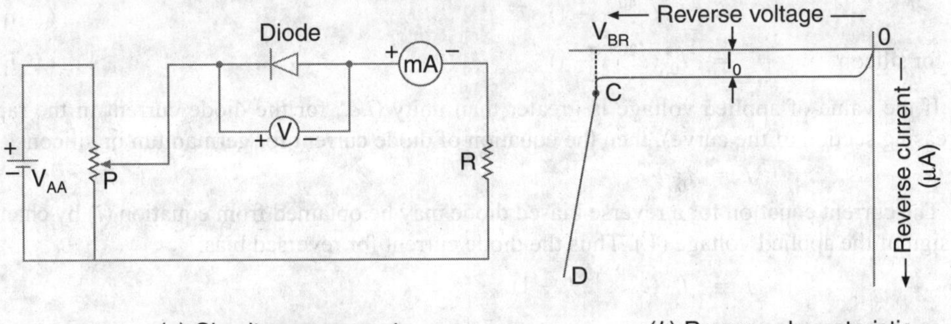

(a) Circuit arrangement. (b) Reverse characteristic.

Fig. 12.5.

It may be noted that negative terminal of the voltage source is connected to the anode of a diode and positive terminal to the cathode. Hence, the diode is reverse biased. The applied reverse voltage is gradually increased above zero in suitable steps and the values of diode current are recorded at each step. Now, if we plot a graph with reverse voltage along the horizontal axis and the diode current along the vertical axis, we shall obtain a curve marked *OCD* as shown in Fig. 12.5 (*b*). The curve *OCD* is called reverse characteristic of the diode.

A careful study of the reverse characteristic indicates that when the applied reverse voltage is below the breakdown voltage (V_{BR}), the diode current is small and remains constant. This value of current is called reverse saturation current (I_O). It is of the order of nanoamperes (1 nA = 10^{-9}A) for silicon diode and microamperes (1μA = 10^{-6} A) for germanium diode. When the reverse voltage is increased to a sufficiently large value, the diode reverse current increases as rapidly as shown by the curve *CD* in the figure. The applied reverse voltage, at which this happens, is known as break down voltage (V_{BR}) of a diode.

12.5. Diode Current Equation

The mathematical equation, which describes the forward and reverse characteristic of a semiconductor diode is called the diode current equation.

Let I I = Forward (or reverse) diode current,

I_0 = Reverse saturation current,

V = External voltage. It is positive for forward bias and negative for reverse bias.

η = A constant called ideality factor, quality factor or emission coefficient, whose value is equal to 1 for germanium diodes and 2 for silicon diodes for relatively low value of diode current (*i.e.*, at or below the knee of the curve) and $\eta = 1$ for germanium and silicon for higher levels of diode current (*i.e.*, in the rapidly increasing section of the curve)

V_T = Volt-equivalent of temperature. Its value is given by the relation, $\frac{T}{11\,600}$, where T is the absolute temperature. At room temperature (*i.e.*, 300 K), V_T is equal to 26 mV.

The current equation for a forward-biased diode is given by the relation,

$$I = I_0\,(e^{V/\eta . V_T} - 1) \quad \text{... } (i)$$

We know that at room temperature, V_T is equal to 26 mV. Substituting the value of V_T (equal to 26 mV or 0.026 V) in equation (*i*),

$$I = I_0\,(e^{40\,V/\eta} - 1)$$

∴ Diode current at or below the knee of the curve for germanium

$$= I_0\,(e^{40\,V} - 1) \qquad \text{... } (\because \eta = 1)$$

and for silicon

$$= I_0\,(e^{20\,V} - 1) \qquad \text{... } (\because \eta = 2)$$

If the value of applied voltage is greater than unity (*i.e.*, for the diode current in the rapidly increasing section of the curve), then the equation of diode current for germanium or silicon

$$I = I_0 \cdot e^{\,20\,V}$$

The current equation for a reverse biased diode may be obtained from equation (*i*) by changing the sign of the applied voltage (V). Thus the diode current for reversed bias,

$$I = I_0\,(e^{-V/\eta . V_T} - 1)$$

If the value of $V >> V_T$, then the term $e^{-V/\eta \cdot V_T} << 1$. Therefore $I = I_0$. Thus the diode current under reverse-bias is equal to the reverse saturation current as long as the external voltage is below its breakdown value.

Example 12.1.

The current flowing in a certain PN junction diode at room temperature is 2×10^{-7} A, when the large reverse voltage is applied. Calculate the current flowing, when 0.1 V forward bias is applied at room temperature.

Solution.

Given: $I_0 = 2 \times 10^{-7}$ A and $V_F = 0.1$ volt.

We know that the current flowing through the diode under forward bias,

$$I = I_0\,(e^{40V_F} - 1) = 2 \times 10^{-7}\,(e^{40 \times 0.1} - 1) = 2 \times 10^{-7}\,(e^4 - 1) = 1.07 \times 10^{-5}\text{ A}$$

$$= 10.7\ \mu\text{A}\ \textbf{Ans.}$$

Example 12.2.

Determine the germanium PN junction diode current for the forward bias voltage of 0.22 V at room temperature 25°C with reverse saturation current equal to 1 mA. Take $\eta = 1$.

Solution.

Given : $V_F = 0.22$ volt; $T = 25°\text{C} = 298$ K; $I_0 = 1$ mA $= 1 \times 10^{-3}$ A and $\eta = 1$.

We know that volt equivalent of the temperature (T),

$$V_T = \frac{T}{11\,600} = \frac{298}{11\,600} = 25.7\text{ mV} = 25.7 \times 10^{-3}\text{ V}$$

∴ The diode current,

$$I = I_0\,(e^{V_F/\eta \cdot V_T} - 1)$$

$$= (1 \times 10^{-3})\,(e^{0.22/(1 \times 25.7 \times 10^{-3})} - 1) \text{ ... } (\because \eta = 1 \text{ for germanium})$$

$$= (1 \times 10^{-3})\,(e^{8.56} - 1) = 5.22\text{ A}\ \textbf{Ans.}$$

Example 12.3.

The current voltage characteristic of a PN junction diode is given by the relation:

$$I = I_0\,(e^{qV/\eta \cdot k \cdot T} - 1)$$

The diode current is 0.5 mA at V = 340 mV and 15 mA at V = 440 mV. Determine the value of η Assume kT/q = 25 mV.

Solution.

Given : $I_1 = 0.5$ mA $= 0.5 \times 10^{-3}$ A; $V_1 = 340$ mV; $I_2 = 15$ mA $= 15 \times 10^{-3}$ A; $V_2 = 440$ mV and $k.T/q = 25$ mV.

We know that the diode current (I_1)

$$0.5 \times 10^{-3} = I_0\,(e^{qV_1/\eta \cdot k \cdot T} - 1) = I_0 \cdot e^{qV_1/\eta \cdot k \cdot T}$$
$$= I_0 \cdot e^{340/25\eta}$$
$$= I_0 \cdot e^{13.6/\eta} \quad \text{... (i)}$$

Similarly, the diode current (I_2)

$$15 \times 10^{-3} = I_0 \cdot e^{440/25\eta} = I_0 \cdot e^{17.6/\eta} \quad \text{... (ii)}$$

Dividing equation (*ii*) by (*i*),

$$30 = e^{(17.6 - 13.6)/\eta} = e^{4/\eta}$$

Taking natural logarithms on both sides,

$$\log_e 30 = \log_e (e^{4/\eta}) = \frac{4}{\eta} \cdot \log_e = \frac{4}{\eta} \quad \text{...}(\because \log_e = 1)$$

$$3.4 = 4/\eta \quad \text{or} \quad \eta = 4/3.4 = 1.18 \ \textbf{Ans.}$$

Example 12.4.

A germanium diode carries a current of 1 mA at room temperature when a forward bias of 0.15 V is applied. Estimate the reverse saturation current at room temperature. (GBTU/MTU., 2006)

Solution.

Given: $V_F = 0.15$ Volts and $I_F = 1$mA $= 1 \times 10^{-3}$ A.

We know that the volt-equivalent of room temperature (T),

$$V_T = \frac{T}{11600} = \frac{300}{11600} = 26\text{mV} = 26 \times 10^{-3}\text{ V}$$

∴ The diode current,

$$I_F = I_0\,(e^{V_F/\eta . V_T} - 1)$$

$$(1 \times 10^{-3}) = I_0\,(e^{0.15/(1 \times 26 \times 10^{-3})} - 1) \quad (\because \eta = 1 \text{ for germanium})$$

$$= \frac{(1 \times 10^{-3})}{(e^{5.8} - 1)} = 3.03 \times 10^{-6} = 3.03\ \mu\text{A} \ \textbf{Ans.}$$

12.6. Effect of Temperature on Diode Characteristic

We have already discussed in the last article that the diode current is a function of temperature. And the temperature appears in the denominator of the exponent term of the diode current equation (*i.e.*, $V/\eta \cdot V_T$ where V_T is equal to $T/11\,600$). *It is thus obvious that with the increase in temperature, the exponent will reduce and hence the diode current should also decrease.* However, it has been found that the variation of saturation current (I_0) is much *greater* than the exponential term. The above fact may be expressed in the form of a mathematical relation as given below:

Let I_{O1} = Saturation current at some temperature (T_1) for either germanium or silicon diode,

and I_{O2} = Saturation current at some other temperature (T_2).

Then $I_{O2} = I_{O1} \cdot 2^{(T_2 - T_1)/10}$ ...(For either germanium or silicon)

As discussed in the last chapter (Art. 11.10), the reverse saturation current (I_0) will be just about double in magnitude for every 10°C increase in temperature. For example, a germanium diode with an I_O in the order of 1 or 2μA at 25°C has a leakage current of 100 μA (= 0.1 mA) at a temperature of 100°C. Current levels of this magnitude in the reverse bias region would certainly question our desired open-circuit condition in the reverse bias region. However, typical values of I_O for silicon diode are much lower than that of germanium for similar power and current levels.

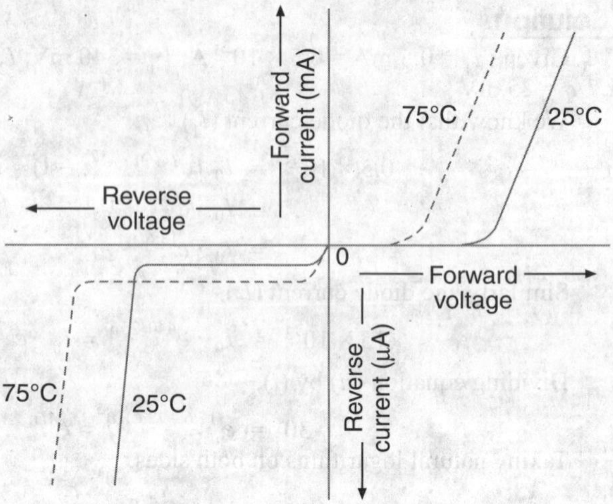

Fig. 12.6. Effect of temperature on the diode characteristic.

The increasing level of I_O with temperature account for the lower threshold voltage as shown in Fig. 12.6. Because of this reason the forward characteristic at 75°C is shown to the left to that of the characteristic at 25°C.

As the temperature increases, the forward characteristics shifts more and more to the left of the characteristic at 25°C (*i.e.*, become more and more "ideal"). However, temperatures beyond the normal operating range can have a very detrimental effect on the diode's maximum power and current levels. We see in Fig. 12.6, that in the reverse bias region, the breakdown voltage is increasing with the increase in temperature.

Example 12.5.

A germanium diode has a saturation current of 10 μA at room temperature (i.e., 300 K). Find the saturation current at 400 K.

Solution.

Given : $I_{300} = 10\mu\text{ A} = 10 \times 10^{-6}$ A; $T_1 = 300$ K and $T_2 = 400$ K.

We know that saturation current at 400 K,

$$I_{400} = I_{300} \times 2^{(T_2 - T_1/10)} = 10 \times 10^{-6} \times 2^{(400-300)/10}$$

$$= 10 \times 10^{-6} \times 2^{10} = 10.2 \text{ mA } \textbf{Ans.}$$

12.7. The Ideal Diode

We know that a PN junction diode conducts well, when it is forward biased and poorly (practically no current) when reverse biased. It will behave as an ideal, if the diode acts as a perfect conductor (with no voltage drop across it) when forward biased and as a perfect insulator (with no current through it) when reverse biased. The *V-I* characteristic of such an ideal diode is shown in Fig. 12.7 (*a*). An ideal diode acts like an automatic switch. The switch is closed when the diode is forward biased and is open when it is reverse biased as shown in Fig. 12.7 (*b*).

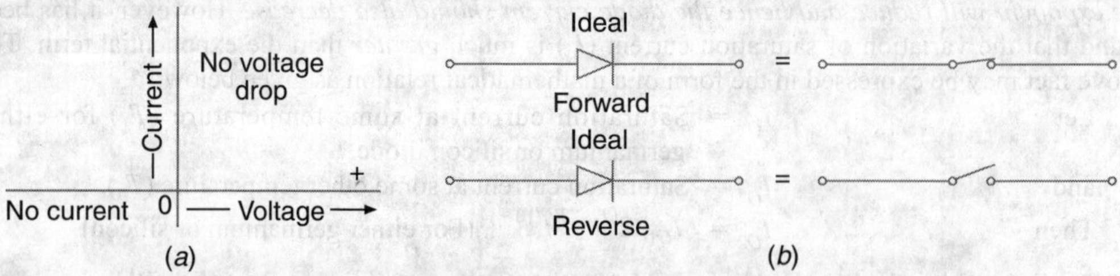

Fig. 12.7. Ideal diode.

The ideal diode.

In other words, an ideal diode is a device, which conducts with zero resistance when forward biased and appears as an infinite resistance when reverse biased. As a matter of fact, an ideal diode cannot be manufactured in actual practice. It is only a theoretical approximation of a real diode. However, in a well-designed electric circuit, a real diode behaves almost like an ideal diode because the forward voltage across the diode is small as compared to the input and output voltages.

12.8. Static and Dynamic Resistance of a Diode

We know that a real diode does not behave as a perfect conductor, when it is forward biased. Similarly, it does not behave as a perfect insulator, when it is reverse biased. In other words, it does not offer zero resistance when forward biased and an infinite resistance when reverse biased. It means that a diode has a definite value of resistance when forward biased. This resistance is known as the d.c or static forward resistance of the diode. It is given by the ratio of the d.c. voltage across the diode to the d.c. current flowing through it. Mathematically, the static forward resistance (in ohms)

$$R_F = V_F / I_F$$

The static forward resistance may be obtained graphically from the diode forward characteristic as shown in Fig. 12.8.

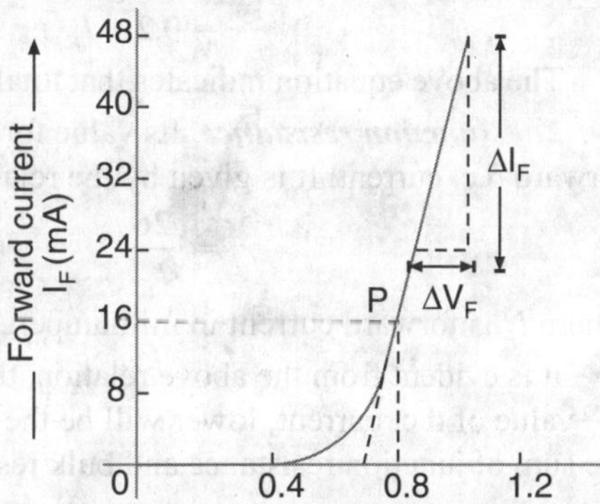

Fig. 12.8. Static and dynamic forwared resistances of a diode from the characteristic curve.

For the operating point *P*, the static forward resistance,

$$R_F = \frac{0.8}{16} = 0.05\ \Omega$$

As seen, the value of R_F is quite low, as it should be. Because of the non-linear shape of the diode characteristic, the value of R_F depends upon the exact location of the point on the curve. In practice, we don't use the static forward resistance. Instead, we use the dynamic or a.c. resistance. The resistance offered by the diode to an *a.c.* signal is called its dynamic or a.c. resistance. The a.c. resistance of a diode, at a particular *d.c.* voltage, is equal to the reciprocal of the slope of the characteristic at that point, *i.e.*, the a.c resistance,

$$r_{ac} = \frac{1}{\Delta I_F / \Delta V_F} = \frac{\Delta V_F}{\Delta I_F} = \frac{\text{Change in voltage}}{\text{Resulting change in current}}$$

Because of the non-linear shape of the forward characteristic, the value of a.c. resistance of a diode is in range of 1 to 25 Ω. Usually, it is smaller than the d.c. resistance of a diode.

In addition to the forward resistance, the diode also possesses another resistance, when it is reverse biased. Such a resistance is known as reverse resistance. It can be either d.c. or a.c. depending upon whether the reverse bias is direct or alternating voltage. It may be noted that, ideally, the reverse resistance of a diode is infinite. However, in actual practice, the reverse resistance is never infinite. It is due to the existance of leakage current in a reverse biased diode. The reverse resistance of a diode is very large as compared to its forward resistance. Its value for germanium and silicon diodes is of several meghoms. The a.c. resistance of a diode may also be determined from the following two resistances:

1. ***Bulk resistance.*** The resistance of the P-and N-semiconductor materials of which the diode is made of, is known as bulk resistance or body resistance. It also includes the resistance introduced by the connection between the semiconductor material and the external metallic conductor also called contact resistance. It is generally designated by r_B. Mathematically, the bulk resistance,

$$r_B = r_P + r_N$$

where r_P = Ohmic resistance of the P-type semiconductor, and

r_N = Ohmic resistance of the N-type semiconductor.

The bulk resistance can range from typically 0.1Ω for high power devices to 2 Ω for some low-power general purpose diodes. The diode current passing through the bulk resistance will develop a voltage equal to (I_F . r_B). Then total voltage drop across the diode is, $V_F = V_B + I_F \,.\, r_B$

$$= 0.6 + I_F \cdot r_B \quad \text{... (For silicon diode)}$$

$$= 0.2 + I_F \; r_B \quad \text{... (For germanium diode)}$$

The above equation indicates that total voltage drop across a diode is directly proportional to I_F.

2. ***Junction resistance.*** Its value for a forward-biased PN junction depends upon the value of forward d.c. current. It is given by the relation,

$$r_j = \frac{26}{I_F}$$

where I_F is forward current in milliamperes.

It is evident from the above relation, that the junction resistance is a variable resistance. Higher the value of d.c. current, lower will be the value of r_j and vice versa. The a.c. resistance is equal to the sum of junction resistance and bulk resistance. Mathematically, the a.c. resistance,

$$r_{ac} = r_j + r_B$$

Example 12.6.

A silicon diode has a bulk resistance of 2 Ω and a forward current of 12 mA. What is the actual value of V_F for the device.

Solution.

Given: r_B = 2 Ω and I_F = 12 mA = 12 × 10^{-3} A.

We know that the voltage drop across a silicon diode,

$$V_F = 0.6 + I_F \,.\, r_B = 0.6 + (12 \times 10^{-3}) \times 2 = 0.624 \text{ V } \textbf{Ans.}$$

Example 12.7.

An ideal germanium diode at a temperature of 125°C has a reverse saturation current of 30 μA. At a temperature of 125°C, find the dynamic resistance for a 0.2 V bias in (a) the forward direction and (b) the reverse direction.

Solution.

Given : T = 125°C = 398 K; I_0 = 30 μA = 30 × 10^{-6} A and V = 0.2 volt.

(a) Dynamic resistance in forward direction

We know that volt equivalent of the temperature,

$$V_T = \frac{T}{11\,600} = \frac{398}{11\,600} = 0.0343$$

$$\therefore \text{ Diode current, } I = I_0 \cdot (e^{V/\eta \cdot V_T} - 1) = I_0 \cdot (e^{V/V_T} - 1) \quad \text{...}(\because \eta = 1 \text{ for germanium})$$

Differentiating the above expression with respect to V,

$$\frac{dI}{dV} = \left(\frac{I_0}{V_T}\right) \times e^{V/V_T}$$

or $$\frac{1}{r_{ac}} = \left(\frac{I_0}{V_T}\right) \times e^{V/V_T} \qquad \ldots\left(\frac{\Delta V}{\Delta I} = \frac{dV}{dI} = r_{ac}\right)$$

$\therefore$ $$r_{ac} = \frac{V_T}{I_0} \cdot e^{-V/V_T} \qquad \ldots (i)$$

$$r_{ac} = \frac{0.0343}{30 \times 10^{-6}} \times e^{-0.2/0.0343}\ 1143.3 \times e^{-5.83} = 1143.3 \times (2.938 \times 10^{-3})$$

$$= 3.36\ \Omega \textbf{ Ans.}$$

(b) Dynamic resistance in reverse direction

Now substituting the given values in equation (i) with V equal to – 0.2 (due to reverse direction), we find that the dynamic resistance in reverse direction,

$$= \frac{0.0343}{30 \times 10^{-6}} \times e^{(-0.2)/0.0343} = 1143.3 \times e^{5.83}$$

$$= 1143.3 \times 340.35 = 0.389 \times 10^6\ \Omega$$

$$= 0.389\ \text{M}\Omega \textbf{ Ans.}$$

Example 12.8.

Determine the d.c resistance levels for the diode in Fig. 12.9.

(a) $I_F = 2\ mA$ *(b)* $I_F = 20\ mA$ *(c)* $V_F = -10\ V$ *(GBTU/MTU., 2007)*

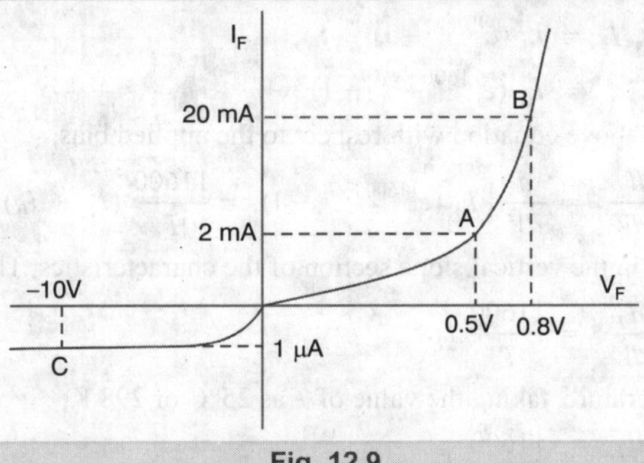

Fig. 12.9

Solution.

(a) DC resistance at $I_F = 2$ mA

Given : $I_F = 2\ \text{mA} = 2 \times 10^{-3}$ A.

From the graph, value of $V_F = 0.5$ V.

We know that for the operating point 'A' d.c resistance,

$$R_F = \frac{V_F}{I_F} = \frac{0.5}{2 \times 10^{-3}} = 250\ \Omega \textbf{ Ans.}$$

(*b*) DC resistance at $I_D = 20$ mA

Given: $I_F = 20$ mA $= 20 \times 10^{-3}$ A.

From the graph, value of $V_F = 0.8$ V.

We know that for the operating point 'B', d.c resistance,

$$R_F = \frac{V_F}{I_F} = \frac{0.8}{20 \times 10^{-3}} = 40\ \Omega\ \textbf{Ans.}$$

(*c*) DC resistance at $V_F = -10$ V

Given: $V_F = -10$ V.

From the graph, value of $I_F = -1\ \mu$A $= -1 \times 10^{-6}$ A

We know that for the operating point 'C' d.c resistance,

$$R_F = \frac{V_F}{I_F} = \frac{-10}{-1 \times 10^{-6}} = 10^7\ \Omega = 10\ \text{M}\Omega\ \textbf{Ans.}$$

Example 12.9.

Show that the junction resistance for a silicon or germanium diode is given by the equation:

$$r_j = \frac{26}{I_F}$$

where I_F is the diode current in milliamperes.

Solution.

We know that the equation for diode current,

$$I = I_0\ (e^{V/\eta V_T} - 1)$$

and the equation for the diode current in the rapidly increasing section of the forward characteristic,

$$I_F = I_0\ (e^{V/V_T} - 1) \qquad \text{... } (\because \eta = 1)$$

$$= I_0\ (e^{11600\, V/T} - 1) \qquad \text{... } (\because = V_T\ T/11600)$$

Differentiating the above equation with respect to the applied bias,

$$\frac{dI_F}{dV} = \frac{d}{dV}\left[I_0\ (e^{11600\, V/T} - 1)\right] = \frac{11600}{T}\ (I_F + I_0)$$

In general, $I_F >> I_0$ in the vertical slope section of the characteristics, Therefore

$$\frac{dI_F}{dV} = \frac{11600}{T}\ I_F$$

And at room temperature, taking the value of T as 25°C or 298 K,

$$\therefore \quad \frac{dI_F}{dV} = \frac{11600}{298}\ I_F$$

$$\text{or} \quad \frac{dV}{dI_F} = r_j = \frac{298}{11600\ I_F} = \frac{0.026}{I_F}$$

$$\therefore \quad r_j \approx \frac{0.026}{I_F} = \frac{26\ \text{mV}}{I_F}$$

Now if I_F is expressed in milliamperes, then the junction resistance,

$$r_j = \frac{26\ \text{mV}}{I_F\ \text{mA}} = \frac{26}{I_F}$$

12.9. The Real Diode

We have already discussed in Art. 12.7 that a diode can be approximated to an automatic switch, which is closed when a diode is forward biased and open when it is reverse biased. But in actual practice, we have to consider two more factors, in addition to a simple switch, for a diode operation as discussed below:

When a diode is forward biased, it does not pass any current until the applied voltage exceeds the knee voltage or the cut-in voltage. Therefore, a diode may be represented by an automatic switch (or an ideal diode) in series with a battery of e.m.f. equal to the knee voltage. The knee voltage has no effect on the diode operation, when it is reverse biased. Therefore, the equivalent circuit of a real diode may be represented as shown in Fig. 12.10 (*a*) and its *V-I* characteristic as shown in Fig. 12.10 (*b*).

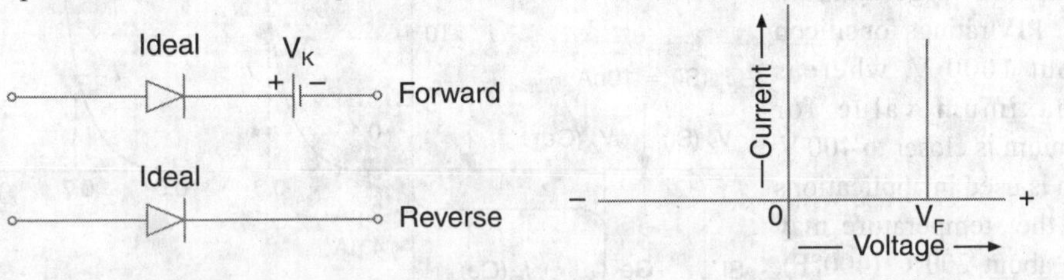

(*a*) Equivalent circuit. (*b*) VI characteristic.

Fig. 12.10.

In addition to a knee voltage (V_K), a diode possesses a.c. or dynamic resistance, which is equal to the sum of bulk resistance and the junction resistance. Thus when a diode is forward biased, it does not conduct, until the applied voltage exceeds the knee voltage. As the applied voltage exceeds further, the current through a diode increases and produces a voltage drop across the a.c. resistance. Since the a.c. resistance is linear, therefore, the voltage drop across a real diode is given by the relation,

$$V_F = V_K + I_F \cdot r_{ac}$$

where V_K = Knee voltage whose value is 0.6 V for silicon and 0.2 V for germanium,

I_F = Forward current, and

r_{ac} = A.C resistance.

Piecewise Linear Diode Model

When a diode is reverse biased, it does not behave like an open switch. In other words, the reverse current is not zero, though it has a small value. It is known as reverse saturation current. This means that a reverse biased diode is equivalent to an ideal diode in series with a high resistance (r_r) which is of the order of MΩ. Therefore, the equivalent circuit of a real diode may be represented as shown in Fig. 12.11 (*a*) and its *V-I* characteristic as shown in Fig. 12.11 (*b*).

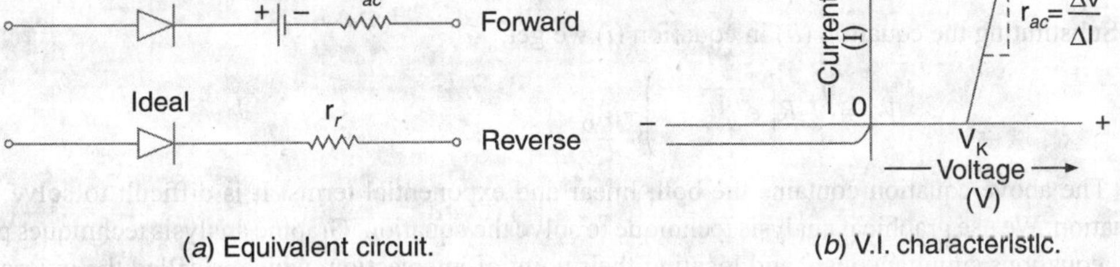

(*a*) Equivalent circuit. (*b*) V.I. characteristic.

Fig. 12.11. The real diode.

12.10. Comparison of Si and Ge Semiconductor Diodes

We have already discussed about semiconductor diode in detail. The silicon and germanium are the two most widely used semiconductors. The PIV rating, current rating and wider temperature ranges of the silicon diode is higher as compared to germanium diodes. PIV ratings for silicon is about 1000 V, whereas the maximum value for germanium is closer to 400 V. Silicon is used in applications where the temperature may rise to about 200°C (400°F), whereas germanium has a much lower maximum rating (100°C). The disadvantage of silicon as compared to germanium is shown in Fig. 12.12. The forward-bias voltage of silicon is 0.7 V whereas for germanium is 0.3 V.

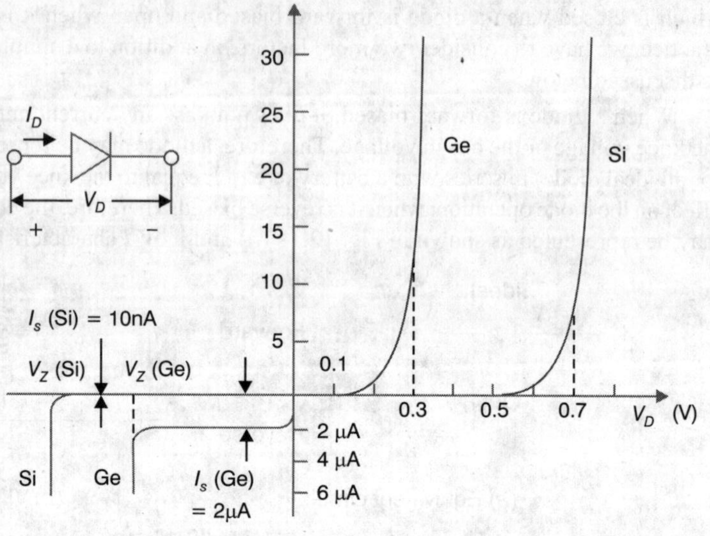

Fig. 12.12.

12.11. The Diode as a Circuit Element

The basic diode circuit is shown in the Fig. 12.13. It consists of the diode *D* in series with a resistance and input voltage source. This circuit is now analyzed to find the diode current and the V_S is applied to the circuit voltage, when the voltage. According to Kirchhoff's voltage law,

$$V_S = IR + V_D \quad \text{...}(i)$$

$$I = \frac{V_S}{R} - \frac{V_D}{R}$$

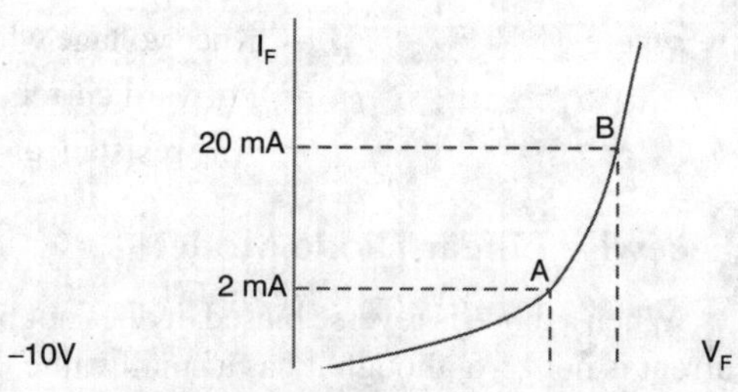

Fig. 12.13.

The ideal diode current equation relating the diode voltage and current is

$$I = I_o\left(e^{\frac{V}{\eta \cdot V_T}} - 1\right) \quad \text{... }(ii)$$

Substituting the equation (*ii*) in equation (*i*) we get,

$$V_S = I_o R\left(e^{\frac{V}{\eta \cdot V_T}} - 1\right) + V_D$$

The above equation contains the both linear and exponential terms. It is difficult to solve the equation. We use graphical analysis technique to solve the equation. Graphic analysis techniques plot two equations simultaneously and locating their point of intersection, which is called the quiescent point, or the *Q*-point.

12.12. Diode Applications

A PN junction diode has an important characteristic that it conducts well in forward direction and poorly in reverse direction. This characteristic makes a diode very useful in a number of applications given below:

1. As rectifiers or power diodes in d.c. power supplies.
2. As signal diodes in communication circuits.
3. As zener diodes in voltage stabilizing circuits.
4. As varactor diodes in radio and TV receivers.
5. As a switch in logic circuits used in computers.
6. As over-voltage protection device in integrated circuits and motor controllers.
7. As a ionizing radiation detector.
9. As a current steering device is uninterruptable power supplies and electronic musical keyboards.

12.13. Power and Current Ratings of a Diode

We know that every diode carrying certain current dissipates some power, whose value is given by the product of voltage and the current through it. Mathematically, the power dissipation, for a forward biased diode,

$$P_D = V_F \times I_F$$

where V_F = Forward voltage drop, and

I_F = Forward current

Similarly, power dissipation for a reverse biased diode,

$$P_D = V_R \times I_R$$

where V_R = Reverse voltage drop, and

I_R = Reverse current.

The maximum value of power, which a diode can dissipate without failure, is called power rating. Thus the power dissipation should not exceed power rating in any case. If it happens, the diode gets destroyed. The manufactures specify power rating of a diode on data sheets. For example, the diode 1 N 914 has a power rating of 250 mW. It means that the diode can dissipate a maximum power of 250 mW.

It will be interesting to know, that the diode manufactures more oftenly list the maximum current, which a device can handle, (called current rating), rather than power rating. It is because of the fact, that it is easy to measure current rating than power rating. For example, the data sheet of 1 N 4003 does not indicate power rating. But it does list a maximum d.c. current of 1 ampere. This means that if we pass a d.c. current of magnitude more than 1 ampere, through 1N 4003, diode it may be destroyed.

As a matter of fact, the data sheets define two categories of diodes namely small signal diodes and rectifiers or power diodes. The small signal diodes are those which have a power rating below 0.5 W whereas the rectifier diodes have a power rating above 0.5 W. Thus 1 N 914 is a small signal diode because its power rating is 0.25 W, whereas 1 N 4003 is a rectifier diode because its power rating is 1 W. These are some rectifier diode that can handle large amount of power (100 W). Such diodes has a special package designed to provide large power.

12.14. Peak Inverse Voltage (PIV)

The peak inverse voltage (PIV) or PRV (peak reverse voltage) rating of the diode is the maximum reverse voltage of diode that can be handle by the diode before breaking down. If the reverse voltage across a P-N junction diode exceeds its PIV, then the reverse current increase sharply and breaks down the junction because of excessive heat generated.

PIV or PRV rating of a P-N junction or diode is also specified by the manufacturer in its data sheet. However, it may be noted that there if we increase the voltage across the junction at reverse bias condition beyond the specified value, the junction will be destroyed.

12.15. Diode Specification Sheets

All manufacturers of the semiconductor device provide data on specific diodes for the users to make proper utilization of the device. This data could be a brief description limited to a one page or more than that. It includes the information arranged in tables, graphs etc. The data is usually for:

1. Forward voltage, V_F (at a specific current and temperature).
2. Maximum forward current, I_F (at a specified temperature).
3. Reverse saturation current I_R or I_O (at a specified voltage and temperature).
4. Reverse voltage rating [*PIV*, *PRV* or *V*(*BR*)], where *BR* implies breakdown at a specified temperature.
5. Maximum power dissipation level at a particular temperature.
6. Capacitance values.
7. Reverse recovery time, $t_{rr.}$
8. Operating temperature range.

Besides this, depending on the type of diode being considered, more data may also be provided such as frequency range, noise level, switching time, thermal resistance level and peak repetitive values. For the application in mind, the significance of the data will usually be self-apparent. If the maximum power or dissipation rating is also provided, it is understood to be equal to the produce,

$$P_{D_{max}} = V_D I_D$$

where I_D and V_D are the diode current and voltage at a particular point of operation.

Fig. 12.14 (*a*) and (*b*) shows the data sheet for 1N 40001 through 1N 4007 silicon diodes the information important from the subject point of view is as follows:

1. From the "maximum ratings" table, we find that peak repetitive voltage or reverse voltage rating for 1N 4001 is 50 *V* and For 1N 4007 is 1000 *V*. The average value of forward current for all the diodes in this series (*i.e.*, 1N 4001 through 1N 4007) is 1.0 A.
2. From the "Electrical characteristics" table we find that average value of forward voltage drop for all diodes is 0.8 V. The maximum value of reverse current at room temperature (25 °C) is 10 μA and it can increase to 50 mA at 100 °C.
3. From the tables on page 192, we find that for handling different values of current, we have different packages for the diode. Fir example if the forward current is in the range of 1.0 to 6.0 A, the small plastic packaging is used. However if the forward current is int the ran ge of 12 to 50 A, the metal packaging is more suitable. The student is advised to carefully observe the shapes of the packages.

MOTOROLA SEMICONDUCTORS

BOX 20912 PHOENIX ARIZONA 85036

Designers 'Data Sheat

1N4001 thru 1N4007

"SURMETIC"▲ RECTIFIERS

... subminiature size, axial lead mounted rectifiers for general- purpose low-power applications.

Designers Dat for "Wirst Case" Conditions

The Designers Data Sheets permit the design of most circuits entirely from the information presented. Limit curves–represented boundaries on device characteristics–are guveb ti facukutate "worst case" design.

LEAD MOUNTED
LILICON RECTIFIERS
50-1000 VOLTS
DIFFUSED JUNCTION

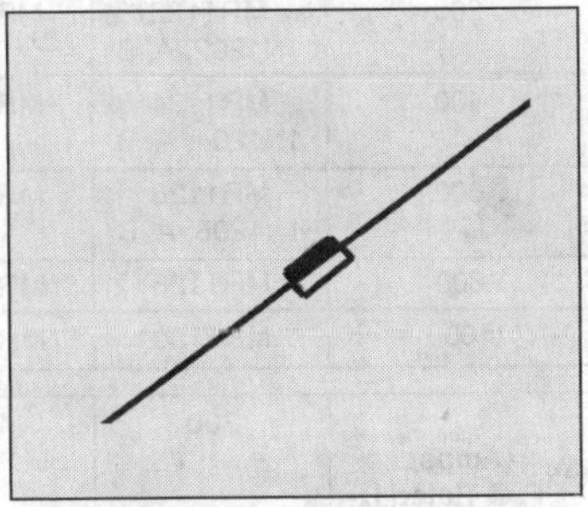

*MAXIMUM RATING

Rating	Symbol	1N4001	1N4002	1N4003	1N4004	1N4005	1N4006	1N4007	Unit
Peak Repetitive Reverse Voltage Working Peak Reverse Voltage DC Blocking Voltage	V_{RRW} V_{RWM} V_R	50	100	200	400	600	800	1000	Volts
Non-Repetitive Peak Reverse Voltage (halfwave, single phase, 60 Hz)	V_{RWM}	60	120	240	480	720	1000	1200	Volts
RMS Reverse Voltage	$V_{R(RMS)}$	35	70	140	280	420	560	700	Volts
Average Rectified Forward Current (single phase, resistive load, 60 Hz, see Figure B, $T_A = 75°C$)	I_O	1.0							Amp
Non-Repetitive Peak Surge Current (surge applied at rated load conditiono, coo Figure 2)	I_{FMS}	30 (for 1 cycle)							Amp
Operating and Storage Junction Temperature Range	$T_J . T_{stg}$	– 65 to + 175							°C

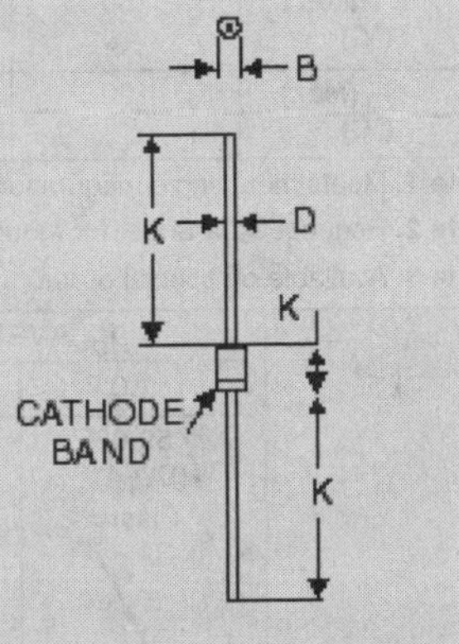

*MAXIMUM RATING

Characteristic and Conditions	Symbol	Typ	Max	Unit
Maximum Instantaneous Forward Voltage Drop ($i_F = 1.0$ Amp, $T_J = 25°C$) Figure 1	V_F	0.93	1.1	Volts
Maximum Full-Cycle Average Forward Votage Drop $I_O = 1.0$ Amp, $T_L = 75°C$, 1 inch leads)	$V_{F(AV)}$	-	0.8	Volts
Maximum Reverse Current (rated dc voltage) $T_J = 25°C$ $T_J = 100°C$	I_R	0.05 1.0	10 50	μA
Maximum Full-Cycle Average Reverse Current ($I_O = 1.0$ Amp, $T_L = 75°C$, 1 inch leads)	$I_{R(AV)}$	-	30	μA

*Indicates JEDEC Registered Data.

DIM	MILLIMETERS MIN	MILLIMETERS MAX	INCHES MIN	INCHES MAX
A	5.97	6.60	0.235	0.260
B	2.97	3.05	0.110	0.120
D	0.76	0.86	0.030	0.034
K	27.94	–	1.100	–

CASE 59-04

Doses Not Conform to Do-41 Outline

MECHANICAL CHARACTERISTICS

CASE: Void free, Transfer Molded

MAXIMUM LEAD TEMPERATURE FOR SOLDERING PURPOSES: 350°C 3/8" from case for 10 seconds at 5 lbs. tension

FINISH: All external surfaces are corrosion-resistant, leads are readily solderable

POLARITY: Cathode indicated by color band

WEIGHT: 0.40 Grams (approximately)

▲Trademark of Motorola Inc.

MOTOROLA INC. 1975 DS 6015 R3

Fig. 12.14 (*a*) Conrtesy of Motorola Inc.

VRRM (Volts)	I_O, AVERAGE RECTIFIED FORWARD CURRENT (Amperes) 12	20	24	25	30		40	50
	245A-02 (DO-203AA) Metal 	Plastic Note 1	339-02 Plastic Note 2 	193-04 (DO-02) Metal 	43-02 (DO-203AB) Metal 		42A-01 Metal 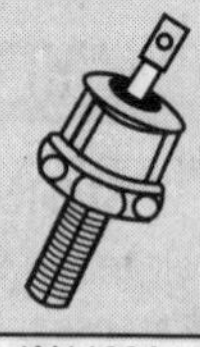	43-04 
50	MR1120 1N1199, A, B	MR2000	MR2400	MR2500	1N3491	1N36659	1N1183A	MR5005
100	MR1121 1N1200, A, B	MR2001	MR2401	MR2501	1N3492	1N3660	1N1184A	MR5010
200	MR1122 1N1202, A, B	MR2002	MR2402	MR2502	1N3493	1N3661	1N1186A	MR5020
400	MR1124 1N1204, A, B	MR2004	MR2404	MR2504	1N3495	1N3663	1N1188A	MR5040
600	MR1126 1N1206, A, B	MR2006	MR2406	MR2506		Note 3	1N1190A	Note 3
800	MR1128	MR2008		MR2508		Note 3	Note 3	Note 3
1000	MR1130	MR2010		MR2510	Note 3	Note 3	Note 3	Note 3
I_{FMS} (Amps) T_A @ Rated I_O (°C)	300	400	400	400	300	400	800	600
T_C @ Rated I_O (°C)	150	150	125	150	130	100	150	150
T_J (Max) (°C)	190	175	175	175	175	175	190	195

Note 1. Meets mounting configuration of TO-220 outline

Note 2. Request Data Sheet for Mounting Information.

Note 3. Available on special order.

VRRM (Volts)	I_O, AVERAGE RECTIFIED FORWARD CURRENT (Amperes) 1.0	1.5		3.0		6.0
	59-03 (DO-41) Plastic 	59-04 Plastic 	60-01 Metal 	267-03 Plastic 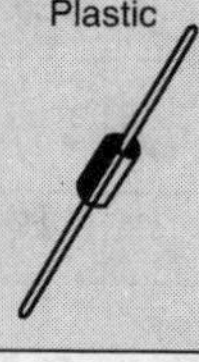	267-02 Plastic 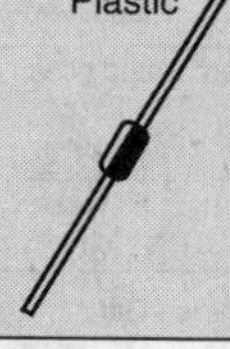	194-04 Plastic 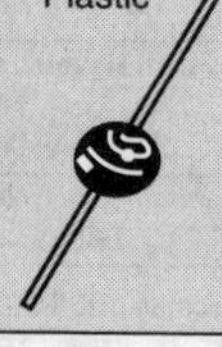
50	†1N4001	**1N5391	1N4719	**MR500	1N5400	MR750
100	†1N4002	**1N5392	1N4720	**MR501	1N5401	MR751
200	†1N4003	1N5393 *MR5059	1N4721	**MR502	1N5402	MR752
400	†1N4004	1N5395 *MR5060	1N4722	**MR504	1N5404	MR754
600	†1N4005	1N5397 *MR5061	1N4723	**MR506	1N5406	MR756
800	†1N4006	1N5398	1N4724	MR508		MR758
1000	†1N4007	1N5399	1N4725		MR510	

(Continued)

IFMS (Amps)	30	50	300	100	200	400
TA @ Rated IO (°C)						
TC @ Rated IO (°C)	150	150	125	150	130	100
TJ (Max) (°C)	190	175	175	175	175	175

† Meets mounting configuration of TO-220 outline

* Request Data Sheet for Mounting Information.

** Available on special order.

Table 12.1 summarizes some common specifications of few rectifier diodes. Notice the maximum current rating (I_F), reverse voltage rating the forward voltage and the catagogry (or technology to which it belongs to. As seen the rectifier diodes are available with a current rating as low as 0.003A and as high as 440. A.

Table : 12.1: Some common specifications of Rectifier diodes.

S. No	*Diode type*	I_f *(A)*	V_{rms} *(V)*	$V_{f(max)}$ *(V)*	*Technology*
1.	IN 4001	1	50	1.1	Standced rectifier
2.	IN 4003	1	200	1.1	____ do ____
3.	IN 4007	1	1000	1.1	____ do ____
4.	BY 8424	0.003	30 000	98	Standard rectifier High voltage
5.	BYV 26C	0.65	600	2.5	Fast rectifier
6.	STPR 520F	5	200	1.25	Ultrafast rectifier
7.	BYT 200 PIV 400	100	400	1.60	Ultrafast rectifier
8.	150K 40A	150	400	1.33	Hyperfast rectifier
9.	MBR 4060WT	40	60	0.77	Schottky barrier rectifier
10.	440 CNQO 30	440	30	0.41	Shottly barrier rectifier

Example 12.10.

A diode has maximum power dissipation of 0.5 W. (a) What is the maximum d.c. current allowed in the forward direction if we allow an approximate forward voltage drop of one volt? (b) What is the approximate breakdown current that burns out the diode, if it has a breakdown voltage of 150 V?

Solution.

Given: $P_{D_{max}} = 0.5$ W, $V_F = 1$ volt and $V_{BR} = 150$ volts.

(a) Maximum forward d.c. current

Let $I_{F(max)}$ = Maximum forward d.c. current.

We know that maximum power dissipation, $P_{D(max)}$

$$0.5 = V_F \times I_{F(max)} = 1I \times I_{F(max)}$$

$$\therefore \quad I_{F(max)} = 0.5/1 = 0.5 \text{ A } \textbf{Ans.}$$

(b) Breakdown current

Let I_R = Breakdown current that burns out the diode.

We also know that maximum power dissipation $P_{D(max)}$

$$0.5 = V_{BR} \times I_R = 150 \times I_R$$

$$\therefore \quad I_R = 0.5/150 = 0.003\,33 \text{ A} = 3.33 \text{ mA } \textbf{Ans.}$$

12.16. Testing of Diode with an Ohmmeter

The internal battery in certain ohmmeters will forward-bias or reverse bias a diode. This permits a simple and quick check for proper functioning. This works for ohmmeters as such VOMS (volt-ohm meter).

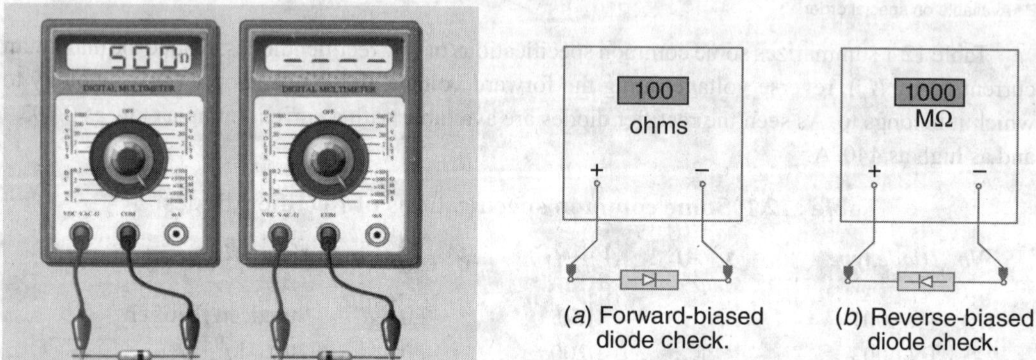

(*a*) Forward-biased diode check. (*b*) Reverse-biased diode check.

Fig. 12.15.

In order to check the diode, in the forward direction, the positive meter lead is connected to the anode and the negative lead to the cathode as shown in Fig. 12.15 (*a*). When a diode is forward biased, its internal resistance is low. Typically, it is around 100 ohms. But it can be more or less also.

When the meter leads are reversed as shown in Fig. 12.15 (*b*), the internal ohmmeter battery reverse biases the diode. The meter indicates a very large resistance (ideally infinite). The PN junction is short-circuited, if a low resistance is indicated in both forward and reverse bias conditions. It is open, if a high resistance is indicated for both the forward and reverse biasing.

12.17. Diode Circuits with a D.C. Voltage Source

Let us consider an ideal diode connected to a voltage source and a resistor as shown in Fig. 12.16 (*a*). Our aim is to determine the various voltages and current through the circuit. The first thing to be noted carefully is that the cathode terminal of the diode is connected to the positive terminal of the source through a resistor (Alternatively diode arrow points towards the positive terminal of the source). This indicates that the diode is reverse biased. Because of this, there will be no current through the circuit (*i.e.*, $I = 0$). Therefore full applied voltages is dropped across the diode (*D*). Further since there is no current through, *R*, there is no voltage drop across this component, *i.e.*, $V_R = 0$.

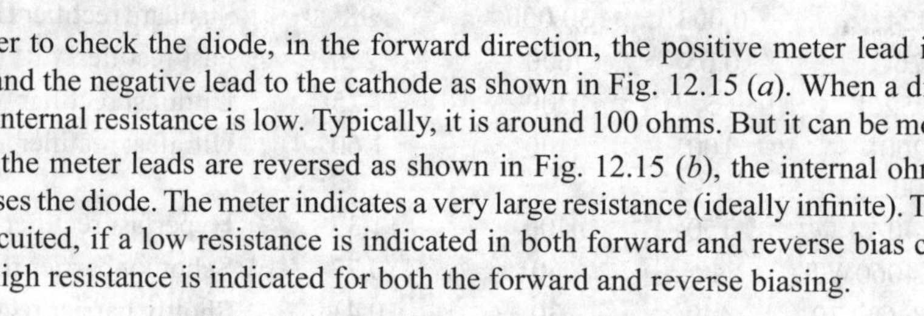

Fig. 12.16.

Now consider the circuit shown in Fig. 12.17 (*b*). Notice that the diode arrow points toward the negative terminal of the source. This indicates that the diode is forward-biased. As a result of this the diode conducts curent throught it, hence there is no voltage drop across it *i.e.*,

$$V_D = 0$$

Further the voltage drop across the resistor is equal to the dc source voltage, *i.e.*,

$$V_R = V_S$$

and the current through the circuit,

$$I = \frac{V_R}{R} = \frac{V_S}{R}$$

Next consider the circuit shown in Fig. 12.18. Notice that this circuit is similar to the one shown in Fig. 12.17 (*b*) except that the diode is a real 'one'. In such a circuit, there will be a voltage drop across the diode whose value for a silicon diode is 0.7*V*. The circuit is given by the equation.

$$I = \frac{V_S - 0.7}{R}\,(A)$$

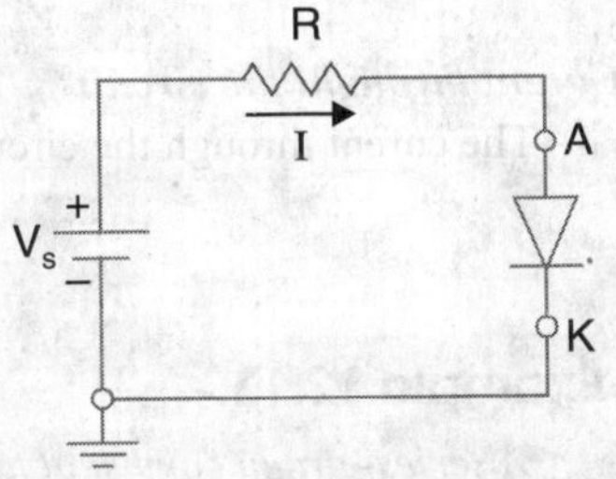

Fig. 12.17.

Example 12.11.

Determine the values of a voltage drop across (i) the diode and (ii) the resistance and (b) the current for the circuit shown in Fig. 12.18. Assume the diode to be an ideal.

Solution.

Given: D = An ideal diode,

$$R = 330\ \Omega$$

(*a*) *Voltage drop across the diode*

Notice that in the cirucit of Fig. 12.18, the cathode (*K*) of the ideal diode is connected to positive terminal of the battery and the andode (*A*) is connected to the negative terminal. Therefore the diode is reverse biased hence the voltage drop across the diode.

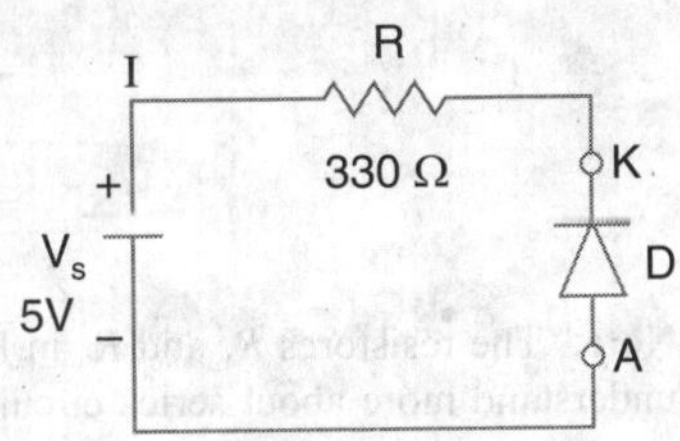

Fig. 12.18.

$$V_D = V_S = 5\ V \ \textbf{Ans.}$$

(*b*) *Voltage drop across the resistance*

Since the diode is reverse biased, therefore, there is no current theough the diode or resistance, hence, no voltage drop across resistance, *i.e.*,

$$V_R = 0 \ \textbf{Ans.}$$

(*c*) *Current through the circuit*

Since the diode is reverse biased, therefore the circuit current is zero, *i.e.*,

$$I = 0 \ \textbf{Ans.}$$

Example 12.12.

A series circuit consist of a 12-V source, a 450 Ω resistor (R) and a diode (D) as shown in Fig. 12.19. Deternine the value of V_D, V_R and I.

Solution.

Given : $V_S = 12\ V$; $R = 470\ \Omega$ and D = An ideal diode

Value of V_D

Notice that the diode, *D* shown in Fig. 12.19 is forward biased because its anode (*A*) is connected to the positive terminal of the battery and the cathode (K) to the negative terminal. Since the diode is assumed to be an ideal.

$$\therefore \quad V_D = 0 \textbf{ Ans.}$$

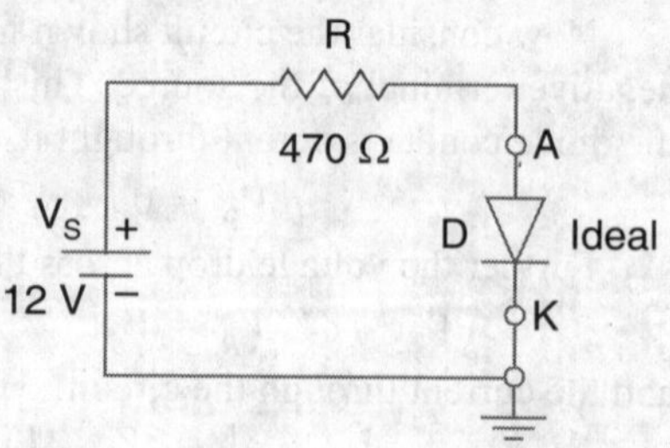

Fig. 12.19.

Value of V_R

Since there is no voltage drop across an ideal diode, the voltage drop across, the resistor.

$$V_R = V_S = 12\ V \textbf{ Ans.}$$

Curent throught the circuit

The curent through the circuit,

$$I = \frac{V_S}{R} = \frac{12}{470} = 2.53 \times 10^{-2}\ \text{A}$$

$$= 22.53 \text{ mA } \textbf{Ans.}$$

Example 12.13.

A series-circuit consist of forward biased diode, and two resistors 330 Ω and 470 Ω connected as shown in Fig. 12.20. Find the value of current through the circuit. Assume the diode to be real diode.

Solution.

Given : $V_S = 6$ V; $R_1 = 330\ \Omega$; $R_2 = 470\ \Omega$; $V_D = 0.7$ V.

We know that total resistance in the circuit shown in Fig. 5-20,

$$R_T = 330 + 470 = 800\ \Omega$$

and the current through the diode,

$$I = \frac{V_S - 0.7}{R_T} = \frac{6 - 0.7}{800}$$

$$= 6.62 \times 10^{-3}\ \text{A}$$

$$= 6.62 \text{ mA } \textbf{Ans.}$$

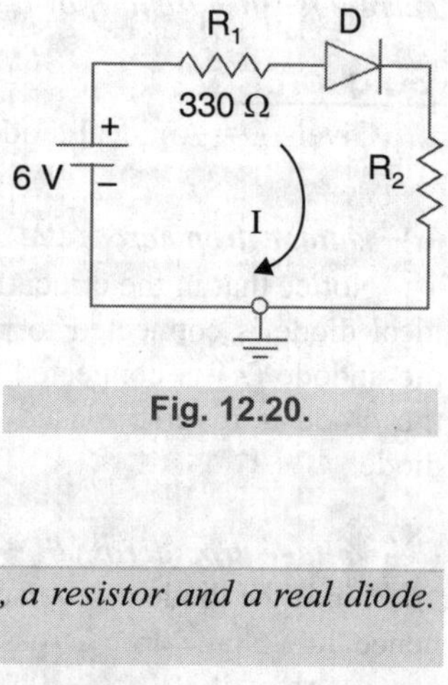

Fig. 12.20.

Note. The resistores R_1 and R_2 in Fig. 5-16 are connected in series. The understand more about series circuit, refer to Appendix *B: BC* rersestive cirucits.

Example 12.14.

Fig. 12.21 shows a series circuit consisting of a voltage source, a resistor and a real diode. Determine the voltage across the resistor and the circuit current.

Solution.

Given: $V_S = 5$ V; $R = 510\ \Omega$; $V_F = 0.7$ V

(*a*) Voltage across *R*

We know that the voltage drop across R,

$$V_R = V_S - 0.7 = 5 - 0.7 = 4.3 \text{ V } \textbf{Ans.}$$

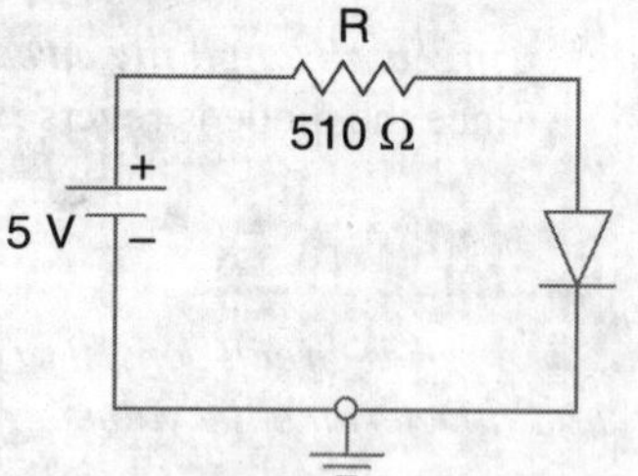

Fig. 12.21.

(*b*) Circuit current

We know that current through the given series circuit,

$$I = \frac{V_R}{R} = \frac{4.3}{510} = 0.00843$$

$$= 8.43 \times 10^{-3}\ \text{A}$$

$$= 8.43 \text{ mA } \textbf{Ans.}$$

Example 12.15.

Determine the value of the total current throught the series circuit shown in Fig. 12.22. Assume both the diodes D_1 and D_2 to be real.

Solution.

Given, $V_S = 6$ V; $V_{D1} = V_{D2} = 0.7$ V; $R = 1.5$ kΩ $= 1.5 \times 10^3$ Ω

We know tha the total series current through the given circuit,

$$I = \frac{V_S - V_{D_1} - D_{D_2}}{R}$$

$$= \frac{6 - 0.7 - 0.7}{1.5 \times 10^3}$$

$$= 3.067 \times 10^{-3} \text{ A}$$

$= 3.067$ mA **Ans.**

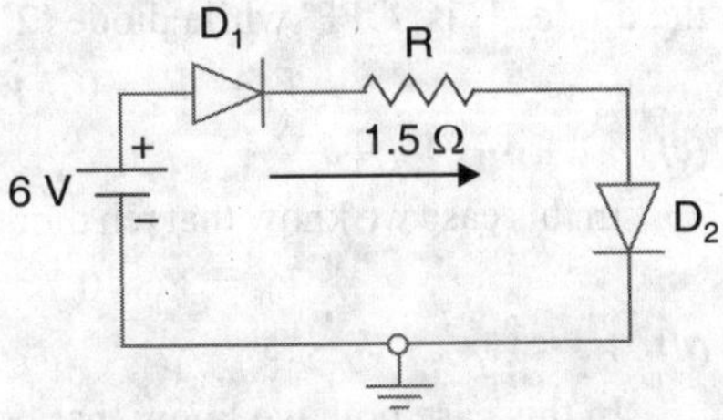

Fig. 12.22.

Example 12.16.

A series circuit consists of two forward biased real diodes, and two resistore connected as shown in Fig. 12.23. Determine the value of total current through the circuit.

Solution.

Given: $V_S = 12$ V; $R_1 = 1.5$ kΩ $= 1.5 \times 10^3$ Ω;

$R_2 = 1.8$ kΩ $= 1.8 \times 10^3$ Ω; $V_{D1} = V_{D2} = 0.7$ V.

We know that current through the given series circuit,

$$I = \frac{V_S - 0.7 - 0.7}{R_T} = \frac{12 - 0.7 - 0.7}{(1.5 \times 10^3) + (1.8 \times 10^3)}$$

$$= \frac{10.6}{3.3 \times 10^3} = 3.212 \times 10^{-3} \text{A}$$

$= 3.212$ mA **Ans.**

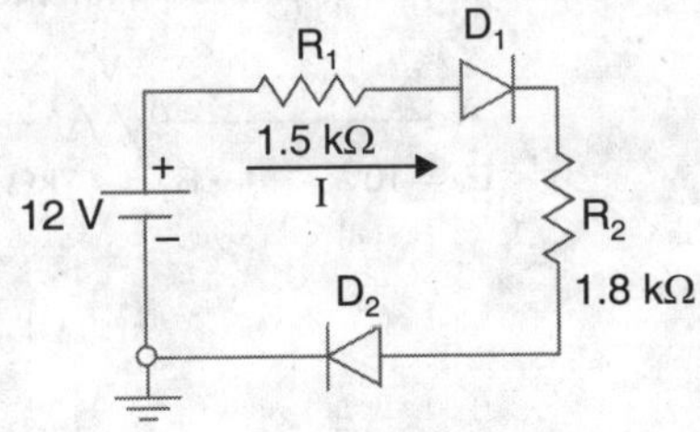

Fig. 12.23.

Example 12.17.

For the circuit shown in Fig. 12.24, determine the output voltage when

(a) $V_1 = V_2 = 0$ *(b)* $V_1 = 0, V_2 = 5$ V

(c) $V_1 = 5$ V, $V_2 = 0$ V *(d)* $V_1 = V_2 = 5$ V

(Banglore Univ. B.Sc. Electronics, May 2001)

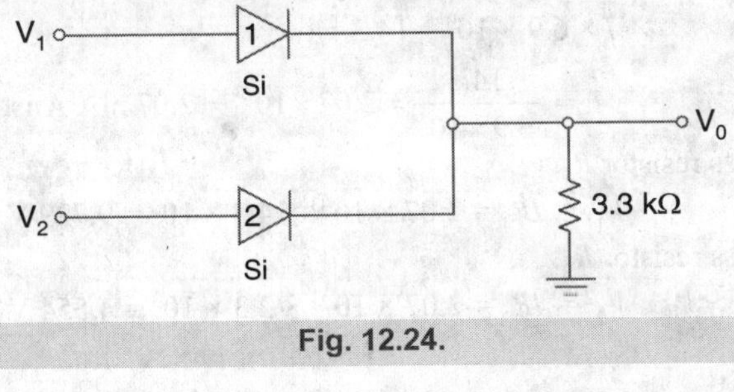

Fig. 12.24.

Solution.

(a) $V_1 = V_2 = 0$

In this case, we know that $V_1 = V_2 = 0$, both the Si diodes are reverse biased. Therefore the output voltage $V_0 = 0$. **Ans.**

(b) $V_1 = 0, V_2 = 5$ V

In this case we know that diode '1' is reverse biased while diode '2' is forward biased. Therefore the diode '1' is 'OFF' while diode '2' is ON. Therefore the output voltage,

$$V_0 = V_2 - 0.7\ V = 5 - 0.7 = 43\ \text{V}\ \textbf{Ans.}$$

(c) $V_1 = 5\ V, V_2 = 0$ V

In this case we know that teh diode '1' is ON while diode '2' is OFF. Therefore the output voltage,

$$V_0 = V_1 - 0.7 = 5 - 0.7 = 4.3\ \text{V}\ \textbf{Ans.}$$

(d) $V_1 = V_2 = 5$ V

In this case both we know that both diodes are ON, therefore the output voltage,

$$V_0 = V_1 - 0.7\ \text{V}\ \text{ or }\ V_2 - 0.7$$
$$= 5 - 0.7\ \text{V}$$
$$= 4.3\ \text{V}\ \textbf{Ans.}$$

Example 12.18.

Determine I, V_1, and V_2 for the series d.c configuration of the following Fig. 12.25.

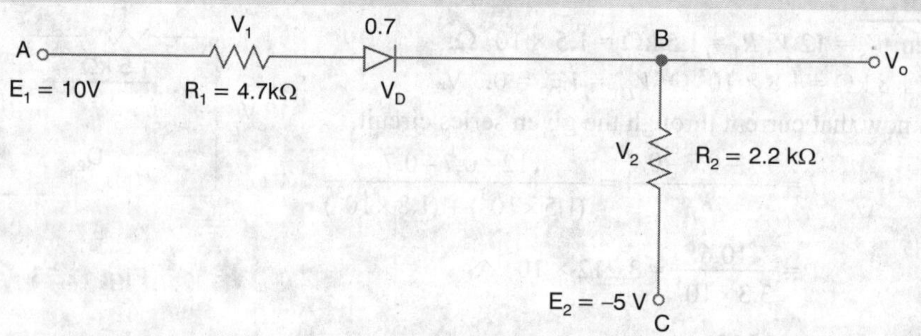

Fig. 12.25.

Solution.

Given: $E_1 = 10$ V; $V_D = 0.7\ V$; $R_1 = 4.7\ \text{k}\Omega = 4.7 \times 10^3\ \Omega$ and $R_2 = 2.2 \times 10^3\ \Omega$.

Applying Kirchhoff's Voltage Law in Loop *ABCA*,

$$IR_1 + V_D + IR_2 - E_2 - E_1 = 0$$
$$4.7 \times 10^3 + 0.7 + I \times 2.2 \times 10^3 - 5 = 0$$
$$I \times 6.9 \times 10^3 - 14.3 = 0$$
$$I = \frac{14.3}{6.9 \times 10^3} = 2.07 \times 10^{-3} = 2.07\ \text{mA}\ \textbf{Ans.}$$

Voltage drop across resistor R_1,

$$V_1 = IR_1 = 2.07 \times 10^{-3} \times 4.7 \times 10^3 = 9.729\ \text{V}\ \textbf{Ans.}$$

Voltage drop across resistor R_2,

$$V_2 = IR_2 = 2.07 \times 10^{-3} \times 2.3 \times 10^3 = 4.554\ \text{V}\ \textbf{Ans.}$$

Example 12.19.

Calculate current through 48 Ω resistance in the circuit shown in Fig. 12.26 below. Consider the diode to be of silicon and forward resistance of diode is 1 Ω. (Nagpur University, Summer 2011)

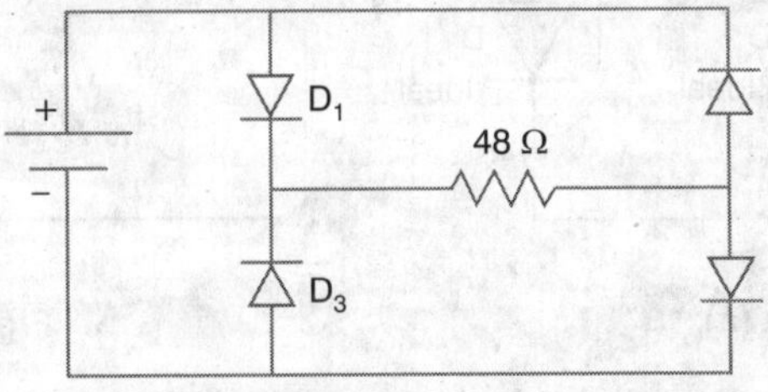

Fig. 12.26.

Solution.

As seen from Fig. 12.27 the diode D_1 and D_4 are forward biased and the diode D_2 and D_3 are reverse biased let us replace. The reverse biased diode open circuit and the forward biased diode with a battery voltage equivalent to knee voltage $V_K = 0.7$ V and 1 Ω resistance. The resulting diagram is as an shown in Fig. 12.27.

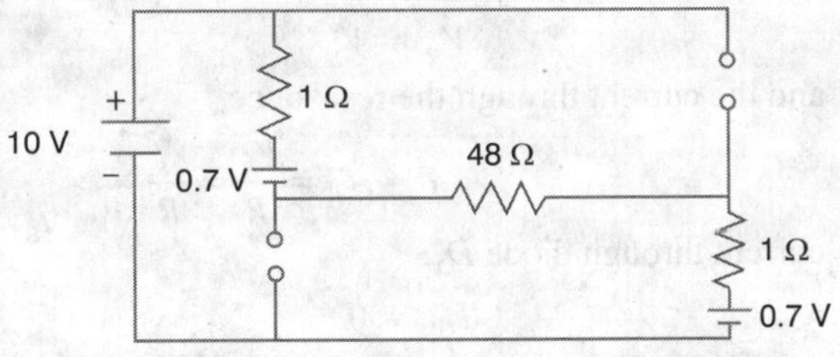

Fig. 12.27.

We know that the current through 48 Ω resistor,

$$= \frac{10 - 0.7 - 0.7}{50} = 172 \times 10^{-3} \text{ A} = 172 \text{ mA} \quad \textbf{Ans.}$$

12.18. Parallel Configuration of Diode Circuits with a DC Voltage Source

We have already discussed about series diode configurations in the last article. Now we shall study the parallel and series –parallel configurations of diode circuits with a DC voltage source. Let us consider two ideal diodes connected in parallel to a voltage source and resistor as shown in Fig. 12.28. We will determine the various voltages and currents through the circuit.

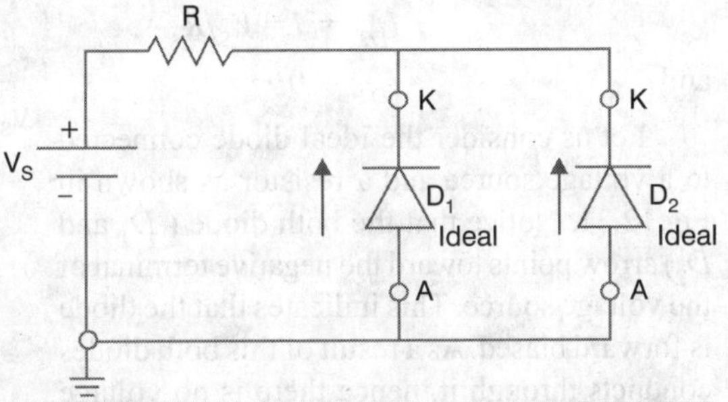

Fig. 12.28.

Now consider the circuit shown in Fig. 12.28. Notice that both the ideal diodes and arrow points toward the positive terminal of the voltage source. This indicates that the diodes are reverse biased. Because of this there will be no current through the circuit (*i.e.*, $I = 0$). Therefore full applied voltage is dropped across the diodes and there is no current through, R, there is no voltage drop across this component, *i.e.*, $V_R = 0$.

Next consider the circuit shown in Fig. 12.29 (*a*) and Fig. 12.29 (*b*). In these circuits the position of one of the diodes is changed. Let us consider Fig. 12.29 (*a*), notice that the diode D_2 arrow points toward the negative terminal and the diode D_1 arrow points toward the positive terminal of the voltage source. Arroios indicate that the diode D_1 is reverse biased and diode D_2 is forward biased.

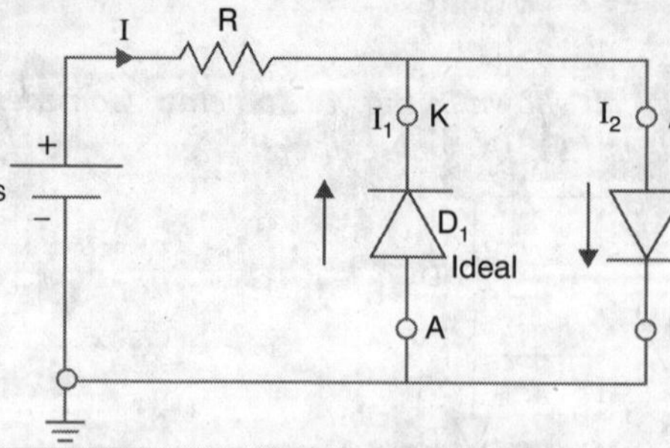

Fig. 12.29 (a)

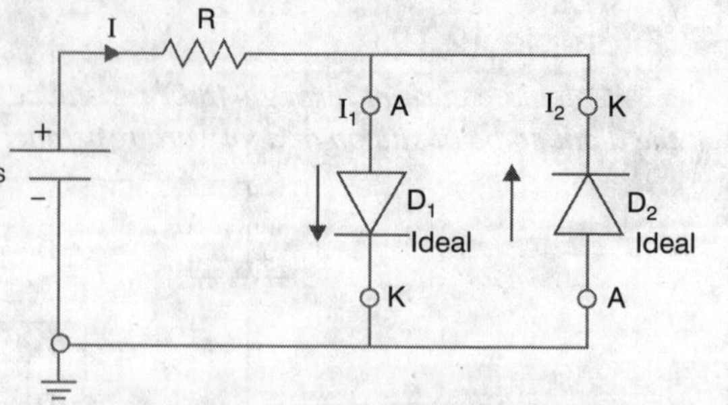

Fig. 12.29 (b)

There will be no current through the diode D_1 and diode D_2 conducts current through it. There is no voltage drop across diode D_2

$$V_{D_2} = 0$$

The voltage drop the resistor is equal to the dc source voltage, i.e.,

$$V_R = V_S$$

and the current through the resistance,

$$I = I_R = \frac{V_R}{R} = \frac{V_S}{R}$$

current through diode D_1,

$$I_{D_1} = 0$$

current through diode D_2,

$$I_{D_2} = I$$

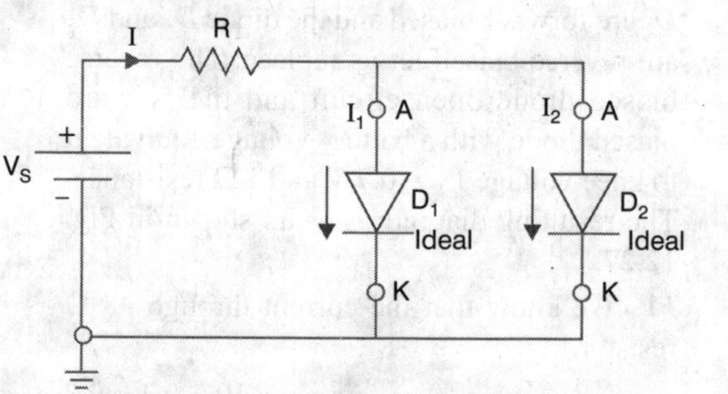

Fig. 12.30

Similarly, for Fig. 12.31 (*b*), we can study in detail and can conclude that,

$$V_{D1} = 0$$

$$I_{D2} = I = V_S/R$$

and $$I_{D2} = 0$$

Let us consider the ideal diode connected to a voltage source and a resistor as shown in Fig.12.31. Notice that the both diode (D_1 and D_2) arrow points toward the negative terminal of the voltage source. This indicates that the diode is forward biased. As a result of this both diodes conducts through it, hence there is no voltage drop across it.

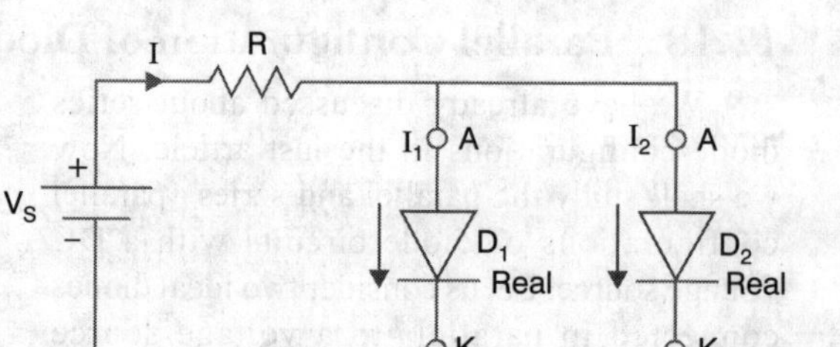

Fig. 12.31

$$V_{D_1} = V_{D_2} = 0$$

The voltage drop across the resistor is equal to the dc source voltage, i.e.,

$$V_R = V_S$$

and the current through the circuit,

$$I = \frac{V_R}{R} = \frac{V_S}{R}$$

Next consider the circuit shown in Fig. 12.31. Notice that this circuit is similar to the circuit shown in Fig. 12.31 assume that the diode is real 'one'. In such a circuit, the voltage drop across the parallel diode,

$$V_{D_1} = V_{D_2} = 0.7$$

and the current through the circuit,

$$I = \frac{V_R - V_{D_1}}{R} = \frac{V_R - V_{D_2}}{R} (A)$$

assuming that the diode having the similar characteristics, we have

$$I_{D_1} = I_{D_1} = \frac{1}{2}$$

Example 12.20.

Determine V_0, V_1, V_{D1} and V_{D2} for parallel diode configuration of Fig. 12.32.

(GBTU/MTU., 2008-09)

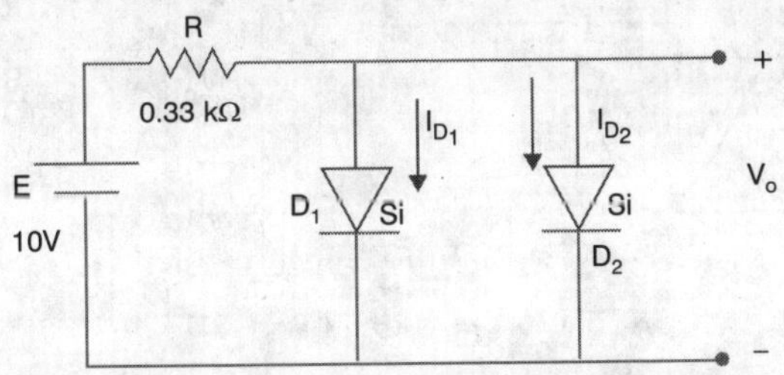

Fig. 12.32.

Solution.

Given: $R = 0.33\ k\Omega = 0.33 \times 103\Omega$; $E = 10V$ and $V_D = 0.7V$.

We know that the voltage across parallel elements is the same and,

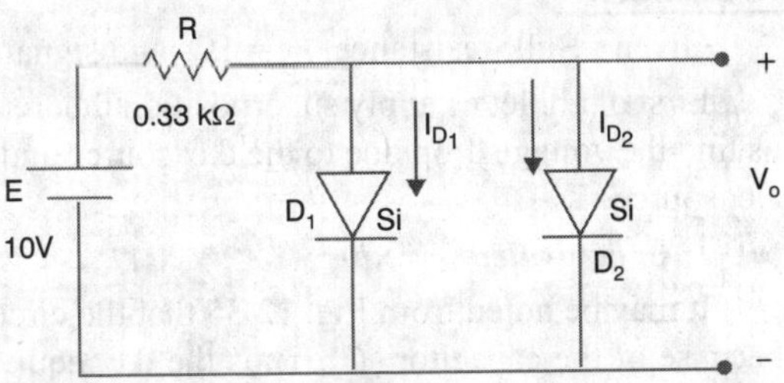

Fig. 12.33.

$$V_D = 0.7V$$

$$\therefore \quad V_o = V_D = 0.7V \textbf{ Ans.}$$

From the Fig. 12.33 the current,

$$I_1 = \frac{V_R}{R} = \frac{E - V_D}{R} = \frac{10 - 0.7}{0.33} = 28.18 \text{ mA } \textbf{Ans.}$$

We know that,

$$I_{D1} = I_{D2} = \frac{I_1}{2} = \frac{28.18}{2} = 14.09 \text{ mA } \textbf{Ans.}$$

12.19. Diode Circuit with D.C. and A.C. Voltage Sources

The electronic circuits used in radios and televisions contain both d.c. and a.c. sources. We have already discussed in Art. 5.9 that such circuits can be easily analysed by using superposition theorem. According to this theorem, first we draw a d.c. equivalent circuit (neglecting a.c. sources) and find the required current and voltage values. Now we draw the a.c. equivalent circuit (neglecting the d.c. sources) and again find the current and voltage values. Finally, we superimpose the two sets of values to obtain the result. While drawing the equivalent circuits, the following two points must be taken care of:

1. Direct current cannot flow 'through' a capacitor. Therefore all capacitors look like an open to a d.c. source.
2. The capacitors used in most of the circuits are large enough to look like a short to an a.c. source. It is particularly true for circuits operating at high frequency. Similarly, d.c. sources would also act as short circuit, unless they have very high internal resistances.

Example 12.21.

The silicon diode shown in Fig. 12.34 has a bulk resistance of 1Ω. The frequency of the 10 mV (peak) signal is so high that the reactance of the coupling capacitor may be taken as zero.

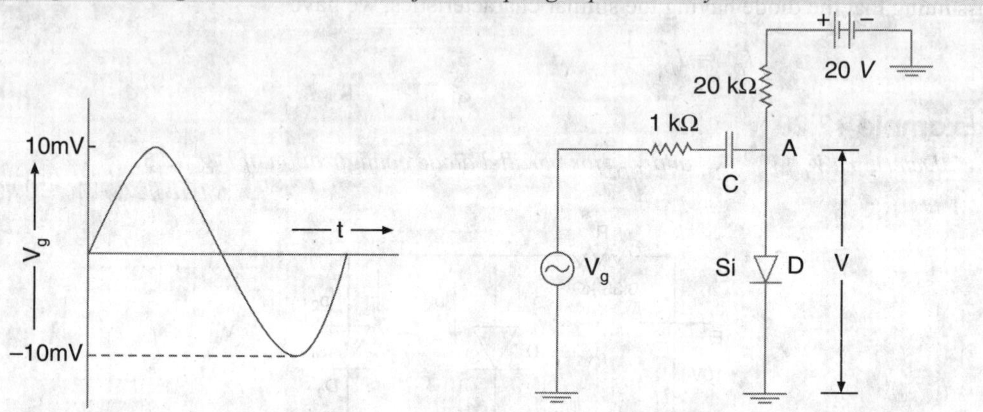

Fig. 12.34.

Sketch the approximate waveform of the total voltage 'V' across the diode.

Solution.

Given : Bulk-resistance, r_B = 1Ω and signal amplitude = 10 mV (peak).

First of all, let us apply superposition theorem to find the voltage drop 'V' across the diode. Let us find the voltage drop due to the d.c. source and then it will be added to the voltage drop due to the a.c. source.

(a) *D.C. equivalent circuit*

It may be noted from Fig. 12.35 that the circuit to the left of the point A is 'open' to d.c. source, because of the capacitor 'C'. Thus the d.c. equivalent circuit is as shown in Fig. 12.34 (*a*). In this circuit, the diode is forward biased by the d.c. source. Under this condition, the voltage drop across the diode is equal to 0.7 V d.c. and current through it,

$$I = \frac{20 - 0.7}{20} = 1 \text{ mA} \quad \textbf{Ans.}$$

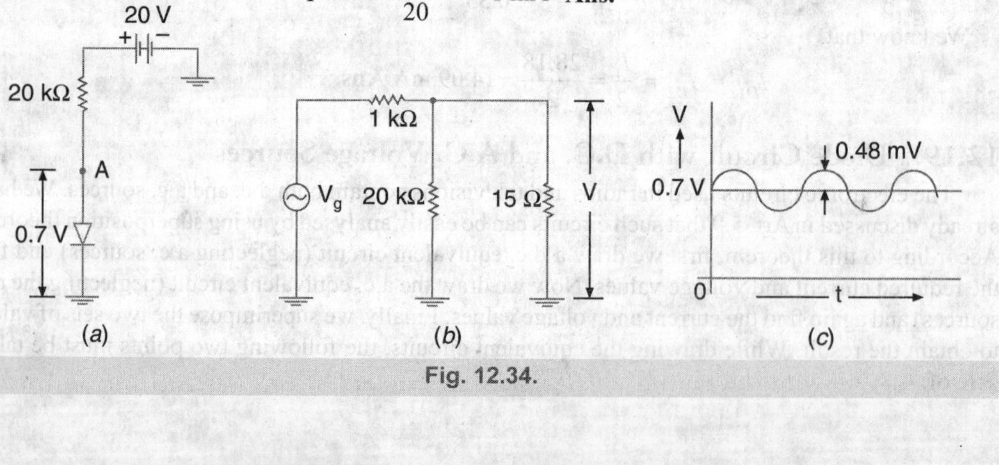

Fig. 12.34.

(b) A.C. equivalent circuit

Here, the capacitor '*C*' and the 20 *V* d.c. source are treated as shorts. This gives us the a.c. equivalent circuit as shown in Fig. 12.35 (*b*). Now, for a given silicon diode, the junction resistance,

$$r_j = \frac{50}{1} = 50\ \Omega$$

and the a.c. resistance, $r_e' = r_B + r_j = 1 + 50 = 51\ \Omega$

It may be noted from Fig. 12.35 (*b*), that for an a.c. signal source, 20 k Ω and 51 Ω are in parallel. The equivalent resistance of this parallel combination

$$= 20\ \text{k}\Omega \parallel 51\ \Omega = \frac{20\ 000 \times 51}{20\ 000 + 51} \approx 51\ \Omega$$

Hence, 1 k Ω and 51Ω are put in series across the a.c. signal source of peak value 10 mV. The peak value of the a.c. voltage drop across 51Ω resistance

$$= 10 \times \frac{51}{51 + 1000} = 0.48\ \text{mV}$$

The total voltage drop across the diode is equal to the sum of d.c. and a.c. voltage drops and is as shown in Fig. 12.35 (*c*). It may be noted that the resulting waveform consists of a d.c. voltage of 0.7 *V* over which an a.c. voltage of peak value 0.48 mV is superimposed. **Ans.**

NUMERICAL PROBLEMS

1. A silicon diode has a saturation current of 0.01 μA at room temperature (300 K). Find the saturation current at 400 K. **(Ans. 10.24 μA)**
2. For silicon diode with $I_0 = 2.5$ μA at 300 K, find the forward voltage at a forward current of 10 mA. **(Ans. 0.4 V)**
3. Calculate the voltage which has to be applied across a PN semiconductor junction to cause a forward current of 5 A/cm², assuming that the saturation current density at 300 K is 25 μA/cm². **(Ans. 0.316 V)**
4. For what voltage will the reverse current in a PN junction germanium diode attains a value of 90% of its saturation value at room temperature? **(Ans. 0.06 V)**
5. A junction diode has a maximum power dissipation of 250 mW and the reverse saturation current of 100 mA. Calculate the value of the maximum forward current at a temperature of 27°C. **(Ans. 2.85 A)**

   [**Hint.** Forward voltage drop across the diode

   $$V_F = 250 \times 10^{-3} / 100 \times 10^{-3} = 2.5\ V$$

   Use this value in the relation $I = I_0 \cdot (e^{V_F/V_T} - 1)$]
6. Determine the voltage across a silicon diode for value for value of $I_F = 5$ mA. Assume that bulk resistance of the diode is 5Ω. **(Ans. 0.625 V)**
7. Determine the applied voltage to achieve a forward current of 0.45 μA in a PN junction silicon diode at *T* = 300 K, if the reverse saturation current is 1.0 nA. (*b*) If the reverse saturation current in a germanium diode is 10 μA, what current would result if the voltage in part (*a*) is applied in the forward direction? **(Ans. 0.3 V and 1.63 A)**
8. Using approximate Boltzmann's diode equation, find the change in forward bias for doubling the forward current of a germanium semiconductor at 290 K.

   (*Basic Electronics, Osmania Uni. 1993*) **(Ans. 17.3 mV)**

9. A certain PN junction diode has a leakage current of 10^{-14} A at room temperature of 27°C and 10^{-9} A at 125°C. The diode is forward biased with a constant current source of 1 mA at room temperature. If the current is assumed to remain constant, calculate the junction barrier voltage at room temperature and at 125°C.

(Ans. At 25°C ⇒ 0.66 *V*; at 125°C ⇒ 0.54 *V*)

10. A silicon diode has a forward voltage drop of 1.2 *V* for a forward dc current of 100 mA. It has a reverse current of 1 mA for a reverse voltage of 10 *V*. calculagte:
(*a*) bulk and reverse registance of a diode.
(*b*) ac resistance at forward dc current of (*ii*) 2.5 mA and (*i*) 25 mA.

(Ans. (*a*) $r_B = 5\ \Omega$; $R_R = 10\ M\Omega$ (*b*) r_{ac} at 2.5 mA = 15 Ω and at 25 mA = 6 Ω)

11. Using analytical expression for diode curent, calculate the dynamic slope resistance of a germanium diode at 290 K, when forward biased at curent of (*i*) 10 mA and (*ii*) 5 mA.

(Ans. (*i*) $r_{ac} = 2500\ \Omega$, (*ii*) $r_{ac} = 5\ \Omega$)

12. A germanium diode draws 40 mA with a forward bias of 0.25 V. The junction is at room temperature of 293 K. Calculate the reverse saturation curent of the diode. **(Ans. 1.82 μA)**

13. Calculate forward curent in Germanium diode at 20°C when forward voltage is 0.3 V. Compare this value with that after a temperature rise of 50°C. Assume that reverse saturation curent doubled for every 10°C rise in temperature.

(Ans. $I_{20°C} = 162, 75\ I_O$; $I_{70°C} = 704, 800\ I_O$; This indicates that $I_{70°C} = 4.3\ I_{20°C}$; *i.e.*, for a rise in temperature of 50°C, I_O has increased 32 times whereas the forward curent has increased only 4.3 times for the same rise in temperature)

14. Determine the value of V_D, V_R and I_T for the circuits shown in Fig. 12.36 (*a*), (*b*) and (*c*) respectively Assume the diode to be an ideal.

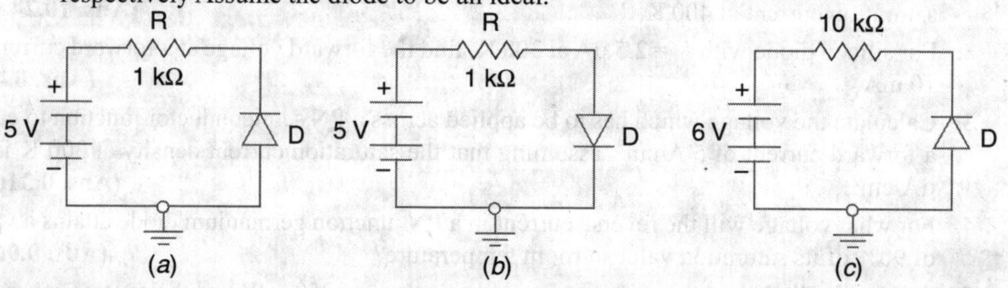

Fig. 12.35.

(Ans. (*a*) $V_D = 5$ V, $V_R = 0$, $I_T = 0$A (*b*) $V_D = 0$; $V_R = 5$ V; $I_T = 5$ mA; (*c*) $V_D = 0$; $V_R = 6$ V; $I_T = 6$ mA)

15. Determine the current through resistor, (*R*) in the circuits shown in Fig. 12.37 (*a*), (*b*) and (*c*) respectively. Assume the diode to be real. **(Ans. (*a*) 0.53 mA; (*b*) 1.26 mA; (*c*) 2.61 mA)**

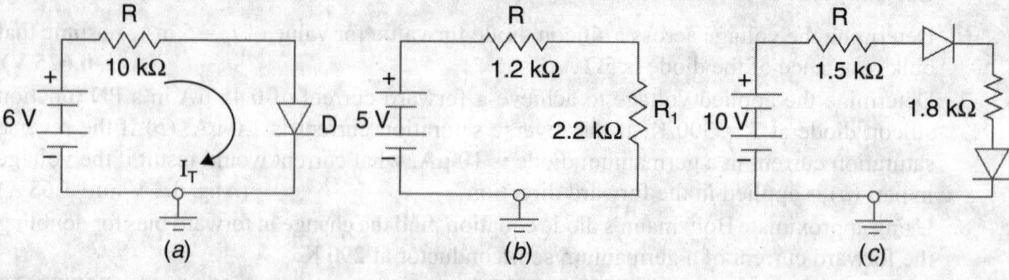

Fig. 12.36.

DESCRIPTIVE QUESTIONS

1. What is a PN junction diode? How its terminals are identified?
2. Draw the *V-I* characteristics of a junction diode when it is (*a*) forward biased and (*b*) reverse biased.
3. For silicon and germanium diode, state approximately, cut in voltages. Also compare their reverse saturation currents.
4. Briefly explain (without derivations) the behaviour of junction in forward bias and reverse bias mode and draw its volt-ampere characteristics.
5. Explain the effect of temperature on the volt-ampere characteristic of a diode.
6. What is an ideal diode and real diode?
7. Explain the following terms: (*a*) static resistance (*b*) bulk resistance (*c*) junction resistance, (*d*) a.c. or dynamic resistance and (*e*) reverse resistance of a diode.
8. What are the important applications of a diode?
9. Draw the symbol and explain the V-I characteristics of the PN junction diode. *(JNTU., Aug/Sep 2006, GBTU/MTU 2003-04)*
10. Draw the V-I characteristic of PN junction diode. Write the current-voltage equation, specify all terms of equation. Discuss its temperature dependence. *(GBTU/MTU., 2006-07)*
11. Sketch and explain the V-I characteristics of a semiconductor diode. *(M.G University., April/May 2004)*
12. With the volt-ampere characteristics, explain the working principle of the diode and also explain the static dynamic resistance of the diodes? *(Anna University., Nov/Dec 2007)*
13. If two PN junction diodes are connected back to back, will it work as a transistor will the terminal designated in the Fig. 12.38 explain. *(GBTU/MNTU., 2003-04)*

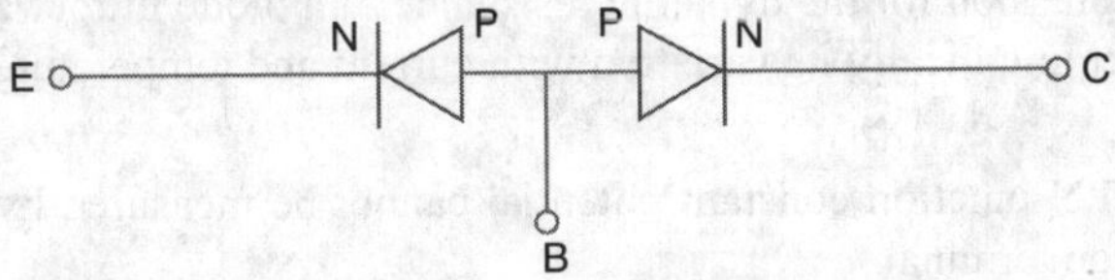

Fig. 12.38.

14. Define Cut in voltage of a diode. *(Anna University., April/May 2008)*
15. Explain the temperature effects on the V-I characteristics of a diode. *(PBUT., 2009)*
16. Explain the working of a PN Junction diode under reverse biased condition. *(GBTU/MNTU., 2004-05)*
17. Compare the V-I characteristics of Si and Ge diodes? *(M.G University., May/June 2006)*
18. Sketch typical forward and reverse characteristics for a germanium diode and for a silicon diode. Compare the characteristics and explain why the reverse saturation current in a silicon diode is much smaller than that in a comparable germanium diode. *(GBTU/MTU., 2008-09)*
19. Which of the two semiconductor material Si or Ge has largest conductivity at room temperature and why ? What happen to the conductivity of the semiconductor with the rise in the temperature? *(GBTU/MNTU., 2002-03)*
20. Differentiate between static and dynamic resistance of a diode. *(GBTU/MTU., 2009-10, PBUT., 2009-10)*

21. Explain how you will determine the static resistance and dynamic resistance of PN diode after defining both of them. *(GBTU/MTU., 2006-07, 2005-06)*
22. Define static and dynamic resistance of diode. How it can be obtained. *(GBTU/MTU., 2006-07)*
23. What is static and dynamic resistance in the forward biasing of the diode? Explain with suitable diagram. *(GBTU/MTU., 2005-06)*
24. Breakdown diode with a breakdown voltage greater than 5 V generally have a positive temperature coefficient but those with lower breakdown voltage have a negative temperature coefficient. Do you agree with statement ? Justify *(GBTU/MNTU., 2003-04)*
25. Explain, how a PN Junction diode can be tested? *(GBTU/MNTU., 2004-5)*
26. Applications of diodes. *(Anna University., Nov/Dec 2007)*
27. Describe the operation of a semiconductor PN junction diode under reverse bias condition. Also draw the V-I characteristics of Si diode and explain the meaning of dynamic resistance of diode. *(Jamia University., 2008)*
28. Draw the characteristics of an ideal diode. How does it differ from practical diode characteristics? Explain *(GBTU/MTU., 2004-05, 2008-09)*
29. Explain the different diode equivalent circuits with necessary approximation if any. *(VTU., Dec 08/Jan 09)*
30. For a reverse-biased diode, does the transition region increase or decrease in width ? Explain *(PBUT., Dec 2005)*
31. Write the analytical equation for the current which describes both the forward and reverse characteristics of a diode. *(PBUT., 2008)*
32. The resistance of a semiconductor material decrease with an increase in temperature – Justify this statement. *(PBUT., April 2005)*
33. Derive the expression for the dynamic resistance of a diode in forward bias condition and hence interpret graphically its variation with current and temperature. *(GBTU/MNTU., 2002-03)*
34. Explain why PN junction constant potential cannot be measured by placing a volt meter across the diode terminal. *(JNTU., Aug/Sep 2006)*
35. Determine the required PIV rating of each diode. *(Jamia University., 2008)*
36. Find the maximum current through each diode during conduction. *(Jamia University., 2008)*
37. What is the required power rating of each diode? *(Jamia University., 2008)*
38. Discuss the working of PN junction diode with the help of V-I characteristics. *(R.G.P.V., June 20009)*
39. Draw voltage current characteristics of a semiconductor diode. Explain the working of diode and current flow mechanism in forward and reverse bias. *(R.G.P.V., June 2005)*
40. Explain the working of the semiconductor diode characteristics. Does the characteristics temperature dependent, if so give reasoning and shape of new characteristics? *(R.G.P.V., May/June 2006)*
41. Draw the piecewise linear volt-ampere characteristics of a PN diode. What is the circuit model for the ON- state and the OFF state? *(R.G.P.V., Dec. 2002)*
42. Discuss piecewise linear volt-ampere diode characteristics with its significance. *(R.G.P.V., Dec. 2003)*
43. Compare Volt-ampere characteristics of Si and Ge diode. Explain why Si preferred over Ge in IC fabrication. *(R.G.P.V., June 2004)*

44. Write the volt ampere equation for a PN diode. Plot the volt ampere curve for PN diode and explain nature of the wave. *(R.G.P.V., June 2007, Dec. 2008)*

45. Discuss how an ideal diode is different from a practical diode. Draw the small signal equivalent circuit of a diode. *(R.G.P.V., June 2003)*

46. Describe in your own words the conditions established by forward and reverse bias condition on a PN junction diode. Describe how diffusion and transition capacitances differ. *(R.G.P.V., Feb 2010)*

47. Draw the piecewise linear volt-ampere characteristics of a PN diode. What is the circuit model for the ON state and the OFF state? *(R.G.P.V., Dec. 2002)*

48. Discuss piece wise linear volt-ampere diode characteristics with its significance. *(R.G.P.V., Dec. 2003)*

49. Derive V-I relation for a forward biased PN junction. *(Nagpur University, Winter 2005)*

50. Why cut in voltage is higher for silicon than that of Germanium? What is the effect of temperature on characteristics of PN junction diode? *(Nagpur University, Summer 2003)*

51. Derive V-I relationship for PN junction diode and briefly discuss its dependence on the temperature. *(Nagpur University, Winter 2001)*

52. Derive V-I relationship of a PN junction diode. Explain assumptions made, if any. *(Nagpur University, Summer 2001)*

53. Explain the effect of temperature on the characteristics of a diode. *(Nagpur University, Summer 2000)*

54. Develop the relationship between applied voltage across the PN junction diode and current flowing through it. *(Nagpur University, Winter 2002)*

55. When a diode is driven from forward condition to reverse condition, draw and explain waveforms for (*a*) minority carrier concertration at the junction of the diode, (*b*) current flowing through the diode circuit and (*c*) voltage aross the diode. Assume resistance (R_L) is present in series with diode. *(GTU., May 2011)*

56. How does the barrier field form in a PN junction? Is it possible to measure this with the help of a voltmeter. *(WBUT., 2011, 12)*

57. Explain in detail the V-I characteristics of a semiconductor diode. *(RGTU., Feb. 2010)*

58. With the help of V-I characteristics explain the working of semiconductor diode. *(RGTU., March, April, 2010)*

59. Define the following pertaining to the diode:

(*a*) Dynamic resistance

(*b*) Diffusion capacitance

(*c*) Reverse recovery time *(VTU., June, July, 2011)*

60. Discuss the approximate and piecewise linear model of a diode. *(VTU., June, July, 2011)*

MULTIPLE CHOICE QUESTIONS

1. The schematic symbol for a PN junction diode is:

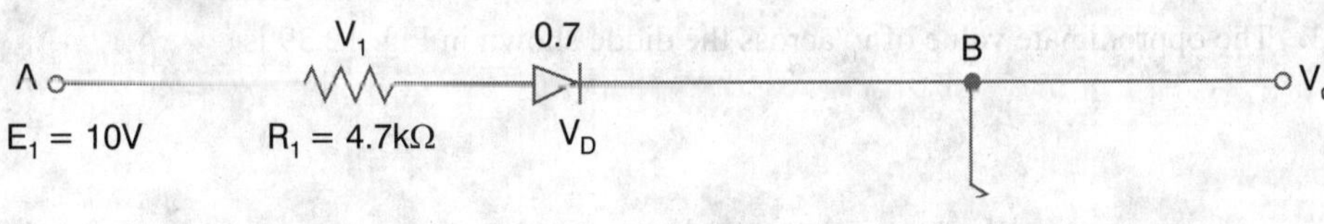

2. The conventional current in a PN junction diode flows:
   (*a*) From positive to negative
   (*b*) From negative to positive
   (*c*) In the direction opposite to the electron flow
   (*d*) Both (*a*) and (*c*) above
3. The cut in voltage (or knee voltage) of a silicon diode is
   (*a*) 0.2 V (*b*) 0.6 V (*c*) 0.8 V (*d*) 1.0 V
4. When a diode is reverse biased, it is equivalent to:
   (*a*) An OFF switch (*b*) An ON switch
   (*c*) A high resistance (*d*) None of the above
5. The resistance of a diode is equal to:
   (*a*) Ohmic resistance of the P-and N-semiconductors
   (*b*) Junction resistance
   (*c*) Reverse resistance
   (*d*) Algebraic sum of (*a*) and (*b*) above
6. For a silicon diode, the value of the forward-bias voltage typically,
   (*a*) must be greater than 0.3 V
   (*b*) must be greater than 0.7 V
   (*c*) depends on the width of the depletion region
   (*d*) depends on the concentration of majority cariesrs
7. When forward biased, a diode
   (*a*) blocks current (*b*) conducts current
   (*c*) has a high resistance (*d*) drops a large voltage
8. A PN junction diode's dynamic conduction is directly proportional to
   (*a*) the applied voltage (*b*) the temperature
   (*c*) the curent (*d*) the thermal voltage

   Hint: Conduction is reciprocal of the resistance.
9. The forward region of a semiconductor diode characteristic curve is where diode appears as
   (*a*) constant current source (*b*) a capacitor
   (*c*) on OFF switch (*d*) an ON switch
10. At room temperature of 25°C, the barrier potential for silicon is 0.7. It value at 125 °C is
    (*a*) 0.6 V (*b*) 0.3 V (*c*) 0.9 V (*d*) 0.7 V
11. Junction break down of a PN junction occurs
    (*a*) with forward bias (*b*) with reverse bias
    (*c*) because of manufacturing defect (*d*) None of these
12. Reverse saturatioin current in a silicon PN junction diode nearly doubles for every
    (*a*) 2 °C rise in temperature (*b*) 5 °C rise in temperature
    (*c*) 6 °C rise in temperature (*d*) 10 °C rise in temperature
13. The opproximate value of v_0 across the diode shown in Fig. 12.39 is

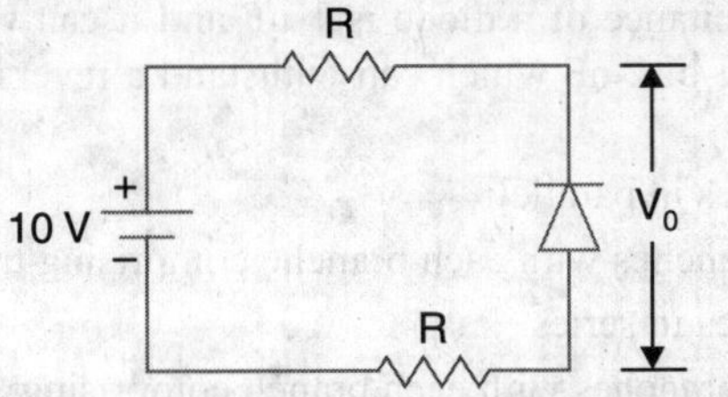

Fig. 12.39.

(*a*) zero (*b*) 10 V

(*c*) 5 V (*d*) dependent on the value of *R*

14. The diode *D* is an ideal in the circuit shown in Fig 12.40. The current, I will be:

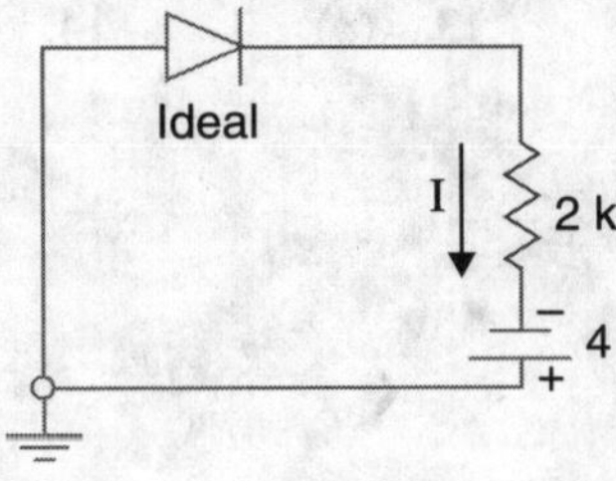

Fig. 12.40.

(*a*) – 2 nA (*b*) zero (*c*) 2 mA (*d*) 2 mA

15. The voltage at V_1 and V_2 of the arragement shown in Fig. 12.41 will be respectively:

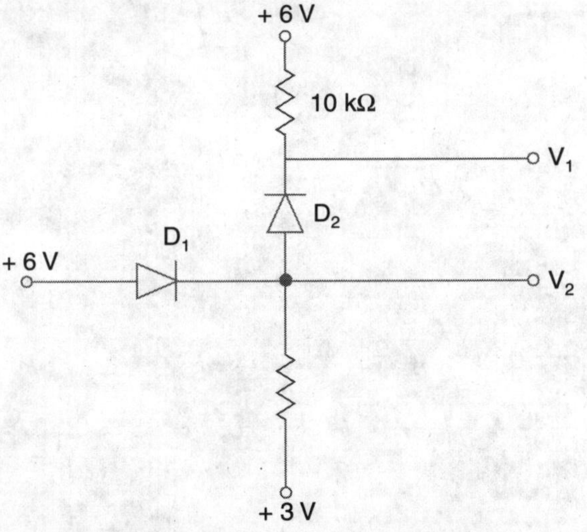

Fig. 12.41.

(*a*) 6 V and 5.4 V (*b*) 5.4 V and 6 V

(*c*) 3 V and 5.4 V (*d*) 6 V and 5 V

16. The transition capacitance of a diode is 1 nF and it can withstand a reverse potential of 400 V.A capacitance of 2 nF which can withstand a reverse potential of 1 kV is obtained by connecting

(*a*) two 1 nF diodes in parallel

(*b*) six parallel branches with each branche comprising three 1 nF diodes in series

(*c*) two 1 nF diodes in series

(*d*) three parallel branches with each branch comprising 1 nF diodes in series

ANSWERS

1. (*b*)	**2.** (*d*)	**3.** (*b*)	**4.** (*a*)	**5.** (*d*)
6. (*b*)	**7.** (*b*)	**8.** (*c*)	**9.** (*d*)	**10.** (*a*)
11. (*b*)	**12.** (*d*)	**13.** (*a*)	**14.** (*c*)	**15.** (*d*)
16. (*b*)				

13 Chapter

Special Purpose Diodes and Opto-Electronic Devices

Outline

1. Introduction.
2. Zener Diode.
3. Reverse Characteristic of a Zener Diode
4. Zener Diode Specifications
5. Zener Diode Equivalent Circuit
6. Zener Diode Applications
7. Tunnel Diode
8. V-I Characteristics of a Tunnel Diode
9. Tunnel Diode Parameters
10. Tunnel Diode Equivalent Circuit
11. Tunnel Diode Applications
12. Varactor Diode
13. Varactor Diode Specifications
14. Varactor Diode Applications
15. Schottky Doide
16. Schottky Diode Applications
17. Step-Recovery Diode
18. Harmonic Generator (or Frequency Multiplier)
19. PIN Diode
20. PIN Diode Applications
21. Light Emitting Diode (LED)
22. LED Voltage Drop and Current
23. LED Applications
24. Multicolour LEDs
25. Seven-Segment Display
26. Dot Matrix Displays
27. Use of LEDs in Facsimile Machines
28. Liquid Crystals Displays
29. Advantages of Liquid Crystal Displays (LCDs)
30. Applications of Liquid Crystal Displays (LCDs)
31. Photodiodes
32. Commercially Available Photodiodes
33. Photodiode Applications
34. Dust Sensor
35. Photoconductive Cell
36. Photovoltaic or Solar Cell
37. Laser Diode
38. Laser Diode Applications
39. Optical Discs
40. Read-only Optical Discs Equipment
41. Printers Using Laser Diodes
42. Hologram Scanners
43. Laser Range Finder
44. Light-activated SCR (LASCR)
45. Optical Isolators
46. Optical Modulators
47. Optical Fibre Communication Systems
48. Optical Fibre Data Links
49. Thermistor

13.1. Introduction

We have already discussed about general purpose diodes in the last chapter. Now we shall discuss several other types of diodes, which are used for specific applications. Such diodes are zener-, varactor-, schottky-, tunnel-, step recovery-, photo- and light emitting diodes. These diodes are also called *special purpose diodes* or simply special diodes. The devices that produce light or responds to the light (opto-electronic devices) are also discussed in this chapter.

13.2. Zener Diode

A zener diode is also called a voltage-reference, voltage regulator or breakdown diode. Like a rectifier diode, it is also important in many power applications. Fig. 13.1 (*a*) shows a schematic symbol for a zener diode. It will be interesting to know that the line at the end of the arrow looks like the letter *Z*.

Zener diode

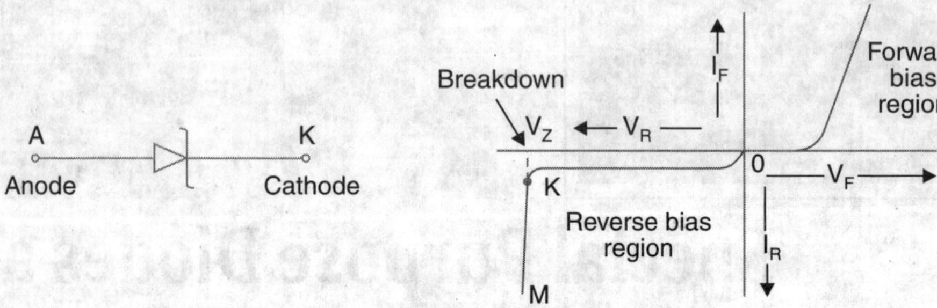

(*a*) Symbol (*b*) V.I. characteristic

Fig. 13.1. Zener diode.

The zener diode is a silicon PN junction device, which differs from a rectifier diode, in the sense, that it is operated in the reverse break down region. The breakdown voltage of a zener diode is set by carefully controlling the *doping level* during manufacture. A zener diode is made by heavily doping P and N sides of the junction. Due to this, the deplection region formed in the diode is very thin (less than 0.000001 m). Consequently, electric field for such a very thin deplection region is very high (about 500,000 v/m) even for a small reverse bias voltage of about 5V. As discussed in the last chapter, when a reverse voltage across a diode is increased, a critical voltage called breakdown voltage is reached at which the reverse current increases sharply as shown by the curve *KM* in Fig. 13.1 (*b*).

The reverse breakdown of a PN junction may occur either due to *avalanche* or *zener effect*. As discussed in chapter on PN-junctions the avalanche breakdown occurs, when the accelerated free electrons acquire sufficient energy to ionize a lattice atom by bombardment. The additional free electrons, created in this way, are accelerated by the reverse field causing more and more ionization. The multiplication of the number of free carriers causes the reverse current to increase rapidly.

The zener breakdown occurs when the electric field across the junction, produced due to reverse voltage, is sufficiently high. This electric field exerts a force on the electrons in the outer most shell. This force is so high that the electrons are pulled away from their parent nuclei and become free carriers. This ionization, which occurs due to the electrostatic force of attraction, is known as zener effect. It causes an increase in the number of free carriers and hence an increase in the reverse current.

The zener diodes, with breakdown voltages of less than 6 V, operate predominantly in zener breakdown. Those with breakdown, voltages greater than 6 V, operate predominantly in avalanche breakdown. Strictly speaking, the first one should be called a zener diode and the second an avalanche diode. But in actual practice, both the types are called zener diodes.

13.3. Reverse Characteristic of a Zener Diode

We have just discussed that a zener diode is operated only in the reverse-bias region. That is why we shall examine its characteristic in this region only. Fig. 13.2 shows the reverse portion of the V-I characteristic of the zener diode. It may be noted from this figure that as the reverse voltage (generally written as V_R) is increased, the reverse current (usually called zener current, I_Z) remains negligibly small up to the 'knee' of the curve (point K in Fig. 13.2). At this point, the effect of breakdown process begins. From the bottom of the knee, the breakdown voltage (also called zener breakdown voltage or simply zener voltage V_Z) remains essentially constant. This ability of a diode is called regulating ability and is an important feature of a zener diode. It maintains, an essentially a constant voltage across its terminals over a specified range of zener current values. Following two points are important from the characteristics of a zener diode:

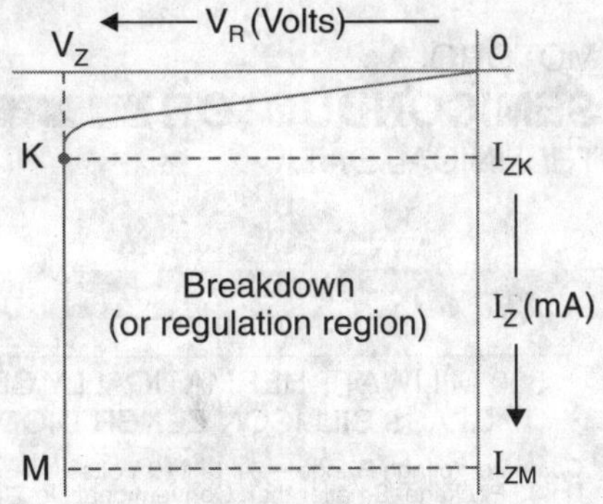

Fig. 13.2. Reverse characteristic of a zener diode.

1. There is a minimum value of zener current called breakover current designated as I_{ZK} or $I_{Z\,(min)}$ which must be maintained in order to keep the diode in breakdown (or regulation) region. When the current is reduced below the knee of the curve, the voltage changes drastically and the regulation is lost.
2. There is a maximum value of zener current designated as I_{ZM} or $I_{Z\,(\text{max})}$ above which the diode may be damaged. The value of this current is given by the maximum power dissipation of the zener diode. As long as the maximum power dissipation is not exceeded, the diode will not be damaged. It will come out of the breakdown region, when the applied reverse voltage is reduced below the breakdown voltage.

13.4. Zener Diode Specifications

The zener diodes are generally specified in terms of four factors namely zener voltage (V_Z) maximum power dissipation (P_{ZM}, $P_{D\,(max)}$ or $P_{Z\,(max)}$), breakover current (I_{ZK}) and zener resistance (r_Z). They are available with breakdown voltages ranging from 1.8 V to 2000 V.

The power dissipation of a zener diode is the product of breakdown voltage (V_Z) and reverse current (I_Z). Mathematically, the power dissipation,

$$P_Z = V_Z \times I_Z$$

The maximum value of power dissipation, which a zener can dissipate, without failure, is called power rating and is designated by P_{ZM}. The zener diodes are available with power rating from 150 mW to 50 W. Generally, the data sheets of zener diodes include the maximum current, a zener can handle without exceeding its power rating. The value of this maximum current is given by the relation,

$$I_{ZM} = \frac{P_{ZM}}{V_Z}$$

where P_{ZM} = Power rating of a zener diode, and

V_Z = Breakdown voltage.

Fig 13.3 (*a*) and (*b*) shows data sheet of silicon zener diodes. Notice that it shows zener diodes with three different number series, (1) 1N 746 through 1N 759, (2) 1N 957 A through 1N 986 A and (3) 1N 4370 through 4372. All these motorola zener diode have a maximum power rating of 500 mW as seen from the table marked "Maximum Ratings". The devices can handle this amount of power until 50°C. If the device is used beyond 50°C temperature, power handling capability reduces by 3.33 mW/°C. From the table marked "Electrical Characteristics", we can find the nominal zener

MOTOROLA
SEMICONDUCTOR
TECHNICAL DATA

1.5KE.8, A thru 1.5KE250,A
See Page 4- 59

500-MILIWATT HERMATICALLY SEALED GLASS SILLICON ZENER DIODES

- Compete Voltage Range – 2.4 to 110 Volts
- Do-35 Package–Smaller than Conventional Do-7 package
- Double slug Type Construction
- Metallurgically Bonded Construction
- Oxide Passivated Die

Designers' Data for 'Worst Case"Conditions

The Designer's Data sheets permit the design of must circuits entirely from the information presented. Limit curves – representing boundaries on device characterisitics – are given to facillitate "worst case" design.

1N1746 thru 1N759
1N957A thru 1N986A
1N4370 thru 1N4372

GLASS ZENER DIODES
500 MILLIWATTS
2.4-110 VOLTS

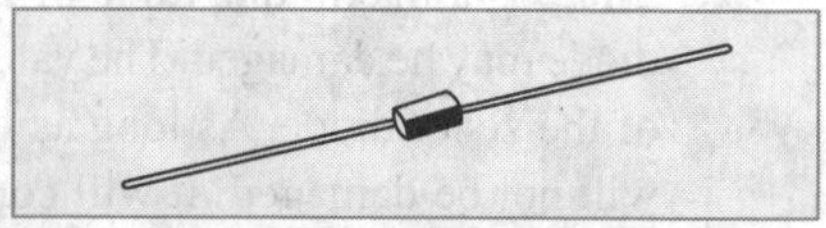

MAXIMUM RATING

Rating	Symbol	Value	Unit
DC Power Dissipation @ T_L < 50°c.	P_D		
•JEDEC Registration		400	mW
•Derate above T_L = 50°c		3.2	mW/°c
Motorola Device Rating		500	mW
Derate above T_L = 50°c		3.33	mW/°c
Operating and Storage Junction Temperature Range	T_J T_{stg}		°c
• JEDEC Registration		-65 to + 175	
Motorola Device Rating		-65 to + 200	

•Indicates JEDEC Registered Date.

MECHANICAL CHARACTERISTICS

MAXIMUM LEAD TEMPERATURE FOR SOLDERING PURPOSES:230 °C, 1/16" from case for 10 seconds

FINISH: All external surface are corrosion resistant with readily solderable leads.

POLARITY: Cathode indicated by color band. When operated in ziner mode, cathode will be possitive with respect to anode

MOUNTING POSITION: Any

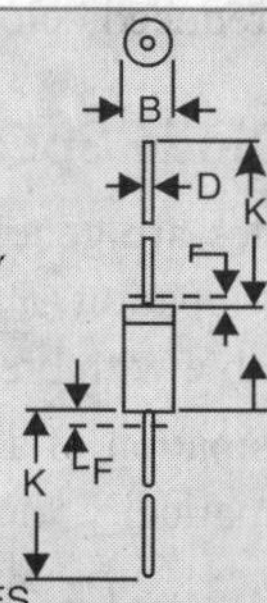

NOTES
1. PACKAGE CONTOUR OPTIONAL WITHIN A AND B. HEAT SLUGS,IF ANY. SHALL BE NOT SUBJECT TO THE MINNIUM LIMIT OF B.
2. LEAD DIAMETER NOT CONTROLLED IN ZONE TO ALLOW FOR FLASS,LEAD FINISH BUILDUP AND MINOR IRRUGULARITIES OTHER THAN HEAT SLUGS.
3. pOLARITY DENDTED BY CATHODE BAND. DIMENSIONNING AND TOLERNCINGPER
4. ANSL Y14.5, 1973.L

DIM	MILLIMETERS		INCHES	
A	3.05	5.08	0.102	0.200
D	1.52	2.29	0.060	0.090
B	0.46	0.56	0.018	0.022
F	—		—	0.050
K	25.40	38.10	1.000	1.500

All JEDEC dimensions and notes apply.

CASE 299-02
DO-204AH
GLASS

STEADY STATE POWER DERATING

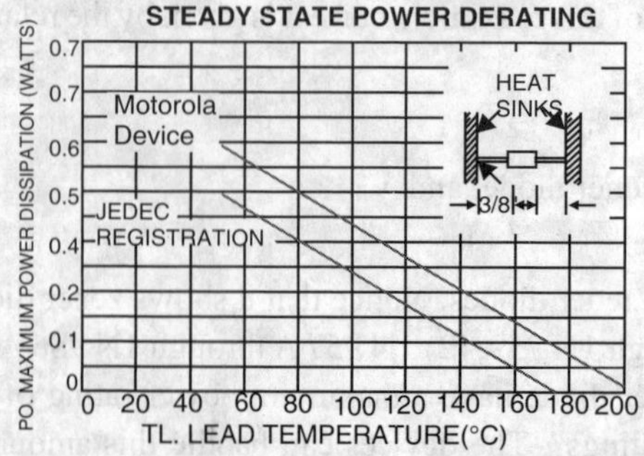

Fig. 13.3 (*a*).

ELECTRICAL CHARACTERISTICS (TA = 25°C, VF = 1.5 V max at 200 mA for all types)

Type Number (Note 1) Volts	Nominal zener voltage (Note 2) mA	Test Current I_{ZT} Ohms	Maximum zoner Impedance Z_{ZT} @ I_{ZT} (Note 3) mA	Maximum DC zenner current I_{ZM} (Note 4) µA		Maximum Reverse Leakage Current $T_A = 25°C$ I_R @ $V_R = 1$ V µA	Maximum Reverse Leakage Current $T_A = 250°C$ I_R @ $V_R = V$ µA
1N 4370	2.4	20	30	150	190	100	200
1N 4371	2.7	200	30	135	165	75	150
1N 4372	3.0	20	29	120	150	50	100
1N 746	3.3	20	28	110	135	10	30
1N 747	3.6	20	24	100	125	10	30
1N 748	3.9	20	23	95	115	10	30
1N 749	4.3	20	22	85	105	2	30
1N 750	4 7	20	19	75	95	2	30
1N 751	5.1	20	17	70	85	1	20
1N 752	5.6	20	11	65	80	1	20
1N 753	6.2	20	7	60	70	0.1	20
1N 754	6.8	20	5	55	65	0.1	20
1N 755	7.5	20	6	50	60	0.1	20
1N 756	8.2	20	8	45	55	0.1	20
1N 757	9.1	20	10	40	50	0.1	20
1N 758	1.0	20	17	35	45	0.1	20
1N 759	1.2	20	30	30	35	0.1	20

Type Number (Note 1)	Nominal zener voltage V_Z (Note 2) Volts	Test Current I_{ZT} mA	Maximum zoner Impedance (Note 3) Z_{ZT} @ I_{ZT} Ohms	Z_{ZK} @ I_{ZT} Ohms	I_{ZK} mA	Maximum DC zenner current I_{ZM} (Note 4) mA	Maximum Reverse Current IR Maximum I_R Maximum µA		Test Voltage Vdc 5%	V_B 10%
1N 957 A	6.8	18.5	4.5	700	1.0	47	61	150	5.2	4.9
1N 958 A	7.5	16.5	5.5	700	0.5	42	55	75	5.7	5.4
1N 959 A	8.2	15	6.5	700	0.5	38	50	50	6.2	5.9
1N 960 A	9.1	14	7.5	700	0.5	35	45	25	6.9	6.6
1N 961 A	10	12.5	8.5	70	0.25	32	41	10	7.6	7.2
1N 962 A	11	11.5	9.5	70	0.25	28	37	4	8.4	8.0
1N 963 A	12	10.5	11.5	700	0.25	26	34	5	9.1	8.6
1N 964 A	13	9.5	13	700	0.25	24	32	5	9.9	9.4
1N 965 A	15	8.5	16	700	0.25	21	27	5	11.4	10.8
1N 966 A	16	7,8	17	700	0.25	29	37	5	12.2	11.5
1N 967 A	18	7.0	21	750	0.25	17	23	5	13.7	13.0
1N 968 A	20	6.2	25	750	0.25	15	20	5	15.2	14.4
1N 969 A	22	5.6	29	750	0.25	14	18	5	16.7	15.8
1N 970 A	24	5.2	33	750	0.25	13	17	5	18.2	17.3
1N 971 A	27	4,6	41	750	0.25	11	15	5	20.6	19.4
1N 972 A	30	4.2	49	1000	0.25	10	13	5	22.8	21.6
1N 973 A	33	3.8	58	1000	0.25	9.2	12	5	25.1	23.8
1N 974 A	36	3.4	70	1000	0.25	8.5	11	5	27.4	25.9
1N 975 A	39	3.2	80	1000	0.25	7.8	10	5	29.7	28.1
1N 976 A	43	3.0	93	150	0.25	7.0	9.6	5	32.7	31.0
1N 977 A	47	2.7	105	1500	0.25	6.4	8.8	5	35.8	33.8
1N 978 A	51	2.5	125	1500	0.25	5.9	8.1	5	38.8	36.7
1N 979 A	56	2.2	150	2000	0.25	5.4	7.4	5	42.6	40.3
1N 980 A	62	2.0	185	2000	0.25	4.9	6.7	5	47.1	44.6
1N 981 A	66	1.8	230	2000	0.25	4.5	6.1	5	51.7	49.0
1N 982 A	75	1.7	270	2000	0.25	1.0	5.5	5	56.0	54.0
1N 983 A	82	1.5	300	3000	0.25	3.7	5.0	5	62.2	59.2
1N 984 A	91	1.4	400	3000	0.25	3.3	4.5	5	69.2	665.2
1N 985 A	100	1.3	500	3000	0.25	3.0	4.5	5	76	72
1N 986 A	110	1.1	750	4000	0.25	2.7	4.1	5	83.6	79.2

Fig. 13.3 (*b*) (Courtesy Motorola Inc.)

voltage (V_Z), maximum value of zener current (I_Z), corresponding to any zener diode given in the list. For example, 1N 4370 has $V_Z = 2.4$ V and $I_{ZM} = 190$ mA, 1N 4372 has $V_Z = 3.0$ V and $I_{ZM} = 150$ mA. Similarly, 1N 746 has $V_Z = 3.3$ V and $I_{ZM} = 135$ mA, 1N 759 has $V_Z = 12$ V and $I_{ZM} = 35$ mA. The zener diode 1N 957 A has $V_Z = 6.8$ V and $I_{ZM} = 61$ mA and 1N 986 A has $V_Z = 110$ V and $I_{ZM} =$ 4.1 mA. The data sheet can help the circuit designer to select the proper device to handle the required zener voltage (V_Z) and the required zener current.

Usually, the zener diodes with maximum power dissipation below 1 W have glass packing while those with 1 W uses metal case and those with power dissipation below 5 W uses Sumetic packaging. It may be carefully noted that zener diodes are available with leads as well as without leads.

Example 13.1.

A IN754 zener diode has a dc power dissipation rating of 500 mW and a zener voltage rating of 6.8 V. What is the value of I_{ZM} for the device?

Solution.

Given: $P_{ZM} = 500$ mW $= 500 \times 10^{-3}$ W and $V_Z = 6.8$ V.

We know that maximum value of zener current,

$$I_{ZM} = \frac{P_{ZM}}{V_Z} = \frac{500 \times 10^{-3}}{6.8} = 73.5 \times 10^{-3} \text{ A} = 73.5 \text{ mA}$$

Note: While designing circuits using zener diodes, we must make sure that the current through the zener does not exceed the I_{ZM} value. For example, in the above situation, we find $I_{ZM} = 73.5$ mA. If the current through the zener exceeds this value, we shall have to replace the diode.

Example 13.2.

A 4.5 V zener is rated at 1.5 watt. What is the maximum safe current of the zener ?

(GBTU., 2011, 2012)

Solution.

Given: $P_{ZM} = 1.5$ watt and $V_Z = 4.5$ V

We know that the maximum current,

$$I_{ZM} = \frac{P_{ZM}}{V_Z} = \frac{1.5}{4.5} = 0.333 \text{ A} \quad \textbf{Ans.}$$

Example 13.3.

A Motorola's IN746 zener diode is being operated at 75°C. What is the maximum power dissipation for the device? Assume that the d.c. power dissipation rating of the device is 500 mW and the derating factor is 3.33 mW/°C above 50°C.

Solution.

Given: $P_{ZM} = 500$ mW and derating factor = 3.33 mW.

We know that so long the zener diode is being operated at or below 50°C, its maximum power dissipation is 500 mW. However, if it is to operated at 75°C, its maximum power dissipated will be below 500 mW, because of the excessive heat generated inside.

We know that total derating value

$$= \text{Derating factor} \times (75 - 50)$$

$$= (3.33) \times (75 - 50) = 83.25 \text{ mW}$$

∴ Maximum power dissipation for the device,

$$P_Z = 500 - 83.25 = 416.75 \text{ mW} \quad \textbf{Ans.}$$

13.5. Zener Diode Equivalent Circuit

Fig. 13.4 (*a*) shows an ideal approximation of a zener diode in reverse breakdown. It shows that a zener diode is simply equivalent to a battery having a voltage equal to the zener voltage (V_Z) Fig. 13.4 (*b*) shows a practical equivalent circuit of a zener diode. This circuit shows that a zener diode is equivalent to a battery with voltage, (V_Z) in series with a resistance (r_Z) called zener resistance. The zener resistance is also known as *dynamic resistance* or *a.c. resistance.*

The presence of resistance (r_Z) indicates that the reserve V-I characteristic of a zener diode is not ideally vertical but it is slightly tilted as shown in Fig. 13.5. The value of zener resistance may be obtained from the V-I characteristic as discussed below:

Consider any two point *P* and *Q* on the reverse V-I curve as shown in Fig. 13.5.

Let ΔV_Z = Change in the values of zener voltage between the points *P* and *Q*, and

ΔI_Z = Change in the values of zener current at the corresponding points.

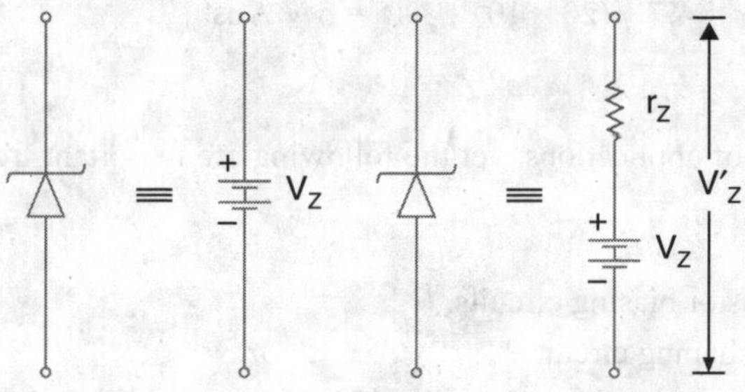

(*a*) Ideal zener equivalent diode circuit.

(*b*) Practical zener diode equivalent circuit.

Fig. 13.4.

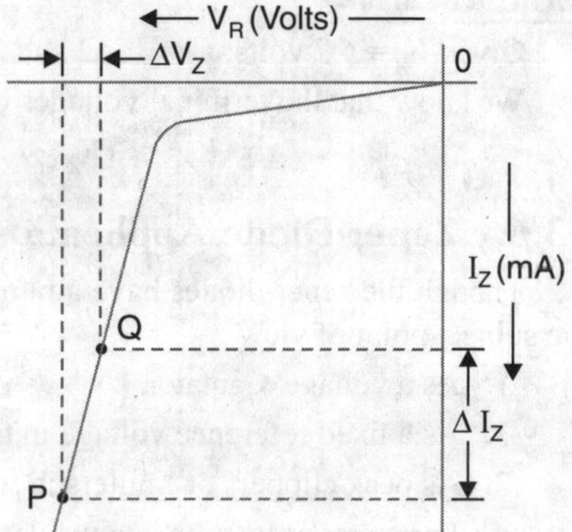

Reverse V.I characteristic curve illustrating zener resistance.

Fig. 13.5.

We know that zener resistance is given by the relation,

$$r_Z = \frac{\Delta V_Z}{\Delta I_Z}$$

For a perfectly vertical breakdown curve, the value of zener resistance is zero. But in actual practice, its value may vary from few ohms to several hundred ohms. The variation in its value depends upon the particular zener voltage and the value of operating current. It may be noted that if the zener current flows through the zener diode in the direction as shown in Fig. 13.6, then voltage drop (V_Z) across the zener resistance has the same polarity as that of a battery. In this case, the voltage across the terminals *A* and *B*,

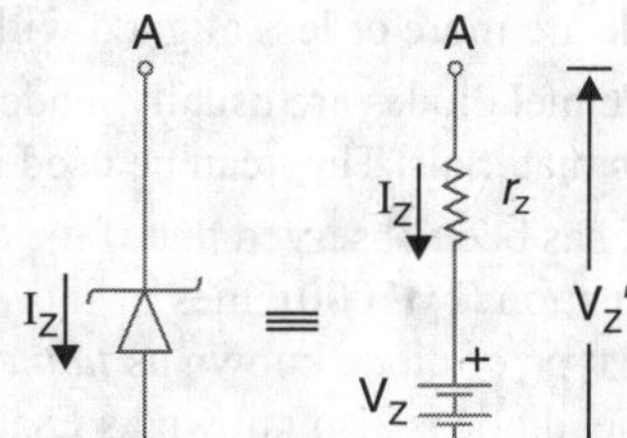

Fig. 13.6. Terminal voltage of a zener diode.

$$V_Z' = I_Z \cdot r_Z + V_Z$$

If the value of r_Z is very small, then the voltage across the terminals *A* and *B*,

$$V_Z' = V_Z$$

Example 13.4.

When the reverse current in a particular zener diode increases from 20 mA to 30 mA, the zener voltage changes from 5.6 V to 5.65 V. What is the resistance of device?

Solution.

Given: $\Delta I_Z = 30 - 20 = 10$ mA = 10×10^{-3}A and $\Delta V_Z = 5.65 - 5.6 = 0.05$ V.

We know that resistance of the zener diode,

$$r_Z = \frac{\Delta V_Z}{\Delta I_Z} = \frac{0.05}{10 \times 10^{-3}} = 5\ \Omega \textbf{ Ans.}$$

Example 13.5.

A 4.7 V zener has a resistance of 15 W . What is the terminal voltage, when the current is 20 mA?

Solution.

Given $V_Z = 4.7$ volts; $r_Z = 15\ \Omega$ and $I_Z = 20$ mA = 20×10^{-3} A.

We know that the terminal voltages of a zener diode,

$$V_Z' = (V_Z + I_Z \cdot r_Z) = 4.7 + (20 \times 10^{-3}) \times 15 = 5 \text{ V } \textbf{Ans.}$$

13.6. Zener Diode Applications

Though the zener diodes have a number of applications, yet the following are important from the subject point of view :

1. As a voltage regulator.
2. As a fixed reference voltage in transistor biasing circuits.
3. As peak clippers or limiters in waveshaping circuits.
4. For meter protection against damage from accidental applications.

13.7. Tunnel Diode

A tunnel diode is a type of semiconductor diode which is capable of very fast operation, well into the microwave frequency region by using quantum mechanical effects.

The tunnel diodes have a heavily doped PN junction. The PN junction is typically about 10 nm wide. The heavy doping results in a broken bandgap where conduction band electron states on the N- side are more or less aligned with valence band hole states on the P-side.

Tunnel diodes are usually made from germanium but can also be made in gallsum arsenide and silicon materials. They can be used as oscillators, amplifiers, frequency converters and detectors.

It has been observed that if the concentration of impurity atoms is greatly increased in a normal PN junction (by 1000 times or more), its characteristics are completely changed. This gives rise to a new type of diode known as *tunnel diode.* It was invented in 1958 by Dr. Leo Esaki. That is why, a tunnel diode is also known as Esaki diode.

In a normally doped PN junction, the impurity concentration (*i.e.*, doping level) is of very small value (*i.e.*, about one part in 10^8 atoms). With this amount of doping, the width of depletion layer is of the order of one micron (which is equal to 10^{-6} m). This constitutes a potential barrier at the junction, which controls the flow of charge carriers (*i.e.*, electrons and holes) across the junction. The charge carriers cannot cross-over the potential barrier, unless they acquire sufficient energy to overcome it. However, when impurity concentration is increased (say about one part in 10^3 atoms), the width of depletion layer reduces to about 10 nanometer (where one nanometer = 10^{-9}m). Under such conditions, the charge carriers will penetrate through the junction at the speed of light, even though they do not have enough energy to overcome the potential barrier. As a result of this, a large to overcome the potential barrier. As a result of this, a large forward current is produced, even if the applied forward voltage is much less than 0.3 V.

The phenomenon of penetrating the charge carriers, directly through the potential barrier, instead of climbing over it, is called tunneling. That is why, highly doped PN junction devices are called tunnel diodes. These diodes are usually made of germanium (Ge) or gallium-assenide (GaAs).

Fig. 13.7 shows two commonly used symbols for a tunnel diode. The tunnel diodes should be handled with a great care, because they are low power devices and can be easily damaged by heat and static electricity.

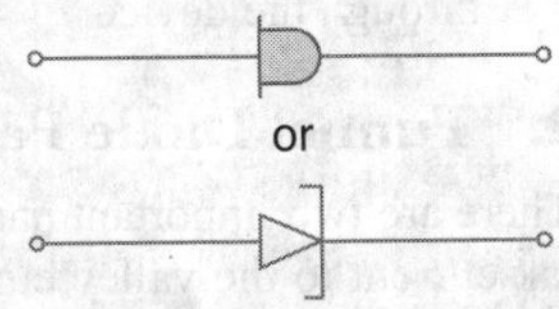

Fig. 13.7. Symbols for tunnel diodes.

13.8. V-I Characteristics of a Tunnel Diode

Fig. 13.8 shows V-I characteristics of a tunnel diode. From the characteristic curve, we see that as the applied forward voltage is increased from zero, the current increases very rapidly, till it reaches its maximum value known as peak current (I_P) as indicated by the point *A*. The corresponding value of the forward voltage is indicated by peak voltage (V_P). The value of this voltage is typically 65 mV for germanium and 160 mV for gallium arsenide tunnel diode.

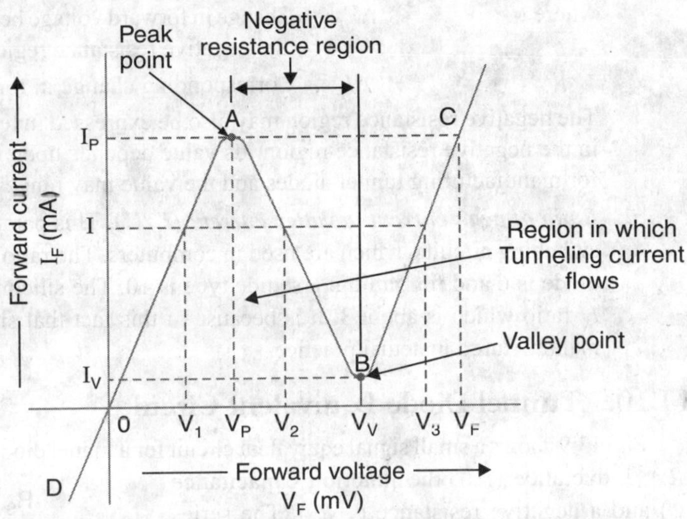

Fig. 13.8. V-I characteristic of a tunnel diode.

It will be interesting to know that if the forward voltage is further increased (*i.e.*, beyond V_P), the current decreases, till it reaches its minimum value known as valley current (I_V) as indicated by the point *B*. As the voltage is further increased, the current increases in a usual manner as in a normal PN junction diode. It has been observed that the current again reaches its peak value (*i.e.*, I_P) as indicated by the point *C*. The corresponding value of the voltage is indicated by V_F as shown in the figure. For larger values of voltages, the current increases beyond these values.

If the tunnel diode is reversed biased (*i.e.*, P side of junction is made negative with respect to its N-side), it acts like an excellent conductor, *i.e.*, the reverse current increases with the increase in reverse voltage. It is indicated by the curve *OD* in the figure. Following are some of the important points of the characteristics of a tunnel diode.

1. Between the peak point *A* and valley point *B*, the current decreases with the increase in voltage. Therefore, the tunnel diode possess a negative resistance in this region as indicated in the figure. This feature makes the tunnel diode useful in high frequency oscillators.
2. For currents, whose values are between I_V and I_P, the curve is triple valued. It means that each current can be obtained at three different applied voltages. It is indicated in the figure by the voltages V_1, V_2 and V_3 for the current *I*. This multivalued feature makes the tunnel diode useful in pulse and digital circuits.

3. The portion *BC* of the characteristic is similar to that of a forward-characteristic of a normal PN junction.
4. The shaded region in the figure indicates the region in which the tunnelling current flows through the device.

13.9. Tunnel Diode Parameters

There are two important parameters of a tunnel diode namely negative resistance and the ratio of peak current to the valley current (*i.e.*, I_P/I_V). These parameters are discussed in more detail as below:

1. ***Negative resistance.*** It is the resistance of a tunnel diode which it offers when operated in a negative resistance region. The negative resistance is given by the relation,

$$R_n = -\frac{\Delta V_F}{\Delta I_F}$$

where ΔV_F = Change in forward voltage between any two points lying within the negative resistance region of the V-I characteristic, and

ΔI_F = Corresponding change in forward current.

The negative resistance region may also be expressed in terms of slope of the characteristic in the negative resistance region. Its value depends upon the semiconductor material used for manufacturing tunnel diodes and the value may range from 10 W to 200 Ω.

2. ***Ratio of peak current to valley current (I_P/I_V).*** This parameter is important in high speed switching circuits, which are used in computers. The ratio I_P/I_V for germanium type tunnel diode is 6 and for gallium arsenide type is 10. The silicon type tunnel diode has a low I_P/I_V ratio which is about 3. It is because of this fact that silicon type tunnel diodes are not manufactured in actual practice.

13.10. Tunnel Diode Equivalent Circuit

Fig. 13.9 shows a small signal equivalent circuit for a tunnel diode. It consists of a series resistance (R_S), inductance (L_S), the junction capacitance (*C*) and a negative resistance ($-R_N$). The series resistance is the resistance due to leads, contacts and semiconductor material. Its typical value may range from 1 to 5 ohms. The series inductance is due to lead lengths. Its typical value may range from 0.1 to 4 *n*H.

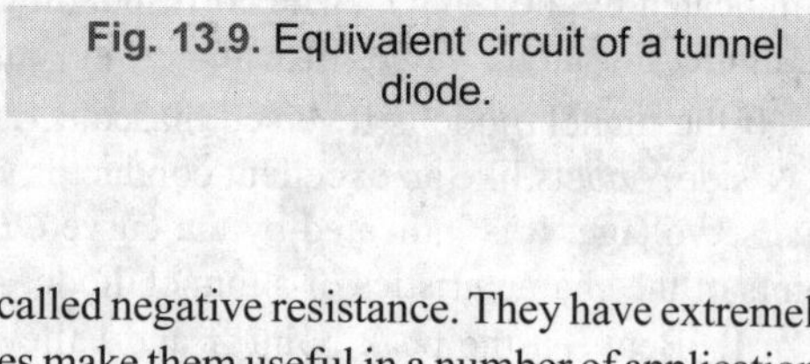

Fig. 13.9. Equivalent circuit of a tunnel diode.

The junction capacitance is due to diffusion capacitance and the applied voltage. Its typical value may range from 0.35 to 100 pF as shown in Fig. 13.9.

13.11. Tunnel Diode Applications

The tunnel diodes exhibit a specific characteristic called negative resistance. They have extremely low values of inductance and capacitance. These features make them useful in a number of applications as discussed below:

Varactors are commonly used in parametric amplifiers, parametric oscillators and voltage controlled oscillators as a part of phase-locked loops and frequency synthesizers. For example, varactors are used in the tuners of television sets to electronically tune the receiver to different stations.

1. As an ultra high-speed switching device. It is possible due to the tunneling mechanism, which takes place at the speed of light. The switching time is of the order of nono-seconds (10^{-9} sec).
2. As a logic memory storage device. It is possible due to the triple valued of the curve for current between I_P and I_V.
3. As a microwave oscillator at frequencies in the order of 10 GHz. It is possible due to extremely low values of inductance and capacitance of the device.
4. In relaxation oscillator. It is possible due to negative resistance of the device.

13.12. Varactor Diode

A varacter diode is, basically, a reverse biased PN junction, which utilizes the inherent capacitance of the depletion layer. It is also known as varicap, voltcap or tunning diode. It is used as a voltage-variable capacitor.

We have already discussed in chapter on PN junction, that the depletion layer created by the reverse bias acts as a capacitor dielectric, whereas the P- and N-regions act as the capacitor plates as shown in the Fig. 13.10 (*a*).

When the reverse bias voltage increases, the depletion layer widens. This increases the dielectric thickness, which in turn, reduces the capacitance. When the reverse bias voltage decreases, the depletion layer narrows down. This decreases the dielectric thickness, which in turn increases the capacitance. Fig. 13.10 (*b*) shows the variation of capacitance with the reverse voltage. This indicates that the variation of capacitance is maximum, when the reverse voltage is equal to zero. It reduces in a non-linear manner, as the value of reverse voltage is increased.

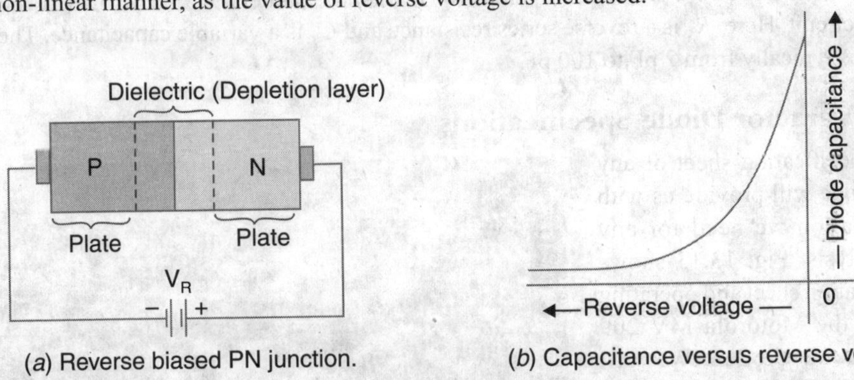

(*a*) Reverse biased PN junction. (*b*) Capacitance versus reverse voltage.

Fig. 13.10.

In a varactor diode, the capacitance parameters are controlled by the method of doping in the depletion layer, or the size and geometry of the diode construction. For example, Fig. 13.11 (*a*) shows the doping profile for an abrupt junction diode (*i.e.,* the normal PN junction). In this type, the doping is uniform on both sides of the junction. The range of capacitance variation (called tuning range) of an abrupt junction diode is 4 : 1. It means that if its maximum transition capacitance is 100 pF and the minimum is 25 pF, then its tuning range is 4 : 1.

The higher tuning range is obtained for varactors, which have hyper abrupt junction. The doping profile of such a junction is as shown in Fig. 13.11 (*b*). In this type, the doping level increases as we approach the junction. The heavy doping at the junction results in a narrower depletion layer and hence a larger capacitance. Moreover, any change in reverse voltage produces a larger variation in capacitance than that of abrupt junction. A hyper abrupt varactor diode has a tuning range of 10 : 1.

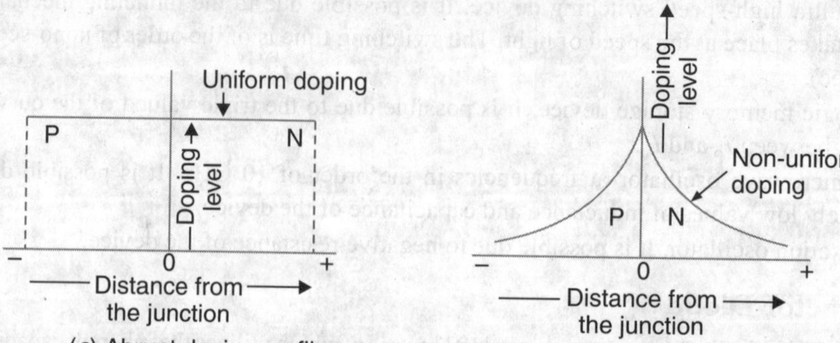

(*a*) Abrupt doping profile. (*b*) Hyper abrupt profile.

Fig. 13.11.

This range is enough to tune a broadcast receiver, over its frequency range of medium wave band (*i.e.*, frequencies from 550 kHz to 1650 kHz).

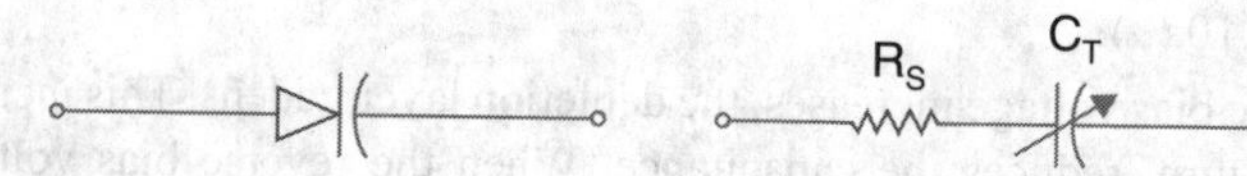

(*a*) Symbol (*b*) Equivalent circuit

Fig. 13.12. Varactor diode.

Fig. 13.12 (*a*) shows a schematic symbol for a varactor diode and Fig. 13.12 (*b*) shows its equivalent circuit. Here, R_S is a reverse series resistance and C_T is a variable capacitance. The value of C_T ranges typically from 2 pF to 100 pF.

13.13. Varactor Diode Specifications

The specification sheet of any varactor diode will provide us with the information we need for any circuit analysis. Fig. 13.13 shows the specification sheet and operating curves for the Motorola MV 209 series varactor diodes. We will use this specification sheet for our discussion on the commonly used varactor parameter and characteristics.

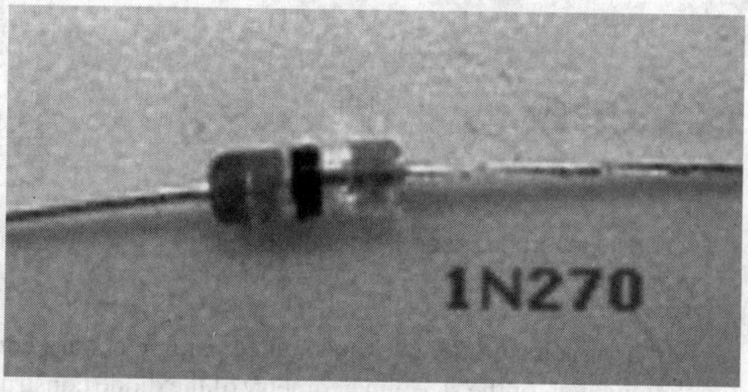

The maximum ratings for the varactor diode are the same as those that are used for the PN junction diode and the zener diode. The same holds true for the reverse breakdown voltage and reverse leakage current ratings that appear in the electrical characteristics section of the specification sheet. We have already discussed these ratings in the last chapter. However, some of the additional ratings are given below:

1. ***The diode capacitance temperature coefficient (TC_C) rating.*** This rating of a varactor diode gives the charge in the component's capacitance for each 1°C rise in temperature above 25°C. The TC_C rating for the MV 209 series varactor diodes is 300 ppm/°C. This means that varactor diode capacitance will increase by 300 parts per million for each 1°C

SILICON EPICAP DIODE

... designed for gernal frequency control and tuning applications; providing solid-state reliability in replacement of mechinical tuning methods.

- High Q with Guranteed minimum Values at VHF Frequencies
- Controlled and Uniform Tuning Ratio
- Available in Surface Mount Package

MMBV109L
MV209

CASE 182-02 STYLE 1
(TO-226 AC)

2 Cathode ⊣◀ 1 Anode

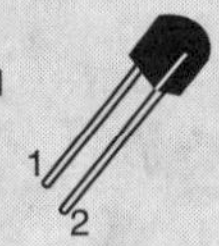

CASE 318-03 STYLE 8
SOTO-23(TO236 AB)

2 Cathode ⊣◀ 1 Anode

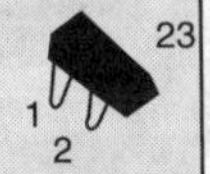

26-32 pF
VOLTAGE VARIABLE
CAPACITANCE DIODES

MAXIMUM RATINGS

Rating	Symbol	MV209 Value	MMBV209,L Value	unit
Reverse Voltage	V_R	30		Volts
Forward Current	I_F	200		mA
Forward Power Dissipation @ TA = 25°c Derate above 25°	P_D	200 2.8	200 2.0	mW mW/°c
Junction Temperature		+125		°c
Storage Temperature Range	T_{stg}	– 55 to + 150		°c

DEVICE MARKING

MMBV109 = 4A

ELECTRICAL CHARACTERISTICS (TA = 25°Unless Otherwise noted.)

Characteristic	Symbol	Min	Typ	Max	Unit
Reverse Breakdown Voltage (I_R = 10 μAdc)	$V_{(BR)R}$	30	—	—	Vdc
Reverse Voltage Leakage Current (V_R = 25 Vdc)	I_R	—	—	o.1	μAdc
Diode Capacltance Temperature Coefficient (V_R = 3.0 Vdc, f = 1.0 MHz)	TC_C	—	300	—	ppm/°C

Device	Ct. Diode. Capacltance VR = 3.0 Vdc,f = 1.0 MHz pF Min	Nom	Max	Q. Figure of Merit V_R = 3.0 Vdc t = 50 MHz (Note 1) Min	C_R, Capacitance Ratio C_2/C25 f = 1.0 MHz (Note 2) Min	Max
MMBV109L, MV209	26	29	32	200	5.0	6.5

FIGURE 1 – DIODE CAPACITANCE

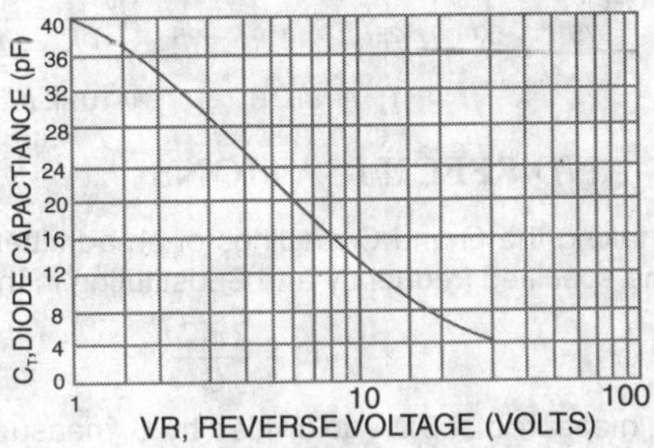

FIGURE 2 – FIGURE OF MERIT

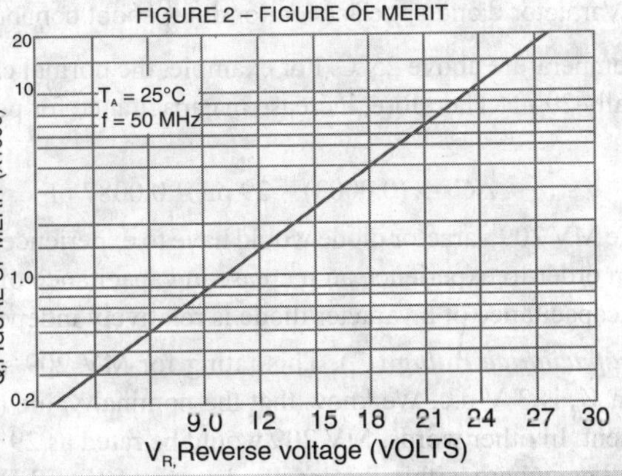

Fig. 13.13 (*b*). Varactor diode specification (or data) sheet. (Courtesy of Motorola Inc.)

FIGURE 3 – LEAKAGE CURRENT

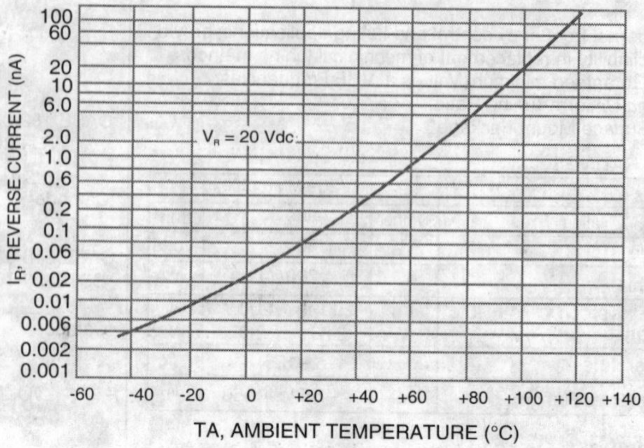

FIGURE 4 – DIODE CAPACITANCE

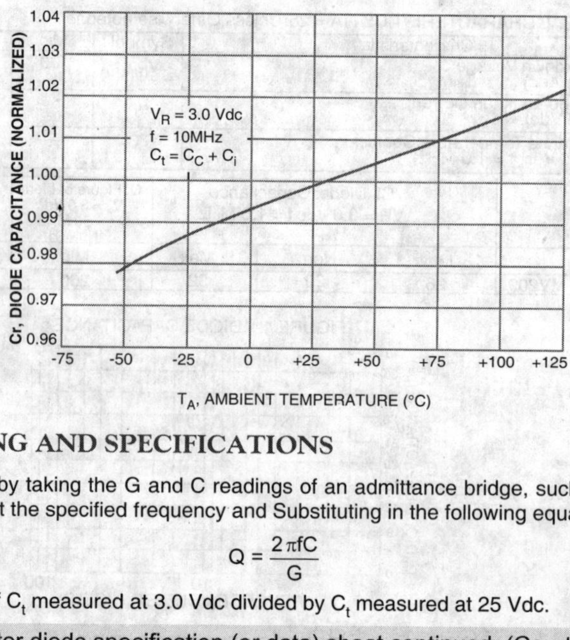

NOTES ON TESTING AND SPECIFICATIONS

1. Q is calculated by taking the G and C readings of an admittance bridge, such as Boonton Electronic Model 33AS8, at the specified frequency and Substituting in the following equation :

$$Q = \frac{2\pi fC}{G}$$

2. C_R is the ratio of C_t measured at 3.0 Vdc divided by C_t measured at 25 Vdc.

Fig. 13.13 (*b*) Varatctor diode specification (or data) sheet continued. (Courtery of Motorola Inc)

rise in temperature above 25°C. For example, the normal capacitance rating of the MV 209 is typically 29 pF. Therefore, increase in capacitance for per 1°C rise in temperature would be,

$$\Delta C = (0.0003) \times 29 \text{ pF} = 0.0087 \text{ pF}$$

Thus, the MV 209 varactor diode would have to experience a temperature increase of nearly 115°C in order to experience an increase in capacitance of 1 pF. This leads us to conclude that the capacitance of a varacter diode is relatively independent of operating temperature.

2. ***Diode capacitance rating*** **(C_t).** This rating for MV 209 series varactor diode is 26 to 32 pF when V_R is 3 V d.c. We know that the nominal value (29 pF) is the rated value of the component. In other words, MV 209 would be rated as 29 pF even though the actual value could fall anywhere within the specified range of V_R = 3 V dc.

3. ***Capacitance ratio*** (C_R). This rating gives us the variation of the junction capacitance over the given range of voltages. For the MV 209 series, the C_R rating ranges from 5.0 to 6.5 for V_R = 3 to 25 V dc.
4. ***Figure of merit*** (Q). It is the quality factor of a junction capacitance. We know that 'Q' of a capacitor is the ratio of energy stored in the capacitor to the energy lost through leakage current. In other words, it is the ratio of power returned to the circuit by the capacitor to the power dissipated by the capacitor. For the MV 209, Q rating is 200 (minimum).

13.14. Varactor Diode Applications

A major application of varactor diode is in *tuning circuits. It is used in electronic tuners in radio, television and other commercial receivers. When a varactor is used in a tuning circuit, it allows a resonant frequency to be adjusted by a variable voltage level as shown in Fig. 13.14. In this circuit, two varactor diodes D_1 and D_2 provide the total variable capacitances in a parallel resonant circuit and V_C is a variable d.c. voltage, which controls the reverse bias and the capacitance of the diodes. The resonant frequency of the tank circuit is given by the relation,

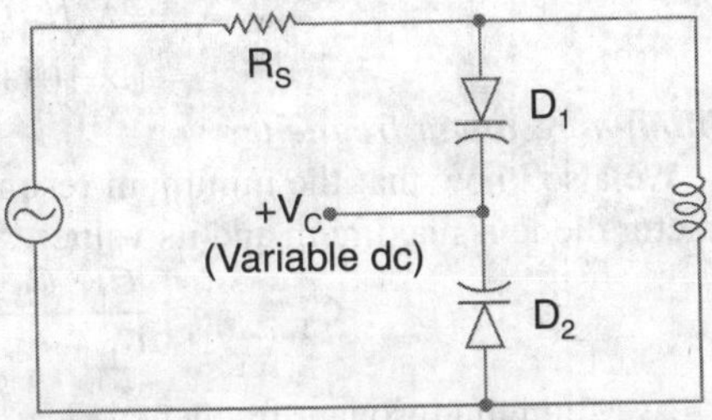

Fig. 13.14.

$$f_0 = \frac{1}{2\pi\sqrt{LC}}$$

where L = The value of the inductance,

C = The value of total capacitance, which is given by the relation $\dfrac{C_1 \cdot C_2}{C_1 + C_2}$.

where C_1 and C_2 are the maximum and minimum values of the capacitances of diode.

The varactor diode is also used in automatic frequency control device, adjustable band-pass filter and parametric amplifier.

Example 13.6.

The capacitance of certain varactor diode can be varied from 5 pF to 50 pF. The diode is used in a tuned circuit of a radio receiver as shown in Fig. 13.15.

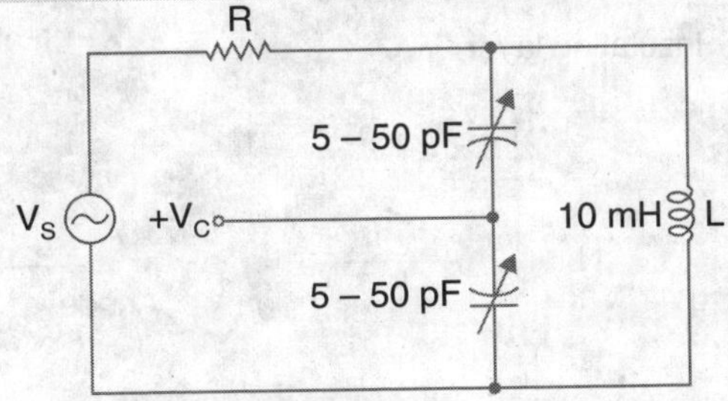

Fig. 13.15.

Determine the tuning range for circuit, if the inductance is 10 mH.

* It is a circuit in which the resonant frequency is altered to accept the desired channel frequency.

Solution.

Given: C_1 (*i.e.*, C_{min}) = 5 pF; C_2 (*i.e.*, C_{max}) = 50 pF and L = 10 mH = 10×10^{-3} H.

Maximum resonant frequency

We know that the maximum resonant frequency is obtained, when the total capacitance of the varactor diode is minimum and its value,

$$C_{T\,(min)} = \frac{C_1 \cdot C_2}{C_1 + C_2} = \frac{5 \times 5}{5 + 5} = 2.5 \text{ pF} = 2.5 \times 10^{-12} \text{ F}$$

∴ Maximum resonant frequency,

$$f_{o\,(max)} = \frac{1}{2\pi\sqrt{L \cdot C_{T(min)}}} = \frac{1}{2\pi\sqrt{(10 \times 10^{-3}) \times (2.5 \times 10^{-12})}} \text{ Hz}$$

$$= 1 \times 10^6 \text{ H}_Z = 1 \text{ MH}_Z \textbf{ Ans.}$$

Minimum resonant frequency

We also know that the minimum resonant frequency is obtained, when the capacitance of the varactor diode is maximum and its value

$$C_{T\,(max)} = \frac{C_1 \cdot C_2}{C_1 + C_2} = \frac{50 \times 50}{50 + 50} = 25 \text{ mF} = 25 \times 10^{-12} \text{ F}$$

∴ Minimum resonant frequency,

$$f_{o\,(min)} = \frac{1}{2\pi\sqrt{L \cdot C_{T(max)}}} = \frac{1}{2\pi \times \sqrt{(10 \times 10^{-3}) \times (25 \times 10^{-12})}} \text{ Hz}$$

$$= 318 \times 103 \text{ Hz} = 318 \text{ kHz} \textbf{ Ans.}$$

13.15. Schottky Doide

The Schottky diode is a semiconductor diode with a low forward voltage drop and a very fast switching action.

When a current flows through a PN junction diode, there is a small voltage drop across the diode terminals. A normal silicon diode has a voltage drop between 0.6–1.1 volts while a Schottky diode voltage drop is between 0.15–0.45 volts. This lower voltage drop can provide high switching speed and better system efficiency.

It is formed by joining a doped semiconductor region (usually N-type) with a metal such as gold, silver or platinum. Thus in a Schottky diode, there is a metal-to-semiconductor junction rather than a simple PN junction as shown in Fig. 13.16 (*a*). A Schottky diode is also called a hot carrier diode or a Schottky barrier diode. Fig. 13.16 (*b*) shows the schematic symbol of a Schottky diode. It will be interesting to know that in the symbol, that the line at the end of a arrow looks like the letter S.

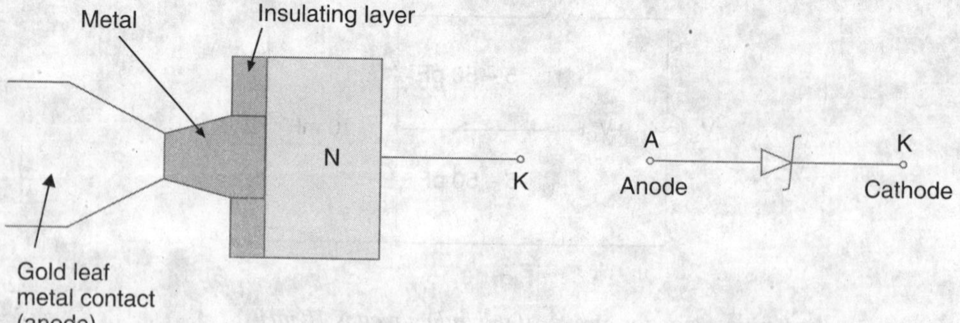

(*a*) Construction (*b*) Symbol

Fig. 13.16. Schottky diode.

Fig. 13.17 shows the characteristic curve of a Schottky diode. As shown in the figure, the Schottky diode has been lower forward voltage drop (V_F) and reverse breakdown voltage (V_{BR}) than those of PN junction diode. Typically, the Schottky diode will have a V_F of approximately 0.3 V and a V_{BR} of about –50 V. These are much lower than the typical PN junction ratings of V_F = 0.7 V and V_{BR} = –150V.

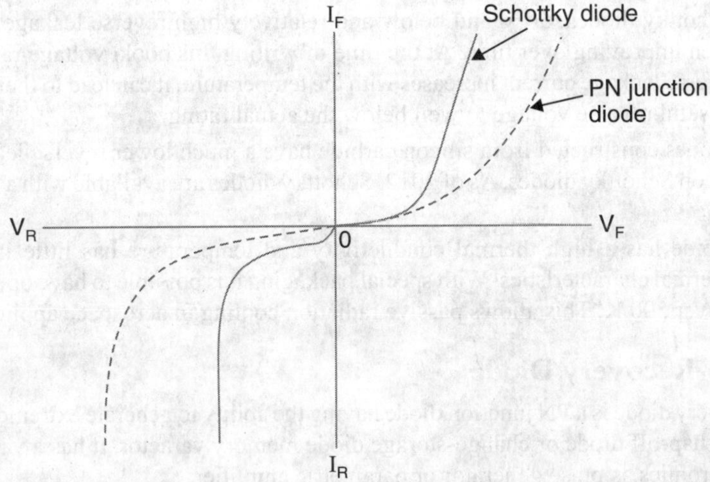

Fig. 13.17. Characteristic curve of a Schottky diode.

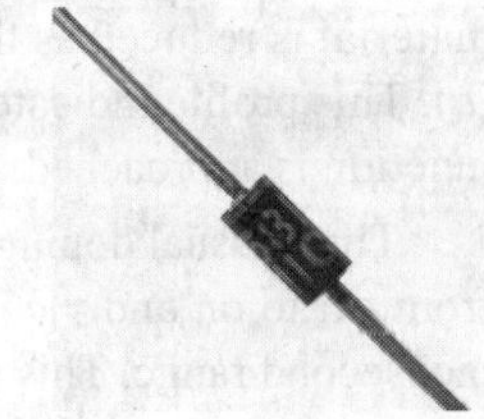

The Schottky diode has a very little junction capacitance. Because of this, it can be operated at much higher frequencies (of 20 GHz or more) than the typical PN junction diode. The reduced junction capacitance also results in a much faster *switching time.* Because of this reason, Schottky devices are being used more and more in digital switching applications.

13.16. Schottky Diode Applications

We know that Schottky diodes have two important features namely it operates only with majority carriers, and there is no charge storage when the device is switched from its 'ON' to 'OFF' state. These features make it useful for a variety of applications as discussed below:

1. To rectify very high frequency (above 300 MHz) signals.
2. As a switching device in digital computers. The speed of computers depends on how fast their diodes and transistors can turn 'ON' and 'OFF'.
3. In clipping and clamping circuits.
4. In low power Schottky transistor-transistor logic (Schottky TTL) circuits.
5. In mixing and detecting circuits used in communication systems.

Schottky barrier diode family expands to meet needs of power related application.

6. In low voltage power supply circuits. The Schottky diode has a threshold voltage of 0.25V only. It is an advantage in low voltage rectifiers because we have to subtract only 0.25V (instead of 0.7 V) for each diode from the low input voltage to get an output.

Schottky diode limitations

The most evident limitation of Schottky diodes are the relatively low reverse voltage rating for silicon-metal Schottky diodes, 50 V and below and relatively high reverse leakage current. Diode designs have been improving over time. At the time of writing this book, voltage ratings can reach 200 V. As the reverse leakage current increases with the temperature, it can lead to thermal instability. This limits the useful reverse voltage to well below the actual rating.

Schottky diodes constructed from silicon carbide have a much lower reverse leakage curent as compared to silicon Schottky diodes. As of 2012, Schottky diodes are available with a reverse voltage up to 1700 V.

Silicon carbide has a high thermal conductivity and temperature has little influence on its switching and thermal characteristics. With special packaging it is possible to have operating junction temperature of over 500 K. This allows passive radiation cooling in aerospace applications.

13.17. Step-Recovery Diode

A step recovery diode is a PN junction diode having the ability to generate extremely short pulses. It is also called snap-off diode or change-storage diode memory varactor. It has a variety of uses in microwave electromics as pulse generator or parametric amplifier.

A step-recovery diode employs a graded doping, where the doping level of the semiconductor material is reduced as the junction is approached. Such a doping profile is as shown in Fig. 13.18 (*a*). This profile indicates that the doping level (*i.e.,* the density) of charge carriers decreases as the junction is approached. Fig.13.18 (*b*) shows the schematic symbol of a step-recovery diode.

The unusual doping of the step-recovery diode affects the time required for the device to switch from off to on and vice versa. The typical switching time for a step-recovery diode is in the low-picosecond range. This makes them ideal for ultra-high frequency switching applications.

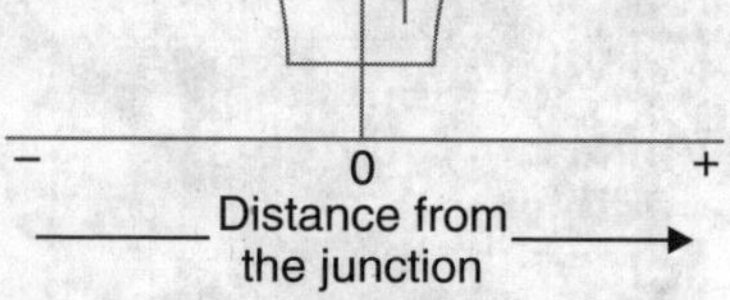

(*a*) Doping profile.

(*b*) Symbol.

Fig. 13.18. Step-recovery diode.

When a step-recovery diode is connected to a sinusiodal a.c. input source, it produces an output as shown in Figure 13.19. It may be noted that during the positive half cycle of the input signal, the diode conducts like a normal silicon diode. But during the negative half cycle, reverse current exists for a short time and then suddenly drops to zero. The reverse current flows because of the stored

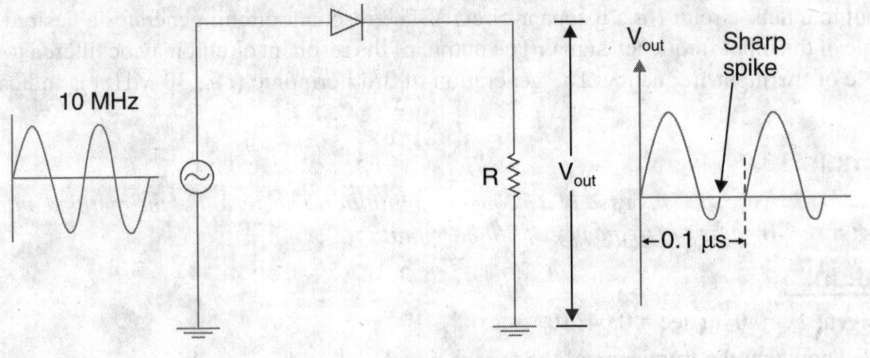

Fig. 13.19. Output showing a sharp spike.

charge in a diode, which takes some finite time for its removal. Since the reverse recovery period is very sharp, therefore, the diode is called step-recovery diode. The step-recovery diode may also be thought of as if the diode is suddenly snapped open. Therefore, the step-recovery diode is also known as a snap-diode.

A fourier analysis of such a transition (*i.e.,* sharp reverse recovery) shows that the output consists of a number of frequency components called harmonics. These harmonics have frequencies of f, $2f$, $3f$, ... nf. The frequency f (10 MHz in the present case) is called fundamental frequency whereas $2f$, $3f$, ..., nf are called second, third ..., harmonics. The output waveform can be filtered to produce a sine wave of higher frequency. Because of this fact, the step-recovery diodes are useful in frequency multipliers (*i.e.,* circuits whose output frequency is a multiple of the input frequency).

13.18. Harmonic Generator (or Frequency Multiplier)

In the last article, we have discussed the step-recovery diode. As a matter of fact, the step-recovery diode has a specific feature that when it is switched from ON to OFF position, it gives a spike at its output. This feature makes it useful in harmonic generator or frequency multiplier circuit. The output frequency of such a circuit is a multiple of the input frequency signal.

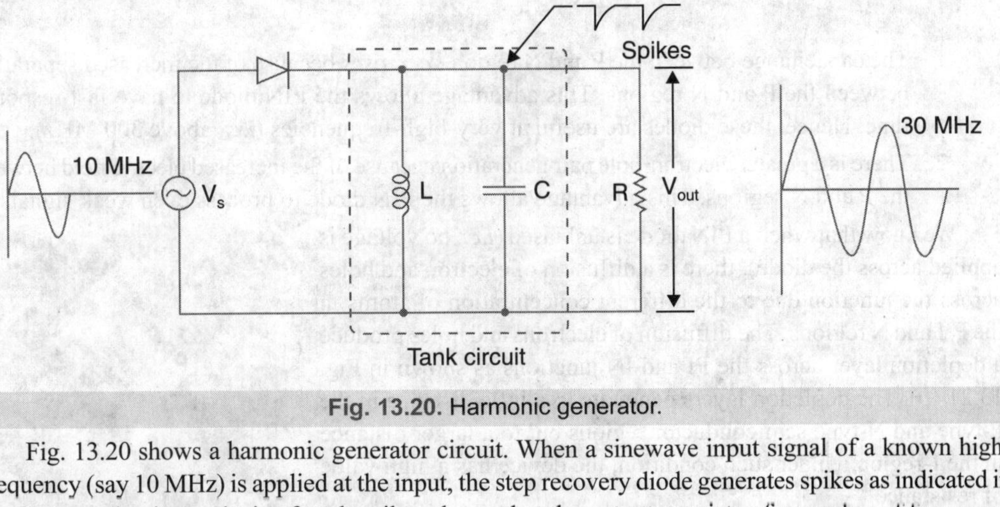

Fig. 13.20. Harmonic generator.

Fig. 13.20 shows a harmonic generator circuit. When a sinewave input signal of a known high-frequency (say 10 MHz) is applied at the input, the step recovery diode generates spikes as indicated in the figure. A fourier analysis of such spikes shows that the output consists of a number of frequency components called harmonics. These harmonics have frequencies of f, $2f$, $3f$ ..... These spikes act as an input to a tank-circuit (or a resonant-circuit). The resonant circuit generates a desired harmonic multiple of the input frequency signal. The output of the resonant circuit may be filtered to get a pure multiple of the input frequency. The generation of 3rd hormonic (*i.e.*, 30 MHz) is indicated in the figure.

Example 13.7.

A step recovery diode is used in a frequency multiplier. The period of the rectified output signal is 0.04 msec. *What is the frequency of 5th harmonic?*

Solution.

Given: $T = 0.04\ \mu$ sec $= 0.04\ \ 10^{-6}$ sec.

We know that the frequency of the output signal,

$$f = \frac{1}{T} = \frac{1}{0.04 \times 10^{-6}} = 25 \times 10^{6}\ \text{Hz} = 25\ \text{MHz}$$

∴ Frequency of the 5th harmonic

$= 5f = 5 \times 25 = 125$ MHz **Ans.**

13.19. PIN Diode

A PIN diode is made up of three semiconductor materials: two *heavily doped* P- and one N-type material separated by an *intrinsic* (*i.e.*, undoped) semiconductor (I) as shown in Fig. 13.19 (*a*). The intrinsic region offers high resistance to the current flowing through it. The PIN diode has the following two advantages over the normal PN junction diode:

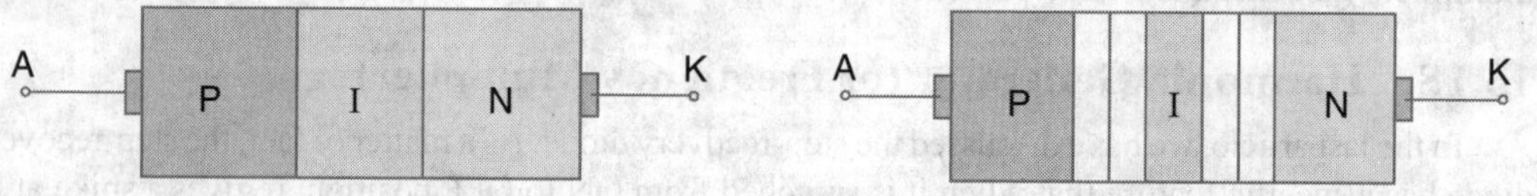

(*a*) Base structure (*b*) Formation of a depletion layer

Fig. 13.21.

1. The capacitance between the P and N region *decreases* because of the increased separation between the P and N regions. This advantage allows the PIN diode to have fast response time. Hence, these diodes are useful at very-high-frequencies (*i.e.*, above 300 MHz).
2. There is a greater electron-hole pair generation because of the increased electric field between the P and N regions. This advantage allows the PIN diode to process even weak signals.

We know that when a PIN diode is unbiased (*i.e.*, no voltage is applied across the diode), there is a diffusion of electron and holes across the junction due to the different concentration of atoms in the P, I and N regions. The diffusion of electrons and holes produce a depletion layer across the PI and IN junctions as shown in Fig. 13.21 (*b*). The depletion layers penetrate to a little distance in the P-type and N-type semiconductor regions but to a larger distance in the I-region. Under such condition, the device has a high value of resistance.

PIN diode

When the PIN diode is forward biased, the width of depletion layers decreases. As a result of this, more carriers are injected into the I-region. This reduces the resistance of the I-region. If the depletion layers are not thick, then the I-region becomes flooded with the carriers at a suitable bias. Thus, when a PIN diode is forward biased, it acts like a variable resistance as shown in Fig. 13.22 (*a*). The forward resistance of an intrinsic region decreases with the increasing current.

On the other, when the PIN diode is reverse biased, the depletion layers become thicker. As the reverse bias is increased, the thickness of the depletion layer increases till the I-region becomes free of mobile carriers. The reverse bias, at which this happens, is called *swept out voltage*. At this stage, the PIN diode acts like an almost constant capacitance as shown in Fig. 13.22 (*b*).

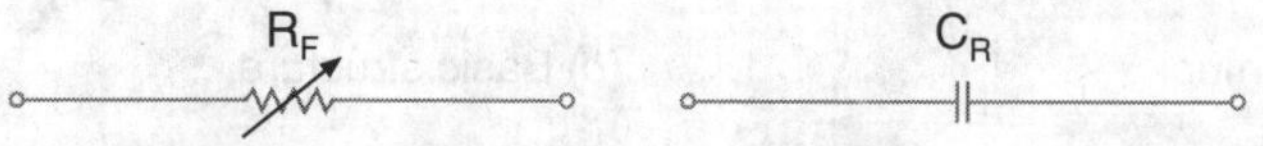

(*a*) Forward biased. (*b*) Reverse biased.

Fig. 13.22. Equivalent circuit of a PIN diode

13.20. PIN Diode Applications

We know that when a PIN diode is forward biased, it acts like a variable resistance which decreases with the increasing current. And when it is reverse biased, the PIN diode acts like an almost constant capacitance. These features make the PIN diode useful for a variety of applications as discussed below:

1. As a d.c. controlled microwave switch operated by rapid changes in bias or as a modulating device. This application takes advantage of the variable forward-resistance characteristic. Modulators are used to combine two signals of different frequencies into a single signal.
2. In attenuater applications because its resistance can be controlled by the current.
3. As photo-detector in fibre optic network card and switches.

13.21. Light Emitting Diode (LED)

Besides tunnel diode, varactor diode, Schottky diode, step-recovery diode, PIN diode, there is a Gunn diode which is used for controlling electronic signals in the GHz range. See the pictures above:

A PN junction diode, which emits light when forward biased, is know as a light emitting diode (abbreviated as LED). The emitted light may be visible or invisible. The amount of light output is directly proportional to the forward current. Thus, higher the forward current, higher is the light output. The schematic symbol of a light emitting diode is shown in Fig. 13.23 (*a*). The arrows, pointing away from the diode symbol represent the light, which is being transmitted away from the junction.

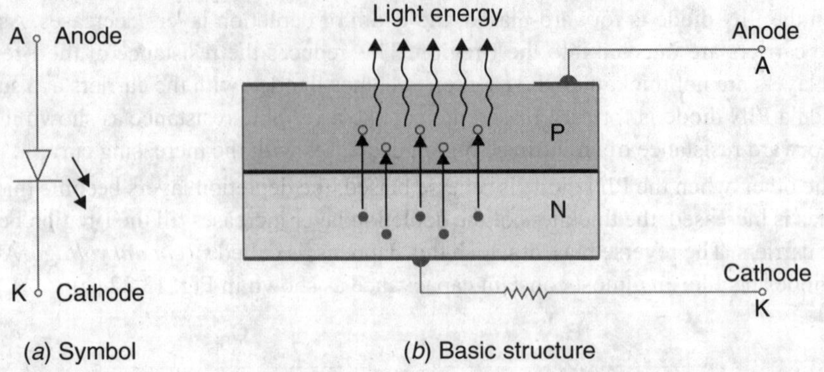

Fig. 13.23. Light emitting diode (LED).

Fig. 13.23 (*b*) shows the basic structure of a light emitting diode. Here, an N-type layer is grown on a P-type substrate (not indicated in figure) by a diffusion process. Then a thin P-type layer is grown on the N-type layer. The metal connections to both the layers make anode and cathode terminals as indicated. The light energy is released at the junction, when the recombination of electrons with holes take place. After passing through the P-region, the light is emitted through the window provided at the top of the surface.

It will be interesting to know that when the LED is forward biased, the electrons and holes move towards the junction and the recombination takes place. After recombination, the electrons, lying in the conduction bands of N-region, fall into the holes lying in the valence band of a P-region. The difference of energy between the conduction band and valence band is radiated in the form of light energy. In ordinary diodes, this energy is radiated in the form of heat.

The semiconducting materials used for manufacturing light emitting diodes are Gallium Phosphide, gallium arsenide phosphide, Zinc selenide and boron nitride. The silicon and germanium is not used for manufacturing light emitting diodes because these are heat producing materials. Moreover, these materials are very poor in emitting light radiations.

The LED's radiate light in different colours such as red, green, yellow, blue, orange etc. Some of the LED's emit infrared (*i.e.,* invisible) light also. The colour, of the emitted light, depends upon the type of the semiconductor used. Thus gallium arsenide emits infrared radiations, gallium arsenide phosphide produces either red or yellow light, gallium phosphide emits red or green light and gallium nitrite produces blue light.

Since LED's have clear (or semiclear) cases, there is normally no label on the cased to identify the leads. The two leads of a LED are identified using one of the several schemes as discussed below.

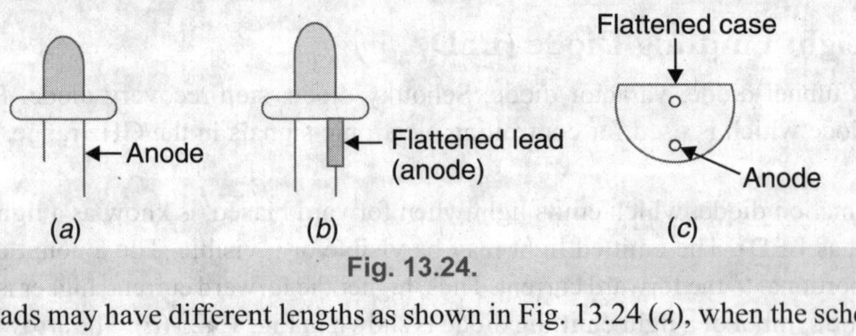

Fig. 13.24.

1. The leads may have different lengths as shown in Fig. 13.24 (*a*), when the schemes is used, the longer of the two leads is usually the anode.
2. One of the leads may be flattened as shown in Fig. 13.24 (*b*). The flattened lead is usually the anode

3. One side of the case is flattened as shown in Fig. 13.24 (*c*). The lead closest to the flattened side is usually the anode.

There is a large variety of light emitting diodes (LDEs) available in the market. They are available in different shapes and sizes. But the more commonly used LEDs are cylindrical in shape and are available in 1.8, 2, 3, 5, 8 and 10 mm diameter sizes. Fig. 13.25 (*a*) through (*f*) shows a variety of LEDs used in different applications.

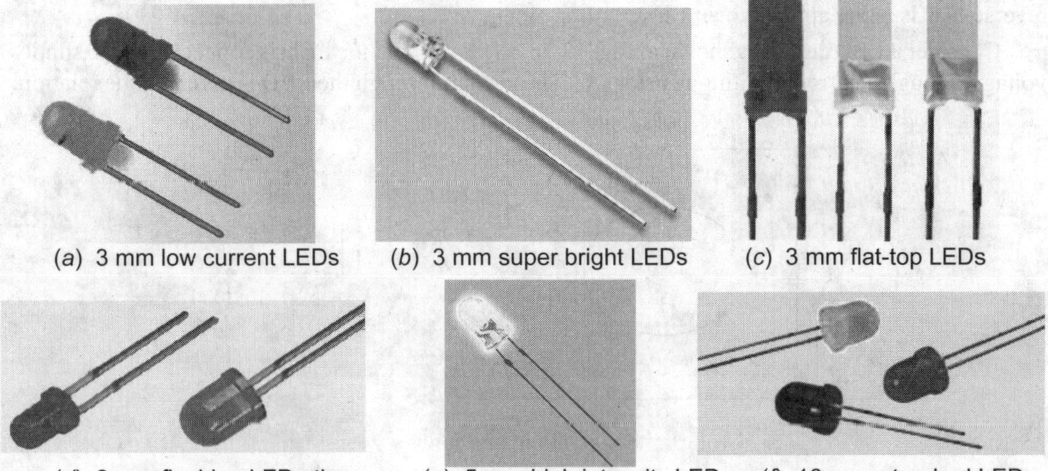

(*a*) 3 mm low current LEDs

(*b*) 3 mm super bright LEDs

(*c*) 3 mm flat-top LEDs

(*d*) 3 mm flashing LEDs-they don't need series resistor

(*e*) 5 mm high intensity LED

(*f*) 10 mm standard LEDs

Fig. 13.25. LED's of different shapes and sizes.

The low current LEDs are useful in low power d.c. circuits, telecommunication indicators, portable equipment and keyboard indicators. The super bright LESs are suitable for applications where normal LEDs are not bright enough or they too restricted viewing angle. The flashing LEDs don't need any external series resistor because they have buit in integrated circuit to give flashing effect.

Special ICs are available for driving LEDs for example the IC BCT32115 device can drive up to 10 LEDs. It can supply 20 mA current to each LED. In other words, it has a total current capability of 200 mA.

The pictures below shows the difference in the quatity of light produced by a bulb and an LED along with their beam patterns.

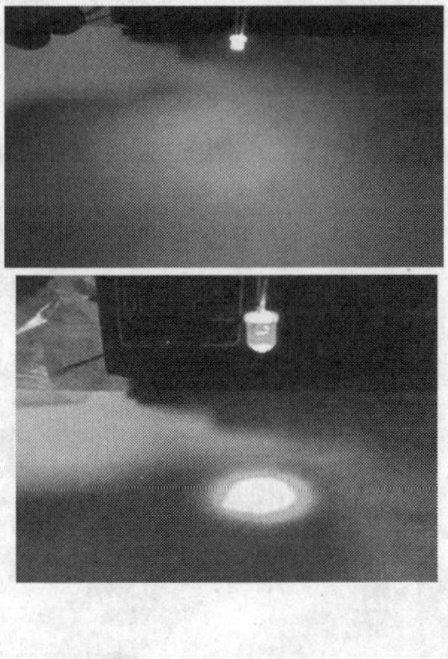

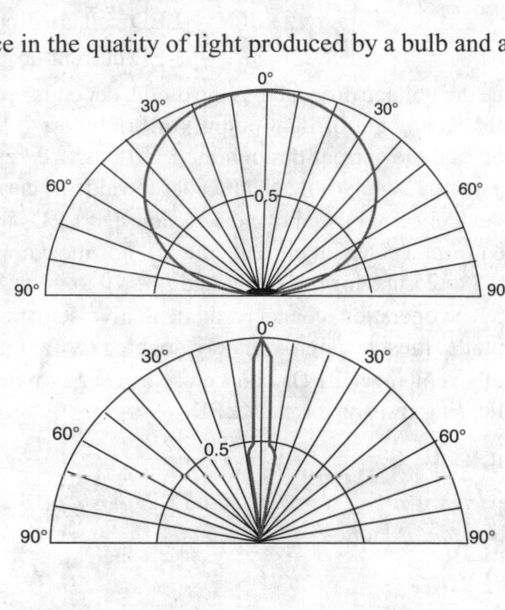

13.22. LED Voltage Drop and Current

The LED's have also a certain forward voltage drop, when they are used in the circuit like an ordinarily diode (which has a voltage drop of 0.7 V). This voltage drop depends upon the LED current, colour of the emitted light etc. Usually, the voltage drop has a typical value that varies from 1.5 to 2.5 V for currents between 10 and 50 mA. It will be interesting to know that LED's are not able to withstand reverse bias of even every small voltages. For this reason, it is necessary to assure that reverse bias is never applied to an LED.

Consider a LED circuit as shown in Fig. 13.26. In this circuit, the LED is connected to the supply voltage through a current limiting resistor (*R*). The current through the LED is given by the relation.

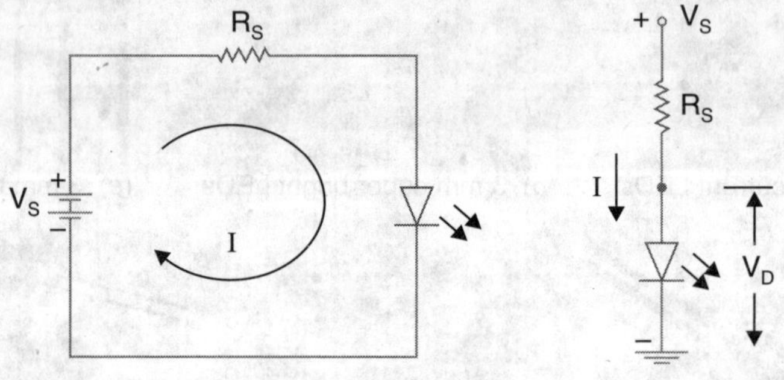

Fig. 13.26. LED circuit.

$$I = \frac{V_S - V_D}{R_S}$$

where V_S = Supply voltage,

V_D = Voltage drop across LED, and

R_S = Current limiting resistor.

The value of current limiting resistor is given by the relation,

$$R_S = \frac{V_{out(pk)} - V_F}{I_F}$$

where $V_{out\ (pk)}$ = Peak output voltage

V_F = LED voltage drop, and

I_F = LED current drop.

Since the voltage drop, across the diode, depends upon the value of current through it, therefore, it controls the amount of light output (or brightness) of the LED. In order to avoid fluctuations in the brightness, the current through the LED must be kept constant. In actual practice, it is done by selecting a large supply voltage and a large value of current limiting resistor.

It is evident from the above discussion that LEDs must be used with an external series resistor (or driven from a constant current source). For most applications, adequate light output is obtained with I_F = 5 to 25 mA for *red* LEDs and I_F = 10 to 40 mA for *green* and *yellow* LEDs.

For an ac operation, connect a diode in inverse parallel with the LED, noting that for equivalent light ouptut I_F must be approximately doubled (within its limits) by reducing the value of R_S.

Finally remember LEDs can be used in series but cannot be configured in parallel without a current limiting resistor for *each* LED.

Example 13.8.

The driving circuit shown in Fig. 13.27 has a peak output voltage of 8 V. The LED has a rating of V_F = 1.8 to 2.0 V and $I_{F\ (max)}$ = 16 mA.

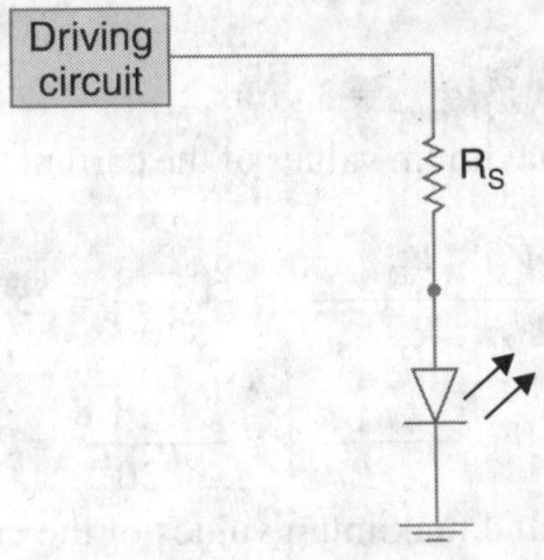

Fig. 13.27.

What is the smallest standard resistor value that can be used as R_S. (Note. A listing of the standard resistor values appears in Appendix A)

Solution.

Given: output voltage $V_{out\,(pk)}$ = 8 V; V_F = 1.8 to 2.0 V and $I_{F\,(max)}$ = 16 mA = 16 × 10^{-3} A.

We know that current limiting resistor,

$$R_S = \frac{V_{out(pk)} - V_F}{I_F} = \frac{8 - 1.8}{16 \times 10^{-3}} = 387.5\ \Omega$$

and if V_F = 2.0 V then R_S = 375 Ω. In either case, the smallest standard-value resistor that has a value greater than 387.5 Ω or 375 Ω is the 390 Ω resistor. This is the component which we have to use in the circuit of Fig. 13.27.

Example 13.9.

Fig. 13.28 shows an LED circuit. The LED has a minimum drop of 1.5 V and a maximum drop of 2.3 V. If the supply voltage is 10 V and R_1 = 470 Ω. What are the minimum and maximum values of LED current?

Solution.

Given: $V_{D\,(min)}$ = 1.5 volts; $V_{D\,(max)}$ = 2.3 volts; V_S = 10 volts and R_1 = 470 Ω.

We know that the maximum value of current,

$$I_{max} = \frac{V_S - V_{D\,(min)}}{R_1} = \frac{10 - 1.5}{470}\ \text{A}$$

$$= 18.1 \times 10^{-3}\ \text{A} = 18.1\ \text{mA}\ \textbf{Ans.}$$

and minimum value of current,

$$I_{min} = \frac{V_S - V_{D\,(max)}}{R_1} = \frac{10 - 2.3}{470}\ \text{A}$$

$$= 16.4 \times 10^{-3}\ \text{A} = 16.4\ \text{mA}\ \textbf{Ans.}$$

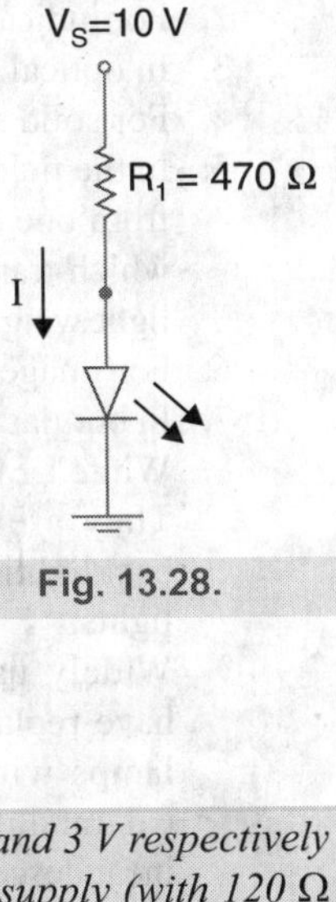

Fig. 13.28.

Example 13.10.

A light emitting diode with minimum and maximum voltage drops of 1.8 V and 3 V respectively is connected alternately with 24 volt supply (with 820 Ω resistor), and 5 volt supply (with 120 Ω resistor). Indicate which supply voltage will keep the brightness of the diode as constant.

Solution.

Given: $V_{D\,(min)} = 1.1$ volts and $V_{D\,(max)} = 3$ volts.

We know that minimum and maximum values of the current through the LED with 24 volt supply (V_S) and 820 Ω resistor (R_S).

$$I_{min} = \frac{V_S - V_{D\,(max)}}{R_S} = \frac{24 - 3}{820} = 25.6 \times 10^{-3}\ \text{A} = 25.6\ \text{mA} \quad ...(i)$$

and

$$I_{max} = \frac{\{V_S - V_{D\,(min)}}{R_S} = \frac{24 - 1.8}{820} = 27.1 \times 10^{-3}\ \text{A} = 27.1\ \text{mA} \quad ...(ii)$$

We also know that minimum and maximum values of the current through the LED with 5 volt supply (V_S) and 120 Ω resistor (R_S).

$$I_{min} = \frac{V_S - V_{D\,(max)}}{R_S} = \frac{5 - 3}{120} = 16.7 \times 10^{-3}\ \text{A} = 16.7\ \text{mA} \quad ...(iii)$$

and

$$I_{max} = \frac{V_S - V_{D\,(min)}}{R_S} = \frac{5 - 1.8}{120} = 26.7 \times 10^{-3}\ \text{A} = 26.7\ \text{mA} \quad ...(iv)$$

It may be noted that the variation in current for the first case (27.1 – 25.6 = 1.5 mA) is small whereas in the second case it is (26.7 – 16.7 = 10.0 mA) large. Therefore, the brightness in the first case will remain constant whereas in the second case it will be changing. Hence in order to get an approximately constant brightness, we use as large a supply voltage as possible.

13.23. LED Applications

The LED's operate at low voltages *i.e.*, from 1.5 V to 2.5 V. They have a long life of about 10 000 hours and can be switched 'ON' and 'OFF' at a very fast speed (≈ 1 n sec). These features make LED's very important electronic device. Following are the important applications of the LED's:

1. In 7-segment, 16-segment and dot matrix displays. Such displays are used to indicate alphanumeric characters and symbols in various systems such as digital clocks, microwave-ovens, stereo tuners, calculators etc.
2. For indicating power ON/OFF conditions, power level indicators or stereo amplifiers.
3. In optical switching applications.
4. For solid state video displays, which are rapidly replacing cathode ray tubes (CRT's).
5. In the field of optical communication, where LED's are used to transfer (or couple) energy from one circuit to another. They are also used to send light energy to fibre optical cable, which transmits energy by means of total internal reflection. The fibre optical cable is of light weight, flexible, often transparent and as small as 0.043 mm in diameter.
6. For image sensing circuits in picturephon.
7. In burglar alarm systems. In such applications, LED's radiating infrared light are preferred.
8. White LEDs are well established in the market these days.
   They are ideal for backlighting of automobile dashboards, backlighting of keypads on cellular phones, and path marking lights.
9. Widely used for traffic signal management. Many countries have replaced traffic control signals using normal incandescent lamps with the LEDs, refer of Fig. 13.29. Even most of the car manufactures have started using LED lights in place of incandescent lamps.
   These day IC's are available which can drive 10 LEDs or more. For example BCT 3211S chip can drive up to 10 LEDs with each LED consuming up to 20 mA current.

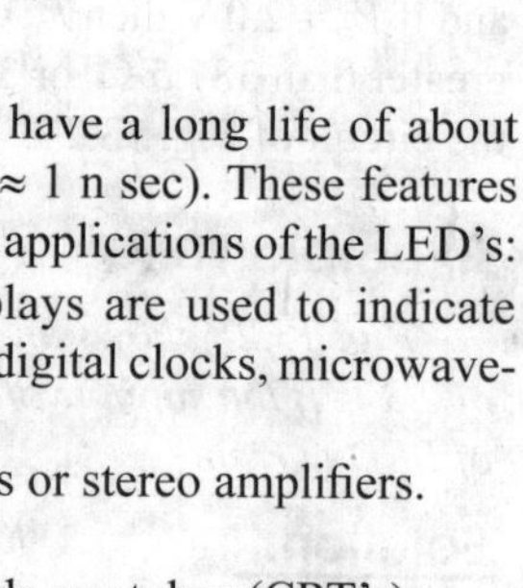

Fig. 13.29.

10. Aviation lighting, Automotive lighting, advertising and general lighting.
11. Infrared LEDs are used in remote control units of many commercial products including television, DVD players and other domestic appliances.

13.24. Multicolour LEDs

The LEDs, which emit one colour of light when forward biased and another when reverse biased, are called multicoloured LEDs. One commonly used symbol for a multicolour LED is as shown in Fig. 13.30.

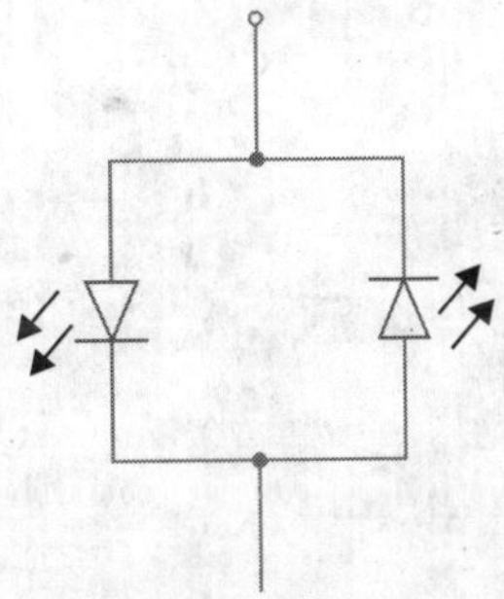

Fig. 13.30. Schematic symbol for multicolour-LED.

Multicolour LEDs actually contain two PN junction that are connected in reverse-parallel, *i.e.*, they are in parallel, with the anode of one being connected to the cathode of the other. Multicolour LEDs are typically red when biased in one direction and green when biased in the other direction. Incidently if a multicolour LED is switched fast, the LED will produced a third colour. For example, a red/green LED will produce a yellow light when rapidly switched back and forth between biasing polarities.

Multicolour LEDs are available in three sizes: 3, 5 and 10 mm diameter. They come as *bicolour* in three didfferent combinations: red-green, red-yellow, green-yellow or tricolour where the third colour (mixed colour) is produced when the other to diodes are ON. Fig. 13.31 shows the picture of a 5 mm bicolour LEDs whereas Fig. 13.32 shows a 10 mm tricolour LED. Multicolour LEDs are used in application such as polarity indication, null indication, logic output status, go/no-go indication etc.

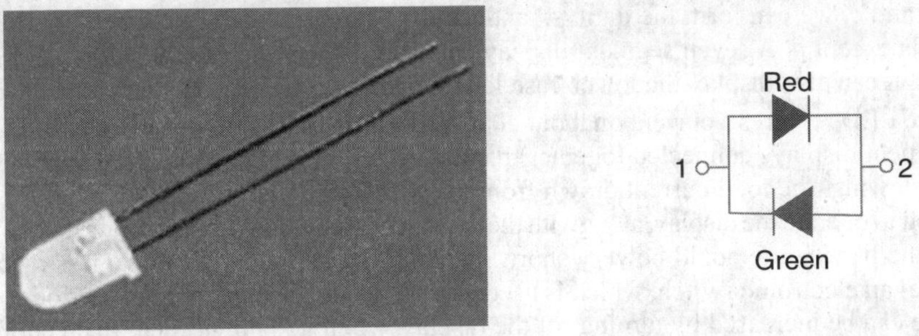

Fig. 13.31. 5 mm bicolour LEDs.

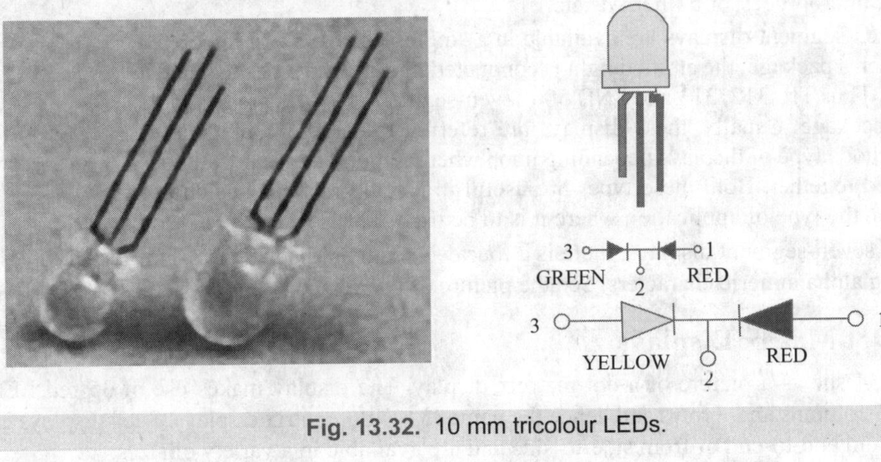

Fig. 13.32. 10 mm tricolour LEDs.

13.25. Seven-Segment Display

Fig. 13.33 (*a*) shows a seven segment display. It is used to display alphanumeric characters. It consist of 7 rectangular light emitting diodes designated by the letters *a, b, c, d, e, f* and *g*. Each LED is called a segment, because it forms a part of the character being displayed.

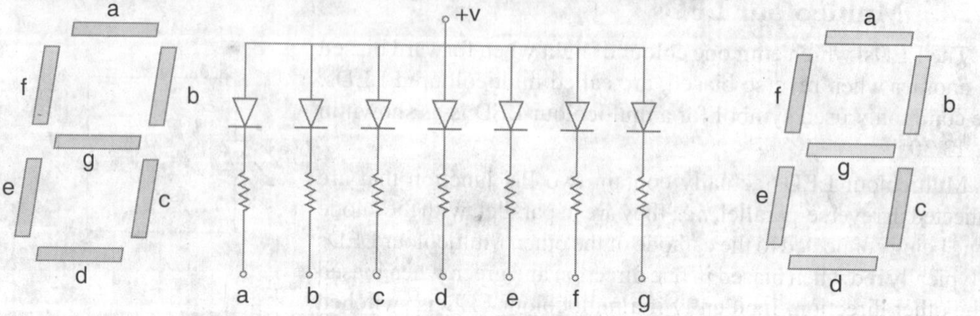

(*a*) 7-segments (*b*) Schematic diagram (*c*) Digit 2 in a 7-segment display.

Fig. 13.33. Seven-segment display.

Fig. 13.33 (*b*) shows a schematic diagram of a seven-segment display. In this circuit, the anodes of all the diodes are connected together to the positive terminal of the dc voltage source. The cathodes are connected to the external resistors. The external resistor is necessary to limit the current through the LED. By grounding the external resistors, we can form any decimal digit from 0 to 9. For example, by grounding *a, b, g, e* and *d*, we can form the digit '2' as shown in Fig. 13.33 (*c*). Similarly, by grounding *f, g* and *c*, we can form the digit '4' and so on. The digit 0 can be formed by grounding all the terminals except *g*. A seven segment display can also display the capital letters, *A, C, E* and *F*. Besides this, it can also display the lower-case letters *b* and *d*.

An average LED requires a current of about 20 mA for its operation. If we have a two-digit (*i.e.*, two seven-segment displays connected together) display in a circuit, then there are 14 LED's. If all the LEDs are glowing, the total current drawn from the supply voltage is 280 mA. With this much current, required to operate the display, alongwith the current, which is needed to operate the remainder of the circuit, the battery life could be very short. In battery powered applications, it is a common practice to instal an electronic switch, which is used to activate the display for short periods of time. The battery life can be increased by turning off the display, when it is not needed. In an equipment, where power consumption is not a problem, the LED's make an excellent display. These days, the seven-segment displays are used in digital clocks, calculators microwave ovens, stereo tuners, digital multimeters, microprocessor trainer kits etc.

The seven-segment displays are available in a single package as an integrated circuit. In such a package, the emitted light is conducted through the pipes to make it more useful. Thus TIL 312, 313 and FND 500 seven-segment displays are available in a single package. Usually, these displays are referred to as common-anode or common-cathode type of displays depending upon whether their anodes or cathodes are connected together. Both these types are useful in actual practice. The choice depends upon the type of application where it is to be used.

Besides seven-segment displays, there is a fourteen-segment display also. This display is used for indicating alphanumeric characters. See the picture of the device below.

13.26. Dot Matrix Displays

Fig. 13.34 shows a picture of a dot matreix display. The display makes use of 35 red LEDs arranged in 5 columns and 7 rows and hence the name (5 × 7 dot matrix display). Such displays can be stacked from end-to-end or from side-to-side and are available in a variety of sizes.

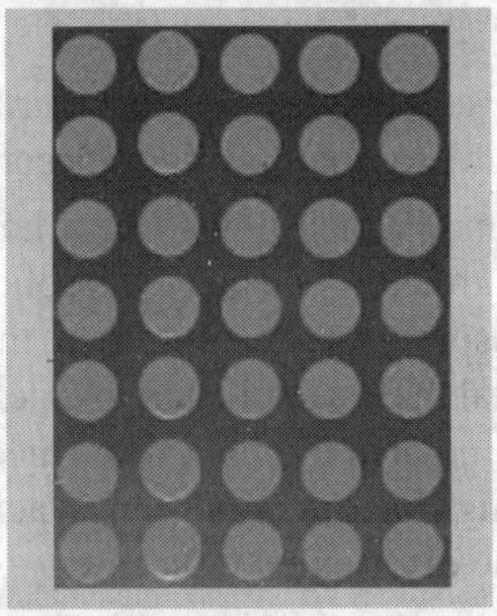

Fig. 13.34. 50 mm 5 × 7 dot matrix display

Fig. 13.35. 4 digit 5 × 7 dot matrix display

These days the dot matrix displays are available in a stack of 4 and 8 (4 being more common) making thereby a 4-digit or 8-digit display. Such displays are TL/CMOS compatible and have on-board decoding, multiplexing circuitry and driverss. These dot xatrix displays find application in instrumentation and process control, test equipment, telecommunications, computer peripherals, portable data entry devices, medical equipment etc. Fig. 13.35 shows a picture of a 4 digit 5 × 7 dot matrix LED display.

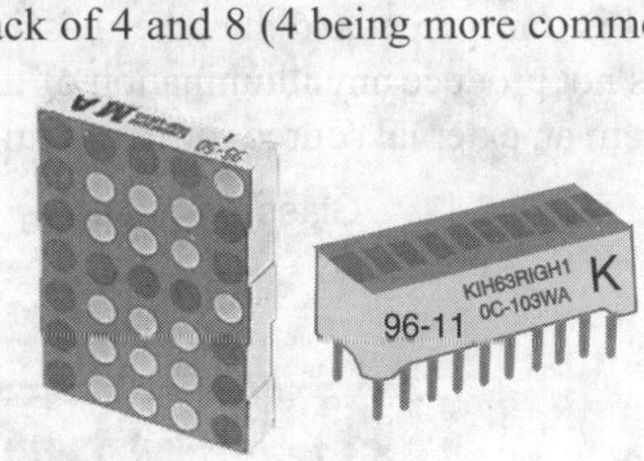

13.27. Use of Leds in Facsimile Machines

Fig. 13-36 shows a simplified schematic diagram of a facsimile (or fax) machine. As seen, the light from the LED array is focussed on the document paper. The light reflected at the paper is focussed on a charge-coupled device (CCD) by a combination. The electrical information is then sent through the data-processing unit to its destination via telephone line.

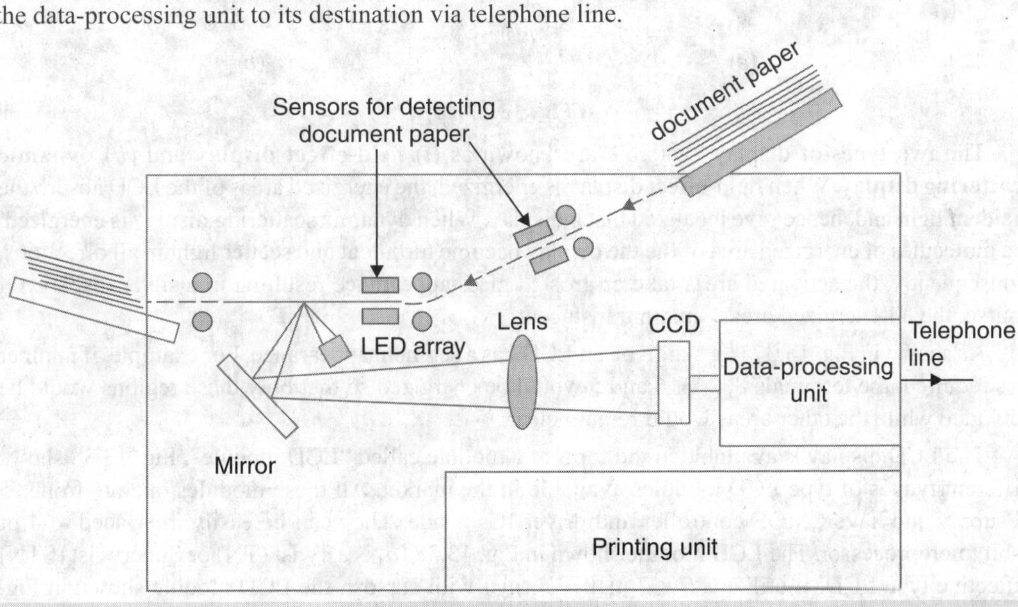

Fig. 13.36.

13.28. Liquid Crystals Displays (LCDs)

A liquid crystal display is a flat panel display, electronic visual display or video display that uses the light modulating proporties of liquid crystals. Liquid crystal do not emit light directly.

A liquid crystal is a material (usually, an organic compound) which flows like a liquid at room temperature but whose molecular structure has some properties normally associated with solids (examples of such compounds are : cholestery l nonanoate and p-azoxyanisole). As is well-known, the molecules in ordinary liquids have random orientation but in a liquid crystal they are oriented in a definite crystal pattern. Normally, a thin layer of liquid crystal is transparent to incident light but when an electric field is applied across it, its molecular arrangement is disturbed causing changes in falls on an activated layer of a liquid crystal, it is either absorbed or else is scattered by the disoriented molecules.

As shown in Fig. 13-37 (*a*), a liquid crystal 'cell' consists of a thin layer about 10 mm) of a liquid crystal sandwiched between two glass sheets with transparent electrodes deposited on their inside faces. With both glass sheets transparent, the cell is known as *transmittive* type cell. When one glass is transparent and the other has a reflective coating, the cell is called *reflective* type. The LCD does not produce any illumination of its own. It, in fact, depends entirely on illumination falling on it from an external source for its visual effect.

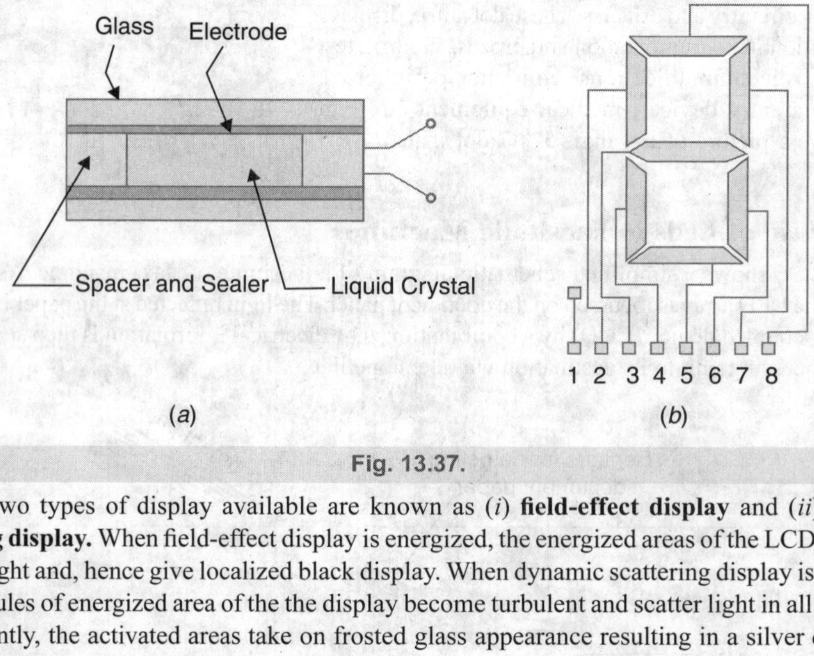

Fig. 13.37.

The two types of display available are known as (*i*) **field-effect display** and (*ii*) **dynamic scattering display.** When field-effect display is energized, the energized areas of the LCD absorb the incident light and, hence give localized black display. When dynamic scattering display is energized, the molecules of energized area of the the display become turbulent and scatter light in all directions. Consequently, the activated areas take on frosted glass appearance resulting in a silver display. Of course, the un-energized areas remain translucent.

As shown in Fig. 13.37 (*b*), a digit on an LCD has a segment appearance. For example, if number 5 is required, the terminals 8, 2, 3, 6 and 5 would be energized so that only these regions would be activated while the other areas would remain clear.

The LCD display is available in the form of a module called "LCD module". Fig. 13.38 shows different tyupes of type LCD modules available in the market. All these modules operate from +5 V supply and have CMOS controller and driver ICs inside. They can be easily interfaced to 4 or 8-bit microprocessor. The LCD module shown in Fig. 13.38 (*a*) is a Twist (TN) or Supertwist (STN) reflective type LCD. It utilize a 5 × 7 matrix format with cursosr, the LCD module shown in Fig. 13.38 (*b*) is of a transmissive type with LED backlighting. This module utilizes a 5 × 8 dot matrix format.

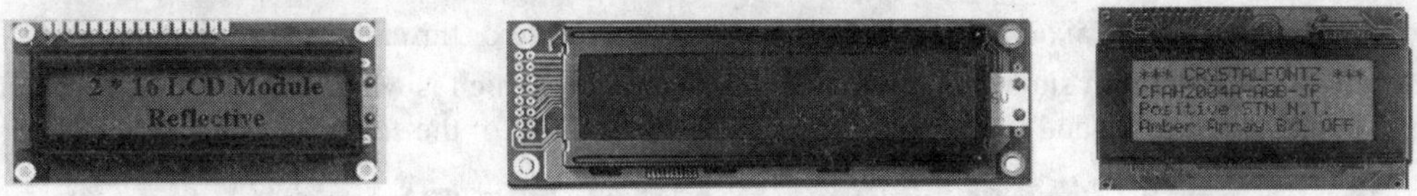

(*a*) TN/STN Reflective alphanumeric dot matrix LCD module
(*b*) STN Transmissive alphanumeric dot matrix LCD module
(*c*) Monochrome STN tranflective graphic LCD module

Fig. 13.38. Different types of LCD modules.

The LCD module shown in Fig. 13.38 (*c*) has a 128 character generator, MOM, allocation of text, user definable graphics can easily be made and strored in RAM and the display window can be freely moved within the allocated mamory range.

To summarizc, the reflective type LCD modules are suitable for use in sunlight and low light applications. The viewing is enhanced by direct sunlight unlike other texhnologies. The transflective types can be backlit making them ideal for use in low or no light conditions.

13.29. Advantages and Disadvantages of Liquid Crystal Displays (LCDs)

Avdantages

Following are some of the main advantages of LCD's as compared to the other display components:

1. Very compact and light weight.
2. Low power consumption.
3. No geometric distortion.
4. Little or no flicker depending on backlight technology.
5. Not affected by screen burn-in.
6. Can be made in almost any size or shape.
7. No limit on thc resolution.
8. Life time of about 50,000 hours.

Note. Screen burn is a permanent discolouration of areas on an electronic display such as CRT display or television stes with plasma display.

Disadvantages

Following are some of the main disadvantages of LCD's as compared to the other display components:

1. Limited viewing angle. This causes the colour, saturation and contrast of LCD to vary.
2. Smearing and ghostly artifacts caused by slow response time (more than 8 ms) and sample and hold operation.
3. Only one native resolution (*i.e.,* single fixed resolution). Display resolutions either require a video-scalar, lower perceptual quality or display at 1:1 pixel mapping in which images will be physically too large or will not fill the whole screen.
4. Fixed bit depth or colour depth. Bit depth is the number of bits used to represent the colour of a single pixel in a bitmapped image or a video frame buffer. Many cheaper LCDs are only able to display 262,000 colours. 8 bit super in-plane suitching (S-IPS) panels can display 16 million colours. But such displays have significantly better black level but are expensive and have slower response time.
5. Low bit depth. This results in images with unnatural or excessive contrast.
6. Input lag. *i.e.,* there is a delay that occurs when a user presses a button to when the screen reacts.

7. Dead or struck pixels may occur during manufacturing or through use.
8. In a constant-on situation, thermalization may occur which is when only part of the screen has overheated and looks discoloured compared to rest of the screen.

13.30. Applications of Liquid Crystal Displays (LCDs)

The liquid crystal Display is a matured technology these days so it is being used in practically all areas where sunlight does not directly on the display. Following are some of the important application of LCDs:

1. Computer monitors.
2. Television.
3. Instrument panels.
4. Aircraft cockpit displays
5. Signages.
6. Consumer devices such as video players, gaming devices, clocks, watches, calculators and telephones.

The liquid crystal display (LCDs) commonly used on notebook computers and handheld PADs are also appearing on desktop. These flat panel displays promise great clarity at increasingly high resolutions and are available in screen sizes upto 15 inches. The LCD monitor offers benefits and drawbacks. The first benefit is size. Because of the need to house the tube itself, cathode-ray tube (CRT) monitors are big and heavy. LCD monitors are expensive than CRTs at present. Another problem is the viewing angle. The optimal viewing angle of an LCD is from straight in front and as you move further to the side the screen becomes harder to read, much more so than with a CRT. Moreover screen resolutions generally reach only as high as 1,024 × 768, which is insufficient for some applications.

13.31. Photodiodes

A photodiode is a type of photodetector capable of converting light into either current or voltage depending upon the mode of operation. The common traditional solar cell used to generate electric solar power is a large area of application of photodiode.

A photodiode is a two terminal PN junction device, which *operates in a reverse bias*. It has a small transparent window, which allows light to strike the PN junction. Fig. 13.39 (*a*) and (*b*) shows the structure and schematic symbol of a photodiode. The basic biasing arrangement of a photodiode is as shown in Figure 13.39 (*c*).

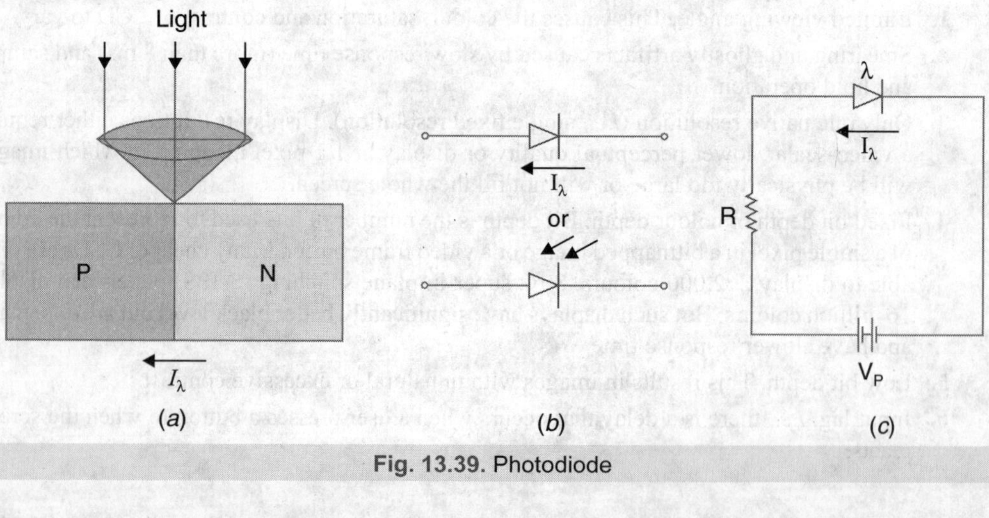

Fig. 13.39. Photodiode

We know that a rectifier diode has a very small reverse-current, when it is reverse biased. The same is true for a photodiode also. The reverse biased current is produced by thermally generated electron-hole pairs in the depletion layer, which are swept across the junction by the electric field created by the reverse voltage. In a rectifier diode, the reverse-current increases with the temperature due to an increase in thc number of electron-hole pairs

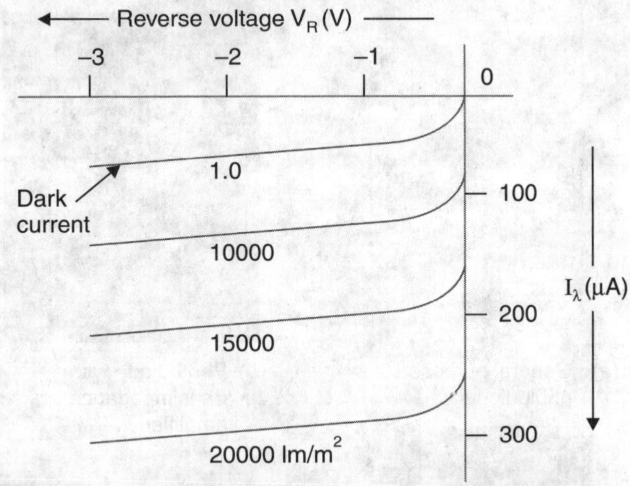

Fig. 13.40. Characteristic curves of a photodiode.

A photodiode differs from a rectifier diode in a sense that its reverse current increases with the light intensity at the PN junction. When there is no incident light, the reverse current is almost negligible and is called the dark current. An increase in the amount of light energy produces an increase in the reverse current for a given value of reverse-bias voltage. Fig. 13.40 shows the characteristic curves for a typical diode.

It may be noted from the characteristic curve that the dark current, for this particular device, is approximately 250 mA at a reverse bias voltage of 3 V. Therefore, resistance of the device with an incident light,

$$r_R = \frac{3}{25 \times 10^{-6}} = 120\ 000\ \Omega = 120\ \text{k}\Omega$$

At 20 000 lm/m², the current is approximately 300 μA at 3 V. Thc rcsistance under this condition,

$$r_R = \frac{3}{300 \times 10^{-6}} = 10\ 000\ \Omega = 10\ \text{k}\Omega$$

These calculations show that the photodiode can be used as a variable resistance device controlled by light intensity. A photodiode can be used to switch on the current at a very fast rate (in nanosecond). Thus it is one of the fastest photo detectors.

13.32. Commercially Available Photodiodes

Fig. 13.41 (*a*) Through (*b*) shows a variety of photodiodes in different types, shapes and sizes. The photodiode shown in (*a*) is a general purpose photodiode. It is suitable for applications in photometers, modulated light detectors, hight-speed punched card readers etc. The photodiode shown in (*b*) finds its application in analytical and medical instrumentation, smoke detectors and photographic processing. The photodiode shown in (*c*) is ideal for low light level applications where a hight signal-to-noise ratio is needed such as light monitoring and control circuit. The photodiode shown in (*d*) has an integral high gain low noise amplifier housed in a lensed housing TO5 package. The lens increases the signal-tonoise ratio making it ideal for electrically noisy environments. The applications for this diode include instrumentation, photometry and light monitoring.

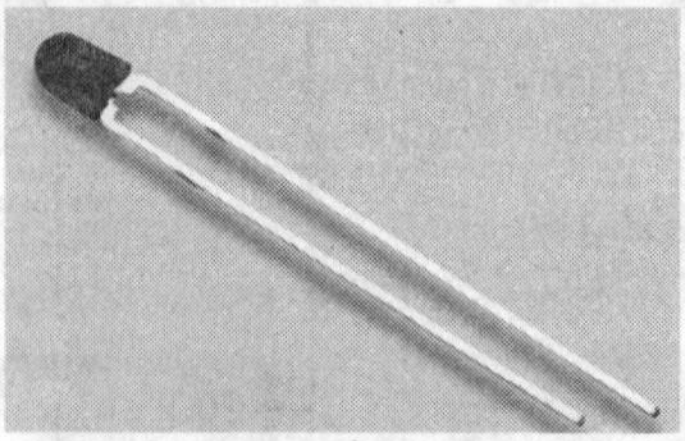

(*a*) General purpose photodiode

(*b*) Photodiode with transimpedance amplifier

(*c*) Photodiode in TO5 package

(*d*) Photodiode with integral amplifier

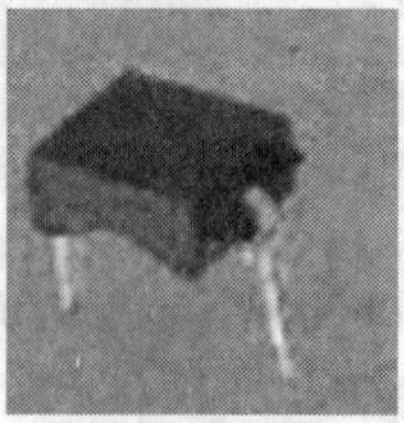

(*e*) A planar silicon PIN photodiode housed in a epoxy package

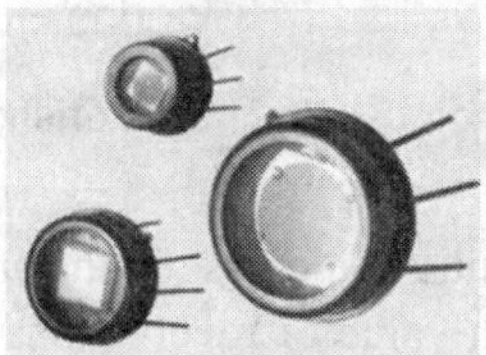

(*f*) Silicon photodiodes

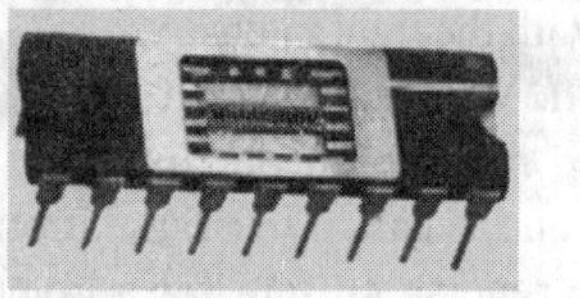

(*g*) A linear silicon PIN photodiode array

(*h*) A planar silicon PIN photodiode in a surface mount package

Fig. 13.41. Photodiodes of different shapes and sizes.

The photodiode shown in (*e*) is 7 mm^2 planar silicon PIN photodiode housed in a two pin epoxy package with a integral daylight cut-off filter. This device is a low noise, high speed and operates over a wide temperature range. The applications for this photodiode include remote controls, light curtains, data transmission and measurement and control. The photodiodes shown in (*f*) are silicon photodiodes ideal for use in general photometric applications. Both these devices have integral colour correction filter. The device shown in (*g*) is a 16-element linear silicon PIN photodiode array housed in a ceramic dual-in-line (DIL) package. This hight speed device consists of 16 individual elements arranged on a 1 mm pitch in common cathode configuration. This array is ideal for linear position sensing, wide aperture detection and edge and hole detection in strip materials. The device shown in (*h*) is a 7 mm^2 planar silicon PIN photodiode housed in a surface mount package. The application of this device are the same as that of the photodiode shown in (*e*), *i.e.*, in remote controls, light curtains, data transmission and measurement and contol.

13.33. Photodiode Applications

Photodiodes are used in similar applications to other photodetectors, such as photoconductors, charge couple devices and photo multiplier tubes. They may be used to generate an analog output which is dependent upon the illumination or to change the state of digital circuiting.

Following are some of the important applications of photodiodes which are important from the subject point of view:

1. Used in consumer electronic devices such as compact disc (CD) players, smoke detectors, and the receivers for infrared remote control equipments from televisions to air-conditioners.
2. Used for accuate measurement of light intensity in science and industry. The photodiodes have a more linear response than photo-conductors.

3. Widely used in medical applications such as detectors for computer tomography, instruments to analyze samples pulse oximeters.
4. Because of their fast switching speed, used for optical communication and in lighting regulation.
5. Optical communication systems
6. Character recognition
7. Encoders *etc.*

Notes : 1. PN-junction photodiodes are not used to measure extremely low light intensities. Instead, if high sensitivity is needed avalanch photodiodes, intensified charge-couple devices or photomultiplier tubes are used for applications such as astronomy, spectroscopy, might vision equipment and laser range finding.

2. A one-dimensional array of hundred or thousands of photodiodes can be used as a position sensor, for example, as a part of angle sensor. One advantage of photodiode arrays is that they allow for high speed parrallel read out since the driving electronics may not be built in like a traditional CMOS or CCD sensor.

13.34. Dust Sensor

Fig. 13.42 shows a combination of an LED and a photodiode used as a dust sensor. As seen, the light emitted from the LED gets reflected by the dust particles. The reflected light is collected by the dust particles. The reflected light is collected by the photodiode and is converted into an electrical signal. The dust sensor is employed in cleaners.

The combination of an LED and a photodiode is also used as : (1) a paper sensor in facsimile machines, (2) as a tape-end sensor in videotape recorders/players, and (3) as a dirt detector for rinsing in washing machines.

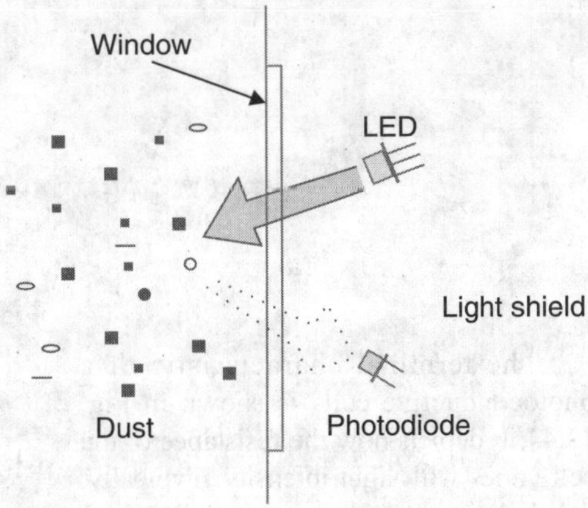

Fig. 13.42.

13.35. Photoconductive Cell

It is a semiconductor device whose resistance varies inversely with the intensity of light falls upon it. It is also known as *photoresistive* cell or *photoresistor* because it operates on the principle of photoresistivity.

The resistivity (and, hence, resistance) of a semiconductor depends on the number of free charge carriers available in it. When the semiconductor is not illuminated, the number of charge carriers is small and, hence, resistivity is high. But when light in the form of photons strikes the semiconductor, each photon delivers energy to it. If the photon energy is greater than the energy band gap of the semiconductor, free mobile charge carriers are liberated and, as a result, resistivity of the semiconductor is decreased.

Photoconductive cells are generally made of cadmium compounds such as cadmium sulphide (CdS) and cadmium selenide (CdSe). Spectral response of CdS cell is similar to the human eye, hence such cells are often used to simulate the human eye. That is why they find use in light metering circuits in photographic cameras.

The construction of a typical photoconductive cell and its two alternative circuit symbols are shown in Fig. 13.43 (*a*) and (*b*) respectively. As seen, a thin layer of photosensitive semiconductor material is deposited in the form of a long strip zig-zagged across a disc-shaped ceramic base with protective sides. For added protection, a glass lens or plastic cover is used. The two ends of the strip are bought out to connecting pins below the base.

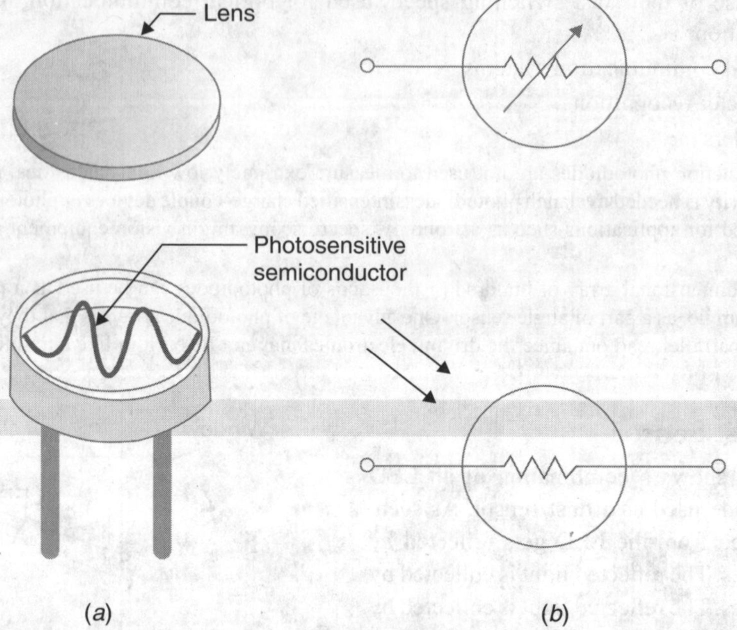

Fig. 13.43.

The terminal characteristic of a photoconductive cells is shown in Fig. 13.44. It depicts how the resistance of the cell varies with light intensity. Typically, the dark resistance of the cells is 1 MΩ or larger. Under illumination, the cell resistance drops to a value between 1 and 100 kΩ depending on surface illumination.

A photoconductive cell is an inexpensive and simple detector which is widely used in OFF/ON circuits, light-measurement and light-detecting circuits.

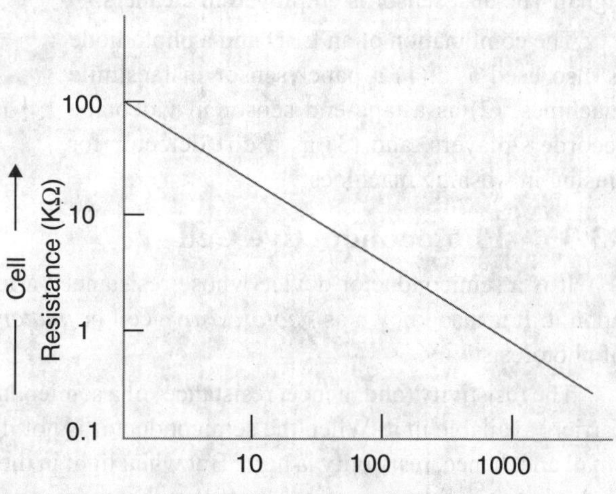

Fig. 13.44.

Example 13.11.

A relay is controlled by a photo-conductive cell which has resistance of 100 k Ω when illuminated and 1 kΩ when in the dark. The realy is supplied with 10 mA from a 30-V supply when cell is in illuminated and is required to be de-energized when the cell is in the dark Sketch a suitable circuit and calculate the required series resistance and value of dark current. (*Optoelectronic Devices, Gujarat Univ. 1993*)

Solution.

The circuit is as shown in Fig. 13.45 where *R* is a current-limiting resistor.

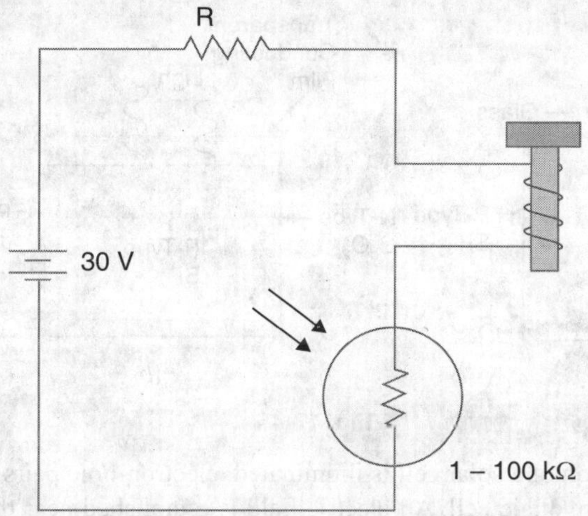

Fig. 13.45.

The current through the circuit,

$$I = \frac{30}{R + r} \quad ...(i)$$

where 'R' is the series resistance and 'r' is the cell resistance. Rearranging the equation (i),

$$R = \left(\frac{30}{I} - r\right) \quad ...(ii)$$

When illuminated, from equation (ii),

$$R = \left(\frac{30}{10 \times 10^{-3}} - (1 \times 10^{3}\right) = 2 \times 10^3\ \Omega = 2\ \text{k}\Omega \quad \textbf{Ans.}$$

From equation (i), the dark current,

$$I_d = \frac{30}{\left((2 \times 10^{3}) + (100 \times 10^{3}\right)} = 0.3 \times 10^{-3}\ \text{A} \quad \text{or} \quad 0.3\ \text{mA} \quad \textbf{Ans.}$$

13.36. Photovoltaic or Solar Cell

Such cells operate on the principle of photovoltaic action *i.e.* conversion of light energy into electrical energy. This action occurs in all semiconductors which are constructed to absorb energy.

As shown in Fig. 13.46 (*a*), a basic solar cell consists of *P*-type and *N*-type semiconductor material (usually, silicon or selenium) forming a *P-N* junction. The bottom surface of the cell (which is always away from light) covered with a continuous conductive contact to which a wire lead is attached. The upper surface has a maximum area exposed to light with a small contact either along the edge or around the perimeter. The surface layer of *P*-type material is extremely thin (0.5 mm) so that light can penetrate to the junction.

Although silicon is commonly used for fabricating solar cells, another construction consists of *P*-type selenium covered with a layer of *N*-type cadmium oxide *P-N* junction as shown in Fig.13.46 (*b*). Two alternative circuit symbols are shown in Fig. 13.46 (*c*). Power solar cells are also fabricated in flat strips to form efficient coverage of available surface area. Incidentally, the maximum efficiency of a solar cell in converting sunlight into electrical energy is nearly 15 per cent at the present.

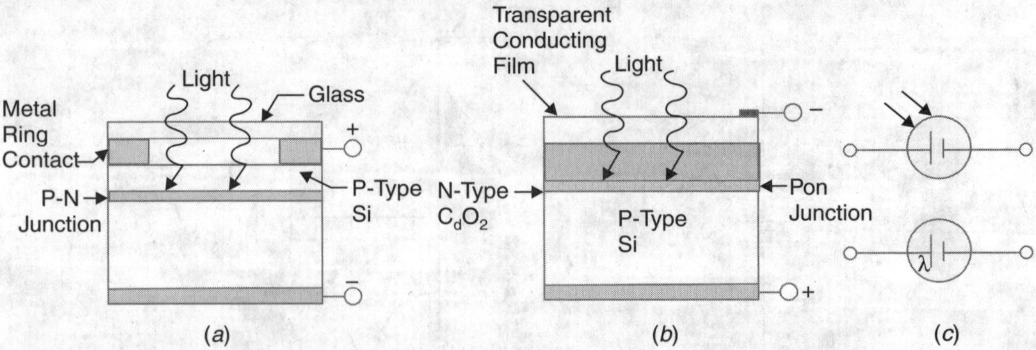

Fig. 13.46.

When the *P-N* junction of a solar cell is illuminated, electron-hole pairs are generated in much the same way, as in photovoltaic cell. An electric field is established near the *P-N* junction by the positive and negative ions created due to the production of electron-hole pairs which leads to the development of potential across the junction. Since the number of electron-hole pairs far exceeds the number exceeds for thermal equilibrium, many of the electrons are pulled across the junction by the force of the electric field. Those that cross the junction contribute to the current in the cell and through the external load. The terminal voltage of the cell is directly proportional to the intensity of the incident light. The voltage may be as high as 0.6 V depending on the external load. Usually a large number of cells are arranged in an array in order to obtained higher voltages and currents as shown in Fig. 13.47 (*a*).

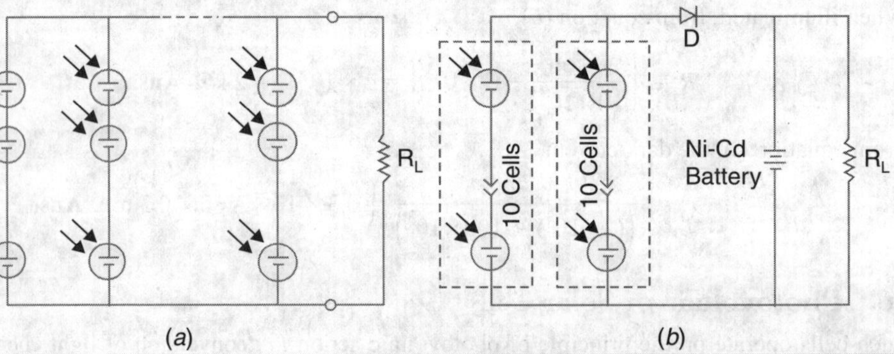

Fig. 13.47.

Solar cells act like a battery when connected in series or parallel. Fig. 13.47 (*b*) show two groups of 10 series cells connected in parallel with each other. If each cell provides 0.5 V at 150 mA, the overall value of the solar bank is 5 V at 150 mA. The two parallel solar banks provides 5 V at 300 mA. This solar power source supplies the load and also charges the Ni-Cd battery. The battery provides power in the absence of light. A blocking diode

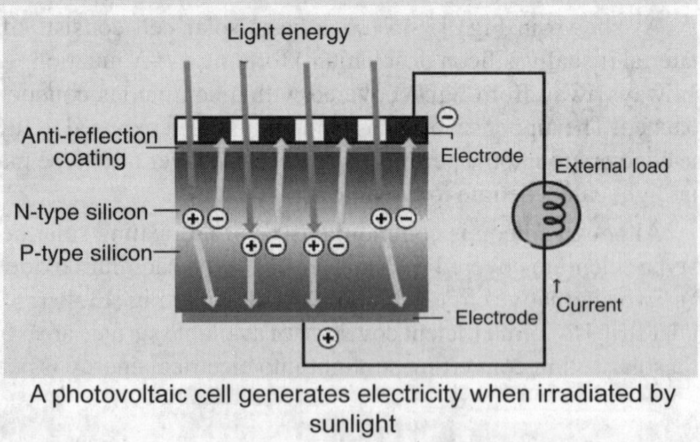

A photovoltaic cell generates electricity when irradiated by sunlight

D is used to isolate the solar cells from Ni-Cd battery otherwise in the absence of light, the battery will discharge through the cells thereby damaging them.

A solar cell operates with fair efficiency, has unlimited. life, can be easily mass-produced and has a high power capacity per weight. It is because of these qualities that it has becomes an important source of power for earth satellites.

Example 13.12.

An earth satellite has on board 12- V battery which supplies a continuous current of 0.5 A. Solar cells are used to keep the battery charged. The solar cells are illuminated by the sun for 12 hours in every 24 hours. If during exposure, each cell gives 0.5 V at 50 mA, determine the number of cells required. (*Optoelectronics Devices, Gujarat Univ. 1994*)

Solution

The solar cell battery-charging circuit is shown in Fig. 13.48. The cells must be connected in series to provided the necessary voltage and such groups must be connected in parallel to provide the necessary current. The charging voltage has to be greater than the battery voltage of 12 V. Allowing for different drops, let the solar bank voltage be 13.5 V.

Number of series connected solar cells = 13.5/0.5 = 27

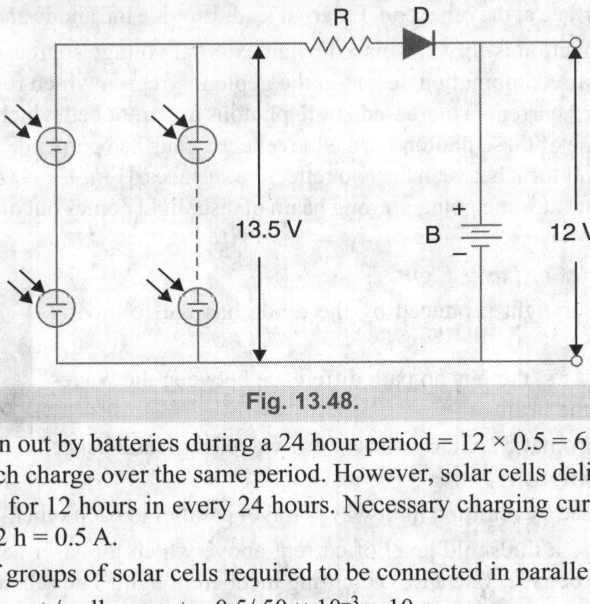

Fig. 13.48.

The charge given out by batteries during a 24 hour period = 12 × 0.5 = 6 Ah. Hence, solar cells must supply this much charge over the same period. However, solar cells deliver current only when they illuminated *i.e.,* for 12 hours in every 24 hours. Necessary charging current required from the solar cells = 6 Ah/ 12 h = 0.5 A.

Total number of groups of solar cells required to be connected in parallel is

= output current / cell current = $0.5/ 50 \times 10^{-3} = 10$

∴ total number of solar cells required for the earth satellite = 27 × 10 = **270**

13.37. Laser Diode

Like LEDs, laser diodes are typical *PN* junction devices used under a forward-bias. The word LASER in an acronym for *Light Amplification* by *Stimulated Emission of Radiation.* The use of laser is (becoming increasing common) in medical equipment used in surgery and in consumer products like compact disk (CD) players, laser printers, hologram scanners etc.

Broadly speaking, the laser diode structure can be divided into two categories :

1. ***Surface-emitting laser diodes :*** These laser diodes emit light in a direction **perpendicular** to the PN junction plane. The vertical cavity surface emitting laser (VCSELs) are of this type of structure.

2. ***Edge-emitting laser diodes*** : These laser diodes emit light in a direction **parallel** to the PN junction plane. Double hetrostructure laser, quantum well lasers are of these types.

Fig. 13.49 (*a*) shows the structure of an edge-emitting laser diode. This type of structure is called Febry-Perot type laser. As seen from the figure, a *P-N* junction is formed by two layers of doped

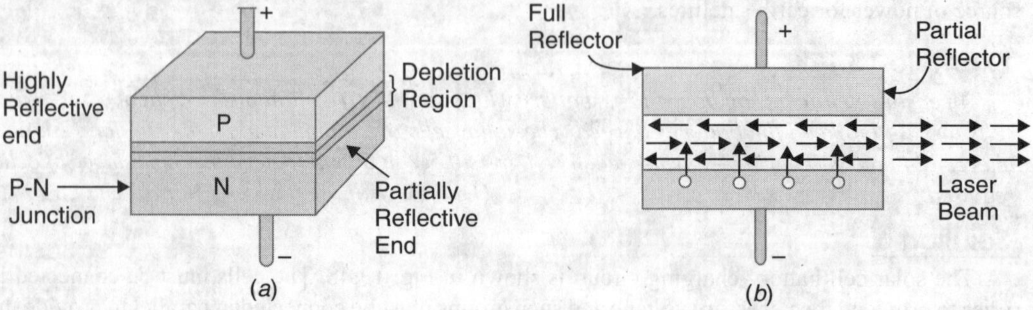

Fig. 13.49.

gallium arsenide (GaAs). The length of the *PN* junction bears a precise relationship with the wavelength of the light to be emitted. As seen, there is a highly reflective surface at one end of the junction and a partially reflective surface at the other end. External leads provide the anode and cathode connections.

When the *P-N* junction is forward-biased by an external voltage source, electrons move across the junction and usual recombination occurs in the depletion region which results in the production of photons. As forward current is increased, more photons are produced which drift at random in the depletion region. Some of these photons strike the reflective surface perpendicularly. These reflected photons move back and forth between the two reflective surfaces. [Fig. 13.49 (*b*)] The photon activity becomes so intense that at some point, a strong beam of laser light comes out of the partially reflective surface of the diode.

Unique Characteristics of Laser Light

The beam of laser light produced by the diode has the following unique characteristics:

Laser diode

1. It is coherent *i.e.* there is no path difference between the waves comprising the beam;
2. It is monochromatic *i.e.* it consists of one wavelength and hence one colour only;
3. It is collimated *i.e.* emitted light waves travel parallel to each other.

Laser diodes have a threshold level of current above which the laser action occurs but below which the laser diode behaves like a LED emitting incoherent light. The schematic symbol of a laser diode is similar to that of LED. Incidentally, a filter or lens is necessary to view the laser beam.

13.38. Laser Diode Applications

Laser diodes have numerous applications. Followings are important from the subject point of view:

1. Used in telecommunications as easily modulated and easily coupled light sources fibre optic communication.
2. Used in various measuring devices such as range finders, bar-code readers.
3. Used in printing industry both as light sources for scanning images and for very high speed and high resolution printing plate manufacturing.
4. Infrared and red laser diodes are common in CD players, CD-ROM and DVD technology. Violet lasers are used in HD-DVD and Blue-ray technology.

5. Used in laser absorption spectroscopy for high-speed low cost assessment or monitoring of the concentration of various species in gas-phase.
6. High power laser diodes are used in industrial applications such as heat treating, cladding, seam welding and for pumping other lasers.
7. Used in laser medicine especially, dentistry.
8. Used in optical data recording, combustion iginition, industrial sorting, industry machining and directed energy weaponery.

Thus laser diodes used in variety of applications ranging from medical equipment used in surgery to consumer products like optical disk equipment, laser printers, hologram scanners etc. Laser diodes emitting visible light are used as pointers. Those emitting visible and infrared light are used to measure range (or distance). The laser diodes are also widely used in parallel processing of information and in parallel interconnections between computers. Some of these applications are discussed in the following articles.

13.39. Optical Discs

The major application field for laser diodes is in optical disc equipment is used for reading or recording information and can be broadly divided into two groups :

1. Reading-only and 2. Recording-and -reading type.

The optical disc equipment of either type make use of a laser diode, lenses and photodiodes. During recording, it changes electrical information into optical information and then records the information on the optical disc. During reading (or playback), the head optically reads the recorded information and changes the optical information into electrical information. Fig. 13.50 shows the different types of optical discs used in practice. The commercial system make use of discs that are 90, 120, 130 and 300 mm in diameter. A mini disk, 64 mm in diameter is also used for digital audio.

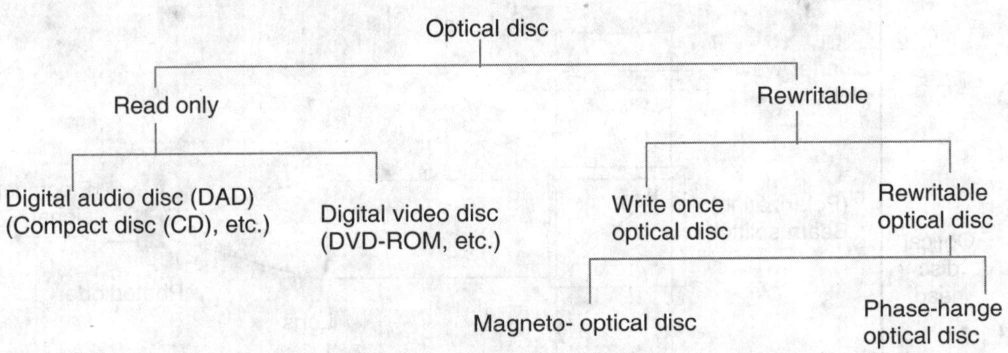

Fig. 13.50.

The optical discs have several advantages over semiconductor memories. Some of these include their larger data storage capacity, shorter access time size. Therefore they are used in terminal equipment of computers as well as in audio visual equipment.

13.40. Read-only Optical Dics Equipment

Fig. 13.51 shows an optical equipment for reading data from digital audio (compact) discs. Compact discs (CDs) which are 120 mm in diameter are typical digital audio discs. Compact discs usually means digital audio compact disk, but it also includes the read-only memory (CD-ROM) for data memory and interactive compact disk (CD-I) for multimedia use.

Audio information (*i.e.* sound) is digitally recorded in stereo on the surface of a CD in the form of microscopic "pits" and "flats". As seen from Fig. 13.51, the light emitted from the laser diode passes through the lens and is focussed to a diameter of about 1 mm on the surface of a disk. As the CD rotates, the lens and beam follow the track under control of a servo motor. The laser light which is altered by the pits and flats along the recorded track is reflected back from the track through the lens and optical system to infrared photodiodes. The signal from the photodiodes in then used to reproduce the digitally recorded sound.

A CD-ROM

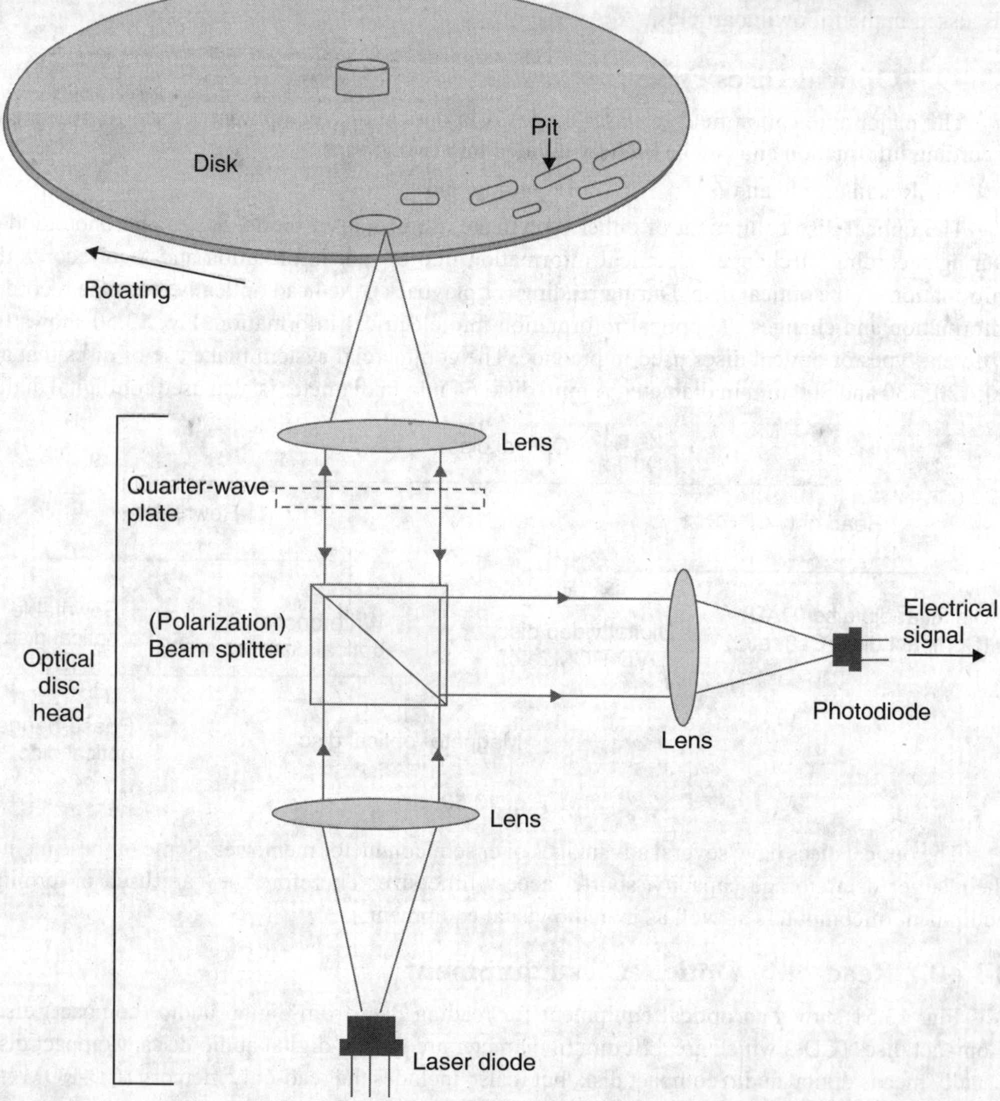

Fig. 13.51.

13.41. Printers Using Laser Diodes

These are two types of optical source usually used in printers; (1) laser diodes and (2) LED arrays. The printers using laser diodes are called laser beam printers (or simply laser printers). These are one of the most attractive type of equipment in office automation in today's world. Words and figures can be printed rapidly and clearly more easily by a laser printer than by other type of printers.

Fig. 13.52 shows a simplified diagram of a laser printer. As seen the laser diode is driven by modulated signals from the computer. The optical beam after passing through the lens is reflected by

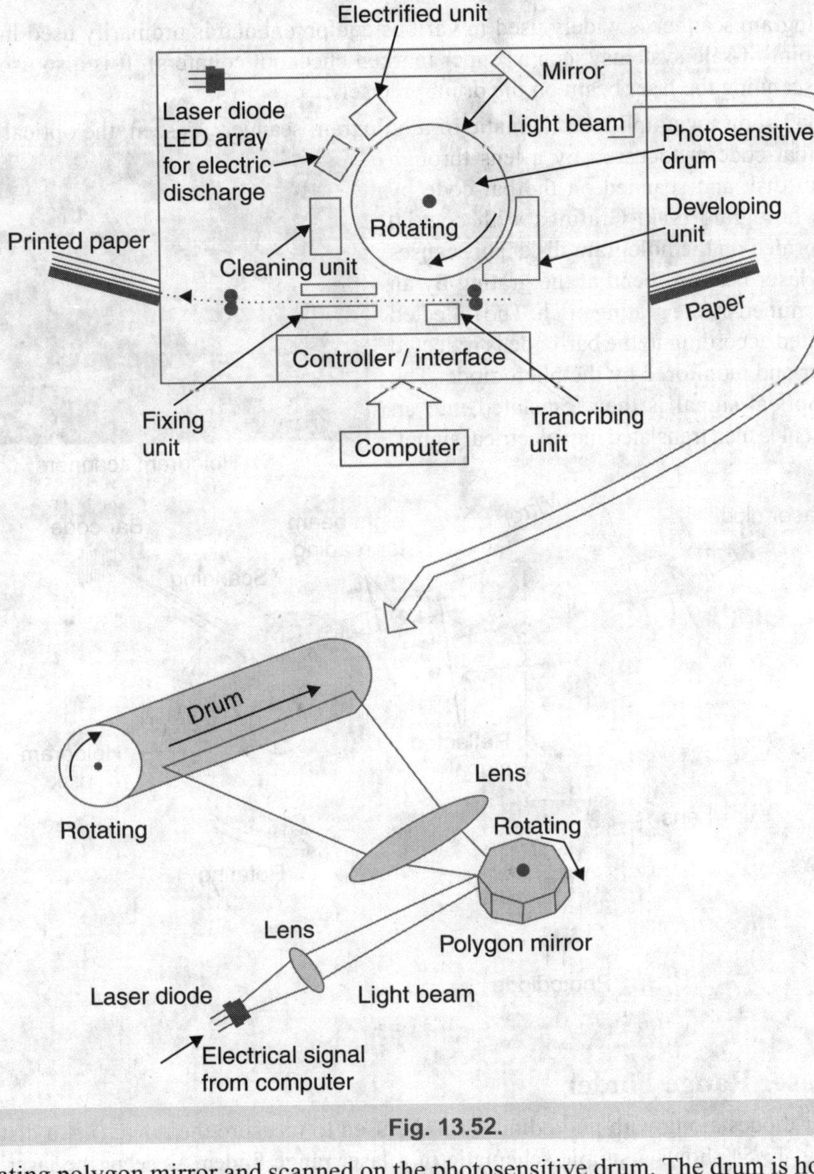

Fig. 13.52.

the rotating polygon mirror and scanned on the photosensitive drum. The drum is homogeneously charged when it passes through the charging unit consisting of an LED array. The homogeneous electrification is partially erased in accordance with the scanned optical beam. This is because of the fact that the electrical resistance at the light-irradiated part decreases and the electric charge is released. This causes the signals (*i.e.* data) from the computer to be written on to the drum. At the developing unit, an electrically charged powder (called toner) is electrostatically attached to the written parts. At the transcribing unit, the powder is transferred to the paper. Next, the transferred pattern is fixed by heating and pressing at the fixing unit. The data from the computer is thus printed on the paper.

(*Courtesy*: Optical Semiconducto Devices by M.Fukuda published by John Weliy & Sons Inc.)

13.42. Hologram Scanners

The hologram scanner is widely used in various equipment and is ordinarily used in bar-code readers in point-of-sale systems (such as super marked checkout counters). It is also used in laser printers for scanning the laser beam on the drum precisely.

Fig. 13.53 shows a simplified schematic of a hologram scanner. As seen, the optical beam for reading the bar-code is focussed by a lens through the hologram disk and scanned on the bar-code by rotating the hologram disk. Gratings with coaxial circles are formed on the hologram disk. This causes the incident laser beam to bend at the grating by an amount determined by the grating pitch. The reflected light modulated according to the bar-code is reflected by the mirror and monitored by the photodiode. The monitored optical signal is then translated into an electrical signal is then translated into electrical signal.

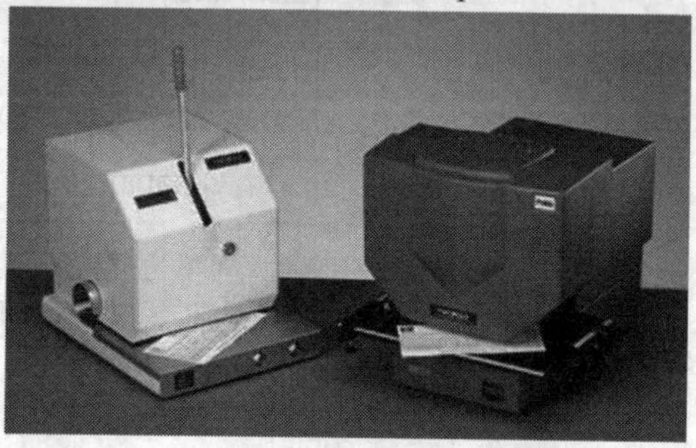

Hologram scanners

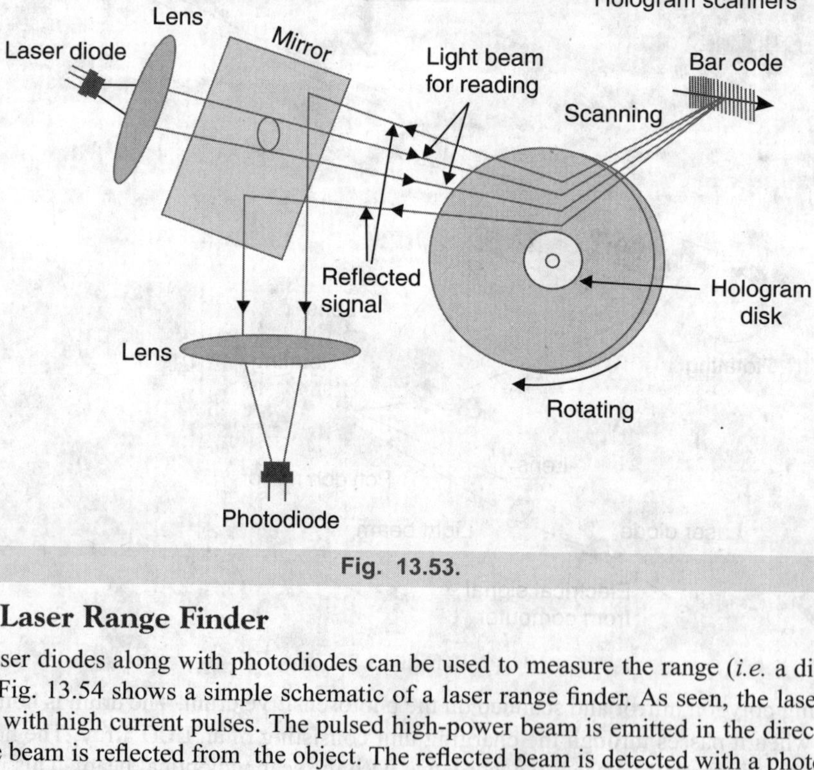

Fig. 13.53.

13.43. Laser Range Finder

The laser diodes along with photodiodes can be used to measure the range (*i.e.* a distance) of an object. Fig. 13.54 shows a simple schematic of a laser range finder. As seen, the laser diode is modulated with high current pulses. The pulsed high-power beam is emitted in the direction of an object. The beam is reflected from the object. The reflected beam is detected with a photo detector (or photodiode). The range is calculated as the difference between the time the light was emitted from the laser diode and the time it was detected by the photodiode.

Let D = distance between the laser range finder and the object.

ΔT = Time difference between the instance when the light was emitted from the laser diode and the instance when it was detected by the photodiode.

Then
$$D = \frac{1}{2} \times \text{speed of light} \times \Delta T$$

A 2-diamensional array of laser diodes and photodetectors can be constructed. Such a system is used to obtain 3-D images of an object.

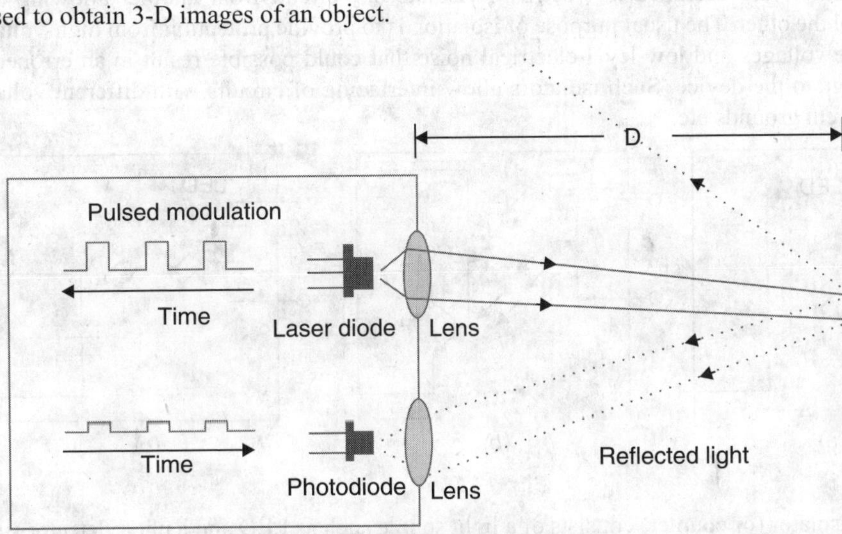

Fig. 13.54.

13.44. Light-activated SCR (LASCR)

The operation of a LASCR is essentially similar to that of a conventional SCR (refer to chapter 17 Art. 17.3) except that it is light-triggered (Fig. 13.55). It has a window and lens which focuses light on the gate junction area. Refer to Fig. 13.55 (*a*). The LASCR operates like a latch. It can be triggered ON by a light input on the gate area but does not turn OFF when light source is removed. It can be turned OFF only be reducing the current through it below its holding current . Depending on its size, a LASCR is capable of handling larger amount of current than be handled by photodiode or a photo-transistor.

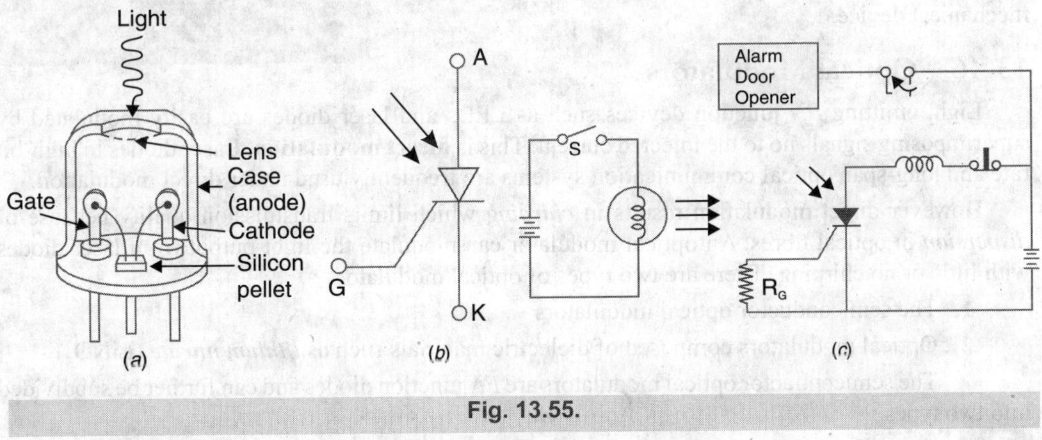

Fig. 13.55.

Fig. 13.55 (*b*) shows the circuit symbol while (*c*) how a LASCR can be used for energizing a latching relay. The input dc source turns on the electric lamp and the resulting incident light triggers the LASCR into conduction. The anode current energizes the relay and closes the contact. It is seen that the input dc source is electrically isolated from the rest of the circuit.

13.45. Optical Isolators

Optical isolators are designed to electrically isolate one circuit from another allowing one circuit to control the other. The usual purpose of isolation is to provide protection from high-voltage transients, surge voltages and low-level electrical noise that could possible result in an erroneous output or damage to the device. Such isolators allow interfacing of circuits with different voltage levels and different grounds etc.

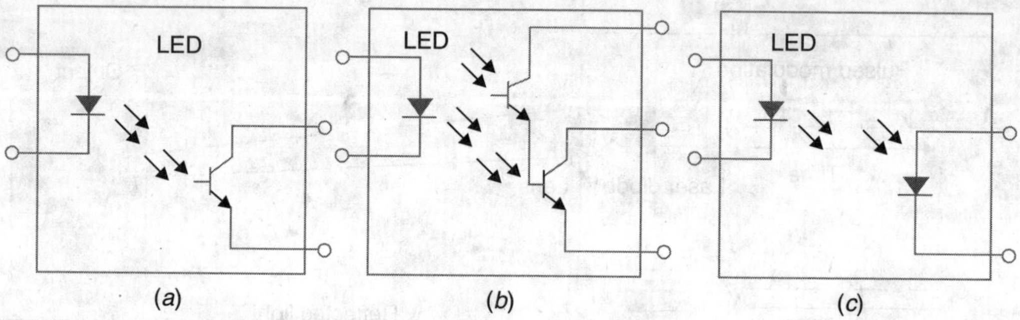

Fig. 13.56.

An optical isolator (or coupler) consists of a light source such as LED and a photodetector such as photo transistor as shown in Fig. 13.56 (*a*) and is available in a standard *IC* package.

A photo transistor is a 3 Terminal device which responds to its input in the same way as a photodiode. In addition it produces larger output.

When LED is forward-biased, the light produced by it is transferred to the photo transistor which is turned ON thereby producing current through the external load.

Fig. 13.56 (*b*) shows a Darlington transistor coupler which is used when increased output current capability is needed beyond that provided by the phototransistor output. The LASCR output coupler of Fig. 13.56 (*c*) can be used in applications where a low-level input voltage is required to latch a high voltage relay for activating some king of electro-mechanical device.

13.46. Optical Modulators

Light emitting *PN* junction devices such as LEDs and laser diodes are easily modulated by superimposing signals no to the injected current. This is **direct modulation.** Laser diodes in high-bit rate and long-span optical communication systems are frequently used under direct modulation.

However direct modulation results in *chirping* which limits transmission quality because of *dispersion* in optical fibres. An optical modulator can modulate the light output from laser diodes with little or no chirping. There are two types of optical modulators :

1. The semiconductor optical modulators
2. Optical modulators composed of dielectric materials such as *lithium nitrate* ($LiNO_3$)

The semiconductor optical modulators are *PN* junction diodes and can further be subdivided into two types :

1. Devices used under forward bias (as LEDs and laser diodes are used). The optical modulation in these devices is carried out by changing gain or loss within the modulators.
2. Devices used under reverse bias (*i.e.,* as photodiodes are used). Most high-performance semiconductor optical modulators are used under reverse bias. The reverse bias is needed to generate strong electric field. Optical modulation is basically performed by modulating the *refractive index* or *optical absorption coefficient* of the modulators. The devices which make use of refractive index phenomenon of modulation are called phase modulation type devices while those that use optical absorption coefficient phenomenon are called intensity modulation type devices.

There are several different of optical modulators available today. But the waveguide type optical modulator is more common in use. Further there are several different waveguide type optical modulator structures possible. Fig. 13.57 shows a mesa type optical modulator structure.

It may be noted that although we have shown the structure making use of a simple *N*-and *P*-layer but in reality each layer (*N*-type or *P*-type) is made up of several different semiconductors.

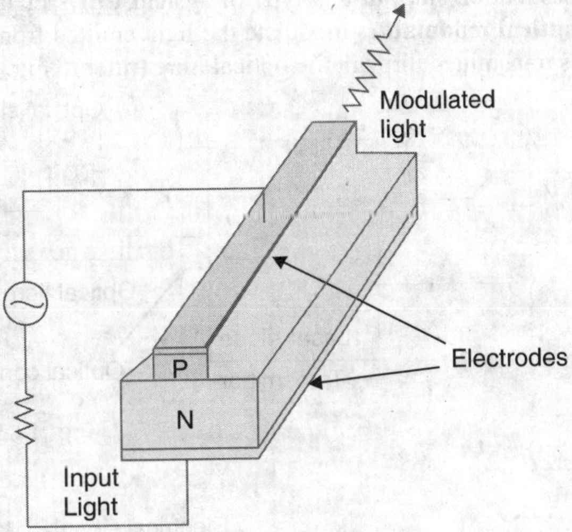

Fig. 13.57.

13.47. Optical Fibre Communication Systems

The optical fibre communication systems (such as public communication networks and data links) are the basic infrastructure of the information hungry society. There are several advantages of the optical fibre system over metallic transmission systems as listed below :

1. Data can be transmitted at a very high-frequency over longer distances without much loss.
2. Electromagnetic induction (EMI) noise is never induced during transmission through optical fibre cables.
3. Optical fibre cable is light, flexible and economical.

Fig. 13.58 shows the public optical fibre communication system broadly divided into two groups: (1) Submarine systems, and (2) Land systems. Submarine systems have already been used to connect countries all over the world. The submarine systems help people to talk overseas without any time delay.

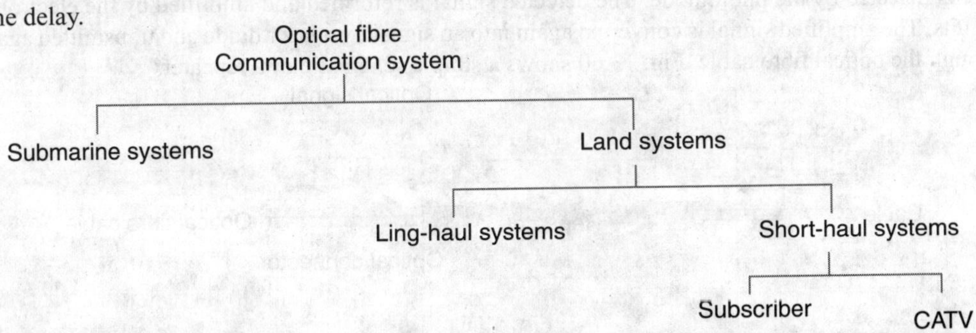

Fig. 13.58.

In land systems, long-haul systems have been connected between large cities. The land systems also include systems such as subscriber systems and CA TV (*i.e.* community or common antenna television, cable and telecommunication television systems, or cable television system).

Fig. 13.59 shows an application of LEDs, laser diodes and photodiodes in a simplified optical fibre communication systems. The LEDs and laser diodes emit light modulated with a signal. The optical signal is then transmitted through the optical fibre and is received with photodiodes on the destination side. In this type of system LEDs or laser diodes emit the light directly through the **optical modulators** modulate the light emitted from the laser diodes and then the modulated light is transmitted through the optical fibre (refer to Fig. 13.60).

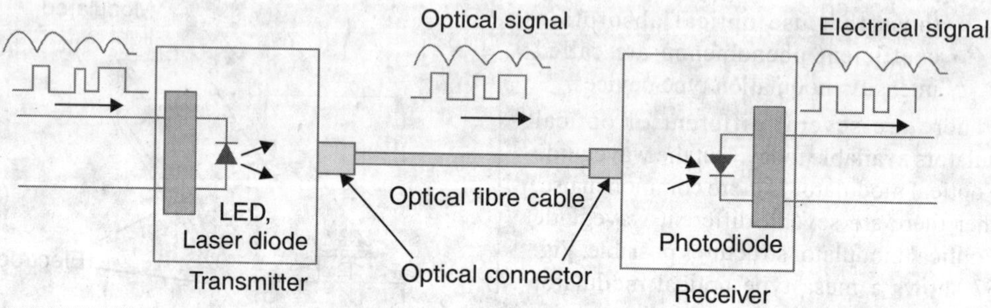

Fig. 13.59.

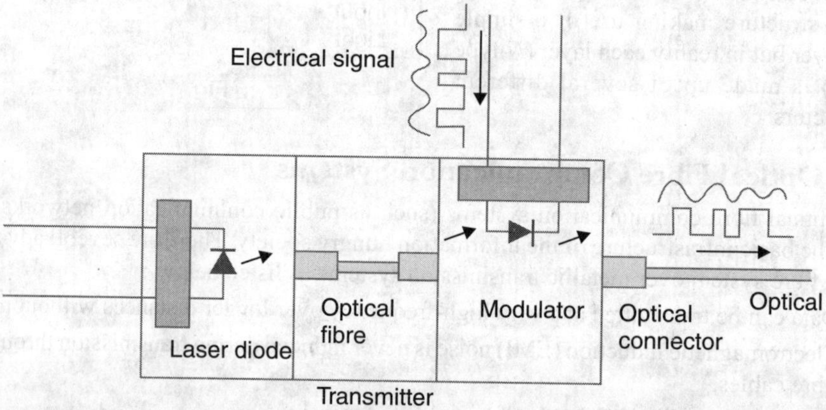

Fig. 13.60.

In long-haul systems, **repeaters** (which include photodiode and laser diodes and electronic circuits) are inserted. In the repeater, the weak optical signal being transmitted through the optical fibre is detected by the photodiode. The detected signal is reformed and amplified by the electronic circuits. The amplified signal is converted again into an signal by a laser diode and transmitted again through the optical fibre cable. Fig. 13.60 shows a simple schematic of a repeater.

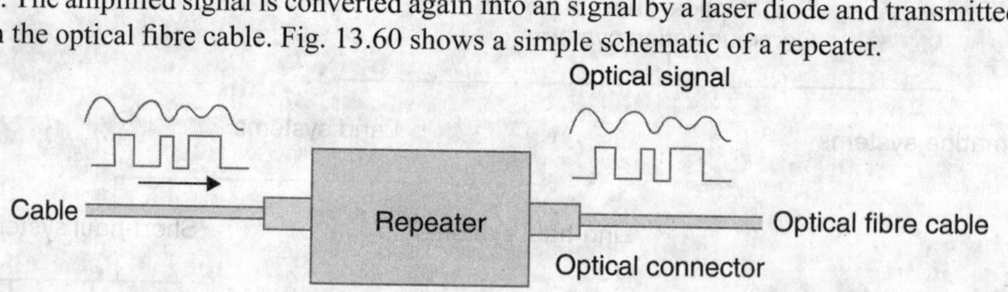

Fig. 13.61.

From the modulation point of view, the optical fibre communication systems can be divided into *digital systems* and *analog systems*. Most long-haul and large capacity optical fibre communication systems are digital systems. The analog systems are used for transmitting information over a short distance.

13.48. Optical Fibre Data Links

The use of optical fibre data links has wide spread in the past few decades. Its application ranges from local area networks (LANs) to the computer, digital audio and mobile fields. Several different types of LEDs and laser diodes emitting light at wavelengths ranging from visible to the infrared are used as optical sources. The transmission data rate is a function of transmission distance and varies from application. For computer links where the distance varies from 1 m to 100 m, the data transmission rate varies from 1 M bits/s to 100 M bits/s. For local-area -networks used in factory, office and building automation, the data transmission rate varies from 10 K bits/s to 5 M bits/s. In digital audio field, where the distance is below 1 m the data transmission rate varies from 500 bits/s. to over 10 M bits/s. Similarly in mobile fields (such as chip, aircraft, train and automotive applications) where the distance could vary from 1 m to 100 m, the data transmission rate varies from 1 K bits/s to 1 M bits/s.

1. Optical fibre local area networks. The optical fibre local area networks (LANs) are similar to the public communication systems. Some of their advantages over the systems using metallic cables are ; (1) high transmission capacity and bit rate and (2) longer transmission distance. However, the range of LANs is restricted. They are more commonly used within factories, offices, buildings etc. Computers, printers, facsimile machines and other office equipment are connected with each other by optical fibre cables as shown in Fig. 13.62.

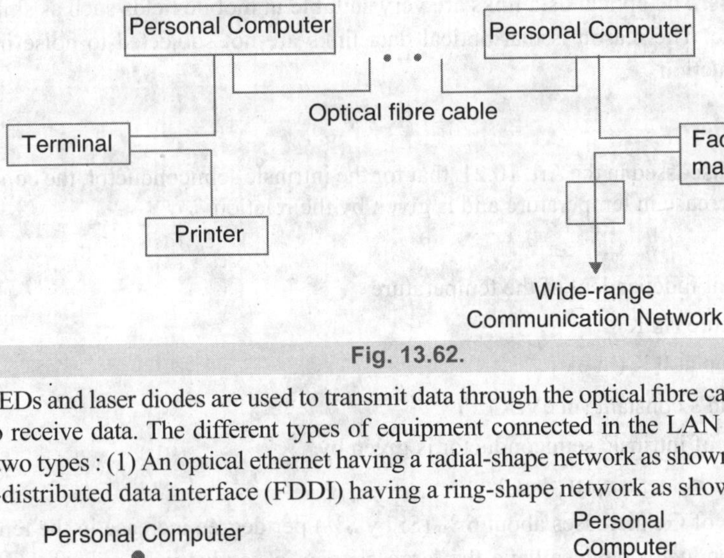

Fig. 13.62.

The LEDs and laser diodes are used to transmit data through the optical fibre cable and photodiodes are used to receive data. The different types of equipment connected in the LAN could be one of the following two types : (1) An optical ethernet having a radial-shape network as shown in Fig. 13.63 (*a*) or (2) a fibre-distributed data interface (FDDI) having a ring-shape network as shown in Fig. 13.63 (*b*).

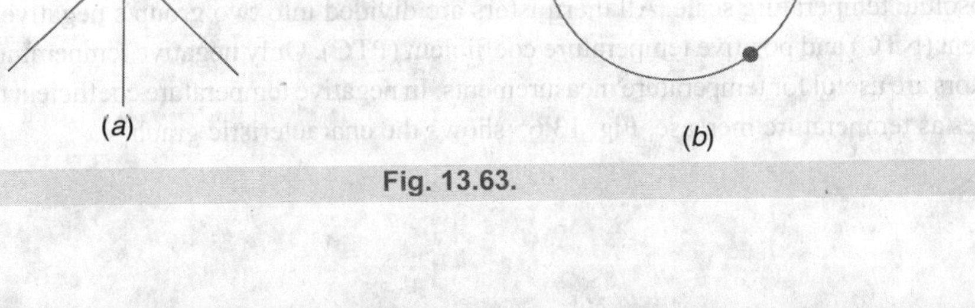

Fig. 13.63.

2. Digital audio field. Fig.13.64 shows an example of a data link in digital audio field. As seen, the optical fibre cable is used to connect compact disk (CD) player, laser disk (LD) player, digital audio type (DAT) and tuner with the amplifier and everything except DAT is unidirectional. The audio digital signals from CD, LD player, DAT and tuner are converted into optical signals by LEDs or laser diodes at one end of the fibre optic cable and then transmitted through the cable to the opposite end. At the opposite end, the signals are received by photodiodes and converted into an electrical signal for amplification and finally speaker for reproduction to a sound.

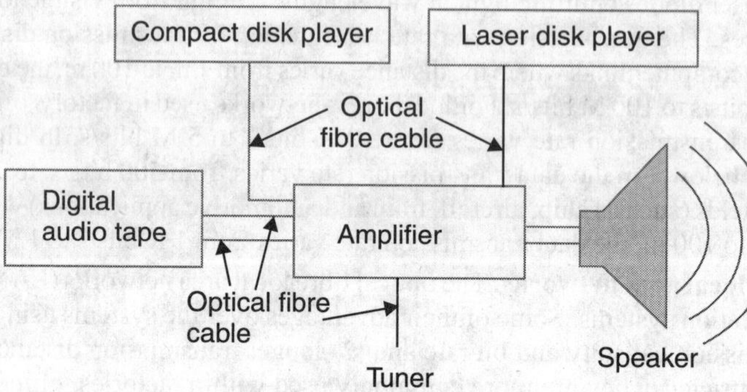

Fig. 13.64.

3. Mobile fields. The optical data links are very suitable in mobile fields such as ship, aircraft, train, automotive etc. The reason is that optical data links are not subjected to noise induced by electromagnetic induction.

13.49. Thermistor

We have already discussed in the Art. 10.21, that for the intrinsic semiconductor, the concentration increase with the increase in temperature and is given by the relation

$$= A_0 . T^3 . e^{-E_{G_0} . k . T}$$

A_0 = A constant independent of the temperature

T = Temperature (in K)

E_{G_0} = Energy gap at 0 K (in eV)

k = Boltzmann's constant (in eV/K)

and the conductivity of intrinsic semiconductor is given by,

$$\sigma_i = q . n_i (\mu_n + \mu_p)$$

The conductivity of Ge increases about 6 % (Si by 8%) per degree increase in the temperature. Some of the semiconductor devices have the large change in conductivity with the temperature change. The device based on this property is called thermistor

The term thermistor is a contraction of words thermal and resistor. The name is usually applied to metal-oxide sensors fabricated in forms of droplets, bars, cylinders, rectangular flakes, and thick films. Thermistors are fabricated of silicon and germanium. It measures the temperature that is referenced to an absolute temperature scale. All thermistors are divided into two group's negative temperature coefficient (NTC) and positive temperature coefficient (PTC). Only negative temperature coefficient thermistors are useful for temperature measurements. In negative temperature coefficient the resistance decreases as temperature increase. Fig. 13.65 shows the characteristic graph.

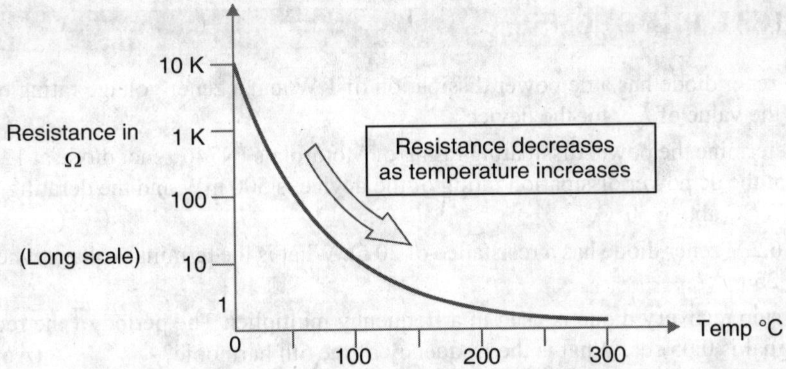

Fig. 13.65.

The resistance of thermistor can be expressed as,

$$R_T = R_1 \, exp\left[\left(\frac{1}{T} - \frac{1}{T_1}\right)\right]$$

R_T = Resistance at T °K

R_1 = Resistance at known temperature T_1 ° K

β = Characteristics temperature.

The thermistors are sensitive and provide large change in resistance for small change in temperature. The temperature range for a thermistor is –60°C to +15°C. Its resistance varies from 0.5 ohm to 0.75 mega ohm. A thermistor is placed in contact with the media whose temperature is to be measured. As the temperature of the media changes, the resistance of the thermistor gets changed.

Advantages

Following are some of the important advantages of thermistor:

1. It has small size and is compact.
2. It gives good response time from fraction of seconds to minutes.
3. It can be installed in small areas like pipeline, inlet or outlet of tanks.
4. There is no need of cold junction compensation.
5. It has good stability.

Disadvantages

Following are the disadvantage of a thermistor:

1. It has non-linear performance, i.e. temperature versus change in resistance graph.
2. It is not suitable for higher temperatures.
3. It requires bridge circuit arrangement for converting the change in resistance to voltage.

Applications

Through the thermistor have a number of applications, yet the following are important from the subject point of view:

1. It useful as temperature transducer.
2. Measurement of conductivity.
3. It used in a thermometry in the measurement of microwave-frequency power and as a thermal relay.
4. Because thermistors have a negative temperature coefficient of resistance thermistors are widely used to compensate for the effects of temperature on circuit performances.

NUMERICAL PROBLEMS

1. A zener diode has a dc power dissipation of 1 W and a zener voltage rating of 27 V. What is the value of I_{ZM} for the device? **(Ans. 37 mA)**
2. Determine the power dissipating rating of Motorola's IN746 zener diode at 125°C. Assume that the dc power dissipation rating of the device is 500 mW and the derating factor is 3.33 mW/°C above 50°C. **(Ans. 250.25 mW)**
3. A 6.2 V zener diode has a resistance of 20 Ω. What is the terminal voltage when the current in 25 mA? **(Ans. 6.7 V)**
4. A step recovery diode is used in a frequency multiplier. The period of the rectified output signal is 0.05 sec. What is the frequency of the 5th harmonic? **(Ans. 100 MHz)**
5. A light emitting diode (LED) is connected across a voltage source of +10 volts through a series resistance of 820 Ω. Calculate the LED current. Assume the voltage drop across an LED of 1.5 volt. **(Ans. 10.4 mA)**

DESCRIPTIVE QUESTIONS

1. What is a zener diode? Draw the equivalent circuit of an ideal zener in the breakdown region.
2. Explain the working of a tunnel diode. Give its two applications.
3. Draw the V-I characteristics of a tunnel diode and indicate the useful region in the curve.
4. What is a varactor diode? Explain briefly, how it can be used in tuning circuits.
5. Briefly explain the PIN diode operation. Give its applications.
6. In what respect is an LED different from an ordinary PN junction diode? State applications of LEDs.
7. Bring out the differences in the reverse bias characteristics of ordinary junction diode and zener diode. Why do the differences exist? *(Calicut University., Dec. 2006)*
8. Discuss the working of Zener diode. *(M.G University., April/May 2003)*
9. Explain the applications of Zener diode. *(Bhartiya University., April 2004)*
10. Describe the construction and characteristics of Zener diode. *(Bhartiya University., April 2004)*
11. Explain V-I characteristics of zener diode *(Anna University., April/May 2008, Calicut University., June 2007)*
12. What do you mean by zener diode ? Explain how zener diode maintains constant voltage across the load. *(GBTU/MNTU., 2009-10)*
13. Draw the characteristics curve for a Zener diode and explain its working. *(M.G University., April/May 2004)*
14. What is zener effect? Explain the function of a zener diode and draw its characteristics. *(Anna University., Nov/Dec 2007)*
15. Explain the principle of an operation of photodiode. *(M.G University., Dec. 2005)*
16. Explain the V-I characteristics of a tunnel diode. *(Bhartiya University., April 2004)*
17. Describe the operating principle and working of Tunnel Diode. *(GBTU/MNTU., 2006)*
18. Draw the V-I characteristics of a Zener diode and explain how the Zener regulates the voltage. *(GBTU/MTU., 2009)*
19. Define tunneling phenomenon. Explain how tunnel diode operates under different operating

conditions. In what way it is different from the conventional diodes? Give the necessary energy level diagrams. *(Anna University., Nov/Dec 2007)*

20. With neat diagrams explain the operation of tunnel diode. *(Anna University., April/May 2008)*

21. With structural diagram explain the V-I characteristics of Schottky diode. *(Anna University., April/May 2008)*

22. What is a backward diode? *(Anna University., Nov/Dec 2007)*

23. What is LDR? *(Anna University., Nov/Dec 2007)*

24. Write short notes on :
(*a*) Light Emitting Diode
(*b*) PN Junction Solar Cell or Photovoltaic *(GBTU/MNTU., 2008)*

25. List the applications of Opto-Electronic devices. *(Anna University., Nov/Dec 2007)*

26. What is LED? Discuss its advantages, disadvantages and applications? *(PTU., Dec 2008)*

27. Describe the working principle of LED. *(M.G University., Aug. 2007)*

28. Discuss the working of LED. *(M.G University., April/May 2003)*

29. Write the short note on LED voltage drop and current. *(JNTU., June 2009)*

30. Write short notes on Multi color LED. *(JNTU., June 2009)*

31. Explain the concept of tunneling with cnergy bond diagrams. *(JNTU., May/Jun. 2006)*

32. Write short notes on LED voltage drop and current. *(JUTN., June. 2009)*

33. Write short notes on Multi colour LED *(JUTN., June. 2009)*

34. State differences between LED and photodiodes *(PTU., May 2008)*

35. What is LCD? Discuss its advantages and applications? *(PTU., May 2009)*

36. What is the physics of photo emissivity? Explain in detail, the construction and working of any one optoelectronic device which uses this principle.*(Anna University., Nov/Dec 2007)*

37. Give a detailed account on Opto -Couplers and their application. *(Anna University., Nov/Dec 2007)*

38. Describe the construction, working, and applications of photo diode. *(Anna University., Nov/Dec 2007)*

39. State the working principle of photodiode. *(PTU., May 2007)*

40. Write any two applications of photodiode. *(PTU., May 2009)*

41. Explain the operation of an Opto-Coupler with a neat circuit. *(Anna University., April/May 2008)*

42. Write short notes no
(*a*) LED
(*b*) Photo Diode *(Anna University., April/May 2008)*

43. Differentiate a PN junction diode and LASER diode. *(Anna University., Nov/Dec 2007)*

44. Explain the V-I characteristics and applications of zener diode. *(M.G University., Dec. 2005, R.G.P.V., June 2009)*

45. Discuss the effect of temperature on Zener and avalanche breakdown. *(R.G.P.V., Dec. 2004, Nov. /Dec. 2007)*

46. Explain zener breakdown. Draw the V-I characteristics of zener diode and explain its operation. *(R.G.P.V., June 2007)*

47. What is a varactor diode? Explain. Give the circuit symbol and circuit model of a varactor diode. *(R.G.P.V., June 2007)*

48. Briefly discuss the working principle and application of varactor diode. *(R.G.P.V., June 2011, 2010)*

49. Describe the characteristics and application of schottky diode, phototransistor and varactor diode. *(R.G.P.V., Dec. 2008)*

50. What is the varactor diode? Explain and give its specific characteristics. *(R.G.P.V., Dec. 2008)*

51. Explain the Schottky diode. *(R.G.P.V., Feb. 2010)*

52. Discuss the characteristics and application of Schottky diode. *(R.G.P.V., June 2009, Dec 2010)*

53. Explain, why Schottky diode can switch – off faster than an ordinary diode. *(R.G.P.V., Dec. 2005, 2003)*

54. With the help of volt-ampere characteristics, explain the working of tunnel diode. *(R.G.P.V., Feb. 2010, Dec. 2010, Dec. 2005)*

55. Explain the characteristics and application of tunnel diode. *(R.G.P.V., June 2009)*

56. Explain the principle of working of tunnel diode. *(R.G.P.V., June 2010)*

57. Briefly discuss the working principle and application of tunnel diode. *(R.G.P.V., June 2011)*

58. Explain tunnel diode and PIN diode with proper characteristics and diagram. *(R.G.P.V., Dec. 2008)*

59. Sketch and explain the volt-ampere characteristic of a tunnel diode. Also draw the small signal model of the tunnel diode. Also draw the small signal model of tunnel diode operating in the negative resistance region. Define each circuit element. *(R.G.P.V., Dec. 2002)*

60. Explain tunnel diode with reference to:

(*a*) V- I characteristics (*b*) Tunnel Theory

(*c*) Advantage (*d*) Disadvantage

(*e*) Applications *(R.G.P.V., Dec. 2004)*

61. Explain the following diodes with their construction, working and application:

(*a*) Varactor diode (*b*) PIN diode

(*c*) Tunnel diode *(R.G.P.V., Dec. 2007)*

62. Give specific characteristics of a tunnel diode also give its applications. *(R.G.P.V., June 2008)*

63. Give V-I characteristics of a tunnel diode and explain the same. *(R.G.P.V., Dec. 2008)*

64. Write a short note on PIN diode. *(R.G.P.V., June 2011)*

65. Explain the construction and working of a PIN Diode and discuss its applications. *(R.G.P.V., June 2002)*

66. Explain the following diodes with their construction, working and applications:

(*a*) Zener Diode (*b*) Varactor Diode

(*c*) PIN Diode (*d*) LED. *(R.G.P.V., May/ June 2006)*

67. What is PIN diode ? Give its working principle and applications. *(R.G.P.V., June 2008)*

68. Write short note on light emitting diodes (LEDs). *(R.G.P.V., June 2002)*

69. Explain the working of LCD. How is it different with LED's ? Comment on the advantage of LCD's over LED's. *(R.G.P.V., Nov. /Dec. 2007, Dec. 2006, Dec. 2001)*
70. What is LED ? Write the advantage and disadvantages of LEDs. How are they different than LCDs ? *(R.G.P.V., June 2007, 2003)*
71. What is photodiode? Draw the volt-ampere characteristics of photodiode. Also, write the applications of photodiode. *(R.G.P.V., Dec 2005)*
72. List down various types of diode available and give at least two applications of each type of diode. *(R.G.P.V., June 2003)*
73. Explain in brief Varactor Diode. *(Nagpur University, Summer 2007, 2000)*
74. Sketch and explain the V-I characteristics of tunnel diode. Indicate the negative resistance region. *(Nagpur University, Summer 2006, Winter 2004)*
75. What are the unique characteristics of a varactor diode? Mention the application of it. *(Nagpur University, Winter 2003)*
76. Draw and explain characteristics of Tunnel diode. *(Nagpur University, Summer 2001)*
77. Explain the construction and working of a photodiode and draw its V-I characteristics. *(Nagpur University, Summer 2000)*
78. Explain the working principle of opto-couplers. *(R.G.P.V., Dec 2010)*
79. Write short notes on :
    (*a*) Effect of temperature on zener and avalanche break-down. *(Nagpur University, Winter 2005)*
    (*b*) Tunnel Diode *(Nagpur University, Summer 2005)*
    (*c*) Breakdown of zener diode *(Nagpur University, Summer 2004)*
    (*d*) Zener Diode *(R.G.P.V., Dec. 2001)*
    (*e*) Solar cells *(R.G.P.V., Dec. 2001)*
    (*f*) Varactor diode. *(R.G.P.V., Dec. 2004, 2002)*
    (*g*) LED and LCD *(R.G.P.V., June 2005)*
    (*h*) PIN Diode *(R.G.P.V., June 2011)*

MULTIPLE CHOICE QUESTIONS

1. A zener diode
    (*a*) has a high forward-voltage rating
    (*b*) has a sharp breakdown at low reverse voltage
    (*c*) is useful as an amplifier
    (*d*) has a negative resistance
2. A 5 V reference is drawn from the circuit shown in Fig. 13.66 if the zener diode is of 5 mA, then R_z will be
    (*a*) 50 Ω
    (*b*) 500 Ω
    (*c*) 50,00 Ω
    (*d*) 50,000 Ω

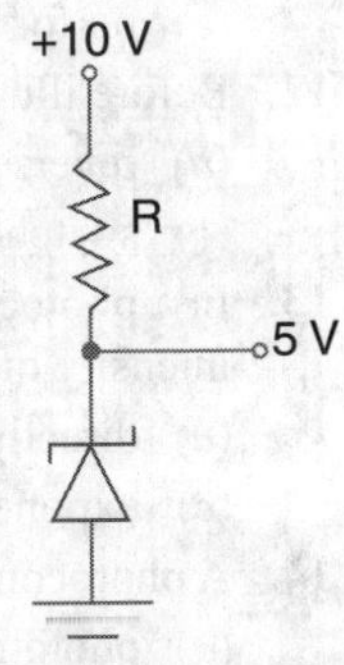

Fig. 13.66

3. A tunnel diode is
   (*a*) a very heavily-doped PN junction diode (*b*) a high resistivity PN junction diode
   (*c*) a slow switching device (*d*) used with reverse bias
4. A PIN diode is frequently used as a
   (*a*) voltage regulator (*b*) peak detector
   (*c*) harmonic generator (*d*) switching diode for very high frequencies
5. A step recovery diode
   (*a*) has an extremely short recovery twice
   (*b*) is an ideal rectifier of high frequency signals
   (*c*) is mainly used as a harmonic generator
   (*d*) conducts equally well in both director
6. The light-emitting diode (LED),
   (*a*) is usually made from silicon
   (*b*) uses a reverse-biased junction
   (*c*) gives a light output which increases with the increase in temperature
   (*d*) depends on the recombination of holes and electrons
7. LED'S do not require
   (*a*) heating (*b*) warm-up time
   (*c*) both (*a*) and (*b*) above (*d*) non of above
8. The sensitivity of a photodiode depends upon
   (*a*) light intensity and depletion region width
   (*b*) depletion region width and excess carrier life time
   (*c*) excess carrier life time and forward bias current.
   (*d*) forward bias current and light intensity.
9. LEDs are commonly fabricated from gallium compounds like gallium arsenide and gallium phosphide because they
   (*a*) are cheap (*b*) are easily available
   (*c*) emit more heat (*d*) emit more light.
10. A LED is basically a ....P-N junction.
    (*a*) forward-biased (*b*) reverse-biased
    (*c*) lightly-doped (*d*) heavily-doped
11. As compared to a LED display, the distinct advantage of an LCD display is that it requires
    (*a*) no illumination (*b*) extremely-bias
    (*c*) no forward-bias (*d*) a solid crystal
12. Before illuminating a P-N junction photodiode, it has to be
    (*a*) reverse-biased (*b*) forward-biased
    (*c*) switched ON (*d*) switched OFF.
13. In a photoconductive cell, the resistance of the semiconductor material varies .....with the intensity of incident light.
    (*a*) directly (*b*) inversely
    (*c*) exponentially (*d*) logarithmically
14. A photoconductive cell is known as ....cell.
    (*a*) phototransistor (*b*) photoresistor
    (*c*) photovoltaic (*d*) both (*a*)and (*b*).

15. A phototransistor excels a photodiode in the matter of
(*a*) faster switching (*b*) greater sensitivity
(*c*) higher current capacity (*d*) both (*a*) and (*b*)
(*e*) both (*b*) and (*c*).

16. A photodarlington comprises of
(*a*) a phototransistor (*b*) a transistor
(*c*) a photodiode (*d*) both (*a*) and (*b*).

17. Solar cell operates on the principle of
(*a*) diffusion (*b*) recombination
(*c*) photo voltaic action (*d*) carrier flow.

18. Solar cells are used as source of power in earth satellites because they have
(*a*) very high efficiency (*b*) unlimited life
(*c*) higher power capacity per weight (*d*) both (*b*) and (*c*)
(*e*) both (*a*) and (*b*).

19. The device possessing the highest sensitivity is a
(*a*) photo conductive cell (*b*) photovoltaic cell
(*c*) photodiode (*d*) phototransistor

20. The unique characteristics of LASER light are that it is
(*a*) coherent (*b*) monochromatic
(*c*) collimated (*d*) all of the above

21. The LASCR operates like a
(*a*) latch (*b*) LED (*c*) photodiode (*d*) phototransistor.

22. Optical couplers are designed to .....one circuit from another.
(*a*) control (*b*) isolate (*c*) disconnect (*d*) protect.

23. The main purpose of using optical isolators is to provide protection to devices from
(*a*) high-voltage transients (*b*) surge voltages
(*c*) low-level noise (*d*) all of the above.

24. A LED emits visible light when its........
(*a*) P-N junction is reverse-biased (*b*) depletion region widens
(*c*) holes and electrons recombine (*d*) P-N junction becomes hot.

25. In LED, light is emitted because
(*a*) recombination of charge carriers takes place
(*b*) diode gets heated up
(*c*) light falling on the diode gets amplified
(*d*) light gets reflected due to lens action.

26. GaAs, LEDs emit radiation in the
(*a*) ultraviolet region
(*b*) violet-blue green range of the visible region
(*c*) visible region
(*d*) infra-red region

27. Phototransistors respond much like a conventional transistor except that, in their case, light energy is used to...

(*a*) alter leakage current (*b*) change base voltage
(*c*) switch it ON (*d*) alter emitter current.

28. Breakdown occurs in Zener diode under
(*a*) forward bias (*b*) reverse bias
(*c*) both forward and reverse bias (*d*) none of the above

(*Bharthiar Univ. April 2004*)

29. For detecting light intensity, we use
(*a*) LED (*b*) zener diode
(*c*) Photodiode (*d*) Tunnel diode

(*Bharthiar Univ. April 2004*)

ANSWERS

1. (*b*)	**2.** (*c*)	**3.** (*a*)	**4.** (*d*)	**5.** (*c*)
6. (*d*)	**7.** (*c*)	**8.** (*a*)	**9.** (*d*)	**10.** (*a*)
11. (*b*)	**12.** (*a*)	**13.** (*b*)	**14.** (*d*)	**15.** (*e*)
16. (*d*)	**17.** (*c*)	**18.** (*d*)	**19.** (*d*)	**20.** (*d*)
21. (*a*)	**22.** (*b*)	**23.** (*d*)	**24.** (*c*)	**25.** (*a*)
26. (*d*)	**27.** (*c*)	**28.** (*b*)	**29.** (*c*)	

14 Chapter

Bipolar Junction Transistors

Outline

1. Introduction.
2. Bipolar Junction Transistor Construction.
3. Bipolar Junction Transistor Manufacturing Techniques.
4. Bipolar Junction Transistor Packaging and Terminal Indentification.
5. Bipolar Junction Transistor Symbols.
6. Unbiased Bipolar Junction Transistor.
7. Bipolar Junction Transistor Biasing.
8. Operation of an NPN Bipolar Junction Transistor.
9. Operation of a PNP Bipolar Junction Transistor.
10. Bipolar Junction Transistor Currents.
11. Bipolar Junction Transistor Circuit Configuration.
12. Current Gain of a Bipolar Junction Transistor in Common-Base Configuration.
13. Current Gain of a Bipolar Junction Transistor in Common Emitter Configuration.
14. Relation between Current Gain α and β.
15. Current Gain of a Bipolar Junction Transistor in Common-Collector Configuration.
16. Leak Current in a Common-Base Bipolar Junction Transistor.
17. Leakage Current in a Common-Emitter Bipolar Junction Transistor.
18. Base Spreading Resistance.
19. Current Gain or β-Curve.
20. Multiple Emitter Transistor.

14.1. Introduction

A transistor is a semiconductor device used to **amplify** and **switch** electronic signals and power. It is composed of a semiconducting material with at least 3 terminals for connection to the external circuit. A voltage (or a current) applied to one pair of terminals charges the current flowing through another pair of terminals. As a result, the controlled (output) power of a transistor can be much more than the controlling (input) power. Hence, the transistor can amplify an input signal.

Today some transistors are packaged individually (called discrete transistors), but many are found embedded in integrated circuits.

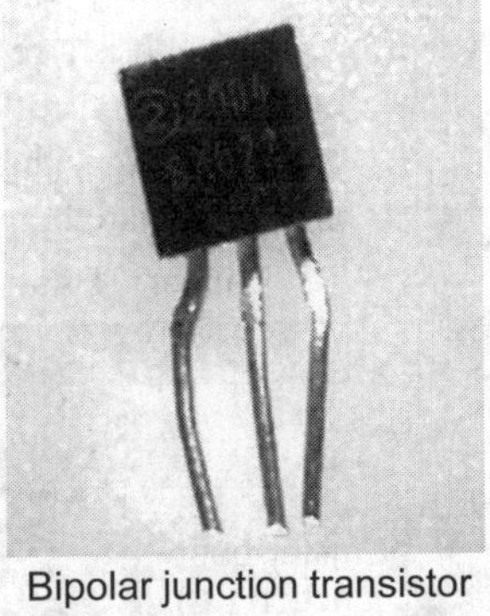

Bipolar junction transistor

Transistor is a basic building block of modern electronic devices. Following its release in 1950's, the transistor has revolutionized the field if electronics. This paved the way for smaller and cheaper radios, calculators, computers, mobile phones, and many more things.

The transistor's low cost, flexibility and reliability have made it a ubiquitous or universal device.

Transistorized mechatronic systems have replaced many electromechanical devices in control engineering and instrumentation industry. It is often easier and cheaper to use a standard microcontroller and write a computer program to carry out a desired control function than to design and equivalent mechanical control function.

Broadly speaking, there are two types of transistors:

(*i*) Biploar junction transistors (or simply biopolar transistors) and

(*ii*) Field-effect transistors. As explained later there is a slight difference in how they are used in the circuit.

Bipolar junction transistor (BJT) is discussed in this chapter while field-effect transistor is covered in chapter 16.

Bipolar junction transistor was developed by Dr. William Shockley and his team in 1947 at Bell Laboratories. The first silicon transistor was produced by Texas instrument in 1954. Since then it has revolutionized the electronic industry.

A bipolar junction transistor consists of two PN junctions. The junctions are formed by sandwiching either P-type or N-type semiconductor layers between a pair of opposite types as shown in Fig. 14.1 (*a*) and (*b*).

Fig. 14.1. Bipolar Junction Transistor.

14.2. Bipolar Junction Transistor Construction

A bipolar junction transistor has, essentially, three regions known as emitter, base and collector. All these three regions are provided with terminals, which are labelled as *E* (for emitter), *B* (for base) and *C* (for collector) respectively. A brief description of the above regions is discussed below:

1. ***Emitter.*** It is a region situated in one side of transistor, which supplies charge carries (*i.e.*, electrons or holes) to the other two regions. The emitter is a heavily doped region.
2. ***Base.*** It is the middle region that forms two P-N junctions in the transistor. The base of transistor is thin, as compared to the emitter and is a lightly doped region.
3. ***Collector.*** It is a region situated in the other side of transistor (*i.e.*, the side opposite to the emitter), which collects charge carriers (*i.e.*, electrons or holes). The collector of a transistor is always larger than the emitter and base of a transistor. The doping level of the collector is intermediate between the heavy doping of emitter and the light doping of the base.

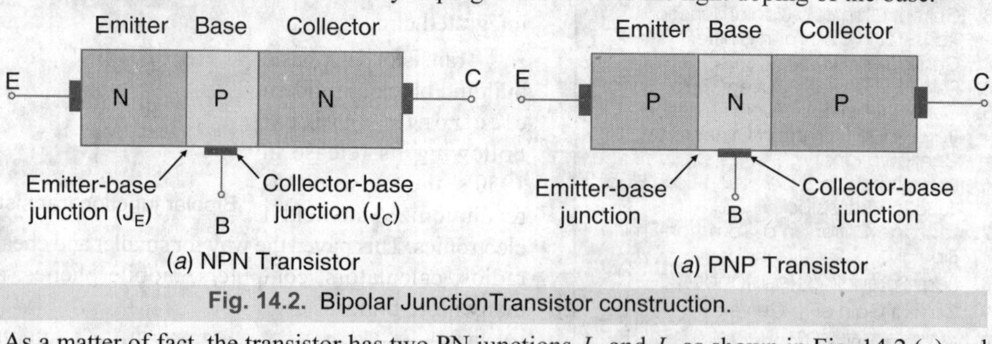

Fig. 14.2. Bipolar JunctionTransistor construction.

As a matter of fact, the transistor has two PN junctions J_C and J_E as shown in Fig. 14.2 (*a*) and (*b*). The junction J_E is a junction between emitter and base regions. Thus it is known as emitter-base junction. Similarly, the junction J_C is a junction between collector and base regions. Thus it is known as collector-base junction.

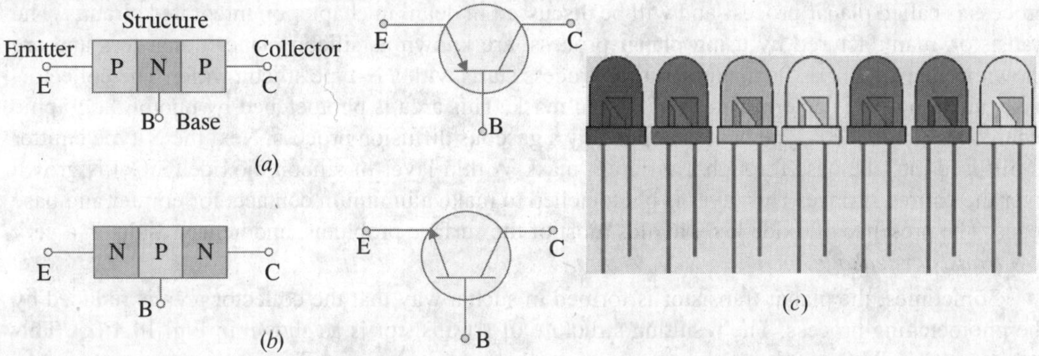

14.3. Bipolar Junction Transistor Manufacturing Techniques

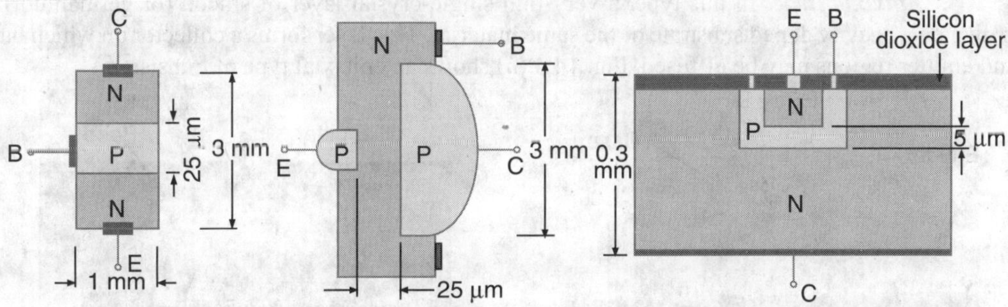

(*a*) Grown-junction type (*b*) Alloy-junction type (*c*) Diffusion type

Fig. 14.3. Transistor manufacturing techniques.

The techniques used for manufacturing the bipolar junction transistor (BJT)are the same as discussed for the PN junctions. Following are the basic techniques used for the BJT. manufacturing:

1. Grown-junctions
2. Alloy of fused-junctions
3. Diffused-junctions
4. Epitaxial

Depending upon the manufacturing technique, a transistor may be called grown-junction type transistor, alloy-junction type transistor, diffused-junction transistor and epitaxial type transistor. Each type of the transistor construction is discussed as below:

1. *Grown-junction type.* In this type, a single crystal is drawn from the molten silicon or germanium. The impurity concentration of single crystal is changed during the crystal drawing operation by adding N-type or P-type impurities as required. Then the grown crystal is cut into small area devices and the contacts are made to each region as shown in Fig. 14.3 (*a*).

2. *Alloy-type.* In this type, two small dots of indium (a P-type impurity) are placed on opposite sides of a thin wafer of N-type material. The whole structure is heated for a short time to a temperature high enough to melt indium into the base material. The alloying of indium, with the original wafer, produces the PN junctions. The leads may be connected to the structure as shown in Fig. 14.3 (*b*). It may be noted, that collector region is made larger than the emitter region. This has on advantage that the collector collects the whole emitter current and thus prevents the current from diffusing into the base region.

3. *Diffused junction type.* In this type, a semiconductor wafer is subjected to gaseous diffusion of both N and P-type impurities to form both the emitter-base and collector-base junctions. This process is called planar process and will be discussed in detail in chapter on integrated circuits. The transistor, manufactured by using planar process, are known as silicon planer transistors and are shown in Fig. 14.3 (*c*). The manufacturing process starts with a N-type silicon wafer. The collector-base junction area is determined by diffusion mask. This area is photoetched by photolithographic techniques. Then the P-type base is applied by a gaseous diffusion process. Next the N-type emitter is diffused into the base through a different mask. A thin layer of silicon dioxide (SiO_2) is grown over the entires surface. This layer is photoetched to make aluminium contacts for emitter and base leads. The presence of oxide layer avoids most of the surface problems, and hence results in a very low *transistor leakage current*.

Sometimes, the planar transistor is formed in such a way that the collector area is reduced by the photoetching process. The resulting structure of a transistor is as shown in Fig. 14.4 (*a*). This structure has a flat topped called mesa, and is called mesa transistor. This structure is very rugged and has large power dissipation capability. It can operate at very high frequencies.

4. *Epitaxial type.* In this type, a very thin single-crystal layer of silicon (or germanium) is grown on a heavily doped substrate of the same material. This layer forms a collector on which base and emitter regions may be diffused. Fig. 14.4 (*b*) shows an epitaxial type of transistor.

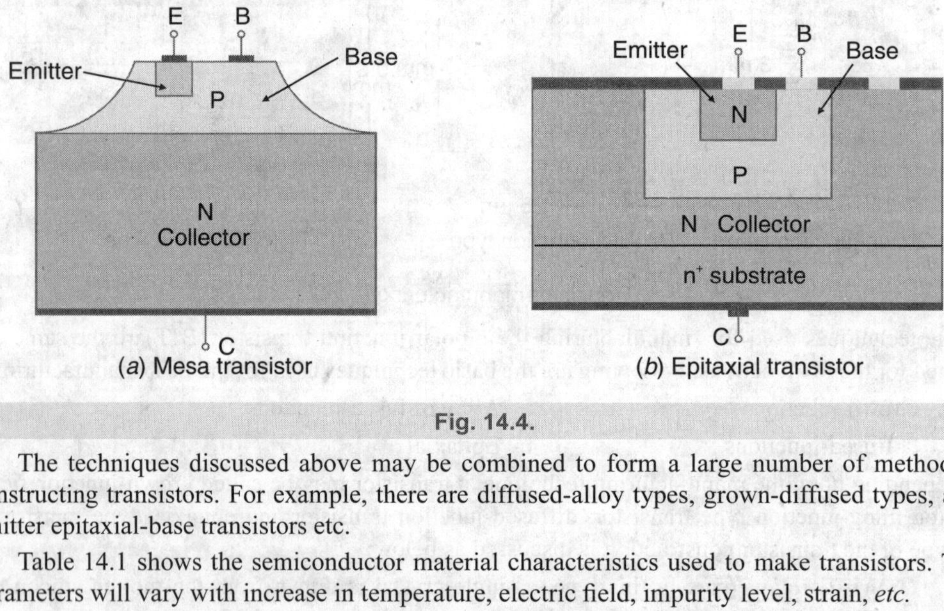

Fig. 14.4.

The techniques discussed above may be combined to form a large number of methods for constructing transistors. For example, there are diffused-alloy types, grown-diffused types, alloy-emitter epitaxial-base transistors etc.

Table 14.1 shows the semiconductor material characteristics used to make transistors. These parameters will vary with increase in temperature, electric field, impurity level, strain, *etc.*

Table 14.1

Material	*Junction forward voltage V @ 25°C*	*Electron mobility $m^2/V.s$ @ 25°C*	*Hole mobility $m^2/V.s$ @ 25°C*	*Max. junction temp. °C*
Ge	0.27	0.39	0.19	70 to 100
Si	0.71	0.14	0.05	150 to 200
GaAs	1.03	0.85	0.05	150 to 200
Al-Si junction	0.3	–	–	150 to 200

The first BJTs were made from germanium (Ge). These days, silicon (Si) transistors predominate in industry. However certain advanced microwave and high performance BJT's use compound semiconductor material like gallium arsenide (Ga As) and the semiconductor alloy silicon germanium (Si Ge).

14.4. Bipolar Junction Transistor Packaging and Terminal Identification

We have already discussed in the last article the different techniques of transistor manufacturing. After the transistor is manufactured, by using any one of the techniques, leads of gold, aluminium or nickel are then attached and the entire structure is packed or encapsulated in a container as shown in Fig. 14.5 (*a*). The transistors with small can (top hat) or plastic body are low-to medium-power devices, while those with studs and heat sinks are high power devices.

First silicon transistor.

(*a*) Package types.

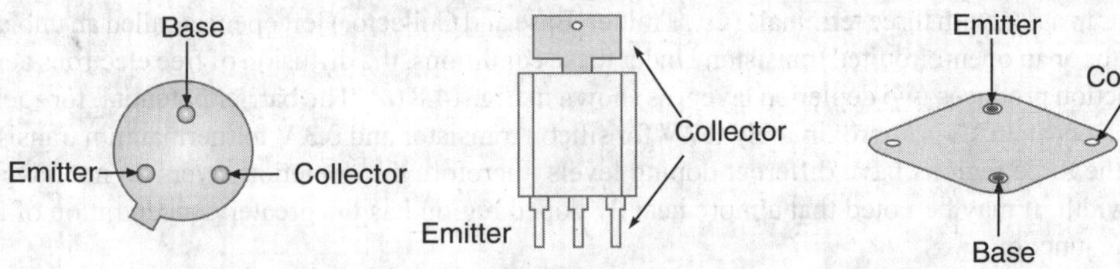

(*b*) Terminal identification methods.

Fig. 14.5.

Most of the transistor packages have some marking to identify its terminals (*i.e.*, to identify which leads are connected to the emitter, base or collector of a transistor). Some of the commonly used methods are as shown in Fig. 14.5 (*b*). In the metal 'top hat' cases (*i.e.*, packages), the tab is closest to emitter lead and the collector is often connected to the case. Generally, the heat sink mounting or stud on the power transistors is the collector terminal.

14.5. Bipolar Junction Transistor Symbols

We have already mentioned in the Art. 14.2 that there are two types of transistors namely NPN and PNP. When the transistor is used as circuit element, it is difficult to represent it with Fig. 14.2 (*a*) and (*b*). In order to make it convenient, both the transistors are represented by their circuit symbols as shown in Fig. 14.6 (*a*) and (*b*) respectively. The transistor symbol carries an arrowhead in the emitter pointing from the P-region towards the N-region. The arrowhead indicates the direction of a conventional current flow in a transistor.

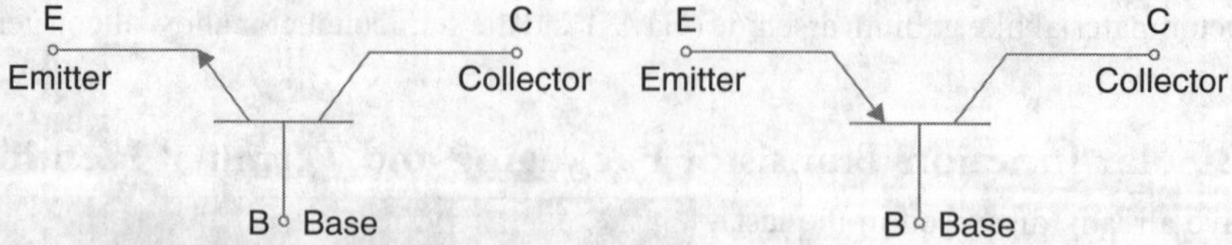

(*a*) Symbol for NPN transistors. (*b*) Symbol for PNP transistors.

Fig. 14.6. Bipolar Junction Transistor symbols.

It may be noted that the direction of arrowheads at the emitter in NPN and PNP transistors is opposite to each other. Sometimes, it is more convenient to represent the symbols for NPN and PNP transistors as shown in Fig. 14.7 (*a*) and (*b*).

(*a*) Symbol for NPN transistor (*b*) Symbol for PNP transistor

Fig. 14.7. Unbaised transistor.

The PNP transistor is a complement of the NPN transistor. Thus in NPN transistor, the majority carries are free electrons, while in PNP, transistor, these are holes. It means that the current and voltages involved in the action of PNP transistor are opposite to those of NPN transistor. In order to avoid confusion, we shall discuss the NPN transistor in our earlier discussion.

14.6. Unbiased Bipolar Junction Transistor

A transistor, with three terminals (*i.e.*, Emitter, Base and Collector) left open, is called an unbiased transistor or an open-circuited transistor. Under these conditions, the diffusion of free electrons across the junction produces two depletion layers as shown in Fig. 14.8 (*a*). The barrier potential, for each of these layers at 25°C, is approximately 0.7 V for silicon transistor and 0.3 V for germanium transistor. Since the three regions have different doping levels, therefore the depletion layers do not have the same width. It may be noted that a more heavily doped region has the greater concentration of ions near the junction.

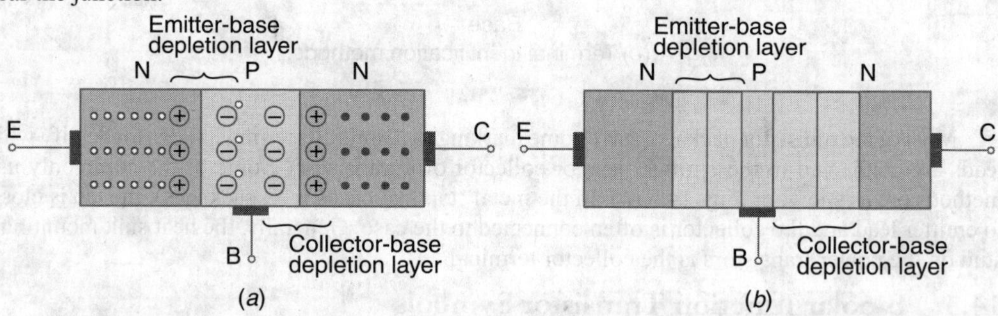

Fig. 14.8. Unbaised Bipolar Junction transistor.

It has been observed that an emitter-base depletion layer penetrates slightly into the emitter, as it is a heavily doped region, whereas it penetrates deeply into the base as its a lightly doped region. Similarly, the collector-base depletion layer penetrates more into the base region and less into the collector region. Both these depletion layers are as shown in Fig. 14.8 (*b*). It may be noted that *emitter-base depletion layer width is smaller than that of collector base depletion layer.*

An unbiased transistor is never used in actual practice. Its terminals are always connected, suitably, to the dc voltage sources (or a battery) for proper transistor action.

14.7. Bipolar Junction Transistor Biasing

The application of a suitable dc voltages, across the transistor terminals, is called biasing. Each junction of a transistor may be forward biased or reverse-biased independently. There are following three different ways of biasing a transistor, which are also known as modes of transistor operation.

1. ***Forward-active.*** In this mode, the emitter-base junction of a transistor is forward biased and the collector base junction is reverse biased as shown in Fig. 14.9 (*a*). In a forward active biasing the negative terminal of a battery is connected to N-side and positive terminal to P-side. The reverse biasing requires all the connections to be opposite to those for forward biasing.

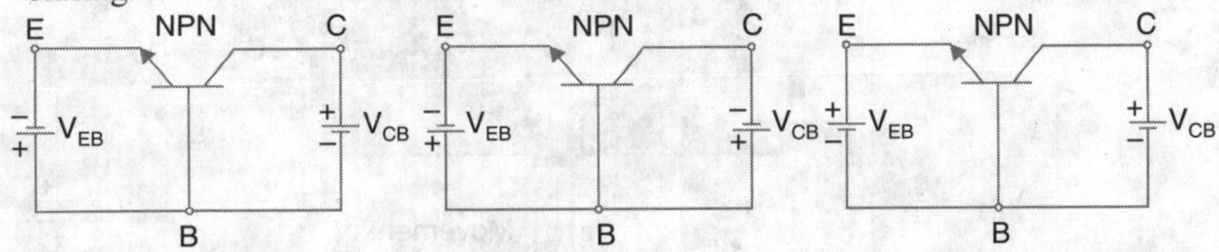

(*a*) Forward-active. (*b*) Saturation. (*c*) Cut-off.

Fig. 14.9. Different modes of transistor biasing.

2. ***Saturation.*** In this mode, both the emitter-base and collector-base junctions of a transistors are forward-biased as shown in Fig. 14.9 (*b*). In this mode, the transistor has a very large value of current. The transistor is operated in this mode, when it is used as a closed switch.
3. ***Cut-off. In this mode***, both the emitter-base and collector-base junctions of a transistor are reverse biased as shown in Fig. 14.9 (*c*). In this mode, the transistor has practically zero current. The transistor is operated in this mode, when it is used as an open switch.
4. ***Reverse Active.*** In this mode the emithr-base junction of a transistor is reverse biased and the collector base junction is forward biased. This mode has very limited application.

The four different modes are summarized in Table 14.2. It must be remembered that for amplifier, a transistor is used in forward-active mode and for switching applications (*i.e.*, ON and OFF applications) a transistor is used alternatively is saturation and cut-off modes.

Table 14.2: Modes of transistor operation

S. No.	*Mode*	*Junction-bias condition*	
		Emitter Base	***Collector Base***
1.	Forward active	Forward	Reverse
2.	Saturation	Forward	Forward
3.	Cut-off	Reverse	Reverse
4.	Reverse active	Reverse	Forward

Note. The modes of operation are the same for both NPN and PNP transistors. However, in PNP transistor biasing, the polarity of battery voltages are opposite to that of NPN transistor.

14.8. Operation of an NPN Bipolar Junction Transistor

Fig. 14.10 (*a*) shows an NPN transistor biased in forward-active mode. .*i.e.*, the emitter-base of a transistor is forward-biased and collector-base junction is reverse-biased. The emitter-base junction

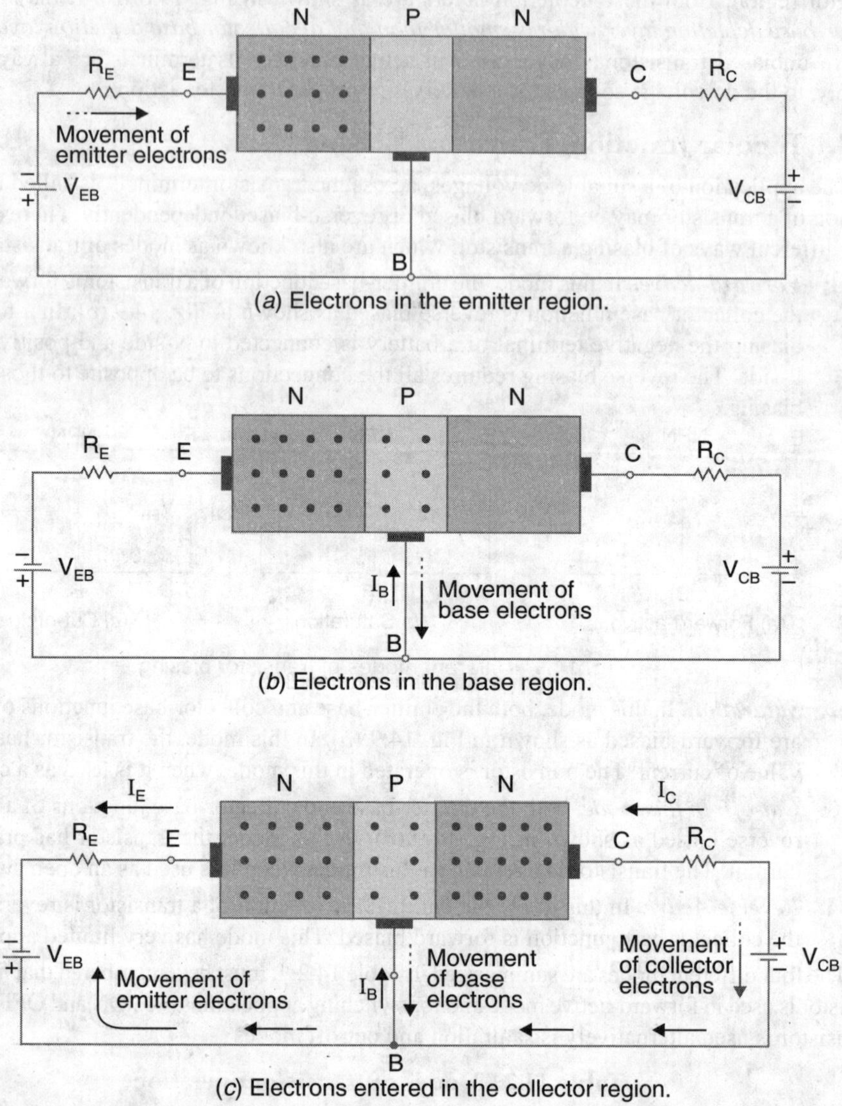

Fig. 14.10.

is forward biased only if V_{EB} is greater than barrier potential which is 0.7 V for silicon and 0.3 V for germanium transistors. The forward bias on the emitter-base junction causes the free electrons in the N-type emitter to flow towards the base region. This constitutes the emitter current (I_E). It may be noted that the direction of conventional current (I_C) is opposite to the flow of electrons. Therefore electrons, after reaching the base region as shown in Fig. 14.10 (*b*), tend to combine with the holes. If these free electrons combine with the holes in the base, they constitute base current (I_B). However, most of the free electrons do not combine with the holes in the base. This is because of the fact that the base width is made extremely small and electrons do not get sufficient holes for recombination. Thus most of the electrons will diffuse to the collector region as shown in Fig. 14.10 (*c*) and constitutes collector current (I_C). This collector current is also called injected current because this current is produced due to electrons injected from the emitter region.

It will be interesting to know that there is another component of collector current due to the thermally generated carriers. This current component is called reverse saturation current and is quite small.

Notes. **1.** The emitter current of a transistor consists of two components namely base current and collector current. The base current is about 2% of the emitter current, while collector current is about 98% of the emitter current.

2. The collector current is mainly due to the electrons injected from the emitter. However, there is another small component of collector current due to thermally generated carriers. This small component of collector current is called reverse saturation current.

3. The resistances R_E and R_C shown in the circuit of Fig. 14.10 are added to limit the magnitude of current in the transistor.

14.9. Operation of a PNP Bipolar Junction Transistor

The operation of a PNP transistor is similar to that of an NPN transistor. However, the current within a PNP transistor is due to the movement of holes, whereas in an NPN transistor it is due to the movement of free electrons. Fig. 14.11 shows a PNP transistor with its emitter-base junction as forward-biased and collector-base junction as reverse biased. The forward bias on the emitter-base junction causes the holes in the emitter region to flow towards the base region. This constitutes the emitter current (I_E). The holes, after reaching in the base region, tend to combine with the electrons. If these holes combine with the electrons in the base, they constitute base current (I_B). However, most of the holes do not combine with the electrons in the base region. It is due to the fact that base width is made extremely small and holes do not get sufficient electrons for recombination. Thus most of the holes diffuse to the collector regions and constitute collector current (I_C). This collector current is called injected current because this current is produced due to the holes injected from the emitter region.

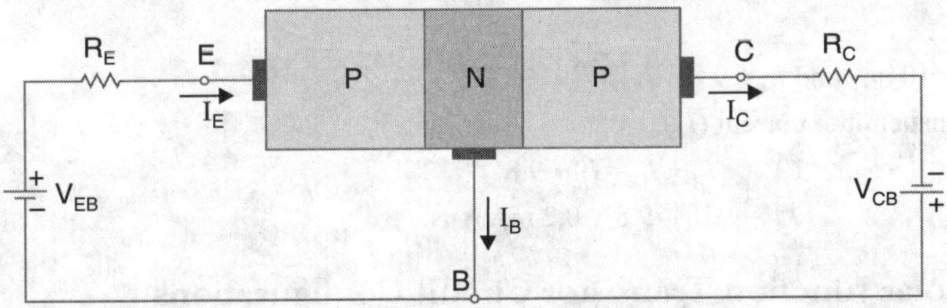

Fig. 14.11. Operation of a PNP transistor.

There is another small component of collector current due to thermally generated carriers. This current component is called reverse saturation current. The conventional directions of emitter, base and collector currents are shown by I_E, I_B and I_C respectively in Fig. 14.11.

14.10. Bipolar Junction Transistor Currents

We know that the direction of a conventional current is always opposite to the electron current in any electronic device. Therefore the conventional currents in a NPN transistor are as shown in Fig. 14.12 (*a*). However, the direction of a conventional current is the same as that of a hole current in a PNP transistor and is as shown in Fig. 14.12 (*b*). In this figure, I_E denotes the emitter current, I_B the base current and I_C the collector current respectively.

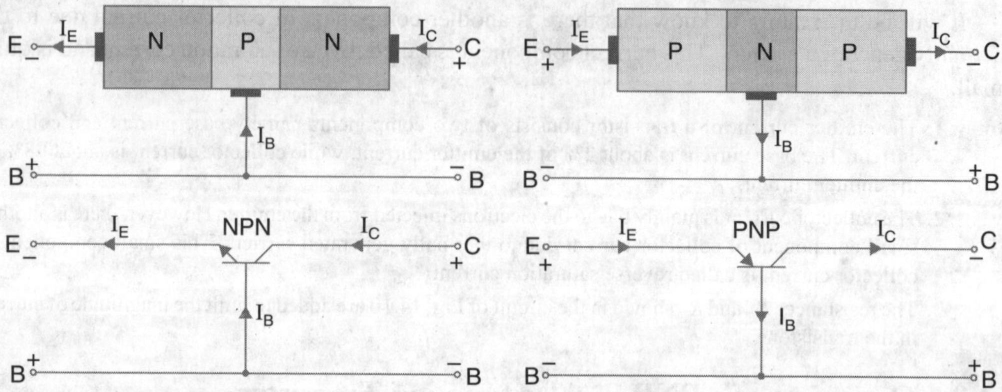

(a) NPN transistor currents. (b) PNP transistor currents.

Fig. 14.12.

It is evident from these diagrams that the *emitter current is the sum of the collector and base currents*. Mathematically, the emitter current,

$$I_E = I_B + I_C$$

Since the base current is very small, therefore

$$I_E \approx I_C$$

Example 14.1.

In a common-base connection, a certain transistor has an emitter current of 10mA and a collector current of 9.8 mA. Calculate the value of the base current.

Solution.

Given: $I_E = 10$ mA and $I_C = 9.8$ mA

We know that emitter current (I_E),

$$10 = I_B + I_C = I_B + 9.8$$

$$\therefore \quad I_B = 10 - 9.8 = 0.2 \text{ mA } \textbf{Ans.}$$

14.11. Bipolar Junction Transistor Circuit Configurations

We know that a transistor has three terminals or leads namely emitter (*E*), base (*B*) and collector (*C*). However, when a transistor is connected in a circuit, we require four terminals *i.e.*, two terminals for input and two for output. This difficulty is overcome by using one of the three terminals as a common terminal to the input and output terminals. Depending upon the terminals, which are used as a common terminals, the transistors can be connected in the following three different connections or configurations:

1. ***Common-base* (*CB*) *connection.*** In this configuration, the transistor is connected with the base as a common terminal as shown in Fig. 14.13 (*a*). The input is applied between the emitter and base terminals. The output is taken between the collector and base terminals. This type of configuration is used to explain the operation of NPN and PNP transistors.

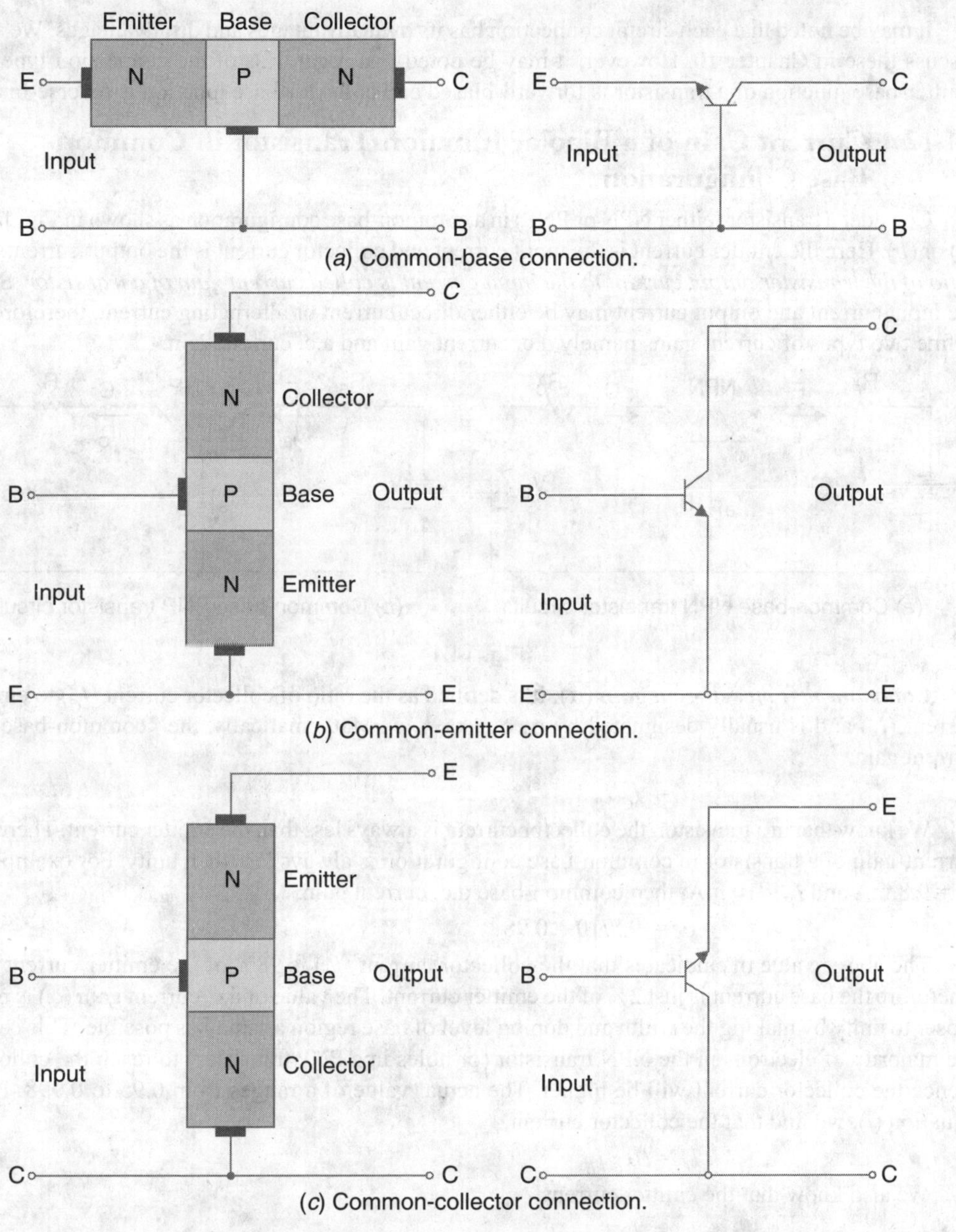

Fig. 14.13. Transistor circuit connections or configurations.

2. ***Common-emitter*** **(*CE*)** ***connection.*** In this configuration, the transistor is connected with the emitter as a common terminal as shown in Fig. 14.13 (*b*). The input is applied between the base and collector terminals. The output is taken between collector and emitter terminals. This is one of the most commonly used connections of a transistor.
3. ***Common-collector*** **(*CC*)** ***connection.*** In this configuration, the transistor is connected with collector as a common terminal as shown in Fig. 14.13 (*c*). The input is applied between the base and collector terminals. The output is taken between emitter and collector terminals.

It may be noted that each circuit connection has its own advantages and disadvantages. We shall discuss these in Chapter 10. However, it may be noted that regardless of the connection type, the emitter-base junction of a transistor is forward-biased and collector-base junction is reverse-biased.

14.12. Current Gain of a Bipolar Junction Transistor in Common-Base Configuration

Consider a transistor (either NPN or PNP) in a common-base configuration as shown in Fig. 14.14 (*a*) or (*b*). Here the emitter current is the input current and collector current is the output current. *The ratio of the transistor output current to the input current is called current gain of a transistor*. Since the input current and output current may be either direct current or alternating current, therefore we define two types of current gains namely d.c. current gain and a.c. current gain.

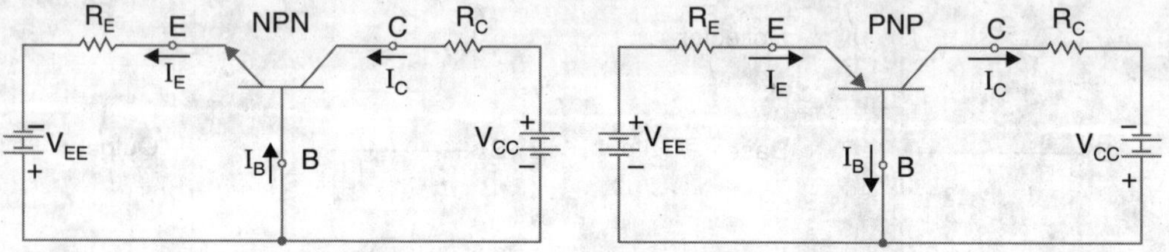

(*a*) Common-base NPN transistor circuit. (*b*) Common-base PNP transistor circuit.

Fig. 14.14.

1. ***Common-base d.c. current gain*** (α). It is defined as the ratio of collector current (I_C) to emitter current (I_E) and is usually designated by α, α_{DC} or h_{FB} , Mathematically, the *common-base d.c. current gain,

$$\alpha = I_C/I_E \quad ...(i)$$

We know that in a transistor, the collector current is always less than the emitter current. Therefore current-gain of a transistor in common-base configuration is always less than unity. For example, if $I_C = 9.8$ mA and $I_E = 10$ mA, then common-base d.c. current gain,

$$\alpha = 9.8/10 = 0.98$$

The above value of indicates that the collector current (I_C) is 98% of the emitter current (I_E). Therefore the base current is just 2% of the emitter current. The value of d.c. current gain (α) is made closer to unity by making the width and doping level of base region as small as possible. This causes the majority of electrons in the NPN transistor (or holes in a PNP transistor) to reach the collector. Hence the collector current will be higher. The actual value of a ranges from 0.95 to 0.998. From equation (*i*), we find that the collector current,

$$I_C = \alpha \cdot I_E$$

We also know that the emitter current,

$$I_E = I_B + I_C$$

$$\therefore \quad I_B = I_E - I_C = I_E - \alpha \cdot I_E = (1 - \alpha)I_E$$

2. ***Common-base a.c. current gain*** **(α_0).** It is defined as the ratio of small change in collector current (ΔI_C) to a small change in emitter current (ΔI_E) for a constant collector-to-base voltage (V_{CB}). It is designated by α_0, α_{ac} or h_{fb}. Mathematically, the common-base a.c. current gain,

$$\alpha_0 = \Delta I_C/\Delta_E$$

The term α_0 is also called common-base short-circuit current gain or small signal current gain. The difference between d.c. current gain (h_{FB}) and a.c. current gain (h_{fb}) should be carefully noted.

* It is also referred to as large-signal common base d.c. current gain.

Here the upper case subscript (*i.e.*, *FB*) designates the d.c. value and lower-case subscript (*i.e.*, *fb*) designates the a.c. value. The values of α_0 is also less than unity and is approximately the same as α. For all practical purposes, d.c. current gain is considered equal to the a.c. current gain. *i.e.*,

$$\alpha = \alpha_0$$

It may be noted that current gain of a transistor in a common-base configuration (α) is less than unity. But still it is called current gain. It is due to the fact that the output resistance of a common-base transistor is much higher than the input resistance. This produces a large voltage gain and hence the large power gain.

Example 14.2.

In a common-base connection, the emitter current is 6.28 mA and the collector current is 6.20 mA. Determine the common-base d.c. current gain.

Solution.

Given: $I_E = 6.28$ mA and $I_C = 6.20$ mA.

We know that common-base d.c. current gain,

$$\alpha = \frac{I_C}{I_E} = \frac{6.20}{6.28} = 0.987 \textbf{ Ans.}$$

Example 14.3.

The common-base d.c. current gain of a transistor is 0.967. If the emitter current is 10 mA, what is the value of base current?

Solution.

Given: $\alpha = 0.967$ and $I_E = 10$ mA.

We know that common-base d.c. current gain (α),

$$0.967 = \frac{I_C}{I_E} = \frac{I_C}{10}$$

$$\therefore \quad I_C = 0.967 \times 10 = 9.67 \text{ mA}$$

We also know that emitter current (I_E),

$$10 = I_B + I_C = I_B + 9.67$$

$$\therefore \quad I_B = 10 - 9.67 = 0.33 \text{ mA } \textbf{Ans.}$$

Example 14.4.

If the emitter current of transistor is 8 mA and I_B is 1/100 of I_C. Determine the levels of I_C and I_B. (*Jamia University., 2008*)

Solution.

Given: $I_E = 8$ mA $= 8 \times 10^{-3}$ and $I_B = 1/100\ I_C$.

We know that, emitter current,

$$I_E = I_B + I_C$$

$$8 \times 10^{-3} = \frac{1}{100} \times I_C + I_C = \frac{101}{100} I_C$$

$$\therefore \quad I_C = \frac{8 \times 10^{-3} \times 100}{101} = \frac{0.8}{101}$$

$$= 0.00792 = 7.92 \times 10^{-3} \text{A} = 7.92 \text{ mA } \textbf{Ans.}$$

We also know that the value of I_B,

$$I_B = I_E - I_C = 8\text{mA} - 7.92 \text{ mA} \textbf{ Ans.}$$

Example 14.5.

The transistor shown in Fig. 14.15 has values of I_E = 10 mA and α = 0.987. Determine the value of I_C and I_B for device.

Solution.

Given: I_E = 10 mA and α = 0.987

Value of I_C

We know that common-base d.c. current gain, (α),

$$0.987 = \frac{I_C}{I_E} = \frac{I_C}{10}$$

$$\therefore \quad I_C = 0.987 \times 10 = 9.87 \text{ mA } \textbf{Ans.}$$

We also know that emitter current (I_E),

$$10 = I_B + I_C = I_B + 9.87$$

$$\therefore \quad I_B = 10 - 9.87 = 0.13 \text{ mA } \textbf{Ans.}$$

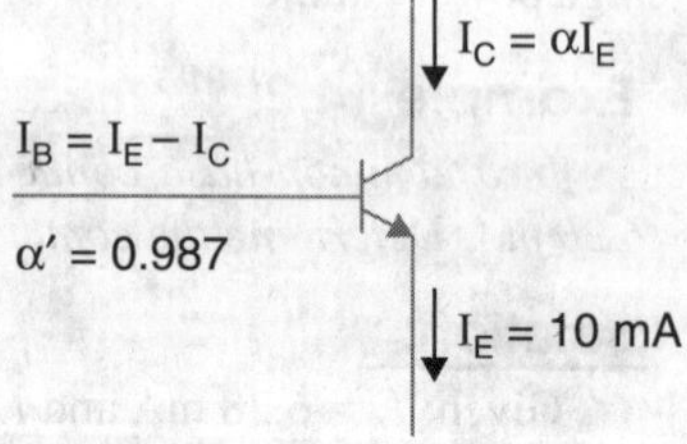

Fig. 14.15.

14.13. Current Gain of Bipolar Junction Transistor in Common-Emitter Configuration

Consider a transistor (either NPN or PNP) in a common-emitter configuration as shown in Fig. 14.16 (*a*) and (*b*). Here, the base current is the input current and the collector current is the output current. The current gain of a transistor in the common-emitter configuration is the ratio of collector current to the base current. Since the base current and collector current may be direct or alternating current, therefore we define two types of current gains namely d.c. current gain and a.c. current gains.

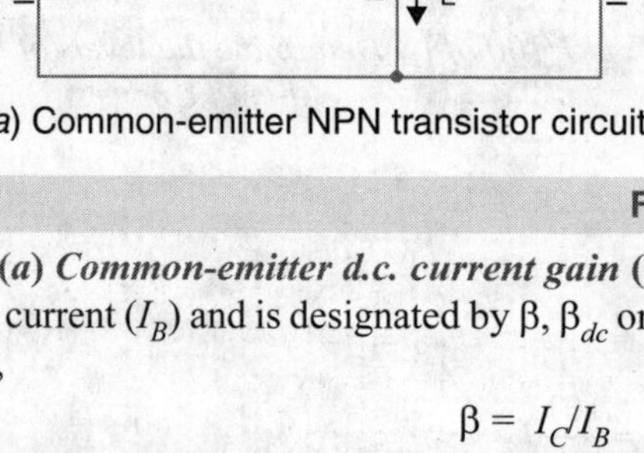

(*a*) Common-emitter NPN transistor circuit.

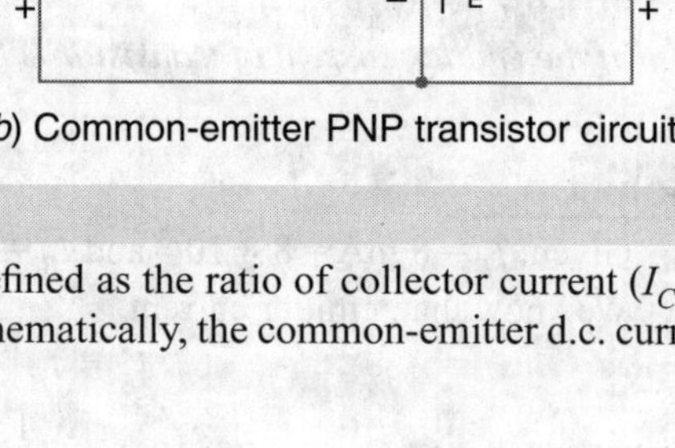

(*b*) Common-emitter PNP transistor circuit.

Fig. 14.16.

(*a*) ***Common-emitter d.c. current gain*** (β). It is defined as the ratio of collector current (I_C) to base current (I_B) and is designated by β, β_{dc} or h_{FE}. Mathematically, the common-emitter d.c. current gain,

$$\beta = I_C/I_B \qquad ...(i)$$

Now we know that collector current of a transistor is much larger than the base current. Therefore the value of β is much greater than unity. For example, if I_C = 5 mA and I_B = 0.05 mA, then common-emitter d.c. current gain,

$$\beta = 5/0.05 = 100$$

The above value of β indicates that collector current is 100 times that of base current. Typical values of β may range from 20 to 250. Sometimes β of a transistor is also known as large-signal common-emitter current gain.

(b) Common emitter a.c. current gain **(β_0).** It is defined as the ratio of small change in collector current (ΔI_C) to the small change in base current (ΔI_B) for a constant collector-to-emitter voltage (V_{CE}). It is designated by β_0. β_{ac} or h_{FE}. Mathematically common-emitter a.c. current gain,

$$\beta_0 = \Delta I_C/\Delta I_B \qquad ...(ii)$$

The term β_0 is also called common-emitter short circuit current gain or small-signal common-emitter current gain. It will be interesting to know that β_0 or h_{fe} is a parameter used in the analysis of small-signal transistor amplifier circuits. β_0 have a range of values similar to that mentioned for β. For all practical purposes, d.c. current gain and a.c. current gain are considered to be equal. *i.e.*,

$$\beta = \beta_0$$

14.14. Relation Between Current Gain α and β

We know that emitter current (I_E) of a transistor is the sum of its base current (I_B) and collector current (I_C). *i.e.*,

$$I_E = I_B + I_C$$

Dividing the above equation on both sides by I_C,

$$\frac{I_E}{I_C} = \frac{I_B}{I_C} + 1$$

Since $I_C/I_E = \alpha$ and $I_C/I_B = \beta$ therefore

$$\frac{1}{\alpha} = \frac{1}{\beta} + 1 = \frac{1 + \beta}{\beta}$$

$$\therefore \qquad \alpha = \frac{\beta}{\beta + 1}$$

The above expression may be written as

$$\alpha(\beta + 1) = \beta$$

$$\alpha \cdot \beta + \alpha = \beta$$

$$\alpha = \beta - \alpha \cdot \beta = \beta(1 - \alpha)$$

$$\therefore \qquad \beta = \frac{\alpha}{1 - \alpha}$$

Example 14.6.

(a) A transistor has an α of 0.975. What is the value of β. (b) if β = 200, what is the value of α ?

Solution.

Given: α = 0.975 and β = 200

Value of β when α is 0.975

We know that,

$$\beta = \frac{\alpha}{1 - \alpha} = \frac{0.975}{1 - 0.975} = 39 \textbf{ Ans.}$$

Value of α when β is 200

We also know that,

$$\alpha = \frac{\beta}{\beta + 1} = \frac{200}{200 + 1} = 0.995 \textbf{ Ans.}$$

Example 14.7.

A transistor has a typical of β = 100. If the collector is 40 mA, what is the value of emitter current?

Solution.

Given: $\beta = 100$ and $I_C = 40$ mA.

We know that (β),

$$100 = \frac{I_C}{I_B} = \frac{40}{I_B}$$

$$I_B = 40/100 = 0.4 \text{ mA}$$

and the emitter current,

$$I_E = I_B + I_C = 0.4 + 40 = 40.4 \text{ mA } \textbf{Ans.}$$

Example 14.8.

A transistor has β = 150. Calculate the approximate collector and base currents, if the emitter current is 10 mA.

Solution.

Given: $\beta = 150$ and $I_E = 10$ mA.

Collector current

Let I_C = Collector current.

We know that common-base current gain,

$$\alpha = \frac{\beta}{\beta + 1} = \frac{150}{150 + 1} = 0.993$$

and common-base current gain (α),

$$0.993 = \frac{I_C}{I_E} = \frac{I_C}{10}$$

$$\therefore \quad I_C = 0.993 \times 10 = 9.93 \text{ mA } \textbf{Ans.}$$

Base current

Let I_B = Base current.

We also know that emitter current (I_E),

$$10 = I_B + I_C = I_B + 9.93$$

$$\therefore \quad I_B = 10 - 9.93 = 0.07 \text{ mA } \textbf{Ans.}$$

Example 14.9.

If the base current in transistor is 30 μA when the emitter current is 7.2 mA, what are the values of α and β.

Solution.

Given: $I_B = 30\ \mu\text{A} = 30 \times 10^{-6}$ A and the $I_E = 7.2 \text{ mA} = 7.2 \times 10^{-3}$ A

We know that the value of α,

$$\alpha = \frac{I_C}{I_E} = \frac{I_E - I_B}{I_E} = \frac{(7.2 \times 10^{-3}) - (30 \times 10^{-6})}{7.2 \times 10^{-3}} = 0.995 \textbf{ Ans.}$$

We also know that the value of β,

$$\beta = \frac{\alpha}{1 - \alpha} = \frac{0.995}{1 - 0.995} = 199 \textbf{ Ans.}$$

Example 14.10.

Determine the value of I_B and I_E for the circuit shown in Fig. 14.17.

Solution.

Given: $\beta = 170$ and $I_C = 80$ mA.

We know that (β),

$$170 = \frac{I_C}{I_B} = \frac{80}{I_B}$$

$$\therefore \quad I_B = \frac{80}{170} = 0.47 \text{ mA } \textbf{Ans.}$$

and emitter current,

$$I_E = I_B + I_C = 0.47 + 80 = 80.47 \text{ mA } \textbf{Ans.}$$

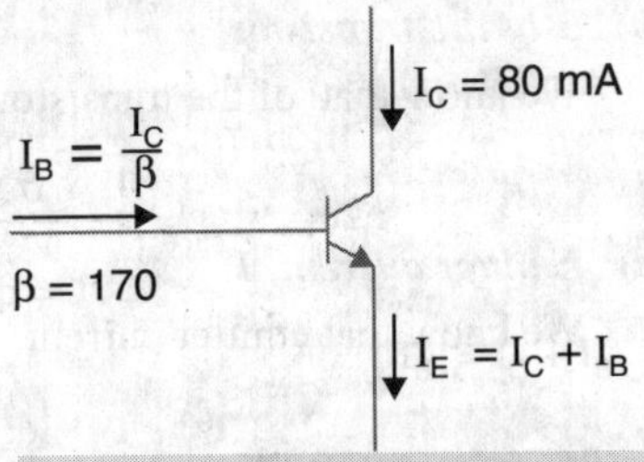

Fig. 14.17.

Example 14.11.

Determine the value of I_C and I_E for the circuit shown in Fig. 14.18.

Solution.

Given: $I_B = 125$ mA = 0.125 mA and $\beta = 200$.

We know that (β),

$$200 = \frac{I_C}{I_B} = \frac{I_C}{0.125}$$

$$\therefore \quad I_C = 200 \times 0.125 = 25 \text{ mA } \textbf{Ans.}$$

and emitter current,

$$I_E = I_B + I_C = 0.125 + 25 = 25.125 \text{ mA. } \textbf{Ans.}$$

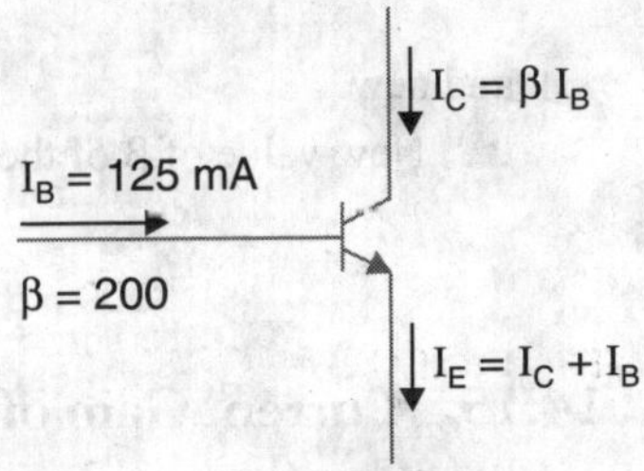

Fig. 14.18.

Example 14.12.

Determine the value of I_C and I_B for the circuit shown in Fig. 14.19.

Solution.

Given: $I_E = 12$ mA and $\beta = 140$.

We know that base current,

$$I_B = \frac{I_E}{1+\beta} = \frac{12}{1+140} = 0.085 \text{ mA } \textbf{Ans.}$$

and collector current,

$$I_C = I_E - I_B = 12 - 0.085 = 11.915 \text{ mA } \textbf{Ans.}$$

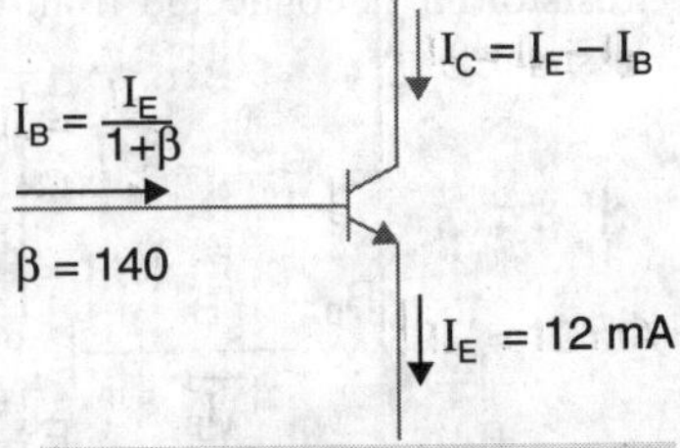

Fig. 14.19.

Example 14.13.

A transistor has I_B = 105 mA and I_C = 2.05 mA. Find: (a) β of the transistor; (b) α of the transistor, (c) emitter current I_E, (d) Now, if I_B changes by + 27 mA and I_C changes by + 0.65 mA, find the new value of β.

Solution.

Given: $I_B = 105$ mA = 105×10^{-6} A and $I_C = 2.05$ mA = 2.05×10^{-3} A.

(a) ***β of the transistor***

We know that β of the transistor,

$$\beta = \frac{I_C}{I_B} = \frac{2.05 \times 10^{-3}}{105 \times 10^{-6}} = 19.5 \textbf{ Ans.}$$

(*b*) *α of the transistor*

We know that of the transistor,

$$\alpha = \frac{\beta}{1+\beta} = \frac{19.5}{1+19.5} = 0.95 \textbf{ Ans.}$$

(*c*) *Emitter current, I_E*

We know that emitter current,

$$I_E = I_B + I_C = 105\ \ 10^{-6} + 2.05\ \ 10^{-3}\ \text{A}$$
$$= (0.105 + 2.05) \times 10^{-3} = 2.155 \times 10^{-3}\ \text{A}$$
$$= 2.155\ \text{mA} \textbf{ Ans.}$$

(*d*) *New value of β when*

$$\Delta I_B = 27\ \text{mA and } \Delta I_C = 0.65\ \text{mA}.$$

We know that new base current,

$$I_B = 105 + 27 = 132\ \text{mA} = 132\ \ 10^{-6}\ \text{A}$$

and new

$$I_C = 2.05 + 0.65 = 2.7\ \text{mA} = 2.7 \times 10^{-3}\ \text{A}$$

∴ New value of β of the transistor,

$$\beta = \frac{I_C}{I_B} = \frac{2.7 \times 10^{-3}}{132 \times 10^{-6}} = 20.5 \textbf{ Ans.}$$

14.15. Current Gain of a Bipolar Junction Transistor in Common-Collector Configuration

Fig. 14.20 (*a*) and (*b*) show a NPN transistor connected in a common-collector configuration. The voltage V_{BB} forward-biases the emitter-base junction and V_{CC} reverse-biases the collector-base junction of the transistor. Here the collector is not at d.c. ground. But it is at a.c. ground. It is due to the fact, that the voltage of d.c. supply source (V_{CC}) has zero resistance to an a.c. signal. An external resistor R_E is connected from the emitter to ground. The output of the circuit is taken across the resistor R_E.

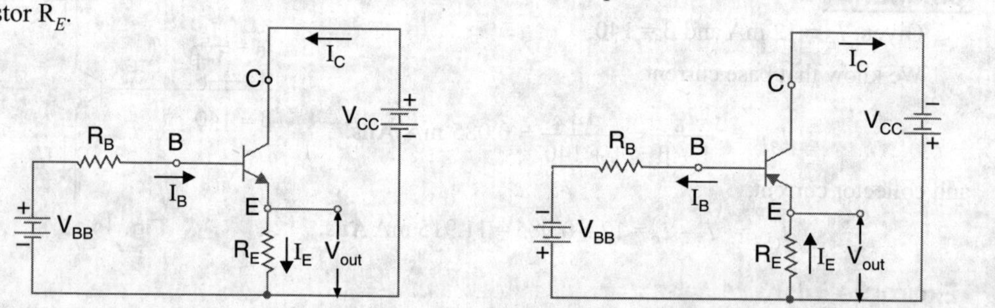

(*a*) Common-collector NPN transistor circuit.

(*b*) Common-collector PNP transistor circuit.

Fig. 14.20.

It may be noted that in a common-collector transistor circuit, the input current is the base current (I_B) and the output current is the emitter current (I_E). Therefore common-collector current gain is given by the relation,

$$= \frac{I_E}{I_B} = \frac{I_E}{I_C} \times \frac{I_C}{I_B} = \frac{1}{\alpha} \times \beta$$

Substituting the value of $\alpha = \dfrac{\beta}{1+\beta}$ in the above equation,common-collector current gain

$$= \frac{\beta}{\beta/(1+\beta)} = 1+\beta$$

It means that the output current of a common-collector transistor circuit is $(1 + \beta)$ times that of the input current. We know that $\beta >> 1$, therefore common-collector current gain $\approx \beta$.

14.16. Leakage Current in a Common-Base Bipolar Junction Transistor

Consider a common base (CB) PNP transistor circuit as shown in Fig. 14.21. When the switch is in its closed position, the forward-bias voltage (V_{EE}) injects the holes in the base region. The reverse-bias voltage (V_{CC}) on the collector-base junction attracts the majority of holes, from the base region, and constitute collector current. A small number of holes combine with the electrons in the base region and constitute base current. It means that the total emitter current splits up into two components namely the base current (I_B) and the collector current (I_C).

Now let the switch *'S'* be opened. It disconnects the emitter from the base and hence the emitter-base junction of the transistor is open-circuited. Thus there is no emitter current and therefore no base current (I_B) or collector current (I_C). However, it may be noted that the collector-base junction of the transistor is reverse-biased due to the holes injected from the emitter. But this junction is forward-biased due to the thermally generated minority carriers (*i.e.*, electrons in the P-type collector region and holes in the N-type base region). The minority carriers diffuse across the collector-base junction and hence produce a certain value of current known as leakage current. This current is called the leakage current from collector to base with emitter open and is designated by I_{CBO}. It is also known as the reverse saturation current or collector cut-off current (I_{CO}). Thus I_{CBO} or I_{CO} is similar to the reverse saturation current in a PN junction. The direction of I_{CBO} or I_{CO} is the same as that of a injected current. It is evident from the above discussion that the total collector current, in a transistor, consists of the following two components :

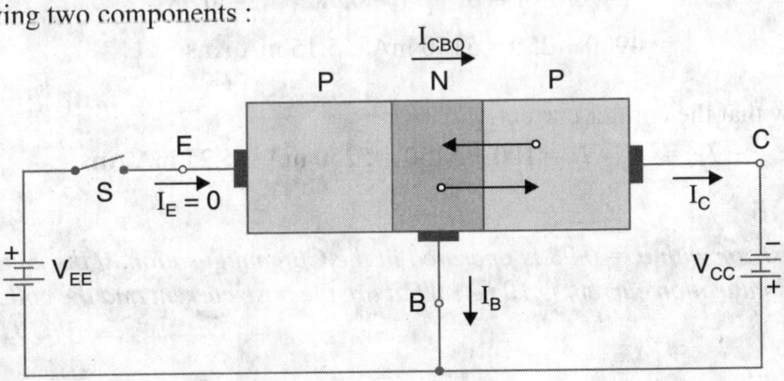

Fig. 14.21. Leakage current in a common-base transistor.

1. The current produced by normal transistor action. This component of a current is called injected current. Its value is equal to $\alpha \cdot I_E$ and is due to majority carriers.
2. The reverse saturation current (I_{CO}) produced by the thermally generated carriers or minority carriers.

Hence the total collector current,

$$I_C = \alpha I_E + I_{CO}$$

The above relation gives us a more accurate definition of common base current gain (α) of a transistor.

$$\therefore \quad \alpha = \frac{I_C - I_{CO}}{I_E}$$

Substituting the value of $I_E = I_B + I_C$ in this equation,

$$\alpha = \frac{I_C - I_{CO}}{I_B + I_C}$$

$$\therefore \quad I_C = \alpha\,(I_B + I_C) + I_{CO} = \alpha \cdot I_B + \alpha \cdot I_C + I_{CO}$$

$$(1 - \alpha)\, I_C = \alpha \cdot I_B + I_{CO}$$

$$\therefore \quad I_C = \frac{\alpha \cdot I_B}{1 - \alpha} + \frac{I_{CO}}{1 - \alpha}$$

It will be interesting to know that at room temperature, for silicon transistor, the value of I_{CO} is usually of the order of nanoamperes and thus is negligible. However, for germanium transistors the value of I_{CO} is of the order of microamperes and thus cannot be neglected. The value of I_{CO} is strongly temperature dependent for both silicon and germanium. It doubles for every 10°C increase in temperature. However, because of the lower magnitude of I_{CO} in silicon, the transistors can be used up to 200°C. On the other hand, germanium transistors are limited to about 100°C only.

Example 14.14.

A certain transistor has $\alpha = 0.98$, $I_{CO} = 5\ \mu A$ and $I_B = 100\ \mu A$. Find the values of collector and emitter currents.

Solution.

Given: $\alpha = 0.98$, $I_{CO} = 5$ mA and $I_B = 100$ mA

Collector current

We know that the collector current,

$$I_C = \frac{\alpha \cdot I_B}{1 - \alpha} + \frac{I_{CO}}{1 - \alpha} = \frac{0.98 \times 100}{1 - 0.98} + \frac{5}{1 - 0.98}\ \mu A$$

$$= 4900 + 250 = 5150 \text{ mA} = 5.15 \text{ mA } \textbf{Ans.}$$

Emitter current

We also know that the emitter current,

$$I_E = I_B + I_C = 100 + 5150 = 5250 \text{ mA} = 5.25 \text{ mA } \textbf{Ans.}$$

Example 14.15.

An NPN transistor with $\alpha = 0.98$ is operated in the CB configuration, if the emitter current is 3 mA and reverse saturation current is 10 μA. What are the base current and the collector current?

(GBTU., 2010–11)

Solution.

Given: $\alpha = 0.98$, $I_{CO} = 10\ \mu A = 10 \times 10^{-6}$ A and $I_E = 3$ mA $= 3 \times 10^{-3}$ A.

Collector current

We know that the collector current,

$$I_C = \alpha I_E + I_{CO} = 0.98 \times (3 \times 10^{-3}) + (10 \times 10^{-6})$$

$$= 2.95 \times 10^{-3} \text{ A} = 2.95 \text{ mA } \textbf{Ans.}$$

Base current

We also know that the emitter current,

$$I_B = I_E - I_C = (3 \times 10^{-3}) - (2.95 \times 10^{-3}) \text{ A}$$

$$= 0.05 \times 10^{-3} = 50 \times 10^{-6} \text{ A} = 50\ \mu A \textbf{ Ans.}$$

14.17. Leakage Current in a Common-Emitter Bipolar Junction Transistor

Consider a common-emitter (CE) PNP transistor circuit as shown in Fig. 14.22. When the switch is in its closed position (*i.e.*, 'ON' position), the current flows in the same way as discussed in the last article. However, if the switch is in open position (*i.e.*, 'OFF' position), then the base is disconnected from the emitter. Hence the base-emitter junction of the transistor is opened and therefore base current is zero. Under this condition, a leakage current flows from emitter to collector terminal. This current is designated by I_{CEO}. This leakage current is not just due to the thermally generated carriers across the collector-base junction, but also due to the movement of electrons across the base-emitter junction. This flow of electrons slightly forward biases the base-emitter junction. It appears as if a base current (equal to I_{CO}) is being supplied to the transistor. This produces an additional collector current (equal to $\beta \cdot I_{CO}$). Thus the total leakage current flowing through the transistor with base open,

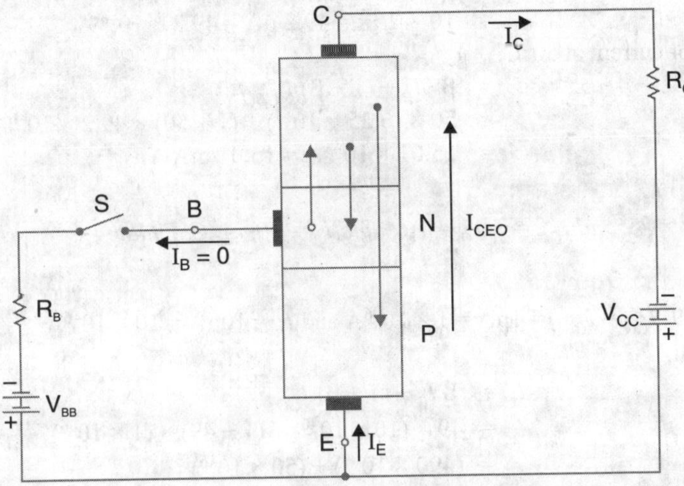

Fig. 14.22. Leakage current in a common-emitter transistor.

$$I_{CEO} = I_{CO} + \beta \cdot I_{CO} = (1 + \beta)\, I_{CO}$$

The total collector current in a transistor consists of the injected current ($\beta \cdot I_B$) and the leakage current (I_{CEO}). Thus the total collector current,

$$I_C = \beta \cdot I_B + I_{CEO} = \beta \cdot I_B + (1 + \beta)\, I_{CO}$$

It is evident from the above equation that leakage current in a common-emitter transistor is β times larger than in a common-base transistor. Since the value of I_{CO} is affected by the temperature, therefore when the temperature changes, there is also a change in the collector current of a common-emitter circuit. For example, let a silicon transistor has I_{CO} at 25°C equal to 0.025 μA and at 150°C, I_{CO} equal to 15 A. The transistor has a β of 100. Thus if this transistor is used in a common-emitter circuit, then its leakage current (I_{CEO}) would be 25 μA at 15°C and 1500 μA at 150°C. This may lead to instability in the common-emitter transistor circuit. Thus we conclude from the above discussion that common-base transistor circuit is much less temperature dependent than the common-emitter transistor circuit.

Example 14.16.

A germanium transistor used in a complementary symmetry amplifier has a collector cut-off current I_{CBO} = 10 mA at a temperature of 27°C and h_{FE} = 50.

(*a*) *What is the collector current, when the base current is 0.25 mA?*

(*b*) *Assuming h_{FE} does not increase with temperature, what would be the value of new collector current, if the transistor's temperature rises to 50°C?*

Solution.

Given: $I_{CBO} = 10\ \mu A = 10\ \ 10^{-6}$ A and $h_{FE}\ (= \beta) = 50$

(a) Value of collector current when $I_B = 0.25$ mA (0.25×10^{-3} A)

We know that collector current,

$$I_C = \beta I_B + (1 + \beta)\, I_{CBO}$$
$$= 50 \times (0.25 \times 10^{-3}) + (1 + 50) \times (10 \times 10^{-6})\text{A}$$
$$= 13.01 \times 10^{-3}\ \text{A} = 13.01\ \text{mA}\ \textbf{Ans.}$$

(b) Value of new collector current if temperature rises to 50°C

We also know that I_{CBO} doubles for every 10°C rise in temperature. Thus I_{CBO} at 50°C,

$$I_{CBO50} = I_{CBO} \times 2^{\,(T2 - T1)/10} = 10 \times 2^{\,(50 - 27)/10}\ \mu\text{A}$$
$$= 10 \times 2^{\,2.3} = 49.2\ \mu\text{A} = 49.2 \times 10^{-6}\ \text{A}$$

∴ Collector current at 50°C,

$$I_C = \beta \cdot I_B + (1 + \beta)\, I'_{CBO}$$
$$= 50 \times (0.25 \times 10^{-3}) + (1 + 50) \times 49.2 \times 10^{-6}\text{A}$$
$$= 15.01 \times 10^{-3}\ \text{A} = 15.01\ \text{mA}\ \textbf{Ans.}$$

Example 14.17.

A BJT with $\beta = 49$ and $I_{CO} = I_{CBO} = 1\mu A$ and $I_B = 10mA$. Calculate I_C? *(PTU., May 2007)*

Solution.

Given: $\beta = 49$; $I_{CO} = I_{CBO} = 1\mu\text{A} = 1 \times 10^{-6}\text{A}$ and $I_B = 10\mu\text{A} = 10 \times 10^{-6}\text{A}$

We know that,

$$I_C = \beta\, I_B + (1 + \beta)\, I_{CO}$$
$$I_C = 49 \times (10 \times 10^{-6}) + (1 + 49) \times (1 \times 10^{-6})$$
$$= (490 \times 10^{-6}) + (50 \times 10^{-6})$$
$$= 540 \times 10^{-6}\text{A or } 540\ \text{mA}\ \textbf{Ans.}$$

14.18. Base Spreading Resistance

We know that the width of the base region of a transistor is extremely small. Therefore the current, which enters the base region, across the emitter-base junction, has a narrow path to reach the base terminal. The resistance, offered by this narrow path region, is called base spreading resistance and is designated by r_b'. The value of this resistance can be increased by increasing the reverse-bias voltage (V_{CB}) across the collector-base junction. It is because of the fact that by increasing the value of V_{CB}, the width of collector-base depletion layer increases. Since the depletion layer penetrates more into the base-region than that of collector, therefore the cross-sectional area the base-region is reduced. The smaller cross-sectional means a large base-spreading resistance. It may be noted that the resistances of emitter and collector regions are very small, as these regions are much larger in size as compared to the base.

Usually, the value of base spreading resistance (r_b') is about 50 to 150 ohms. But in some cases, it may be as high as 1000 ohms. When a current passes through base spreading resistance, it produces a voltage drop across it. It has been observed that the effect of base spreading resistance is significant in high frequency circuits. At low frequencies, it has practically no effect.

14.19. Current Gain or β-Curve

We have already discussed in Art. 14.13 that common-emitter current gain (β) is very important transistor parameter. Therefore its study is very essential for every electronic engineer and technician. It has been experimentally found that the value of β varies with the collector current (I_C) and the junction temperature (T_j).

Fig. 14.23 shows a graph of β versus I_C at different junction temperatures (*i.e.*, – 55°C, 25°C and 125°C). Such a graph is called current gain graph, (or β curves). It may be noted from the graph that for a constant junction temperature, β has a maximum value at certain value of I_C. At a fixed value of I_C, the value of β increases with the increase in junction temperature. The variation in β over a large temperature range may be as high as 3 : 1. The manufacturer always specifies the value of β for the fixed values of I_C and junction temperature. It will be interesting to know that for a fixed value of I_C and junction temperature, the value of β varies from device to device for a given transistor.

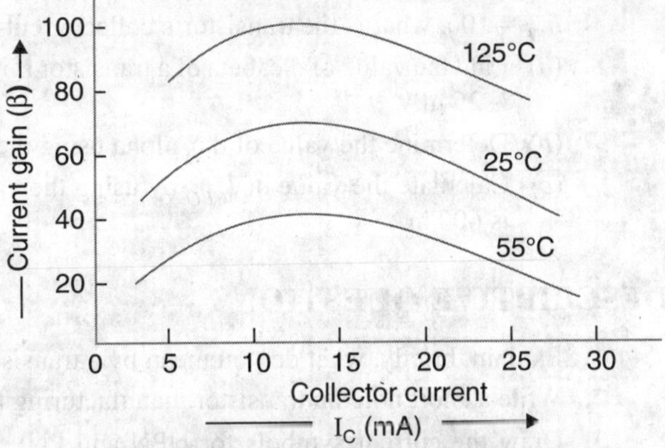

Fig. 14.23. Current gain β curves.

14.20. Multiple Emitter Transistor

Fig. 14.24 illustrates a Multiple Emitter Transistor. It is a specialized bipolar transistor mostly used at the input of transistor-transistor logic (TTL) NAND logic gates. The input signals are applied to the emitters. The collector current through the transistor flows only when the input is at logic low state. There is no collector current when the emitters are driven by the logic HIGH voltage. The multiple emitter transistor can be used to perform AND operation using a signal transistor. The advantage of multiple emitter transistor is that it allows reduction of switching time and power dissipation.

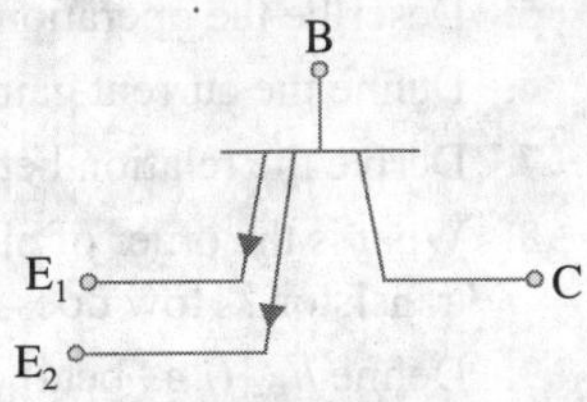

Fig. 14.24.

NUMERICAL PROBLEMS

1. A certain transistor has α of 0.98 and a collector leakage current I_{CO} of 1 mA. Calculate the collector and the base currents, when I_E = 1 mA. **(Ans. 0.981 mA and 19 mA)**
2. In a *CB* connection, the emitter current is 5.80 mA and the collector current is 5.6 mA. Find the value of α. **(Ans. 0.966)**
3. The beta of a transistor is 100. If the collector current is 50 mA, find the value of base current. **(Ans. 0.5 mA)**
4. A germanium transistor has I_{CBO} = 10 mA, = 0.98 and I_C = 1 mA. Determine the emitter current and β of the transistor. **(Ans. 1.01 mA; 98)**
5. The transistor shown in Fig. 14.25 has values of I_B = 50 mA and β = 400. Determine the value of I_C and I_E for the device. **(Ans. 20 mA, 20.05 mA)**
6. Determine the value of I_B and I_E for the circuit shown in Fig. 14.26.

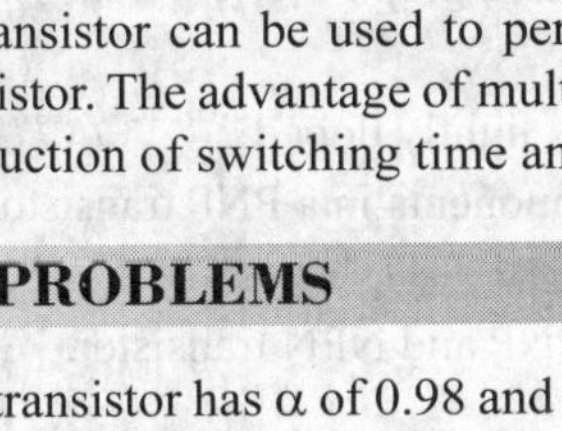

Fig. 14.25.

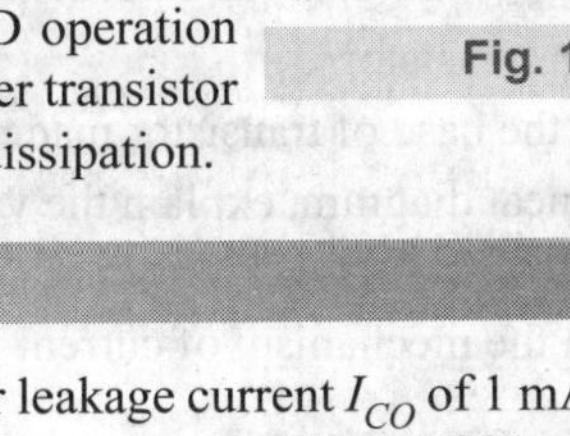

Fig. 14.26.

7. A germanium transistor's collector current is 50 mA, when the base current is 0.49 mA. If h_{FE} = 100, what is the transistor's collector cut-off current I_{CEO}. **(Ans. 1 m)**
8. (*a*) Find the value of d.c. beta of a transistor if its collector current is 3 mA and base current is 25 mA.

   (*b*) Determine the value of d.c. alpha using the value of d.c. beta obtained in part (*a*).

   (*c*) Calculate the value of I_{CBO} by using the value of d.c. beta obtained in part (*a*) if I_{CBO} is 600 mA. **(Ans. (*a*) 120 (*b*) 0.992 and (*c*) 4.96 μA)**

DESCRIPTIVE QUESTIONS

1. Explain, briefly, what do you mean by a transistor.
2. Write a short note on transistor manufacturing techniques.
3. Draw the current symbols for NPN and PNP transistors. Also indicate the conventional directions for three currents.
4. What are the different ways of biasing a transistor? Which method of biasing is used for operating transistor as an amplifier ?
5. Describe the operation of a NPN transistor.
6. Define the current gain and current gain β of a transistor.
7. Derive the relation between alpha and beta of a transistor.
8. What is the order of magnitude of the reverse collector saturation current I_{CBO} for a silicon transistor ? How does I_{CBO} vary with the temperature ?
9. Define h_{FE} (*i.e.*, beta) of a transistor and sketch a graph of h_{FE} versus I_C.
10. Define a transistor. *(JNTU., June 2009)*
11. Why is the base of transistor made thin and is lightly doped ? *(JNTU., Aug./Sep. 2006)*
12. With a neat diagram, explain the various components in a PNP transistor. *(JNTU., May/Jun. 2006)*
13. Explain the mechanism of current flow in a PNP and NPN transistor. *(JNTU., Aug/Sep. 2006)*
14. Write note on transistor construction. *(Anna University., Nov/Dec 2007)*
15. Explain with necessary figures the principle of operation of a PNP transistor. Explain how a transistor could be operated as an amplifier. *(M.G University., Aug. 2007)*
16. Explain the working principle of transistors. *(M.G University., April/May 2004)*
17. Compare different types of transistor configuration with necessary circuit diagrams using transistor. *(JNTU., Aug./Sep. 2006, PTU., May 2005)*
18. What is the width of the base in a BJT in comparison with either the emitter or the collector region ? Why ? *(PBUT., June 2005)*
19. Explain why in the active operation the base current I_B is much smaller than I_C or I_E. What is the relation among the three currents? *(GBTU/MTU., 2008-09)*
20. Which of the transistor currents is always the greatest? Which is always the smallest? Which two currents are relatively closer in magnitude? *(Jamia University., 2008, GBTU/MTU., 2006-07, 2005-06)*
21. Explain the working principle of PNP transistor. Mention 3 important uses of a transistor? *(M.G University., May/June 2006)*
22. Derive the relationship between of a BJT. *(M.G University., Dec. 2007)*

23. Define α and β. Express the relationship between them. (*M.G University., Aug. 2007, GBTU/MTU 2004-05, 2000-01, PBUT., Dec. 2004, R.G.P.V., June 2004*)
24. Define I_{CBO} and I_{CEO}. How are they different? How are they related? Are they typically close in magnitude. (*Jamia University., 2008*)
25. Explain how a Bipolar junction transistor can be used as an amplifier. Derive the relationship between the beta and alpha of the transistor. (*GBTU/MTU., 2008-09*)
26. Which of the transistor is always the largest ? Which one is the smallest? (*GBTU/MTU., 2009-10*)
27. Draw the power flow diagram of the transistor. (*Anna University., April/May 2008*)
28. If a transistor has its collector-base junction forward biased and the other junction reverse biased, will it work ? Explain. (*GBTU/MNTU., 2003-04*)
29. Discuss the leakage currents in a common collector circuits. Find the relation between leakage current I_{CO} and total outputs current I_C of a common emitter circuit. (*Bhartiya University., April 2004*)
30. The common base amplification factor $\alpha < 1$, why this called amplification factor ? Justify. (*GBTU/MNTU., 2001-02*)
31. List the three factors that contribute to thermal instability in transistor amplifiers. Also draw a circuit for *CE* amplifier with fixed, bias and show that the stability factor "S" for the circuit is (β + 1). (*GBTU/MNTU., 2002-03*)
32. Explain why an ordinary junction transistor called bipolar. (*GBTU/MNTU., 2004-05*)
33. Derive the expression for the collector current for a *CE* transistor. (*PBUT., 2009*)
34. Derive the relation between alpha and beta of a transistor. (*JNTU., Aug./Sep. 2006*)
35. Define α and β with respect to BJT and drive the relationship between them. (*GBTU/MTU., 2008-09*)
36. Show that $I_C = \alpha I_E + I_{CBO}$ and discuss the parameters given in the equation. (*Anna University., Nov/Dec 2007*)
37. What is 'β' of a transistor ? Draw a graph showing the variation of 'α' with emitter current in a BJT. (*JNTU., Aug./Sep. 2006*)
38. Write any two applications of transistor. (*JNTU., June 2006*)
39. Analyses the operation of BJT graphically. (*Calicut University., Dec. 2006*)
40. Explain the different regions of operation of transistor. (*R.G.P.V., Feb. 2010*)
41. Explain the working of transistor in common emitter configuration. (*R.G.P.V., June 2010*)
42. Explain the construction and basic operation of CC configuration of transistor. (*R.G.P.V., June 2009*)
43. Explain amplification mechanism of a transistor. Also elaborate the term "Large signal current gain", "α " with reference to a BJT. (*R.G.P.V., Dec. 2003*)
44. Explain the working principle of PNP transistor. (*R.G.P.V., Dec. 2001*)
45. Draw the circuit diagram of NPN transistor in CB and CE configuration with dc batteries. Compare the typical values of current gain in both CB and CE configurations. (*R.G.P.V., June 2007, 2005*)
46. Show the current components in NPN transistor, when biased in active region. (*Nagpur University, Summer 2008, 2000, 2006*)
47. Derive the relation between α and β. (*Nagpur University, Winter 2006*)

48. What is necessary and sufficient condition for a transistor to operate in cut-off region? *(Nagpur University, Summer 2003)*
49. Comment on the junction areas and doping levels in a transistor. *(Nagpur University, Summer 2003)*
50. Explain the transistor action and with neat diagram shown different current component in the transistor. *(Nagpur University, Summer 2003)*
51. Explain minority concentration variation in a PNP transistor operating in active region of common base configuration. *(Nagpur University, Winter 2002)*
52. Comment on the junction areas and doping levels in a transistor. *(Nagpur University, Summer 2000)*
53. Explain the working of transistor in common emitter configuration. *(RGTU.,Dec.2009)*
54. Give the structure of a transistor and the circuit symbol. Compare the physical size doping levels and the functions of the three layers in the transistor. *(RGTU., Feb. 2010)*
55. Explain the following with reference to BJT:
    (*a*) Reverse saturation current (I_{CBO})
    (*b*) Base width modulation (Early effect) *(RGTU., Feb. 2010)*
56. Derive the current equation $I_C = \beta\, I_B + (\beta + 1)\, I_{CBO}$ for the transistor. *(RGTU., Feb. 2010)*

MULTIPLE CHOICE QUESTIONS

1. In a PNP transistor with normal bias
   (*a*) Only holes cross the collector junction
   (*b*) Only majority carriers cross the collector junction
   (*c*) The collector junction has a low resistance
   (*d*) The emitter junction is forward-biased and collector junction is reverse biased
2. For transistor action
   (*a*) The collector must be more heavily doped than the base region
   (*b*) The base region must be N-type material
   (*c*) The base region must be very narrow
   (*d*) The collector-base junction must be forward biased
3. In a PNP transistor, the electrons flow
   (*a*) Into the transistor at the emitter and base leads
   (*b*) Out of the transistor at the emitter and base leads
   (*c*) Into the transistor at the collector and base leads
   (*d*) Out of the transistor at the collector and base leads
4. The leakage current I_{CBO} flows in
   (*a*) The emitter, base and collector leads (*b*) The emitter and base leads
   (*c*) The emitter and collector leads (*d*) The base and collector leads
5. A small increase in collector-base reverse bias will cause
   (*a*) A large increase in emitter current
   (*b*) A large increase in collector current
   (*c*) Very little change in collector reverse saturation current
   (*d*) A large decrease in collector current

6. Eary effect in BJT refers to

(*a*) avalanche breakdown (*b*) thermal breakdown

(*c*) base narrowing (*d*) Zener breakdown

(UPSC Engg. Sciences 2002)

7. A PNP transistor has

(*a*) only donor ions (*b*) only acceptor ions

(*c*) two P-regions and one N-region (*d*) three PN junction

8. The emitter of a transistor is generally doped the heaviest because it

(*a*) has to dissipate maximum power (*b*) has to supply the charge carriers

(*c*) is the first region of the transistor (*d*) must possess low resistance.

9. For current working of an *NPN* bipolar junction transistor, the different electrodes should have the following polarities with respect to emitter.

(*a*) collector + *ve*, base – *ve* (*b*) collector – *ve*, base + *ve*

(*c*) collector – *ve*, base – *ve* (*d*) collector + *ve*, base + *ve*

10. Select the CORRECT alternative. In a bipolar transistor

(*a*) emitter region is of low/high resistivity matterial which is lightly/heavily-doped.

(*b*) collector region is of lower/higher conductivity than emitter region

(*c*) base region is of high/low resistivity material which is only lightly/heavily-doped.

11. In a properly-biased *NPN* transistor, most of the electrons from the emitter

(*a*) recombine with holes in the base (*b*) recombine in the emitter itself

(*c*) pass through the base to the collector (*d*) are stopped by the junction barrier.

12. The following relationship between α and β are correct EXCEPT

(*a*) $\beta = \frac{\alpha}{1-\alpha}$ (*b*) $\alpha = \frac{\beta}{1-\beta}$

(*c*) $\alpha = \frac{\beta}{1+\beta}$ (*d*) $1-\alpha = \frac{1}{1+\beta}$.

13. The value of total collector current in a *CB* circuit is

(*a*) $I_C = \alpha I_E$ (*b*) $I_C = \alpha I_E + I_{CO}$

(*c*) $I_C = \alpha I_E - I_{CO}$ (*d*) $I_C = \beta I_E$.

14. In a junction transistor, the collector cut off current I_{CBO} reduces considerably by doping the

(*a*) emitter with high level of impurity (*b*) emitter with low level of impurity

(*c*) collector with high level of impurity (*d*) collector with low level of impurity

15. In a transistor amplifier, the reverse saturation current I_{CO}

(*a*) doubles for every 10°C rise in temperature

(*b*) doubles for every 1°C rise in temperature

(*c*) increase linearly with the temperature

(*d*) doubles for every 5°C rise in temperature

16. In the case of a bipolar transistor, α is

(*a*) positive and > 1 (*b*) positive and < 1

(*c*) negative and > 1 (*d*) negative and < 1

17. The emitter base of a given transistor is forward-biased and its collector-base reverse-biased. If the base current is increased, then its

(a) I_C will decrease (b) V_{CE} will increase
(c) I_C will increase (d) V_{CC} will increase

18. The collector characteristics of a common/emitter connected transistor may be used to find its
(a) input resistance (b) base current
(c) output resistance (d) voltage gain

19. Which of the following approximations is often used in electronic circuits ?
(a) $I_C \cong I_E$ (b) $I_B \cong I_C$ (c) $I_B \cong I_E$ (d) $I_E \cong I_B + I_C$

20. If in a bipolar junction transistor, $I_B = 100\ \mu A$ and $\mu_C = 10$ mA, in what range does the value of its beta lie ?
(a) 0.1 to 1.0 (b) 1.01 to 10 (c) 10.1 to 100 (d) 100.1 to 1000

21. In a BJT, largest current flow occurs
(a) in the emitter (b) in the collector
(c) in the base (d) through CB junction

22. Which of the following transistor configuration circuits is much less temperature dependent?
(a) common-base (b) common-emitter
(c) common-collector (d) none of the above

(Bharthiar Univ. April 2004)

ANSWERS

1. (d)	**2.** (c)	**3.** (c)	**4.** (d)	**5.** (d)
6. (c)	**7.** (c)	**8.** (b)	**9.** (d)	**10.** (a)
11. (c)	**12.** (c)	**13.** (b)	**14.** (d)	**15.** (a)
16. (b)	**17.** (a)	**18.** (c)	**19.** (a)	**20.** (a)
21. (a)	**22.** (a)			

15

Chapter

BJT Characteristics

Outline

1. Introduction.
2. Characteristics of a Common-base Configuration.
3. Input Characteristics of a Transistor in a Common-base Configuration.
4. Base Width Modulation or Early Effect.
5. Output Characteristics of a Transistor in Common-Base Configuration.
6. Characteristic of a Transistor in Common-Emitter Configuration.
7. Input Characteristics of a Transistor in Common-Emitter Configuration.
8. Output Characteristics of a Transistor in Common-Emitter Configuration.
9. Characteristics of a common-collector configuration
10. Input characterics of a Transistor in common-collector configuration
11. Output characteristics of a transistor in common-collector configuration
12. Transistor Ratings.
13. Maximum Collector Current.
14. Maximum Power Dissipation.
15. Maximum Output Voltage.
16. Bipolar Junction Transistor Specification Sheet.
17. Transistor Testing.
18. Integrated Transistors.
19. High-Voltage Transistors.
20. High-current Transistors
21. High-power Transistors.
22. Thermal Runaway and Heat Sink.
23. Phototrancictor.
24. Photodarlington

15.1. Introduction

We have already discussed in the last chapter about the construction of a bipolar junction transistor and its current gains α and β. As a matter of fact, the values of α and β do not completely describe the behaviour of a transistor. Many more details can be studied with the help of curves, which relate the transistor currents and voltages. These curves are known as characteristic curves. Following are the two sets of characteristic curves, which are important from the subject point of view :

1. ***Input characteristics.*** These curves give the relationship between the input current and input voltage for a given output voltage.
2. ***Output characteristics.*** These curves give the relationship between the output current and the output voltage for a given input current.

These two sets of characteristic curves completely describe the static operation of a transistor. Depending upon the configuration, the transistor characteristics may be studied under the following two heads:

1. Characteristics of a transistor in a common-base configuration.
2. Characteristics of a transistor in a common-emitter configuration.
3. Characteristics of a transistor in a common-collector configuration.

That the characteristics of a transistor, in a common-collector configuration, are not needed as they can be treated as a special case of common-emitter configuration (with feedback applied).

15.2. Characteristics of a Transistor in a Common-Base Configuration

Following are two important characteristics of a transistor in a common-base (CB) configuration:

1. ***Input characteristics.*** These curves give the relationship between the emitter current (I_E) and the emitter-to-base voltage (V_{EB}) for a constant collector-to-base voltage (V_{CB}). These curves are also known as base curves of a transistor.
2. ***Output characteristics.*** These curves give the relationship between the collector current (I_C) and the collector-to-base voltage (V_{CB}) for a constant emitter current (I_E). These curves are also known as collector curves of a transistor.

These characteristics may be obtained by means of a circuit arrangement as shown in Fig. 15.1. In this circuit, the NPN transistor is connected in a common-base configuration. The collector-to-base voltage can be varied by changing the position of potentiometer R_2, while the emitter-to-base voltage can be varied by changing the position of potentiometer R_1. The d.c. milliammeters and d.c. voltmeters are connected in the emitter and collector circuits of a transistor to measure the voltages and currents.

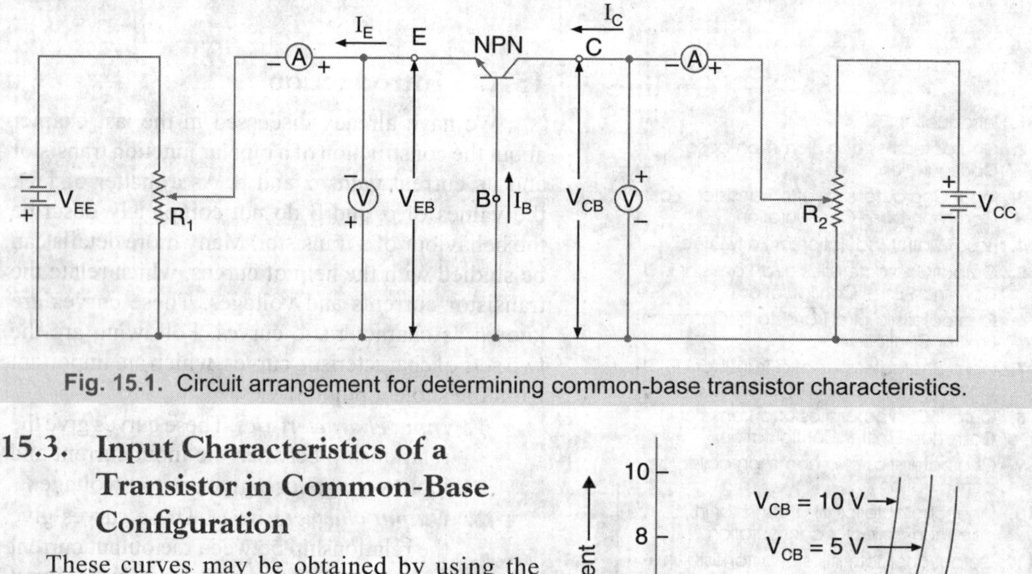

Fig. 15.1. Circuit arrangement for determining common-base transistor characteristics.

15.3. Input Characteristics of a Transistor in Common-Base Configuration

These curves may be obtained by using the circuit arrangement as shown in Fig. 15.1. First of all, we adjust the collector-to-base voltage (V_{CB}) to one volt. Then increase the emitter-to-base voltage (V_{EB}) in small suitable steps (*i.e.*, of the order of 0.1 V) and record the corresponding values of emitter current (I_E) at each step. Now, if we plot a graph with emitter-to-base voltage (V_{EB}) along the horizontal axis and the emitter current (I_E) along the vertical axis, we shall obtain a curve marked V_{CB} = 1 V as shown in Fig. 15.2. A similar procedure may be used to obtain curves at different collector-to-base voltage 5 V and 10 V as shown in the Fig. 15.2.

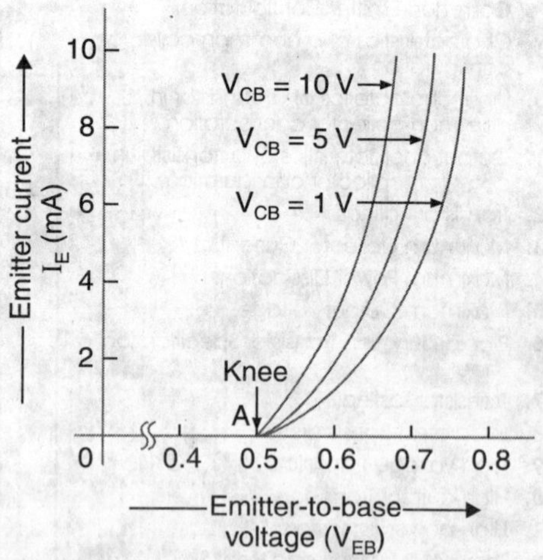

Fig. 15.2. Input characteristics of a common-base transistor.

It will be interesting to know that the input characteristics give us the information about the following important points:

1. There exists a threshold voltage (also called offset voltage, cut-in voltage or knee voltage) as indicated by the region *OA*, below which the emitter current is negligibly small. The knee-or threshold voltage is designated by V_K (orV_g). The value of knee voltage (V_K) is approximately 0.5 V for silicon and 0.1 V for germanium transistors.
2. Beyond the point *A*, for a fixed collector-to-base voltage the emitter current (I_E) increases rapidly with a small increase in emitter-to-base voltage (V_{EB}). It means that input resistance (R_i) of a transistor in common-base configuration is very small.
3. As the collector-to base voltage (V_{CB}) is increased above one volt, the curves shift upwards. It occurs due to a phenomenon called base-width modulation or early effect. It will be explained in the next article.
4. The input characteristic may be used to determine the value of *a.c. input resistance. Its value at any point on the curve is given by the ratio of a change in emitter-to-base voltage (ΔV_{EB}) to the resulting change in emitter current (ΔI_E) for a constant collector-to-base voltage (V_{CB}). Mathematically the a.c. input resistance,

$$R_i = \frac{\Delta V_{EB}}{\Delta I_E}$$

It may be noted that input characteristic is linear in the upper region, but non-liner (*i.e.*, curved) in the lower region. Therefore *the a.c. input resistance depends upon the location of the operating point selected along the curve*. Its value, in the linear region of the curve, is about 50 ohms.

15.4. Base Width Modulation or Early Effect

We have already discussed in the last article that the emitter-base diode characteristic shifts upwards by increasing the value of collector-to-base voltage (V_{CB}). This occurs due to the phenomenon called base width modulation or early effect, which may be explained as follows:

We know that in a PN junction, the width of a depletion region increases as the reverse-bias voltage increase. In a transistor, since the emitter-base junction is forward-biased, therefore this has no effect on the width of depletion region. However, the collector-base junction is reverse biased. Therefore as the reverse-bias voltage across the collector-base junction increases, the width of depletion region also increases. Since the base is lightly doped, as compared to the collector, therefore depletion region penetrates deeper into the base region. This reduces the effective width of the base region. This variation or modulation of the effective base width, by the collector voltage, is known as base width modulation or early effect. The decrease in base width by the collector voltage, has the following three effects:

1. It reduces the chances of recombination of electrons with the holes in the base region. Hence the common base current gain (α) increases with the increase of collector-to-base voltage (V_{CB}).
2. The concentration gradient of minority carriers within the base increases. This, in turn, increases the emitter current.
3. For extremely large collector voltage, the effective base width may be reduced to zero, causing voltage breakdown of a transistor. This phenomenon is known as *punch through*.

* The d.c. input resistance of a transistor is not of any use in actual practice.

15.5. Output Characteristics of a Transistor in Common-Base Configuration

These characteristics may be obtained by using the circuit shown in Fig. 15.1. First of all, we adjust, the emitter-to-base voltage (V_{EB}) to get a suitable value of emitter current (say 2 mA). Keeping the emitter current constant, we increase the collector-to-base voltage (V_{CB}) from zero in a number of suitable steps and record the corresponding values of the collector current (I_C) at each step. If we plot a graph with collector-to-base voltage (V_{CB}) along the horizontal axis and the collector current (I_C) along the vertical axis, we shall obtain a curve marked I_E = 2 mA as shown in Fig. 15.3. A similar procedure may be used to the obtain the characteristics at different values of emitter current *i.e.*, I_E = 4, 6, and 8 mA. The output characteristic gives us the information about the following important points:

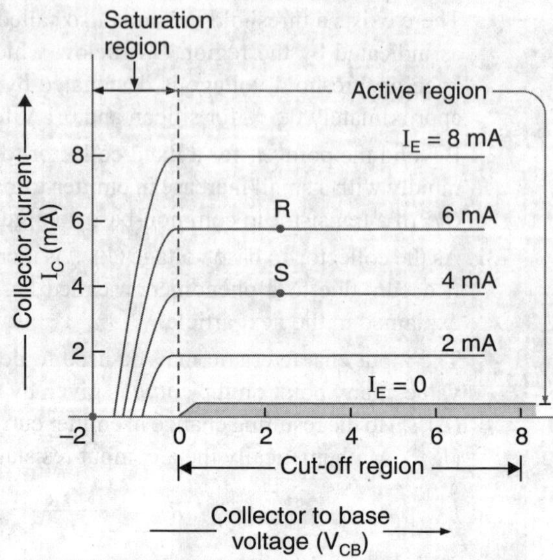

Fig. 15.3. Output characteristics of a common-base transistor.

1. The curve may be divided into three important regions namely saturation region, active region and cut-off region. The saturation region is the region to the left of the vertical dashed line. It maybe noted that in this region, collector-to-base voltage (V_{CB}) is negative for a NPN transistor. It means that collector-base junction of a transistor is also forward biased in the saturation region. In this region, a small change in V_{CB} results in a large value of current. The active region is the region between the vertical dashed line and The horizontal axis. In the active region, the collector current is constant and is equal to the emitter current. The cut-off region is the region along the horizontal axis as shown by a shaded region in the Fig. 15.3. It corresponds to the curve marked I_E = 0. In the cut-off region, both the junctions of a transistor are reverse-biased.
2. The collector current flows even when the collector-to-base voltage (V_{CB}) is zero. This is due to the fact that electrons are injected into the base under the action of a forward-biased emitter-base junction. These electrons are collected by the collector due to the internal junction voltage (*i.e.*, the barrier potential) at the collector-base junction.
3. A small collector current flows even when emitter current (I_E) is zero. This current is called collector leakage current and is designated as I_{CBO}.
4. The collector current is practically independent of collector-to-base voltage (V_{CB}) in the active region. However, if V_{CB} is increased beyond a certain large value, the collector current increases rapidly due to avalanche breakdown (not shown in the figure) and the transistor action is lost.
5. The characteristic may be used to determine the common-base transistor a.c. output resistance. Its value at any point is given by the ratio of a change in collector-to-base voltage (ΔV_{CB}) to the resulting change in collector current (ΔI_C) for a constant emitter current (I_E). Mathematically, the a.c. output resistance,

$$R_O = \frac{\Delta V_{CB}}{\Delta I_C}$$

It may be noted that the characteristic curves of common-base transistor are almost horizontal. This indicates that the value of output resistance (R_0) is very high. Its typical value is about 500 kΩ.

6. The characteristic may be used to determine small-signal common-base current gain or a.c. alpha (α_0) of transistor. This can be done by selecting two point *R* and *S* on the characteristics and note down the corresponding values of ΔI_C and I_C. Thus if $\Delta I_C = 6 - 4 = 2$ mA and $I_E = 6 - 4 = 2$ mA, then small signal common-base current gain.

$$\alpha_0 = \frac{\Delta I_C}{\Delta I_E} = \frac{2}{2} = 1$$

15.6. Characteristics of a Transistor in Common-Emitter Configuration

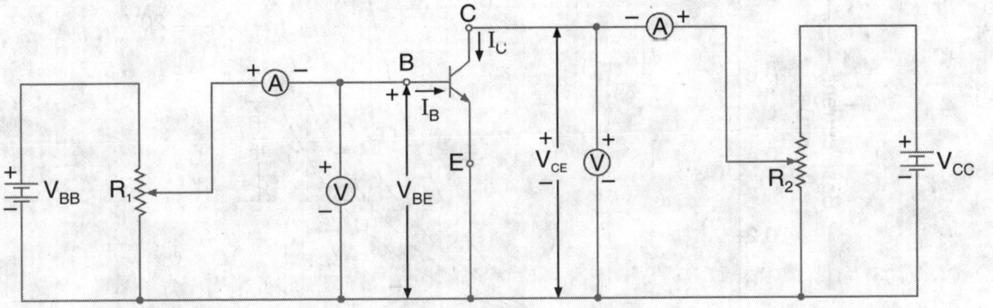

Fig. 15.4. Circuit arrangement for determining common-emitter transistor characteristics.

A transistor, in a common-emitter configuration, has also two important characteristics namely input characteristics and output characteristics as discussed below:

1. ***Input characteristics.*** These curves give the relationship between the base current (I_B) and the base-to-emitter voltage (V_{BE}) for a constant collector-to-emitter voltage (V_{CE}).
2. ***Output characteristics.*** These curves give the relationship between the collector current (I_C) and collect-to-emitter voltage (V_{CE}) for a constant base current (I_B).

These characteristics may be obtained by means of a circuit arrangement as shown in Fig. 15.4. In this circuit, an NPN transistor is connected in a common-emitter configuration. The collector-to-emitter voltage (V_{CE}) can be varied by changing the position of potentiometer , while the base-to-emitter voltage (V_{BE}) can be varied by changing the position of potentiometer R_1. The dc milliammeters and dc voltmeters are connected in the base and the collector circuit of a transistor to measure the voltages and currents.

15.7. Input Characteristics of a Transistor in Common-Emitter Configuration

These characteristics may be obtained by using the circuit arrangement shown in Fig. 15.4. First of all, we adjust the collector-to-emitter voltage (V_{CE}) to 1 volt. Then we increase to base-to-emitter voltage (V_{BE}) in small suitable steps and record the corresponding values of base current (I_B) at each step. If we plot a graph with base-to-emitter voltage (V_{BE}) along the horizontal axis and the base current (I_B) along the vertical axis, we shall obtain a curve marked $V_{CE} = 1$ V as shown in Fig. 15.5. A similar procedure may be used to obtain characteristics at different values of collector-to-emitter voltage *i.e.*, $V_{CE} = 2$, 10 and 20V. The input characteristic gives us the information about the following important points:

1. There exists a threshold or knee voltage (V_K) below which the base current is very small. The value of knee voltage is 0.5 V for silicon and 0.1 V for germanium transistors.

2. Beyond the knee, the base current (I_B) increases with the increase in base-to-emitter voltage (V_{BE}) for a constant collector-to-emitter voltage (V_{CE}). However, it may be noted that the value of base current does not increase as rapidly as that of the input characteristic of a common-base transistor. It means that input resistance of a transistor in common-emitter configuration is higher as compared to the common base configuration.
3. As the collector-to-emitter voltage (V_{CE}) is increased above 1 V, the curves shift downwards. It occurs because of the fact, that as V_{CE} is increased, the depletion width in the base-region increases. The reduces the effective base width, which in turn reduces the base current.

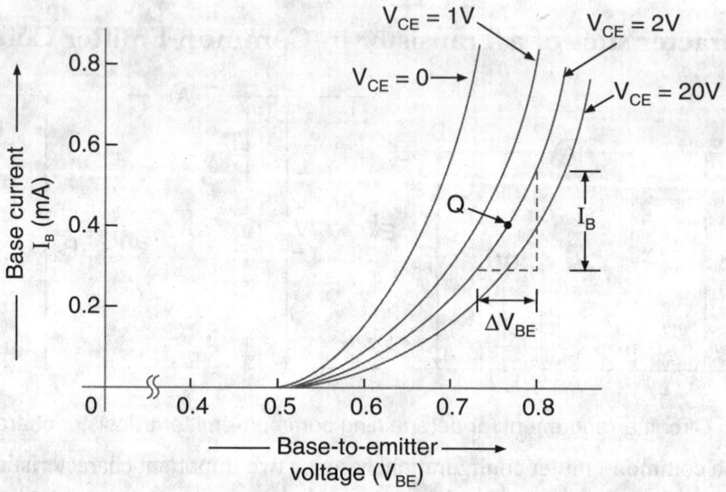

Fig. 15.5. Input characteristics of a common-emitter transistor.

4. The input characteristic may be used to determine the value of common-emitter transistor a.c. input resistance (R_i). Its value is given by the ratio of change in base-to-emitter voltage (ΔV_{BE}) to the resulting change in base current (I_B) at a constant collector-to-emitter voltage (V_{CE}). Mathematically, the a.c. input resistance,

$$R_i = \frac{\Delta V_{BE}}{\Delta I_B}$$

It may be noted that the input characteristic is not linear in the lower region of the curve. Therefore, the input resistance varies with the location of the operating point. The value of a.c. input resistance ranges from 600 Ω to 4000 Ω.

15.8. Output Characteristics of a Transistor in Common-Emitter Configuration

These characteristics may be obtained by using the circuit shown in Fig. 15.4. To begin with, adjust the base current (I_B) to 40 mA value. Then increase the collector-to-emitter voltage (V_{CE}) in a number of steps and record the corresponding values of collector current (I_C) at each step. If we plot a graph with collector-to-emitter voltage (V_{CE}) along the horizontal axis and collector current (I_C) along the vertical axis, we shall obtain a curve marked I_B = 40 μA as shown in Fig. 15.6. A similar procedure may be used to obtain characteristics at I_B = 80 μA, 120 μA, 160 μA and so on. The output characteristics give us the information about the following important points:

1. The output characteristics may be divided into three important regions namely saturation region, active region and cut-off region. The saturation and cut-off regions are shown by

the shaded areas, while the active region is the region between the saturation and cut-off region.

2. As the collector-to-emitter voltage (V_{CE}) is increased above zero, the collector current (I_C) increases rapidly to a saturation value, depending upon the value of base current. It may be noted that collector current (I_C) reaches to a saturation value when V_{CE} is about 1 V.
3. When collector-to-emitter voltage (V_{CE}) is increased further, the collector current slightly increases. This increase in collector current is due to the fact that increased value of collector-to-emitter voltage (V_{CE}) reduces the base current and hence the collector current increases. This phenomenon is called an early effect.
4. When the base current is zero, a small collector current exists. This is called leakage current. However, for all practical purposes, the collector current (I_C) is zero, when the base current (I_B) is zero. Under this condition the transistor is said to be cut-off.

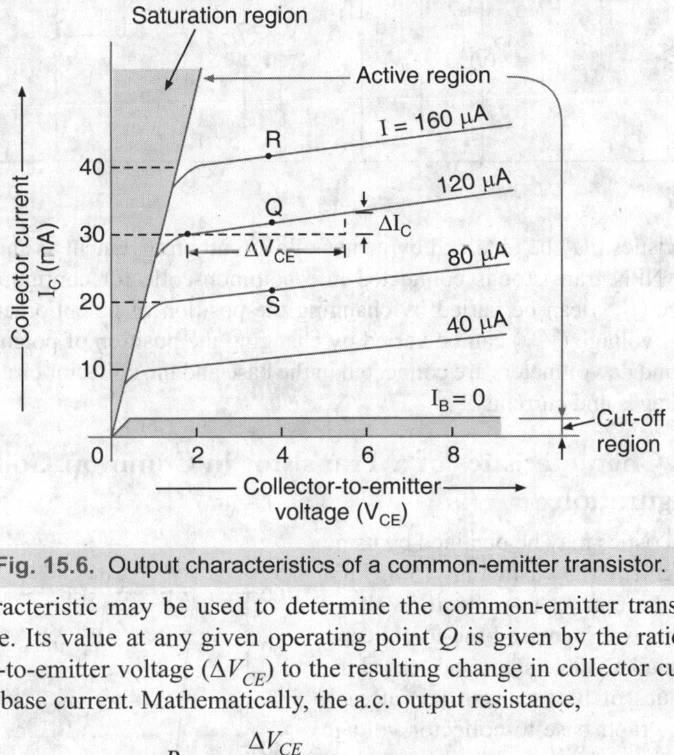

Fig. 15.6. Output characteristics of a common-emitter transistor.

5. The characteristic may be used to determine the common-emitter transistor a.c. output resistance. Its value at any given operating point Q is given by the ratio of a change in collector-to-emitter voltage (ΔV_{CE}) to the resulting change in collector current (ΔI_C) for a constant base current. Mathematically, the a.c. output resistance,

$$R_0 = \frac{\Delta V_{CE}}{\Delta I_C}$$

The common-emitter output resistance of a transistor range from 10 kΩ to 50 kΩ.

6. The characteristic may be used to determine the small-signal common-emitter current gain or a.c. beta (β_0) of a transistor. This can be done by selecting two points R and S on the characteristics and note down the corresponding valued of ΔI_C and ΔI_B. Thus if $\Delta I_C = 40 - 20 = 20$ mA $= 20 \times 10^{-3}$ A and $\Delta I_B = 160 - 80 = 80$ mA $= 80 \times 10^{-6}$ A then small signal common emitter current again,

$$\beta_0 = \frac{\Delta I_C}{\Delta I_B} = \frac{20 \times 10^{-3}}{80 \times 10^{-6}} = 250$$

15.9. Characteristics of a Transistor in Common-Collector Configuration

We have already discussed that the common-collector configuration is treated as a special case of common-emitter configuration. Following are two important characteristics of a transistor in a common-collector (CC) configuration:

1. ***Input Characteristics.*** These curves give the relation between base current (I_B) and base-to-collector voltage (V_{BC}) for a constant emitter-to-collector voltage (V_{EC}).
2. ***Output Characteristics.*** These curves give the relation between the emitter current (I_B) and emitter –to-collector (V_{BC}) voltage for a constant base current (I_B).

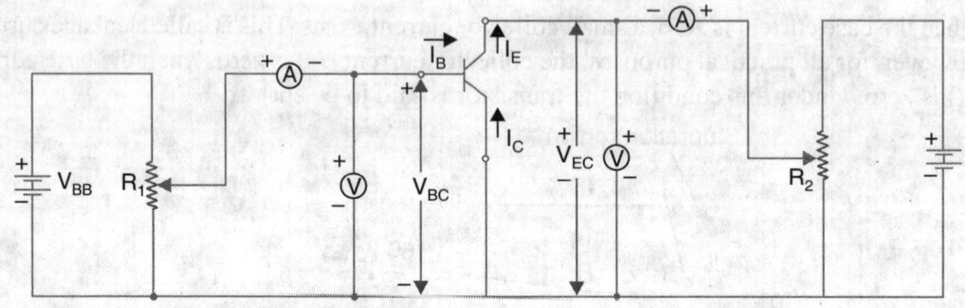

Fig. 15.7

These characteristics may be obtained by means of a circuit arrangement as shown in Fig. 15.7. In this circuit, an NPN transistor is connected in a common-collector configuration. The base-to-collector voltage (V_{BC}) can be varied by changing the position of potentiometer R_1, while the emitter-to-collector voltage (V_{EC}) can be varied by changing the position of potentiometer R_2. The dc milliammeters and dc voltmeters are connected in the base and the collector circuit of a transistor to measure the voltages and currents.

15.10. Input Characteristics of a Transistor in Common-Collector Configuration

These characteristics may be obtained by using the circuit arrangement shown in Fig. 15.8. First of all, we adjust the emitter-to-collector voltage (V_{EC}) to 2 volt. Then we increase base-to-collector voltage (V_{BC}) in small suitable steps and record the corresponding values of base current (I_B) at each step. If we plot a graph base-to-collector voltage (V_{BC}) along the horizontal axis and the base current (I_B) along the vertical axis, we shall obtain a curve marked emitter-to-collector voltage (V_{EC}) = 2 V as shown in Fig15.8. A similar procedure may be used to obtain characteristics at different values of emitter-to-collector voltage (V_{EC}) i.e., V_{EC} = 4 V.

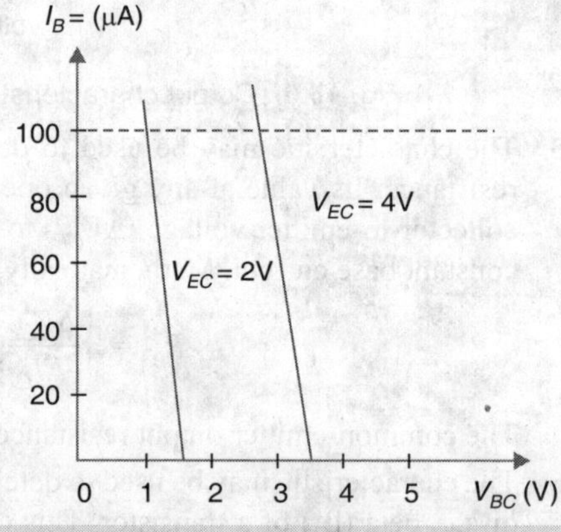

Fig. 15.8. Input characteristics of a common-collector transistor.

15.11. Output Characteristics of a Transistor in Common-Collector Configuration

These characteristics may be obtained by using the circuit shown in Fig. 15.9. To begin with, adjust the base current (I_B) to 20 μA value. Then increase the emitter –to-collector (V_{EC}) in a number of steps and record the corresponding values of emitter current (I_E) at each step. If we plot a graph with emitter –to-collector (V_{EC}) along the horizontal axis and emitter current (I_E) along the vertical axis, we shall obtain a curve marked $I_B = 20$ μA as shown in Fig. 15.9. A similar procedure may be used to obtain characteristics at $I_B = 40$ μA , 60 μA, 80 μA and so on.

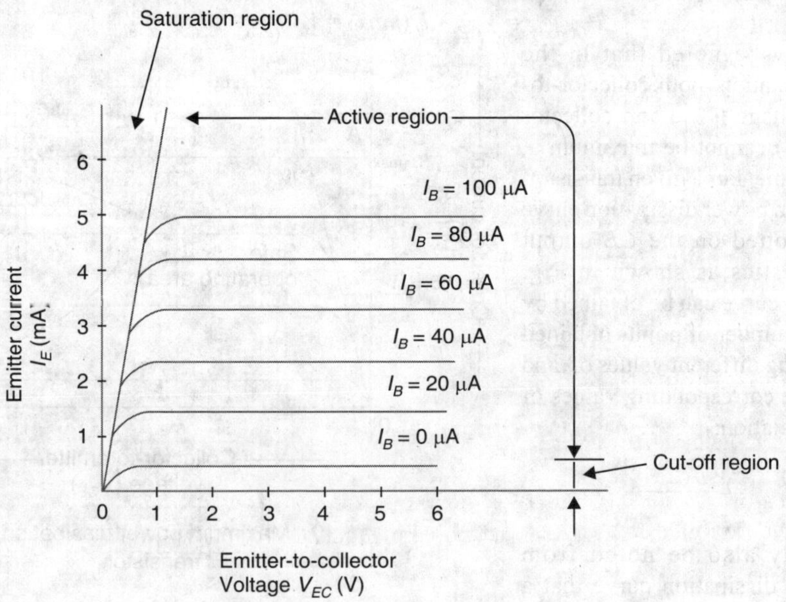

Fig.15.9. Output characteristics of a common-collector transistor.

15.12. Transistor Ratings

A transistor, like any other electronic device, has limitations on its operation. These limitations are stated in terms of maximum ratings. If these ratings are exceeded, it may cause either permanent damage to transistors or temporarily change their operating characteristics. The maximum ratings are normally specified on the manufacturer's data sheet. Following are some of the important maximum ratings of a transistor:

1. Maximum collector current.
2. Maximum power dissipation.
3. Maximum output voltage.

15.13. Maximum Collector Current

This is the absolute maximum value of the current, which a collector can handle safely and is designated by $I_{C(max)}$. This limitation is due to the localized hot spots, which can occur within the transistor due to high current densities at unavoidable current path recstrictions or high resistance points within the transistor. A common source of difficulty is the junction area and the wire bonds that connect the transistor terminals to the external leads. For example, for a general purpose NPN silicon transistor, 2 N 2222 A, $I_{C(max)}$ is 800 mA and for 2 N 3904, $I_{C(max)}$ is 200 mA.

15.14. Maximum Power Dissipation

It is a rating used to indicate the limits on the power handling capability of a transistor. The power dissipation in a transistor is mainly the power dissipated in its junctions. Under d.c. conditions, the junction power is found from the product of junction voltage and junction current. In normal operation, two junctions carry approximately the same current. But the collector-base junction has a much higher voltage across it than emitter-base junction. As a result of this, the power dissipation of a collector-base junction is of more importance. Mathematically, the maximum power dissipation of a transistor is given by the relation,

$$P_{D(max)} = V_{CE} \times I_C(max) = V_{CE(max)} \times I_C$$

It may be noted that in the equation relation, both collector-to-emitter voltage (V_{CE}) and collector current (I_C) cannot be maximum at the same time. For a given transistor, a maximum power dissipation curve can be plotted on the *CE* output characteristics as shown in Fig. 15.10. The curve can be obtained by joining a number of points obtained by assuming different values of and finding the corresponding values of from the relation,

$$V_{CE} = \frac{P_{D\,(max)}}{I_C}$$

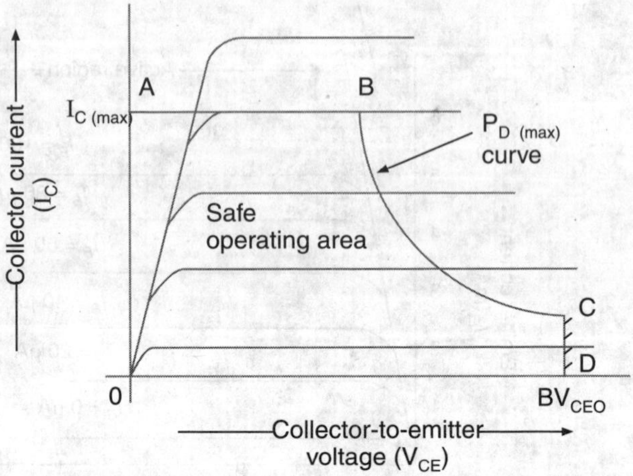

Fig. 15.10. Maximum power dissipation curve for a transistor.

It may also be noted from the power dissipation curve that a transistor cannot be operated in the shaded portion of the graph. Thus the maximum collector current $I_{C(max)}$ is the limiting rating between the points *A* and *B*, is the limiting rating between the points *B* and *C* and break- down voltage from collector-to-emitter with base open (BV_{CEO}) is the limiting rating between the points *C* and *D*. The region inside the curve *ABCD* (or $P_{D(max)}$ curve) is called a safe operating area.

The value of maximum power dissipation $P_{D(max)}$ is, usually, specified at an ambient temperature of 25°C. For example, $P_{D(max)}$ of 2 N 2222 A is 0.5 W at 25°C. For higher temperatures, its value is less. The data sheets often give derating factors for determining the value of $P_{D(max)}$ at any temperature above 25°C. For example, a derating factor of 2 mW / °C means that the maximum power dissipation is reduced 2 mW for each °C increase in temperature. The maximum power dissipation at any temperature (*T*) may be obtained from the relation,

$$P_{D(max)}\,(T) = P_{D(max)} - DF \times (T - 25)$$

where $P_{D(max)}$ = Maximum power dissipation at 25°C,

DF = Derating factor, and

T = Any temperature in °C.

Example 15.1.

A 2 N 2222 A transistor has a $P_{D(max)}$ of 500 mW at 25°C. The derating factor is 2.28 mW/°C. What is the at a temperature of 70°C ?

Solution.

Given: $P_{D(max)}$ = 500 mW; DF = 2.28 mW/°C and T = 70°C.

We know that maximum power dissipation at a temperature of 70°C,

$$P_{D(max)}\,(70) = P_{D(max)} - DF\,(T - 25) = 500 - 2.28 \times (70 - 25)$$

$$= 500 - 102.6 = 397 \text{ mW} = 0.397 \text{ W}$$ **Ans.**

15.15. Maximum Output Voltage

In most of the applications, the collector-base junction is subjected to the highest voltages. Since there is a possibility of voltage breakdown at high voltages, therefore it is necessary to specify the maximum allowable collector-base voltage or collector-emitter voltage or both. The voltage breakdown, which can occur in a transistors is of two types namely avalanche breakdown and punch-through (also called *reach-through*).

1. ***Avalanche breakdown.*** It is a form of voltage breakdown that arises, when the electrons and holes in the semiconductor are accelerated enough by the applied voltage. The accelerated electrons and holes collide with the bound electrons and produce more free electrons. This causes even more collisions due to which current increases to a large value.
2. ***Punch-through.*** It is a form of voltage breakdown that arises due to the increased width of collector-or-base junction depletion region with increased collector-base junction voltage (*i.e.*, early effect). The depletion region at a junction is the region of fixed ions on both sides of the junction. As the applied voltage across the junction increases, the depletion region penetrates deeper into the base. Since the base is very thin, therefore it is possible that at moderate voltages, the depletion region will have spread completely across the base to reach the emitter-base junction. As a result of this, the current increases to a very large value. The punch-through differs from the avalanche breakdown, in the sense, that it takes place at fixed voltage between the collector and base. It is not dependent on the circuit configuration.

The avalanche breakdown in a transistor may take place because of avalanche multiplication of the reverse saturation current (I_{CO}) that crosses the collector junction. The manufacturer, usually, specifies two voltage ratings namely the maximum reverse voltage, (BV_{CEO}) which may be applied between the collector and emitter terminals under the condition that the base lead is open (*i.e.*, when base current, $I_B = 0$) and the maximum reverse voltage (V_{CBO}) which may be applied between the collector and base terminals under the condition that the emitter lead is open (*i.e.*, when emitter current ($I_E = 0$). Usually, the value of BV_{CBO} is double than that of the value of BV_{CEO}. The collector-to-emitter breakdown voltage is affected by the condition of emitter-base circuit. Consequently, additional ratings are given either for a specified resistance connected from base to emitter (designated by BV_{CER}) or for the base shorted to emitter (designated by B_{CEB}). If the base is connected to the emitter through a resistance R, the breakdown voltage will lie between and BV_{CBO}.

Fig. 15.11. shows the *CE* output characteristics extended into the breakdown region. It may be noted that the breakdown voltage decreases for the larger collector currents. The data sheets list the minimum value of BV_{CEO}, which is approximately 40 V as shown in the figure. Transistors are available with breakdown voltage ranging from 20 V to 500 V. In actual practice, the transistor is never operated near the breakdown voltage.

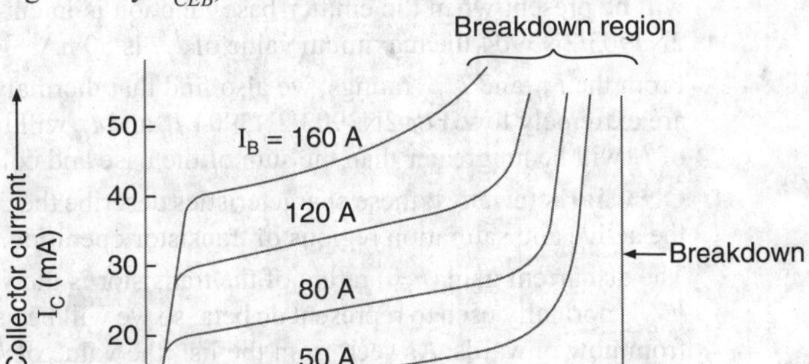

Fig. 15.11. CE output characteristics extended into the breakdown region.

Therefore the collector-to-emitter voltage (V_{CE}) of a transistor is kept well below the breakdown voltage.

15.16. Bipolar Junction Transistor Specification Sheet

As a matter of fact the transistor specification sheet (or datasheet) contains a wide variety of *dc* and *ac* operating characteristics. But in this article we will focus only on some of the commonly used dc maximum ratings and electrical characteristics.

Fig. 15.12 this shows the datasheet for the Motorola 2N3903 and 2N3904 transistors. As seen from this datasheet, we find that the right top corner box shows the transistor numbers, *i.e.*, 2N3903 and 2N3903 and 2N3904. A box below that shows that both these transistors are NPN silicon type are NPN silicon type devices. Both the transistor can be used in switching and amplifier circuits. The box below that shows the transistor package. Besides the three boxes, the specification sheet also shows four lists namely *Maximum Ratings, Thermal Characteristics, OFF Characteristics and ON Characteristics*.

1. **Maximum Ratings :** This list gives us the information about the maximum values of the junction voltage which the transistor can handle safely. From this list we find that, 2N3903/2N3904 has maximum *collector—base voltage* (V_{CB}) of 60 V, the maximum *collector-emitter voltage with base open* (V_{CEO}) of 40 V and maximum *emitter-base voltage* (V_{EB}) of 6.0 V.
2. **Thermal Characteristics:** These are used in circuit development applications and will not be discussed here.
3. **OFF Characteristics:** These characteristics define the operation of the transistor when it is being operated in cut-off. The cut-off means that the collector current through the transistor, I_C is nearly zero. The first three ratings in this list, *i.e.*, BV_{CBO}, BV_{CEO} and BV_{CBO} are the same ratings that appear in the **Maximum Ratings** list. They are repeated here for the convenience of engineer who must rapidly find the ratings.

   The collector cut-off current (I_{CEX}) rating indicates the maximum value of I_C when the device is cut-off. From the list, we find that for 2N3903/2N3904 the maximum value of I_{CEX} is 50 nA. The base cut-off current (I_{BL}) ratings, indicates the maximum amount of base current that will be present when the emitter-base junction is in cut-off. From the list, we find that for 2N3903/2N3904 the maximum value of I_{BL} is 50 nA.

   From the I_{BL} and I_{CEX} ratings, we also find that thermal currents of the transistor at cut-off are extremely low. For 2N3903/2N3904 I_B and I_C will be no greater than 50 nA. the value of I_E will be no greater than the sum of the base and collect currents (*i.e.*, 100 nA).
4. **ON Characteristics:** these characteristics describe the dc operating characteristics for both the active and saturation regions of transistor operation.

   The dc current gain (h_{FE}) rating of the transistor is the value of dc beta. Note that the label h_{FE} is normally used to represent dc beta, so we will be using this label in all our discussions from now onwards. As seen from the list, the value of h_{FE} is measured at different values of I_C. this goes along with the discussion we had in the last article on current gain or beta curve.

   The collector-to-emitter saturation voltage [$V_{CE(sat)}$] rating indicates the maximum value of VCE when the device is in saturation. Recall that saturation means that I_C is maximum. For the 2N3904, the value of [$V_{CE(sat)}$] is 0.2 V when I_C is 10 mA and 0.3 V when I_C is 50 mA. The base-emitter saturation voltage [$V_{BE(sat)}$] ratings indicates the maximum value of V_{BE} when the device is in saturation. As seen from the list, we find that this value is 0.85 or 0.95 depending upon the rated value of I_B.

Note: Specification of another bipolar junction transistor 2N4123 is given in Appendix F.

MAXIMUM RATINGS

Rating	Symbol	Value	Unit
*Collector-Base Voltage	V_{CB}	60	Vdc
*Collector-Emiter Voltage	V_{CEO}	40	Vdc
*Emitter-Base Voltage	V_{EV}	6.0	Vdc
*Collector Current	I_C	200	mAdc
Total Power Dissipation @ T_A = 60°C	P_D	250	mW
**Total Power Dissipation @ T_A = 25°C	P_D	350	mW
Derate above 25°C		2.8	mW/°C
**Total Power Dissipation @ T_C = 25°C	P_D	1.0	Watts
Derate above 25°C		8.0	mW/°C
**Junction Operating Temperature	T_J	150	°C
**Storage Temperature Range	T_{stg}	– 55 to + 150	°C

2N3903
2N3904

NPN SILICON
SWITCHING & AMPLIFIER
TRANSISTORS

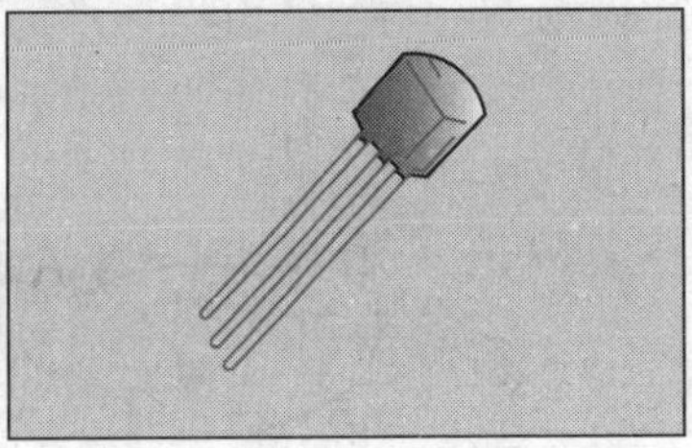

THERMAL CHARACTERISTICS

Characteristics	Symbol	Max	Unit
Thermal Resistance, Junction to Ambient	$R_{\theta JA}$	357	°C/W
Thermal Resistance, Junction to Case	$R_{\theta JC}$	125	°C/W

OFF CHARACTERISTICS (T_A = 25°C unless otherwise noted)

Characteristic	Fig. No.	Symbol	Min	Max	Unit
Collector-Base Breakdown Voltage (I_C = 10 μAdc, I_E = 0)		BV_{CBO}	60	-	Vdc
Collector-Emitter Breakdown Voltage (1) (I_C = 10 mAdc, I_B = 0)		BV_{CEO}	40	-	Vdc
Emitter-Base Breakdown Voltage (I_E = 10 μAdc, I_C = 0)		BV_{EBO}	6.0	-	Vdc
Collector-Cutoff Current (V_{CE} = 30 Vdc, $V_{EB(off)}$ = 0.3 Vdc)		I_{CEX}	-	50	nAdc
Base-Cutoff Current (V_{CE} = 30 Vdc, $V_{EB(off)}$ = 0.3 Vdc)		I_{BL}	-	50	nAdc

ON CHARACTERISTICS

Characteristic		Fig. No.	Symbol	Min	Max	Unit
DC Current Gain (1)		15	h_{FE}			-
(I_C = 0.1 mAdc, V_{CE} = 1.0 Vdc)	2N3903			20	-	
	2N3904			40	-	
(I_C = 1.0 mAdc, V_{CE} = 1.0 Vdc)	2N3923			35	-	
	2N3904		70	-		
(I_C = 10 mAdc, V_{CE} = 1.0 Vdc)	2N3903			50	150	
	2N3904		100	300		
(I_C = 50 mAdc, V_{CE} = 1.0 Vdc)	2N3903			30	-	
	2N3904		60	-		
(I_C = 100 mAdc, V_{CE} = 1.0 Vdc)	2N3903			15	-	
	2N3904		30	-		
Collector-Emitter Saturation Voltage (1)		16, 17	$V_{CE(sat)}$			Vdc
(I_C = 10 mAdc, I_B = 1.0 mAdc)				-	0.2	
(I_C = 50 mAdc, I_B = 5.0 mAdc)				-	0.3	
Base-Emitter Saturation Volatage (1)		17	$V_{CE(sat)}$			Vdc
(I_C = 10 mAdc, I_B = 1.0 mAdc)				0.65	0.85	
(I_C = 50 mAdc, I_B = 5.0 mAdc)				-	0.95	

Fig. 15.12. Bipolar junction transistor specification sheet (Courtery, Motorola Inc.)

15.17. Transistor Testing

You may recall from the discussion in the last chapter that diode can be tested with an ohmmeter. The transistor can be checked in the same basic manner. However, there are many more measurements required to verify whether the transistor under test is good or bad.

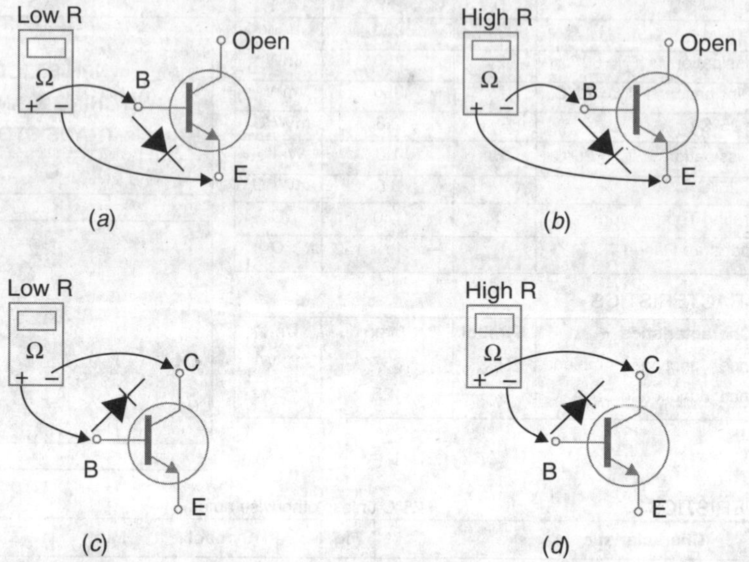

Fig. 15.13.

Fig. 15.13 shows the process for testing the transistor. The emitter-base junction is tested for a low forward resistance and a high reverse resistance. When connected to the ohmmeter as shown in Fig. 15.13 (*a*), the meter should read a low resistance, typically 1 kΩ or less. When connected as shown in Fig. 15.13 (*b*), the resistance, should be high, typically 100 kΩ or higher. If both of these tests show the emitter-base junction to be good, the collector-base junction should be tested in the same manner. These tests illustrated in Fig. 15.13 (*c*) and (*d*) should yield the same results. The forward resistance of the collector-base junction should be 1 kΩ and the reverse resistance should be 100 kΩ or higher.

The last test involves checking the collector-emitter terminals. These terminals should have extremely high resistance regardless of polarity of the meter. If the transistor passes all the resistance checks, it is a good transistor. If it fails any one of the checks mentioned above, transistor if faulty and must be replaced.

Note: The polarities shown in Fig. 15.13 are for an NPN transistor. When testing a PNP transistor, the ohmmeter leads must be reversed.

15.18. Integrated Transistors

Integrated transistors are single devices that contain more than one transistor in integrated from. You will study in chapter on Integrated Circuits (ICs), that such devices contain more than one device in a single casing. Fig. 15.14 (*a*) shows a typical IC transistor casing. It may be noted from the pin listing shows in Fig. 15.14 (*a*) or the internal connection shows in Fig. 15.14 (*b*) that the package contains four individual transistors.

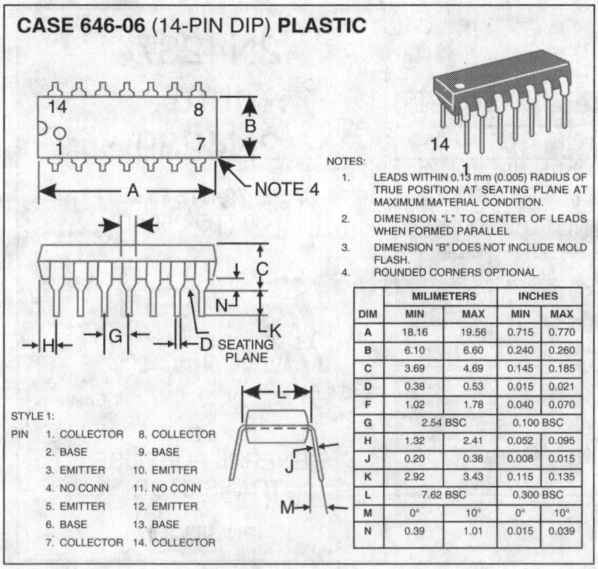

DIM	MILIMETERS MIN	MILIMETERS MAX	INCHES MIN	INCHES MAX
A	18.16	19.56	0.715	0.770
B	6.10	6.60	0.240	0.260
C	3.69	4.69	0.145	0.185
D	0.38	0.53	0.015	0.021
F	1.02	1.78	0.040	0.070
G	2.54 BSC		0.100 BSC	
H	1.32	2.41	0.052	0.095
J	0.20	0.38	0.008	0.015
K	2.92	3.43	0.115	0.135
L	7.62 BSC		0.300 BSC	
M	0°	10°	0°	10°
N	0.39	1.01	0.015	0.039

(a)

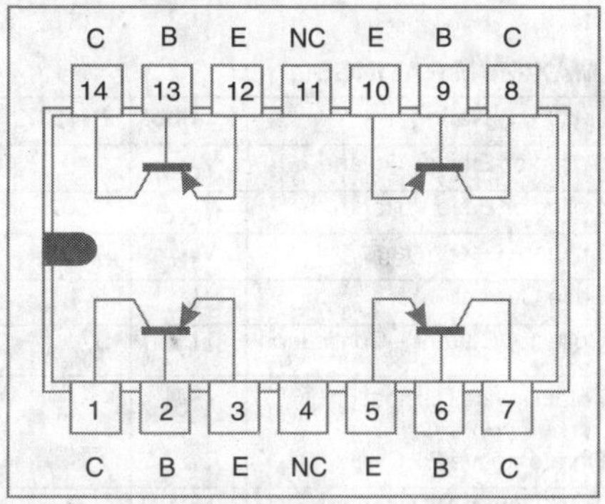

(b)

Fig. 15.14. Integrated transistor (Courtery : Motorola Inc).

15.19. High Voltage Transistors

High-voltage transistors have unusually high reverse breakdown voltage ratings. For example, the maximum ratings for the Motorola BFW 43 transistor are shown in Fig. 15.15 As seen from the "maximum ratings" list, the V_{CEO} and V_{CBO} are much higher than those for the 2N3904 transistor. High-voltage transistors are used in circuits that are extremely high supply voltages, such as television CRT

15.20. High Current Transistors

High-current transistors have very high maximum current ratings. Fig. 15.16 shows the maximum ratings of the Motorola 2N4237 transistor. As seen from the list, this device has an I_C rating of 1A. this is much higher as compared to the maximum ratings of 2N3904 (200 mA only). These devices are used in applications that have high current demands such as current regulator circuits, motor drives etc.

MAXIMUM RATINGS

Rating	Symbol	Value	Unit
Collector-Emitter Voltage	V_{CEO}	150	Vdc
Collector-Base Voltage	V_{CBO}	150	Vdc
Emitter-Base Voltage	V_{EBO}	6.0	Vdc
Collector Current – Continuous	I_C	0.1	Adc
Total Device Dissipation @ T_A = 25°C Derate above 25°C	P_D	0.4 2.66	Watt mW
Total Device Dissipation @ T_C = 25°C Derate above 25°C	P_D	1.4 8.0	Watts mW/°C
Operating and Storage Junction Temperature Range	T_J, T_{stg}	– 65 to + 200	°C

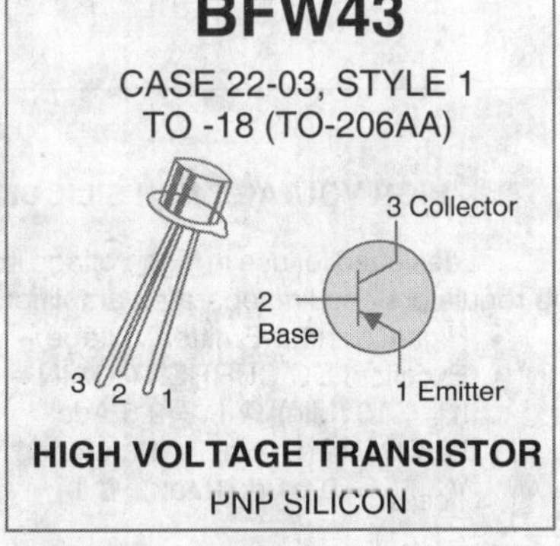

Fig. 15.15. High-Voltage Transistors (Courtery: Motorola Inc.)

MAXIMUM RATINGS

Rating	Symbol	2N4237	2N4238	2N4239	Unit
Collector-Emitter Voltage	V_{CEO}	40	60	80	Vdc
Collector-Base Voltage	V_{CBO}	50	80	100	Vdc
Emitter-Base Voltage	V_{EBO}		6.0		Vdc
Base Current	I_B		500		mA
Collector Current – Continuous	I_C		1.0		Adc
Total Device Dissipation @ $T_A = 25°C$ Derate above 25°C	P_D		1.0 5.3		Watt mW/°C
Total Device Dissipation @ $T_C = 25°C$ Derate above 25°C	P_D		6.0 34		Watts mW/°C
Operating and Storage Junction Temperature Range	T_J, T_{stg}		– 65 to + 200		°C

2N4237
thru
2N4239

CASE 79-04STYLE 1
TO-39 (TO-250AD)

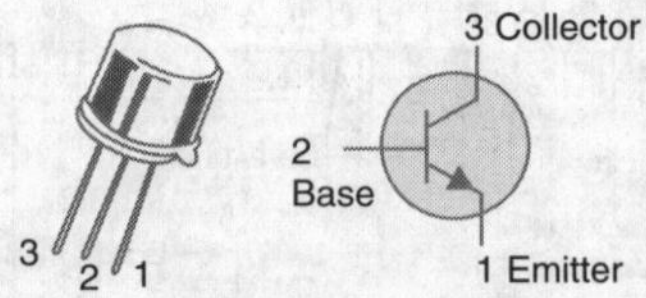

GENERAL PURPOSE
TRANSISTOR

NPN SILICON

Fig. 15.16. High current transistors (Courtery: Motorola Inc.)

15.18. High-Power Transistors

These transistors are designed for use in high-power circuits, such as regulated and switch-mode power supplies. The high-power transistors are capable of dissipating large amount of power. Fig. 15.14 shows the specifications sheets of Motorola 2N3902. As seen from the list, this transistor can dissipate 100 W of power Also notice from the picture that the casing of power transistors is different from the low and medium power transistors. This device has V_{CEO} of 400 V, V_{CBO} of 700 V, V_{EB} of 5.0 V and I_C of 3.5 A.

The high-power transistors are used in regulated power supply applications. The regulated power supply circuits are discussed in Chapter 33.

MOTOROLA
SEMICONDUCTOR
TECHNICAL DATA

NPN
2N3902

HIGH VOLTAGE NPN SILICON TRANSISTORS

. . . designed for use in high-voltage inverters, converters, switching regulators and line operated amplifiers.

- High Collector-Emitter Voltage – V_{CEX} = 700 Vdc
- Excellent DC CURRENT GAIN – h_{FE} = 10 (Min) @ I_C = 2.5 Adc
- Low Collector-Emitter Saturation Voltage – $V_{CE(sat)}$ = 0.8 Vdc (Max). @ I_C = 1.0 Adc

3.5 AMPERE
POWER TRANSISTORS
NPN SILICON
400/VOLTS
100 WATTS

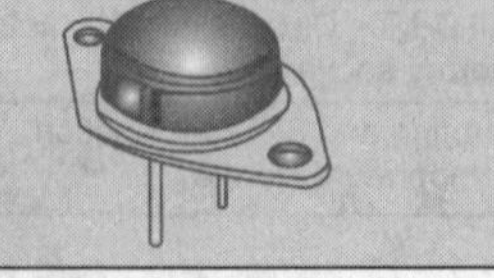

*MAXIMUM RATINGS			
Rating	Symbol	2N3902	Unit
Collector-Emitter Voltage	V_{CEO}	400	Vdc
Collector-Emitter Voltage	V_{CEX}	700	Vdc
Emitter-Base Voltage	V_{EB}	5.0	Vdc
Collector Current – Continuous	I_C	3.5	Adc
Base Current	I_B	2.0	Adc
Total Device Dissipation @ T_C = 75°C Derate above 75°C	P_D	100 1.33	Watts W/°C
Operating Junction Temperature Range	T_J	– 65 to + 150	°C
Storage Temperature Range	T_{stg}	– 65 to + 200	°C

THERMAL CHARACTERISTICS			
Characteristic	Symbol	Max	Unit
Thermal Resistance, Junction to Case	q_{JC}	0.75	°C/W

*Indicates JEDEC Registered Data

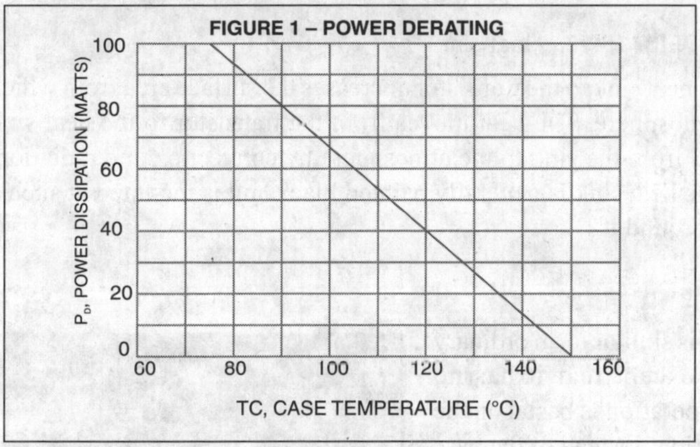

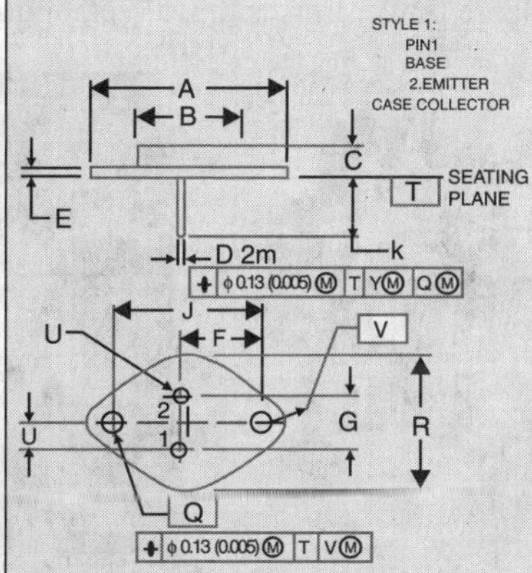

NOTES:
1. DIMENSIONING AND TOLERANCING PER ANSI Y14.5M, 1982.
2. CONTROLLING DIMENSION: INCH.
3. ALL RULES AND NOTES ASSOCIATED WITH REFERENCED TO-204AA OUTLINE SHALL APPLY.

DIM	MILIMETERS		INCHES	
	MIN	MAX	MIN	MAX
A	–	39.37	–	1.550
B	–	21.08	–	0.830
C	6.35	8.25	0.250	0.325
D	0.97	1.09	0.038	0.043
E	1.40	1.77	0.055	0.070
F	30.15 BSC		1.187 BSC	
G	10.92 BSC		0.430 BSC	
H	5.46 BSC		0.215 BSC	
J	16.89 BSC		0.665 BSC	
K	11.18	12.19	0.440	0.480
Q	3.84	4.19	0.151	0.165
R	–	26.67	–	1.050
U	4.83	5.33	0.190	0.210
V	3.84	4.19	0.151	0.165

CASE 1-06
TO-204AA
(TO-3)

Fig. 15.17. Specification sheet of high power transistor 2N3902 (Courtery, Motorola Inc.)

15.22. Thermal Runaway and Heat Sink

We have already discussed that the reverse saturation current in semiconductor devices, changes with temperature. Moreover, the reverse saturation current approximately doubles for every 10°C rise in temperature. This fact may cause considerable practical difficulty in using a transistor as an amplifier. It is because of the fact, that if the temperature of the collector-base junction increases, the leakage current of the transistor increases. And the collector current also increases. The increase in collector current produces an increase in the power dissipation at the collector-base junction. This, in turn, further increases the temperature of the collector-base junction causing the collector current to further increase. This process may become cumulative and it is possible that the ratings of the transistor are exceeded. If it happens, the device gets burnt out. This process is described as the thermal runaway of the transistor. In actual practice, a thermal runaway is avoided by using a stabilization circuitry or a heat sink with the transistor.

A heat sink is a metallic heat-conducting device placed in close contact with a transistor to increase the power dissipation capability of the transistor. The heat sink is, usually, not required for the transistors handling small signals. However, for power transistors, its use is essential. For low-power transistors, a fin-type heat sink is used as shown in Fig. 15.18 (left). On the other hand, for large power transistors, a rectangular heat sink is used as shown in Fig. 15.18 (right).

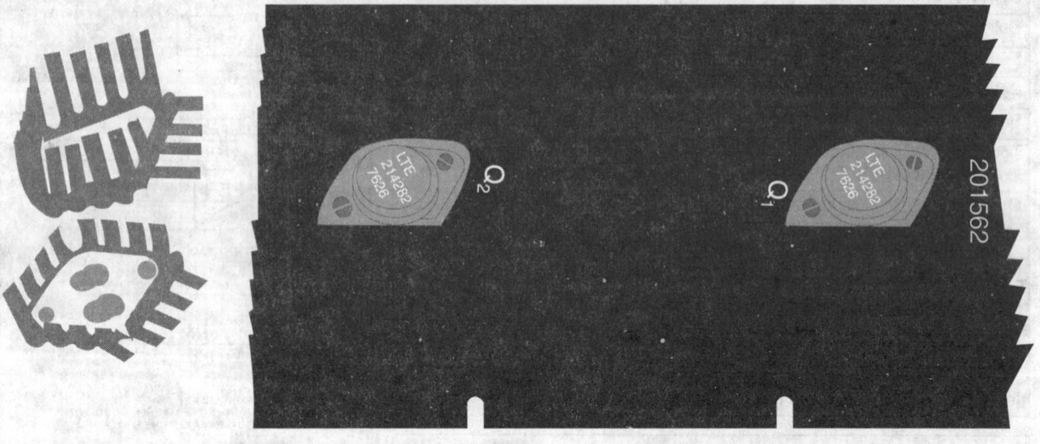

Fig. 15.18. Heat sink.

It may be noted that connecting a heat sink to the transistor, increases the surface area from which the heat is to be dissipated to the atmosphere. The heat moves from the transistor to the heat sink by conduction and then it is removed from the sink to the atmosphere by convection and radiation. In order to radiate maximum heat, the heat sink is generally painted black and is mounted at such a position where the free air can flow around it.

15.23. Phototransistor

It is light-sensitive transistor and is similar to an ordinary bipolar junction transistor (BJT) except that it has no connection to the base terminal. Its operation is based on the photodiode that exists at the *CB* junction. Instead of the base current, the input to the transistor is provided in the form of light as shown in the schematic symbol of Fig. 15.19(*a*).

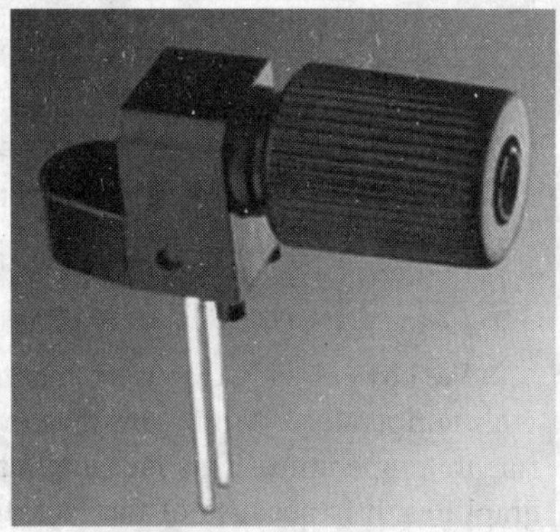

Phototransistor

Silicon NPNs are mostly used as phototransistor. The device is usually packed in a TO-type can with a lens on top although it is sometimes encapsulated in clear plastic. When there is no incident light on the *CB* junction, there is a small thermally-generated collector-to-emitter leakage current I_{CEO} which, in this case, is called dark current and is in the *nA* range.

When light is incident on the *CB* junction, a base current I_λ is produced which is directly proportional to the light intensity. Hence, collector current $I_C = \beta\, I_\lambda$.

Typical collector characteristic curves of a phototransistor are shown in Fig. 15.19 (*b*). Each individual curve corresponds to a certain value of light intensity expressed in mW/cm^2. As seen, I_C increases with light intensity.

The phototransistor has applications similar to those of a photodiode. Their main differences are in the current and response time. The photo-transistor has the advantages of greater sensitivity and current capacity than photodiodes. However, photodiodes are faster of the two, switching in less than a nanosecond.

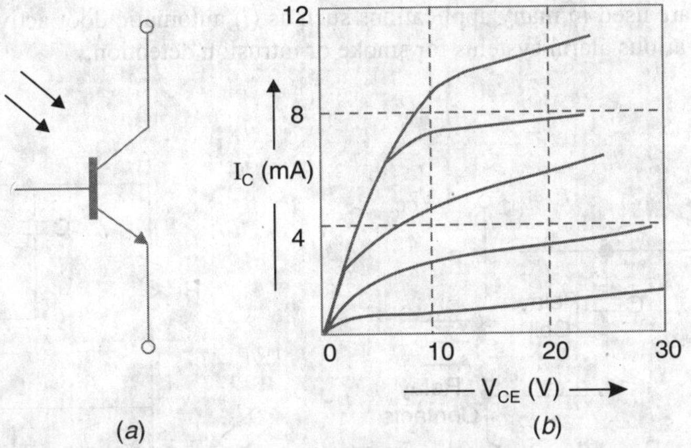

Fig. 15.19.

15.24. Photodarlington

As shown in Fig. 15.20 a photodarlington consists of a phototransistor in a Darlington arrangement with a common transistor. It has a much greater sensitivity to incident radiant energy than a phototransistor because of higher current gain. However, its switching time of 50 ms is much longer than the phototransistor (2 μs) or the photodiode (1 ns). Its circuit symbol is shown in Fig. 15.20.

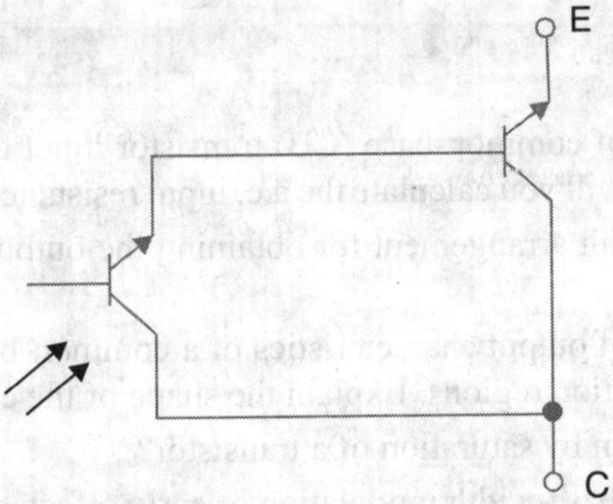

Fig. 15.20.

Photodarlington are used in a variety of applications some of which are given below.

A light-operated relay is shown in Fig. 15.21 (*a*). The phototransistor Q_1 drives the bipolar transistor Q_2. When sufficient light falls on Q_1, it is driven into saturation so that I_C is increased manifold. This collector current while passing through the realy coil energizes the relay.

Fig. 15.21 (*b*) shows a dark-operated relay circuit *i.e.*, one in which relay is deenergized when light falls on the phototransistor. Here, Q_1 and R form a potential divider across V_{CC}. With insufficient light incident on Q_1, transistor Q_2 is biased ON thereby keeping the relay energized. However, when there is sufficient light, Q_1 turns ON and pulls the base of Q_2 low thereby turning Q_2 OFF and hence, deenergizing the relay.

Such relays are used in many applications such as (*i*) automatic door activators, (*ii*) process counters and (*iii*) various alarm systems for smoke or intrusion detection.

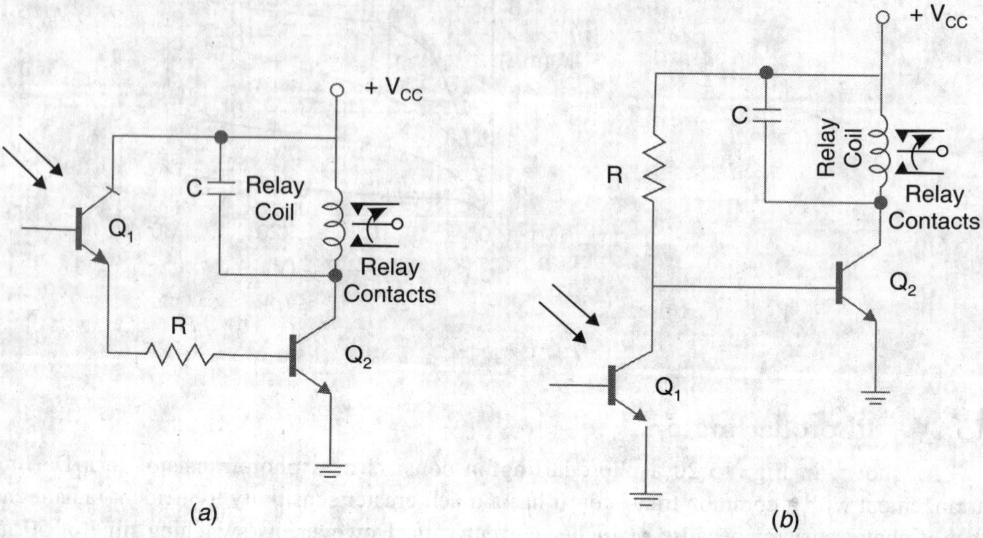

Fig. 15.21.

DESCRIPTIVE QUESTIONS

1. Sketch a family of common-base (CB) transistor input characteristics. Label all variables and explain how will you calculate the a.c. input resistance of a transistor from these curves.
2. Describe the circuit arrangement for obtaining the output characteristics of common-base transistor.
3. Sketch a family of output characteristics of a common-base transistor. Indicate the active, cut-off and saturation regions. Explain the shape of the curves quantitatively.
4. What do you mean by saturation of a transistor?
5. Explain the term base-width modulation or early effect.
6. Give the circuit arrangement for obtaining the input and output characteristics of transistor in common-emitter configuration.
7. Sketch the V-I characteristics of an NPN transistors in common emitter operation and indicate the different regions of importance.
8. Explain theterms (*a*) Maximum collector current (*b*) Maximum power dissipation and (*c*) Maximum output voltage.
9. Draw a derating curve (*i.e.*, power dissipation curve) for a power transistor.
10. Discuss thermal runaway. How it can be prevented in a high power transistor?
11. What is a hat-sink ?
12. What is a heat-sink compound ?
13. Compare the characteristics of CB, CE and CC configurations.

(*M.G University., Aug. 2007, PTU., Dec 2006*)

14. Write two distinct applications each of CB and CC configurations. *(M.G University., May 2005)*
15. Sketch the common emitter configuration. *(M.G University., April/May 2004)*
16. Sketch and explain the output characteristics of a PNP transistor. *(M.G University., April/May 2004)*
17. Sketch and neatly label the base and collector characteristics of the CE. *(M.G University., May/June 2006)*
18. Draw the output characteristic curve of NPN transistor in CE configuration level the parameter and indicate different regions of operations. *(GBTU/MTU., 2006-07)*
19. Show the current components in PNP transistor, when biased in active region. *(Nagpur University., 2008)*
20. Draw the circuit diagram of transistor in the CE configuration. Sketch the output characteristics and indicate the active, saturation and cut off region. *(GBTU/MTU., 2005-06, 2004-05, 2002-03)*
21. Draw the output characteristic for a common base transistor amplifier. Mark and define the various region of operations. *(Jamia University., 2008, GBTU/MTU 2002-03)*
22. Draw the input and output characteristics of transistor in common emitter configuration. How h-parameter can be obtained from these characteristics? *(GBTU/MTU., 2009, 05-06, 04-05, 01-02)*
23. Draw the Common-Basc configuration of transistor and explain its input and output characteristics. *(Calicut University., June 2007)*
24. For common emitter-bipolar junction transistor configuration, analyze the input and output characteristics. Also, give the inferences and necessary analytical expression for the same. *(Anna University., Nov/Dec 2007)*
25. Draw the input and output characteristics of a transistor in CE configuration and mark the cutoff, saturation and active regions. *(Anna University., April/May 2008)*
26. Draw and explain the input and output characteristics of a common base configuration. *(Anna University., Nov/Dec 2007)*
27. Which of the BJT configuration is suitable for impedance matching applications? Why? *(Anna University., Nov/Dec 2007)*
28. With necessary diagram explain the input & output characteristics of common emitter configuration. *(JNTU., June 2009)*
29. Draw the output characteristic of CE configuration showing different operating regions along with necessary conditions. *(PBUT., 2008)*
30. Explain active region, saturation region and cut-off region in transistor characteristics. *(JNTU., Aug./Sep. 2006)*
31. Explain the input and output characteristics of a transistor in CB Configuration. *(JNTU., May/Jun. 2006, GBTU/MTU 2004-05.)*
32. Indicate the various regions of operation on the output characteristics. *(JNTU., May./Jun. 2006)*
33. Why does the CE configuration provide large current amplification while CB does not? *(JNTU., Aug./Sep. 2006)*
34. Explain the input and output characteristics of the transistor in CC configuration with diagrams. What is the inference form these characteristics? *(JNTU., Aug./Sep. 2006)*
35. With the necessary diagram explain the input & output characteristics of common emitter configuration. *(JNTU., June 2009)*

36. Draw circuit diagram for CC and CE configuration of transistor. Which is having higher gain and why. *(PTU., Dec 2007)*
37. Compare three configuration CB, CC and CE in terms their of output resistances. *(PTU., May 2008)*
38. Draw the input and output characteristic curves of a CB configuration for a transistor. Explain the shape of curves. Clearly indicate cut off, active and saturation regions. *(PTU., May 2008)*
39. Explain with equivalent circuit working of Photo transistor. *(PTU., May 2009)*
40. Define Heat sink. *(Anna University., April/May 2008)*
41. Draw the common emitter configuration for BJT transistor and draw for input and output characteristics. Discuss the different regions of operation. *(R.G.P.V., June 2011)*
42. Explain the early effect in BJT with load the aid of necessary plots. *(R.G.P.V., Dec 2005)*
43. Explain what is early effect? *(R.G.P.V., June 2009, 2010, Dec. 2010)*
44. Explain the working of CB configuration of transistor supported by input, output characteristics and current equations. *(R.G.P.V., Feb. 2010)*
45. Draw and explain CB, CE and CC configuration input and output characteristics including active, cut-off and saturation region. *(R.G.P.V., Dec. 2008)*
46. Draw CE configuration of BJT amplifier. Give V-I characteristics of this CE configuration and explain the same. *(R.G.P.V., Dec. 2008)*
47. Explain the characteristics and application of phototransistor. *(R.G.P.V., June 2009)*
48. How thermal runway is initiated in BJT ? How it can be avoided? Is it possible in JFET ? *(R.G.P.V., June 2011)*
49. Draw the common emitter output characteristics of a transistor and show the cut-off, active and saturation region on it. *(R.G.P.V., June 2003)*
50. Discuss thermal runaway. *(R.G.P.V., Dec. 2002)*
51. What is a heat sink? How does it contribute to increasing power dissipation as far as thermal resistance is concerned ? *(R.G.P.V., Dec. 2006)*
52. Explain the input and output characteristics of a transistor in CB configuration. *(R.G.P.V., June 2007)*
53. Draw and explain the input and output characteristics for CE configuration. *(Nagpur University, Summer 2006)*
54. Compare CC, CE and CB configuration of BJT. With reference give the application areas of the configuration. *(Nagpur University, Summer 2004)*
55. What are the maximum ratings of a transistor? Explain in short. *(Nagpur University, Winter 2003)*
56. Discuss the differences between various configurations of transistor and explain why CE configuration is useful?
57. Draw power derating curve for a power transistor and explain its use. *(Nagpur University, Winter 2001)*
58. Draw the input-output characteristics of CE transistor configuration and explain. *(Nagpur University, Summer 2007, 2004)*
59. What is early effect? Discuss the impact of early effect on input and output characteristics of CE amplifier. *(Nagpur University, Winter 2005)*
60. Explain Base width modulation in a BJT. *(Nagpur University, Summer 2007)*
61. What is Base modulation? *(Nagpur University, Summer 2008, 2006, 2004)*
62. Explain base-width modulation with the help of plots of potential and minority concentration. *(Nagpur University, Winter 2004)*

63. Explain Early Effect, Discuss its consequences. *(Nagpur University, Summer 2003)*
64. What are the three transistor configurations? Compare and contrast the characteristics of these configurations. Also state any one application of each of these configuration. *(RGTU., Dec. 2010)*
65. Draw and explain input and output characteristics of common emitter configuration. *(RGTU., June 2011)*
66. Draw the circuit diagram to obtain the characteristics of BJT in CE configuration. Draw output characteristics curves and indicate the different regions of operation. *(RGTU., June 2011)*
67. Discuss the characteristics of BJT. *(RGTU., June 2009)*
68. Explain the working of BJT with diagram. Discuss its VF characteristics. *(RGTU., June 2010)*
69. Discuss the operation characteristics and possible applications of a bipolar junction transistor (BJT) clearly distinguishing between cut-off linear and saturation regions. *(RGTU., June 2011)*
70. What is thermal runaway? *(RGTU., June 2011)*
71. Explain the output characteristic of NPN transistor in CE configuration. Also indicate different regions. *(GTU., Dec. 2011)*

MULTIPLE CHOICE QUESTIONS

1. The "cut-in" voltage of a silicon small-signal transistor is
   (*a*) 0 V (*b*) 0.2 V (*c*) 0.5 V (*d*) 0.8 V
2. When the collector junction in transistors is biased in the reverse direction and the emitter junction in the forward direction, the transistor is said to be in the
   (*a*) active region (*b*) cut-off region (*c*) saturation region (*d*) none of them.
3. The transistor is said to be in saturation region when
   (*a*) both collector and emitter junctions are forward biased
   (*b*) both collector and emitter junctions are reversed biased
   (*c*) emitter junction is forward biased, but the collector junction is reverse biased
   (*d*) emitter junction is reverse biased, but the collector junction is forward biased.
4. For a silicon transistor in the common emitter configuration the cut-off condition is achieved by applying a minimum reverse voltage across the emitter junction of the order of
   (*a*) 0 V (*b*) 0.7 V (*c*) 1.5 V (*d*) 5 V
5. A transistor connected in common-base configuration has
   (*a*) a high input resistance and a low output resistance.
   (*b*) a low input resistance and high output resistance
   (*c*) a low input resistance and a low output resistance
   (*d*) a high input resistance and a high output resistance

ANSWERS

1. (*c*) 2. (*a*) 3. (*a*) 4. (*a*) 5. (*b*)

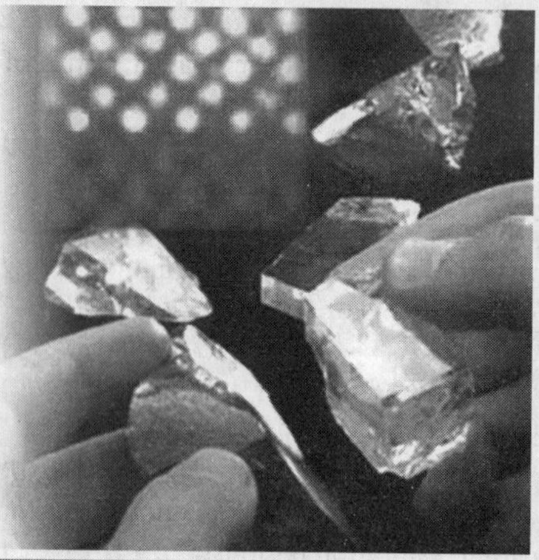

16

Chapter

Field-Effect Transistors

Outline

16.1. Introduction

The field-effect transistor (FET) is a transistor that relies on electric field to control the shape and hence conductivity of a (N-or P-channel) in a semiconductor. FETs are sometimes called unipolar transistors in order to differentiate them from bipolar transistors.

The field-effect transistor (FET) is a three terminal device used for a variety of applications that match, to a large extent, those of the BJT described in Chapter 14. Although there are lost of difference between the two types of devices, there are also many similarities that will be pointed out later.

The primary difference between the two types of transistors is highlighted in Fig. 16.1. It may be carefully noted that the BJT transistor is a *current-controlled* device (Fig. 16.1 (*a*)), while the JFET transistor is a *voltage-controlled* device (Fig. 16.1 (*b*)). In other words, the current I_C in Fig. 16.1 (*a*) is adirect function of the level of I_B. For the FET, the current I will be a function of the voltage VGS applied to the input circuit as shown in Fig. 16.1 (*b*). In each case the current of the output circuit is being controlled by a parameter of the input circuit-in one case a current level and in the other an applied voltage.

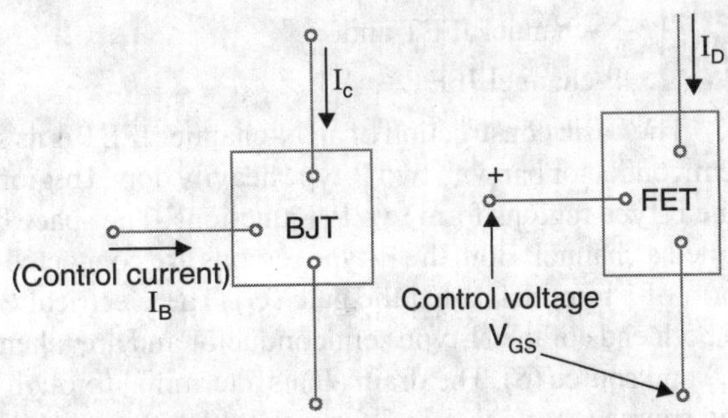

Fig. 16.1.

Field effect transistors offer many attractive feature for applications in analog switching, high input impendance amplifiers, microwave amplifications and integrated circuits,, The FETs have considerably higher input impedance than bipolar transistors which allows the input of a FET to be more readily matched to the standard microwave system. The FET has a negative temperature at high current lives; that is the current decreases as temperature increases. This characteristic leads to a more uniform temperature distribution over the device area and prevents the FET from thermal runway or second breakdown which occurs in the bipolar transistor. The device is thermally stable, even when the active area is large or when many devices are connected in parallel.

Because FETs are unipolar devices, they do not suffer from minority carrier storage effects and consequently have higher switching speeds and higher cut-off frequencies. In addition the devices are basically square-law or linear devices; inter-modulation and cross-modulation products are much smaller.

16.2. Types of Field-effect Transistor

Broadly speaking there are two main types of field effect transistors :

1. Junction field-effect transistor (JFET).
2. Metal oxide semiconductor field-effect transistor (MOSFET) or insulated gate field-effect transistor (IGFET).

Both of these types of FET's can be further sub-divided as shown in Fig. 16.2.

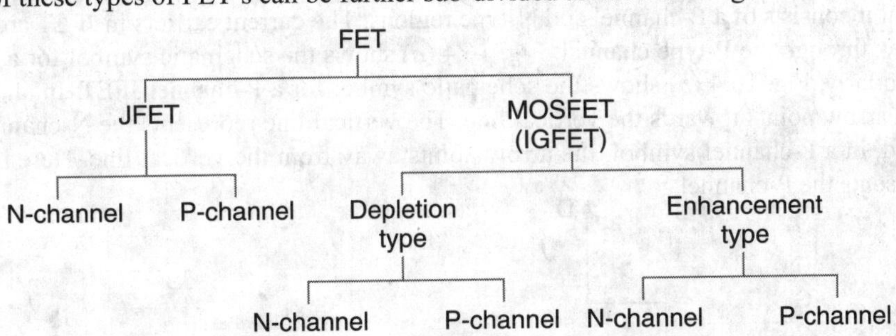

Fig. 16.2.

Now we shall discuss the construction, operation and characteristic of all the above mentioned field-effect transistors in the following pages.

16.3. Junction Field-Effect Transistor

The junction field-effect transistors (JFET's) can be divided depending upon their structure into the following two categories:

1. N-channel JFET and
2. P-channel JFET.

The basic construction of an N-channel JFET is as shown in Fig. 16.3 (*a*). It consists of an N-type semiconductor bar with two P-type heavily doped regions diffused on opposite sides of its middle part. The P-type regions form two PN junctions. The space between the junctions (*i.e.*, N-type regions) is called a channel. Both the P-type regions are connected internally and a single wire is taken out in the form of a terminal called the gate (*G*). The electrical connections (called ohmic contacts) are made to both ends of the N-type semiconductor and are taken out in the form of two terminals called drain (*D*) and source (*S*). The drain (*D*) is a terminal through which electrons leave the semiconductor bar and source (*S*) is a terminal through which the electrons enter the semiconductor.

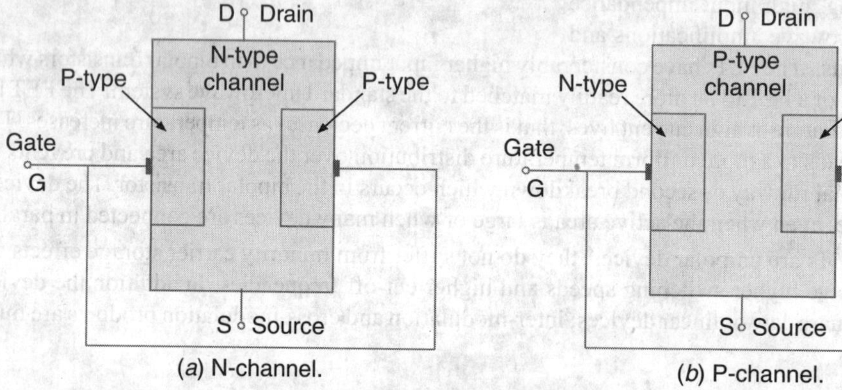

Fig. 16.3. JFET's.

Whenever a voltage is applied across the drain and source terminals, a current flows through the N-channel. The current consists of only one type of carriers (*i.e.*, electrons) Therefore the field-effect transistor (FET) is called a unipolar device. This distinguishes an FET from a BJT (*i.e.*, a bipolar junction transistor) where the current consists of the flow of both the electrons and holes.

A P-channel JFET is shown in Fig. 16.3 (*b*). Its construction is similar to that of N-channel JFET, except that it consists of a P-channel and N-type regions. The current carriers in JFET are the holes, which flow through the P-type channel. Fig. 16.4 (*a*) shows the schematic symbol for a N-channel JFET. Similarly, Fig. 16.4 (*b*) shows the schematic symbol for a P-channel JFET. In an N-channel JFET, the arrow points towards the vertical line. The vertical line represents the N-channel. On the other hand, in a P-channel symbol, the arrow points away from the vertical line. Here the vertical line represents the P-channel.

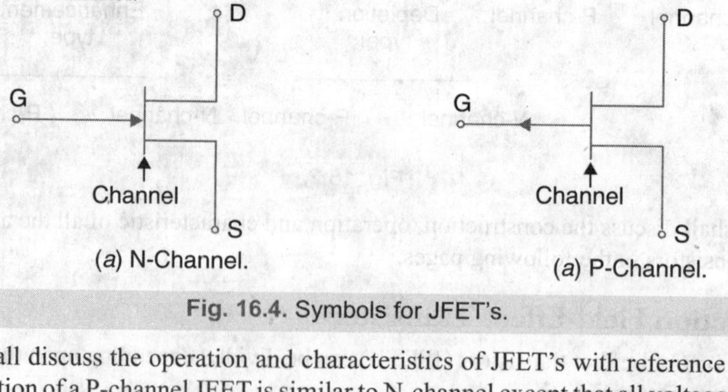

Fig. 16.4. Symbols for JFET's.

Now we shall discuss the operation and characteristics of JFET's with reference to N-channel JFET. The operation of a P-channel JFET is similar to N-channel except that all voltages and currents are reversed.

16.4. Formation of Depletion Region in JFET

Consider an N-channel JFET as shown in Fig. 16.5 (*a*). Here the P-type gate and N-type channel constitute PN junction. This PN junction is always reverse biased in JFET operation. The reverse bias is applied by a voltage V_{GG} connected between the gate and source terminals as shown in Fig. 16.5 (*b*). It may be noted that positive terminal of the voltage (V_{GG}) is connected to source (*S*) and negative terminal to gate (*G*).

We know that whenever a PN junction is reverse biased, the electrons and holes diffuse across the junction and leave behind the positive ions on N-side and negative ions on P-side. The region, containing these immobile ions, is known as depletion region. As the reverse bias voltage across the junction is increased, thickness of the depletion region is also increased. If both the P-side and N-side of the junctions are equally doped, the depletion region will extend equally in both the regions. However, if one side of the junction is heavily doped, as compared to the other side, the depletion region extends more into the region of lower doping.

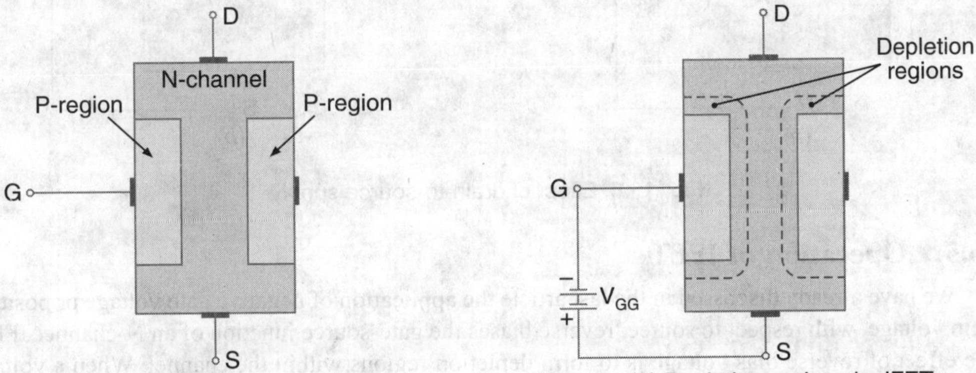

(*a*) N-channel JFET. (*b*) Depletion regions in JFET.

Fig. 16.5.

A similar action takes place in JFET. Since the *P-region of a N-channel JFET is heavily doped, as compared to the N-channel, the depletion region extends less into the P-region and deeper into the N-channel* as shown in Fig. 16.4 (*b*). Incidentally, when there is no applied voltage between the gate and source, the depletion region is symmetrical around the junction. The conductivity of depletion region is zero because there are no mobile charge carries in this region. Hence the effective width of the N-channel is reduced. It further reduces with the increased reverse-bias voltage applied across the gate and source terminals of the JFET.

The reverse-bias across the gate source junction of a JFET may also be achieved by applying a voltage across the drain and source terminals as shown in Fig. 16.5 (*a*). It may be noted that drain (*D*) is connected to the position terminal of the dc supply (V_{DD}) and source (*S*) is connected to the negative terminal.

Now we shall discuss as to how the drain-to-source voltage produces a reverse-bias across the gate-source junction in the same way as the gate-to-source voltage. In order to do so, we represent the channel resistances r_a and r_b as shown in the Fig. 16.6 (*a*). These resistances are shown variable, as their values depends upon the magnitude of the drain-to-source voltage (V_{DS}) and gate-to-source voltage (V_{GS}). In the presence of positive supply voltage V_{DD} (with gate open), the electrons flow from source to drain through the N-channel and constitutes a current known as drain current (I_D). The conventional direction of drain current is indicated from drain-to-source through the device. The drain current causes a voltage drop across the resistance r_b, which has the effect of reverse biasing the gate-to-source junction. Thus even if the gate is open, the gate-to-source junction is reverse biased by the drain-to-source voltage. It creates depletion regions within the channel as shown in Fig. 16.6 (*b*).

It may be noted from Fig. 16.6 (*b*) that depletion region is not symmetrical around the gate-to-source junction. As a matter of fact, it is extended more deeper into the channel near the drain terminal and less on the source terminal. It is because of the fact, *that voltage drop across r_a is greater than that across r_b. Due to this, reverse bias voltage is higher near the drain end of the channel as compared to the source end.*

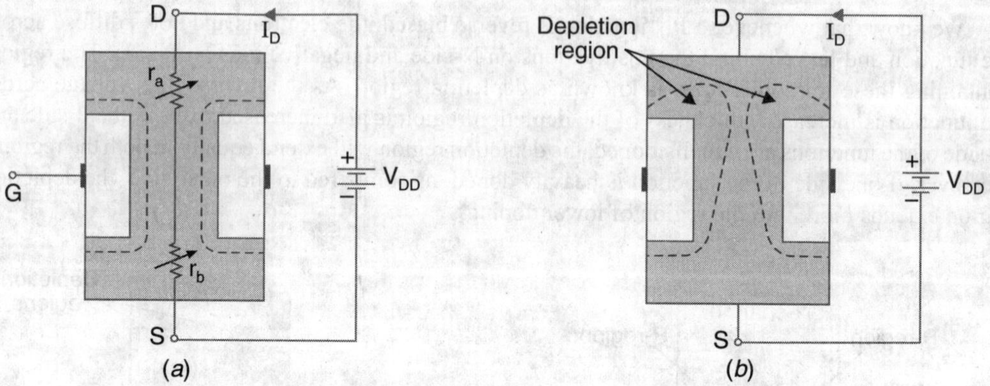

Fig. 16.6. Effect of drain-to-source supply.

16.5. Operation of JFET

We have already discussed in the last article the application of negative gate voltage or positive drain voltage, with respect to source, reverse biases the gate-source junction of an N-channel JFET. The effect of reverse bias voltage is to form depletion regions within the channel. When a voltage is applied between the drain and source with a dc supply voltage (V_{DD}), the electrons flows from source to drain through the narrow channel existing between the depletion regions. This constitutes the drain current (I_D) and its conventional direction is indicated from drain-to-source. The value of drain current is maximum, when no external voltage is applied between the gate and source and is designated by the symbol I_{DSS}.

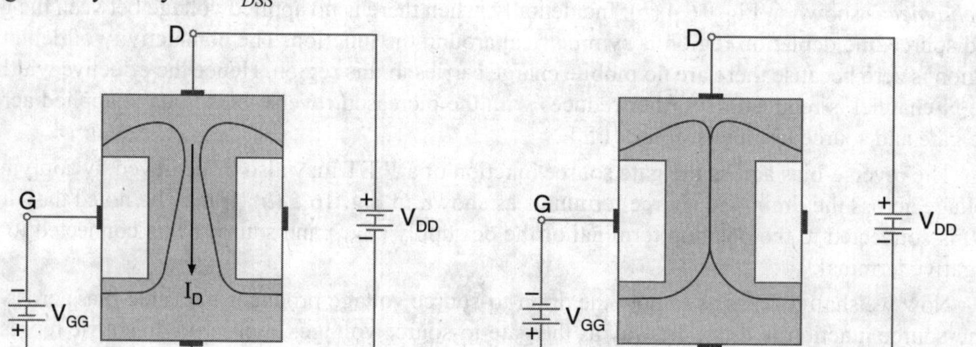

Fig. 16.7. Operation of JFET.

When the gate-to-source voltage (V_{GS}), if applied by a dc supply, (V_{GG}) and increased above zero as shown in Fig. 16.7 (*a*), the reverse-bias voltage across the gate-source junction is increased. As a result of this, the depletion regions are widened. This reduces the effective width of the channel and therefore controls the flow of drain current through the channel. When the gate-to-source voltage (V_{GS}) is increased further, a stage is reached at which two depletion regions touch each other as shown in Fig. 16.7 (*b*). At this gate-to-source voltage, the channel is completely blocked or pinched off and drain current is reduced to zero. The gate-to-source voltage (V_{GS}) at which the drain current is zero (or completely cut-off) is called pinch off voltage. It is designated by the symbol V_P or V_{GS} (off). The value of pinch off voltage V_P is negative for N-channel JFET's. It depends on (*i*) doping of the N and P regions of the device and (*ii*) width of the original channel structure.

The operation of P-channel JFET is exactly similar to N-channel JFET, except that current carriers are holes and polarities of the dc supply, voltage V_{GG} and V_{DD} are reversed.

16.6. Characteristics of JFET

We know that a family (or a set) of curves which relate device current and voltages are known as characteristic curves. Followings are the two important characteristics of a JFET:

1. ***V-I or drain characteristics.*** These curves give relationship between the drain current (I_D) and drain-to-source voltage (V_{DS}) for different values of gate-to-source voltage (V_{GS}).
2. ***Transfer characteristics.*** These curves give relationship between drain current (I_D) and gate-to-source voltage (V_{GS}) for different values of drain-to-source (V_{DS}) voltage.

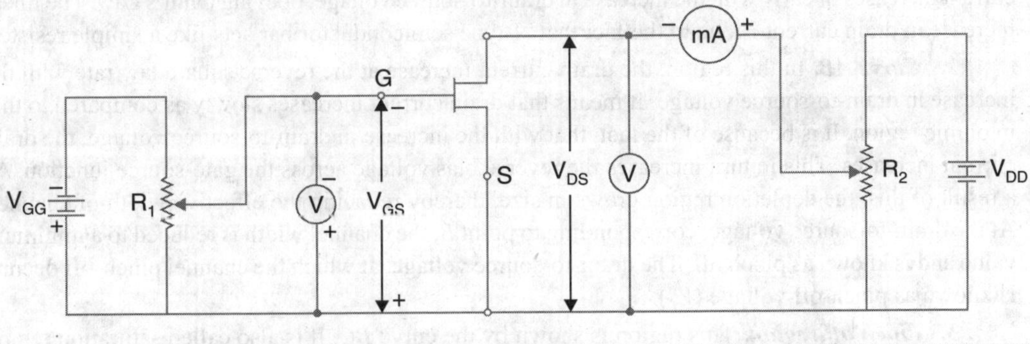

Fig. 16.8. Circuit arrangement for plotting JFET characteristics.

The drain and transfer characteristics of JFET may be obtained by using an N-channel JFET connected in the common source mode as shown in Fig. 16.8. Here the potentiometers R_1 and R_2 are used to vary the voltages V_{GS} and V_{DS} respectively. The voltages V_{DS} and V_{GS} may be measured by the voltmeters connected across the JFET terminals. The drain current (I_D) can be measured by the milliammeter connected in series with the JFET and the supply voltage V_{DD}.

16.7. Drain Characteristics

These curves may be obtained by using the circuit arrangement shown in Fig. 16.8. First of all, we adjust the gate-to-source voltage (V_{GS}) to zero volt. Then increase the drain-to-source voltage (V_{DS}) in small suitable steps and record the corresponding values of drain current (I_D) at each step. Now if we plot a graph with drain-to-source voltage along the horizontal axis and drain current along the vertical axis, we shall obtain a curve marked $V_{GS} = 0$ as shown in Fig. 16.9. A similar procedure may be used to obtain curves for different values of gate-to-source voltage *i.e.*, V_{GS} = 1, 2, 3 and 4 V.

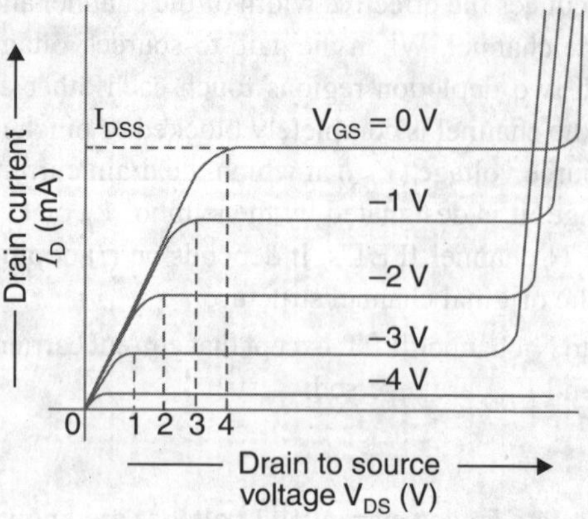

Fig. 16.9. Drain characteristics of JFET.

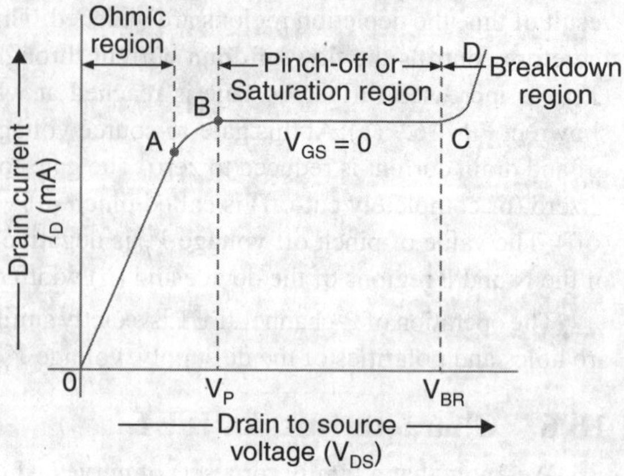

Fig. 16.10. Drain characteristics with V_{GS} = 0 volt.

In order to explain, the typical shape of drain characteristics, let us select the curve with V_{GS} = 0 volt as shown in Fig. 16.10. The curve may be sub-divided into the following regions:

1. ***Ohmic region.*** This region is shown as a curve *OA* in the figure. In this region, the drain current increases linearly with the increase in drain-to-source voltage, obeying Ohm's Law. The linear increase in drain current is due to the fact that N-type semiconductor bar acts like a simple resistor.

2. ***Curve AB.*** In this region, the drain current increase at the reverse square law rate with the increase in drain-to-source voltage. It means that drain current increases slowly as compared to that in ohmic region. It is because of the fact, that with the increase in drain-to-source voltage, the drain current increases. This in turn increases the reverse bias voltage across the gate-source junction. As a result of this, the depletion region grows in size, thereby reducing the effective width of channel. At the drain-to-source voltage, corresponding to point *B*, the channel width is reduced to a minimum value and is known as pinch off. The drain-to-source voltage, at which the channel pinch-off occurs, is known as pinch-off voltage (V_P).

3. ***Pinch off region.*** This region is shown by the curve *BC*. It is also called saturation region or constant current region. In this region, the drain current remains constant at its maximum value (*i.e.*, I_{DSS}). The drain, current in the pinch off region, depends upon the gate-to-source voltage and is given by the relation,

$$I_D = I_{DSS}\left(1 - \frac{V_{GS}}{V_P}\right)^2$$

The above relation is known as *Shockly's equation.* The pinch off region is the normal operating region of JFET, when used as an amplifier.

4. ***Breakdown region.*** This region is shown by the curve *CD*. In this region, the drain current increases rapidly as the drain-to-source voltage is also increased. It happens because of the breakdown of gate-to-source junction due to avalanche effect. The drain-to-source voltage corresponding to point *C* is called breakdown voltage.

16.8. Effect of Gate-to-Source Voltage on Drain Characteristics

We have already discussed in the last article about the drain characteristics of JFET. The effect of increasing gate-to-source voltage on the drain characteristics may be noted from Fig. 16.9. As the gate-to-source voltage (V_{GS}) is increased above zero to the 1, 2, 3 and 4 volts, following two effects take place:

1. The value of pinch off voltage is reached at a smaller value of drain current as compared to that when $V_{GS} = 0$, and
2. The value of drain-to-source voltage (V_{DS}) is decreased as compared to that when $V_{GS} = 0$.

The first effect may be understood from the fact that reverse bias voltage across the gate-source junction of JFET is the sum of gate-to-source voltage (V_{GS}) and drain-to-source voltage (V_{DS}). Thus for fixed V_{DS}, the increased value of V_{GS} will increase the reverse bias across the junction. For example, if V_{GS} of – 1V is applied between the gate and source of JFE, then the gate-to-source junction becomes reverse biased, even when the drain current is zero. This causes the depletion regions to be formed within the channel. As the voltage V_{DS} is increased above zero, the drain current increases. This produces voltage drop across channel resistance, which in turn produces reverse bias voltage across the gate-source junction. The amount of reverse bias voltage required to be produced by drain current would obviously be decreased by one volt. In other words, a smaller voltage drop along the channel (*i.e.*, smaller than that when $V_{GS} = 0$) will increase the depletion regions to the point where they will pinch off the drain current. As a result of this, the pinch-off voltage (V_P) is reached at a lower value of drain current than that when $V_{GS} = 0$.

The second effect is caused due to the fact that the increased value of gate-to-source voltage (V_{GS}) keeps adding to the reverse bias at the junction produced by the drain current. Therefore its lower value is required to reach the breakdown voltage.

It may be noted that with zero gate-to-source voltage, the drain current saturates at I_{DSS} value and the characteristic indicates the pinch off voltage (V_P) equal to 4 V. However, if V_{GS} of – 1 V, is applied, the channel still requires – 4 V to achieve pinch off condition. It means that now the channel requires only 3 V drop (instead of previously 4 V) from the drain-to-source voltage. This 3 V drop may be achieved with the lower value of drain current. Similarly, when V_{GS} is – 2 and –3 V, pinch off is achieved with 2 V and 1 V drops respectively along the channel. These voltage drops are obtained with further reduced values of the drain current. It may be noted that when V_{GS} is – 4 V (*i.e.*, equal to pinch off voltage), no channel drop is required and hence the drain current is zero.

16.9. Transfer Characteristics

These are also called transconductance curves, which give us the relationship between drain current (I_D) and gate-to-source voltage (V_{GS}) for a constant value of drain-to-source voltage (V_{DS}). The transfer characteristics may be obtained by using the circuit arrangement shown in Fig. 16.8. First of all, we adjust the drain-to-source voltage to some suitable value and increase the gate-to-source voltage in small suitable steps. Now record the corresponding values of drain current at each step. If we plot a graph with gate-to-source voltage (V_{GS}) along the horizontal axis and the drain current (I_D) along the vertical axis, we shall obtain a curve as shown in Fig. 16.11. A similar procedure may be used to obtain curves at different values of gate-to-source voltage (V_{GS}).

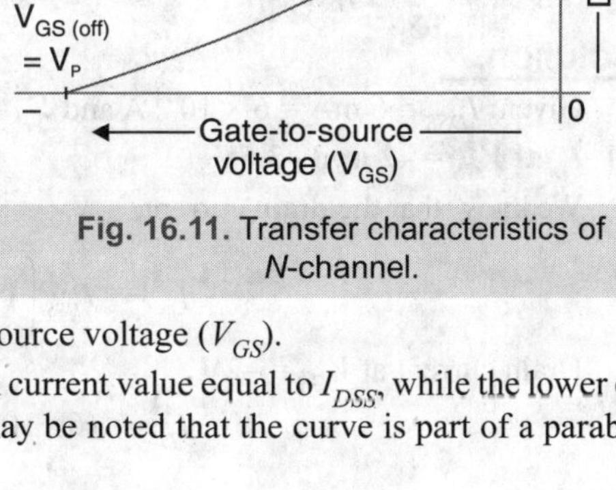

Fig. 16.11. Transfer characteristics of *N*-channel.

The upper end of the curve is shown by the drain current value equal to I_{DSS}, while the lower end is indicated by a voltage equal to $V_{GS(off)}$ or V_P. It may be noted that the curve is part of a parabola and therefore may be expressed by the equation,

$$I_D = I_{DSS}\left(1-\frac{V_{GS}}{V_{GS(off)}}\right)^2 = I_{DSS}\left(1-\frac{V_{GS}}{V_P}\right)^2$$

There is one important point which must be mentioned. Most JFET specifications sheet will list more than one value of $V_{GS(\text{off})}$ and I_{DSS}. For example, the specifications sheet for the JFET 2N5457 lists the following:

$$V_{GS(\text{off})\,(\text{minimum})} = -0.5\text{ V and } I_{DSS(\text{min})} = 1\text{ mA}$$

$$V_{GS(\text{off})\,(\text{maximum})} = -6\text{ V and } I_{DSS(\text{max})} = 5\text{ mA}$$

In such a situation, we must use the two minimum values to plot one curve and the two maximum values to plot another curve on the same graph.

Example 16.1.

The data sheet of a certain JFET indicates that I_{DSS} equal to 15 mA and V_{GS} (off) equal to –5V. Determine the drain current for V_{GS} equal to 0 V, – 1 V and – 4 V.

Solution.

Given: I_{DSS} = 15 mA and $V_{GS\,(off)}$ = – 5 V

We know that drain current,

$$I_D = I_{DSS}\left(1-\frac{V_{GS}}{V_{GS(off)}}\right)^2$$

∴ Drain current when (V_{GS} = 0,)

$$I_D = 15\left(1-\frac{0}{-5}\right)^2 = 15\text{ mA } \textbf{Ans.}$$

Similarly, drain current when V_{GS} = – 1 V,

$$I_D = 15\left(1-\frac{-1}{5}\right)^2 = 9.6\text{ mA } \textbf{Ans.}$$

and drain current when V_{GS} = – 4 V,

$$I_D = 15\left(1-\frac{-4}{-5}\right)^2 = 0.6\text{ mA } \textbf{Ans.}$$

Example 16.2.

Given, I_{DSS} = 6mA and V_P = –4.5V.

(a) Determine I_D at V_{GS} = –2 and –3.6V

(b) Determine V_{GS} at I_D = 3 and 5.5 mA. *(R.G.P.V., Feb. 2010)*

Solution.

Given: I_{DSS} = 6 mA = 6 × 10^{-3} A and V_P = –4.5V.

(*a*) I_D at V_{GS} = –2 and –3.6V

We know that the drain current,

$$I_D = I_{DSS}\left(1-\frac{V_{GS}}{V_P}\right)^2$$

Drain current at V_{GS} = –2V,

$$I_D = (6\times10^{-3})\times\left(1-\frac{-2}{-4.5}\right)^2 = 1.85\times10^{-3}\text{ A} = 1.85\text{ mA } \textbf{Ans.}$$

(b) V_{GS} at $I_D = 3$ and 5.5 mA

We know that the drain current,

$$I_D = I_{DSS}\left(1-\frac{V_{GS}}{V_P}\right)^2$$

V_{GS} at $I_D = 3$mA,

$$3 \times 10^{-3} = (6\times10^{-3})\times\left(1-\frac{V_{GS}}{-4.5}\right)^2$$

$$V_{GS} = -1.318 \times 10^{-3} = -1.318\text{V } \textbf{Ans.}$$

V_{GS} at 5.5 mA,

$$3 \times 10^{-3} = (6\times10^{-3})\times\left(1-\frac{-3.6}{-4.5}\right)^2$$

$$V_{GS} = -0.192 \text{ V } \textbf{Ans.}$$

Example 16.3.

For the given value $V_P = -2V$, $I_{DSS} = 1.55$ mA. It is desired to bias the circuit at $I_D = 0.8$ mA, and $V_{DD} = 24$ V.

Find : (a) V_{GS} (b) g_m *(R.G.P.V., June 2011)*

Solution.

Given: $V_P = -2V$; $I_{DSS} = 1.55 \times 10^{-3}A$; $I_D = 0.8mA = 0.8 \times 10^{-3}$ A and $V_{DD} = 24V$.

(a) V_{GS}

We know that the drain current,

$$I_D = I_{DSS}\left(1-\frac{V_{GS}}{V_P}\right)^2$$

$$0.8 \times 10^{-3} = (1.55\times10^{-3})\times\left(1-\frac{V_{GS}}{-2}\right)^2$$

$$V_{GS} = -0.56\text{V } \textbf{Ans.}$$

(b) g_m

We know that transconductance,

$$g_m = -\frac{2I_{DSS}}{V_P} = -\frac{2\times1.55\times10^{-3}}{-2} = 1.12\times10^{-3}\text{ A/V}$$

$$= 1.12 \text{ mA/V } \textbf{Ans.}$$

Example 16.4.

For an N-channel JFET $I_{DSS} = 8$ mA, $V_P = -4$ V, $V_{GS} = -1$ V. Estimate I_D, g_{mo} and g_m.

(M.G. University., May/June 2006)

Solution.

Given: $I_{DSS} = 8mA$; $V_P = -4$ Volts; $V_{GS} = -1$ Volts

We know that the drain current,

$$I_D = I_{DSS}\left(1-\frac{V_{GS}}{V_P}\right)^2 = 8\left(1-\frac{-1}{-4}\right)^2 = 4.5\text{ mA } \textbf{Ans.}$$

Example 16.6.

A JFET has a drain current of 5 mA. If I_{DSS} is 10 mA and V_{GS} is –6 V, find the values of V_P

(RGTU., June 2011)

Solution.

Given: $I_D = 5\ mA = 5 \times 10^{-3}$ A; $I_{DSS} = 10$ mA $= 10 \times 10^{-3}$ A and $V_{GS} = -6$ V

We know that the drain current,

$$I_D = I_{DSS}\left(1 - \frac{V_{GS}}{V_P}\right)^2$$

$$5 \times 10^{-3} = 10 \times 10^{-3}\left(1 - \frac{-6}{V_p}\right)^2$$

$$V_p = 20.6 \text{ V } \textbf{Ans.}$$

Example 16.6.

A JFET has parameters of $V_{GS\ (off)}$ equal to – 20 V and I_{DSS} equal to 12 mA. Plot the transconductance curve for the device using V_{GS} values of 0 V, – 5 V, –10 V, – 15 V and – 20 V.

Solution.

Given: $V_{GS\ (off)} = -20$ V and $I_{DSS} = 12$ mA

First of all, let us find the value drain currents for the given values of V_{GS} (0 V,– 5 V, – 10 V, – 15 V and – 20 V) as discussed below:

We know that drain current,

$$I_D = I_{DSS}\left(1 - \frac{V_{GS}}{V_{GS\ (off)}}\right)^2$$

∴ Drain current when $V_{GS} = -5$ V,

$$I_D = 12\left(1 - \frac{-5}{-20}\right)^2 = 6.75 \text{ mA}$$

Similarly, drain current when $V_{GS} = -10$ V,

$$I_D = 12\left(1 - \frac{-10}{-20}\right)^2 = 3 \text{ mA}$$

and drain current when $V_{GS} = -15$ V,

$$I_D = 12\left(1 - \frac{-15}{-20}\right)^2 = 0.75 \text{ mA}$$

The value of V_{GS} and the corresponding values of I_D obtained above are shown in Table 16.1.

Table 16.1:

V_{GS} (*V*)	I_D (*mA*)
– 20	0
– 15	0.75
– 10	3
– 5	6.75
– 0	12

These points are now plotted and joined together to form a curve shown in Fig. 16.12.

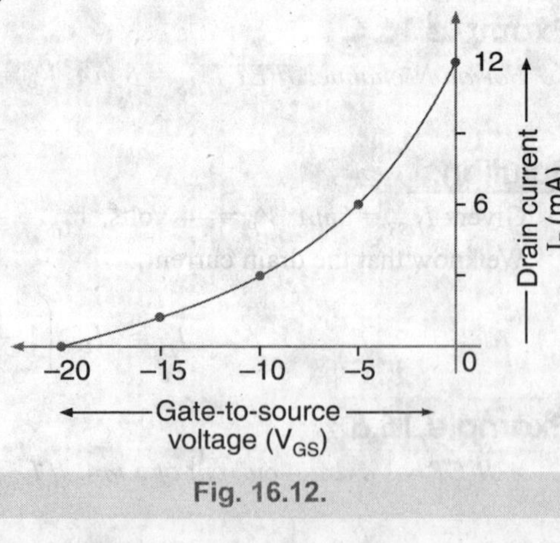

Fig. 16.12.

Example 16.7.

A 2 N 5486 JFET has values of $V_{GS(off)}$ equal to – 2 V to – 6 V and I_{DSS} equal to 8 mA to 20 mA. Plot the minimum and maximum transconductance curves for the device.

Solution:

Given: $V_{GS\ (OFF)}$ = –2 V to – 6 V and I_{DSS} = 8 mA to 20 mA.

First of all, let us find the values of drain currents for the given values of V_{GS}.

We know that drain current,

$$I_D = I_{DSS}\left(1 - \frac{V_{GS}}{V_{GS(off)}}\right)^2 \quad ...(i)$$

Substituting the values of V_{GS} = – 0.5, – 1.0 and – 1.5 with $V_{GS\ (off)}$ = – 2.0 V, we get the corresponding values of I_D = 4.5 mA, 2 mA and 0.5 mA. With these values, we have the set of V_{GS} versus I_D value, shown in Table 16.2

Table 16.2:

For minimum transconductance curve

$V_{GS}(V)$	$I_D\ (mA)$
– 2.0	0
– 1.5	0.5
– 1.0	2.0
– 0.5	4.5
0	8

Table 16.3:

For maximum transconductance curve

$V_{GS}(V)$	$I_D\ (mA)$
– 6	0
– 4	2.2
– 2	8.9
0	20

Similarly, using the maximum values of $V_{GS(off)}$ = – 6 V and I_{DSS} = 20 mA, we can use equation (*i*) again for V_{GS} = – 4 V and – 2 V to obtain the values of I_D in between the end points. The end points of the transconductance curve are (– 6 V, 0 mA) and (0 V, 20 mA). The values of I_D for in between points are 2.2 mA and 8.9 mA. The values obtained are shown in Table 16.3.

Now if we plot the values corresponding to the sets shown in Tables 16.2 and 16.3 and joining them, we get the minimum and maximum trans conductance curve as shown in Fig. 16.13.

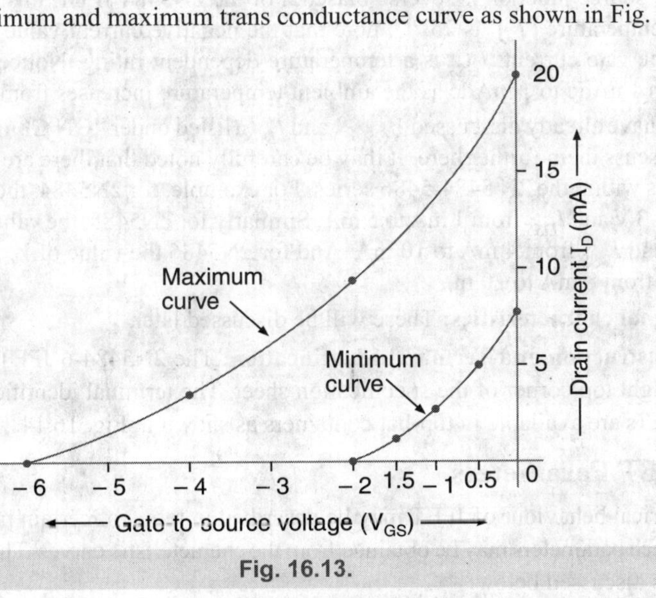

Fig. 16.13.

16.10. Specifications Sheet of JFET

Although the general content of specification sheets may very from the absolute minimum to an extensive display of graph and charts, there are a few fundamental parameters that will be provided by all manufactures. The specifications sheet for 2N5484 through 2N5486 JFETs as provided by the Motorola is as shown in Fig. 16.14.

Maximum Ratings: Most of the maximum ratings for these devices are self-explanatory at this point. In out discussion for the BJT specifications sheet, we covered all the commonly used maximum voltage, current and power ratings. The JFET maximum ratings are no different. As seen form the table for maximum ratings, we find that for 2N5484, the maximum ratings for V_{DG} and V_{GSR} is 25 V.

The next maximum ratings in the list is for the drain current which is indicated as 30 mA dc. Then it is the maximum value of forward gate current. This rating is indicated as 10 mAdc. This rating may produce a through in your mind that the FET could be operated in the forward region. But the answer to such a thought is NO. Recall that the contorl of drain current in FET depends upon the amount of *reverse* bias on the JFET. The maximum drain current (I_{DSS}) is reached when the reverse bias reaches 0 V. any increase in V_{GS} above 0 V will not cause any increase in drain current. However, should the gate accidentally becomes forward biassed; I_G must be greater than 10 mAdc, for the device to be destroyed. As long as I_G is below 10 mAdc, the device will be safe.

The maximum power dissipation which the device under consideration can safely handle at 25°C is 310 mW. As the temperature exceeds 25°C temperature, the maximum power dissipation reduces at the rate of 2.82 mW for each 1°C increase in temperature above 25°C.

Off Characteristics : In this list three values are indicated:

(*i*) gate-source breakdown voltage, $V_{(BR)GSS}$,

(*ii*) gate reverse current, I_{GSS}, and

(*iii*) gate-source cut-off voltage, $V_{GS(off)}$. The gate-source breakdown voltage defines the limit on V_{GS}. If V_{GS} is allowed to exceed this value (–25 V for 2N5484), the JFET will have to be replaced.

The gate reverse current rating indicates the maximum value of gate current that will occur when the gate-source junction is reverse biased. For the 2N5484 JFET, this rating is 1 nAdc when the ambient temperature (T_A) is 25 °C. note that the negative current value is used to indicate the direction of the gate current. I_{GSS} is a temperature dependent rating. Notice that the value of I_{GSS} increases form 1 nAdc to 1 mAdc as the ambient temperature increases from 25 ° C to 100°C

Since we have already disscussed $V_{GS(off)}$ and I_{DSS} (listed under "**ON Characteristics**") in detail, we will not discuss them further here. It may be carefully noted that there are wide ranges of $V_{GS(off)}$ and I_{DSS} values within the 2N 5484-5486 series. For example, for 2N5484, the values of $V_{GS(off)}$ vary from – 0.3 to – 3 V and I_{DSS} from 1 mA to 5 mA. Similarly for 2N5485, the values of $V_{GS(off)}$ vary from –0.5 to –4 V and I_{DSS} from 4 mA to 10 mA. And for 2N5486 the value of $V_{GS(off)}$ vary from –2 V to –6 V and I_{DSS} from 8 mA to 20 m..

Small signal characteristics: These will be discussed later.

Case Construction and Terminal Identification: The 2N5484-6 JFETs have the appearance as shown of right top corner of the specification sheet. The terminal identification is also provided there. The JFETs are available in top-hat containers as shown in Fig. 16.14.

16.11. JFET Parameters

The electrical behaviour of JFET may be described in terms of certain parameters called JFET parameters. Such parameters may be obtained from the characteristic curves. The important parameters of JFET are as discussed below:

1. ***D.C drain resistance*** **(R_{DS}).** It is also called the static or ohmic resistance of the channel and is given by the ratio of voltage (V_{DS}) to the drain current (I_D). Mathematically d.c. drain resistance,

$$R_{DS} = \frac{V_{DS}}{I_D}$$

2. ***A.C. drain resistance*** **(r_d).** It is also called dynamic drain resistance and is the a.c. resistance between the drain and source terminal when the JFET is operating in the pinch-off or saturation region. It is given by the ratio of small change in drain-to-source voltage (ΔV_{DS}) to the corresponding change in drain current (ΔI_D) for a constant gate-to-source voltage (V_{GS}). Mathematically, the a.c. drain resistance,

$$r_d = \frac{\Delta V_{DS}}{\Delta I_D}$$

It is sometimes written as r_{ds}, which indicates that it is the resistance from drain-to-source terminals. Since r_d is the output resistance of JFET, it may also be expressed as an output admittance y_{DS}.(*i.e.*, equal to $1/r_d$).

3. ***Transconductance*** **(g_m).** It is also called forward transconductance or forward transmittance. It is given by the ratio of small change in drain current (ΔI_D) to the corresponding change in gate-to-source voltage (ΔV_{GS}) for a constant drain-to-source voltage (V_{DS}). Mathematically, the transconductance,

$$g_m = \frac{\Delta I_D}{\Delta V_{GS}}$$

The value of transconductance (g_m) is expressed in siemens (*S*) or mhos. We know that transfer characteristic of a JFET is a part of a parabola *i.e.*, it is non-linear. Therefore the value of g_m varies depending upon the location on the curve as set by the gate-to-source voltage (V_{GS}). It has a greater value near the top of the curve than near the bottom.

4. ***Amplication factor*** **(μ).** It is given by the ratio of small change in drain-to-source voltage (ΔV_{DS}) to the corresponding change in gate-to-source voltage (ΔV_{GS}) for a constant drain current (I_D). Mathematically, the amplification factor,

$$\mu = \frac{\Delta V_{DS}}{\Delta V_{GS}}$$

The amplification factor (μ) may also be expressed as a product of transconductance (g_m) and a.c. drain resistance (r_d).

Thus
$$\mu = \frac{\Delta V_{DS}}{\Delta V_{GS}} = \frac{\Delta V_{DS}}{\Delta I_D} \times \frac{\Delta I_D}{\Delta V_{GS}} = r_d \times g_m$$

5. ***Input Resistance*** **(R_i).** We know that a JFET operation with its gate-source junction reverse biased. As the reverse current of a junction is, usually, very small. Therefore gate-reverse current of a JFET is also very small. Since the gate-source junction of a JFET is used as an input, therefore the input resistance of a JFET is very high. This high input resistance is an advantage of JFET over the bipolar transistor. Mathematically, the input resistance,

$$R_i = \frac{V_{GS}}{I_{GSS}}$$

where V_{GS} = Gate-to-source voltage and

I_{GSS} = Gate reverse current. It is of the order of nano-amperes.

2N5484
thru
2N5486

CASE 29-04, STYLE 5
TO-92 (TO-226AA)

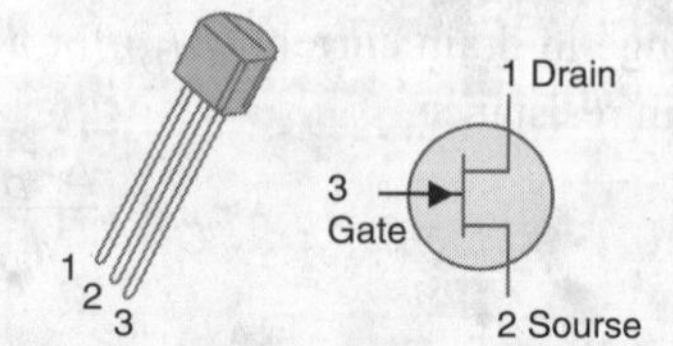

JFET
VHF/UHF AMPLIFIERS

N-CHANNEL– DEPLETION

Refer to 2N4416 for graphs.

MAXIMUM RATINGS

Rating	Sympol	Value	Unit
Drain-Gate Voltage	V_{DG}	25	Vdc
Reverse Gate-Source Voltage	V_{GSR}	25	Vdc
Drain Current	I_D	30	mAdc
Forward Gate Current	$I_{G(f)}$	10	mAdc
Total Device Dissipation @ T_C = 25°C	P_D	310	mW
Derate above 25°C		2.82	mW/°C
Operating and Storage Junction Temperature Range	T_J, T_{stg}	– 65 to + 150	°C

ELECTRICAL CHARACTERISTICS (T_A = 25°C unless otherwise noted.)

Characteristic		Symbol	Min	Typ	Max	Unit
OFF CHARACTERISTICS						
Gate-Source Breakdown Voltage (I_G = – 1.0 μAdc, V_{DS} = 0)		$V_{(BR)GSS}$	– 25	—	—	Vdc
Gate Reverse Current		I_{GSS}				
(V_{GS} = – 20 Vdc, V_{DS} = 0)			—	—	– 1.0	nAdc
(V_{GS} = – 20 Vdc, V_{DS} = 0, T_A = 100°C)			—	—	– 0.2	μAdc
Gate Source Cutoff Voltage		$V_{GS(off)}$				Vdc
(V_{DS} = 15 Vdc, I_D = 10 nAdc)	2N5484		– 0.3	—	– 3.0	
	2N5485		– 0.5	—	— 4.0	
	2N5486		– 2.0	—	– 6.0	
ON CHARACTERISTICS						
Zero-Gate-Voltage Drain Current		I_{DSS}				mAdc
(V_{DS} = 15 Vdc, V_{GS} = 0)	2N5484		1.0	—	5.0	
	2N5485		4.0	—	10	
	2N5486		8.0	—	20	
SMALL-SIGNAL CHARACTERISTICS						
Forward Transfer Admittance		$\|y_{fs}\|$				μmhos
(V_{DS} = 15 Vdc, V_{GS} = 0, f = 1.0 kHz)	2N5484		3000	—	6000	
	2N5485		3500	—	7000	
	2N5486		4000	—	8000	
Input admittance		$Re(y_{is})$				μmhos
(V_{DS} = 15 Vdc, V_{GS} = 0, f = 100 MHz)	2N5484		—	—	100	
(V_{DS} = 15 Vdc, V_{GS} = 0, f = 400 MHz)	2N5485, 2N5486		—	—	1000	
Output Admittance		$\|y_{os}\|$				μmhos
(V_{DS} = 15 Vdc, V_{GS} = 0, f = 1.0 kHz)	2N5484		—	—	50	
	2N5485		—	—	60	
	2N5486		—	—	75	
Output Conductance		$Re(y_{os})$				μmhos
(V_{DS} = 15 Vdc, V_{GS} = 0, f = 100 MHz)	2N5484		—	—	75	
(V_{DS} = 15 Vdc, V_{GS} = 0, f = 400 MHz)	2N5485, 2N5486		—	—	100	
Forward Transconductance		$Re(y_{fs})$				μmhos
(V_{DS} = 15 Vdc, V_{GS} = 0, f = 100 MHz)	2N5484		2500	—	—	
(V_{DS} = 15 Vdc, V_{GS} = 0, f = 400 MHz)	2N5485		3000	—	—	
	2N5486		3500	—	—	

Fig. 16.14. (Courtesy : Motorola Inc.)

(Continued)

ELECTRICAL CHARACTERISTICS (TA = 25°C unless otherwise noted)

Substrate connected to source.

Characteristic	Fig. No.	Symbol	Min	Max	Unit
OFF CHARACTERISTICS					
Drain-Source Breakdown Voltage (I_D = 10 μAdc, V_{GS} = 0)	–	$V_{(BR)DSS}$	25	–	Vdc
*Gate Leakage Current	–	I_{GSS}			pAdc
(V_{GS} = – 35 Vdc, V_{DS} = 0)			–	10	
(V_{GS} = – 35 Vdc, V_{DS} = 0, T_A = 125°C			–	100	
*Zero-Gate-Voltage Drain Current	–	I_{DSS}			nAde
(V_{DS} = 10 Vdc, V_{GS} = 0)			–	10	
(V_{DS} = 10 Vdc, V_{GS} = 0, T_A = 125°C)			–	1.0	μAdc
*ON CHARACTERISTICS					
Gate-Source Threshold Voltage (V_{DS} = 10 Vdc, I_D = 10 μAdc) 3N169	–	$V_{GS(th)}$	0.5	1.5	Vdc
3N170			1.0	2.0	
3N171			1.5	3.0	
"ON" Drain Current (V_{GS} = 10 Vdc, V_{DS} = 10 Vdc)	3	$I_{D(on)}$	10	–	mAdc
Drain-Source "ON" Voltage (I_D = 10 mAdc, V_{GS} = 10 Vdc)	–	$V_{DS(on)}$	–	2.0	Vdc
SMALL SIGNAL CHARACTERISTICS					
*Drain-Source Resistance (V_{GS} = 10 Vdc, I_D = 0, f = 1.0 kHz)	4	$r_{ds(on)}$	–	200	Ohms
Forward Transfer Admittance (V_{DS} = 10 Vdc, I_D = 2.0 mAdc, f = 1.0 kHz)	1	$\|y_{fs}\|$	1000	–	μmhos
*Reverse Transfer Capacitance (V_{DS} = 0, V_{GS} = 0, f = 1.0 MHz)	2	C_{rss}	–	1.3	pF
*Input Capacitance (V_{DS} = 10 Vdc, V_{GS} = 0, f = 1.0 MHz)	2	C_{iss}	–	5.0	pF
*Drain-Substrate Capacitance ($V_{D(SUB)}$ = 10 Vdc, f = 1.0 MHz	–	$C_{d(sub)}$	–	5.0	pF

* SWITCHING CHARACTERISTICS

Characteristic	Conditions	Fig. No.	Symbol	Min	Max	Unit
Turn-On Delay Time	(V_{DD} = 10 Vdc, $I_{D(on)}$ = 10 mAdc, $V_{GS(on)}$ = 10 Vdc, $V_{GS(off)}$ = 0, R_G = 50 Ohms)	6,10	$t_{d(on)}$	–	3.0	ns
Rise Time		7,10	t_r	–	10	ns
Turn-Off Delay Time		8,10	$t_{d(off)}$	–	3.0	ns
Fall Time		9,10	t_f	–	15	ns

*Indicates JEDEC Registered Data.

FIGURE 1 – FORWARD TRANSFER ADMITTANCE

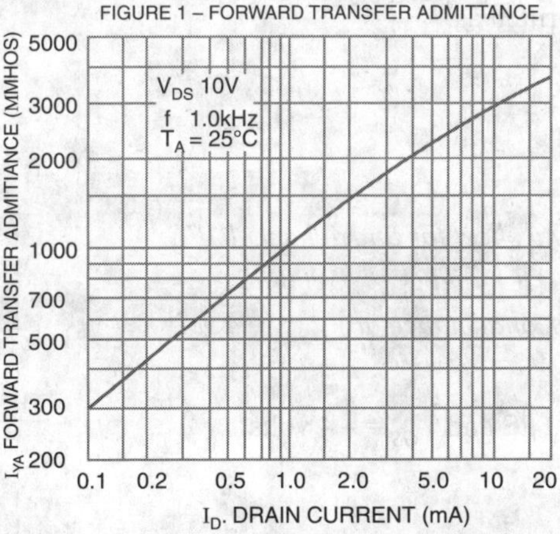

FIGURE 2 – CAPPACITANCE

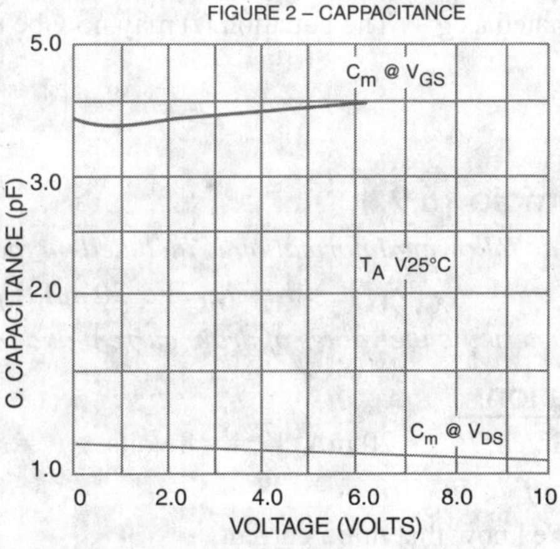

MOTOROLA Semiconductor Products Inc.

Fig. 16.14. (*a*) (Contd.) (Courtesy : Motorola Inc.)

Example 16.8.

When V_{GS} of a FET changes from –3.1 V to –3 V, the drain current charges from 1 mA and 1.3 mA. What is the value of transconductance ?

Solution.

Given: V_{GS} of a FET changes from –3.1 V to 3 V and the drains current, 1 mA to 1.3 mA.

We know that change in V_{GS},

$$\Delta V_{GS} = -3(-3.1) = 0.1.$$

and the change in drain current,

$$\Delta I_D = 1.3 - 1 = 0.3 \text{ mA.}$$

∴ The value of transconductance,

$$g_m = \frac{\Delta I_D}{\Delta V_{GS}} = \frac{0.3 \text{ mA}}{0.1 \text{ V}} = 30 \text{ mA/V} \quad \text{or} \quad 30 \text{ mS} \quad \textbf{Ans.}$$

16.12. Mathematical Expression for Transconductance

We have already discussed in Art 16.6 that the drain current is given by the relation,

$$I_D = I_{DSS}\left(1 - \frac{V_{GS}}{V_P}\right)^2$$

Differentiating on both sides with respect to V_{GS},

$$\frac{dI_D}{dV_{GS}} = 2I_{DSS}\left(1 - \frac{V_{GS}}{V_P}\right)^2 \times \left(\frac{0-1}{V_P}\right)$$

$$g_m = -\frac{2\,I_{DSS}}{V_P}\left(1 - \frac{V_{GS}}{V_P}\right) \qquad \text{...}(i)$$

Substituting $V_{GS} = 0$ in the above expression,

$$g_m = -\frac{2I_{DSS}}{V_P} = g_{mo}$$

The value of g_m at $V_{GS} = 0$ is the maximum and is equal to $-2I_{DSS}/V_P$. This value of g_m is designated as g_{mo}. The equation (*i*) may now be rewritten as:

$$g_m = g_{mo}\left(1 - \frac{V_{GS}}{V_P}\right)$$

Example 16.9.

The following information is included on the data sheet for an N-channel JFE:

$$I_{DSS} = 20 \text{ mA}, V_P = -8 \text{ V and } g_{mo} = 5000\ \mu\text{S}.$$

Determine the values of drain current and transconductance at $V_{GS} = -4$ volts.

Solution.

Given: $I_{DSS} = 20$ mA, $V_P = -8$ volts; $g_{mo} = 5000$ µS and $V_{GS} = -4$ volts.

Value of drain current

We know that drain current,

$$I_D = I_{DSS}\left(1 - \frac{V_{GS}}{V_P}\right)^2 = 20\left(1 - \frac{-4}{-8}\right)^2 = 5 \text{ mA} \quad \textbf{Ans.}$$

Value of transconductance

We also know that transconductance,

$$g_m = g_{mo}\left(1 - \frac{V_{GS}}{V_P}\right) = 5000\left(1 - \frac{-4}{-8}\right) = 2500\ \mu\text{S} \quad \textbf{Ans.}$$

Example 16.10.

As V_{GS} changed from –1 V to –1.5 V keeping V_{DS} constant, I_D of the FET drops from 7 mA to 5 mA. What is the transconductance of FET? If the a.c. drain resistance is 200 kΩ, find also the amplification factor of the FET. *(WBTU., 2011)*

Solution.

Given: $r_d = 200\ \text{k}\Omega = 200 \times 10^3\ \Omega$; Change in value of V_{GS} of a FET = –1 V to 1.5 V and the change in value of drains current = 7 mA to 5 mA

Transconductance

We know that the change in V_{GS},

$$\Delta V_{GS} = (-1) - 1.5 = 0.5$$

and the change in drain current,

$$\Delta I_D = 7 - 5 = 2\ \text{mA}.$$

∴ The value of transconductance,

$$g_m = \frac{\Delta I_D}{\Delta V_{GS}} = \frac{2\ \text{mA}}{0.5\ \text{V}} = 10\ \text{mA/V} \text{ or } 10\ \text{mS} \quad \textbf{Ans.}$$

Amplification factor

We know that the amplification factor,

$$\mu = r_d \times g_m = (200 \times 10^3) \times (10 \times 10^{-3})$$

$$= 2000\ \Omega = 2 \times 10^3\ \text{k}\Omega \quad \textbf{Ans.}$$

Example 16.11.

For JFET if I_{DSS} = 20 mA, $V_{GS(off)}$ = –5 V and g_{mo} = 4 mS. Determine the transconductance for V_{GS} = –4 V and I_D at this point. *(Nagpur University., Summer 2011)*

Solution.

Given: $V_{GS(off)}$ = –5 V; I_{DSS} = 20 mA; g_{mo} = 4 mS = 4×10^{-3} S and V_{GS} = –4 V

We also know that the transconductance,

$$g_m = g_{mo}\left(1 - \frac{V_{GS}}{V_{GS}(off)}\right)$$

$$= 4\left(1 - \frac{-4}{-5}\right) = 0.8\ \text{mS} \quad \textbf{Ans.}$$

We know that the drain current,

$$I_D = I_{DSS}\left(1 - \frac{V_{GS}}{V_{GS(off)}}\right)^2$$

$$= 20\left(1 - \frac{-4}{-5}\right)^2 = 0.8\ \text{mA} \quad \textbf{Ans.}$$

16.13. Comparison Between Field-Effect Transistor and Bipolar Junction Transistor

Table 16.4 shows some of the important points of comparison between the field-effect transistor and bipolar junction transistor.

Table 16.4: Comparison between field-effect transistor and bipolar junction transistor

S.No	*Field-Effect Transistor (FET)*	*Bipolar junction transistor (BJT)*
1.	It is a unipolar device. *i.e.*, current in the device is carried either by electrons or holes.	It is a bipolar device. *i.e.* current in the device is carried by both electrons and holes.
2.	It is a voltage-controlled device *i.e.*, voltage at the gate (or drain) terminal controls the amount of current flowing through the device.	It is a current-controlled device. *i.e.* the base current controls the amount of collector current.
3.	Its input resistance is very high and is of the order of several megaohms.	Its input resistance is very low as compared to FET and is of the order of few kilohms.
4.	It has a negative temperature coefficient at high current levels. It means that current decreases as the temperature increases. This characteristic prevents the FET from thermal breakdown.	It has a positive temperature coefficient at high current level. It means that collector current increases with the increase in temperature. This characteristic leads the BJT to thermal breakdown.
5.	It does not suffer from minority-carrier storage effects and therefore has higher switching speeds and cut-off frequences.	It suffers from minority carrier storage effects and therefore has lower switching speed and cut-off frequencies than that of FET's.
6.	It is less noisy than a BJT or vacuum tube and is thus more suitable as an input amplifier for low-level signals. It is used extensively in high fidelity frequency modulated receivers for quiet reception.	It is comparatively more noisy than a field-effect transistor.
7.	It is much simpler to fabricate as a integrated circuit (IC) and occupies a less space on IC chip than that of BJT.	It is comparatively difficult to fabricate as an integrated circuit and occupies more space on IC chip than that of FET.
8.	It is relatively immune to radiation and hence makes excellent signal chopper.	It is susceptible to radiations. With the radiation exposure, B of the transistor decreases. As a result, input impedance, gain, bandwidth also reduced. Overall affect is that stability of the transistor circuit gets disturbed.
9.	It has a relatively lower gain-bandwidth product as compared to BJT.	It has a relatively higher gain-bandwidth product as compared to FET.
10.	It has a drawback of being very susceptible to overload voltage, thus requiring special handling during installation.	BJT does not require any special handling installation like FETS.

16.14. MOSFETs

The MOSFET is an abbreviation for metal-oxide semiconductor field-effect transistor. Like JFET, it has a source, gate and drain. However, unlike JFET, the gate of a MOSFET is insulated from the channel. Because of this, the MOSFET is sometimes known as an IGFET which stands for insulated-gate field effect transistor.

16.15. Types of MOSFET

There are two basic types of MOSFETs; depletion-type MOSFETs and enhancement-type MOSFETs. The depletion type MOSFETs are also called D-type MOSFETs and enhancement-type MOSFETs. as E-type MOSFETs. The primary difference between the two types of MOSFETs is the difference between the constructions. Both the D-type and E-type MOSFETs are disscussed one by one in the following pages.

16.16. Depletion-Type MOSFET

Fig. 16.15 shows the basic structure of an N-channel depletion type MOSFET. It consists of a conducting bar of N-type material with an insulated gate on the left and P-region on the right. Free electrons can flow from source to drain through the N-type material. The P-region is called substrate (or body). It physically reduces the conducting path to a narrow channel. A thin layer of silicon dioxide is deposited on the left side of the channel. This layer insulates the gate from the channel. Because of this, a negligible gate current flows even when the gate voltage is positive. It will be interesting to know that a PN junction, which exists in a JFET, has been eliminated in the MOSFET.

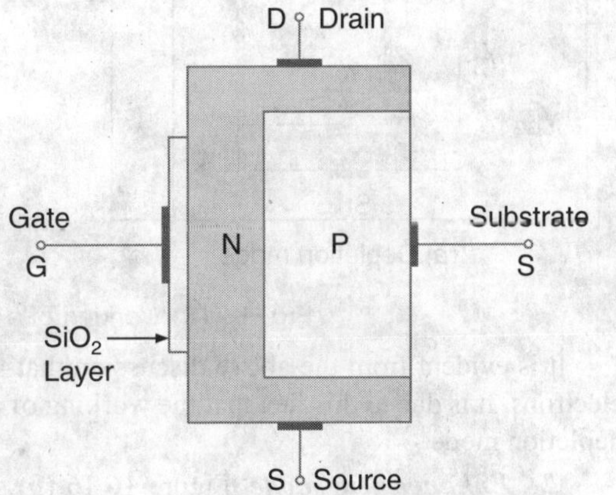

Fig. 16.15. Basic structure of an N-channel depletion-type MOSFET.

The basic construction of a depletion-type P-channel MOSFET is similar to that of N-channel except that the conducting bar is of P-type material and the substrate is of N-type material.

16.17. Working of a Depletion-Type MOSFET

The depletion-type MOSFET can be operated in two different modes as given below:

1. ***Depletion mode.*** The device operates in this mode, when the gate voltage in negative.
2. ***Enhancement mode.*** The device operates in this mode, when the gate voltage is positive.

Since the depletion-type MOSFET can be operated in either depletion or enhancement mode, therefore this device is commonly known as depletion-enhancement (DE) type MOSFET. The working of a MOSFET may be explained easily, if we visualize the entire structure of the device as a parallel plate capacitor. One of the plates is formed by the gate and the other by the semiconductor channel. The plates are separated by a dielectric (SiO_2 layer). We know that if one plate of a capacitor is made negative, it induces a positive charge on the opposite plate and vice versa. This principle is used below in explaining the working of MOSFET's in the depletion and enhancement modes.

1. ***Depletion mode.*** Fig. 16.16 (*a*) shows a MOSFET with a negative gate-to-source voltage. The negative voltage, on the gate, induces a positive charge in the channel. Because of this, free electrons in the vicinity of positive charge are repelled away in the channel. As a result of this, the channel is depleted of free electrons. This reduces the number of free electrons (which constitute the drain current) passing through the channel. Thus as the value of negative gate-to-source voltage is increased, the value of drain voltage, called $V_{GS(\text{off})}$, the channel is totally depleted of free electrons and therefore the drain current reduces to zero. Thus with the negative gate voltage, the operation of MOSFET is similar to that of a JFET

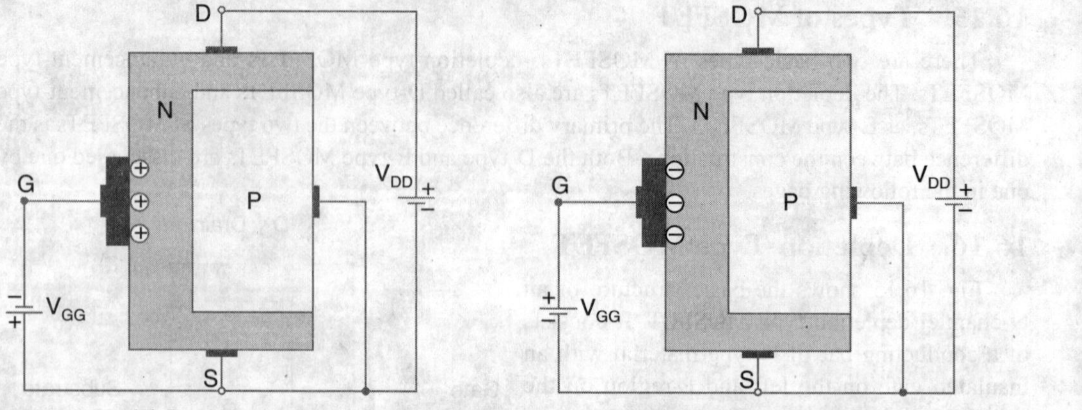

(a) Depletion mode. (b) Enhancement mode.

Fig. 16.16. Working of a depletion-type MOSFET.

It is evident from the above discussion that negative gate voltage depletes the channel of free electrons. It is due to this fact that the working of a MOSFET, with a negative gate voltage, is called depletion mode.

2. ***Enhancement mode.*** Figure 16.16 (*b*) shows a MOSFET with a positive gate-to-source voltage. The positive gate voltage increases the number of free electrons passing through the channel. The greater the gate voltage, greater is the number of free electrons passing through the channel. This increases *i.e.*, enhances the conducting of the channel. Because of this fact, positive gate operation is called enhancement mode.

It will be interesting to know that depletion-type MOSFET can conduct even if the gate-to-source voltage (V_{GS}) is zero. Because of this, it is commonly known as Normally-ON MOSFET.

16.18. Drain Characteristic of Depletion-Type MOSFET

Fig. 16.17 shows the drain characteristic for the N-channel depletion-type MOSFET in the common source configuration. These curves are plotted for both negative and positive values of gate-to-source voltage (V_{GS}). The curves shown above the curve for $V_{GS} = 0$ have a positive zero whereas those below it have a negative value of V_{GS}. When V_{GS} is zero and negative, the MOSFET operates in the depletion-mode. On the other hand, if VGS is zero and positive, the MOSFET operates in the enhancement-mode.

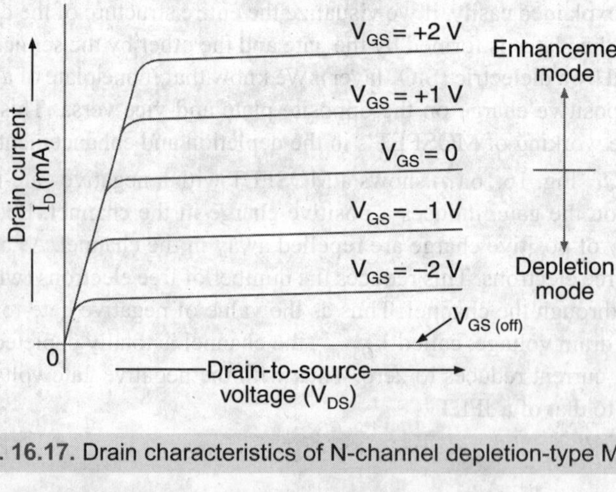

Fig. 16.17. Drain characteristics of N-channel depletion-type MOSFET.

It may be noted that the drain characteristics of depletion-type MOSFET's are similar to that of JFET. The only difference is that JFET does not operate for positive values of gate-to-source voltage (V_{GS}).

16.19. Transfer Characteristic of Depletion-Type MOSFET

Fig. 16.18 shows the transfer characteristic (also called transconductance curve) for an N-channel depletion-type MOSFET. It may be noted from this curve that the region *AB* of the characteristic is similar to that of JFET. But here, this curve extends for the positive values of gate-to-source voltage (V_{GS}) also. The value I_{DSS} represents the current from drain-to-source with $V_{GS} = 0$. The drain current at any point along the transfer characteristic (*i.e.*, the curve *ABC*) is given by the relation,

$$I_D = I_{DSS}\left(1 - \frac{V_{GS}}{V_{GS(off)}}\right)^2$$

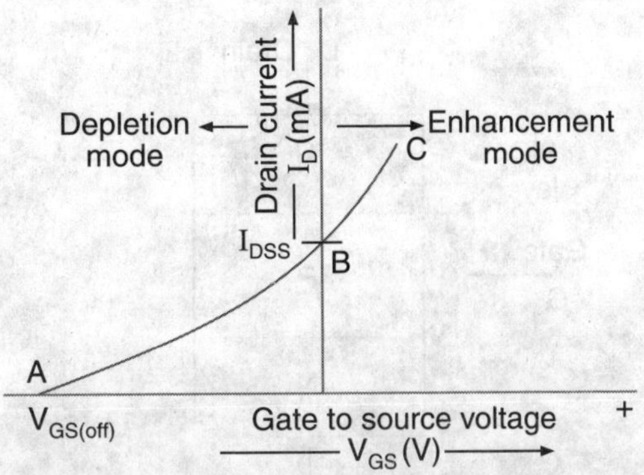

Fig. 16.18. Transfer characteristics of a N- channel depletion-type MOSFET.

It may be noted that even if $V_{GS} = 0$, the device has a drain current equal to I_{DSS}. Due to this fact, it is called normally-ON MOSFET.

16.20. Circuit Symbol for Depletion-Type MOSFET

Fig. 16.19 (*a*) and (*b*) shows the circuit symbols for the N-channel depletion-type MOSFET.

In these figures, a thin vertical line (just right to the gate) represents the channel. The drain and source terminals are connected to the top and bottom of the channel as shown. The arrow, on the P-type substrate, points towards the channel. This indicates that the channel is N-type. In some MOSFET's, a connection from the substrate is also taken out. Such MOSFET's have 4-terminals as indicated in the Fig. 16.19 (*a*). But in most of the MOSFET's the substrate is internally connected to the source. This results in a three terminal device, whose circuit symbol is as shown in Fig. 16.19 (*b*).

Fig. 16.19 (*c*) shows the circuit symbol for a P-channel depletion type MOSFET. It may be noted that the symbol is similar to that of N-channel, except the direction of the arrow on the substrate. Its direction is away from the channel, which indicates that the channel is of P-type material.

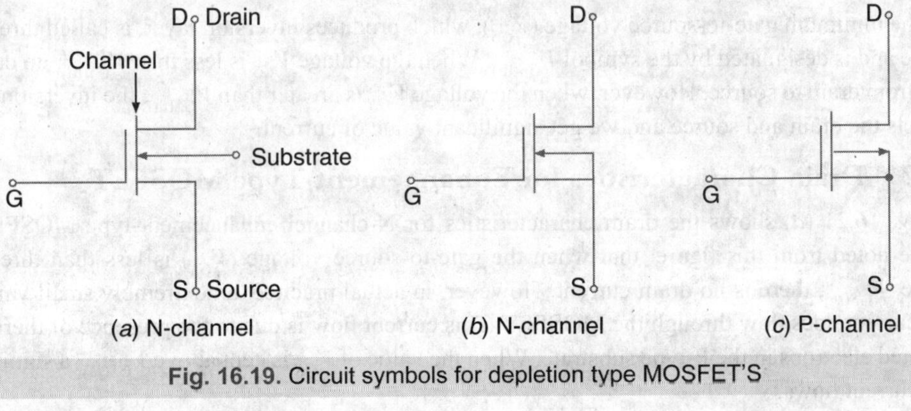

Fig. 16.19. Circuit symbols for depletion type MOSFET'S

16.21. Enhancement-Type MOSFET

The enhancement-type MOSFET has no depletion mode and it operates only in enhancement mode. It differs in construction from the depletion-type MOSFET in the sense that it has no physical channel. Fig. 16.20 (*a*) shows the basic structure of the N-channel enhancement-type MOSFET. It may be noted, that the P-type substrate extends the silicon dioxide layer completely.

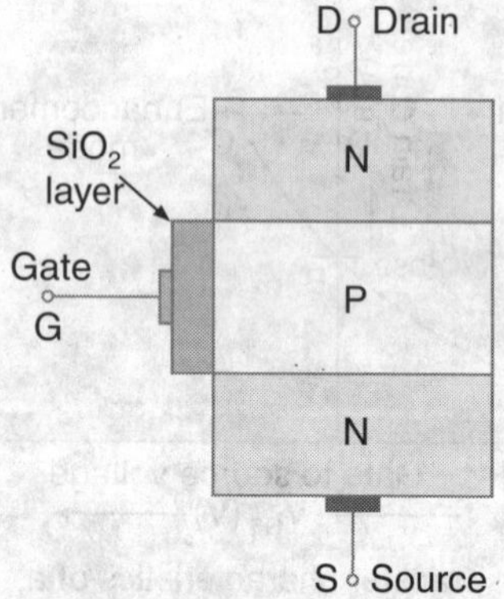

(*a*) Basic construction.

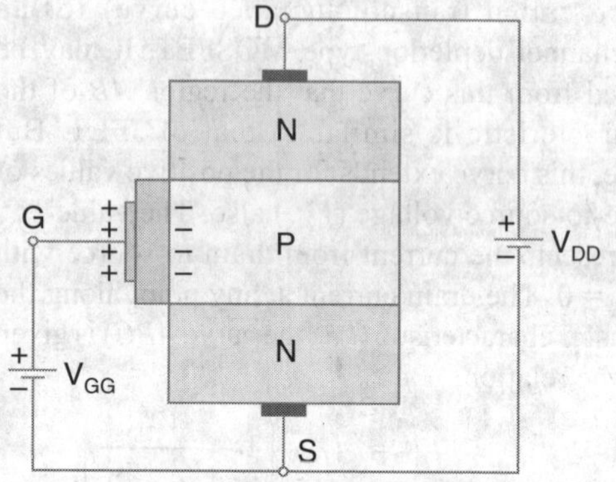

(*b*) Formation of invertion layer.

Fig. 16.20. Enhancement type MOSFET.

Fig. 16.20 (*b*) shows the normal biasing polarities for the N-channel enhancement-type MOSFET. It must be remembered that this M OSFET is always operated with the positive gate-to-source voltage (V_{GS}). When the gate-to-source voltage is zero, the V_{DD} supply tries to force free electrons from source-to-drain. But the presence of P-region does not permit the electrons to pass through it. Thus there is no drain current for $V_{GS} = 0$. Due to this fact, the enhancement type MOSFET is also called normally-OFF MOSFET.

Now, if some positive voltage is applied at the gate, it induces a negative charge in the P-type substrate just adjacent to the silicon dioxide layer. The induced negative charge is produced by attracting the free electrons from the source. When the gate is positive enough, it can attract a number of free electrons. This forms a thin layer of electrons, which stretches from source to drain. This effect is equivalent to producing a thin layer of N-type channel in the P-type substrate. This layer of free electrons is called N-type inversion layer.

The minimum gate-to-source voltage (V_{GS}), which produces inversion layer, is called threshold voltage and is designated by the symbol $V_{GS(th)}$. When the voltage V_{GS} is less than $V_{GS(th)}$, no current flows from drain to source. However, when the voltage V_{GS} is greater than $V_{GS(th)}$, the inversion layer connects the drain and source and we get significant value of current.

16.22. Drain Characteristics for Enhancement-Type MOSFET

Fig. 16.21 (*a*) shows the drain characteristics for N-channel enhancement-type MOSFET. It may be noted from this figure, that when the gate-to-source voltage (V_{GS}) is less than threshold voltage, $V_{GS(th)}$, there is no drain current. However, in actual practice, an extremely small value of drain current does flow through the MOSFET. This current flow is due to the presence of thermally generated electrons in the P-type substrate. When the value of V_{GS} is kept above $V_{GS(th)}$, a significant drain current flows.

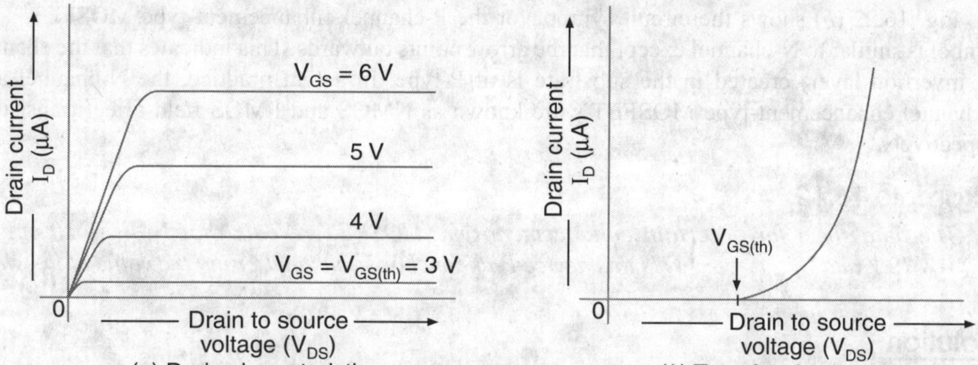

(*a*) Drain characteristic. (*b*) Transfer characteristic.

Fig. 16.21. Characteristics for N-channel enhancement-type MOSFET.

The values of drain current increases with the increase in gate-to-source voltage. It is because of the fact that the width of inversion layer widens for increased values of V_{GS} and therefore allows more number of free electrons to pass through it. The drain current reaches its saturation value above a certain value of drain-to-source voltage (V_{DS}).

16.23. Transfer Characteristic for Enhancement-Type MOSFET

Fig. 16.21 (*b*) shows the transfer characteristic for N-channel enhancement type MOSFET. It may be noted from this figure that there is no drain current when the gate-to-source voltage, $V_{GS} = 0$. However, if V_{GS} is increased above the threshold voltage, $V_{GS(\text{th})}$, the drain current increases rapidly as shown in figure. The drain current at any point along the curve is given by the relation,

$$I_D = K\left[V_{GS} - V_{GS(th)}\right]^2$$

where *K* is a constant, whose value depends on the type of MOSFET. Its value can be determined from the data sheet by taking specified value of drain current called $I_{D(\text{ON})}$ at the given value of V_{GS} and then substituting these values in the above equation. Incidently, it may be noted that enhancement-type MOSFET does not have an I_{DSS} parameter like JFET and depletion-type MOSFET.

16.24. Circuit Symbol for Enhancement-Type MOSFET

Fig. 16.22 (*a*) shows the circuit symbol for *N*-channel enhancement-type MOSFET. In this figure, the broken line indicates that there is no conducting channel between drain and source, when $V_{GS} = 0$. Due to this fact, this device is also known as "Normally-OFF MOSFET". The drain and source terminals are shown at the top and bottom end of the broken line. The substrate is internally connected to the source as shown. The arrow points in the direction of channel (or inversion layer), which is created when V_{GS} is increased above $V_{GS(\text{th})}$.

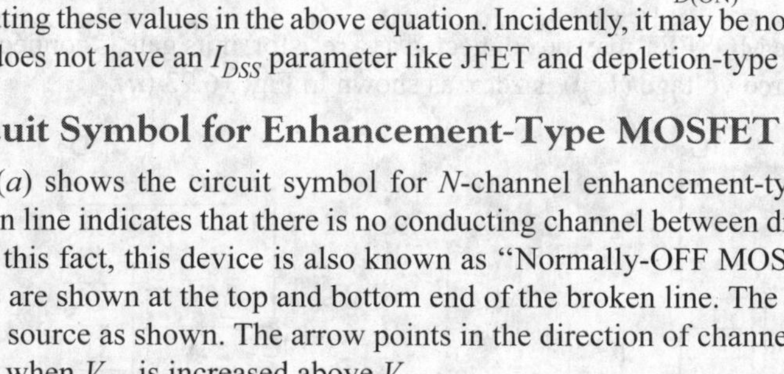

(*a*) N-channel. (*b*) P-channel.

Fig. 16.22. Circuit symbols for enhancement-type MOSFET.

Fig. 16.22 (*b*) shows the circuit symbol for the P-channel enhancement-type MOSFET. This symbol is similar to N-channel except that the arrow points outwards. This indicates that the channel (or invertion layer) created in the substrate is of P-type. In actual practice, the N-channel and P-channel enhancement-type MOSFET's are known as NMOS and PMOS field-effect transistors respectively.

Example 16.12.

The data sheet for a certain enhancement-type MOSFET reveals that $I_{D(on)} = 10$ *mA at* $V_{GS} = -12$ *V and* $V_{GS(th)} = -3$ *V. Is this device P-channel or N-channel? Find the value of* I_D, *when* $V_{GS} = -6$ *V.*

Solution.

Given: $I_{D(on)} = 10$ mA; $V_{GS} = -12$ volts and $V_{GS(th)} = -3$ volts.

Since the value of V_{GS} is negative for the enhancement-type MOSFET, this indicated that the device is P-channel. We know that the drain current,

$$I_D = K[V_{GS} - V_{GS(th)}]^2 \quad ...(i)$$

or

$$I_{D(on)} = K[(V_{GS} - V_{GS(th)})]^2$$

$$10 = K[-12-(-3)]^2 = 81\ K$$

$$\therefore \quad K = 10/81 = 0.12 \text{ mA/V}$$

Substituting this value of K and V_{GS} (equal to -6) in the equation (*i*),

$$I_D = 0.12[-6-(-3)]^2 = 1.08 \text{ mA} \quad \textbf{Ans.}$$

16.25. The MOSFET as a Resistor

The MOSFET's have an important property that they can be used as a resistor, capacitor, amplifier and a switch. This makes the design of electronic circuits very simple, because the entire circuit consists of only MOSFETs and no other component. The examples of such circuits are microprocessors and memory circuits.

The depletion type MOSFET may be connected as a resistor, if its gate is connected to the source so that the gate-to-source voltage (V_{GS}) is zero, as shown in Fig. 16.23 (*a*).

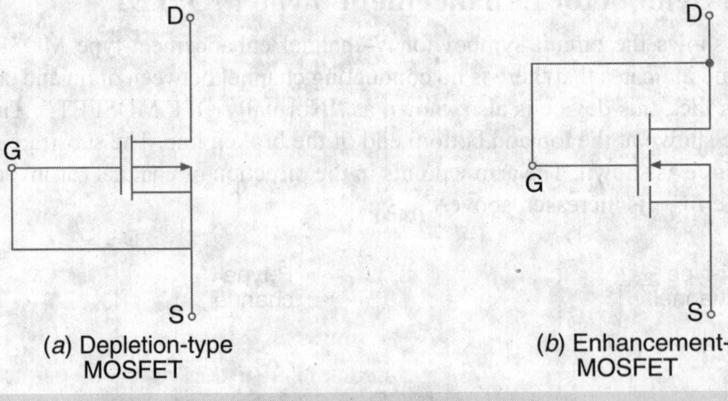

(*a*) Depletion-type MOSFET

(*b*) Enhancement-type MOSFET

Fig. 16.23. MOSFET as a resistor.

The enhancement-type MOSFET may be connected as a resistor, if its gate is connected to drain so that $V_{GS} = V_{DS}$ as shown in Fig. 16.23 (*b*).

16.26. Advantages of N-Channel MOSFETs Over P-Channel

These days N-channel MOSFETs have become much popular than P-channel for fabricating MOS circuits and systems. The P-channel MOSFETs find their use only in complementary metal-oxide semiconductor (CMOS) integrated circuits. The main advantage of N-channel MOSFET, over P-channel, is due to the fact that charge carriers in N-channel devices are the electrons, which have a mobility of about 1300 $cm^2/V.s$. On the other hand, the charge carriers in P-channel devices are the holes which have a mobility of about 500 $cm^2/V.s$. Since the current, in a semiconductor, is directly proportional to the mobility, therefore the current in a N-channel MOSFET is more than two times that of P-channel for the same dimension.

The ON resistance of the N-channel MOSFET is one-third of that for P-channel. It means that in order to achieve the same value of current and ON resistance, the P-channel MOSFET requires three times the area of an equivalent N-channel MOSFET. Thus electronic currents using N-channel MOSFETs are much smaller in size than those of P-channel. This results in higher packing density.

16.27. Complementary MOSFETS (CMOS)

MOSFETs are used extensively in the field of digital electronics to manufacture logic gates and many other synchronous and asynchronous circuits. The logic gates include AND, OR, NOT gates and their various combinations. The synchronous circuits include flip-flops, counters, memories, A/D and D/A converters. The asynchronous circuits include combination of logic gates such as decoders, encoders, multiplexers and demultiplexers.

A CMOS logic gate consists of a combination of NMOS and PMOS devices. The detailed discussion of CMOS devices is beyond the scope of this book.

16.28. Daul-gate MOSFETs

As a matter of fact, the operation of MOSFETs is limited at high frequencies because of their high gate to-channel capacitance. The cause of this high capacitance can be seen in Fig. 16.24. The metal

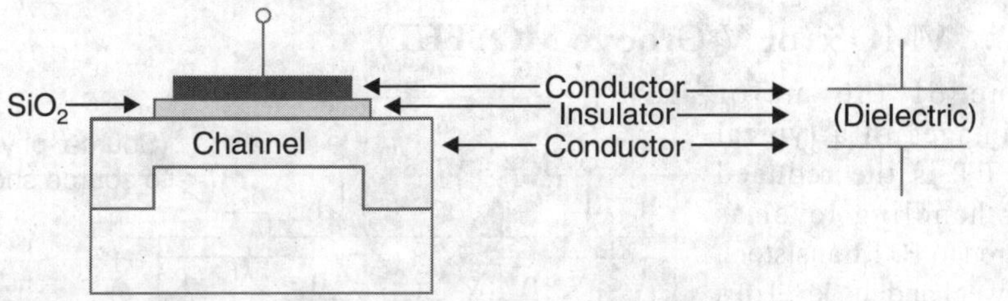

Fig. 16.24.

plate that is used for the gate is a conductor. The silicon dioxide between the gate and the channel in an insulator. The channel itself can be viewed as being a conductor to the silicon dioxide layer. Thus the combination of the three forms a parallel-plate capacitor.

The capacitance offered by this parallel plate capacitor can be reduced by employing a dual gate MOSFET structure. The physical construction and schematic symbols for the dual-gate MOSFET are shown in Fig. 16.25.

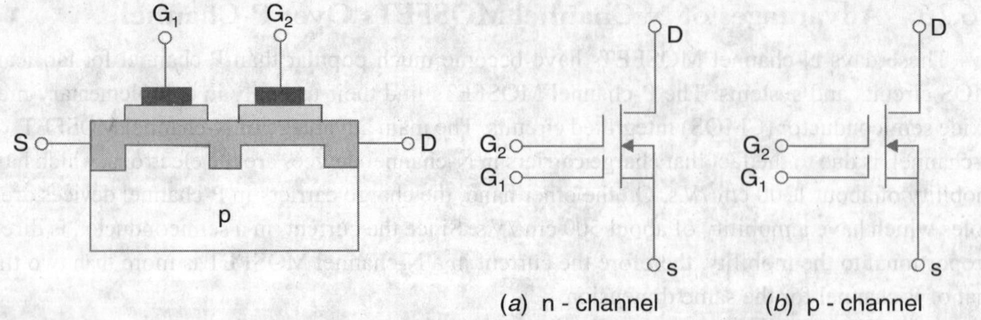

Fig. 16.25.

The reduced capacitance of the dual-gate MOSFET is the result of the way in which the component is used. Normally the component is used in such a way as toact as two series-connected MOSFETs. When the dual gate MOSFET is used as series MOSFETs, the effect is similar to connecting two capacitors in series Recall that the total capacitance in a sries connection is lower than either individual component value. Thus by connecting the two gates in a series configuration, the overall capacitance is reduced. This allows the dual-gate MOSFET to have a better high-frequency response as compared to a normal MOSFET.

16.29. Power MOSFETs

With the advancement of technology, the engineers have produced a wide variety of MOSFETs that are designed specifically for high current, high voltage and high power applications. Some applications for these components are found in cascode amplifiers, RF amplifier, power MOSFET derivers. The detail of these applications is beyond the scope of this book.

Following are some of the power MOSFETs which are important from the subject point of view.

1. Vertival MOSFET (VMOS) **2.** Lateral Double-Diffused MOSFET(LDMOS)

Both there power MOSFETs are discussed below one by one in the following pages.

16.30. VMOS (or V-Groove MOSFET)

One of the major disavantages of a typical MOSFET is the reduced power handling level as compared to BJT transistors. The power handling level of a typical MOSFET is less than 1W. This drawback of the MOSFET can be overcome by changing the construction mode from one of the planar nature to one with a vertical structure as shown in Fig. 16.26.

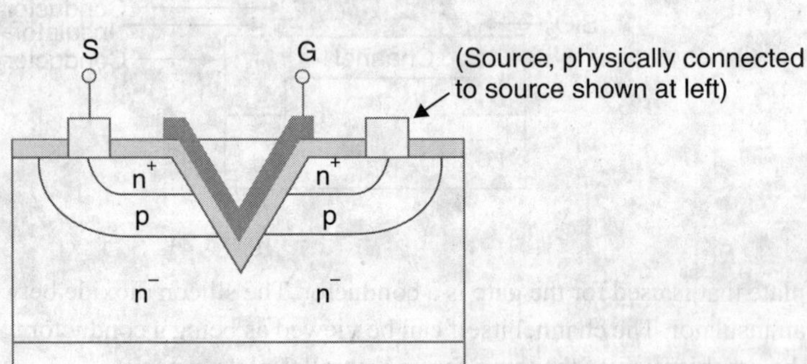

Fig. 16.26. Illustrating the construction of VMOS.

As seen form this figure, all the elements of the planar MOSFET are present in the vertical metal-oxide silicon FET (or simply VMOS) the metallic surface connection to the terninalnsis a component designed to handle much larger

The *vertival*-MOSFET (or simply VMOS) is a component disigned to handle much larger drain currents than the standard MOSFET. The current handling capability of the VMOS is a result of its physical consturction which is illustrated in Fig.16.26. As seen from this figure, the component has materials that are labeled as P, N^+ and N^-. The N- material labels indicate the differences in doping levels. Also notice that there is no physical channel connecting the source (at top) and the drain (at bottom). Thus VMOS is an *enchancement type* MOSFET.

With the V-shaped gate, a larger channel is formed by a positive gate voltage. With a large channel, the device is capable of handling a large amount of drain current. The operation of the VMOS is illustrated in Fig. 16.27 below.

VMOS Operation : When a positive gate voltage is applied to the device, an N-type channel forms in the P-type region. This effective channel connects the source to the drain. As seen from Fig. 16.27 the shape of the gate causes a wider channel to form than is created in the standard MOSFET. Because of this, the amount of drain current is much higher for this component. Moreover, the VMOS can exhibit a higher transconductance and a lower turn-on resistance than the conventional planer MOSFET.

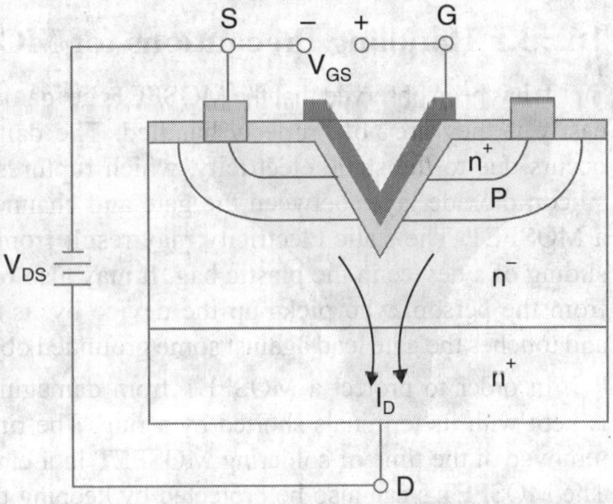

Fig. 16.27. Illustrating VMOS Opertor.

Another advantage of using VMOS is the fact that it is not susceptible to thermal runway. The VMOS has a *positive* temperature coefficient. This means that the resistance of the component *increases* when temperature increases. Thus an increases in temperature will cause a *decrease* in drain current.

The VMOS device can be fabricated with more than one V-groove to increase amount of drain current and some other performance characteristics.

16.31. LDMOS

The LDMOS (*i.e.*, lateral double-diffused MOSFET) is another type of power MOSFET. This MOSFET uses a very small channel region and a heavily doped N-type region (N^+) to obtain a high drain current and low channel resistance [$r_{d(ON)}$]. Fig. 16.28 shows the basic construction of LDMOS. As seen from this figure, the narrow channel (shaded region) is made up of the P-type material that lies between the N^-substrate (lightly doped) and the N^+ (heavily doped) source region. Since only the N-type material lies between the channel and the drain, the effecitve length of the channel is extrenely short. This coupled with the N-type matterial in the channel-to-drain path

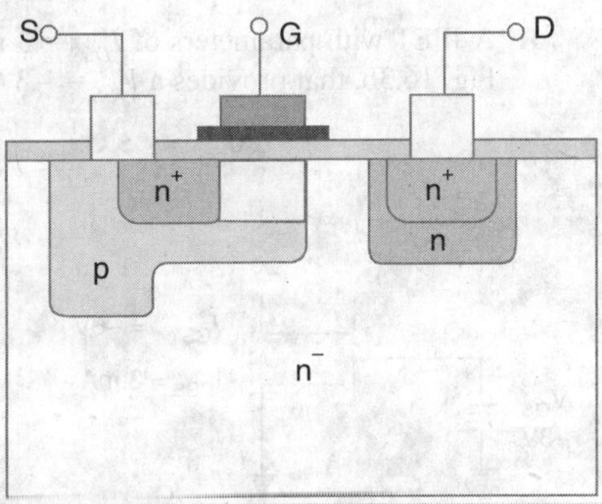

Fig. 16.28. LDMOS

provides an extermely low value of $r_{d(ON)}$. With a low channel resistance, the LDMOS device can handle very high amounts of current without generaing any damaging amount of heat.

The LDMOS has typical values of $r_{d(ON)}$ that are in the range of 2 Ω or less. With this low channel resistance, it is typically capable of handling currents as high as 20 A.

16.32. Multichannel FETs

To improve the FET's power handling capability, Multichannel devices have been stuied. The multichannel FET is basically many JFETs connected in parallel. The current-voltage characteristic of such devices is similar to a single-channel JFET except the drain current is substantially higher.

16.33. Handling Precautions for MOSFETs

It has been observed that the MOSFETs get damaged easily if they are not properly handled. The damage occurs due to the static electricity, which ruptures the silicon dioxide layer between the gate and channel of a MOSFET. The static electricity may result from the sliding of a device in the plastic bag. It may also result from the person, who picks up the device by its case and touches the gate lead against some grounded object.

In order to protect a MOSFET from damaging, it is kept with its terminals shorted by a ring. The ring is removed at the time of soldering MOSFET in a circuit. The MOSFET's can also be protected by keeping them in a conducting foam. The MOSFET's should never be inserted or removed from the circuits with power ON.

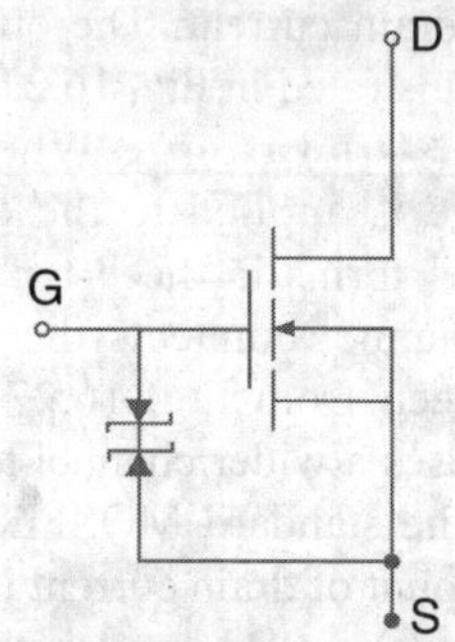

Fig. 16.29. Pair of zener diodes as a protection against static electricity.

These days, some of the MOSFETs have a built-in protection against static electricity and high voltages. This is done by connecting a pair of zener diodes back-to-back between the gate and source of a MOSFET as shown in Fig. 16.29. These diodes are designed in such a way that if the gate-to-source voltage exceeds say by + 10 V (or – 10 V) typically, one of the zener diodes will conduct and the other will breakdown. This provides a path for the excessive charge from gate-to-source. But this protection provides a minor disadvantage that it lowers the input resistance of the MOSFET.

NUMERICAL PROBLEMS

1. A JFET with parameters of I_{DSS} = 3 mA and $V_{GS(off)}$ = – 6 V is used in a circuit shown in Fig. 16.30, that provides a V_{GS} = – 3 V. Determine the value of I_D in the circuit.

(Ans. 1.33 mA)

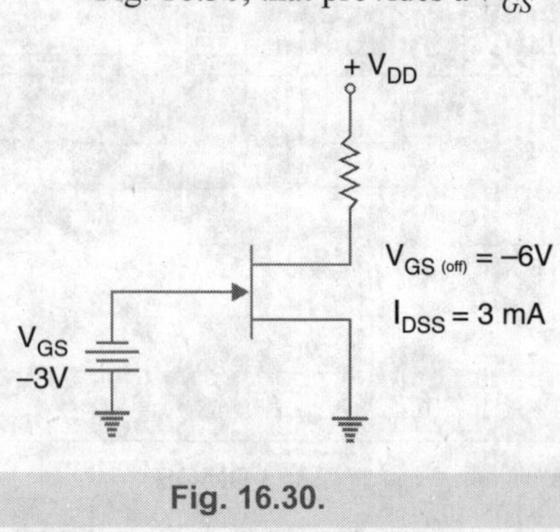

Fig. 16.30.

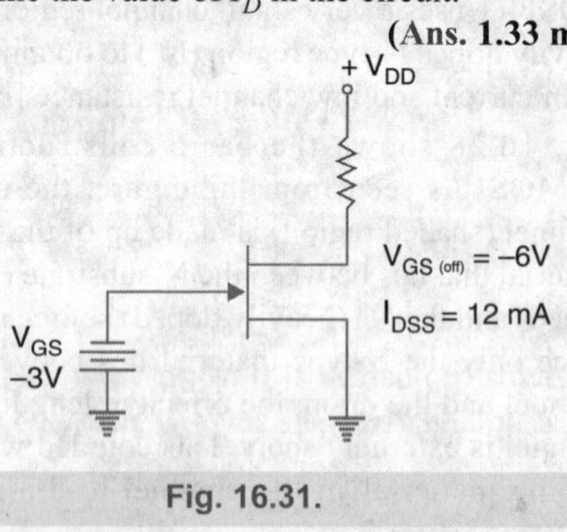

Fig. 16.31.

2. Determine the value of drain current for the circuit shown in Fig. 16.31. **(Ans. 6.32 mA)**
3. A JFET has parameters of $V_{GS(off)} = -6$ V and $I_{DSS} = 3$ mA. Plot the transconductance curve for the device using V_{GS} values of 0 V, – 1 V, – 3 V, – 5 V and – 6 V.
4. Plot the transconductance curve for a JFET with $V_{GS\ (off)} = -20$ V and I_{DSS} is 12 mA. Using V_{GS} values of 0 V, – 5 V, –10 V, – 15 V and – 20 V.

DESCRIPTIVE QUESTIONS

1. (*a*) Explain the working of a JFET (junction field-effect transistor).
   (*b*) Draw the V-I characteristics of an N-channel JFET.
   (*c*) Define dynamic drain resistance (r_d) and amplification factor (μ) for a JFET.
2. Define the following terms for a JFET:
   (*a*) The pinch-off voltage
   (*b*) Channel ohmic region
   (*c*) Drain resistance
   (*d*) Transconductance
   (*e*) I_{DSS}
   (*f*) I_{GSS}
3. Explain with the help of neat diagrams, the structure of a N-channel FET, and its volt-ampere characteristic. In what ways it is different from a bipolar junction transistor?
4. List the advantages and disadvantages of FET over bipolar transistors.
5. Bring out a neat comparison between a JFET and a MOSFET.
6. Describe the construction and explain the operation of depletion enhancement IGFET. Also draw the static characteristics.
7. Discuss with the aid of diagrams, the essential differences between enhancement mode FET's and depletion mode FETs.
8. Explain in details the construction and operation of Field effect transistor (FET). Also draw the small signal model for the FET. *(GBTU/MTU., 2008-09)*
9. With the help of the constructional diagram explain the working of a JFET. Sketch the two characteristics and explain them. *(M.G University., May 2005, PTU., 2007)*
10. Explain the construction, working and drain characteristics of a JFET. *(Bhartiya University., April 2004)*
11. Sketch the structure of a P-type channel depletion type MOSFET and explain its principle of operation with neat diagrams. Also sketch its V-I characteristics and circuit symbol for it. *(GBTU/MTU., 2008-09)*
12. Sketch the cross-section of a P-channel depletion type MOSFET and explain the operation of the device. Discuss and draw drain and transfer characteristics of it. *(GBTU/MTU., 2008-09)*
13. What is MOSFET? With the help of a suitable diagram, explain the working of different MOSFET. *(PTU., May 2005)*
14. How is a JFET used as a voltage variable resistance? Explain. *(Anna University., Nov/Dec 2007)*
15. Explain how FET is used as a VVR. *(M.G University., May 2005)*
16. How is an FET used as a voltage variable resistor? Explain. *(GBTU/MTU., 2008-09)*
17. Give the difference between the following:
    (*a*) JFET and MOSFET and
    (*b*) Enhancement MOSFET and depletion MOSFET. *(Bhartiya University., April 2004)*

18. What is the significant difference between the construction of an enhancement type MOSFET and depletion type MOSFET? Explain with suitable diagram.
(GBTU/MTU., 2005-06, 2004-05)

19. What are the advantages of the FET over a conventional bipolar junction transistor? Define pinch-off voltage, amplification factor and drain resistance of FET. Explain with the help of circuit diagram, how an FET is used as a voltage dependent resistor.
(GBTU/MTU., 2005-06)

20. Sketch the constructional diagram of a JFET biased for amplification, bringing out the shape of the deflection region. *(M.G University., May 2008)*

21. Explain the working of JFET along with its construction with neat sketch. Also draw its transfer characteristic curve with explanation. *(GBTU/MTU., 2005-06, 2006-07)*

22. Draw the structure of n channel JFET its drain and transfer characteristics. Indicate the different regions of operation in the drain characteristics. Why the name is "Field Effect Transistor"? *(GBTU/MTU., 2006-07)*

23. What is the major difference between a bipolar and a unipolar device?
(GBTU/MTU., 2006-07)

24. What is the major difference between bipolar and unipolar devices?
(Jamia University., 2008)

25. What are the advantages of FET over bipolar transistor? How is an FET used as a voltage variable resistance? Explain. *(GBTU/MTU., 2008-09)*

26. Explain the N-channel JFET working, drain characteristic and transfer characteristics.
(GBTU/MTU., 2009)

27. Explain hoe an FET can be used as a voltage controlled resistor
(GBTU/MNTU., 2004-05)

28. With an neat sketch explain the working of an N-channel JFET, Draw the transfer characteristics and explain pinch off. *(GBTU/MNTU., 2004-05)*

29. Explain the transfer and output characteristics of a depletion type MOSFET.
(Bhartiya University., April 2004)

30. With suitable explanation discuss the difference between the enhancement and depletion type of MOSFET. Also draw the characteristic and transfer characteristic curves
(GBTU/MTU., 2008-09)

31. Define the pinch-off voltage. *(Anna University., Nov/Dec 2007)*

32. Compare the performance of FET with MOSFET? *(PTU., May 2007)*

33. Describe construction, working and characteristic of MOSFET. *(PTU., Dec 2008)*

34. Explain how an FET is used as VVR? *(PTU., May 2007)*

35. State advantages of JFET over BJT. *(PTU., Dec 2007)*

36. Explain the VI characteristics of JFET. Label its different regions of operation. And also explain their physical significance. *(PTU., Dec 2006)*

37. What is physical significance of high input impedance of an FET? *(PTU., Dec 2006)*

38. How is a JFET used as a voltage variable resistance? Explain.
(Anna University., Nov/Dec 2007)

39. Differentiate an E-MOSFET from D-MOSFET. Explain their characteristics.
(M.G University., July/August 2003)

40. Explain the phenomenon of "Pinch off" in JFET. *(GBTU/MTU., 2006-07)*

41. Define Transconductance amplification factor and drain resistance of FET. *(GBTU/MNTU., 2006)*

42. Why FET is called voltage controlled device? Draw the circuit diagram of JFET common drain amplifier and drive the expression for its voltage gain. *(GBTU/MTU., 2006-07)*

43. Write the basic difference between enhancement and depletion MOSFET. Draw the structure of P-channel depletion MOSFET and explain working with drain and transfer characteristics. *(GBTU/MTU., 2006-07, 2004-05)*

44. How does MOSFET differ from JFET? *(PTU., Dec 2008, May 2006)*

45. Describe construction, working and characteristics on N-channel JFET? *(PTU., May 2009, May 2005)*

46. Explain in detail with diagram that how voltage applied at gate of FET controls the flow of carrier through channel. *(PTU., May 2008)*

47. What happens to the drain current of a P-channel JFET when a positive voltage is applied on its gate? *(PTU., Dec 2008)*

48. In what respect FETs better than BJT's. *(PTU., May 2009)*

49. What happens to the drain current of N-channel JFET when a negative voltage is applied on its gate? *(PTU., May 2009)*

50. Draw diagram showing constructional details of N-channel MOSFET. Explain the operating of N-channel MOSFET in depletion mode. *(PTU., May 2008)*

51. What do you understand by terms Channel and Drain in JFET? *(PTU., May 2008)*

52. What is meant by 'pinch of voltage' 'Pinch of Locus' of JFET? Mark Pinch of locus from drain characteristic *(JNTU., Aug./Sep. 2006)*

53. Explain giving illustrative diagrams how the pinch-off condition occurs in a MOSFET. *(JNTU., Aug./Sep. 2008)*

54. What is the parameters that control the pinch-off voltage of a JFET. *(M.G University., May/June 2006)*

55. Draw labeled diagram showing constructional features of N-channel MOSFET. Explain principle and working of N-channel MOSFET in briefly explains. *(PTU., Dec 2007)*

56. Draw and explain the VI characteristics of Depletion and enhancement type MOSFETs? *(PTU., Dec 2006)*

57. Give the drain current equation of JFET. *(Anna University., Nov/Dec 2007)*

58. State the advantages of FET over the BJT. *(Anna University., April/May 2008, PTU., May 2008)*

59. Describe the operation of N channel JFET with diagrams and characteristic curves. Explain the important parameters of JFET. *(Anna University., April/May 2008)*

60. Mention the applications of FETs. *(Anna University., Nov/Dec 2007)*

61. With neat diagram, Explain the operation, V-I and transfer characteristics of N-channel depletion type MOSFET. *(Anna University., Nov/Dec 2007)*

62. Compare depletion and enhancement type of MOSFETs. *(Anna University., Nov/Dec 2007)*

63. What is the difference between Bipolar Junction transistor & Field effect transistor? *(JNTU., June 2009)*

64. Define DC grain resistance. *(JNTU., June 2009)*

65. Define AC drain resistance. *(JNTU., June 2009)*

66. Define transconductance. *(JNTU., June 2009)*
67. Define amplification factor. *(JNTU., June 2009)*
68. Explain the principle of MOSFET in depletion mode with neat sketches and output characteristics. *(JNTU., May/Jun. 2006)*
69. Use proper diagrams to explain the structure of enhancement only type MOSFETs. Why are the devices so named? Can they be operated in the depletion mode? *(JNTU., Aug./Sep. 2008)*
70. Compare JFET and MOSFET with respect to various features. *(JNTU., Aug./Sep. 2006)*
71. List the advantages and disadvantages of FET over bipolar transistors. *(JNTU., June. 2009)*
72. Write with the help of static drain characteristics; explain the principle of a junction FET. *(Nagpur University., 2008)*
73. Show that the transconductance gain of a JFET is related to the drain current by
$$g_m = \frac{2}{V_p}\sqrt{I_{dss} \times I_{dss}}$$
(Nagpur University., 2008, Summer 2011)
74. Discuss Triac and power MOSFET with proper diagram and applications. *(R.G.P.V., Dec. 2008)*
75. Discuss basic characteristics and application of MOSFET. *(R.G.P.V., Dec. 2008)*
76. Compare an enhancement type MOSFET and depletion type MOSFET. Briefly describe operation of enhancement type MOSFET. *(R.G.P.V., Dec. 2007)*
77. Compare BJT and FET in their construction and working. How MOSFET can be compared with FET? *(R.G.P.V., May/ June 2006)*
78. What is the significant difference between the construction of an enhancement type MOSFET and a depletion type MOSFET? Briefly describe operation of enhancement type MOSFET. *(R.G.P.V., Dec. 2004)*
79. Discuss the construction details of a JFET. How is it different than of MOSFET? Draw their normal operating model circuit. *(R.G.P.V., June 2003)*
80. Derive the relation for calculating the threshold voltage of MOSFET considering zero voltage at gate. *(R.G.P.V., Dec. 2008)*
81. Explain the work of MOSFET. *(R.G.P.V., Dec. 2006)*
82. What is the significant difference between the construction of an enhancement type MOSFET and depletion type MOSFET ? Briefly describe operation of enhancement type MOSFET. *(R.G.P.V., Dec. 2004)*
83. Discuss the construction and principle of working of enhancement MOSFET. *(R.G.P.V., June 2010)*
84. Discuss the working principle of enhancement MOSFET. *(R.G.P.V., June 2009)*
85. Discuss the working of depletion type MOSFET. *(R.G.P.V., June 2009)*
86. Discuss the construction and working of depletion MOSFET. *(R.G.P.V., June 2010)*
87. In what ways in the construction of depletion type MOSFET similar to that of a JFET. In what ways is it different? Explain why the application of a positive voltage to the gauge of an N channel depletion type MOSFET will result in a drain current exceeding I_{DSS}. *(R.G.P.V., Feb 2010)*
88. Write short note on transfer characteristic of MOSFET. *(R.G.P.V., Dec. 2002)*
89. Discuss the characteristic of depletion type MOSFET. *(R.G.P.V., Dec. 2010)*
90. Draw the drain characteristics of depletion type MOSFET .Discuss with reasoning, its various region of operation. *(R.G.P.V., June 2011)*

91. Discuss the characteristics and working of enhancement type MOSFET. *(R.G.P.V., Dec. 2010)*

92. Explain enhancement MOSFET with working characteristics and necessary diagrams. *(R.G.P.V., Dec. 2008)*

93. Explain the basic operation and characteristics of N-channel depletion type MOSFET. *(R.G.P.V., Feb. 2010)*

94. Discuss the main application of the MOSFET. *(R.G.P.V., Dec. 2005)*

95. Derive the expression for V_{Dsat} and I_{Dsat} of MOSFET with diagram and curve. *(R.G.P.V., Dec. 2008)*

96. Discuss the construction of N -channel FET. *(R.G.P.V., June 2009)*

97. Explain the construction and working of N-channel FET. *(R.G.P.V., June 2010)*

98. Explain the principle of working of FET. What is voltage controlled resistance region ? *(R.G.P.V., Feb. 2010)*

99. Draw the circuit diagram and briefly discuss the drain characteristics of JFET. In the drain characteristics plot, mark various regions of operation and discuss their characteristics. *(R.G.P.V., June 2010)*

100. Explain the field effect phenomena in JFET. *(R.G.P.V., Dec. 2001)*

101. Draw the symbol and explain the characteristics of N-channel JFET and mark linear, saturation and break down regions. *(R.G.P.V., June 2002)*

102. Define the pinch-off voltage, V_P, sketch the depletion region before and after pinch-off. How FET does behaves in the region before pinch-off and after pinch-off. *(R.G.P.V., Dec. 2002)*

103. Explain the operation of a JFET. Also give its characteristics curves and enlist its merits and demerits over BJT. *(R.G.P.V., Dec. 2003)*

104. Explain the working of a JFET and give its characteristics curves. *(R.G.P.V., June 2004)*

105. What are the advantages of an FET over a bipolar transistor? Explain how depletion layer in FET formed. *(R.G.P.V., June 2005)*

106. Explain the working of FET with the help of equivalent model. *(R.G.P.V., June 2009)*

107. With the help of equivalent model explain the working of FET. *(R.G.P.V., June 2010)*

108. Draw and explain common source drain characteristics on an N-channel JFET. *(Nagpur University, Summer 2007)*

109. Explain how an FET is used as a voltage variable resistance. *(Nagpur University, Summer 2007, Winter 2004)*

110. Draw and explain common drain characteristics of an N-channel JFET. *(Nagpur University, Summer 2006)*

111. What do you understand by enhancement and depletion mode? *(Nagpur University, Summer 2006)*

112. Compare BJT and FET. *(Nagpur University, Winter 2005)*

113. What is pinch-off voltage. *(Nagpur University, Winter 2005)*

114. Draw and explain common source drain characteristics of an N-channel JFET. *(Nagpur University, Winter 2004)*

115. With the help of static drain characteristics. Explain the principle of junction FET. *(Nagpur University, Summer 2004, 2001)*

116. Write short note on biasing of JFET. *(Nagpur University, Summer 2004)*

117. Explain the enhancement effect of MOSFET. *(Nagpur University, Winter 2003)*

118. Give the order of magnitude of g_m, r_d and μ for MOSFET. *(Nagpur University, Winter 2003)*

119. Comment on the biasing of BJT and FET. *(Nagpur University, Winter 2003)*

120. What are the advantages of MOSFET over BJT and JFET? *(Nagpur University, Summer 2003)*

121. Explain the difference between JFET and BJT. *(Nagpur University, Winter 2001, Summer 2000)*

122. What do you understand by enhancement and depletion mode. Draw their symbols and output characteristics. *(Nagpur University, Winter 2001)*

123. Draw and explain the static output characteristics of a FET. Show that, FET exhibits voltage dependent resistance property in certain region of these characteristics. *(Nagpur University, Winter 2002)*

124. With the help of neat sketch, explain the working of JFET. Define pinch off voltage and I_{DSS}. *(Nagpur University, Summer 2000)*

125. Write short notes on :

(*a*) Enhancement mode MOSFET. *(Nagpur University, Winter 2006)*

(*b*) Compare BJT and FET. *(Nagpur University, Summer 2007, Winter 2006)*

(*c*) MOSFET *(R.G.P.V., June 2005, Dec. 2001)*

(*d*) Transfer characteristics of MOSFET *(R.G.P.V., Dec. 2002)*

126. Draw and explain the characteristics of JFET. *(RGTU., June 2009, 2010)*

127. Explain pinch-off voltage and threshold voltage. *(RGTU., June 2010)*

128. Explain the construction and operation of P-channel enhancement type MOSFET. *(RGTU., June 2010)*

129. Draw and explain the drain characteristics for N-channel deplection MOSFET. *(RGTU., June 2010)*

130. Using illustrations explain the operations of a P-channel JFET. Show the internal depletion region and explain their shapes. *(RGTU., June 2011)*

131. Draw and explain static drain and transfer characteristics of a JFET. *(Nagpur University, Summer 2011)*

132. Write the types of MOSFET and their two major differences. *(GBTU., 2011, 2012)*

133. Draw the structure of an N-channel JFET and explain its principle of operation. Also draw its drain and transfer characteristics with the help of suitable circuit. Describe how an FET can be used as a voltage controlled resistor. *(GBTU., 2011, 2012)*

134. Explain the working of deplection type MOSFET. *(RGTU., Dec. 2009)*

MULTIPLE CHOICE QUESTIONS

1. A field-effect transistor (FET) operates on

   (*a*) Majority carriers only (*b*) Minority carriers only

   (*c*) Positively charged ions only

2. In JFET operating above pinch-off voltage, the

   (*a*) Drain current remains practically constant

   (*b*) Drain current starts decreasing

   (*c*) Drain current increases rapidly

   (*d*) Depletion region becomes smaller

3. The JFET is oftenly called square-law device because its
   (a) Transconductance curve is parabolic
   (b) A.C. resistance from drain-to-source varies inversely as square of the drain current
   (c) Drain current varies as square of drain voltage for a fixed gate-to-source voltage
   (d) Reverse gate leakage current varies as a square of the reverse gate voltage
4. For the operation of depletion-type MOSFET, the gate voltage has to be
   (a) Low positive (b) High positive
   (c) High negative (d) Zero
5. The N-channel MOSFET devices are preferred more than P-channel's because
   (a) N-channel devices are faster than P-channel devices
   (b) N-channel devices consumes less power then P-channel devices
   (c) N-channel devices have higher packing density than P channel devices
   (d) Both (a) and (c) above
6. As compared to N-channel MOS switch, the P-channel MOS switch has
   (a) Same ON resistance (b) Less ON resistance
   (c) More ON resistance (d) either 'a' or 'b'
7. Thermal runway is not possible in FET because as the temperature of the FET increases:
   (a) the modify decreases (b) the transconductance increases
   (c) the drain current increases (c) the mobility increases

   (UPSC Engg. Service 2001)
8. The trans conductance 'g_m' of a JFET is equal to
   (a) $\dfrac{-2\,I_{DSS}}{V_P}$ (b) $\dfrac{2}{|V_P|}\sqrt{I_{DSS}\cdot I_D}$
   (c) $-\dfrac{2\,I_{DSS}}{V_P}\left\{1-\dfrac{V_{GS}}{V_P}\right\}$ (d) $\dfrac{I_{DSS}}{V_P}\left\{1-\dfrac{V_{GS}}{V_P}\right\}$
9. An FET is a better chopper than a BJT because it has :
   (a) lower off-set voltage (b) higher series ON resistance
   (c) lower input current (d) higher input impedance.
   (e) both (c) and (d)

   (UPSC Engg. Secvice 2000)
10. In a biased JFET, the shape of the channel is as shown in Fig. 16.32, because

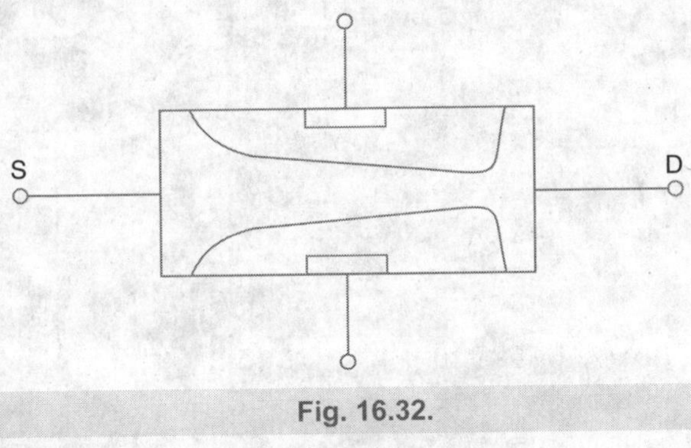

Fig. 16.32.

   (a) it is the property of the material used.
   (b) the drain end is more reverse biased than source end.

(*c*) the drain end is more forward biased than source end.
(*d*) the impurity profile varies with the distnce from the source.

(*UPSC Engg. Services 1999*)

11. Which of the following statement is not trave in case of FET.
(*a*) It has high input impedance
(*b*) It is less noisy than bipolar transistor.
(*c*) It has a large gain × band width
(*d*) all of the above.

12. The JFET is a
(*a*) current controlled device with high input resistance
(*b*) voltage controlled device with high input resistance
(*c*) voltage controlled device with low input resistance
(*d*) current controlled device with low input reisistance

13. The input impedance of a JFET is in the range of
(*a*) above 2 MΩ
(*b*) 200 to 400 kΩ
(*c*) 20 to 40 kΩ
(*d*) below 2 kΩ

(*Bharthiar Univ. April 2004*)

14. FET is
(*a*) current controlled device
(*b*) voltage controlled device
(*c*) resistance controlled device
(*d*) reactance controlled device

(*Chatrapati Shahuji Maharaj Univ. 2004*)

15. In a FET, 10 volts recerse voltage is applied. If gate current is 0.1 μA, the input resistance is,
(*a*) 1000 MΩ
(*b*) 10 MΩ
(*c*) 100 MΩ
(*d*) none of these

(*Chatrapati Shahuji Maharaj Univ. 2004*)

ANSWERS

1. (*a*)	**2.** (*a*)	**3.** (*a*)	**4.** (*c*)	**5.** (*d*)	**6.** (*c*)
7. (*b*)	**8.** (*a*)	**9.** (*e*)	**10.** (*b*)	**11.** (*c*)	**12.** (*b*)
13. (*a*)	**14.** (*b*)	**15.** (*c*)			

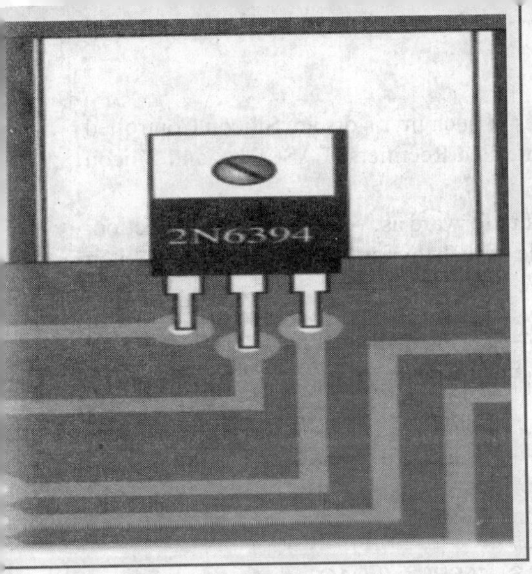

17

Chapter

Thyristors

Outline

1. Introduction.
2. Types of Thyristors.
3. Silicon Controlled Rectifier. (SCR)
4. SCR Biasing.
5. SCR Operation.
6. SCR Equivalent Circuit.
7. Turning ON (or Triggering) SCR.
8. Turning OFF SCR.
9. V-I Characteristics of SCR.
10. Forward Characteristic of SCR.
11. Reverse Characteristic of SCR.
12. SCR Ratings.
13. Series and Parallel combination of SCR
14. Series combination of SCR.
15. Parallel combination of SCR.
16. Applications of SCR.
17. Protection circuits
18. SCR crowbar circuit
19. Snubber circuit
20. Triac.
21. Triac Operation.
22. V-I Characteristic of Triac.
23. Triac Ratings.
24. Applications of Triac.
25. Difference between SCR and Triac.
26. Unijunction Transistor (UJT).
27. Basic Construction of UJT.
28. Equivalent Circuit of UJT.
29. Intrinsic Stand-off Ratio.
30. UJT Operation.
31. V-I Characteristic of UJT.
32. Applications of UJT.
33. Programmable Unijunction Transistor.
34. Diac.
35. V-I Characteristic of Diac.
36. Silicon Controlled Switch (SCS).
37. SCS Operation.
38. Applications of SCS.
39. Silicon Unilateral Switch (SUS).
40. Silicon Bilateral Switch (SBS).

17.1. Introduction

A thyristor is a semiconductor device with few layers of N and P type material. They act as bistable switches, conducting when there gate receives a current trigger. The switches continue to conduct until they are forward biased.

In other words thyristor is a four-layer PNPN device, which has three PN junctions. It has two stable switching states namely the ON or conducting state and the OFF or non-conducting state. There is no other state in between these two, as is in bipolar and field-effect transistors. The thyristors are used specifically for high power switching applications such as control of a.c. power to the load, motor speed control, light dimmers etc. Unlike BJT's and operational amplifiers (Op-amps), which can also be used as switches, most thyristors are not designed to be used as linear amplifying devices.

17.2. Types of Thyristors

Following are two main types of thyristors:

1. ***Unidirectional.*** The thyristors, which conduct in forward direction only, are

known as unidirectional thyristors. The examples of such thyristors are Silicon Controlled Rectifiers (SCR's), Light Activated Silicon Controlled Rectifiers (LASCR's), and Silicon Controlled Switch (SCS).

2. ***Bidirectional.*** The thyristors, which can conduct in forward as well as in reverse direction, are known as bidrectional thyristors. The example of such thyristors is a Triode A.C. Switch (TRIAC).

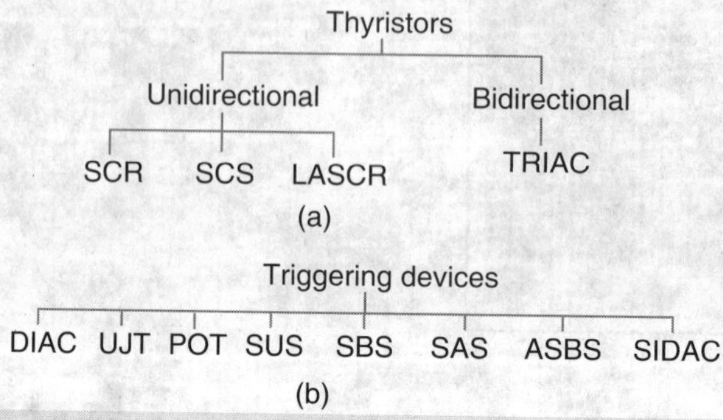

Fig. 17.1.

The thyristors require a control signal to switch from the non-conducting to the conducting state. The devices, which generate such signals, are called triggering devices.

Fig. 17.1 (*b*) shows the various types of triggering devices, which are available in these days. It includes Unijunction Transistor (UJT), Diode A.C. Switch (DIAC). Silicon Unilateral Switch (SUS), Silicon Bilateral Switch (SBS), silicon Asymmetrical swith (SAS), Silicon Axisymmetrical Silicon Bilateral Switch (ASBS), and SIDAC.

17.3. Silicon Controlled Rectifier (SCR)

These days, the Silicon Controlled Rectifier briefly written as (SCR) is one of the most important semiconductor device in the industrial or power electronics field. It is used as a controlled switch to perform a variety of functions such as rectification, d.c. to a.c. inversion and power control. It is an important element in the control of electrical motor speed, electrical furnace heat, lighting and many other uses.

Silicon controlled rectifier

A Silicon Controlled Rectifier (SCR) consists of four semiconductor layers forming a PNPN structure as shown in Fig. 17.2 (*a*). It has three PN junctions namely J_1, J_2 and J_3. There are three terminals called anode (*A*), cathode (*K*) and the gate (*G*) taken out as shown in the figure. The anode (*A*) terminal is taken out from the P_1-layer. The cathode (*K*) terminal is taken out from the N_2-layer, and the gate (*G*) terminal from the P_2-layer. The schematic symbol of SCR is shown in Fig. 17.2 (*b*). It may be noted that the SCR symbol is quite similar to that of a diode. In fact, the SCR resembles the diode electrically, since it conducts the current in one direction only, when forward biased. However, the SCR's different from a diode because it has an additional gate terminal. This gate is used to turn 'ON' the device.

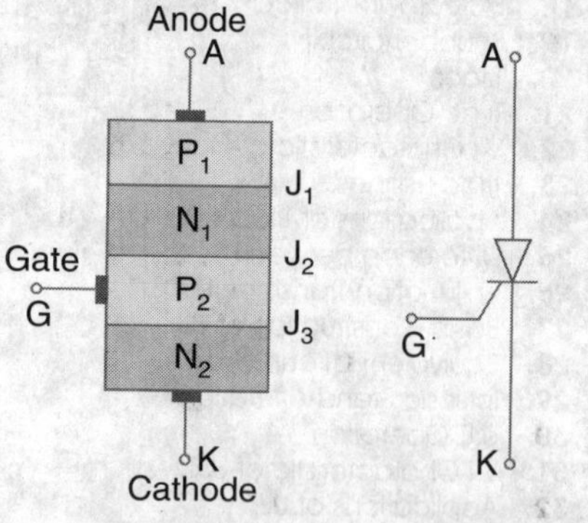

(*a*) Construction. (*b*) Symbol.

Fig. 17.2. Silicon controlled rectifier.

Fig. 17.3 (*a*) throngh (*f*) shown pictures of several SCRs. Notice the different shapes, sizes and the packages.

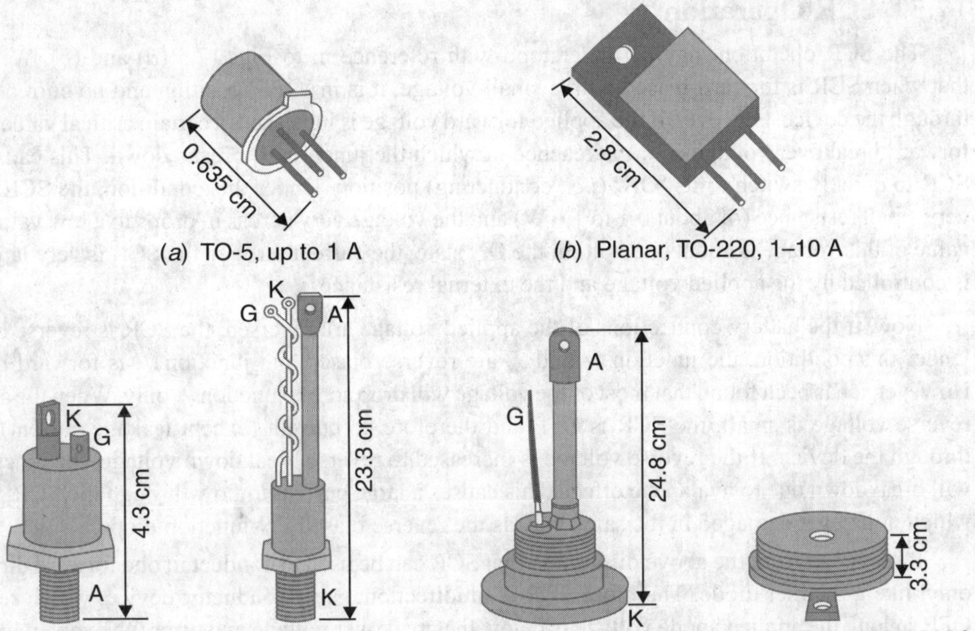

Fig. 17.3.

17.4. SCR Biasing

The SCR can be biased in two modes depending upon the polarity of the applied voltage across anode and cathode terminals. When the anode is positive, with respect to the cathode, the SCR is said to be forward-biased as shown in Fig. 17.4 (*a*).

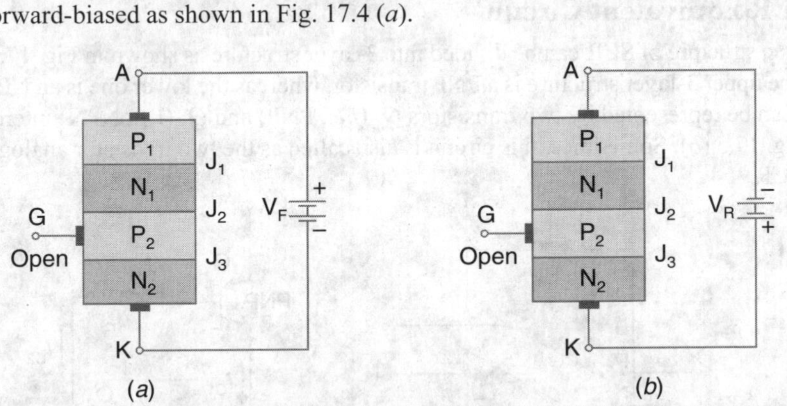

Fig. 17.4. SCR biasing.

In this mode, the junctions J_1 and J_3 are forward biased and the junction J_2 is reverse biased. There is no current (except leakage current) through the SCR. Therefore the SCR is in 'OFF' (non-conducting) state. Under this condition, the device offers a very high resistance. The value of this resistance is several megohms.

On the other hand, if the anode is negative, with respect to cathode, the SCR is said to be reverse biased as shown in Fig. 17.4 (*b*). In this mode, the junctions J_1 and J_3 are reverse biased and the junction J_2 is forward biased. Again there is no flow of current (except leakage current) through the SCR. Therefore the SCR is in 'OFF' (non-conducting) state.

17.5. SCR Operation

The SCR operation may be understood with reference in to Fig. 17.5 (*a*) and (*b*). We know that when SCR is forward-biased with a small voltage, it is in 'OFF' position and no current flows through the device. However, if the applied forward voltage is increased, a certain critical value called forward breakover voltage (V_{BO}) is reached at which the junction J_2 breaks down. This causes the SCR to quickly switch to its 'ON' (*i.e.*, conducting) position. Under this condition, the SCR offers very small resistance (of about 0.1 to 1.0 W) and the voltage very across it, drops to a low value. The value of this voltage is about one volt. In the ON state, the current through the SCR is very large and is controlled by the applied voltage and the external resistance.

Now, if the battery connections of the applied voltage are reversed, the SCR is reverse biased. Under this condition, the junction J_1 and J_3 are reverse biased and junction J_2 is forward biased. However, it has been found that most of the voltage will drop across junction J_1 only. When the applied reverse voltage is small, the SCR is OFF and therefore no current (except leakage current) flows through the device. If the reverse voltage is increased to reverse breakdown voltage, the junction J_1 will breakdown due to avalanche effect. This causes a large current to flow through the SCR, due to which it may get damaged in the same way as the reverse biased PN-junction diode.

It is evident from the above discussion that SCR can be used to conduct in one forward direction only, like a rectifier diode. Therefore SCR is unidirectional semiconductor device, which remains OFF so long the applied anode voltage is below the breakover voltage and turns ON when it exceeds the breakover voltage.

It will be interesting to know that SCR is never operated with the anode to cathode voltage (*i.e.*, supply voltage across SCR) equal to the forward breakover voltage. In fact, it is operated at a supply voltage much smaller than the forward breakover voltage. In that case, SCR is turned ON by the gate voltage and gate current.

17.6. SCR Equivalent Circuit

The basic structure of SCR can be divided into 3-layer structure as shown in Fig. 17.5 (*a*). It may noted that the upper 3-layer structure is a PNP transistor, whereas the lower one is an NPN transistor. Thus SCR can be represented by two transistors Q_1 (*i.e.*, PNP) and Q_2 (*i.e.*, NPN) interconnected as shown in Fig. 17.5 (*b*). Sometimes, this circuit is also called as the two transistor analogy or an ideal latch of a SCR.

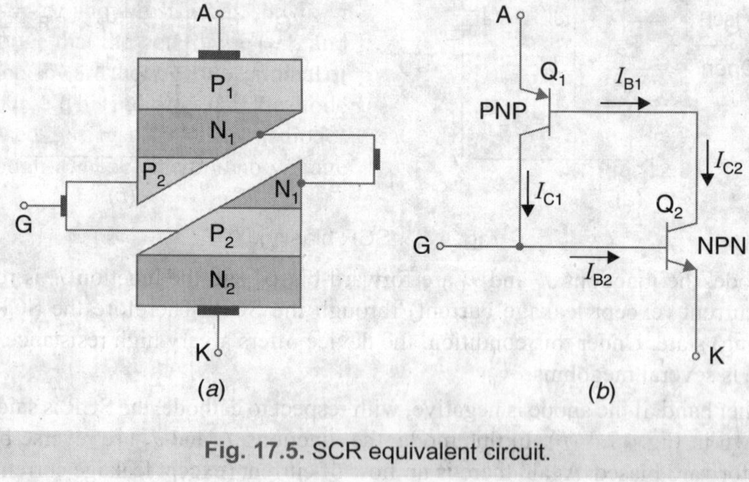

Fig. 17.5. SCR equivalent circuit.

It is evident from the figure that the collector of each transistor is connected to the base of other transistor. Therefore the collector current of transistor Q_1, is the base current of transistor Q_2, and the base current of transistor Q_1 is the collector current of transistor Q_2. For this circuit, we have an action of positive feedback or regeneration. It means that if there is a change in current, at any point in the loop (formed by transistors Q_1 and Q_2), it is amplified and returned to the starting point with the same phase. For example, if the base current of transistor Q_2 increases, the collector current of transistor. Q_2 will also *increase. It causes more base current through transistors Q_1 due to which the collector current of transistor Q_1 increases. This action will continue till both the transistors are driven into saturation. In this case, the SCR acts like a ON switch and it will pass the current from anode to cathode.

On the other hand, if the base current of transistor Q_2 decreases, the collector current of transistor Q_2 will also decrease. It causes the reduced base current through transistor Q_1 due to which the collector current of transistor Q_1 decreases. This action will continue till both the transistors are driven into cut-off. In this case, the SCR acts like a OFF switch and hence it will block the current from anode to cathode.

17.7. Turning ON (or Triggering) SCR

The SCR can be turned ON, from its OFF position, by several methods as discussed below:

1. *Forward breakover voltage.* If the voltage across the SCR exceeds the rated forward breakover voltage, the SCR will start conducting due to avalanche breakdown.

2. *Gate triggering.* This is the most commonly used method to trigger the SCR. In this method, the SCR is operated with an anode voltage slightly less than the rated forward breakover voltage and is triggered into conduction by a low-power gate pulse. It may be noted that once the SCR is switched ON, the gate has no further control on the device current. The gate pulse signals can be supplied either from a d.c. source or an a.c. source.

Fig. 17.6 (*a*) shows an SCR connected to the d.c. source through a load. In this case, the gate signal is generated by a push button switch (*S*). When the switch (*S*) is pressed, momentarily, a positive voltage is applied at the gate. As a result of this, the SCR is turned ON, and the current flows through the load. The SCR will remain in its ON position, until the supply voltage is removed or reversed.

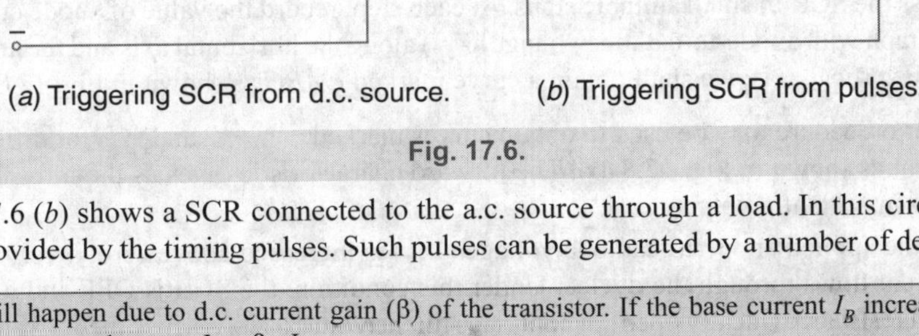

(*a*) Triggering SCR from d.c. source. (*b*) Triggering SCR from pulses.

Fig. 17.6.

Fig. 17.6 (*b*) shows a SCR connected to the a.c. source through a load. In this circuit, the gate signal is provided by the timing pulses. Such pulses can be generated by a number of devices, called

* This will happen due to d.c. current gain (β) of the transistor. If the base current I_B increases, then the collector current become $I_C = \beta \cdot I_B$.

triggering devices. Such devices are unijunction transistor, diac, silicon unilateral switch etc.

3. *Rate-effect or dV/dt triggering.* In this method, the SCR is turned ON by rapidly increasing the anode-to-cathode voltage. The rapidly increasing anode-to-cathode voltage produces a charging current, which triggers the SCR to conduction.

4. *Light triggering.* In this method, the SCR is triggered by irradiating it with light. When the light falls on the middle junction (J_2) of the SCR, the device turns ON. Such a device is called Light-Activated Silicon Controlled Rectifier (LASCR).

17.8. Turning OFF SCR

We have already discussed in the last article, that once the SCR turns ON (*i.e.*, starts conducting), it continues to conduct even when the gate signal is removed. This ability of the SCR to remain conducting, even when the gate signal is removed, is known as latching. It means that SCR cannot be turned OFF simply by removing the gate signal. A number of methods are used for turning OFF the SCR as mentioned below:

1. Reversing polarity of anode-to-cathode voltage.
2. Interrupting anode current by means of momentarily series or parallel switching arrangement. This method is known as *anode current interruption.*
3. Reducing the current through SCR below the holding current. This method is known as *forced commutation.*

17.9. V-I Characteristics of SCR

It gives the relationship between the anode current and the anode-to-cathode voltage of SCR for different values of gate current. The SCR has two types of V-I characteristics namely forward characteristic and reverse characteristics.

The forward characteristic may be obtained by using the SCR circuit having adjustable d.c. supplies V_{AA} and V_{GG} as shown in Fig. 17.7. The reverse characteristic may be obtained by reversing the connections of both the d.c. supplies.

An ammeter is used to measure the value of anode current, while the voltmeter is used to measure the anode-to-cathode voltage.

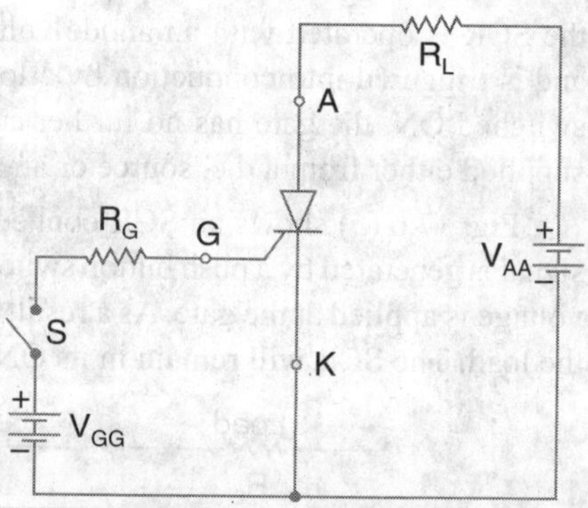

Fig. 17.7. SCR circuit.

17.10. Forward Characteristic

The forward characteristic of SCR may be obtained by using the circuit shown in Fig. 17.6. First of all, adjust the gate current to zero value by keeping the switch (*S*) open. Then increase the applied voltage across the SCR in small suitable steps. At each step, record the value of anode current. Now, if we plot a graph with anode-to-cathode voltage (V_{AK}) along the horizontal axis and the anode current (I_A) along the vertical axis, we shall obtain a curve marked *OABC* as shown in Fig. 17.8 (*a*).

A similar procedure may be used to obtain curves marked I_{G_1}, I_{G_2} and I_{G_3} for different values of gate-current as shown in Fig. 17.8 (*b*). The forward characteristic gives us the information about the following important points:

1. As the applied anode-to-cathode voltage (V_{AK}) is increased above zero, a very little anode current flows through the device. Under this condition, the SCR is OFF and it offers very high resistance (ideally open). It continues till the voltage V_{AK} reaches the forward breakover voltage marked by point *A* in Fig. 17.8 (*a*).

2. As the anode-to-cathode voltage exceeds the breakover voltage the SCR turns ON and the anode-to-cathode voltage decrease quickly to a value marked by point *B*. The value of breakover voltage is about one volt. At this stage, the current through the SCR increases rapidly to a large value, which is determined by the supply voltage and the value of load resistance in the circuit.

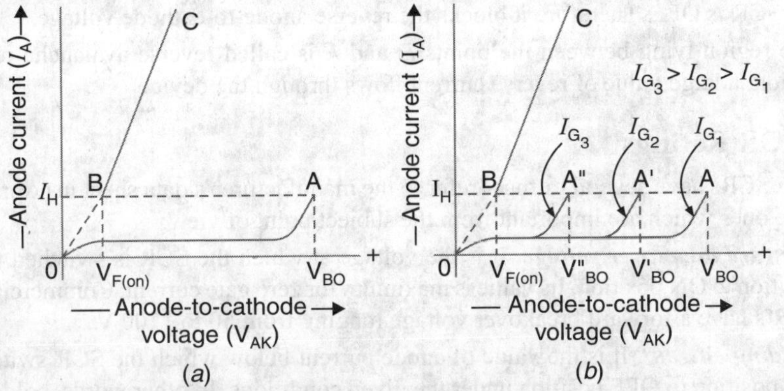

Fig. 17.8. Forward characteristic of SCR.

3. The current corresponding to the point *B* is called the holding current and is designated by the symbol I_H. It is the minimum value of anode current to keep the SCR is ON. If the anode current falls below the value of holding current, the SCR turns OFF as shown by the broken line *OB*.
4. As the value of gate current (I_G) is increased above zero, the SCR turns ON at lower breakover voltages as indicated by the points marked *A'* and *A''* in Fig. 17.8 (*b*) respectively.
5. The region lying between the points *O* and *A* is called forward blocking region. In this region, the SCR is OFF and blocks the forward anode-to-cathode voltage. This voltage is also known as forward blocking voltage.
6. The region lying between the points *B* and *C* is called forward conduction region. In this region, the SCR is ON and conducts current.

17.11. Reverse Characteristic of SCR

The reverse characteristic of SCR may be obtained by using the circuit as shown in Fig. 17.7 and reversing the connections of the d.c. supplies V_{AA} and V_{GG}. First of all, adjust the gate current to any suitable value. Then increase the reverse applied voltage in suitable steps and record the value of reverse anode current. Now, if we plot a graph with applied reverse anode-to-cathode voltage (V_R) along the horizontal axis and anode current (I_R) along the vertical axis, we shall obtain a curve marked *ODE* as shown in Fig. 17.9. The reverse characteristic of SCR gives us the information about the following important points:

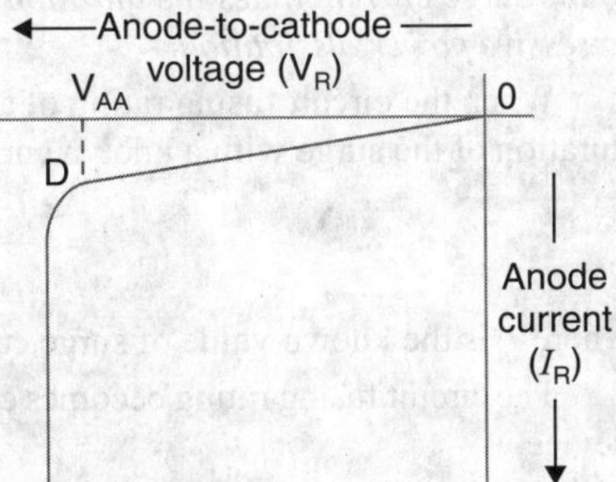

Fig. 17.9. Reverse characteristic of SCR.

1. As the applied reverse voltage (V_R) is increased above zero, a very little current (called leakage current) flows through the SCR. Under this condition, the SCR is OFF and it offers a very high resistance. It continues till the applied reverse voltage reaches breakdown voltage (V_{BR}).

2. As the applied reverse voltage is increased, above the breakdown voltage, the reverse current increase more rapidly as shown by the curve *DE*. This rapid increase is because of the breakdown of SCR (due to avalanche effect) and may damage the device if the current exceeds the rated value.
3. The region lying between the points *O* and *D* is called reverse blocking region. In this region, the SCR is OFF. Therefore it blocks the reverse anode-to-cathode voltage.
4. The region lying between the points *D* and *E* is called reverse avalanche region. In this region, a large value of reverse current flows through the device.

17.12. SCR Ratings

Athough SCRs have several ratrngs (refer to the manufacturer's data sheet in (Appendex B) yet following are ones which are important from the subject point of view:

1. ***Forward breakover voltage.*** It is the voltage at which the SCR is switched from its OFF position to ON position. Its value is maximum for zero gate current. Commercially available SCR's have a forward breakover voltage ranging from 50 to 1200 V.
2. ***Holding current.*** It is the value of anode current below which the SCR switches from its ON position to OFF position under the given conditions. In other words, holding current is the minimum value of current required to maintain conduction.
3. ***Gate trigger current.*** It is the value of anode current necessary to switch SCR from its OFF position to ON position under specified conditions.
4. ***Average forward currents.*** It is the maximum continuous value of anode current, which the SCR can handle in its ON position under specified conditions. It is also called current rating. Commercially available SCR's have a current rating ranging from 1 to 1800 A.
5. ***Reverse breakdown voltage.*** It is the value of reverse voltage from cathode-to-anode at which the device breaks into avalanche region and starts conducting heavily.

Other parameters, usually, included on the specification sheet of an SCR are the turn-on-time, turn-off-time, junction temperature and case temperature, circuit fusing (I^2t) rating, gate non-trigger voltage (V_{GD}) rating and critical rise (dv/dt) rating.

The circuit fusing (I^2t) rating indicates the maximum forward surge current capability of the SCR device. For example, the I^2t rating of 2N 682-92 series SCR's is 103 A^2s. *If the product of the square of the surge current times the duration (time) of the surge exceeds 103 A^2s, the SCR is destroyed by excessive power dissipation.*

When the circuit fusing rating of an SCR is known, we can determine the maximum allowable duration of the surge with a known current value as follows:

$$t_{max} = \frac{I^2t \text{ (rated)}}{I_s^2}$$

where I_s is the known value of surge current.

The circuit fusing rating becomes critical in applications where SCR is used as a surge protection device.

The gate nontrigger voltage (V_{GD}) rating indicates the maximum gate voltage that can be applied without triggering the SCR into conduction. If V_G exceeds this rating, the SCR will be triggered into the ON state. The V_{GD} rating is important because it points out one of the potential reason of false triggering of SCR. False triggering is a situation where the SCR is accidentally triggered into conduction, usually by some type of electrical noise. For example, 2N 682-92 series of SCRs has a V_{GD} rating of 250 mV. This means that if a noise signal with a peak value greater than 250 mV appears at the gate, the device may be triggered into conduction.

Another common reason of false triggering is a rise in anode voltage that exceeds the critical rise (*dv*/*dt*) rating of the SCR. This rating indicates the maximum rate of increase in anode-to-cathode voltage that the SCR can handle without false triggering occurring. The *dv*/*dt* rating of 2N 682-92 series SCRs is 30 V/ms. This means that a noise signal in V_{AK} with a rate of rise equal to 30 V/ms may cause false triggering.

Example 17.1.

The 2N 682 SCR is subjected to 40 A surge that lasts for 15 ms. Determine whether or not this surge will destroy the device. The circuit fusing rating of the device is 93 A^2s.

Solution.

Given: $I = 40$ A and $t = 15$ ms $= 15 \times 10^{-3}$ sec.

We know that I^2t value for the SCR,

$$= (40^2) \times (15 \times 10^{-3}) = 24 \text{ A}^2\text{s}.$$

Since this value is well below the maximum rating of 93 A^2s, therefore the device will not be destroyed.

Example 17.2.

The C35 series SCR have a circuit fusing rating of 75 A^2s. Determine the maximum allowable duration of a 100 A surge that passes through one of these devices.

Solution.

Given: $I^2t = 75$ A^2s and $I_s = 100$ A

We know that maximum allowable duration,

$$t_{max} = \frac{I^2t \text{ (rated)}}{I_s^2} = \frac{75}{(100)^2} = 7.5 \times 10^{-3} \text{ s} = 7.5 \text{ ms} \quad \textbf{Ans.}$$

17.13. Series and Parallel Combination of SCR

We have already discussed the characteristics of SCR in the previous article. In Industrial applications the required voltage and current rating are so high such that one SCR cannot fulfill the requirements.

The rated blocking voltage and rated forward current of an SCR decide the maximum power which can be controlled by a single device. A series combination of SCR is used where the voltage is higher than the capability of an individual device and the same is applicable for parallel combination of SCR for application of currents. High voltage convertors require high voltage and high current applications which ensure the series-parallel combination of devices.

17.14. Series Combination of SCR

When the input voltage is higher than the voltage rating of a single SCR, then the two or more SCRs are connected in series to increase their forward blocking capability as shown in Fig. 17.10. In which two SCRs, SCR_1 and SCR_2 are connected in series. Consider the series connection of SCR of the same class and rating are connected in series. As in any other device the characteristics of two SCRs are different. Two SCRs in series divide the voltage between them is inverse proportion to their leakage currents. This leads to an unequal distribution of voltage across the two series SCRs. Efficiency of SCR in series

$$\eta_s = \frac{\text{Total voltage rating of the series combination}}{\text{Individual voltage of thyristor} \times \text{Number of thyristors in series}}$$

When two SCRs are connected in series, the same current flows through the SCRs. But the voltage across SCR_1 V_1 is higher than that across SCR_2 is V_2 because the leakage current for SCR_2 is smaller than that of for the same voltage. Therefore, the two SCRs do not share the supply voltage equally. The maximum voltage that the SCRs can block is only $V_1 + V_2$ not 2 V_{BO}. To use the full forward blocking capabilities of each SCR, the forward blocking voltage must be equally distributed.

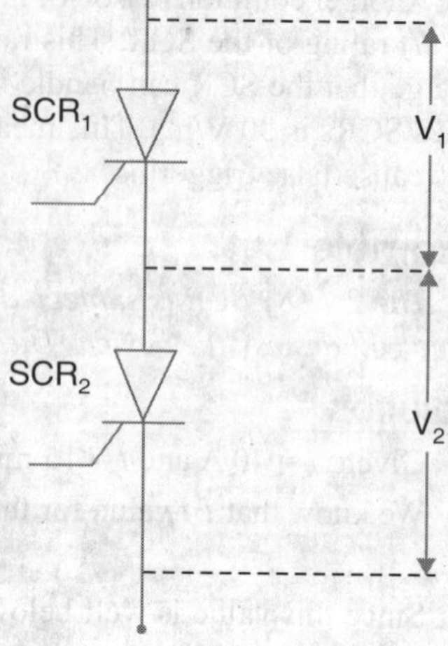

Fig.17.10

The forward V-I characteristic of SCRs is shown in Fig.17.11. Voltage V_{FBO1} and V_{FBO2} is the forward breakover voltage for SCR_1 and SCR_2 respectively. The leakage resistance V_1/I_0 for SCR_1 is very low compared to the leakage resistance V_2/I_0 for SCR_2. For small leakage current I_0, SCR_2 supports voltage V_2 which is nearly the point of breakdown whereas SCR_1 supports voltage V_1 which is less compared to V_2. Hence SCR with high-leakage resistance will share a larger portion of the applied voltage.

The equal sharing of voltage, forced by connecting a resistance across each SCR, called equalizing resistor or static equalizing circuit. The resistance is designed in such a way that the excess leakage current flows through the resistance. The static equalizing circuit is shown in Fig. 17.11. A nearly equal distribution of voltage during blocking is easily accomplished by connecting voltage-equalizing resistors R_1 and R_2 in parallel with each SCR such that each parallel combination has the same resistance as shown in Fig. 17.12(a). When the large numbers of SCRs are connected in series, we used the other method, which permits a reasonable uniform distribution of voltages. In this we use the same value of resistance in parallel with each SCR as shown in Fig. 17.12(b). This allows a different but fixed voltage to appear across each SCR. In this arrangement the SCR with the lower leakage current will have a greater portion of the blocking voltage than the SCR with the higher leakage current.

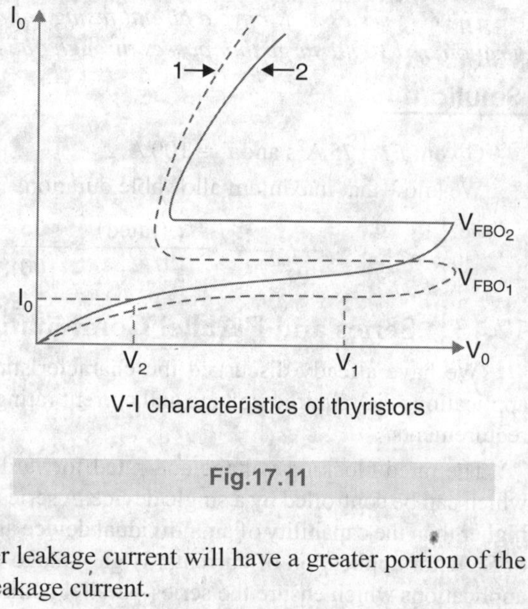

V-I characteristics of thyristors

Fig.17.11

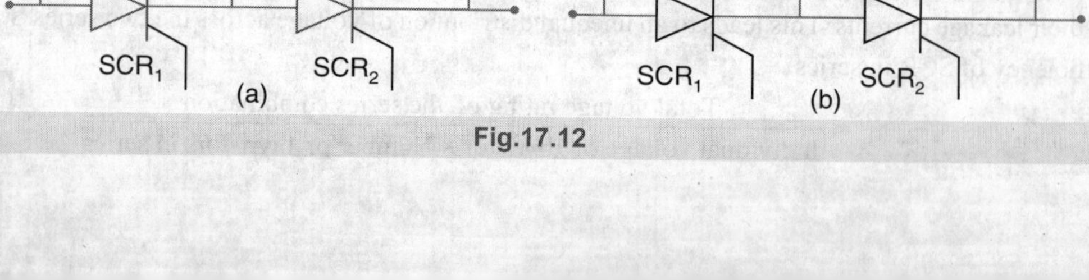

Fig.17.12

17.15. Parallel Combination of SCR

We know that a parallel combination of SCR is used where the current is higher than the capability of an individual device. When the load current exceeds the rating of a single SCR, then the SCRs are connected in parallel to increase their common current capability as shown in Fig. 17.13. When the SCRs are not perfectly matched, then the result of unequal sharing between them occurs. The efficiency of SCRs is parallel is given by,

Efficiency of SCRs in parallel

$$\eta_p = \frac{\text{Total current rating of the series combination}}{\text{Individual current of thyristor} \times \text{Number of thyristors in parallel}}$$

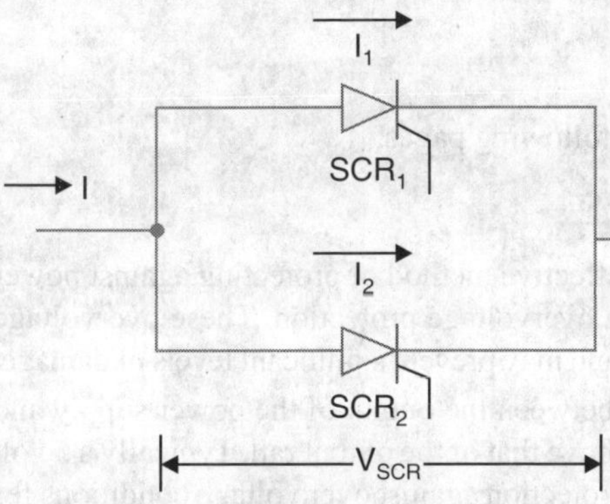

Fig. 17.13.

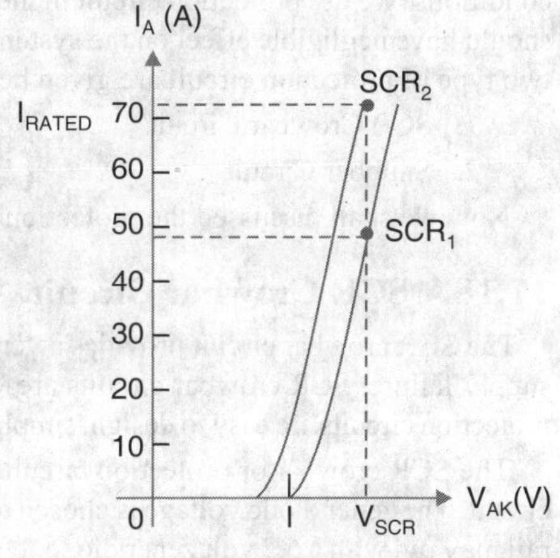

Fig. 17.14.

Fig. 17.14 shown the V-I characteristics of two SCRs, SCR_1 and SCR_2. The rating of the SCRs is the same. When these SCRs are connected in parallel they will have equal voltage drop across them. However due to their mismatch in characteristics SCR_2 is carrying the rated current I_2, while SCR_1 is carrying a current I_1. Current rating of SCR_2 is which is much less than SCR_1. The total rated current of the parallel connections is only $I_1 + I_2$ instead of $2I_2$.

Matched-pair SCRs are generally available for parallel connection but they are every expensive. With unmatched SCRs, equal current sharing is enforced by adding a low-values resistor or inductor in series with each SCR forced current sharing using equal-values resistors is shown in Fig. 17.15. The basic requirement is to make current I_1 close to I_2, a maximum difference of 20 % is acceptable. If we assume the voltage V_1 across SCR_1 to be greater than the voltage V_2 across SCR_2, the value of R can be obtained from

$$I_1R + V_1 = I_2R + V_2$$

$$R = \frac{V_1 - V_2}{I_2 - I_1}$$

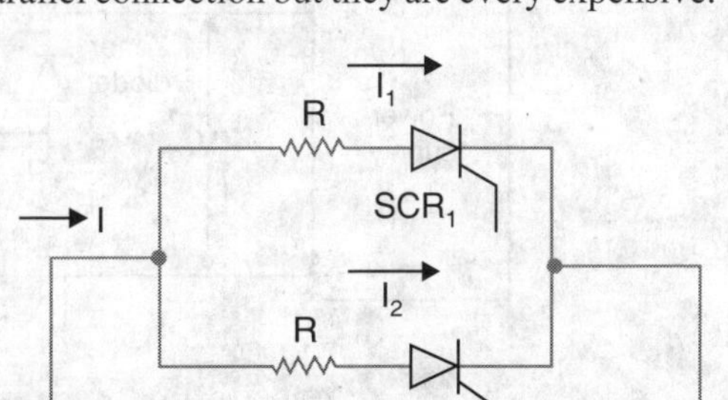

Fig. 17.15.

17.16. Applications of SCR

The SCR has a number of applications, yet the following industrial applications are important from the subject point of view:

1. Motor speed control **2.** Light-dimming control **3.** Heater control **4.** Phase control **5.** Battery charger **6.** Inverters **7.** Static switches **8.** Rectifier power supplies **9.** Relay control

17.17. Protection Circuits

A thyristor may be subjected to overvoltage's or over currents, high rates of change of current (di/dt) or rates of change of voltage (dV/dt) due to which the components may get damaged. In order to avoid these damages and for reliable operation of thyristor circuit it must be protected against such conditions. We use protection circuit in the SCR circuit to prevent the damage. The protection circuit should have negligible effect on the system during normal operation. It must be fast responding. The two type of protection circuit are given below:

1. SCR Crowbar Circuit
2. Snubber Circuit

Now we shall discussed the protection circuit in the following pages.

17.18. SCR Crowbar Circuit

The SCR crowbar circuit provides a very simple but effective method of protecting against power supply failure. SCR crowbar circuits are widely used in overvoltage protection. These overvoltage protection circuits are easy to design, simple to construct and may prevent significant levels of damage.

The SCR crowbar or protection circuit is connected between the output of the power supply and circuit. The zener diode voltage is chosen to be slightly above that of the output rail. Typically a 5 volt rail may run with a 6.2 volt zener diode. It provides the protection against over voltage conditions for entire circuit. The Fig. 17.16 shows SCR crowbar circuit. The zener diode in the circuit is selected such that at normal output voltage it acts as an open switch. Hence voltage across R is zero and SCR remains open.

When output voltage of the supply increase than the normal supply voltage by any reason, the zener diode conducts and a voltage appears across R. This voltage is sufficient to turn ON the SCR. The conduction of SCR reduces the voltage drop across it and thus protects the circuit from large over voltage.

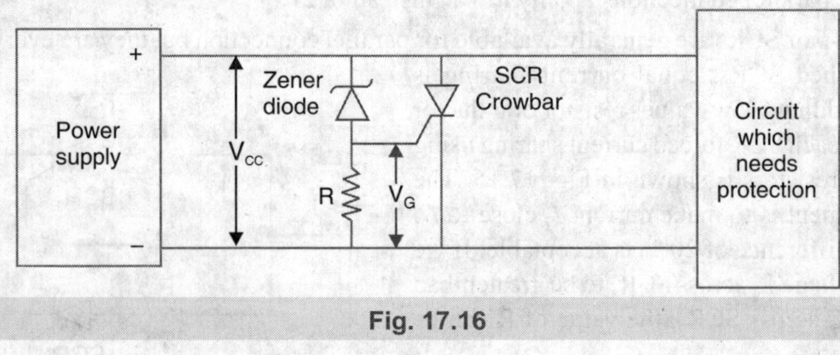

Fig. 17.16

17.19. Snubber Circuit

A snubber circuit consists of a series combination of resistance R_S and capacitance C_S in parallel with the SCR as shown in Fig. 17.17. A capacitor C_S in parallel with the device is sufficient to protect unwanted voltage rise of the SCR. When switch S is closed, then a sudden voltage appears across the circuit. Capacitor C_S behaves like a short circuit, therefore voltage across SCR is zero. After some time voltage across C_S builds up at a slow rate. Before SCR is fired by gate pulse, capacitor C_S is fully charged to voltage V_S. When the SCR is turned ON, capacitor discharges through the SCR and sends a current equal to local path formed by C_S and SCR. As this load resistance is quite low, the turn-on *di/dt* will tend to be excessive and as a result, SCR may be destroyed.

In order to limit the magnitude of discharge current, a resistance R_S is inserted in series with C_S as shown in Fig .17.17. Now when SCR is turned ON, initial discharge current (V_S/R_S) is relatively small and turn-on *di/dt* is reduced.

R_S, C_S and the load circuit parameters should be such that *dv/dt* across C_S during its charging is less than the specified *dv/dt* rating of the SCR and discharge current at the turn-on of SCR is within reasonable limits.

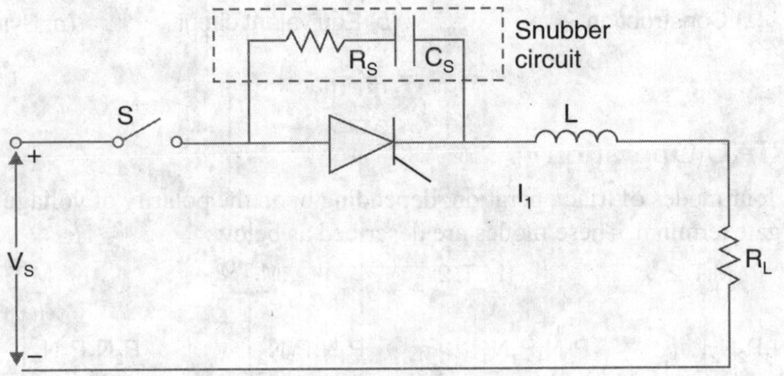

Fig. 17.17

The design of snubber circuit parameters is quite complex. In practice, designed snubber parameters are adjusted up or down in the final assembled power circuit so as to obtain a satisfactory performance of the power electronics system.

17.20. TRIAC

A triac is a three terminal device, which can conduct in either direction, when triggered either by a positive or a negative pulse irrespective of the polarity of the voltage across its main terminals. The triac behaves like two SCR's connected in parallel but in opposite directions, with a common gate terminal. *i.e.*, the anode of one SCR is connected to the cathode of the other and their gates are connected together. Thus anode and gate voltage applied in either direction will trigger the triac. It is due to the fact that the applied voltage will trigger at least one of the SCR's connected in opposite direction. The triacs are available with current ratings up to 25 A and voltage ratings up to 500 volts.

The basic construction of a triac is shown in Fig. 17.18 (*a*). It may be noted from this figure that the triac consists of two four layer switches in parallel. These switches are $P_1N_1P_2N_2$ and $P_2N_1P_1N_4$ as shown by the broken lines in the figure. Since the triac conducts in both directions, therefore the terminals are designated by numbers instead of anode and cathode as in SCR. The triac has two main terminals namely main terminal 1 [*i.e.*, (M.T. 1)] and main terminal 2 (*i.e.*, M.T.2) and one gate terminal (*G*). Fig. 17.18 (*b*) shows the equivalent circuit of a triac, which consists of two SCR's connected in parallel but in opposite directions with a common gate terminal. Fig. 17.18 (*c*) shows the schematic symbol of a triac.

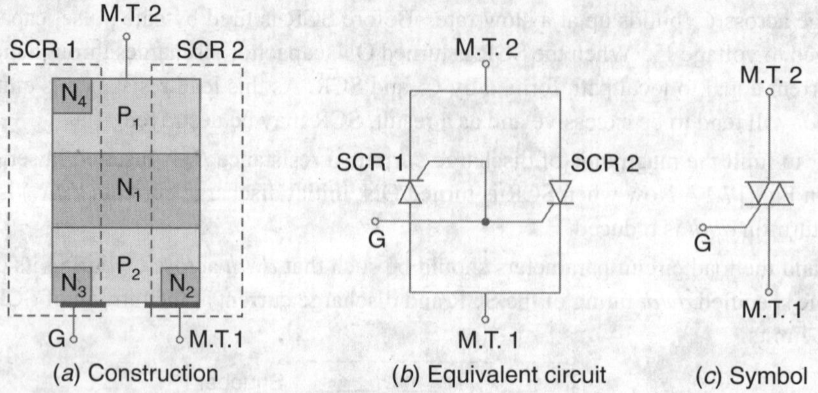

Fig. 17.18. Triac.

17.21. TRIAC Operation

There are four modes of triac operation, depending upon the polarity of voltage across its main terminals and gate terminal. These modes are described as below:

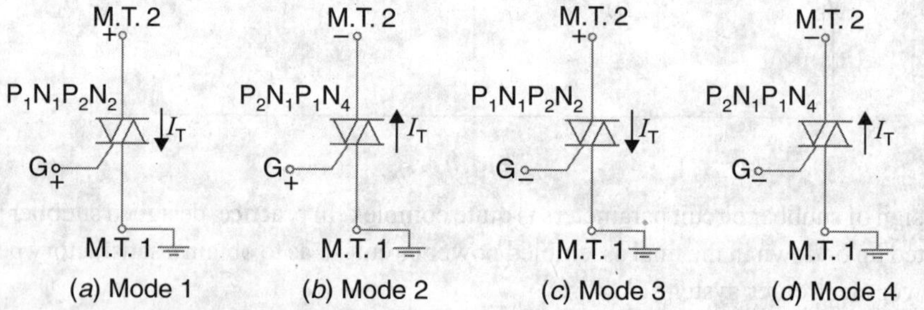

Fig. 17.19. Triac operation.

1. *M.T 2 is positive and G is positive*. In this mode, the operation of the triac is identical to SCR. The current flows through the switch $P_1N_1P_2N_2$ from M.T. 2 to M.T. 1 as shown in Fig. 17.19 (*a*).
2. *M.T. 2 is negative and G is positive*. In this mode, the current flows through the switch $P_2N_1P_1N_4$ from M.T. 1 to M.T. 2 as shown in Fig. 17.19 (*b*). This is an inefficient mode and must be avoided.
3. *M.T. 2 is positive and G is negative*. In this mode, the current flows through the switch $P_1N_1P_2N_2$ from M.T. 2 to M.T 1 as shown in Fig. 17.19 (*c*). This mode is less efficient than mode 1 but not as poor as mode 2.
4. *M.T. 2 is negative and G is negative*. In this mode, the current flows through the switch $P_2N_1P_1N_4$ from M.T. 1 to M.T. 2 as shown in Fig. 17.19 (*d*). This mode is slightly less efficient than mode 1.

The mode 1 and mode 4 are efficient modes in triac operation. Therefore these two modes are called normal modes of triac operation. However, as discussed for SCR's the triac is triggered with the gate pulses of proper polarity of d.c. voltages. The triac can be turned OFF only by reducing the device current below the holding value of the current.

17.22. V-I Characteristics of a TRIAC

It gives the relationship between the triac current and the voltage applied across its two main terminals. We have already discussed in the last article that a triac is operated usually in two ways (*i*) main terminal 2 (*i.e.*, M.T. 2) and gate (*G*) both are positive with respect to main terminal 1 (*i.e.*, M.T. 1) and (*ii*) main terminal 2 (*i.e.*, M.T. 2) and gate (*G*) both are negative with respect to main terminal 1 (*i.e.*, M.T. 1). In the first case, the triac current flows from the main terminal 2 to the main terminal 1. And in the second case, the triac current flows from the main terminal 1 to the main terminal 2 (*i.e.*, in a direction opposite to that of the first case).

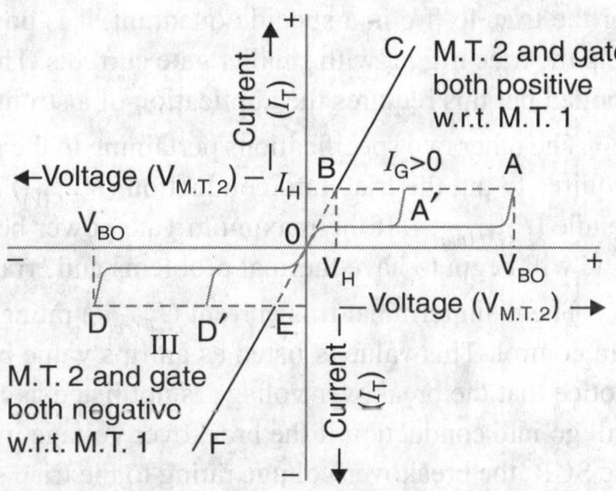

Fig. 17.20. *V-I* characteristics of a triac.

When the triac is operated, with its main terminal 2 and gate both positive with respect to main terminal 1, the V-I characteristic obtained is as shown in Fig. 17.20 by the curve marked *OABC*. Similarly, when the Triac is operated with its main terminal 2 and gate both negative with respect to main terminal 1, the V-I characteristic obtained is shown in Fig. 17.20 by the curve marked *ODEF*.

The V-I characteristic of a triac gives us the information about the following important points:

1. The curves OABC and ODEF are symmetrical and identical to the forward characteristic of silicon controlled rectifier (SCR).
2. The Triac is OFF until the applied voltage of either polarity (*i.e.*, whether M.T 2 is positive with respect to M.T. 1 or M.T. 2 is negative with respect to M.T. 1) exceeds the breakover voltage.
3. As the applied voltage of either polarity exceeds the breakover voltage, the triac turns ON and the voltage drop across the triac decreases to a low value (indicated by V_H). The triac current increases to a value determined by the supply voltage and load resistance.
4. As the value of gate current (I_G) is increased above zero, the breakover voltage is lowered, (indicated by points *A'* and *D'*). Like SCR, the triac is never operated with the zero gate current. When the gate current of a suitable value is applied (usually in the form of pulses), the triac turns ON at much lower breakover voltage.

17.23. TRIAC Ratings

The specification sheets for triacs (refer to Appendix B) lists several useful characteristics in circuit design and troubleshooting. These characteristic include:

1. Latching current (I_L)
2. Holding current (I_H)
3. Gate-trigger current (I_{GT})
4. Main terminal RMS current ($I_{T(RMS)}$)
5. Reverse current (I_R)
6. Breakover voltage (V_{DROM})
7. Gate-to-cathode voltage or gate voltage (V_{GT})
8. Maximum peak gate power ($P_{GM(max)}$)
9. Maximum average gate power ($P_{G(AV)}$)
10. On-state voltage across MT1 and MT2 (V_{TH}) also called forward voltage (V_T)

The latching current (I_L) is the amount of current the triac gate needs to stay in conduction after it has been initially turned on. The holding current (I_H) is the main terminal current required after the triac has been turned ON. The gate-trigger current (I_{GT}) is the amount of current that is necessary to go into conduction initially. It may be carefully noted that since the gate must be a specific polarity for the triac to fire in a specific quadrant, it is possible to apply bias voltage of similar polarity to help the triac trigger with smaller gate currents. The bias voltage of the opposite polarity can also be applied but this requires the application of a stronger gate current to get the triac into conduction.

The other two specifications pertaining to the gate refer to the average amount of power the gate requires to put the triac into conduction ($P_{G(AV)}$) and the maximum amount of power the gate can handle ($P_{GM(max)}$). If the maximum gate power begins to exceed the limit in the specifications, the triac will begin to have thermal problems and erratic triggering will occur.

The main terminal rms current ($I_{T(rms)}$) rating is the maximum amount of current that the triac can control. This value is listed as an rms value because the triac is primarily used in AC circuits. Notice that the breakover voltage is not listed as forward breakover like the SCR because the triac will go into conduction if the breakover voltage in either direction exceeds the specified level. Like the SCR, the breakover voltage rating to the triac should exceed the maximum peak circuit voltage, so it does not go into conduction if a spike of voltage occurs. Triacs are available commercially with typical breakover voltage ratings (V_{DROM}) that range from 100 to 1000 volts in 100 volts increments. The on-state voltage is the voltage measured across MT1 and MT2 when the triac is in conduction.

17.24. Applications of TRIAC

The triac has an important property that it can conduct current in either forward or reverse direction, depending upon the polarity of the voltage across its terminals. This property makes the triac very useful in a large number of industrial applications. Some of the triac application are given below:

1. Phase control **2.** Motor speed control **3.** Heater control **4.** Light-dimming control **5.** Static switch to turn a.c. power ON and OFF

17.25. Difference Between SCR and TRIAC

Table 17.1 gives some of the important points of difference between SCR and a triac.

Table 17.1 : Difference between SCR and Triac

S.No.	*SCR*	*Triac*
1.	It is unidirectional *i.e.*, it can conduct current in forward direction only.	It is bidirectional *i.e.*, it can conduct in forward as well as reverse direction.
2.	It is triggered by a narrow positive pulse applied at the gate terminal.	It is triggered by a narrow pulse of either positive or negative polarity applied at the gate terminal.
3.	SCR's are available with a large current rating.	Triacs are available for lower current rating as compared to that of SCR's.
4.	It has a fast turn-off and therefore can be used to switch a.c. supply frequencies up to few kHz.	It has a comparatively longer turn-off time than SCR. Therefore its use is limited to a.c. supply frequencies up to 400 Hz.

17.26. Unijunction Transistor (UJT)

Unijunction transistor

A unijunction transistor (abbreviated as UJT) is a three terminal silicon semiconductor device. As the name indicates, the UJT has only one PN junction like an ordinary diode. However, it is different from the ordinary diode in the sense that it has three terminals. It will be interesting to know that the behaviour of unijunction transistor is quite different from the other transistors like bipolar junction transistor (BJT) and the field-effect transistor (FET).

17.27. Basic Construction of UJT

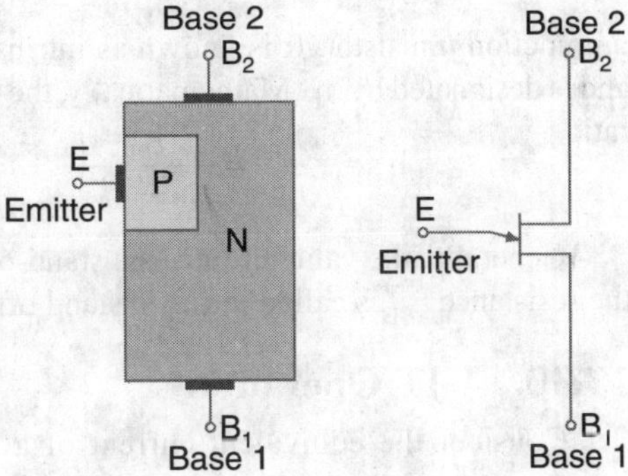

(a) Construction (b) Symbol

Fig. 17.21. Unijunction transistor.

The basic construction of a unijunction transistor (UJT) is as shown in Fig. 17.21 (*a*). It consists of an N-type silicon semiconductor bar and a P-type silicon region. The N-type bar is called a base and the P-type region as the emitter. Thus a PN junction is formed between the emitter and base regions. The emitter region is heavily doped, while the base region is lightly doped. Due to this reason, the resistivity of the base material is very high. Three terminals are taken out of the whole structure : one from the emitter region and two form the ends of the base region. These terminals are labelled as emitter (E), base 1 (B_1) and base 2 (B_2). It may be noted that the emitter region is shown closer to base 2 terminal than base 1. Fig. 17.21 (*b*) shows the schematic symbol of UJT.

It may be noted that the schematic symbol of UJT is different from that of JFET (junction field-effect transistor). The difference is that the arrow is at some angle in the schematic symbol of UJT.

17.28. Equivalent Circuit of UJT

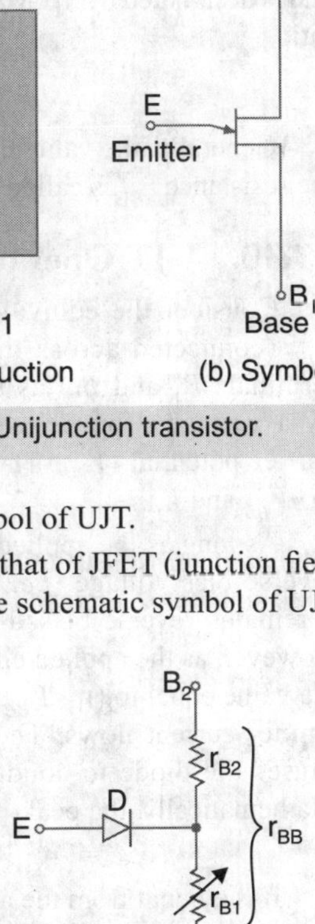

Fig. 17.22. UJT equivalent circuit.

Figure 17.22 shows the equivalent circuit of a unijunction transistor (UJT). It consists of a diode and a resistance. The diode (D) represents the PN junction, while the resistance (r_{BB}) is the internal bulk resistance of the silicon bar from one end to the other. In other words, the resistance r_{BB} represents the total resistance between the base terminals and is called the interbase resistance. The resistance r_{BB} is represented by the sum of two separate resistances r_{B1} and r_{B2} in the equivalent circuit of UJT. The resistances r_{B1} represents the bulk resistance between the emitter (E) and base 1 (B_1), whereas resistance r_{B2} is the bulk resistance between the emitter (E) and base 2 (B_2). Mathematically, the resistance

$$r_{BB} = r_{B1} + r_{B2}$$

When there is no voltage applied to the UJT, the value of resistance, r_{BB} is typically 5 to 10kW. The resistance r_{B1} is shown as a variable resistance in the UJT equivalent circuit. It is because of the fact that the value of resistance r_{B1} varies inversely with the emitter current (I_E). Depending upon the value of emitter current, the value of resistance r_{B1} can vary typically from 4 kΩ to 40Ω.

17.29. Intrinsic Stand-off Ratio

Consider the equivalent circuit of a unijunction transistor (UJT) with a battery voltage V_{BB} applied across its base terminals B_1 and B_2 as shown in Fig. 17.23. As the emitter is open, the applied voltage V_{BB}, divides itself across resistance r_{B1} and r_{B2}. The voltage across the resistance r_{B1},

$$V_1 = \frac{r_{B1}}{r_{B1} + r_{B2}} \times V_{BB} = \frac{r_{B1}}{r_{BB}} \times V_{BB}$$

The resistance ratio $\frac{r_{B1}}{r_{BB}}$ is an important characteristic of unijunction transistor. It is known as intrinsic stand-off ratio and is designated by η. Mathematically, the intrinsic stand-off ratio,

$$\eta = \frac{r_{B1}}{r_{BB}} = \frac{r_{B1}}{r_{B1} + r_{B2}}$$

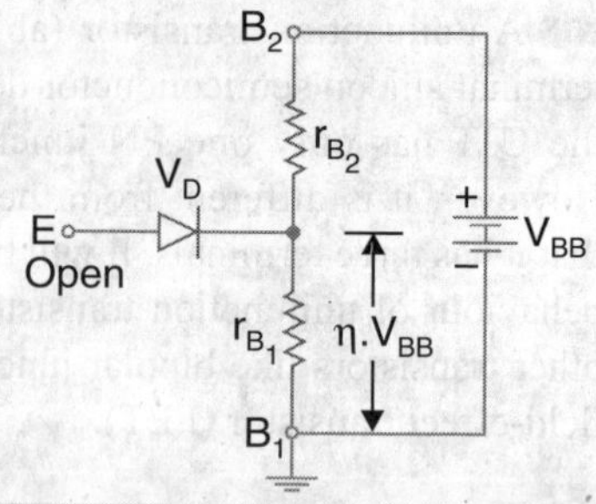

Fig. 17.23. UJT equivalent circuit with battery V_{BB}

Generally, the value of intrinsic stand-off ratio is between 0.5 and 0.8. The voltage drop across the resistance r_{B1} is called intrinsic stand off voltage. It reverse biases the emitter diode.

17.30. UJT Operation

Consider the equivalent current of a unijunction transistor (UJT) with the voltage source V_{EE} (connected across the emitter and base 1 terminal) and V_{BB} (connected across the base terminals B_1 and B_2) as shown in Fig. 17.24. Hence, the emitter diode (*D*) is reverse biased by a voltage drop across the resistance r_{B1} (whose value is equal to $\eta \cdot V_{BB}$) and its own barrier potential (V_D). Thus the total reverse bias voltage across a diode is equal to the sum of $\eta \cdot V_{BB}$ and V_D.

As long as the applied emitter voltage is below the total reverse bias voltage (*i.e.*, $\eta \cdot V_{BB} + V_D$) across the diode, it remains reverse biased. And there is no emitter current. However, as the applied emitter voltage reaches (or exceeds) the value equal to ($\eta \cdot V_{BB} + V_D$), the diode conducts and the emitter current flows. The value of emitter voltage, which causes the diode to conduct, is called peak-point voltage. Mathematically, the peak-point voltage,

$$V_P = \eta \cdot V_{BB} + V_D$$

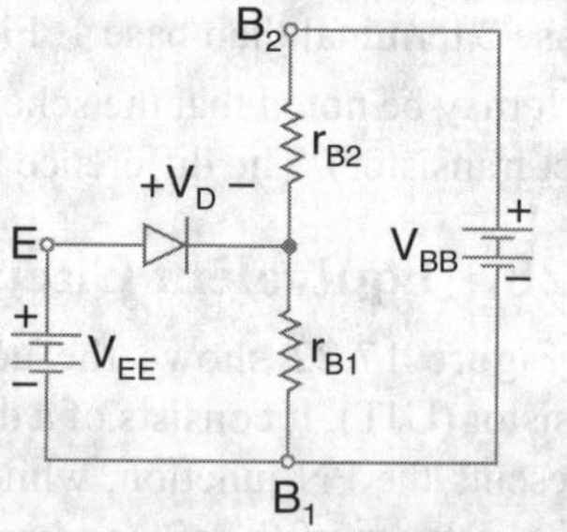

Fig. 17.24. UJT equivalent circuit with V_{EE} and V_{BB} supplies.

It is evident from the above discussion that as the emitter voltage reaches the peak-point voltage, the diode conducts (*i.e.*, it becomes forward biased) and the emitter current begins to flow. Under this condition, the uninjunction transistor is said to be fired, triggered or turned ON. At this instant, the holes from the P-type emitter region are injected into the base region and are swept by the electric field towards the base terminals B_1. The presence of excess holes, slightly reduces the resistance r_{B1} which in turn reduces the intrinsic stand-off voltage ($\eta \cdot V_{BB}$). This action is called conductivity modulation because the conductivity of the material between the emitter and base of terminals increases as the holes are injected into it. It is, obviously, a regenerative process because a smaller value of intrinsic stand-off voltage results in a stronger forward bias across the diode. Due to this, more holes are injected and the intrinsic stand-off voltage ($\eta \cdot V_{BB}$) is further reduced. As a result of this, the emitter current increases, while the voltage at the emitter (V_E equal to $\eta \cdot V_{BB} + V_D$) decreases. It produces a negative resistance region in the V-I characteristic of UJT. And the UJT switches from its OFF position to ON position.

Notes: 1. When the diode is forward biased, the current entering from the positive terminal of the supply voltage V_{BB} in the base 2 terminal also increases. Therefore it is necessary to place an external resistor (called current limiting resistor) to the UJT in series with base 2 terminal. The value of current limiting resistor for the UJT's 2 N 2646 and 2 N 2647 is given by the relation,

$$R = \frac{1000}{\eta \cdot V_{BB}}$$

The value of resistor (R) ranges from 100 to 200 Ω.

2. Sometimes, it is required to obtain the output voltage from base of terminal of UJT. This can be done by placing a resistor of small value (ranging from 39 Ω to 47 Ω) in the base 1 lead.

3. A UJT can be switched ON from its OFF position by applying a positive trigger at its emitter terminal. It can be brought back to its OFF position by applying a negative trigger at the emitter terminal.

4. The output of UJT can be taken at anyone of the three terminals *i.e.*, emitter, base 1 or base 2.

Example 17.3.

Determine the peak-point voltage (V_P) value for the 2N 4870 UJT shown in Fig. 17.25. Take $V_D = 0.7$ V.

Solution.

Given: $V_D = 0.7$ volts; $\eta = 0.75$ (max) and $V_{BB} = 12$V

We know that peak-point voltage of the circuit,

$$V_P = \eta \cdot V_{BB} + V_D$$
$$= (0.75 \times 12) + 0.7 = 9.7 \text{ V } \textbf{Ans.}$$

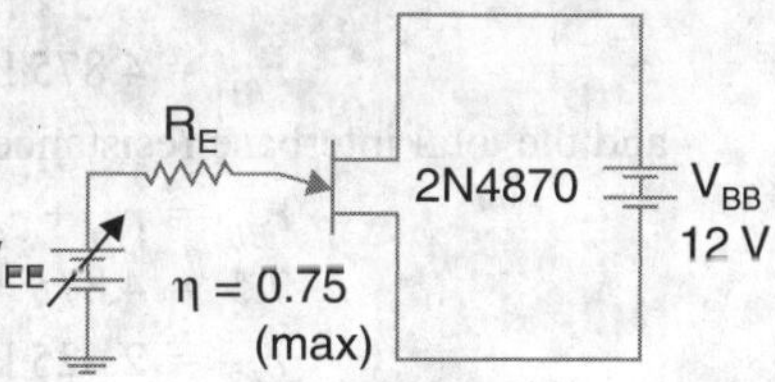

Fig. 17.25.

Example 17.4.

A unijunction transistor used for triggering the silicon controlled rectifier has $r_{B1} = 4$ kΩ and $r_{B2} = 2.5$ kΩ. Find (a) the value of intrinsic stand-off ratio, and (b) the peak-point voltage, if $V_{BB} = 15$ V. Take the barrier potential as 0.7 V.

Solution.

Given: $r_{B1} = 4$ kΩ $\quad r_{B2} = 2.5$ kΩ $V_{BB} = 15$ V and $V_D = 0.7$ V

(*a*) *Intrinsic stand-off ratio*

We know that intrinsic stand-off ratio,

$$\eta = \frac{r_{B1}}{r_{B1} + r_{B2}} = \frac{4}{4 + 2.5} = 0.62 \textbf{ Ans.}$$

(*b*) *Peak-point voltage*

We also know that the peak-point voltage,

$$V_P = \eta \cdot V_{BB} + V_D = (0.62 \times 15) + 0.7 \text{ V}$$
$$= 10 \text{ V } \textbf{Ans.}$$

Example 17.5.

The intrinsic stand-off ratio for a UJT is determined to be 0.60. A measurement of its interbase resistance indicates 7 kΩ. What are the UJT's static values of r_{B1} and r_{B2}?

Solution.

Given: $\eta = 0.60$ and $r_{BB} = 7$ kΩ.

We know that intrinsic stand-off ration (η)

$$0.60 = \frac{r_{B1}}{r_{BB}} = \frac{r_{B1}}{7}$$

$\therefore \quad r_{B1} = 0.60 \times 7 = 4.2 \text{ k}\Omega$ **Ans.**

and the total interbase resistance (r_{BB}),

$$\therefore \quad 7 = r_{B1} + r_{B2} = 4.2 + r_{B2}$$

$$r_{B2} = 7 - 4.2 = 2.8 \text{ k}\Omega \textbf{ Ans.}$$

Example 17.6.

The intrinsic standoff ratio of UJT is 0.65. If the interbase resistance is 7.5 kΩ, determine r_{B1} and r_{B2}? (*Anna University, Nov/Dec. 2007*)

Solution.

Given: $\eta = 0.65$ and $r_{BB} = 7.5 \text{ k}\Omega = 7.5 \times 10^3 \ \Omega$

We know that intrinsic stand-off ration (η),

$$\eta = \frac{r_{B1}}{r_{BB}}$$

$$0.65 = \frac{r_{B1}}{7.5 \times 10^3}$$

$$\therefore \quad r_{B1} = 4.875 \text{ k}\Omega \textbf{ Ans.}$$

and the total interbase resistance (r_{BB}),

$$r_{BB} = r_{B1} + r_{B2}$$

$$7.5 = 4.875 + r_{B2}$$

$$r_{B2} = 2.625 \text{ k}\Omega \textbf{ Ans.}$$

Example 17.7.

If I_{B1} and I_{B2} through the base resistors of differential amplifier are 1.65μA and 1.75μA respectively, calculate the input offset current and input bias current of the differential amplifier.

(*Anna University., April May 2008*)

Solution.

Given: $I_{B1} = 1.6$ mA, $I_{B2} = 1.75$ mA.

We know that the value of input offset current,

$$I_{i0} = I_{B1} - I_{B2}$$

$$= 1.65 - 1.75 = 0.10 \text{ mA} \textbf{ Ans.}$$

We also know that the value of input bias current,

$$I_B = \frac{(I_{B1} + I_{B2})}{2} = \frac{(1.65 + 1.75)}{2} = 1.70 \ \mu\text{A} \textbf{ Ans.}$$

17.31. V-I Characteristic of UJT

Fig. 17.26 shows the V-I characteristic of UJT. There are two important points on the characteristic curve namely the peake-point and the valley-point. These points divide the curve into three important regions *i.e.*, cut-off region, negative resistance region and saturation region. These regions are explained below:

1. ***Cut-off region.*** The region, to the left of peak-point, is called cut-off region. In the region, the emitter voltage is below the peak-point voltage (V_P) and the emitter current is approximately zero. The UJT is in its OFF position in this region.
2. ***Negative resistance region.*** The region, between the peak-point and the valley-point, is called negative-resistance region. In this region, the emitter voltage decreases from V_P to V_V

and the emitter current increases from I_P to I_V. The increase in emitter current is due to the decrease in resistance r_{B1}. It is because of this fact that this region is called negative-resistance region. It is the most important region from the application point of view. For example, when the UJT is operated as an oscillator, it works in the negative-resistance region.

3. ***Saturation region.*** The region, beyond the valley point, is called saturation region. In this region, the device is in its ON position. The emitter voltage (V_E) remains almost constant with the increasing emitter current.

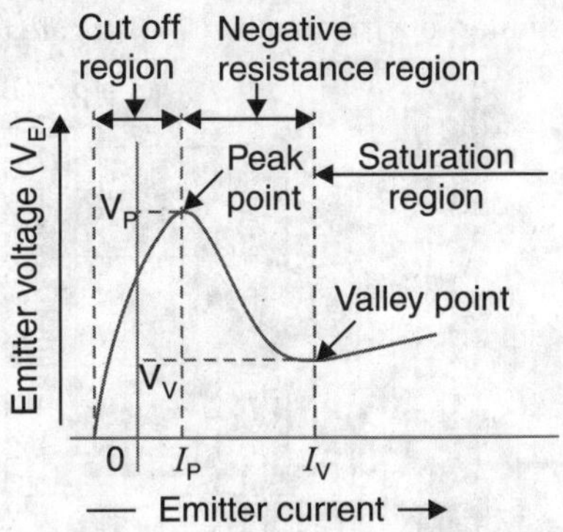

Fig. 17.26. V-I characteristics of UJT.

17.32. UJT Applications

The UJT is not classified as a thyristor but rather as a trigger device. This device is mainly used to control the triggering of SCRs and triacs. Followings are its important from the subject point of view:

1. Trigger device for SCR's and TRIACs
2. Non-sinusiodal oscillators.
3. Saw-tooth generators
4. Timing circuits

17.33. Programmable Unijunction Transistor

The programmable unijunction transistor (PUT) is a thyristor that has a function similar to that of a UJT. It is commonly used in relaxation oscillators. Unlike the UJT, the PUT is a thyristor with a gate connected to anode PN junction.

Fig. 17.27 (*a*) shows the physical model, (*b*) the electrical equivalent circuit and (*c*), the schematic symbol of PUT. Noice that PUT is more like an SCR than a UJT. The PUT has an anode gate instead of a cathode gate. Thus to triger the PUT, we need a negative voltage on the gate with respect to the cathode.

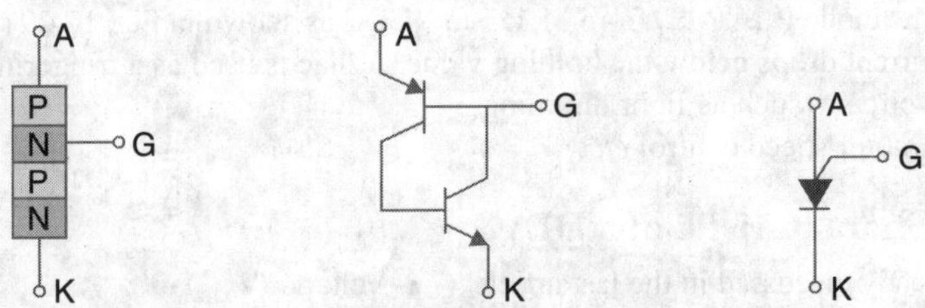

(*a*) Physical PNPN Model (*b*) Electrical Equivalent (*c*) Schematic Symbol

Fig. 17.27.

The functional difference between the PUT and UJT is that the PUT has an intrinsic stand-off ratio that is variable or programmable whereas the UJT does not.

17.34. DIAC

The diac is a two terminal device, which can pass current in either direction when the breakover voltage is reached in either polarity across the two terminals. The basic construction of a diac is similar to a triac, but without a gate terminal as shown in Fig. 17.28 (*a*).

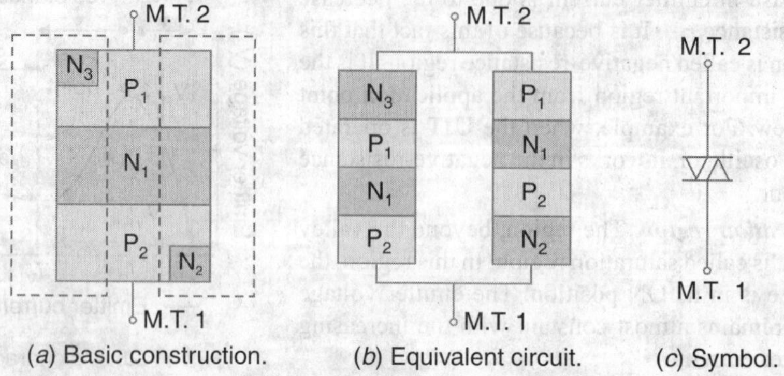

Fig. 17.28. Diac

It may be noted from the figure that a diac consists of two 4-layer diodes connected in parallel but in opposite directions. The four layer diodes are $P_1N_1P_2N_2$ and $P_2N_1P_1N_3$ as shown by the broken lines in the figure. Since the diac conducts in both directions, therefore the terminals are designated by numbers instead of anode and cathode. The diac has two main terminals namely M.T. 1 and M.T. 2. Fig. 17.28 (*b*) shows the equivalent circuit of a diac, which consists of two four-layer diodes connected in parallel but in opposite directions. Fig. 17.28 (*c*) shows the schematic symbol of the diac.

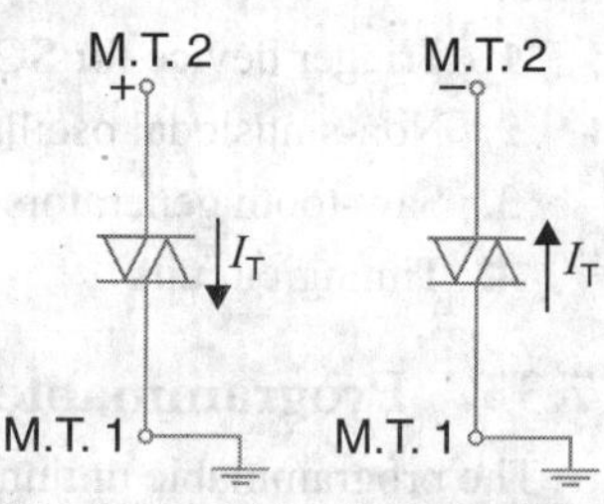

Fig. 17.29.

The diac can pass current in either direction depending upon the polarity of voltage across its main terminals. It can be turned ON only when the applied voltage across its main terminals reaches the breakover voltage. When the applied voltage makes the M.T. 2 positive with respect to M.T. 1, the diac passes current through the diode $P_1N_1P_2N_2$ from M.T. 2 to M.T. 1 as shown in Fig. 17.29 (*a*).

However, if the applied voltage makes M.T. 2 negative with respect to the M.T. 1, the diac passes current through the diode $P_2N_1P_1N_3$ from M.T. 1 to M.T. 2 as shown in Fig. 17.29 (*b*). The diac turns OFF when the current drops below the holding value. A diac is used as a triggering device for triac in phase control circuits such as light dimming, heat control and motor speed control *etc.*

17.35. V-I Characteristic of a DIAC

We have already discussed in the last article that a diac can conduct in either direction, depending upon the polarity of the applied voltage across its main terminals. When a diac is operated with M.T. 2 positive with respect to M.T 1, the V-I characteristic obtained is as shown in Fig. 19.30 by the curve marked OAB. Similarly, when the diac is operated with its M.T. 2 negative with respect to M.T. 1, the V-I characteristic obtained is as shown in Fig. 17.30 by the curve marked OCD.

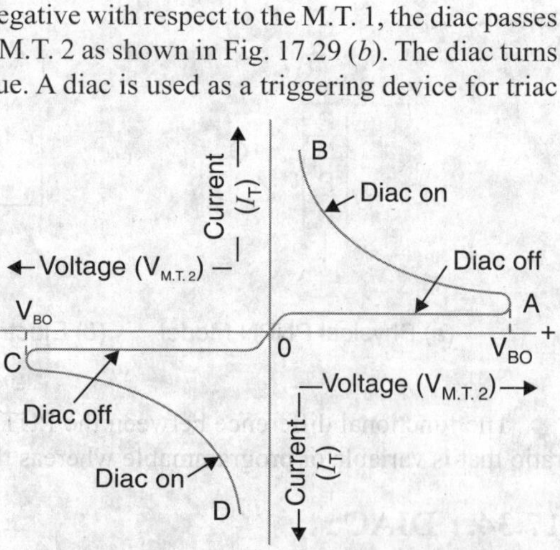

Fig. 17.30. V-I characteristic of a disc.

The V-I characteristic of a diac gives us an important information about the following points:

1. The curves OAB and OCD are symmetrical and identical. It is true because the operation of a diac is identical in both directions of conduction.
2. The diac does not pass any current, until the applied voltage of either polarity reaches the breakover voltage.
3. At the breakover voltage, the diac turns ON and the current through the diac increases rapidly as shown in the figure.
4. The operating voltage and currents are the same in either direction. Therefore the diac is a symmetrical bilateral diode.

The diacs are available with a breakdown voltage ranging from 28 to 36 V and maximum pulse current of 2 A.

17.36. Silicon Controlled Switch (SCS)

A silicon controlled switch (abbreviated as SCS) is similar in construction to a silicon controlled rectifier as shown in Fig. 17.31 (*a*). It has four terminals namely anode, cathode and two gate terminals called anode gate and cathode gate. Fig. 17.31 (*b*) shows the schematic symbol of silicon controlled switch.

The silicon controlled switch has an advantages that it can be turned ON and OFF using either gate terminal. On the other hand, an SCR can be turned ON only using its gate terminal. But it cannot be turned off using the same gate terminal.

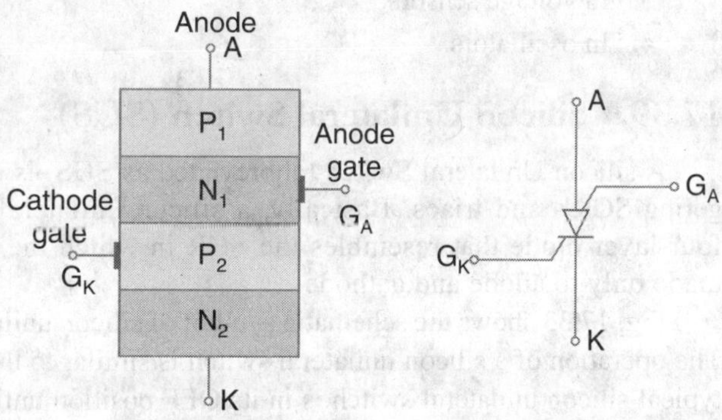

(*a*) Construction. (*b*) Symbol

Fig. 17.31. Silicon controlled switch.

17.37. SCS Operation

The operation of silicon controlled switch (SCS) may be understood by referring to its transistor equivalent circuit as shown in Fig. 17.32. To start with, let us assume that the transistors Q_1 and Q_2 are OFF. Therefore the silicon controlled switch is also OFF. If a positive pulse is applied on the cathode gate, the transistor Q_2 will be driven into conduction and thus provides a path for the base current of transistor Q_1. When the transistor Q_1 turns ON, its collector current provides the base drive to the transistor Q_2. This action drives both the transistors Q_1 and Q_2 into saturation and hence the silicon controlled switch turns ON. The silicon controlled switch can also be turned ON by applying a negative pulse at the anode gate terminal as shown in the figure.

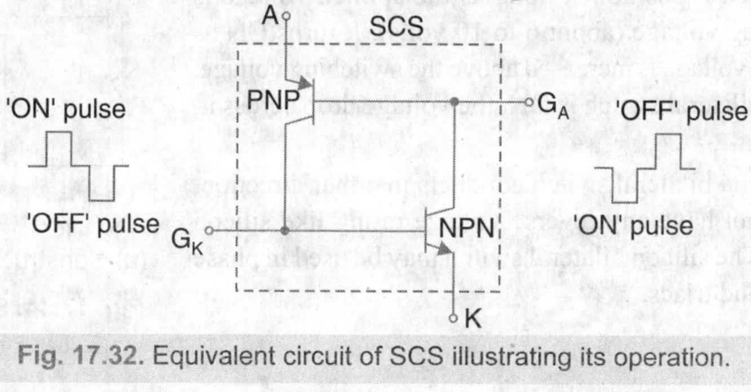

Fig. 17.32. Equivalent circuit of SCS illustrating its operation.

The silicon controlled switch can be turned OFF by applying a positive pulse at the anode gate or a negative pulse at the cathode gate. The applied anode gate voltage pulse reverse biases the base-emitter junction of the transistor Q_1 and turns it OFF. This, in turn, cuts-off the transistor Q_2 due to which the silicon controlled switch is also turned OFF. It will be interesting to know that the silicon controlled switch can also be turned OFF by reducing the anode current below the holding current value.

17.38. SCS Applications

The silicon controlled switch (SCS) has two major advantages over SCR, namely it has a fast turn OFF and it can be turned OFF with pulses on either gate terminal. But the SCS has one disadvantage that it is available only in lower power ratings. Because of this fact, SCS is used only in low power circuits. Some of the important applications of SCS are given below:

1. In counters, registers and timing circuits.
2. In pulse generators.
3. In voltage sensors.
4. In oscillators.

17.39. Silicon Unilateral Switch (SUS)

A silicon Unilateral Switch (abbreviated as SUS) is used for triggering SCR's and triacs. Basically, a silicon unilateral switch is a four-layer diode that resembles the SCR in which the contacts are made only to anode and cathode.

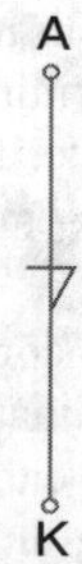

Fig. 17.33. SUS symbol.

Fig. 17.33 shows the schematic symbol of silicon unilateral switch. The operation of a silicon unilateral switch is similar to that of SCR. A typical silicon unilateral switch is in its OFF position until the applied voltage is below the switching voltage (about 6 to 10 V). The voltage drop across, a silicon unilateral switch when it is ON, is about 1.5 V. The maximum forward current capability is about 200 mA. The holding current ranges from about 0.5 mA to 1.5 mA.

A silicon unilateral switch can be turned OFF by reversing the anode voltage or disconnecting the supply. The maximum reverse voltage rating of a silicon unilateral switch is about 30 V.

17.40. Silicon Bilateral Switch (SBS)

The silicon bilateral switch (abbreviated as SBS) is used for triggering SCR's and triacs. Basically, it consist of two silicon unilateral switches, connected in parallel but in opposite directions as shown in Fig. 17.34 (*a*). The schematic symbol of silicon bilateral switch is shown in Fig. 17.34 (*b*).

The operation of a silicon bilateral switch is similar to that of a triac. It is in its OFF position so long as the applied voltage is below the switching voltage (about 6 to 10 volts). It turns ON as soon as the applied voltage is increased above the switching voltage. When the silicon bilateral switch is ON, the voltage drop across it, is about 1.7 volts.

Since the silicon bilateral switch conducts in either direction, therefore it does not have any reverse voltage rating like silicon unilateral switch. The silicon bilateral switch may be used in phase control of SCR's and triacs.

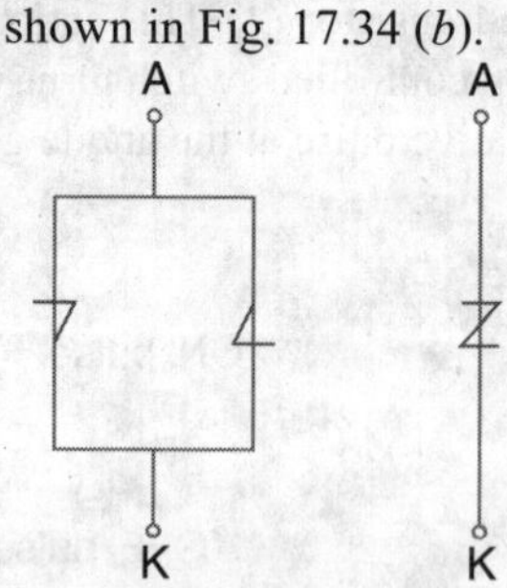

(*a*) Construction (*b*) Symbol

Fig. 17.34. Silicon bilateral switch.

17.41. Silicon Asymmetrical Switch (SAS)

A *silicon asymmetrical switch* (SAS) is similar to silicon bilateral switch (SBS). The major difference between the two is that as SAS has a different firing voltage in the positive quadrant (*i.e.*, quadrant I) than it does in the negative quadrant. (*i.e.*, quadrant IV). This feature is useful in those circuits that are affected by hysteresis, *e.g.* circuits with large inductors, motor circuits and transformers.

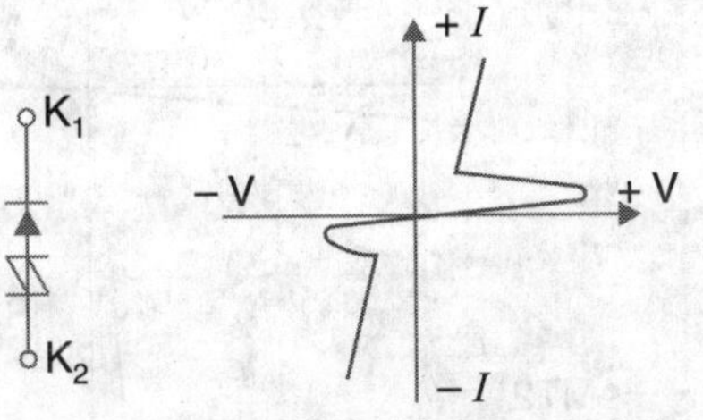

(*a*) Symbol (*b*) V-I characteristic

Fig. 17.35.

We know that when an inductor (coil) charges with current, it will become magnetized at a specific rate. When the current flow is stopped or reversed, (such as at the end of each half-cycle when ac voltage is used), the rate at which the original magnetic field collapses and the reverse magnetic field builds will not be the same rate as the original signal. This effect is called *hysteresis* and it may carry over to the electronic components used to provide the firing (or triggering) pulse to the thyristors that are controlling the current flow. When hysteresis occurs in electronic devices, it will cause then to heat up more than the normal. The asymmetrical characteristic of the SAS will help limit the effects of hysteresis.

Fig. 17.35 shows the symbol and the V-I characteristic of SAS. As seen from this diagram, the terminals on SAS are identified by K_1 and K_2.

17.42. Asymmetrical Silicon Bilateral Switch (ASBS)

An asymmetrical silicon bilateral switch (ASBS) is similar to the silicon bilateral switch (SBS) except it has two different firing (or triggering) levels. This feature is similar to the silicon asymmetrical switch (SAS) and provides the same function. The ASBS provides positive and negative pulses that are nearly identical except for the difference in voltages.

Fig. 17.36 (*a*) and (*b*) shows the schematic symbol and the V-I characteristic of the ASBS. As seen from the ASBS symbol, its terminals are also called cathode 1 (K_1) and cathode 2 (K_2).

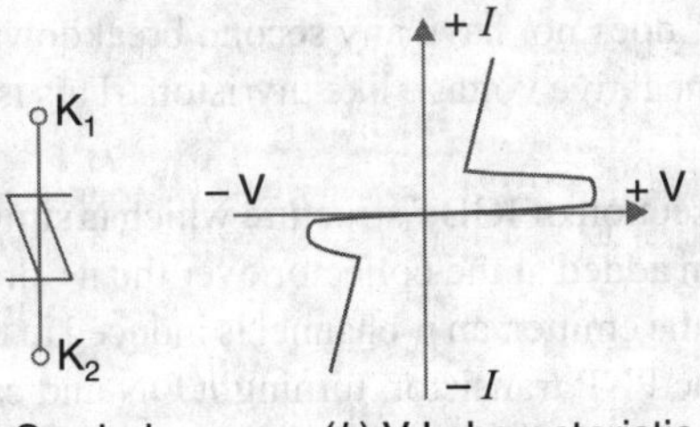

(*a*) Symbol (*b*) V-I characteristic

Fig. 17.36. ASBS

17.43. SIDAC

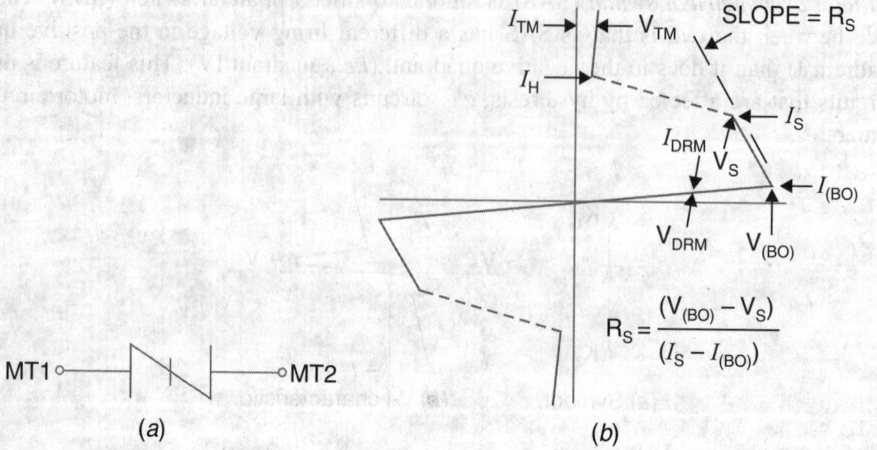

Fig. 17.37.

It is a newer triggering device. SIDAC is a high voltage bilateral trigger device. This device will switch from the blocking state to the conducting state anytime the voltage exceeds the breakover voltage (V_{BO}) level of the device in either the positive or negative direction. Typical breakover voltage for this device range from 104 to 280 V.

Fig. 17.37 (*a*) shows the schematic symbol and (*b*), V-I characteristics of SIDAC.

17.44. IGBT (Insulated Gate Bipolar transistor)

We have already discussed about BJT and MOSFET. To make use of the advantages of both MOSFET and BJT, the IGBT has been introduced. The Insulated Gate Bipolar Transistor (IGBT) is a minority-carrier device with high input impedance and large bipolar current-carrying capability. IGBT is like a device with MOS input characteristics and bipolar output characteristic that is a voltage-controlled bipolar device.

The insulated gate bipolar transistor is a bipolar transistor with an insulated base terminal. This insulated base terminal is called a gate. An IGBT is a faster device than a MOSFET. IGBT is a three terminal device having a collector (C), emitter (E), and the gate (G). The collector and emitter are the power terminals. Gate ad emitter is the control terminals.

An IGBT has the merits of both BJT and MOSFET. Like a MOSFET it has very high impedance and draws very small gate current, and like a BJT its ON state resistance is lower even at higher blocking voltage rating. IGBT does not have any second breakdown problem like a BJT. An IGBT can also be designed to block negative voltage like thyristors. This is not possible in the case of BJTs and MOSFETs.

Fig. 17.38. shows the cross section of IGBT structure which is similar to that of a MOSFET, except an additional p+ layer has been added at the collector over the n+ drain layer of the MOSFET. If the gate is positive with respect to the emitter, an n-channel is induced in the p-region. This forward biases the base-emitter junction of the PNP transistor, turning it ON and causing conductivity modulation of the N-region. This gives a signification reduction of the conduction drop when compared to a MOSFET. At the ON conduction the driver MOSFET in the equivalent circuit of IGBT carries most of the total terminal current.

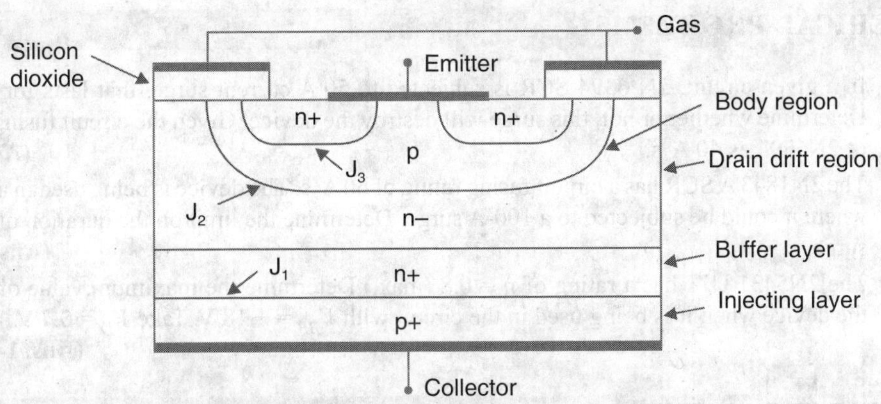

Fig. 17.38

An important point to note is the presence of a parasitic SCR in the device structure. This can lead to latch-up of the device in the ON state. Then it becomes a semi-controlled device like an SCR. This latching action caused by the parasitic NPN transistor with the PNP transistor is prevent by sufficient reducing the resistivity of the p+ layer and diverting most of the current through the MOSFET shown in the equivalent circuit. The device is turned OFF by reducing the gate voltage to zero or negative, which shuts OFF the conducting channel in the p region. The device has higher current density than of a BJT or MOSFET.

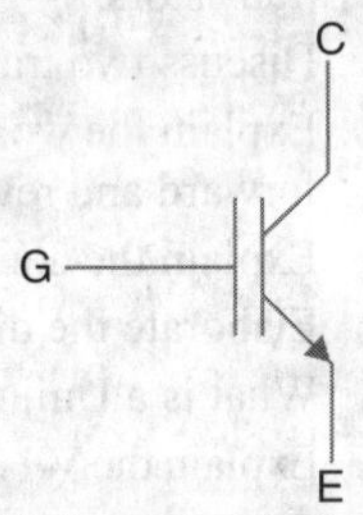

Fig. 17.39. IGBT Circuit Symbol

The n+ buffer layer between the p+ collector contact and the n- drift layer, with proper doping density and thickness, can significantly improve the operation of the IGBT in two important respects. It lowers the ON-state voltage drop of the device and shortens the turn- OFF time. On the other hand, the presence of this greatly reduced the reverse blocking capability of the IGBT. A circuit symbol for the IGBT is shown in Fig. 17.39. It has three terminals called Collector (C), Gate (G) and Emitter (E).

Advantages of IGBT over a MOSFET and a BJT

Following are some of the important advantages of IGBT over MOSFET and BJT:

1. It has a very low on-state voltage drop due to conductivity modulation and has superior on-state current density. So smaller chip size is possible and the cost can be reduced.
2. Low driving power and a simple drive circuit due to the input MOS gate structure. It can be easily controlled as compared to current controlled devices (thyristor, BJT) in high voltage and high current applications.
3. It has superior current conduction capability compared with the bipolar transistor. It also has excellent forward and reverse blocking capabilities.

Disadvantages of IGBT

Following are the disadvantage of IGBT:

1. Switching speed is inferior to that of a Power MOSFET and superior to that of a BJT. The collector current tailing due to the minority carrier causes the turnoff speed to be slow.
2. There is a possibility of latchup due to the internal PNPN thyristor structure.

NUMERICAL PROBLEMS

1. In a given circuit, 2N 6394 SCR is subjected to 50 A current surge, that lasts for 10 ms. Determine whether or not, this surge will destroy the device. Given the circuit fusing rating of 2N6394 as 40 A^2s. **(Ans. No)**
2. The 2N1843 A SCR has a circuit fusing rating of 60 A^2s. The device is being used in a circuit where it could be subjected to a 100-A surge. Determine the limit on the duration of such a surge. **(Ans. 6 ms)**
3. The 2N5431 UJT has a rating of $\eta = 0.8$ (max.) Determine the maximum value of V_P for the device when it is being used in the circuit with $V_{BB} = +18$ V. Take $V_D = 0.7$ V. **(Ans. 144.7 V)**

DESCRIPTIVE QUESTIONS

1. Discuss briefly, the construction, working, characteristics and applications of silicon controlled rectifier.
2. Give a schematic diagrams of the SCR and derive its equivalent representation in terms of transistors.
3. Discuss two transistor analogy of a SCR.
4. Explain the working of SCR with the help of its two transistor equivalent circuits. Draw its forward and reverse characteristics.
5. Explain the construction and working of a Triac.
6. Elaborate the difference between SCR and TRIAC.
7. What is a Unijunction Transistor (UJT)? How does it differ from a JFET?
8. Explain the V-I characteristics of UJT. Why it is called a current controlled negative resistance device?
9. Explain the construction and working of a diac.
10. What is SCS? How does it differ from SCR?
11. Explain briefly about the difference between SUS and SBS.
12. Give the characteristic features of devices belonging to the thyristors family. *(Anna University., Nov/Dec 2007)*
13. Draw the transistor equivalent circuit of SCR. *(Anna University., Nov/Dec 2007)*
14. Draw the two transistor version of a SCR and explain its firing characteristics with the circuit. *(JUTN., May/Jun. 2006)*
15. Draw the two transistor model of SCR. *(Anna University., April/May 2008)*
16. How the forced turn-off of an SCR is different from natural turn-off? *(Anna University., Nov/Dec 2007)*
17. Describe construction and working UJT? Also draw its characteristics. *(PTU., Dec 2008)*
18. Discuss the function of a UJT applied device with a circuit diagram. *(Bhartiya University., April 2004)*
19. What do you understand by UJT? Explain its operation. *(PTU., May 2008)*
20. What is negative resistance region in UJT? Explain its physical significance. *(PTU., Dec 2006)*
21. Explain the working of UJT. *(Bhartiya University., April 2004, R.G.P.V., June 2009, R.G.P.V., Dec. 2001)*

22. With neat diagram explain the construction, working characteristics of Unijunction transistor. Give its equivalent circuit *(Anna University., Nov/Dec 2007)*
23. Define intrinsic standoff ratio of UJT. *(Anna University., April/May 2008)*
24. Describe the operation of UJT with its equivalent circuit and its emitter characteristic. *(Anna University., April/May 2008)*
25. Explain the terms peak voltage and valley voltage with respect to UJT. *(Anna University., Nov/Dec 2007)*
26. Explain the working principle of UJT with its characteristics. *(JUTN., May/Jun. 2006, Aug./Sep. 2008)*
27. Describe a UJT. Draw its equivalent circuit and hence define the intrinsic standoff ratio. Draw its characteristic curve and explain the various parameters. *(JNTU., Aug./Sep. 2008)*
28. Diac is a bidirectional device. Explain. *(Anna University., Nov/Dec 2007)*
29. Write a note on PUT. *(Anna University., Nov/Dec 2007)*
30. Discuss briefly the operation of a TRIAC. *(GBTU/MNTU., 2008)*
31. Mention the applications of UJT, Diac, and Triac. *(Anna University., Nov/Dec 2007)*
32. Explain the operation of DIAC. *(Anna University., April/May 2008, GBTU/MNTU., 2008)*
33. Explain the principle of working of UJT. *(R.G.P.V., Feb. 2010, June 2010)*
34. Discuss the construction working and characteristics of unijunction transistor. List few application of the UJT. *(R.G.P.V., June 2011)*
35. DIAC is a bidirectional device. Explain. *(R.G.P.V., June 2011)*
36. Discuss the working of the following:
    (*a*) SCR
    (*b*) IGBT *(R.G.P.V., Dec 2010)*
37. Discuss the working of the following:
    (*a*) Triac
    (*b*) GTO *(R.G.P.V., Dec 2010)*
38. Explain the operation of SCR using two transistor analogy. Plot the V-I Characteristics and mark various currents. *(R.G.P.V., June 2011)*
39. Describe various mechanisms of Turn-on and Turn-off for SCR. *(R.G.P.V., June 2011)*
40. Explain SCR with working principle and characteristics. *(R.G.P.V., Dec. 2008)*
41. Describe the following items:
    (*a*) Diac
    (*b*) IGBT and GTO. *(R.G.P.V., June 2009, Dec. 2008)*
42. Explain series and parallel combination of SCR and draw its characteristics. *(R.G.P.V., Dec. 2008)*
43. With the help of characteristics explain the working of SCR. *(R.G.P.V., June 2009)*
44. How is SCR is used in protection circuit? *(R.G.P.V., June 2009)*
45. Write short note on the following:
    (*a*) SCR
    (*b*) GTO *(R.G.P.V., Feb. 2010)*
46. Write short note on the following:
    (*a*) Diac
    (*b*) IGBT *(R.G.P.V., Feb. 2010, June 2010)*

47. Draw the V-I characteristics of SCR and explain its working. *(R.G.P.V., June 2010)*
48. Discuss the series and parallel combination of SCR. *(R.G.P.V., June 2010)*
49. What is a UJT? What are its characteristics? Give a typical circuit for a practical application of UJT.Explain. *(M.G University., July/August 2007)*
50. Write short notes on :
    (*a*) UJT characteristics and applications *(R.G.P.V., Dec. 2002)*
    (*b*) UJT *(R.G.P.V., June 2005)*
    (*c*) TRIAC *(R.G.P.V., June 2011, 2010)*

MULTIPLE CHOICE QUESTIONS

1. An SCR is a semiconductor device which consists of
   (*a*) four PN junctions (*b*) three PN junctions
   (*c*) two PN junctions (*d*) one PN junction.
2. After triggering an SCR, the gate pulse is removed. The current in the SCR will
   (*a*) immediately fall to zero (*b*) rise up
   (*c*) remain the same (*d*) immediately rise a little and then falls to zero
3. An SCR may be turned OFF by
   (*a*) reversing polarity of the anode-to-cathode voltage
   (*b*) interrupting its anode current
   (*c*) low-current drop out
   (*d*) all of the above.
4. A triac can be triggered into conduction by applying
   (*a*) only positive voltage at the gate
   (*b*) positive or negative voltage at the gate
   (*c*) positive or negative voltage at either main terminal
   (*d*) both (*a*) and (*c*) above.
5. Which of the following acts like a diode and two resistors
   (*a*) SCR (*b*) triac (*c*) diac (*d*) UJT
6. A four layer diode is sometime called
   (*a*) UJT (*b*) PNPN diode (*c*) DIAC (*d*) Switch
7. Which of the following PNPN device has two gates:
   (*a*) SCR (*b*) DIAC (*c*) TRIAC (*d*) TRANSISTOR
8. When a SCR is conducting, it has
   (*a*) infinitely large resistance (*b*) resistance of few megohms
   (*c*) low resistance (*d*) zero resistance

ANSWERS

1. (*b*)	2. (*c*)	3. (*d*)	4. (*d*)
5. (*d*)	6. (*c*)	7. (*c*)	8. (*c*)

18 Chapter

Integrated Devices and Circuits

Outline

18.1. Introduction

Electronic circuitry has undergone tremendous changes since the invention of a triode by Lee De Forest in 1907. In those days, the active components (like triode) and passive components (like resistors, inductors and capacitors etc.) of the circuits were **separate and distinct units** connected by soldered leads.

With the invention of the transistor in 1948 by W.H. Brattain and I. Bardeen, the electronic circuits became considerably reduced in size. It was due to the fact that a transistor was not only cheaper, more reliable and less power consuming but was also much

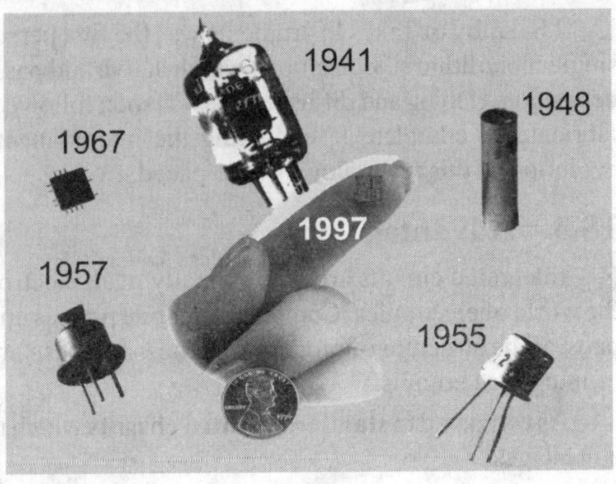

smaller in size than an electron tube. To take advantage of small transistor size. the passive components too were greatly reduced in size thereby making *the entire circuit very small.* Development of printed circuit boards (PCBs) further reduced the size of electronic equipment by eliminating bulky wiring and tie points.

In the early 1960s, a new field of **Microelectronics** was born primarily to meet the requirements of the Military which wanted to reduce the size of its electronic equipment to approximately one-tenth of its then existing volume. This drive for extreme reduction in the size of electronic circuits has led to the development of microelectronic circuits called **integrated circuits (ICs)** which are so small that their actual construction is done by technicians using high powered microscopes.

18.2. Integrated Circuit

An integrated circuit (refered to as an IC, chip or microchip) is an electronic circuit manufactured by the patterned diffusion of trace elements into the surface of a then substrate of a semiconductor. Additional materials are deposited and patterned to form interconnection between semiconductor devices.

An integrated circuit (IC) **is just a packaged electronic circuit.** In other words it is a complete electronic circuit in which both the *active* and *passive* components are fabricated on a tiny single chip of silicon. **Active components are** those which have the ability to produce gain. Examples are : transistors and FETs. **Passive components** or devices are those which do not have this ability. Examples are : transistors capacitors and inductors.

Integrated circuit

ICs are produced by the same process as are used for manufacturing individual transistors and diodes etc. In such circuits, different components are isolated from each other by isolation diffusion within the crystal chip and are interconnected by an aluminium layer that serves as wires.

A discrete circuit, on the other hand, is one *that is built by connecting separate components.* In this case, each component is produced separately and then all are assembled together to make the electronic circuit.

J.S. Kilby of Taxes Instruments was the first person to develop (in 1959) an integrated circuit—a single monolithic silicon chip in which active and passive components were fabricated by successive deposition, etching and diffusion. He was soon followed by Robert Noyce of Fairchild who successfully fabricated a complete IC including the interconnections on a single silicon chip. Since then the evolution of this technology is fast-paced.

18.3. Advantage of ICs

Integrated circuits are used virtually in all electronic equipments today and have revolutionized the world of electronics. Computers, mobile phones and other digital applicances are now inextricable parts of the structure of modern societies. All these are made possible by the low cost of production of integrated circuits.

As compared to **standard printed circuits** which use discrete components, ICs have the following advantages :

1. ***Extremely small physical size***

Often the size is thousands of times smaller than a discrete circuit. The various components and their interconnections are distinguishable only under a powerful microscope.

2. ***Very small weight***

Since many circuit functions can be packed into a small space, complex electronic equipment can be employed in many applications where weight and space are critical, such as in aircraft or space-vehicles.

3. ***Reduced cost***

It is major advantage of ICs. The reduction in cost per unit is due to the fact that many identical circuits can be built simultaneously on a single wafer—this process is called **batch fabrication.** Although the processing steps for the wafer are complex and expensive, the large number of resulting integrated circuits make the ultimate cost of each IC fairly low.

4. ***Extremely high reliability***

It is perhaps the **most important** advantage of an IC and is due to many factors. Most significant factor is the absence of soldered connections. Another is the need for fewer interconnections—the major cause of circuit failures. Small temperature rise due to low power consumptions of ICs also improves their reliability. In fact, an IC logic gate has been found to be 100,000 times more reliable than a vacuum tube logic gate and 100 times more reliable than a transistor logic gate.

Obviously, higher reliability means that ICs will work for longer periods without giving any trouble—something most desirable from both military and consumer applications point of view.

5. ***Increased response time and speed***

Since various components of an IC are located close to each *in* or *on* a silicon wafer, the time delay of signals is reduced. Moreover, because of the short distances, the chance of stray, electrical pickup (called parasitic capacitance) is practically nil. Hence it makes them very suitable for small signal operation and high frequency operation. As a result, the response time or the operating speed of the system is improved.

6. ***Low power consumption***

Because of their small size, ICs are more suitable for low power operation than bulky discrete circuits.

7. ***Easy replacement***

ICs are hardly ever repaired because in case of failure, it is more economical to replace them than to repair them.

8. ***Higher yield***

The *yield* is the percentage of usable devices. Because of the batch fabrication, the yield is very high. Faulty devices usually occur because of some defect in the silicon wafer or in the fabrication steps. Defects in silicon wafer can occur because of lattice imperfection and strains introduced in crystal growth, cutting and handling of the wafers. Usually such defects are extremely small, but their presence can ruin devices built on or around. Reducing the size of each device greatly increases the chance for a given device to be free of such defects. The same is true for fabrication defects such as the presence of a dust particle on the photolithographic mask.

18.4. Limitations of ICs

The integrated circuits suffer from the following limitations :

1. Coils or inductors cannot be fabricated,
2. ICs function at fairly low voltages,

3. They handle only limited amount of power,
4. They are quite delicate and cannot withstand rough handling or excessive heat.

However, the advantages of ICs far outweigh their disadvantages or limitations.

18.5. Scale of Integration

Level of integration in ICs has been increasing ever since they were developed some three–and–a–half decades back. The number of electronic circuits or components that can be fitted into a standard size ICs has been dramatically increasing with each passing year. In fact, whole **electronic systems** rather than just **a circuit** are incorporated in one package.

An approximate method of classifying the amount of circuit or component density is as follows:

1. **SSI**— *Small Scale Integration*

   In this case, the number of **circuits** contained in one IC package is less than 12 (or number or **components** is less than 50).

2. **MSI**—*Medium Scale Integration*

   Here, number of circuits per package is between 13 and 99 (or number or components is between 50 and 5000)

3. **LSI**—*Large Scale Integration*

   In this case, circuit density is between 100 and 9,999 (or component density is between 5000 and 100,000). Integrated circuits such as 1 K bit RAMs, calculators chips, first and second generation microprocessors were LSI devices.

4. **VLSI**—*Very Large Scale Integration*

   Here the number of circuits per package is between 10,000 to 99,999 (or number of components is between 100,000 —10,000,000).

5. **ULSI**—*Ultra Large Scale Integration*

   In this case, the circuit density is between 100,000 to 999,999 (or component density is between 1,000,000—10,000,000).

6. **GSI**—*Giga Scale Integration*

   Here the number of circuits per package is 1,000,00 or more (or number of components are over 100,000,000).

7. *SOC and 3D-IC:*

   A system-on-a-chip (SOC) is an integrated circuit in whcih all components needed for a computer or other system are included on a single chip. The design of a such a device can be complex and costly. Moreover building different types of components on a single piece of silicon may compromise efficiency of some elements. However these drawbacks are offset by lower manufacturing and assembly costs and greatly reduced power consumption.

   A three-dimensional integrated circuit (3D - 2C) has two or more layers of active electronic components that are integrated both vertically and horizontally into a single circuit. Communication between the layers use on-die signalling. As a result, power consumption is much lower than in equivalent separate circuits.

In summary,

SSI	<	12	
MSI		12 – 99	
LSI		100 – 9,999	Circuit per chip
VLSI		10,000 – 99,999	
ULSI		100,000 – 999,999	
GSI	>	1,000,000	

18.6. Classification of ICs by Structure

Structurally speaking ICs can be classified into the following three types :

1. Monolithic Integrated Circuits

The word 'monolithic' means 'single stone' or more appropriately 'a single-solid structure'. In these ICs, all circuit components (both active and passive) are fabricated inseparably within a single continuous piece of silicon crystalline material called **wafer (or substrate).** All components are atomically part of the same chip. Transistors, diodes and other passive components are fabricated at appropriate spots in the substrate using epitaxial diffusion technique.

Monolithic IC equipment

Component interconnections are provided on the surface of the structure and external connecting wires are taken out to the terminals. It is a complete circuit requiring no 'glad ons'

Despite some of its distinct disadvantages, monolithic ICs are in wide use because for mass production, monolithic process has been found to be the most economical.

2. Thick and Thin-Film ICs

The essential difference between thick-film and thin-film ICs *is not their relative thickness but* the method of depositing the film. Both have similar appearance, properties and general characteristics though they both differ in many respects from monolithic ICs. These ICs are not formed within a silicon wafer but on the surface of an insulating substance such as glass or a ceramic material. Moreover, **only passive components** (resistor, capacitors) are formed through thick or thin-film techniques on the insulating surface. The active elements (transistor, diodes) are added externally as **discrete elements** to complete a functional circuit. These discrete active components are frequently produced by using the monolithic process.

As stated above, the primary difference between the thick and thin film techniques is the process used for forming passive components and the metallic conduction pattern.

(*a*) Thin-film ICs

Such circuits are constructed by depositing films (typically 0.1 to 0.5 mm) of conducting material through a mask on the surface of a substrate made of glass or ceramic. Resistors and conductors are formed by varying the width and thickness of the film and by using materials of different resistivity. Capacitors are produced by sandwiching an insulating oxide film between two conducting films. Small inductors can be made by depositing a spiral formation of film. The active components like transistors and diodes etc. are externally added and inter-connected by wire bonds.

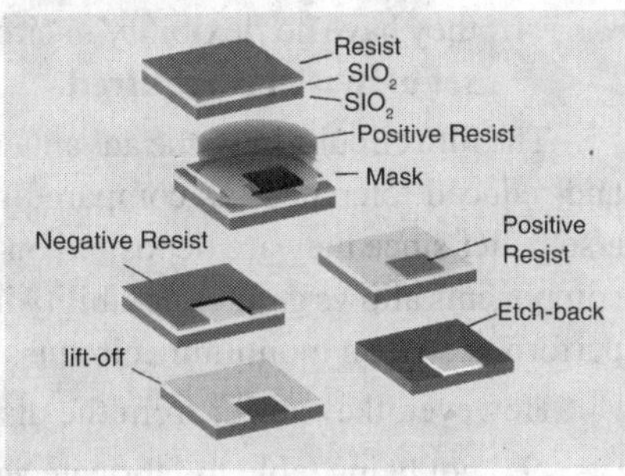

Thin film ICs

Following two methods are used to produced thin films :

(*i*) vacuum evaporation

In this method, the vaporised material is deposited through a set of masks on the glass or ceramic substrate contained in vacuum.

(*ii*) cathode sputtering

In this method, atoms from a cathode made of the desired film material are deposited on the substrate which is located between the cathode and the anode.

(b) Thick-film ICs

Hybrid or Multichip ICs.

Such type of integrated circuits are sometimes referred to as **printed** thin-film circuits. They are so called because silk-screen printing techniques are employed to create the desired circuit pattern on the surface of the substrate. The screens are made of fine stainless steel wire mesh and the 'inks' are pastes (of pulverised glass and aluminium) which have conductive. resistive or dielectric properties. After printing, the circuits are high-temprature fired in a furnace to fuse the films to the insulating substrate. As with thin-film ICs, active elements are added extermally as discrete components.

3. Hybrid or Multichip ICs

As the name implies, such circuits are formed either by inter-connecting a number of individual chips or by a combination of film and monolithic IC techniques. In such ICs, active components are first formed within a silicon wafer (using monolithic) which is subsequently covered with an insulating layer such as SiO_2. Film techniques are then employed to form passive components on the SiO_2 surface. Connections are made from the film to the monolithic structure through 'windws' cur in the SiO_2 layer.

18.7. Comparison Between Different ICs

Each type of IC has its own advantages and disadvantage.

Monolithic circuits have *the advantage of lowest cost and highest reliability.* However they have the following disadvantages :

1. isolation between components is poorer,
2. range of values of passive components used in the circuits is comparatively small,
3. inductors cannot be fabricated,
4. they have no flexibility in circuit design because for making any changes in the circuit, **new set of masks is required.**

The film circuits have the advantage of forming passive components with broader range of values and reduced tolerances as compared to monolithic circuits. Isolation between their components is also better since they are deposited on a substrate that is an insulator. Use of external discrete active components allows greater flexibility in circuit design. These circuits also give better high frequency performance than monolithic circuits.

However, they suffer from the disadvantages of

1. not being able to fabricate active components
2. comparatively higher cost and
3. larger physical size.

The chief advantage of multichip IC*s* is their greater flexibility but they are **too expensive** for mass production and **have least reliability.** Hence such circuits are generally used prototypes for monolithic IC*s*.

The various integrated circuits are arranged in the form of an 'IC' tree of Fig. 18.1.

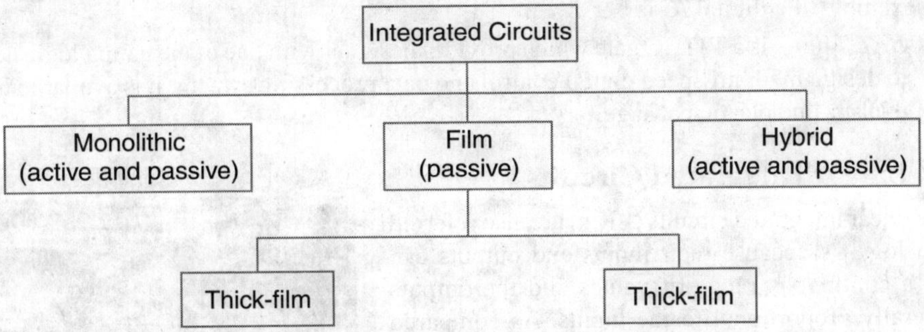

Fig. 18.1.

18.8. Classification of ICs Function

The earlier classification of ICs was based on *their method of construction.* However, the integrated circuits can also be classified according to their general function. The two most important categories are:

1. Analog (or Linear)
2. Digital
3. Mixed singal (both analog and digit on the same chip)

Analog ICs work by processing continuous signals. They perform functions like amplification, active filtering, demodulation and mixing. The examples of analog ICs are operational amplifiers, sensors and power management circuits.

Digital ICs can contain anything from logic gates, flip-flops, multiplexers microcontrollers, digital signal processors in few square millimeters. The small size of these circuits allows high-speed, low power dissipation and reduced manufacturing cost compared with board level integration.

The mixed-signal ICs can combine analog and digital circuits on the same chip to create functions such as A/D converters and D/A converters, radio subsystem of a cell phone, the read data path and laser sled control logic of a DVD player digitally controlled sound chips LAN and WAN router. They often contain an entire system on a chip.

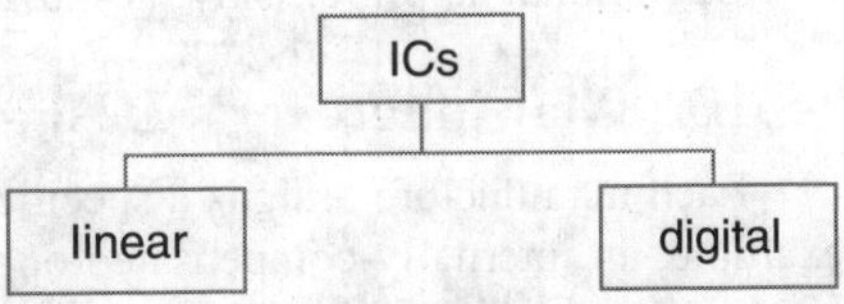

Fig. 18.2.

1. linear and
2. digital

The same fact has been shown in Fig. 18.2.

Examples of linear *ICs* are :

1. ***BEL CA*** **-3020**—used as multipurpose wide-band power amplifier.
2. ***BEL CA*** **-3065** — it is a monolithic IC which combines a multistage *IF* amplifier, limiter, an *FN* detector, an electronic attenuator, a Zener diode regulated power supply and an audio amplifier. In fact, this IC provides a high performance multistage sub-system of a TV reveiver. It is available in 14-pin dual-in-line package.

Both are manufactured by Bharat Electronic Ltd., Bangalore.

3. **SSD 710**—is a linear IC used as Differential Comparator it is manufactured by Solid State Devices, Syed Abdullah Road, Bombay.

An example of a digital *IC* is

1. ***BEL* 7400**—is a TTL IC gate which provides designer with one of the gating logic necessary to design medium-speed digital control and data processing system. It is available in 14-pin dual-in-line plastic package.

18.9. Linear Integrated Circuits

The linear integrated circuits (LICs) are also referred to as analog ICs because their inputs and outputs can take on a continuous range of values and the outputs are generally proportional to the inputs. As compared to digital ICs, LICs are used much less. But LICs are quickly replacing their discrete circuit counterparts in many applications as their cost becomes competitive. They also possess much higher reliability because so many external connections (major source of circuit failure) are eliminated. LICs find wide use in military and industrial applications as well as in consumer products. They are frequently used in

Linear ICs.

1. Operational amplifiers,
2. Small-signal amplifiers,
3. Power amplifiers,
4. *RF* and *IF* amplifiers,
5. Microwave amplifiers,
6. Multipliers
7. Voltage comparators,
8. Voltage regulators *etc.*

Operational amplifier is by far the most versatile form for an LIC and is discussed separately.

18.10. Manufacturer's Designation of LICs

Each manufacture assigns a specific code and type number to the LIC*s* produced by him. For example, an internally-compensated op-amp 741 produced by Fairchild is designated as mA 741. Here, mA is the identifying code used by Fairchild. Many other manufacturers produce LIC*s* similar to 741 but use their own code while retaining the same type number. For example, Fiarchild's original mA 741 is manufactured by other manufacturers with the following codes :

1. National Semiconductor — *LM* 741
2. Motorola — *MC* 1741
3. RCA — *CA* 3741
4. Texas Instruments— *SN* 52741

It is seen that the last three digits in each manufacturer's designation are the same *i.e.,* 741. All these op-amps have the same specifications. Hence, in practice, the manufacturer's code number is often ignored and all such LIC*s* are referred to as 741.

Many LIC*s* are available in different classes such as A, B, C, E, S and SC. For example, main classes of 741 are as under :

741	—	Military grade of-amp
741 C	—	Commercial grade op-amp

741 A	—	Improved version of 741
741 E	—	Improved version of 741 C
741 S	—	Military grade op-amp with higher slew rate
741 SE	—	Commercial grade op-amp with higher slew rate

18.11. Digital Integrated Circuits

About 80 per cent of the IC market has been captured by digital IC*s* which are mostly utilized by the computer industry. Digital IC*s* lend themselves easily to monolithic integration because a computer *uses a large number of identical circuits.* Moreover, such circuits employ relatively few capacitors and values of resistances, voltage and currents are low.

Digital IC*s* contain circuits whose input and output voltages are limited to two possible levels— low or high. It is so because *digital signals are usually binary.* somethings, digital circuits are referred to as switching circuits. Digital IC*s* include circuits such as

1. logic gates
2. flip-flops
3. counters
4. clock-chips
5. calculator chips
6. memory chips
7. microprocessors (mP) etc.

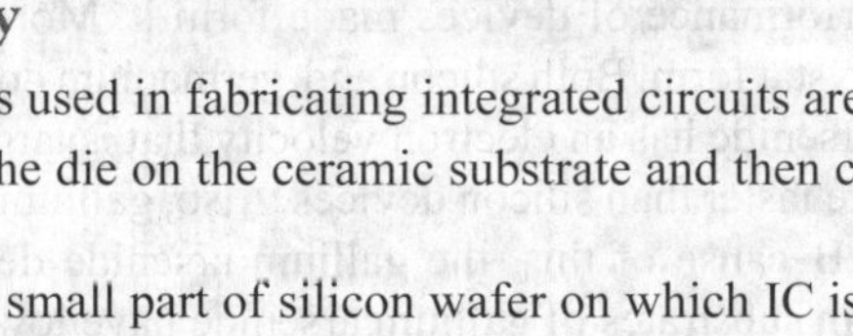

18.12. IC Terminology

Some of the common terms used in fabricating integrated circuits are defined below :

1. **Bonding** – attaching the die on the ceramic substrate and then connecting the leads to the package.
2. **Chip** — an extremely small part of silicon wafer on which IC is fabricated. A photograph of the wafer containing hundreds of chips (or dice) and a drawing of a chip are shown in

Fig. 18-3. The identical chips, each of which may vary in area from 10 to over 100 mm^2, may contain up to several million devices.

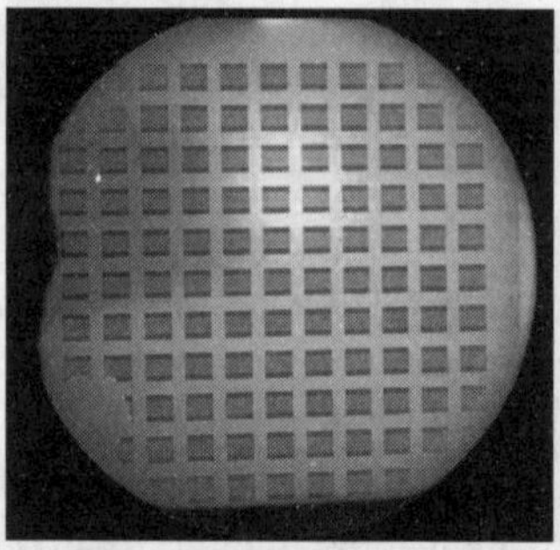

Dish silicon water in DHD

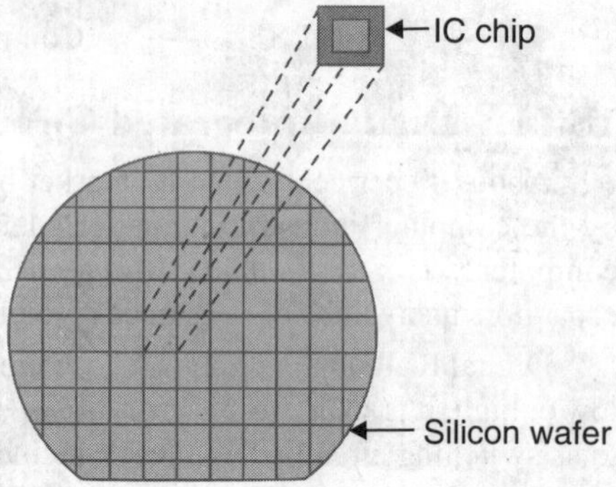

Fig. 18.3.

3. **Circuit probing** — testing the electrical performance of each IC chip with the help of a microscope and multi-point probe.
4. **Die** — same as a chip.
5. **Diffusion** — a process that consist of the introduction of impurities into selected regions of a wafer to form junctions.
6. **Encapsulation** — putting a cap over the IC and sealing it in an inert atmosphere.
7. **Epitaxy** — a process of the controlled growth of a crystalline doped layer of silicon on a single crystal substrate.
8. **Etching** — a process of selective removal of regions of a semiconductor, metal or silicon dioxide.
9. **Mask** — a glass plate with desired pattern of diffusion or metallization. Usually a single mask is not sufficient to fabricate an IC.
10. **Metallization** — a process for providing ohmic contacts and interconnections by evaporating aluminium over the chip.
11. **Photolithography** — a process to transfer geometrical pattern from the mask to the surface of the wafer.
12. **Photoresist** — a light-sensitive material that hardens when exposed to ultraviolet light.
13. **Wafer** — a thin disk of semiconductor in which number of ICs are fabricated simultaneously.

18.13. Semiconductors Used in Fabrication of ICs and Devices

The fabrication of IC*s* has been based on the use of silicon (Si) as the premier semiconductor. Two other semiconductors used for IC fabrication are germanium and gallium arsenide (GaAs). But these semiconductors present special problems for device fabrication as discussed below.

Gallium arsenide has very attractive electrical properties but its crystals have a high density of defects which limit the performance of devices made form it. Moreover gallium arsenide is more difficult to grow in single crystal form. Both silicon and germanium do not suffer from these problems. On the plus side, gallium arsenide has an electron velocity that is larger than silicon. Because of this gallium arsenide devices are faster than silicon devices. Also, gallium arsenide has a lower saturation electric field than silicon. Because of this, the gallium arsenide devices have lower power-delay product. Devices made from substrates of gallium arsenide have lower parasitic capacitances.

This property contributes to their speed advantage over the silicon devices. Another advantage of gallium arsenide results from direct band gap which makes it possible to provide certain functions not possible in silicon such as *coherent* and *incoherent light emission*.

A major advantage of silicon, in addition to its abundant availability in the form of sand, is that it is possible to form a superior stable oxide (SiO_2). This oxide has superb insulating properties and provides an essential and excellent ingredient in the fabrication and protection of devices or ICs. On the other hand, germanium oxide is unsuited for device applications. The intrinsic resistivity of germanium is 47 Ω-cm, while that of silicon is 230,000 Ω-cm. The low resistivity of germanium would have precluded the fabrication of rectifying devices with high breakdown voltages. Thus high-voltage rectifying devices and certain infrared sensing devices are practical with silicon. Finally, there is an economic consideration and that is electronic grade gernanium is now more costly than silicon. Thus, at present time, silicon remains the major semiconductor in the fabrication of ICs.

Some other semiconductors used in semiconductor industry are gallium phosphide (GaP), gallium nitride (GaN), zinc sulphide (ZnS), indium antimonide (InSb), compound of cadmium and selenium (CdSe). GaAs, GaP and GaN are used in high-speed devices and devices requiring emission and absorption of light such as lasers and light emitting diodes (LEDs). ZnS is used as fluorescent material such as those in television screens. InSb and CdSe are used as light detectors.

18.14. IC Design and Manufacturing

The IC design and manufacturing is a long process. We shall study these two processes separately in the following pages.

18.15. Integrated Circuit Design

Integrated circuit design or simply IC design can be divided into two broad categories :

1. Digital IC design

2. Analog IC design and mixed signal IC design. The digital IC design is to produce components such as microprocessors, field programmable gate arrays (FPGAs), memories (RAM, ROM and flash) and digital application specific integrated circuits (ASICs).

Analog IC design is to produce components such as op-Amps, linear regulators, phase-locked loops, oscillators and active filters.

Mixed signal IC design includes components such as data converters using delta-sigma modulation, analog-to-digital converters, digital to analog-converters, digital radio chips, cellular telephone, software radio, Local Area Network (LAN) and Wide Area Network (WAN) route ICs.

Modern ICs are enormously complicated. A long IC may have over 1 billion transistors. An IC process might have more than 700 rules. Moreover, since the manufacturing process itself is not completely predictable, designers must accounts for its statistical nature. The complexity of modern IC design, as well as the market pressure to produce the designs rapidly, has led to extensive use of automated design tools in the IC design process. In short the design of an IC using electronic design automation (EDA) software is the design, test and verification of the instructions that the IC is to carryout.

Steps involved in IC design

IC design involves the creation of electronic components such as resistors, capacitors, diodes, transistors and the metallic interconnect on a piece of silicon. A method to isolate the individual components formed in the substrate is necessary since the substrate is condcutive and often forms active region of the individual components. The two common methods for isolation are : (1) P-N junction isolation and (2) dielectric isolation. Attention must be give to power dissipation of transistors and interconnect resistances and current density of interconnect, contacts and vias. Electromigration in metallic interconnect and electrostatic discharge (ESD) damage to the tiny components are also of concern.

Finally the physical layout of certain circuit sub-blocks is critical in order to achieve the desired speed operation, to segregate noisy portions of an IC from quiet portions, to balance the effects of heat generation across the IC or to facilitate the placement of connections to circuiting outside IC.

A typical IC design cycle as shown in Fig. involves several step. These steps are listed below.

1. Feasibility study and die size estimate
2. Functional verification
3. Circuit/RTL design
4. Circuit/RTL simulation Logic simulation
5. Floorplanning
6. Design review
7. Layout
8. Layout verification
9. Static timing analysis
10. Layout review
11. Design for test and Automatic test pattern generation
12. Design for manufacturability (IC)
13. Mask data preparation

} IC design

14. Wafer fabrication
15. Die test
16. Packaging
17. Post silicon validation and integration
18. Device characterization
19. Tweak (if necessary)
20. Datasheet generation Portable Document Format
21. Ramp up
22. Production
23. Yield Analysis / Warranty Analysis Reliability (semoiconductor)
24. Failure analysis on any returns
25. Plan for next generation chip using production information if possible

} IC fabrication testing and packaging

Digital IC design

A digital IC design can be broadly classified into 3 steps:

1. ***Electroic system level design.*** This step creates user functional specification. The designer may use a variety of computer languages and tools to create this description. Example include C/C++ model, System C, System Verilog Transaction level Models, Simulink and MATLAB.

2. ***RTL Design.*** This step is the hardest part. It converts the user specification into register transfer level (RTL) description. The RTL describes the exact behaviour of the digital circuits on the chip as well as the inter connections to inputs and outputs.

3. ***Physical design.*** This step takes the RTL and a library of available logic gates and creates a chip design. This involves figuring out which logic gates to use, defiring places for them and wining them together. The steps from 5 to 13 is a part of physical design.

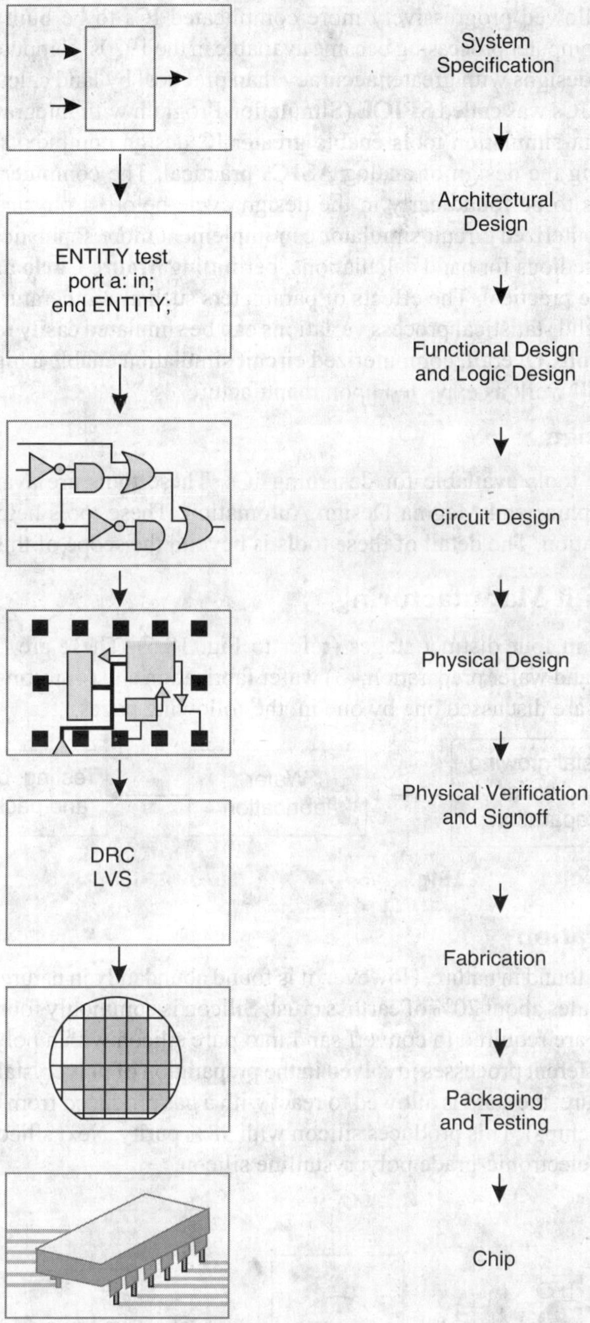

Fig. 18.4. Courtesy on-line Wikipedia

Analog IC Design

Before the advent of the microprocessor and software based design tools, analog ICs were designed using hand calculations. These ICs were basic circuits, or op-amps. An interative trial-and-error process and "*overengineering*" of device size was often necessary to achieve a manufacturable

IC. Reuse of proven designs allowed progressively more complicated ICs to be built upon prior knowledge. When inexpensive computer processing became available in the 1970s, computer programs were written to simulate circuit designs with greater accuracy than practical by land calculation. The first circuit simulator for analog ICs was called **SPICE** (Simulation Program with Integrated Circuits Emphasis). Computerized circuit simulation tools enable greater IC design complexity than hand calculations can achieve, making the design of analog **ASICs** practical. The computerized circuit simulators also enable mistakes to be found early in the design cycle before a physical device is fabricated. Additionally, a computerized circuit simulator can implement more sophisticated device models and circuit analysis too tedious for hand calculations, permitting **Monte Carlo analysis** and process sensitivity analysis to be practical. The effects of parameters such as temperature variation, doping concentration variation and statistical process variations can be simulated easily to determine if an IC **design is manufacturable.** Overall, computerized circuit simulation enable a higher degree of confidence that the circuit will work as expected upon manufacture.

Software Tools for IC Design

There are several software tools available for designing ICs. These tools are available from synopsis, Cadence, Mentor graphics and Magma Design Automation. These tools help in design, simulation, analysis and verification. The detail of these tools is beyond the scope of this book.

18.16. Integrated Circuit Manufacturing

The ICs are manufactured in four distinct stages (refer to Fig. 18.5). These are (1) material preparation, (2) crystal growing and wafer preparation, (3) wafer fabrication and (4) testing, bounding and packaging. All these stages are discussed one by one in the following pages.

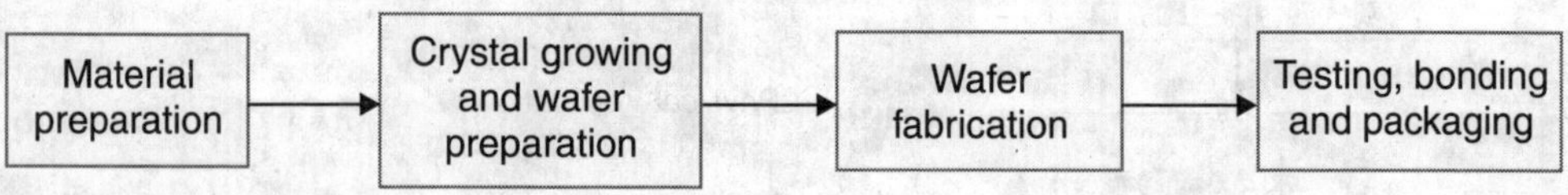

Fig. 18.5.

18.17. Material Preparation

Silicon, as an element is not found in nature, However, it is found abundantly in nature in the form of silicon dioxide, which constitutes about 20% of earth' s crust. Silicon is commonly found as quartz or sand. A number of processes are required to convert sand into pure silicon with a polycrystalline structure. Fig. 18.6 shows the different processes involved in the preparation of polycrystalline silicon from sand. As seen from this figure, the sand is allowed to react with a gas produced from the burning of carbon (coal, coke and wood chips). This produces silicon with 98% purity. Next silicon is further purified in a reactor to produce electronic-grade polycrystalline silicon.

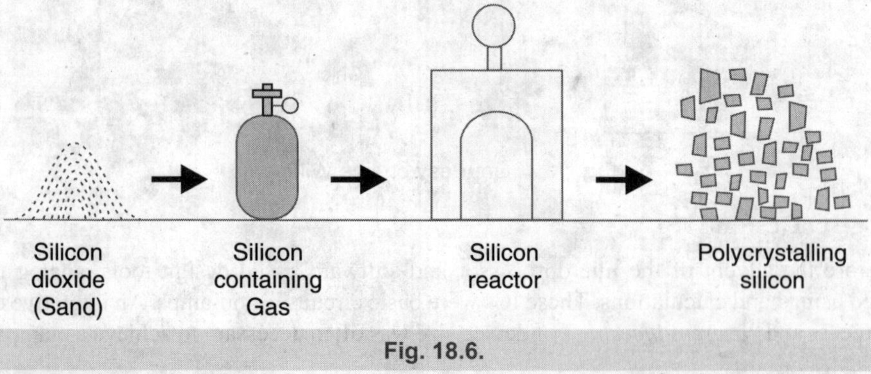

Fig. 18.6.

18.18. Crystal Growing and Wafer Preparation

The polycrystalline silicon is composed of many small crystals having random orientation and containing many defects. For silicon to be used in the fabrication of ICs, it must be nearly perfect and crystalline in nature. We, therefore, now need to produce single. crystals of silicon. This is done by a process called crystal growth. There are two methods to carryout the crystal growth : (*i*) the *Czochralski* and (*ii*) the *flat zone* process. The Czochralski process prepares virtually all the silicon used for IC fabrication. The flat zone process is used to prepare crystals for fabricating high power, high voltage semiconductor devices.

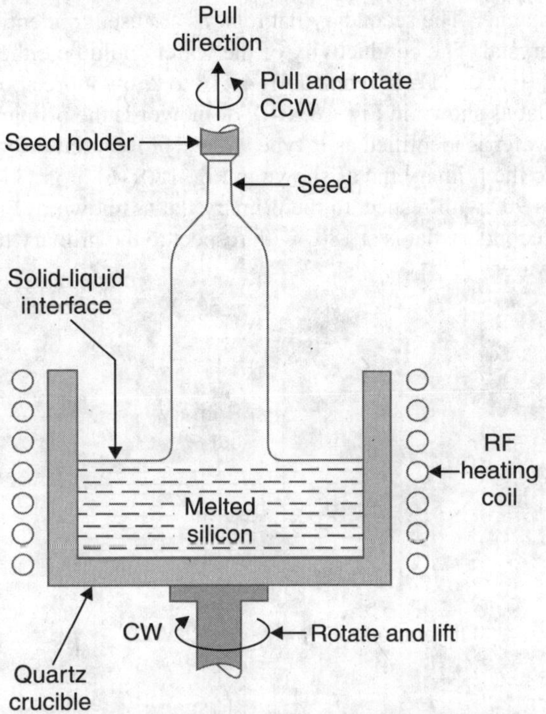

Fig. 18.7.

The Czochralski process. The equipment used for single crystal growth (called puller) is as shown in Fig. 18.7 The puller has three main components : (*i*) a furnace which includes quartz crucible, a rotation mechanism (clockwise as shown), and a radio frequency (RF) hating element, (*ii*) a crystal pulling mechanism which includes a seed holder and a rotation which includes a seed holder and a rotation mechanism (counter-clockwise), and (*iii*) an ambient control which includes an argon gas source, a flow control and an exhaust system. In addition the puller has a computer system to contorl process parameters such as temperature, crystal diameter, pull rate and rotationn speed.

To grow crystals, the polycrystalline sillicon is placed in the cruvible. The furnace is heated to a temperature of 1690 K which is slightly greater than melting point (1685 K) of silicon. A precisely contorlled amount of dopant (boron or phosphours) is added to the mleting make the silicon as *P*-type or *N*-type. A suitable oriented seed crystal (*i.e.*, a small highly perfect crystal) is suspended over the crucible in a seed holder. The seed is inserted into the melt and a small portion of it is allowed to melt. The seet is rotated and pulled up very slowly, while at the same tiime, the crucible is rotated in the opposite direction. The molten silicon attaches itself to the seed and it becomes indentical to the seed in structure and orienation. As the seed is pulled up, the material that is attached to the seed solidifies. (*i.e.,* freezes). It crystal structure becomes the same as that of the seed and a larger crystal is formed. Thus using this method. cylindrical single crystal bars (called ingots) of silicon are produced.

The desired diameter of the silicon ingot is obtained by controlling both the temperature and the pulling speed. In the final step, when the bulk of the melt has been grown, the crystal diameter is decreased until there is a point contact with the melt. The resulting ingot is cooled and is removed to be made into thin discs called *wafers.* The ingots have diameters as large as 200 mm with the latest ones approaching 300 mm. The ingot length is of the order of 1000 mm.

Wafer preparation : In this stage the ingot surface is grounded throughout to an exact diameter and the top bottom portions are cut off. After that one or more flat regions are ground along the length of the ingot. These flat regions mark the specific crystal orientation of the ingot and coductivity type (*i.e., P*-type or *N*-type) of silicon material. Refer to Fig. 18.8.

Notice that the ingot is marked with two flats regions : (*i*) the larger flat (called primary flat) and (*ii*) the smaller flat or secondary flat regions. The primary flat allows a mechanical locator in automatic processing equipment to position the wafer and to orient the devices relative to the crystal in a specific manner. The secondary flat regions are used to identify the orientation and conductivity type of the crystal. The conductivity of the wafer could be either *P*-type or *N*-type and the crystal orientation, {100} or {111} (refer to Art 48.16 to know more about crystal orientation). If there is no secondary flat as shown in Fig. 18.8 (*a*), or in words the primary and the secondary flat are superimposed, the wafer is identified as *P*-type with {111}-crystal orienation. If the secondary flat is 45° with respect to the primary flat as shown in Fig. 18.8 (*b*), it is {111} *N*-type wafer. However, if the secndary flat is 90° with respect to the primary flat as shown in Fig. 18.8 (*c*), it is {100} *P*-type wafer and if the secondary flat is at 180° with respect to the primary flat as shown in Fig. 18.8 (*d*), it is {100} *N*-type wafer.

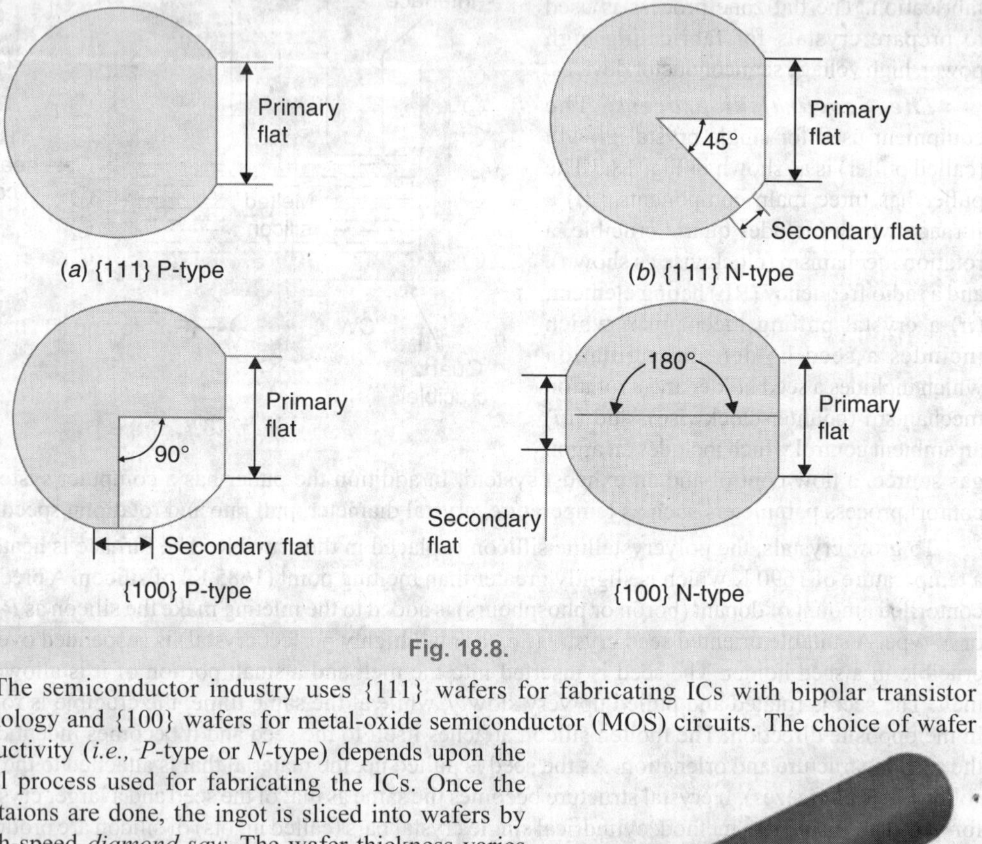

Fig. 18.8.

The semiconductor industry uses {111} wafers for fabricating ICs with bipolar transistor technology and {100} wafers for metal-oxide semiconductor (MOS) circuits. The choice of wafer conductivity (*i.e., P*-type or *N*-type) depends upon the actual process used for fabricating the ICs. Once the orientaions are done, the ingot is sliced into wafers by a high-speed *diamond saw.* The wafer thickness varies from 0.4 to 1.0 mm.

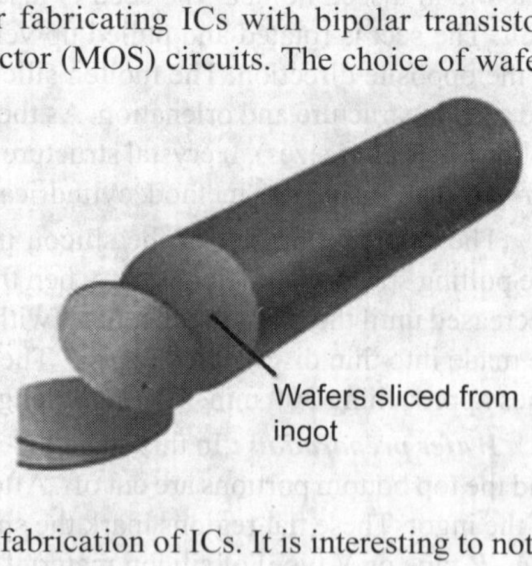

After slicing, both sides of the wafers are lapped to produce typical flantess uniformity with in 2 mm. The lapped orientation usually leaves the surface ane edges of the wafer, damaged and contaminated. These can be removed by a process called chemical etching. The next step is to polish the wafers suface to a mirror-like finish. Finally the wafers are cleaned, rinsed and dried for use in fabrication of ICs. It is interesting to note that the final wafer thickness is about one-third less than after the slicing.

18.19. Wafer Fabrication

Following is the category of the processes that are used in the fabrication of *ICs* :

(*i*) oxidation (*ii*) etching
(*iii*) diffusion (*iv*) ion implantation
(*v*) photolithography (*vi*) epitaxy
(*vii*) metallization and interconnections.

We will study each process separately and apply some of these later to the formation of a diode bipolar transistor and metal oxide semiconductor field effect transistor. The basic fabrication process is called *planar process, i.e.,* a process in which the introduction of impurities and metallic interconnections is carried out from the top of the wafer. A major advantage of the planer process is that each fabrication step is applied to all identical circuits and each of the many wafers at the same time.

It is important to initially emphasize that the fabriation requires an extermely clean environment in addition to the precise contorl of temperature and humidity.

18.20. Oxidation

The process of oxidation consists of growing a thin film of silicon dioxide (SiO_2) on the surface of a silicon wafer. Silicon dioxide. has several uses :

1. to serve as a mask against implant or diffusion of dopant into silicon,
2. to provide surface passivation,
3. to isolate one device from another,
4. to act as a component in MOS structures.

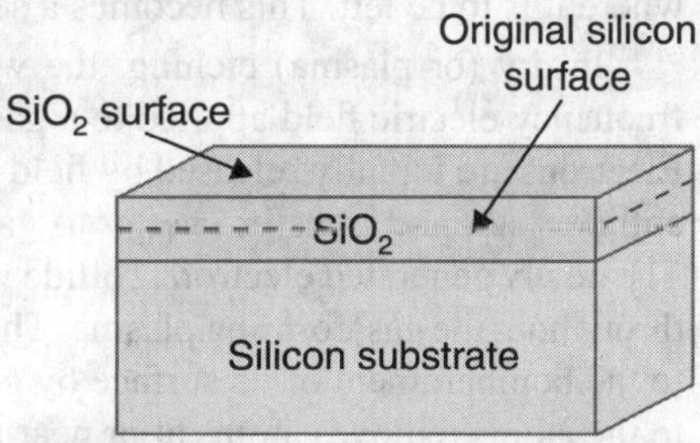

Fig. 18.9.

Fig. 18.9 shows oxide layer grown on the surface of silicon substrate. The commonly used silicon dopants, such as boron, phosphorus, arsenic, and antimony, have very low diffusion coefficients (*i.e.,* they diffuse with great difficulty) in silicon dioxide Because of this reason, silicon dioxide. is used as a shield against infiltration of these dopants. On the other hand, these dopants diffuse easily if the surface is silicon.

Several techniqes have been developed for forming oxide layers. Some of these are thermal oxidation, vapour phase technique [chemical vapour deposition (CVD)], and plasma oxidation. However, thermal oxidation is the more commonly used technique in IC processing.

Fig. 18.10 shows a thermal oxidation system. As seen, the oxidation is accomplished by placing the silicon wafers vertically into a quartz boat in a quartz tube. The quartz boat is slowly passed through a resistance heated furnace, in the presence of oxyzen, operating at a temperature of about 1000°C. The oxidizing agent may be *dry* by using a mizture of water vapour and oxyzen. Acomputer contorls the whole operation in the thermal oxidation system. The operation include regulating the gas flow sequence, automatic insertion and removal of wafers and the furnace temperature.

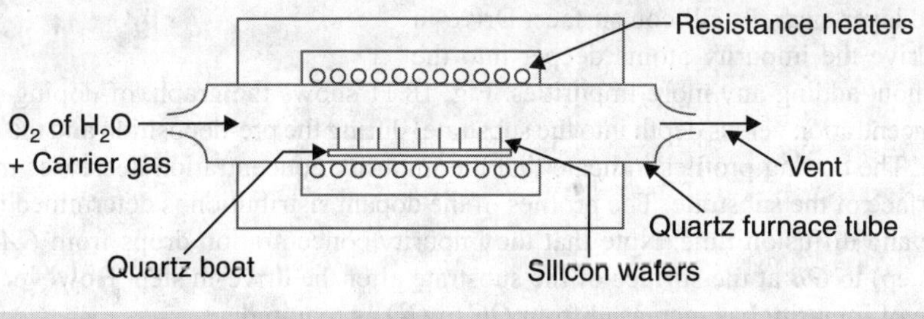

Fig. 18.10.

It will be interesting to note that the oxide growth in the dry process is much *slower* but it produces an oxidized layer that has *excellent* electrical properties. For relatively thin oxides such as gate oxide in a MOSFET (typically 10 mm), dry oxidation is used. However, for thicker oxides (^{3}500 nm) such as field oxides in MOS integrated circuits and for bipolar devices, wet oxidation is used to provide both isolation and passivation.

Once the silicon dioxide layer has been formed on the surface of the wafer, it is selectively removed (etched) from those surface where impurities are to be introduced and kept as a shield, for the underlying silicon surface, where no dopants are to be allowed.

Oxide layers are relatively free from defects and provide stable and reliable electrical properties.

18.21. Etching

Etching is the process of selective removal of regions of a semiconductor, metal or silicon dioxide. There are two types of etching : *wet* and *dry*. In wet etching, the wafers are immersed in a chemical solution at a predeteermind temperature. In this process, the material to be etched is removed equally in all directions (*i.e.,* etching is isotropic). Because of this, some material is etched from the regions where it is to be left. This becomes a serious problem when dealling with small dimension.

In dry (or plasma) etching, the wafers are immersed in gaseous plasma created by a radio-frequency electric field applied to a gas such as argon. The gas breaks down and becomes ionized. Electrons are initially released by field emission. These electrons gain kinetic energy from the field, collide with, and transfer energy to gas molecules, which result in generating ions and electrons. The newly generated electrons collide with other gas molecules and the avalanche process continues throughout the gas, forming plasma. The wafer to be etched is placed on an electrode and is subjected to the bombardment of its surface by gas ions. As a result, the transfer of momentum from the ions to the atoms removes atoms at or near the surface to be etched.

Dry etching also called reactive ion etching (RIF) is directional (or anisotropic). In other words, the material is removed only from those regions where it is required. Most modern processes use only dry etching to produce fine line patterns needed for VLSI integrated circuits.

18.20. Diffusion

This process consists of the intorduction of impurities into selected regions of a wafer to form junctions. Diffusion occurs in two steps : the *pre-deposition* and the *drive-in* diffusion. In the predeposition step, a high concentration of dopant atoms is intorduced at the sillicon by a vapour that contains the dopant at a temperature of 1000°C. More recently, a more accurate method of pre-depo-sition known as *ion implantation* is used.

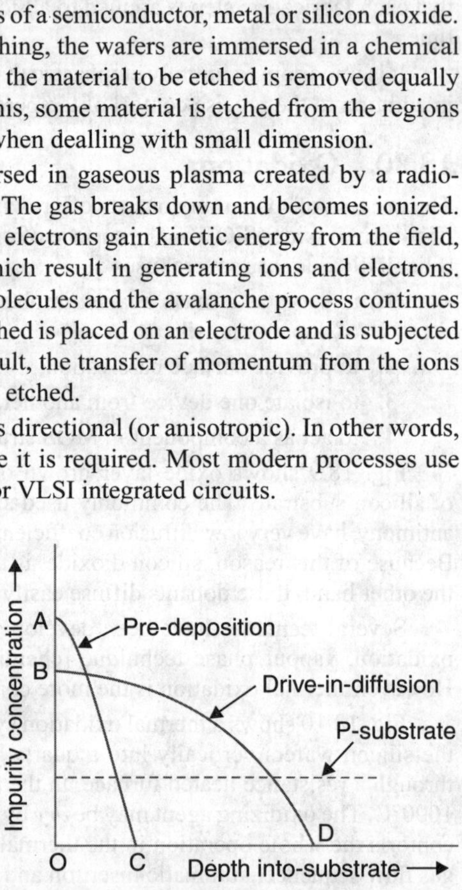

Fig. 18.11.

Pre-deposition tends to produce, a shallow but havily deped layer, near the sillicon surface. Drive-in is used to drive the impurity atoms deeper into the surface, without adding any more impurities. Fig. 18.11 shows the graph, of doping profiles, (*i.e.,* impurity concentration versus depth into the substrate) during the pre-deposition and the drive-in steps of diffusion. The doping profiles indicate that the impurity concentration decreases monotonically from the surface of the substrate. The profiles of the dopant sistribution is determined mainly by the temperature and diffusion time. Note that the impurity concentration drops from *OA* (during pre-deposition step) to *OB* at the surface of the substrate after the drive-in step. However the path into the substrate of impurity has increased from *OC* to *OD* as required.

Common depants are boron for *P*-type layers and phosphorus, antimony and arsenic for *N*-type layers. However, diffusion is rarely performed using pure elements themselves. Rather, compounds of elements are used and impurities may be introduced from either solid, liquid, or gaseous substance.

Fig. 18.12 shows the equipment used in diffusion. The wafers and placed in a quartz boat within a quartz furnance tube. The furnace is heated by resistance heaters surrounding it. To intorduce as impurity, phosphorus for example, phosphorus oxychloride ($POCl_3$) is placed in a container either inside the quartz tube in a region of relatively low temperature or in container outside the furnace at a temperature that maintains its liquid form. For a *P*-type dopant, boron is used. The proper vapour pressure is maintained by a control of the temperature. Nitrogen and oxygen gas are made to pass over the container. These gases react with the silicon, forming a layer on the surface of the wafer that contains silicon, oxygen and phosphorus. At the high temperature of the furnace, phosphorus easily diffuse into the silicon.

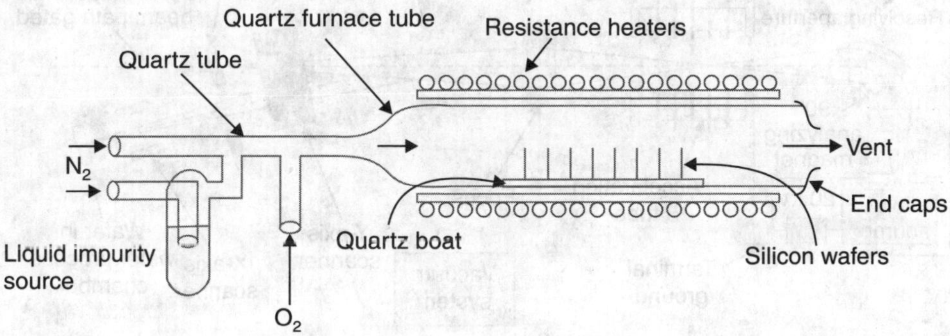

Fig. 18.12.

In order that the dopant may be diffused deeper into silicon, the drive-in step follows. This is done at a higher temperature of about 1100°C inside a furnace similar to that used for pre-deposition, except that no dopant is introduced into the furnace. The higher temperature, causes the dopant atoms to move into silicon more quickly. Diffusion depth is controlled by the time and temperature of the drive-in process. By precise control of the time and temperature, accurate junction depths of fraction of a micron can be obtained. Fig. 18.13 shows the diffusion of dopant into silicon.

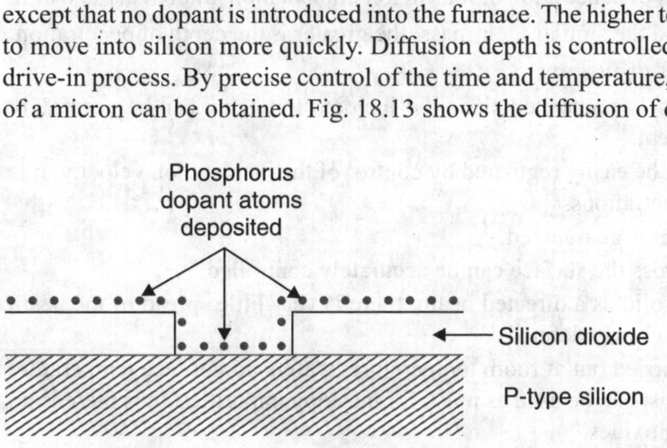

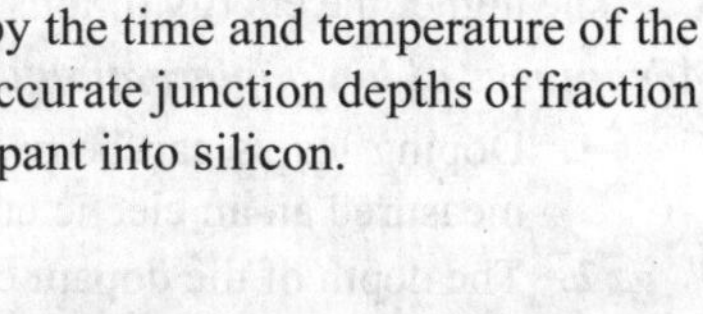

The photolithograph is used to add circuit elements to the wafer. The wafer is coated with layer of a chemical called photoresist.

Fig. 18.13.

18.23. Ion Implantation

This is a process of intorducing dopants into selected areas of the surface of the wafer by bombarding the surface with high-energy ions of the particular dopant.

Fig. 18.14 shows a typical ion implantation equipment. To generate ions, such as those of phosphorus, an arc discharge is made to occur in a gas, such as phosphine (PH_3), that contains the dopant. The ions are then accelerated in an electirc field. The ions are further accelerated so that their energy reaches several hundred ke V or Me V, whereupon they are focused on and strike the surface of the silicon wafer.

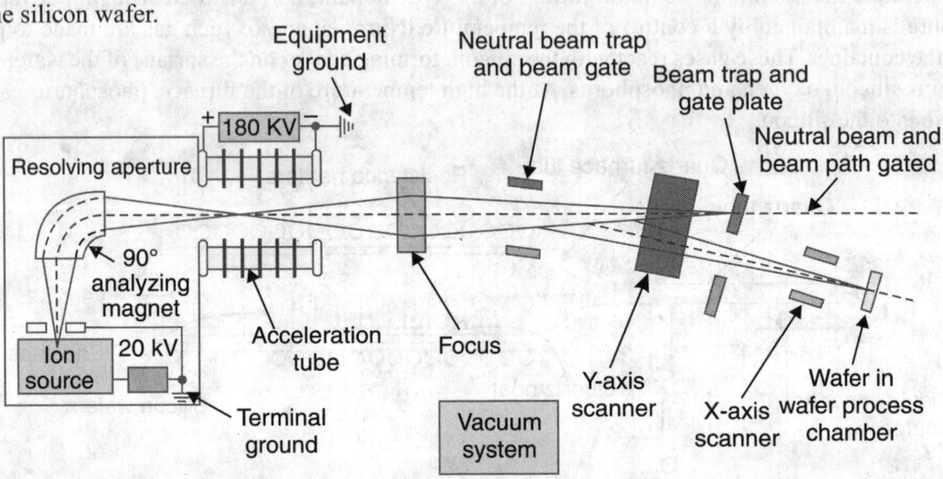

Fig. 18.14.

As is the case with diffusion, the ion beam is made to penetrate only into selected regions of the wafer by a process of making (discussed later). On entering the wafer, the ions collide with silicon atoms and lose their energy. The depth of penetration of ions in ion implantation is about 0.1 to 0 mm.

The higher the energy of ions and the similar their mass, the greater is the depth of penetration.

***Advantages of ion implantation over diffusion* :**

1. Doping levels can be precisely controlled since the incident ion beam can be accurately measured an an electic current.
2. The depth of the dopant can be easily regulated by control of the incident ion velocity. It is capable of very shallow penetrations.
3. Extreme purity of the dopant is guaranteed.
4. The dopiong uniformity across the surface can be accurately controlled.
5. Because the ions enter the solid as a directed beam, there is very little spread of the beam thus the doping area can be clearly defined.
6. Since ion implantation is carried out at room temperature, wafers do not face temperature stress. In addition photo-resist can be used as mask for masking impurities, thus there is no need to grow thick masking oxides.

***Disadvantages of ion implantation over diffusion* :**

1. The ion implantation may create considerable damage to the crystal structure because of the collisons of the high-energy ions with the silicon. Such damage results in inferior performance of *ICs* made by this process. If the damage is not extensive, the process of *annealing* restores the structure.
2. High initial investment and operational cost of the equipment (> US 1 million).
3. Uses very toxic gases for some of the depants such as phosphorus and arsenic.

18.24. Photomask Generation

The whole process of IC fabrication consists of identifying selected regions of each circuit of the wafer surface into which identical dopant or metallic interconnections are made, while protecting other regions of the wafer surface. To carry out one of the many fabrication process, a separate mask is required for each operation whose function is to expose the selected regions and protect the others. There may be hunreds of identical dies (or IC*s*) on a wafer with each circuit containing hundreds of thousands or millions of devices. Identacal steps are carried out simultaneously for each process. For each process, separate mask is needed.

The mask production starts with a drawing using a computer-assisted graphics system with all the information about the drawing stored in digital form. Commands from the computer are prepared that drive a pattern generator, which uses an electron beam to write the partiular pattern, for one or several dies, on a glass plate covered with thin chromium film. When the glass plate is prepared, it is called *recticle.* A mask usually refers to a glass plate that contains a pattern for the whole wafer. The reticle pattern is projected onto the wafer and a *wafer stepper* is used that reduces the reticle circuit onto the photo-resist covered wafer that steps across over the surface until the entire array of circuits is built up.

It may be noted that the use of a single mask for all the circuits on a wafer is not feasible for printing very small (<1 μm) features because of alignment problems. However, the use of single mask is still in use for fabricating simple digital and analog circuits such as light emitting diodes (LEDs).

18.25. Photolithography

It is a process in which the geometrical pattern on the glass plate (called reticle) is transferred to the surface of the wafer. This is done to open identical windows so that the diffusion (or the ion implantation) process may take place in all identical regions of the same *IC* and for all *ICs* on the wafer. As an illustration we assume that the first reticle is used over an oxidized surface as shown in Fig. 18.15 (*a*).

To transfer the pattern, the wafer is coated with light-sensitive material called *photoresist.* By applying small amount (about 1 ml) of the photoresist to the wafer surface and spinning the wafer very rapidly, a uniform film of about 1 mm thick is formed over the oxidized surface of the wafer [refer to Fig. 18.15 (*b*)]. After this the following steps are taken to open a window on the surface of the wafer :

1. The wafer is baked at 100°C to solidify the photoresist on the wafer.
2. The glass plate is placed on the wafer and aligned by the computer control.
3. The glass plate is exposed to vltraviolet (UV) light with the transparent parts of the glass plate passing the light on the wafer. The photoresist under the opaque regions of the glass plate is unaffected [refer to Fig. 18.15 (*c*)].
4. The exposed photoresist is chemically removed by dissolving it in an organic solvent and exposing the silicon dioxide underneath [refer to Fig. 18.15 (*d*)].
5. The exposed silicon dioxide is then etched away using hydrofluoric acid. The hydrofluoric acid dissolves silicon dioxide and not silicon dioxide and the photoresist [refer to Fig. 18.15 (*e*)].
6. The photoresist. under the opaque regions of the glass plate is removed using a proper solvent and the silicon dioxide is exposed [refer ot Fig. 18.15 (*f*)].

All surface in Fig.18.15 (*f*) are protected except those covered by silicon only in which diffusion or ion implantation is to take place. It may be noted that the surface covered bysilicon dioxide do not permit any entry of dopants.

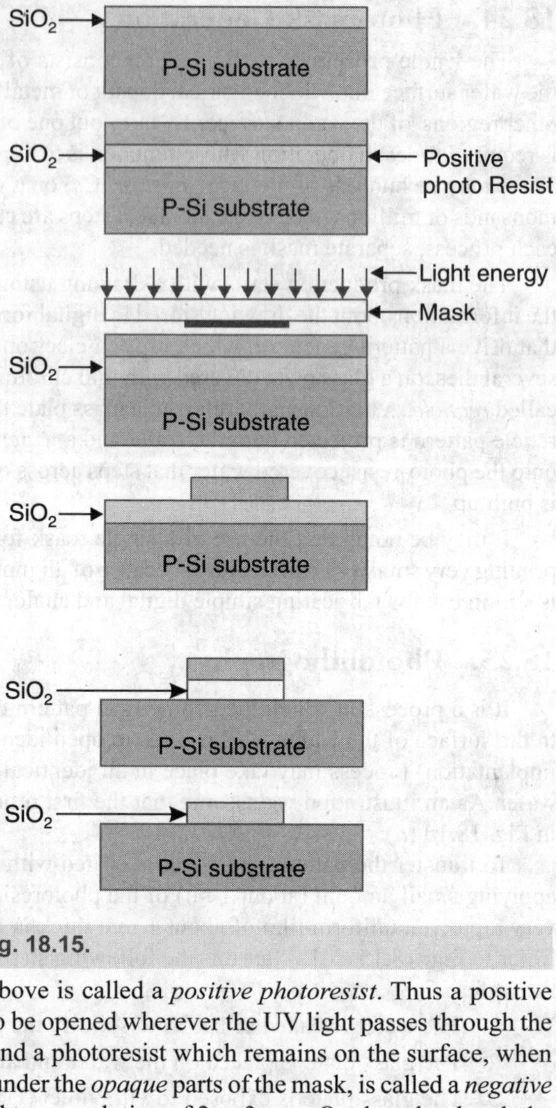

(*a*) P silicon substrate and oxide film

(*b*) Photoresist applied

(*c*) Mask placed above photoresist and UV light directed at it

(*d*) Photoresist etched away under transparent regions of mask

(*e*) SiO_2 etched away

(*f*) Remaining photoresist etched away

Fig. 18.15.

The photoresist used in the explanation above is called a *positive photoresist*. Thus a positive photoreisst is that which allows the windows to be opened wherever the UV light passes through the *transparent* parts of the mask. On the other hand a photoresist which remains on the surface, when exposed to UV light, and windows are opened under the *opaque* parts of the mask, is called a *negative photoresist*. The negative photoresist is limited to a resolution of 2 to 3 mm. On the other hand, the positive photoresist sacrifice adhersion and simplicity of development to achieve higher resolution.

The method of using UV light has a practical limit for transferring patterns of the silicon. It can be used for the patterns where the linewidths are above 1 to 2 mm. In order to transfer small patterns (*i.e.,* those with below 1 mm linewidth), very short wavelength radiation such as electron-beam or X-rays are used.

18.26. Epitaxy

Epitaxy or **Epitaxial growth** is the proess of the *conrolled growth* of a crystalline doped layer of silicon on a single crystal substrate. The process of diffusion and ion implantation, which were earlier described, produce a layer at the surface that is of higher doping density than that which existed before the dopant was added. It is not possible by these methods to produce, at the surface a layer of lower conecntration than exists there. This can however, be accomplished by the method of epitaxy. In the processes of diffusion and ion implantation, a dopant is driven into a substrate of doped silicon. In epitaxy, a layer of doped silicon is deposited on top of the surface of the substrate. Normally this single crystal layer has different type of doping from that of the substrate.

Epitaxy is used to deposit N or N^+ (*i.e.,* heavily doped *N* type) silicon, which is impossible to accomplish by diffusion. It is also used in isolation between bipolar transistors where N^- (*i.e.,* lightly doped *N*-type) is deposited on *P*-type layer. It may also be used to improve the surface quality of an *N*-type substrate by depositing *N*-type material over it. A variety of methods are used to grow the epitaxial layer. These methods include vapour-phase epitaxy (VPE), liquid-phase epitaxy (LPE), and molecular beam epitaxy (MPE). The system for growing an epitaxial layer using vapur-phase epitaxy (VPE) is shown in Fig. 18.16.

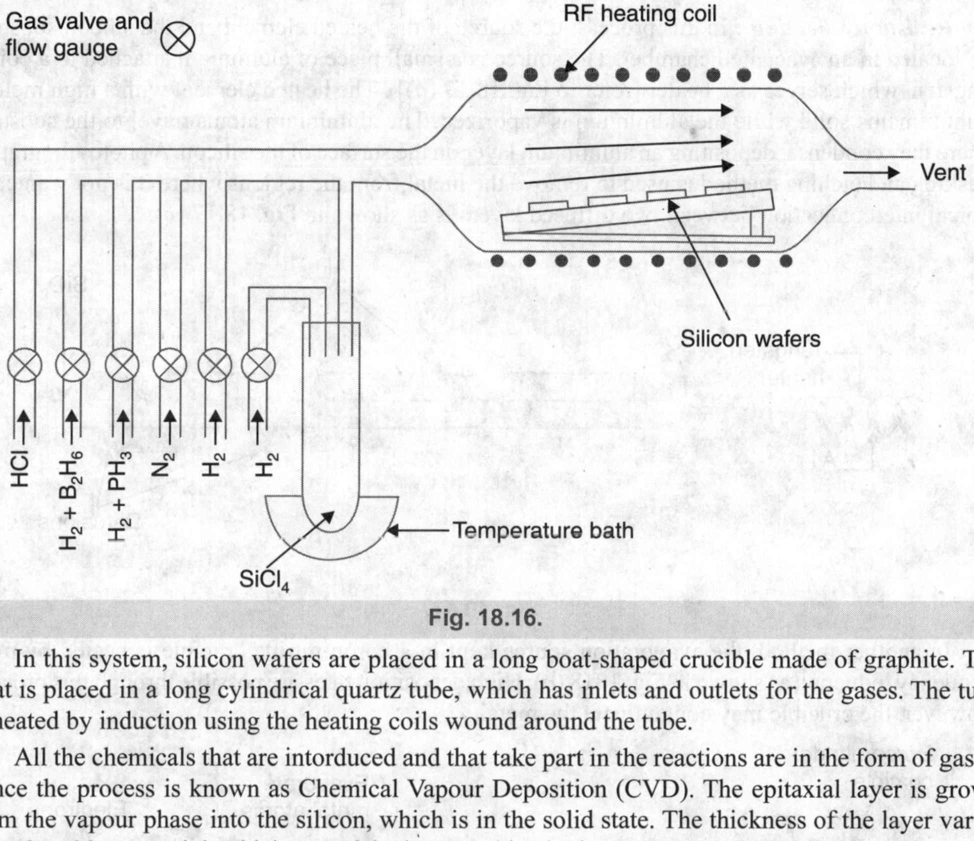

Fig. 18.16.

In this system, silicon wafers are placed in a long boat-shaped crucible made of graphite. The boat is placed in a long cylindrical quartz tube, which has inlets and outlets for the gases. The tube is heated by induction using the heating coils wound around the tube.

All the chemicals that are intorduced and that take part in the reactions are in the form of gases, hence the process is known as Chemical Vapour Deposition (CVD). The epitaxial layer is grown from the vapour phase into the silicon, which is in the solid state. The thickness of the layer varies from 3 to 30 mm and the thickness of the layer and its doping content are controlled to an accuracy of less than 2 percent. The reactions are carried out at a temperature of approximately 1200°C. The high temperature is necessary so that the dopant atoms can acquire a sufficient amount of energy to allow them to move into the crystal to form covalent bonds and bocome extension of the single crystal. Because the layer is grown on the substrate, epitaxy is a growth technique where the crystal is formed without reaching the meling point of silicon.

18.27. Metallization and Interconnections

After all the fabrication steps of an IC are completed, it becomes necessary to provide metallic interconnections for the IC and for external connections to the IC.The requirement that must be met by the interconnections is that they have low resistance to minimize both the voltage drops on the lines as well as the capacitances between the lines so as to reduces delay times. The connections must also make *ohmic contracts* to semiconductors in the devices such as *P* and *N* regions of a *PN* junction diode. An ohmic contact is one that exhibits a very low resistance, allowing currents to pass easily in both direction through the contact.

The high conductivity of aluminium makes it the metal of obvious choice, particularly in silicon-based devices. It also has the following advantages :

1. easy to evaporate
2. can be easily etched
3. not expensive,
4. adheres well to silicon dioxide.

There are variety of processes for depositing aluminium on silicon substrates but the following three are important from the subject point of view : (1) resistance heating, (2) electron beam heating and (3) suputtering.

Resistance heating : In this process, the source of the heated element and the silicon substrate are located in an evacuated chamber. The source is a small piece of aluminium attached to a coil of tungsten, which serves as a heater [refer to Fig. 18.17 (*a*)] . The heated element with a high melting point remains solid while the aluminium is vaporized. The aluminium atoms travel to the substrate where they condense, depositing an aluminium layer on the surface of the silicon. A photolithographic masking and etching method is used to remove the metal from the regions where it is not wanted. A typical interconnection between two diffused layers is as shown in Fig. 18.17 (*b*).

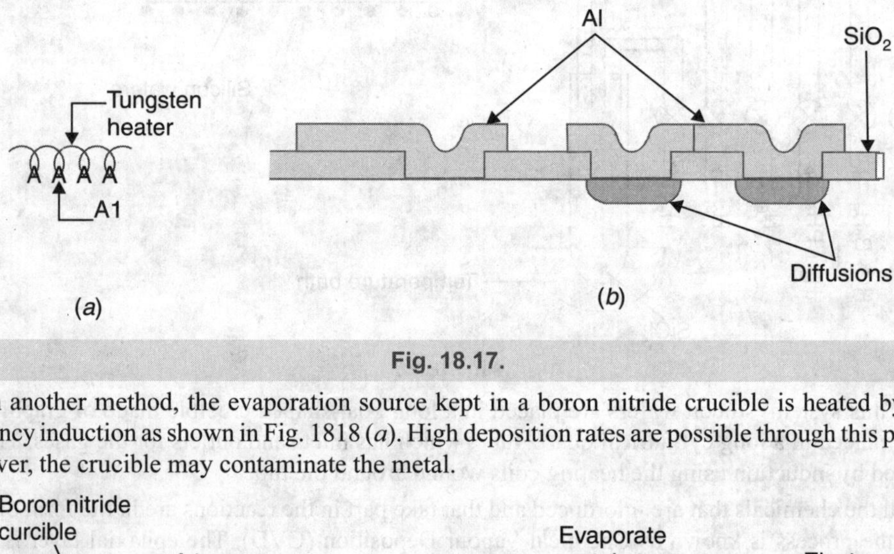

Fig. 18.17.

In another method, the evaporation source kept in a boron nitride crucible is heated by radio frequency induction as shown in Fig. 1818 (*a*). High deposition rates are possible through this process. However, the crucible may contaminate the metal.

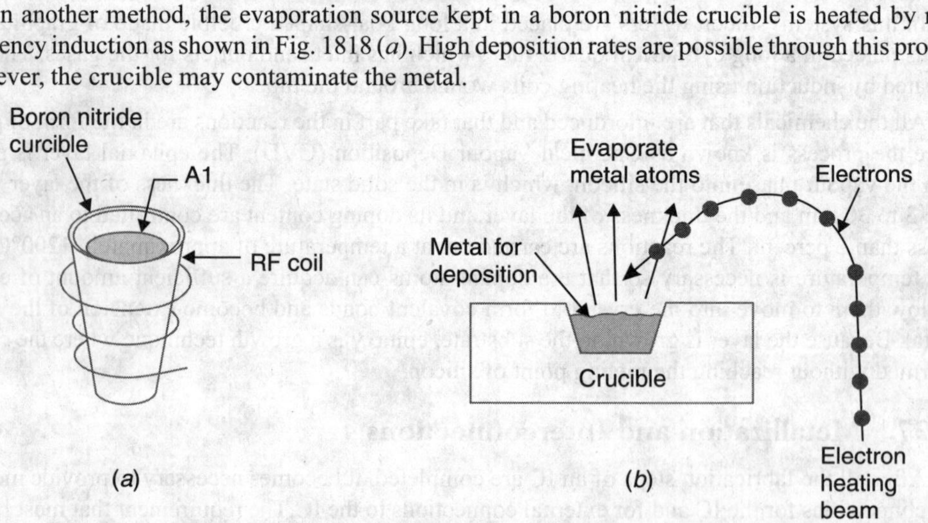

Fig. 18.18.

Electron-beam heating : In this process, aluminium in a crucible is placed into a vacuum chamber together with the substrate. The aluminium is subjected to a high intensity electron beam formed by an electron gun, which vaporizes the aluminium as shown in Fig. 18.18 (*b*). This causes the aluminium to travels to the wafer. By the use of mask and photolithography, the aluminium is deposited on the indenified regions on the wafer surface.

Sputtering : In this process, the material to be deposited is placed in a container maintained at low pressure in the vicinity of the substrate. The material to be deposited is labelled the cathode or the target, while the anode is the substrate. A DC or a radio-frequency high voltage is applied between anode and cathode. This high voltage ionizes the inert gas in the chamber. The ions are accelerated to the cathode. (in this operation the anode is usually grounded) where, by impact with the aluminium target, atoms of aluminium are vaporized. A gas of aluminium atoms is generated and deposited on the surface of the wafer.

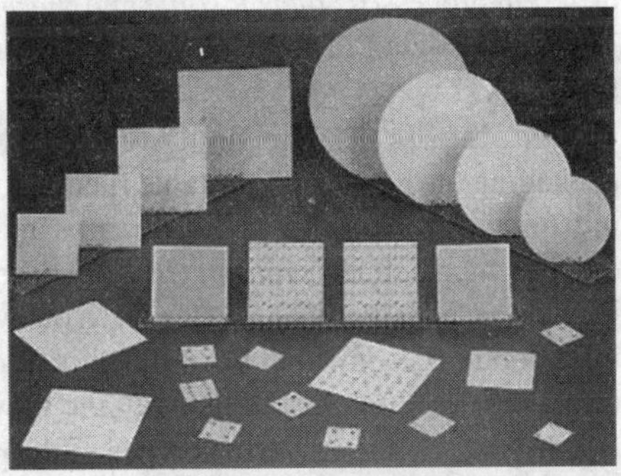

Several cycles of photolithography, etching and doping are performed producing multiple layers of circuit elements on the wafer.

Follwing the deposition of aluminium, the silicon wafers are placed in a furnace to solidify the connections so that low resistance metallic contacts are made.

The interconnections between the elements of an ICare made by aluminium lines having a thichkness of about 0.5 μm. These are laid on top of the silicon dioxide layer, which covers the surface of the wafer. By using photollithography, openings are made in silicon dioxide so that the aluminium layer is connected to the silicon or to the ohmic contact on the silicon. In very compicated integrated circuits, it is necessary to have two or three vertically stacked layers of interconnections separated by silicon dioxide layers. The interconnecting lines terminate at aluminium pads (called bonding pads) from which connections to the outside are made.

18.28. Testing, Bonding and Packaging

The individual IC chip must be connected properly ot outside leads and packaged in a way that is convenient for use in a larger circuit or a system. Since the devices are handled individually once they are separated form the wafer, bonding and packaging are expensive process.

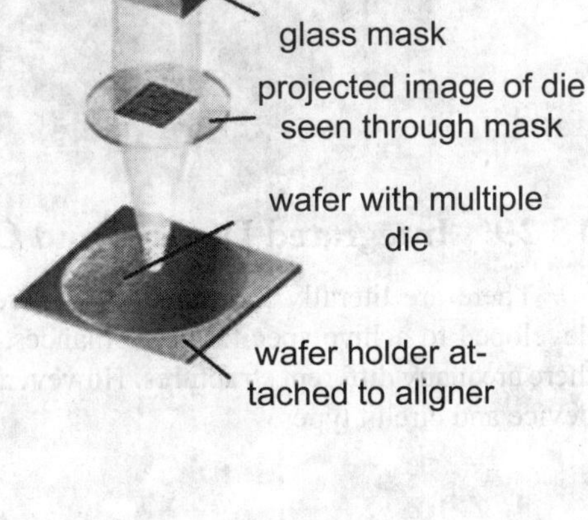

Testing — After the wafer of monolithic circuits has been processed and the final metallization pattern defined, it is placed in a holder under a microscope and is aligned for testing by a machine called *multiple-point probe*. The probe contacts the various pads on an individual circuit, and a series of tests are made to verify the electrical properties of the device in a very short time. After all the circuits are tested, the wafer is removed from the testing machine, sawed between the circuits, and broken apart. Then each die that passed the test is picked up and placed in the package.

Bonding — It consists of two steps. In the first step the back of the die is mechanically attached to an appropriate mount medium such as ceramic substrate, multi-layer-ceramic package or metal lead frame. The two common die bonding methods are *hard solders* and *polymers*. In the second step, the bond pads on the circuit side of the die are connected by wires to the package as shown in Fig. 18.19. The three common schemes of interconnection to the chip bond pads are (*i*) *wire bonding*,(*ii*) *tape-outomated bonding* (*TAB*), and (*iii*) *flip-chip solder bonding*. Wire bonding is further split into two different processes : *thermosonic* and *thermocompression*. The further detail of these process is beyond the scope of this book.

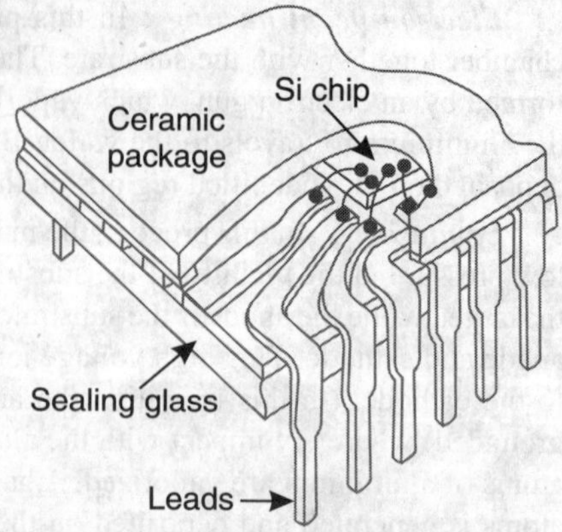

Fig. 18.19.

Packaging — The final step in IC fabrication is packaging the device in a suitable medium that can protect in from the moisture and contaminants and the bonds and other elements must be protected from corrosion and mechnical shocks. The modern ICs are mounted in package with many output leads. In one version, chip is mounted on a stamped metal lead frame and wire bonding is done between the chip and the leads. The package is formed by applying a ceramic or plastic case and trimming away the unwanted parts of the lead frame.

Fig. 18.19 shows various types of the packaging of ICs used these days. As seen from this figure, the packages can be broadly classifed into the follwing two categories :

1. *Through-hole mount* that involve inserting the package pins through holes on the printed circuit board (PCB) before soldering [Fig. 18.20 (*a*) through (*d*)].
2. *Surface mount type* where the leads do not pass through holes in the PCB. Instead, surface-mounted package leads are aligned to electrical contacts on the PCB, and are connected simultaneously by solder reflow [Fig.18.20 (*e*) through (*m*)].

Most packages can be made using ceramic or plastic. The ICs are hermetically sealed for protection from the environment. The pins can be on one side (single inline or zigzag pattern of leads), two sides (dual inline package or DIP), or four sides of the package (quad package). Most advanced packages have leads distributed over a large portion of the surface of the package as in through-hole pin grid arrays (PGAs) or surface-mounted ball grid arrays (BGAs).

18.29. Integrated Devices and Circuits Formation

There are literally thousands of different semiconductor device structures. They have been developed to achive specific performances, either as discrete components or in ICs. Accrodingly there are many different structures. However, there are basic structures required for each of the major device and circuit types.

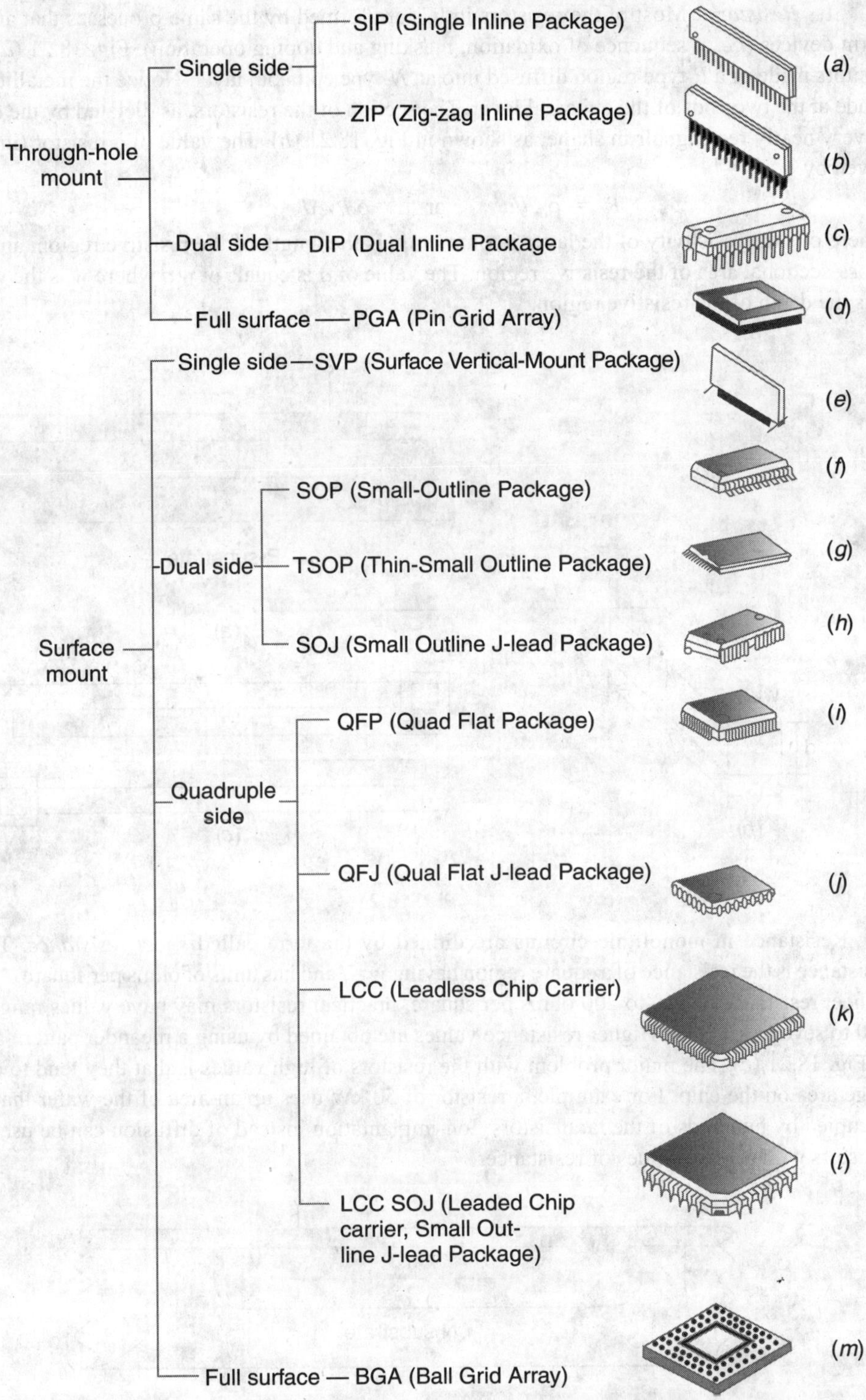
Through-hole mount
Single side
SIP (Single Inline Package)
ZIP (Zig-zag Inline Package)
Dual side
DIP (Dual Inline Package
Full surface
PGA (Pin Grid Array)
Surface mount
Single side
SVP (Surface Vertical-Mount Package)
Dual side
SOP (Small-Outline Package)
TSOP (Thin-Small Outline Package)
SOJ (Small Outline J-lead Package)
Quadruple side
QFP (Quad Flat Package)
QFJ (Qual Flat J-lead Package)
LCC (Leadless Chip Carrier)
LCC SOJ (Leaded Chip carrier, Small Out-line J-lead Package)
Full surface
BGA (Ball Grid Array)
(a)
(b)
(c)
(d)
(e)
(f)
(g)
(h)
(i)
(j)
(k)
(l)
(m)

Fig. 18.20.

1. ***Resistors*** **:** Most of the resistors in ICs are formed by the same processes that are used to form devices (*i.e.,* a sequence of oxidation, masking and doping operation). Fig. 18.21 (*a*) shows a resitors made of a *P*-type region diffused into an *N*-type epitaxial layer. Notice the metallic contacts made at the two ends of the epitaxial layer. The section of the resistors, as dictated by the diffusion, is very nearly rectangualr in shape, as shown in Fig. 18.21 (*b*). The value of a resistor (in ohms) is given by

$$R = \rho \, . \, l/a \qquad \text{or} \qquad \rho.l/w.d$$

where ρ is the resistivity of the layer in ohm-cm, *l* is the length of the resistive region, and *a* is the cross-sectional area of the resistive region. The value of *a* is equals ot *w.d.*where *w* is the width and *d* is the depth of the resistive region.

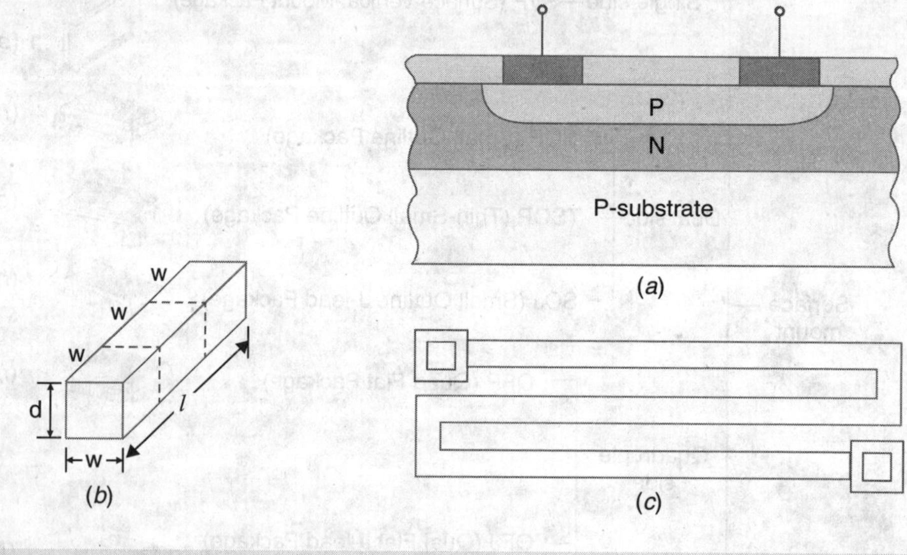

Fig. 18.21.

Resistance in monolithic circuits are difined by the term called *sheet resistance.* The sheet resistance is the resistance of a square region having $w = l$ and has units of ohms per square. Assuming a sheet resistance of 100 to 200 ohms per square, practical resistors may have values ranging form 100 to several kilohms. Higher resistance values are obtained by using a meander pattern as shown in Fig. 18.21 (*c*). The major problem with the resistors of high values is that they tend to occupy a large area on the chip. For example, a resistor of 50 kW uses up an area of the wafer that may be occupied by hundreds of the taransistors. Ion-implantation instead of diffusion can be use to make resistors with precise values of resistance.

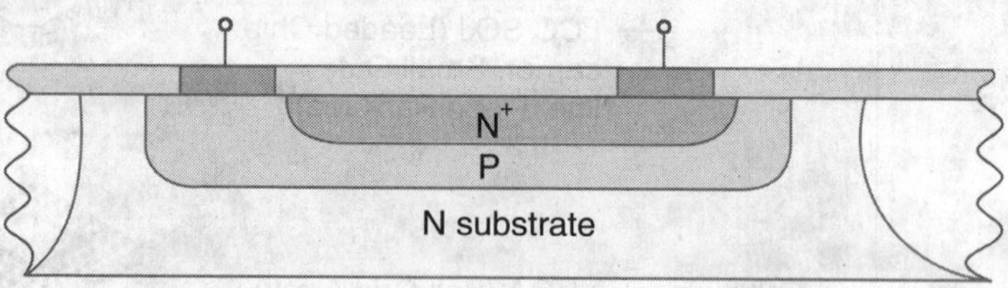

Fig. 18.22.

2. ***Capacitors*** : One type of capacitor in monolithic circuits is made by using the capacitance formed between the *P* and *N* regions of a reverse-baised diode. Such a capacitors is shown in Fig. 18.23. These capacitors are formed by using the same diffusion processes that are used to from devices. Bipolar transistors are made of three regions in which either of the two *PN* junctions may be used as capacitors whereby the breakdown voltage of the capacitors may very considerably from one to the other. The disadvantage of the junction capacitors is the dependence of the capacitance on the voltage applied to the junciton. Capacitors that are voltage-indipendent can be formed from metal insulator N^+ (*i.e.*, heavily-doped) semiconductor layers as used in MOS structures.

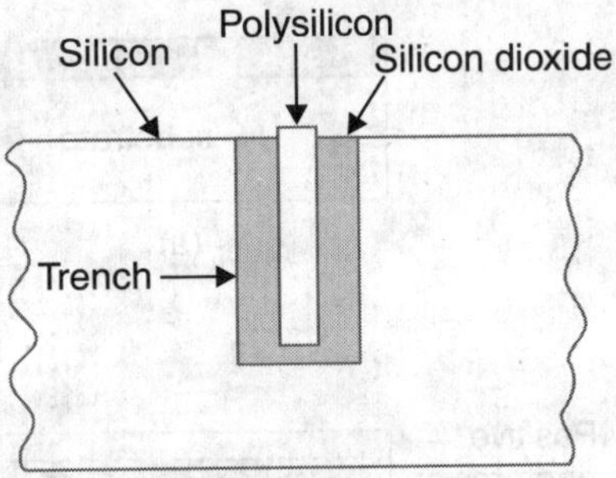

Fig. 18.23.

In dense circuits, an oxide/nitride/oxide dielectric sandwich is used. The combination film has a lower dielectric constant, allowing a capacitor area smaller than a vonventional silicon dioxide capacitor.

Capacitors can also be fabricated by creating a trench etched vertically into the wafer surface (refer to Fig. 18.23). The trenches are etched either isotropically with wetetching or anisotropically with dry etching techniques. The trench side walls are oxidised and the centre of the trench is filled with deposited polysilicon. The final structure is "wired" from the surface, with the silicon and polysilicon serving as two electrodes and silicon dioxide as a dielectric material.

The trenched capacitors are useful when preservation of wafer surface is main criteria. Another alternative to conserve wafer surface area is to build stacked capacitors on the wafer surface. This effort has driven by the need for small high dielectric capacitors for dynamic randon access memory (DRAM)

3. ***Diodes*** : In integrated circuits, where all the interconnections and device terminals are made at the surface, a diode is formed in the following tow ways :

(a) In bipolar circuits from a bipolar transistor by placing a *short-circuit* between two of the three terminals of the transistor (*i.e.*, collector to base or emitter to base). Thus there are emitter-base diodes and base-collector diodes.

(b) In MOS circuits, most of the diodes are formed with the source-drain doping step.

However, in order to make a discrete *PN* junction diode, we will discuss the various steps involved in the fabrication.

The starting material is a heavily doped *N*-type (N^+) substrate about 150 μm thick. A layer of *N*-type silicon (1 to 5 μm) is grown on the substrate by epitaxy as shown in Fig. 18.24 (*a*). Then a layer of silicon dioxide (SiO_2) is deposited by oxidation as shown in Fig. 18.24 (*b*). Next, the surface is coated with positive photoresist as shown in Fig. 18.24 (*c*). Then mask is palced on the surface of silicon, aligned and exposed to ultraviolet light (UV) as shown in Fig. 18.24 (*d*). Next, mask is removed, resist is removed and silicon dioxide layer under the exposed resist is etched as shown in Fig. 18.24 (*e*). Then boron is diffused to form P-type region a shown in Fig. 18.24 (*f*). Note that boron diffuses easily in silicon but not in silicon dioxide. Next, a thin aluminium film is deposited over the surface as shown in Fig. 18.24 (*g*). Then the metallized area is covered with resist another mask is used to identify areas where metal is to be preserved. Wafer surface is etched to remove unwanted metal. Resist is then dissolved and we get a structure as shown in Fig. 18.24 (*h*). Finally, contact metal is deposited on the back surface and ohmic contacts are made by heat treatment.

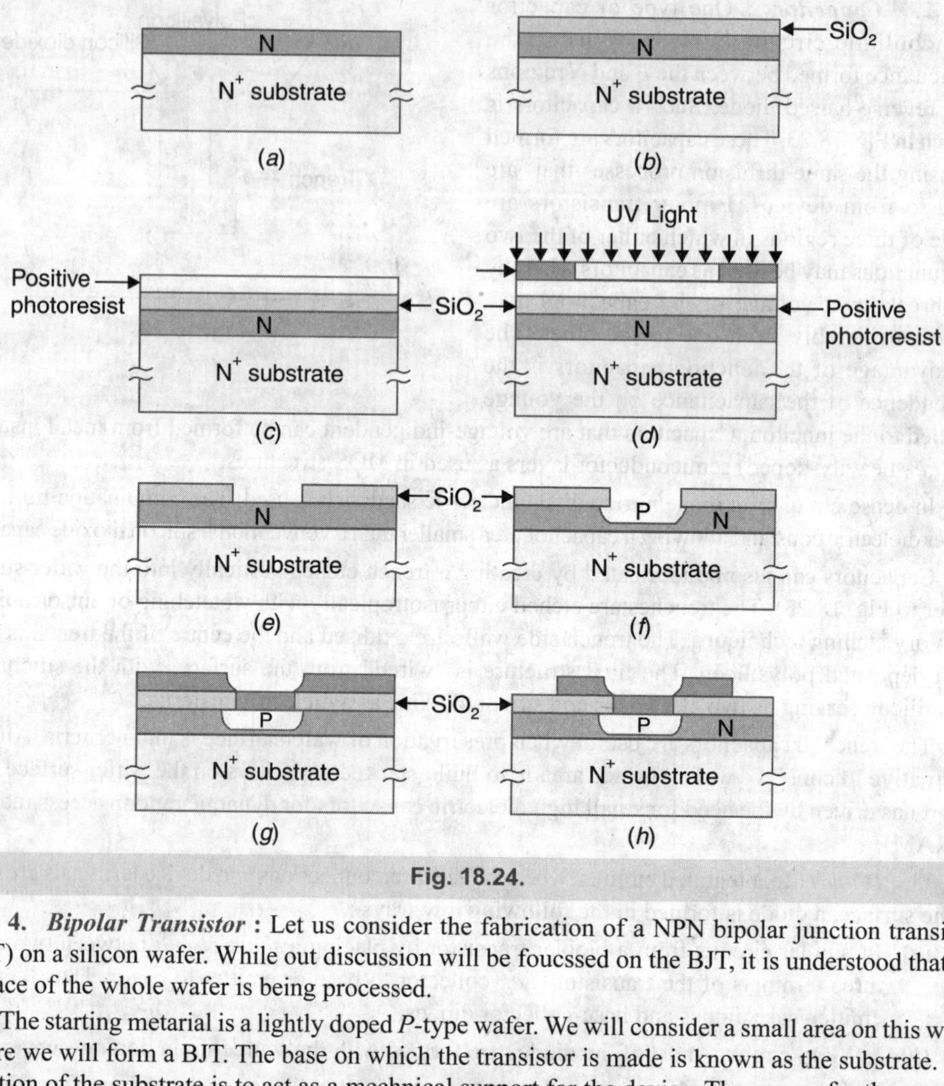

Fig. 18.24.

4. ***Bipolar Transistor*** : Let us consider the fabrication of a NPN bipolar junction transistor (BJT) on a silicon wafer. While out discussion will be foucssed on the BJT, it is understood that the surface of the whole wafer is being processed.

The starting metarial is a lightly doped *P*-type wafer. We will consider a small area on this wafer where we will form a BJT. The base on which the transistor is made is known as the substrate. The function of the substrate is to act as a mechnical support for the device. The reason for the use of a *P*-type substrate for the NPN transistor will be clarified when the strem isolation is discussed. The substrate has a resistivity of 3-10 Ω cm with a thickness between 500 and 700 μm for wafers having diameters over 100 mm.

Fig. 18.25 shows the fabrication steps up to, and including the metal contacts to the three regions (*i.e.,* emitter, base and collector).

First a layer of silicon dioxide (SiO_2), about 0.5 μm thick, is deposited on the surface of the substrate by thermal oxidation as shown in Fig. 18.25 (*a*). Using the *first mask* and the photolithographic process, widows are opened in the oxide for the *buried layer.* The N^+ (*i.e.,* heavily doped *N*-type) buried layer is diffused to a depth of about 3mm as shown in Fig. 18.25 (*b*). This layer serves to collect the carriers that have crossed the base on their way to the collector terminal. It serves as a sub-collector and is used to reduce the collector ohmic-resistance. After the buried layer is difused, the wafer is stripped of all oxide to permit the next deposition. It is to be noted that during the subsequent high-temperature processes, the buried layer tends to diffuse out.

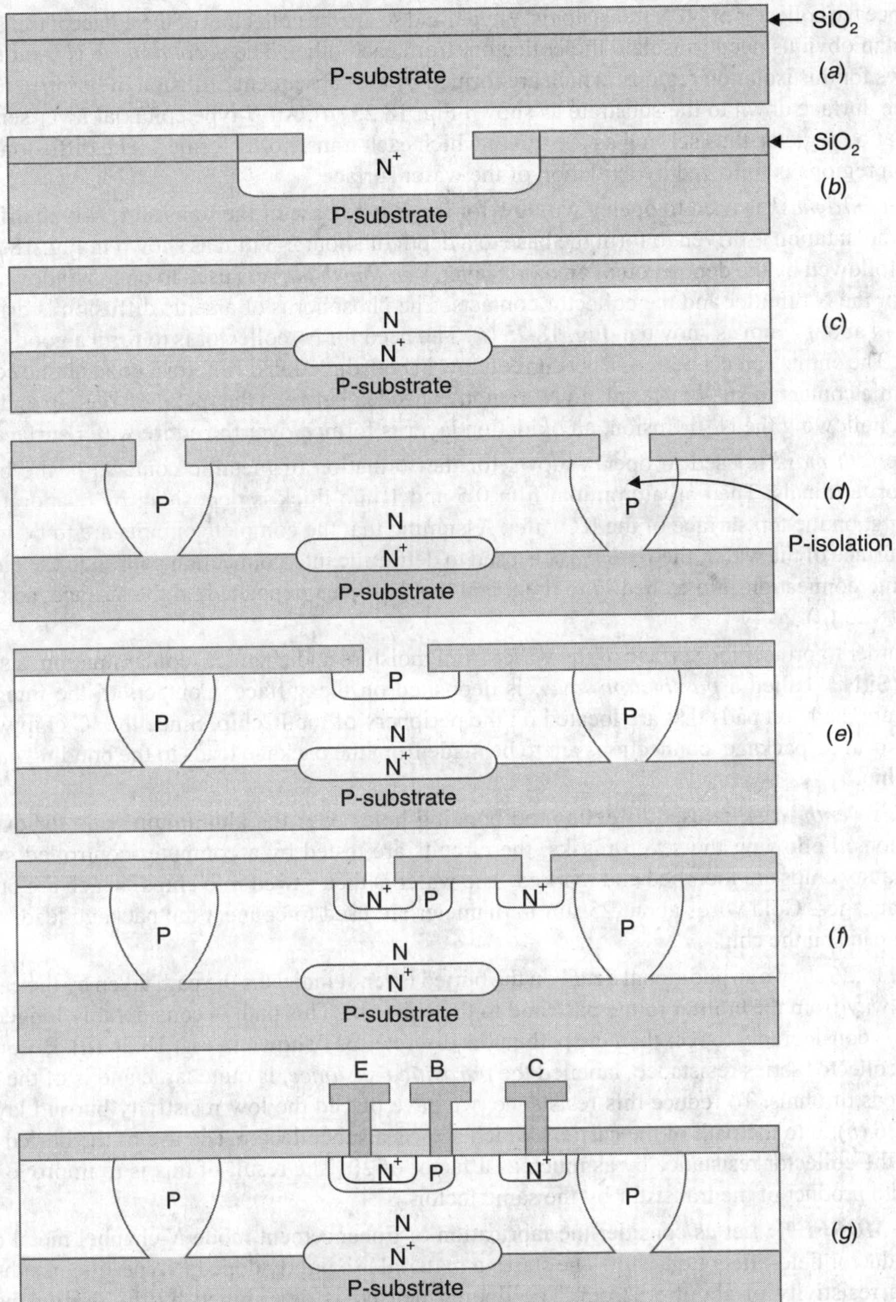

Fig. 18.25.

The next operation is the deposition of a phosphorus-doped *N*-type *epitaxial layer* on the whole wafer. This layer has a resistivity of 0.1 to 1Ω cm. The thickness of this layer is 0.5 to 5 μm for high-speed digital circuit applications and 10-20 μm for linear analog circuits. A layer of silicon dioxide (SiO_2) about 0.5 to 1 μm thick is grown thermally on the surface of the epitaxial as shown in Fig. 18.25 (*c*).

Since the collector of NPN transistor is *N*-type, and so are the collectors of the adjacent transistors, there is an obvious need to isolate the collectors from each other. The *second mask* is used to etch windows for this isolation regions, which are formed by the subsequent diffusion of boron extending from the durface down to the substrate as shown Fig. 18.25 (*d*). An *N*-type epitaxial layer separates the isolation regions, thus serving as the tub in which each transistor is formed. The diffusion of the isolation regions is followed by oxidation of the wafer surface.

The *third mask* is used to open a window for the *P*-type base of the transistor, *P*-type diffusion or ion implantation is driven to form the base to a depth of about 2-3 mm as shown in Fig. 18.25 (*e*). This is followed by the deposition of an oxide layer. The *fourth mask* is used to open windows in the oxide for the N^+ emitter and the collector contacts. The phosphorus or arsenic diffusion is driven to a depth of about 2 mm as shown in Fig. 18.25 (*f*). The need for N^+ collector is to form a good ohimic contact. The ohmic contact permits easy current flow in both directions. To form a good ohmic contact. The ohmic contact to an *N* material, an N^+ region is needed between the metal on the top and the *N* region . Following the N^+ diffusion, an oxidation layer is formed over the entire wafer surface.

The *fifth mask* is used to open widows for the formation of metallic contacts ot the bipolar transistor terminals. Then an aluminium film 0.5 and 1 mm thick is deposited, by evaporation or sputtering, on the top surface of the IC wafer. Assuming that the complete circuits are to be formed on the surface of the wafer, the *sixth mask* is used to define the interconnection pattern in the circuits. These interconnections are etched into the metal that has been deposited on the surface, as shown in Fig. 18.25 (*g*).

In order to protect the surface of the wafer from moisture and chemical contamination, a silicon nitride (Si_3N_4), called a *passivation layer,* is deposited on the surface. Contacts to the integrated circuits are made on pads that are located on the periphery of the ICchip. Since the IC chip will be bonded to an IC package, connections are to be made from the package leads to the bonding pads on the IC chip.

The *seventh* mask is used to define the bonding holes over the aluminium pads for external connections. Following the seven masks, the circuits are tested by a computer-controlled system and all faulty chips are idenified and marked. The wafer is then sawed into chips, which are bonded on *IC* packages. Gold wires about 25 μm in diameter are used to connect the package leads to the bonding pads on the chip.

Fig. 18.26 shows an additional detail on the buried layer. It indicates the path taken by the carriers on theie way from the emitter to the base and to the collector. This path is considerably longer than the path is considerably longer than the path in the discrete. BJT shown in Fig. 18.26 (*a*). Because of this the collector series resistance, labelled the *parasitic resistance,* is quite large and is of the order if hundreds of ohms. To reduce this resistance, we have placed the low resistivity burried layer in Fig. 18.26 (*b*). into the path of the carriers which acts as a subcollector. The use of the buried layer reduces the collector resistance by as much as a factor of 20. The result of this is to improve gain-bandwidth product of the transistor by the same factor.

5. *MOSFET* : Let us consider the fabrication of Enhancement mode *N*-channel metal oxide semiconductor field-effect transistor. The starting material is a lightly doped *P*-type silicon substrate having a resistivity of about 5 Ω cm. The doping density is determined by the drain-substrate breakdown voltage of 20-30 V. A thin (about 20 nm) layer of silicon dioxide is formed over the substrate to provide stress relief to the wafer. Next a thiin (about 20 nm) layer of silicon nitride (Si_3N_4) is deposited by the chemical vapour deposition (CVD) process on top of the silicon dioxide. The silicon dioxide permits selective oxidation so that a thick oxide (about 500 nm) can be formed in the field region. These operations result in Fig. 18.27 (*a*).

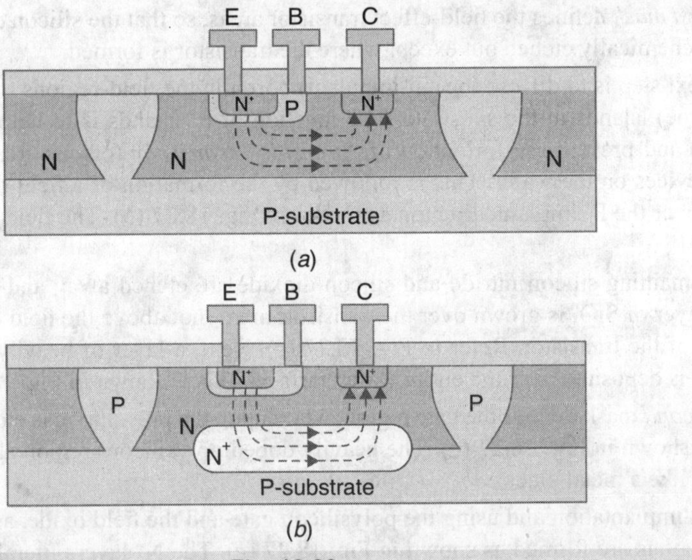

Fig. 18.26.

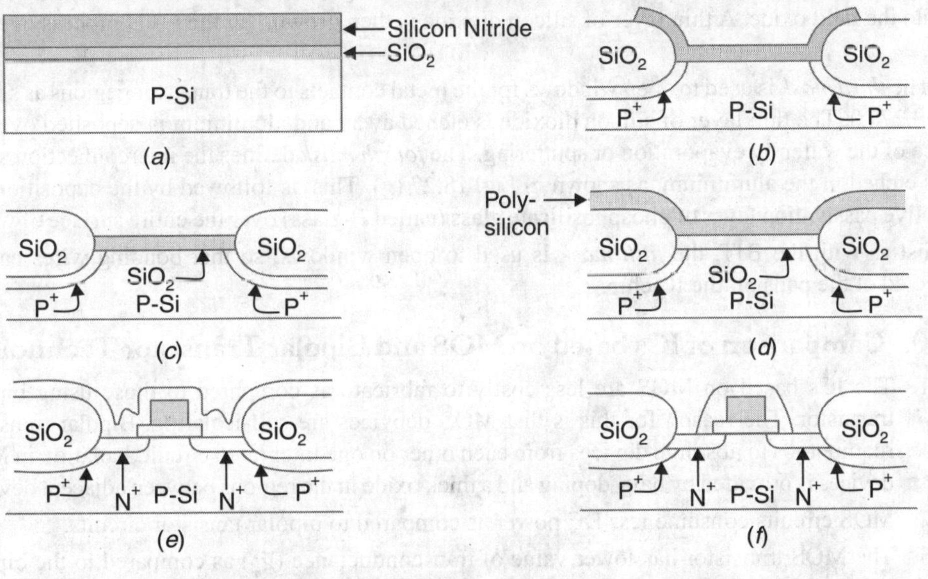

Fig. 18.27.

The *first mask* defines the field-effect transistor areas, so that the silicon dioxide and the silicon nitride are chemically etched out except where the transistor is formed.

The next step is to diffuse (or ion-impulant) boron in the field-regions to form P^+ (*i.e.,* heavily doped *P*-type) islands in the substrate. The function of P^+ islands is to help increas the threshold voltage, V_T and prevent the formation of "*prarsitic*" *transistor* (or electircal cross talk) between adjacent devices on the wafer. This is followed by the formation of a field oxide layer (about 500 nm thick) over the P^+ implanted region as shown in Fig. 18.27 (*b*). The field oxide layer also helps to increase V_T.

The remaining silicon nitride and silicon dioxide are etched away and an ultra thin (about 5 to10 nm) layer of SiO_2 is grown over the transisotr area (not above the field oxide). This forms the gate-oxide of the transistor. Refer to Fig. 18.27 (*c*). Next, a layer of heavily doped (typically N^+) polysilicon is deposited over the entire wafer surface. This is shown in Fig. 18.27 (*d*).

The *second* mask defines the gate region, whereupon the polysilicon is etched away except over the gate as shown in Fig. 18.27 (*e*). The heavily doped polysilicon region above the gate behaves electrically like a metal electrode.

By ion-implantation and using the polysilicon gate and the field oxide, as the mask, the source drain N^+ regions are formed as shown in Fig. 18.27 (*e*). The N^+ layers literally diffuse a sufficient substance ot ensure proper alignment so that the channel length is well difined. The dopants do not centrate the field oxide. A thin layer of silicon dioxide is then grown., by the CVD process over the wafer.

The *third mask* is used to open windows for the metal contacts to the transistor gegions as shown Fig. 18.27 (*f*). The thin layer of silicon dioxide is etched away and aluminium is deposited over the surface of the wafer by evaporation or sputtering. The *fourth mask* defines the interconnection strem that is etched in the aluminium, as shown in Fig. 18.27 (*g*). This is followed by the deposition if a protective passivation layer of phosphosilicate glass (called *P*-glass) over the entire surface of wafer.

Just as with the BJT, the *fifth mask* is used to open windows, so that bonding wires can be connected ot the pads on the IC chip.

18.30. Camparison of ICs based on MOS and Bipolar Transistor Technology

1. The ICs based on MOS are less costly to fabricate as compared to those using bipolar transistor, The reason for this is that MOS debvices are self-isolating. Bipolar transistor require tubs to absolute devices from each other on one integrated circuit. Isolation in MOS devices is provided by havy doping and a thick oxide in the regions between adjacent devices.
2. MOS circuits consume less DC power as compared to bipolar tansistor circuits.
3. The MOS transistor has lower value of transconductance (g_m) as compared to the bipolar transistor. This feature makes the bipolar transistor ICs superior to MOS circuits for analog circuit applications.
4. For the same channel length and the width, the limiting cut-off frequency of the MOS transistor is better than a bipolar transistor. Therefore the MOS transistor has a higher bandwidth than the bipolar transistor.
5. The packing density of MOS ICs is at least 10 times more than that for bipolar ICs. Also, a MOS resistors occupies less than 1% of the area of a conventional diffused resistor. This high packing density makes MOS ICs especially suited for LSI, VLSI and ULSI circuits.

The main disadvantage of MOS ICs is their slower speed as compared to bipolar ICs. Hence they do not compete with bipolar ICs in ultrahigh-speed applications.

However, due to their (*i*) low cost, (*ii*) low power consumption and (*iii*) high packing density, MOS ICs are widely produced by semiconductor manufacturing industry. The MOS ICs are available as calculator chips, momory chips, microprocessors (μP), single-chip computers etc.

18.31. Popular Applications of ICs

There are plenty of electronic products in the market, which used integrated circuits extensively. One of the widely accepted applications is the digital watch, which can display hours, minutes, seconds, day and month. Another popular application is electronic calculator which can perform various functions like addition, subtraction, mulitiplication and division. There are also advanced scientific calculators that are programmable and can display the graphs.

Fig. 18.28. Digital Camera
(*Courtesy* : Sony Corporation, Japan)

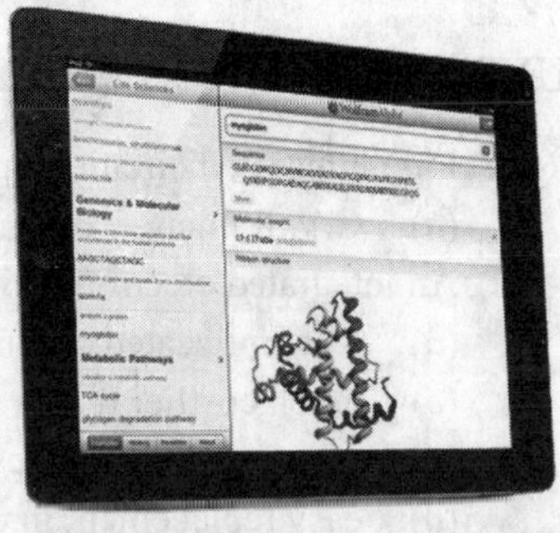

Fig. 18.29. ipad
(*Courtesy* : Apple Inc.)

Modern electronic products such as pocket PC, personal digital assistant (PDA), MP3 players, digital cameras, digital camcorders, mobile phones, digital dictionaries and digital translators, CD compact disk) players, DVD (digital versatile disk) players etc. also make use of ICs extensively. with a Pocket PC, you can read and send e-mail, edit Word, Excel and Outlook files, listen to digital music, read electronic books (eBooks), track your finances, and browse the web. The digital camera shown in FIg. 18.28 can take pictures and store on the floppy diskette or memory stick. A personal digital assistant (PDA) as shown in Fig. 18.29 is a handheld computer which allows the user to aganise calendar, contacts or tasks. These little electronic wonders might put office secretary out of a job.

DVD-player.

Electronic games have evolved from microelectronics and computer knowhow. Video games that have to be hooked up to home TV have become very popular. Hand held PCs and mobile phones also come with games that requires no TV hook-up.

DESCRIPTIVE QUESTIONS

1. What is an integrated circuit ? What are its advantages and limitations over the discrete circuits ?
2. Explain the different types of integrated circuits.
3. What are thick film integrated circuits ? How the films are deposited on such circuits?

4. Explain the difference between the analog and digital integrated circuits. Give suitable examples of both types.
5. Name and explain the different processes involved in preparing the surface of a semiconductor wafer for fabricating an integrated circuit.
6. Explain in detail, how using photolithography, a window is opened in the oxide layer diffusing impurites in the specified regions
7. What are the lateral and vertical PNP transistors ? Explain them by giving their constructional details.
8. Outline the process sequence of fabrications of MOSFET's.
9. Explain, how a bipolar transistor can be used as a diode.2

MULTIPLE CHOICE QUESTIONS

1. First integrated circuit chip was developed by
   (*a*) C.V. Raman (*b*) W.H. Brattain (*c*) J.S. Kilby (*d*) Robert Noyce.
2. An integrated electronic circuits is
   (*a*) a compliecated circuit (*b*)an integrating device
   (*c*) much costlier than a single transistor (*d*) fabricated on a tiny silicon chip.
3. Most important advantages of an IC is its
   (*a*) easy replacement in case of circuit failure
   (*b*) extremely high reliability (*c*) reduced cost
   (*d*) low power consumption.
4. In monolithic ICs, all components are fabricated by process.
   (*a*) evaporation (*b*) sputtering (*c*) diffusion (*d*) oxidisation.
5. Monolithic ICs are fabricated within a
   (*a*) soft stone (*b*) single stone (*c*) silicon layer. (*d*) oxidisation.
6. As compared to monolithic ICs, film ICs have the advantage of
   (*a*) better high-frequency response (*b*) much reduced cost
   (*c*) smaller size (*d*) less flexibility in circuit design.
7. Monolithic technique is ideally suited for fabricating ....ICs.
   (*a*) thin-film (*b*) digital (*c*) linear (*d*) thick-film
8. In the contex of IC fabrication, metallisation means
   (*a*) connecting metallic wires
   (*b*) forming interconnecting conduction pattern and bonding pads
   (*c*) depositing SiO_2 layer
   (*d*) convering with a metallic cap.
9. Monollithic transistors are formed in the epitaxial *N*-layer
   (*a*) in one operation (*b*) by evaporation
   (*c*) by successive impurity diffusions (*d*) by oxidisation.
10. The major component of a MOS IC is a/an
    (*a*) FET (*b*) MOSFET (*c*) BJT (*d*) SCR.
11. Processing of MOS ICs is less expensive that bipolar ICs primarily because they
    (*a*) use cheaper components (*b*) need no component isolation
    (*c*) require much less steps (*d*) have very high packing density.

12. In an integrated circuit, the SiO_2 layer provedes
    (*a*) electrical connection to the external circuit
    (*b*) physical strength
    (*c*) isolation
    (*d*) conducting path.
13. The foundation on which an IC is built is called an
    (*a*) insulator (*b*) base (*c*) wafer (*d*) plate
14. Monollithic IC construction uses
    (*a*) discrete components (*b*) high-value resistors
    (*c*) connecting wires (*d*) extensive number of components
15. Microelectronics refers to
    (*a*) circuits using miniature discrete components
    (*b*) only monolithic integrated circuits
    (*c*) small circuits made by evaporation silk-screen or semiconductor
    (*d*) electric circuits using subminiature tubes
16. In a monolithic integrated circuits.
    (*a*) all components are fabricated in a single wafer of semiconductor material
    (*b*) each component is formed in a separate wafer of semiconductor material
    (*c*) active components assembled in a separate wafer of semiconductor material
    (*d*) the components are not isolated from each other
17. Microelectronics integrated circuits
    (*a*) are not made with silicon
    (*b*) are made using the diffused planer process
    (*c*) have an internonnection pattern formed under the final oxide
    (*d*) are made on ceramic substrates
18. A multichip circuit
    (*a*) consists of several interconnected thin film circuits
    (*b*) is a monolithic silicon wafer with thin film components
    (*c*) cosists of several interconnected monolithic circuit wafers
    (*d*) is normally made with thick film techniques
19. In thin film circuits
    (*a*) active components are diffused into the substrate
    (*b*) passive components are formed on the surface of the substrate by evaporation
    (*c*) components are formed by silk-screen techniques
    (*d*) active components are not used
20. Thick film circuits are formed
    (*a*) on a thick semiconductors substrate
    (*b*) by sputtering and diffusing tantalum into a glass substrate
    (*c*) with conducing ink and a glass-metal slurry silk screened into a cermic substrate
    (*d*) by fabricating active and passive components simulataneously
21. The photomasking
    (*a*) provides control of the regions of silicon oxide to be etched
    (*b*) controls the depth of diffusion
    (*c*) prevents ambient light from shining on the silicon wafer during processing
    (*d*) reduces component size

22. In a necessary because silicon is electrically conducting
    (*a*) necessary because silicon is electrically conducting
    (*b*) separates components mechanically
    (*c*) is not necessary because the silicon substrate has a high resistivity
    (*d*) produces a high resistance but only between active components
23. In a monolithic integrated circuit, the isolation can be obtained by
    (*a*) a layer of photoresist
    (*b*) mounting a separating chip on the insulating base
    (*c*) scribing
    (*d*) a reverse-biased PN junction
24. A diffused resistor is an integrated circuit
    (*a*) can be of any value
    (*b*) is formed after the transistor diffusion
    (*c*) formed at the same time as one of the transistor regions
    (*d*) can be only of -type.
25. Capacitors for integration circuits
    (*a*) cannot be made using different techniques
    (*b*) can be made with very high calues of capacitance
    (*c*) can be made using silicon dioxide as the dilectiric
    (*d*) are always discreet components cannected extermally
26. Integrated circuits are probe tested
    (*a*) after separation into individual chips
    (*b*) to check small-singal as characteristics
    (*c*) to measure individual components
    (*d*) on the complete wafer for dc operation

ANSWERS

1. (*c*)	2. (*d*)	3. (*b*)	4. (*c*)	5. (*c*)
6. (*a*)	7. (*b*)	8. (*b*)	9. (*c*)	10. (*b*)
11. (*c*)	12. (*c*)	13. (*c*)	14. (*d*)	15. (*c*)
16. (*a*)	17. (*b*)	18. (*c*)	19. (*b*)	20. (*c*)
21. (*a*)	22. (*a*)	23. (*d*)	24. (*c*)	25. (*c*)
26. (*d*)				

19 Chapter

Rectifiers and Filters

Outline

19.1. Introduction

The rectifiers and filters are the circuits, which are used in the initial stages of a d.c. power supply. The d.c. power supply is essential for the operation of many electronic devices and circuits. A rectifier is a circuits, which uses one or more diodes to convert a.c. voltage into pulsating d.c. voltage. It may be broadly categorized in the following two types:

1. Half-wave rectifier, and
2. Full-wave rectifier.

19.2. Single-phase Half-wave Rectifier

Fig. 19.1 (*b*) shows a half-wave rectifier circuit. It consists of a single diode in series with a load resistor. The input to the half-wave rectifier is supplied from the 50 Hz a.c. supply, whose waveform is shown in Fig. 19.1 (*a*). The working of a half-wave rectifier circuit may be studied by considering separately the positive and negative half cycles of the a.c. input voltage.

During the positive half cycle of the a.c. input voltage, the diode is forward biased and conducts for all instantaneous voltages greater than the offset voltage (0.7 V for silicon and 0.3 V for germanium diodes). However, for all practical purposes, we assume that the diode is forward biased, whenever the a.c. input voltage goes above zero. While conducting, the diode acts as a short-circuit, so that the circuit current flows and produces a voltage across the load resistor (R_L). The voltage produced across the load resistor has the same shape as that of the positive input half cycle of a.c. input voltage as shown in Fig. 19.1 (*c*). The waveform of diode current (which is equal to the load current) is also shown in Fig. 19.1 (*c*).

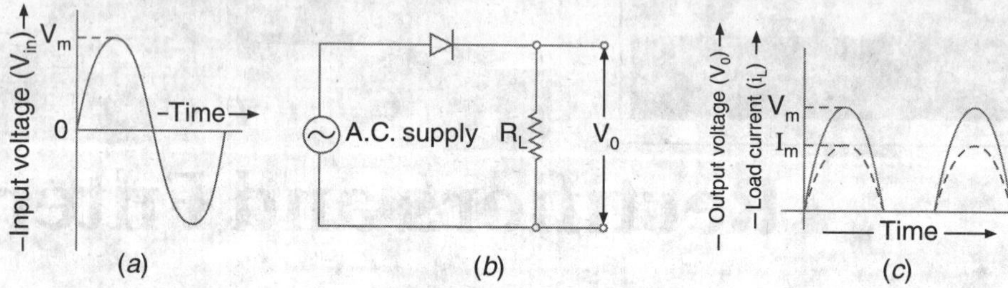

Fig. 19.1. Half-wave rectifier.

During the negative half-cycle, the diode is reverse biased and hence it does not conduct. Thus there is no current flow or voltage drop across load resistor (R_L) *i.e.*, $i_D = 0$ and $v_0 = 0$. The net result is that only the positive half cycle of the a.c. input voltage appears across R_L. It means that only the positive half cycle of the a.c. input voltage is utilized for delivering a.c. power. It is evident that output voltage (v_0) is not a steady d.c. But it is a pulsating d.c. wave having a ripple frequency equal to the input voltage frequency. It will be interesting to know that we can see the waveform of output voltage on an oscilloscope connected across R_L. If we measure the output voltage (or output current) by a d.c. meter, it will indicate some average positive value for both voltage and current.

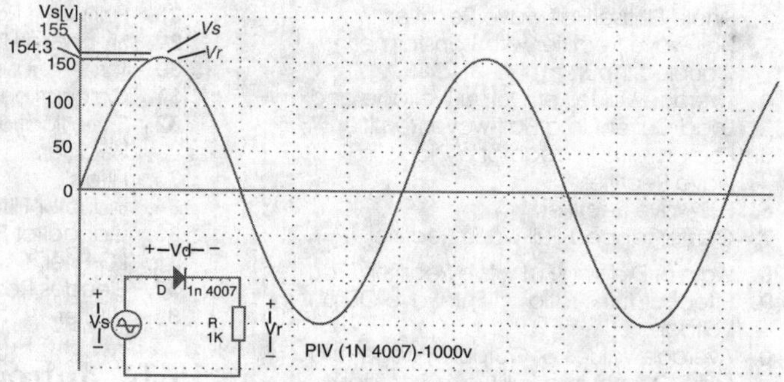

It is evident from the above discussion, that *as the circuit uses only one-half cycle of the a.c. input voltage, therefore it is popularly known as a half-wave rectifier.*

19.3. Half-wave Rectifier with Transformer Coupled Input

These days, most of the electronic circuits use a transformer to couple a.c. input voltage from source to the rectifier circuit as shown in Fig. 19.2. The transformer coupling provides the following two advantages:

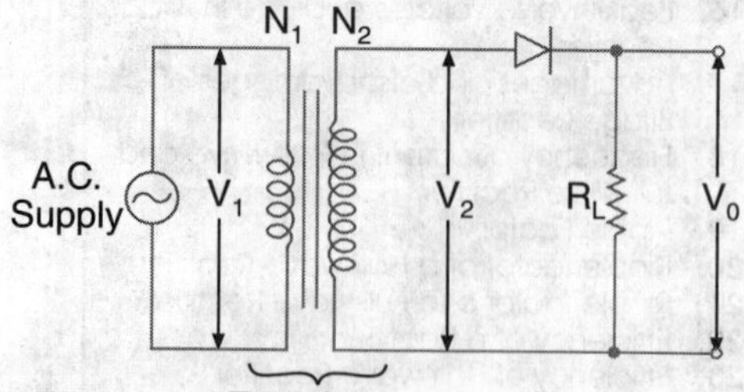

Fig. 19.2. Half-wave rectifier with transformer coupled input.

1. It allows the source voltage to be stepped up or stepped down.
2. The a.c. power source is electrically isolated from the rectifier circuit. It reduces the chances of getting a shock.

The primary and secondary voltages of an ideal transformer are related by the expression,

$$\frac{V_2}{V_1} = \frac{N_2}{N_1}$$

where V_1 = The r.m.s. value of primary voltage.
V_2 = The r.m.s. value of secondary voltage.
N_1 = Number of turns in primary winding, and
N_2 = Number of turns in secondary winding.

19.4. Average Values of Output Voltage and Load Current in a Half-wave Rectifier

Consider a half-wave rectifier circuit with a sinusoidal a.c. input voltage. We know that instantaneous value of the sinusoidal a.c. input voltage is given by the relation,

$$v = V_m \sin \omega t = V_m \sin \theta \qquad ...(\because \theta = \omega t)$$

where V_m = Maximum value of the a.c. input voltage

It may be noted that in case of a transformer coupling, the equation of a.c. input voltage would remain the same as given above. However, the maximum value of a.c. input voltage (V_m) will represent the maximum voltage across the secondary winding of the transformer.

Now let I_m = Maximum value of diode current or load current. It is the current through the resistor R_L.
V_{dc} = Average or d.c. value of output (or load) voltage across the load resistor, and
I_{dc} = Average or d.c. value of load current.

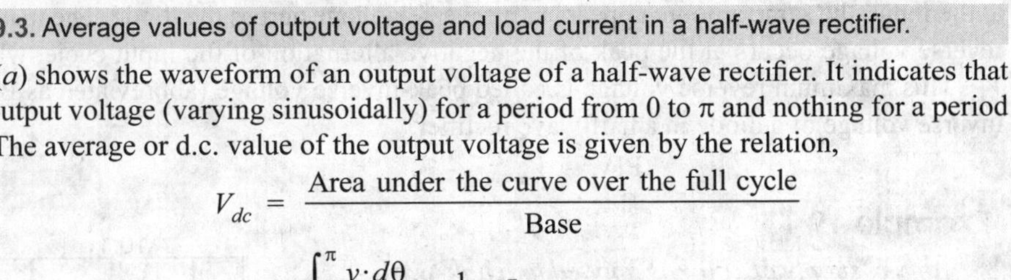

Fig. 19.3. Average values of output voltage and load current in a half-wave rectifier.

Fig. 19.3 (*a*) shows the waveform of an output voltage of a half-wave rectifier. It indicates that there is as an output voltage (varying sinusoidally) for a period from 0 to π and nothing for a period from π to 2π. The average or d.c. value of the output voltage is given by the relation,

$$V_{dc} = \frac{\text{Area under the curve over the full cycle}}{\text{Base}}$$

$$= \frac{\int_0^{\pi} v \cdot d\theta}{2\pi} = \frac{1}{2\pi} \int_0^{\pi} V_m \cdot \sin\theta \cdot d\theta$$

$$= \frac{V_m}{2\,\pi} \left| (-\cos\theta) \right|_0^{\pi} = \frac{V_m}{2\pi} (+1 - (-1))$$

$$= \frac{V_m}{\pi} = 0.318\, V_m$$

The above expression indicates that the average or d.c. value of the output voltage is 31.8 per cent of the maximum a.c. input voltage. This value is indicated by a broken horizontal line in Fig. 19.3 (*a*).

Let us now find the average or d.c. value of load current. It may be found by dividing the d.c. value of output voltage (V_{dc}) by the value of load resistor (R_L). Mathematically, the average or d.c. value of load current,

$$I_{dc} = \frac{V_{dc}}{R_L} = \frac{V_m}{\pi \cdot R_L} = \frac{I_m}{\pi} \quad \ldots\ldots \left(\because I_m = \frac{V_m}{R_L}\right)$$

$$= 0.318\, I_m$$

The above expression indicates that the average or d.c. value of load current is 31.8% of the maximum load current. This value is indicated by a broken horizontal line in Fig. 19.3 (*b*).

Notes. 1. In deriving the average values of output voltage and load current, we have assumed that the diode used in the rectifier circuit, is an ideal one *i.e.*, it has zero forward resistance. However, In actual practice, it is not so. If R_f is the diode forward resistance, then the average output voltage across the load,

$$V_{dc} = \frac{V_m}{\pi} - I_{dc} \cdot R_f$$

and the average value of load current,

$$I_{dc} = \frac{V_m}{\pi(R_f + R_L)}$$

2. If the a.c. input voltage is applied through the transformer, then it is necessary to include the secondary winding resistance (R_s) in the relation of average values of output voltage and current. In this case, average value of output voltage,

$$V_{dc} = \frac{V_m}{\pi} - I_{dc}(R_f + R_s)$$

and the average value of load current,

$$I_{dc} = \frac{V_m}{\pi\,(R_f + R_s + R_L)}$$

19.5. Peak Inverse Voltage of a Diode in Half-wave Rectifier

We have already discussed in Art. 19.2 that during negative half-cycle of the a.c. input voltage, the diode in a half-wave rectifier of Fig. 19.1 is reverse biased. And it does not conduct. As a result of this, there is no current in the circuit and hence no voltage drop across the load resistance R_L. Now, if we apply Kirchoff's Voltage Law to the circuit, then we find that whatever be the negative voltage at the input, the same amount appears as a reverse voltage across the diode. The maximum value of reverse voltage occurs at the peak of the negative alternation of the input cycle, which is equal to V_m. This maximum reverse voltage is called peak-inverse voltage (abbreviated as PIV). Thus peak inverse voltage of a diode in a half-wave rectifier,

$$\text{PIV} = V_m$$

Example 19.1.

An a.c. supply of 200 V is applied to a half-wave rectifier circuit through a transformer of turns ratio 10 : 1 as shown in Fig. 19.4.

Find (a) d.c. output voltage and (b) PIV of a diode. Assume the diode to be an ideal one.

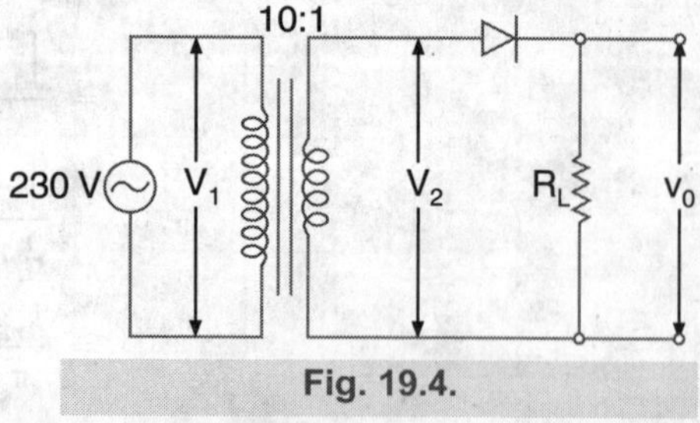

Fig. 19.4.

Solution.

Given: $V_1 = 230$ volts; $N_2/N_1 = 1/10$

(*a*) ***D.C. output voltage***

We know that the secondary voltage,

$$V_2 = V_1 \times \frac{N_2}{N_1} = 230 \times \frac{1}{10} = 23 \text{ V}$$

and maximum value of secondary voltage,

$$V_m = \sqrt{2} \times V_2 = \sqrt{2} \times 23 = 32.5 \text{ V}$$

∴ D.C. output voltage,

$$V_{dc} = 0.318\, V_m = 0.318 \times 32.5 = 10.3 \text{ V } \textbf{Ans.}$$

(*b*) ***PIV of a diode***

We also know that peak-inverse voltage of a diode,

$$PIV = V_m = 32.5 \text{ V } \textbf{Ans.}$$

Example 19.2.

Fig. 19.5 shows a half-wave rectifier circuit with a transformer coupled input. Determine the value of dc load current.

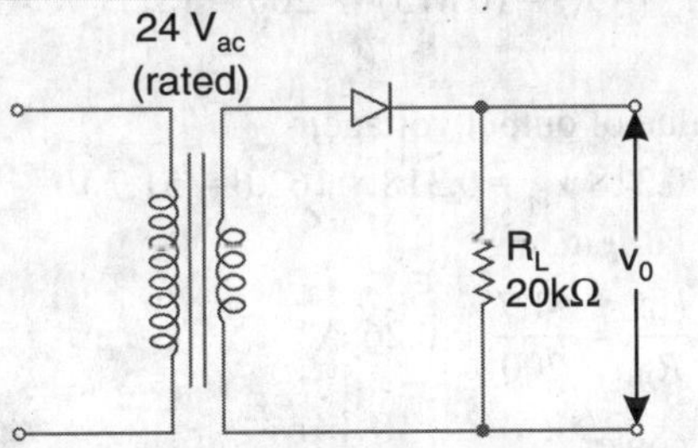

Fig. 19.5.

Solution.

Given: $R_L = 20 \text{ k}\Omega = 20000\ \Omega$.

We know that maximum value of secondary voltage,

$$V_m = \sqrt{2} \times V_2 = \sqrt{2} \times 24 = 33.94 \text{ V}$$

∴ Maximum value of load current,

$$I_m = \frac{V_m}{R_L} = \frac{33.94}{20000} = 1.697 \times 10^{-3} \text{ A} = 1.697 \text{ mA}$$

and dc or average value of load current,

$$I_{dc} = 0.318\, I_m = 0.318 \times (1.697 \times 10^{-3}) = 0.54 \times 10^{-3} \text{ mA } \textbf{Ans.}$$

Example 19.3.

Fig. 19.6 shows a half-wave rectifier circuit with a transformer coupled input.

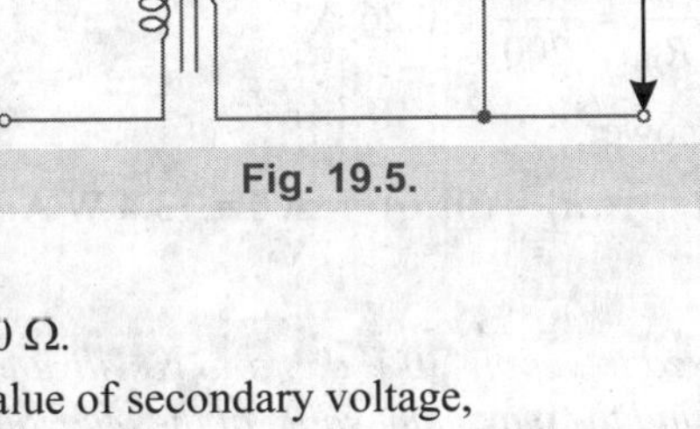

Fig. 19.6.

Determine the maximum and average values of power delivered to the load. Take R_L equal to 200 Ω.

Solution.

Given: $V_1 = 230$ volts; $N_2/N_1 = 1/2 = 0.5$ and $R_L = 200\ \Omega$.

Maximum value of load power

We know that r.m.s. value of the secondary voltage.

$$V_2 = V_1 \times \frac{N_2}{N_1} = 230 \times \frac{1}{2} = 115 \text{ V}$$

and maximum value of secondary voltage,

$$V_m = \sqrt{2} \times V_2 = \sqrt{2} \times 115 = 162.6 \text{ V}$$

We also know that maximum value of load current,

$$I_m = \frac{V_m}{R_L} = \frac{162.6}{200} = 0.813 \text{ A}$$

and maximum value of load power,

$$P_m = I_m^2 \times R_L = (0.813)^2 \times 200 = 132.3 \text{ W } \textbf{Ans.}$$

Average value of load power

We know that average value of output voltage,

$$V_{dc} = 0.318\ V_m = 0.318 \times 162.6 = 51.7 \text{ V}$$

and the average value of load current,

$$I_{dc} = \frac{V_{dc}}{R_L} = \frac{51.7}{200} = 0.26 \text{ A}$$

∴ Average value of load power,

$$P_{dc} = I^2_{dc} \times R_L = (0.26)^2 \times 200 = 13.4 \text{ W } \textbf{Ans.}$$

Example 19.4.

A half-wave rectifier is used to supply 30 V d.c. to a resistive load of 600 Ω. The diode has a forward resistance of 25 Ω. Find the maximum value of the a.c. voltage required at the input.

Solution.

Given: $V_{dc} = 30$ volt; $R_L = 600\ \Omega$ and $R_f = 25\ \Omega$

We know that the average value of load current,

$$I_{dc} = \frac{V_{dc}}{R_L} = \frac{30}{600} = 0.05 \text{ A}$$

and maximum value of load current,

$$I_m = \pi \times I_{dc} = \pi \times 0.05 = 0.157 \text{ A}$$

∴ Maximum a.c. voltage required at the input,

$$v_{in\,(max)} = I_m \times (R_f + R_L) = 0.157 \times (25 + 600) \text{ V} = 98.125 \text{ V } \textbf{Ans.}$$

Example 19.5.

A diode having internal resistance 20 Ω is used for half-wave rectification. If the applied voltage V = 50 sin ωt and the load resistance R_L = 800 Ω, find :

(a) I_m, I_{dc}, I_{rms} (b) d.c output voltage (c) Efficiency of rectification

Solution.

Given: $V_m = 50$ V; $R_L = 800\ \Omega$ and $R_f = 20\ \Omega$.

(a) I_m, I_{dc}, I_{rms}

We know that average value of load current,

$$I_{dc} = \frac{V_m}{\pi(R_f + R_L)} = \frac{50}{3.14 \times (20 + 800)} = 0.0194 = 19.4 \text{ mA} \quad \textbf{Ans.}$$

and the maximum value of load current,

$$I_m = \pi \times I_{dc} = \pi \times 0.0198 = 0.0609 \text{ A} = 60.9 \text{ mA} \quad \textbf{Ans.}$$

We also know that the r.m.s value of load current,

$$I_{rms} = \frac{I_m}{2} = \frac{0.0609}{2} = 0.0304 \text{ A} = 30.4 \text{ mA} \quad \textbf{Ans.}$$

(b) *d.c. output voltage*

We know that the d.c. output voltage (V_{dc}),

$$V_{dc} = \frac{V_m}{\pi} - I_{dc}R_f = \frac{50}{\pi} - (0.0194) \times 20$$

$$= 15.92 - 0.388 = 15.532 \text{ V} \quad \textbf{Ans.}$$

(c) *Efficiency of rectification*

We know that the efficiency,

$$\eta = \frac{0.406}{1 + \frac{R_f}{R_L}} = \frac{0.406}{1 + \frac{20}{800}} = \frac{0.406}{1.025} = 0.396 \text{ or } 39.6\% \quad \textbf{Ans.}$$

19.6. Full-wave Rectifier

A full-wave rectifier is a circuit, which allows a unidirectional current to flow through the load during the entire input cycle as shown in Fig. 19.7. The result of full-wave rectification is a d.c. output voltage that pulsates every half-cycle of the input. On the other hand, a half-wave rectifier allows the current to flow through the load during positive half-cycle only.

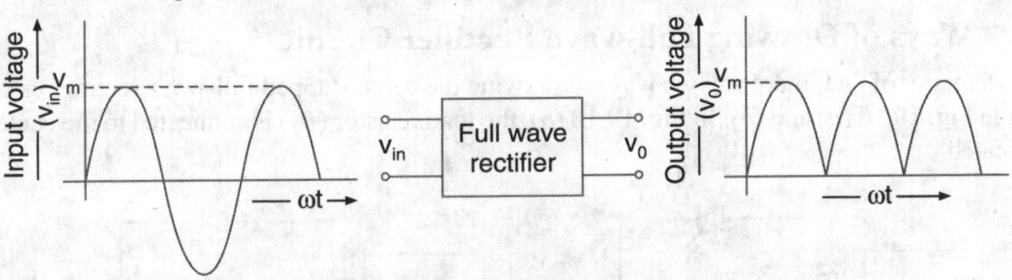

Fig. 19.7. Full wave rectifier, with input and output voltage waveforms.

There are two types of full-wave rectifiers namely center-tapped and bridge rectifier. Now we shall discuss both these types of rectifiers in the following pages.

19.7. Center-tapped Full-wave Rectifier

Fig. 19.8 shows the circuit of a center-tapped full-wave rectifier. The circuit uses two diodes, which are connected to the center-tapped secondary winding of the transformer. The input signal is applied to the primary winding of the transformer. The center-tap on the secondary winding of a transformer is, usually, taken as the ground or zero voltage reference point. It may be

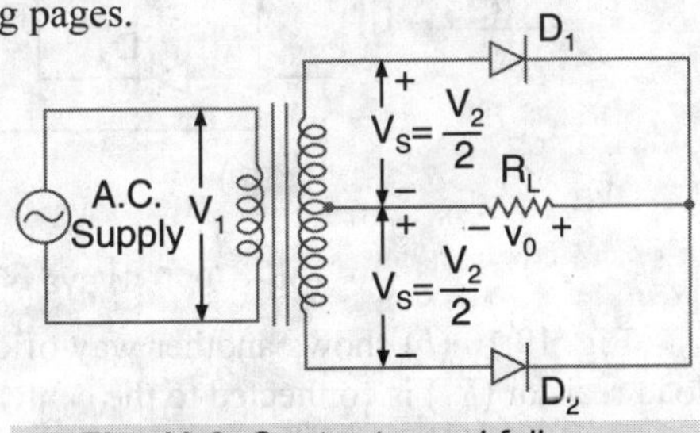

Fig. 19.8. Center-tapped full-wave rectifier

noted that the voltage between the center-tap and either end of the secondary winding is half of the secondary voltage, *i.e.*,

$$V_S = \frac{V_2}{2}$$

The operation of a center-tapped full-wave rectifier circuit may be studied as follows:

During the positive input half-cycle, the polarities of the secondary voltage are as shown in Fig. 19.9 (*a*). This forward biases the diode D_1 and reverse-biases the diode D_2. As a result of this, the diode D_1 conducts some current whereas the diode, D_2 is OFF. The current through load R_L is as indicated in the Fig.

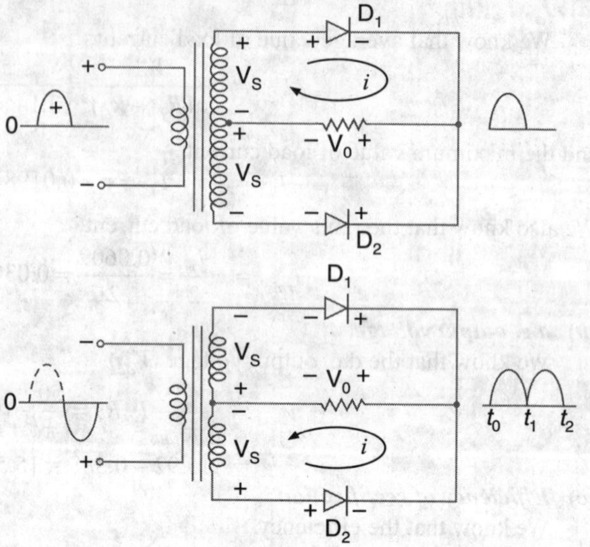

Fig. 19.9. Operation of centre-tapped full-wave rectifier.

During the negative input half-cycle, the polarities of the secondary voltage are as shown in Fig. 19.9 (*b*). This reverse-biases the diode D_1 and forward-biases the diode, D_2. As a result of this, the diode D_1 is OFF and the diode D_2 conducts some current. The current through the load R_L is as indicated in the figure.

It may be noted that current through the load flows in the same direction, during both the positive and negative portions of the input cycle. Therefore, the output voltage developed across load R_L is full-wave rectified d.c. voltage as shown in Fig. 19.9 (*b*).

19.8. Ways of Drawing Full-wave Rectifier Circuit

As a matter of fact, there are two ways of drawing the center-tapped full-wave rectifier circuit as shown in Fig. 19.10 (*a*) and (*b*). In Fig. 19.10 (*a*), the load resistor (R_L) is connected to the center-tap as indicated,

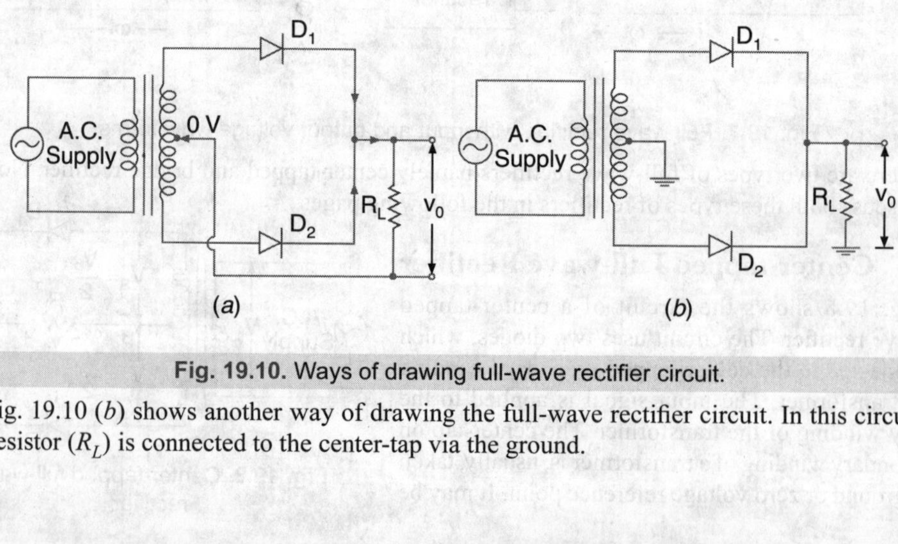

Fig. 19.10. Ways of drawing full-wave rectifier circuit.

Fig. 19.10 (*b*) shows another way of drawing the full-wave rectifier circuit. In this circuit, the load resistor (R_L) is connected to the center-tap via the ground.

19.9. Effect of Turns Ratio on Full-wave Rectifier Output Voltage

We have already discussed in Art. 19.7, that in a center-tapped full-wave rectifier, the voltage between center tap and either end of secondary winding is half of the secondary voltage. However, the value of secondary voltage depends upon the turns ratio (*i.e.*, N_2/N_1) of the transformer. Thus if the transformer turns ratio is '1' (*i.e.*, $N_2 = N_1 = 1$), the primary and secondary voltages are equal. Therefore, output voltage is equal to half of the primary voltage (or secondary voltage).

It is possible to obtain an output voltage equal to the input, by using a step-up transformer with a turns ratio of 2 : 1 (*i.e.*, $N_2 = 2$ and $N_1 = 1$). In this case, the total secondary voltage (V_2) is twice the primary voltage. Therefore the voltage between center-tap and either end of the secondary winding is V_1 (which is equal to $V_2/2$). Mathematically,

$$V_S = V_1 = \frac{V_2}{2}$$

19.10. Average Values of Output Voltage and Load Current in a Full-wave Rectifier

Consider a center-tapped full-wave rectifier with a sinusoidal a.c. input voltage.

Let V_m = Maximum value of the voltage across each half of the secondary winding.

V_S = The r.m.s. value of the voltage across each half of the secondary winding,

I_m = Maximum value of the load current,

V_{dc} = Average or d.c. value of the output voltage across the load resistor, and

I_{dc} = Average or d.c. value of the current through the load resistor (R_L).

We know that the equation for the voltage across each half of the secondary winding,

$$v_S = V_m \sin wt = V_m \sin\theta \qquad ... (\because \theta = wt)$$

where v_S = Instantaneous value of the voltage across each half of the secondary winding.

We also know that the average or d.c. value of the output voltage,

$$V_{dc} = \frac{\text{Area under the curve over a * half-cycle}}{\text{Base}}$$

$$= \frac{\int_0^\pi v_S \cdot d\theta}{\pi} = \frac{\int_0^\pi V_m \sin\theta\, d\theta}{\pi}$$

$$V_{dc} = \frac{1}{\pi}\int_0^\pi V_m \cdot \sin\theta\, d\theta$$

$$V_{dc} = \frac{V_m}{\pi}\left|(-\cos\theta)\right|_0^\pi = \frac{V_m}{\pi}[+1-(-1)]$$

$$= \frac{2\,V_m}{\pi} = 0.636\,V_m.$$

* Here, we have taken into account only the half cycle (and not full cycle) of the output voltage.

It is evident from the above relation that the average value of a full-wave rectifier is 0.636 V_m and is as shown in Fig. 19.11 (*a*). It may be noted that this value is twice that of a half-wave rectifier.

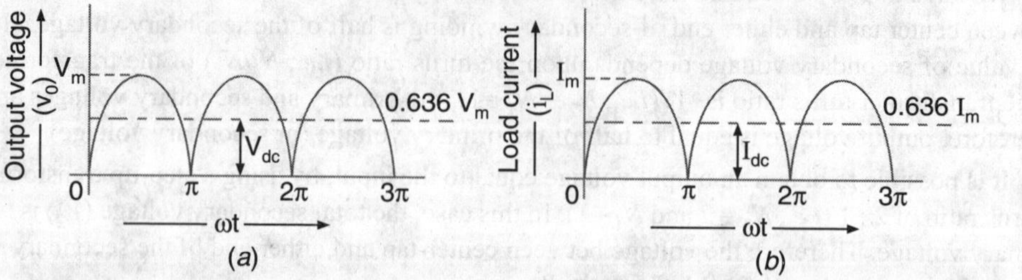

Fig. 19.11. Average values of output voltage and load current in a full-wave rectifier.

We also know that the average or d.c. value of load current,

$$I_{dc} = \frac{V_{dc}}{R_L} = \frac{2V_m}{\pi \cdot R_L} = \frac{2I_m}{\pi} = 0.636\, I_m$$

It is evident from the above relation that average value of load current is 0.636 I_m and is as shown in Fig. 19.11 (*b*). It may be noted that this value is twice that of a half-wave rectifier.

Note. In deriving the average values of output, voltages and load current, we have assumed that the diodes and the transformer used in the rectifier circuit is an ideal one. However, in actual practice it is not so. Thus if R_f is the diode forward resistance and R_S is the secondary winding resistance, then average value of output voltage,

$$V_{dc} = \frac{2\,V_m}{\pi} - I_{dc}\,(R_f + R_S)$$

and the average value of load current,

$$I_{dc} = \frac{2\,I_m}{\pi} = \frac{2\,V_m}{\pi\,(R_f + R_S + R_L)}$$

Example 19.6.

Fig. 19.12 shows the circuit of a center-tapped full-wave rectifier circuit.

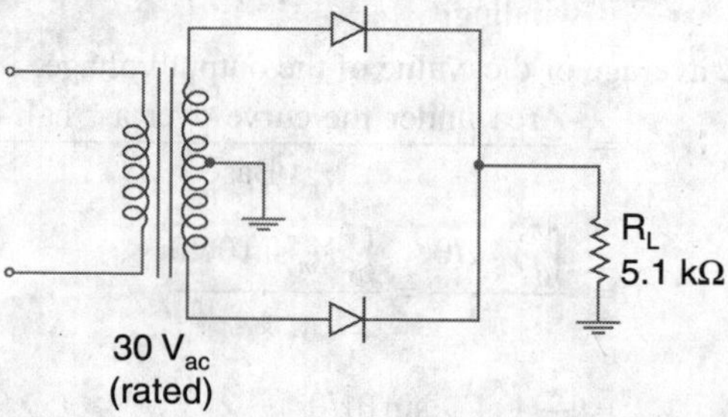

Fig. 19.12.

Determine the dc output voltage and the average (or dc) load current.

Solution.

Given: $V_2 = 30\ V_{ac}$ and $R_L = 5.1\ \text{k}\Omega = 5100\ \Omega$.

DC output Voltage

We know that voltage between the center-tap and either end of secondary winding of a transformer,

$$V_S = \frac{V_2}{2} = \frac{30}{2} = 15\ V_{ac}.$$

and maximum value of this voltage,

$$V_m = \sqrt{2} \times V_S = \sqrt{2} \times 15 = 21.2 \text{ V}$$

∴ The dc output voltage,

$$V_{dc} = 0.636\, V_m = 0.636 \times 21.2 = 13.5 \text{ V } \textbf{Ans.}$$

Average or dc load current

We also know that the average or dc load current,

$$V_{dc} = \frac{V_{dc}}{R_L} = \frac{13.5}{5100} = 2.647 \times 10^{-3} \text{ A} = 2.647 \text{ mA } \textbf{Ans.}$$

19.11. Peak Inverse Voltage of a Diode in Center-tapped Full-wave Rectifier

We know that each diode in a full-wave rectifier, is alternately forward biased and reversed biased. The maximum value of reverse voltage, which each diode must with stand, is equal to the maximum secondary voltage (*i.e.*, 2 V_m) as discussed.

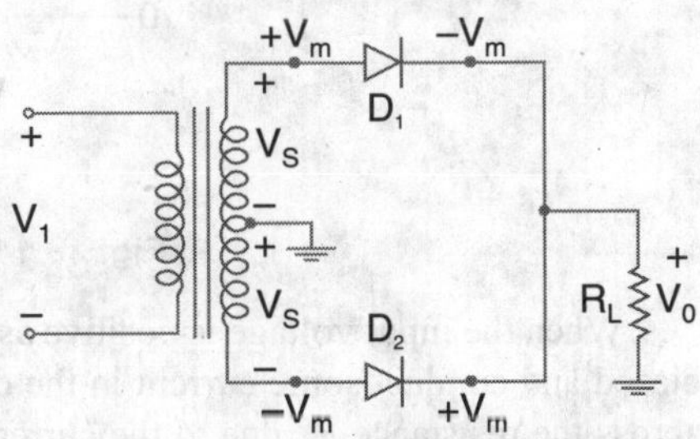

Fig. 19.13. Peak-inverse voltage of a diode in center-tapped full-wave rectifier.

Consider the circuit of a center-tapped full-wave rectifier as shown in Fig. 19.13. It may be noted that when the secondary voltage (v_S) has the polarity as shown in the figure, the anode of the diode D_1 is + V_m (where V_m is the maximum half secondary voltage) and the anode of D_2 is − V_m. Since D_1 is forward biased, its cathode is at the same voltage as its anode (neglecting barrier potential) *i.e.*, + V_m. This is also the voltage on the cathode of the diode D_2. Therefore the total reverse voltage across the diode D_2

$$= V_m - (-V_m) = 2\, V_m$$

∴ Peak-inverse voltage of each diode in a center-tapped full-wave rectifier,

$$PIV = 2\, V_m$$

19.12. Advantages and Disadvantages of Center-tapped Full-wave Rectifier

A center-tapped full-wave rectifier has the following advantages and disadvantages:

Advantages

1. The d.c. output voltage and load current values are twice than those of a half-wave rectifier.
2. The *ripple factor is much less (0.482) than that of half-wave rectifier (1.21).
3. The **efficiency is twice that of half-wave rectifier. For a full-wave rectifier, the maximum possible value of efficiency is 81.2%, while that of half-wave rectifier is 40.6%.

Disadvantages

1. The output voltage is half of the secondary voltage.
2. The peak-inverse voltage (PIV) of a diode is twice that of the diode used in the half-wave rectifier.
3. It is expensive to manufacture a center-tapped transformer, which produces equal voltages on each half of the secondary winding.

* Please see Art. 19.19.

** Please see Art. 19.22.

19.13. Full-wave Bridge Rectifier

It uses four diodes connected across the main supply, as shown in Fig. 19.14 (*a*). The operation of the circuit may be studied as follows:

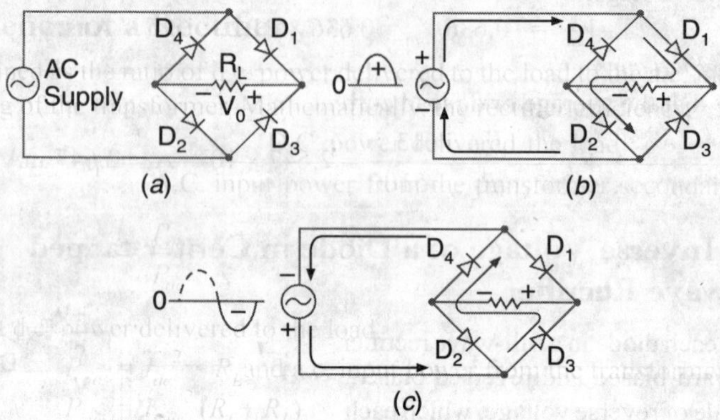

Fig. 19.14. Bridge rectifier operation.

When the input voltage is positive as shown in Fig. 19.14 (*b*), the diodes D_1 and D_2 are forward biased and conduct some current in the direction as indicated in the figure. A voltage is developed across the resistance R_L due to the current flow through it. The voltage looks like the positive half of the input cycle. At this time the diodes D_3 and D_4 are reverse biased. When the input voltage is negative as shown in Fig. 19.14 (*c*), the diodes D_3 and D_4 are forward biased and conduct some current in the same direction through R_L as during the positive half-cycle. During this time, the diodes D_1 and D_2 are reverse biased. As a result of this action, a full-wave rectified output voltage is developed across the resistance R_L.

The bridge rectifier shown in Fig. 19.14 (*a*) is directly connected across the mains supply. But in actual practice, the mains supply is applied through a transformer as shown in Fig. 19.15.

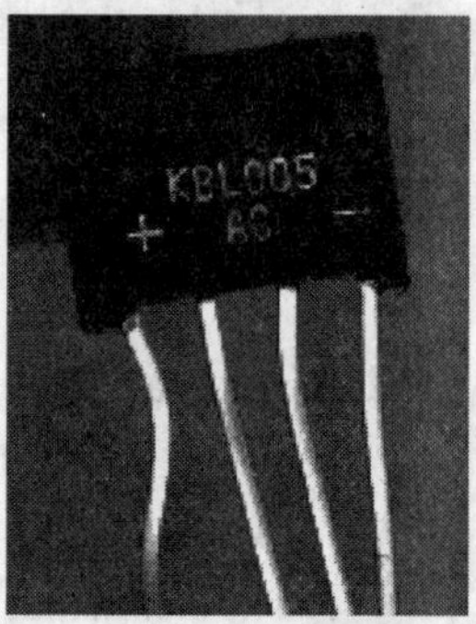

Full wave rectifier.

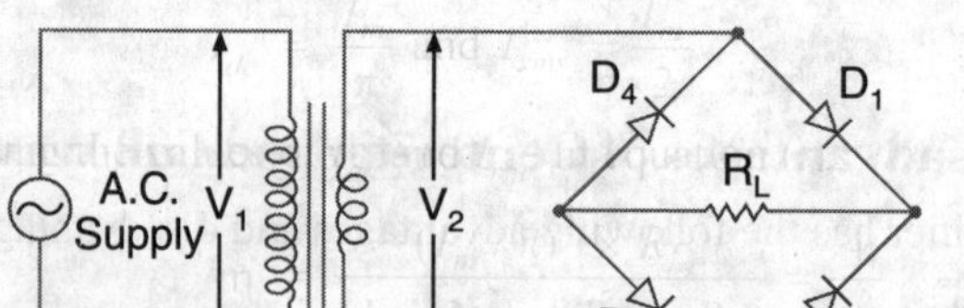

Fig. 19.15. Bridge rectifier with transformer.

19.14. Ways of Drawings a Bridge Rectifier Circuit

As a matter of fact, there are two ways of drawing a bridge rectifier circuit as shown in Fig. 19.16 (*a*) and (*b*). In Fig. 19.16 (*a*), the load resistor (R_L) is connected to the points *B* and *C* of the diode bridge as indicated.

Fig. 19.16 (*b*) shows another way of drawing the bridge rectifier circuit. In this circuit, the load resistor (R_L) is connected to the point *C* of the diode bridge *via* the ground.

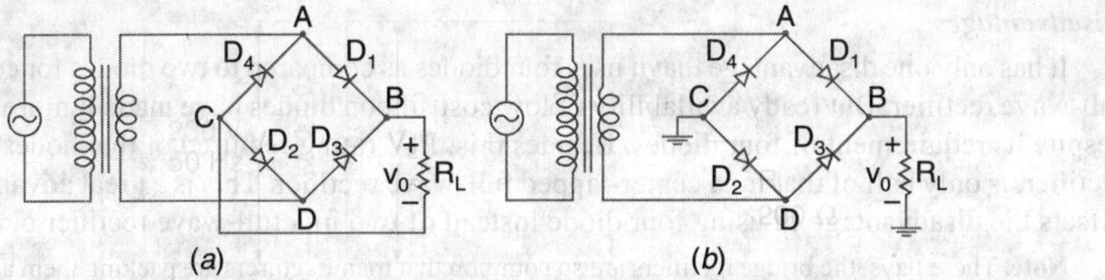

Fig. 19.16. Ways of drawing bridge rectifier circuit.

19.15. Average Values of Output Voltage and Load Current in a Bridge Rectifier

We have already discussed these values in Art. 19.10. The average values of output voltage and load current are the same as for a center-tapped full wave rectifier. Thus these values in a bridge rectifier may also be obtained in the same way as in Art. 19.10. The average values of output voltage and load current for bridge rectifier are:

$$V_{dc} = \frac{2\,V_m}{\pi} \text{ and } I_{dc} = \frac{2\,I_m}{\pi} = \frac{2\,V_m}{\pi \cdot R_L}$$

where V_m = Maximum value of secondary voltage.

Note: If the values of diode forward resistance (R_F) and the secondary winding resistance (R_S) are also taken into account, then the average value of output voltage.

$$V_{dc} = \frac{2\,V_m}{\pi} - I_{dc}\,(R_S + R_F)$$

and the average value of load current,

$$I_{dc} = \frac{2I_m}{\pi} = \frac{2V_m}{\pi(R_S + R_F + R_L)}$$

19.16. Peak Inverse Voltage of a Diode in a Bride Rectifier

Consider a bridge rectifier circuit as shown in Fig. 19.17. The polarities of primary and secondary windings correspond to the positive half cycle of the input. Under this condition, the diodes D_1 and D_2 are forward biased. Therefore, these diodes are shown as shorted points in the figure.

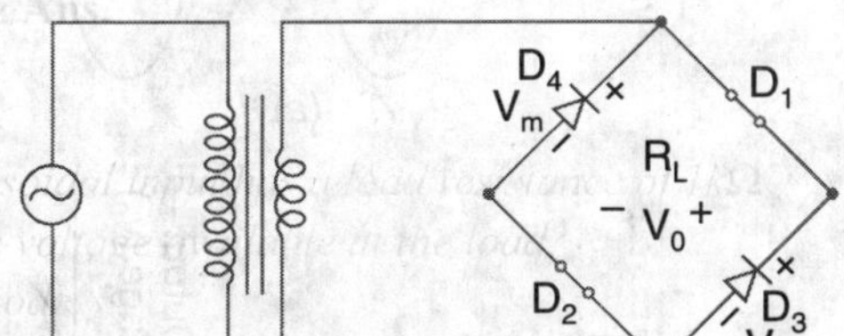

Fig. 19.17. Peak-inverse voltage of a diode in bridge rectifier.

On the other hand, the diodes D_3 and D_4 are reverse biased and have a maximum reverse voltage equal to the maximum secondary voltage (V_m). Thus, peak-inverse voltage of a diode in a bridge rectifier,

$$\text{PIV} = V_m$$

19.17. Advantages and Disadvantages of Bridge Rectifier

A bridge rectifier has the following advantages and disadvantages:

Advantages

1. It can be used, with advantage, in applications allowing floating output terminals *i.e.*, no output terminal is grounded.
2. The transformer is less costly as it is required to provide only half the voltage of an equivalent center-tapped transformer used in a full wave rectifier circuit.
3. No center-tap is required on the transformer.

Disadvantages

It has only one disadvantage that it uses four diodes as compared to two diodes for center-tapped full-wave rectifier. But ready availability of low cost silicon diodes have made it more economical despite its requirement of four diodes. Besides this, PIV rating required of the diodes, in a bridge rectifier, is only half of that for a center-tapped full-wave rectifier. This is a great advantage, which offsets the disadvantage of using four diode instead of two in a full-wave rectifier circuit.

Note. These days, the bridge rectifiers are so common that manufacturers are packing them as a single unit with bakelite or some other plastic encapsulation with external connections brought out.

(*a*) (*b*) (*c*) (*d*) (*e*)

19.18. Frequency of Output in Half-wave and Full-wave Rectifiers

Fig 19.18 (*a*) shows the waveform of an a.c. input voltage while Fig. 19.18 (*b*) and (*c*) shows the waveforms of a half-wave and full-wave rectifiers respectively.

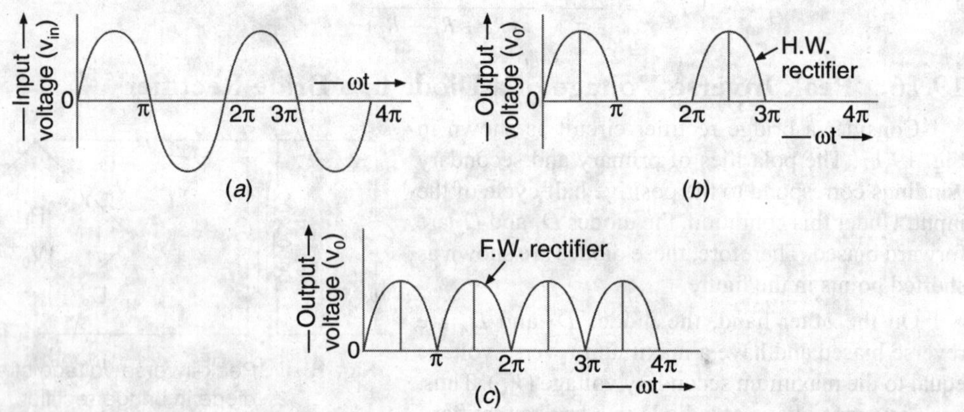

Fig. 19.18. Frequency of output in half-wave and full-wave rectifiers.

It may be noted that frequency of half-wave rectifier output is the same as of its input, *i.e.*, $f_{out} = f_{in}$.

However, the frequency of full-wave rectifier is twice that of input, *i.e.*, $f_{out} = 2 f_{in}$.

Thus in power supply applications, the a.c. power line voltage is of 50 Hz frequency. Therefore d.c. output voltage of a half-wave rectifier is 50 Hz. But d.c. output voltage of a full-wave rectifier is 100 Hz.

Example 19.7.

Fig. 19.19 shows a bridge rectifier circuit.

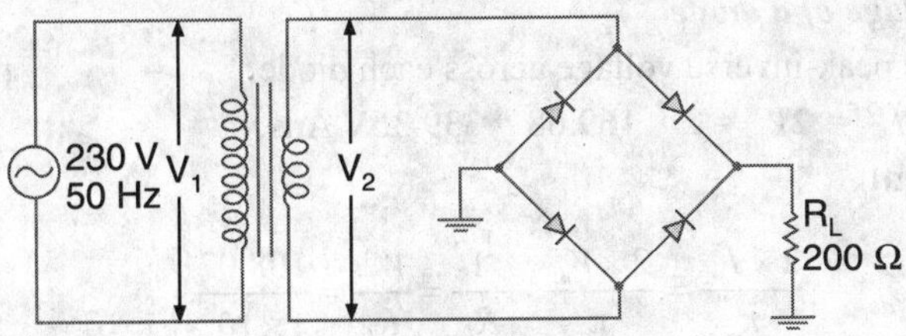

Fig. 19.19.

Assuming the diodes to be ideal, determine the quantities (a) d.c. output voltage, (b) peak-inverse voltage of a diode and (c) output frequency.

Solution.

Given: $V_1 = 230$ volt; $N_2/N_1 = 1/4 = 0.25$; $R_L = 200\ \Omega$ and $f_{in} = 50$ Hz.

(a) D.C. output voltage

We know that r.m.s. value of secondary voltage,

$$V_2 = V_1 \times \frac{N_2}{N_1} = 230 \times 0.25 = 57.5 \text{ V}$$

and maximum value of the secondary voltage,

$$V_m = \sqrt{2} \times V_2 = \sqrt{2} \times 57.6 = 81.3$$

∴ D.C. output voltage,

$$V_{dc} = 0.636 \text{ V}_m = 0.636 \times 81.3 = 51.7 \text{ V } \textbf{Ans.}$$

(b) Peak-inverse voltage of a diode

We know that peak-inverse voltage of a diode,

$$PIV = V_m = 81.3 \text{ V } \textbf{Ans.}$$

(c) Output frequency

We also know that output frequency,

$$f_{out} = 2 f_{in} = 2 \times 50 = 100 \text{ Hz } \textbf{Ans.}$$

Example 19.8.

A full wave bridge rectifier with a 120 V rms sinusoidal input has a load resistance of 1kΩ

(a) if silicon diodes are employed ,what is the dc voltage available at the load?

(b) Determine the required PIV rating of each diode.

(c) Find the maximum current through each diode during conduction.

(d) What is the required power rating of each diode?(Jamia Millia Islamia University., 2008)

Solution.

Given: $V_{rms} = 120$ V and $R_L = 1\text{k}\Omega = 1 \times 103\Omega$.

(a) D.C. voltage

We know that,

$$V_{dc} = \frac{2 \times V_m}{\pi}$$

$$V_m = V_{rms}\sqrt{2} = 120 \times \sqrt{2} = 169.68\ V$$

$$\therefore \quad V_{dc} = \frac{2 \times V_m}{\pi} = \frac{2 \times 169.68}{\pi} = 108.07\ V \quad \textbf{Ans.}$$

(b) *Peak inverse voltage of a diode*

We know that the peak-inverse voltage across each diode,

$$PIV = 2V_m = 2 \times 169.68 = 339.36\text{V Ans.}$$

(c) Maximum current

We know that,

$$V_{dc} = \frac{2 \times I_m}{\pi} = \frac{2 \times V_m}{\pi} \times \frac{1}{R_L} = \frac{V_{dc}}{R_L} = \frac{108.07}{1 \times 10^3}$$

$$= 108.07 \times 10\text{–}3 = 108.07 \text{ mA}$$

$$\therefore \quad I_m = \frac{V_m}{R_L} = \frac{169.68}{1 \times 10^3} = 169.68 \times 10^{-3} = 169.68 \text{ mA } \textbf{Ans.}$$

(d) Power rating of each diode

Required power rating,

$$P = V \times R = 169.68 \times 1000 = 169.68 \text{ kW Ans.}$$

$$P = V_m . I_m = (169.68 \times 10^{-3}) \times (169.68 \times 10^{-3})\text{W}$$

$$= 28791.3 \times 10^{-3} \text{ W}$$

$$= 28.8 \text{ mW}$$

19.19. Ripple Factor

The output voltage (or load current) of a rectifier consists of two components namely d.c. component and a.c. component. The a.c. component present in the output is called a ripple. As a matter of fact, the ripple is undesirable and accounts for pulsations in the rectifier output. The effectiveness of a rectifier depends upon the magnitude of ripple in the output. Smaller the ripple, more effective will be the rectifier. Mathematically, the ripple factor,

$$\gamma = \frac{\text{The r.m.s value of a.c. component of output voltage}}{\text{The d.c. component of output voltage}}$$

$$= \frac{V_{r(rms)}}{V_{dc}} = \frac{I_{r(rms)}}{I_{dc}}$$

where $V_{r(rms)}$ = The r.m.s. value of the a.c. component of the output voltage.

V_{dc} = The average or d.c. value of the output voltage.

$I_{r\,(rms)}$ = The r.m.s. value of the a.c. component of current, and

I_{dc} = The average or d.c. value of the load current.

We know that the r.m.s. value of the rectified load current,

$$I_{rms} = \sqrt{I_{dc}^2 + I_{r(rms)^2}}$$

Dividing the above equation on both sides by I_{dc}.

$$\frac{I_{rms}}{I_{dc}} = \frac{\sqrt{I_{dc}^2 + I_{r(rms)^2}}}{I_{dc}} = \sqrt{1 + \left(\frac{I_{r(rms)}}{I_{dc}}\right)^2}$$

Squaring and rearranging the above expression,

$$\frac{I_{r(rms)}}{I_{dc}} = \sqrt{\left(\frac{I_{rms}}{I_{dc}}\right)^2 - 1}$$

or

$$\gamma = \sqrt{\left(\frac{I_{rms}}{I_{dc}}\right)^2 - 1}$$

Now we shall use the above relation to find out the ripple factor for half-wave and full-wave rectifier circuits.

19.20. Ripple Factor of a Half-wave Rectifier

We have already discussed in Art. 19.4 that the average value of load current in a half-wave rectifier,

$$I_{dc} = \frac{I_m}{\pi} \qquad ...(i)$$

where I_m is the maximum value of load current. The r.m.s. value of the load current for a half-wave rectifier may be found by using *calculus and its value,

$$I_{rms} = \frac{I_m}{2} \qquad ...(ii)$$

Substituting these values of I_{dc} and I_{rms} in the expression for ripple factor (Art. 19),

$$\gamma = \sqrt{\left(\frac{I_{rms}}{I_{dc}}\right)^2 - 1} = \sqrt{\left(\frac{I_m / 2}{I_m / \pi}\right)^2 - 1} = 1.21$$

If we express the ripple factor in terms of a percentage, its value is 121%. This indicates that the amount of a.c. component present in the output of a half-wave rectifier is 121% of d.c. output voltage. Hence the half-wave rectifier is not very successful in converting the current from a.c. to d.c.

19.21. Ripple Factor of a Full-wave Rectifier

We have already discussed in Art. 19.10 that the average value of load current in a full-wave rectifier (either center-tapped or bridge rectifier) is given by the relation, $I_{dc} = 2\,I_{ms}/\pi$ ...(i)

The r.m.s. value of the load current may be found by using the **calculus and its value is found to be,

$$I_{rms} = \frac{I_m}{\sqrt{2}} \qquad ...(ii)$$

Substituting these values of I_{dc} and I_{rms} in the expression for ripple factor (Art. 19),

$$\gamma = \sqrt{\left(\frac{I_{rms}}{I_{dc}}\right)^2 - 1} = \sqrt{\left(\frac{I_m / \sqrt{2}}{2I_m / \pi}\right)^2 - 1} = 0.482$$

* We know that expression for the load current,

$$iL = I_m \cdot \sin\theta$$

$$\therefore \quad I_{rms} = \sqrt{\frac{1}{2\pi}\int_0^{2\pi} i_L^2 \cdot d\theta} = \sqrt{\frac{1}{2\pi}\left[\int_0^{\pi} I_m^2 \sin^2\theta \cdot d\theta + \int_{\pi}^{2\pi} (0 \cdot d\theta)\right]}$$

$$\sqrt{\frac{I_m^2}{2\pi}\int_0^{\pi}\left(\frac{1-\cos 2\theta}{2}\right) d\theta} = \sqrt{\frac{I_m^2}{4\pi}\left(\theta - \frac{\sin 2\theta}{2}\right)_0^{\pi}} = \frac{I_m}{2}$$

** We know that expression for load current is given by the relation,

$$i_L = I_m \cdot \sin\theta$$

$$\therefore \quad I_{rms} = \sqrt{\frac{1}{2\pi}\int_0^{\pi} i_L^2 \cdot d\theta} = \sqrt{\frac{1}{\pi} I_m^2 \sin^2\theta \cdot d\theta}$$

$$= \sqrt{\frac{I_m^2}{\pi}\int_0^{\pi}\left(\frac{1-\cos 2\theta}{2}\right) d\theta} = \sqrt{\frac{I_m^2}{2\pi}\left(\theta - \frac{\sin 2\theta}{2}\right)_0^{\pi}} = \sqrt{\frac{I_m^2}{\pi} \times \frac{\pi}{2}} = \frac{I_m}{\sqrt{2}}$$

It is evident from the above result that ripple factor of a full-wave rectifier is 0.482 and is much smaller than that of a half-wave rectifier. Because of this reason, full-wave rectifier is used more commonly in actual practice.

19.22. Efficiency of a Rectifier

It may be defined as the ratio of d.c. power delivered to the load to the a.c. input power from the secondary winding of the transformer. Mathematically, the rectifier efficiency,

$$\eta = \frac{\text{D.C. power delivered the load}}{\text{A.C. input power from the transformer secondary}}$$

$$= \frac{P_{dc}}{P_{ac}} \quad \text{... (i)}$$

We know that d.c. power delivered to the load,

$P_{dc} = I_{dc}^{\ 2} \times R_L$ and a.c. input power from the transformer secondary,

$P_{ac} = I_{rms}^{\ 2} (R_f + R_L)$

where R_L = Load resistance, and

R_f = Forward resistance of a diode.

Substituting the values of P_{dc} and P_{ac} in equation (*i*),

$$\eta = \frac{I_{dc}^2 \times R_L}{I_{rms}^2 \ (R_f + R_L)}$$

19.23. Efficiency of a Half-wave Rectifier

We have already discussed in the last article that efficiency of a rectifier is given by the expression,

$$\eta = \frac{I_{dc}^2 \times R_L}{I_{rms}^2 \ (R_f + R_L)} \quad \text{... (i)}$$

We also know that for a half-wave rectifier,

$$I_{dc} = \frac{I_m}{\pi} \text{ and } I_{rms} = \frac{I_m}{2}$$

Substituting the values of I_{dc} and I_{rms} in equation (*i*),

$$\eta = \frac{(I_m / \pi)^2 \times R_L}{(I_m / 2)^2 \times (R_f + R_L)} = \frac{4}{\pi^2} \times \frac{R_L}{R_f + R_L} = \frac{0.406}{1 + \dfrac{R_f}{R_L}}$$

Now efficiency will be maximum, if $R_L >> R_f$. Thus

$$\eta_{max} = 0.406 \text{ or } 40.6\%.$$

It shows that efficiency of a half-wave rectifier is 40.6% under the condition that the value of load resistance is very large as compared to the forward resistance of a diode (*i.e.*, $R_L >> R_f$). However, in actual practice, the efficiency is always less then 40.6%.

19.24. Efficiency of a Full-wave Rectifier

We have already discussed in Art. 19.22 that rectifier efficiency is given by the relation,

$$\eta = \frac{I_{dc}^2 \times R_L}{I_{rms}^2 \ (R_f + R_L)} \quad \text{... (i)}$$

We know that for a full-wave rectifier,

$$I_{dc} = 2\frac{I_m}{\pi} \text{ and } I_{rms} = \frac{I_m}{\sqrt{2}}$$

Substituting the values of I_{dc} and I_{rms} in equation (*i*),

$$\eta = \frac{\left(\frac{2\,I_m}{\pi}\right)^2 \times R_L}{\left(\frac{I_m}{\sqrt{2}}\right)^2 \times (R_f + R_L)} = \frac{8}{\pi^2} \times \frac{R_L}{R_f + R_L} = \frac{0.812}{1 + \frac{R_f}{R_L}}$$

Now efficiency will be maximum if $R_L >> R_f$. Thus $h_{max} = 0.812$ or 81.2%

It is evident from the above result, that maximum efficiency of a full-wave rectifier is twice that of half-wave rectifier. It means that a full-wave rectifier is twice as effective as a half-wave rectifier.

Example 19.9.

Fig. 19.20 shows a center-tapped full-wave rectifier circuit:

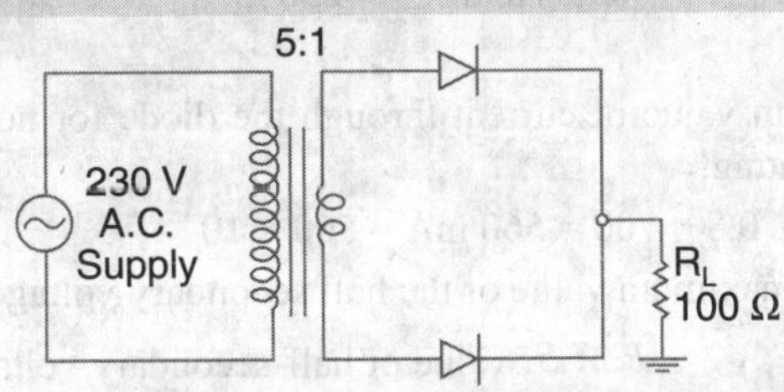

Fig. 19.20.

Assuming both the diodes to be ideal, determine (a) d.c. output voltage; (b) peak-inverse voltage, and (c) rectification efficiency. *(J.N.T.U., 2009)*

Solution.

Given: $V_1 = 230$ volts; $N_2/N_1 = 1/5 = 0.2$ and $R_L = 100$ W.

(*a*) *D.C. output voltage*

We know that the r.m.s. value of secondary voltage,

$$V_2 = V_1 \times \frac{N_2}{N_1} = 230 \times 0.2 = 46 \text{ V}$$

and half of the secondary voltage,

$$V_S = \frac{V_2}{2} = \frac{46}{2} = 23 \text{ V}$$

∴ Maximum value of half of the secondary voltage,

$$V_m = \sqrt{2} \times V_S = \sqrt{2} \times 23 = 32.5 \text{ V}$$

and d.c. output voltage,

$$V_{dc} = \frac{2\,V_m}{\pi} = \frac{2 \times 32.5}{\pi} = 20.7 \text{ V } \textbf{Ans.}$$

(*b*) *Peak-inverse voltage of a diode*

We know that peak-inverse voltage of a diode,

$$\text{PIV} = 2\,V_m = 2 \times 32.5 = 65 \text{ V } \textbf{Ans.}$$

(c) ***Rectification efficiency***

We also know that rectification efficiency of a full-wave rectifier,

$$\eta = 0.812 \text{ or } 81.2\% \textbf{ Ans.}$$

Example 19.10.

A full-wave rectifier, using a center-tap transformer of 200-0-200 V, has two rectifying diodes. Each one of them is rated at I_{max} = 700 mA and I_{avg} 250 mA. Assuming ideal rectifying diodes, determine:

(a) *The value of load resistor that can be connected to give largest d.c. power output without exceeding any diode rating.*

(b) *The d.c. load voltage and current for the condition of part (a) above.*

(c) *The peak-inverse voltage of each diode that can be safely used in this rectifier.*

(A.M.I.E., Summer, 1983)

Solution.

Given: $V_S = 200$ volts; $I_{max} = 700$ mA and $I_{avg} = 250$ mA.

(a) ***Value of load resistor***

We know that maximum value of current through the diode for normal operation should not exceed 80% of its current rating.

$$\therefore \quad I_{max} = 0.8 \times 700 = 560 \text{ mA} = 560 \times 10^{-3}$$

We also know that the maximum value of the half secondary voltage,

$$V_m = \sqrt{2} \times R.M.S. \text{ value of half-secondary voltage}$$

$$= \sqrt{2} \times V_S = \sqrt{2} \times 200 = 282.8 \text{ V}$$

∴ Value of load resistor that gives the largest d.c. power output,

$$R_L = \frac{\text{Maximum value of half-secondary voltage}}{\text{Maximum current through the diode}}$$

$$= \frac{V_m}{I_{max}} = \frac{282.8}{560 \times 10^{-3}} = 505\ \Omega \textbf{ Ans.}$$

(b) ***D.C. load voltage and current***

We know that d.c. load voltage (or output voltage),

$$V_{dc} = \frac{2V_m}{\pi} = \frac{2 \times 282.8}{\pi} = 180 \text{ V} \textbf{ Ans.}$$

and the d.c. load current,

$$I_{dc} = \frac{V_{dc}}{R_L} = \frac{180}{505} = 0.36 \text{ A} \textbf{ Ans.}$$

(c) ***Peak-inverse voltage of a diode***

We also know that peak-inverse voltage across each diode,

$$\text{PIV} = 2\,V_m = 2 \times 282.8 = 565.6 \textbf{ Ans.}$$

19.25. Transformer Utilization Factor

It may be defined as the ratio of d.c. power delivered to the load and the a.c. rating of the transformer secondary. Mathematically, transformer utilization factor,

$$\text{T.U.F.} = \frac{\text{D.C. power delivered to the load}}{\text{A.C. rating of the transformer secondary}}$$

$$= \frac{P_{dc}}{P_{ac(rated)}}$$

For a half-wave rectifier, the value of transformer utilization factor is found to be equal to 0.287. For a center-tapped full-wave rectifier, its value is equal to 0.693 and for a bridge rectifier, its value is equal to 0.812. The transformer utilization factor is very useful in determining the rating of a transformer to be used with rectifier circuit.

19.26. Comparison of Rectifiers

The characteristics of the three rectifiers, which we have discussed, in the previous articles are summarized in table 19.1.

Table 19.1 : Comparison of rectifiers

S. No.	*Item*	*Half-wave*	*Full-wave*	
			Center-tapped	*Bridge*
1.	Number of diodes	1	2	4
2.	Peak-inverse voltage of diode	V_m	$2\ V_m$	V_m
3.	D.C. Output voltage	$0.318\ V_m$	$0.636\ V_m$	$0.636\ V_m$
4.	Ripple factor	1.21	0.482	0.482
5.	Ripple frequency	$2f_{in}$	$2f_{in}$	$2f_{in}$
6.	Transformer utilization factor	0.287	0.693	0.812

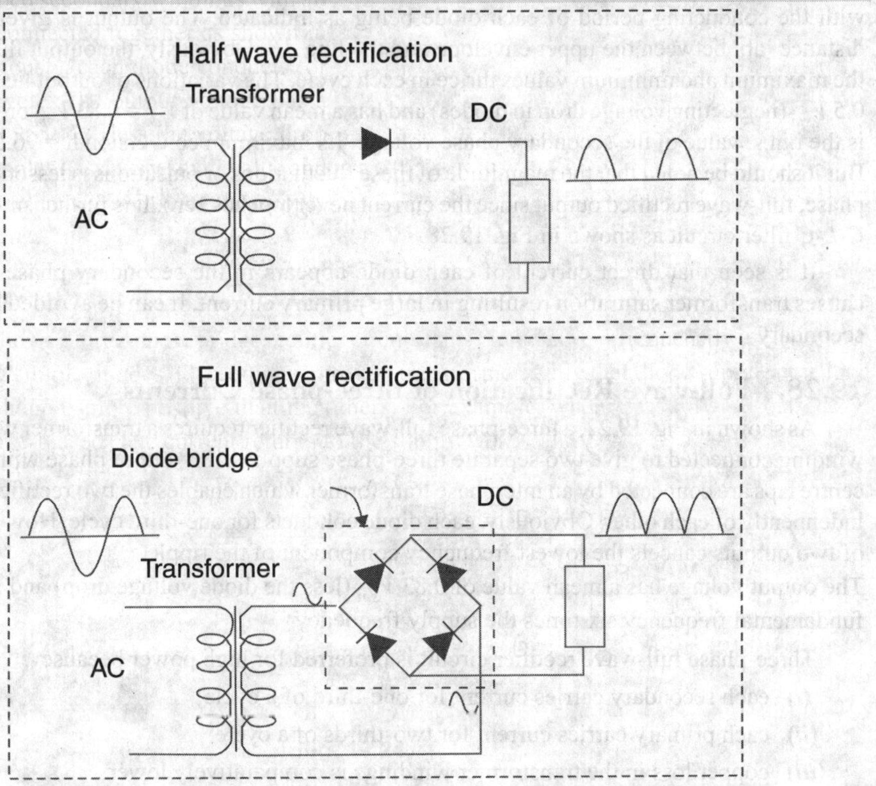

19.27. Three-phase Half-wave Rectifier

Rectification of a three-phase supply with the help of diodes is shown in Fig. 19.21(*a*) along with a smoothing circuit. The three diodes are connected to the three phases of star-connected secondary of a three-phase transformer. Neutral point N of the secondary is the negative terminal for the rectified output and is earthed as shown.

The shape of the output is shown in Fig. 19.21(*b*). The horizontal line *AB* represents the potential of the negative d.c. terminal output and the sine waves 1, 2 and 3 each represent the anode potentials of the three diodes.

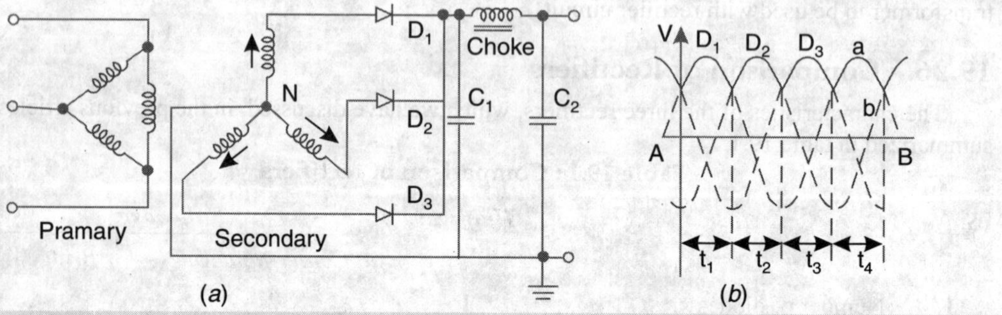

Fig. 19.21.

During one-third of the cycle *i.e.*, during time *t*1, only diode D_1 will conduct. It will cease conducting at t_2 and then D_2 will conduct current upto t_2 after which D_2 will take over and will supply anode current till t_3. When one diode conducts, the other two remain inactive because then their anodes are more –ve than their cathodes. This process repeats itself during each ensuing cycle, with the conducting period of each diode being as indicated. The output is given by the vertical distance `ab' between the upper envelope and the line *AB*. Obviously, the output fluctuates between the maximum and minimum values thrice in each cycle. The variations of output lie between V_{sm} and 0.5 V_{sm} (neglecting voltage drop in diodes) and has a mean value of $V_{dc} = 0.83\ V_{sm}$ or 1.17 V_S where V_S is the r.m.s. value of the secondary phase voltage. Its maximum conversion h = 96.5% and g = 0.17. But it should be noted that the magnitude of these fluctuations or pulsations is lesser than for a single-phase, full-wave rectified output since the current never touches zero. It is further smoothened up by a *C-L-C* filter circuit as shown in Fig. 19.28.

It is seen that direct current of each diode appears in the secondary phase winding and so causes transformer saturation resulting in large primary current. It can be avoided by using zig-zag secondary.

19.28. Full-wave Rectification of three-phase Currents

As shown in Fig. 19.22, a three-phase full-wave rectifier requires a transformer with six secondary winding connected to give two separate three-phase supplies 180° out of phase with each other. The centre taps are connected by an interphase transformer which enables the two rectifier units to operate indenpently of each other. Obviously, each diode conducts for one-third cycle. However, the addition of two outputs cancels the lowest frequency component of the ripple.

The output voltage has a mean value of 0.83 V_{sm} (less the diode voltage drop) and a ripple having a fundamental frequency six times the supply frequency.

Three-phase full-wave rectifier circuit is preferred for high power because

(*i*) each secondary carries current for one-third of a cycle;

(*ii*) each primary carries current for two-thirds of a cycle;

(*iii*) copper loss in the transformer windings is comparatively lower.

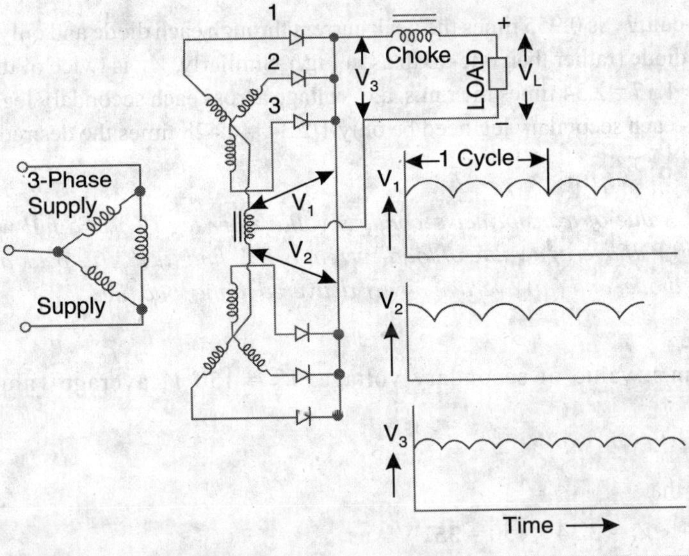

Fig. 19.22.

19.29. Six-phase Half-wave Reactifier

Such a rectifier can be operated from a three-phase supply using a transformer with three centre-tapped secondary windings with all the centre taps connected together as shown in Fig. 19.23. Obviously, each diode conducts for one-sixth of a cycle. The output voltage has a mean value of 0.955 V_{sm} (less the voltage drop in the diode). Ripples has a very small value and a fundamental frequency six times the supply frequency.

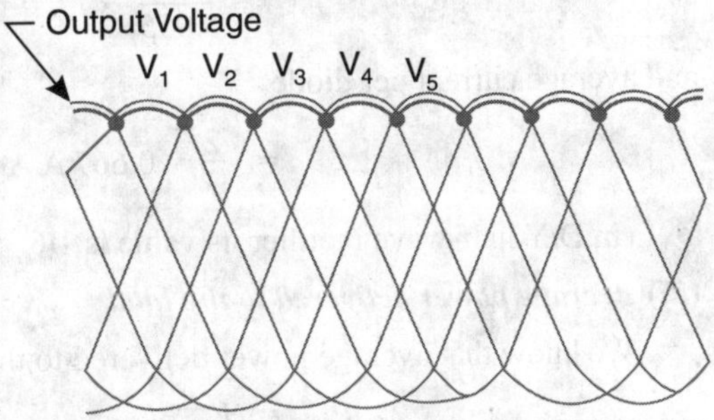

Fig. 19.23.

19.30. Three-phase Bridge Rectifier Circuit

Fig. 19.24 shows a three-phase bridge rectifier circuit. This circuit is very frequenctly used because apart from being simple, it does not require centre-tap transformer. Only that diode supplies the load whose phase voltage is more positive than the others. For example, when D_1 is forward-biased, D_2 and D_3 are reverse-biased. The current returns to the supply *via* D_5 and D_6.

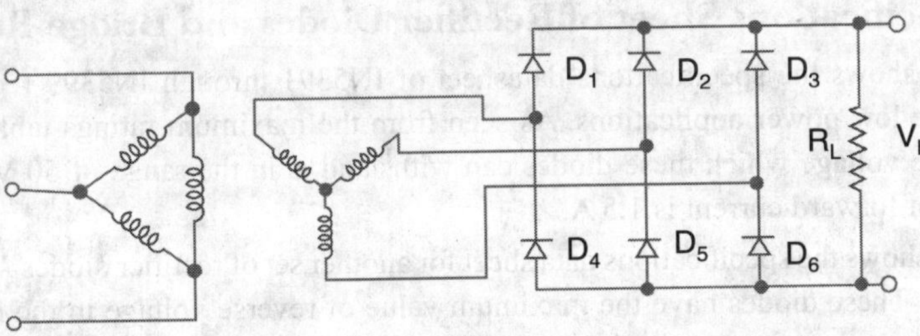

Fig. 19.24.

In this circuit, I_{dc} is 0.955 times the peak current through each diode and only one-third of it flows through each diode (rather than one-sixth as in Fig.). Similarly, V_{dc} is twice of the V_{in} 3-ϕ, half-wave rectifier or $2 \times 1.17 = 2.34$ times the r.m.s. a.c. voltage across each secondary leg. Accordingly, r.m.s. voltage across each secondary leg need be only $1/2.34 = 0.428$ times the desired d.c. output voltage.

Example 19.11.

The r.m.s. value of transformer secondary voltage per leg (V_S) in a full-wave. D/Y six-phase rectifier is 150 V. If average value of load current is 2 A, find (i) V_{dc} (ii) peak and average current through each diode, and (iii) average power delivered to the load, i.e., P_{dc}.

Solution.

Given r.m.s. value of secondary voltage, $V_S = 150\ V$, average value of load current, $I_{dc} = 2\ A$.

(i) Value of V_{dc}

We know that,

$$V_{dc} = 2.34 \times V_S = 2.34 \times 150 = \mathbf{351\ V\ Ans.}$$

(ii) Peak and average current through each diode

We know that peak current per diode,

$$= \frac{1}{0.955} \times I_{dc} = 1.05 \times 2 = 2.1\ \text{A}\ \mathbf{Ans.}$$

and average current per diode,

$$= \frac{2}{3} = 0.667\ \text{A}\ \mathbf{Ans.}$$

(In D/Y half-wave rectifier its value is $1/6^{th}$ rather than $1/3^{rd}$)

(iii) Average power delivered to the load

We know that average power delivered to the load,

$$P_{dc} = V_{dc} \cdot I_{dc}$$

$$= 351 \times 2 = 701\ \text{W}\ \mathbf{Ans.}$$

19.31. Calculations with Resistive Load

Let us suppose that no filter is connected across the rectifier but only a load of resistance *R*. Further, we will neglect transformer resistance and leakage and internal diode drops. In that case,

$$V_{dc} = 0.827\ V_{sm}\ ;\ I_{dc} = 0.827\ V_{sm}/R$$

19.32. Specifications Sheet of Rectifier Diodes and Bridge Rectifier

Fig. 19.25 shows the specifications datasheet of IN5391 through IN5399 rectifier diodes for general purpose low power applications. As seen from the maximum ratings table, the maximum value of reverse voltage which these diodes can withstand is in the range of 50 V to 1000 V. The average value of forward current is 1.5 A.

Fig. 19.26 shows the specifications data sheet for another set of rectifier diodes *MUR* 5005, 5010, 5015 and 5020. These diodes have the maximum value of reverse voltage in the range of 50 V to 200 V. However, the average value of current which all these diodes can handle is 50 A. Therefore, these diodes are more suitable for rectifiers with high curent applications.

MOTOROLA
SEMICONDUCTOR
TECHNICAL DATA

1N5391 thru 1N5399

Designers Data Sheet

"SURMETIC" RECTIFIERS

...subminiature size, axial lead-mounted rectifiers for general-purpose, low-power applications.

Designers Data for "Worst Case" Conditions

The Designers Data Sheets permit the design of most circuits entirely from the information presented. Limits curves–representing boundaries on device characteristics–are given to facilitate "worst-case" design.

LEAD-MOUNTED SILICON RECTIFIERS

50-1000 VOLTS DIFFUSED JUNCTION

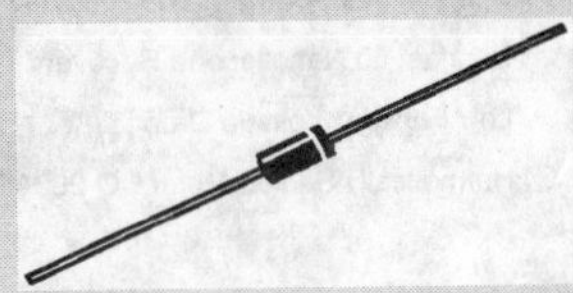

***MAXIMUM RATINGS**

Rating	Symbol	1N5391	1N5392	1N5393	1N5395	1N5397	1N5398	1N5399	Unit
Peak Repetitive Reverse Voltage Working Peak Reverse Voltage DC Blocking Voltage	V_{RRM} V_{RWM} V_R	50	100	200	400	600	800	1000	Volts
Nonrepetitive Peak Reverse Voltage (Halfwave, Single Phase, 60 Hz)	V_{RSM}	100	200	300	525	800	1000	1200	Volts
RMS Reverse Voltage	$V_{R(RMS)}$	35	70	140	280	420	560	700	Volts
Average Rectified Forward Current (Single Phase, Resistive Load, 60 Hz, $T_L = 70°C$, 1/2" From Body)	I_O	← 1.5 →							Amp
Nonrepetitive Peal Surge Current (Surge Applied at Rated Load Conditions, See Figure 2)	I_{FSM}	← 50 (for 1 cycle) →							Amp
Storage Temperature Range	T_{stg}	← – 65 to + 175 →							°c
Operating Temperature Tange	T_L	← – 65 to + 170 →							°c
DC Blocking Voltage Temperature	T_L	← 150 →							°c

***ELECTRICAL CHARACTERISTICS**

Characteristic and Conditions	Symbol	Typ	Max	Unit
Maximum Instantaneous Forward Voltage Drop ($i_F = 4.7$ Amp Peak, $T_L = 170°C$, 1/2 Inch Leads)	V_F	–	1.4	Volts
Maximum Reverse Current (Rated dc Voltage) ($T_L = 150°C$)	I_R	250	300	µA
Maximum Full-Cycle Average Reverse Current (1) ($I_O = 1.5$ Amp, $T_L = 70°C$, 1/2 Inch Leads)	$I_{R(AV)}$	–	300	µA

*Indicates JEDEC Registered Data.

NOTE 1: Measured in a single-phase, halfwave circuit such as shown in Figure 6.25 of EIA RS-282, November 1963. Operated at rated load conditions $I_O = 1.5$ A, $V_r = V_{RWM}$, $T_L = 70°C$.

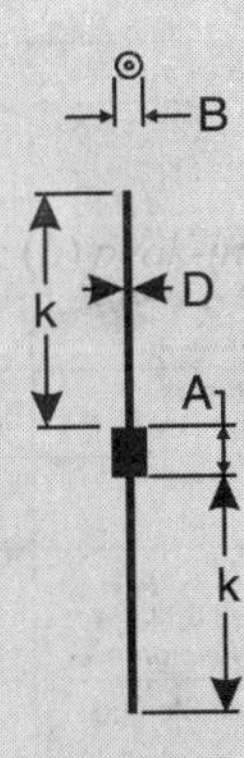

NOTES:

1. ALL RULES AND NOTES ASSOCIATED WITH JEDEC DO-41 OUTLINE SHALL APPLY.
2. POLARITY DENOTED BY CATHODE BAND.
3. LEAD DIAMETER NOT CONTROLLED WITHIN "F" DIMENSION.

DIM	MILLIMETERS MIN	MILLIMETERS MAX	INCHES MIN	INCHES MAX
A	5.97	6.60	0.235	0.260
B	2.79	3.05	0.110	0.120
D	0.76	0.86	0.030	0.034
K	27.94	–	1.100	–

CASE 59-04 PLASTIC

MECHANICAL CHARACTERISTICS

CASE: Transfer molded plastic

MAXIMUM LEAD TEMPERATURE FOR SOLDERING PURPOSES: 240°C, 1/8" form case for 10 seconds at 5 lbs. tension

FINISH: All external surfaces are corrosion-resistant, leads are readily solderable

POLARITY: Cathode indicated by color band

WEIGHT: 0.40 grams (approximately)

Fig. 19.25. (Courtery of Motorola Inc.)

MOTOROLA
SEMICONDUCTOR
TECHNICAL DATA

MUR5005
MUR5010
MUR5015
MUR5020

ULTRAFAST RECTIFIERS
50 AMPERES
50 to 200 VOLTS

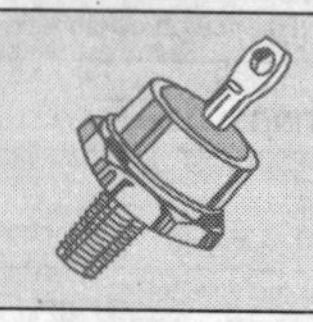

SWITCHMODE POWER RECTIFIERS

...designed for use in switching power supplies, inverters and as free wheeling diodes, these state-of-the-art devices have the following features:

- Ultrafast 50 Nanosecond Recovery Time
- Low Forward Voltage Drop
- Hermetically Sealed Metal DO-203AB Package

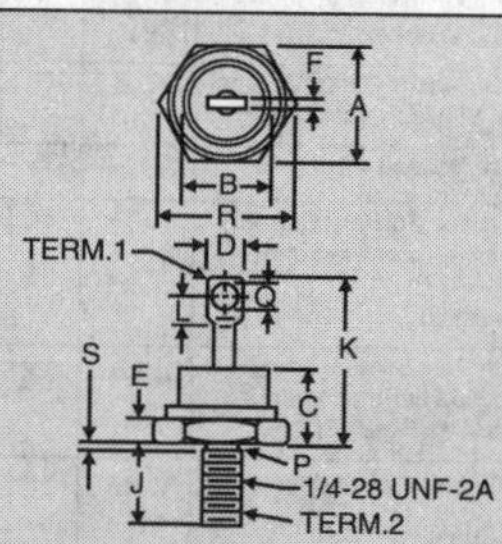

NOTES:
1. DIM "P" IS DIA.
2. CHAMFER OR UNDERCUT ON ONE OR BOTH ENDS OF HEXAGONAL BASE IS OPTIONAL
3. ANGULAR ORIENTATION AND CONTOUR OF TERMINAL ONE OPTIONAL.
4. THREADS ARE PLATED.
5. DIMENSIONING AND TOLERANCING PER ANSI Y14.5, 1973.

DIM	MILLIMETERS		INCHES	
	MIN	MAX	MIN	MAX
A	16.94	17.45	0.669	0.687
B	—	16.94	—	0.667
C	—	11.43	—	0.450
D	—	9.53	—	0.375
E	2.92	5.08	0.115	0.200
F	—	2.03	—	0.080
J	10.72	11.51	0.422	0.453
K	—	25.40	—	1.000
L	3.86	—	0.156	—
P	5.59	6.32	0.220	0.249
Q	3.56	4.45	0.140	0.175
R	—	20.16	—	0.794
S	—	2.26	—	0.089

CASE 257-01
DO-203AB
METAL

MAXIMUM RATINGS

Rating	Symbol	MUR				Unit
		5005	5010	5015	5020	
Peak Repetitive Reverse Voltage Working Peak Reverse Voltage DC Blocking Voltage	V_{RRM} V_{RWM} V_R	50	100	150	200	Volts
Nonrepetitive Peak Reverse Voltage	V_{RSM}	55	110	165	220	Volts
Average Forward Current TC = 150°C	$I_{F(AV)}$	50				Amps
Nonrepetitive Peak Surge Forward Current (half cycle, 60 Hz, Sinusoidal Waveform)	I_{FSM}	600				Amps
Operating Junction and Storage Temperature	T_J, T_{stg}	– 55 to + 175				°C

THERMAL CHARACTERISTICS

Tating	Symbol	All Devices	Unit
Thermal Resistance, Junction to Case	$R_{\theta JC}$	1.0	°C/W

ELECTRICAL CHARACTERISTICS

Maximum Instantaneous Forward Voltage Drop (i_F = 50 Amp, T_J = 25°C) (i_F = 50 Amp, T_J = 125°C) (i_F = 100 Amp, T_J = 125°C)	v_F	 1.15 0.95 1.10	Volts
Maximum Reverse Current @ DC Voltage (T_J = 25°C) (T_J = 125°C)	I_F	 10 1.0	 μA mA
Maximum Reverse Recovery Time (I_F = 1.0 Amp, di/dt = 50 Amp/ms, V_R = 30 V, T_J = 25°C)	t_{rr}	50	ns

MECHANICAL CHARACTERISTICS

CASE: Welded, hermetically sealed

FINISH: All external surface correction resistant and terminal leads are readily solderable

POLARITY: Cathode to Case

MOUNTING POSITIONS: Any

MOUNTING TORQUE: 25 in-lb max

Fig. 19.26. (Courtesy of Motorola, Inc.)

MOTOROLA
SEMICONDUCTOR
TECHNICAL DATA

MDA2500 Series

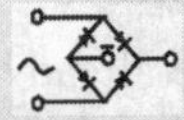

SINGLE-PHASE FULL-WAVE BRIDGE

25 AMPERES
50-600 VOLTS

RECTIFIER ASSEMBLY

...utilizing individual viod-free molded rectifiers, interconnected and mounted on an electrically isolated aluminum heat sink by a high thermal-conductive epoxy resin.

- 400 Ampere Surge Capability
- Electrically Isolated Base
- UL Recognized
- 1800 Volt Heat Sink Isolation

MAXIMUM TATINGS

Rating (Per Diode)	Symbol	MDA 2500	2501	2502	2504	2506	2508	2510	Unit
Peak Repetitive Reverse Voltage Working Peak Reverse Voltage DC Blocking Voltage	V_{RRM} V_{RWM} V_R	50	100	200	400	600	800	1000	Volts
DC Output Voltage Resistive Load Capacitive Load	Vdc	 30 50	 62 100	 124 200	 250 400	 380 600	 500 600	 620 1000	Volts
Sine Wave RMS Input Voltage	$V_{R(RMS)}$	35	70	140	280	420	560	700	Volts
Average Rectified Forward Current (Single phase bridge resistive load, 60 Hz, T_C = 55°C)	I_O	25							Amp
Nonrepetitive Peak Surge Current (Surge applied at rated load conditions)	I_{FSM}	400							Amp
Operating and Storage Junction Temperature Range	T_J, T_{stg}	– 65 to + 175							°C

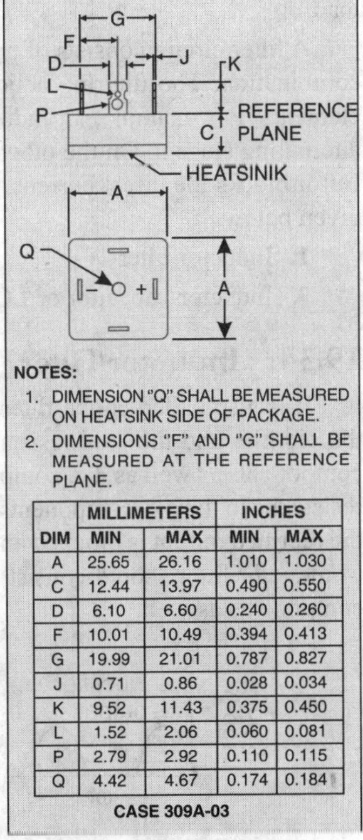

THERMAL CHARACTERISTICS

Characteristic	Symbol	Typ	Max	Unit
Thermal Resistance, Junction to Case	$R_{\theta JC}$			°C/W
Each Die		4.5	6.0	
Total Bridge		2.0	2.8	

ELECTRICAL CHARACTERISTICS (TC = 25°C unless otherwise noted)

Characteristic	Symbol	Min	Typ	Max	Unit
Instantaneous Forward Voltage (Per Diode) (i_F = 40 A)*	v_F	–	0.95	1.05	Volts
Reverse Current (Per Diode) (Rated V_R)	I_F	–	–	10	μA

NOTES:
1. DIMENSION "Q" SHALL BE MEASURED ON HEATSINK SIDE OF PACKAGE.
2. DIMENSIONS "F" AND "G" SHALL BE MEASURED AT THE REFERENCE PLANE.

DIM	MILLIMETERS MIN	MAX	INCHES MIN	MAX
A	25.65	26.16	1.010	1.030
C	12.44	13.97	0.490	0.550
D	6.10	6.60	0.240	0.260
F	10.01	10.49	0.394	0.413
G	19.99	21.01	0.787	0.827
J	0.71	0.86	0.028	0.034
K	9.52	11.43	0.375	0.450
L	1.52	2.06	0.060	0.081
P	2.79	2.92	0.110	0.115
Q	4.42	4.67	0.174	0.184

CASE 309A-03

MECHANICAL CHARACTERISTICS

CASE: Plastic case with an electrically isolated aluminum base.

POLARITY: Terminal designation embossed on case:
+ DC output
– DC output
AC not marked

MOUNTING POSITION: Bolt down. Highest heat transfer efficiency accomplished through the surface opposite the terminals. Use silicone heat sink compound on mounting surgace for maximum heat transfer.

WEIGHT: 25 grams (approx.)

TERMINALS: Suitables for fast-on connections. Readily solderable, corrosion resistant. Soldering recommended for applications greater than 15 amperes.

MOUNTING TORQUE: 20 in-lb max

* Pulse Width = 100 ms, Duty Cycle ≤ 2%.

Fig. 19.27. (Courtesy of Motorola, Inc.)

Fig. 19.27 shows the specifications data sheet of MDA 2500 series single-phase full-wave bridge rectifiers. As seen from the maximum ratings table, the maximum value of reverse voltage for MDA 2500 series devices ranges from 50 V to 1000 V whereas the dc output voltage ranges from 30 V to 620 volts. The average value of rectified current for all the devices in MDA 2500 series in 25 A.

19.33. Filters

We have already discussed in Art. 19.19 that the output of a rectifier contains d.c. component as well as an a.c. component. The presence of an a.c. component is most undesirable and therefore must be removed from the rectifier output. It is achieved by a circuit called filter.

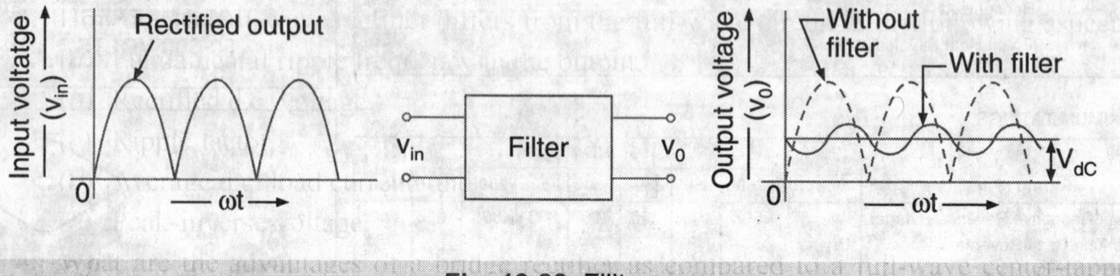

Fig. 19.28. Filiter.

Fig. 19.28 shows the concept of a filter, where the full-wave rectified output voltage is applied at its input. The output of a filter is not exactly a constant d.c. level. But it contains a small amount of a.c. component as well. Thus a filter circuit may be defined as a circuit, which removes (or minimizes) the unwanted a.c. component of the rectifier output and allows only the d.c. component to reach the load.

A filter circuit consists of passive circuit elements *i.e.*, inductors, capacitors, resistors and their combination. The filtering action depends upon the basic electrical properties of passive circuit elements. For example, an inductor allows the direct current to pass through it. But it blocks the alternating current. On the other hand, a capacitor allows the alternating current to pass through it. But it blocks the direct current. Some of the important filters used in the field of electronics are as given below:

1. Inductor filter
2. Capacitor filter
3. Inductor-capacitor or LC filter
4. π-filter

19.34. Inductor Filter

It is also called choke filter. It consists of an inductor (L), which is suitably inserted between the rectifier and the load resistance (R_L) as shown in Fig. 19.29. the rectifier output contains a.c. components as well as d.c. components. When the output passes through an inductor, it offers a high resistance to the a.c. components and no resistance to d.c. components. Therefore, a.c. component of the rectified output is blocked and only d.c. component reaches at the load. Theoretically, the output of inductor filter should consist of only d.c. voltage. But in actual practice, it contains a small a.c. components as well.

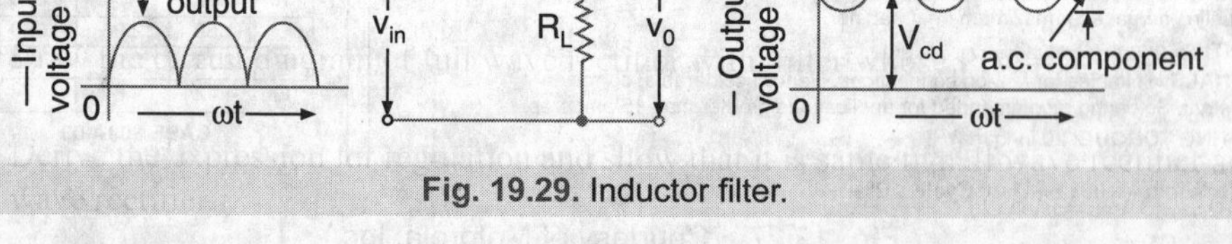

Fig. 19.29. Inductor filter.

The ripple factor of an inductor filter (for 50 Hz supply) is given by the expression,

$$\gamma = \frac{R_L}{3\sqrt{2}\cdot\omega.L} = \frac{R_L}{1330\,L} \quad ...(i)$$

where R_L is in ohms and L in Henrys. It is evident from equation (*i*) that the ripple will decrease as the value of R_L is decreased and L is increased.

The inductor filter is more effective only for heavy load currents *i.e.*, when the load resistance (R_L) is small. The a.c. component, present at the output of a filter, can be reduced if we use an inductor of a large value. But it may be noted that an inductor of a large value will also have a higher d.c. resistance, which will result in a lower d.c. output voltage.

Example 19.12.

(*a*) *What should be the value of inductance to use in an inductor filter connected to a full-wave rectifier operating at 50 Hz, if the ripple is not to exceed 5% for a 100 Ω load.*

(*b*) *Repeat the above problem for the standard aircraft power frequency of 400 Hz.*

Solution.

(*a*) Given: $f = 50$ Hz; $\gamma = 5\% = 0.05$ and $R_L = 100\ \Omega$.

We know that ripple factor (*g*),

$$0.05 = \frac{R_L}{3\sqrt{2}\cdot\omega.L} = \frac{100}{3\sqrt{2}\times(2\pi\times 50)\times L} = \frac{0.075}{L}$$

$\therefore \quad L = 0.075/0.05 = 1.5$ H **Ans.**

(*b*) Given: $f = 400$ Hz; $\gamma = 0.05$ and $R_L = 100\ \Omega$.

We also know that ripple factor (γ),

$$0.05 = \frac{R_L}{3\sqrt{2}\cdot\omega\cdot L} = \frac{100}{3\sqrt{2}\times(2\pi\times 400)\times L} = \frac{1}{106.6\,L}$$

$\therefore \quad L = 1/(106.6 \times 0.05) = 0.188$ H **Ans.**

19.35. Capacitor Filter

We have already discussed in the last article that an inductor filter is suitable only for heavy loads (*i.e.*, a load with smaller values of resistance). An inexpensive filter for light loads (*i.e.*, a load with larger value of resistance) is available in the form of a capacitor filter. The operation of a capacitor filter depends upon the property of a capacitor to oppose any change in voltage, when it is connected across a pulsating d.c. voltage. The action of a capacitor filter is to be smoothed out (or filter out) the voltage ripples or pulsations.

Fig. 19.30. shows a half-wave rectifier with a capacitor filter. The operation of this circuit may be explained in terms of the waveforms as shown in Fig. 19.30 (*a*) (*b*) and (*c*) respectively. Fig. 19.31 (*a*) shows the waveform of the a.c. input voltage (indicated by the broken curve) and the half-wave rectifier output. Fig. 19.31 (*b*) shows the waveform of a voltage across a capacitor and Fig. 19.31 (*c*) shows the waveform of an output voltage. These waveforms may be better understood from the following discussion.

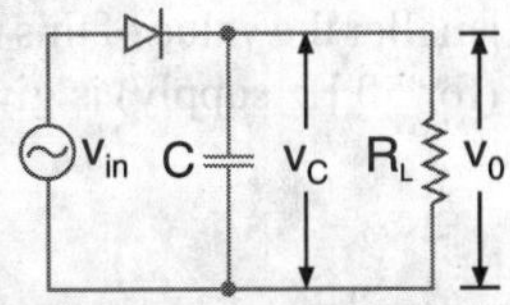

Fig. 19.30. Half-wave rectifier with capacition.

During the positive half-cycle of the a.c. input voltage, the diode is forward biased. It allows the capacitor to charge quickly to its maximum value of voltage (V_m) as indicated by the curve *oa* in Fig. 19.31 (*b*). Since there is no resistance in the charging path, except diode forward resistance (which is negligible), the charging time is almost zero. Therefore, the capacitor follows the charging voltage. When the a.c. input voltage begins to decrease below its maximum value, the capacitor returns its charge. We know that during the negative half-cycle of the input, the capacitor tries to discharge through the load resistance as shown by the curve *ab* in Fig. 19.31 (*b*). It may be noted that the capacitor cannot discharge through the diode, because it is reverse biased. The discharging time constant may be determined by the product of the capacitance value of the capacitor (*C*) and the value of load resistance (R_L). Mathematically, time constant,

$$\tau = C \cdot R_L$$

It will be interesting to know that larger the time-constant, lesser will be the capacitor discharge. In actual practice, the discharging time constant is kept about 100 times larger than that of the charging time. Therefore, the capacitor does not have enough time to discharge appreciably. As a result of this, the capacitor maintains a large voltage across the load resistance (R_L).

We also know that during the next half cycle, when the a.c. input voltage exceeds the capacitor voltage, as indicated by point '*b*' in Fig. 19.31 (*a*), the diode is again forward biased and the capacitor charges to its maximum value of the voltage (*i.e.*, V_m). It is indicated by the curve *dc* in Fig. 19.31 (*b*). During the negative half-cycle, the diode is reverse biased and the capacitor tries to discharge through the load resistance. The process continues for the other cycles and we get an output voltages as shown in Fig. 19.31 (*c*).

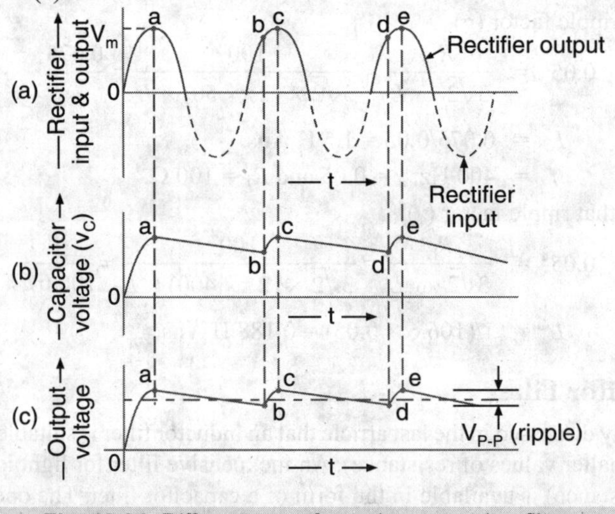

Fig. 19.31. Different waveforms in a capacitor filter.

It may be noted from Fig. 19.31 (*c*) that due to charging and discharging of a capacitor, there is a ripple (*i.e.*, a.c. component) present in the output voltage of a capacitor filter. As a matter of fact, smaller the value of this ripple, better will be the filtering action. The ripple factor of a capacitor filter (for 50 Hz supply) is given by the relation,

$$\gamma = \frac{1}{4\sqrt{3} \cdot f \cdot C \cdot R_L} = \frac{2890}{C \cdot R_L}$$

where *C* is in μF and R_L in ohms.

Example 19.13.

A full-wave rectifier, using a capacitor filter, has to supply 30 V d.c. to a load resistance of 1 kΩ.. Assuming the diode and transformer winding resistance to be negligible, estimate the value of capacitor filter for a ripple factor of 0.01.

Solution.

Given: $V_{dc} = 30$ volts; $R_L = 1\ \text{k}\Omega = 1 \times 10^3\ \Omega$ and $\gamma = 0.01$.

Let C = Value of capacitor in μF

We know that ripple factor for a 50 Hz supply (γ)

$$0.01 = \frac{2890}{C \cdot R_L} = \frac{2890}{C \times (1 \times 10^3)} = \frac{2.89}{C}$$

$$C = 2.89/0.01 = 289\ \mu\text{F}\ \textbf{Ans.}$$

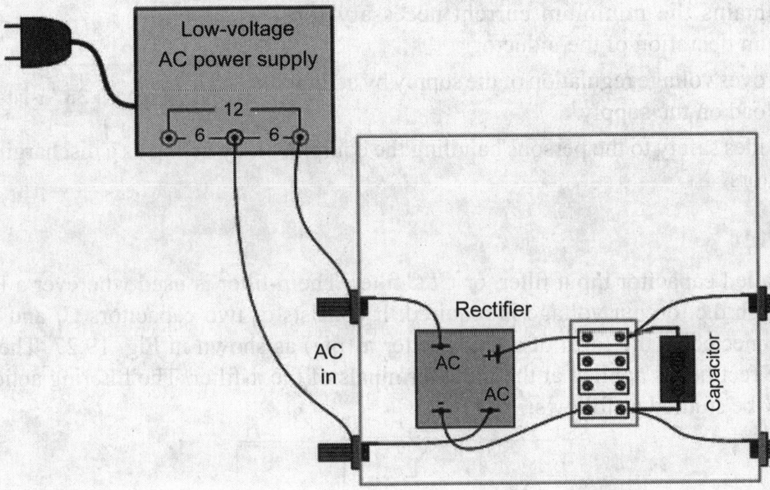

19.36. LC Filter

We have already discussed that in an inductor filter, the ripple factor is directly proportional to the load resistance. On the other hand in a capacitor filter, it is varying inversely with the load resistance. Hence, if we combine the inductor (*L*) filter and the capacitor (*C*), the ripple factor will become almost independent of the load resistance. The combination of *L* and *C* filter is called *LC* filter and is as shown in Fig. 19.32. The *LC* filter is also known as inductor input filter, choke input filter, *L* input filter or *LC*-section.

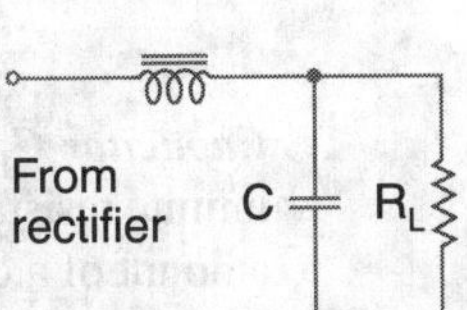

Fig. 19.32. LC filter.

The ripple factor for 50 Hz supply of a *LC* filter is given by the relation,

$$\gamma = \frac{\sqrt{2}}{12\,\omega^2 \cdot LC} = \frac{1.195}{LC}$$

where *L* is in Henry and *C* in μF.

Example 19.14.

A full-wave rectifier circuit, with LC filter supplies of 12 V d.c. at 100 mA with a maximum ripple constant of 1%. If the value of L = 1 H, estimate the size of the capacitor.

Solution.

Given: $V_{dc} = 12$ volts; $I_{dc} = 100$ mA; $\gamma = 1\% = 0.01$ and $L = 1$ H

Let C = Size or value of the capacitor in μF.

We know that ripple factor (γ)

$$0.01 = \frac{1.195}{LC} = \frac{1.195}{1 \times C} = \frac{1.195}{C}$$

$$\therefore \quad C = 1.195/0.01 = 119.5\ \mu\text{F} \textbf{ Ans.}$$

19.37. Bleeder Resistor

It is a common practice to use a resistor (called bleeder resistor) across the filter output as shown in Fig. 19.33. Such a resistor provides the following three advantages:

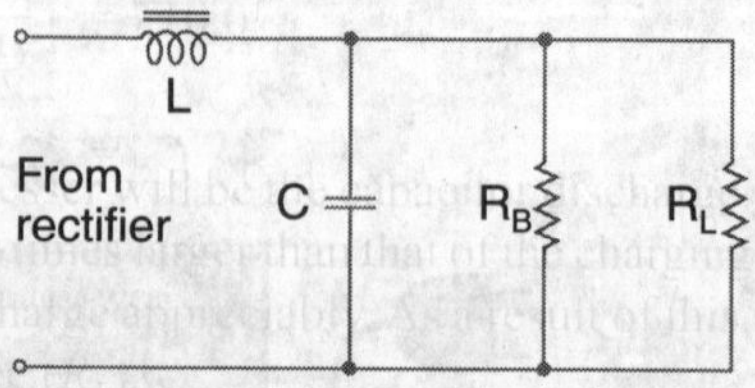

Fig. 19.33. π-filter

1. It maintains the minimum current necessary for optimum operation of the inductor.
2. It improves voltage regulation of the supply by acting as preload on the supply.
3. It provides safety to the persons handling the equipment, by acting as a discharging path for capacitors.

19.38. π-Filter

It is also called capacitor input filter, or *CLC* filter. The p-filter is used wherever a low output current and a high d.c. output voltage is required. It consists of two capacitors C_1 and C_2 and an inductor (L) connected in the form of a greek letter π (pie) as shown in Fig. 19.27. The pulsating output from the rectifier is applied at the input terminals of the π-filter. The filtering action of each component may be studied as follows:

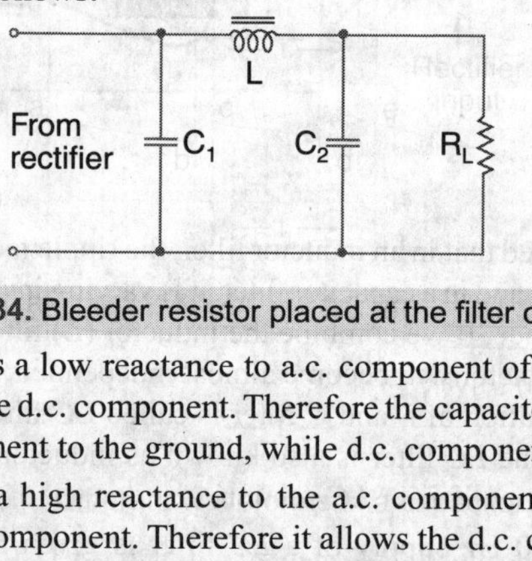

Fig. 19.34. Bleeder resistor placed at the filter output.

1. ***Capacitor C_1.*** It offers a low reactance to a.c. component of rectifier output. But it offers infinite resistance to the d.c. component. Therefore the capacitor C_1 bypasses an appreciable amount of a.c. component to the ground, while d.c. component moves towards L.
2. ***Inductor L.*** It offers a high reactance to the a.c. component of rectifier output but zero resistance to the d.c. component. Therefore it allows the d.c. component to pass through it, and blocks the a.c. component, which could not be bypassed by the capacitor C_1.
3. ***Capacitor C_2.*** Its behaviour is similar to the capacitor C_1. It bypasses the a.c. component of rectifier output, which could not be blocked by an inductor L. As a result, only the d.c. component is available at the output.

It has been found that the ripple factor in a π-filter is dependent only on the product of the capacitance values of the capacitors C_1 and C_2 (and not on the individual values of capacitors C_1 and C_2). But the d.c. output voltage will be high only if the value of capacitor C_1 is large. Hence for a given ripple factor, the -filter is designed with a large value of capacitor C_1. The ripple factor (for 50 Hz supply) of α -filter is given by the relation,

$$\gamma = \frac{1}{4\sqrt{2} \cdot \omega^3 \cdot C_1 \cdot C_2 L \cdot R_L} = \frac{5700}{C_1 \cdot C_2 \cdot L \cdot R_L}$$

where C_1 and C_2 are in μF, L in Henrys and R_L in ohms.

Example 19.15. *A full-wave rectifier supplies 0.2 A current at 30 V d.c. Find the ripple factor to be expected when two 100 mF, capacitors and a 5 H inductor are used in a p-filter with a 50 Hz supply.*

Solution.

Given: $I_{dc} = 0.2$ A; $V_{dc} = 30$ volt; $C_1 = C_2 = 100$ mF; $L = 5$ H and $f = 50$ Hz

We know that the load resistance,

$$R_L = \frac{V_{dc}}{I_{dc}} = \frac{30}{0.2} = 150\ \Omega$$

and the ripple factor for 50 Hz supply,

$$\gamma = \frac{5700}{C_1 \cdot C_2 \cdot R_L} = \frac{5700}{100 \times 100 \times 5 \times 150}$$

$$= 7.6 \times 10^{-4} = 0.076\%\ \textbf{Ans.}$$

NUMERICAL PROBLEMS

1. A half-wave rectifier circuit supplies 100 mA d.c. to a 250 Ω load. Find the d.c. output voltage, the *PIV* rating of a diode and the r.m.s. voltage for the transformer supplying the rectifier. **(Ans. 25 V; 35.4 V and 17.7 V)**
2. An a.c. supply of 230 V is applied to a half-wave rectifier circuit through a transformer of turns ratio 20 : 1. If the load resistance is 1kΩ, find (*a*) the d.c. output voltage, (*b*) maximum value of the output voltage, (*c*) d.c. load current, and (*d*) maximum value of load current. **(Ans. 5.17 V; 16.3 V; 5.17 mA; 16.3 mA)**
3. A half-wave rectifier uses a diode having a forward resistance of 2 Ω. If the transformer secondary voltage is 10 V (r.m.s.) and has a winding resistance of 5 Ω. Find (*a*) the d.c. output voltage at a load of 100 mA and (*b*) *PIV* of a diode. **(Ans. 3.8 V and 14.1 V)**
4. A full-wave rectifier with a center-tapped transformer (10 – 0 – 10 V) supplies a load current of 100 mA. Neglecting the diode forward resistance and secondary winding resistance, find (*a*) the d.c. output voltage, (*b*) *PIV* of each diode, and (*c*) ripple frequency. **(Ans. 9 V; 28.3 V and 100 Hz)**
5. A full-wave rectifier is fed by a 24 V_{ac} (rated) center-tapped transformer. What is the d.c. output voltage of the circuit. If the rectifier is connected to a load resistance of 2.2 k Ω load, what is the d.c. value of load current. **(Ans. 10.8 V; 4.9 mA)**
6. A full-wave rectifier uses two diodes, the forward resistance of each diode is 20 Ω. The transformer r.m.s. secondary voltage from center-tapped to each end of the secondary is 50 V and the load resistance is 980 Ω. Determine (*a*) the average value of load current and (*b*) the r.m.s. value of load current. **(Ans. 45 mA and 25 mA)**
7. A transformer with 20 V r.m.s., 50 Hz secondary, supplies a bridge rectifier having a load of 200 Ω. Determine the d.c. output voltage, d.c. load current and *PIV* rating of the diode. **(Ans. 18 V; 0.09 A and 28.3 V)**
8. A 50 Hz transformer having 50 V r.m.s. on each side of the center-tap supplies a full-wave rectifier circuit. The circuit load is 150 Ω with a shunt capacitor filter of 1000 μF. Find the ripple factor. **(Ans. 1.9%)**

9. A π-filter in a full-wave rectifier uses $C_1 = C_2 = 100\ \mu F$ and $L = 10$ H. The load current is 0.5 A at 100 V d.c. Find the ripple factor? **(Ans. 0.0285%)**

DESCRIPTIVE QUESTIONS

1. Draw the circuit diagram of half-wave rectifier. Explain its working. What is the frequency of ripple in its output?

2. Draw the circuit diagram of a full-wave rectifiers (*a*) with center-tap connection and (*b*) bridge connection. Explain their working. What is the peak-inverse voltage of a diode in each case?

3. How does a half-wave rectifier differs from the full-wave circuit in the following aspects?
(*a*) Fundamental ripple frequency in the output,
(*b*) Rectified d.c. voltage,
(*c*) Ripple factor,
(*d*) Average d.c. load current, and
(*e*) Peak-inverse voltage.

4. What are the advantages of a bridge rectifier as compared to a full-wave center-tapped rectifier?

5. What is efficiency of rectification? Compare half-wave and full-wave rectifiers from the point of view of rectification efficiency.

6. Define (*i*) a ripple factor, (*ii*) rectification efficiency, and (*iii*) transformer utilization factor for a rectifier.

7. Why do we need filters in a power supply? Under what condition we shall prefer a capacitor filter?

8. What is the function of a bleeder resistor?

9. Describe the terms capacitor-input filter and inductor input filter.

10. Describe a half wave rectifier. *(M.G University., April/May 2004)*

11. Describe a half wave rectifier using a junction diode. Derive expressions for ripple factor and efficiency for half wave rectifier circuit. *(GBTU/MTU., 2008-09, 2004-05)*

12. Derive all the necessary parameters of HWR *(JNTU., June 2009)*

13. Derive an expression for I_{dc} and I_{rms} for half–wave rectifier circuit and fine its maximum efficiency. *(PBUT., 2008)*

14. Show that the maximum rectification efficiency of HWR is 40.6 % and that of FWR is 81.2 %. *(JNTU., Aug./Sep. 2008)*

15. Define the following terms of a half-wave rectifier with a resistive load: (i) Ripple factor, (ii) Peak inverse voltage, (iii) Rectification efficiency. *(JNTU., Aug./Sep. 2006)*

16. Draw the circuit of full wave rectifier and derive ripple factor, rectification efficiency and PIV. *(Calicut University., June 2007)*

17. Explain the advantages and disadvantages of a bridge rectifier as compared with centre tapped circuit. *(M.G University., May 2005)*

18. For a two diode full wave rectifier derive the value of ripple factor? *(M.G University., May/June 2006)*

19. Draw the circuit diagram of full wave rectifier with filter whose PIV = V_m *(GBTU/MTU., 2006-07)*

20. Derive the expression for regulation and show that it is same in half wave rectifier and full wave rectifier. *(GBTU/MTU., 2006-07)*

21. Draw a neat diagram of full wave bridge rectifier circuit. Explain it's working in details, clearly making the direction of flow of currents for positive and negative cycles. *(GBTU/MTU., 2005-06)*

22. Explain the working of a full wave bridge rectifier. Derive the expression for ripple factor. *(PTU., May 2006)*

23. Explain with figures the principle of operation of a bridge rectifier? What are the advantages of this circuit? *(M.G University., Dec. 2005)*

24. What is the need for filters in rectifiers? Explain .How filter circuit can be classified. *(M.G University., July/August 2007)*

25. Different type of filters used in rectifiers. *(GBTU/MTU., 2006-07)*

26. Compare the performance of half wave rectifier and full wave rectifier. *(Anna University., Nov/Dec 2007)*

27. Write four advantages of Bridge rectifier compared to a centre-tapped Full-wave rectifier. *(M.G University., Dec. 2007)*

28. Show that rectification efficiency for a full wave rectifier is 81.2%? *(PTU., Dec 2008)*

29. Define the ripple factor of a rectifier. Give the values for Half wave and Full wave rectifier. *(M.G University., May 2008)*

30. Draw the circuit diagram of a bridge rectifier and explain .Derive the expression for its:

(*i*) r.m.s current (*ii*) average current

(*iii*) ripple factor (*iv*) rectification efficiency

(*v*) TUF. *(M.G University., May 2008)*

31. What are the advantages of full wave rectifier? *(Bhartiya University., April 2004)*

32. Derive an expression for efficiency and ripple factor for a full wave rectifier. *(Bhartiya University., April 2004)*

33. Draw circuit diagram of Bridge rectifier and explain its working. Draw output waveforms with and without filters. *(Nagpur University., 2008)*

34. Draw and explain with the help of waveforms a bridge rectifier with and without C filter and derive the expressions for the ripple factor in both cases. *(M.G University., May 2005)*

35. Explain bridge rectifier with capacitor filter. What is the expression for ripple factor for this circuit? *(M.G University., May/June 2006)*

36. Draw the circuit diagram of full wave Bridge rectifier. Explain its action with input and output wave forms. Derive an expression for its efficiency. *(Bhartiya University., April 2004)*

37. Derive expression for I_{dc}, V_{dc}, I_{rms} and V_{rms} for half-wave rectifier. *(Nagpur University 2008)*

38. What will be the PIV of non-conducting diode if is the peak voltage across secondary of the transformer in a half - wave rectifier? *(PTU., Dec 2008)*

39. Draw the circuit diagram of a full wave rectifier using Bridge connection. Explain its working and derive expressions for the RMS value and average value of current. *(PTU., May 2008)*

40. Explain the working of a two diode full-wave rectifier? *(M.G University., May/June 2006)*

41. Draw and explain the working of a centre-tapped full-wave rectifier. Derive its average current, ripple factor, efficiency, TUF, PIV. *(M.G University., May/June 2006)*

42. Define ripple factor and TUF. *(M.G University., July/Aug.2007)*

43. Show that full-wave rectifier in twice as efficient as a half-wave rectifier. *(M.G University., July/Aug.2007)*

44. Define ripple factor. Write its value for the HWR without any filter. *(M.G University., May/June 2006)*
45. Draw circuits diagrams to show two methods of producing a negative output voltage from a half wave rectifier. Explain briefly the circuit operations. *(GBTU/MTU., 2008-09)*
46. Drive an expression for ripple factor in case of half wave rectifier. *(GBTU/MTU., 2009)*
47. Why transformer utility factor of half wave rectifier and center tapped full wave rectifier lesser then bridge type rectifier? *(Calicut University., Dec. 2006)*
48. What is a capacitor filter? What are its characteristics? *(M.G University., Aug. 2007)*
49. What is the drawback of C filter? How is it rectified? *(Calicut University., Dec. 2006)*
50. Define and compare the following for a HWR and center tapped FWR, Ripple factor, PIV, efficiency and rectification factor. *(Calicut University., Dec. 2006)*
51. What is π- filter? Emphasize its importance. *(Calicut University., Dec. 2006)*
52. Draw and explain the function of inductive filter. *(Calicut University., June 2007)*
53. Explain the operation of full wave voltage doubler circuit. *(VTU., June /July 2008, GBTU/MTU 2005-06)*
54. Derive the ripple factor for FWR with capacitor filter. *(Anna University., April/May 2008)*
55. Derive the expression for ripple factor of FWR with π section filter. *(Anna University., April/May 2008)*
56. Define Transformer Utilization Factor. *(Anna University., Nov/Dec 2007)*
57. Draw the circuit diagram of a FWR with capacitor input filter. With suitable waveform explain its working. Derive the expression for ripple factor. *(Anna University., Nov/Dec 2007)*
58. What is a switched capacitor filter? *(Anna University., April/May 2008)*
59. Draw the circuit diagram of a FWR with centre tap connection and Bridge connection and explain its operation. *(JNTU, June 2009)*
60. Write the short notes on multiple L- Section and Multiple π-Section filter. *(JNTU., June 2009)*
61. Compare all the filter circuits from the point of view of ripple factor. *(JNTU., June 2009)*
62. Why silicon diodes are used as rectifiers ? *(PBUT., June 2005)*
63. Draw the circuit diagram of a full wave rectifier and derive expression for dc output voltage, dc output current and ripple factor. *(PBUT., Dec. 2004)*
64. Compare the advantages and disadvantages between centre-tapped and bridge-type full-wave rectifier. *(PBUT., 2009)*
65. Define the following terms as referred to full-wave rectifier circuit.
    (*i*) PIV (*ii*) Average dc voltage
    (*iii*) RMS current (*iv*) Ripple factor *(JNTU., May/Jun. 2006)*
66. Bridge connection and explain its operation. *(JNTU., June 2009)*
67. Define the following terms: (*i*) Ripple factor (*ii*) form factor (*iii*) PIV *(JNTU., Aug./Sep.2006)*
68. In a full-wave single-phase bridge rectifier circuit, can the transformer and load be interchanged ? Justify your statement with reasons. *(JNTU., Aug./Sep. 2006)*
69. Explain why a bridge rectifier is preferred over a centre-tap rectifier. *(JNTU., May/Jun. 2006, Aug./Sep. 2008)*

70. What are the advantages and disadvantages of C filters? At what circumstances they are used. *(M.G University., May 2008)*
71. Compare all the filter circuits from the point of view of ripple factor. *(JNTU., June 2009)*
72. Compare various filter circuits in terms of their circuits, ripple factor and voltage waveforms. *(JNTU., Aug./Sep. 2008)*
73. Write the short notes on multiple L- section and multiple π -section filter. *(JNTU., June 2009)*
74. What is the necessity of having filters in power supplies? Explain with diagrams how RL-filter improves the output of a rectifier. *(PTU., May 2008)*
75. Explain the necessity of a bleeder resistor. *(JNTU., May/Jun 2006, Aug./Sep. 2008)*
76. What is bleeder resistance? *(Calicut University., June 2007)*
77. Write a short note on rectifier circuits. *(R.G.P.V., May/June 2006)*
78. Define a rectifier with types and explain bridge full wave rectifier with diagram and analysis. *(R.G.P.V., Dec. 2008)*
79. Draw the circuit diagram of full-wave bridge rectifier and briefly explain its operation. For the full-wave rectifier, reduce the expression for:
    (*i*) Output dc voltage
    (*ii*) Ripple Factor
    (*iii*) Efficiency. *(R.G.P.V., June 2011)*
80. Explain the application of diode in a full-wave bridge rectifier.*(R.G.P.V., June 2009, 2010)*
81. Explain the working of bridge rectifier. Also, calculate PIV for the same. *(R.G.P.V., Dec. 2009)*
82. Explain the working of full-wave bridge rectifier. *(R.G.P.V., Feb. 2010)*
83. What are the function of rectifier? Draw the circuit and explain different types of rectifiers. *(R.G.P.V., June 2002)*
84. Explain the function of the filtering capacitor in rectifiers. *(R.G.P.V., Dec. 2003)*
85. Draw the circuit diagram of full wave centre tapped rectifier and bridge rectifier. Sketch the input and output wave forms for full wave centre tapped rectifier. *(R.G.P.V., Dec. 2004)*
86. Draw the circuit diagram of bridge rectifier and explain its working. Draw output waveforms with and without filters. *(Nagpur University, Summer 2008)*
87. Derive the expression for I_{dc}, V_{dc}, I_{rms} and V_{rms} for half wave rectifier. *(Nagpur University, Summer 2008)*
88. Derive the expression for ripple factor for Half wave rectifier. *(Nagpur University, Summer 2007)*
89. Draw the circuit diagram of Bridge rectifier. Explain the working of capacitor filter. *(Nagpur University, Winter 2005)*
90. Derive and explain the operation of Bridge rectifier with the help of neat waveform for input voltage, output voltage and diode current. Derive the expression for V_{rms}. *(Nagpur University, Summer 2003)*
91. Mention the advantages of four diode bridge rectifier over a full-wave rectifier using two diodes. Give the necessary proof. *(Nagpur University, Winter 2001)*
92. Calculate the peak inverse voltage, if V_m in the peak value of an applied voltage in a half wave rectifier circuit with a large capacitor connected across the load. *(Nagpur University, Winter 2001)*

93. With neat circuit diagram explain the working of bridge rectifier.
(Nagpur University, Winter 2002)

94. Explain the operation of a three phase bridge rectifier and draw the output voltage waveforms.
(Nagpur University, Summer 2000)

95. Compare different filters as regards to ripple factor and the output voltage.
(Nagpur University, Summer 2000)

96. Write short notes on :

(*i*) Comparison of bridge rectifier over centre tapped rectifier.
(Nagpur University, Summer 2000)

(*ii*) Peak Inverse Voltage *(R.G.P.V., Dec. 2005, 2003)*

(*iii*) Rectifiers *(R.G.P.V., Dec. 2001)*

(*iv*) Bridge Rectifier *(Nagpur University, Summer 2001)*

97. Explain the center tapped full wave rectifier. *(R.G.T.U., March, April 2010)*

98. With the help of circuit diagram explain the working of half wave rectifier. Show the relevant waveforms. *(R.G.T.U., Feb. 2010)*

99. Obtain expression for efficient of rectification and ripple factor for half wave rectifier.
(R.G.T.U., Feb. 2010)

100. What are the disadvantages of half wave rectifier? How are they overcome in full wave rectifier? *(R.G.T.U., Feb. 2010)*

101. Draw the neat sketch of Bridge rectifier with capacitor filter and explain. Derive the expression for ripple factor and efficiency. *(Nagpur University, Summer 2011)*

MULTIPLE CHOICE QUESTIONS

1. In a half-wave rectifier, the load current flows for

(*a*) Complete cycle of the input signal,

(*b*) Less than half-cycle of the input signal,

(*c*) More than half-cycle but less than complete cycle of the input signal.

(*d*) Only for the positive half-cycle of the input signal.

2. In a full-wave rectifier, the current in each of the diodes flows for

(*a*) Complete cycle of the input signal,

(*a*) Complete cycle of the input signal.

(*b*) Half cycle of the input signal.

(*c*) Less than half of the input signal.

(*d*) None of the above.

3. The ripple factor of a bridge rectifier is

(*a*) 0.482 (*b*) 0.812 (*c*) 1.11 (*d*) 1.21

4. A bridge rectifier is preferable to an ordinary two-diode full-wave rectifier because

(*a*) Fundamental ripple frequency in the output,

(*a*) It needs much smaller transformer for the same output.

(*b*) It uses four diodes.

(*c*) Its transformer does not require center-tap.

(*d*) None of the above.

5. The basic purpose of a filter is to:
   (*a*) Minimise variations in a.c. input signal.
   (*b*) Suppress harmonics in rectified output.
   (*c*) Remove ripples from the rectified output.
   (*d*) Stabilise d.c. output voltage.
6. The use of a capacitor filter in a rectifier circuit gives satisfactory performance only when the load
   (*a*) current is high (*b*) current is low
   (*c*) voltage is high (*d*) voltage is low (*UPSC Engg. Services 2001*)
7. A half-wave rectifier is equivalent to
   (*a*) a clamper circuit (*b*) a clipper circuit
   (*c*) a clamper circuit with negative bias (*d*) a clamper circuit with positive bias
8. Bridge rectifiers are preferred because
   (*a*) they require small transformer
   (*b*) they have less peak-inverse voltage
   (*c*) they need small transformer and also have less peak-inverse voltage
   (*d*) they have low ripple factor.
9. If V_m is the peak value of an applied voltage in a half-wave rectifier with a large capacitor across the load, then the peak-inverse voltage will be :
   (*a*) $V_m/2$ (*b*) V_m (*c*) $\sqrt{2}\,V_m$ (*d*) $2\,V_m$
10. A voltage of 200 cos 100 *t* is applied to a half-wave rectifier with a load resistance of 5 kW. The rectifier is represent by an ideal diode in series with a resistance of 1 kW. The maximum value of curent, *dc* component of current and r.m.s. value of current will be respectively,
   (*a*) 33.33 mA, 10.61 mA and 16.67 mA (*b*) 2.22 mA, 8.61 mA and 12.38 mA
   (*c*) 28.33 mA, 14.61 mA and 13.33 mA (*d*) 40 mA, 20 mA and 25 mA
   [**Hint :** Refer to Fig. 19.35.)

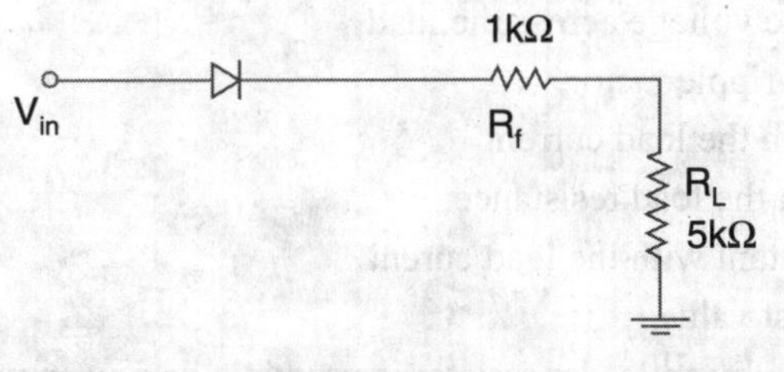

Fig. 19.35.

$$I_m = \frac{V_m}{R_f + R_L} = \frac{200}{1\text{k} + 5\text{k}} = 33.33 \text{ mA } \textbf{Ans.}$$

$$I_{dc} = \frac{I_m}{\pi} = \frac{33.33 \text{ mA}}{\pi} \; 10.61 \text{ mA } \textbf{Ans.}$$

$$I_{rms} = \sqrt{I_{dc}^2 + I_{L(ac)}^2}$$

where $$I_{L(ac)} = 0.385\, I_m$$

Substituting the values, we get, I_{rms} = 16.67 mA]

11. The basic reason why a full-wave rectifier has a twice the efficiency of a half-wave rectifier is that

(*a*) it makes use of transformer

(*b*) its ripple factor is much less

(*c*) it utilizes both half-cycle of the input

(*d*) its output frequency is double the line frequency.

12. The output of a half-wave rectifier is suitable only for

(*a*) running car radios (*b*) running ac motors

(*c*) charging batteries (*d*) running tape-recorders.

13. The ripple factor of a bridge rectifier is

(*a*) 0.406 (*b*) 0.812 (*c*) 1.21 (*d*) 1.11.

14. The ripple factor of a power supply is given by (symbols have the usual meaning).

(*a*) $\frac{P_{dc}}{P_{ac}}$ (*b*) $\sqrt{\left(\frac{I_{rms}}{I_{dc}}\right)^2 - 1}$ (*c*) $\sqrt{\left(\frac{I_{dc}}{I_{rms}}\right)^2 - 1}$ (*d*) $\frac{I_{dc}}{I_{rms}}$

15. The PIV of a half-wave rectifier circuit with a shunt capacitor filter is

(*a*) $2V_{sm}$ (*b*) V_{sm} (*c*) $V_{sm}/2$ (*d*) $3V_{sm}$

16. The primary function of a rectifier filter is to

(*a*) minimise ac input variations

(*b*) suppress odd harmonics in the rectifier output

(*c*) stabilise dc level of the output voltage

(*d*) remove ripples from the rectified output.

17. In a rectifier, larger the value of shunt capacitor filter

(*a*) larger the peak-to-peak value of ripple voltage

(*b*) larger the peak curent in the rectifying diode

(*c*) longer the time that current pulse flows through the diode

(*d*) smaller the dc voltage across the load.

18. In a LC filter, the ripple factor,

(*a*) increases with the load current

(*b*) increase with the load resistance

(*c*) remains constant with the load curent

(*d*) has the lowest value.

19. The main reason why a bleeder resistor is used in a dc power supply is that it

(*a*) keeps the supply ON (*b*) improves voltage regulation

(*c*) improves filtering action (*d*) both (*b*) and (*c*).

20. Which rectifier requires four diodes ?

(*a*) half-wave voltage doubler (*b*) full-wave voltage doubler

(*c*) full-wave bridge circuit (*d*) voltage quadrupler.

21. If, by mistake, ac source in a bridge rectifier, is connected across the dc terminals, it will burn out and hence short .... diodes.

(*a*) one (*b*) two

(*c*) three (*d*) four

22. The dc output polarity from a half-wave rectifier can be reversed by reversing
    (*a*) the diode (*b*) transformer primary
    (*c*) transformer secondary (*d*) both (*b*) and (*c*).
23. In a half-wave rectifier if a resistance equal to load resistance is connected in parallel with the diode, then
    (*a*) output voltage would be halved
    (*b*) output voltage would be doubled
    (*c*) circuit will stop rectifying
    (*d*) output voltage will remain unchanged.
24. If the input supply frequency is 50 Hz, the output ripple frequency of a bridge rectifier is .... Hz.
    (*a*) 100 (*b*) 75 (*c*) 50 (*d*) 25
25. The average dc power obtained from a load resistor (R_L) of a half-wave rectifier is :
    (*a*) $I_m^{\,2} R_L$ (*b*) $\dfrac{I_m^2 R_L}{\pi^2}$
    (*c*) $\dfrac{I_m R_L}{\pi}$ (*d*) $I_m R_L$
    (*Bharthiar Univ. April 2004*)
26. Which type of rectifier requires four diodes :
    (*a*) half-wave rectifier
    (*b*) full-wave bridge rectifier
    (*c*) full-wave rectifier
    (*d*) none of the above. (*Bharthiar Univ. April 2004*)
27. The efficiency of a full-wave rectifier is :
    (*a*) 40.6% (*b*) 81.2%
    (*c*) 50% (*d*) 100%
    (*Bharthiar Univ. April 2004*)

ANSWERS

1. (*d*)	**2.** (*b*)	**3.** (*a*)	**4.** (*a*)
5. (*c*)	**6.** (*b*)	**7.** (*b*)	**8.** (*c*)
9. (*b*)	**10.** (*a*)	**11.** (*c*)	**12.** (*b*)
13. (*c*)	**14.** (*b*)	**15.** (*a*)	**16.** (*d*)
17. (*b*)	**18.** (*c*)	**19.** (*d*)	**20.** (*b*)
21. (*d*)	**22.** (*a*)	**23.** (*c*)	**24.** (*a*)
25. (*b*)	**26.** (*b*)	**27.** (*b*)	

20

Chapter

Regulated Power Supplies

Outline

20.1. Introduction

It is a well known fact, that all the electronic devices (*i.e.*, diodes, transistors, integrated circuits etc.) and circuits require a constant d.c. voltage for their operation. The d.c. voltage can be supplied from dry cells or batteries. But these are expensive as compared to conventional regulated d.c. power supplies. The regulated d.c. power supply is made by converting the domestic a.c. supply to d.c. supply. The essential elements, which constitute a regulated d.c. power supply, are given below :

1. Rectifier,
2. Filter, and
3. Voltage regulator.

The rectifier and filter has already been discussed in the last chapter. Now we shall study the voltage regulator in the following pages.

20.2. Voltage Regulator

We have already discussed in the last article that a regulated d.c. power supply consists of three elements namely rectifier, filter and voltage regulator. The rectifier is the front or the first element of a d.c. power supply, the filter is the second (or intermediate element) and the voltage regulator is the last element. The concept of a voltage regulator may be understood from the following discussion. (Refer to Fig. 20.1.)

A rectifier is used to convert the ac output of the transformer to pulsating d.c. (Fig. 20.2 (*a*)). This pulsating dc is then applied to a filter which reduces the variations in the rectifier ouput (Fig. 20.2 (*b*).

Fig. 20.1.

Consider a combination of full-wave rectifier and a filter circuit connected to an a.c. input voltage source. When the a.c. input voltage, at the rectifier input, changes above or below its normal value, it will cause a change in d.c. voltage produced at the filter output. A similar change in d.c. voltage may also occur, when the load resistance connected at the filter output changes above or below its normal value. It means that d.c. output voltage fluctuates (or changes) whenever the a.c. input voltage or the load resistance varies above or below the normal values.

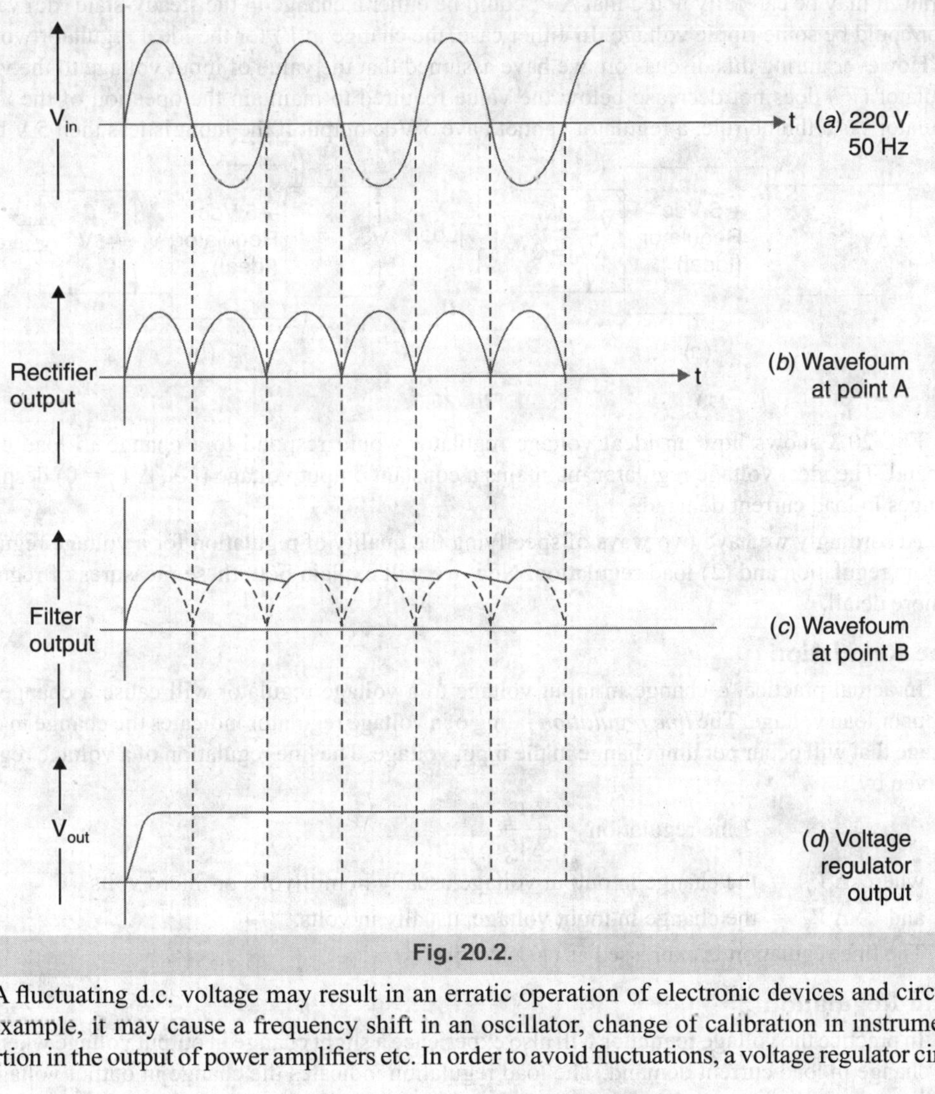

Fig. 20.2.

A fluctuating d.c. voltage may result in an erratic operation of electronic devices and circuits. For example, it may cause a frequency shift in an oscillator, change of calibration in instruments, distortion in the output of power amplifiers etc. In order to avoid fluctuations, a voltage regulator circuit is used. The voltage regulator is connected between the filter and the load. Its function is to maintain a constant output d.c. voltage inspite of the a.c. input voltage fluctuations or changes in load resistance values.

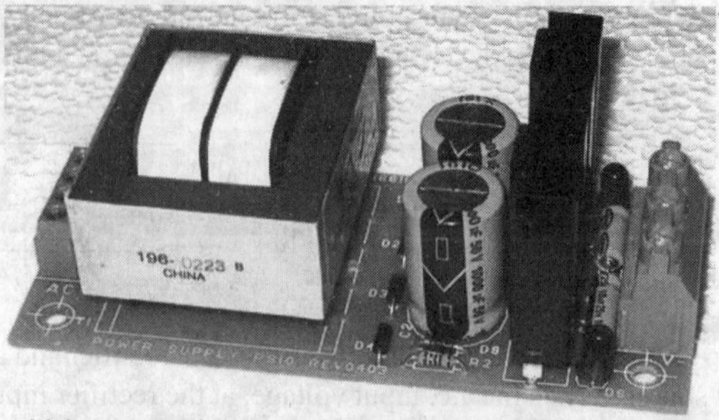

Voltage regulator provides a constant DC output voltage that is essentially independent of the input voltage, output load current and temperature.

20.3. Voltage Regulation

An ideal voltage regulator will maintain a constant dc output voltage regardless of changes in either its input voltage or its load current demand. For example consider a +5 *Vdc* regulator shown in Fig 20.3. A change in the regulator input voltage shown as ΔV_S in Fig. 20.3 (*a*) would not appear at the regular output. It may be carefully noted that ΔV_S could be either a change in the steady-state (dc) value of V_S or could be some ripple voltage. In either case, the change in V_L for the ideal regulator would be 0V. However during this discussion, we have assumed that the value of input voltage to the voltage regulator (V_S) does not decrease below the value required to maintain the opeation of the voltage regulator. As a thumb rule, a regulator cannot have 5 Vdc output if the input is less then 5 V d.c.

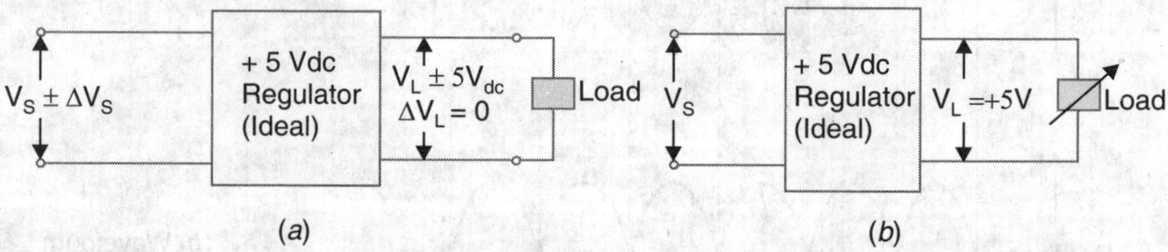

Fig. 20.3.

Fig. 20.3 shows how an ideal voltage regulator would respond to a change in load current demand. The ideal voltage regulator maintains a constant output voltage (*i.e.*, $\Delta V_L = 0$) despite the changes in load current demand.

Accordingly we have two ways of specifying the quality of regulation for a voltage regulator : (1) line regulation and (2) load regulation. Now we will explain both these measures of regulation in more detail.

Line Regulation

In actual practice, a change in input voltage to a voltage regulator will cause a change in its output or load voltage. The *line regulation* rating of a voltage regulator indicates the change in output voltage that will occur per unit change in the input voltage. The line regulation of a voltage regulator is given by,

$$\text{Line regulation} = \frac{\Delta V_L}{\Delta V_S}$$

where ΔV_L = the change in output voltage usually in millivolts or micro volts

and ΔV_S = the change in input voltage, usually in volts.

The line regulation is expressed as (V/V), or $\mu V/V$.

Load Regulation

In practice the voltage regulator will also experience a slight change in output voltage when there is a change in load current demand. The load regulation indicates the change in output voltage that will occur per unit change in load current. Mathematically, the load regulation may be expressed as,

$$\text{Load regulation} = \frac{V_{NL} - V_{FL}}{\Delta I_L}$$

where V_{NL} = no load output voltage.
V_{FL} = the full-load output voltage and
ΔI_L = the change in load current demand

The load regulation of a voltage regulator is expressed in terms of μ*V*/μA.

Ideally speaking, whether it is line regulation or load regulation, the lower the two values are, the better is the quality of regulation.

Some people prefer another way of expressing voltage regulation as given below. Here the voltage regualtion is expressed as the ability of the regulator to maintain a constant output voltage inspite of a.c. input voltage fluctuations and changes in load resistance.

Mathematically,

$$\% \text{ regulation} = \frac{V_{NL} - V_{FL}}{V_{FL}} \times 100$$

where V_{NL} = no load voltage and
V_{FL} = full-load voltage

Example 20.1.

The change in output voltage of a voltage regulator is measured at 100 μV when the input voltage changes by 5 V. Determine the value of line regulation for this regulator.

Solution.

Given : $\Delta V_L = 100\ \mu V = 100 \times 10^{-6}$ V and $\Delta V_{in} = 5$ V

We know that live regulation,

$$\text{Line regulation} = \frac{\Delta V_L}{\Delta V_S} = \frac{100 \times 10^{-6}}{5} = 20 \times 10^{-6} \text{ V/V}$$

$$= 20\ \mu\text{V/V} \textbf{ Ans.}$$

Example 20.2.

A voltage regulator has a line regulator of 1.4 μV/V. Determine the change in its output voltage of the input voltage changes byu 10 V.

Solution.

Given : Line regualtion = 1.4 μV/V = 1.4×10^{-6} V/V and $\Delta V_S = 10$ V.

Let ΔV_L = change in output voltage of the regulator,

We know that live regulation,

$$1.4 \times 10^{-6} = \frac{\Delta V_L}{\Delta V_S} = \frac{\Delta V_L}{10}$$

$$\therefore \quad \Delta V_L = (1.4 \times 10^{-6}) \times 10 = 14 \times 10^{-6} \text{ V or } 14\ \mu\text{V} \textbf{ Ans.}$$

Example 20.3.

A voltage regualtor is rated for an output (or load) current of I_L = 0 to 40 mA. Under no load conditions, the output voltage from the circuit is 8V. Underfull load conditions, the output voltage drops to 7.995 V. What is the value of load regulation for this voltage regulator.

Solution.

Given : $\Delta I_L = 40$ mA; $V_{NL} = 8$ V and $V_{FL} = 7.995$ V

We know that load regulation of a voltage regulator,

$$\text{load regualtion} = \frac{V_{NL} - V_{FL}}{\Delta I_L} = \frac{8 - 7.995}{40 \text{ mA}} = \frac{0.005}{40 \text{ mA}}$$

$$= \frac{5 \text{ mV}}{40 \text{ mA}} = 0.125 \text{ mV/mA}$$

$$= 125 \ \mu\text{V/mA. } \textbf{Ans.}$$

Example 20.4.

A voltage regulator has a load regulation of 10 μV/mA. Under no-load conditions, the output voltage is 5 V. Calculate the voltage under full-load condition, if the load current is 20 mA.

Solution.

Given : Load regulation = 10 μV/mA: $V_{NL} = 5$ V, $\Delta I_L = 20$ mA.

Let V_{FL} = output voltage at full-load.

We know that the load regulation of a voltage regulation,

$$10 \ \mu\text{V/mA} = \frac{V_{NL} - V_{FL}}{\Delta I_L} = \frac{5 - V_{FL}}{20 \text{ mA}}$$

$$\therefore \quad V_{FL} = 5 \text{ V} - (10 \ \mu\text{V/mA}) \times 20 \text{ mA}$$

$$= 5 \text{ V} - 200 \ \mu\text{V} = 4.9998 \text{ V } \textbf{Ans.}$$

Example. 20.5.

A 10 V regulated d.c. power supply has a regulation of 0.002 per-cent. Find the magnitude of variation in output voltage.

Solution.

Given $V_0 = 10$ volts; Regulation = 0.002% = 0.00 002.

We know that a regulation of 0.00 002 means, that the output voltage will vary within the limits of 0.002% of 10 V. Therefore,

magnitude of variation = 0.00 002 × 10 = 0.00 002 V = 0.2 mV

This indicates that the variation in output voltage is 0.2 mV. However, since the variation is equal on the increasing or decreasing side, therefore the magnitude of variation may also be expressed as ± 0.1 mV. **Ans.**

20.4. Types of Voltage Regulators

Although there are many types of voltage regulators, yet the following ones are important from the subject point of view :

1. Zener diode shunt regulator.
2. Transistor shunt regulator.
3. Transistor series regulator.
4. Controlled transistor series regulator.
5. Transistor current regualtor
6. Op-amp series regualtor
7. Op-amp shunt regulator
8. Switching regulators
9. Monolithic regulator.

20.5. Zener Diode Shunt Regulator

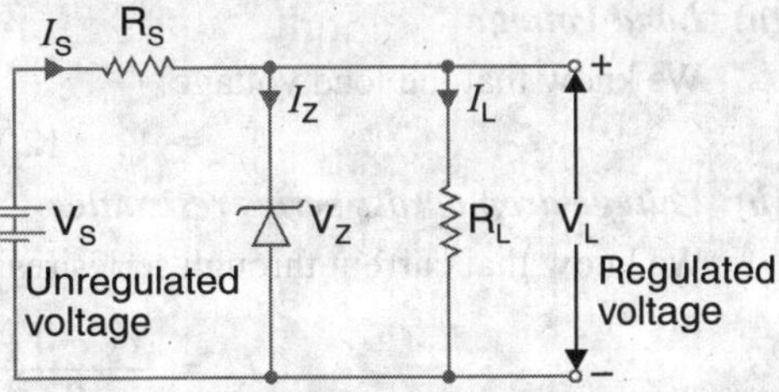

Fig. 20.4. Zener diode shunt regulator.

Fig. 20.4 shows a circuit of zener diode shunt regulator. Since the zener is connected in parallel (or shunt) with the load, therefore the circuit is known as shunt regulator. A resistance, (R_S) is connected in series with the zener to limit current in the circuit. Therefore the resistance R_S is also known as series current limiting resistor.

The output voltage (V_L) is taken across the load resistance (R_L). For proper operation, the input voltage (V_S) must be greater than the zener voltage (V_Z). This ensures that zener operates in the reverse breakdown region.

The input current (*i.e.*, current through the limiting resistor) is given by the relation,

$$I_S = \frac{V_S - V_Z}{R_S}$$

where V_S = The d.c. input voltage to the regulator circuit, and

V_Z = The zener voltage.

The ideal zener diode acts as a constant voltage source of voltage (V_Z). However, a practical zener diode has a finite value of resistance called zener resistance (r_Z). Because of the zener resistance, there is a voltage drop across it which is equal to $I_Z \cdot r_Z$. Therefore the voltage across the terminals of the zener diode (which is equal to load voltage),

$$V_L = V_Z + I_Z \cdot r_Z \qquad ...(i)$$

If the zener resistance is negligible, then the load voltage,

$$V_L = V_Z \qquad ...(ii)$$

and the current through the load resistance, is given by the relation,

$$I_L = \frac{V_L}{R_L} \qquad ...(iii)$$

We know that input current is the sum of zener current and load current *i.e.*,

$$I_S = I_Z + I_L \quad \text{or} \quad I_Z = I_S - I_L$$

Example 20.6.

Fig. 20.5 shows the circuit of a zener diode shunt regulator.

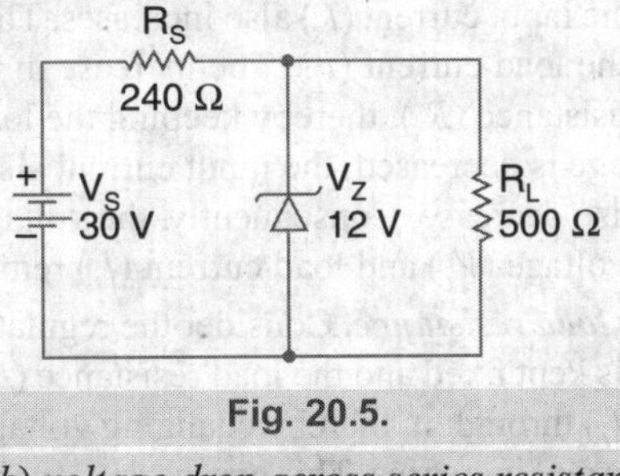

Fig. 20.5.

Find (a) the load voltage (b) voltage drop across series resistance and (c) current through the zener diode. (R.G.P.V., June 2011)

Solution.

Given : V_S = 30 volts; R_S = 240 Ω; V_Z = 12 volts and R_L = 500 Ω.

(*a*) ***Load Voltage***

We know that the load voltage

$$V_L = V_Z = 12 \text{ V } \textbf{Ans.}$$

(*b*) ***Voltage drop across series resistance***

We know that current through series resistance R_S.

$$I_S = \frac{V_S - V_Z}{R_S} = \frac{30 - 12}{240} = 0.075 \text{ A}$$

and voltage drop across series resistance R_S.

$$= I_S \cdot R_S = 0.075 \times 240 = 18 \text{ V } \textbf{Ans.}$$

(*c*) ***Current through zener diode***

We also know that current through R_L,

$$= I_L = \frac{V_L}{R_L} = \frac{12}{500} = 0.024 \text{ A}$$

and current through zener diode,

$$I_Z = I_S - I_L = 0.075 - 0.024 = 0.051 \text{ A } \textbf{Ans.}$$

20.6. Working of Zener Diode Shunt Regulator

It may be discussed under the two heads, namely regulation with varying input voltage and regulation with varying load resistance.

1. ***Regulation with varying input voltage.*** Consider the regulator circuit as shown in Fig. 20.6(*a*). Here the load resistance (R_L) is kept fixed and the input voltage (V_S) varies within the limits.

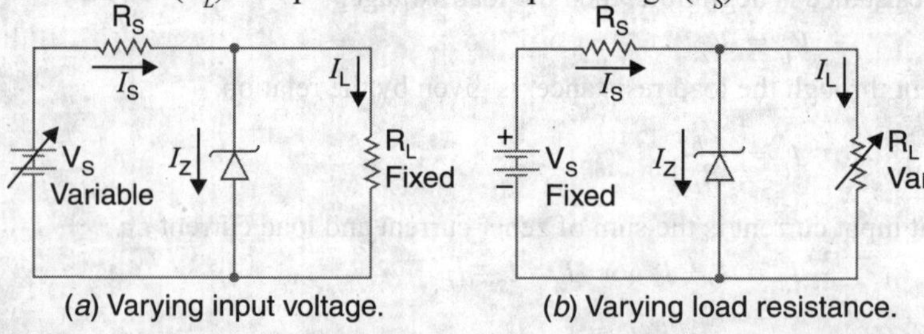

Fig. 20.6. Working of zener diode shunt regulator.

As the input voltage increases, the input current (I_S) also increases. This increases the current through zener diode, without affecting the load current (I_L). The increase in input current will also increase the voltage drop across series resistance (R_S), thereby keeping the load voltage (V_L) as constant. On the other hand, if the input voltage is decreased, the input current also decreases. As a result of this, the current through zener will also decrease. Consequently, the voltage drop across series resistance will be reduced. Thus the load voltage (V_L) and load current (I_L) remains constant.

2. ***Regulation with varying load resistance.*** Consider the regulator circuit as shown in Fig. 20.6 (*b*). Here the input voltage (V_S) is kept fixed and the load resistance (R_L) varies. The variation of load resistance changes the current (I_L) through it, thereby changing voltage (V_L) across it. When the load resistance decreases, the load current increases. This causes the zener current to decrease. As a result of this, the input current and the voltage drop across series resistance remains constant. Thus the load voltage (V_L) is also kept constant. On the other hand, if the load resistance increases, the load current decreases. As a result of this, the zener current increases. This again keeps the values of input current and voltage drop across series resistance as constant. Thus the load voltage remains constant.

Example 20.7.

A zener diode shown in Fig. 20.7 is used in the circuit has a breakdown voltage of 5.1 V, the zener resistance (r_z) of 10 Ω and the minimum and maximum values of current through it are 1 mA and 15 mA respectively.

Determine the minimum and maximum values of the input voltage, which can be regulated by the zener shunt regulator.

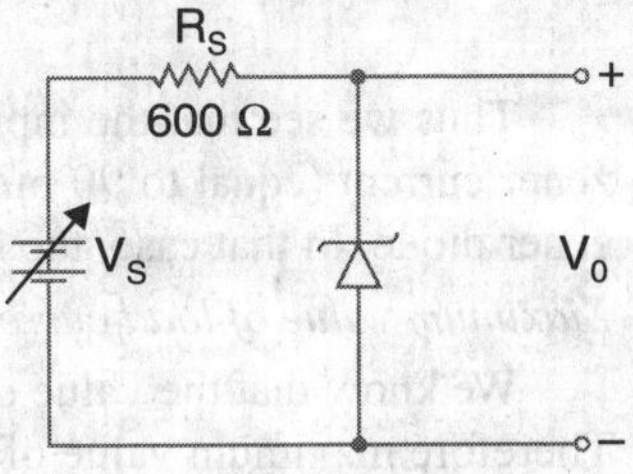

Fig. 20.7.

Solution.

Given : $V_Z = 5.1$ volts; $r_Z = 10$ Ω; $I_{Z(min)} = 1$ mA $= 1 \times 10^{-3}$ A; $I_{Z(max)} = 15$ mA $= 15 \times 10^{-3}$ A and $R_S = 600$ Ω.

Fig. 20.8 shows the equivalent circuit of the given zener diode shunt regulator.

It may be noted that the zener diode is replaced by a battery of 5.1 V in series with a resistance of 10 Ω.

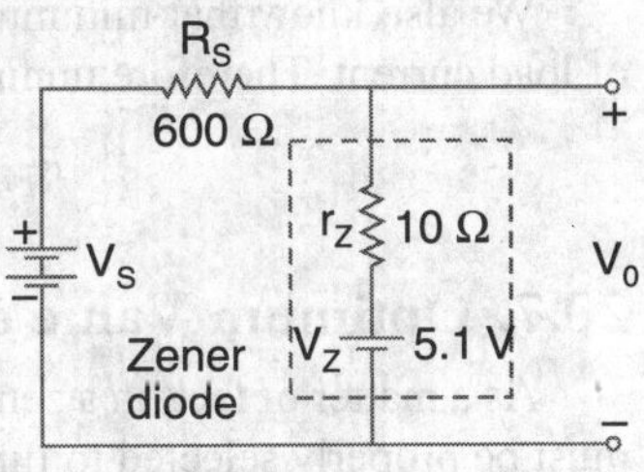

Fig. 20.8.

Minimum value of input voltage

We known that minimum value of output voltage,

$$V_{O\,(min)} = V_Z + I_{Z(min)} \cdot r_Z = 5.1 + (1 \times 10^{-3})\ 10 = 5.11 \text{ V}$$

and the minimum value of input voltage,

$$V_{S\,(min)} = I_{Z\,(min)} \cdot R_S + V_{O\,(min)} = (1 \times 10^{-3}) \times 600 + 5.11 = 5.71 \text{ V } \textbf{Ans.}$$

Maximum value of input voltage

We know that maximum value of output voltage,

$$V_{O\,(max)} = V_Z + I_{Z\,(max)} \cdot r_Z = 5.1 + (15 \times 10^{-3}) \times 10 \text{ V}$$
$$= 5.25 \text{ V}$$

and the maximum value of the input voltage,

$$V_{S\,(max)} = I_{Z\,(max)} \cdot R_S + V_{O\,(max)} = (15 \times 10^{-3}) \times 600 + 5.25 \text{ V}$$
$$= 14.25 \text{ V } \textbf{Ans.}$$

Example 20.8.

Determine the minimum and maximum values of load current for which the zener diode shunt regulator shown in Fig. 20.9 will maintain regulation.

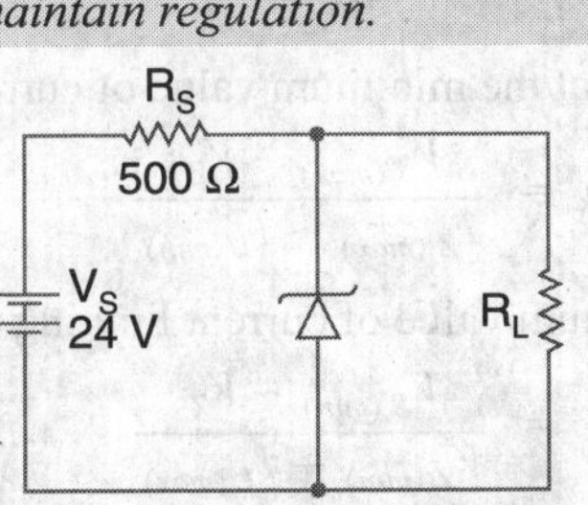

Fig. 20.9.

What is the minimum value of load resistance (R_L) that can be used in the circuit. The zener diode has a V_Z equal to 12 V, $I_{Z\,(min)}$ of 3 mA, $I_{Z\,(max)}$ of 90 mA and r_Z is zero.

Solution.

Given : $V_S = 24$ volts; $R_S = 500$ Ω; $V_Z = 12$ volts; $I_{Z\,(min)} = 3$ mA; $I_{Z\,(max)} = 90$ mA and $r_Z = 0$.

Minimum value of load current

We know that current through series resistance,

$$I_S = \frac{V_S - V_Z}{R_S} = \frac{24 - 12}{500} = 0.024 \text{ A} = 24 \text{ mA}$$

Thus we see that the input current (equal to 24 mA) is much less than the maximum value of zener current (equal to 90 mA). Therefore, we can assume that all the input current flows through zener diode. In that case, the minimum value of load current is zero. **Ans.**

Maximum value of load current

We know that the value of load current will be maximum, when the zener current is minimum. Therefore maximum value of load current,

$$I_{L\,(max)} = I_S - I_{Z\,(min)} = 24 - 3 = 21 \text{ mA } \textbf{Ans.}$$

Minimum value of load resistance

We also know that minimum value of load resistance is obtained by using the maximum value of load current. Therefore minimum value of load resistance,

$$R_{L\,(min)} = \frac{V_Z}{I_{L\,(max)}} = \frac{12}{21} = 0.571 \text{ k}\Omega = 571\ \Omega \textbf{ Ans.}$$

20.7. Optimum Value of Current Limiting Resistor

As a matter of fact, in a zener diode shunt regulator circuit, the value of current limiting resistor must be properly selected to fulfil the following two requirements :

1. When the input voltage is minimum and the load current is maximum, sufficient current must be supplied to keep the zener diode within its breakdown region (or regulating region).
2. When the input voltage is maximum and the load current is minimum, the zener current must not increase the maximum rated value.

The optimum value of current limiting resistor can be determined by using the following two equations for maximum and minimum values of zener current :

$$I_{Z\,(max)} = \frac{V_{S\,(max)} - V_Z}{R_S} - I_L\,(min) \qquad \text{...}(i)$$

$$I_{Z\,(min)} = \frac{V_{S\,(min)} - V_Z}{R_S} - I_L\,(max) \qquad \text{...}(ii)$$

From equation (*i*), we find that the minimum value of current limiting resistor,

$$R_{S\,(min)} = \frac{V_{S\,(max)} - V_Z}{I_{Z\,(max)} + I_{L\,(min)}}$$

and from equation (*ii*), the maximum value of current limiting resistor,

$$R_{S\,(max)} = \frac{V_{S\,(min)} - V_Z}{V_{Z\,(min)} + I_{L\,(max)}}$$

The value of current limiting resistor (R_S) should be chosen in such a way that its value is between $R_{S\,(max)}$ and $R_{S\,(min)}$ *i.e.*,

$$R_{S\,(min)} < R_S < R_{S\,(max)}$$

Example 20.9.

For a zener shunt regulator if $V_Z = 10$ V, $R_S = 1$ kΩ, $R_L = 2$ kΩ and the input voltage varies from 22 to 40 V, find the maximum and minimum values of zener current.

Solution.

Given : $V_Z = 10$ volts; $R_S = 1\ k\Omega = 1 \times 10^3\ \Omega$; $R_L = 2\ k\Omega = 2 \times 10^3\ \Omega$; $V_{S\,(min)} = 22$ volts and $V_{S\,(max)} = 40$ volts.

Maximum value of zener current

We know that load current,

$$I_L = \frac{V_Z}{R_L} = \frac{V_Z}{R_L} = \frac{100}{2 \times 10^3} = 5 \times 10^{-3}\ \text{A} = 5\ \text{mA}$$

and maximum value of zener current,

$$I_{Z\,(max)} = \frac{V_{S\,(max)} - V_Z}{R_S} - I_{L\,(min)} = \frac{40 - 10}{1 \times 10^3} - 5 \times 10^{-3} \quad ...(\because I_{L\,(min)} = I_L)$$

$$= 25 \times 10^{-3}\ \text{A} = 25\ \text{mA}\ \textbf{Ans.}$$

Minimum value of zener current

We also know that minimum value of zener current,

$$I_{Z\,(min)} = \frac{V_{S(min)} - V_Z}{R_S} - I_{L(max)} = \frac{22 - 10}{1 \times 10^3} - (5 \times 10^{-3}) \quad ...(\because I_{L\,(max)} = I_L)$$

$$= 7 \times 10^{-3}\ \text{A} = 7\ \text{mA}\ \textbf{Ans.}$$

Example 20.10.

A 10 V zener diode is used to regulate the voltage across a variable load resistor. The input voltage varies between 13 and 16 V. The load current (I_L) varies between 10 and 85 mA. The minimum zener current is 15 mA. Find (i) the maximum value of R_S and (ii) the maximum power dissipated by the zener diode, using this value of R_S. *(A.M.I.E., Summer, 1987)*

Solution.

Given : $V_Z = 10$ volts; $V_{S\,(min)} = 13$ volts; $V_{S\,(max)} = 16$ volts; $I_{L\,(min)} = 10$ mA $= 10 \times 10^{-3}$ A; $I_{L\,(max)} = 85$ mA $= 85 \times 10^{-3}$ A; and $I_{Z\,(min)} = 15$ mA $= 15 \times 10^{-3}$ A.

(i) Maximum value of R_S

We know that maximum value of R_S,

$$R_{S\,(max)} = \frac{V_{S\,(min)} - V_Z}{I_{Z\,(min)} + I_{L\,(max)}} = \frac{13 - 10}{(15 \times 10^{-3} + 85 \times 10^{-3})} = 30\ \Omega\ \textbf{Ans.}$$

(ii) Maximum power dissipated

We also know that maximum value of zener current,

$$I_{Z\,(max)} = \frac{V_{S\,(max)} - V_Z}{R_{S\,(max)}} - I_{L\,(min)}$$

$$= \frac{16 - 10}{30} - (10 \times 10^{-3})$$

$$= 0.2 - 0.01 = 0.19\ \text{A}$$

and maximum power dissipated by the zener,

$$P_{Z\,(max)} = I_{Z\,(max)} \times V_Z = 0.19 \times 10 = 1.9\ \text{W}\ \textbf{Ans.}$$

Example 20.11.

A zener regulator has to handle supply voltage variation from 19.5 volts to 22.5 volts. Find the magnitude of regulating resistance, if the load resistance is 6 kΩ. The zener diode has the following specifications :

Breakdown voltage = 18 volts

Minimum zener diode current = 2 μA

Maximum power dissipation = 60 mW

Zener resistance = 20 Ω. *(A.M.I.E., Winter, 1986)*

Solution.

Given : $V_{S(min)} = 19.5$ volts; $V_{S(max)} = 22.5$ volts; $R_L = 6$ kΩ; $V_Z = 18$ volts; $I_{Z(min)} = 2\ \mu A = 2 \times 10^{-6}$ A; $P_{Z(max)} = 60$ mW $= 60 \times 10^{-3}$ W and $r_Z = 20\ \Omega$.

Let $I_{Z(max)}$ = Maximum value of zener current.

We know that maximum value of power dissipation $P_{Z(max)}$,

$$60 \times 10^{-3} = I^2_{Z(max)} \times r_Z = I_{Z(max)}{}^2 \times 20$$

$$\therefore \quad I_{Z(max)}{}^2 = \frac{60 \times 10^{-3}}{20} = 3 \times 10^{-3}$$

and $$I_{Z(max)} = \sqrt{(3 \times 10^{-3})} = 0.0548 \text{ A} = 54.8 \times 10^{-3} \text{ mA}$$

We also know that load current,

$$I_L = \frac{V_L}{R_L} = \frac{V_Z}{R_L} = \frac{18}{6 \times 10^3} = 3 \times 10^{-3} \text{ A} = 3 \text{ mA}$$

∴ Maximum value of regulating (or series resistance),

$$R_{S(max)} = \frac{V_{S(min)} - V_Z}{I_{Z(min)} + I_{L(max)}} = \frac{19.5 - 18}{(2 \times 10^{-6}) + (3 \times 10^{-3})} \quad ...(\because I_{L(max)} = I_L)$$

$$= 500\ \Omega$$

and minimum value of regulating resistance,

$$R_{S(min)} = \frac{V_{S(max)} - V_Z}{I_{Z(max)} + I_{L(min)}} = \frac{22.5 - 18}{(54.8 \times 10^{-3}) + 3 \times 10^{-3}} \quad ...(\because I_{L(min)} = I_L)$$

$$= 77.8\ \Omega \quad ...(i)$$

∴ Magnitude of regulating resistance should be between 77.8 Ω and 500 Ω, *i.e.*, $77.8 < R_S < 500\ \Omega$. **Ans.**

Example 20.12.

Fig. 20.10 shows the circuit of a zener diode voltage regulation.

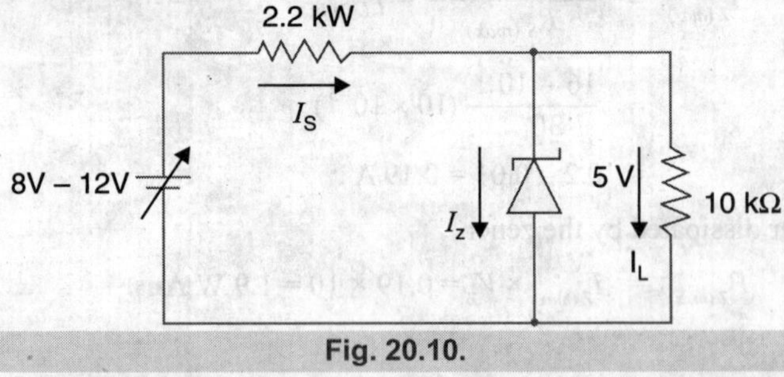

Fig. 20.10.

Find the maximum and minimum value of zener current in the circuit.

(Banglore Univ. B.sc. Electronics, May 2001)

Solution.

Given : $V_{S(min)} = 8$ V; $V_{S(max)} = 12$ V; $R_S = 2.2$ kΩ; $V_Z = 5$ V and $R_L = 10$ kΩ.

We know that minimum value of input current,

$$I_{S(min)} = \frac{V_{S(min)} - V_Z}{R_S} = \frac{8-5}{2.2 \text{ k}} = 1.364 \text{ mA}$$

and maximum value of input current,

$$I_{S(max)} = \frac{12-5}{2.2 \text{ k}} = 3.182 \text{ mA}$$

We also know that the value of load current,

$$I_L = \frac{V_L}{R_L} = \frac{5}{10 \text{ k}} = 0.5 \text{ mA}$$

∴ The minimum value of zener current,

$$I_{Z(min)} = I_{S(min)} - I_L = 0.364 - 0.5 = 0.864 \text{ mA } \textbf{Ans.}$$

and maximum value of zener current,

$$I_{Z(max)} = I_{S(max)} - I_L = 3.182 - 0.5 = 2.682 \text{ mA } \textbf{Ans.}$$

Example 20.13.

Design the zener regulator for the following specifications :

Output voltage = 5 V

Load current = 20 mA

Zener power dissipation $P_{Z(max)}$ = 500 mW

Input voltage = 12 ± 3 V *(Banglore Univ. Jan. 2004)*

Solution:

Given: $V_O = V_L = 5$ V; $I_L = 20$ mA; $P_{Z(max)} = 500$ mW and $V_S = 12$ V ± 3 V.

Let $I_{Z(max)}$ = Max. value of current through the zener.

We know that maximum power dissipation of a zener, $P_{Z(max)}$,

$$500 \times 10^{-3} = V_Z \cdot I_{Z(max)} = 5\, I_{Z(max)}$$

$$\therefore \quad I_{Z(max)} = \frac{500 \times 10^{-3}}{5} = 100 \times 10^{-3} = 100 \text{ mA}$$

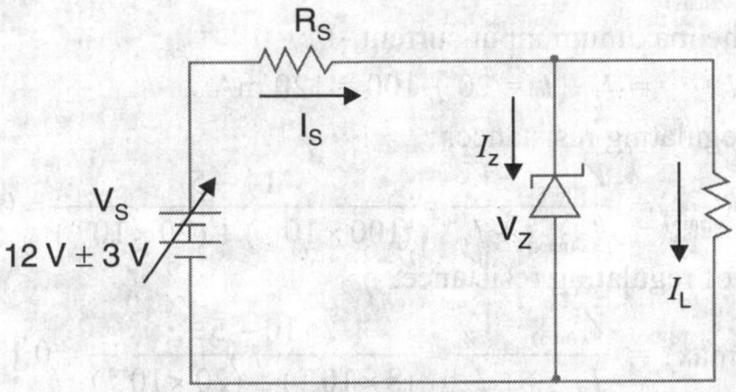

Fig. 20.11.

We also know that the maximum input current, (refer to Fig. 20.11),

$$I_{S(max)} = I_L + I_Z$$
$$= 20 + 100 = 120 \text{ mA}$$

and the minimum value of regulating resistance,

$$R_{S(min)} = \frac{V_{S(max)} - V_Z}{I_{Z(max)} + I_{L(min)}} = \frac{15 - 5}{120 \text{ mA}} = 83.3\ \Omega \textbf{ Ans.}$$

Notice that if we fix the value of $R_S = 83.3\ \Omega$, the minimum value of zener current,

$$I_Z = \frac{V_{S(min)} - V_Z}{R_S} - I_{L(max)} = \frac{9 - 5}{83.3} - (20 \times 10^{-3}) = 28.02 \text{ mA}$$

So when the input voltage varies from the normal 12 V within the range of ± 3 V, the zener current will vary from 28.02 mA to maximum 100 mA if the current limiting resistor is at least 83.3 Ω. But to be more on the safeside, you can select $R_S = 100\ \Omega$. **Ans.**

Note : In the present case, the statement does not give any value for $I_{Z(min)}$. That value will put a maximum limit on the value of R_S. So the chosen valve of R_S should not be very high as well.

Example 20.14.

Design a Zener regulator shown in Fig. 20.12. Already there as (Ex. 20.13) for following specifications: load current $I_L = 20$ mA, output voltage $V_o = 5$ V, Zener wattage $P_Z = 500$ mW, Input voltage $V_i = 12 \pm 2$ V and $I_{Z(min)} = 8$ mA. *(Nagpur Univ. 2004)*

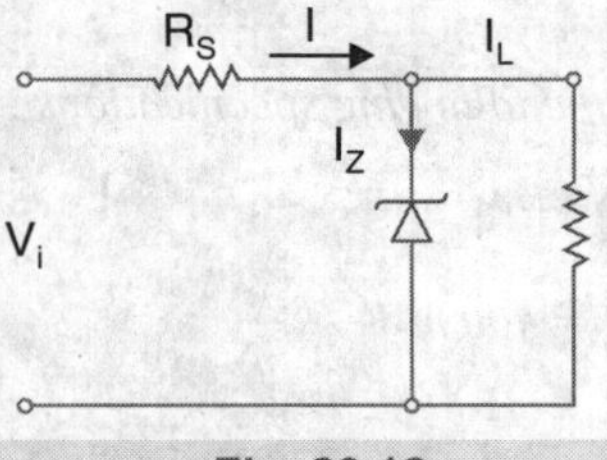

Fig. 20.12.

Solution.

Given : $V_o = 5$ V; $I_L = 20$ mA $= 20 \times 10^{-3}$ A; $P_Z = 500$ mW $= 500 \times 10^{-3}$ W; $V_s = 12 \pm 2$ and $I_{Z(min)} = 8$ mA $= 8 \times 10^{-3}$ A.

Let $I_{Z(max)}$ = Max. value of current through the zener

We know that maximum power dissipation of a zener, $P_{Z(max)}$

$$500 \times 10^{-3} = V_Z I_{Z(max)} = 5\, I_{Z(max)}$$

$$\therefore \quad I_{Z(max)} = 100 \times 10^{-3} \text{ A} = 100 \text{ mA}$$

We also know that the maximum input current,

$$I_{S(max)} = I_L + I_Z = 20 + 100 = 120 \text{ mA}$$

The minimum value of regulating resistance,

$$R_{s(min)} = \frac{V_{s(max)} - V_Z}{I_{Z(max)} + I_L} = \frac{14 - 5}{(100 \times 10^{-3}) + (20 \times 10^{-3})} = 0.075 \times 10^3 = 75\ \Omega$$

and the maximum value of regulating resistance,

$$\text{Rs(max)} = \frac{V_{s(min)} - V_Z}{I_{Z(min)} + I_L} = \frac{10 - 5}{(8 \times 10^{-3}) + (20 \times 10^{-3})} = 0.1785 \times 10^3 = 178.5\ \Omega$$

The value of regulating resistance R_s must be greater than 75 Ω and less than 178.5 Ω **Ans.**

Example 20.15.

A load of 1 kΩ is connected across a 10 V zener regulator as shown in Fig. 20.13. The zener current can vary between 5 mA to 55 mA while maintaining the voltage constant. Find the minimum and maximum voltage level at the input. (*Nagpur Univ. 2004*)

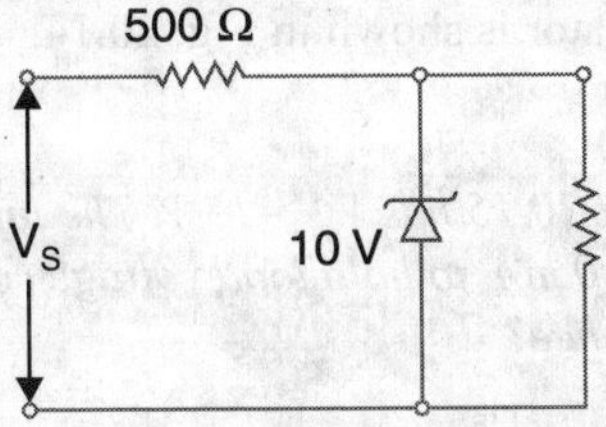

Fig. 20.13.

Solution.

Given : $R_S = 500\ \Omega$; $V_Z = 10$ V; $R_L = 1$ kΩ; $I_{Z(min)} = 5$ mA; $I_{Z(max)} = 55$ mA.

We know that the load current,

$$I_L = \frac{V_Z}{R_L} = \frac{10}{1 \times 10^3} = 10 \times 10^{-3} \text{ A or } 10 \text{ mA}$$

Let $V_{S(min)}$ = Minimum value of input voltage

$V_{S(max)}$ = Maximum value of input voltage

We know that minimum value of input voltage,

$$V_{S(min)} = [\{I_L + I_{Z(min)}\} \times R_S] + V_Z$$
$$= [\{(10 \times 10^{-3}) + (5 \times 10^{-3})\} \times 500] + 10$$
$$= 17.5 \text{ V}$$

and the maximum value of input voltage,

$$V_{S(max)} = [\{(I_L + I_{Z(max)}\} \times 500] + 10$$
$$= [\{(10 \times 10^{-3}) + (55 \times 10^{-3})\} \times 500] + 10$$
$$= 42.5 \text{ V}$$

∴ The minimum value of voltage level at input is 17.5 V and the maximum, 42.5 V **Ans.**

Example 20.16.

Draw and design a zener voltage regulator circuit to provide output of 12 V for a maximum load current of 0.5 A, when the input voltage variation is 15 to 18 V. (*PTU., May 2007*)

Solution.

Given: $V_{out} = 12$ V; $I_{L(max)} = 0.5$ A; $V_{S(min)} = 15$ V and $V_{S(max)} = 18$ V.

We know that Zener voltage regulator,

$$I_{S(min)} = \frac{V_{s(\min)} - V_{out}}{R_s}$$

$$R_S = \frac{V_{s(\min)} - V_{out}}{I_{s(\min)}} = \frac{15 - 12}{0.506} = 5.9\ \Omega$$

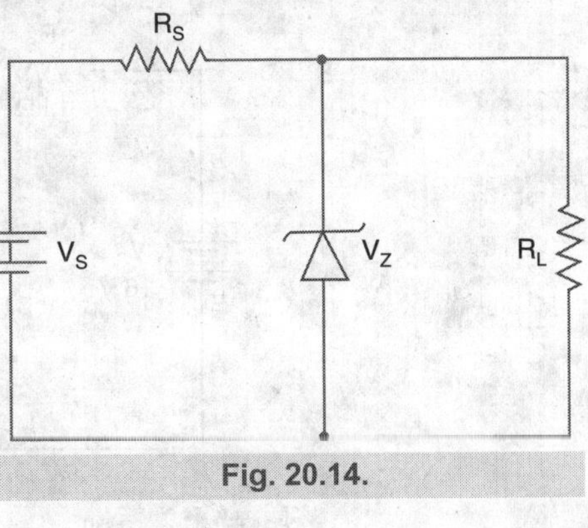

Fig. 20.14.

We also know that,

$$R_L = \frac{V_{out}}{I_L} = \frac{12}{0.5} = 24\ \Omega$$

And $\quad V_Z = V_{out} = 12\text{V}$

Design of a zener voltage regulator is shown in Fig. 20.14.

Example 20.17.

A zener regulator shown in Fig. 20.15 has V_Z =15 V. The input voltage may vary from 22 V to 40 V and load current from 20 to 100 mA to hold load voltage constant under all conditions. What should be the value of series resistance? (PTU., May 2005)

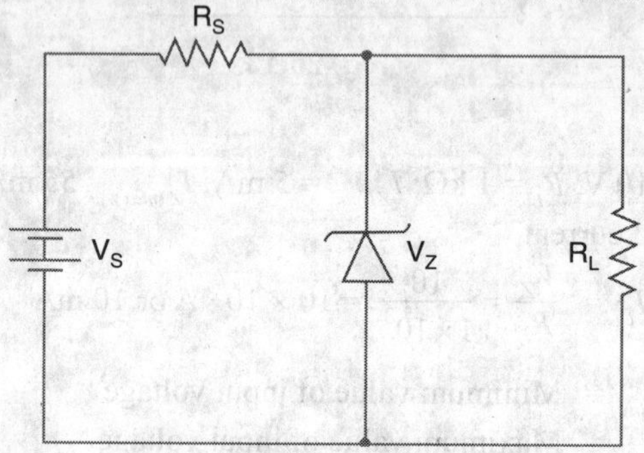

Fig. 20.15.

Solution.

Given: $V_Z = 12$ V; $V_{S(\min)} = 22$ V; $V_{S(\max)} = 40$ V; $I_{L(\min)} = 20$ mA; $I_{L(\max)} = 100$ mA and consider $I_Z = 6$ mA.

We know that the value of series resistance in case of maximum voltage variation,

$$R_S = \frac{V_{S(\max)} - V_{out}}{I_Z + I_{L(\max)}}$$

$$= \frac{40 - 15}{6 \times 10^{-3} + 100 \times 10^{-3}} = \frac{25}{106 \times 10^{-3}} = 236\ \Omega \quad \textbf{Ans.}$$

Example 20.18.

For the circuit shown below in Fig. 20.16, determine V_L, V_R, I_Z and P_Z. (GBTU., 2010–2011)

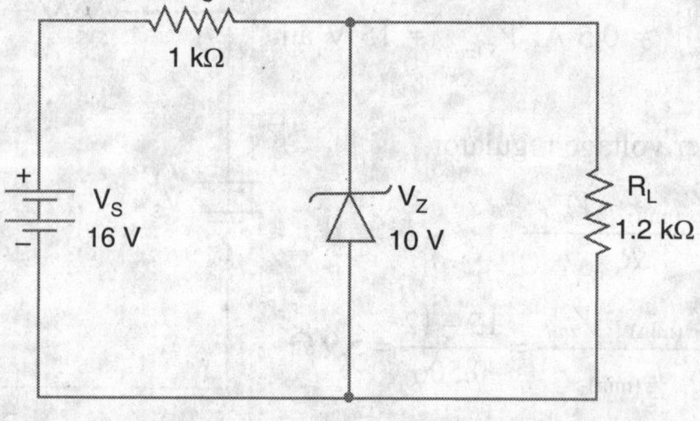

Fig. 20.16.

Solution.

Given: $V_S = 16$ V; $R_S = 1$ kΩ $= 1 \times 10^3$ Ω; $V_Z = 10$ V and $R_L = 1.2$ kΩ $= 1.2 \times 10^3$ Ω.

We know that the load voltage,

$$V_L = V_Z = 10 \text{ V Ans.}$$

and the current through the series resistances,

$$I_S = \frac{V_S - V_Z}{R_S} = \frac{16 - 10}{1 \times 10^3} = 6 \times 10^{-3} \text{ A}$$

We know that the voltage drop across the series resistances,

$$V_R = I_S R_S = (6 \times 10^{-3}) \times (1 \times 10^3) = 6 \text{ V } \textbf{Ans.}$$

and the current through the load resistance,

$$I_L = \frac{V_L}{R_L} = \frac{10}{1.2 \times 10^3} = 8.33 \times 10^{-3} \text{ A}$$

We also known that current through the zener diode,

$$I_Z = I_S - I_L = (6 \times 10^{-3}) - (8.33 \times 10^{-3})$$
$$= -2.33 \times 10^{-3} \text{ A} = -2.33 \text{ mA Ans.}$$

and the power dissipated by zener diode,

$$P_Z = V_Z I_Z = 10 \times (2.33 \times 10^{-3}) = 2.33 \times 10^{-3} \text{ W} = 23.3 \text{ mW } \textbf{Ans.}$$

Example 20.18.

For the zener diode network of following Fig. 20.17, determine V_L, V_R, I_Z and P_Z

(RGPV, Dec. 2008)

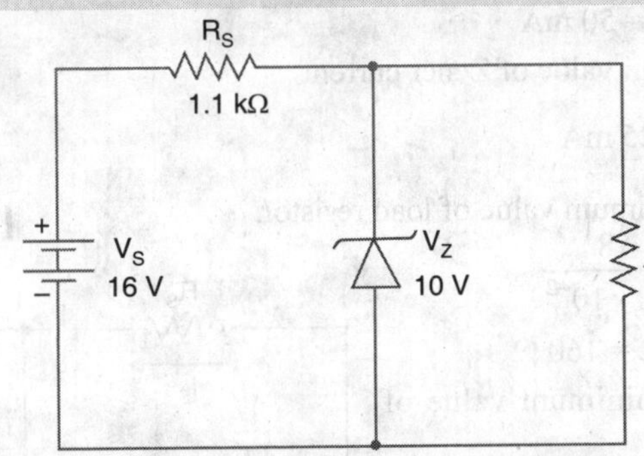

Fig. 20.17.

Solution.

Given: $V_S = 16$ V; $R_S = 1.1$ kΩ; $V_Z = 10$ volts and $R_L = 1.8$ kΩ.

We know that the load voltage

$$V_L = V_Z = \frac{R_L . V_S}{R_S + R_L} = \frac{(1.8 \times 10^3) \times 16}{(1.1 \times 10^3) + (1.8 \times 10^3)} = 10 \text{ V } \textbf{Ans.}$$

as $V_S < V_Z$, therefore zener is in OFF state as shown in Fig. 20.18.

$$I_S = \frac{V_S - V_Z}{R_S} = \frac{16 - 10}{1.1 \times 10^2} = 5.45 \text{ mA}$$

$$I_L = \frac{V_Z}{R_L} = \frac{10}{1.8 \times 10^2} = 5.56 \text{ mA.}$$

Notice that $I_L > I_S$, thus zener is OFF as shown in Fig. 20.18

then,

$$I_S = I_L \text{ and } I_Z = 0$$

from the circuit,

$$V_R = V_S - V_L = 16 - 10 = 6 \text{ V} \quad \textbf{Ans.}$$

We know that the,

$$P_Z = V_Z I_Z = 10 \times 0 = 0 \quad \textbf{Ans.}$$

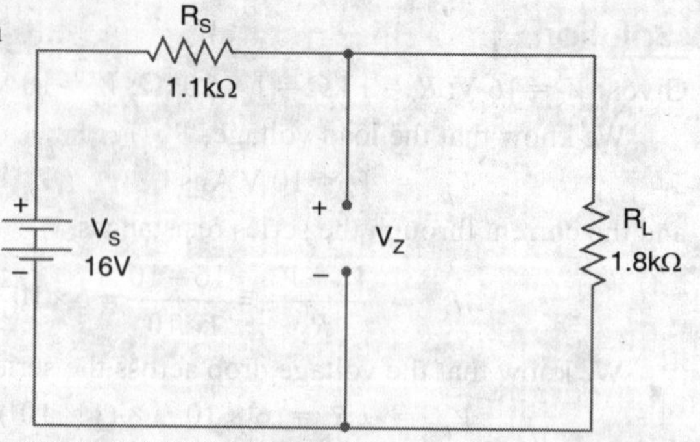

Fig. 20.16.

Example 20.20.

Design and draw a Zener regulator circuit to meet the following specification:
Load voltage =8 V
Input voltage = 30 V
Load current = 0-50 mA
$I_{Z(MIN)}$ = 5 mA
PZ = 1 watts

(Anna University., Nov/Dec 2007)

Solution.

Given: $V_L = 8$ V; $V_{in} = 30$ V; $I_L = 0 - 50$ mA; $I_{Z\,(min)} = 5$ mA $= 5 \times 10^{-3}$ A and $P_Z = 1$ W

Input voltage is constant,

$$V_{in(min)} = V_{in(max)} = 30 \text{ V}$$

$$I_{L(min)} = 0 \text{ A}, \quad I_{L(max)} = 50 \text{ mA}$$

We know that the maximum value of Zener current

$$I_{Z(max)} = \frac{P_Z}{V_o} = \frac{1}{8} = 125 \text{ mA}$$

We also know that the minimum value of load resistor,

$$R_{L(min)} = \frac{V_o}{I_{L(max)}} = \frac{8}{50 \times 10^{-2}}$$

$$= 0.16 \times 10^3 \,\Omega = 160 \,\Omega$$

We also know that the minimum value of current limiting resistor,

$$R_{S(max)} = \frac{V_{in(min)} - V_o}{I_{L(max)} + I_{Z(min)}}$$

$$= \frac{30 - 8}{50 \times 10^{-3} + 5 \times 10^{-3}} = 400\Omega$$

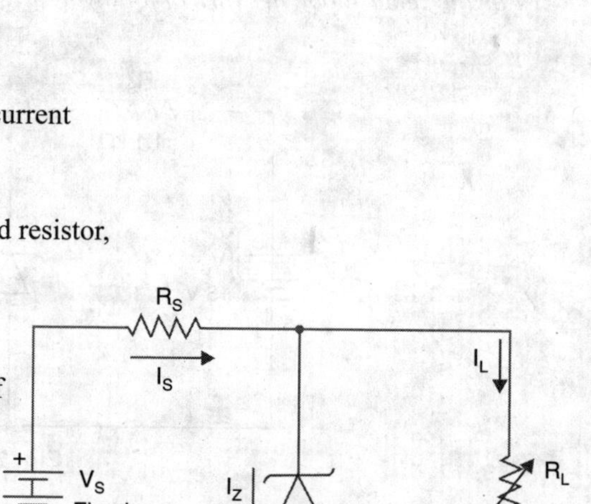

Fig. 20.19.

We also know that the minimum value of current limiting resistor,

$$R_{S(min)} = \frac{V_{in(max)} - V_o}{I_{L(min)} + I_{Z(max)}}$$

$$= \frac{30 - 8}{0 + (125 \times 10^{-3})} = 176\,\Omega$$

For any value of R_S between 176 Ω to 400 Ω, the circuit shown in Fig. 20.19 will act as a regulator circuit.

Example 20.21.

For the network of Fig. 20.20, determine the range of R_L and I_L that will result in V_Z being maintained at 10 V. Also determine the maximum wattage rating of diode. (GBTU., 2011, 2012)

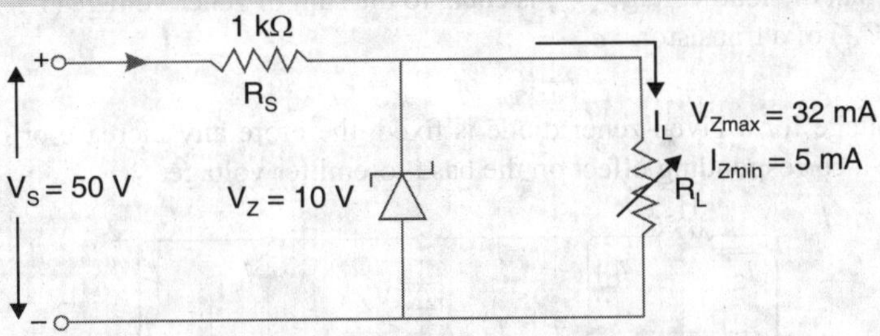

Fig. 20.20.

Solution.

Given: V_S = 50 V; R_S = 1 kΩ = 1 × 10³ Ω; V_Z = 10 V; $I_{Z(max)}$ = 32 mA = 32 × 10⁻³ A and $I_{Z(min)}$ = 32 mA = 32 × 10³ A

We know that the current through the series resistance,

$$I_S = \frac{V_S - V_Z}{R_S} = \frac{50 - 10}{1 \times 10^3} = 40 \times 10^{-3}\text{ A} = 40\text{ mA}$$

and the minimum value of load current,

$$I_{L(min)} = I_S - I_{Z(min)} = 40 - 32 = 8\text{ mA} \textbf{ Ans.}$$

We also know that the maximum value of load resistance,

$$R_{L(max)} = \frac{V_Z}{I_{L(min)}} = \frac{10}{8 \times 10^{-3}} = 1.25 \times 10^3\,\Omega = 1.25\text{ k}\Omega \textbf{ Ans.}$$

and the maximum value of load current,

$$I_{L(max)} = I_S - I_{Z(max)} = 40 - 5 = 35\text{ mA Ans.}$$

We also know the minimum value of load resistance,

$$R_{L(min)} = \frac{V_Z}{I_{L(min)}} = \frac{10}{35 \times 10^{-3}} = 285.71\,\Omega \textbf{ Ans.}$$

We also know that the maximum wattage rating of diode,

$$P_{D(max)} = V_Z \times I_{Z(max)} = 10 \times (32 \times 10^{-3})$$

$$= 320 \times 10^{-3} = 320\text{ mW} \textbf{ Ans.}$$

20.8. Disadvantages of Zener-diode Shunt Regulator

Following are the disadvantages of a zener-diode shunt regulator :

1. The maximum load current, which can be supplied to load resistor (R_L) is limited to $I_{Z\,(max)} - I_{Z\,(min)}$, which is usually of few milliamperes.
2. A large amount of power is wasted in the zener-diode and the series resistance (R_S) in comparison with the load power.
3. The regulation factor and the output resistance are not very low.

These drawbacks are overcome in a transistorized voltage regulators, which will be discussed in the following pages.

20.9. Transistor Shunt Regulator

Fig. 20.21 shows a circuit of a transistor shunt regulator. Since the transistor is connected in shunt with the load resistance, therefore the circuit is known as shunt regulator. It may be noted from the regulator circuit, that the load voltage (V_L) is equal to the sum of zener voltage (V_Z) and the base-to-emitter voltage (V_{BE}) of a transistor, *i.e.*,

$$V_L = V_Z + V_{BE} \text{ or } V_{BE} = V_L - V_Z \qquad ...(i)$$

Since the voltage for a given zener diode is fixed, therefore any increase or decrease in load voltage will have a corresponding effect on the base-to-emitter voltage.

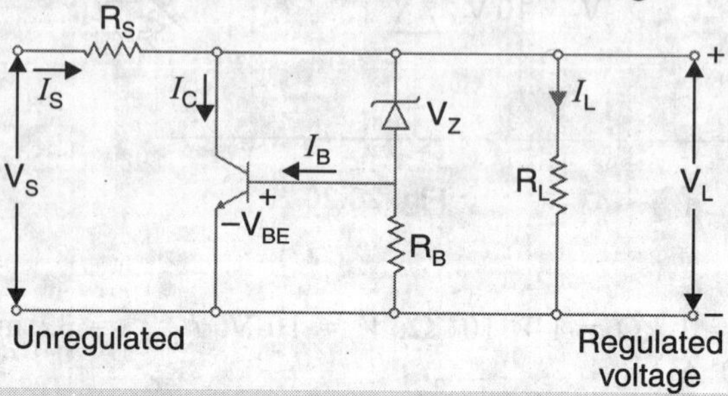

Fig. 20.21. Transistor shunt regulator.

The working of a transistor shunt regulator may be understood by supposing that the unregulated input voltage (V_S) increases. Because of this, the load voltage also increases. As a result of this and from equation (*i*) above, we find that base-to-emitter voltage also increases. And the base current of the transistor (I_B) increases. Due to this, the collector current of a transistor (I_C) also increases. This causes the input current (I_S) to increase, which in turn increases the voltage drop across series resistance (V_{RS}). Consequently, the load voltage decreases. It is true because sum of the voltage drop across series resistance (V_{RS}) and the load voltage (V_L) is equal to input voltage (V_S) at all times. *i.e.*,

$$V_S = V_{R_S} + V_L \text{ or } V_L = V_S - V_{R_S} \qquad ...(ii)$$

The above discussion may be summarized in the form of an equation as give below :

$$V_S \uparrow \rightarrow V_L \uparrow \rightarrow V_{BE} \uparrow \rightarrow I_C \uparrow V_{R_S} \uparrow \rightarrow V_L \downarrow$$

Here arrows in the upward direction show the increase in the value, while the arrow in the downward direction indicates the decrease in value. The transistor shunt regulator is suitable for the loads in which the load current varies from a finite minimum value to a finite maximum value.

Example 20.22.

Determine the regulated voltage and circuit currents for the shunt regulator of Fig. 20.22.

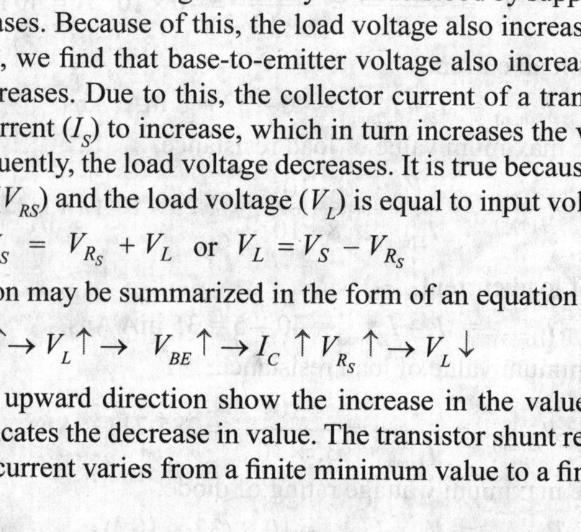

Fig. 20.22.

Solution.

We know that load voltage,

$$V_L = V_Z + V_{BE} = 10 + 0.7 = 10.7 \text{ V}$$

and the value of load current,

$$I_L = \frac{V_L}{R_L} = \frac{10.7}{100} = 0.107 \text{ A} = 107 \text{ mA}$$

The current through, R_S,

$$I_S = \frac{V_S - V_L}{R_S} = \frac{15 - 10.7}{33} = 0.1303 \text{ A} = 130.3 \text{ mA}$$

∴ The collector current,

$$I_C = I_S - I_L = 130.3 - 107 = 23.3 \text{ mA}$$

The base current for the transistor (or current through the zener),

$$I_B = I_Z = \frac{I_C}{\beta} = \frac{23.3 \text{ mA}}{100} = 0.233 \text{ mA} = 233 \text{ μA } \textbf{Ans.}$$

20.10. Transistor Series Regulator

Fig. 20.23 shows a circuit of a transistor series regulator. Since the transistor is connected in series with the load, therefore the circuit is known as a series regulator. This circuit is also called as emitter follower, because its transistor behaves as an emitter follower. In other words, the transistor behaves as a variable resistance whose value is determined by the amount of base current. It may be noted from the circuit that the load voltage (V_L) is equal to the difference of zener voltage (V_Z) and the base-to-emitter voltage (V_{BE}), *i.e.*,

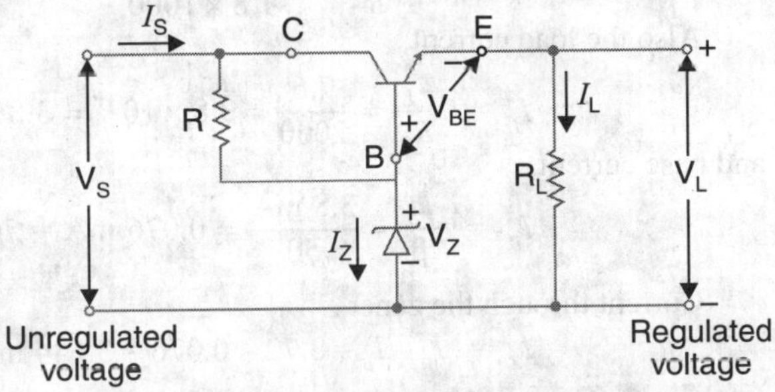

Fig. 20.23. Transistor series regulator.

$$V_L = V_Z - V_{BE} \quad \text{or} \quad V_{BE} = V_Z - V_L \qquad ...(i)$$

The working of a transistor series regulator may be understood by supposing that value of load resistance is increased. Because of this, the load current decreases and the load voltage tends to increase. However, as seen from the equation (*i*), that any increase in the load voltage will decrease the base-to-emitter voltage (V_{BE}) because the zener voltage is fixed. As a result of this, the forward bias of the transistor is reduced, which reduces its level of conduction. This increases the collector-to-emitter voltage of the transistor. This in turn will slightly decrease the input current to compensate for the increase in the value of load resistance, so that load voltage may remain at a constant value. The circuit action may be summarized in the form of the following equation :

$$V_L \uparrow \rightarrow V_{BE} \downarrow \rightarrow V_{CE} \uparrow \rightarrow V_Z \downarrow$$

The output of a transistor series regulator is approximately equal to zener voltage (V_Z). This regulator can also be used for larger load currents.

Example 20.23.

Calculate the output voltage and Zener current in the regulator circuit shown in Fig. 20.24.

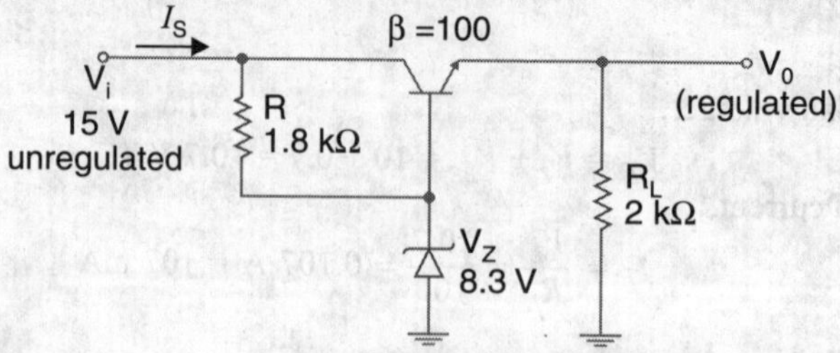

Fig. 20.24.

Solution.

Given : $V_i = V_S = 15$ V; $V_Z = 8.3$ V; $\beta = 100$, $R = 1.8$ kΩ and $R_L = 2$ k.Ω

We know that the load voltage,

$$V_L = V_Z - V_{BE} = 8.3 - 0.7 = 7.6 \text{ V}$$

and collector-to-emitter voltage,

$$V_{CE} = V_S - V_L = 15 - 7.6 = 7.4 \text{ V}$$

∴ Current through resistor, R

$$I_R = \frac{V_S - V_Z}{R} = \frac{15 - 8.3}{1.8 \times 1000} = 3.72 \times 10^{-3} \text{ A} = 3.72 \text{ mA}$$

Also the load current,

$$I_L = \frac{V_L}{R_L} = \frac{7.6}{2000} = 3.8 \times 10^{-3} = 3.8 \text{ mA}$$

and base current,

$$I_B = \frac{I_L}{\beta} = \frac{3.8 \text{ mA}}{50} = 0.076 \text{ mA} = 76 \text{ μA} \qquad (\because I_C \approx I_E = I_L)$$

∴ Current through the zener,

$$I_Z = I_R - I_B = 3.72 - 0.076 = 3.644 \text{ mA } \textbf{Ans.}$$

Example 20.24.

A transistor series regulator is to supply a load current of 0 to 2 A at 12 V. The unregulated d.c. input varies from 15 V to 20 V. Assuming the transistor $\beta = 100$, $V_{BE} = 0.5$ V, zener voltage, $V_Z = 12.5$ V and $I_{Z\,(min)} = 1$ mA, find the maximum value of the resistance (R) and power rating of resistance R, zener diode and transistor.

Solution.

Given : $I_{L(min)} = 0$; $I_{L\,(max)} = 2$ A; $V_L = 12$ volts ; $V_{S\,(min)} = 15$ volts; $V_{S\,(max)} = 20$ volts; $\beta = 100$; $V_{BE} = 0.5$ volts; $V_Z = 12.5$ volts; and $I_{Z\,(min)} = 1$ mA.

Fig. 20.25 shows the circuit diagram of the transistor series regulator.

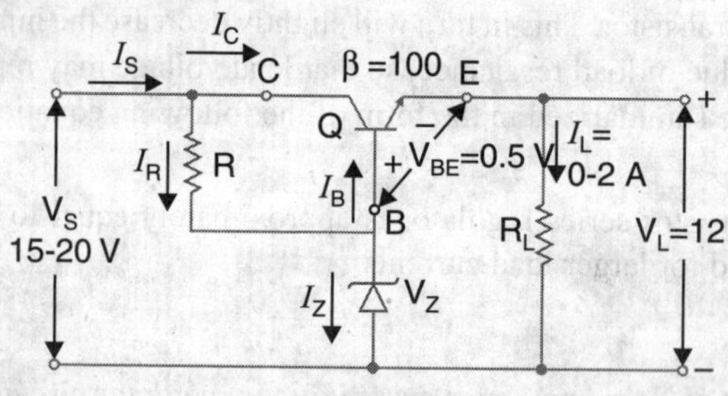

Fig. 20.25.

Maximum value of resistance R

We know that the maximum value of base current of a series transistor,

$$I_{B\,(max)} = \frac{I_{C\,(max)}}{\beta} = \frac{I_{L\,(max)}}{\beta} = \frac{2}{100} = 0.02 \text{ A} = 20 \text{ mA}$$

and current through the resistance R,

$$I_R = I_{B\,(max)} + I_{Z\,(min)} = 20 + 1 = 21 \text{ mA} = 21 \times 10^{-3} \text{ A}$$

$\therefore$ Maximum value of resistance R,

$$R_{max} = \frac{V_{S\,(min)} - V_Z}{I_R} = \frac{15 - 12.5}{21 \times 10^{-3}} = 119 \ \Omega \textbf{ Ans.}$$

Maximum power dissipation (or power rating) of the zener diode

We know that maximum value of zener current,

$$I_{Z\,(max)} = \frac{V_{S\,(max)} - V_Z}{I_R} = \frac{20 - 12.5}{119} = 63 \text{ mA} = 63 \times 10^{-3} \textbf{ Ans.}$$

and the maximum power dissipation (or power rating) of a zener diode,

$$P_{Z\,(max)} = V_Z \times I_{Z\,(max)} = 12.5 \times 63 \times 10^{-3} = 0.79 \text{ W} \textbf{ Ans.}$$

Maximum power dissipation (or power rating) of resistance R

We know that maximum power dissipation of resistance R,

$$P_{R\,(max)} = (V_{S\,(max)} - V_Z) \cdot I_{Z\,(max)} = (20 - 12.5) \times 63 \times 10^{-3} \text{ W} = 0.47 \text{ W} \textbf{ Ans.}$$

Maximum power dissipation (or power rating) of the transistor

We also know that maximum value of collector-to-emitter voltage of a transistor.

$$V_{CE\,(max)} = V_{S\,(max)} - V_L = 20 - 12 = 8 \text{ V}$$

and maximum power dissipation of the transistor,

$$P_{D\,(max)} = V_{CE\,(max)} \times I_{L\,(max)} = 8 \times 2 = 16 \text{ W} \textbf{ Ans.}$$

Example 20.25.

Draw and design a series transistor voltage regulator circuit to provide output of 9 V for a maximum load current of 1.0 A, when the input voltage variation is 15 to 18 V and transistor gain is 50. *(PTU., Dec 2006)*

Solution.

Given : $V_L = 9$ V; $I_L = 1.0$ A; $V_{S\,(min)} = 15$ V; $V_{S\,(max)} = 18$ V and $\beta = 50$

We know that,

$$I_B = \frac{I_L}{\beta} = \frac{1.0}{50} = 0.02$$

and that output voltage,

$$V_Z = V_L + V_{BE} = 9 + 0.5 = 9.5 \text{ V} \qquad (\because V_{BE} = 0.5 \text{ V})$$

and the voltage drop in resistor,

$$V_R = V_{S(max)} - V_Z = 18 - 9.5 = 8.5 \text{ V}$$

and the current through the resistance R,

$$I_R = I_B + I_Z = 0.02 + 0.01 = 0.03 \text{ A}$$

and

$$R = \frac{V_R}{I_R} = \frac{8.5}{0.03} = 283 \ \Omega$$

Zener diode of breakdown voltage 9.5 V and resistor of 283 Ω is required. Circuit of a series transistor voltage regulator is shown in Fig. 20.26.

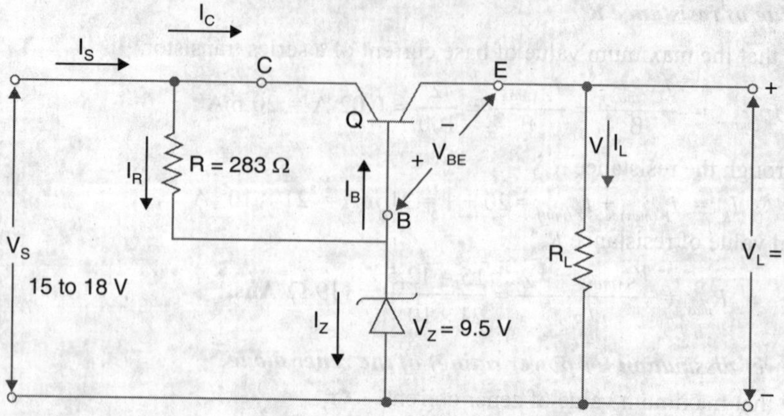

Fig. 20.26.

20.11. Controlled Transistor Series Regulator

Fig. 20.27 shows a controlled transistor series regulator circuit. It is also known as feedback voltage regulator circuit. This circuit employs two transistors namely Q_1 and Q_2. The Q_1 transistor is a series regulator transistor. It is also called control transistor or pass transistor. And the Q_2 transistor functions both as a voltage comparator and d.c. amplifier. Therefore it is known as an amplifier transistor.

The action of a controlled transistor series regulator may be studied by supposing that there is an increase in load voltage either due to input voltage variation or change of load (V_{BE}) of Q_2 transistor. Because of this, the collector current of Q_2 transistor (I_{C_2}) increases. As a result of this, total current (I_{ST}) flowing through the resistance (R_S) increases. This, in turn decreases the collector current of Q_2 transistor. Since the base of Q_1 transistor is tied to the collector of Q_2 transistor,

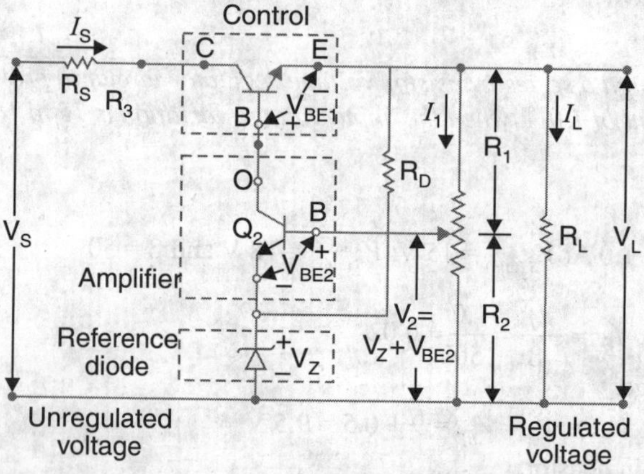

Fig. 20.27. Controlled transistor series regulator.

therefore base voltage of Q_1 transistor, with respect to ground, decreases. This in turn decreases the base-to-emitter voltage (V_{BE1}) of the transistor. Therefore the collector-to-emitter voltage (V_{CE1}) of Q_1 transistor has to increase in order to maintain the same emitter current. If the change in collector-to-emitter voltage of a transistor can be made equal to the variation in input voltage (*i.e.*, ΔV_S), the load voltage will remain constant. Since the collector-to-base voltage of Q_1 transistor is equal to its collector-to-emitter voltage, we can assume that if the value of ΔV_S is dropped across the resistance

R_3, the load voltage will remain constant. It may be noted that by changing the tapping on the potentiometer ($R_1 - R_2$), the d.c. load voltage of the regulator circuit can be varied. The circuit action may be summarized in the form of following equation :

$$V_L \uparrow \rightarrow V_{BE_2} \uparrow \rightarrow I_{C_2} \uparrow \rightarrow V_{BE1} \downarrow \rightarrow V_{CE1} \uparrow \rightarrow V_L \downarrow$$

The regulated load voltage for a controlled transistor series regulator is given by the relation,

$$V_L = (V_Z + V_{BE_2})\left(1 + \frac{R_1}{R_2}\right) = V_2 \times \left(1 + \frac{R_1}{R_2}\right)$$

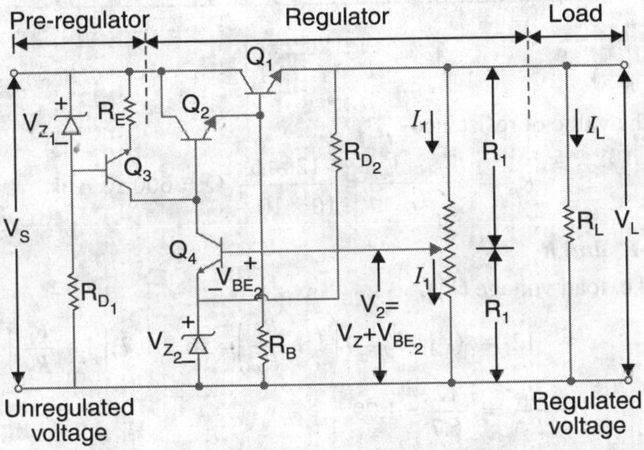

Fig. 20.28 A series regulator using pre-regulator and Darlington pair.

The current gain of the series transistor can be increased by using a Darlington Pair. This can be increased further by increasing the value of the resistance R_3, which is achieved by using a pre-regulator. A pre-regulator is a constant-current source and therefore its equivalent resistance (R_3) is very large. Such a circuit, which incorporates both the above features is shown in Fig. 20.28.

Example 20.26.

Design a series regulated power supply to provide a nominal output voltage of 12 volts and supply load current of 200 milliamperes. The nominal unregulated input voltage is 30 volts and source resistance is 10 ohms. Ratings of the active devices used are as follow:

Control transistor (silicon) : $h_{fe} = 150$; $h_{ie} = 1.5$ kW ; $V_{BE} = 0.7$ volt; $I_C = 10$ mA

Amplifier transistor (silicon) : $h_{fe} = 100$; $h_{ie} = 250$ W ; $V_{BE} = 0.7$ volt

Reference zener diode : $V_R = 6$ volts; $R_Z = 10$ ohms at $I_Z = 20$ mA

Sketch the complete circuit and obtain reasonable values of the components you use.

Solution.

Given : $V_L = 12$ volts; $I_L = 200$ mA; $V_S = 30$ volts; $R_S = 10\ \Omega$; $\beta_1 (= h_{fe1}) = 150$; $I_{C1} = 10$ mA; $V_{BE1} = 0.7$ volts; $\beta_2 (= h_{fe2}) = 100$; $V_{BE2s} = 0.7$ volts; $V_Z (= V_R) = 6$ volts; $R_Z = 10\ \Omega$ and $I_Z = 20$ mA; $I_D = 10$ mA $= 10 \times 10^{-3}$ A; $I_1 = 10$ mA $= 10 \times 10^{-3}$ A.

Fig. 20.29 shows the complete circuit of a transistor series regulator power supply. Let us choose the collector current of Q_2 transistor (I_{C_2})and the current I_D equal to 10 mA each. Now we shall use these values to determine the values of the resistances R_D, R_1 and R_2.

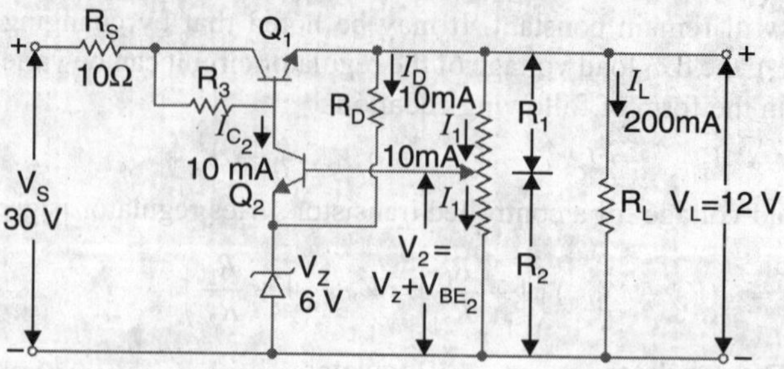

Fig. 20.29.

Value of resistance R_D

We know that the value of resistance

$$R_D = \frac{V_L - V_Z}{I_D} = \frac{12 - 6}{10 \times 10^{-3}} \Omega = 600 \ \Omega \ \textbf{Ans.}$$

Value of resistance R_1 and R_2

We know that d.c. load voltage (V_L),

$$12 = (V_Z + V_{BE_2})\left(1 + \frac{R_1}{R_2}\right) = (6 + 0.7)\left(1 + \frac{R_1}{R_2}\right) = 6.7\left(1 + \frac{R_1}{R_2}\right)$$

$$\therefore \quad 1 + \frac{R_1}{R_2} = \frac{12}{6.7} = 1.79$$

$$\text{or} \quad \frac{R_1}{R_2} = 1.79 - 1 = 0.79 \qquad \text{... } (i)$$

We also know that base current of Q_2 transistor,

$$I_{B2} = \frac{I_{C_2}}{\beta_2} = \frac{10}{100} = 0.1 \text{ mA}$$

Since we require the value of current through potentiometer ($R_1 - R_2$) to be very large than that of base current (*i.e.*, I_1 I_{B2}), therefore we select the current I_1 equal to 10 mA. Then the voltage drop across resistance R_2,

$$V_2 = V_Z + V_{BE_2} = 6 + 0.7 = 6.7 \text{ V}$$

$$\therefore \quad \text{Value of resistance, } R_1 = \frac{V_Z - V_2}{I_1} = \frac{12 - 6.7}{10 \times 10^{-3}} \Omega = 530 \ \Omega \ \textbf{Ans.}$$

Now substituting the value of R_1 in equation (*i*),

$$\frac{530}{R_2} = 0.79$$

$$\text{or} \quad R_2 = \frac{530}{0.79} = 670 \ \Omega \ \textbf{Ans.}$$

Value of resistance R_3

We know that base current of Q_1 transistor,

$$I_{B_1} = \frac{I_L + I_1 + I_D}{h_{fe_1}} = \frac{200 + 10 + 10}{150} = 1.5 \text{ mA} = 1.5 \times 10^{-3} \text{ A}$$

and current through resistance R_3,

$$I = I_{B_1} + I_{C_2} = 1.5 + 10 .5 \text{ mA} = 11.5 \times 10^{-3} \text{ A}$$

$$\therefore \quad \text{Value of resistance } R_3 = \frac{V_S - (V_{BE1} + V_L)}{I} = \frac{30 - (0.7 + 12)}{11.5 \times 10^{-3}} \ \Omega$$

$$= 1500 \ \Omega \text{ or } 1.5 \text{ k}\Omega \quad \textbf{Ans.}$$

20.12. Short-Circuit Protection Against Overload

As a matter of fact, a practical regulator circuit has to be always protected against short-circuit overload. A typical short-circuit overload protection circuit is shown in Fig. 20.30. It employs two diodes D_1 and D_2, and a sensing resistance R connected as shown in the figure. The diodes D_1 and D_2 normally remain open.

When the output terminals of a regulator circuit are shorted, then current through Q_1-transistor increases. This increases voltage drop across sensing resistance R. When the drop across resistance R is such that two diodes start conducting, then current flowing through resistance R is given by the relation,

$$I = \frac{2V_D - V_{BE_1}}{R} \qquad \text{...}(i)$$

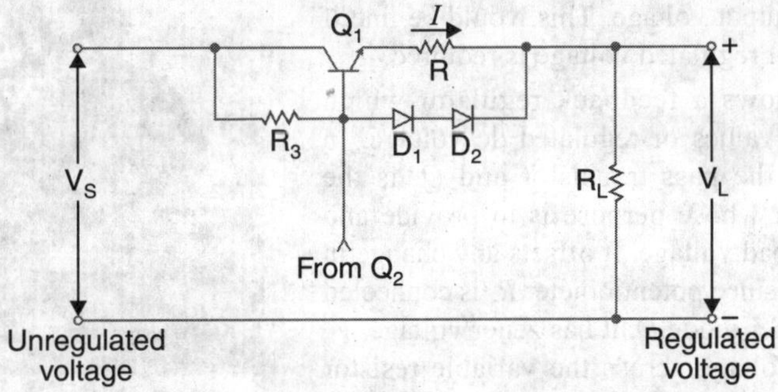

Fig. 20.30. Short-circuit protection against overload.

where V_D = Voltage drop across each diode, and

V_{BE_1} = Base-to-emitter voltage of Q_1-transistor.

There can be no further increase in current through resistance R or the Q_1-transistor. Thus the current I given by the equation (i) is the maximum current, which will flow through Q_1-transistor under short-circuit condition. Thus the series transistor (Q_1) is protected from being damaged.

20.13. Transistor Current Regulator

The main function of a current regulator is to maintain a fixed current through the load despite variations in the terminal voltage. Such a circuit employing a zener diode and a PNP transistor is shown in Fig. 20.31.

Suppose due to drop in V_L, current I_L $(= I_C)$ is decreased, then I_E $(\approx I_C)$ will also decrease. Hence drop across R_E, *i.e.*, V_{BE} will decrease, As per kirchoff's Voltage law,

$$-V_{RE} - V_{BE} + V_Z = 0 \text{ or } V_{BE} = V_Z - V_{RE}$$

(Fixed)

Hence a decrease in V_{RE} will decrease V_{BE} and therefore the conductivity of the transistor will also decrease. This process keeps I_L at a fixed level.

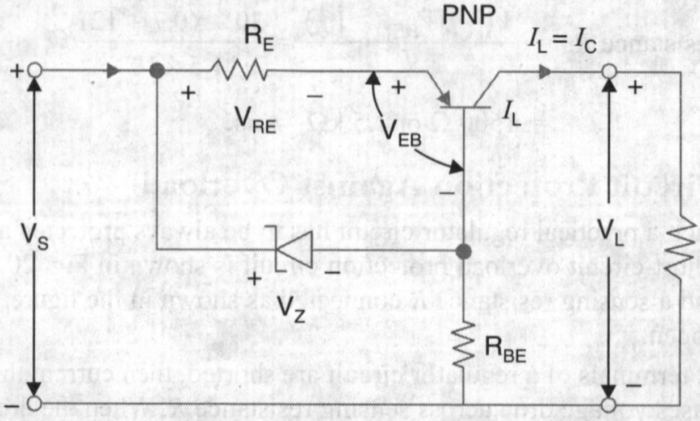

Fig. 20.31.

20.14. Variable Feedback Regulator

The voltage regulators considered so far provide a non-adjustable output voltage. This would be fine if only single value of regulated voltage is required.

Transistor current regulator.

Fig. 20.32 shows a feedback regulator which provides different values of regulated dc voltage. In Fig. 20.32, Q_1 is the pass transistor and Q_2 is the feedback transistor whose purpose is to provide and sample output or load voltage. It offsets any change in the output voltage. since potentiometer R_3 is connected in parallel with zener diode D, it has zener voltage, V_Z aplied across it. Voltage across the variable resistor varies from O to V_Z. Capacitor C ensures that voltage across D and R_3 does not change suddenly.

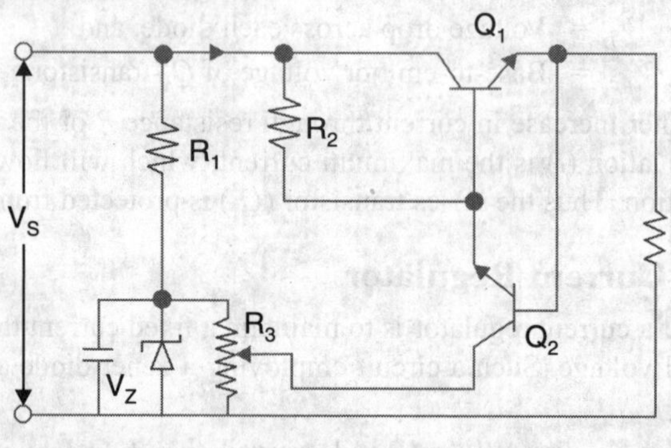

Fig. 20.32.

Voltage at the base of Q_2 is 0.7 V more positive than the voltage at the emitter. Its emitte voltage an hence the base voltage can be changed with the help of R_3. Since the base of Q_2 is tied to the output, it is responsible for providing output or load voltage. The voltage V_{CE_1} across the pass transistor is given by the difference of input voltage and output voltage. The current through Q_1 is equal to the load current. R_2 prevents saturation of the transistors where as R_1, limits the current flowing through D.

The working of feedback regulator may be explained as follows :

Since base voltage of Q_2 is directly related to V_{out} it will change if V_{out} changes. The base and collector of Q_2 are 180° out of phase with each other. If base voltage increases due to increase in V_{out} collector voltage would decrease. Now, collector of Q_2 controls base of Q_1. As the bas voltage of Q_1 decreases, its collector-emitter resistance increases which lowers the load current. This, in turn, lowers the output voltage thereby offsetting the attempted increases in V_{out}. The opposite of these steps provides the action of an attempted decrease in output voltage.

Example 20.27.

In the variable feedback regulator circuit of Fig. 20.32 $V_S = 25$ *V,* $V_Z = 15$ *V and* $R_L = 1$ *kΩ. If the wiper of* R_3 *is adjusted half-way and assuming silicon transistor, compute (i)* V_{out} *(ii)* I_L *(iii)* I_{E1} *(iv)* P_1.

Solution.

Given $V_S = 25$ V; $V_Z = 15$ V and $R_L = 1$ kΩ

(i) V_{out}

We know that output voltage,

$$V_{out} = \text{Voltage at the wiper} + V_{BE_2} = (15/2) + 0.7 = 8.2 \text{ V } \textbf{Ans.}$$

(ii) I_L

We know that load current,

$$I_L = V_{out}/R_L = 8.2/1 \text{ k}\Omega = 8.2 \text{ mA}$$

(iii) I_{E_1}

We know that, $I_{E_1} = I_L = 8.2$ mA **Ans.**

(iii) P_1

We know that, $I_{E_1} = I_L = 8.2$ mA **Ans.**

We know that $V_{CE_1} = V_S - V_{out} = 25 - 8.2 = 16.8$ V;

∴ $P_1 = 16.8 \text{ V} \times 8.2 \text{ mA} = 140$ mW **Ans.**

20.15. Basic Op-amp Series Regulator

Its circuit is shown in Fig. 20.33 and its operation may be explained as follows :

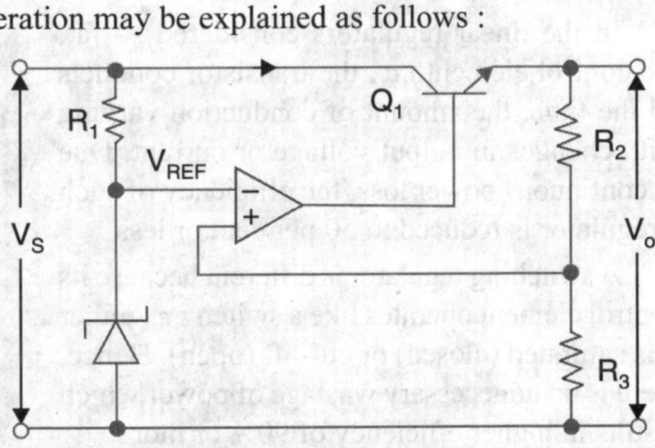

Fig. 20.33.

The potentiometer R_2-R_3 senses any change in output voltage Vout. When Vout attempts to decrease because of decrease in V_S or because of the increase in I_L, a proportional voltage decrease is applied to the inverting output of the op-amp by the potentiometer. Since, the other op-amp input is held by the Zener voltage at a fixed reference voltage V_{REF}, a small difference voltage (called error voltage) is developed across the two inputs of the op-amp. This differenece voltage is amplified and op-amp's output voltage increases. This increases in voltage is applied to the base of Q_1 casuing the emitter voltage (= V_{out}) to increase till the voltage to the inverting input again equals the reference (Zener) voltage. This action offsets attempted decrease in the output voltage thus keeping it almost constant. The opposite action occurs if the output voltage tries to increase.

It will be seen that the op-amp of Fig. is actually connected as a non-inverting amplifier where V_{REF} is the input at the noninverting terminal and the R_2/R_3 voltage divider forms the negative feedback network. The colsed-loop voltage gain is given by $A = 1 + (R_2/R_3)$. Neglecting base-emitter voltage of Q_1, we get

$$V_{out} = V_{REF}\,(1 + R_2/R_3)$$

It is seen that V_{out} depends on Zener voltage and potential divider resistors R_2 and R_3 but is independent of input voltage V_S.

20.16. Basic Op-amp Shunt Regulator

Such a shunt type linear regulator is shown in Fig. 20.34. Here, the control element is a series resistor R_2 and a transistor Q_1 in parallel with the load. In such a regulator, regulation is achieved by controlling the current through Q_1.

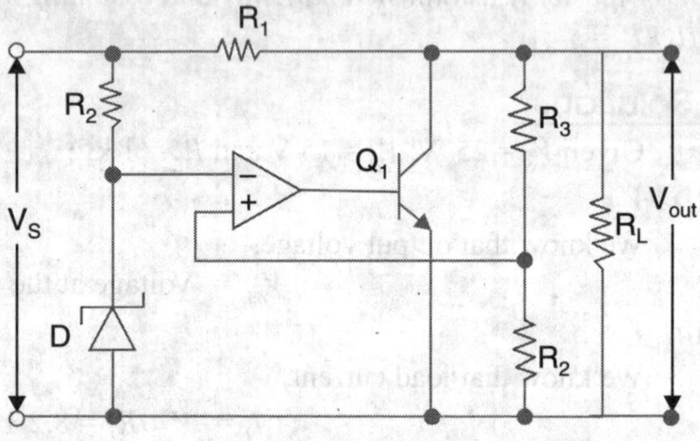

Fig. 20.34.

The circuit opearation may be explained as follows. When output voltage tries to decrease due to change in either the input voltage or load current or temperature, the attempted decrease is sensed by R_3 and R_4 and applied to the non-inverting input of the op-amp. The resulting difference in voltage reduces the op-amp's output, driving Q_1 less thus reducing its collector current (shunt current), and increasing its collector-to-emitter resistance. Since collector-to-emitter resistance acts as a voltage divider with R1, this action offsets the attempted decrease in output voltage and hence, maintains it at a constant value. The opposite action occurs when output voltage tries to increase. The shunt regulator is less efficient than the series type but offers inherent short-circuit protection.

20.17. Switching Regulators

In the linear regulators considered so far, the control element *i.e.*, the transistor conducts all the time, the amount of conduction varying with changes in output voltage or current. Due to continuous power loss, the efficiency of such a regulator is reduced to 50 per cent or less.

A switching regulator is different because its control element operates like a switch *i.e.*, either it is saturated (closed) or cut-off (open). Hence, there is no unnecessary wastage of power which results in higher efficiency of 90% or more.

Switching regulators are of three basic types : (*i*) step-down regulator, (*ii*) step-up regulator and (*iii*) inverting regulator.

Step up/down switching regulator

20.18. Step-down Switching Regulator

In this regulator (Fig. 20.35), V_{out} is always less than V_S. An unregulated positive dc voltage is applied to the collector of the *NPN* transistor. A series of pulses from oscillator is sent ot the base of transistor *T* which gets saturated (closed) on each of the positive pulses. It is so because *NPN* transistor needs a positive voltage pulse on its base in order to turn ON. A saturated transistor acts as a closed switch, hence it allows V_S to send current through *L* and charge *C* to the value of output voltage during the on-time (T_{ON}) of the pulse. The diode D_1 is reverse-biased at this point and hence, does not conduct.

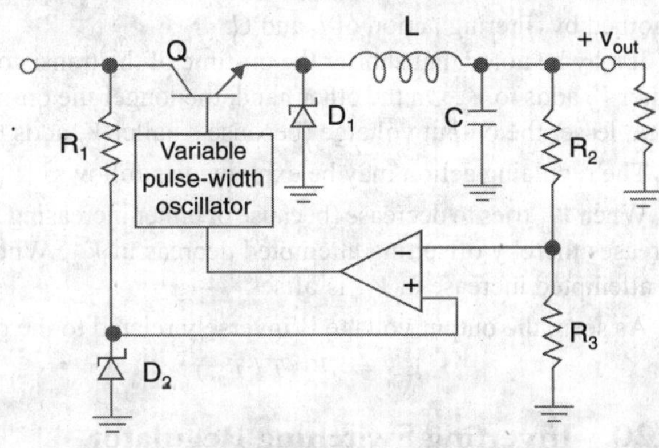

Fig. 20.35.

Eventually when positive pulse turns to zero, *Q* is cut-off and acts like an open switch during the off period (T_{OFF}) of the pulse. The collapsing magnetic field of the coil produces self-induced voltage and keeps thecurrent flowing by returning energy to the circuit.

The value of output voltage depdends on input voltage and pulse width *i.e.*, on-time of the transistor. When on-time is increased relative to off-time, *C* charges more thus increasing V_{out}. When T_{ON} is decreased, *C* discharges more thus decreasing V_{out}. By adjusting the duty cycle (T_{ON}/T) of the transistor, V_{out} can be varied.

$$\therefore \quad V_{out} = V_S (T_{ON}/T)$$

where *T* is the period of the ON-OFF cycle of the transistor and is related to frequency by $T - 1/f$. Also, $T = T_{ON} + T_{OFF}$ and the ratio (T_{ON}/T) is called the *duty cycle*.

The regulating action of the circuit may be explained as follows:

When V_{out} tries to decrease, on-time of the transistor is increased causing an additional charge on the capacitor *C* to offset the attempted decrease. When V_{out} tries to increase, T_{ON} of the transistor is decreased causing *C* to discharge enough to offset the attempted increase.

20.19. Step-up Switching Regulator

The circuit is shown in Fig. 20.36. When transistor *Q* turns ON on the arrival of the positive pulse at its base, voltage across *L* increases quickly to $V_S - V_{CE(sat)}$ and magnetic field of *L* expands quickly. During on-time of the transistor, V_L keeps decreasing from its initial maximum value. The longer transistor is ON, the smaller V_L becomes.

When transistor turns OFF, magnetic field of *L* collapses and its polarity reverses so that its voltge adds to the input voltage thus producing an output

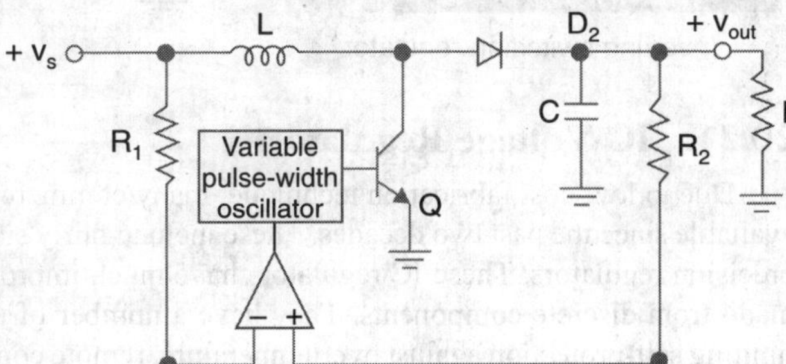

Fig. 20.36.

voltage greater than the input voltage. During off-time of the transistor, D_2 is forward-biased and allows C to charge. The variations in V_{out} due to charging and discharging action are sufficiently smoothed by filtering action of L and C.

It may be noted that shorter the on-time of the transistor, greater the inductor voltage and hence greater V_L adds to V_S. On the other hand, the longer the on-time, the smaller the inductor voltage and hence, lesser the output voltatge (because smaller V_L adds to V_S).

The regulaing action may be explained as follows:

When V_{out} tries to decrease (because of either increasing load or decreasing V_S), transistor on time) decreases thereby offsetting attempted decreas in V_{out}. When V_{out} tries to increase, on-time increase and attempted increase in V_{out} is offset.

As seen, the output voltage is inversely related to the duty cycle.

$$\therefore \quad V_{out} = V_S (T / T_{ON})$$

20.20. Inverting Switching Regulator

The basic diagram of such a regulator is shown in Fig. 20.37. This regulator provides an output voltages that is opposite in polarity to the input voltages.

When transistor Q turns ON by the positive pulse, the inductor voltage V_L jumps to $V_S - V_{CE\,(sat)}$ and the magnetic field of the inductor expands rapidly. When transistor is ON, the diode D_2 is reverse-biased and V_L decreases from its initial maximum value. When transistor turns OFF, the magnetic field collapses and inductor's polarity reverses. This forward-biases D_2, charges C and produces a negative output voltage. This repetitive ON-OFF action of the transistor produces a repetitive charging and discharging that is smoothed by LC filter action. As in the case of a step-up regulator, lesser the time for which transistor is ON, greater the output voltage and *vice versa*.

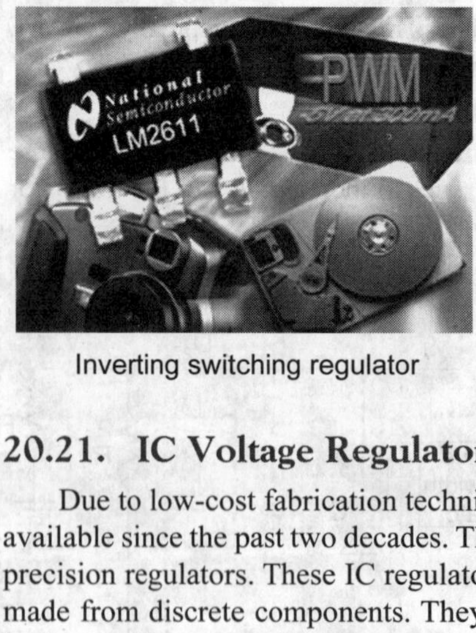

Inverting switching regulator

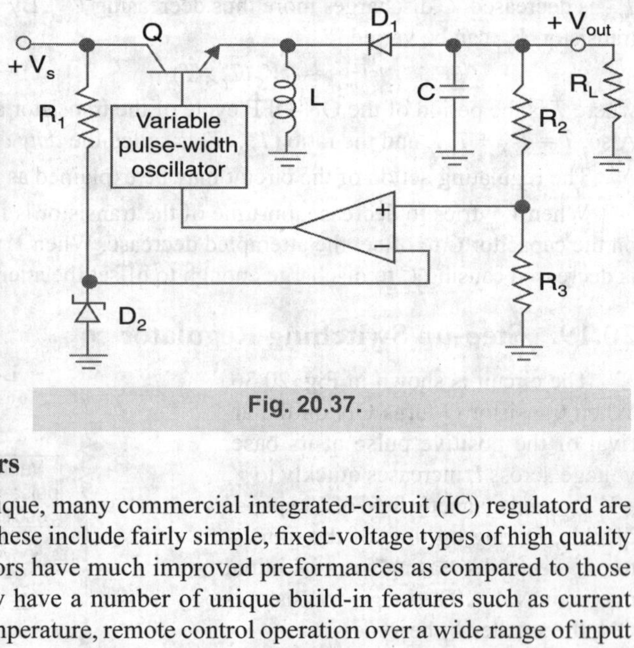

Fig. 20.37.

20.21. IC Voltage Regulators

Due to low-cost fabrication technique, many commercial integrated-circuit (IC) regulatord are available since the past two decades. These include fairly simple, fixed-voltage types of high quality precision regulators. These IC regulators have much improved preformances as compared to those made from discrete components. They have a number of unique build-in features such as current limiting self-protection against overtemperature, remote control operation over a wide range of input voltages and foldback current limiting.

Now we will study the following types of IC voltage regulators : (1) fixed positive linear voltage regulators, (2) fixed negative linear voltage regulators, (3) adjustable positive linear voltage regulators, and (4) adjustable negative linear voltage regulators.

20.22. Fixed Positive Linear Voltage Regulators

There are many IC regulators available in the market that produces a fixed positive output voltage. But 7800 series of IC regulators is representative of three teminal devices that are available with several fixed positive output voltages making them useful in a wide range of applications. Fig. 20.38(a) shows a standard configuration of a fixed positive voltage IC regulator of 7800 series. Notice that it has three terminals labelled as input, output and ground. The last two digits (marked xx) in the part number designate the input voltage. For example, IC 7805 is a + 5V regulator. Similarly IC 7812 is a +12 V regulator and IC 7815 is a +15 V regulator. The capacitor C_1 (typically 0.33 μF) is required only if the power supply filter is located more than 3 inches from the IC regulator. The capacitor C_2 (typically 0.01 μF) acts basically as a line filter to improve transient response.

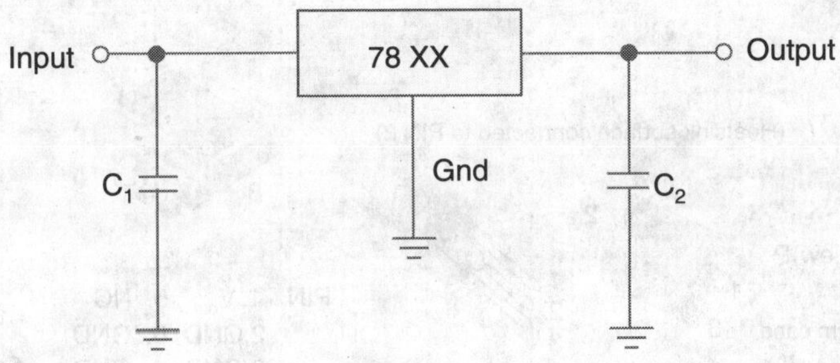

(a)

Part Number	Output Voltage
7805	+5.0 V
7806	+6.0 V
7808	+8.0 V
7809	+9.0 V
7812	+12.0 V
7815	+15.0 V
7818	+18.0 V
7824	+24.0 V

(b)

Fig. 20.38.

Fig. 20.38 (b) shows the part number and the output voltage of 7800 series IC voltage regulators. As seen from this figure, the 7800 series has IC regualtors tha can produce output voltages ranging from +5.0 to +24.0 volt. It may be carefully noted that although these regulators are designed primarily to produce fixed output voltage but they can be used with external components to obtain adjustable output voltage and current.

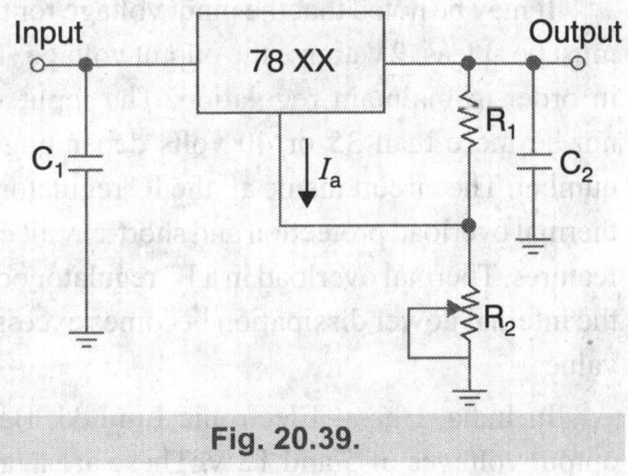

Fig. 20.39.

Fig. 20.39 shows the circuit indicating the use of 78 XX as an adjustable voltage regulator. The output voltage is given by the equation,

$$V_{out} = V_{fixed} + \left(\frac{V_{fixed}}{R_1} + I_Q \right) R_2$$

For example, for a 7805 IC regulator, V_{fixed} = 5V. Let $R_1 = R_2$ = 1 kΩ and I_Q = 2 mA, then its output voltage is,

$$V_{out} = 5 + \left(\frac{5}{1\ \text{k}\Omega} + 5\ \text{mA} \right) \times 1\ \text{k}\Omega$$

Thus output voltage of IC 7805 regulator can be adjusted anywhere between 5 V to 15 V.

This example indicates that the output of IC 7805 regulator is adjusted to 15 V using external resistances R_1 and R_2.

The standard 7800 series can produce output current input in excess of A when used with adequate heat sink. It is available in aluminimum can package TO-3 (indicated by K in the part number) and plastic package TO-220 (indicated by *T* in the part number), The 78L00 series can provide up to 100 mA and is available in TO-92 and metal TO-39 low profile packages. The 78M00 series can provide upto 0.5 A and is available in plastic TO-202 package. Fig. 20.40 shows the typical metal and plastic packages for the IC voltages regulators.

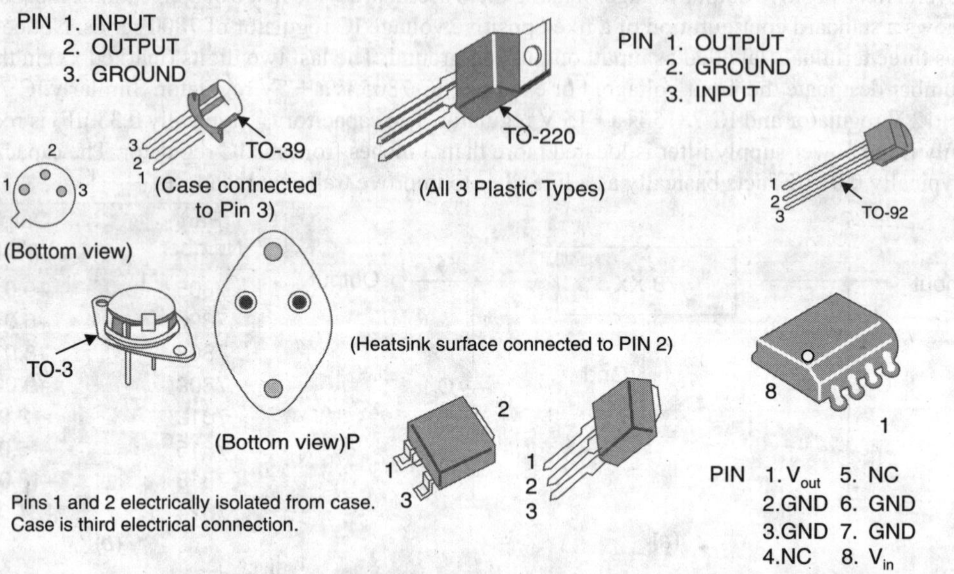

Fig. 20.40. Typical metal and plastic packages.

It may be noted that the input voltage for the IC regulator must be at least 2V above the ouptut voltage. This is required in order to maintain regulation. The input voltage should not be more than 35 or 40 volts depending upon the part number. The circuit inside all the IC regulators have internal thermal overload protection and short-circuit current-limiting features. Thermal overload in a IC regulator occurs whenever the internal power dissipation becomes excessive and the temperature of the device exceeds a certain value.

IC 7805 regulator

In India, Bharat Electronic Limited, Bangalore manufactures the IC voltage regulators with output voltages of 5 and 12 V. These are available in the market with part numbers BEL 7805 and BEL 7812 respectively.

20.23. Fixed Negative Linear Voltage Regulators

The 7900 series is typical of three-terminal *IC* regulators that provide a fixed negative output voltage. This series is a negative-voltage counterpart of the 7800 series and shares most of the same features, charateristics and package types. Fig. 20.41 (*a*) indicates the standard configuration and Fig. 20.41 (*b*), the part numbers with corresponding output voltages that are available in 7900 series.

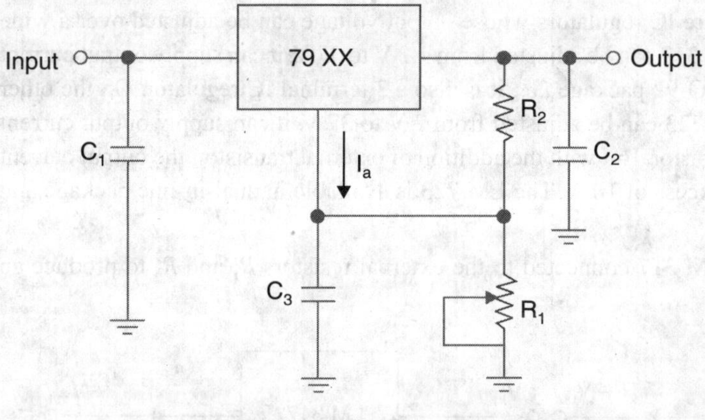

(a)

Part Number	Output Voltage
7905	– 5.0 V
7905.2	– 5.2 V
7906	– 6.0 V
7908	– 8.0 V
7912	– 12.0 V
7915	– 15.0 V
7918	– 18.0 V
7924	– 24.0 V

(b)

Fig. 20.41.

The capacitor C_1 (typically 0.22 μF) is required only if the power supply filter is located more than 3 inches away from the *IC* regulator. The capacitor C_2 (typically 1 μF) is required for stability of the output voltage. Both capacitors C_1 and C_2 must be solid tantalum capacitors.

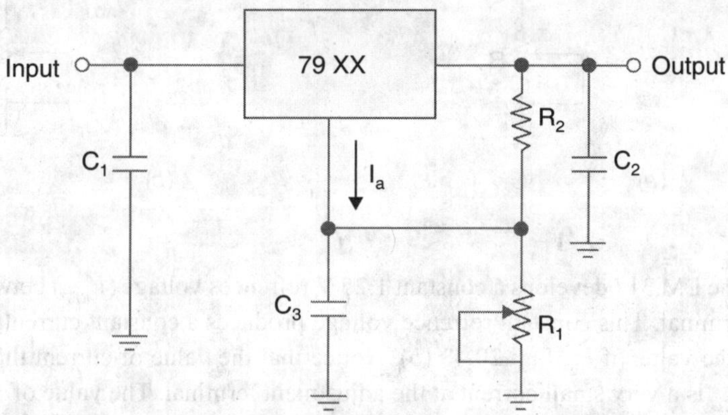

Fig. 20.42.

Fig. 20.42 shows the use of 79XX to produce an adjustable output voltage. The capacitor C_3 (typically 25 μF) improves the transient response of the output voltage. The output voltage is given by the equation,

$$V_{out} = V_{fixed}\left(\frac{R_1 + R_2}{R_2}\right)$$

The recommended value of R_2, for 7905 is 300 Ω, for 7915, its value is 750 Ω and for 7915 is 1 kΩ.

In india, BEl manufactures IC regulators with output voltage of –5 V and –12 V. These are available in the market with part numbers BEL 7905 and BEL 7915 respectively.

20.24. Adjustable Positive Output Linear Voltage Regulators

We have already seen in Art. 20.22 that by adding external resistors, we can adjust the output voltage of 7800 series IC regulators higher than their fixed (or set) voltages. For example, the output voltage of 7805 can be adjusted higher than 5 V. But the performance and reliability of 7800 series to produce voltage higher than its fixed value is not considered to be good.

The LM 317 and LM 723 are IC regulators whose output voltage can be adjusted over a wide range. The output voltage of LM 317 can be adjusted from 1.2 V to 37 V, it can supply output current of 100 mA and is available in TO-92 package *i.e.*, it is also a 3 terminal IC regulator. On the other hand, the output voltage of LM 723 can be adjusted from 2 V to 37 V, it can supply output current of 150 mA without external transistor. But with the addition of external transistor, the output current capability can be increased in excess of 10 A. The LM 723 is available in dual-in-line package and in a metal can package.

Fig. 20.43 (*a*) shows the LM 317 connected to the external resistors R_1 and R_2 to produce an adjustable output voltatge.

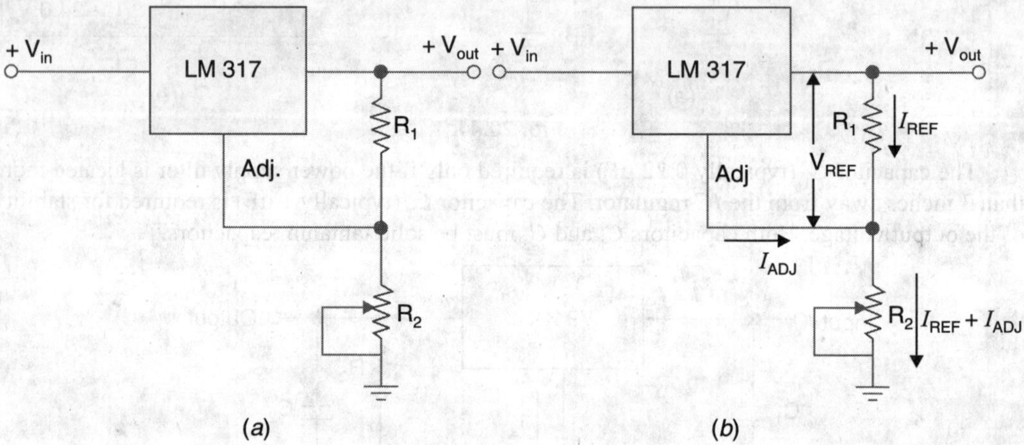

Fig. 20.43.

In opeation, the LM 317 develops a constant 1.25 V refrences voltage (V_{REF}) between the output and adjustment terminal. This constant refrence voltage produces a constant current, (I_{REF}) through R_1, regardless of the value of R_2, Fig. 20.43 (*b*). Notice that the value of current through R_2 is the sum of I_{REF} and I_{ADJ} is a very small current at the adjustment terminal. The value of I_{ADJ} is typically around 100 mA. It can be shown that the output voltage.

$$V_{out} = V_{REF}\left(1+\frac{R_2}{R_1}\right)+I_{ADJ}\cdot R_2$$

It is evident from the above equation that the output voltate is a function of R_1 and R_2. Usually the value of R_1 is recommended to be around 220 Ω. Once the value of R_1 is set, the output voltage is adjusted by varying R_2.

In order to study LM 317 in more detail, refer to the device specification sheet given in Appendix B.

Example 20.28.

Calculate the minimum and maximum output voltages for the IC voltage regulator shown in Fig.20.44. Assume I_{ADJ} *= 100 mA,* V_{in} *= + 35 V.*

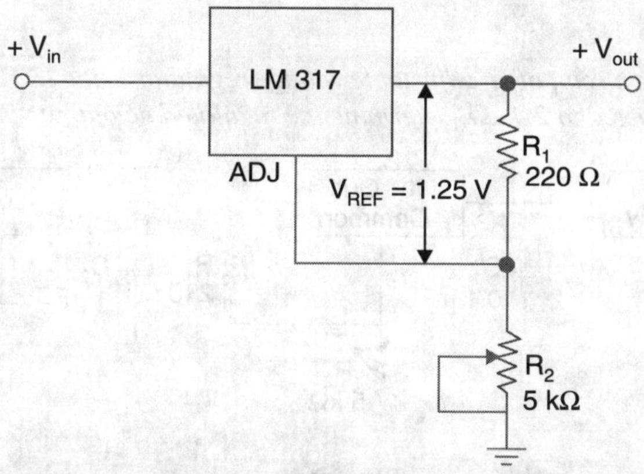

Fig. 20.44.

Solution.

Given : $I_{ADJ} = 100\ \mu A$, $V_{in} = +35$ V.

The equation for output voltage of the IC voltage regulator is given by,

$$V_{out} = V_{REF}\left(1+\frac{R_2}{R_1}\right)+I_{ADJ}\cdot R_2$$

When R_2 is set at its minimum value of (*i.e.* 0 Ω), the output voltage,

$$V_{out\,(min)} = 1.25\left(1+\frac{0}{220}\right)+(100\times10^{-6})\times 0$$

$$= \mathbf{1.25\ V}$$

When R_2 is set at its maximum values (*i.e.*, 5 kΩ), the output voltage,

$$V_{out\,(max)} = 1.25\left(1+\frac{5000}{220}\right)+(100\times10^{-6})\times 5000$$

$$= 29.66 + 0.5 = 30.16\ \text{V.}\ \textbf{Ans.}$$

Example 20.29.

Determine the regulated voltage in the circuit of Fig. 20.45 with $R_1 = 220\ \Omega$ and $R_2 = 1.5\ k\Omega$.

Solution.

Given : $R_1 = 220\ \Omega$ and $R_2 = 1.5\ k\Omega = 1500\ \Omega$.

We know that output voltage of an adjustable voltage regulator,

$$V_o = 1.25\left(\frac{R_2}{R_1}+1\right) = 1.25\left(\frac{1500}{220}+1\right)$$

= 7.82 V **Ans.**

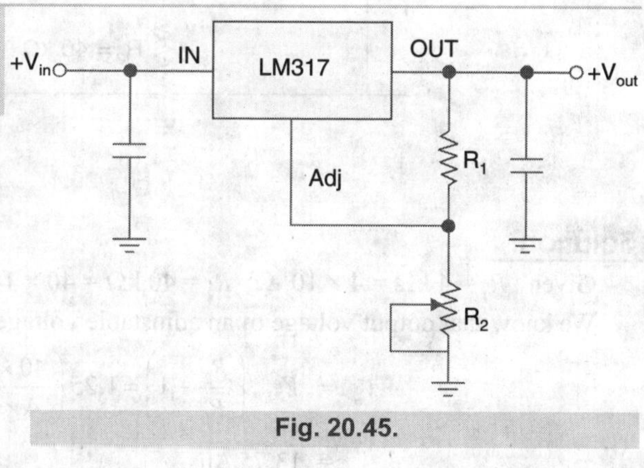

Fig. 20.45.

Example 20.30.

Fig. 20.46 shows the circuit of an adjustable voltage regulator using LM 317 integrated circuit. If the value of R_2 is adjusted to 2.4 kΩ, determine the regulated dc output voltage for the circuit.

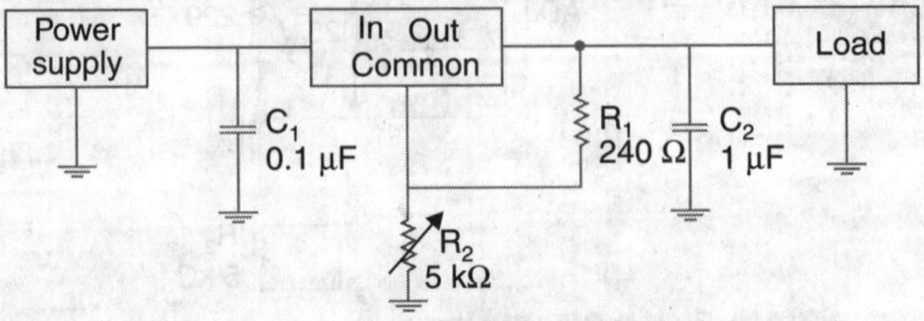

Fig. 20.46.

Solution.

Given : $R_1 = 240\ \Omega$ and $R_2 = 2.4\ k\Omega = 2400\ \Omega$

We know that the regulated d.c. output voltage,

$$V_o = 1.25\left(\frac{R_2}{R_1}+1\right) = 1.25\left(\frac{2400}{240}+1\right) = 13.75 \text{ V } \textbf{Ans.}$$

Example 20.31.

Calculate the approximate output voltage and the maximum, minimum efficiency, if the input voltage V_{in} varies between 30 V and 48 V. What should be the output voltage if R_2 value is doubled? Assume V_{REF} = 1.25 V. (VTU., June/July 2011)

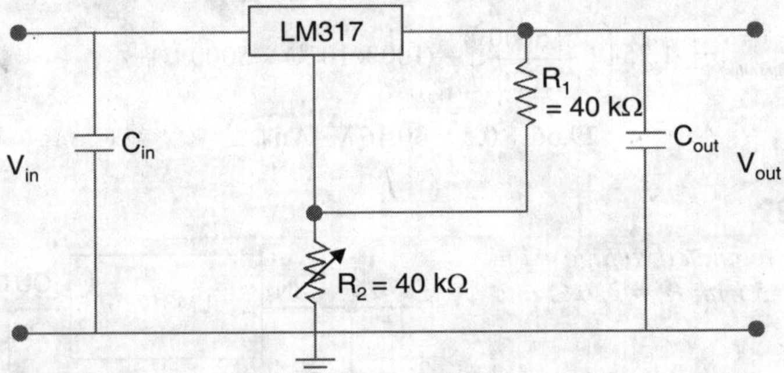

Fig. 20.47.

Solution.

Given : $R_1 = 4\ k\Omega = 4 \times 10^3\ \Omega$; $R_2 = 40\ k\Omega = 40 \times 10^3\ \Omega$ and $V_{in} = 30$ V to 48 V.

We know that output voltage of an adjustable voltage regulator,

$$V_o = V_{REF}\left(\frac{R_2}{R_1}+1\right) = 1.25\left(\frac{40\times10^3}{4\times10^3}+1\right)$$

$$= 13.75 \text{ Ans.}$$

V_o when R_2 value is doubled

In this case, the value of $R_2 = 80 \times 10^3\ \Omega$

We know that output voltage of an adustable voltage regulator,

$$V_o = V_{REF}\left(\frac{R_2}{R_1}+1\right) = 1.25\left(\frac{80\times10^3}{4\times10^3}+1\right)$$

$$= 26.25 \text{ V } \textbf{Ans.}$$

Note that the output voltage also doubles when the value of R_2 doubles.

20.25. Adjustable Negative Output Linear Voltage Regulators

A good example of this type of regulators is LM 337. The regulator is a negative output counter part of LM 317. The LM 337 (like LM 317) requires two external resistors for adjustment of output voltage as shown in Fig. 20.48. The output voltage can be adjusted anywhere from –1.2 V to – 37 V depending upon the external resistor values.

The LM 723 can also be used as an adjustable negative output voltage regulator. The output voltage of this IC regulator can be adjusted anywhere from – 2.0 V to – 37 V depending upon the external resistor values.

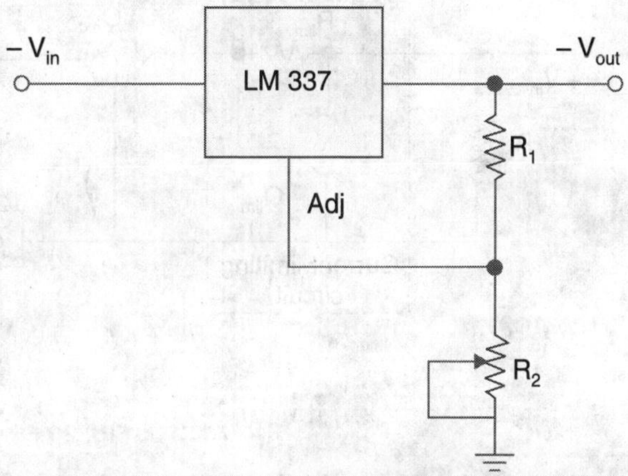

Fig. 20.48.

20.26. Use of External Pass Transistor with Linear Voltage Regulators

We have already mentioned in the last two articles that a linear voltage regulator (7800 and 7900 series) is capable of delivering only a certain amount of output current to a load. For example, the 7800 series regulators can handle a maximum output current of at least 1.3 A and typical 2.5 A. If the load current exceeds the maximum allowable value, there will be a **thermal overload** and the regulator will **shut down.** *A thermal overload condition means that there is excessive power dissipation inside the regulator.*

If an application requires a larger value of load current than the maximum current that the regulator can deliver, we will have to use an external pass transistor as shown in Fig. 20.50. The value of R_{ext} (current-sensing resistor) determines the value of current as which the external pass transistor (Q_{ext}) begins to conduct because it sets the base-to-emitter voltage of the transistor.

As long as the current is less than the value set by R_{ext}, the transistor Q_{ext} is OFF and the regulator operates normally. This is because the voltage drop across R_{ext} is less than 0.7 V (*i.e.*, the base to emitter voltage required to turn Q_{ext} ON). The value of R_{ext} is determined by the equation $R_{ext} = 0.7\ V/I_{max}$ where I_{max} is the value of current that the voltage regulator is to handle **internally.**

Fig. 20.49.

When the current is sufficient to produce at least a 0.7 drop across R_{ext}, the transistor Q_{ext} turns on and conducts any current in excess of I_{max}. The transistor Q_{ext} will conduct current depending on the load requirement. For example, if the total load current is 5 A and I_{max} was selected to be 1 A, then the external pass transistor (Q_{ext}) will conduct 4 A of current through it.

It may be noted that the external pass transistor is a power with heat sink that must be capable of handling a maximum power given by the equation,

$$P_{ext} = I_{ext}(V_{in} - V_{out})$$

One major drawback of the circuit shown in Fig. 20.42 is that the external pass transistor is not protected from excessive current, such as would result from a shorted output. This drawback can be overcome by using an additional current limiting circuit as shown in Fig. 20.50.

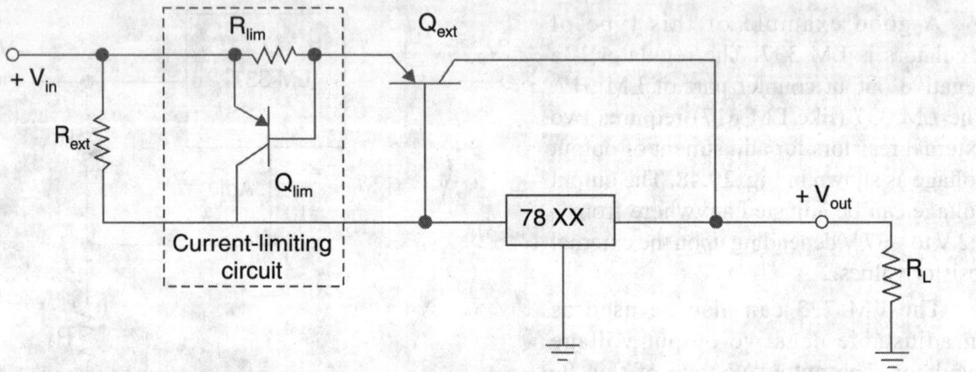

Fig. 20.50.

The operation of this circuit may be explained as follows. The current sensing resistor, R_{lim} sets the base-to-emitter voltage of Q_{lim}. The base-to-emitter voltage of Q_{ext} is now determined by $\left(V_{R_{ext}} - V_{R_{lim}}\right)$ because they have opposite polarities. So for normal operation, the drop across R_{ext} must be sufficient to oyercome the opposing drop across R_{lim}.

If the current through T_{ext} exceeds a certain maximum value, ($I_{ext\,(max)}$) because of a shorted output or a faulty load, the voltage across R_{lim}, reaches 0.7 V and turns Q_{lim} now conducts current away from Q_{ext} and through the regulator. This forces a thermal overload to occur and shut down the regulator. Remember, the *IC* voltage regulator circuitry is internally protected from thermal overload as part of its design. This way the external pass transistor is protected from excessive current.

20.27. Use of Linear Voltage Regulators as a Current Regulator

The 3-terminal linear voltage regulator can be used as a current source when an application requires that a constant current be supplied to a variable load. The basic circuit is shown in Fig. 20.51. Here R_1 is the current-setting resistor. The regualtor provides a fixed output voltage, V_{out} between the ground terminal and the output terminal. However, it may be noted that the ground pin of the regulator is not connected to the circuit ground. The constant current suplied to the load, is given by the equation,

$$I_L = \frac{V_{out}}{R_L} + I_Q$$

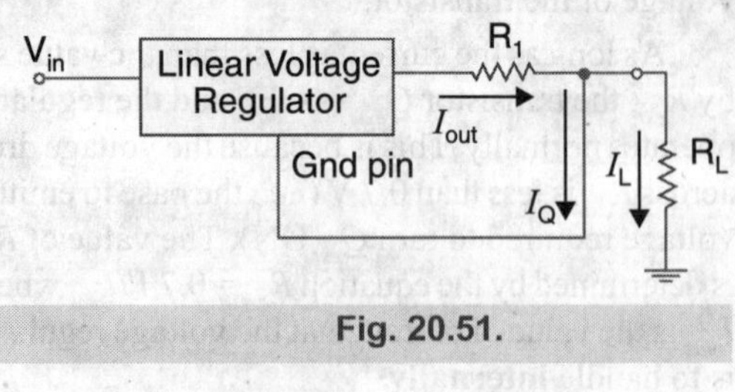

Fig. 20.51.

Usually the current, I_Q is very small as compared to the output current and hence can be neglected, therefore

$$I_L \approx \frac{V_{out}}{R_L}$$

For example, if we use 7805 regulator to provide a constant current of 1 A to a variable load, then

$$R_1 = \frac{5}{1} = 5\ \Omega$$

Please note that input voltage must be at least 2 V greater than the output voltage. Thus for 5 V regulator, V_{in} must be greater than 7 V.

20.28. Switching Voltage IC Regualtors

There are several switching voltage *IC* regulators available in the market. The choice depends upon the desired application and the cost. However, we will illustrate it with the IC 78S40. This device is a universal device that can be used with external components to provide step-up, step down and inverting operation.

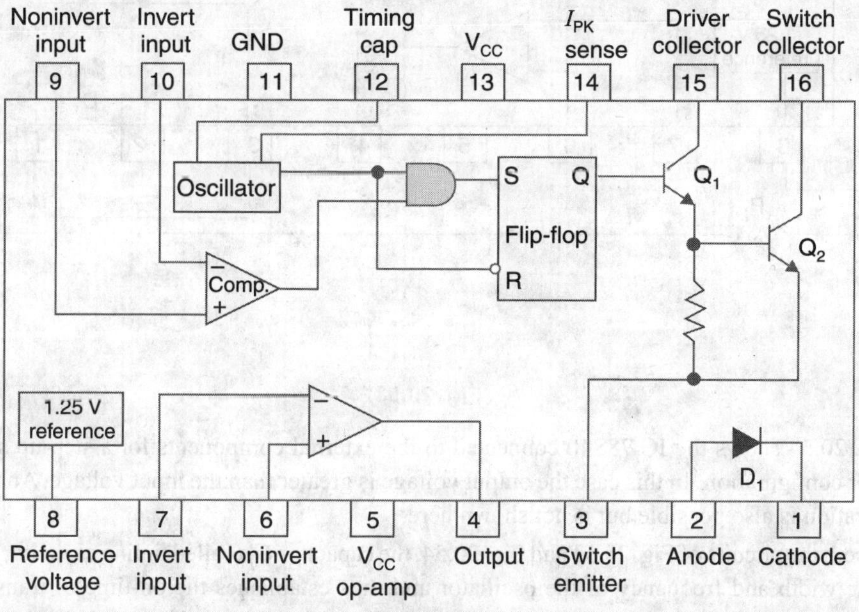

Fig. 20.52.

Fig. 20.53 shows that internal circuitry of the IC 78C40. The circuitry can be compared to the basic switching regulators discussed in Art. As seen from this diagram, the oscillator and comparator functions are directly comparable. The logic gate and flip-flop in the 78S40 we not included in the basic circuit of Fig. 20.51, but they provide additional regulation. Transistor Q_1 and Q_2 perform the same function as Q in the basic circuit. The 1.25 V reference block in the 78S40 has the same purposes as that of the zener diode in the basic circuit and diode D_1 in the 78S40 corresponds to D_1 in the basic circuit.

The circuitry of 78S40 has also an uncommitted or unused OP-AMP. We require external circuitry to operate this device as a regulator.

Fig. 20.53 shows the external connections of the IC 78S40 for a step-down switching regualtor configuration. Note that in this configuration the circuit produces an output voltage which is less than input voltage.

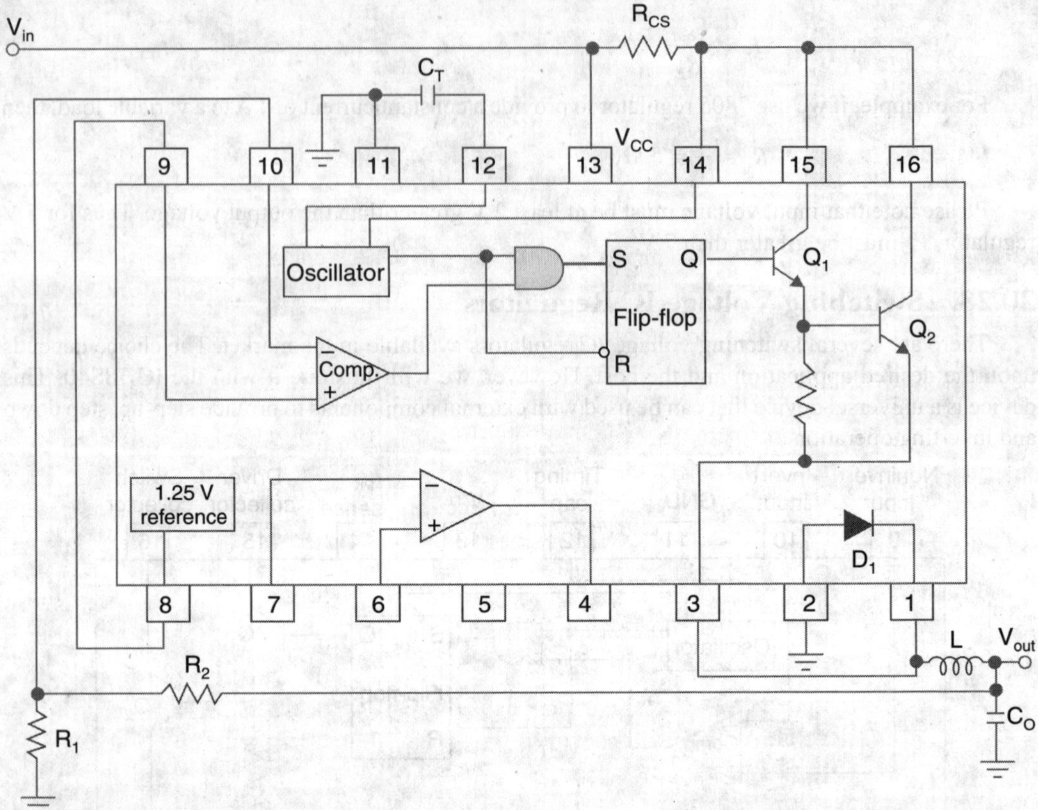

Fig. 20.53.

Fig. 20.54 shows the IC 78S40 connected to the external components for a step-up switching regulator configuration. In this case the output voltage is greater than the input voltage. An inverting configuration is also possible but is not shown here.

In both the circuits of Fig. 20.53 and Fig. 20.54, the capacitor C_T (called timing capacitor) controls the pulse width and frequency of the oscillator and thus establishes the on-time of transistor Q_1. The voltage across the resistor R_{CS} (called current-sensing resistor) is used internally by the oscillator to vary the duty cycle based on the desired peak current. The voltage divider made up of R_1 and R_2 reduces the output voltage to a value equal to the reference voltage. If the output voltage (V_{out}) exceeds its set value, the output of the comparator switches to the low state, disabling the gate to turn off unitl the output decreases. This regulating action is in addition to that produced by the duty cycle variation of the oscillator.

20.29. Complete D.C. Power Supply Circuit

Fig. 20.55 shows a complete d.c. power supply circuit. It consists of a step down transformer, full-wave bridge rectifier, a π-filter, a transistor series voltage regulator. The unregulated a.c. voltage voltage is fed from the transformer through a bridge rectifier. It is then filtered by a π-filter (CLC filter) and finally regulated by a transistor series regulator.

It is possible to take either smaller or greater d.c. output voltage than that produced by the d.c. power supply circuit. The smaller ouptut can be taken by using a voltage divider at the output terminals of the d.c. supply. A d.c. output voltage, greater than the normal value, may be obtained by using voltage multiplier. A voltage multiplier produces a d.c. voltage equal to a multiple of the maxium input voltage (*i.e.*, 2 V_m, 3 V_m and so on).

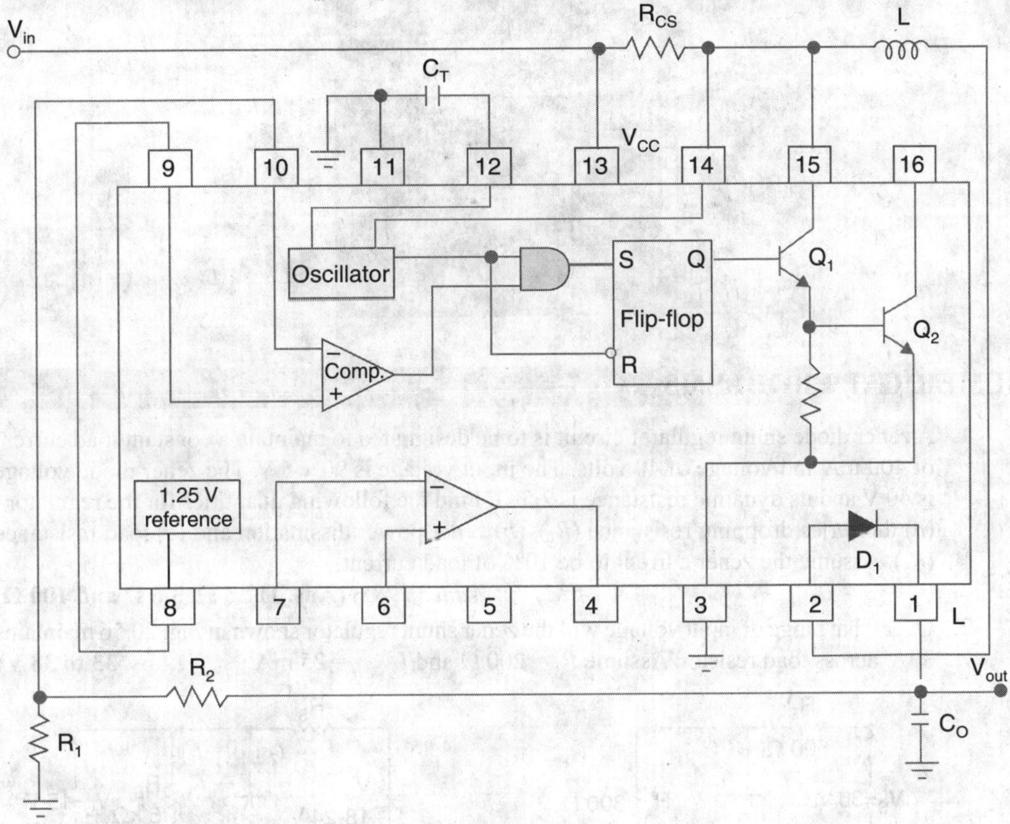

Fig. 20.54.

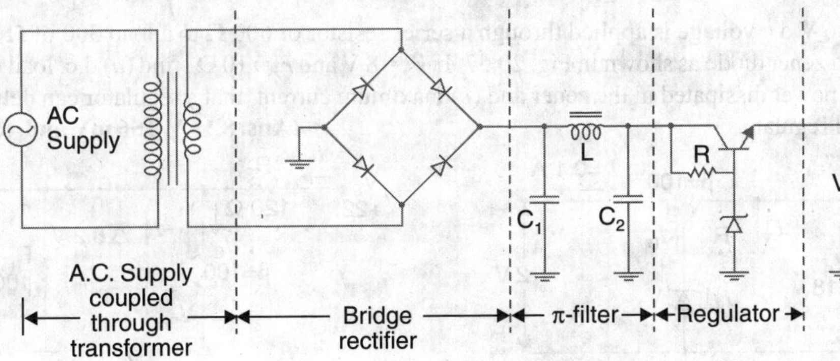

Fig. 20.55.

Such power supplies are used for high-voltage/low current devices like cathode-ray tubes (*i.e.*, the picture tubes in TV receivers, oscilloscopes and computer displays).

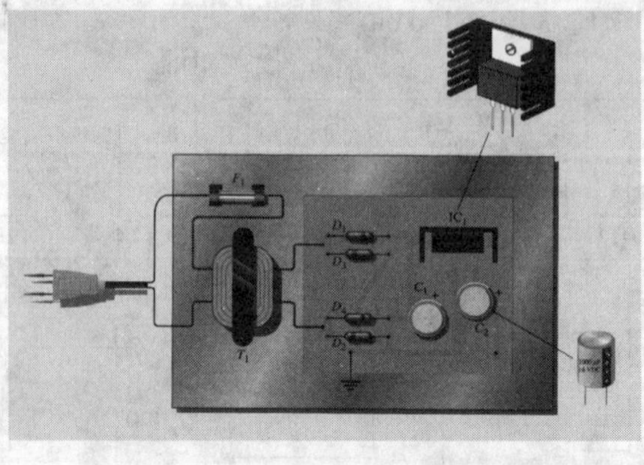

NUMERICAL PROBLEMS

1. A zener diode shunt regulator circuit is to be designated to maintain a constant load current of 400 mA and voltage of 40 volts. The input voltage is 90 ± 5 V. The zener diode voltage is 40 V and its dynamic resistance is 2.5 Ω. Find the following quantities for the regulator : (*a*) the series dropping resistance (R_S), (*b*) zener power dissipation and (*c*) load resistance (R_L). Assume the zener current to be 10% of load current.
(*A.M.I.E., Winter 1985*) **(Ans. 112.5 Ω, 3.6 Ω and 100 Ω)**

2. Over what range of input voltage will the zener shunt regulator shown in Fig. 20.56 maintains 30 V across load resistor. Assume R_S = 200 Ω and $I_{Z(max)}$ = 25 mA. **(Ans. 33 to 38 V)**

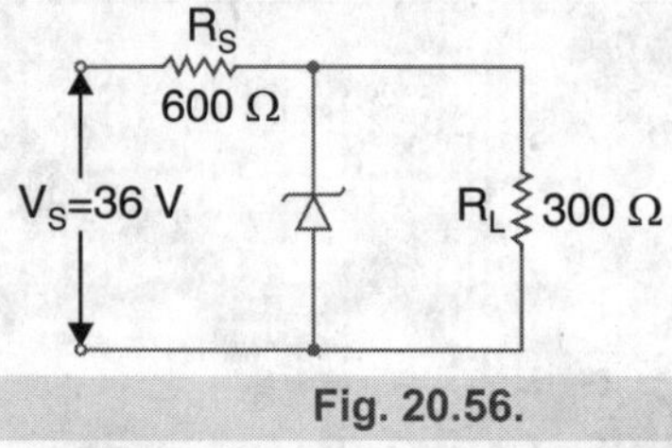

Fig. 20.56.

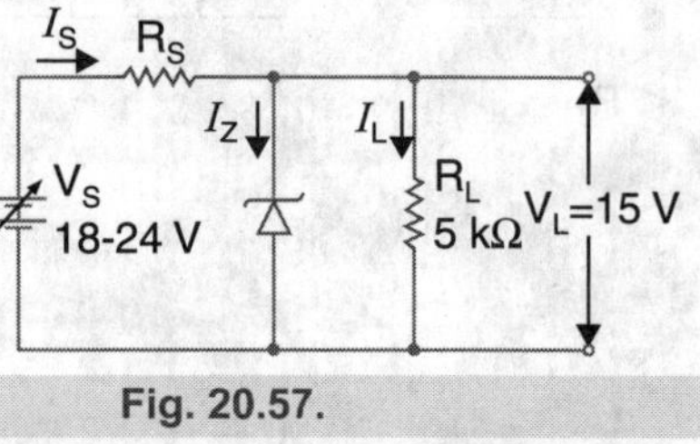

Fig. 20.57.

3. A 36 V d.c voltage is applied through a series resistor of 600 Ω to a load 300 of Ω shunted by a zener diode as shown in Fig. 20.57. If V_Z = 8 V and r_Z = 10 Ω, find (*a*) d.c. load voltage; (*b*) power dissipated in the zener and (*c*) maximum current, that a regulator can deliver and still regulate. **(Ans. 8.19 V, 156 mV and 46.5 mA)**

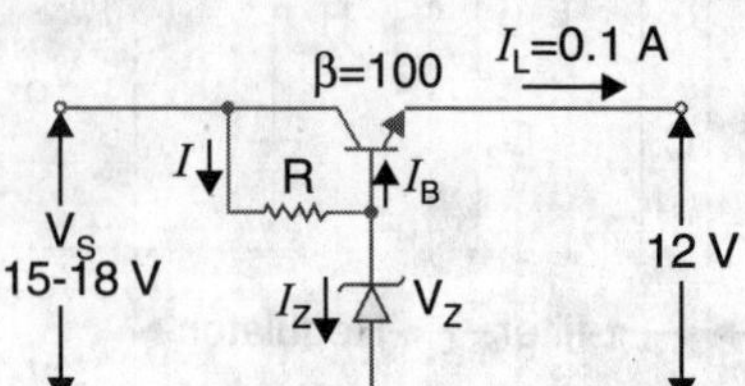

Fig. 20.58.

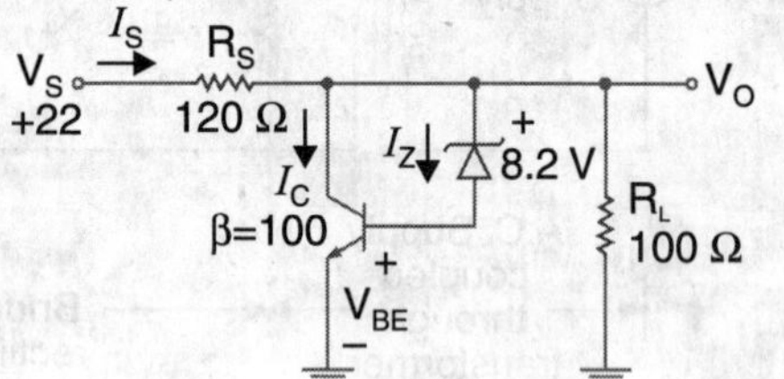

Fig. 20.59.

4. A zener diode voltage regulating circuit is as shown in Fig. 20.58. The zener diode used has a zener voltage (V_Z) of 15 V and minimum current $I_{Z\,(min)}$ of 2 mA, a power dissipation of 120 mΩ and a zener resistance of 40 W. If the load resistance is 5 kW and the input voltage varies from 18 to 24 V, find the value of R_S. **(Ans. 158 Ω < R_S < 1000 Ω)**

5. Assuming the base current (I_B) to be 20% of for V_S = 15 volts and V_{EB} = 0.6 volts, estimate the values of R, power in the transistor, and zener diode for the circuit shown in Fig. 20.59.

(Ans. 400 Ω, 0.17 W and 0.6 W)

6. Determine the regulated voltage and circuit currents for the shunt regulator of Fig. 20.60.

(Ans. V_L = 8.9 V; I_L = 89 mA; I_S = 109 mA; I_C = 20 mA and I_Z = 0.2 mA)

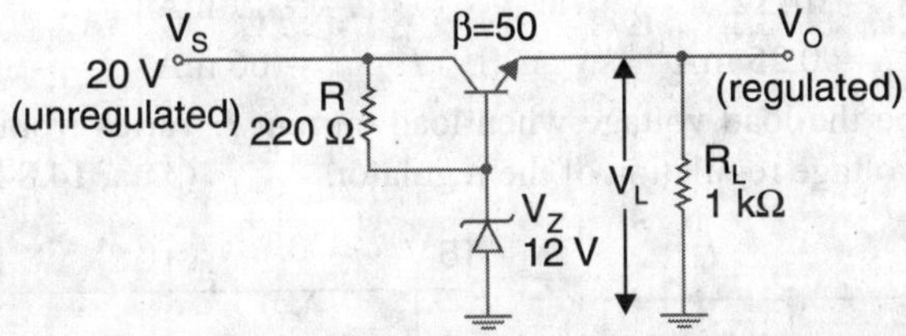

Fig. 20.60.

7. Calculate the output voltage and zener current in the regulator circuit of Fig. 20.61.

(Ans. V_0 = 11.3 V; I_Z = 36 mA)

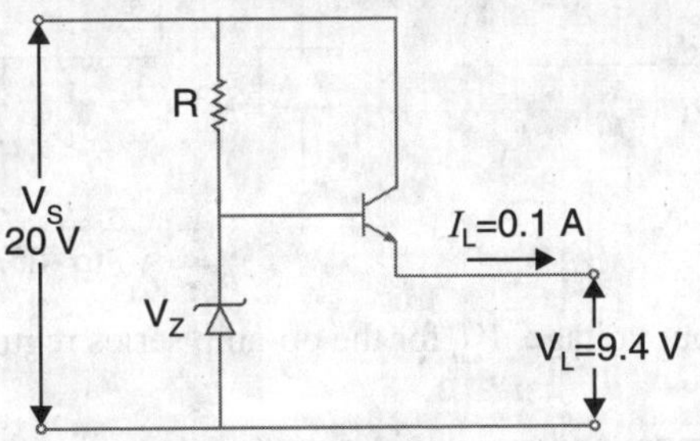

Fig. 20.61.

8. For the circuit shown in Fig. 20.62, find the ratings of R and zener, assuming the zener current to be 5 mA under normal full load condition.

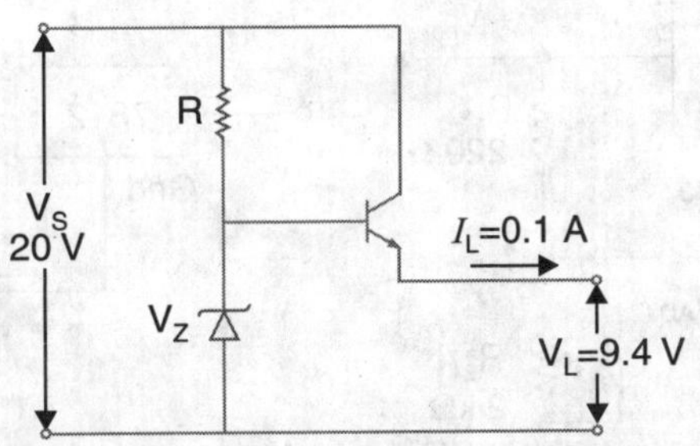

Fig. 20.62.

The available transistors are of β = 50; V_{BE} = 0.6 V and having the following specifications :

Type 1: V_{CEO} = 30 V, $I_{C\,(max)}$ = 2.5 A, P = 30 W

Type 2: V_{CEO} = 25 V $I_{C\,(max)}$ = 3A, P = 60 W

Type 3: V_{CEO} = 15 V, $I_{C\,(max)}$ = 5 A, P = 50 W

Choose the type of transistor suitable for the above circuit. *(A.M.I.E., Summer, 1984)*

(Ans. 400 Ω; 0.25 W and transistor type 1)

9. The LM 317 is being used in a circuit with adjustment resistance values of $R_1 = 510\ \Omega$ and $R_2 = 6.834\ k\Omega$ (adjusted potentiometer value). Determine the output voltage for the circuit. **(Ans. 14.4 V)**

10. The Zener diode of Fig. 20.63 has the following rating :

$$V_z = 15\text{ V} \quad \text{at} \quad I_z = 17\text{ mA}$$
$$r_z = 14\ \Omega \quad \text{at} \quad I_z = 17\text{ mA}$$
$$I_{z\,(min)} = 0.25\text{ mA} \quad I_{z\,(max)} = 66\text{ mA}$$

What would be the load voltage when load current I_L varies from 1 mA to 60 mA. Also calculate the voltage regulation of the regulator. **(Ans. 14.846 V, 15.672 V, 5.27%)**

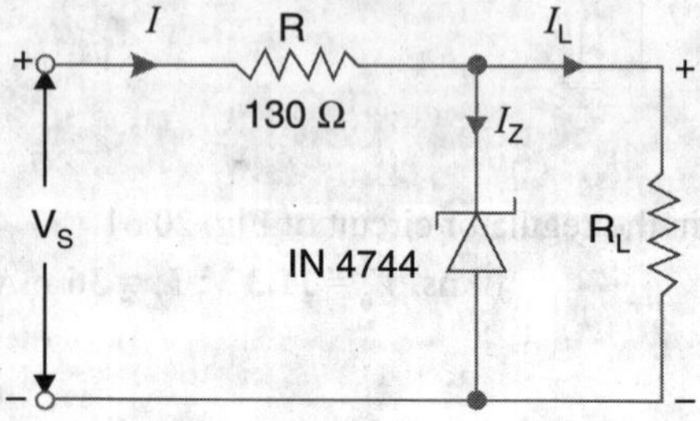

Fig. 20.63.

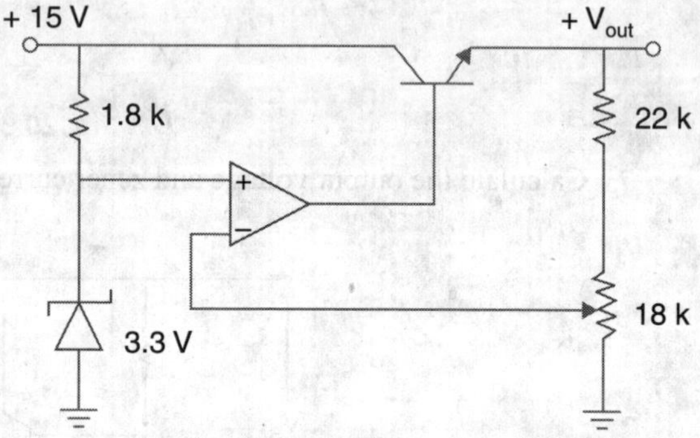

Fig. 20.64.

11. Determine the output voltage, V_{out} for the op-amp series regulator shown in Fig. 20.64. **(Ans. 7.3 V)**

12. Calculate the minimum and maximum output voltages for the IC voltage regulator shown in Fig. 20.65. Assume $I_{ADJ} = 50$ mA, $V_{in} = +35$ V, $V_{REF} = 1.2$ V **(Ans. 1.25 V, 12.71 V)**

13. Fig. 20.66 shows the circuit of a current regulator. What value of R_1 is necessary to provide a constant current of 1 A. **(Ans. 12 Ω)**

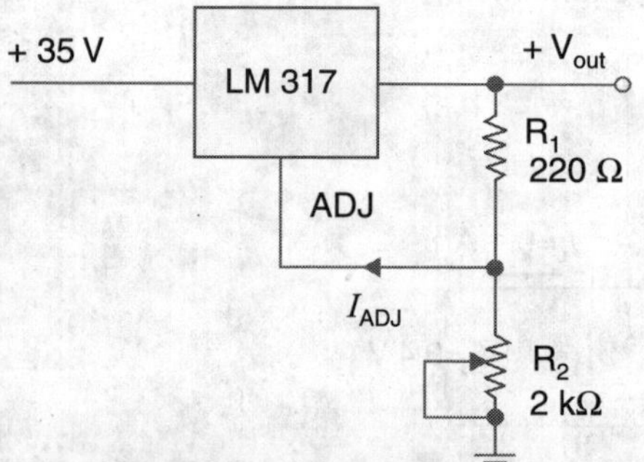

Fig. 20.65.

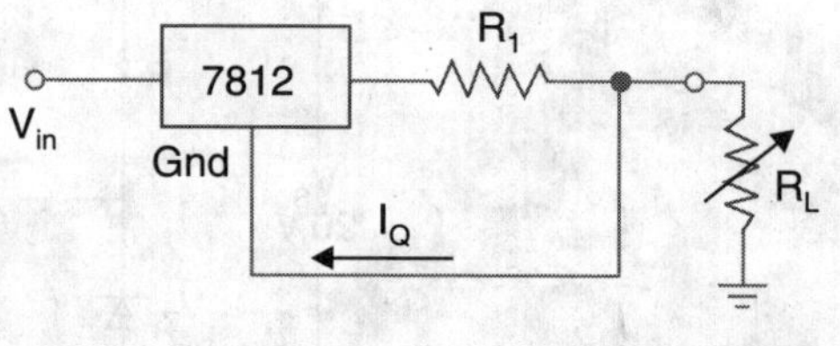

Fig. 20.66.

DESCRIPTIVE QUESTIONS

1. Describe a zener shunt regulator circuit.
2. What are the disadvantages of zener shunt regulator ?
3. Draw circuit diagram of transistor shunt regulator. Explain it briefly.
4. Draw circuit diagram of a transistorized series voltage regulator.

5. Describe the monolithic IC voltage regulators.
6. Define voltage regulation. *(Anna University., April/May 2008, GBTU/MTU., 2006-07)*
7. Define Line & Load Regulation. *(PTU., Dec. 2009, May 2008, Dec. 2008, May 2006)*
8. Define Regulation and Ripple factor. *(Nagpur University., 2008)*
9. Draw the circuit of simple zener voltage regulator and explain its working principle. *(Calicut University., June 2007)*
10. What is Zener diode? Draw the voltage regulator circuit using Zener diode and explain its working. *(Jamia University., 2008)*
11. Draw the circuit diagram of voltage regulator using zener diode. *(GBTU/MTU., 2006-07)*
12. Explain how zener diode acts as a regulator. *(Anna University., April/May 2008)*
13. Zener diode can be used as voltage regulator-Justify it. *(Anna University., Nov/Dec 2007)*
14. Sketch and explain the circuit diagram of shunt and series voltage regulators. Compare and contrast their performances. *(M.G University., May 2008)*
15. Draw a series pass voltage regulator circuit with sort circuit protection and explain the working. *(M.G University., May. 2005)*
16. Draw the circuit diagram of a series voltage regulator and explain its operation. Why and how in short circuit protection included in this. *(M.G University., July/Aug. 2007)*
17. Draw and explain the working of a series pass voltage regulator with feedback. How does the circuit provide regulation against the (1) line voltage variation; (2) load current variation *(M.G University., May/June 2006)*
18. Sketch the circuit diagram of a simple Zener shunt voltage regulator and explain how it regulates the output voltage against variations in (i) input line voltage and (ii) in load current. *(M.G University., Dec 2007)*
19. How does Zener diode maintain constant voltage across load in the breakdown region? *(PTU., May 2005)*
20. Draw the equivalent circuit of Zener diode as voltage regulator and explain? *(PTU., May 2006)*
21. What is the need of voltage regulator circuit in a power supply? Discuss the application of zener diode as voltage regulator. *(PTU., Dec 2009)*
22. Draw the circuit of transistor shunt voltage regulator and explain. *(Calicut University., June 2007)*
23. What are the various blocks that form a Basic Voltage Regulator? Explain the series shunt voltage regulator. List advantages of Voltage Regulator? *(Anna University., April/May 2008)*
24. Explain the procedure for designing a shunt type voltage regulator. *(Anna University., April/May 2008, GBTU/MTU 2004-05)*
25. Draw a circuit diagram which uses a breakdown diode to regulate the output voltage across a load and explain the working. *(GBTU/MTU., 2005-06)*
26. What is a switching regulator? What are the types of switching regulator? Explain them in detail with suitable figures. *(Anna University., April/May 2008)*
27. Write short notes on switching mode power supply. *(Anna University., April/May 2008)*
28. Draw and explain the working principle of a SMPS circuit with its output waveforms. *(Anna University., April/May 2008)*
29. With neat sketch explain the switched Mode Power supplies. *(Anna University., Nov/Dec 2007)*

30. Draw the functional block diagram of switching regulator and explain. *(Anna University., April/May 2008)*
31. Draw the circuit of shunt voltage regulator and explain its working. *(JNTU., Aug./Sep. 2008)*
32. What is the practical importance of voltage regulation in power supplier? *(PTU., Dec 2005)*
33. Write a short note on transistor series and shunt regulator? *(PTU., Dec 2005)*
34. Write a note on transistor series regulators. *(PTU., May 2005)*
35. With the help of circuit diagram Explain the working of a transistor series regulator. *(PTU., May 2009)*
36. Explain the circuit diagram of a 3 pin voltage regulator, specifying the input voltage range. *(M.G University., May. 2005)*
37. Draw the characteristics of zener diode and explain how it can be used as a voltage regulator. *(R.G.P.V., June 2011)*
38. What are voltage regulators? Classify them and explain in brief? *(R.G.P.V., June 2002)*
39. What are the reasons needs of a voltage regulator? Enlist various types of voltage regulators. *(R.G.P.V., June 2003)*
40. Explain the term "Voltage regulation". Discuss the working of a simple series regulator. Draw it circuit. *(R.G.P.V., June 2005)*
41. How the regulator voltage may be achieved. *(R.G.P.V., Dec. 2006)*
42. Give the classification of voltage regulators. Explain in the working of adjustable voltage regulators. *(R.G.P.V., June 2007)*
43. Define all the characteristics parameters of voltage regulator device. *(R.G.P.V., June 2008)*
44. Draw a current regulator circuit using IC-7805 and discuss the regulation mechanism. Also give its merits and demerits. *(R.G.P.V., June 2008)*
45. Explain the function of zener diode as regulator. *(Nagpur University, Winter 2006)*
46. Draw the circuit of a series regulator and explain its action in brief. *(Nagpur University, Winter 2002)*
47. Write short notes on :
    (*a*) Series voltage regulator *(Nagpur University, Summer 2005)*
    (*b*) Voltage regulators *(R.G.P.V., Dec. 2002)*
48. Explain the types of switching regulators and enlist their characteristics. Why do switching regulator have better efficiency than the series regulator? *(GBTU., 2010, 2011)*
49. Explain the significance of line and load regulation. *(VTU., June/July 2011)*

MULTIPLE CHOICE QUESTIONS

1. Which of the following is not an essential element of a d.c. power supply.
   (*a*) Rectifier (*b*) Filter
   (*c*) Voltage regulator (*d*) Voltage amplifier
2. A voltage regulator is a circuit which
   (*a*) Converts the a.c. voltage to d.c. voltage
   (*b*) Smoothens the a.c. variation in d.c. output voltage

(*c*) Maintains a constant d.c. output voltage inspite of the fluctuations in a.c. input voltage or load current

(*d*) None of the above

3. The percentage voltage regulation of voltage supply providing 100 V unloaded and 95 V at full load is

(*a*) 5.3% (*b*) 5.0%

(*c*) 0.53% (*d*) None of the above

4. Which of the following voltage regulator is preferred for providing large values of load current.

(*a*) Zener diode shunt regulator (*b*) Transistor series regulator

(*c*) Transistor shunt regulator (*d*) None of the above

5. A pre-regulator is used with a transistor series regulator in order to

(*a*) Increase the input resistance of the series transistor

(*b*) Decrease the input resistance of the series transistor

(*c*) Increase the output resistance of the series transistor

(*d*) Increase the current gain of the series transistor

6. The main job of a voltage regulator is to provide a nearly—output voltage.

(*a*) sinusoidal (*b*) constant (*c*) smooth (*d*) fluctuating

7. A 10-V dc regulator power supply has a regulation of 0.005 per cent. Its output voltage will vary within an envelope of .... millivolt.

(*a*) ± 2.5 (*b*) ± 0.5 (*c*) ± 5 (*d*) ± 0.05

8. An ideal voltage regulator has a voltage regulation of

(*a*) 1 (*b*) 100 (*c*) 50 (*d*) 0.

9. In a Zener diode shunt voltage regulator, the diode regulates so long as it is kept in .... condition.

(*a*) forward (*b*) reverse (*c*) loaded (*d*) unloaded

10. The power efficiency of a switching voltage regulator is much higher than that of a linear regulator because it operates.

(*a*) in saturation (*b*) in cut-off

(*c*) like a switch (*d*) on high duty cycle.

11. A transistor series voltage regulator is called emitter-follower regulator because the emitter of the pass transistor follows the ....voltage.

(*a*) output (*b*) input (*c*) base (*d*) collector

12. In an op-amp series voltage regulator, output voltage depends on

(*a*) Zener voltage (*b*) voltage divider resistors

(*c*) output voltage (*d*) both (*a*) and (*b*)

13. In a feedback series regulator circuit, the output voltage is regulated by controlling the

(*a*) magnitude of input voltage

(*b*) gain of the feedback transistor

(*c*) reference voltage

(*d*) voltage drop across the series pass transistor

14. An op-amp shunt regulator differs from the series regulator in the sense that its control element is connected in

(*a*) series with line resistor (*b*) parallel with the resistor

(*c*) parallel with load resistor (*d*) parallel with input voltage.

15. A switching voltage regulator can be of the following type :
(*a*) step-down (*b*) step-up (*c*) inverting (*d*) all of the above

16. In an inverting tupe switching regulator, output voltage is .... input voltage.
(*a*) lesser than (*b*) greater than (*c*) equal to (*d*) opposite to.

17. The output voltage of a step-down type switch ing voltage regulator depends on
(*a*) input voltage (*b*) duty cycle
(*c*) transistor on time (*d*) all of the above.

18. As compared to voltage regulators made up of discrete components, IC regulators have the inherent advantage/s of
(*a*) self protection against over-temperature (*b*) remote control
(*c*) current limiting (*d*) all of the above.

19. A 12 V monolithic regulator is adjusted to obtain a higher output voltages as shown in Fig. 20.67. The v_o will be
(*a*) 12 V (*b*) 17 V (*c*) 24 V (*d*) 29 V

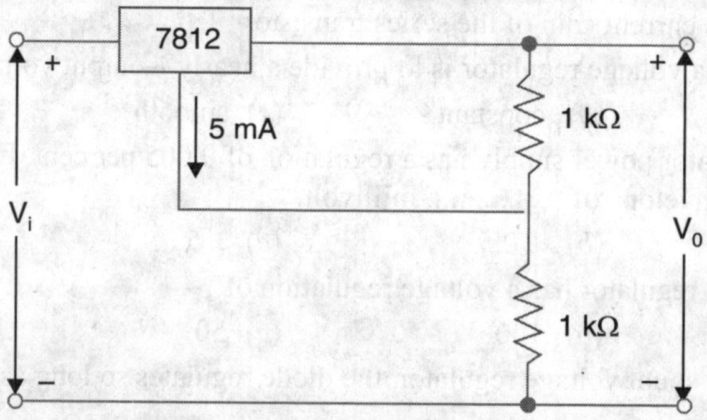

Fig. 20.67.

20. A three terminal monolithic IC regulator can be used as
(*a*) an adjustable output voltage regulator alone
(*b*) an adjustable output voltage regulator and a current regulator
(*c*) a current regulator and a power switch
(*d*) a current regulator alone

21. In a switched-mode power supply (SMPS), after conversion of ac supply to a highly filtered dc voltage, a switching transistor is switched ON and OFF at a very high speed by a pulse width modulator (PWM) which generates very high frequency square pulses. The frequency of the pulses is typically in the range of
(*a*) 100 Hz-200 Hz (*b*) 500 Hz-1 kHz
(*c*) 2 kHz-5 kHz (*d*) 20 kHz-50kHz

(*UPSC Engg. Services. 2002*)

ANSWERS

1. (*d*)	**2.** (*c*)	**3.** (*a*)	**4.** (*b*)	**5.** (*d*)
6. (*b*)	**7.** (*a*)	**8.** (*d*)	**9.** (*b*)	**10.** (*c*)
11. (*c*)	**12.** (*d*)	**13.** (*d*)	**14.** (*a*)	**15.** (*d*)
16. (*d*)	**17.** (*d*)	**18.** (*d*)	**19.** (*d*)	**20.** (*d*)
21. (*d*)				

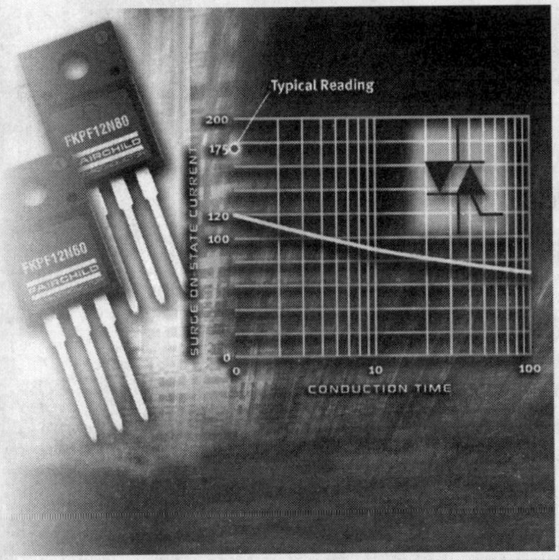

21 Chapter

Controlled Rectifiers

Outline

1. Introduction.
2. Output Waveforms for Different Firing Angles.
3. Values of Output Voltage and Current in Controlled Rectifiers.
4. Average values for Full-wave Controlled Rectifier.
5. Silicon Controlled Rectifier (SCR).
6. Power Control Using SCR.
7. SCR Half-wave Rectifier.
8. Average Value of Load Voltage and Current.
9. 90° Variable Half-wave Rectifier.
10. 180° Variable Half-wave Rectifier.
11. SCR Full-wave Rectifier.
12. UJT Triggered SCR Phase Control.
13. Triac Power Control.
14. Diac-Triac Phase Control Circuit.
15. Chopper.
16. Inverters.
17. Single-phase Inverter.
18. Push-Pull Inverter.

21.1. Introduction

We have already discussed in chapter on rectifiers and filters about diodes, half-wave and full-wave rectifiers. In such rectifier circuits, the current flows through the diode, whenever the instantaneous value of the a.c. input voltage is greater then the voltage across the load at that instant. This point is made clear in the half-wave diode rectifier circuit as shown in Fig. 21.1. In such case, the load current flows during all times of the positive half cycles of the input voltage.

We shall now introduce a new class of rectifier called *controlled rectifier*. In such a rectifier, special semiconductor device such as *silicon controlled rectifier* abbreviated as SCR or a *triac* is used instead of an ordinary PN junction diode. Such a device has three terminals, namely anode, cathode and gate. The load current, in a controlled rectifier, flows only when a control signal is applied at the gate terminal to turn 'ON' the rectifier.

Fig. 21.2 (*a*) shows the half-wave controlled rectifier circuit and Fig. 21.2 (*b*) shows the input voltage waveform and load current waveform. It may be noted that the control signal is applied to turn on the rectifier in such a way that the conduction starts at a specific point marked A in Fig. 21.2 (*b*). The point A (called *firing point*) corresponds to the angle

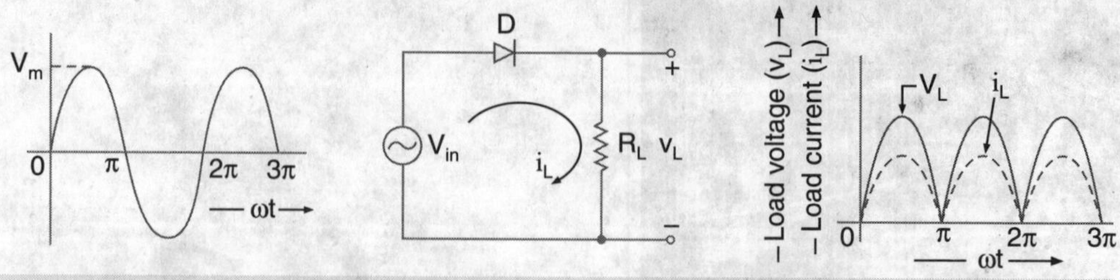

Fig. 21.1. Diode half-wave rectifier.

α, so that the conduction is delayed by this much period. It is because of this delay that the amount of load current has reduced in a controlled rectifier as compared to an ordinary rectifier.

It will be interesting to know that once the rectifier is turned 'ON' it remains in conduction for the rest of the positive half-cycle *i.e.*, up to 180° (or π-radians). Thus the point *A* (at which the conduction starts) is determined by the angle of delay in applying control signal to the rectifier circuit. As the angle α increases, conduction is further delayed and the load current is also reduced.

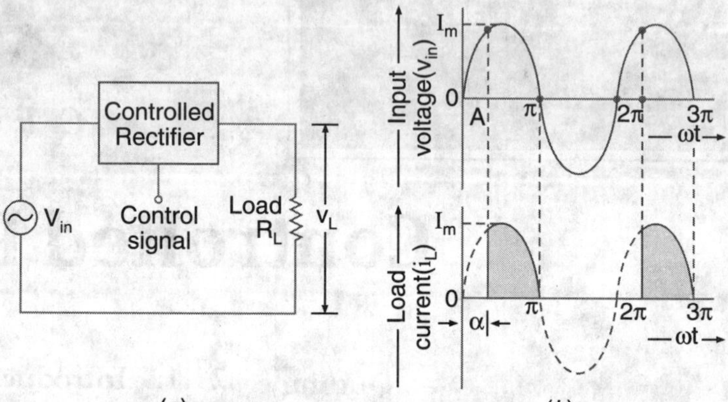

(*a*) (*b*)

Fig. 21.2. Half-wave controlled rectifier.

It is evident from the above discussion, that load current in a controlled rectifier may be controlled easily by selecting the firing point. This feature (*i.e.*, control over the load current) is of great advantage in controlled rectifiers.

Let the equation of the input voltage be,

$$v_i = V_m \sin \omega t$$

$$= V_m \sin \theta$$

For a half-wave rectifier, the average value is found by taking the average over the whole cycle (0 to 2π) even though the conduction takes place only from 0 to π in the entire cycle.

$$\therefore \quad V_L = \frac{1}{2\pi}\int_0^{\pi} v_{in}\cdot d\theta$$

$$= \frac{1}{2\pi}\int_0^{\pi} v_m \sin\theta \cdot d\theta$$

$$= \frac{V_m}{2\pi}\int_0^{\pi} \sin\theta \cdot d\theta$$

$$= \frac{V_m}{2\pi}\Big|-\cos\theta\Big|_{\alpha}^{\pi}$$

$$= \frac{V_m}{2\pi}\left|(-\cos\pi) - (-\cos\alpha)\right|$$

$$= \frac{V_m}{2\pi}(1 + \cos\alpha)$$

and dc output voltage,

$$V_{dc} = V_L = \frac{V_m}{2\pi}(1 + \cos\alpha)$$

For a resistive load,

$$I_{dc} = I_L = \frac{V_m}{2\pi R_L}(1 + \cos\alpha)$$

Notice that if $\alpha = 0$, then

$$V_{dc} = \frac{V_m}{\pi} \text{ and } I_{dc} = \frac{V_m}{\pi R_L}$$

Recall that these are the same values as obtained for a uncontrolled half-wave rectifier.

21.2. Output Waveforms for Different Firing Angles

Fig. 21.3 (*a*), (*b*), (*c*) and (*d*) shows the output waveforms for different firing angles. (0°, 45°, 90° and 135°) along with the waveforms for input voltage and voltage drop across the controlled rectifier.

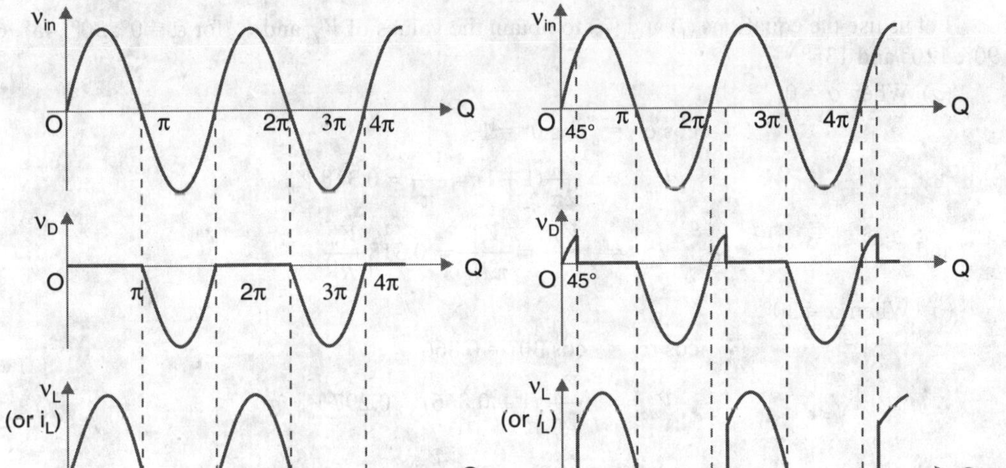

Fig. 21.3.

For all these waveforms, we have assumed that ac supply voltage is sinusoidal ($V_{in} = V_m \sin \theta$). Also a resistive load is assumed. Further, the diode is assumed to be an ideal *i.e.*, when it conducts, the voltage drop across it is 0. When the diode doesnot conduct, it is 'open' circuit.

It is evident from the waveforms that as the firing angle 'α' is increased from 0 through 135°, the output voltage keeps decreasing till for α = 180°, the output is zero.

21.3. Values of Output Voltage and Current in Controlled Rectifiers

We have already mentioned in Art. 21.1. that the dc output voltage for a half-wave rectifier is given by the equation,

$$V_{dc} = \frac{V_m}{2\pi}(1 + \cos\alpha) \qquad ...(i)$$

and the dc output current,

$$I_{dc} = \frac{V_m}{2\pi R_L}(1 + \cos\alpha) = \frac{V_{dc}}{R_L} \qquad ...(ii)$$

Let us use the equations (*i*) and (*ii*) to obtain the values of V_{dc} and I_{dc} for α = 0°, 30°, 45°, 60°, 90°, 120° and 135°.

(*a*) When α = 0,

$$\cos\alpha = \cos 0° = 1$$

$$\therefore \quad V_{dc} = \frac{V_m}{2\pi}(1 + 1) = \frac{V_m}{\pi} = 0.318\, V_m$$

$$\text{and} \quad I_{dc} = \frac{V_{dc}}{R_L} = \frac{V_m}{\pi R_L} = 0.318 \frac{V_m}{R_L}$$

(*b*) When α = 30°

$$\cos\alpha = \cos 30° = 0.866$$

$$\therefore \quad V_{dc} = \frac{V_m}{2\pi}(1 + 0.866) = 0.297\, V_m$$

$$\text{and} \quad I_{dc} = \frac{V_{dc}}{R_L} = 0.297 \frac{V_m}{R_L}$$

(*c*) When α = 45°

$$\cos\alpha = \cos 45° = 0.707$$

$$\therefore \quad V_{dc} = \frac{V_m}{2\pi}(1 + 0.707) = 0.27\, V_m$$

$$\text{and} \quad I_{dc} = \frac{V_{dc}}{R_L} = 0.27 \frac{V_m}{R_L}$$

(*d*) When α = 60°

$$\cos\alpha = \cos 60° = 0.5$$

$$\therefore \quad V_{dc} = \frac{V_m}{2\pi}(1 + 0.5) = 0.239\, V_m$$

$$\text{and} \quad I_{dc} = \frac{V_{dc}}{R_L} = 0.239 \frac{V_m}{R_L}$$

(*e*) When α = 90°

$$\cos\alpha = \cos 90° = 0$$

$$\therefore \quad V_{dc} = \frac{V_m}{2\pi}(1+0) = \frac{V_m}{2\pi} = 0.159\, V_m$$

$$\text{and} \quad I_{dc} = \frac{V_{dc}}{R_L} = \frac{V_m}{2\pi R_L} = 0.159\,\frac{V_m}{R_L}$$

(*f*) When $\alpha = 120°$

$$\cos\alpha = \cos 120° = -0.5$$

$$\therefore \quad V_{dc} = \frac{V_m}{2\pi}(1-0.5) = 0.08\, V_m$$

$$\text{and} \quad I_{dc} = \frac{V_{dc}}{R_L} = 0.08\,\frac{V_m}{R_L}$$

(*g*) When $\alpha = 135°$

$$\cos\alpha = \cos 135° = -0.707$$

$$\therefore \quad V_{dc} = \frac{V_m}{2\pi}(1-0.707) = 0.0466\, V_m$$

$$\text{and} \quad I_{dc} = \frac{V_{dc}}{R_L} = 0.0466\,\frac{V_m}{R_L}$$

Example 21.1.

A 100 Ω load is connected to a peak supply of 300 V through a controlled half-wave diode rectifier. The load power is to varied from 25 W to 80 W. What is the angular firing control required ? Neglect forward voltage drop of the diode. (*Pune Univ. 1990*)

Solution.

Given : $R_L = 100\ \Omega$ and $V_m = 300$ V.

We know that load power,

$$P = V_{dc} \cdot I_{dc}$$

$$= \left[\frac{V_m}{2\pi}(1+\cos\alpha)\right] \times \left[\frac{V_m}{2\pi R_L}(1+\cos\alpha)\right]$$

$$= \left(\frac{V_m}{2\pi}\right)^2 \cdot \frac{1}{R_L} \cdot (1+\cos\alpha)^2$$

$$= \left(\frac{300}{2\pi}\right)^2 \cdot \frac{1}{100} \cdot (1+\cos\alpha)^2$$

$$= 22.8\,(1+\cos\alpha)^2$$

Angular firing control when load power P = 25 W,

$$25 = 22.8\,(1+\cos\alpha)^2$$

$$\text{or} \quad 1+\cos\alpha = \sqrt{\left(\frac{25}{22.8}\right)} = 1.096$$

$$\therefore \quad \cos\alpha = 0.096 \text{ or } \alpha = 89.4°$$

Angular firing control when load power P = 80 W,

$$80 = 22.8\,(1+\cos\alpha)^2$$

or $$1 + \cos\alpha = \sqrt{\frac{80}{22.8}} = 1.873$$

$\therefore$ $$\cos\alpha = 0.873 \text{ or } \alpha = 29.2°$$

This shows that when the load power is varied from 25 W to 80 W, the firing angle is reduced form 89.4° to 29.2°. **Ans.**

Example 21.2.

In a controlled half-wave rectifier, peak supply voltage is 200 V and the value of load resistor is 1 kΩ. Calculate the power delivered to the load circuit for firing angles of (i) 0°, (ii) 45°, (iii) 90° and (iv) 135°. *(BHU, 1998)*

Solution.

Given : $V_m = 200$, and $R_L = 1$ kΩ

(i) $\alpha = 0°$

We know that dc output voltage,

$$V_{dc} = 0.318\, V_m = 0.318 \times 200 = 63.6 \text{ V}$$

and the dc output current,

$$I_{dc} = \frac{V_{dc}}{R_L} = \frac{63.6}{1000} = 63.6 \text{ mA}$$

$\therefore$ Load power, $$P = 63.6 \text{ V} \times 63.6 \text{ mA} = V_{dc} \cdot I_{dc}$$
$$= 4045 \text{ mW or } 4.045 \text{ W } \textbf{Ans.}$$

(ii) $\alpha = 45°$

We know that dc output voltage,

$$V_{dc} = 0.27\, V_m = 0.27 \times 200 = 54 \text{ V}$$

and the dc output current,

$$I_{dc} = \frac{V_{dc}}{R_L} = \frac{54}{1000} = 54 \text{ mA}$$

$\therefore$ Load power,

$$P = V_{dc} \cdot I_{dc}$$
$$= 54 \text{ V} \times 54 \text{ mA}$$
$$= 2916 \text{ mW or } 2.916 \text{ W } \textbf{Ans.}$$

(iii) $\alpha = 90°$

We know that dc output voltage,

$$V_{dc} = 0.159\, V_m = 0.159 \times 200 = 31.8 \text{ V}$$

and $$I_{dc} = \frac{V_{dc}}{R_L} = \frac{31.8}{1000} = 31.8 \text{ mA}$$

$\therefore$ $$P = V_{dc} \cdot I_{dc} = 31.8 \text{ V} \times 31.8 \text{ mA} = 1011 \text{ mW or } 1.011 \text{ W } \textbf{Ans.}$$

(iv) $\alpha = 135°$

We know that dc output voltage,

$$V_{dc} = 0.0466\, V_m = 0.0466 \times 200 = 9.3 \text{ V}$$

and $$I_{dc} = \frac{V_{dc}}{R_L} = \frac{9.3}{1000} = 9.3 \text{ mA}$$

$\therefore$ $$P_{dc} = V_{dc} \cdot I_{dc} = 9.3 \text{ V} \times 9.3 \text{ mA} = 86.49 \text{ mW } \textbf{Ans.}$$

21.4. Average Values for Full-wave Controlled Rectifier

Fig. 21.4 shows the full-wave rectified waveform to be controlled by the rectifier for θ = 60°. As seen from this diagram, the average values will be doubled because there are two half sinusoids to be arranged.

Thus the dc output voltage,

$$V_{dc} = 2 \times \left[\frac{V_m}{2\pi}(1+\cos\theta_1)\right] = \frac{V_m}{\pi}(1+\cos\theta_1)$$

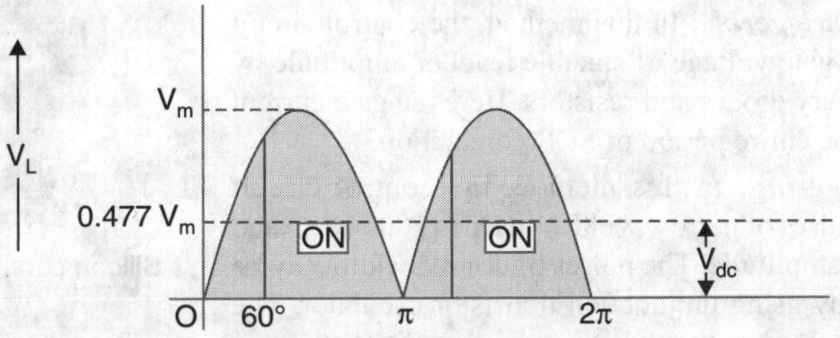

Fig. 21.4.

In the present case, θ = 60°,

$$\therefore \quad V_{dc} = \frac{V_m}{\pi}(1+\cos 60°) = \frac{V_m}{\pi}(1+0.5) = 0.477\,V_m$$

and the dc output current,

$$I_{dc} = \frac{V_{dc}}{R_L}$$

21.5. Silicon Controlled Rectifier (SCR)

It is a three junction, four-layer and three terminal semiconductor device. The three terminals are anode, cathode and gate. The basic construction and circuit symbol for the silicon controlled rectifier (SCR) is as shown in Fig. 21.5 (*a*) and (*b*).

When the anode is made positive, with respect to cathode, the junctions J_1 and J_3 are forward biased, while the junction J_2 is reverse biased. As a result of this, the SCR does not conduct. However, if a positive voltage of suitable amplitude is applied at the gate, with respect to the cathode the junction J_2 is forward biased and the SCR is triggered *i.e.*, it starts conducting like an ordinary diode.

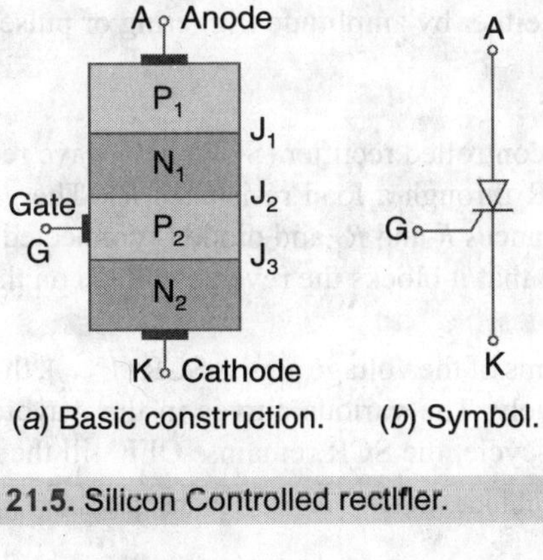

(*a*) Basic construction. (*b*) Symbol.

Fig. 21.5. Silicon Controlled rectifier.

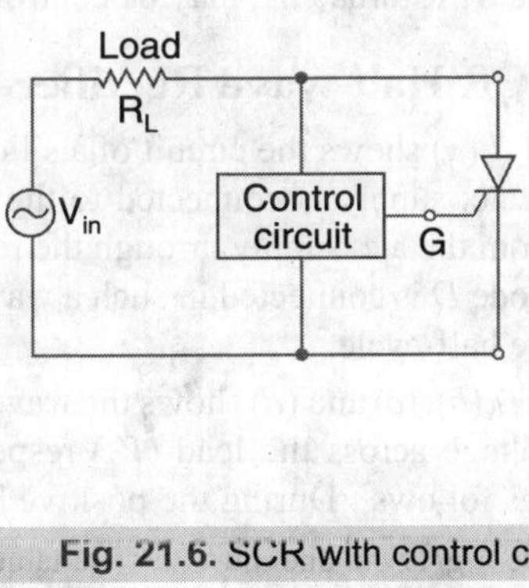

Fig. 21.6. SCR with control circuit.

It will be interesting to know that once the SCR turns 'ON' the gate has no control over the SCR. The conduction can only be terminated by removing the applied voltage from the anode.

As mentioned above, the SCR needs positive voltage at the gate terminal to turn 'ON'. This voltage is provided by a control circuit as shown in Fig. 21.6. Although there are a number of methods of turning SCR in its 'ON' position, yet the following two are important from the subject point of view :

Silicon controlled rectifier.

1. ***Amplitude triggering.*** In this method, the control circuit produces a gate voltage of suitable level or amplitude by using ordinary diodes and resistors. Here the gate current flows for the entire period of SCR conduction.
2. ***Pulse triggering.*** In this method, the control circuit produces pulses of narrow width (1 μsec to 1000 μsec) and of suitable amplitude. The pulses of desired width may be generated by using unijunction transistor oscillator. The pulse triggering has an advantage over amplitude triggering that in case of pulse triggering the gate current is supplied only for a short duration.

21.6. Power Control Using SCR

There are large number of applications in industry, which require some control over the amount of current supplied to the load. Some of these applications are *electric welding*, *lighting control*, *motor speed control* etc. We know that a simple way to vary the load current is either to control the transformer secondary voltage or insert a resistor in the load circuit. But in actual practice, neither of these two methods is desirable. The control of transformer secondary voltage requires an expensive equipment, whereas the insertion of resistor in the load circuit results in a loss of power. Therefore, the use of silicon controlled rectifier (SCR)* is an efficient way of controlling the power. When an a.c. voltage is applied across an SCR, it will turn 'ON' during each positive half cycle and 'OFF' during each negative half cycle. The average load current can be varied over a wide limit by controlling the point in each positive half-cycle at which the SCR is turned ON. The point, at which the SCR turns ON, may be controlled either by amplitude triggering or pulse triggering.

SCR Power Control

21.7. SCR Half-wave Rectifier

Fig. 21.7 (*a*) shows the circuit of a silicon controlled rectifier (SCR) half-wave rectifier. In this circuit, the a.c. supply is connected to the SCR through a load resistance R_L. The gate current is obtained from the a.c. supply through the resistances R and R_1 and diode D connected in series.

The diode D is connected in such a way, so that it blocks the reverse voltage on the gate during the negative half-cycle.

Fig. 21.7 (*b*), (*c*) and (*d*) shows the waveforms of the voltage across SCR (V_{SCR}), the load current (I_L) and voltage across the load (V_L) respectively. The various curves in the waveforms may be explained as follows : During the positive half-cycle, the SCR remains 'OFF' till the input voltage

* The gate trigger voltage must be of suitable amplitude, which can provide the necessary gate current for SCR triggering.

reaches the gate trigger voltage *i.e.*, point A in Fig. 21.7 (*b*). As a result, there is no current through R_L. Hence load current (I_L) and load voltage (V_L) are zero. At point *A*, SCR is fired into conduction. It acts like a short and voltage across it drops to zero *i.e.*, curve *AB* in Fig. 21.7 (*b*). Under this condition, whole of the applied voltage drops across the load resistance. During the negative input-half cycle, the SCR is reverse biased and hence does not conduct. As a result, all the applied voltage appears across SCR and none across the load resistance (R_L).

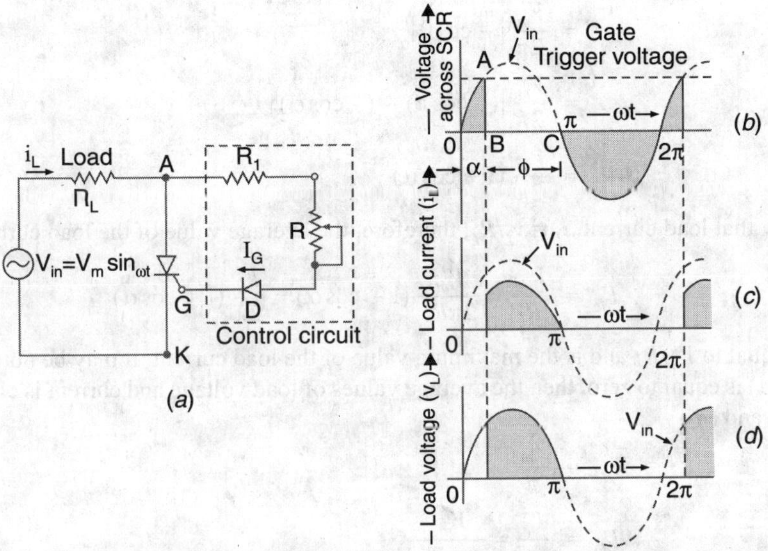

Fig. 21.7. SCR half-wave rectifier and its waveforms.

The horizontal distance between the points *O* and *A* in Fig. 21.7 (*b*) represents the time in the positive half-cycle, when SCR is not conducting. This distance in degrees is called the firing angle, phase angle or delay angle (α). This angle gives us an idea about the position at which the SCR starts conduction with respect to the origin. It may be noted from the figure that the SCR is conducting between the points *B* and *C*. This angle is called conduction angle (ϕ) and its value is equal to (π – α).

21.8. Average Values of Load Voltage and Current

We have already discussed in the last article, about the SCR half-wave rectifier. Now we shall determine its average values of load voltage (V_L) and load current (i_L). We know that the equation of a.c. supply voltage is

$$v_{in} = V_m \cdot \sin \omega t = V_m \cdot \sin \theta \qquad \text{... } (\because \theta = \omega t)$$

We also know that for a half-wave rectifier, the average value of a load voltage is determined by the relation,

$$V_{dc} = \frac{\text{Area under the curve over a complete cycle}}{\text{Base}}$$

$$= \frac{\int_0^{2\pi} v_{in} \cdot d\theta}{2\pi} = \frac{1}{2\pi} \int_0^{2\pi} V_m \cdot \sin \theta \, d\theta \qquad \text{... } (i)$$

Now the load voltage is developed only during the period SCR* conducts. This period lies

* Here we have neglected the forward voltage drop across the device. Its value is about 1 volt.

between the angles α and π. Therefore, we must take the average over the limits α to π instead of 0 to 2π in equation (*i*).

$$\therefore \quad V_{dc} = \frac{1}{2\pi}\int_{\alpha}^{\pi} V_m \cdot \sin\theta \, d\theta$$

$$= \frac{V_m}{2\pi}\int_{\alpha}^{\pi} \sin\theta \cdot d\theta$$

$$= \frac{V_m}{2\pi}\left| -\cos\theta \right|_{\alpha}^{\pi}$$

$$= \frac{V_m}{2\pi}\left|(-\cos\pi) - (-\cos\alpha)\right|$$

$$= \frac{V_m}{2\pi}(1 + \cos\alpha) \quad \text{... (ii)}$$

We know that load current, $i_L = v_L/R_L$, therefore, the average value of the load current,

$$I_{dc} = \frac{V_{dc}}{R_L} = \frac{V_m}{2\pi R_L}(1 + \cos\alpha) = \frac{I_m}{2\pi}(1 + \cos\alpha) \quad \text{...(iii)}$$

where I_m is equal to V_m/R_L and is the maximum value of the load current. It may be noted that if the firing angle (α) is equal to zero, then the average values of load voltage and current is obtained from equations (*ii*) and (*iii*)

$$\therefore \quad V_{dc} = \frac{V_m}{\pi}$$

$$\text{and} \quad I_{dc} = \frac{I_m}{\pi} = \frac{V_m}{\pi \cdot R_L}$$

It may be noted that the above results are the same as obtained for an ordinary diode half-wave rectifier.

Example 21.3.

An SCR is used for converting a.c. to d.c. The anode supply is 220 volts, 50 Hz and the firing angle is adjusted at 60°. Find out the d.c. output voltage.

(U.P.S.C., Engg. Services, 1985)

Solution.

Given : $V_{rms} = 220$ V and $\alpha = 60°$.

We know that maximum value of the applied voltage *i.e.*, anode supply,

$$V_m = \sqrt{2} \times V_{rms} = \sqrt{2} \times 220 = 311 \text{ V}$$

and the d.c. output voltage,

$$V_{dc} = \frac{V_m}{2\pi}(1 + \cos\alpha) = \frac{311}{2\pi}(1 + \cos 60°) = 74.2 \text{ V } \textbf{Ans.}$$

Example 21.4.

Find the value of resistance to limit the average current to 0.5 A, when 100 V r.m.s., is applied to an SCR with a firing angle of 45°.

Solution.

Given : $I_{dc} = 0.5$ A; $V_{rms} = 100$ volts and $\alpha = 45°$

Let R_L = Value of resistance, which limits the anode current to 0.5 A

We know that maximum value of applied voltage,

$$V_m = \sqrt{2} \times V_{rms} = \sqrt{2} \times 100 = 141.4 \text{ V}$$

and the average current (I_{dc})

$$0.5 = \frac{V_m}{2\pi \cdot R_L}(1 + \cos\alpha) = \frac{141.4}{2\pi \cdot R_L}(1 + \cos 45°)$$

$$= \frac{141.4}{2\pi \cdot R_L}(1 + 0.707) = \frac{38.42}{R_L}$$

$$\therefore \quad R_L = 38.42/0.5 = 76.84 \text{ W } \textbf{Ans.}$$

21.9. 90° Variable Half-wave Rectifier

The SCR half-wave rectifier circuit shown in Fig. 21.7(*a*) is known as 90° variable half-wave rectifier (or 90° variable phase control circuit). The term 90° variable implies that we can make SCR to conduct anywhere from the origin to the maximum value of the positive half-cycle *i.e.*, the firing angle can be varied from 0° to 90°. The firing angle depends upon the value of variable resistance (*R*). If the resistance *R* is set at a low value, the SCR will start conducting near the origin. On the other hand, if *R* is set at a large value, the SCR may not conduct until the maximum value of positive half-cycle.

21.10. 180° Variable Half-wave Rectifier

Fig. 21.8 (*a*) shows that circuit of a 180° variable half-wave rectifier. It is also called 180° variable phase control circuit. We can see the difference between this circuit and that of a 90° variable phase control circuit. In this circuit, a capacitor (*C*) has been added between the variable resistance (*R*) and the cathode of SCR. The combination of resistance (*R*) and capacitor (*C*) forms a phase-shift network. This network delays the point at which SCR will trigger on the positive half of the applied voltage. The delay is provided by charging of the capacitor through the resistance *R*. This allows the firing angle to be controlled from 0° up to a maximum of 180°.

Fig. 21.8 (*b*) shows the waveforms of input voltage (V_{in}), capacitor voltage (V_C) and load voltage (V_L). It may be noted that capacitor voltage lags behind the input voltage. Their phase relationship is given by the relation,

$$V_C = \frac{V_{in}}{1 + \omega \cdot C \cdot R}$$

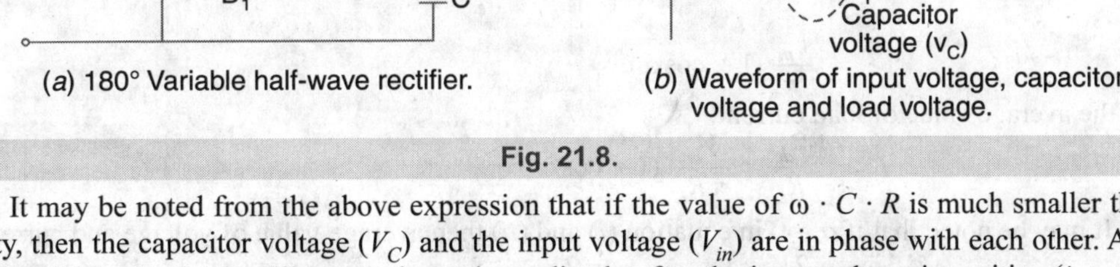

(*a*) 180° Variable half-wave rectifier.

(*b*) Waveform of input voltage, capacitor voltage and load voltage.

Fig. 21.8.

It may be noted from the above expression that if the value of $\omega \cdot C \cdot R$ is much smaller than unity, then the capacitor voltage (V_C) and the input voltage (V_{in}) are in phase with each other. As a result of this, the firing will occur almost immediately after the input voltage is positive (*i.e.*, $\alpha = 0°$, $\phi = 180°$). On the other hand, if the value of $\omega \cdot C \cdot R$ is much larger than unity, then capacitor voltage is delayed by practically 90° from the input voltage and is much reduced in amplitude. This allows the firing to be delayed to almost the end of half a cycle (*i.e.*, $\alpha = 180°$, $\phi = 0°$).

The diode D_2 has been added to reset capacitor (C) by discharging it during the negative half cycle. This allows the firing to occur at the same point in each cycle.

21.11. SCR Full-wave Rectifier

Fig. 21.9 (*a*) shows one of the circuits used for SCR full-wave rectifier. It consists of a diode bridge in series with SCR. The gate control input is left open, because a number of different firing circuits can be used.

The working of the SCR full-wave rectifier circuit may be understood from the following discussion. When the terminal A is positive, the current will flow through the diode D_1, through SCR, then through load resistance and then finally through diode D_2 it passes to terminal B. However, if the terminal B is positive, the current will flow through diode D_3, SCR, load resistance, diode D_4 and then finally to terminal A. Thus the four diodes forming a bridge provide full-wave rectification, while the SCR controls the amount of voltage applied to the load.

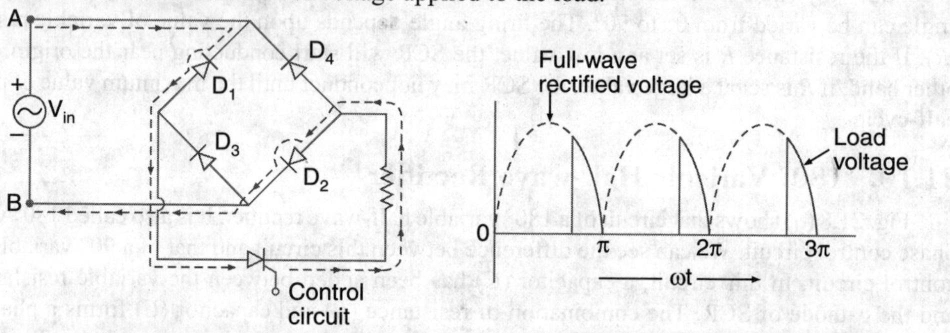

(*a*) SCR full-wave rectifier. (*b*) The full-wave rectified and load voltage waveforms.

Fig. 21.9.

Fig. 21.9 (*b*) shows the full-wave rectified and load voltage waveforms. The amount of voltage applied to the load may be controlled by adjusting the firing angle of the SCR. This can be achieved by adjusting the gate current supplied by the control circuit.

The average value of load voltage and load current may be obtained from the equation of the applied voltage, *i.e.*,

$$V_{in} = V_m \sin \omega t = V_m \sin \theta$$

Then average value of load voltage,

$$V_{dc} = \frac{1}{\pi}\int_{\alpha}^{\pi} V_m \sin\theta \cdot d\theta = \frac{V_m}{\pi}\int_{\alpha}^{\pi} \sin\theta \, d\theta$$

$$= \frac{V_m}{\pi}\left| -\cos\theta \right|_{\alpha}^{\pi} = \frac{V_m}{\pi}[(-\cos\pi) - (-\cos\alpha)]$$

$$= \frac{V_m}{\pi}(1 + \cos\alpha) \quad \text{... (i)}$$

and the average value of load current,

$$I_{dc} = \frac{V_{dc}}{R_L} = \frac{V_m}{\pi R_L}(1 + \cos\alpha) \quad \text{...(ii)}$$

It may be noted that if $\alpha = 0$ in equation (*i*) and (*ii*) then average value of voltage and current,

$$V_{dc} = \frac{2V_m}{\pi} \text{ and } I_{dc} = \frac{2V_m}{\pi R_L}$$

These values are the same as in an ordinary diode full-wave rectifier.

21.12. UJT Triggered SCR Phase Control

We have already discussed in previous articles that SCR's are used in power control circuits. In these circuits, a control circuit is needed to vary the firing or the phase angle at which the SCR is triggered. The value of phase angle controls the power delivered to the load. If we want to deliver the maximum power to the load, then the SCR must be triggered as soon as the a.c. voltage across it goes positive *i.e.*, the firing angle must be zero. Therefore, the SCR will conduct for 180° after triggering. If the power requirement is below the maximum value, the firing angle must be between 0° and 180°. In order to adjust the firing angle between 0° and 180°, we need a control circuit. We have already seen that a circuit shown in Fig. 21.8 (*a*) can be used to accomplish the control. But this circuit is affected by loading and supply voltage variation. Therefore, this circuit does not give a better phase control. However, a better control can be achieved by using a unijunction transistor (UJT) relaxation oscillator circuit.

Fig. 21.10 shows a UJT oscillator controlling a SCR circuit. The SCR is in it 'OFF' state until the UJT fires. When the UJT fires, the capacitor will discharge quickly through the resistor R_4. This positive pulse will turn 'ON' the SCR. The SCR will remain 'ON' until the SCR line voltage approaches zero. At this point, the SCR will turn 'OFF' throughout the entire negative circle. A similar circuit can also be made for triac phase control.

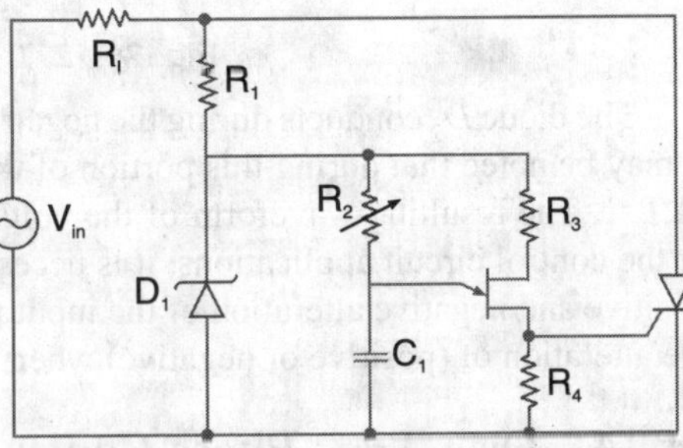

Fig. 21.10. UJT triggered SCR phase control.

21.13. Triac Power Control

The triacs, like SCR's are also used to control average power delivered to the load by the method of phase control. The triac can be triggered such that the a.c. power is supplied to the load for a controlled position of each half-cycle.

The triac is in its 'OFF' position during each positive half-cycle* of the a.c. for a certain interval called the delay angle or firing angle (measured in degrees). Then triac is turned 'ON' and passes current through the load for the remaining portion of the positive half-cycle, except that the current passes in the opposite direction through the load. The action during both half-cycles is illustrated in Fig. 21.11 (*a*). The resulting waveform of load current is as shown Fig. 21.11 (*b*).

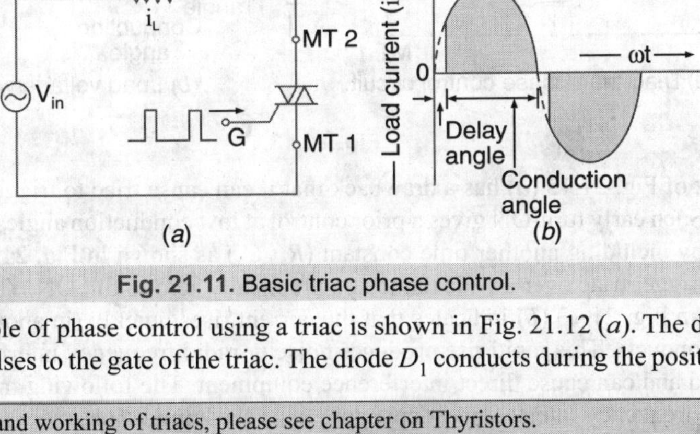

Fig. 21.11. Basic triac phase control.

A simple example of phase control using a triac is shown in Fig. 21.12 (*a*). The diodes are used to provide trigger pulses to the gate of the triac. The diode D_1 conducts during the positive half-cycle.

* For construction and working of triacs, please see chapter on Thyristors.

The value of resistance R_1 sets the point on the positive half-cycle at which the triac triggers. It may be noted that during this portion of the a.c. cycle, M.T. 2 and *G* are positive with respect to M.T. 1.

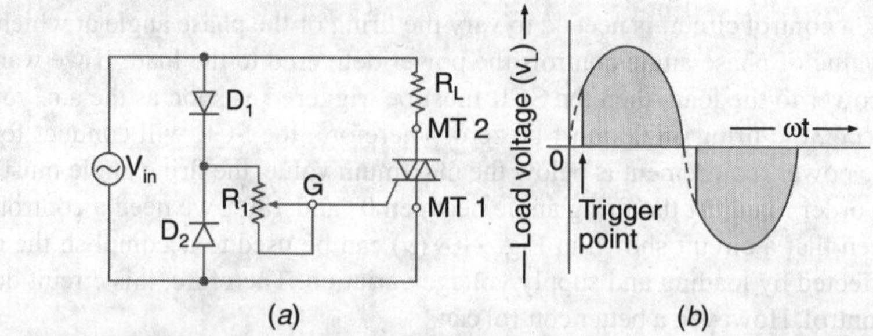

Fig. 21.12. Triac phase control circuit.

The diode D_2 conducts during the negative half-cycle and the resistance R_1 sets the trigger point. It may be noted that during this portion of the a.c. cycle, M..T 2 and G are negative with respect to M.T. 1. The resulting waveform of the voltage across the load resistor is shown in Fig. 21.12 (*b*). In the control circuit applications, it is necessary that the triac turn 'OFF' occurs at the end of each positive and negative alteration of the input voltage. But the triac turn 'OFF'occurs, usually, before the alteration of (positive or negative) when the load current falls below the holding current value.

21.14. Diac-Traic Phase Control Circuit

Fig. 21.13 (*a*) shows a basic diac-traic phase control circuit, which may be used for light dimming, heat control or universal motor speed control. The operation of the circuit may be explained as follows: As the input voltage increases, either positively or negatively, the capacitor C_2 charges at the rate determined by setting of resistance R_2. When the voltage across capacitor C_2 exceeds the breakover voltage of diac, it turns 'ON'. As a result of this, the capacitor C_2 discharges through the conducting diac into the gate of triac. Hence the triac turns ON and it conducts current through the load. The waveform of voltage across the load in as shown is Fig. 21.13 (*b*).

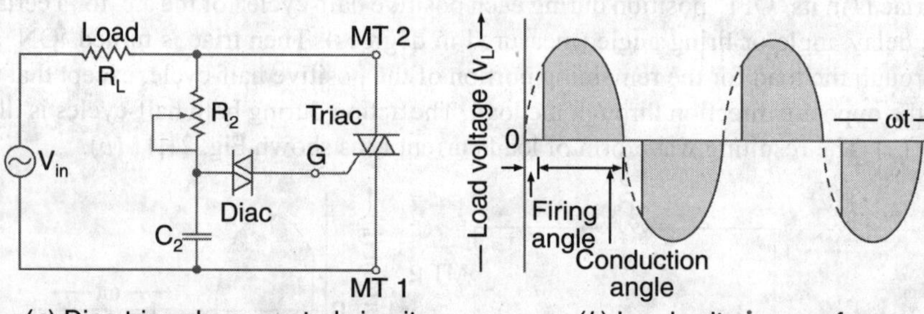

(*a*) Diac-triac phase control circuit. (*b*) Load voltage waveform.

Fig. 21.13.

The circuit of Fig. 21.13 (*a*) has a drawback that it can cause triac to trigger before the desired turn-ON time. Such early turn-ON gives a prior control at low conduction angles. This drawback can be minimized by including another time constant ($R_3 \cdot C_3$) as shown in Fig. 21.14. This circuit can be adjusted to trigger triac over a wide range *i.e.*, practically zero to full ON. The waveform of load voltage shown in Fig. 21.13 (*b*) indicates that the current rises quickly (in about 1 or 2 msec) each time the triac conducts. This rapid rise produces noise or radio frequency interference (RFI), which may be radiated and can cause direct interference equipment. The following methods may be used to reduce radio frequency interference, depending upon the application :

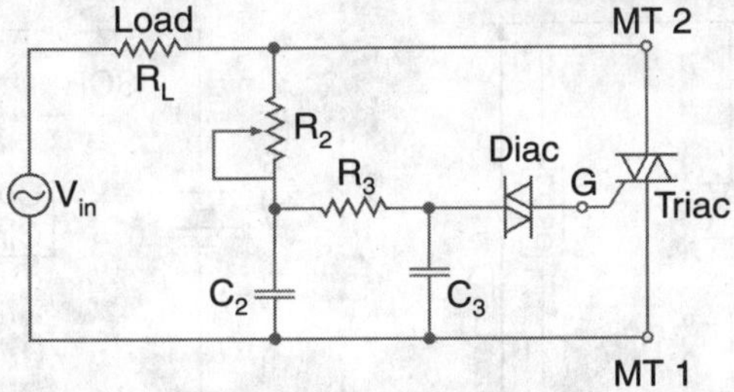

Fig. 21.14. Phase control circuit using double time constant.

1. In light dimming and speed controls, a RFI filter may be used as shown in Fig. 21.15.
2. In heat controls, the current through an electrical heater may be turned ON for a few cycles and then turned OFF for a few cycles to maintain a desired temperature. If the turn ON is only allowed to take place when the voltage is zero, no rapid increase in current takes place as it does if the switching takes place during the cycle.

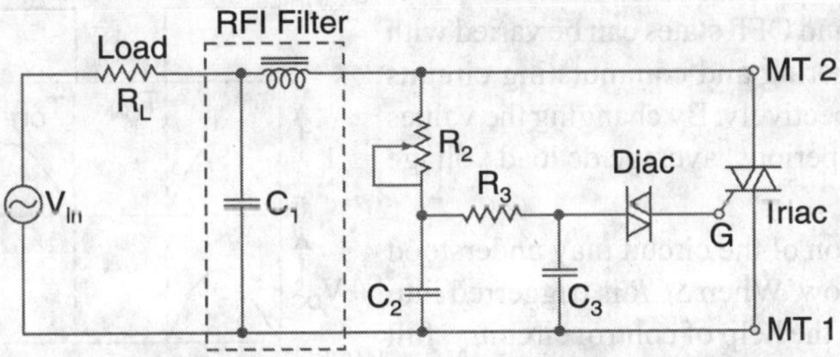

Fig. 21.15. Phase control circuit using RFI filter.

Several manufactures provide IC's that accomplish the above mentioned type of switching called zero-voltage switching (ZVS). One such circuit is the CA 3059 manufactured by RCA. This device is a silicon monolithic integrated circuit, which is designed to act as a triggering circuit for SCR's and triacs. This circuit provides an output pulse (typically 100 msec duration) each time the applied line voltage crosses zero on each half cycle. The pulses are suitable for frequencies ranging from 50 Hz to 400 Hz. This circuit may also be used with light dimming circuits, if the a.c. frequency is 400 Hz as is found in aircraft applications. But it is not very useful in controlling the speed of motors.

21.15. Chopper

A chopper is a *dc-to-dc converter.* It converts a given constant dc voltage into a variable average dc voltage across a load by placing a high-speed static switch between the dc source and the load. This high-speed static switch is called a chopper because it chops off the dc supply into ON and OFF periods of flow.

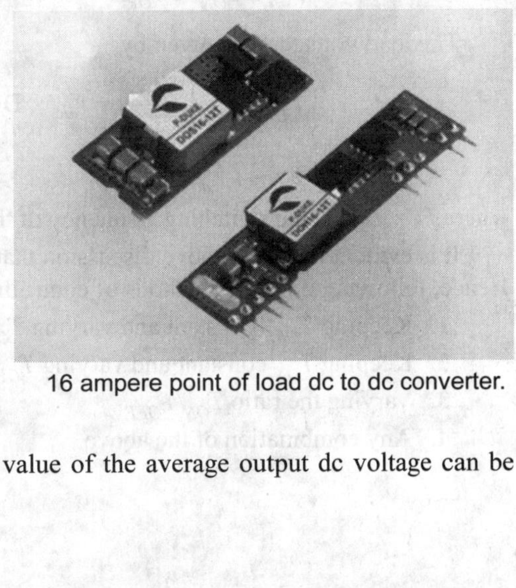

16 ampere point of load dc to dc converter.

A basic chopper circuit is shown in Fig. 21.16. When switch *S* is closed, the dc supply voltage V_{dc} is applied across the load and when it is open, the load is disconnected from the supply. By varying the ratio of the switch-closed time (T_{ON}) to the switch-open time (T_{OFF}) at a fixed frequency, the value of the average output dc voltage can be controlled.

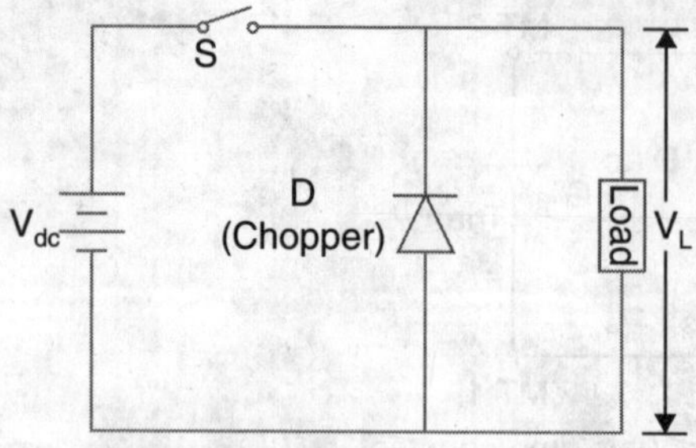

Fig. 21.16.

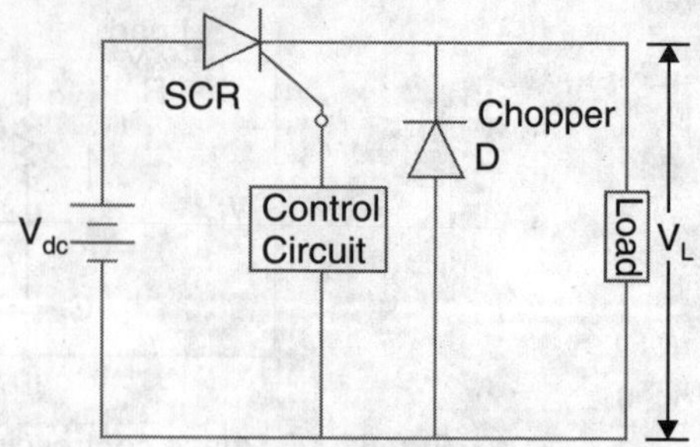

Fig. 21.17.

The switch *S* in Fig. 21.16 could be either a transistor or an *SCR* depending on the amount of power involved. An *SCR* is used in high-power applications whereas transistors are used when power involved is low. A dc chopper circuit using an *SCR* is shown in Fig. 21.17. The *SCR* acts like a static switch and has two states of ON and OFF. The duration of ON and OFF states can be varied with the help of triggering and commutating circuits (not shown) respectively. By changing the values of ON and OFF periods, average dc load voltage can be changed.

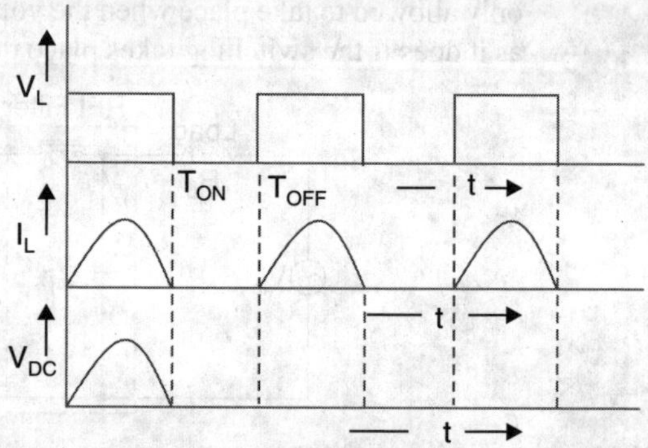

Fig. 21.18.

The operation of the circuit may understood as explained below. When *SCR* is triggerred into conduction with the help of control circuitry, full dc voltage V_{dc} is applied across the load for a period of T_{ON}. When the *SCR* is switched OFF by the control circuitry, there is no voltage across the load. The output load voltage is in the form of a square wave as shown in Fig. 21.18. The waveshapes of the load current and chopper current for an inductive load are as shown. The freewheeling diode *D* provides path for the stored inductive energy to flow.

If T_{ON} is the ON time and T_{OFF} time of the chopper, the duty cycle of the chopper is given by

$$\text{duty cycle} = \frac{T_{ON}}{T} = \frac{T_{ON}}{T_{ON} + T_{OFF}}$$

The load voltage V_L is given by

$$V_L = V_{dc}\left(\frac{T_{ON}}{T}\right) = V_{dc} \times \text{Duty cycle}$$

$$= f\,V_{dc}\,T_{ON}$$

where $f\,(= 1/T)$ is the switching frequency of the chopper.

It is evident from the above discussion that V_L depends on the duty cycle since V_{dc} is constant. Hence, following different methods of controlling V_L are available.

1. Keeping T_{OFF} constant and varying T_{ON}.
2. Keeping T_{ON} constant and varying T_{OFF}.
3. Varying the ratio T_{ON}/T_{OFF}
4. Any combination of the above.

Example 21.5.

A dc chopper has ON time of 30 ms and OFF time of 10 ms. Calculate (i) chopper duty cycle and (ii) chopping frequency.

Solution.

Given : $T_{ON} = 30$ ms and $T_{OFF} = 10$ ms.

(i) ***Chopper duty cycle***

We know that chopper duty cycle,

$$= \frac{T_{ON}}{T} = \frac{T_{ON}}{T_{ON} + T_{OFF}}$$

$$= \frac{30\ \mu s}{30\ \mu s + 10\ \mu s} = \frac{30}{40} = 0.75 \text{ or } 75\% \textbf{ Ans.}$$

(ii) ***Chopping frequency***

We know that chopping frequency,

$$f = \frac{1}{T} = \frac{1}{T_{ON} + T_{OFF}} = \frac{1}{30\ \mu s + 10\ \mu s}$$

$$= \frac{1}{40\ \mu s} = 0.025 \times 10^6 \text{ Hz or } 25 \text{ kHz } \textbf{Ans.}$$

Example 21.6.

A chopper supplied by a 200 V d.c. has ON time of 30 ms and OFF time of 10 ms. Determine the value of average dc output voltage. (*Mysore Univ. 1998*)

Solution.

Given : $T_{ON} = 30$ ms and $T_{OFF} = 10$ ms.

We know that the duty cycle,

$$= \frac{T_{ON}}{T_{ON} + T_{OFF}} = \frac{30 \text{ ms}}{30 \text{ ms} + 10 \text{ ms}} = 0.75$$

∴ Average value of dc output voltage,

$$V_L = V_{dc} \times \text{duty cycle}$$

$$= 200 \times 0.75 = 150 \text{ V } \textbf{Ans.}$$

21.16. Inverters

An inverter is a device that *changes d.c. power into a.c. power* (just the opposite of converters). The inversion process can be achieved with the help of transistors, *SCR*s and tunnel diodes etc. For low and medium outputs, transistorised inverters are suitable but for high power outputs, SCR inverters are essential. For very low voltage and high current requirements, tunnel diode invertors are used.

For inverter applications, transistors have definite advantages over *SCR*s regarding the switching speed, simplicity of control circuitry, high efficiency and greater reliability. It is mainly due to this fact that *SCR* inverters require complicated circuitry for triggerring and commutation.

The basic working principle of an inverter may be explained with the help of circuit shown in Fig. 21.19. It is called voltage-driven inverter because a dc voltage source is connected through semiconductor switches directly to the primary of a transformer.

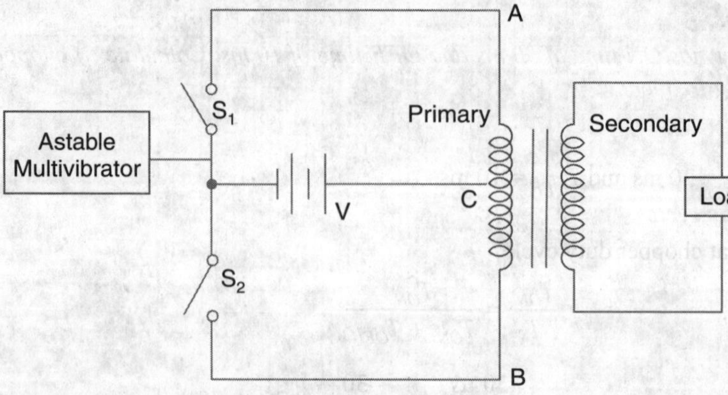

Fig. 21.19.

In Fig. 21.19, S_1 and S_2 are switching devices (transistors or *SCR*s) which open and close alternately at regular intervals of time. The two switching devices are generally driven by an astable multi-vibrator operating at the desired frequency. When S_1 is closed, the entire dc source voltage *V* is applied across points *A* and *B* of the transformer primary. S_1 remains closed for a certain period of time after which it is cut off and S_2 closes. It also remains closed for the same period of time during which the source voltage *V* is impressed across points *B* and *C* of the primary. S_2 then opens out and S_1 closes. In this way, an alternating voltage is applied across the primary which induces an ac voltage in the secondary. Since dc supply voltage is connected directly across the primary, the output waveform of the secondary voltatge is a square wave (Fig. 21.20.) irrespective of the type of load and load power factor. However, the waveforms of both the primary and secondary currents depend on the type of load whether resistive, induictive or capacitive.

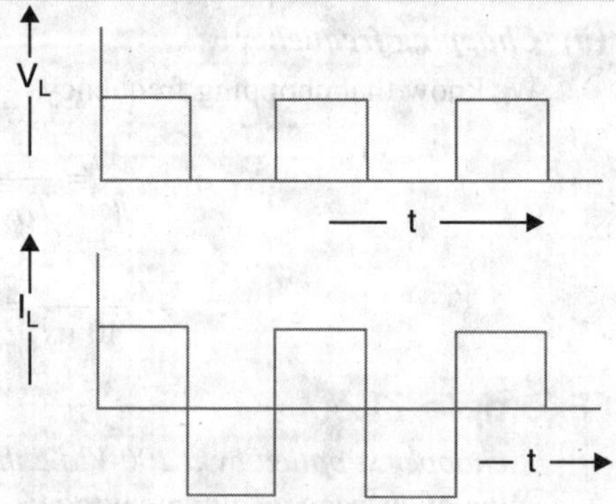

Fig. 21.20.

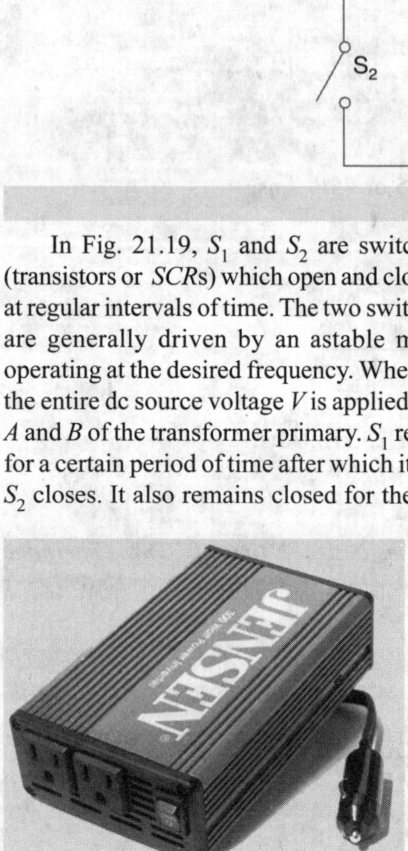

Power Inverter.

21.17. Single-phase Inverter

Fig. 21.21 (*a*) shows a single-phase inverter with a load resistor using 4 *SCR*s working in pairs. The triggerring and commutating circuitry of the *SCR*s has not been shown in the figure. The two thyristors SCR_1 and SCR_4 are triggerred simultaneously so that load current passes through R_L from left to right. Exactly when these two *SCR*s are switched off by the commutating circuitry, thyristors SCR_2 and SCR_3 are triggerred into conduction thereby sending current through R_L from right to left. Hence, an ac voltage is developed across the load whose waveform is as shown in Fig. 21.21.

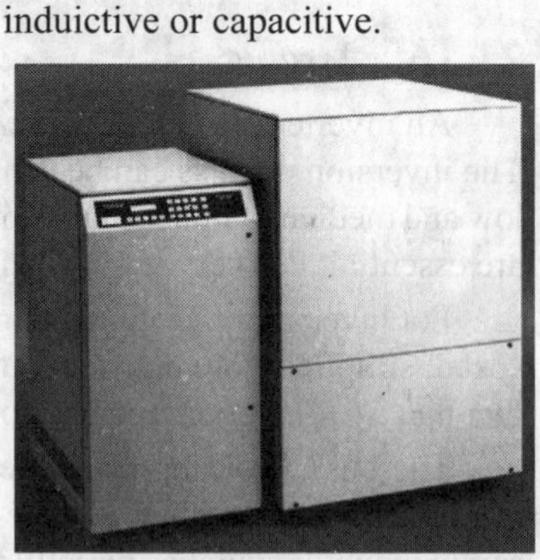

Single-phase inverter.

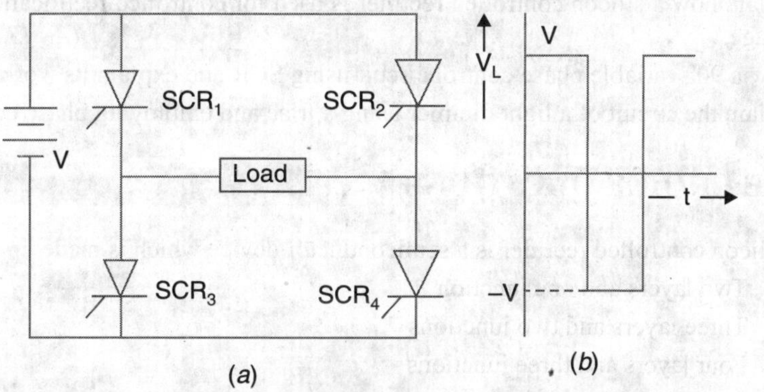

Fig. 21.21.

21.18. Push-Pull Inverter

Fig. 21.22 shows an inverter which employes two *SCR*s and one tranformer. These two *SCR*s are triggerred into conduction alternately for the same period of time. As a result, current through the primary becomes alternating which induces an ac voltages across the secondary and hence the load. As explained earlier in Art. 21.17, the secondary ac voltage has a square waveform. The capacitor *C* is connected across the anodes of the two *SCR*s and provides commutation *i.e.*, switching off of the *SCR*s. The capacitor charges to double the supply voltage as a result of transformer action between the two halves of the primary winding. This large voltage is sufficient to reverse-bias the *SCR*s and drive the holding current below its rated value.

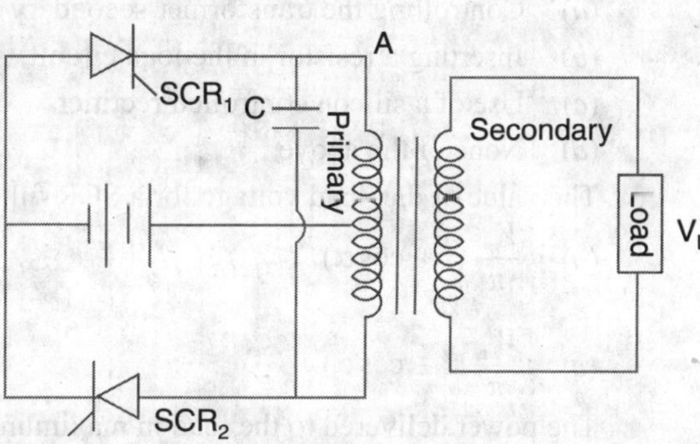

Fig. 21.22.

NUMERICAL PROBLEMS

1. An SCR is rated at 50 A peak and 10 A average. What is the greatest possible delay in the trigger angle, if the direct current is at the rated value ? **(Ans. 75°)**
2. Find the value of load resistance, which can limit the average current to 0.8 A, if 220 V r.m.s. is applied to a SCR with a firing angle 60°. **(Ans. 92.8 Ω)**
3. A silicon controlled rectifier with anode supply of 220 V r.m.s. 50 Hz, applied through a 100 W resistor is fired at 50° of the anode cycle. Assuming the internal drop to be zero; find

   (*a*) the average current and (*b*) average power delivered to the load.

   (Ans. 0.74 A and 54.9 W)

DESCRIPTIVE QUESTIONS

1. What is a controlled rectifier ? How it is different from an ordinary diode rectifier ?
2. Write a short note on the use of SCR for power control. (*Grad. I.E.T.E., June 1985*)

3. Explain how a silicon controlled rectifier is used for controlled rectification of a.c. supply voltage. *(A.M.I.E., Winter 1986)*
4. Draw a 90° variable phase control circuit using SCR and explain its working.
5. Explain the circuit of a light dimmer using a triac and employing phase control.

MULTIPLE CHOICE QUESTIONS

1. A silicon controlled rectifier is a semiconductor device which is made up of
   (*a*) Two layers and one junction
   (*b*) Three layers and two junctions
   (*c*) Four layers and three junctions
   (*d*) None of the above
2. Which of the following is preferred for controlling the amount of power supplied to the load
   (*a*) Controlling the transformer secondary voltage
   (*b*) Inserting a resistor in the load circuit
   (*c*) Use of a silicon controlled rectifier
   (*d*) None of the above
3. The value of d.c. load voltage for a SCR full-wave rectifier circuit is
   (*a*) $\frac{V_m}{\pi}(1 + \cos\alpha)$ (*b*) $\frac{V_m}{2\pi}(1 + \cos\alpha)$
   (*c*) $\frac{V_m}{\pi}(1 - \cos\alpha)$ (*d*) $\frac{V_m}{2\pi}(1 - \cos\alpha)$
4. The power delivered to the load in maximum when the SCR firing angle is
   (*a*) zero (*b*) 90°
   (*c*) 180° (*d*) Between 90° and 180°
5. The triac is a device which conducts during
   (*a*) Positive half cycle of the input voltage
   (*b*) Negative half cycle of the input voltage
   (*c*) Both positive and negative half cycles of the input voltage
   (*d*) None of the above

ANSWERS

1. (*c*) **2.** (*c*) **3.** (*a*) **4.** (*a*) **5.** (*c*)

22 Chapter

BJT Biasing & Stabilization

Outline

22.1. Introduction

It is a well known fact, these days, that bipolar junction transistors (BJTs) are used in a large variety of applications and in many different ways. To use a transistor in any application, it is necessary first to bias the transistor. The usual reason for this biasing is to turn the device ON and to place it in a region where it operates linearly and provides a constant amount of voltage gain.

The biasing deals with setting a fixed level of the current, which should flow through the transistor with a desired fixed voltage drop across the transistor junctions. Usually, the collector current, base current, collector-to-emitter voltage and the base-to-emitter voltage are the currents and voltages, which are required to be set by the biasing circuit. The proper values of these currents and voltages allow a transistor to amplify the weak signals faithfully. It means that the transistor must increase the magnitude of the signal without changing its shape.

Fig. 22.1 (*a*) shows a properly biased transistor amplifier. Here the signal variations at the input of a transistor are accurately reproduced at the input. However, Fig. 22.1 (*b*) and (*c*) shows an improperly biased transistor amplifiers. It may be noted that in

Fig. 22.1 (*b*), the positive portion of the output signal is clipped. It is because of the fact, that the transistor is biased near its cut-off point. Similarly, in Fig. 22.1 (*c*), the negative portion of the output signal is clipped. In this case, the transistor is biased near its saturation-point.

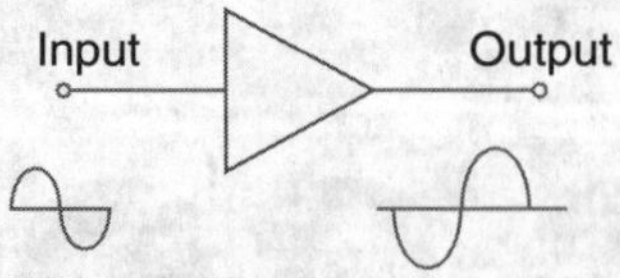

(*a*) Amplifier linear operation.

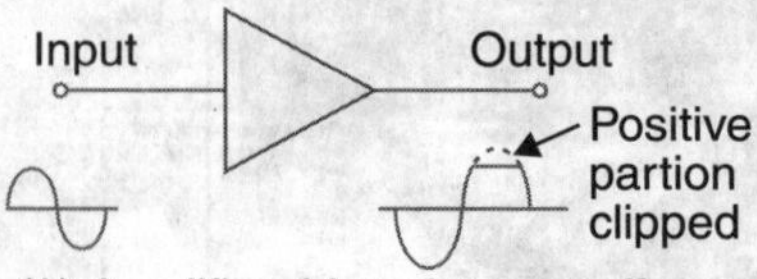

(*b*) Amplifier driven into cut off.

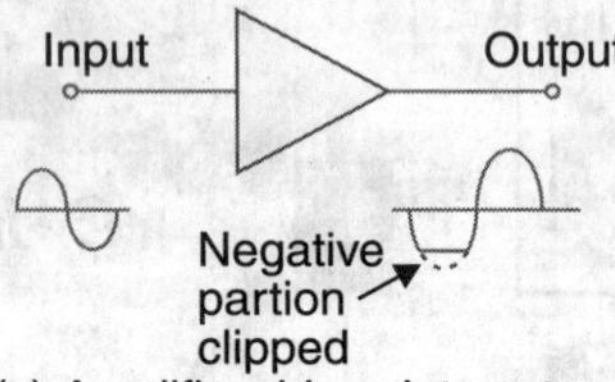

(*c*) Amplifier driven into saturation.

Fig. 22.1.

22.2. The D.C. Operating Point and Load Line

We have already discussed in the last article that of proper operation of a transistor, in any application, we set a fixed level of certain currents and voltages in a transistor. These values of currents and voltages define the point, at which the transistor operates. This point is called operating point. It is also known as quiescent point or simply *Q*-point. Since the level of currents and voltages are fixed, therefore the operating point is also called d.c. operating point.

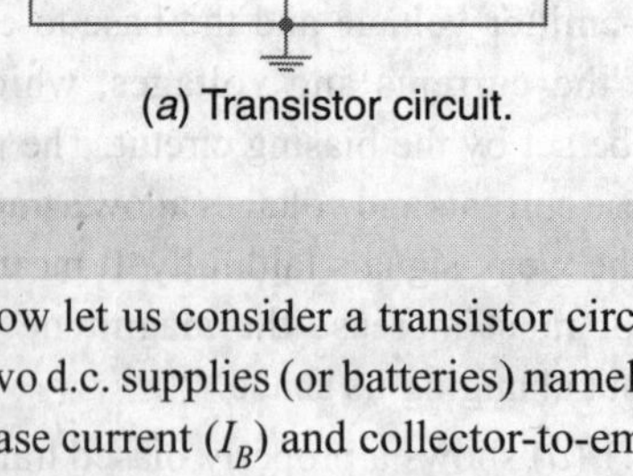

(*a*) Transistor circuit.

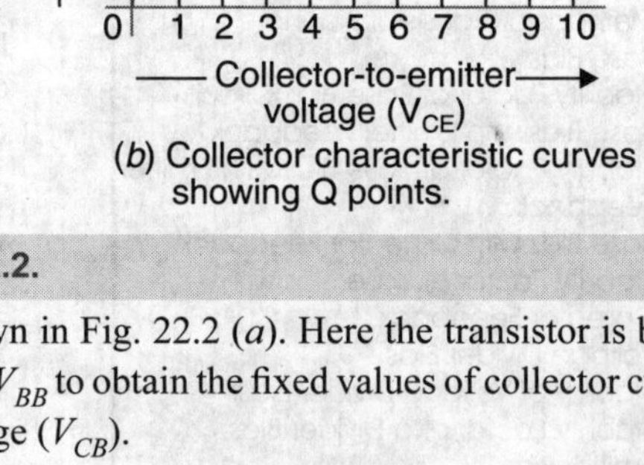

(*b*) Collector characteristic curves showing Q points.

Fig. 22.2.

Now let us consider a transistor circuit as shown in Fig. 22.2 (*a*). Here the transistor is biased with two d.c. supplies (or batteries) namely V_{CC} and V_{BB} to obtain the fixed values of collector current (I_C), base current (I_B) and collector-to-emitter voltage (V_{CB}).

Fig. 22.2 (*b*) shows the collector characteristic curves for this particular transistor. We shall use these characteristics to graphically illustrate the effects of d.c. biasing. In order to draw these curves, we set various values of base current (I_B) say from 0.1 mA to 0.6 mA, by varying the V_{BB} supply and then observe the changes in collector current and collector-to-emitter voltage.

First of all, let us adjust the V_{BB} supply to produce a base current of 0.3 mA. The corresponding value of collector current,

$$I_C = \beta \cdot I_B = 100 \times 0.3 = 30 \text{ mA} = 30 \times 10^{-3} \text{ A}$$

and the collector-to-emitter voltage,

$$V_{CE} = V_{CC} - I_C \cdot R_C = 10 - (30 \times 10^{-3}) \times 200 = 10 - 6 = 4 \text{ V}$$

These values of collector current and collector-to-emitter voltage define a Q-point marked as Q_1 on the characteristic curves of Fig. 22.3 (*a*). Now the V_{BB} supply is increased to produce a base current of 0.4 mA. This gives a collector current (I_C) of 40 mA and collector-to-emitter voltage (V_{CE}) of 2 V. These values of collector current and collector-to-emitter voltage define another Q-point marked as Q_2 on the characteristic curves of Fig. 22.3 (*a*). Finally, the V_{BB} supply is reduced to produce a base current of 0.2 mA. This gives a collector current of 20 mA and collector-to-emitter voltage of 6 V. These values of collector current and collector-to-emitter voltage define Q-point marked Q_3 on the characteristic curves as shown in Fig. 22.3 (*a*).

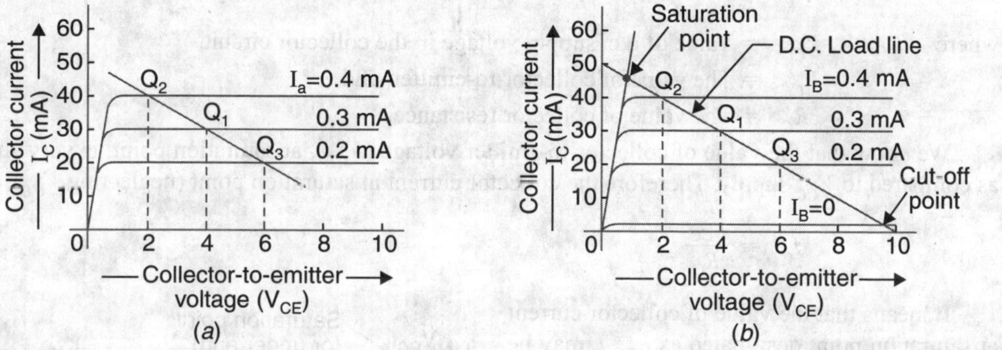

Fig. 22.3. D.C. load line

It is evident from the above discussion that if the value of base current increases, the collector current also increases, while the collector-to-emitter voltage decreases. On the other hand, if the value of base current decreases, the collector current also decreases while the collector-to-emitter voltage increases. It means that as the V_{BB} supply is adjusted up or down, the d.c. operating point of a transistor moves up and down along a sloping straight line connecting the Q-points as shown in Fig. 22.3 (*a*).

If we extend the line joining the operating points Q_1, Q_2 and Q_3 on both sides, we shall obtain a straight line as shown in Fig. 22.3 (*b*). This line is called a *d.c. load line.* The values of base current, collector current and collector-to-emitter voltage can be read at any point along this line.

The d.c. load line gives us the information about the following two important points :

1. The load line intersects the horizontal axis at a point marked V_{CC}, which is equal to 10 V. This point is called *transistor cut-off point* or *lower end of the load line*. At this point, the values of base current and collector current are *zero* (ideally). But actually, there is a small value of reverse saturation current (I_{CBO}) at cut off point as indicated in the figure. As a result of this, the collector-to-emitter voltage (V_{CE}) is slightly less than the value of V_{CC}.

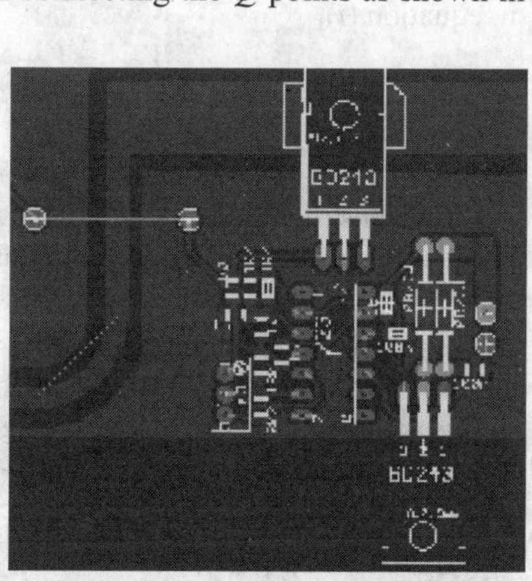

DC biasing is used to establish a steady level of transistor current and voltage called the dc operating point or quescent point (Q-point).

2. The load line intersects the vertical axis at a point marked I_C equal to 50 mA. This point is called transistor *saturation point* or *upper end of the load line*. At this point, the collector current is maximum and the collector-to-emitter voltage is very small equal to $V_{CE\,(sat)}$.

It may be noted that if the upper and lower ends of the load line are known, the load line can be obtained by joining the upper and lower ends.

22.3. Simpler Way of Drawing a D.C. Load Line

We have already discussed in the last article about the d.c. load line and the way it is obtained from the collector characteristic curves. A much simpler way of drawing a d.c. load line is possible, if we know the values of V_{CC} supply and the collector resistance (R_C). For example, consider again the transistor circuit shown in Fig. 22.2 (*a*). For this circuit, we know that the value of collector current is given by the relation,

$$I_C = \frac{V_{CC} - V_{CE}}{R_C} \qquad ...(i)$$

where V_{CC} = Value of d.c. supply voltage in the collector circuit,

V_{CE} = The value of collector-to-emitter, and

R_C = The value of collector resistance.

We know that the value of collector-to-emitter voltage (V_{CE}), at saturation point, is very small as compared to V_{CC} supply. Therefore the collector current at saturation point (neglecting V_{CE}),

$$I_C = \frac{V_{CC}}{R_C} \qquad ...(ii)$$

It means that the value of collector current at saturation point designated as $I_{C\,(sat)}$ may be obtained by dividing the value of V_{CC} supply by the value of collector resistance (R_C). This value gives us the upper end of the load line as shown in Fig. 22.4.

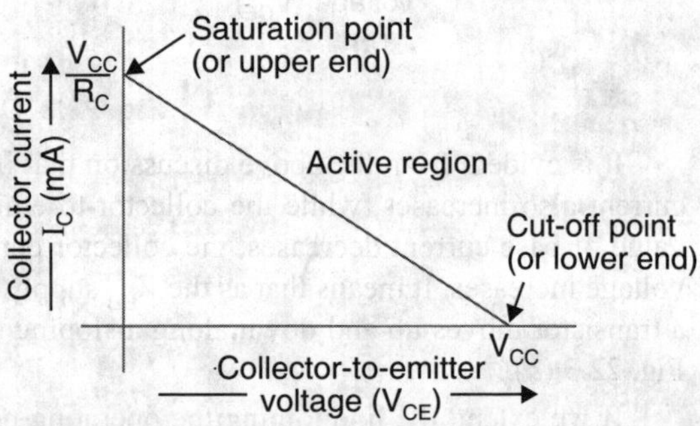

Fig. 22.4. D.C. load line.

We also know that at cut-off point, the value of collector current is zero. Substituting $I_C = 0$ in equation (*i*),

$$0 = \frac{V_{CC} - V_{CE}}{R_C}$$

or $$V_{CE} = V_{CC} = V_{CE\,(cut\text{-}off)}$$

It means that the value of collector-to- emitter voltage at cut-off point designated as $V_{CE\,(cut\text{-}off)}$ is equal to the value of V_{CC} supply. This value gives us the lower end of the load line as shown in the figure.

It may be noted that region lying in between the saturation and cut-off points of the load line is called *active region* (or linear region), of the transistor's operation. As long as the transistor is operated in this active region, the output voltage is a *linear reproduction* of the input. It will be interesting to know that for normal operation, the *Q*-point of a transistor is selected *midway* between the saturation and cut-off point *i.e.*, where the collector-to-emitter voltage (V_{CE}) is equal to half of the V_{CC} supply.

Example 22.1.

A transistor circuit of a pulse amplifier is shown Fig. 22.5.

What is the value of saturation current and cut-off voltage? Also draw the d.c. load line.

Solution.

Given : $V_{BB} = 10$ V; $R_B = 47$ kΩ; $V_{CC} = 20$ V; $R_C = 10$ kΩ $= 10 \times 10^3$ Ω and β = 100.

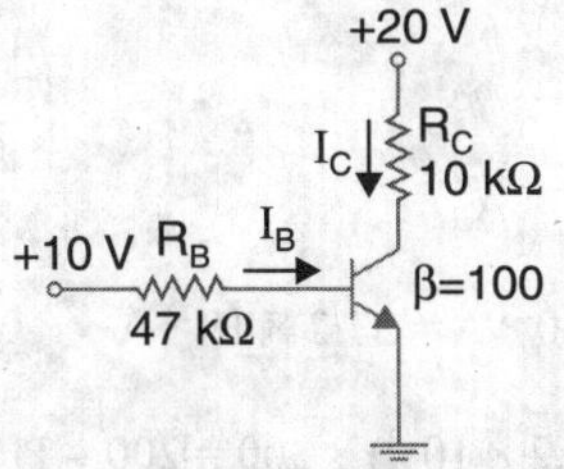

Fig. 22.5. D.C. load line

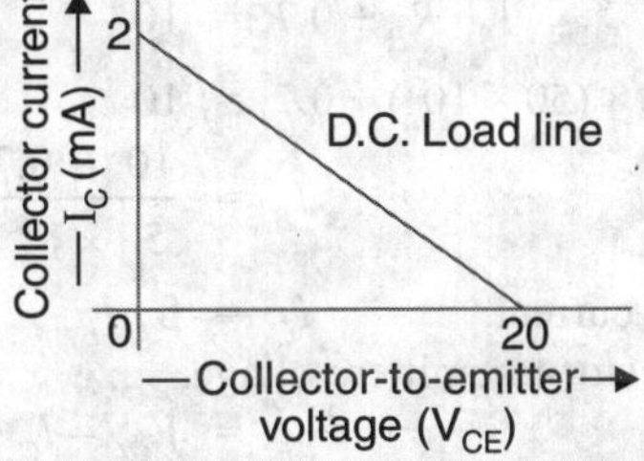

Fig. 22.6.

Value of saturation current

We know that the value of saturation current,

$$I_{C\,(sat)} = \frac{V_{CC}}{R_C} = \frac{20}{10 \times 10^3} = 0.002 \text{ A or } 2 \text{ mA } \textbf{Ans.}$$

Value of cut-off voltage

We also know that the value of cut-off voltage,

$$V_{CE\,(cut\text{-}off)} = V_{CC} = 20 \text{ V } \textbf{Ans.}$$

D.C. load line

The values of $I_{C\,(sat)}$ and $V_{CC\,(cut\text{-}off)}$ gives the upper and lower ends of d.c. load line. By using these two points, we get a straight line (called d.c. load line) as shown in Fig. 22.6.

Example 22.2.

Find the upper and lower ends of the d.c. load line for the circuit shown in Fig. 22.7. Also find the Q-point and locate it on the d.c. load line.

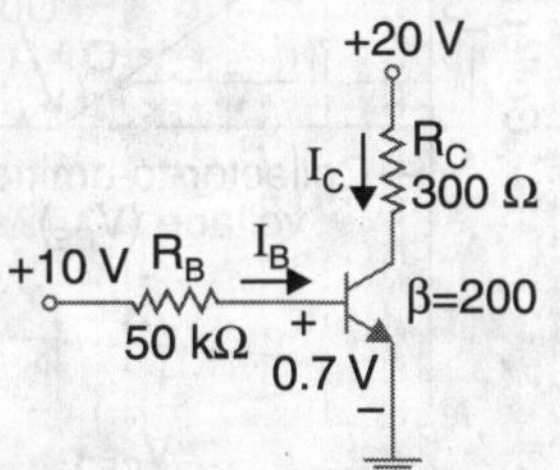

Fig. 22.7.

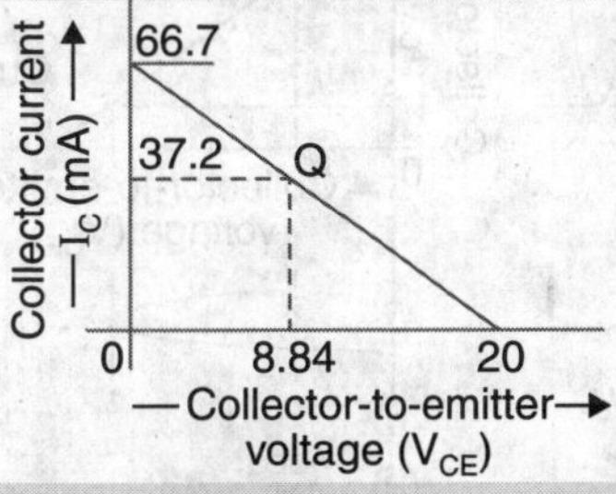

Fig. 22.8.

Solution.

Given : $V_{CC} = 20$ volts; $R_C = 300$ Ω; $V_{BB} = 10$ volts; $R_B = 50$ kΩ $= 50 \times 10^3$ Ω and β = 200.

Upper and lower ends of d.c. load line

We know that the upper end of the load line is at a point, where the collector current is maximum and its value,

$$I_{C\,(sat)} = \frac{V_{CC}}{R_C} = \frac{20}{300} = 0.0667 \text{ A} = 66.7 \text{ mA}$$

Similarly, the lower end of the load line is at a point, where the collector-to-emitter voltages maximum and its value,

$$V_{CE\,(cut\text{-}off)} = V_{CC} = 20 \text{ V } \textbf{Ans.}$$

Now if we join the upper and lower ends, we get a d.c. load line as shown in Fig. 22.8.

Q-point

Applying Kirchoff's Voltage Law to the base loop of the given circuit,

$$I_B \cdot R_B + 0.7 = 10$$

or $$I_B \times (50 \times 10^3) + 0.7 = 10$$

$$\therefore \quad I_B = \frac{10 - 0.7}{50 \times 10^3} = 0.186 \times 10^{-3} \text{ A}$$

and collector current, $$I_C = \beta \cdot I_B = 200 \times 0.186 \times 10^{-3} \text{ A} = 37.2 \times 10^{-3} \text{ A}$$

∴ Collector-to-emitter voltage,

$$V_{CE} = V_{CC} - I_C \cdot R_C = 20 - (27.2 \times 10^{-3}) \times 300 = 200 - 11.16 = 8.84 \text{ V}$$

Now let us locate the *Q*-point on the d.c. load line by drawing a vertical line through the 8.84 V as shown in Fig. 22.8. From the figure, we find that the *Q*-point corresponds to 37.2 mA **Ans.**

22.4. Q-Point and Maximum Undistorted Output

It has been observed that under certain input signal conditions, the location of *Q*-point, on the load line, may cause one peak of the output signal to be clipped. For example consider *Q*-point located near the saturation point as shown in Fig. 22.9 (*a*). In this case, during the negative half cycle of the input, the transistor is driven into saturation. As a result of this, the negative peak of the input signal is clipped at the output.

Now consider a *Q*-point located near the cut-off point as shown in Fig. 22.9 (*b*). In this case, the transistor is driven into cut-off. Therefore the positive peak of the input signal is clipped at the output. It may be noted that if the *Q*-point is located at the centre of the load line as shown in Fig. 22.9 (*c*), we get an undistorted signal at the output.

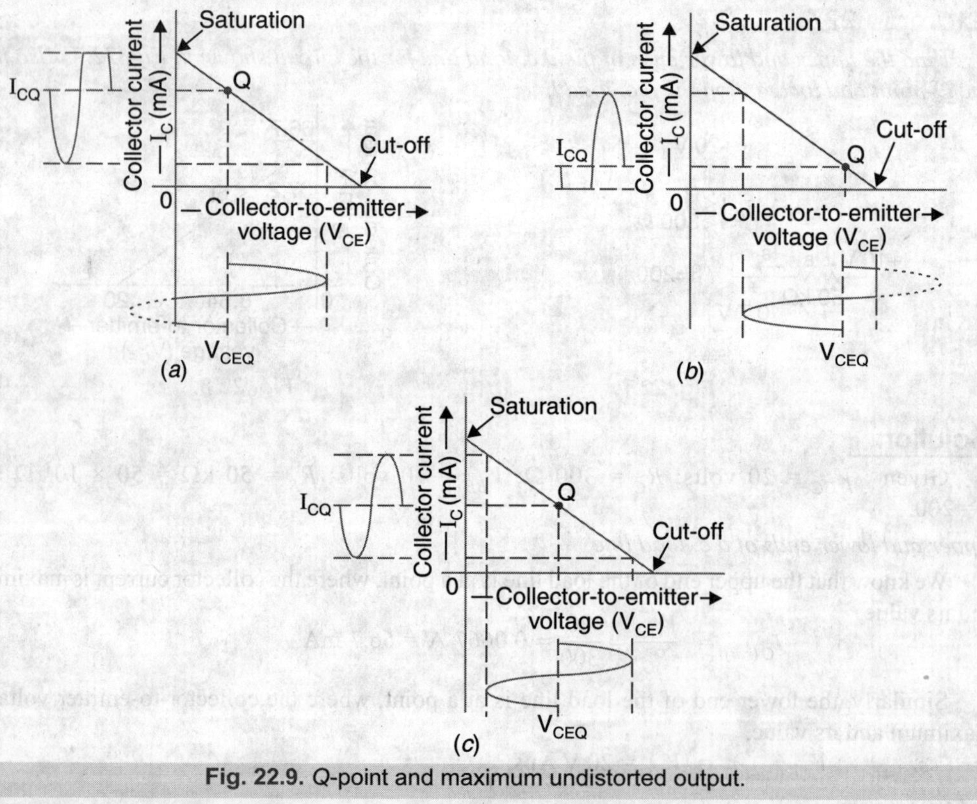

Fig. 22.9. *Q*-point and maximum undistorted output.

22.5. Factors Affecting Stability of Q-Point

It will be interesting to know that even after the suitable selection of Q-point, it tends to shift from its position. It happens because of the following two factors :

1. ***Inherent variations of transistor parameters.*** We know that the collector current for a common-emitter transistor amplifier is given by the relation,

$$I_C = \beta \cdot I_B + (I + \beta) \cdot I_{CO} \qquad ...(i)$$

where β = Common-emitter current gain,

I_B = Base-current and

I_{CO} = Reverse saturation current.

The above equation contains three variables namely, β_2, I_B and I_{CO}. All the three variables are found to be strongly dependent upon temperature. As the temperature, increases, all the three variables also increase. It results in the increase of collector current. This causes the operating point to *shift* towards the saturation point, resulting in clipping and bad distortion of the output. In some cases, a transistor may even *burn* out because of the excessive collector current.

2. ***Variation in parameter values of transistors of the same type.*** It has been observed that inspite of the tremendous advancement, made in the semiconductor device manufacturing technology, some transistors of a particular type with a large variation in parameter values are manufactured. This variability is provided by the manufacturer in the data sheet, where minimum, maximum and typical values of parameters are mentioned. This gives rise to an instability of the Q-point.

We know that if the temperature increases, the reverse saturation current also increases. This causes the collector current to increase by a factor of $(1 + \beta)$ as seen from equation (i). However if by some means, the base current is made to decrease with the increase in temperature, then the collector current will decrease by a factor of β. This will compensate for the increase in $(1 + \beta)I_{CO}$ and hence the collector current will remain nearly constant. This principle is used to provide stabilization of collector current by designing it as biasing circuit of different types and later providing compensation techniques.

22.6. Stability Factor

The stability factor may be defined as the rate of change of collector current with respect to the reverse saturation current keeping the common-emitter current gain (β) and base current (I_B) as constant. Mathematically, the stability factor,

$$S = \frac{dI_C}{dI_{CO}}$$

The stability factor is a measure of *bias stability* of a transistor circuit. It will be interesting to know that a *higher value of stability factor indicates poor stability, whereas a lower value indicates good stability*. In this sense, the stability factor should be called as an instability factor.

It may be noted that if the rate of change of collector current (dI_C) is equal to the rate of change of reverse saturation current (dI_{CO}), then the value of stability factor is unity. The unity is the lowest value of stability factor. It may be noted that if the value of stability factor is closer to the unity, the variation of collector current with the temperature will be less. Therefore there will be less variation in the Q-point.

The stability factor may also be expressed alternatively by using the relationship between base current (I_B), collector current (I_C) and reverse saturation current (I_{CO}). We know that the value of collector current in a transistor is given by the relation,

$$I_C = \beta \cdot I_B + (1 + \beta)\, I_{CO}$$

Differentiating the above expression with respect to I_C,

$$1 = \frac{d\,(\beta \cdot I_B)}{dI_C} + \frac{d\,(1+\beta)\,I_{CO}}{dI_C}$$

$$= \beta \times \frac{dI_B}{dI_C} + (1+\beta)\,\frac{dI_{CO}}{dI_C} \qquad \text{... (Assuming as constant)}$$

$$= \beta \times \frac{dI_B}{dI_C} + (1+\beta)\,\frac{1}{S} \qquad \text{...}\left(\because S = \frac{dI_C}{dI_{CO}}\right)$$

$$\therefore \quad S = \frac{1+\beta}{1-\beta\left(\dfrac{dI_B}{dI_C}\right)} \qquad \text{...}(i)$$

The above expression is a general expression, which is very useful for determining the stability factor (S) of any biasing circuit. It can be done first by finding the relationship between the base current and collector current and then using the above equation to determine the value of stability factor.

22.7. Stability Factor of Common Base Circuit

We have already discussed in the chapter on transistors that the collector current of a transistor, in common base configuration, is given by the relation,

$$I_C = \alpha \cdot I_E + I_{CO}$$

Differentiating this equation with respect to I_{CO},

$$\frac{dI_C}{dI_{CO}} = 0 + 1$$

$$\therefore \quad S = 1$$

The value of stability factor indicates that the *common base circuits are highly stable*. Thus there is no need of bias stabilization in these circuits.

22.8. Stability Factor of Common Emitter Circuit

We have already discussed in chapter on transistors that the collector current of a transistor in common emitter configuration is given by the relation,

$$I_C = \beta \cdot I_B + (1+\beta)\,I_{CO}$$

Differentiating this equation with respect to I_{CO}

$$\frac{dI_C}{dI_{CO}} = 0 + (1+\beta)$$

$$\therefore \quad S = 1+\beta$$

It will be interesting to know that if $\beta = 100$, then the value of $S = 1 + 100 = 101$. It means that the collector current changes 101 times the change in reverse saturation current. Thus in a common emitter circuit, the collector current is highly dependent upon the reverse saturation current and hence upon the temperature. Thus *there is a strong need to provide bias stabilization in common emitter circuits to improve the stabilization factor*.

22.9. Conditions for Proper Biasing of a Transistor

We have already discussed that for faithful amplification of an input signal, a transistor must be biased properly. Following are some of the important conditions for biasing the transistor properly:

1. Proper d.c. value of the collector current.
2. Proper value of V_{BE} (0.7 V for silicon and 0.3 V for germanium) at any instant.
3. Proper value of V_{CE} (1 V for silicon and 0.5 V for germanium) at any instant.

The conditions 1 and 2 make sure that base-emitter junction of a transistor shall remain forward biased, during all parts of the signal. The condition 3 makes sure that collector-base junction of input a transistor shall remain reverse biased at all instants.

22.10. Methods of Transistor Biasing

In the transistor amplifier circuits, drawn so far, the biasing is done with two power supplies V_{BB} and V_{CC}. The V_{BB} supply is used for biasing of the emitter-base junction and V_{CC} supply for biasing the collector base junction. However, in actual practice, one power supply is used for biasing both the junctions of a transistor. Following are the most commonly used methods for biasing the transistors :

1. Base bias (also called *fixed bias*).
2. Base bias with emitter feedback (also called *emitter feedback bias*).
3. Base bias with collector feedback (also called *collector feedback bias*).
4. Voltage divider bias (also called *self bias*).
5. Emitter bias.

Now we shall discuss the above mentioned methods of biasing in the following pages.

22.11. Base Bias

Fig. 22.10 (*a*) shows a base bias circuit for a NPN transistor. The base bias circuit is also known as fixed bias circuit. In this case, the circuit uses two d.c. supplies namely V_{BB} and V_{CC}. But a more practical method is to use a single d.c. supply (V_{CC}) as shown in Fig. 22.10 (*b*). In this case, both the base and collector resistance are connected to the positive side of the V_{CC} supply. To simplify the circuit diagram, we may omit the battery symbol and replace it by a line termination with a voltage indicated as shown in Fig. 22.10 (*c*).

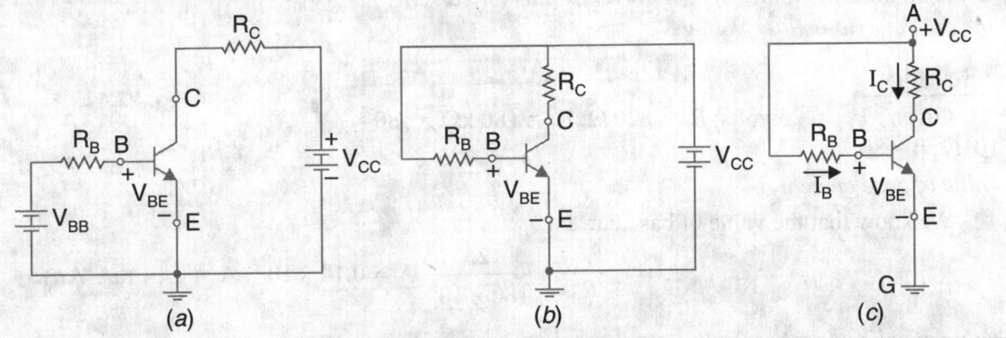

Fig. 22.10. Base bias circuit.

Now we shall determine the d.c. bias currents and voltages in the base and collector of the transistor. First of all, consider the base-emitter circuit loop (AFBEGA) in the base bias circuit. Applying Kirchoff's Voltage Law for the given loop,

$$I_B \cdot R_B + V_{BE} - V_{CC} = 0$$

or the base current,
$$I_B = \frac{V_{CC} - V_{BE}}{R_B} \quad ...(i)$$

Since the supply voltage V_{CC} and the base-emitter voltage (V_{BE}) have fixed values of voltage, the selection of base bias resistor (R_B) fixes the value of the base current. The equation (*i*) may be simplified, if we neglect the value of base-emitter voltage, because it is usually very small as compared to the value of V_{CC}. Thus the approximate value of base current,

$$I_B = \frac{V_{CC}}{R_B} \quad ...(\because V_{CC} >> V_{BE})$$

Now consider the collector-emitter circuit loop (AFCEGA) in the base bias circuit. Applying Kirchoff's Voltage Law for this loop,

$$I_C \cdot R_C + V_{CE} = V_{CC}$$

$$\therefore \quad V_{CE} = V_{CC} - I_C \cdot R_C \quad ...(ii)$$

The above equation gives the voltage drop across the collector-emitter terminals of the transistor. The value of collector current is given by the relation,

$$I_C = \beta \cdot \frac{V_{CC}}{R_B} = \frac{V_{CC}}{R_B / \beta} \quad ...(iii)$$

The above relation shows that the collector current is β times greater than the base current and is not at all dependent on the resistance of the collector circuit.

It may be noted from the equations (*ii*) and (*iii*) that the values of collector current (I_C) and collector-to-emitter voltage (V_{CE}) are dependent on β. But β is strongly dependent upon temperature. It means that collector current and collector-to-emitter voltage of a base bias circuit (which sets the *Q*-points of a transistor) will vary with the change in value of β due to variation in temperature. It means that it is impossible to obtain a stable *Q*-point in a base-bias circuit. Because of this fact, the base bias is never used in amplifier circuits. However, it is used in digital circuits, where the transistor is used as a switch between saturation and cut-off-regions.

Example 22.3.

Fig. 22.11 shows a fixed bias circuit using an NPN transistor circuit.

Determine the values of base current, collector current and the collector-to-emitter voltage.

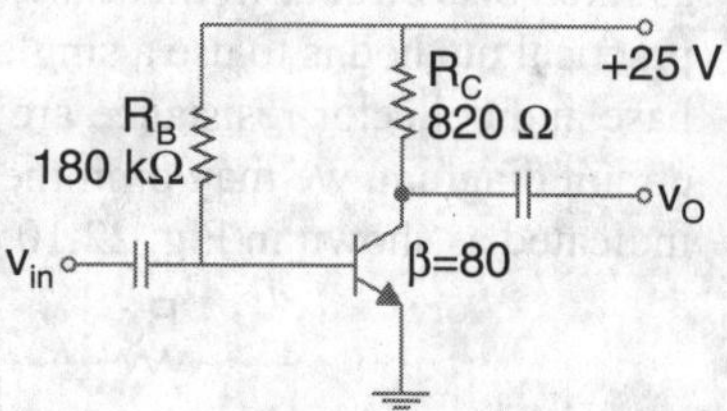

Fig. 22.11.

Solution.

Given : V_{CC} = 25 volts; R_C = 820 Ω; R_B = 180 kΩ = 180 × 10^3 Ω; β = 80.

Value of base current

We know that the value of base current,

$$I_B = \frac{V_{CC}}{R_B} = \frac{25}{180 \times 10^3} \text{ A} = 0.14 \times 10^{-3} \text{ A} = 0.14 \text{ mA } \textbf{Ans.}$$

Value of collector current

We know that the value of collector current,

$$I_C = \beta \cdot I_B = 80 \times 0.14 \times 10^{-3} = 11.2 \times 10^{-3} \text{ A}$$
$$= 11.2 \text{ mA } \textbf{Ans.}$$

Value of collector-to-emitter voltage

We know that the value of collector-to-emitter voltage,

$$V_{CE} = V_{CC} - I_C \cdot R_C = 25 - (11.2 \times 10^{-3}) \times 820 \text{ V}$$
$$= 25 - 9.2 = 15.8 \text{ V } \textbf{Ans.}$$

Example 22.4.

In the circuit shown in Fig. 22.12, calculate the value of collector current.

(UPSC Engg. Serviecs 2002)

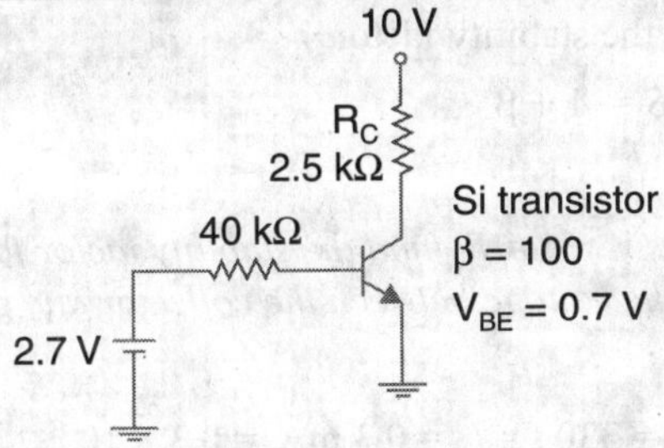

Fig. 22.12.

Solution.

Given: $V_{BB} = 2.7$ V; $R_B = 40$ kΩ; $V_{CC} = 10$ V; $R_C = 2.5$ kΩ; β = 100; $V_{BE} = 0.7$ V.

We know that the base current,

$$I_B = \frac{V_{BB} - V_{BE}}{R_B} = \frac{2.7 - 0.7}{40\,k} = 0.05 \times 10^{-3} \text{ A} \quad \text{or} \quad 50\ \mu\text{A}.$$

and the collector current,

$$I_C = \beta\, I_B = (0.05 \times 10^{-3}) \times 100 = 0.005 \text{ A} \quad \text{or} \quad 5 \text{ mA } \textbf{Ans.}$$

Example 22.5.

Find whether the transistor shown in Fig. 22.13 is operating in saturation active or cut-off region. Justify your answer with reasons.

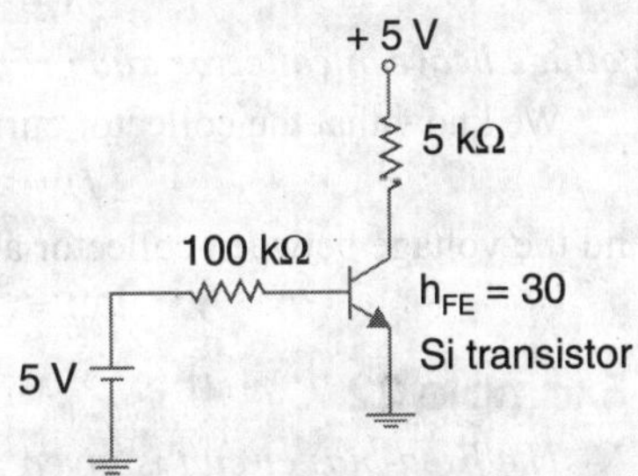

Fig. 22.13.

Solution.

Given: $V_{BB} = 5$V; $R_B = 100$ kΩ; $V_{CC} = 5$ V; $R_C = 5$ kΩ and h_{FE} (or β) = 30.

We know that the base current,

$$I_B = \frac{V_{BB} - V_{BE}}{R_B} = \frac{5 - 0.7}{100 \times 10^3}$$

...(∵ $V_B = 0.7$ V for Si)

$$= \frac{4.3}{100 \times 10^3} = 0.043 \times 10^{-3} \text{ A}$$

$$= 43 \times 10^{-6} \text{ A} \quad \text{or} \quad 43\ \mu\text{A}$$

and the collector current,

$$I_C = \beta\, I_B = 30 \times (43 \times 10^{-6}) = 1290 \times 10^{-6} \text{ A} \quad \text{or} \quad 1.29 \text{ mA.}$$

For the transistor to operate in saturation, the collector current, should at least be,

$$I_C = \frac{V_{CC}}{R_C} = \frac{5}{5 \times 10^3} = 1 \times 10^{-3} \text{ A} \quad \text{or} \quad 1 \text{ mA}$$

Since the actual current in the circuit (1.29 mA) is more than the minimum required value (T.C. 1 mA), therefore the transistor shown in Fig. 22.13 is operating in saturation. **Ans.**

22.12. Stability Factor of Base-Bias Circuit

We know that the basic circuit used in the base bias (or fixed bias) current is that of common-emitter configuration. We have already discussed in Art. 22.8 that stability factor, is equal to (1+ β). Therefore for base-bias circuit, the stability factor.

$$S = 1 + \beta$$

Example 22.6.

Determine the value of bias resistor R_B and the stability factor for the fixed bias circuit shown in Fig. 22.14. Also determine the voltage between the collector and ground.

Solution.

Given : V_{CC} = 12 volts; R_C = 330 Ω; I_B = 0.3 mA = 0.3×10^{-3} A and β = 100.

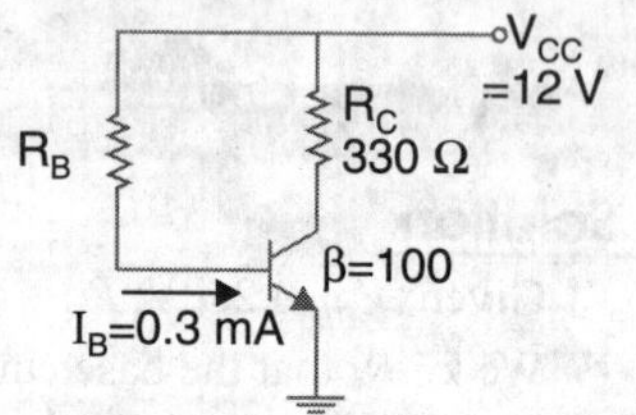

Fig. 22.14.

Value of bias resistor

Let R_B = Value of base resistor.

We know that the bias current (I_B),

$$0.3 \times 10^{-3} = \frac{V_{CC}}{R_B} = \frac{12}{R_B}$$

$$R_B = 12/(0.3 \times 10^{-3}) = 40 \text{ k}\Omega \textbf{ Ans.}$$

Stability factor

We know that the stability factor for the given circuit,

$$S = 1 + \beta = 1 + 100 = 101 \textbf{ Ans.}$$

Voltage between collector and ground

We know that the collector current,

$$I_C = \beta \cdot I_B = 100 \times 0.3 \times 10^{-3} = 30 \times 10^{-3} \text{ A}$$

and the voltage between collector and ground (or emitter),

$$V_{CE} = V_{CC} - I_C \cdot R_C = 12 - (30 \times 10^{-3}) \times 330 = 2.1 \text{ V } \textbf{Ans.}$$

Example 22.7.

The fixed-bias circuit is given in Fig. 20.15, and it is subjected to an increase in temperature from 25°C to 75°C. If β = 100 at 25°C and β = 125 at 75°C. Determine the percentage change in Q-point value (V_{CE}, I_C) over the temperature range. Neglect any change in V_{BE}. Take V_{BE} = 0.7 V.

(GTU., May 2011)

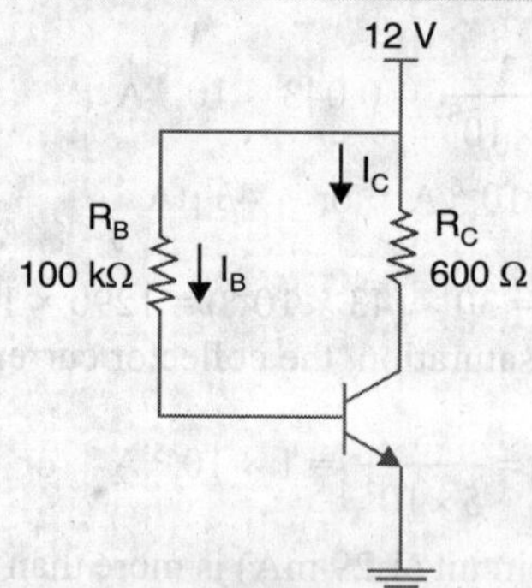

Fig. 22.15.

Solution.

Given : $V_{CC} = 12$ V; $R_B = 100$ kΩ $= 100 \times 10^3$ Ω; $R_C = 600$; β = 100 at 25°C and β = 125 at 75°C.

We know that the base current,

$$I_B = \frac{V_{CC} - V_{BE}}{R_B} = \frac{12 - 0.7}{100 \times 10^3} = 113 \times 10^{-6} \text{ A}$$

Values at 25°C

We know that the collector current,

$$I_C = \beta I_B = 100 \times (113 \times 10^{-6}) = 11.3 \times 10^{-3} \text{ A} = 11.3 \text{ mA}$$

and the collector current,

$$V_C = V_{CC} - I_C R_C = 12 - (11.3 \times 10^{-3}) \times (600) = 5.22 \text{ V}$$

Value at 75°C

We know that the collector current,

$$I_C = \beta I_B = 125 \times (113 \times 10^{-6}) = 14.12 \times 10^{-3} \text{ A} = 14.12 \text{ mA}$$

and the collector to emitter voltage

$$V_C = V_{CC} - I_C R_C = 12 - (14.12 \times 10^{-3}) \times (600) = 3.52 \text{ V}$$

We also know that the percentage change in collector current,

$$\% \Delta I_C = \frac{I_C(75°C) - I_C(25°C)}{I_C(25°C)} \times 100 \%$$

$$= \frac{(14.12 \times 10^{-3}) - (11.3 \times 10^{-3})}{(11.3 \times 10^{-3})} \times 100 \% = 25\% \textbf{ Ans.}$$

and the percentage change in collector-to-emitter voltage,

$$\% \Delta V_{CE} = \frac{V_{CE}(75°C) - V_{CE}(25°C)}{V_{CE}(25°C)} \times 100 \%$$

$$= \frac{3.52 - 5.22}{5.22} \times 100\% = -32.47 \% \textbf{ Ans.}$$

Example 22.8.

For the circuit shown in the Fig. 22.16, determine the operating point with β =100.

(Anna University., April/May 2008)

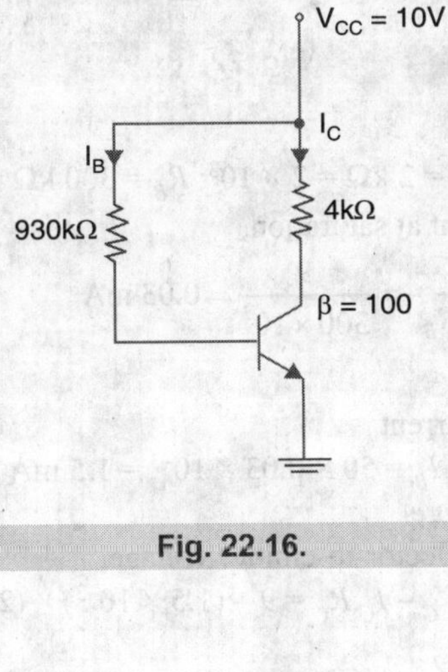

Fig. 22.16.

Solution.

Given : $V_{CC} = 10$ V; $R_C = 4$ kΩ $= 4 \times 10^3$ Ω; $R_B = 930$ kΩ $= 930 \times 10^3$ Ω and β = 100.

We know that collector current at saturation,

$$I_{C(sat)} = \frac{V_{CC}}{R_C} = \frac{10}{4 \times 10^3} = 2.5 \text{ mA}$$

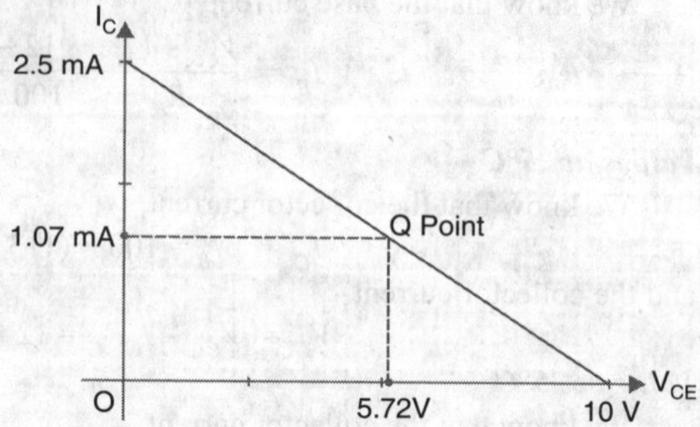

Fig. 22.17.

and the value of cut-off voltage,

$$V_{CE(cut\text{-}off)} = V_{CC} = 30 \text{ V}$$

We also know that base current,

$$I_B = \frac{V_{CC}}{R_B} = \frac{10}{930 \times 10^3} = 10.7 \ \mu\text{A}$$

and the collector current,

$$I_C = \beta I_B = 100 \times 10.7 \times 10^{-6}$$
$$= 1.07 \times 10^3 \text{ A} = 1.07 \text{ mA}$$

The collector-to-emitter voltage,

$$V_{CE} = V_{CC} - I_C R_C$$
$$= 10 - (1.07 \times 10^{-3}) \times (4 \times 10^3)$$
$$= 10 - 4.28 \text{ V} = 5.72 \text{ V}$$

∴ The *Q*-point is located on the d.c. load line as shown in Fig. 22.17.

Example 22.9.

Find the collector current and collector- to- emitter voltage for the given circuit as shown in Fig. 22.18.

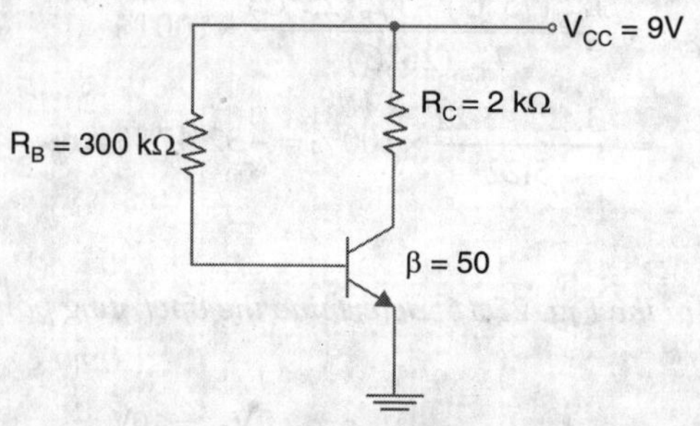

Fig. 22.18.

Solution.

Given : $V_{CC} = 9$ V; $R_C = 2$ kΩ; $R_C = 2$ kΩ $= 2 \times 10^3$; $R_B = 300$ kΩ $= 300 \times 10^3$ Ω and β = 50.

We know that collector current at saturation,

$$I_B = \frac{V_{CC}}{R_B} = \frac{9}{300 \times 10^3} = 0.03 \text{ mA}$$

Value of collector current

We know that the collector current

$$I_C = \beta I_B = 50 \times 0.03 \times 10^{-3} = 1.5 \text{ mA } \textbf{Ans.}$$

Value of collector-to-emitter voltage

We know that the value of collector-to-emitter voltage,

$$V_{CE} = V_{CC} - I_C R_C = 9 - (1.5 \times 10^{-3}) \times (2 \times 10^3) = 6 \text{ V } \textbf{Ans.}$$

Example 22.10.

In a fixed bias circuit of transistor, V_{CC} = 15 V, R_B = 300 kΩ, R_L = 2 kΩ, β = 100, I_{CO} = 20 mA and V_{BE} = 0.7 V. Find the stability factor of q factor with respect to I_{CO}. (PTU., May 2007)

Solution.

Given : V_{CC} = 15 V; R_B = 300 kΩ; R_L = 2 kΩ; β = 100; I_{CO} = 20 mA and V_{BE} = 0.7 V.

We know that the stability factor (*S*),

$$S = 1 + \beta$$
$$= 1 + 100 = 101 \textbf{ Ans.}$$

22.13. Base Bias with Emitter Feedback

Fig 22.19 shows the circuit of a base bias with emitter resistor. The emitter resistor provides better bias stability than the base bias circuit, as discussed in Art. 22.12.

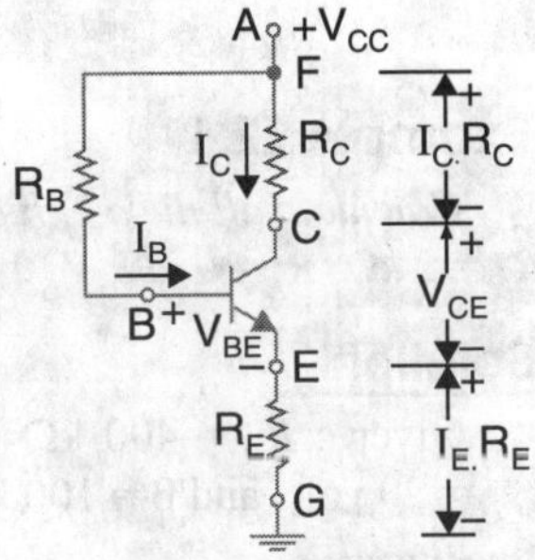

Fig. 22.19. Base bias with emitter feedback.

When the current gain (β) increases (say due to temperature), the collector currents in the circuit also increases. It produces more voltage across the emitter resistor (R_E), which reduces the voltage drop across the base resistor. This causes the base current to decrease. As a result of this, the collector current also decreases. Thus the increase in collector current, due the increase in value of current gain, is compensated.

It is evident from the above discussion that the emitter resistor provides compensation for the variation in the value of current gain (β) by reducing the base current. Thus the emitter resistor is called emitter feedback resistor. The emitter feedback circuit is effective only if the emitter resistor is as large as possible. But in actual practice, it is not possible because the emitter resistor is kept relatively small to avoid the transistor to operate in saturation.

Let us now determine the d.c. bias currents and voltages in the base, collector and emitter of the transistor. First of all, consider the base-emitter circuit loop (AFBEGA) in the d.c. bias circuit. Applying Kirchoff's Voltage Law to the base-emitter circuit loop,

$$V_{BE} + I_E \cdot R_E - V_{CC} + I_B \cdot R_B = 0$$

Substituting I_E equal to (β + 1) I_B in the above equation, and solving it for the base current,

$$I_B = \frac{V_{CC} - V_{BE}}{R_B + (\beta + 1) R_E} = \frac{V_{CC} - V_{BE}}{R_B + \beta \cdot R_E} \quad \text{...(Taking } \beta + 1 \approx \beta)$$

∴ Collector current,

$$I_C = \beta \cdot I_B = \frac{V_{CC} - V_{BE}}{R_E + R_B / \beta} = \frac{V_{CC}}{R_E + R_B / \beta} \quad \text{...}(\because V_{CC} >> V_{BE})$$

It is evident from the above equation that if the value of emitter resistor (R_E) is made very large as compared to the value of (R_B/β), the bias circuit is independent of variations in current gain (β). But in actual practice, it is not possible to make the emitter resistor large enough to compensate for the effects of current gain (β) without saturating the transistor. Thus some compromise will have to be made between the bias stability and the performance of the circuit. Now consider the collector-emitter circuit loop (AFCEGA) in the d.c. bias circuit. Applying Kirchoff's Voltage Law to this loop.

$$V_{CE} + I_E \cdot R_E - V_{CC} + I_C \cdot R_C = 0$$

Replacing I_E with I_C in the above equation and solving it for the voltage across collector-emitter of a transistor,

$$V_{CE} = V_{CC} - I_C (R_C + R_E) \quad ...(i)$$

The voltage measured from emitter to ground,

$$V_E = I_E \cdot R_E = R_C \cdot R_C \quad ...(\because I_E \approx I_C)$$

The voltage measured from collector to ground,

$$V_C = V_{CC} - I_C \cdot R_C$$

Notes: 1. The voltage, at which the transistor is biased, may also be calculated by the relation,

$$V_{CE} = V_C - V_E$$

2. The saturation value of collector current may be obtained by using the equation (*i*) and its value is given by the relation,

$$I_{C\,(sat)} = \frac{V_{CC} - V_{CE}}{R_C + R_E} = \frac{V_{CC}}{R_C + R_E} \quad ...(\text{if } V_{CC} >> V_{CE})$$

Example 22.11.

Calculate all the d.c. bias voltages and currents in circuit shown in Fig. 22.16, if $R_B = 400$ kΩ, $R_C = 2$ kΩ; $R_E = 1$ kΩ, $V_{CC} = 20$ V and β = 100. Neglect V_{BE} of the transistor. (*JNTU, June 2009*)

Solution.

Given : R_B = 400 kΩ = 400 × 10^3 Ω; R_C = 2 kΩ = 2 × 10^3 Ω; R_E = 1 kΩ = 1 × 10^3 Ω; V_{CC} = 20 volts and β = 100.

Base current

We know that the base current,

$$I_B = \frac{V_{CC}}{R_B + \beta \cdot R_E} = \frac{20}{(400 \times 10^3) + [100 \times (1 \times 10^3)]} \text{ A}$$

$$= 0.04 \times 10^{-3} \text{ A} = 0.04 \text{ mA } \textbf{Ans.}$$

Collector current

We know that the collector current,

$$I_C = \beta \cdot I_B = 100 \times 0.04 = 4 \text{ mA } \textbf{Ans.}$$

Collector-to-emitter voltage

We also know that the collector-to-emitter voltage,

$$V_{CE} = V_{CC} - I_C \cdot (R_C + R_E)$$

$$= 20 - (4 \times 10^{-3}) [(2 \times 10^3) + (1 \times 10^3)] \text{ V}$$

$$= 20 - 12 = 8 \text{ V } \textbf{Ans.}$$

Example 22.12.

In the circuit shown in Fig. 22.20, a silicon transistor with β = 50 is used. Find I_C and V_{CE}. Draw the D.C. load line on output characteristics and indicate the Q-point. Take V_{BE} = 0.7 V.

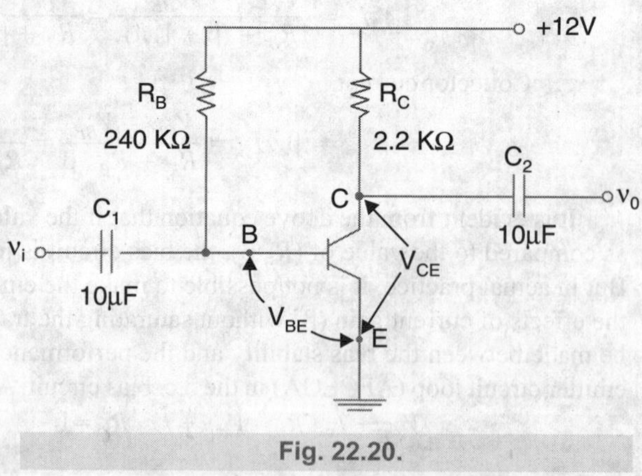

Fig. 22.20.

Solution.

Given : V_{CC} = 12 V, R_C = 2.2 kΩ; R_B = 240 kΩ; β = 50 and V_{BE} = 0.7 V.

We know that the value of collector current,

$$I_C = \frac{V_{CC} - V_{BE}}{R_E + \frac{R_B}{\beta}}$$

$$= \frac{12 - 0.7}{0 + \left(\frac{240 \times 10^3}{50}\right)} = 2.35 \text{ mA}$$

and collector-to-emitter voltage,

$$V_{CE} = V_{CC} - I_C \cdot R_C$$

$$= 12 - (2.35 \times 10^{-3}) \times (2.2 \times 10^3) \text{ V}$$

$$= 12 - 5.17 \text{ V}$$

$$= 6.83 \text{ V}$$

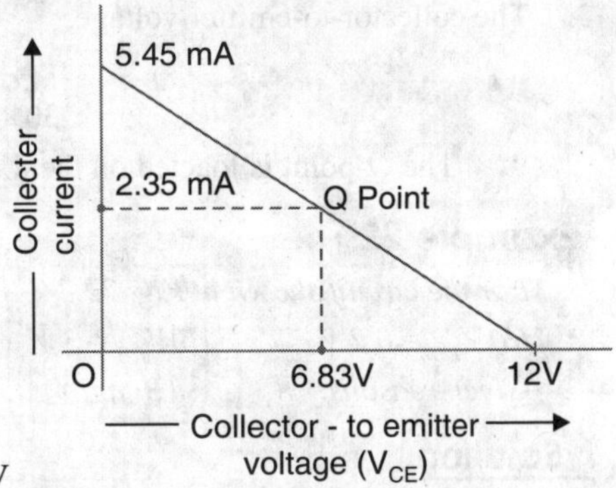

Fig. 22.21.

We also know that,

$$I_{C(sat)} = \frac{V_{CC}}{R_C} = \frac{12}{2.2 \times 10^3} = 5.45 \text{ mA}$$

and the cut-off voltage,

$$V_{CE(cut\text{-}off)} = V_{CC} = 12 \text{ V}$$

Using the values of $I_{C(sat)}$, I_C, V_{CE} and V_{CC} the d.c. load line can be obtained as shown in Fig. 22.21.

Example 22.13.

For the common-emitter circuit shown in Fig. 22.22, draw the dc load line and mark the dc operating point on it. Assume, β = 100 and neglect V_{BE}. (*Bangalore Univ. July 2003*)

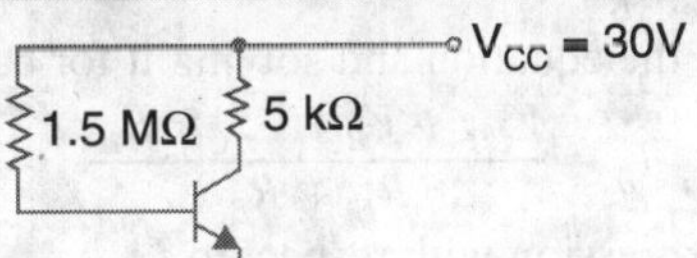

Fig. 22.22.

Solution.

Given : $V_{CC} = 30$ V, $R_C = 5$ kΩ $= 5 \times 10^3$ Ω; $R_B = 1.5$ MΩ $= 1.5 \times 10^6$ Ω; β = 100.

We know that collector current at saturation,

$$I_{C(sat)} = \frac{V_{CC}}{R_C} = \frac{30}{5 \times 10^3} = 6 \text{ mA}$$

and the value of cut-off voltage,

$$V_{CE(cut\text{-}off)} = V_{CC} = 30 \text{ V}$$

We also know that base current,

$$I_B = \frac{V_{CC}}{R_B} = \frac{30}{1.5 \times 10^6} = 20 \times 10^{-6} \text{ A} = 20 \text{ μA}$$

and the collector current,

$$I_C = \beta I_B = 100 \times 20 \times 10^{-6}$$

$$= 20 \times 10^{-4} \text{ A} = 2 \text{ mA}$$

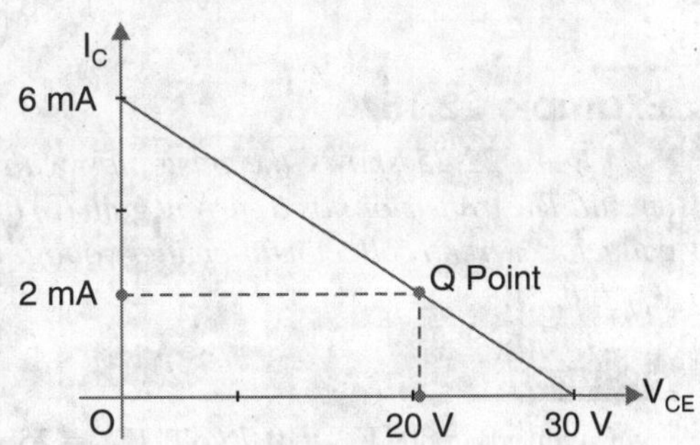

Fig. 22.23.

The collector-to-emitter voltage,

$$V_{CE} = V_{CC} - I_C R_C = 30 - (2 \times 10^{-3}) \times (5 \times 10^3)$$
$$= 30 - 10 \text{ V} = 20 \text{ V}$$

∴ The Q-point is loacted on the d.c. load line as shown in Fig. 22.23.

Example 22.14.

For the circuit shown in Fig. 22.24. What will be the state of LED when
(a) $V_{BB} = 0.2$ V (b) $V_{BB} = 3$ V
Given reasons. (Banglore Univ. B.Sc. Electronics, May 2001)

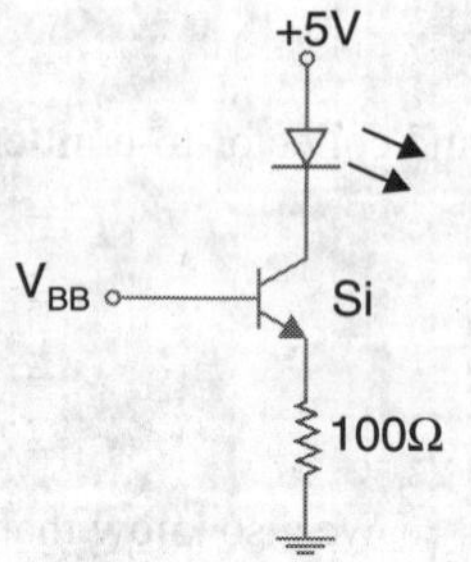

Fig. 22.24.

Solution.

Given : $V_{CC} = 5$ V, $R_E = 100\ \Omega$

We know that the state of LED will be determined by the fact whether the transistor is ON or OFF.

(a) $V_{BB} = 0.2$ V

Since the transistor in the given circuit is Si, it needs 0.7 V between the base-emitter junction to turn ON. Therefore, the LED is OFF. **Ans.**

(b) $V_{BB} = 3$ V

Since $V_{BB} = 3$ V is more that 0.7 V, the transistor will turn ON and therefore the LED will be ON. **Ans.**

22.14. Stability Factor of Base Bias with Emitter Feedback

We have already discussed in the last article that if we apply Kirchoff's Voltage Law to the base-emitter circuit loop shown in Fig. 22.16, then we get

$$V_{BE} + I_E \cdot R_E - V_{CC} + I_B \cdot R_B = 0$$

Replacing I_E with $I_B + I_C$ in the equation and solving it for base current,

$$I_B = \frac{V_{CC} - V_{BE} - I_C \cdot R_E}{R_E + R_B}$$

Differentiating the above expression with respect to I_C,

$$\frac{dI_B}{dI_C} = \frac{0 - 0 - 1 \cdot R_E}{R_E + R_B} = -\frac{R_E}{R_E + R_B}$$

Substituting the value of dI_B/dI_C in the general expression for stability factor,

$$S = \frac{1+\beta}{1-\beta\left(\frac{dI_B}{dI_C}\right)} = \frac{1+\beta}{1-\beta\left(-\frac{R_E}{R_E+R_B}\right)} = \frac{1+\beta}{1+\beta \cdot \frac{R_E}{R_E+R_B}}$$

Example 22.15.

Figure 22.25 shows the base-bias with emitter feedback resistor circuit. The transistor has a current gain (β) of 80. Determine the value of collector current, collector-to-emitter voltage and stability factor. Assume $V_{BE} = 0.7$ V.

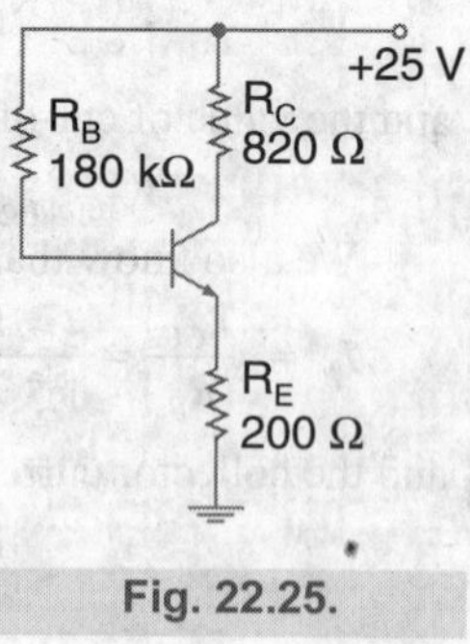

Fig. 22.25.

Solution.

Given : $\beta = 80$; $V_{BE} = 0.7$ volt; $V_{CC} = 25$ volts; $R_C = 820\ \Omega$; $R_B = 180\ \text{k}\Omega = 180 \times 10^3\ \Omega$ and $R_E = 200\ \Omega$.

Value of collector current

We know that the value of collector current,

$$I_C = \frac{V_{CC} - V_{BE}}{R_E + \frac{R_B}{\beta}} = \frac{25 - 0.7}{200 + \frac{180 \times 10^3}{80}} \text{ A}$$

$$= 9.9 \times 10^{-3} \text{ A or } 9.9 \text{ mA } \textbf{Ans.}$$

Value of collector-to-emitter voltage

We also know that the value of collector-to-emitter (or ground) voltage,

$$V_{CE} = V_{CC} - I_C \cdot R_C = 25 - (9.9 \times 10^{-3}) \times 820 \text{ V}$$

$$= 25 - 8.12 = 16.88 \text{ V } \textbf{Ans.}$$

22.15. Base Bias with Collector Feedback

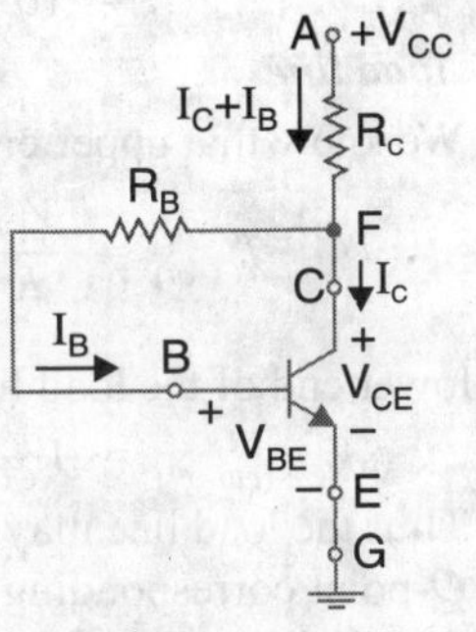

Fig. 22.26. Base bias with collector feedback.

Fig. 22.26 shows a d.c. bias circuit in which the base resistor (R_B) is connected to the collector terminal of a transistor rather than to the V_{CC} supply as in a base bias circuit. Here, the collector voltage provides the bias-emitter junction. The resistor R_B acts as a feedback resistor. It provides a very stable *Q*-point by reducing the effect of variations in current gain (β).

Let us now determine the d.c. bias currents and voltages in the base and collector of the transistor. First of all, consider the base-emitter circuit loop (AFBEGA) in the d.c. bias circuit. Applying Kirchoff's Voltage Law to the base-emitter circuit loop,

$$V_{BE} - V_{CC} + (I_B + I_C)\, R_C + I_B \cdot R_B = 0$$

Substituting I_C equal to β in the above equation and solving for the base current,

$$I_B = \frac{V_{CC} - V_{BE}}{R_B + (\beta + 1)R_C} = \frac{V_{CC} - V_{BE}}{R_B + \beta \cdot R_C} \quad \text{...(Taking } \beta + 1 - \beta)$$

and the collector current,

$$I_C = \beta \cdot I_B = \frac{V_{CC} - V_{BE}}{R_C + R_B / \beta} = \frac{V_{CC}}{R_C + R_B / \beta} \quad ...(\because V_{CC} >> V_{BE})$$

Now consider the collector-emitter circuit loop (*i.e.*, AFCEGA) in d.c. bias circuit. Applying Kirchoff's Voltage Law to this loop,

$$V_{CE} - V_{CC} + (I_C + I_B)\, R_C = 0$$

Substituting $I_C + I_B$ equal to I_C in the above equation and solving it for collector-to-emitter voltage,

$$V_{CE} = V_{CC} - I_C \cdot R_C$$

Example 22.16.

Calculate the Q-point values i.e., the values of collector current and collector-to-emitter voltage for the d.c. bias circuit shown in Fig. 22.27. Also draw the load line and locate Q-point on it. Assume $V_{BE} = 0.7$ V.

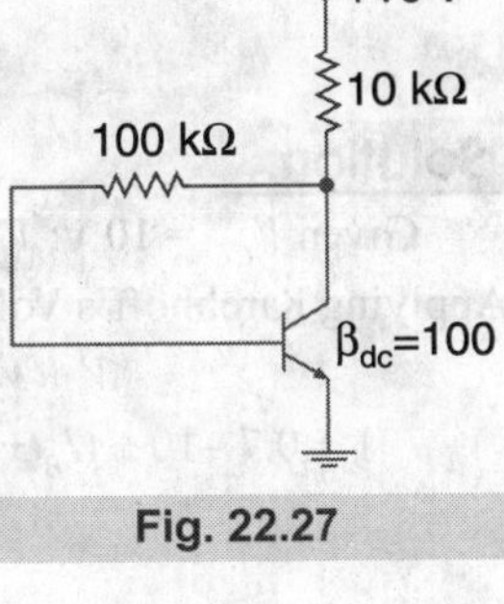

Fig. 22.27

Solution.

Given : $V_{CC} = 10$ volts; $R_C = 10$ kΩ $= 10 \times 10^{-3}$ Ω; $R_B = 100$ kΩ $= 100 \times 10^3$ Ω; $\beta_{dc} = 100$ and $V_{BE} = 0.7$ volt.

Q-point values

We know that the value of collector current,

$$I_C = \frac{V_{CC} - V_{BE}}{R_C + \frac{R_B}{\beta}} = \frac{10 - 0.7}{10 \times 10^3 + \frac{100 \times 10^3}{100}}$$

$$= 0.845 \times 10^{-3} \text{ A} = 0.845 \text{ mA } \textbf{Ans.}$$

and the collector-to-emitter voltage,

$$V_{CE} = V_{CC} - I_C \cdot R_C$$
$$= 10 - [(0.845 \times 10^{-3}) \times (10 \times 10^3)]$$
$$= 10 - 8.45 = 1.55 \text{ V } \textbf{Ans.}$$

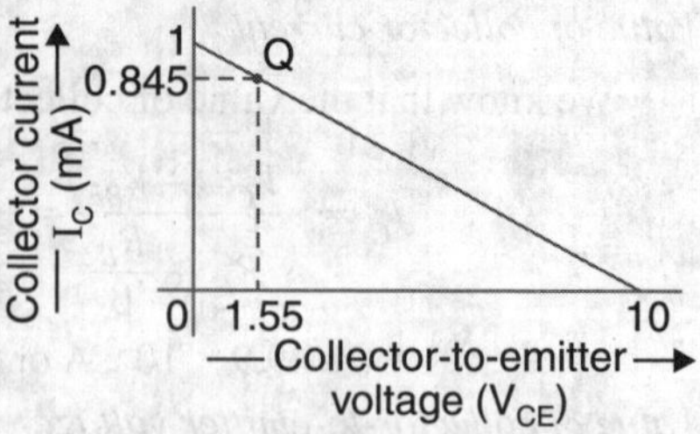

Fig. 22.28.

D.C. load line

We know that upper end of the load line

$$I_{C(sat)} = \frac{V_{CC}}{R_C} = \frac{10}{10 \times 10^3} = 1 \times 10^{-3} \text{ A} = 1 \text{ mA}$$

and lower end of the load line,

$$V_{CE\,(cut\text{-}off)} = V_{CC} = 10 \text{ V}$$

Thus the load line may be obtained by joining the upper and lower ends as shown in Fig. 22.28. The *Q*-point corresponding to the collector current of 0.845 mA and collector-to-emitter voltage of 1.55 V is indicated on the load line.

Example 22.17.

For the circuit shown in Fig. 22.29, determine the minimum value of I_B to saturate the transistor. Assume $\beta = 100$.

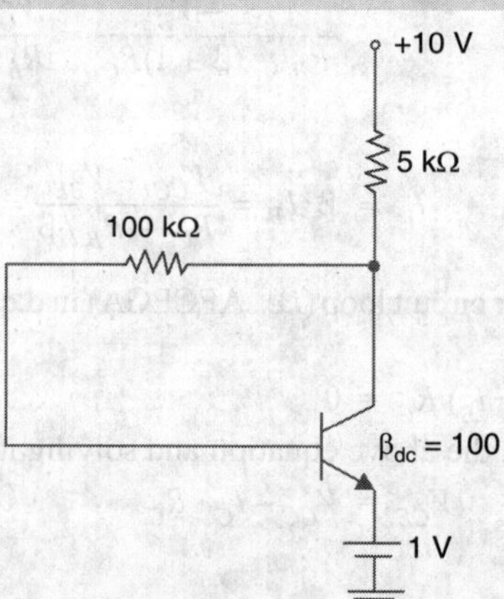

Fig. 22.29.

Solution.

Given $V_{CC} = 10$ V; $V_E = 1$V; $V_{BE} = 0.7$ V; $R_C = 5$ kΩ; $= 5 \times 10^3$ Ω; $R_B = 100$ kΩ and $\beta = 100$.

Applying Kirchhoff's Voltage Law to the base-emitter circuit loop,

$$V + V_{BE} - V_{CC} + (I_B + I_C)R_C + I_B \cdot R_B = 0$$
$$1 + 0.7 - 10 + (I_B + I_C)(5 \times 10^3) + I_B \times 100 \times 10^3 = 0$$

Since $I_C > I_B$, then

$$1 + 0.7 - 10 + I_C (5 \times 10^3) + I_B \times 100 \times 10^3 = 0$$

Substituting in the above equation, we get, $I_C = \beta I_B$

$$1 + 0.7 - 10 + \beta I_B (5 \times 10^3) + I_B \times 100 \times 10^3 = 0$$

or $$(600 \times 10^3) I_B = 8.3$$

or $$I_B = \frac{8.3}{(600 \times 10^3)} = 13.83 \times 10^{-6}\ \text{A} = 13.83\ \text{mA}\ \textbf{Ans.}$$

Example 22.18.

For the circuit shown in Fig. 22.29, Calculate I_B, I_C and I_E if Si transistor is used with $\beta = 50$ and stability factor 5. *(Nagpur Univ. Winter 2003)*

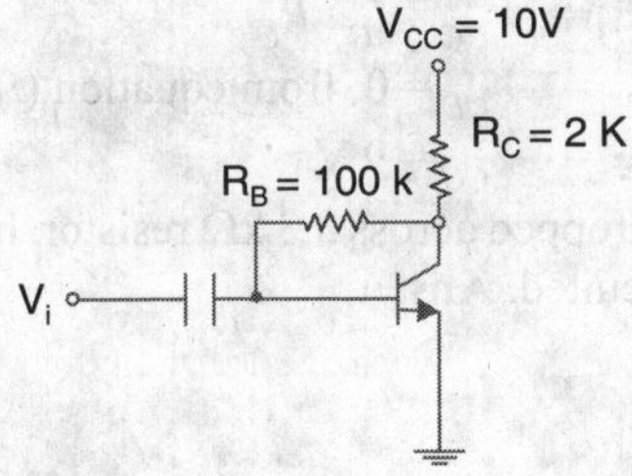

Fig. 22.30.

Solution.

Given $V_{CC} = 10$ V; $R_C = 2$ kΩ; $R_B = 100$ kΩ and $\beta = 50$.

We know that base current,

$$I_B = \frac{V_{CC} - V_{BE}}{R_B + \beta \cdot R_C} = \frac{10 - 0.7}{100\text{k} + (50 \times 2\text{k})}$$ ...(For *Si* transistor $V_{BE} = 0.7$ V)

$= 0.0465$ mA **Ans.**

and the collector current,

$$I_C = \beta \cdot I_B = 50 \times 0.0465\ \mu\text{A} = 2.325\ \text{mA}\ \textbf{Ans.}$$

We also know that emitter current,

$$I_E \approx I_C = 2.325\ \text{mA}\ \textbf{Ans.}$$

Example 22.19.

For the circuit shown in Fig. 22.31, what are the possible causes if (a) $V_C = 0$; (b) $V_C = 9$ V. *(Banglore Univ. B.sc. Electronics, May 2001)*

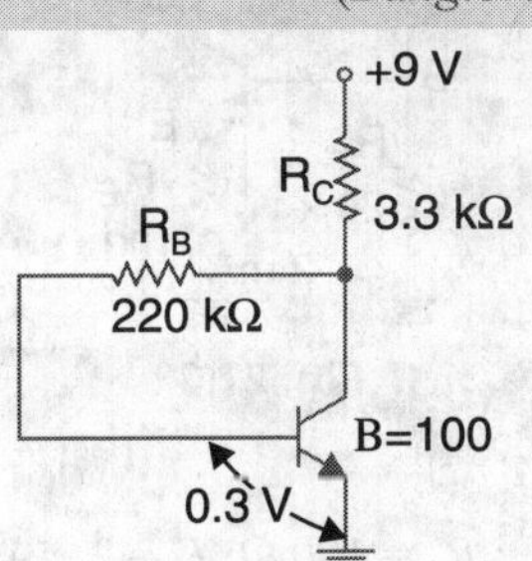

Fig. 22.31.

Solution.

Given : $V_{CC} = 9$ V; $R_B = 220$ kΩ $= 220 \times 10^3$ Ω; $R_C = 3.3$ kΩ $= 3.3 \times 10^3$ Ω; $V_{BE} = 0.3$ V and $\beta = 100$.

(a) *If $V_C = 0$*

We know that

$$I_B = \frac{V_{CC} - V_{BE}}{R_B + \beta \cdot R_C} = \frac{9 - 0.3}{(220 \times 10^3) + (100 \times 3.3 \times 10^3)}$$

$$= 0.016 \times 10^{-3} \ \Omega$$

$$= 16 \ \mu A$$

and the collector current,

$$I_C = \beta \cdot I_B = 100 \times 16 \ \mu A = 1600 \ \mu A = 1.6 \ \mu A.$$

Also $\qquad V_{CE} = V_{CC} - I_C \cdot R_C$

Since $\qquad V_C = V_{CE} = 0$; from equation (*i*) we know that

$$I_C R_C = V_{CC} = 9 \text{ V}.$$

This means the entire V_{CC} is dropped across 3.3 kΩ resistor. In other words the collector-emitter unction of a transistor is short-circuited. **Ans.**

(b) *If $V_C = 9$ V*

We know that the base current,

$$I_B = 16 \ \mu A$$

and the collector current,

$$I_C = \beta \cdot I_B = 1.6 \ \mu A$$

But also notice that, the collector-to-emitter voltage (V_{CE}),

$$9 = V_{CC} - I_C \cdot R_C = 9 - I_C \cdot R_C$$

$$\therefore \qquad I_C \cdot R_C = 0$$

Since $\quad I_C \neq 0$, therefore $R_C = 0$.

This means the collector resistance (R_C) is short-circuited. **Ans.**

Example 22.20.

If $\alpha = 0.98$, and $V_{BE} = 0.7$ V, find R_1 in the circuit shown in Fig. 22.32 for an emitter current $I_E = 2$ mA. Neglect reverse saturation current. (*Nagpur Univ. Winter 2003*)

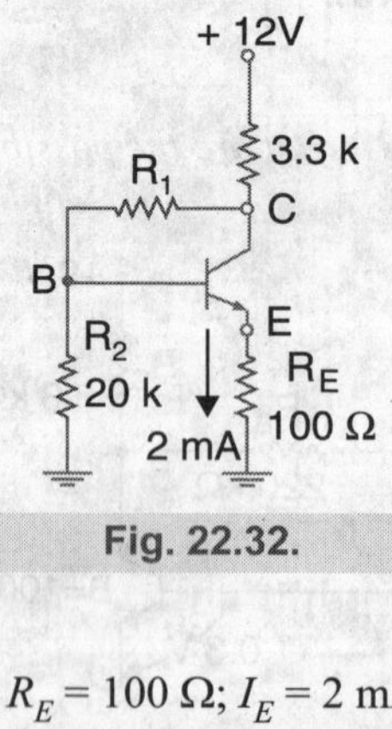

Fig. 22.32.

Solution.

Given : $V_{CC} = 12$ V; $R_C = 3.3$ kΩ; $R_E = 100$ Ω; $I_E = 2$ mA; $V_{BE} = 0.7$ V, $\alpha = 0.98$.

We know that collector current,

$$I_C = \alpha \cdot I_E = 0.98 \times 2 \text{ mA} = 1.96 \text{ mA}$$

We know that the base voltage, (measured with respect to ground)

$$V_B = V_{BE} + I_E \cdot R_E$$
$$= 0.7 \text{ V} + 2 \text{ mA} \times 100 \ \Omega$$
$$= 0.9 \text{ V}$$

and the collector voltage (measured with respect to ground),

$$V_C = V_{CC} - I_C \cdot R_C = 12 - (1.96 \text{ mA}) \times (3.3 \text{ k}\Omega)$$
$$= 12 - 6.488 = 5.532 \text{ V}$$

We also know that current through R_2,

$$I_{R_2} = \frac{V_B}{R_2} = \frac{5.532}{20 \text{ k}} = 0.277 \text{ mA}$$

Also the base current,

$$I_B = I_E - I_C$$
$$= 2 \text{ mA} - 1.96 \text{ mA}$$
$$= 0.04 \text{ mA}$$

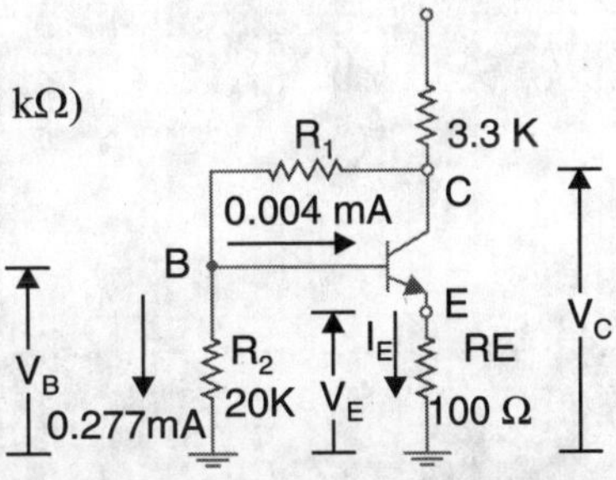

Fig. 22.33.

∴ The current through R, (refer to Fig. 22.33),

$$I_{R_1} = I_{R_2} + I_B = 0.277 \text{ mA} + 0.04 \text{ mA} = 0.317 \text{ mA}$$

and the value of the resistance,

$$R_1 = \frac{V_C - V_B}{I_{R_1}} = \frac{5.532 - 0.9}{0.317 \text{ mA}} = 14.6 \text{ k}\Omega \textbf{ Ans.}$$

Example 22.21.

In a circuit shown in Fig. 22.34, V_{CC} = 24 V, R_C = 10 k and R_E = 270 Ω. If a Si transistor is used with β = 45 and if V_{CE} = 5 V. Find R_B, neglect reverse saturation current.

(Nagpur Univ. Winter 2003, Summer 2008)

Solution.

Given V_{CC} = 24 V; R_C = 10 kΩ; R_E = 270 Ω; For a transistor, V_{BE} = 0.7 V; β = 45 and V_{CE} = 5 V.

We know that collector-to emitter voltagte, (V_{CE})

$$5 = V_{CC} \cdot I_C \cdot R_C = 24 - I_C$$

$$\therefore \quad I_C = \frac{24 - 5}{10 \times 10^3} = 1.9 \text{ mA}$$

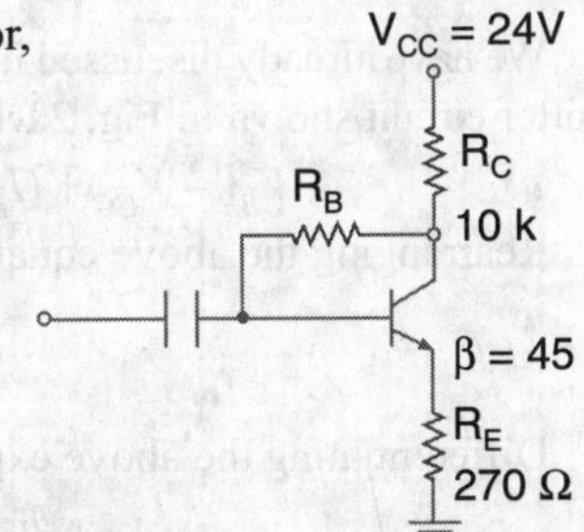

Fig. 22.34.

We also knwo that collector current (I_C)

$$1.9 \times 10^{-3} = \frac{V_{CC} - V_{BE}}{R_C + \dfrac{R_B}{\beta}} = \frac{24 - 0.7}{(10 \times 10^3) + \dfrac{R_B}{45}}$$

$$\therefore \quad \frac{R_B}{45} = \frac{23.93}{1.9 \times 10^{-3}} - (10 \times 10^3) = (12.6 \times 10^3) - (10 \times 10^3)$$

$$= 2.6 \times 10^3$$

or $$R_B = (2.6 \times 10^3) \times 45 = 117 \text{ k}\Omega \textbf{ Ans.}$$

Example 22.22.

For the circuit shown in Fig. 22.35, find I_B, I_C, V_C. (GBTU/MTU., 2009)

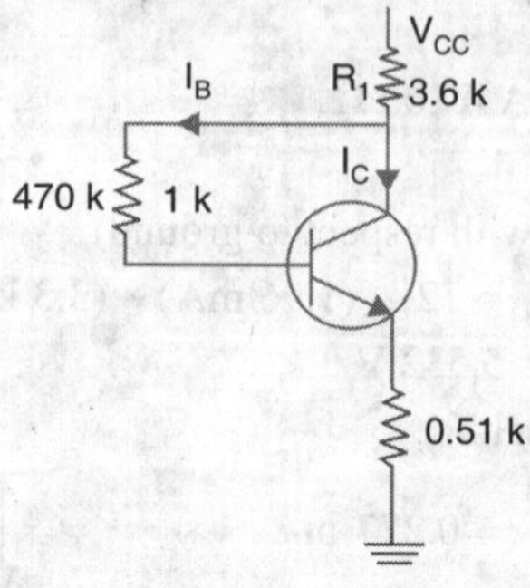

Fig. 22.35.

Solution.

Given $R_B = 470\ k\Omega = 470 \times 10^3\ \Omega$; $R_C = 3.6\ k\Omega = 3.6 \times 10^3\ \Omega$; $V_{CC} = +16$ V and $R_E = 0.51 k\Omega = 0.51 \times 10^3 \Omega$.

We know that,

$$I_B = \frac{V_{CC} - V_{BE}}{R_B + \beta R_C}$$

$$= 1.7738 \times 10^5 A = 17.73\ \mu A \textbf{ Ans.}$$

We also know that,

$$I_C = \beta . I_B = 120 \times 17.73\ \mu A = 2127.6\ \mu A = 2.1276 \text{ mA} \textbf{ Ans.}$$

and the value of V_C given by,

$$V_C = V_{CC} - I_C R_C = 16 - (2.1276 \times 10^{-3}) \times (3.6 \times 10^3)$$

$$= 8.34 \text{ V} \textbf{ Ans.}$$

22.16. Stability Factor of Base Bias with Collector Feedback

We have already discussed in the last article that if we apply Kirchoff's Voltage Law to the base-emitter circuit shown in Fig. 22.19, then we get

$$V_{BE} - V_{CC} + (I_B + I_C) R_C + I_B \cdot R_B = 0$$

Rearranging the above equation and solving for base current,

$$I_B = \frac{V_{CC} - V_{BE} - I_C \cdot R_C}{R_C + R_B}$$

Differentiating the above expression with respect to I_C,

$$\frac{dI_B}{dI_C} = \frac{0 - 0 - 1 \cdot R_C}{R_C + R_B} = -\frac{R_C}{R_C + R_B}$$

Substituting the value of dI_B/dI_C in the general expression for stability factor,

$$S = \frac{1 + \beta}{1 - \beta\left(\dfrac{dI_B}{dI_C}\right)} = \frac{1 + \beta}{1 - \beta\left(-\dfrac{R_C}{R_C + R_B}\right)} = \frac{1 + \beta}{1 + \beta \times \dfrac{R_C}{R_C + R_B}}$$

It is evident from the above expression that the stability factor (*S*) of a collector feedback bias is smaller than $(1 + \beta)$. Therefore the biasing arrangement of collector feedback resistor is certainly an improvement over the fixed bias circuit.

Example 22.23.

Determine the d.c. bias current and voltage for the d.c. bias circuit shown in Fig. 22.36.

Also determine the stability factor of the bias circuit. Assume $V_{BE} = 0.7$ V.

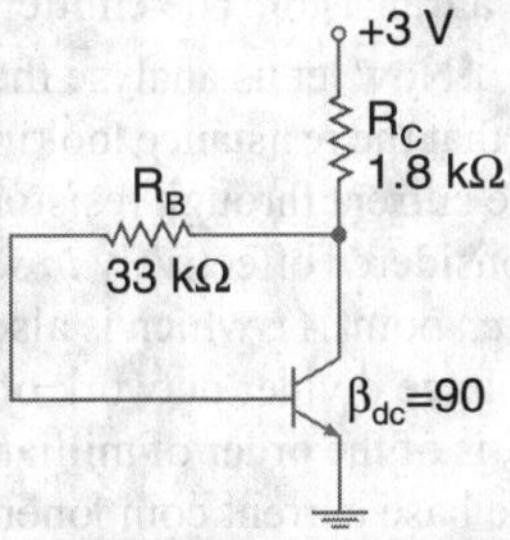

Fig. 22.36.

Solution.

Given : $V_{CC} = 3$ volts; $R_C = 1.8\ k\Omega = 1.8 \times 10^3\ \Omega$; $R_B = 33\ k\Omega = 33 \times 10^3\ \Omega$; $\beta = 90$ and $V_{BE} = 0.7$ V

D.C bias current and voltage

We know that the base current,

$$I_B = \frac{V_{CC} - V_{BE}}{R_B + \beta \cdot R_C} = \frac{3 - 0.7}{(33 \times 10^3) + 90\ (1.8 \times 10^3)}\ \text{A}$$

$$= 0.0118 \times 10^{-3}\ \text{A} = 0.0118\ \text{mA}$$

and the collector current,

$$I_C = \beta \cdot I_B = 90 \times 0.0118 = 1.06\ \text{mA}\ \textbf{Ans.}$$

$\therefore$ ***Collector-to-emitter voltage,***

$$V_{CE} = V_{CC} - I_C \cdot R_C = 3 - (1.06 \times 10^{-3}) \times (1.8 \times 10^3)\ \text{V}$$

$$= 3 - 1.9 = 1.1\ \text{V}\ \textbf{Ans.}$$

Stability factor

We also know that stability factor,

$$S = \frac{1 + \beta}{1 + \beta \times \dfrac{R_C}{R_C + R_B}} = \frac{1 + 90}{1 + 90 \times \dfrac{1.8 \times 10^3}{(1.8 \times 10^3) + (33 \times 10^3)}}$$

$$= \frac{91}{5.65} = 16.1\ \textbf{Ans.}$$

22.17. Voltage Divider Bias

In all the d.c. bias circuits, discussed in the previous articles, we have found that the values of d.c. bias current and voltage of the collector depends upon the current gain (β) of the transistor. But we know that the value of current gain (β) is temperature sensitive especially for silicon transistors. Since the nominal value of current gain (β) is not well defined, therefore it would be desirable to provide a d.c. bias circuit which is independent of the transistor current gain (β). The d.c. bias circuit shown in Fig. 22.37 (*a*) meets this condition and is thus a very popular bias circuit. It is commonly known as *voltage divider bias* or *self bias circuit.*

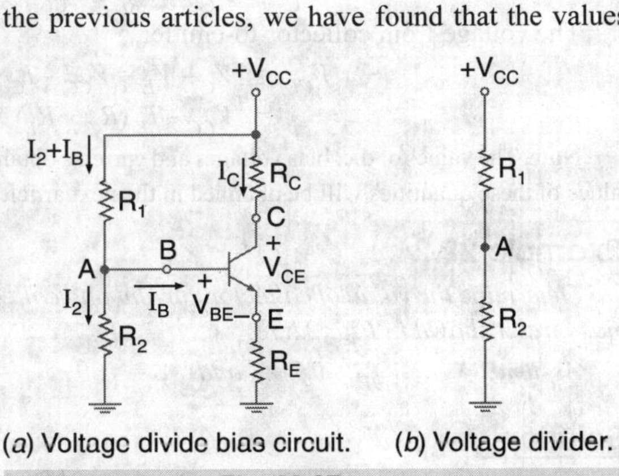

(*a*) Voltage divide bias circuit. (*b*) Voltage divider.

Fig. 22.37.

The name voltage divider is derived from the fact that resistors R_1 and R_2 form a potential divider across the V_{CC} supply. The voltage drop across resistor R_2 forward biases the base-emitter junction of a transistor. The emitter resistor (R_E) providers the d.c. stability.

Now let us analyse the base-emitter loop of the voltage divider bias circuit. A basic assumption is that the resistance looking into the base is much larger than that of the resistor R_2. If this is so, then the current through resistor R_1 flows almost completely into resistor R_2 and the two resistors may be considered effectively *in series* as shown in Fig. 22.37 (*b*). The voltage at the junction of the resistor (*i.e.*, point *A*), which is also the voltage at the base of the transistor, is then determined simple by the voltage divider network of R_1 and R_2 and the supply voltage. If the current through resistors R_1 and R_2 is of the order of milliamperes and that through the base is of the order of the microamperes, then the base current component can be neglected.

It is evident from Fig. 22.37 (*b*), that the voltage at the transistor base (due to the voltage divider network of resistors R_1 and R_2)

$$V_B = V_{CC} \times \frac{R_2}{R_1 + R_2}$$

Since the voltage drop across the base-emitter junction (V_{BE}) when forward biased is very small, as compared to the voltage at the base (V_B), therefore the voltage at the emitter is almost equal to the voltage at the base *i.e.*,

$$V_E = V_B \quad \text{... (Neglecting } V_{BE}\text{)}$$

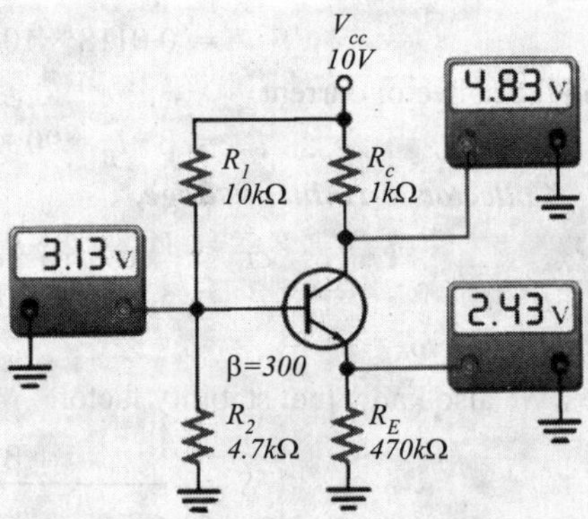

A voltage divider biased transistor with correct voltages.

∴ Value of emitter current,

$$I_E = \frac{V_E}{R_E}$$

and the value of collector current,

$$I_C \approx I_E$$

The voltage drop across the collector resistor,

$$V_{RC} = I_C \cdot R_C$$

and the voltage at the collector (measured with respect to the ground)

$$V_C = V_{CC} - V_{RC} = V_{CC} - I_C \cdot R_C$$

The voltage from collector-to-emitter,

$$V_{CE} = V_C - V_E = V_{CC} - I_C \cdot R_C - I_E \cdot R_E$$

$$= V_{CC} - I_E (R_C + R_E) \quad ...(\because I_C = I_E)$$

Note. The values of d.c. bias voltages and currents obtained above are only approximate. The more accurate values of these quantities will be obtained in the next article.

Example 22.24.

Determine the value of collector current and collector-to-emitter voltage for the voltage divider bias circuit shown in Fig. 22.38.

Assume $V_{BE} = 0.7\ V$ *and* $\beta = 100.$

Solution.

Given : $V_{CC} = 10$ volts; $R_C = 1\ k\Omega = 1 \times 10^3\ \Omega$; $R_1 = 10\ k\Omega$; $R_2 = 5\ k\Omega$; $R_E = 500\ \Omega$; $V_{BE} = 0.7$ volt and $\beta = 100$.

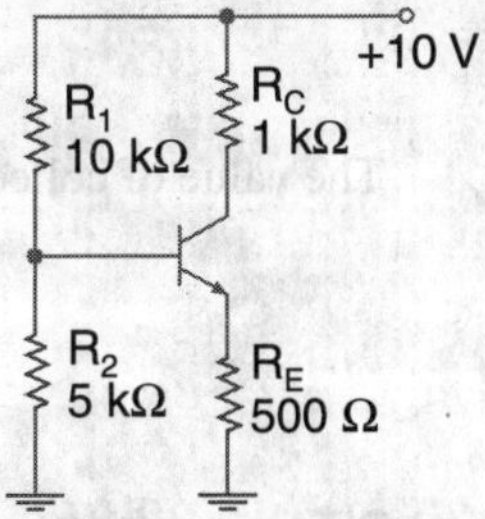

Fig. 22.38.

Collector current

We know that voltage at the base,

$$V_B = V_{CC}\left(\frac{R_2}{R_1+R_2}\right) = 10\left(\frac{5}{10+4}\right)\text{ V}$$

$$= 3.33 \text{ V}$$

and voltage at the emitter,

$$V_E = V_B - V_{BE} = 3.33 - 0.7 = 2.63 \text{ V}$$

∴ Emitter current,

$$I_E = \frac{V_E}{R_E} = \frac{2.63}{500} = 5.26 \times 10^{-3} \text{ A} = 5.26 \text{ mA}$$

and collector current,

$$I_C = I_E = 5.26 \text{ mA } \textbf{Ans.}$$

Collector-to-emitter voltage

We also know that the collector-to-voltage,

$$V_{CE} = V_{CC} - (R_C + R_E) = 10 - 5.26 \times 10^{-3}\,(1 \times 10^3 + 500) \text{ V}$$

$$= 10 - 7.89 = 2.11 \text{ V } \textbf{Ans.}$$

Example 22.25.

Calculate the exact value of emitter current and V_{CE} in the circuit shown in Fig. 22.39. (*Nagpur Univ. 2004*)

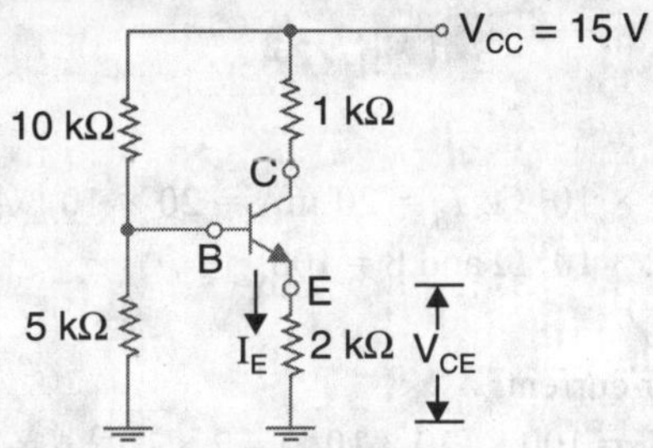

Fig. 22.39.

Solution.

Given $V_{CC} = 15$ V; $R_C = 1$ kΩ; $R_E = 2$ kΩ; $R_1 = 10$ kΩ; $R_2 = 5$ kΩ.

We know that the Thevenin's voltage,

$$V_{TH} = V_{CC}\left(\frac{R_2}{R_1+R_2}\right) = 15 \times \left(\frac{5k}{10k+5k}\right) = 5 \text{ V}$$

and the Thevenin's equivalent resistance,

$$R_{TH} = R_1 \parallel R_2 = \frac{R_1 \times R_2}{R_1 + R_2} = \frac{10\,k \times 5k}{10k + 5k} = 3.33 \text{ k}\Omega$$

We also know that emitter current,

$$I_E = \frac{V_{TH} - V_{BE}}{R_E + \dfrac{R_H}{\beta}} \approx \frac{V_{TH} - V_{BE}}{R_E}$$

$$= \frac{5 - 0.7}{2k} = 2.15 \text{ mA } \textbf{Ans.}$$

The value of collector-to-enitter voltage,

$$V_{CE} = V_{CC} - I_E (R_C + R_E) \qquad ...(\because I_C \approx I_E)$$

$$= 15 - (2.15 \times 10^{-3}) \times (1000 + 2000)$$

$$= 8.55 \text{ V } \textbf{Ans.}$$

Example 22.26.

Determine the following values of the voltage divider configuration shown in Fig. 22.40.

(a) I_C (b) V_E (c) V_{CC} (d) V_{CE} (e) V_B (f) R_1 (GBTU., 2011–2012)

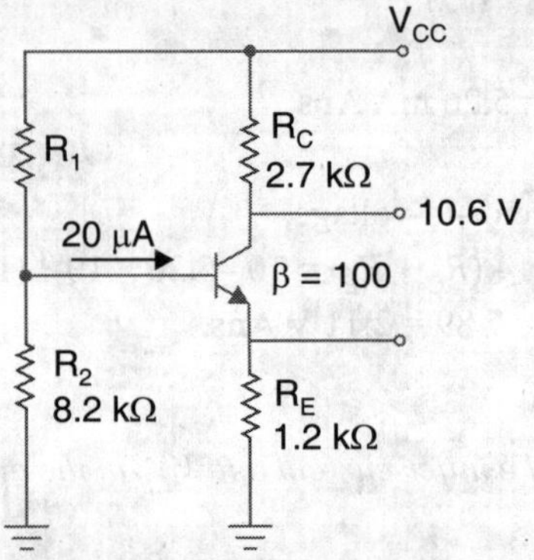

Fig. 22.40.

Solution.

Given: $R_2 = 8.2 \text{ k}\Omega = 8.2 \times 10^3 \ \Omega$; $I_B = 20 \text{ mA} = 20 \times 10^{-6}$ A; $R_C = 2.7 \text{ k}\Omega = 2.7 \times 10^3 \ \Omega$; $V_C = 10.6$ V; $R_E = 1.2 \text{ k}\Omega = 1.2 \times 10^3 \ \Omega$ and $\beta = 100$.

(a) $\boldsymbol{I_C}$

We know that the collector current,

$$I_C = \beta I_B = 100 \times (20 \times 10^{-6}) = 2 \times 10^{-3} \text{ A} = 2 \text{ mA } \textbf{Ans.}$$

(b) $\boldsymbol{V_E}$

We know that the emitter voltage,

$$V_E = I_E R_E = (I_C + I_B) R_E$$

$$= [(2 \times 10^{-3}) + (20 \times 10^{-6})] \times (1.2 \times 10^3) = 2.42 \text{ V } \textbf{Ans.}$$

(c) $\boldsymbol{V_{CC}}$

We also know that,

$$V_{CC} = V_C + I_C R_C = [10.6 + (2 \times 10^{-3}) \times (2.7 \times 10^3)]$$

$$= 10.6 + 5.4 = 16 \text{ V } \textbf{Ans.}$$

(d) $\boldsymbol{V_{CE}}$

We know that the collector-to-emitter voltage,

$$V_{CE} = V_C - V_E = 16 - 2.42 = 13.58 \text{ V } \textbf{Ans.}$$

(e) $\boldsymbol{V_B}$

We know that the emitter voltage is almost equal to the base voltage,

$$V_E = V_B = 2.42 \text{ V } \textbf{Ans.}$$

(*f*) R_1

We know that the,

$$V_B = V_{CC} \times \frac{R_2}{R_1 + R_2}$$

$$2.42 = 16 \times \frac{(8.2 \times 10^3)}{R_1 + (8.2 \times 10^3)}$$

Showing the above equations, we get

$$R_1 = 46 \times 10^3 \text{ W} = 46 \text{ k}\Omega \text{ **Ans.**}$$

22.18. Stability of Voltage Divider Bias

Consider again the voltage divider bias circuit shown in Fig. 22.37 (*a*). Looking out from the base terminal, the bias circuit can be redrawn as shown in Fig. 22.41 (*a*).

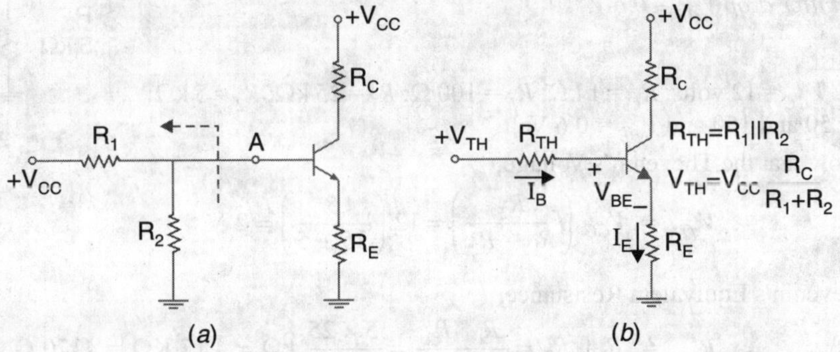

Fig. 22.41. Thevenizing the voltage divider bias circuit.

Now applying the Thevenin's Theorem to the circuit lying on the left of the point *A*, we get the Thevenin's Voltage,

$$V_{TH} = \left(\frac{R_2}{R_1 + R_2}\right) V_{CC}$$

and the Thevenin's Equivalent Resistance

$$R_{TH} = R_1 \parallel R_2 = \frac{R_1 \cdot R_2}{R_1 + R_2}$$

We know that the Thevenin's Voltage (V_{TH}) is a voltage drop across resistor R_2. As a matter of fact, it is the voltage at the base (V_B) measured with respect to the ground. Fig. 22.41 (*b*) shows the voltage divider bias circuit in which the series combination of resistors R_1 and R_2 is replaced by a Thevenin's equivalent circuit. Applying, Kirchoff's Voltage Law to the base-emitter loop of this circuit,

$$I_B \cdot R_{TH} + V_{BE} + I_E \cdot R_E - V_{TH} = 0$$

Replacing I_B with I_E / β (equal to I_C / β) in the above relation and solving for emitter current,

$$I_E = \frac{V_{TH} - V_{BE}}{R_E + \frac{R_{TH}}{\beta}} \quad ...(i)$$

$$= \frac{V_{TH} - V_{BE}}{R_E} \quad ...(R_E >> R_{TH} / \beta)$$

It is evident from the above relation that *the emitter current is independent of current gain* (*b*). In actual practice, it is achieved by making the emitter resistor 10 times greater than R_{TH} / β.

Mathematically, emitter resistance,

$$R_E \geq 10\, R_{TH}/\beta$$

or

$$R_{TH} \leq 0.1\, \beta \cdot R_E \qquad ...(ii)$$

Usually, the value of resistor R_2 is smaller than the value of resistor R_1. Therefore the Thevenin's equivalent resistance R_{TH} (equal to $R_1 \parallel R_2$) is approximately equal to R_2. In that case, the [equation (*ii*)] may be written as

$$R_2 \leq 0.1\, \beta \cdot R_E \qquad ...(iii)$$

Example 22.27.

Fig. 22.42 shows that d.c. bias circuit of a common emitter transistor amplifier.

Find the percentage change in collector current, if the transistor with $h_{FE} = 50$ is replaced by another transistor with $h_{FE} = 150$. It is given that the base emitter drop $V_{BE} = 0.6$ V.

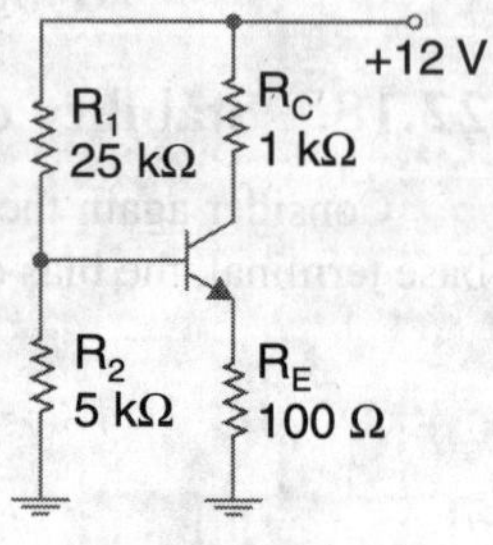

Fig. 22.42.

Solution.

Given : $V_{CC} = 12$ volts; $R_C = 1$ kΩ; $R_E = 100$ Ω; $R_1 = 25$ kΩ; $R_2 = 5$ kΩ; h_{FE} $(= \beta) = 50$ and 150 and $V_{BE} = 0.6$ Volt.

We know that the Thevenin's Voltage,

$$V_{TH} = V_{CC}\left(\frac{R_2}{R_1 + R_2}\right) = 12\left(\frac{5}{25+5}\right) = 2 \text{ V}$$

and the Thevenin's Equivalent Resistance,

$$R_{TH} = R_1 \parallel R_2 = \frac{R_1 \cdot R_2}{R_1 + R_2} = \frac{5 \times 25}{5 + 25} \text{ k}\Omega = 4.17 \text{ K}\Omega = 4170\ \Omega$$

We also know that the emitter current,

$$I_E = \frac{V_{TH} - V_{BE}}{R_E + \dfrac{R_{TH}}{\beta}}$$

For $\beta = 50$, the emitter current,

$$I_E = \frac{2 - 0.6}{100 + \dfrac{4170}{50}} = 7.63 \times 10^{-3} \text{ A} = 7.63 \text{ mA}$$

and for $\beta = 150$, the emitter current,

$$I_E = \frac{2 - 0.6}{100 + \dfrac{4170}{150}} = 10.95 \times 10^{-3} \text{ A} = 10.95 \text{ mA}$$

∴ Change in the collector current, when the transistor with h_{FE} equal to 50 is changed with another transistor with h_{FE} equal to 150,

$$\frac{10.95 - 7.63}{7.63} = 0.435 = 43.5\% \textbf{ Ans.}$$

Example 22.28.

A Si transistor is biased using voltage divider method as shown in Fig. 22.43. Determine the operating point. (*Banglore Univ. B.sc. Electronics, May 2001*)

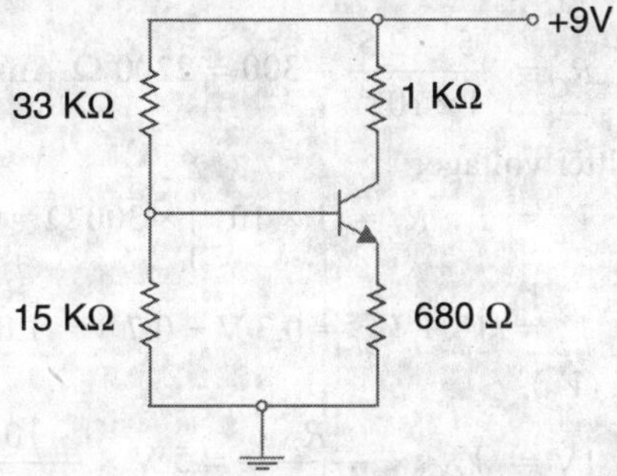

Fig. 22.43.

Solution.

Given : $V_{CC} = 9$ V; $R_C = 1$ kΩ $= 1 \times 10^3$ Ω; $R_E = 680$ Ω; $R_1 = 33$ kΩ and $R_2 = 15$ kΩ.

We know that the base voltage,

$$V_B = V_{CC} \times \frac{R_2}{R_1 + R_2} = 9 \times \frac{15 \text{ k}}{33 \text{ k} + 15 \text{ k}} = 2.81 \text{ V}$$

Since the voltage drop across the base emitter junction for a Si transistor is 0.7 V, therefore,

$$V_E = V_B - 0.7 \text{ V} = 2.81 - 0.7 = 2.11 \text{ V}$$

and the value of emitter current,

$$I_E = \frac{V_E}{R_E} = \frac{2.11}{680} = 3.1 \times 10^{-3} \text{ A} = 3.1 \text{ mA}$$

∴ Collector current,

$$I_C \approx I_E = 3.1 \text{ mA}$$

We know that voltage across the collector resistor,

$$V_{RC} = I_C \cdot R_C = 3.1 \text{ mA} \times 1 \text{ k}\Omega = 3.1 \text{ V}$$

and the voltage at the collector (measured with respect to the ground),

$$V_C = V_{CC} - V_{R_C} = 9 - 3.1 = 5.9 \text{ V}$$

∴ The voltage from collector-to-emitter,

$$V_{CE} = V_C - V_E = 5.9 - 2.11 = 3.79 \text{ V.}$$

Hence the operating point values are,

$$I_C = 3.1 \text{ mA and } V_{CE} = 3.79 \text{ V} \quad \textbf{Ans.}$$

Example 22.29.

The BJT in amplifier of Fig. 22.44, has $h_{fe} = 100$, $V_{BE} = 0.7$ V, $I_{CO} = 0$. Calculate the value of R_1 and R_C such tha its $I_C = 1$ mA and $V_{CE} = 2.5$ V. (UPSC Engg. Services 2001)

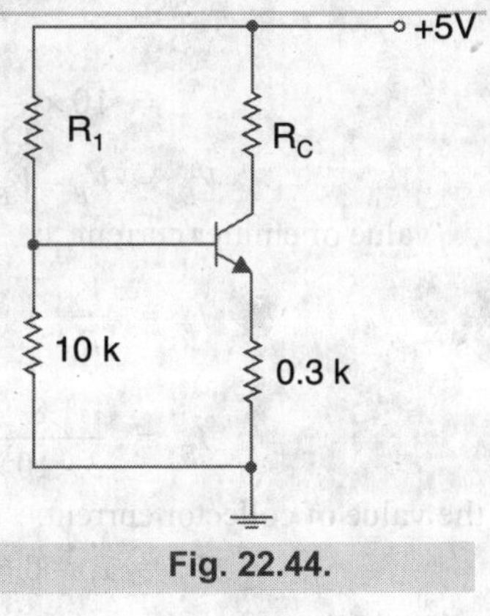

Fig. 22.44.

Solution.

Given : $V_{CC} = 5$ V; $R_E = 0.3$ kΩ $= 300$ Ω; $I_C = 1$ mA; $V_{CE} = 2.5$ V; $h_{fe} = \beta = 100$; $V_{BE} = 0.7$ V; $I_{CO} = 0$.

We know that the emitter current,

$$I_E \approx I_C = 1 \text{ mA}$$

and the voltatge from collector-to-emitter, (V_{CE})

$$2.5 = V_{CC} - I_E (R_C + R_E)$$
$$= 5 - (1 \times 10^{-3})(R_C + 300)$$

or $$R_C = \frac{5 - 2.5}{1 \times 10^{-3}} - 300 = 2200\ \Omega \textbf{ Ans.}$$

We also know that the emitter voltage,

$$V_E = I_E \cdot R_E = (1 \times 10^{-3}) \times 300\ \Omega = 0.3\ \text{V}$$

and the base voltage,

$$V_B = V_E + V_{BE} = 0.3\ \text{V} + 0.7\ \text{V} = 1.0\ \text{V}$$

For a voltage-divider bias, (V_B),

$$1\text{V} = V_{CC} \times \frac{R_2}{R_1 + R_2} = 5\ \text{V} \times \frac{10\ \text{k}}{R_1 + 10\ \text{k}}$$

$$\therefore \quad R_1 = 50\ \text{k} - 10\ \text{k} = 40\ \text{k}\Omega \textbf{ Ans.}$$

Example 22.30.

For the circuit shown in Fig. 22.45, below, draw the d.c load line and mark the Q-point. Assume silicon transistor. *(Calicut University., June 2007)*

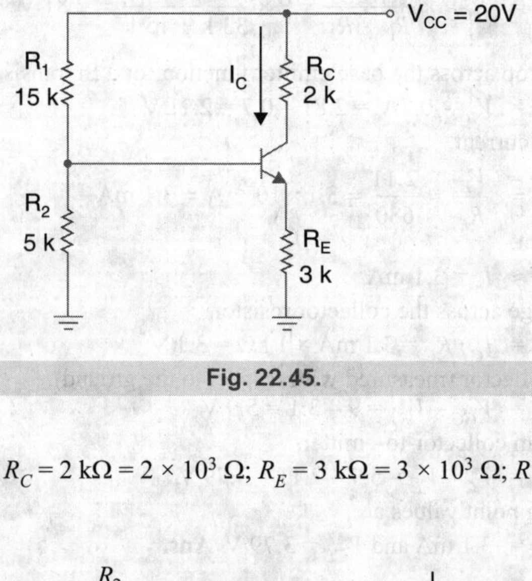

Fig. 22.45.

Solution.

Given : $V_{CC} = 20$ V; $R_C = 2\ \text{k}\Omega = 2 \times 10^3\ \Omega$; $R_E = 3\ \text{k}\Omega = 3 \times 10^3\ \Omega$; $R_1 = 15\ \text{k}\Omega = 15 \times 10^3\ \Omega$; $R_2 = 5\ \text{k}\Omega = 5 \times 10^3\ \Omega$.

We know that,

$$V_B = V_{CC} \times \frac{R_2}{R_1 + R_2}$$

$$= 10 \times \frac{5}{15 + 5} = 2.5\text{V}$$

$$V_E = V_B - V_E = 2.5 - 0.7 = 1.8\ \text{V}$$

$\therefore$ value of emitter current,

$$I_E = \frac{V_E}{R_E}$$

$$I_E = \frac{1.8}{3 \times 10^3} = 0.6 \times 10^{-3}\ \text{A} = 0.6\ \text{mA}$$

the value of collector current,

$$I_C = I_E = 0.6\ \text{mA}$$

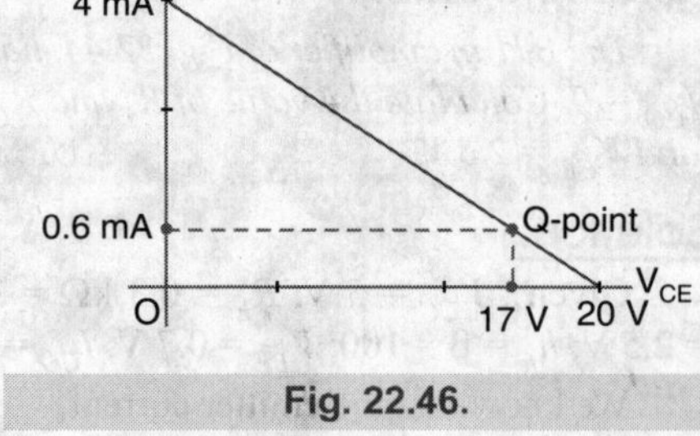

Fig. 22.46.

the voltage from collector-to-emitter,

$$V_{CE} = V_{CC} - I_E(R_C + R_E) = 20 - 0.6\,(2 + 3)$$
$$= 17 \text{ V}$$

D.C. load line and Q-point of transistor is shown in Fig. 22.46.

Example 22.31.

Calculate the value of emitter current in the voltage divider circuit shown in Fig. 22.47, Aslo find the value of V_{CE} and collector potential V_C. *(Dip. IETE., June 2001)*

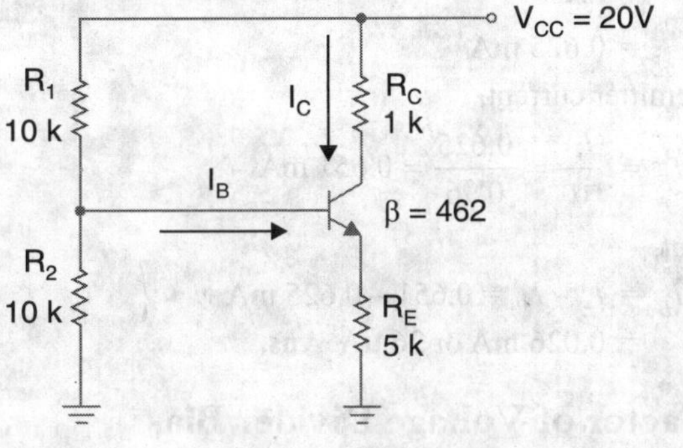

Fig. 22.47.

Solution.

Given : $V_{CC} = 20$ V; $R_C = 1$ kΩ ; $R_E = 5$ kΩ; $R_1 = 10$ kΩ; $R_2 = 10$ kΩ. β = 0.7; $V_{BE} = 0.7$ V.

We know that voltage at the base,

$$V_B = V_{CC} \times \frac{R_2}{R_1 + R_2} = 20 \times \frac{10 \text{ k}}{10 \text{ k} + 10 \text{ k}} = 10 \text{ V}$$

and voltage at the emitter,

$$V_E = V_B - V_{BE} = 10 - 0.7 = 9.3 \text{ V}$$

∴ Emitter current,

$$I_E = \frac{V_E}{R_E} = \frac{9.3}{5 \text{ k}} = 1.86 \text{ mA } \textbf{Ans.}$$

and collector current,

$$I_C \approx I_E = 1.86 \text{ mA}$$

We also know that collector-to-emitter voltage,

$$V_{CE} = V_{CC} - I_C \cdot R_C = 20 - (1.86 \text{ mA}) \times (1 \text{ k}\Omega)$$
$$= 18.14 \text{ V } \textbf{Ans.}$$

Example 22.32.

A transistor is connected in common-emitter (CE) configuration in which collector suply is 8 V and the voltage drop across resistance R_C connected in the collector circuit is 0.5 V. The value of R_C is 800 W. If a = 0.96 determine (a) Collector-emitter voltage (V_{CE}) and (b) base current. *(Dip, IETE,. June 2001)*

Solution.

Given : $V_{CC} = 8$ V; $V_{R_C} = 0.5$ V; $R_C = 800$ Ω; α = 0.96.

Fig. 22.48 shows the arrangement.

We know that the collector-to-emitter voltage,

$$V_{CE} = V_{CC} - I_C \cdot R_C$$
$$= 8 - 0.5$$
$$= 7.5 \text{ V } \textbf{Ans.}$$

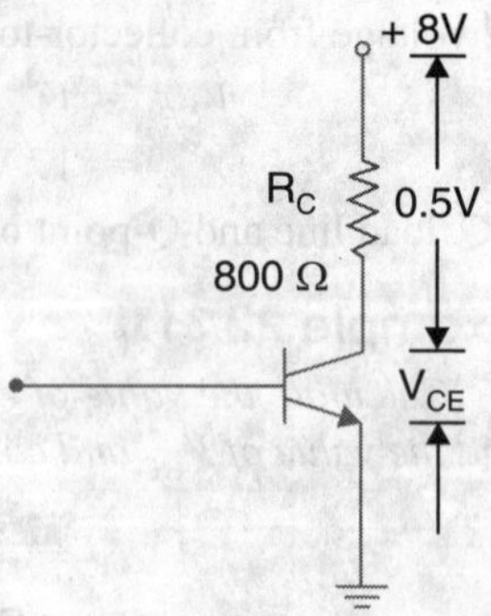

Fig. 22.48.

and the collector current,

$$I_C = \frac{V_{R_C}}{R_C} = \frac{0.5}{800} = 625 \times 10^{-4} \text{ A}$$
$$= 0.625 \text{ mA}$$

We also know that emitter current,

$$I_E = \frac{I_C}{\alpha} = \frac{0.625}{0.96} = 0.651 \text{ mA}$$

∴ The base current,

$$I_B = I_E - I_C = 0.651 - 0.625 \text{ mA}$$
$$= 0.026 \text{ mA or } 26 \text{ µA } \textbf{Ans.}$$

22.19. Stability Factor of Voltage Divider Bias

We have already discussed in the last article that if we apply Kirchoff's Voltage Law to the base-emitter loop of the voltage divider bias circuit shown in Fig. 22.37 (*b*), we get,

$$I_B \cdot R_{TH} + V_{BE} + I_E \cdot R_E - V_{TH} = 0 \qquad ...(i)$$

Substituting the value of I_E (equal to $I_B + I_C$) in the above equation and solving it for base current,

$$I_B = \frac{V_{TH} - V_{BE} - I_C \cdot R_E}{R_E + R_{TH}}$$

Differentiating the above expression with respect to I_C,

$$\frac{dI_B}{dI_C} = \frac{0 - 0 - 1 \cdot R_E}{R_E + R_{TH}} = -\frac{R_E}{R_E + R_{TH}}$$

Substituting the above value of dI_B/dI_C in the general expression for stability factor,

$$S = \frac{1+\beta}{1-\beta\left(\frac{dI_B}{dI_C}\right)} = \frac{1+\beta}{1-\beta\left(-\frac{R_E}{R_E - R_{TH}}\right)} = \frac{1+\beta}{1+\beta\left(\frac{R_E}{R_E + R_{TH}}\right)}$$

Rearranging the above expression,

$$S = \frac{(1+\beta)\left(1 + \frac{R_{TH}}{R_E}\right)}{1 + \beta + \frac{R_{TH}}{R_E}}$$

Substituting the value of R_{TH} equal to $\frac{R_1 \cdot R_2}{(R_1 + R_2)}$ in the above expression and rearranging,

$$S = \frac{R_E (R_1 + R_2) + R_1 \cdot R_2}{R_E (R_1 + R_2) + R_1 \cdot R_2 (1 - \alpha)}$$

Example 22.33.

Show that the stability factor $(S = \dfrac{dI_C}{dI_{CO}})$ *for the d.c. bias circuit shown in Fig. 22.49 is given by the expression :*

$$S = (1+\beta) \times \frac{1 + \dfrac{R_B}{R_E}}{1 + \beta + \dfrac{R_B}{R_E}}$$

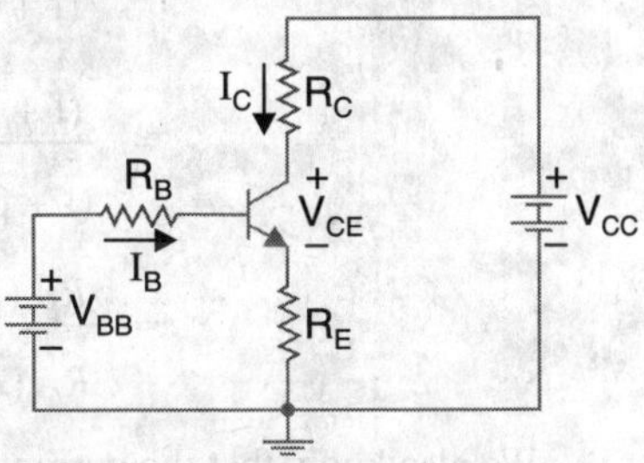

Fig. 22.49.

Solution.

The given circuit is similar to the voltage divider bias circuit and may by compared with the Thevenin's Equivalent Circuit of a voltage divider bias circuit. By comparing, we find that the V_{BB} supply is the same as the Thevenin's Voltage (V_{TH}) and R_B is the same as the Thevenin's Equivalent Resistance (R_{TH}).

We know that the stability factor,

$$S = \frac{(1+\beta)\left(1 + \dfrac{R_{TH}}{R_E}\right)}{1 + \beta + \dfrac{R_{TH}}{R_E}}$$

Replacing R_{TH} with R_B in the above expression,

$$S = \frac{(1+\beta)\left(1 + \dfrac{R_B}{R_E}\right)}{1 + \beta + \dfrac{R_B}{R_E}} \quad \textbf{Ans.}$$

Example 22.34.

Show that in a voltage divider single battery bias type common emitter voltage amplifier using NPN transistor, the stability factor is given by the expression,

$$S = \frac{\delta I_C}{\delta I_{CBO}} = \frac{R_E\ (R_1 + R_2) + R_1 \cdot R_2}{R_E\ (R_1 + R_2) + R_1 \cdot R_2\ (1 - \alpha)}$$

where R_1, R_2, R_E *and* α *have their usual meanings.*

Solution.

We know that the stability factor for a voltage divider bias circuit,

$$S = \frac{(1+\beta)}{1 + \beta\left(\dfrac{R_E}{R_E + R_{TH}}\right)}$$

Substituting the value of R_{TH} (equal to $R_1 \cdot R_2/(R_1 + R_2)$ in the above expression,

$$S = \frac{(1+\beta)}{1 + \beta \cdot \dfrac{R_E}{R_E + \dfrac{R_1 \cdot R_2}{R_1 + R_2}}} = \frac{(1+\beta)\left(R_E + \dfrac{R_1 \cdot R_2}{R_1 + R_2}\right)}{R_E + \dfrac{R_1 \cdot R_2}{R_1 + R_2} + \beta \cdot R_E}$$

$$= \frac{(1+\beta)[R_E(R_1+R_2)+R_1\cdot R_2]}{R_E(R_1+R_2)+R_1\cdot R_2+\beta\cdot R_E(R_1+R_2)}$$

$$= \frac{(1+\beta)[R_E(R_1+R_2)+R_1\cdot R_2]}{(1+\beta)R_E(R_1+R_2)+R_1\cdot R_2}$$

$$= \frac{(1+\beta)[R_E(R_1+R_2)+R_1\cdot R_2]}{(1+\beta)\left[R_E(R_1+R_2)+\dfrac{R_1\cdot R_2}{1+\beta}\right]}$$

$$= \frac{R_E(R_1+R_2)+R_1\cdot R_2}{R_E(R_1+R_2)+\dfrac{R_1\cdot R_2}{1+\beta}} \quad \text{... } (i)$$

We also know that the current gain,

$$\beta = \frac{\alpha}{1-\alpha}$$

Adding unity on both sides of the above expression,

$$1+\beta = 1+\frac{\alpha}{1-\alpha} = \frac{1-\alpha+\alpha}{1-\alpha} = \frac{1}{1-\alpha}$$

Replacing $1+\beta$ with $1/(1-\alpha)$ in equation (*i*),

$$S = \frac{R_E(R_1+R_2)+R_1\cdot R_2}{R_E(R_1+R_2)+R_1\cdot R_2(1-\alpha)} \quad \textbf{Ans.}$$

Example 22.35.

A transistor with $\beta = 50$; $V_{BE} = 0.7$ V, $V_{CC} = 22.5$ V and $R_C = 5.6$ kΩ is used in a biasing circuit as shown in Fig. 22.50. It is designed to establish the quiescent point at $V_{CE} = 12$ V, $I_C = 1.5$ mA and stability factor $S \leq 3$.

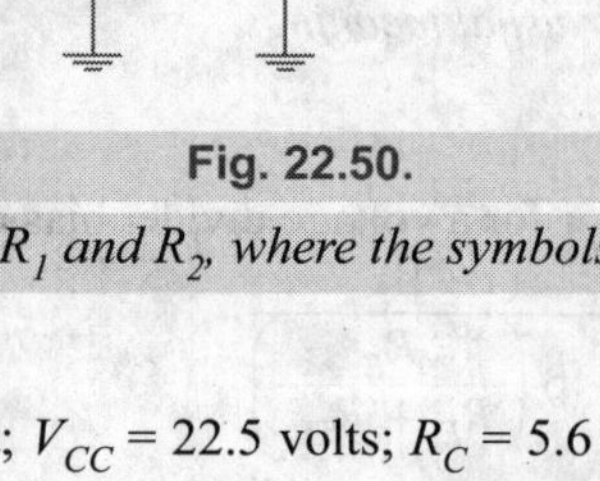

Fig. 22.50.

Find the values of resistor R_E, R_1 and R_2, where the symbols have their there usual meanings.

Solution.

Given : $\beta = 50$; $V_{BE} = 0.7$ volt; $V_{CC} = 22.5$ volts; $R_C = 5.6$ kΩ $= 5.6 \times 10^3$ Ω, $V_{CE} = 12$ volts; $I_C = 1.5$ mA $= 1.5 \times 10^{-3}$ *A* and $S \leq 3$.

Emitter resistance (R_E)

We know that the value of collector-to-emitter voltage (V_{CE}),

$$12 = V_{CC} - I_C \cdot (R_C + R_E)$$

$$= 22.5 - (1.5 \times 10^{-3})[(5.6 \times 10^3) + R_E]$$

$$= 14.1 - 1.5 \times 10^{-3} R_E$$

$$\therefore \quad R_E = 1.4 \times 10^3\ \Omega = 1.4\ \text{k}\Omega\ \textbf{Ans.}$$

Resistances R_1 and R_2

We know that the stability factor (S)

$$3 = \frac{\beta + 1}{1 + \beta \times \dfrac{R_E}{R_{TH} + R_E}} = \frac{50 + 1}{1 + 50 \times \dfrac{1.4 \times 10^3}{R_{TH} + 1.4 \times 10^3}}$$

$$= \frac{51}{1 + \dfrac{70 \times 10^3}{R_{TH} + 1.4 \times 10^3}}$$

$$3\left[\frac{(R_{TH} + 1.4 \times 10^3) + 70 \times 10^3}{R_{TH} + 1.4 \times 10^3}\right] = 51$$

$$3\ [[(R_{TH} + (1.4 \times 10^3)] + (70 \times 10^3)] = 51\ [R_{TH} + (1.4 \times 10^3)]$$

$$48\ [R_{TH} + (1.4 \times 10^3)] = 210 \times 10^3$$

or $$R_{TH} + (1.4 \times 10^3) = (210 \times 10^3)/48 = 4375$$

$$\therefore \quad R_{TH} = 4375 - (1.4 \times 10^3) = 2975\ \Omega = 2.98\ \text{k}\Omega$$

We also know that for a good voltage divider, the value of resistor

$$R_2 = 0.1\ \beta \cdot R_E = 0.1 \times 50 \times (1.4 \times 10^3)\ \Omega$$

$$= 7 \times 10^3\ \Omega = 7\ \text{k}\Omega\ \textbf{Ans.}$$

and the Thevenin's Resistance (R_{TH}),

$$2.98 = R_1 \parallel R_2 = \frac{R_1 \cdot R_2}{R_1 + R_2} = \frac{7 R_1}{R_1 + 7}$$

or $$2.98\ (R_1 + 7) = 7 R_1$$

$$4.02\ R_1 = 20.86$$

$$\therefore \quad R_1 = 20.86/4.02 = 5.2\ \text{k}\Omega\ \textbf{Ans.}$$

Example 22.36.

In a self biased potential divider CE amplifier circuit with R_1 = 100 kΩ, R_2 = 90 kΩ, V_{CC} = 22.5 V, R_C = 5.6 kΩ, R_E = 10 kΩ and a BJT with β = 55, V_{BE} = 0.7 V. Calculate the operating point and the stability factor S of the circuit. *(M. G. University., May/June 200)*

Solution.

Given : V_{CC} = 22.5 V; R_1 = 100 kΩ = 100 × 10^3 Ω; R_2 = 90 kΩ = 90 × 10^3 Ω; R_C = 5.6 kΩ = 5.6 × 10^3 Ω; R_E = 10 kΩ = 1 × 10^3 Ω; β = 55 and V_{BE} = 0.7 V.

We know that the base voltage,

$$V_B = V_{CC} \times \frac{R_2}{R_1 + R_2} = 22.5 \times \frac{90 \times 10^3}{(100 \times 10^3) + (90 \times 10^3)} = 10.66\ \text{V}$$

and the voltage across the base emitter junction,

$$V_E = V_B - V_{BE} = 10.66 - 0.7 = 9.96\ \text{V}$$

the value of emitter current,

$$I_E = \frac{V_E}{R_E} = \frac{9.96}{10 \times 10^3} = 0.996 \times 10^{\ 3}\ \text{A} = 0.996\ \text{mA}$$

$\therefore$ *Collector current,*

$$I_C = I_E = 0.996 \text{ mA}$$

We know that the voltage across the collector resistor,

$$V_{RC} = I_C . R_C = 0.996 \times 10^{-3} \times 5.6 \times 10^3 = 5.576 \text{ V}$$

and the voltage at the collector current (measured with respect to ground),

$$V_C = V_{CC} - V_{R_C} = 22.5 - 5.576 = 16.924 \text{ V}$$

∴ The voltage from collector to emitter,

$$V_{CE} = V_C - V_E = 16.924 - 9.96 = 6.694 \text{ V}$$

Hence the operating point values are,

$$I_C = 0.996 \text{ mA and } V_{CE} = 6.694 \text{ V } \textbf{Ans.}$$

We know that the stability factor is given by,

$$S = \frac{(1+\beta)}{1+\beta\left(\dfrac{R_E}{R_E + \left(\dfrac{R_1 . R_2}{R_1 + R_2}\right)}\right)}$$

$$= \frac{(1+55)}{1+55\left(\dfrac{10\times 10^3}{10\times 10^3 + \left(\dfrac{100\times 10^3 \times 90\times 10^3}{100\times 10^3 + 90\times 10^3}\right)}\right)} = \frac{56}{10.587} = 5.194 \textbf{ Ans.}$$

Example 22.37.

A silicon transistor uses voltage divider biasing as shown in Fig. 22.51, V_{CC} = 12 V, R_1 = 10 kΩ, R_2 = 5 kΩ, R_L = 1 kΩ and R_E = 3 kΩ. Determine the operating point using Thevenin's theorem.

(*Anna University., April/May 2008*)

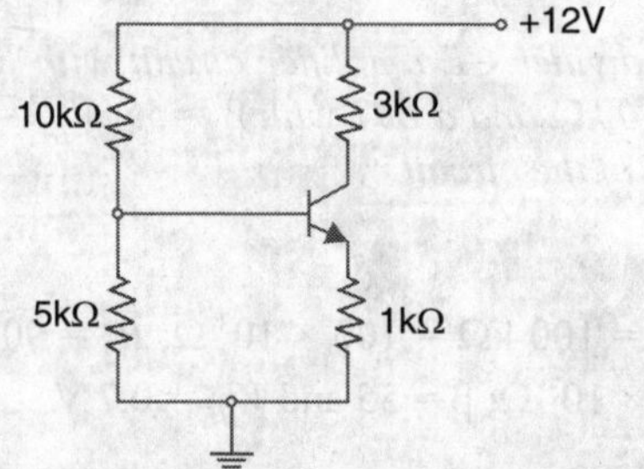

Fig. 22.51.

Solution.

Given : V_{CC} = 12 V; R_1 = 10 kΩ = 10 × 10³ Ω; R_2 = 5 kΩ = 5 × 10³ Ω; R_L = 5 kΩ = 5 × 10³ Ω; R_E = 3 kΩ = 3 × 10³ Ω; and V_{BE} = 0.7 V.

We know that the base voltage,

$$V_{TH} = V_{CC}\left(\frac{R_2}{R_1 + R_2}\right)$$

$$= 12\left(\frac{5}{10+5}\right) = 4\text{V}$$

and the Thevenin's Resistance,

$$R_{TH} = R_1 \parallel R_2 = \frac{R_1 . R_2}{R_1 + R_2} = \frac{10 \times 5}{10+5} \text{ k}\Omega = 3.333 \text{ k}\Omega = 3333 \ \Omega$$

We also know that the emitter current,

$$I_E = \frac{V_{TH} - V_{BE}}{R_E} = \frac{4 - 0.7}{3333} = 0.9 \times 10^{-3}\text{A}$$

∴ Collector current,

$$I_C \approx I_E = 0.9 \text{ mA}$$

We know that voltage across the collector resistor,

$$V_{RC} = I_C . R_C = 0.9 \times 10^{-3} \times 3 \times 10^3 = 2.7 \text{ V}$$

and the voltage at the collector (measured with respect to the ground),

$$V_C = V_{CC} - V_{RC} = 12 - 2.7 = 9.3 \text{ V}$$

Hence the operating point values are,

$$I_C = 0.9 \text{ mA and } V_{CE} = 9.3 \text{ V } \textbf{Ans.}$$

Example 22.38.

A Ge arrangement with β = 49 has self biasing arrangement as given in the Fig. 22.52. Given V_{CC} = 10 V, R_L = 1 kΩ, V_{CE} = 5 V, I_C = 4.9 mA and V_{BE} = 0.2 V the stability factor S is desired to be 10. Obtain the values of R_1, R_2 and R_E. (*GBTU., 2010–2011*)

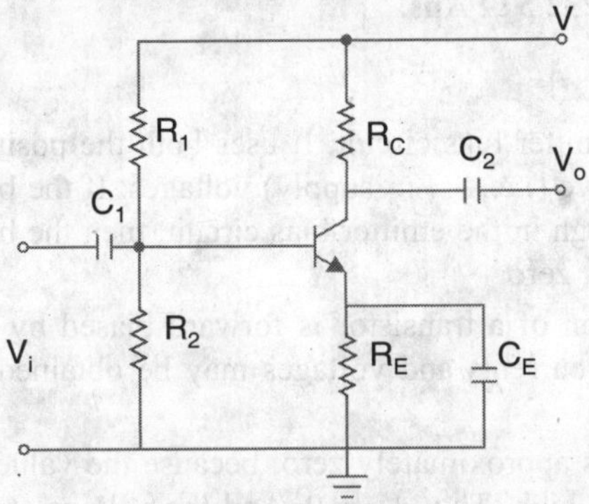

Fig. 22.52.

Solution.

Given : V_{CC} = 10 V; R_C = 1 kΩ = 1 × 10^3 Ω; V_{CE} = 5 V; I_C = 4.9 mA = 4.9 × 10^{-3} A; V_{BE} = 0.2V; S = 10 and β = 49.

We know that the base current,

$$I_B = \frac{I_C}{\beta} = \frac{4.9}{49} = 0.1 \text{ mA}$$

and the value of R_E,

$$R_E = \frac{V_{CC} - V_{CE} - I_C R_C}{I_C + I_B}$$

$$= \frac{10 - 5 - [(4.9 \times 10^{-3}) \times (1 \times 10^3)]}{(4.9 \times 10^{-3}) + (0.1 \times 10^{-3})} = 20 \times 10^3 \Omega = 20 \text{ k}\Omega \textbf{ Ans.}$$

We also know that the stability factor,

$$S = \frac{(1+\beta)\left(1 + \dfrac{R_{TH}}{R_E}\right)}{1 + \beta + \dfrac{R_{TH}}{R_E}}$$

$$10 = \frac{(1+49)\left(1 + \dfrac{R_{TH}}{(20 \times 10^3)}\right)}{1 + 49 + \dfrac{R_{TH}}{(20 \times 10^3)}} = \frac{50\left(1 + \dfrac{R_{TH}}{(20 \times 10^3)}\right)}{50 + \dfrac{R_{TH}}{(20 \times 10^3)}}$$

$$R_{TH} = 225 \times 10^3 \ \Omega \text{ or } 225 \text{ k}\Omega$$

and the Thevenin's equivalent resistance,

$$R_{TH} = \frac{R_1 . R_2}{R_1 + R_2} \quad \text{... } (i)$$

and the Thevenin's voltage,

$$V_{TH} = \left(\frac{R_2}{R_1 + R_2}\right) V_{CC} \quad \text{... } (ii)$$

solving equation (*i*) and (*ii*) we get the value of R_1 and R_2,

$$R_1 = 6.97 \times 10^3 \ \Omega = 6.97 \ \Omega \textbf{ Ans.}$$

$$R_2 = 232.5 \ \Omega \textbf{ Ans.}$$

22.20. Emitter Bias

Fig. 22.53 shows the emitter bias circuit. It uses both the positive (*i.e.*, V_{CC} supply) and negative (*i.e.*, $-V_{EE}$ supply) voltages. If the base resistance (R_B) is small enough in the emitter bias circuit, then the base voltage (V_B) is approximately zero.

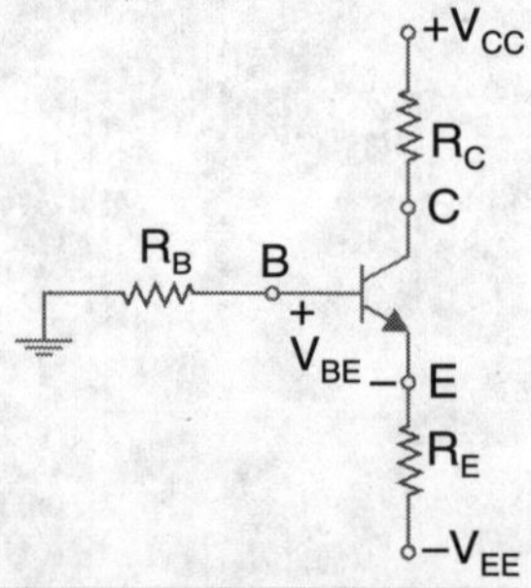

Fig. 22.53.

The base-emitter junction of a transistor is forward biased by the $-V_{EE}$ supply. The d.c. bias currents and voltages may be obtained as given below :

The base voltage (V_B) is approximately zero, because the value of resistance R_B is usually very high. Thus $V_B = 0$. And the voltage at the emitter (measured with respect to ground),

$$V_E = -V_{BE}$$

$\therefore$ Value of emitter current,

$$I_E = \frac{V_E - V_{EE}}{R_E} = \frac{-V_{BE} - V_{EE}}{R_E} \quad \text{...}(\because V_E = -V_{BE})$$

Since V_{EE} supply is negative, therefore, the emitter current,

$$I_E = \frac{-V_{BE} - (-V_{EE})}{R_E} = \frac{V_{EE} - V_{BE}}{R_E}$$

We know that the value of collector current is approximately the same as that of emitter current, *i.e.*, $I_C = I_E$.

∴ Collector voltage (measured with respect to ground),

$$V_C = V_{CC} - I_C \cdot R_C$$

and the collector-to-emitter voltage,

$$V_{CE} = V_C - V_E$$

Example 22.39.

Find the values of emitter current, collector current, and collector-to-emitter voltage for the emitter bias circuit shown in Fig. 22.54. Assume $V_{BE} = 0.7$ V.

Solution.

Given : $V_{CC} = 10$ volts; $V_{EE} = 10$ volts; $R_C = 1\ \text{k}\Omega = 1 \times 10^3\ \Omega$; $R_E = 5\ \text{k}\Omega = 5 \times 10^3\ \Omega$ and $R_B = 50\ \text{k}\Omega$.

Emitter current

We know that the emitter voltage,

$$V_E = -V_{BE} = -0.7\text{ V}$$

and the value of emitter current,

$$I_E = \frac{V_{EE} - V_{BE}}{R_E} = \frac{10 - 0.7}{5 \times 10^3}\text{ A}$$

$$= 1.86 \times 10^{-3}\text{ A} = 1.86\text{ mA } \textbf{Ans.}$$

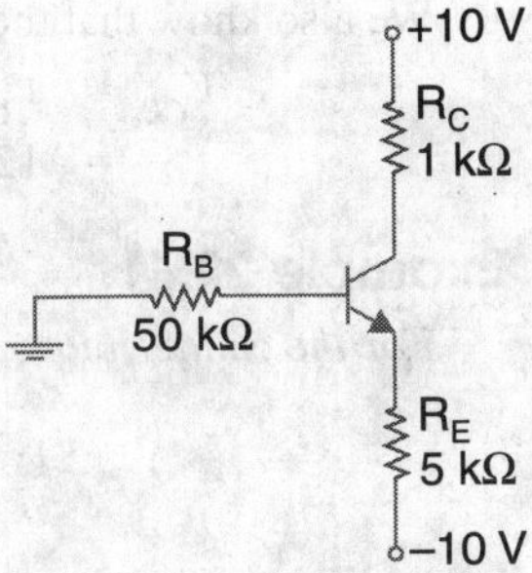

Fig. 22.54.

Collector current

We know that the value of collector current,

$$I_C \approx I_E = 1.86 \times 10^{-3} = 1.86\text{ mA } \textbf{Ans.}$$

Collector-to-emitter voltage

We also know that the collector voltage,

$$V_C = V_{CC} - I_C \cdot R_C = 10 - (1.86 \times 10^{-3}) \times (1 \times 10^3) = 8.14\text{ volts}$$

and the collector-to-emitter voltage,

$$V_{CE} = V_C - V_E = 8.14 - (-0.7) = 8.84\text{ volts } \textbf{Ans.}$$

Example 22.40.

Find I_C and V_{CE} for the circuit shown in Fig. 22.55. Assume, β = 100 for the BJT.

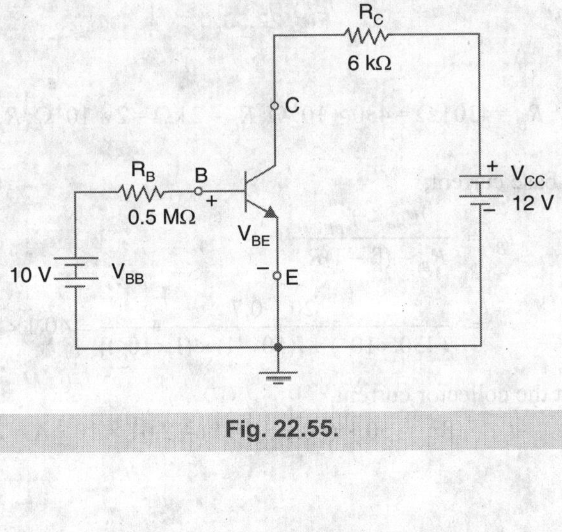

Fig. 22.55.

Solution.

Given : $V_{CC} = 12$ V; $V_{BB} = 10$ V; $R_B = 0.5$ MΩ $= 0.5 \times 10^6$ Ω; $R_C = 6$ kΩ $= 6 \times 10^3$ Ω and β = 100.

We know that the base current,

$$I_B = \frac{V_{BB} - V_{BE}}{R_B}$$

and $V_{BE} = 0.7$ V

$$\therefore \quad I_B = \frac{10 - 0.7}{0.5 \times 10^6} = 18.6 \times 10^{-6} = 18.6\ \mu\text{A}\ \textbf{Ans.}$$

and the collector current,

$$I_C = \beta I_B = 100 \times 18.6 \times 10^{-6} = 18.6 \times 10^{-4} = 1.86 \times 10^{-3}\ \text{A or } 1.86\ \text{mA}$$

We also know that the voltage between collector and emitter (V_{CE}),

$$V_{CE} = V_{CC} - I_C R_C = 12 - (18.6 \times 10^{-3}) \times (6 \times 10^3)$$
$$= 12 - 11.16 = 0.84\ \text{V}\ \textbf{Ans.}$$

Example 22.41.

For the emitter bias network of Fig. 22.56 determine I_C, V_{CE}, V_C, V_E, V_B, V_{BC}.

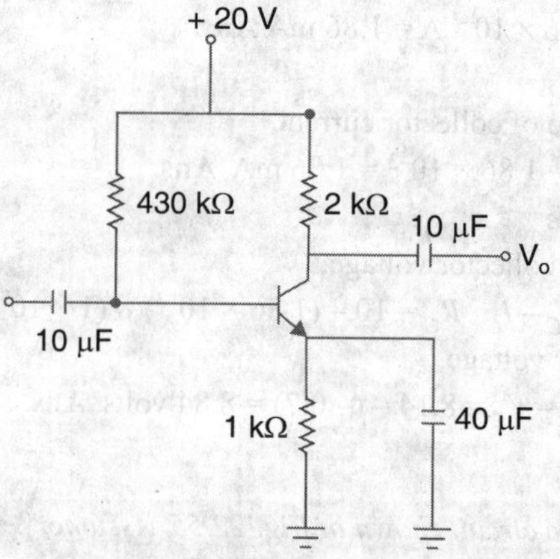

Fig. 22.56.

Solution.

Given : $V_{CC} = 20$ V; $R_B = 430$ kΩ $= 430 \times 10^3$ Ω; $R_C = 2$ kΩ $= 2 \times 10^3$ Ω; $R_E = 1$ kΩ $= 1 \times 10^3$ Ω and β = 50.

We know that the base current,

$$I_B = \frac{V_{CC} - V_{BE}}{R_B + (\beta - 1)R_E}$$

$$= \frac{20 - 0.7}{(430 \times 10^3) + [(50 - 1) \times (1 \times 10^3)]} = 40.1 \times 10^{-6}\ \text{A} = 40.1\ \mu\text{A}$$

We also know that the collector current,

$$I_C = \beta I_B = 50 \times (40.1 \times 10^{-6}) = 2.01 \times 10^{-3}\ \text{A} = 2.01\ \text{mA}\ \textbf{Ans.}$$

and the collector to emitter voltage,

$$V_{CE} = V_{CC} - I_C (R_C + R_E)$$
$$= 20 - (2.01 \times 10^{-3}) [(2 \times 10^3) + (1 \times 10^3)] = 13.9 \text{ V } \textbf{Ans.}$$

We know that the base voltage,

$$V_B = V_{BE} + V_E = 0.7 + 2.05 = 2.75 \text{ V } \textbf{Ans.}$$

and the collector voltage,

$$V_C = V_{CC} - I_C R_C = 20 - [(2.01 \times 10^{-3}) \times (2 \times 10^3)] = 15.98 \text{ V } \textbf{Ans.}$$

22.21. Stability of Emitter Bias

Consider again the emitter bias circuit as shown in Fig. 22. 50. Applying Kirchoff's Voltage Law to the base-emitter loop of this circuit,

$$I_B \cdot R_B + V_{BE} + I_E \cdot R_E - V_{EE} = 0$$

Substituting the value of base current I_B (equal to $I_C / \beta = I_E / \beta$) in the above equation,

$$I_E \left(\frac{R_B}{\beta} + R_E \right) = V_{EE} - V_{BE}$$

$$\therefore \quad I_E = \frac{V_{EE} - V_{BE}}{\dfrac{R_B}{\beta} + R_E}$$

$$= \frac{V_{EE} - V_{BE}}{R_E} \qquad \text{...(If } R_E >> R_B / \beta)$$

$$= V_{EE} / R_E \qquad \text{...(If } V_{EE} >> V_{BE})$$

The above relation indicates that the emitter current is independent of current gain (β) and base-to-emitter voltage (V_{BE}) as long as the value of resistance R_E and V_{EE} supply are very large. In actual practice, the value of R_E is selected in such a way that it is 100 times greater than R_B / β. Mathematically, it is equivalent to,

$$R_E \geq 100 \, R_B / \beta$$

or

$$R_B \leq 0.01 \, \beta \cdot R_E$$

Note. Since the emitter current is independent of β and V_{BE}, therefore the *Q*-point is not affected appreciably by the variations in these parameters. Thus the emitter bias provides a reasonably stable bias point.

22.22. Stability Factor of Emitter Bias

We have already discussed in the last article that if we apply Kirchoff's Voltage Law to the base-emitter loop of the emitter bias circuit shown in Fig. 22.25, we get

$$I_B \cdot R_B + V_{BE} + I_E \cdot R_E - V_{EE} = 0$$

Substituting the value of I_B (equal to $I_B + I_C$) in the above equation and solving it for base current,

$$I_B = \frac{V_{EE} - V_{BE} + I_C \cdot R_C}{R_E + R_B}$$

Differentiating the above expression with respect to I_C

$$\frac{dI_B}{dI_C} = \frac{0 - 0 - 1 \cdot R_E}{R_E + R_B} = -\frac{R_E}{R_E + R_B}$$

Substituting the above value of dI_B / dI_C in the general expression for stability factor,

$$S = \frac{1+\beta}{1-\beta\left(\dfrac{dI_B}{dI_C}\right)} = \frac{1+\beta}{1-\beta\left(-\dfrac{R_E}{R_E+R_B}\right)} = \frac{1+\beta}{1+\beta\times\dfrac{R_E}{R_E+R_B}}$$

Example 22.42.

Fig. 22.57 shows the emitter bias current using NPN transistor. Determine how much the Q-point will change over a temperature range, where b increases from 50 to 100 and V_{BE} decreases from 0.7 V to 0.6 V.

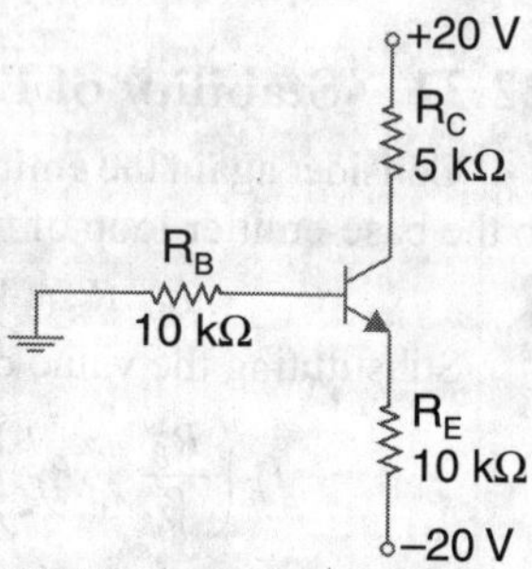

Fig. 22.57.

Solution.

Given : V_{CC} = 20 volts; V_{EE} = 20 volts; $R_C = 5\text{ k}\Omega = 5\times10^3\ \Omega$; $R_E = 10\text{ k}\Omega = 10\times10^3\ \Omega$; $R_B = 10\text{ k}\Omega = 10\times10^3\ \Omega$; β = 50 and 100; V_{BE} = 0.7 volt and 0.6 volt.

We know that the *Q*-point (*i.e.*, operating point) is defined by the values of collector current (I_C) and collector-to-emitter voltage (V_{CE}). Therefore in order to determine the change in *Q*-point, we shall find the variation in collector current and the collector-to-emitter voltage in both the cases.

We know that the value of emitter current, when β = 50 and V_{BE} = 0.7 V,

$$I_E = \frac{V_{EE}-V_{BE}}{R_E+\dfrac{R_B}{\beta}} = \frac{20-0.7}{10\times10^3+\dfrac{10\times10^3}{50}}$$

$$= 1.892\times10^{-3}\text{ A} = 1.892\text{ mA}$$

and the collector current,

$$I_C = I_E = 1.892\times10^{-3}\text{ A} = 1.892\text{ mA} \qquad ...(i)$$

We also know that the collector voltage,

$$V_C = V_{CC} - I_C\cdot R_C = 20-(1.892\times10^{-3})\times(5\times10^3)$$

$$= 20-9.46 = 10.54\text{ volts}$$

and the collector-to-emitter voltage,

$$V_{CE} = V_C - V_E = 10.54-(-0.7) = 11.24\text{ V} \qquad ...(ii)$$

Similarly, value of the emitter current, when β = 100 and V_{BE} = 0.6,

$$I_E = \frac{V_{EE}-V_{BE}}{R_E+\dfrac{R_B}{\beta}} = \frac{20-0.6}{10\times10^3+\dfrac{10\times10^3}{100}} = 1.921\times10^{-3}\text{ A} = 1.921\text{ mA}$$

and the collector current,

$$I_C = I_E = 1.921\times10^{-3} = 1.921\text{ mA}$$

We also know that the collector voltage,

$$V_C = V_{CC} - I_C\cdot R_C = 20-(1.921\times10^{-3})\times(5\times10^3) = 20-9.605 = 10.395\text{ V}$$

and the collector to emitter voltage,

$$V_{CE} = V_C - V_E = 10.395-(-0.6) = 10.995\text{ V}$$

∴ Change in collector current,

$$\Delta I_C = \frac{1.921 - 1.892}{1.892} = 0.0153 = 1.53\% \textbf{ Ans.}$$

and change in collector to emitter voltage,

$$\Delta V_{CE} = \frac{11.24 - 10.995}{11.24} = 0.0218 = 2.18\% \textbf{ Ans.}$$

22.23. Additional Stability Factors

We have already discussed a number of d.c. biasing circuits, which provide stabilization of collector current (I_C) against the variation in reverse saturation current (I_{CO}). However, there are two other factors, which affect the stability of the d.c. biasing circuit, namely *variation in the base current* (I_B) *and the current gain* (β) *with temperature*. The stabilities, due to these two quantities, are measured by the stability factors S' and S'' respectively.

The stability factor S' may be defined as *the rate of change of collector current* (I_C) *with respect to the base current* (I_B) *keeping the common-emitter current gain* (β) *and reverse saturation current* (I_{CO}) *as constant*. Mathematically, the stability factor,

$$S' = \frac{dI_C}{dI_B}$$

In actual practice, the variation in base current occurs because of the change in base-to-emitter voltage (V_{BE}) with the change in temperature. We know that the base-to-emitter voltage (V_{BE}) decreases at a rate of 2.5 m V/°C for a constant collector current. Therefore it is more convenient to define the stability factor S' as the rate of change of collector current with respect to the base-to-emitter voltage *i.e.*,

$$S' = \frac{dI_C}{dV_{BE}}$$

For a voltage divider bias, the stability factor,

$$S' = -\frac{S}{R_{TH} + R_E} \times \frac{\beta}{\beta + 1} \qquad ...(i)$$

where R_{TH} = Thevenin's Equivalent Resistance

$$= \frac{R_1 \cdot R_2}{R_1 + R_2}$$

R_E = Emitter resistance, and

S = Stability factor (equal to dI_C/dI_{CO}).

Similarly, the stability factor, S'' may be defined as *the rate of change of collector current* (I_C) *with respect to current gain* (β) *keeping the base current* (I_B) *and reverse saturation current* (I_{CO}) *as constant*. Mathematically, the stability factor,

$$S'' = \frac{dI_C}{d\beta}$$

For a voltage divider bias, the stability factor,

$$S'' = \frac{I_C \cdot S}{\beta_1 (\beta_2 + 1)} \qquad ...(ii)$$

where S = Stability factor (equal to dI_C / dI_{CO}),

β_1 = Current gain at temperature T_1, and

β_2 = Current gain at temperature T_2.

From the expressions (*i*) and (*ii*) above, we find that the stability factors *S'* and *S"* increase as the value of stability factor *S* is increased. The total change in collector current as a function of all the three stability factors,

$$\Delta I_C = S \cdot \Delta I_{CO} + S' \cdot \Delta V_{BE} + S'' \cdot \Delta\beta$$

This variation may be caused either due to temperature variation or due to transistor replacement. It will be interesting to know that for a silicon transistor, the collector current (I_C) is least dependent on reverse saturation current (I_{CO}). Therefore the manufacturer specifies the average range of operation for a silicon transistor as – 55°C to + 175°C. However, for a germanium transistor, the effect of variation in the reverse saturation current on collector current is large. Therefore the manufacturer specifies the average range of operation for a germanium transistor as –55°C to + 75°C.

22.24. Biasing Circuits for PNP Transistors

A PNP transistor is considered as a complement of an NPN transistor. It means that all voltages and currents in a PNP transistor are opposite to those of the NPN transistor. Thus the d.c. biasing circuit for a PNP transistor may be obtained by the following two actions :

1. Replace the NPN transistor by a PNP transistor.
2. Reverse all voltages and currents.

A positive V_{CC} supply in a PNP transistor biasing circuit will become the negative supply with positive terminal of the supply grounded. It is true because to forward bias the base-emitter junction of a PNP transistor, the base-emitter voltage (V_{BE}) has a negative value. Similarly, to reverse bias the collector-base junction, the collector-to-base voltage (V_{CB}) has a negative value. Fig. 22.58 (*a*), (*b*) and (*c*) shows the d.c. biasing circuits for PNP transistor. Fig. 22.58 (*a*) is a collector feedback bias, (*b*) is a emitter feedback bias and (*c*) is a voltage divider bias.

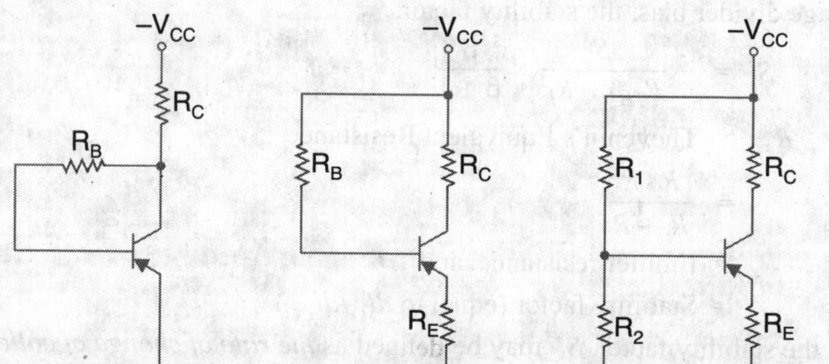

(*a*) Collector feed-back bias. (*b*) Emitter feed-back bias. (*c*) Voltage divider bias.

Fig. 22.58. Biasing circuits for *PNP* transistors.

Sometimes, the power supply to be used is positive. In such a case, the PNP transistors are, usually, drawn upside down as shown in Fig. 22.59 (*a*), (*b*) and (*c*). It may be noted that there is no difference in the working of a transistor, if it is oriented in any direction in the space. It is just for convenience that we use this kind of convention.

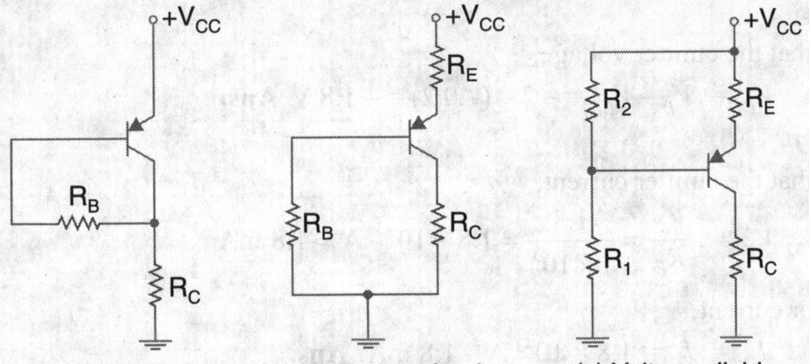

(a) Collector feed-back bias. (b) Emitter feed-back bias. (c) Voltage divider bias.

Fig. 22.59. Upside down conversion for *PNP* transistor.

The current and voltages for any PNP transistor biasing circuit may be determined in the same way as for NPN transistor biasing circuits.

Now consider a voltage divider bias circuit using PNP transistor. We know that the voltage at the base of a PNP transistor is given by the relation,

$$V_B = V_{EE}\left(\frac{R_1}{R_1 + R_2}\right)$$

and the voltage at the emitter,

$$V_E = V_B - V_{BE}$$

∴ Value of emitter current,

$$I_E = \frac{V_{EE} - V_E}{R_E}$$

and the value of collector current,

$$I_C = I_E$$

∴ Voltage at the collector,

$$V_C = I_C \cdot R_C$$

and the collector-to-emitter voltage,

$$V_{CE} = V_C - V_E = I_C \cdot R_C - (V_B - V_{BE})$$
$$= -[(V_B - V_{BE}) - I_C \cdot R_C]$$

Note. If a d.c. biasing circuit for PNP transistor is shown using a V_{CC} supply, the expressions for calculating the d.c. biasing currents and voltages are in the same as that of a NPN transistor.

Example 22.43.

Calculate the d.c. bias voltages and currents for a voltage divider bias circuit using PNP transistor. The circuit has V_{CC} = 12 V; R_C = 2 kΩ; R_E = 1 kΩ; R_1 = 100 kΩ; and R_2 = 20 kΩ. The transistor has β = 100 and V_{BE} = – 0.2 V.

Solution.

Given : V_{CC} = 12 volts; R_C = 2 kΩ = 2 × 10³ Ω; R_E = 1 kΩ = 1 × 10³ Ω; R_1 = 100 kΩ; R_2 = 20 kΩ; β = 100 and V_{BE} = – 0.2 volt.

Base voltage

We know that the base voltage,

$$V_B = -V_{CC} \times \frac{R_1}{R_1 + R_2} = -12 \times \frac{20}{100 + 20} = -2 \text{ V } \textbf{Ans.}$$

Emitter voltage

We know that the emitter voltage,

$$V_E = V_B - V_{BE} = -2 - (-0.2) = -1.8 \text{ V} \quad \textbf{Ans.}$$

Collector current

We know that the emitter current,

$$V_E = \frac{V_E}{R_E} = \frac{1.8}{1 \times 10^3} = 1.8 \times 10^{-3} \text{ A} = 1.8 \text{ mA}$$

and the collector current,

$$I_C = I_E = 1.8 \times 10^{-3} \text{ A} = 1.8 \text{ mA} \quad \textbf{Ans.}$$

Collector voltage

We know that the collector voltage,

$$V_C = -[V_{CC} - I_C \cdot R_C] = -[12 - (1.8 \times 10^{-3}) \times (2 \times 10^3) \text{ V}]$$
$$= -(12 - 3.6) = -8.4 \text{ V} \quad \textbf{Ans.}$$

Collector-to-emitter voltage

We know that the collector-to-emitter voltage,

$$V_{CE} = V_C - V_E = -8.4 - (-1.8) = -6.6 \text{ V} \quad \textbf{Ans.}$$

Example 22.44.

For the circuit shown in Fig. 22.60, calculate the values of base and emitter voltages.

(*UPSC Engg. Services 2000*)

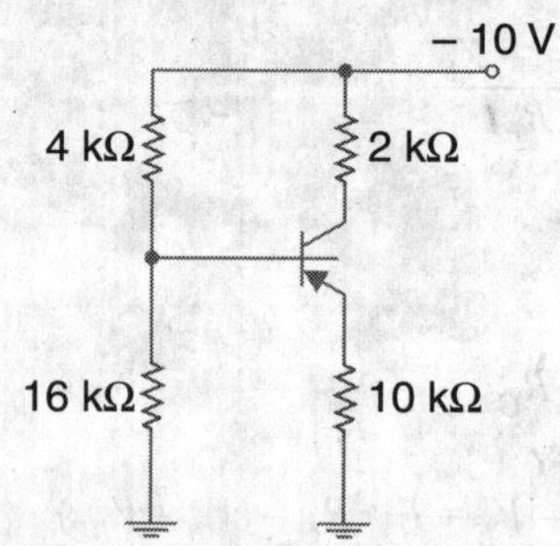

Fig. 22.60.

Solution.

Given: $V_{CC} = 10$ V; $R_C = 10$ kΩ; $R_E = 2$ kΩ; $R_1 = 16$ kΩ and $R_2 = 4$ kΩ. Assusing Si transistor, $V_{BE} = 0.7$ V and $\beta = 100$.

We know that the base voltage,

$$V_B = V_{CC}\left(\frac{R_2}{R_1 + R_2}\right) = 10 \times \left(\frac{4k}{16k + 4k}\right) = 2 \text{ V} \quad \textbf{Ans.}$$

The ernitter voltage,

$$V_E = V_B - V_{BE} = 2 - 0.7 = 1.3 \text{ V} \quad \textbf{Ans.}$$

Example 22.45.

A PNP Ge trnasistor is used in the self-biasing arrangement shown in Fig. 22.61. The circuit component values are $V_{CC} = 4.5$ V; $R_C = 1.5$ kΩ; $R_E = 0.27$ kΩ; $R_2 = 2.7$ kΩ and $R_1 = 27$ kΩ. If $\beta = 44$, find the quiescent points.

(*Nagpur Univ. Winter 2003*)

Solution.

Given : $V_{CC} = 4.5$ V; $R_C = 1.5$ kΩ; $R_E = 0.27$ kΩ; $R_2 = 2.7$ kΩ; $R_1 = 27$ kΩ; $\beta = 44$; V_{BE} for a Ge transistor = 0.3 V.

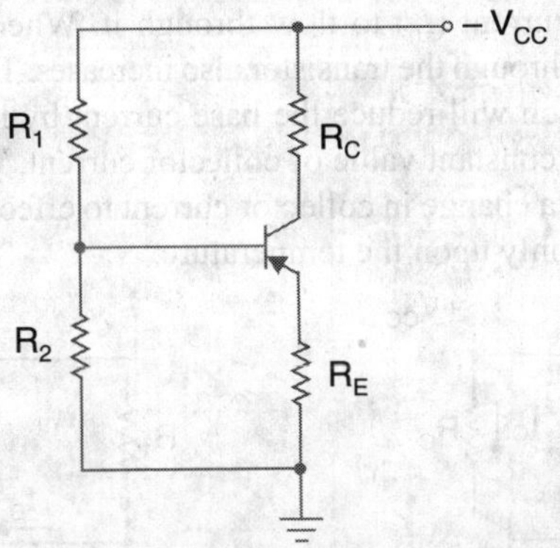

Fig. 22.61.

Quiescient Point

We need to find out the values of collector current (I_C) and collector-emitter voltage (V_{CE}).

We know that base voltage,

$$V_B = -V_{CE} \times \frac{R_2}{R_1 + R_2} = 4.5 \times \frac{2.7\text{ k}}{27\text{ k} + 2.7\text{ k}} = -0.409\text{ V}$$

and the voltage at the emitter

$$V_E = V_B - V_{BE} = -0.409 - (-0.3) = -0.109\text{ V}$$

∴ Value of emitter current,

$$I_E = \frac{V_E}{R_E} = \frac{0.109}{0.27\text{ k}\Omega} = 0.404\text{ mA}$$

and the collector current,

$$I_C \approx I_E = 0.404\text{ mA}$$

We also know that voltage across the collector resistor,

$$V_{R_C} = I_C \cdot R_C = 0.404\text{ mA} \times 1.5\text{ k} = 0.606\text{ V}$$

and the voltatge at the collector (measured with respect to the ground),

$$V_C = -[V_{CC} - V_{R_C}] = -[4.5 - 0.606] = -3.894\text{ V}$$

∴ The voltage from collector to emitter,

$$V_{CE} = -V_C - (-V_E) = -3.894 - (-0.109) = -3.785\text{ V.}$$

∴ The operating points values are,

$$I_C = 0.404 \text{ and } V_{CE} = -3.785\text{ V} \textbf{ Ans.}$$

22.25. Bias Compensation

The d.c. biasing circuits, which we have discussed in the previous articles, use a *negative feedback* to stabilize the operating point. The negative feedback means that a fraction of the collector voltage (or collector current) is supplied back at the input of the transistor. We know that the use of negative feedback reduces the gain of amplifier. However, in some cases, it is not tolerable that the gain of an amplifier should reduce. In such cases, we use *compensation techniques* to stabilize the *Q*-point instead of d.c. biasing circuits. The compensation technique is the use of temperature sensitive devices such as diodes, thermistors, transistors etc. Sometimes, a combination of the biasing circuits and temperatures sensitive devices is also used.

Fig. 22.62 (*a*) shows a diode of the same material as that of the transistor for providing bias compensation. The diode is reverse biased by the base-emitter junction voltage. It allows the reverse saturation current or leakage current (I_0) to flow through it. When the temperature increases, the reverse saturation current (I_{CO}) through the transistor also increases. This, in turn, increases the leakage current through the diode, which will reduce the base current by keeping the current (I) constant. This action is required to keep constant value of collector current. It may be noted that this method of compensation does not need a change in collector current to effect the change in base current. The collector current is dependent only upon the temperature.

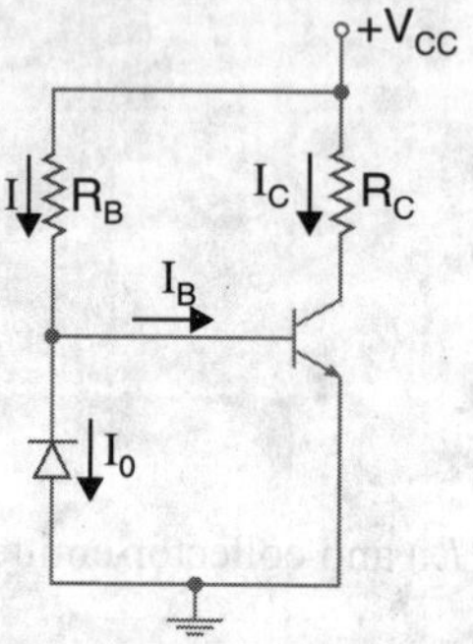

(*a*) Use of a diode.

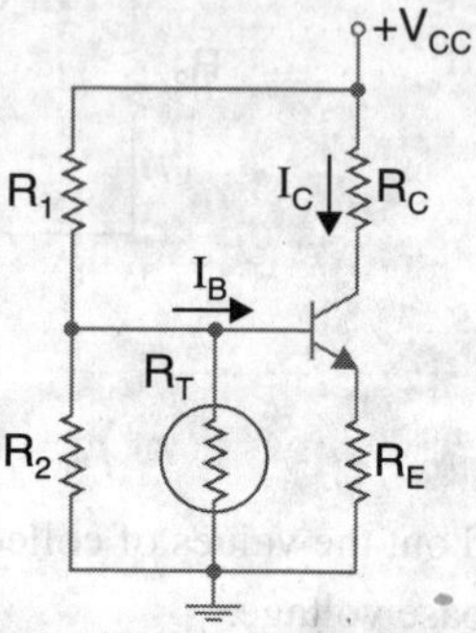

(*b*) Use of a thermistor.

Fig. 22.62. Bias compensation.

Fig. 22.62 (*b*) shows a thermistor (R_T) for providing bias compensation. The thermistor has a *negative temperature coefficient i.e.*, its resistance decreases with the increase in temperature. It is placed in parallel with the resistance R_2. The resistance values are selected in such a way so that a correct bias is applied at the normal temperature. As the temperature increases, the drop across thermistor decreases. This, in turn, reduces the forward bias across the base emitter junction, which reduces the base current and hence the collector current.

NUMERICAL PROBLEMS

1. Find the collector current and collector-to-emitter voltage for the circuit shown in Fig. 22.63. **(Ans. 1.5 mA and 6 V)**

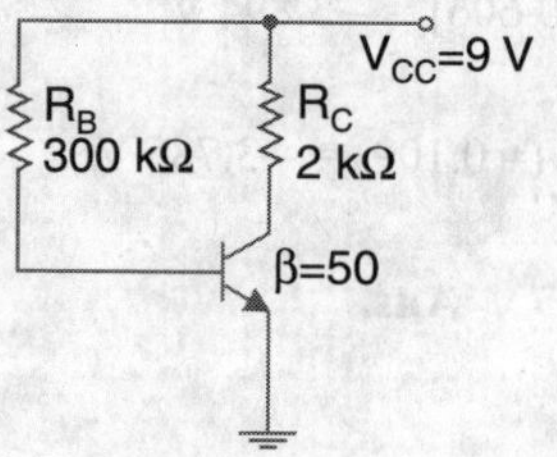

Fig. 22.63.

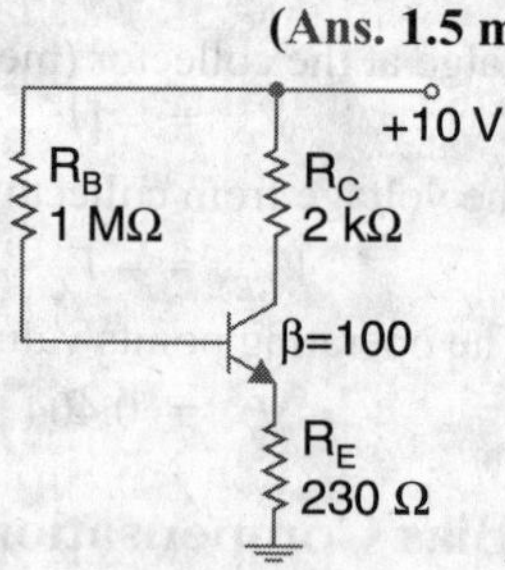

Fig. 22.64.

2. Find the d.c. load line for problem 1 and also locate the *Q*-point on the d.c load line. **(Ans. Upper end: 4.5 mA; Lower end: 9 V)**
3. Calculate the values of emitter, base and collector currents for the circuit shown in Fig. 22.64. **Ans. 0.909 mA; 9.09 mA; 0.909 mA)**

4. For the collector feed backbias circuit shown in Fig. 22.65, find $I_{C\ (sat)}$, V_{CE} and stability factor (S). Take $\beta = 100$ and neglect V_{BE}. **(Ans. 1 mA 5 V, 34.1)**

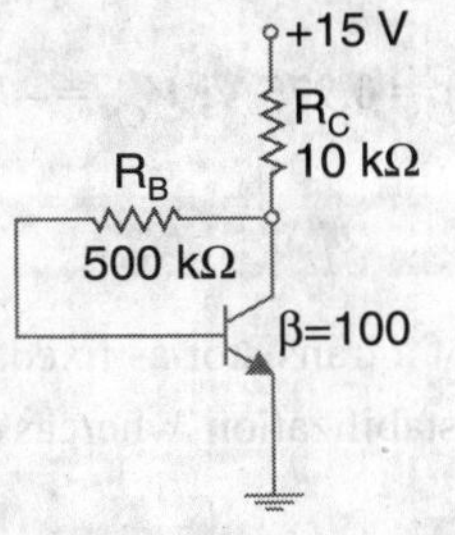

Fig. 22.65.

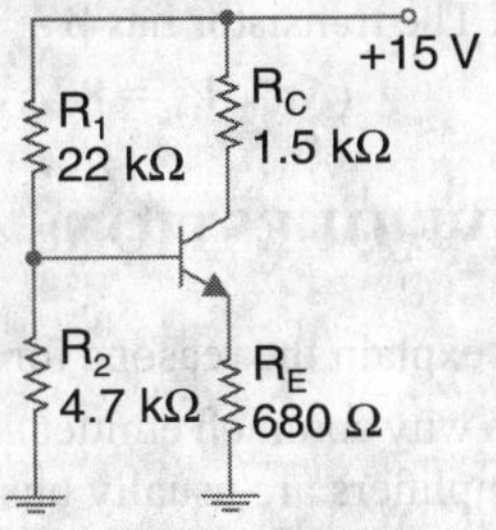

Fig. 22.66.

5. Find collector current and collector-to-emitter voltage for the voltage divider bias circuit shown in Fig. 22.66. **(Ans. 2.75 mA and 9 V)**

6. What is the collector current in a circuit shown in Fig. 22.67 ? Also determine the following voltages to ground : V_B, V_E and V_C. **(Ans. 1.43 mA; 5 V; 4.3 V and 17.14 V)**

7. Consider a common emitter NPN circuit with voltage divider bias as shown in Fig. 22.68. Compute the percentage change in the collector current (I_C), if h_{FE} of the transistor varies from 50 and 200 in this circuit. Neglect V_{BE} and I_{CBO}. **(Ans. 5.7%)**

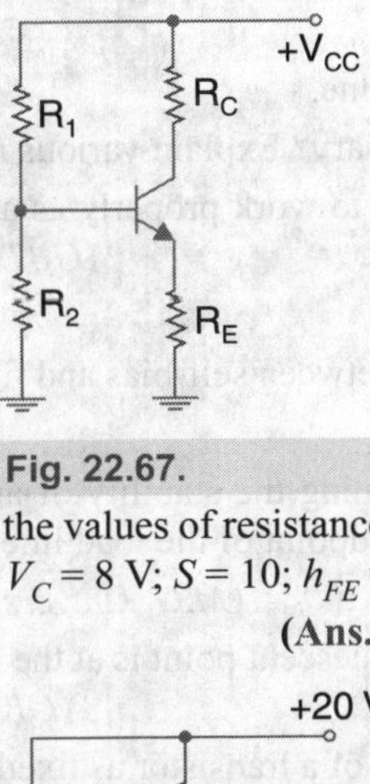

Fig. 22.67.

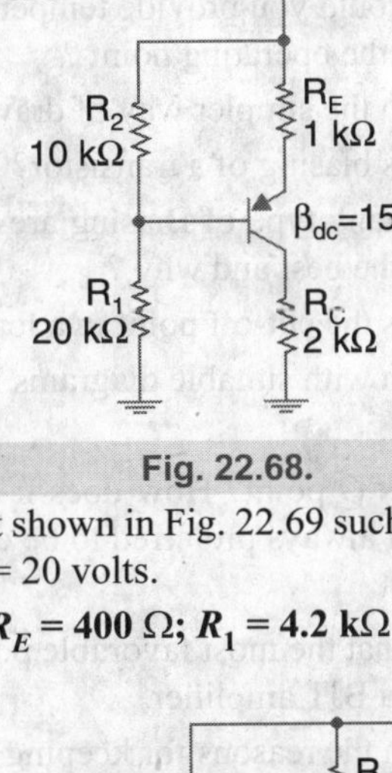

Fig. 22.68.

8. Determine the values of resistances for the circuit shown in Fig. 22.69 such that $I_C = 5$ mA; $V_{CE} = 6$ V; $V_C = 8$ V; $S = 10$; $h_{FE} = 200$ and $V_{CC} = 20$ volts.

(Ans. $R_C = 2.4$ kΩ; $R_E = 400$ Ω; $R_1 = 4.2$ kΩ and $R_2 = 2$ kΩ)

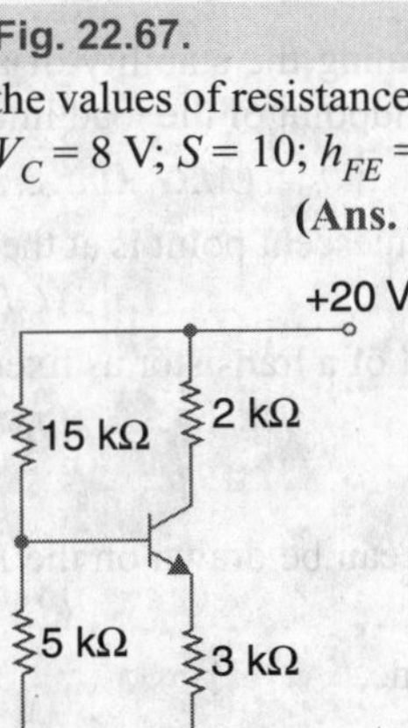

Fig. 22.69.

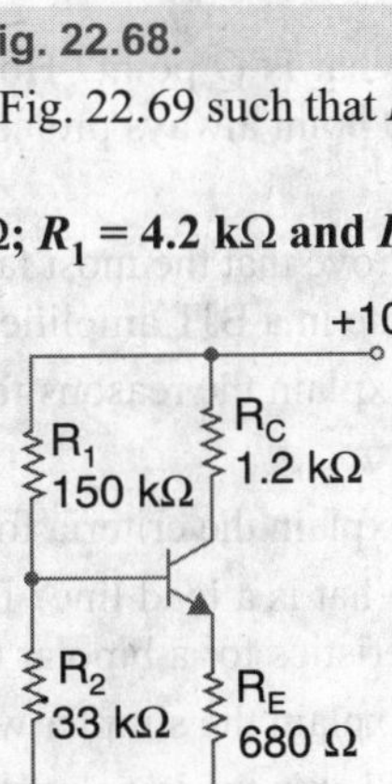

Fig. 22.70.

9. Find the Q-point values (*i.e.*, the values of collector current and collector-to-emitter voltage) for the voltage divider bias circuit shown in Fig. 22.70. Also draw a load line and locate Q-point on it. **(Ans. $I_C = 2.63$ mA and $V_{CE} = -2.11$ V)**

10. Calculate the d.c. bias currents and voltages for a voltage divider bias circuits using *PNP* transistor. The circuit has R_C = 1.5 kΩ; R_E = 2 kΩ; R_1 = 40 kΩ; R_2 = 10 kΩ and V_{EE} = 10 V. The transistor has β = 100 and V_{BE} = 0.7 V.

(Ans. V_B = 8 V; V_E = 8.7 V; I_C = 0.65 mA; V_C = 0.975 V; V_{CE} = – 7.725 V)

DESCRIPTIVE QUESTIONS

1. Briefly explain the reasons for keeping the operating point of a transistor as fixed.
2. Explain why common emitter amplifier require a form of d.c. stabilization, whereas common base amplifiers are usually unstabilized.
3. Can the value of stability factor be less than unity ? Explain it briefly.
4. Name different types of biasing circuits and give three circuit configurations.
5. Compare the relative stability of (*a*) emitter bias and fixed bias circuit, (*b*) emitter bias and voltage divider bias circuits.
6. Derive an expression for the stability factor of a fixed bias current.
7. Draw a voltage divider bias circuit and derive an expression for its stability factor.
8. List the three sources of instability of collector current and hence define the three stability factors.
9. How would you provide temperature compensation for the variations of V_{BE} and stabilization of the operating point ?
10. Explain the simpler way of drawing D.C. load line. *(JNTU., June 2009)*
11. What is biasing of a transistor? Why is it necessary? Explain various method of biasing.
12. How many type of Biasing are done on a BJT, to work properly as an amplifier ? Which one is the best and why ? *(GBTU/MNTU., 2005-06)*
13. What is the cut-off point in a load line of a BJT. *(PBUT., Dec. 2004)*
14. Explain with suitable diagrams the difference between self-bias and fixed-bias *(PBUT., 2009)*
15. What is *Q* point? How does it help in determining the stability of an amplifier? Why is Q point always preferred to be chosen as the midpoint of the load line. *(M.G. University., July/Aug. 2007)*
16. Prove that the most favorable position for the quiescent point is at the midpoint of the load line in a BJT amplifier. *(GBTU/MTU., 2008-09)*
17. Explain the reasons for keeping operating point of a transistor as fixed. *(JNTU., Aug./Sep. 2008)*
18. Explain the criteria for fixing operating point. *(JNTU., June 2009)*
19. What is a load line? Discuss how the load live can be drawn on the I_C versus V_{CE} characteristics for a bipolar transistor amplifier. *(JNTU., Aug./Sep. 2006)*
20. Explain the simpler ways of drawing dc load line. *(JNTU., June 2009)*
21. What is the important of D.C. load line. *(JNTU., June 2009)*
22. Can the value of stability factor be less than unity ? Explain briefly. *(JNTU., Aug./Sep. 2008)*
23. List out the different types of biasing methods. *(JNTU., June 2009)*
24. Explain with the help of load line the effect of variation of V_{CC}, I_B, on *Q*- point of a transistor. *(VTU., Dec 08/Jan 09)*

25. Define and give expressions of the three stability factors of a BJT amplifier. *(M.G University., May/June 2006)*
26. Define three stability factors of BJT amplifier giving their basic expressions. *(M.G University., May 2008)*
27. What is meant by biasing of transistors? Sketch and explain a common base amplifier. *(M.G University., April/May. 2004)*
28. What are the factors affecting the stability of Q-point? Explain them. *(Bhartiya University., April 2004)*
29. Why biasing is necessary in BJT amplifier and explain the concept of D.C. load line with neat diagram. *(Anna University., Nov/Dec 2007)*
30. Determine the stability factor '*S*' for the circuit shown below. *(Nagpur University., 2008)*
31. In how many modes the BJT works? Also explain the biasing pattern for each of them. *(GBTU/MTU., 2006-07)*
32. Draw various biasing technique and explain. *(Calicut University., June 2007)*
33. Explain *Q*-point and explain. *(Calicut University., June 2007)*
34. Explain the need for biasing of a transistor. *(Calicut University., June 2007)*
35. What is meant by the term quiescent operation point? How it should located to get distortion less amplification. *(GBTU/MTU., 2009)*
36. What is a *Q*-point? Explain the d.c. load line. Find the upper and lower end points of the load line for the circuit given below. Draw the d.c. load line using these end points. *(Bhartiya University., April 2004)*
37. What is meant by fixed bias? How it is provided. *(M.G University., July/Aug. 2007)*
38. What are the various possible configurations in BJT? Explain in detail the potential divider biasing and compare it with other biasing with reference to the stability factor *(GBTU/MTU., 2008-09)*
39. Discuss the cause for bias instability in a transistor. *(VTU., June /July 2008)*
40. With neat circuit diagram, explain how compensation for V_{BE} can be obtained using diode in emitter circuit of a transistor. *(VTU., June /July 2008)*
41. Why it is always considered better to fix *Q*-point in the middle of active region? *(PTU., May 2009)*
42. Draw the neat diagram of voltage divider configuration and perform the dc analysis. *(Jamia University., 2008)*
43. Prove that self bias is the best bias, by way of derivations. *(M.G University., July/Aug. 2007)*
44. Draw and explain the circuit of a fixed bias CE amplifier. Derive the current stability factor expression. Draw the D.C. load line of the circuit. *(M.G University., May/June 2006)*
45. Sketch the circuit diagram of a simple negative clamping circuit and describe its operation for a sine-wave input. *(M.G University., May 2008)*
46. Draw and explain the circuit of a potential divider biased CE-BJT amplifier. Sketch the AC and DC load lines of this circuit and explain the same. *(M.G University., May 2008)*
47. Draw and explain the circuit of fixed biased BJT amplifier and give the design steps. Discuss the merits and demerits of this circuit. *(M.G University., May2008)*
48. What do you understand by "transistor biasing"? Mention the important points to be considered for the selection of operating point. *(GBTU/MTU., 2006-07)*

49. Explain any one compensation circuit used in transistor amplifier. *(Calicut University., June 2007)*

50. What is the different compensation circuits used in BJT amplifiers? Explain any two. *(Calicut University:, Dec 2006)*

51. Sketch a voltage divider bias circuit using an NPN transistor. Show all voltage polarities and current directions. Explain the operation of the circuit and write the approximate equation for V_B, I_E, I_C and V_{CE}. *(GBTU/MTU., 2008-09)*

52. Define stability factor. Explain the self-bias circuit with diagrams and derive expression for stability factor. *(PTU., May 2008)*

53. What is operating point? Why is it necessary to stabilize operating point of a transistor amplifier? *(PTU., Dec 2007)*

54. What is the requirement of good biasing circuit? *(PTU., May 2007)*

55. Define stability factor. Explain with circuit diagram of potential divider method of biasing in amplifiers. Explain with Circuit diagram of potential divider method of biasing in amplifiers. *(PTU., Dec 2007)*

56. Explain the different bias compensation methods of transistor biasing? *(PTU., Dec 2006)*

57. What is operation point? Explain its physical significance? *(PTU., Dec 2006)*

58. Explain why fixed bias circuit, inspite of its simplicity, is not much used in practice? *(PTU., May 2009)*

59. Why there is a need for bias stabilization? *(PTU., May 2009, Dec. 2008)*

60. Explain working of voltage divider bias circuit. *(PTU., May 2009)*

61. What do you understand by transistor Biasing? Why is it necessary to bias a transistor? *(PTU., May 2008)*

62. Explain working of emitter bias circuit. *(PTU., Dec 2008)*

63. What are the limitations in selecting operating region in a transistor circuit? *(PTU., Dec 2008)*

64. Give the D.C. load line of the circuit and find Q point. *(PTU., May 2007)*

65. Discuss the biasing techniques used for linear integrated circuits. *(VTU., June /July 2008)*

66. Define three stability factors. *(Anna University., April/May 2008)*

67. Define quiescent point. *(Anna University., April/May 2008)*

68. What is bias compensation using thermistor? *(Anna University., April/May 2008)*

69. Define stability factor. Derive and explain the condition to avoid thermal runaway. *(Anna University., April/May 2008)*

70. Write short notes on temperature compensation methods. *(Anna University., April/May 2008)*

71. Why biasing is necessary n BJT amplifier and Explain the concept of DC load line with neat diagram. *(Anna University., Nov/Dec 2007)*

72. What is the important of D.C. load line? *(JNTU., June 2009)*

73. Explain the criteria for fixing operating point. *(JNTU., June 2009)*

74. Sketch the circuit, which compensates for VBE and explain. (M.G University., May 2008)

75. Explain the simple way of drawing dc load line. *(JNTU., June 2009)*

76. Can the value of stability factor be less than unity? Explain briefly. *(JNTU., June 2009)*

77. Draw a BJT fixed bias circuit and derive the expression for the stability factor 'S'. *(JNTU., May/Jun. 2006)*

78. Draw a fixed bias circuit and explain why this circuit is unsatisfactory if the transistor is replaced by another of the same type. *(JNTU., Aug./Sep. 2006)*

79. Draw the circuit diagram of a collector-to-base bias circuit of CE amplifier and derive expression for 'S'. *(JNTU., May/Jun. 2006)*

80. Draw the collector to base bias circuit and derive the expression for the stability factor S. *(JNTU., Aug./Sep. 2008)*

81. Explain bias compensation using thermistor. *(JNTU., Aug./Sep. 2008)*

82. With the help of necessary circuits and equations show how bias compensation is achieved for I_{CO} and V_{BE}. *(M.G University., May. 2005)*

83. Draw the circuit of I_{CO} bias compensation and explain how it works. *(M.G University., May/June 2006)*

84. Draw a self bias circuit for a transistor. Why this circuit has an improvement on fixed-bias circuit? How load line is draw in self-bias circuit? *(R.G.P.V., Dec. 2004, June 2003)*

85. Write short note on biasing of transistors. *(R.G.P.V., June. 2002)*

86. Give analysis of voltage-divider bias and derive design equations. *(R.G.P.V., June. 2004)*

87. Discuss voltage divider biasing (self bias) technique for BJT. Also give its merits and demerits. *(R.G.P.V., June. 2008)*

88. Write short note on voltage divider biasing of BJT. *(R.G.P.V., Dec. 2008)*

89. What do you understand by transistor biasing ? What is its need ? Describe the various used for transistor biasing. State their advantage and disadvantage? *(R.G.P.V., Nov. /Dec. 2007)*

90. What is bias stabilization? Discuss fixed and self biasing circuits. *(R.G.P.V., June 2005)*

91. List the different methods of biasing of transistor. Explain emitter feedback bias (self bias) in detail. *(R.G.P.V., Dec. 2005)*

92. Explain BJT. Why biasing is required? Explain different biasing techniques. *(R.G.P.V., May/ June 2006)*

93. Prove that self bias circuit has better stability than that of a collector-to-base bias circuit. *(R.G.P.V., Dec. 2006)*

94. In a transistor define stabilization. Explain the condition for thermal stability. *(Nagpur University, Summer 2007)*

95. Define stabilization and compensation techniques. Explain what is the condition for thermal stability. *(Nagpur University, Summer 2006)*

96. Define stabilization and compensation techniques. Explain, what is the condition for thermal stability. *(Nagpur University, Winter 2004)*

97. Comment on choice of operation point and its stability. *(Nagpur University, Winter 2005)*

98. What do you understand by stabilization of operating point? Discuss point? Discuss different biasing circuits? *(Nagpur University, Summer 2004)*

99. Explain the compensation of V_{BZ} in case of silicon transistor. *(Nagpur University, Winter 2003)*

100. Explain, what do you mean by transistor biasing and stabilization of Q- point. *(Nagpur University, Summer 2003)*

101. Define:
(*a*) Stabilisation techniques
(*b*) Compensation techniques. (*Nagpur University, Summer 2000*)

102. Draw a circuit which uses a diode to compensate for changes
(*a*) V_{BZ}
(*b*) I_{CO} (*Nagpur University, Summer 2000*)

103. Enumerate the factors responsible for instability in a transistor. (*Nagpur University, Summer 2000*)

104. Explain in brief "Thermistor Compensation". (*Nagpur University, Summer 2000*)

105. How is the collector current stability obtained in case of a feedback bias? (*Nagpur University, Winter 2003*)

106. Write short notes on :
(*a*) D. C load Line (*R.G.P.V., Dec. 2001*)
(*b*) Load Line (*R.G.P.V., Dec. 2004*)
(*c*) Necessity of transistor biasing (*Nagpur University, Summer 2001*)

107. Draw the circuit diagram of fixed bias circuit. Why is it called a fixed bias circuit? Derive the expression for stability factor. (*RGTU., June 2011*)

108. Define operating point. (*RGTU., June 2011*)

109. Explain any two types of biasing and also determine their stability factor. (*RGTU., June 2010*)

110. Explain various types of biasing used for BJT. Also discuss bias stability. (*RGTU., June 2009*)

111. What is biasing of a transistor? Explain the factors that affect selection of Q-point anywhere in the active region for the transistor to operate as an amplifier. (*VTU., June, July 2011*)

112. What is bias stabilization? Discuss the stability of the operating point against variation in its parameters. (*VTU., June, July 2011*)

113. Define stabilization factors: S, S′, and S″. Derive expressions for S and S′ for self-bias transistor circuit.. (*GTU., May 2011*)

114. Explain DC load line and Q-point for any transistor configuration. Also state the necessity of biasing and list biasing method for transistor. (*GTU., Dec. 2011*)

115. With a neat diagram explain the concepts of "load line" and Q-point of a transistor. (*WBUT., 2011*)

116. Draw the circuit of self biasing arrangement of a PNP transistor. Explain physically how this areagement provides a good stability against temperature variation. Also derive the expression for stability factor. (*WBUT., 2011, 2012*)

117. Draw the circuits of fixed bias and cellector to base bias areagements of transistor. (*WBUT., 2011, 2012*)

MULTIPLE CHOICE QUESTIONS

1. The d.c. load line of a transistor circuit
(*a*) Is a graph between I_C and V_{CE} (*b*) Is a graph between I_C and I_B
(*c*) Does not contain the locating point (*d*) Is a curved line

2. The positive part of the output signal in a transistor circuit starts clipping, if Q-point of the circuit moves
   (*a*) Towards the saturation point (*b*) Towards the cut-off point
   (*c*) Towards the center of the load line (*d*) None of the above
3. The negative part of the output signal in a transistor circuit starts clipping, if Q-point of the circuit moves
   (*a*) Towards the saturation point (*b*) Towards the cut-off point
   (*c*) Towards the center of the load line (*d*) None of the above
4. The voltage divider biasing circuit is used in amplifiers quite often because it
   (*a*) Limits the a.c. signal going to the base
   (*b*) Makes the operating point almost independent of β
   (*c*) Reduces the d.c. base current (*d*) Reduces the cost of the circuit
5. Which of the following components are used for bias compensation in transistor circuits
   (*a*) Resistors (*b*) Rectifier diodes
   (*c*) Thermistors (*d*) Both (*b*) and (*c*) above
6. To avoid thermal run away in the design of analog circuit, the operating point of the BJT should be such that it satisfies the condition
   (*a*) $V_{CE} = 1/2\ V_{CE}$ (*b*) $V_{CE} \leq 1/2\ V_{CC}$
   (*c*) $V_{CE} > 1/2\ V_{CC}$ (*d*) $V_{CE} \leq 0.78\ V_{CC}$
7. In the case of a BJT amplifier, bias stability is achiveved by
   (*a*) keeping the base current constant
   (*b*) changing the base current in order to keep the I_C and V_{CB} constant
   (*c*) keeping the temperature constant
   (*d*) keeping the temperature and the base current constant.
8. For a transistor amplifier with self-biasing network, the following components are used : $R_1 = 4$ kΩ, $R_2 = 4$ kΩ and $R_E = 1$ kΩ the approximate value of the stability factor 'S' will be
   (*a*) 4 (*b*) 3 (*c*) 2 (*d*) 1.5
9. A transistor circuit employing base bias with collector feedback has greater stability than the one without feedback because
   (*a*) I_C decrease in magnitude (*b*) V_{BE} is decreased
   (*c*) of negative feedback effect (*d*) I_C beocmes independent of b.
10. The universal bias stabilization circuit is most popular because
    (*a*) I_C does not depend on transistor characteristics
    (*b*) its β-sensitive is high
    (*c*) voltage divider is heavily loaded by transistor base
    (*d*) I_C equals to I_E.
11. Improper biasing of a transistor circuit leads to
    (*a*) excessive heat production at collector terrminal
    (*b*) distortion in output signal
    (*c*) faulty location of load line
    (*d*) heavy loading of emitter terminal.
12. The negative output swing in a transistor circuit starts clipping first when Q-point
    (*a*) has optimum value (*b*) is near saturation point
    (*c*) is near cut-off point (*d*) is in the active region of the load line.

13. When a *BJT* is employed as an amplifier, it operates

(*a*) in cut-off (*b*) in saturation

(*c*) well into saturation (*d*) over the active region.

14. Which of the following method used for biasing a *BJT* in integrated circuits is considered independent of transistor beta ?

(*a*) fixed biasing (*b*) voltage divider bias

(*c*) collector feedback bias (*d*) base bias with collector feedback.

15. The voltage V_0 of the circuit shown in Fig. 22.71 is,

(*a*) 5.1 V (*b*) 3.1 V (*c*) 2.5 V (*d*) zero

16. The collector voltage V_C of the circuit shown in Fig. 22.72, is approximately

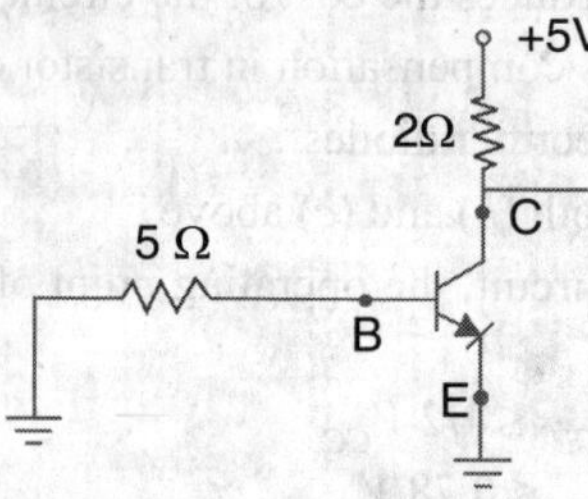

Fig. 22.71.

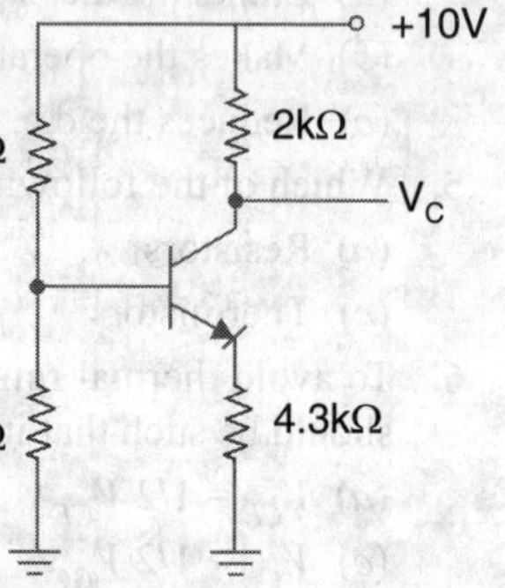

Fig. 22.72.

(*a*) 2 V (*b*) 4.6 V (*c*) 8 V (*d*) 8.6 V

17. In the circuit shown in Fig. 22.73. The transistor is biased at :

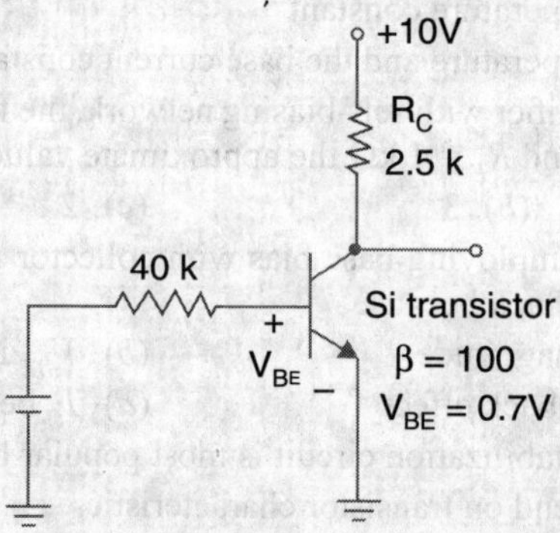

Fig. 22.73.

(*a*) 0 mA (*b*) 5 mA (*c*) 3.9 mA (*d*) ∞

(*UPSC Engg. Services 2002*)

Hint. [$I_B = \dfrac{2.7 - 0.7}{40\text{ k}} = 0.05\text{ mA}; I_C = \beta I_B = 100 \times 0.05 = 5\text{ mA}$]

18. Assume $V_{BE} = 0.7$ V and $\beta = 50$ for the transistor shown in Fig. 22.74. For $V_{CE} = 2$ V, the value of R_B is :

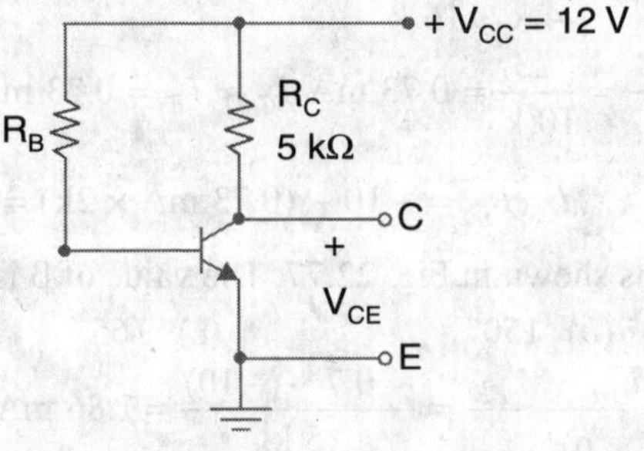

Fig. 22.74.

(*a*) 200 kΩ (*b*) 242 kΩ (*c*) 283 kΩ (*d*) 300 kΩ

Hint. [$V_{CE} = V_{CC} - I_C \cdot R_C \Rightarrow 2\text{V} = 12 - I_C = 2$ mA]

$\therefore\ I_B = I_C/\beta = 2 \text{ mA}/50 = 0.04$ mA;

$$R_B = \frac{12 - 0.7}{0.04 \text{ mA}} = 282.5 \text{ k}\Omega \approx 283 \text{ k}\Omega$$

19. The collector voltage VC of the circuit shown in Fig. 22.75, is approximately,

(*a*) 2 V (*b*) 4.6 V (*c*) 8 V (*d*) 8.6 V

(*UPSC Engg. Services 1999*)

Hint. [$V_B = 10 \times \dfrac{20 \text{ k}}{20 \text{ k} + 20 \text{ k}} = 5$ V; $V_E = V_B - 0.7 = 5 - 0.7 = 4.3$ V

$$I_E = \frac{V_E}{R_E} = \frac{4.3}{4.3 \text{ k}} = 1 \text{ mA}\ I_C \approx I_E = 1 \text{ mA};$$

$$V_C = V_{CC} - I_C \cdot R_C = 10 - (1\text{mA} \times 2\text{k}) = 8 \text{ V}$$

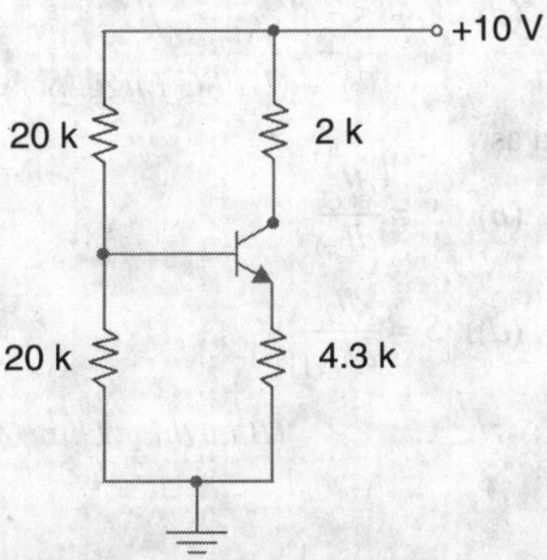

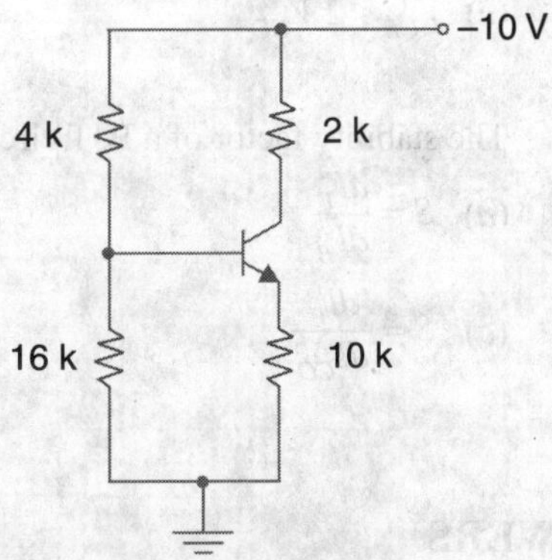

Fig. 22.75. Fig. 22.76.

20. In the circuit shown in Fig. 22.76, the approximate voltages atr the transistor,

(*a*) base and emitter are – 8 V and – 7.3 V

(*b*) base and collector respectively are – 8 V and – 7.3 V

(*c*) collector and emitter respectively are – 5 V and – 7.3 V

(*d*) base, emitter and collector respectively are – 8 V, – 7.3 V and – 5 V.

(*UPSC Engg. Services 2000*)

Hint. $V_B = -10 \times \dfrac{16\text{ k}}{4\text{ k} + 16\text{ k}} = -8\text{V}; V_E = V_B - 0.7 = -7.3\text{V}$

$$I_E = \frac{V_E}{R_E} = \frac{7.3}{10\text{ k}} = 0.73\text{ mA } I_C \approx I_E = 0.73\text{ mA};$$

$$\therefore \quad V_C = V_{CC} - I_C \cdot R_C = -10 - (0.73\text{ mA} \times 2\text{k}) = -8.54\text{ V}]$$

21. A circuit using BJT is shown in Fig. 22.77. The value of β is

(*a*) 120 (*b*) 150 (*c*) 165 (*d*) 186

Hint. $I_E = \dfrac{-V_{BE} - V_{EE}}{R_E} = \dfrac{-0.7 - (-10)}{5\text{ k}} = 1.86\text{ mA}, I_C \approx I_B = 1.86\text{ mA}$

$\beta = I_C / I_B = 1.86/0.01 = 1.86$ **Ans.**

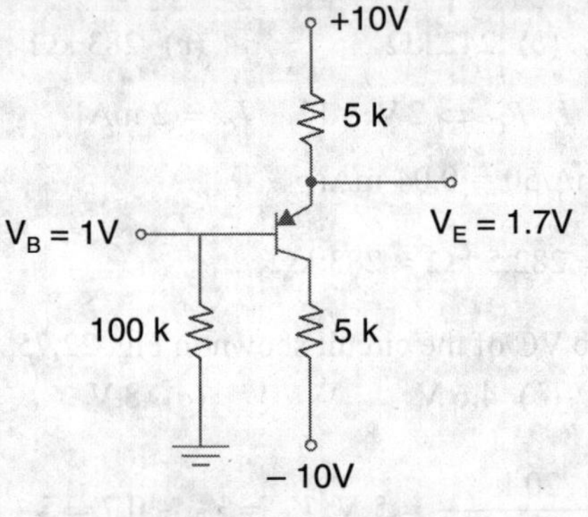

Fig. 22.77.

22. Thermal runway will take place if the Q-point is such that

(*a*) $V_{CE} > \dfrac{1}{2} V_{CC}$ (*b*) $V_{CE} < V_{CC}$

(*c*) $V_{CE} < 2\,V_{CC}$ (*d*) $V_{CE} < \dfrac{1}{2} V_{CC}$

(*UPSC Engg. Services 1999*)

23. The stability factor of a BJT circuit is defined as

(*a*) $S = \dfrac{dI_E}{dI_B}$ (*b*) $S = \dfrac{dI_C}{dI_E}$

(*c*) $S = \dfrac{dI_C}{dI_{CO}}$ (*d*) $S = \dfrac{dI_C}{dI_{CO}}$

(*Bharthiar Univ. April 2004*)

ANSWERS

1. (*a*)	**2.** (*b*)	**3.** (*a*)	**4.** (*b*)	**5.** (*d*)
6. (*c*)	**7.** (*a*)	**8.** (*b*)	**9.** (*b*)	**10.** (*d*)
11. (*b*)	**12.** (*b*)	**13.** (*d*)	**14.** (*b*)	**15.** (*d*)
16. (*a*)	**17.** (*b*)	**18.** (*c*)	**19.** (*c*)	**20.** (*a*)
21. (*d*)	**22.** (*d*)	**23.** (*c*)		

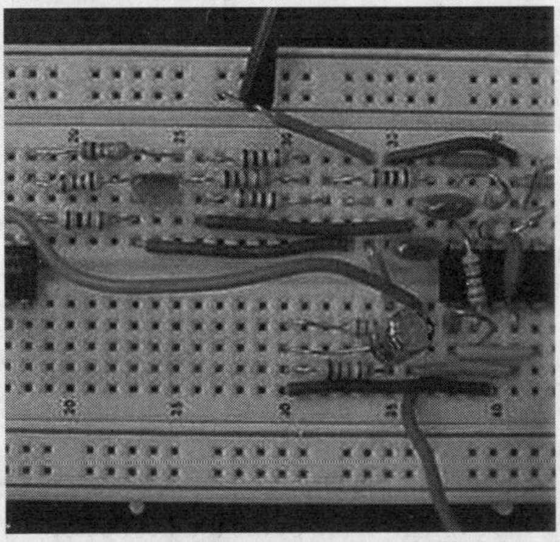

23
Chapter

Low and High Frequency BJT Models

Outline

23.1. Introduction

A model is a combination of a circuit element, properly chosen, that best approximates the actual behaviour of a semiconductor device under specific operating conditions.

The bipolar junction *transistor models* are useful for analysing amplifier circuits. The models are more easy to use than graphical procedures for the analysis purpose. When the signal is small, a transistor behaves as a linear device. However, when the signal is large, the transistor behaves as a non-linear device. Therefore, separate models are required for transistors in small-signal operation and large-signal operation. The large-signal model is also known as a d.c. model.

On the other hand, a small signal model is known as an a.c. model. Following are the two different approaches of developing transistor models:

1. In terms of current gain and circuit components.
2. In terms of hybrid parameters.

In this chapter, we shall discuss transistor models in terms of the first approach. The transistor models, in terms of hybrid parameters will be discussed in the chapter on hybrid parameters of a transistor.

23.2. Notations for Transistor Currents and Voltages

It is important to introduce a standard terminology for the symbols representing specific current and voltage components to avoid confusion between a.c. and d.c. components. This terminology is taken from the list recommended by the Institution of Electrical and Electronic Engineers (IEEE) of U.S.A. The system underlying the notation is as follows:

1. Lower case letters (such as v and i) designate instantaneous voltage and current respectively.
2. Upper case letters (such as V and I) designate the r.m.s. or peak values of a.c. components or average (d.c.) values of total quantities.
3. Lower case subscripts are used only with time varying components of voltage and current. For example, the instantaneous value of collector voltage is represented by v_c and r.m.s. value by V_c.
4. Upper case subscripts refer to total or d.c quantities. For example, the instantaneous value of the total collector voltage is enoted by v_C. Similarly, the average value of the total collector voltage is denoted by V_C.
5. Repeated upper case subscripts refer to the magnitudes of supply quantities. For example, collector voltage supply is designated by V_{CC} and base voltage supply by V_{BB}.
6. The currents are taken as positive, when entering a terminal from the external circuit. Fig. 23.1 (*a*) and (*b*) shows the positive directions for the currents in a NPN and PNP transistors respectively. It may be noted that the actual direction of the emitter current in a NPN transistor is opposite to that indicated by the arrow (I_E) in Fig. 23.1 (*a*). Similarly, for a PNP transistor, the actual directions of base current and collector current are opposite to that indicated by the arrows I_B and I_C in Fig. 23.1 (*b*) respectively.

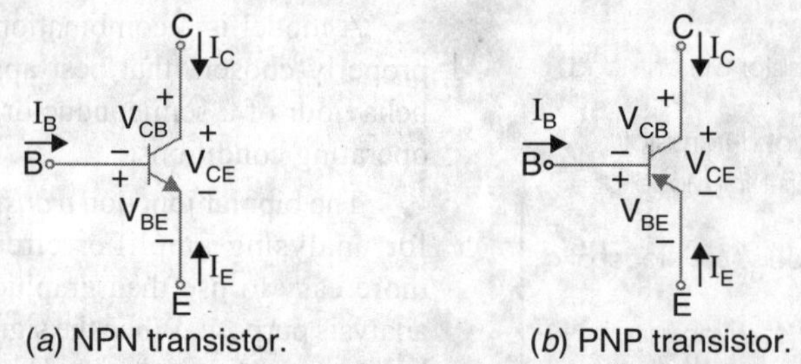

Fig. 23.1. Standard terminology for voltages and currents in a transistor.

7. The voltages are taken as positive, when measured with respect to a reference node (usually ground). For voltages measured between a pair of terminal (not the reference) an additional subscript is used as shown in Fig. 23.1 (*a*) and (*b*). For example, the instantaneous collector-to-emitter voltage is indicated by v_{CE}.
8. The additional subscript Q is added to circuit variables to indicate quiescent or operating point values. For example, the quiescent value of collector current is denoted by I_{CQ}.

Table 23.1 summarize the notations for transistor voltage and currents. It may be noted that total quantity (voltage or current) is the sum of d.c. and a.c. components. For example, the total value of collector voltage may be expressed as,

$$v_C = \underset{\text{d.c. value}}{V_C} + \underset{\text{a.c. value}}{v_c}$$

where V_C = D.C. or the quiescent collector voltage, and

v_c = Instantaneous value of a.c. collector voltage.

Table 23.1 : Notations for transistor voltages and currents

S. No.	*Item*	*Supply*	*Quiescent (static)*	*A. C. or time varying component*		*Total d.c. + a.c.*	
				Instaneous	*r.m.s.*	*Instaneous*	*Avg.*
1.	Collector voltage	V_{CC}	V_{CQ}	v_c	V_c	v_C	V_C
2.	Collector current	I_{CC}	I_{CQ}	i_c	I_c	i_C	I_C
3.	Base voltage	V_{BB}	V_{BQ}	v_b	V_b	v_B	V_B
4.	Base current	I_{BB}	I_{BQ}	i_b	I_b	i_B	I_B
5.	Emitter voltage	V_{EE}	V_{EQ}	v_e	V_e	v_E	V_E
6.	Emitter current	I_{EE}	I_{EQ}	i_e	I_e	i_E	I_E

23.3. Ebers-Moll Model of a Transistor

We know that a transistor consists of two coupled PN junctions. The base region is common to both junctions and forms a coupling between them. Since the transistors are constructed with very narrow base regions, therefore, a significant electrical interaction exists between the junctions. This interaction is called transistor action. Fig. 23.2 (*a*) shows the current components of a PNP transistor, which constitute the terminals current *i.e.*, emitter current (I_E) and collector current (I_C). The voltages V_{EB} and V_{CB} are the voltage drops from emitter-to-base and collector-to-base respectively. Assuming that there is no voltage drop across all the semiconductors forming the emitter, base and collector regions (*i.e.*, $r_{ee'} = r_{bb'} = r_{cc'} = 0$), the voltages V_{EB} and V_{CB} appear across the respective junctions. It may be noted in Fig. 23.2 (*a*) that emitter current has two components namely the current associated with the emitter-base diode indicated by I_{ED} and the current associated with the collector-base diode indicated by I_{CD}. The component $a_R \cdot I_{CD}$ is the portion of collector-base diode current, which is coupled through the base to emitter. Similarly, $a_F \cdot I_{ED}$ is the portion of emitter-base diode current, which is coupled through the base to the collector.

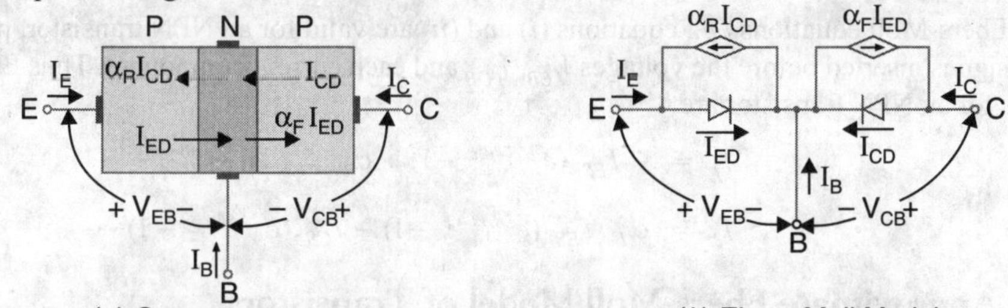

(*a*) Current components. (*b*) Ebers-Moll Model.

Fig. 23.2. PNP transistor.

On the basis of above discussion, we can construct Ebers Moll model, as shown in Fig. 23.2 (*b*). From this figure, the following important points may be carefully noted:

1. The model consist of two diodes connected back-to-back (*i.e.*, the cathodes are connected together). These diodes represent the emitter-base junction and collector-base junction of the transistor.
2. There are two controlled current sources indicated by $\alpha_R \cdot I_{CD}$ and $\alpha_F \cdot I_{CD}$. These sources indicate coupling between the two junctions of the transistor.
3. The currents I_{ED} and I_{CD} are related to V_{EB} and V_{CB} by the following equations:

$$I_E = I_{ED} - \alpha_R \cdot I_{CD}$$

$$= I_{ES}\,(e^{V_{EB}/V_T} - 1) - \alpha_R \cdot I_{CS}\,(e^{V_{CB}/V_T} - 1) \quad \text{... (i)}$$

and

$$I_C = -\alpha_F \cdot I_{ED} + I_{CD}$$

$$= -\alpha_F \cdot I_{ES}\,(e^{V_{EB}/V_T} - 1) + I_{CS}\,(e^{V_{CB}/V_T} - 1) \quad \text{... (ii)}$$

where I_{ES} = Reverse saturation current of the emitter base junction,

I_{CS} = Reverse saturation current of the collector-base junction.

α_F = Forward transmission from emitter to collector (*i.e.*, I_C/I_B). It is also called common-base current gain (α) of transistor. Its value is in the range of 0.98 to 0.998.

α_R = Reverse transmission from collector to emitter (*i.e.*, I_E/I_C). It is also called inverse alpha (α_1) of a transistor. Its value is in the range of 0.4 to 0.8.

The equations (*i*) and (*ii*) are known as Ebers-Moll equations and give us the values of emitter current and collector current respectively. The four quantities I_{ES}, I_{CS}, α_E and α_R are functions of doping densities and transistor geometry. These quantities are related by the equation,

$$\alpha_F \cdot I_{ES} = \alpha_R \cdot I_{CS} \qquad ...(iii)$$

The above equation is, sometimes, called reciprocity condition for the transistor. Fig. 23.3(*a*) shows the current components in an NPN transistor and Fig. 23.3 (*b*) shows its Ebers-Moll model. It may be noted that for an NPN transistor, the direction of all current components and junction voltages are reversed from those of a PNP transistor.

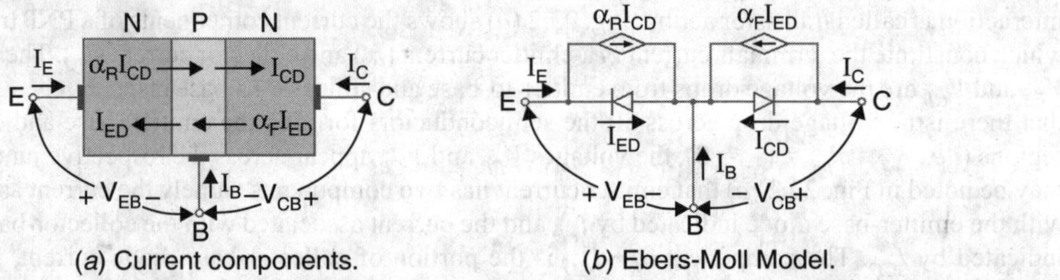

(*a*) Current components. (*b*) Ebers-Moll Model.

Fig. 23.3. NPN transistor.

The Ebers-Moll equations *i.e.*, equations (*i*) and (*ii*) are valid for an NPN transistor, provided a negative sign is inserted before the voltages V_{EB}, V_{CB} and each current component. Thus Ebers-Moll equations for an NPN transistor are:

$$I_E = -I_{ES}\,(e^{-V_{EB}/V_T} - 1) + \alpha_R \cdot I_{CS}\,(e^{-V_{CB}/V_T} - 1)$$

and

$$I_C = \alpha_F \cdot I_{ES}\,(e^{-V_{EB}/V_T} - 1) - I_{CS}\,(e^{-V_{CB}/V_T} - 1)$$

23.4. Approximate Ebers-Moll Model of Transistors

The Ebers-Moll model of a transistor can be simplified, if the junction diodes are replaced by an ideal one. An ideal diode is one, which acts as an open circuit when it is reverse biased; and as a short circuit when it is forward biased. Now we shall discuss the behaviour of this approximate model in the saturation, active and cut-off regions of a PNP transistor.

1. ***Saturation region.*** In this region, both the emitter-base and collector-base junctions are forward biased. Therefore, both the diodes act as short circuits. Under this condition, the Ebers-Moll model in saturation region is as shown in Fig. 23.4 (*a*). It may be noted that in saturation region the transistor acts as a short-circuit.

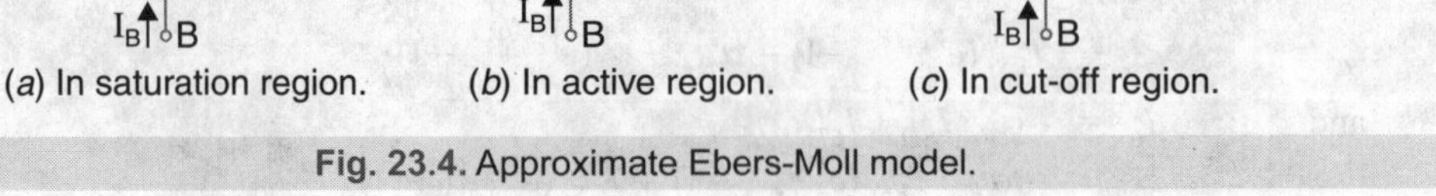

(*a*) In saturation region. (*b*) In active region. (*c*) In cut-off region.

Fig. 23.4. Approximate Ebers-Moll model.

2. ***Active region.*** In this region, the emitter-base junction is forward biased and collector-base junction is reverse biased. Therefore, the emitter diode acts as a short-circuit and collector diode as an open circuit. Under this condition, the Ebers-Moll model in active region is as shown in Fig. 23.4 (*b*).
3. ***Cut-off region.*** In this region, both the emitter-base and collector-base junctions are reverse biased. Therefore, both the diodes and current sources act as open circuits. Under this condition, the Ebers-Moll model in cut-off region is as shown in Fig. 23.4 (*c*). It may be noted that in the cut-off region, the transistor acts as an open-circuit.

23.5. More Accurate Ebers-Moll Model of Transistor

As a matter of fact, the Ebers-Moll models discussed in the last article are very simple. Although these models provide accurate results in most of the times, yet in some other times we require complex models to get more accurate results. We shall now discuss as to how these complex models can be obtained in the three regions.

1. ***More accurate saturation region model.*** The model shown in Fig. 23.4 (*a*) indicates that voltages V_{EB} and V_{CB} are zero in the saturation region. But in actual practice, it is not the case. As a matter of fact, there is always some voltage drop between the transistors terminals. The values of voltages V_{EB} and V_{CB} are essentially those of forward biased diodes. Above the cut-in or knee voltage (V_K), the diode current increases very rapidly with the applied voltage. Thus if the external circuitry limits the current, as it often does, the forward biased PN junction voltage will remain essentially constant. Therefore, saturation models, which are more accurate, are shown in Fig. 23.5 (*a*) and (*b*) for PNP and an NPN transistors respectively. The values of battery voltages $V_{EB\,(sat)}$ and $V_{CB\,(sat)}$ are 0.8 V and 0.6 V respectively for both the transistors.

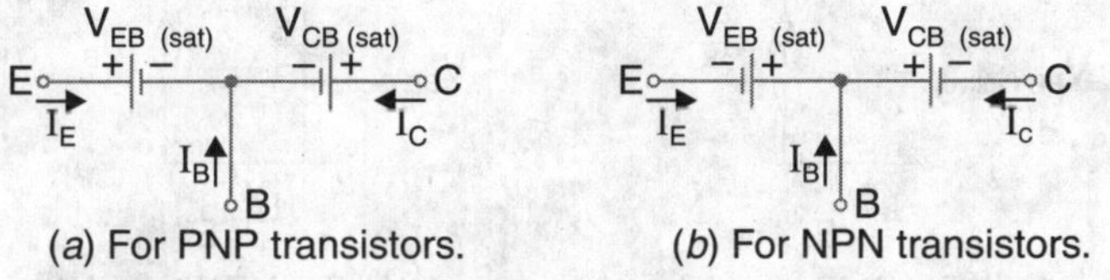

Fig. 23.5. More accurate saturation region model.

2. ***More accurate active region model.*** The active region model shown in Fig. 23.4 (*b*) can be made, somewhat, more accurate by adding a battery voltage $V_{EB\,(active)}$. The value of this battery voltage is 0.7 V.

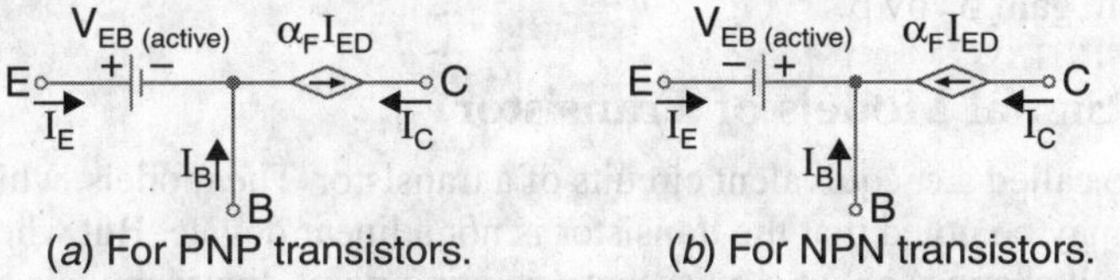

Fig. 23.6. More accurate active region model.

Figure 23.6 (*a*) and (*b*) shows, somewhat, more accurate model for a PNP and NPN transistor respectively.

3. ***More accurate cut-off region model.*** Generally, the model shown in Fig. 23.4 (*c*) is sufficient in the cut-off region *i.e.*, the transistor in cut-off region can be considered to be an open-circuited. But sometimes we are interested in few additional details of the transistor in this region. Consider the model shown in Fig. 23.2 (*b*). Since the diodes are not ideal, therefore, they have an effective cut-in or knee voltage (V_K). For instance, if in Fig. 23.2 (*b*) the voltage

V_{EB} is less than V_K, then the emitter diode will be cut-off and the emitter current will be zero. If the collector diode is cut-off, then the transistor will also be cut-off. However, if the external voltages are such that the collector diode is cut-off, then the transistor will be cut off only if the emitter-base voltage is not of the proper polarity and its magnitude is less than V_K. This defines the edges of cut-off region.

23.6. Ebers-Moll Model of Transistor in Common Emitter Configuration

The Ebers-Moll model of a transistor discussed in the last article is that of a common-base configuration. However, the transistors are used more commonly in a common-emitter configuration. Therefore, it is important to discuss the Ebers-Moll Model also called d.c. model of a transistor in common emitter configuration.

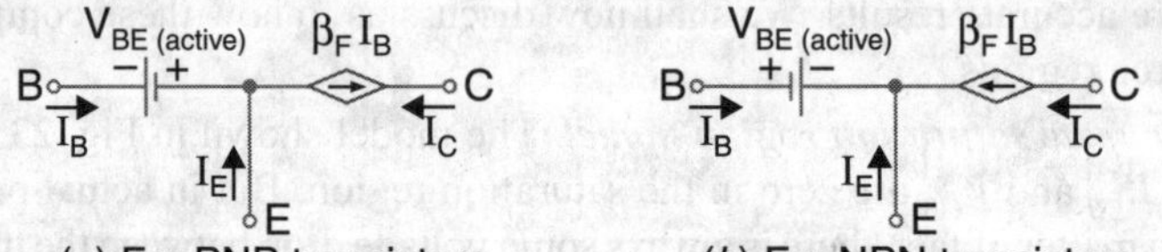

(*a*) For PNP transistors. (*b*) For NPN transistors.

Fig. 23.7. D.C. models for a transistor in common emitter configuration.

Fig. 23.7 shows d.c. model of a transistor in the active region. Here β_F (equal to β) is a common emitter current gain. It is the ratio of collector current to base current. The term $V_{BE(active)}$ is the barrier potential across the base-emitter diode and its value is 0.7 V for silicon transistor.

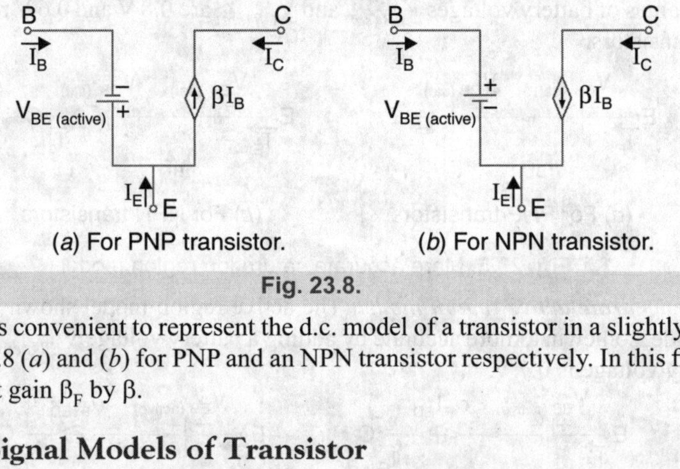

(*a*) For PNP transistor. (*b*) For NPN transistor.

Fig. 23.8.

Sometimes, it is convenient to represent the d.c. model of a transistor in a slightly different way as shown in Fig. 23.8 (*a*) and (*b*) for PNP and an NPN transistor respectively. In this figure, we have replaced the current gain β_F by β.

23.7. Small Signal Models of Transistor

These are also called a.c. equivalent circuits of a transistor. The models, which we shall discuss here are linear. It may be noted that the transistor is not a linear device. But when we are interested in determining small changes about a quiescent operating point, linear models can be accurate. We know that signals can be represented by small changes. Thus linear models can be used in the analysis of small signal circuits for a variety of amplifier applications. It has been observed that the behaviour of transistor is different for low-frequencies and high-frequencies. Therefore, we shall discuss the small signal models of a transistor under the following two heads:

1. Low-frequency small signal model.
2. High-frequency small signal model.

23.8. Low-Frequency Small Signal Model of Transistor

A low frequency small signal model can be obtained from the Ebers-Moll model of a transistor as shown in Fig. 23.2 (*a*). Since in the normal operation of a transistor, the emitter-base junction is forward biased and collector-base junction is reverse biased, therefore, the Ebers-Moll model shown in Fig. 23.2 (*a*), can be reduced to a circuit as shown in Fig. 23.9 (*a*).

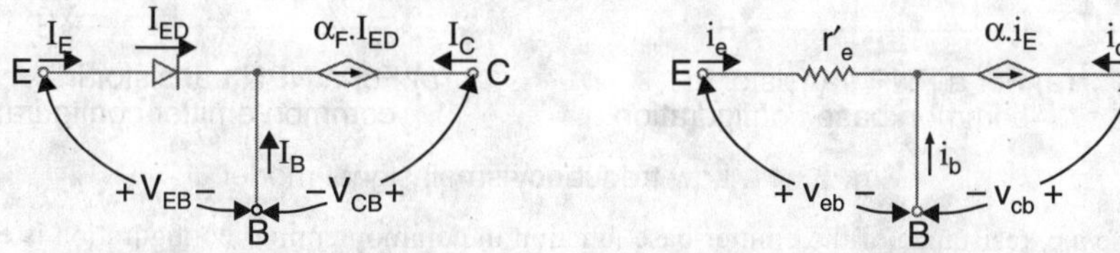

(*a*) Ebers-Moll model. (b) Small signal model.

Fig. 23.9. PNP transistor.

From this circuit, we can obtain a small signal model of a transistor as shown in Fig. 23.9 (*b*). This model is for a PNP transistor in a common base configuration. Following are some of the importants points, which may be carefully noted from the Fig. 23.9 (*b*):

1. The emitter diode is replaced by an a.c. resistance r'_e. The resistance r'_e represents the a.c. resistance (or dynamic resistance) of a forward-biased emitter base junction. Its value is givcn by thc rclation,

$$r'_e = \frac{25}{I_E}$$

where I_E is the d.c. or quiescent operating current in milliamperes.

2. All currents and voltages of a transistor are indicated by lower case letter *i* and *v* respectively.
3. The diode current is represented by i_e instead of i_{ed}, as it does not make any difference. Similarly, α_F is shown as α.

The small-signal model of a PNP transistor in common-base configuration can be represented in a more convenient way as shown in Fig. 23.10 (*a*).

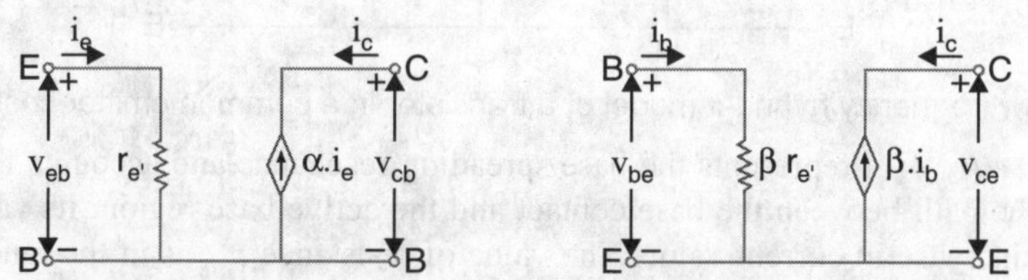

(*a*) For a PNP transistor in common base configuration. (*b*) For a NPN transistor in common emitter configuration.

Fig. 23.10. Low-frequency small signal model.

The small signal model of a PNP transistor in common emitter configuration may by obtained in the same way as discussed above and is as shown in Fig. 23.10 (*b*).

It may be noted that in a common emitter model, the a.c. resistance of a base-emitter junction is represented by $\beta \cdot r'_e$. The current of a controlled source is indicated by $\beta \cdot i_b$. The small signal models for an NPN transistor in a common base and common emitter configurations are shown in Fig. 23.11 (*a*) and (*b*) respectively.

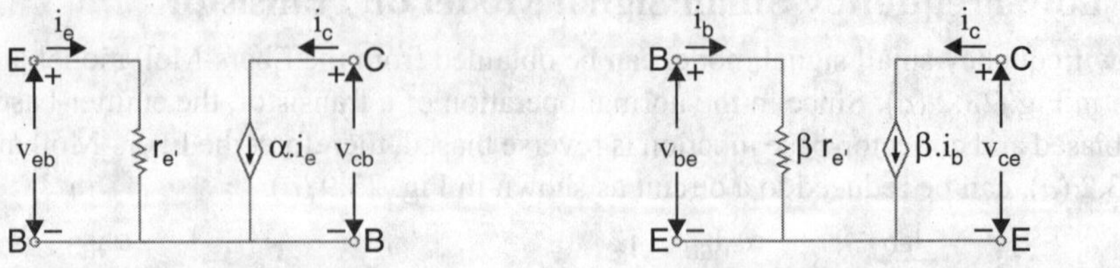

(*a*) For a PNP transistor in common base configuration.

(*b*) For a NPN transistor in common emitter configuration.

Fig. 23.11. Low frequency small signal model.

Note. The a.c. resistance of the emitter-base junction in common emitter configuration is equal to $\beta \cdot r_e'$ (and not equal to r_e'). It is because of the fact that the a.c. resistance, looking into the base in common emitter configuration, is given by the relation,

$$r_{ac} = 25/I_B$$

where I_B is the base current in milliamperes. Since the base current

$$I_B = I_C/\beta = I_E/\beta$$

$$\therefore \quad r_{ac} = \frac{25}{I_B} = \frac{25}{I_E/\beta} = \beta\frac{25}{I_E} = \beta \cdot r_e'$$

23.9. Low Frequency Hybrid-π Model of Transistor

Fig. 23.12 shows the low frequency hybrid-π model of a transistor with common emitter configuration. It is more accurate than that of the Ebers-Moll model discussed in the last article. Following are the components of a hybrid-π model:

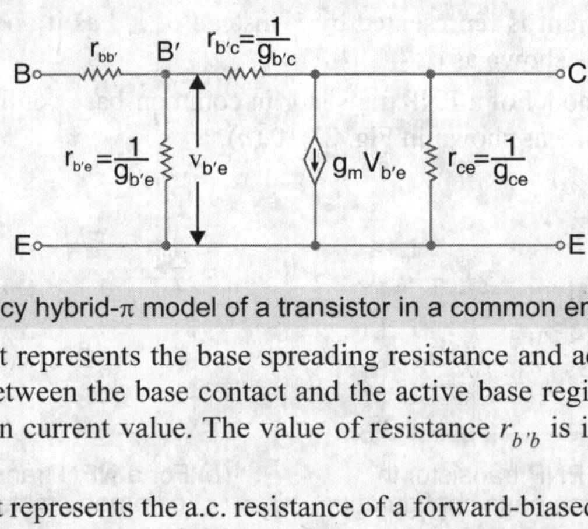

Fig. 23.12. Low frequency hybrid-π model of a transistor in a common emitter configuration.

1. ***Resistance*** ($r_{bb'}$). It represents the base spreading resistance and accounts for the voltage drop in the path between the base contact and the active base region. Its value decreases with the increase in current value. The value of resistance $r_{b'b}$ is in the range of 40 Ω to 400 Ω.
2. ***Resistance*** ($r_{b'e}$). It represents the a.c. resistance of a forward-biased emitter-base junction. Its value is in the range of about few hundred ohms to several thousand ohms. The reciprocal of resistance $r_{b'e}$ is called a.c. conductance and is designated by $g_{b'e}$. Mathematically, the a.c. conductance

$$g_{b'e} = 1/r_{b'e}$$

3. ***Resistance*** ($r_{b'c}$). It represents the effect of feedback between the emitter-base junction and collector-base junction due to an early effect. Its value is of the order of several megaohms. Because of its extremely high value, it is neglected in computation.
4. ***Controlled current source*** ($g_m \cdot v_{b'e}$). It represents the coupling between the junctions. Its value (equal to $g_m \cdot v_{b'e}$) is proportional to the base current (i_b).

5. ***Resistance (r_{ce}).*** It is called output resistance. Its typical value is in the range of about few kiloohms to hundreds of kiloohms. The reciprocal of resistance r_{ce} is called output conductance and is designated by g_{ce}. Mathematically, the output conductance, $g_{ce} = 1/r_{ce}$. *Sometimes, the resistance r_{ce} is also represented by a symbol r_o.

The hybrid-π model may by simplified, if we omit $r_{bc'}$ from the model. The simplified low frequency model is as shown in Fig. 23.13.

The parameter g_m is called transconductance (or mutual conductance) and is expressed in Siemens (*S*) or mhos (Ω). The transconductance represents the small changes in collector current about the operating point produced by the small changes in the base-emitter voltage. Its value is given by the relation,

$$g_m = \frac{I_c}{25}$$

where I_c is the collector current in milliamperes.

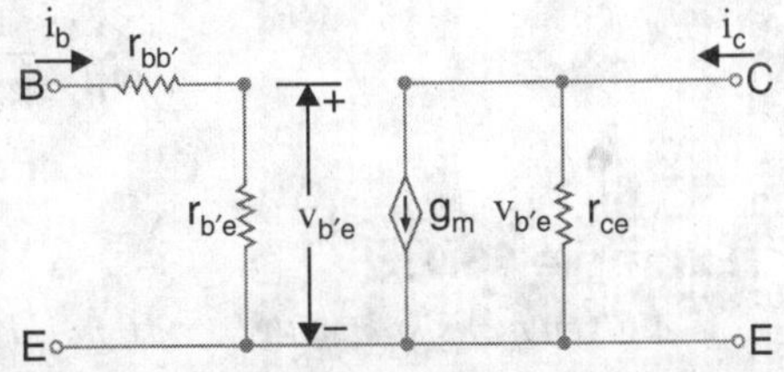

Fig. 23.13. Simplified hybrid-p low frequency model of a common emitter transistor.

It may be noted from Fig. 23.13 that the voltage drop across resistance $r_{be'}$ is given by the relation,

$$v_{b'e} = i_b \cdot r_{b'e}$$

where i_b is the base current.

It may also be noted that if the collector-to-emitter voltage (v_{ce}) is equal to zero, then there is no current through resistance r_{ce}. In this case, the whole current will pass through the controlled current generator. The value of collector current,

$$i_c = g_m \cdot v_{b'e}$$

Substituting the value of $v_{b'e}$ (equal to $i_{b'} \cdot r_{b'e}$) in the above equation, the collector current,

$$i_c = g_m \cdot i_b \cdot r_{b'e}$$

or

$$i_c/i_b = g_m \cdot r_{b'e}$$

The ratio i_c/i_b is defined as a.c. current gain (β_{ac}) of a common emitter transistor. It is also called small signal short circuit current gain, and is designated by h_{fe}. Thus a.c. current gain,

$$h_{fe} = i_c/i_b = g_m \cdot r_{b'e.}$$

$$\therefore \quad r_{b'e} = h_{fe}/g_m \text{ and } g_{b'e} = \frac{1}{r_{b'e.}} = g_m/h_{fe}$$

The parameter h_{fe} is supplied by the manufacturer. Knowing the value of quiescent collector current, the value of transconductance may be determined by using the relation $g_m = I_c/25$. And then may be calculated by using the relation $r_{b'e} = h_{fe}/g_m$. The value of base spreading resistance may be obtained by using the relation,

$$r_{bb'} = h_{ie} - r_{b'e}.$$

where h_{ie} is the input resistance of a common emitter transistor and is usually specified by the manufacturer. The value of output conductance may be obtained by the relation,

$$g_{ce} = h_{oe} - (1 + h_{fe})\, g_{b'e}$$

where g_{ce} = Output conductance with input as short-circulated. Its value is usually supplied by the manufacturer.

$g_{b'c}$ = Feedback conductance. Its value is given by the relation $h_{re} \cdot g_{b'e,}$. The parameter h_{re} is called reverse voltage gain and is usually specified by the manufacturer.

* According to this phenomenon, the width of the depletion region of collector-base junction increases as the magnitude of the reverse bias voltage (across the collector base junction) increases.

In actual practice, the value of $h_{fe} >> 1$. In that case, the output conductance,

$$g_{ce} = h_{oe} - h_{fe} \cdot g_{b'e'}$$

Substituting the value of $g_{b'c'}$ (equal to $h_{re} \cdot g_{b'c'}$) in the above expression, the value of output conductance,

$$g_{ce} = h_{oe} - h_{fe} \cdot h_{re}\, g_{b'c'}$$

$$= h_{oe} - h_{re} \cdot g_m \qquad \ldots (\because g_{b'e'} = g_m/h_{fe})$$

$\therefore$ Value of output resistance,

$$r_{ce} = \frac{1}{g_{ce}} = \frac{1}{h_{oe} - (1 + h_{fe})\, g_{b'c}}$$

$$= 1/(h_{oe} - h_{re}\, g_m)$$

Example 23.1.

A given transistor with $I_C = 10$ mA, $V_{CE} = 10$ V and at room temperature has the following set of low frequency parameters:

$$h_{ie} = 500\ \Omega\ h_{oe} = 10^{-5}\ A/V$$

$$h_{fe} = 100\ h_{re} = 10^{-4}$$

Find the values of all the hybrid - π parameters of a low frequency model.

Solution.

Given: $I_C = 10$ mA; $V_{CE} = 10$ volt; $h_{ie} = 500\ \Omega$; $h_{oe} = 10^{-5}$ A/V; $h_{fe} = 100$ and $= h_{re} = 10^{-4}$.

Resistance $r_{b'e}$

We know that the value of transconductance,

$$g_m = \frac{I_C}{25} = \frac{10}{25} = 0.4\ \Omega$$

and the value of resistance, $r_{b'e'}$

$$= \frac{h_{fe}}{g_m} = \frac{100}{0.4} = 250\ \Omega\ \textbf{Ans.}$$

Base spreading resistance

We know that the value of base spreading resistance,

$$r_{bb'} = h_{ie} - r_{b'e} = 500 - 250 = 250\ \textbf{Ans.}$$

Output resistance

We know that the value of feedback conductance,

$$g_{b'c} = \frac{h_{re}}{r_{b'e}} = \frac{10^{-4}}{250} = 4 \times 10^{-7}$$

and the value of output resistance,

$$r_{ce} = \frac{1}{h_{oe} - (1 + h_{fe})\, g_{b'c}} = \frac{1}{10^{-5} - (1 + 100) \times (4 \times 10^{-7})}$$

$$= 3.3 \times 10^4\ \Omega = 33\ k\Omega\ \textbf{Ans.}$$

23.10. High Frequency Hybrid-π Model of a Transistor

Fig. 23.14 shows a high frequency hybrid-π model of a transistor in common emitter configuration. This model is similar to the low frequency hybrid-π model, except the addition of two capacitances, c_e and c_c respectively. The values of resistance $r_{bb'}$, $r_{b'c}$, $r_{b'e}$ and r_{ce} may be obtained in the same way as we discussed for the low-frequency hybrid -π model.

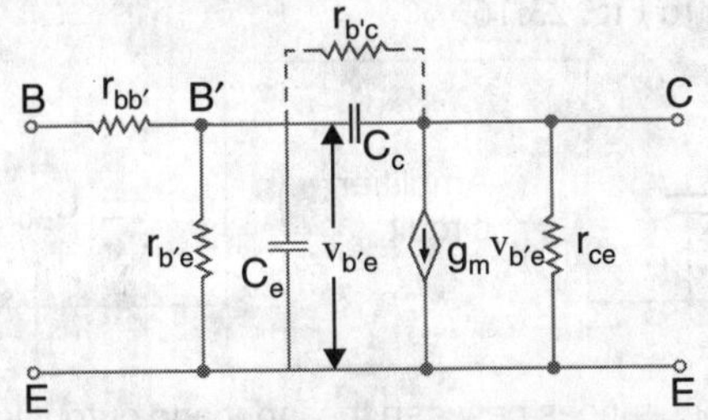

Fig. 23.14. High frequency hybrid-p model of a transistor in a common emitter configuration.

The capacitance c_e is the sum of the emitter diffusion capacitance (C_{D_e}) and the emitter junction capacitance (C_{T_e}). Mathematically, the capacitance,

$$C_e = C_{D_e} + C_{T_e}$$

For a forward biased emitter-base junction, the emitter diffusion capacitance (C_{D_e}) is, usually, much larger than the emitter junction capacitance (C_{T_e}). Therefore, the capacitance,

$$C_e = C_{D_c}$$

It has been observed that the capacitance C_{D_e} is proportional to the emitter bias current, I_E and is almost independent of temperature. The value of C_c may be determined experimentally from measurement of frequency (f_T) at which the common emitter short-circuit current gain drops to unity. Its value is given by the relation,

$$C_e = \frac{g_m}{2\pi f_T}$$

where g_m is the transconductance.

The capacitance C_e is the depletion region capacitance of the reverse biased collector-base junction. It varies as $(V_E)^{-n}$, where n is 1/2 or 1/3 for an abrupt or graded junction.

23.11. Miller Effect Capacitance

Strictly speaking, in the high frequency region, the capacitive elements of importance are the interelectrode capacitances and the wiring capacitances between the leads of the circuit. The inter-electrode capacitances are the internal capacitances between the terminals of the active device (BJT or FET). Refer to Fig. 23.15(*a*) and (*b*).

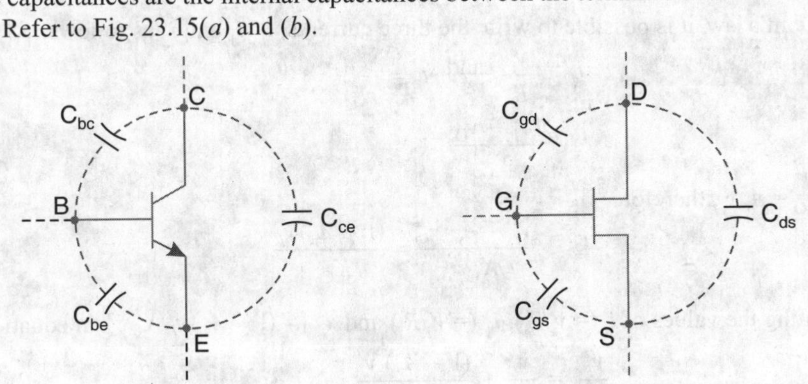

(*a*) Inter electrode capacitances between the terminals of BJT

(*b*) Inter electrode capacitances between the terminals of FET

Fig. 23.15.

The wiring capacitances are the capacitances between the leads of the circuit on the input and output sides of the circuit. Refer to Fig. 23.16.

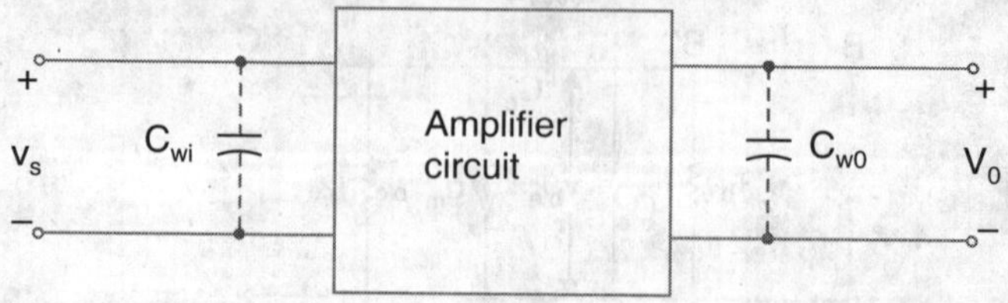

Fig. 23.16. Illustrating wiring capacitances between the input and output leads of the amplifier circuit.

The large capacitors of the amplifiers circuit that controlled the low frequency response have all been replaced by their short-circuit equivalent due to their very low reactance at high frequences.

It may be carefully noted that inverting amplifiers introduce a phase shift of 180° between input and output. This results in a negative value of voltage gain (A_v). Because of this phase shift, the input and output capacitance is increased by a capacitance level sensitive to (1) the interelectrode capacitance between the input and output terminals of the active device (BJT or FET) and (2) the gain of the amplifier. This feedback capacitance as illustrated in Fig. 23.17 is designated by C_f.

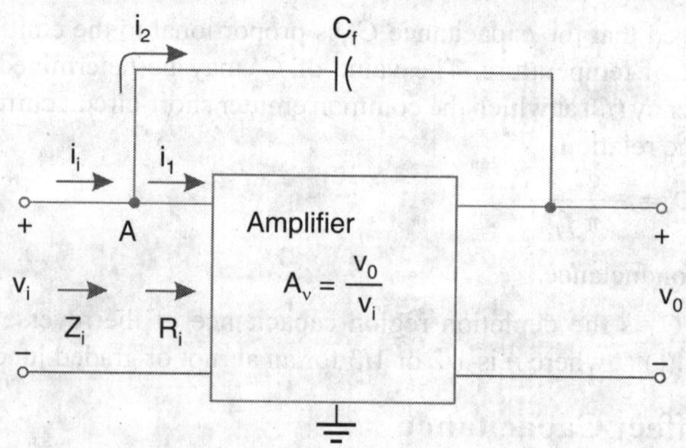

Fig. 23.17.

Applying Kirchoff's current Law at point A, we get,

$$i_i = i_1 + i_2 \qquad \text{...}(i)$$

Using Ohm's law, it is possible to write the three currents, i_i, i_1 and i_2 as follows,

$$i_i = \frac{v_i}{R_i} \text{ and } i_1 = \frac{v_i}{r_i}$$

Also,

$$i_2 = \frac{v_i - v_O}{X_{C_f}}$$

Since, $v_O = A_v\, v_i$, therefore,

$$i_2 = \frac{v_i - A_v \cdot v_i}{X_{C_f}} = \frac{(1 - A_v)\, v_i}{X_{C_f}}$$

Substituting the values of i_i (= v_i/Z_i), i_1 (= V_i/R_i) and i_2 [= $(1 - A_v)v_i/X_{C_f}$] in equation (*i*), we get,

$$\frac{v_i}{R_i} = \frac{v_i}{R_i} + \frac{(1 - A_v)\, v_i}{X_{C_f}}$$

Taking out v_i common from R.H.S. and cancelling it out from two sides we get,

$$\frac{1}{Z_i} = \frac{1}{R_i} + \frac{1 - A_v}{X_{C_f}}$$

$$= \frac{1}{R_i} + \frac{1}{\dfrac{X_{C_f}}{1 - A_v}} \quad ...(ii)$$

But
$$\frac{X_{C_f}}{1 - A_v} = \frac{1}{\omega C_f (1 - A_v)} = \frac{1}{\omega (1 - A_v) C_f} \quad ...(iii)$$

Let $(1 - A_v) = C_M$ (*i.e.*, Miller effect capacitance),

Then equation (*iii*) can be rewritten as,

$$\frac{X_{C_f}}{1 - A_v} = \frac{1}{\omega C_M} = X_{C_M}$$

where, X_{C_M} = capacitive reactance due to the Miller effect capacitance C_M.

Substituting the value of in equation (*iii*) we get,

$$\frac{1}{Z_i} = \frac{1}{R_i} + \frac{1}{X_{C_M}} \quad ...(iv)$$

The above equation helps us to establish an equivalent circuit of Fig. 23.18. The resulting equivalent circuit is as shown in Fig. 23.18.

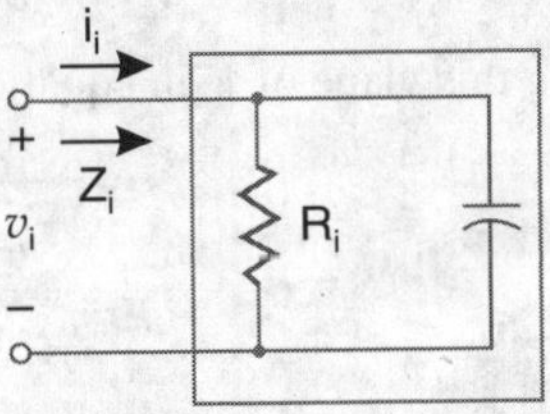

Fig. 23.18.

Notice that the equivalent circuit of Fig. 23.18, includes the some R_i that we have dealt with in previous chapters (especially chapter-16), with the addition of feedback capacitor magnified by the gain of the amplifier. Any interelectrode capacitance at the input terminals to the amplifier will simply be added in parallel with the elements of Fig. 23.18.

In general, the Miller effect input capacitance is given by the equation,

$$C_{M_i} = (1 - A_v)\, C_f \quad ...(iv)$$

It is evident from the above equation that for any inverting amplifier, the input capacitance will be increased by a Miller effect capacitance sensitive to the gain of the amplifier and the interelectrode capacitance connected between the input and output terminals of the active device (BJT or FET).

At high frequencies, the gain A_v will be a function of the level of C_{M_i}. However, since the maximum gain is the midband value, using the midband value will result in the highest level of C_M. This represents the worst-case scenario. Hence, remember the value of A_v to be used in equation (*iv*) is midband gain value.

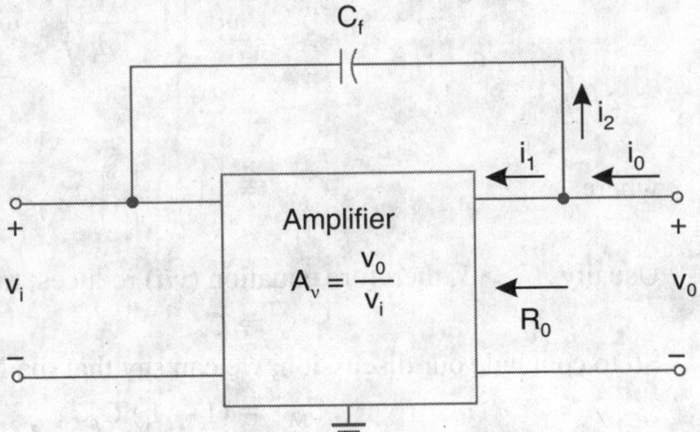

Fig. 23.19.

If you examine equation (*iv*) carefully, you will find that if we use a positive value for A_v, it would result in a negative capacitance. It is because of this reason that we use an inverting amplifier configuration in out discussion.

Let us now focus on the output capacitance of the amplifier. We will use the circuit shown in Fig. 23.19 to determine the output of Miller effect capacitance.

Applying Kirchoff's current Law at point B (refer to Fig. 23.19), we get,

$$i_o = i_1 + i_2$$

With $$i_1 = \frac{v_o}{r_o} \text{ and } i_2 = \frac{v_o - v_i}{X_{C_f}}$$

$$\therefore \quad i_o = \frac{v_o}{R_o} + \frac{v_o - v_i}{X_{C_f}}$$

Since, $R_o >> X_{C_f}$, therefore the term v_o/R_o is negligible as compared to the term $(v_o - v_i) / X_{C_f}$. Hence,

$$i_o \approx \frac{v_o - v_i}{X_{C_f}}$$

Since, $$\frac{v_o}{v_i} = A_v$$

$$\therefore \quad v_i = v_o/A_v$$

Substituting this vlaue of V_i in equation (v), we get,

$$i_o = \frac{v_o - \dfrac{v_o}{A_v}}{X_{C_f}} = \frac{v_o\left(1 - \dfrac{1}{A_v}\right)}{X_{C_f}}$$

or $$\frac{i_o}{v_o} = \frac{1 - \dfrac{1}{A_v}}{X_{C_f}}$$

or $$\frac{v_o}{i_o} = \frac{X_{C_f}}{1 - \dfrac{1}{A_v}}$$

Substituting $X_{C_f}\ (= 1/\omega c_f)$ in the above equation, we get,

$$\frac{v_o}{i_o} = \frac{1}{\omega c_f\left(1 - \dfrac{1}{A_v}\right)} = \frac{1}{\omega \underbrace{\left(1 - \dfrac{1}{A_v}\right) C_f}_{C_{M_o}}} = \frac{1}{\omega C_{M_o}} \quad \text{...(vi)}$$

where, $$C_{M_o} = \left(1 - \frac{1}{A_v}\right) C_f \quad \text{...(vii)}$$

Usually $A_v >> 1$, therefore equation (vii) reduces, to,

$$C_{M_o} = C_f$$

So to conclude our discussion, we can say that the Miller effect capacitance,

$$C_{M_i} = (1 - A_v)\, C_f$$

and $$C_{M_o} = C_f.$$

DESCRIPTIVE QUESTIONS

1. For a PNP transistor in the active region, what is the sign (positive or negative) of emitter current, collector current, base current, collector-to-base voltage and emitter-to-base voltage?
2. Draw the Ebers-Moll model for an NPN transistor. Also give the equations for emitter current and collector current.
3. Draw an approximate d.c. model of a PNP transistor in each of the three regions of operation.
4. Draw the low-frequency hybrid-π model and explain the meaning of each component of a model.
5. (*a*) Define the transconductance g_m. (*b*) write an equation, which relates g_m to current gain β (or h_{fe}).
6. Draw a neat diagram of a high-frequency hybrid-π transistor equivalent circuit. Explain the meaning of each component.
7. Draw the small signal high frequency common emitter model of a transistor. *(A.M.I.E., Winter, 1987)*
8. Explain the high frequency model of a bipolar transistor. *(Grad. I.E.T.E., Dec. 1987)*
9. Derive an Ebers-Moll model for a PNP transistor. *(Anna University., April/May 2008)*
10. With the neat sketch explain hybrid - pi (π) Common Emitter transistor model. *(Anna University., Nov/Dec 2007)*
11. With the neat sketch π common emitter transistor model. *(Anna University., Nov/Dec 2007)*
12. Derive the expression for transistor conductance g_m for hybrid pi common emitter transistor model. *(Anna University., Nov/Dec 2007)*
13. Derive the expression for transistor conductance (g_m) for hybrid common emitter transistor model. *(Anna University., Nov/Dec 2007)*
14. Sketch the small signal high frequency circuit of a common source amplifier and derive the expression for a voltage gain. *(Anna University., Nov/Dec 2007)*
15. Draw the equivalent circuit for a CE amplifier at high frequencies and obtain the current gain expression at high frequencies. *(M.G University., May/June 2006)*
16. Explain what will happen to the voltage gain of amplifier. If the bypass capacitor is open circuited. *(GBTU/MNTU., 2004-05)*
17. What will the affect of an emitter bypass capacitor on frequency response of an amplifier? *(PTU., May 2009)*
18. Draw the hybrid π model of CE transistor and explain the origin of various conductance and resistances. *(Nagpur University., 2008)*
19. Write short note on Ebers- Moll model of BJT. *(R.G.P.V., June 2011)*
20. Explain the Ebers-Moll model of transistor. *(R.G.P.V., Dec. 2010)*
21. Derive the expression for I_E and I_C for BJT using Ebers-Moll model with proper necessary diagram. *(R.G.P.V., Dec. 2008)*
22. Explain Ebers-Moll model. *(R.G.P.V., Feb 2010)*
23. Write the Elbers-Moll equations for BJT and sketch the circuit model which satisfies these equations. *(Nagpur University, Summer 2007)*
24. Draw the Elbers-Moll model for a PNP transistor. Also elaborate why we cannot construct a transistor using two diodes. *(Nagpur University, Winter 2005)*

25. Write the Elbers- Moll equation and sketch the circuit model which satisfies these equations. *(Nagpur University, Winter 2004)*

26. Show that at low frequencies the hybrid π model with r_{bc} and r_{ce} taken as infinite reduces to the approximate CE h parameter model. *(Nagpur University, Winter 2001)*

27. Give the significance of two capacitors and all resistive components of hybrid model and give their typical values. *(Nagpur University, Summer 2000)*

28. Write short notes on :

(*a*) Hybrid π model parameters *(Nagpur University, Summer 2006, 2005, 2004)*

(*b*) Eber Moll model of BJT *(R.G.P.V., June 2011, Dec. 2004)*

MULTIPLE CHOICE QUESTIONS

1. Which of the following is not a time varying quantity

(*a*) v_{ce} (*b*) V_{CE} (*c*) v_{CE} (*d*) V_{ce}

2. In the Ebers-Model of a bipolar transistor, the parameter α_F is the

(*a*) Forward transmission from emitter to collector

(*b*) Reverse transmission from collector to emitter

(*c*) Common base current gain

(*d*) Both (*a*) and (*c*) above

3. The value of transconductance of a bipolar transistor for a collector current of 1.5 mA is

(*a*) 0.05 Ω (*b*) 0.05×10^{-3} Ω

(*c*) 37.5 Ω (*d*) None of the above

4. The resistance $r_{b'c}$ in the low frequency hybrid- π model of a bipolar transistor represents

(*a*) Base spreading resistance

(*b*) A.C. resistance for forward biased emitter-base junction

(*c*) The effect of feedback between the emitter-base junction and collector-base junction due to Early effect

(*d*) None of the above

5. The capacitance C_e in the high frequency hybrid-π model of a bipolar transistor represents the

(*a*) Depletion region capacitance (*b*) Emitter diffusion capacitance

(*c*) Emitter-base junction capacitance (*d*) Sum of the (*b*) and (*c*) above

ANSWERS

1. (*b*) **2.** (*d*) **3.** (*a*) **4.** (*c*)
5. (*d*)

24 Chapter

Single-Stage BJT Amplifiers

Outline

24.1. Introduction

When we cannot hear a stereo system, we increase the volume, when the picture on our television is too dark, we increase the brightness control. In both of these cases, we are taking a relatively weak signal and making its stronger (*i.e.*, increasing of its power). The process of increasing the power of an a.c. signal is called *amplification*. The circuits used to performs this function are called *amplifiers*.

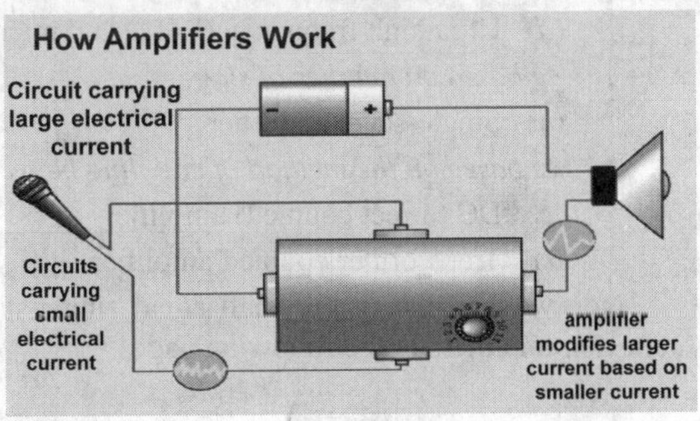

An amplifier may also be defined as a device, which produces a larger electrical output of similar characteristics than that of the input parameters. It increases the magnitude of the current or voltage input by means of energy drawn from an external source. The input signal may be obtained from a phonograph cartridge, tape head, or a transducer such as thermocouple, pressure gauge, etc. The output signal may be supplied to a loud speaker in an audio amplifier, a motor in a servo amplifier, a relay in a control application etc.

As mentioned above, an amplifier needs a source of energy for amplification. This may be provided by a battery (usually a dry battery) or a d.c. source resulting from a rectifier and filter combination. The amplifier also contains atleast one active device. This may be an electron tube, bipolar transistor or a field-effect transistor, which can provide a control function. Basically, the active device converts the energy from the d.c. source into energy at the output of the amplifier, which is proportional to the input signal. The a.c. input signal merely supplies a means of controlling the d.c. to a.c. convertion, which takes place in the tube or transistor. The control is, usually, accomplished with comparatively little input signal power.

The amplifier, in which the instantaneous output signal is directly proportional to the corresponding input signal, is called a *linear amplifier*. On the other hand, an amplifier, in which the output is not directly proportional to the input, is called a *non-linear amplifier*.

24.2. Classification of Amplifiers

The linear amplifiers may be classified according to their mode of operation *i.e.*, the way they operate on the predetermined set of values. Their descriptions are based on the following factors:

1. *As based on the input*
   (*a*) Small-signal amplifier (*b*) Large-signal amplifier
2. *As based on the output*
   (*a*) Voltage amplifier (*b*) Power amplifier
3. *As based on transistor configuration*
   (*a*) Common-emitter (CE) amplifier (*b*) Common-base (CB) amplifier
   (*c*) Common-collector (CC) amplifier
4. *As based on biasing conditions*
   (*a*) Class-A (*b*) Class-B (*c*) Class-AB (*d*) Class-C
5. *As based on nature of load resistance*
   (*a*) Untuned amplifiers (Wide-band amplifier)
   (*b*) Tuned amplifier (Narrow-band amplifier)
6. *As based on frequency response*
   (*a*) Direct coupled (DC) amplifier (*b*) Audio frequency (AF) amplifier
   (*c*) Radio frequency (RF) amplifier
   (*d*) Ultra high frequency (UHF) and microwave frequency amplifier
7. *As based on number of stages*
   (*a*) Single-stage amplifier (*b*) Multistage amplifier
8. *As based on the method of coupling between the stages*
   (*a*) DC (direct coupled) amplifier (*b*) R-C coupled amplifier
   (*c*) Transformer coupled amplifier

Now we shall discuss the small signal, single-stage common-emitter, common base and common collector amplifiers in the following pages.

24.3. Transistor as an Amplifier

Fig. 24.1 shows a basic circuit of an amplifier. In this circuit, an NPN transistor is used in the common emitter (CE) configuration. Therefore the basic circuit is known as a basic common emitter amplifier. Here V_{BB} supply forward-biases the emitter-base junction and V_{CC} supply reverse-biases the collector-base junction. This biases the transistor to operate in the active region V_S is a sinusoidal a.c. input signal source. It has a source resistance R_S. The magnitude of signal source voltage is such that it always forward-biased the emitter-base junction regardless of the polarity of the signal. A brief description of the operation of amplifier circuit is given below:

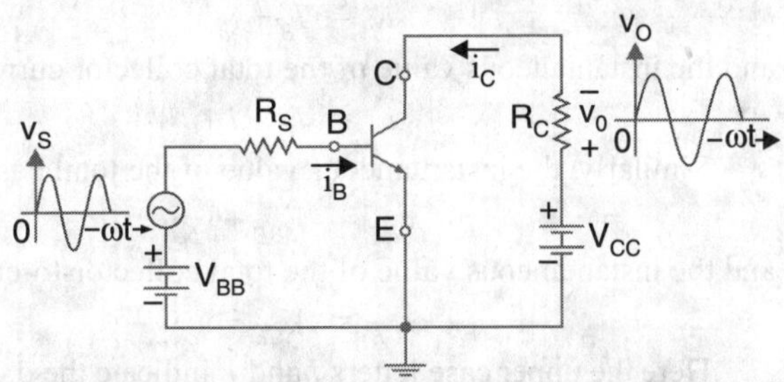

Fig. 24.1. Basic common emitter amplifier.

First of all, let us assume that there is no a.c. signal source. Under this condition, a d.c collector current (I_C) flows through the collector load (R_C). This is called zero signal current or quiescent operating current. Now let an a.c. signal be applied between the emitter-base junction. During the positive half-cycle of the input signal, the forward-bias across the emitter-base junction is increased. As a result of this, more electrons are injected into the base and reach the collector, which increase the collector current. The increased collector current produces greater voltage drop across the resistance R_C. However, during the negative half-cycle, the forward-bias across the emitter-base junction is decreased. Due to this, the collector-current decreases. The decreased collector current produces smaller voltage drop across resistance R_C.

It is evident from the above discussion, that a small a.c. signal at the input produces a large a.c. signal at the output or load resistance. Thus the transistor acts as an amplifier. The ratio of output voltage to input voltage is called voltage gain of the amplifier.

24.4. Common-Emitter Transistor Amplifier

Fig. 24.2 shows a circuit of a common-emitter amplifier using NPN transistor. It may be noted from this circuit that the input a.c. signal is applied across the base-emitter terminals and the output signals is taken across the collector and emitter terminals. The emitter-base junction of a transistor is forward-biased by the V_{BB} supply. The collector-base junction is reverse-biased by the V_{CC} supply. The Q-point is determined by the V_{CC} supply alongwith the resistance R_B and R_C. The capacitors C_1 and C_2 are called blocking capacitors (or coupling capacitors). Each capacitor acts like a switch, which is open to a direct current but shorted to an alternating current. Because of this, a blocking capacitor blocks the direct current, but passes the alternating current. This action isolates d.c. bias from an a.c. signal in the circuit. The function of capacitor C_1 is to connect the a.c. signal source to the input circuit (or base-emitter circuit) of the amplifier. The function of capacitor C_2 is to connect the amplifier to its load resistance or to the next stage of the amplifier.

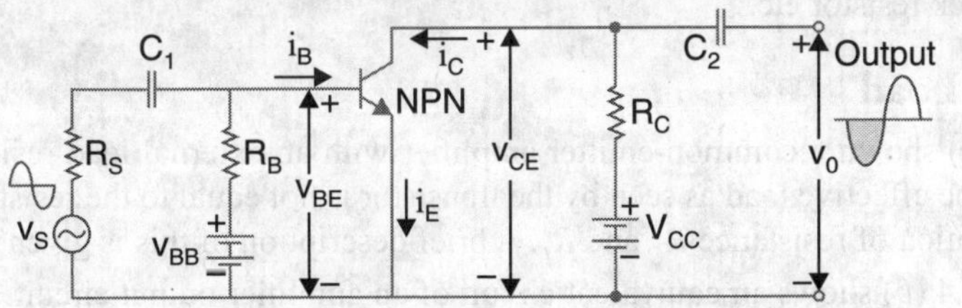

Fig. 24.2. Common-emitter transistor amplifier.

All the voltages and currents of a transistor are indicated by the instantaneous total values (*i.e.*, d.c + a.c.). Thus the instantaneous value of the total base current is given by the relation,

$$i_B = I_B + i_b$$

and the instantaneous value of the total collector current,

$$i_C = I_C + i_c$$

Similarly, the instantaneous value of the total base-to-emitter voltage,

$$v_{BE} = V_{BE} + v_{be}$$

and the instantaneous value of the total collector-to-emitter voltage,

$$v_{CE} = V_{CE} + v_{ce}$$

Here the upper case letters *I* and *V* indicate the d.c. (or quiescent-operating) current and voltage values respectively. The lower case letters *i* and *v* represent the a.c. current and voltage values respectively. It may be noted that emitter terminal of a transistor is shown grounded, therefore common emitter amplifier is also known as grounded emitter amplifier.

24.5. Different Ways of Drawing Common Emitter Amplifier Circuit

In most of the commonly used amplifier circuits, the polarity of bias voltage required for the base, is the same as that required for a collector. Consequently, the circuit shown in Fig. 24.2 may be simplified by eliminating the V_{BB} supply and simply connecting it to the V_{CC} supply as shown in Fig. 24.3 (*a*).

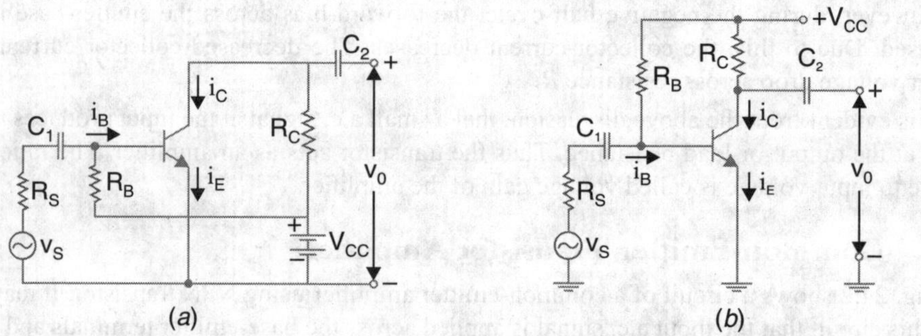

Fig. 24.3. Different ways of drawing common emitter amplifier circuit.

Fig. 24.3 (*b*) shows a circuit drawn in another form. It is more practical way of drawing an amplifier circuit. It may be noted that the battery symbol is not shown in this circuit. It is always understood to be connected from the terminal marked + V_{CC} to the ground terminal, even though it may not be actually shown. Moreover, the grounded terminals of the transistor and the signal source are shown separately for the sake of convenience.

The common emitter amplifier circuit shown in Fig. 24.3 has a base bias or fixed bias circuit. The biasing circuit can be a voltage-divider circuit, voltage-divider with feedback emitter resistor, collector feedback resistor etc.

24.6. A.C. Load

Fig. 24.4 (*a*) shown a common-emitter amplifier with an external load resistance (R_L). Under this condition, the effective load as seen by the transistor is not equal to the resistance R_C. But it is a parallel combination of resistance R_C and R_L. A brief description of this is given as below:

The Fig. 24.4 (*b*) shows an equivalent circuit of an amplifier output circuit. This circuit shows a capacitor across V_{CC} supply. This capacitor exists within the electronic power supplies. Because of this capacitor, the d.c. supply *i.e.*, (V_{CC}) supply presents a very low resistance to the a.c. signal currents. Thus an a.c. signal current, flowing through resistance R_C and into the supply, produces a very little a.c. voltage drop within the power supply. In other words, the entire a.c. voltage is dropped across the resistance R_C. Hence the resistance R_C may be considered as connected to ground as for as a.c. signal is concerned. It is for this reason that the equivalent circuit shown in Fig. 24.4 (*c*) does not show power supply.

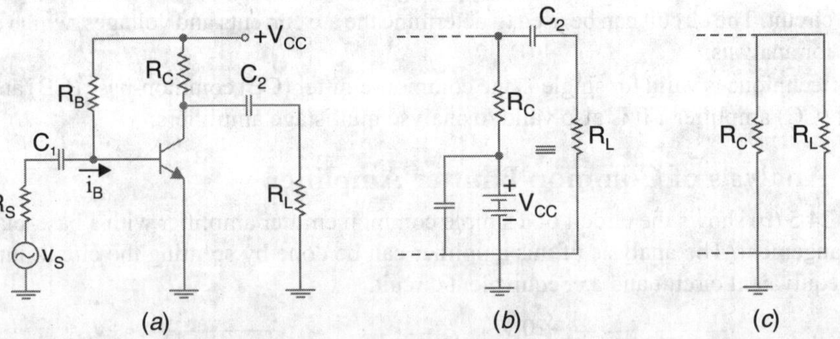

Fig. 24.4. A.C. load.

Now consider the effect of coupling the capacitor C_2. We know that for a.c. current, the capacitor C_2 acts as short-circuit. Therefore, the a.c. voltage drop across the resistance R_L must be equal to that across the resistance R_C. It means that the two resistances R_C and R_L may be considered in parallel as far as a.c. currents are concerned. It is illustrated in the circuit shown in Fig. 24.4 (*c*).

It is evident from the above discussion, that the actual signal load, on the amplifier is a parallel combination of resistances R_C and R_L. This load is called a.c. load or effective load and is designated by r_L. Thus the a.c. load,

$$r_L = R_C \parallel R_L = \frac{R_C \times R_L}{R_C + R_L}$$

If $R_L = \infty$ or $R_L >> R_C$, then the a.c. load is equal to the collector resistance, *i.e.*, $r_L = R_C$.

Sometimes, it is very convenient to call resistance R_C as the d.c. load and the parallel combination of resistance R_C and R_L (designated as r_L) as a.c. load on an amplifier. It is the a.c. load, which determines the voltage gain of an amplifier.

24.7. Analysis of Transistor Amplifier

We know that in a transistor amplifier, the d.c. sources set up d.c. biasing currents and voltages, which are known as quiescent operating currents and voltages. The a.c. input signal source produces fluctuations in these currents and voltages. As a result of this, we get an amplifier signal at the output. We have to analyse the amplifier circuit in order to determine the a.c. and d.c. currents and voltage gain, input resistance and output resistance of an amplifier. The simplest way to analyse the circuit is to split the analysis into the following parts:

1. ***D.C. analysis.*** The following procedure is adopted for the d.c. analysis:
   (*a*) First of all, reduce the a.c. sources to zero. It means that a voltage source is replaced by a short-circuit and a current source by an open-circuit.
   (*b*) Now open all the capacitors, as they block direct current.

   The circuit, which remains after applying the above two steps, is called a d.c. equivalent circuit. This circuit can be used to determine the d.c. currents and voltages, which are required for the analysis.

1. ***A.C. analysis.*** The following procedure is adopted for the a.c. analysis:
   (*a*) First of all, reduce the d.c. sources to zero. It means that a voltage source is replaced by a short-circuit and a current source by an open-circuit.
   (*b*) Now short all the capacitors as they pass alternating current.
   (*c*) Then replace the transistor with its small-signal equivalent circuit.
   The circuit, which remains after applying the above three steps, is called an a.c. equivalent circuit. The circuit can be used to determine the a.c currents and voltages, which are required for analysis.

This technique is valid for single-stage common-emitter (CE) common-base (CB) and common-collector (CC) amplifiers. It is also valid to analyse multistage amplifiers.

24.8. Analysis of Common Emitter Amplifier

Fig. 24.5 (*a*) shows the circuit of a simple common emitter amplifier with a base-bias (or fixed-bias) arrangement. The analysis of this amplifier can be done by splitting the circuit into two parts *i.e.*, d.c. equivalent circuit and a.c. equivalent circuit.

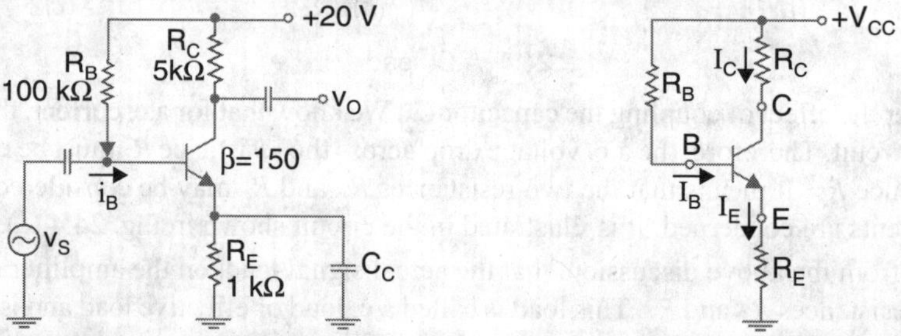

(*a*) Common emitter amplifier with emitter bypass capacitor.

(*b*) D.C. equivalent circuit of common emitter amplifier.

Fig. 24.5.

(*a*) ***D.C. equivalent circuit.*** It may be obtained from the amplifier circuit by opening (or disconnecting) the capacitors. The d.c. equivalent circuit is shown in Fig. 24.5 (*b*). We have already discussed in chapter on transistor biasing that base current in a base-bias circuit is given by the relation,

$$I_B = \frac{V_{CC} - V_{BE}}{R_B} = \frac{V_{CC}}{R_B} \qquad ...(\because V_{CC} >> V_{BE})$$

and the collector current,

$$I_C = \beta \cdot I_b$$

$\therefore$ Emitter current,

$$I_E \approx I_C \approx \beta \cdot I_B$$

(*b*) ***A.C. equivalent circuit.*** It may be obtained from the amplifier circuit by replacing the d.c. supply (V_{CC}) by a ground and short-circuiting the capacitors. The a.c. equivalent circuit is shown in Fig. 24.6.

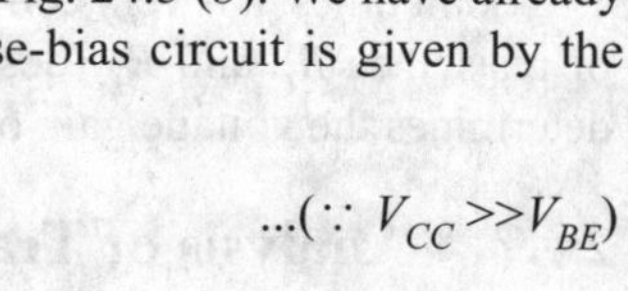

Fig. 24.6. A.C. equivalent circuit.

We shall use d.c. and a.c. equivalent circuits to determine the amplifier parameters in the next article.

24.9. Common Emitter Amplifier Parameters

We have already discussed in the last article a simple a.c. equivalent circuit of a common emitter amplifier with base-bias. A more detailed a.c. equivalent circuit may be obtained by replacing NPN transistor with a small-signal model as discussed in the last chapter. Fig. 24.7 shows such an a.c. equivalent circuit. This circuit is used to determine amplifier parameters such as input resistance (R_i), output resistance (R_o), current gain (A_i), voltage gain (A_v) and power gain (A_p). It may be noted that we have not shown the signal source resistance R_s in the figure. Therefore, we shall take resistance R_s as zero. The effect of signal source resistance shall be discussed later in this chapter.

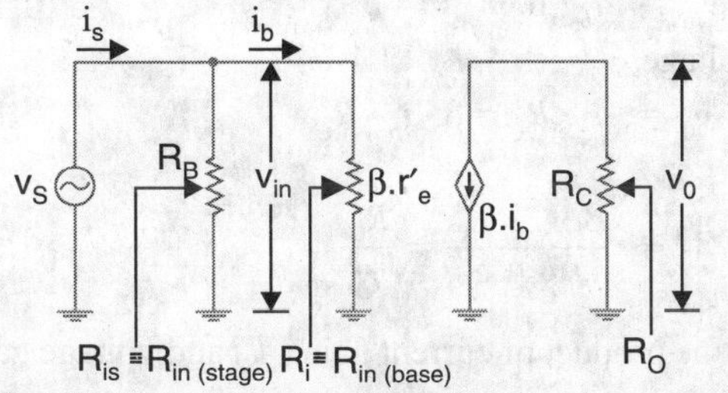

Fig. 24.7. Detailed a.c. equivalent circuit of common emitter.

It may be noted from Fig. 24.7 that $\beta \cdot r'_e$ represents the a.c. resistance of the emitter-base junction as seen by the input signal. It is known as input resistance, looking directly into the base and is designated by R_i or R_{in} (base). Here β is common emitter d.c. current gain (or h_{FE}) of a transistor. The input resistance, (R_i) does not include the effect of external biasing resistors connected to the base. The input resistance, which does include this, is called an input resistance of amplifier stage. It is designated by R_{is} or R_{in} (stage). It may be noted that in the absence of signal source resistance (R_s), all the source voltage appears as the input voltage across the base-emitter junction. Another important point is that since $\beta \cdot r'_e$ is much smaller than resistance R_B, therefore, a major part of source current (i_s) passes through $\beta r'_e$ and a negligible part through the resistance R_B. Following parameters of a common emitter amplifier are important from the subject point of view:

1. ***Input resistance.*** It is the resistance looking directly into the base and is given by the ratio of base voltage to base current. Thus input resistance,

$$R_i = \frac{v_b}{i_b} = \frac{v_{in}}{i_b}$$

Since $v_{in} = \beta \cdot r'_e \cdot i_b$, therefore, input resistance,

$$R_i = \frac{\beta \cdot r'_e \cdot i_b}{i_b} = \beta \cdot r'_e$$

and the input resistance of the amplifier stage,

$$R_{is} = R_B \,||\, (\beta \cdot r'_e)$$

Since $R_B >> \beta \cdot r'_e$, therefore,

$$R_{is} = \beta \cdot r'_e = R_i$$

2. ***Output resistance.*** It is the resistance, looking into collector, and is approximately equal to the collector resistance (R_C). Mathematically the output resistance,

$$R_0 = R_C$$

3. ***Current gain.*** It is the ratio of collector current (i_c) to the base current (i_b). Mathematically, the current gain,

$$A_i = \frac{i_c}{i_b} = \beta$$

4. ***Voltage gain.*** It is the ratio of output voltage (v_0) to the input voltage (v_{in}). Since the output voltage is the same as collector voltage and input voltage is the same as base voltage, therefore

it is also known as voltage gain from base to collector. Mathematically the voltage gain,

$$A_v = \frac{v_0}{v_{in}}$$

We also know that the input voltage,

$$v_{in} = i_b \cdot \beta \cdot r'_e$$

and the output voltage,

$$v_0 = i_e \cdot R_C = \beta \cdot i_b \cdot R_C \qquad ...(\because (i_e = \beta \cdot i_b))$$

∴ Voltage gain,

$$A_v = \frac{\beta \cdot i_b \cdot R_C}{i_b \cdot \beta\, r'_e} = \frac{R_C}{r'_e}$$

5. ***Power gain.*** It is the product of current gain (A_i) and voltage gain (A_v). Mathematically, the power gain,

$$A_p = A_i \cdot A_v = \beta \cdot R_C / r'_e$$

and the power gain in decibels,

$$G_p = 10 \log_{10} A_p$$

Example 24.1.

Fig. 24.8 shows a common emitter amplifier with a base-bias arrangement.

Determine the values of (a) input resistance looking into the base; (b) input resistance of the stage; (c) output resistance; and (d) voltage gain.

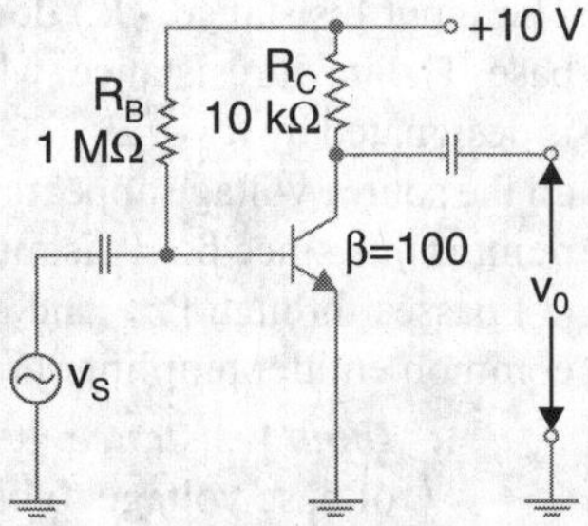

Fig. 24.8.

Solution.

Given: V_{CC} = 10 volts; R_C = 10 kΩ = 10 × 10^3 Ω; R_B = 1 MΩ = 1 × 10^6 Ω and β = 100.

Fig. 24.9 shows the d.c. equivalent circuit for the common emitter amplifier.

(*a*) *Input resistance looking into the base*

We know that the base current,

$$I_B = \frac{V_{CC} - V_{BE}}{R_B} = \frac{10 - 0.7}{1 \times 10^6} = 9.3 \times 10^{-6} \text{ A}$$

$$= 9.3\ \mu\text{A}$$

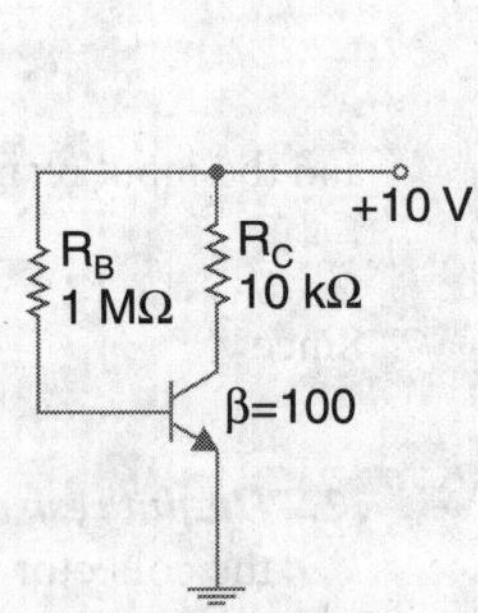

Fig. 24.9.

and the collector current,

$$I_C = \beta \cdot I_B = 100 \times (9.3 \times 10^{-6})$$

$$= 0.93 \times 10^{-3} \text{ A} = 0.93 \text{ mA}$$

∴ Emitter current,

$$I_E \approx I_C = 0.93 \text{ mA}$$

and a.c. resistance of emitter diode,

$$r'_e = \frac{25}{I_E} = \frac{25}{0.93} = 26.9\ \Omega$$

∴ Input resistance looking directly into the base,

$$R_i = \beta \cdot r'_e = 100 \times 26.9 = 2690\ \Omega = 2.69\ \text{k}\Omega \textbf{ Ans.}$$

(*b*) ***Stage input resistance***

We know that stage input resistance,

$$R_{is} = R_B \parallel (\beta \cdot r'_e) = (1 \times 10^6) \parallel (2.69 \times 10^3)$$

$$= \frac{(1 \times 10^6) \times (2.69 \times 10^3)}{(1 \times 10^6 + 2.69 \times 10^3)} = 2.69 \text{ k}\Omega \textbf{ Ans.}$$

(*c*) ***Output resistance***

We know that the output resistance,

$$R_0 = R_C = 10 \text{ k}\Omega \textbf{ Ans.}$$

(*d*) ***Voltage gain***

We also know that the voltage gain,

$$A_v = \frac{R_C}{r'_e} = \frac{10 \times 10^3}{26.9} = 372 \textbf{ Ans.}$$

Example 24.2.

A common emitter amplifier has an input resistance R_i = 2.5 kΩ and a voltage gain of 200. If the input signal voltage is 5 mV, find (a) the base current (b) the collector current (c) the power gain, and (d) dB power gain. Take β = 50.

Solution.

Given: $R_i = 2.5 \text{ k}\Omega = 2.5 \times 10^3\ \Omega$; $A_v = 200$; $V_s = 5 \text{ mV} = 5 \times 10^{-3}$ V and β = 50.

(*a*) ***Base current***

Let i_b = Base current.

We know that the input resistance (R_i),

$$2.5 \times 10^3 = \frac{V_s}{i_b} = \frac{5 \times 10^{-3}}{i_b}$$

$$\therefore \quad i_b = 2 \times 10^{-6} \ A = 2 \text{ mA} \textbf{ Ans.}$$

(*b*) ***Collector current***

We know that the collector current,

$$i_c = \beta \cdot i_b = 50 \times 2 = 100 \text{ mA} \textbf{ Ans.}$$

(*c*) ***Power gain***

We know that the current gain,

$$A_i = \beta = 50$$

and power gain,

$$A_p = A_i \cdot A_v = 50 \times 200 = 10{,}000 \textbf{ Ans.}$$

(*d*) ***dB Power gain***

We also know that *dB* power gain,

$$G_p = 10 \log_{10} A_p = 10 \log_{10} 10000$$

$$= 10 \log 10^4 = 40 \text{ dB. } \textbf{Ans.}$$

24.10. Effect of Emitter Feedback Resistor on Common Emitter Amplifier Parameters

Fig. 24.10 (*a*) shows a common emitter amplifier employing base bias with emitter feedback resistor. The d.c. equivalent circuit for the given amplifier circuit is shown in Fig. 24.10 (*b*). It is obtained from the amplifier circuit by disconnecting the capacitors.

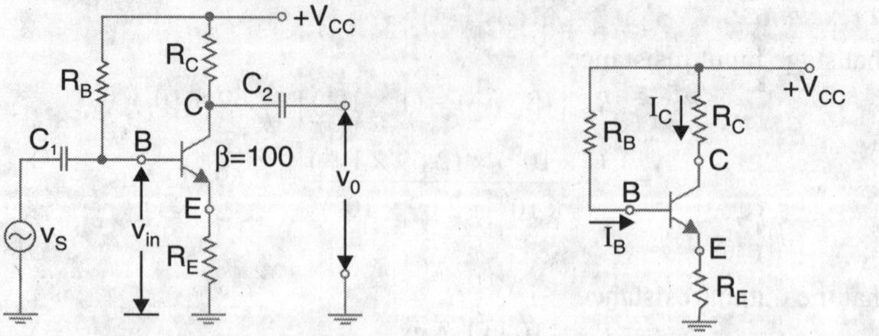

(*a*) Common emitter amplifier. (*b*) D.C. equivalent circuit.

Fig. 24.10.

We have already discussed in chapter on transistor biasing that for such a biasing circuit, the collector current is given by the relation,

$$I_C = \frac{V_{CC} - V_{BE}}{R_E + \frac{R_B}{\beta}} = \frac{V_{CC}}{R_E + \frac{R_E}{\beta}} \quad ...(\because V_{CC} >> V_{EE})$$

The a.c. equivalent circuit for the amplifier circuit is shown in Fig. 24.11. This circuit is obtained from the amplifier circuit by short circuiting the capacitors and replacing the d.c. supply by a ground.

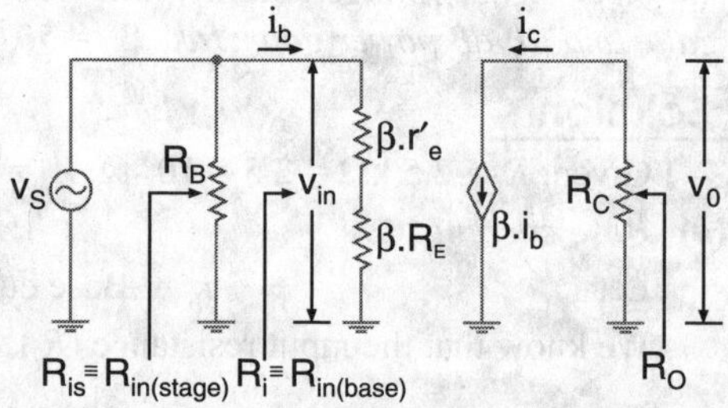

Fig. 24.11. A.C. equivalent circuit.

It may be noted that the emitter resistance, when referred to base circuit, becomes $\beta \cdot R_E$. Therefore, the total resistance in the base circuit is $\beta\,(r'_e + R_E)$

The input resistance looking into the base,

$$R_i = \beta\,(r'_e + R_E)$$

If $R_E >> r'_e$, then $R_i \approx \beta \cdot R_E$

The input resistance of amplifier stage (*i.e.*, the total input resistance as seen by the source),

$$R_{is} = R_B \,\|\, (\beta \cdot R_E) = R_B \,\|\, R_i$$

This indicates that input resistance of an amplifier increases with the addition of emitter resistance. Now we shall see the effect of emitter resistance on the voltage gain. As seen from Fig. 24.11, the input voltage,

$$v_{in} = i_b \cdot \beta\,(r'_e + R_E) = i_b \cdot \beta \cdot R_E$$

and the output voltage,

$$v_0 = i_c \cdot R_C = \beta \cdot i_b \cdot R_C$$

$\therefore$ Voltage gain,

$$A_v = \frac{v_o}{v_{in}} = \frac{\beta \cdot i_b \cdot R_C}{i_b \cdot \beta \cdot (r'_e + R_E)} = \frac{R_C}{r'_e + R_E}$$

$$= \frac{R_E}{R_B} \quad ...(\because R_E >> r'_e)$$

We know that voltage gain of a common emitter amplifier without emitter feedback resistor is given by the relation,

$$A_v = \frac{R_C}{r'_e}$$

Since the resistance R_E is much larger than a.c resistance r'_e, therefore, the addition of emitter resistance in the common emitter amplifier reduces the voltage gain. Incidently, it may be noted that the emitter feedback resistor has no effect on the output resistance.

Example 24.3.

A common emitter amplifier is shown in Fig. 24.12.

Determine the values of (a) Input resistance looking directly into the base: (b) Input resistance of the stage; (c) Voltage gain; and (d) dB voltage gain.

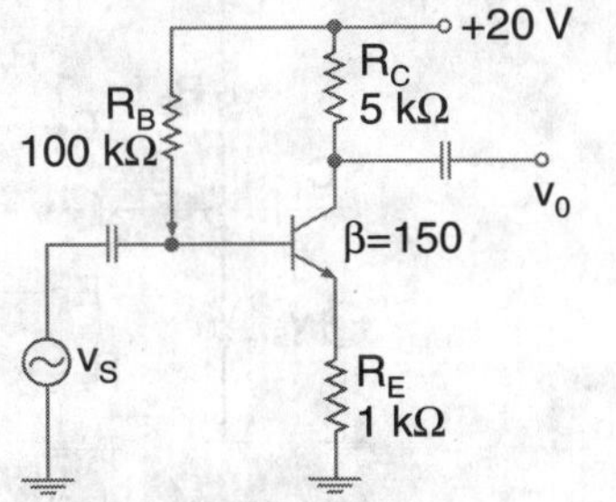

Fig. 24.12.

Solution.

Given: $V_{CC} = 20$ volts; $R_C = 5\ k\Omega = 5 \times 10^3\ \Omega$; $R_E = 1\ k\Omega$, $= 1 \times 10^3\ \Omega$; $R_B = 100\ \Omega = 100 \times 10^3\ \Omega$ and $\beta = 150$.

(a) Input resistance looking directly into the base

We know that the collector current,

$$I_C = \frac{V_{CC}}{R_E + \dfrac{R_B}{\beta}} = \frac{20}{1 \times 10^3 + \dfrac{100 \times 10^3}{150}} = 12 \text{ mA}$$

and emitter current,

$$I_E \approx I_C = 12 \text{ mA}$$

$$\therefore \qquad r'_e = \frac{25}{I_E} = \frac{25}{12} = 2.1\ \Omega$$

and input resistance looking directly into the case,

$$R_i = \beta\,(r'_e + R_E) = 150\,[2.1 + (1 \times 10^3)]\ \Omega = 150 \times 10^3\ \Omega = 150\ k\Omega \textbf{ Ans.}$$

(b) Input resistance of the stage

We know that input resistance of the stage,

$$R_{is} = R_B \,\|\, R_i = \frac{(R_B \cdot R_i)}{(R_B + R_i)} = \frac{(100 \times 150)}{(100 + 150)} = 60\ k\Omega \textbf{ Ans.}$$

(c) Voltage gain

We know that voltage gain,

$$A_v = \frac{R_C}{R_E} = \frac{5}{1} = 5 \textbf{ Ans.}$$

(d) dB Voltage gain

We also know that *dB* voltage gain,

$$G_p = 10 \log_{10} A_v = 10 \log_{10} 5 = 7 \text{ dB} \textbf{ Ans.}$$

24.11. Effect of Emitter Bypass Capacitor on Common Emitter Amplifier Parameters

Fig. 24.13 (*a*) shows a common emitter amplifier with a capacitor connected across the resistance R_E. This capacitor is called a **bypass capacitor*, because it connects the emitter terminal to ground for a.c. signal current. The emitter bypass capacitor acts as an open switch to the direct current.

* The action of a bypass capacitor depends upon its value of capacitance (*c*). Usually, a capacitor is selected with a large value. The advantage of this is that the reactance offered by the capacitor to a.c. current is much less than resistance R_E. Therefore, resistance R_E effectively appears as a short for a.c. currents.

Therefore it does not affect the d.c. bias of the circuit. The d.c. equivalent circuit for the given amplifier is as shown in Fig. 24.13 (*b*). It may be noted that it is the same circuit as shown in Fig. 24.10 (*b*).

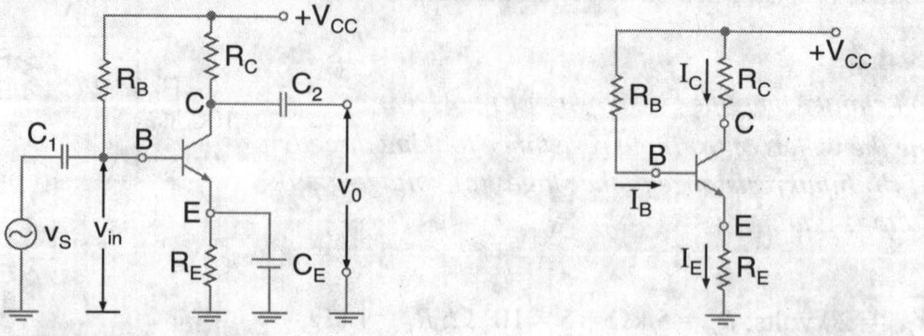

(*a*) Common emitter amplifier with emitter feedback. (*b*) D.C. equivalent circuit.

Fig. 24.13.

The a.c. equivalent circuit for the given amplifier is shown in Fig. 24.14. It may be noted, in this circuit, that input is directly looking into the base is $\beta \cdot r'_e$ only because resistance R_E is assumed to be shorted by the capacitor for a.c. signal current. Thus

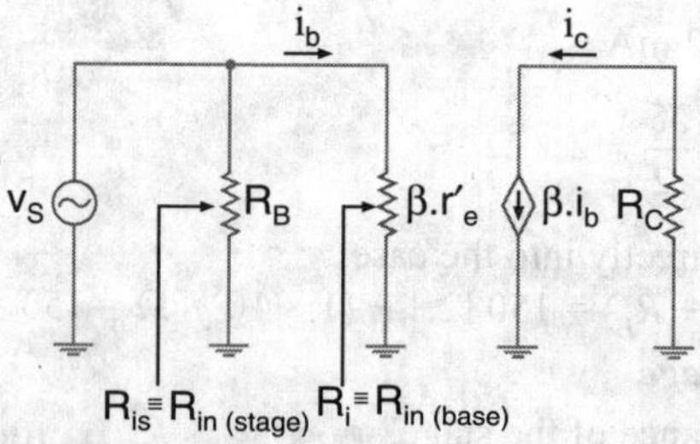

Fig. 24.14. A.C. equivalent circuit of common emitter amplifier with emitter bypass capacitor.

$$R_i = \beta \cdot r'_e \qquad \text{...}(i)$$

and $$R_{is} = R_B \,||\, (\beta \cdot r'_e) \qquad \text{...}(ii)$$

It is evident from equations (*i*) and (*ii*) that the input resistance is lower than that obtained without emitter bypass capacitor in the last article. It means that emitter bypass capacitor reduces the input resistance of a common emitter amplifier than that of an amplifier without emitter capacitor. Now let us study the effect of emitter bypass capacitor on the voltage gain. We know that the input voltage,

$$v_{in} = i_b \cdot \beta \cdot r'_e$$

and the output voltage,

$$v_0 = i_c \cdot R_C = \beta \cdot i_b \cdot R_C \qquad \text{...}(\because i_e = \beta \cdot i_b)$$

$\therefore$ Voltage gain,

$$A_v = \frac{v_0}{v_{in}} = \frac{\beta \cdot i_b \cdot R_C}{i_b \cdot \beta \cdot r'_e} = \frac{R_C}{r'_e}$$

The above value of voltage gain is the same as obtained for a common emitter amplifier without feedback resistor. However, the voltage gain with feedback resistor has been found to be equal to R_C/R_E. As the value of the resistance (R_E) is very large as compared to that of r'_e, therefore, we say that a common emitter amplifier with emitter bypass capacitor provides a higher voltage gain than that of an amplifier without emitter bypass capacitor.

Example 24.4.

A transistor used in a single-stage common emitter amplifier shown in Fig. 24.15 has β = 50. Determine (a) the input resistance looking directly into the base; (b) the total input resistance; (c) the output resistance; and (d) the voltage gain. What will be the voltage gain, if the emitter bypass capacitor is removed?

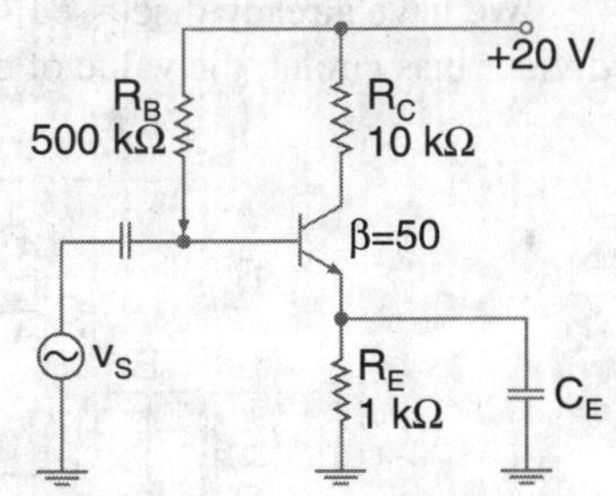

Fig. 24.15.

Solution.

Given: V_{CC} = 12 volts; R_C = 10 kΩ = 10 × 10^3 Ω; R_E = 1 kΩ = 1 × 10^3 Ω; R_B = 500 kΩ = 500 × 10^3 Ω and β = 50.

(a) Input resistance looking directly into the base

We know that collector current,

$$I_C = \frac{V_{CC}}{R_E + \frac{R_B}{\beta}} = \frac{12}{1 \times 10^3 + \frac{500 \times 10^3}{50}} \text{ A}$$

$$= 1.09 \times 10^{-3} \text{ A} = 1.09 \text{ mA}$$

$$\therefore \quad I_E \approx I_C = 1.09 \text{ mA}$$

and
$$r'_e = \frac{25}{I_E} = \frac{25}{1.09} = 22.9 \ \Omega$$

∴ Input resistance looking directly into the base,

$$R_i = \beta \cdot r'_e = 50 \times 22.9 = 1146 \ \Omega \textbf{ Ans.}$$

(b) Total input resistance

We know that total input resistance,

$$R_{is} = R_B \parallel R_i = (500 \times 10^3) \parallel 1146 = 1146 \ \Omega$$

(c) Output resistance

We know that output resistance,

$$R_0 = R_C = 10 \text{ k}\Omega \textbf{ Ans.}$$

(d) Voltage gain

We know that the voltage gain,

$$A_v = \frac{R_C}{r'_e} = \frac{(10 \times 10^3)}{22.9} = 436.7 \textbf{ Ans.}$$

Voltage gain if the emitter bypass capacitor is removed

We also know that voltage gain, if the emitter bypass capacitor is removed,

$$A_v = \frac{R_C}{R_E} = \frac{10}{1} = 10 \textbf{ Ans.}$$

24.12. Effect of A.C. Load on Common Emitter Amplifier Parameters

Fig. 24.16 (*a*) shows a common emitter amplifier with a voltage divider-bias and connected to an external load resistance (R_L). The external load may be a loud speaker, an electric motor or the next stage of an amplifier. The voltage divider bias has been selected to illustrate change in a.c. equivalent circuit from base-bias of the amplifier. The d.c. equivalent circuit for the given amplifier is as shown in Fig. 24.16 (*b*). This has been obtained by opening (or disconnecting) all the capacitors.

We have already discussed in chapter on transistor biasing and stabilization that for a voltage divider bias circuit, the value of emitter current (or collector current) is given by the relation,

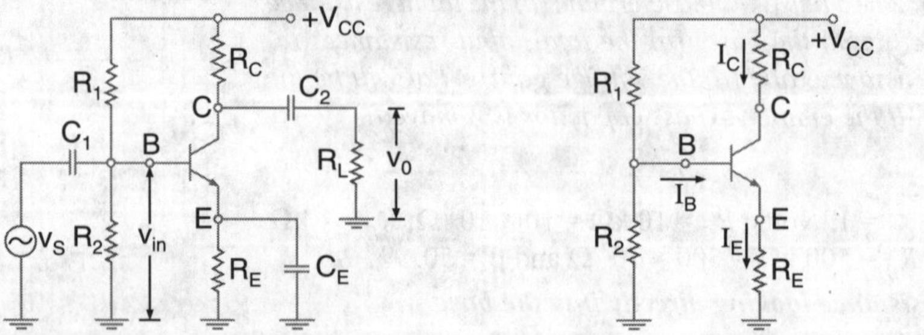

(*a*) Common emitter amplifier with load resistance.

(*b*) D.C. equivalent circuit.

Fig. 24.16.

$$I_E = \frac{V_{TH} - V_{BE}}{R_E + \dfrac{R_{TH}}{\beta}} = \frac{V_{TH} - V_{BE}}{R_E} \qquad ...(\because R_E >> R_{TH}/\beta)$$

where V_{TH} = Thevenin's Voltage and its value is equal to $V_{CC}\left(\dfrac{R_2}{R_1 + R_2}\right)$ and

R_{TH} = Thevenin's Equivalent Resistance and its value is equal to $\dfrac{(R_1 \cdot R_2)}{(R_1 + R_2)}$

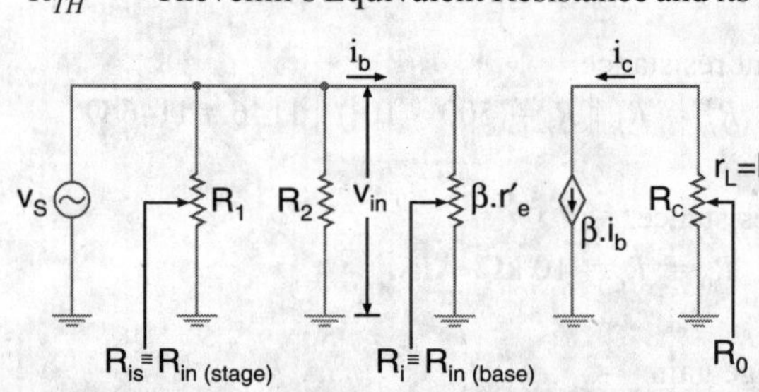

Fig. 24.17. A.C. equivalent circuit of a common emitter amplifier with load resistance.

The a.c. equivalent circuit of the given amplifier is shown in Fig. 24.17. It has been obtained by replacing the d.c. supply by ground and capacitor by a short-circuit. It may be noted that the input resistance directly looking into the base,

$$R_i = \beta \cdot r'_e$$

The total input resistance as seen by the source (*i.e.*, the input resistance of the amplifier stage),

$$R_{is} = (R_1 \parallel R_2) \parallel (\beta \cdot r'_e)$$

Now let us discuss the effect of an a.c. load on the voltage gain. We have already discussed in Art 24.5 that the a.c. load (or effective load) as seen by the amplifier, at its output, is a parallel combination of R_C and R_L. Thus a.c. load,

$$r_L = R_C \parallel R_L$$

and input voltage,

$$v_{in} = i_b \cdot \beta \cdot r'_e$$

Output voltage,

$$v_0 = i_c \cdot r_L = \beta \cdot i_b \cdot r_L$$

and voltage gain,

$$A_v = \frac{v_0}{v_{in}} = \frac{\beta \cdot i_b \cdot r_L}{i_b \cdot \beta \cdot r'_c} = \frac{r_L}{r'_e}$$

Since the value of a.c. load is smaller than the value of collector resistance, therefore, the addition of load resistance reduces the voltage gain of common emitter amplifier. However, if $R_L >> R_C$, then the a.c. load,

$$r_L \approx R_C$$

and the voltage gain,

$$A_v = \frac{R_C}{r'_e}$$

It means that if the load resistance is very high, it does not effect the voltage gain of the amplifier. It may be noted that the a.c. load (or effective load) seen by the amplifier is r_L. Therefore the output resistance of an amplifier,

$$R_0 = r_L$$

Example 24.5.

Draw the a.c. equivalent circuit for the common emitter amplifier shown in Fig. 24.18. Given β = 200, V_{BE} = 0.7 V and a.c. resistance of the emitter diode is of the form,

$$r'_e = \frac{25}{\text{D.C emitter current in mA}} \quad (\Omega)$$

Also find out the a.c. output voltage and input impedance for this circuit.

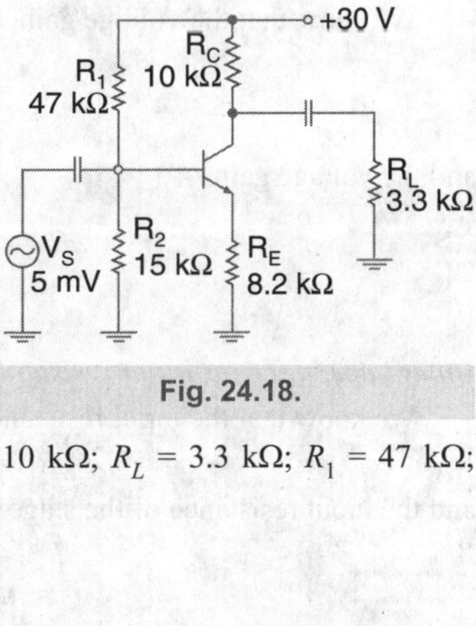

Fig. 24.18.

Solution.

Given: β = 200; V_{BE} = 0.7 volts; V_{CC} = 30 volts; R_C = 10 kΩ; R_L = 3.3 kΩ; R_1 = 47 kΩ; R_2 = 15 kΩ; R_E = 8.2 kΩ and V_S = 5 mV.

A.C. equivalent circuit

We know that the Thevenin's Voltage,

$$V_{TH} = V_{CC}\left(\frac{R_2}{R_1 + R_2}\right) = 30 \times \left(\frac{15}{47 + 15}\right) = 7.26 \text{ V}$$

and the equivalent resistance,

$$R_{TH} = R_1 \parallel R_2 = \frac{R_1 \cdot R_2}{R_1 + R_2} = \frac{47 \times 15}{47 + 15} = 11.4 \text{ k}\Omega = 11.4 \times 10^3 \ \Omega$$

∴ Value of the emitter current,

$$I_E = \frac{V_{TH} - V_{BE}}{R_E + \dfrac{R_{TH}}{\beta}} = \frac{7.26 - 0.7}{8.2 \times 10^3 + \dfrac{11.4 \times 10^3}{200}} = 0.794 \text{ mA} = 0.8 \text{ mA}$$

and the value of a.c. resistance of the emitter diode,

$$r'_e = \frac{25}{(I_E)(\text{in } mA)} = \frac{25}{0.794} = 31.5 \ \Omega$$

We also know that the a.c. load seen by the amplifier,

$$r_L = R_C \| R_L = \frac{R_C \cdot R_L}{R_C + R_L} = \frac{10 \times 3.3}{10 + 3.3} = 2.5 \text{ k}\Omega$$

Now let us draw the a.c. equivalent circuit for the given common emitter amplifier as shown in Fig. 24.19. **Ans.**

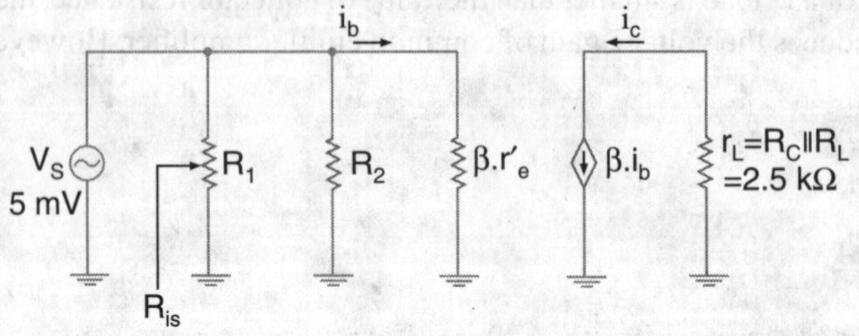

Fig. 24.19.

Output voltage

Let v_0 = Output voltage,

We know that the voltage gain,

$$A_v = \frac{r_L}{r'_e} = \frac{2.5 \times 10^3}{31.5} \approx 79 \textbf{ Ans.}$$

and the voltage gain (A_v)

$$79 = \frac{v_0}{v_{in}} = \frac{v_0}{5}$$

$$\therefore \quad v_0 = 79 \times 5 = 395 \text{ mV } \textbf{Ans.}$$

Input impedance (or input resistance)

We know that the input resistance looking directly into the base,

$$R_i = \beta \cdot r'_e = 200 \times 31.5 = 6.3 \times 10^3 \ \Omega = 6.3 \text{ k}\Omega$$

and the input resistance of the stage (*i.e.*, total input resistance) of the circuit,

$$R_{is} = (R_1 \| R_2) \| R_i = R_{TH} \| R_i$$
$$= 11.4 \| 6.3 = (11.4 \times 6.3) / (11.4 + 6.3)$$
$$= 4 \text{ k}\Omega \textbf{ Ans.}$$

24.13. Effect of Signal Source Resistance on Common Emitter Amplifier Parameters

Fig. 24.20 shows a common emitter amplifier connected to a signal source. The signal source has an internal resistance designated as R_S. If the internal resistance of the source is zero, then all the input signal voltage appears at the base terminal. However, if the a.c. source has a non-zero internal resistance, then the actual signal voltage at the base will be less by a factor equal to the voltage drop across resistance R_S. Thus the actual signal voltage,

$$v_{in} = v_S = \text{Drop across the resistance } R_S$$

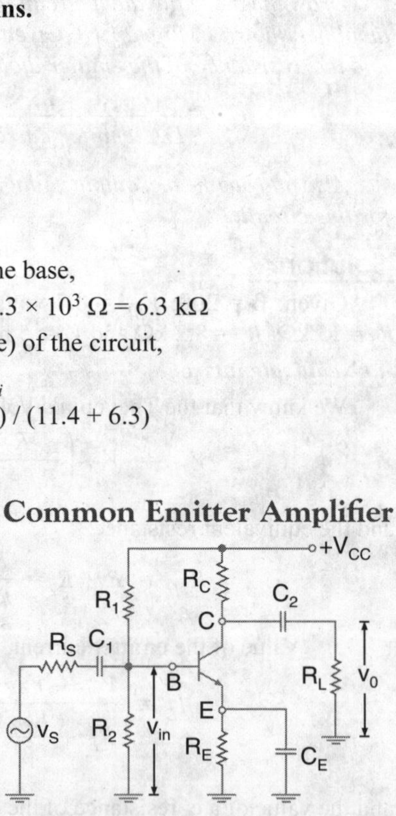

Fig. 24.20. Common emitter amplifier with signal source resistance.

It is evident from the above relation, that greater the internal resistance of an a.c. signal source, greater the voltage drop across the resistance R_S and hence lesser the input voltage.

The a.c. equivalent circuit of the given amplifier is shown in Fig. 24.21. The total input resistance of the amplifier as seen by the signal source,

$$R_{is} = (R_1 \parallel R_{2)} \parallel (\beta \cdot r'_{e)}$$

It may be noted from the a.c. equivalent circuit that the signal source resistance (R_S) is in series with the total input resistance. Therefore, the a.c. source voltage is divided between resistances R_S and R_{is}. The voltage at the base is given by the relation,

$$v_{in} = \left(\frac{R_{is}}{R_S + R_{is}}\right) v_S$$

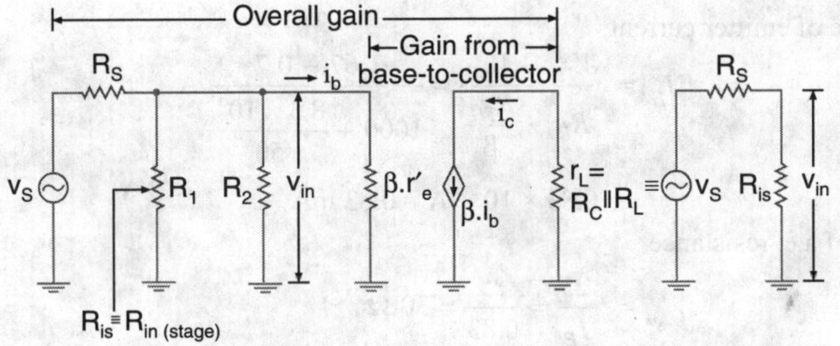

Fig. 24.21. A.C. equivalent circuit.

If $R_S << R_{is}$, then the voltage at the base,

$$v_{in} \approx v_S$$

We know that voltage gain from base to collector (or output) is given the relation,

$$A_v = \frac{r_L}{r'_e} = \frac{R_C \parallel R_L}{r'_e}$$

The overall voltage gain (called stage gain) of an amplifier is given by the relation,

$$A_{vs} = A_v \cdot \frac{v_{in}}{v_S}$$

Example 24.6.

For the amplifier shown in Fig. 24.22, determine (a) output voltage and (b) the overall voltage gain,

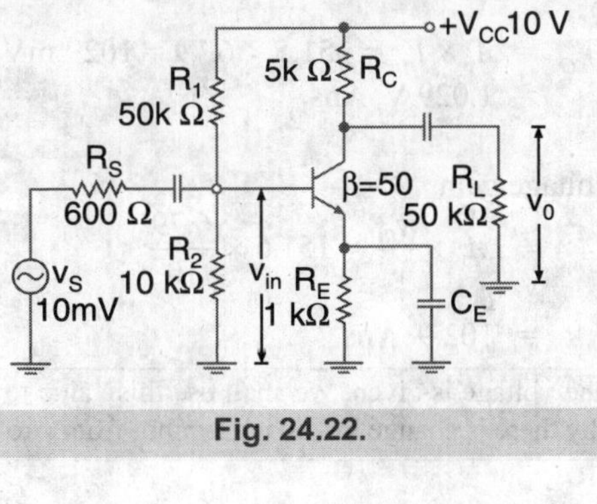

Fig. 24.22.

Solution.

Given: V_{CC} = 10 volts; R_C = 5 kΩ; R_E = 1 kΩ; R_1 = 50 kΩ; R_2 = 10 kΩ; R_S = 600 Ω; R_L = 50 kΩ; β = 50 and *V_S = 10 mV.

(a) Output voltage

We know that Thevenin's Voltage,

$$V_{TH} = V_{CC}\left(\frac{R_2}{R_1 + R_2}\right) = 10 \times \left(\frac{10}{50 + 10}\right) = 1.67 \text{ V}$$

and the Thevenin's Resistance,

$$R_{TH} = R_1 \parallel R_2 = (R_1 \cdot R_2)/(R_1 + R_2)$$
$$= (50 \times 10)/(50 + 10) = 8.3 \text{ k}\Omega = 8.3 \times 10^3 \ \Omega$$

∴ Value of emitter current,

$$I_E = \frac{V_{TH} - V_{BE}}{R_E + \dfrac{R_{TH}}{\beta}} = \frac{1.67 - 0.7}{1000 + \dfrac{8.3 \times 10^3}{50}} \text{ A}$$
$$= 0.83 \times 10^{-3} \text{ A} = 0.83 \text{ mA}$$

and the value of a.c. resistance,

$$r'_e = \frac{25}{I_E} = \frac{25}{0.83} = 30 \ \Omega$$

We also know that the total input resistance,

$$R_{is} = (R_1 \parallel R_2) \parallel (\beta \cdot r'_e) = R_{TH} \parallel (\beta \cdot r'_e)$$
$$= 8.3 \times 10^3 \parallel (50 \times 30) = 1270 \ \Omega$$

and the value of a.c. load resistance,

$$r_L = R_C \parallel R_L = 5 \parallel 50 = 4.545 \text{ k}\Omega$$
$$= 4.545 \times 10^3 \ \Omega$$

∴ Voltage gain,

$$A_v = \frac{r_L}{r'_e} = \frac{4.545 \times 10^3}{30} = 151.5$$

and the input voltage,

$$V_{in} = V_S\left(\frac{R_{is}}{R_{is} + R_s}\right) = 10\left(\frac{1270}{1270 + 600}\right)$$
$$= 6.79 \text{ mV}$$

Now the output voltage,

$$V_O = A_v \times V_{in} = 151.5 \times 6.79 = 1029 \text{ mV}$$
$$= 1.029 \text{ V } \textbf{Ans.}$$

(b) Overall voltage gain

We know that overall voltage gain,

$$A_{vs} = A_v \times \frac{v_{in}}{v_s} = 151.5 \times \frac{6.79}{10}$$
$$= 102.9 \ \textbf{Ans.}$$

* Here the r.m.s. value of the voltage is given. We shall use this value to calculate the r.m.s. value of the output voltage. That is why there is change in voltage symbol from *v* to V.

Example 24.7.

Determine the voltage given (V_o/V_s) for the amplifier circuit shown in Fig. 24.23.

(UPSC Engg. Services 2000)

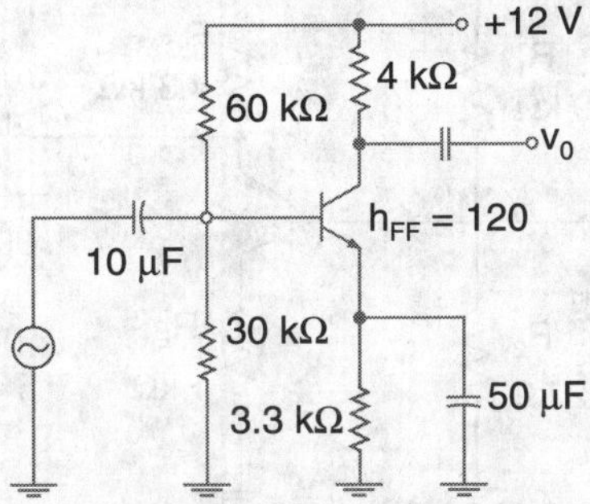

Fig. 24.23.

Solution.

Given: $V_{CC} = 12$ V; $R_1 = 60$ kΩ; $R_2 = 30$ kΩ; $R_C = 4$ kΩ; $R_E = 3.3$ kΩ.

We know that the Thevevin's equivalent voltage at the base,

$$V_{TH} = V_{CC} \times \left(\frac{R_2}{R_1 + R_2}\right) = 12 \times \left(\frac{30k}{60k + 30k}\right) = 4 \text{ V}$$

and the Thevenin's equivalent resistance,

$$R_{TH} = 60\,k \parallel 30\,k = \frac{60k \times 30k}{60k + 30k} = 20 \text{ k}\Omega$$

and the emitter current,

$$I_E = \frac{V_{TH} - V_{BE}}{R_E + \dfrac{R_{TH}}{\beta}} = \frac{4 - 0.7}{(3.3 \times 10^{-3}) + \left(\dfrac{20 \times 10^3}{120}\right)} = \frac{3.3}{3.5 \times 10^3} = 0.943 \times 10^{-3} \text{ A.}$$

$$= 0.949 \text{ mA}$$

We know that the resistance,

$$r'_e = \frac{25\,mV}{I_E\,(mA)} = \frac{25}{0.949} = 26.34\ \Omega$$

and the a.c. load,

$$r_L = R_C = 4 \text{ k}\Omega = 4 \times 10^3\ \Omega.$$

∴ The voltage gain,

$$A_V = \frac{R_C}{r'_e} = \frac{4 \times 10^3}{26.34} = 151.84\ \Omega \textbf{ Ans.}$$

Example 24.8.

For a single-style transistor amplifier shown in Fig. 24.24, find the voltage gain A_v, input impedance Z_{in} and V_{CE}. Take $\beta = 200$ and $r_e = 30\ mV/\ I_E$.

(U.P.S.C. Engg. Services 2002)

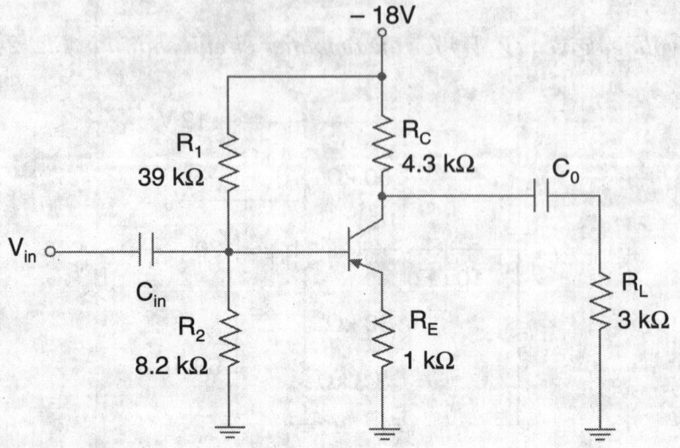

Fig. 24.24.

Solution.

Given : $V_{CC} = -18\ V$, $R_C = 4.3\ k\ \Omega$; $R_E = 1\ k\ \Omega$; $R_1 = 39\ k\Omega$, $R_2 = 8.2\ k\Omega$; $R_L = 3\ k\Omega$; $\beta = 200$.

We know that Thevenin's voltage,

$$V_{TH} = V_{CC}\left(\frac{R_2}{R_1 + R_2}\right) = -18 \times \frac{8.2\ k}{39\ k + 8.2\ k}$$

and the Thevenin's resistance,

$$R_{TH} = R_1 \| R_2 = (R_1 \cdot R_2) / R_1 + R_2$$

$$= (39\ \text{k} \times 8.2\ \text{k}) / (39\ \text{k} + 8.2\ \text{k}) = 6.78\ \text{k}\ \Omega$$

∴ Value of emitter current,

$$I_C = \frac{V_{TH} - V_{BE}}{R_E + \dfrac{R_{TH}}{\beta}} = \frac{-3.13 - (-0.7)}{1000 + \dfrac{6780}{200}}$$

$$= -2.35\ \text{mA}$$

and value of a.c. resistance,

$$r'_e = \frac{30\ \text{mV}}{I_E} = \frac{30\ \text{mV}}{2.35\ \text{mA}} = 12.77\ \Omega$$

We also know that the total input resistance, (or impedence),

$$R_{is} = (R_1 \| R_2) \| (\beta \cdot r'_e) = R_{TH} \| (\beta . r'_e)$$

$$= (6.78\ \text{k}) \| (200 \times 12.77\ \Omega)$$

$$= 6.78\ \text{k} \| 2.55\ \text{k}$$

$$= 1.85\ \text{k}\Omega$$

and the value of a.c. load resistance,

$$r_e = R_C \| R_2 = 4.3 \| 8.2 = 1.77\ \text{k}\ \Omega$$

$$= 1.77 \times 10^3\ \Omega$$

∴ Voltage gain,

$$A_v = \frac{r_L}{r_e'} = \frac{1.77 \times 10^3}{12.77} = 138.6\ \textbf{Ans.}$$

Example 24.9.

For a self-biased circuit show in Fig. 24.25. Calculate A_V, R_i and R_o. Also calculate A_{VS}.

(Nagpur Univ. Winter 2003)

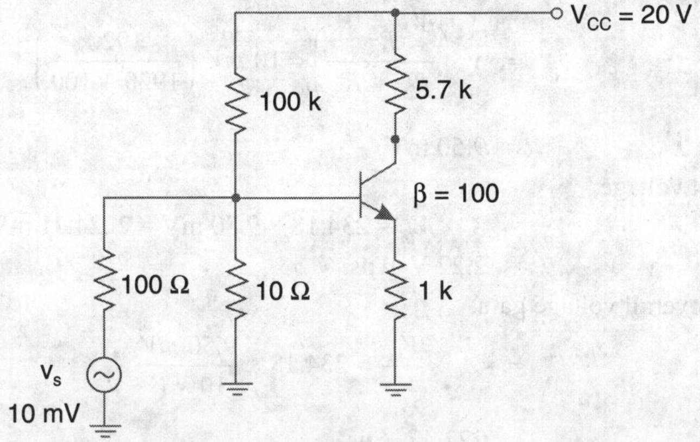

Fig. 24.25.

Solution.

Given : $V_{CC} = 20$ V; $R_C = 5.7$ kΩ; $R_E = 1$ k Ω; $R_1 = 100$ kΩ, $R_2 = 10$ kΩ; $R_S = 100$ Ω; β = 100

We know that Thevenin's voltage,

$$V_{TH} = V_{CC}\left(\frac{R_2}{R_1 + R_2}\right) = 20 \times \frac{10\,k}{100\,k + 10\,k} = 1.82 \text{ V}$$

and the Thevenin's resistance,

$$R_{TH} = R_1 \parallel R_2 = (R_1 \cdot R_2) / (R_1 + R_2)$$
$$= (100 \text{ k} \times 10 \text{ k})\,(100 \text{ k} + 10 \text{ k}) = 9.09 \text{ k}\Omega$$

∴ Voltage of emitter current,

$$I_E = \frac{V_{TH} - V_{BE}}{R_E + \dfrac{R_{TH}}{\beta}} = \frac{1.82 - 0.7}{1000 + \dfrac{9.09 \times 10^3}{100}} = 1.027 \text{ mA}$$

and the value of a.c. resistance,

$$r'_e = \frac{25}{I_E} = \frac{25}{1{\cdot}027} = 24.34\ \Omega$$

We also known that the total input resistance,

$$R_{is} = (R_1 \parallel R_2) \parallel (\beta . r'_e) = R_{TH} \parallel (\beta\, r'_e)$$
$$= (9.09 \text{ k}) \parallel (100 \times 24.34)$$
$$= (9.09 \text{ k}) \parallel (2.434 \text{ k})$$
$$= 1.92 \text{ k}\Omega$$
$$= 1920\ \Omega$$

and the value of ac load resistance,

$$r_L = R_C = 5.7 \text{ k}\Omega = 5700\ \Omega$$

∴ Voltage gain,

$$A_v = \frac{r_L}{r'_e} = \frac{5700}{2434} = 234.18 \textbf{ Ans.}$$

and the input voltage,

$$v_{in} = v_s \left(\frac{R_{is}}{R_{is} + R_S}\right) = 10\ mV \left(\frac{1920}{1920 + 100}\right)$$

$$= 9.50 \text{ mV}$$

Now the output voltage,

$$V_o = A_v \times V_{in} = 234.18 \times 9.50 \text{ mV} = 2224.71 \text{ mV}$$
$$= 2.22 \text{ V } \textbf{Ans.}$$

We know that overall voltage gain,

$$A_{vs} = A_v \times \frac{V_{in}}{V_S} = 234.18 \times \frac{9.50\ mV}{10\ mV}$$

$$= 222{\cdot}47 \textbf{ Ans.}$$

Example 24.10.

For a single-stage transistor amplifier circuit shown in Fig. 24.26, find the voltage gain A_v the input impedance R_{in} and V_{CE}. Take $\beta = 200$ and $r_e = 30\ mV/I_E$. (*UPSC Engg. Services 2002*)

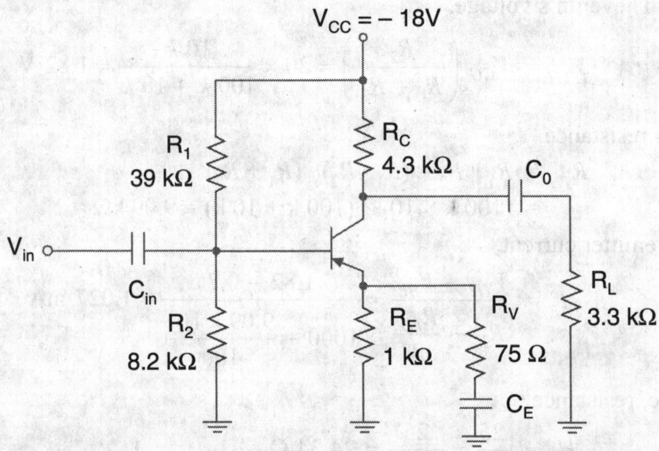

Fig. 24.26.

Solution.

Given: $V_{CC} = -18$ V; $R_C = 4.3$ kΩ $= 4.3 \times 10^3$ Ω; $R_1 = 39$ kΩ $= 39 \times 10^3$ Ω; $R_2 = 8.2$ kΩ $= 8.2 \times 10^3$ Ω; $R_E = 1$ kΩ $= 1 \times 10^3$ Ω; $R_V = 75$ Ω; $r'_e (= r_e) = 30$ mV/I_E; β = 200.

We know that Thevenin's voltage,

$$V_{TH} = V_{CC}\left(\frac{R_2}{R_1 + R_2}\right) = -18 \times \left(\frac{8.2}{39 + 8.2}\right) = -3.13 \text{ V}$$

and the Thevenin's resistance,

$$R_{TH} = R_1 \parallel R_2 = R_1 \cdot R_2 / (R_1 + R_2)$$
$$= (39 \times 8.2) / (39 + 8.2) = 6.78 \text{ kΩ} = 6.78 \times 10^3 \text{ Ω}.$$

∴ Value of enitter current,

$$I_E = \frac{V_{TH} - V_{BE}}{R_E + R_{TH}} = \frac{-3.13 - (-0.7)}{1000 + \dfrac{6.78 \times 10^3}{200}}$$

$$= -2.35 \times 10^{-3} \text{ A} = -2.35 \text{ mA}$$

We know that collector-to-enitter voltage,

$$V_{CE} = V_{CC} - I_C (R_C + R_E) \approx V_{CC} - I_E (R_C + R_E)$$

$$= -18 - (-2.35 \times 10^{-3}) \{(4.3 \times 10^3) + (1 \times 10^3)\}$$

$$= -19 + 12.455 = -5.545 \text{ V} \quad \textbf{Ans.}$$

and the value of a.c. resistance,

$$r'_e = \frac{30 \text{ mV}}{I_E} = \frac{30}{2.35} = 12.77 \ \Omega$$

We also know that the total input resistance or impedance,

$$R_{is} = (R_1 \parallel R_2) \parallel (\beta \cdot r'_e + R_v) = R_{TH} \parallel (\beta \cdot r'_e + R_v)$$

$$= (6.78 \times 10^3) \parallel (200 \times 12.77 + 75)$$

$$= (6.78 \times 10^3) \parallel 2629 = 1894.4 \ \Omega \text{ or } 1.894 \text{ k}\Omega \quad \textbf{Ans.}$$

and the value of a.c. load resistance,

$$r_L = R_C \parallel R_L = 4.3 \parallel 3.3 = 1.867 \text{ k}\Omega$$

$$= 1.867 \times 10^3 \ \Omega$$

∴ Voltage given

$$A_v = \frac{r_L}{r'_e + R_v} = \frac{1.894 \times 10^3}{12.77 + 75} = 21.56 \quad \textbf{Ans.}$$

Note: It is evident from the example that the addition of 75 Ω resistance has contributed to the input resistance as well as the gain. If you compare the total input resistance and voltage gain of this example with the Example 24.8, it is clear that the addition of 75 resistance has increased the input reistance by a small value (1.894 kΩ – 1.85 kΩ ≈ 0.044 kΩ) but reduced the voltage gain drastically to 21.56 (instead of 138.6).

24.14. Gain Stability in Common Emitter Amplifier

We have already discussed that the voltage gain of a common emitter amplifier is given by the ratio of r_L/r'_e, where r_L is the a.c. load resistance and r'_e is the emitter diode a.c. resistance. It indicates that the voltage given depends upon to value of a.c resistance r'_e. We have also discussed that the a.c. resistance depends upon the temperature. Thus as the temperature increases, the a.c. resistance also increases and the voltage gain decreases. The gain variations may be acceptable in some applications, like a transistor radio receiver where we can adjust the volume control. But there are many applications where we need as stable a voltage gain as possible. This can be achieved by inserting a resistor (r_E) in series with the emitter as shown in Fig. 24.27. Now the emitter is no longer at the a.c. ground. Because of this, a.c. current flows through resistor (r_E) and produces voltage at the emitter. If the value of resistor (r_E) is much greater than a.c. resistance (r'_e), almost all the a.c. input signal appears at the emitter. Therefore the emitter is said to be bootstrapped for alternating current as well as direct current. Because of this, the d.c. emitter voltage is within 0.7 V of the d.c. base voltage, and the a.c. emitter voltage is approximately equal to the a.c. base voltage.

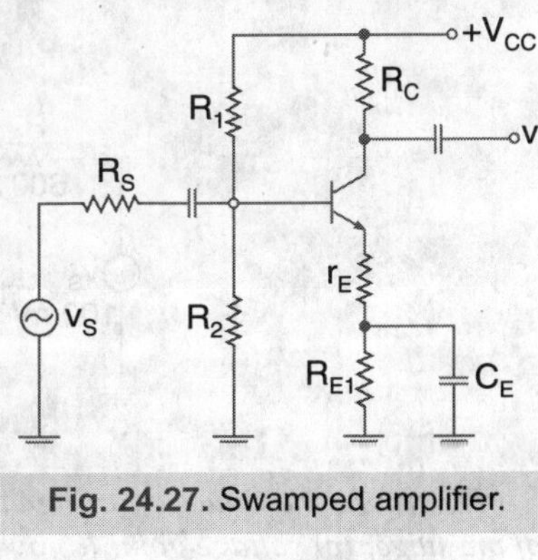

Fig. 24.27. Swamped amplifier.

The voltage gain for this amplifier circuit is given by the relation,

$$A_v = \frac{R_C}{r'_e + r_E}$$

If $r_E >> r'_e$, then,

$$A_v = R_C/r_E$$

It means that if the value of resistor r_E is made much larger than that of a.c. resistance (r'_e), the voltage gain is independent of r'_e. Therefore any variation in r'_e due to temperature change will have, practically, no effect on the voltage gain and hence the gain is stable. In other words, the resistor r_E has swamped the variations in r'_e. Because of this, the circuit shown in Fig. 24.27 is called swamped amplifier.

24.15. A.C. Equivalent Circuit of a Swamped Amplifier

Fig. 24.28 shows an a.c. equivalent circuit of a swamped amplifier. The input resistance of a swamped amplifier, when looking directly into the base, is given by the relation,

$$R_i = \beta\,(r_E + r'_e)$$

It is evident from the above relation, that the addition of resistor (r_E) increases the input resistance of the amplifier as compared to the input resistance of the unswamped amplifier (whose value is equal to $\beta \cdot r'_e$).

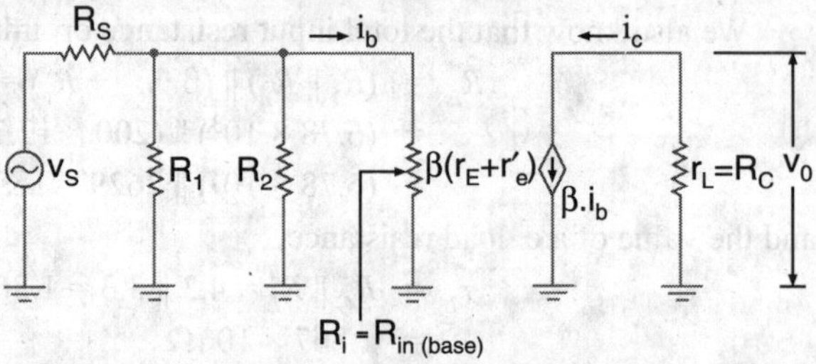

Fig. 24.28. A.C. equivalent circuit of a swamped amplifier.

24.16. Advantages of a Swamped Amplifier

Following are some of the important advantages of a swamped amplifier over an unswamped common emitter amplifier:

1. Stable voltage gain.
2. Higher input resistance.
3. Produces less distortion in the output signal.

Example 24.11.

Fig. 24.29 shows the circuit of a swamped amplifier.

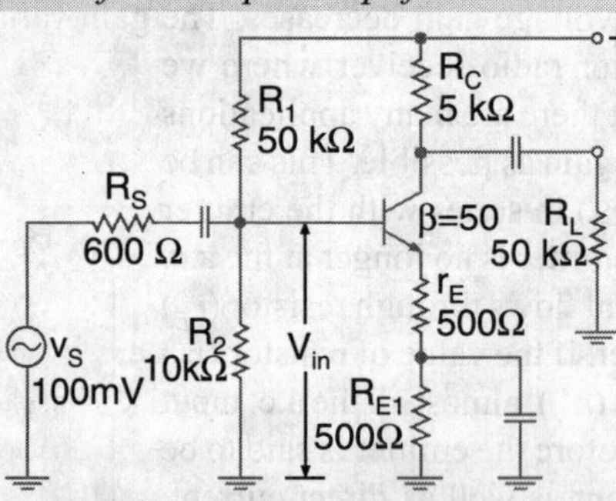

Fig. 24.29.

Determine the values of (a) input resistance looking directly into the base; (b) input resistance of the stage; (c) voltage gain; (d) overall voltage gain; and (e) the output voltage.

Solution.

Given: $V_{CC} = 10$ volt; $R_C = 5$ kΩ; $R_L = 50$ kΩ; $\beta = 50$; $R_1 = 50$ kΩ; $R_2 = 10$ kΩ; $r_E = 500$ Ω; $R_{E1} = 500$ Ω; $R_S = 600$ Ω and $V_S = 100$ mV.

(a) *Input resistance looking directly into the base*

We know that the Thevenin's Voltage,

$$V_{TH} = V_{CC}\left(\frac{R_2}{R_1 + R_2}\right) = 10 \times \left(\frac{10}{50+10}\right) \text{V} = 1.67 \text{ V}$$

and the value of Thevenin's Equivalent Resistance,

$$R_{TH} = R_1 \parallel R_2 = 50 \parallel 50 = 8.3 \text{ k}\Omega = 8.3 \times 10^3 \ \Omega$$

We also know that emitter resistance,

$$R_E = R_{E1} + r_E = 500 + 500 = 1000 \ \Omega$$

∴ Value of emitter current,

$$I_E = \frac{V_{TH} - V_{BE}}{R_E + \frac{R_{TH}}{\beta}} = \frac{1.67 - 0.7}{1000 + \frac{8.3 \times 10^3}{50}} = 0.83 \times 10^{-3} \text{ A} = 0.83 \text{ mA}$$

and the value of a.c. resistance,

$$r'_e = \frac{25}{I_E} = \frac{25}{0.83} = 30 \ \Omega$$

We also know that the input resistance looking directly into the base,

$$R_i = \beta\,(r_E + r'_e) = 50\,(500 + 30) = 26.5 \times 10^3 \ \Omega = 26.5 \text{ k}\Omega \textbf{ Ans.}$$

(b) *Input resistance of the stage*

We know that input resistance of the stage,

$$R_{is} = (R_1 \parallel R_{2)} \parallel [\beta\,(r_E + r'_e)]$$
$$= 8.3 \parallel 26.5 = 6.32 \text{ k}\Omega = 6320 \ \Omega \textbf{ Ans.}$$

(c) *Voltage gain*

We know that the value of a.c. load resistance,

$$r_L = R_C \parallel R_L = 5 \parallel 50 = 4.55 \text{ k}\Omega = 4.55 \times 10^3 \ \Omega$$

∴ Voltage gain

$$A_v = \frac{r_L}{r_E + r'_e} = \frac{4.55 \times 10^3}{500 + 30} = 8.58 \textbf{ Ans.}$$

(d) *Overall voltage gain*

We know that the ratio of base voltage to the source voltage,

$$\frac{v_{in}}{v_s} = \frac{R_{is}}{R_s + R_{is}} = \frac{6320}{600 + 6400} = 0.913$$

∴ Overall voltage gain,

$$A_{vs} = A_v \times \frac{v_{in}}{v_s} = 8.58 \times 0.913 = 7.83 \textbf{ Ans.}$$

(e) *Output voltage*

We also know that the value of output voltage,

$$V_o = A_{vs} \times V_s = 7.83 \times 100 = 7.83 \text{ mV} \textbf{ Ans.}$$

24.17. Characteristics and uses of Common Emitter Amplifier

A common emitter amplifier has the following important characteristics:

1. Its input resistance is in the range of 1 kΩ to 2 kΩ, which is considered to be moderately low.
2. Its output resistance is about 50 kW and is considered to be moderately large.
3. Its current gain (β) is very high and is in the range of 50 to 300.
4. Its voltage gain is of the order of 1500.
5. It produces very large power gain and is of the order of 10,000 (equal to 40 dB) or so.
6. It produces phase reversal of the input signal *i.e.*, output is 180° out of phase with respect to the input signal.

The common emitter amplifier is the most widely used amplifier in actual practice because of its large voltage and power gains. In addition to this, its input and output resistances are suitable for most of the applications.

24.18. Common Base Amplifier

Fig. 24.30 shows the circuits of a common-base (CB) amplifier using NPN transistor. Since the base is grounded, therefore, this circuit is also called a grounded-base amplifier. In this circuit, the input is applied across the emitter-base terminals and the output is taken across collector-base terminals. The emitter-base junction is forward biased by V_{EE} supply and the collector-base junction is reverse biased by V_{CC} supply. The capacitors C_1 and C_2 are called blocking capacitors (or coupling capacitors). These capacitors block the direct current, but pass the alternating current. Therefore, they are used to isolate d.c. biase from a.c. signal current in the circuit.

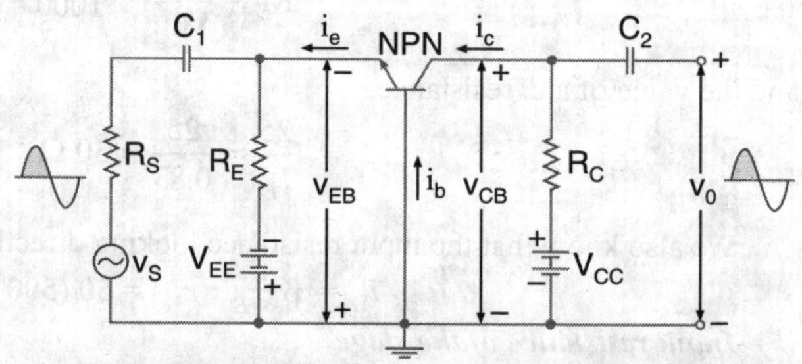

Fig. 24.30. Common-base transistor.

The *Q*-point or d.c. working conditions in the amplifier circuit are obtained by the d.c. batteries (V_{EE} and V_{CC}) alongwith the resistances R_E and R_C. The d.c. collector-to-base-voltage is given by the equation,

$$V_{CB} = V_{CC} - I_C \cdot R_C \qquad ...(i)$$

It will be interesting to know that when there is no a.c. signal source (*i.e.*, no v_s), the current flowing through the load resistor (R_L) is the quiescent operating current. Now suppose an a.c. signal is applied across the emitter-base terminals. Then during positive half-cycle of the input signal, the forward-bias across the emitter-base junction decreases because the voltage V_{EB} is negative with respect to the ground. As a result of this, a small number of electrons are injected into the base and reach the collector. Therefore, the base and collector currents decrease. Now the decrease in collector current produces smaller voltage drop across the resistor R_C. Therefore the voltage V_{CB} is increased as seen from the equation (*i*). During the negative half cycle, the forward-bias across emitter-base junction increases. As a result of this, more number of emitter electrons are injected into the base, due to which collector current increases. The increased collector current produces a larger voltage drop across the resistance R_C and hence the voltage V_{CB} decreases.

It is evident from the above discussion, that the positive-going input signal produces a positive going output signal. Therefore, there is no phase reversal between the input and output signals. It may be noted that the collector current is less than emitter current. Therefore, current gain of the common base amplifier (α) is less than unity. But still we get a voltage gain from the amplifier circuit. It is because of the fact that the output resistance of the common-base amplifier is very large as compared to the input resistance. Thus even though the changes in input and output currents are same, yet the voltage drop across output resistance is very large. Therefore, the changes in output voltage are much larger than the input a.c. signal.

24.19. Different Ways of Drawing Common Base Amplifier Circuit

Fig. 24.31 (*a*) and (*b*) shows a common base a amplifier circuit drawn in a different form. In Fig. 24.31 (*a*), the battery symbol is not indicated. It is always understood to be connected from the terminal marked + or – to the ground terminal, even though it may not be actually shown.

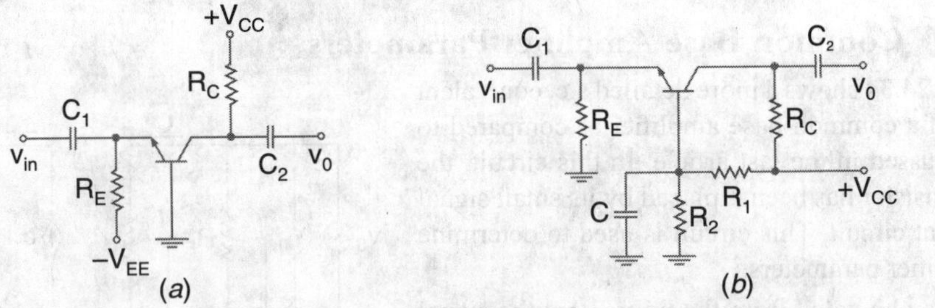

Fig. 24.31. Different ways of drawing common base amplifier circuit.

This circuit has a emitter-bias arrangement. A common base amplifier can have a voltage-divider bias arrangement as shown in Fig. 24.31 (*b*). This circuit uses a single d.c. supply. Here the base is at a.c. ground through a capacitor (*C*).

24.20. Analysis of a Common Base Amplifier

Fig. 24.32 shows a simple common base amplifier with a emitter-bias arrangement and connected to an external load resistance (R_L). The analysis of this amplifier circuit can be done by splitting it into two parts namely d.c. equivalent circuit and a.c. equivalent circuit. Both the d.c. and a.c. equivalent circuits may be obtained exactly in the same way as discussed for a common emitter amplifier.

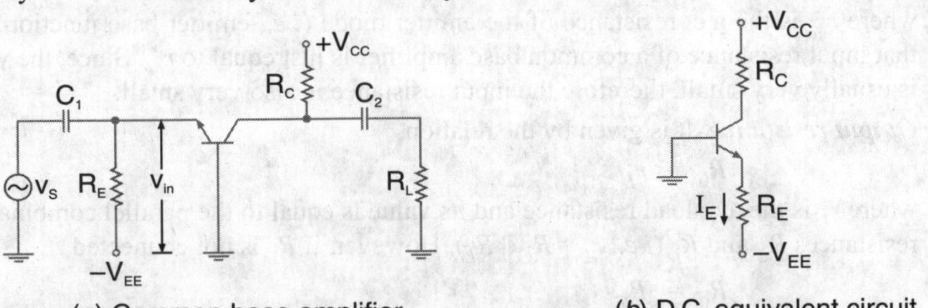

(*a*) Common base amplifier. (*b*) D.C. equivalent circuit.

Fig. 24.32.

Fig. 24.32 (*b*) shows the d.c. equivalent circuit of the common base amplifier. We have already discussed in chapter on transistor bias that such a circuit is called emitter-bias arrangement.

The emitter current in such a circuit is given by the relation,

$$I_E = \frac{V_{EE} - V_{BE}}{R_E}$$

and the collector current,

$$I_C = I_E$$

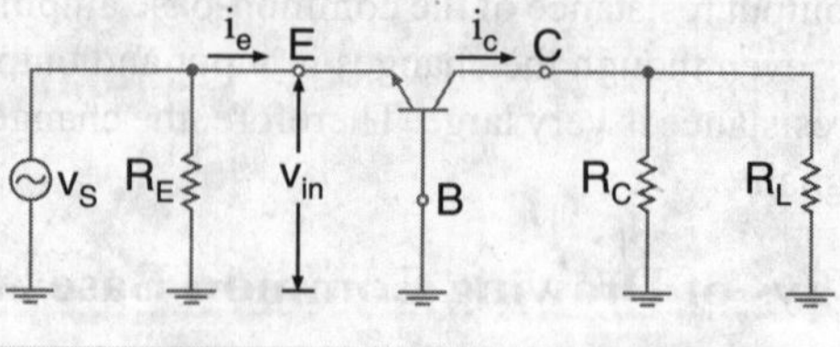

Fig. 24.33. A simple a.c. equivalent circuit for CB amplifier.

Fig. 24.33 shows a simple a.c. equivalent circuit for a common base amplifier. The input signal drives the emitter terminal and the output is taken from the collector.

24.21. Common Base Amplifier Parameters

Fig. 24.34 shows a more detailed a.c. equivalent circuit of a common-base amplifier as compared to that discussed in the last article. In this circuit, the NPN transistor has been replaced by its small signal equivalent circuit. This circuit is used to determine the amplifier parameters.

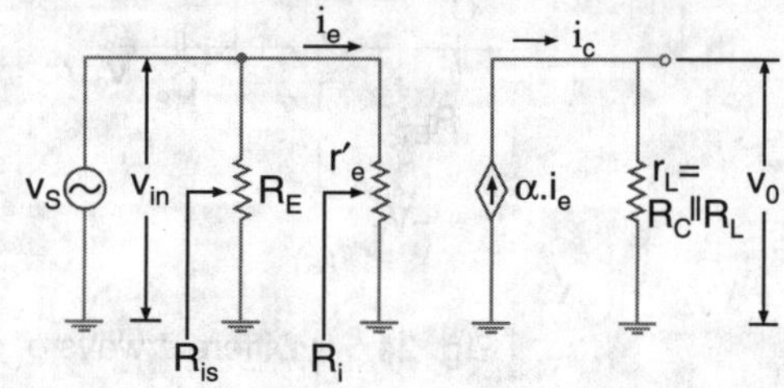

Fig. 24.34.

It may be noted from the figure, that the signal source has no resistance *i.e.*, R_s is equal to zero. Therefore all of the signal voltage appears as the input voltage (*i.e.*, $v_{in} = v_s$). Now we shall determine the voltage of amplifier parameters in terms of circuit components.

1. ***Input resistance.*** The input resistance, looking directly into the emitter, is given by the relation,

$$R_i = r'_e$$

and the input resistance of the stage,

$$R_{is} = R_E \| r'_e = r'_e \quad \text{... (if } R_E >> r'_e\text{)}$$

where r'_e is the a.c. resistance of the emitter diode (*i.e.*, emitter-base junction). It means that input resistance of a common base amplifier is just equal to r'_e. Since, the value of r'_e is usually very small, therefore the input resistance is also very small.

2. ***Output resistance.*** It is given by the relation,

$$R_0 = r_L$$

where r_L is the a.c. load resistance and its value is equal to the parallel combination of the resistances R_C and R_L (*i.e.*, $r_L = R_C \| R_L$). However, if R_L is not connected

$$R_0 = R_C$$

3. ***Current gain.*** It is given by the ratio of output current to the input current or the ratio of collector current to the emitter current. Mathematically the current gain,

$$A_i = \frac{i_c}{i_e} = \alpha$$

4. ***Voltage gain.*** It is given by the ratio of output voltage to the input voltage. Mathematically, the voltage gain.

$$A_v = \frac{v_0}{v_{in}} \quad \text{...}(i)$$

We know that the input voltage,

$$v_{in} = i_e \cdot r'_e$$

and the output voltage,

$$v_0 = i_c \cdot r_L = i_e \cdot r_L \qquad ...(\because i_c = i_e)$$

Substituting the values of v_{in} and v_0 in equation (*i*) the voltage gain,

$$A_v = \frac{i_e \cdot e_L}{i_e \cdot r'_e} = \frac{r_L}{r'_e}$$

It indicates that voltage gain of a common-base amplifier is the ratio of a.c. load resistance to the a.c. resistance of emitter diode. It is also called the voltage gain from emitter to collector.

5. ***Power gain.*** It is given by the product of voltage gain and current gain. Mathematically, the power gain,

$$A_p = A_v \cdot A_i$$

The power gain, in decibels, is given by the relation,

$$G_p = 10 \log_{10} A_p$$

Example 24.12.

For the single-stage common base amplifier shown in Fig. 24.35, find the values of (a) input resistance, (b) current gain, (c) voltage gain and (d) power gain.

What would be the r.m.s. value of the signal voltage across the load, if the input signal voltage is 10 mV? Assume the transistor to be made of silicon with current gain = 0.98.

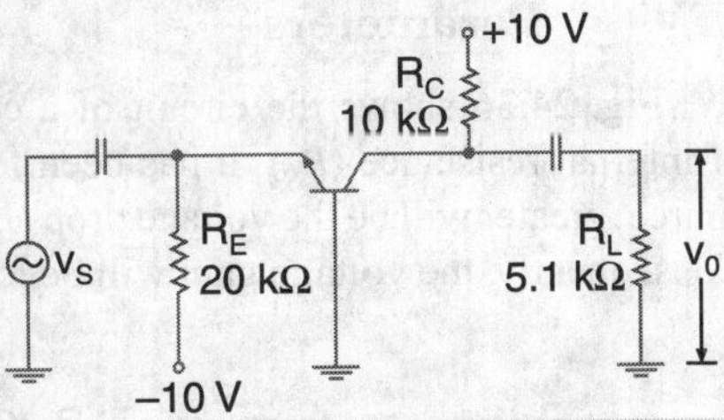

Fig. 24.35.

Solution.

Given: $V_S = 10$ mV; $\alpha = 0.98$; V_{BE} (for silicon transistor) = 0.7 volt; $V_{CC} = 10$ volts; $R_C = 10$ kΩ; $R_L = 5.1$ kΩ; $R_E = 20$ kΩ = 20×10^3 Ω and $V_{EE} = 10$ volts.

Input resistance

We know that the value of emitter current,

$$I_E = \frac{V_{EE} - V_{BE}}{R_E} = \frac{10 - 0.7}{20 \times 10^3} = 0.465 \times 10^3 \text{ A} = 0.465 \text{ mA}$$

and the value of a.c. emitter resistance,

$$r'_e = \frac{25}{I_E} = \frac{25}{0.465} = 53.8\ \Omega$$

∴ Value of input resistance, looking directly into the emitter,

$$R_i = r'_e = 53.8\ \Omega$$

and the input resistance of the stage,

$$R_{is} = R_E \,||\, r'_e = (20 \times 10^3) \,||\, 53.8 = 53.8\ \Omega \textbf{ Ans.}$$

Current gain

We know that the value of current gain,

$$A_i = \alpha = 0.98 \textbf{ Ans.}$$

Voltage gain

We know that the value of a.c. load resistance,

$$r_L = R_C \,||\, R_L = 10 \,||\, 5.1 = 3.38 \text{ k}\Omega = 3.38 \times 10^3\ \Omega$$

∴ Voltage gain,

$$A_v = \frac{r_L}{r'_e} = \frac{3.38 \times 10^3}{53.8} = 62.8 \textbf{ Ans.}$$

Power gain

We know that the value of power gain,

$$A_p = A_v \cdot A_i = 62.8 \times 0.98 = 61.5 \textbf{ Ans.}$$

and the power gain in decibels,

$$G_P = 10 \log_{10} A_P = 10 \log_{10} 61.5 = 17.9 \text{ dB} \textbf{ Ans.}$$

Output voltage

We know that the value of input voltage,

$$v_{in} = v_s = 10 \text{ mV} \qquad \ldots (R_s = 0)$$

∴ Value of output voltage,

$$V_0 = A_v \times v_{in} = 62.8 \times 10 = 628 \text{ mV} \textbf{ Ans.}$$

24.22. Effect of Signal Source Resistance on Common Base Amplifier Parameters

Fig. 24.36 shows the circuit of a common base amplifier connected to an a.c. signal source of internal resistance (R_S). It has been found that greater value of internal resistance of a.c. signal source, greater will be the voltage drop across internal resistance, and lesser will be the input voltage. Consequently, the voltage gain will be lower with a large value of internal resistance.

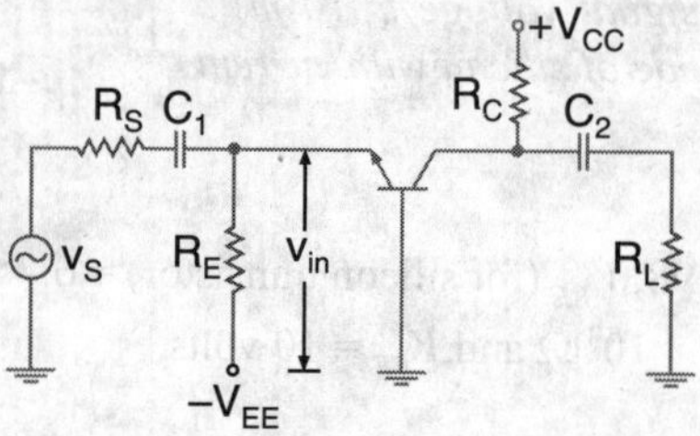

Fig. 24.36. Common base amplifier connected to an a.c. signal source with internal resistance (R_s).

The a.c. equivalent circuit of the given amplifier is shown in Fig. 24.37. The total resistance of the amplifier as seen by the signal source,

$$R_{is} = R_E \parallel r'_e$$

$$= r'_e \qquad \ldots (\text{If } R_E >> r'_e)$$

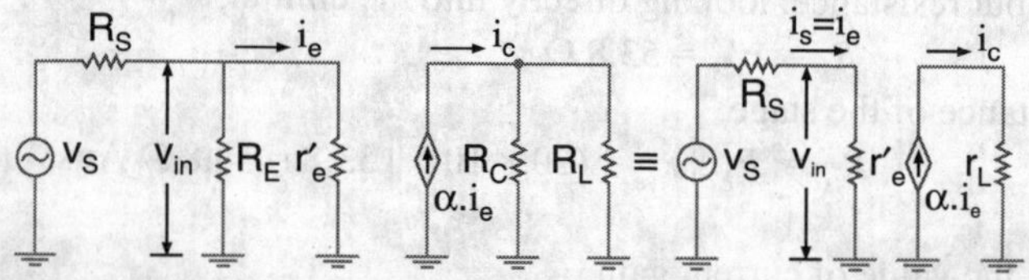

Fig. 24.37. A.C. equivalent circuit of a common base amplifier.

It may be noted from the a.c. equivalent circuit, that the internal resistance of a signal source (R_S) is in series with the total input resistance. Therefore the a.c. source voltage is divided between the internal resistance (R_S) and the total input resistance (R_{is}). The voltage at the base is given by the relation,

$$v_{in} = v_s\left(\frac{R_{is}}{R_s + R_{is}}\right)$$

$$= v_s\left(\frac{r'_e}{R_s + r'_e}\right) \quad \text{... (If } R_{is} = r'_e\text{)}$$

$$= v_s\left(\frac{1}{1 + R_s/r'_e}\right)$$

It is evident from the above relation that as the value of R_s increase, in comparison to r'_e, the ratio R_s/r'_e also increases. It decreases the value of voltage at the base (v_{in}). Now as seen from the source, the value of input resistance,

$$= R_s + r'_e$$

∴ Voltage gain of the amplifier stage,

$$A_{vs} = \frac{v_0}{v_{in}} = \frac{r_L}{R_s + r'_e} \approx \frac{r_L}{R_s} \quad \text{... (If } R_S >> r'_e\text{)}$$

The above relation indicates that if the value of R_s is much larger than r'_e the voltage gain of the amplifier stage becomes independent of r'_e. In other words, the voltage gain is independent of the transistor to be used in the amplifier circuit. This technique is called swamping out the emitter diode resistance of a transistor. It may be noted that the voltage gain of the stage is different from the voltage gain from emitter to collector, whose value is calculated from the ratio r_L/r'_e.

Example 24.13.

Consider the common base amplifier circuit as shown in Fig. 24.35 of Example 24.12. Assuming, the signal sources resistance (R_S) = 50 Ω, find (a) the voltage gain from source to output, (b) the voltage gain from emitter to output, and (c) approximate value of V_{in}.

Solution.

Given: $R_S = 50\ \Omega$.

We have already obtained the following in Example 24.12.

$I_E = 0.465$ mA; $r'_e = 53.8\ \Omega$; $R_i = 53.8\ \Omega$; $R_{is} = 52.4\ \Omega$ and $R_L = 3.38$ kΩ. Now we shall use the above values in this example.

(a) Voltage gain from source to output

We know that the voltage gain from source to output (called stage gain) is given by the relation,

$$A_{vs} = \frac{r_L}{R_s + r'_e} = \frac{3.38 \times 10^3}{50 + 53.8} = 32.6 \textbf{ Ans.}$$

(b) Voltage gain from emitter to output

We know that voltage gain from emitter to output,

$$A_v = \frac{r_L}{r'_e} = \frac{3.38 \times 10^3}{53.8} = 62.8 \textbf{ Ans.}$$

(c) Value of v_{in}

We know that the voltage gain from source to output, (A_{vs}),

$$32.6 = \frac{v_0}{v_s} = \frac{v_0}{10}$$

$\therefore \quad v_0 = 32.6 \times 10 = 326$ mV

We also know that the voltage gain from emitter to collector (A_v),

$$62.8 = \frac{v_0}{v_{in}} = \frac{326}{v_{in}}$$

$\therefore \quad v_{in} = 326/62.8 = 5.2$ mV **Ans.**

24.23. Characteristics and Uses of Common Base Amplifier

A common base amplifier has the following important characteristics:

1. It has very low input resistance and is of the order of 30 to 150 ohms.
2. It has very high output resistance and is of the order of 500 kΩ.
3. Its current gain (α) is less than unity.
4. Its voltage gain is of the order of 1500.
5. Its power gain is of the order of 30 dB.
6. Its input and output are in phase with each other *i.e.*, there is no phase reversal between the input and output signals.

A common base amplifier has a very important application as it is used for matching circuits with low output resistance to those with a high input resistance.

24.24. Common-Collector Amplifier

Fig. 24.38 shows a circuit of common-collector (CC) amplifier using *NPN* transistor. Since the collector resistance (R_C) is zero, therefore the collector is at a.c. ground. That is why this circuit is also known as grounded collector amplifier. In this circuit, the input is applied between the base and collector terminals and the output is taken across the emitter and collector terminals. The capacitors C_1 and C_2 block the direct current; but pass the a.c. signal current in the circuit.

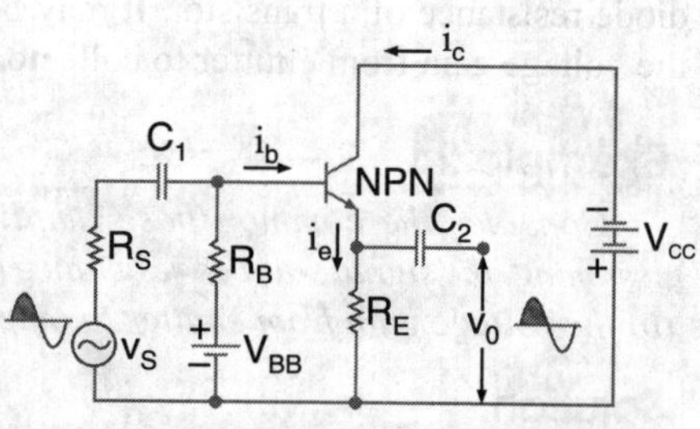

Fig. 24.38. Common-collector transistor amplifier.

The *Q*-point in the circuit is obtained by the d.c. batteries V_{BB} and V_{CC} alongwith the resistors R_E and R_B. The d.c. collector to emitter voltage is given by the equation,

$$V_{CE} = V_{CC} - I_E \cdot R_E = V_{CC} - V_0 \quad \text{...(where } V_0 \text{ is the output voltage)}$$

It will be interesting to know that when there is no a.c. signal source (*i.e.*, v_s is equal to zero), the current flowing through the emitter (or load) resistor is the quiescent operating current. Now let an a.c. signal be applied across the input terminals. Then during the positive half cycle of the input a.c. signal, the forward bias is increased because the voltage V_B is positive with respect to the collector (*i.e.*, ground). This causes the base current to increase, due to which the emitter current is also increased. As a result of this, the voltage drop across resistor R_E is increased and hence the output voltage (v_0) is also increased. Consequently, we get positive half cycle of the output. It means that a positive going input signal results in a positive going output signal. Thus the input and output signals are in phase with each other as shown in Figure 24.34. In other words, the emitter voltage follows the base voltage. That is why the common collector amplifier is known as emitter follower. The voltage gain of an emitter follower is less than unity. But it has a large current gain and large power gain.

24.25. Different Ways of Drawing Common Collector Amplifier Circuit

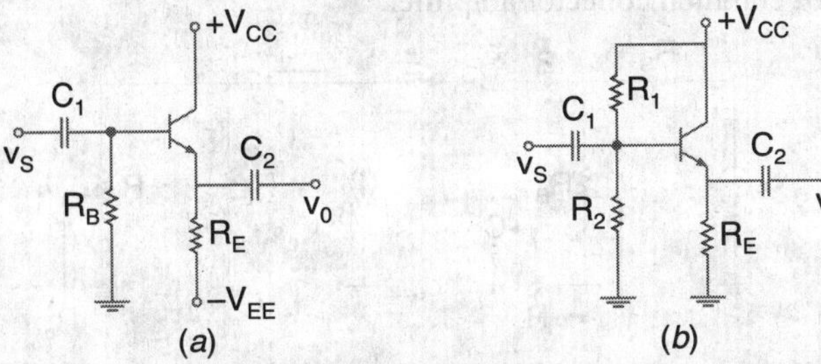

Fig. 24.39. Different ways of drawing common collector amplifier circuit.

Fig. 24.39 (*a*) and (*b*) indicates a common collector amplifier drawn in different forms. In Fig. 24.39 (*a*) the battery is indicated with the emitter terminal. This circuit has a base bias arrangement. A common collector amplifier may have a voltage divider bias arrangement as shown in Fig. 24.39 (*b*). This circuit uses a single d.c. supply. It may be noted that in all the circuits, collector is at a.c. ground.

24.26. Analysis of Common Collector Amplifier

Fig. 24.40 (*a*) shows a simple common collector amplifier with a base-bias arrangement. The amplifier is connected to a load resistance (R_L). The d.c. equivalent circuit for the given amplifier is as shown in Fig. 24.40 (*b*). The value of emitter current for such a circuit is given by the relation,

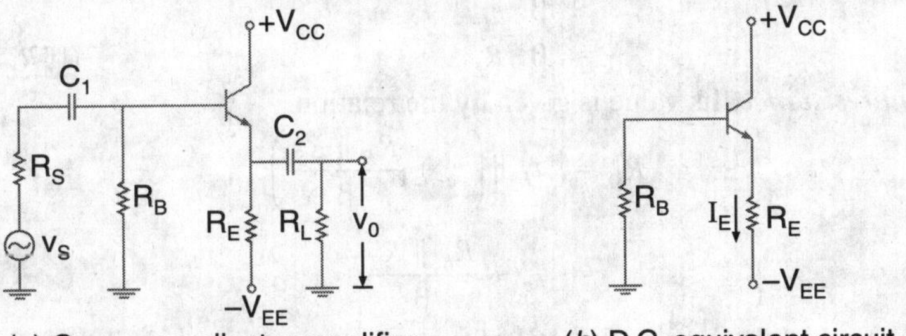

(*a*) Common collector amplifier. (*b*) D.C. equivalent circuit.

Fig. 24.40.

$$I_E = \frac{V_{EE} - V_{BE}}{R_E - R_B/\beta} \approx \frac{V_{EE}}{R_E + R_B/\beta}$$

Fig. 24.41 shows a simple a.c. equivalent circuit for a common collector amplifier. The input signal drives the base terminal and the output is taken from the emitter.

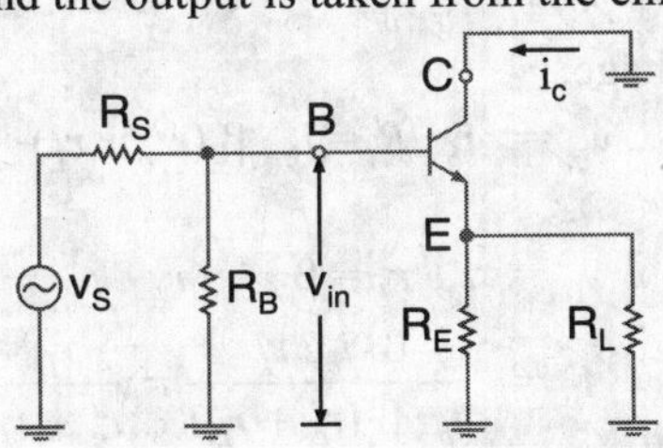

Fig. 24.41. A.C. equivalent circuit.

24.27. Common Collector Amplifier Parameters

Fig. 24.42 shows a more detailed a.c. equivalent circuit of a common collector amplifier as compared to that of Fig. 24.41. In this circuit, an NPN transistor is replaced by its small-signals equivalent circuit. This circuit is used to determine input and output resistances, current gain, voltage gain and power gain of common collector amplifier.

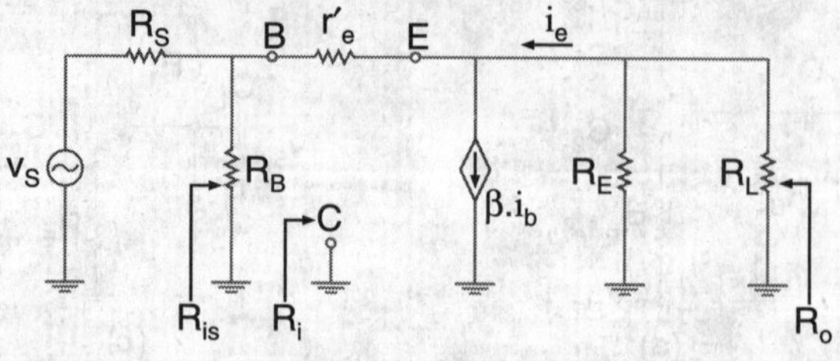

Fig. 24.42. Detailed a.c. equivalent circuit of a common collector amplifier.

It may be noted that if the value of signal source resistance (R_S) is zero, then all the signal voltage appears as the input voltage (*i.e.*, $v_{in} = v_S$). Now we shall determine the values of common-collector amplifier parameters in terms of circuit components.

1. ***Input resistance.*** The input resistance of the stage is given by the parallel combination of R_E and R_i. Now R_i is the input resistance looking into the base. Its value is equal to $\beta\,(r'_e + r_L) \approx \beta \cdot r_L$ · *i.e.*, β times the a.c. load resistance seen by the emitter.

$$\therefore \quad R_{is} = R_B \,\|\, \beta\,(r'_e + r_L)$$

$$\approx \beta\,(r'_e + r_L) \qquad \text{...[If } R_B >> \beta\,(r'_e + r_L)]$$

$$\approx \beta \cdot r_L \qquad \text{...(If } r_L >> r'_e)$$

$$= \beta \cdot R_E \qquad \text{(If } R_L \text{ is equal to zero)}$$

2. ***Output resistance.*** Its value is given by the relation,

$$R_O = r_L \,\|\, \left(r'_e + \frac{R_B \,\|\, R_S}{\beta} \right)$$

$$= r'_e + \frac{R_B \,\|\, R_S}{\beta} \qquad \text{...(If } r_L >> r'_e)$$

3. ***Current gain.*** It is given by the ratio of output current to input current *i.e.*, the ratio of emitter current to base current. Thus current gain,

$$A_i = \frac{i_e}{i_b} \approx \frac{i_c + i_b}{i_b} = 1 + \frac{i_c}{i_b} = 1 + \beta \approx \beta$$

4. ***Voltage gain.*** It is given by the ratio of output voltage to the input voltage. Thus voltage gain

$$A_v = v_O / v_{in}$$

We know that input voltage,

$$v_{in} = i_b \cdot R_i = i_b \cdot \beta\,(r'_e + r_L)$$

and output voltage,

$$v_O = i_e \cdot r_L = \beta \cdot i_b \cdot r_L$$

$$\therefore \quad A_v = \frac{\beta \cdot i_b \cdot r_L}{i_b \cdot \beta\,(r'_e + r_L)} = \frac{r_L}{r'_e + r_L} \approx \frac{r_L}{r_L} = 1 \qquad \text{(If } r'_e << r_L)$$

It means that output signal from the emitter has the same magnitude as the input signal at the base.

5. ***Power gain.*** It is given by the product of voltage and current gain. Mathematically, the power gain.

$$A_p = A_v \times A_i = 1 \times \beta = \beta$$

and the power gain in decibels,

$$G_p = 10 \log_{10} A_p \; dB$$

It is evident from the above discussion that the voltage gain of emitter follower is unity and current gain is β. However, input resistance is b times greater than the output resistance. It means that emitter follower steps up the resistance level of the output to a much higher value at its input. It is the most useful characteristic of an emitter follower (or common collector amplifier). This characteristic makes the emitter follower very useful as a resistance matching device *i.e.*, it is used to match circuits with high output resistance to those with low input resistance.

Example 24.14.

Fig. 24.43 shows the circuit of an emitter follower. The transistor used in the circuit has current gain (β) of 50 and $V_{BE} = 0.7$ V

Determine the values of input resistance, and the exact voltage gain.

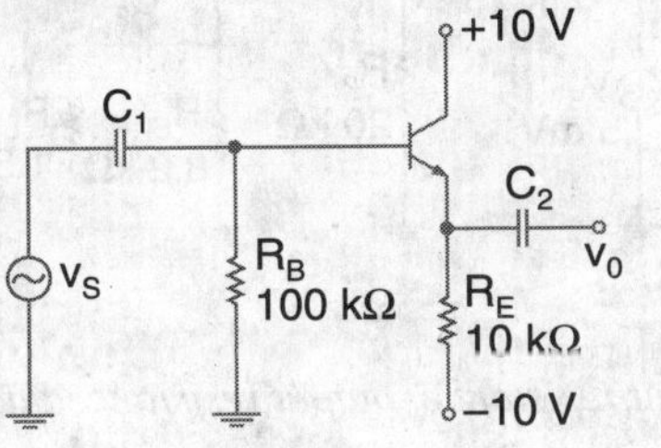

Fig. 24.43.

Solution.

Given: $V_{EE} = 10$ volts ; $R_E = 10\ k\Omega = 10 \times 10^3\ \Omega$; $R_B = 100\ k\Omega = 100 \times 10^3\ \Omega$; $\beta = 50$ and $V_{BE} - 0.7$ volt.

Input resistance

We know that the value of emitter current,

$$I_E = \frac{V_{EE} - V_{BE}}{R_E + \dfrac{R_B}{\beta}} = \frac{10 - 0.7}{(10 \times 10^3) + \dfrac{100 \times 10^3}{50}} = 0.775 \times 10^{-3}\ A = 0.775\ mA$$

and a.c. resistance of emitter diode,

$$r'_e = \frac{25}{I_E} = \frac{25}{0.775} = 32.3\ \Omega$$

∴ Input resistance looking directly into the base,

$$R_i = \beta\,(R_E + r'_e) = 50\,[(10 \times 10^3) + 32.3] = 501.6\ k\Omega \textbf{ Ans.}$$

and input resistance of the stage,

$$R_{is} = R_B \parallel R_i = 100 \parallel 501.6 = 83.4\ k\Omega \textbf{ Ans.}$$

Output resistance

We know that the output resistance,

$$R_0 = r'_e + \frac{R_B \parallel R_S}{\beta} = r'_e = 32.3\ \Omega \textbf{ Ans.} \quad ...(\because R_s = 0)$$

Voltage gain

We also know that the voltage gain,

$$A_v = \frac{R_E}{r'_e + R_E} = \frac{10 \times 10^3}{32.3 + (10 \times 10^3)} = 0.997 \textbf{ Ans.}$$

Example 24.15.

The emitter follower circuit shown in Figure 24.44 uses a voltage divider bias arrangement. The transistor has a current gain (β) of 80 and V_{BE} = 0.7 V.

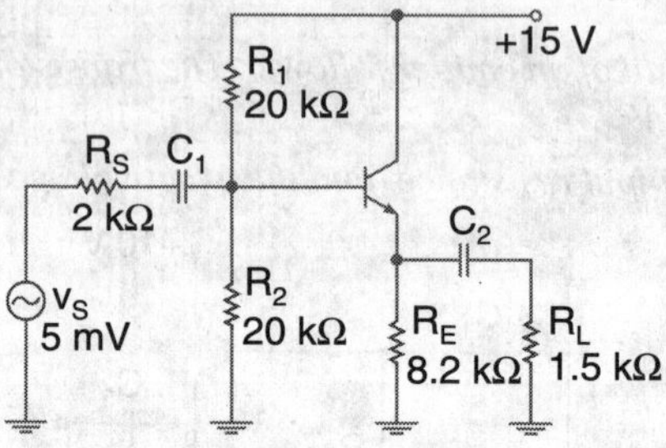

Fig. 24.44.

Calculate the value of input resistance, output resistance and the a.c. output voltage.

Solution.

Given: β = 80; V_{BE} = 0.7 volt; V_{CC} = 15 volt; R_1 = 20 kΩ; R_2 = 20 kΩ R_S = 2 kΩ; V_S = 5 mV; R_E = 8.2 kΩ = 8.2 × 10^3 Ω and R_L = 1.5 kΩ.

Input resistance

We know that Thevenin's voltage

$$V_{TH} = V_{CC} \times \frac{R_2}{R_1 + R_2} = 15 \times \frac{20}{20 + 20} = 7.5 \text{ V}$$

and the value of equivalent resistance,

$$R_{TH} = R_1 \parallel R_2 = 20 \parallel 20 = 10 \text{ k}\Omega = 10 \times 10^3 \ \Omega$$

∴ Value of emitter current,

$$I_E = \frac{V_{TH} - V_{BE}}{R_E + \dfrac{R_{TH}}{\beta}} = \frac{7.5 - 0.7}{8.2 \times 10^3 + \dfrac{(10 \times 10^3)}{80}}$$

$$= 0.817 \times 10^{-3} \text{ A} = 0.817 \text{ mA}$$

and a.c. resistance of emitter diode,

$$r'_e = \frac{25}{I_E} = \frac{25}{0.817} = 30.6 \ \Omega$$

We also know that the a.c. load resistance,

$$r_L = R_E \parallel R_L = 8.2 \parallel 1.5 = 1.27 \text{ k}\Omega = 1.27 \times 10^3 \ \Omega$$

∴ Value of input resistance, looking directly into the base,

$$R_i = \beta (r_L + r'_e) = 80 (1.27 \times 10^3 + 30.6) = 104 \times 10^3 \ \Omega = 104 \text{ k}\Omega$$

and input resistance of the stage,

$$R_{is} = R_{TH} \parallel R_i = 10 \parallel 104 = 9.1 \text{ k}\Omega \textbf{ Ans.}$$

Output resistance of the stage,

We know that the output resistance of the stage

$$R_0 = r'_e + \frac{R_S \parallel R_{TH}}{\beta} = 30.6 + \frac{(2\times10^3 \parallel 10\times10^3)}{80} = 51.4\ \Omega \textbf{ Ans.}$$

A.C. output voltage

We know that the input voltage,

$$V_{in} = V_s \times \left(\frac{R_{is}}{R_s + R_{is}}\right) = 5 \times \left(\frac{9.1}{2 + 9.1}\right) \text{ mV} = 4.1 \text{ mV}$$

and the output voltage,

$$v_0 = v_{in} = 4.1 \text{ mV} \textbf{ Ans.}$$

24.28. Characteristics and Uses of Common Collector Amplifier

A common collector (or emitter follower) has the following important characteristics:

1. It has a high input resistance, which is of the order of 20 to 500 kΩ.
2. It has a low input resistance, which is of the order of 50 to 1000 Ω.
3. It has a high current gain. Its value depends upon the current gain (β) of the transistor, which is in the range of 50 to 500.
4. Its voltage gain is less than unity.
5. Its power gain is much smaller than that of common emitter and common base amplifiers.
6. Its output signal is in phase with the input signal.

The main application of common collector amplifier is for resistance matching. It is used to transfer energy from a high output resistance circuit to that of a low input resistance circuit. The common collector amplifier can also be used as a two-way amplifier, because it can pass a signal in either direction.

24.29. Comparison of Three Amplifier Configurations

Table 24.1 show the comparison of important characteristics of three amplifier configurations, namely common emitter, common base and common collector.

Table 24.1: Comparison of three amplifier configurations.

S. No.	*Characteristic*	*Type of configuration*		
		Common Emitter	*Common Base*	*Common Collector*
1.	Input resistance (R_i)	$\beta \cdot r'_e$ (Moderate)	r'_e (Lowest)	$\beta \cdot R_E$ (Highest)
2.	Output resistance (R_0)	R_c (Moderate)	R_c (Highest)	$\approx r'_e$ (Lowest)
3.	Current gain (A_i)	β (High)	≈ 1 (Lowest)	1 + β (Highest)
4.	Voltage gain (A_v)	r_L/r'_e (Highest)	r_L/r'_e (High)	≈ 1
5.	Power gain (A_p)	Highest	Moderate	Lowest
6.	Phase reversal	Yes	No	No

24.30. Nonlinear Distortion

Any unwanted change in the shape of an a.c. signal is called distortion. One problem that commonly occurs in common-emitter amplifiers is called *nonlinear distortion*. The output from an waveform from an amplifier experiencing nonlinear distortion is shown in Fig. 24.45 (*a*).

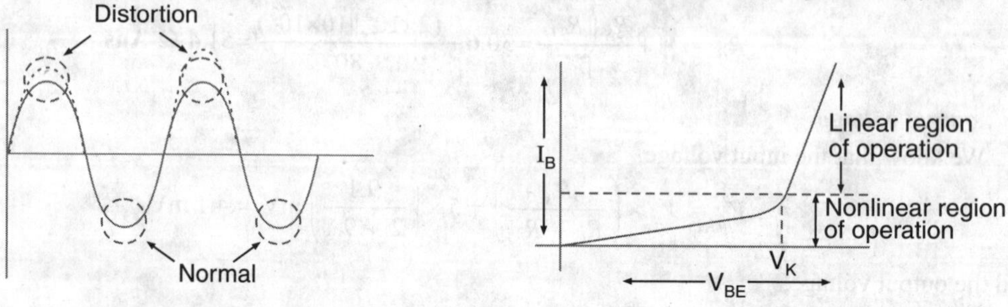

(*a*) Output waveform illustrating nonlinear distortion.

(*b*) Base characteristic curve.

Fig. 24.45.

Notice the difference between the shape of the negative alteration of the signal (normal) and that of its positive alteration (distorted). Nonlinear distortion is caused by driving the base-emitter junction of the transistor into its nonlinear operating region. This point is illustrated with the help of Fig. 24.45 (*b*) which shows the basic characteristic curve of a transistor.

Normally a transistor is operated so that the base-emitter junction stays in the linear region of operation. In this region, a change in V_{BE} causes a linear (*i.e.*, constant rate) change in I_B. If the transistor is operated in the nonlinear operating region, a change in V_{BE} will cause a change in I_B that is not linear. It is evident if we compare the slope of the curve at the points immediately above and below V_K. Since the slope of the curve changes, the rate of change in I_B also changes.

The base-emitter junction of a transistor may be driven into its non linear operating region in one of two sets of circumstances:

1. A transistor is normally biased so that I_B will have a value well within the linear region of operation. If an amplifier is poorly biased so that the *Q*-point value of I_B is near the nonlinear region of operation, a relatively small a.c. input signal will drive the amplifier into the nonlinear region of operation, and nonlinear distortion will result.
2. The amplifier input signal may be sufficient to drive even a well-biased amplifier into nonlinear operation, causing nonlinear distortion.

The solution to the first problem is to redesign the amplifier for a higher value of I_B. While for the second problem, the solution is to reduce the amplitude of the amplifier input signal.

Nonlinear distortion can cause a variety of problems in communications electronics. In stereos, it may cause the audio to sound "grainy" or may even cause the tones of various musical instruments to change. In televisions, crossover distortion in the video circuitry can cause distortion in the picture on the CRT. To avoid these types of problems, amplifier design normally emphasizes avoiding *crossover distortion*. Since amplifiers are normally designed to avoid nonlinear distortion problems, the most common cause of nonlinear distortion is *overdriving* an amplifier.

NUMERICAL PROBLEMS

1. For the single-stage common emitter amplifier shown in Fig. 24.42, find the values of input resistance looking directly into the base, output resistance, voltage and dB voltage gain.

(Ans. 415 Ω; 5 kΩ; 602.4 and 55.6 dB)

2. If for the common emitter amplifier shown in Fig. 24.46, the input voltage $v_s = 5$ mV, find the value of output voltage and the overall voltage gain. **(Ans. 681 mV and 136)**

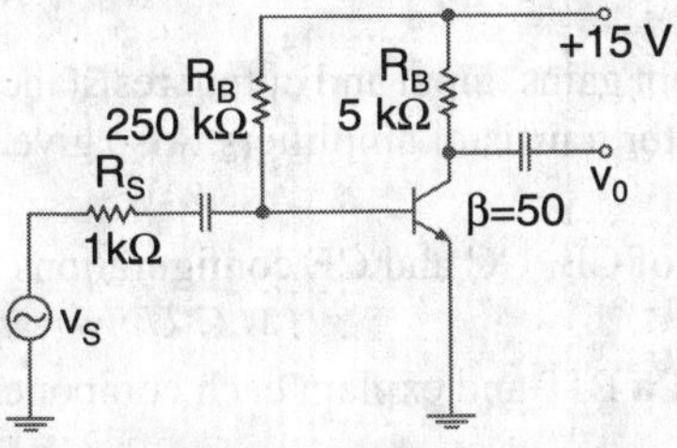

Fig. 24.46.

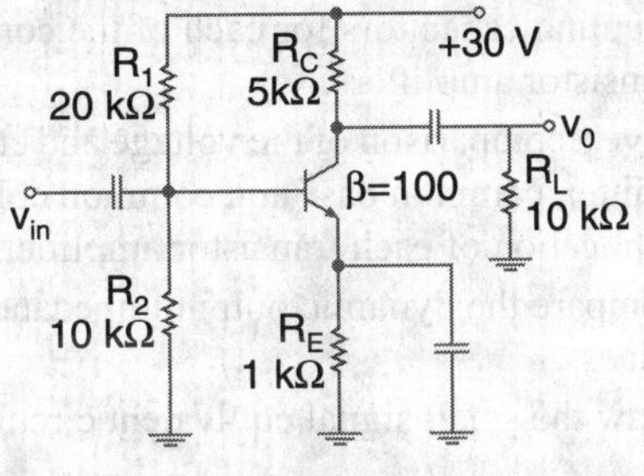

Fig. 24.47.

3. Draw the a.c. equivalent circuit for the single-stage common emitter amplifier shown in Fig. 24.47. Also find the values of R_i, R_{is}, R_0 and A_v. **(Ans. 270 Ω; 270 Ω; 3.3 kΩ, 1222)**
4. Fig. 24.48 shows the circuit of a single stage common base amplifier. Find the values of (*a*) stage input resistance (*b*) stage output resistance, (*c*) current gain and (*d*) stage voltage gain. **(Ans. 25 Ω; 5 kΩ; 1 and 200)**
5. For the single stage common collector amplifier circuit shown in Fig. 24.49, calculate the values of R_i, R_0, A_i and A_v. Take transistor β = 100 and $r'_e = 25/I_E$ (mA). **(Ans. 500 kΩ; 29.4 Ω; 100 and 0.99)**

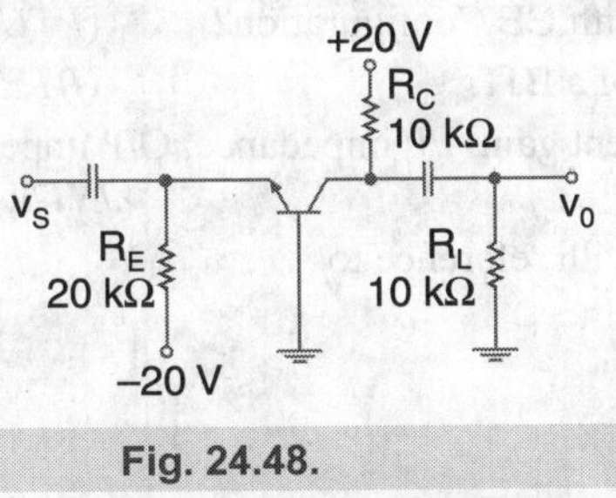

Fig. 24.48.

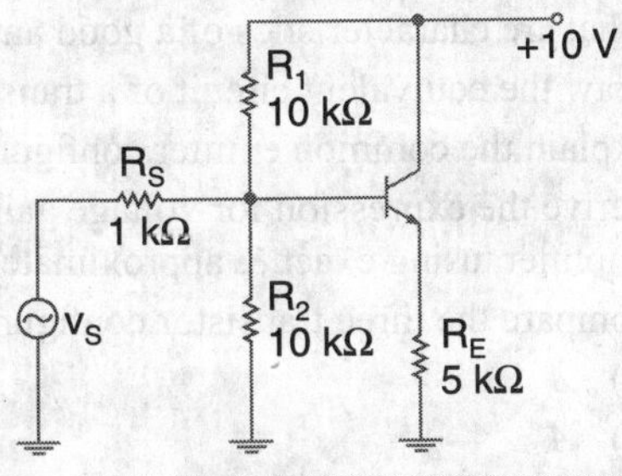

Fig. 24.49.

DESCRIPTIVE QUESTIONS

1. With a neat circuit diagram, explain the working of a transistor amplifier.
2. Draw the circuit diagram of a common emitter amplifier with emitter and voltage divider biasing circuit. Why normally an emitter bypass capacitor is used?
3. Derive the expressions for input resistance, output resistance and voltage gain of a common emitter amplifier.
4. With a neat circuit diagram, explain the working of a transistor amplifier in which phase inversion of the input signal does not take place. Obtain the expressions for input resistance for such an amplifier.
5. Discuss the merits and limitations of common base amplifier.
6. Draw the circuit diagram of common base amplifier and derive expressions for current gain and voltage gain.
7. With suitable derivations, bring out the salient features of an emitter follower and indicate its applications.

8. Derive the expressions for input resistance, output resistance and voltage gain of an emitter follower circuit.
9. Give neat circuit diagrams showing complete biasing arrangements, including bypass and coupling capacitors for each of the common emitter, common base and common collector transistor amplifiers.
10. Give a comparison of the voltage and current gains, input and output resistances of common emitter, common base and common collector transistor amplifiers. Also give one important application of each transistor amplifier.
11. Compare the dynamic output impedances of CB, CC and CE configurations giving typical values. *(M.G University., May 2008)*
12. Draw the small signal equivalent circuit of a BJT and explain each component. *(GBTU/MTU., 2009-10)*
13. Explain the operation of BJT as an amplifier. Also explain in detail the various possible combinations of BJT amplifiers with their corresponding input and output characteristic curves *(GBTU/MTU., 2008-09)*
14. Give reasons why common emitter configuration is widely used in amplifier circuits. *(PTU., May 2008)*
15. Draw circuit diagram of a practical amplifier with transistor in common base configuration. *(PTU., May 2008)*
16. Why common collector amplifier called emitter follower? *(PTU., May 2009)*
17. Which amplifier configuration is most popularly used and why? *(PTU., Dec 2008)*
18. What are characteristics of a good amplifier? *(PTU., May 2009)*
19. Draw the equivalent circuit of a transistor with CE Configuration? *(PTU., Dec 2006)*
20. Explain the common emitter configuration of a BJT. *(PTU., Dec 2006)*
21. Derive the expression for voltage gain, current gain, I/P impedance, O/P impedance of CE amplifier, using exact & approximate model. *(JNTU., June 2009)*
22. Compare the three transistor configuration with reference to
    (*a*) A_e
    (*b*) A_v
    (*c*) R_i
    (*d*) R_o
    (*e*) Applications *(R.G.P.V., Dec. 2004)*
23. Explain how a bipolar junction transistor works as an amplifier. *(Nagpur University, Summer 2007, 2005)*
24. Compare CE, CB and CC amplifier configurations. *(Nagpur University, Summer 2005)*
25. Explain the transmitter amplifying action in case of a CB circuit. *(Nagpur University, Winter 2003)*
26. Compare CE and CB configuration with reference to A_p, A_v and R_i. *(Nagpur University, Winter 2005)*
27. Compare the following parameters of the three transistor configurations *i.e.*, CE, CC and CB:
    (*a*) Input impedance
    (*b*) Output impedance
    (*c*) Current gain
    (*d*) Voltage gain *(R.G.P.V., June 2004)*

MULTIPLE CHOICE QUESTIONS

1. For a common emitter amplifier having a small unbypassed emitter resistance (R_E) the input resistance is approximately equal to
   (*a*) R_E (*b*) h_{fe} (*c*) $h_{fe} \cdot R_E$ (*d*) R_E/h_{fe}
2. The voltage gain of a common base amplifier is
   (*a*) zero (*b*) less than unity
   (*c*) unity (*d*) greater than unity
3. For a common base transistor amplifier having input resistance (R_i) and output resistance (R_0), which of the following statements holds good
   (*a*) R_i is low, R_0 is high (*b*) R_i is high, R_0 is low
   (*c*) R_i and R_0 are both medium
4. The current gain of an emitter follower is
   (*a*) zero (*b*) greater than unity
   (*c*) less than unity (*d*) all of them
5. Which of the following transistor amplifiers has the highest voltage gain
   (*a*) common-base (*b*) common-collector
   (*c*) common-emitter (*d*) none of them
6. In an ac amplifier, larger the internal resistance of the ac signal source
   (*a*) greater the overall voltage gain (*b*) greater the input impedance
   (*c*) smaller the current gain (*d*) smaller the circuit voltage gain.
7. The main use of an emitter follower is as
   (*a*) power amplifier (*b*) impedance matching device
   (*c*) low-input impedance circuit (*d*) follower of base signal.
8. An ideal amplifier is one which
   (*a*) has infinite voltage gain
   (*b*) responds only to signal at its input terminals
   (*c*) has positive feedback
   (*d*) gives uniform frequency response.
9. The voltage gain of a single-stage amplifier is increased when
   (*a*) its ac load is decreased (*b*) resistance of signal source is increased
   (*c*) emitter resistance R_E is increased. (*d*) as load resistance is increased.
10. When emitter bypass capacitor in a common-emitter amplifier is removed, its.................... is considerably reduced.
    (*a*) input resistance (*b*) output load resistance
    (*c*) emitter current (*d*) voltage gain
11. Unique features of a *CC* amplifier circuit is that it
    (*a*) steps up the impedance level
    (*b*) does not increases signal voltage
    (*c*) acts as an impedance matching device
    (*d*) all of the above.

12. An amplifier circuit is shown in Fig. 24.50. The voltage gain (V_o / V_s) is

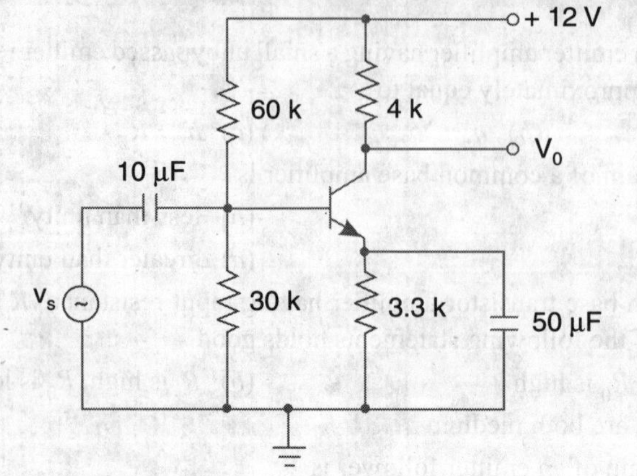

Fig. 24.50.

(*a*) 4/3.33 (*b*) 100
(*c*) 150 (*d*) 160

(*UPSC Engg. Services* 2000)

Hint. $V_B = 12 \times \dfrac{30}{30k + 160\ k} = 4V;\ I_E = \dfrac{V_B - V_{BE}}{R_E} = \dfrac{4 - 0{\cdot}7V}{3{\cdot}3k}$

$$\therefore \quad r'_e = \frac{25\text{mV}}{I_E} = \frac{25\text{ mV}}{1 - \text{mA}} = 25\ \Omega$$

And gain $\quad A_v = \dfrac{R_C}{r'_e} = \dfrac{4k}{25\ \Omega} = 160\ \Omega$

13. The amplifier configuration which gives maximum gain is:

(*a*) common emitter (*b*) common collector
(*c*) common base (*d*) all-of-these

(*Bharthiar Univ. April 2004*)

ANSWERS

1. (*c*) **2.** (*d*) **3.** (*a*) **4.** (*b*) **5.** (*c*)
6. (*d*) **7.** (*b*) **8.** (*b*) **9.** (*d*) **10.** (*d*)
11. (*d*) **12.** (*d*) **13.** (*a*)

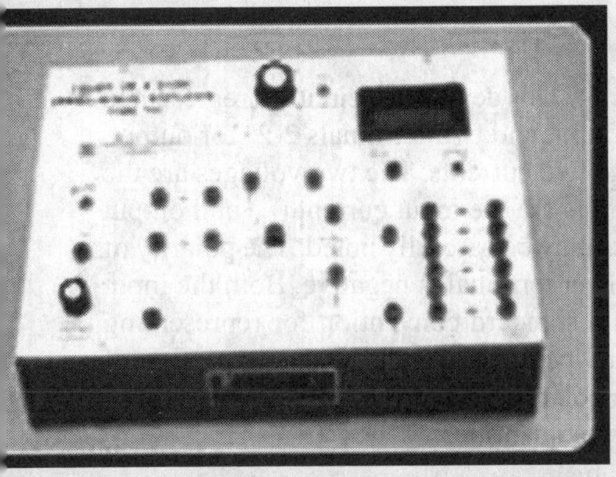

25

Chapter

Hybrid Parameters

Outline

25.1. Introduction

The hybrid parameters are commonly known as *h*-parameters. These are generally used to determine amplifier characteristic parameters such as voltage gain, input and output resistances etc. We have already discussed in the last chapter that amplifier characteristic parameters may be determined by using current gain (β) and values of other circuit components. This approach had the following advantages:

1. Value of circuit components are easily available, and
2. The procedure followed is quite simple and easy to understand.

However, this method has a drawback that the results obtained are not very accurate. It is mainly because of the fact that base and collector circuits of a transistor are not completely independent of each other. For example, we know that collector current should be β-times of the base current. Moreover, it should remain constant at this value. However, in actual practice, it is not so, as it is affected by the value of load resistance. Similarly, changes in the output voltage also cause changes in input voltage. These drawbacks are removed by the use of hybrid parameters. The hybrid parameters give very accurate results in transistor amplifier circuit analysis.

25.2. The *h*-parameters of a Linear Circuit

Fig. 25.1 (*a*) shows a model of any linear device or a circuit. This device or a circuit is represented by a box and has four terminals *i.e.*, two-terminals 1-1′ for input and two-terminals 2-2′ for output. The behaviour of this circuit is specified by two voltages and two currents. The two voltages are the input voltage (v_1) and the output voltage (v_2). The two currents are the input current (i_1) and output current (i_2). The polarity of voltages and direction of currents may be carefully noted. The polarity of voltage is such that the upper terminal is positive and the lower terminal is negative. Both the input and output currents are assumed to flow into the box. It is a standard convention for representing currents and voltages and do not correspond with the actual polarity of voltages and direction of currents. Thus if in any circuit, the voltages are of opposite polarity and currents are flowing out of the box; then these voltages and currents are taken as negative quantities.

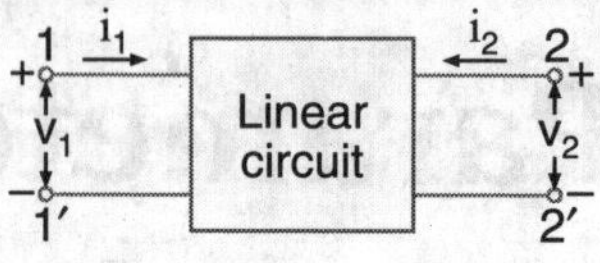

(*a*) Linear circuit.

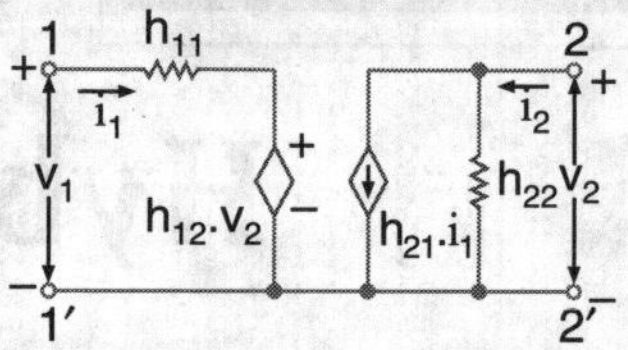

(*b*) Hybrid model for a linear circuit.

Fig. 25.1.

The linear circuit may be replaced by an equivalent circuit as shown in Fig. 25.1 (*b*). The equivalent circuit is called hybrid model of a linear circuit. In such a circuit, the input and output voltages and the input and output currents (called variables) may be related by the set of the following two equations:

$$v_1 = h_{11}\cdot i_1 + h_{12}\cdot v_2 \quad \text{... (i)}$$

$$i_2 = h_{21}\cdot i_1 + h_{22}\cdot v_2 \quad \text{... (ii)}$$

where v_1 = Input voltage, v_2 = Output voltage,
i_1 = Input current, i_2 = Output current, and
h_{11}, h_{12}, h_{21} and h_{22} = The hybrid or *h*-parameters.

The equations (*i*) and (*ii*) have been obtained by applying Kirchhoff's Voltage Law to the input and output circuits of the hybrid model. Thus a linear circuit has a set of four parameters namely h_{11}, h_{12}, h_{21} and h_{22}. These parameters completely describe the behaviour of the circuit and are constant for a given circuit. They are expressed in different units. The parameter h_{11} is expressed in ohms (Ω), h_{12} and h_{21} are dimensionless and h_{22} is expressed in mhos (Ω) or (*S*). Since these parameters have mixed units, therefore, these are called hybrid (means mixed) or *h*-parameters.

25.3. Determination and Meaning of *h*-parameters

We have already discussed in the last article that every linear circuit may be represented by a set of four *h*-parameters namely h_{11}, h_{12}, h_{21} and h_{22}. The parameters h_{11} and h_{21} may be determined by short-circuiting the output terminals of a given circuit. On the other hand, h_{12} and h_{22} may be determined by open-circuiting the input terminals of the given circuits.

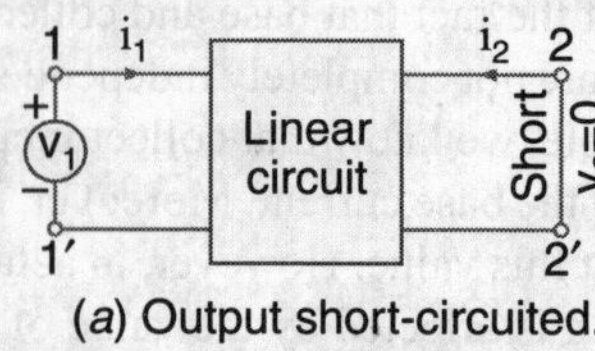

(*a*) Output short-circuited.

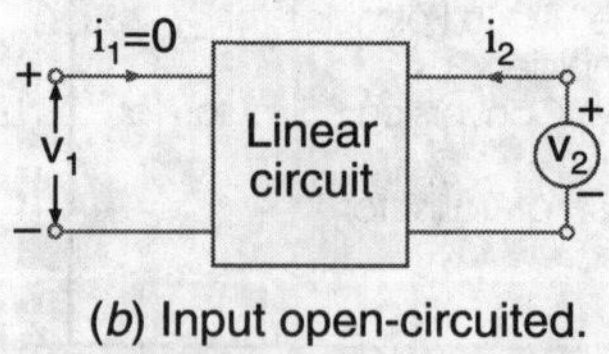

(*b*) Input open-circuited.

Fig. 25.2.

1. ***Determination of h_{11} and h_{21}.*** These are determined by short-circuiting the output terminals of a given circuit as shown in Fig. 25.2 (*a*). A short-circuit at the output terminals makes the voltage v_2 equal to zero. We know that the input voltage is given by the relation,

$$v_1 = h_{11} \cdot i_1 + h_{12} \cdot v_2$$

Substituting the value of v_2 (equal to zero) in the above equation, the input voltage,

$$v_1 = h_{11} \cdot i_1 \text{ or } h_{11} = v_1/i_1$$

Thus h_{11} may be determined from the ratio v_1/i_1. The value of i_1 is obtained by applying a voltage at the input and then measuring the value of input current (i_1). Since h_{11} is the ratio of voltage to current, therefore, it has the units of ohms *i.e.*, the same unit as that of a resistance. Because of this fact, h_{11} is also called input resistance of the circuit with output short-circuited. Similarly, we know that the output current is given by the relation,

$$i_2 = h_{21} \cdot i_1 + h_{22} \cdot v_2$$

Again substituting the value of v_2 (equal to zero) in the above equation, the output current,

$$i_2 = h_{21} \cdot i_1 \text{ or } h_{21} = i_2/i_1$$

Thus h_{21} may be determined from the ratio i_2/i_1. The values of i_1 and i_2 may be obtained by applying a voltage at the input and then measuring the input current (i_1) and output current (i_2). Since h_{21} is the ratio of currents, therefore, it has no units. The parameter h_{21} is called the forward current gain of the circuit with output short-circuited.

2. ***Determination of h_{12} and h_{22}.*** These are determined by open circuiting the input terminals of a given circuit as shown in Fig. 25.2 (*b*). An open circuit, at the input terminals, makes the current (i_1) equal to zero. We also know that the input voltage is given by the relation,

$$v_1 = h_{11} \cdot i_1 + h_{12} \cdot v_2$$

Substituting the value of i_1 (equal to zero) in the above equation, the input voltage,

$$v_1 = h_{12} \cdot v_2 \text{ or } h_{12} = v_1/v_2$$

Thus h_{12} may be determined from the ratio v_1/v_2. The value of v_1 may be obtained by applying a voltage v_2 at the output and then measuring the input voltage (v_1). Since, h_{12} is a ratio of voltages, therefore, it has no units. As h_{12} is the ratio of input voltage (v_1) to the output voltage (v_2), therefore, its value is known as the reverse voltage gain in order to distinguish it from to forward voltage gain, whose value is equal to v_2/v_1.

Similarly, we know that the output current is given by the relation,

$$i_2 = h_{21} \cdot i_1 + h_{22} \cdot v_2$$

Again substituting the value of i_1 (equal to zero) in the above equation, the output current,

$$i_2 = h_{22} \cdot v_2 \quad \text{or} \quad h_{22} = i_2/v_2$$

Thus h_{22} may be determined from the ratio i_2/v_2. The value of current (i_2) may be obtained by applying a voltage at the output (v_2) and then measuring the output current with input open. Since h_{22} is the ratio of current to voltage, therefore, it has the units of mhos (Ω) or Siemens (*S*). The parameter h_{22} is also called output conductance with input open.

Example 25.1.

Determine the value of h-parameters for a circuit shown in Fig. 25.3.

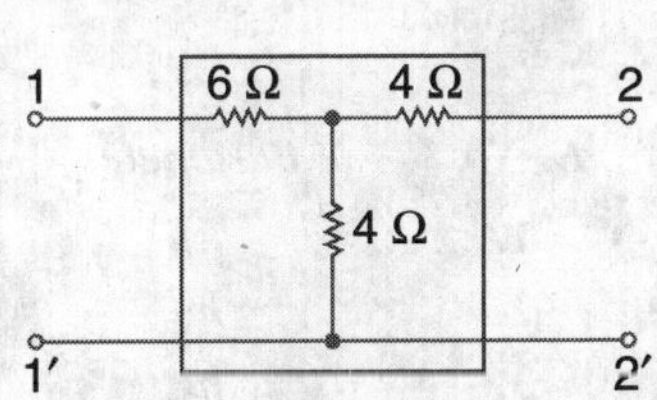

Fig. 25.3.

Solution.

(a) Parameters h_{11} and h_{21}

We know that the values of h_{11} and h_{21} can be determined by putting a short-circuit across the output terminals of the given circuit as shown in Fig. 25.4 (*a*). And h_{11} (*i.e.*, input resistance looking from the input terminals 1-1′) is given by the relation,

$$h_{11} = 6 + (4 \| 4) = 6 + \frac{4 \times 4}{4+4} = 8\ \Omega \quad \textbf{Ans.}$$

It may be noted from Figure 25.4 (*a*), that the input current is divided into two equal parts at point N. A current $i_1/2$ flows outwards and $i_1/2$ flows outwards and $i_1/2$ flows downward. The output current,

$$i_2 = -i_1/2$$

The negative sign indicates that the current is actually flowing out of the box.

$$\therefore \quad h_{21} = \frac{i_2}{i_1} = \frac{-i_1/2}{i_1} = -\frac{1}{2} = -0.5 \quad \textbf{Ans.}$$

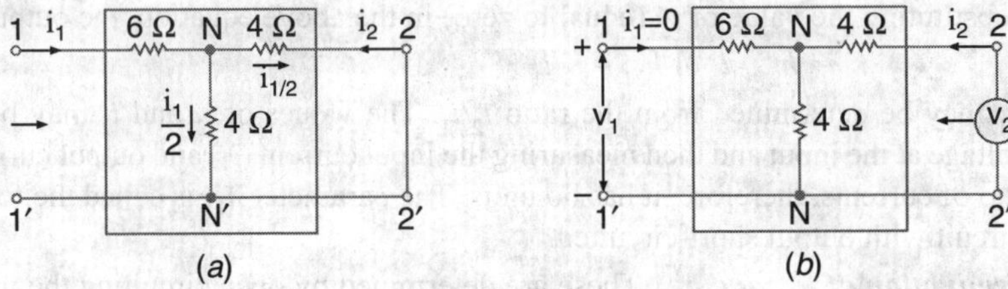

Fig. 25.4.

(b) Parameters h_{12} and h_{22}

We know that the values of h_{12} and h_{22} may be determined by keeping the input terminals open and applying a voltage source v_2 across the output terminals as shown in Fig. 25.4 (*b*). Due to the open input, $i_1 = 0$. The current flowing through 6 Ω resistor is zero, and hence there is no voltage drop across it. The voltage (v_2) produces a current (i_2) in the circuit. This current will produce equal voltage drops across the two 4 Ω resistors. The voltage drop across the 4 Ω resistor connected between *N* and *N'* appears as the voltage v_1 across the input terminals 1 : 1′. Thus $v_1 = v_2/2$.

$$\therefore \quad h_{12} = \frac{v_1}{v_2} = \frac{v_2/2}{2} = 0.5 \quad \textbf{Ans.}$$

The output resistance (*i.e.*, the resistance looking from the output terminals 2 : 2′ with input terminals open)

$$= 4 + 4 = 8\ \Omega$$

∴ Output conductance,

$$h_{22} = 1/8 = 0.125 \text{ mhos} \quad \textbf{Ans.}$$

25.4. Another Representation for *h*-parameters

Table 25.1 summaries the meaning of each *h*-parameter and the required condition.

Table 25.1 : *h*-parameters

S. No.	*Parameter*	*Meaning*	*Condition*
1.	h_{11}	Input resistance	Output shorted
2.	h_{12}	Reverse voltage gain	Input open
3.	h_{21}	Forward current gain	Output shorted
4.	h_{22}	Output conductance	Input open

It will be interesting to know that the parameters h_{11} and h_{21} are called forward parameters h_{12} and h_{22} are called reverse parameters. The forward and reverse parameters indicate the way they represent themselves. For example, h_{12} is a reverse parameter because it is a voltage gain of v_1/v_2 (and not a forward gain v_2/v_1). Similarly, h_{22} is a reverse parameter because it is an output conductance. Sometimes, it is more convenient to represent the parameters h_{11}, h_{12} and h_{22} as h_i, h_r, h_f and h_o respectively. In other words,

$h_i = h_{11}$ = Input resistance with output shorted,

$h_r = h_{12}$ = Reverse voltage gain with input open,

$h_f = h_{21}$ = Forward current gain with output shorted, and

$h_o = h_{22}$ = Output conductance with input open.

These notations for *h*-parameters may be easy to remember by knowing the meaning of *i*, *r*, *f* and 0, which are given as below:

i = Input r = Reverse

f = Forward o = Output

25.5. The *h*-parameter Notations for Transistors

The *h*-parameters of a transistor depend upon the types of the connection (or configuration) used *i.e.*, common-emitter (CE), common-collector (CC) or common-base (CB). Because of this, each of the four h-parameters carries a second subscript letter *e*, *b* or *c*. The letter '*e*' is used to designate common-emitter; '*c*' for common collector and '*b*' for common base configuration. Table 25.2 shows the notations commonly used for transistor *h*-parameters.

Table 25.2.

S. No.	*General Parameter*	*Transistor configuration*		
		Common emitter	*Common base*	*Common collector*
1.	h_{11}	h_{ie}	h_{ib}	h_{ic}
2.	h_{12}	h_{re}	h_{rb}	h_{rc}
3.	h_{21}	h_{fe}	h_{fb}	h_{fc}
4.	h_{22}	h_{oe}	h_{ob}	h_{oc}

25.6. *h*-parameters of a Transistor

It has been observed that for small signal operation, the behaviour of a transistor is like a linear device (*i.e.*, a device whose output is directly proportional to the input and varies continuously with the input signal). Thus for small signal operation, every transistor has a set of four *h*-parameters. The values of these parameters depend upon the following factors:

1. Transistor type.
2. Transistor configuration.
3. Operating point.
4. Frequency.
5. Temperature.

The values of *h*-parameters may be obtained experimentally or graphically from the transistor characteristics. The procedure for determining *h*-parameters will be discussed later in this chapter.

25.7. Hybrid Equivalent Circuit of a Transistor

Fig. 25.5. (*a*) shows the transistor with its input and output terminals. The amplifier can be formed by connecting a signal source at its input terminals and a load resistance at its output terminals.

Here the transistor represents in any one of the three possible configurations. Fig. 25.5(b) shows the general form of a h-parameter equivalent circuit of such a transistor. The equivalent circuit is known as small-signal low-frequency hybrid model of a transistor.

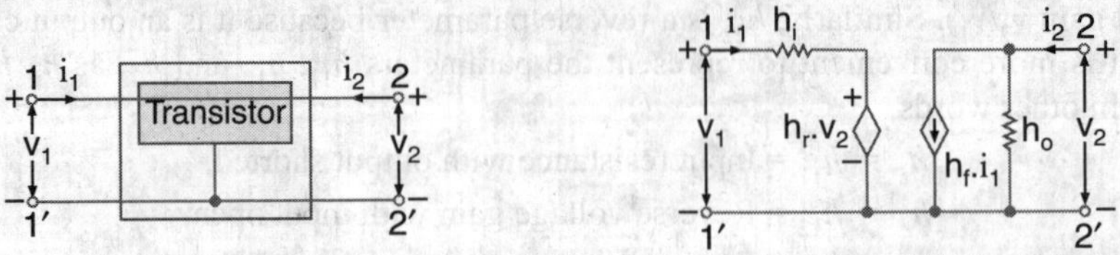

(a) Transistor. (b) Hybrid equivalent circuit of a transistor.

Fig. 25.5.

It may be noted in Fig. 25.5 (b) that the input resistance of a transistor (h_i) appears in series at the input. The reverse voltage gain (h_r) is multiplied by the output voltage to produce an equivalent voltage source (equal to $h_r \cdot v_2$) in series with the input. The equivalent voltage source indicates that in a practical transistor, some of the output voltage is fedback to the input. This represents an unavoidable interaction between the input and output circuits. The forward current gain (h_f) is multiplied by the input current and appears as an equivalent current source ($h_f \cdot i_1$) in the output. The output resistance ($i.e.$, reverse of output conductance, which is equal to $1/h_o$) appears across the output terminals. Since the transistor can be connected in any one of the three different ways namely common emitter, common base and common collector, therefore, there are three hybrid equivalent circuits, which will be discussed in the following pages.

25.8. Hybrid Equivalent Circuit for Common Emitter Transistor

Fig. 25.6 (a) shows the transistor connected in common-emitter (CE) configuration and Fig. 25.6 (b) shows the hybrid equivalent circuit of such a transistor.

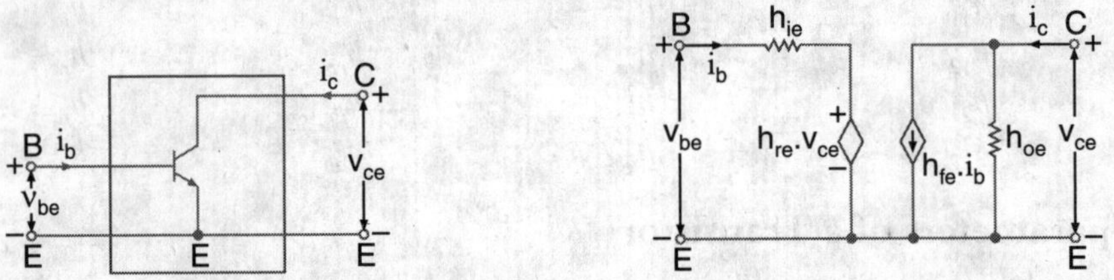

(a) Common emitter transistor. (b) Hybrid equivalent circuit of a common emitter transistor.

Fig. 25.6.

In a common emitter transistor configuration, the input signal is applied between the base and emitter terminals of a transistor and output appears between the collector and emitter terminals. The input voltage (v_{be}) and the output current (i_c) are given by the following equation:

$$v_{be} = h_{ie} \cdot i_b + h_{re} \cdot v_{ce}$$
$$i_c = h_{fe} \cdot i_b + v_{ce}$$

25.9. Hybrid Equivalent Circuit for Common Base Transistor

Fig. 25.7 (a) shows the transistor connected in a common-base (CB) configuration and Fig. 25.7 (b) shows its hybrid equivalent circuit. In a common-base configuration, the input signal is applied between emitter and base terminals and output appears between collector and base terminals.

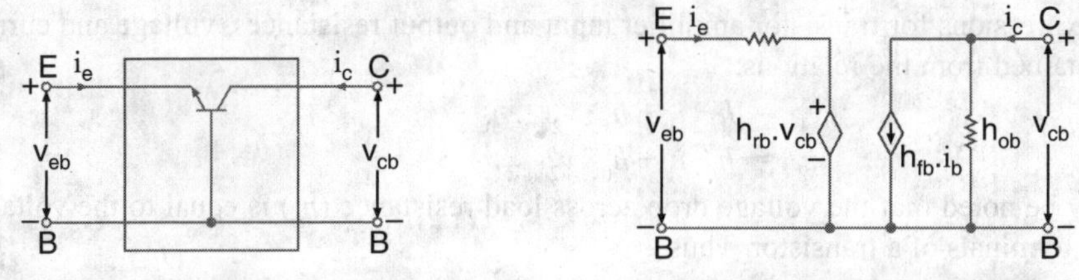

(*a*) Common base transistor. (*b*) Hybrid equivalent circuit of a common base transistor.

Fig. 25.7.

The input voltage (v_{eb}) and the output current (i_c) are given by the following equations:

$$v_{eb} = h_{ib} \cdot i_e + h_{rb} \cdot v_{cb}$$
$$i_e = h_{fb} \cdot i_c + h_{ob} \cdot v_{cb}$$

25.10. Hybrid Equivalent Circuit for Common Collector Transistor

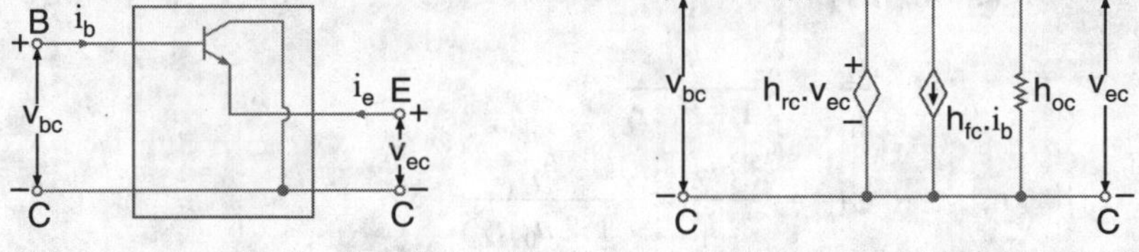

(*a*) Common collector transistor. (*b*) Hybrid equivalent circuit of a common collector transistor.

Fig. 25.8.

Fig. 25.8 (*a*) shows the transistor connected in common-collector (CC) configuration and Fig. 25.8 (*b*) shows its hybrid equivalent circuit. In a common collector configuration, the input signal is applied between base and collector terminals and the output appears between emitter and collector terminals. The input voltage (v_{bc}) and the output current (i_e) are given by the following equations:

$$v_{bc} = h_{ic} \cdot i_b + h_{rc} \cdot v_{ec}$$
$$i_e = h_{fc} \cdot i_b + h_{oc} \cdot v_{ec}$$

25.11. Amplifier Expressions

Fig. 25.9 (*a*) shows a general amplifier circuit. In this circuit, a transistor is connected in any one of the three possible configurations (*i.e.*, common emitter, common base or common collector) to a voltage source (v_s) and load resistance (r_L). The voltage source has an internal resistance (R_s) as shown in the figure. The load resistance (r_L) is the effective or a.c. load resistance seen by the transistor at its output. Fig. 25.9 (*b*) shows the hybrid equivalent circuit of a general transistor amplifier circuit. Now we shall derive the expressions (called hybrid formulas) for input resistance, output resistance, current gain and voltage gain of a transistor.

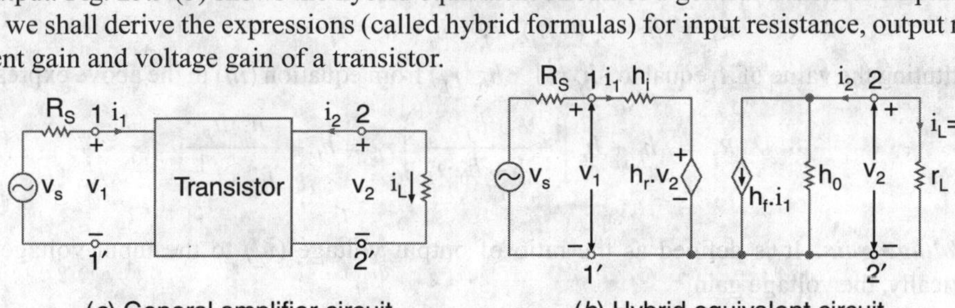

(*a*) General amplifier circuit. (*b*) Hybrid equivalent circuit.

Fig. 25.9.

The expressions for transistor amplifier input and output resistances, voltage and current gains may be obtained from the relations:

$$v_1 = h_i \cdot i_1 + h_r \cdot v_2 \quad ...(i)$$

$$i_2 = h_f \cdot i_1 + h_0 \cdot v_2 \quad ...(ii)$$

It may be noted that the voltage drop across load resistance (r_L) is equal to the voltage across the output terminals of a transistor. Thus

$$v_2 = i_L \cdot r_L = -i_2 \cdot r_L \quad ...(\because i_2 = -i_L)$$

(*a*) ***Current gain.*** It is the ratio of output current (i_L) to input currents (i_1). Mathematically, the current gain,

$$A_i = \frac{i_L}{i_1} = -\frac{i_2}{i_1}\alpha$$

Substituting the value of v_2 (equal to $-i_2 \cdot r_L$) in equation (*ii*),

$$i_2 = h_f \cdot i_1 + h_0(-i_2 \cdot r_L) = h_f \cdot i_1 - h_0 \cdot i_2 . r_L$$

$$i_2 + h_0 \cdot i_2 \cdot r_L = h_f \cdot i_1$$

$$i_2(1 + h_0 \cdot r_L) = h_f \cdot i_1$$

$$\frac{i_2}{i_1} = \frac{h_f}{1 + h_0 . r_L}$$

$$\therefore \quad A_i = -\frac{i_2}{i_1} = -\frac{h_f}{1 + h_0 \cdot r_L} \quad ...(iii)$$

(*b*) ***Input resistance.*** The resistance, which we see looking into the amplifier input terminals (1,1′) is called the amplifier input resistance. It is also known as the input resistance looking into the base of a transistor for common emitter and common collector amplifier and is designated by R_i. Its value is given by the ratio of input voltage (v_1) to the input current (i_1). Mathematically, the input resistance looking into the input terminals,

$$R_i = \frac{v_1}{i_1}$$

Again substituting the value of v_2 (equal to $-i_2 \cdot r_L$) in equation (*i*)

$$v_1 = h_i \cdot i_1 + h_r(-i_2 \cdot r_L)$$

$$= h_i \cdot i_1 - h_r \cdot i_2 \cdot r_L$$

Dividing the above equation by i_1 on both sides,

$$\frac{v_1}{i_1} = \frac{h_i \cdot i_1 - h_r \cdot i_2 \cdot r_L}{i_1} = h_i - h_r\left(\frac{i_2}{i_1}\right) r_L$$

Replacing v_1/i_1 with R_i and i_2/i_1 by $-A_i$ in the above expression,

$$R_i = h_i - h_r(-A_i) \cdot r_L = h_i + h_r \cdot A_i \cdot r_L \quad ...(iv)$$

Substituting the value of A_i equal to $-h_f/(1 + h_0 \cdot r_L)$ from equation (*iii*) in the above expression,

$$R_i = h_i + h_r\left(-\frac{h_f}{1 + h_0 \cdot r_L}\right) r_L = h_i - \frac{h_r \cdot h_f}{h_0 + \frac{1}{r_L}} \quad ...(v)$$

(*c*) ***Voltage gain.*** It is defined as the ratio of output voltage (v_2) to the input voltage (v_1). Mathematically, the voltage gain,

$$A_v = \frac{v_2}{v_1} = -\frac{i_2 \cdot r_L}{v_1} \quad ...(\because v_2 = -i_2 \cdot r_L)$$

Substituting the value of i_2 (equal to $-A_i \cdot i_1$) from equation (*iii*) the above expression,

$$A_v = \frac{A_i \cdot i_1 \cdot r_L}{v_1} = A_i \cdot r_L \left(\frac{i_1}{v_1}\right)$$

Replacing i_1/v_1 by R_i in the above expression,

$$A_v = \frac{A_i \cdot r_L}{R_i}$$

Substituting the value of A_i (equal to $-h_f/(1 + h_0 \cdot r_L)$) and R_i (equal to $\frac{h_i - h_r \cdot h_L}{(h_0 + 1/r_L)}$) in the above expression and rearranging,

$$A_v = -\frac{h_f \cdot r_L}{h_i + (h_i \cdot h_0 - h_r \cdot h_f)\, r_L}$$

If we replace($h_i \cdot h_0 - h_r \cdot h_f = \Delta h$) in the above expression, then the voltage gain,

$$A_v = -\frac{h_f \cdot r_L}{h_i + \Delta h \cdot r_L}$$

(*d*) ***Output resistance.*** It is obtained by setting the source voltage (V_S) to zero and the load resistance (r_L) to infinity and by driving the output terminals from a generator v_2 as shown in Fig.25.10.

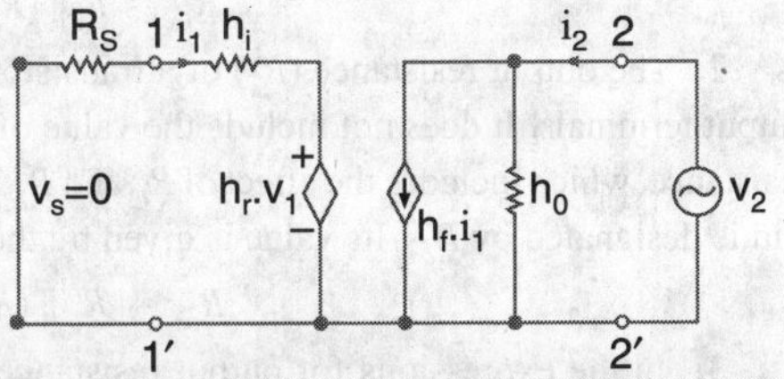

Fig. 25.10. Hybrid equivalent circuit of a general transistor amplifier with voltage source connected across output terminals.

Then output resistance is the ratio of voltage (v_2) to the current drawn from the voltage source (i_2). Mathematically, the output resistance,

$$R_0 = \frac{v_2}{i_2} = \frac{v_2}{h_f \cdot i_1 + h_0 \cdot v_2} \quad \text{...(vi)}$$

The value of current (i_1) may be obtained by applying Kirchhoff's Voltage Law to the input side of transistor amplifier circuit. Thus

$$R_s \cdot i_1 + h_i \cdot i_1 + h_r \cdot v_2 = 0$$

$$\therefore \quad i_1 = -\frac{h_r \cdot v_2}{R_s + h_i}$$

Substituting this value of i_1 in equation (*vi*),

$$R_0 = \frac{v_2}{h_f\left(-\frac{h_r \cdot v_2}{R_s + h_i}\right) + h_0 \cdot v_2}$$

$$= \frac{v_2}{-\frac{h_f \cdot h_r \cdot v_2}{R_s + h_i} + h_0 \cdot v_2}$$

Rearranging the above equation,

$$R_0 = \frac{R_s + h_i}{(R_s + h_i)\, h_0 - h_f \cdot h_r} = \frac{R_s + h_i}{R_s \cdot h_0 + (h_i \cdot h_0 - h_f \cdot h_r)}$$

$$= \frac{R_s + h_i}{R_s \cdot h_0 + \Delta h}$$

It may be noted from the above relation that if source resistance (R_s) is zero, then output resistance,

$$R_0 = \frac{h_i}{\Delta h}$$

(*e*) ***Power gain.*** It is defined as the product of voltage gain (A_v) and the current gain (A_i). Mathematically, the power gain,

$$\Delta P = A_v \cdot A_i$$

Notes: 1. The input resistance (R_i) of a transistor is the value of resistance looking directly into the transistor from the input terminals. It does not include the values of resistor connected externally for the biasing purpose. The input resistance, which takes into account the biasing resistors, is called resistance of the stage (or stage input resistance) and is designated by R_{is}. For a fixed bias current, the value of stage input resistance,

$$R_{is} = R_i \,\|\, R_B$$

and for voltage divider bias circuit,

$$R_{is} = R_i \,\|\, (R_1 \| R_2)$$

2. The output resistance (R_o) of a transistor is the resistance looking directly into the transistor from the output terminals. It does not include the value of collector resistance (R_C) and load resistance (R_L). The output resistance, which includes the effect of R_C and R_L, is called output resistance of a stage (or stage output resistance) and is designated by R_{os}. Its value is given by the relation,

$$R_{os} = R_o \,\|\, (R_C \,\|\, R_L) = R_O \,\|\, r_L \qquad \text{... } (\because r_L = R_C \,\|\, R_L)$$

3. In the expressions for output resistance, current and voltage gains, we require the value of a.c load resistance (*i.e.*, r_L). The value of r_L depends upon the type of connection used for transistor. For common emitter and common base connections, the a.c. load resistance,

$$r_L = R_C \qquad \text{... (If } R_L \text{ is not connected)}$$

$$= R_C \,\|\, R_L \qquad \text{... (If } R_L \text{ is connected)}$$

However, for a common collector (or emitter follower connection) the value of a.c. load resistance,

$$r_L = R_E \qquad \text{... (If } R_L \text{ is not connected)}$$

$$= R_E \,\|\, R_L \qquad \text{... (If } R_L \text{ is connected)}$$

25.12. Effect of Source Resistance on Voltage and Current Gains

We have already discussed in the last article the current and voltage gains of a transistor amplifier. The expressions for current and voltage gains derived there are from the amplifier input terminals (1, 1′) to the output terminal (2, 2′). In other words, these values do not take into consideration the source resistance (R_S). However, it is important to take into account the source resistance (R_S), while calculating the voltage and current gains of the transistor amplifier. In this case, the voltage and current gains are known as overall gains or stage gains.

The overall gains of the transistor amplifier is the ratio of output voltage (v_2) to the source voltage (v_s). Mathematically, the overall voltage gain,

$$A_{vs} = \frac{v_2}{v_s}$$

Multiplying and dividing the right hand side of the above equation by v_1,

$$A_{vs} = \frac{v_2}{v_s} \times \frac{v_1}{v_1} = \frac{v_2}{v_1} \times \frac{v_1}{v_s} = A_v \times \frac{v_1}{v_s}$$

where A_v (equal to v_2/v_1) is the voltage gain of amplifier from the input terminals (1, 1′) to the output terminals (2, 2′). In order to determine the voltage ratio (v_1/v_s), consider the circuit shown in Fig. 25.11 (*a*).

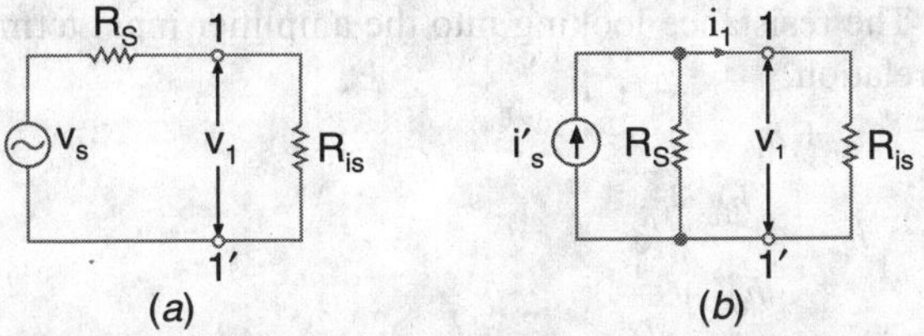

Fig. 25.11. Effect of source resistance on amplifier voltage and current gains.

In this figure, the resistance R_{is} represents the amplifier input resistance. It is evident from this figure that the voltage drop across input resistance,

$$v_1 = v_s \times \frac{R_{is}}{R_s + R_{is}}$$

$$\therefore \quad \frac{v_1}{v_s} = \frac{R_{is}}{R_s + R_{is}}$$

$$\text{and} \quad A_{vs} = A_v \times \frac{R_{is}}{R_s + R_{is}}$$

Similarly, the overall current gain of the transistor amplifier is the ratio of output current (i_L) to the current delivered by the source (i_s). Mathematically, the overall current gain,

$$A_{is} = \frac{i_L}{i_s} = -\frac{i_2}{i_s} \qquad \ldots (\because i_L = -i_2)$$

Multiplying and dividing the right hand side of the above equation by *i*,

$$A_{is} = \frac{i_2}{i_s} \times \frac{i_1}{i_1} = -\frac{i_2}{i_1} \times \frac{i_1}{i_s} = -A_i \times \frac{i_1}{i_s}$$

where A_i (equal to (i_2 / i_1) is the current gain of the amplifier from the input terminals (1, 1′) to the output terminals (2,2′). In order to determine the current ratio, consider the circuit shown in Fig. 25.11 (*b*). Here the voltage source has been replaced by its equivalent current source. It may be noted that the current through the amplifier input resistance,

$$i_1 = i_s \times \frac{R_s}{R_s \times R_{is}}$$

$$\therefore \quad \frac{i_1}{i_s} = \frac{R_s}{R_s + R_{is}}$$

$$\text{and} \quad A_{is} = A_i \times \frac{i_1}{i_s} = A_i \times \frac{R_s}{R_s + R_{is}}$$

25.13. Hybrid Formulas for Common Emitter Amplifier

These formulas may be obtained from the general hybrid formulas derived in Art. 25.11 by adding a second subscript letter '*e*' (which stands for common emitter) with the *h*-parameters and are as discussed below:

(a) Current gain. It is given by the relation,

$$A_i = -\frac{h_f \cdot r_L}{h_i + \Delta h \cdot r_L}$$

where r_L is the a.c. load resistance. Its value is equal to the parallel combination of resistances R_C and R_L. Since h_{fe} of a transistor is a positive number, therefore, A_i of a common emitter amplifier is negative.

(*b*) ***Input resistance.*** The resistance looking into the amplifier input terminals (*i.e.*, base of a transistor) is given by the relation,

$$R_i = h_{ie} + h_{re} \cdot A_i \cdot r_L$$

$$= h_{ie} - \frac{h_{re} \cdot h_{fe}}{h_{oe} + \dfrac{1}{r_L}}$$

The input resistance of the amplifier stage (called stage input resistance R_{is}) depends upon the biasing arrangement. For a fixed bias circuit, the stage input resistance,

$$R_{is} = R_i \parallel R_B$$

If the circuit has no biasing resistances, then $R_{is} = R_i$.

(*c*) ***Voltage gain.*** It is given by the relation,

$$A_v = \frac{A_i \cdot r_L}{R_i}$$

Since the current gain (A_i) of a common emitter amplifier is negative, therefore, the voltage gain (A_v) is also negative. It means that there is a phase difference of 180° between the input and output. In other words, the input signal is inverted at the output of a common emitter amplifier. The voltage gain, in terms of h-parameters, is given by the relation,

$$A_v = -\frac{h_{fe} \cdot r_L}{h_{ie} + \Delta h \cdot r_L}$$

where $\Delta h = h_{ie} \cdot h_{oe} - h_{re} \cdot h_{fe}$

(*d*) ***Output resistance.*** The resistance looking into the amplifier output terminals is given by the relation,

$$R_0 = \frac{R_s + h_{ie}}{R_s \cdot h_{oe} + \Delta h}$$

where R_s = Resistance of the source, and

$$\Delta h = h_{ie} \cdot h_{oe} - h_{re} \cdot h_{fe}$$

The output resistance of the stage,

$$R_{os} = R_o \parallel r_L$$

(*e*) ***Overall voltage gain.*** It is given by the relation,

$$A_v = \frac{A_v \cdot R_s}{R_s + R_{is}}$$

(*f*) ***Overall current gain.*** It is given by relation,

$$A_{is} = \frac{A_i \cdot R_s}{R_s + R_{is}}$$

Example 25.2.

The h-parameters of a transistor used in a common emitter circuit are h_{ie} = *1.0 kΩ,* $h_{re} = 10 \times 10^{-4}$, h_{fe} = *50 and* h_{oe} = *100* m *mhos. The load resistor for the transistor is 1 kΩ in the collector circuit. The transistor is supplied from a signal source of resistance 1000 Ω. Determine the value of input and output impedances (or resistances), voltage and current gains in the amplifier stage.*

Solution.

Given: $h_{ie} = 1.0\ k\Omega = 1000\ \Omega$; $h_{re}\ 1.0 \times 10^{-4}$; $h_{oe} = 100\ \mu$ mhos $= 100 \times 10^{-6}$ mhos; $R_C = 1k\Omega$ $= 1000\ \Omega$ and $R_s = 1000\ \Omega$.

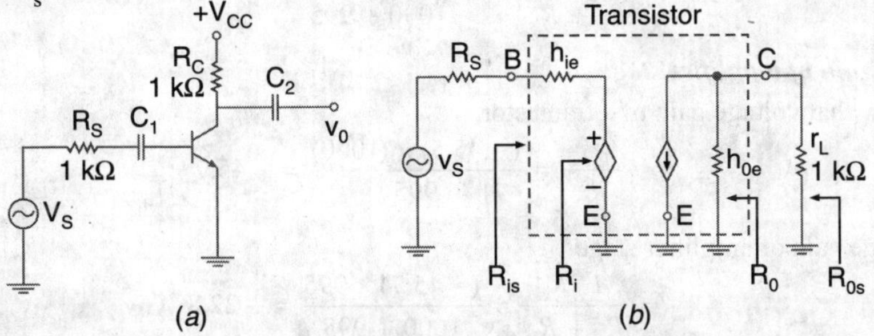

Fig. 25.12.

Fig. 25.12. (*a*) shows the common emitter transistor amplifier circuit and (*b*) shows its hybrid equivalent circuit.

(*a*) *Input resistance of the amplifier stage*

We know that there is no load connected at the output of the amplifier (*i.e.*, $R_L = 0$), therefore, the value of a.c. load resistance,

$$r_L = R_C = 1000\ \Omega$$

We also know that current gain of a transistor,

$$A_i = \frac{h_{fe}}{1 + h_{oe} \cdot r_L} = -\frac{50}{1 + (100 \times 10^{-6}) \times 1000} = -45.5$$

and the input resistance of a transistor, looking directly into the base,

$$R_i = h_{ie} + h_{re} \cdot A_i \cdot r_L$$
$$= 1000 + [(1.0 \times 10^{-4}) \times (-45.5) \times 1000]\ \Omega$$
$$= 995\ \Omega$$

$\therefore$ Input resistance of the amplifier stage,

$$R_{is} = R_i = 995\ \Omega \textbf{ Ans.}$$

(*b*) *Output resistance of amplifier stage*

We know that

$$\Delta h = h_{ie} \cdot h_{oe} - h_{re} \cdot h_{fe}$$
$$= [1000 \times (100 \times 10^{-6})] - (1.0 \times 10^{-4}) \times 50$$
$$= 95 \times 10^{-3} = 0.095$$

$\therefore$ Output resistance of the transistor looking directly into collector,

$$R_0 = \frac{R_s + h_{ie}}{R_s \cdot h_{oe} + \Delta h}$$

$$= \frac{1000 + 1000}{[1000 \times (100 \times 10^{-6})] + 0.095}\ \Omega$$

$$= \frac{2000}{0.1 + 0.095} = 10.256\ \Omega$$

and output resistance of the amplifier stage,

$$R_{os} = R_o \parallel r_L = 10256 \parallel 1000 = 911\ \Omega \textbf{ Ans.}$$

(c) ***Current gain of amplifier stage***

We know that the current gain of amplifier stage,

$$A_{is} = \frac{A_i \cdot R_s}{R_s + R_{is}} = \frac{(-45.5) \times (1000)}{1000 + 995} = -22.8 \textbf{ Ans.}$$

(d) ***Voltage gain of amplifier stage***

We know that voltage gain of a transistor,

$$A_v = \frac{A_i \cdot r_L}{R_i} = \frac{(-45.5) \times (1000)}{995} = -45.7$$

∴ Voltage gain of amplifier stage,

$$A_{vs} = \frac{A_v \cdot R_{is}}{R_s + R_{is}} = \frac{(-45.7) \times 995}{1000 + 995} = -22.8 \textbf{ Ans.}$$

Example 25.3.

A BJT has h_{ie} = 2kΩ, $h_{re} = 16 \times 10^{-5}$, h_{fe} = 50, h_{oe} = 50 μS. Calculate the current gain, voltage gain input resistance and output resistance, if the load resistance is 10 kΩ.

(M. G. University., May 2005)

Solution.

Given: $h_{ie} = 2\text{ k}\Omega = 2 \times 10^3\Omega$; $h_{re} = 16 \times 10^{-5}$; $h_{fe} = 50$; $h_{oe} = 50\ \mu\text{S} = 50 \times 10^{-6}$ S; and $r_L = 10\text{ k}\Omega = 10 \times 10^3\ \Omega$

Current Gain

We know that current gain,

$$A_i = \frac{h_{fe}}{1 + h_{oe}.r_L} = \frac{50}{1 + (50 \times 10^{-6}) \times (10 \times 10^3)}$$

$$= 33.33 \textbf{ Ans.}$$

Input Impedance

We know that input impedance,

$$R_i = h_{ie} + h_{re}.A_i.r_L$$
$$= (2 \times 10^3) + (16 \times 10^{-5}) \times (33.33) \times (10 \times 10^3)$$
$$= 106{,}656\ \Omega \textbf{ Ans.}$$

Voltage Gain

We know that the voltage gain,

$$A_v = \frac{A_i.r_L}{R_i} = \frac{33.33 \times (10 \times 10^3)}{106{,}656}$$

$$= 3.125 \textbf{ Ans.}$$

Example 25.4.

A transistor connected in CE configuration has the following h-parameters.
h_{ie} = 1·1 kΩ; $h_{re} = 2{\cdot}5 \times 10^{-4}$; h_{fe} = 50, h_{oe} = 25 μ siemens and $r_S = r_L$ = 1 kΩ
Calculate the current gain, input impedance and voltage gain.

(Bangalore University, B.Sc. April/May 2001)

Solution.

Given: h_{ie} = 1·1 kΩ = 1·1 × 10³ Ω; h_{re} = 2·5 × 10⁻⁴; h_{fe} = 50, h_{oe} = 25 μ_S = 25 × 10⁻⁶ *S*; and $r_S = r_L$ = 1 kΩ = 1 × 10³ Ω

Current Gain

We know that current gain,

$$A_i = \frac{h_{fe}}{1 + h_{oe} \cdot r_L} = \frac{50}{1 + (25 \times 10^{-6}) \times (1 \times 10^3)}$$

$$= 48.78 \textbf{ Ans.}$$

Input Impedance

We know that input impedance,

$$R_i = h_{ie} + h_{re} \cdot A_i \cdot r_L$$

$$= (1{\cdot}1 \times 10^3) + (2.5 \times 10^{-4}) \times (48{\cdot}78) \times (1 \times 10^3)$$

$$= 1112.2\ \Omega \textbf{ Ans.}$$

Voltage Gain

We know that voltage gain,

$$A_v = \frac{A_i \cdot r_L}{R_i} = \frac{48{\cdot}78 \times (1 \times 10^3)}{1112.2} = 43.86 \textbf{ Ans.}$$

25.14. Hybrid Formulas for Common Emitter Amplifier with Collector Feedback Bias

Consider a common emitter amplifier with collector feedback bias as shown in Fig. 25.13 (*a*). Such a circuit can be analysed by using Miller's Theorem. According to this theorem, the collector feedback bias resistance (R_B) can be disconnected from its position and can be shown as an equivalent of two resistances R_{B_1} and R_{B_2} as shown in Fig. 25.13 (*b*).

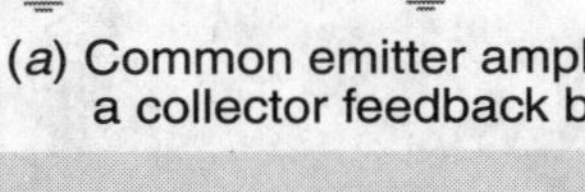

(*a*) Common emitter amplifier with a collector feedback bias.

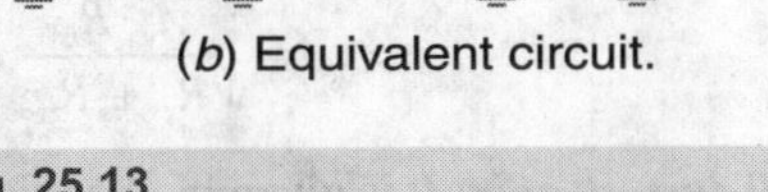

(*b*) Equivalent circuit.

Fig. 25.13.

The resistance R_{B_1} is shown to be connected between the base and ground and its value is given by the relation,

$$R_{B_1} = \frac{R_B}{1 - A_v}$$

where A_v = Voltage gain

Since the value of voltage gain of common emitter amplifier is very large, as compared to unity (*i.e.*, $A_v >> 1$), therefore, the value of resistance R_{B_1} is much smaller than that of resistance R_B. The resistance R_{B_2} is shown to be connected between the collector and ground and its value is given by the relation,

$$R_{B_2} = \frac{R_B}{1 - \frac{1}{A_v}}$$

Since the voltage gain is very large, as compared to unity (*i.e.*, $A_v >> 1$), therefore, value of resistance R_{B_2} is approximately equal to resistance R_B. The circuit shown in Figure 25.13 (*b*) can now be analysed easily as discussed earlier in Art. 25.13. Thus a.c. equivalent load resistance of the circuit,

$$r_L = R_C \parallel R_{B_2} = \frac{R_C \times R_{B_2}}{R_C + R_{B_2}}$$

(*a*) ***Current gain.*** It is given by the relation,

$$A_i = \frac{h_{fe}}{1 + h_{oe} \cdot r_L}$$

(*b*) ***Input resistance.*** The resistance looking into the amplifier input terminal (*i.e.*, base of the transistor) is given by the relation,

$$R_i = i_{ie} + h_{re} \cdot A_i \cdot r_L$$

and input resistance of the amplifier stage,

$$R_{is} = R_i \parallel R_{B_1}$$

(*c*) ***Voltage gain.*** It is given by the relation,

$$A_v = \frac{A_i \cdot r_L}{R_i}$$

(*d*) ***Output resistance.*** The resistance looking into the amplifier output terminals is given by the relation,

$$R_0 = \frac{R_s + h_{ie}}{R_s \cdot h_{oe} + \Delta h}$$

where R_s = Resistance of the source, and

$$\Delta h = h_{ie} \cdot h_{0e} - h_{re} \cdot h_{fe}$$

The output resistance of the amplifier stage,

$$R_{0s} = R_0 \parallel r_L$$

(*e*) ***Overall voltage gain.*** It is given by the relation,

$$A_{vs} = \frac{A_v \cdot R_{is}}{R_s + R_{is}}$$

(*f*) ***Overall current gain.*** It is given by the relation,

$$A_{is} = \frac{A_i \cdot R_s}{R_s + R_{is}}$$

Example 25.5.

Consider a common emitter transistor amplifier with collector load resistance $R_C = 4$ kΩ, collector-to-base feedback resistance $R_B = 40$ kΩ and a source resistance $R_S = 10$ kΩ. The transistor parameters are $h_{ie} = 1100$ Ω; $h_{fe} = 50$; $h_{re} = h_e = 0$. Calculate the voltage gain, the input resistance looking from the source terminals and the output resistance. (*Grad. I.E.T.E., June 1987*)

Solution.

Given: $R_C = 4\text{ k}\Omega$; $R_B = 40\text{ k}\Omega$; $R_S = 10\text{ k}\Omega = 10 \times 10^3\ \Omega$; $h_{ie} = 1100\ \Omega$; $h_{fe} = 50$ and $h_{re} = h_{oe} = 0$.

Fig. 25.14 (*a*) shows the circuit with the given resistance value, while Fig. 25.14 (*b*) shows its equivalent circuit.

Input resistance

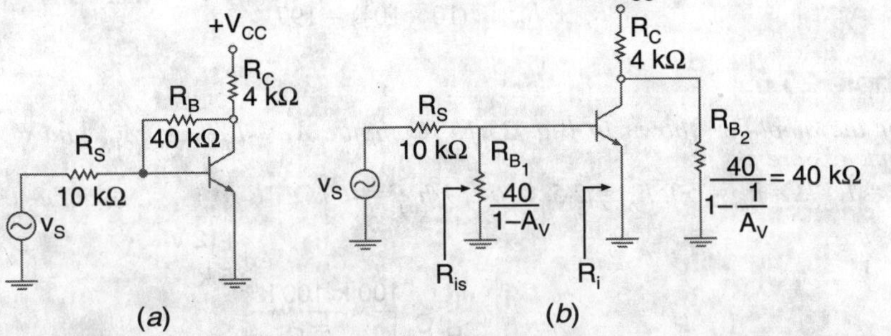

Fig. 25.14.

We know that the resistance,

$$R_{B_2} = \frac{40}{1 - \frac{1}{A_v}} = 40\text{ k}\Omega \quad \text{....}(\because A_v >> 1)$$

∴ A.C. load resistance,

$$r_L = R_C \,\|\, R_{B_2} = 4 \,\|\, 40\text{ k}\Omega = 3.64\text{ k}\Omega = 3640\ \Omega$$

We also know that current gain,

$$A_i = \frac{h_{fe}}{1 + h_{oe} \cdot r_L} = -h_{fe} = -50 \quad \text{...}(\because h_{oe} = 0)$$

and the input resistance of the amplifier looking into the base,

$$R_i = h_{ie} + h_{re} \cdot A_i \cdot r_L = h_{ie} = 1100\ \Omega \quad \text{...}(\because h_{re} = 0)$$

We also know that the value of voltage gain,

$$A_v = \frac{A_i \cdot r_L}{R_i} = \frac{(-50) \times 3640}{1100} = -166$$

and the value of resistance,

$$R_{B_1} = \frac{40}{1 - A_v} = \frac{40}{1 - (-166)}\text{ k}\Omega$$

$$= 0.240\text{ k}\Omega = 240\ \Omega$$

∴ Input resistance (*i.e.*, resistance looking from the source terminals),

$$R_{is} = R_i \,\|\, R_{B_1} = 1100 \,\|\, 240 = 197\ \Omega \textbf{ Ans.}$$

Output resistance

We know that output resistance,

$$R_O = \frac{R_s + h_{ie}}{R_s \cdot h_{oe} + \Delta h} = \infty \quad \text{... }(\because h_{oe} - \Delta h - 0)$$

∴ Output resistance of the stage,

$$R_{os} = R_0 \| r_L = r_L = 3640\ \Omega\ \textbf{Ans.} \qquad ...(\because R_0 = \infty)$$

Voltage gain of the stage

We know that voltage gain of the stage (*i.e.*, voltage gain of the amplifier looking from the source terminals called stage gain),

$$A_{vs} = \frac{A_v \cdot R_{is}}{R_s + R_{is}} = \frac{(-166) \times 197}{(10 \times 10^3) + 197} = -3.2\ \textbf{Ans.}$$

Example 25.6.

For the amplifier shown in Fig. 25.15. Compute $A_i = I_o/I_i$, A_v, A_{VS} and R_i. The values of h-parameters are :

$h_{ie} = 1.1\ k\Omega$; $h_{fe} = 50$; $h_{re} = 2.5 \times 10^{-4}$, $1/h_{oe} = 40\ k\Omega$. *(Nagpur University 2004)*

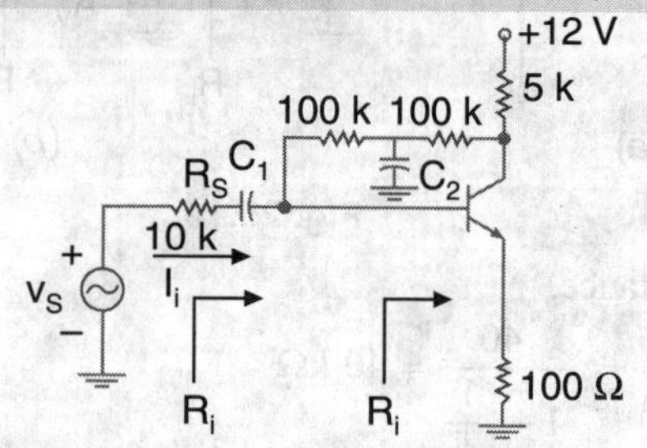

Fig. 25.15.

Solution.

Given : $h_{ie} = 1.1\ k\Omega$, $h_{fe} = 50$; $h_{re} = 2.5 \times 10^{-4}$, $1/h_{oe} = 40\ k\Omega$ or $h_{oe} = 1/40\ k\Omega = 0.025 \times 10^{-3}$ or $25 \times 10^{-6}\ \Omega$.

We know that the circuit shown in Fig. has a collector feedback bias. The value of feedback resistor $R_B = 100\ k + 100\ k = 200\ k\Omega$. The capacitor ($C_2$) in between the two 100 *k* resistors has no useful role accept that it will filter out any ac signal compled from base to collector or collector to base.

The collector feedback resistor (R_B) = 200 kΩ is equivalent of two resistances R_{B_1} and R_{B_2} (refer to Fig. 25.13)

We know that the ac load resistance,

$$r_L = R_c \| R_{B_2} = \frac{R_c \times R_{B_2}}{R_c + R_{B_2}}$$

$$= \frac{5\ \text{k} \times 200\ \text{k}}{5\ \text{k} + 200\ \text{k}} = 4.88\ \text{k}\Omega \qquad (\because R_{B_2} \simeq R_B)$$

and the current gain,

$$A_i = \frac{h_{fe}}{1 + h_{oe} \cdot r_L}$$

$$= \frac{50}{1 + (25 \times 10^{-6}) \times (4.88 \times 10^3)}$$

$$= 44.52$$

The input resistance,

$$R_i = h_{ie} + h_{re} \cdot A_i \cdot r_L$$

$$= (1.1 \times 10^3) + (2.5 \times 10^{-4}) \times 44.52 \times (4.88 \times 10^3)$$
$$= 1.154 \times 10^3\ \Omega \text{ or } 1.154\ k\Omega$$

We also know that voltage gain,

$$A_v = \frac{A_i \cdot r_L}{R_i}$$

$$= \frac{44.68 \times (4.88 \times 10^3)}{(1.154 \times 10^3)} = 188.94 \textbf{ Ans.}$$

Note. The voltage for a common emitter amplifier is negative.

We also know that the value of resistance,

$$R_{B_1} = \frac{R_B}{1 - A_v} = \frac{200k}{1-(-17.4)} = 10.87\ k\Omega$$

∴ Input resistance looking from the source terminal,

$$R_{is} = R_i \,||\, R_{B_1} = 1.154\ k \,||\, 10.87 = 1.04\ k\Omega$$

and the voltage gain of the stage,

$$A_{vs} = \frac{A_v \cdot A_{is}}{R_s + R_{is}}$$

$$= \frac{188.94 \times 1.04k}{10k + 1.04k} = 17.8 \textbf{ Ans.}$$

25.15. Hybrid Formulas for Common Base Amplifier

These formulas may be obtained from the general hybrid formulas derived in Art. 25.11 by adding a second subscript letter '*b*' (which stands for common base) with *h*-parameters and are as discussed below:

(a) Current gain. It is given by the relation,

$$A_i = -\frac{h_{fb}}{1 + h_{ob} \cdot r_L}$$

where r_L is the a.c. load resistance. Its value is equal to the parallel combination of resistances R_C and R_L. Since h_{fb} of a transistor is a negative number, therefore, the value of current gain (A_i) of a common base amplifier is positive.

(b) Input resistance. The resistance looking into the input terminals (*i.e.*, emitter of a transistor) is given by the relation,

$$R_i = h_{ib} + h_{rb} \cdot A_i \cdot r_L$$

$$= h_{ib} - \frac{h_{rb} \cdot h_{fb}}{h_{ob} + \dfrac{1}{r_L}}$$

The input resistance of the amplifier stage (called stage input resistance) R_{is} depends upon the biasing arrangement. If the amplifier circuits has no biasing resistance, then $R_{is} = R_i$.

(c) Voltage gain. It is given by the relation,

$$A_v = \frac{A_i \cdot r_L}{R_i}$$

Since current gain (A_i) of a common base amplifier is positive, therefore, the voltage gain (A_v) is also positive. It means that there is no phase difference between the input and output signals of a common base amplifier. The voltage gain, in terms of *h*-parameters, is given by the relation,

$$A_v = -\frac{h_{fb} \cdot r_L}{h_{ib} + \Delta h \cdot r_L}$$

where $\Delta h = h_{ib} \cdot h_{ob} - h_{rb} \cdot h_{fb}$

(*d*) ***Output resistance.*** The resistance looking into the amplifier output terminals is given by the relation,

$$R_0 = \frac{R_s + h_{ib}}{R_s \cdot h_{ob} + \Delta h}$$

where R_s = Resistance of the source, and

$\Delta h = h_{ib} \cdot h_{ob} - h_{rb} \cdot h_{fb}$

The output resistance of the amplifier stage,

$$R_{os} = R_o \parallel r_L$$

(*e*) ***Overall voltage gain.*** It is given by the relation,

$$A_{vs} = \frac{A_v \cdot R_{is}}{R_s + R_{is}}$$

(*f*) ***Overall current gain.*** It is given by the relation,

$$A_{is} = \frac{A_i \cdot R_s}{R_s + R_{is}}$$

Example 25.7.

A transistor used in a common base amplifier has the following values of h-parameters:

$$h_{ib} = 28\ \Omega;\ h_{fb} = -0.98;\ h_{rb} = 5 \times 10^{-4};\ \text{and}\ h_{ob} = 0.34 \times 10^{-6}\ S$$

Calculate the values of input resistance, output resistance, current gain and voltage gain, if the load resistance is 1.2 kΩ. Assume source resistance as zero.

Solution.

Given: $h_{ib} = 28\ \Omega$; $h_{fb} = -0.98$; $h_{rb} = 5 \times 10^{-4}$; $h_{ob} = 0.34 \times 10^{-6}$ S; $r_L = 1.2$ kΩ = 1200 Ω and $R_s = 0$.

Current gain

We know that current gain,

$$A_i = -\frac{h_{fb}}{1 + h_{ob} \cdot r_L}$$

$$= -\frac{(-0.98)}{1 + [(0.34 \times 10^{-6}) \times 1200]}$$

$$= 0.98 \textbf{ Ans.}$$

Input resistance

We know that input resistance,

$$R_i = h_{ib} + h_{rb} \cdot A_i \cdot r_L$$

$$= 28 + [(5 \times 10^{-4}) \times 0.98 \times 1200]\ \Omega$$

$$= 28 + 0.6 = 28.6\ \Omega \textbf{ Ans.}$$

Output resistance

We know that

$$\Delta h = h_{ib} \cdot h_{ob} - h_{rb} \cdot h_{fb}$$

$$= [28 \times (0.34 \times 10^{-6})] - [(5 \times 10^{-4}) \times (-0.98)]$$
$$= 5 \times 10^{-4}$$

∴ Output resistance,

$$R_o = \frac{R_s + h_{ib}}{R_s \cdot h_{ib} + \Delta h} = \frac{h_{ib}}{\Delta h} \quad \text{... } (\because R_s = 0)$$

$$= \frac{28}{5 \times 10^{-4}} = 5.6 \times 10^4 \Omega = 56 \text{ k}\Omega \textbf{ Ans.}$$

Voltage gain

We know that voltage gain,

$$A_v = \frac{A_i \cdot r_L}{R_i} = \frac{0.98 \times 1200}{28.6} = 41 \textbf{ Ans.}$$

Example 25.8.

The CB h-parameters of a transistor are h_{ib} = 30 Ω, h_{rb} = 4 × 10^{-4}, h_{fb} = – 0.99 and h_{ob} = 0.9 × 10^{-6} S for a suitable operating point. The amplifier is used in the CB mode with a load resistance of 6 kW. Calculate the current gain and input resistance. (WBTU., 2001–2012)

Solution.

Given: h_{ib} = 20 Ω; h_{rb} = 4 × 10^{-4}; h_{fb} = –0.99; h_{ob} = 0.9 × 10^{-6} S; and r_L = 6 kΩ = 6 × 10^3 Ω.

Current gain

We know that current gain,

$$A_i = -\frac{h_{fb}}{1 + h_{ob} \cdot r_L}$$

$$= -\frac{(-0.99)}{1 + [(0.9 \times 10^{-6}) \times (6 \times 10^3)]} = 0.98 \textbf{ Ans.}$$

Input resistance

We know that input resistance,

$$R_i = h_{ib} + h_{rb} \cdot A_i \cdot r_L$$
$$= 30 + [(4 \times 10^{-4}) \times (0.98) \times (6 \times 10^3)]$$
$$= 30 + 2.35 = 32.35 \ \Omega \textbf{ Ans.}$$

25.16. Hybrid Formulas for Common Collector Amplifier

These formulas may be obtained from the general hybrid formulas derived in Art. 25.11 by adding a second subscript letter '*c*' (which stands for common collector) with *h*-parameters and are as discussed below:

(a) Current gain. It is given by the relation,

$$A_i = -\frac{h_{fc}}{1 + h_{oc} \cdot r_L}$$

where r_L is the a.c. load resistance. Its value is equal to the parallel combination of resistances R_E and R_L. Since h_{fc} of a transistor is a negative number, therefore, current gain A_i is positive.

(b) Input resistance. The resistance looking into the amplifier input terminals (*i.e.*, base of a transistor) is given by the relation,

$$R_i = h_{ic} + h_{rc} \cdot A_i \cdot r_L$$

$$= h_{ic} - \frac{h_{rc} \cdot h_{fc}}{h_{oc} + \dfrac{1}{r_L}}$$

The input resistance of the amplifier stage R_{is} (called stage input resistance) depends upon the biasing arrangement, if the amplifier circuit has no biasing resistance, then $R_{is} = R_s$.

(c) ***Voltage gain.*** It is given by the relation,

$$A_v = \frac{A_i \cdot r_L}{R_i}$$

Since current gain (A_i) of a common collector amplifier is positive, therefore, the voltage gain (A_v) is also positive. It means that there is no phase difference between the input and output signals of a common collector amplifier. The voltage gain, in terms of h-parameters, is given by the relation,

$$A_v = -\frac{h_{fc} \cdot r_L}{h_{ic} + \Delta h \cdot r_L}$$

where $\Delta h = h_{ic} \cdot h_{oc} - h_{rc} \cdot h_{fc}$

(d) ***Output resistance.*** The resistance looking into the amplifier output terminals is given by the relation,

$$R_0 = \frac{R_s + h_{ic}}{R_s \cdot h_{oc} + \Delta h}$$

where R_s = Resistance of the source, and

$\Delta h = h_{ic} \cdot h_{oc} - h_{rc} \cdot h_{fc}$

(e) ***Overall voltage gain.*** It is given by the relation,

$$A_{vs} = \frac{A_v \cdot R_{is}}{R_s + R_{is}}$$

(f) ***Overall current gain.*** It is given by the relation,

$$A_{is} = \frac{A_i + R_s}{R_i + R_{is}}$$

Example 25.9.

A transistor used in a common collector (or emitter follower) circuit as shown in Fig. 25.16 has the following set of h-parameters:

$h_{ic} = 2\ k\Omega;\ h_{fc} = -51;\ h_{rc} = 1$ *and* $h_{oc} = 25 \times 10^{-6}$ *mhos*

Find the values of input and output resistances, current and voltage gains of the amplifier stage. Use the approximate or exact hybrid formulas.

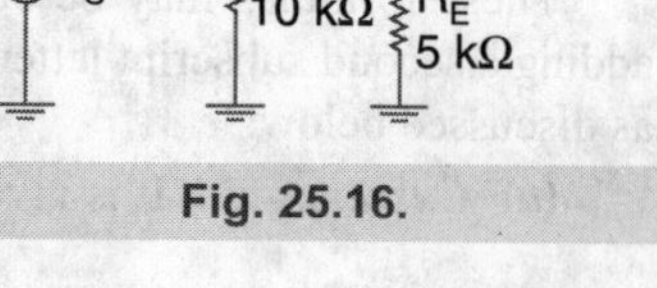

Fig. 25.16.

Solution.

Given: $h_{ic} = 2\ k\Omega = 2000\ \Omega$; $h_{fc} = -51$; $h_{rc} = 1$; $h_{oc} = 25 \times 10^{-6}$ mhos; $r_L = R_E = 5\ k\Omega = 5 \times 10^3\ \Omega$; $R_s = 1\ k\Omega = 1000\ \Omega$ and $R_1 = R_2 = 10\ k\Omega$

Fig. 25.16 shows the hybrid equivalent circuit of the given common collector amplifier.

Input resistance of the amplifier stage

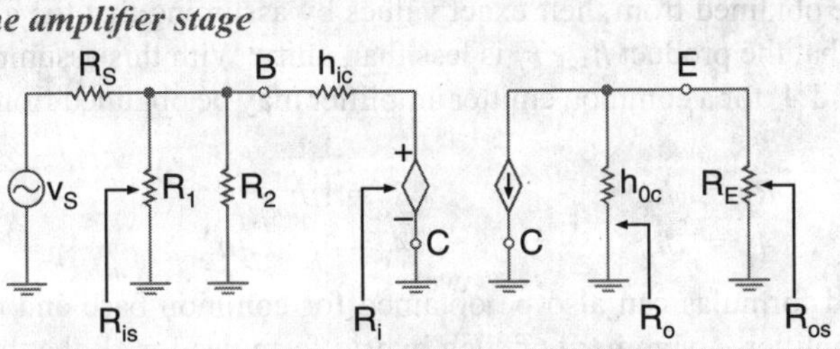

Fig. 25.17.

We know that current gain,

$$A_i = \frac{h_{fc}}{1 + h_{oe} \cdot r_L} = \frac{(-51)}{1 + [(25 \times 10^{-6}) \times (5 \times 10^3)]}$$

$$= 45.3$$

and input resistance, $R_i = h_{ic} + h_{rc} \cdot A_i \cdot r_L$

$$= 2000 + (1 \times 45.3 \times 5 \times 10^3)\ \Omega$$

$$= 228 \times 10^3\ \Omega = 228\ \text{k}\Omega$$

∴ Input resistance of the amplifier stage,

$$R_{is} = R_i \,\|\, (R_1 \,\|\, R_2) = 228 \,\|\, (10 \,\|\, 10) = 4.9\ \text{k}\Omega = 4900\ \Omega \textbf{ Ans.}$$

Output resistance of the amplifier stage

We know that output resistance,

$$R_o = -\frac{R_s + h_{ic}}{h_{fc}} = -\frac{1000 + 2000}{(-51)} = 58.8\ \Omega$$

and output resistance of the amplifier stage,

$$R_{os} = R_o \,\|\, R_E = 58.8 \,\|\, (5 \times 10^3) = 58.1\ \Omega \textbf{ Ans.}$$

Current gain of the amplifier stage

We know that current gain of the amplifier stage,

$$A_{is} = \frac{A_i \cdot R_s}{R_s + R_{is}} = \frac{45.3 \times 1000}{1000 + 4900} = 7.7 \textbf{ Ans.}$$

Voltage gain of the amplifier stage

We know that voltage gain,

$$A_v = \frac{A_i \cdot r_L}{R_i} = \frac{45.3 \times (5 \times 10^3)}{228 \times 10^3} = 1$$

and voltage gain of the amplifier stage,

$$A_{vs} = \frac{A_v \cdot R_{is}}{R_s + R_{is}} = \frac{1 \times 4900}{1000 + 4900} = 0.83 \textbf{ Ans.}$$

25.17. Approximate Hybrid Formulas for Transistor Amplifier

The hybrid formulas derived in the previous articles for common emitter, common base and common collector amplifiers are called exact formulas. These formulas are used to calculate the exact input resistance (R_i), output resistance (R_o), current gain (A_i) and voltage gain (A_v). As a matter of fact, the calculation of exact values is time consuming and liable to some calculation error. Therefore, we can use approximate values of R_i, R_o, A_i and A_v in actual practice. The approximate values of R_i,

R_o, A_i and A_v may be obtained from their exact values by assuming that the a.c. load resistance (r_L) is small enough, so that the product $h_{oe} \cdot r_L$ is less than unity. With this assumption, the approximate values of R_i, R_o, A_i and A_v for a common emitter amplifier may be obtained from the following hybrid formulas:

$$R_i = h_{ic} \qquad R_o = 1/h_{oe}$$
$$A_i = h_{fe} \qquad A_v = h_{fe} \cdot r_L / h_{ie}$$

A similar hybrid formulas can also be obtained for common base and common collector (or emitter follower) amplifier. A summary of such hybrid formulas for all the three types of transistor amplifiers is given in Table 25.3.

Table 25.3 : Approximate hybrid formulas for common emitter, common base and common collector amplifiers

S. No.	*Item*	*Symbol*	*Common emitter*	*Common base*	*Common collector*
1.	Input resistance	R_i	h_{ie}	h_{ib}	$h_{ic} + h_{fc} \cdot r_L$
2.	Output resistance	R_o	$\dfrac{1}{h_{oe}}$	$\dfrac{1}{h_{ob}}$	$\dfrac{R_s + h_{ic}}{h_{fc}}$
3.	Current gain	A_i	$-h_{fe}$	$h_{fb} = 1$	h_{fc}
4.	Voltage gain	A_v	$\dfrac{h_{fe} \cdot r_L}{h_{ie}}$	$\dfrac{h_{fb} \cdot r_L}{h_{ib}}$	1

Example 25.10.

In a single stage common emitter amplifier as shown in Fig. 25.18, the transistor h-parameters are:

h_{ie} = 1500 Ω; h_{fe} = 50; h_{re} = 50 × 10^{-4} and h_{oe} = 20 × 10^{-6}

Calculate the values of input and output resistances, current and voltage gains of the stage. Use approximate hybrid formulas for calculations.

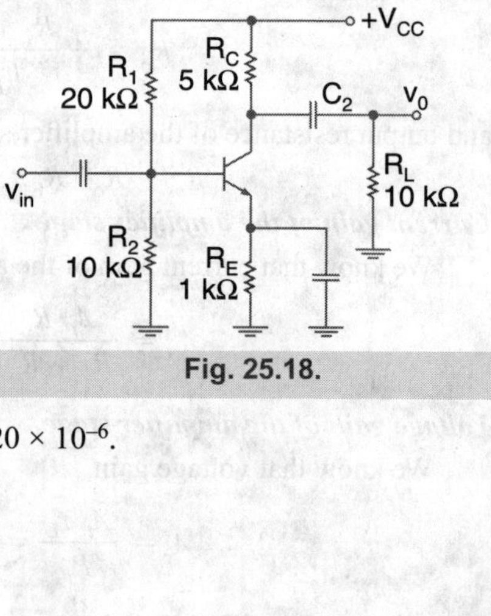

Fig. 25.18.

Solution.

Given h_{ie} = 1500 Ω; h_{fe} = 50; h_{re} = 50 × 10^{-4} and h_{oe} = 20 × 10^{-6}.

Input resistance of the voltage

We know that the value of a.c. load resistance,

$$r_L = R_C \parallel R_L = 5 \parallel 10 = 3.3 \text{ k}\Omega$$
$$= 3.3 \times 10^3 \ \Omega$$

and current gain, $\quad A_i = -h_{fe} = -50$

We also know that the input resistance,

$$R_i = h_{ie} = 1500 \ \Omega = 1.5 \text{ k}\Omega$$

∴ Input resistance of the stage,

$$R_{is} = R_i \parallel (R_1 \parallel R_2) = 1.5 \parallel (20 \parallel 10) \text{ k}\Omega \ 1.22 \text{ k}\Omega \textbf{ Ans.}$$

Output resistance of the stage

We know that output resistance,

$$R_o = \frac{1}{h_{oe}} = \frac{1}{20 \times 10^{-6}} = 0.05 \times 10^6 \ \Omega = 50 \text{ k}\Omega$$

and output resistance of the stage,

$$R_{os} = R_o \| R_L = 50 \| 3.3 = 3.1 \text{ k}\Omega \textbf{ Ans.}$$

Voltage gain of the stage

We know that the voltage gain,

$$A_v = \frac{A_i \cdot r_L}{R_i} = \frac{(-50) \times (3.3 \times 10^3)}{1500} = -110$$

and voltage gain of the stage,

$$A_{vs} = \frac{A_v \cdot R_{is}}{R_s + R_{is}} = A_s = -110 \textbf{ Ans.} \qquad ... (\because R_s = 0)$$

Current gain of the stage

We know that current gain of the stage,

$$A_{is} = \frac{A_i \cdot R_s}{R_s + R_{is}} = A_v = -50 \textbf{ Ans.} \qquad ... (\because R_s = 0)$$

Example 25.11.

A common-emitter amplifier with voltage divider bias has values of $R_C = 12$ kΩ, $R_L = 4.7$ kΩ, $R_1 = 33$ kΩ, $R_2 = 4.7$ kΩ and $I_C = 1$ mA. At 1 mA, the transistor has h-parameter values of [1]h_{ie} = 1 kΩ to 5 kΩ and [2]h_{fe} = 70 to 350. Determine the value of (a) input impedance and (b) the overall voltage gain for the circuit.

Solution.

Given: $R_C = 12$ kΩ, $R_L = 4.7$ kΩ, $R_1 = 33$ kΩ, $R_2 = 4.7$ kΩ, $I_C = 1$ mA; $h_{ie} = 1$ kΩ to 5 kΩ and $h_{fe} = 70$ to 350.

With these values, the circuit for the common-emitter amplifier is as shown in Fig. 25.19.

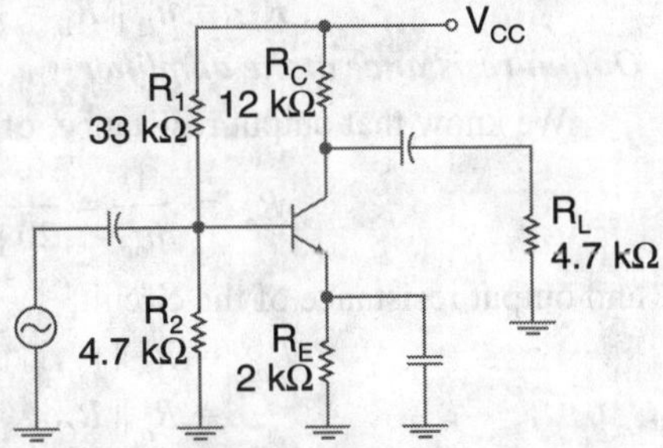

Fig. 25.19.

We know that, $$^*h_{ie} = \sqrt{h_{ie(min)} \times h_{ie(max)}}$$

$$= \sqrt{1 \text{ k}\Omega \times 5 \text{k}\Omega}$$

$$= 2.24 \text{ k}\Omega$$

and $$h_{fe} = \sqrt{h_{fe(min)} \times h_{fe(max)}} = \sqrt{70 \times 350}$$

$$= 104.67$$

(a) Input resistance (or impedance) of the amplifier

We know that input resistance of a transistor is,

$$R_i = h_{ie} = 2.24 \text{ kW}$$

and input resistance of the amplifier,

$$R_{is} = R_i \| R_1 \| R_2 = 2.24\text{k } \Omega \| 33 \text{ k}\Omega \| 4.7\text{k}\Omega \approx 1.45 \text{ k}\Omega \textbf{ Ans.}$$

(b) Overall voltage gain

We know that current gain of a transistor,

$$A_i = h_{fe} = 104.67$$

* In commercially available transistors, the value of h_{ie} and h_{fe} varies over a certain range, $h_{ie(min)}$ to $h_{ie(max)}$, and $h_{fe(max)}$ to $h_{fe(min)}$ respectively. In such a case, the value of $h_{ie} = \sqrt{h_{ie(min)} \times h_{ic(max)}}$ and $h_{fe} = \sqrt{h_{fe(min)} \times h_{fe(max)}}$

We also know that a.c. load resistance,

$$r_c = R_C \parallel R_L = 12 \text{ k}\Omega \parallel 4.7 \text{ k}\Omega = 3.38 \text{ k}\Omega,$$

and the overall voltage gain,

$$A_{vs} = A_v = \frac{A_i \cdot r_L}{R_i} = \frac{104.67 \times 3.38 \text{ k}\Omega}{2.24 \text{ k}\Omega}$$

$$= 157.94 \textbf{ Ans.} \qquad (\because R_S = 0)$$

Example 25.12.

Fig. 25.20 shows the circuit of a common-emitter amplifier with fixed bias. Determine the value of input resistance, output resistance, current gain and voltage gain of the amplifier circuit.

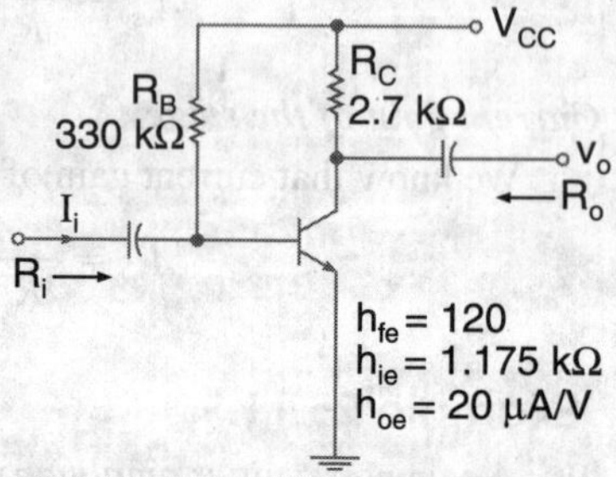

Fig. 25.20.

Solution.

Given: $R_B = 330 \text{ k}\Omega$; $R_C = 2.7 \text{ k}\Omega$; $h_{fe} = 120$; $h_{ie} = 1.175 \text{ k}\Omega$ and $h_{oe} = 20\ \mu\text{A/V}$.

Input resistance of the amplifier

We know that input resistance of the transistor,

$$R_i = h_{ie} = 1.175 \text{ k}\Omega$$

and input resistance of the circuit,

$$R_{is} = h_{ie} \parallel R_B = 1.175 \text{ k}\Omega \parallel 330 \text{ k}\Omega = 1.17 \text{ k}\Omega \textbf{ Ans.}$$

Output resistance at the amplifier

We know that output resistance, of the transistor,

$$R_o = \frac{1}{h_{oe}} = \frac{1}{20\ \mu \text{ A/V}} = \frac{1}{20 \times 10^{-6}} = 5 \times 10^4 \Omega = 50 \text{ k}\Omega$$

and output resistance of the circuit,

$$R_{os} = R_o \parallel r_L$$

$$= R_o \parallel R_C \qquad (\because r_L = R_C)$$

$$= 50 \text{ k}\Omega \parallel 2.7 \text{ k}\Omega.$$

$$= 2.56 \text{ k}\Omega \textbf{ Ans.}$$

Current gain of the amplifier

We know that current gain of the circuit,

$$A_i = h_{fe} = 120 \textbf{ Ans.}$$

Voltage gain of the amplifier

We know that voltage gain of the amplifier,

$$A_{vs} = \frac{A_i \cdot r_L}{R_i} = \frac{h_{fe} \times R_C}{R_i} = \frac{120 \times 2.7 \text{ k}\Omega}{1.175 \text{ k}\Omega} = 275.7 \textbf{ Ans.}$$

25.18. Approximate Conversion Formulas for h-parameters

These days, transistors are used in most of the circuits in common emitter configuration. Therefore, the manufacturers list only the common emitter *h*-parameters in their data sheets. However, if the transistor is to be used in a configuration, other than the common emitter, then we must use the specified conversion formulas for determining its *h*-parameters. The conversion formulas for common base and common collector configuration are as given below.

1. ***Convertion formulas for common base configuration.*** The convertion formulas for determining the values of h-parameters of a transistor in common base configuration, from the common emitter h-parameter values, are as given below:

$$h_{ib} = \frac{h_{ie}}{1 + h_{fe}}$$

$$h_{rb} = \frac{h_{ie} \cdot h_{oe}}{1 + h_{fe}} - h_{re}$$

$$h_{fb} = -\frac{h_{fe}}{1 + h_{fe}}$$

$$h_{ob} = \frac{h_{oe}}{1 + h_{fe}}$$

2. ***Conversion formulas for common collector configuration.*** The conversion formulas for determining the values of h-parameters of a transistor in common collector configuration from the common emitter h-parameter value, are as given below:

$$h_{ic} = h_{ie}$$

$$h_{rc} = 1 - h_{re} \approx 1 \qquad ...(\because h_{rc} << 1)$$

$$h_{fc} = -(1 + h_{fe})$$

$$h_{oe} = h_{oe}$$

Example 25.13.

Find the values of h_{fb} and h_{fc}, if the value of h_{fe} of a transistor is 50.

Solution.

Given: $h_{fe} = 50$

We know that $h_{fb} = -\dfrac{h_{fe}}{1 + h_{fe}} = -\dfrac{50}{1 + 50} = -0.98$ **Ans.**

and $h_{fc} = -(1 + h_{fe}) = -(1 + 50) = -51$ **Ans.**

Example 25.14.

A transistor is connected in common collector configuration and its h-parameters are:
$h_{ie} = 1100\ \Omega$; $h_{re} = 2.5 \times 10^{-4}$ $h_{fe} = 50$; $h_{oe} = 24\ \mu A/V$
The circuit uses $R_L = 10\ k\Omega$ and $R_S = 1\ k\Omega$. Calculate gain A_i, the input resistance R_i and the voltage gain, A_v of this amplifier. (A.M.I.E., Winter, 1988, R.G.P.V. Dec., 2003, J.N.T.U., 2009)

Solution.

Given: $h_{ie} = 1100\ \Omega$; $h_{re} = 2.5 \times 10^{-4}$; $h_{fe} = 50$; $h_{oe} = 24\ \mu$ A/V $= 24 \times 10^{-6}$ A/V; $r_L\ (= R_L) = 10\ k\Omega = 10 \times 10^3\ \Omega$ and $R_S = 1\ k\Omega$.

Current gain

First of all, let us find the values of h-parameters for a transistor in common collector configuration because the given h-parameter values are for a transistor in common emitter configuration. We know that

$$h_{ic} = h_{ie} = 1100\ \Omega$$

and $h_{rc} = 1 - h_{re} = 1 - (2.5 \times 10^{-4}) \approx 1$

Similarly we know that

$$h_{fc} = -(1+h_{fe}) = -(1+50) = -51$$

and $$h_{oe} = h_{oe} = 24 \times 10^{-6}\ A/V$$

∴ Current gain,

$$A_i = -\frac{h_{fc}}{1+h_{oe}\cdot r_L} = \frac{(-51)}{1+(24\times10^{-6})(10\times10^{3})} = 41.1 \textbf{ Ans.}$$

Input resistance

We know that input resistance,

$$R_i = h_{ic} + h_{rc}\cdot A_i \times r_L = 1100 + [1 \times 41.1 \times (10 \times 10^3)]$$
$$= 1100 + (41.1 \times 10^4) = 412.1 \times 10^3\ \Omega$$
$$= 412.1\ k\Omega \textbf{ Ans.}$$

Voltage gain

We know that voltage gain, $$A_v = \frac{A_i \cdot r_L}{R_i} = \frac{41.1\times(10\times10^3)}{412.1\times10^3} = 0.997 \textbf{ Ans.}$$

Example 25.15.

A CE transistor amplifier is characterized by h_{ie} = 2 kohm, $h_{re} = 2 \times 10^{-4}$, h_{fe} = 50 and $h_{oe} = 20 \times 10^{-6}$ A/V if the load resistance is 4 kΩ and the source resistance is 200 Ω. Determine the R_i, R_o, A_v, and A_i.

(GBTU., 2010–2011)

Solution.

Given: $h_{ie} = 2\ k\Omega = 2 \times 10^3\ \Omega$; $h_{re} = 2 \times 10^{-4}$; $h_{fe} = 50$; $h_{oe} = 20 \times 10^{-6}$ A/V; $r_L = 4\ k\Omega = 4 \times 10^3\ \Omega$ and $R_S = 200\ \Omega$.

We know that the current gain of transistor,

$$A_i = -\frac{h_{fe}}{1+h_{oe}.r_L}$$

$$= -\frac{50}{1+(20\times10^{-6})\times(4\times10^3)} = -46.29 \textbf{ Ans.}$$

We know that,

$$R_i = h_{ie} + h_{re}.A_i.r_L$$
$$= (2 \times 10^3) + [(2 \times 10^{-4}) \times (-46.29) \times (4 \times 10^3)]$$
$$= 1.9 \times 10^3 = 1.9\ k\Omega \textbf{ Ans.}$$

We also know that,

$$\Delta_h = h_{ie}.h_{oe} - h_{re}.h_{fe}$$
$$= [(2 \times 10^3) \times (20 \times 10^{-6})] - [(2 \times 10^{-4}) \times (50)] = 0.03$$

and $$R_o = \frac{R_S + h_{ie}}{R_S.h_{oe} + \Delta h} = \frac{200 + (2\times10^3)}{200\times(20\times10^{-6}) + 0.03}$$

$$= \frac{2200}{0.034} = 64705\ \Omega \textbf{ Ans.}$$

We also know that,

$$A_v = \frac{A_i.r_L}{R_i}$$

$$= \frac{(-46.29)\times(4\times10^3)}{1.9\times10^3} = -97.45\ \Omega \textbf{ Ans.}$$

25.19. Determination of Transistor *h*-parameters

The transistor *h*-parameters may be determined by any one of the following two methods:

1. Graphical determination of the *h*-parameters; and
2. Experimental determination of the *h*-parameters.

Now we shall discuss both these methods one by one in the following pages.

25.20. Graphical Determination of *h*-parameters

We have already discussed in Art. 25.2 that the input and output voltages and currents of any linear circuit are related by the following two equations:

$$v_1 = h_i \cdot i_1 + h_r \cdot v_2 \quad ...(i)$$

and

$$i_2 = h_f \cdot i_1 + h_o \cdot v_2 \quad ...(ii)$$

where v_1 = Input voltage; v_2 = Output voltage

i_1 = Input current; i_2 = Output current and

h_f, h_i, h_r, h_o = Hybrid or *h*-parameters

These equations may also be used to relate input and output voltages and currents for a transistor. For example, in a common emitter transistor, the input voltage (v_1) is the base-to-emitter voltage (v_{BE}) and the output voltage (v_2) is the collector-to-emitter voltage (v_{CE}). Similarly, the input current (i_1) is the base current (i_B) and the output current (i_2) is the collector current (i_C). Thus, for a common emitter transistor, the equations (*i*) and (*ii*) may be written as:

$$v_{BE} = h_{ie} \cdot i_B + h_{re} \cdot v_{CE}$$

$$i_C = h_{fe} \cdot i_B + h_{oe} \cdot v_{CE}$$

With the help of partial derivatives for the above two equations, it can be shown that the *h*-parameters for the common emitter transistor can be found by using the following equations:

$$h_{ie} = \frac{\delta v_{BE}}{\delta i_B} = \frac{\Delta v_{BE}}{\Delta i_B} \quad \text{... (for a constant } v_{CE}\text{)}$$

$$h_{re} = \frac{\delta v_{BE}}{\delta v_C} = \frac{\Delta v_{BE}}{\Delta v_{CE}} \quad \text{... (for a constant } i_B\text{)}$$

$$h_{fe} = \frac{\delta i_C}{\delta i_B} = \frac{\Delta i_C}{\Delta i_B} \quad \text{... (for a constant } v_{CE}\text{)}$$

$$h_{oe} = \frac{\delta i_C}{\delta v_{CE}} = \frac{\Delta i_C}{\Delta v_{CE}} \quad \text{... (for a constant } i_B\text{)}$$

It is evident from equation (*i*) that parameter h_{ie} is determined by the ratio of a small change in base-to-emitter voltage (Δv_{BE}) to the corresponding change in base current (Δi_B) for a constant value of collector-to-emitter voltage (v_{CE}). Similarly, parameter h_{re} is given by the ratio of a small change in base-to-emitter voltage (Δv_{BE}) to the corresponding change in collector-to-emitter voltage (Δv_{CE}) for a constant base current (i_B). On the other hand, the parameter h_{fe} is determined by the ratio of a small change in collector current (Δi_C) to the corresponding small change in base current (Δi_B) for a constant collector to emitter voltage (v_{CE}). Similarly, the parameter h_{oe} is given by the ratio of small change in collector current (Δi_C) to the corresponding small change in collector-to-emitter valtage (ΔV_{CE}) for a constant base current (i_B). It will be interesting to know that the values of h_{ie} and h_{re} parameters are determined graphically from the input or basc characteristics, while the values of h_{fe} and h_{oe} parameters are determined from the output or collector characteristics.

25.21. Graphical Determination of h_{ie} and h_{re} Parameters

We have already discussed in the last article that the value of h_{ie} and h_{re} parameters are determined from the input or base characteristics. In order to determine the values of these parameters, the first step is to locate a point on the input or base characteristics as shown in Fig. 25.21. Now to find the value of h_{ie}, we draw a line which is tangent to the curve marked v_{CE_2} and passes through the *Q*-point.

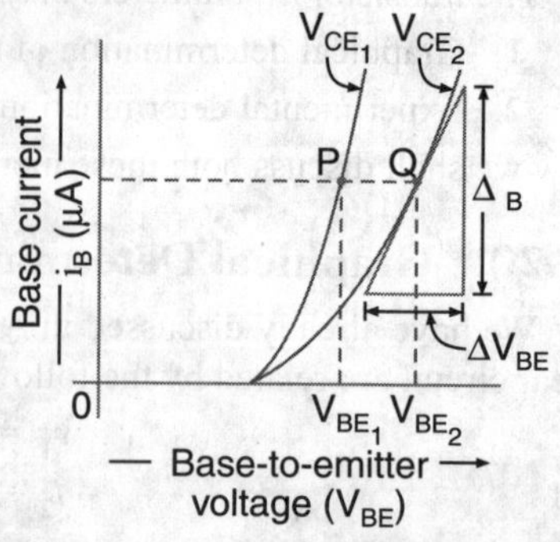

Fig. 25.21. Determination of h_{ie} and h_{re}, parameters.

A small change in base-to-emitter voltage (Δv_{BE}) is then chosen and the corresponding change in base current (i_B) is measured. Substituting these values in the relation $\Delta v_{BE}/\Delta i_B$, we get the value of h_{ie}.

The parameter h_{re} can be found by selecting another point *P* on the curve marked v_{CE_1} in such a way so that the point *P* and *Q* have the same value of base current. Then measure the values of changes in collector-to-emitter voltage Δv_{CE} (*i.e.*, $v_{CE_2} - v_{CE_1}$) and the resulting change in base-to-emitter valtage ΔV_{BE}. (equal to $v_{BE_2} - v_{BE_1}$). Substituting these values in the relation $\Delta v_{BE}/\Delta v_{CE}$, we get the value of h_{re} .

25.22. Graphical Determination of h_{fe} and h_{oe} Parameters

We have already discussed in Art. 25.20 that the values of h_{fe} and h_{oe} parametess are determined from the output or collector characteristics as shown in Fig. 25.22. Now in order to find the value of h_{fe}, we select two points *R* and *S* (above and below the *Q*-point) in such a way that these points have the same collector-to-emitter voltage (*i.e.*, v_{CE_2}). Then measure the values of selected change in base current Δi_B (equal to ($i_{B_2} - i_{B_1}$) and the corresponding change in collector current Δi_C (equal to $i_{C_2} - i_{C_1}$). Substituting these values in relation $\Delta i_C / \Delta i_B$, we get the value of h_{fe}.

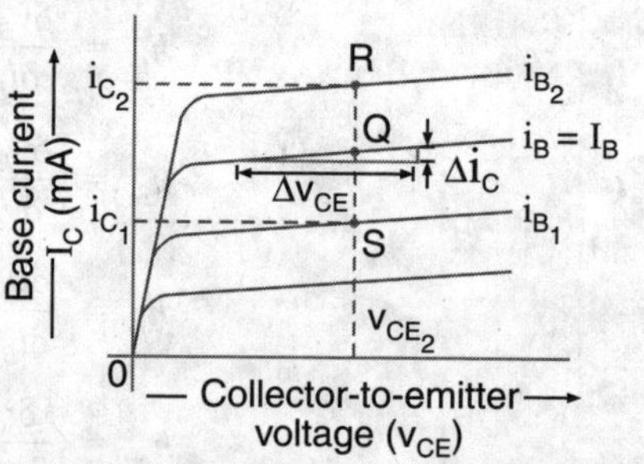

Fig. 25.22. Determination of h_{fe} and h_{oe} parameter.

The parameter h_{oe} can be found by drawing a line, which is tangent to the curve marked $i_B = I_B$ and passes through the *Q*-point as shown in the Fig. A small change in collector-to-emitter voltage (Δv_{CE}) is then chosen and the corresponding change in collector current (Δi_C) is measured. Substituting these values in the relation $\Delta i_C / v_{CE}$, we get the value of h_{oe}.

25.23. Experimental Determination of *h*-parameters

The *h*-parameters may be determined experimentally by constructing a circuit, which will satisfy the conditions specified in the definition of *h*-parameters. Although there are four parameters, which are to be determined, yet only two circuits are needed for the parameters. The circuit shown in Fig. 25.23 (*a*) may be used to determine h_{ie} and h_{fe}, while the circuit shown in Fig. 25.23 (*b*) may be used to determine h_{re} and h_{oe}.

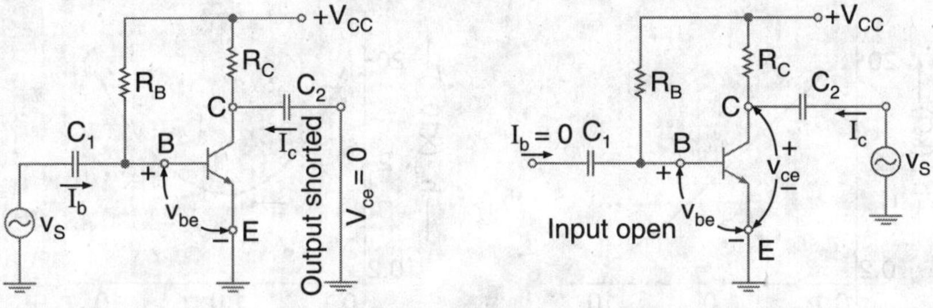

Fig. 25.23. Experimental determination of *h*-parameters.

The transistor in the circuits shown above is connected in a common emitter configuration. The capacitors C_1 and C_2 are the coupling capacitors *i.e.*, they act as a short circuit for the a.c. signal. But they prevent any d.c. level associated with the signal from upsetting the biasing conditions set up by the V_{CC} supply and the resistors R_B and R_C. It may be noted that all voltages and currents shown in the figures are the r.m.s. values.

The values of h_{ie} and h_{fe} may be obtained by measuring the values of base-to-emitter voltage (V_{be}),base current (I_b) and collector current (I_c) from the Fig. 25.23 (*a*). Then,

$$h_{ie} = \frac{V_{be}}{I_b} \quad \text{and} \quad h_{fe} = \frac{I_c}{I_b}$$

Similarly, the values of h_{re} and h_{oe} may be obtained by measuring the values of collector-to-emitter voltage (V_{ce}), base-to-emitter voltage (V_{be}) and collector current (I_c) from Fig. 25.22 (*b*). Then,

$$h_{re} = \frac{V_{be}}{V_{ce}} \quad \text{and} \quad h_{oe} = \frac{I_c}{V_{ce}}$$

25.24. Variation of Transistor *h*-parameters

We have already discussed the procedures for determining transistor *h*-parameters from the characteristic curves. The parameters can be obtained at the operating point from the slopes and spacing between the curves at that point. Since the characteristic curves are not in general straight lines and not equally spaced for equal changes in base current (I_B) or collector-to-emitter voltages (V_{CE}) therefore, the values of *h*-parameters varies depending upon the position of the operating point of the curves. The variation of *h*-parameters for a typical transistor depends upon the following factors:

1. Collector current (I_C).
2. Junction temperature (T_j).
3. Collector-emitter voltage (V_{CE}).

However, the variation of *h*-parameters due to the collector current is most important from subject point of view and is discussed as below:

(*a*) ***Variation of h_{ie}.*** Fig. 25.24 (*a*) shows the variation of h_{ie} with the operating value of collector current (I_C). It may be noted that the value of h_{ie} decreases as the operating value of collector current increases.

(*b*) ***Variation of h_{re}.*** Fig. 25.24 (*b*) shows the variation of h_{re} with the collector current. It may be noted that first h_{re} decreases with the increase in collector current. It reaches to a minimum value at a certain collector current. Beyond that, the parameter h_{re} increases with the increase in collector current.

(*c*) ***Variation of h_{fe}.*** Fig. 25.24 (*c*) shows the variation of h_{fe} (a.c. beta) with the operating value of collector current. It may be noted that h_{fe} increases with the increase in collector current and reaches a maximum at certain collector current. Beyond that value, h_{fe} decreases with

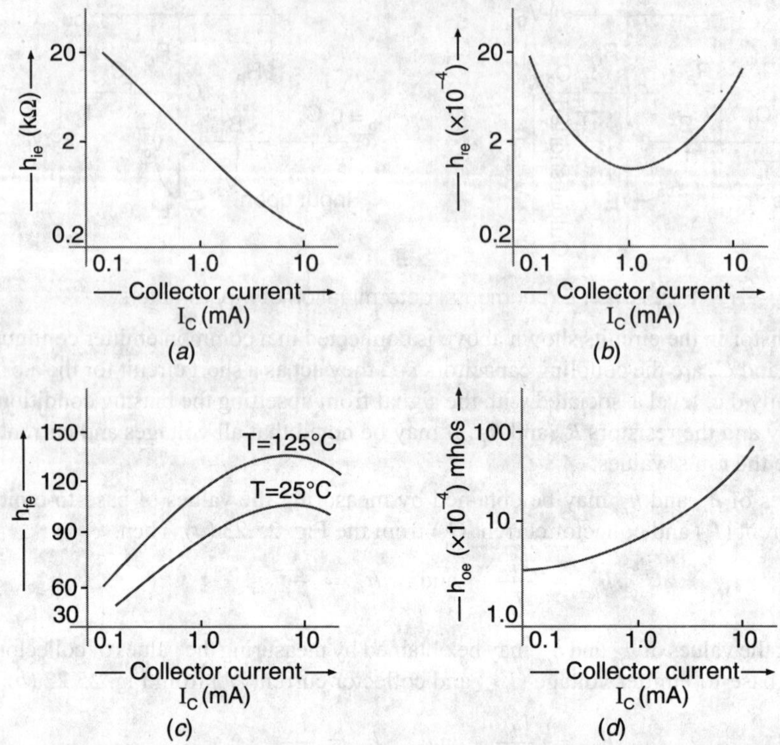

Fig. 25.24. Variation of transistor *h*-parameters.

the increase in collector current. The parameter h_{fe} is very sensitive to temperature. Its value increases with the increase in temperature as shown in figure by the curves market $T = 25°C$ and $T = 125°C$.

(*d*) *Variation of* h_{oe}. Fig. 25.24 (*d*) shows the variation of h_{oe} with the operating value of collector current. It may be noted that h_{oe} increases with the increase of collector current.

It is evident from the above discussion that *h*-parameters vary over a wide range with the change in collector current. Therefore, the data sheet of a transistor, usually, specifies the values of *h*-parameters at a certain value of collector current. For example, for a transistor 2 N 3905, the data sheet specifies the following values of *h*-parameters at a collector current of $I_C = 1$ mA:

$$h_{ie} = 3.5\ \text{k}\Omega;\ h_{re} = 1.3 \times 10^{-4};\ h_{fe} = 120 \text{ and } h_{oe} = 8.5\ \mu \text{ mhos.}$$

NUMERICAL PROBLEMS

1. The hybrid parameters for a transistor used in common emitter configuration are:

$$h_{ie} = 1000\ \Omega; \qquad h_{fe} = 150$$

$$h_{re} = 1.2 \times 10^{-4}, \qquad h_{oe} = 25 \times 10^{-6} \text{ mhos}$$

The transistor has a load resistance of 10 kΩ in the collector and is supplied from a signal source of resistance 5 kΩ. Compute the values of input resistance, output resistance, current gain and voltage gain of the stage. **(Ans. 8.56 Ω; 8.2 kΩ; 102.5 and 205)**

2. A single common emitter amplifier circuit is shown in Fig. 25.25.

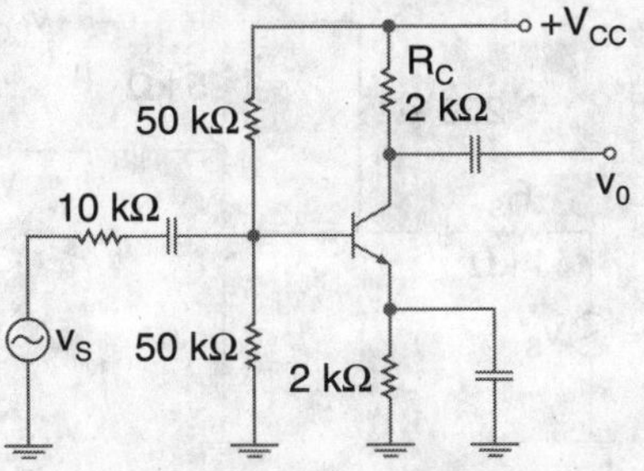

Fig. 25.25.

Calculate the values of R_i, R_o, Δ_x, Δ_v, A_{vs}, A_{is}. The transistor has a set of following parameters:

h_{fe} = 50; h_{ie} = 1.1 kΩ

h_{re} = 0; h_{oe} = 0 **(Ans. 1100 Ω; ∞; 50; 90.9; 9 and 45)**

3. Find the values of voltage gain, current gain, input resistance and power gain for a common emitter transistor amplifier, with r_L = 1600 Ω and R_S = 1000 Ω. The transistor has a following set of *h*-parameters: h_{ie} = 1100 Ω; h_{re} = 2.5 × 10^{-4}; h_{fe} = 25 μ A/V.

(*U.P.S.C. Engg. Service, 1987*)

(Ans. 69.8, 48, 1084 Ω; 50 kΩ and 3350)

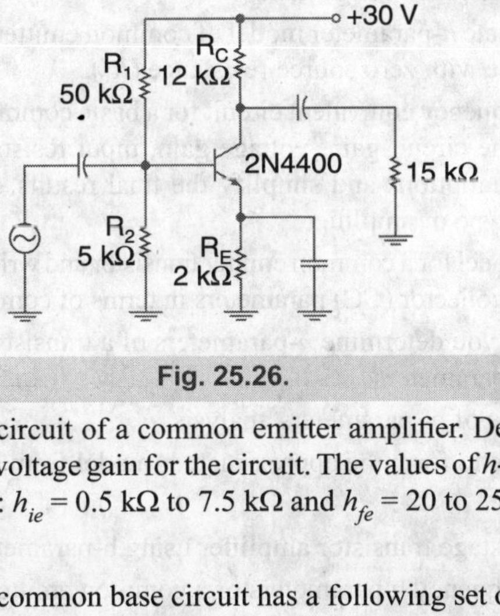

Fig. 25.26.

4. Fig. 25.26 shows the circuit of a common emitter amplifier. Determine the value of input resistance and overall voltage gain for the circuit. The values of *h*-parameter for the transistor 2N 4400 are as below: h_{ie} = 0.5 kΩ to 7.5 kΩ and h_{fe} = 20 to 250.

(Ans. 1.35 kΩ and 244)

5. A transistor used in a common base circuit has a following set of parameters: h_{ib} = 20 Ω; h_{fb} = 0.98; h_{rb} = 3 × 10^{-4} and h_{ob} = 0.5 × 10^{-6}. Find the values of R_o, k_o, A_i, A_v and A_{vs}; if R_s = 600 Ω, and r_L = 1.5kΩ. **(Ans. 20.4 Ω; 1 MΩ; 0.98; 72 and 2.3)**

6. Design an emitter follower circuit having R_i = 500 kΩ, R_o = 20Ω , which uses a transistor with h_{fe} = 50; h_{ie} = 1 kΩ and h_{oe} = 25μ A/V. Find the values of A_i and A_v for this circuit.

(*A.M.I.E., Summer, 1984*) **(Ans. R_E = 10 kΩ; A_i = 51 and A_v = 0.998)**

7. A transistor used in common emitter configuration as shown in Fig. 25.27 has the following set of *h*-parameters h_{ie} = 1 kΩ; h_{fe} = 50; h_{re} = 5 × 10^{-4} and h_{oe} = 2 × 10^{-5} mhos;

Determine R_i, R_{is}, R_o. A_i and A_v. Use the approximate hybrid formulas.

(Ans. 1 kΩ; 996 Ω; 50 kΩ; 50 and 250)

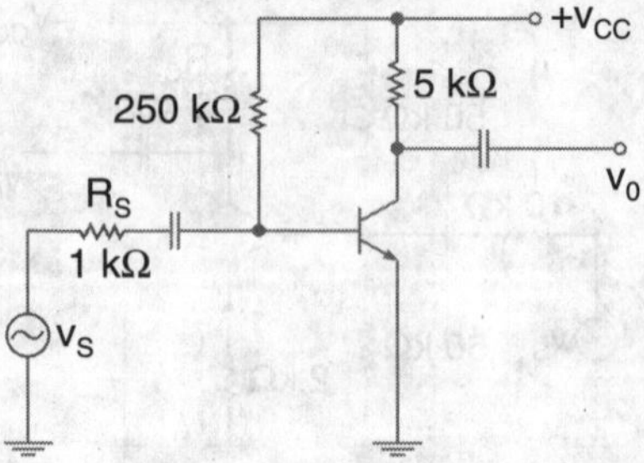

Fig. 25.27.

8. For I_E = 1 mA, give the order of magnitudes (including the sign) of (*a*) h_{ib} (*b*) h_{ie} (*c*) h_{re} (*d*) h_{rc} (*e*) h_{fb} (*f*) h_{fe} (*g*) h_{fc} (*h*) h_{oe} (*i*) h_{ob} and (*j*) h_{oc}.

(Ans. (*a*) 29.4 (*b*) 4 kΩ (*c*) 2 × 10^{-4} Ω (*d*) 1 (*e*) – 0.99 (*f*) 120 (*g*) – 120 (*h*) 50 × 10^{-6}Ω (*i*) 0.5 × 10^{-6} Ω and (*j*) 50 × 10^{-6}Ω)

DESCRIPTIVE QUESTIONS

1. Draw the equivalent circuit of a common emitter amplifier in terms of *h*-parameter and obtain the expression for voltage gain.
2. Draw an approximate *h*-parameter model of common emitter transistor configuration driven by a voltage source with zero source resistance (R_S).
3. Develop a low-frequency equivalent circuit for a basic common collector amplifier and drive the relations for the circuit gain, voltage gain, input resistance in terms of *h*-parameters. Make suitable assumptions and simplify the final results. Also justify the name 'emitter follower' for this type of amplifier.
4. Draw the hybrid model for a common emitter transistor and write the input and output equations. Find the common collector (CC) parameters in terms of common emitter *h*-parameters.
5. Explain, how will you determine *h*-parameters of a transistor experimentally?
6. Define the four h parameters. *(Anna University., April/May 2008)*
7. Give the advantage of h- parameters analysis. *(JNTU., June 2009)*
8. Give the procedure to form the approximate h-model from exact h-model of amplifier. *(JNTU., June 2009)*
9. Analyses a single stage transistor amplifier using h-parameters. *(JNTU., June. 2009)*
10. Define the low frequency h parameters of a common emitter transistor. *(M.G University., May 2007)*
11. Write the h-parameter notations for CE and CB transistor circuits. *(Bhartiya University., April 2004)*
12. Discuss determination of h-parameters from transistor characteristics. *(PTU., May 2009)*
13. Define

$$h_{11} = \frac{v_i/i_1}{v_2} = 0$$ *(PTU., Dec 2008)*

14. Discuss h-parameter equivalent circuit of transistor? *(PTU., Dec 2008)*
15. Define h-parameters. Derive expression for

(*a*) Voltage gain

(*b*) Current gain
(*c*) Input resistance
(*d*) Output resistance of CE amplifiers using h-parameters. (*PTU., Dec 2007*)

16. Derive h-parameters for a transistor in CB configuration? (*PTU., Dec 2006*)
17. Draw its equivalent circuit of common emitter configuration using h-parameters. Also discuss its important features. (*PTU., Dec 2006*)
18. From fundamentals derive the h parameter equations and model of a common base BJT amplifier. Give typical values for the parameters and define them. (*M.G University., May 2008*)
19. Draw the hybrid model of CE amplifier and obtain its gain, input impedance and output impedance. Compare the performance of this CE amplifier with CC and CB configuration. (*Anna University., April/May 2008*)
20. Draw the circuit and low frequency equivalent circuit using h-parameter for CE amplifiers. Use the equivalent circuit to find current gain and voltage gain to the amplifier. (*Bhartiya University., April 2004*)
21. Sketch and neatly label the base and collector characteristics of the CE and clearly show how the four h-parameters are evaluated from the characteristics. (*M.G University., May/June, 2006*)
22. Draw the hybrid equivalent circuit for common base configuration and find the expression of current gain and input impedance. (*GBTU/MTU., 2005-06*)
23. Voltage gain and current gain expressions for CB Configuration using transistor hybrid model. (*Anna University., Nov/Dec 2007*)
24. Draw the small signal h-parameter model of CE-BJT amplifier and explain. Derive expression for voltage gain, input impedance and output impedance. (*M.G University., July/Aug. 2007*)
25. Discuss analysis of transistor amplifier using h-parameter in CB configuration. (*PTU., May 2009*)
26. Define hybrid parameters. Derive equations for current gain and voltage gain of a CE amplifier using h-parameters. (*PTU., May 2008*)
27. Discuss analysis of transistor amplifier using h-parameter in CE configuration. (*PTU., Dec 2008*)
28. Define CB, CC, and CE configurations of a transistor amplifier with the help of sketches. Compare their four properties (R_i, R_o, A_v, and A_i). Write one applications each from the above configurations. (*M.G University., Dec. 2007*)
29. Draw the low frequency hybrid equivalent circuit for CE & CB amplifier. (*JNTU., June 2009*)
30. Give the approximate h-parameters conversion for CB and CC configuration in terms of CE. (*JNTU., June 2009*)
31. Draw the low frequency small signal model of a transistor in CB and CE configurations and explain significance of each model. (*JNTU., Aug./Sep. 2008*)
32. Give the approximate h-parameter conversion formulae for CB and CC configuration in terms of CE. (*JNTU., June. 2009*)
33. Draw and explain the input and output characteristics of a transistor in CE configuration with regions of operation. Show how h-parameters can be determined graphically. (*JNTU., Aug/Sep 2006, Aug./Sept. 2008*)

34. Draw a low-frequency equivalent circuit for a CC amplifier and derive the relations for the current gain, voltage gains and input resistance in terms of h-parameter. *(JNTU., May/Jun 2006, Aug./Sep. 2008)*
35. Draw the circuit diagram of CE amplifier with emitter resistance and obtain its equivalent hybrid model and derive expressions for A_i, R_i and A_v. Use approximate analysis. *(JNTU., Aug./Sep. 2006)*
36. Derive the expression for voltage gain, current gain, I/P impedance, O/P impedance of CE amplifier, using exact for approximate model. *(JNTU., June. 2009)*
37. Draw the circuit diagram of CC amplifier using hybrid parameters and derive expressions for A_i, A_v, R_i and R_o. *(JNTU., Aug./Sep. 2006, GBTU/MTU 2002-03)*
38. What are the h-parameters? Find the h_{re} in terms of the common base (CB) h-parameters. *(R.G.P.V., Dec. 2001)*
39. Write short note on h-parameters of BJT. *(R.G.P.V., June 2004, Nov. / Dec. 2007)*
40. Draw the h-parameter equivalent of an amplifier and explain different h-parameters. *(R.G.P.V., May / June 2006)*
41. Analyze and draw the h-parameter model for common base mode. *(R.G.P.V., June 2002)*
42. Draw the circuit of a CC transistor configuration and give its h-parameter model. How will you determine:
    (*a*) Current Gain (*b*) Input resistance
    (*c*) Voltage Gain (*d*) Output Impedance *(R.G.P.V., Dec. 2002)*
43. Give the h-parameter model for CE configuration. Also define each of the terms used. *(R.G.P.V., Dec. 2003)*
44. Give low frequency h-parameter model of BJT and derive expressions for voltage and current gain for CE configuration. *(R.G.P.V., June 2008)*
45. Give low frequency h-parameter model of BJT and explain the same. *(R.G.P.V., Dec. 2008)*
46. Draw the transistor characteristics in common emitter configuration and discuss the method for determination of various hybrid parameter from these characteristics. Determine CC h-parameter in terms of CE h-parameters. *(R.G.P.V., Dec. 2005)*
47. Give high frequency h-parameter model of BJT. Derive expressions for voltage and current gain for CE configuration using low frequency h-parameter model of BJT. *(R.G.P.V., Dec. 2008)*
48. Draw the hybrid-π model off CE transistor and explain the origin of various conductance and resistances. *(Nagpur University, Summer 2008, Winter 2003)*
49. Derive the overall h-parameter of cascaded amplifiers. *(Nagpur University, Winter 2005, 2002)*
50. In terms of *h*-parameter and load impedance derive the expresion for current gain (A_i), input impedance (R_i) and voltage gain (A_v). *(R.G.P.V., June 2003)*
51. Explain hybrid π transfer model for transistor in CE configuration. *(R.G.P.V., June 2010)*
52. Write down the *h*-parameter of a bipolar junction transistor. *(GBTU., 2011, 2012)*
53. Draw the hybrid equivalent circuit of CB configuration and write the expression for A_i, R_i, A_v and R_i. *(GBTU., 2011, 2012)*
54. Derive the expression for A_i, R_i, A_v and R_i in terms of CE *h*-parameters for emitter follows circuit. *(GBTU., May, 2011)*

55. Represent/Derive CC *h*-parameters (h_{ie} and h_{fe}) in terms of CE *h*-parameters. *(GTU., May 2011)*

56. Determine *h*-parameters for two port network. Also draw the hybrid model for CE, CB and CC configuration. *(GTU., Dec. 2011)*

MULTIPLE CHOICE QUESTIONS

1. The *h*-parameters are called hybrid parameters because they
(*a*) Are different from *y*- and *z*-parameters.
(*b*) Are mixed with other parameters
(*c*) Apply to circuits contained in a box
(*d*) Are defined by using both open-circuit and short-circuit terminations

2. Which of the following statement is not correct regarding the *h*-parameters of a transistor
(*a*) The values of *h*-parameters can be obtained from transistor characteristics.
(*b*) Their values depends upon the transistor configuration
(*c*) Their values depend on operating point
(*d*) They are four in number

3. Which of the following four *h*-parameters of a transistor has a greatest value?
(*a*) h_i (*b*) h_r (*c*) h_o (*d*) h_f

4. Which of the following four *h*-parameters of a transistor has a smallest value?
(*a*) h_i (*b*) h_r (*c*) h_o (*d*) h_f

5. The typical value of h_{ic} is
(*a*) 1 kΩ (*b*) 40 kΩ (*c*) 100 kΩ (*d*) 2 mΩ

6. The input impedance h_{11} of a network with output shorted is given by the ratio
(*a*) v_1/i_1 (*b*) v_1/v_2 (*c*) i_2/i_1 (*d*) i_2/v_2

7. The *h*-parameters of a transistor depend on its
(*a*) configuration (*b*) operating point
(*c*) temperature (*d*) all of the above

8. The output admittance ho of an ideal transistor connected is common-base configuration is .... siemens
(*a*) 0 (*b*) $1/r$ (*c*) $1/\beta\, r_e$ (*d*) – 1

9. A transistor has $h_{fe} = 100$, $h_{ie} = 5.2$ kΩ, and $r_{bb} = 0$.
At room temperature, $V_T = 26$ mV. The collector current, I_C will be
(*a*) 10 mA (*b*) 5 mA (*c*) 1 mA (*d*) 0.5 mA

ANSWERS

1. (*d*)	**2.** (*c*)	**3.** (*a*)	**4.** (*c*)
5. (*a*)	**6.** (*a*)	**7.** (*a*)	**8.** (*a*)
9. (*a*)			

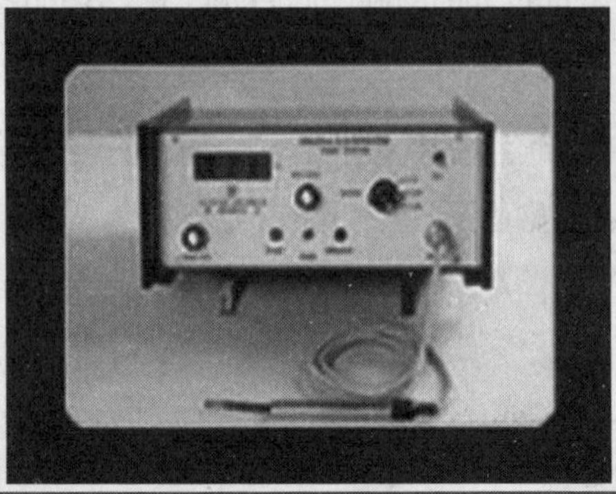

26 Chapter

Multistage BJT Amplifiers

Outline

1. Introduction.
2. Gain of a Multistage Amplifier.
3. Different Coupling Schemes Used in Amplifiers.
4. RC Coupled Amplifier.
5. Calculation of Voltage Gain for RC Coupled Amplifier.
6. Frequency Response of RC Coupled Amplifier.
7. Advantages and Disadvantages of RC Coupled Amplifier.
8. Application of RC Coupled Amplifier.
9. Impedance Coupled Amplifier.
10. Transformer Coupled Amplifier.
11. Calculations of Voltage Gain for Transformer Coupled Amplifier.
12. Frequency Response of Transformer Coupled Amplifier.
13. Advantages and Disadvantages of Transformer Coupled Amplifier.
14. Applications of Transformer Coupled Amplifier.
15. Direct Coupled Amplifier.
16. Calculations of Voltage Gain for Direct Coupled Amplifier.
17. Direct Coupled Amplifier Using Complementary Transistors.
18. Frequency Response of Direct Coupled Amplifier.
19. Advantages and Disadvantages of Direct Coupled Amplifier.
20. Applications of Direct Coupled Amplifier.
21. Darlington Amplifier.
22. Darlington Amplifier Characteristics.
23. Darlington Pair.
24. Darlington Connection of BJTs
25. Parallel Darlington Transistor
26. D C Bias operation of Darlingtion Circuit
27. A C Analysis of Darlingtion Pair
28. Feedback Pair
29. DC Bias Operation of Feedback Pair
30. AC Analysis of Feedback Pair

26.1. Introduction

It has been observed that the voltage (or power) gain, obtained from a single stage small signal amplifier, is limited. Moreover, it is not sufficient for all practical applications. Therefore in order to achieve greater voltage and power gain, we have to use more than one stage of amplification. Such an amplifier is called a *multistage amplifier*. It will be interesting to know that when we use multistage amplifier, then we feed the output of one stage to the input of the next as shown in Figure 26.1. Such a connection of amplifiers is called cascading. The amplifier used in radio and television receivers is, usually, a multistage amplifier.

A multistage amplifier, using two or more single stage common emitter amplifiers, is called a *cascade amplifier*. The term cascade denotes the type of connection used for coupling two amplifier stages. On the other hand, a multistage amplifier with common emitter amplifier as the first stage and common base (or common collector) amplifier as the second stage, is called a *cascode amplifier*. It will be interesting to know that cascade and cascode connections are not only used in transistor amplifiers but also in field-effect transistors.

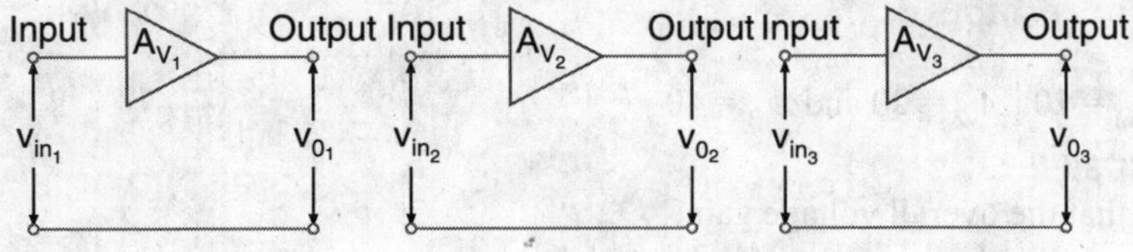

Fig. 26.1. A general multistage amplifier.

26.2. Gain of a Multistage Amplifier

We have already discussed in the last article that in a multistage amplifier the output voltage of the first stage acts as an input voltage to the second stage. And the output voltage of second stage as an input to the third stage and so on. *The voltage gain of a multistage amplifier is equal to the product of the gains of the individual stages*. Thus if A_{v_1}, A_{v_2} and A_{v_3} are the individual stage gains, then overall voltage gain,

$$A_v = A_{v_1} \times A_{v_2} \times A_{v_3}$$

The amplifier voltage gain may also be expressed in decibels (dB). However, in that case the overall decibel voltage gain is the sum of the decibel gains of the individual stages. Thus if G_{v1}, G_{v2} and G_{v_3} are the individual stage decibel gains, then overall decibel voltage gain,

$$G_v = G_{v_1} \times G_{v_2} \times G_{v_3}$$

$$= 20 \log_{10} A_{v_1} + 20 \log_{10} A_{v_2} + 20 \log_{10} A_{v_3}$$

$$= 20 \log_{10} A_v \text{ (dB)}$$

Similarly, the overall current amplification,

$$A_i = A_{i_1} \times A_{i_2} \times A_{i_3}$$

Then the overall power gain,

$$A_p = A_v \cdot A_i$$

and the overall decibel power gain,

$$G_p = 10 \log_{10} A_p$$

Cascade Connection

A Connection scheme in which the output of one amplifier stage is connected to the input of the second stage is called cascade connection (Refer to Fig. 26.2 below).

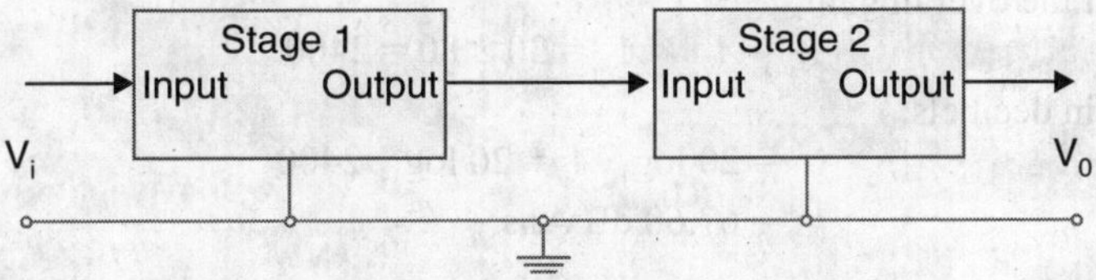

Fig. 26.2. A general multistage amplifier.

Example 26.1.

A given amplifier arrangement has the following voltage gains. $A_{v1} = 10$, $A_{v2} = 20$ and $A_{v3} = 40$. What is the overall voltage gain? Also express each gain in dB and determine the total dB voltage gain.

Solution.

Given: $A_{v1} = 10$; $A_{v2} = 20$ and $A_{v3} = 40$

Overall voltage gain

We know that the overall voltage gain,

$$A_v = A_{v1} \times A_{v2} \times A_{v3} = 10 \times 20 \times 40 = 8000 \textbf{ Ans.}$$

Total dB voltage gain

We know that dB voltage gain of the first stage

$$G_{v_1} = 20 \log_{10} A_{v_1} = 20 \log_{10} 10 = 20 \text{ dB}$$

Similarly, $G_{v_2} = 20 \log_{10} A_{v_2} = 20 \log_{10} 20 = 26 \text{ dB}$

and $G_{v_3} = 20 \log_{10} A_{v_3} = 20 \log_{10} 40 = 32 \text{ dB}$

∴ Total dB voltage gain,

$$G_v = G_{v_1} + G_{v_2} + G_{v_3} = 20 + 26 + 32 \text{ dB} = 78 \text{ dB } \textbf{Ans.}$$

Note: Check $G_v = 20 \log_{10} A_v = 20 \log_{10} 8000 = 78$ dB.

Example 26.2.

A multistage amplifier consists of three stages. The voltage gains of stages are 60, 100 and 160. Calculate the overall gain in decibels? (*PTU., May 2005*)

Solution.

Given: $A_{v_1} = 60$; $A_{v_2} = 100$ and $A_{v_3} = 160$

We know that the overall gain.

$$Av = A_{v_1} \times A_{v_2}$$
$$= 60 \times 100 \times 60 = 960000$$

Overall gain in decibels,

$$= 20 \log_{10} A_v = 20 \log_{10} 960000$$
$$= 119.6 \text{ dB } \textbf{Ans.}$$

Example 26.3.

A multistage amplifier consists of two stages. The voltage gain of the stages are 30 and 80. Calculate the overall gain in dB. (*PTU., May 2007*)

Solution.

Given: $A_{v_1} = 30$ and $A_{v_2} = 80$

We know that the overall gain,

$$A_v = A_{v_1} \times A_{v_2} = 30 \times 80 = 2400$$

Overall gain in decibels,

$$= 20 \log_{10} A_v = 20 \log_{10} 2400$$
$$= 67.60 \text{ dB } \textbf{Ans.}$$

Example 26.4.

Three amplifier stages are working in cascade with 0.05 V peak-to-peak input providing 150 V peak-to-peak output. If the voltage gain of the first stage is 20 and input to the third stage is 15 V peak-to-peak, determine,

(*i*) *the overall voltage gain*

(*ii*) *voltage gain of 2nd and 3rd stages*

(*iii*) *input voltage of the 2nd stage.* (*Banglore Univ. Jan. 2004*)

Solution.

Gives : No. of amplifier stages = 3. Input to the first stage, $v_{in1} = 0.05\ V_{pk-pk}$, Output of the final stage $(v_{out})_3 = 150\ V_{pk-pk}$. Voltage gain of the first stage $A_{v1} = 20$; Input to the third stage $v_{in3} = 15\ V_{pk-pk}$. Refer to Fig. 26.3 for clear explanation.

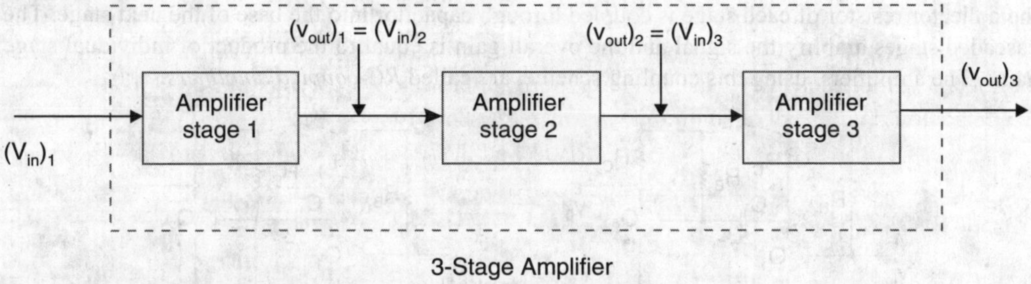

Fig. 26.3.

(i) ***Overall Voltage gain***

We know that overall voltage gain,

$$A_v = \frac{(v_{out})_3}{(v_{in})_1} = \frac{150}{0.05} = 3000 \textbf{ Ans.}$$

(ii) ***Voltage gain of 2nd and 3rd stage***

We know that voltage gain of the 3rd stage,

$$A_{v_3} = \frac{(v_{out})_3}{(v_{in})_3} = \frac{150}{15} = 10 \textbf{ Ans.}$$

Let A_{v_2} = Gain of the 2nd stage,

We know that the overall voltage gain (A_v),

$$3000 = A_{v_1} \cdot A_{v_2} \cdot A_{v_3} - 20 \times A_{v_2} \times 10 = 200\ A_{v_2}$$

$$\therefore \quad A_{v_2} = \frac{3000}{200} = 15 \textbf{ Ans.}$$

(iii) ***Input voltage of the second stage***

We know that voltage gain of the 2nd stage (A_{v2}),

$$15 = \frac{(v_{out})_2}{(v_{in})_2} = \frac{(v_{in})_3}{(v_{in})_2} = \frac{15}{(v_{in})_2}$$

$$\therefore \quad (v_{in})_2 = \frac{15}{15} = 1\ V_{kp-pk} \textbf{ Ans.}$$

26.3. Different Coupling Schemes Used in Amplifiers

As a matter of fact, all amplifiers, need some kind of *coupling* network. Even a single stage amplifier, needs coupling to the input source and output load. The multistage amplifiers need coupling between their individual stages. This type of coupling is called *interstage coupling*. It serves the following two purposes:

1. It transfers a.c. output of one stage to the input of the next stage.
2. It isolates the d.c. conditions of one stage to the next. It is necessary to prevent the shifting of *Q*-points.

The coupling network (or coupling device) must ensure that both the above purposes are fulfilled, when an a.c. signal is to be amplified. Following are the four coupling schemes used in amplifiers.

1. ***Resistance-capacitance (RC) coupling.*** It is the most important method of coupling the signal from one stage to the next and is shown in Fig. 26.4 (*a*). In this method, the signal developed across the collector resistor of each stage is coupled through capacitor into the base of the next stage. The cascaded stages amplify the signal and the overall gain is equal to the product of individual stage gains. The amplifiers, using this coupling scheme, are called *RC-coupled amplifiers*.

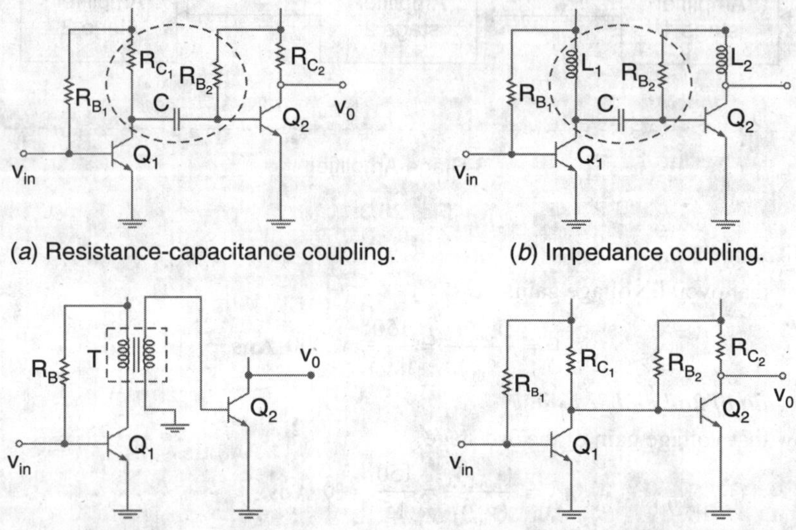

(*a*) Resistance-capacitance coupling.

(*b*) Impedance coupling.

(*c*) Transformer coupling.

(*d*) Direct coupling.

Fig. 26.4. Coupling schemes in amplifiers.

2. ***Impedance coupling***. This coupling is shown in Fig. 26.4 (*b*). It may be noted that the collector resistor (R_C) is replaced by an inductor (L). As the frequency increases, inductive reactance X_L (equal to $\omega \cdot L$) approaches infinity and each inductor appears open. In other words, the inductors pass direct current but block alternating current. The amplifiers, using this coupling scheme, are called *impedance-coupling amplifiers*.

3. ***Transformer coupling.*** This coupling is shown in Fig. 26.4 (*c*). In this method, primary winding of the transistor acts as a collector load and the secondary winding conveys the a.c. output signal directly to the base of the next stage. It may be noted that there is no need of coupling capacitor in the transformer coupling. The amplifiers, using this coupling scheme, are called *transformer-coupled amplifiers*.

Modern coupling transformer

4. ***Direct-coupling.*** This coupling is shown in Fig. 26.4 (*d*). In this method, the a.c. output signal is fed directly to the next stage. This type of coupling is used where low frequency signals are to be amplified. The coupling devices such as capacitors, inductors and transformers cannot be used at low frequencies because their size becomes very large. The amplifiers, using this coupling scheme, are called *direct coupled amplifiers* or *d.c. amplifiers*.

26.4. RC Coupled Amplifier

Fig. 26.5 shows a two-stage RC coupled transistor amplifier. The circuit consists of two single-stage common emitter transistor amplifiers. The resistors R_C, R_B and capacitor C_C form the coupling network. The capacitor C_1 is used to couple the input signal to the base of Q_1, while the capacitor C_2 is used to couple the output signal from the collector of Q_2 to the load. The capacitor C_E, connected at the emitters of Q_1 and Q_2, are needed because they *bypass* the emitter to the ground. Without these capacitors, the voltage gain of each stage will be lost.

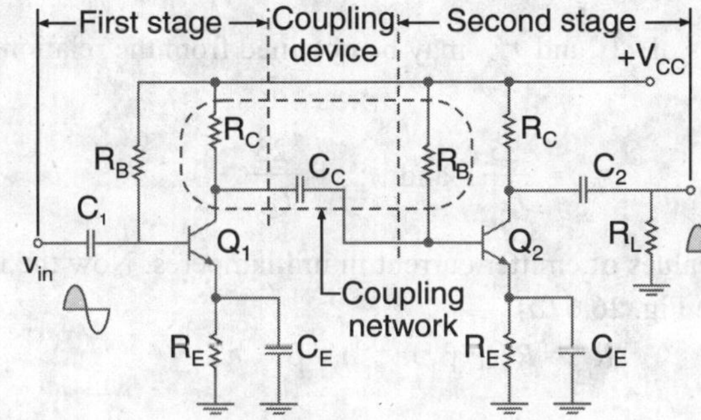

Fig. 26.5. RC coupled amplifier.

The operation of the above circuit may be understood as follows: When an a.c. signal is applied to the input of the first stage, it is amplified by a transistor and appears across the collector resistor (R_C). This signal is given to the input of the second stage through a coupling capacitor C_C. The second stage does further amplification of the signal. In this way, the cascaded stages amplify the signal and the overall gain is equal to the product of the individual stage gains.

It will be interesting to know that the *output signal of a two-stage RC coupled amplifier is in phase with the input signal*. It is because of the fact that its phase has been reversed twice by the amplifier.

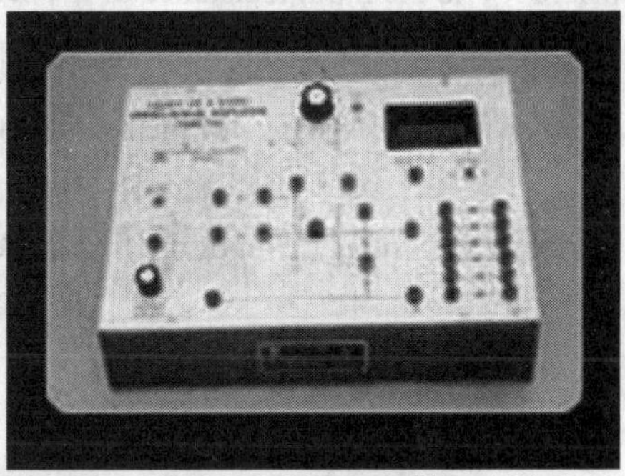

RC coupled amplifier

26.5. Calculation of Voltage Gain for RC Coupled Amplifier

Consider a two stage RC coupled transistor amplifier circuit as shown in Fig. 26.5. The a.c. equivalent circuit for each amplifying stage of that circuit is shown in Fig. 26.6 (*a*) and (*b*) respectively. Here we have not shown the complete small signal a.c. equivalent circuit of the transistor. Rather, it is represented by an a.c. emitter diode resistance ($\beta \cdot r'_e$) and the three points marked *E* (for emitter), *B* (for base) and *C* (for collector). It has been done because a.c. equivalent circuit of a transistor is slightly different at low and high-frequencies.

The parameters R_{i_1} and R_{o_1} in the equivalent circuits shown below, represent the input resistance and output resistance of the first stage respectively. Similarly, the parameters R_{i_2} and R_{o_2} represent the input resistance and output resistance of the second stage.

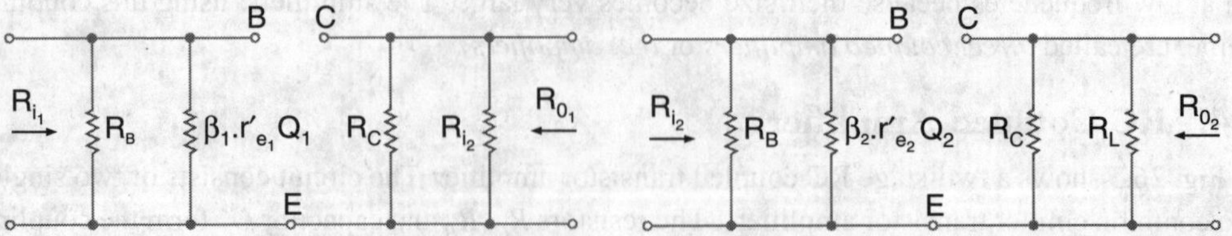

(*a*) A.C. equivalent circuit for first stage. (*b*) A.C. equivalent circuit for second stage.

Fig. 26.6. A.C. equivalent circuit of a two-stage RC coupled amplifier.

The parameters r'_{e_1} and r'_{e_2} represent the a.c. emitter diode resistances of the transistors Q_1 and Q_2 respectively. The values of r'_{e_1} and r'_{e_2} may be obtained from the relations,

$$r'_{e_1} = \frac{25}{I_{E_1}} \quad \text{and} \quad r'_{e_2} = \frac{25}{I_{E_2}}$$

where I_{E_1} and I_{E_2} are the values of emitter current in milliamperes. Now the input resistance of the first stage as seen from the Fig. 26.6 (*a*),

$$R_{i_2} = R_B \,||\, (\beta_1 \cdot r'_{e_1}) = \beta_1 \cdot r'_{e_1} \qquad \text{... (If } R_B >> \beta_1 \cdot r'_{e_1}\text{)}$$

and the output resistance,

$$R_{o_1} = R_C \,||\, R_{i2}$$

where R_{i2} is the input resistance of the second stage and its value is given by the relation,

$$R_{i_1} = R_B \,||\, (\beta_2 \cdot r'_{e_2}) = \beta_2 \cdot r'_{e_2} \qquad \text{... (If } R_B >> \beta_2 \cdot r'_{e_2}\text{)}$$

The output resistance of the second stage,

$$R_{o_2} = R_C \,||\, R_L$$

Now the voltage gain of the first stage,

$$A_{v_1} = \beta_1 \times \frac{R_{o_1}}{R_{i_1}} = \beta_1 \times \frac{R_{o_1}}{\beta_1 \cdot r'_{e_1}} = \frac{R_{o_1}}{r'_{e_1}}$$

Similarly, voltage gain of the second stage,

$$A_{v_2} = \frac{R_{o_2}}{r'_{e_2}}$$

∴ Overall voltage gain of the amplifier,

$$A_v = A_{v_1} \cdot A_{v_2} = \frac{R_{o_1}}{r'_{o_1}} \times \frac{R_{o_2}}{r'_{o_2}}$$

If the transistors used in both the stages are identical, then current gain of the transistors Q_1 and Q_2 (*i.e.*, β_1 and β_2) will be equal. In that case, the emitter currents will also be equal. Thus $r'_{e_1} = r'_{e_2} = r'_e$. Then overall voltage gain,

$$A_v = R_{o_1} \cdot R_{o_2} / r_e^2$$

Example 26.5.

Fig. 26.7 shows the circuit of a two stage RC coupled amplifier.

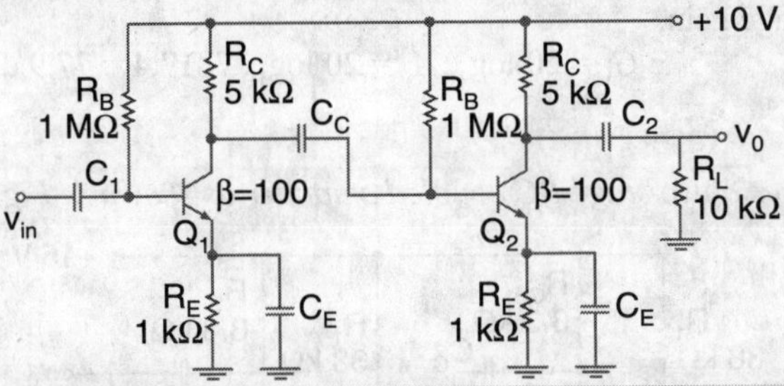

Fig. 26.7.

Determine the values of input resistance, output resistance, current gain, voltage gain for the first and second stages. Also determine the overall voltage gain and overall dB voltage gain.

Solution.

Given: $V_{CC} = 10$ volts; $R_C = 5\ \text{k}\Omega = 5000\ \Omega$; $R_B = 1\ \text{M}\Omega = 1 \times 10^6\ \Omega$; $R_E = 1\ \text{k}\Omega = 1 \times 10^3\ \Omega$; $R_L = 10\ \text{k}\Omega = 10\ 000\ \Omega$ and $\beta_1 = \beta_2 = 100$.

Input resistance of the first and second stage

We know that the value of emitter current for the first as well as second stage,

$$I_E = \frac{V_{CC}}{R_E + \dfrac{R_B}{\beta}} = \frac{10}{1 \times 10^3 + \dfrac{1 \times 10^4}{100}}$$

$$= 0.91 \times 10^{-3}\ \text{A} = 0.91\ \text{mA}$$

and the value of a.c. emitter diode resistance,

$$r'_e = \frac{25}{I_E} = \frac{25}{0.91} = 27.5\ \Omega$$

∴ Input resistance of the first stage,

$$R_{i_1} = \beta \cdot r'_e = 100 \times 27.5 = 2750\ \Omega \ \textbf{Ans.}$$

and input resistance of the second stage,

$$R_{i_2} = \beta \cdot r'_e = 100 \times 27.5 = 2750\ \Omega \ \textbf{Ans.}$$

Output resistance of the first and second stage

We know that output resistance of the first stage,

$$R_{o_1} = R_C \,\|\, R_{i_2} = 5000 \,\|\, 2750 = 1774\ \Omega \ \textbf{Ans.}$$

and output resistance of the second stage,

$$R_{o_2} = R_C \,\|\, R_L = 5000 \,\|\, 10\ 000 = 3333\ \Omega \ \textbf{Ans.}$$

Voltage gain of the first and second stage,

We know that the voltage gain of the first stage,

$$A_{v_1} = \frac{R_{o_1}}{r'_e} = \frac{1774}{27.5} = 64.5 \ \textbf{Ans.}$$

and voltage gain of the second stage,

$$A_{v_2} = \frac{R_{o_2}}{r'_e} = \frac{3333}{27.5} = 121.2 \ \textbf{Ans.}$$

Overall voltage gain and dB voltage gain

We also know that the overall voltage gain,

$$A_v = A_{v_1} \times A_{v_2} = 64.5 \times 121.2 = 7817.4 \textbf{ Ans.}$$

and the overall dB voltage gain,

$$G_v = 20 \log_{10} A_v = 20 \log_{10} 7817.4 = 77.9 \text{ dB} \textbf{ Ans.}$$

Example 26.6.

A radio receiver uses a two stage RC coupled amplifier as shown in Fig. 26.8.

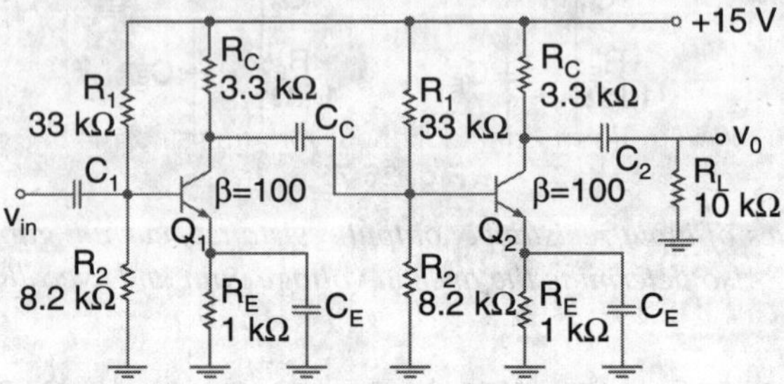

Fig. 26.8.

Determine the values of voltage gain of each stage and overall voltage gain. Also express the overall voltage gain in decibels.

Solution.

Given: V_{CC} = 15 volts; R_C = 3.3 kΩ = 3300 Ω; R_E = 1 kΩ = 1000 Ω; R_1 = 33 kΩ; R_2 = 8.2 kΩ; R_L = 10 kΩ = 10 000 Ω and $\beta_1 = \beta_2 = 100$.

Voltage gain of each stage

We know that voltage drop across the resistor R_2,

$$V_{TH} = V_{CC}\left(\frac{R_2}{R_1 + R_2}\right) = 15\left(\frac{8.2}{33 + 8.2}\right) \text{ V} = 2.985 \text{ V}$$

and the parallel combination of resistances R_1 and R_2,

$$R_{TH} = R_1 \parallel R_2 = 33 \parallel 8.2 = 6.57 \text{ k}\Omega = 6570 \ \Omega$$

∴ Emitter current,

$$I_E = \frac{V_{TH} - V_{BE}}{R_E + \dfrac{R_{TH}}{\beta}} = \frac{2.985 - 0.7}{1000 + \dfrac{6570}{100}} \text{ A}$$

$$= 2.14 \times 10^{-3} \text{ A} = 2.14 \text{ mA}$$

and the a.c. emitter diode resistance,

$$r'_e = \frac{25}{I_E} = \frac{25}{2.14} = 11.7 \ \Omega$$

We also know that input resistance of the second stage,

$$R_{i_2} = \beta \cdot r'_e = 100 \times 11.7 = 1170 \ \Omega$$

and the output resistance of the first stage,

$$R_{o_1} = R_C \parallel R_{i_2} = 3300 \parallel 1170 = 864 \ \Omega$$

Similarly, output resistance of the second stage,

$$R_{o_2} = R_C \parallel R_L = 3300 \parallel 10\,000 = 2481 \ \Omega$$

∴ Voltage gain of the first stage,

$$A_{v_1} = \frac{R_{o_1}}{r'_e} = \frac{864}{11.7} = 73.8 \textbf{ Ans.}$$

and voltage gain of the second stage,

$$A_{v_2} = \frac{R_{o_2}}{r'_e} = \frac{2481}{11.7} = 212 \textbf{ Ans.}$$

Overall voltage gain

We know that the overall voltage gain,

$$A_v = A_{v_1} \cdot A_{v_2} = 73.8 \times 212 = 15646 \textbf{ Ans.}$$

Overall voltage in decibels

We also know that the overall voltage gain in decibels (called the dB voltage gain),

$$G_v = 20 \log_{10} A_v = 20 \log_{10} 15\,646 = 83.9 \text{ dB } \textbf{Ans.}$$

26.6. Frequency Response of RC Coupled Amplifier

The frequency response of an amplifier is a graph, which indicates the relationship between the voltage gain as a function of frequency. Usually, the voltage gain (in decibels) is plotted along the vertical axis and the frequency (in hertz or kilohertz) along the horizontal axis of the frequency response graph. Fig. 26.9 shows the frequency response of RC coupled amplifier.

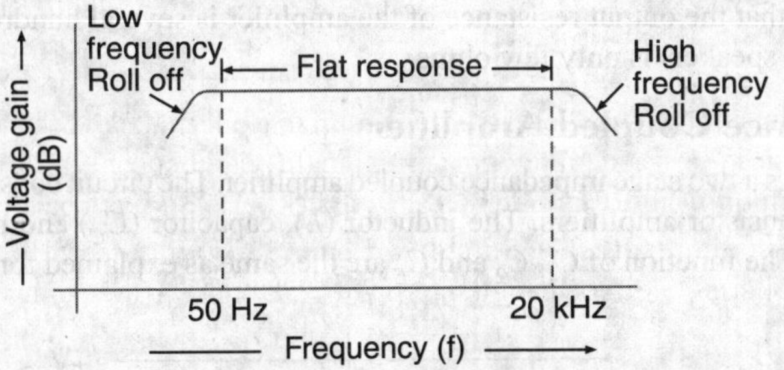

Fig. 26.9. Frequency response of RC coupled amplifier.

It is evident from the graph that the voltage gain *drops off* (or rolls off) at low frequencies (*i.e.*, frequencies below 50 Hz) and at high frequencies (*i.e.*, frequencies above 20 kHz), while it remains *constant* in the mid-frequency range. The behaviour is discussed in more detail as follows:

1. *At low frequencies (i.e., below 50 Hz).* We know that the capacitive reactance (X_C) is inversely proportional to the frequency. Thus at low frequencies, the reactance of the capacitor C_C is quite large. Therefore it will allow only a small part of the signal to pass from one stage to the next stage. In addition to this, the emitter bypass capacitor (C_E) cannot shunt the emitter resistor effectively, because of its large reactance at low frequencies. As a result of these two factors, the voltage gain rolls off at low frequencies.

2. *At high frequencies (i.e., above 20 kHz).* In this frequency range, the reactance of capacitor C_C becomes quite small, therefore it behaves like a short-circuit. As a result of this, the loading effect of the next stage increases, which reduces the voltage gain. In addition to this, the capacitance of the emitter diode plays an important role at high frequencies. It increases the base current of the transistor due to which the *current gain (β) reduces*. Hence the voltage gain rolls off at high frequencies.

3. *At mid-frequencies (i.e., 50 Hz to 20 kHz).* The effect of coupling capacitor, in this frequency range, is such that it maintains a constant voltage gain. Thus as the frequency increases, the reactance of capacitor C_C decreases, which tends to increase the gain. However, at the same time, the lower capacitive reactance increases the loading effect of the next stage due to which the gain reduces. These two factors almost cancel each other. Thus a constant gain is maintained throughout this frequency range.

26.7. Advantages and Disadvantages of RC Coupled Amplifier

Advantages

1. It is the most convenient and least expensive multistage amplifier.
2. It has a wide frequency response *i.e.*, its gain versus frequency curve remains constant over a wide frequency range.
3. It provides less frequency distortion.

Disadvantages

1. The overall gain of the amplifier is comparatively small because of the loading effect of successive stages.
2. It has a tendency to become noisy with age, especially in moist climates.
3. It provides poor resistance (or impedance) matching between the stages.

26.8. Applications of RC Coupled Amplifiers

We have already discussed in Art. 26.6 that an RC coupled amplifier has an excellent frequency response from 50 Hz to 20 kHz. This property makes it very useful in the initial stages of all the *public address systems*. However, it may be noted that a coupled amplifier cannot be used as a final stage of the amplifier because of its poor resistance matching characteristics. The resistance mismatching is because of the fact that the output resistance of the amplifier is several hundred ohms, whereas that of the load (such as speaker) is only few ohms.

26.9. Impedance Coupled Amplifier

Fig. 26.10 shows a two stage impedance coupled amplifier. The circuit consists of two single-stage common emitter transistor amplifiers. The inductor (L), capacitor (C_C) and resistor (R_B) forms the coupling network. The function of C_1, C_2 and C_E are the same as explained for RC coupled amplifier in Art. 26.4.

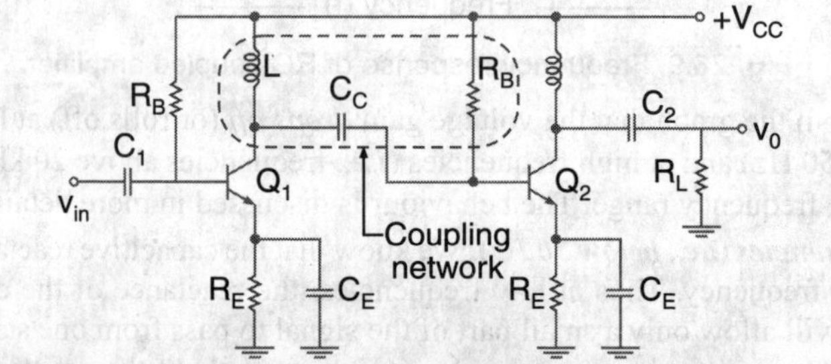

Fig. 26.10. Two-stage impedance coupled amplifier.

The expression for overall voltage gain and individual stage gains may be obtained in the same way as obtained for RC coupled amplifier. The only change to be made there is that R_C should be replaced by inductive reactance X_L (= ωL). Thus voltage gain of the first stage,

$$A_{v_1} = \frac{R_{o_1}}{r'_{e_1}} = \frac{X_L \parallel R_{i_2}}{r'_{e_1}} = \frac{\omega L \parallel R_{i_2}}{r'_{e_1}}$$

$$\approx \frac{R_{i_2}}{r'_{e_1}} \qquad \text{...(If } \omega_L >> R_{i2}\text{)}$$

where R_{i_2} is the input resistance of the second stage and is given by the expression,

$$R_{i_2} = R_B \parallel (\beta_2 \cdot r'_{e_2}) = \beta_2 \cdot r'_{e_2} \qquad \text{...(If } R_B >> \beta_2 \cdot r'_{e_2}\text{)}$$

and voltage gain of the second stage,

$$A_{v_2} = R_{o_2} / r'_{e_2}$$

∴ Overall voltage gain,

$$A_v = A_{v_1} \times A_{v_2}$$

The main advantage of an impedance coupled amplifier is that it *can be operated at low collector supply voltage*, as there is hardly any drop across inductor (L). The disadvantage of impedance coupled amplifier is that it *is relatively expensive because of the use of inductors*. Moreover, the impedance of the inductor drops off at lower frequencies. Therefore as a rule, *an impedance coupled amplifier is suitable only at frequency above 20 kHz (called radio frequencies)*.

26.10. Transformer Coupled Amplifier

Fig. 26.11 shows a two-stage transformer coupled amplifier. The circuit consists of two single-stage common emitter transistor amplifiers. The function of transformer (T_1) is to couple the a.c. output signal from the output of the first stage to the input of the second stage, while transformer (T_2) couples the output signal to the load. The input coupling capacitor is C_1, while the emitter bypass capacitor is C_E,

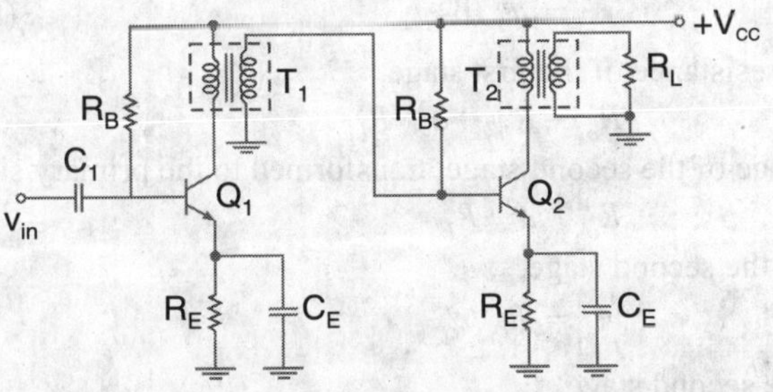

Fig. 26.11. Two stage transformer coupled amplifier.

The operation of the above circuit may be understood from the condition that when an a.c. input signal is applied to the base transistor Q_1, it appears in the amplified form across primary winding of the transformer (T_1). The voltage developed across the primary winding is then transferred to the input of the next stage by the secondary winding of the transformer (T_1). The second stage does amplification in an exactly similar manner.

In actual practice, a bypass capacitor is used on the bottom of each primary winding to produce an a.c. ground. This avoids the load inductance of the connecting line that returns to the d.c. supply point. Similarly, a bypass capacitor is used on the bottom of each secondary winding to get an a.c. ground. This prevents the signal power loss in the biasing resistors.

26.11. Calculations of Voltage Gain for Transformer Coupled Amplifier

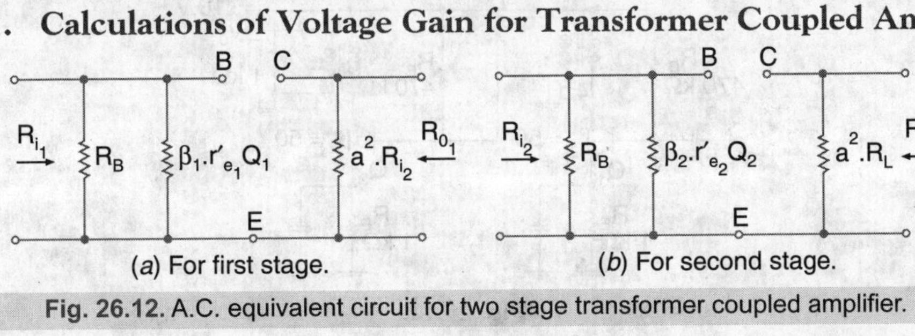

Fig. 26.12. A.C. equivalent circuit for two stage transformer coupled amplifier.

Consider the circuit of a two-stage transformer coupled amplifier shown in Fig. 26.11. The a.c. equivalent circuit for the amplifier is shown in Fig. 26.12 (*a*) and (*b*). In these figures, the a.c. equivalent circuit of the transistor is shown by the a.c. emitter diode resistance and three points marked *E*, *B* and *C*. Assuming the transformer to be an ideal (*i.e.*, neglecting its leakage current, winding resistance etc.), whatever be the load on its secondary winding side, it transforms the same on the primary side. The value of the transformed load,

$$R'_L = \left(\frac{N_1}{N_2}\right)^2 R_L = a^2 \cdot R_L \qquad \ldots \left(\text{where } a = \frac{N_1}{N_2}\right)$$

where N_1 = Number of primary turns, and

N_2 = Number of secondary turns.

Similarly, if R_{i_2} is the input resistance of the second stage, then its value transformed to the primary side,

$$R'_{i_2} = a^2 \cdot R_{i_2}$$

We know that the input resistance of the first stage is given by the relation,

$$R_{i_1} = R_B \parallel (\beta_1 \cdot r'_{e_1}) \qquad \ldots(\text{If } R_B >> \beta_1 \cdot r'_{e_1})$$

and the input resistance of the second stage, transformed to the primary side,

$$R'_{i_2} = a^2 \cdot R_{i_2} = a^2 \cdot R_B \parallel (\beta_2 \cdot r'_{e_1})$$

$$= a^2 (\beta_2 \cdot r'_{e_2}) \qquad \ldots (\text{If } R_B >> \beta \cdot r'_{e_1}\, \beta_2 \cdot r'_{e_2})$$

Similarly, output resistance of the first stage,

$$R_{o_1} = R'_{i_2} = a^2 \cdot R_{i_2}$$

and the output resistance of the second stage, transformed to the primary side,

$$R_{o_2} = a^2 \cdot R_L$$

∴ Voltage gain of the second stage,

$$A_{v_1} = R_{o_1} / r'_{e_1}$$

and voltage gain of the second stage,

$$A_{v_2} = \frac{R'_{o_2}}{r'_{e_2}} = \frac{a^2 \cdot R_L}{r'_{e_2}}$$

∴ Overall voltage gain,

$$A_v = A_{v_1} \cdot A_{v_2} = \frac{R_{o_1}}{r'_e} \times \frac{a^2 \cdot R_L}{r'_{e_2}}$$

Example 26.7.

Fig. 26.13 shows the circuit diagram of a two-stage transformer coupled transistor amplifier.

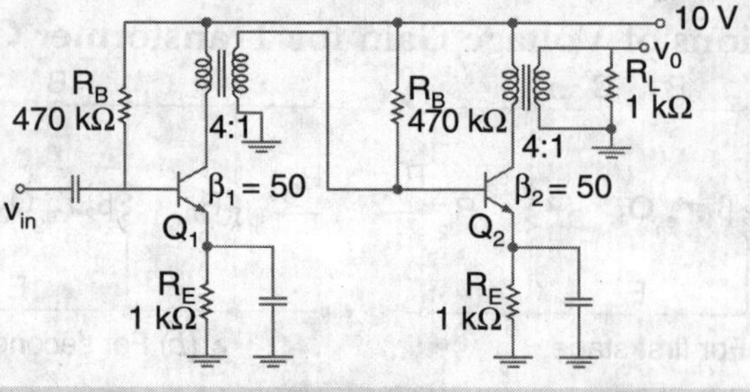

Fig. 26.13.

Calculate the values of individual stage gains and the overall voltage gain in decibels. Neglect V_{BE} and take $r'_e = 25\ mV/I_E\ (mA)$.

Solution.

Given: $V_{CC} = 10$ volts; $R_E = 1\ k\Omega = 1 \times 10^3\ \Omega$; $R_B = 470\ k\Omega = 470 \times 10^3\ \Omega$; $a = 4$; $\beta_1 = \beta_2 = 50$ and $R_L = 1\ k\Omega = 1000\ \Omega$.

Individual stage gains

We know that the value of emitter current,

$$I_E = \frac{V_{CC}}{R_E + \frac{R_B}{\beta}} = \frac{10}{(1 \times 10^3) + \frac{470 \times 10^3}{50}}\ A$$

$$= 0.96 \times 10^{-3}\ A = 0.96\ mA$$

and the value of a.c. emitter diode resistance,

$$r'_e = \frac{25\ mV}{I_E\ (mA)} = \frac{25}{0.95} = 26\ \Omega$$

We also know that input resistance of the first stage,

$$R_{i_1} = R_B \parallel (\beta_1 \cdot r'_{e_1}) = (470 \times 10^3) \parallel (50 \times 26)\ \Omega$$

$$= 1296\ \Omega$$

and input resistance of the second stage,

$$R_{i_1} = R_B \parallel (\beta_2 \cdot r'_{e_2}) = (470 \times 10^3) \parallel (50 \times 26)\ \Omega$$

$$= 1296\ \Omega$$

The input resistance of the second stage, transformed to the primary side,

$$R'_{i_2} = a^2 \cdot R_{i_2} = (4)^2 \times 1296 = 20736\ \Omega$$

Similarly, output resistance of the first stage,

$$R_{o_1} - R'_{i_2} = 20736\ \Omega$$

and output resistance of the second stage, transformed to the primary side,

$$R'_{i_2} = a^2 \cdot R_L = (4)^2 \times 1000 = 1600\ \Omega$$

∴ Voltage gain of the first stage,

$$A_{v_1} = \frac{R_{o_1}}{r'_{e_1}} = \frac{20736}{26} = 797.5 \textbf{ Ans.}$$

and voltage gain of the second stage,

$$A_{v_2} = \frac{R'_{o_2}}{r'_{e_2}} = \frac{16000}{26} = 615.4 \textbf{ Ans.}$$

Overall voltage gain in decibels

We know that the overall voltage gain,

$$A_v = A_{v_1} \times A_{v_2} = 797.5 \times 615.4 \textbf{ Ans.}$$

and the overall voltage gain in decibels (called dB voltage gain),

$$G_v = 20 \log_{10} A_v = 20 \log_{10} 490\ 782 = 114\ dB \textbf{ Ans.}$$

26.12. Frequency Response of Transformer Coupled Amplifier

Fig. 26.14 shows the frequency response graph (*i.e.*, a graph of voltage gain *versus* frequency) for a transformer coupled transistor amplifier.

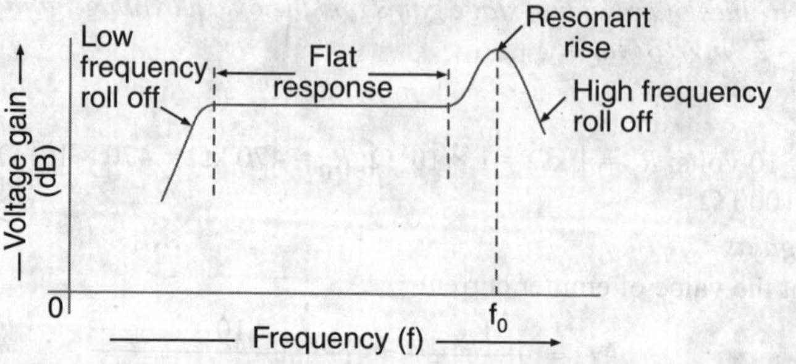

Fig. 26.14. Frequency response of a transformer coupled amplifier.

It is evident from this figure, that the voltage gain drops off (or rolls off) at low as well as at high frequencies , whereas it remains constant in mid-frequency range. Another noticeable feature is that *at one particular frequency* (f_0) *the voltage gain increases and then rolls off continuously*. This typical behaviour may be explained as follow:

We know that output voltage of a transformer coupled amplifier is equal to the product of collector current and the reactance of the primary winding of coupling transformer. At low frequencies, the reactance of primary winding ($X_L = \omega \cdot L$) begins to decrease and hence the voltage gain reduces. At high frequencies, the effect of leakage inductance and distributed capacitance (*i.e.*, the capacitance between the turns of the winding) becomes significant and hence the voltage gain reduces. *The peak gain results due to the resonance (or turning) effect of inductance and distributed capacitance, which forms a resonant circuit*. The frequency, at which the peak occurs, is called resonant frequency (f_0).

It has been found that flat part of frequency response curve of transformer coupled amplifiers is small as compared to that of RC coupled amplifiers. As a result of this, *these amplifiers can not be used over a wide range of frequencies*. Moreover, if they are used, *they produce frequency distortion*, which means that all frequency components in a complete input signal (such as music, speech signal) are not equally amplified. However, the transformers can be suitably designed to provide a fairly flat response curve and excellent fidelity over the entire audio frequency range (*i.e.*, 20 Hz to 20 kHz).

26.13. Advantages and Disadvantages of Transformer Coupled Amplifier

Advantages

1. No signal power is lost in the collector or base resistors, because of the low winding resistance of the transformer.
2. It provides a higher voltage gain than the RC coupled amplifier.
3. It provides an excellent resistance (or impedance) matching between the stages. The resistance matching is desirable for maximum power transfer.

Disadvantages

1. The coupling transformer is expensive and bulky, particularly when operated at audio frequencies.
2. At radio frequencies, the winding inductance and distributed capacitance produces reverse frequency distortion.
3. It tends to produce '*hum*' in the circuit.

26.14. Applications of Transformer Coupled Amplifier

We have already discussed in the last article that transformer coupled amplifiers provide excellent impedance matching between the individual stages. This ability makes it very useful in a multistage amplifier, where it is used as a final stage. *It is used to transfer power to the low impedance load (such as speaker)*. The impedance of a speaker varies from 4 Ω to 16 Ω, whereas the output resistance of a transistor amplifier is several hundred ohms. In order to match the load impedance, with that of the amplifier output, a step-down transformer of proper turns ratio is used. The resistance of the secondary winding of the transformer is made equal to the speaker impedance, while that of the primary winding is made equal to the output resistance of the amplifier.

26.15. Direct Coupled Amplifier

It is also called as *DC amplifier* and is used to *amplify very low frequency* (*i.e.*, below 10 Hz) signals including direct current or zero frequency. It may be noted that the capacitors, inductors and transformers can not be used as a coupling network at very low frequencies because the electrical size of these devices, at low frequencies, becomes very large.

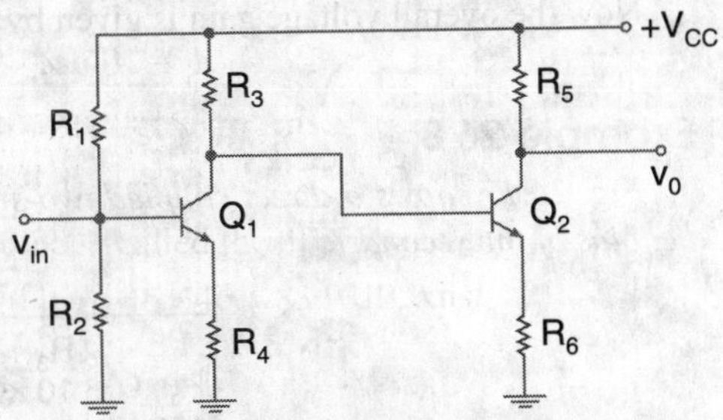

Fig. 26.15. Two stage direct coupled amplifier.

Fig. 26.15 shows a two-stage direct coupled transistor amplifier. It may be noted that the output of the first stage is directly connected to the base of the next transistor. Moreover, there are no input or output coupling capacitors. The operation of this circuit is discussed below:

The signal to be amplified is applied directly to the input of the first stage. Due to the transistor action, it appears in the amplified form across the collector resistor or transistor Q_1. This voltage then drives the base of the second transistor Q_2 and the amplified output is obtained across the collector resistor of transistor Q_2.

26.16. Calculation of Voltage Gain of Direct Coupled Amplifier

Consider the circuit diagram of a direct coupled amplifier shown in Fig. 26.15. The a.c. equivalent circuit for such an amplifier is shown in Fig. 26.16.

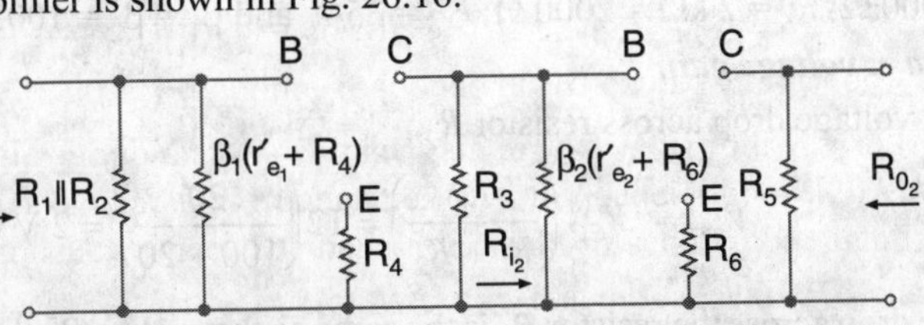

Fig. 26.16. A.C. equivalent circuit of a two stage direct coupled amplifier.

It is evident from the figure shown above that the input resistance of the first stage,

$$R_{v_1} = (R_1 \| R_2) \| \beta_1(r'_{e_1} + R_4)$$

and input resistance of the second stage,

$$R_{i_2} = \beta_2(r'_{e_2} + R_6)$$

Similarly, the output resistance of the first stage,

$$R_{o_1} = R_3 \| R_{i_2}$$

and output resistance of the second stage,

$$R_{o_2} = R_5$$

∴ Voltage gain of the first stage,

$$A_{v_1} = \beta \times \frac{R_{o_1}}{R_{i_1}} = \frac{R_{o_1}}{r'_{e1} + R_4}$$

and voltage gain of the second stage,

$$A_{v_2} = \beta_2 \times \frac{R_{o_2}}{R_{i_3}} = \frac{R_{o_2}}{r'_{e_2} + R_4}$$

Now the overall voltage gain is given by the relation,

$$A_v = A_{v_1} \cdot A_{v_2}$$

Example 26.8.

Fig. 26.17 shows a direct coupled two-stage amplifier. The d.c. bias of the first stage sets the d.c. bias of the second.

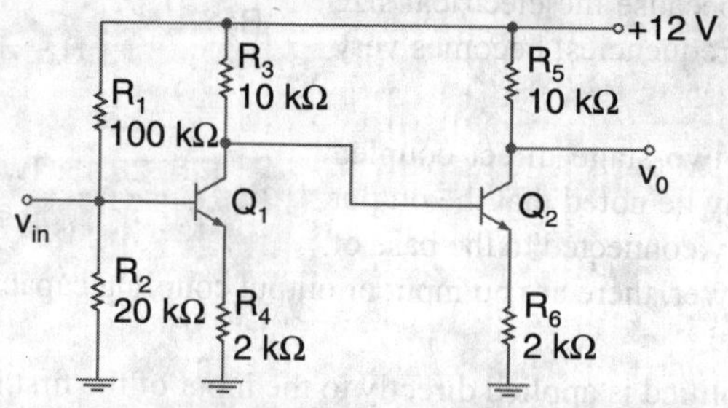

Fig. 26.17.

Determine the individual stages a.c. voltage gain and the overall voltage gain. Neglect V_{BE} and use $r'_e = 25/I_E$ in mA. Take $\beta_1 = \beta_2 = 100$.

Solution.

Given: $V_{CC} = 12$ volts; $R_1 = 100$ kΩ; $R_2 = 20$ kΩ; $R_3 = 10$ kΩ = 10 000 Ω; $R_4 = 2$ kΩ = 2000 Ω; $R_5 = 10$ kΩ = 10 000 Ω; $R_6 = 2$ kΩ = 2000 Ω; $r'_e = 25/I_E$ and $\beta_1 = \beta_2 = 100$.

Individual stages a.c. voltage gain

We know that voltage drop across resistor R_2,

$$V_{TH} = V_{CC}\left(\frac{R_2}{R_1 + R_2}\right) = 12\left(\frac{20}{100 + 20}\right) = 2 \text{ V}$$

As the voltage drop across the resistor R_4 is the same as that across R_2. Therefore emitter current of Q_1 transistor,

$$r'_{e_1} = \frac{V_{TH}}{R_4} = \frac{2}{2000} = 0.001 \text{ A} = 1 \text{ mA}$$

and the a.c. emitter diode resistance,

$$V_{R_6} = \frac{25}{I_{E_1}} = \frac{25}{1} = 25 \; \Omega$$

Again neglecting V_{BE}, the voltage drop across resistor, R_e is the same as the collector voltage of Q_1 transistor and its value,

$$I_{E_2} = V_{C_1} = V_{CC} - I_{E_6} \cdot R_3$$
$$= 12 - (0.001 \times 10\,000) = 12 - 10 = 2 \text{ V}$$

∴ Emitter current of transistor,

$$I_{E_2} = \frac{V_{R_6}}{R_6} = \frac{2}{2000} = 0.001 \text{ A} = 1 \text{ mA}$$

and the a.c. emitter diode resistance,

$$r'_{e_2} = \frac{25}{I_{E_2}} = \frac{25}{1} = 25\ \Omega$$

We also know that input resistance of the second stage

$$R_{i_2} = \beta_2\,(r'_{e_2} + R_6) = 100\ (25 + 2000) = 202{,}500$$

and output resistance of the first stage,

$$R_{o_1} = R_3 \parallel R_{i_2} = 10\ 000 \parallel 202\ 500 = 10{,}000\ \Omega$$

Similarly, output resistance of the second stage,

$$R_{o_1} = R_5 = 10{,}000\ \Omega$$

∴ Voltage gain of the first stage.

$$A_{v_1} = \frac{R_{o_1}}{r'_{e_1} + R_4} = \frac{10\ 000}{25 + 2000} = 4.9 \textbf{ Ans.}$$

and voltage gain of second stage,

$$A_{v_2} = \frac{R_{o_2}}{r'_{e_2} + R_6} = \frac{10000}{25 + 2000} = 4.9 \textbf{ Ans.}$$

Overall voltage gain

We know that overall voltage gain,

$$A_v = A_{v_1} \cdot A_{v_2} = 4.9 \times 4.9 = 24 \textbf{ Ans.}$$

26.17. Direct Coupled Amplifier Using Complementary Transistor

Fig. 26.18 shows a two-stage direct coupled amplifier using *complementary transistors*. The term '*complementary*' means that one stage uses *NPN* transistor, while the second stage uses *PNP* transistor. The operation of the circuit may be understood from the condition that when the positive going input signal is applied to the base of transistor Q_1, its collector current increases. This increases the voltage drop across resistor R_3 due to which the collector voltage of transistor Q_1 decreases. Consequently, the base of transistor Q_2 becomes less positive (or more negative). It means that a negative going signal is applied to the base of Q_2 transistor. It causes the forward bias across emitter-base junction of Q_2 (*i.e.*, PNP transistor) to increase, which in turn increases the collector current. As a result of this, the voltage drop across resistor R_6 increases and we get a positive going amplified output. Similarly, during the positive going input signal, we get a negative going amplified output across the resistor R_6.

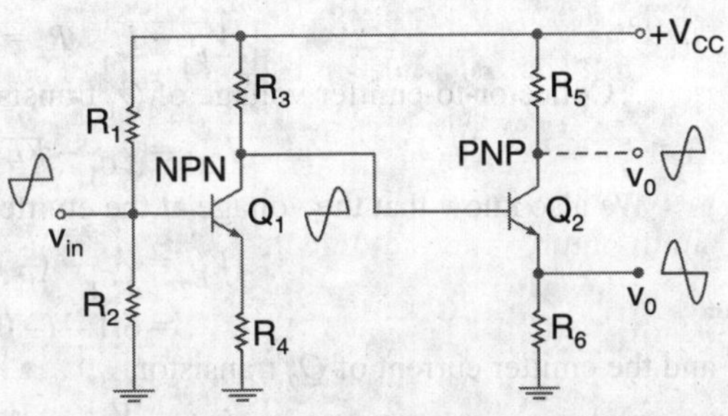

Fig. 26.18. Direct coupled amplifier using complementary transistors.

It may be noted that if we take an output from the emitter terminal of Q_2 transistor, its magnitude and polarity will be the same as that of its input (*i.e.*, at the base of Q_2 transistor).

Example 26.9.

For the circuit shown in Fig. 26.19, determine the d.c. collector currents and collector-to-emitter voltages of transistors Q_1 and Q_2.

Also determine the approximate overall a.c. voltage gain of the cascaded amplifier. Assume h_{FE} (i.e., β) of both the transistors to be 100 and C_1 and C_2 to be very large. Assume V_{BE} = 0.7 V.

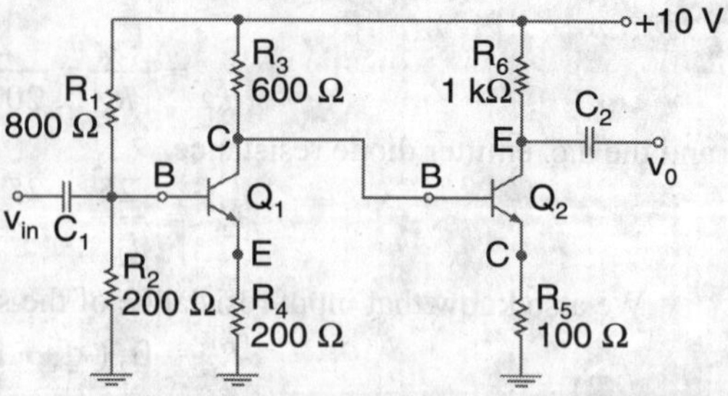

Fig. 26.19.

Solution.

Given: V_{CC} = 10 volts; R_1 = 800 Ω; R_2 = 200 Ω; R_3 = 600 Ω; R_4 = 200 Ω; R_5 = 100 Ω; R_6 = 1 kΩ = 1 × 10^3 Ω, $\beta_1 = \beta_2$ = 100 and V_{BE} = 0.7 volt.

D.C. collector currents and collector-to-emitter voltages

We know that the voltage drop across resistor R_2,

$$V_{R_2} = V_{CC}\left(\frac{R_2}{R_1 + R_2}\right) = 10\left(\frac{200}{800 + 200}\right) = 2 \text{ V}$$

and the emitter current of Q_1 transistor,

$$I_{E_1} = \frac{V_{R_2} - V_{BE_1}}{R_E} = \frac{2 - 0.7}{200} dx = 6.5 \times 10^{-3} \text{ A}$$

∴ Collector current of Q_1, transistor,

$$I_{C_1} = I_{E_1} = 6.5 \times 10^{-3} \text{ A} = 6.5 \text{ mA } \textbf{Ans.}$$

We also know that the voltage at the collector of Q_1 transistor,

$$V_{C_1} = V_{CC} - I_{C_1} \cdot R_3 = 10 - (6.5 \times 10^{-3}) \times 600 = 6.1 \text{ V}$$

and the voltage at the emitter of Q_1 transistor,

$$V_{E_1} = I_{E_1} \cdot R_4 = (6.5 \times 10^{-3}) \times 200 = 1.3 \text{ V}$$

∴ Collector-to-emitter voltage of Q_1 transistor,

$$I_{CE_1} = V_{C_1} - VE_1 = 6.1 - 1.3 = 4.8 \text{ V } \textbf{Ans.}$$

We also know that the voltage at the emitter of Q_2 transistor,

$$V_{E_2} = V_{C_1} - (-V_{BE}) \qquad (\because V_{BE} \text{ for a PNP Transistor is negative})$$

$$= 6.1 - (-0.7) = 6.8 \text{ V}$$

and the emitter current of Q_2 transistor,

$$I_{E_2} = \frac{V_{CC} - V_{E_2}}{R_6} = \frac{10 - 6.8}{1 \times 10^3} = 3.2 \times 10^{-3} \text{ A}$$

∴ Collector current of Q_2 transistor,

$$I_{C_2} = I_{E_2}$$

and the voltage at the collector of Q_2 transistor,

$$V_{C_2} = I_{C_1} \cdot R_5 = (3.2 \times 10^{-3}) \times 100 = 0.32 \text{ V}$$

∴ Collector-to-emitter voltage of Q_2 transistor,

$$V_{CF_2} = V_{C_2} - V_{E_2} = 0.32 - 6.8 = -6.48 \text{ V } \textbf{Ans.}$$

Overall voltage gain

We know that a.c. emitter diode resistance of Q_1 transistor,

$$r'_{e_1} = \frac{25}{I_{E_1}} = \frac{25}{6.5} = 3.85\ \Omega$$

and a.c. emitter diode resistance of Q_2 transistor,

$$r'_{e_2} = \frac{25}{I_{E_2}} = \frac{25}{3.2} = 7.8\ \Omega$$

We also know that the input resistance of the second stage,

$$R_{i_2} = \beta_2\,(r'_{e_2} + R_6) = 100\,(7.8 + 1000) = 100.8 \times 10^3\ \Omega$$

and output resistance of the first stage,

$$R_{o_1} = R_3 \parallel R_{i_2} = 600 + (100.85 \times 103) = 596.3\ \Omega$$

$\therefore$ Voltage gain of the first stage,

$$A_{v_1} = \frac{R_{o_1}}{r'_{e_1} + R_4} = \frac{596.3}{3.85 + 200} = 2.93$$

and voltage gain of the second stage,

$$A_{v_2} \approx 1 \quad \text{...}(\because \text{ It is an emitter follower stage})$$

$\therefore$ Overall voltage gain,

$$A_v = A_{v_1} \cdot A_{v_2} = 2.93 \times 1 = 2.93 \textbf{ Ans.}$$

26.18. Frequency Response of Direct Coupled Amplifier

Fig. 26.20 shows the frequency response (*i.e.*, a graph of dB voltage gain versus frequency) of a direct coupled amplifier. It is evident from this figure, the *gain is uniform up to a certain frequency denoted by f_2. Beyond this frequency, the gain rolls off slowly*. The gain rolls off at high frequencies due to the increased emitter diode capacitance and stray wiring capacitance.

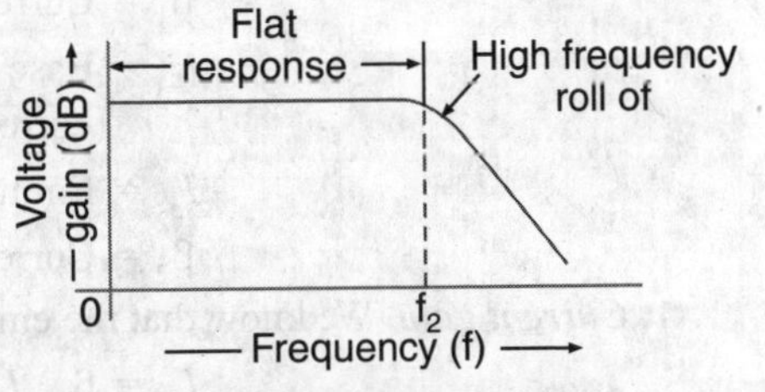

Fig. 26.20. Frequency response of direct coupled amplifier.

26.19. Advantages and Disadvantages of Direct Coupled Amplifier

Advantages

1. The circuit arrangement is very simple because it uses a minimum number of resistors.
2. The circuit cost is low because of the absence of expensive coupling devices.
3. It can amplify very low frequency signals down to zero frequency.

Disadvantages

1. It cannot amplify high frequency signals.
2. It has a poor temperature stability. Because of this, its, *Q*-point shifts. In a multistage direct coupled amplifier, the *Q*-point shifts are amplified in succeeding stages. Thus a small d.c. shift, in the first stage can cause the final stage to be either saturated or cut-off. All integrated circuit amplifiers are direct coupled because of the difficulty of fabricating large integrated capacitors. It leads to special problems in their design.

26.20. Applications of Direct Coupled Amplifiers

The direct coupled amplifierr are used in many electronic systems that handle signals, which change very slowly with time. Some of the important applications are as given below:

1. Analog computation.
2. Power supply regulators.
3. Bioelectric measurements.
4. Linear integrated circuits.

26.21. Darlington Amplifier

The Darlington Amplifier consists of two cascaded emitter followers as shown in Fig. 26.21. The configuration results in a set of improved amplifier characteristics.

The Darlington amplifier has a high input resistance, low output resistance and high current gain. These characteristics make it very useful as a current amplifier. The voltage gain of a Darlington amplifier is less than unity.

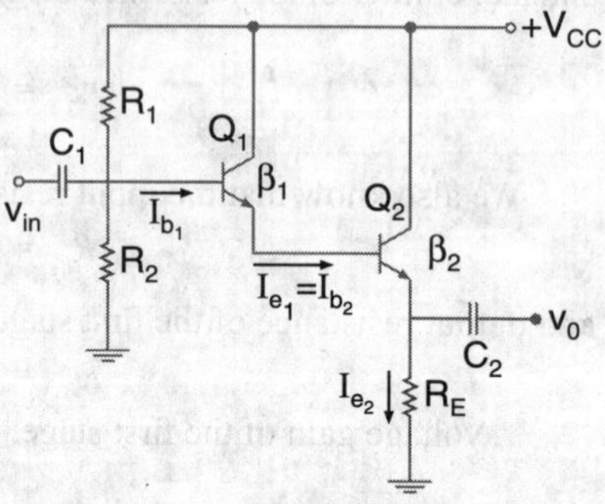

Fig. 26.21. Darlington amplifier.

26.22. Darlington Amplifier Characteristics

Consider the Darlington amplifier circuit shown in Fig. 26.20.

Let I_{b_1} = Base current of Q_1 transistor,

I_{e_1} = Emitter current of Q_1 transistor,

β_1 = Current gain of Q_1 transistor,

I_{b_2} = Base current of Q_2 transistor. Its value is equal to the emitter current of Q_1 transistor,

I_{e_2} = Emitter current of Q_2 transistor, and

β_2 = Current gain of Q_2 transistor.

1. *Current gain.* We know that the emitter current of Q_2 transistor,

$$I_{e_1} = \beta_1 \cdot I_{b_1}$$

and emitter current Q_2 of transistor,

$$I_{e_2} = \beta_2 \cdot I_{b_2} = \beta_2 \cdot I_{e_1} \qquad ...(\because I_{b_2} = I_{e_2})$$

$$= \beta_2 (\beta_1 \cdot I_{b_1}) \qquad ...(\because I_{e_1} = \beta_1 \cdot I_{b_1})$$

$$= \beta_1 \cdot \beta_2 \cdot I_{b_1} = \beta^2 \cdot I_{e_1}$$

Now let both the transistors used in the amplifier be identical. Then current gain β_1 and β_2 will be equal. In that case, emitter current of Q_2 transistor,

$$I_{e_2} = \beta_2 \cdot I_{b_1} \qquad ...(\because \beta_1 \cdot \beta_2 = \beta^2)$$

$\therefore$ Overall current gain,

$$A_i = \frac{I_{e2}}{I_{b_1}} = \beta^2$$

2. *Input resistance.* We know that input resistance of the second stage is given by the relation,

$$R_{i_2} = \beta_2 (r'_{e_2} + R_E)$$

$$= \beta_2 \cdot R_E \qquad ...(\text{If } R_E\ r_{e_2})$$

It is the value of resistance seen by the emitter of Q_1 transistor and is given by the relation,

$$R'_{i_1} = \beta_1 (r'_{e_1} + R_{i_2})$$
$$= \beta_1 \cdot R_{i_2} \qquad \text{... (If } R_{i_2} >> r'_{e_1})$$
$$= \beta_1 \cdot \beta_2 \cdot R_E \qquad \text{...}(\because R_{i_2} = \beta_2 \cdot R_E)$$

For identical transistor, we know that current gain β_1 and β_2 are also equal. Therefore input resistance,

$$R'_{i_1} = \beta^2 \cdot R_E$$

and input resistance of the amplifier stage,

$$R_{i_1} = (R_1 \parallel R_2) \parallel R'_{i_1} = (R_1 \parallel R_2) \parallel \beta^2 \cdot R_E$$
$$= R_1 \parallel R_2 \qquad \text{...}(\because (R_1 \parallel R_2) << \beta^2 \cdot R_E)$$

3. *Output resistance.* The output resistance of the first stage is given by the relation,

$$R_{o_1} = r'_{e_1} + \frac{R_1 \parallel R_2}{\beta_1}$$

and output resistance of the second stage,

$$R_{o_2} = r'_{e_2} + \frac{R_{o_1}}{\beta_2}$$

$$= r'_{e_2} + \frac{r'_{e_1} + \dfrac{R_1 \parallel R_2}{\beta_1}}{\beta_2}$$

$$= r'_{e_2} + \frac{r'_{e_1}}{\beta_2} + \frac{R_1 \parallel R_2}{\beta_1 \cdot \beta_2}$$

$$= r'_{e_2} \qquad \text{...}\left(\frac{r'_{e_1}}{\beta_2} \text{ and } \frac{(R_1 \parallel R_2)}{\beta_1 \cdot \beta_2} \text{ are negligible}\right)$$

4. *Voltage gain.* The voltage gain of Darlington amplifier is given by the relation,

$$A_v = \frac{v_o}{v_{in}} \qquad \text{...}(i)$$

The value of output voltage of a Darlington amplifier,

$$v_o = I_{e_2} \cdot R_E$$

and the input voltage,

$$v_{in} = I_{e_1} \cdot r'_{e_1} + (r'_{e_2} + R_E) \qquad \text{...}(ii)$$

The above expression is obtained by applying Kirchoff's Voltage Law to the input circuit of the Darlington amplifier. Substituting the value of I_{e_1} (equal to I_{e_2}/β_2) in equation (*ii*)

$$v_{in} = \frac{I_{e_2}}{\beta_2} r'_{e_1} + I_{e_2} (r'_{e_2} + R_E)$$

$\therefore$ Voltage gain,

$$A_v = \frac{I_{e_2} \cdot R_E}{\dfrac{I_{e_2}}{\beta_2} \cdot r'_{e_1} + I_{e_2} (r'_{e_2} + R_E)} = \frac{R_E}{\dfrac{r'_{e_1}}{\beta_2} + (r'_{e_2} + R_E)}$$

$$\approx 1 \qquad \text{...}(\because R_E >> r'_{e_2} \text{ or } r'_{e_1}/\beta_2)$$

26.23. Darlington Pair

These days, the semiconductor device manufacturers are supplying a pair of identical transistors, connected suitably, in a single package. It is known as a *Darlington pair*. The two transistors are connected in such a way that the emitter of one transistor is connected to the base of the other as shown in Fig. 26.22 (*a*) and (*b*). The Darlington pairs shown in Fig. 26.22 (*a*) and (*b*) are called NPN Darlington pair and PNP Darlington pair respectively

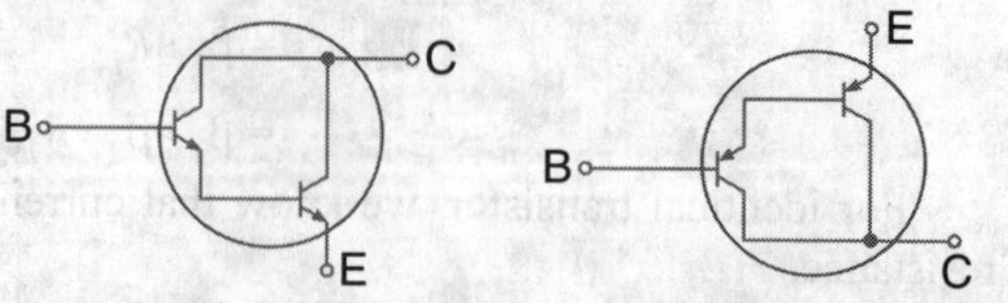

(*a*) NPN transistor Darlington pair. (*b*) PNP transistor Darlington pair.

Fig. 26.22.

A Darlington pair is a three terminal device namely base (*B*) emitter (*E*) and collector (*C*). It acts like a transistor, with an extremely high current gain (β). The Darlington pairs are available for both NPN and PNP transistors. Even photo-transistor Darlington pairs are also available. Such a device is b times more sensitive than that of a photo-transistor.

Example 26.10.

Fig. 26.23 shows the circuit diagram of a Darlington amplifier.

Determine (a) the overall gain; (b) the a.c. emitter diode resistance for each transistor; (c) total input resistance; and (d) the overall voltage gain.

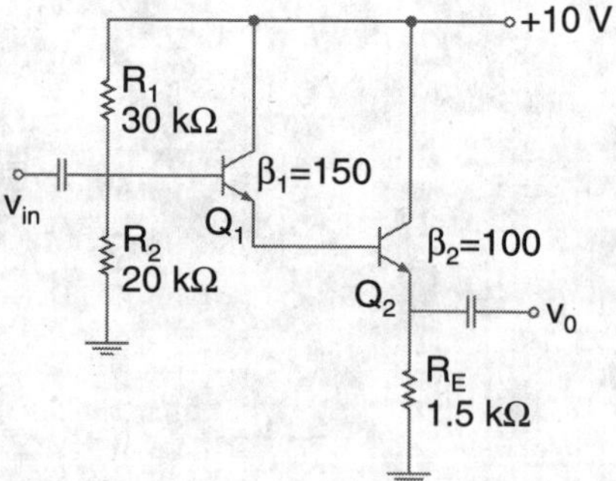

Fig. 26.23.

Solution.

Given V_{CC} = 10 volt; R_1 = 30 kΩ; R_2 = 20 kΩ; R_E = 1.5 kΩ = 1.5 × 10³ Ω; β_1 = 150 and β_2 = 100.

(*a*) *Overall current gain*

We know that overall current gain,

$$A_i = \beta_1 \cdot \beta_2 = 150 \times 100 = 15\,000 \textbf{ Ans.}$$

(*b*) *A.C. emitter diode resistance for each transistor*

We know that voltage drop across resistor R_2

$$V_{R_2} = V_{CC} \times \frac{R_2}{R_1 + R_2} = 10\left(\frac{20}{30+20}\right) = 4 \text{ V}$$

and the voltage at the base of Q_2 (or emitter of Q_1),

$$V_{B_2} = V_{R_2} - V_{BE_2} = 4 - 0.7 = 3.3 \text{ V}$$

Similarly, voltage at the emitter of Q_2,

$$V_{E_2} = V_{B_2} - V_{BE_2} = 3.3 - 0.7 = 2.6 \text{ V}$$

and the value of emitter current of Q_2,

$$I_{E_2} = \frac{V_{E_2}}{R_E} = \frac{2.6}{1.5 \times 10^3} = 1.73 \times 10^{-3} \text{ A} = 1.73 \text{ mA}$$

∴ A.C. emitter diode resistance of Q_2 transistor

$$r'_{e_2} = \frac{25}{I_{E_2} \text{ (in } mA)} = \frac{25}{1.73} = 14.5\ \Omega \textbf{ Ans.}$$

Now the base current of Q_2,

$$I_{b_2} = \frac{I_{E_2}}{\beta_2} = \frac{1.73}{100} = 0.0173 \text{ mA}$$

and the emitter current of Q_1,

$$I_{E_1} = I_{b_2} = 0.0173 \text{ mA}$$

∴ A.C. emitter diode resistance of Q_1,

$$r'_{e_1} = \frac{25}{I_{E_1} \text{ (in } mA)} = \frac{25}{0.0173} = 1445\ \Omega \textbf{ Ans.}$$

(c) ***Total input resistance***

We know that total input resistance (*i.e.*, input resistance of the amplifier stage),

$$R_{i_1} = R_1 \parallel R_2 = 30 \parallel 20 = 12 \text{ k}\Omega \textbf{ Ans.}$$

(d) ***Overall voltage gain***

We also know that overall voltage gain,

$$A_v = \frac{R_E}{\dfrac{r'_{e_1}}{\beta_2} + (r'_{e_2} + R_E)} = \frac{1.5 \times 10^3}{\dfrac{1445}{100} + [14.4 + (1.5 \times 10^3)]}$$

$$= 0.98 \textbf{ Ans.}$$

26.24. Darlington Connection of BJTs

This connection is a very popular connection of two bipolar junction transistors for operation as one "*Superbeta*" transistor. Fig. 26.24 shows a Darlington connection of two NPN transistors. The main feature of Darlington connection is that the composite transistor acts as a single unit with a current gain that is equal to the product of the current gains of the individual transistors.

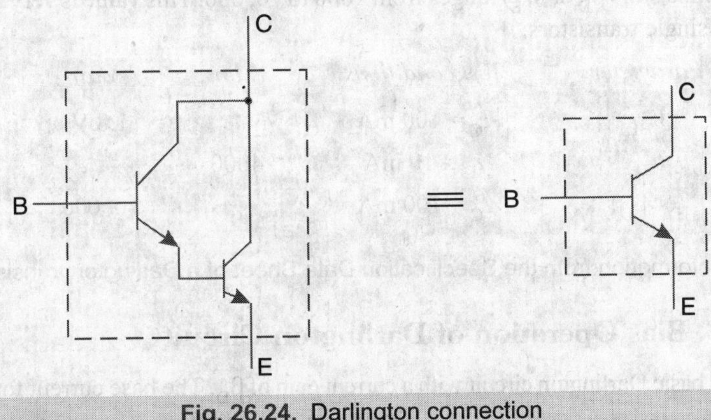

Fig. 26.24. Darlington connection

Let β_1 = Current gain of transistor, Q_1 and

β_2 = Current gain of transistor, Q_2.

Then, current gain of a darlington transistor (or a composite transistor),

$$\beta = \beta_1 \cdot \beta_2$$

If the two transistors are fabricated in such a way they are "*well matched*", then $\beta_1 = \beta_2 = \beta$. In such a case, the current gain of a composite transistor,

$$\beta_D = \beta \cdot \beta = \beta^2$$

Example 26.11.

A certain IC has a Darlington Connection of two transistors each having a current gain, 160 find the current gain of the composite transistor.

Solution.

Given : $\beta_1 = \beta_2 = 160$

We know that current gain of a composite transistor with $\beta_1 = \beta_2$ is,

$$\beta_D = \beta^2 = 160^2 = 25600 \text{ Ans.}$$

Example 26.12.

Calculate the value of current gain provided by each Darlington connection of two identical transistors. Given : overall current gain, $\beta_D = 6000$.

Solution.

Given β_D of two identical transistors = 6000

Let $\beta_1 = \beta_2 = \beta$ = current gain each transistor.

We know that current gain of a Darlington (or composite) transistor (β_D),

$$6000 = \beta_1 \times \beta_2 = \beta^2$$

$$\therefore \quad \beta = \sqrt{6000} = 77.46 \textbf{ Ans.}$$

26.25. Packaged Darlington Transistor

Several Darlington transistors are available commercially. Fig. 26.25 shows a part of specification data sheet of a typical Darlington transistor (2N999). As seen from the Fig. 26.25, the current gain of the Darlington transistor (β_D or h_{FE}) ranges from 4000 to 70, 000. This value is very high as compared to other typical single transistors.

Parameter	*Test conditions*	*Min*	*Max*
V_{BE}	$I_C = 100$ mA		1.8 V
h_{FE}	$I_C = 10$ mA	4000	
(or β_D)	$I_C = 100$ mA		70, 000

Fig. 26.25 Information from the Specification Data Sheet of a Darlington transistor (2N999).

26.26. D.C. Bias Operation of Darlington Circuit

Fig. 26.26 a basic Darlington circuit with a current gain of β_D. The base current for this circuit may be obtained in the same way as discussed earlier in Chapter 22. It is given by the equation :

$$I_B = \frac{V_{CC} - V_{BE}}{R_B + \beta_D R_E} \quad ...(i)$$

Notice that equation (*i*) is the same as for a regular transistor. But the value of β_D is much higher and V_{BE} is much larger. The emitter current,

$$I_E = (\beta_D + 1)\, I_B \approx \beta_D I_B$$

and the collector current,

$$I_C \approx I_E$$

In order to find the values of V_E, V_B, V_C and V_{CE}, consider the collector-emitter loop circuit (AFCEGA). Applying Kirchoff's Voltage Law to this loop,

$$V_{CE} + I_E \, . \, R_E - V_{CC} + I_C \, . \, R_C = 0$$

Replacing I_C with I_E in the above equation and solving it for the voltage across collector-emitter of a transistor,

$$V_{CE} = V_{CC} - I_C\,(R_C + R_E),$$

$$V_E = I_E\,R_E \approx I_C \, . \, R_E$$

$$V_C = V_{CC} - I_C \, . \, R_C$$

$$V_{CE} = V_C - V_E$$

and

$$V_B = V_E + V_{BE}$$

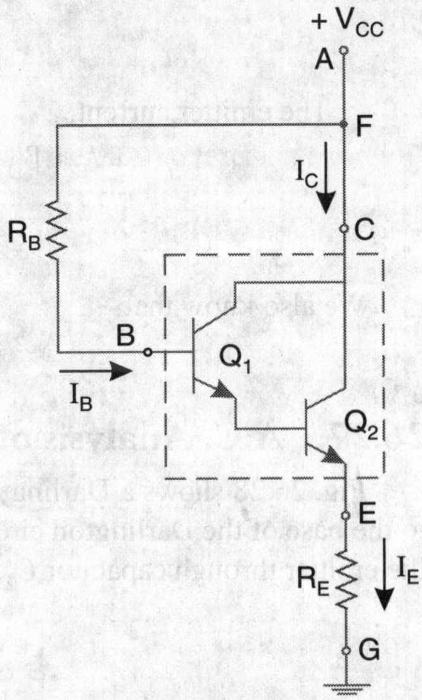

Fig. 26.26.

Example 26.13.

Fig. 26.27 shows a circuit of a Darlington connection. Given : β_D = 6000 and V_{BE} = 10.6 V.

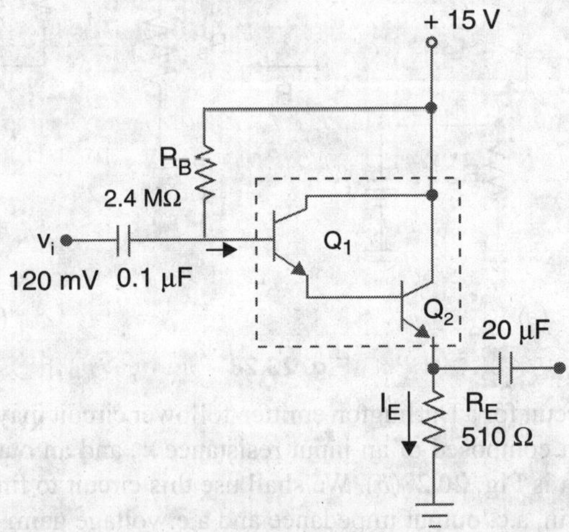

Fig. 26.27.

Calculate the d.c. bias voltage, V_{E2}, emitter current I_{E2}, and the amplifier voltage gain.

Solution.

Given : V_{CC} = 15V R_B = 2.4 MΩ = 2.4 × 10^6 Ω; β_D = 6000; R_E = 510 Ω; V_i = 120 mV

We know that the base current,

$$I_B = \frac{V_{CC} - V_{BE}}{R_B + \beta_D\,R_E}$$

$$= \frac{15 - 1.6}{2.4 \times 10^{+6} + (6000 \times 510)}$$

$$= 2.45 \times 10^{-6}\text{ A}$$
$$= 2.45\ \mu\text{A}$$

∴ The emitter current,

$$I_E \approx \beta_D \cdot I_B = 6000 \times (2.45 \times 10^{-6}) = 14700 \times 10^{-6}\text{ A}$$
$$= 4.7\text{ mA}$$
$$I_{E_2} = I_E = 14.7\text{ mA. } \textbf{Ans.}$$

We also know that,

$$V_{E_2} = I_{E_2} \cdot R_E = 14.7\text{ mA} \times 510\ \Omega = 7.5\text{ V } \textbf{Ans.}$$

26.27. A.C. Analysis of Darlington Pair

Fig. 26.28 shows a Darlington emitter-follower circuit. As seen, the a.c. input signal is applied to the base of the Darlington circuit through capacitor C_1. While the a.c. output, v_o is obtained from the emitter through capacitor C_2.

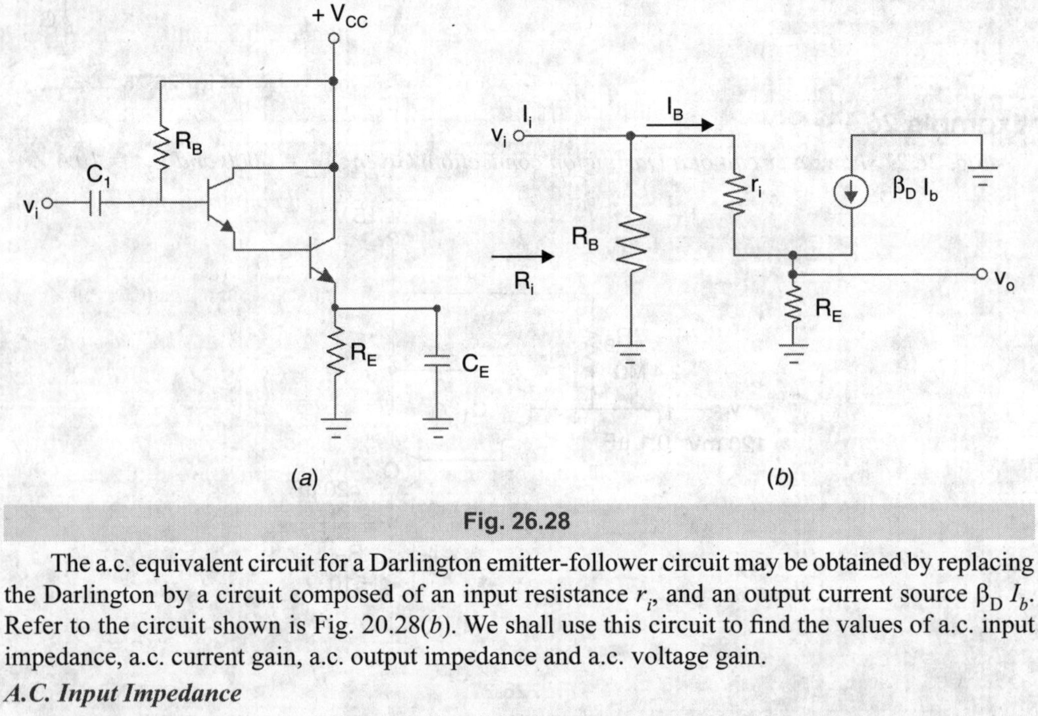

Fig. 26.28

The a.c. equivalent circuit for a Darlington emitter-follower circuit may be obtained by replacing the Darlington by a circuit composed of an input resistance r_i, and an output current source $\beta_D I_b$. Refer to the circuit shown is Fig. 20.28(*b*). We shall use this circuit to find the values of a.c. input impedance, a.c. current gain, a.c. output impedance and a.c. voltage gain.

A.C. Input Impedance

We know that the a.c. base current through the input resistance r_i is given by the equation.

$$i_b = \frac{v_i - v_o}{r_i} \qquad ...(i)$$

and the output voltage,

$$v_o = (I_b + \beta_D I_b)\, R_E \qquad ...(ii)$$

Re-arranging the equation (*i*), we get,

$$i_b \cdot r_i = v_i - v_o$$

or

$$v_i = i_b \cdot r_i + v_o \qquad ...(iii)$$

Substituting the value of v_o from equation (*ii*) to equation (*iii*), we get,

$$v_i = i_b \cdot r_i + (i_b + \beta_D\, i_b)\, R_E$$

or
$$v_i = i_b\, [r_i + (1 + \beta_D)\, R_E]$$

Now the a.c. input impedance looking into the transistor base,

$$\frac{v_i}{i_b} = r_i + (1 + \beta_D)\, R_E$$

$$\approx r_i + \beta_D\, R_E \qquad ...(\because \beta_D >> 1)$$

And the a.c. input impedance looking into the circuit,

$$R_i = R_B \,||\left(\frac{V_i}{i_b}\right)$$

or
$$R_i = R_B \,||\, (r_i + \beta_D\, R_E)$$

A.C. Current Gain

We know that a.c. output current through R_E (refer to Fig. 26.28 (*b*)) is given by the equation,

$$i_o = i_b = \beta_D\, i_b$$

$$= (1 + \beta_D)\, i_b$$

$$= \beta_D\, i_b \qquad ...(\beta_D >> 1)$$

We also that transistor gain, is

$$\frac{i_o}{i_b} = \beta_D \qquad ...(i)$$

and the a.c. current gain of the circuit,

$$A_i = \frac{i_o}{i_i}$$

Multiplying and dividing the right hand side of the above equation by i_b, we get,

$$A_i = \frac{i_o}{i_b} \cdot \frac{i_b}{i_i} \qquad ...(ii)$$

Now the ratio i_b/i_i can be determined by using the current division rule from circuit shown in Fig. 26.28 (*b*)

$$i_b = \frac{R_B}{(r_i + \beta_D\, R_E) + R_B}\, i_i$$

$$= \frac{R_B}{\beta_D\, R_E + R_B}\, i_i \qquad ...(\because r_i << \beta_D\, R_E \text{ or } R_B)$$

or
$$\frac{i_b}{i_i} = \frac{R_B}{\beta_D\, R_E + R_B} \qquad ...(iii)$$

Substituting the values of i_o/i_b $(= \beta_D)$ and i_b/i_i $(= \beta_D/(\beta_D\, R_E + R_B))$ in equation (*ii*), the a.c. circuit current gain,

$$A_i = \beta_D \cdot \frac{R_B}{\beta_D\, R_E + R_B} = \frac{\beta_D\, R_B}{R_B + \beta_D\, R_E}$$

A.C. Output Impedance

The a.c. output impedance of a Darlington emitter follower circuit shown in Fig. 26.29 can be determined by using the a.c. circuit shown in Fig. 26.29 (*a*) and (*b*).

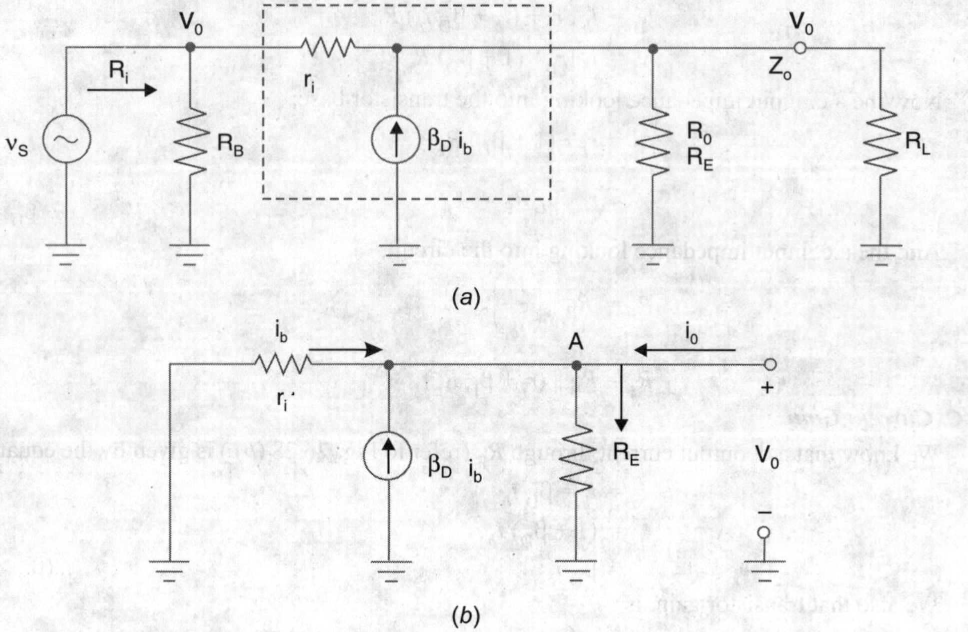

Fig. 26.29. AC, equivalent circuit for calculation of a.c. output impedance

We know that as seen by load R_L is determined by applying a voltage, v_o and measuring the current i_o (with input v_S set to zero).

Applying Kirchoff's Current Law at node *A*,

$$I_o = \frac{v_o}{R_E} + \frac{v_o}{r_i} - \beta_D \, . \, i_b$$

Substituting for i_b (= v_o/r_i) in the above equation,

$$i_o = \frac{v_o}{R_E} + \frac{v_o}{r_i} - \beta_D\left(-\frac{v_o}{r_i}\right)$$

$$= v_o\left(\frac{1}{R_E} + \frac{1}{r_i} + \frac{\beta_D}{r_i}\right)$$

Therefore, the a.c. output impedance,

$$R_o = \frac{v_o}{i_o} = \frac{1}{\dfrac{1}{R_E} + \dfrac{1}{r_i} + \dfrac{\beta_D}{r_i}}$$

$$= R_E \parallel r_i \parallel \left(\frac{r_i}{\beta_D}\right)$$

$$\approx \frac{r_i}{\beta_D}$$

A.C. Voltage Gain

The a.c. voltage gain of a Darlington emitter follower circuit shown in Fig. 26.30 G can be determined by using the a.c. equivalent circuit shown in Fig. 26.30.

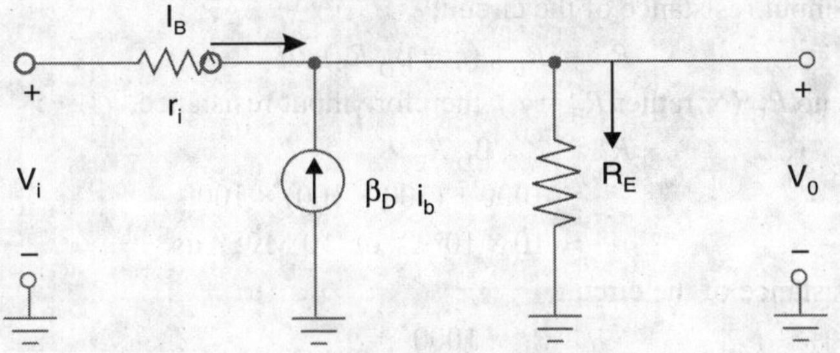

Fig. 26.30.

As seen from this circuit, the a.c. output voltage,

$$v_o = (i_b + \beta_D \cdot i_b)\, R_E$$
$$= i_b\,(R_E + \beta_D\, R_E) \qquad ...(iv)$$

and the a.c. input voltage,

$$v_i = i_b\, r_i + (i_b + \beta_D\, i_b)\, R_E$$
$$= i_b\,(r_i + R_E + \beta_D\, R_E)$$

or

$$i_b = \frac{v_i}{r_i + R_E + \beta_D\, R_E} \qquad ...(v)$$

Substituting the value of i_b from equation (*v*) to equation (*iv*), the ac output voltage,

$$v_o = \left(\frac{v_i}{r_i + R_E + \beta_D\, R_E}\right)(R_E + \beta_D\, R_E)$$

or

$$A_v = \frac{v_o}{v_i} = \frac{R_E + \beta_D\, R_E}{r_i + R_E + \beta_D\, R_E} \approx 1$$

Example 20.14.

For the Darlington amplifier shown in Fig. 26.31, obtain its Z-matrix for indentical Q_1 and Q_2 having h_{fe} = 100, $R_E = h_{ie}$ = 1 kΩ, (*UPSC Engg. Services*. 2001)

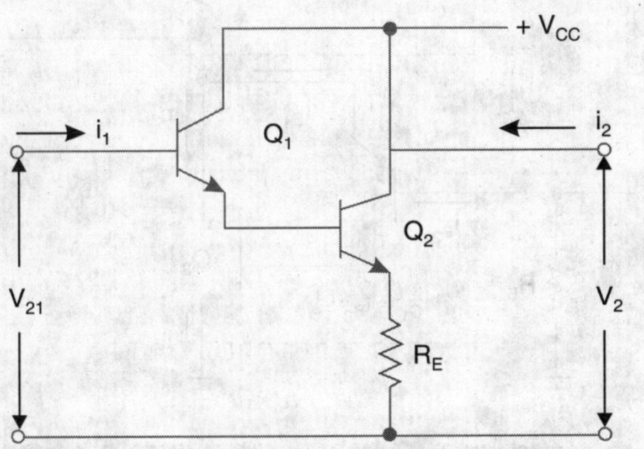

Fig. 26.31.

Solution.

Given : $h_{fe} = \beta = 100$; R_E = 1 kΩ and $h_{ie} = r_i$ = 1 kΩ

We know that input resistance of the circuit,

$$R_i = R_B \parallel (r_i + \beta_D R_E)$$

Since, there is no R_B (or rather $R_B = \infty$), therefore input resistance,

$$R_i = r_i + \beta_D R_E$$
$$= 1000 + (100 \times 100) \times 1000$$
$$\approx 10 \times 10^6\ \Omega \text{ or } 10\ \text{M}\Omega \textbf{ Ans.}$$

The output resistance of the circuit,

$$R_o = \frac{r_i}{\beta_D} = \frac{1000}{(100)^2} = 0.1\ \Omega \textbf{ Ans.}$$

26.28. Feedback Pair

Fig. 26.32 shows a feedback pair connection, As seen, this connection is a two transistor circuit that operater like a Darlington circuit. Notice that the feedback pair uses a PNP transistor driving an NPN transistor. Overall, the feedback pair acts like one PNP transistor which provides very high current gain.

The feedback pair is used along with a Darlington connection to provide a complementary transistor operation in Power Amplifiers. This is covered later in Chapter 22, Art 22-21.

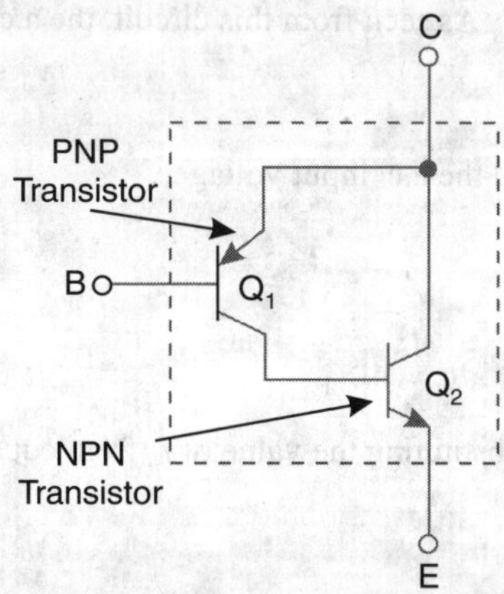

Fig. 26.32. Illustrating a feedback pari connection

26.29. D.C. Bias Operation of Feedback Pair

Fig. 26.33 shows a basic circuit of a feedback pair.

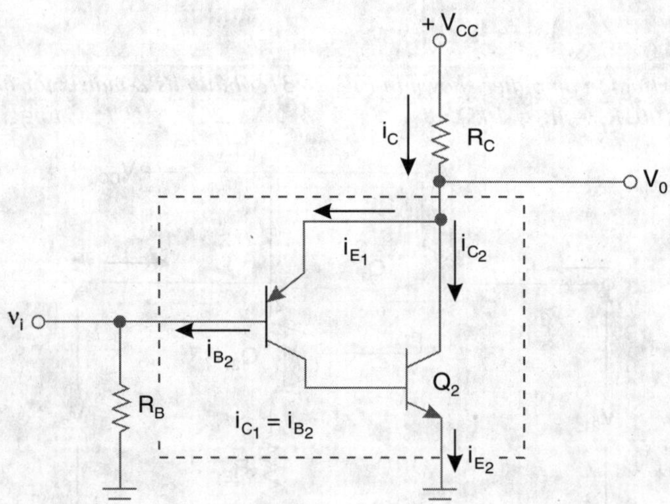

Fig. 26.33. Illustrating the operation of a feedback pair

From the base-emitter loop of transistor, Q_1, we obtain,

$$V_{CC} - I_C R_C - V_{BE_1} - I_{B_1} - \cdot R_B = 0$$

Substituting, $I_C = \beta_1 . \beta_2 . I_B$ in the above equation.

$$V_{CC} - \beta_1 \cdot \beta_2 \cdot I_B \cdot R_C - V_{BE_1} - I_{B_1} \cdot R_B = 0$$

Solving for, I_{B1} we get,

$$I_{B_1} = \frac{V_{CC} - V_{BE_1}}{R_B + \beta_1 . \beta_2 . R_C}$$

The collector current of transistor, Q_1,

$$I_{C_1} = \beta_1 \cdot I_{B_1}$$

This is also the base current of transistor, Q_2

$$\therefore \quad I_{C_1} = I_{B_2}$$

Next, the collector current of transistor, Q_2,

$$I_{B_1} = \beta_2 \cdot I_{B_2} \approx I_{E_2}$$

The current through R_C,

$$I_C = I_{E_1} + I_{C_2}$$

Since, $I_{E_1} \approx I_{C_1}$, therefore,

$$I_C = I_{C_2} + I_{C_2}.$$

Further since, $I_{C_1} = I_{B_2}$ (usually a small value because it is a base current and β_1 is high), therefore,

$$I_C = I_{C_2}.$$

Example 26.15.

Calculate the d.c. bias currents and voltages for the circuit shown in Fig. 26.34 to provide V_o at one-third the supply voltage.

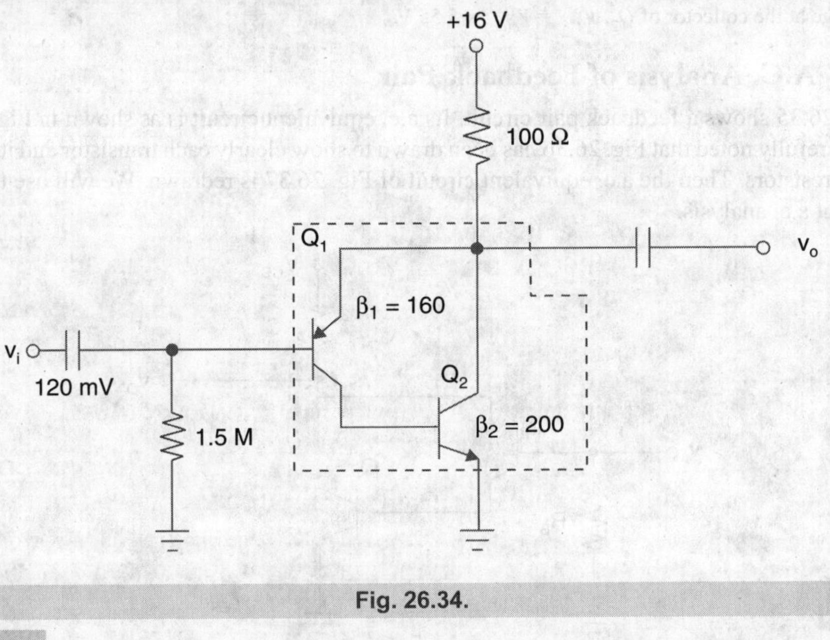

Fig. 26.34.

Solution.

Given : $V_{CC} = 16$ V; $\beta_1 = 160$, $\beta_2 = 200$; $R_B = 1.5$ MΩ; $V_i = 120$ mV

We know that the base current,

$$I_B = \frac{V_{CC} - V_{EB_1}}{R_B + \beta_1 \beta_2 R_C} = \frac{16 - 0.7}{(1.5 \times 10^6) + (160 \times 200 \times 100)} \text{ A}$$

$$= 3.255 \times 10^{-6} \text{ A or } 3.255 \text{ μA}$$

and the base current of Q_2,

$$I_{B_1} = I_{C_2} - I_{B_1} = 160 \times (3.225 \times 10^{-6})$$
$$= 0.520 \times 10^{-3} \text{ or } 0.520 \text{ mA}$$

We also know that collector current of Q_2,

$$I_{C_2} = \beta_2 I_{B_1} - 200 \times (0.520 \times 10^{-3}) \text{ A}$$
$$= 0.104 \text{ A or } 104 \text{ mA}$$

and the current through R_C,

$$I_C = I_{E_1} + I_{C_1} = 0.520 + 104 = 104.5 \text{ mA}$$

The d.c. voltage at the output,

$$V_{o(dC)} = V_{CC} - I_C R_C$$
$$= 16 - (104.5 \times 10^{-3}) \times 100 \text{ V}$$
$$= 5.55 \text{ V}$$

and the d.c. voltage at the input,

$$V_{i(dc)} = V_{o(dc)} - V_{BE} = 5.55 - 0.7 = 4.85 \text{ V Ans.}$$

Notes : Voltage at the base of Q_1, $V_{B_1} = V_{i(dc)} = 4.85$ V; the

Voltage at the collector of Q_2, $V_{C_2} = V_{o(dc)} = 5.55$ V.

26.30. A.C. Analysis of Feedback Pair

Fig. 26.35 shows a feedback pair circuit. Its a.c. equivalent circuit is as shown in Fig. 26.36. It may be carefully noted that Fig. 26.36 has been drawn to show clearly each transistor and its base and collector resistors. Then the a.c. equivalent circuit of Fig. 26.37 is redrawn. We will use this circuit to carryout a.c. analysis.

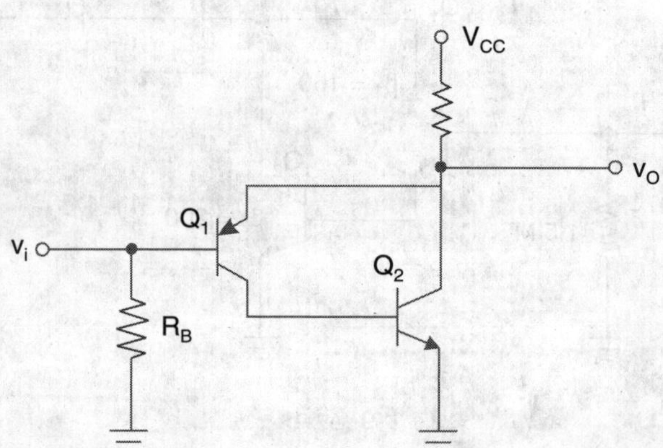

Fig. 26.35

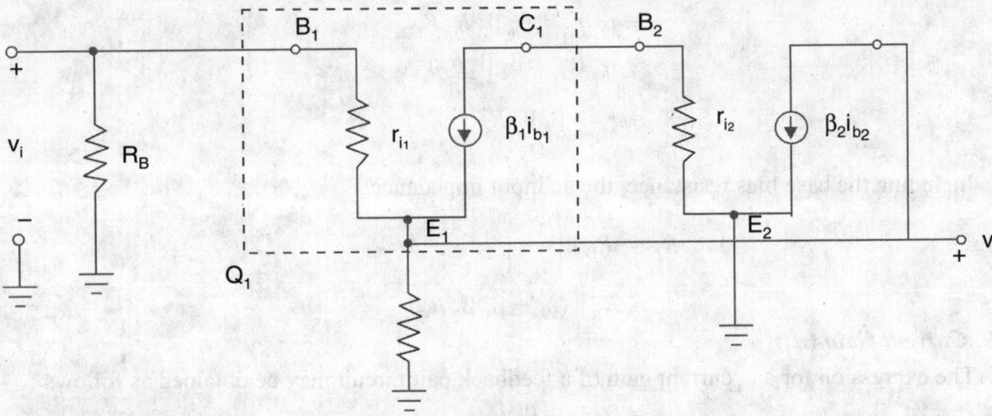

Fig. 20.36.

A.C. Input resistance (R_i)

Refer to circuit shown in Fig. 26.37. The a.c. input resistance seen looking into the base of transistor Q_1 is obtained as follows.

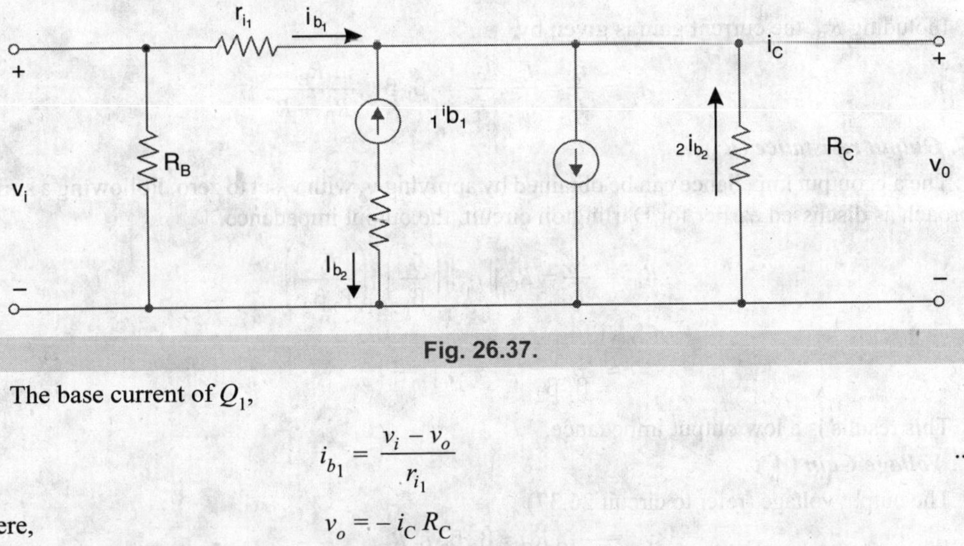

Fig. 26.37.

The base current of Q_1,

$$i_{b_1} = \frac{v_i - v_o}{r_{i_1}} \qquad \text{...}(i)$$

where,

$$v_o = -\,i_C\,R_C$$
$$\approx (-\,\beta_2\,i_{b_1} + \beta_2\,i_{b_2})\,R_C$$
$$\approx (-\,\beta_2\,i_{b_1})\,R_C$$

Substituting the value of v_o from above to equation (i) we get,

$$i_{b_1} = \frac{v_i - (\beta_2\,i_{b_2})\,R_C}{r_{i_1}}$$

Rearranging the above equation,

$$i_{b_1} \cdot r_{i_1} = v_i - \beta_2\,i_{b_2}\,R_C$$

or

$$v_i = i_{b_1}\,r_{i_1} - \beta_1\,i_{b_2}\,R_C$$

Substituting for i_{b_2} ($= i_{c_1} = \beta_1\,i_{b_1}$), we get,

$$V_i = i_{b_1}\,r_{i_1} + \beta_2\,(\beta_1\,i_{b_1})\,R_C$$

$$= i_{b_1} r_{i_1} + \beta_1 \beta_2 i_{b_1} R_C$$
$$= i_{b_1} (r_{i_1} + \beta_1 \beta_2 R_C)$$
$$\therefore \quad \frac{v_i}{i_{b_1}} = r_{i_1} + \beta_1 \beta_2 R_C$$

Including the base bias resistance, the ac input impedance,

$$R_i = R_B \| \left(\frac{v_i}{i_{b_1}}\right)$$
$$= R_B \| (r_{i_1} + \beta_1 \beta_2 R_C)$$

A.C. Current Gain (A_i)

The expression for a.c. current gain of a feedback pair circuit may be obtained as follows :

$$i_o = \beta_2 i_{b_1} - \beta_2 i_{b_1} - i_{b_1}$$
$$= \beta_2 (\beta_2 i_{b_1}) - (1 + \beta_2) i_{b_1}$$
$$\approx \beta_1 \beta_2 i_{b_1}$$
$$\therefore \quad \frac{i_o}{i_{b_1}} = \beta_1 \beta_2$$

Including R_B, the current gain is given by,

$$A_i = \frac{i_o}{i_i} = \frac{i_o}{i_{b_1}} \cdot \frac{i_{b_1}}{i_i} = \beta_1 \beta_2 \cdot \frac{R_B}{R_B + R_i}$$

A.C. Output resistance (R_0)

The a.c. output impedance can be obtained by applying v_o with v_i set to zero. Following a similar approach as discussed earlier for Darlington circuit, the output impedance,

$$R_o = \frac{v_o}{i_o} = R_C \| r_{i_1} \| \left(\frac{r_{i_1}}{\beta_1}\right) \| \left(\frac{r_{i_1}}{\beta_1 \beta_2}\right)$$
$$\approx \frac{r_{i_1}}{\beta_1 \beta_2}$$

This results is a low output impedance.

A.C. Voltage Gain (A_v)

The output voltage (refer to circuit 26.37)

$$v_o = - i_C R_C \approx \beta_1 \beta_2 i_{b_1} R_C$$

Since,
$$= \frac{v_i - v_o}{r_{i_1}}$$

$$\therefore \quad = v_i - i_{b_1} \cdot r_{i_1} = v_i - \left(\frac{v_o}{\beta_1 \beta_2 R_C}\right) r_{i_1}$$

$\therefore$ The voltage gain,

$$A_v = \frac{v_o}{v_i} = \frac{1}{1 + \dfrac{r_{i_1}}{\beta_1 \beta_2 R_C}}$$
$$= \frac{\beta_1 \beta_2 R_C}{\beta_1 \beta_2 R_C + r_{i_1}}$$

Example 20.14.

In a Darlington amplifier R_E = 100. If β_1 = 100 and β_2 = 200, calculate the overall input impedance.
(*Anna University., April/May* 2008)

Solution.

Given : $\beta_1 = 100$; $\beta_2 = 200$ and $R_E = 100$.

We know that over all input impedance,

$$= \beta_1 \beta_2 R_E$$

$$= 100 \times 100 \times 200 = 2000 \text{ k}\Omega \text{ or } 2 \text{ M}\Omega$$ **Ans.**

NUMERICAL PROBLEMS

1. For the two-stage RC coupled amplifier shown in Fig. 26.38, determine the individual stage gains, input resistance and the overall voltage gain,
(Ans. $A_{v_1} = 66$; $A_{v_2} = 193$; $R_i = 1.98$ kΩ; $A_v = 82.1$ dB)

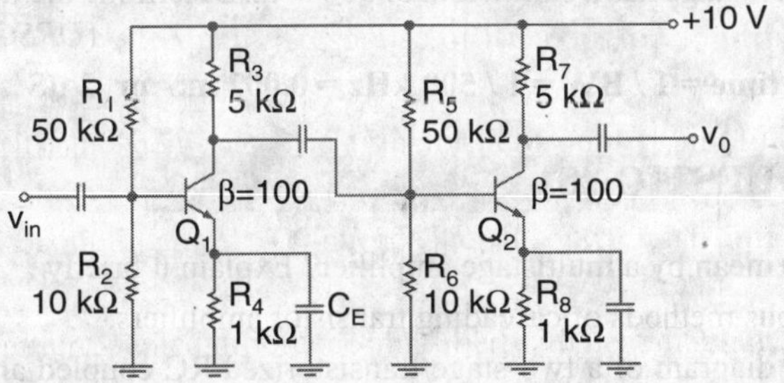

Fig. 26.38.

2. Fig. 26.39 shows a two-stage RC coupled amplifier consisting of common emitter stage followed by an emitter follower. Determine A_{v1}, A_{v2}, A_v, R_{i1} and the a.c. output voltage.
(Ans. $A_{v_1} = 159$; $A_{v_2} = 1$; $A_v = 159$; $R_{i_1} = 3.6$ kΩ and $V_0 = 100$ mV)

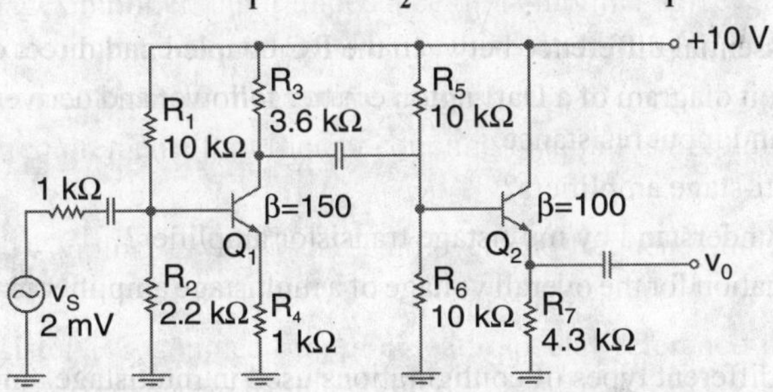

Fig. 26.39.

3. Fig. 26.40 shows a direct coupled two-stage amplifier. The d.c. bias of the first stage sets the d.c. base for the second stage. Find the individual stage voltage gain and the overall voltage gain. Neglect V_{BE} and use $r'_e = 25 \text{ mV}/I_E$ (mA). Take $\beta_1 = \beta_2 = 100$.
(Ans. $A_{v_1} = 259$, $A_{v_2} = 1$; $A_v = 259$)

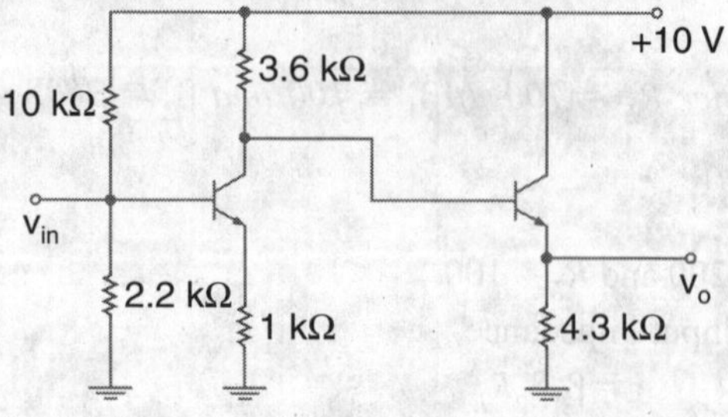

Fig. 26.40.

4. The power input to an amplifier is 2 μW. The power given of the amplifier is 40 dB. Calculate the output power of the amplifier. (*UPSC Engg. Services 2002*)

   (Hint: 40 dB = 10 $\log_{10} A_P \Rightarrow A_P$ = 10,000,

   P_{out} = 10,000 × 2 μW = 20,000 μW or 20 mW Ans.)

5. An RC amplifier stage has a bandwidth of 500 kHz. Determine the rise time of the amplifier stage. (*UPSC Engg. Services 2002*)

   (Hint: Rise time = 1 / BW = 1 / 500 kHz = 0.002 mS or 2 μS Ans.)

DESCRIPTIVE QUESTIONS

1. What do you ℓmean by a multistage amplifier? Explain it briefly.
2. State the various methods of cascading transistor amplifiers.
3. Give a circuit diagram of a two-stage transistorized RC coupled amplifier. Also draw the frequency response of the amplifier.
4. With the help of a suitable circuit diagram, explain the working of a RC coupled amplifier. Derive the expression for voltage gain of the amplifier.
5. Draw the circuit diagram of a RC coupled amplifier using *PNP* transistor.
6. Explain with suitable circuit diagram, the operation of transformer coupled transistorized amplifier.
7. Explain the essential difference between the RC coupled and direct coupled amplifier.
8. Draw the circuit diagram of a Darlington emitter follower and derive the expressions for its voltage gain and input resistance.
9. What are multi-stage amplifiers? (*PTU., May 2009*)
10. What do you understand by multistage transistor amplifier? (*PTU., Dec 2005*)
11. Derive the equation for the overall voltage of a multistage amplifier in terms of the individual voltage gains (*PTU., Dec 2008*)
12. What are the different types of configurations used in multistage amplifier circuits? (*PTU., Dec 2007*)
13. What are the different types of configurations used in multistage amplifier circuits? (*PTU., May 2008*)
14. State and explain atleast one typical application of each type of coupling (*PTU., May 2007*)
15. Give the equation of overall upper and lower cut off frequencies of multistage amplifier. (*Anna University., April/May 2008*)

16. Derive for the upper and lower cut off frequencies of RC coupled BJT amplifier.
(Anna University., April/May 2008)
17. Draw a RC coupled amplifier and justify the presence of each component. Draw its frequency response. *(M.G University., July/Aug. 2007)*
18. Why a capacity coupling is used to connect a signal source to an amplifier? For a coactively coupled load, is the dc load larger or smaller than ac load? Explain?
(GBTU/MTU., 2005-06)
19. Draw the frequency response of a single stage R-C coupled amplifier and explain for its different slopes with equivalent circuits for each one of the slope.
(M.G University., July/Aug. 2007)
20. Sketch the frequency response of a RC coupled amplifier and mark the three regions in it.
(M.G University., May/June 2006)
21. Draw an RC differentiating circuit and prove that it can differentiate the input. Derive the condition for best differentiation. Sketch the frequency response of the circuit.
(M.G University., May/June 2006)
22. What are the advantages of RC coupled amplifier. *(M.G University., Dec. 2005)*
23. Draw and explain the frequency response of RC coupled amplifier.
(M.G University., Dec. 2005)
24. Write the merits and demerits of a transformer coupled amplifiers. Give notes on its applications. *(Bhartiya University., April 2004)*
25. Why capacitive coupling used to connect a signal source to an amplifier?
(Anna University., Nov/Dec 2007)
26. Explain the emitter coupled difference amplifier with neat diagram.
(Anna University., Nov/Dec 2007)
27. Explain how the characteristics are modified with transformer coupling.
(Anna University., Nov/Dec 2007)
28. Why capacitive coupling used to connect a signal source to an amplifier?
(Anna University., Nov/Dec 2007)
29. "In a multi stage amplifier, input impedance of a amplifier stage should be very high, and output impedance must be very low" Justify this statement. *(PTU., Dec. 2006)*
30. Give applications of transformer coupled amplifier. *(PTU., Dec. 2009)*
31. What are the requirements for efficient coupling in a multistage amplifier?
(PTU., Dec. 2009)
32. Why does RC Coupling give constant gain over mid frequency range.
(PTU., May 2005)
33. Explain transistor RC coupled amplifier with special reference to frequency response, advantages, disadvantages and applications. *(PTU., Dec 2005)*
34. Why is coupling capacitor provided in RC coupled amplifier also called blocking capacitor?
(PTU., Dec 2008)
35. Draw and label the frequency response of an RC coupled amplifier. *(PTU., Dec 2009)*
36. Explain the essential difference between the RC coupled and direct coupled amplifier.
(PTU., Dec 2009)

37. Write short note on RC coupled amplifier.
(R.G.P.V., Nov. / Dec. 2007June 2004)

38. Explain cascading. Also discuss the effect of cascading in bandwidth.
(R.G.P.V., June 2007, Nov. /Dec. 2007)

39. Draw circuit diagram of a DC coupled amplifier and derive for its voltage and current gain.
(R.G.P.V., Dec. 2008)

40. What is a Darlington amplifier? Explain in brief. *(R.G.P.V., Dec. 2008)*

41. What is Darlington amplifier? How high input impedance can be obtained from it? Explain.
(R.G.P.V., June 2008)

42. Explain the Darlington amplifier. *(R.G.P.V., Dec. 2001)*

43. Why cascading of amplifier is needed? What are the effect of cascading in gain and bandwidth? *(R.G.P.V., June 2003)*

44. Write short notes on :
(*a*) Cascade Amplifiers *(R.G.P.V., Dec. 2004)*
(*b*) RC coupled amplifier *(R.G.P.V., Dec. 2007)*

MULTIPLE CHOICE QUESTIONS

1. The gain of a cascaded amplifier is equal to the
(*a*) product of individual gains (*b*) sum of individual gains
(*c*) ratio of stage gains (*d*) none of these

2. Two stages of a multistage amplifier have a gains of 50 and 20. The dB voltage gain is
(*a*) 3 (*b*) 30 (*c*) 300 (*d*) 1000

3. The resistance-capacitance coupling is popular in audio amplifiers because
(*a*) it provides an output signal in phase with the input signal
(*b*) it needs low voltage collector supply
(*c*) it has better audio frequency response
(*d*) none of these

4. The major advantage of a direct coupled amplifier is that
(*a*) it uses less number of components
(*b*) it has a very good temperature stability
(*c*) it does not use frequency sensitive components
(*d*) it can amplify direct current and low frequency signals

5. In a Darlington pair,
(*a*) the two transistors are connected in parallel
(*b*) the emitter of the first transistor feeds the base of the second
(*c*) the collector of the first transistor feeds the base of the second

6. Cascading two amplifiers will result in
(*a*) reduction in overall gain and increase in overall bandindth
(*b*) reduction in overall gain and reduction in overall bandwidth
(*c*) increase in overall gain and increase in overall bandwidth
(*d*) increase in overall gain and reduction in overall bandwidth

7. The overall bandwidth of two identical voltage amplifiers connected in cascade will
   (*a*) remain the same as that of a single stage
   (*b*) be worse than that of a single stage
   (*c*) be better than that of a single stage
   (*d*) be better if stage gain is low and worse if stage gain is high.
8. RC coupling is popular in low-level audio amplifiers because it
   (*a*) has better low frequency response
   (*b*) is inexpensive and needs no adjustments
   (*c*) provides an output signal in phase with the input signal
   (*d*) needs low voltage battery for collector supply
9. Frequency response characteristic at a single-stage RC coupled amplifier is shown in Fig. 26.41. The fall in gain at both ends of the characteristic is due to

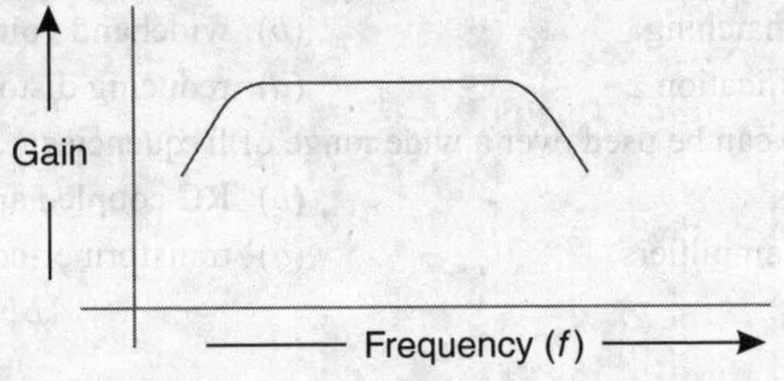

Fig. 26.41.

   (*a*) transistor shunt capacitances
   (*b*) bypass and coupling capacitances of the circuit
   (*c*) transistor shunt capacitances of the lower end and bypass and coupling capacitances at the higher end
   (*d*) transistor shunt capacitances at the higher end and bypass and coupling capacitances at the lower end.
10. The most desireable feature of a transformer coupling is
   (*a*) higher voltage gain
   (*b*) wide frequency range
   (*c*) ability to provide impedance matching between the stages
   (*d*) ability to eliminate hum from the output
11. A transformer coupled amplifier would give
   (*a*) maximum voltage gain (*b*) impedance matching
   (*c*) maximum current gain (*d*) larger bandwidth
12. In multistage amplifiers, direct coupling is especially suited for amplifying
   (*a*) high frequency a.c. singals (*b*) changes in d.c. voltages
   (*c*) high-level voltages (*d*) sinusoidal singals.
13. The outstanding characteristic of a direct coupled amplifier is its
   (*a*) utmost economy
   (*b*) temperature stability
   (*c*) avoidance of frequency-sensitive components
   (*d*) ability to amplify direct current and low frequency singals.

14. The coupling that produces minimum interferance with frequency response is

(*a*) direct coupling (*b*) impedance coupling

(*c*) RC coupling (*d*) transformer coupling

15. A singal has frequency components lying in the range of 0.001 Hz to 10 Hz. Which one of the following types of couplings should be chosen in a multistage amplifier designed to amplify this signal ?

(*a*) RC coupling (*b*) transformer coupling

(*c*) direct coupling (*d*) double-tuned transformer

16. The darlington pair consists of the two stages :

(*a*) CE & CC (*b*) both CE

(*c*) both CC (*d*) CE & CB

17. The Darlington pair is mainly used for

(*a*) impedance matching (*b*) wideband voltage amplification

(*c*) power amplification (*d*) reducing distortion.

18. These amplifiers can be used over a wide range of frequencies:

(*a*) dc amplifiers (*b*) RC coupled amplifiers

(*c*) LC coupled amplifiers (*d*) transformer-coupled amplifiers

(*Bharthiar Univ. April 2004*)

ANSWERS

1. (*a*)	**2.** (*d*)	**3.** (*c*)	**4.** (*d*)	**5.** (*b*)
6. (*c*)	**7.** (*c*)	**8.** (*b*)	**9.** (*d*)	**10.** (*c*)
11. (*b*)	**12.** (*b*)	**13.** (*d*)	**14.** (*a*)	**15.** (*c*)
16. (*c*)	**17.** (*c*)	**18.** (*b*)		

27

Chapter

Power Amplifiers

Outline

27.1. Introduction

A *power amplifier* is an amplifier, which is capable to providing a large amount of power to the load such as *loudspeaker*, or motor etc. It is essential in almost all electronic systems, where a large amount of power is required to be supplied to the load. The power amplifier, is used as a last stage in as electronic system. For example, a *public address system* (PAS) consists of a microphone, a multistage amplifier, a power amplifier and a loudspeaker. The microphone converts the sound waves into electrical signal, which is of very low voltage (usually of few millivolts). This signal is *insufficient* to drive the loudspeaker. Therefore this signal is first raised to a sufficiently high value (a few volts) by passing it through a multistage small-signal (or voltage) amplifier. This signal is then used to drive the power amplifier, because it is incapable of delivering a large amount of power to the loudspeakers.

A power amplifier is more commonly known *as audio amplifier*. The audio amplifiers are used in public address systems, tape recorders, stereo systems, television receivers, radio receivers, broadcast transmitters etc. It will be interesting to know that a *power amplifier does not actually amplify the power*. As a matter of fact, *it takes power from the d.c. power supply connected to the output circuit and converts it into useful a.c. signal power*. The power is fed to the load. The type of a.c. power developed, at the output of a power amplifier, is controlled by the input signal. Thus we can say that actually a power amplifier is a *d.c. to a.c. power converter,* whose action is controlled by the input signal.

The power amplifiers, are also known as *large signal amplifiers*. The term 'large signal' for the power amplifiers arises because these amplifiers use a large part of their a.c. load line for operation. It is in contrast to the small signal amplifiers, which use only 10% of their a.c. load line for operation. The small signal amplifiers are commonly known as voltage amplifiers.

Alternating current power amplifier – 60 Hz to 25 Hz, 3000 W

27.2. Difference Between Voltage Amplifier and Power Amplifier

We have already discussed in the last article that a small signal amplifier is called voltage amplifier and a large signal amplifier as a power amplifier. It will be interesting to know that a voltage amplifier produces an increased voltage into a high resistance load. On the other hand, power amplifier produces a large increase in output power or an appreciable amount of current flow into a relatively low resistance load. Following are some of the important differences in the construction of a voltage and power amplifier circuit :

1. The transistor used in voltage amplifier has a large value of current gain (β) – typically 100 as compared to that of a power amplifier. The current gain (β) of a transistor used in power amplifier is, usually, between 20 and 50.
2. In a voltage amplifier, the input resistance of a transistor is required to be quite low as compared to its output resistance. On the other hand, in a power amplifier, the input resistance of a transistor is generally very large as compared to its output resistance.
3. The *physical size* of the transistor used in voltage amplifiers is usually small and is known as low or medium-power transistor. On the other hand, power amplifiers employ larger size transistors called *power transistors*. Such transistors are needed because a considerable amount of heat is dissipated within the transistor and hence a large surface area is required for heat dissipation.
4. The voltage amplifiers, usually employ *RC coupling* for interstage connection, whereas power amplifiers use *transformer coupling*.

27.3. Performance Parameters

As a matter of fact, there are three parameters namely *circuit efficiency*, *distortion and power dissipation* capability of a transistor, which determines the performance of power amplifiers. These are discussed as below :

1. ***Circuit efficiency***. It is also called the overall efficiency or convertion efficiency of the power amplifier. The major criterion for a power amplifier is not the power gain, but the maximum a.c. power output. We have already, discussed that an amplifier is basically an energy convertion device *i.e.,* it converts d.c. power supply into a.c. power output. Therefore the ability of a power amplifier, to convert d.c. power supply from the source into a.c. power, is a measure of its effectiveness and is called its *circuit efficiency*.

Thus circuit efficiency may be defined as the maximum a.c. output power to the d.c. input power. Mathematically, the circuit efficiency,

$$\eta = \frac{\text{Maximum a.c. output power}}{\text{d.c. input power}}$$

The maximum theoretical values for circuit efficiency depend upon the way in which the load is coupled to the transistor and the class of operation of the amplifier. Its value may lie, anywhere, from 25% to 90%.

2. *Distortion.* A good amplifier should produce an output, which does not differ from the input in any respect, except amplitude. In other words, the amplifier output is expected to be enlarged but faithful reproduction of the input. However, in actual practice, it is not possible to construct such an ideal amplifier, whose output is exact reproduction of the input. The output is always found to be different from the input either in its waveform or frequency content. This difference between the output and input of an amplifier is called distortion.

Two types of distortions namely *amplitude* (or harmonic) *distortion* and *crossover distortion* are considered to be important in power amplifiers. The amplitude distortion results from the *non-linearity* of the transistor. It occurs because of the fact that transistor output may not increase, equally, for all portions of the input signal during positive and negative half cycles. On the other hand, crossover distortion (which usually exists in class *B* amplifiers) occurs when transistors *do not operate in the correct phase* with each other.

3. *Power dissipation capability.* The ability of power transistor to dissipate heat is known as power dissipation capability or power rating. The heat is generated within the transistor due to excessive current passing through it during amplification. If the heat generated, within the transistor, is greater than its power dissipation capability, it is likely to burn out. The power dissipation capability of a transistor may be increased by connecting a mass of metal (called *heat sink*) to the transistor case. The increased surface area at the transistor case allows the heat to escape more easily into the surrounding air. It keeps the case temperature within permissible limits.

27.4. A.C. Load Line

We know that every amplifier experiences two types of loads namely a d.c. load and an a.c. load. Consequently, the amplifier has two types of load lines *i.e.*, d.c. load line and a.c. load line. We have already discussed about d.c. load line in chapter on transistor biasing circuits. In this chapter, we shall discuss about a.c. load line.

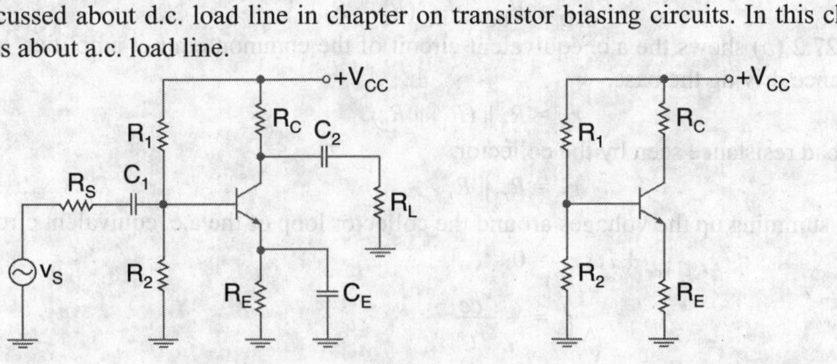

(*a*) Common emitter amplifier circuit. (*b*) d.c. equivalent circuit.

Fig. 27.1.

Consider a common emitter amplifier circuit as shown in Fig. 27.1 (*a*). Its d.c. equivalent circuit is shown in Fig. 27.1 (*b*). With the help of this circuit, we can determine the upper and lower ends of the d.c. load line. The upper end is given by the relations,

$$V_{CE} = 0 \text{ and } I_C = \frac{V_{CC}}{R_C + R_E}$$

It is shown in Fig. 27.2 (*a*) as point *A*. Similarly, the lower end is given by the relations,

$$I_C = 0 \text{ and } V_{CE} = V_{CC}$$

It is shown in Fig. 27.2 (*a*) as point *B*. Joining the points *A* and *B*, by a straight line, we get the d.c. load line.

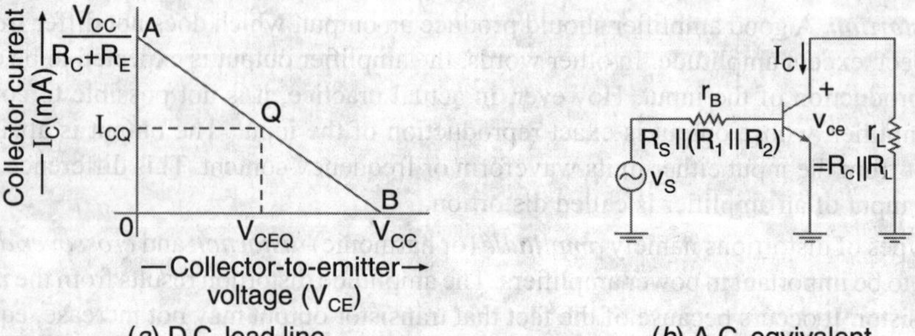

(*a*) D.C. load line. (*b*) A.C. eqvivalent.

Fig. 27.2.

The d.c. equivalent circuit may also be used to determine the *Q*-point values of collector current (I_{CQ}) and collector-to-emitter voltage (V_{CEQ}). We know that voltage drop across the resistor R_2,

$$V_{R_2} = V_{CC}\left(\frac{R_2}{R_1 + R_2}\right)$$

∴ *Q*-point emitter current,

$$I_{EQ} = \frac{V_{R2} - V_{BE}}{R_E}$$

and the *Q*-point collector current,

$$I_{CQ} = I_{EQ}$$

The *Q*-point collector-to-emitter voltage,

$$V_{CEQ} = V_{CC} - I_{CC}\,(R_C + R_E)$$

Fig. 27.2 (*b*) shows the a.c. equivalent circuit of the common emitter amplifier. In this figure, the resistance driving the base,

$$r_B = R_S \parallel (R_1) \parallel R_2)$$

and a.c. load resistance seen by the collector,

$$r_L = R_C \parallel R_L$$

Now summing up the voltages around the collector loop of the a.c. equivalent circuit,

$$v_{ce} + i_c \cdot r_L = 0$$

$$\therefore \quad i_c = -\frac{v_{ce}}{r_L} \qquad \text{...(i)}$$

where i_c = A.C. collector current, and

v_{ce} = A.C. collector-to-emitter voltage,

Now let I_{CQ} = The d.c. collector current at *Q*-point

V_{CEQ} = The d.c. collector-emitter voltage at *Q*-point.

$I_{C\,(sat)}$ = The d.c. saturation current, and

$V_{CE\,(\text{cut-off})}$ = The d.c. cut-off voltage.

Then the a.c. collector current,

$$i_c = \Delta I_C = I_C - I_{CQ}$$

and the a.c. collector-to-emitter voltage,

$$v_{ce} = \Delta V_{CE} = V_{CE} - V_{CEQ}$$

Substituting the values of i_c and v_{ce} in equation (*i*),

$$I_C - I_{CQ} = -\frac{V_{CE} - V_{CEQ}}{r_L} = -\frac{V_{CE}}{r_L} + \frac{V_{CEQ}}{r_L}$$

or

$$I_C = I_{CQ} + \frac{V_{CEQ}}{r_L} - \frac{V_{CE}}{r_L} \qquad ...(ii)$$

The above equation is an equation of a straight line called a c. load line. We know that when the transistor goes into saturation, the collector-to-emitter voltage (V_{CE}) becomes zero. In that case, the equation (*ii*) may be written as

$$I_C = I_{CQ} + \frac{V_{CEQ}}{r_L} = I_{C(sat)}$$

The above expression gives the upper end of the a.c. load line. Similarly, when the transistor goes into cut-off, the collector current (I_C) becomes zero. In that case, the equation (*ii*) may be written as

$$0 = I_{CQ} + \frac{V_{CEQ}}{r_L} - \frac{V_{CE}}{r_L}$$

$$= I_{CQ} + \frac{V_{CEQ}}{r_L} - \frac{V_{CE(cut-off)}}{r_L} \qquad ...(\because V_{CE} + V_{CE\,(cut\text{-}off)} \text{ at } I_C = 0)$$

or

$$V_{CE\,(cut\text{-}off)} = V_{CEQ} + I_{CQ} \cdot r_L$$

The above equation gives us the lower end of a.c. load line. By joining the upper and lower ends, we get a complete a.c. load line as shown in Fig. 27.3 (*a*). The locus of a.c. load line gives all the possible a.c. operating points. At any instant, during a.c. cycle, the operating point of the transistor is, somewhere, along the a.c. load line. The exact location may be determined by the amount of change from the *Q*-point.

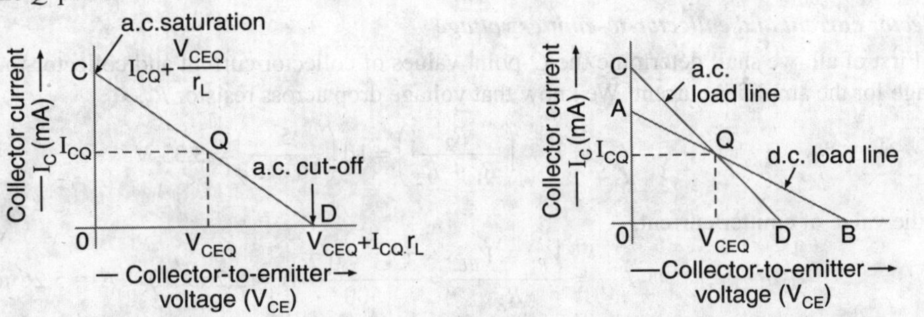

(*a*) A.C. load line.

(b) A.C. and D.C. load lines.

Fig. 27.3.

Fig. 27.3 (*b*) shows the a.c. load line superimposed on the d.c. load line. It may be noted that the slope of a.c. load line (equal to $1/r_L$) is greater than that of d.c. load line. Both the load lines intersect at *Q*-point.

Notes : 1. The a.c. load line for emitter follower (*i.e.,* common collector amplifier), common base amplifier and swamped amplifier may be obtained by a derivation almost identical to that given for the common emitter amplifier.

2. For emitter follower, the saturation value of current,

$$I_{C\,(sat)} = I_{CQ} + \frac{V_{CEQ}}{r_E}$$

where r_E is the a.c. equivalent load resistance at the emitter (*i.e.*, equal to $R_E \parallel R_L$). The a.c. collector-to-emitter cut-off voltage,

$$V_{CE\,(cut\text{-}off)} = V_{CEQ} + I_{CQ} \cdot r_E$$

3. For a common base amplifier, the values of $I_{C(sat)}$ and $V_{CE(cut\text{-}off)}$ are approximately the same as those given for a common emitter amplifier.

4. For swamped amplifier, the value of saturation current is given by the relation.

$$I_{C(sat)} = I_{CQ} + \frac{V_{CEQ}}{r_E + r_L}$$

where r_L = A.C. equivalent load resistance and

r_E = Unbypassed emitter resistance.

The a.c. collector-to-emitter cut-off voltage is given by the relation,

$$V_{CE\,(cut\text{-}off)} = V_{CEQ} + I_{CQ}(r_E + r_L)$$

Example 27.1.

Fig. 27.4 shows the circuit of a common emitter amplifier.

Determine the values of collector current and collector-to-emitter voltage at the points of saturation and cut-off under a.c. signal operation. Also draw the a.c.load line. Assume V_{BE} = 0.7 V.

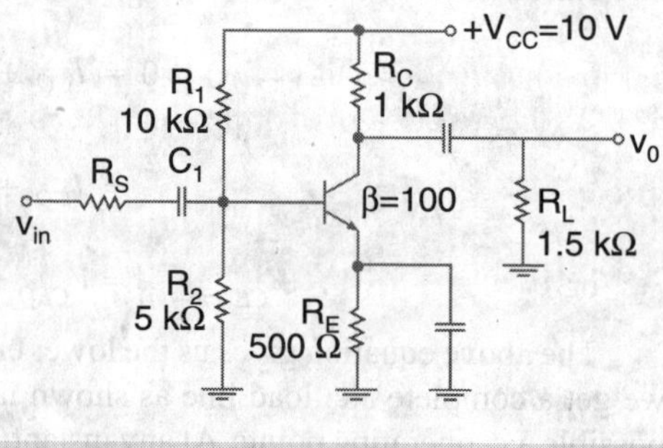

Fig. 27.4.

Solution.

V_{CC} = 10 volts; R_1 = 10 kΩ; R_2 = 5 kΩ; R_C = 1 kΩ = 1000 Ω; R_E = 500 Ω; R_L = 1.5 kΩ = 1500 Ω; β = 100 and V_{BE} = 0.7 V.

Collector current and collector-to-emitter voltage

First of all, we shall determine the *Q*-point values of collector current and collector-to-emitter voltage for the amplifier current. We know that voltage drop across resistor R_2,

$$V_{R_2} = V_{CC}\left(\frac{R_2}{R_1 + R_2}\right) = 10\left(\frac{5}{10+5}\right) = 3.33 \text{ V}$$

and the value of emitter current,

$$I_{EQ} = \frac{V_{R_2} - V_{BE}}{R_E} = \frac{3.33 - 0.7}{500} \text{ A} = 5.26 \times 10^{-3} \text{ A} = 5.26 \text{ mA}$$

∴ Value of collector current,

$$I_{CQ} = I_{EQ} = 5.26 \times 10^{-3} \text{ A} = 5.26 \text{ mA}$$

We also know that the collector-to-emitter voltage,

$$V_{CEQ} = V_{CC} - I_{CQ}(R_C + R_E)$$

$$= 10 - (5.26 \times 10^{-3}) \times (1000 + 500) = 2.11 \text{ V}$$

Thus the *Q*-point is located at I_C = 5.26 mA and V_{CE} = 2.11 V as shown in Fig. 27.5.

Now we shall find the values of collector current and collector-to-emitter voltage at saturation and cut-off points. We know that at saturation point, the collector current,

$$I_{C\,(sat)} = I_{CQ} + \frac{V_{CEQ}}{r_L}$$

$$= 5.26 \times 10^{-3} + \frac{2.11}{1000 \,||\, 1500} \quad ...(\because r_L = R_C \,||\, R_L)$$

$$= 8.78 \times 10^{-3}\ \text{A} = 8.78\ \text{mA}\ \textbf{Ans.}$$

and the collector-to-emitter voltage,

$$V_{CE\,(sat)} = 0\ \textbf{Ans.}$$

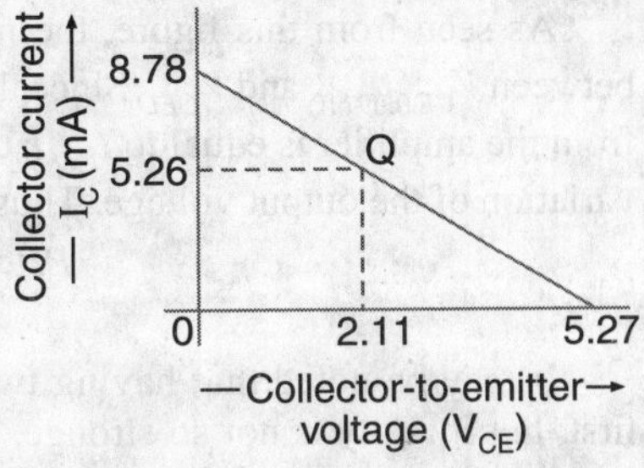

Fig. 27.5. A.C. load line.

Similarly, the collector current at cut-off point

$$I_{C\,(cut\text{-}off)} = 0\ \textbf{Ans.}$$

and the collector-to-emitter voltage,

$$V_{CE\,(cut\text{-}off)} = V_{CEQ} + I_{CQ} \cdot r_L$$

$$= 2.11 + (5.26 \times 10^{-3}) \times 600\ \text{V}$$

$$= 5.27\ \text{V}\ \textbf{Ans.}$$

A.C. load line

The a.c. load line may be obtained by plotting the points of a.c. saturation and a.c. cut-off and then joining them by a straight line as shown in Fig. 27.5.

27.5. What Does the A.C. Load Line Tell Us ?

The ac load line is used to tell us the maximum output voltage swing for a gives common-emitter amplifier. In other words, the ac load line will tell us the maximum possible peak (V_{PP}) output voltage from a given amplifier The maximum V_{PP} is know as *compliance* of the amplifier.

The compliance of an amplifier is found by obtaining the maximum values of I_C and V_{CE} from there respective values of I_{CQ} and V_{CEQ} The maximum possible variations are illustrated in Fig. 27.6.

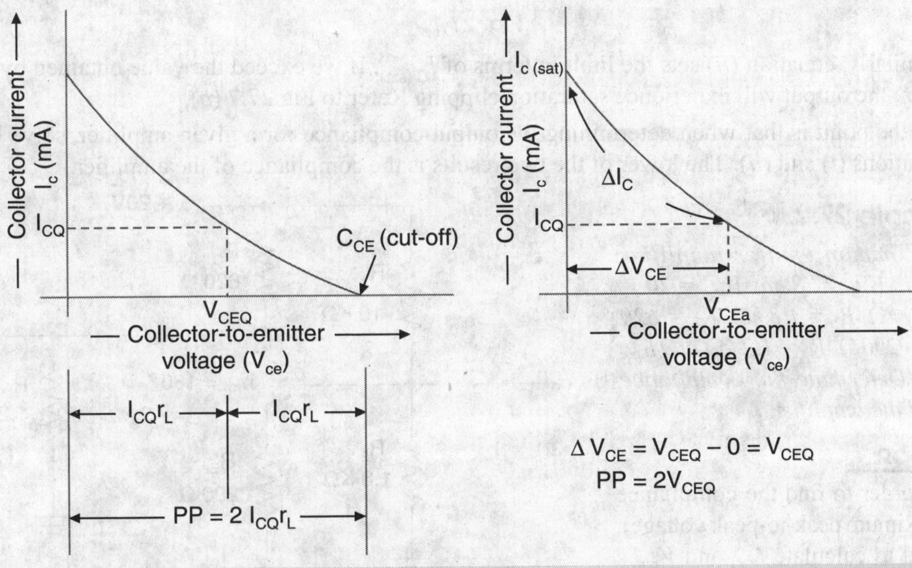

Fig. 27.6.

As seen from this figure, the maximum possible variation for V_{CE} is equal to the difference between $V_{CE\,(cut\text{-}off)}$ and V_{CEQ} Since this variation is equal to $I_{CQ}\, r_L$, the maximum peak output voltage from the amplifier is equal to $I_{CQ}\, r_L$. Two times this value will gives us the maximum peak-to-peak variation of the output voltage. Thus,

$$PP = 2\, I_{CQ}\, r_L \qquad ....(i)$$

$$= 2\, V_{CEQ} \qquad ....(ii)$$

It may be noted that having two different equations (*i*) and (*ii*), may seam a bit confusing at first, but it really is not so stronge. The equation (*i*) sets the limit in terms of $V_{CE\,(cut\text{-}off)}$. In other words, of we expected the value obtained by this equation, the output voltage will try to expected $V_{CE\,(cut\text{-}off)}$ which it cannot do. The result of exceeding the value obtained by equation (*i*) is shown in Fig. 27-7 (*a*), this is called cut-off clipping, because the output voltage is clipped off at the value of $V_{CE\,(cut\text{-}off)}$.

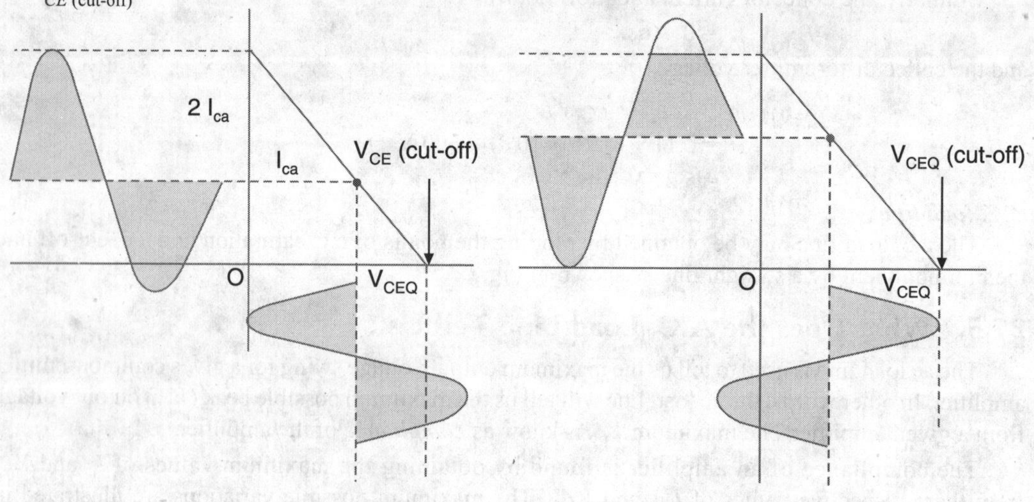

(*a*) Cut - off clipping (*a*) Saturation clipping

Fig. 27.7.

Similarly, equation (*ii*) sets the limit in terms of $I_{C\,(sat)}$. If we exceed the value obtained by this equation, the output will experience saturation clipping Refer to Fig 27.7 (*b*).

So the point is that when determining the output compliance for a given amplifier, solve both the equations (*i*) and (*ii*). The lower of the two results is the compliance of the amplifier.

Example 27.2.

A common emitter amplifier shown in Fig. 27.8 has V_{CC} = 20 V, R_1 = 10 kΩ, R_2 = 1.8 kΩ, R_C = 620 Ω, R_E = 200 Ω, R_L = 1.2 kΩ and h_{FE} = 180 Determine the compliance (PP) of the amplifier.

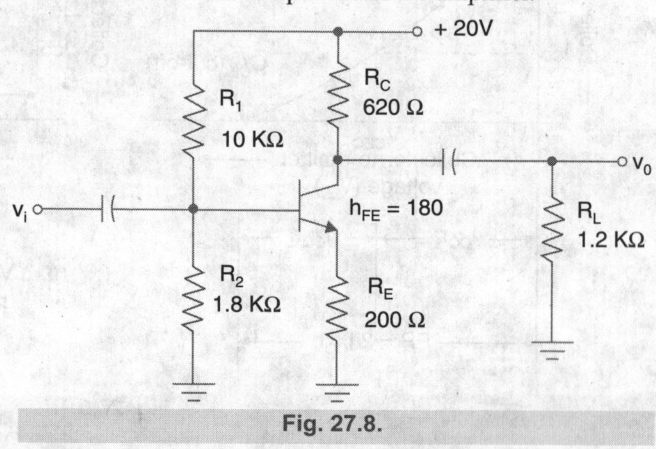

Fig. 27.8.

Solution.

In order to find the compliance (*i.e.* maximum peak-to-peak voltage), we need to calculate I_{CQ} and V_{CEQ}. We can follow the same approach as given in Example 25.4. Alternatively follow as below.

We know that base voltage,

$$V_B = V_{CC} \times \frac{R_2}{R_1 + R_2} = 20 \times \frac{1.8\ k}{10\ k + 1.8\ k} = 3.05\ \text{V}$$

and voltage at the emitter,

$$V_E = V_B - V_{BE} = 3.05 - 0.7 = 2.35\ \text{V}$$

∴ Value of emitter current,

$$I_E = \frac{V_E}{R_E} = \frac{2.35}{200} = 0.01175\ \text{A or } 11.75\ \text{mA}$$

and collector current,

$$I_C \approx I_E = 11.75\ \text{mA}$$

We also known that the voltage from collector-to-emitter,

$$V_{CE} = V_{CC} - I_E\,(R_C + R_E) = 20 - (0.01175) \times (620 + 200)\ \text{V}$$
$$= 20 - 9.635\ \text{V}$$
$$= 10.365\ \text{V}$$

Thus $I_{CQ} = 0.01175$ A or 11.75 mA

and $V_{CEQ} = 10.36$ V

We also know that the ac load resistance,

$$r_L = R_C \,||\, R_L = 620 \,||\, 1.2\ \text{k}$$
$$= 408.8\ \Omega$$

Now let us solve for both the compliance values,

$$PP = 2\, I_{CEQ} = r_L = 2 \times 10.365\ 20.73\ \text{V}$$

Since the overall compliance is always equal to smaller of the two values obtained, therefore the maximum *PP* for the amplifier show in Fig. 27.8 is 9.61 V. If the value is exceeded, cut-off clipping will occur.

27.6. Classification of Power Amplifiers

The classification of power amplifiers is based on transistor biasing condition and amplitude of the input signal. It also takes into account the portion of the cycle for which the transistor conducts. On the above basis, following are the three main classifications of power amplifiers :

1. ***Class-A amplifier.*** It is an amplifier in which the transistor bias and amplitude of the input signal is such that the output current flows for the complete cycle (*i.e.*, 360°) of the input signal. Fig. 27.9 (*a*) shows output for a class-A amplifier.

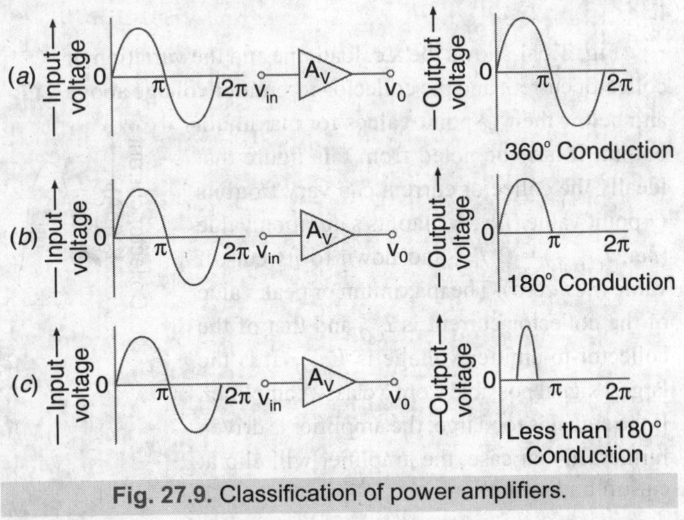

Fig. 27.9. Classification of power amplifiers.

2. ***Class-B amplifier.*** It is an amplifier in which the transistor bias and amplitude of the input signal are such that the output current flows for only one half cycle (*i.e.*, 180°) of the input signal. Fig. 27.9 (*b*) shows the output for a class-B amplifier.
3. ***Class-C amplifier.*** It is an amplifier in which the transistor bias and amplitude of the input signal are such that output current flows for less than half cycle (*i.e.*, less than 180°) of the input signal. Fig. 27.9 (*c*) shows the output for a class-C amplifier.

In addition to the above three main types, there is another amplifier called class-AB amplifier. The characteristics of such an amplifier lies in-between those of class-A and class-B amplifiers. Thus in a class-AB amplifier, the output current flows for more than half cycle but less than the complete cycle (*i.e.*, current flows for more than 180° but less than 360°). The class-A and class-B amplifiers are called audio power amplifiers because they are capable of amplifying the signals with frequencies ranging from 20 Hz to 20 kHz.

27.7. Class-A Amplifier

The class-A amplifier is, basically, a common emitter amplifier as shown in Fig. 27.10. This circuit is called direct coupled class-A power amplifier, in order to distinguish it from a transformer coupled power amplifiers. The only difference between this circuit and the small signal version, considered earlier, is that the signals handled by the power amplifier circuit are in the range of volts. And the transistor used is a power transistor, capable of operating in the range of few watts. In the class-A amplifier, the transistor bias and amplitude of the input signal are such that the output current flows for the complete cycle (*i.e.*, 360°) of the input signal. This condition is achieved by locating the *Q*-point somewhere near the centre of the load line. However, in order to obtain the maximum output signal, the *Q*-point is set at the centre of the load line.

Class-A amplifier.

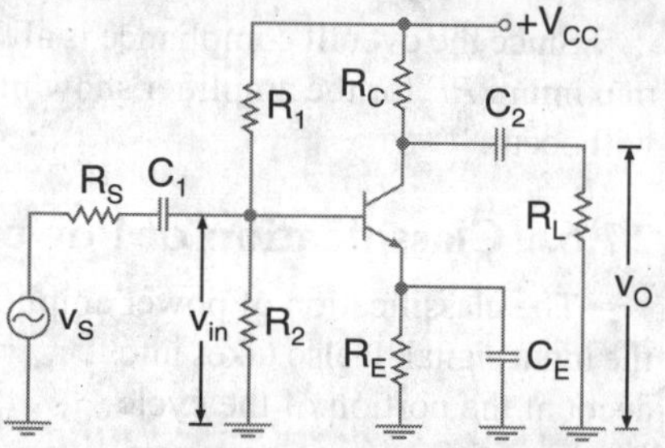

Fig. 27.10. Circuit for class-A amplifier.

Fig. 27.11 shows the a.c. load line and the variation of collector current and the collector-to-emitter voltage above and below their *Q*-point values for maximum output. It may be noted from this figure that ideally the collector current can vary from its *Q*-point value (*i.e.*, I_{CQ}) to its saturation value (*i.e.*, $I_{C\,(sat)}$) = 2 I_{CQ} and down to its cut-off value (*i.e.*, zero). The maximum or peak value of the collector current is I_{CQ} and that of the collector-to-emitter voltage is V_{CEQ}. It is the largest signal possible from a class-A amplifier. If the signal is too large, the amplifier is driven further. In that case, the amplifier will clip at cut-off and saturation.

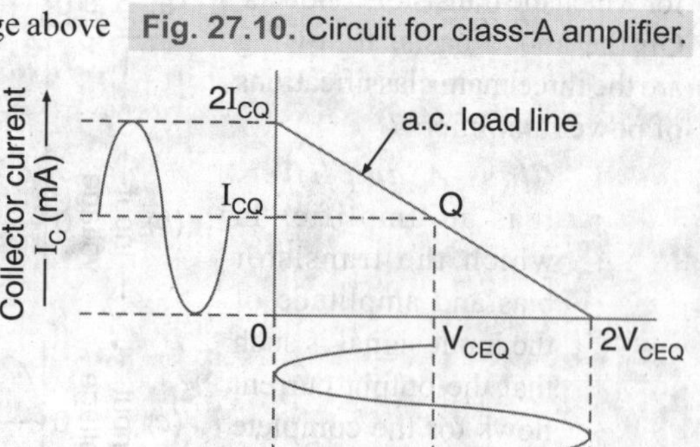

Fig. 27.11. Maximum output for class-A amplifier.

Now we shall obtain the expressions for amplifier parameters such as voltage gain, current gain and power gain for class-A amplifier.

Voltage gain. The voltage gain for a class-A power amplifier may be obtained in the same way as discussed for a small-signal amplifier in chapter 24. It is given by the relation,

$$A_v = \frac{r_L}{r'_e}$$

where r_L = A.C. load resistance whose value is equal to the parallel combination of collector resistance (R_C) and load resistance (R_L) *i.e.*, $r_L = R_C \parallel R_L$. If R_L is not connected then $r_L = R_C$.

r'_e = A.C. emitter diode resistance.

It will be interesting to know that the formula for a.c. emitter diode resistance, $r'_e = 25/I_E$ where I_E is in milliamperes is not void for large signal or power amplifier.

The large signal a.c. emitter diode resistance is determined graphically from the transconductance curve as shown in Fig. 27.12. It is done by using the relation,

$$r'_e = \frac{\Delta V_{BE}}{\Delta I_C}$$

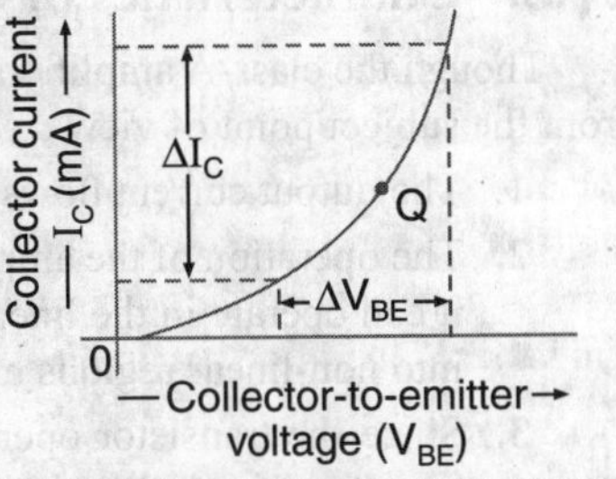

Fig. 27.12. Transconductance curve.

Current gain. The current gain of a transistor is the ratio of a.c. collector current (i_c) to the a.c base current (i_b). Mathematically, the current gain,

$$A_i = \frac{i_c}{i_b} = \beta$$

Power gain. We know that the a.c. input power to the base of transistor,

$$p_{in} = v_{in} \cdot i_b$$

and the a.c. output power from the collector,.

$$p_o = -v_o \cdot i_c$$

The negative sign in the above equation indicates that the phase of input signal is reversed at the output.

$\therefore$ Power gain,

$$A_p = \frac{p_o}{p_{in}} = \frac{-v_o \cdot i_c}{v_{in} \cdot i_b} = -\frac{v_o}{v_{in}} \times \frac{i_c}{i_b}$$

$$= -A_v \cdot A_i = -\left(\frac{r_L}{r'_e}\right) \times \beta$$

where A_v = Voltage gain, and

A_i = Current gain.

Example 27.3.

Find the large-signal voltage gain and power gain of the amplifier shown in Fig. 27.13. Assume that the value of r'_e has been found to be 8 Ω from the graphical data.

Solution.

Given : $r'_e = 8\ \Omega$; $R_C = 220\ \Omega$; $R_E = 47\ \Omega$; $R_1 = 4.7\ \text{k}\Omega$; $R_2 = 470\ \Omega$ and $\beta = 50$.

We know that a.c. load resistance,

$$r_L = R_C \parallel R_L = R_C = 220\ \Omega \text{ and}$$

voltage gain,

$$A_v = \frac{r_L}{r'_e} = \frac{200}{8} = 27.5$$

We also know that current gain,

$$A_i = \beta = 50$$

$\therefore$ Power gain,

$$A_p = A_v \cdot A_i = 27.5 \times 50 = 1375 \textbf{ Ans.}$$

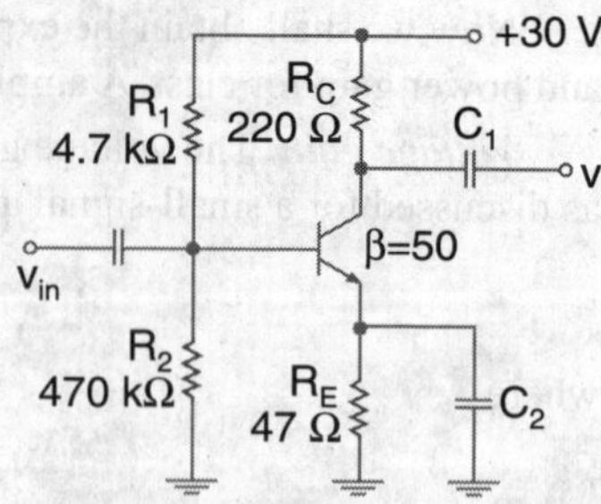

Fig. 27.13.

27.8. Characteristics of Class-A Amplifier

Though the class-A amplifiers have a number of characteristics, yet the following are important from the subject point of view :

1. The output current flows during the entire cycle of the a.c. input signal.
2. The operation of the amplifier is restricted to smaller central region of the load line, so that it can operate in the linear region of the load line. The large-signals may shift the *Q*-point into non-linear regions near saturation or cut-off and hence produce amplitude distortion.
3. Since the transistor operates over the linear region of the load line, therefore the output waveform is almost similar to the input waveform.
4. The a.c. power output per active device (or transistor) is smaller than that of class-B or class-C amplifier.
5. The overall efficiency or circuit efficiency of the amplifier circuit is an important parameter. It is defined as the ratio of a.c. power delivered to the load to the total power supplied by the d.c. source. Mathematically, the overall efficiency,

$$\eta_o = \frac{\text{A.C. power delivered to the load}}{\text{Total power supplied by the d.c. source}}$$

$$= \frac{\text{Average a.c. output power}}{\text{Average d.c. input power}}$$

   The maximum possible overall efficiency of a class-A amplifier with series fed resistive load is 25%.
6. The collector efficiency of the amplifier circuit is another important parameter. It is defined as the ratio of a.c. power delivered to the load, to the power supplied by the d.c. source to the transistor. Mathematically, collector circuit efficiency,

$$\eta_C = \frac{\text{A.C. power delivered the load}}{\text{Power supplied by the d.c. source to the transistor}}$$

$$= \frac{\text{Average a.c. output power}}{\text{Average d.c. input power to the transistor}}$$

   The maximum possible value of collector efficiency for a series fed resistive load is 50%.
7. If we use a transformer coupled load instead of a direct coupled resistive load, the maximum possible overall efficiency increases to 50%.

27.9. Power Relations for Class-A Amplifier

Fig. 27.14 (*a*) shows the output circuit of a single stage Class-A amplifier. In this circuit, the load itself is shown connected as a collector resistor. When the load is connected to the amplifier circuit in this configuration, it is referred to as direct coupled load.

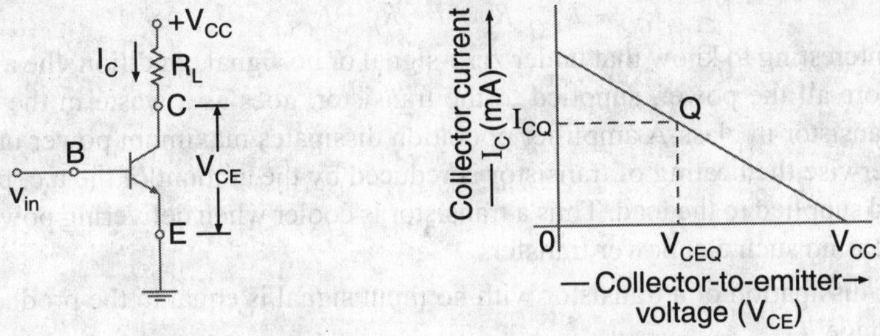

(*a*) Output-circuit of a class-A amplifier.

(*b*) A.C. load line.

Fig. 27.14.

Fig. 27.14 (*b*) shows its a.c. load line with *Q*-point located at its centre. We know that when an a.c. input signal is applied, the *Q*-point shifts up and down from its centre position. Accordingly, the output current will increase or decrease from its quiescent value (*i.e.*, I_{CQ}). Similarly, the collector-to-emitter voltage (V_{CE}) will also increase or decrease from its quiescent value (*i.e.*, V_{CEQ}). It may be noted that a long as the signal variations are limited to the linear region of the load line, average value of the collector current is I_{CQ}. It is true because the positive and negative input signal swings produce equal changes in I_{CQ}.

The total average power supplied to the output circuit by the collector (or V_{CC}) supply is given by the relation,

$$P_{in\,(dc)} = V_{CC} \cdot I_{CQ} = 2\,V_{CEQ} \cdot I_{CQ} \qquad ...(\because V_{CC} = 2\,V_{CEQ})$$

This power is utilized within the load resistor and transistor in the following ways :

1. The power consumed in the load resistor (R_L) whose value is given by the relation,

$$P_{RL\,(dc)} = (I_{CQ})^2 \cdot R_L$$

It is dissipated in the form of heat.

2. The power consumed within the transistor, whose value is given by the relation,

$$P_{tr\,(dc)} = P_{in\,(dc)} - P_{RL\,(dc)}$$

The power consumed by the transistor is further utilized in the following two parts :

1. A part of power is utilized to develop a.c. power across the load resistor. It constitutes the useful a.c. power output and its value is given by the relation,

$$P_{O\,(ac)} = I^2 \cdot R_L = \frac{V^2}{R_L}$$

$$= \frac{V_p^2}{2R_L} \qquad ...(\because V_p = \sqrt{2}\,V)$$

where V = The r.m.s. value of a.c. output voltage,

V_p = The peak or maximum value of a.c. output voltage, and

I = The r.m.s. value of a.c. output current.

2. The remaining part is dissipated in the form of heat within the transistor collector region. Its value is designated as $P_{C\,(dc)}$ and it goes as a waste.

It is evident from above, that the complete equation may be written as follows.

$$P_{in\,(dc)} = P_{RL\,(dc)} + \underbrace{P_{O(ac)} + P_{C(dc)}}_{P_{tr\,(dc)}}$$

or $$V_{CC} \cdot I_{CQ} = I^2_{CQ} \cdot R_L + I^2 \cdot R_L + P_{C\,(dc)}$$

It will be interesting to know that under zero-signal or no signal condition, the a.c. output power is zero. Therefore all the power, supplied to the transistor, goes as a waste in the form of heat. It means that a transistor in class-A amplifier operation dissipates maximum power under zero-signal condition. Otherwise the heating of transistor is reduced by the amount of the a.c. power converted by the stage and supplied to the load. Thus a transistor is cooler when delivering power to a load than that when there is no such a.c. power transfer.

The power dissipation of a transistor with no input signal is equal to the product of its Q-point current and voltage, *i.e.*,

$$P_{tr\,(dc)} = V_{CEQ} \cdot I_{CQ} \quad \text{...(Under zero signal condition)}$$

27.10. Overall and Collector Efficiency of Class-A Amplifier

We have already discussed in Art. 27.7 that overall efficiency is defined as the ratio of a.c. power delivered to the load to the total power supplied by the d.c. source. Mathematically, the overall efficiency,

$$\eta_o = \frac{\text{A.C. Power delivered to the load}}{\text{Total power supplied by the d.c. source}}$$

$$= \frac{\text{Average a.c. output power}}{\text{Average d.c. input power}} = \frac{P_{o\,(ac)}}{P_{in\,(dc)}}$$

We know that average a.c. output power,

$$P_{o\,(ac)} = \frac{V_p^2}{2R_L} = \frac{V_p \cdot I_p}{2} \quad \text{...}(\because V_p = I_p \cdot R_L)$$

and the average d.c. input power,

$$P_{in\,(dc)} = V_{CC} \cdot I_{CQ}$$

$$\therefore \quad \eta_o = \frac{V_p \cdot I_p / 2}{V_{CC} \cdot I_{CQ}} = \frac{V_p \cdot I_p}{2V_{CC} \cdot I_{CQ}} \quad \text{...}(i)$$

Now for the largest possible output signal, from a class-A amplifier, the Q-point must be located at the centre of a.c. load line. As a result of this, the maximum or peak value of the output voltage (V_p) is equal to V_{CEQ} and the peak value of output current (I_p) is equal to I_{CQ} *i.e.*,

$$V_p = V_{CEQ} \quad \text{and} \quad I_p = I_{CQ}$$

Substituting these values of V_p and I_p in equation (*i*)

$$\eta_o = \frac{V_{CEQ} \cdot I_{CQ}}{2V_{CC} \cdot I_{CQ}} = \frac{V_{CEQ} \cdot I_{CQ}}{2\,(2V_{CEQ}) \cdot I_{CQ}} = 0.25 = 25\%$$

The above value is the maximum possible value of overall efficiency and is designated as $h_{o\,(max)}$. It means that the maximum value of overall efficiency for class-A amplifier is 25%. However, in actual practice, it is always found to be less than this value. The collector-efficiency for a class-A amplifier is given by the relation,

$$\eta_o = \frac{\text{Average a.c. output power}}{\text{Average d.c. input power to the transistor}}$$

$$= \frac{P_{o\,(ac)}}{P_{tr\,(dc)}}$$

The maximum value of collector efficiency is obtained when the output signal is the largest possible. Under this condition, the average a.c. output power,

$$P_{o\,(ac)} = \frac{V_p \cdot I_p}{2} = \frac{V_{CEQ} \cdot I_{CQ}}{2}$$

The d.c. input power to the transistor is maximum under zero signal condition and its value is,

$$P_{tr\,(dc)} = V_{CEQ} \cdot I_{CQ}$$

$$\therefore \quad \eta_{C\,(max)} = \frac{V_{CEQ} \cdot I_{CQ}}{2\,(V_{CEQ} \cdot I_{CQ})} = 0.5 = 50\%$$

It means that maximum collector efficiency is 50%. In other words, the maximum a.c. power, which is transistor can deliver is $P_{tr\,(dc)}/2$. Incidentally, the power dissipation of a transistor under zero signal condition, is maximum and is equal to $P_{in\,(dc)}/2$. It is evident from the above discussion that a class-A amplifier, with a direct coupled load, can supply only one-half the amount of a.c. power to the load, which its transistor can dissipate. The a.c. power input is four times the amount of a.c. signal power in the load.

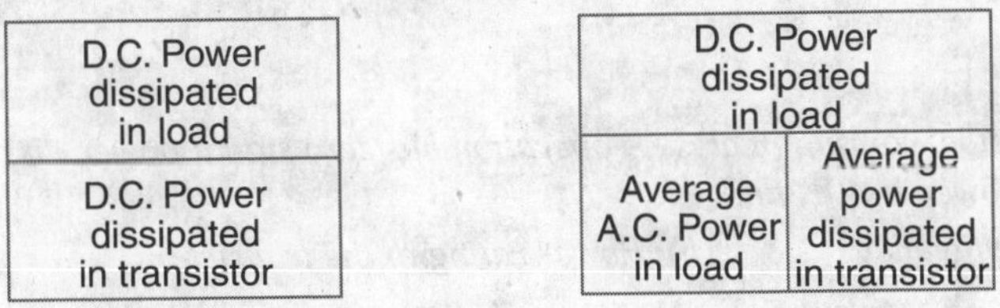

(*a*) No input signal. (*b*) Full input signal.

Fig. 27.15.

These facts are shown in Fig. 27.15 (*a*) and (*b*). Fig. 27.15 (*a*) shows that power relationships in a class-A amplifier, when there is no input signal. It is evident from the figure that the d.c. input power is utilized is two ways. One-half is dissipated in the load resistor, while the other half is dissipated in the transistor. Similarly, Fig. 27.15 (*b*) shows the power relationships, when there is a full input signal. Here it indicates that one-fourth (or 25%) of the total d.c. input power is converted into a.c. output power, while the remaining power is dissipated as a heat in load resistor and transistor.

Example 27.4.

A certain power transistor meant for class-A operation has a zero signal power dissipation of 20 watts. If the a.c. output power is 5 W, find (a) collector efficiency and (b) power rating of the transistor.

Solution.

Given : $P_{tr\,(dc)} = 20$ W and $P_{o\,(ac)} = 5$ W

(*a*) ***Collector efficiency***

We know that collector efficiency of class-A amplifier,

$$\eta_e = \frac{P_{o\,(ac)}}{P_{tr\,(dc)}} = \frac{5}{20} = 0.25 = 25\% \textbf{ Ans.}$$

(*b*) ***Power rating of the transistor***

We know that zero signal condition represents the condition under which the transistor dissipates maximum power. Therefore power rating of the transistor

= 20 W **Ans.**

Example 27.5.

The maximum collector dissipation of a transistor used in a class A amplifier is 10 W. When a signal is applied, the collector efficiency of the circuit is 32%. Calculate the a.c. power output.

Solution.

Given : $P_{c\,(dc)} = 10$ W and $\eta_c = 32\% = 0.32$

Let $P_{o\,(ac)}$ = The a.c. power output.

We know that collector efficiency, (η_c),

$$0.32 = \frac{P_{o\,(ac)}}{P_{tr\,(dc)}} = \frac{P_{o\,(ac)}}{P_c + P_{o\,(ac)}} = \frac{P_{o\,(ac)}}{10 + P_{o\,(ac)}}$$

or
$$0.32 \times (10 + P_{o\,(ac)}) = P_{o\,(ac)}$$
$$3.2 + 0.32 \times P_{o\,(ac)} = P_{o\,(ac)}$$

$\therefore$
$$P_{o\,(ac)} = \frac{3.2}{(1 - 0.32)} = 4.7 \text{ W } \textbf{Ans.}$$

Example 27.6.

A power transistor working in class A operation has zero signal power dissipation of 10 Watts. If the a.c. output power is 4 W, find

(a) Collector efficiency *(b) Power rating of the transistor.* *(PTU., Dec 2005)*

Solution.

Given: $P_{tr(dc)} = 10$ W and $P_{o(ac)} = 4$ W.

(a) Collector efficiency

We know that collector efficiency of class – A amplifier,

$$\eta_e = \frac{P_{0(ac)}}{P_{tr(dc)}} = \frac{4}{10} = 0.4 = 40\% \textbf{ Ans.}$$

(b) Power rating of the transistor

We know that zero signal condition represents the condition under which the transistor dissipates maximum power. Therefore power rating of the transistor,

$$= 10 \text{ W } \textbf{Ans.}$$

Example 27.7.

An amplifier has a collector efficiency of 50% and operates from a 24 V supply. If the output power is 3.5 W, what is the total power dissipated within the circuit? What is most likely source for most of this power loss ?

Solution.

Given : $\eta_c = 50\% = 0.5$; $V_{CC} = 24$ V; $P_{o\,(ac)} = 3.5$ W.

We know that collector efficiency (η_c),

$$0.5 = \frac{P_{o\,(ac)}}{P_{tr\,(dc)}} = \frac{3.5}{P_{tr\,(dc)}}$$

$\therefore$
$$P_{tr\,(dc)} = \frac{3.5}{0.5} = 7 \text{ W}$$

We also know that,

$$P_{tr\,(dc)} = P_{c\,(dc)} + P_{o\,(ac)}$$

or $$7 = P_{c\,(dc)} + 3.5$$

or $$P_{c\,(dc)} = 7 - 3.5 = 3.5 \text{ W}$$

The power $P_{c\,(dc)} = 3.5$ W is dissipated in the form of heat within the transistor collector region.

Example 27.8.

A single-stage, class-A amplifier has $V_{CC} = 20$ volt, $V_{CEQ} = 10$ volt, $I_{CQ} = 600$ mA and collector load resistor, $R_L = 16\ \Omega$. The a.c. output current varies by $\pm$ 300 mA with the a.c. input signal.

Determine (a) the power supplied by the d.c. source to the amplifier circuit, (b) d.c. power consumed by the load resistor, (c) a.c power developed across the load resistor, (d) d.c. power delivered to the transistor, (e) d.c. power wasted in transistor collector, (f) overall efficiency, and (g) collector efficiency.

Solution.

Given : $V_{CC} = 20$ volt; $V_{CEQ} = 10$ volt; $I_{CQ} = 600$ mA $= 0.6$ A; $R_L = 16\ \Omega$ and $I_p = 300$ mA $= 0.3$ A.

(a) *Power supplied by the d.c. source to the amplifier circuit*

We know that power supplied by the d.c. source to the amplifier circuit,

$$P_{in\,(dc)} = V_{CC} \cdot I_{CQ} = 20 \times 0.6 = 12 \text{ W } \textbf{Ans.}$$

(b) *D.C. power consumed by the load resistor*

We know that d.c. power consumed by the load resistor,

$$P_{RL\,(dc)} = (I_{CQ})^2 \times R_L = (0.6)^2 \times 16 \text{ W} = 5.76 \text{ W } \textbf{Ans.}$$

(c) *A.C. power developed across the load resistor*

We know that r.m.s. value of the collector current,

$$I = \frac{I_p}{\sqrt{2}} = \frac{0.3}{\sqrt{2}} = 0.212 \text{ A}$$

∴ A.C. power developed across the load resistor,

$$P_{o\,(ac)} = I^2 \cdot R_L = (0.212)^2 \times 16 = 0.72 \text{ W } \textbf{Ans.}$$

(d) *D.C. power delivered to the transistor*

We know that d.c. power delivered to the transistor,

$$P_{tr\,(dc)} = P_{in\,(dc)} - P_{RL\,(dc)} = 12 - 5.76 \text{ W} = 6.24 \text{ W } \textbf{Ans.}$$

(e) *D.C. power wasted in the transistor collector*

We know that d.c. power wasted in the transistor collector,

$$P_{c\,(dc)} = P_{tr\,(dc)} - P_{o\,(ac)} = 6.24 - 0.72 \text{ W} = 5.52 \text{ W } \textbf{Ans.}$$

(f) *Overall efficiency*

We know that overall efficiency,

$$\eta_o = \frac{P_{o\,(ac)}}{P_{in\,(ac)}} = \frac{0.72}{12} = 0.06 = 6\% \textbf{ Ans.}$$

(g) *Collector efficiency*

We know that collector efficiency,

$$\eta_o = \frac{P_{o\,(ac)}}{P_{tr\,(ac)}} = \frac{0.72}{6.24} = 0.115 = 11.5\% \textbf{ Ans.}$$

Example 27.9.

A class A CE amplifier operates from V_{CC} = 20 V draws a current I_{CQ} = 5 A and feeds a load of 40 Ω through a step up transformer of n_1/n_2 = 3.16. Find the efficiency of the amplifier when it is properly matched for maximum power supply. (*Anna University., April/May 2008*)

Solution.

Given: $V_{CC} = 20\ V;\ I_{CQ} = 5\ A;\ n_1/n_2 = 3.16$ and $R_L = 40\ \Omega$.

We know that the transformer ratio,

$$n = \frac{n_2}{n_1}$$

and effective load resistance (R'_L),

$$R'_L = \frac{R_L}{(n)^2} = \frac{40}{(3.16)^2} = 4\ \Omega$$

We also know that, average input d.c. power

$$P_{dc} = V_{CC}\, I_{CQ} = 20 \times 5 = 100\ \text{W}$$

and the average a.c. power

$$P_{ac} = \frac{1}{2}\frac{V_{CC}^2}{R'_L} = \frac{1}{2}\frac{(20)^2}{4} = 50\ \text{W}$$

We know that efficiency,

$$\%\eta = \frac{P_{ac}}{P_{dc}} \times 100 = \frac{50}{100} \times 100 = 50\%\ \textbf{Ans.}$$

Example 27.10.

A class A power amplifier has a transformer as the load. If the transformer has a turns ratio (N_1/N_2) of 10 and the secondary Load is 100 Ω, find the maximum a.c. power output. Given that zero signal collector current is 100 mA. (*PTU., Dec 2005*)

Solution.

Given: $a = 10$; $R_L = 100\ \Omega$ and $I_{CQ} = 100$ mA.

We know that the effective load, as seen by the primary of the transformer,

$$R'_L = a^2 . R_L = (10)^2 \times 100 = 10^4\ \Omega$$

and the maximum power delivered to the load,

$$P_{o(ac)} = \frac{1}{2} I_{CQ}^2 \times R'_L$$

$$= \frac{1}{2} I_{CQ}^2 \times R'_L\ (100 \times 10^{-3})^2 \times 10^4 = 50\ \text{W}\ \textbf{Ans.}$$

Example 27.11.

Calculate the collector efficiency and overall efficiency of a class-A, CE amplifier circuit shown in Fig. 27.16 and with the following parameters

$(V_{ce})_Q$ = 10 volts

$(I_C)_Q$ = 500 mA

$(I_{out})_{peak}$ = ± 250 mA. (*R.G.P.V., June 2008*)

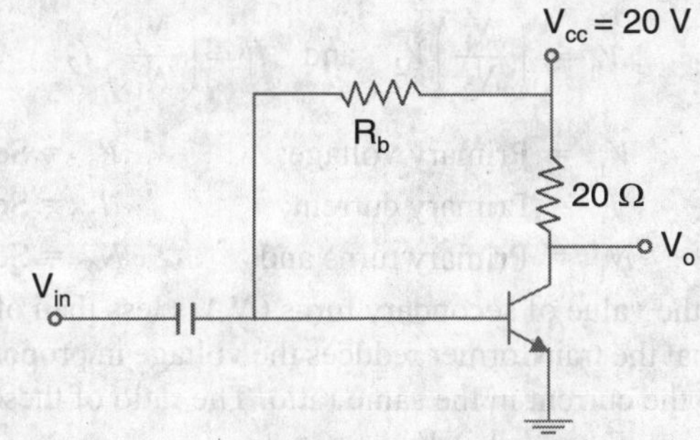

Fig. 27.16

Solution.

Given : $(V_{ce})_Q = 10$ volts; $(I_C)_Q = 500$ mA $= 500 \times 10^{-3}$; $(I_{out})_{peak} = \pm 250$ mA $= 250 \times 10^{-3}$ A; $V_{CC} = 20\ V$ and $R_L = 20\ \Omega$.

We know that the r.m.s value of output current,

$$I = \frac{I_P}{\sqrt{2}} = \frac{250 \times 10^{-3}}{\sqrt{2}} = 0.177 \text{ A}$$

We also know that the average power delivered to the load,

$$P_{o(ac)} = I^2 R_L = (0.177)^2 \times 20 = 0.63 \text{ W}$$

and,

$$P_{in(dc)} = P_{in(ac)} = V_{CC} \times I_{CQ} = 20 \times 0.5 = 10 \text{ W}$$

$$P_{RL(dc)} = (I_{CQ})^2 \times R_L = (0.5)^2 \times 20 = 5 \text{ W}$$

then,

$$P_{tr(dc)} = P_{in(dc)} - P_{RL(dc)} = 10 - 5 = 5 \text{ W}$$

We know that the value of collector efficiency,

$$\eta_o = \frac{P_{o(ac)}}{P_{tr(dc)}} = \frac{0.63}{5} = 0.126 = 12.6 \% \textbf{ Ans.}$$

27.11. Transformer Coupled Class-A Amplifier

We have already discussed that the overall efficiency of a direct coupled class-A amplifier does not exceed 25%. The main reason for this is that load resistance is directly connected in the output circuit of the amplifier. Due to this, the quiescent current (I_{CQ}) passes through the resistance.

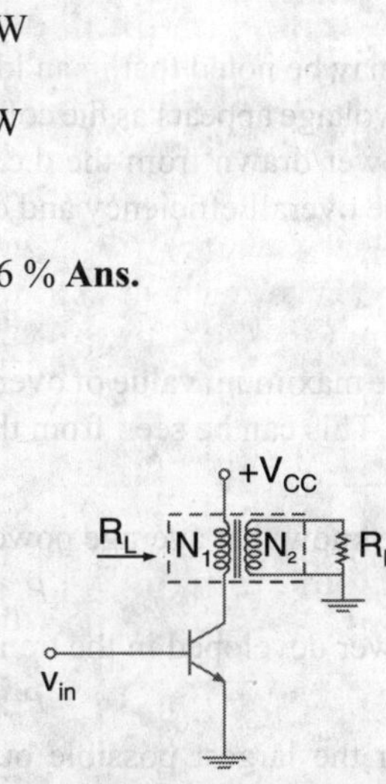

Fig. 27.17. Transformer coupled class-A amplifier.

This current represents a considerable waste of power, whose value is equal to $I^2_{CQ} \cdot R_L$ because it does not contribute to a.c. output power. This problem is solved by using a suitable transformer for coupling the load to the amplifier stage as shown in Fig. 27.17. Since the primary winding of a transformer has a low resistance, therefore the power absorbed in the winding is negligible as compared to the resistive load.

The transformer coupled class-A amplifier is also known as single-ended class-A amplifier. The term 'single-ended' is used to differentiate this type from the push-pull amplifiers. The function of a transformer is to match the low impedance load (such as loudspeaker) to that of the output impedance of the amplifier. The impedance matching property follows from the relations :

$$V_1 = \left(\frac{N_1}{N_2}\right) V_2 \text{ and } I_1 = \left(\frac{N_2}{N_1}\right) I_2$$

where V_1 = Primary voltage; V_2 = Secondary voltage

I_1 = Primary current; I_2 = Secondary current

N_1 = Primary turns and N_2 = Secondary turns

It may be noted that if the value of secondary turns (N_2) is less than of primary turns (N_1), then the above equations show that the transformer reduces the voltage in proportion to the turns ratio N_2/N_1 (equal to n) and steps up the current in the same ratio. The ratio of these two equations gives,

$$\frac{V_1}{I_1} = \left(\frac{1}{n^2}\right)\frac{V_2}{I_2}$$

Now V_1/I_1 represents the effective load resistance or the resistance seen looking into the primary of the transformer and is designated as R'_L. On the other hand, V_2/I_2 represents the value of load resistance (*i.e.*, R_L). Therefore the equation (*i*) may be rewritten as,

$$R'_L = \left(\frac{1}{n^2}\right) R_L$$

or

$$R'_L = a^2 \cdot R_L$$

where

$$a = \frac{1}{n} \text{ or } \left(\frac{N_1}{N_2}\right)$$

and is called step down turns ratio. Usually, the value of 'a' is made much larger than unity. It is necessary to make the load resistance appear as a large effective resistance seen from the transformer primary.

It may be noted that in an ideal transformer, there is no primary drop. As a result of this, all the supply voltage appears as the collector-to-emitter voltage of the transistor (*i.e.*, $V_{CC} = V_{CEQ}$). Therefore total power drawn from the d.c. supply is equal to the d.c. power developed within the transistor. Thus the overall efficiency and collector efficiency becomes equal, *i.e.*,

$$\eta_o = \frac{P_{o\,(ac)}}{V_{CC} \cdot I_{CQ}} = \frac{P_{o\,(ac)}}{V_{CEQ} \cdot I_{CQ}} = \eta_c$$

The maximum value of overall or collector efficiency of a transformer coupled class-A amplifier is 50%. This can be seen from the proof given below :

Proof

We know that average power delivered by the d.c. supply,

$$P_{in\,(dc)} = V_{CC} \cdot I_{CQ}$$

and power developed in the transistor under zero-signal condition,

$$P_{tr\,(dc)} = V_{CEQ} \cdot I_{CQ} = V_{CC} \cdot I_{CQ}$$

For the largest possible output signal, the peak (or maximum) value of the output voltage developed across the load resistor,

$$V_p = V_{CEQ}$$

and the peak value of output current,

$$I_p = I_{CQ}$$

$\therefore$ *Overall efficiency*,

$$\eta_o = \frac{P_{o\,(ac)}}{P_{in\,(dc)}} = \frac{V_p \cdot I_p / 2}{V_{CC} \cdot I_{CQ}} = 0.5 = 50\%$$

and *collector efficiency*,

$$\eta_c = \frac{P_{o\,(ac)}}{P_{tr\,(dc)}} = \frac{V_p \cdot I_p / 2}{V_{CC} \cdot I_{CQ}} = 0.5 = 50\%$$

Example 27.12.

Calculate the effective resistance seen looking into the primary of a 15 : 1 transformer connected to an output load of 8 Ω.

Solution.

Given : $a = N_1/N_2 = 15$ and $R_L = 8\ \Omega$.

We know that the value of effective resistance,

$$R'_L = a^2 \cdot R_L = (15)^2 \times 8 = 1800\ \Omega = 1.8\ \text{k}\Omega\ \textbf{Ans.}$$

Example 27.13.

What turns ratio is required to match a 16 Ω speaker load to an amplifier, so that the effective load resistance is 10 kΩ ? (P.T.U., Dec. 2006)

Solution.

Given : $R_L = 16\ \Omega$ and $R'_L = 10\ \text{k}\Omega = 10 \times 10^3\ \Omega$

Let a = Required turns ratio

We know that the value of effective resistance (R'_L)

$$10 \times 10^3 = a^2 \cdot R_L = a^2 \cdot 16$$

or

$$a = \sqrt{\frac{(10 \times 10^3)}{16}} = \sqrt{625} = 25$$

Thus we need a transformer with turns ratio of 25 : 1. **Ans.**

Example 27.14.

A transformer coupling is used in the final stage of a multistage amplifier. If the output impedance of a transistor is 1 kΩ, find the turns ratio (N_1/N_2) of the transformer so that maximum power is transferred to the load? Speaker has a resistance of 10 Ω? (PTU., May 2005)

Solution.

Given: $R_O = 1\ k\Omega = 1 \times 10^3\ \Omega$ and $R_L = 1\ \Omega$.

We know that the turns ratio of transformer,

$$a = \sqrt{\frac{R_0}{R_L}} = \sqrt{\frac{1 \times 10^3}{1}} = 31.62\ \textbf{Ans.}$$

Example 27.15.

A single-ended class-A amplifier has a transformer coupled load of 8 Ω. If the transformer turns ratio (N_1/N_2) is 10, find the maximum power output delivered to the load. Take the zero signal collector current of 500 mA.

Solution.

Given : $R_L = 8\ \Omega$; $a = N_1/N_2 = 10$ and $I_{CQ} = 500\ \text{mA} = 0.5\ \text{A}$.

We know that the effective load, as seen by the primary of the transformer,

$$R'_L = a^2 \cdot \text{R}_L = (10)^2 \times 8 = 800\ \Omega$$

and the maximum power delivered to the load,

$$P_{o\,(ac)} = \frac{1}{2} I_{CQ}^2 \times R'_L = \frac{1}{2} \times (0.5)^2 \times 800 \text{ W} = 100 \text{ W } \textbf{Ans.}$$

Example 27.16.

A transistor rated for a maximum collector dissipation of 100 mW operates a single-ended class-A output stage from a 10 volt supply. Calculate the approximate values of (a) maximum undistorted a.c. output power, (b) the quiescent current, and (c) turns-ratio of the output transformer, if the load resistance is 16 Ω. Assume overall and collector efficiency = 0.5.

Solution.

Given : $P_{tr\,(dc)} = 100$ mW = 0.1 W; $V_{CC} = 10$ volt; $R_L = 16\ \Omega$ and $\eta_o = \eta_c = 0.5$.

(a) *Maximum undistorted a.c. output power*

We know that maximum collector efficiency (η_c)

$$0.5 = \frac{P_{o\,(ac)}}{P_{tr\,(dc)}} = \frac{P_{o\,(ac)}}{0.1}$$

$$\therefore \quad P_{o\,(ac)} = 0.5 \times 0.1 = 0.05 \text{ W} = 50 \text{ mW } \textbf{Ans.}$$

(b) *Quiescent collector current*

We know that a.c. output power, ($P_{o\,(ac)}$),

$$0.05 = \frac{1}{2} V_{CC} \cdot I_{CQ} = \frac{1}{2} \times 10 \times I_{CQ} = 5\, I_{CQ}$$

$$\therefore \quad I_{CQ} = 0.05/5 = 0.01 \text{ A } \textbf{Ans.}$$

(c) *Transformer turns ratio*

Let a = Transformer turns ratio

We know that effective load resistance *i.e.*, the load resistance seen looking into the primary of the transformer,

$$R'_L = \frac{V_{CC}}{I_{CQ}} = \frac{10}{0.01} = 1000\ \Omega$$

and effective load resistance (R'_L),

$$1000 = a^2 \cdot R_L = a^2 \times 16$$

$$a = \sqrt{1000/16} = 8 \textbf{ Ans.}$$

Example 27.17.

A power transistor working in class-A amplifier is supplied from a.c. supply of 10 V. If the maximum change in collector current is 50 mA, find the power transferred to a 4 Ω speaker when (a) speaker is directly connected; and (b) speaker is connected through the transformer to the transistor circuit. *(U.P.S.C., Civil Services, 1986)*

Solution.

Given : $V_{CC} = 10$ volt; $I_P = 50$ mA = 0.05 A and $R_L = 4\ \Omega$.

(a) *Speaker directly connected*

Fig. 27.17 (*a*) shows the speaker directly connected to the transistor circuit. Now we know that the peak (or maximum) value of the change in collector-to-emitter voltage is equal to V_{CC} (*i.e.*, 10 V). Moreover the r.m.s. value of output current,

$$I = \frac{I_P}{\sqrt{2}} = \frac{0.05}{\sqrt{2}} = 0.035 \text{ A}$$

∴ Average power delivered to the speaker,

$$P_{o\,(ac)} = I^2 \cdot R_L = (0.035)^2 \times 4 = 4.9 \times 10^{-3}\text{ W} = 4.9\text{ mW } \textbf{Ans.}$$

(b) Speaker connected through the transformer

Fig. 27.17 (*b*) shows the speaker connected through the transformer to the transistor circuit. In this case, also the maximum or peak change in collector-to-emitter voltage is equal to V_{CC} (*i.e.*, 10 V). We know that the effective load resistance seen looking directly into the primary of a transformer.

$$R'_L = \frac{V_1}{I_1} = \frac{V_{CC}}{I_{CQ}} = \frac{10}{0.05} = 200\ \Omega$$

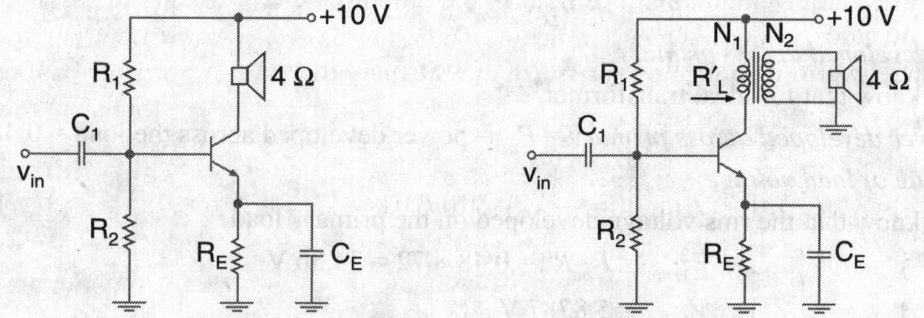

(*a*) Speaker directly connected. (*b*) Speaker connected through a transformer.

Fig. 27.18.

We also know that the effective load resistance (R'_L),

$$200 = a^2 \cdot R_L = a^2 \times 4 \quad \text{or} \quad a = \sqrt{200/4} = \sqrt{50} = 7$$

∴ Output transformer has a turns ratio (N_1/N_2) equal to 7 : 1. The transformer secondary voltage,

$$V_2 = \frac{V_1}{a} = \frac{10}{7} = 1.43\text{ V}$$

and peak value of secondary current,

$$I_{2P} = \frac{V_2}{R_L} = \frac{1.43}{4} = 0.36\text{ A}$$

∴ r.m.s. value of secondary current,

$$I_2 = 0.36/\sqrt{2} = 0.255\text{ A}$$

and the average power transferred to the speaker,

$$I_2^2 \times R_L = (0.255)^2 \times 4 = 0.26\text{ W} = 260\text{ mW } \textbf{Ans.}$$

Example 27.18.

The loudspeaker of 8 Ω is connected to the secondary of the output transformer of a class A amplifier circuit. The quiescent collector current is 140 mA. The turns ratio of the transformer is 3: 1. The collector supply voltage is 10 V. If ac power delivered to the loudspeaker is 0.48 W, Assuming the transformer to be an ideal, calculate,

(a) power developed across primary
(b) value of load voltage
(c) value of primary voltage
(d) value of load current
(e) value of primary current
(f) the dc power input

(g) the efficiency

(h) the power dissipation *(Anna University., Nov./Dec. 2006)*

Solution.

Given: $R_L = 8\ \Omega$; $I_{CQ} = 140$ mA $= 140 \times 10^{-3}$ A; $V_C = 10$ V and $P_{ac} = 0.48$ W.

We know that the transformer ratio,

$$a = \frac{N_1}{N_2} = \frac{3}{1} = 3$$

We also know that the value of effective resistance,

$$R'_L = a^2 R_L = 32 \times 8 = 72\ \Omega$$

(a) *Power developed across primary*

We know that for ideal transformer,

Power developed across primary $= P_{ac} =$ power developed across the load $= 0.48$ W **Ans.**

(b) *rms value of load voltage*

We know that the rms voltage developed on the primary load,

$$V_{1rms}^2 = P_{ac} R'_L = 0.48 \times 72 = 34.56 \text{ V}$$

$$V_{1rms} = 5.8787 \text{ V}$$

We also know that,

$$\frac{V_{1rms}}{V_{2rms}} = \frac{N_1}{N_2} = 3$$

$$V_{2rms} = \frac{V_{1rms}}{3} = \frac{5.8787}{3} = 1.9595 \text{ V } \textbf{Ans.}$$

(c) *rms value of primary voltage*

We know that the rms voltage developed on the primary load,

$$V_{1rms}^2 = P_{ac} R'_L = 0.48 \times 72 = 34.56 \text{ V}$$

$$V_{1rms} = 5.8787 \text{ V } \textbf{Ans.}$$

(d) *rms value of load current*

We know that the load current,

$$V_{2rms}^2 = \frac{P_{ac}}{R_L} = \frac{0.48}{8} = 0.6$$

$$I_{2rms} = \sqrt{0.6} = 0.2449 \text{ A } \textbf{Ans.}$$

(e) *rms value of primary current*

We know that,

$$\frac{I_{1rms}}{I_{2rms}} = \frac{N_2}{N_1} = 0.3333$$

$$I_{1rms} = I_{2rms} \times 0.3333 = 0.2449 \times 0.3333$$

$$= 0.0816 \text{ A} = 81.64 \text{ mA Ans.}$$

(f) *The d.c. power input*

We know that,

$$P_{DC} = V_{CC}\, I_{CQ} = 10 \times 140 \times 10^{-3} = 1.4 \text{ W } \textbf{Ans.}$$

(g) *The efficiency*

We know that,

$$\% \eta = \frac{P_{ac}}{P_{dc}} \times 100 = \frac{0.48}{1.4} \times 100 = 34.28 \ \% \quad \textbf{Ans.}$$

(h) *The power dissipation*

We know that,

$$P_d = P_{dc} - P_{ac} = 1.4 - 0.48 = 0.92 \text{ W} \quad \textbf{Ans.}$$

27.12. Class-B Amplifier

We have already discussed in Art. 27.5 that in a class-B, amplifier, the transistor bias and amplitude of the input signal is such that the output current flows only for one-half cycle (*i.e.*, 180°) of the input signal. This condition is achieved by setting the *Q*-point at cut-off (or lower end) of the load line.

Class B- amplifier

Fig. 27.19 (*a*) illustrates the biasing condition on the transconductance curve, while Fig. 27.19 (*b*) shows the biasing condition on the a.c. load line. The variation of collector current with the input signal is also shown in both these figures. It may be noted that with no input signal, the transistor is biased at cut-off. At this point, there is no current flow through the transistor and hence no power is dissipated by the transistor. However, when the signal is applied, the transistor handles an average current, which increases it for large input signals.

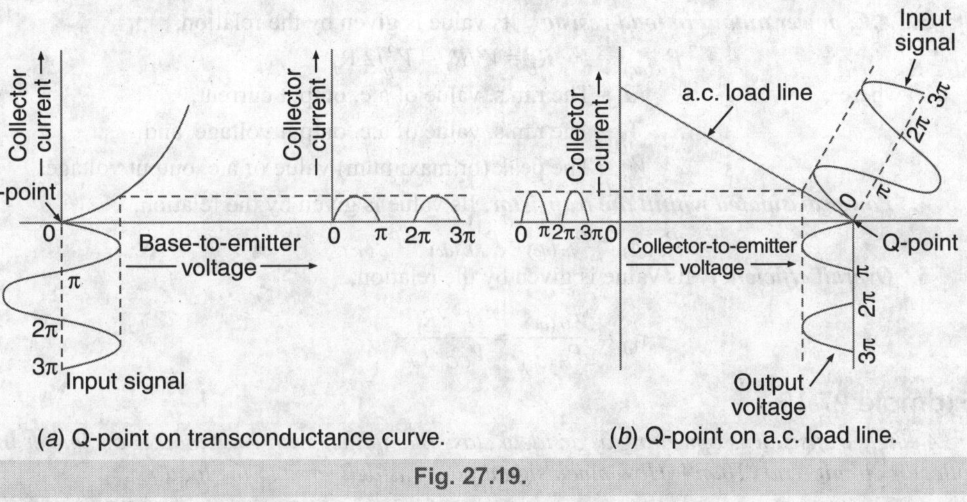

(*a*) Q-point on transconductance curve. (*b*) Q-point on a.c. load line.

Fig. 27.19.

Thus contrary to class-A amplifier operation, in which the power dissipation of a transistor is maximum with no input signal and minimum with largest input signal, the operation of class-B amplifier circuit is to increase transistor dissipation for increased input signal. Since the average current in class-B amplifier is less than that in class-A, therefore the amount of power dissipated by the transistor is less in class-B amplifier.

27.13. Characteristics of Class-B Amplifier

Following are some of the important characteristics of class-B amplifier :

1. The output current flows only for one-half of the cycle (*i.e.*, 180°) of the input signal.
2. The transistor dissipates no power with zero input signal. However, it increases with the increase in the amplitude of input signal. It is contrary to class-A amplifier operation, where the transistor dissipation is maximum with no input signal and minimum with the largest input signal.
3. The average current drawn by the circuit in class-B operation is smaller than that in class A. As a result of this, the amount of power dissipated by the transistor is less in class-B. Thus the overall efficiency of the circuit is higher than that of class-A. Its maximum value has been found to be equal to 78.5%.

27.14. Power Relations for Class-B Amplifier

Following are some of the power relations for class-B amplifier which are important from the subject point of view :

1. ***D.C input power.*** The input power comes from the d.c. source (*i.e.*, the V_{CC} supply) and is given by the relation,

$$P_{in\,(dc)} = V_{CC} \cdot I_{dc}$$

where I_{dc} is the average value of current drawn from the V_{CC} supply. If I_{CP} is the peak (or maximum) value of average value of current, then

$$I_{dc} = I_{CP}/\pi$$

$$\therefore \quad P_{in\,(dc)} = V_{CC} \cdot I_{CP}/\pi$$

2. ***D.C. power loss in load resistor.*** Its values is given by the relation,

$$P_{RL\,(dc)} = I_{dc}^2 \cdot R_L = \left(\frac{I_{CP}}{\pi}\right)^2 \cdot R_L \qquad \ldots (\because I_{dc} = I_{CP}/\pi)$$

3. ***A.C. power output in load resistor.*** Its value is given by the relation,

$$P_{o\,(ac)} = I^2 \cdot R_L = V^2/R_L = V_p^2/2\,R_L$$

where I = The r.m.s. value of a.c. output current,

V = The r.m.s. value of a.c. output voltage, and

V_p = The peak (or maximum) value of a.c. output voltage.

4. ***Power dissipated within the transistor.*** Its value is given by the relation,

$$P_{c\,(dc)} = P_{in\,(dc)} - P_{RL\,(dc)} - P_{o\,(ac)}$$

5. ***Overall efficiency.*** Its value is given by the relation,

$$\eta_o = \frac{P_{o\,(ac)}}{P_{in\,(dc)}} = \frac{I^2 \cdot R_L}{V_{CC} \cdot I_{dc}}$$

Example 27.19.

A single transistor is operating as an ideal class B amplifier with 1 kΩ load. A d.c. meter in the collector circuit reads 10 mA. How much signal power is delivered to the load?

(*Nagpur Univ. 2004*)

Solution.

Given: $R_L = 1\text{ k}\Omega = 1 \times 10^3\ \Omega$; $I_C = 10$ mA. $= 10 \times 10^{-3}$ A.

Let $P_L = P_{O\,(ac)}$. The power delivered to the load.

We know that power delivered to the road,

$$P_L = I^2 \cdot R_L = (10 \times 10^{-3})^2 \times (1 \times 10^3) = 0.1 \text{ W } \textbf{Ans.}$$

27.15. Maximum Value of A.C. Output Power in Class-B Amplifier

Fig. 27.20 shows the location of a Q-point for class-B amplifier alongwith the maximum variation of collector-to-emitter voltage (V_{CE}) and collector current (I_C) with the largest possible signal.

It may be noted from this figure that the peak or maximum value of the output current is I_{CP} and that of the output voltage is V_{CEQ} or V_{CC}.

We know that the a.c. output power in a class-B amplifier is developed only during one-half cycle of the input signal. Therefore its value is given by the relation,

$$P_{o\,(ac)} = \frac{1}{2}\left(\frac{V_{CC}}{\sqrt{2}}\right)\cdot\left(\frac{I_{CP}}{\sqrt{2}}\right) = \frac{1}{4} \times V_{CP} \cdot I_{CP}$$

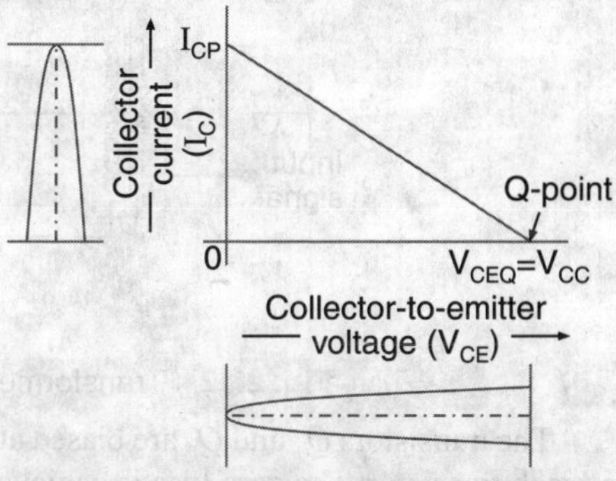

Fig. 27.20. Illustrating maximum a.c. output power in class-B amplifier.

27.16. Maximum Overall Efficiency for Class-B Amplifier

We have already discussed in Art. 27.13 that the overall efficiency for a class-B amplifier,

$$\eta_o = \frac{P_{o\,(ac)}}{P_{in\,(dc)}}$$

We also know that maximum power developed, $P_{o\,(ac)} = \frac{1}{4} V_{CP} \cdot I_{CP}$

∴ Maximum value of overall efficiency,

$$\eta_{o\,(max)} = \frac{\frac{1}{4} \cdot V_{CP} \cdot I_{CP}}{V_{CC} \cdot I_{dc}} = \frac{\frac{1}{4} \cdot V_{CP} \cdot I_{CP}}{V_{CC} \cdot I_{CP} / \pi} \qquad ...\left(I_{dc} = \frac{I_{CP}}{\pi}\right)$$

$$= \frac{\frac{1}{4} \cdot V_{CC} \cdot I_{CP}}{V_{CC} \cdot I_{CP} / \pi} = \frac{\pi}{4} \qquad ...(\because V_{CP} = V_{CC})$$

$$= 0.785 = 78.5\%$$

27.17. Class-B Push-Pull Amplifier

We have already discussed in previous articles that in a class-B amplifier, the transistor conducts only for one-half-cycle and remains off during the other half cycle. It means that we obtain an output signal, whose either positive or negative half cycles are missing. This type of output signal gives a large distortion. In order to avoid this, we use two transistors connected in a push-pull arrangement. In this arrangement, one of the transistors conducts during one half cycle and the other conducts during the second half cycle. With push-pull arrangement, it is possible to build class-B amplifier, which have low distortion, large a.c. output power and high efficiency.

Fig. 27.21 shows a transformer coupled class-B push-pull amplifier. The circuit consists of two centre-tapped transformers T_1 and T_2 and two identical transistors Q_1 and Q_2. The transformer (T_1) is an input transformer and is called phase splitter. It is required to produce two signal voltages, which are 180° out of phase with each other. These two signal voltages, with opposite polarity, drive the input of transistors Q_1 and Q_2. The transformer (T_2) is an output transformer and is required to couple the a.c. output signal from the collector to the load.

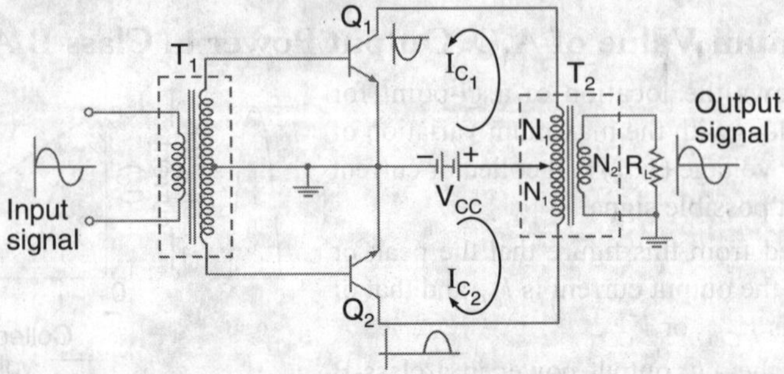

Fig. 27.21. Transformer coupled class-*B* push-pull amplifier.

The transistors Q_1 and Q_2 are biased at cut-off, which is done by providing zero bias on the base of each transistor. The zero bias is achieved by connecting the emitter and base terminals together. In order to get a balanced circuit, the two emitters are connected to the centre tap of transformer T_1 secondary and the V_{CC} supply, to the centre tap of transformer T_2 secondary.

Now we shall discuss the working of push-pull amplifier circuit. When there is no input signal, both the transistors Q_1 and Q_2 are cut-off. Hence no current is drawn from the $V_{(CC)}$ supply. Thus there is no power wasted on standby ant the power dissipation in both the transistors is, practically, zero. It will be interesting to know that even if both the transistors Q_1 and Q_2 are biased, little above their cut-off points, the transformers still operate near zero flux. It is because of the fact that d.c. bias current, for each transistor, flows in opposite directions through the secondary of transformer T_2. The magnetic flux set up by each of these currents results in opposite flux through the magnetic core, so that the net flux in the perfectly matched case is zero. Thus the transformer is not required to handle a large flux due to d.c. bias currents. As a result of this, the transformer can have a small size core, biased to operate near zero flux.

Now consider only the positive half cycle of the input signal. During this, the base of transistor Q_1 is positive and that of Q_2 is negative. As a result of this, the Q_1 conducts, while the transistor Q_2 is OFF. When the input signal goes negative, Q_1 turns off and Q_2 conducts. Thus at any instant, only one transistor in the circuit is conducting. In other words, each transistor handles one-half of the input signal. Then the output transformer joins these two halves and produces a full-sine wave in the load resistor. However, it has been found that joining of outputs of two transistors (*i.e.*, Q_1 and Q_2) is never perfect and hence the output is distorted.

The class-B push-pull amplifier is known as double ended class-B amplifier. This amplifier is frequently employed in the output (or final) stages of the amplifier circuits. These are used in public address systems (PAS 's) tape recorders and stereo amplifiers.

27.18. Advantages of Class-B Push-Pull Amplifier

Though the class-B push-pull amplifier has a number of advantages over class-A amplifier, yet the following are important from the subject point of view :

1. The circuit efficiency of a class-B push-pull amplifier is 78.5%, which is much higher than that of class-A whose value is 25%. It is mainly due to the reason that no power is drawn from the d.c. supply under no signal condition.
2. The use of push-pull system in the class-B amplifier eliminates even order harmonics in the a.c. output signal.
3. Because of the absence of even harmonics, the circuit gives more output, per device, for a given amount of distortion.

4. There is no d.c. component in the output signal. It is because of the fact, that d.c. components of two collector currents, through the two-halves of the primary of the T_2-transformer flow in opposite directions. As a result of this, there is no possibility of the core saturation of the T_2-transformer, even at the peak value of the signals. Thus we can use smaller sized cores in the transformers, without affecting the circuit performance.

27.19. Crossover Distortion

We know that transistors in a class-B push-pull amplifier are biased at cut-off. It means that when the d.c. bias voltage is zero, the input signal voltage must exceed the barrier voltage before a transistor conducts. In other words, a transistor does not conduct, until the input signal voltage exceeds 0.7 V for silicon and 0.3 V for germanium transistors. Because of this, there is a time interval between the positive and negative alternations of the input signal when neither transistor is conducting as shown in Fig. 27.22. The resulting distortion in the output signal is quite common and is called crossover distortion.

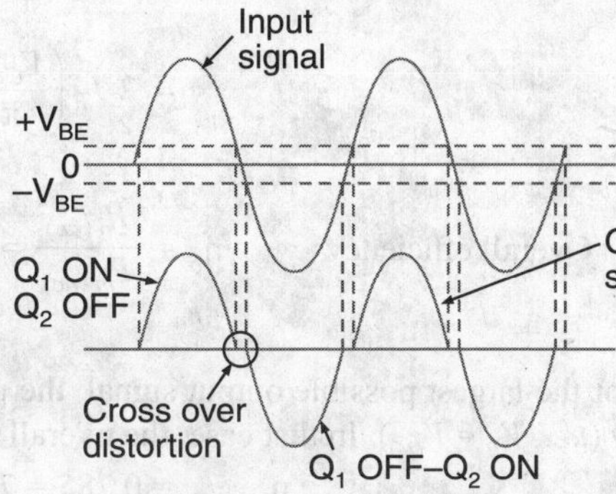

Fig. 27.22. Illustrating crossover distortion.

The crossover distortion may be avoided by applying a slight forward bias (equal to 0.7 V for silicon and 0.3 V for germanium transistors) to the base-emitter junction of both the transistors of the amplifier circuit. It causes the transistor to conduct immediately, when the a.c. input signal is applied. The application of slight forward bias shifts the Q-point slightly above the cut-off. In that case, each transistor operates for more than one half cycle. The resulting operation of the transistor is called class-AB operation.

27.20. Efficiency of Class-B Push-Pull Amplifier

Consider a class-B push-pull amplifier circuit as shown in Fig. 27.21. The a.c. load line, the operating point and the output current and voltage for the push-pull circuit are as shown in Fig. 27.23. In this figure, the a.c. load line for Q_2-transistor has been plotted upside down with respect to that of Q_1-transistor in order to get the combined output.

Now let I_C = Average (or d.c.) value of output current of Q_1 or Q_2 transistor.

I_P = Peak or maximum value of output current,

V_C = Average (or d.c.) value of *output voltage, and

V_P = Peak (or maximum) value of the output voltage.

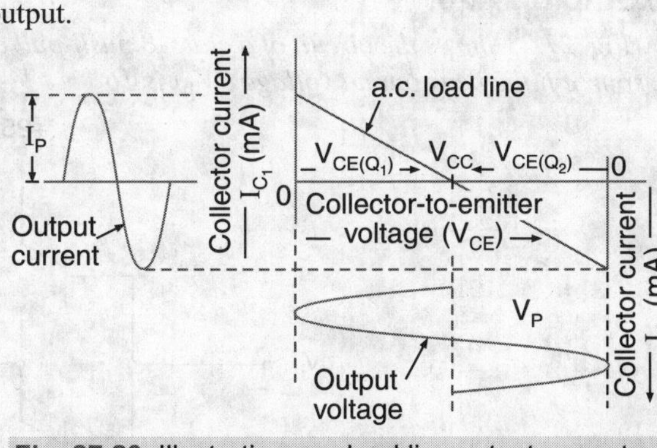

Fig. 27.23. Illustrating a.c. load line output current and voltage waveforms for the class-B push-pull amplifier.

Then a.c. power delivered to the load resistor,

* Here we have assumed that the transformer turns ratio is $(N_1 + N_1) : N_2$. It means that turns ratio for each transistor (Q_1 or Q_1) is $N_1 = N_2$. In other words, each transistor uses half the secondary of the output transformer.

$$P_{o\,(ac)} = V_C \cdot I_C = \left(\frac{V_P}{\sqrt{2}}\right)\cdot\left(\frac{V_P}{\sqrt{2}}\right) = \frac{V_P \cdot I_P}{2}$$

and total power dissipation of the two transistors,

$$2P_{C\,(dc)} = P_{in\,(ac)} - P_{o\,(ac)} = V_C \cdot I_C - V_P \cdot I_P / 2$$

$$= V_{CC} \cdot \frac{2I_P}{\pi} - \frac{V_P \cdot I_P}{2} \qquad \ldots\left(I_C = \frac{2I_P}{\pi}\right)$$

$$= 2\,I_P\left(\frac{V_{CC}}{\pi} - \frac{V_P}{4}\right)$$

$\therefore$ Overall efficiency,

$$\eta_o = \frac{P_{o\,(ac)}}{P_{in\,(dc)}} = \frac{\dfrac{V_P \cdot I_P}{2}}{V_{CC}\left(\dfrac{2I_P}{\pi}\right)} = \frac{\pi}{4} \cdot \frac{V_P}{V_{CC}} = 0.785\,\frac{V_P}{V_{CC}}$$

For the largest possible output signal, the peak value of the output voltage is equal to the V_{CC} supply (*i.e.*, $V_P = V_{CC}$). In that case, the overall efficiency is maximum, and its value,

$$\eta_{o\,(max)} = 0.785 = 78.5\%$$

The value of collector efficiency is equal to the overall efficiency, whose maximum value is also 78.5%.

Note. The class B cmplifier has the same output power characteristic as the class A amplifier, By formoula, ac power out,

$$P_{o(ac)} = \frac{V_P^2}{2\,R_L} \text{ or } \frac{V_{PP}^2}{8\,R_L}$$

To calculate the maximum possible output power, we need to determine its compliance the compliance of a class *B* amplifier is found as

$$PP = 2\,V_{CEQ}$$

$$= V_{CC} \qquad \ldots(\because V_{CEQ} = V_{CC}/2)$$

Example 27.20.

Fig. 27.23 shows the circuit of a class-B push-pull amplifier. What is the power drawn from the dc source it the peak output voltage ,(V_P) is 16 V.

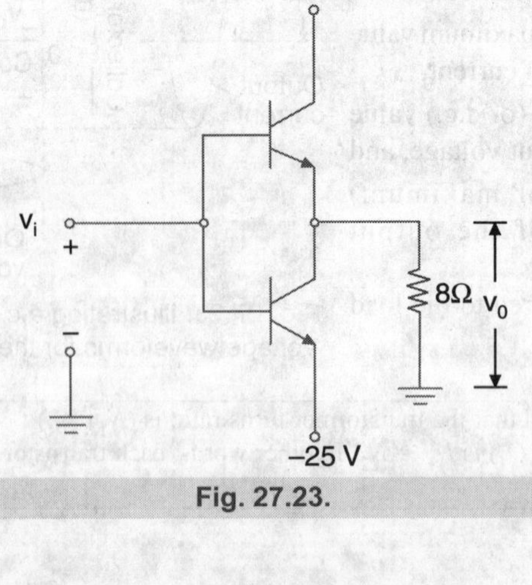

Fig. 27.23.

Solution.

Given $R_L = 8\ \Omega$; peak-output voltage $(V_P) = 16$ V.

We know that power drawn from the source

$$\approx \text{ac power output}$$

$$= \frac{V_P^2}{2R_L} = \frac{16^2}{2 \times 8} = 16 \text{ W } \textbf{Ans.}$$

Example 27.21.

A class-B push-pull amplifier with transformer coupled load uses two transistors rated 10W each. What is the maximum power output one can obtain at the load from they circuit?

(UPSC Engg. Services 2002)

Solution.

Given : Transistor rating in class-B push-pull amplifier = 10 W

Let $P_{o(ac)}$ = The maximum power output of a push-pull amplifier

We know that total power dissipation of two transistors.

$$= 2\,P_{c(dc)} = 2 \times 10 = 20 \text{ W}$$

and the maximum overall efficiency,

$$\eta = 0.785 = \frac{P_{o\,(ac)}}{P_{in\,(dc)}} = \frac{P_{o\,(ac)}}{2P_{c(dc)} + P_{o(ac)}} = \frac{P_{o(ac)}}{20 + P_{o(ac)}}$$

$$\therefore \quad 0.785\,(20 + P_{o(ac)}) = P_{o(ac)}$$

$$15.70 + 0.785\,P_{c(ac)} = P_{o(ac)}$$

or $$(1 - 0.785)\,P_{o\,(ac)} = 15.70$$

$$\therefore \quad P_{o(ac)} = \frac{15.70}{1 - 0.785} = 73.02 \text{ W } \textbf{Ans.}$$

Example 27.22.

A class B complementary A.F. power amplifier shown in Fig. 27.25. Calculate.

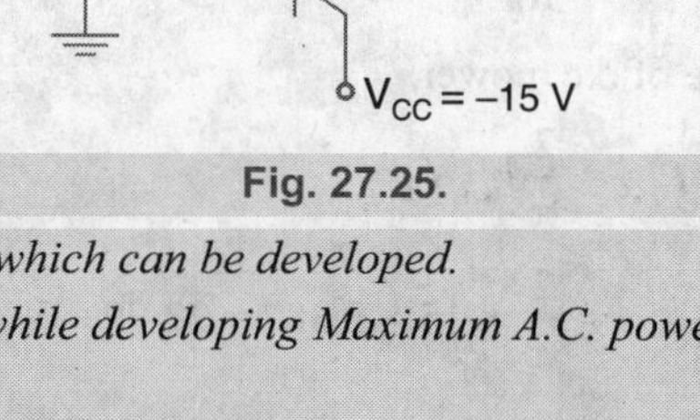

Fig. 27.25.

(a) Maximum A.C. power which can be developed.

(b) Collector dissipation while developing Maximum A.C. power.

(c) Efficiency

(d) Maximum power dissipation per transistor

(e) Efficiency under maximum power dissipation condition.(Anna University., Nov/Dec. 2007)

Solution.

Given: $V_{CC} = \pm 15\ V$ and $R_L = 4\ \Omega$.

(*a*) ***Maximum AC power***

We know that the value of maximum AC power,

$$(P_{ac})_{max} = \frac{1}{2}\frac{V_{CC}^2}{R_L} = \frac{1}{2} \times \frac{(15)^2}{4} = 28.125 \text{ W} \quad \textbf{Ans.}$$

(*b*) ***Collector dissipation while developing AC power,***

When the power is maximum $V_m = V_{CC}$

$$R_L = \frac{V_m}{I_m}$$

$$I_m = \frac{V_m}{R_L} = \frac{V_{CC}}{R_L} = \frac{15}{4} = 3.75 \text{ A}$$

We know that,

$$P_{dc} = \frac{2}{\pi} V_{CC}\, I_m$$

$$= \frac{2}{\pi} \times 15 \times 3.75 = 35.809 \text{ W}$$

and,

$$P_d = P_{dc} - P_{ac}$$

$$= 35.809 - 28.125 = 7.684 \quad \textbf{Ans.}$$

(*c*) ***Efficiency***

We know that the value of efficiency (η),

$$\%\,\eta = \frac{P_{ac}}{P_{dc}} \times 100$$

$$= \frac{28.125}{35.809} \times 100 = 78.5\% \quad \textbf{Ans.}$$

(*d*) ***Maximum Power dissipation per transistor***

We know that the maximum value,

$$V_m = \frac{2}{\pi} V_{CC} = \frac{2}{\pi} \times 15 = 9.549 \text{ V}$$

and,

$$I_m = \frac{V_m}{R_L} = \frac{9.5492}{4} = 2.387 = 2.387 \text{ A}$$

and we also know that the value of d.c. power,

$$P_{dc} = \frac{2}{\pi} V_{CC} I_m$$

$$= \frac{2}{\pi} \times 15 \times 2.387 = 22.797 \text{ W}$$

and the value of a.c. power,

$$P_{ac} = \frac{1}{2}\frac{V_m^2}{R_L} = \frac{1}{2}\frac{(9.549)^2}{4} = 11.398 \text{ W}$$

We also know that,

$$(P_d)_{max} = P_{dc} - P_{ac} = 22.797 - 11.398 = 11.39 \text{ W}$$

and the maximum power dissipation per transistor,

$$(P_d)_{max} = \frac{(P_d)_{max}}{2} = \frac{11.39}{2} = 5.699 \text{ W per transistor } \textbf{Ans.}$$

(e) *Efficiency under maximum power dissipation condition*

We know that,

$$\% \eta = \frac{P_{ac} \text{ under } (P_d)_{max}}{P_{dc} \text{ under } (P_d)_{max}} \times 100$$

$$= \frac{11.398}{22.797} \times 100 = 50\ \% \textbf{ Ans.}$$

Example 27.23.

A complementary symmetry push pull amplifier is operated using V_{CC} = ± 10 V and deliver power to a load R_L = 5 Ω. Calculate.

(a) Maximum power output

(b) Power rating of transistor

(c) D.C. input at maximum power output. (*VTU., June, July 2011*)

Solution.

Given : $V_{CC} = \pm 10$ V and RL = 5 Ω

(a) *Maximum power output*

We know that the maximum output power,

$$P_{o(max)} = \frac{V_{CC}^2}{2R_L} = \frac{(10)^2}{2 \times 5} = 10 \text{ W } \textbf{Ans.}$$

(b) *Power rating of transistor*

We know that,

$$V_m = \frac{2}{\pi} V_{CC} = \frac{2}{\pi} \times 10 = 6.36 \text{ V}$$

We also know that the total collector power rating in two transistor

$$P_{c(dc)} = P_{in(dc)} - P_{o(ac)} = V_{CC}\left(\frac{2V_m}{\pi R_L}\right) - \left(\frac{V_m^2}{2R_L}\right)$$

$$= \left[10 \times \left(\frac{2 \times 6.36}{\pi \times 5}\right)\right] - \left[\frac{(6.36)^2}{2 \times 5}\right] = 4.052 \text{ W}$$

Power rating of each transistor,

$$= \frac{P_{c(dc)}}{2} = \frac{4.052}{2} = 2.026 \text{ W } \textbf{Ans.}$$

(c) *D.C. input at maximum power output*

We know that the d.c. input at maximum power input,

$$P_{in(dc)max} = V_{CC} \times \frac{2\,V_{CC}}{\pi R_L} = 10 \times \left(\frac{2 \times 10}{\pi \times 5}\right) = 12.73 \text{ W } \textbf{Ans.}$$

Example 27.24.

A class-B output stage has an efficiency of 60%. If the maximum collector dissipation of each transistor is 2.5 W, calculate the d.c. input power and the a.c. output power.

Solution.

Given : $\eta_o = 60\% = 0.6$; Maximum collector dissipation of each transistor $P_{c\,(dc)} = 2.5$ W

Let $P_{in\,(dc)}$ = The d.c. input power, and

$P_{o\,(ac)}$ = The a.c. output power.

We know that total collector (or power) dissipation in two transistors

$$2P_{c\,(dc)} = 2 \times 2.5 = 5 \text{ W}$$

We also know that overall efficiency, (η),

$$0.60 = \frac{P_{o\,(ac)}}{P_{in\,(dc)}}; \quad \therefore \quad P_{o\,(ac)} = 0.6 \times P_{in\,(dc)} \quad ...(i)$$

and the d.c. input power,

$$P_{in\,(dc)} = 2P_{c\,(dc)} + P_{o\,(ac)} = 5 + 0.6 \times P_{in\,(dc)}$$

or $$(1 - 0.6)\,P_{in\,(dc)} = 5; \quad \therefore \quad P_{in\,(dc)} = \frac{5}{1 - 0.6} = 12.5 \text{ W } \textbf{Ans.}$$

From equation (*i*), $$P_{o\,(ac)} = 0.6 \times P_{in\,(dc)} = 0.6 \times 12.5 = 7.5 \text{ W } \textbf{Ans.}$$

27.21. Transformerless Class-B Push-Pull Amplifier

The push-pull amplifier discussed in Art. 27.16 has two major drawbacks namely it requires a bulky and expensive transformer and also input signal voltage 180° out of phase with each other. The second requirement necessitates the use of an input centre-tapped transformer called phase splitter. It makes the driver circuitry of the amplifier complicated. The transformerless amplifier removes both the above mentioned drawbacks while retaining all the advantages of push-pull configuration.

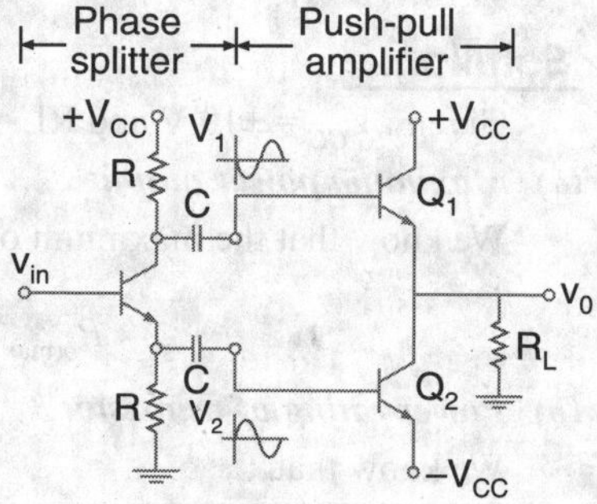

Fig. 27.26. Transformerless class-B push-pull amplifier.

Fig. 27.26 shows a transformerless class-B push-pull amplifier. In this figure, the input centre tapped transformer is removed by an input driver called phase splitter or phase inverter. It consists of an NPN transistor with equal collector and emitter resistances.

It may be noted that the emitter output of phase splitter produces an inphase signal, while that of collector output produces an out of phase signal. Therefore the output signals are equal in amplitude, but opposite in phase. The load is directly connected to the amplifier circuit. The removal of output transformer does not create any problem of impedance mismatch between the output impedance of the circuit and the load resistance. These days, the high-power transistors available have an output impedance of the order of 4 to 8 ohms.

The split supply (*i.e.*, positive and negative V_{CC}) used in the circuit gives us an advantage of making the d.c. component of output power as zero. It is due to this fact that we have connected the load directly with the amplifier circuit.

27.22. Complementary Symmetry Class-B Push-Pull Amplifier

Fig. 27.27 shows another class-B push-pull amplifier, which requires neither an input nor an output transformer. This arrangement uses transistors having complementary symmetry in the emitter follower configuration. The term 'complementary' means that the circuit uses two identical transistors – one NPN and the other PNP. Both these transistors have identical input and output characteristics. The term 'symmetry' means that the biasing resistors are equal. As a result of this, the emitter-base junction of each transistor is biased with the same voltage. It may be noted that the input signal is capacitively coupled, whereas the output signal is direct coupled.

The circuit operation may be understood that with zero input signal, none of the transistor conducts. Thus the output voltage is zero. During the positive input half-cycle, Q_1 conducts and Q_2 is off. Similarly, during the negative input half-cycle, Q_1 is driven into cut-off and Q_2 conducts.

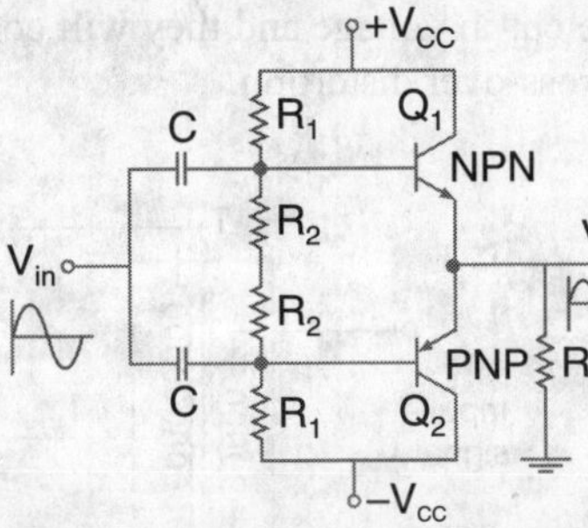

Fig. 27.27. Complementary symmetry Class-B push-pull amplifier.

It may be noted that amplifier circuit has a unity gain because of the emitter follower configuration. Moreover, there is no phase inversion of the output signal. The split supply (*i.e.*, equal positive and negative V_{CC}) used in the circuit gives us an advantage that the d.c. component of output voltage can be mode zero. Thus only a.c. component of the power is available across the load resistor (R_L).

27.23. Class AB Amplifier

If the amplifying device is biased in such a way that current flows in the device for 51% - 99% of the input signal, the amplifier is operating class-AB. A simple class-AB amplifier is shown in Fig. 27.28. Notice that the output signal is distorted. The output signal no longer has the same shape as the input signal. The portion of the output signal that appears to be cut off is caused by the lack of current through the transistor. When the emitter becomes positive enough, the transistor cannot conduct because the base- to-emitter junction is no longer forward biased. Any further increase in input signal will not cause an increase in output signal voltage. Class-AB amplifiers have better efficiency and poorer fidelity than class-*A* amplifiers. They are used when the output signal need not be a complete reproduction of the input signal, but both positive and negative portions of the input signal must be available at the output. Class-AB amplifiers are usually defined as amplifiers operating between class-A and class-B because class-A amplifiers operate on 100% of input signal and class-B amplifiers (discussed next) operate on 50% of the input signal. Any amplifier operating between these two limits is operating class-AB.

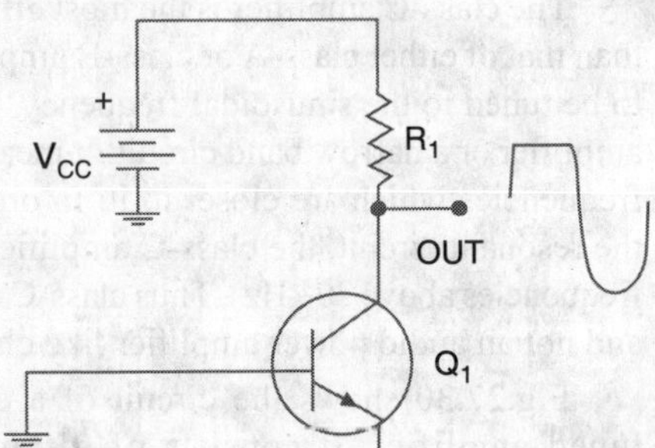

Fig. 27.28

27.24. Class AB Push Pull Amplifier

We have already discussed the Class-B push-pull amplifier is Art. 27.17. The working of class-AB push-pull amplifier is same as Class-B push-pull amplifier. The class-AB push-pull amplifier is obtained by using class-B push-pull amplifier connected with a resistor and diode in the circuit.

Fig. 27.29 shows a class-AB push-pull amplifier. The circuit consists of two centre-tapped transformers T_1 and T_2, two identical transistors Q_1 and Q_2, resistor R and diode D. The dc voltage developed across the diode D is connected to the bases of both the transistors through the secondary winding of the input transformer. This voltage acts as a dc bias for the transistors because it is equal to the cut-in voltage and they will conduct for complete half cycle period of the input to eliminate the cross-over distortion.

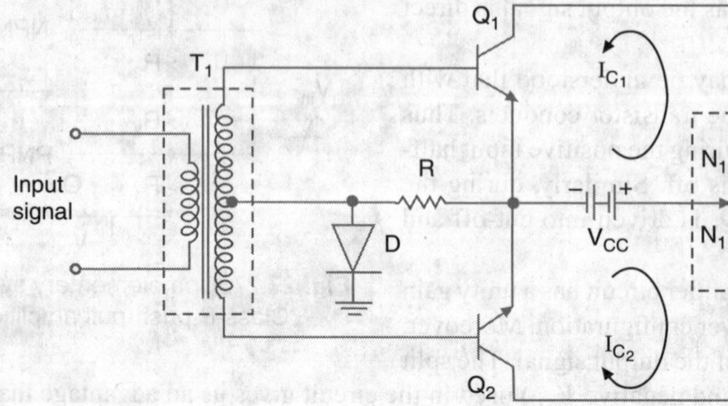

Fig. 27.29.

27.25. Class-C Amplifier

The class-C amplifier is the most efficient power amplifier, which can produce more load power than that of either class-A or class-B amplifier. To amplify a sinusoidal signal, a class-C amplifier has to be tuned to the sinusoidal frequency. Because of this fact, the class-C amplifier is called a tuned amplifier or a narrow band circuit. It means that it can amplify only the resonant frequency and those frequencies which are closer to it. In order to avoid the need for large inductors and capacitors, in the resonant circuit, the class-C amplifiers are used to amplify the signals at radio frequencies (*i.e.*, frequencies above 20 kHz). Thus class-C amplifier is basically a radio-frequency (RF) power amplifier and not an audio power amplifier like class-A and class-B amplifier.

Fig.27.30 shows the circuit of a class-C tuned amplifier. It consists of a transistor amplifier with a tuned circuit. The input voltage of the amplifier must have sufficiently large amplitude so that the base emitter junction is forward biased for less than 1800 of the input signal cycle. The transistor conducts during this portion of the input signal cycle. The collector current is consists of a series of pulses. One pulses for each cycle of the input signal. The peaks of these pulses resemble the peaks of the input sine wave and the duration of each is less than the half the period of the input signal.

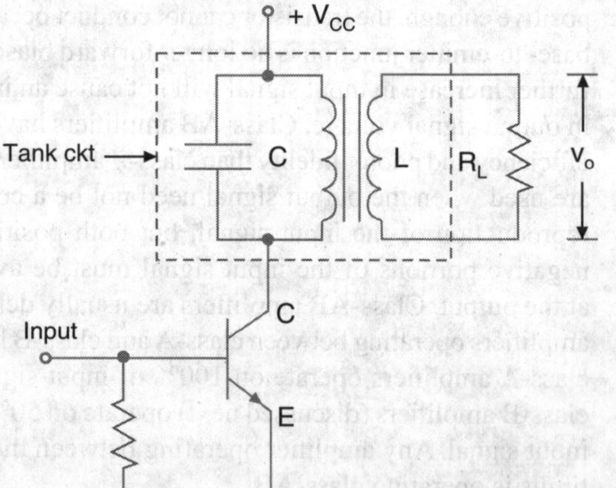

Fig. 27.30.

The parallel resonant circuit provides the load impedance is tuned to the input frequency and has a high Q. The output voltage across the tuned circuit is sinusoidal and has the same frequency as that of the input signal.

A class-C tuned amplifier operates at the resonant frequency fo of the tank circuit, where

$$f_o = \frac{1}{2\pi\sqrt{LC}}$$

The bandwidth is given by B

$$B = \frac{f_o}{Q_o}$$

where Q_o is the effective Q of the tank circuit.

A class-C tuned amplifier is used as a frequency multiplier if the resonant circuit is tuned to a harmonic of the input signal.

27.26. Characteristics of Class-C Amplifier

Following are some of the important characteristics of class-C amplifier :

1. The output current flows only during a part of the positive (or negative) half cycle of the input signal. This condition is achieved by biasing the transistor below cut-off.
2. The output signal does not resemble the input signal because it consists of narrow pulses.
3. The class-C amplifier is the most efficient power amplifier and its overall efficiency, under certain conditions, may approach even 100%.

27.27. Class-D Amplifier

The class-D amplifier is also known as *switching amplifier*. It is an electronic amplifier, where all the power devices like MOSFET are operated as binary switches Fig. 27.31 shows the circuit diagram of a class-D amplifier. The MOSFET switch are either fully ON or fully OFF. As seen from this Fig. the input is converted into a square wave. This square wave signal is fed to a switching controller which produces high amplitude (between V_{CC} and ground levels) square wave output.

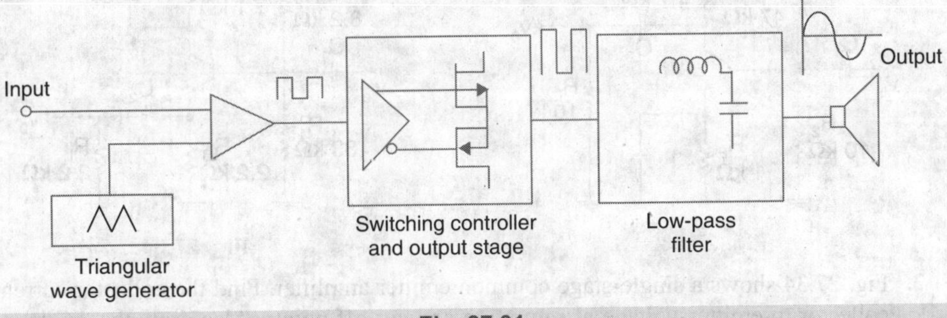

Fig. 27.31.

A low pass filter is inserted between the switching controller and the load to supress harmonics. Since the MOSFET switches, there are no losses in ideal inductors and capacitors. Ideally the efficiency of class D amplifier is 100%. All the power supplied to it is delivered to the load. This is because an ideal switch in its ON state will conduct all current but has no voltage across it, hence no heat is dissipated.

Whenever the stage switches, the capacitances at the input and output are charged and discharged which takes energy. The higher the switching frequency the more power is last. The class D amplifier are mostly used in home theatre systems and mobile phones, powered speakers, bars amplifiers, *etc.*

27.28. Class-E Amplifier

The class-E amplifier has in improved high frequency efficiency compared to class-D amplifier. Fig. 27.32 show the class-E amplifier, here the output capacitance of the MOSFET is incorporated in the output network.

The input signal is either square wave or sine wave. The drain voltage of the MOSFET is closet to zero, when the MOSFET is ON. There is no switching losses in ON state, however in OFF state it still causing losses.

The class emplifiers is used extensively as a RF power amplifier.

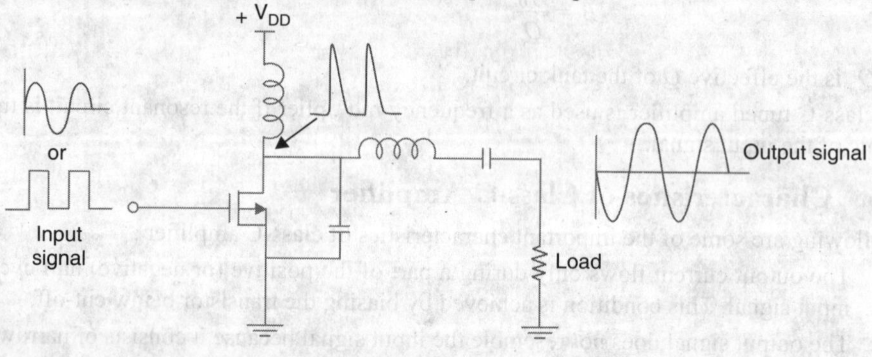

Fig. 27.32. Class-E amplifier.

NUMERICAL PROBLEMS

1. Determine the collector current and the collector-to-emitter voltage at the points of saturation and cut-off in Fig. 27.33. under a.c. signal operation and hence draw the AC load line. Assume $V_{BE} = 0.7$ V. **(Ans. 1.4 mA; 4 V)**

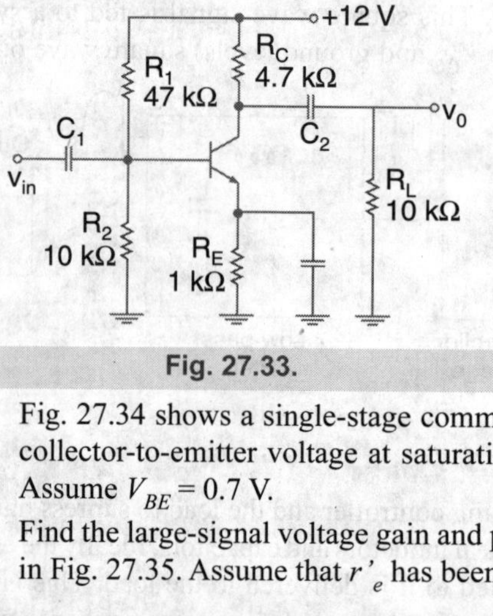

Fig. 27.33.

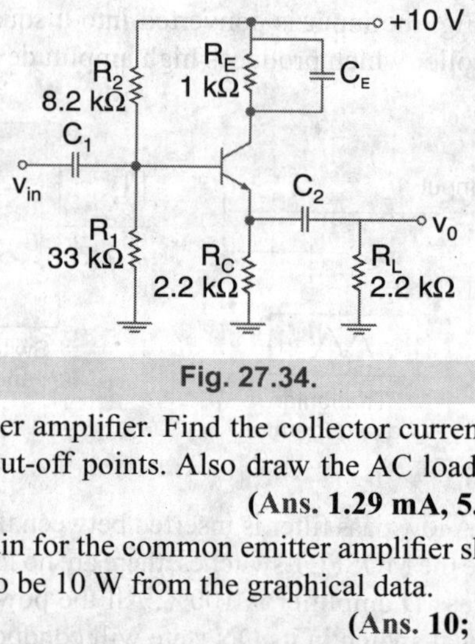

Fig. 27.34.

2. Fig. 27.34 shows a single-stage common emitter amplifier. Find the collector current and collector-to-emitter voltage at saturation and cut-off points. Also draw the AC load line. Assume $V_{BE} = 0.7$ V. **(Ans. 1.29 mA, 5.8 V)**
3. Find the large-signal voltage gain and power gain for the common emitter amplifier shown in Fig. 27.35. Assume that r'_e has been found to be 10 W from the graphical data. **(Ans. 10; 250)**
4. Fig. 27.36 shows a single-stage common emitter amplifier. Find the large signal voltage gain and power gain for this circuit. Assume that r'_e has been found to be 50 Ω from the graphical data. **(Ans. 5.28; 2.64)**

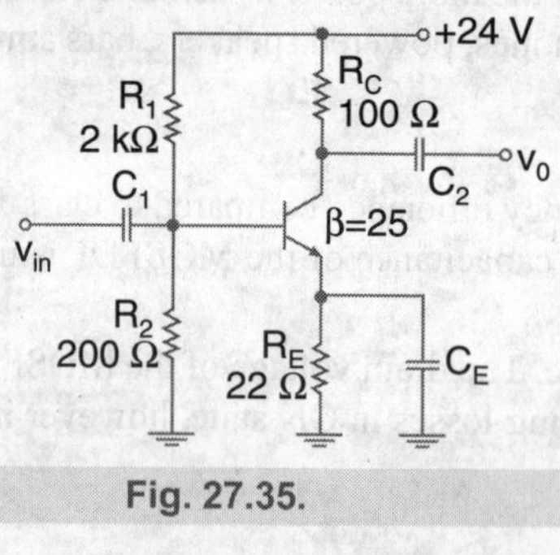

Fig. 27.35.

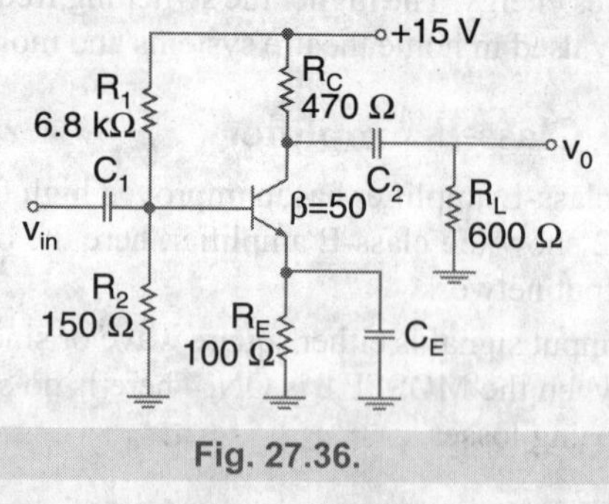

Fig. 27.36.

5. A certain power transistor meant for Class-A operation has a zero signal power dissipation of 10 watts. If the a.c. output power is 4 W , find (*a*) collector efficiency and (*b*) power rating of the transistor. **(Ans. 25%; 10 W)**
6. An amplifier has an efficiency of 31% and power output of 12 W. Is this likely to be a class-A or class-B amplifier and what is the dc input power to the circuit ? **(Ans. Class A, 38.7 W)**
7. An amplifier has an efficiency of 32% and a collector dissipation of 0.8 W. Calculate the d.c. power input and a.c. power output of the circuit. **(Ans. 1.176 W, 0.376 W)**
8. A class-A transformer coupled amplifier uses a 25 : 1 transformer to drive 4 Ω load. Calculate the effective a.c load (seen by the transistor connected to the larger turns side of the transformer). **(Ans. 2500 W)**
9. What turns ratio transformer is needed to couple to an 8 Ω load so that it appears as a 10 kW effective load ? **(Ans. 35 : 1)**
10. A power transistor of a transformer coupled class-A amplifier has to deliver a maximum of 4 W of low frequency power to a 4 Ω load with a power supply 24 V. Find the values of (*a*) d.c. input power, (*b*) quiescent voltage and current, (*c*) peak value of collector current, and (*d*) transformer turns ratio. Take $\mu = 0.4$. **(Ans. (*a*) 8 W; (*b*) 24 V, 0.33 A; (*c*) 0.4 A; and (*d*) 4.2)**
11. A power transistor of a class-A amplifier is supplied from a d.c. supply of 10 V. If the maximum collector current is 100 mA, find the power transferred to a 4 Ω load, if load is (*a*) directly coupled, and (*b*) transformer coupled. **(Ans. (*a*) 20 mW; (*b*) 500 mW)**

DESCRIPTIVE QUESTIONS

1. Why a power amplifier is called a large signal amplifier ? Explain it briefly.
2. What is the difference between the voltage amplifier and power amplifier ?
3. How do you classify the power amplifiers ? With the help of suitable sketches or otherwise compare class-A, class-B and class-C amplifiers.
4. Explain the meaning of class-A, class-B and class-C as applied to power amplifiers.
5. Draw a neat circuit diagram of a class-A direct coupled amplifier using NPN transistor and explain clearly its operation.
6. Draw the circuit diagram of a single ended class-A power amplifier. Explain its operation.
7. Draw neatly the configuration of push-pull amplifier and explain its working. Derive an expression for its efficiency.
8. Show that the maximum collector efficiency of class-B amplifier is 78.6 per cent.
9. Give the advantages and disadvantages of push-pull configuration in power amplifiers.
10. (*a*) With a neat circuit diagram, explain the principle of operation of a push-pull amplifier circuit using transistors.
    (*b*) Derive the equation for conversion efficiency of class-B power amplifier.
    (*c*) Clearly explain the reason why most of the power amplifier used in practice are designed to operate in class AB stage.
11. Explain the classification of power amplifiers. (*Bhartiya University., April 2004*)
12. Write short note on large signal amplifier. (*JNTU., June 2009*)
13. Explain the difference between voltage and power amplifier. (*PTU., May 2005*)
14. Explain class-A RC coupled power amplifier. (*Anna University., Nov/Dec 2007*)
15. With a neat circuit diagram, explain the operation of transformer coupled class-A power amplifier. (*Anna University., April/May 2008*)

16. What is the function of transformer in power amplifier? *(PTU., Dec 2009)*

17. Give applications of transformer coupled amplifier. *(PTU., Dec 2009)*

18. Show that maximum collector efficiency of class-A transformer coupled power amplifier is 50%. *(PTU., Dec 2005, May 2005)*

19. Derive the expression for conversion efficiency of a transformer coupled class-A power amplifier. *(PTU., Dec 2009)*

20. Derive an expression for the maximum conversion efficiency of a class-B push-pull amplifier. *(VTU., June /July 2008)*

21. Discuss the complementary symmetry class-B amplifier and obtain its efficiency. *(Anna University., April/May 2008)*

22. Explain the operations of a push-pull power amplifier with a circuit. Find its efficiency. *(Bhartiya University., April 2004)*

23. Explain working of transformer coupled Class-B push-pull power amplifier. What are the advantages of push pull configuration? *(Nagpur University., 2008)*

24. Derive expressions for efficiency of class-A and class-B power amplifier. *(Nagpur University., 2008)*

25. Draw the class-B push-pull emitter follower circuit and explain its operation. *(Anna University., April/May 2008)*

26. What is called as crossover distortion and how to minimize this distortion? *(Anna University., Nov/Dec 2007)*

27. What is cross over distortion? How it can be eliminated? *(Anna University., April/May 2008)*

28. Discuss crossover distortion? *(PTU., May 2009, 2006, 2005, Nagpur University, Winter 2002)*

29. Describe the operation of class-AB amplifier to avoid cross over distortion. *(Anna University., April/May 2008)*

30. Explain the operation of Class-C amplifier and derive its efficiency and figure of merit. *(Anna University., April/May 2008)*

31. What are the factors affecting thermal stability of power amplifiers? Explain its significance. *(Anna University., April/May 2008)*

32. With neat circuit diagram, explain amplifier and derive its efficiency. *(Anna University., May/June 2007)*

33. What is meant by distortion in an amplifier? *(PTU., Dec 2006)*

34. What do you understand by class-A, class-B and class-C power amplifier? Define and explain the following terms as applied to power amplifiers

(*a*) Collector efficiency

(*b*) Distortion

(*c*) Power dissipation Capability *(PTU., May 2005)*

35. Why is class C amplifier more efficient than class-A amplifier? *(PTU., Dec 2009)*

36. Show that the maximum efficiency of a class-B operation will not exceed 78.5 percent. *(PTU., Dec 2006)*

37. Explain the working of a class-B push-pull Amplifier using Transformer. Also show the efficiency of such as amplifier ≤ 78.5%. *(PTU., Dec 2008)*

38. Discuss the advantages of push-pull amplifier. *(PTU., May 2006)*

39. Explain the push-pull amplifier circuit in detail with a neat diagram? *(PTU., May 2005)*

40. Although a class-B single ended power amplifier has high efficiency, yet it is open circuited. *(PTU., May 2007)*

41. Draw and explain the working of a push-pull amplifier circuit. *(PTU., May 2007)*

42. Draw and explain the working of push-pull class-B Amplifier. What are its advantages & disadvantages? *(PTU., Dec 2007)*

43. Sketch the circuit of a push-pull class-B transistor amplifier in the common Collector configuration with an output Transformer, and explain its working. *(PTU., May 2009)*

44. With a suitable circuit diagram and waveform, explain the operation of a class-B power amplifier using complementary symmetry transistors. *(PTU., Dec 2009)*

45. Write a note on complementary symmetry amplifier. *(PTU., Dec 2007)*

46. Draw a circuit using complementary symmetry and discuss its advantages. *(PTU., May 2009)*

47. What are the advantages of complimentary symmetry Amplifier? *(PTU., May 2008)*

48. Compare the efficiency of class A, B, C, AB *(Anna University., Nov/Dec 2007)*

49. Give short notes on class-C power amplifiers. *(Bhartiya University., April 2004)*

50. How do you classify the operation of the amplifiers? *(R.G.P.V., June 2003)*

51. Derive the collector efficiency of a collector of a class-A power amplifier with resistive load and transformer coupled class-A power amplifier. *(R.G.P.V., June 2002)*

52. Derive expression for the conversion efficiency of class-A amplifier assuming distortion terms to be negligible small. *(R.G.P.V., June 2007)*

53. What is class-A amplifier? Explain. Derive an expression for its power conversion efficiency. *(R.G.P.V., Dec. 2008)*

54. Show that the maximum theoretical efficiency of the class B amplifier is $\pi/4 \times 100$ percent. *(R.G.P.V., June 2003)*

55. Show that the maximum conversion efficiency of an idealized class-B amplifier is 78.5 %. *(R.G.P.V., Dec. 2006, Dec. 2003)*

56. Explain the working of push-pull amplifier and its basic principle of operation. *(R.G.P.V., June 2002)*

57. Explain the working of a class-B push-pull amplifier. *(R.G.P.V., June 2004)*

58. Draw circuit diagram of a class-B push-pull amplifier and explain its working. *(R.G.P.V., Dec. 2008)*

59. Give working principle, merits and demerits of a push-pull amplifier circuit. *(R.G.P.V., June 2008)*

60. Write short note on class-AB amplifier. *(R.G.P.V., Dec. 2008, June 2004)*

61. List the types of power amplifier. Explain with the help of circuit diagram of the working of class-C amplifier. *(R.G.P.V., Nov. /Dec. 2007)*

62. Draw the circuit of a push-pull amplifier using complementary symmetry without transformer and discuss its working. *(R.G.P.V., Nov. /Dec. 2007, June 2005)*

63. Derive a simple expression for the output power of an idealized class-B push-pull power amplifier. *(R.G.P.V., Dec. 2002)*

64. Explain why even harmonics are not present in push-pull amplifier. Give other additional advantages of the circuit over that of a single transistor amplifier. *(R.G.P.V., Dec. 2004)*

65. Define the following modes of operation of an amplifier.
(*a*) Class-A
(*b*) Class -C. (*R.G.P.V., Dec. 2005*)

66. Explain how even harmonics get eliminated in the output of a push-pull amplifier. (*R.G.P.V., Dec. 2005*)

67. Give the classification of power amplifiers on the basic of location of Q-point. Compare the performance on efficiency and distortion. (*R.G.P.V., May/ June 2006*)

68. Find the efficiency of class-A power amplifier. (*R.G.P.V., May/ June 2006*)

69. What are the advantages of push-pull amplifier? (*R.G.P.V., Dec. 2006*)

70. Draw the circuit of a push-pull amplifier stage suing two transistors and explain its working. (*R.G.P.V., June 2007*)

71. What is meant by the term cross over distortion? How it is reduced? (*R.G.P.V., June 2007*)

72. Explain the working of transformer coupled class-B push-pull power amplifier. What are the advantages of push-pull configuration? (*Nagpur University, Summer 2008*)

73. Derive the expression for efficiency of class-A and class-B power amplifier. (*Nagpur University, Summer 2008*)

74. Differentiate A, B, AB and C type of power amplifier. (*Nagpur University, Summer 2007, 2005*)

75. Show that the efficiency of class-B push-pull amplifier is 78.53 %. Also derive the relation for power dissipated in each transistor and obtain the condition for maximum power dissipated. (*Nagpur University, Winter 2006*)

76. Draw the maximum efficiency expression for transformer coupled class-A amplifier. (*Nagpur University, Summer 2006*)

77. With the circuit diagram discuss the working of the push-pull class-A power amplifier. Discuss advantages of push-pull system. (*Nagpur University, Winter 2005*)

78. Derive equation of efficiency of transformer coupled class A power amplifier. What is maximum possible efficiency? (*Nagpur University, Winter 2005*)

79. Show that the maximum conversion efficiency of an idealized class-B push pull circuit is 78.5 percentages. (*Nagpur University, Winter 2004*)

80. Draw schematic diagram of class-B push-pull amplifier and show that O/P consists of odd harmonics only and the maximum efficiency is 78.5 %. (*Nagpur University, Summer 2004, 2001, Winter 2000*)

81. Explain impedance matching in case of a power amplifier. How does it effect the power output and efficiency of the circuit. (*Nagpur University, Winter 2003*)

82. Why conversion efficiency of class-A. RC coupled amplifier is less than class A transformer coupled amplifier? (*Nagpur University, Summer 2003*)

83. What is crossover distortion in amplifier and how to eliminate it? (*Nagpur University, Summer 2003*)

84. Draw class-B push-pull amplifier and derive expression for maximum efficiency. (*Nagpur University, Summer 2003*)

85. Explain the origin of crossover distortion and describe a method to minimize this distortion. (*Nagpur University, Winter 2002*)

86. With neat circuit diagram explain the working of complementary symmetry push-pull amplifier. Discuss the harmonic distortion components in the output. (*Nagpur University, Winter 2002*)

87. Derive equation for efficiency of class-A amplifier. *(Nagpur University, Winter 2002)*

88. What reasons can you give for the improved efficiency of a class-B power amplifier compared to class A stage? What is disadvantage, if any, occurs with class-B stage? How it can be eliminated? *(Nagpur University, Winter 2001)*

89. Draw a circuit of complementary symmetrical class-AB push-pull amplifier. How cross over distortion is eliminated? Why this circuit is popular than the transformer coupled power amplifier. *(Nagpur University, Winter 2001)*

90. What are the different possible distortions in a power amplifier? Which is the most significant? Why? *(Nagpur University, Summer 2000)*

91. Explain how amplifier is classified. *(Nagpur University, Winter 2000)*

92. Explain the push-pull amplifier. *(R.G.P.V., Dec. 2001)*

93. Explain the biasing of power amplifier and write its application. *(R.G.P.V., Nov. /Dec. 2007, Dec. 2001)*

94. Discuss crossover distortion in power amplifiers. *(R.G.P.V., Dec. 2003)*

95. Explain the different between voltage and power amplifier. Give the classification of power amplifiers. *(Nagpur University, Winter 2002)*

96. Write short notes on :
    (*a*) Crossover Distortion. *(Nagpur University, Summer 2006)*
    (*b*) Crossover distortion in power amplifier. *(Nagpur University, Summer 2007)*
    (*c*) Class-B push-pull power amplifier. *(Nagpur University, Winter 2006)*
    (*d*) Push-pull amplifier. *(Nagpur University, Summer 2005)*
    (*e*) Classification of power amplifier *(R.G.P.V., Dec. 2005)*
    (*f*) Complementary symmetry push-pull circuit *(Nagpur University, Summer 2000)*

97. Explain the working of class A push-pull amplifier and derive its efficiency. *(RGTU., June 2010)*

98. Compare class A, B AB and C power amplifier. *(RGTU., June 2009, 2010)*

99. Explain the following :
    (*a*) Class-B amplifier and its efficieny *(Nagpur University, Summer 2006)*
    (*b*) Type of distortion. *(RGTU., June 2010)*

100. Give classification of power amplifier. *(RGTU., June 2011)*

101. Draw the circuit diagram of a class-B push-pull amplifier and explain its working. *(RGTU., June 2011)*

102. Show that the maximum efficiency of class-A amplifier is 25%. *(RGTU., Dec. 2009)*

103. Explain the principle of push-full power amplifier. *(RGTU., Dec. 2009)*

104. What are the types of distortion in power amplifier. *(RGTU., Dec. 2009, July 2009)*

105. Explain the operation of transformer coupled class-A power amplifier. *(VTU., June/July 2011)*

106. Show that maximum efficiency of the push-pull class-B amplifier circuit is 78.5%. *(VTU., June/July 2011)*

MULTIPLE CHOICE QUESTIONS

1. In a class-A amplifier, the current in the output circuit flows for :
   (*a*) 360° (*b*) 180° (*c*) 90° (*d*) less than 90°

2. The maximum collector circuit efficiency of class-A amplifier with a transformer coupled load is
   (*a*) 25% (*b*) 50% (*c*) 78.6% (*d*) 95.4%
3. Which of the following is a main function of the transformer used in the output of a power amplifiers :
   (*a*) To increase the output power
   (*b*) To increase the voltage gain
   (*c*) To match the load resistance with the output resistance of a transistor
   (*d*) None of the above
4. Class-A power amplifiers, as compared to class-B amplifiers
   (*a*) Have more efficiency
   (*b*) Have less distortion
   (*c*) Are not susceptible to power supply hum
   (*d*) None of the above
5. In which of the following amplifiers, does the amplifying device becomes cooler when the input device is changed from no signal to a rated value.
   (*a*) Class-A amplifier (*b*) Class-B amplifier
   (*c*) Class-AB amplifier (*d*) Class-C amplifier
6. High-power efficiency of a push-pull amplifier is due to the fact that
   (*a*) each transistor conducts on different cycles of input
   (*b*) transistors are placed in CE configuration
   (*c*) there is no quiscient collector current
   (*d*) low forward biasing voltage is required.
7. The circuit efficiency of a class-A amplifier can be increased by using
   (*a*) low dc power input (*b*) direct-coupled load
   (*c*) low-rating transistor (*d*) transformer-coupled load.
8. In a class-A amplifier, worst-case condition occurs with
   (*a*) zero signal input (*b*) maximum signal input
   (*c*) high load resistance (*d*) transformer coupling.
9. The output of a class-B amplifier
   (*a*) is distortion-free
   (*b*) consists of positive half-cycle only
   (*c*) is like the output of a full-wave rectifier
   (*d*) comprises short-duration current pulses.
10. The maximum overall efficiency of a transformer-coupled class-A amplifier is — per cent.
   (*a*) 78.5 (*b*) 25 (*c*) 50 (*d*) 85
11. A transistor audio amplifier is found to have an overall efficiency of 70 per cent. Most probably, it is a .....amplifier.
   (*a*) class-B push-pull (*b*) single-stage class-C
   (*c*) transformer-coupled class-A (*d*) direct-coupled class-A
12. The main purpose of using transformer coupling in a class-A amplifier is to make it more
   (*a*) distortion-free (*b*) bulky (*c*) costly (*d*) efficient.
13. A class-B push-pull amplifier has the main advantage of being free from
   (*a*) any circuit imbalances (*b*) unwanted noise
   (*c*) even-order harmonic distortion (*d*) dc magnetic saturation effects.

14. Crossover distortion occurs in........amplifiers.
(*a*) push-pull (*b*) class-A (*c*) class-B (*d*) class AB
15. The maximum overall efficiency of a class-B push-pull amplifier cannot exceed —per cent.
(*a*) 100 (*b*) 78.5 (*c*) 50 (*d*) 85
16. The dissipation at the collector is zero in the quiescent. state and increases with excitation in the case of a
(*a*) class A series-fed amplifier (*b*) class A transistor coupled amplifier
(*c*) class AB amplifier (*d*) class B amplifiers
17. Class AB operation is often used in power (large signal) amplifiers in order to,
(*a*) get maximum efficiency (*b*) remove even harmonics
(*c*) overcome cross-over distortion (*d*) reduce collector dissipation
18. In a "public address system" amplifier, the input stage is,
(*a*) a current amplifier (*b*) power amplifier
(*c*) a voltage amplifier (*d*) either (*a*), (*b*) or (*c*)
(*Bharthier Univ. April 2004*)

ANSWERS

1. (*a*)	**2.** (*b*)	**3.** (*c*)	**4.** (*b*)	**5.** (*a*)
6. (*c*)	**7.** (*d*)	**8.** (*a*)	**9.** (*b*)	**10.** (*c*)
11. (*a*)	**12.** (*d*)	**13.** (*c*)	**14.** (*a*)	**15.** (*b*)
16. (*d*)	**17.** (*a*)	**18.** (*c*)		

28

Chapter

Tuned Amplifiers

Outline

28.1. Introduction

In our day to day work, sometimes it is desired that an amplifier should be a selective one *i.e.*, the amplifier should select a desired frequency (or a narrow band of frequencies) and amplify it to the desired levels. For example, radio and television stations transmit at a particular radio frequency (*i.e.*, a frequency in the range of 30 kHz to 300 MHz) assigned to them. The radio or a television receiver is required to pick up and amplify the desired radio frequency signal, while rejecting all others. This job cannot be done by the audio amplifiers discussed in the last chapter. It is because of the fact that such amplifiers have resistive loads and consequently their gains is independent of the signal frequency over a large bandwidth. In other words, since the audio amplifier amplifies a wide range of frequencies equally well, therefore it cannot be used for the selection of a specific desired frequency, while rejecting all other frequencies.

In order to pick up and amplify the desired radio frequency signal, the resistive load in the audio amplifier is replaced by a tuned circuit (also called a parallel resonant circuit) as shown in Fig. 28.1. The tuned circuit is capable of selecting as particular frequency and rejecting the others.

Thus the use of tuned circuit in the transistor amplifier circuit, makes possible the selection and amplification of a particular desired radio frequency. Such an amplifier is called a tuned voltage amplifier.

Thus an amplifier, which amplifies a specific frequency (or a narrow band of frequencies), is called a *tuned voltage amplifier* or *simply tuned amplifier.* It serves the following two purposes:

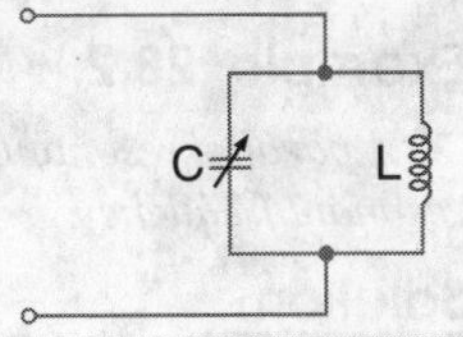

Fig. 28.1. A tuned circuit.

1. Selection of a desired radio frequency signal.
2. Amplification of the selected signal to a suitable voltage level.

28.2. Parallel Resonant Circuit

A parallel resonant circuit consists of an inductor (L) and a capacitor (C) connected in parallel to each other with respect to a supply source. Fig. 28.2 shows a parallel resonant circuit connected across an a.c. supply source of variable frequency. Here the resistance R represents the coil resistance. Its value is, usually, of the order of few ohms and therefore may be neglected as compared to the impedance of the resonant circuit.

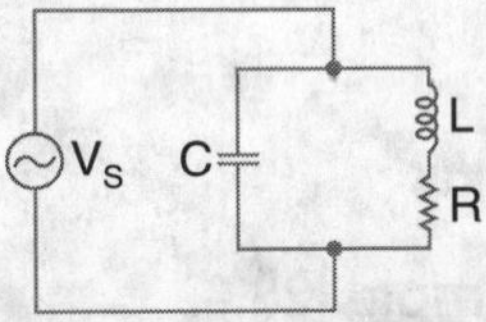

Fig. 28.2. A parallel resonant circuit connected to an a.c. supply source.

Now consider the frequency of the a.c. supply source to be varied suitably. As a result of this, the circuit will encounter different impedance at different frequencies. As the frequency is increased, the inductive reactance (X_L) is also increased and the capacitive reactance (X_C) is decreased. There is a certain frequency of the applied a.c. voltage at which the inductive reactance is equal to the capacitive reactance. This frequency is called resonance frequency and is designated by the symbol f_o. At resonance frequency, the circuit is said to be in electrical resonance. Under resonance condition, the *impedance of the resonance circuit becomes maximum and the line current (i.e., the current drawn from the source) is minimum.*

The expression for resonant frequency may be obtained from the condition,

$$X_L = X_C$$

Substituting the values of X_L equal to $2\pi f_o \cdot L$ and X_C equal to $1/2\pi f_o \cdot C$ in this equation,

$$2\pi f_o \cdot L = 1/\pi f_o \cdot C$$

$$\therefore \quad f_0 = \frac{1}{2\pi\sqrt{L \cdot C}} = \frac{0.159}{\sqrt{L \cdot C}}$$

If the value of inductance is in henrys and the capacitance is by farads, then the resonant frequency is expressed in hertz.

Example 28.1.

A parallel resonant circuit has an inductance of 150 μH and a capacitor of 100 pF. Find the value of resonant frequency. *(Anna University, May/June 2007)*

Solution.

Given: $L = 150\ \mu\text{H} = 150 \times 10^{-6}$ H and $C = 100\ \text{pF} = 100 \times 10^{-12}$ F.

We know that the resonant frequency,

$$f_o = \frac{0.159}{\sqrt{L \cdot C}} = \frac{0.159}{\sqrt{(150 \times 10^{-6}) \times (100 \times 10^{-12})}}\ \text{Hz}$$

$$= 1.3 \times 10^6\ \text{Hz} = 1.3\ \text{MHz}\ \textbf{Ans.}$$

Example 28.2.

A parallel resonant circuit has an inductance of 9 mH and a capacitance of 0.001 μF. Find the resonant frequency.

Solution.

Given: $L = 9 \text{ mH} = 9 \times 10^{-3}$ H and $C = 0.001$ μF.

We know that the value of resonant frequency,

$$f_o = \frac{1}{2\pi\sqrt{LC}}$$

$$= \frac{1}{2\pi\sqrt{(9\times10^{-3})\times(0.01\times10^{-6})}}$$

$$= 0.053 \times 10^6 \text{ Hz or } 53 \text{ kHz} \quad \textbf{Ans.}$$

Example 28.3.

Calculate the resonant frequency of a tuned amplifier whose capacitor value is 10 pF and inductor value is 1 mH. (*Anna University., May/June 2007*)

Solution.

Given: $C = 10 \text{ pF} = 10 \times 10^{-12}$ F and $L = 1 \text{ mH} = 10^{-3}$ H

We know that resonant frequency

$$f_o = \frac{1}{2\pi\sqrt{LC}}$$

$$= \frac{1}{2\pi\sqrt{(1\times10^{-3})\times(10\times10^{-12})}} = 1.59\times10^6$$

$$= 1.59 \text{ kHz} \quad \textbf{Ans.}$$

Example 28.4.

A parallel resonant circuit has a capacitor of 100 pF and an inductor of 100 mH. The inductor has a resistance of 5 ohms. Find the value of frequency at which the circuit resonates and the circuit impedance at resonance. (*Anna University., Nov./Dec. 2008*)

Solution.

Given: $C = 100 \text{ pF} = 100 \times 10^{-12}$ F; $L = 100 \text{ μH} = 100 \times 10^{-6}$ and $R = 5\ \Omega$.

We know that the resonant frequency,

$$f_o = \frac{1}{2\pi\sqrt{LC}}$$

$$= \frac{1}{2\pi\sqrt{(100\times10^{-6})\times(100\times10^{-12})}} = 1.59\times10^6 \text{ Hz}$$

$$= 1.59 \text{ MHz} \quad \textbf{Ans.}$$

We also know that impedance at resonance,

$$= \frac{L}{RC} = \frac{100\times10^{-6}}{(5)\times(100\times10^{-12})} = 200 \times 10^3\ \Omega = 200 \text{ k}\Omega \quad \textbf{Ans.}$$

28.3. Resonance Curve

It is a curve, which shows the variation of circuit impedance (or circuit current) with the change in frequency of the applied voltage.

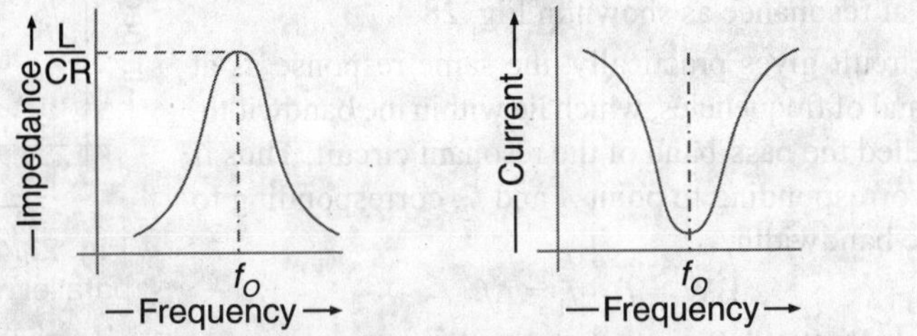

(*a*) Impedance versus frequency. (*b*) Current versus frequency.

Fig. 28.3. Resonance curves of a parallel tuned circuit.

Fig. 28.3 (*a*) shows the variation of circuit impedance (Z_p) with the change in frequency of the applied voltage. It is evident from this curve, that the impedance is maximum at the resonance and its value is equal to *L*/ *CR i.e.*,

$$Z_p = L/CR$$

As the frequency is changed, above or below the resonance, the value of impedance decreases rapidly. Figure 28.3 (*b*) shows the resonance curve, in terms of the variation of circuit current with the change in frequency of the applied voltage. It is evident from the curve that the circuit current is minimum at resonance frequency and its value is equal to *V*/(*L*/*CR*). As the frequency is changed, above or below the resonance, the value of circuit current increases rapidly.

Example 28.5.

A parallel resonant circuit consists of a capacitor of 100 pF and an inductor of 100 mH. The inductor has a resistance of 5 Ω. Find the value of frequency at which the circuit will resonate and the circuit impedance at resonance.

Solution.

Given : $C = 100 \text{ pF} = 100 \times 10^{-12}$ F; $L = 100$ μH $= 100 \times 10^{-6}$ H and R = 5 Ω.

Resonant frequency

We know that the resonant frequency,

$$f_o = \frac{0.159}{\sqrt{L \cdot C}} = \frac{0.159}{(100 \times 10^{-6}) \times (100 \times 10^{-12})}$$

$$= 1.59 \times 10^6 = 1.59 \text{ MHz } \textbf{Ans.}$$

Circuit impedance at resonance

We also know that the circuit impedance at resonance,

$$Z_p = \frac{L}{CR} \frac{100 \times 10^{-6}}{(100 \times 10^{-12}) \times 5} = 2 \times 10^5 \ \Omega = 200 \text{ k}\Omega \textbf{ Ans.}$$

28.4. Bandwidth of a Parallel Resonant Circuit

We have already discussed in the last article about the variation of circuit impedance (or current) of a parallel resonant circuit. This variation gives the resonance circuit an ability to discriminate between the frequency at resonance and those at off resonance. This discriminating ability of a resonance circuit is expressed in terms of its *bandwidth*.

The bandwidth of a resonant circuit is given by the band of frequencies, which lie between two points (*A* and *B*) on either side of the resonance curve where the impedance drops to $1/\sqrt{2}$ or 0.707 of its maximum value at resonance as shown in Fig. 28.4.

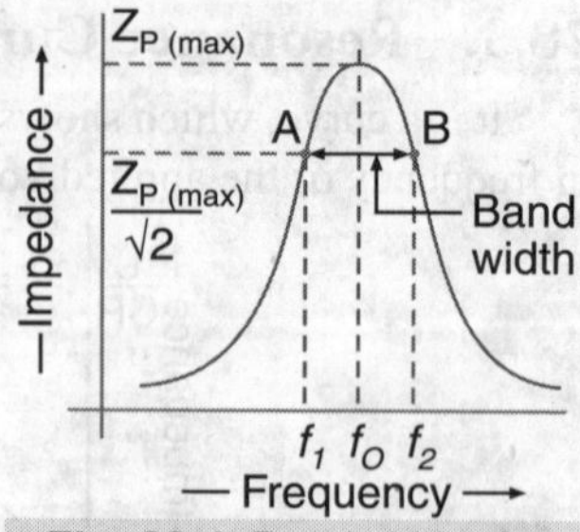

Fig. 28.4. Bandwidth of a parallel resonance circuit.

The resonant circuit gives practically, the same response as at resonance, for the band of frequencies, which lie within the bandwidth. This band is also called the pass-band of the resonant circuit. Thus if f_1 is the frequency corresponding to point *A* and f_2, corresponding to the point *B*, then the bandwidth,

$$BW = f_2 - f_1 = \Delta f$$

The points marked *A* and *B* on the resonance curve shown in Figure 28.4 are know as half power points, because the power developed at each of these points in half the power developed at resonance. Consequently, the frequencies f_1 and f_2 are known as lower half-power and upper half-power frequencies respectively.

28.5. Sharpness of Resonance

The resonance curve, of a resonant circuit, is required to be as sharp as possible in order to provide a high selectivity. The sharp resonance curve means that the impedance falls off rapidly as the frequency is varied above and below the resonant frequency. Mathematically, the sharpness of a resonance curve is defined as ratio of the bandwidth of the circuit to its resonant frequency *i.e.*,

Sharpness of resonance

$$= \frac{\text{Bandwidth}}{\text{Resonant frequency}} = \frac{BW}{f_o} = \frac{f_2 - f_1}{f_o} = \frac{1}{Q_o} \quad \text{... (i)}$$

where Q_o is called the quality factor or simply *Q-factor* of the circuit at resonance. Its value is equal to the ratio of inductive reactance (or capacitive reactance) at resonance to the circuit resistance. Mathematically, the *Q-factor,*

$$Q_o = \frac{X_L}{R} = \frac{\omega_o \cdot L}{R} = \frac{2\pi f_o \cdot L}{R}$$

where L = Value of circuit inductance, and

R = Value of circuit resistance or coil resistance.

It is evident from equation (*i*) that the sharpness of resonance (*i.e.*, selectivity) depends upon the value of *Q*-factor. Higher the value of Q_o, more selective is the resonant circuit. The value of Q_o, in turn, depends upon the value of coil resistance (*R*). Smaller the value of coil resistance, higher is the value Q_o.

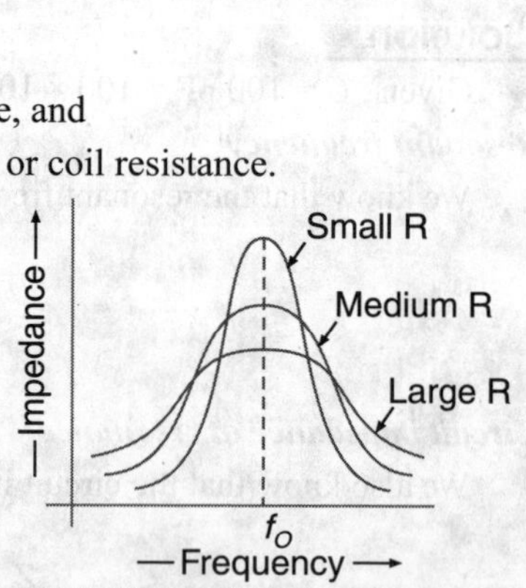

Fig. 28.5. Effect of coil resistance (*R*) on the sharpness of resonance curves.

Fig. 28.5 shows the effect of coil resistance (*R*) on the sharpness of the resonance curve. It is evident from this curve that when the coil resistance is small, the resonance curve is very sharp. On the other hand, if the coil resistance is large, the resonance curve is less sharp. It may be noted from equation (*i*), that the bandwidth,

$$BW = \frac{f_o}{Q_o} \quad \text{... (ii)}$$

or

$$f_o = BW \times Q_o \quad \text{... (iii)}$$

It is evident from equation (*ii*) that higher the value of quality factor (Q_o), the lower will be the value of bandwidth.

Thus we conclude that a higher value of quality factor provides a higher selectivity, but a smaller bandwidth, whereas a lower value of quality factor provides a poor selectivity but a larger bandwidth.

Example 28.6.

Calculate the bandwidth between the half-power points of circuit which resonates at 1MHz and has a Q of 100. (UPSC Engg. Services 2002)

Solution.

Given $f_o = 1$ MHz and $Q_o = 100$

We know that bandwidth of a circuit,

$$\text{BW} = \frac{f_o}{Q_o} = \frac{1 \text{ MHz}}{100} = 0.01 \text{ MHz} = 10 \text{kHz} \textbf{ Ans.}$$

Example 28.7.

A tuned circuit has resonant frequency of 1600 kHz and a bandwidth of 10 kHz. What is the value of its Q-factor?

Solution.

Given: $f_o = 1600$ kHz and $BW = 10$ kHz.

Let Q_o = Value of quality factor.

We know that the bandwidth (BW),

$$10 = \frac{1600}{Q_o}$$

or $Q_o = 1600/10 = 160$ **Ans.**

Example 28.8.

A tuned amplifier has its maximum gain at a frequency of 2 MHz and has a bandwidth of 50 kHz. Calculate the Q-factor.

Solution.

Given: $f_o = 2 \text{ MHz} = 2 \times 10^6$ Hz and $BW = 50 \text{ kHz} = 50 \times 10^3$ Hz.

Let Q_o = The value of Q-factor,

We know that the bandwidth (BW_o),

$$50 \times 10^3 = \frac{f_o}{Q_o} = \frac{2 \times 10^6}{Q_o}$$

or

$$Q_o = \frac{2 \times 10^6}{50 \times 10^3} = 40 \textbf{ Ans.}$$

Example 28.9.

A circuit is resonant at 455 kHz and has a 10 kHz bandwidth. The inductive reactance is 1255 Ω. What is the parallel impedance of the circuit at resonance?

Solution.

Given: $f_o = 455 \text{ kHz} = 455 \times 10^3$ Hz; $\text{BW} = 10 \text{ kHz} = 10 \times 10^3$ Hz and $X_L = 1255\ \Omega$.

Let Z_P = Value of impedance at resonance

We know that the value of bandwidth (BW),

$$10 \times 10^3 = \frac{f_o}{Q_o} = \frac{455 \times 10^3}{Q_o}$$

$$\therefore \quad Q_o = \frac{455 \times 10^3}{10 \times 10^3} = 45.5$$

and value of Q-factor (Q_o),

$$45.5 = \frac{X_L}{R} = \frac{1255}{R}$$

$$\therefore \quad R = 1255/45.5 = 27.6\ \Omega$$

We also know that the inductive reactance (X_L),

$$1255 = 2\pi f_o \cdot L = 2\pi \times (455 \times 10^3) \times L = 2859 \times 10^3\ L$$

$$L = 1255/\,(2859 \times 10^3) = 0.439 \times 10^{-3}\ \text{H}$$

and value of capacitive reactance at resonance,

$$X_C = X_L\ 1255\ \Omega$$

or

$$\frac{1}{2\pi f_o \cdot C} = 1255$$

$$\frac{1}{2\pi \times (455 \times 10^3) \times C} = 1255$$

or

$$\frac{1}{(2859 \times 10^3)\ C} = 1255$$

$$\therefore \quad C = \frac{1}{(1255 \times 2859 \times 10^3)} = 278.7 \times 10^{-12}\ \text{F}$$

and value of circuit impedance at resonance,

$$Z_P = \frac{1}{(1255 \times 2859 \times 10^3)} = 278.7 \times 10^{-12}\ \text{F}$$ **Ans.**

28.6. Single-tuned Voltage Amplifier

Fig. 28.6 shows the circuit of a single-tuned voltage amplifier using a bipolar junction transistor (BJT). The circuit shown in Fig. 28.6 (*a*) is called capacitively coupled tuned amplifier because the output is taken through a coupling capacitor (C_c). Similarly, the circuit shown in Fig. 28.6 (*b*) is called

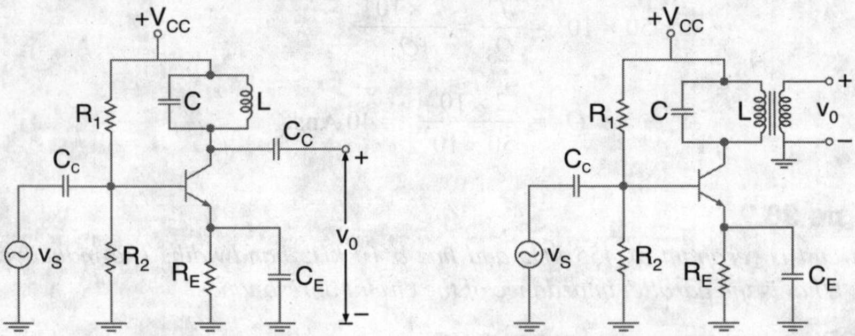

(*a*) Capacitively coupled. (*b*) Inductively coupled.

Fig. 28.6. Single-tuned voltage amplifier.

inductively coupled tuned amplifier because the output is taken across an inductor. Both these circuits consist of a transistor amplifier and tuned circuit as the load. The values of capacitance (*C*) and inductance (*L*) of the tuned circuit are selected in such a way that the resonant frequency of the tuned circuit is equal to the frequency to be selected and amplified. The resistors R_1, R_2 and R_E are called biasing resistors. These resistors provide the d.c. operating currents and voltages for the transistor.

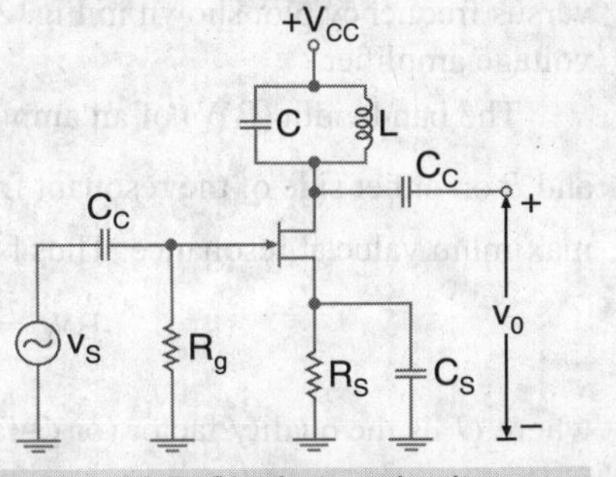

Fig. 28.7. Single-tuned voltage amplifier using field-effect transistor.

It will be interesting to know that the other active devices like field-effect transistor (FET's) and metal-oxide semiconductor field-effect transistors (MOSFET's) may also be used in tuned voltage amplifier circuit instead of bipolar junction transistor. Fig. 28.7 shows the circuit of single-tuned voltage amplifier using a field effect transistor.

The working of tuned voltage amplifier may be understood by considering a radio frequency signal, to be amplified, applied at the input of the amplifier. The resonant frequency, of the tuned circuit, is made equal to the frequency of the input signal by changing the value of capacitance (*C*) or inductance (*L*). When the frequency of the tuned circuit becomes equal to that of the input signal, a large signal appears across the output terminals. It may be noted that if the input signal is a complex wave (*i.e.*, it contains many frequency components), in that case the signal with frequency equal to the resonant frequency will be amplified. And all the other frequencies will be rejected by the tuned circuit.

28.7. Frequency Response of Single-tuned Voltage Amplifier

We have already discussed in the chapter on single stage transistors amplifier that the voltage gain of an amplifier depends upon current gain (β), input resistance (R_i) and effective or a.c. load resistance. The voltage gain is given by the relation,

$$A_v = \beta \times \frac{r_L}{R_i}$$

We know that the a.c. load resistance of a parallel resonant circuit (*i.e.*, tuned circuit) is given by the relation,

$$r_L = Z_P = \frac{L}{CR}$$

where L = Value of inductance,

C = Value of capacitance, and

R = Value of effective resistance of the inductor.

∴ Voltage gain of a tuned voltage amplifier is given by the relation,

$$A_v = \beta \times \frac{\frac{L}{CR}}{R_i}$$

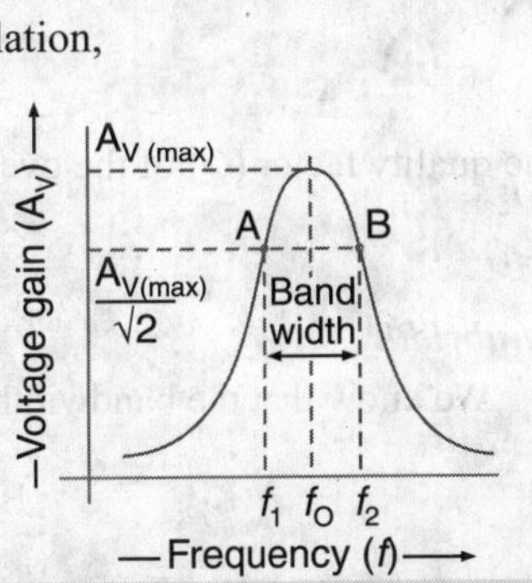

Fig. 28.8. Frequency response curve.

We also know that the value of the quantity *L*/*CR* (called impedance of the tuned circuit) is very high at the resonant frequency and it decreases as the frequency changes above or below the resonant frequency. Therefore voltage gain of a tuned amplifier is very high at the resonant frequency and decreases as the frequency changes above or below the resonant frequency.

The above facts are shown in the form of a voltage gain

versus frequency plot shown in Fig. 28.8. Such a plot is called frequency response curve of a tuned voltage amplifier.

The bandwidth (BW) of an amplifier is equal to the frequency difference between the points A and B on either side of the resonant frequency, where the value of voltage gain drops to $1/\sqrt{2}$ of its maximum value at resonance. Thus bandwidth,

$$\text{BW} = \Delta f = f_2 - f_1 = \frac{f_o}{Q_o}$$

where Q_o is the quality factor (or Q-factor) of the tuned circuit.

Example 28.10.

Calculate the bandwidth, maximum dissipated power in the transistor and maximum output power for the tuned amplifier circuits shown in Fig. 28.9. (*UPSC Engg. Services 2002*)

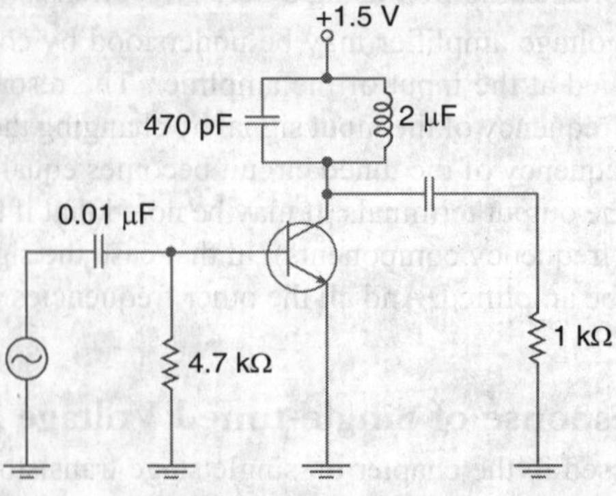

Fig. 28.9.

Solution.

Given: $V_{CC} = 15$ V; $C = 470$ pF $= 470 \times 10^{-12}$ F; $L = 2$ µH $= 2 \times 10^{-6}$ H; $C_1 = 0.01$ µF $= 0.01 \times 10^{-6}$ F; $C_2 = 1000$ pF $= 1000 \times 10^{-12}$ F; $R_B = 4.7$ kΩ $= 4.7 \times 10^3$ Ω; $R_L = 1$ kΩ $= 1 \times 10^3$ Ω and $Q_L = 100$.

We know that the resonant frequency,

$$f_o = \frac{0.159}{\sqrt{L \cdot C}} = \frac{0.159}{\sqrt{(2 \times 10^{-6}) \times (470 \times 10^{-12})}} = 5.19 \times 10^6 \text{ Hz}$$

and the a.c. collector resistance of the tuned amplifier,

$$\text{rc} = Q_L \omega_o L \,\|\, R_L = Q_L 2\pi f_o L \,\|\, R_L$$

$$= (100 \times 2\pi \times 5.19 \times 10^6 \times 2 \times 10^{-6}) \,\|\, (1 \times 10^3) = 867\ \Omega$$

The quality factor (Q) of the overall circuit is given by,

$$Q = \frac{r_c}{2\pi f_o L} = \frac{867}{2\pi \times (5.19 \times 10^6) \times (2 \times 10^{-6})} = 13.29$$

Bandwidth

We know that the bandwidth (BW),

$$\text{BW} = \frac{f_o}{Q} = \frac{5.19 \times 10^6}{13.29} = 390 \times 10^3 \text{ Hz} = 390 \text{ kHz } \textbf{Ans.}$$

Power dissipated in transistor,

We know that the maximum dissipated power in transistor,

$$p_{tr} = \frac{V_{PP}^2}{40\, r_c} = \frac{(30)^2}{40 \times 867} = 26 \times 10^{-3}\ \text{W} = 26\ \text{mW}\ \textbf{Ans.}$$

Maximum output power

We know that the maximum output power for the tuned amplifier,

$$P_{o(\max)} = \frac{V_{rms}^2}{R_L} = \frac{V_{PP}^2}{8 \times R_L} = \frac{(2 \times V_{CC})^2}{8 \times R_L}$$

$$= \frac{(2 \times 15)^2}{8 \times (1 \times 10^3)} = 112.5 \times 10^{-3}\ \text{W} = 112.5\ \text{mW}\ \textbf{Ans.}$$

28.8. Limitations of Single-tuned Voltage Amplifier

The tuned voltage amplifier are used in the wireless communication systems. Generally, the communication transmitters do not transmit the audio (*i.e.*, voice or music) signal directly. Instead the audio signal is raised to some high frequency level. This high frequency is in the radio frequency (RF) range and it serves as a carrier of the audio signal. The process of raising the audio signal to the radio frequency level is called modulation. The communication transmitters transmit the modulated wave.

The modulated wave has a relatively narrow band of frequencies centred around the carrier frequency (f_c) as shown in Fig. 28.10 (*a*). The frequencies (f_1 and f_2) in the narrow band actually represent the information and hence is an important part of the transmitted signal.

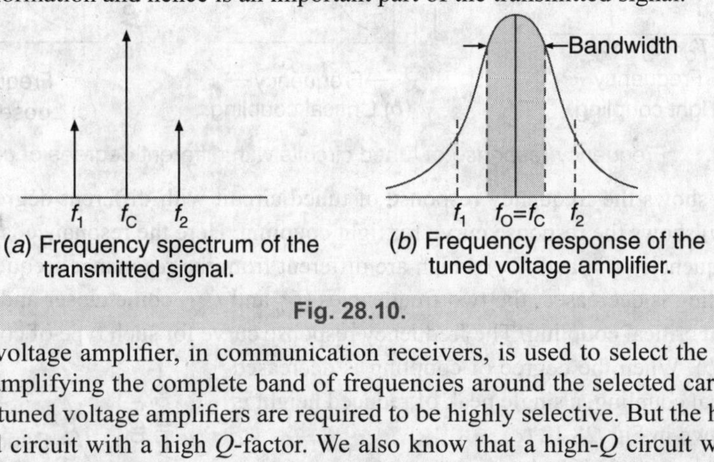

(*a*) Frequency spectrum of the transmitted signal. (*b*) Frequency response of the tuned voltage amplifier.

Fig. 28.10.

The tuned voltage amplifier, in communication receivers, is used to select the desired carrier frequency and amplifying the complete band of frequencies around the selected carrier frequency. In other words, tuned voltage amplifiers are required to be highly selective. But the high selectivity requires a tuned circuit with a high *Q*-factor. We also know that a high-*Q* circuit will give a high voltage gain. But, at the same time, it will give reduced bandwidth (because bandwidth is reciprocal to the *Q*-factor). It means that a tuned voltage amplifier with reduced bandwidth may not be able to amplify equally the complete band of the transmitted signal as shown in Fig. 28.10 (*b*). In other words, narrow bandwidth or smaller pass-band of the amplifier will result in a poor reproduction of the audio signal. It is the major limitation of a single tuned, voltage amplifier and is overcome by using a double tuned circuit. In double-tuned voltage amplifiers, we use two tuned circuits which are inductively coupled to each other.

28.9. Double-tuned Voltage Amplifier

Fig. 28.11 shows the circuit of a double tuned voltage amplifier. It consists of a transistor amplifier with two tuned circuits. One of the tuned circuits (*i.e.*, L_1 C_1) is shown as the collector load and other (*i.e.*, L_2C_2) as the output. The resistor R_1, R_2 and R_E are used to provided d.c. currents and voltages for the transistor operation.

The signal to be amplified is applied at the input terminal through the coupling capacitor (C_C).

The resonant frequency of the tuned circuit L_1C_1 is made equal to that of the signal (by adjusting L_1 or C_1). Under these conditions, the tuned circuit offers a very high impedance to the input signal. As a result of this, a large output appears across the tuned circuit L_1C_1. The output from this tuned circuit is inductively coupled to the L_2C_2 tuned circuit.

The double tuned voltage amplifiers are extensively used in intermediate frequency (IF) amplifiers in radio and television receivers.

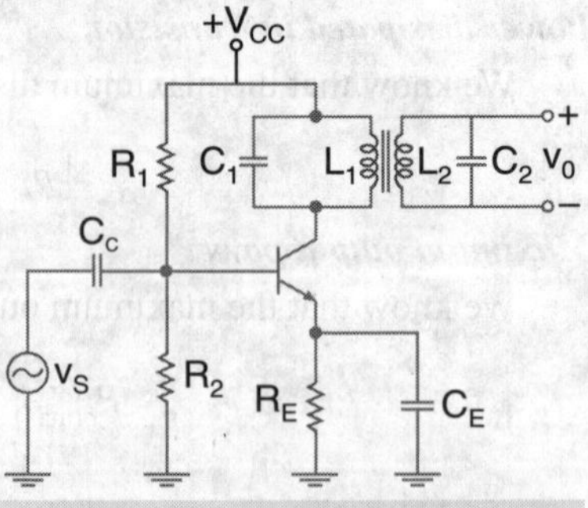

Fig. 28.11. Double tuned voltage amplifier.

28.10. Frequency Response of Double-tuned Voltage Amplifier

The frequency response of a double tuned voltage amplifier depends upon its degree of coupling (also called coefficient of coupling) between the two tuned circuits. The degree of coupling gives us an idea of the amount of energy transferred between the two tuned circuits.

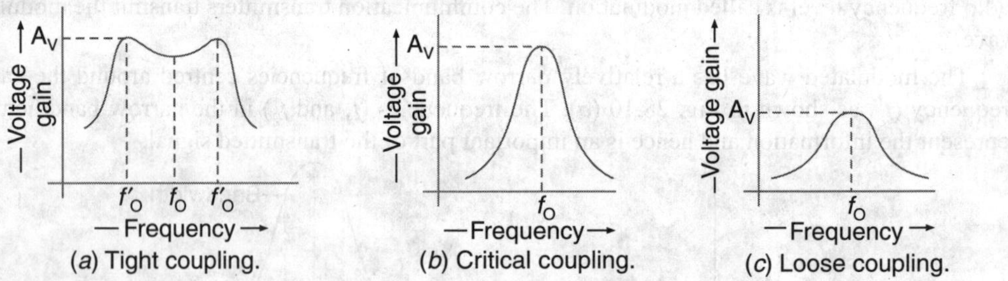

Fig. 28.12. Frequency response of tuned circuits with different degrees of coupling.

Fig. 28.12 shows the frequency response of tuned circuit with different degrees of coupling. Figure 28.12 (*a*) shows the response curve for tight coupling. Here the resonance is shown to occur at two new frequencies (f'_o and f''_o) which are different from the resonance frequency (f_o). As the degree of coupling is decreased, the two frequencies (f'_o and f''_o) come closer and merge into one frequency (f_o) at critical coupling. The frequency response curve for such type of coupling is shown in Fig. 28.12 (*b*). When the degree of coupling is decreased below the critical coupling, a single peak of reduced height is obtained as shown in Fig. 28.12 (*c*).

The tuned circuits of a double-tuned voltage amplifier are tightly coupled to each other. Therefore the frequency response of a double-tuned voltage amplifier is shown in Fig. 28.13. This type of response curve provides a high selectivity, high gain and relatively large bandwidth to the tuned voltage amplifier.

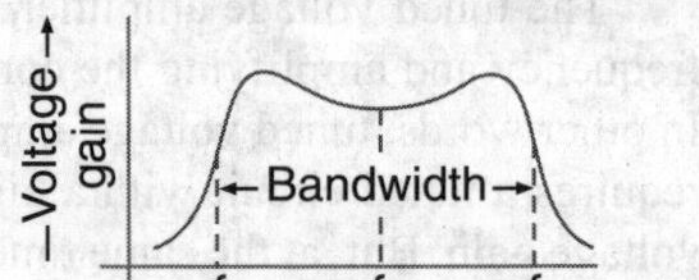

Fig. 28.13. Frequency response of a double-tuned voltage amplifier.

28.11. Stagger-tuned Voltage Amplifier

It has been observed that if two or more tuned circuits, which are synchronously tuned are cascaded, the overall bandwidth decreases. However, if the different tuned circuits which are cascaded, are tuned to slightly different frequencies, it is possible to obtain an increased bandwidth with a flat pass-band with steep sides. This technique is known as *stagger-tuning*.

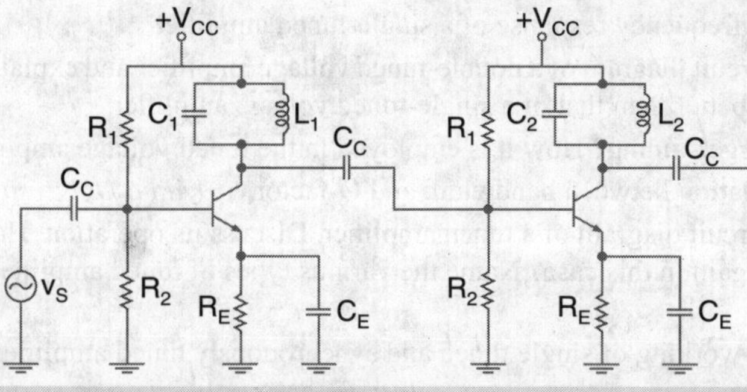

Fig. 28.14. A two-stage stagger tuned voltage amplifier.

Fig. 28.14 shows a two-stage tuned voltage amplifier. The stagger tuning in this circuit, may be achieved by resonating the tuned circuits L_1C_1 and L_2C_2 to slightly different frequencies.

Fig. 28.15 shows the frequency response of a stagger-tuned amplifier. In this figure, curve '*a*' shows the gain versus frequency response of the L_1C_1 tuned circuit. Similarly, the curve '*b*' shows the response of the L_2C_2 tuned circuit. The curve '*c*' indicates the combined (or the overall) response of the circuit. It is evident from this curve that the amplifier has a greater bandwidth and flatter passband. It has been found that more the number of stages used, the flatter will be the pass-band and steeper will be the gain fall-off outside the pass-band. It may be noted that because of stagger-tuning, there is a loss of voltage gain (the gain reduces from A_v to A'_v). If an optimum stagger-tuning is employed, the response curve of the amplifier is very close to a rectangular response curve. Such a response curve is known as Butterworth response.

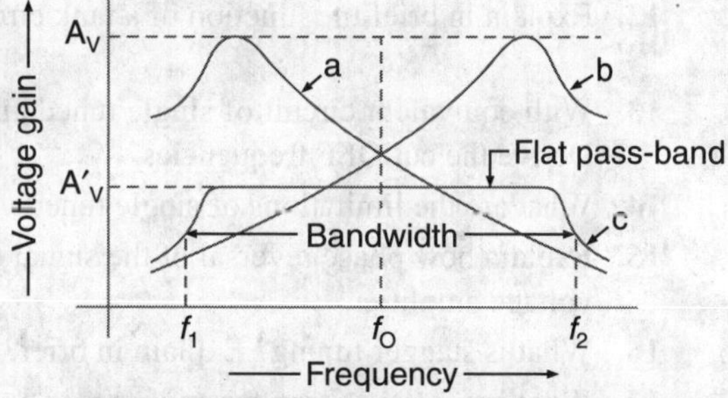

Fig. 28.15. Frequency response of a stagger-tuned voltage amplifier.

NUMERICAL PROBLEMS

1. A parallel resonant circuit has an inductance of 100 mH and a capacitor of 100 pF. Find the value of resonant frequency. **(Ans. 1.59 MHz)**
2. A parallel *LC* circuit consists of a capacitor of 3000 pF and an inductor of 100 mH. The inductor has a resistance of 5 W. Find the value of resonant frequency and the circuit impedance at resonance. **(Ans. 291 kHz and 6.66 kW)**
3. A tuned circuit has resonant frequency of 800 kHz and a band width of 10 kHz. What is the value of its *Q*-factor? **(Ans. 80)**
4. A parallel resonant circuit has a band width of 15 kHz and *Q*-factor of 120. What is the resonant frequency of the circuit? **(Ans. 1800 kHz)**

DESCRIPTIVE QUESTIONS

1. What is a tuned voltage amplifier? Explain it briefly.
2. Derive an expression for the resonance frequency of a parallel resonant circuit.
3. Explain the meaning of a resonance curve. Plot a resonance curve for a parallel tuned circuit.

4. Discuss the frequency response of a single-tuned amplifier.
5. Draw the circuit diagram of a double-tuned voltage amplifier and explain how its frequency response is better than that of a single-tuned voltage amplifier.
6. What is stagger-tuning? How it is employed in the tuned voltage amplifier?
7. Brief the relation between bandwidth and Q-factor. *(Anna University., May/June 2007)*
8. Draw the circuit diagram of a tuned amplifier. Discuss its operation. How current gain and voltage are gain in this case? Name the various types of tuned amplifier. *(PTU., May 2009)*
9. Discuss the working of single tuned and synchronously tuned amplifiers. *(Anna University., May/June 2007, Nov/Dec 2007)*
10. Explain single tuned voltage amplifier and discuss its frequency response. *(Anna University., Nov/Dec 2006)*
11. Mention two applications of tuned amplifiers. *(Anna University., Nov/Dec 2007)*
12. Explain in brief the function of a tank circuit in a tuned- voltage amplifier. *(PTU., May 2007, Dec 2009)*
13. With equivalent circuit of single tuned amplifier derive the gain as function of frequency. Derive the cut OFF frequencies. *(Anna University., May/June 2006)*
14. What are the limitations of single tuned voltage amplifier? *(PTU., Dec 2009)*
15. Explain how phase reversal of the signal takes place when it is amplified by a single-stage voltage amplifier. *(PTU., Dec 2006)*
16. What is stagger tuning? Explain in brief? *(PTU., Dec 2005)*
17. What are the synchronous and stagger tuned amplifier circuits. *(Anna University., May/June 2007)*
18. Write shorts notes on stagger tuned amplifier. *(Anna University., May/June 2007)*
19. What is the advantage of Stagger tuned Amplifier? *(PTU., Dec 2007, May 2008)*
20. Explain the working and frequency response of single tuned amplifier circuit. *(Anna University., May/June 2007)*
21. Draw the circuit of a Single tuned amplifier. Draw its frequency response. Using small signal equivalent circuit derives expressions for gain at resonance and 3 dB bandwidth. *(Anna University., Nov/Dec 2007)*
22. Draw the equivalent circuit of capacitance coupled single tuned amplifier and derive the expression fir the bandwidth. *(Anna University., May/June 2007)*
23. Explain class 'C' tuned amplifier derives its efficiency. *(Anna University., May/June 2007)*
24. Draw the circuit of class C tuned amplifier and explain its operation with relevant waveforms. *(Anna University., April/May 2008)*
25. Discuss instability of tuned amplifier. *(Anna University., April/May 2008)*
26. Mention applications of tuned amplifiers. *(Anna University., Nov/Dec 2008)*
27. Explain the single and double tuned voltage amplifier. *(RGTU., June 2011)*
28. What do you meant by critical coupling in double tuned amplifier? *(RGTU., June 2011)*
29. Why do we used tuned amplifiers in the IF and RF range? *(RGTU., June 2011)*
30. Explain the working of the following:
    (*a*) Double tuned amplifier
    (*b*) Stagger tuned amplifier *(RGTU., June 2010)*

31. Compair single tuned and double tuned circuits. *(RGTU., June 2010)*

32. Discuss the characteristics and frequency response of tunned amplifiers. *(RGTU., June 2009)*

33. Discuss the characteristics and frequency response of stagger tuned amplifier. Compare it with rest of the tuned amplifiers. *(RGTU., Dec. 2009)*

MULTIPLE CHOICE QUESTIONS

1. Which of the following amplifiers should be used to amplify a radio frequency signal?
(*a*) Audio-frequency amplifier (*b*) Push-pull amplifier
(*c*) Video amplifier (*d*) Tuned voltage amplifier

2. An amplifier has a pass-band from 745 kHz to 755 kHz, and a resonant frequency of 750 kHz. Its bandwidth is
(*a*) 5 kHz (*b*) 10 kHz (*c*) 20 kHz (*d*) None of the above

3. If the *Q*-factor of a resonant circuit is doubled, its bandwidth is
(*a*) Halved (*b*) Doubled (*c*) Quadrupled (*d*) Remains the same

4. A tuned circuit has resonant frequency of 1 MHz. To get a bandwidth of 20 kHz, its *Q*-factor must be equal to
(*a*) 20 (*b*) 50 (*c*) 100 (*d*) 500

5. Stagger-tuning in tuned amplifiers is achieved by using
(*a*) Double tuned circuit
(*b*) Tuned circuits which are tuned to same frequency
(*c*) Tuned circuits which are tuned to slightly different frequencies
(*d*) None of the above

6. If a *Q* of a single-stage tuned amplifier is doubled, than it bandwidth will
(*a*) remain same (*b*) become half (*c*) become double (*d*) become four times
(UPSC Engg. Services 2000)

ANSWERS

1. (*d*) **2.** (*b*) **3.** (*a*) **4.** (*b*)
5. (*c*) **6.** (*b*)

29

Chapter

Feedback Amplifiers

Outline

29.1. Introduction

An ideal linear amplifier, in the midfrequency region, provides an output signal, which is an *exact replica* of the input signal. However, the practical amplifiers depart from the ideal for many reasons, *e.g.*, non-linearity of transistor characteristics, parameter variation, temperature effects etc. Some of these factors can be minimized by improving the involved basic devices. But there is a limit, beyond which this approach does not work. The best approach is to use the *principle of feedback* to achieve a desired degree of improvement. The *feedback* is a process of injecting some energy from the output and then return it back to the input. The amplifiers, which use the feedback principle, are called *feedback amplifiers*.

A feedback amplifier, basically, consists of two parts namely amplifier circuit and feedback circuit (or network). Depending upon whether the feedback signal increases or decreases the input signal, there are two basic types of feedback in amplifiers:

1. *Positive feedback.* If the feedback signal (voltage or current) is applied in such a way that it is *in phase with the input signal* and thus increases it, then it is called a positive feedback. Sometimes, it is also known as regenerative feedback or direct feedback.

The positive feedback increases the gain of amplifier. However, if produces excessive distortion due to which it is seldom used in amplifiers. The positive feedback is used in oscillators.

2. *Negative Feedback.* If the feedback signal (*i.e.*, voltage or current) is applied in such a way that it is *out of phase with the input signal* and thus decreases it, then it is called a negative feedback. Sometimes, it is also called as degenerative feedback or inverse feedback.

The negative feedback reduces gain of the amplifier. However, it improves the amplifier performance in many other respects. Thus a negative feedback is frequently used in small-signal as well as with large-signal amplifier circuits.

29.2. Principle of Feedback Amplifiers

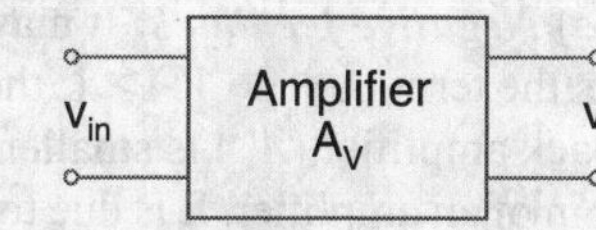

Fig. 29.1. An ordinary amplifier with voltage gain.

Fig. 29.1 shows an ordinary amplifier *i.e.*, an amplifier which does not employ any feedback.

We know that the voltage gain of such an amplifier is given by the ratio of the output voltage (V_o) to the input voltage (V_{in}). Mathematically, the voltage gain,

$$A_v = \frac{V_o}{V_{in}}$$

The voltage gain (A_v) is also known as open loop gain. Now let us add a feedback circuit to the amplifier as shown in Figure 29.2. The addition of feedback circuit changes the values of amplifier gain, its input as well as the output voltages.

Let V'_o = Output voltage of feedback amplifier.

β = Fraction of the output voltage fed to the input. It is called feedback ratio and is different from common emitter transistor current gain, and

A'_v = Voltage gain of feedback amplifier.

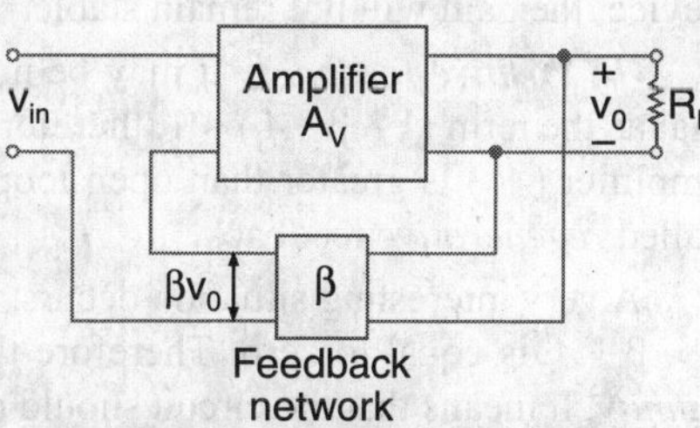

Fig. 29.2. Block diagram of a feedback amplifier.

The feedback circuit injects a fraction (β) of the output voltage (V'_o) and return it to the input voltage, whose value changes to $V_{in} \pm \beta \cdot V'_o$. It will be interesting to know that the value at input voltage is $V_{in} + \beta V'_0$ if the feedback voltage is in phase with the input voltage. And it is equal to $V_{in} - \beta \cdot V'_o$, if the feedback voltage is out of phase with the input voltage. This input voltage is amplified by the circuit and its value at the output,

$$A_v (V_{in} + \beta \cdot V'_o) = V'_o$$

Rearranging the above equation,

$$V'_o (1 - \beta \cdot A_v) = A_v \cdot V_{in}$$

or

$$\frac{V'_o}{V_{in}} = \frac{A_v}{1 - \beta \cdot A_v} = A'_v$$

where A'_v is equal to V'_o/V_{in} and is the voltage gain of the feedback amplifier.

For a positive feedback, the feedback ratio (β) is positive and for negative feedback, it is negative. Therefore voltage gain of the feedback amplifier,

$$A'_v = \frac{A_v}{1 - \beta \cdot A_v} \quad \text{... (For positive feedback)}$$

$$= \frac{A_v}{1 - (-\beta \cdot A_v)} = \frac{A_v}{1 + \beta \cdot A_v} \quad \text{... (For negative feedback)}$$

The term ($\beta \cdot A_v$) is called *loop gain*. The difference between unity and the loop gain (*i.e.*, $1 - \beta \cdot A_v$) is called return difference (or *desentivity*). The amount of feedback introduced into an amplifier may also be expressed in decibels by the expression,

Feedback in dB,

$$= 20 \log_{10}\left(\frac{A'_v}{A_v}\right) = 20 \log_{10}\left(\frac{1}{1 + \beta \cdot A_v}\right)$$

Now we shall discuss in more details the effect of negative feedback and positive feedback on the amplifiers.

(*a*) *Negative feedback.* It may be noted from the voltage gain expression of negative feedback that as the term $(1 + \beta \cdot V_v) > 1$, the value of A'_v is smaller that A_v. In other words, gain of a negative feedback amplifier (A'_v) is smaller than the open loop gain (A_v). Thus the negative feedback reduces the gain of an amplifier. It is due to this fact that a negative feedback is called *degenerative* feedback. The value of closed loop gain (A'_v) may also be looked from the following angle:

We know that the term $\beta \cdot A_v >> 1$. Therefore $1 + \beta \cdot A_v \approx \beta \cdot A_v$.

Thus
$$A'_v = \frac{A_v}{\beta \cdot A_v} = \frac{1}{\beta}$$

The above relation indicates that the voltage gain of a negative feedback amplifier (A'_v) depends only upon the feedback ratio (β). Though the value of voltage gain of a negative feedback amplifier (A'_v) is quite low as compared to (A_v), yet it is found to be very stable. It means that the voltage gain (A'_v) is not affected by any change in device parameters, supply voltage and from aging of circuit components etc.

It may be noted that stability of voltage gain (A'_v) comes only if the feedback network is made up of stable elements such as resistors, inductors and capacitors. If the feedback network has an active device, the gain will not remain stable.

(*b*) *Positive feedback.* It may be noted from the voltage gain expression of positive feedback that as the term $(1 - \beta \cdot A_v) < 1$, therefore $A'_v > A_v$. It shows that voltage gain of a positive feedback amplifier (A'_v) is greater than open loop gain (A_v). It is due to this fact that a positive feedback is called *regenerative* feedback.

A very interesting situation occurs, when the product b $\cdot A_v$ becomes equal to unity. In that case $1 - \beta \cdot A_v$ is equal to zero. Therefore the voltage gain of a positive feedback amplifier becomes-*infinity*. It means that the circuit should give an output without any input. But in actual practice, it is not possible. As a matter of fact, the amplifier becomes an oscillator, which supplies its own input.

29.3. Advantages and Disadvantages of Negative Feedback

Though there are numerous advantages of employing negative feedback in the amplifier, yet some of the important are given below:

1. Increased stability.
2. Increased bandwidth.
3. Less amplitude and harmonic distortion.
4. Decreased noise.
5. Less frequency distortion.
6. Less phase distortion.
7. Input and output resistances can be modified as desired.

The only disadvantage of negative feedback is that it reduces the amplifier gain. However, this disadvantage is outweighed due to numerous advantages obtained through negative feedback.

Example 29.1.

The voltage gain of certain amplifier without feedback is 400. If the feedback ratio (β) is equal to 0.1, find the voltage gain of the amplifier with negative feedback.

(Anna University, Nov/Dec. 2000)

Solution.

Given: $A_v = 400$ and $\beta = 0.1$

We know that the voltage gain of an amplifier with negative feedback,

$$A'_v = \frac{A_v}{1 + \beta \cdot A_v} = \frac{400}{1 + (0.1 \times 400)} = \frac{400}{41} = 9.76 \textbf{ Ans.}$$

Example 29.2.

Calculate the gain of a negative feedback amplifier having A = – 2000 and β = – 1/10.

(Jamia University., 2008)

Solution.

Given $A = -2000$ and $\beta = -1/10$.

We know that for negative feedback amplifier,

$$A' = \frac{A}{1+\beta A} = \frac{-2000}{1+\left(\frac{1}{10}\times(-2000)\right)} = 9.95 \textbf{ Ans.}$$

Note: We need not consider the negative sign with the value of β.

Example 29.3.

The voltage gain of an amplifier without feedback is 3000. Calculate the voltage gain of the amplifier if negative feedback is introduced in the circuit. Given the feedback fraction β = 0.01.

(PTU., May 2006)

Solution.

Given: $A_v = 3000$ and $\beta = 0.01$.

We know that the voltage gain,

$$A'_v = \frac{A_v}{1+\beta . A_v} = \frac{3000}{1+(0.01\times 3000)} = 96.77 \textbf{ Ans.}$$

Example 29.4.

Calculate the gain of negative feedback amplifier with an internal gain A_v = 75 and feedback fraction β = 1/15. What will be the gain if A_v doubles? *(PTU., Dec 2005)*

Solution.

Given: $A_v = 75$ and $\beta = 1/15$.

We know that the voltage gain,

$$A'_v = \frac{A_v}{1+\beta . A_v} = \frac{75}{1+\left(\frac{1}{15}\right)\times 75} = 12.5 \textbf{ Ans.}$$

if A_v is doubled, $A_v = 150$

$$A'_v = \frac{A_v}{1+\beta . A_v} = \frac{75}{1+\left(\frac{1}{15}\right)\times 150} = 13.63 \textbf{ Ans.}$$

There is a small increase in feedback gain from 12.5 to 13.63.

Example 29.5.

The voltage gain of an amplifier without feedback and negative feedback respectively are 100 and 20. Calculate the percentage of negative feedback β. (*UPSC Engg. Sources 2000*)

Solution.

Given : $A_v = 100; A'_v = 20$

We know that voltage gain of an amplifier with feedback (A'_v)

$$20 = \frac{A_v}{1+\beta \,.\, A_v} = \frac{100}{1+\beta \times 100}$$

$$\therefore \quad 20\,(1 + \beta \times 100) = 100$$

$$1 + 100\,\beta = \frac{100}{20} = 5$$

or $$100\,\beta = 5 - 1 = 4$$

$$\therefore \quad \beta = \frac{4}{100} = 0.04 \text{ or } 4\% \textbf{ Ans.}$$

Example 29.6.

An amplifier has a voltage gain of 1000. With negative feedback, the voltage gain reduces to 10. Calculate the fraction of the output that is feedback to the input.

Solution.

Given: $A_v = 1000$ and $A'_v = 10$

Let β = Feedback ratio.

We know that the voltage gain with feedback (A'_v)

$$10 = \frac{A_v}{1+\beta \cdot A_v} = \frac{1000}{1+(\beta \times 1000)}$$

or $$1 + 1000\,\beta = \frac{1000}{10} = 100$$

$$\therefore \quad \beta = \frac{99}{1000} = 0.099 \textbf{ Ans.}$$

Example 29.7.

An amplifier with negative feedback gives an output of 12.5 V with an input of 1.5 V. When feedback is removed, it requires 0.25 V input for the same output. Find (a) value of voltage gain without feedback; and (b) value of β, if the input and output are in phase and β is real.

Solution.

Given: $V'_o = V_o = 12.5$ volt; $V'_{in} = 1.5$ volt and $V_{in} = 0.25$ volt.

(*a*) ***Voltage gain without feedback***

We know that voltage gain of an amplifier without negative feedback

$$A_v = \frac{V_o}{V_{in}} = \frac{12.5}{0.25} = 50 \textbf{ Ans.}$$

(*b*) ***Value of β when input and output are in phase***

We know that the voltage gain with negative feedback,

$$A'_v = \frac{V'_o}{V'_{in}} = \frac{12.5}{1.5} = 8.33 \text{ V}$$

and the voltage gain with negative feedback, (A'_v)

$$8.33 = \frac{A_v}{1 + \beta \cdot A_v} = \frac{50}{1 + (\beta \times 50)}$$

or $$8.33\,(1 + 50\,\beta) = 50$$

$$1 + 50\,\beta = \frac{50}{8.33} = 6$$

$\therefore$ $$\beta = \frac{(6 - 1)}{50} = 0.1 \textbf{ Ans.}$$

Example 29.8.

With a negative feedback, an amplifier gives an output of 10 V with an input of 0.5V. When the feedback is removed, it requires 0.25 V input for the same output. Calculate (a) gain without feedback (b) feedback fraction β. *(Anna University., May/June 2007)*

Solution.

Given: With negative feedback, $V_o = 10$ V; $V_{in} = 0.5$ V; without feedback, $V_o = 10$ V, $V_{in} = 0.25$ V

(a) Calculation of Voltage gain without feedback

We know that voltage gain of an amplifier without negative feedback

$$A_v = \frac{V_o}{V_{in}} = \frac{10}{0.25} = 40 \textbf{ Ans.}$$

(b) Calculation of voltage gain and feedback fraction β with feedback

We know that the voltage gain with negative feedback,

$$A_v = \frac{V'_o}{V'_{in}} = \frac{10}{0.5} = 20 \textbf{ Ans.}$$

and the voltage gain with negative feedback, (A_v')

$$20 = \frac{A_v}{1 + \beta . A_v} = \frac{40}{1 + \beta \times 40}$$

or $$20\,(1 + \beta \times 40) = 40$$

$\therefore$ $$\beta = 0.025 \textbf{ Ans.}$$

Example 29.9.

An engineer designs an amplifier to have a voltage gain of 80, but when constructed it gives only a voltage gain of 60. A technician suggests that a positive feedback may be introduced to increase the gain to the required level, (a) How much feedback should be given to increase the voltage gain to the required level? (b) What disadvantage may arise due to this type of operation? (c) What percentage of feedback will cause oscillation?

Solution.

Given: $A_v = 60$ and $A'_v = 80$

(a) Value of feedback ratio

Let β = Value of feedback ratio required to increase the voltage gain from 60 to 80.

We know that voltage gain with positive feedback (A'_v)

$$80 = \frac{A_v}{1 - \beta \cdot A_v} = \frac{60}{1 - \beta \cdot 60} = \frac{60}{1 - 60\,\beta}$$

$$80\,(1 - 60\,\beta) = 60$$

$$1 - 60\,\beta = \frac{60}{80} = 0.75$$

$$\therefore \quad 60\,\beta = 1 - 0.75 = 0.25$$

or $$\beta = \frac{0.25}{60} = 0.004 \textbf{ Ans.}$$

(*b*) *Disadvantage of employing positive feedback*

We know that the amplifier with positive feedback cannot have a stable voltage gain, because its gain increases with feedback. The increased voltage gain will further increase the feedback. This process will continue till the feedback ($\beta \cdot A_v$) becomes equal to unity, Under this condition, the amplifier will start oscillating. In other words, the amplifier becomes an oscillator. **Ans.**

(*c*) *Percentage of feedback which causes oscillation*

We know that the condition for oscillation,

$$\beta \cdot A_v = 1$$

or $$\beta \times 60 = 1$$

$$\therefore \quad \beta = 1/60 = 0.017 = 1.7\% \textbf{ Ans.}$$

Example 29.10.

An amplifier with negative feedback has a voltage gain of 100. It is found that without feedback, an input signal of 50 mV is required to produce a given output; whereas with feedback, the input signal must be 0.6 V for the same output. Calculate the value of voltage gain without feedback and feedback ratio. *(Bangalore Univ. B.Sc. April/May 2001)*

Solution.

Given $A'_v = 100;\ v_{in} = 50$ mV

Let A_v = Voltage gain without feedback

v_o = Output voltage without feedback

v'_o = Output voltage with feedback

v'_{in} = Input voltage with feedback (= 0.6 V given)

We know that voltage gain with feedback, (A'_v),

$$100 = \frac{v'_o}{v'_{in}} = \frac{v'_o}{0.6}; \qquad \therefore v'_o = 60 \text{ V}$$

and the voltage gain without feedback,

$$A_v = \frac{v_o}{v_{in}} = \frac{v_o}{50 \text{ mV}} = \frac{60}{50 \text{ mV}} = 1200 \textbf{ Ans.}$$

We also know that voltage gain with feedback (A'_v),

$$100 = \frac{A_v}{1 + \beta . A_v} = \frac{1200}{1 + (\beta + 1200)}$$

$$\therefore \quad \beta = \left(\frac{1200}{100} - 1\right) \Big/ 1200$$

$$= 9.1 \times 10^{-3} \text{ or } 0.91\% \textbf{ Ans.}$$

29.4. Gain Stability

We have already discussed in the last article that the use of negative feedback reduces that voltage gain of an amplifier. Moreover, the voltage gain of a feedback amplifier (A'_v) depends on the feedback network (*i.e.*, $A'_v = 1/\beta$). It means that the voltage gain of a feedback amplifier does not change with the variation in transistor parameters due to temperature, aging etc. Thus the use of negative feedback provides a greater stability to the voltage gain of the amplifier.

We know that the voltage gain of an amplifier with negative feedback,

$$A'_v = \frac{A_v}{1 + \beta \cdot A_v}$$

Taking natural logarithms on both sides of the above expression,

$$\log_e A'_v = \log_e \left(\frac{A_v}{1 + \beta \cdot A_v} \right)$$

$$= \log_e A_v - \log_e (1 + \beta \cdot A_v)$$

Differentiating both side,

$$\frac{dA'_v}{A'_v} = \frac{dA_v}{A_v} - \left(\frac{\beta \cdot dA_v}{1 + \beta \cdot A_v} + \frac{A_v \cdot d\beta}{1 + \beta \cdot A_v} \right)$$

The derivative of β (*i.e.*, $d\beta$) is zero, because the feedback network, usually, consists of stable elements such as resistor, capacitors and inductors. Therefore,

$$\frac{dA'_v}{A'_v} = \frac{dA_v}{A_v} - \frac{\beta \cdot dA_v}{1 + \beta \cdot A_v}$$

$$= dA_v \left(\frac{1}{A_v} - \frac{\beta}{1 + \beta \cdot A_v} \right)$$

$$= \frac{1}{1 + \beta \cdot A_v} \times \frac{dA_v}{A_v}$$

The termd dA'_v/A'_v represents the fractional change in amplifier voltage gain with feedback and dA_v/A_v the fractional change in voltage gain without feedback. The term $1/(1 + \beta \cdot A_v)$ is called *sensitivity. It indicates the ratio of percentage change in voltage gain with feedback to the percentage change in voltage gain without feedback.*

Mathematically, the sensitivity

$$= \frac{\dfrac{dA'_v}{A'_v}}{\dfrac{dA_v}{A_v}} = \frac{1}{1 + \beta \cdot A_v}$$

The reciprocal of the term sensitivity is called densensitivity. Thus densensitivity = $1 + \beta \cdot A_v$

The term 'densensitivity' indicates the factor by which the voltage gain has been reduced due to feedback. It may be proved that

Densensitivity $\approx \dfrac{A_v}{A'_v}$ and

Sensitivity $\approx \dfrac{A'_v}{A_v}$

Example 29.11.

An amplifier has an open loop gain of 800 and a feedback ratio of 0.05. If the open loop gain changes by 20% due to temperature, find the percentage change in closed loop gain.

Solution.

Given: $A_v = 800$; $\beta = 0.05$ and $dA_v/A_v = 20$

We know that the percentage change in closed loop gain,

$$\frac{dA'_v}{A'_v} = \frac{1}{1+\beta \cdot A_v} \times \frac{dA_v}{A_v} = \frac{1}{1+0.05\times 800} \times 20 = 0.5\ \%\ \textbf{Ans.}$$

Example 29.12.

An amplifier has an open loop gain of 1000 and feedback ratio of 0.04. If the open loop gain changes by 10 % due to temperature, find the percentage change in gain of the amplifier with feedback. (*JNTU., 2009*)

Solution.

Given: $A_v = 1000$; $\beta = 0.04$; $dA_v/A_v = 10\ \%$

We know that the percentage change in closed loop gain,

$$\frac{dA_v'}{A_v'} = \frac{1}{1+\beta A_v} \times \frac{dA_v}{A_v} = \frac{1}{1+0.04\times 1000} \times 10 = 0.244\ \%\ \textbf{Ans.}$$

Example 29.13.

An amplifier with a closed loop gain of 100 is required and this gain should not vary more than 1%, when the inherent gain of the amplifier without feedback varies by 20%. Find the values of A_v and β.

Solution.

Given: $A'_v = 100$; $dA'_v/A'_v = 1\% = 0.01$ and $dA_v/A_v = 20\% = 0.20$.

We know that,

$$\frac{dA'_v}{A'_v} = \frac{1}{1+\beta \cdot A_v} \times \frac{dA_v}{A_v}$$

$$0.01 = \frac{1}{1+\beta \cdot A_v} \times 0.2$$

or $$1+\beta \cdot A_v = \frac{0.2}{0.01} = 20$$

We also know that the closed loop gain (A'_v),

$$100 = \frac{A_v}{1+\beta \cdot A_v} = \frac{A_v}{20}$$

$\therefore$ $$A_v = 100 \times 20 = 2000\ \textbf{Ans.}$$

and $$\beta = (1+\beta \cdot A_v)/A_v = 20/2000 = 0.01\ \textbf{Ans.}$$

Example 29.14.

An amplifier with open loop voltage gain $A_v = 1000 \pm 100$ is available. It is required to have an amplifier whose voltage gain varies by no more than $\pm$ 0.1%. Find the reverse transmission factor β.

(*Anna University., April/May. 2008*)

Solution.

Given : $A_v = 1000 \pm 100$ and $\dfrac{dA'_v}{A'_v} = \pm 0.1\%$

We know that,

$$\frac{dA_v'}{A_v'} = \frac{1}{1+\beta.A_v} \times \frac{dA_v}{A_v}$$

$$\frac{0.1}{100} = \left(\frac{1}{1\times\beta.A_v}\right) \times \frac{100}{1000}$$

$$1 + \beta.A_v = \frac{100\times100}{0.1\times1000}$$

$$1 + \beta.A_v = 100$$

$$\beta = \frac{99}{A_v} = \frac{99}{1000} = 0.099 \textbf{ Ans.}$$

Example 29.15.

An amplifier has voltage gain with feedback of 100. If the gain without feedback changes 20% and the gain with feedback should not vary more than 2 %, determine the values of open loop gain A and feedback ratio β. (PTU., May 2009)

Solution.

Given : $A_v = 100$; $\frac{dA}{A} = 20\% = 0.2$ and $\frac{dA_v'}{A_v'} = 2\% = 0.02.$

Case 1: *Gain with feedback is maximum i.e. 2 %*

We know that gain with feedback,

$$\frac{dA_v'}{A_v'} = \frac{1}{1+\beta.A_v} \times \frac{dA}{A}$$

$$0.02 = \frac{1}{1+\beta.A_v} \times 0.2$$

$$1 + \beta.A_v = 10$$

We also know that the closed loop gain,

$$A_v' = \frac{A_v}{1+\beta.A_v}$$

$$100 = \frac{A_v}{10}$$

$$\therefore \quad A_v = 1000 \textbf{ Ans.}$$

and $$1 + \beta.A_v = 10$$

$$1 + \beta \times 1000 = 10$$

$$\therefore \quad \beta = 0.009 \textbf{ Ans.}$$

Case 2 : *Gain with feedback is maximum i.e., 1%*

We know that gain with feedback,

$$\frac{dA_v'}{A_v'} = \frac{1}{1+\beta.A_v} \times \frac{dA}{A}$$

$$0.01 = \frac{1}{1+\beta.A_v} \times 0.2$$

$$1 + \beta.A_v = 20$$

We also know that closed loop gain,

$$A_v' = \frac{A_v}{1+\beta.A_v}$$

$$100 = \frac{A_v}{20}$$

$$A_v = 2000 \textbf{ Ans.}$$

and $$1 + \beta.A_v = 20$$

$$1 + \beta \times 1000 = 20$$

$$\beta = 0.019 \textbf{ Ans.}$$

29.5. Increased Bandwidth

Consider an amplifier, whose bandwidth is required to be found out when it is subjected to negative feedback.

Let BW = Bandwidth of an amplifier without feedback,

and BW′ = Bandwidth of an amplifier with feedback,

A_v = Voltage gain of the amplifier without feedback, and

β = Feedback ratio.

We have already discussed that the negative feedback increases the bandwidth of an amplifier by a factor equal to $(1 + \beta \cdot A_v)$.

$$\therefore \quad BW' = (1 + \beta \cdot A_v)\, BW \qquad \ldots (i)$$

It may be noted that voltage gain of a feedback amplifier reduces by the same factor by which the bandwidth is increased (*i.e.*, $1 + \beta \cdot A_v$). Thus the product of voltage gain and bandwidth of an amplifier without feedback and with feedback remains the same. We know that bandwidth of an amplifier is given by the separation between the upper and lower 3 *d*B frequencies.

Now let f_1 = the lower 3 dB frequency, and

f_2 = the upper 3 dB frequency.

∴ Bandwidth of the amplifier,

$$BW = f_2 - f_1$$

It can be proved easily that with negative feedback the upper (f_2') and lower (f_1') 3 dB frequencies of an amplifier are given by the relation,

$$f_2' = (1 + \beta \cdot A_v) f_2$$

and $$f_1' = \frac{f_1}{(1 + \beta \cdot A_v)}$$

It is evident from the above relations that the upper 3 dB frequency (f_2') is greater than f_2 by a factor of $(1 + \beta \cdot A_v)$. Similarly, f_1' is smaller than f_1 by a factor of $(1 + \beta \cdot A_v)$. Since the product of voltage gain and bandwidth is the same without feedback and with feedback, therefore

$$A'_v \times BW' = A_v \times BW$$

or $$A'_v (f_2' - f_1') = A_v (f_2 - f_1)$$

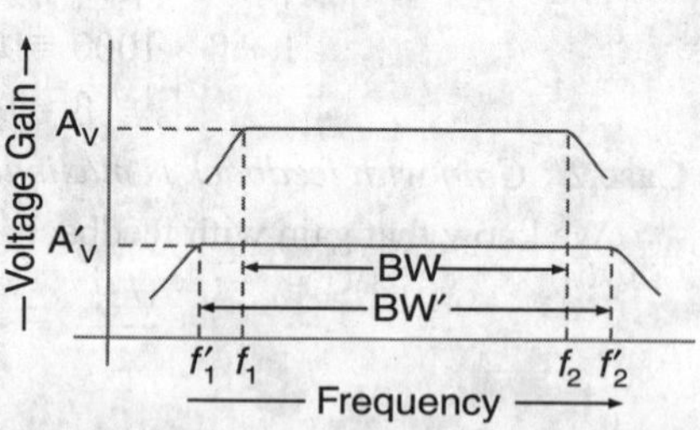

Fig. 29.3. Gain versus frequency graph of an amplifier with and without feedback.

The above facts are illustrated in the frequency response graph (*i.e.*, gain versus frequency graph) as shown in Fig. 29.3.

In this figure, the voltage gain is plotted along the vertical axis and the frequency along the horizontal axis. It is evident from the plot that the negative feedback has increased the bandwidth from BW to BW′.

Example 29.16.

An amplifier has a mid-frequency gain of 100 and a bandwidth of 200 kHz. (a) What will be the new bandwidth and gain, if 5% negative feedback is introduced. (b) What should be the amount of feedback, if the bandwidth is to be restricted to 1 MHz.

Solution.

Given: $A_v = 100$; BW = 200 kHz = 200×10^3 Hz and $\beta = 5\ \% = 0.05$.

(a) New bandwidth and gain with 5% negative feedback

We know that the new bandwidth (*i.e.*, the bandwidth with 5% or 0.05 negative feedback),

$$\begin{aligned} BW' &= (1 + \beta \cdot A_v)\ BW \\ &= [1 + (0.05 \times 100)] \times 200 \times 10^3 \text{ Hz} \\ &= 1200 \times 10^3 \text{ Hz} = 1200 \text{ kHz } \textbf{Ans.} \end{aligned}$$

and gain with feedback,

$$A'_v = \frac{A_v}{1 + \beta \cdot A_v} = \frac{100}{1 + (0.05 \times 100)} = 16.7 \textbf{ Ans.}$$

(b) Amount of feedback with BW' equal to 1 MHz (1 × 10⁶ Hz)

Let β = Amount of feedback to restrict the bandwidth to 1 MHz.

We know that bandwidth with feedback (BW′),

$$\begin{aligned} 1 \times 10^6 &= (1 + \beta \cdot A_v)\ BW \\ &= (1 + \beta \times 100) \times (200 \times 10^3) \end{aligned}$$

$$\therefore \quad 1 + 100\,\beta = (1 \times 10^6)/(200 \times 10^3) = 5$$

$$\text{or} \quad \beta = 4/100 = 0.04 = 4\% \textbf{ Ans.}$$

Example 29.17.

An amplifier has a midband gain of 1500 and a bandwidth of 4 MHz. The midband gain reduces to 150 when a negative feedback is applied. Determine the value of feedback factor and the bandwidth. *(UPSC Engg. Services 1998)*

Solution.

Given : $A_v = 1500$; BW = 4 MHz and $A'_v = 150$.

Let, β = feedback factor and

BW′ = bandwidth with feedback

We know that the voltage gain with feedback (A'_v),

$$150 = \frac{A_v}{1 + \beta \cdot A_v} = \frac{1500}{1 + (\beta \times 1500)}$$

$$\therefore \quad \beta = \frac{\left(\dfrac{1500}{150} - 1\right)}{1500} = 6 \times 10^{-3} \text{ or } 0.6\ \% \textbf{ Ans.}$$

We also know that bandwidth with feedback,

$$BW' = (1 + \beta \cdot A_v)\, BW = (1 + (6 \times 10^{-3}) \times 1500)\,(4 \times 10^6)\text{Hz}$$
$$= 10 \times 4 \times 10^6 \text{ Hz} = 40 \times 10^6 \text{ Hz}$$
$$= 40 \text{ MHz } \textbf{Ans.}$$

Example 29.18.

The midband gain of the amplifier is 100 and the lower cut off frequency is 1 kHz. Find the gain of the amplifier at a frequency of 20 Hz. (*Anna University., Nov./Dec. 2006*)

Solution.

Given : $A_{mid} = 100$; $f_1 = 1$ kHz = 1 × 103 Hz and $f = 20$ Hz.

We know that the gain of the amplifier,

$$A = \frac{A_{mid}}{\sqrt{1+\left(\frac{f_1}{f}\right)^2}} = \frac{100}{\sqrt{1+\left(\frac{1\times10^3}{20}\right)^2}} = 2 \text{ Ans.}$$

Example 29.19.

Gain of an amplifier is 100. If 1/10 of the output is feedback to the input out of phase by 180, find the gain with feedback. If the bandwidth of amplifier was 10 kHz, what will be the new bandwidth?

(*R.G.P.V., May/ June 2006*)

Solution.

Given : $A_v = 100$, and Bandwidth = 10 kHz, $\beta = 1/10$.

We know that the voltage gain with feedback,

$$A_v' = \frac{A_v}{1+\beta A_v} = \frac{100}{1+\frac{1}{10}\times 100} = 9.09$$

We know that the bandwidth with feedback,

$$BW' = (1 + \beta A_v)\, BW = \left(1+\frac{1}{10}\times 100\right)(10\times10^3)$$
$$= 110 \times 10^3 \text{ Hz} = 110 \text{ kHz } \textbf{Ans.}$$

Example 29.20.

An amplifier has a midband gain 125 and bandwidth 250 kHz. If 4 % negative feedback is introduced, find the new bandwidth and gain. (*PTU., Dec 2008*)

Solution.

Given : $A_v = 125$; BW = 250 kHz = 250 × 10^3 and β = 4% = 0.04.

We know that the new bandwidth,

$$BW' = (1 + \beta.A_v)BW$$
$$= (1 + 0.04 \times 125)250$$
$$= 1500 \text{ kHz } \textbf{Ans.}$$

and the gain,

$$A_v' = \frac{A_v}{1+\beta.A_v} = \frac{125}{1+0.04\times125} = 20.8 \textbf{ Ans.}$$

Example 29.21.

(a) An amplifier has a value of R_{in} = 4.2 kΩ, A_v = 220, and β = 0.01. Determine the value of input resistance of the feedback amplifier.

(b) The amplifier in part (a) had cut-off frequencies of f_1 = 1.5 kHz and f_2 = 501.5 kHz before the feedback path was added. What are the new cut-off frequencies for the circuit?

Solution.

Given: R_{in} = 4.2 kΩ, A_v = 220 and β = 0.01

(a) ***Input resistance of a feedback amplifier***

We know that input resistance of a feedback amplifier

$$R'_i = (1 + \beta \cdot A_v)\, R_i = (1 + 0.01 \times 220) \times 4.2 \text{ k}\Omega$$

$$= 13.44 \text{ k}\Omega \textbf{ Ans.}$$

(b) ***New cut-off frequencies (f_1' and f_2')***

We know that lower 3 dB frequency,

$$f_1' = \frac{f_1}{1 + \beta \cdot A_v} = \frac{1.5 \text{ kHz}}{1 + 0.01 \times 220} = 0.469 \text{ kHz} = 469 \text{ Hz } \textbf{Ans.}$$

and the upper 3 dB frequency,

$$f_2' = (1 + \beta\, A_v) f_2 = (1 + 0.01 \times 220)\; 501.5 \text{ kHz} = 1604.8 \text{ kHz } \textbf{Ans.}$$

29.6. Decreased Distortion

Consider a large amplitude signal applied to a stage of an amplifier, so that the operation of an active device (*i.e.*, transistor) extend slightly beyond its range of linear operation. As a result of this, the output signal is slightly distorted. Now if a negative feedback is introduced to the amplifier stage, the voltage gain reduces. But if the input signal is increased, by the same amount by which the gain is reduced, the output signal amplitude remains the same (*i.e.*, as it was without feedback). Now if we measure the distortion in both the cases, it will be found that distortion has reduced due to feedback by a factor (1 + β · A_v). Mathematically, amplifier distortion with feedback,

$$D' = \frac{D}{1 + \beta \cdot A_v}$$

where D is the amplifier distortion without feedback.

It may be noted that the input signal to the feedback amplifier may be the actual signal available from an external source or it may be an output of an amplifier preceding the feedback stage. To increase the input of the feedback amplifier by a factor (1 + β · A_v), we can either increase the nominal gain of the preamplifying stages or add a new stage. It will be interesting to know that the full benefit of the feedback amplifier, in reducing distortion, can be obtained. This can be done if the preamplifying stages do not introduce any additional distortion because of the increased output.

Example 29.22.

An RC coupled amplifier has a voltage gain of 1000, f_1 = 50 Hz, f_2 = 200 kHz and a distortion of 5%, without feedback. Find the amplifier voltage gain, f_1', f_2' and distortion when a negative feedback is applied with feedback ratio of 0.01. (Anna University, April/May. 2008)

Solution.

Given: A_v = 1000; f_1 = 50 Hz; f_2 = 200 kHz; D = 5% and β = 0.01.

Voltage gain with feedback

We know that voltage gain with feedback,

$$A'_v = \frac{A_v}{1 + \beta \cdot A_v} = \frac{1000}{1 + (0.01 \times 1000)} = 90.9 \textbf{ Ans.}$$

f_1' and f_2'

We know that the lower 3 dB frequency,

$$f_1' = \frac{f_1}{1 + \beta \cdot A_v} = \frac{50}{1 + (0.01 \times 1000)} = 4.5 \text{ Hz} \textbf{ Ans.}$$

and the upper 3 dB frequency,

$$f_2' = (1 + \beta \cdot A_v) f_2 = [1 + (0.01 \times 1000)] \times 200 \text{ kHz}$$
$$= 2200 \text{ kHz} = 2.2 \text{ MHz} \textbf{ Ans.}$$

Distortion with feedback

We know that distortion with feedback,

$$D' = \frac{D}{1 + \beta \cdot A_v} = \frac{5}{1 + (0.01 \times 1000)} = 0.45\% \textbf{ Ans.}$$

29.7. Decreased Noise

It has been observed that there are many sources of noise in an amplifier depending upon the active device used in the circuit. It will be interesting to know that with the use of feedback, the magnitude of noise (N) is reduced by a factor of $(1 + \beta \cdot A_v)$ in a similar manner as distortion. Thus noise with feedback in given by the relation,

$$N' = \frac{N}{1 + \beta \cdot A_v}$$

and decrease in noise,

$$= \frac{N - N'}{N} = 1 - \frac{N'}{N}$$

$$= 1 - \frac{\dfrac{N}{1 + \beta \cdot A_v}}{N} = 1 - \frac{1}{1 + \beta \cdot A_v}$$

Example 29.23.

A negative feedback is used to reduce the noise from an amplifier by 80%. (a) What must be the percentage of negative feedback to accomplish this, if the input voltage gain is 100? (b) What will be the voltage gain with feedback? *(Anna University, Nov./Dec. 2008)*

Solution.

Given: $A_v = 100$; Reduction of noise = 80% = 0.8.

(*a*) ***Percentage of negative feedback***

Let β = Value of negative feedback to reduce the noise of amplifier by 80%.

We know that the reduction of noise of an amplifier with feedback,

$$0.8 = 1 - \frac{1}{1 + \beta \cdot A_v} = 1 - \frac{1}{1 + \beta \times 100}$$

$$\frac{1}{1 + 100\beta} = 1 - 0.8 = 0.2$$

$$1 + 100\beta = 1/0.2 = 5$$

$$\therefore \quad \beta = 4/100 = 0.04 \textbf{ Ans.}$$

(b) *Voltage gain with feedback*

We also know that voltage gain with feedback,

$$A'_v = \frac{A_v}{1 + \beta \cdot A_v} = \frac{100}{1 + (0.04 \times 100)} = 20 \textbf{ Ans.}$$

Example 29.24.

The gain and the distortion of an amplifier are 150 and 5% respectively without feedback. If the stage has 10 % of its output voltage applied as negative feedback, find the distortion of the amplifier with feedback. (*PTU., May 2005*)

Solution.

Given : $A_v = 100$; $D = 5\% = 0.05$ and $\beta = 10\% = 0.1$.

We know that the distortion with negative feedback,

$$D' = \frac{D}{1+\beta A_v} = \frac{0.05}{[1+(0.1\times 100)]} = 3.125\times 10^{-3}$$

$$= 0.3125\% \textbf{ Ans.}$$

29.8. Types of Feedback Connections

There are four basic types (or ways) of connecting the feedback signal from an amplifier output to its input. These types are given below:

1. Voltage-series feedback connection.
2. Voltage-shunt feedback connection.
3. Current-series feedback connection.
4. Current-shunt feedback connection.

It means that both voltage and current can be feedback to the input either in series or parallel. In the feedback connection types, the term 'voltage' refers to connecting the output voltage as input to the feedback network. The term 'current' refers to tapping off some output current through the feedback network. The term 'series' refers to connecting the feedback signal in series with the input signal voltage, and the term 'shunt' refers to connecting the feedback signal in shunt (parallel) with an input current source.

It has been observed that the *series feedback connections tend to increase the input resistance, while the shunt feedback connections tend to decrease the input resistance*. Moreover, the *voltage feedback will tend to decrease the output resistance, while the current feedback tends to increase the output resistance*. As a matter of fact, higher input resistance and lower output resistance is desired for most cascade amplifiers. Both of these characteristics are obtained by using the voltage-series feedback connection.

29.9. Voltage-series Feedback Connection

It is also called a *shunt-derived series-fed feedback* connection. A block diagram of a voltage-series feedback connection is as shown in Fig. 29.4. In this connection, a fraction of the output voltage is applied in series with the input voltage through the feedback network. However, the input to the feedback network is in parallel with output of the amplifier.

It can be shown easily that the voltage-series feedback connection increase the input resistance and decreases the output resistance of the feedback amplifier.

The input resistance of a feedback amplifier is given by the relation,

$$R'_i = (1 + \beta \cdot A_v)\, R_i$$

where β = The feedback fraction,

A_v = The voltage gain of a amplifier without feedback, and

R_i = The input resistance of an amplifier without feedback.

Similarly, the output resistance of a feedback amplifier is given by the relation,

$$R'_o = \frac{R_o}{1 + \beta \cdot A_v}$$

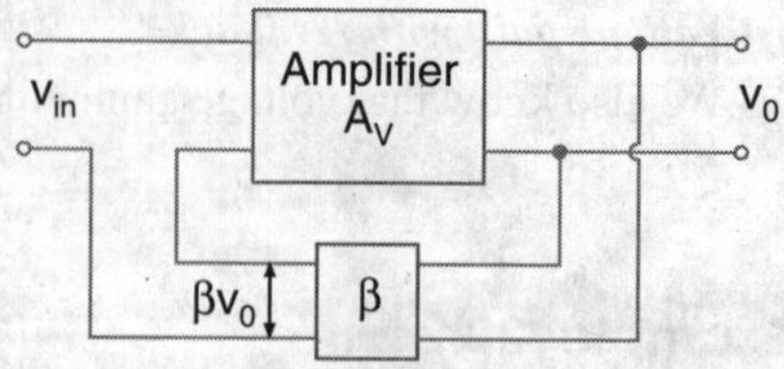

Fig. 29.4. Voltage-series feedback connections.

where R_o is the output resistance of an amplifier without feedback.

As usual, the voltage-series feedback connection reduces the voltages gain of a feedback amplifier and its value is given by the relation,

$$A'_v = \frac{A_v}{1 + \beta \cdot A_v}$$

Example 29.25.

Calculate the voltage gain, input and output resistances of a voltage-series feedback amplifier having $A_v = 300$, $R_i = 1.5$ kΩ, $R_o = 50$ kΩ and β = 1/15.

Solution.

Given: $A_v = 300$; $R_i = 1.5$ kΩ; $R_o = 50$ kΩ and β = 1/15.

Voltage gain

We know that voltage gain of a voltage-series feedback amplifier,

$$A'_v = \frac{A_v}{1 + \beta \cdot A_v} = \frac{300}{1 + \left(\frac{1}{15}\right) \times 300} = 14.3 \textbf{ Ans.}$$

Input resistance

We know that input resistance,

$$R'_i = (1 + \beta \cdot A_v)R_i = \left[1 + \left(\frac{1}{15}\right) \times 300\right] 1.5 = 31.5 \text{ k}\Omega \textbf{ Ans.}$$

Output resistance

We also know that output resistance,

$$R'_i = \frac{R_o}{1 + \beta \cdot A_v} = \frac{50}{1 + \left(\frac{1}{15}\right) \times 300} = 2.4 \text{ k}\Omega \textbf{ Ans.}$$

Example 29.26.

Determine the voltage gain, input and output resistances with feedback for voltage series feedback having $A_v = -100$, $R_i = 10$ kΩ, $R_o = 20$ kΩ for the feedback of (a) β = –0.1 (b) β = –0.5.

(GTU, June 2011)

Solution.

Given: $A_v = -100$; $R_i = 10$ kΩ = 10 × 10^3 Ω and $R_o = 20$ kΩ = 20 × 103Ω.

(a) Values with β = –0.1

Voltage gain

We know that voltage gain of a voltage-series feedback amplifier,

$$A'_v = \frac{A_v}{1 + \beta \cdot A_v} = \frac{-100}{1 + (-0.1) \times (-100)} = -9.09 \textbf{ Ans.}$$

Input resistance

We know that input resistance,

$$R_i = (1 + \beta . A_v)\, R_i = [1 + (-0.1) \times (-100)] \times (10 \times 10^3)$$

$$= 110 \times 10^3\ \Omega = 110 \text{ k}\Omega \textbf{ Ans.}$$

Output resistance

We know that the output resistance,

$$R'_o = \frac{R_o}{1+\beta.A_v} = \frac{20\times10^3}{1+(-0.1)\times(-100)} = 1.81\times10^3\,\Omega = 1.81\,\text{k}\Omega \textbf{ Ans.}$$

(b) Values with β = –0.5

Voltage gain

We know that voltage gain of a voltage-series feedback amplifier,

$$A'_v = \frac{A_v}{1+\beta\cdot A_v} = \frac{-100}{1+(-0.5)\times(-100)} = -1.96 \textbf{ Ans.}$$

Input resistance

We know that the input resistance,

$$R_i = (1+\beta.A_v)\,R_i = [1+(-0.5)\times(-100)]\times(10\times10^3)$$
$$= 510\times10^3\,\Omega = 510\,\text{k}\Omega \text{ Ans.}$$

Output resistance

We know that the output resistance,

$$R'_o = \frac{R_o}{1+\beta.A_v} = \frac{20\times10^3}{1+(-0.5)\times(-100)} = 392\,\Omega = 0.392\,\text{k}\Omega \textbf{ Ans.}$$

29.10. Voltage-shunt Feedback Connection

It is also called a *shunt-derived shunt-fed feedback* connection. A block diagram of a voltage-shunt feedback connection is shown in Fig. 29.5. In this connection, a fraction of the output voltage is applied in parallel with the input voltage through the feedback network.

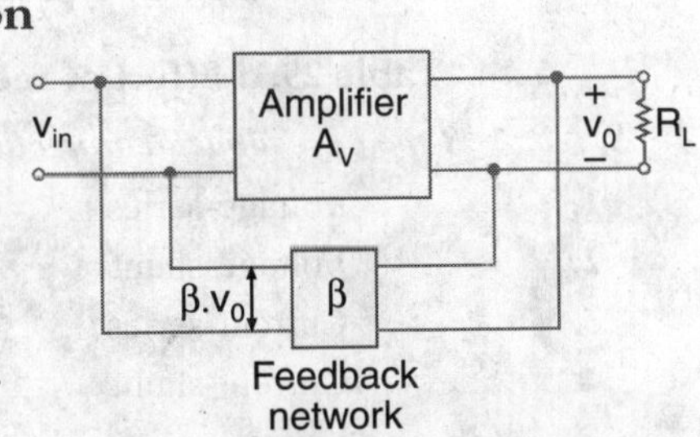

Fig. 29.5. Voltage-shunt feedback connection.

It can be shown easily that the voltage-shunt feedback connection decreases both input and output resistances of the feedback amplifier by a factor equal to $(1 + \beta \cdot A_v)$. Thus

$$R'_i = \frac{R_i}{1+\beta\cdot A_v} \text{ and } R'_o = \frac{R_o}{1+\beta\cdot A_v}$$

29.11. Current-series Feedback Connection

It is called a *series-derived series-fed feedback* connection. A block diagram of a current-series feedback connection is shown in Fig. 29.6. In this connection a fraction of the output current is converted into a proportional voltage by the feedback network and then applied in series with the input.

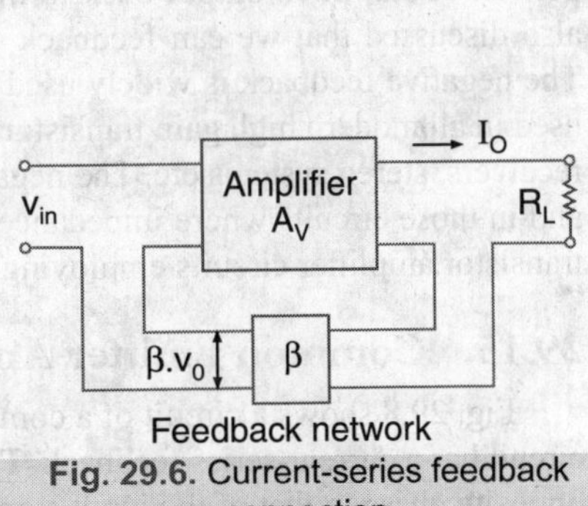

Fig. 29.6. Current-series feedback connection.

It can be shown easily that the current-series feedback connection increases both the input resistance and the output resistance of a feedback amplifier by a factor equal to $(1 + \beta \cdot A_v)$. Thus

$$R'_i = (1+\beta\cdot A_v)\,R_i \text{ and } R'_o = (1+\beta\cdot A_v)\,R_o$$

29.12. Current-shunt Feedback Connection

It is also called a *series-derived shunt-fed feedback* connection. A block diagram of a current-shunt feedback connection is shown in Fig. 29.7. In this connection, a fraction of the output current is converted into a proportional voltage by the feedback network and then applied in parallel with the input voltage.

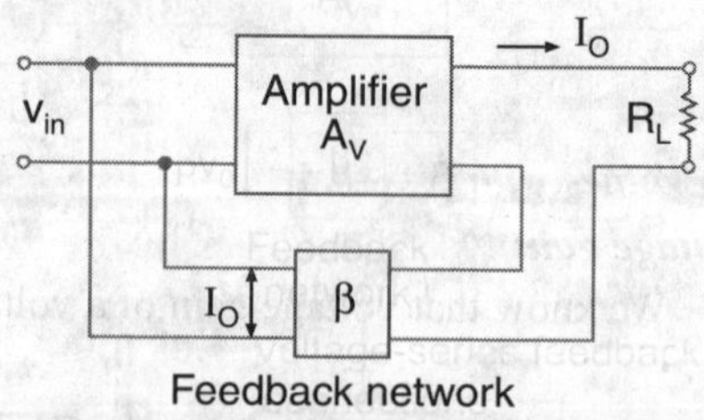

Fig. 29.7. Current-shunt feedback connection.

It can be shown easily that the current-series feedback connection decreases the input resistance but increase the output resistance by a factor equal to $(1 + \beta \cdot A_v)$. Thus

$$R'_i = \frac{R_i}{1 + \beta \cdot A_v} \text{ and } R'_o = (1 + \beta \cdot A_v) \cdot R_o$$

29.13. Comparison of Feedback Connections

We have already discussed in the last articles that a series feedback always increase the input resistance, while the shunt feedback decreases its value. Similarly, a voltage feedback (either series or shunt type) always decreases the output resistance, whereas a current feedback (either series or shunt type) always increases its value. All these facts are summarized in Table 29.1. It may be noted that the four types of feedback connectional decrease the voltage gain, distortion and noise. However, the bandwidth increases in all the types of connections.

Table 29.1: Effect of feedback on input and output resistances

S. No.	*Type of feedback connection*	*Input resistance*	*Output resistance*
1.	Voltage-series	increases	decreases
2.	Voltage-shunt	decreases	decreases
3.	Current-series	increases	increases
4.	Current-shunt	decreases	increases

29.14. Negative Feedback in Transistor Amplifiers

We have already discussed in the last articles that negative feedback is a method for feeding a portion of amplified output back to the input so as to *decrease* the value of input signal. We have also discussed that we can feedback either voltage or current (series or shunt) back to the input. The negative feedback is widely used in electronic amplifiers to achieve desired performance. It is used in all modern high gain transistor (and integrated circuit) amplifier used in radio and television receivers, stereo systems etc. The negative feedback is also used in regulated power supply circuits and in those circuits where impedance matching is required. We shall now discuss some important transistor amplifier circuits employing negative feedback.

29.15. Common Emitter Amplifier Without Emitter Bypass Capacitor

Fig. 29.8 shows a circuit of a common emitter amplifier without emitter bypass capacitor. This circuit has a current-series feedback. The current through resistor R_E results in a voltage, which is in opposite phase to that of the input signal. As a result of this, the output voltage (v_o) is reduced.

In order to remove current-series feedback, the emitter resistor must be removed or bypassed by a capacitor (as is usually done).

It can be shown easily that voltage gain of a common emitter amplifier without feedback,

$$A_v = \frac{R_C}{r'_e}$$

and voltage gain of a common emitter feedback amplifier,

$$A'_v = \frac{R_C}{(r'_e + R_E)}$$

Similarly, the feedback ratio,

$$\beta = \frac{R_E}{R_C}$$

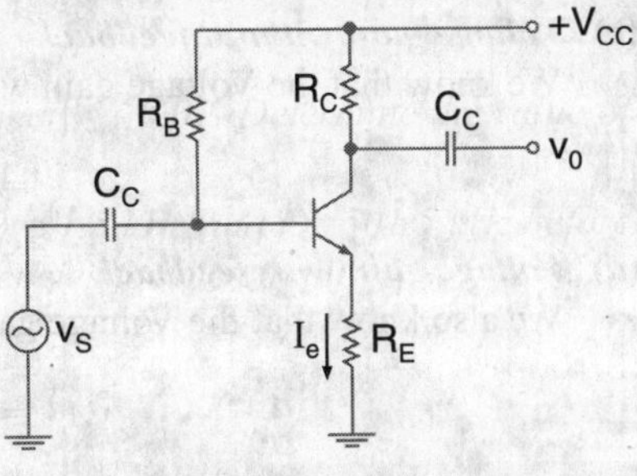

Fig. 29.8. Common-emitter amplifier without emitter bypass capacitor.

Example 29.27.

Find (a) Feedback ratio (b) feedback factor (c) Voltage gain without feedback (d) Voltage gain with feedback for a circuit given in Fig. 29.9. Assume transistor $\beta = 200$ and neglect V_{BE}.

(PTU., Dec 2007, May 2008)

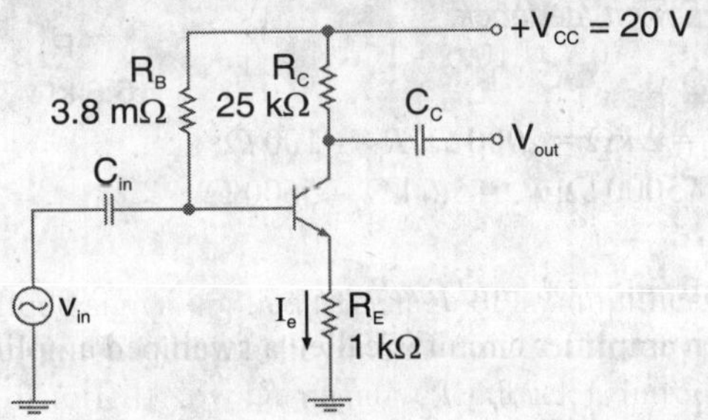

Fig. 29.9

Solution.

Given : $V_{CC} = 20$ V; $R_B = 3.8$ MΩ $= 3.8 \times 10^6$ Ω, $R_C = 25$ kΩ $= 2.5 \times 10^3$ Ω; $R_E = 1$ kΩ $= 1 \times 10^3$ Ω and $\beta = 200$.

We know that emitter current,

$$I_E = \frac{V_{CC}}{(R_B / \beta) + R_E} = \frac{20}{\left(\frac{3.8 \times 10^6}{200}\right) + (1 \times 10^3)} = 1\,\text{mA}$$

and emitter resistance

$$r'_e = \frac{25\,mV}{I_E} = \frac{25\,mV}{1\,mA} = 25\,\Omega$$

(a) *Feedback ratio*

We know that the feedback ratio,

$$\beta = \frac{R_E}{R_C} = \frac{1}{25} = 0.04 \textbf{ Ans.}$$

(b) *Feedback factor*

We know that the feedback factor,

$$\beta A_v = \beta\left(\frac{R_C}{r'_e}\right) = \frac{2.5 \times 10^3 \times 0.04}{25} = 40 \textbf{ Ans.}$$

(c) *Voltage gain without feedback*

We know that the Voltage gain without feedback,

$$A_v = \frac{R_C}{r_e'} = \frac{2.5 \times 10^3}{25} = 1000 \textbf{ Ans.}$$

(d) *Voltage gain with feedback*

We also know that the Voltage gain with feedback,

$$A_v' = \frac{A_v}{1 + \beta . A_v} = \frac{1000}{1 + 40} = 24.39 \textbf{ Ans.}$$

Example 29.28.

An amplifier with current series feedback has the form shown in Fig. 29.10. The transistor has $h_{fe} = 100$, $h_{ie} = 2$ kΩ, $R_c = 470$ Ω and R_e consists of two 100 Ω resistors, one is bypassed by capacitor C_E.

Calculate the values of voltage gain and input resistance of the amplifier with and without feedback.

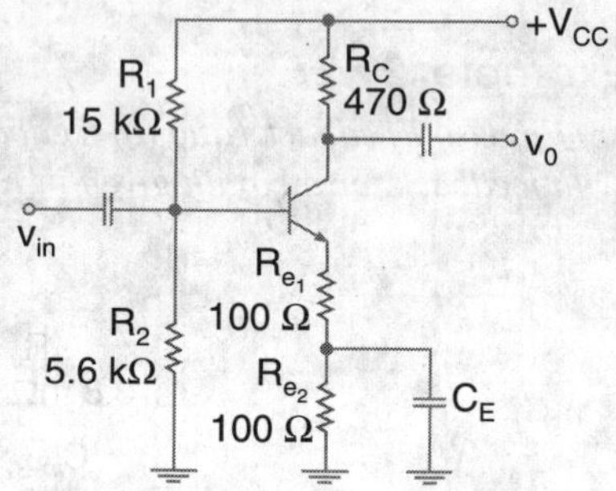

Fig. 29.10.

Solution.

Given: $h_{fe} = 100$ Ω; $h_{ie} = 2$ kΩ = 2000 Ω; $R_{e_1} = 100$ Ω; $R_{e_2} = 100$ Ω; $R_1 = 15$ kΩ = 15000 Ω; $R_2 = 5.6$ kΩ = 5600 Ω and $R_c = 470$ Ω.

Voltage gain and input resistance without feedback

We know that the given amplifier circuit is called a swamped amplifier and its voltage gain,

$$A_v = \frac{A_i \cdot R_c}{R_i} = \frac{h_{fe} \cdot R_c}{h_{ie}} \qquad \text{... } (\because A_i = h_{fe};\ R_i = h_{ie})$$

$$= \frac{100 \times 470}{2000} = 23.5 \textbf{ Ans.}$$

and input resistance,

$$R_i = (R_1 \parallel R_2) \parallel h_{ie} = (15000 \parallel 5600) \parallel 2000 \ \Omega$$

$$= 1342 \ \Omega \textbf{ Ans.}$$

Voltage gain and input resistance with feedback

We also know that the feedback ratio,

$$\beta = \frac{R_e}{R_c} = \frac{R_{e_1}}{R_c} = \frac{100}{470} = \frac{1}{4.7}$$

∴ Voltage gain of the feedback amplifier,

$$A'_v = \frac{A_v}{1 + \beta \cdot A_v} = \frac{23.5}{1 + \left(\frac{1}{4.7} \times 23.5\right)} = 3.92 \textbf{ Ans.}$$

and the input resistance,

$$R'_i = R_i (1 + \beta \cdot A_v)$$

$$= 1342 \left[1 + \left(\frac{1}{4.7} \times 23.5\right)\right] = 8052 \ \Omega \textbf{ Ans.}$$

29.16. Emitter Follower

Fig. 29.11 shows the circuit of an emitter follower or a common collector amplifier. This circuit is also a case of current-series feedback.

The negative feedback is produced in the same way as discussed for the common emitter amplifier without emitter bypass capacitor. In an emitter follower, the feedback ratio is equal to unity, (*i.e.*, $\beta = 1$). Therefore, the overall voltage gain of an emitter follower is approximately unity (*i.e.*, $A_v = A'_v = 1$).

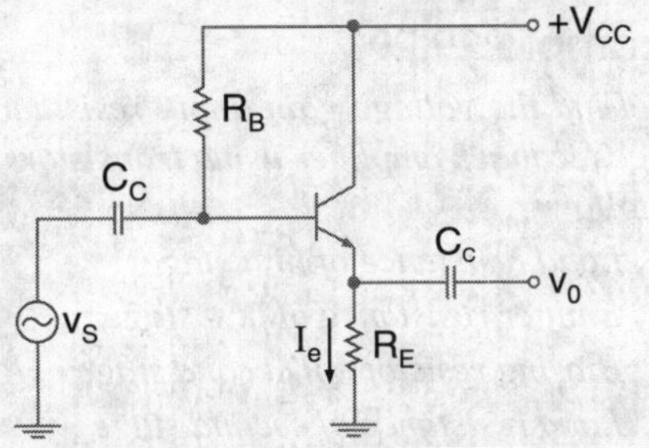

Fig. 29.11. Emitter follower.

29.17. Collector Feedback Biased Common Emitter Amplifier

Fig. 29.12 shows the circuit, of a collector feedback biased common emitter amplifier. This circuit is also a case of voltage-shunt feedback.

The feedback is obtained by coupling a portion of the output voltage to the base through the resistor R_B. The voltage-shunt feedback connection provides a stabilized amplifier overall gain. But it reduces the input and output resistances. It can be shown that the feedback ratio,

$$\beta = \frac{R_C}{R_B}$$

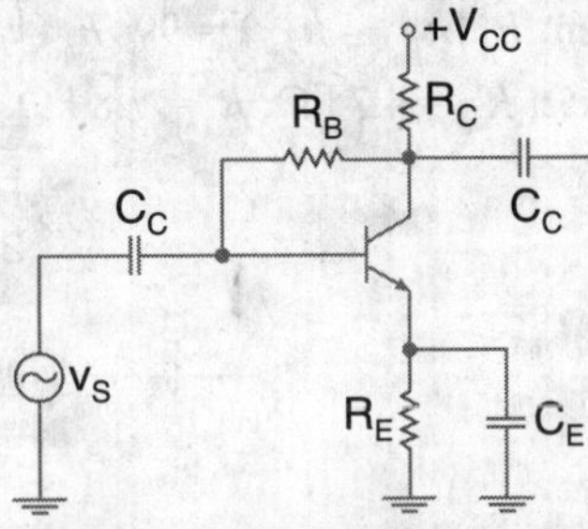

Fig. 29.12. Collector feedback biased common emitter amplifier.

29.18. Negative Feedback in Multistage Amplifiers

It will be interesting to know that the negative feedback is very useful in multistage transistor amplifiers. Fig. 29.13 shows the two cascaded stages of a transistor amplifier with voltage-series feedback. The cascaded amplifier stages, with transistor Q_1 and Q_2, provides an overall voltage gain. A feedback network consists of resistors R_3 and R_4 coupled by a capacitor C_6 (to block d.c.) between output and input.

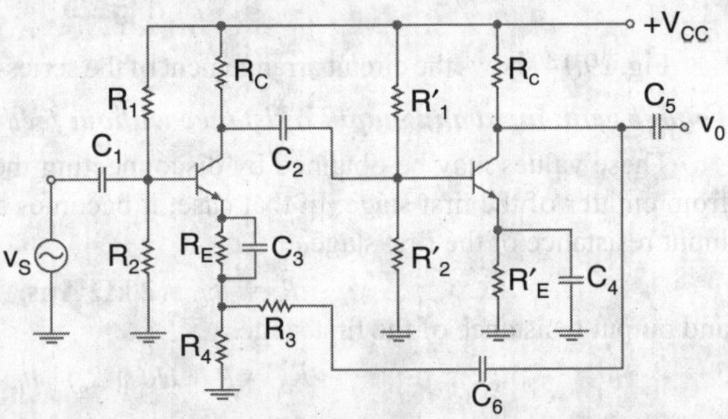

Fig. 29.13. Negative feedback in multistage amplifier.

The feedback signal is taken from the collector of the second stage and is connected to the emitter (neglecting the bypassed R_E resistor) of the first amplifier stage. The negative feedback results from the connection, since the inphase output signal of the second stage connected through the feedback network to the emitter opposes the input signal between the base-emitter junction of the transistor of the first stage.

It is also possible to obtain the current-shunt feedback in a multistage amplifier. In order to obtain current-shunt feedback, the feedback signal is taken from the emitter of the second stage (through a feedback network consisting of a resistor and capacitor) and is connected to the base of the first stage.

Example 29.29.

Find the voltage gain, input resistance and output resistance of a series-shunt pair type two stage feedback amplifier using transistors with h_{fe} = 99 and h_{ie} = 2 kΩ. The circuit components are as follows:

Load resistance of first stage $(R_c) = 22\ k\Omega$
Emitter resistance of first stage $(R_4) = 100\ \Omega$
Biasing resistors of second stage $(R_1\ and\ R_2) = 220\ k\Omega\ and\ 22\ k\Omega$
Load resistance of second stage $(R'_c) = 4.7\ k\Omega$
Feedback resistor from collector of second stage to emitter of first stage
$(R_3) = 7.8\ k\Omega$

Solution.

Given: $h_{fe}\,(h_{fe_1} = h_{fe_2}) = 99$; $h_{ie}\,(h_{ie_1} = h_{ie_2}) = 2$ kΩ; $R_c = 22$ kΩ; $R_4 = 100$ Ω; $R_1 = 220$ kΩ; $R_2 = 22$ kΩ; $R'_c = 4.7$ kΩ; $R_3 = 7.8$ kΩ.

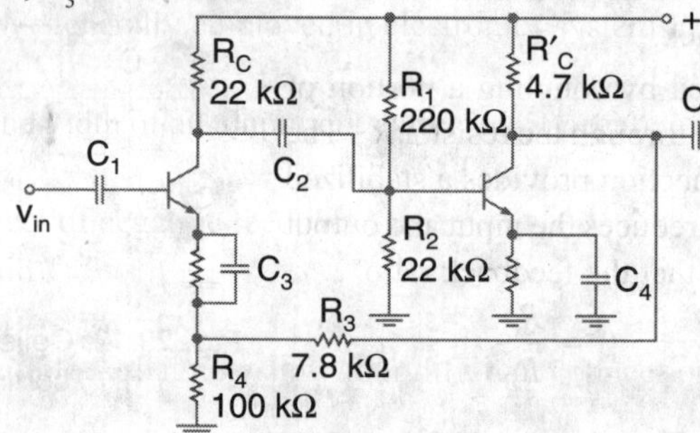

Fig. 29.14.

Fig. 29.14 shows the circuit arrangement of the series-shunt pair type two stage feedback amplifier.

Voltage gain, input and output resistance without feedback

These values may be obtained by disconnecting the feedback network *i.e.*, resistors R_3 and R_4 from emitter of the first stage. In that case, it becomes a simple two stage amplifier. We know that input resistance of the first stage,

$$R_i = h_{ie_1} = 2\ \text{k}\Omega\ \textbf{Ans.}$$

and output resistance of the first stage,

$$R_{o_1} = R_c \parallel (R_1 \parallel R_2) \parallel h_{ie_2} = 22 \parallel (220 \parallel 20) \parallel 2 = 1.67\ \text{k}\Omega$$

Similarly, input resistance of the second stage,

$$R_{i_2} = h_{ie} = 2\ \text{k}\Omega$$

and output resistance of the second stage,

$$R_{o_2} = R'_c \parallel (R_3 + R_4) = 4.7 \parallel (7.8 + 0.1) = 2.95\ \text{k}\Omega\ \textbf{Ans.}$$

Voltage gain, input and output resistance with feedback

We know that the voltage gain of the first stage without feedback,

$$A_{v_1} = \frac{h_{fe} \cdot R_{o_1}}{h_{ie}} = \frac{99 \times 1.67}{2} = 82.7$$

and voltage gain of second stage without feedback,

$$A_{v_2} = \frac{h_{fe} \cdot R_{o_2}}{h_{ie}} = \frac{99 \times 2.95}{2} = 146$$

∴ Overall gain of the cascaded amplifier, without feedback,

$$A_v = A_{v_1} \cdot A_{v_2} = 82.7 \times 146 = 13155$$

Now let us connect the feedback network back at its place. In that case, the value of feedback ratio,

$$\beta = \frac{R_4}{R_3 + R_4} = \frac{0.1}{7.8 + 0.1} = \frac{1}{79}$$

and the input resistance of the amplifier with feedback,

$$R_{i_1} = R_i (1 + \beta . A_v) \left[1 + \left(\frac{1}{79} \times 12074\right)\right] = 308. \text{ k}\Omega \textbf{ Ans.}$$

Similarly, output resistance of the amplifier with feedback,

$$R'_{o_2} = \frac{R_{o_2}}{1 + \beta \cdot A_v} = \frac{2.95}{1 + \left(\frac{1}{79} \times 12074\right)} = 0.0192 \text{ k}\Omega$$

$$= 19.2 \ \Omega \textbf{ Ans.}$$

and the overall voltage gain of the amplifier with feedback,

$$A'_v = \frac{A_v}{1 + \beta \cdot A_v} = \frac{12074}{1 + \left(\frac{1}{79} \times 12074\right)} = 78.5 \textbf{ Ans.}$$

NUMERICAL PROBLEMS

1. An amplifier has a gain of 2000. If the feedback ratio is 0.04, then find the voltage gain of the amplifier with negative feedback. **(Ans. 24.7)**
2. An amplifier has a value of $A_v = 1000$ and $\beta = 0.22$. Determine the value of A'_v for the circuit. **(Ans. 4.52)**
3. An amplifier has a value of $A_v = 588$ and $A'_v = 10.6$. Determine the value of feedback fraction β. **(Ans. 0.093)**
4. An amplifier has a gain of 4000. With negative feedback, the gain reduces to 25. Calculate the fraction of the output that is feedback to the input. **(Ans. 0.04)**
5. An amplifier with negative feedback has a gain of 50. It is found that without feedback, an input signal of 0.1 V is required to produce a given output, whereas with feedback, the input signal must be 0.8 V for the same output. Calculate the amount of voltage gain and feedback ratio. **(Ans. $A_v = 400$; $\beta = 0.018$)**
6. An amplifier has a voltage gain of 500. A technician decides that 10% negative feedback should be employed to reduce distortion. (*a*) What will be the voltage gain with feedback? (*b*) What should be the feedback in order to double the gain that existed in (*a*)? **(Ans. (*a*) 9.8; (*b*) 4.9%)**
7. An amplifier has an open loop gain of 200 and feedback ratio of 0.1. If the open loop gain changes by 20% due to variation of transistor parameters, find the percentage change in closed loop gain. **(Ans. 0.95%)**
8. The midband voltage gain of a certain amplifier is 500. Its upper half-power frequency (f_2) is 20 kHz and the lower half-power frequency (f_1) is 20 Hz. What will be the upper and lower half power frequencies and voltage gain, if 2% negative feedback in introduced? **(Ans. $f_2 = 220$ kHz; $f_1 = 1.8$ Hz, $A_v = 45.5$)**
9. An amplifier with 2.5 kΩ input resistance and 50 kΩ output resistance has a voltage gain of 100. The amplifier is now modified to provide 5% negative voltage feedback in series

with the input. Calculate (*a*) the voltage gain, (*b*) the input resistance, and (*c*) the output resistance with feedback. **(Ans. (*a*) 16.67, (*b*) 15 kΩ, and (*c*) 8.3 kΩ)**

10. Fig. 29.15 shows a transistor feedback amplifier. If the transistor has h_{fe} = 100 and h_{ie} = 2 kΩ, find the voltage gain, and input impedance of the amplifier with feedback and without feedback. **(Ans. A_v = 83.3; R_i = 12 kΩ; A'_v = 500 and R_i = 2 kΩ)**

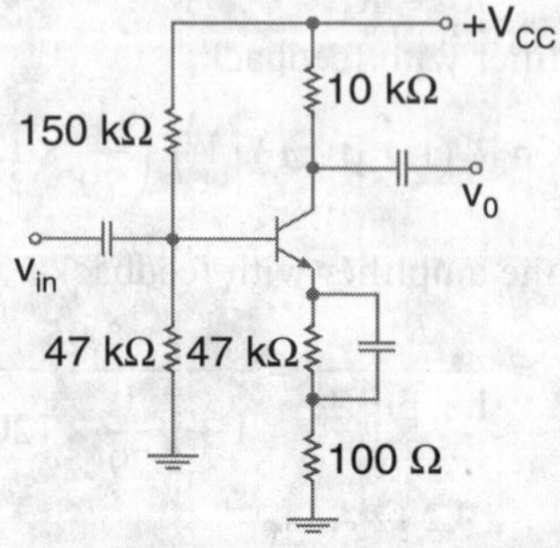

Fig. 29.15.

11. Find the voltage gain, input resistance and output resistance of a series-shunt pair type two stage feedback amplifier shown in Fig. 29.16. The amplifier uses transistors with h_{fe} = 99 and h_{ie} = 2 kΩ. **(Ans. A'_v = 96.5; R'_{i1} = 45.4 kΩ and R'_{o2} = 130 kΩ).**

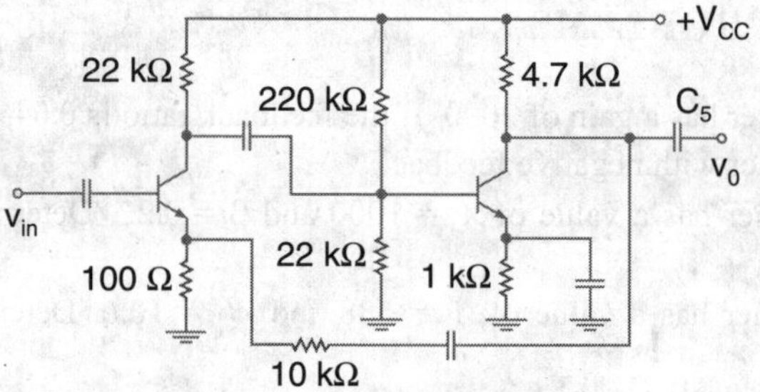

Fig. 29.16.

DESCRIPTIVE QUESTIONS

1. What do you understand by the term 'negative feedback' in amplifiers?
2. Compare the negative feedback and positive feedback.
3. Enumerate the effects of negative feedback on the various characteristics of the amplifier.
4. Describe with necessary derivations, the effect of negative feedback on the bandwidth and distortion in an amplifier.
5. What is the effect of a voltage series negative feedback in the following performance measures of a BJT amplifier:
   (*i*) Input resistance (*ii*) Output resistance
   (*iii*) Bandwidth (*iv*) Distortion and noise
   (*v*) Gain stability
6. Using the block diagram, derive the closed loop form transfer ratio of a feedback system in terms of the open-loop gain. (*Anna University., June 2009*)
7. What is feedback concept? Discuss the basic ways of connecting the feedback signal. (*Jamia University., 2008*)

8. Draw the circuit of an emitter follower .Why it is called so. *(M.G University., May/June 2006)*
9. Draw the circuit of an emitter follower. Identify the type of negative feedback. Calculate the gain, input and output resistance with and without feedback. *(Anna University., May/June 2007)*
10. Identify the type of feedback in the circuit. *(Anna University., May/June 2006, April/May 2008)*
11. Name the four types of feedback topologies. *(Anna University., May/June 2006)*
12. Explain the advantages of negative feedback in amplifiers. *(Nagpur University, Summer 2008, 2004, Nagpur University, Winter 2001, R.G.P.V., Dec. 2005)*
13. Mention the advantage of negative feedback. *(Anna University., Nov/Dec 2006, VTU., June /July 2008, Nagpur University., Winter 2004, Summer 2003)*
14. List the characteristic of negative feedback amplifier. *(Anna University., Nov/Dec 2007)*
15. Define negative and positive feedback. *(Anna University., Nov/Dec 2007)*
16. Draw the circuit of a current series feedback amplifier and explain. Derive expressions for input and output impedance. How does it improve the stability of the amplifier? *(Anna University., Nov/Dec 2007)*
17. Give the block diagram of feedback amplifier and discuss the effect of negative feedback with respect to closed loop gain, bandwidth and distortion. *(Anna University., Nov/Dec 2006)*
18. Draw the block diagrams of the four possible feedback topologies and explain. *(Anna University., Nov/Dec 2007)*
19. Prove that the bandwidth of the amplifier increases with negative feedback. *(Anna University., Nov/Dec 2007)*
20. Compare negative and positive feedback. *(Anna University., May/June 2007)*
21. With neat circuit diagram, explain the current series feedback amplifier and also derive the expression for feedback ratio, input resistance, output resistance and voltage gain. *(Anna University., May/June 2007)*
22. Compare the four types of feedback topologies with respect to basic amplifier, R_{if} and, R_{of}. Draw example circuit for each type of feedback. *(Anna University., May/June 2006)*
23. Draw the feedback topology block diagram with voltage series feedback and derive the input impedance. *(Anna University., May/June 2006)*
24. Explain the effect of negative feedback on the input resistance for a voltage shunt feedback amplifier. *(Anna University., Nov/Dec 2007)*
25. Draw the circuit diagram of an amplifier with current series feedback. Derive the expression for gain, input and output resistance with feedback. *(Anna University., Nov/Dec 2007)*
26. Discusses the classification of feedback amplifiers with schematic, how is impedance level modified in each type. *(Anna University., April/May 2008)*
27. Derive expression of voltage gain with positive feedback and negative feedback and state condition for stability in negative feedback amplifiers.*(Anna University., April/May 2008)*
28. With example prove that negative feedback increases stability. *(Anna University., April/May 2008)*

29. What is the effect of negative feedback on the bandwidth and distortion of an amplifier ? *(Anna University., May/June 2007)*

30. What is the effect of a voltage series negative feedback in the following performance measures of a BJT amplifier:
(*a*) Input resistance
(*b*) Output resistance
(*c*) Bandwidth
(*d*) Distortion and noise
(*e*) Gain stability *(PTU., Dec 2005)*

31. How does negative feedback increase bandwidth of an amplifier? *(PTU., Dec 2005)*

32. With negative feedback voltage gain reduces, explain why? *(Anna University., Nov/Dec 2007, PTU., Dec 2006)*

33. Explain in detail, the effect of negative feedback on gain, bandwidth, distortion, noise, input resistance and output resistance. *(Anna University., May/June 2007)*

34. Derive an expression for input and output resistance of voltage series feedback amplifier. *(VTU., June /July 2008)*

35. What is an emitter follower? Where it is used *(M.G University., July/Aug. 2007)*

36. Draw a block diagram of voltage shunt feedback amplifier and give its input and output resistance. *(Anna University., May/June 2007)*

37. Discuss the effects of negative feedback on gain, bandwidth, distortion and noise. *(Anna University., May/June 2007)*

38. How the distortion is decrease in negative feedback? *(JNTU., June 2009)*

39. Draw the practical circuit for current shunt feedback and find the voltage gain, input impedance & output impedance. *(JNTU., June 2009)*

40. Discuss the effect of negative feedback on the frequency of an amplifier. *(Anna University., June 2009)*

41. Draw the circuit of an emitter follower. Derive an expression for its output impedance. List two applications of an emitter follower. *(PBUT., 2008)*

42. How emitter bypass capacitor in a CE amplifier prevents signal degeneration ? *(PBUT.,2008)*

43. What type of feedback exists in an emitter follower ? Why ? *(PBUT., Dec 2005)*

44. Explain the concept of feedback *(PTU., May 2007)*

45. Compare the negative feedback and positive feedback. *(PTU., Dec 2008)*

46. Name the feedback-amplifier topologies. *(PTU., Dec 2008)*

47. Why negative feedback is employed in high gain amplifiers? *(PTU., May 2005)*

48. What are the different types of negative feedback? Explain each with block diagram. *(PTU., Dec 2007, May 2008)*

49. What are the different types of negative feedback? Explain how the input and output impedances of an amplifier are affected by the different types of negative feedback. *(PTU., Dec 2008)*

50. Describe with necessary derivations, the effect of negative feedback on the bandwidth and distortion in an amplifier. *(PTU., Dec 2009)*

51. Explain how the negative feedback amplifier improves stability, reduced noise and increases the input impedance. *(PTU., Dec 2009)*

52. How the negative feedback changes the impedances? Explain with mathematical expression. *(PTU., May 2006)*
53. Explain types of negative feedback in transistor circuits. *(PTU., May 2007)*
54. Derive the expression for the gain of negative voltage feedback amplifier. *(PTU., May 2005)*
55. Draw a voltage-series feedback circuit. How are its input and output resistances calculated? *(PTU., Dec 2008)*
56. Explain the current series feedback circuit in detail with a neat diagram and perform a suitable analysis *(PTU., Dec 2005)*
57. Explain the current shunt feedback circuit and perform a suitable analysis? *(PTU., May 2005)*
58. Write short note on positive and negative feedback. *(R.G.P.V., Dec. 2008)*
59. Compare negative feedback with positive feedback. *(R.G.P.V., Dec. 2006)*
60. What are the advantages of negative feedback? Explain. *(R.G.P.V., Dec. 2003)*
61. Discuss the principle of negative feedback in amplifiers with a neat diagram. Derive an expression for the gain and bandwidth. *(R.G.P.V., Nov. /Dec. 2007)*
62. Explain different kinds of negative feedback with circuit diagrams. *(R.G.P.V., May/ June 2006)*
63. Write short note on types of feedback. *(R.G.P.V., June 2003)*
64. What is the relationship between transfer gain with feedback A_f and that without feedback A?
    (*i*) Define negative feedback.
    (*ii*) Define positive feedback.
    (*iii*) Define the amount of feedback in decibels. *(R.G.P.V., Dec. 2002)*
65. What is the condition of feedback in amplifier to work it as an oscillator? Show how these conditions can be realized? *(R.G.P.V., June 2003)*
66. Draw the circuit diagram of emitter follower. Identify the feedback is employed in amplifiers. *(R.G.P.V., June 2004)*
67. What is the relation between gain with feedback, A_f and the without feedback A? Define negative feedback and positive feedback. *(R.G.P.V., Dec. 2004)*
68. Explain in positive and negative feedback. Give merits of negative feedback in terms of BW, gain and sensitivity. *(R.G.P.V., June 2005)*
69. Find the expression for input impedance for an amplifier circuit in which voltage series feedback has been employed. *(R.G.P.V., Dec. 2005)*
70. Explain the following types of feedback circuit:
    (*i*) Voltage-series feedback
    (*ii*) Current-series feedback *(R.G.P.V., Dec. 2006)*
71. What do you understand by feedback in amplifiers? Explain the term feedback factor and open-loop gain. What are the types of feedback? *(R.G.P.V., June 2007)*
72. Draw the equivalent circuit of voltage amplifier with voltage series feedback and derive expressions for the output resistance with feedback R_{of} *(R.G.P.V., June 2007)*
73. What are different types of feedbacks? Explain with merits, demerits and application. *(R.G.P.V., June 2008)*
74. Draw the circuit of an emitter follower and analyse its feedback mechanism. *(R.G.P.V., June 2008)*

75. With a neat sketch, explain the working principle of an emitter follower. *(Nagpur University, Winter 2000)*
76. Define negative and positive feedback. Draw the block schematic of amplifier with negative feedback. What type of feedbacks are used in amplifiers and oscillators? *(Nagpur University, Summer 2000)*
77. Draw the circuit of emitter follower regulator using PNP transistor and explain its operation. *(Nagpur University, Summer 2000)*
78. Explain in brief the comparison between negative feedback voltage and current amplifiers as far as the impedance level and gain is concerned. Draw block schematic for the above mentioned amplifiers. *(Nagpur University, Summer 2001)*
79. List the five characteristics of an amplifier which modified by negative feedback. *(Nagpur University, Winter 2002)*
80. What are the four possible topologies of a feedback amplifier? Draw the block schematics of each. *(Nagpur University, Winter 2002)*
81. Why feedback is generally employed in electronics systems? *(Nagpur University, Summer 2003)*
82. Explain the bandwidth of an amplifier is affected, if negative feedback is in corporate. *(Nagpur University, Winter 2003)*
83. Explain the effect of negative feedback on
    (*i*) Voltage gain
    (*ii*) Bandwidth *(Nagpur University, Summer 2005)*
84. With the help of block diagram show four different connections of negative feedback. *(Nagpur University, Winter 2005)*
85. Show block schematically the different connections in an amplifier and practical circuit also and explain the effect of each type of feedback on input and output impedance. *(Nagpur University, Winter 2006)*
86. Explain the effect of negative feedback on
    (*i*) Voltage Gain
    (*ii*) Bandwidth
    (*iii*) Distortion
    (*iv*) Stability *(Nagpur University, Summer 2007)*
87. Explain voltage shunt, voltage series, current shunt and current series feedback system in transistor amplifiers. *(Nagpur University, Summer 2007)*
88. Write short notes on :
    (*i*) Advantage of negative feedback *(Nagpur University, Summer 2001)*
    (*ii*) Comparison between positive and negative feedback amplifiers. *(Nagpur University, Summer 2007)*

MULTIPLE CHOICE QUESTIONS

1. For amplifiers using negative feedback, the bandwidth
    (*a*) lowers its input and output resistances
    (*b*) increases its input and output resistances
    (*c*) lowers its input resistance and increases the output resistance

2. In a current-series feedback amplifier, the input resistance
(*a*) increases (*b*) decreases (*c*) remains unchanged
3. In an amplifier, if you employ voltage-shunt negative feedback
(*a*) the gain increases (*b*) the input resistance decreases
(*c*) the output resistance decreases (*d*) the bandwidth decreases
4. A common-emitter (CE) transistor amplifier employing an n-bypassed series emitter resistance is a case of
(*a*) current-series feedback (*b*) voltage-shunt feedback
(*c*) voltage-series feedback (*d*) current-shunt feedback
5. The advantage of using negative feedback in an amplifier is that its gain can be made practically independent of
(*a*) temperature changes (*b*) age of components
(*c*) frequency (*d*) all of the above.
6. Feedback in an amplifier always helps to
(*a*) control its output (*b*) increase its gain
(*c*) decrease its input impedance (*d*) stabilize its gain.
7. The only drawback of using negative feedback in amplifiers is that it involves
(*a*) gain sacrifice (*b*) gain stability
(*c*) temperature senstivity (*d*) frequency dependence.
8. Closed-loop gain of a feedback amplifier is the gain obtained when
(*a*) its output terminals are closed (*b*) negative feedback is applied
(*c*) feedback loop is closed (*d*) feedback factor exceeds unity.
9. A large sacrifice factor in a negative feedback amplifiers leads to
(*a*) inferior performance
(*b*) increased output impedance
(*c*) characteristics impossible to achieve without feedback
(*d*) precise control over output.
10. Negative feedback in an amplifier
(*a*) lowers its lower 3-dB frequency (*b*) raises its upper 3-dB frequency
(*c*) increases its bandwidth (*d*) all of the above.
11. Regarding negative feedback in amplifiers which statement is WRONG ?
(*a*) it widens the separation between 3-dB frequencies
(*b*) it incerases the gain-bandwidth product
(*c*) it improves gain stability
(*d*) it reduces distortion.
12. Negative feedback reduces distortion in an amplifier only when it
(*a*) comes as part of input signal (*b*) is part of its output.
(*c*) is generated within the amplifier (*d*) exceed a certain safe level.
13. An amplifier with no feedback has a gain-bndwidth product of 4 MHz. Its closed-loop gain is 40. The new band-width is
(*a*) 100 kHz (*b*) 160 MHz
(*c*) 10 MHz (*d*) 20 kHz.

14. The shunt-derived series-fed feedback in an amplifier

(*a*) increases its output impedance (*b*) decreases its output impedance

(*c*) increases its input impedance (*d*) both (*b*) and (*c*).

15. A feedback amplifier has a closed gain of –200. It should not vary more than 50% despite 25% variation in amplifier gain A without feedback. The value of A is

(*a*) 800 (*b*) – 800

(*c*) 1000 (*d*) – 1000

16. The gain of a negative feedback amplifier is 40 dB. If the attenuation of the feedback path is 50 dB, then the gain of the amplifier without feedback is

(*a*) 78.92 (*b*) 146.32

(*c*) 215.51 (*d*) 317.23

17. In a common emitter amplifier, the unbypassed emitter resistor provides,

(*a*) voltage-shunt feedback (*b*) current-series feedback

(*c*) negative-voltage feedback (*d*) positive-current feedback.

ANSWERS

1. (*b*)	**2.** (*a*)	**3.** (*a*)	**4.** (*b*)	**5.** (*d*)
6. (*a*)	**7.** (*a*)	**8.** (*c*)	**9.** (*c*)	**10.** (*d*)
11. (*b*)	**12.** (*c*)	**13.** (*a*)	**14.** (*d*)	**15.** (*d*)
16. (*b*)	**17.** (*b*)			

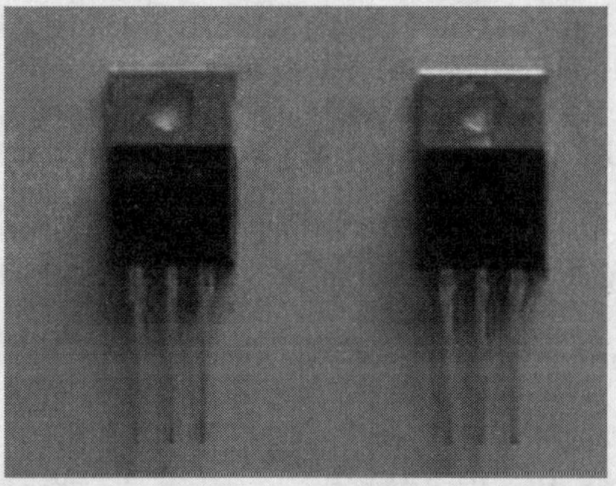

30 Chapter

Field-Effect Transistor Amplifiers

Outline

1. Introduction.
2. Biasing the FET.
3. Biasing the JFET.
4. Gate Bias.
5. Self Bias.
6. Setting a Q-Point.
7. Setting a Q-Point Using Load Line.
8. Biasing Against Device Parameters Variation.
9. Voltage Divider Bias.
10. Source Bias.
11. Current Source Bias.
12. Biasing the Enhancement Type MOSFET's.
13. Biasing the Depletion Type MOSFET's.
14. Small-Signal FET Models.
15. Small-Signal Low-Frequency FET Models.
16. Small Signal High-Frequency FET Models.
17. The Field-Effect Transistor Amplifier.
18. Common Source Amplifier.
19. Analysis of Common Source Amplifier.
20. Effect of A.C. Load on Amplifier Parameters.
21. Effect of External Source Resistance on Voltage Gain.
22. Common Drain Amplifier.
23. Analysis of Common Drain Amplifier.
24. Common Gate Amplifier.
25. Analysis of Common Gate Amplifier.

30.1. Introduction

We have already discussed the various types of amplifiers using bipolar junction transistors. Now we shall discuss some amplifiers, using field-effect transistors (abbreviated as FET's). The field-effect transistor is a three terminal unipolar device in which the current is controlled by an electric-field. A field-effect transistor has a number of advantages over a bipolar junction transistor namely *extremely high input resistance, less noisy, better thermal stability etc.* Its only disadvantage is that it has a relatively *small gain-bandwidth product.* However, this drawback is being improved considerably by using advanced manufacturing techniques. Because of the above mentioned advantages, the field-effect transistor amplifier has become more popular.

30.2. Biasing the FET

To use a FET in any application, it is necessary first to bias the device. The usual reason for this biasing is to turn the device ON and to place it in a region where it operates properly and provide a constant amount of voltage gain. The biasing deals with setting a fixed level of current which should flow through the FET with a desired fixed voltage drop across the FET junctions.

30.3. Biasing the JFET

There are several methods of biasing the JFET. These are (*a*) Gate bias, (*b*) Self-bias, (*c*) Voltage divider bias, and (*d*) Source bias. Whatever be the biasing method, the gate-source junction of a JFET will always be reverse biased. Even when the gate and source terminals are shorted, the drain current (I_D) will cause the junction to be reverse biased. As a matter of fact, the gate-source junction of a JFET must never be allowed to become forward biased. It is because of the reason that the gate material is not constructed to handle any significant amount of current. If the junction is allowed to become forward biased, current will start to flow in the gate circuit. Under most circumstances, this current will destroy the JFET.

Now we shall discuss the JFET biasing methods one by one in the following pages.

30.4. Gate Bias

The gate bias for JFET is similar to the base bias of bipolar transistor Figure 30.1 (*a*) shows the gate bias circuit for N-channel, JFET (junction field-effect transistor). In this circuit, the gate supply voltage ($-V_{GG}$) is used to ensure that the gate-source junction is reverse biased. Since there is no gate current, there is no voltage drop across R_G.

It will be interesting to know that gate bias cannot provide a stable *Q*-point from one JFET to another. The reason for this is that for a given JFET, there is a range of values for $V_{GS(off)}$ and I_{DSS} which can lead to a major problem with the gate bias circuit. It may also be noted that even though there is no gate current or any voltage drop across R_G, but still the circuit contains R_G. Actually, R_G is used for ac operation purpose. Thus R_G resistor has little to do with the dc operation of the amplifier, but it is a vital part of the ac operation of the circuit.

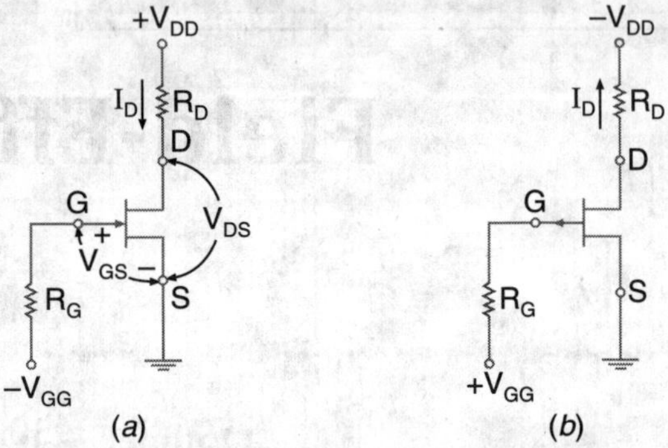

Fig. 30.1. Gate Bias.

Fig. 30.1 (*b*) shows the gate bias circuit for P-channel JFET. This circuit is similar to the one for a N-channel JFET, except that the polarity of V_{DD} and V_{GG} supplies are reversed.

30.5. Self-Bias

Fig. 30.2 (*a*) shows the self-bias circuit for an N-channel junction field-effect transistor (JFET). In this circuit, there is only one drain supply and no gate supply. The gate terminal is connected through a resistor (R_G) to the ground. The source terminal is connected through a resistor (R_S) to the ground.

When the drain voltage is applied, a drain current flows even if there is no gate current (V_G). The drain current produces a voltage drop across resistor R_S (equal to $I_D \cdot R_S$). This voltage drop produces the gate-to-source reverse voltage required for an FET operation. The resistor R_S is called a feedback resistor. Its function is to prevent any variation in the FET drain current, which may be understood from the following discussion.

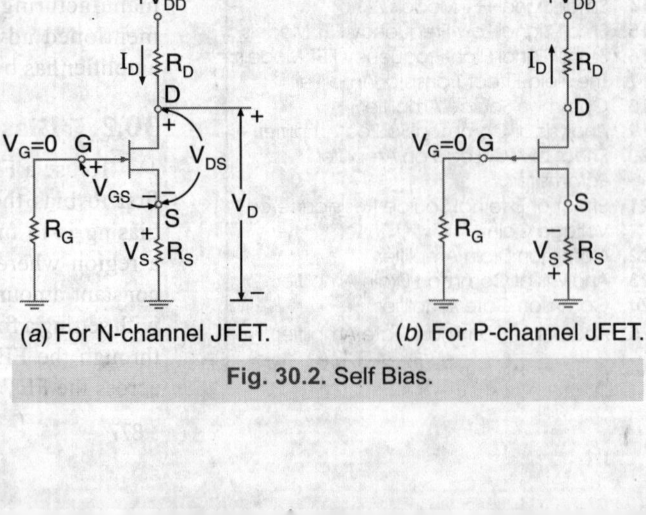

(*a*) For N-channel JFET. (*b*) For P-channel JFET.

Fig. 30.2. Self Bias.

First of all, consider an increase in drain current, which will increase the voltage drop across the resistor (R_S). The increased voltage drop increases the reverse gate-to-source voltage, which decreases the effective width of the channel. It reduces the value of drain current. Now, if the drain current decreases, then the reverse action taken place, *i.e.*, the reduced drain current decreases the gate-to-source voltage, which in turn increases the effective width of channel, thereby increasing the of value of drain current.

Fig. 30.2 (*b*) shows the self-bias circuit for P-channel JFET. It is similar to N-channel JFET circuit, except that the polarity of drain supply voltage (*i.e.*, V_{DD}) is reversed. We know that in a JFET, no current flows through the gate terminal to the ground because the gate-source junction is reverse biased. In actual practice, a small current called reverse leakage current (I_{GSS}) does flow through the gate terminal. Usually, its value is very small and can be neglected. Therefore we shall assume that the gate terminal is at zero voltage. In other words, the gate voltage with respect to ground *i.e.*, $V_G = 0$. We know that the source voltage with respect to ground,

$$V_S = I_D \cdot R_S$$

and the drain voltage,

$$V_D = V_{DD} - I_D \cdot R_D$$

The drain-to-source voltage (V_{DS}) is equal to the difference between the drain voltage (V_D) and the source voltage (V_S).

i.e.,
$$V_{DS} = V_D - V_S = (V_{DD} - I_D \cdot R_D) - I_D \cdot R_S$$
$$= V_{DD} - I_D (R_D + R_S)$$

The gate-to-source voltage (V_{GS}) is equal to the difference between the gate voltage (V_G) and the source voltage (V_S).

i.e.,
$$V_{GS} = V_G - V_S$$
$$= 0 - I_D \cdot R_S$$
$$= - I_D \cdot R_S \quad \text{... } (i)$$

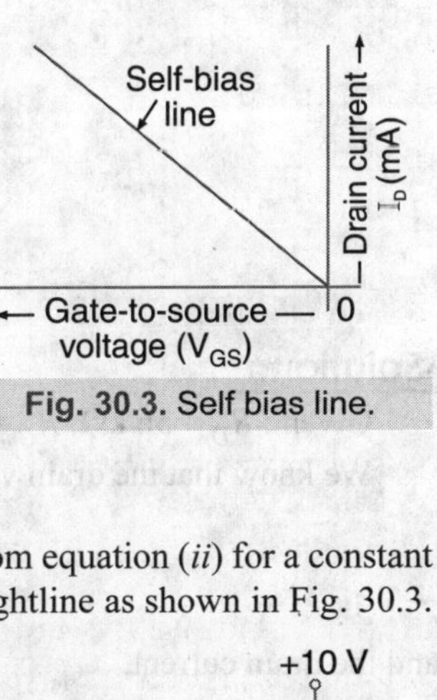

Fig. 30.3. Self bias line.

It is evident from the above relation that the gate-to-source voltage is equal to the negative of the voltage across the source resistor. It is thus obvious that greater the value of drain current, more negative will be the gate-to-source voltage. Rearranging equation (*i*),

$$I_D = -\frac{V_{GS}}{R_S} \quad \text{... } (ii)$$

Now it is plot a graph using the value of drain current obtained from equation (*ii*) for a constant value of resistor (R_S) against the gate-to-source voltage we get a straightline as shown in Fig. 30.3. This straight-line is called a self-bias line.

Example 30.1.

Find the values of drain-to-source voltage (V_{DS}) and the gate-to-source voltage (V_{GS}) for the circuit shown in Fig. 30.4. Given $I_D = 5$ mA.

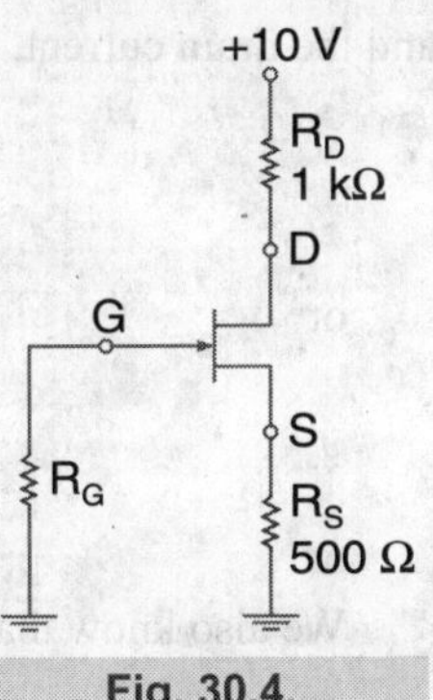

Fig. 30.4.

Solution.

Given: $I_D = 5$ mA $= 5 \times 10^{-3}$ A; $V_{DD} = 10$ volts; $R_D = 1$ kΩ $= 1 \times 10^3$ Ω and $R_S = 500$ Ω

Drain-to-source voltage

We know that the source voltage,

$$V_S = I_D \cdot R_S = (5 \times 10^{-3}) \times 500 = 2.5 \text{ V}$$

and the drain voltage,

$$V_D = V_{DD} - I_D \cdot R_D$$
$$= 10 - (5 \times 10^{-3}) \times (1 \times 10^3) \text{ V}$$
$$= 10 - 5 = 5 \text{ V}$$

∴ Drain-to-source voltage,

$$V_{DS} = V_D - V_S = 5 - 2.5 \text{ V} = 2.5 \text{ V} \textbf{ Ans.}$$

Gate-to-source voltage

We know that gate-to-source voltage,

$$V_{GS} = -V_S = -2.5 \text{ V} \textbf{ Ans.}$$

Example 30.2.

The N-channel FET shown is Fig. 30.5 has I_{DSS} = 1.5 mA, V_P = – 1.5 V. If the quiscient drain to ground voltage is 10 volts, find R_1. (*Nagpur Univ. Winter 2003*)

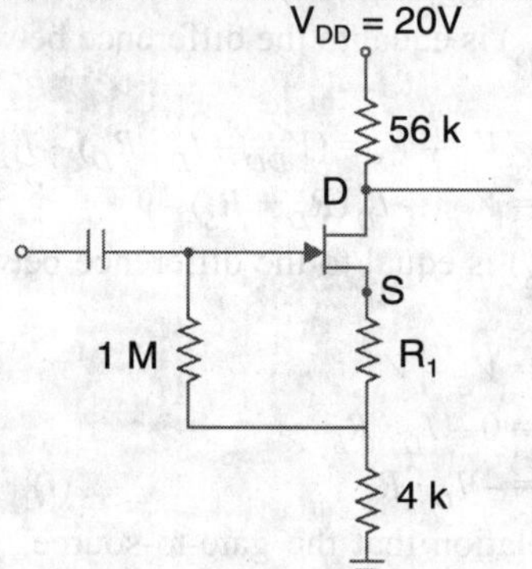

Fig. 30.5.

Solution.

Given : R_D = 56 kΩ; R_G = 1 MΩ; I_{DSS} = 1.5 mA; V_P = – 1.5 V, V_D = 10 V and V_{DD} = 20 V

We know that the drain voltage (measured with respect to ground), (V_D)

$$10 = V_{DD} - I_D \,.\, R_D = 20 - I_D \times 56 \text{ k.}$$

$$\therefore \quad I_D = \frac{10}{56\,k} = 0.179 \text{ mA}$$

and the drain current,

$$0.179 \times 10^{-3} = I_{DSS}\left[1 - \frac{V_{GS}}{V_P}\right]^2 = (1.5 \times 10^{-3}) \times \left[1 - \frac{V_{GS}}{1.5}\right]^2$$

$$\text{or} \quad \left[1 - \frac{V_{GS}}{1.5}\right]^2 = \frac{0.179 \times 10^{-3}}{1.5 \times 10^{-3}} = 0.119$$

$$\therefore \quad 1 - \frac{V_{GS}}{1.5} = \sqrt{0.119} = 0.345$$

$$\text{or} \quad V_{GS} = -1.5 \times (1 - 0.345) = -0.98 \text{ V}$$

We also know that the source voltage,

$$V_S = V_{GS} = -0.98 \text{ V}$$

and voltage drop across the source resistor (V_S),

$$0.98 = I_D \,.\, (R_1 + 4\,k) = 0.179 \times 10^{-3}\,(R_1 + (4 \times 10^3))$$

$$R_1 + (4 \times 10^3) = \frac{0.98}{0.179 \times 10^{-3}} = 5.475 \times 10^3 \approx 5.5 \times 10^3\ \Omega$$

$$\therefore \quad R_1 = (5.5 - 4) \times 10^3 = 1.5\ \text{k}\Omega\ \textbf{Ans.}$$

Example 30.3.

In a self-bias N-channel JFET shown in Fig. 30.6, the operating point is to be set at $I_D = 1.5$ mA and $V_{DS} = 10$ V. The JFET parameters are $I_{DSS} = 5$ mA and $V_P = -2$ V. Find the values of R_S and R_D. Given that $V_{DD} = 20$ V. *(Dip IETE., June 2001)*

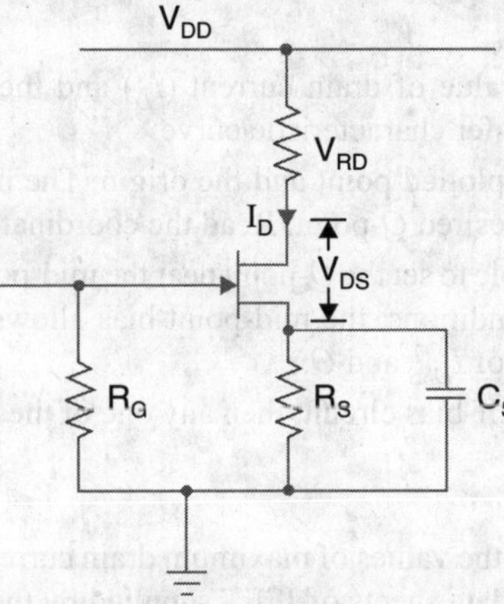

Fig. 30.6.

Solution.

Given : $I_D = 1.5$ mA; $V_{DS} = 10$ V; $I_{DSS} = 5$ mA, $V_P = -2$ V and $V_{DD} = 20$ V.

We know that the drain current, (I_D),

$$1.5\text{ mA} = I_{DSS}\left(1 - \frac{V_{GS}}{V_P}\right) = 5\text{ mA}\left(1 - \frac{V_{GS}}{-2}\right)$$

$$\therefore \quad 1 + \frac{V_{GS}}{2} = \frac{1.5}{5} \Rightarrow \frac{V_{GS}}{2} = \frac{1.5}{5} - 1 = -0.7$$

or $\quad V_{GS} = -0.7 \times 2 = -1.4$ V

We also know that the dc voltage from gate-to-source,

$$V_{GS} = V_G - V_S = -V_S$$

$$\therefore \quad V_S = +1.4\text{ V.}$$

The drain current (I_D),

$$1.5\text{ mA} = \frac{1.4\text{ V}}{R_S};\quad R_S = \frac{1.4\text{ V}}{1.5\text{ mA}} = 0.933\text{ k}\Omega = 933\ \Omega\ \textbf{Ans.}$$

We also know that drain-to-source voltage (V_{DS}),

$$10 = V_{DD} - I_D\,(R_D + R_S)$$

$$= 20 - 1.5 \times 10^{-3}\,(R_D + 933)$$

$$\therefore \quad R_D = \left(\frac{20 - 10}{1.5 \times 10^{-3}}\right) - 933 = 5733.1\ \Omega = 5.7\text{ k}\Omega\ \textbf{Ans.}$$

30.6. Setting a *Q*-point

We know that the *Q*-point (*quiescent operating point* or bias point) for a self-biased JFET is established by determining the value of drain current (I_D) for a desired value of gate-to-source voltage (V_{GS}) or vice versa. However, if the data sheet of a JFET includes a transfer characteristic curve (also called *transconductance curve*), then we can find the *Q*-point by using the procedure as given below:

1. First of all, select a convenient value of drain current. Usually, its value is taken half of the maximum possible value of drain current (*i.e.*, I_{DSS}). Then find the voltage drop across source resistor (R_S) by the relation,
$$V_S = I_D \cdot R_S$$
and the gate-to-source voltage from the relation,
$$V_{GS} = -V_S$$
2. Now plot the assumed value of drain current (I_D) and the corresponding gate-to-source voltage (V_{GS}) on the transfer characteristic curve.
3. Draw a line through the plotted point and the origin. The point of intersection of the line and the curve gives the desired *Q*-point. Read the coordinates of *Q*-point.

As a matter of fact, it is desirable to set the *Q*-point near the mid-point of the transfer characteristic curve of a JFET. Under signal conditions, the mid-point bias allows a maximum amount of drain current swing between the values of I_{DSS} and *O*.

Now if we want to design a self-bias circuit, then any one of the following two methods is used for this purpose:

1. *Analytical method*

In this method, we note down the values of maximum drain current (I_{DSS}) and the gate-to-source cut-off voltage ($V_{GS(off)}$) from the data sheets of JFET supplied by the manufacturer. Now the value of drain current is obtained by the relation,

$$I_D = I_{DSS}\left[1 - \frac{V_{GS}}{V_{GS(off)}}\right]^2$$

For example, if we select the gate to source voltage (V_{GS}) equal to $V_{CS(off)}/4$, then the value of drain current,

$$I_D = I_{DSS}\,[1 - 0.25]^2 = I_{DSS}\,(0.75)^2 = 0.56\,I_{DSS}$$

It indicates that the drain current is slightly more than one-half of I_{DSS}. But it will bias the JFET close to the mid-point of the curve. The value of drain resistor (R_D) is selected in such a way that the drain voltage (V_D) is equal to half the drain supply voltage (V_{DD}). The value of gate resistor (R_G) is chosen arbitrarily large, so that it prevents loading on the driving stages.

2. *Graphical method*

In this method, we draw a self-bias line in such a way that it intersects the transfer characteristic curve near its mid-point. The point of intersection of the self-bias line and the transfer characteristic curve gives us the required *Q*-point. Then read the coordinates of the *Q*-point. The value of source resistance (R_S) is given by the ratio of gate-to-source voltage (V_{GS}) to the drain current (I_D). Mathematically, the source resistance,

$$R_S = \frac{V_{GS}}{I_D}$$

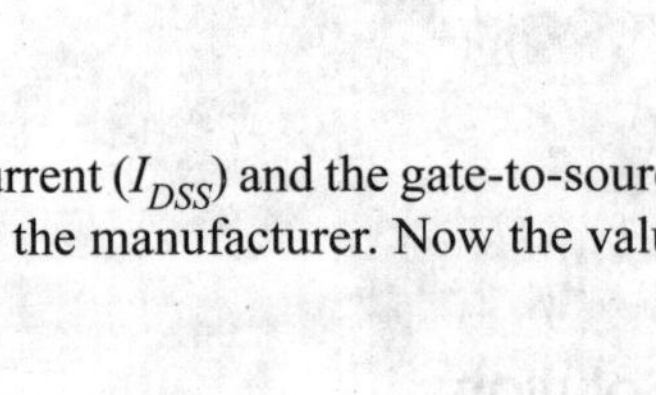

Fig. 30.7. Setting a *Q*-point.

However, a more accurate method is to draw a self-bias line through the coordinates of I_{DSS} and $V_{GS(off)}$ as shown in Fig. 30.7. Then the point of intersection of self-bias line and the transfer characteristic curve locates the Q-point. The value of source resistor is given by the relation,

$$R_S = \frac{V_{GS(off)}}{I_{DSS}}$$

The value of drain resistor (R_D) and gate resistor (R_G) are selected in the same way as discussed above for the analytical method.

Example 30.4.

Fig. 30.8 (a) shows that circuit of a self-biased JFET amplifier and Fig. 30.8 (b) indicates the transfer characteristic curve of the JFET.

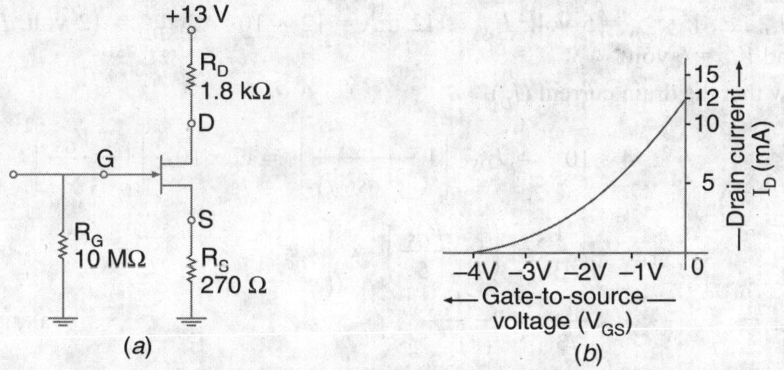

Fig. 30.8.

Find the quiescent values of I_D and V_{GS}. Calculate the values of d.c. voltage between the drain and ground.

Solution.

Given: $R_D = 1.8\ k\Omega = 1.8 \times 10^3\ \Omega$; $R_S = 270\ \Omega$; $R_G = 10\ M\Omega$ and I_{DSS} (from transfer characteristic curve) = 12 mA.

(a) Quiescent values of I_D and V_{GS}

1. First of all, assume the drain current (I_D) equal to the half of I_{DSS} (*i.e.*, 6 mA or 6×10^3 A). Using this value of drain current, the voltage drop across resistor (R_S),

$$V_S = I_D \cdot R_S = (6 \times 10^{-3}) \times 270 = 1.62\ V$$

and the gate-to-source voltage,

$$V_{GS} = -V_S = -1.62\ V$$

2. Now plot the values of drain current (I_D = 6 mA) and the corresponding value of voltage V_{GS} (*i.e.*, – 1.62 V) on the graph of transfer characteristic curve as shown in Fig. 30.9.

3. Draw a line through the plotted point (P) and the origin as shown in the figure and read the coordinates of Q-point.

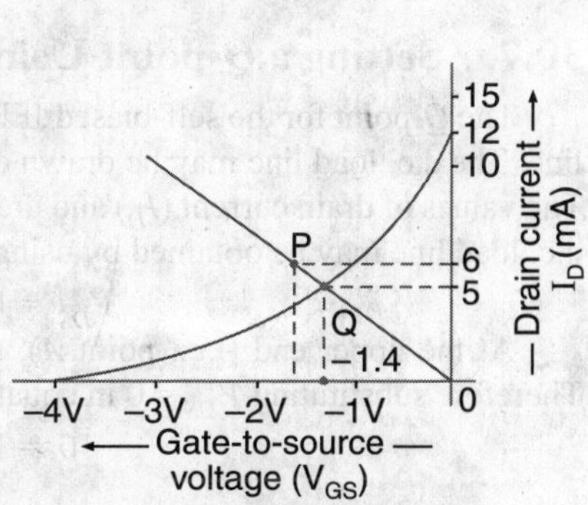

Fig. 30.9.

It is evident from this graph that the coordinates of Q-point are

$$I_D = 5\ mA$$
$$V_{GS} = -1.4\ V\ \textbf{Ans.}$$

(b) D.C. voltage between drain and ground

We know that the d.c. voltage between drain and ground,

$$V_D = V_{DD} - I_D \cdot R_D$$

$$= 15 - (5 \times 10^{-3}) \times (1.8 \times 10^3)\text{ V}$$

$$= 15 - 9 = 6\text{ V } \textbf{Ans.}$$

Example 30.5.

A self-biased P-channel JFET has a pinch off voltage (V_P) = 5 V and I_{DSS} = 12 mA. The supply voltage available is 12 V. Determine the values of resistors R_D and R_S, so that I_D = 5 mA and V_{DS} = 6 V.

Solution.

Given: V_P *i.e.*, $V_{GS(off)}$ = 5 volt; I_{DSS} = 12 mA = 12 × 10^{-3}A; V_{DD} = 12 volt, I_D = 4 mA = 4 × 10^{-3} A and V_{DS} = 6 volt.

We know that the drain current (I_D),

$$4 \times 10^{-3} = I_{DSS}\left[1 - \frac{V_{GS}}{V_{GS(off)}}\right]^2 = 12 \times 10^{-3}\left[1 - \frac{V_{GS}}{5}\right]^2$$

or
$$\frac{1}{3} = \left(1 - \frac{V_{GS}}{5}\right)^2$$

or
$$\frac{1}{\sqrt{3}} = 1 - \frac{V_{GS}}{5} \qquad \text{... (Taking square root)}$$

$$\therefore \quad V_{GS} = \left(1 - \frac{1}{\sqrt{3}}\right) \times 5 = (1 - 0.58) \times 5\text{ V} = 2.1\text{ V}$$

We also know that the source voltage,

$$V_S = V_{GS} = 2.1\text{ V}$$

and voltage drop across the source resistor (V_S),

$$2.1 = I_D \cdot R_S = (4 \times 10^{-3})\, R_S$$

$$\therefore \quad R_S = \frac{2.1}{(4 \times 10^{-3})} = 525\ \Omega\ \textbf{Ans.}$$

and drain-to-source voltage (V_{DS}),

$$6 = V_{DS} - I_D \cdot R_D = 12 - (4 \times 10^{-3}) \times R_D$$

$$R_D = 1500\ \Omega = 1.5\text{ k}\Omega\ \textbf{Ans.}$$

30.7. Setting a *Q*-point Using D.C. Load Line

The *Q*-point for the self-biased JFET circuit may also be obtained graphically by using d.c. load line. The d.c. load line may be drawn easily on the drain characteristics as shown in Fig. 30.10 (*a*). The values of drain current (I_D) and drain-to-source voltage (V_{DS}) at the upper and lower ends of the d.c. load line may be obtained by using the relation,

$$V_{DS} = V_{DD} - I_D \cdot (R_D + R_S) \qquad \text{... }(i)$$

At the upper end (*i.e.*, point *A*), the value of drain-to-source voltage (V_{DS}) is equal to zero. Therefore substituting V_{DS} = 0 in equation (*i*),

$$0 = V_{DD} - I_D \cdot (R_D + R_S)$$

$$\therefore \quad I_D = \frac{V_{DD}}{(R_D + R_S)}$$

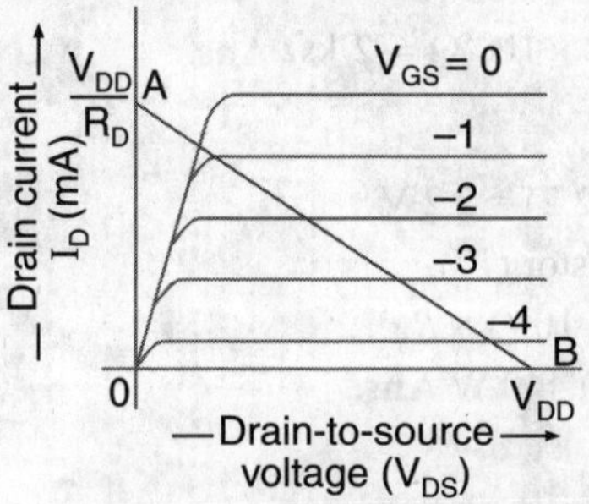

(a) Drawing d.c. load line.

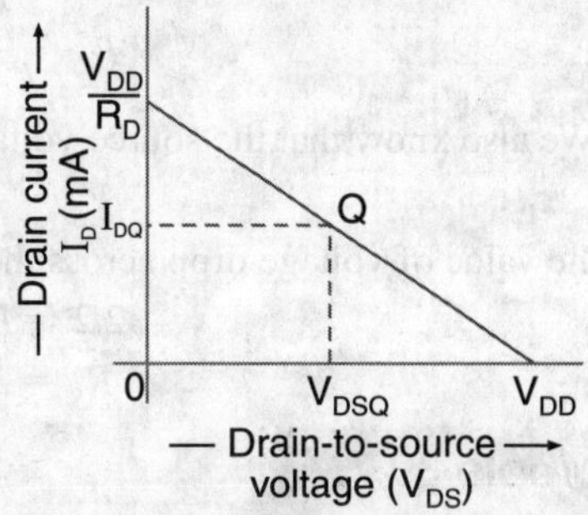

(b) Setting a Q-point.

Fig. 30.10.

Similarly, at the lower end (*i.e.*, point *B*) the value of drain current (I_D) is equal to zero. Therefore substituting, $I_D = 0$ in equation (*i*),

$$V_{DS} = V_{DD} - 0\,(R_D + R_S) = V_{DD}$$

Thus the upper and lower ends of the d.c. load line are given by the points,

$$I_D = V_{DD}/(R_D + R_S) \text{ and } V_{DS} = 0 \quad \text{... (Upper end)}$$

$$I_D = 0 \text{ and } V_{DS} = V_{DD} \quad \text{... (Lower end)}$$

The *Q*-point is set at the mid-point of the d.c. load line as shown in Fig. 30.10 (*b*). In that case, the value of drain current at *Q*-point,

$$I_{DQ} = \frac{V_{DD}}{2(R_D + R_S)}$$

and the drain-to-source voltage at the *Q*-point,

$$V_{DSQ} = \frac{V_{DD}}{2}$$

Example 30.6.

Calculate the self-bias operation point for the FET circuit shown in Fig. 30.11.

Also calculate the values of resistors R_D and R_S to obtain this bias condition. Given the maximum value of drain current as 10 mA and V_{GS} = − 2.2 V at I_D = 5 mA.

Solution.

Given: I_{DSS} = 10 mA and V_{DD} = 20 volt.

(*a*) *Operating point*

We know that the value of drain current at *Q*-point,

$$\therefore \quad I_{DQ} = \frac{I_{DSS}}{2} = \frac{10}{2} = 5 \text{ mA}$$

and the value of drain-to-source voltage,

$$V_{DSQ} = \frac{V_{DD}}{2} = \frac{20}{2} = 10 \text{ V}$$

∴ Operating point,

$$I_D = 5 \text{ mA and } V_{DS} = 10 \text{ V } \textbf{Ans.}$$

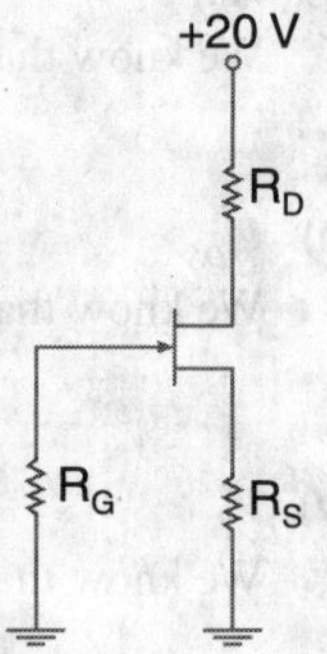

Fig. 30.11.

(*b*) *Values of R_D and R_S*

Here V_{GS} = − 2.2 volt and $I_D = 5 \text{ mA} = 5 \times 10^{-3}$ A

We know that the drain-to-source voltage (V_{DS}),

$$10 = V_{DD} - I_D \cdot R_S = 20 - (5 \times 10^{-3})\,R_D$$

$$\therefore \qquad R_D = \frac{20 - 10}{5 \times 10^{-3}} = 2 \times 10^3 \ \Omega = 2 \ \text{k}\Omega \ \textbf{Ans.}$$

We also know that the source voltage,

$$V_S = - V_{GS} = - (- 2.2) = 2.2 \text{ V}$$

and the value of voltage drop across the source resistor (V_S),

$$2.2 = I_D \cdot R_S = (5 \times 10^{-3}) \times R_S$$

$$\therefore \qquad R_S = 2.2/(5 \times 10^{-3}) \ 440 \text{ W} \ \textbf{Ans.}$$

Example 30.7.

Determine the following for the network for Fig. 30.12.

(a) V_{GSQ} *(b)* I_{DQ} *(c)* V_{DS} *(d)* V_D *(e)* V_G *(f)* V_S

(GBTU., 2011–2012)

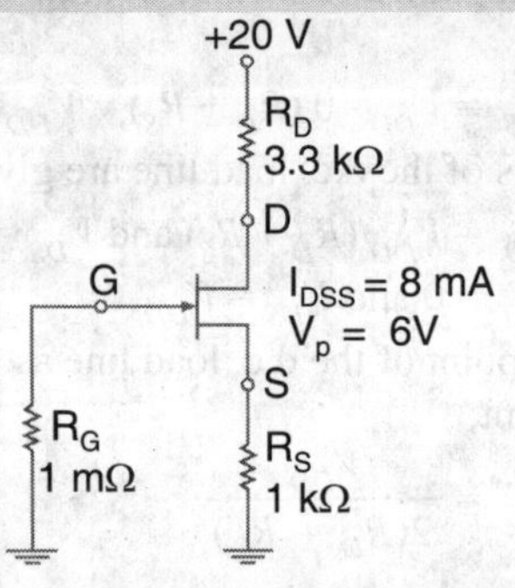

Fig. 30.12.

Solution.

Given: V_{DD} = 20 V; R_D = 3.3 kΩ = 3.3 × 10^3 Ω; I_{DSS} = 8 mA = 8 × 10^{-3} A; V_P = –6 V; R_G = 1 MΩ = 1 × 10^6 Ω and R_S = 1 kΩ = 1 × 10^3 Ω

(a) V_{DSQ}

We know that the value of drain-to-source voltage at *Q*-point,

$$V_{DSQ} = \frac{V_{DD}}{2} = \frac{20}{2} = 10 \text{ V} \ \textbf{Ans.}$$

(b) I_{DQ}

We know that the value of drain current at Q-point,

$$I_{DQ} = \frac{I_{DSS}}{2} = \frac{8 \times 10^{-3}}{2} = 4 \times 10^{-3} = 4 \text{ mA} \ \textbf{Ans.}$$

(c) V_{DS}

We know that,

$$V_{DS} = V_{DD} - I_D (R_D + R_S)$$

$$= 20 - (4 \times 10^{-3}) [(3.3 \times 10^3) + (1 \times 10^3)] = 2.8 \text{ V} \ \textbf{Ans.}$$

(d) V_S

We know that the value of source voltage,

$$V_S = I_D R_S = (4 \times 10^{-3}) \times (1 \times 10^3) = 4 \text{ V} \ \textbf{Ans.}$$

(e) V_G

We also know that,

$$V_G = 0 \ \textbf{Ans.}$$

(*f*) V_D

We know that,

$$V_D = V_{DD} - I_D R_D = 10 - [(4 \times 10^{-3}) \times (3.3 \times 10^3)] = -3.2 \text{ V } \textbf{Ans.}$$

30.8. Biasing Against Device Parameter Variation

We have already discussed that for a self-biased JFET, the *Q*-point is set at the mid-point of the transfer characteristic curve. However, it has been found that the JFET parameters *like maximum drain current (I_{DSS}) and the gate-to-source cut-off ($V_{GS(off)}$) vary with the temperature. In that case, the Q-point will also vary with the temperature and hence will not remain stable.* This degrades the performance of the FET amplifier circuit.

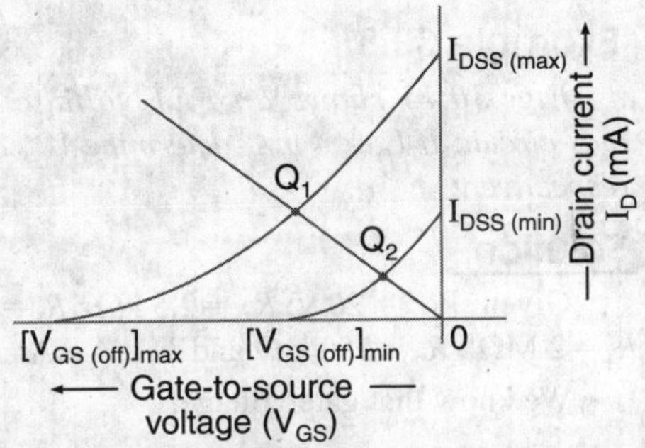

Fig. 30.13. Use of two transfer characteristic curves for setting a Q-point.

In order to provide compensation for the device parameter variation, the manufacturers provide two transistor characteristic curves as shown in Fig. 30.13. The first curve corresponds to the maximum values of I_{DSS} and $V_{GS(off)}$. In such a case, we draw a bias line through both the curves, which is near the mid-point of either curve. Then read the values of gate-to-source voltage (V_{GS}) and the drain current (I_D) for either *Q*-point. The optimum value of source resistor (R_S) is given by the ratio of gate-to-source voltage (V_{GS}) and drain current (I_D).

Sometimes, the manufacturers do not supply the transfer characteristic curves, but provide the maximum and minimum values of the I_{DSS} and $V_{GS(off)}$. In that case, we calculate the maximum and minimum values of source resistor (R_S) and then take the average of these two values. This average value will provide a *Q*-point near the mid-point of the either transfer characteristic curve.

30.9. Voltage Divider Bias

Fig. 30.14 shows the voltage divider bias circuit. The name 'voltage divider' is derived from the fact that resistor R_1 and R_2 are connected on the gate side (between the V_{DD} supply and ground) form a voltage divider. Assuming the current to be zero, the voltage is given by the relation,

$$V_G = \left(\frac{R_2}{R_1 + R_2}\right) V_{DD}$$

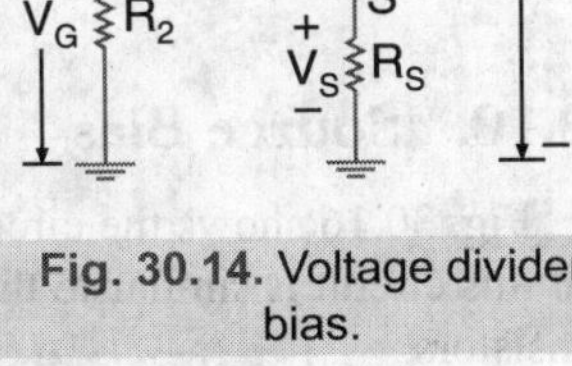

Fig. 30.14. Voltage divider bias.

and the voltage from source to ground,

$$V_S = V_G - V_{GS}$$

∴ Value of drain current,

$$I_D = \frac{V_S}{R_S} = \frac{V_G - V_{GS}}{R_S} \quad ... (i)$$

and the d.c. voltage from drain to ground,

$$V_D = V_{DD} - I_D \cdot R_D$$

Now, if the gate voltage is very large, as compared to gate-to-source voltage, then the drain current is approximately constant for any JFET. It has been found that in actual practice the voltage divider bias is less effective with JFET's than bipolar transistors. It is because of the fact that in bi-polar transistors the base-to-emitter voltage (V_{BE}) is approximately 0.7 V with only minor variations from one transistor to next. However, in a JFET, the gate-to-source voltage (V_{GS}) can vary several volts from one JFET to another. As a result of this, it is difficult to make gate voltage (V_G) large enough than the gate-to-source voltage (V_{GS}).

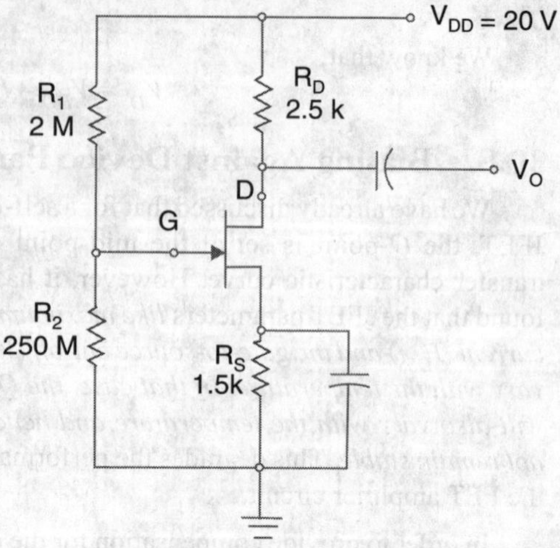

Fig. 30.15.

Example 30.8.

Fig. 30.15 shows an FET voltage divider bias circuit. If I_D = 4 mA, determine V_{GS} and V_{DS} respectively.

Solution.

Given: $V_{DD} = 20$ V; $R_D = 2.5$ kΩ ; $R_S = 1.5$ kΩ; $R_1 - 2$ MΩ ; $R_2 = 250$ k Ω and $I_D = 4$ mA.

We know that gate voltage,

$$V_G = V_{DD}\left(\frac{R_2}{R_1 + R_2}\right)$$

$$= 20\times\left(\frac{250\times10^3}{\{(2\times10^6)+(250\times10^3)\}}\right) = 2.22 \text{ V}$$

and the voltage from source to ground,

$$V_S = V_G - V_{GS}$$

The value of V_{GS} is obtained as follows. We know that drain current (I_D),

$$4\times10^{-3} = \frac{V_S}{R_S} = \frac{V_S}{1.5\times10^3}$$

$$V_S = (4\times10^{-3})\times(1.5\times10^3) = 6.0 \text{ V}$$

Using equation (i), we find that,

$$V_{GS} = V_G - V_S = 2.22 - 6.0 = -3.8 \text{ V } \textbf{Ans.}$$

We know that dc voltage from drain to ground,

$$V_D = V_{DD} - I_D \cdot R_D$$

$$= 20 - \{(4\times10^{-3})\times(2.5\times10^3)\}$$

$$= 4.0 \text{ V } \textbf{Ans.}$$

30.10. Source Bias

Fig. 30.16 shows the circuit of a source bias for JFET. It may be noted that this circuit is similar to the emitter bias circuit used for biasing bipolar transistors.

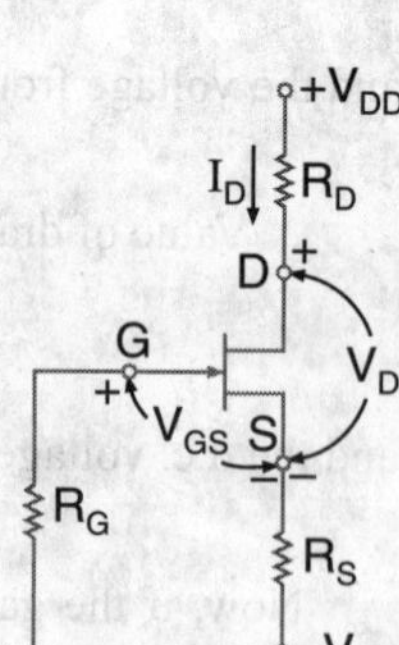

Fig. 30.16. Source bias.

The value of drain current is given by the relation,

$$I_D = \frac{V_{SS} - V_{GS}}{R_S}$$

Since, $V_{SS} >> V_{GS}$, therefore drain current,

$$I_D \approx \frac{V_{SS}}{R_S}$$

30.11. Current Source Bias

Fig. 30.17 shows the current-source bias circuit for a JFET. The circuit of Fig. 30.16 (*a*) uses two supplies (positive and negative), while that of Fig. 30.16 (*b*) employs only one (*i.e.*, positive) supply.

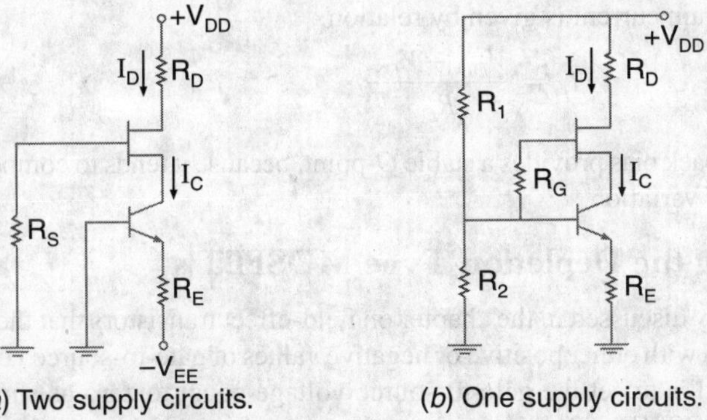

Fig. 30.17. Current-source bias circuit.

First of all, consider the two supply current-source bias circuit. Here the bipolar transistor is emitter biased. Therefore its collector current is given by the relation,

$$I_C = \frac{V_{EE} - V_{BE}}{R_E} = \frac{V_{EE}}{R_E} \quad \text{... (If } V_{EE} >> V_{BE}\text{)}$$

The drain current of the JFET is equal to the collector current because the bipolar transistor acts like a d.c. current-source, *i.e.*,

$$I_D = I_C$$

Now consider the one supply current-source bias circuit. Here the bipolar transistor is voltage divider biased. Its collector current is given by the relation,

$$I_C = \frac{V_B - V_{BE}}{R_E}$$

and the drain current,

$$I_D = I_C$$

The current-source bias circuit provides a solid *Q*-point with JFET. Hence this circuit is very useful in actual practice.

30.12. Biasing the Enhancement Type MOSFET's

We have already discussed in the chapter on field-effect transistors that the operation of an enhancement types MOSFET requires a gate-to-source voltage greater than its threshold value (*i.e.*, $V_{GS} > V_{GS(th)}$). This eliminates the use of self-bias because it makes the gate voltage negative with respect to source. There are two methods for biasing the enhancement type MOSFET's namely the voltage divider bias and the drain feedback bias. The voltage divider bias has been already discussed in Art. 30.9. The drain feedback bias is similar to the collector feedback bias of bipolar transistors and is discussed as below:

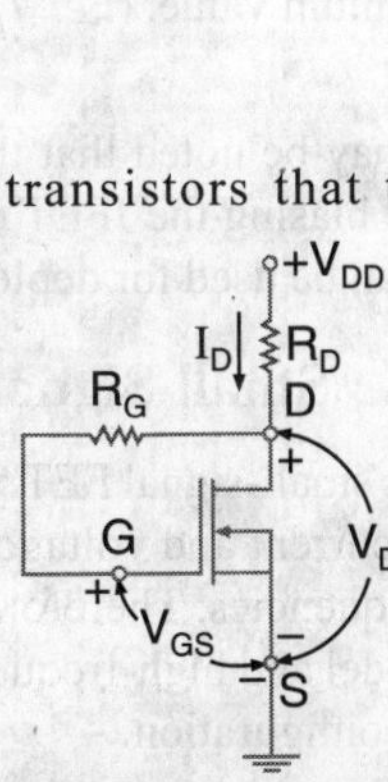

Fig. 30.18. Drain feedback bias.

Fig. 30.18 shows the circuit of a drain feedback bias. It may be noted that this type of bias is used only for enhancement type

MOSFET's. Since the value of gate current in FET's is approximately zero, therefore there is no voltage drop across gate resistor (R_G). This makes the gate-to-source voltage equal to the drain-to-source voltage, *i.e.*,

$$V_{GS} = V_{DS}$$

The value of drain current is given by relation,

$$I_D = \frac{V_{DD} - V_{DS}}{R_D}$$

The drain feedback bias provides a stable *Q*-point, because it tends to compensate for any changes in device parameter variation.

30.13. Biasing the Depletion Type MOSFET's

We have already discussed in the chapter on field-effect transistors that the depletion type MOSFET can be operated with either positive or negative values of gate-to-source voltage (V_{GS}). Therefore the simplest method is to set the gate-to-source voltage equal to zero as shown in Fig. 30.18 (*a*). This will bias the MOSFET at a point lying on the vertical axis of the transfer characteristic curve as shown in Fig. 30.19 (*b*). If an a.c. signal is applied at the gate, it will vary the gate-to-source voltage above and below the *Q*-point.

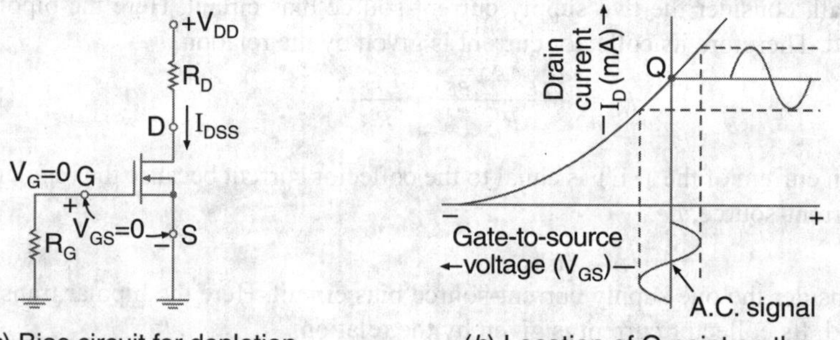

(*a*) Bias circuit for depletion type MOSFET.

(*b*) Location of Q-point on the transfer characteristic curve.

Fig. 30.19.

It may be noted that the biasing circuit of a depletion type MOSFET has no applied gate or source voltage. Therefore the gate-to-source voltage is equal to zero and the drain current is equal to its maximum value, (*i.e.*, $I_D = I_{DSS}$). The drain-to-source voltage is given by the relation,

$$V_{DS} = V_{DD} - I_D \cdot R_D = V_{DD} - I_{DSS} \cdot R_D$$

It may be noted that the zero bias is possible only with depletion type MOSFET. It cannot be used for biasing the JFET or bipolar transistor. However, all the biasing methods, discussed for the JFET, can be used for depletion type MOSFET in the depletion mode but not in enhancement mode.

30.14. Small-Signal FET Models

The small-signal FET *model (valid both for JFET and MOSFET) is used to relate small changes in FET current and voltages about the quiescent operating point. The model is different at low and high-frequencies. Therefore we shall study the small-signal models separately as the low-frequency FET model and high-frequency model. In both these models, the FET will be considered in common source configuration.

* It is also called an equivalent circuit.

30.15. Small-Signal Low-Frequency FET Model

Fig. 30.20 shows the small-signal low-frequency model of a field-effect transistor. In this model, the gate-to-source junction is represented by an open circuit and no current is drawn by the input terminal of the field-effect transistor. It is because of the fact, that the input resistance (*i.e.*, the resistance of gate-source junction) is very large. Its value at d.c. or zero frequency is typically 10^8 to 10^{10} Ω for JFET's and 10^{10} to 10^{14} Ω for MOSFET's. For all practical purposes, these values can be considered to be so large that the input of FET can be considered as an open circuit.

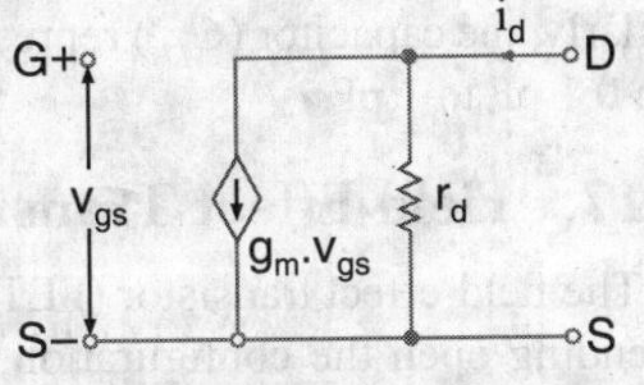

Fig. 30.20. Low-frequency FET model.

It will be interesting to know that although the gate-source junction appears as an open circuit, yet the gate-to-source voltage affects the value of drain current. It is indicated by a voltage-controlled current source ($g_m \cdot v_{gs}$) whose value is proportional to the gate-to-source voltage. The FET transconductance (g_m) is measured in milliamperes per volt (mA/V), milli siemens (mS) or milli mhos (mΩ). Typical values of transconductance are from 0.5 mA/V to 10 mA/V for JFET's and 0.5 mA/V to 20 mA/V for MOSFET's.

The FET drain resistance (also called FET output resistance) is represented by the resistance (r_D). Typical values of drain resistance are from 100 kΩ to 1 MΩ for JFET's and 1 kΩ to 50 kΩ for MOSFET's.

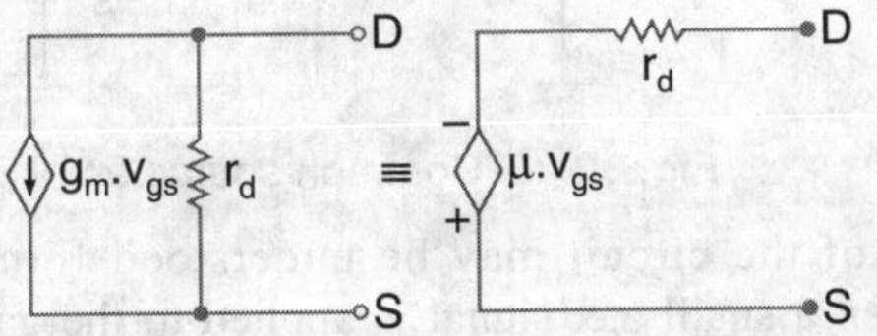

(*a*) Voltage-controlled current source with its equivalent voltage-controlled voltage source.

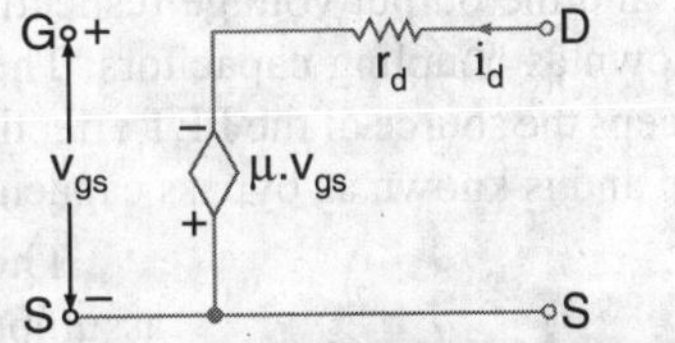

(*b*) Another form of low-frequency FET model.

Fig. 30.21.

It is possible to convert the voltage-controlled current source ($g_m \cdot v_{gs}$) into its equivalent voltage-controlled voltages source as shown in Fig. 30.21 (*a*).

The equivalent circuit consists of a voltage-controlled voltage source ($\mu \cdot v_{gs}$) in series with the drain resistance (r_D). The parameter (μ) is the amplification factor of the field effect transistor and its value is equal to the product of transconductance (g_m) and the drain resistance (r_D). Mathematically, the amplification factor,

$$\mu = g_m \cdot r_D$$

Fig. 30.21 (*b*) shows another form of the low frequency model of a field effect transistor using the voltage source representation. Sometimes this representation is also very useful in analysing the FET amplifiers.

30.16. Small-Signal High-Frequency FET Model

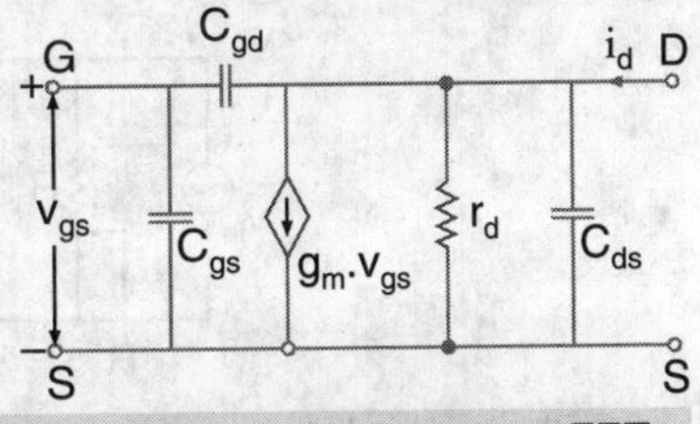

Fig. 30.22. High-frequency FET model.

Fig. 30.22 shows the small-signal high-frequency model of a field-effect transistor. It is identical to the low-frequency model, except the addition at capacitances between each pair of terminals. The capacitor, (c_{gs}) represents the barrier capacitance between the gate and source. Its typical value is from 1 pF to 10 pF for both JFET's and MOSFET's. The capacitor (C_{gD}) represents the barrier capacitance between the gate and drain. Its typical value is also from 1 pF to 10 pF for both JFET's and MOSFET's. Similarly, the capacitor (C_{Ds}) represents the drain-to-source capacitance. The typical value of C_{Ds} is from 0.1 pF to 1 pF.

30.17. Field-Effect Transistor Amplifier

The field-effect transistor (FET) has a capability to amplify a.c. signals like a bipolar transistor. Depending open the configuration, the FET amplifiers may be studied under the following three heads:

1. Common source amplifier.
2. Common drain amplifier, and
3. Common gate amplifier.

30.18. Common Source Amplifier

Fig. 30.23 shows the circuit of a common source N-channel JFET amplifier. It is similar to a common emitter amplifier. Here the resistors R_1 and R_2 (called a voltage divider) are used to bias the field-effected transistor. The capacitor (C_1) and (C_2) are used to couple the a.c. input voltage source and the output voltage respectively, these are known as coupling capacitors. The capacitor (C_S) keeps the source of the FET effectively at a.c. ground and is known as bypass capacitor.

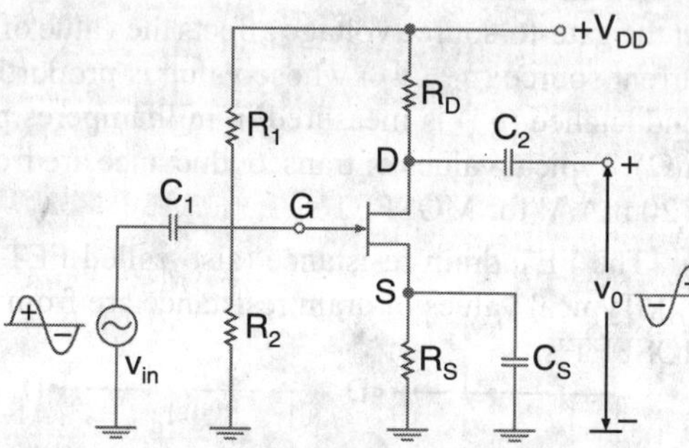

Fig. 30.23. Common source amplifier.

Silicon N-channel MOSFET.

The operation of the circuit may be understood from the assumption that when a small a.c. signal is applied to the gate, it produces variations in the gate-to-source voltage. This produces variations in the drain current. As the gate-to-source voltage increases, the drain current also increases. As a result of this, the voltage drop across the resistor (R_D) also increases. This causes the drain voltage to decrease. It means that the positive half cycle of the input voltage produces the negative half cycle of the output voltage. In other words, the output voltage (at the drain) is 180° out-of-phase with the input voltage (at the gate). This phenomenon of *phase inversion* is similar to that exhibited by a common emitter bipolar transistor amplifier.

30.19. Analysis of Common Source Amplifier

Fig. 30.24 shows the a.c. equivalent circuit of a common source amplifier. This circuit has been obtained from the amplifier circuit shown in Fig. 30.23 by short-circuiting the capacitors and

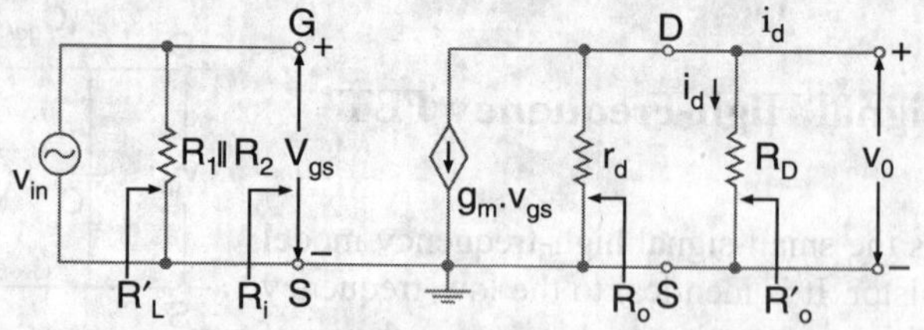

Fig. 30.24. A.C. equivalent circuit of a common source amplifier.

the d.c. voltage supplies. The field-effect transistor is also replaced by its low-frequency model (or equivalent circuit). Now we shall use this circuit to find the expressions for amplifier voltage gain, *input resistance and output resistance.

1. *Voltage gain.* It is the ratio of the output voltage (v_o) to the input voltage (v_{in}). Mathematically the voltage gain,

$$A_v = \frac{v_o}{v_{in}}$$

It may be noted that the current from the current source splits between the resistors r_D and R_D. The current through resistor R_D (as per current divider rule) is given by the relation,

$$i_D = \frac{r_d}{R_D + r_d}(g_m \cdot v_{gs})$$

and the output voltage,

$$v_o = -i_d \cdot R_D = -\left(\frac{r_d}{R_D + r_d}\right)(g_m \cdot v_{gs})\,R_D$$

$$= -g_m\left(\frac{r_d \cdot R_D}{R_D + r_d}\right)v_{gs}$$

$$= -g_m\,(r_D \parallel R_D) \cdot v_{gs}$$

$$= -g_m \cdot r_L \cdot v_{gs} \qquad \dots (\because r_L = r_D \parallel R_D)$$

$$= -g_m \cdot r_L \cdot v_{in} \qquad \dots (\because v_{in} = v_{gs})$$

The minus sign indicates that the output voltage is 180° out-of-phase with the input voltage.

∴ Voltage

$$A_v = \frac{v_o}{v_{in}} = -g_m \cdot r_L$$

It may be noted that if the drain resistance (r_D) is sufficiently greater than resistor R_D, (*i.e.*, $r_D \geq 10\,R_D$) then the equivalent resistance of the resistors r_D and R_D in parallel,

$$r_L = R_D$$

and the voltage gain,

$$A_v = g_m \cdot R_D$$

Note. If the resistance R_D is much larger than r_D (*i.e.*, $R_D >> r_D$) then a.c. load resistance (r_L) is approximately equal to r_D. In that case, the amplifier gain is equal to the FFT amplification factor (μ), which is the theoretical maximum voltage gain of the FET. But in actual practice, we cannot select R_D to be much larger than r_D because it makes the d.c. bias current in the range of microamperes. The FET transconductances (*i.e.*, g_m) is near zero at this low bias current value.

2. *Input resistance.* It is the ratio of the input voltage (v_{in}) to the input current (i_{in}). Mathematically, the input resistance,

$$R_i = \frac{v_{in}}{i_{in}}$$

We know that input resistance (R_i) of a field-effect transistor is very high and hence can be considered to be infinity (*i.e.*, open-circuit). However, the input resistance of the amplifier stage (R'_i) is equal to the parallel combination of resistors R_1 and R_2 and the FET input resistance (R_i). Thus

* In some books, the input and output resistances are referred as the input and output impedance respectively. But as we are studying the amplifier performance at low frequencies, at which the effect of capacitances is negligible, therefore it is more appropriate to refer the impedances as resistances.

$$R'_i = (R_1 \parallel R_2) \parallel R_i$$
$$= R_1 \parallel R_2 \qquad \text{... (When } R_i \text{ is infinite)}$$

It may be noted that if we use a self-bias circuit (instead of a voltage divider bias), then the input resistance of the amplifier stage,

$$R'_i = R_G$$

3. ***Output resistance.*** It is the ratio of the output voltage (v_o) to the output current (i_D). Mathematically, the output resistance,

$$R'_O = \frac{v_o}{i_d}$$

We know that the output resistance (R_O) of a field-effect transistor is r_D. But for the amplifier stage, the output resistance is the parallel combination of resistors R_D and r_D. *i.e.,*

$$R'_o = R_D \parallel r_D$$

However, if $r_D >> R_D$, then the output resistance of amplifier stage,

$$R_o = R_D$$

Example 30.9.

A certain JFET has a g_m of 4 mS. With an external drain resistance of 1.5 kΩ, find the value of ideal voltage gain.

Solution.

Given: $g_m = 4 \text{ mS} = 4 \times 10^{-3}$ S and $R_D = 1.5 \text{ k}\Omega = 1.5 \times 10^3\ \Omega$

We know that the voltage gain,

$$A_v = -g_m \cdot R_D = -(4 \times 10^{-3}) \times (1.5 \times 10^3) = -6 \textbf{ Ans.}$$

Example 30.10.

A JFET amplifier has g_m = 2.5 mA/V and r_D = 500 kΩ. The load resistance is 10 kΩ. Find the value of voltage gain. (*Grad. I.E.T.E., Dec. 1987*)

Solution.

Given: $g_m = 2.5 \text{ mA/V} = 2.5 \times 10^{-3}$ A/V;

$$r_d = 500 \text{ k}\Omega \text{ and } R_D = 10 \text{ k}\Omega$$

We know that the a.c. equivalent resistance,

$$r_L = \frac{R_D \times r_d}{R_D + r_d} = \frac{10 \times 500}{10 \times 500} \text{ k}\Omega = 9.8 \text{ k}\Omega = 9.8 \times 10^3\ \Omega$$

∴ Voltage gain,

$$A_v = -g_m \cdot r_L = -(2.5 \times 10^{-3}) \times (9.8 \times 10^3)$$
$$= -24.5 \textbf{ Ans.}$$

Example 30.11.

The input and output resistances of the FET amplifier are shown in Fig. 30.25. Calculate the value of voltage gain. The FET amplifier has g_m = 2 mA/V and r_d = 40 kΩ.

Solution.

Given: $g_m = 2 \text{ mA/V} = 2 \times 10^{-3}$ A/V; $r_d = 40$ kΩ; $R_D = 20$ kΩ and $R_G = 100$ MΩ.

Voltage gain

We know that the a.c. equivalent resistance,

$$r_L = \frac{R_D \times r_d}{R_D + r_d} = \frac{20 \times 40}{20 \times 40} = 13.3 \text{ k}\Omega = 13.3 \times 10^3\ \Omega$$

∴ *Voltage gain*

$$A_v = -g_m \cdot r_L = -(2 \times 10^{-3}) \times (13.3 \times 10^3)$$
$$= -26.7 \textbf{ Ans.}$$

Input resistance

We know that the input resistance,

$$R'_i = R_G = 100 \text{ M}\Omega \textbf{ Ans.}$$

Outpur resistance

We also know that the output resistance,

$$R'_o = r_L$$
$$= 13.3 \text{ k}\Omega \textbf{ Ans.}$$

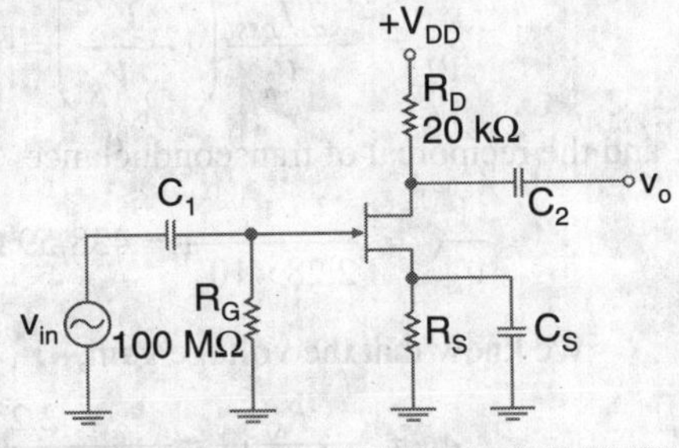

Fig. 30.25.

Example 30.12.

Calculate the voltage gain, $A_v = V_o / V_i$ at 1 kHz for the circuit in Fig. 30.26. The FET parameters are g_m = 2 mA/V and r_d = 10 k. Neglect capacitances. (Nagpur Univ. Winter 2003)

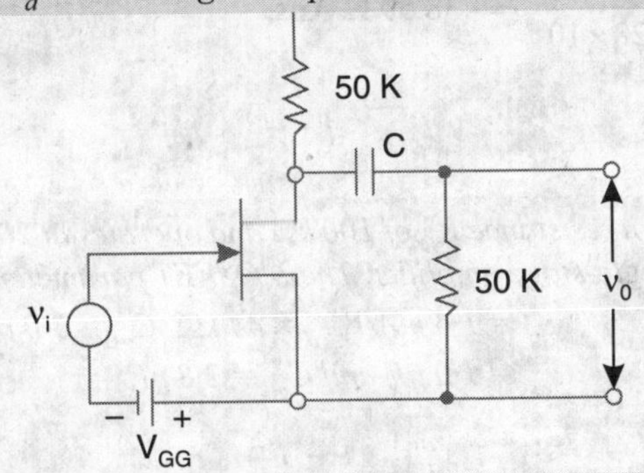

Fig. 30.26.

Solution.

Given : g_m = 2 mA/V; r_d = 10 kΩ; R_D = 50 kΩ.

We know that ac load resistance,

$$r_L = r_d \parallel R_D = 10\ k \parallel 50\ k$$
$$= \frac{10\ k \times 50\ k}{10\ k + 50\ k} = 8.33 \text{ k}\Omega$$

and voltage gain of the FET,

$$A_v = -g_m \cdot r_L = -(2 \text{ mA/V}) \times (8{\cdot}33 \text{ k}\Omega)$$
$$= -16{\cdot}67 \textbf{ Ans.}$$

Example 30.13.

For the Fig. 30.27 V_{GS} = –2.86 V and I_D = 4.56 mA. Find g_m, r_d, R'_o and A_V. Assume I_{DSS} = 16 mA and V_p = –4 V and Y_{os} = 25 μ℧. (VTU., June/July 2011)

Solution.

Given : V_{GS} = –2.86 V; I_D = 4.56 mA = 4.5 × 10^{-3} A; I_{DSS} = 16 mA = 16 × 10^{-3} A; V_p = –4 V; R_G = 1 mA = 1 × 10^{-6} A and R_S = 2.2 kΩ = 2.2 × 10^3 Ω.

We know that the transconductance,

$$g_m = -\frac{2\,I_{DSS}}{V_p}\left(1-\frac{V_{GS}}{V_p}\right) = -\frac{2\times(16\times10^{-3})}{(-4)}\left(1-\frac{-2.86}{-4}\right) = 2.28 \text{ mA/V}$$

and the reciporcal of transconductance,

$$\frac{1}{g_m} = \frac{1}{2.28\times10^{-3}} = 438.59\ \Omega$$

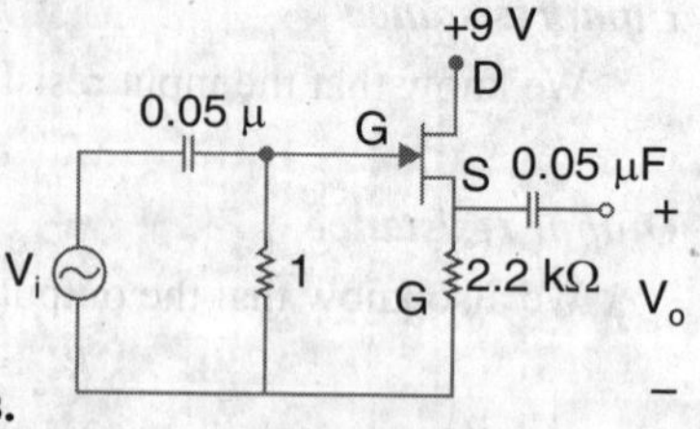

Fig. 30.27.

We know that the voltage gain,

$$A_v = \frac{R_S}{R_S+\frac{1}{g_m}} = \frac{2.2\times10^3}{(2.2\times10^3)+(438.59)} = 0.833 \textbf{ Ans.}$$

We also know that the output resistance,

$$R'_o = \frac{1}{g_m} = \frac{1}{2.28\times10^{-3}} = 438.59\ \Omega \textbf{ Ans.}$$

Example 30.14.

A MOSFET has a drain resistance R_D of 100 kΩ and operates at 20 kHz. Calculate the voltage gain of this device as a single-stage amplifier. The MOSFET parameters are :

$$g_m = 1{\cdot}6 \text{ mA/V},\ r_d = 44 \text{ k}\Omega,\ C_{gs} = 3{\cdot}0 \text{ pF}$$
$$C_{ds} = 1{\cdot}0 \text{ pF and } C_{gd} = 2{\cdot}8 \text{ pF}$$

(*UPSC Engg. Services* 1996)

Solution.

Given : R_D = 100 kΩ; g_m = 1·6 mA/V, r_d = 44 kΩ, C_{gs} = 3·0 pF; C_{ds} = 1·0 pF and C_{gd} = 2·8 pF.

Since the operating frequency is 20 kHz (a low frequency), interelectrode capacitances, C_{gs}, C_{ds} and C_{gd} are not be considered.

We know that the ac load resistance,

$$r_L = R_D \parallel r_d = (100 \text{ k}\Omega) \parallel (44 \text{ k}\Omega)$$
$$= \frac{100\ k \times 44\ k}{100\ k + 44\ k} = 30.56 \text{ k}\Omega$$

and the voltage gain,

$$A_v = -g_m \cdot r_L$$
$$= 1{\cdot}6 \text{ mA/V} \times 30{\cdot}56 \text{ k}\Omega$$
$$= 48{\cdot}9 \textbf{ Ans.}$$

30.20. Effect of A.C. Load on Amplifier Parameters

Consider a common source amplifier shown in Fig. 30.28. Here a load resistance (R_L) is connected to the amplifier's output through a coupling capacitor. In this case, the total or effective drain resistance is the parallel combination of resistors R_D and R_L *i.e.*,

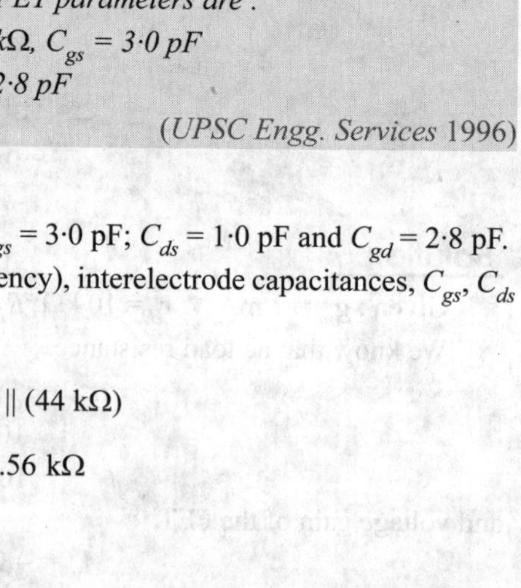

Fig. 30.28. Common source amplifier connected to a load resistor.

$$r_L = R_D \parallel R_L$$

It may be noted that in the above expression, if we also consider the FET drain resistance (r_D), then the effective drain resistance,

$$r_L = (R_D \parallel R_L) \parallel r_d$$

The effect of load resistor (R_L) is to reduce the voltage gain as well as the output resistance. The value of voltage gain with load resistance is given by the relation,

$$A_v = -g_m \cdot r_L$$
$$= -g_m (R_D \parallel R_L) \parallel r_d$$
$$= -g_m (R_D \parallel R_L) \qquad \text{... (If } r_d >> (R_D \parallel R_L))$$

and the output resistance, of the amplifier stage

$$R'_o = (R_D \parallel R_L) \parallel r_d$$
$$= R_D \parallel R_L \qquad \text{...(If } r_d >> (R_D \parallel R_L))$$

Example 30.15.

Fig. 30.29 shows the circuit of a common source FET amplifier.

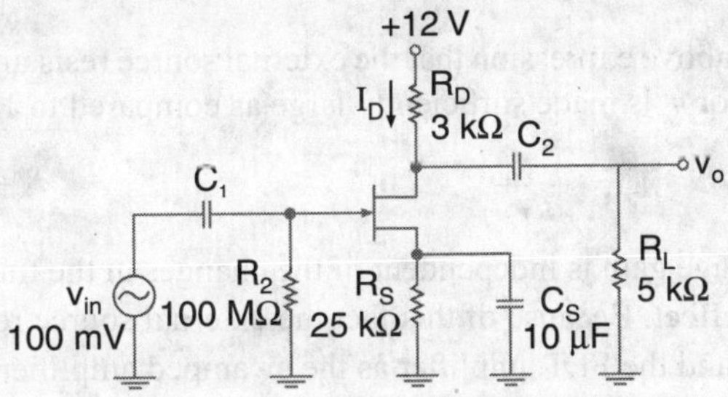

Fig. 30.29.

Find the value of r.m.s. output voltage. The g_m is 4500 µS and I_D is 2 mA. Neglect the FET drain resistance (r_d).

Solution.

Given: $g_m = 4500\ \mu S = 4500 \times 10^{-6}\ S$; $R_D = 3\ k\Omega$, $R_L = 5\ k\Omega$; $v_{in} = 100\ mV = 100 \times 10^{-3}$ V and $I_D = 2$ mA.

We know that the effective value of a.c. drain resistance,

$$r_L = R_D \parallel R_L = \frac{R_D \times R_L}{R_D + R_L} = \frac{3 \times 5}{3 + 5}\ k\Omega = 1.875\ k\Omega = 1875\ \Omega$$

∴ Output voltage,

$$v_o = -g_m \cdot r_L \cdot v_{in}$$
$$= -(4500 \times 10^{-6}) \times (1875) \times (100 \times 10^{-3})$$
$$= 0.844 \text{ V } \textbf{Ans.}$$

30.21. Effect of External Source Resistance on Voltage Gain

We have already discussed a field-effect transistor amplifier in which the source resistor is completely bypassed for a.c. signals. It means that for a.c. signals, the source is grounded. Now consider the amplifier circuit in which we include an external resistor (V_S) to the source resistance (R_S) as shown in Fig. 30.30. The source is no longer at a.c. ground. The drain current through resistor (r_S) produces an a.c. voltage between the source and ground.

It may be noted that the total input voltage between the gate and ground,

$$v_{in} = v_{gs} + i_d \cdot r_s$$

and the output voltage,

$$v_o = -\, i_d \cdot r_L$$

where r_L is the parallel combination of resistor R_D and the FET output resistance (r_D). Therefore voltage gain,

$$A_v = \frac{v_o}{v_{in}} = -\frac{i_d \cdot r_L}{v_{gs} + i_d \cdot r_s}$$

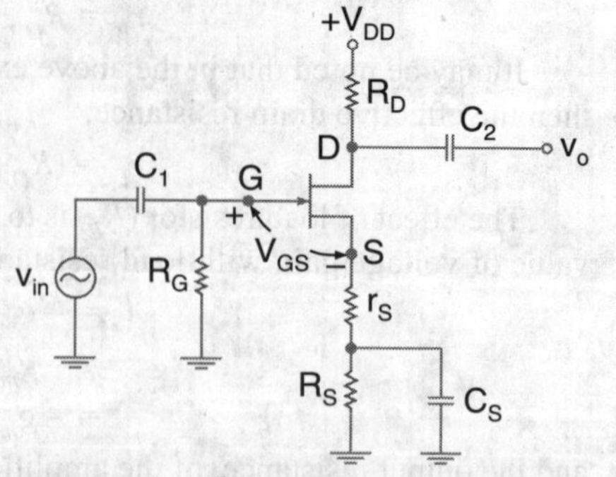

Fig. 30.30. JFET amplifier with external source resistance.

Substituting the value of i_D (equal to $g_m \cdot v_{gs}$) in the above expression,

$$A_v = -\frac{g_m \cdot v_{gs} \cdot r_L}{v_{gs} + g_m \cdot v_{gs} \cdot r_s} = -\frac{g_m \cdot v_{gs} \cdot r_L}{v_{gs}\,(1 + g_m \cdot r_s)} = -\frac{g_m \cdot r_L}{1 + g_m \cdot r_s} = \frac{r_L}{r_e + \dfrac{1}{g_m}}$$

It is evident from the above expression that the external source resistance (r_s) reduces the voltage gain. If the value of resistor r_s is made sufficiently large as compared to $1/g_m$, then the voltage gain,

$$A_v = -\frac{r_L}{r_s}$$

It means that the voltage gain is independent of the changes in the transconductance value. It is known as the swamping effect. Because of this fact the external source resistance (r_s) is referred to as the swamping resistor and the FET amplifier as the swamped amplifier.

Example 30.16.

Fig. 30.31 shows the circuit of a swamped FET amplifier.

Determine the voltage gain (a) when $R_L = 0$ and (b) when R_L is 100 kΩ. Neglect the FET output resistance (r_D) Take g_m = 4 mS.

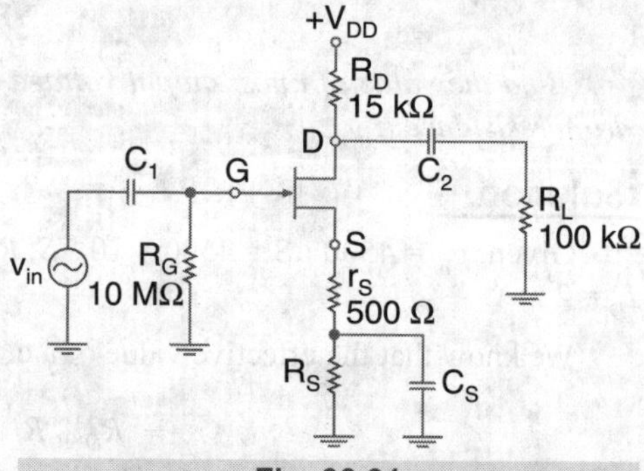

Fig. 30.31.

Solution.

Given: $g_m = 4\text{ mS} = 4 \times 10^{-3}\ S$; $R_D = 1.5\ \text{k}\Omega$; $R_G = 10\ \text{M}\Omega$ and $r_s = 500\ \Omega$.

(a) Voltage gain when R_L is zero

We known that the effective a.c. drain resistance,

$$r_L = R_D = 1.5 \times 10^3\ \Omega$$

and the voltage gain,

$$A_v = -\frac{g_m \cdot r_L}{1 + g_m \cdot r_s}$$

$$= -\frac{(4 \times 10^{-3}) \times (1.5 \times 10^3)}{1 + (4 \times 10^{-3}) \times 500}$$

$$= -\ 2$$ **Ans.**

(b) *Voltage gain when R_L is 100 kΩ*

We know that the effective a.c. drain resistance,

$$r_L = \frac{R_D \times R_L}{R_D + R_L} = \frac{1.5 \times 100}{1.5 + 100} = 1.48 \text{ k}\Omega = 1.48 \times 10^3 \ \Omega$$

and the voltage gain,

$$A_v = -\frac{g_m \cdot r_L}{1 + g_m \cdot r_s} = -\frac{(4 \times 10^3) \times (1.48 \times 10^3)}{1 + (4 \times 10^{-3}) \times 500}$$

$$= -1.97 \textbf{ Ans.}$$

Example. 30.17.

Determine the voltage gain of the amplifier shown in Fig. 22.32, with (i) unbypassed R_S and (ii) with bypassed R_S. Assume the output persistance (r_d) of the FET is infinity. The JFET parameters :

$$I_{DSS} = 10 \text{ mA and } V_p = -3{\cdot}5 \text{ V. The Q-point}$$

$$I_{DQ} = 2{\cdot}3 \text{ mA and } V_{GSQ} = -1{\cdot}8 \text{ V.}$$

(Dip. IETE., Dec. 2000)

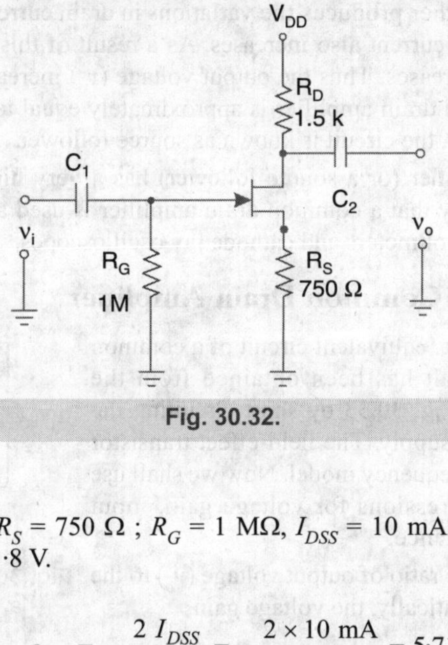

Fig. 30.32.

Solution.

Given : $R_D = 1{\cdot}5$ kΩ ; $R_S = 750$ Ω ; $R_G = 1$ MΩ, $I_{DSS} = 10$ mA; $V_p = -3{\cdot}5$ V. The Q-point $I_{DQ} = 2{\cdot}3$ mA and $V_{GSQ} = -1{\cdot}8$ V.

We know that,

$$g_{mo} = -\frac{2\, I_{DSS}}{V_p} = -\frac{2 \times 10 \text{ mA}}{-3{\cdot}5 \text{ V}} = 5{\cdot}71 \text{ mA/V}$$

At the Q-point,

$$g_m = g_{mo}\left(1 - \frac{V_{GS}}{V_p}\right) = 5{\cdot}71\left(1 - \frac{-1{\cdot}8}{-3{\cdot}5}\right)$$

$$= 2{\cdot}77 \text{ mA/V}$$

The ac load resistance,

$$r_L = r_d \,\|\, R_D = R_D$$

(i) *Unbypassed R_S*

We know that the Voltage gain with unbypassed R_S

$$A_v = -\frac{g_m \cdot r_L}{1 + g_m R_S} = \frac{2{\cdot}77 \text{ mA/V} \times 1{\cdot}5\, k}{1 + 2{\cdot}77 \text{ mA/V} \times 0{\cdot}750\, k}$$

$$= -\frac{4{\cdot}155}{1 + 2{\cdot}08} = -1{\cdot}35 \textbf{ Ans.}$$

(ii) Bypassed R_S

We know that the voltage gain with bypassed R_S,

$$A_v = -g_m \cdot r_L = -2.77 \text{ mA/V} \times 1.5 \text{ k} = -4.155 \textbf{ Ans.}$$

30.22. Common Drain Amplifier

Fig. 30.33 shows the circuit of a common drain amplifier. It is similar to common collector (or emitter follower) amplifier. Self-biasing is used in the circuit. The input signal is applied to the gate through a coupling capacitor (C_1). And the output is taken from the source terminal through the coupling capacitor (C_2).

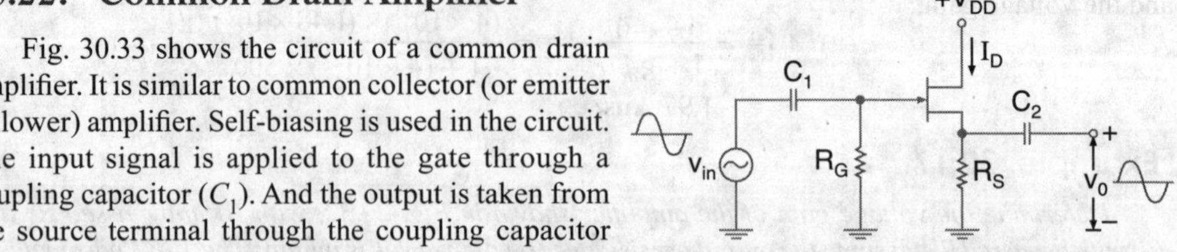

Fig. 30.33. Common drain amplifier.

The operation of the amplifier may be understood by assuming that when a small a.c. signal is applied to the gate, it produces variations in the gate-to-source voltage. This further produces the variations in drain current (I_D). As the gate-to-source voltage increases, the drain current also increases. As a result of this, the voltage drop across the source resistor (R_S) also increases. Thus the output voltage (v_o) increases. It may be noted that the output voltage of a common drain amplifier is approximately equal to and in phase with the input voltage. Because of this fact, the circuit is known as source follower.

A common drain amplifier (or a source follower) has a very high value of input resistance. It is because of this property that a common drain amplifier is used at the front end of measuring instruments like electronic voltmeters and cathode ray oscilloscopes.

30.23. Analysis of a Common Drain Amplifier

Fig. 30.34 shows the a.c. equivalent circuit of a common drain amplifier. This circuit has been obtained from the amplifier circuit shown in Fig. 30.33 by short circuiting the capacitors and d.c. voltage supply. The field-effect transistor is also replaced by its low-frequency model. Now we shall use this circuit to find the expressions for voltage gain, input resistance and output resistance.

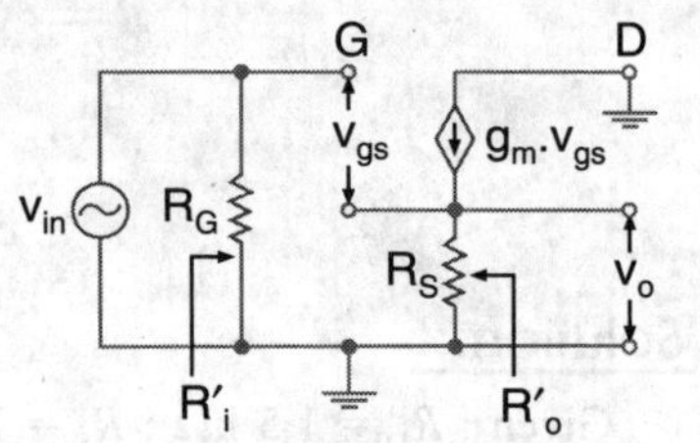

Fig. 30.34. A.C. equivalent circuit of a common drain amplifier.

1. ***Voltage gain.*** It is the ratio of output voltage (v_o) to the input voltage (v_{in}). Mathematically, the voltage gain,

$$A_v = \frac{v_o}{v_{in}}$$

For the common drain amplifier, the input voltage,

$$v_{in} = v_{gs} + i_d \cdot R_s$$

and the output voltage,

$$v_o = i_d \cdot R_S$$

∴ Voltage gain,

$$A_v = \frac{v_o}{v_{in}} = \frac{i_d \cdot R_s}{v_{gs} + i_d \cdot R_s}$$

Substituting the value of $i_d \, (= g_m \cdot v_{gs})$ in the above expression,

$$A_v = \frac{(g_m \cdot v_{gs})\, R_s}{v_{gs} + g_m \cdot v_{gs} \cdot R_s}$$

$$= \frac{g_m \cdot v_{gs} \cdot R_s}{v_{gs}\,(1 + g_m \cdot R_s)}$$

$$= \frac{g_m \cdot R_s}{1 + g_m \cdot R_s} = \frac{R_s}{R_s + \dfrac{1}{g_m}}$$

It may be noted that when R_s is much greater than $1/g_m$, the value of voltage gain approaches unity.

2. *Input resistance.* We know that the input signal is applied to the gate of the amplifier. As a result of this, the input resistance seen by the input signal source is extremely high just as in the common source configuration. The total input resistance (or the input resistance of the amplifier),

$$R'_i = R_G \,\|\, R_i$$

where R_i is the input resistance of the field effect transistor. Since the value of R_i is extremely high, therefore the total input resistance,

$$R'_i = R_G$$

3. *Output resistance.* We know that voltage gain of the common drain amplifier,

$$A_v = \frac{v_o}{v_{in}} = \frac{R_s}{R_s + \dfrac{1}{g_m}}$$

∴ Output voltage,

$$v_o = \left(\frac{R_s}{R_s + \dfrac{1}{g_m}}\right) \times v_{in}$$

The above expression is a voltage-divider equation. It implies that the input voltage (v_{in}) drives two resistors R_s and $1/g_m$ with output voltage taken across resistor (R_s). It means that the output side of the amplifier appears as shown in Fig. 30.35 (*a*). It is evident from this figure, that the source resistor is driven by an a.c. source with an output resistance *i.e.*,

$$R_D = 1/g_m$$

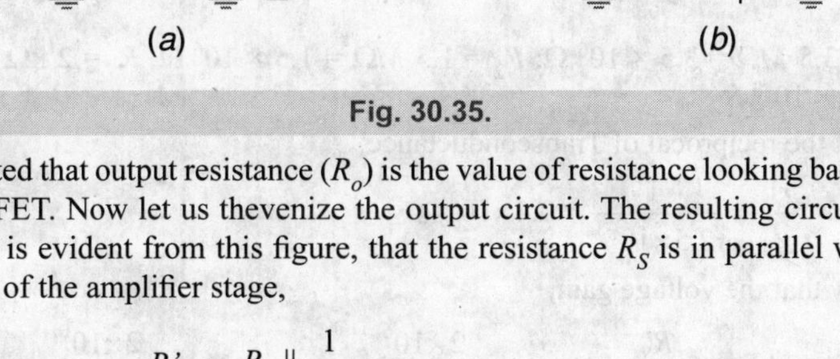

Fig. 30.35.

It may be noted that output resistance (R_o) is the value of resistance looking back into the source terminal of the JFET. Now let us thevenize the output circuit. The resulting circuit is as shown in Fig. 30.35 (*b*). It is evident from this figure, that the resistance R_S is in parallel with $1/g_m$ and the output resistance of the amplifier stage,

$$R'_o = R_s \,\|\, \frac{1}{g_m}$$

If the value of source resistance (R_s) is much larger than $1 / g_m$, then the output resistance of the amplifier stage,

$$R'_o = \frac{1}{g_m}$$

Example 30.18.

Fig. 30.36 shows the circuit of a source follower. Determine the voltage gain of the amplifier.

Also determine the input and output resistance of the amplifier. Assume g_m = 800 μS, infinite input resistance and neglect FET output resistance.

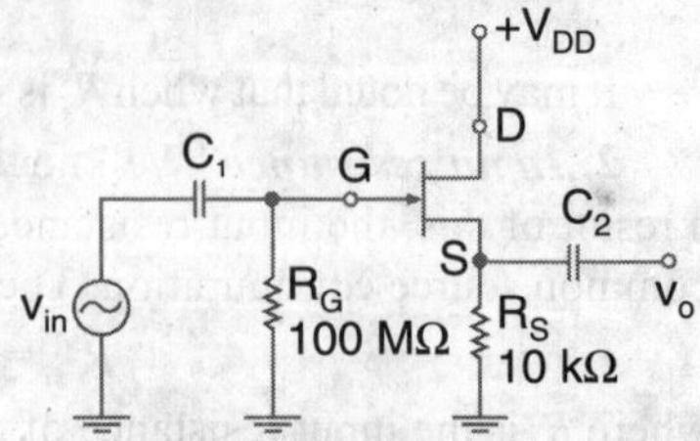

Fig. 30.36.

Solution.

Given: g_m = 8000 μS = 8000 × 10^{-6}S; R_G = 10 kΩ = 10 × 10^3 Ω and R_G = 100 MΩ.

Voltage gain

We know that the reciprocal of transconductance,

$$\frac{1}{g_m} = \frac{1}{8000 \times 10^{-6}} = 125\ \Omega$$

and the voltage gain,

$$A_v = \frac{R_s}{R_s + \dfrac{1}{g_m}} = \frac{10 \times 10^3}{(10 \times 10^3) + 125} = 0.988 \textbf{ Ans.}$$

Input resistance

We know that the input resistance

$$R'_i = R_G = 100\ \text{M}\Omega \textbf{ Ans.}$$

Output resistance

We also know that the output resistance,

$$R'_o = \frac{1}{g_m} = \frac{1}{8000 \times 10^{-6}} = 125\ \Omega \textbf{ Ans.}$$

Example 30.19.

The Common Drain FET amplifier circuit has R_1 = 3.5 MΩ, R_2 = 1.5 MΩ, R_S = 2kΩ and g_m = 2.5 μS Determine A_v (M. G. University, May/June 2006)

Solution.

Given : R_1 = 3.5 *M*Ω = 3.5 × 10^6 Ω; R_2 = 1.5 *M*Ω = 1.5 × 10^6 Ω; R_S = 2 kΩ = 2 × 10^6 Ω and g_m = 2.5 *mS* = 2.5 × 10^{-3} *S*

We know that the reciprocal of Transconductance,

$$\frac{1}{g_m} = \frac{1}{2.5 \times 10^{-3}} = 0.4 \times 10^3\ \Omega$$

We also know that the voltage gain,

$$A_v = \frac{R_S}{R_S + \dfrac{1}{gm}} = \frac{2 \times 10^6}{(2 \times 10^6) + (0.4 \times 10^3)} = \frac{2 \times 10^6}{(2 \times 10^6) + (0.4 \times 10^3)} = 0.999 \textbf{ Ans.}$$

Example 30.20. *Find the voltage gain for the source follower shown in Fig. 30.35.*

Also find the input and output resistances. If the input voltage is 2 mV, find the value of the output voltage. Assume g_m = 5500 μS.

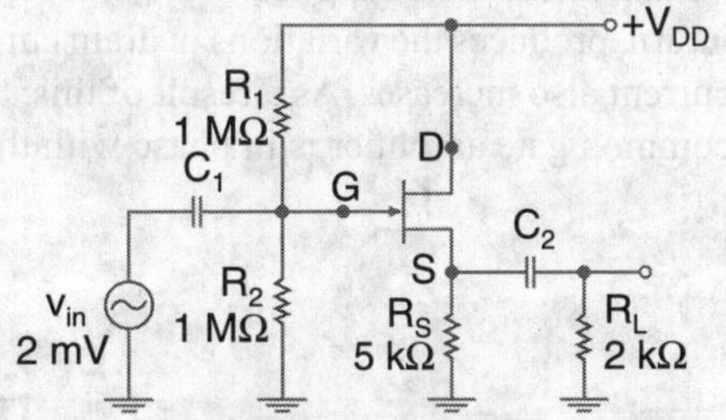

Fig. 30.37.

Solution.

Given: v_{in} = 2 mA; g_m = 5500 μS = 5500 × 10^{-6} S; $R_1 = R_2$ = 1 MΩ; R_S = 5 kΩ = 5000 Ω and R_L = 2 kΩ = 2000 Ω .

Voltage gain

We know that the reciprocal of transconductance,

$$\frac{1}{g_m} = \frac{1}{5500 \times 10^{-6}} = 181.8\ \Omega$$

and the voltage gain,

$$A_v = \frac{R_s}{R_s + \frac{1}{g_m}} = \frac{5000}{5000 + 181.8} = 0.965 \textbf{ Ans.}$$

Input resistance

We know that input resistance,

$$R'_i = R_1 \,\|\, R_2 = \frac{R_1 \cdot R_2}{R_1 + R_2} = \frac{1 \times 1}{1 \times 1}\ \text{M}\Omega = 0.5\ \text{M}\Omega \textbf{ Ans.}$$

Output resistance

We know that the input resistance,

$$R'_o = R_s \,\|\, \frac{1}{g_m} = 5000 \,\|\, 181.8 = 175.4\ \Omega \textbf{ Ans.}$$

Output voltage

Let v_o = Value of the output voltage.

Fig. 30.38 shows the a.c. equivalent circuit of the output side of the source follower. It is evident from this figure, that an a.c. source of voltage $A_v \cdot v_{in}$ (equal to 0.965 × 2 or 1.93 V) is in series with an output resistance of 175.4 Ω.

We know that the a.c. voltage across the load resistor,

$$v_o = \left(\frac{R_L}{R_L + R'_o}\right) \times A_v \cdot v_{in}$$

$$= \left(\frac{2000}{2000 + 175.4}\right) \times 1.93 \text{ mV}$$

$$= 1.77 \text{ mV} \textbf{ Ans.}$$

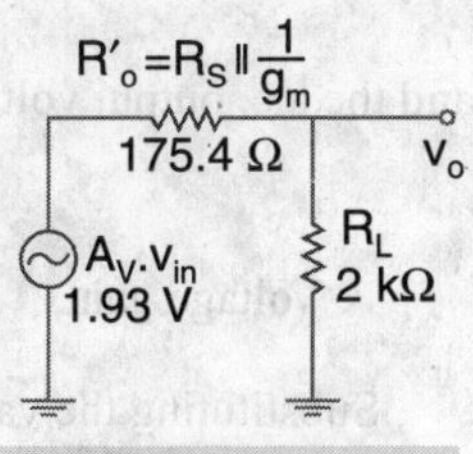

Fig. 30.38.

30.24. Common Gate Amplifier

Fig. 30.39 shows the circuit of a common gate amplifier. It is similar to a common base amplifier. The input signal is applied at the source through a coupling capacitor (C_1) and the output is taken from the drain through the coupling capacitor (C_2). The gate is effectively at a.c. ground because of the capacitor, (C_G).

The operation of the common gate amplifier may be understood from the assumption that when a small a.c. signal is applied to the source, it produces variation in gate-to-source voltage (V_{GS}). This inturn, produces the variations in drain current (I_D). As the gate-to-source voltage increases, the drain current also increases. As a result of this, the output voltage also increases. Thus output voltage of a common gate amplifier is in phase with the input voltage.

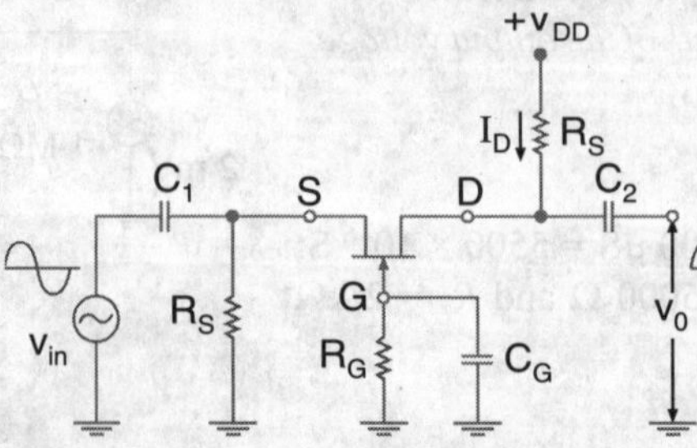

Fig. 30.39. Common gate amplifier.

A common gate amplifier has a low input resistance, high output resistance, high voltage gain and no phase reversal.

30.25. Analysis of a Common Gate Amplifier

Fig. 30.40 shows an a.c. equivalent circuit of a common gate amplifier. This circuit has been obtained from the amplifier circuit shown in Figure 30.34 by short-circuiting the capacitors and the d.c. supply. The field-effect-transistor is also replaced by its low-frequency model.

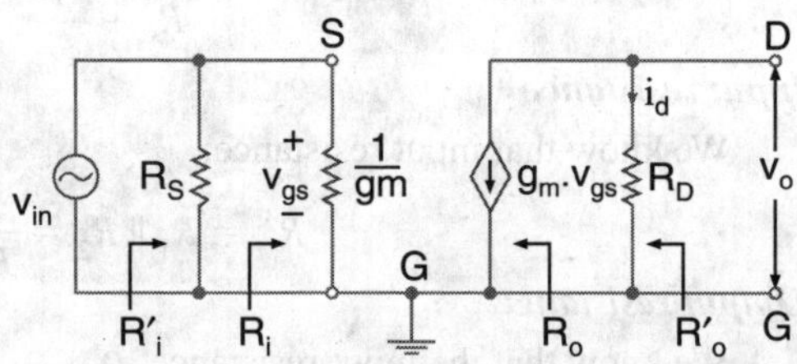

Fig. 30.40. Common gate amplifier.

Now we shall use this circuit to find the expressions for amplifier voltage gain, input resistance and output resistance.

1. *Voltage gain.* It is the ratio of a.c. output voltage (v_o) to the a.c. input voltage (v_{in}). Mathematically, the voltage gain,

$$A_v = \frac{v_o}{v_{in}}$$

We know that the a.c. input voltage,

$$v_{in} = v_{gs}$$

and the a.c. output voltage,

$$v_o = i_d \cdot R_D$$

∴ Voltage gain,
$$A_v = \frac{i_d \cdot R_D}{v_{gs}}$$

Substituting the value of i_d (equal to $g_m \cdot v_{gs}$) in the above expression,

$$A_v = \frac{(g_m \cdot v_{gs}) \cdot R_D}{v_{gs}} = g_m \cdot R_D$$

It may be noted that the above expression for voltage gain is true only when there is no load resistor. If there is a load resistor (R_L) connected to the amplifier output, then the voltage gain.

$$A_v = g_m (R_D \parallel R_L)$$

2. *Input resistance.* It is the ratio of input voltage (v_{in}) to the input current (i_{in}). Mathematically, the input resistance,

$$R_i = \frac{v_{in}}{i_{in}}$$

For a common gate amplifier, the input voltage (v_{in}) is equal to the gate-to-source voltage (v_{gs}) and the input current (i_{in}) is equal to the drain current (i_d). Therefore the input resistance,

$$R_i = \frac{v_{gs}}{i_d} = \frac{v_{gs}}{g_m \cdot v_{gs}} = \frac{1}{g_m} \quad \text{... } (\because i_d = g_m \cdot v_{gs})$$

It may be noted that the resistance (R_i) is the value of input resistance looking directly into the source terminal of the JFET. The input resistance of the amplifier,

$$R'_i = R_s \,||\, \frac{1}{g_m}$$

However, if R_s is much larger than the reciprocal of tranconductance ($1/g_m$) then the input resistance of the amplifier,

$$R'_i = R_s$$

It is evident from the above relation that a common gate amplifier has a low input resistance (equal to the value of external source resistance only). It is in contrast to the common source and common drain amplifier, which have extremely high input resistances.

3. *Output resistance.* It is the ratio of a.c. output voltage (v_o) to the a.c. output current (i_o). Mathematically, the output resistance of the amplifier,

$$R'_o = \frac{v_o}{i_o} = \frac{i_d \cdot R_D}{i_d} \quad \text{... } (\because v_o = i_D \cdot R_D \text{ and } i_o = i_D)$$

$$= R_D$$

It means that the output resistance of the common gate amplifier is equal to the external drain resistance (R_D). If there is a load resistor (R_L) connected to the amplifier output, then the output resistance of the amplifier,

$$R'_o = R_D \,||\, R_L$$

Example 30.21.

Fig. 30.41 shows the circuit of a common gate JEET amplifier. The JFET has transconductance g_m = 2500 μS.

Determine the amplifier voltage gain and input resistance.

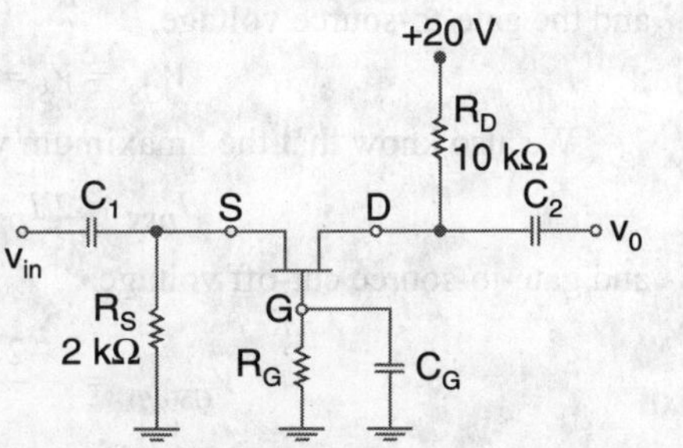

Fig. 30.41.

Solution.

Given: g_m = 2500 μS = 2500 × 10^{-6}S; R_D = 2 kΩ = 2000 Ω and R_D = 10 kΩ = 10,000 Ω.

Voltage gain

We know that voltage gain,

$$A_v = g_m \cdot R_D = (2500 \times 10^{-6}) \times 10{,}000 = 25 \textbf{ Ans.}$$

Input resistance

We know that reciprocal of transconductance,

$$\frac{1}{g_m} = \frac{1}{2500 \times 10^{-6}} = 400\ \Omega$$

and the amplifier input resistance,

$$R'_i = R_i \,||\, \frac{1}{g_m} = 2000 \,||\, 400\ \Omega = \frac{2000 \times 400}{2000 + 400} = 333\ \Omega \textbf{ Ans.}$$

Example 30.22.

For a JFET, the transconductance at zero gate-to-source voltage, g_{mo} = 5 mS. Determine the amplifier input resistance (R'_i) and a.c. voltage gain (v_o / v_{in}) for the circuit shown in Fig. 30.42. Assume the capacitors to be short circuit for a.c. (*Grad. I.E.T.E., Dec. 1987*)

Solution.

Given: $g_{mo} = 5 \text{ mS} = 5 \times 10^{-3}$ S; $R_D = 1 \text{ k}\Omega = 1 \times 10^3\ \Omega$; $R_S = 200\ \Omega$ and $I_D = 5 \text{ mA} = 5 \times 10^{-3}$ A.

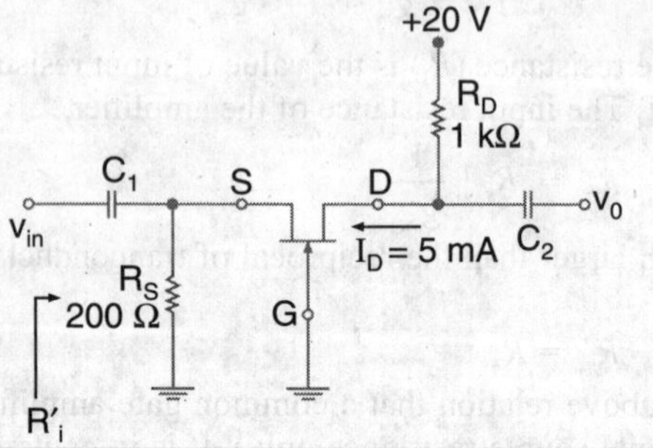

Fig. 30.42.

Input resistance

We know that reciprocal of transconductance

$$\frac{1}{g_{mo}} = \frac{1}{5 \times 10^{-3}} = 200\ \Omega$$

and the input resistance of the amplifier,

$$R'_i = R_s \parallel \frac{1}{g_{mo}} = 200 \parallel 200\ \Omega = 100\ \Omega \textbf{ Ans.}$$

Voltage gain

We know that the d.c. source voltage, $V_S = I_D \cdot R_S = (5 \times 10^{-3}) \times 200 = 1$ V

and the gate-to-source voltage,

$$V_{GS} = V_S = 1\text{V}$$

We also know that the *maximum value of drain current,

$$I_{DSS} = 2\,I_D = 2 \times (5 \times 10^{-3}) = 10 \times 10^{-3} \text{ A}$$

and gate-to-source cut-off voltage

$$V_{GS(off)} = -\frac{2\,I_{DSS}}{I_D} = -\frac{2 \times (10 \times 10^{-3})}{5 \times 10^{-3}} = -4\text{V}$$

and the transconductance (g_m) $= g_{mo}\left[1 - \frac{V_{GH}}{V_{GS(off)}}\right] = (5 \times 10^{-3})\left[1 - \frac{1}{4}\right] = 3.75 \times 10^{-3}$ S

∴ The voltage gain, $A_v = g_m \cdot R_D = (3.75 \times 10^{-3}) \times (1 \times 10^3) = 3.75$ **Ans.**

* In actual practice, the amplifiers are biased in such a way that the Q-point is located at the mid-point of the trans-conductance curve. In that case, the maximum value of drain current (I_{DSS}) is twice the value of drain current.

NUMERICAL PROBLEMS

1. Find the values of drain-to-source voltage (V_{DS}) and the gate-to-source voltage (V_{GS}) for the circuit shown in Fig. 30.43. **(Ans. 7.5 V and – 1.5 V)**

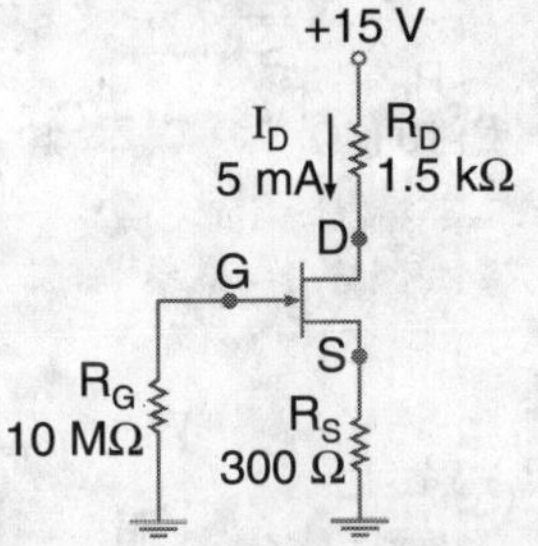

Fig. 30.43.

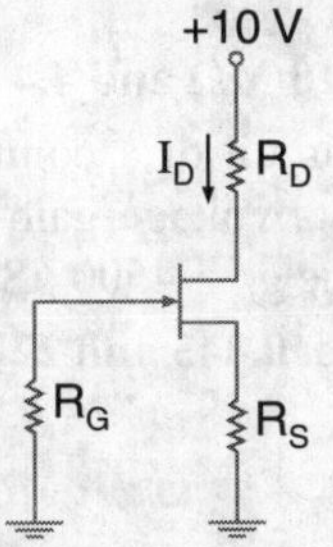

Fig. 30.44.

2. Calculate the self-bias operating point for the FET shown in Fig. 30.44. Also calculate the values of resistors R_D and R_S to obtain this bias condition. Given the maximum value of drain current as 5 mA and V_{GS} = 0.6 V or I_D = 2.5 mA.
(Ans. I_{DQ} = 2.5 mA : V_{DSQ} = 5 V: R_D = 2 kΩ and R_S = 240 Ω)
3. Fig. 30.45 (*a*), shows the circuit of a self-biased JFET amplifier and Fig. 30.45 (*b*) indicates the transfer characteristic curve of the JFET.

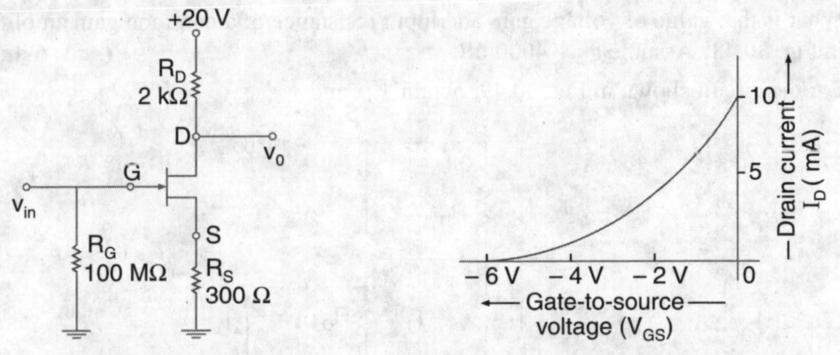

Fig. 30.45.

Find the quiescent values of I_D and V_{GS}. Also find the value of d.c. voltage between the drain and ground. **(Ans. I_{DQ} = 5.5 mA and V_{GSQ} = – 1.6 V: V_D = 9 V)**
4. A certain JFET has a transconductance (g_m) of 2500 μS. With an external drain resistance of 2 kΩ, find the value of ideal voltage gain. **(Ans. 5)**
5. The gain of a certain JFET amplifier with a source resistance of zero is 25. Determine the value of drain resistance, if the transconductance is 5500 μS. **(Ans. 4.55 kΩ)**
6. A certain JFET amplifier has g_m of 4 mS, r_d = 10 kΩ and R_D = 5 kΩ. What is the voltage gain? Assume the source resistance to be zero. **(Ans. 13.3)**

7. Calculate the value of voltage gain, input resistance and output resistance of the amplifier shown in Fig. 30.46.

   The FET has a transconductance of 4.2 mA/V and $r_D = 400$ kΩ.

   (Ans. 18.6 : 10 MΩ and 4.4 kΩ)

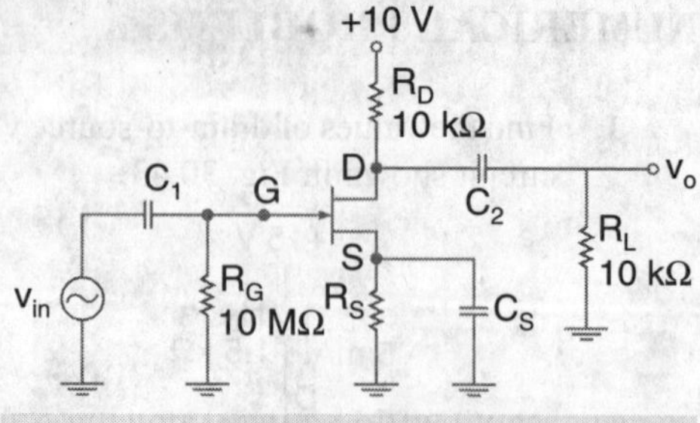

Fig. 30.46.

8. Fig. 30.47 shows the circuit of a common source amplifier. Find its voltage gain and output resistance. Assume $g_m = 4500$ μS.

   (Ans. 0.845 and 220 Ω)

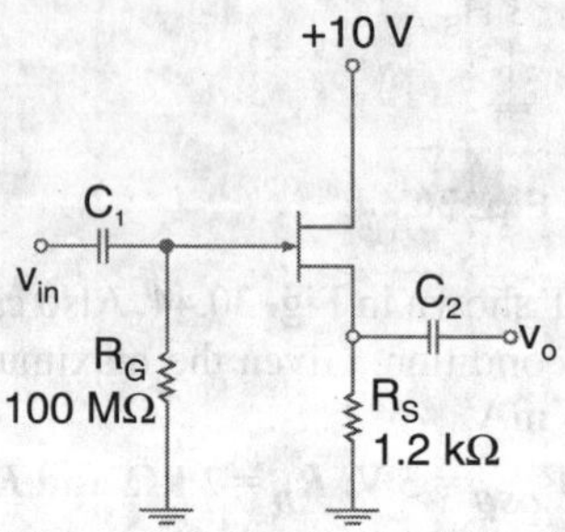

Fig. 30.47.

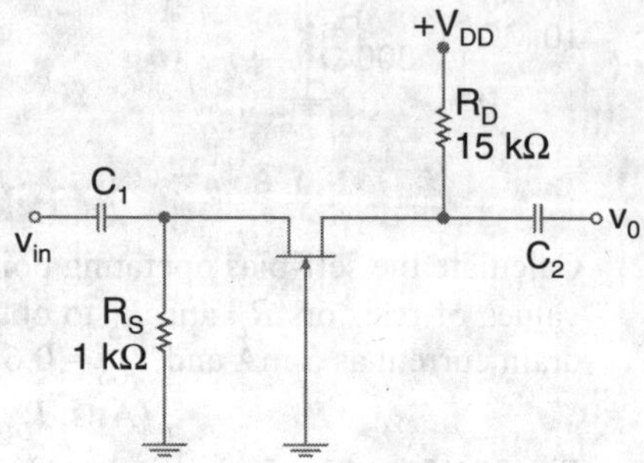

Fig. 30.48.

9. What is that value of voltage gain and input resistance of a common gain amplifier shown in Fig. 30.48? Assume $g_m = 4000$ μS. **(Ans. 6 and 200 Ω)**

10. For the circuit shown in Fig. 30.49, obtain V_{o_1} and V_{o_2} . *(UPSC Engg. Services 2002)*

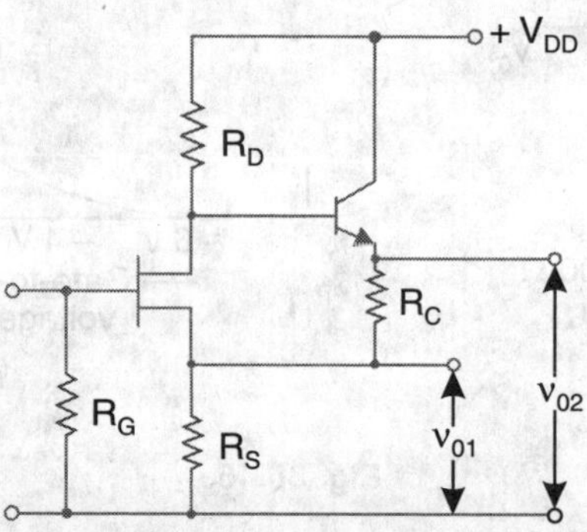

Fig. 30.49.

[**Hint :** State 1 : Self-bias JFET circuit; $V_{o_1} = V_S = I_D R_S$ Stage 2 : $V_{o_2} = V_{o_1} - I_C R_C = I_D R_S - I_C R_C$]

DESCRIPTIVE QUESTIONS

1. What are the biasing schemes available to achieve the required bias in a junction field-effect transistor (JFET)? Explain any one of the biasing schemes.
2. How do you set a *Q*-point in a self-biased JFET?
3. Draw a small signal low-frequency model of a field-effect transistor and explain its various elements. Also give the approximate range of the elements.

4. How a small-signal high frequency model is different from a low-frequency model? Explain it briefly.
5. Draw neatly the circuit diagram of a common source JFET amplifier and explain its working.
6. Derive the expression for voltage gain and output resistance for a common source JFET amplifier.
7. What is the effect of external source resistance on the voltage gain of a common source amplifier? Explain with necessary derivation.
8. Draw the circuit diagram of a common gate amplifier. Derive the expression for its voltage gain and input resistance.
9. Discuss different biasing methods of JFET. *(GBTU/MTU., 2006-07)*
10. With the help of a neat diagram explain the voltage divider biasing method for JFET *(Anna University., April/May 2008)*
11. Write the construction, operation and characteristics behavior of JFET under various biasing conditions. Give the necessary figures. *(Anna University., Nov/Dec 2007)*
12. Draw the circuit of fixed biased FET and explain it. *(GBTU/MNTU., 2004-05)*
13. Draw the two biasing circuits for an enhancement type MOSFET and explain. *(JNTU., May./Jun. 2006)*
14. Draw the biasing circuit suitable for JFET and if the JFET is replaced by a MOSFET, for what mode of operation it is valid and explain about the function of each component used in the circuit. *(JNTU., May./Jun. 2006)*
15. What specific capacitance has the greatest effect on the high frequency response of a cascade of FET amplifier? Explain. *(Anna University., Nov/Dec 2007)*
16. Draw the high frequency equivalent circuit of a JFET. *(M.G University., May 2005)*
17. What specific capacitance has the greatest effect on the high frequency response of a cascade of FET amplifier? Explain. *(Anna University., Nov/Dec 2007)*
18. Draw and explain voltage divider bias using FET and derive for its stability factor. *(Anna University., April/May 2008)*
19. Sketch and explain the circuit diagrams of a self biased CS-FET amplifier. *(M.G University., May 2008)*
20. Draw the input output characteristic of CD configuration. *(GBTU/MNTU., 2004-05)*
21. Draw the high frequency equivalent circuit for a FET amplifier and derive the values of all the parameters. *(Anna University., April/May 2008)*
22. Draw the circuit for common drain amplifier and obtain the expression for voltage gain. *(GBTU/MTU., 2009)*
23. For common source FET amplifier derive the expression for voltage gain at high frequency. *(Nagpur University., 2008)*
24. Derive the expression for input and output impedances of a signal FET amplifier. *(GBTU/MNTU., 2004-05)*
25. Derive the expression for A_v, R_i and of a JFET source follower. *(GBTU/MNTU., 2004-05)*
26. Draw the circuit diagram of common source amplifier and derive expressions for voltage gain and output resistance. *(JNTU., Aug./Sep. 2008)*
27. Draw the circuit diagram of a common drain amplifier and derive expressions for voltage gain and output resistance. *(JNTU., May/Jun. 2006)*
28. Determine the input output impedance and voltage gain of fixed bias configuration of JFET. *(R.G.P.V., Dec. 2010)*

29. Discuss the working of common gate configuration of JFET and analyse the different parameter. *(R.G.P.V., Dec. 2010)*

30. Draw the circuit of
(*i*) CS amplifier
(*ii*) Source follower using FET at low frequencies.
Derive the expression for the voltage gain at low frequencies in each case. *(R.G.P.V., Dec. 2005)*

31. For a common source FET amplifier, derive the expression for voltage gain at high frequency. *(Nagpur University, Winter 2006)*

32. Sketch the circuit of common source amplifier. Derive the expression for the voltage gain at low frequencies. Also sketch the small-signal high frequency circuit of above amplifier and derive the expression for the voltage gain. *(Nagpur University, Winter 2002)*

33. For common source FET amplifier derive the expression for voltage gain at high frequency. With the help of static drain characteristics. Explain the principle of a junction FET. *(Nagpur University, Summer 2008)*

MULTIPLE CHOICE QUESTIONS

1. The best location for setting a *Q*-point on d.c. load line of an FET amplifier is at
(*a*) saturation point (*b*) cut-off point
(*c*) mid-point (*d*) none of these

2. Which of the following bias methods provides a solid *Q*-point in JFET, amplifiers
(*a*) Gate bias (*b*) Self-bias
(*c*) Voltage divider bias (*d*) Current source bias

3. Which of the following technique is used for biasing the enhancement type MOSFET's?
(*a*) Voltage divider bias (*b*) Collector feedback bias
(*c*) Current source bias (*d*) Self-bias

4. The voltage gain of a common source JFET amplifier depends upon its
(*a*) transconductance (g_m) (*b*) amplification factor (μ)
(*c*) external load resistance (R_D) (*d*) both (*a*) and (*c*) above

5. A common gate amplifier has
(*a*) high input resistance and high output resistance
(*b*) low input resistance and high output resistance
(*c*) low input resistance and low output resistance
(*d*) high input resistance and low output resistance

6. The transconductance 'g_m' of a JFET is equal to
(*a*) $-\dfrac{2I_{DSS}}{V_P}$ (*b*) $\dfrac{2}{|V_P|}\sqrt{I_{DSS}\,I_D}$
(*c*) $-\dfrac{2I_{DSS}}{V_P}\left(1-\dfrac{V_{GS}}{V_P}\right)$ (*d*) $\dfrac{I_{DSS}}{V_P}\left(1-\dfrac{V_{GS}}{V_P}\right)$

7. A transconductance amplifier has
(*a*) high input impedance and low output impedance
(*b*) low input impedance and high output impedance

(*c*) high input and output impedances
(*d*) low input and output impedances

8. The threshold voltage of an *n*-channel enhancement mode MOSFET is 0.5 V, when the device is biased at a gate voltage of 3 V, pinch-of would occur at a drain voltage of

(*a*) 1.5 V (*b*) 2.5 V
(*c*) 3.5 V (*d*) 4.5 V

9. The zero gate bias channel resistance of a junction field-effect transistor is 750 and the pinch-off voltage is 3 V. For a gate bias of 1.5 V and very low drain voltage, the device would behave as a resistance of

(*a*) 320 Ω (*b*) 816 Ω
(*c*) 1000 Ω (*d*) 1270 Ω

10. If properly biased, JFET with act as a

(*a*) current controlled current source (*b*) voltage controlled voltage source
(*c*) voltage controlled current source (*d*) current controlled voltage source

ANSWERS

1. (*c*)	**2.** (*d*)	**3.** (*b*)	**4.** (*d*)	**5.** (*b*)
6. (*c*)	**7.** (*a*)	**8.** (*c*)	**9.** (*b*)	**10.** (*b*)

31

Chapter

Sinusoidal Oscillators

Outline

31.1. Introduction

We have already discussed in the chapter on Feedback Amplifiers that if the feedback signal is applied in such a way that it is *in phase* with the input signal and thus increases it, then it is called a *positive feedback, degenerative feedback* or *direct feedback*.

The use of positive feedback results in a feedback amplifier having closed-loop gain (A'_v) greater than the open-loop gain (A_v). It results in *instability* and operates as an *oscillatory circuit*. An oscillator circuit provides a, constantly, varying amplified output signal of any desired frequency.

31.2. Comparison Between an Amplifier and an Oscillator

Fig. 31.1 (*a*) show a block diagram of an amplifier, while (*b*) shows that of an oscillator. We know that an amplifier is a device, which produces an output signal with a similar waveform as that of the input. But its power level is generally high. This additional power is supplied by an external d.c. source. Thus an amplifier is essentially an energy convention device, *i.e.*, a device, which gets energy from the d.c. source and converts it into an a.c. energy at the same frequency as that of the input signal. The d.c. to a.c. conversion is controlled by the input signal. It means that *if there is no input signal then no energy convertion take place. Thus there is not output signal.*

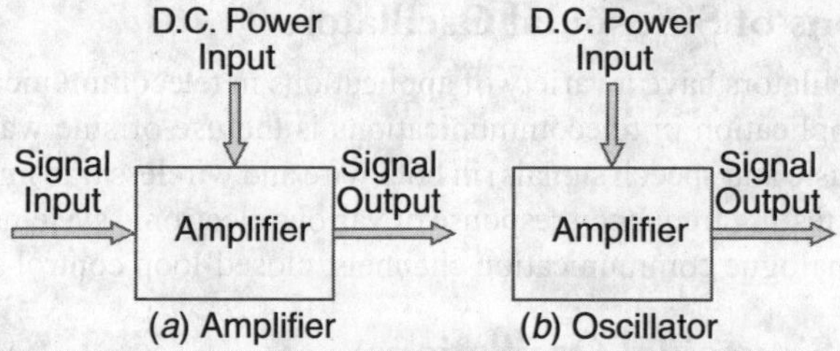

Fig. 31.1.

On the other hand, *an oscillator is a device, which produces an output signal, without any input signal of any desired frequency*. It keeps producing an output signal, so long as the d.c. power is supplied. It is evident from the above discussion that *an oscillator does not require any external signal to start or maintain energy convertion process*.

31.3. Classification of Oscillators

The electronic oscillators may be broadly classified into the following two categories:

1. ***Sinusoidal or harmonic oscillators.*** The oscillators, which provide an output having a *sine waveform*, are called sinusoidal or harmonic oscillators. Such oscillators can provide output at frequencies ranging from 20 Hz to 1 GHz (1 GHz = 10^{12} Hz).
2. ***Non-sinusoidal or relaxation oscillators.*** The oscillators, which provide an output having a square, rectangular or sawtooth wavform, are called non-sinusoidal or relaxation oscillators. Such oscillators can provide output at frequencies ranging from zero to 20 MHz. (1 MHz = 10^6 Hz).

In this chapter, we shall discuss only the sinusoidal oscillators. The non-sinusoidal oscillators will be discussed separately in the next chapter. The sinusoidal oscillators may be further subs-divided into the following types:

(*a*) ***Tuned circuit oscillators.*** These oscillators use a tuned-circuit consisting of inductors (L) and capacitors (C) and are used to generate high-frequency signals. Thus they are also known as *radio frequency* (R.F.) *oscillators*. Such oscillators are *Hartley, Colpitts, Clapp-oscillators* etc.

(*b*) ***RC oscillators.*** There oscillators use resistors and capacitors and are used to generate low-or audio-frequency signals. Thus they are also known as *audio-frequency* (A.F.) *oscillators*. Such oscillators are *phase-shift* and *Wien-bridge* oscillators.

(*c*) ***Crystal oscillators.*** These oscillators use *quartz crystals* and are used to generate highly stabilized output signal with frequencies up to 10 MHz. The *pierce oscillator* is an example of a cystal oscillator.

(*d*) ***Negative-resistance oscillators.*** These oscillators use *negative-resistance characteristic* of the devices such as tunnel diodes. A tuned diode oscillator is an example of negative resistance oscillator.

Crystal oscillator.

31.4. Applications of Sinusoidal Oscillators

The sinusoidal oscillators have a variety of applications in telecommunications and electronics. The most important application in telecommunications is the use of sine wave as a *carrier signal* (used in modulating music and speech signals) in both wire and wireless transmissions. The sine wave signals are also used in testing frequency response of various electronic systems and equipments such as amplifiers, filters, analogue communication channels, closed-loop control systems etc.

31.5. Nature of Sinusoidal Oscillations

The nature of oscillations, produced by sinusoidal oscillators, may be of two types, namely damped oscillations and undamped oscillations.

1. *Damped oscillations.* The electrical oscillations, whose amplitude goes on decreasing with time are called *damped oscillations*. The waveform of such oscillations is shown in Fig. 31.2 (*a*). The damped oscillations are produced by those oscillator circuits in which power losses take place continuously during each oscillation without any means for compensating the same. However, it may be noted that frequency of oscillations remains unchanged because is depends upon the circuit parameters.

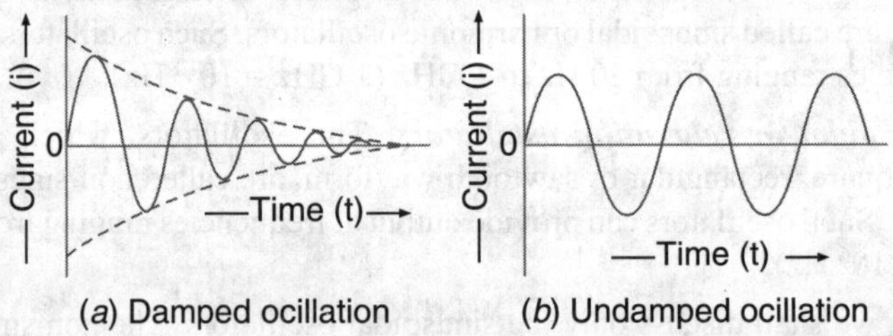

(*a*) Damped ocillation (*b*) Undamped ocillation

Fig. 31.2. Nature of sinosoidal oscillations.

2. *Undamped oscillation.* The electrical oscillations, whose amplitude remains constant with time, are called *undamped* oscillations. The waveform of such oscillations is shown in Fig. 31.2 (*b*). The undamped oscillations are produced by those oscillator circuits in which either there are no power losses or if they have any power loss then they have provision for compensating those losses. It may be noted that an oscillator is required to produce undamped oscillations for utilizing in various telecommunication and electronics applications.

31.6. Oscillatory Circuit

A circuit, which produces electrical oscillations of any desired frequency, is known as an oscillatory circuit. This circuit consists of two reactive elements namely inductor (L) and a capacitor (C) connected in parallel with each other as shown in Fig. 31.3. Thus such a circuit is also known as LC circuit or *tank circuit*. Both the reactive devices are capable of storing electrical energy. It is a well known fact that an inductor stores energy in its magnetic field, whenever a current flows through it. Similarly, a capacitor stores energy in its electric field, whenever a voltage (or

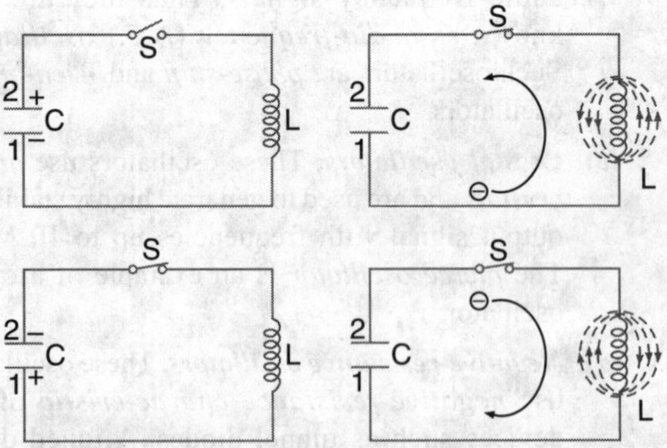

Fig. 31.3. LC circuit acting as an oscillatory circuit.

potential difference) is applied across its plates. Both the reactive elements are supposed to be an ideal one (*i.e.*, they do not have any power losses). Now we shall discuss as to how an LC circuit generates oscillations.

Suppose a capacitor has been charged from a d.c. voltage source with a polarity as shown in Fig. 31.3 (*a*). As the switch (S) is open, it cannot discharge through the coil (L).

When the switch (S) is closed as shown in Figure 31.3 (*b*), the capacitor discharge through the coil (L). The discharging the due to the flow of electrons from plate 1 to plate 2 through the coil (L) as indicated by the arrow (note that conventional current direction is from plate 2 to plate 1). This current flow sets up *magnetic field* around the coil. This magnetic field stores the energy released by the electric field. Because of the inductive effect, the current grows up slowly towards the maximum value. This situation occurs, when the capacitor is fully discharged. At this instant, the *electrical energy stored in the capacitor becomes zero and the energy stored in magnetic field is maximum*. In other words, we say that electrical energy across the capacitor is completely converted into magnetic field energy around the coil.

Once the capacitor is discharged completely, the magnetic field around the coil begins to collapse and produces a *counter e.m.f.* According to Lenz's Law, the counter e.m.f., keeps the electrons moving in the same direction as shown in Fig. 31.3 (*b*). This again charges the capacitor, though in the opposite direction, as shown in Fig. 31.3 (*c*). It may be noted that plate 1 has become positive whereas the plate 2, is still negative. When the capacitor is charged completely, in the opposite direction, the magnetic field around the coils is also collapsed completely. *At this instant, the energy previously stored in the magnetic field, has now converted into the energy stored in the electric field of the capacitor*.

After this, the capacitor starts discharging in the opposite direction, so that the electrons now move from plate 2 to plate 1 as shown in Fig. 31.3 (*d*). The electric field starts collapsing whereas magnetic field starts building up again, but in the reverse direction. Fig. 31.3 (*d*) shows the condition when the capacitor has fully discharged once again.

The above mentioned sequence of charging and discharging of a capacitor results in an alternating motion of electrons or an alternating current. As a result of this, the energy is alternatively stored in the electric field of the capacitor and the magnetic field of the coil. *This interchange of energy between the capacitor and coil continues to be repeated, and results in the production of electrical oscillations.*

If there are no power losses in the tank circuit, the waveform of the electrical oscillations will be undamped (*i.e.*, of constant amplitude). However, in a practical tank circuit, the waveform of the oscillations is damped (*i.e.*, of decreasing amplitude). It is because of the energy lost by the L and C element, which may take place because of the following two reasons:

1. Some energy is lost in the form of heat produced in the resistance of the coil, connecting wires and within the dielectric material of the capacitor.
2. Some energy is lost from the circuit due to radiation in the form of electromagnetic waves.

As a result of these losses, the amplitude of oscillating current in the tank circuit decreases gradually and eventually it becomes zero. The waveform of such an oscillating current will be of the form as shown in Fig. 31.2 (*a*).

31.7. Frequency of Oscillatory Circuit

The frequency (*f*) of the oscillatory current, produced in the tank circuit, is determined by the following two factors:

1. ***Capacitance of the capacitor.*** The frequency of oscillation (f_o) is inversely proportional to the square root of the capacitance of a capacitor. Thus, larger the value of capacitance, greater is the time required for removal of the discharge current and hence lower is its frequency.

Mathematically, the frequency,

$$f_0 \;\alpha\; 1/\sqrt{C} \qquad ...(i)$$

2. ***Self-inductance of the coil.*** The frequency of oscillation (f_o) is proportional to the square root of the self-inductance of the coil. Thus larger the inductance, greater is the inductance effect and hence longer the time required by the current to stop flowing during the discharge of the capacitor. Mathematically, the frequency,

$$f_0 \;\alpha\; 1/\sqrt{L} \qquad ...(ii)$$

Combining the equations (*i*) and (*ii*) we get

$$f_o \;\alpha\; \frac{1}{\sqrt{L \cdot C}} = \frac{1}{2\pi\sqrt{L \cdot C}}$$

$$= \frac{0.159}{\sqrt{L \cdot C}} \text{ in Hertz}$$

If L is in μH and C in μF, then the frequency,

$$f_o = \frac{159}{\sqrt{L \cdot C}}$$

It may be noted that the damped oscillations, so produced, are not useful in actual practice. We need continuous undamped oscillations for practical applications such as radio and cable transmission. The *undamped oscillations can be produced if some additional energy is supplied in correct phase and correct direction to the tank circuit for making up the power losses continuously occurring in the circuit.*

31.8. Frequency Stability of an Oscillator

The frequency stability of an oscillator is a measure of its ability to maintain a constant frequency, over a long time interval. However, it has been found that if an oscillator is set at some particular frequency, it does not maintain it for a longer period. In other words, the frequency of an oscillator changes slowly (or drifts away) from the initially set value. Sometimes, the change in frequency is uniform in one direction (*i.e.*, either increasing or decreasing). But at some other times, it may be changing quite erratically. The change in oscillation frequency may arise due of the following factors:

1. ***Operating point of the active device.*** The operating point of the active device (*i.e.*, bipolar transistor or FET) is selected in such a way so that its operation in non-linear region, changes the values of device parameters which, in turn, affects the frequency stability of the oscillator.

2. ***Circuit components.*** The values of circuit components (*i.e.*, resistor, inductors and capacitors) change with the variation in temperature. Since such changes take place slowly, they also cause a drift in oscillator frequency.

3. ***Supply voltage.*** The changes in d.c. supply voltage applied to the active device, shift the oscillator frequency. This problem can be avoided by using a highly regulated power supply.

4. ***Output load.*** A change in the output load may cause a change in the Q-factor of the tank circuit, thereby causing a change in oscillator output frequency.

5. ***Interelement capacitances and stray capacitances.*** Any change in the interelement capacitances of a transistor (particularly collector-to-emitter capacitance), cause changes in the oscillator output frequency and thus affect the frequency stability. Similarly, the stray capacitances also affect the frequency stability of an oscillator. The effect of change in interelement capacitances can be neutralized by putting an additional capacitor across the corresponding elements. However, it is difficult to avoid the effect of stray capacitances.

31.9. The Barkhausen Criterion

We have already discussed in the chapter of Feedback Amplifiers, that the overall voltage gain of a positive feedback amplifier is given by the expression,

$$A'_v = \frac{A_v}{1 - \beta \cdot A_v}$$

A_v = Gain of an amplifier without feedback also called open loop-gain, and

* $\beta \cdot A_v$ = Product of feedback fraction and open loop-gain. It is called loop-gain

It will be interesting to know that if the product ($\beta \cdot A_v$) is made equal to unity, then the denominator of equation (*i*) is zero. It means that the gain (A'_v) will increase to infinity. Of course, the output of the feedback amplifier, in actual practice, cannot be infinite. Therefore the condition (*i.e.*, $1 - \beta \cdot A_v = 0$) represents that there will be an output voltage, whose frequency is completely different from the input signal. In other words, the circuit has stopped amplifying and started oscillating. It may be noted that the oscillations will not be maintained, if the value of $\beta \cdot A_v$ is less than unity.

We know that an amplifier reverses the phase of an input signal at its output. It means that an amplifier causes a phase shift of 180° between the input and output signals. In order to provide positive feedback, the feedback network must provide a phase shift of 180°, so as to provide a signal with a phase shift of 360° or 0° at the amplifier input.

The above two conditions for positive feedback (*i.e.*, $\beta \cdot A_v = 1$ *and the net phase shift around the loop equal to 360° or 0°) are called Barkhausen criterion for oscillation.* Mathematically, the Barkhausen criterion may be stated as follows:

$$\beta \cdot A_v = 1 \quad \text{...}(i)$$

and

$$\angle \beta \cdot A_v = 0°$$

Thus if $\beta \cdot A_v$ is a complex quantity (*i.e.*, a quantity which contains both real and imaginary parts) then its real part must be equal to unity and imaginary part equal to zero, *i.e.*,

$$\beta \cdot A_v = 1 + j_0$$

However, an oscillator in which the quantity $\beta \cdot A_v$ is exactly unity, is not realizable in practice. In every practical oscillator, the quantity $\beta \cdot A_v$ is slightly larger than unity. Thus an oscillator circuit must satisfy the two conditions mentioned above to produce sustained oscillations. The conditions are (*a*) loop gain ($\beta \cdot A_v$) ≥ 1 and (*b*) phase-shift between the input and output signals must be equal to 360° or 0°. It may be noted that if a circuit does not satisfy any one of these conditions, the oscillations will not be produced.

31.10. Tuned Circuit Oscillators

These are also known as *LC oscillators, resonant circuit oscillators or tank circuit oscillators*. These oscillators are used to produce an output with frequencies ranging from 1 MHz to 500 MHz. Hence they are also known as R.F. oscillators. A **bipolar transistor or an FET is used as an amplifier with tuned circuit oscillators. With an amplifier and an LC tank circuit, we can feedback a signal with right amplitude and phase to maintain oscillations.

It will be interesting to know that most of the oscillators used in radio transmitters and receivers are of LC oscillators type. Depending upon the way, the feedback is used in the circuit, the LC oscillators are of the following five types:

* $\beta \cdot A_v$ is actually a complex quantity in positive feedback amplifiers, because of the use of inductive elements.

** The operational amplifiers (Op Amps) cannot be used as an amplifier with LC oscillators, because its unity gain frequency (f_{unity}) is much smaller than the frequency range of LC oscillators.

1. ***Tuned-collector or Armstrong oscillator.*** It uses inductive feedback from the collector of a transistor to the base. The LC circuit is in the collector circuit of the transistor.
2. ***Tuned base oscillator.*** It also uses inductive feedback. But the LC circuit is in the base circuit.
3. ***Harley oscillator.*** It also uses inductive feedback.
4. ***Colpitts oscillator.*** It uses capacitive feedback.
5. ***Clapp oscillator.*** It also uses capacitive feedback.

We shall now, discuss all the above mentioned tuned circuit oscillators in detail in the following pages.

31.11. Tuned-Collector Oscillator

It is also called *Armstrong Oscillator*. Fig. 31.4 shows the circuit of such an oscillator. In this circuit, the collector drives an LC tank circuit. The feedback signal is taken from a small secondary winding (L_1) and feedback to the base. There is a phase shift of 180° in the transformer and another 180° phase shift is provided by the transistor amplifier. It means that the total phase shift around the loop is zero and hence the feedback is positive. Ignoring the loading effect of base, the feedback fraction,

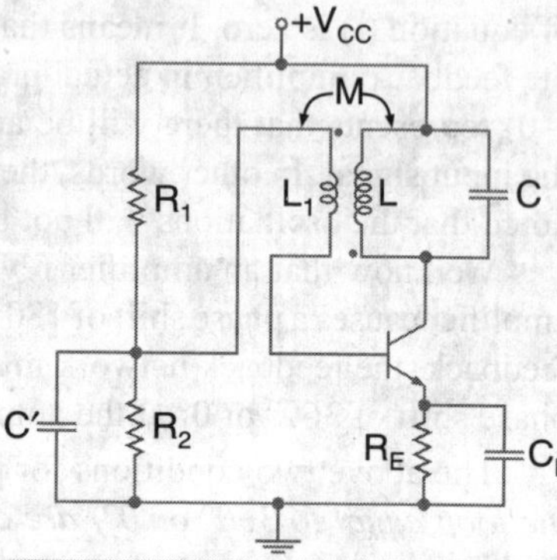

Fig. 31.4. Tuned-collector oscillator.

$$\beta \approx M/L$$

where M = The mutual inductance between the primary and secondary windings, and

L = Self-inductance of the primary winding.

It may be noted that for oscillations to start, the voltage gain must be greater then $1/\beta$.

The resistors R_1, R_2 and R_E are used to provide d.c. bias to the transistor. The capacitor C' and C_E act as bypass capacitors for the resistors R_2 and R_E respectively. The capacitor C¢ also provides a.c. ground for the secondary transformer (L_1). The operation of the circuit may be explained as follows.

When the circuit is energied by switching on the supply, the collector current starts increasing to the quiescent value. This current charges the capacitor (C). When this capacitor is fully charged, it discharges through the primary winding (L) producing magnetic field around the primary coil. When the capacitor is fully discharged, the magnetic field collapses and charges the capacitor in the reverse direction. Then again the capacitor discharges in the reverse direction. This process continues again and again. As a result of this, the oscillations are produced.

The frequency of oscillations produced by the tank circuit depends upon the values of primary inductance (L) and capacitance (C). Its value given by the expression,

$$f_o = \frac{1}{2\pi\sqrt{L \cdot C}} = \frac{0.159}{\sqrt{L \cdot C}} \text{ Hz}$$

These oscillations induce some voltage in secondary winding (L_1) by mutual induction. The magnitude of the voltage induced depends upon the number of turns of secondary winding (L_1) and the coefficient of coupling (k) between the coils L and L_1. However, the frequency of the induced voltage is the same as that of the oscillations produced in the tank circuit. The voltage across L_1 winding appears between the base and emitter terminals of the transistor. This voltage is amplified at the collector of a transistor and is partially used for overcoming the losses, which take place in the tank circuit. The remaining voltage may be utilized for any other purpose.

Example 31.1.

A tuned collector oscillator is tuned to operate at 22 kHz by a variable capacitor set to 2 nF. What is the value of the tuned circuit inductance.

Solution.

Given $f_o = 22$ kHz $= 22 \times 10^3$ Hz and C = 2 nF $= 2 \times 10^{-9}$ F

We know that for a tuned collector oscillator the frequency of oscillations (f_o),

$$22 \times 10^3 = \frac{0.159}{\sqrt{L \cdot C}} = \frac{0.159}{\sqrt{L \times (2 \times 10^{-9})}}$$

$$\therefore \quad L = \frac{\left(\dfrac{0.159}{22 \times 10^3}\right)^2}{(2 \times 10^{-9})} = 0.026 \text{ H } \textbf{Ans.}$$

Example 31.2.

A tuned collector oscillator operates at 2.2 MHz. At what frequency will it work if its tuned circuit capacitance is reduced by 50%.

Solution.

Given: $f_o = 2.2$ MHz $= 2.2 \times 10^6$ Hz.

Let f_o' = The frequency at which the tuned circuit will work when the capacitance is reduced by 50%.

We know that frequency of oscillations, for a tuned collector oscillator,

$$f_o = \frac{0.159}{\sqrt{L \cdot C}} \text{ or } f_o \propto \frac{1}{\sqrt{C}} \quad ...(i)$$

and when the capacitance is reduced by 50%, the frequency of oscillations,

$$f_o' \propto \frac{1}{\sqrt{C/2}} = \frac{\sqrt{2}}{\sqrt{C}} = \frac{1.414}{\sqrt{C}} \quad ...(ii)$$

Dividing equation (*ii*) by (*i*), we get

$$\frac{f_o'}{f_o} = \frac{1.414/\sqrt{C}}{1/\sqrt{C}} = 1.414;$$

$$\therefore \quad f_o' = 1.414 f_o = 1.414 \times 2.2 \times 10^6 = 3.11 \times 10^6 \text{ Hz} = 3.11 \text{ MHz } \textbf{Ans.}$$

31.12. Tuned-Drain (FET) Oscillator

Fig. 31.5 shows the tuned drain (FET) oscillator. It is similar to the tuned collector transistor oscillator shown in Fig. 31.4. This FET has a high input resistance and amplification factor than that of a bipolar transistor. Therefore FET oscillators are more commonly used in actual practice. The frequency of this oscillator is given by the relation.

$$f_o = \frac{1}{2\pi\sqrt{L \cdot C}} \quad ...(i)$$

and the condition (*i.e.*, $\beta \cdot A_v = 1$) is equivalent to

$$\mu = L/M \quad ...(ii)$$

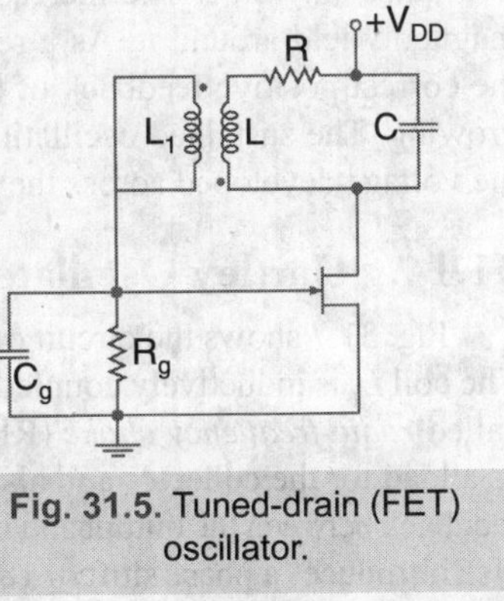

Fig. 31.5. Tuned-drain (FET) oscillator.

where μ = Amplification factor for the FET, and

M = Multual inductance between the coils L_1 and L.

If we take into account, the coil resistance (R), then the above expression for frequency modifies to

$$f'_o = \frac{1}{2\pi\sqrt{L \cdot C}}\left(1 + \frac{R}{r_D}\right) \quad ...(iii)$$

where r_D = The a.c. drain resistance.

and the condition ($\beta \cdot A_v = 1$) is equivalent to

$$g_m = \frac{\mu \cdot R \cdot C}{\mu \cdot M - L} \quad ...(iv)$$

where g_m is the value of transoconductance for an FET.

The equation (*iv*) gives the minimum value of transconductance required for sustained oscillations. The circuit operation may be understood from the following discussion.

When the circuit is first energized by switching on the supply, the gate bias (provided by the parallel combination of the $R_g \cdot C_g$) is zero and the FET operates with a large value of transconductance (g_m). The loop gain is therefore greater than unity and the amplitude of oscillations starts growing. As the oscillations grow, the bias becomes negative and the value of transconductance (g_m) decreases. Finally, the amplitude stabilizes itself at that value for which the loop gain for oscillation frequency is reduced to unity.

31.13. Tuned-Base Oscillator

It is also known as *tickler oscillator*. Fig. 31.6 shows the circuit of such an oscillator. In this circuit, the tuned, circuit is put in the base circuit. The feedback signal is supplied to the base through the coil L_1 (called tickler coil). There is a phase shift of 180° by the transistor amplifier and another 180° phase shift is provided by the transformer. It means that the total phase shift, around the loop, is zero and hence the feedback is positive.

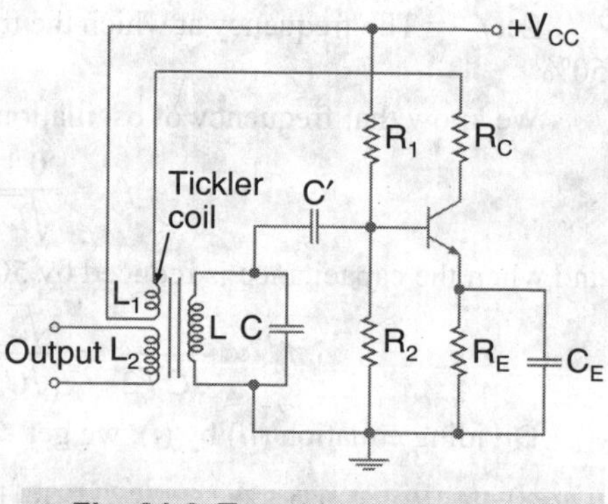

Fig. 31.6. Tuned-base oscillator.

The resistors R_1, R_3 and R_E are used to provide d.c. bias to the transistor. The capacitors C' and C_E act as coupling and bypass capacitors. The operation of the circuit may be understood from the following discussion.

When the circuit is energized, by switching on the supply, the collector current starts increasing to a quiscient value. The increasing collector current, through the tickler coil, creates a changing magnetic field around it. As a result of this, a voltage is induced in the tuned circuit. Because of the correct positive feedback of the coils, and sufficient gain of the amplifier, the oscillations start growing. The sustained oscillations are produced, when the transistor is driven into cut-off due to the voltage developed across the capacitor (C).

31.14. Hartley Oscillator

Fig. 31.7 shows the circuit of a Hartley oscillator. It tank circuit consists of two coils L_1 and L_2. The coil L_1 is inductively coupled to coil L_2 and the combination works as an autotransformer. A coil called *radio frequency choke* (RFC) is connected between the collector and the V_{CC} supply. It acts as a load for the collector and also permits an easy flow of d.c. current. But blocks a.c. current. The feedback between the output and input circuits is accomplished through autotransformer action, which also introduces a phase shift of 180°. The phase reversal between the output and input voltage occurs because they are taken from the opposite ends of the coils (L_1 and L_2) with respect to the tap, which is grounded. It may be noted that the tap on the combination of L_1 and L_2 coils is actually connected to the transistor emitter terminal via ground and through the capacitor (C_E). Since the transistor also introduces a phase shift of 180°, therefore, the total phase shift is 360° and hence the feedback is positive. Ignoring the loading effects of the base, the feedback fraction is given by the relation,

$$\beta = L_2/L_1$$

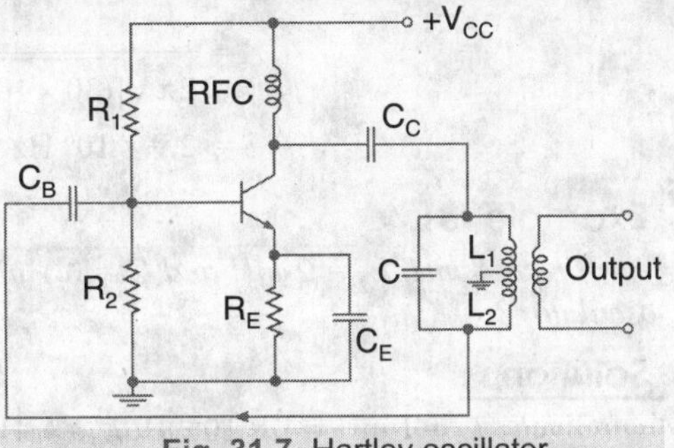

Fig. 31.7. Hartley oscillator.

For oscillations to start, the voltage gain must be greater than 1/b which is equal to L_1/L_2.

The capacitor C_C, connected between the collector and the tuned circuit, is called *coupling capacitor*. It permits only the a.c. currents to pass to the tank circuit. In other words, the capacitor C_C blocks the d.c. currents. The capacitor C_B, called *blocking capacitor*, further blocks the d.c. currents reaching at the base. The resistors R_1, R_2 and R_E are used to provide d.c. bias to the transistor.

A 1929 vintage transmitter utilizing a Hartley oscillator and 45 vacuum tube.

When the circuit is energised, by switching on the supply, the collector current flows. The oscillations are produced because of positive feedback from the tank circuit. The frequency of oscillations is given by the relation,

$$f_o = \frac{1}{2\pi\sqrt{L \cdot C}}$$

where $L = L_1 + L_2 + 2M$

$= L_1 + L_2$

(If the mutual inductance M is neglected)

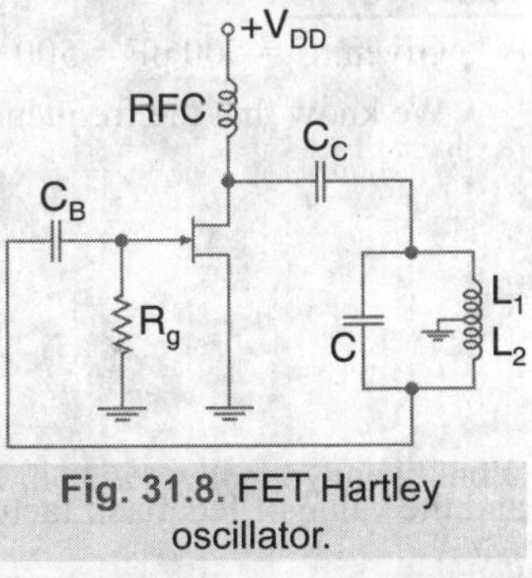

Fig. 31.8. FET Hartley oscillator.

The output from the tank may be taken out by means of another coil coupled either to L_1 or L_2 coil as shown in the figure. Sometimes, a field-effect transistor (FET) may also be used as an amplifying device instead of a bipolar transistor. Fig. 31.8 shows the circuit of a Hartley oscillator using the FET.

Example 31.3.

In a transistorized Hartley oscillator, the tank circuit has the capacitance of 100 pF. The value of inductance between the collector and tapping point is 30 μH and the value of inductance between the tapping point and the transistor base is 1 × 10^{-8} henry. Determine the frequency of oscillations. Neglect the mutual inductance.

Solution.

Given: $C = 100$ pF $= 100 \times 10^{-12}$ F; $L_1 = 30$ μH $= 30 \times 10^{-6}$ H and $L_2 = 1 \times 10^{-8}$ H.

We know that the frequency of oscillations

$$f_o = \frac{1}{2\pi\sqrt{L \cdot C}} = \frac{1}{2\pi\sqrt{(L_1 + L_2)\, C}}$$

$$= \frac{1}{2\pi\sqrt{[(30\times10^{-6})]+(1\times10^{-8})\times(100\times10^{-12})}}$$

$= 2.9 \times 10^6$ Hz = 2.9 MHz **Ans.**

Example 31.4.

If $L_1 = 1$ mH, $L_2 = 2$ mH and C = 0.1 nF, what is the frequency of oscillation of the Hartley oscillator ?
(Anna University., Nov. /Dec. 2007)

Solution.

Given : $L_1 = 1$ mH = 1×10^{-3} H; L_2 = mH = 2×10^{-3} H; and $C = 0.1$ nF = 0.1×10^{-9} F.

We know that the frequency of oscillations

$$f_0 = \frac{1}{2\pi\sqrt{(L_1+L_2)C}}$$

$$= \frac{1}{2\pi\sqrt{[(1\times10^{-3})+(2\times10^{-3})]\times(0.1\times10^{-9})}} = 290.58\times10^6 \text{ Hz}$$

= 290.58 MHZ **Ans.**

Example 31.5.

In a Hartley Oscillator, the value of the capacitor in the tuned circuit is 500 pF and two sections of coil have inductance 38 μH and 12 μ H. Find the frequency of oscillations and the feedback factor β.
(PTU., May 2009)

Solution.

Given: $C = 500$ pF = 500×10^{-12} F; $L_1 = 38$ μH = 38×10^{-6} H and $L_2 = 12$ μH = 12×10^{-6} H.

We know that the frequency of oscillations

$$f_0 = \frac{1}{2\pi\sqrt{L.C}} = \frac{1}{2\pi\sqrt{(L_2+L_2)C}}$$

$$= \frac{1}{2\pi\sqrt{[(38\times10^{-6})+(12\times10^{-6})]\times500\times10^{-12}}}$$

= 1 MHz **Ans.**

and the value of feedback factor,

$$\beta = \frac{L_2}{L_1} = \frac{12}{38} = 0.315 \textbf{ Ans.}$$

Example 31.6.

Determine the frequency of oscillation of a transistorized Hartely Oscillator having $L_1 = 150$ μH, $L_2 = 1.5$ μH, M = 75 μH and C = 150 pF.
(M. G. University., April/May 2009)

Solution.

Given : $L_1 = 150$ μH = 150×10^{-6} H; $L_2 = 1.5$ μH = 1.5×10^{-6} H; $M = 75$ μH = 75×10^{-6} H and $C = 150$ pF = 150×10^{-12} F.

We know that the frequency of oscillations

$$f_0 = \frac{1}{2\pi\sqrt{(L_1+L_2+2M)C}}$$

$$= \frac{1}{2\pi\sqrt{([150\times10^{-6}) + (1.5\times10^{-6}) + (75\times10^{-6})]\times(150\times10^{-12})}}$$

$$= \frac{1}{2\pi\sqrt{(226.5\times10^{-6})\times(150\times10^{-12})}} = 864 \text{ kHz } \textbf{Ans.}$$

Example 31.7.

Calculate the (i) operating frequency, and (ii) feedback fraction for the hartley oscillator shown in Fig. 31.9 the mutual inductance between the coils, M = 20 μH, and L_1 = 1000 μH and L_2 = 100 μH. (Dip. IETE, June 2001)

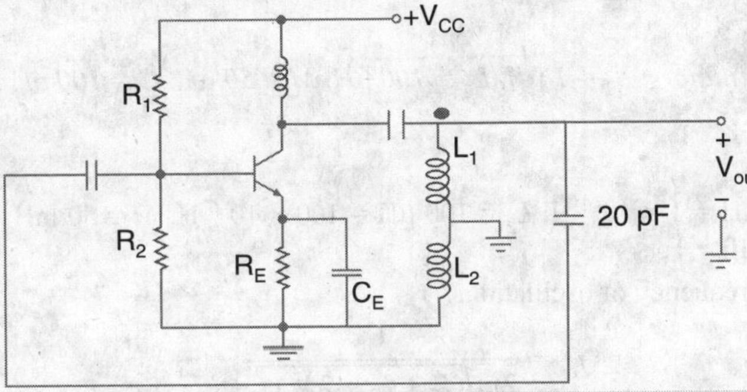

Fig. 31.9.

Solution.

Given: L_1 = 1000 μH; L_2 = 100 μH; M = 20 μH, and . C - 20 pF

We know that frequency of oscillations of a given circuit (a Hartley oscillator),

$$f_o = \frac{1}{2\pi\sqrt{LC}} = \frac{1}{2\pi\sqrt{(L_1 + L_2 + 2M)C}}$$

$$= \frac{1}{2\pi\sqrt{[(1000\times10^{-6}) + (100\times10^{-6}) + 2\times(20\times10^{-6})]\times 20\times10^{-12}}}$$

$$= 1.05\times10^6 \text{ or } 1.05 \text{ MHz } \textbf{Ans.}$$

Example 31.8.

For the Hartley oscillator shown in Fig. 31.10, find the frequency oscillations of the circuit.

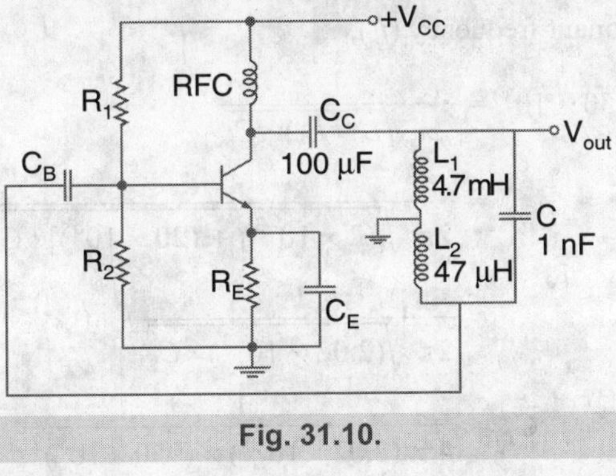

Fig. 31.10.

Solution.

Given: $C = 1$ nF $= 1 \times 10^{-9}$ F; $L_1 = 4.7$ mH $= 4.7 \times 10^{-3}$ H; and $L_2 = 47$ μH $= 47 \times 10^{-6}$ H

We know that frequency of oscillations,

$$f_o = \frac{1}{2\pi\sqrt{L \cdot C}} = \frac{1}{\sqrt{2\pi\,(L_1 + L_2) \cdot C}}$$

$$= \frac{1}{2\pi\sqrt{[(4.7 \times 10^{-3}) + (47 \times 10^{-6})] \times (1 \times 10^{-9})C}}$$

$$= \frac{1}{2\pi \times 2.179 \times 10^{-6}} = 0.0731 \times 10^{6} \text{ Hz} = 73.1 \text{ kHz } \textbf{Ans.}$$

Example 31.9.

For Hartley oscillators, $L_1 = 1$ μH, $L_2 = 100$ μH, $M = 50$ μH, $C = 100$ pF. Find its frequency.
(GTU., June 2011)

Solution.

Given: $L_1 = 1$ μH $= 1 \times 10^{-6}$ H; $L_2 = 100$ μH $= 100 \times 10^{-6}$ H; $M = 50$ mH $= 50 \times 10^{-6}$ H and $C = 100$ pF $= 100 \times 10^{-12}$ F.

We know that frequency of oscillations,

$$f_o = \frac{1}{2\pi\sqrt{(L_1 + L_2 + 2M)\,C}}$$

$$= \frac{1}{2\pi\sqrt{[(1 \times 10^{-6}) + (100 \times 10^{-6}) + 2(50 \times 10^{-6})]\,(100 \times 10^{-12})}}$$

$$= 1.123 \times 10^{6} \text{ Hz} = 1.123 \text{ MHz } \textbf{Ans.}$$

Example 31.10.

A Hartley oscillator is designed with $L_1 = 2$ mH, $L_2 = 20$ μH and a variable capacitance. Determine the range of capacitance values, if the frequency of oscillation is varied between 950 and 2050 k Hz.

Solution.

Given: $L_1 = 2$ mH $= 2 \times 10^{-3}$ H; $L_2 = 20$ μH $= 20 \times 10^{-6}$ H; $f_o = 950$ to 2050 kHz $= 950 \times 10^{3}$ Hz to 2050×10^{3} Hz.

Let C_1 = Value of capacitance required for a frequency of 950 kHz, and

C_2 = Value of capacitance required for a frequency of 2050 kHz.

We know that the resonant frequency (f_o),

$$950 \times 10^{3} = \frac{1}{2\pi\sqrt{(L_1 \cdot L_2) \cdot C_1}}$$

$$= \frac{1}{2\pi\sqrt{[(2 \times 10^{-3}) + (20 \times 10^{6})] \cdot C_1}}$$

$$= \frac{1}{2\pi\sqrt{(2.02 \times 10^{-3}) \cdot C_1}}$$

or

$$C_1 = \frac{1}{4\pi^2 (2.02 \times 10^{-3}) \times (950 \times 10^{3})^2}$$

$$= 13.9 \times 10^{-12} \text{ F} = 13.9 \text{ pF}$$

and
$$2050 \times 10^3 = \frac{1}{2\pi\sqrt{(L_1 + L_2) \cdot C_2}}$$

$$= \frac{1}{2\pi\sqrt{[(2 \times 10^{-3}) + (20 \times 10^{-6})] \cdot C_2}}$$

$$= \frac{1}{2\pi\sqrt{(2.02 \times 10^{-3}) \cdot C}}$$

or
$$C_2 = \frac{1}{4\pi^2\sqrt{(2.02 \times 10^{-3}) \times (2050 \times 10^3)^2}}$$

$$= 2.98 \times 10^{-12} \text{ F} = 2.98 \text{ pF}$$

Hence the range of capacitance values required is 2.98 pF to 13.9 pF **Ans.**

Example 31.11.

In an Hartley oscillator if L_1 = 0.1 mH, L_2 = 10 μH and mutual inductance between the coils equal to 20 μH. calculate the value of capacitor C of the oscillatory circuit to obtain frequency of 4110 kH and also find the condition for sustained oscillations.

(Bangalore Univ B Sc Electronics May 2001)

Solution.

Given : $L_1 = 0.1$ μH; $L_2 = 10$ μH; $M = 20$ μH, $f_o = 4110$ kHz.

Let C = value of the capacitor required.

We know that the frequency of oscillation of a Hartley oscillator circuit (f_o),

$$4110 \times 10^3 = \frac{1}{2\pi\sqrt{LC}} = \frac{1}{2\pi\sqrt{(L_1 + L_2 + 2M)\, C}}$$

or
$$4110 \times 10^3 = \frac{1}{2\pi\sqrt{[(0.1 \times 10^{-3}) + (10 \times 10^{-6}) + (2 \times 10^{-6})] \times C}}$$

$$\therefore \quad C = 1.154 \times 10^{-11} \text{ or } 11.54 \text{ pF}$$

The condition for sustained oscillation is that voltage gain

$$A_V = \frac{1}{\beta} = \frac{1}{L_1 / L_2} = \frac{L_1}{L_2} = \frac{0.1 \times 10^{-3}}{10 \times 10^{-6}} = 10 \textbf{ Ans. Ans.}$$

31.15. Colpitts Oscillator

It is similar to the Hartley oscillator, except for one difference. The colpitts oscillator uses tapped capacitance instead of tapped inductance used in Hartley oscillator. Fig. 31.11 (*a*) shows the colpitts oscillator circuit, whereas Fig. (*b*) shows its a.c. equivalent circuit. The tank circuit is made up of two capacitors C_1 and C_2 connected in series with each other across a fixed inductance (L). The resistorss (R_1, R_2, and R_E), capacitors (C_C and C_B) and RFC have the function as mentioned for Hartley oscillator. The feedback between the output and input circuit is accomplished by the voltage developed across the capacitor C_2. Ignoring the loading effect of the base, the feedback fraction,

$$\beta = C_1 / C_2$$

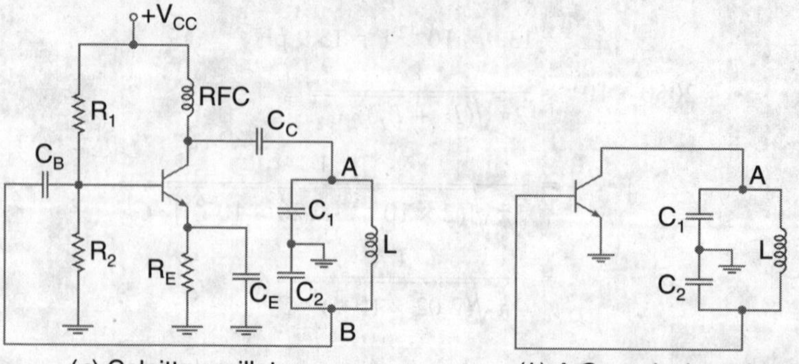

(a) Colpitts oscillator. (b) A.C. equivalent circuit.

Fig. 30.11.

For oscillation to start, the voltage gain (A_v) must be greater than $1/\beta$ (or C_2/C_1). Or in other words, $A_v > C_2/C_1$. The frequency of oscillations (neglecting mutual inductance), is given by the relation,

$$f_o = \frac{1}{2\pi\sqrt{L \cdot C}}$$

(Where C is the effective value of capacitance and equal to $\frac{C_1 \cdot C_2}{C_1 + C_2}$)

$$\therefore \quad f_o = \frac{1}{2\pi\sqrt{L \times \frac{C_1 \cdot C_2}{C_1 + C_2}}}$$

It may be noted that in a colpitts oscillator, the capacitors C_1 and C_2 act as a simple alternating voltage divider. Therefore points A and B are 180° out of phase with each other. Another phase shift of 180° is provided by the transistor itself. Thus there is a total phase shift of 360° between the emitter-base and collector-base circuits.

When the circuit is energized by switching on the supply, the capacitors C_1 and C_2 are charged. These capacitors discharge through the coil (L), which sets up the oscillations of frequency $f_o = 1/2\pi\sqrt{L \cdot C}$ where $C = C_1 \cdot C_2/(C_2 + C_2)$. The oscillations across the capacitor C_2 are feedback to the base-emitter junction and appear in an amplified form at the collector. Because of the positive feedback, the oscillations of constant amplitude are produced. The colpitts oscillators are widely used as a signal generator for frequencies between 1 MHz and 500 MHz.

Example 31.12.

In a transistor colpitts oscillator, $C_1 = 0.001$ μF; $C_2 = 0.01$ μF and $L = 5$ μH. Find the required gain for oscillation and the frequency of oscillation.

Solution.

Given: $C_1 = 0.0001$ μF $= 0.001 \times 10^{-6}$ F; $C_2 = 0.01$ μF $= 0.01 \times 10^{-6}$ F and $L = 5$ μH $= 5 \times 10^{-6}$ H.

We know that voltage gain required to produce oscillations,

$$A_v > \frac{C_2}{C_1} = \frac{0.01 \times 10^{-6}}{0.001 \times 10^{-6}} = 10 \textbf{ Ans.}$$

and the capacitance,

$$C = \frac{C_1 \cdot C_2}{C_1 + C_2} = \frac{(0.001 \times 10^{-6}) \times (0.01 \times 10^{-6})}{(0.001 \times 10^{-6}) + (0.01 \times 10^{-6})} = 9.091 \times 10^{-10} \text{ F}$$

∴ Frequency of oscillations,

$$f_o = \frac{1}{2\pi\sqrt{L \cdot C}}$$

$$= \frac{1}{2\pi \times \sqrt{(5 \times 10^{-6}) \times (9.091 \times 10^{-10})}} \text{ Hz}$$

$= 2.37 \times 10^6$ Hz = 2.37 MHz **Ans.**

Example 31.13.

If C_1 and C_2 are 200 pF and 50 pF respectively. Calculate the value of inductance for producing oscillations at 1 MHz in the Colpitt's Oscillator circuit. (Anna University., Nov. /Dec. 2007)

Solution.

Given : $C_1 = 200\ pF = 200 \times 10^{-12}$ F; $C_2 = 50$ pF $= 50 \times 10^{-12}$ F and $f_0 = 1$ MHz $= 1 \times 10^6$ Hz.

We know that the frequency of oscillation,

$$f_0 = \frac{1}{2\pi\sqrt{L\left(\frac{C_1C_2}{C_1 + C_2}\right)}}$$

$$1 \times 10^6 = \frac{1}{2\pi\sqrt{L\frac{(200\times10^{-12})\times(50\times10^{-12})}{(200\times10^{-12})+(50\times10^{-12})}}}$$

$L = 6.635 \times 10^{-3}$ H = 0.633 mH **Ans.**

Example 31.14.

In Colpitts oscillator, the values of the inductors and capacitors in the tank circuit are L = 40, C_1 = 100pF and C_2 = 500pF. (a) Find the frequency of oscillations (b) If the output voltage is 10 V, find the feedback voltage. (PTU., Dec 2008)

Solution.

Given: $L = 40\ mH = 40 \times 10^{-3}$ H, $C_1 = 100$ pF $= 100 \times 10^{-12}$ F and $C_2 = 500$ pE $= 500 \times 10^{-12}$ F.

(a) Find the frequency of oscillations

We know that the frequency of oscillations of a Colpitts oscillator,

$$f_0 = \frac{1}{2\pi\sqrt{\left[L\left(\frac{C_1C_2}{C_1 + C_2}\right)\right]}}$$

$$= \frac{1}{2\pi\sqrt{(40\times10^{-3})\times\frac{(100\times10^{-12})\times(500\times10^{-12})}{(100\times10^{-12})+(500\times10^{-12})}}}$$

= 87.21 MHz **Ans.**

(b) The value of feedback voltage with output voltage = 10 V

We know that the feedback voltage,

$$V_f = \beta V_{out}$$

and

$$\beta = \frac{C_1}{C_2} = \frac{100 \times 10^{-12}}{500 \times 10^{-12}} = 0.2$$

$$V_f = 0.2 \times 10 = 2 \text{ V } \textbf{Ans.}$$

Example 31.15.

In a Colpitts Oscillator, C_1 = 0.001 μF, C_2 = 0.01 μF and L = 10μH. Find the frequency of oscillation, feedback factor and voltage gain. *(Anna University., June 2009)*

Solution.

Given: C_1 = 0.001 μF = 0.001 × 10^{-6} F; C_2 = 0.01 μF = 0.01 × 10^{-6} F and L = 10 μH = 10 × 10^{-6} H.

Frequency of oscillation

We know that the frequency of oscillator,

$$f_0 = \frac{1}{2\pi\sqrt{L\left(\frac{C_1 C_2}{C_1 + C_2}\right)}}$$

$$= \frac{1}{2\pi\sqrt{(10\times10^{-6})\times\left(\frac{0.001\times10^{-6}\times0.01\times10^{-6}}{0.001\times10^{-6}+0.01\times10^{-6}}\right)}} \text{Hz}$$

Feedback factor

$$= \frac{1}{2\pi\times0.0953\times10^{-6}} \text{Hz} = 1.67\times10^{6} \text{Hz} = 1.67 \text{ MHz } \textbf{Ans.}$$

We know that the value of feedback factor,

$$\beta = \frac{C_1}{C_2} = \frac{0.001}{0.01} = 0.1 \textbf{ Ans.}$$

Voltage gain

We know that the voltage gain,

$$A = \frac{C_2}{C_1} = \frac{0.01}{0.001} = 10 \textbf{ Ans.}$$

Example 31.16.

In Colpitt's Oscillator, the value of the inductors and capacitors in the tank are L = 40 mH, C_1 = 100 pF and C_2 = 500 pF

(a) Find the frequency oscillation.
(b) If the input voltage is 10V, Find the feedback voltage.
(c) Find the minimum gain, if the frequency is changed by changing 'L' alone.
(d) Find the value of C_1 for a gain of 10.
(e) Also find the new frequency of oscillation. *(Anna University., Nov/Dec. 2007)*

Solution.

Given: L = 40 mH = 40 × 10^{-3} H; C_1 = 100 pF = 100 × 10^{-12} C and C_2 = 500 × 10^{-12} = 100 × 10^{-12} F

(a) Frequency of Oscillation

We know that the value of frequency is given by,

$$f = \frac{1}{2\pi\sqrt{L\,C_{eq}}}$$

$$C_{eq} = \frac{C_1 + C_2}{C_1 + C_2} = \frac{(100 \times 10^{-12}) \times (500 \times 10^{-12})}{(100 \times 10^{-12}) + (500 \times 10^{-12})} = 83.333 \times 10^{-12} \text{ F}$$

$$\therefore \qquad f = \frac{1}{2\pi \sqrt{40 \times 10^{-3} \times 83.333 \times 10^{-12}}} = 87.173 \text{ kHz } \textbf{Ans.}$$

(b) feedback voltage if V_0 = 100 V

We know that the value of gain,

$$\text{Gain} = \frac{C_2}{C_1} = \frac{500}{100} = 5$$

and the value of feedback voltage,

$$= \frac{V_0}{Gain} = \frac{10}{5} = 2 \text{ V } \textbf{Ans.}$$

(c) Minimum gain, if the frequency is changed by changing 'L' alone

We know that the minimum gain,

$$\frac{C_2}{C_1} = \frac{500}{100} = 54 \textbf{ Ans.}$$

(d) Value of C_1 for a gain of 10

We know that the value of gain,

$$Gain = 10 = \frac{C_2}{C_1}$$

$$10 = \frac{500 \times 10^{-12}}{C_1}$$

$$C_1 = 50 \text{ pF } \textbf{Ans.}$$

(e) New Frequency of Oscillation,

$$C_1 = 50pF = 50 \times 10^{-12} \text{ F and } C_2 = 500\ pF = 500 \times 10^{-12} \text{ F}$$

We know that the C_{eq},

$$C_{eq} = \frac{C_1 C_2}{C_1 + C_2} = \frac{(50 \times 10^{-12}) \times (500 \times 10^{-12})}{(50 \times 10^{-12}) + (500 \times 10^{-12})} = 45.455 \times 10^{-12} \text{ F}$$

and the frequency of oscillation,

$$f = \frac{1}{2\pi \sqrt{L\, C_{eq}}}$$

$$f = \frac{1}{2\pi \sqrt{(40 \times 10^{-3}) \times (45.45 \times 10^{-12})}} = 0.118 \times 10^{6} \text{ Hz or } 118 \text{ kHz } \textbf{Ans.}$$

Example 31.17.

In a transistor Colpitt's oscillator we have L= 100 μH, L_{RFC} = 0.6 μH, C_1 = 0.001 μF, C_2 = 0.01 μF. Find (a) Operating Frequency (b) Feedback fraction (c) Minimum gain to sustain oscillations & Emitter resistance if R_C = 2.5 kΩ. *(PTU., May 2008, Dec 2008)*

Solution.

Given: $L = 100\ \mu H = 100 \times 10^{-6}$ H, $L_{RFC} = 0.6\ \mu H$, $C_1 = 0.001 \times 10^{-6}$ and $C_2 = 0.01\ \mu F = 0.01 \times 10^{-6}$ F.

(a) Frequency of oscillations

We know that the frequency of oscillations of a Colpitts oscillator,

$$f_0 = \frac{1}{2\pi\sqrt{\left[L \times \left(\frac{C_1 \cdot C_2}{C_1 + C_2}\right)\right]}}$$

$$= \frac{1}{2\pi\sqrt{(100 \times 10^{-6}) \times \frac{(0.001 \times 10^{-6}) \times (0.01 \times 10^{-6})}{(0.001 \times 10^{-6}) + (0.01 \times 10^{-6})}}} = 528 \text{ kHz } \textbf{Ans.}$$

(b) Feedback fraction

We know that the value of feedback fraction,

$$\beta = \frac{C_1}{C_2} = \frac{0.001 \times 10^{-6}}{0.01 \times 10^{-6}} = 0.1 \textbf{ Ans.}$$

(c) Minimum gain to sustain oscillations

We know that the minimum gain,

$$A_{min} = \frac{1}{\beta} = \frac{1}{0.1} = 10$$

$$R_E \cong \frac{R_C}{A_{min}} = \frac{2.5 \text{ k}\Omega}{10} = 250\ \Omega \textbf{ Ans.}$$

Example 31.18.

In a Colpitts oscillator L = 100 μH, C_1 = 0.001 μF, C_2 = 0.02 μH. Determine the frequency of oscillation. (M. G. University., Dec. 2007)

Solution.

Given : $L = 100\ \mu H = 100 \times 10^{-6}$ H; $C_1 = 0.001\ \mu F = 0.001 \times 10^{-6}$ F and $C_2 = 0.02\ \mu F = 0.02 \times 10^{-6}$ F.

We know that the frequency of oscillations

$$f_0 = \frac{1}{2\pi\sqrt{L.C}} = \frac{1}{2\pi\sqrt{L \times \frac{C_1 \cdot C_2}{C_1 + C_2}}}$$

$$= \frac{1}{2\pi\sqrt{(100 \times 10^{-6}) \times \frac{(0.001 \times 10^{-6}) \times (0.02 \times 10^{-6})}{(0.001 \times 10^{-6}) \times (0.02 \times 10^{-6})}}}$$

$$= \frac{1}{2\pi\sqrt{(100 \times 10^{-6}) \times \frac{0.00002 \times 10^{-6}}{0.021}}}$$

$= 515.7$ kHz **Ans.**

Example 31.19.

For the transistor Colpitts oscillator shown in Fig. 31.12, find the frequency of oscillations.

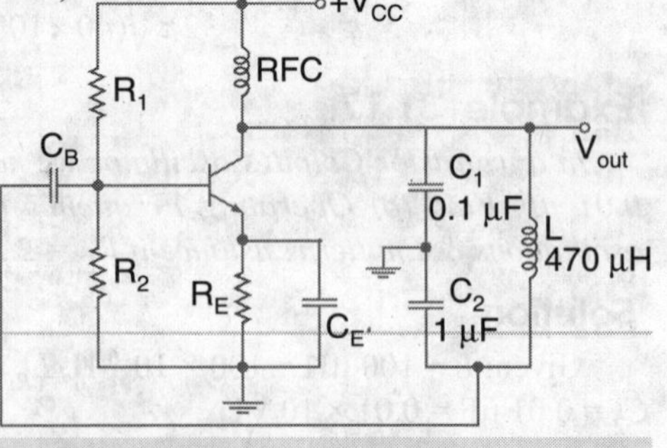

Fig. 31.12.

Solution.

Given: $C_1 = 0.1\ \mu F = 0.1 \times 10^{-6}\ F$; $C_2 = 1\ \mu F = 1 \times 10^{-6}$ F and $L = 470\ \mu H = 470 \times 10^{-6}$ H.

We know that frequency of oscillations,

$$f_o = \frac{1}{2\pi\sqrt{L \cdot C}} = \frac{1}{2\pi\sqrt{L \times \dfrac{C_1 \cdot C_2}{C_1 + C_2}}}$$

$$= \frac{1}{2\pi\sqrt{(470 \times 10^{-6}) \times \dfrac{(0.1 \times 10^{-6}) \times (1 \times 10^{-6})}{(0.1 \times 10^{-6}) + (1 \times 10^{-6})}}}$$

$$= \frac{1}{2\pi\sqrt{(470 \times 10^{-6}) \times \dfrac{0.1 \times 10^{-6}}{1.1}}} = \frac{1}{2\pi \times 6.537 \times 10^{-6}}$$

$$= 0.0244 \times 10^6 \text{ Hz} = 24.4 \text{ kHz } \textbf{Ans.}$$

Example 31.20.

A Colpitts oscillator is designed with $C_1 = 100$ pF and $C_2 = 7500$ pF. The inductance is variable. Determine the range of inductance values, if the frequency of oscillation is to vary between 950 kHz and 2050 kHz.

Solution.

Given: $C_1 = 100$ pF $= 100 \times 10^{-12}$ F; $C_2 = 7500$ pF $= 7500 \times 10^{-12}$ F and $f_o = 950$ kHz to 2050 kHz $= 950 \times 10^3$ Hz to 2050×10^3 Hz.

Let L_1 = Value of inductance required for a frequency of 950×10^3 Hz,

and L_2 – Value of inductance required for a frequency of 2050×10^3 Hz.

We know that the effective value of capacitance,

$$C = \frac{C_1 \cdot C_2}{C_1 + C_2} = \frac{(100 \times 10^{-12}) \times (7500 \times 10^{-12})}{(100 \times 10^{-12}) + (7500 \times 10^{-12})} \text{ F}$$

$$= 98.68 \times 10^{-12} \text{ F}$$

and the resonant frequency (f_o),

$$950 \times 10^3 = \frac{1}{2\pi\sqrt{L_1 \cdot C}} = \frac{1}{2\pi \times \sqrt{L_1 \times (98.68 \times 10^{-12})}}$$

or
$$(950 \times 10^3)^2 = \frac{1}{4\pi^2 \cdot L_1 \times (98.68 \times 10^{-12})}$$

$\therefore$
$$L_1 = \frac{1}{4\pi^2 \times (98.68 \times 10^{-12}) \times (950 \times 10^3)^2}$$

$$= 284 \times 10^{-6} \text{ H} = 284\ \mu\text{H}$$

and
$$2050 \times 10^3 = \frac{1}{2\pi\sqrt{L_2 \cdot C}}$$

$$= \frac{1}{2\pi\sqrt{L_2 \times (98.68 \times 10^{-12})}}$$

$$\text{or} \qquad (2050 \times 10^3)^2 = \frac{1}{4\pi^2 \times L_2 \times (98.68 \times 10^{-12})}$$

$$\therefore \qquad L_2 = \frac{1}{4\pi^2 \times (98.68 \times 10^{-12}) \times (2050 \times 10^3)^2}$$

$$= 61 \times 10^{-6}\ \text{H} = 61\ \mu\text{H}$$

Thus the range of inductance required for the variation of the desired frequency is from 61 μH to 284 μH. **Ans.**

Example 31.21.

The frequency of oscillation of a Colpitts oscillator is given by,

$$f_o = \frac{1}{2\pi\sqrt{\left[L\left(\frac{C_1 C_2}{C_1 + C_2}\right)\right]}}$$

where L, C_1 and C_2 are the frequency - determining components. Such a circuit operates at 450 kHz with $C_1 = C_2$. What will be the oscillation frequency if the value of C_2 is doubled?

Solution.

Given: $f_o = 450$ kHz $= 450 \times 10^3$ Hz; and $C_1 = C_2 = C$ (say)

Let $\qquad f_o$ = The frequency when the value of C_2 is doubled.

We know that the frequency of oscillations of a Colpitts oscillator,

$$f_o = \frac{1}{2\pi\sqrt{L\left(\frac{C_1 C_2}{C_1 + C_2}\right)}} = \frac{1}{2\pi\sqrt{L \cdot \frac{C \cdot C}{C + C}}}$$

$$= \frac{1}{2\pi\sqrt{\frac{L . C}{2}}} = \frac{1}{2\pi} \cdot \sqrt{\frac{3}{L \cdot C}} \qquad ...(i)$$

when the value of C_2 is doubled, the frequency of oscillation,

$$f_o' = \frac{1}{2\pi\sqrt{\left[L\left(\frac{C_1 \cdot 2C_2}{C_1 + 2C_2}\right)\right]}} = \frac{1}{2\pi\sqrt{L \cdot \frac{2C \cdot C}{3C}}}$$

$$= \frac{1}{2\pi\sqrt{\frac{2\ LC}{3}}} = \frac{1}{2\pi} \cdot \sqrt{\frac{3}{2\ LC}} \qquad ...(ii)$$

Dividing equation (*ii*) by (*i*),

$$\frac{f_o'}{f_o} = \frac{\frac{1}{2\pi} \cdot \sqrt{\frac{3}{2LC}}}{\frac{1}{2\pi} \cdot \sqrt{\frac{2}{LC}}} = \sqrt{\frac{3}{2LC} \times \frac{LC}{2}} = \sqrt{\frac{3}{4}} = 0.866$$

$\therefore\ f_o' = 0.866 \times f_o = 0.866 \times 450 \times 10^3 = 389 \times 10^3$ Hz $= 389.7$ kHz **Ans.**

31.16. FET Collpits Oscillator

Fig. 31.13 shows the circuit of a field-effect transistor (FET) oscillator. In this circuit, the feedback signal is applied to the gate terminal of FET. Since the gate has a high input resistance, therefore the loading effect on the tank circuit is much less than that with a bipolar transistor.

The operation of the FET colpitt's oscillator circuit may be explained in the same way as discussed for transistor colpitt's oscillator.

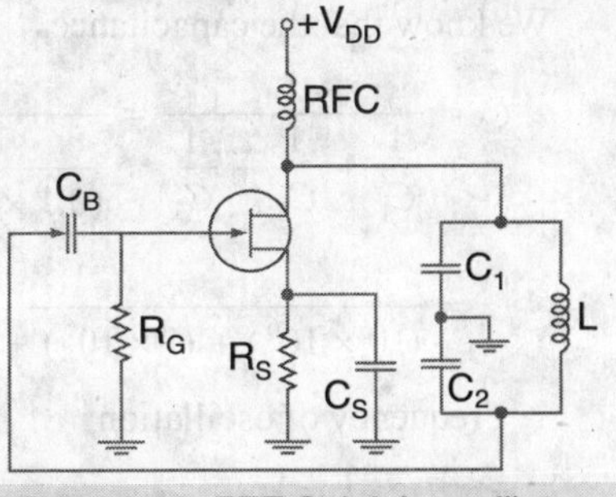

Fig. 31.13. FET Colpitt's oscillator.

31.17. Clapp Oscillator

It is an improved version of the colpitts oscillator. Fig. 31.14 shows the circuit of a clapp oscillator. The circuit differs from the colpitts oscillator only in one respect, that it contains one additional capacitor (C_3) connected in series with the inductor. The addition of capacitor (C_3) improves the frequency stability and eliminates the effect of transistor parameters and stray capacitances.

The operation of clapp oscillator circuit is in the same way as that of colpitts oscillator. The frequency of oscillator is given by the relation,

$$f_o = \frac{1}{2\pi\sqrt{L \cdot C}}$$

where
$$C = \frac{1}{\dfrac{1}{C_1} + \dfrac{1}{C_2} + \dfrac{1}{C_3}}$$

Usually, the value of C_3 is much smaller than C_1 and C_2. As a result of this, C is approximately equal to C_3. Therefore the frequency of oscillation,

$$f_o = \frac{1}{2\pi\sqrt{L \,.\, C_3}}$$

Example 31.12.

Calculate the frequency of oscillations for the Clapp Oscillator shown in Fig. 31.15.

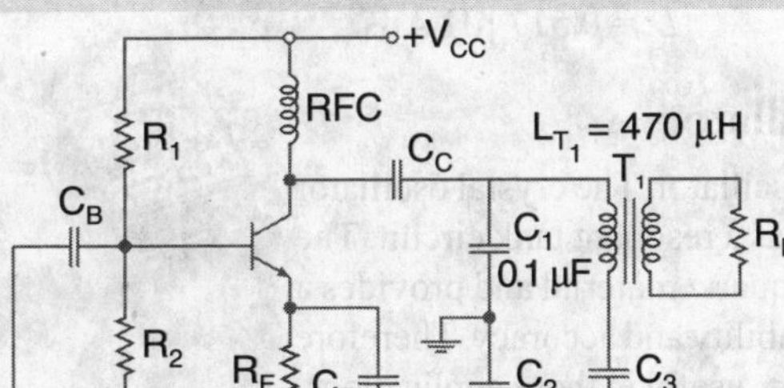

Fig. 31.15.

Solution.

Given: $C_1 = 0.1\ \mu\text{F} = 0.1 \times 10^{-6}$ F; $C_2 = 1\mu$ F $= 1 \times 10^{-6}$ F; $C_3 = 100$ pF $= 100 \times 10^{12}$ F and $L_{T_1} = L = 470\ \mu\text{H} = 470 \times 10^{-6}$H

We know that the capacitance,

$$C = \frac{1}{\frac{1}{C_1}+\frac{1}{C_2}+\frac{1}{C_3}} = \frac{1}{\frac{1}{(0.1\times10^{-6})}+\frac{1}{(1\times10^{-6})}+\frac{1}{(100\times10^{-12})}}$$

$$= \frac{1}{(10\times10^6)+(1\times10^6)+(10{,}000\times10^6)} = \frac{1}{10011\times10^6} = 99.9\times10^{-12}\ \text{F} = 99.9\ \text{pF}$$

$\therefore$ Frequency of oscillation

$$f_o = \frac{1}{2\pi\sqrt{LC}} = \frac{1}{2\pi\times\sqrt{(470\times10^{-6})\times(99.9\times10^{-12})}} = 734.5\ \text{kHz}\ \textbf{Ans.}$$

Example 31.23.

Calculate the inductance value to produce 734.5 kHz frequency of oscillation in Clapp Oscillator having C_1 = 0.1 μF, C_2 = 1 μF, and C_3 = 100 μF. *(Anna University., April/May. 2006)*

Solution.

Given: C_1 = 0.1 μF = 0.1 × 10^{-6} F; C_2 = 1 μF = 1 × 10^{-6} F; C_3 = 100 μF = 100 × 10^{-6} F and f_0 = 734.5 kHz = 734.5 × 10^3 Hz.

We know that capacitance,

$$C = \frac{1}{\frac{1}{C_1}+\frac{1}{C_2}+\frac{1}{C_3}} = \frac{1}{\frac{1}{0.1\times10^{-6}}+\frac{1}{1\times10^{-6}}+\frac{1}{100\times10^{-6}}}$$

$$= \frac{1}{10\times10^6+1\times10^6+0.01\times10^6} = 90.826\times10^{-9} = 90.826\ \text{nF}.$$

and the frequency of oscillation (f_o),

$$734.5 = \frac{1}{2\pi\sqrt{LC}}$$

or
$$734.5\times10^3 = \frac{1}{2\pi\times\sqrt{L\times90.826\times10^{-9}}}$$

$\therefore$
$$L = 0.517\ \mu\text{H}\ \textbf{Ans.}$$

31.18. Crystal Oscillator

It is basically a tuned oscillator. The crystal oscillator uses a piezoelectric crystal as a resonant tank circuit. The crystal is, usually, made of quartz material and provides a high degree of frequency stability and accuracy. Therefore the crystal oscillators are very useful in those applications, where frequency stability is very essential. The crystal oscillators are widely used in communication transmitters, digital watches and clock etc.

Voltage controlled crystal oscillator

31.19. Quartz Crystal

A quartz crystal has a very peculiar property known as piezoelectric effect. According to this effect, when an a.c. voltage is applied across a quartz crystal, it vibrates at the frequency of the applied voltage. Conversely, if a mechanical force is applied to vibrate a quartz crystal, it generates an a.c. voltage. The other materials, which exhibit the piezoelectric effect are Rochelle salt and tourmaline.

The quartz material is preferred because it is inexpensive and readily available in nature. Moreover, its properties lie in between those of Rochelle salt and tourmaline.

31.20. Characteristics of Quartz Crystal

The natural shape of a quartz crystal is a hexogonal prism with pyramids at the ends as shown in Fig. 31.16 (*a*). In order to get a usable crystal out of it, we have to cut a rectangular slab out of a natural crystal as shows in Fig. 31.16 (*b*). There are a number of ways to cut the natural crystal namely *X*-cut, *Y*-cut, *XY*-cut, AT-cut etc. Different cuts have different piezoelectric properties (*i.e.*, resonant frequency, temperature coefficient etc.).

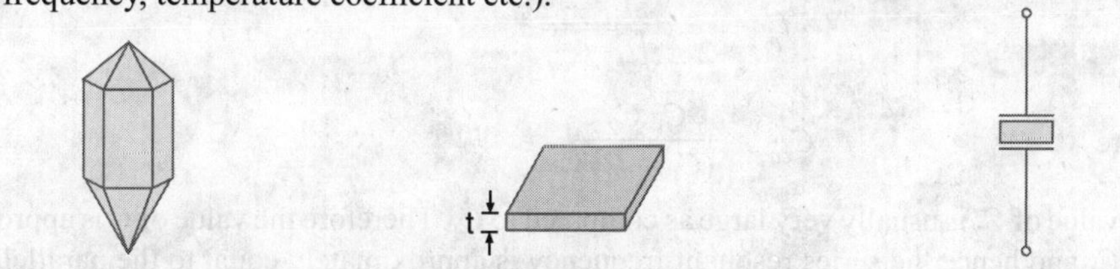

(*a*) Natural shape of a Quartz crystal. (*b*) Rectangular slab. (*c*) Crystal symbol.

Fig. 31.16.

For use in electronic circuits, the slab is mounted between two metal plates and housed in a package, which is equal to the size of postage stamp. In actual practice, the package as a whole is known as crystal and its symbol is shown in Fig. 31.16 (*c*). When a crystal is placed across an a.c. source, it starts vibrating. The amount of vibration depends upon the frequency of the applied voltage. By changing the frequency, we can find a frequency at which the crystal vibrations reach its maximum value. The frequency, at which it happens, is called resonant frequency of the crystal.

The quartz crystals are available with resonant frequencies ranging from 15 kHz to 10 MHz. For higher frequencies (*i.e.*, frequencies up to 100 MHz), the crystal can be made to vibrate on overtones (*i.e.*, multiples of fundamental frequency). However, for frequencies greater than 10 MHz, the crystals are not useful.

31.21. Electrical Equivalent Circuit of a Crystal

Fig. 31.17 (*a*) and (*b*) shows the symbol and electrical equivalent circuit of a crystal respectively. It consists of a series R-L-C_1 circuit in parallel with a capacitance (C_2). When the crystal mounted across the a.c. source is not vibrating, it is equivalent to the capacitance C_2. However, when the crystal is vibrating, it acts like a tuned circuit R-L-C_1.

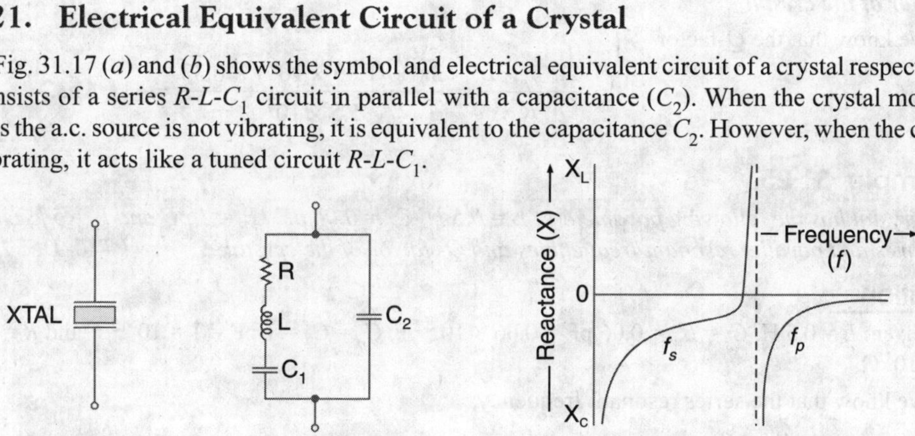

(*a*) Symbol. (*b*) Equivalent circuit. (*c*) Reactance versus frequency.

Fig. 31.17.

Fig. 31.17 (*c*) shows the graph of a reactance (X_L or X_C) versus frequency (*f*). It is evident from this graph that the crystal has two closely spaced resonant frequencies. The first one is the series resonant frequency (f_s), which occurs when reactance of the inductance (L) is equal to the reactance of the capacitance C_1. In that case, the impedance of the equivalent circuit is equal to the resistance R and the frequency of oscillation is given by the relation,

$$f_s = \frac{1}{2\pi\sqrt{L \cdot C_{eq}}}$$

The second one is the parallel resonant frequency (f_p), which occurs when the reactance of R-L- C_1 branch is equal to the reactance of capacitor C_2. At this frequency, the crystal offers a very high impedance to the external circuit and the frequency of oscillation is given by the relation.

$$f_p = \frac{1}{2\,\pi\sqrt{L \cdot C_{eq}}}$$

where $$C_{eq} = \frac{C_1\,C_2}{(C_1 + C_2)}$$

The value of C_2 is usually very large as compared to C_1. Therefore the value of C is approximately equal to C_1 and hence the series resonant frequency is approximately equal to the parallel resonant frequency (*i.e.*, $f_s = f_p$).

Example 31.24.

A crystal has the following parameters: L = 0.33 H, C = 0.065 pF, C_2 = 1.0 pF and R = 5.5 kΩ. Find the series resonant frequency and Q-factor of the crystal. (*Anna University., Nov/Dec. 2008*)

Solution.

Given: $L = 0.33$ H; $C_1 = 0.065$ pF $= 0.065 \times 10^{-12}$ F; $C_2 = 1.0$ pF and $R = 5.5$ kΩ $= 5.5 \times 10^3$ Ω

Series resonant frequency

We know that the series resonant frequency,

$$f_s = \frac{1}{2\pi\sqrt{L \cdot C_1}} = \frac{1}{2\pi \times \sqrt{0.33 \times (0.065 \times 10^{-12})}}$$

$$= 1.09 \times 10^6 \text{ Hz } \textbf{Ans.}$$

Q-factor of the crystal

We know that the Q-factor

$$= \frac{2\pi\, f_s \cdot L}{R} = \frac{2\pi \times (1.09 \times 10^6) \times 0.33}{5.5 \times 10^3} = 411 \textbf{ Ans.}$$

Example 31.25.

A crystal has the following parameters: L = 0.5 H, C_S = 0.06 pF, C_P = 1 pF and R = 5 kΩ. Find the series and parallel resonant frequencies and Q- factor of the crystal. (*PTU., Dec 2008*)

Solution.

Given: $L = 0.5$ H, $C_s = C_1 = 0.06$ pF $= 0.06 \times 10^{-12}$ F, $C_p = C_2 = 1$ pF $= 1 \times 10^{-12}$ F and $R = 5$ kΩ $= 5 \times 10^3$ Ω.

We know that the series resonant frequency,

$$f_s = \frac{1}{2\pi\sqrt{LC_1}}$$

$$= \frac{1}{2\pi\sqrt{0.5 \times 0.056 \times 10^{-12}}} = 920.43 \text{ kHz } \textbf{Ans.}$$

and the parallel resonant frequency,

$$f_p = \frac{1}{2\pi\sqrt{L\,C_{eq}}}$$

$$C_{eq} = \frac{0.06\times10^{-2}\times1\times10^{-12}}{0.06\times10^{-12}+1\times10^{-12}} = 0.057\times10^{-12}\,\text{F}$$

$$\therefore \quad f_p = \frac{1}{2\pi\times\sqrt{0.5\times0.057\times10^{-12}}} = 942.8\text{ kHz } \textbf{Ans.}$$

We also know that the Q-factor of the crystal,

$$Q = \frac{2\pi\, f_s\,.\,L}{R}$$

$$= \frac{2\pi\times(920.43\times10^{6})\times0.05}{5\times10^{3}} = 577 \textbf{ Ans.}$$

Example 31.26.

A crystal has L = 2 H, C = 0.01 pF and R = 2 kΩ. Its mounting capacitance is 2 pF. Calculate its series and parallel resonating frequency. (*VTU., June/July 2011*)

Solution.

Given: $L = 2$ H, $C = 0.01$ pF $= 0.01\times10^{-12}$ F, $R = 2$ kΩ $= 2\times10^{3}$ Ω and $C_m = 2$ pF $= 2\times10^{-12}$ F.

We know that the series resonant frequency,

$$f_s = \frac{1}{2\pi\sqrt{LC}} = \frac{1}{2\pi\sqrt{(2)\times(0.01\times10^{-12})}}$$

$$= 1.125\times10^{6}\text{ Hz} = 1.125\text{ MHz } \textbf{Ans.}$$

the effective value of capacitance,

$$C_{eq} = \frac{C\,.\,C_m}{C+C_m} = \frac{(0.01\times10^{-12})\times(2\times10^{-12})}{(0.01\times10^{-12})+(2\times10^{-12})} = 9.95\times10^{-15}\,\text{F}$$

We know that the parallel resonant frequency,

$$f_p = \frac{1}{2\pi\sqrt{L.C_{eq}}} = \frac{1}{2\pi\sqrt{2\times(9.95\times10^{-15})}}$$

$$= 1.128\times10^{6}\text{ Hz} = 1.128\text{ MHz } \textbf{Ans.}$$

31.22. Q-Factor and Frequency Stability of a Crystal

The crystal has a very high value of Q-factor as compared to the discrete LC tank circuits. For example, the Q-factor of a typical crystal is 20,000 as compared to a maximum of about 100 for high quality LC tank circuits. The extremely high values of the Q-factor of a crystal produces oscillators with very stable frequency values.

The frequency of an oscillator tends to change slightly with time. This change in frequency with time is called frequency drift. The frequency drift is produced by temperature aging etc. In a crystal oscillator, the frequency drift is very small *i.e.*, less than 0.0001% (*i.e.*, 1 part in 10^{6}) per day. It is because of this fact that the *crystals are used as a basic timing device in digital watches.*

31.23. Crystal Oscillator Circuit

A variety of crystal oscillator circuits is possible such as crystal-controlled tuned collector-oscillator, colpitts crystal oscillator clapp crystal oscillator etc. But the pierce crystal oscillator is the most commonly used circuit in actual practice. Fig. 31.18 shows the transistor pierce crystal oscillator. In this circuit, the crystal in connected as a series element in the feedback path from collector to the base.

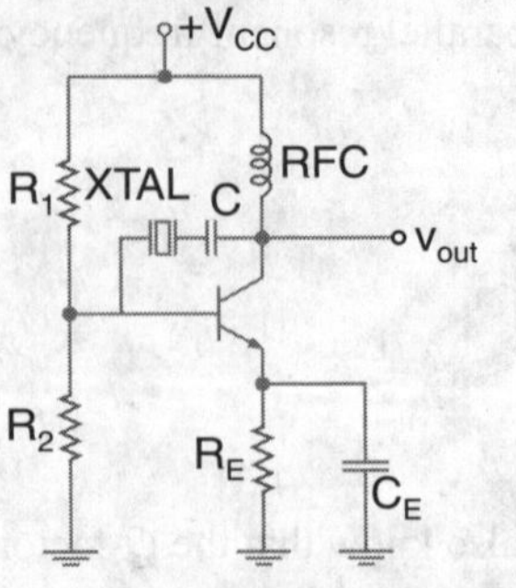

Fig. 31.18. Transistor pierce crystal oscillator.

The resistors R_1, R_2 and R_3 provide a voltage-divider stabilized d.c. bias circuit. The capacitor C_E provides a.c. bypass of the emitter resistor and RFC (radio frequency choke) coil provides for d.c. bias while decoupling any a.c. signal on the power lines from affecting the output signal. The coupling capacitor *C* has negligible impedance at the circuit operating frequency. But it blocks any d.c. between collector and base.

The circuit frequency of oscillation is set by the series-resonant frequency of the crystal and its value is given by the relation,

$$f_o = \frac{1}{2\pi\sqrt{L \cdot C_1}} \qquad ...(i)$$

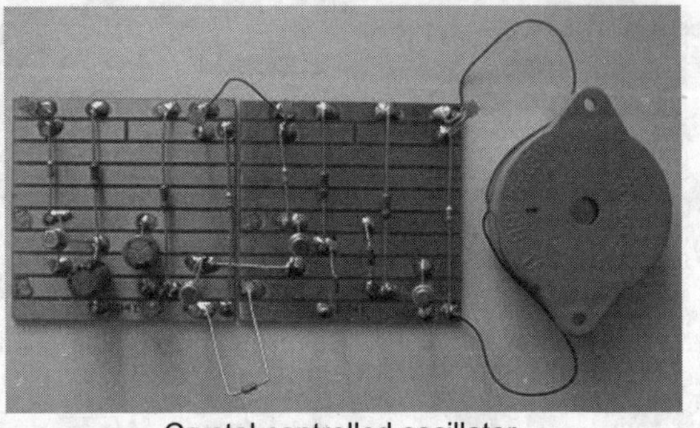

Crystal controlled oscillator.

It may be noted that the changes in supply voltage, transistor device parameters etc. have no effect on the circuit operating frequency, which is held stabilized by the crystal.

Fig. 31.19 shows the FET version of the pierce crystal oscillator. The advantage of using a FET is that it provides a higher input resistance to the crystal. However, the circuit frequency of oscillation is the same as given by equation (*i*).

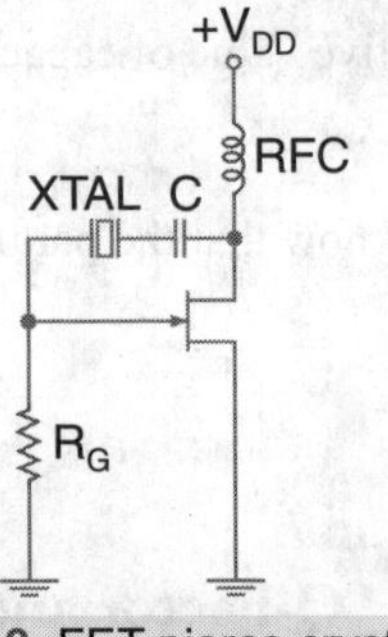

Fig. 31.19. FET pierce crystal oscillator.

31.24. RC Oscillators

These oscillators employ resistors and capacitors and are used to generate low-or audio-frequency signals. Hence they are also known as audio-frequency (A.F.) oscillators. The tuned (or *LC*) oscillators are not suitable at low-frequencies because the size of inductors (*L*) and capacitors (*C*) becomes very large and therefore are not practicable to manufacture in actual practice.

31.25. Basic Principle of RC Oscillators

We know that a single stage of common-emitter (CE) amplifier not only amplifies the input signal, but also introduces a phase shift of 180°. If we take a fraction of the output and feed it back to the input, without considering the phase shift, the negative feedback takes place. Due to the negative feedback, the output voltage of the amplifiers decreases. However, for producing oscillations, we must have positive feedback of sufficient magnitude. This occurs only when the fraction of output voltage is feedback with the same phase as that of the input signal.

Model 4402B Ultra low distoration RC oscillator.

This condition may be achieved in any one of the following two ways:

1. A fraction of the output of a single-stage amplifiers is passed through a phase-shift network, before feeding back to the input. The phase-shift network gives another phase-shift of 180° in addition to the phase-shift of 180° introduced by the amplifier. Thus there is a total phase-shift of 360°, which is also equal to 0°. The RC oscillator, utilizing this principle, is known as phase-shift oscillator.
2. A second-stage of the amplifier is used for producing another 180° phase shift in addition to the phase-shift of 180° produced by the first stage. Thus there is a total phaseshift of 360°, which is also equal to 0°. A fraction of the output from the second stage is feedback to the input of the first stage without producing any further phase-shift. The RC oscillator, utilizing this principle, is known as Wien-bridge oscillator.

31.26. Phase-Shift Oscillator

Fig. 31.20 shows the circuit of a phase shift oscillator using field-effect transistor (FET). The circuit is drawn to show clearly the amplifier and feedback network. The amplifier stage is self-biased with a capacitor by-passed source resistor (R_S) and drain bias resistor (R_D). The expression for voltage gain of the amplifier is given by their relation,

$$A_v = g_m \cdot r_L$$

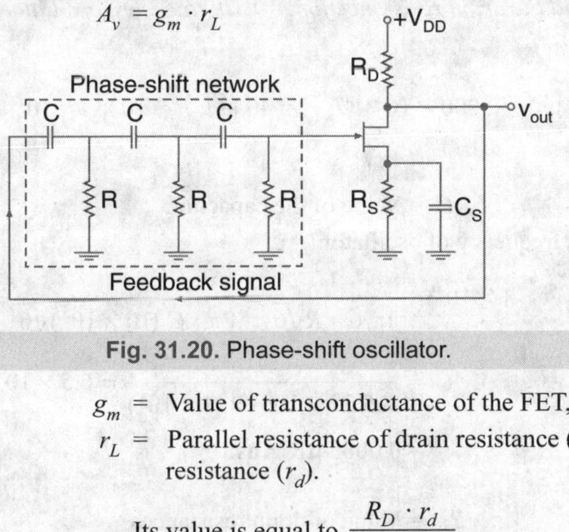

Fig. 31.20. Phase-shift oscillator.

where g_m = Value of transconductance of the FET,

and r_L = Parallel resistance of drain resistance (R_D) and FET output resistance (r_d).

Its value is equal to $\dfrac{R_D \cdot r_d}{(R_D + r_d)}$

The feedback network consists of three *identical RC sections. Each section produces a phase-shift of 60°. Therefore the net phase shift of the feedback network is 180°. Since the amplifier stage also introduces a phase-shift of 180°, therefore the total phase shift between the input and output circuit is 360° or 0°.

When the circuit is energised, by switching on the supply, the circuit starts oscillating. The oscillations may start due to the minor variation in d.c. supply or the inherent noise in the FET. The variation in the gate current is amplified in the drain circuit. Then it is fedback through the phase shift network and finally applied to the gate. The oscillations will be maintained if the loop gain (*i.e.*, $A_v \cdot \beta$) is at least equal to unity. However, to start the oscillations, the loop gain must be greater than

* In actual practice, it is difficult to get a feedback network of three identical RC sections, which produces a total phase shift of 180°. It is because of the fact, that each RC section in the feedback network loads down the previous one.

unity. Using classical network analysis, it can be shown that circuit frequency of oscillation is given by the relation,

$$f_o = \frac{1}{2\pi\ CR\sqrt{6}}$$

and the feedback fraction of the RC phase-shift network,

$$\beta = 1/29$$

Using the value of β (equal to 1/29) we find that for the loop gain ($A_v \cdot \beta$) to be greater than unity, the gain of the amplifier stage must be greater than 1/β or 29.

The phase shift oscillator is particularly useful for generating signals in the audio frequency range. Usually, it is used for generating fixed-spot frequencies, which can be selected by switching in preset RC networks. For the variable frequency oscillator, the three capacitors are ganged and varied simultaneously.

Example 31.27.

It is desired to design a phase shift oscillator using an FET having g_m = 5000 μmhos, r_d = 40 kΩ and feedback circuit resistor (R) = 10 kΩ. Select the value of capacitor for oscillator operation at 1 kHz and the drain resistance for A_v > 29 to ensure oscillator action.

Solution.

Given: $g_m = 5000\ \mu℧ = 5000 \times 10^{-6}\ ℧$; $r_d = 40$ kΩ; $R = 10$ kΩ $= 10 \times 10^3$ Ω and $f_o = 1$ kHz $= 1 \times 10^3$ Hz.

Value of capacitor

Let C = Value of the capacitor.

We know that the frequency of oscillator (f_o),

$$1 \times 10^3 = \frac{1}{1\pi\ C \cdot R\sqrt{6}} = \frac{1}{2\pi \cdot C\ (10 \times 10^3)\sqrt{6}} = \frac{1}{1.539 \times 10^5\ C}$$

or

$$C = \frac{1}{(15.39 \times 10^5) \times (1 \times 10^3)} = 6.5 \times 10^{-9}\ \text{F}$$

$$= 0.0065\ \mu\text{F}\ \textbf{Ans.}$$

Value of drain resistance

Let R_D = Value of drain resistance.

We know that to start oscillations, the amplifier voltage gain (A_v) must be greater than 29. Therefore let us assume that the value of amplifier voltage gain is 40. We also know that the amplifier voltage gain (A_v),

$$40 = g_m \cdot r_L = (5000 \times 10^{-6})\ r_L$$

$$\therefore \quad rL = \frac{40}{(5000 \times 10^{-6})} = 8 \times 10^3\ \Omega = 8\ k\Omega$$

We also know that the parallel resistance (r_L),

$$\therefore \quad 8 = \frac{R_D \cdot r_d}{R_D + r_d} = \frac{R_D \times 40}{R_D + 40}$$

$$8\ (R_D + 40) = 40\ R_D$$

$$\text{or} \quad 32\ R_D = 320$$

$$\therefore \quad R_D = 320/32 = 10\ \text{k}\Omega\ \textbf{Ans.}$$

Example 31.28.

In a RC phase shift, if $R_1 = R_2 = R_3 = 200$ kΩ and $C_1 = C_2 = C_3 = 100$ pF, find frequency of oscillations. (PTU., May 2009)

Solution.

Given: $R = R_1 = R_2 = R_3 = 200 \text{ k}\Omega = 200 \times 10^3 \ \Omega$ and $C = C_1 = C_2 = C_3 = 100 \text{ pF} = 100 \times 10^{-12}$ F

We know that the frequency of oscillations,

$$f_0 = \frac{1}{2\pi \, CR\sqrt{6}}$$

$$= \frac{1}{2\pi \times (100 \times 10^{-12}) \times (200 \times 10^3)\sqrt{6}}$$

$$= 3248.079 \text{ Hz} \approx 3.248 \text{ kHz} \textbf{ Ans.}$$

Example 31.29.

Design a phase shift oscillator to operate at a frequency of 2 kHz using a JFET. Assume that $\mu = 50$ and $r_d = 5$ kΩ and the phase shift network does not load the amplifier.

(a) Find the minimum value of R_D to be used in the drain circuit.

(b) Find the value of the RC product.

(c) Chose reasonable values of R and C to make this oscillator work satisfactorily.

Solution.

Given: $f_o = 2 \text{ kHz} = 2 \times 10^3$ Hz; $\mu = 50$ and $r_d = 5$ kΩ.

(a) Maximum value of R_D

We know that the FET amplification factor (μ),

$$50 = g_m \cdot r_d = g_m \times 5$$

$$\therefore \quad g_m = 50/5 = 10 \text{ m℧}$$

We also know that in order to start oscillations, the amplifier voltage gain (μ) is 50, therefore the circuit will oscillate. Let us assume that the amplifier voltage gain (A_v) is 40.

Now the amplifier voltage gain (A_v),

$$40 = g_m \cdot r_L = 10 \times r_L$$

$$\therefore \quad r_L = 40/10 = 4 \text{ k}\Omega$$

We also know that a.c. load resistance (r_L)

$$4 = R_D \parallel r_d = \frac{R_D \times r_d}{R_D + r_d} = \frac{R_D \times 5}{R_D + 5}$$

or

$$4R_D + 20 = 5R_D$$

$$\therefore \quad R_D = 20 \text{ k}\Omega \textbf{ Ans.}$$

(b) Value of the RC product

We know that the frequency of oscillation (f_o),

$$2 \times 10^3 = \frac{1}{2\pi \, C \cdot R\sqrt{6}} = \frac{0.065}{RC}$$

$$\therefore \quad RC = \frac{0.065}{(2 \times 10^3)} = 3.25 \times 10^{-5} \textbf{ Ans.}$$

(c) ***Value of R and C***

Let us chose R equal to 10 kΩ or 10×10^3 Ω. Then the value of capacitor,

$$C = \frac{(3.25 \times 10^{-5})}{(10 \times 10^3)} = 3.25 \times 10^{-9}\,\text{F} = 3.25\ \text{nF}$$

∴ The values of R and C, which make the oscillator to work satisfactorily are

$$R = 10\ \text{k}\Omega \text{ and } C = 3.25\ \text{nF}$$ **Ans.**

Example 31.30.

Prove that the feedback 'β' of the phase-shift network is given by:

$$-\beta = \frac{1}{1 - 5\alpha^2 - j(6\alpha - \alpha^3)}$$

where $$\alpha = \frac{1}{\omega CR}$$

Assume that the network does not load the amplifier. (*A.M.I.E., Summer, 1986; Winter 1985*)

Solution.

Fig. 31.21 shows the phase-shift network of an *RC* phase-shift oscillator. In this circuit, V_o represents the output voltage and V'_f represents the negative of feedback voltage (V_f); *i.e.*, $V'_f = -V_f$. The negative sign with V_f indicates that it is 180° out of phase with respect to V_o or V'_f.

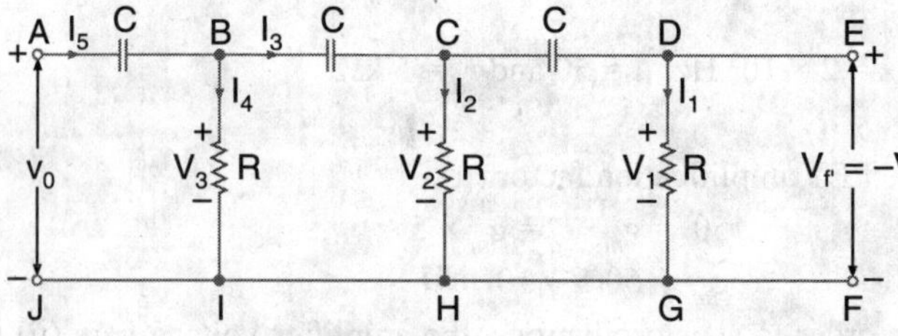

Fig. 31.21.

In order to find the value of feedback 'β', consider the distribution of voltages and currents within the circuit as shown in the figure. Then apply Kirchoff's Voltage Law (*i.e.*, sum up the voltages around the various meshes) and determine the ratio $V'_f / V_o = -V_f / V_o = \beta$

Summing up the voltages around the mesh *DEFG*,

$$V'_f = I_1 \cdot R$$

or $$I_1 = \frac{V'_f}{R} \qquad \ldots(i)$$

and around the mesh *CDGH*,

$$V_2 = \frac{I_1}{j\,\omega\,C} + I_1 \cdot R = I_1\left(\frac{1}{j\,\omega\,C} + R\right)$$

Now let $$\alpha = \frac{1}{\omega\,C\,R} \text{ or } \frac{1}{\omega\,C} = R \cdot \alpha$$

∴ $$V_2 = I_1\left(\frac{R \cdot \alpha}{j} + R\right) = I_1 \cdot R\left(\frac{\alpha}{j} + 1\right)$$

We also know that

$$I_2 = \frac{V_2}{R} = I_1\left(\frac{\alpha}{j} + 1\right) \qquad ...(ii)$$

It may be noted that the current,

$$I_3 = I_2 + I_1 = I_1\left(\frac{\alpha}{j} + 1\right) + I_1$$

$$= I_1\left(\frac{\alpha}{j} + 2\right) \qquad ...(iii)$$

Similarly, summing up the voltages around the mesh *BCHI*,

$$V_3 = \frac{I_3}{j\,\omega\,C} + I_2 \cdot R = I_3 \cdot \frac{R \cdot \alpha}{j} + I_2 \cdot R \qquad ...(\because 1/\omega\,C = R \cdot \alpha)$$

Substituting the values I_2 and I_3 in the above equation from equations (*ii*) and (*iii*),

$$V_3 = \left[I_1\left(\frac{\alpha}{j} + 2\right)\right]\left(\frac{R \cdot \alpha}{j}\right) + I_1\left(\frac{\alpha}{j} + 1\right) R$$

$$= I_1\left(\frac{R \cdot \alpha^2}{j^2} + \frac{2R \cdot \alpha}{j} + \frac{R \cdot \alpha}{j} + R\right)$$

$$= I_1 \cdot R\left(\frac{\alpha^2}{j^2} + \frac{3\alpha}{j} + 1\right)$$

and

$$I_4 = \frac{V_3}{R} = I_1\left(\frac{\alpha^2}{j^2} + \frac{3\alpha}{j} + 1\right) \qquad ...(iv)$$

It may also be noted that the current,

$$I_5 = I_3 + I_4$$

$$= I_1\left(\frac{\alpha}{j} + 2\right) + I_1\left(\frac{\alpha^2}{j^2} + \frac{3\alpha}{j} + 1\right)$$

$$= I_1\left(\frac{\alpha^2}{j^2} + \frac{4\alpha}{j} + 3\right) \qquad ...(v)$$

Now summing up the voltages around, the mesh *ABIJ*,

$$V_o = \frac{I_5}{j\,\omega\,C} + I_4 \cdot R = I_5 \cdot \frac{R \cdot \alpha}{j} + I_4 \cdot R$$

Substituting the values of I_4 and I_5 in the above equation from equations (*iv*) and (*v*)

$$V_o = I_1\left(\frac{\alpha^2}{j^2} + \frac{4\alpha}{j} + 3\right) \cdot \left(\frac{R \cdot \alpha}{j}\right) + I_1 \cdot \left(\frac{a^2}{j^2} + \frac{3\alpha}{j} + 1\right) R$$

$$= I_1 \cdot R\left(\frac{\alpha^3}{j^3} + \frac{4\alpha^2}{j^2} + \frac{3\alpha}{j} + \frac{\alpha^2}{j^2} + \frac{3\alpha}{j} + 1\right)$$

$$= I_1 \cdot R\left(\frac{\alpha^3}{j^3} + \frac{4\alpha^2}{j^2} + \frac{3\alpha}{j} + \frac{\alpha^2}{j^2} + \frac{3\alpha}{j} + 1\right) \qquad ...(\because I_1 \cdot R = V'_f)$$

$$= V'_f(j\alpha^3 - 5\alpha^2 - j\,6\alpha + 1) \qquad ...(\because j^2 = -1)$$

$$= V'_f[1 - 5\alpha^2 - j\,(6\alpha - \alpha^3)]$$

$$\therefore \qquad \frac{V'_f}{V_o} = \frac{1}{1 - 5\alpha^2 - j\,(6\alpha - \alpha^3)}$$

Now, since the actual feedback factor is the ratio of V_f/V_o, therefore

$$\frac{V_f}{V_o} = \beta = \frac{-V'_f}{V_o}$$

and
$$-\beta = \frac{V'_f}{V_o} = \frac{1}{1 - 5\alpha^2 - j\,(6\alpha - \alpha^3)}$$

Example 31.31.

Obtain the expression for the frequency of oscillations for a phase-shift oscillator and show that gain of an amplifier must exceed 29 in this oscillator. Given the feedback factor,

$$-\beta = \frac{1}{1 - 5\alpha^2 - j\,(6\alpha - \alpha^3)}$$

where $\alpha = \dfrac{1}{\omega R C}$ *(A.M.I.E., Summer, 1986, JNTU., 2009)*

Solution.

Given: The feedback factor,

$$-\beta = \frac{1}{1 - 5\alpha^2 - j\,(6\alpha - \alpha^3)} \qquad \text{... (i)}$$

The expression for the frequency of oscillations may be obtained from the above expression by equating the reactive part equal to zero *i.e.*,

$$6\alpha - \alpha^3 = 0$$

$$\therefore \qquad \alpha = \sqrt{6}$$

or
$$\frac{1}{\omega C R} = \sqrt{6}$$

$$\omega = \frac{1}{C R \sqrt{6}}$$

or
$$2\pi f = \frac{1}{C R \sqrt{6}} \qquad \text{...}(\because \omega = 2\pi f)$$

$$\therefore \qquad f = \frac{1}{2\pi\, C R \sqrt{6}} \quad \textbf{Ans.}$$

Substituting the value of a (equal to) in equation (*i*),

$$-\beta = \frac{1}{1 - 5\,(\sqrt{6})^2 - j\,(6\sqrt{6} - (\sqrt{6})^3)}$$

$$= \frac{1}{1 - 30} = -\frac{1}{29}$$

or
$$\beta = \frac{1}{29}$$

Now, we know that the Barkhauscn condition for positive feedback is that the loop gain ($A_v \cdot \beta$) must be equal to or greater than unity *i.e.*,

$$A_v \cdot \beta \geq 1$$

or $$A_v \geq 1/\beta = 29 \textbf{ Ans.}$$

31.27. Transistor Phase Shift Oscillator

It has been found that by using a transistor as the active element of the amplifier stage, the output of the feedback network is loaded appreciably by the relatively low input resistance (*i.e.*, h_{ie}) of the transistor. This problem can be removed by using an emitter follower input stage followed by a common-emitter amplifier stage. But if the single transistor stage is required, then the use of voltage-shunt feedback is more suitable.

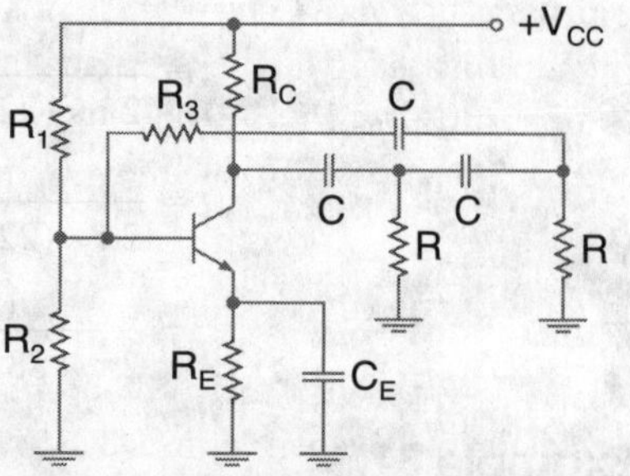

Fig. 31.22. Transistor phase shift oscillator.

Fig. 31.22 shows the circuit of a transistor phase shift oscillator using voltage-shunt feedback connection. In this connection, the feedback signal is coupled through the feedback resistor (R_s) in series with the amplifier stage input resistance R_i (whose value is equal to the parallel combination of resistors R_1, R_2 and h_{ie}). In order to make the calculations simpler, the value of resistor R_3 is made equal to the difference of resistor R_2 and R_i *i.e.*,

$$R_3 = R - R_i$$

where R is the value of resistor in the phase shift network. The frequency of oscillation for the transistor phase shift oscillators is given by relation,

$$f = \left(\frac{1}{2\pi\, RC}\right) \times \frac{1}{\sqrt{6 + 4\,(R_C/R)}}$$

For the loop gain to be greater than unity, the current gain of the transistor,

$$h_{fe} > 23 + \frac{R_C}{29\,R} + \frac{R}{4R_C}$$

Example 31.32.

Draw the circuit of RC phase shift oscillator, with R = 6 kΩ, C = 1500 pF and R_C = 18 kΩ, obtain its operating frequency. (*Anna University., Nov/Dec 2007*)

Solution.

Given: $R = 6$ kΩ, $C = 1500$ pF $= 1500 \times 10^{-12}$ F and $R_C = 18$ kΩ.

We know the frequency of oscillators

$$f_0 = \frac{1}{2\pi\, CR\sqrt{6}}$$

$$= \frac{1}{2\pi \times 1500 \times 10^{-12} \times 6 \times 10^3 \times 2.45} = 7.22 \text{ kHz } \textbf{Ans.}$$

Example 31.33.

A transistorized RC phase oscillator has the following values. Calculate the frequency of oscillations:

$R = 400$ kΩ, $C = 50$ pF, $R_1 = 18$ kΩ, $R_2 = 100$ kΩ, $R_e = 1$ kΩ, $R_c = 4.74$ kΩ and $C_e = 220$ μF.

(*M. G. University., April/May 2009*)

Solution.

Given: $R = 400\ k\Omega = 400 \times 10^3\ \Omega$; $C = 50\ pF = 50 \times 10^{-12}\ F$; $R_1 = 18\ k\Omega = 18 \times 10^3\ \Omega$; $R_2 = 100\ k\Omega = 100 \times 10^3\ \Omega$ $R_e = 1\ k\Omega = 1 \times 10^3\ \Omega$; $R_c = 4.7\ k\Omega = 4.7 \times 10^3\ \Omega$ and $C_e = 220 \times 10^{-12}$ F.

We know that the frequency of oscillation (f_0),

$$f_0 = \frac{1}{2\pi\ RC} \times \frac{1}{\sqrt{6 + 4\ R_C/R}}$$

$$= \frac{1}{2\pi \times (220 \times 10^{-12})\,(400 \times 10^3)} \times \frac{1}{\sqrt{6 + 4\,(4.7 \times 10^3)/400 \times 10^3}}$$

$$= \frac{1}{2\pi \times (220 \times 10^{-12})\,(400 \times 10^3)} \times \frac{1}{\sqrt{6 + 4\,(4.7 \times 10^3)/400 \times 10^3}}$$

$$= \frac{1}{2\pi \times 88 \times 10^{-9}} \times \frac{1}{\sqrt{6.047}} = 735.47\ \text{kHz}$$

Example 31.34.

Select the value of capacitor C and transistor gain h_{fe} to provide an oscillator frequency of $f_o = 2$ kHz. The circuit values are $h_{ie} = 2\ k\Omega$; $R_1 = 20\ k\Omega$; $R_2 = 80\ k\Omega$; $R_C = 10\ k\Omega$ and $R = 8\ k\Omega$. Also find the value of feedback resistor R_3.

Solution.

Given: $f_o = 2\ kHz = 2 \times 10^3$ Hz; $h_{ie} = 2\ k\Omega$; $R_1 = 20\ k\Omega$; $R_2 = 80\ k\Omega$; $R_C = 10\ k\Omega = 10 \times 10^3\ \Omega$ and $R = 8\ k\Omega = 8 \times 10^3\ \Omega$.

Value of capacitor C

We know that frequency of oscillation (f_o)

$$2 \times 10^3 = \frac{1}{2\pi\ RC} \times \frac{1}{\sqrt{6 + 4\ R_C / R}}$$

$$= \left(\frac{1}{2\pi \times (8 \times 10^3) \times C}\right) \times \frac{1}{\sqrt{6 + 4 \times (10 \times 10^3)/(8 \times 10^3)}}$$

$$= \left(\frac{1}{16\pi \times 10^3 \times C}\right) \times \frac{1}{\sqrt{6 + 5}}$$

$$= \left(\frac{1}{16\pi \times 10^3 \times C}\right) \times \frac{1}{3.32}$$

$$\therefore \quad C = \frac{1}{(16\pi \times 10^3) \times 3.32 \times (2 \times 10^3)}$$

$$= 3.0 \times 10^{-9}\ \text{F} = 0.003\ \mu\text{F}\ \textbf{Ans.}$$

Value of transistor gain h_{fe}

We know that current gain of a transistor,

$$h_{fe} \geq 23 + 29\frac{R}{R_C} + 4\frac{R_C}{R}$$

$$\geq 23 + 29 \times \left(\frac{8 \times 10^3}{10 \times 10^3}\right) + 4\left(\frac{10 \times 10^3}{8 \times 10^3}\right)$$

$$\geq 23 + 23.2 + 5 = 51.2$$

Thus a transistor with $h_{fe} \geq 51.2$ will provide sufficient loop gain for the circuit to operate as an oscillator. However, in actual practice, a transistor with h_{fe} greater than 60 should be selected. **Ans.**

Value of feedback resistor R_3

We know that the input resistance of the amplifier stage,

$$R_i = (R_1 \parallel R_2) \parallel h_{ie} = (20 \parallel 80) \parallel 2 = 1.8 \text{ k}\Omega = 1.8 \times 10^3 \ \Omega$$

and the value of feedback resistor,

$$R_3 = R - R_i = (8 \times 10^3) - (1.8 \times 10^3) \ \Omega$$

$$= 6.2 \times 10^3 \ \Omega = 6.2 \text{ k}\Omega \ \textbf{Ans.}$$

31.28. Wien-bridge Oscillator

In the previous articles, we have discussed, phase shift oscillators. They have one drawback *i.e.*, when they are used to have a variable frequency circuit, it is difficult to adjust equally the capacitor values of the phase-shift network simultaneously. An oscillator circuit, which is more useful for variable frequency operation, is the Wein-bridge oscillator.

Fig. 31.23 shows the circuit of a Wein-bridge oscillator using transistors. However, we may also use FET's instead of bipolar transistors. The circuit consists of a two-stage RC coupled amplifier, which provides an approximately 360° or 0° phase-shift. So the feedback network has no need to introduce any additional phase-shift. The feedback network consists of $C_1 - R_1$, $C_2 - R_2$ (called a lead-lag network) and $R_3 - R_4$ (called a voltage-divider). The lead-lag network provides a positive feedback to the input of the first stage and the voltage-divider, the negative feedback to the emitter of Q_1 transistor.

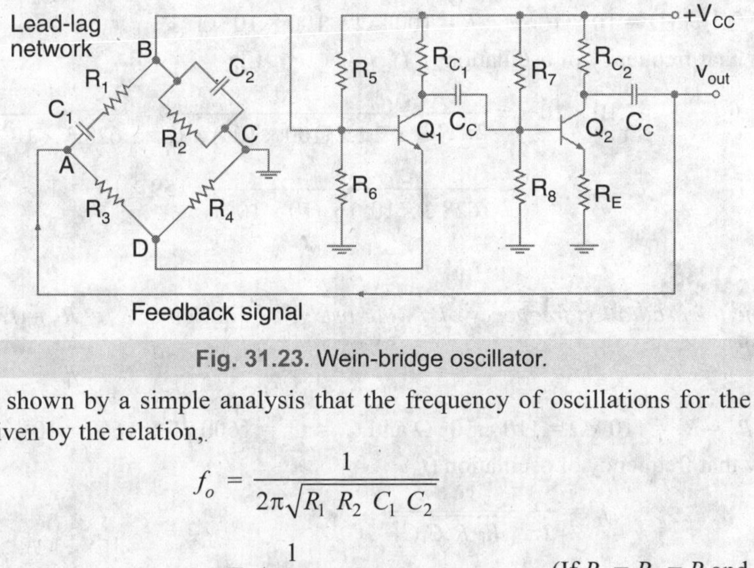

Fig. 31.23. Wein-bridge oscillator.

It can be shown by a simple analysis that the frequency of oscillations for the Wein-bridge oscillator is given by the relation,

$$f_o = \frac{1}{2\pi\sqrt{R_1 \ R_2 \ C_1 \ C_2}}$$

$$= \frac{1}{2\pi \ RC} \qquad \text{... (If } R_1 = R_2 = R \text{ and } C_1 = C_2 = C\text{)}$$

and the ratio of resistor R_3 to R_4,

$$R_3/R_4 = 2$$

Thus a ratio of R_3 to R_4 greater than 2 will provide a sufficient gain for the circuit to oscillate at the desired frequency.

It has been observed that in a FET Wein-bridge oscillator, the oscillator frequency can be varied by varying two capacitors C_1 and C_2 simultaneously. These are variable air gang capacitors. The frequency range of the oscillator can be changed by switching the different preset values of resistors

R_1 and R_2 into the circuit. On the other hand, in a transistor Wein-bridge oscillator, because of the low input resistance of the transistor, the range changes are made by switching in the preset values of capacities C_1 and C_2. For continuous frequency variation, the values of resistors R_1 and R_2 are changed by using a dual potentiometer. The operation of a Wein-bridge oscillator may be explained as follows:

When the circuit is energised by switching on the supply, a small random oscillations appearing at the base of Q_1 transistor are amplified, at its collector. These oscillations are further amplified at the collector of Q_2 transistor. Since the oscillations at the collector of Q_2 transistor have been inverted twice, therefore these oscillations are in phase with the input signal. A part of the output signal from the collector of Q_2 transistor is fedback to the Wein-bridge, which is further amplified. The process continues, till sustained oscillations are produced.

The Wein-bridge oscillator is a standard oscillator circuit for generating low frequencies in the range of 20 Hz to about 1 MHz. It is used in all commercial audio signal generators. The Wein-bridge oscillator has an advantage that it gives an extremely pure sine wave output, good frequency stability and a highly stabilized amplitude. A greater amplitude stability may be obtained by using a thermistor (a device with negative temperature coefficient) in place of the resistor R_3.

Example 31.35.

A Wein-bridge oscillator is used for operation at f_o = 10 kHz. If the value of R is 100 kΩ, find the value of the capacitor C. (*Anna University., Nov/Dec. 2008*)

Solution.

Given: $f_o = 10 \text{ kHz} = 10 \times 10^3 \text{ Hz}$; $R = 100 \text{ k}\Omega = 100 \times 10^3\ \Omega$

We know that frequency of oscillation (f_o)

$$10 \times 10^3 = \frac{1}{2\pi RC} = \frac{1}{2\pi \times (100 \times 10^3) \times C} = \frac{1}{628.3 \times 10^3 C}$$

$$\therefore \quad C = \frac{1}{(628.3 \times 10^3) \times (10 \times 10^3)} = 159 \times 10^{-12} \text{ F} = 159 \text{ pF } \textbf{Ans.}$$

Example 31.36.

Determine the oscillation frequency of a wien-bridge oscillation of $R_1 = R_2 = 110$ kΩ and C_1 = V2 1600 pF. (*M. G. University., Nov. 2008*)

Solution.

Given : $R_2 = R_2 = 110 \text{ k}\Omega = 110 \times 10^3\ \Omega$ and $C_1 = C_2 = 1600 \text{ pF} = 1600 \times 10^{-12}$.

We know that frequency of oscillation (f_0),

$$f_0 = \frac{1}{2\pi\sqrt{R_1 R_2 C_1 C_2}}$$

$$= \frac{1}{2\pi\sqrt{(110 \times 10^3) \times (110 \times 10^3) \times (1600 \times 10^{-12}) \times (1600 \times 10^{-12})}}$$

$$= \frac{1}{2\pi \times 176000 \times 10^9} = 904 \text{ Hz } \textbf{Ans.}$$

31.29. Tunnel Diode Oscillator

A tunnel diode may be used to generate very high frequency signal (*i.e.*, freuencies of above 10 G Hz). Fig. 31.24 shows the circuit of a tunnel diode oscillator. The circuit consists of a tunnel diode in series with a resistor (R) and a switch (S) connected across a supply source (V). The value of resistor (R) is selected in such a way that it biases the tunnel diode in the midway of the negative resistance region.

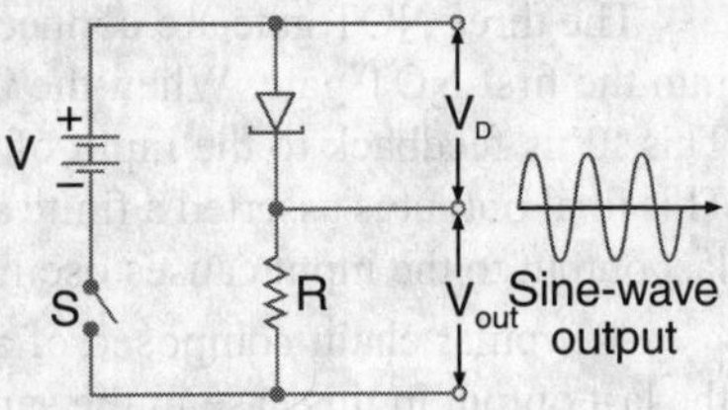

Fig. 31.24. Tunnel diode oscillator.

Fig. 31.25 shows the V-I characteristic of a tunnel diode. The curve AB is the negative-resistance region. It is evident that the Q-point is set at the middle of the curve AB. The Q-point can move between the points A and B during the circuit operation. The point A is called the peak-point and the point B as the valley point. The values of currents and voltages corresponding to the points A and B is indicated in the figure.

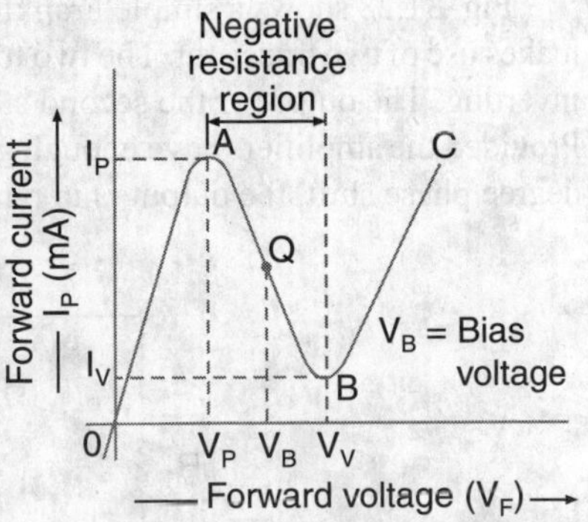

Fig. 31.25. V-I characteristic of tunnel diode.

When the switch (S) is closed, the circuit current rises immediately towards the constant value, whose value is determined by the value of resistor (R) and the diode resistance. However, as the voltage drop across the tunnel diode (V_D) exceeds the peak-point voltage (V_P), the tunnel diode is driven into negative resistance region. In this region, the current starts decreasing, till the voltag V_D becomes equal to the valley-point voltage (V_V). At this point, a further increase in voltage V_D drives the diode into the positive resistance region (*i.e.*, BC curve). As a result of this, the circuit current tends to increase. This increase in circuit current will increase the voltage drop across resistor (R) which will reduce the voltage (V_D).

This brings the diode back into the negative resistance region. The decrease in voltage V_D increases the value of current in the circuit, till point A is reached where the voltage V_D is equal to the voltage V_P. This completes one cycle of operation. In this way, the circuit will continue to oscillate back and fourth in the negative-resistance region and will produce a sine wave output as shown in the figure.

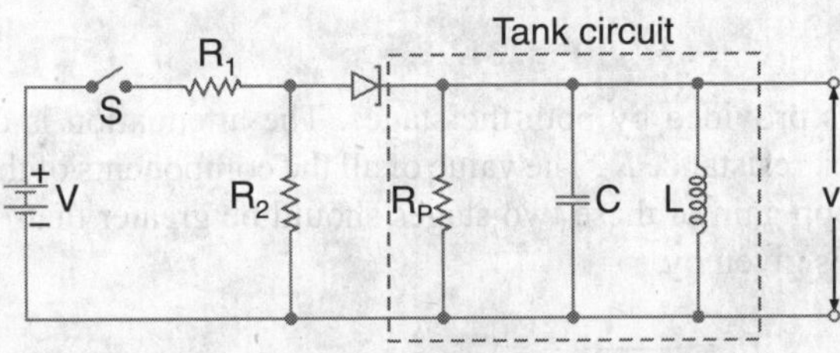

Fig. 31.26. Practical tunnel diode oscillator.

The tunnel diode oscillator discussed above is not used in practice. In fact a circuit shown in Fig. 31.26 is more commonly used. In this circuit, the resistor, R_1 sets the proper biasing for the diode, whereas resistor R_2 sets the proper current level for the tank circuit. The parallel combination of resistor, (R_P), inductor, (L) and a capacitor (C) forms a tank circuit, which resonates at the selected frequency.

31.30. Ring Oscillator

A ring oscilator is composed of an odd number of INVERTER (NOT) gates whase output oscillates between two voltage levels either true or false. The diagram of a simple three NOT ring oscillator is shown in Fig. 31.27.

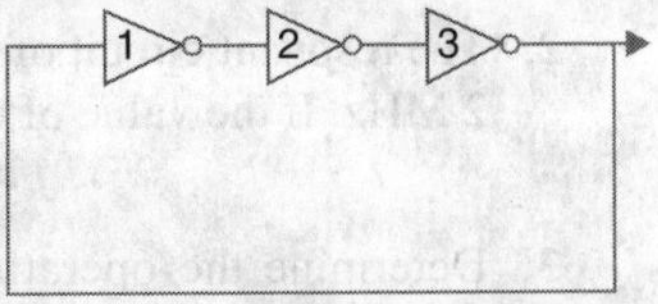

Fig. 31.27.

The three NOT gate are connected in the form of ring. The output of last NOT gate is feedback into the first NOT gate. When the input of the first gate is '1', then the output of the circuit is '0'. This '0' is feedback to the input of first gate, then the output of the circuit changes from '0' to '1'. This final output is asserted a finite amout of time after the first input is asserted, the feedback of this last output to the input causes oscillation.

A circular chain composed of an even number of inverters cannot be used as a ring oscillator, the last output in this case in the same as the input.

31.30. Franklin Oscillator

Fig. 31.28 shows a simple Franklin Oscillator. As seen from the diagrams the franklin oscillator makes use of two transistor. The two transistors are connected as a two stage amplifier, both stages are inverting. The output of the second stage is coupled to the input of the first stage by a tuned circuit. Provided the amplifiers have enough gain, the circuit will oscillate. The two amplifiers each give 180 degree phase shift, the output is in phase with in input.

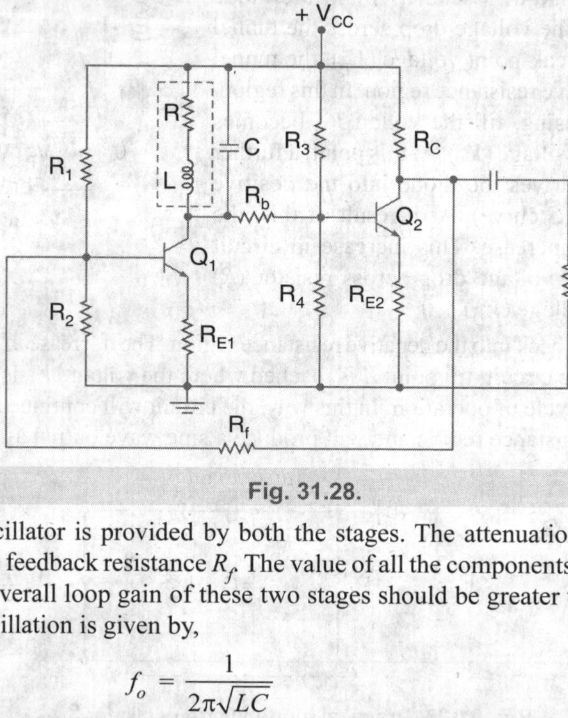

Fig. 31.28.

The gain of oscillator is provided by both the stages. The attenuation is caused by coupling resistance R_b and the feedback resistance R_f. The value of all the components of the circuit are chosen in such a way that overall loop gain of these two stages should be greater than 1 (1 $A\beta_1 > 1$). Here the frequency of oscillation is given by,

$$f_o = \frac{1}{2\pi\sqrt{LC}}$$

NUMERICAL PROBLEMS

1. A tuned-collector oscillator circuit used in a local oscillator of radio receiver has a fixed inductance of 100 μH and a fixed capacitor of 1000 pF. Calculate the frequency of oscillations. **(Ans. 503.3 kHz)**
2. The resonant circuit of a tuned-collector transistor oscillator has a resonance frequency of 2 MHz. If the value of capacitance is 820 pF, determine the value of inductance required. **(Ans. 7.7 H)**
3. Determine the operating frequency of a transistor Hartley Oscillator, if L_1= 100 μH; L_2 = 1 mH; M = 20 μH and C = 20 pF. **(Ans. 1.05 MHz)**

4. The frequency of oscillation of a Hartley oscillator is given by:

$$f_o = \frac{1}{2\pi\sqrt{(L_1 + L_2)\,C}}$$

Calculate the oscillation frequency if $C = 0.01$ μF and $L_1 = L_2 = 25$ mH. **(Ans. 7.117 kHz)**

5. Calculate the frequency of oscillations for the Clapp oscillator shown in Fig. 31.29. **(Ans. 87.7 kHz)**

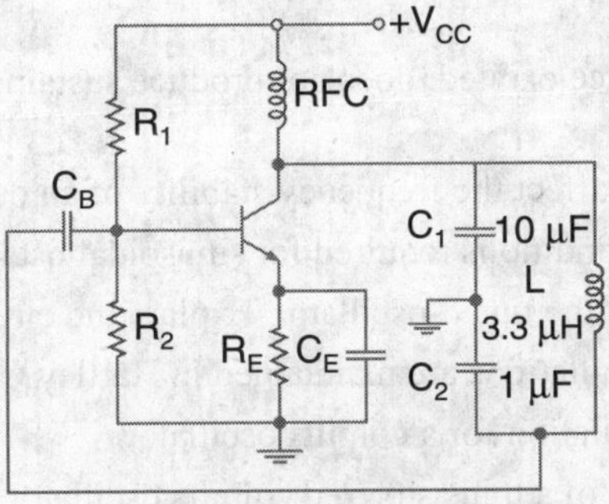

Fig. 31.29.

6. Calculate the value of A_v and the frequency of oscillations for the circuit shown in Fig. 31.30. **(Ans. 5.53 MHz)**

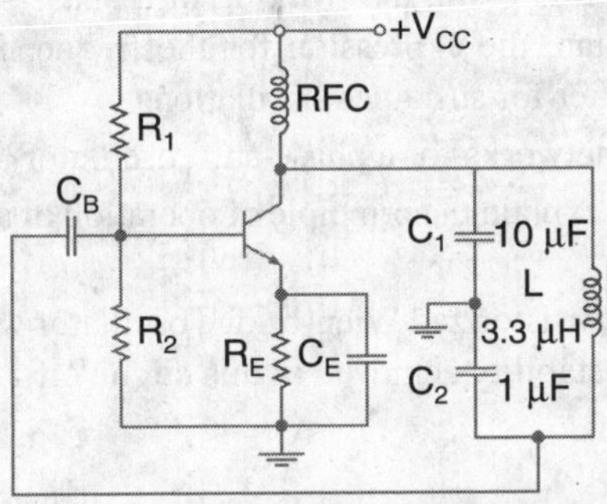

Fig. 31.30.

7. A transistor Colpitts oscillator has $C_1 = 250$ pF; $C_2 = 100$ pF and $L = 60$ μH. Find the value of gain needed and the frequency of oscillation. **(Ans. 2.5 and 2.43 MHz)**

8. A crystal has the following parameters: $L = 0.5$ H, $C_1 = 0.06$ pF, $C_2 = 1$ pF and $R = 5$ kΩ. Find the series and parallel resonant frequencies and Q of the crystal. **(Ans. 918.9 kHz and 946 kHz)**

9. An FET phase-shift oscillator uses three identical RC sections in the feedback network. The values of the components are $R = 100$ kΩ and $C = 0.01$ μF. Determine the frequency of oscillations. **(Ans. 64.97 Hz)**

10. A transistor RC phase-shift oscillator has a frequency of 100 kHz. If the value of R is 10 kHz, find the value of C and gain necessary for sustained oscillations. **(Ans. 41.67 pF, 29)**

11. A Wein-bridge oscillator has a frequency of 500 Hz. If the value of C is 1000 pF, determine the value of R. **(Ans. 318.3 kΩ)**

DESCRIPTIVE QUESTIONS

1. Explain briefly, what do you mean by an electronic oscillator? Give two applications of an oscillator.
2. What are the different types of oscillators?
3. Explain the meaning of the terms
   (*a*) damped oscillations, and
   (*b*) undamped oscillations.
4. Why an LC tank circuit, once excited, does not produce sustained oscillations? Explain it briefly.
5. What are the factors which affect the frequency stability of an oscillator?
6. Give the two Barkhausen conditions required for sinusoidal oscillations to be sustained.
7. Draw the circuit diagram of the tuned-oscillator. Explain the circuit operation.
8. Explain briefly, how the oscillations are maintained in Hartley Oscillator.
9. Discuss the criteria for oscillation for a colpitts oscillator.
10. Draw a neat circuit diagram of a transistorized colpitts oscillator with a suitable equivalent circuit. Give the expressions for the following:
    (*i*) Frequency of oscillations, and
    (*ii*) Minimum gain for sustained oscillations.
11. Explain with the neat circuit diagram, the working of a phase-shift oscillator using three sections of RC network. State the expression for the frequency of oscillations and the minimum gain of the amplifier for sustained oscillations.
12. Why do we need three RC networks for a phase-shift oscillator? Can it be two or four?
13. With a neat circuit diagram, explain the principle of operation of a RC phase-shift oscillator using FET.
14. Explain the operation of a transistorized Wien-bridge oscillator with the help of neat circuit diagram. How is amplitude stability achieved in this circuit?
15. Describe oscillator operation. *(Jamia University., 2008)*
16. What is an oscillator? *(PTU., Dec 2006)*
17. What is an oscillator? How does it differ from an Alternator? *(PTU., May 2008)*
18. What type of feedback is applied for oscillators? *(Anna University, Nov/Dec 2007)*
19. Define 'Q' of tank Circuit. *(Anna University, Nov/Dec 2007)*
20. Mention the essential conditions to be satisfied by an oscillator circuit. *(Anna University., Nov/Dec 2006)*
21. Define quality factor. Obtain the quality factor for a parallel resonant circuit. *(Anna University., May/June 2007)*
22. What are the factors which affect the frequency stability of an oscillator? *(JNTU, June 2009)*
23. Explain briefly about frequency and amplitude stability of oscillators. *(JNTU, June 2009)*
24. What type of feedback a sinusoidal oscillator uses ? Why? *(PBUT., June 2005)*
25. Why is positive feedback necessary to produce oscillations? *(PTU., Dec 2009)*

26. Give the two Barkhansen condition for sinusoidal oscillations to be sustained. *(JNTU, June 2009)*

27. What is Barkhausen criteria? What do you mean by frequency stability and stability criterion in an oscillator? Explain Wien Bridge Oscillator in detail. *(PTU., May 2006)*

28. State Barkhausen's criterion for oscillation. *(Anna University., Nov/Dec 2007, Nagpur University., 2008, PTU., May 2009, R.G.P.V., Dec. 2005, 2008)*

29. What are the Barkhausen conditions of oscillations in electronic systems? What is their significance? *(PTU., May 2008)*

30. What is the figure of merit of a crystal oscillator? How does it affect the stability of oscillators? *(Nagpur University., 2008)*

31. In detail discuss Hartley Oscillator? *(PTU., May 2006)*

32. Explain the circuit of Hartley Oscillator in detail? *(PTU., Dec 2005)*

33. Draw the circuit of a Colpitts Oscillator. Derive expressions for frequency of oscillation and condition for oscillation. *(Anna University, Nov/Dec 2007)*

34. Explain the colpitt oscillator. Derive its function of oscillation. *(Anna University., May/June 2006)*

35. With a circuit diagram, explain the function of Colpitt's oscillator. Give its theory. *(Bhartiya University., April 2004)*

36. Explain the working of a Colpitts Oscillator with a neat diagram. *(PTU., May 2005)*

37. Explain Clapp oscillator and derive its frequency of oscillation. *(Anna University., May/June 2007)*

38. Give the equivalent circuit of quartz crystal and mention its series and parallel resonance frequency *(Anna University., Nov/Dec 2006, April/May 2008)*

39. Give reasons why quartz crystal is used in crystal oscillator. State its advantages and applications. *(Anna University., April/May 2008)*

40. Why crystal oscillator is used in radio transmitter? *(PTU., May 2005)*

41. What is the principle behind operation of a crystal oscillator. *(Anna University., Nov/Dec 2007)*

42. Draw the electrical equivalent circuit of the quartz crystal oscillator and write the expression for series and parallel resonant frequencies. *(Anna University., May/June 2007)*

43. What is piezoelectric effect? Draw the equivalent circuit of quartz. Describe the crystal oscillator with neat circuit diagram. *(Anna University., May/June 2007)*

44. Crystal oscillators possess high degree of frequency stability- Justify. *(Anna University., Nov/Dec 2007)*

45. What are the advantages of LC oscillator and RC oscillator? *(Anna University., May/June 2007)*

46. Draw the circuit of RC phase shift oscillator and explain its working. Also derive the expression for frequency of oscillator and condition for maintenance of oscillator. *(Anna University., May/June 2007, April/May 2008)*

47. Explain the RC phase shift oscillator with neat circuit diagram. Derive its frequency of oscillation. Give the amplifier gain and feedback network gain to sustain oscillator operation. *(Anna University., Nov/Dec 2006)*

48. Explain Colpitt's and Hartley oscillator and derive their frequency of oscillation *(Anna University., Nov/Dec 2006)*

49. Draw a neat circuit diagram of RC phase shift oscillator using BJT and explain its working principle. *(JNTU., June 2009)*

50. Sketch the circuit and explain the operation of a RC phase shift oscillator. Derive the expression for frequency and condition for sustained oscillations for the circuits. *(Anna University., May/June 2007)*

51. Why an LC tank circuit, once excited, does not produce sustained oscillation? Explain it briefly? *(JNTU., June 2009)*

52. What is the basic principle of producing sustained oscillations in LC oscillators? *(PTU., Dec 2009)*

53. Draw and explain the working of RC phase shift oscillator and also derive expression for its frequency of oscillations. *(PTU., May 2008, Dec 2008, Nagpur University, Summer 2004)*

54. What do you use three RC section in RC phase shift oscillator? *(PTU., Dec 2005)*

55. Compare RC phase shift and crystal oscillator. *(PTU., May 2006)*

56. Draw and explain the circuit of Wien bridge oscillator. Derive the expression for frequency of oscillation. Will oscillation take place if the bridge is balanced? Explain *(PTU., May 2007)*

57. Describe the working of Wien bridge oscillator. *(PTU., Dec 2009)*

58. What are two transistor used in a Wien bridge Oscillator? *(PTU., May 2009)*

59. Draw the circuit diagram of a Wien bridge oscillator and briefly explain its operation. How is its frequency of oscillation determined? *(Anna University., Nov/Dec 2007, PTU., Dec 2008)*

60. Draw the circuit diagram of Wein bridge oscillator. Explain the need for providing both positive and negative feedback. *(Anna University., May/June 2007)*

61. Explain Wein bridge oscillator circuit and derive its frequency of oscillator. *(Anna University., May/June 2007)*

62. Why is amplifier circuit necessary in an oscillator? *(R.G.P.V., Nov. / Dec. 2007)*

63. State and briefly explain Barkhausen criterion of oscillation. Draw the circuit diagram of general oscillator. *(R.G.P.V., June 2007)*

64. Give the classification and working of different kinds of oscillators. *(R.G.P.V., May/ June 2006)*

65. List the different types of oscillator with their applications. *(R.G.P.V., Nov. /Dec. 2007)*

66. Derive the expression for the frequency of oscillations of a phase shift oscillator. *(R.G.P.V., June 2002)*

67. Draw the circuit phase shift oscillator using bipolar junction transistor. At what frequency will the circuit oscillate? *(R.G.P.V., Dec. 2005)*

68. Draw circuit diagram of an LC oscillator and explain its working. *(R.G.P.V., June 2004)*

69. Discuss LC oscillator. *(R.G.P.V., June 2005)*

70. Draw circuit diagram of an LC oscillator and explain the working of the same. *(R.G.P.V., Dec. 2008)*

71. Write short note on Hartley and Colpitts oscillators. *(R.G.P.V., June 2002)*

72. Explain the following oscillators in detail

(*a*) Hartley Oscillator

(*b*) Colpitt's Oscillator. *(R.G.P.V., Dec. 2006)*

73. Draw the circuit diagram of Hartley oscillator and explain its working. Derive the expression for frequency of operation. *(R.G.P.V., June 2007)*
74. Sketch the circuit of Wein bridge oscillator. Explain its working. Determine the frequency of oscillation. Will oscillation take place if the bridge is balanced? Explain. *(R.G.P.V., Dec. 2004, 2002)*
75. Derive the expression for frequency of oscillation for Wein bridge oscillator. *(R.G.P.V., Dec. 2008, Dec. 2003)*
76. Draw the circuit of Wein bridge oscillator and explain its working. *(R.G.P.V., June 2005)*
77. Draw the circuit of a Wein bridge oscillator. Discuss its working will oscillation take place if bridge is balanced. *(R.G.P.V., Nov. / Dec 2007)*
78. Explain working principle and give analysis of Wein bridge oscillator. *(R.G.P.V., June 2008)*
79. Prove that the gain of the amplifier used in Wein bridge oscillator must be greater than 3 times for sustained oscillations. *(R.G.P.V., Dec. 2006)*
80. What is oscillator circuit? Write different type of oscillator (Name only). Explain the working of any one of them. *(R.G.P.V., Dec. 2001)*
81. Draw and explain the circuit of Colpitts oscillator. *(R.G.P.V., Dec. 2005)*
82. Draw and explain the principle of operation of clap oscillator. *(R.G.P.V., June 2007)*
83. What do you understand by tank circuit? How can you relate with oscillators? Discuss the advantage of oscillators. *(R.G.P.V., Dec. 2007)*
84. Explain the Barkhausen criterion for the oscillators. *(Nagpur University, Summer 2008 2007, Winter 2004, 2001)*
85. What is the figure of merit of a crystal oscillator? How does it affect the stability of oscillators? *(Nagpur University, Summer 2008)*
86. Draw the circuit diagram of RC phase shift oscillator and explain the working. Why RC phase shift oscillator is preferred for the generation of low frequency. *(Nagpur University, Summer 2007)*
87. With neat circuit diagram of crystal oscillator, derive the expression of series and parallel resonant frequency. *(Nagpur University, Winter 2006)*
88. Draw the circuit diagram for Hartley oscillator and explain its working. *(Nagpur University, Summer 2007)*
89. Discuss the frequency stability of a RC and LC oscillator. *(Nagpur University, Summer 2006)*
90. Derive an expression for frequency of oscillation of RC phase shift oscillation and explain why RC phase shift oscillator is preferred for the generation of low frequency. *(Nagpur University, Winter 2004)*
91. Draw the neat sketch of a wein-bridge oscillator and obtain the minimum gain and the frequency of oscillation. *(Nagpur University, Winter 2003)*
92. Derive the expression for frequency of oscillations for RC Wein Bridge Oscillator. *(Nagpur University, Summer 2003)*
93. State the frequency stability criterion for a sinusoidal oscillator. *(Nagpur University, Winter 2002)*

94. Draw the circuit for wein bridge oscillator. Give brief explanation. *(Nagpur University, Summer 2000)*

95. Write short notes on :

(*a*) RC phase-shift oscillator. *(Nagpur University, Winter 2006)*

(*b*) Crystal oscillator *(Nagpur University, Summer 2007, 2001)*

96. Write the disadvantages of RC phase shift oscillator. Draw the circuit diagram of RC phase shift oscillator and derive the expression for frequency.. *(GBTU., 2010, 2011)*

97. What is piezolectic crystal? Derive the expression of the ratio of frequency in series and parallel resonance in a piezoelectric crystal. *(GBTU., 2010, 2011)*

98. Explain Barkhausen criterion for oscillation. *(Nagpur University., Summer 2011)*

99. Draw and explain wein bridge oscillator. *(Nagpur University., Summer 2011)*

100. Obtain the Brakhausan criterion for operation of the oscillator using basic feedback circuit and hence, explain the operator of the oscillator. *(VTU., June/July 2011)*

101. State Barkhausen in criteria for oscillation and also explain stability criteria for sinusoidal oscillator. *(GTU., June 2011)*

102. Explain wien bridge oscillator with neat sketch. What determines the frequency of oscillation? Will oscillations take place if the bridge is balanced? *(GTU., Nov./Dec. 2011)*

103. Draw the cricuit of a wien-bridge oscillator and derive the expression for its frequency of oscillation. *(WBUT., 2011)*

104. Derive the expression for the frequency of oscillation and the minimum gain required for sustained oscillations of the RC phase shift oscillators. *(GTU., Nov./Dec. 2011)*

MULTIPLE CHOICE QUESTIONS

1. An electronic oscillator is
   (*a*) just like an alternator (*b*) nothing but an amplifier
   (*c*) an amplifier with amplifier (*d*) a converter of ac to dc energy
2. The frequency of oscillation of an elementary *LC* oscillatory circuit depends on
   (*a*) coil resistance (*b*) coil inductance
   (*c*) capacitance (*d*) both (*a*) and (*c*)
3. For sustaining oscillations in an oscillator
   (*a*) feedback factor should be unity (*b*) phase shift should be unity
   (*c*) feedback should be negative (*d*) both (*a*) and (*b*).
4. If Barkhausen criterion is not fulfilled by an oscillator circuit, it will
   (*a*) stop oscillating (*b*) produce damped waves continuously
   (*c*) becomes an amplifier (*d*) produce high-frequency whistles.
5. For sustained oscillations in an oscillator, the loop gain should be
   (*a*) zero (*b*) less than unity
   (*c*) greater than unity
6. Which of the following factors do not affect the frequency stability of an oscillator?
   (*a*) output load
   (*b*) inter-element capacitances and stray capacitances
   (*c*) temperature variation
   (*d*) coil size

7. A transistor Hartley Oscillator uses
   (*a*) resistive feedback (*b*) inductive feedback
   (*c*) capacitive feedback (*d*) none of the above
8. Crystal oscillators are superior to tuned *LC* oscillators mainly because of their
   (*a*) exceptionally high degree of frequency stability
   (*b*) high value of *Q*-factor
   (*c*) size of the crystal (*d*) both (*a*) and (*b*) above
9. In a RC phase-shift oscillator, the expression for frequency of oscillation is:
   (*a*) $f = 1/2\pi\sqrt{LC}$ (*b*) $f = 1/2\pi\, RC\sqrt{6}$
   (*c*) $f = 1/\pi\, CR$ (*d*) None of the above
10. In a transistor Hartley oscillator
   (*a*) inductive feedback is used (*b*) untapped coil is used
   (*c*) entire coil is in the output circuit (*d*) no capacitor is used
11. A Hartley oscillators used for generating
   (*a*) very low frequency oscillation (*b*) radio-frequency oscillation
   (*c*) microwave oscillation (*d*) audio-frequency oscillation
12. A Colpitts oscillator uses
   (*a*) tapped coil (*b*) inductive feedback
   (*c*) tapped capacitance (*d*) no tuned *LC* circuit
13. In *RC* phase-shift oscillator circuits.
   (*a*) there is no need for feedback
   (*b*) feedback factor is less than unity
   (*c*) pure sine wave output is possible
   (*d*) transistor parameters determine oscillation frequency
14. Wien bridge oscillator is most often used whenever
   (*a*) wide range of high purity sine waves is to be generated
   (*b*) high feedback ratio is needed
   (*c*) square output waves are required
   (*d*) extremely high resonant frequencies are required.
15. The primary advantage of a crystal oscillator is that
   (*a*) it can oscillate at any frequency
   (*b*) it gives a high output voltage
   (*c*) its frequency of oscillation remains almost constant
   (*d*) it operates on a very low dc supply voltage
16. The RC network shown in Fig. 31. 31 can provide a maximum theoretical phase-shift of
   (*a*) 90° (*b*) 180°
   (*c*) 270° (*d*) 360° (*UPSC Engg. Services 2001*)

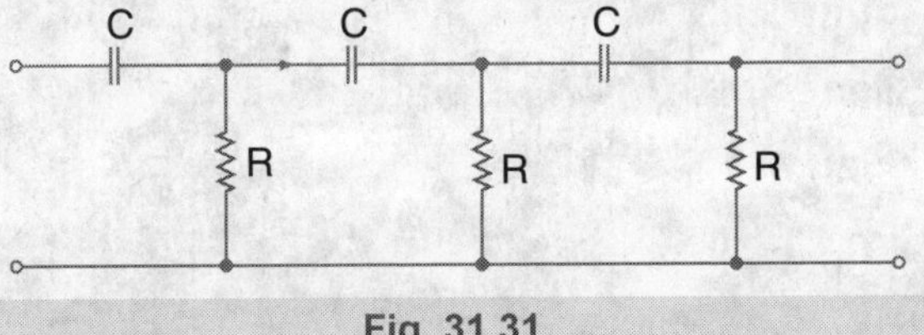

Fig. 31.31.

17. Which one of the following circuit is most suitable as an oscillator at a frequency of 100 Hz.

(*a*) Hartley oscillator (*b*) Colpitts oscillator
(*c*) Crystal oscillator (*d*) Wein bridge oscillator

(*UPSC Engg. Services 2001*)

18. The Bark hausen criterion for sustained oscillation is given by

(*a*) $A\beta = 1$ (*b*) $|A\beta| \geq 1$
(*c*) $|A\beta| < 1$ (*d*) $\angle A\beta = 180°$

19. An oscillator always needs an amplifier with

(*a*) positive feedback (*b*) negative feedback
(*c*) both types of feedback (*d*) an *LC* tank circuit.

20. Crystal oscillators are often used for providing

(*a*) square wave output
(*b*) triangular wave output
(*c*) stable carrier wave in radio transmitters
(*d*) frequency multiplication.

21. A *CE* amplifier can be converted into an oscillator by

(*a*) providing adequate positive feedback
(*b*) phase shifting the output by 180° and feeding this phase shifted output to the input.
(*c*) using only a series tuned circuit as a load on the amplifier
(*d*) using a negative resistance device as a load on the amplifier.
(*e*) all of above

(*UPSC Engg Services*. 2002)

22. Colpitts oscillator is also known as

(*a*) LC oscillator (*b*) Resonant circuit oscillator
(*c*) Tank-Circuit oscillator (*d*) all-of-the above

(*Bharthiar Univ. April 2004*)

ANSWERS

1. (*c*)	**2.** (*d*)	**3.** (*d*)	**4.** (*a*)	**5.** (*c*)	**6.** (*d*)
7. (*b*)	**8.** (*d*)	**9.** (*b*)	**10.** (*a*)	**11.** (*b*)	**12.** (*c*)
13. (*c*)	**14.** (*a*)	**15.** (*c*)	**16.** (*b*)	**17.** (*d*)	**18.** (*a*)
19. (*a*)	**20.** (*c*)	**21.** (*e*)	**22.** (*d*)		

32

Chapter

Non-Sinusoidal Oscillators

Outline

32.1. Introduction

The oscillators, which generate waveforms other than sine waveform, are called nonsinusoidal oscillators. The waveforms generated by nonsinusiodal oscillators include *square, triangular rectangular and sawtooth waveforms.* These oscillators use one or two bipolar junction transistors (or field-effect transistors). The transistor is used as a switch, which operates in two states namely saturation and cut-off state. When the switch operates, there is a sharp transition from one state to another. Usually, a transistor remains in the cut-off state for a certain period of time, which is followed by another interval of time during which it remains in saturation. Since the transistor alternately supplies power to the load, and relaxes when it is in cut-off state, therefore the nonsinusoidal oscillators are also known as *relaxation oscillators*.

32.2. Classification of Nonsinusoidal Oscillators

The nonsinusoidal oscillators may be classified into the following three types:

1. Multivibrators
2. Blocking oscillators and
3. Sawtooth or sweep generators.

The multivibrators and blocking oscillators are

discussed in this chapter, while the sawtooth or sweep generators will be discussed in the chapter on Time-Base Circuits.

32.3. Transistor as a Switch

We have already discussed in Art 32.2 that nonsinusoidal oscillators employ one or two transistors, which are used as a switch. The switch operates between two states namely *saturation* and cut-off state. The saturation state occurs when both the junctions (*i.e.*, emitter-base junction and collector base junction) of a transistor are forward biased. On the other hand, the cut-off state occurs, when both the junctions are reverse biased.

Consider a common emitter transistor switch as shown in Fig. 32.1. A pulse input (v_{in}) as shown in the figure is applied at the based of the transistor. It is used to control the voltage (or state of the switch) between collector and emitter of the transistor. It may be noted that at time $t < T_1$, the input voltage (v_{in}) is equal to $-V_1$ and the emitter-base diode is reverse biased. Since the collector-base diode is also reverse biased, therefore the transistor is cut-off, and practically no current exists anywhere in the circuit. As a result of this, the collector-to-emitter or the output voltage (v_o) is approximately equal to the supply voltage (V_{CC}). Thus when the collector current (i_C) is zero and $v_o = V_{CC}$, the transistor acts as an open switch.

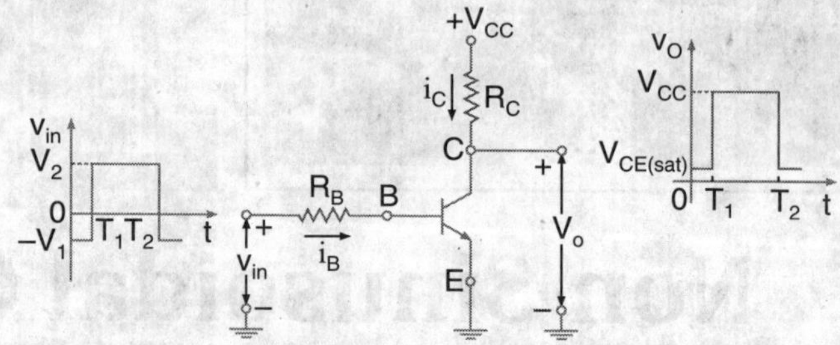

Fig. 32.1. A transistor switch with input and output waveforms.

It will be interesting to know that when the input voltage becomes equal to *V_2 for the time interval $T_1 < t < T_2$, both the emitter-base and collector-base diodes are forward biased and the transistor is saturated. In this interval (*i.e.*, $T_1 < t < T_2$) the output voltage, $v_o = V_{CE(sat)}$. The value of $V_{CE(sat)}$ for silicon transistor is 0.2 V and the collector current is maximum. Its value is approximately equal to V_{CC}/R_C. These values indicate that the transistor acts as a closed switch and the current in the closed switch is determined by the supply voltage (V_{CC}) and the load resistor (R_C).

At time $T = T_2$, the input waveform returns back to $-V_1$.

This causes the transistor to switch back to cut-off state. We may summarize the above two results as follows:

$$i_C = 0 \quad \text{and} \quad v_{CE} = V_{CC} \qquad \text{... (Open switch)}$$

$$i_C = V_{CC}/R_C \quad \text{and} \quad v_{CE} = V_{CE(sat)} \qquad \text{... (Closed switch)}$$

Example 32.1.

Find the upper and lower end points of the load line for the circuit shown in Fig. 32.2. Draw the dc load line using these end points.

(*Bharthiar Univ. 2004*)

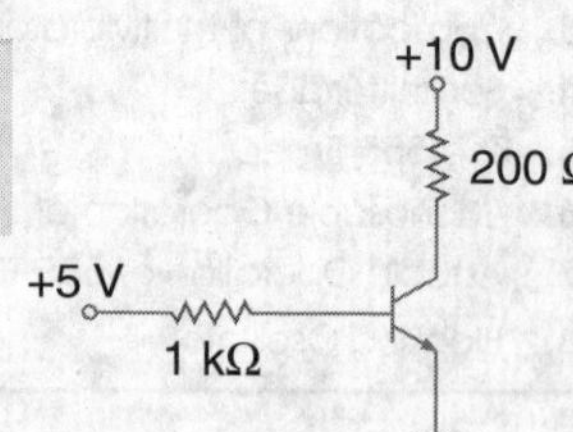

Fig. 32.2.

Solution.

Given: $V_{CC} = 10$ V ; $V_{BB} = 5$ V ; $R_C = 200\ \Omega$ and $R_B = 1000\ \Omega$

The upper cut-off point of the load line for the circuit shown in Fig. 32.2 is given by,

* The value of V_2 is selected to ensure that transistor is in saturation. It is achieved by setting the value of base current (I_B) greater than $I_{B(sat)}$. The value of $I_{B(sat)}$ is equal to $V_{CC}/\beta \cdot R_C$ where β is the current gain of the transistor.

$$i_C = \frac{V_{CC}}{R_C} \text{ and } v_{CE} = V_{CE\,(sat)} \quad ...(i)$$

and the lower cut-off point is given by,

$$i_C = 0 \quad \text{and} \quad v_{CE} = V_{CC} \quad ...(ii)$$

Using equation (*i*), we find that,

$$i_C = \frac{V_{CC}}{R_C} = \frac{10}{200}$$

$$= 0.05 \text{ A or } 50 \text{ mA},$$

and using equation (*ii*)

$$i_C = 0 \quad \text{and} \quad V_{CE} = 10 \text{ V}.$$

Using the values of $i_C = 50$ mA and $v_{CE} = 10$ V, the upper and lower cut-off points of the load line can be sketched as shown in Fig. 32.3.

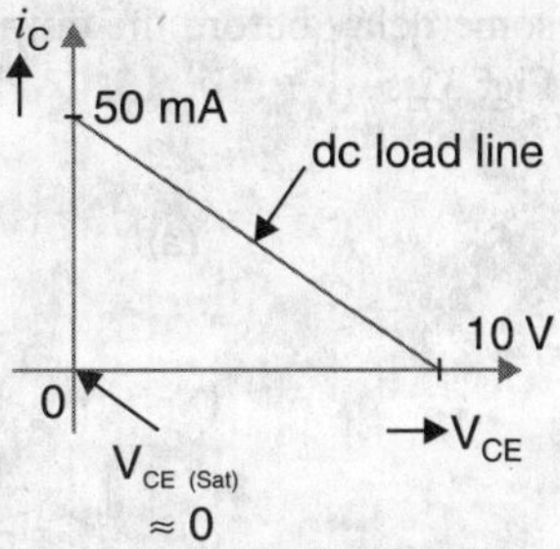

Fig. 32.3.

Example 32.2.

A silicon transistor with $V_{BE(sat)} = 0.8$ V, $\beta = h_{FE} = 100$, $V_{CE(sat)} = 0.2$ V is used in the circuit shown in Fig. 32.4. Find the minimum value of R_C for which the transistor remains in saturation.

(GTU., May 2011)

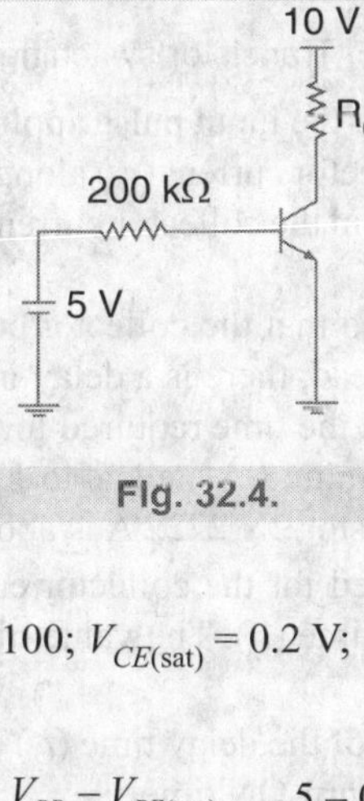

Fig. 32.4.

Solution.

Given: $V_{BE(sat)} = 0.8$ V; $\beta = h_{FE} = 100$; $V_{CE(sat)} = 0.2$ V; $V_{BB} = 5$ V; $V_{CC} = 10$ V and $R_B = 200$ kΩ $= 200 \times 10^3\ \Omega$

We know that the base current,

$$I_B = \frac{V_{BB} - V_{BE(sat)}}{R_B} = \frac{5 - 0.8}{200 \times 10^3}$$

$$= 0.021 \times 10^{-3} \text{ A} = 21 \times 10^{-6} \text{ A} = 21 \text{ mA}$$

and the collector current,

$$I_C = \beta I_B = 100 \times (21 \times 10^{-6}) = 2.1 \times 10^{-3} \text{ A} = 2.1 \text{ mA}$$

We also know that,

$$V_{CC} - I_C R_C - V_{CE(sat)} = 0$$

$$10 - (2.1 \times 10^{-3}) R_C - 0.2 = 0$$

or $$R_C = 9.8/(2.1 \times 10^{-3}) = 4.66 \times 103\ \Omega = 4.66 \text{ k}\Omega$$

The minimum value of $R_C = 4.66$ kΩ to kept the transistor in saturation. **Ans.**

32.4. Transistor Switching Times

It has been observed that the application of a pulse input at the input (*i.e.*, base-emitter junction of a transistor is not followed directly by a change in output voltage. In order words, there is always some delay before the output changes as compared to the input. This phenomenon is illustrated in Fig. 32.5.

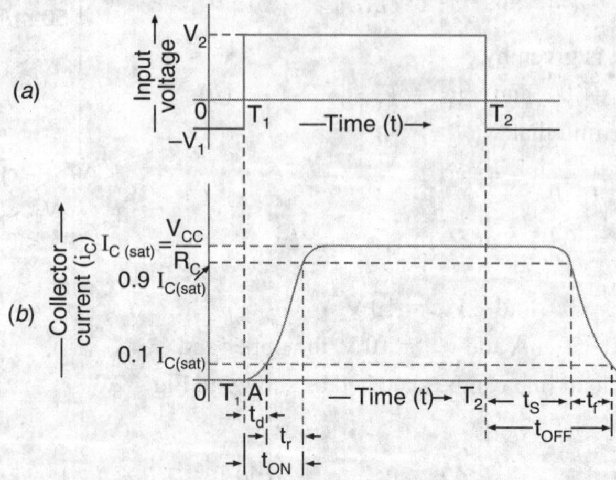

(a) Input pulse waveform

(b) Collector current waveform with various time delays

Fig. 32.5. Transistor switching times.

Fig. 32.5 (*a*) shows the waveform of the input pulse applied to the transistor and Figure 32.5 (*b*) indicates the resulting waveform of collector current (i_C) alongwith the various time delays involved. Following are the important time delays in the collector current waveform, which may be noted from the figure.

1. ***Delay time*** **(t_d).** It may be noted that the collector current does not quickly respond to the input signal at time $t = T_1$. Instead, there is a delay indicated by T_1A. The time that elapses during this delay, together with the time required for the collector current to rise to 10% of its maximum (*i.e.*, saturation) value $I_{C(sat)}$ equal to V_{CC}/R_C is called the delay time (t_d). The value of delay time for a transistor 2N 2222 A is about 10 nanosecond.
2. ***Rise time*** **(t_r).** The time required for the collector current to rise from 10% to 90% of the maximum value is called rise time (t_r). The value of rise time for 2 *N* 2222 A transistor is about 25 nanosecond.
3. ***Time-ON time*** **(t_{ON}).** The sum of the delay time (t_d) and the rise time (t_r) is called the turn-ON time. Mathematically, the turn-ON time,
$$t_{ON} = t_d + t_r$$
For 2N2222 A transistor, the total turn-ON time is equal to 35 nanosecond.
4. ***Storage time*** **(t_s).** It may be noted from the figure that when the input signal returns back to its initial state (*i.e.*, $-V_1$) at time $t = T_2$, the collector current again fails to respond immediately. The interval, which elapses between the transition of the input voltage waveform and the time when collector current has dropped to 90% of its maximum value is called the storage time. The value of storage time for 2 N 2222. A transistor is about 225 nanosecond.
5. ***Fall-time*** **(t_f).** The storage interval is fallowed by full time (l.p.). It is the time required for the collector current to fall from 90% to 10% of its maximum value. Its value for 2N2222 A transistor is about 60 nanosecond.
6. ***Turn-OFF time*** **(t_{OFF}).** It is the sum of storage time (t_s) and the fall time (t_f). Mathematically, the turn-OFF time,
$$t_{OFF} = t_s + t_f$$
For 2N2222 A transistor the total turn-OFF time is equal to 285 nanosecond.

32.5. Factors Contributing to Various Time Delays in a Transistor

We have already discussed in the last article about various time delays like delay time, rise time, storage time and fall time. Now we shall discuss the factors, in more detail, which contribute to these time delays.

(*a*) ***Delay time.*** There are three factors, which contribute delay time (t_d) and are as given below:

1. When the driving signal is applied to the transistor input, a certain time is required to charge-up the emitter-base junction transistor capacitance so that the transistor may be brought from cut-off region to the active regain.
2. Even when the transistor has been brought to the point, where minority carriers have begun to cross the emitter-base junction, a time interval is required before these carriers can cross the base region to the collector-base junction and the recorded as a collector current.
3. Some time is always required for the collector current to rise to 10% of its maximum value.

(*b*) ***Rise and fall time.*** These are due to the fact that if a base-current step is used to saturate the transistor or return it from saturation to cut-off, the transistor collector current must traverse the active region.

It has been observed that the collector current increases or decreases along an exponential curve as shown in Fig. 32.6.

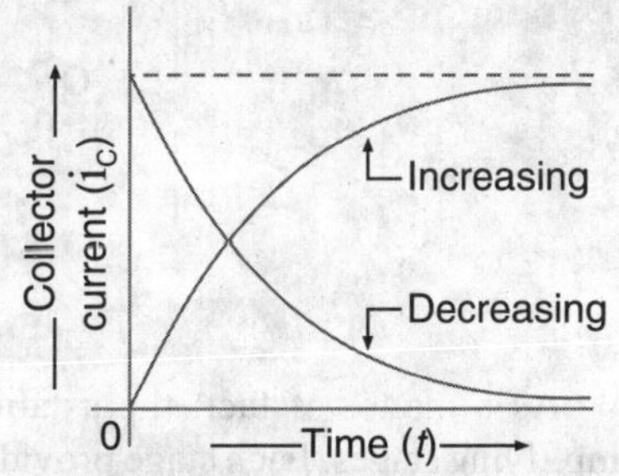

Fig. 32.6. Increase or decrease of collector current.

(*c*) ***Storage time.*** It results from the fact that a transistor, in saturation, has a saturation charge of excess minority carriers stored in the baseregion. A transistor cannot respond, until this saturation charge has been removed. The storage time can be reduced by adding a capacitor (C) across the base resistor (R_B). It can also be eliminated by not allowing the transistor to enter into saturation by adding a Schottky diode across the collector and base terminals of the transistor.

32.6. Multivibrators

These are two stage amplifiers, which are coupled in such a way that the output of one amplifier provides an input for the other. The amplifiers use positive feedback (also called regenerative feedback) which is supplied in such a manner that the transistor of one stage is driven into saturation (*i.e.*, ON state) and the other to cut-off (*i.e.*, OFF state). The action is reversed after a certain time, depending upon the circuit conditions *i.e.*, the saturated transistor is driven into cut-off and the cut-off transistor is driven into saturation.

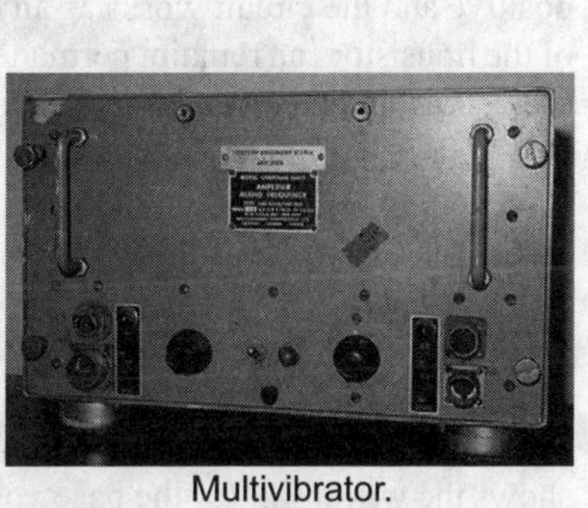
Multivibrator.

32.7. Types of a Multivibrators

There are three basic types of multivibrators depending upon the type of coupling network used.

1. Astable multivibrator.
2. Monostable multivibrator.
3. Bistable multivibrator.

An astable multivibrator uses capacitive coupling and monostable multivibrator uses resistive-capacitive coupling, while the bistable multivibrator uses resistive coupling. Now we shall discuss all the above three types of multivibrators one by one in the following pages.

32.8. Astable Multivibrator

It is also called a *free-running relaxation oscillator* and is commonly used to generate square waveform. Fig. 32.7 shown the circuit of a collector-coupled astable multivibrator using two identical NPN transistors Q_1 and Q_2. It is possible to have $R_{C1} = R_{C2} = R_C$, $R_1 = R_2 = R$ and $C_1 = C_2 = C$. In that case, the circuit is known as symmetrical astable multivibrator. The transistor Q_1 is forward biased by the V_{CC} supply through resistor R_1. Similarly, the transistor Q_2 is forward biased by the V_{CC} supply through resistor R_2. The output of transistor Q_1 is coupled to the input of transistor Q_2 through the capacitor C_1. Similarly, the output of transistor Q_2 is coupled to the input of transistor Q_1 through the capacitor C_2.

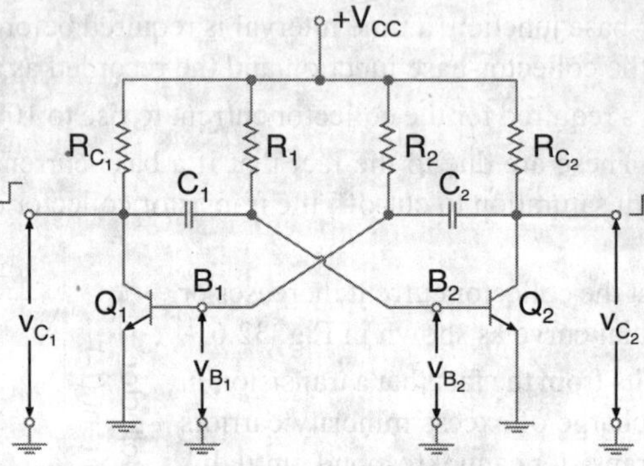

Fig. 32.7. Astable multivibrator.

As a matter of fact, the astable multivibrator may be thought of as two common-emitter amplifying stages. Each stage provides a feedback through a capacitor at the input of the other. Since the amplifying stage introduces a 180° phase-shift and another 180° phase-shift is introduced by a capacitor, therefore the feedback signal has a total phase-shift of 360° or 0°. Thus the feedback is positive and the circuit works as an oscillator. In other words, because of capacitive coupling, none of the transistor can remaint permanently cut-off or saturated. Instead the circuit has two quasi-stable states (ON and OFF) and it makes periodic transition between these two states.

The output of an astable multivibrator is available at the collector terminal of either transistor (*i.e.*, Q_1 and Q_2) as shown in the figure. However, the two outputs are 180° out of phase with each other. Therefore one of the output is said to be the complement of the other.

32.9. Circuit Operation of Astable Multivibrator

The circuit operation of an astable multivibrator is easy to understand with the waveforms shown in Fig. 32.8. These waveforms are at the base and collector of transistors Q_1 and Q_2. Fig. 32.8 (*a*) shows the waveform for the base voltage of transistor Q_1 (*i.e.*,v_{B1}) and Fig. 32.8 (*b*) for the collector voltage of transistor Q_1 (*i.e.*, v_{C1}). Similarly Fig. 32.8 (*c*) shows the waveform for the base voltage of transistor Q_2 (*i.e.*, v_{B2}) and Fig. 32.8 (*d*) for the collector voltage of transistor Q_2 (*i.e.*, v_{C2}). The circuit operation may be explained as follows:

1. When the d.c. power supply (V_{CC}) is switched ON, (say at $t = 0$) one of the transistor will start conducting more than the other due to some imbalance in the circuit. Let us suppose that transistor Q_1 start conducting more than that of transistor Q_2. Then because of positive feedback, the transistor Q_1 will be driven into saturation and transistor Q_2 to cut-off. Thus at $t > 0$, the transistor Q_1 is ON and Q_2 is OFF. Thus at $t > 0$, $v_{B1} = V_{BE(sat)}$ (*i.e.*, 0.7 V for silicon transistor), $= V_{C2} = V_{CE\,(sat)}$ (*i.e.*, 0.3 V for silicon transistor), v_{B1} is negative and $V_{C2} = V_{CC}$.

2. During the time $t > 0$ (*i.e.*, when Q_1 is ON and Q_2 is OFF), the capacitor C_1 is charging towards the voltage V_{CC} through R_1. The charging takes place exponentially with time constant $\tau_1 = R_1 . C_1$. Since the base of transistor Q_2 is directly connected to capacitor C_1, as shown in Fig. 32.8. Therefore the voltage V_{B2} also increases exponentially towards V_{CC}.
3. As soon as the voltage V_{B2} increases, above the cut-in voltage (*i.e.*, 0.5 V for silicon transistor), the transistor Q_2 starts conducting. It occurs at $t = t_1$. As the transistor Q_2 goes into saturation, its collector voltage (v_{C1}) falls to $v_{CE(sat)}$. The fall in voltage v_{C2} causes an equal fall, (*i.e.*, $V_{CC} - V_{CE(sat)} = V_{CC}$) in voltage v_{B1} because the two are capacitively coupled. The fall in voltage v_{B2} cuts-off the transistor Q_1 and its collector voltage (v_{C1}) starts rising towards V_{CC} with a time constant $\tau' = R_C . C_1$. The rise in voltage V_{C1} is coupled through capacitor C_1 to the base of the transistor Q_2, causing a small overshoot in voltage v_{B2}. Soon the voltage v_{B_2} settles at $V_{BE(sat)}$ *i.e.*, 0.7 V level. Thus at $t \geq t_1$, the transistor Q_1 is OFF and Q_2 is ON. The voltage levels at this instant are v_{B1} is negative, $v_{C1} = V_{CC}$, $v_{B2} = V_{BE} = V_{BE(sat)}$ and $v_{C_2} = V_{CE(sat)}$.

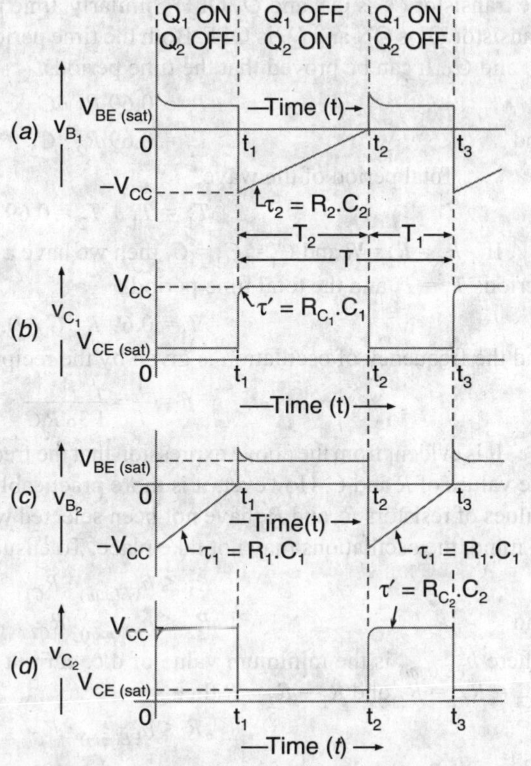

Fig. 32.8. Waveforms at the base and collector of transistors Q_1 and Q_2 in astable multivibrator.

4. During the time $t > t_1$ (*i.e.*, when Q_1 is OFF and Q_2 is ON) the voltage v_{B_1} rises exponentially with time constant $\tau_2 = R_2 \cdot C_2$ towards V_{CC}. At $t = T_2$, the voltage v_{B_1} reaches the cut-in level (*i.e.*, 0.5 V) and a reverse transition takes place (*i.e.*, Q_1 turns ON and Q_2 turns OFF). The voltage levels for $t > t_2$ are $v_{B_1} = v_{BE(sat)}$, $v_{C_1} = v_{CE(sat)}$; v_{B_2} is negative and $v_{C_2} = V_{CC}$. Thus the voltage levels for $t > t_2$ are the same as for $t > 0$.

It may be noted that for an astable multivibrator, using PNP transistors, the voltages and currents are negative to those for a circuit with NPN transistors. Thus all the waveform shown in Fig. 32.7 must be inverted, if PNP transistors are used.

32.10. Switching Times and Frequency of Oscillation

We have already discussed in the last article the waveforms at the base and collector of the transistors Q_1 and Q_2 in the circuit astable multivibrator. Each cycle of the waveform consists of time period T_1 and T_2 as shown in Fig. 32.8. The time period T_1 represents the duration for which the transistor Q_1 is ON and Q_2 OFF. Similarly, time period T_2 represents the duration for which the transistor Q_2 is ON and Q_1 is OFF. Both the time periods T_1 and T_2 depend upon charge of capacitors C_1 and C_2. It can be proved that the time periods,

$$T_1 = 0.69\ R_1 \cdot C_1 \qquad \text{... (i.e., ON time for } Q_1\text{)}$$

and

$$T_2 = 0.69\ R_2 \cdot C_2 \qquad \text{... (i.e., ON time for } Q_2\text{)}$$

$\therefore$ Total period of the wave,

$$T_1 = T_1 + T_2 = 0.69\ (R_1 \cdot C_1 + R_2 \cdot C_2)$$

If $R_1 = R_2 = R$ and $C_1 = C_2 = C$, then we have a symmetrical astable multivibrator, whose time periods $T_1 = T_2$ and the total time period,

$$T = 0.69\ R \cdot C + 0.69\ R \cdot C = 1.38\ R \cdot C$$

and the frequency of oscillation is given by the reciprocal of the period, *i.e.*,

$$f = \frac{1}{T} = \frac{1}{1.38\ R.C}$$

It is evident from the above expression that the frequency of oscillation may be varied by adjusting the values of R and C. However, it is more practicable to adjust R than C. It may be noted that if the values of resistors R_1 and R_2 have not been selected with d.c. current gain of the transistor (h_{FE} or β) in mind, the oscillations may not take place. To ensure oscillations, the value of resistor

$$R_1 \le h_{FE(min)} \cdot R_{C1}$$

and

$$R_2 \le h_{FE(min)} \cdot R_{C2}$$

where $h_{FE\ (min)}$ is the minimum value of d.c. current gain of transistors Q_1 and Q_2. If the values of $R_{C1} = R_{C2} = R_C$ and $R_1 = R_2 = R$ then

$$R \le h_{FE(min)} \cdot R_C$$

and

$$f = \frac{1}{1.38\ R \cdot C}$$

Example 32.3.

An astable multivibrator has component values, $R_1 = R_2 = 20\ k\Omega$ and $C_1 = C_2 = 100\ pF$. Calculate the frequency of oscillator.

Solution.

Given: $R_1 = R_2 = R = 20\ \text{k}\Omega = 20 \times 10^3\ \Omega$ and $C_1 = C_2 = C = 100\ \text{pF} = 100 \times 10^{-12}\ \text{F}$.

We know that the frequency of oscillator,

$$f = \frac{1}{1.38\ R\ C}$$

$$= \frac{1}{1.38\ (20 \times 10^3) \times (100 \times 10^{-12})}\ Hz$$

$$= 0.362 \times 10^6\ \text{Hz} = 362\ \text{kHz}\ \textbf{Ans.}$$

Example 32.4.

Determine the period and frequency of oscillation for an astable multivibrator with component values $R_1 = 2\ k\Omega$; $R_2 = 20\ k\Omega$ and $C_1 = 0.01\ \mu F$, $C_2 = 0.05\ \mu F$.

Solution.

Given: $R_1 = 2\ \text{k}\Omega = 2 \times 10^3\ \Omega$; $R_2 = 2t0\ \text{k}\Omega = 20 \times 10^3\ \Omega$; $C_1 = 0.01\ \text{mF} = 0.01 \times 10^{-6}\text{F}$ and $C_2 = 0.05\ \mu\text{F} = 0.05 \times 10^{-6}\ \text{F}$.

Period of oscillation

We know that period of oscillation,

$$T = 0.69\,(R_1 \cdot C_1 + R_2 \cdot C_2)$$
$$= 0.69\,[(2 \times 10^3) \times (0.01 \times 10^{-6}) + (20 \times 10^3) \times (0.05 \times 10^{-6})]$$
$$= 0.69\,[(0.02 \times 10^{-3}) + (1 \times 10^{-3})]$$
$$= 0.7\times 10^{-3} = 0.7 \text{ ms } \textbf{Ans.}$$

Frequency of oscillation

We know that frequency of oscillation,

$$f = \frac{1}{T} = \frac{1}{0.7 \times 10^{-3}}$$
$$= 1.43 \times 10^3 \text{ Hz} = 1.43 \text{ kHz } \textbf{Ans.}$$

Example 32.4.

Determine the value of capacitors to be used in an astable multivibrator to provide a train of pulse 1 μ sec wide at a repetition rate of 100 kHz. Given $R_1 = R_2 = 10$ kΩ.

Solution.

Given: Pulse width (T_1) = 1 μsec = 1×10^{-6} sec; f = 100 kHz = 100×10^3 Hz and $R_1 = R_2 = 10$ kΩ = 10×10^3 Ω.

Fig. 32.9 shows the waveform of the required pulse train.

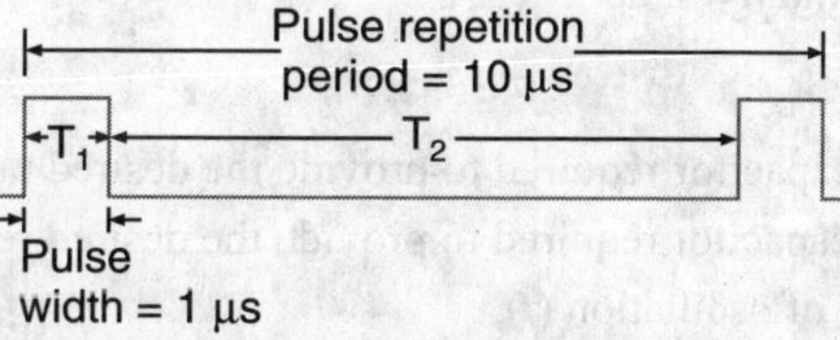

Fig. 32.9.

Let C_1 = Value of the capacitor required to provide the desired pulse width, and

C_2 = Value of the capacitor required to provide the desired repetition rate.

We know that frequency of oscillation (f),

$$100 \times 10^3 = \frac{1}{T}$$

or $$T = \frac{1}{(100 \times 10^3)} = 10 \times 10^{-6} \text{ sec}$$

We also know that pulse width (T_1),

$$1 \times 10^{-6} = 0.69\, R_1 \cdot C_1 = 0.69 \times (10 \times 10^3) \times C_1$$

$$\therefore \quad C_1 = \frac{1 \times 10^{-6}}{0.69 \times (10 \times 10^3)} = 145 \times 10^{-12} \text{ F}$$
$$= 145 \text{ pF } \textbf{Ans.}$$

and the time period (T),

$$10 \times 10^{-6} = T_1 + T_2 = 1 \times 10^{-6} + T_2$$

$$\therefore \quad T_2 = (10 \times 10^{-6}) - (1 \times 10^{-6}) = 9 \times 10^{-6} \text{ sec}$$

We also know that the time period (T_2),

$$9 \times 10^{-6} = 0.69\, R_2 \cdot C_2 = 0.69 \times (10 \times 10^3) \times C_2$$

$$\therefore \quad C_3 = \frac{9 \times 10^{-6}}{0.69 \times (10 \times 10^3)} = 1304 \times 10^{-12}\ \text{F}$$

$$= 1304\ \text{pF}\ \textbf{Ans.}$$

Example 32.6.

Determine the value of capacitors to be used in an astable multivibrator to provide a train of pulse 2 μs wide at a repetition rate of 100 kHz if $R_1 = R_2 = 20k\Omega$.

(Anna University., Nov/Dec 2007)

Solution.

Given: Pulse width (T_1) = 2 μsec = 2×10^{-6} sec; f = 100 kHz = 100×10^3 Hz and $R_1 = R_2 = 20\ \text{k}\Omega = 20 \times 103\ \Omega$.

Fig. 32.10 shows the waveform of the required pulse train.

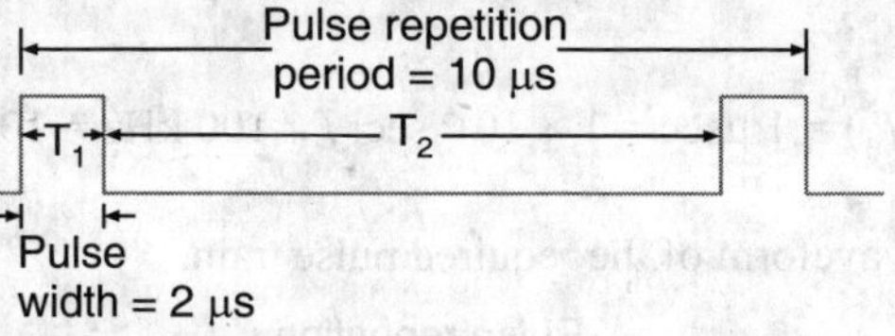

Fig. 32.10.

Let C_1 = Value of the capacitor required to provide the desired pulse width, and
C_2 = Value of the capacitor required to provide the desired repetition rate.

We know that frequency of oscillation (f),

$$100 \times 10^3 = \frac{1}{T}$$

or
$$T = \frac{1}{100 \times 10^3} = 10 \times 10^{-6}\ \text{sec}$$

We also know that pulse width (T_1),

$$2 \times 10^{-6} = 0.69\, R_1 \cdot C_1 = 0.69 \times (20 \times 10^3) \times C_1$$

$$\therefore \quad C_1 = \frac{2 \times 10^{-6}}{0.69 \times (20 \times 10^3)}$$

$$= 145\ \text{pF}\ \textbf{Ans.}$$

and the time period (T),

$$10 \times 10^{-6} = T_1 + T_2 = 2 \times 10^{-6} + T_2$$

$$T_2 = (10 \times 10^{-6}) - (2 \times 10^{-6}) = 8 \times 10^{-6}\ \text{sec}$$

We also know that the time period (T_2),

$$8 \times 10^{-6} = 0.69\, R_2 \cdot C_2 = 0.69 \times (20 \times 10^3)\, C_2$$

$$\therefore \quad C_2 = \frac{8 \times 10^{-6}}{0.69 \times (20 \times 10^3)} = 579.\ 5\ \text{pF}\ \textbf{Ans.}$$

Example 32.7.

A collector coupled astable multivibrator is to be used to produce an unsymmetrical square wave of durations $\tau_A = 310$ msec and $\tau_B = 250$ msec, $V_{CC} = 15$ V; $I_{C(sat)} = 5$ mA, $h_{E(min)} = 20$.

Find the values of circuits components and output waveform time constant.

(A.M.I.E., Summer, 1987)

Solution.

Given: $T_2\ (\tau_A) = 310$ msec $= 310 \times 10^{-6}$ sec; $T_2\ (\tau_B) = 250$ msec $= 250 \times 10^{-6}$ sec; $V_{CC} = 15$ volt; $I_{C(sat)} = 5$ mA $= 5 \times 10^{-3}$ A and $h_{FC(min)} = 20$.

Value of circuit components

We know that the value of collector resistance,

$$R_C = \frac{V_{CC}}{I_C} = \frac{15}{5 \times 10^{-3}} = 3000\ \Omega$$

∴ Collector resistance of transistors Q_1 and Q_2,

$$R_{C_1} = R_{C_2} = R_C = 3000\ \Omega \textbf{ Ans.}$$

We also know that the value of base current required to saturate the transistor,

$$I_{B(sat)} = \frac{I_{C\,(sat)}}{h_{FE\,(min)}} = \frac{5 \times 10^{-3}}{20}\ \text{A} = 0.25 \times 10^{-3}\ \text{A}$$

Assuming a base over drive factor of 2, the value of base current required,

$$I_B = 2\, I_{B(sat)} = 2 \times 0.25\ 10^{-3}\ \text{A} = 0.5 \times 10^{-3}\ \text{A}$$

$$\therefore \quad R_1 = R_2 = R = \frac{V_{CC}}{I_B} = \frac{15}{0.5 \times 10^{-3}}$$

$$= 30 \times 10^3\ \Omega = 30\ \text{k}\Omega \textbf{ Ans.}$$

We know that the time period (T_1)

$$310 \times 10^{-6} = 0.69\ R_1 \cdot C_1 = 0.69 \times (60 \times 10^3) \times C_1$$

$$\therefore \quad C_1 = \frac{310 \times 10^{-6}}{0.69 \times (30 \times 10^3)} = 14.98 \times 10^{-9}\ F$$

$$= 14980\ \text{pF} \textbf{ Ans.}$$

Similarly, the time period (T_2),

$$250 \times 10^{-6} = 0.69\ R_2 \cdot C_2 = 0.69 \times (30 \times 10^3) \times C_2$$

$$\therefore \quad C_2 = \frac{250 \times 10^{-6}}{0.69 \times (60 \times 10^3)} = 1208 \times 10^{-9}\ F = 12080\ pF \textbf{ Ans.}$$

Output waveform time constants

We know that the time constant,

$$\tau_1 = R_1 \cdot C_1 = (30 \times 10^3) \times (14980 \times 10^{-12})\ \text{sec}$$
$$= 449.4 \times 10^{-6}\ \text{sec} = 449.4\ \text{msec} \textbf{ Ans.}$$

and the time constant,

$$\tau_2 = R_2 \cdot C_2 = (30 \times 10^3) \times (12080 \times 10^{-12})\ \text{sec}$$
$$= 362.4 \times 10^{-6}\ \text{sec} = 362.4\ \text{msec} \textbf{ Ans.}$$

Similarly, the time constant,

$$\tau' = R_{C1} \cdot C_1 = 3000 \times (7.49 \times 10^{-9})\ \text{sec}$$
$$= 22.5 \times 10^{-6}\ \text{sec} = 22.5\ \text{msec} \textbf{ Ans.}$$

and the time constant,

$$\tau'' = R_{C1} \cdot C_2 = 3000 \times (6.04 \times 10_{-9}) \text{ sec}$$
$$= 18.1 \times 10^{-6} \text{ sec} = 18.1 \text{ msec } \textbf{Ans.}$$

32.11. Monostable Multivibrator

It is also called *one shot or univibrator* and can be used to generate a gating pulse, whose width can be controlled. The monostable moltivibrator provides a single pulse of desired duration in response to an *external trigger. The external trigger causes the circuit to go to the quasi-stable state. After a certain interval of time, the circuit returns to its original stable state.

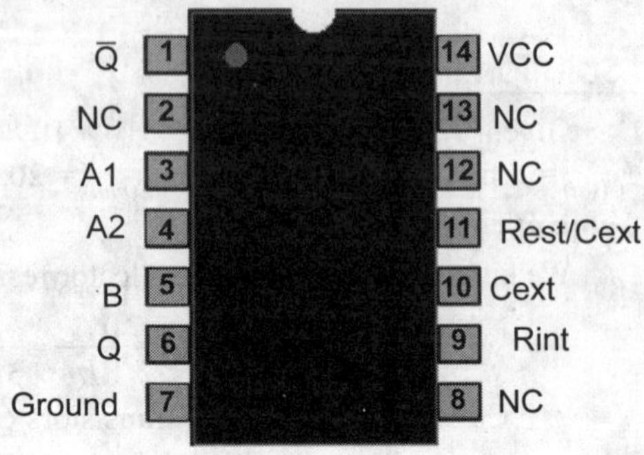

Monostable multivibrator

Fig. 32.11 shows the circuit of a monostable multivibrator using NPN transistors. Here, the output of transistor Q_2 is coupled to the base of transistor Q_1 through the resistance R_1. On the other hand, the output of transistor Q_1 is coupled to the base of transistor Q_2 through the capacitor C_2. The capacitor C_1 is known as commutating capacitor or speed up capacitor. Its function is to speed up the circuit in making abrupt transitions between the ON and OFF states. The base of transistor Q_2 is returned to the V_{CC} supply through a resistor R_3, while the base of transistor Q_1 is connected to the negative supply through a resistor R_2. The advantage of this biasing is that it keeps the transistor Q_1 OFF and Q_2 ON. This state is known as a stable state of the monostable multivibrator.

The output of a monostable multivibrator is available at the collector terminal of either transistor (*i.e.*, Q_1 or Q_2) as shown in the figure. However, the two outputs are the complement of each other *i.e.*, when one of the output is at V_{CC} level, the other is at $V_{CE(sat)}$ level.

The circuit operation may be explained by keeping in mind that initially the circuit is in its stable state *i.e.*, Q_1 is OFF and Q_2 is ON. Now Let us see as to what happens when the trigger is applied:

1. When a positive trigger pulse of sufficient amplitude is applied to the base of transistor Q_1, it overrides the reverse bias provided by the V_{BB} supply and gives it a forward bias. Because of this, the transistor Q_1 starts conducting.
2. As the transistor Q_1 conducts, its collector voltage falls due to the voltage drop across resistor R_{C1}. This fall in voltage is coupled through capacitor C_2, which decreases the forward bias of transistor Q_2.
3. Because of the reduced forward bias, the collector current of transistor Q_2 starts decreasing and its collector voltage rises exponentially towards $V_{CC} \cdot R_1/(R_1 + R_{C2})$ with a time constant $\tau_2 = C_1 \cdot (R_1 \parallel R_{C2})$.
4. The rising collector voltage of transistor Q_2 is coupled to the base of

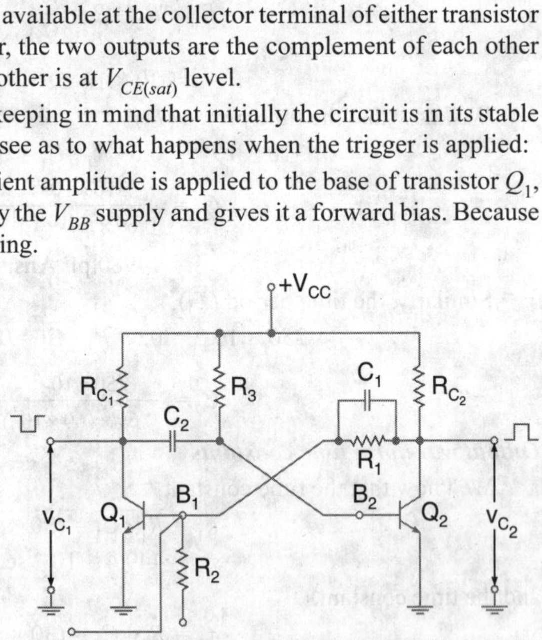

Fig. 32.11. Monostable multivibrator.

* It is short duration pulse of suitable amplitude to initiate operation of an electronic circuit. Such a pulse me be generated by a differeniator (R-C circuit) or by a monostable multivibrator.

transistor Q_1 through resistor R_1, where it further increases its forward bias. Because of the increased forward bias, the transistor Q_1 conducts more. This action is cummulative because of the positive feedback, and the collector voltage of harristor Q_1 falls to $V_{CE\,(sat)}$.

5. Immediately, the capacitor C_2 starts charging exponentially towards V_{CC} with a time constant $\tau_1 = R_3 \cdot C_2$. As C_1 charges, the voltage at the base of the transistor Q_2 decreases. As C_1charges further, the transistor Q_2 is pulled out from the cut-off and the reverse transition takes place *i.e.*, Q_2 turns ON and Q_1 turns OFF.
6. When the transistor Q_2 starts conducting, its collector voltage falls because of the drop across resistor R_{C2}. This drop is coupled to the base of the transistor Q_1, whose collector voltage rises towards V_{CC} with time constant $\tau_3 = R_{C1} \cdot C_2$. Finally, the transistor Q_1 turns fully ON and the transistor Q_1 goes OFF. The circuit remains in this stable state till another pulse is applied.

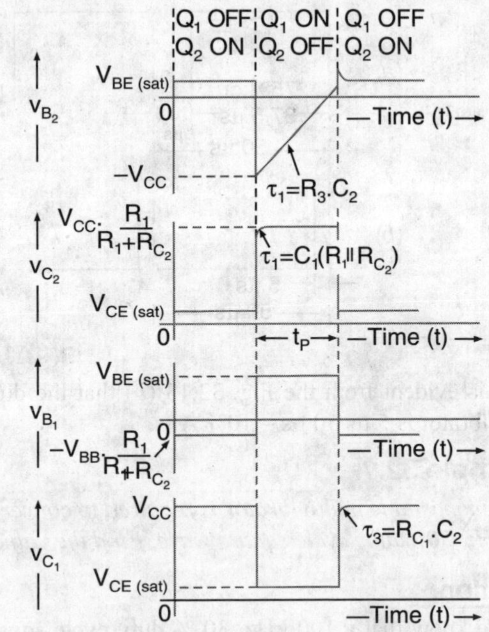

Fig. 32.12. Waveforms at the base and collector of transistors. Q_1 and Q_2 of a monostable multivibrator.

Fig. 32.12 shows the waveforms at the base and collector of the transistor Q_1 and Q_2 of a monostable multivibrator. The width or duration of the pulse obtained at the collector or output of either transistor (Q_1 or Q_2) of the monostable multivibrator is given by the expression,

$$t_p = 0.69\, R_3 \cdot C_1$$

Example 32.8.

A 20 kHz, 75% duty cycle square wave is used to trigger continuously, a monostable multivibrator with a triggered pulse duration (t_p) of 5 μs. What will be the duty cycle of the waveform at the output (Q) of the monostable?

Solution.

Given: Frequency of square wave, $f = 20$ kHz $= 20 \times 10^3$ Hz.

We know that time period of the square wave,

$$T = \frac{1}{f} = \frac{1}{20 \times 10^3} = 0.05 \times 10^{-3} s = 50\ \mu s.$$

Since the duty cycle of the square wave is 75%, therefore the waveform for this is as shown in Fig. 32.13 (*a*). Now the monostable multivibrator is triggered once each time a new pulse arrives. The monostable multivibrator remains triggered only for a duration, $t_p = 5$ μs. The waveform of the output Q of the monostable multivibrator is as shown in Fig. 32.13 (*b*).

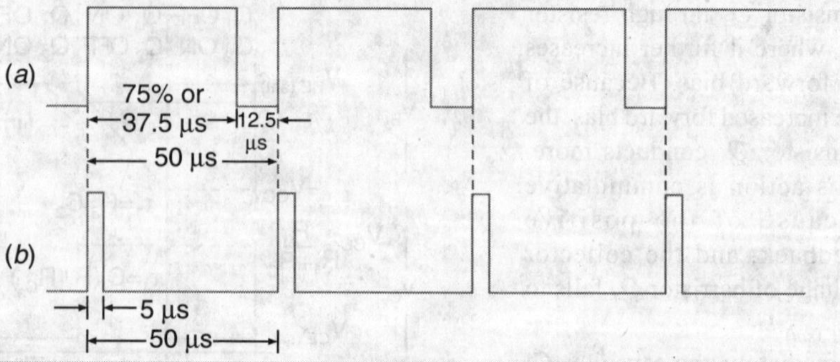

Fig. 32.13.

It is evident from the Fig. 32.13 (*b*) that the duty cycle of the output (*Q*) of the monostable multivibrator is 5 µs/50 µs = 10% **Ans.**

Example 32.9.

A monostable multivibrator is required to convert a 100 kHz, 30% duty cycle square wave to a 100 kHz, 50% duty cycle square wave. Find the values of R_3 and C_1.

Solution.

We know that a 100 kHz, 30 % duty cycle square wave is applied at the trigger input of the monostable multivibrator. This waveform is shown in Fig. 32.14 (*a*). In order to convert it to a square waveform with 50% duty cycle, we want that the monostable multivibrator should have,

$$t_p = 50\% \text{ of the time period of 100 kHz waveform.}$$

Now time period of 100 kHz square wave,

$$T = \frac{1}{f} = \frac{1}{100\ kHz} = \frac{1}{100 \times 10^{-3}} = 10 \times 10^{-6}\ s = 10\ \mu s.$$

$$\therefore \quad t_p = 50\% \text{ of } 10\ \mu s\ \textit{i.e.,}\ \frac{1}{2} \times 10\ \mu s = 5\ \mu s$$

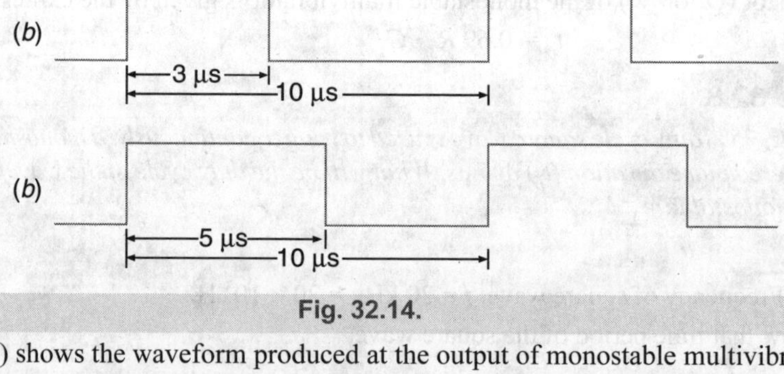

Fig. 32.14.

Fig. 32.14 (*b*) shows the waveform produced at the output of monostable multivibrator.

We know that pulse width (t_p),

$$5 \times 10^{-6} = 0.69\ R_3 \cdot C_1 \qquad \text{... } (i)$$

Let us select $C_1 = 0.001$ mF (*i.e.*, 1 nF), then, from equation (*i*)

$$5 \times 10^{-6} = 0.69 \times R_3 \times (0.001 \times 10^{-6})$$

$$\therefore \quad R_3 = \frac{5 \times 10^{-6}}{0.69 \times (0.001 \times 10^{-6})} = 7246.4\ \Omega$$

$$= 7.246 \text{ kW } \textbf{Ans.}$$

Note. In the above example, the choice of 'C' should be such that the value of R_3 turns out to be in kΩ. The reason for this is that we want the current in the circuit to be in milliamperes.

32.12. Bistable Multivibrator

It is also called *flip-flop, Eccles-Jordan circuit, trigger-circuit or binary.* The bistable multivibrator is used for counting and storing of binary information in computer circuits. It also finds extensive application in the generation and processing of pulse type waveforms. The bistable multivibrator has *two stable states* and can stay in any one of these two states, indefinitely, as long as the power is supplied. It changes to other state only when it receives a trigger (a short direction pulse) from outside.

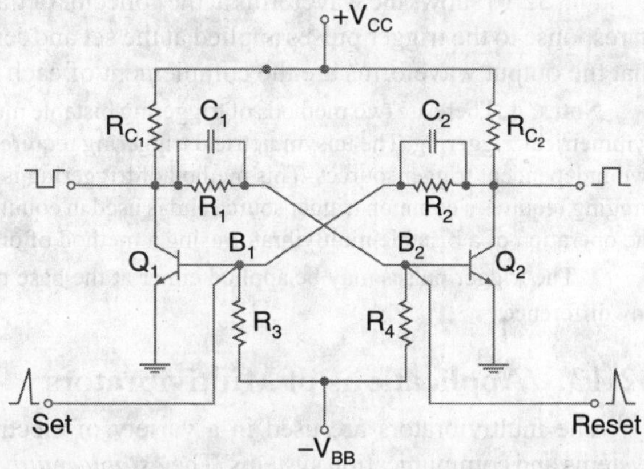

Fig. 32.15. Bistable multivibrator.

Fig. 32.15 shows the circuit of a bistable multivibrator using two NPN transistors. Here, the output of a transistor Q_2 is coupled to the base of transistor Q_1 through a resistor R_1. Similarly, the output of a transistor Q_1 is coupled to the base of transistor Q_2 through a resistor R_2. The capacitors C_1 and C_2 are known as *speed up* capacitors. Their function is to increase the speed of the circuit in making abrupt transition from one stable state to another stable state. The base resistors (R_3 and R_4) of both the transistors are connected to a common source ($-V_{BB}$). The output of a bistable multivibrator is available at the collector terminal of both the transistors Q_1 and Q_2. However, the two outputs are the complements of each other.

The operation of the circuit may be explained as given below:

1. When V_{CC} supply is switched on one of the transistor will start conducting more than the other. Then because of the feedback action, this transistor will be driven into saturation and the other to cut-off. Let us assume that the transistor Q_1 is in saturation (*i.e.*, ON) and Q_2 is cut-off (*i.e.*, OFF). It is a stable state of the circuit and will remain in this state, till a trigger pulse is applied from outside. A negative pulse applied to the set input will turn OFF the transistor Q_1 and Q_2 to ON. A similar action can also be achieved by applying a positive pulse at reset input.
2. Suppose a positive pulse is applied at the reset input. It will cause the transistor Q_2 to conduct. As the collector voltage of Q_2 falls, it cuts-off the transistor Q_1. Thus the circuit switches to the other stable state *i.e.*, a state where Q_1 is OFF and Q_2 is ON.

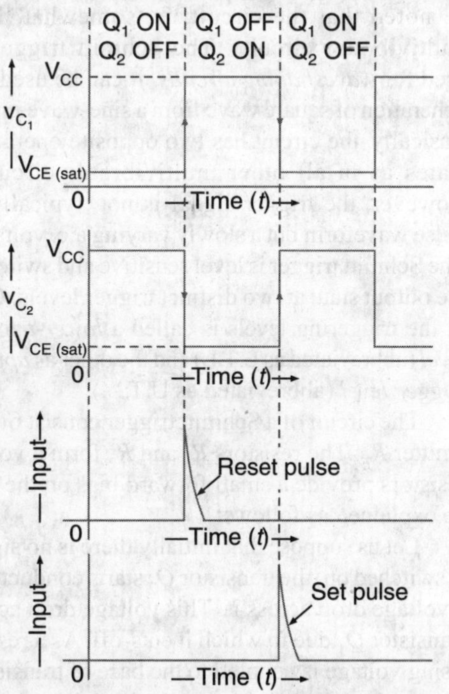

Fig. 32.16. Waveforms at the collector of transistor Q_1 and Q_2.

3. Now if a positive pulse is applied at the set input, it will switch the circuit back to its original stable state *i.e.*, Q_1 ON and Q_2 OFF.

Fig. 32.16 shows the waveforms at the collector of transistor Q_1 and Q_2 of the bistable multivibrator in response to the trigger pulses applied at the set and reset input. It is evident from these waveforms, that the output waveforms are the complement of each other.

Notes: **1.** There are two methods of triggering bistable multivibrators namely unsymmetrical triggering and symmetrical triggering. The unsymmetrical triggering requires two trigger inputs called set and reset. These are two independent trigger sources. This method of triggering is used in computer logic circuits. The symmetrical trigging requires a common trigger source and is used in counter circuits. It may be noted that we have explained the operation of a bistable multivibrator using a method of unsymmetrical triggering.

2. The trigger pulses may be applied either at the base or collector of the transistors, as it does not make any difference.

32.13. Applications of Multivibrators

The multivibrators are used in a variety of electronic circuits such as in computers, control systems and communication systems. The *astable multivibrators are used as square-wave generators, voltage-to-frequency converters and in pulse synchronization, as clock for binary logic signals etc.* The monostable multivibrators are used for the generation of *well defined pulses,* the logic design of pulse delay, variable pulse width etc. The bistable multivibrators are used as memory elements in shift registers, counters, etc.

32.14. Schmitt Trigger

Fig. 32.17 shows the circuit of a Schmitt trigger (after the name of its inventor). It may be noted that the circuit, is, somewhat, like multivibrator circuits. The Schmitt trigger is used for *waveshaping circuits.* It can be used for generation of square wave from a sine-wave input. Basically, the circuit has two opposite operating states as in all other multivibrator circuits. However, the trigger signal is not, typically, a pulse waveform but a slowly varying a.c. voltage. The Schmitt trigger is level sensitive and switches the output state at two distinct trigger levels. One of the triggering levels is called a *lower-trigger level* (abbreviated as L.T.L) and the other as *upper-trigger level* (abbreviated as U.T.L.).

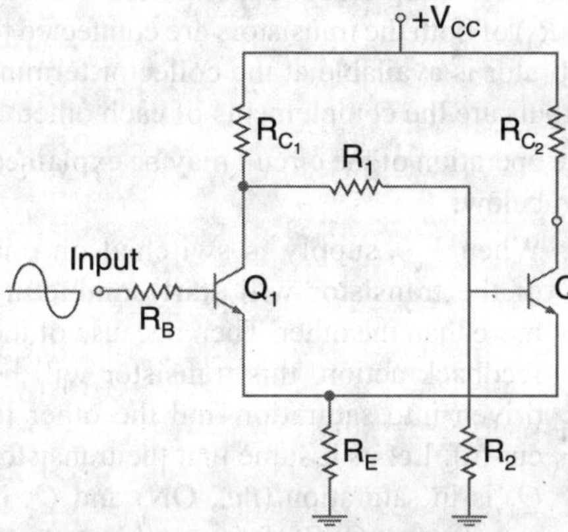

Fig. 32.17. Schmitt trigger.

The circuit of a Schmitt trigger consist of two identical transistors Q_1 and Q_2 coupled through an emitter R_E. The resistors R_1 and R_2 form a voltage divider across the V_{CC} supply and ground. These resistors provide a small forward bias on the base of transistor Q_2. The operation of the circuit may be explained as follows:

Let us suppose that initially there is no signal at the input. Then as soon as the power supply V_{CC} is switched on, the transistor Q_2 starts conducting. The flow of its current through resistor R_E produces a voltage drop across it. This voltage drop acts as a reverse bias across the emitter-base junction of transistor Q_1 due to which it cuts-off. As a result of this, the voltage at its collector rises to V_{CC}. This rising voltage is coupled to the base of transistor Q_2 through the resistor R_1. It increases the forward bias at the base of transistore Q_2 and therefore drives it into saturation and holds it there. At this instant, the collector voltage, levels are $v_{C_1} = V_{CC}$ and $v_{C_1} = v_{CE(sat)}$ as shown in Fig. 32.18.

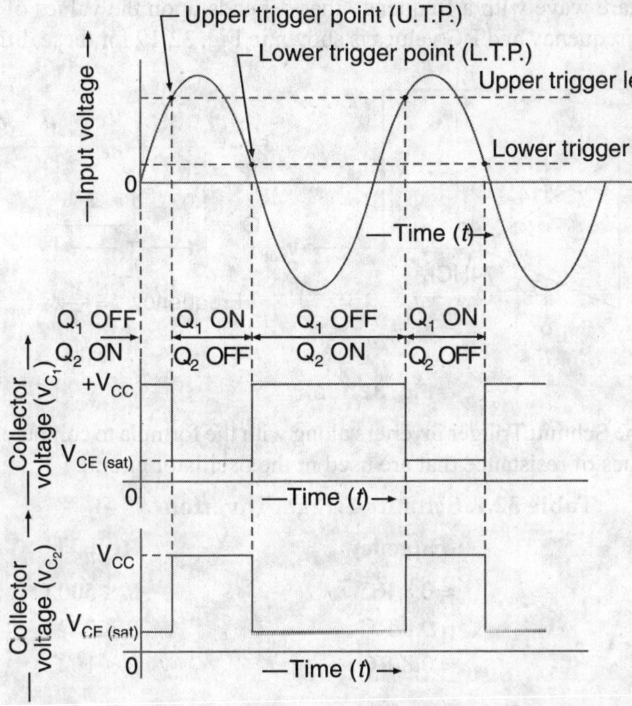

Fig. 32.18. Waveforms at the input and collector of transistors Q_1 and Q_2.

Now suppose an a.c. signal is applied at the input of the Schmitt trigger (*i.e.*, at the base of the transistor Q_1). As the input voltage increases above zero, nothing will happen until it crosses the upper trigger level (U.L.T.). As the input voltage increases, above the upper-trigger level, the transistor Q_1 conducts. *The point, at which it starts conducting, is known as upper-trigger point* (U.T.P.). As the transistor Q_1 conducts, its collector voltage falls below V_{CC}. This fall is coupled through resistor R_1 to the base of transistor Q_2 which reduces its forward bias. This in turn reduces the current of transistors Q_2 and hence the voltage drop across the resistor R_E. As a result of this, the reverse bias of transistor Q_1 is reduced, and it conducts more. As the transistor Q_1 conducts more heavily, its collector current further reduces due to which the transistor Q_2 conducts near cut-off. This process continues till the transistor Q_1 is driven into saturation and Q_2 into cut-off. At this instant, the collector voltage levels are $V_{C1} = V_{CE\,(sat)}$ and $V_{C2} = V_{CC}$ as shown in the figure.

The transistor Q_1 continue to conduct *till the input voltage falls below the lower-trigger level* (L.T.L.). It will be interesting to know that when the input voltage becomes equal to the lower-trigger level, the emitter-base junction of transistor Q_1 becomes reverse biased. As a result of this, its collector voltages starts rising towards V_{CC}. This rising voltage increases the forward bias across transistor Q_2, due to which it conducts. *The points, at which transistor Q_2 starts conducting is called lower-trigger point* (L.T.P.). Soon the transistor Q_2 is driven into saturation and Q_1 to cut-off. This completes one cycle. The collector voltage levels at this instant are $V_{C1} = V_{CC}$ and $V_{C2} = V_{CE(sat)}$. No change in state will occur during the negative half cycle of the input voltage.

It is evident from the above discussion that the output of a Schmitt trigger is a positive going pulse, whose width depends upon the time during which transistor Q_1 is conducting. The conduction time is set by the upper- and lower-trigger levels.

These days Schmitt trigger is also available in the form of integrated circuit (IC) Fig. 32.19 shows a schmitt trigger inverter connected as an oscillate (*i.e.* as an astable multivibractor) The signal at

V_{out} is an aproximate square wave with a frequency that depends upon the values of R and C. The relationship between the frequency and RC values is shown in Fig. 32.19 for these different Schmitt trigger integral circuits.

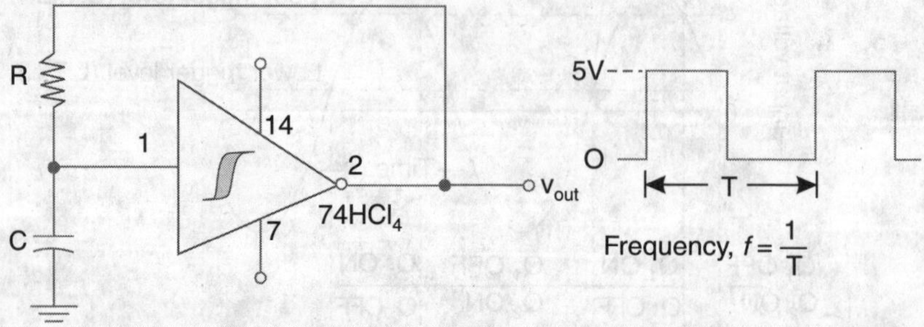

Fig. 32.19.

Table 32.1 shows some Schmitt Trigger inverters along with the formula to compute the frequency of oscillation and the values of resistance that are used in the oscillator circuit.

Table 32.1. Schmitt Trigger Inverters

IC	Frequency	Remark
7414	≈ 0.8/RC	R ≤ 500 Ω
74LS14	≈ 0.8/RC	R ≤ 2 Ω
74HC14	≈ 0.8/RC	R ≤ 10 Ω

As seen from Table 32.1 for 7414, Schmitt trigger invertier, the output frequency is given by $f \approx 0.8 / RC$. Notice that for this IC, the value of $R \leq 500\ \Omega$. For 74 LS 14 IC, the output frequency is again $0.8 / RC$ but the value of $R \leq 2\ k\Omega$. On the other hand for 74 HC 14, the square were output frequency is given by $1.2/RC$ where the value of $R \leq 10\ m\Omega$.

Example 32.10.

Design a Schmitt trigger circuit to have $V_{CC} = 12$ V, UTP = 5 V, LTP = 3 V and I = 2 mA using two silicon NPN transistors with $h_{FE\,(min)} = 100$ and $I_2 = 0.1.\ I_{C2}$.

(Anna University., Nov/Dec 2007)

Solution.

Given: $UTP = 5$ V; $LTP = 3$ V and $V_{CC} = 12$.V.

We know that, $V_i = V_{B_2} = UTP = 5$ V when Q_2 is "ON"

$$V_E = V_i - V_{BE_1} = V_{B_2} - V_{BE_2} = 5 - 0.7 = 4.3 \text{ V}$$

Let $I_{C_2} = I_{E_3} = 1$ mA ...in "ON" state

Now $I_{C_2} R_{C_2} = V_{CC} - V_E - V_{CE2(sat)}$ ...Let $V_{CE2(sat)} = 0.2$ V

$$(1 \times 10^{-3}) \times R_{C_2} = 12 - 4.3 - 0.2$$

$$R_{C_2} = 7.5\ k\Omega$$

Now $I_2 = 0.1\, I_{C_2} = 0.1 \times (1 \times 10^{-3}) = 1 \times 10^{-4}$ A

$$R_2 = \frac{V_{B_2}}{I_2} = \frac{5}{1 \times 10^{-4}} = 50\ k\Omega$$

$$I_{B_2} = \frac{I_{C_2}}{h_{fe\,(min)}} = \frac{1 \times 10^{-3}}{100} = 10\ \mu A$$

$$I_2 + I_{B_2} = \frac{V_{CC} - V_{B_2}}{R_{C_1} + R_1}$$

$$R_{C_1} + R_1 = \frac{12 - 5}{1 \times 10^{-4} + 10 \times 10^{-6}} = 63.64 \times 10^3 \quad ...(i)$$

Now $V_{B_2} = V_{B_1} = LTP = 3$ V and Q_1 is "ON"

$$I_1 = \frac{V_{B_2}}{R_2} = \frac{3}{50 \times 10^3} = 60\ \mu\text{A}$$

and $I_{C_1} \approx I_{E_1} = \frac{V_{B_1} - V_{BE_1}}{R_E} = \frac{3 - 0.7}{4.3 \times 10^3} = 5.348 \times 10^{-4}$ A

$$V_{CC} = R_{C_1}(I_{C_1} + I_1) + I_1(R_1 + R_2) \quad ...(ii)$$

or $V_{CC} = I_{C_1}R_{C_1} + I_1(R_{C_1} + R_1) + I_1R_2 \quad ...(iii)$

Substituting the values of $R_{C1} + R_1$ from equation (*i*) into equation (*ii*), we get,

$$12 = 5.348 \times 10^{-4} \times R_{C_1} + (60 \times 10^{-6}) \times (63.64 \times 10^3) + (60 \times 10^{-6}) \times (50 \times 10^3)$$

Simplifying the above equation, we get,

$$R_{C1} = 9.69\ \text{k}\Omega$$

$$R_1 = 63.64 \times 10^3 - 9.69 \times 10^3 = 53.95\ \text{k}\Omega$$

Thus when is "ON" $V_o = V_{CC} - I_{C2(ON)}R_{C_2} = 12 - 1 \times 10^{-3} \times 7.5 \times 10^3 = 4.5$ V

and when Q_2 is "OFF", $V_o = V_{CC} = 12$ V

The design of Schmitt trigger circuit is shown in Fig. 32.20 below :

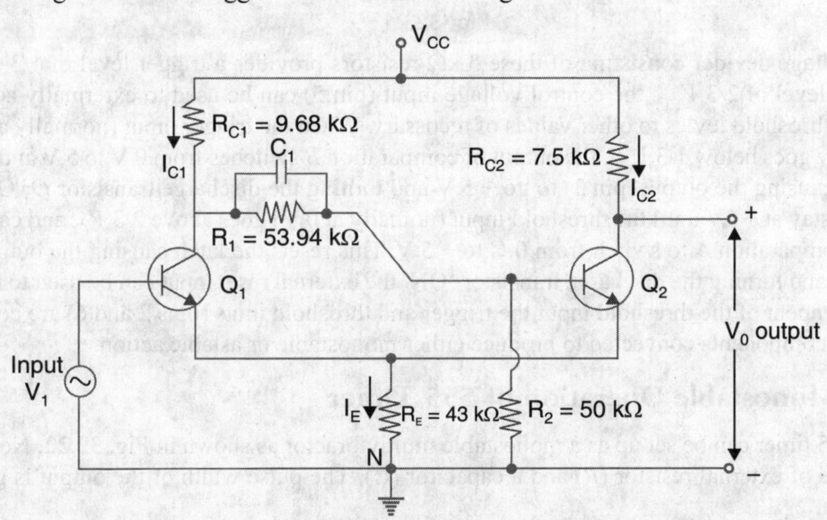

Fig. 32.20.

32.15. The 555 Timer

The 555 Timer is a very popular IC used in electronic industry. This IC can be used in the monostable multivibratior made as well as astable multivibrator made. let us study this device in detail.

Fig. 32.21 shows the functional block diagram of a 555 timer. As seen from this diagram, it has two comparators labelled A and B, an R - S flip-flop labelled as latch, an invated output buffer and these resistors of 5 kW each. the name 555 timer has come from the three resistor of 5 kΩ each within the timer circuits. The comparators are the devices whose output are + 5 V when the voltage on the positive (+) input is greater than the voltage on the negative (–) input and 0 V when the negative (–) input is greater than the voltage on the positive (+) input.

A timer 555 IC

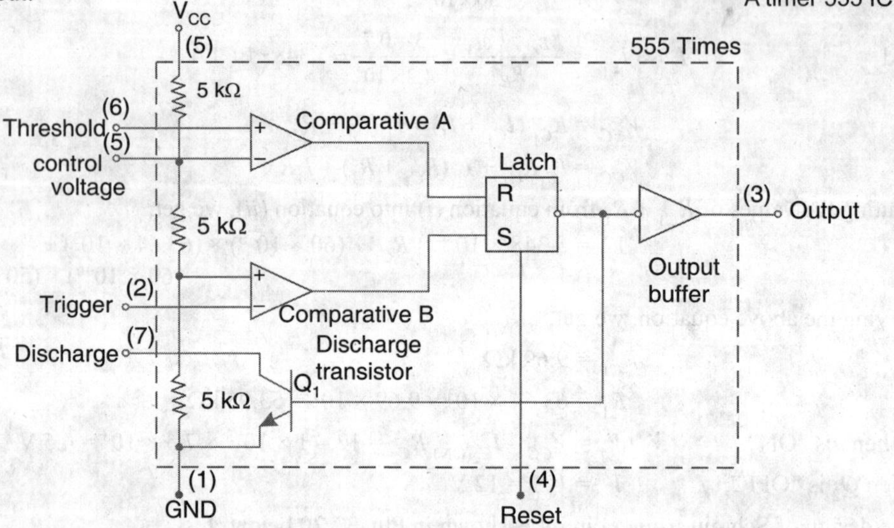

Fig. 32.21.

The voltage devider consisting of these 5 kΩ resistors provides a trigger level of 1/3 V_{CC} and a threshold level of 2/3 V_{CC}. the control voltage input (pin 5) can be used to extermally adjust the trigger and throshold levles to other values of recossary. When the trigger input (normally at + 5 V) momenatrily goes below 1/3 V_{CC}, the output of comparation B switches from 0 V to 5 V and sets the S–R latch, causing the output (pin 3) to go + 5 V and turning the discharge transistor Q_1 OFF, The output will stay at + 5V until the threshold input (nomally at 0 V) goes above 2/3 $V_{CC}$and causes the output of comparation A to switch from 0 C to + 5 V. This resets the latch causing the output to go back to 0 V and turning the discharge transistor 'ON' the external reset input can be used to reset the latch independent of the threshold input the trigger and throshold inuts (pins 2 and 6) are controlled by external components convected to produce either monostable or astable action.

32.16. Monostable Operation of 555 Timer

The 555 timer can be set up as a monostable multivibractor as shown in Fig. 32.22. Notice that it makes use of external resistor (R) and a capacitor (C). The pulse width of the output is given by the equation,

$$t_p = 1.1\, R_1\, C_1$$

As seen from Fig. 32.22, the control voltage input is not used and is connected to a decopling capacitor, C_2 to prevent noise from affenting the trigger and throshold levels.

The operation of this circuit may be explained as follows. Before a trigger pulse is applied, the output is 0 V and the discharge transistor Q_1 is ON, keeping C_1 discharged as shown in Fig. 32.22. When a negative going trigger pulse is applied at to, the output goes to + 5 V and the discharge.

Transistor turns OFF, allowing capacitor C_1 to being charging through R_1 as shown in Fig. 32.24.

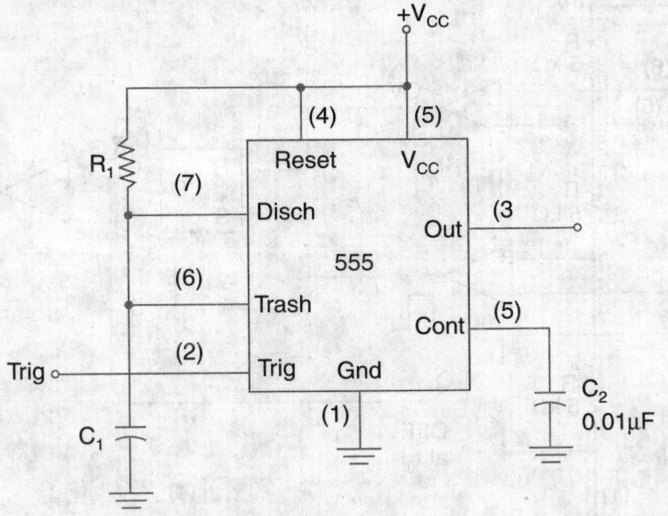

Fig. 32.22. 555 Timer prior to triggering.

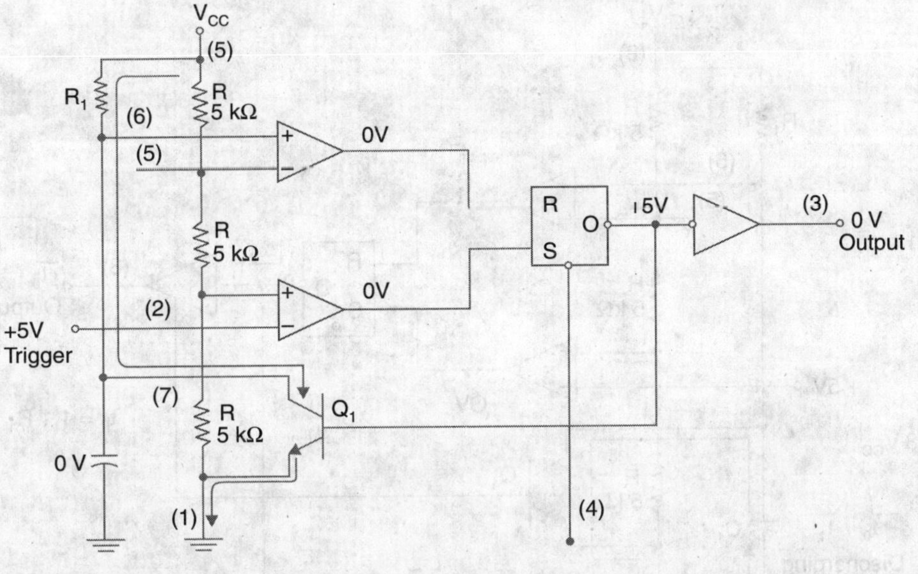

Fig. 32.23. 555 Timer when triggering.

When C_1 charges to 1/3 V_{CC}, the output goes back to 0 V at t_1 and Q_1 turns ON immediately discharging C_1 as shown in Fig. 32.25 it is obvions that the charging rate of C_1 determines long the output is at + 5 V level.

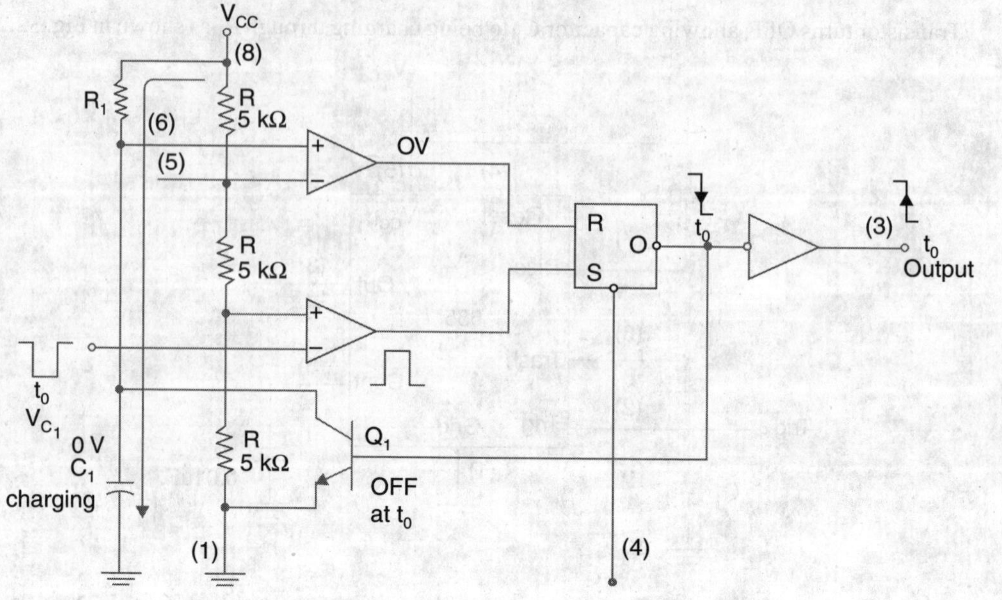

Fig. 32.24. 555 Timer when triggered.

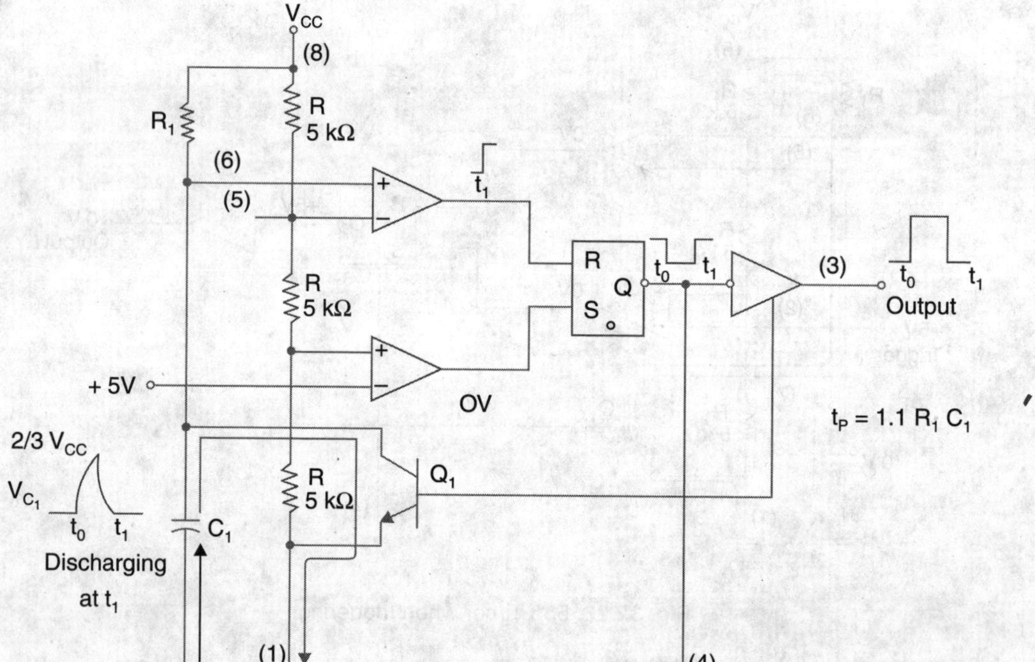

Fig. 32.25. 555 Timer at the end of charging interval.

Example 32.11.

Determine the value of pulse width for a 555 timer operating in a monostable mode with $R_1 = 2.2\ k\Omega$ and $C_1 = 0.01 \mu F$.

Solution.

Given: $R_1 = 2.2 \text{ k}\Omega$ and $C_1 = 0.01 \ \mu\text{F}$

We know that the pulse width

$$t_p = 1.1 \ R_1 \ C_1 = 1.1 \times (2.2 \times 10^3) \times (0.01 \times 10^{-6}) \text{ s}$$

$$= 24.2 \times 10^{-6} \text{ s} = 24.2 \text{ ms } \textbf{Ans.}$$

Example 32.12.

Design a circuit using 555 timer that will stretch a 1 μs negative going pulse that occurs every 60 μs into a 10 μs negative going pulse. Select C = 1000 pF.

Solution.

Given: A negative going pulse of 1 μs duration that occurs every 60 μs. Refer to Fig. 32.25.

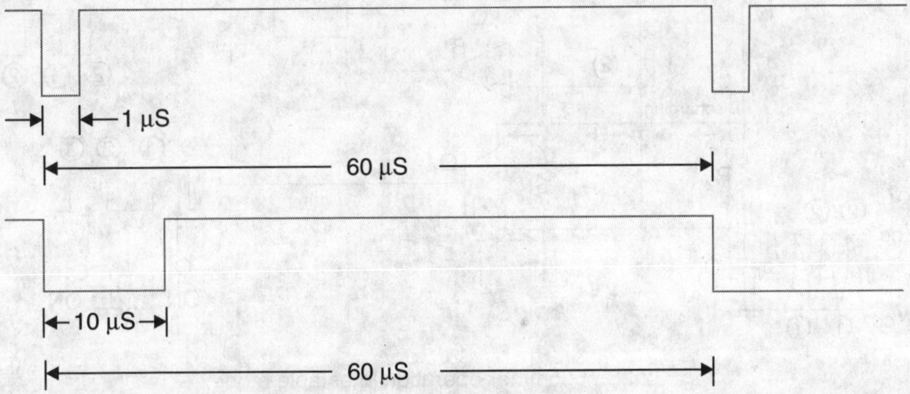

Fig. 32.26.

We know that the pulse width (t_p)

$$10 \times 10^{-6} = 1.1 \times R_1 \times (1000 \times 10^{-12})$$

$$\therefore \quad R_1 = \frac{10 \times 10^{-6}}{1.1 \times (1000 \times 10^{-12})} = 9.09 \times 10^3 \ \Omega$$

or $9.09 \text{ k}\Omega$ **Ans.**

32.17. Astable Operation of 555 Timer

Fig. 32.27 shows an arrangement of 555 Timer connected as an astable multivibrator. Notice that the throshold input (pin 6) is non connected to the trigger input (pin 2). The external components R_1, R_2 and C_1 form the tuning network that sets the frequency of oscillation. The 0.01 μF capacitor, C_2 connected to the control (pin 5) input is strictly for decoupling and has no effect on the circuit. operation In same cases, it can be left off even.

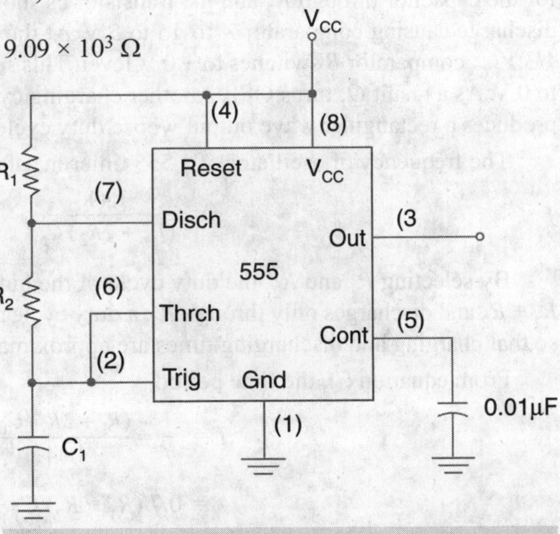

Fig. 32.27.

The circuit operation may be explained as follows. Initially when the power is turned ON, the capacitor (C_1) is uncharged and thus the trigger voltage (Pin 2) is at OV. This causes the output of comparator B to be at + 5 V and the output of comparation A to be at OV forcing the output of the latch and the base of Q_1 to be at OV. As a result, the transistor Q_1 is OFF Now C_1 begins charging through R_1 and R_2 this situation is shown in Fig. 32.28.

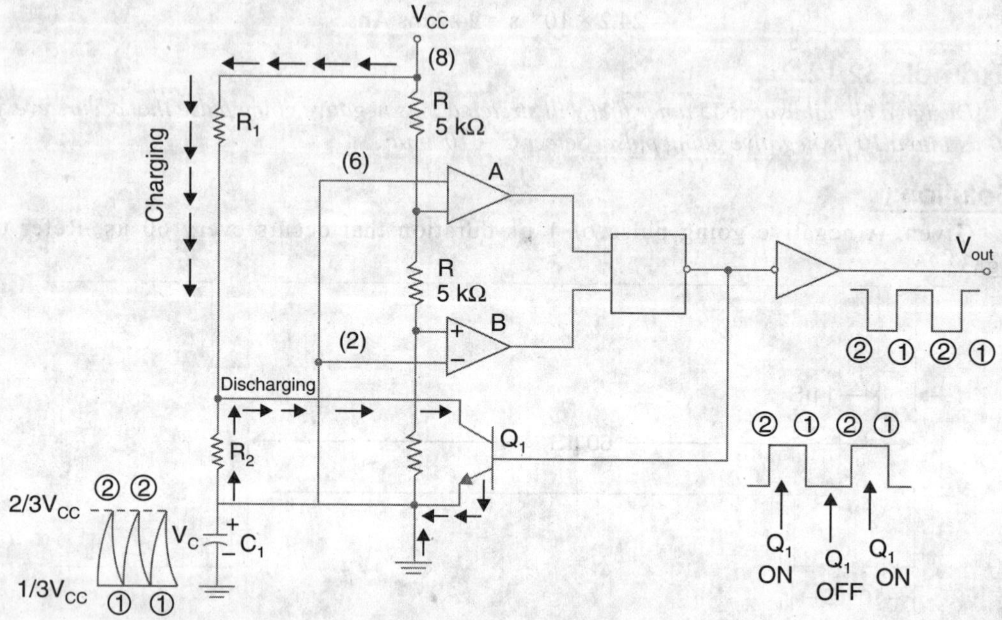

Fig. 32.28. 555 timer operation in astable mode.

When the capacitor voltage reaches 1/3 V_{CC}, comparator B switches to its O output state and when the capacitor reaches 2/3 V_{CC}, comparator A switches to its + 5 V state. This resets the latch causing the base of Q_1 to go + 5 V. As a result Q_1 turns ON. This sequence creates a discharge path for the capacitor through R_2 and the transistor as shown in Fig. 32.28. The capacitor now begins to discharge causing comparator A to go to 5 V. At the point where the capacitor discharges down to 1/3 V_{CC}, comparator B switches to + 0 V level. This sets the latch, making the base of Q_1 to ge down to 0 V. As a result Q_1 turns OFF. Another charging cycle begins and the entire process repeats. This produces a rectangular wave output whose duty cycle depends on the values of R_1 and R_2.

The frequency of oscillator of a 555 timer in astable mode is given by the expression,

$$f = \frac{1.44}{(R_1 + 2R_2)\, C_1} \qquad ...(i)$$

By selecting R_1 and R_2, the duty cycle of the output can be adjusted Since C_1 charges through $R_1 + R_2$ and discharges only through R_2, a duty cycle approaching 50% can be achieved of $R_2 >> R_1$ so that charging and discharging times are approximately equal.

From equation (*i*), the time period,

$$T = \frac{1}{f} = \frac{(R_1 + 2R_2)\, C_1}{1.44} \frac{(R_1 + R_2)\, C_1}{1.44} + \frac{R_2\, C_1}{1.44}$$

$$= 0.7\,(R_1 + R_2)\, C_1 + 0.7\, R_2\, C_1$$

$$= t_1 + t_2$$

Notice that t_1 is the charging time and t_2 is the discharging time. These timings are shown in Fig. 32.29.

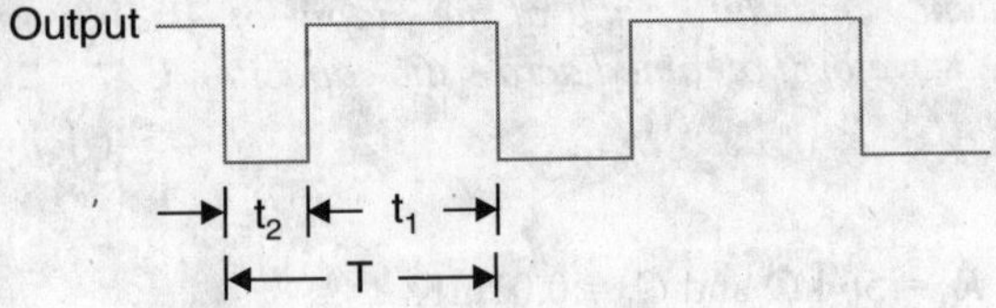

Fig. 32.29.

Example 32.13.

Fig. 32.30 shows the 555 timer in an astable mode. Determine the values of t_1, t_2 and duty cycle of the output waveform.

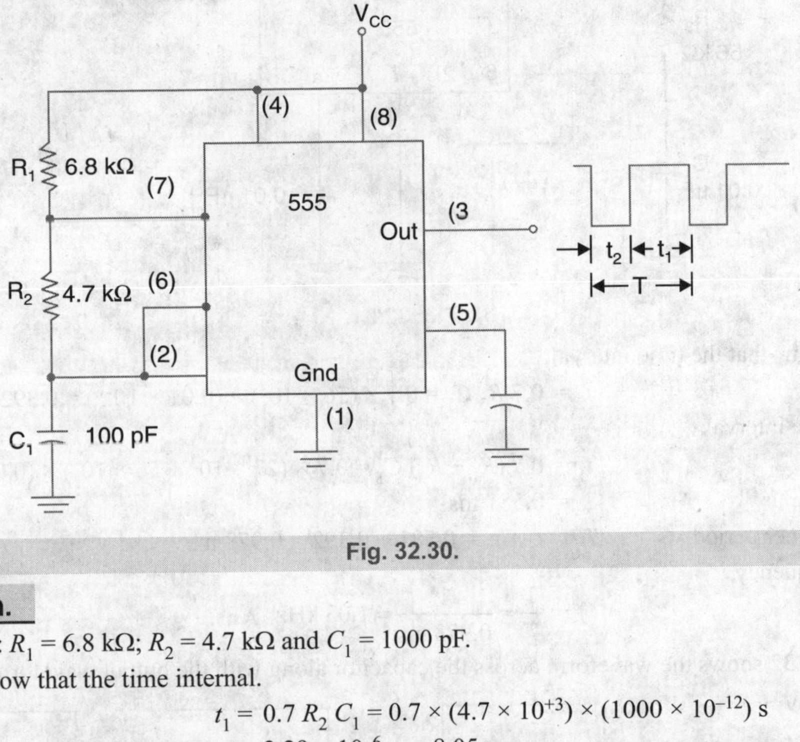

Fig. 32.30.

Solution.

Given: $R_1 = 6.8$ kΩ; $R_2 = 4.7$ kΩ and $C_1 = 1000$ pF.

We know that the time internal.

$$t_1 = 0.7\ R_2\ C_1 = 0.7 \times (4.7 \times 10^{+3}) \times (1000 \times 10^{-12})\ \text{s}$$
$$= 3.29 \times 10^{-6}\ \text{s or } 8.05\ \text{ms.}$$

and the time interval

$$t_1 = 0.7\ (R_1 + R_2)\ C_1$$
$$= 0.7\ [(6.8 \times 10^3) + (4.7 \times 10^3)] \times (1000 \times 10^{-12})$$
$$= 8.05 \times 10^{-6}\ \text{s or } 8.05\ \text{ms.}$$

We also known that duty cycle of the waveform,

$$\text{duty cycle} = \frac{t_1}{T} \times 100 = \frac{8.05\ \mu s}{(t_1 + t_2)} \times 100$$

$$= \frac{8.05\ \mu s}{8.05 + 3.29\ \mu s} \times 100 = 71\%\ \textbf{Ans.}$$

Example 32.14.

Calculate the frequency and duty cycle of the output of Astable multivibractor shown in Fig. 32.31 and draw the waveform obtained across the capacitor, C_1.

(*Bangalore University 2000*)

Solution.

Given: $R_1 = 27$ kΩ; $R_2 = 56$ kΩ and $C_1 = 0.01$ μF.

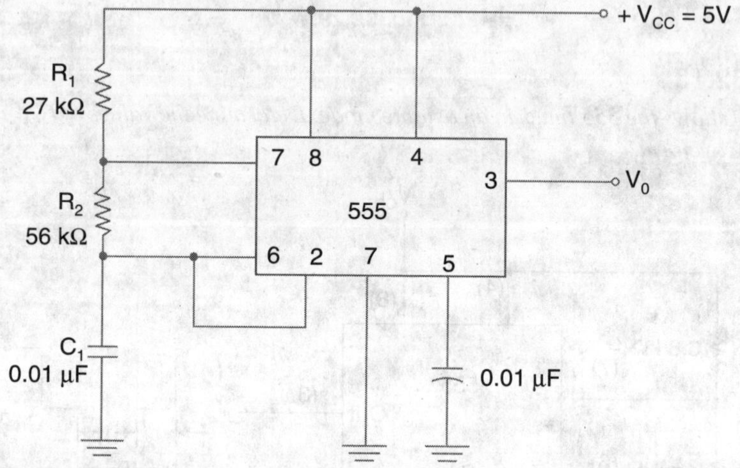

Fig. 32.31.

We know that the time interval,

$$t_2 = 0.7\,R_2\,C_1 = 0.7 \times (56 \times 10^3) \times (0.01 \times 10^{-6}) = 0.392 \text{ ms}$$

and the time interval,

$$t_1 = 0.7\,(R_1 + R_2)\,C_1 = 0.7 \times (27 \times 10^3 + 56 \times 10^3) \times (0.01 \times 10^{-6})$$
$$= 0.581 \text{ ms}$$

∴ Timer period, $T = t_1 + t_2 = 0.581 + 0.392 = 0.973$ ms

and the frequency,

$$f = \frac{1}{T} = \frac{1}{0.973 \text{ ms}} = 1.03 \text{ kHz} \quad \textbf{Ans.}$$

Fig. 32.32 shows the waveform across the capacitor along with the output waveform.

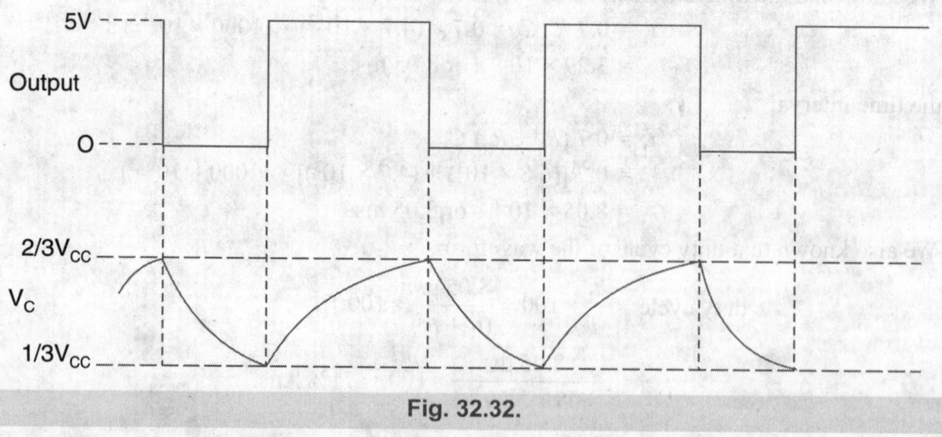

Fig. 32.32.

Example 32.15.

Design an astable multivibractor that will oscillate at 50 kHz, 60% duty cycle . Select C = 0.002 μF.

Solution.

Given: $f = 50$ kHz $= 50 \times 10^3$ Hz; duty cycle = 60%, C = 0.0022 μ*F*.

We know that the time period,

$$T = \frac{1}{F} = \frac{1}{50 \times 10^3} = 0.02 \times 10^{-3} \text{ } s \text{ or } 0.02 \text{ } ms$$

and the duty cycle,

$$0.6 = \frac{t_1}{T} = \frac{t_1}{0.02 \times 10^{-3}}$$

$$\therefore \quad t_1 = 0.012 \times 10^{-3} \text{ s or } 0.012 \text{ ms}$$

We also know that the time period, (T)

$$0.02 \text{ ms} \times t_1 + t_2 = 0.012 + t_2$$

$$\therefore \quad t_2 = 0.02 - 0.012 = 0.008 \text{ ms}$$

Now the time duration (t_2)

$$0.008 \times 10^{-3} = 0.7 R_2 C_1 = 0.7 \times R_2 \times (0.0022 \times 10^{-6})$$

$$\therefore \quad R_2 = \frac{0.008 \times 10^{-3}}{(0.7 \times 0.0022 \times 10^{-6})} = 5.19 \text{ k}\Omega \textbf{ Ans.}$$

The time duration (t_1),

$$0.012 \times 10^{-3} = 0.7 (R_1 + R_2) C_1 = 0.7 (R_1 + 5.19 \text{ k}) \times (0.0022 \times 10^{-6})$$

$$\therefore \quad R_1 = \left(\frac{0.012 \times 10^{-3}}{0.7 \times (0.002 \times 10^{-6})} - 5.19 \, k \right) = (7.79 \, k - 5.19 \, k)$$

$$= 2.6 \text{ k}\Omega \textbf{ Ans.}$$

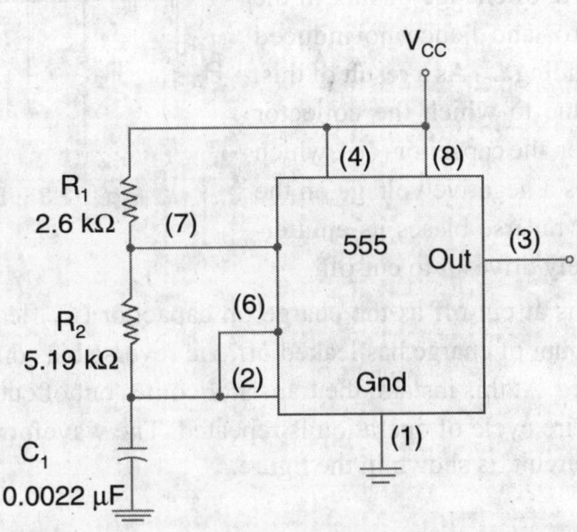

Fig. 32.33.

Fig. 32.33 shows the circuit to produce so kHz square wave with 60% duty cycle.

Note : The 555 timer can be used to generate a timing signal for a stepper motor shown in the picture.

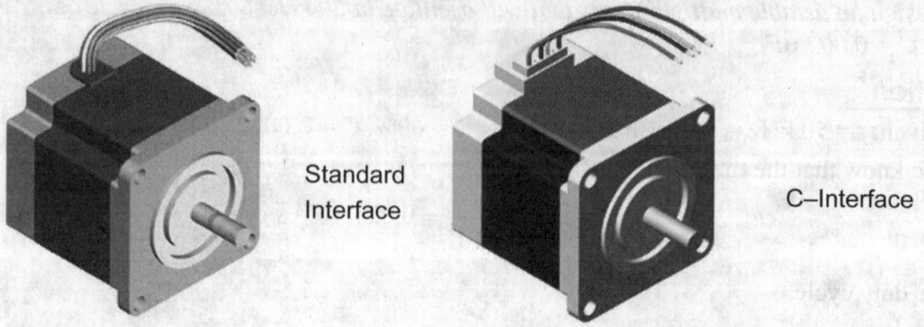

32.18. Blocking Oscillator

It is an oscillator, which is used to *produce high-current pulses of short duration.* Fig. 32.34 shows the circuit of a blocking oscillator. This circuit consists of bipolar junction transistor (*Q*), a *pulse transformer (*T*), a diode (*D*), a capacitor (C_1) and resistors R_1, R_2 and R_3. The pulse transformer has been used to couple the output back to the input through a capacitor (C_1). The resistors R_1 and R_2 provide a small forward bias on the emitter-base junction of the transistor.

The operation of the **blocking oscillator may be understood from the position that when the power supply V_{CC} is switched on, the base voltage rises rapidly. When it becomes higher than the cut-off value (0.5 V for silicon transistor), it causes the collector current to increase. The rising collector current, through the winding L_1, induces a voltage in winding L_2 of the pulse transformer. This makes the emitter-base junction of the transistor more forward biased due to which it conduct more. Therefore the collector current further increases. Due to the regenerative action, the transistor is driven into saturation. At this point, the collector current is maximum and no further increase in its value is possible. Therefore the rate of change of flux in the winding L_1 becomes zero, and hence no induced voltage is produced in winding L_2. As a result of this, the base current drops due to which the collector current also falls. Moreover, the capacitor (C_1), which had charged earlier, places a negative voltage on the base of the transistor and reverse biases its emitter-base junction and ultimately drives it to cut-off.

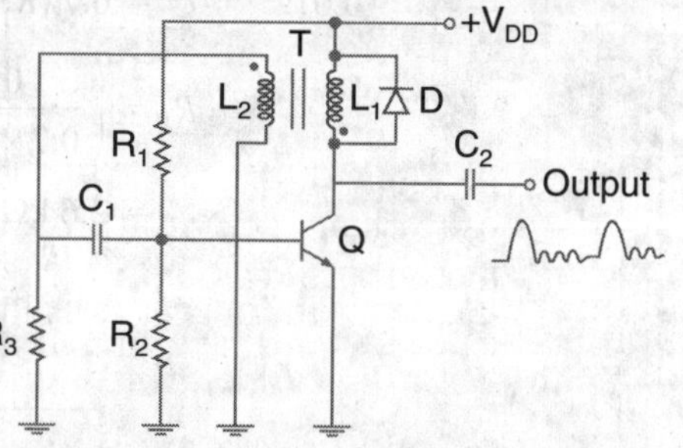

Fig. 32.34. Blocking oscillator.

The transistor remains at cut-off as the charge on capacitor (C_1) leaks away through resistor R_3. When a sufficient amount of charge has leaked off, the reverse bias on the transistor is removed and is again forward biased. At this instant, the transistor comes out of cut-off and starts conducting again. In this way, the entire cycle of operation is repeated. The waveform of the narrow and short pulses, generated by the circuit, is shown in the figure.

* A pulse transformer is ferrite cored transformer to handle fast waveforms.

** The blocking oscillator can also be used to operate in a monostable mode.

NUMERICAL PROBLEMS

1. What is the frequency of an astable multivibrator having the component values of $R_1 = R_2 = R = 10$ kΩ and $C_1 = C_2 = C = 1000$ pF? **(Ans. 72.5 kHz)**
2. Determine the value of resistors to be used to provide a symmetrical oscillation of 100 kHz, if $C_1 = C_2 = C = 330$ pF. **(Ans. 22 kΩ)**
3. Determine the period and frequency of oscillation for an astable multivibrator with component values of $R_1 = R_2 = 4.7$ kΩ and $C_1 = C_2 = 270$ pF. **(Ans. 1.75 ms and 57.1 kHz)**
4. Select the value of R and C for the monostable multivibrator to give an output pulse width of 400 ns. **(Ans. Choose C = 100 pF, then R = 5.78 kΩ)**
5. Refer to Fig. 32.35. For the self-biased binary NPN transistor, worst case values of

$$V_{CC(sat)} = 0.4 \text{ V}, V_{BE(sat)} = 0.8 \text{ V}$$

and zero base -to-emitter cut off. The circuit parameters are $V_{CC} = 20$ V, $R_1 = 30$ kΩ, $R_e = 390$ Ω, $R_C = 4.7$ k, $R_2 = 15$ k.

(*a*) Find stable state current and voltages.

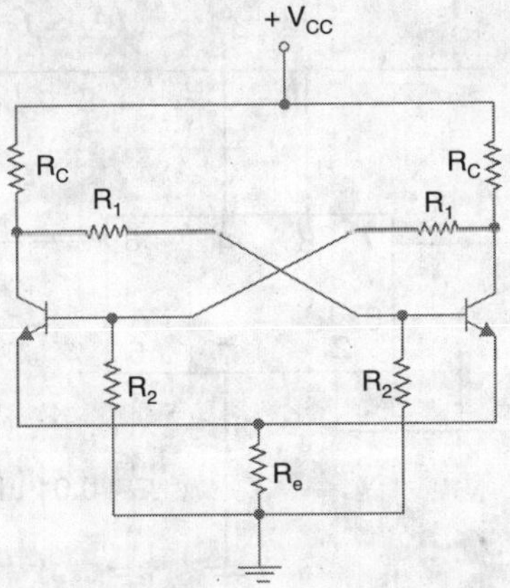

Fig. 32.35.

(*b*) Find min. value of h_{FE} required to give the values in part (*a*).

(*c*) As the temperature is increased, what is the max. value to which I_{CBO} can increase before the condition is reduced where neither transistor is off ?

(*Nagpur Univ. Winter 2003*)

6. The fixed bias binary uses PNP Si transistors with worst case (maximum) values of $V_{CE(sat)} = -0.5$ V, $V_{BE(sat)} = -1.0$ V, $I_{CBO} = -10$ nA at 25°C and zero base to emitter voltage at cut-off. The circuit parameters are $V_{CC} = V_{BB} = 6$ V, $R_C = 1.2$ k, $R_1 = 4.7$ k and $R_2 = 27$ k.

(*a*) Find $h_{FE(min)}$ and verity that one transistor is OFF and the other is ON. Find the stable-state currents and voltages.

(*b*) If the reverse saturation current doubles every 10° C, what is the maximum temperature at which one transistor will remain OFF ? (*Nagpur University. Winter 2003*)

7. The emitter coupled monostable multivibractor uses PNP transistor with $h_{FE} = 40$. Neglect the voltage of a forward biased junction and assume $r_{bb}' = 0$. The circuit parameters are : $V_{CC} = R_{C_1}$ 12 V, = 3.6 k, R_{C_2} = 2.2 k, $R_e = 1$ k and $R = 68$ k.
The bias V is adjusted to be midway between V_{min} and V_{max} A trigger is applied at $t = 0$ Calculate all the voltage levels and plot the waveforms. (*Nagpur Univ. Winter 2003*)
8. The self-biased binary used NPN silicon transistors having worst case (maximum) values of $V_{CE(sat)} = 0.4$ V and $V_{BE(sat)} = 0.8$V and zero base-to-emitter voltage for cut-off. The usual circuit parameters of $V_{CC} = 20$ V, $R_C = 4.7$ kΩ; $R_1 = 30$ kΩ; $R_2 = 15$ kΩ and $R_C = 390$ Ω.

(a) Draw a neat circuit and insert the parameters indicated and find the stable state currents and voltages.
(b) Find the minimum values of h_{FE} required to give the values in part (a).
(c) As the temperature is increased, what is the maximum values to which I_{CBO} can increases before the condition is reached where neither transistor is OFF?

(AMIE, Winter 1993)

9. Two monostables A and B triggered simultaneously produces pulse widths T_A and T_B respectively. Give a suitable circuit which will give the output pulse width $|T_A - T_B|$. Also give a circuit which will indicate through LEDs which monostable has a longer pulse width. *(AMIE., Summer 2000)*

10. A schmitt trigger circuit is to be designed with a VTP of 5V and LTP of 3 V. The NPN silicon transistors employed have $h_{FE(min)} = 100$ and I_C to be 2 mA. The available supply voltage is 12 V. Design a suitable circuit. *(AMIE., Summer 1999)*

11. Calculate the frequency and duty cycle of the output of Astable multivibractor and drawn the waveform obtained across the capacitor, C_1 Refer to Fig. 32.36.

(Bangalore Univ. B.Sc April/May 2001)

(Ans. f = 85 kHZ, duty cycle = 59.5%.)

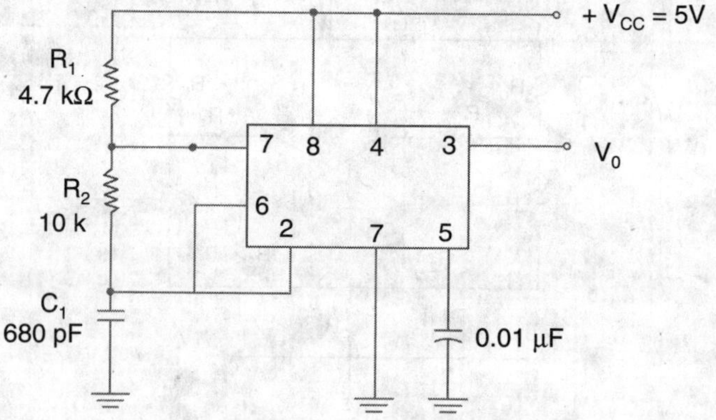

Fig. 32.36.

12. Explain the working of the astable circuit show in Fig. 32.37 Assume that the saturation voltages for the Op-Amp are $\pm V_{CC}$ Derive an expression for the output pulse width of the monostable. Drawn the waveform when triggers are applied at times t_1 and $t_2 < t$. Suggest a suitable trigger circuit of an NPN transistor is used for the switch, S.

(AMIE., Summer 2000)

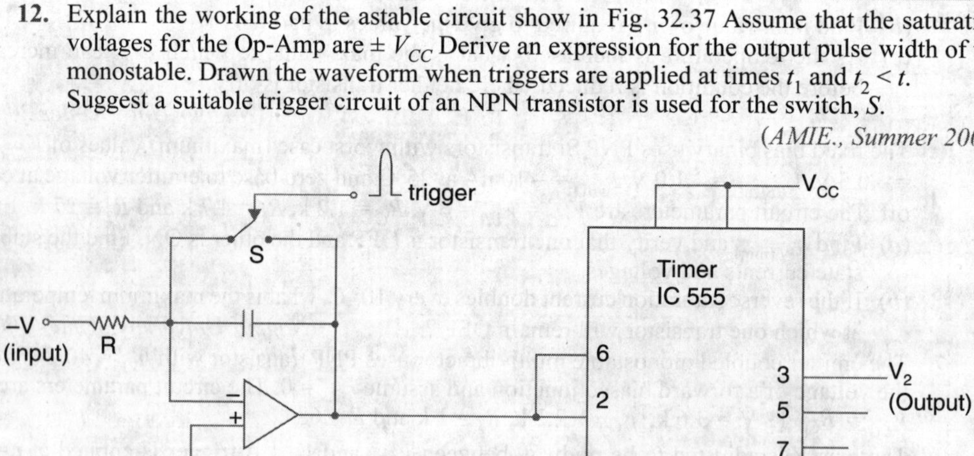

Fig. 32.37.

DESCRIPTIVE QUESTIONS

1. What is a nonsinusoidal oscillator? Explain it briefly.
2. Explain the switching action of a transistor.
3. What is a multivibrator? Explain the difference between the three types of multivibrators.
4. With a neat sketch, explain the working of an astable multivibrator.
5. Draw the switching waveforms for the astable multivibrator.
6. What is monostable multivibrator? Explain its working with the help of waveforms.
7. With a neat sketch, explain the operation of a bistable multivibrator.
8. How a Schmitt trigger is different from a multivibrator?
9. What is a blocking oscillator? Explain how it can be used to generate high-current pulses of short duration.
10. Draw any one type of a Schmitt trigger circuit and explain with waveforms.
11. Drawn the circuit of an emitter coupled transistor monostable multivibrator and explain its operation. Show how this multivibractor can be modified to control the width of the output pulse. *(AMIE., Summer 1999)*
12. Drawn the circuit of an astable multivibractor and explain its working. *(AMIE., 1996)*
13. Drawn the circuit of a Schmitt trigger and explain its working Show how the hysteresis loop in the circuit stabiliges the operation of schmitt trigger. *(AMIE., Winter 1996)*
14. Explain the working of a self-biased transistor bistable multivibractor circuit. What is the purpose of the speed-up capacitors ? *(AMIE., Winter 1998)*
15. Explain the working of a collector-coupled monostable multivibractor. Give expression for its gate width. *(AMIE., Winter 1998)*
16. Draw the circuit of a Schmitt trigger and explain its principle of operation. Discuss how the hysteresis loop in the circuit stabiliges the operation of the schmitt trigger. *(AMIE., Winter 1997)*
17. (*a*) Distinguish between a bistable and a monostable multivibractor
    (*b*) Discuss the principle of operation of a transistor bistable multivibractor
    (*c*) Discuss the principle of transistor bistable multivibractor. *(AMIE., Winter 1997)*
18. Explain the operation of an astable multivibractor. *(AMIE., Summer 1995)*
19. Explain the hysteresis loop and hysteresis range associated with schmitt trigger. How can the hysteresis voltage be reduced in the emitter coupled schmitt trigger? *(AMIE., Summer 1995)*
20. Explain with the circuit diagram, the collector coupled monostable multivibractor. *(AMIE., Summer 1995)*
21. Sketch an emitter-coupled transistor Schmitt trigger circuit using NPN transistor and briefly explain its operation. Define the terms : upper trigger point (UTP), lower-trigger point (LTP) and hysteresis related to the Schmitt trigger circuit. *(AMIE., Summer 1997)*
22. Draw the circuit diagram of a collector-coupled mono and explain its working. *(AMITE., Summer 1996)*
23. Explain the working of self-biased transistor bistable multivibractor. What is a retriggeiable monoshot ? *(AMIE., Summer 1996)*
24. (*a*) Explain the operation of a bistable multivibractor
    (*b*) Give a typical triggering circuit by bistable multivibractor. *(AMIE., Summer 1994)*
    (*c*) Draw the circuit diagram of a collector coupled astable multivibractor using transistors and explain its principle of operation. Draw the waveforms at the base and collector of one transistor. *(AMIE., Winter 2001)*

25. (*a*) Draw the schematic of DC based monostable circuit. Explain how the output pulse width is controlled Also examine the concept of positive and negative triggering in respect of input-output relations in a monostable multivibractor.
26. (*a*) Compare a Schmitt trigger with a binary circuit.
    (*b*) Draw a neat circuit diagram of a schmitt trigger and explain the various methods of minimizing the hysteresis in the circuit.
    (*c*) What are delay lines ? Describe their application areas. (*AMIE., Winter 1994*)
27. (*a*) Draw the internal operational circuit diagram of an IC timer 555.
    (*b*) Explain the operation of an IC 555 as an astable multivibractor.
    (*c*) Derive an expression for frequency of the output pulse waveform of the circuit.
    (*d*) What are the two comparator reference voltage of IC 555 if $V_{CC} = +9$ volts. (*AMIE., Winter 1997*)
28. (*a*) Sketch a transistor Schmitt trigger circuit and briefly explain its operation giving input output waveforms.
    (*b*) Draw the circuit diagram of a 555 based astable multivibractor. Explain its operation. Also discuss about the changes in the circuit to generate a 50% duty cycle square waveform. (*AMIE., Winter 2000*)
29. (*a*) Draw an operational internal circuits of a timer IC 555.
    (*b*) Show that the timer works as a Schmitt trigger when pins 2 and 5 are tied-together to an input voltage (v_i) and the output (v_o) is taken at pin 3.
30. (*a*) Use a suitable integrator circuit in the feedback path of the above schmitt trigger to make a pulse generator.
    (*b*) Derive an expression for the frequency of the output pulse waveform. (*AMIE., Summer 1997*)
31. Explain with the aid of a neat circuit diagram, the working of a triggered blocking oscillator with base timing. Derive an expression for the pulse duration, t_p. Enlist the drawbacks of the circuit. (*AMIE., Winter 1997, 1998*)
32. (*a*) Explain with a neat circuit diagram, how a blocking oscillator can be used to produce high current pulses of short duration.
    (*b*) Enumerate the various methods of triggering a bistable multivibractor and describe one such method fully.
    (*c*) Explain how a BJT can be used as a switch. (*AMIE., Winter 1994*)
33. Write short notes on transistor switching times. (*Anna University., April/May 2008*)
34. Draw the response of a transistor to a pulse and mark all the delays which affect the switching speed. (*Anna University., May/June 2006*)
35. Discuss "Junction Diode Switching Time" to justify diode reverse recovery time, storage time, and transition time in brief. Elaborate these definitions by labeling appropriate waveforms for diode. (*GBTU/MNTU., 2002-03*)
36. Define rise time. (*Anna University., Nov/Dec 2007*)
37. Define rise time. (*Anna University., Nov/Dec 2007*)
38. Define duty cycle. (*Anna University., May/June 2007*)
39. Give the relation between rise time and bandwidth. (*Anna University., May/June 2007*)
40. What is an astable multivibrator? What is its application? (*M.G University., July/Aug. 2007*)
41. With the circuit, explain the working of astable multi-vibrator. Give its final output wave form. (*Bhartiya University., April 2004*)
42. Draw and explain the circuit of a BJT astable multivibrator with the help of waveforms (*M.G University., May 2007*)

43. What is the other name of Astable multivibrator. *(Anna University., Nov/Dec 2007)*
44. Explain the working of Emitter coupled Astable multivibrator. *(Anna University., April/May 2008)*
45. Draw and explain the circuit of astable multivibrator and hence draw the waveforms at salient points. Derive expressions for frequency of oscillation. *(Anna University., Nov/Dec 2007)*
46. With neat circuit diagram explain the working of astable multivibrator. Derive its time period. Draw the base and collector waveforms. *(Anna University., Nov/Dec 2006)*
47. List the various applications of an astable multivibrator. *(M.G University., May/June 2006)*
48. Explain the operation of collector coupled astable multivibrator and also derive the expression of T for a collector astable multivibrator. *(Anna University., May/June 2007)*
49. Draw the circuit of emitter coupled astable multivibrator for one of the quasi stable states and show the direction of currents. Mention the terminal voltage in the circuit. *(Anna University., May/June 2006)*
50. Draw the circuit of a collector coupled transistor monostable multivibrator. Sketch the waveform at base and collector for each transistor when the circuit is triggered and explain its working *(Anna University., May/June 2006)*
51. Explain circuit used to trigger a transistor monostable multivibrator circuit. *(Anna University., May/June 2008)*
52. Draw the circuit of a collector coupled transistor astable multivibrator. Explain its operation and mention the disadvantages of the circuit. *(Anna University., Nov/Dec 2006)*
53. Draw the circuit of a collector coupled transistor astable multivibrator. Sketch the base and collector waveform and explain the circuit operation. *(Anna University., Nov/Dec 2007)*
54. Explain the methods of base triggering of a bistable multivibrator. *(Anna University., Nov/Dec 2006)*
55. State the role of commutating capacitors in bistable multivibrator *(Anna University., April/May 2008)*
56. Explain the operation of a mono stable multivibrator. *(Bhartiya University., April 2004)*
57. Draw and explain the circuit of a Monostable multivibrator with suitable triggering arrangements. Give the waveforms at different modes. *(Anna University., Nov/Dec 2007)*
58. Derive an expression for the pulse width of a Monostable multivibrator. *(Anna University., Nov/Dec 2007)*
59. State any two applications of Monostable multivibrator. *(Anna University., Nov/Dec 2007)*
60. What is the necessity for damping in a monostable-blocking oscillator? *(Anna University., May/June 2007)*
61. With neat circuit diagram, explain monostable blocking oscillators with base and emitter timing. Draw necessary wave forms. *(Anna University., May/June 2007)*
62. With neat circuit diagram explain the working of monostable multivibrator. Derive its on time. Draw the base and collector signals. *(Anna University., May/June 2006)*
63. Give the application of bistable multivibrator. *(Anna University., May/June 2007)*
64. State the use of Schmitt trigger. *(Anna University., Nov/Dec 2007)*
65. With a neat circuit, explain the operation of Schmitt trigger. *(Anna University., April/May 2008)*

66. With circuit discuss Schmitt operation. Obtain expression for UTP and LTP. *(Anna University., April/May 2008)*
67. What is Schmitt trigger? Explain the operation of Schmitt using neat circuit diagram. Give the mathematical analysis of the Schmitt trigger. *(Anna University., May/June 2007)*
68. With neat diagram, explain the working of a Schmitt trigger circuit. Draw the corresponding wave forms. *(Anna University., May/June 2007)*
69. Describe pin diagram of 555 IC. *(PTU., May 2008)*
70. What is meant by IC timer ? How is IC 555 used to generate square waveform, explain ? *(PTU., Dec 2005)*
71. Discuss the operation of IC 555 as a monostable multivibrator. Draw the waveform and explain. *(Anna University., April/May 2008)*
72. Discuss the working of a self biased bistable multivibrator. *(Anna University., April/May 2008)*
73. Explain the different methods of triggering monostable multivibrator. *(Anna University., April/May 2008)*
74. Define sweep time with reference to a time base signal. *(Anna University., Nov/Dec 2006)*
75. With neat circuit diagram, explain the working of a monosatable blocking oscillator using emitter timing. Draw the equivalent circuit. Also derive the expression for pulse width. *(Anna University., May/June 2007)*
76. Give the circuit schematic OFF of 555 timer connected as an astable multivibrator. Describe its operation. *(VTU., June /July 2008)*
77. How is IC 555 used as a frequency divider ? Explain. *(PTU., Dec 2004)*
78. What is Blocking Oscillator? Give one application. *(Anna University., Nov/Dec 2007, April/May 2008)*
79. Draw a bistable multivibrator and give the design procedure for the same *(R.G.P.V., June 2008)*
80. Draw the circuit diagram of an astable multivibrator and explain in the working of the same. Also give its designing procedural steps. *(R.G.P.V., Dec. 2008)*
81. Explain the design considerations which are necessary to be considered while designing a bistable multivibrator. *(R.G.P.V., Dec. 2006)*
82. With a neat sketch, explain the working of astable multivibrator. *(R.G.P.V., Dec. 2007)*
83. Explain multivibrator circuits with their circuit diagram. *(R.G.P.V., May/ June 2006)*
84. Draw the circuit of bistable multivibrator. Explain its working with the help of wave forms. *(R.G.P.V., Dec. 2002)*
85. What are different types of multivibrator? Explain monostable multivibrator and its wave forms. *(R.G.P.V., June 2002)*
86. Draw the diagram and explain the working of monostable multivibrator. *(R.G.P.V., Dec. 2006, Dec. 2001)*
87. Explain the utility of 555 timer as a monostable multivibrator. *(RGTU., June 2009)*
88. Explain the working of monostable multivibrator using 555 timer. *(RGTU., March, April 2010)*
89. Explain the working of 555 timer as bistable multivibrator with the help of circuit diagram and waveforms. *(RGTU., Dec. 2010)*
90. Discuss the working principle of Bistable multivibrator using 555 timer. *(RGTU., Dec. 2009)*

MULTIPLE CHOICE QUESTIONS

1. When a transistor is driven into saturation, its collector voltage is
   (*a*) Zero (*b*) $V_{CE(sat)}$
   (*c*) V_{CC} (*d*) None of the above
2. An astable multivibrator is also called free-running relaxation oscillator because it
   (*a*) oscillates continuously between two unstable states
   (*b*) requires no d.c. power for its operation
   (*c*) runs and relaxes alternately
   (*d*) none of the above
3. The frequency of oscillation of an astable multivibrator depends on
   (*a*) transistor current gain (h_{FE})
   (*b*) value of collector load resistors
   (*c*) duration of the input pulse
   (*d*) values of *R* and *C*
4. A monostable multivibrator
   (*a*) gives one output pulse for one input trigger pulse
   (*b*) gives two output pulses for one input trigger pulse
   (*c*) has one stable state
   (*d*) both (*a*) and (*c*) above
5. A bistable multivibrator
   (*a*) has two stable states (*b*) has two stable states
   (*c*) oscillates between two stable states without any trigger pulse
   (*d*) is used for generating square wave
6. Non-sinusoidal waveforms
   (*a*) are departments from sine waveform (*b*) have low mark-to-space ratio
   (*c*) are much easier to generate (*d*) are unfit for digital operation
7. A relaxation oscillator is one which
   (*a*) has two stable states (*b*) relaxes indefinitely
   (*c*) produces non-sinusoidal output (*d*) oscillates continuously
8. A square pulse has a mark-to-space ratio of
   (*a*) 1 : 1 (*b*) 1 : 2 (*c*) 2 : 1 (*d*) 1 : 4
9. Which of the following statement is WRONG ? In a multivibrator
   (*a*) output is available continuously
   (*b*) feedback between two stages is 100%
   (*c*) positive feedback is employed
   (*d*) when one transistor is ON, the other is OFF
10. An monostable multivibractor circuit
    (*a*) has no stable state
    (*b*) gives two output pulsos for one input trigger pulse
    (*c*) returns to its stand-by states automatically
    (*d*) has no energy-storage element
11. A bistable multivibractor circuit
    (*a*) has two unstable states
    (*b*) has one energy-storage element
    (*c*) switches between its two states automatically
    (*d*) is not an oscillator

12. The frequency of oscillation of an astable multivibrator depends mainly on the
(*a*) value of collector load resistors (*b*) *RC* values of the circuit
(*c*) value of transistor β (*d*) width of the input pulse

13. An monostable multivibractor is frequently used
(*a*) in memory and timing circuits
(*b*) for producing triangular waves
(*c*) in counting circuits
(*d*) for regeneration of distorted waves

14. A bistable multivibractor is a .... oscillator.
(*a*) triggered (*b*) free-runnig (*c*) sine-wave (*d*) sawtooth

15. A blocking oscillator
(*a*) is a triggered oscillator
(*b*) is an amplifier with negative feedback
(*c*) generates sinusoidal waves
(*d*) produces sharp and narrow pulse

16. In an astable multivibractor
(*a*) $\beta = 1$ (*b*) $\beta A = 1$ (*c*) $\beta > 1$ (*c*) $\beta < 1$

17. In order to generate a square wave from a sinusoidal input signal one can use
(*a*) schmitt trigger circuit (*b*) clipper and amplifiers
(*c*) monostable multivibrators (*d*) both (*a*) and (*b*)
(*UPSC Engg. Services 2002*)

18. In every practical oscillator, the loop-gain is slightly larger than unity and the amplitude of the oscillations is limited by the
(*a*) magnitude of the loop gain
(*b*) onset of non-linearity
(*c*) magnitude of the gain of the amplifier
(*d*) feedback transmission factor. (*UPSC Engg. Services 2000*)

19. In an astable multivibrator, the base resistor of 100 kW and capacitors of 100 pF are used. The frequency of output will be
(*a*) 1.2 kHz (*b*) 5.5 kHz (*c*) 6.9 kHz (*d*) 12.8kHz
(*AMIE., Winter 1999*)

20. For a Schmitt trigger invertor the upper and lower trip voltage are 3 V and 1 V and higher and lower states are 15 V and 2 V. The output for a sinusoidal input of 10 V peak will be between
(*a*) 1 V and 3 V (*b*) 2V and 15 V (*c*) 3 V and 15 V (*d*) 10 V and 15 V
(*AMIE., Summer 1999*)

21. A transistor used as a switches is operated in
(*a*) active region
(*b*) cut-off region
(*c*) saturation region
(*d*) cut-off and saturation region. (*AMIE., Summer 2001*)

22. A 1 ms pulse can be stretched to 1 s pulse by using.
(*a*) an astable multivibractor
(*b*) a monostable multivibractor
(*c*) a monostable multivibractor
(*d*) a Schmitt trigger circuit (*UPSC Engg. Services 2000*)

23. A monostable multivibractor circuit show in Fig. 32.38. The value of C would be nearly.
(*a*) 0.001 mF (*b*) 0.01 mF (*c*) 0.1mF (*d*) 1 mF

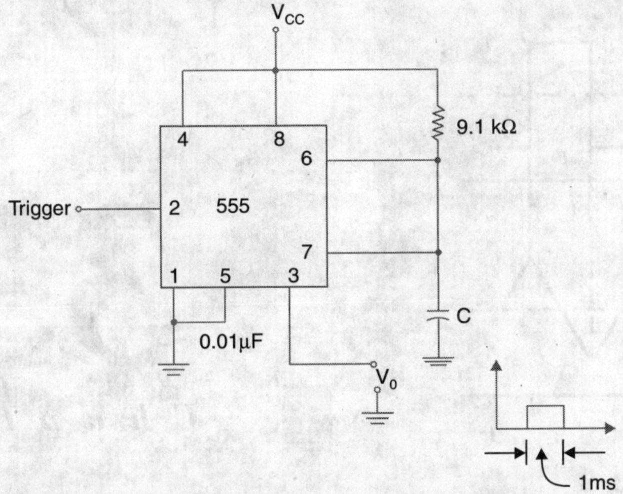

Fig. 32.38.

24. The circuit shown in Fig. 32.39 represent

(*a*) an astable multivibractor
(*b*) a monostable multivibrator
(*c*) voltage-controlled oscillator
(*d*) ramp generator

(*UPSC Engg Services 1999*)

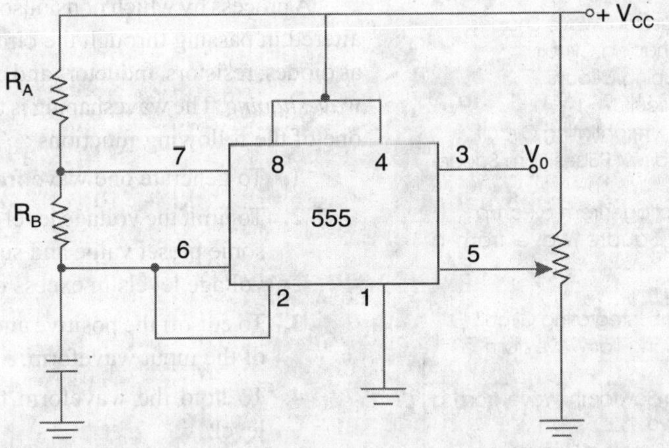

Fig. 32.39.

25. A Schottky diode is connected between the base and collector of the transistor to

(*a*) prevent cut-off
(*b*) prevent saturation
(*c*) decrease power dissipation
(*d*) none of these.

ANSWERS

1. (*b*)	**2.** (*c*)	**3.** (*d*)	**4.** (*d*)	**5.** (*a*)
6. (*a*)	**7.** (*c*)	**8.** (*a*)	**9.** (*a*)	**10.** (*c*)
11. (*d*)	**12.** (*b*)	**13.** (*d*)	**14.** (*a*)	**15.** (*d*)
16. (*a*)	**17.** (*a*)	**18.** (*c*)	**19.** (*c*)	**20.** (*b*)
21. (*d*)	**22.** (*b*)	**23.** (*c*)	**24.** (*c*)	**25.** (*b*)

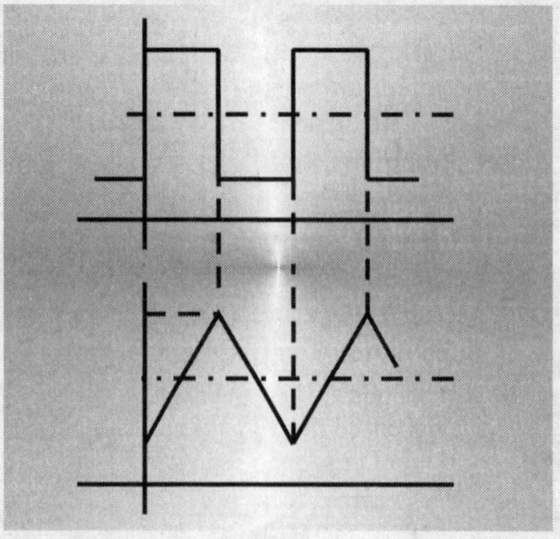

33

Chapter

Wave Shaping

Outline

33.1. Introduction

A process by which nonsinusoidal waveforms are altered in passing through the circuit elements (such as diodes, resistors, inductors and capacitors) is called *waveshaping*. The waveshaping is used to perform any one of the following functions :

1. To generate one wave from the other.
2. To limit the voltage level of the waveform to some preset value and suppressing all other voltage levels in excess of the preset level.
3. To cut-off the positive and negative portions of the input waveform.
4. To hold the waveform to a particular d.c. level.

The waveshaping is important in most of the signal process systems and is performed by the circuits known as **differentiators, integrators, limiters, clippers and clampers.**

The differentiator circuits are used to generate sharp narrow pulses either from distorted pulse waveforms or from rectangular waveforms. The integrator circuits are required to generate a voltage, which increases or decreases linearly with time (called a *ramp*). The limiters (also called amplitude limiters) are used to limit the voltage level of the waveform to some preset value and suppressing all other voltage levels in excess of this preset level. The limiters can also be used to protect a circuit against any possible damage due to excess voltage. These are an important part of signal processing circuit in television and radio receivers. The clipper circuit is required to cut-off the positive and negative portions of an input waveform. On the other hand, a clamper circuit is used to hold (or restore) the waveform to a particular d.c. level. The clamper circuits are also very useful in television receivers.

33.2. Different Non–sinusoidal Wave Shapes

We know that any waveform, whose shape is different from that of a standard sinusoidal wave, is called a nonsinusoidal waveform. Some of the important nonsinusoidal waveforms are *pulse*, *square*, *symmetrical square*, *triangular and sawtooth* etc. All these waveforms are a part of standard voltage or current signals such as step, ramp, and exponential and are discussed as below :

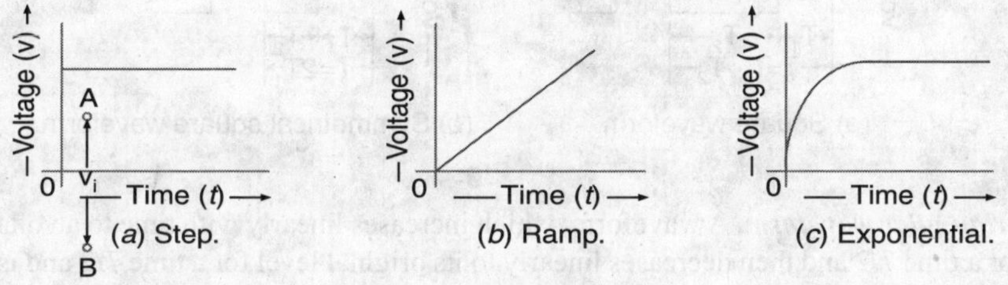

Fig. 33.1. Standard voltage signals.

1. ***Step voltage.*** A voltage, which maintains the zero value for all times $t < 0$ and maintains the value V for all times $t > 0$ is called a step voltage. The transition between the two voltage levels takes place at $t = 0$ and is accomplished in a very short interval. Thus as shown in Fig. 33.1 (*a*), $v = 0$ immediately before $t = 0$ and $v = V$ immediately after $t = 0$.
2. ***Ramp voltage.*** A voltage, which is zero for $t < 0$ and increases linearly with time for $t > 0$, is called a ramp voltage. It is also called a *sweep voltage*. Thus as shown in Fig. 33.1 (*b*), $v = 0$ immediately before $t = 0$ and $v = \alpha \cdot t$ after $t = 0$. Here is a constant of proportionality.
3. ***Exponential voltage.*** A voltage, which is zero for $t < 0$ and increases nonlinearily with time as shown in Fig. 33.1 (*c*), is called an exponential voltage.

Now we shall discuss the pulse, square, symmetrical square, triangular and sawtooth waveforms in more details.

(*a*) ***Pulse waveform.*** Fig. 33.2 (*a*) shows the waveform of an ideal pulse. The pulse amplitudes is V and the pulse duration is t_p. It is evident from Fig. 33.2 (*b*) and (*c*) that the pulse may be considered as the sum of the step voltage $+ V$, whose discontinuity occurs at $t = 0$ and a step voltage $- V$, whose discontinuity occurs at $t = t_p$. The pulse waveforms find extensive use in almost every field of electronics such as communication, computers, defence equipments etc.

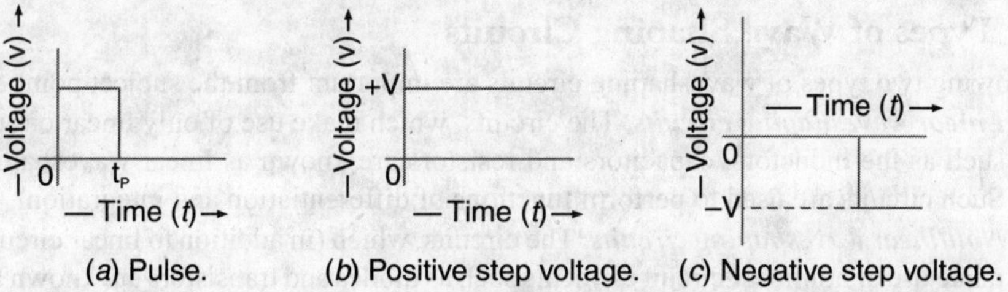

Fig. 33.2.

(*b*) ***Square waveform.*** A waveform, which maintains itself at one constant voltage level V_1 for time T_1 and at another constant level V_2 for time T_2 and is repetitive with a period $T = T_1 + T_2$ as shown in Fig. 33.3 (*a*) is called a square waveform. The square waveforms are used in digital electronic circuits, radars and as synchronising pulses in television.

(*c*) ***Symmetrical square waveform.*** A square waveform for which $T_1 = T_2 = T/2$ as shown in Fig. 33.3. (*b*) is called a square waveform. It may be noted that because of the symmetry, the voltage levels V_1 and V_2 are equal and opposite *i.e.*, $V_1 = - V_2$. The symmetrical square waveforms are very useful in digital electronic circuits.

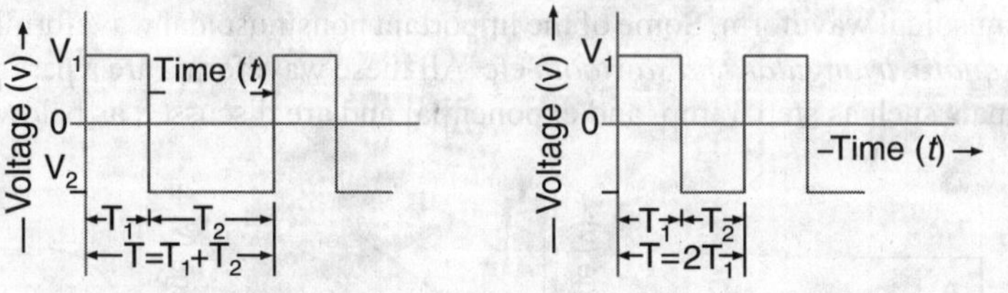

(*a*) Square waveform. (*b*) Symmetrical square waveform.

Fig. 33.3.

(*d*) ***Triangular waveform.*** A waveform, which increases linearly with time to a voltage level V for a time $T/2$ and then decreases linearly to its original level for a time $T/2$ and is repetitive with a period T as shown in Fig. 33.4 (*a*) is called triangular waveform. It may be noted from the figure, that a triangular wave may be considered as the sum of ramp voltage, which increases at a rate of $2\,V/T$ for a time $T/2$ and the ramp voltage which decreases at a rate of $-\,2V/T$ for the remaining time $T/2$. The triangular waveforms are used in scanning circuits, where a uniform left-to-right scan is required as in computer displays. These are also used in timing circuits for electronic applications.

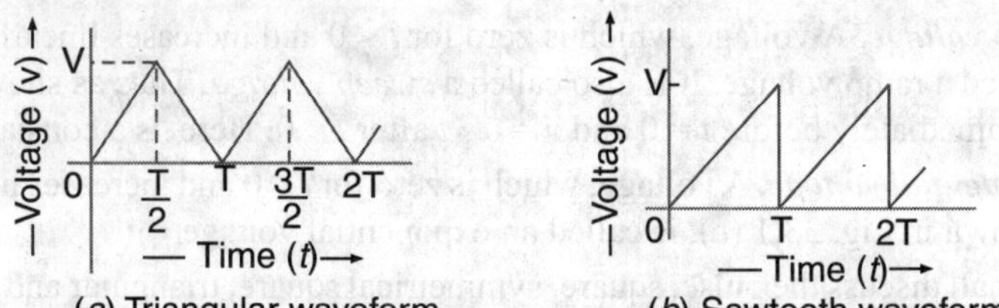

(*a*) Triangular waveform. (*b*) Sawtooth waveform.

Fig. 33.4.

(*e*) ***Sawtooth waveform.*** A waveform, which increases linearly with time to a voltage level V for a time T and then changes abruptly to its original level and is repetitive as shown in Fig. 33.4 (*b*) is called sawtooth waveform. It is also called *sweep waveform* or *time-base waveform*. The sawtooth waveforms are used in the scanning circuits of cathode ray oscilloscopes and televisions.

33.3. Types of Wave Shaping Circuits

Following two types of waveshaping circuits are important from the subject point of view :

1. ***Linear waveshaping circuits.*** The circuits, which make use of only linear circuit elements such as the inductors, capacitors and resistors are known as linear waveshaping circuits. Such circuits are used to perform functions of differentiation and integration.
2. ***Nonlinear waveshaping circuits.*** The circuits, which (in addition to linear circuit elements) make use of nonlinear circuit elements such as diodes and transistors are known as nonlinear wave shaping circuits. Such circuits are used to perform functions of amplitude limiting, clipping and clamping.

Now we shall discuss both the above mentioned types of wave shaping circuits one by one in the following pages.

33.4. Linear Wave Shaping Circuits

We have already discussed in the last article that the linear wave shaping circuits are those which make use of linear circuit elements such as the inductors, capacitors and resistors. Such circuits are used to perform functions of differentiation and integration. The linear wave shaping circuits use the transient properties of resistors, inductors and capacitors to realise the mathematical operations of differentiation and integration.

33.5. Differentiating Circuit

A circuit in which the output voltage is directly proportional to the derivative of the input voltage is called a differentiating circuit. Mathematically, the output voltage of a differentiating circuit,

$$v_o \propto \frac{dv_i}{dt}$$

$$= k.\frac{dv_i}{dt}$$

where k is a constant of proportionality.

A differentiating circuit may be realised by a simple RC series circuit as shown in Fig. 33.5. In this circuit, the output is taken across the resistor (R). The circuit is designed in such a way that reactance of the capacitor ($X_C = 1/2\pi f_C$) is very large as compared to the value of resistor (R) *i.e.*,

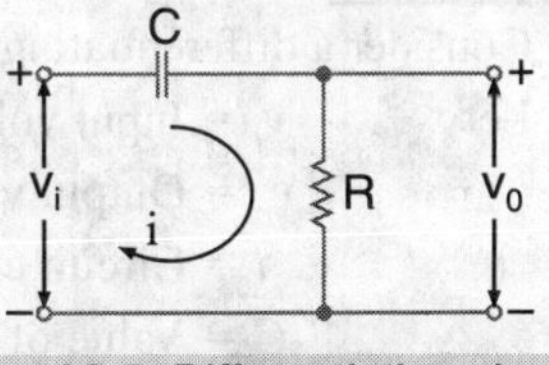

Fig. 33.5. Differentiating circuit.

$$X_C >> R$$

$$\frac{1}{2\pi f_C} >> R$$

or $$\frac{1}{2\pi RC} >> f \qquad \text{... } (i)$$

where f = Frequency of the input signal, and

C = Value of capacitance of the capacitor.

We know that the time period of the input signal,

$$T = \frac{1}{f}$$

∴ Equation (*i*) may be rewritten as

$$\frac{1}{2\pi RC} << \frac{1}{T}$$

$$2\pi RC << T$$

or $$\tau << T \qquad \text{... } (ii)$$

where $\tau = RC$ is called time constant of the circuit.

It is evident from equation (*ii*) that if time constant of the RC circuit (*i.e.*, $\tau = R.C$) is made much smaller than the time period of the input signal waveform, then the output of the circuit.

$$v_o = R.\,C.\,\frac{dv_i}{dt}$$

where $C \cdot dv_i/dt$ is the value of current through the circuit.

In actual practice, it has been found that a time constant of about 1 to 10 per cent of the time period of the input signal waveform is usually sufficient to provide good differentiation, *i.e.*,

$$\tau \leq 0.1\, T \qquad \text{...(For good differentiation)}$$

It will be interesting to know that in a differentiating circuit, the average (or d.c.) level of the output signal is always zero and it does not depend on the average level of the input signal. As a result of this, the output must extend in positive and negative directions, with respect to the zero-voltage axis, and must be equal to the area which is below the zero axis.

Sometimes, the differentiating circuit is also known as a high-pass filter. The name 'high-pass' filter is designated because the circuit blocks the low-frequencies and allows high-frequencies to pass through it. It is due to the reason that reactance of the capacitor decreases with the increasing frequency. As a result of this, at very high frequencies, the capacitor acts as a short circuit and virtually all the input appears at the output.

Example 33.1.

Prove that the average (or d.c.) value of the output of a differentiating circuit is zero.

(A.M.I.E., Summer, 1987)

Solution.

Consider a differentiating circuit as shown in Fig. 33.6.

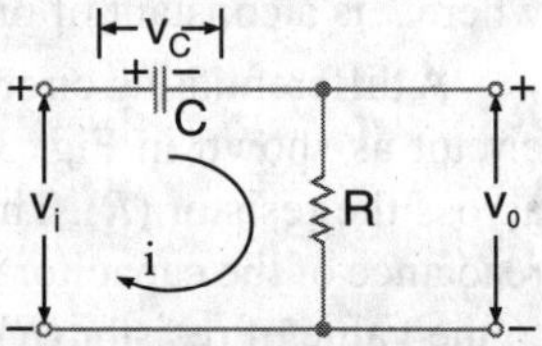

Fig. 33.6.

Let v_i = Input voltage,

v_o = Output voltage,

i = Circuit current,

C = Value of capacitance of the capacitor,

q = Charge on the capacitor, and

R = Value of resistor.

We know that voltage across the capacitor,

$$v_C = q/C$$

Applying Kirchoff's Voltage Law to the differentiating circuit,

$$v_i = v_C + v_o$$

$$v_i = \frac{q}{C} + v_o \qquad \text{...}\left(\because v_c = \frac{q}{C}\right)$$

Differentiating the above equation with respect to τ,

$$\frac{dv_i}{dt} = \frac{1}{C} \times \frac{dq}{dt} + \frac{dv_o}{dt}$$

The term dq/dt in the above equation represents the rate of flow of charge with time *i.e.*, the current i. Therefore replacing dq/dt by i in the above equation,

$$\frac{dv_i}{dt} = \frac{i}{C} + \frac{dv_o}{dt}$$

$$= \frac{v_o}{RC} + \frac{dv_o}{dt} \qquad \text{...}(\because i = v_o/R)$$

Multiplying the above equation on both sides by dt and integrating it over one period T,

$$\int_{t=0}^{T} dv_i = \int_{t=0}^{T} \left(\frac{v_o}{RC}\right) dt + \int_{t=0}^{T} dv_o$$

$$v_i(T) - v_i(0) = \frac{1}{RC}\int_{t=0}^{T} v_o \cdot dt + v_o(T) - v_i(0)$$

Under steady state conditions, the output waveform as well as the input signal is repetitive with a period T, so that $v_o(T) = v_o(0)$ and $v_i(T) = v_i(0)$. Therefore,

$$\frac{1}{RC}\int_0^T v_o \cdot dt = 0$$

or
$$\int_0^T v_o \cdot dt = 0$$

The above integral represents the area under the output waveform over one cycle. Since the value of the integral is zero, therefore the average level of the output signal is always zero.

33.6. Applications of a Differentiating Circuit

Following are some of the important applications of a differentiating circuit, which are important from the subject point of view :

1. To generate a series of narrow pulses (called pips) from the rectangular or square waveforms. The pips are used as trigger pulses or synchronization pulses in circuits used in television and cathode ray oscilloscopes.
2. To generate a step from a ramp input.
3. To generate a square wave from a triangular wave input.

All the applications are discussed in more details one by one in the following pages.

33.7. Generation of Narrow Pulses from Square Wave

Fig. 33.7 (*a*) shows the square waveform, which acts as input to the differentiating circuit. It is assumed, that time constant ($\tau = R \cdot C$) of the differentiating circuit is much smaller than the time period of the input waveform (*i.e.*, $\tau << T$). We know that a square waveform may be considered to be the sum of positive steps and negative voltage steps. Now, as a matter of fact a differentiating circuit for a step voltage input, should produce a sharp spike of infinite amplitude for an abrupt change and zero for a constant part of the step voltage at its output. It is true because, the exact derivative of a discontinuity is infinite and the derivative of a constant is zero. Thus under ideal conditions, the output of a differentiating circuit should be as shown in Fig. 33.7 (*b*). But in actual practice, we get an output as shown in Fig. 33.7 (*c*). Here, the output is shown to rise instantaneously to an amplitude of 2V and then decays down to zero. The difference in the ideal and the real outputs may be explained as below :

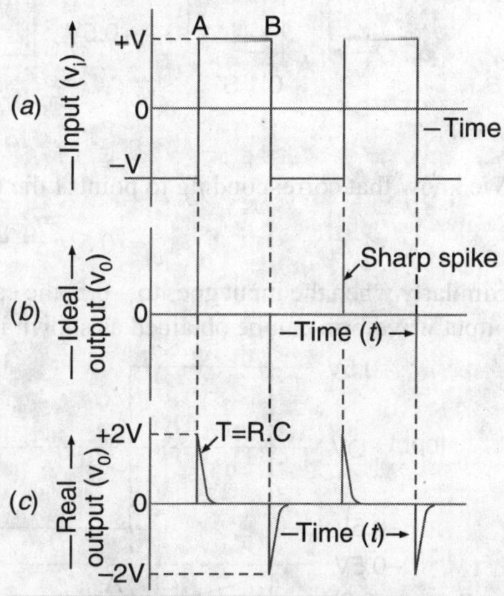

Fig. 33.7. Generation of narrow pulses from a square wave.

It has been observed that whenever, there is an abrupt change (*i.e.*, discontinuity) in voltage at the input of the capacitor, it acts as a short-circuit instantaneously and a full change in the input voltage appears across the output. It is because of the fact, that a capacitor requires an infinite charging current for the voltage across it to change instantaneously. Since the charging current remains finite, therefore it act as short-circuit and passes the step change (or discontinuity) in the input voltage as such to its output. Secondly, the capacitor blocks any constant (or d.c.) input voltage and hence does not allow it to reach at the output.

Thus with square wave input, when the input changes abruptly from a voltage level $-V$ to $+V$ (the total change being 2 V), the differentiating circuit produces it at its output as such. During the constant part *AB*, the output decays exponentially with a time constant τ (equal to $R \cdot C$). At a time $\tau = 5\,\tau$, the output decays to 0.6% of its initial value. Thus we can assume that the output has reached its steady state in a time interval equal to 5 τ.

Note. If a square wave is applied to a differentiating circuit whose time constant (t) is not much smaller as compared to the time period of the input waveform, the output is not a train of narrow pules. But it consists of a square wave whose upper and lower portions are tilted towards the zero-voltage axis. And if a differentiating circuit has $\tau \geq T$, the square wave will be passed through the circuit without any change.

Example 33.2.

A square-wave whose peak-to-peak value is 1V extends ± 0.5 V w.r.t. ground. The half period is 0.1 sec. This voltage is impressed upon an RC differentiating circuit whose time constant is 0.2 sec. What are the steady state max. and min. values of the output voltage.

(Nagpur Univ. Winter 2003)

Solution.

Given: Square-wave input with peak-to-peak value = 1V. Half-period ($T/2$) = 0.1 s. Time constant of RC circuit (τ) = 0.25. Refer to Fig. 33.8 for better view.

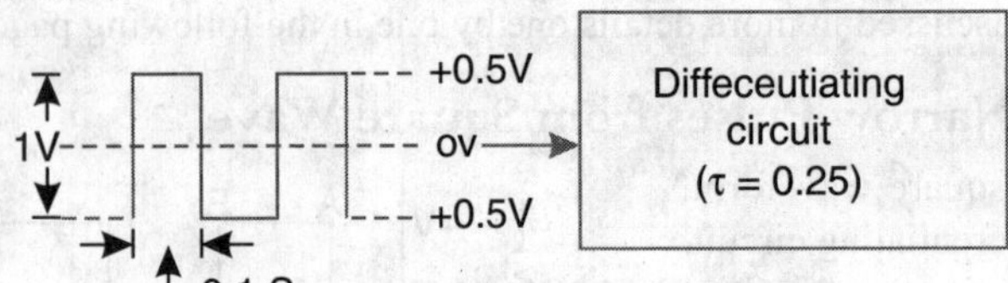

Fig. 33.8.

We know that corresponding to point *A* the capacitor will charge to a voltage,

$$v_c = 0.5 \cdot e^{-t/\tau} = 0.5 \times e^{-0.1/0.2} = 0.5 \times 0.606 = 0.3\,\text{V}$$

Similarly when the input goes to – 0.5, the capacitor will charge to a voltage of – 0.3. Therefore, the output waveform can be obtained as shown in Fig. 33.9.

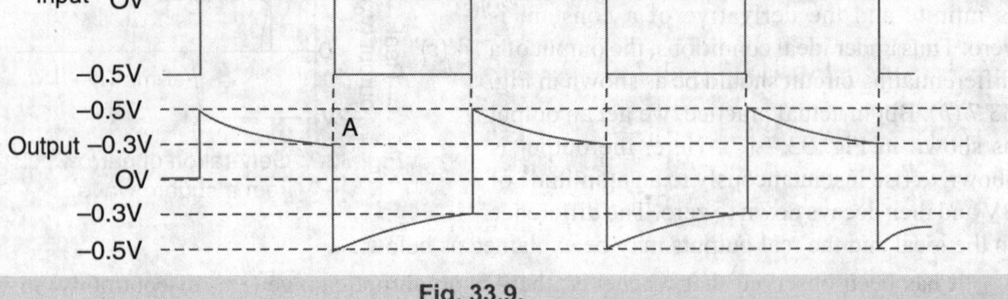

Fig. 33.9.

33.8. Generation of Voltage-step from a Ramp

Fig. 33.10 (*a*) show a ramp waveform, which acts as an input to the differentiating circuit. As a matter of fact, that circuit should produce an output waveform as shown in Fig. 33.10 (*b*). It indicates that under ideal conditions, the output should be a step with amplitude equal to $\alpha \cdot \tau$ where α is a constant which indicates the slope of ramp voltage and τ is a time constant of the circuit (equal to the product $R \cdot C$).

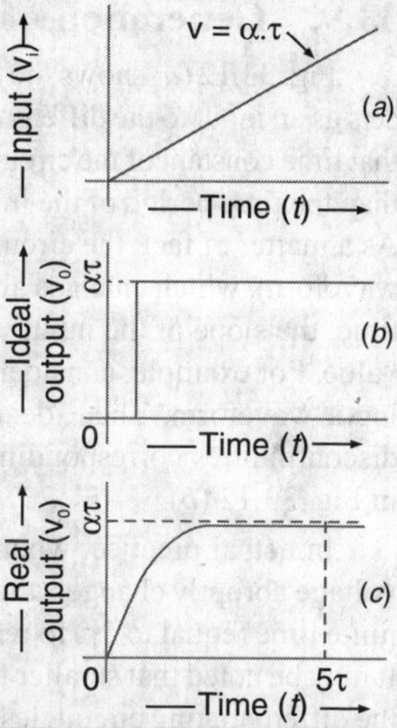

Fig. 33.10. Generation of voltage-step from a ramp.

In actual practice, the output of the circuit increases exponentially in accordance with the relation,

$$v_o = \alpha \cdot \tau\,(1 - e^{-t/\tau})$$

If t ³ 5 τ (*i.e.*, 5 times the time constant of the circuit), the term $e^{-t/\tau}$ is negligible as compared to unity. Therefore the circuit output,

$$v_o = \alpha \cdot \tau$$

It means that output reaches its steady state value (equal to $\alpha \cdot \tau$) after a time $t = 5\,\tau$. The waveform of the actual or real output is as shown in Fig. 33.10 (*c*).

Example 33.3.

The pulse from a high-voltage generator is as shown in Fig. 33.11. This is applied to the RC differentiation with RC = 250 ps. If the maximum value observed at the output of the differentiator is 50 V, what is the peak value of the input voltage ? (*AMIE., Summer 1997*)

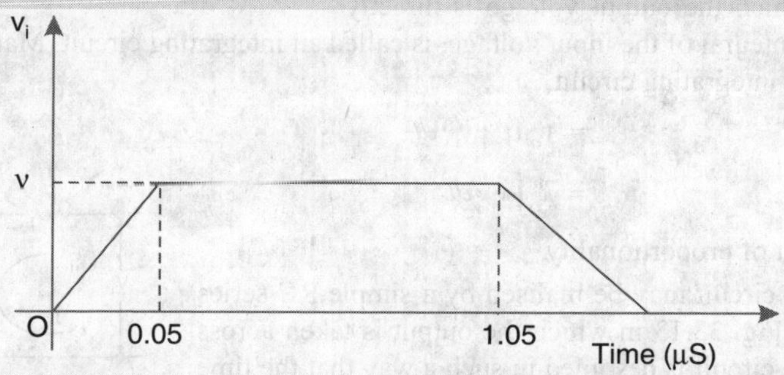

Fig. 33.11.

Solution.

Given time constant of the differentiator = 250 ps and v_o = 5 V.

Let V_P = peak value of the input voltage.

We know that steady state value of the output, (v_o)

$$50\text{ V} = \alpha \cdot \tau = \alpha \cdot (250 \times 10^{-12})$$

$$\therefore \quad \alpha = \frac{50}{250 \times 10^{-12}} = 0.2 \times 10^{12}\ \text{V/s}$$

$\therefore$ At $\tau = 0.05\ \mu$s, the peak value of input voltage

$$V_P = 0.2 \times 10^{12} \times (0.05 \times 10^{-6}) = 0.01 \times 10^{6}\ \text{V}$$

$$= 10\ \text{kV}$$ **Ans.**

33.9. Generation of a Square Wave from a Triangular Wave

Fig. 33.12 (*a*) shows the triangular waveform, which acts as an input to the differentiating circuit. It is assumed that time constant of the circuit ($\tau = R \cdot C$) is much smaller than the time period of the input waveform (*i.e.*, $\tau << T$). As a matter of fact, the circuit should produce an output waveform, which changes abruptly to a new value, each time, the slope of the input waveform changes to a new value. For example, consider points *A*, *B*, *C* and *D* of the input waveform. Then ideally the circuit should have discontinuities corresponding to these points as shown in Fig. 33.12 (*b*).

In actual precice, whenever the slope of the input voltage abruptly changes, the circuit output takes certain finite time (equal to 5τ) to reach to its steady-state level. It may be noted that smaller the value of time constant of the differentiating circuit, lesser will be the taken by its output to reach to its steady-state level. However, a s maller time constant reduces the amplitude of the output voltage.

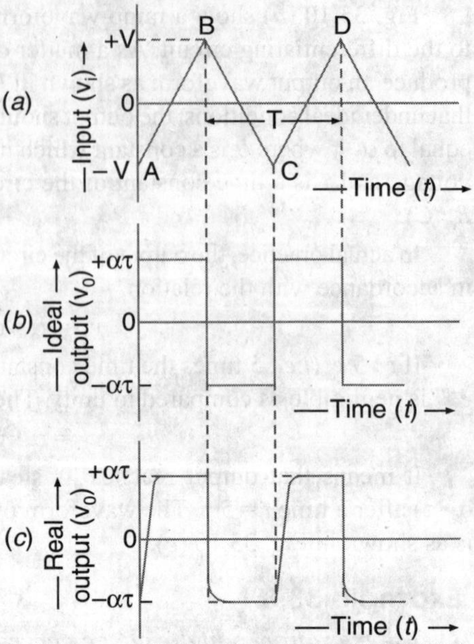

Fig. 33.12. Generation of a square wave from a trinagular wave.

33.10. Integrating Circuit

A circuit, in which the output voltage is directly proportional to the integral of the input voltage, is called an integrating circuit. Mathematically, the output voltage of an integrating circuit,

$$= v_o \; \alpha \int v_i \cdot dt$$

$$= k \int v_i \cdot dt$$

where *k* is a constant of proportionality.

The integrating circuit may be realised by a simple RC series circuit as shown in Fig. 33.13 in which the output is taken across a capacitor (*C*). The circuit is designed in such a way that the time constant of the circuit ($\tau = R \cdot C$) is greater than or equal to 10 times the period of the input signal waveform *i.e.*,

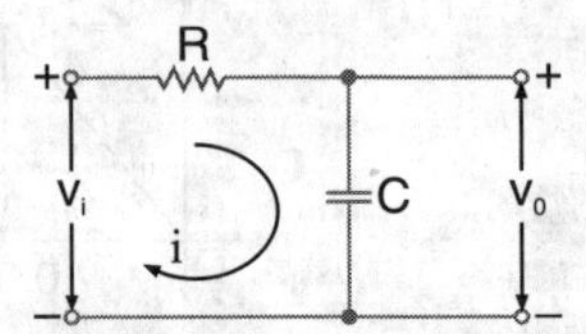

Fig. 33.13. Integrating circuit.

$$\tau \geq 10\,T \qquad \text{... (For good integration)}$$

The output of the integrated circuit is given by the relation,

$$v_o = \frac{1}{RC} \int v_i \cdot dt$$

It will be interesting to know that in an integrating circuit, the average (or d.c.) level of the output signal is the same as that of the input signal (*i.e.*, it is not zero as in a differentiating circuit). Sometimes, the integrating circuit is also called a low-pass filter. The name 'low-pass' filter is designated because of the fact that the circuit passes low-frequencies readily but attenuates high-frequencies. The attenuation of high frequencies is due to the reactance of capacitor, which decreases with the increasing frequency. At very high frequencies, the capacitor acts as a short circuit and therefore the output drops to zero.

33.11. Applications of an Integrating Circuit

Following are some of the important applications of an integrating circuit, which are important from the subject point of view :

1. To perform mathematical integration in analog computers.
2. To generate a triangular wave from a square wave.
3. To generate a sawtooth wave from a rectangular wave.

Now we shall discuss the use of integrating circuit to generate a triangular and a sawtooth in the following pages.

33.12. Generation of a Triangular Wave from a Square Wave

We know that a square wave as shown in Fig. 33.14 (*a*) may be considered as a sum of two voltage steps, of opposite polarities, which are displaced in time by $T/2$ where T is the time period of a square wave. It has been found that the response of an integrating circuit to a voltage step is exponential with a time constant $\tau = RC$. However, if the value of time constant is very large, as compared to time period of the input waveform (*i.e.*, $\tau >> T$), the response is almost a straight line as shown in Fig. 33.14 (*b*). Here the line *LM* is the response to a voltage step *AB*. Similarly, the line *MN* represents the response corresponding to the voltage step *CD*, and so on. In that case, the slope of the line is given by the relation,

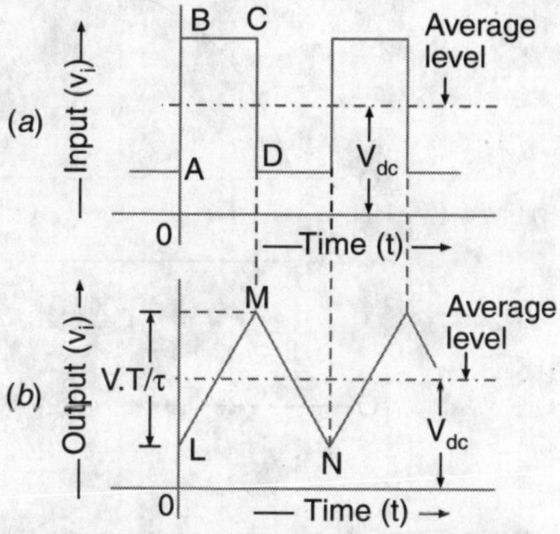

Fig. 33.14. Generation of a triangle wave from a square wave.

$$\text{slope} = \frac{V_f - V_i}{\tau}$$

where V_i = Initial value, and

V_f = Final value.

Now suppose, the input initial value is $-V$ and the final value is $+V$, then the slope of the response curve is equal to,

$$\text{slope} = \frac{V - (-V)}{\tau} = \frac{2V}{\tau}$$

Then in a time equal to $T/2$, the total change of the output voltage is given by the relation,

$$v_o = \frac{2V}{\tau} \times \frac{T}{2} = \frac{V \cdot T}{\tau}$$

However, in actual practice, it has been found that the amplitude of the output voltage is always less than VT/τ.

Example 33.4.

Obtain the response from a cascaded network as shown in Fig. 33.15.

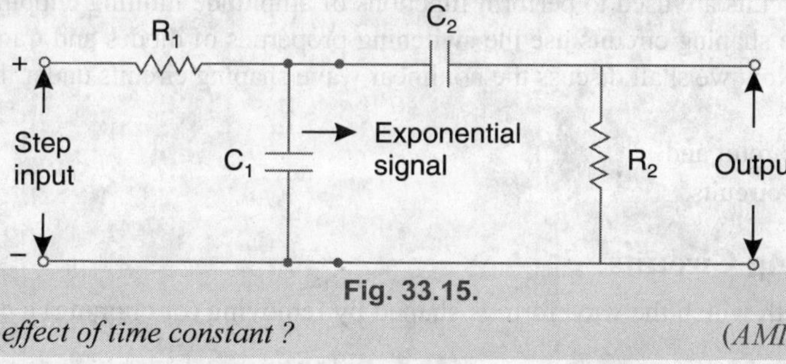

Fig. 33.15.

What is the effect of time constant ? (*AMIE., Summer 1994*)

Solution.

Given : Circuit shown in Fig. 33.15. Notice that the circut consists of an integrator followed by a differentiator circuit. The cascaded response is as shown in Fig. 33.16.

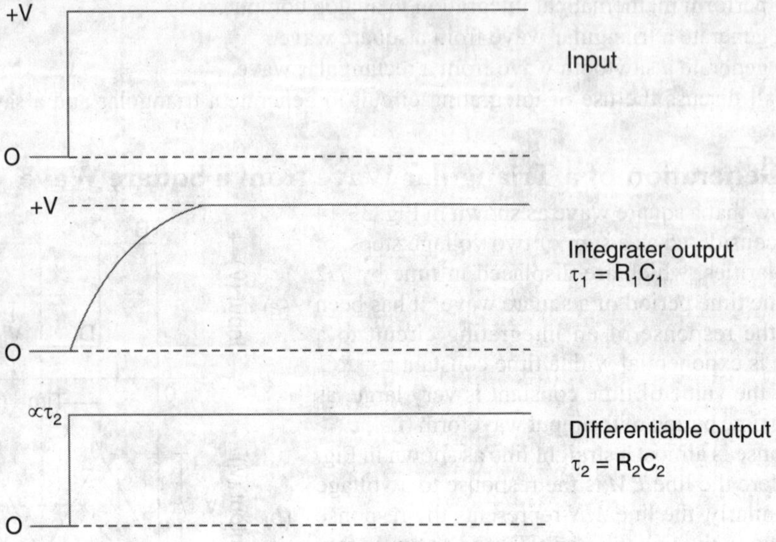

Fig. 33.16.

33.13. Generation of a Sawtooth Wave from a Rectangular Wave

Fig. 33.17 (*a*) shows the waveform of a rectangular wave. It is a modified version of a square wave. The response of an integrating circuit to a rectangular wave is similar to that discussed for a square wave in the last article, except the output waveform, which is a sawtooth wave (instead of a triangular wave) as shown in Fig. 33.17 (*b*),

Note: The average levels in the waveforms in Fig. 33.14 and 33.17 are indicated to remind the reader that the integrating circuit does not disturb these at the output.

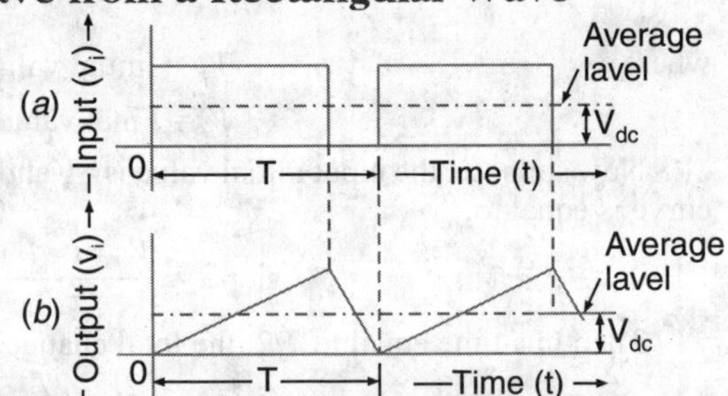

Fig. 33.17. Generation of a sawtooth wave from a rectangular wave.

33.14. Nonlinear Wave Shaping Circuits

We have already discussed in Art. 33.3 that non-linear wave shaping circuits are those which (in addition to resistors, inductors and capacitors) make use of nonlinear elements such as diodes and transistors. Such circuits are used to perform functions of amplitude limiting clipping and clamping. The nonlinear wave shaping circuits use the switching properties of diodes and transistors to realise various functions. Now we shall discuss the nonlinear wave shaping circuits under the following two heads :

1. Clipping circuit and
2. Clamping circuits.

33.15. Clipping Circuits

The circuit, with which the waveform is shaped by removing (or clipping) a certain portion of the input signal voltage above or below a certain level, is called as clipping circuit or simply clipper. Such circuits are used in digital computers, radars, radio and television receivers etc. A clipping circuit may also be used to limit the amplitude of the input signal voltage as required in several application. In that case, a clipping circuit is known as amplitude limiter or simply limiter. The clipping circuits are of the following four types :

1. Positive clipper
2. Negative clipper
3. Biased clipper
4. Combination clipper

33.16. Positive Clipper

Fig. 33.18 (*a*) shows the circuit of a positive clipper. It consists of a diode (*D*) and a resistor (*R*) with output taken across the resistor. The diode acts as an ideal switch between the source and the load. It acts as a closed switch, when the input voltage is negative (*i.e.*, $v_i < 0$) and as an open switch when the input valtage is positive or zero (*i.e.*, $v_i \geq 0$). The purpose of resistor (*R*) is to limit the current through diode, when it acts as closed switch. The operation of the circuit may be explained as follows.

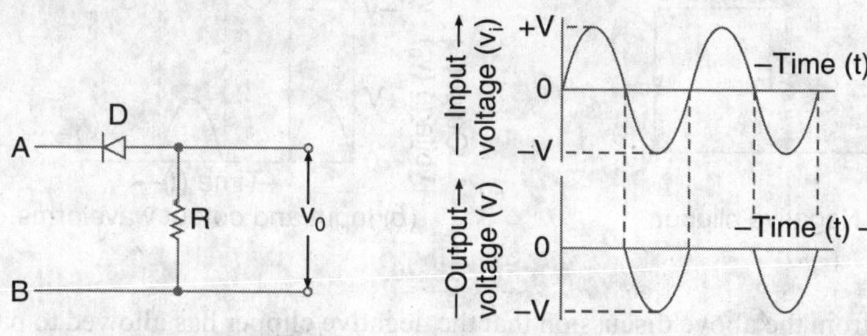

(*a*) Positive clipper. (*b*) Input and output waveforms.

Fig. 33.18.

Fig. 33.18 (*b*) shows that waveform of an input voltage. During the positive half cycle of the input voltage, the terminal *A* is positive with respect to *B*. This reverse biases the diode and it acts as an open switch. Therefore, all the applied voltage drops across the diode and none across the resistor. As a result of this, there is no output voltage during the positive half cycle of the input voltage as shown in the figure. During the negative half cycle of the input voltage, the terminal *B* is positive with respect to *A*. Therefore it forward biases the diode and it acts as a closed switch. Thus there is no voltage drop across diode during the negative half cycle of the input voltage. All the input voltage is dropped across the resistor as shown in the output waveform.

It is evident from the above discussion as well as the output waveform that the positive clipper has allowed to pass the negative half cycle of the input voltage and clipped the positive half cycle completely. Hence the circuit of Fig. 33.18 (*a*) is called a positive clipper. It may be noted in Fig. 33.18 (*a*) that the diode acts as a series switch between the source and the load. Due to this reason, the circuit is called series positive clipper. It is also possible to connect a diode as a shunt switch between the source and the load as shown in Fig. 33.19. Such a circuit is known as shunt positive clipper. In this circuit, the diode acts as a closed switch when the input voltage is positive (*i.e.*, $v_i \geq 0$) and as an open switch when the input voltage is negative (*i.e.*, $v_i < 0$). The output waveform, is the same as that of a series positive clipper.

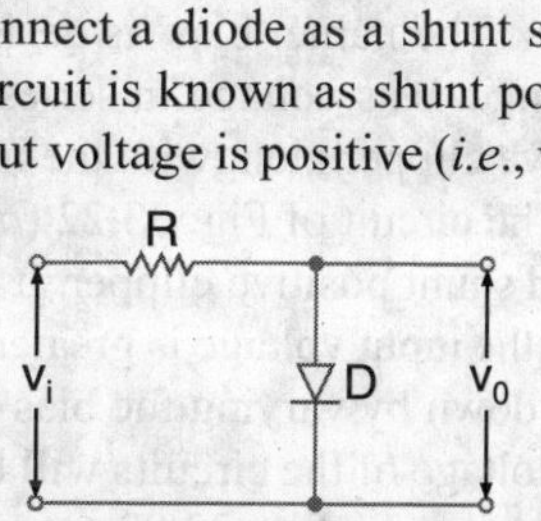

Fig. 33.19. Shunt positive clipper.

33.17. Negative Clipper

Fig. 33.20 (*a*) shows the circuit of a negative clipper. Here the diode is shown connected in a direction opposite to that of a positive clipper. The function of the resistor (R) is to limit the current when the diode is forward biased (*i.e.*, when it acts as a closed switch). The diode acts as a closed switch for a positive input voltage (*i.e.*, $v_i \geq 0$) and as an open switch for a negative input voltage (*i.e.*, $v_i < 0$). The operation of the circuit may be explained in the same way as discussed for a positive clipper and is as given below.

Fig. 33.20 (*b*) shows the waveform of the input voltage. During the positive half cycle of the voltage, the terminal A is positive with respect to the terminal B. Therefore the diode is forward biased and it acts as a closed switch. As a result, all the input voltage appears across the resistor as shown in Fig. 33.20 (*b*). During the negative half cycle of the input voltage, the terminal B is positive with respect to the terminal A. Therefore the diode is reverse biased and it acts as an open switch. Thus, there is no voltage drop across the resistor during the negative half cycle as shown in the output waveform.

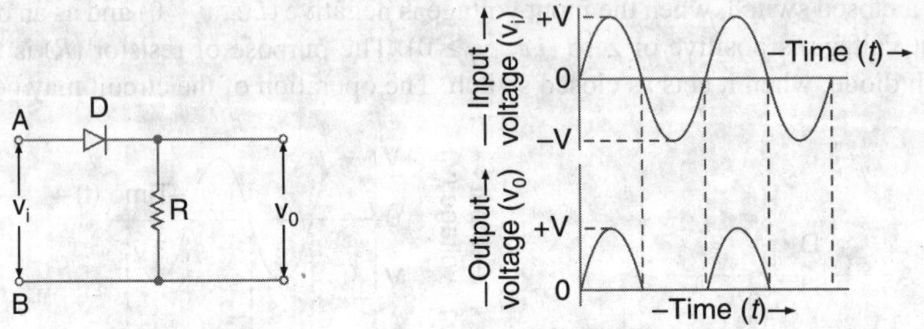

(*a*) Negative clipper. (*b*) Input and output waveforms.

Fig. 33.20.

It is evident from the above discussion that the negative clipper has allowed to pass the positive half cycle of the input voltage and clipped the negative half cycle completely. The negative clipper circuit of Fig. 33.20 (*a*) is called series negative clipper, because the diode acts as a series switch between the source and the load. It is also possible to connect a diode as a shunt switch between the source and the load as shown in Fig. 33.21. Such a circuit is known as shunt negative clipper. In this circuit, the diode acts as a closed switch for a negative input voltage (*i.e.*, $v_i < 0$) and as an open switch for a positive input voltage (*i.e.*, $v_i \geq 0$). The output waveform of the circuit is the same as that of series negative clipper.

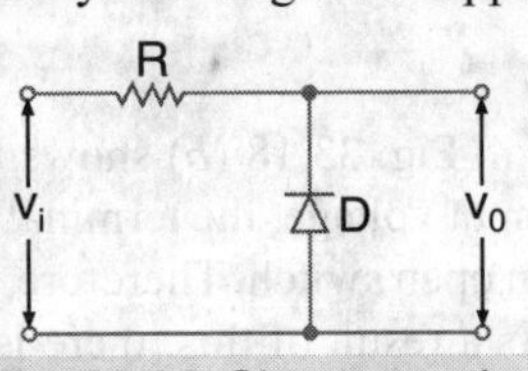

Fig. 33.21. Shunt negative clipper.

33.18. Biased Clipper

We have already discussed in the last articles that a positive clipper clips off the entire positive portions of the input signal. Similarly, a negative clipper clipps off the entire negative portions off the input signal. In such circuits, there is no provision for adjustment of the clipping level.

A clipping circuit, which has a provision for the adjustment of a clipping level is called a biased clipper. The name 'bias' is designated because the adjustment of the clipping level is achieved by additing a bias voltage in series with the diode or resistor. Fig. 33.22 shows the circuit of a biased positive clipper alongwith the input and output voltage waveforms.

The circuit of Fig. 33.22 (*a*) is a biased series positive clipper, while that of Fig. 33.22 (*b*) is a biased shunt positive clipper. It may be noted that the clipping takes place during the positive cycle when the input voltage is greater than battery voltage (*i.e.*, $v_i \geq V_B$). The clipping level can be shifted up or down by varying the bias voltage (V_B). It will be interesting to know that if the polarity of the bias voltage of the circuits will be as shown in Fig. 33.22 is reversed, then the resulting circuits will be as shown in Fig. 33.23 (*a*) and (*b*). Here the input signal lying above the voltage level $-V_B$ is clipped. The resulting waveforms of the output voltage are shown in figures.

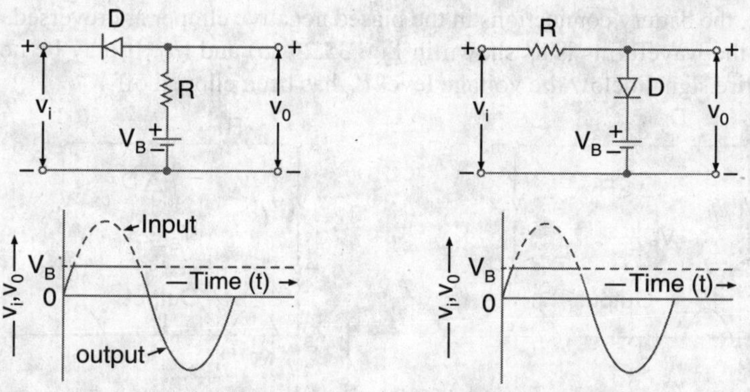

(*a*) Biased series positive clipper. (*b*) Biased shunt positive clipper.

Fig. 33.22. Biased clipper.

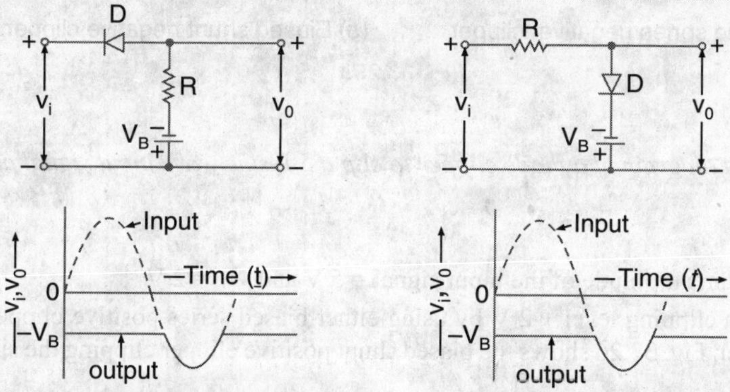

(*a*) Biased series positive clipper. (*b*) Biased shunt positive clipper

Fig. 33.23. Biased clipper with reversed polarity of battery V_B.

Fig. 33.24 shows the circuit of a biased negative clipper. The circuit of Fig. 33.24 (*a*) corresponds to the biased series negative clipper, while that of Fig. 33.24 (*b*) a biased shunt negative clipper. In these circuits, clipping takes place during the negative half cycle only when the input voltage $v_i \leq - V_B$. The clipping level can be shifted up or down by varying the bias-voltage $(- V_B)$.

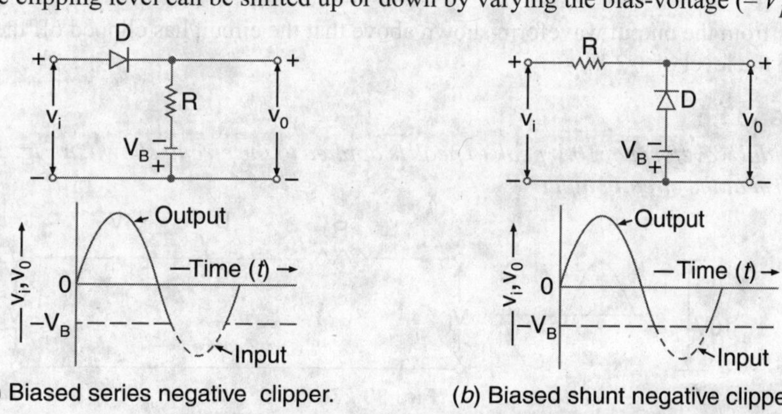

(*a*) Biased series negative clipper. (*b*) Biased shunt negative clipper.

Fig. 33.24.

Now suppose, the battery connections in the biased negative clipper are reversed. The resulting circuit alongwith the waveforms is as shown in Fig. 33.25 (*a*) and (*b*). It may be noted from this circuit that the entire signal below the voltage level V_B has been clipped off.

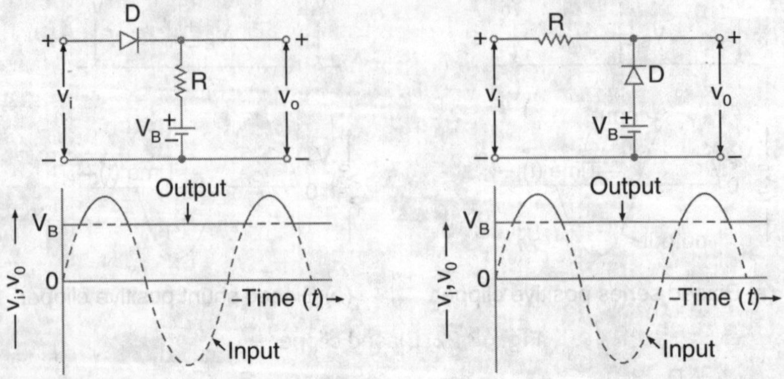

(*a*) Biased series negative clipper. (*b*) Biased shunt negative clipper.

Fig. 33.25.

Example 33.5.

Draw the circuit diagram of a diode clipper to clip a 5 V sine wave input signal at + 2 V.

(R.G.P.V., June. 2005)

Solution.

Given : Maximum amplitude of the input signal = 5 V and V_B = 2 V.

We can obtain a clipping level + 2 V by using either biased series positive clipper or a biased shunt positive clipper. Fig. 33.26 shows the biased shunt positive clipper clipping the signal at + 2 V.

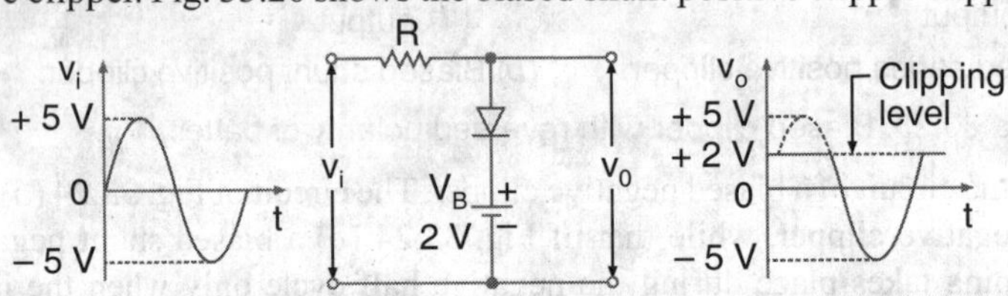

Fig. 33.26.

It is evident from the output waveform shown above that the circuit has clipped off the input signal lying above the level of + 2 V.

Example. 33.6.

Sinusoidal waveform of 10 V peak to peak is applied to the circuit shown in Fig. 33.27. Sketch the waveform at the circuit output.

(AMIE, Summer 1996)

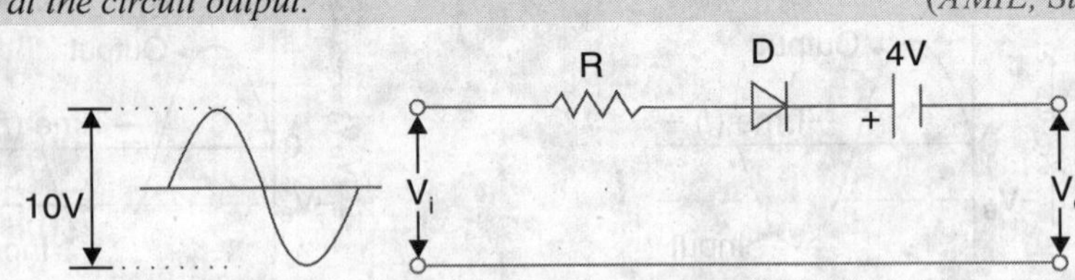

Fig. 33.27.

Solution.

Fig. 33.27 (*a*) shows the given input voltage waveform. The output voltage waveform can be obtained as discussed below.

Let us assume the diode, *D* as an ideal diode (*i.e.*, a diode with no barrier voltage and zero forward resistance).

The diode, *D*, acts like an open circuit for an input voltage below (+ 4 V). Therefore, the output voltage for an input voltage for an input greater than or equal to 4 V can be calculated as $v_0 = (v_i - 4)$ V. The output remains at zero for input voltages less than or equal to 4 V. Therefore the output voltage waveform can be sketched as shown in Fig. 33.28.

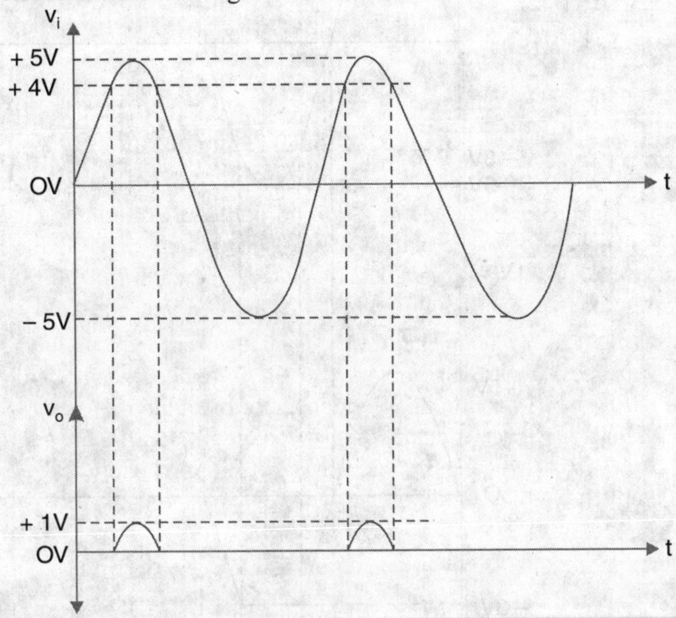

Fig. 33.28.

Note. If we assume the diode to be real *i.e.*, with foreword voltage drop of 0.7 V. Then it will turn on only when the input is greater than 4.7 V.

Example. 33.7.

Sinusoidal waveform of 10 V peak-to-peak is applied to the circuit shown in Fig. 33.29. Sketch the waveform at the circuit output. (*AMIE, Summer 1996*)

Solution.

Given, circuit shown in Fig. 33.29. It resembles that of a biased negative shunt chipping circuit.

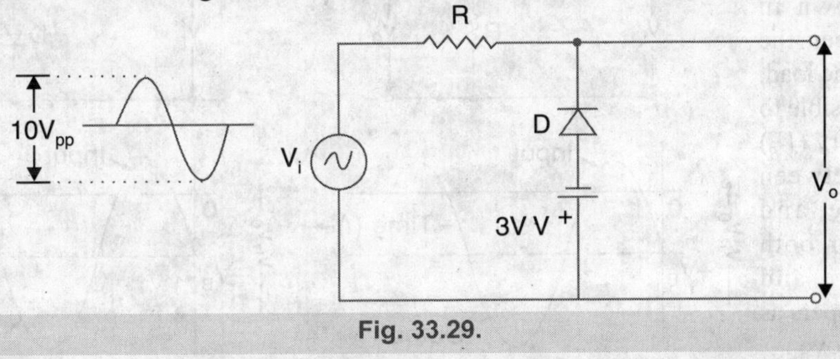

Fig. 33.29.

When the input signal is below – 3 V, it is clear from the circuit, that the diode, *D*, turns 'ON' and starts conducting. So, the output is clipped at –3 V for any input lower than thus value. Therefore the output waveform is as shown in Fig. 33.30.

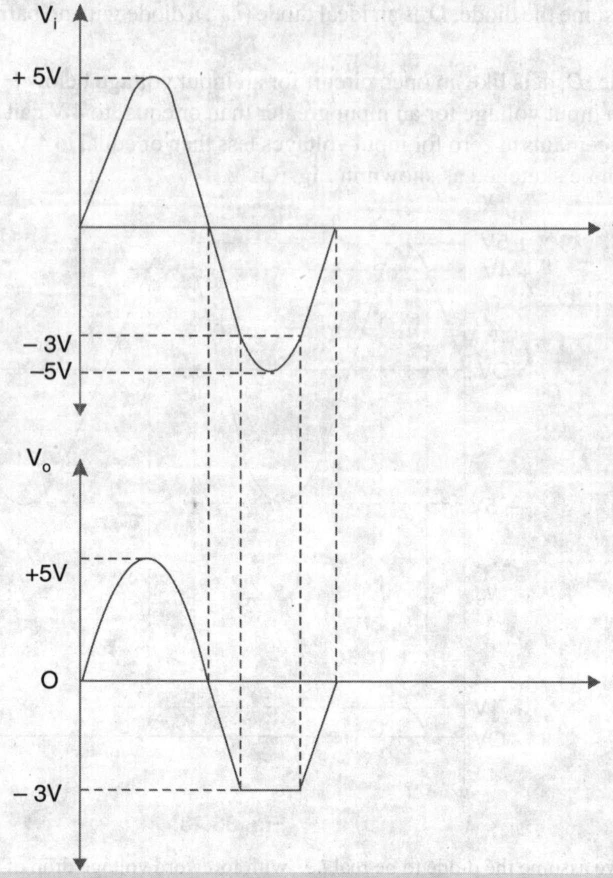

Fig. 33.30.

Note. If the diode is assumed to be a 'real' or practical; the output will be clipped at – 3.7 V.

33.19. Some Other Biased Clippers

We have already discussed the circuit for biased positive and negative clippers. In all these circuits, the battery for the bias voltage (V_B) is shown in shunt between the source and the load. It is also possible to use the battery (V_B) in series between the source and the load for both the positive and negative clippers as discussed above :

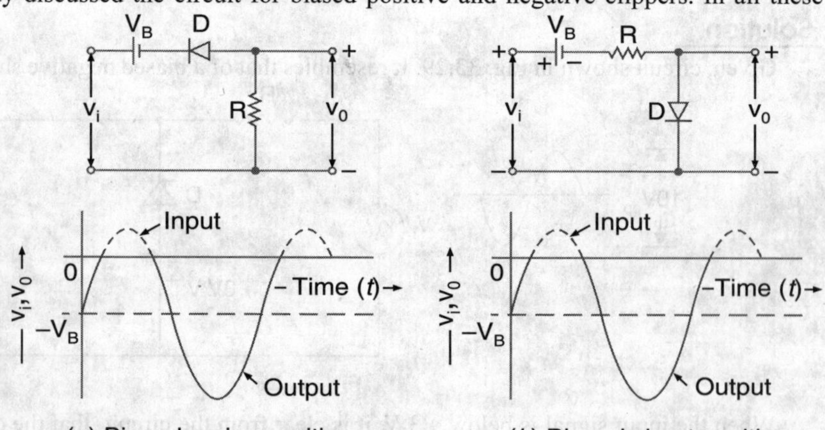

(*a*) Biased series positive clipper.

(*b*) Biased shunt positive clipper.

Fig. 33.31.

Fig. 33.31 (*a*) shows the circuit

of a biased series positive clipper alongwith its output waveform. Here the addition of bias voltage V_B (with polarity of the battery as shown in the figure) shifts the entire input signal below the zero-voltage axis by an amount equal to the bias voltage (V_B). Then the circuit clipps off the part of the positive half cycle lying above the zero-voltage axis. Fig. 33.31 (*b*) shows the circuit of a biased shunt positive clipper. It produces an output similar to that of a series positive clipper as shown in the figure.

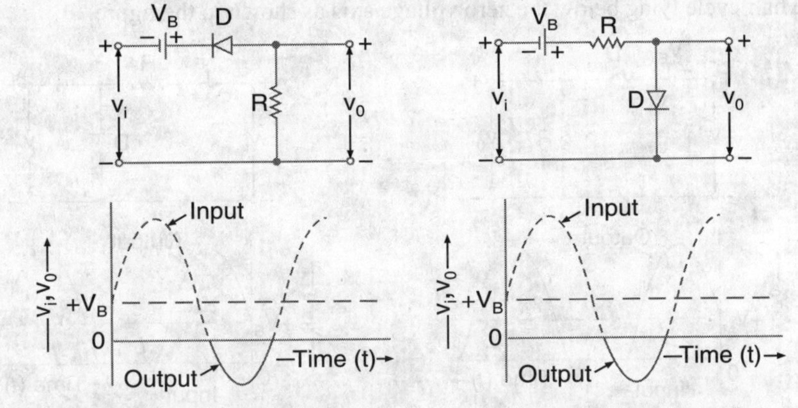

(*a*) Biased series positive clipper. (*b*) Biased shunt positive clipper.

Fig. 33.32.

Now suppose the connections of the battery V_B in a biased positive clipper Fig. 33.31 (*a*) and (*b*) are reversed. Then the resulting circuits for the series and shunt-type positive clipper are as shown in Fig. 33.32 (*a*) and (*b*) respectively. It may be noted that the addition of bias voltage V_B (with polarity of the battery as shown in the figure) shifts the entire input signal above the zero voltage axis by an amount equal to the bias voltage (*i.e.*, V_B). Then the circuit clipps off the part of the negative cycle lying below the zero voltage axis as shown in the figure.

Fig. 33.33 (*a*) shows the circuit of a biased series negative clipper alongwith its output waveform. Here the addition of bias voltage V_B (with the polarity of the battery as shown in the figure) shifts the entire input signal below the zero-voltage axis by an amount equal to the bias voltage (V_B). Then the circuit, clipps off the part of the negative half cycle lying below the zero valtage axis. Fig. 33.33 (*b*), it shows the circuit of a bissed shunt negative clipper. It produces an output similar to that of a series negative clipper as shown in the figure.

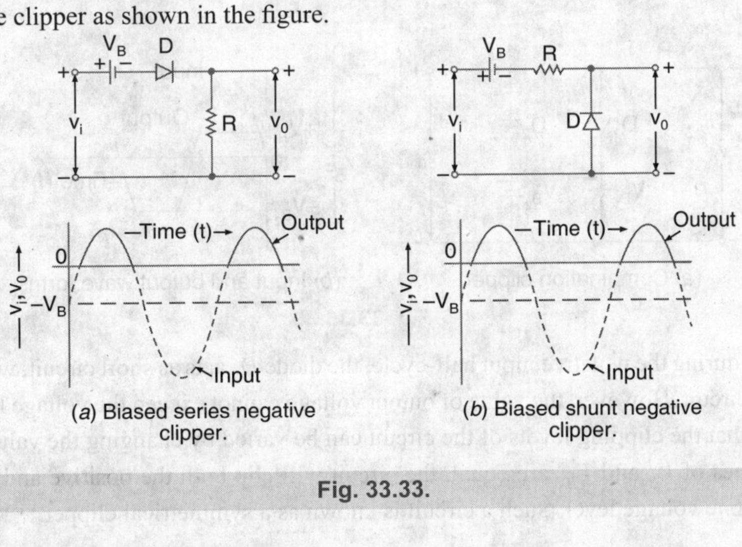

(*a*) Biased series negative clipper. (*b*) Biased shunt negative clipper.

Fig. 33.33.

Now suppose the connections of the battery V_B in a biased negative clipper [Fig. 33.33 (*a*) and (*b*)] are reversed. Then the resulting circuits for the series and shunt-type negative clipper are as shown in Fig. 33.34 (*a*) and (*b*) respectively. It may be noted that the addition of a bias voltage V_B (with polarity of the battery as shown in the figure) shifts the entire input signal above the zero voltage axis by an amount equal to the bias voltage (V_B). Then the circuit clipps off the part of the negative half cycle lying below the zero voltage axis as shown in the figure.

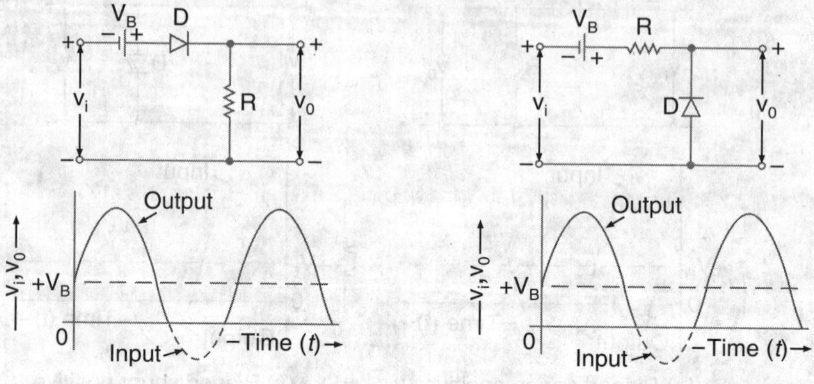

(*a*) Biased series negative clipper.

(*b*) Biased shunt negative clipper.

Fig. 33.34.

33.20. Combination Clipper

The combination of a biased positive clipper and a biased negative clipper is called combination clipper. Such a clipper circuit can clip at two-independent levels depending upon the bias voltages. Fig. 33.35 (*a*) shows the circuit of a combination clipper. The operation of the circuit may be explained as discussed below.

Suppose a sinusoidal a.c. voltage is applied at the input terminals of the circuit. Then during the positive half-cycle, the diode D_1 is forward biased, while diode D_2 is reverse biased. Therefore the diode D_1 will conduct and will act as short-circuit. On the other hand, diode D_2 will act as an open-circuit. However, the value of output voltage cannot exceed the voltage level of V_{B_1} as shown in Fig. 33.35 (*b*).

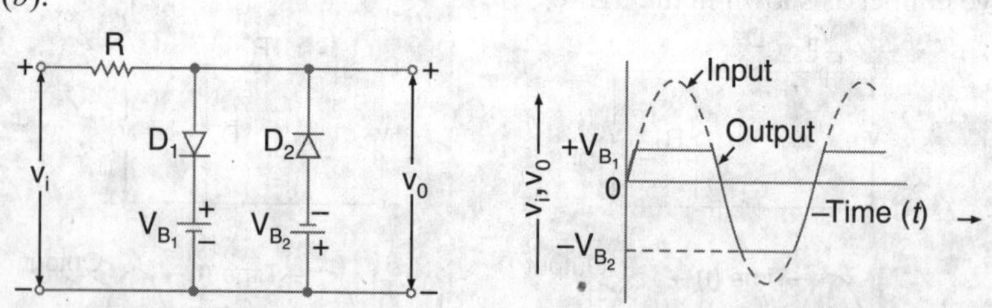

(*a*) Combination clipper.

(*b*) Input and output wave forms.

Fig. 33.35.

Similarly, during the negative input half-cycle, the diode D_2 acts as short circuit, while the diode D_1 as an open circuit. However, the value of output voltage cannot exceed the voltage level of V_{B_2}. It may be noted that the clipping levels of the circuit can be varied by changing the values of V_{B_1} and V_{B_2}. If the values of V_{B_1} and V_{B_2} are equal, the circuit will clip both the positive and negative half cycles at the same voltage level. Such a circuit is known as a symmetrical clipper.

Example 33.8.

Draw the circuit diagram of a clipper to clip a 5 V sine wave input signal at +3 V and – 2 V.

Solution.

Fig. 33.36 shows the circuit of a combination clipper alongwith the input and output waveforms.

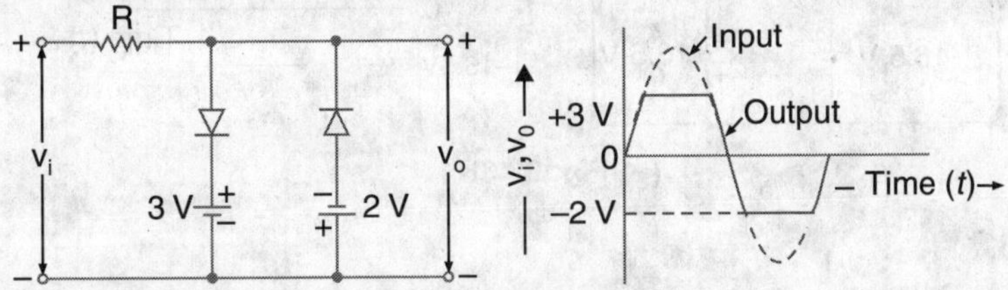

Fig. 33.36.

It is evident from the waveforms, that the positive half cycle of the input voltage is clipped at + 3 V and the negative cycle is clipped at – 2 V.

Example 33.9.

Sinusoidal waveform of 10 V peak-to-peak is applied to the circuit shown in Fig. 33.37 (a). Sketch the waveform at the circuit output. (*AMIE, Summer 1996, RGPV, Dec. 2004*)

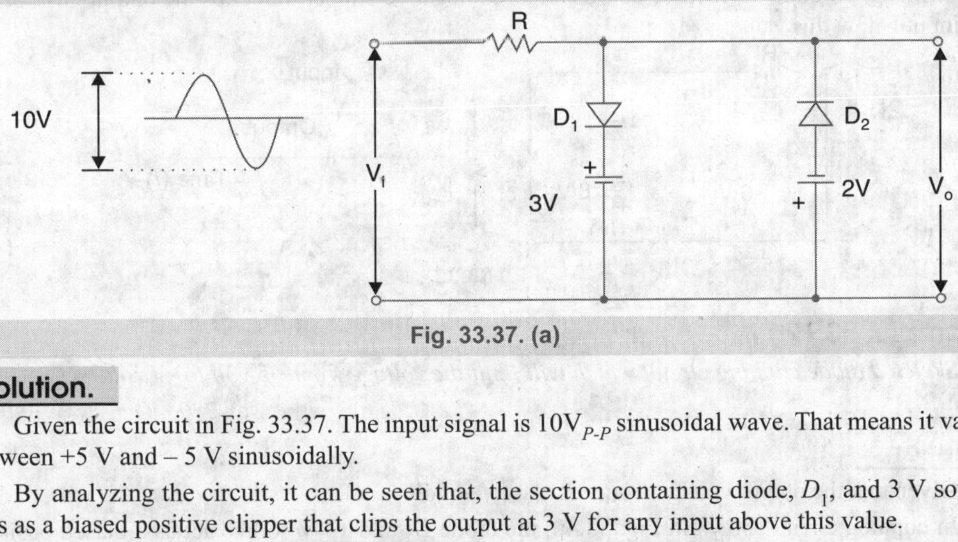

Fig. 33.37. (a)

Solution.

Given the circuit in Fig. 33.37. The input signal is $10V_{P\text{-}P}$ sinusoidal wave. That means it varies between +5 V and – 5 V sinusoidally.

By analyzing the circuit, it can be seen that, the section containing diode, D_1, and 3 V source acts as a biased positive clipper that clips the output at 3 V for any input above this value.

Similarly, the section containing the diode, D_2, and 2 V source cuts as a biased negative clipper that clips the output 2 V at for any input below this value.

Therefore, the output waveform will be as shown in Fig. 33.37 (*b*).

Example 33.10.

Draw the diagram of a diode clipping circuit which clips a voltage at ± 16.5 volts.

(*M. G. University., Dec. 2007*)

Solution.

Given that the output voltage at ±16.5 volts.

Fig. 33.38 shows the clipping circuit. Here the diode D_1 and +16.5 V source act as a biased positive clipper that clips the output from 16.5 V for any input above the value.

Similarly D_2 and 16.5 V source acts as a biased negative clipper that clips the output 16.5 V at for any input below this value

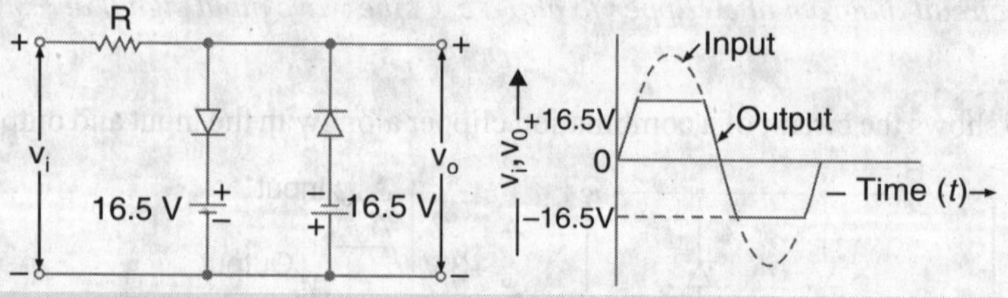

Fig. 33.38

Example 33.11.

Draw and explain the working of a clipping circuit which limits the voltage at + 6 V and – 8 V.

(M. G. University., May 2005)

Solution.

Given that the output voltage at + 6 V and – 8 V.

In clipping circuit shown in Fig. 33.39, the diode D_1 and + 6 V source act as a biased positive clipper that clips the output from 6 V for any input above the value.

Similarly D_2 and –8 V source cuts as a biased negative clipper that clips the output –8 V at for any input below this value.

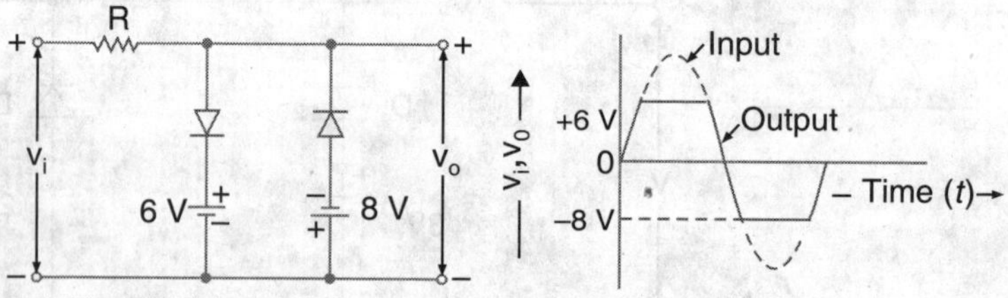

Fig. 33.39

Example 33.12.

Sketch a diode clipping circuit which will limit the voltage from +5 volt to –4 volt.

(M. G. University., May 2005, Oct. /Nov. 2004)

Solution.

Given that the output voltage from +5 volt to –4 volt.

In clipping circuit shown in Fig. 33.40, the diode D_1 and +5 V source act as a biased positive clipper that clips the output from 5 V for any input above the value.

Similarly D_2 and –4 V source cuts as a biased negative clipper that clips the output 4 V at for any input below this value.

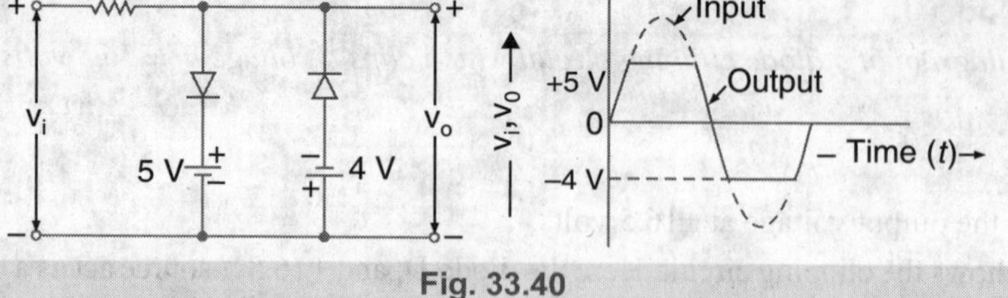

Fig. 33.40

Example 33.13.

Sketch the output voltage wavefrom for the circuit shown in Fig. 33.41. Sketch the output voltage wave from: (*GBTU/MTU., 2008-09*)

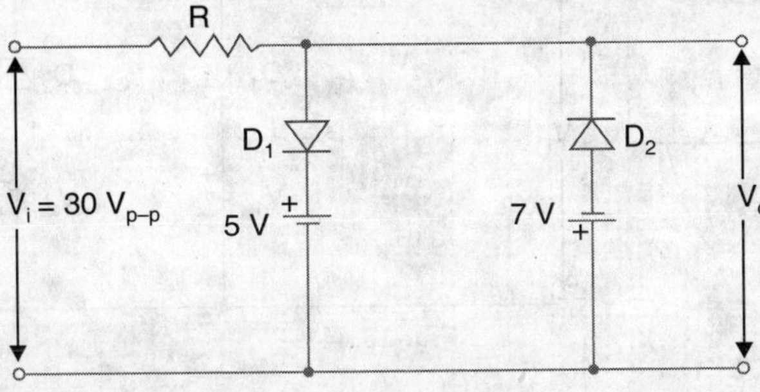

Fig. 33.41

Solution.

Given: The circuit in Fig. 33.41 and reproduced again in 33.42 also. The input signal is 30 V_{P-P} triangular waves. That means it varies between +15 V and – 15 V triangular.

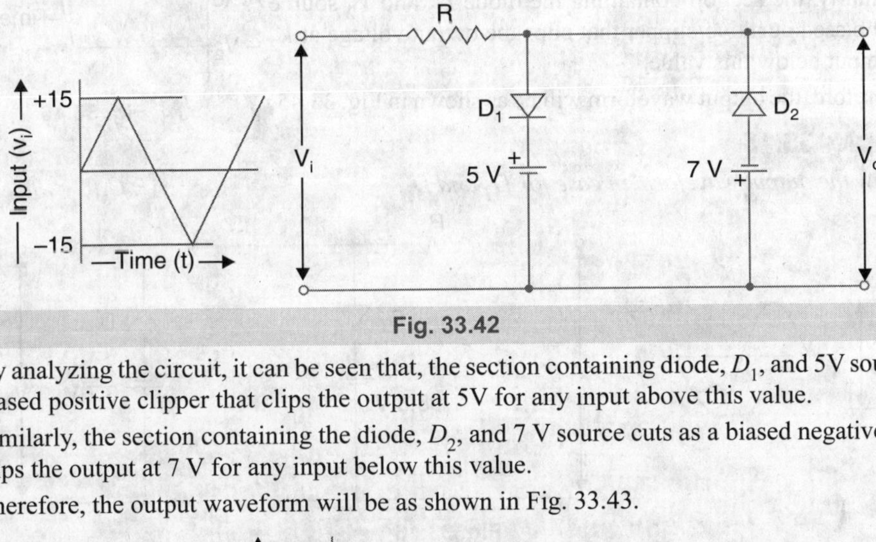

Fig. 33.42

By analyzing the circuit, it can be seen that, the section containing diode, D_1, and 5V source acts as a biased positive clipper that clips the output at 5V for any input above this value.

Similarly, the section containing the diode, D_2, and 7 V source cuts as a biased negative clipper that clips the output at 7 V for any input below this value.

Therefore, the output waveform will be as shown in Fig. 33.43.

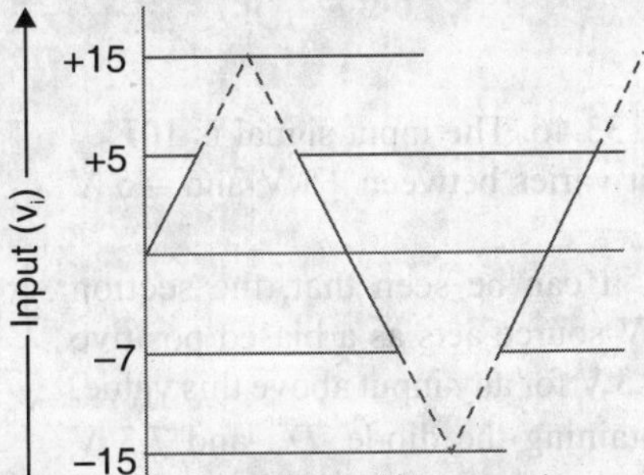

Fig. 33.43 Output Waveform

Example 33.14.

Draw the output waveform for the following Fig. 33.44. *(GBTU/MTU., 2004-05)*

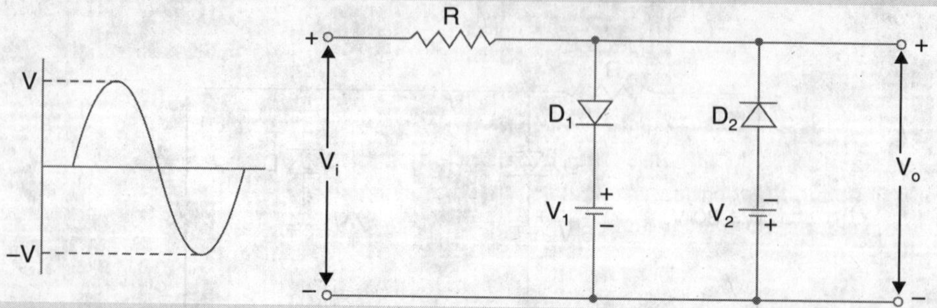

Fig. 33.44

Solution.

Given: The circuit in Fig. 33.44. The input signal varying sinusoidally between +V and –V.

By analyzing the circuit, it can be seen that, the section containing diode, D_1 and V_1 source acts as a biased positive clipper that clips the output voltage at V_1 for any input above this value.

Similarly, the section containing the diode D_2 and V_2 source cuts as a biased negative clipper that clips the output voltage at V_2 for any input below this value.

Therefore, the output waveform will be as shown in Fig. 33.45.

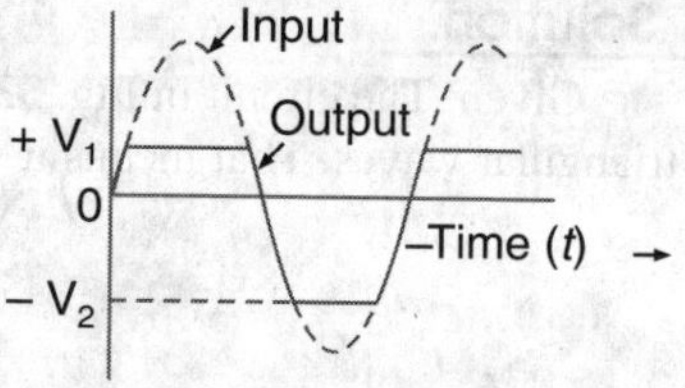

Fig. 33.45

Example 33.15.

Draw the output waveform in case for D_1 and D_2. *(R.G.P.V., June 2011)*

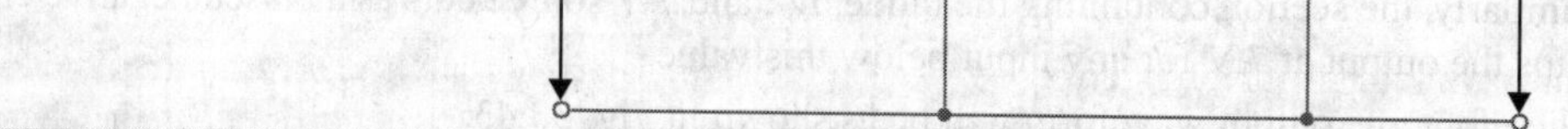

Fig. 33.46

Solution.

Given : The circuit in Fig. 33.46. The input signal is $10V_{p-p}$ sinusoidal wave. That means it varies between +5 V and – 5 V sinusoidally.

By analyzing the circuit, it can be seen that, the section containing diode, D_1, and 5.3 V source acts as a biased positive clipper that clips the output at 5.3 V for any input above this value.

Similarly, the section containing the diode, D_2, and 7.3 V source cuts as a biased negative clipper that clips the output 7.3 V at for any input below this value.

Therefore, the output waveform will be as shown in Fig. 33.47.

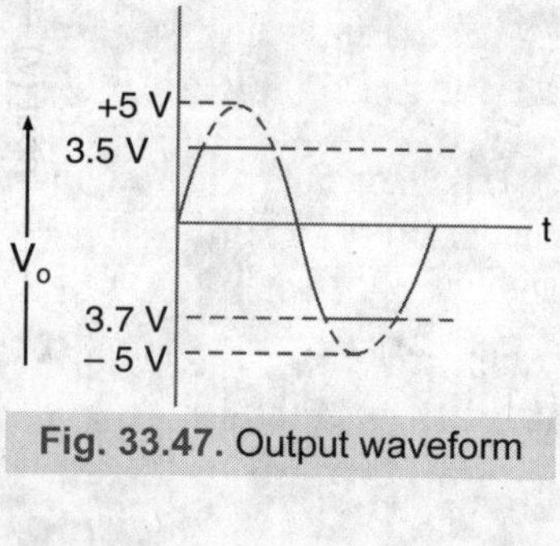

Fig. 33.47. Output waveform

Example 33.16.

A biased shunt clipper circuit as shown in Fig. 33.48 is to be designed to protect a circuit which cannot accept voltage axceeding, $V_0 = \pm 2.7$ V. The input to the clipper is a ± 8 V square wave and output current is to be a maximum of 1 mA.

Calculate the value of R_1 and specify the diode features required. (*AMIE, Summer 2002*)

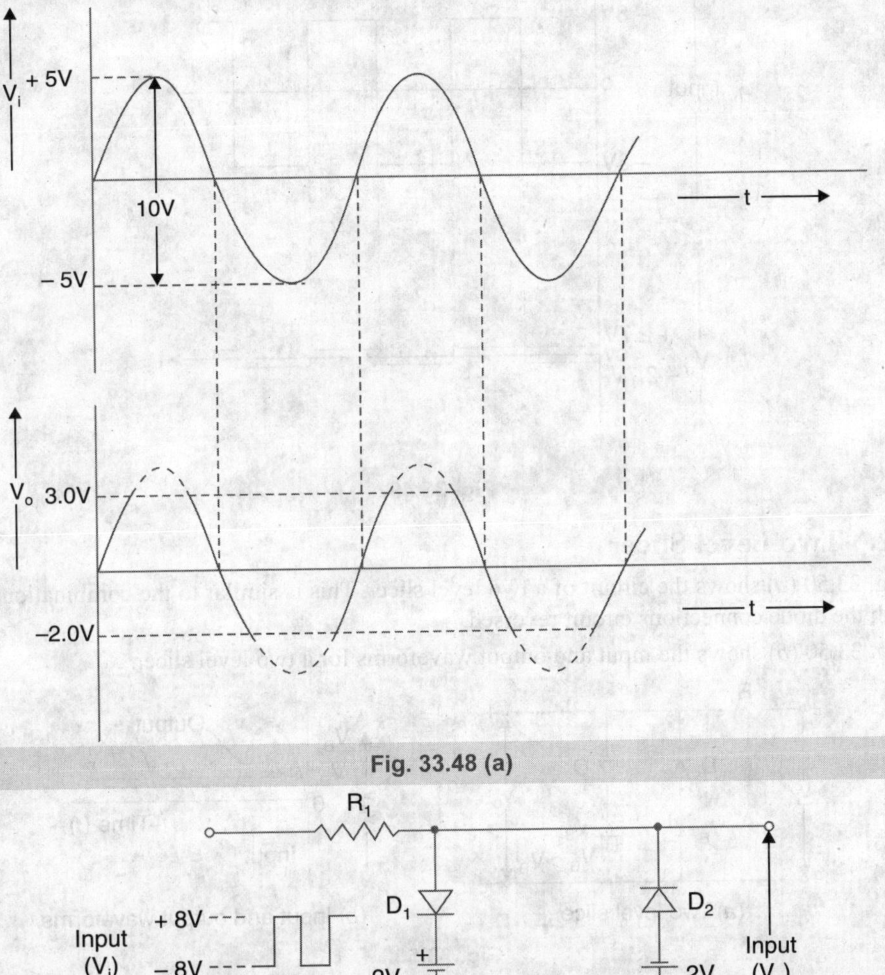

Fig. 33.48 (a)

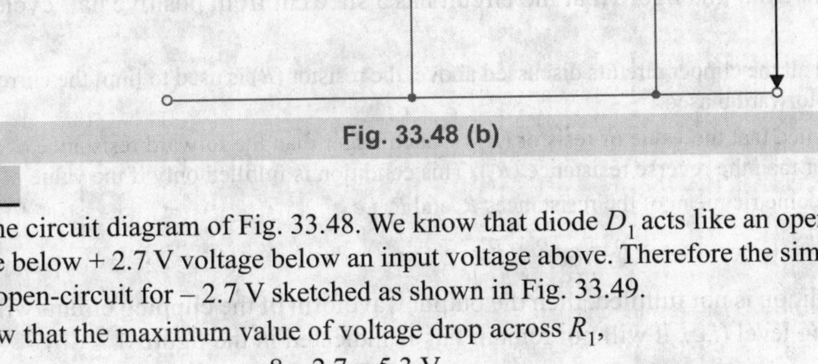

Fig. 33.48 (b)

Solution.

Given the circuit diagram of Fig. 33.48. We know that diode D_1 acts like an open-circuit for an input voltage below + 2.7 V voltage below an input voltage above. Therefore the similarly diode D_2 acts like-an open-circuit for – 2.7 V sketched as shown in Fig. 33.49.

We know that the maximum value of voltage drop across R_1,

$$= 8 - 2.7 = 5.3 \text{ V}$$

∴ Value of R_1, so that the current does not exceed 1 mA

$$R_{1(min)} = \frac{5.3}{1 \times 10^{-3}} = 5.3 \times 10^3\,\Omega \text{ or } 5.3\text{ k}\Omega \textbf{ Ans.}$$

The diodes D_1 and D_2 has to be silicon diodes.

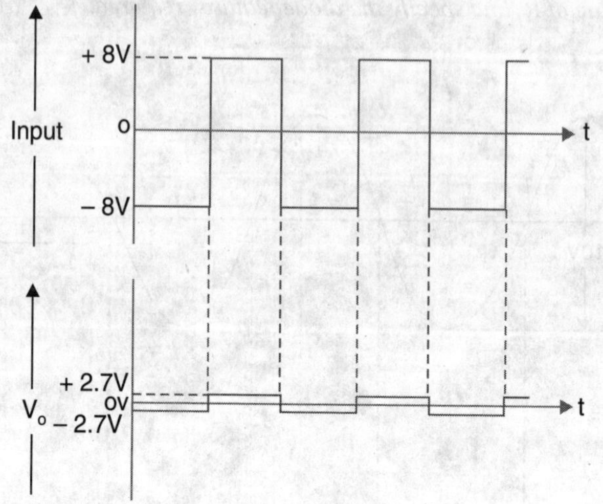

Fig. 33.49.

33.21. Two Level Slicer

Fig. 33.50 (*a*) shows the circuit of a two level slicer. This is similar to the combination clipper but with the diode connections circuit reversed.

Fig. 33.50 (*b*) shows the input and output waveforms for a two level slicer.

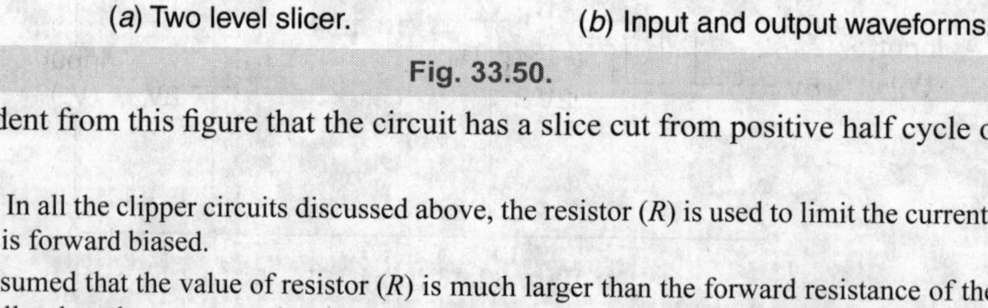

(*a*) Two level slicer. (*b*) Input and output waveforms.

Fig. 33.50.

It is evident from this figure that the circuit has a slice cut from positive half cycle of the input signal.

Notes. 1. In all the clipper circuits discussed above, the resistor (R) is used to limit the current through the diode when it is forward biased.

2. It is assumed that the value of resistor (R) is much larger than the forward resistance of the diode (R_f), and much smaller than the reverse resistance (R_r). This condition is fulfilled only if the value of resistor (R) is selected as the geometric mean of the resistances R_f and R_r *i.e.*,

$$R = \sqrt{R_f \cdot R_r}$$

If this condition is not fulfilled, then the output waveform of the clipping circuit will be rounded near the clipping level (*i.e.*, it will not remain flat as indicated in the figures).

Example 33.17.

A sinusoidal waveform of 10 V peak- to-peak. is applied to the circuit shown in Fig. 33.51. Sketch the waveform at the circuit output.

Solution.

Given a sinusoidal waveform of 10 V peap-to-peak applied to the circuit shown in Fig. 33.51.

In order to determine the output waveform, let us assume both the diodes D_1 and D_2 to be to be an ideal diodes. The input voltage waveform is sketchad in Fig. 33.52 (*a*). notice that the diode D_1 is off as long as the input voltage is below 3 V and D_2 is off as long as the input voltage is below 2 V.

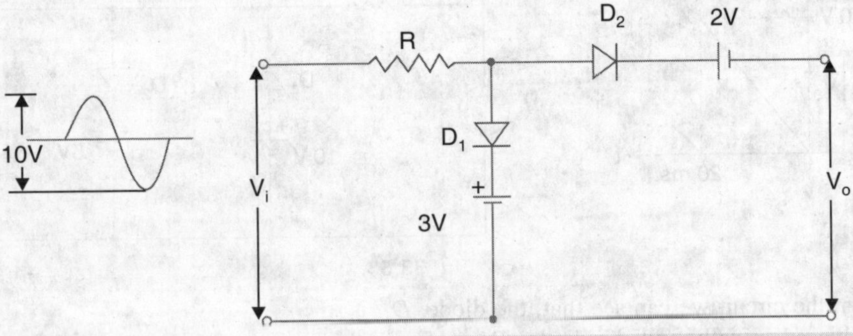

Fig. 33.51.

For an input voltage below 2 V, the output 3 V, D_2 is ON, therefore the output voltage $V_0 = V_i - 2$ V. When the input is equal to or greater than, 3 V, both the diodes D_1 and D_2 one ON and the output voltage. $V_0 = -2\text{ V} + 3\text{ V} = 1$ V. The output voltage waveform is sketched as shown in Fig. 33.52 (*b*).

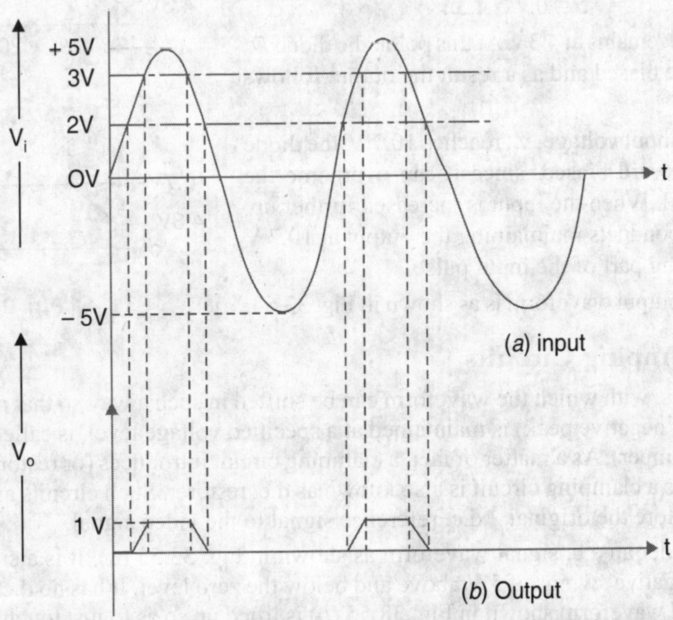

Fig. 33.52. Input and output waveforms of circuit shown in this Fig.

Example 33.18.

Sketch the output of the clipper circuit shown Fig. 33.53 for the given input. Assume diodes are made of silicon and practical. (*Dip. IETE, Dec. 2000*)

Solution.

Given, The input voltage waveform and a clipper circuit shown in Fig. 33.53. The diodes D_1 and D_2 are silicon and practical. It means, when the diode D_1 or D_2 is forward biased (> 0.7 V), it acts like a closed switch and the voltage drop D_1 or D_2 is reversed biased, it acts like an open switch with no voltage drop across it.

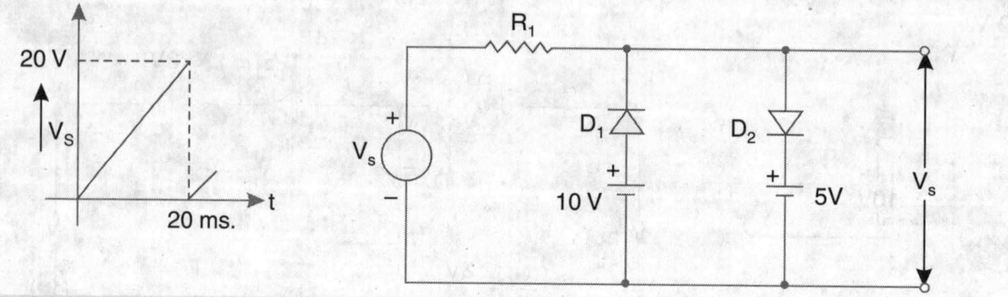

Fig. 33.53.

From the circuit we can see that, the diode, D_1, is reverse biased initially and the diode, D_2, is forward biased, due to the biasing sources connected in series with them. Therefore the output can be calculated as,

$$v_0 = 5 \text{ V} - \text{voltage drop across the diode } D_2$$

$$= 5 - 0.7 = \mathbf{4.3\ v}$$

The output remains at 4.3 v. At this point, the diode D_2 become reverse biased and as a result the output follows the input.

When the input voltage, v_s, reaches 10.7 V, the diode D_1 become forward biased, since it can overcome the barrier potential. When the input is increased further up, the diode, D_1, conducts maintaining the output at 10.7 V for the remaining part of the input pulse.

Therefore, the output waveform is as shown in Fig. 33.54.

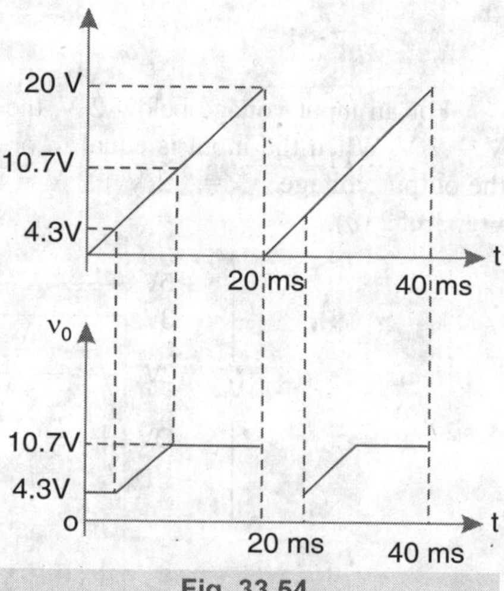

Fig. 33.54.

33.22. Clamping Circuits

The circuits, with which the waveform can be shifted in such a way so that a particular part of it (say positive or negative peak) is maintained at a specified voltage level, is called a clamping circuit (or simply a clamper). As a matter of fact, a clamping circuit introduces (or restores) a d.c. level to an a.c. signal. Thus a clamping circuit is also known as d.c. restorer. Such circuits are used in television receivers to restore the original. *d.c. reference signal to the video signal.

Now consider an a.c. signal waveform as shown in Fig. 33.55 (*a*). It is a sine wave with equal positive and negative swings of 5 V above and below the zero level. It has no d.c. level. On the other hand, the signal waveform shown in Fig. 33.55 (*b*) is lifted up so as to just touch the horizontal axis.

* In television systems, the d.c. reference level corresponding to the brightness level of the picture is not transmitted with video signal.

It is said to have a d.c. level of + 5 V and is known in Fig. 33.55 (*c*) has a d.c. level of – 5 V and is known as waveform negatively clamped at 0 V.

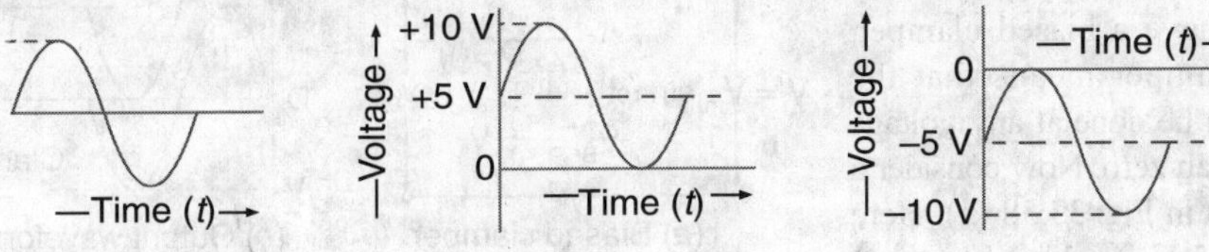

(*a*) A.C. signal. (*b*) Positively clamped A.C. signal. (*c*) Negatively clamped A.C. signal.

Fig. 33.55.

It is possible to clamp a waveform positively at 0 V by connecting a d.c. battery in series with the source, whose e.m.f is equal to the peak value of the sine wave ($v_i = V_m \sin \omega t$) as shown in Fig. 33.56 (*a*). However, it is a very crude way of achieving the required clamping action and therefore is not used in actual practice.

A correct way of clamping the input signal, positively at 0 V, is to use circuit as shown in Fig. 33.56 (*b*). This circuit consists of a diode and capacitor and is popularly known as a diode clamper. The clamping action of the circuit may be understood from the fact that during the negative half cycle of the input voltage (v_i), the diode is forward biased and the current flows through the circuit. As a result of this, the capacitor (*C*) is charged to a voltage equal to the negative peak value (*i.e.*, – V_m). Once the capacitor is fully charged to – V_m, it cannot discharge because the diode cannot conduct in the reverse direction. It means that this capacitor acts as a battery with an e.m.f, equal to the – V_m. The polarity of this voltage is such that it adds to the input signal. Therefore the output voltage is equal to the sum of the a.c. input signal and the capacitor voltage V_m *i.e.*,

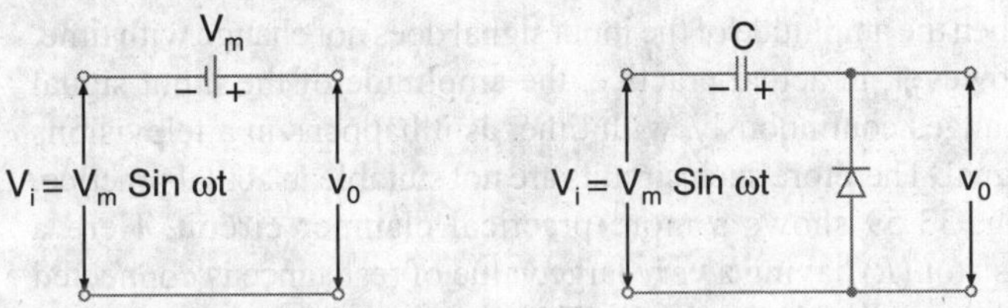

(*a*) Battery connected in series with the source. (*b*) Diode clamper.

Fig. 33.56. Circuits for clamping input signal positively at 0 V.

$$v_o = v_i + V_m = V_m \cdot \sin \omega t + V_m$$

It is evident from the above relation, that a d.c. voltage equal to V_m is added to input signal. This causes the waveform to clamp positively at 0 V.

Now suppose it is required to clamp the input signal waveform ($v_i = V_m \sin \omega t$) negatively at 0 V. This can be done by reversing the diode connections as shown in Fig. 33.57. Here the diode is forward biased during the positive half-cycle of the input signal. Therefore the capacitor is charged to a voltage equal to the positive peak of the input signal (*i.e.*, V_m), the polarity of this voltage is such that it is subtracted from the input signal. Hence, the output voltage,

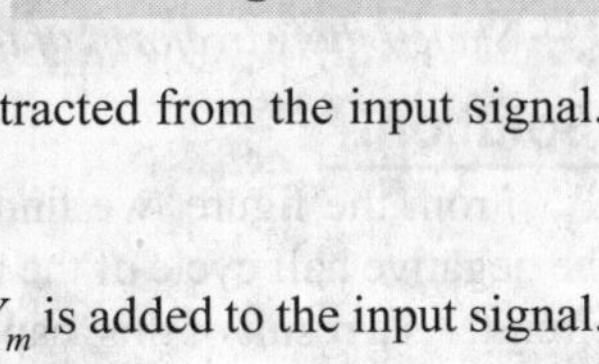

Fig. 33.57.

$$v_o = v_i - V_m = V_m \sin \omega t - V_m$$

It is evident from the above relation that a d.c. voltage equal to – V_m is added to the input signal. This causes the waveform to clamp negatively at 0 V.

The diode clamper circuits shown in Fig. 33.56 (*b*) and 33.57 are also known as positive clamper and negative clamper respectively.

33.23. Biased Clampers

Like a biased clipper, it is also possible to have a biased clamper. A biased clamper means that the clamping can be done at any voltage level other than zero. Now consider a circuit shown in Fig. 33.58 (*a*). Here a battery of 5 V is added in such a way that the clamping takes place positively at – 5 V as shown in Fig. 33.58 (*b*).

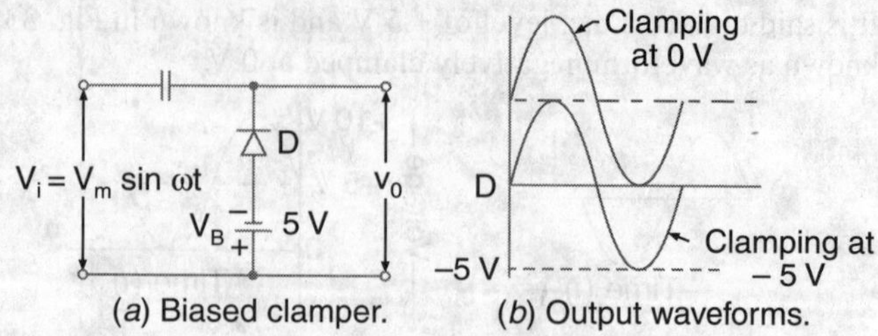

(*a*) Biased clamper. (*b*) Output waveforms.

Fig. 33.58.

Similarly, it is also possible to clamp the input waveform positively at + 5 V by reversing the battery connections.

33.24. Practical Clamper Circuit

The clamper circuits used in last articles work properly, only when the amplitude of the input signal does not change with time. However, in actual practice, the amplitude of the input signal changes continuously, with time, as it happens in a televisions signal. Therefore such circuits are not suitable in actual practice. Fig. 33.59 shows a more practical clamper circuit. Here a resistor (*R*) having a very large value of resistance is connected in parallel with the diode. The purpose of this resistor is to provide a discharge path for the capacitor. The large value for this resistor is required to produce a time constant ($\tau = R \cdot C$) which is large enough than the period of the input signal waveform. For good clamping action, the value of time constant should be at least 10 times the period of the input signal waveform.

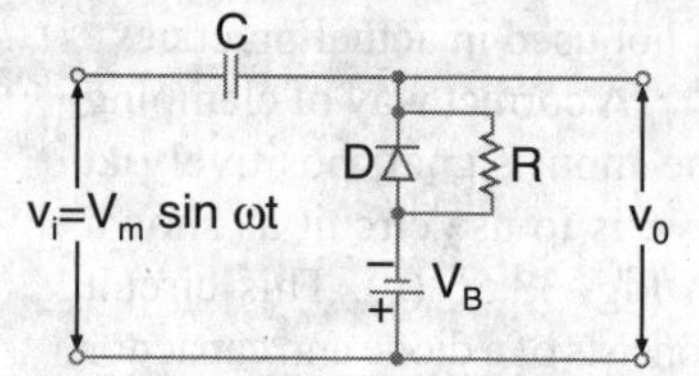

Fig. 33.59. Practical clamper circuit.

Example 33.19.

The input signal with waveform shown in Fig. 33.60 (a) is applied to the clamper circuit shown in Fig. 33.60 (b).

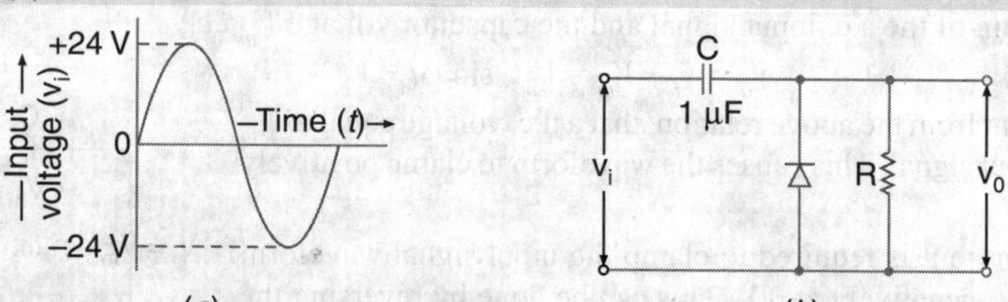

(*a*) (*b*)

Fig. 33.60.

Sketch the waveform of the output voltage.

Solution.

From the figure, we find the diode is forward biased during the negative half cycle of the input signal. Therefore it acts as short circuit. As a result of this, capacitor is charged to a voltage equal to the negatives peak of the input signal (*i.e.*, 24 V). The polarity of this voltage is such that it gets added to the input signal. Therefore the output voltage $v_o = 24 \sin \omega t + 24$. This causes the voltage to clamp positively at zero volt as shown in Fig. 33.61.

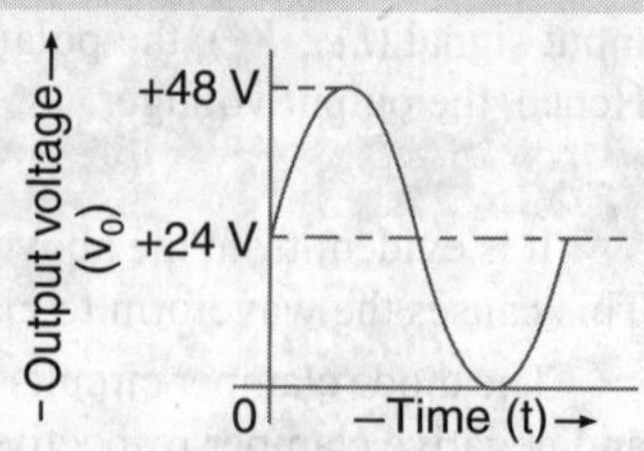

Fig. 33.61.

Example 33.20.

Sketch V_o for the circuit show in Fig. 33.62. *(GBTU/MTU., 2006-07)*

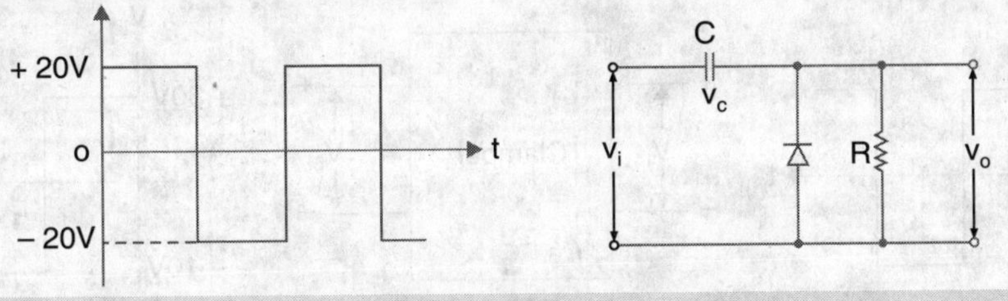

Fig. 33.62.

Solution.

From Fig. 33.62, we find the diode is forward biased during the negative half cycle of the input signal. Therefore it acts as short circuit. As a result of this, capacitor is charged to a voltage equal to the negatives peak of the input signal (*i.e.,* 20 V). The polarity of this voltage is such that it gets added to the input signal. Therefore the output voltage $v_o = 20 + 20 = 40$ V. This causes the voltage to clamp positively at zero volt as shown in Fig. 33.63.

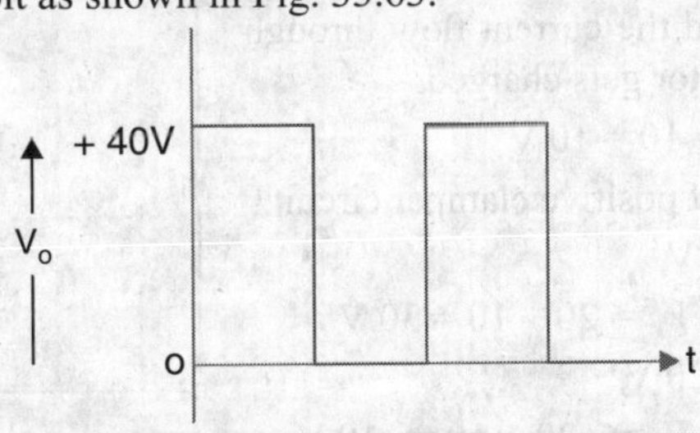

Fig. 33.63.

Example 33.21.

Draw the output waveform of circuit shown in Fig. 33.64, in case, Assume, diode D is ideal. *(R.G.P.V., June 2011)*

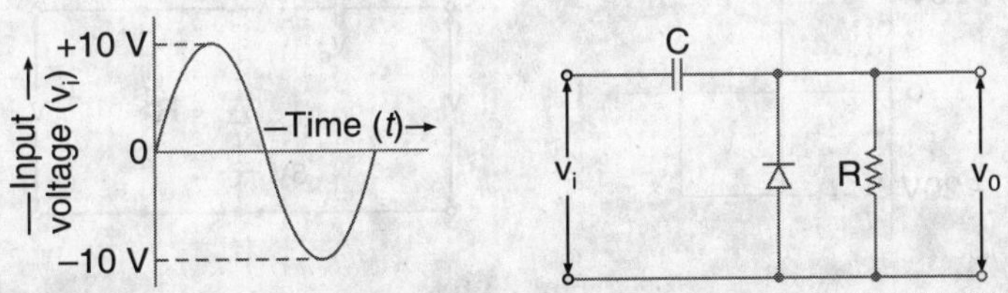

Fig. 33.64.

Solution.

From the figure, we find the diode is forward biased during the negative half cycle of the input signal. Therefore it acts as short circuit. As a result of this, capacitor is charged to a voltage equal to the negatives peak of the input signal (*i.e.,* 10 V). The polarity of this voltage is such that it gets added to the input signal. Therefore the output voltage $v_o = 10 \sin \omega t + 10$. This causes the voltage to clamp positively at zero volts as shown in Fig. 33.65.

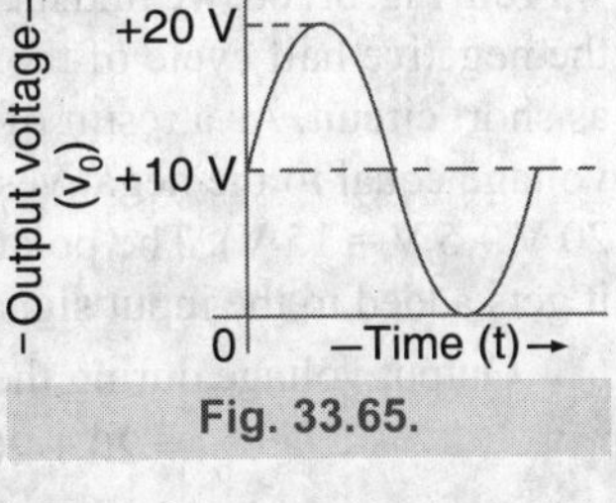

Fig. 33.65.

Example 33.22.

Design a clamper to perform the function indicated in the Fig. 33.66. (*GBTU.,2011,2012*)

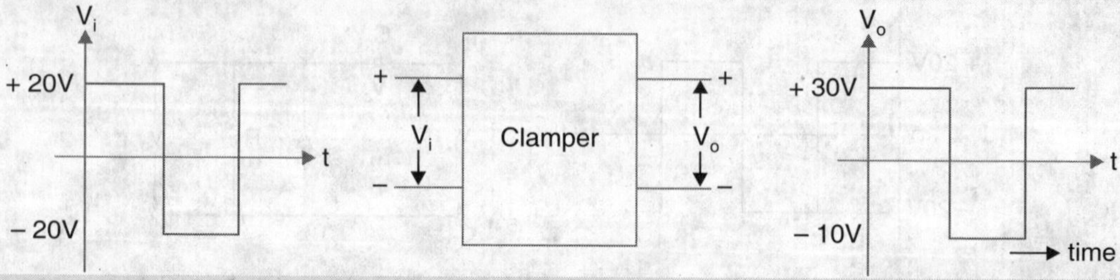

Fig. 33.66.

Solution.

From the given input and output waveform, we see that the output is same as input waveform but a d.c level of +10 V is added to the input voltage so the V_B = 10 V. The clamper circuit is positive clamper.

During the negative half cycle of the input voltage the diode is forward bias and the current flow through the circuit. As a result capacitor gets charged.

$$V_C = 20 - 10 = 10 \text{ V}$$

Fig. 33.67 show the design of positive clamper circuit.

for the positive cycle V_i = 20 V,

$$V_o = V_i + V_C = 20 + 10 = 30 \text{ V}$$

for the negative cycle V_i = – 20 V,

$$V_o = V_i + V_C = -20 + 10 = -10 \text{ V}$$

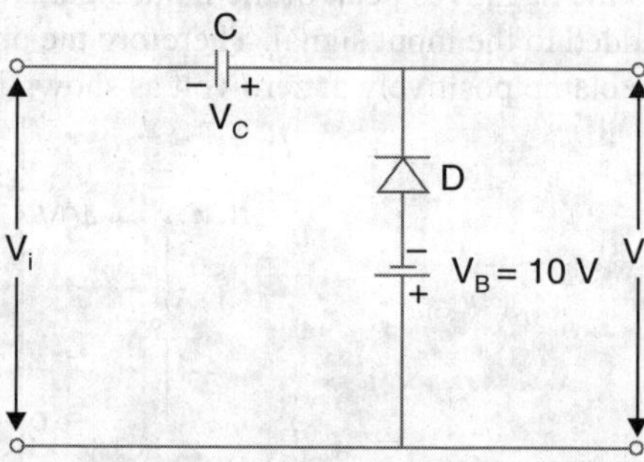

Fig. 33.67.

Example 33.23.

Determine V_o for the following circuit shown in Fig. 33.68. (*GBTU/MTU.,2006-07*)

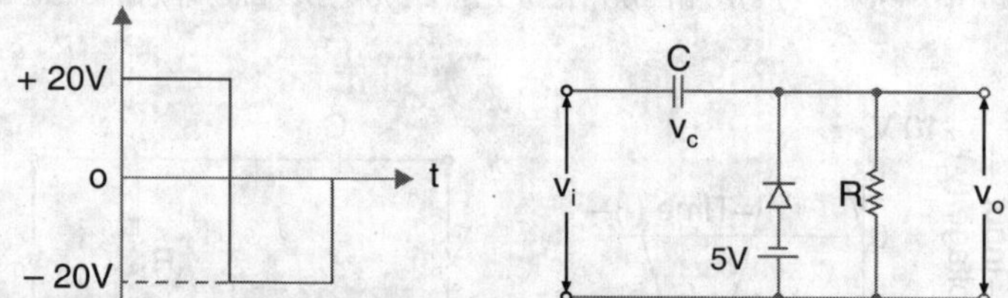

Fig. 33.68.

Solution.

From Fig. 33.68, we find the diode is forward biased during the negative half cycle of the input signal. Therefore it acts as short circuit. As a result of this, capacitor is charged to a voltage equal to the negatives peak of the input signal (*i.e.,* 20 V – 5 V = 15 V). The polarity of this voltage is such that it gets added to the input signal.

Output voltage during the positive half cycle,

$$V_o = 20 + 20 + (-5) = 35 \text{ V}$$

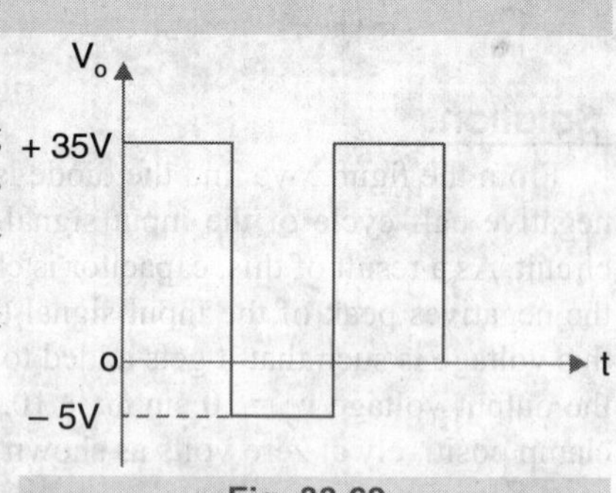

Fig. 33.69.

Output voltage during the negative half cycle,

$$V_o = -5\text{ V}$$

Therefore, the output waveform will be as shown in Fig. 33.69.

Example 33.24.

A ± 10 V square wave is applied (a) to the input terminals of negative voltage clamping circuit and (b) to the input of a negative shunt clipper. Sketch the output waveform that result in each case.

(AMIE, Summer 1999)

Solution.

(*a*) The output of a negative voltage clamping circuit remains at 0 V, during positive half cycle of the input wave. During negative half cycle, the output is given as 2 V_{in}.

Therefore, the resulting output waveform is as shown in Fig. 33.70.

(*b*) The output of a negative shunt clipper remains at 0 V, during negative half cycle of the input wave. During positive half cycle, the output follows the input. Therefore, the resulting output waveform is as shown in Fig. 33.71.

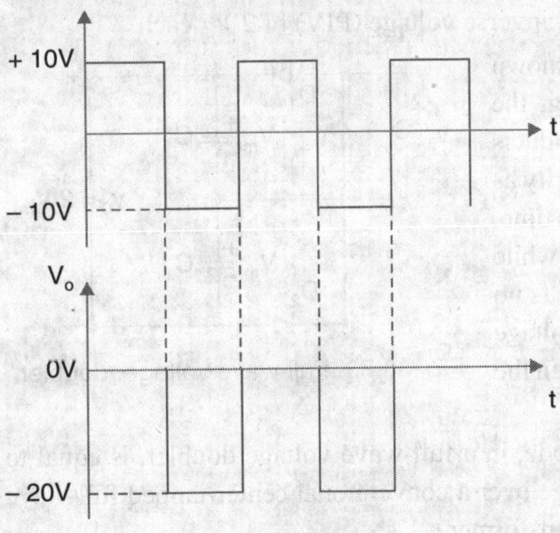

Fig. 33.70.

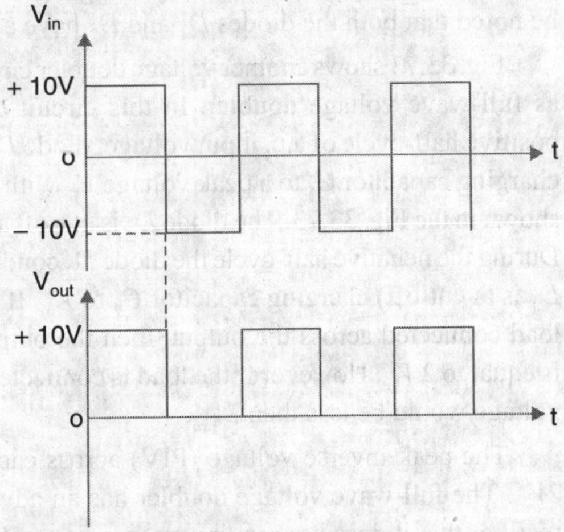

Fig. 33.71.

33.25. Applications of Clamping Circuits

We have already discussed that the clamping circuits are used to shift any part of the input signal waveform and can be maintained at a specified voltage level. Such circuits are used in television receivers to restore the original d.c. reference signal (corresponding to the brightness level of the picture) to the video signal. The clamping circuits are also used to produce a d.c. voltage whose value is a multiple of peak a.c. input voltage (*i.e.*, $2V_m$, $3V_m$, $4V_m$ etc.). Such circuits are known as voltage multipliers. These circuits are used to supply power to high voltage/low current devices like cathode ray tubes used in television receivers, oscilloscopes and computer displays.

33.26. Voltage Multipliers

We have already discussed in the last article that a voltage multiplier is a circuit, which produces an output d.c. voltage whose value is multiple of peak a.c. input voltage (*i.e.*, $2V_m$, $3V_m$, $4V_m$ and so on). Such circuits are used as a power supply for high voltage/low current device like cathode ray tubes.

33.27. Voltage Doubler

A voltage multiplier, whose output d.c. voltage is double the peak a.c. input voltage, is called a voltage doubler Fig. 33.72 shows the circuit, of a half-wave voltage doubler. In this circuit, each section of a diode and capacitor (say D_1 and C_1 or D_2 and C_2) is called a peak rectifier. The operation of the circuit may be explained as given below :

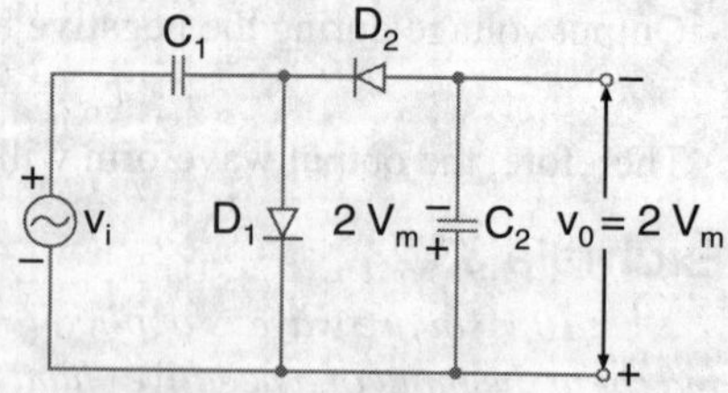

Fig. 33.72. Half-wave voltage. doubler.

During the positive half-cycle of the input signal, the diode D_1 conducts (and diode D_2 is cut-off), charging the capacitor C_1 up to the peak rectified voltage (*i.e.*, V_m). During the negative half-cycle, diode, D_1 is cut-off and diode D_2 conducts charging capacitor C_2. It may be noted that during negative half-cycle, the voltage across capacitor C_1 is in series with the input voltage. Therefore the total voltage presented to capacitor C_2 is equal to $2\ V_m$. As a result of this, the capacitor C_2 is charged to a voltage of $2\ V_m$ during the negative half-cycle. On the next positive half-cycle, the diode D_2 is non-conducting and the capacitor will discharge through the load. If no load is connected across capacitor C_2, both capacitors stay charged at their full values (*i.e.*, C_1 to V_m and C_2 to $2\ V_m$). It may be noted that both the diodes D_1 and D_2 have a peak inverse voltage (PIV) of $2\ V_m$ each.

Fig. 33.73 shows another voltage doubler circuit known as full-wave voltage doubler. In this circuit during, the positive half-cycle of a.c. input voltage, diode D_1 conducts charging capacitor C_1 to a peak voltage V_m with polarity as shown in the Fig. 33.73. The diode D_2 is cut-off at this time. During the negative half-cycle the diode D_2 conducts (while D_1 is at cut-off) charging capacitor C_2 to V_m. If there is no load connected across the output, then the output voltage is equal to $2\ V_m$. However if the load is connected, then the voltage would be less then $2V_m$.

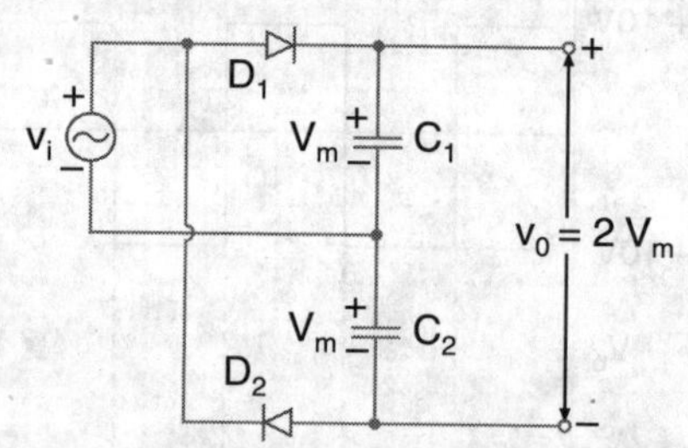

Fig. 33.73. Full-wave voltage. doubler.

The peak inverse voltage (PIV) across each diode, in a full-wave voltage doubler, is equal to $2V_m$. The full-wave voltage doubler has an advantage over a conventional center-tapped full-wave rectifier that it does not require any center-tapped transformer.

33.28. Voltage Trippler and Quadrupler

Fig. 33.74 shows the extension of a half-wave voltage doubler. This circuit develops an output voltage equal to 3 and 4 times the peak input voltage. It is obvious from the pattern of the circuit connection as to how additional diodes and capacitors may be connected for obtaining higher multiplications of the basic peak voltage. Theoretically, there is no upper limit for the amount of voltage multiplication. But it has been observed that if we increase the voltage multiplication beyond four, the regulation is very poor. It is because of this fact, that voltage multipliers are not used in low-voltage supplies.

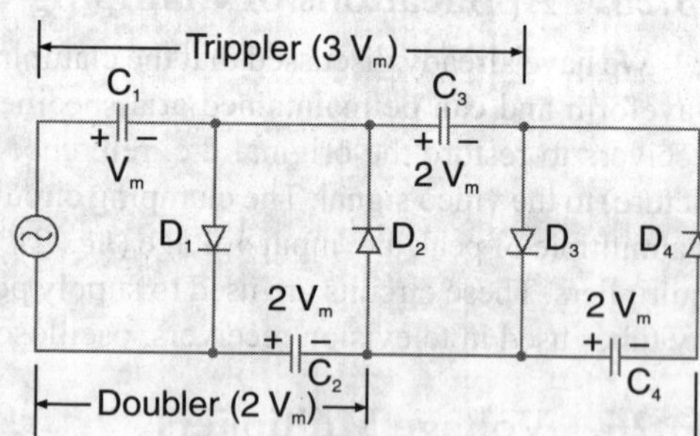

Fig. 33.74. Voltage tripper and quadrupler.

The operation of the voltage trippler and quadrupler circuit may be understood from the fact, that during the first positive half-cycle, the capacitor C_1 charges through diode D_1 to a peak voltage V_m. During the negative half-cycle, capacitor C_2 charges through diode D_2 to twice the peak voltage $2V_m$ developed by the sum of the voltages across capacitor C_1 and the input signal.

During the second positive half-cycle, the diode D_3 conducts and the voltage across capacitor C_2 charges the capacitor C_3 to the same 2 V_m peak voltage. And during the negative half-cycle, diodes D_2 and D_4 conduct allowing capacitor C_3 to charge capacitor C_4 to the same 2 V_m peak voltage.

It is evident from Fig. 33.72, that the voltage across capacitor C_2 is 2 V_m, across C_1 and C_3 is 3 V_m and across C_2 and C_4 is 4 V_m. It means that the voltage quadrupler circuit can provide three different voltages *i.e.*, 2 V_m, 3 V_m and 4 V_m. It will be interesting to know that if we use additional sections of diode and capacitor with the voltage quadrupler circuit, then each capacitor will be charged to 2 V_m. The peak inverse voltage for each diode in the circuit is equal to 2 V_m.

NUMERICAL PROBLEMS

1. Find the output voltage shape for the clipping circuit shown in Fig. 33.75 (*a*) for the inputs shown in Fig. 33.75 (*b*).

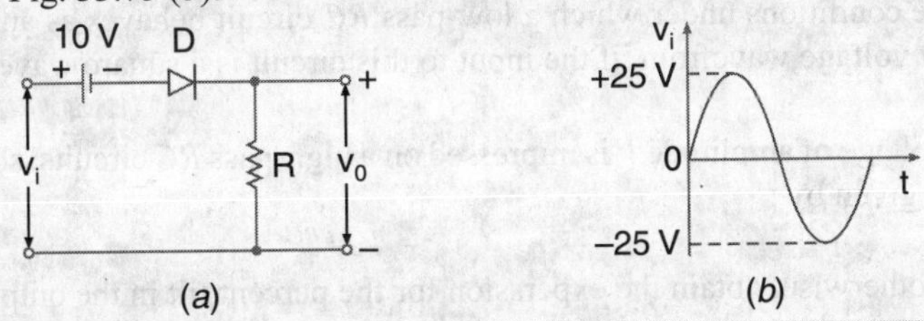

Fig. 33.75.

2. Sketch the output waveform for a diode clamper circuit shown in Fig. 33.76 (*a*) for a sine wave input with a peak value of 30 V.

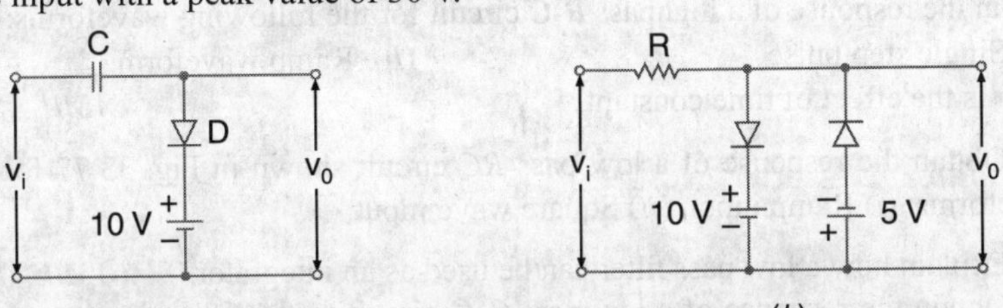

Fig. 33.76.

3. Draw neatly the output waveform for a diode clamper circuit shown in Fig. 33.76 (*b*) for a sine wave input with a peak voltage of 20 V.
4. The pulse from a high voltage generator reises linearly for 0.04 ms and than remains constant for 2 μs. The rate of rise of pulse of the pulse is measured with RC differentiating circuit whose time constant is 400 ps. If the positive output voltage from the differentiator has a maximum value of 100 V, what is the peak value of the generator ? Derive clearly the expressions used. (*AMIE., Winter 1996*)
5. The pulse from a high voltage generator 'a magnetron' rises linearly for 0.05 ms and then remians constant for 1 μs. The rate of rise of pulse is measured with an *RC* differentiating circuit whose time constant is 250 ps. If the positive output voltage from the differentiator

has a maximum value of 100 V, what is the peak voltage of generator ? Develop clearly, any expression employed ? (*AMIE., Winter 1993*)

6. A sinusoidal input, $v_i(t) = V_m \sin Wt$ is fed to a passive *RC* differentiator. If $\omega RC = 0.01$, determine (*i*) the phase error and (*ii*) the magnitude reduction factor. (*AMIE., Summer 1997*)

DESCRIPTIVE QUESTIONS

1. What do you mean by the term 'wave shaping'? Also discuss, why it is needed in practical applications?
2. Discuss the necessary criteria for a good differentiating circuit. Give some applications of this circuit.
3. Show by drawing the neat waveforms as to how we can generate
   (*i*) narrow pulses from square wave, and
   (*ii*) square wave from triangular wave
4. Enumerate the conditions under which an RC circuit behaves as an integrator.
5. Sketch the output waveforms for an integrating circuit when the input signal is
   (*i*) a square wave and (*ii*) a rectangular wave
6. Obtain the conditions under which a low-pass *RC* circuit behaves as an integrator. Sketch the output voltage wavefrom, if the input to this circuit is a square wave. (*AMIE., Summer 1993*)
7. If a step voltage of amplitude *V* is impressed on a high-pass *RC* circuits, show that the output voltage is given by :
   $$v_o = Ve^{-t/RC}$$
   Hence or otherwise, obtain the experssion for the percent tilt in the output. (*AMIE., Summer 1993*)
8. Show how an *RC* circuit can be used as (*i*) an integrator and (*b*) a differentiator. (*AMIE., Winter 1993*)
9. Obtain the response of a highpass *R-C* circuit for the following waveforms:
   (*a*) Single step pulse (*b*) Ramp waveform.
   What is the effect of time constant. (*AMIE., Summer 1994*)
10. (*a*) Obtain the response of a low pass *RC* circuit shown in Fig. 33.77 for the following waveforms: (*i*) Ramp input, (*ii*) Square wave input.
    (*b*) Explain how a low pass filter can be used as an integrator. (*AMIE., Summer 1995*)
11. (*a*) Obtain the response of a low-pass R-C circuit to the following inputs (*i*) single step pulse (*ii*) ramp waveform
    (*b*) Obtain the conditions under which a low-pass *R-C* circuit behaves as an integrator. Sketch the output voltage waveform if the input to this circuit is a symmetrical square wave. (*AMIE., Winter 1994*)

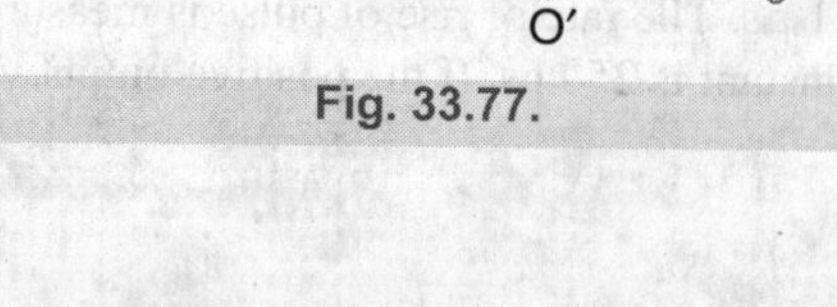

Fig. 33.77.

12. Plot the capacitor voltage and current versus time for a series *RC* circuit with a switch and a dc voltage source. Explain the shape of the graph. Derive equations for the capacitor voltage when $t = RC$ and when $t = 5\ RC$. State any assumptions made by you. *(AMIE., Winter 1998)*
13. A pulse is applied to a low-pass *RC* circuit. Prove by direct integration that the area under the pulse is the same as the area under the output waveform across capacitor. Explain this result physically. *(Nagpur Univ., Winter 2003)*
14. Show how the response of a lowpass *RC* circuit for ramp input for different time constant. How can it be modified for square wave input ? *(AMIE., Winter 1996)*
15. Explain the working of a positive, diode clipping circuit. *(M.G University., May 2005)*
16. Draw the circuit of a double clipper. Show the input and corresponding output. *(M.G University., July/Aug. 2007)*
17. Draw the types of biased-clippers and explain their operation. *(Calicut University., Dec. 2006)*
18. Draw the circuit of negative clipper with input and output waveforms. *(Anna University., May/June 2007)*
19. Draw and explain the negative clamping circuit using a diode. *(M.G University., May/June 2006)*
20. Draw the various clipper and clamper configurations along with input output waveforms. *(GBTU/MTU., 2009)*
21. Draw and explain the working of positive clamper *(VTU., June /July 2008)*
22. Draw the circuit diagram of a negative voltage clamper and plot its output for a sinusoidal input. *(Anna University., Nov/Dec 2007)*
23. Give the circuit of combination clipper. Draw its input –output waveform for sinusoidal input. *(Anna University., Nov/Dec 2006)*
24. What is a clamper circuit ? Draw the circuit diagram of a positive clamper and a negative clamper showing their output waveforms *(PBUT., 2009)*
25. Explain how diodes are useful for clamping operation Draw a negative clamping circuit. Find the relation between the time constant of the circuit and the time period of the input signal. What should be the full discharge time, in terms of the time constant of the circuit, for the capacitor you have used. *(PBUT., 2008)*
26. What are clippers and clampers? Explain in detail the various types of clippers and clampers with suitable diagrams *(GBTU/MTU., 2008-09)*
27. Write short notes on any two of the following:
    (*a*) Voltages multiplier circuit using diodes.
    (*b*) Clipping circuits *(GBTU/MTU., 2006-07)*
28. What is the difference between clipping circuit and clamping circuit? Determine for the network. *(GBTU/MNTU., 2002-03)*
29. Draw the voltage double circuit and the working of the circuit. *(GBTU/MTU., 2005-06)*
30. What is a voltage multiplier? *(Anna University., April/May 2008)*
31. What are Multipliers circuits? Explain and draw the circuit required for obtaining an output voltage 10 times that of applied input voltage. *(GBTU/MTU., 2008-09)*
32. With neat diagram circuit diagram, explain the working of voltage trippler. *(GBTU/MNTU., 2004-05)*

33. What is a clamper? Explain with the help of neat diagram. *(R.G.P.V., Dec. 2004)*
34. Write a short note on clamping circuits. *(R.G.P.V., June 2005)*
35. Write a short note on clipper and clamper circuits. *(R.G.P.V., May/ June 2006)*
36. Differentiate between clippers and clampers. *(R.G.P.V., June 2011, Dec. 2006)*
37. What is the function of a clipping circuit? Draw the circuit diagram of a diode clipper that limits the positive peak of the input voltage. Explain how the circuit works. *(R.G.P.V., Dec. 2006)*
38. Explain and draw the circuit of a simple positive level diode clipper using resistor R in the series arm and shunt arm consisting of reference voltage in series with the diode D. *(R.G.P.V., June 2007)*
39. What is clamper? Explain with the help of neat diagram. *(R.G.P.V., Dec. 2004)*
40. Explain the working of a clamper circuit with input/output wave forms. *(R.G.P.V., June 2004)*
41. Give circuit diagram of a clamper with input and output wave forms. *(R.G.P.V., Dec. 2008, 2003)*
42. Draw the diagram of clipping and clamping circuits and explain in working principle. *(R.G.P.V., Dec. 2001)*
43. Write short note on half-wave voltage doublers. *(Nagpur University, Winter 2002)*
44. Draw and explain the voltage doublers circuit. *(Nagpur University, Summer 2000)*
45. Write short notes on :
    (*a*) Clamping circuit *(R.G.P.V., June 2005)*
    (*b*) Clipper and clamper *(R.G.P.V., Dec. 2007)*

MULTIPLE CHOICE QUESTIONS

1. The criteria for a good differentiating circuit is that its time constant ($\tau = R \cdot C$) should be
   (*a*) much larger than the period of the input waveform
   (*b*) less than or equal to the period of the input waveform
   (*c*) much smaller than the period of the input waveform
2. If the input to a differentiating circuit is a triangular wave, then the output wave will be
   (*a*) square wave
   (*b*) rectangular wave
   (*c*) sawtooth wave
   (*d*) none of the above
3. The criteria for good integrating circuit is that its time constant ($\tau = R \cdot C$) should be
   (*a*) equal to the period of the input waveform
   (*b*) much larger than the period of the input waveform
   (*c*) much smaller than the period of the input waveform
4. The main function of a clipping circuit is to
   (*a*) remove a certain portion of the input signal above or below a certain level
   (*b*) restore d.c. level to the signal
   (*c*) suppress amplitude variations in the input signal voltage
   (*d*) both (*a*) and (*c*) above
5. The clamper circuit is used to
   (*a*) introduce a d.c. level to a.c. signal
   (*b*) suppress variations in amplitude of the input signal
   (*c*) obtain an output which is integral of the input signal

6. A capacitor of 100 μ*F* is charged to 10 V through a resistance of 10 kΩ. It will be fully charged in

   (*a*) 5 s (*b*) 0.1 s (*c*) 1 s (*d*) 0.5 s

7. An integrator has a transfer function whose

   (*a*) amplitude increases linearly with the frequency
   (*b*) amplitude varies inversely with the frequency
   (*c*) amplitude remains constant
   (*d*) amplitude decreases linearly with the frequency (*AMIE., Summer 2000*)

8. A triangular waveform can be generated by

   (*a*) differentiating a square waveform
   (*b*) integrating a square waveform
   (*c*) differentiating a sinusoidal waveform
   (*d*) integrating a sinusoidal waveform. (*AMIE., Winter 2001*)

ANSWERS

1. (*c*)	**2.** (*a*)	**3.** (*b*)	**4.** (*d*)
5. (*a*)	**6.** (*a*)	**7.** (*a*)	**8.** (*b*)

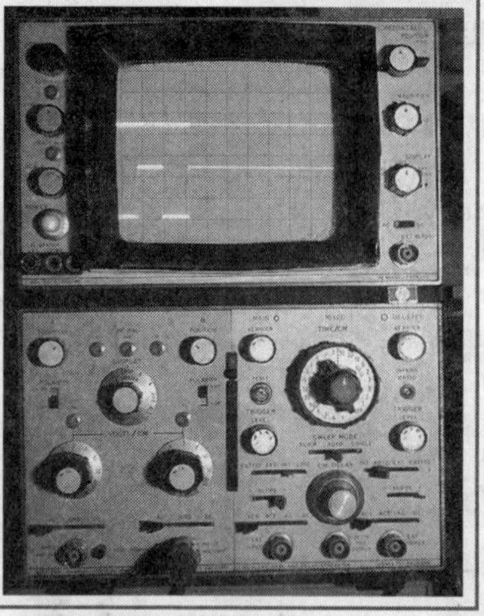

34

Chapter

Time Base Circuits

Outline

34.1. Introduction

The circuits, which provide a linear time scale, are called time base circuits or time base generators. In such circuits, either the output voltage (or the output current) is a linear function of time, over a specified time interval. These circuits are extremely important in most of the electronic systems. For example, a linear time base voltage is required on the deflection plates of cathode ray oscilloscope, to sweep the electron beam, from left to right across the screen. Similarly, a linear time base current waveform is required in the deflection coils of a television receiver. Because of the sweep application, the circuits are also sometimes known as sweep circuits (or sweep generators).

34.2. General Features of a Time Base Signal

Fig. 34.1 (*a*) shows a typical time base signal. Here the voltage starts from some initial value as indicated by point *A* in the figure. Then it increases linearly with time to a maximum value indicated by point *B* in the figure. After this, the voltage returns back to its initial value as indicated by point *C* in the figure.

The time, required by the voltage from its initial value to reach its maximum value, is known as sweep time and is designated by the symbol T_s. Similarly, the time taken by the voltage to its initial value is called a restoration time and is designated by the symbol T_r. The restoration time is also known as return time or the flyback time.

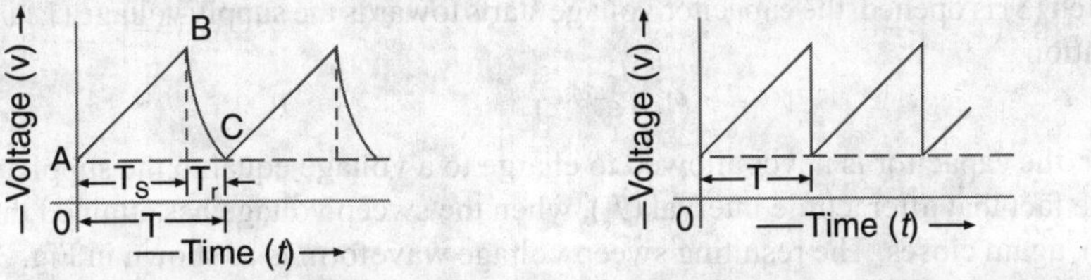

(a) General sweep voltage. (b) Sawtooth voltage waveform.

Fig. 34.1. Sweep waveforms.

In most of the applications, the shape of the waveform during the restoration time, as well as the restoration time itself, is of no importance. However, in certain applications, it is desired that the restoration time is very short in comparison with the time occupied by the linear portion of the waveform (*i.e.*, $T_r << T_s$). In that case, the restoration time is extremely short and a new voltage is initiated at the instance the previous one is terminated. The waveform, in such a case, is as shown in Fig. 34.1 (*b*) and is known as sawtooth waveform. It is customary to refer to waveforms shown in Fig. 34.1 (*a*) and (*b*) as sweep waveforms even in applications not involving the deflection of an electron beam.

It will be interesting to know that the circuits, which generate time base signals, do not ordinarily provide sweep voltages that are precisely linear. A most useful way of expressing the deviation from linearity is the slope error or sweep-speed error which is given by the relation,

$$e_s = \frac{\text{Difference in slope at the beginning and end of sweep}}{\text{Initial value of slope}}$$

There is one another term called 'sweep speed', which is used to compare the performance of different sweep generating circuits. In all these circuits, a capacitor is an essential element. The charging and discharging of a capacitor generates a sweep voltage. Suppose a capacitor (C) is charged by a constant current (I). Then the voltage across the capacitor at any time (v_C) is given by the relation,

$$v_c = \left(\frac{I}{C}\right) t$$

The rate of change of capacitor voltage with time is known as sweep speed. Thus

$$\text{Sweep speed} = \frac{dv_c}{dt} = \frac{d}{dt}\left(\frac{I}{C}\right) t = \frac{I}{C}$$

The sweep speed is required to be constant in most of the applications.

34.3. Types of Time Base Circuits

Following two types of time base circuits are important from the subject point of view:

1. ***Free running time base generator.*** A circuit in which the periodic sawtooth waveform is generated, without the application of any signal, is called a free running time base generator. Such a circuit is required to display a periodic waveform. It may be noted that in a free running time base generator, the sweep time (T_s) must be larger than the period of the waveform to be displayed.
2. ***Triggered time base generator.*** A circuit in which the linear waveform with a prescribed duration of time is generated by the application of a trigger signal, is called a triggered time base generator. Such a circuit is required to display widely separated narrow-width pulser or a waveform which may not be periodic but may occur at irregular intervals.

The operation of the circuit may be understood from the condition, that the switch (*S*) is initially assumed to be closed. At this instant, the capacitor voltage and hence the output voltage (v_o) is zero, when the switch (*S*) is opened, the capacitor voltage starts towards the supply voltage (*V*) in accordance with the equation,

$$V_s = V(1 - e^{-t/RC}) \qquad ...(i)$$

However, the capacitor is never allowed to charge to a voltage equal to the supply voltage. It is because of the fact that after a time interval (T_s), when the sweep voltage has attained the value (V_s), the switch (*S*), again closes. The resulting sweep voltage waveform is as shown in Fig. 34.2 (*b*). The sweep-speed error for the exponential sweep circuit is given by the relation,

$$e_s = \frac{V_s}{V}$$

It is evident from the above relation that smaller the value of sweep voltage (V_s), lower will be the value of sweep-speed error and hence higher is the value of linearity. Due to this reason, the exponential charging sweep circuit is used only in those applications, which require lower sweep voltages (usually of the order of volts or tens of volts). If $t/RC << 1$, then we can expand equation (*i*) as follows:

$$v_s = \frac{V \cdot t}{R \cdot C}\left(1 - \frac{t}{2\,RC} + \frac{t^2}{6\,R^2C^2} \cdots\right)$$

$$= \frac{V \cdot t}{R \cdot C} \qquad \text{...(Retain first term and neglect others)}$$

Since $v_s = V_s$ when $t = T_s$, therefore

$$V_s = \frac{V \cdot T_s}{R \cdot C}$$

or

$$\frac{V_s}{V} = \frac{T_s}{R \cdot C}$$

∴ Sweep-speed error,

$$e_s = \frac{V_s}{V} = \frac{T_s}{R \cdot C} = \frac{T_s}{\tau}$$

where τ is the time constant and is equal to ($R \cdot C$). It is obvious from this relation that for good linearity (*i.e.*, smaller sweep-speed error), the time constant (τ) must be very large as compared to sweep duration (T_s).

34.4. Methods of Generating a Time Base Waveform

Although there are a number of methods of generating a time base waveform, yet the following are important from the subject point of view:

1. ***Exponential charging.*** In this method, a capacitor is charged through a resistor to a voltage, which is small in comparison with the supply voltage.
2. ***Constant current charging.*** In this method, a capacitor is charged linearly from a constant-current source.
3. ***Miller integration.*** In this method, an operational amplifier (Op Amp) based integrator is used to convert a step into a ramp waveform.
4. ***Boot strap.*** In this method, a constant current is approximated by maintaining nearly constant voltage across a fixed resistor in series with a capacitor.

34.5. Exponential Sweep Circuit

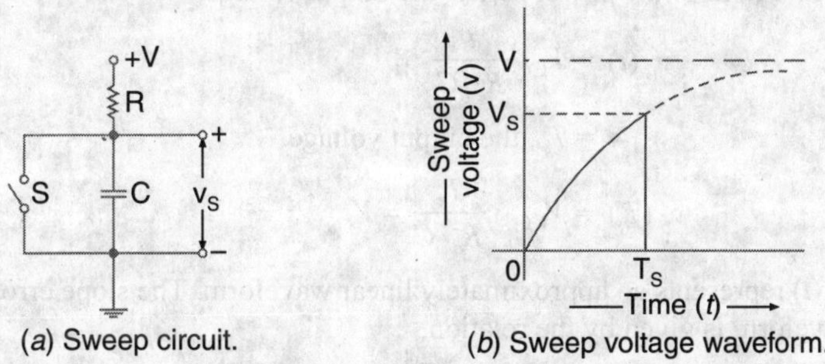

Fig. 34.2.

Fig. 34.2 shows a simplest circuit to generate a time base signal. It is known as exponential sweep circuit. This circuit consists of a series RC circuit connected to a supply voltage *V*. *A* switch *S* (which is usually is transistor switch) is connected across the capacitor *C*. The output voltage is taken across the capacitor (*C*).

34.6. Sweep Circuit Using Transistor Switch

We have already discussed in the last article about the exponential sweep circuit. We know that the circuit requires an electronic switch *S*. Many devices are available to serve as a switch *S* such as bipolar junction transistor, field-effect transistor, unijunction transistor, thyratron etc.

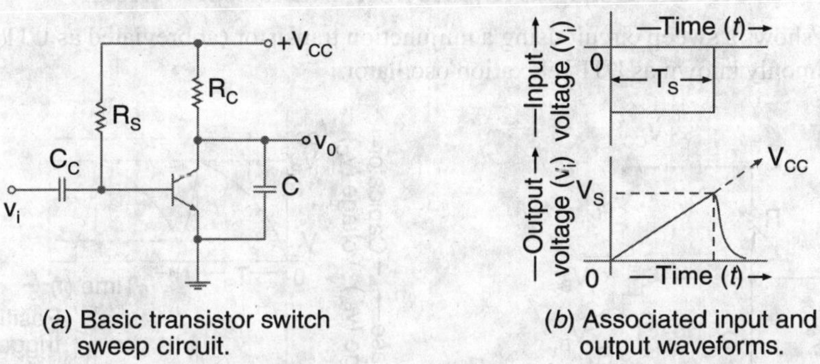

Fig. 34.3.

Fig. 34.3 (*a*) shows the basic circuit of a transistor switch sweep circuit. This circuit requires a gating waveform (v_i) as shown in Fig. 34.3 (*b*). It may be obtained from a monostable multivibrator (*i.e.*, one shot) or an astable multivibrator. If the 'gating' waveform is obtained from a monostable multivibrator, the sweep generated by the circuit is called *triggered sweep* and the circuit is known as *triggered voltage time base generator*. However, if the 'gating' waveform is obtained from an astable multivibrator, the sweep generated is called *free running sweep* and the circuit is known as *free running voltage time base generator.*

Initially, the transistor is biased ON and operates in the saturation region. Thus when there is no input (*i.e.*, $v_i = 0$), the output voltage is zero. (Actually its value is equal to $V_{CE\,(sat)}$). When the 'gating' pulse (*i.e.*, a negative pulse) is applied, the transistor turns OFF. As a result of this, the capacitor voltage rises to a target value V_{CC} with a time constant $R_C \cdot C$. The charging curve ignoring $V_{CE(sat)}$ is governed by relation,

$$v_o\,(t) = V_{CC}\,(1 - e^{-t/Rc.C})$$

If $t/R_c \cdot C << 1$, the above equation may be expanded into a power series in $t/R_c \cdot C$. Then retaining only the first term of the power series, the output voltage,

$$v_o(t) = V_{CC} \frac{t}{R_c \cdot C} \quad ...(i)$$

At $t = T_s$, the output voltage,

$$v_o(T_s) = V_{CC} \cdot \frac{T_s}{R_c \cdot C} = V_s \quad ...(ii)$$

The equation (*i*) represents an approximately linear waveform. The slope error or sweep-speed error or such a waveform is given by the relation,

$$e_s = \frac{V_s}{V_{CC}} = \frac{T_s}{R_c \cdot C}$$

It is evident from the above relation that the output or sweep voltage level (V_s) must be a small fraction of the supply voltage (V_{CC}) and this is not a desirable feature, if a certain sweep voltage level is desired. This shortcoming can be remedied by the Miller sweep and the Bootstrap sweep circuits.

It may be noted that the transistor switch if OFF only for the gating time (T_s). At the end of time T_s, the capacitor discharges, and the voltage is again zero. It is also possible to generate a negative-going sweep using transistor switch. But in that case, we have to use a PNP transistor instead of a NPN transistor in the sweep circuit.

34.7. Sweep Circuit Using Unijunction Transistor Switch

Fig. 34.4 shows a sweep circuit using a unijunction transistor (abbreviated as UJT) switch. The circuit is commonly known as UJT relaxation oscillator.

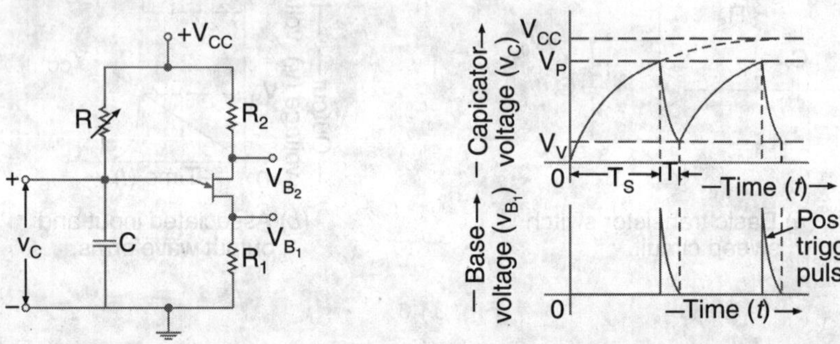

(*a*) UJT relaxation oscillator. (*b*) Associated waveforms.

Fig. 34.4.

The operation of the circuit may be understood form the condition that when the supply voltage (V_{CC}) is switched ON, the capacitor (*C*) charges through resistor (*R*), till the capacitor voltage reaches the voltage level (V_P) which is called peak-point voltage. At this voltage, the unijunction transistor turns ON. As a result of this, the capacitor (*C*) discharges rapidly through resistor (R_1). When that capacitor voltage drops to level V_v (called valley-point voltage) the unijunction transistor switches OFF allowing the capacitor (*C*) to charge again.

Fig. 34.4 (*b*) shows the collector voltage (v_c) and the base 1 voltage waveforms during the capacitor charging and discharging interval. It may be noted that the voltage developed at base 1 terminal is in the form of narrow pulses commonly known as trigger pulses. The similar pulses (but with opposite polarity) are also available at the base terminal of the unijunction transistor.

It will be interesting to know that the sweep period, also called period of oscillation, depends upon time constant ($R \cdot C$) and the type of unijunction transistor used in the circuit. However, it is independent of supply voltage (V_{CC}) and the temperature. The sweep period is given by the relation,

$$T = R \cdot C \cdot \log_e \frac{1}{1-\eta}$$

$$= 2.3\ \text{R} \cdot \text{C} \cdot \log_{10} \frac{1}{1-\eta}$$

where η is a intrinsic stand-off ratio. Its value is approximately equal to the ratio of peak-point voltage to the supply voltage *i.e.*, $\eta = V_P / V_{CC}$. The sweep frequency (or the frequency of oscillation) is the reciprocal of sweep period *i.e.*,

$$f = \frac{1}{T} = \frac{1}{2.3\ R \cdot C \cdot \log_{10} \dfrac{1}{1-\eta}}$$

The sweep frequency can be varied by changing values of either resistance (R) or capacitor (C). However, it is more convenient to change the value of (R) than (C). It is due to this fact that the resistor (R) is shown as a variable resistor in Fig. 34.4 (*a*).

Example 34.1.

In a simple UJT sweep circuit, the resistance and capacitance are 100 kΩ and 0.4 μF. The ratio of the peak-point voltage to the supply voltage is 0.57. Find the frequency of sweep.

(*Anna University Nov/Dec. 2007*)

Solution.

Given: $R = 100\ \text{k}\Omega = 100 \times 10^3\ \Omega$; $C = 0.4\ \mu\text{F} = 0.4 \times 10^{-6}$ F and $\eta = V_P/V_{CC} = 0.57$.

We know that the frequency of sweep,

$$f = \frac{1}{2.3\ R \cdot C \cdot \log_{10} \dfrac{1}{1-\eta}}$$

$$= \frac{1}{2.3 \times (100 \times 10^3) \times (0.4 \times 10^{-6}) \log_{10} \dfrac{1}{1-0.57}}$$

$$= \frac{1}{0.92 \times \log_{10} 2.33} = \frac{1}{0.092 \times 0.367}\ \text{Hz}$$

$= 29.6$ Hz **Ans.**

Example 34.2.

A unijunction transistor with η = 0.62 is used in a relaxation oscillator circuit with R = 5 kΩ and C = 0.05 μF.

(*a*) *Determine the period and frequency of oscillation.*

(*b*) *Determine the new value of R, which must be changed in order to obtain a frequency of oscillation of 50 Hz.*

(*c*) *If C is increased to 0.5* mF*, what should be the value of R, if the frequency is to be 50 Hz?*

Solution.

Given: $\eta = 0.62$; $R = 5\ \text{k}\Omega = 5 \times 10^3\ \Omega$ and $C = 0.05\ \mu\text{F} = 0.05 \times 10^{-6}$ F.

(a) *Period and frequency of oscillation*

We know that period of oscillation,

$$T = 2.3\ R \cdot C \log_{10} \frac{1}{1-\eta}$$

$$= 2.3 \times (5 \times 10^3) \times (0.05 \times 10^{-6}) \log_{10} \frac{1}{1-0.62}$$

$$= (5.75 \times 10^{-4}) \times \log_{10} 2.63$$

$$= 2.4 \times 10^{-4} \text{ sec} = 0.24 \text{ msec } \textbf{Ans.}$$

and the frequency of oscillation,

$$f = \frac{1}{T} = \frac{1}{2.4 \times 10^{-4}} = 4140 \text{ Hz } \textbf{Ans.}$$

(b) *New value of R*

Here $f = 50$ Hz

Let R = Value of resistance which gives 50 Hz frequency.

We know that period of oscillation,

$$= \frac{1}{f} = \frac{1}{50} = 0.02 \text{ sec}$$

and the period (T),

$$0.02 = 2.3\ R \cdot C \log_{10} \frac{1}{1-\eta}$$

$$= 2.3 \times R \times (0.05 \times 10^{-6}) \times \log_{10} \frac{1}{1-0.62}$$

$$= 2.3 \times R \times (0.05 \times 10^{-6}) \times \log_{10} 2.36$$

$$= 2.3 \times R \times (0.05 \times 10^{-6}) \times 0.42$$

$$= 4.83 \times 10^{-8}\ R$$

$$\therefore \quad R = 0.02/(4.83 \times 10^{-8}) = 4.14 \times 10^3\ \Omega$$

$$= 414 \text{ k}\Omega \textbf{ Ans.}$$

(c) *Value of R with C = 0.5 μF*

Here $f = 50$ Hz and $C = 0.5\ \mu F = 0.5 \times 10^{-6}$ F

Let R = Value of resistance which gives 50 Hz frequency.

We know that period of oscillation,

$$T = \frac{1}{f} = \frac{1}{50} = 0.02 \text{ sec}$$

and the period of oscillation (T),

$$0.02 = 2.3 \times R \cdot C \log_{10} \frac{1}{1-\eta}$$

$$= 2.3 \times R \times (0.5 \times 10^{-5}) \times \log_{10} \frac{1}{1-0.62}$$

$$= 2.3\ \text{R} \times (0.5 \times 10^{-6}) \times 0.42 \times 4.83 \times 10^{-7}\ \text{R}$$

$\therefore \quad R = 0.02/(4.83 \times 10^{-7}) = 4.14 \times 10^4 \ \Omega$

$= 41.4 \text{ k}\Omega$ **Ans.**

The value of R can also be determined straight forward by looking at the expression for the period, when the capacitance is increased by a factor of 10 (*i.e.*, from 0.05 µF to 0.5 µF) then to keep frequency of 50 Hz as constant. The value of R will have to reduced by the same factor *i.e.*, from 414 kΩ to 41.4 kΩ. **Ans.**

34.8. A Transistor Constant Current Sweep

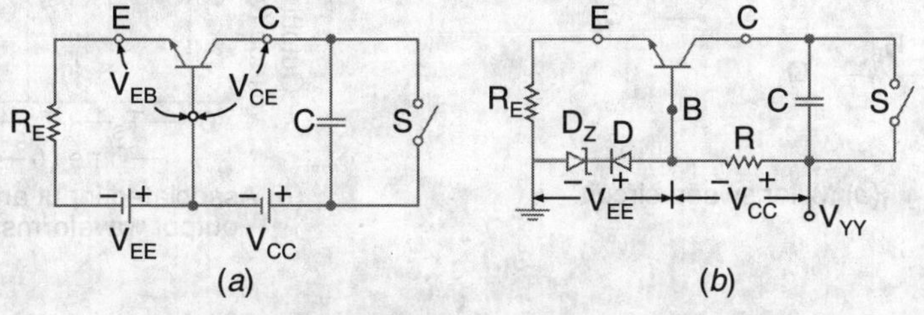

Fig. 34.5. A constant current sweep circuit.

Fig. 34.5 (*a*) shows a circuit to generate a sweep using constant current obtained from a common base transistor. We know that except for very small values of collector-to-base voltage, the collector current of a transistor in the common base configuration is very nearly constant, when the emitter current is held fixed. This characteristic may be used to generate a quite linear sweep by causing a constant current to flow into a capacitor. The value of emitter current in Fig. 34.5 (*a*) is given by the expression,

$$I_E = \frac{V_{EE} - V_{EB}}{R_E}$$

If the emitter-to-base voltage (V_{BE}) remains constant with times after the switch (S) opened, then the collector current will be a constant whose nominal value,

$$I_C = h_{FB} \cdot I_B = \alpha \cdot I_E$$

where h_{FB} (or a) is the common base current gain of the transistor.

The constant current sweep circuit discussed above has one drawback, that it makes the sweep rate as a function of temperature. Since the emitter-base junction voltage (V_{BE}) for a fixed current decreases by about 2 mV/°C, therefore the *sweep speed increases with the temperature.

Fig. 34.5 (*b*) shows a constant current sweep circuit, which employs a single supply V_{YY} and provides a considerable temperature compensation. The emitter supply voltage is now equal to the voltage (V_Z) across the zener diode (D_Z) plus the drop (V_D) across the diode D. The collector supply voltage V_{CC} is now equal to the drop across resistance R *i.e.*, $V_{CC} = V_{YY} - V_{EE}$. The diode D is made of the same material as that of the transistor (*i.e.*, both silicon or germanium). Hence the diode (D) serve to compensate for the temperature dependent emitter-top-base voltage (V_{BE}). If the drops across diode and emitter-base junction are always the same, then the voltage across resistor (R_E) remains equal to (V_Z) and the emitter current is equal to V_Z/R_E, which is constant. Thus the sweep speed is essentially independent of temperature.

34.9. Miller Sweep Circuit

It is also called *Miller integrator.* Figure 34.6 shows the circuit of a Miller integrator or sweep

* We know that expression for sweep speed is I/C. In the present case, $I = I_C = \alpha \cdot I_E$. Therefore sweep speed is $\alpha \cdot I_E/C$. Since $I_E = (V_{EE} - V_{EB})/R_E$, therefore sweep speed $= \alpha\ (V_{EE} - V_{EB})/R_E$. It is evident from this relation that as the voltage V_{EB} decreases, the sweep speed increases.

circuit. Here the transistor Q_1 acts as a switch and transistor Q_2 is a common emitter amplifier *i.e.*, a high gain amplifier. The operation of the circuit may be understood from the condition that initially, the transistor Q_1 is ON and Q_2 is OFF. At this instant, the voltage across the capacitor and the output voltage is equal to V_{CC}.

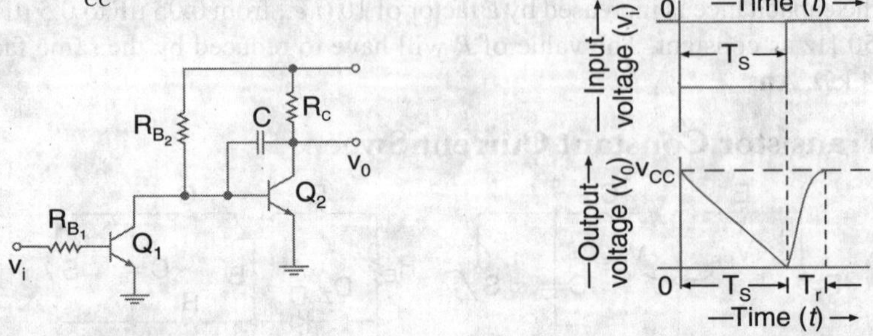

(*a*) Miller sweep circuit. (*b*) Associated input and output waveforms.

Fig. 34.6.

Now suppose a pulse of negative polarity as shown in Fig. 34.6 (*b*) is applied at the base of the transistor Q_1. As a result of this, the emitter-base junction of the transistor Q_1 is reverse biased and it turns OFF. This causes the transistor Q_2 to turn ON. As the transistor Q_2 conducts, the output voltage begins to decrease towards zero. Since the capacitor (C) is coupled to the base of transistor Q_2, therefore the rate of decrease of the output voltage is controlled by the rate of discharge of capacitor (C). The time constant of the discharge is given by the relation,

$$\tau = R_B \cdot C$$

Since the value of time constant is very large, therefore the discharge current practically remains constant. As a result of this, the run down of the collector voltage is linear. When the input pulse is removed, the transistor Q_1 turns ON and Q_2 turns OFF. It will be interesting to know that as the transistor Q_1 turns OFF, the capacitor (C) charges quickly, through resistor R_C to V_{CC} with a time constant (τ) equal to $R_C.C$. The waveform of the generated sweep or the output voltage (v_0) is shown in Fig. 34.6 (*b*).

It may be noted from the circuit that we have shown only one stage of amplification. But in actual practice more than one stage of amplification is used. Usually, the amplifier stage uses emitter followers at the input and output to increase the low-frequency input and output resistances (*i.e.*, R_i and R_o). Sometimes, the resistor R_C inthe Miller sweep cicuit is replaced by a diode. The diode forward resistance (R_f) helps the capacitor to charge quickly from V_{CC}. This reduces the flyback time (T_r) of the generated sweep. The Miller sweep circuit provides an excellent sweep linearity as compared to other sweep circuits.

34.10. Bootstrap Sweep Circuit

Fig 34.7 (*a*) shows a practical form of a bootstrap sweep circuit. Here the transistor Q_1 acts as a switch and Q_2 as an emitter follower (*i.e.*, a unity gain amplifier). The operation of the circuit may be explained as follows.

Initially, the transistor Q_1 is ON and Q_2 is OFF. Therefore the capacitor C_1 is charged to V_{CC} through the diode forward resistance (R_F). At this instant, the output voltage (v_o) is zero. When a negative pulse as shown in Fig. 34.7 (*b*) is applied to the base of transistor Q_1, it turns OFF. Since transistor Q_2 is an emitter follower, therefore the output voltage (v_o) is the same as the base voltage of transistor Q_2. Thus as the transistor Q_1 turns OFF, the capacitor (C_1) starts charging this capacitor (C) through resistor (R). As a result of this, both the base voltage of Q_2 and the output voltage begins to increase from zero. As the output voltage increases, the diode D becomes reverse biased. It is because of the fact that the output voltage is coupled through the capacitor (C_1) to the diode. Since the value of capacitor (C_1) is much larger than that of capacitor (C), therefore the voltage across capacitor (C_1) practically remains constant. Thus the voltage drop across the resistor (R) also remains constant. Because of this, the current i_R through the resistor R also remains constant. It means that capacitor C is charged with a constant current. This causes the voltage across the capacitor C (and hence the output voltage) to increase linearly with time.

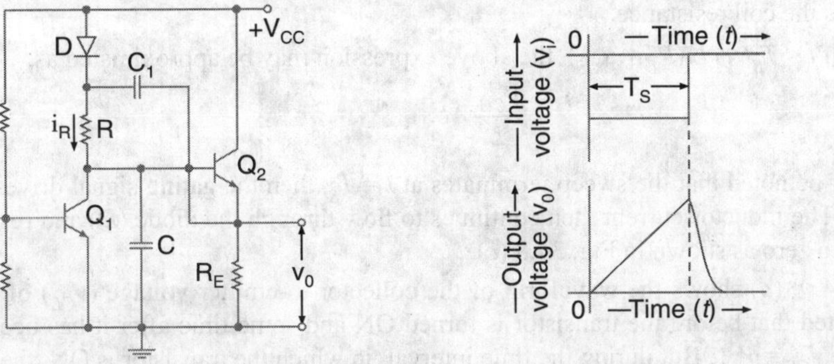

(a) Bootstrap sweep circuit. (b) Associated waveforms.

Fig. 34.7.

It is evident from the above discussion that the circuit pulls itself up by its own bootstrap and hence it is known as bootstrap sweep circuit. When the negative pulses removed from the input of transistor Q_1, the capacitor C discharges rapidly through Q_1 and the output voltage returns to zero. Then capacitor C_1 again charges to voltage V_{CC} through diode D. The waveform of the output voltage is as shown in Fig. 34.7 (b).

34.11. Current Time Base Generator

We have already discussed the methods and circuits by means of which we can generate a voltage that varies linearly with the time. Such a voltage may be applied to a resistor, thereby generating a current that increases linearly with time. However, the more accurate method is to cause a linearly time varying current to flow through an inductor. Such circuits are used in radar and television displays.

Fig. 34.8 (a) shows a simple circuit of a current time base generator. Here an inductor (L) in series with a transistor is connected across the V_{CC} supply. The transistor operates as a switch in this circuit. The gating waveform at the base operates between two levels V_1 and V_2 as shown on Fig. 34.8 (b). The lower level (V_1) keeps the transistor in cut-off, while the upper level drives the transistor into saturation.

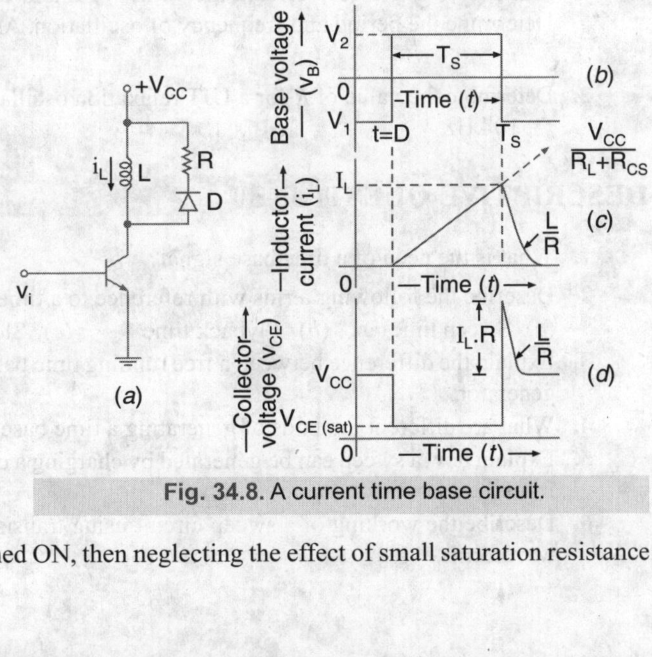

Fig. 34.8. A current time base circuit.

The operation of the circuit may be understood from the condition that when the transistor switch is turned ON, then neglecting the effect of small saturation resistance

(R_{CS}), the current through an inductor (i_L) increases linearly with the time. The diode D does not conduct during the sweep, because it is reverse biased. The growth of current through an inductor is given by the relation,

$$i_L = \frac{V_{CC}}{R_L + R_{CS}} 1 - e^{\left(\frac{R_L + R_{CS}}{L}\right) t}$$

where R_L is the coil resistance.

Thus if $t < T_S << L/(R_L + R_{CS})$, the above expression may be approximated as,

$$i = \frac{V_{CC}}{L} \cdot t$$

It may be noted that the sweep terminates at $t = T_S$, then the gating signal drives the transistor to cut-off. The inductor current then continues to flow through the diode (D) and resistor (R), until it reduces to zero as shown in Fig. 34.8 (c).

Fig. 34. 8 (d) shows the waveform of the collector-to-emitter voltage (v_{CE}) of a transistor. It may be noted that before the transistor is turned ON and some time after it has been turned OFF, the voltage $v_{CE} = V_{CC}$. But during the time interval, in which the transistor is ON, the voltage v_{CE} is quite low *i.e.*, equal to $V_{CE\,(sat)}$. It may also be noted that at the moment the transistor is turned OFF, a large spike of amplitude $I_L \cdot R$ appears across the inductor. This voltage must be kept below the breakdown voltage, so that it does not damage the transistor.

The sweep-speed error for the current time base circuit is given by the relation,

$$e_S = \frac{T_S\,(R_L + R_{CS})}{L}$$

It is evident from the above expression that to maintain linearity the time constant $L\,/\,(R_L + R_{CS})$ must be kept large enough as compared to the sweep time (T_S).

NUMERICAL PROBLEMS

1. In a simple UJT sweep circuit, the values of R and C are 47 kW and 0.2 mF respectively. Determine the period and frequency of oscillation. Assume h = 0.56.

(Ans. 7.7 ms and 129.6 Hz)

2. Determine the value of R for a UJT relaxation oscillator having $C = 0.01$ μF, $\eta = 0.60$ and $f = 10$ kHz. **(Ans. 10.9 kΩ)**

DESCRIPTIVE QUESTIONS

1. What is the need of a time base signal?
2. Describe the following terms with reference to a time base signal
   (*a*) Sweep time (*b*) Flyback time (*c*) Slope error (*d*) Sweep speed
3. Explain the difference between a free running time base generator and a triggered time base generator.
4. What are different methods of generating a time base signal?
5. Explain how a sweep can be generated by charging a capacitor C through resistor R directly from the source V.
6. Describe the working of a sweep circuit using transistor switch. Also draw the associated waveforms.

7. Draw a circuit for generating sweep using unijunction transistor switch. Give expressions for sweep period and frequency.
8. Explain the working of a Miller integrator type of sawtooth voltage generator.
9. Give the simple Miller sweep circuit using NPN transistor.
10. Explain the working of a Bootstrap sweep circuit. Also give the expression for sweep-speed error.
11. Draw the circuit of a current time base generator. Explain the working of the circuit with the associated waveforms.
12. Draw and explain the working of UJT relaxation sweep generator circuit. *(M.G University., May/June 2006)*
13. Describe the working of UJT relaxation oscillator. *(Anna University., May/June 2007)*
14. Draw and relaxation oscillator circuit using UJT. *(M.G University., May 2005)*
15. Explain the use of UJT as a relaxation oscillator? *(PTU., Dec 2006)*
16. With the help of neat circuit diagram, characteristics and waveforms, explain the working of UJT relaxation oscillator. *(M.G University., May 2008)*
17. Explain with suitable diagrams and waveforms the working of a UJT relaxation oscillator. Also derive its frequency of oscillation. *(Anna University., Nov/Dec 2007, Nov/Dec 2006)*
18. Discuss briefly the operation of a current time base generator with necessary diagram. *(Anna University., May/June 2007, Nov/Dec 2006)*
19. Draw the simple current time base circuit and give its expression relating current and time *(Anna University., Nov/Dec 2006)*
20. Define time base generator. *(Anna University., May/June 2007)*
21. Explain with circuit diagram and waveforms the Transistor voltage time base generator that uses Bootstrapping. *(Anna University., Nov/Dec 2007)*
22. Explain the operation of bootstrap ramp generator circuit. Discuses the advantage of this circuit with other ramp generators *(Anna University., Nov/Dec 2007)*
23. Explain bootstrap time base sweep generator with circuit diagram and waveform. Derive for its output voltage. *(Anna University., May/June 2006)*
24. Draw the circuit for generating sweep using UJT switch and explain its operation. Give expression for sweep period and frequency. *(Anna University., Nov/Dec 2006)*
25. What does bootstrapping mean? *(Anna University., April/May 2008)*
26. Explain Miller and Bootstrap sweep generatiors with necessary circuits and waveforms *(Anna University., Nov/Dec 2006)*
27. Explain miller generator and triggered based oscillator. *(Anna University., April/May 2008)*
28. Explain the working of Miller type oscillator with circuit. Give two applications *(Anna University., April/May 2008)*
29. Draw the circuit of a transistor bootstrap ramp generator and explain how the circuit operates *(Anna University., Nov/Dec 2006)*
30. Draw the circuit of an oscillator using an UJT. Discuss its working. *(R.G.P.V., June 2003)*

MULTIPLE CHOICE QUESTIONS

1. Which the following time base circuit is suitable to display widely separated narrow-width pulses?
   (*a*) Voltage time base generator (*b*) Current time base generator
   (*c*) Free running time base generator (*d*) Triggered time base generator
2. The sweep speed of a simple RC circuit is given by
   (*a*) I/C (*b*) $I \cdot t/C$ (*c*) T_S/RC (*d*) $V \cdot T_S/RC$
3. The sweep period in a sweep circuit using unijunction transistor is
   (*a*) directly proportional to R (*b*) inversely proportional to R
   (*c*) directly proportional to C (*d*) directly proportional to the product $R \cdot C$
4. If you require good linearity, which type of sweep generator you will use?
   (*a*) An RC circuit with SCR (*b*) A UJT sweep circuit
   (*c*) A Miller integrating type circuit (*d*) A 555 time circuit
5. Which of the following sweep circuit is used in television receivers for display?
   (*a*) Voltage time base generator (*b*) Current time base generator
   (*c*) Bootstrap sweep circuit (*d*) None of the above
6. In the circuit shown in Fig. 34.9., if $R_1 >> R_P$ and the pulses can satuate the transistor Q, then the output voltage, V_o will be :

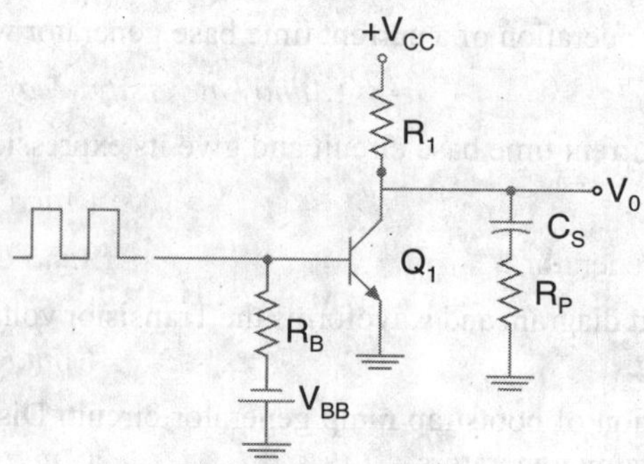

Fig. 34.9.

(*a*)  (*b*) 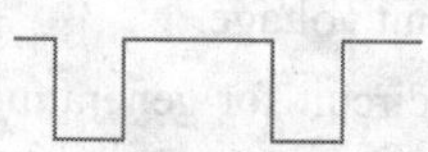

(*c*) 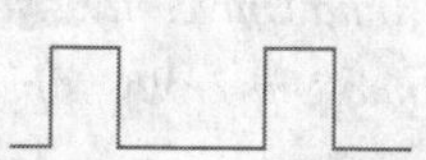 (*d*) 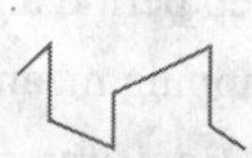

ANSWERS

1. (*d*) 2. (*a*) 3. (*d*) 4. (*c*) 5. (*b*) 6. (*a*)

35

Chapter

Operational Amplifiers (Op-Amps)

Outline

35.1. Introduction

An operational amplifier (or Op-Amp in short) is a linear integrated circuit (IC) that has a very high voltage gain, a high input impedance and a low output impedance. The internal circuit of a very popular Op-Amp (μA 741) is shown in Fig. 35.1 (*a*). Its schematic symbol is shown in Figure 35.1 (*b*) and its pin diagram in Fig. 35.1 (*c*).

As seen from Fig. 35.1 (*b*), Op-Amp is provided with two input terminals. One of the terminals, labeled '-' is known as *inverting input terminal*. The word "inverting" implies that if a signal is applied to the '-' terminal of the Op-Amp, it appears with the opposite polarity at the output, *i.e.*, a sinusoidal signal will experience a phase shift of 180°. The other input terminal, labeled "+" is called the *non-inverting input terminal*. In other words, a signal applied to this terminal of the Op-Amp is amplified without inversion. The Op-Amp actually amplifies the difference between the voltages applied to its two input terminals.

Two further terminals are provided for the connection of positive and negative power supply voltages. Both polarities are necessary so that the output voltage can vary either side of zero volts. Most Op-Amps operate satisfactory from a wide range of supply voltages and a few designed to operate from a single-polarity supply.

It may be carefully noted that the schematic diagram of the μA 741 Op-Amp shown in Fig. 35.1 (*a*), that it makes use of all bipolar transistors. Such Op-Amps are called bipolar type Op-Amps. Some of the Op-Amps have a FET input circuit with the remainder of the circuit using bipolar transistors. Such Op-Amps are called FET type Op-Amps.

Operational-amplifier.

35.2. Operational Overview

The input circuit of an Op-Amp, which is driven by inverting and non-inverting inputs, is called a *differential amplifier*. This means that the Op-Amp amplifies the difference between the voltages applied at the non-inverting input and inverting input. This point is illustrated in Fig. 35.2. In this figure, v_1 is the voltage applied at the inverting input of the Op-Amp and v_2 is the voltage applied at the non-inverting input. The difference between the two voltages *i.e.*, $v_2 - v_1$ (or $v_2 - v_1$) acts as an input to the Op-Amp. Mathematically, the difference voltage that will be applied is,

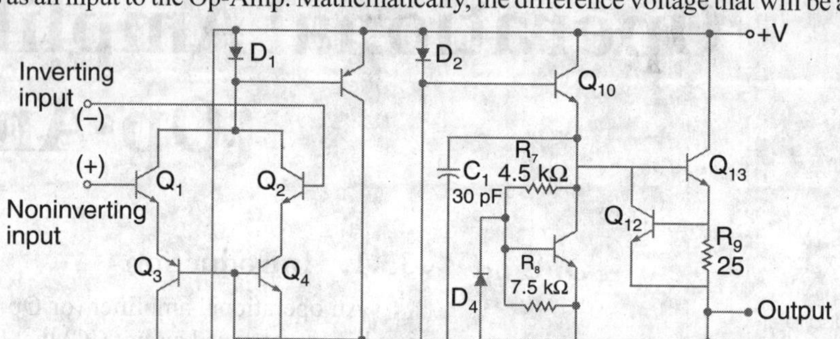

(*a*) Schematic diagram of a 741.

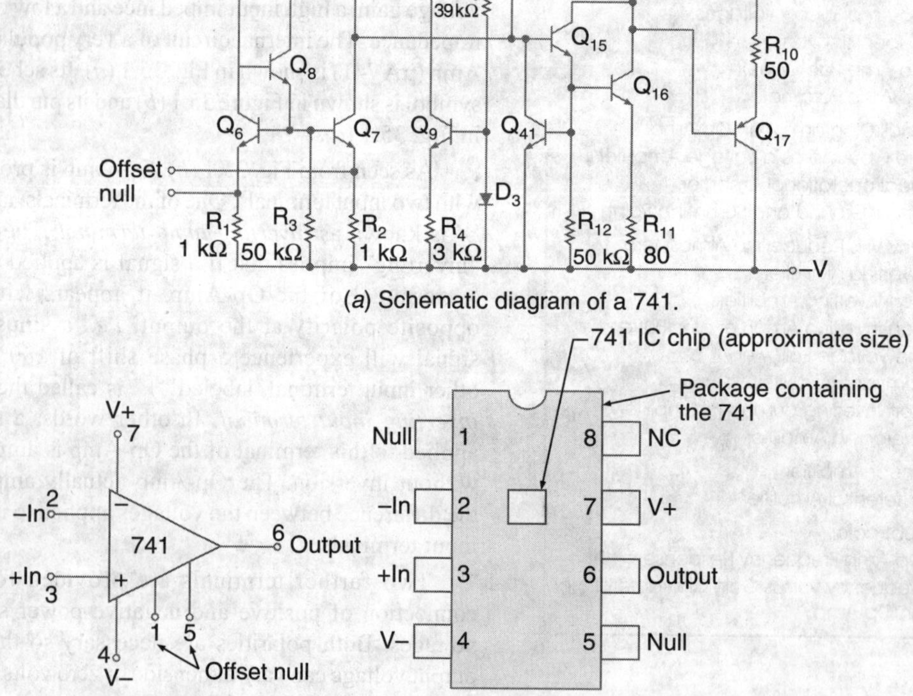

(*b*) Circuit symbol for a 741. (*c*) The 741 chip is shown packaged in an 8-pin dual-in-line package (top view).

Fig. 35.1. Op-Amp.

$$v_{in} = v_2 - v_1$$

The output (v_o) from the Op-Amp for a given pair of input voltages depends upon the factors such as (1) the gain of the amplifier, (2) the polarity relationship between v_1 and v_2 and values of supply voltage + V and – V.

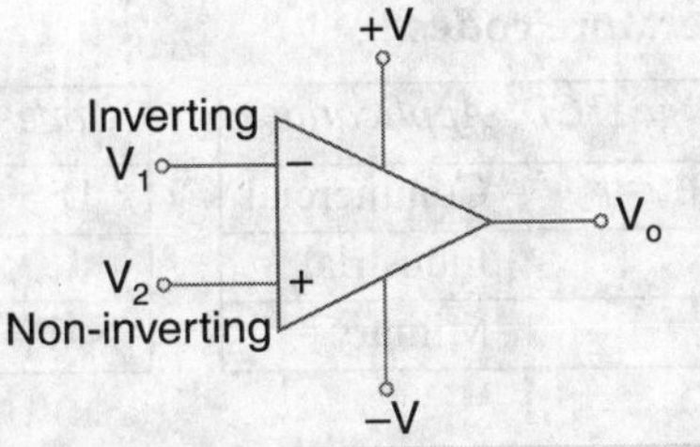

Fig. 35.2. Illustrating the overview of an Op-Amp.

35.3. Op-Amp Supply Voltages

The Op-Amp has two supply voltage inputs, labeled as + V and – V. These two supply pins are normally connected in any one of the following three ways as shown in Fig. 35.3 (*a*), (*b*), and (*c*).

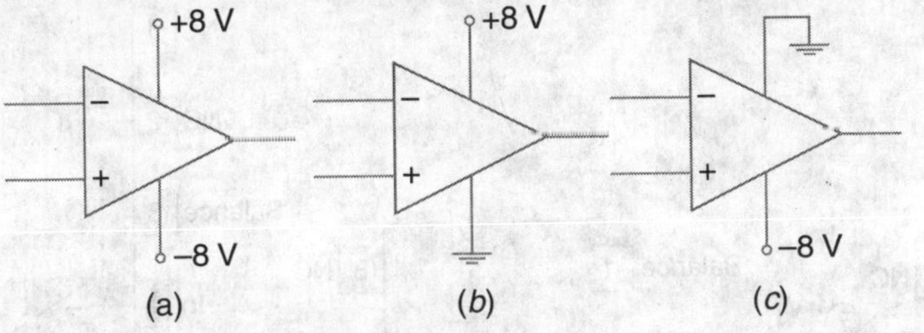

Fig. 35.3. Op-Amp supply voltages.

As seen from the figure, one way is to have + V and – V set to equal voltages (say 8 volts) that are of opposite polarities as shown in Fig. 35.3 (*a*). The other way is to provide a single supply voltage to one of the supply pins and grounding the other as shown in Fig. 35.3 (*b*) and (*c*). In this figure, the supply voltage is connected to + V pin while the – V pin is grounded. Whereas in Fig. 35.3 (*c*), the supply voltage is connected to –V pin and the + V is grounded. The choice among the three methods of connecting supply voltage to an Op-Amp depends on the particular application.

35.4. IC Identification

There are several types of Op-Amps produced in the form of integrated circuit (IC) by different manufacturers. These Op-Amps are identified using a seven character identification (ID) code. This code is shown in Fig. 35.4. The code has three parts namely *prefix*, *designator* and *suffix*. The prefix is used to identify the particular manufacturer. A listing of the most common prefixes is given in Table 35.1.

Prefix	**Designator**	**Suffix**
MC	741 C	N

Fig. 35.4. An Op-Amp ID code.

Table 35.1. Manufacturer Prefixes

Prefix	*Manufacturer*
AD	Analog Devices
CA	RCA
LM	National Semiconductor
MC	Motorola
NE/SE	Signetics
OP	Precision Monolithics
RC/RM	Ratheon
SG	Silicon General
TI	Texas Instruments
UA	Fairchild

The designator code indicates two things: (1) the three-digit number indicates the specific type of

Op-Amp, (2) the final letter indicates the operating temperature range. Table 35.2 gives the commonly used temperature codes for the Op-Amps. The designator code is used not only to determine the specific type of Op-Amp you are dealing with, but also to help you to determine which Op-Amps can be substituted for each other.

Table 35.2. Temperature codes

Code	*Temperature range (°C)*	*Application*
C	0 to 70	Commercial
I	– 25 to 85	Industrial
M	– 55 to 125	Military

Table 35.3. Suffix codes

Code	*Package type*
D	Plastic Dual-in-line (DIP)
J	Ceramic DIP
N, P	Plastic DIP with longer lead

35.5. Op-Amp Packages

The Op-Amps are available in two packages: (1) dual-in-line (DIP) and metal can as shown in Fig. 35.5. Metal cans (type TO-5) are available with 8, 10 or 12 leads. The DIP packages of Op-Amps have 8 or 14 pins. The DIP package is more common than metal can. The type of package is an important factor when considering whether or not the Op-Amp may be substituted for another.

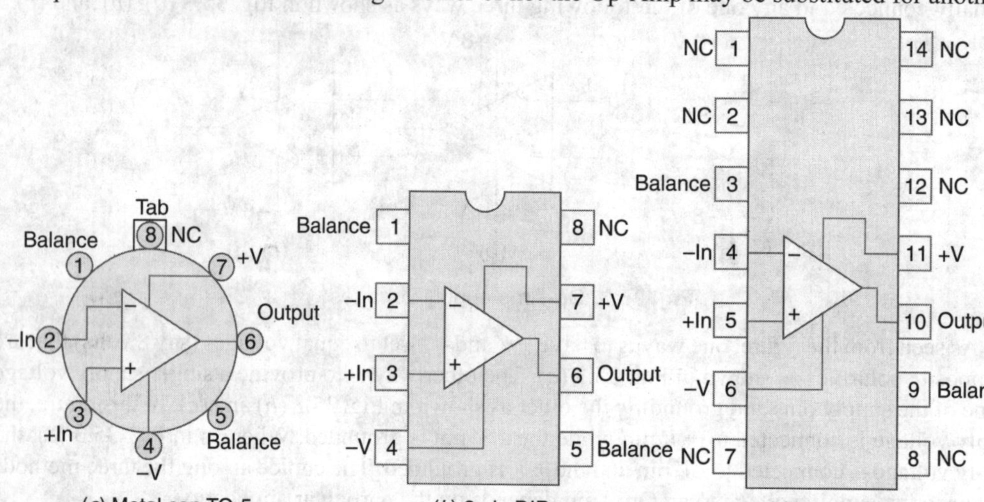

(*a*) Metal can TO-5. (*b*) 8 pin DIP. (*c*) 14 pin DIP.

Fig. 35.5. Op-Amp packages.

35.6. Op-Amp Parameters

Following are some of the important Op-Amp parameters which are necessary to understand the fundamental concepts of Op-Amp operation.

1. *Gain.* The ideal Op-Amp would have an infinite open-loop differential gain. But in all practical circuits using Op-Amp, the gain falls short of the infinite value. Practical Op-Amps have open-loop gains which vary considerably from one type to another but may be somewhere between 25,000 to 3,00,000. The gain is often expressed in decibels (dB). If v_1 is the input and v_2 is the output-voltage of the Op-Amp, then gain in decibels.

$$dB = 20 \log_{10} \frac{v_2}{v_1}$$

2. *Input Resistance.* Ideally the input resistance of an Op-Amp is infinitely high. But in practice, the input resistance may be any value between 250 k Ω and 40 MΩ for the Op-Amps with bipolar transistor input and 10^{12} Ω for the Op-Amps with field-effect transistor (FET) input.

3. *Output Resistance*. Basically an Op-Amp is a voltage amplifier. Therefore it should have its input resistance as low as possible. Practical Op-Amps have the output resistance of the order of 100 Ω.

4. *Common-mode Rejection Ratio (CMRR)*. The output voltage of an Op-Amp is proportional to the difference between the voltages applied to the inverting and non-inverting terminals. Ideally when the two voltages are equal, the output voltage should be zero. A signal applied to both the input terminals is called a common-mode signal. It is nearly always an unwanted noise voltage. The ability of an Op-Amp to suppress common-mode signals is expressed in terms of its common-mode rejection ratio (CMRR). Mathematically, the common-mode rejection ratio is given by the relation,

$$CMRR = \frac{\text{Differential voltage gain}}{\text{common mode voltage gain}} = \frac{A_{dm}}{A_{cm}}$$

where A_{dm} = Differential voltage gain, and

A_{cm} = Common-mode voltage gain.

The common-mode rejection ratio can also be expressed in decibels. Thus,

$$CMRR = 20 \log_{10} \frac{A_{dm}}{A_{cm}} \text{ dB}$$

Typically an Op-Amp might have a common-mode rejection ratio of 90 dB.

5. *Slew rate*. The slew rate of an Op-Amp is the maximum rate at which its output voltage is capable of changing. It is expressed in volts per second (V/μs). For example, μA 741 Op-Amp has a slew rate of 0.5 V/μs. This means that the output from μA 741 Op-Amp can change by 0.5 V every micro-second. When a signal at a given frequency is applied to the Op-Amp, the maximum permissible output voltage is determined by the slew rate. And if we try to develop the output voltage higher than the maximum permissible value, the signal waveform will be distorted. This fact is illustrated in Fig. 35.6. In this figure we have shown the input (a square waveform) and output waveforms of an Op-Amp with inadequate slew rate.

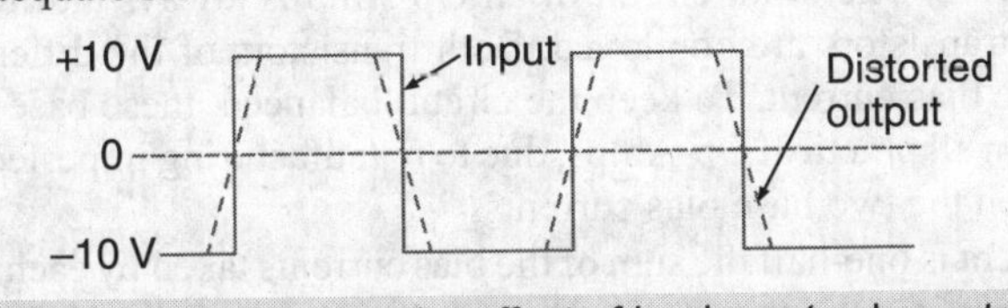

Fig. 35.6. Illustrating the effect of inadequate slew rate.

Typically the slew rate for Op-Amps is in the range of 0.3 – 12 V/μs. Since frequency is a function of time, the slew rate can be used to determine the maximum operating frequency of the Op-Amp as follows :

$$f_{max} = \frac{\text{slew rate}}{2\pi V_{pk}}$$

where V_{pk} is the maximum or peak value of the applied input signal.

6. *Bandwidth*. The open-loop voltage gain of an Op-Amp is not constant at all frequencies but falls off at high frequencies because of the capacitive effects. The gain frequency characteristics of Op-Amps are specified by the manufacturers in two ways: (1) the bandwidth for large signals over which less than 5% distortion is obtained and (2) the frequency at which the gain falls to unity.

Fig. 35.6 shows the gain-frequency characteristic of a μA 741 Op-Amp. As seen from this figure, we find that at frequencies below 10 Hz, the voltage gain is constant (approximately 106 dB). But at higher frequencies, the gain falls off at the constant rate of 6 dB/octave. Its value becomes 0 dB (*i.e.*, = 1) at a frequency (f_t) of 1 MHz. Other Op-Amps have similar gain-frequency characteristics but with different values of low frequency gain and frequency (f_t) at which the gain reduces to 0 dB.

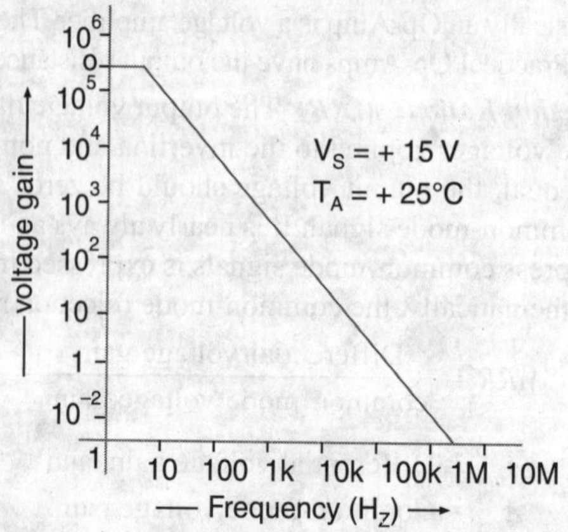

Fig. 35.7. Open-loop voltage gain of Op-Amp as a function of frequency.

7. ***Input Offset Voltage.*** Ideally speaking, the output voltage of an Op-Amp should be zero when the value of an applied voltage at both the input terminals is zero. However for any practical Op-Amp, it is found that the output voltage does exist for zero input voltage. This voltage arises because of small imbalances within the Op-Amp. But for the convenience of discussion, we assume that the output is processed by an input offset voltage (V_{OS}).

The input offset voltage of an Op-Amp is equal to the output voltage for zero input voltage divided by the open-loop voltage gain of the amplifier. The input offset voltage is in the rage of 10 μV to 7 mV for Op-Amps with bipolar transistor input type and 0.5 to 15 mV for Op-Amps with field-effect transistor (FET) input type. Although the offset voltage is small, yet it is amplified by the circuit and may not be of negligible amplitude at the output. Most Op-Amps are provided with some means of nulling out the effects of input offset voltage.

8. ***Input Bias Current.*** The input circuit of an Op-Amp is always a differential amplifier and in many cases bipolar transistors are employed. Both transistors of the differential amplifier must be provided with a base bias current. To keep the circuit balanced, these base currents should be of equal value. However, in all practical Op-Amps, due to manufacturing imperfections, there is always some difference between the two base bias currents.

The input bias current is one-half the sum of the bias currents taken by each input of the Op-Amp. These are the bias currents for the input transistors. Typically, the input bias current is 10 to 50 nA for a bipolar transistor input Op-Amp and as low as 10 to 100 pA for the FET input Op-Amp.

9. ***Input Offset Current.*** The difference between the bias currents is known as the input offset current. Typical values are in the range of 3-20 nA for bipolar transistor input Op-Amps and a few pico-amperes (pA) for FET input Op-Amps.

In order to study more about Op-Amp parameters, refer to the μA 741 device specifications sheet given in Appendix B.

Example 35.1.

Calculate the common-mode rejection ratio (CMRR) of an Op-Amp, that has a differential gain of 2 00 000 and a common-mode gain of 6.33.

Solution.

Given : $A_{dm} = 200\ 000$ and $A_{cm} = 6.33$

We know that common-mode rejection ratio,

$$CMRR = 20 \log_{10} \left(\frac{A_{dm}}{A_{cm}} \right) \text{ in dB}$$

$$= 20 \log_{10}\left(\frac{200000}{6.33}\right) \text{ in dB} = 90 \text{ dB } \textbf{Ans.}$$

Example 35.2.

An Op-Amp has a CMRR of 90 dB. If its differential voltage gain is 30 000, calculate its common-mode gain.

Solution.

Given : $CMRR = 90$ dB and $A_{dm} = 30\ 000$

Let A_{cm} = The common-mode voltage gain of the Op-Amp.

We know that common-mode rejection ratio (CMRR)

$$90 = 20 \log_{10}\left(\frac{A_{dm}}{A_{cm}}\right) = 20 \log_{10}\left(\frac{30\ 000}{A_{cm}}\right)$$

or

$$4.5 = \log_{10}\left(\frac{30\ 000}{A_{cm}}\right)$$

or

$$10^{4.5} = \frac{30\ 000}{A_{cm}}; \quad \therefore A_{cm} = \frac{30\ 000}{10^{4.5}} = 0.95 \textbf{ Ans.}$$

Example 35.3.

Determine output voltage of differential amplifier for input voltage of 300 μV and 240 μV. Differental gain is 5000 and CMRR is 100. (GTU., June 2011)

Solution.

Given : $V_1 = 300$ μV = 300×10^{-6} V; $V_2 = 240$ mV = 240×10^{-6} V; $A_{dm} = 5000$ and CMRR = 100.

We know that, CMRR,

$$100 = \frac{A_{dm}}{A_{cm}} = \frac{5000}{A_{cm}}$$

$$\therefore \quad A_{cm} = 50$$

We also know that the output voltage of differential amplifier,

Vo = output voltage in differential mode ($A_{dm}V_{dm}$) + Output voltage in common mode ($A_{cm}V_{cm}$)

$$V_o = (A_{dm}V_{dm}) + (A_{cm}V_{cm})$$

$$= A_{dm} \times (V_1 - V_2) + A_{cm} \times \left(\frac{V_1 + V_2}{2}\right)$$

$$= \{5000 \times [(300 \times 10^{-6}) - (240 \times 10^{-6})]\} + \left\{50 \times \left(\frac{(300 \times 10^{-6}) + (240 \times 10^{-6})}{2}\right)\right\}$$

$$= 0.3 + 0.0135 = 0.3135 \text{ V} = 313.5 \times 10^{-3} \text{ V} = 313.5 \text{ mV } \textbf{Ans.}$$

Example 35.3.

The amplifier shown in Fig. 35.8 is being used to amplify an input signal to a peak output voltage of 100 mV. What is the maximum operating frequency of the amplifier? The slew rate for μA 741 Op-Amp is 0.5 V/μs.

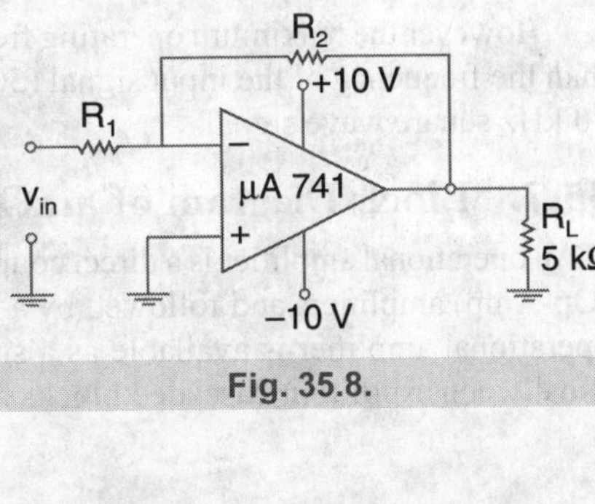

Fig. 35.8.

Solution.

Given : Slew rate = 0.5 V/μs = 0.5×10^6 V/s; $V_{pk} = 100$ mV = 0.1 V.

We know that maximum operating frequency for the amplifier,

$$f_{max} = \frac{\text{Slew rate}}{2\pi V_{pk}} = \frac{0.5 \times 10^6}{2\pi \times 0.01} = 796 \times 10^3 \text{ Hz} = 796 \text{ kHz } \textbf{Ans.}$$

Example 35.5.

A 50 Hz square wave is to be amplified by an Op-Amp to have an output voltage swing ± 10 V. Two Op-Amps are available: (a) the μA 741 which has a slew rate of 0.5 V/μs and (b) the TLO 81 with a slew rate of 13 V/μs. Determine if either or both of the Op-Amps are suitable for the purpose.

Solution.

Given : The frequency of the square wave (f) = 50 kHz = 50×10^3 Hz; Output voltage swing = ± 10 V.

We know that time-period of the square waveform,

$$T = \frac{1}{f} = \frac{1}{50 \times 10^3} = 20 \times 10^{-6} = 20 \text{ μs}$$

Now in order to amplify the input signal the Op-Amp must change its output voltage from 10 V to – 10 V (from – 10 V to + 10 V) *i.e.*, a total voltage change of 20 V in one half-period *i.e.*, 20/2 = 10 μs.

(*a*) *μA 741*

For this Op-Amp, the slew rate is 0.5 V/μs. Therefore in 10 μs, the output voltage of μA 741 can only change by 10 × 0.5 = 5 V. But the requirement is of 20 V. Therefore μA 741 Op-Amp is not suitable.

(*b*) *TLO 81*

For this Op-Amp, the slew rate is 13 V/μs. Therefore in 10 μs, the output voltage of TLO 81 can change 10 × 13 = 130 V. But the requirement is only of 20 V. Therefore TLO 81 Op-Amp is suitable.

Alternative solution. Given V_{pk} = 10 V.

The maximum operating frequency for the μA 741 Op-Amp,

$$f_{max} = \frac{\text{Slew rate}}{2\pi V_{pk}} = \frac{0.5 \text{ V/μs}}{2\pi \times 10}$$

$$= \frac{0.5 \times 10^6}{20\pi} = 7957.7 \text{ Hz} = 7.957 \text{ kHz}$$

Similarly, the maximum operating frequency for the TLO 81 Op-Amp,

$$f_{max} = \frac{\text{Slew rate}}{2\pi V_{pk}} = \frac{13 \text{ V/μs}}{2\pi \times 10} = \frac{13 \times 10^6}{20\pi} = 206.9 \text{ kHz}$$

Comparing the maximum operating frequency of the μA 741 (7.957 kHz) with the frequency of the input signal (50 kHz) we find that the value is much lower than that of the input signal. Hence the μA 741 is not a suitable Op-Amp.

However the maximum operating frequency of the TLO 81 Op-Amp is much higher (206.9 kHz) than the frequency of the input signal (50 kHz), therefore, TLO 81, Op-Amp can be used to amplify 50 kHz square wave signal.

35.7. Block Diagram of an Op-Amp

An operational amplifier is a direct coupled high gain amplifier consisting of one or more differential (Op-Amp) amplifiers and followed by a level translator and an output stage. As mention earlier, the operational amplifier is available as a single integrated circuit package. Commercial IC Op-Amps usually consist of four cascaded blocks as shown in Fig. 35.9.

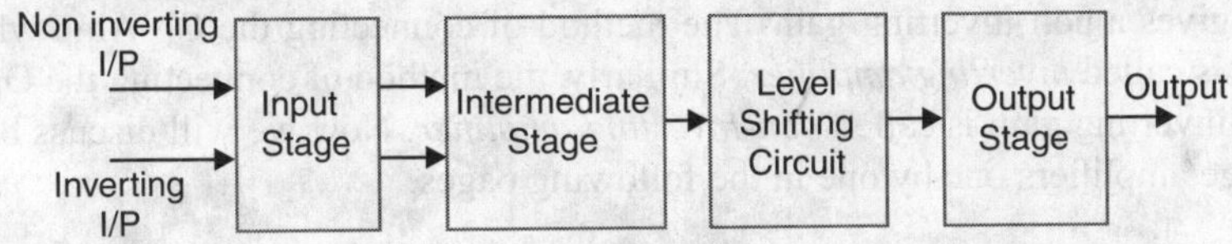

Fig. 35.9

The first two stages are cascaded differential amplifiers used to provide high gain and high input resistance. The main purpose of the differential amplifier stage is to provide high gain to the difference mode signal and cancel the common-mode signal. Thus it is able to suppress any noise which is common to both of the input terminals. The intermediate stage of Op-Amp is another differential amplifier which is driven by the output of the first stage. This is usually dual input unbalanced output.

The third stage acts as a buffer as well as a level shifter. The buffer is usually an emitter follower whose input impedance is very high so that it prevents loading of the high gain stage. The level shifter adjusts the d.c. voltage so that output voltage is zero for zero inputs. The adjustment of d.c. level is required as the gain stages are direct coupled. As it is not possible to fabricate large value of capacitors, all IC's are direct coupled usually. The output stage is designed to provide low output impedance as demanded by the ideal Op-Amp characteristics.

The output stage is generally a push pull complementary amplifier. The output stage increases the output voltage swing and raises the current supplying capability of the Op-Amp. To allow such symmetrical swing the amplifier is provided with both positive and negative supply voltages. Power supply voltages of 15 V are common. Additionally, an Op-Amp generally incorporates circuitry to provide drift compensation and frequency compensation. Fig. 35.10. shows a complete Op-Amp circuit having input different amplifiers with balanced output, intermediate stage with unbalanced output, level shifter and an output amplifier. We will discuss each block in the following pages.

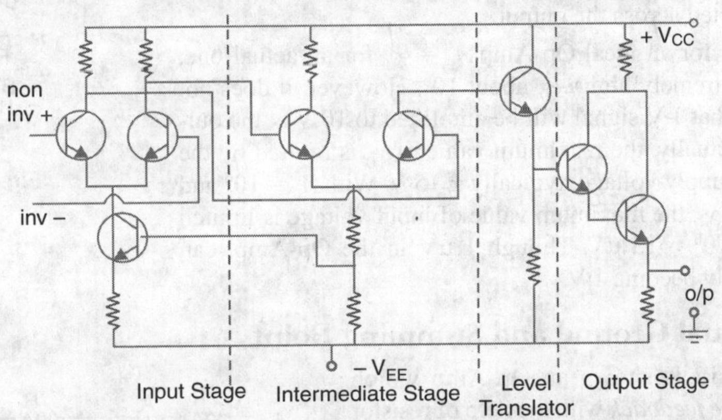

Fig. 35.10

35.8. The Op-Amp as a Voltage Amplifier

High voltage operational-amp.

We have already mentioned in Art 35.1, that Op-Amp has a very high voltage gain. But it will be interesting to know that an Op-Amp is never used without the application of negative or positive feedback. When an Op-Amp is used as a voltage amplifier, a large amount of negative feedback must be applied to specify accurately the voltage gain. There are two ways in which an Op-Amp can be connected to act as an amplifier–one method gives an inverted gain and the other gives a non-inverting gain. The method of connecting the Op-Amp which produces inverting gain is called *inverting amplifier*. Similarly, the method of connecting the Op-Amp which produces non-inverting gain is called *non-inverting amplifier*. Now we will discuss both the above types of voltage amplifiers one by one in the following pages.

35.9. Ideal Operational Amplifier

When an *Op-Amp* is operated without connecting any resistor or capacitor from its output to any one of its inputs (*i.e.,* without feedback), it is said to be in the open-loop condition. The word 'open loop' means *that feedback path or loop is open.* The specifications of Op-Amp under such condition are called open-loop specifications.

An ideal Op-Amp (Fig. 35.11) has the following characteristics:

1. its open-loop gain A_v is *infinite i.e.,* $A_v = -\infty$
2. its input resistance R_i (measured between inverting and non-inverting terminals) is *infinite i.e.,* $R_i = \infty$ ohm
3. its output resistance R_0 (seen looking back into output terminals) is *zero i.e.,* $R_0 = 0\ \Omega$
4. it has *infinite b and with i.e.,* it has flat frequency response from d.c. to infinity.

Though these characteristics cannot be achieved in practice, yet an ideal Op-Amp serves as a convenient reference against which real Op-Amps may be evaluated.

Following additional points are worth noting :

1. infinte input resistance means that input current $i = 0$ as indicated in Fig. 35.11. It means that an ideal Op-Amp is a voltage-controlled device.
2. $R_0 = 0$ means that v_0 is not dependent on the load resistance connected across the output.
3. though for an ideal Op-Amp $A_v = \infty$, for an actual one, it is extremely high *i.e.,* about 10^6. However, it does not mean that 1 V signal will be amplified to 10 V at the output. Actually, the maximum value of v_0 is limited by the basis supply voltage, typically ± 15 V. With $A_v = 10^6$ and v_0– 15 V_2 the maximum value of input voltage is limited to $15/10^6 = 15\ \mu$V. Though 1 μV in the Op-Amp, can certainly become 1 V.

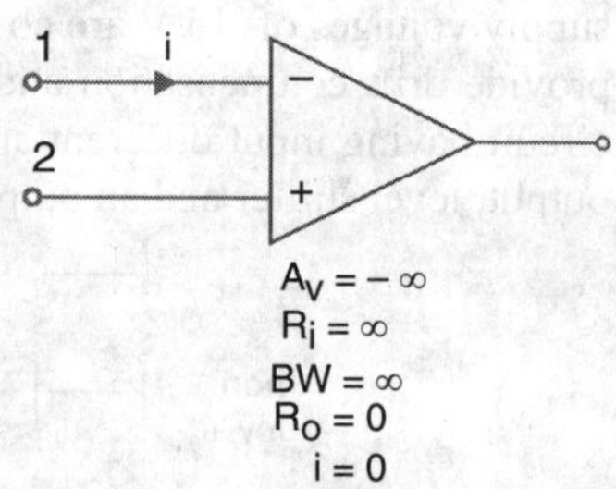

Fig. 35.9.

35.10. Viriual Ground and Summing Point

In Fig. 35.12 is shown an Op-Amp which *employs negative feedback* with the help of resistor R_f which feeds a portion of the output to the input.

Since input and feedback currents are algebraically added at point *A*, it is called the summing point.

The concept of virtual ground arises from the fact that input voltage v_i at the inverting terminal of the Op-Amp is forced to such a small value that, for all practicall purposes, it may be assumed to be zero. Hence, point *A* is essentially at ground voltage and is referred to as *virtual ground.* Obviously, *it is not the actual ground,* which, as seen from , is situated below.

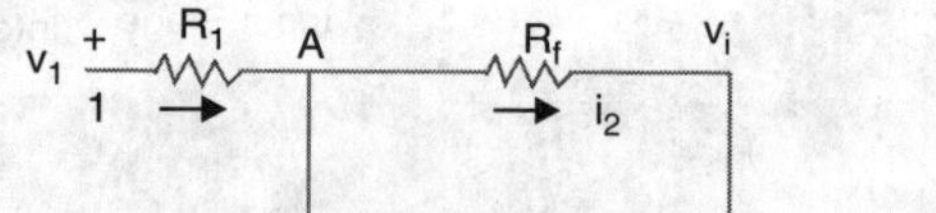

Fig. 35.12.

35.11. Why V_i is Reduced to Almost Zero ?

When v_1 is applied, point A attains some positive potential and at the same time V_0 is brought into existence. Due to negative feedback, some fraction of the output voltage is fed back to point *A* antiphase with the voltage already existing there (due to V_1).

The algebraic sum of the two voltages is almost zero so that $v_i = 0$. Obviously, will become exactly zero when negative feedback voltage at *A* is exactly equal to the positive voltage produced by v_1 at *A*.

Another point worth considering is that there exists a virtual short between the two terminals of the Op-Amp because $v_i = 0$. It is virtual because no current flows (remember $i = 0$) despite the existence of this short.

35.12. Inverting Amplifier

Fig. 35.13 shows the connection method for producing the inverted gain using Op-Amp. Such a configuration is called inverting amplifier. The Op-Amp circuit makes use of a single resistor (R_1) and a single feedback resistor (R_2). Notice that resistor R_1 is connected between the input terminal of the amplifier circuit and inverting input terminal (–) of the Op-Amp. While the resistor R_2 is connected between the inverting input terminal (–) and the output terminal of the Op-Amp. The non-inverting terminal (+) is connected to the ground. The inverting amplifier produces a 180° phase shift in voltage from input to output. Thus the input and output signals of the inverting amplifier are not in-phase with each other.

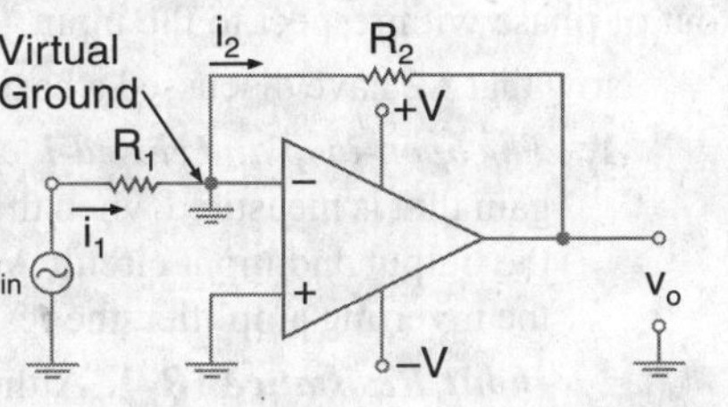

Fig. 35.13. Inverting amplifier.

Now we know that the Op-Amp gain without any feedback (called open-loop gain) is very high. This means that the voltage at the inverting (*i.e.*, '–') terminal must be very small. As a matter of fact, the input voltage at the inverting terminal will be very nearly at the same potential as the non-inverting terminal. Now since the non-inverting input is grounded, the inverting input of an Op-Amp is also at the ground potential and is referred to as *virtual ground.*

Since the inverting terminal is virtual ground, therefore the entire input voltage appears across R_1 and the input current, $i_1 = v_{in}/R_1$. Moreover, the input resistance (R_{in}) of an Op-Amp is very high (ideally infinite), so very little current flows into the Op-Amp. Practically, the current drawn by the Op-Amp is negligible or nearly zero. Thus all the input current flows through R_2 (*i.e.*, $i_1 = i_2$). And the voltage developed across R_2 is equal to the output voltage (v_o) of the circuit. Using this knowledge, let us develop expressions for the voltage gain, input resistance, output resistance and common mode rejection ratio of the inverting amplifier.

We have already mentioned above that the input voltage of an Op-Amp,

$$v_{in} = i_1 \cdot R_1, \quad \text{or} \quad i_1 = v_{in}/R_1$$

And output voltage,

$$v_o = -i_2 \cdot R_2, \quad \text{or} \quad i_2 = -v_o/R_2$$

The negative sign indicates that output is at a lower potential than that of a virtual ground. It is obvious because current through resistor R_2 is flowing from the side connected to the inverting input to the side connected to the output (see Figure 35.7). Since the input- and output-currents are equal (*i.e.*, $i_1 = i_2$), because Op-Amp input current is negligible.

$$\therefore \qquad \frac{v_{in}}{R_1} = \frac{v_o}{R_2}$$

or
$$v_o = \frac{R_2}{R_1} V_{in} \quad \text{or} \quad \frac{v_o}{v_{in}} = -\frac{R_2}{R_1} \qquad ...(i)$$

Now recall that voltage gain (A_v) of an amplifier is defined as the ratio of output voltage to the input voltage. Mathematically,

$$A_v = \frac{\text{Output voltage}}{\text{Input voltage}} = \frac{v_o}{v_{in}}$$

Substituting the value of (v_o/v_{in}) from equation (*i*) into the above equation, we get,

$$A_v = -\frac{R_2}{R_1}$$

The above equation indicates that voltage gain of an inverting amplifier can be determined by dividing the value of R_2 by R_1. The negative sign indicates that the inverting amplifier output is 180° out of phase with respect to the input.

Now that we have discussed a basic Op-Amp circuit, let us give practical meaning to few terms.

1. ***The open-loop and closed-loop gain.*** The open-loop voltage gain (A_{OL}) of an Op-Amp is the gain that is measured when there is no feedback path (*i.e.*, no physical connection) between the output and input circuit. When a feedback path is present, such as the R_2 connection in the inverting amplifier, the resulting circuit gain is referred to as the closed-loop gain (A_{CL}).
2. ***Input Resistance*** (***R_{in}***). Although, the Op-Amp has an extremely high input resistance (ideally infinite) yet the inverting amplifier does not have that high value. Referring to Fig. 35.14, we find that the voltage source "*sees*" the overall resistance to be as the parallel combination of input resistance (R_1) that is going to virtual ground and the Op-Amp input resistance. But since the input resistance of an Op-Amp is very high, therefore, input resistance of an inverting amplifier is,

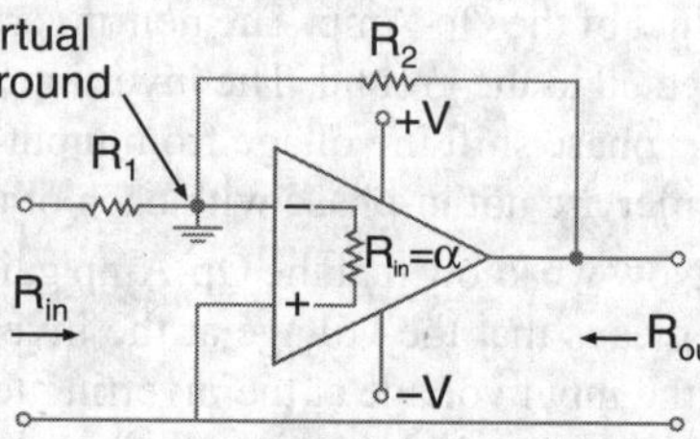

Fig. 35.14. Illustrating input and output resistance.

$$R_{in} \approx R_1$$

It may be noted carefully that the value of R_1 will always be much less than the input resistance of the Op-Amp, therefore, the overall input resistance of an inverting amplifier will be much lower than the Op-Amp input resistance.

3. ***Output Resistance.*** Referring to Fig. 35.14, again, we find that output resistance of an inverting amplifier is the parallel combination of R_2 and the output resistance of Op-Amp itself. Since the output resistance of an Op-Amp is usually very low (of the order of 100 Ω), the presence of R_2 reduces the overall output resistance of the inverting amplifier below 100 Ω.
4. ***Common-mode Rejection ratio.*** We have already defined common-mode rejection ratio (CMRR) of the Op-Amp as the ratio of differential gain to common-mode gain. Since the differential gain of the inverting amplifier (A_{CL}) is much less than the differential gain of the Op-Amp (A_{OL}), therefore, CMRR of the inverting amplifier will be much lower than that of the Op-Amp. Mathematically, CMRR of the inverting amplifier is,

$$CMRR = \frac{A_{CL}}{A_{CM}}$$

where A_{CL} = The closed loop voltage gain of the inverting amplifier, and

A_{CM} = The common-mode gain of the Op-Amp.

Example 35.6.

Determine the maximum allowable value of v_{in} for the circuit shown in Fig. 35.15. Assume that the gain of the amplifier is 200.

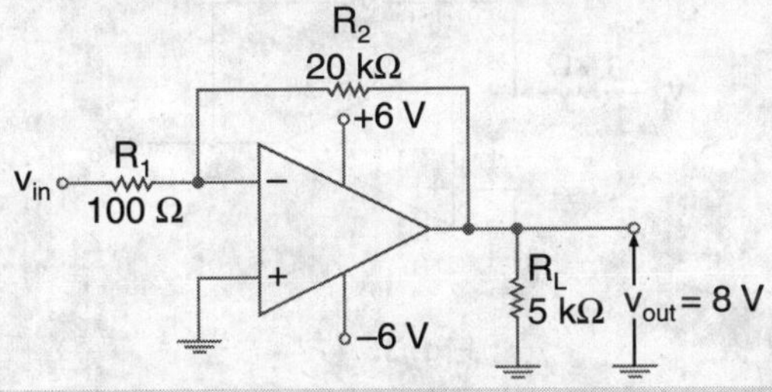

Fig. 35.15.

Solution.

Given $A_{CL} = 200$ and $v_{out} = 8$ V

We know that voltage gain of an inverting amplifier, (A_{CL}),

$$200 = -\frac{v_o}{v_{in}} = -\frac{8}{v_{in}}$$

$$v_{in} = -\frac{8}{200} = 0.04 = 40 \text{ mV } \textbf{Ans.}$$

Example 35.7.

Determine the peak-to-peak output voltage for the circuit shown in Fig. 35.16. Also determine the maximum possible output values for the amplifier. Assume the gain of the amplifier is 150.

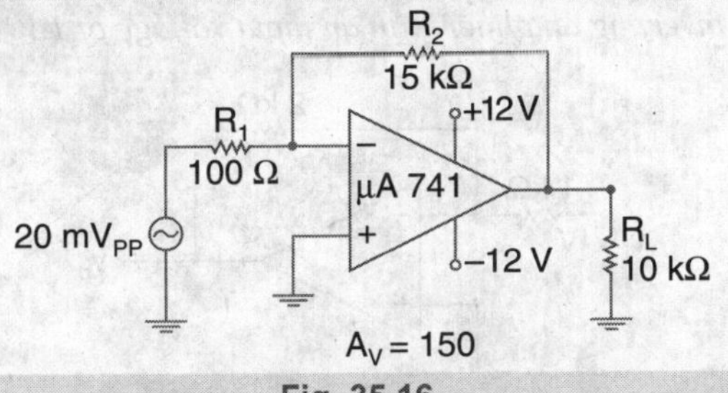

Fig. 35.16.

Solution.

Given : $A_{CL} = 150$

We know that voltage gain of an inverting amplifier, (A_{CL}),

$$150 = \frac{v_{out}}{v_{in}} = \frac{v_{out}}{200 \text{ m } V_{pk}}$$

$$\therefore \quad v_{out} = 150 \times 20 \text{ m } V_{pk} = 3000 \text{ m } V_{pk} = 3\ V_{pk} \ \textbf{Ans.}$$

According to the guidelines when dealing with 741, if the load resistance is between 2 kΩ to 10 kΩ, the maximum possible output values are

$$V_{pk\,(+)} = (+\text{ V}) - 2 \text{ V} = 12 - 2 = 10 \text{ V (maximum)}$$

and $$V_{pk\,(-)} = (-\text{ V}) + 2 \text{ V} = -12 + 2 - 10 \text{ V (maximum)}$$

Thus maximum possible output value could be between + 10 and – 10 V. **Ans.**

Example 35.8.

Fig. 35.17 shows an inverting amplifier with an input voltage v_i.

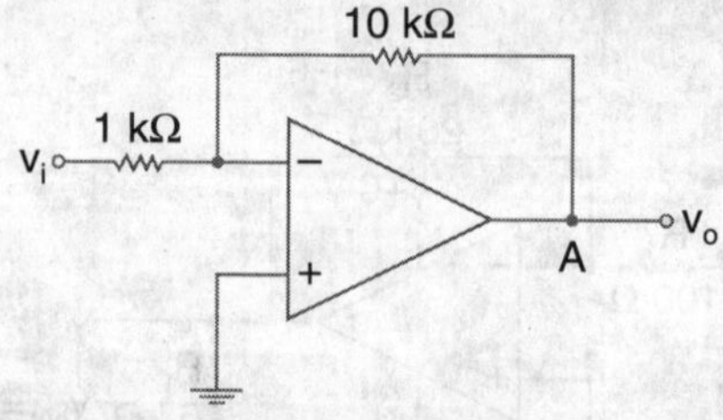

Fig. 35.17.

The input voltage is increased from 0.1 V to 0.4 V. what will be the voltage at node A in each case. why ? *(Bangalore Univ. B.Sc. Electronics. May 2001)*

Solution.

Given : The input voltage increased from 0.1 V to 0.4V.

$$R_1 = 1\text{ k}\Omega;\ R_2 = 10\text{ k}\Omega;$$

We know that voltage gain of the inverting amplifier,

$$A_{CL} = \frac{R_2}{R_1} = \frac{10\text{ k}}{1\text{ k}} = 10$$

Thus when the input voltage is increased from 0.1 V to 0.4 V, the voltage at mode A increases from 1 V to 4 V **Ans.**

Example. 35.9.

Fig. 35.18 shows an inverting amplifier with an input voltage of 1 V.

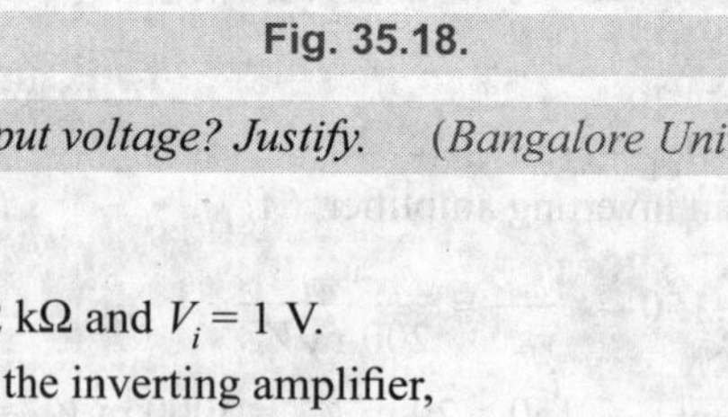

Fig. 35.18.

What will be value of output voltage? Justify. *(Bangalore Univ. B.Sc Electronics May 2001)*

Solution.

Given : $R_1 = 1$ kΩ; $R_2 = 2$ kΩ and $V_i = 1$ V.

We know that the gain of the inverting amplifier,

$$A_{CL} = \frac{R_2}{R_1} = \frac{2\text{ k}}{1\text{ k}} = 2$$

∴ Output voltage of the inverting amplifier,

$$v_o = A_{CC} \times V_i = 2 \times 1 = 2\text{ V } \textbf{Ans.}$$

Example. 35.10.

Design an inverting amplifier with a gain of –5 and input resistance of 10 kΩ.

(GBTU., 2010–2011)

Solution.

Given : $A_v = -5$ and $R_1 = 10$ kΩ.

We know that the gain of the inverting amplifier,

$$A_v = -\frac{R_2}{R_1}$$

and the feedback resistance,

$$R_2 = -A_v R_1 = -(-5) \times (10) = 50 \text{ k}\Omega$$

An inverting amplifier with the required value of R_2 is as shown in Fig. 35.19.

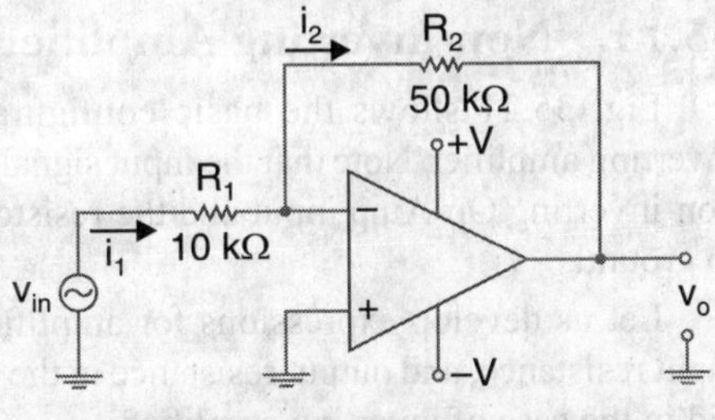

Fig. 35.19.

Example 35.11.

For the circuit shown in Fig. 35.20, find the closed-loop gain, input impedance, common-mode rejection ratio, and maximum operating frequency.

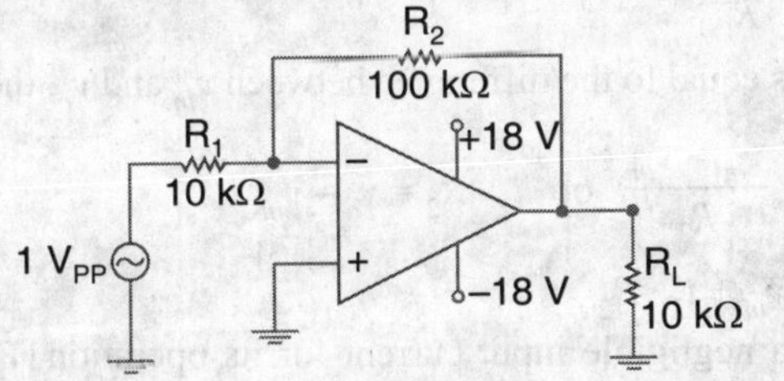

Op-Amp Paramaters.

$A_{CM} = 0.001$

$A_{OL} = 180{,}000$

$Z_{in} = 1 \text{ M}\Omega$

$Z_{out} = 80\ \Omega$ (max)

Slewrate = 0.5 V/μs

Fig. 35.20.

Solution.

We know that closed-loop gain,

$$A_{CL} = \frac{R_2}{R_1} = \frac{100\,\text{k}\Omega}{10\,\text{k}\Omega} = 10 \textbf{ Ans.}$$

and input impedance of the circuit,

$$Z_{in} \approx R_1 = 10 \text{ k}\Omega \textbf{ Ans.}$$

We also know that output impedance of the circuit,

$$Z_{out} \approx Z_{out} \text{ of the Op-Amp} = 80\ \Omega \textbf{ Ans.} \quad (\because Z_{in} \text{ of Op-Amp} >> R_1)$$

and the common-mode rejection ratio,

$$CMRR = \frac{A_{CL}}{A_{CM}} = \frac{10}{0.001} = 10{,}000 \textbf{ Ans.}$$

To calculate the maximum operating frequency for the inverting amplifier circuit, we need to determine its peak output voltage.

With values of $v_{in} = 1\ V_{PP}$ and $A_{CL} = 10$, the peak-to-peak output voltage,

$$v_{out} = 1 \times 10 = 10\ V_{PP}$$

Thus the peak output voltage is 5 V_{PK} and the maximum operating frequency

$$f_{max} = \frac{\text{Slew rate}}{2\pi V_{pk}} = \frac{0.5 \text{ V/}\mu\text{s}}{2\pi \times 5} = \frac{0.5/10^{-6} \text{ (V/s)}}{2\pi \times 5} = \frac{0.5 \times 10^6}{10\pi}$$

$$= 15.9 \times 10^3 \text{ Hz} = 15.9 \text{ kHz } \textbf{Ans.}$$

35.13. Non-inverting Amplifier

Fig. 35.21 shows the basic configuration of a non-inverting amplifier. Note that the input signal is applied to the non-inverting Op-Amp input and the resistor R_1 is returned to ground.

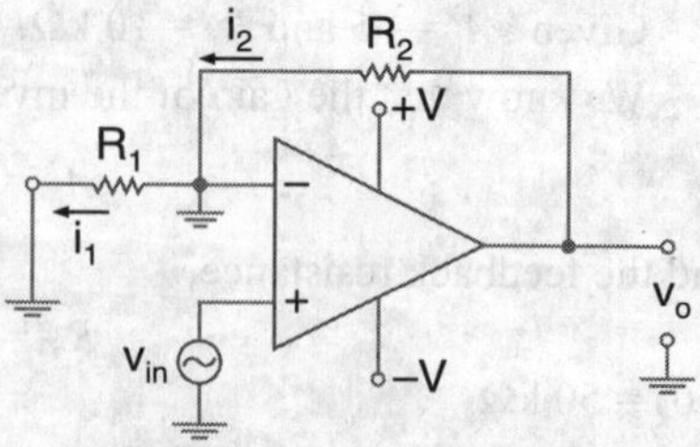

Fig. 35.21. A non-inverting amplifier.

Let us develop expressions for amplifier voltage gain, input resistance, and output resistance in the same way as we did in the case of inverting amplifier.

1. *Voltage gain.* We know that the Op-Amp gain without feedback (called open-loop gain) is very high. This means that voltage at the inverting terminal should nearly be the same as that at the non-inverting terminal. Since the non-inverting input is at a potential of v_{in}, therefore inverting input is also at the same potential. If i_1 is the current through resistor R_1, then its value,

$$i_1 = \frac{v_{in}}{R_1} \quad \text{or} \quad v_{in} = i_1 \cdot R_1 \qquad ...(i)$$

Since the voltage drop across R_2 is equal to the difference between v_{in} and v_o, therefore, current through resistor R_2,

$$i_2 = \frac{v_o - v_{in}}{R_2} \quad \text{or} \quad i_2 \cdot R_2 = v_o - v_{in}$$

$$\therefore \quad v_o = v_{in} + i_2 \cdot R_2 \qquad ...(ii)$$

Now recall that Op-Amp draws a negligible input current for its operation. For all practical purposes, its value is considered to be zero, therefore, $i_1 = i_2$.

Replacing i_2 by i_1 in equation (*ii*), we get

$$v_o = v_{in} + i_1 \cdot R_2$$

The amplifier voltage gain,

$$A_v = \frac{\text{Output voltage}}{\text{Input voltage}} = \frac{v_o}{v_{in}}$$

Substituting the values of v_o and v_{in} from equation (*ii*) and (*i*) into the above equation,

$$A_v = \frac{v_{in} + i_1 \cdot R_2}{i_1 \cdot R_1} = \frac{i_1 \cdot R_1 + i_1 \cdot R_2}{i_1 \cdot R_1} = \frac{R_1 + R_2}{R_1}$$

or
$$A_v = 1 + \frac{R_2}{R_1}$$

The above equation shows that voltage gain of a non-inverting amplifier will always be greater than the equivalent gain of the inverting amplifier by a value of 1. For example, if an inverting amplifier has a gain of 100, the equivalent non-inverting amplifier will have a gain of 101.

2. *Input resistance.* The non-inverting amplifier has extremely high input resistance because the input signal is applied directly to the Op-Amp (*i.e.*, not through a resistor R_1 as it happens in the case of inverting amplifier). As a matter of fact, the presence of feedback network causes the amplifier input resistance to be greater than the input resistance of the Op-Amp itself.

3. *Output Resistance.* The output resistance of a non-inverting amplifier is lower than the output resistance of the Op-Amp itself. This characteristic is similar to that of the inverting amplifier.

4. *Common-mode Rejection Ratio (CMRR).* The common-mode rejection ratio (CMRR) for a non-inverting amplifier is determined in the same way as we do it for the inverting amplifier. Thus

$$CMRR = \frac{A_{CL}}{A_{CM}}$$

where A_{CL} = The closed-loop voltage gain, and

A_{CM} = The common-mode voltage gain.

Example 35.12.

For the non-inverting amplifier shown in Fig. 35.22, find the values of closed-loop gain (A_{CL}), common mode rejection ratio (CMRR), and the maximum operating frequency.

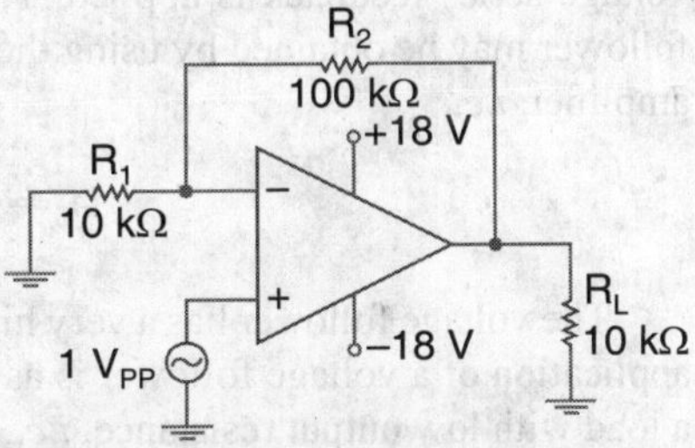

Fig. 35.22

Solution.

We know that closed-loop gain of a non-inverting amplifier,

$$A_{CL} = \left(1 + \frac{R_2}{R_1}\right) = \left(1 + \frac{100\ \text{k}\Omega}{10\ \text{k}\Omega}\right) = (1 + 10) = 11 \text{ Ans.}$$

and the common-mode rejection ratio,

$$CMRR = \frac{A_{CL}}{A_{CM}} = \frac{11}{0.001} = 11\,000 \textbf{ Ans.}$$

To calculate the maximum operating frequency for the mon-inverting amplifier circuit, we need to determine its peak output voltage.

With values of $v_{in} = 1\ V_{PP}$ and $A_{CL} = 11$, the peak-to-peak output voltage,

$v_{out} = 1 \times 11 V_{PP}$

Thus, the peak output voltage is $5.5 V_{pk}$ and the maximum operating frequency,

$$f_{max} = \frac{\text{Slew rate}}{2\pi\ V_{pk}} = \frac{0.5\ \text{V}/\mu\text{s}}{2\pi \times 5.5}$$

$$= \frac{0.5 \times 10^6}{2\pi \times 5.5} = 14.47 \times 10^3 \text{ Hz}$$

$$= 14.47 \text{ kHz} \textbf{ Ans.}$$

35.14. Non-inverting Amplifier as a Buffer

The extremely high input resistance and extremely low output resistance of the non-inverting amplifier makes it very useful as a buffer. You may recall that the emitter follower and source follower are circuits that can be used to match a source with high input resistance to a load with low resistance. The non-inverting amplifier can be used for the same purpose. However, the major difference is that the basic non-inverting amplifier is capable of producing high voltage gain values while the emitter follower and source follower have voltage gain values which are less than '1'.

35.15. The Voltage Follower

The voltage follower circuit can be obtained from the non-inverting amplifier by removing resistors R_1 and R_2. The resulting circuit is as shown in Fig. 35.23. Notice that the input voltage is applied at the non-inverting input of an Op-Amp, and the inverting input is connected to the Op-Amp output. Because of the direct connection between the inverting input and output terminals, a 100% voltage-series feedback is applied. The voltage gain of a voltage follower may be obtained by using the equation of a non-inverting amplifier, *i.e.*,

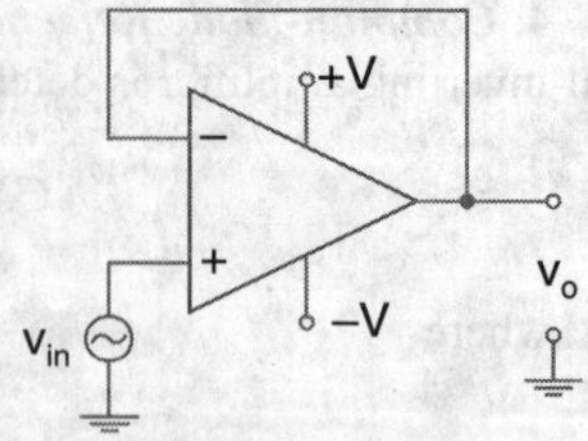

Fig. 35.23. The Voltage follower.

$$A_v = 1 + \frac{R_2}{R_1} = 1, \qquad (\because \text{ for a voltage follower, } R_1 = R_2 = 0)$$

The voltage follower has a very high input resistance and a very low output resistance. The main application of a voltage follower is as a means of connecting a source with high input resistance to a load with low output resistance, *i.e.*, to act as a buffer amplifier.

The common-mode rejection ratio (CMRR) of a voltage follower circuit is given by the relation,

$$CMRR = \frac{A_{CL}}{A_{CM}}$$

Since the closed loop gain (A_{CL}) is equal to '1', therefore, the common-mode rejection ratio,

$$CMRR = \frac{1}{A_{CM}}$$

Example 35.13.

For the voltage follower shown in Fig. 35.24, find the values of A_{CL}, CMRR and f_{max}.

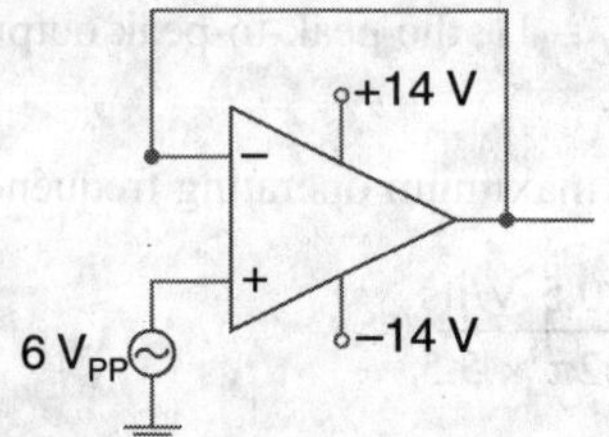

Op-Amp parameters.

$A_{CM} = 0.001$
$A_{OL} = 180,000$
$Z_{in} = 1\ M\Omega$
$Z_{out} = 80\ \Omega$ (max)
Slew rate = 0.5 V/μs.

Fig. 35.24.

Solution.

We know that for a voltage follower,

$$A_{CL} = 1 \textbf{ Ans.}$$

and the common-mode rejection ratio,

$$CMRR = \frac{A_{CL}}{A_{CM}} = \frac{1}{0.001} = 1000 \textbf{ Ans.}$$

To calculate the maximum operating frequency for the voltage follower, we need to determine its peak output voltage.

With values of $v_{in} = 6\ V_{PP}$ and $A_{CL} = 1$, the peak-to-peak output voltage,

$$v_{out} = 6 \times 1 = 6\ V_{PP}$$

Thus the peak output voltage is 3 V_{pk} and the maximum operating frequency,

$$f_{max} = \frac{\text{Slew rate}}{2\pi V_{pk}} = \frac{0.5 \text{ V/}\mu\text{s}}{2\pi \times 3} = \frac{0.5 \times 10^6}{2\pi \times 3} = 26.5 \text{ kHz } \textbf{Ans.}$$

35.16. Summing Amplifier

The summing amplifier is an Op-Amp circuit that accepts several inputs and provides an output that is proportional to the sum of inputs. The basic summing amplifier is shown in Fig. 35.25.

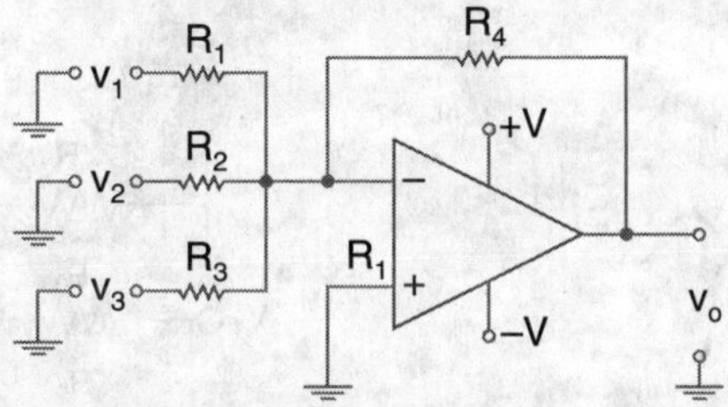

Fig. 35.25. A Basic summing amplifier.

Inverting Summing Amplifiers

The operation of the circuit may be understood as follows. Let us suppose we have an input voltage only at V_1, then output voltage,

$$v_o = -\frac{R_4}{R_1} \cdot v_1$$

Similarly, if v_2, is the only input, the output voltage is,

$$v_o = -\frac{R_4}{R_2} \cdot v_2$$

And if v_3 is the only input, the output voltage is,

$$v_o = -\frac{R_4}{R_3} \cdot v_3$$

However if all the three voltages are present, the output voltage would be,

$$v_o = \left(-\frac{R_4}{R_1} \cdot v_1\right) + \left(-\frac{R_4}{R_2} \cdot v_2\right) + \left(-\frac{R_4}{R_3} \cdot v_3\right)$$

And if $R_1 = R_3 = R_4 = R$, then output voltage,

$$v_o = -(v_1 + v_2 + v_3)$$

It is evident from the above equation that the output voltage is equal to the sum of three input voltages. The negative sign indicates that the output voltage is 180° out of phase with respect to the three input voltages. It may be noted carefully, that there is no practical limit on the number of summing inputs. As long as the amplifier is capable of providing the correct output voltage, the circuit can have any number of inputs. The equation (*i*) can be expanded (or reduced). Moreover the resistor used at the amplifier input do not necessarily have the same value. In other words, the inputs can be weighted so that certain inputs have more of effect on the output.

35.17. Non-inverting Summing Amplifier

Fig. 35.26 shows the basic configuration of a non-inverting summing amplifier. The operation of the circuit may be understood as follows. Let the voltage at the negative input is V_a, then the voltage at positive input is also V_a.

Let us support we have an input voltage only at V_1, V_2, and V_3 at negative input, then the nodal equation at node '*a*' is given by

$$\frac{V_1 - V_a}{R_1} + \frac{V_2 - V_a}{R_2} + \frac{V_3 - V_a}{R_3} = 0$$

$$V_a = \frac{\dfrac{V_1}{R_1} + \dfrac{V_2}{R_2} + \dfrac{V_3}{R_3}}{\dfrac{1}{R_1} + \dfrac{1}{R_2} + \dfrac{1}{R_3}} \quad \text{... } (i)$$

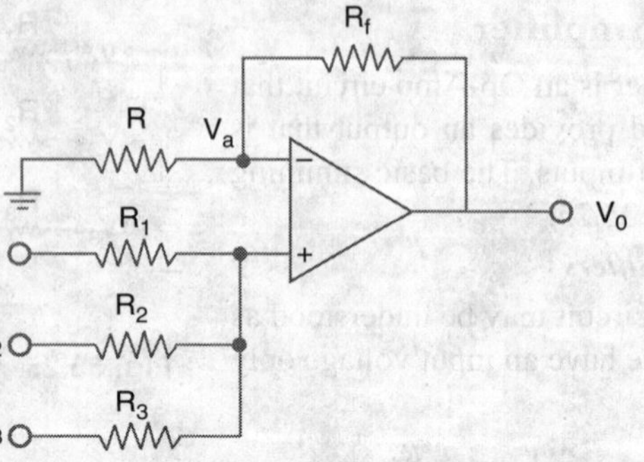

Fig. 35.26

The output voltage of the amplifier is given by,

$$V_0 = \left(1 + \frac{R_f}{R}\right) V_a$$

Substituting the value of from equation (*i*), we get

$$V_0 = \left(1 + \frac{R_f}{R}\right) \frac{\left(\dfrac{V_1}{R_1} + \dfrac{V_2}{R_2} + \dfrac{V_3}{R_3}\right)}{\dfrac{1}{R_1} + \dfrac{1}{R_2} + \dfrac{1}{R_3}}$$

35.18. Averaging Amplifier

A summing amplifier can be used to produce the mathematical average of input voltages. This is obtained by setting the ratio of the feedback resistor to the input resistor equal to the reciprocal of the number of inputs. This is because of the reason that the average of several numbers is obtained by adding the numbers and then dividing by the quantity of numbers we have. Such a summing amplifier is known as an *averaging amplifier*.

A summing amplifier can be used to produce the mathematical average of input voltages. This is obtained by setting the ratio of the feedback resistor to the input resistor equal to the reciprocal of the number of inputs. Fig. 35.27 shows the averaging amplifier the input voltages V_1, V_2 and V_3 are applied across the input resistor *R*.

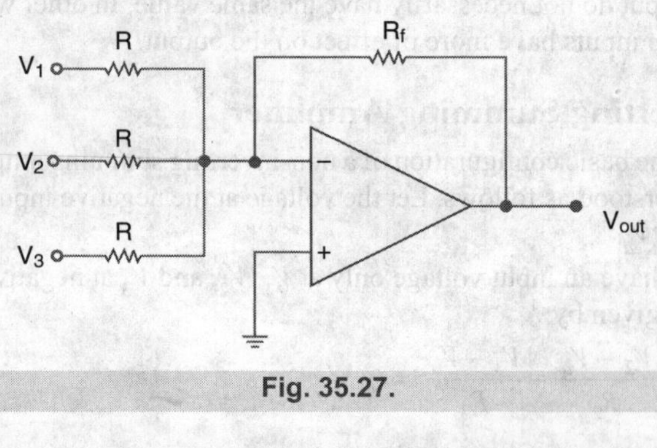

Fig. 35.27.

This is because of the reason that the average of several numbers is obtained by adding the numbers and then dividing by the quantity of numbers we have. Such a summing amplifier is known as an averaging amplifier.

35.19. Linear Amplifier

We will consider the functioning of an Op – Amp as constant-gain amplifer both in the inverting configurations.

Inverting Amplifier or Negative Scale.

As shown in Fig. 35.28 inverting terminal has been grounded, where as R_1 connected the input signal v_1 to the inverting input. A feefback resistor R_f has been connected from the output to the inverting input.

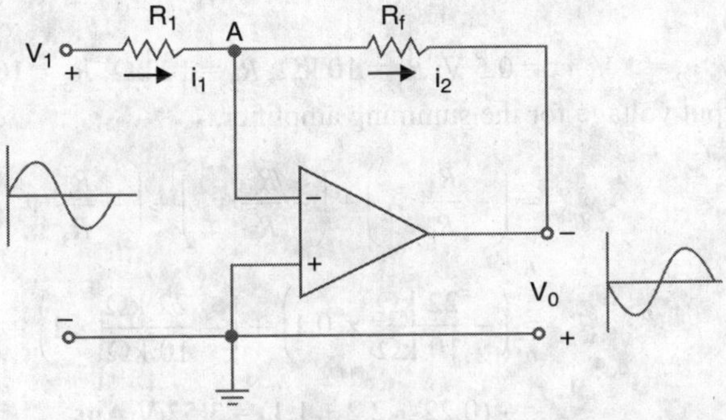

Fig. 35.28

Gain

Since point A is at ground potential,

$$i_1 = \frac{v_{in}}{R_f} = \frac{v_1}{R_1}$$

$$i_2 = \frac{-v_0}{R_f} \qquad \text{Please note – ve sign}$$

Using Kirchoff's current law at point A,

$$i_1 + (-i_2) = 0 \text{ or } \frac{v_1}{R_1} + \frac{v_0}{R_f} = 0$$

or $$\frac{v_0}{R_f} = -\frac{v_1}{R_1} \text{ or } \frac{v_o}{v_1} = -\frac{R_f}{R_1} \text{ or } A_v = -\frac{R_f}{R_1} \text{ or } \therefore v_0 = -k\,V_{in}$$

It is seen from above, that closed-loop gain of the inverting amplifier depends on the ratio of the two external resistors R_i and R_f and is independent of the amplifier parameters.

Also notice that Op – Amp works as a negative scales. It scales the input, *i.e.*, it multiplies the input by a minus constant factor *k*.

Example. 35.14.

Fig. 35.29 shows the circuit of a summing amplifier.

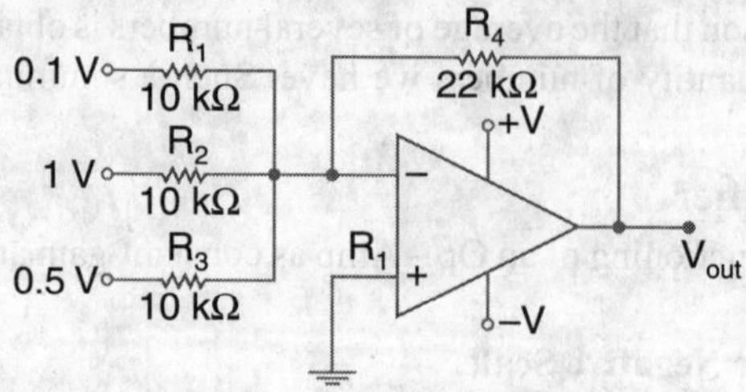

Fig. 35.29.

Determine the value of output voltage for this circuit.

Solution.

Given : $v_1 = 0.1$ V; $v_2 = 1$ V; $v_3 = 0.5$ V; $R_1 = 10$ kΩ; $R_2 = 10$ kΩ; $R_3 = 10$ kΩ and $R_4 = 22$ kΩ.

We know that output voltage for the summing amplifier,

$$v_{out} = \left(-\frac{R_4}{R_1}\cdot v_1\right) + \left(-\frac{R_4}{R_2}\cdot v_2\right) + \left(-\frac{R_4}{R_3}\cdot v_3\right)$$

$$= \left(-\frac{22\text{ k}\Omega}{10\text{ k}\Omega}\times 0.1\right) + \left(-\frac{22\text{ k}\Omega}{10\text{ k}\Omega}\times 1\right) + \left(-\frac{22\text{ k}\Omega}{10\text{ k}\Omega}\times 0.5\right)$$

$$= -(0.22 + 2.2 + 1.1) = 3.52 \text{ V } \textbf{Ans.}$$

Example. 35.15.

Find the output of the following Op-Amp. Circuit shown in Fig. 35.30.(GBTU/MTU, 2008, 09)

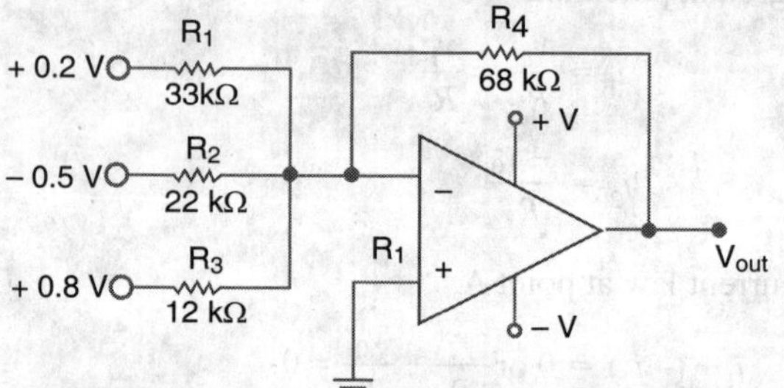

Fig. 35.30.

Solution.

Given : $v_1 = 0.2$ V; $v_2 = -0.5$ V; $v_3 = 0.8$ V; $R_1 = 33$ kΩ $= 33 \times 10^3 \Omega$ $R_2 = 22$ kΩ $= 22 \times 10^3$ Ω; $R_3 = 12$ kΩ $= 12 \times 10^3 \Omega$ and $R_4 = 68$ kΩ. $= 60 \times 10^3 \Omega$.

We know that output voltage for the summing amplifier.

$$V_{\text{out}} = \left(\frac{R_4}{R_1}\cdot v_1\right) + \left(\frac{R_4}{R_2}\cdot v_2\right) + \left(\frac{R_4}{R_3}\cdot v_3\right)$$

$$= \left(\frac{68}{33}\times 0.2\right) + \left(-\frac{68}{22}\times 0.5\right) + \left(\frac{68}{12}\times 0.8\right) = 3.4 \text{ V } \textbf{Ans.}$$

Example. 35.16.

Design an Op Amp circuit that will produce an output equal to $-(4\ V_1 + V_2 + 0.1\ V_3)$. Write an expression for the output and sketch its output waveform when $V_1 = 2w$, $V_2 = +5\ V$ d.c. and $V_3 = -100\ V$ d.c. *(Bangalore Univ. B.Sc. April/May 2001)*

Solution.

Given : $V_1 = 2\ \sin \omega t$, $V_2 = +5$ d.c. and $V_3 = -100\ V$ d.c.

The Op-Amp circuit that will produce an output equal to $-(4\ V_1 + V_2 + 0.1\ V_3)$ is shown in Fig. 35.31. Notice that V_1 bias a given of 4, V_2 has a unity gain and V_3 a gain of 0.1.

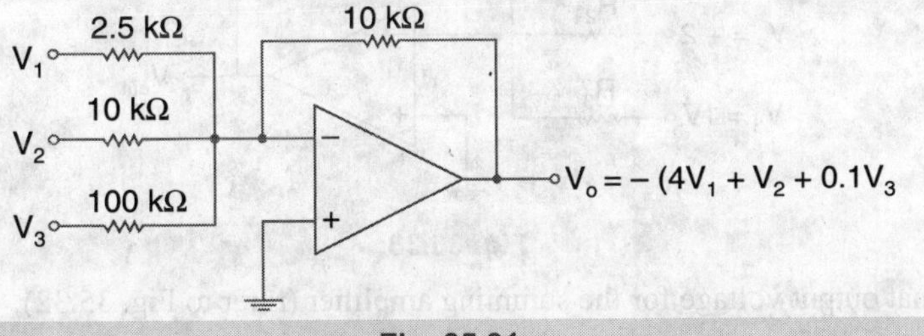

Fig. 35.31.

Now it $V_1 = 2\sin \omega t$

$V_2 = +5$ V d.c.

and $V_3 = -100$ V d.c.

Then output voltage,

$$V_o = -(4\ V_1 + V_2 + 0.1\ V_3)$$
$$= -(4 \times 2\sin \omega t + 5 + 0.1) \times (-100\ V)$$
$$= -(8\sin \omega t + 5 + 10\ V)$$
$$= -(15 + 8\sin \omega t)$$

Fig. 35.32. shows the sketch of the output waveform.

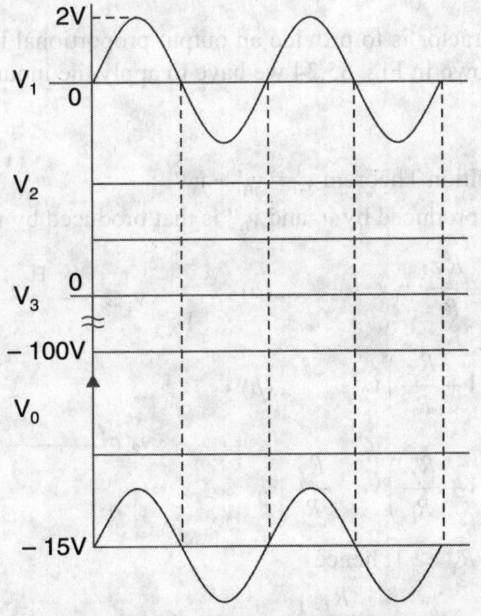

Fig. 35.32.

Example 35.17.

Calculate the output voltage of a these input summing amplifier. Given : $R_1 = 200\ k\Omega$, $R_2 = 250\ k\Omega$; $R_3 = 500\ k\Omega$; $R_f = 1\ m\Omega$, $V_1 = -2\ V$, $V_2 = +2V$, $V_3 = +1\ V$. (Bangalore Univ. July 2003)

Solution.

Given : $V_1 = -2$ V, $V_2 = +2$ V, $V_3 = +1$ V; $R_1 = 200$ k Ω, $R_2 = 250$ kΩ, $R_3 = 500$ kΩ, $R_f = 1$ mΩ.

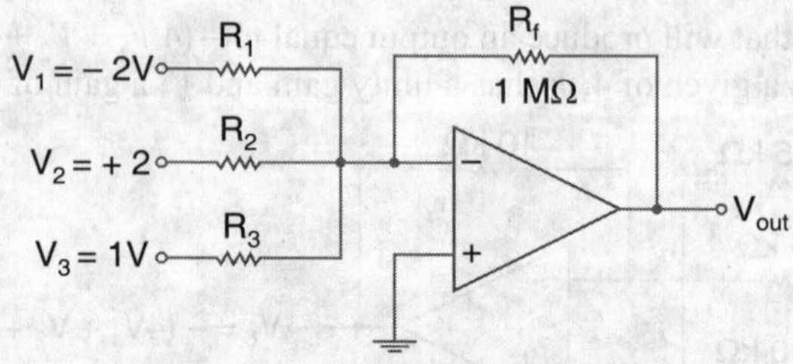

Fig. 35.33.

We know that output voltage for the summing amplifier (refer to Fig. 35.32),

$$V_{out} = \left(-\frac{R_f}{R_1}\right) v_1 + \left(-\frac{R_f}{R_2}\right) v_2 + \left(-\frac{R_f}{R_3}\right) v_3$$

$$= \left(-\frac{1\ m\Omega}{200\ k\Omega}\right)(-2V) + \left(-\frac{1\ m\Omega}{250\ k\Omega}\right)(+2V) + \left(\frac{-1\ m\Omega}{500\ k\Omega}\right)(-1\ V)$$

$$= (-5) \times (-2\ V) + (-4)(+2\ V) + (-2)(-1\ V)$$

$$= 10\ V - 8\ V + 2\ V$$

$= 4$ V **Ans.**

35.21. Subtractor

The function of a subtractor is to provide an output proportional to or equal to the difference of two input signals. As shown in Fig. 35.34 we have to apply the inputs at the inverting as well as non-inverting terminals.

Calculations

According to Superposition Theorem $v_0 = v_0' + v_0''$

where v_0' is the output produced by v_1 and v_0'' is that produced by v_2.

Now, $$v_0' = -\frac{R_f}{R_1} \cdot v_1 \quad ...(a)$$

$$v_0'' = \left(1 + \frac{R_f}{R_1}\right) v_2 \quad ...(b)$$

$$v_0 = \left(1 + \frac{R_f}{R_1}\right) v_2 - \frac{R_f}{R_1} \cdot v_1$$

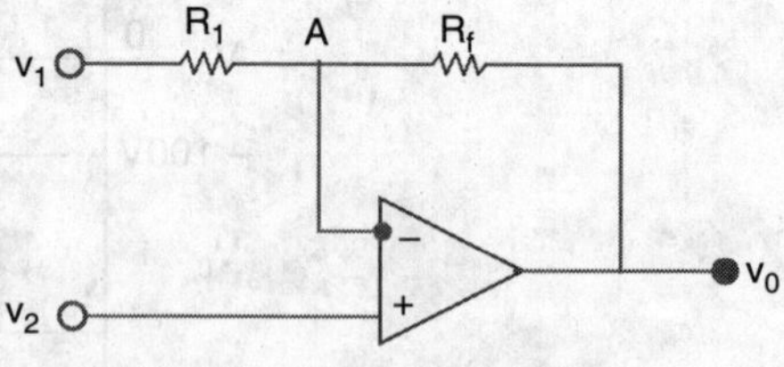

Fig. 35.34.

Since $R_f >> R_1$ and $R_f/R_1 >> 1$, hence

$$v_0 \approx \frac{R_f}{R_1}(v_2 - v_1) = K\,(v_2 - v_1)$$

Further, If $R_f = R_1$, then

$$V_0 = (v_2 - v_1) = \text{difference of the two input voltages}$$

Obviously, if $R_f \neq R_1$, then a scale factor is introduced.

Example 35.18.

Find the output voltages of an Op-Amp inverting adder for the following sets of input voltages and resistors. In all cases, $R_f = 1\ M\Omega$.

$$v_1 = -3\text{ V}, v_2 = +3\text{ V}, v_3 = +2\text{ V}; R_1 = 250\text{ k}\Omega, R_2 = 500\text{ k}\Omega, R_3 = 1\text{ M}\Omega$$

(Electronic Engg. Nagput Univ. 1991)

Solution.

Given : $v_1 = -3$ V, $v_2 = +3$V, $v_3 = +2$ V, $R_1 = 250$ kΩ, $R_2 = 500$ kΩ, $R_3 = 1$ MΩ and $R_f = 1$ MΩ

We know that output voltage,

$$v_0 = -(K_1 v_1 + K_2 v_2 + K_3 v_3)$$

$$K_1 = \frac{R_f}{R_1} = \frac{1000\text{ K}}{250\text{ K}} = 4, K_2 = \frac{1000}{500} = 2, K_3 = \frac{1\text{ M}}{1\text{ M}} = 1$$

$$v_0 = -[(4 \times -3) + (2 \times 3) + (1 \times 2) + (1 \times 2)] = +4\text{ V}$$

Example 35.19.

In the subtractor circuit of Fig. 35.33, $R_1 = 5\ k\Omega$, $R_f = 10\ k\Omega$, $v_1 = 4\ V$ and $v_2 = 5\ V$. Find the value of output voltage.

Solution.

Given : $R_1 = 5$ kΩ, $R_f = 10$ kΩ, $v_1 = 4$ V and $v_2 = 5$ V

We know that output voltage,

$$v_0 = \left(1 + \frac{R_f}{R_1}\right) v_1 - \frac{R_f}{R_1} v_2 = \left(1 + \frac{10}{5}\right) 4 - \frac{10}{5} \times 5 = +2\text{ V}$$

Example 35.20.

Desgin on Op-Amp circuit that will produce an output equal to $-(4\ v_1 + v_2 + 0.1\ v_3)$. Write an expression for the output and sketch its output and sketch its ouptut waveform when $v_1 = 2\sin\omega t$, $v_2 = +5\ V$ d.c. and $v_3 = -100$ d.c. *(Banglore University 2001)*

Solution.

Given : $v_1 = 2\sin\omega t$, $v_2 = +5$ V, $v_3 = -100$ V and $V_o = -(4v_1 + v_2 + 0.1\ v_3)$

We know that output voltage,

$$v_0 = \left[-\frac{R_f}{R_1} v_1 + \frac{R_f}{R_2} v_2 + \frac{R_f}{R_3} v_3\right] \quad ...(i)$$

and also

$$v_0 = -(4\ v_1 + v_2 + 0.1\ v_3) \quad ...(ii)$$

Comparing equations (*i*) and (*ii*), we find,

$$\frac{R_f}{R_1} = 4, \frac{R_f}{R_2} = 1, \frac{R_f}{R_3} = 0.1$$

Therefore if we assume $R_f = 100$ kΩ, then $R_1 = 25$ kΩ, $R_2 = 100$ kΩ and $R_3 = 10$ kΩ. With the values of R_1, R_2 and R_3, the Op-Amp circuit is as shown in Fig. 35.35 (*a*)

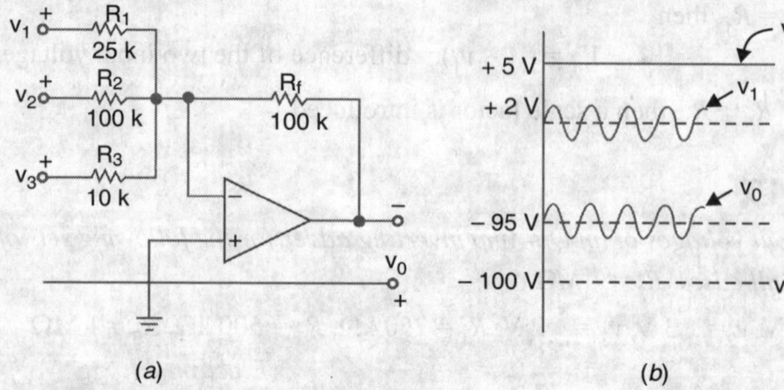

Fig. 35.35

With the given galues of v_1 = 2 sin ω*t*, v_2 = + 5 V, v_3 = – 100 V d.c. the output voltage, v_0 = 2 sin ω*t* + 5 – 100 = 2 sin ω*t* –95 V. The waveform of the output voltage is sketched as shown in Fig. 35.35 (*b*)

35.20. Differential Amplifier

Fig. 35.36 shows the circuit of a differential amplifier. Notice that the voltages are applied at both the Op-Amp inputs (*i.e.*, inverting and non-inverting inputs) through the resistors. And the difference between these two voltages is amplified.

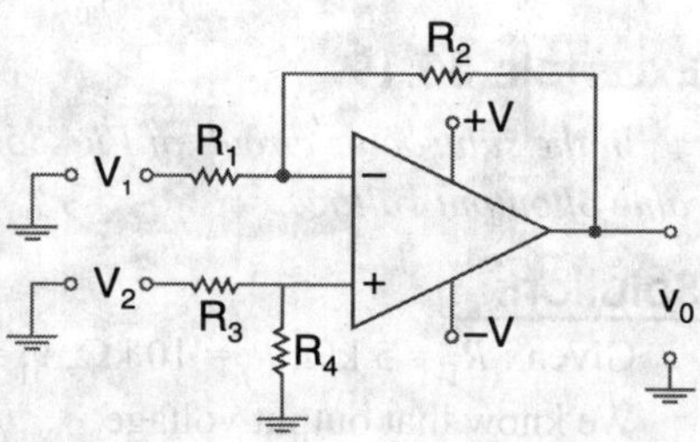

Fig. 35.36. A differential amplifier.

Suppose that voltage '*v*' volts is applied to terminal 1 and zero volts to terminal 2. The difference in the potentials at the inverting and non-inverting Op-Amp terminals is very nearly zero. Therefore the inverting terminal must be at zero potential. This means that the input voltage v_1 is developed across resistor R_1.

∴ Input current, $$i_1 = \frac{v_1}{R_1}$$

Since the input resistance of the Op-Amp is high, the current, (*i.e.*, i_1) flows through resistor R_2. the voltage drop across R_2 which is the output voltage V_{out} of the circuit,

$$= -v_1 \cdot \left(\frac{R_2}{R_1}\right)$$

And the voltage gain of the circuit,

$$A_v = \frac{v_{out}}{v_1} = \frac{-v_1 \cdot (R_2/R_1)}{v_1} = -\frac{R_2}{R_1}$$

Conversely if the voltage applied to input terminals 1 and 2 are respectively zero and v_2 volts, the voltage appearing at the non-inverting terminal

$$= v_2 \cdot \left(\frac{R_4}{R_3 + R_4}\right)$$

This voltage will also appear at the inverting terminal. So the voltage across resistor R_1,

$$= -v_2 \cdot \left(\frac{R_4}{R_3 + R_4}\right)$$

In this case, the output voltage of the circuit,

$$v_{out} = v_2 \cdot \left(\frac{R_4}{R_3 + R_4}\right) + \left\{\left(-v_2 \cdot \frac{R_4}{R_3 + R_4}\right) \times \left(-\frac{R_2}{R_1}\right)\right\}$$

And the voltage gain of the circuit,

$$A_v = \frac{v_{out}}{v_2} = \left(\frac{R_4}{R_3 + R_4}\right)\left(1 + \frac{R_2}{R_1}\right)$$

Lastly, if the voltages applied to the terminals 1 and 2 are v_1 and v_2 volts respectively, the difference between the two voltages is amplified. If $v_1 > v_2$, a voltage $(v_1 - v_2)$ is amplified $(-R_2/R_1)$ times. On the other hand, if $v_2 > v_1$, a voltage $(v_2 - v_1)$ is amplified by $[R_4/(R_3 + R_4)] \cdot [1 + (R_2/R_1)]$.

If $R_1 = R_3$ and $R_2 = R_4$, then the voltage gain,

$$A_v = \frac{v_{out}}{v_1 - v_2} = \frac{R_2}{R_1}; \therefore v_{out} = \frac{R_2}{R_1}(v_2 - v_1)$$

35.21. Op-Amp Frequency Response

We have already discussed in Art. 35.6 that slew rate of an Op-Amp is a measure of how fast the output voltage can change. It is expressed in volts/microseconds (V/μs). The maximum operating frequency of an Op-Amp is,

$$f_{max} = \frac{\text{Slew rate}}{2\pi v_{pk}}$$

Thus the peak output voltage (v_{pk}) limits the maximum operating frequency. When the maximum output frequency of an Op-Amp is exceeded, the result is distorted output waveform. It will be interesting to know that if the operating frequency of an Op-Amp is increased beyond a certain point, it will,

1. Decrease the maximum output voltage swing
2. Decrease the open-loop voltage gain
3. Decrease the input resistance and
4. Increase the output resistance.

35.22. Frequency versus Gain Characteristic of an Op-Amp

Fig. 35.37 shows the frequency versus gain characteristic of a typical Op-Amp (such as 741). As seen from this figure, the Op-Amp gain remains stable from 0 Hz to a certain cut-off frequency (which is less than 10 Hz). Then the gain reduces by a factor of 100 when the frequency is increased by a factor of 10 (*i.e.*, the gain drops at the rate of 20 dB per decade).

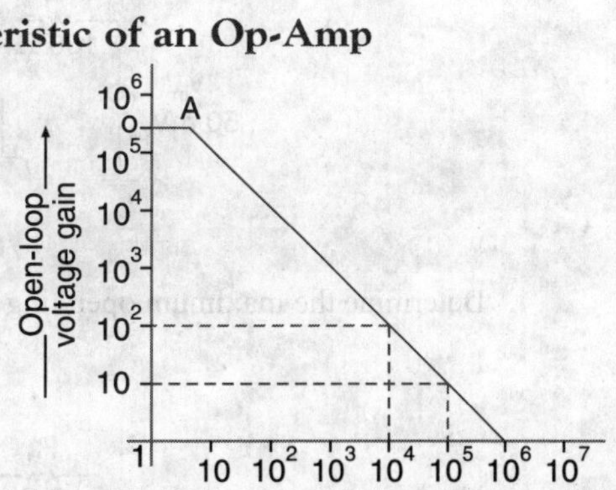

Fig. 35.37. Op-Amp frequency versus gain characteristic.

Since the Op-Amp is a d.c. amplifier, it will exhibit mid-band voltage gain at 0 Hz. As the frequency of operation is increased above 0 Hz, a point (such as 'A' shown in Fig. 35.36) is reached where the gain starts dropping. This drop in gain is due to internal values of input and output capacitance. It may be noted that as the frequency is increased to 10^6 Hz (*i.e.*, 1 MHz), the Op-Amp gain is reduced to '1' (*i.e.*, unity). This means that the 741 Op-Amp cannot be operated above 10^6 Hz. The frequency range from 0 to 10^6 Hz is referred to as the bandwidth of an Op-Amp. If we wish to operate the Op-Amp at a gain of 100, the bandwidth is

10^4 Hz (or 10 kHz). And if the gain required is 10, then the bandwidth is 10^5 Hz (or 100 kHz). This indicates that the gain and bandwidth of an Op-Amp are inversely proportional to each other. Based on this fact, we can make the following two statement.

1. The higher the gain of an Op-Amp, the narrower is its bandwidth, and
2. The lower the gain of an Op-Amp, the wider is its bandwidth.

It is evident from the above discussion, that if we want a wider bandwidth, we have to settle for less gain. On the other hand, if we want a high gain, we have to settle for a narrower bandwidth.

NUMERICAL PROBLEMS

1. A certain Op-Amp has a open-loop gain of 100,000 and a common-mode gain of 0.25. Determine the CMRR and express it in decibels. **(Ans. 400,000 and 112 dB)**
2. Fig. 35.38 shows the circuit of an inverting amplifier. Determine the value of maximum operating frequency for this circuit. **(Ans. 79.6 kHz)**

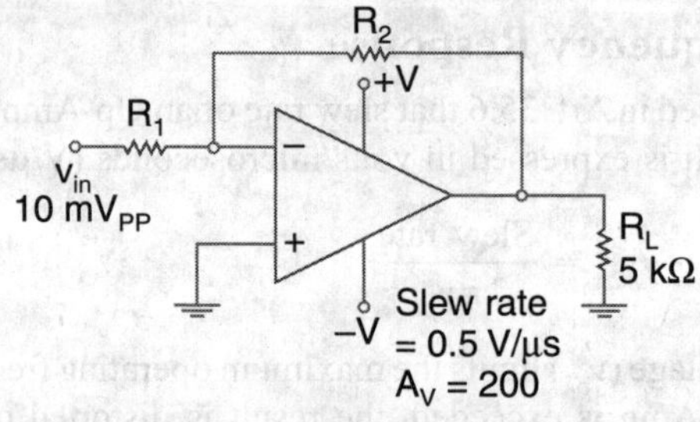

Fig. 35.38.

3. Determine the peak-to-peak output voltage value for the circuit shown in Fig. 35.39. Also determine the maximum possible output values for the amplifier. Assume that the gain of the amplifier is 150. **(Ans. 15 V_{PP}, + 9 V and – V)**

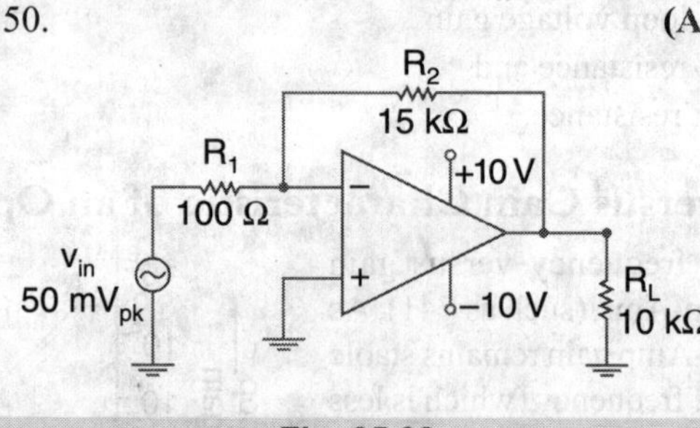

Fig. 35.39.

4. Determine the maximum operating frequency for the amplifier shown in Fig. 35.40. **(Ans. 19.8 kHz)**

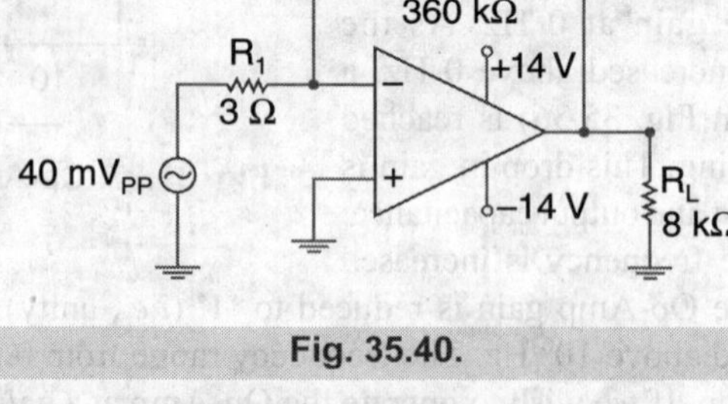

Fig. 35.40.

5. Fig. 35.41 shows the Op-Amp configuration. Determine the value of R_2 required to produce a gain of 100. **(Ans. 220 kΩ)**

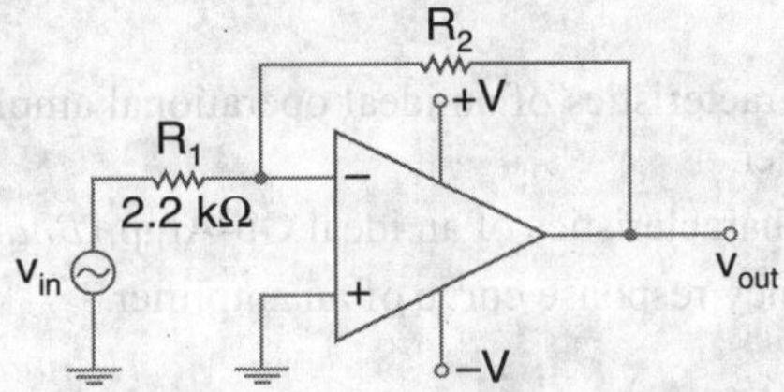

Fig. 35.41.

6. Determine the gain of the amplifier in Fig. 35.42. The open-loop voltage gain is 100,000. **(Ans. 22.3)**

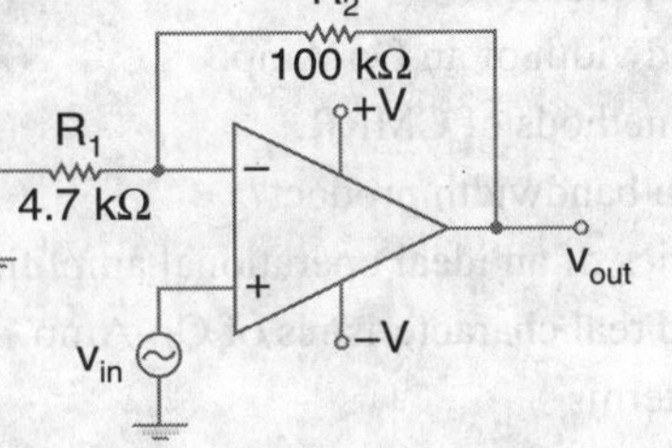

Fig. 35.42.

7. The Op-Amp shown in Fig. 35.43. has the following parameters : A_{OL} = 50,000, Z_{in} = 4 MΩ, and Z_{out} = 50 Ω. (*a*) Find the values of input and output impedance, (*b*) Determine the closed-loop voltage gain. **(Ans. (*a*) 1 kΩ and 50 Ω (*b*) – 100)**

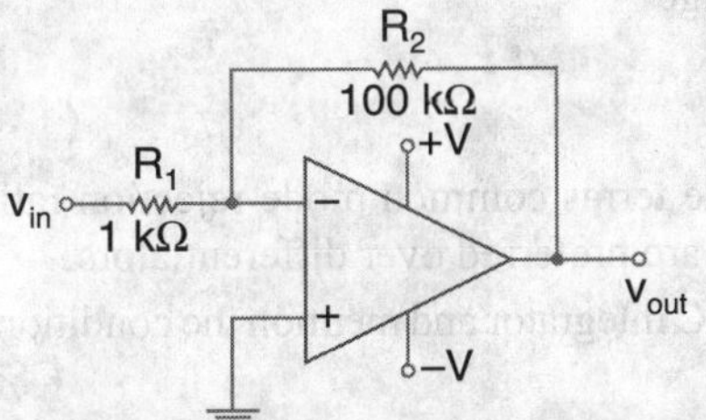

Fig. 35.43.

DESCRIPTIVE QUESTIONS

1. Describe some of the characteristics of a practical Op-Amp.
2. Define the term "common-mode rejection ratio".
3. For a given value of differential gain, does a higher CMRR results in a higher or lower common-mode gain? Explain briefly.
4. How does the input impedance of a non-inverting amplifier compare to the input impedance of the Op-Amp itself.
5. What is a voltage follower? Describe its main characteristics.
6. Draw the circuit of a summing amplifier using inverting amplifier configuration. Write an equation for the output voltage for this circuit.
7. What is a differential amplifier? Can it be used in a single-ended input configuration ?

8. Draw the pin diagram of 741 Op-Amp and mention the function of each pin. What are the characteristics of an ideal Op-Amp? Why it called operational amplifier? *(GBTU/MTU., 2006-07)*
9. Explain the ideal characteristics of an ideal operational amplifier. Also draw the circuit of non-inverting amplifier. *(GBTU/MTU., 2006-07)*
10. Lists the open loop characteristics of an ideal Op-Amp. *(Bhartiya University., April 2004)*
11. Draw general frequency response curve of an amplifier. *(Anna University., Nov/Dec. 2007)*
12. Write the improving methods of CMRR. *(Anna University., Nov/Dec. 2007)*
13. Define CMRR. *Anna University., Nov/Dec. 2007)*
14. Define slew rate and CMRR of an Op-Amp. *(PBUT., 2006)*
15. Define slew rate. What causes it? *(Anna University, April/May 2008)*
16. Define unity gain bandwidth of an Op-Amp. *(Anna University., April/May 2008)*
17. Write the improving methods of CMRR. *(Anna University., Nov/Dec 2007)*
18. What is meant by gain bandwidth product? *(Anna University., April/May 2008)*
19. Enlist the characteristics of an ideal operational amplifier (Op-Amp) *(PBUT., 2009-10)*
20. Describe the ideal and real characteristics of Op-Amp. *(PBUT., 2009)*
21. Define the following terms
    (*a*) Slew rate of an Op-Amp
    (*b*) Unity gain bandwidth *(PBUT., 2009)*
22. Explain the term sin Op-Amp.
    (*a*) Input offset current
    (*b*) Input offset voltage
    (*c*) Slew rate
    (*d*) CMRR *(GBTU/MTU., 2005-06)*
23. Define and explain the terms common mode rejection ratio and virtual ground in an Op-Amp. Why integrator are preferred over differentiators. *(GBTU/MTU., 2008-09)*
24. Draw the circuit of a RC integrator and mention the condition under which the circuit behaves as a integrator. *(Anna University., Nov/Dec 2008)*
25. Define slew rate and CMRR of an Op-Amp. *(PBUT., 2006)*
26. Enlist the characteristics of an ideal operational amplifier (Op-Amp) *(GBTU/MTU., 2009-10)*
27. Explain in detail the characteristic of ideal Op-Amp? Also derive the closed loop gain factors for an inverting and non- inverting amplifier. *(GBTU/MTU., 2008-09)*
28. What are the characteristics of an ideal operation amplifier? Explain an inverting amplifier. *(GBTU/MTU., 2008-09)*
29. Draw the circuit diagram of a non-inverting amplifier. *(Anna University., April/May 2008)*
30. Derive expressions of output voltages for following inverting and non-inverting amplifier. *(PBUT., 2009)*
31. With the circuit diagram explain the working of an Op-Amp, as inverting and non-inverting amplifier. *(Bhartiya University., April 2004)*
32. Draw the circuit diagram of voltage follower using an Op-Amp. *(PBUT., 2009-10, June 2005)*

33. What are the characteristics of an ideal a Op-Amp? Draw its equivalent circuit along with voltage transfer curve. *(GTU., Nov, Dec. 2011)*
34. Write short note on: Measurement of Op-Amp parameters. *(GTU., Nov./Dec. 2011)*
35. Draw the circuit diagram of inverting weighted summer. Calculate the output voltage for the same. *(GBTU., 2010, 2011)*
36. Derive the expression for closed loop gain of non-inverting configuration of Op-Amp. *(GBTU., 2010, 2011)*
37. Explain the following :
    (*a*) CMRR (*b*) Slew rate
    (*c*) Offset voltage (*d*) Offset current.
    (RGTU., Dec. 2009 June 2011, GBTU., 2010, 2011)
38. Explain the working of an Op-Amp as inverting amplifier in close loop and derive an expression for the output. *(RGTU., June 2011)*
39. What is an Op-Amp? List four basic building blocks of an Op-Amp and explain them. *(RGTU., Dec. 2010, June, 2009)*

MULTIPLE CHOICE QUESTIONS

1. Which of the following characteristics do not necessarily apply to the Op-Amp :
   (*a*) high gain (*b*) low power
   (*c*) high input impedance (*d*) low output impedance
2. The output voltage of a particular Op-Amp increases 8 mV in 12 μs in response to a step voltage on the input. The slew rate is
   (*a*) 0.667 V/μs (*b*) 0.75 V/μs (*c*) 1.5 V/μs (*d*) 96 V/μs
3. The voltage gain of a voltage follower is :
   (*a*) unity (*b*) less than unity (*c*) greater than unity (*d*) variable
4. The highest possible impedance is achieved with the :
   (*a*) inverting amplifier (*b*) non-inverting amplifier
   (*c*) differential amplifier (*d*) voltage follower
5. The voltage gain of a non-inverting Op-Amp amplifier :
   (*a*) greater than unity (*b*) less than unity
   (*c*) always greater than unity (*d*) both (*a*) & (*c*)
6. An Op-Amp can be classified as........ amplifier.
   (*a*) linear (*b*) low-r_m (*c*) positive feedback (*d*) *RC*-coupled
7. An ideal Op-Amp has
   (*a*) infinite A_v (*b*) infinite R_i (*c*) zero R_0 (*d*) all the above.
8. Op-Amp have become very popular in industry mainly because
   (*a*) they are very cheap
   (*b*) their external characteristics can be changed to suit any application
   (*c*) of their extremely small size
   (*d*) they are available in different packages.
9. Since input resistance of an ideal Op-Amp is infinite
   (*a*) its output resistance is zero
   (*b*) its output becomes independent of load resistance
   (*c*) its input current is zero
   (*d*) it becomes a current -controlled device.

10. The gain of an actual Op-Amp is around
(*a*) 1,000,000 (*b*) 1000 (*c*) 100 (*d*) 15 V

11. When an input voltage of 1 V is applied to an Op-Amp having $A_v = 10^6$ and bias supply of + 15 V, the output voltage available is
(*a*) 15×10^6 V (*b*) 10^6V (*c*) 15 μ V (*d*) 15 V.

12. An inverting amplifier has $R_f = 2$ M and $R_1 = 2$ k, Its scale factor is
(*a*) 1000 (*b*) – 1000 (*c*) 10^{-3} (*d*) -10^{-3}

13. In an inverting amplifier, the two input terminals of an ideal Op-Amp are at the same potential because
(*a*) the two input terminals are directly shorted infinity
(*b*) the input impedance of the Op-Amp is infinity
(*c*) common-mode rejection ratio is infinity
(*d*) the open-loop gain of the Op-Amp is infinity

14. The open -loop gain of an operational amplifier is 105. An input signal of 1 mV is applied to the inverting input with the non-inverting input connected to the ground. The supply voltages are ± 10 V. The output of the amplifier will be
(*a*) + 100 V (*b*) – 100 V
(*c*) + 10 V (approximately) (*d*) – 10 V (approximately)

15. An ideal Op-Amp has
(*a*) infinite input and output impedance
(*b*) very low input and output impedance
(*c*) low input and infinite output impedance
(*d*) infinite input and zero output impedance.

16. For an ideal difference amplifier, the common mode Rejection Ratio (CMRR) should be:
(*a*) As high as possible (*b*) As low as possible
(*c*) Constant (*d*) unity

17. The input stage of an Op-Amp is usually a
(*a*) differential amplifier (*b*) class-B push-pull amplifier
(*c*) CE amplifier (*d*) Level shifter

18. The unity gain frequency of an Op-Amp equal
(*a*) 10^3 Hz (*b*) 10^6 Hz (*c*) 10^2 Hz (*d*) 10 Hz

19. An Op-Amp possesses
(*a*) very large input resistance and very large output resistance
(*b*) very large input resistance and very small output resistance.
(*c*) very small input resistance and very small output resistance.
(*d*) very small input resistance and very large output resistance

(*UPSC Engg. Service 2002*)

20. In a 741 Op-Amp, there is 20 dB/decade fall-off starting at a relatively ion frequency. This is due to :
(*a*) applied load (*b*) internal compensation
(*c*) impedance of the source (*d*) power dissipation in the chip.

(*UPSC Engg. Service 2001*)

21. The input differential stage of Op-Amp 741 is biased at about 10 mA current. Such a low current of the input stage gives
(*a*) high CMRR (*b*) high differential gain
(*c*) low CMRR (*d*) high input impedance.

(*UPSC Engg. Services 2002*)

22. A differential amplifier is invariably used in the input stage of all Op-Amps. This is done basically to provide the Op-Amp with a very high

(*a*) CMRR (*b*) bandwidth (*c*) slew-rate (*d*) open-loop gain

(*UPSC Engg. Services 2001*)

23. The slew rate of an Op-Amp is 0.5 $V/_{\mu s}$ The maximum frequency of a sinusoidal input of 2 V_{rms} that can be handled without excessive distortion is

(*a*) 3 kHz (*b*) 30 kHz (*c*) 200 kHz (*d*) 2 MHz

24. In a circuit if the open loop gain is 10^6 and output voltage is 10 V, the differential voltage should be

(*a*) 10 μV (*b*) 0.1 V (*c*) 100 μV (*d*) 1 μV.

(*UPSC Engg. Services 2002*)

25. An Op-Amp has a common-mode gain of 0.01 and a differential mode gain of 10^5 Its common mode rejection ratio would be :

(*a*) 10^{-7} (*b*) 10^{-3} (*c*) 10^3 (*d*) 10^7

26. The voltage gain versus frequency curve of an Op-Amp is shown in Fig. 35.44.

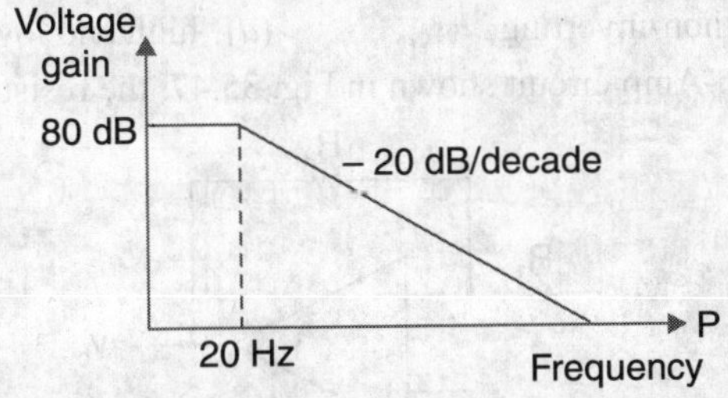

Fig. 35.44.

The gain-based width product of the Op-Amp is

(*a*) 200 Hz (*b*) 200 MHz (*c*) 200 kHz (*d*) 2 MHz.

(*UPSC Engg. Services 1999*)

27. Active load is used the collector of the differential amplifier of an Op-Amp to

(*a*) increase the output resistance
(*b*) increase the differential gain, A_d
(*c*) increase maximum peak-to-peak output voltage
(*d*) eliminate load resistance from the circuit.

28. The output voltage of the circuit shown in Fig. 35.45 is

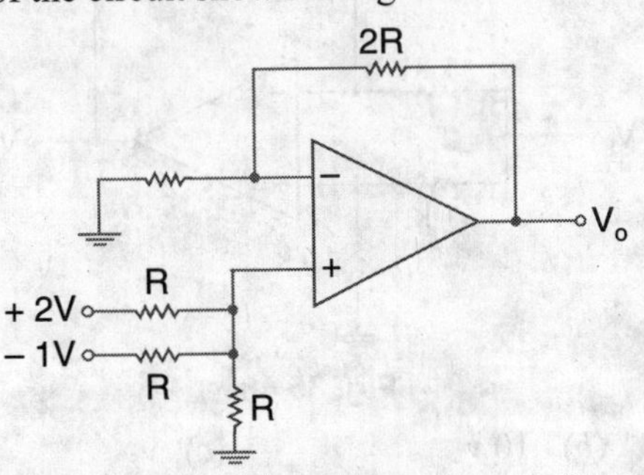

Fig. 35.45.

(*a*) 1.0 V (*b*) 1.5 V (*c*) 2.0 V (*d*) 2.5 V

29. In the circuit hown in Fig 35.46, the value of output, v_0 is

(*a*) + 3 V (*b*) – 3 V (*c*) – 7 V (*d*) + 7 V

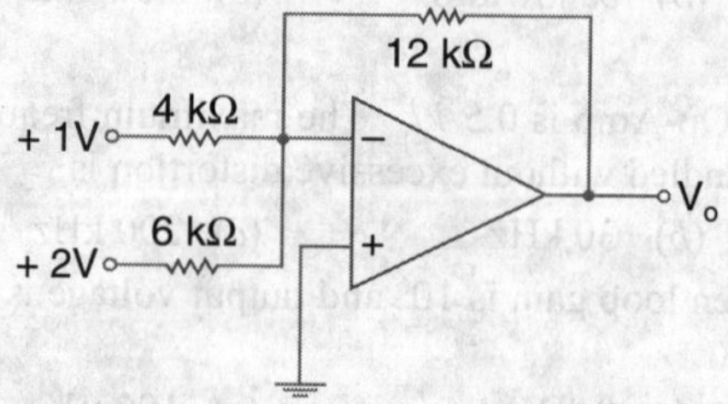

Fig. 35.46.

30. When in a negative scaler, both R_1 and R_f are reduced to zero, the circuit functions as

(*a*) integrator (*b*) subtractor (*c*) comparator (*d*) unity follower.

31. The two input terminals of an Op-Amp are known as

(*a*) positive and negative (*b*) differential and non-differential

(*c*) inverting and non-inverting (*d*) high and low.

32. In the inverting Op-Amp circuit shown in Fig. 35.47, the resistance

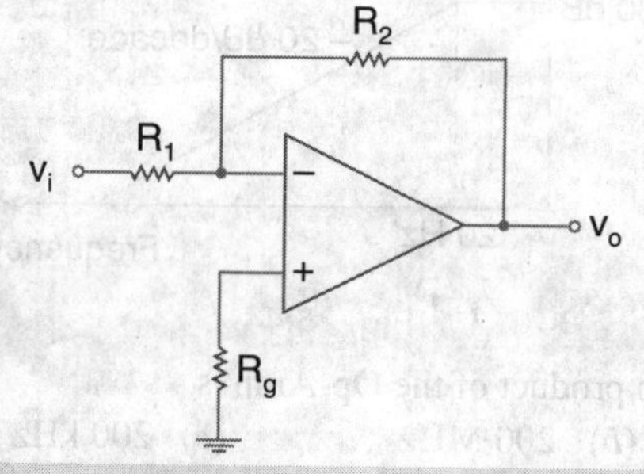

Fig. 35.47.

R_g is chose as $R_1 \| R_2$ in order to

(*a*) increase the gain (*b*) reduce offset voltage

(*c*) reduce offset current (*d*) increase CMRR.

33. The v_0 of the Op-Amp circuit shown in Fig. 35.48, is :

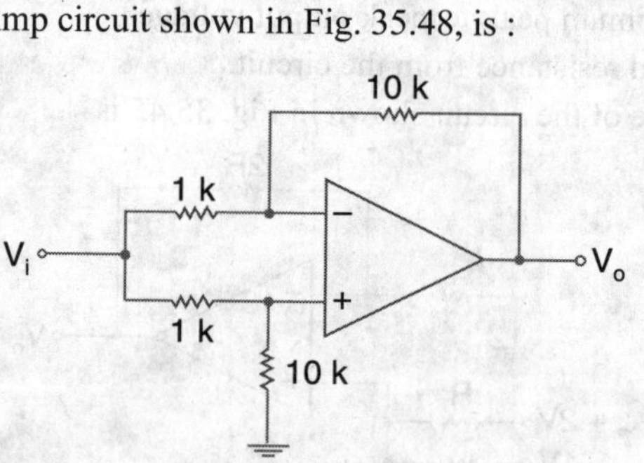

Fig. 35.48.

(*a*) 11 v_i (*b*) 10 v_i (*c*) v_i (*d*) zero

(*UPSC Engg. Services, 1999*)

34. A circuit using an Op-Amp is shown in Fig. 35.49. It has :

(*a*) voltage-series feedback (*b*) voltage-shunt feedback

(*c*) current-shont feedback (*d*) current-series feedback.

(*UPSC Engg. Services, 2001*)

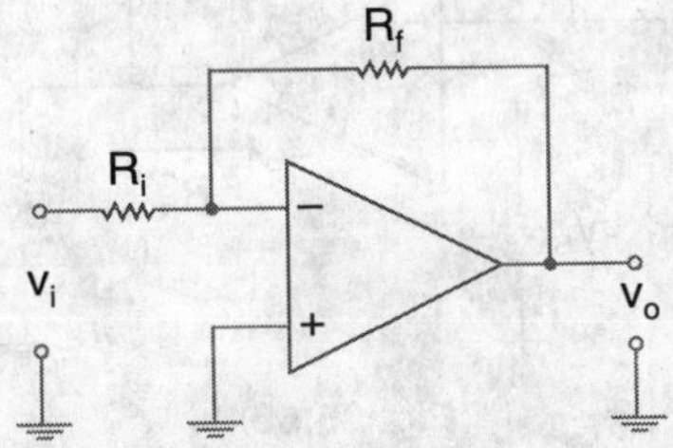

Fig. 35.49.

35. A non-inverting Op-Amp is shown in Fig. 35.50.

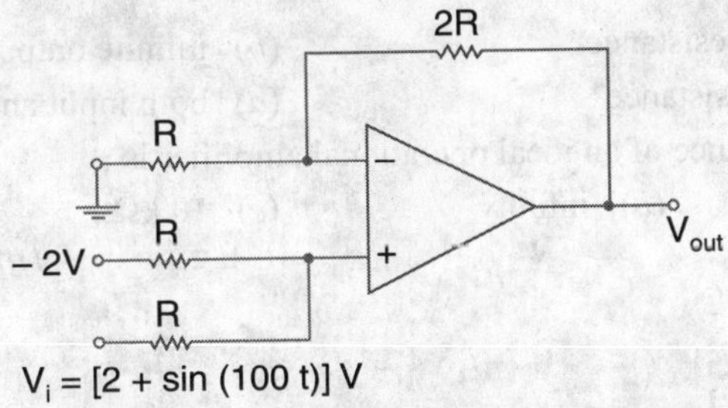

Fig. 35.50.

The output voltage v_0, for an input, $v_i = [2 + \sin(100\,t)]$ V is

(*a*) $\frac{3}{2}\sin(100t)$ (*b*) $3\sin(100\,t)$

(*c*) $2\sin(100\,t)$ (*d*) $3\sin(100\,t) + \frac{1}{2}$

(*UPSC Engg. Services 2002*)

36. The Op-Amp circuit shown in Fig. 35.51 has an input impedance of 1 MΩ and an open-loop gain of 10^5. The output impedance seen by source V_S is

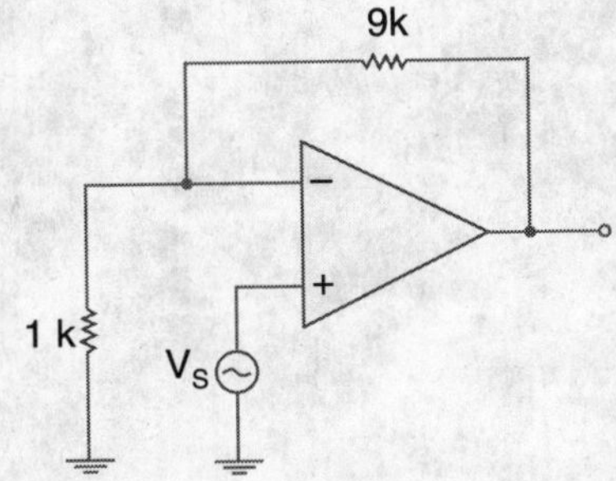

Fig. 35.51.

(*a*) 10^{11} Ω (*b*) 10^{11} Ω (*c*) 10 k Ω (*d*) – 1000

37. An Op-Amp with an open-loop gain of 10,000, R_{in} = 2 k Ω and R_0 = 500 Ω is used in the non-inverting configuration shown in Fig. 35.51. The output resistance R_0 is (*a*) 250.5 Ω (*b*) 21 Ω (*c*) 2 Ω (*d*) 0.998 Ω

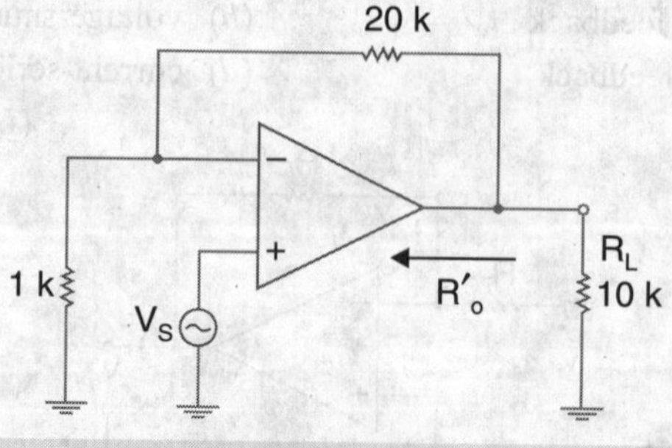

Fig. 35.52.

38. The voltage gain of an open-circuit ideal Op-Amp is

(*a*) infinity (*b*) around 10,000 (*c*) around 300 (*d*) 0

39. An ideal Op-Amp has

(*a*) zero output resistance (*b*) infinite output resistance

(*c*) zero input resistance (*d*) both input and output zero resistance

40. The input impedance of an ideal operational amplifier is

(*a*) 0 (*b*) infinity (*c*) 10 kΩ (*d*) none of these

(*Bharthiar Univ. April 2004*)

ANSWERS

1. (*b*)	**2.** (*a*)	**3.** (*a*)	**4.** (*d*)	**5.** (*d*)
6. (*a*)	**7.** (*d*)	**8.** (*b*)	**9.** (*c*)	**10.** (*a*)
11. (*d*)	**12.** (*b*)	**13.** (*b*)	**14.** (*b*)	**15.** (*d*)
16. (*a*)	**17.** (*a*)	**18.** (*b*)	**19.** (*b*)	**20.** (*b*)
21. (*d*)	**22.** (*d*)	**23.** (*b*)	**24.** (*a*)	**25.** (*d*)
26. (*c*)	**27.** (*d*)	**28.** (*c*)	**29.** (*d*)	**30.** (*d*)
31. (*c*)	**32.** (*a*)	**33.** (*b*)	**34.** (*a*)	**35.** (*b*)
36. (*a*)	**37.** (*d*)	**38.** (*a*)	**39.** (*a*)	**40.** (*b*)

36 Chapter

Basic Op-Amp Applications

Outline

36.1. Introduction

We have already discused in the last chapter about an Op-Amp that it is an integrated circuit with very high gain high input impedance and low output impedance. In this chapter we shall study about the applications of an Op-Amp. An Op-Amp can be used as a constant gain multiplier, voltage summing amplifier voltage buffer amplifier, basic amplifier in instrumentation circuits, comparator, in oscillator circuits and active filters. We shall discuss some of these applications in the following pages.

36.2. Comparators

The comparator is a circuit that is used to compare two voltages and provide an output indicating the relationship between those two voltages. Generally speaking, comparators are used to compare either,

1. Two changing voltages to each other, as in comparing two sine waves, or
2. A changing voltage to a set dc reference voltage.

Non-Inverting Comparator

Fig. 36.1 (*a*) shows the circuit of an Op-Amp comparator. It may be noted that there is no feedback path in the circuit. In this circuit, the input voltage is applied to the non-inverting input terminal and a set dc reference voltage (V_{ref}) if applied to the inverting terminal of the Op-Amp. The operation of the circuit is as follows.

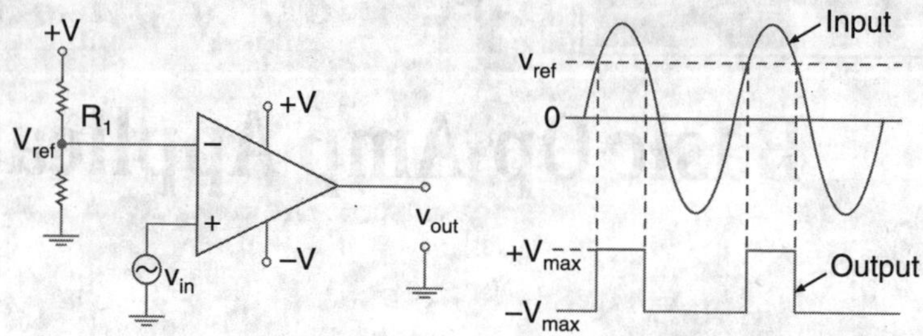

(*a*) Circuit. (*b*) Input and output waveforms.

Fig. 36.1. Op-Amp comparator.

As long as the input voltage is below V_{ref}, the comparator output is approximately "$-V_{max}$" volts. But if the input voltage equals V_{ref} or exceeds it, the comparator output changes to "$+V_{max}$" volts. Thus depending upon the value of input voltage, the comparator produces a dc voltage that indicates the polarity (or magnitude) relationship between the two input voltages as shown in Figure 36.1 (*b*). A special case arises if the inverting input of the Op-Amp is grounded and the input signal voltage is applied to the non-inverting input. In such a situation, $V_{ref} = 0$, and the output is equal to $-V_{max}$, so long the input signal is below 0. As soon as, the input signal crosses zero the output switches to $+V_{max}$. Such a circuit is called zero-level detector. This can be used as a squaring circuit to produce a square wave from a sine wave.

36.3. Inverting Comparator

Fig. 36.2 shows as Op-Amp configured as an inverting Op-Amp comparator is shown in the Fig. 36.2. In this circuit the input voltage is applied to the inverting input terminal and a set d.c. reference voltage V_{ref} is applied to the non-inverting input terminal. The operation of the circuit is as follows.

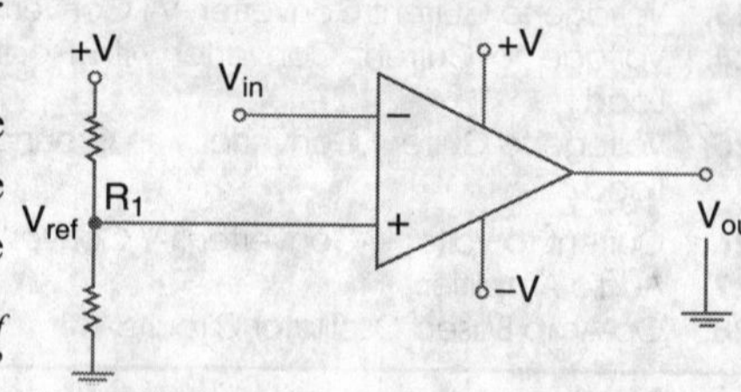

Fig. 36.2. Inverting Comparator

As long as the input voltage is below V_{ref}, the comparator output is approximately "$+V_{max}$" volts because the voltage at the inverting input is less than that at the non-inverting input. But if the input voltage equals V_{ref} or exceeds it, the comparator output changes to "$-V_{max}$" volts. Thus depending upon the value of input voltage, the comparator produces a d.c. voltage that indicates the polarity (or magnitude) relationship between the two input voltages as shown in Fig. 36.3.

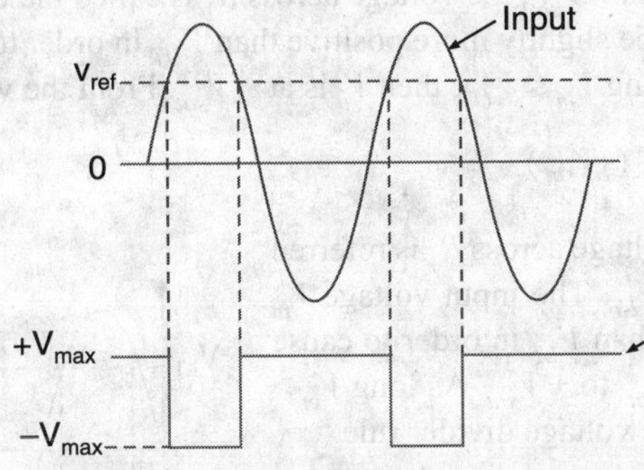

Fig. 36.3.

36.4. Practical Comparator

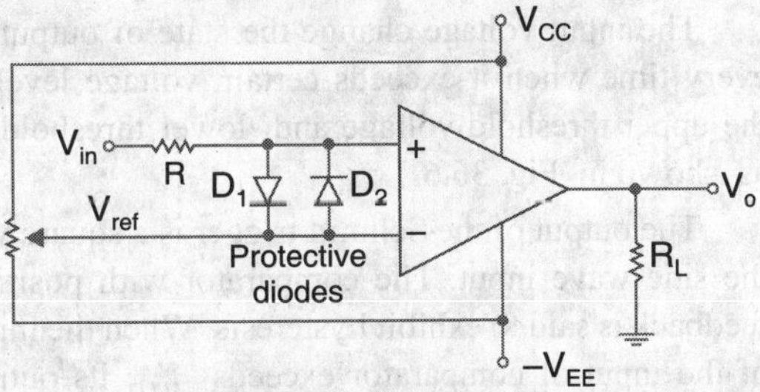

Fig. 36.4. Practical Comparator

Fig. 36.4 shows the circuit of practical non-inverting comparator. The input voltage V_{in} is applied to the non-inverting input terminal and reference voltage V_{ref} is applied to the inverting input terminal. It consists of protective diodes and potentiometer to adjust the reference voltage. Diodes D_1 and D_2 are provided in the circuit to protect the Op-Amp against the excessive input voltage. Due to these diodes the difference input voltagc V_{id} of thc Op-Amp is clamped to either 0.7 V or –0.7 V. These diodes are known as clamp diodes. There are some Op-Amps with built-in input protection so input diodes are unnecessary.

In this circuit of Fig. 36.4, V_{ref} is obtained by using a 10 kΩ potentiometer that forms a voltage divider with the supply voltages $+V_{CC}$ to $-V_{EE}$ and the wiper is connected to the non-inverting input terminal. As the wiper moved toward $-V_{EE}$, the voltage V_{ref} becomes more negative, while if it moves toward $+V_{CC}$, the voltage V_{ref} becomes more positive. Thus V_{ref} of a desired amplitude and polarity can be obtained by simply adjusting the 10 kΩ potentiometer.

The resistance R in series with input voltage V_{in} is used to limit the current through D_1 and D_2. The potentiometer acts as a voltage divider and allows reference voltage to set any value between $+V_{CC}$ to $-V_{EE}$.

The comparator is a type of analog-to-digital converter. At any time the output voltage V_o shows whether input voltage V_{in} is greater or less than V_{ref}. The comparator is sometimes also called a voltae-level detector because, for a desired value of V_{ref}, the voltage level of the input V_{in} can be detected.

36.5. Schmitt Trigger

We have already discussed about comparator circuit. The comparator circuit with positive feedback work as a Schmitt trigger. Fig. 36.5 shows the circuit diagram of Schmitt trigger. This circuit converts an irregular-shaped waveform to a square wave or pulse. This circuit is also known as squaring circuit.

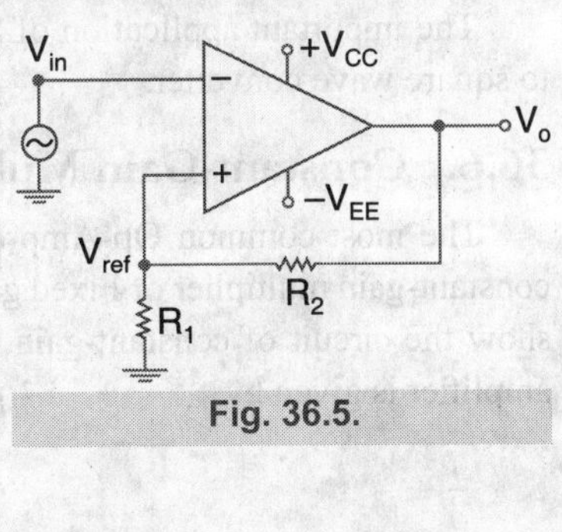

Fig. 36.5.

These threshold voltages are obtained by using the voltage divider. The voltage across R_1 is a variable reference threshold voltage that depends on the value and polarity of the output voltage V_o.

When the output voltage V_o is $+V_{sat}$, the voltage across R_1 is called the upper threshold voltage V_{UT}. The input voltage V_{in} must be slightly more positive than V_{UT} in order to cause the output V_o to switch from $+V_{sat}$ to $-V_{sat}$. As long $V_{in} < V_{UT}$, then V_o is at $+V_{sat}$. From the voltage divider rule,

$$V_{UT} = \frac{R_1}{R_1 + R_2}(+V_{sat})$$

When V_o is $-V_{sat}$, then the voltage across R_1 is referred as the lower threshold voltage V_{LT}. The input voltage V_{in} must be slightly more negative than V_{LT} in order to cause the output V_o to switch from $-V_{sat}$ to $+V_{sat}$. As long $V_{in} > V_{LT}$, then V_o is at $-V_{sat}$. From the voltage divider rule

$$V_{LT} = \frac{R_1}{R_1 + R_2}(-V_{sat})$$

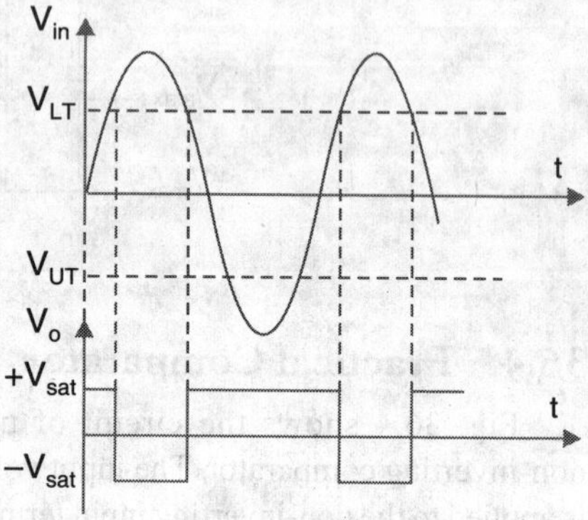

Fig. 36.6.

The input voltage change the state of output voltage every time when it exceeds certain voltage levels called the upper threshold voltage and lower threshold voltage as shown in Fig. 36.6.

The output of the Schmitt trigger is a square wave for the sine wave input. The comparator with positive feedback is said to exhibit hysteresis. When the input of the input of comparator exceeds V_{UT} its output switches from $+V_{sat}$ to $-V_{sat}$. Its revert back to its original state, $+V_{sat}$, when input goes below V_{LT} as shown in Fig. 36.7. The hysteresis voltage V_{HY} is equal to the difference between V_{UT} and V_{LT},

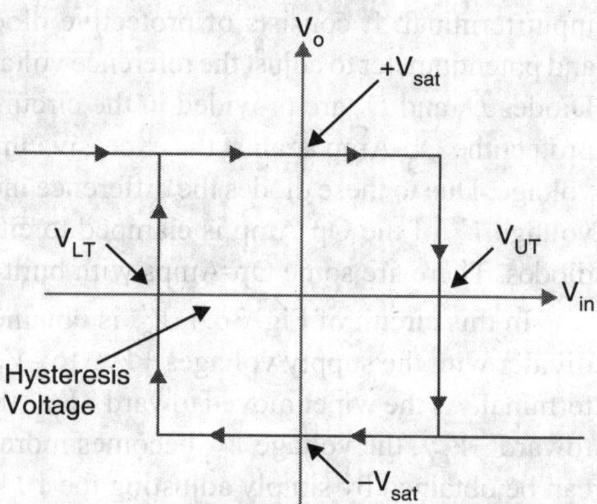

Fig. 36.7

$$V_{HY} = V_{UT} - V_{LT}$$

$$= \left[\frac{R_1}{R_1 + R_2}(+V_{sat})\right] - \left[\frac{R_1}{R_1 + R_2}(-V_{sat})\right]$$

$$= \frac{R_1}{R_1 + R_2}[(+V_{sat}) - (-V_{sat})]$$

$$= \frac{2R_1}{R_1 + R_2}V_{sat}$$

The important application of Schmitt trigger as sine to square wave converter.

36.6. Constant-Gain Multiplier

The most common Op-Amp circuit is the inverting constant-gain multiplier or Fixed gain amplifier. Fig. 36.8 show the circuit of constant-gain multiplier, the gain of amplifier is given by,

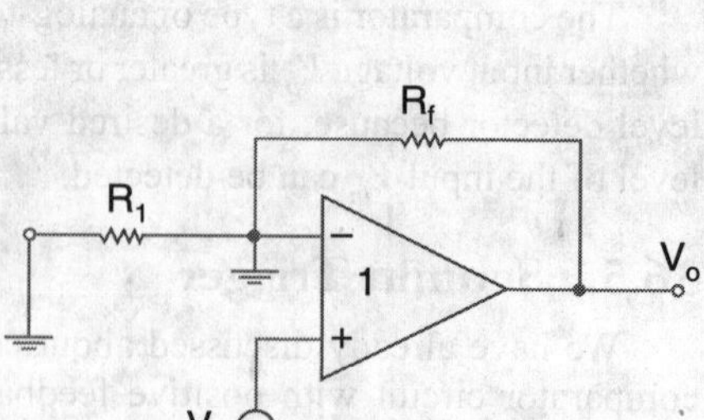

Fig. 36.8.

$$A = -\frac{R_f}{R_1}$$

36.7. Multiple-Stage Gain Amplifier

When a number of amplifiers are connected in series, the overall gain is the product of the individual stage gains. Fig. 36.9 shows a connection of three stage amplifier. The first stage is connected to provide non-inverting gain and next two stages provide an inverting gain.

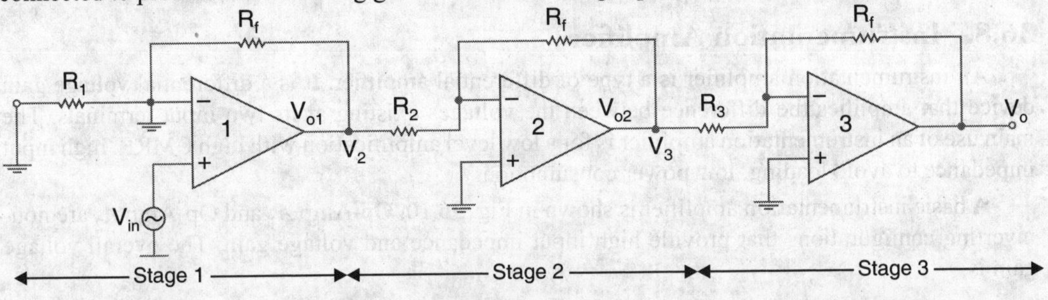

Fig. 36.9.

Voltage gain of the first stage non-inverting amplifier,

$$A_1 = \frac{V_{01}}{V_{in}} = \left(1 + \frac{R_f}{R_1}\right)$$

$$V_{01} = \left(1 + \frac{R_f}{R_1}\right) V_{in}$$

Voltage gain of the second stage inverting amplifier,

$$A_2 = \frac{V_{02}}{V_2} = -\frac{R_f}{R_2}$$

$$V_{02} = -\left(\frac{R_f}{R_2}\right) V_2$$

Since output voltage of first stage amplifier is equal to the input voltage of second stage amplifier, therefore the output voltage of the second stage amplifier,

$$V_{02} = \left(-\frac{R_f}{R_2}\right)\left(1 + \frac{R_f}{R_1}\right) V_{in}$$

Voltage gain of the third stage inverting amplifier,

$$A_3 = \frac{V_o}{V_3} = -\frac{R_f}{R_3}$$

$$\therefore \quad V_o = \left(-\frac{R_f}{R_3}\right) V_3$$

Since output voltage of second stage amplifier is equal to the input voltage of third stage amplifier, therefore the output voltage of the third stage amplifier,

$$V_o = \left(-\frac{R_f}{R_3}\right)\left(-\frac{R_f}{R_2}\right)\left(1 + \frac{R_f}{R_1}\right) V_{in}$$

or $$\frac{V_o}{V_{in}} = A_v = \left(1+\frac{R_f}{R_1}\right)\left(-\frac{R_f}{R_2}\right)\left(-\frac{R_f}{R_3}\right)$$

From the above equation we see that the voltage gain of the multiple stage amplifiers is equal to the product of the voltage gains of the individual stages of the amplifier.

$$A_V = \left(1+\frac{R_f}{R_1}\right)\left(-\frac{R_f}{R_2}\right)\left(-\frac{R_f}{R_3}\right) = A_1 \times A_2 \times A_3$$

36.8. Instrumentation Amplifier

An instrumentation amplifier is a type of differential amplifier. It is a differential voltage gain device that amplifies the difference between the voltages existing at its two input terminals. The main use of an instrumentation amplifier is for a low level amplification with high CMRR, high input impedance to avoid loading, low power consumption.

A basic instrumentation amplifier is shown in Fig. 36.10. Op-Amp A_1 and Op-Amp A_2 are non-inverting configurations that provide high input impedance and voltage gain. The overall voltage gain is

$$A_V = \frac{V_o}{(V_1 - V_2)} = 1 + \frac{2R}{R_P}$$

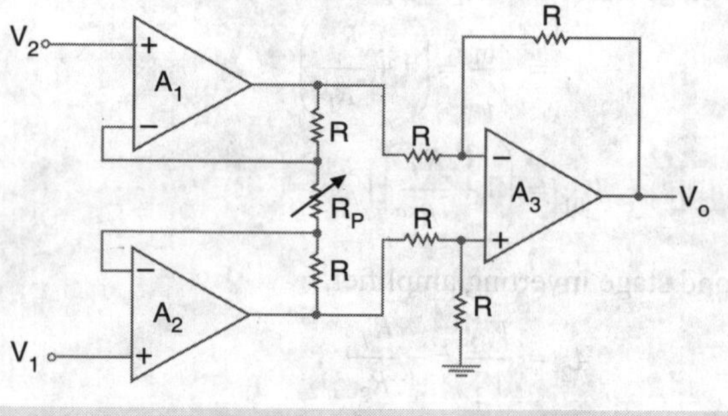

Fig. 36.10.

36.9. Instrumentation amplifier using two amplifiers

We have already discussed that the voltage follower amplifier has a very high input resistance. We can achieve the high input resistance with a variable voltage by using two Op-Amp instrumentation amplifier circuits. Fig. 36.11 shows the instrumentation amplifier with two Op-Amps.

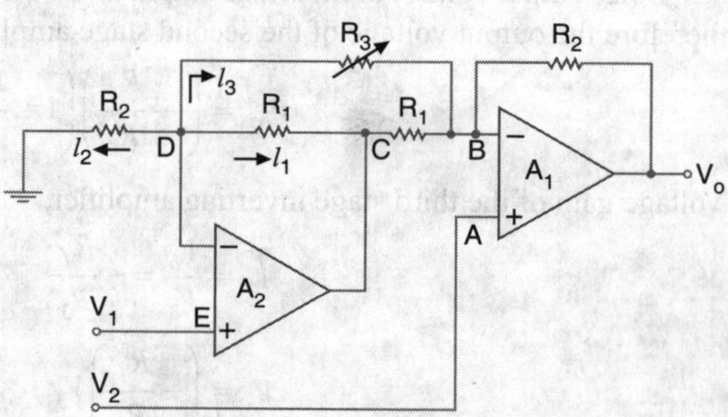

Fig. 36.11.

For amplifier 1 the voltage at node A is same as the node B, then

$$V_A = V_B = V_2$$

Similarly for the amplifier, voltage at node E is same as the node D.

$$V_E = V_D = V_1$$

$$I_2 = \frac{V_D}{R_2} = \frac{V_1}{R_2} \qquad \text{... (i)}$$

$$I_1 = \frac{V_D - V_C}{R_1} = \frac{V_1 - V_C}{R_1} \quad \text{... (ii)}$$

$$I_3 = \frac{V_D - V_B}{R_3} = \frac{V_1 - V_2}{R_3} \quad \text{... (iii)}$$

Applying Kirchoff's current law (KCL) in node D we get,

$$I_1 + I_2 + I_3 = 0 \quad \text{... (iv)}$$

Substituting the value of I_1, I_2 and I_3 from equations (*i*), (*ii*) and (*iii*) to equation (*iv*), we get

$$\frac{V_1 - V_C}{R_1} + \frac{V_1}{R_2} + \frac{V_1 - V_2}{R_3} = 0 \quad \text{...(v)}$$

Similarly applying Kirchoff's current law (KCL) node B we get,

$$\frac{V_B - V_D}{R_3} + \frac{V_B - V_C}{R_1} + \frac{V_B - V_o}{R_2} = 0 \quad \text{...(vi)}$$

From Equation (*v*) and (*vi*) we get,

$$\frac{V_o}{V_2 - V_1} = 1 + \frac{R_2}{R_1} + \frac{2R_2}{R_3}$$

Here $(V_2 - V_1)$ is the difference of input voltage and V_o is the output voltage. Thus the voltage gain of this circuit is,

$$A_V = \frac{R_2}{R_1} + \frac{2R_2}{R_3}$$

Here the resistance R_3 is the variable resistance.

36.10. Instrumentation amplifier using three amplifiers

Instrumentation amplifier using three Op-Amps is shown in Fig. 36.12. This provides high input resistance for accurate measurement of signals. In this circuit a non-inverting amplifier is added to each of the basic difference amplifier inputs.

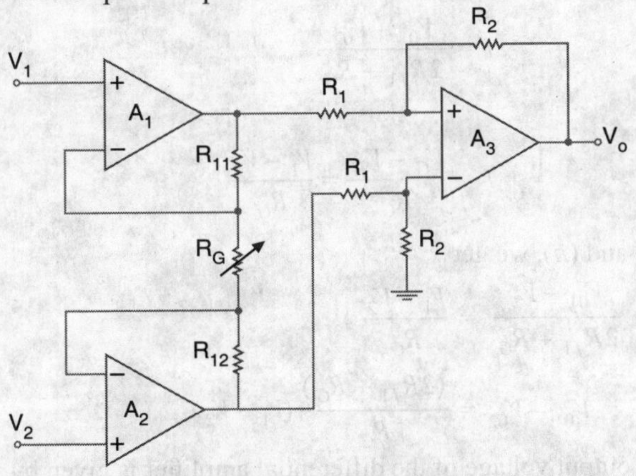

Fig. 36.12.

Here Op-Amps A_1 and A_2 are the non-inverting amplifiers forming the input or first stage of the instrumentation amplifier. The op-amp A_3 is the difference amplifier forming an output stage of the amplifier. The block diagram representation of the three Op-Amp instrumentation amplifier is shown in Fig. 36.13.

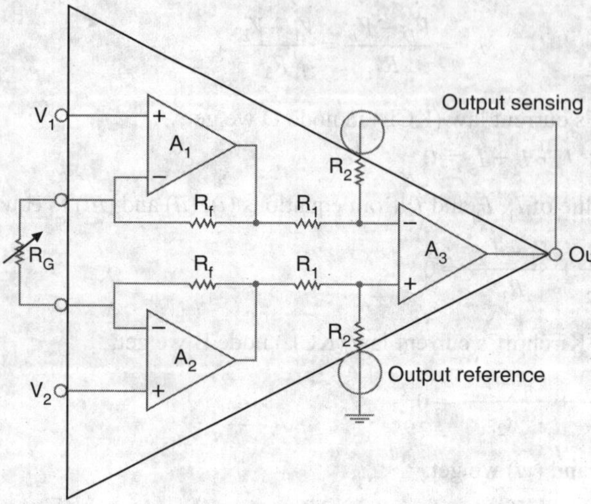

Fig. 36.13

The output of the Op-Amp A_1 is V_{o1} and the output of the Op-Amp A_2 is V_{02}. The output stage of a standard difference amplifier is,

$$V_o = \frac{R_2}{R_1}(V_{02} - V_{01})$$

Let us find out the values of V_{02} and V_{01} from the instrumentation amplifier circuit shown in Fig. 36.14. The voltage at node A is equal to voltage V_1 and node voltage B is equal to node voltage A. And hence voltage of G is also equal to V_1. Similarly the voltage of H is V_2. Applying Ohm's Law between the node E and F we get,

$$I = \frac{V_{01} - V_{02}}{R_{f1} + R_G + R_{f2}}$$

Let $R_{f1} = R_{f2} = R_f$ then,

$$I = \frac{V_{01} - V_{02}}{2R_{f1} + R_G} \quad \ldots(i)$$

In node G and H,

$$I = \frac{V_G - V_H}{R_G} = \frac{V_1 - V_2}{R_G} \quad \ldots(ii)$$

From equation (*i*) and (*ii*), we get

$$\frac{V_{01} - V_2}{2R_{f1} + R_G} = \frac{V_1 - V_2}{R_G}$$

$$\therefore \quad V_{01} - V_{02} = \frac{(2R_{f1} + R_G)}{R_G}(V_1 - V_2)$$

We know that the output voltage of the differential amplifier is given by

$$V_o = \frac{R_2}{R_1}(V_{02} - V_{01})$$

Substituting the value of $V_{o2} - V_{01}$ we get,

$$V_o = \frac{R_2}{R_1}\cdot\left(1+\frac{2R_f}{R_G}\right)(V_{o2}-V_{01})$$

This is the overall gain of the amplifier. If R_1 equal to R_2 then the overall gain of amplifier is given by,

$$V_o = \left(1+\frac{2R_f}{R_G}\right)(V_{o2}-V_{01})$$

With the help of variable resistance the gain can be easily varied without disturbing the symmetry of the circuit. The gain depends in external resistances and hence can be adjusted accurately and made stable by selecting high quality resistances.

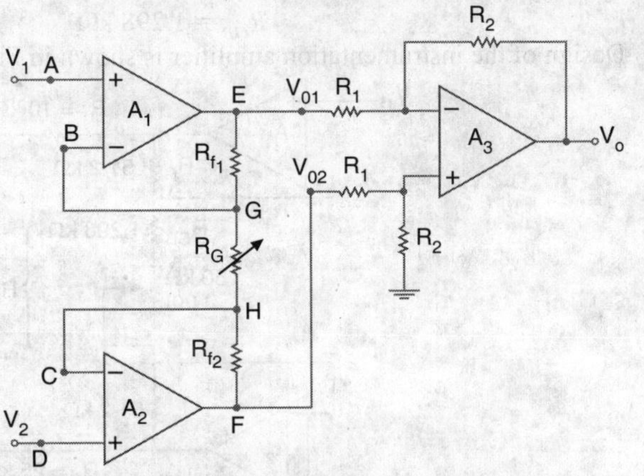

Fig. 36.14.

Example 36.1.

Design the instrumentation amplifier to have a variable gain in the range 5-200. Use a 50 kW potentiometer. *(GBTU., 2010, 2011)*

Solution.

We know that the gain of the instrumentation amplifier,

$$A = \left(1+\frac{2R_f}{R_G}\right)\left(\frac{R_2}{R_1}\right) = A_1 A_2$$

Let us choose $A_2 = 2.5$ then,

$$\frac{R_2}{R_1} = 2.5$$

Now consider,

$R_1 = 10\text{ k}\Omega$ then $R_2 = 2.5\ R_1 = 2.5 \times 10\text{ k}\Omega$

and implement R_G using 50 kΩ potentiometer with a base resistance of R_{GB} in series

We know that on zero resistance positions, the gain of the instrumentation amplifier,

$$\left(1+\frac{2R_f}{R_{GB}}\right)(2.5) = 200$$

$$1+\frac{2R_f}{R_{GB}} = 80$$

or
$$\frac{2R_f}{R_{GB}} = 39.5 \quad \text{... }(i)$$

and on the maximum resistance position, the gain equation can be written as,

$$\left(1+\frac{2R_f}{(50\times10^3)+R_{GB}}\right)(2.5) = 5$$

or
$$\frac{R_f}{(50\times10^3)+R_{GB}} = 1 \quad \text{... }(ii)$$

Solving equation (*i*) and (*ii*) we get,

$$R_f = 51.29\text{ k}\Omega$$

$$R_{GB} = 1.298 \text{ k}\Omega$$

Design of the instrumentation amplifier is shown in Fig. 36.15.

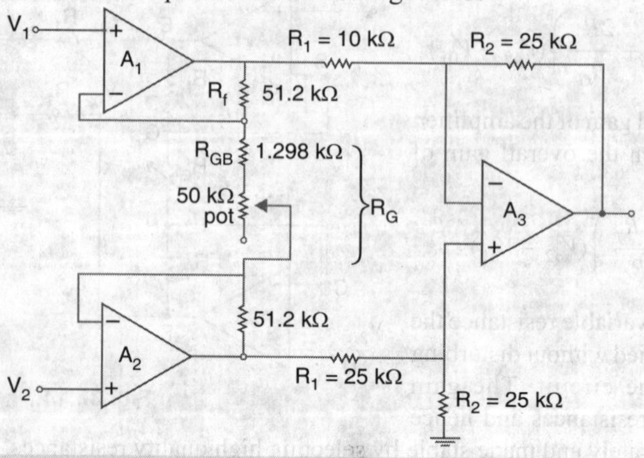

Fig. 36.15.

36.11. The Integrator

Technically speaking, integrator is a circuit whose output is proportional to the area of its input waveform. Fig. 36.16 shows the circuit of an Op-Amp integrator. The RC circuit itself also acts as a simple integrator. But the problem with such a simple circuit is that the output voltage is not a linear triangular output as it should be. The function of the Op-Amp is to linearize the output. It may be noted from the diagram that the inverting input to the Op-Amp is held at a virtual ground by the differential amplifier in the Op-Amp input circuit. Therefore the input current, $I_1 = V_{in}/R_1$. Because of the high input impedance of the Op-Amp, virtually all of I_1 will flow to the capacitor.

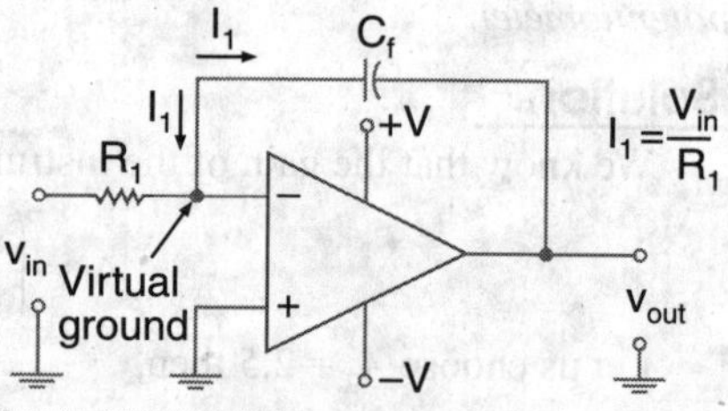

Fig. 36.16. Op-Amp integrator.

Assuming that V_{in} is constant for a given period of time R_1 and is a fixed value, I_1 can also be assumed to be a constant value. Since virtually all of I_1 flows to the capacitor, C is being charged by a constant current source. Thus as long as V_{in} is constant, the capacitor will charge/discharge at a linear rate. This produces a linear ramp output as shown in Fig. 36.17.

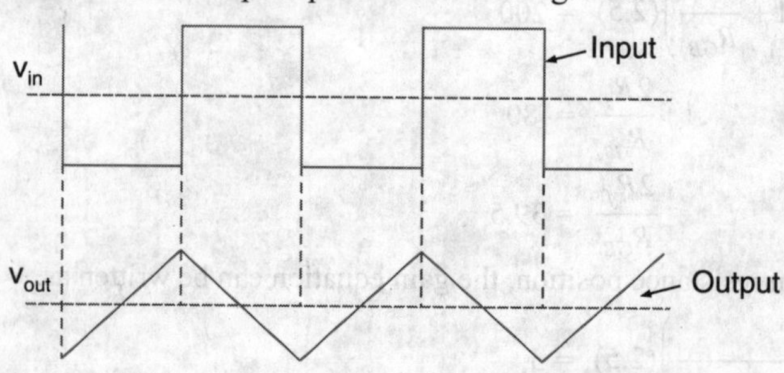

Fig. 36.17. Input versus output of an Op-Amp integrator.

It may be noted from the figure shown above that when the input goes positive (*i.e.*, from – *V* to + *V*), the output is a negative ramp. On the other hand, when the input goes negative (*i.e.*, from + *V* to – *V*), the output is a positive ramp. This indicates that the circuit output is 180° out of phase with the input.

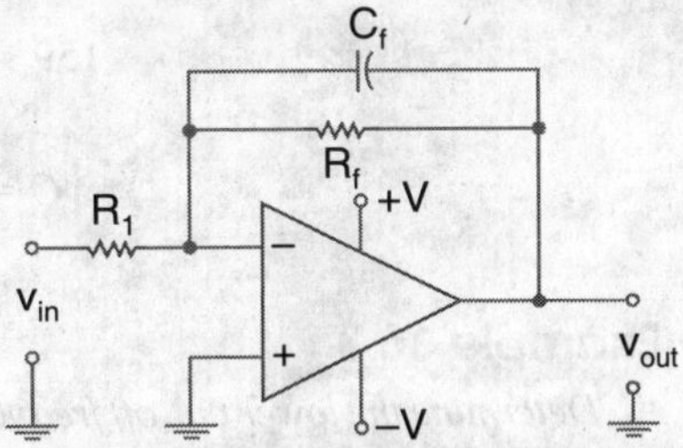

Fig. 36.18.

The output wave from shown in Fig. 36.17 lead us to a potential problem with the Op-Amp integrator As seen from this waveform the mean value of the output is 0V. However, in actual practice this may not be true. As matter of fact, if the Op-Amp has not been properly compersated for the input offset voltage, the output may not cauter it self around 0V. This problem can be rectified by adding a parallel resistor in the feedback path as shown in Fig 36.18. But the addition of this resistor has a disadvantage. The RC feedback circuit like any RC circuit will have a cut-off frequency. In this case it is the lower cut-off frequency with which we are concerned with. As the frequency decreases, the reactance of the capacitor will increase. At some point, the reactance of the capacitor will be greater than the value of R_f. As a result, the integration action of the circuit will stop.

The cut-off frequency of an integrator is given by the equation,

$$f_1 = \frac{1}{2\pi R_f C_f}$$

where R_f = resistance in ohms and
C_f = capacitance in farads

It will be interesting to known that the integrator will start to lose its linear output characteristics before the frequency (f_1) given by equation (*i*) is reached. thus for an optimum output, the value of X_C should always be less than 0.1 R_f, *i.e.*

$$X_C < 0.1 R_f$$

This leads to the point that the integrator circuit should not be operated below the frequency,

$$f_{min} = 10 f_1 = \frac{10}{2\pi R_f C_f}$$

The output voltage of integrator circuit in given by

$$V_o = -\frac{1}{R_1 C_f}\int V_{in}\, dt$$

Example 36.2.

Fig. 36.19 shows an integrator with a lower cut-off frequency of 159 Hz. Determine the value of the capacitance required in the circuit.

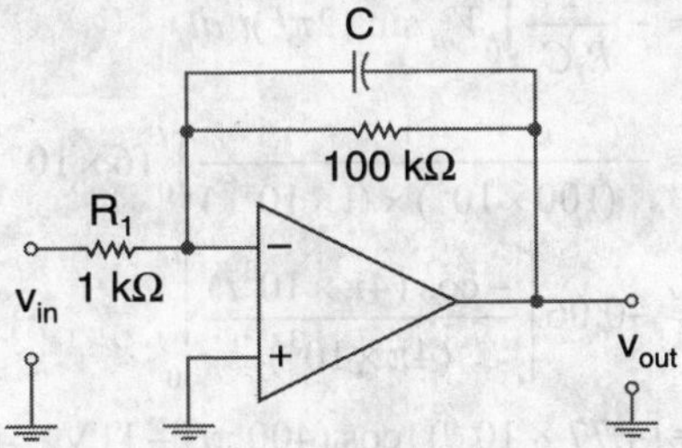

Fig. 36.19.

Solution.

Given : R_1 = 1kΩ; R_2 = 100 kΩ and f_1 = 159 Hz

Let C = value of capacitance (F).

We know that the cut-off frequency of an integrator circuit (f_1),

$$159 = \frac{1}{2\pi\, R_f\, C_f} = \frac{1}{2\pi \times (100 \times 10^3) \times C}$$

$$\therefore \qquad C = \frac{1}{2\pi \times (100 \times 10^3) \times 159} = 1.0 \times 10^{-8} \text{ or } 0.01\ \mu\text{F} \textbf{ Ans}$$

Example 36.3.

Determine the lower cut-off frequency of an integrator circuit shown in Fig. 36.20.

Also determine the frequency at which the integrator will start to lose its linear characteristics.

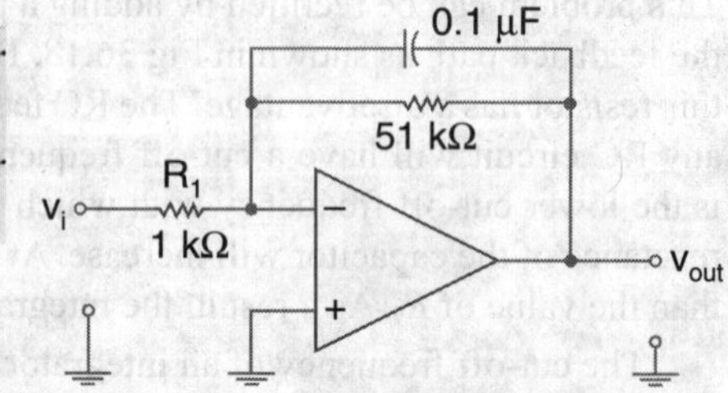

Fig. 36.20.

Solution.

Given : $R_1 = 1\text{ k}\Omega$; $R_f = 51\text{k}\Omega$ and $C_f = 0.1\ \mu f$.

We known that the cut-off frequency of an integrator circuit,

$$f = \frac{1}{2\pi\, R_f\, C_f} = \frac{1}{2\pi \times (51 \times 10^3) \times (0.1 \times 10^{-6})} = 312 \text{ Hz } \textbf{Ans.}$$

The minimum frequency at which the integrator will start to lose its linearly,

$$f_{min} = 10\, f_1 = 10 \times 312 = 3120 \text{ Hz } \textbf{Ans.}$$

Example 36.4.

A sinusoidal signal with peak value 6 mV and 2 kHz frequency is applied to the input of an ideal Op-Amp integrator with R_m = 100 kΩ and C_F = 1 mF. Find the output voltage?

(*GBTU., 2010, 2011*)

Solution.

Given : $V_m = 6\text{ mV} = 6 \times 10^{-3}$ V; $f = 2\text{ kHz} = 2 \times 10^3$ Hz; $R_1 = 100\text{ k}\Omega = 100 \times 10^3\ \Omega$ and $C_1 = 1\ \mu\text{F} = 1 \times 10^{-6}$ F.

We know that the output voltage,

$$V_o = -\frac{1}{R_1 C_1}\int_0^t V_{in}\, dt = -\frac{1}{R_1 C_1}\int_0^t V_{in} \sin \omega t\, dt$$

$$= -\frac{1}{R_1 C_1}\int_0^t V_{in} \sin(2\pi f)t\, dt$$

$$= -\frac{1}{(100 \times 10^3) \times (1 \times 10^{-6})}\int_0^t (6 \times 10^{-3}) \sin[2 \times \pi \times (2 \times 10^3)]t\, dt$$

$$= -0.06 \left[\frac{-\cos(4\pi \times 10^3 t)}{(4\pi \times 10^3)}\right]_0^t$$

$$= (4.77 \times 10^{-6})[\cos(400\,\pi t) - 1]\text{ V} = 4.77[\cos(4000\pi t) - 1]\ \mu\text{V } \textbf{Ans.}$$

Let $\qquad C$ = value of capacitance (F)

36.12. The Miller Integrator

The integrator circuit shown in Fig. 36.21 can also be represented as shown in Fig. 36.21. Notice that the feedback capacitor may be broken into two capacitors, one at the input and the other at the output. The circuit shown in Fig. 36.22 is more commonly known as Miller integrator.

As the operating frequency increases, the feedback capacitor will act more and more like a short circuit. Eventally a stage will come when the capacitor would short out the feedback resistor completely. In such a situation the closed loop gain of the circuit reduces to zero.

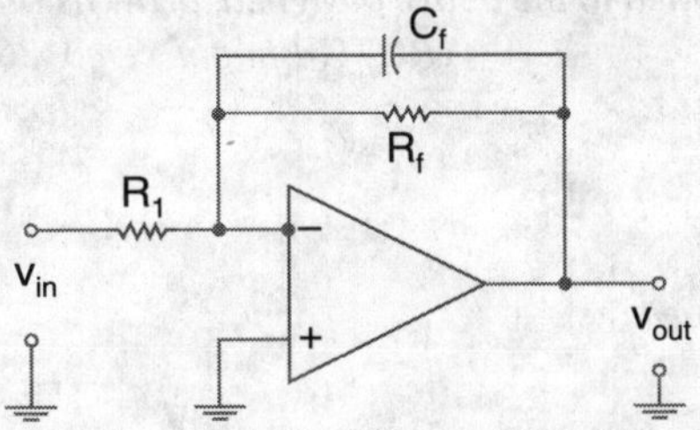

Fig. 36.21.

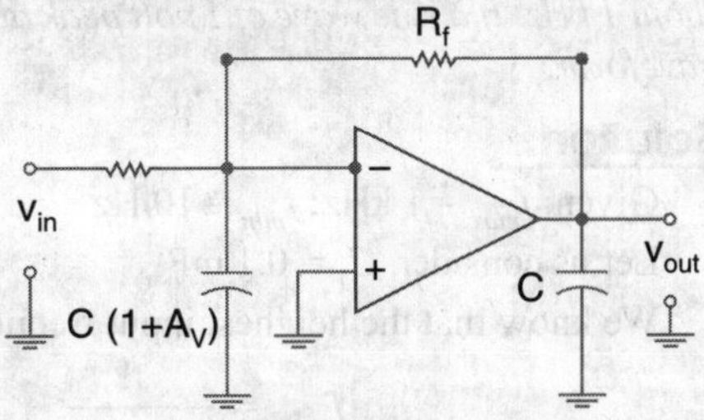

Fig. 36.22.

36.5. The Differentiator

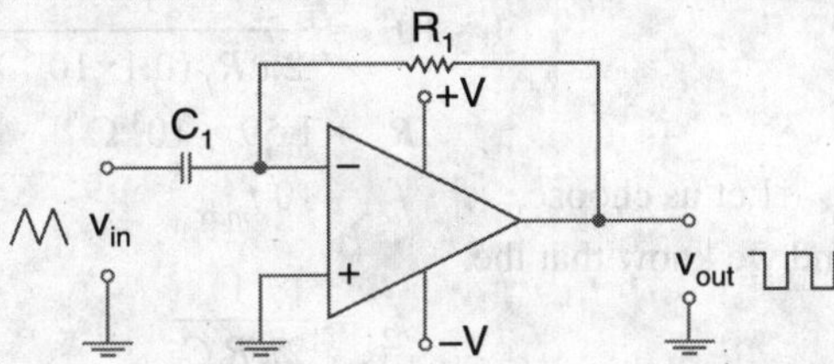

Fig. 36.23. The differentiator circuit.

Technically speaking, the differentiator is a circuit whose output is proportional to the rate of change of its input signal. The circuit for a differentiator is as shown in Fig. 36.23. Notice that the position of capacitor and resistor is reversed in a differentiator in comparison to the integrator. The input signal for the differentiator is shown to be a triangular waveform. The resulting output is a square waveform.

The cut-off frequency of a differentiate is given by the equation,

$$f_2 = \frac{1}{2\pi \, R_1 \, C_1}$$

There is a limit on the high frequency operation of the differentiation. The circuit will start to lose its operating characteristics at one-tenth the value of f_2, *i.e.,*

$$f_{max} = \frac{f_2}{10} = \frac{1}{20\,\pi \, R_1 \, C_1}$$

The output voltage of differentiator is given by,

$$V_o = -R_1 C_1 \frac{dV_{in}}{dt}$$

Example. 36.5.

A differentiator has $R_1 = 10$ kΩ and $C_1 = 0.01$ μF. Determine its cut-off frequency and its maximum linear operating frequency.

Solution.

Given: $R_1 = 10$ kΩ and $C_1 = 0.01$ μF.

We know that cut-off frequency of a differentiator,

$$f_2 = \frac{1}{2\pi \, R_1 \, C_1} = \frac{1}{2\pi \times (10 \times 10^3) \times (0.01 \times 10^{-6})} = 159 \text{ Hz}$$

and the maximum linear operating frequency,

$$f_{max} = \frac{f_2}{10} = \frac{1.59 \text{ kHz}}{10} = 159 \text{ Hz} \quad \textbf{Ans.}$$

Example. 36.6.

Design a differentiator to differentiate an input signal that varies in frequency from 10 Hz to about 1 kHz. If a sine wave of 1 volt peak at 100 Hz is applied to the above designed, draw its output waveform. (*RGTU Bhopal., Dec. 2010*)

Solution.

Given: $f_{max} = 1$ kHz; $f_{min} = 10$ Hz

Let us consider, $C_1 = 0.1$ mF

We know that the heighest input frequency,

$$f_{max} = \frac{1}{2\pi R_f C_1}$$

$$1 \times 10^3 = \frac{1}{2\pi R_f (0.1\times10^{-6})}$$

$$\therefore \quad R_f = 1.59 \times 10^3 \ \Omega$$

Let us choose, $f_2 = 10 f_{max}$

and we know that the,

$$f_2 = \frac{1}{2\pi R_1 C_1}$$

$$10 \times (1 \times 10^3) = \frac{1}{2\pi R_1 (0.1\times10^{-6})}$$

$$R_1 = 159.15 \ \Omega$$

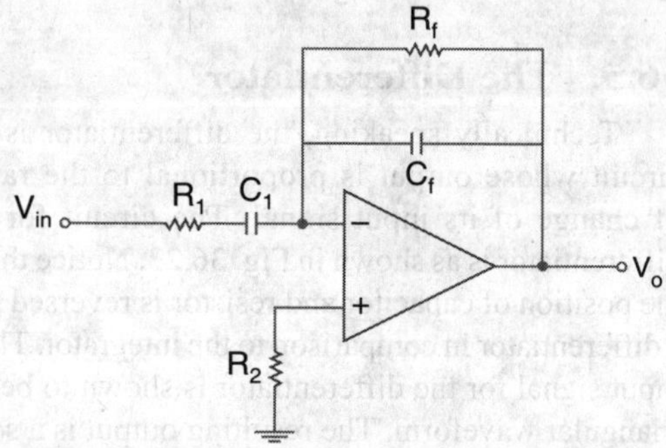

Fig. 36.24.

We also know that,

$$R_1 C_1 = R_f C_f$$

$$\therefore \quad C_1 = \frac{R_f C_f}{R_1} = \frac{(1.59\times10^3)\times(0.1\times10^{-6})}{159.15} = 0.01\times10^{-6}\ \text{F} = 0.01\ \mu\text{F}$$

and $\quad R_2 = R_1 \parallel R_f = (159.15) \parallel (1.59 \times 10^3) = 144.67 \ \Omega$

Fig. 36.24 show the differentiator Op-Amp,

The output voltage of the differentiator is given by,

$$V_o = -R_f C_1 \frac{dV_{in}}{dt}$$

$$= -(1.59 \times 10^3) \times (0.01 \times 10^{-6}) \frac{d}{dt}[1 \sin 2\pi (1000) t]$$

$$= -0.9424 \cos (2\pi \times 10^3 t)$$

The output waveform is shown in Fig. 36.25.

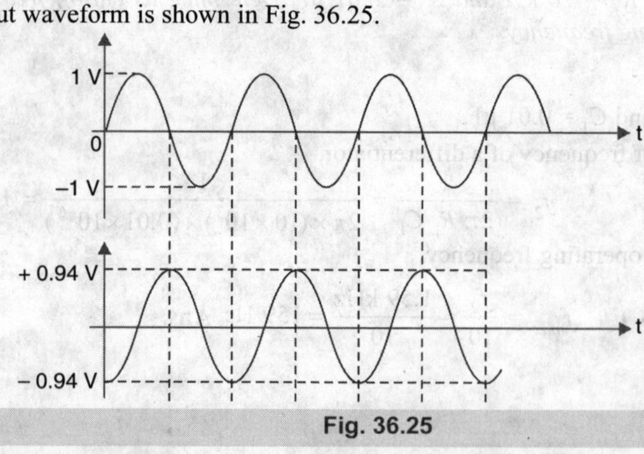

Fig. 36.25

36. 14. Precision Rectifier

The major limitation of ordinary diode is that it cannot rectify voltages below the cut –voltage (V_T= 0.6 V) of the diode. A circuit that acts like an ideal diode can de designed by placing a diode D in the feedback loop of an Op-Amp as shown in Fig. 36.26. This circuit is known as precision diode and used as a rectifier.

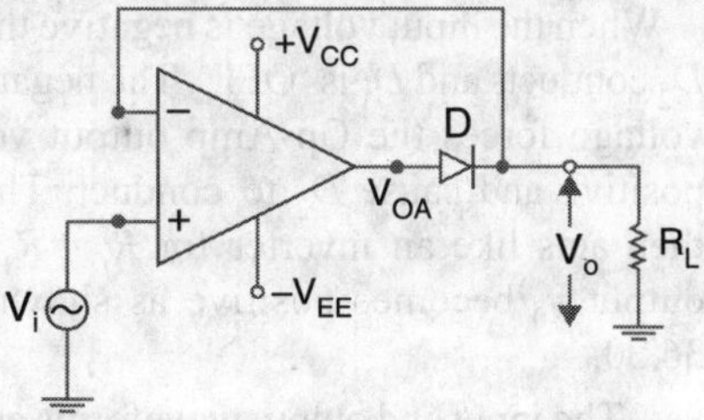

Fig. 36.26.

Here the cut-in voltage is divided by the open loop gain (A_{OL}) of the Op-Amp. The open loop gain of the Op-Amp is approximate10^4. Then the cut-in voltage is virtually eliminated. Thus the circuit acts as a voltage follower for input when $v_i > V_T/A_{OL}$.

During the positive half cycle, when the input voltage (v_i) is positive or greater than V_T/A_{OL}, the diode is 'ON'. When the input voltage (v_i) is negative or less than V_T/A_{OL}, the diode is 'OFF' and no current delivered to the load resistor except for small bias current of the Op-Amp and the reverse saturation current of the diode. The input output waveform is shown in Fig. 36.27.

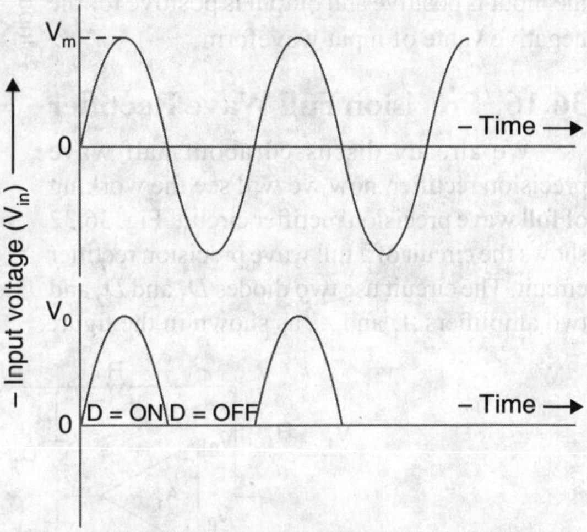

Fig. 36.27.

36.15. Precision Half-Wave Rectifier

An inverting amplifier can be converted into an ideal half-wave rectifier by adding two diodes as shown in Fig. 36.28.

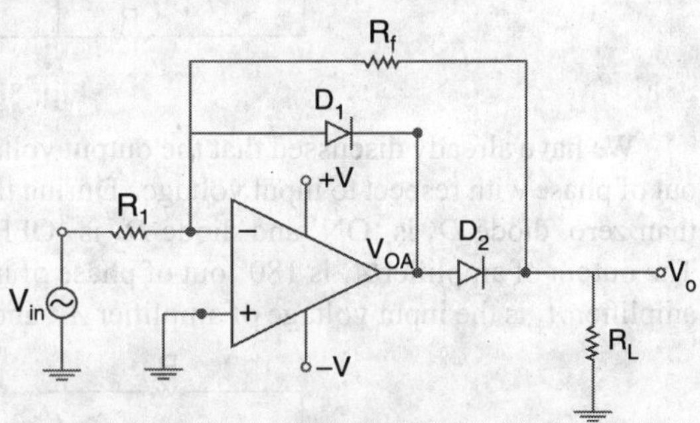

Fig. 36.28.

During the positive half cycle when the input voltage v_i is positive, diode D_1 conducts causing the Op-Amp output voltage V_{OA} to go negative by one diode drop. Hence diode D_2 is reverse biased. The output voltage v_O is zero. No current flows through R_f and the input current flows through D_1 as shown in Fig. 36.29.

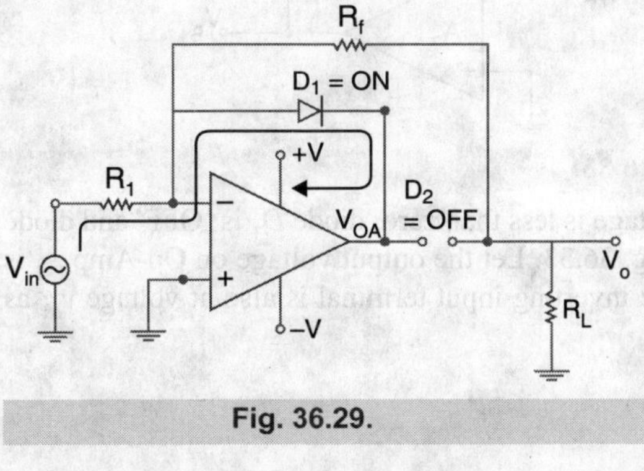

Fig. 36.29.

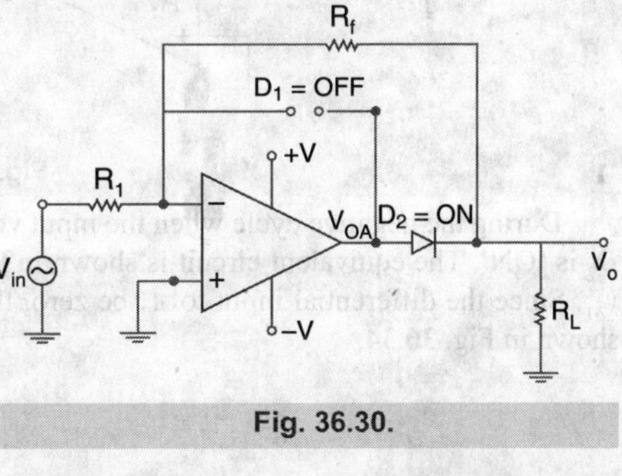

Fig. 36.30.

When the input voltage is negative then diode D_2 conducts and D_1 is 'OFF'. The negative input voltage forces the Op-Amp output voltage to positive and cause D_2 to conduct. The circuit then acts like an inverter for $R_f = R_1$ and the output v_0 becomes positive as shown in Fig. 36.30.

The input and output waveforms are shown in Fig. 36.31. The output waveform is zero when the input is positive and output is positive for the negative value of input waveform.

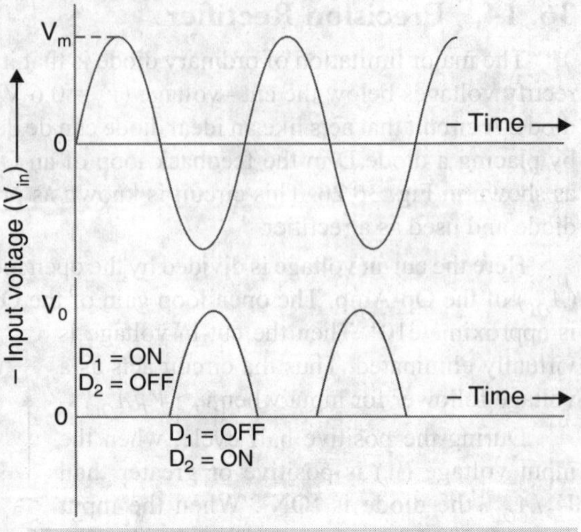

Fig. 36.31.

36.16. Precision Full Wave Rectifier

We already discussed about half wave precision rectifier, now we will see the working of full wave precision rectifier circuit. Fig. 36.32 shows the circuit of a full wave precision rectifier circuit. The circuit use two diodes D_1 and D_2 and two amplifiers A_1 and A_2 as shown in the figure.

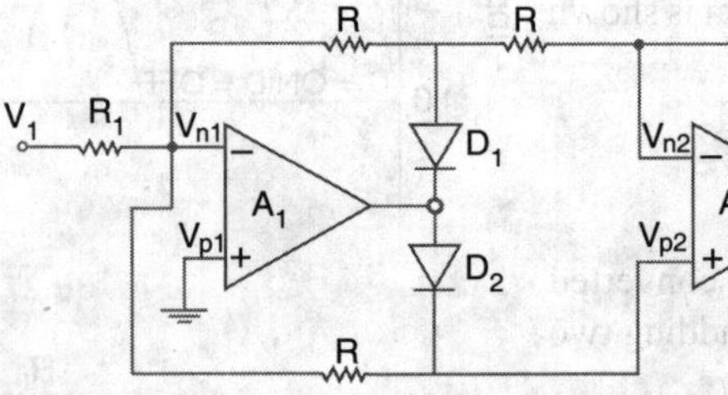

Fig. 36.32.

We have already discussed that the output voltage of the inverter is equal in magnitude but 180° out of phase with respect to input voltage. During the positive cycle when the input voltage is greater than zero, diode D_2 is 'ON' and diode D_2 is 'OFF'. Both the amplifier A_1 and A_2 acts as inverter. The output of amplifier A_1 is 180° out of phase of input voltage $(-v_i)$. The output voltage $(-v_i)$ of the amplifier A_1 is the input voltage of amplifier A_2, and then the output voltage (v_o) of amplifier A_2 is v_i.

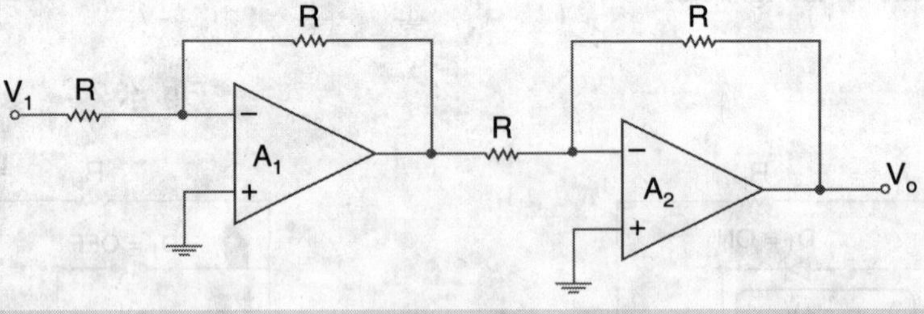

Fig. 36.33.

During the negative cycle when the input voltage is less than zero, diode D_1 is 'OFF' and diode D_2 is 'ON'. The equivalent circuit is shown in Fig. 36.33. Let the output voltage on Op-Amp A_1 is v_{o1}. Since the differential input to A_2 be zero, the inverting input terminal is also at voltage v_{o1} as shown in Fig. 36.34.

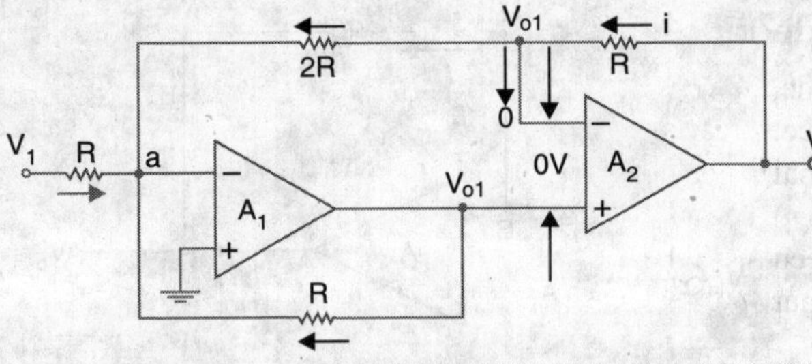

Fig. 36.34.

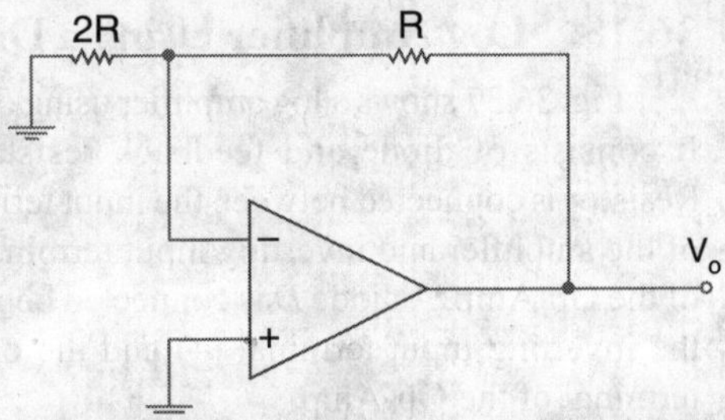

Fig. 36.35.

Applying Kirchoff's current law (KCL) at node 'a' we get

$$\frac{v_i}{R} + \frac{v_{o1}}{2R} + \frac{v_{o2}}{R} = 0$$

$$v_{o1} = -\frac{2}{3}\, v_i$$

To find v_o in term of v_{o1}, the equivalent circuit if amplifier A_2 is shown in Fig. 36.35.

$$v_o = \left(1 + \frac{R}{2R}\right) v_{o1} = \frac{3}{2}\, v_{o1}$$

Substituting the value of v_{o1} we get,

$$v_o = \frac{3}{2}\left(-\frac{2}{3}\, v_i\right) = -\, v_i$$

Hence from the above equation we see that the output is positive from the negative input. The input and output waveforms are shown in Fig. 36.36.

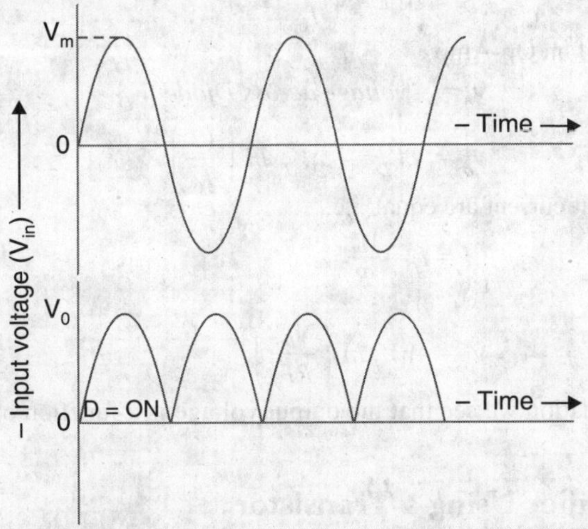

Fig. 36.36.

36.17. Log amplifier

Log amplifier (logarithmic amplifiers) has a wide range of applications ranging from signal compression, linearization of outputs of certain transducers. Log amplifiers are nonlinear circuits in which the output voltage is proportional to the logarithmic of the input. There are two basic circuits for logarithmic amplifiers

1. Log amplifier using a Diode
2. Log Amplifier using a Transistor

Now we shall discussed both these the log amplifiers in the following pages.

36.18. Log Amplifier Using a Diode

Fig. 36.37 shows a log amplifier using diode. It consists of diode and feedback resistance. Resistor is connected between the input terminal of the amplifier and inverting input terminal (–) of the Op-Amp. A diode D is connected between the inverting input terminal (–) and the output terminal of the Op-Amp.

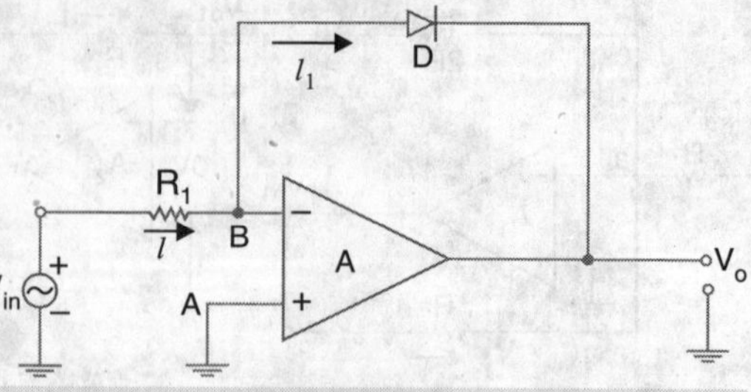

Fig. 36.37.

The node A is grounded and hence the node B is at virtual ground. Hence voltage at node B is also zero. The current flowing through the resistor R is $\boldsymbol{I}$, then the input voltage of an Op-Amp,

$$v_{in} = I R$$

$$I = \frac{v_{in}}{R}$$

The current flowing through the diode is $\boldsymbol{I_f}$ and the voltage across diode is V_D, then the diode current equation

$$I_f = I_o\left(e^{\frac{V_D}{\eta V_T}} - 1\right) \approx I_0\left(e^{\frac{V_D}{\eta V_T}}\right)$$

or
$$\frac{V_D}{\eta V_T} = ln\left(\frac{I_f}{I_o}\right)$$

$\therefore$
$$V_D = \eta V_T \; ln\left(\frac{I_f}{I_0}\right)$$

and the output voltage of an Op-Amp,

$$v_o = -\; voltage\; across\; Diode\; V_D$$

$$v_o = -\,V_D = -\,\eta V_T \; ln\left(\frac{I_f}{I_0}\right) \qquad \ldots(i)$$

The input and output current are equal, i.e.,

$$I = I_f = \frac{v_{in}}{R}$$

Thus equation (*i*),

$$v_o = -\eta V_T \; ln\left(\frac{v_{in}}{RI_o}\right)$$

From the above equation we see that the output voltage is a function of logarithm of the input function.

36.19. Log Amplifier Using a Transistor

This type of log amplifier is obtained by using a transistor in place of diode in the Fig. 36.38. Log amplifier using a transistor is shown in Fig. 36.38. Resistor is connected between the input terminal of the amplifier and inverting input terminal (–) of the Op-Amp. The transistor is connected between the inverting input terminal (–) and the output terminal of the Op-Amp. The base of the transistor in grounded while the collector of transistor is on virtual ground.

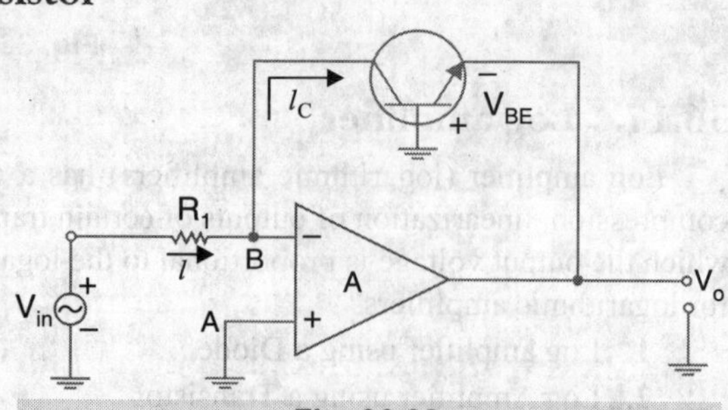

Fig. 36.38.

The node B is at virtual ground. Hence voltage at node B is also zero. The current flowing through the resistor R is **I**, then the input voltage of an Op-Amp,

$$v_{in} = IR$$

$$\therefore \quad I = \frac{v_{in}}{R}$$

The input and output current are equal, then

$$I = I_C = \frac{v_{in}}{R} \quad \text{... } (i)$$

The Voltage $V_{CB} = 0$ because the collector is at virtual ground and base is grounded. Hence we can write the equation for I_C as,

$$I_C = I_o \left(e^{\frac{v_{BE}}{v_T}} \right)$$

$$v_{BE} = V_T \ln \left(\frac{I_C}{I_o} \right)$$

And the output voltage,

$$v_o = -v_{BE} = -v_T \ln \left(\frac{I_C}{I_o} \right) \quad \text{... } (ii)$$

Substituting the value of I_C from equation (i) in the above equation we get,

$$v_o = -v_T \ln \left(\frac{v_{in}}{RI_o} \right)$$

The above equation gives the output voltage v_o proportional to the logarithm of the input voltage v_{in}.

36.20. Antilog amplifier

We have already discussed about log amplifier in the previous article. Log and antilog amplifier circuits include the same elements but arranged in different feedback configurations. The basic antilog amplifier can be obtained by using a diode or a transistor. In antilog amplifier the Antilog amplifier is also known as exponential amplifiers. There are two basic circuits for antilog amplifiers

1. Antilog amplifier using a Diode
2. Antilog Amplifier using a Transistor

Now we shall discussed these antilog amplifiers in the following pages.

36.21. Antilog Amplifier Using a Diode

Fig. 36.39 shows an antilog amplifier using diode. A diode D is connected between the input terminal of the amplifier and inverting input terminal (–) of the Op-Amp. A resistor R is connected between the inverting input terminal (–) and the output terminal of the Op-Amp

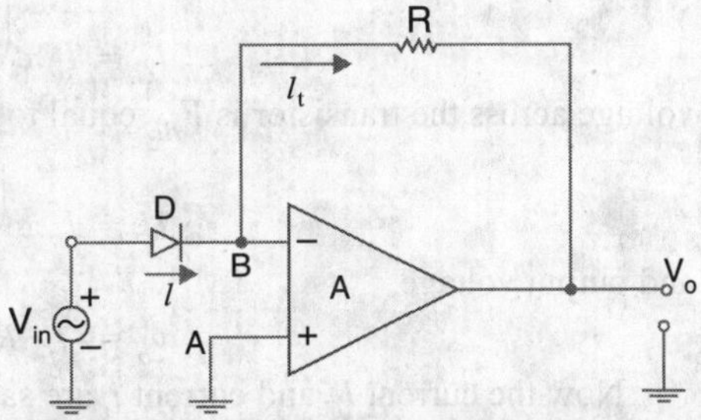

Fig. 36.39.

The node A is grounded and hence the node B is at virtual ground. Hence voltage at node B is also zero. Thus the current flowing through the diode is I_f and the voltage across diode is V_D, then the diode current equation

$$I = I_o \cdot \left(e^{\frac{V_D}{\eta V_T}} - 1 \right) \approx e^{\left(\frac{V_D}{\eta V_T} \right)}$$

voltage across diode is $V_D = v_{in}$.

$$I = I_o.e^{\left(\frac{v_{in}}{\eta V_T}\right)}$$

And output voltage,

$$v_o = -i_f \cdot R$$

$$\therefore \quad i_f = -\frac{v_o}{R}$$

Since the input and output currents are equal *i.e.*, $I = i_f$,

$$I_o.e^{\left(\frac{v_{in}}{\eta V_T}\right)} = -\frac{v_o}{R}$$

$$v_o = -RI_o.e^{\left(\frac{v_{in}}{\eta V_T}\right)}$$

Lets the product RI_o assume to be v_{ref}, so we get

$$v_o = -v_{ref}.e^{\left(\frac{v_{in}}{\eta V_T}\right)}$$

From the above equation we see that the output voltage is proportional to the exponential function of v_{in}. The exponential function is the antilog and thus circuit works as an antilog amplifier.

36.22. Antilog Amplifier Using a Transistor

This type of antilog amplifier is obtained by using a transistor in place of diode in the Fig. 36.39. Antilog amplifier using transistor is shown in Fig. 36.40, the transistor is connected between the input terminal of the amplifier and inverting input terminal (–) of the Op-Amp. The resistor is connected between the inverting input terminal (–) and the output terminal of the Op-Amp. The base of the transistor in grounded while the collector of transistor is on virtual ground.

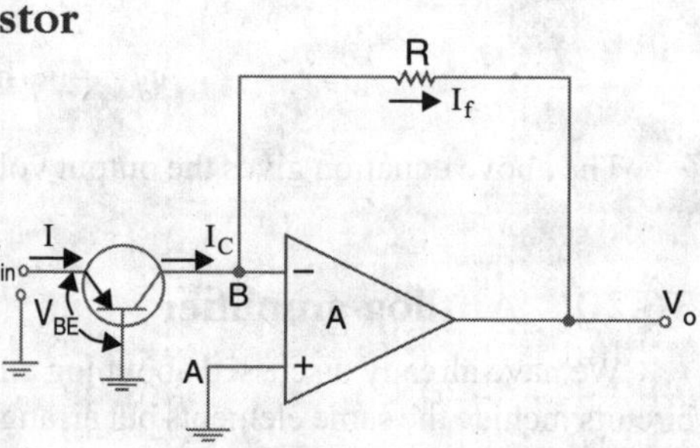

Fig. 36.40.

The node B is at virtual ground. Hence voltage at node B is also zero. Thus both collector and base of the transistor are at ground potential and V_{CB} = 0. Hence the voltage across the transistor is V_{BE} and we can write the expression for its collector current as,

$$I_C = I_o.e^{\left(\frac{v_{Be}}{V_T}\right)}$$

voltage across the transistor is V_{BE} equal to the input voltage v_{in},

$$I_C = I_o.e^{\left(\frac{v_{in}}{V_T}\right)}$$

and output voltage,

$$v_o = -i_f \cdot R$$

Now the current I_C and current i_f are same as Op-Amp input current is zero,

$$I_C = i_f$$

$$I_o.e^{\left(\frac{v_{in}}{V_T}\right)} = -\frac{v_o}{R}$$

$$v_o = -RI_o.e^{\left(\frac{v_{in}}{V_T}\right)}$$

Let the product RI_o assume to be v_{ref}, so we get

$$v_o = v_{ref} . e^{\left(\frac{v_{in}}{V_T}\right)}$$

Thus from the above equation we see that the output voltage is proportional to the exponential of v_{in}. Thus circuit works as basic antilog amplifier.

36.23. Voltage to Current Converter (V-I Converter)

It also called a Transconductance amplifier. In many applications, we have to convert a voltage signal to a proportional output current. In a voltage to current converter, the output load current is proportional to the input voltage. According to the connection of load there are two types of voltage to current converters. They are given below:

1. Voltage to Current Converter with Floating Load
2. Voltage to Current Converter with Grounded Load

In floating type of voltage to current converter the load resistor is not connected to the ground where as in grounded load type of V to I converter the load resistor is grounded.

36.24. Voltage to Current Converter with Floating Load

Fig. 36.41 shows a voltage-to-current converter, in which load resistor R_L is floating. It is not connected to ground. The input voltage is applied to the non-inverting input terminal. The feedback voltage across resistor R_1 drives the inverting input terminal. This circuit is also called a current series negative feedback amplifier. The feedback voltage across R_1 depends on the output current and is in series with the input difference voltage v_{id}.

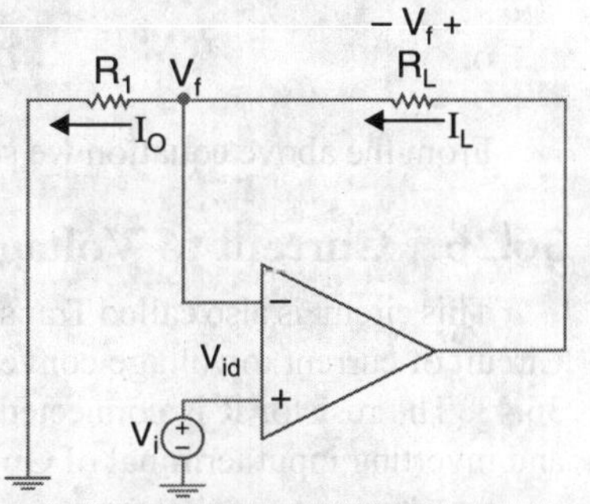

Fig. 36.41

From Kirchhoff's Voltage equation we get,

$$v_i = v_{id} + v_f$$

$v_{id} \cong 0V$, since A is very large, therefore

$$v_i = v_f = R_1 \, i_o$$

$$i_o = \frac{v_{in}}{R_1}$$

we know that $i_o = i_L$,

$$i_L = \frac{v_{in}}{R_1}$$

From the above equation we see that an input voltage v_i is converted into an output current of v_i/R_1. The voltage-to-current converter can be used in such applications as low-voltage *dc* and *ac* voltmeters, diode match finders, light-emitting diodes (LEDs) and zener diode testers.

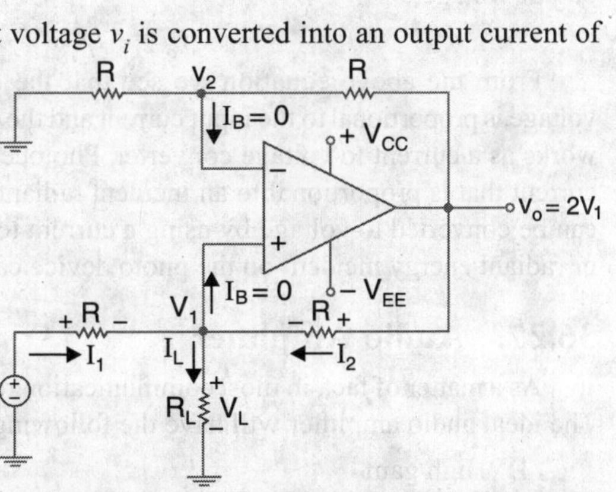

Fig. 36.42.

36.25. Voltage to Current Converter with Grounded Load

This circuit is also called Howland Current Converter from the name of inventor. In this circuit one terminal of the load is grounded and load current is controlled by an input voltage as shown in Fig. 36.42.

First we determined the voltage v_1 at the non-inverting input terminal and then establish the relationship between v_1 and the load current.

From the Kirchhoff's Current equation we get,

$$i_1 + i_2 = i_L$$

$$\frac{v_i - v_1}{R} + \frac{v_o - v_1}{R} = i_L$$

$$v_i - v_1 + v_o - v_1 = i_L R$$

$$v_i - 2v_1 + v_o = i_L R$$

$$v_1 = \frac{v_i + v_o - i_L R}{2} \quad \text{...}(i)$$

The gain of Op-Amp in non-inverting mode is given as $A = 1 + R_f/R_1$. For this circuit $R_f = R_1 = R$, then $A = 1 + R/R = 2$. Hence the output voltage,

$$A = 2 = \frac{v_o}{v_1}$$

or $$v_o = 2v_1$$

substituting the value v_1 of in equation (*i*) we get,

$$v_o = 2v_1 = v_i + v_o - i_L R$$

$$0 = v_i - i_L R$$

$$v_i = i_L R$$

or $$i_L = \frac{R}{v_i}$$

From the above equation we see that the load current depends on the input voltage and resistor.

36.26. Current to Voltage Converter (I-V Converter)

This circuit is also called Transresistance Amplifier. Circuit of current to voltage converter is shown in Fig. 36.43. The resistor R is connected between the output and inverting input terminal of Op-Amp.

The inverting terminal is at virtual ground, the current I_i flows through the feedback resistor. Thus the output voltage,

$$v_o = -I_i R$$

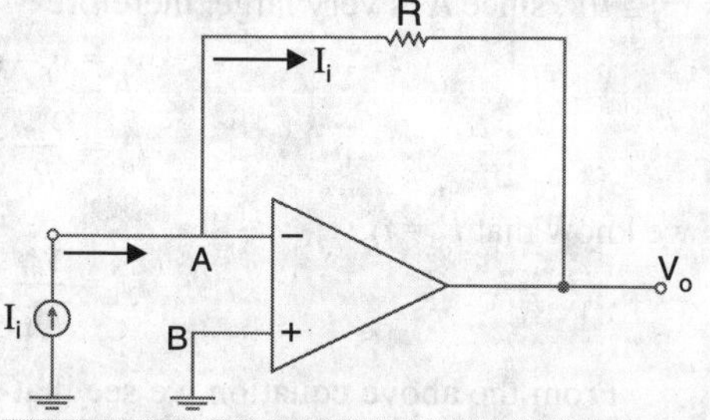

Fig. 36.43.

From the above equation we see that the output voltage is proportional to the input current and the circuit works as a current to voltage converter. Photocell, photodiode and photovoltaic cell give an output current that is proportional to an incident radiant energy or light. The current through there devices can be converted to voltage by using a current to voltage converter and thereby the amount of light or radiant energy incident on the photo device can be measured.

36.27. Audio Amplifier

As a matter of fact, in most communication receivers, the final output stage is the audio amplifier. The ideal audio amplifier will have the following characteristics:

1. High gain
2. Minimum distortion in the audio frequency range (*i.e.*, 20 Hz to 20 kHz range).

3. High input resistance (or impedance).
4. Low output resistance (or impedance) to provide optimum coupling to the speaker.

The use of Op-Amp in an audio amplifier will fulfill the requirements listed above very nicely. An Op-Amp audio amplifier is shown in Fig. 36.44.

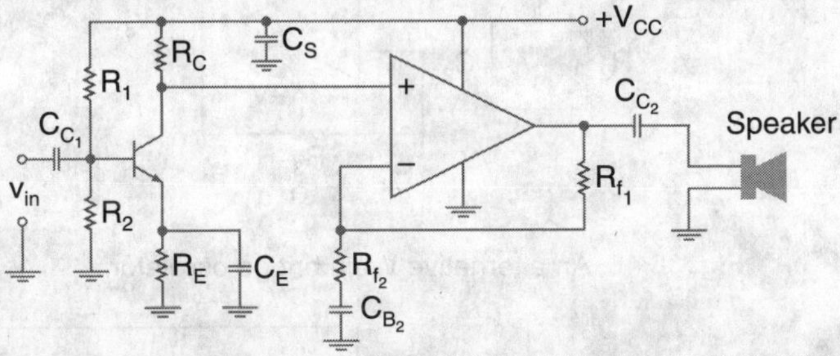

Fig. 36.44. An Op-Amp audio amplifier.

Notice that the Op-Amp is supplied only from + V volt power supply, the – V terminal is grounded. Because of this, the output will be between the limits of (+ V – 1) volt and + 1 volt approximately. Also notice the use of a coupling capacitor (C_{C_2}) between the Op-Amp and speaker. This capacitor is necessary to reference the speaker signal around ground. The capacitor C_S, is included in the V_{CC} line to prevent any transient current caused by the operation of Op-Amp from being coupled back to Q_1 through the power supply. The high gain requirement is accomplished by the combination of two amplifier stages. The high R_{in}/Low R_{out} of the audio amplifier is accomplished by the Op-Amp itself, as the low distortion characteristic.

36.28. Op-Amp Based Oscillator Circuits

We have already discussed about sinusoidal oscillators in Chapter 33. There we defined the oscillator as a circuit that produces an output waveform without any external signal source. The only input to an oscillator is the dc power supply. As such, the oscillator can be viewed as being a signal generator. We have also discussed in the same chapter, about the different types of oscillator circuits (like Wein bridge oscillator, Colpitts Oscillator and Crystal Oscillator) using bipolar junction transistor and field-effect transistor. Now we will study these oscillator circuits using Op-Amp.

36.29. Op-Amp Based Wein-bridge Oscillator

Fig. 36.45 shows the basic version of a Wein bridge oscillator. The circuit uses an Op-Amp and RC bridge circuit. Note the basic bridge connection carefully. Resistors R_1, R_2, and capicitors C_1 and C_2 form the frequency adjustment elements while resistors R_3 and R_4 form part of the feedback path.

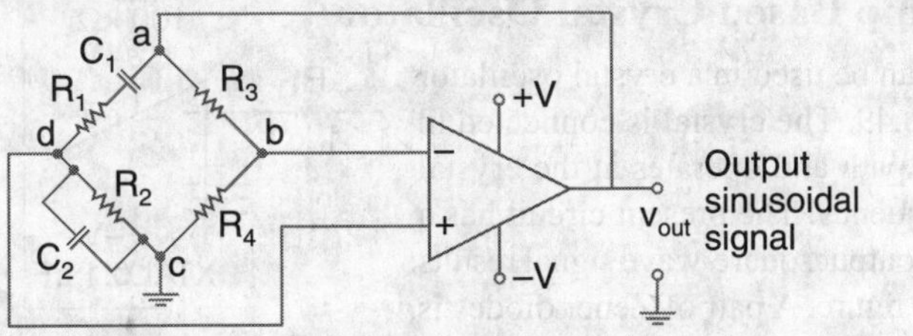

Fig. 36.45. An Op-Amp based Wein bridge oscillator.

The Op-Amp output is connected as the bridge input as points 'a' and 'c'. The bridge circuit output at points 'b' and 'd' is the input to the Op-Amp. Fig. 36.46 shows an alternative way of connecting the RC bridge circuit to the Op-Amp. In a typical Wein bridge oscillator, $R_1 = R_2 = R$, and $C_1 = C_2 = C$.

This means that the two RC circuits will have the same cut-off frequency. The cut-off frequency is given by $(1/2\pi RC)$.

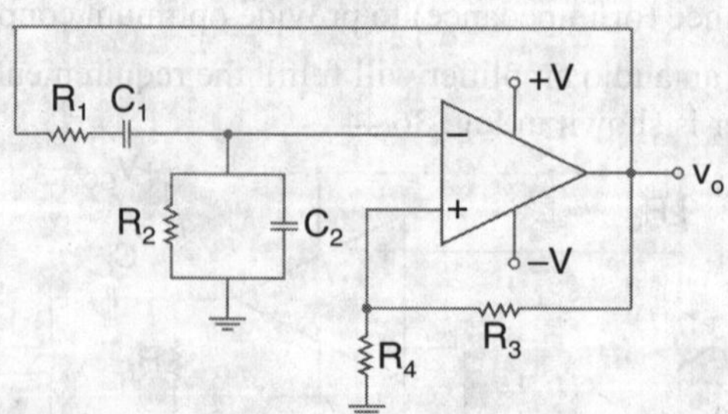

Fig. 36.46. An alternative Wein bridge oscillator.

Example 36.7.

Fig. 36.47 shows the circuit of a Wein-bridge Oscillator using Op-Amp as an amplifier. Notice the components R_1, C_1, R_2, C_2, R_3 and R_4 are connected in the bridge configuration in the same way as shown in Fig. 36.43.

Calculate the frequency of the Wein-bridge oscillator.

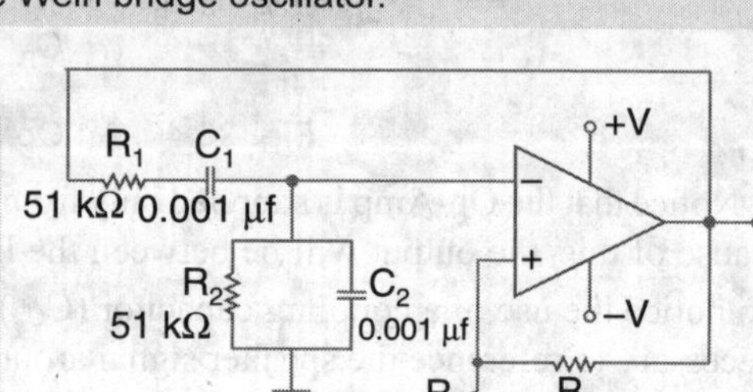

Fig. 36.47.

Solution.

Given: $R_1 = R_2 = 51\ k\Omega = 51 \times 10^3\ \Omega$ and $C_1 = C_2 = C = 0.001\ \mu F = 0.001 \times 10^{-6}$ F.

We know that the frequency of oscillations,

$$f_0 = \frac{1}{2\pi\ RC} = \frac{1}{2\pi \times (51 \times 10^3) \times (0.001 \times 10^{-6})}$$

$= 3120.7$ Hz **Ans.**

36.30. Op-Amp Based Colpitts Oscillator

An Op-Amp based Colpitts oscillator is as shown in Fig. 36.48. Here the Op-Amp provides the basic amplification needed while a LC feedback network of Colpitts configuration sets the oscillator frequency.

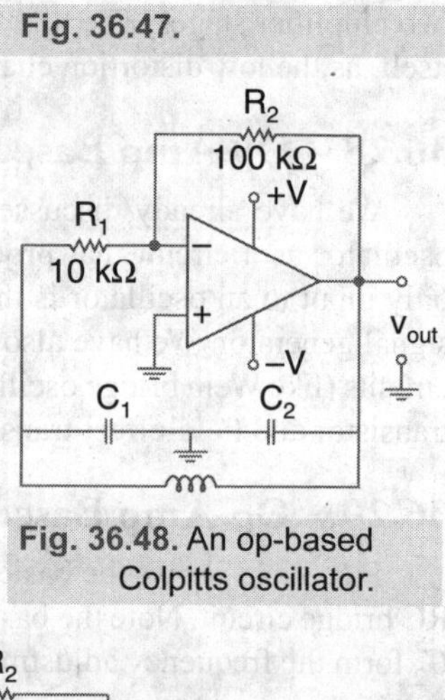

Fig. 36.48. An op-based Colpitts oscillator.

36.31. Op-Amp Based Crystal Oscillator

An Op-Amp can be used in a crystal oscillator as shown in Fig. 36.49. The crystal is connected in the series resonant path and operates at the crystal series resonant frequency. The present circuit has a high gain so that an output square-wave signal results are as shown in the figure. A pair of Zener diodes is shown at the output to provide output amplitude at exactly the Zener voltage (V_Z).

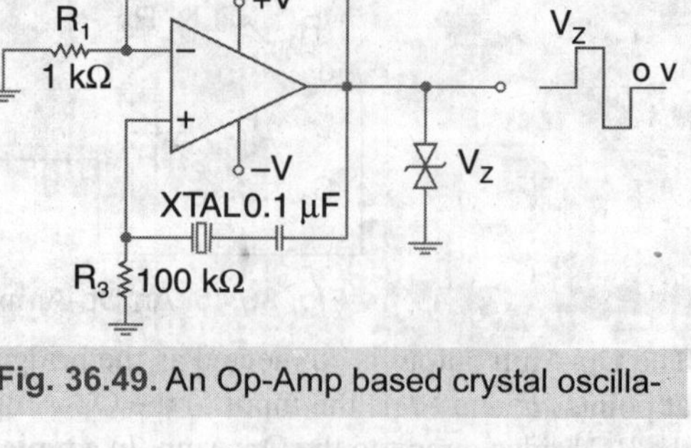

Fig. 36.49. An Op-Amp based crystal oscilla-

36.32. A Triangular-Wave Oscillator

We have already discussed in Chapter 33 a method to generate a triangular-wave from a square-wave. Fig. 36.50 shows an Op-Amp circuit to implement that method. The circuit makes use of two Op-Amps: one of them is used as a comparator and the other as an integrator. The operation of the circuit is as given below:

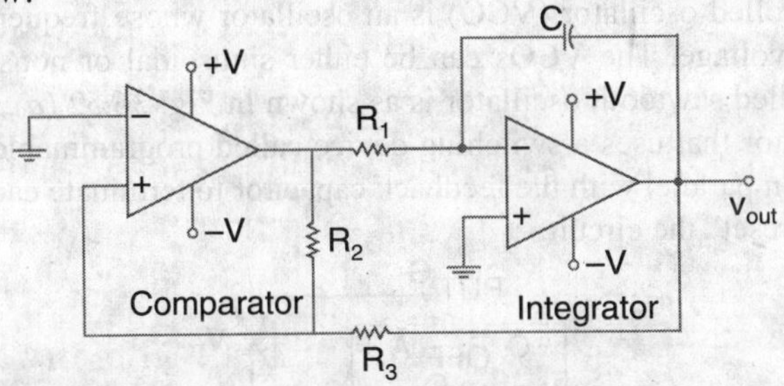

Fig. 36.50. A triangular-wave oscillator.

To begin with, let us assume that the output voltage of the comparator is at its maximum negative level. This output is connected to the inverting input of the integrator through resistor, R_1. This produces a positive-going ramp on the output of the integrator. When the ramp voltage reaches the upper trigger point (UTP), the comparator switches to its maximum positive level. This positive level causes the integrator ramp to change to a negative going direction. The ramp continues in this direction until the lower trigger point (LTP) of the comparator is reached. At this point, the comparator output switches back to its maximum negative level and the cycle repeats. This action is shown in Fig. 36.51.

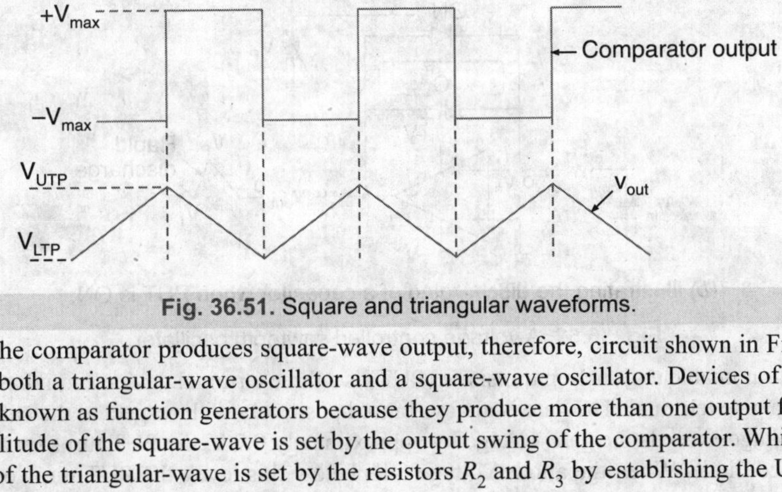

Fig. 36.51. Square and triangular waveforms.

Since the comparator produces square-wave output, therefore, circuit shown in Fig. 36.49 can be used as both a triangular-wave oscillator and a square-wave oscillator. Devices of this type are commonly known as function generators because they produce more than one output function. The output amplitude of the square-wave is set by the output swing of the comparator. While the output amplitude of the triangular-wave is set by the resistors R_2 and R_3 by establishing the UTP and LTP voltages according to the following formulas:

$$V_{\text{UTP}} = +V_{max}\left(\frac{R_3}{R_2}\right)$$

$$V_{\text{LTP}} = -V_{max}\left(\frac{R_3}{R_2}\right)$$

It may be noted that the comparator output levels, $+V_{max}$ are equal. The frequency of both waveforms depends on the $R_1 C$ time constant as well as the amplitude-setting resistor R_2 and R_3. By varying R_1, the frequency of oscillation can be adjusted without changing the output amplitude.

$$f = \frac{1}{4\,R_1 C}\left(\frac{R_2}{R_3}\right)$$

36.33. A Voltage-Controlled Sawtooth Oscillator (VCO)

The voltage-controlled oscillator (VCO) is an oscillator whose frequency can be changed by a variable dc control voltage. The VCOs can be either sinusoidal or nonsinusoidal. One way to build a voltage-controlled sawtooth oscillator is as shown in Fig. 36.52 (*a*). This circuit makes use of an Op-Amp integrator that uses a switching device called programmable unijunction transistor (abbreviated as PUT) in parallel with the feedback capacitor to terminate each ramp at a prescribed level and effectively "reset" the circuit.

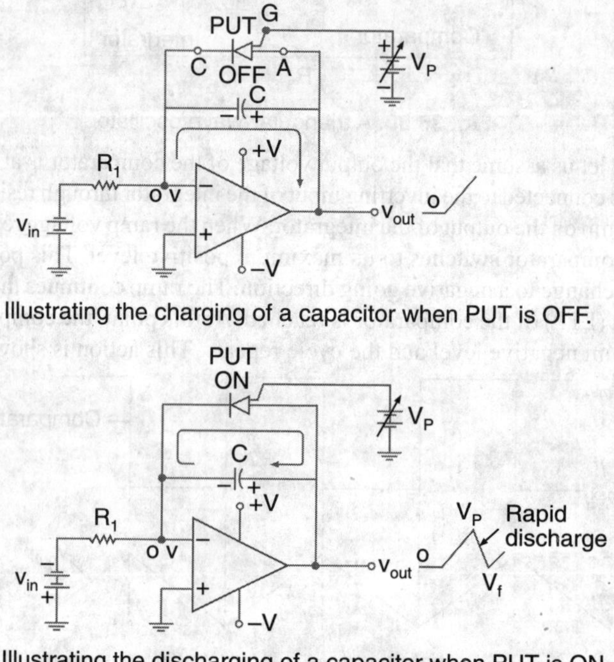

(*a*) Illustrating the charging of a capacitor when PUT is OFF.

(*b*) Illustrating the discharging of a capacitor when PUT is ON.

Fig. 36.52. A voltage controlled sawtooth oscillator.

The programmable unijunction transistor (PUT) is a three terminal device *i.e.*, it has an anode, a cathode and a gate terminal. The gate is always biased positively with respect to the cathode. When the anode voltage exceeds the gate voltage by approximately 0.7 V, the PUT turns on and acts as a forward biased diode. When the anode voltage falls below this level, the PUT turns off. Also, the value of current must be above the holding value to maintain conduction.

The operation of the circuit may be explained as below. The negative d.c. input voltage, $-V_{IN}$, produces a positive-going ramp on the output. During the time that ramp is increasing the circuit acts as a regular integrator. When the ramp voltage (*i.e.*, voltage at PUT anode) exceeds the gate voltage by 0.7 V, the PUT turns on. This forces the capacitor to discharge rapidly as shown in Fig. 36.52 (*b*). However, the capacitor does not discharge completely to zero because of the PUT's forward voltage, V_F. Discharging of the capacitor continues until the current through PUT drops below the holding value. At this point, PUT turn off and capacitor begins to charge again, thus generating a new output ramp. The cycle repeats and the resulting output is a repetitive sawtooth waveform as shown in the figure. It may be noted that the sawtooth amplitude and period can be adjusted by varying the PUT gate voltage.

The frequency of smooth curve is given by the relation:

$$f = \frac{|V_{IN}|}{R_1 C}\left(\frac{1}{V_P - V_F}\right)$$

It is evident from the above equation that the frequency depends upon the time constant "R_1C" of the integrator and the peak voltage set by the PUT. The time period of the sawtooth wave is the reciprocal of the frequency, *i.e.*,

$$T = \frac{V_P - V_F}{|V_{IN}|/R_1C}$$

36.34. A Square-wave Relaxation Oscillator

Fig. 36.53 shows the circuit of a basic relaxation oscillator. Its operation depends upon charging and discharging of a capacitor. Notice that the Op-Amp's inverting input is the capacitor voltage and the noninverting input is a portion of the output fed-back through resistors R_2 and R_3.

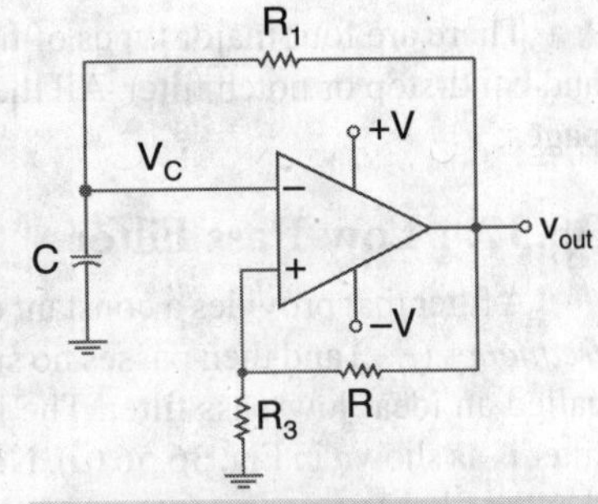

Fig. 36.53. A square-wave relaxation oscillator.

When the circuit is first turned on, the capacitor is uncharged. Because of this, the inverting input is at 0 V. This makes the output a positive maximum, and capacitor begins to charge towards V_{out} through R_1. When the capacitor voltage reaches a value equal to the feedback voltage on the non-inverting input, the Op-Amp switches to the maximum negative stage. At this point, the capacitor begins to discharge from + V_f toward – V_f. When the capacitor voltage reaches – V_f, the Op-Amp switches back to the maximum positive state. This action continues to repeat and a square wave output is obtained as shown in Fig. 36.54.

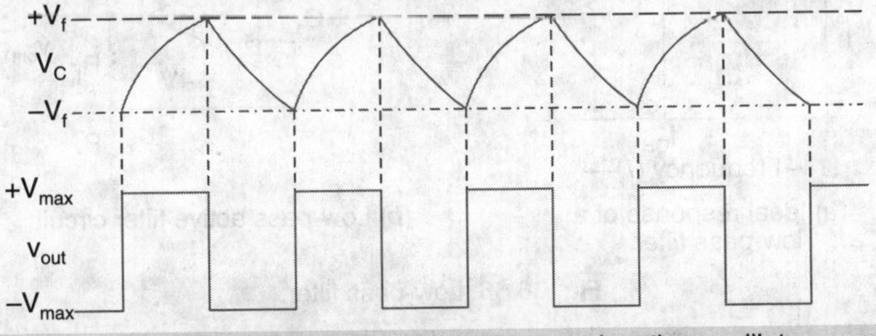

Fig. 36.54. Waveforms for a square-wave relaxation oscillator.

36.35. High-impedance Voltmeter

Fig. 36.55 shows the circuit of a high impedance voltmeter. In such a circuit, the closed loop-gain depends on the internal resistance of the meter, R_M. The input voltage will be amplified and the output voltage will cause a proportional current to flow through the meter. By adding a small series potentiometer in the feedback loop, the meter can be calibrated to provide a more accurate reading.

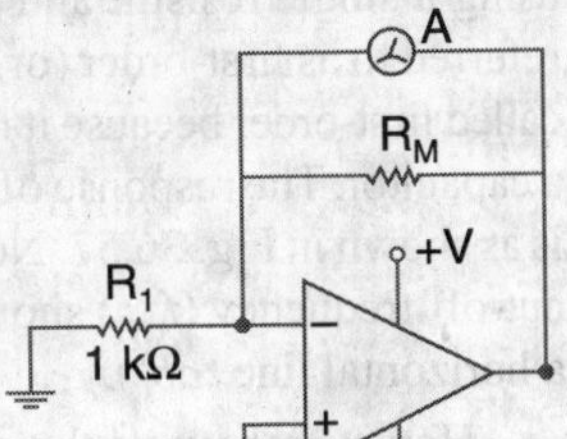

Fig. 36.55. A high impedance voltmeter.

The high input impedance of the Op-Amp reduces the circuit loading that is caused by the use of the meter. Although this type of circuit would cause some circuit loading, it would be much more accurate than a VOM (Volt-Ohm-Meter) with an input impedance of 20 kΩ/V.

36.36. Active Filters

We have already discussed in Chapter 30, about tuned voltage amplifiers (or simply tuned amplifiers). Such amplifiers are designed to amplify only those frequencies that are within certain range. As long as the input signal is within the specified range, it will be amplified. If it goes outside of this frequency range, amplification will be drastically reduced. The tuned amplifier circuits using Op-Amp are generally referred to as active filters. Such circuits do not require the use of inductors. The frequency response of the circuit is determined by resistor and capacitor values.

We have already mentioned above that Op-Amps can be used to build active filter circuits. A filter circuit can be constructed using passive components like resistors and capacitors. But an active filter, in addition to the passive components makes use of an Op-Amp as an amplifier. The amplifier in the active filter circuit may provide voltage amplification and signal isolation or buffering.

There are four major types of filters namely, low-pass filter, high-pass filter, and band-pass filter and band-stop or notch filter. All these four types of filters are discussed one by one in the following pages.

36.37. Low-Pass Filter

A filter that provides a constant output from dc up to a *cut-off frequency* (f_{OH}) and then passes no signal above that frequency is called an ideal low-pass filter. The ideal response of a low-pass filter is as shown in Fig. 36.56 (*a*). Notice that the response shows that the filter has a constant output (indicated by a horizontal line '*ab*') from dc or zero frequency up to a cut-off frequency (f_{OH}). And beyond f_{OH}, the output is zero as indicated by the vertical line '*bc*' in the figure.

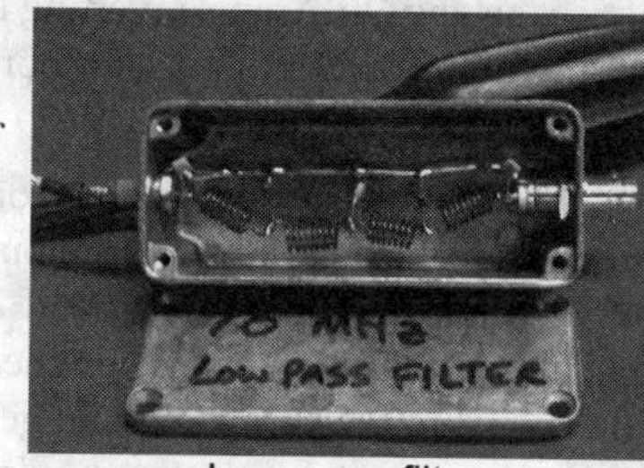

Low-pass filter

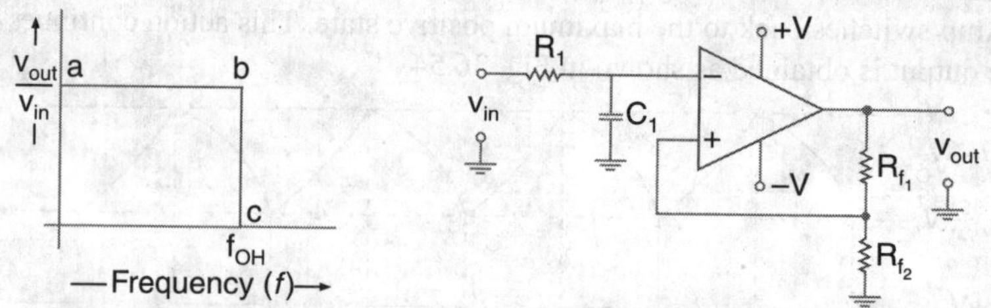

(*a*) Ideal response of a low-pass filter.

(*b*) Low-pass active filter circuit.

Fig. 36.56. Low-pass filter.

Fig. 36.56 (*b*) shows the circuit of a low-pass active filter using a single resistor and capacitor. Such a circuit is also referred to as first-order (or single-pole) low-pass filter. It is called first-order because it makes use of a single resistor and a capacitor. The response of such a first-order low pass filter is as shown in Fig. 36.57. Notice that the response below the cut-off frequency (f_{OH}) shows a constant gain (indicated by a horizontal line '*ab*').

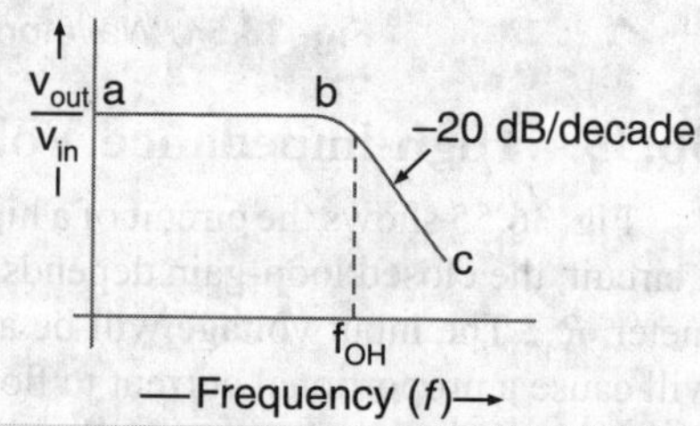

Fig. 36.57. Response of a first-order low-pass active filter.

However, beyond the cut-off frequency, the gain does not reduce immediately to zero as expected in Fig. 36.56 (*a*) but reduces with a slope of 20 dB/decade (means that the output voltage reduces by a factor of 100 when the frequency increases by a factor of 10). The voltage gain for a low-pass filter below the cut-off frequency (f_{OH}) is given by the relation,

$$A_v = 1 + \frac{R_3}{R_1}$$

And the cut-off frequency is determined by the relation,

$$f_{OH} = \frac{1}{2\pi R_1 C_1}$$

It is possible to connect two sections of the filter together as shown in Fig. 36.58 (*a*). Such a circuit is called second-order (or two-pole) low pass filter. Fig. 36.58 (*b*) shows another configuration of the second-order low-pass filter.

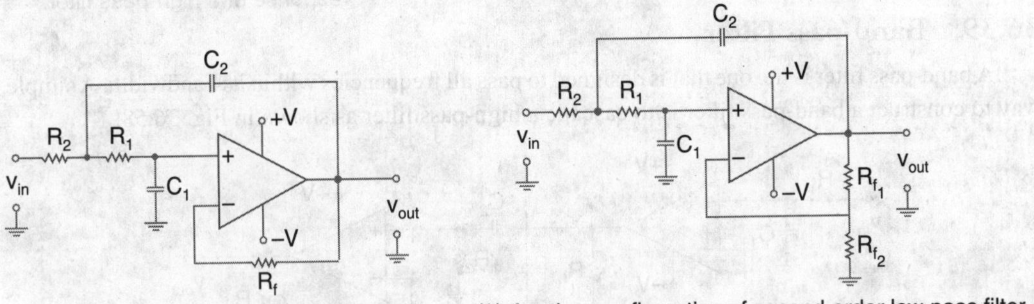

(*a*) Second order low-pass filter. (*b*) Another configuration of second-order low-pass filter.

Fig. 36.58.

Each circuit shown in Fig. 36.58 has two RC circuits, $R_1 - C_1$ and $R_2 - C_2$. As the operating frequency increases beyond f_2, each circuit will be dropping the closed-loop gain by 20 dB, giving a total roll-off rate of 40 dB/decade when operated above f_2. The cut-off frequency for each of the circuit is given by,

$$f_2 = \frac{1}{2\pi\sqrt{R_1 R_2 C_1 C_2}}$$

36.38. High-pass Filter

As a matter of fact, there is very little difference between the high-pass filter and the low-pass filter. Fig. 36.59 (*a*) shows the circuit of a first order (or single-pole) high-pass filter and Fig. 36.59 (*b*),

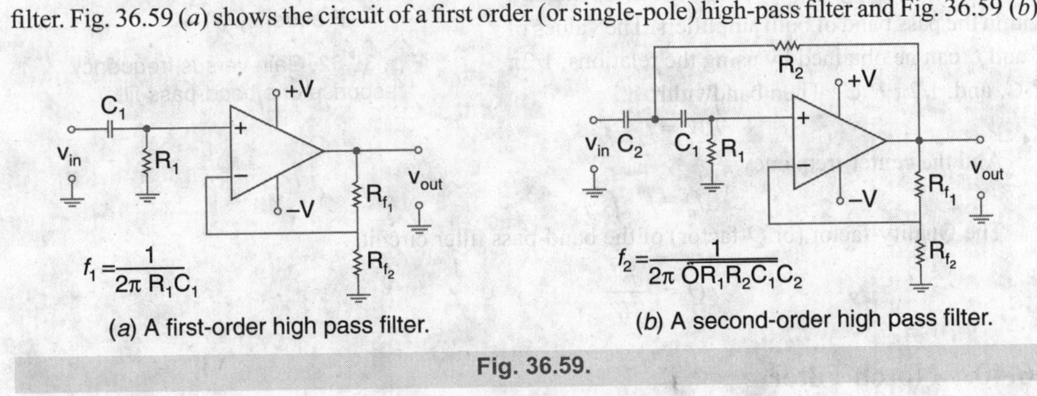

(*a*) A first-order high pass filter. (*b*) A second-order high pass filter.

Fig. 36.59.

the circuit of a second-order (or two-pole) high-pass filter. Notice that the only thing that has changed is the positions of the capacitors and resistors. The value of cut-off frequencies f_1 and f_2 is obtained by using the same equations we used for low-pass filter.

Fig. 36.60 shows the gain versus frequency response of a high-pass filter. Notice that the solid line indicates the ideal response while the dashed line, corresponds to the actual response of the filter circuit. The ideal curve indicates that the filter has a zero output for the frequencies below f_{OL} (indicated by the line '*ab*'). And beyond f_{OL}, it has a constant output. The actual response curve may correspond to the roll-off gain by 20 dB/decade for first order or 40 dB/decade for second-order low-pass filters.

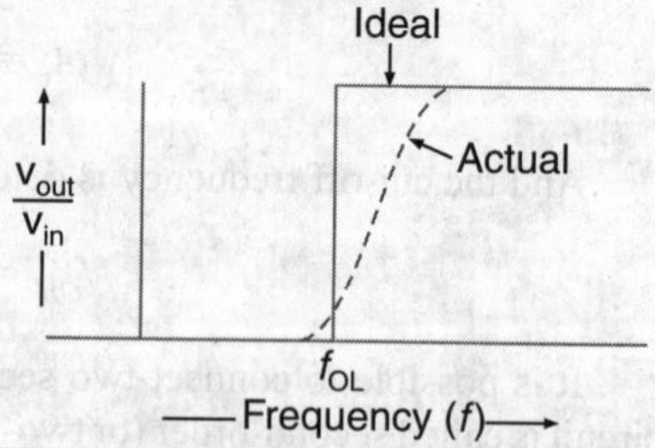

Fig. 36.60. Gain versus frequency response of a high-pass filter.

36.39. Band-pass Filter

A band-pass filter is the one that is designed to pass all frequencies within its bandwidth. A simple way to construct a band-pass filter is to cascade a high-pass filter as shown in Fig. 36.58.

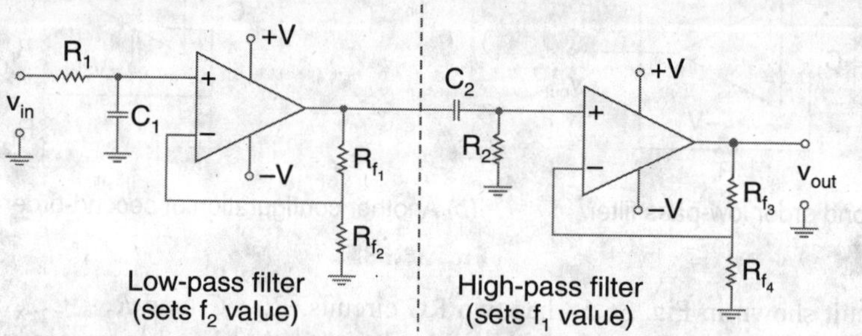

Fig. 36.61. A band-pass filter circuit.

The first stage of the band-pass filter will pass all frequencies that are below its cut-off value, f_2. All the frequencies passed by the first stage will head into the second stage. This stage will pass all frequencies above its value of f_1. The result of this circuit action is as shown in Fig. 36.62. Note that the only frequencies that will pass through the amplifier are those that fall within the pass band of both amplifiers. The values of f_1 and f_2 can be obtained by using the relations, $1/2\pi\, R_1C_1$ and, $1/2\pi\, R_2C_2$. Then bandwidth,

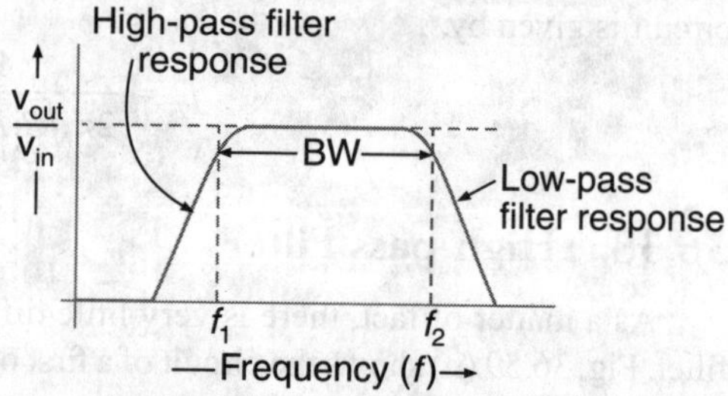

Fig. 36.62. Gain versus frequency response of a band-pass filter.

$$BW = f_2 - f_1$$

And the center frequency,

$$f_o = f_1 \cdot f_2$$

The Quality-factor (or *Q*-factor) of the band-pass filter circuit,

$$Q = \frac{f_o}{BW}$$

36.40. Notch Filter

The notch filter is designed to block all frequencies that fall within its bandwidth. Fig. 36.63 (*a*) shows a block diagram and 36.63 (*b*), the gain versus frequency response curve of a multistage notch filter.

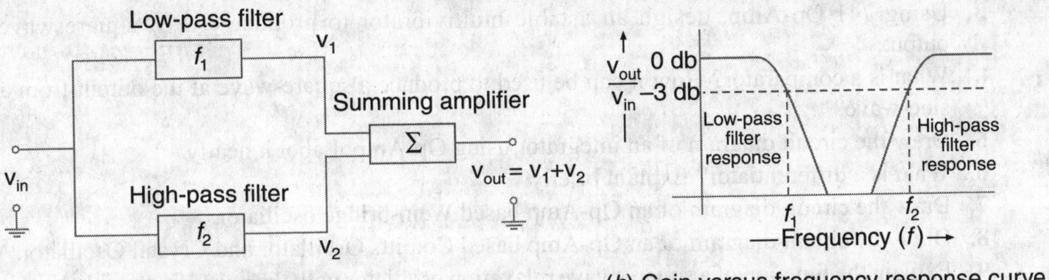

(a) Block diagram. (b) Gain versus frequency response curve.

Fig. 36.63. Notch filter.

The block diagram shows that the circuit is made up of a high-pass filter, a low-pass filter and a summing amplifier. The summing amplifier produces an output that is equal to a sum of the filter output voltages. The circuit is designed in such a way so that the cut-off frequency, f_1 (which is set by a low-pass filter) is lower in value than the cut-off frequency, f_2 (which is set by high-pass filter). The gap between the values of f_1 and f_2 is the bandwidth of the filter.

When the circuit input frequency is lower than f_1, the input signal will pass through low-pass filter to the summing amplifier. Since the input frequency is below the cut-off frequency of the high-pass filter, v_2 will be zero. Thus the output from the summing amplifier will equal the output from the low-pass filter. When the circuit input frequency is higher than f_2, the input signal will pass through the high-pass filter to the summing amplifier. Since the input frequency is above the cut-off frequency of the low-pass filter, v_1 will be zero. Now the summing amplifier output will equal the output from the high-pass filter.

It is evident from the above discussion that frequencies below f_1 and those above f_2, have been passed by the notch filter. But when the circuit frequency between f_1 and f_2, neither of the filters will produce an output. Thus v_1 and v_2 will be both zero and the output from the summing amplifier will also be zero.

The frequency analysis of the notch filter is identical to the band-pass filter. First, determine the cut-off frequencies of the low-pass and the high-pass filters. Then using these calculated values, determine the bandwidth, center frequency and Q values of the circuit.

DESCRIPTIVE QUESTIONS

1. Determine the lower cut-off frequency and the frequency at which the circuit shown in Fig. 36.64 will start to lose its linear output characteristics. **(Ans. 79.6 kHz, 796 kHz)**

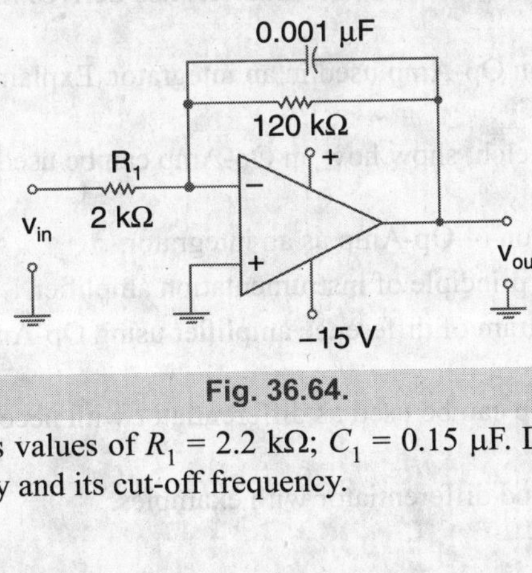

Fig. 36.64.

2. A differentiator has values of R_1 = 2.2 kΩ; C_1 = 0.15 μF. Determine its maximum linear operating frequency and its cut-off frequency. **(Ans. 48.2 Hz and 482 Hz)**

3. Using 741 Op-Amp, design an astable multivibrator to produce ± 9 V, square wave output. *(AMIE., Winter 1998)*
4. What is a comparator? How it can be used to produce a square-wave at the output from a sine-wave?
5. Draw the circuit diagram of an integrator using Op-Amp. Label it neatly.
6. What is a differentiator? Explain briefly.
7. Draw the circuit diagram of an Op-Amp based Wein-bridge oscillator.
8. Draw the circuit diagram of an Op-Amp based Colpitts Oscillator and Crystal Oscillator.
9. Explain the function of a square-wave relaxation oscillator with the help of a suitable circuit diagram.
10. What is the difference between a low-pass and a high-pass filter? Explain it with the help of gain-versus-frequency response graph.
11. Sketch the gain-versus-frequency curves for a band-pass and notch-filter.
12. Give any four applications of a comparator. *(Anna University., April/May 2008)*
13. Explain in detail the application of operational amplifier as an Integrator and Differentiator with suitable circuit and expressions involved. *(GBTU/MTU., 2008-09)*
14. Draw the differentiator circuit using an Op-Amp and derive an expression for its output voltage in terms of input voltage. *(PBUT., 2008)*
15. Draw the input and output waveform of miller integrator circuit. *(Anna University., May/June 2007)*
16. Draw a RC integrator and differentiator circuit. Draw their outputs for a square input. How the output varies with respect to variation in the R and C value .Derive the conditions for ideal integration and differentiation. *(M.G University., July/Aug. 2007)*
17. Define low pass and high pass filters. Explain by drawing frequency response characteristics. *(R.G.P.V., June 2005)*
18. List the applications of passive filter. *(R.G.P.V., Dec. 2006)*
19. What are the concepts of passive filters? Enlist various types of filters. Show how the inductors and capacitors are best suitable elements for designing a filter? *(R.G.P.V., June 2003)*
20. Define low pass and high pass filters. Explain by drawing frequency response characteristics. *(R.G.P.V., Dec. 2007)*
21. Draw and explain the commonly used three Op-Amp instrumentation amplifier circuits. Derive expressions for its gain. *(GBTU., 2010, 2011)*
22. Explain the working of pratical differentiator. Also derives its frequency response. *(GBTU., 2010, 2011)*
23. Draw the circuit of an Op-Amp used in an integrator. Explain its working. *(RGTU., Feb. 2010)*
24. With the help of a sketch, show how an Op-Amp can be used as a differentiator. *(GBTU., Feb. 2010)*
25. Explain the application of Op-Amp as an integrator. *(RGTU., Dec. 2009)*
26. Explain the working principle of instrumentation amplifier. *(RGTU., Dec. 2009)*
27. Draw the circuit diagram of difference amplifier using Op-Amp. Derive the expression for output voltage. *(GBTU., 2010, 2011)*
28. Explain how Op-Amp can be used as differentiator with necessary diagrams. *(RGTU., Dec. 2010)*
29. Describe integrator and differentiator with examples. *(RGTU., June 2009)*

MULTIPLE CHOICE QUESTIONS

1. The purpose of a comparator is to
   (*a*) amplify an input voltage
   (*b*) detect the occurrence of a changing input voltage
   (*c*) maintain a constant output when the dc input voltage changes
   (*d*) produce a change in output when an input voltage equals the reference voltage
2. The Op-Amp comparator circuit uses
   (*a*) positive feedback (*b*) negative feedback
   (*c*) regenerative feedback (*d*) no feedback
3. The feedback path in an Op-Amp differentiator consists of
   (*a*) a resistor (*b*) a capacitor
   (*c*) a resistor and a capacitor in series (*d*) a resistor and a capacitor in parallel
4. The feedback path in an Op-Amp integrator consists of
   (*a*) a resistor (*b*) a capacitor
   (*c*) a resistor and a capacitor in series (*d*) a resistor and a capacitor in parallel
5. For the high-pass circuit to act as a differentiator, the time constant must be
   (*a*) small
   (*b*) very small in comparison to the time period of the input signal
   (*c*) very high in comparison to the time period of the input signal
   (*d*) of modeiate value
6. In the circuit shown in Fig. 36.65.

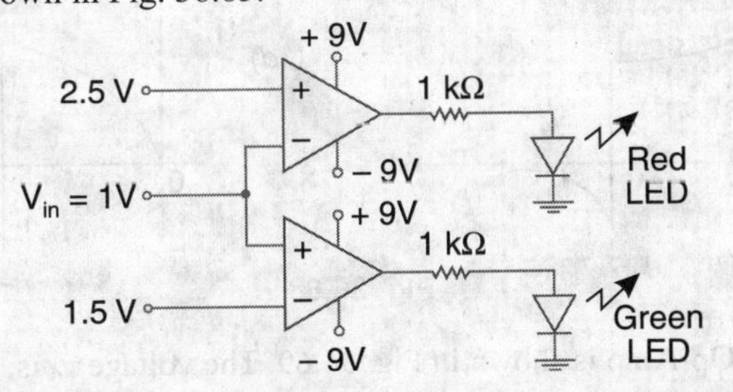

Fig. 36.65.

   (*a*) only red will glow (*b*) only green will glow
   (*c*) both red and green will glow (*d*) neither red nor green will glow
7. The low frequency gain of the low pass filter shown in Fig. 36.66. at an input frequency one decade above the cut-off is

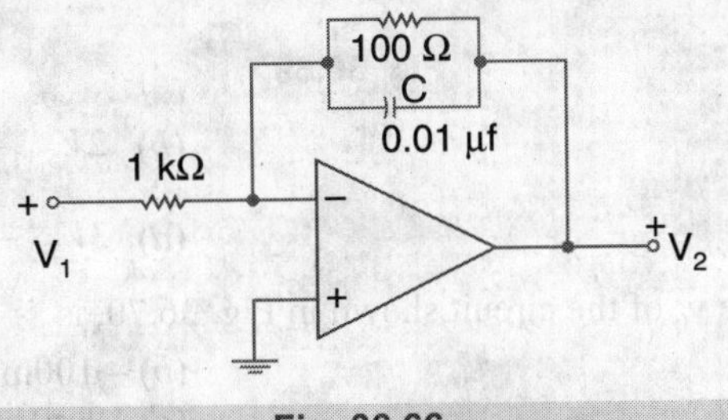

Fig. 36.66.

(*a*) 10dB
(*b*) 20 dB
(*c*) 30 dB
(*d*) 40 dB

8. A unit positive step is applied at the input of the circuit shown in Fig. 36.67. After 20 seconds, the output v_0 will be
(*a*) + 20 V
(*b*) + 10 V
(*c*) – 10 V
(*d*) – 20 V

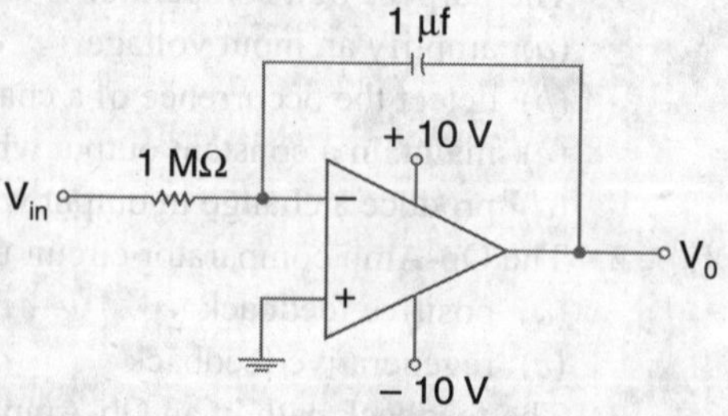

Fig. 36.67.

9. Which one of the following correctly depicts the output of a low-pass filter to a pulse input.

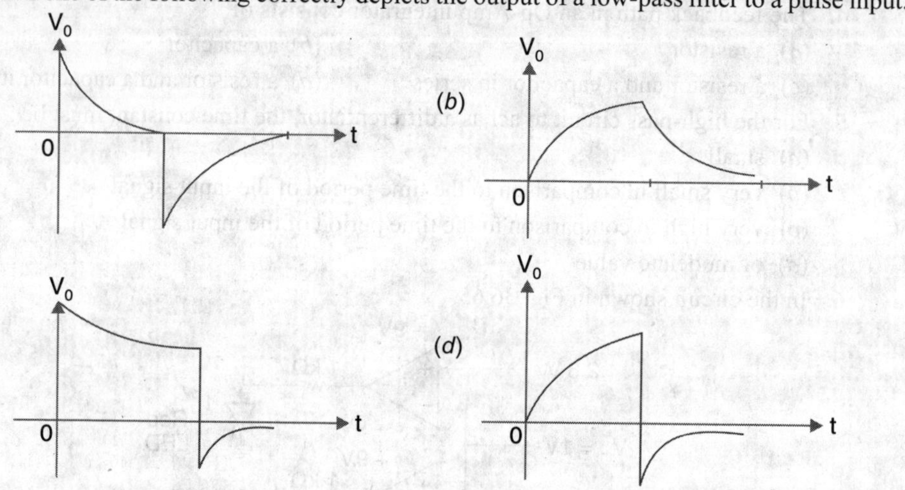

Fig. 36.68.

10. A circuit with Op-Amp is shown in Fig. 36.69. The voltage v_0 is,

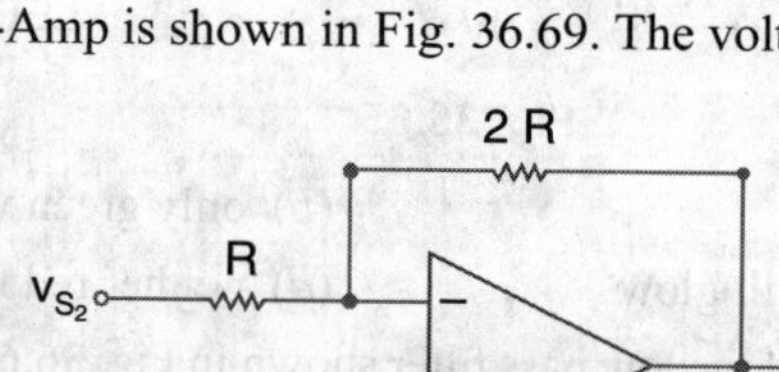

Fig. 36.69.

(*a*) $3V_{S_1} - 6V_{S_2}$
(*b*) $2V_{S_1} - 3V_{S_2}$
(*c*) $2V_{S_1} - 2V_{S_2}$
(*d*) $3V_{S_1} - 2V_{S_2}$

11. The output voltage v_0 of the circuit shown in Fig. 36.70, is
(*a*) – 100 V
(*b*) – 100mV
(*c*) 10 V
(*d*) 10 mV

(*UPSC Engg. Services, 1999*)

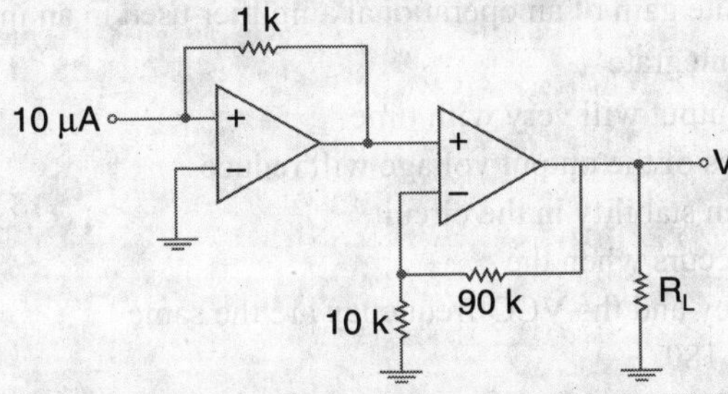

Fig. 36.70.

12. An Op-Amp circuit is shown in Fig. 36.71. The current 'I' is :
(*a*) zero (*b*) 6 mA (*c*) 10 mA (*d*) 12 mA

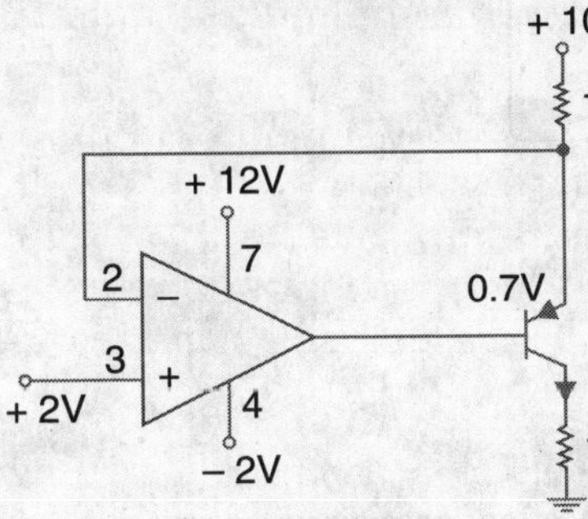

Fig. 36.71.

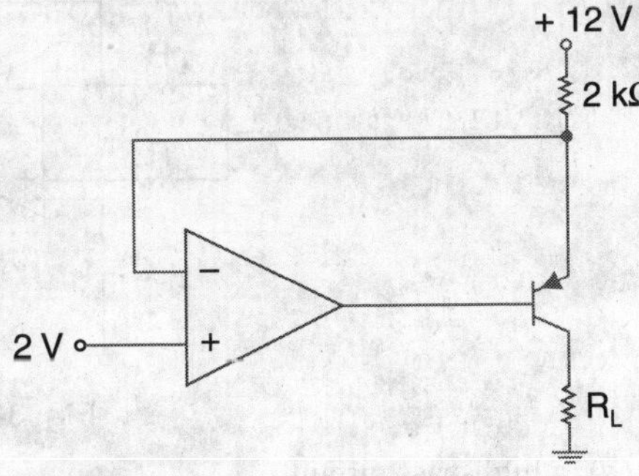

Fig. 36.72.

13. A circuit is shown in Fig 33.72. The largest value of R_L that can be used in this circuit is
(*a*) 100 Ω (*b*) 400 Ω (*c*) 2 kΩ (*d*) 20 kΩ

(*UPSC Engg. Services 2001*)

14. An Op-Amp circuit shown in Fig. 36.73, the output v_o, will be (Assume ideal Op-Amp)

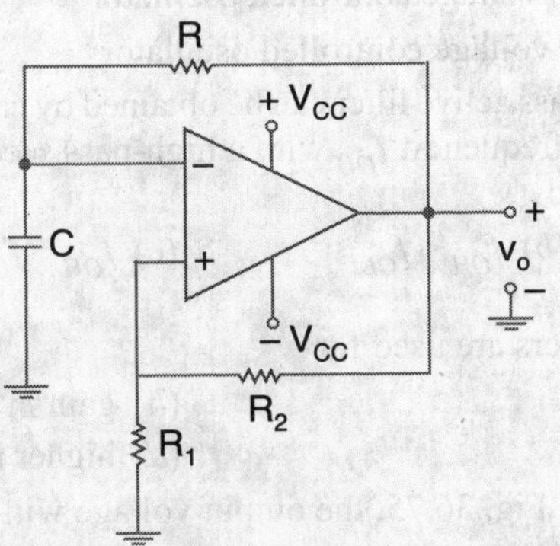

Fig. 36.73.

(*a*) equal to zero because the input is zero
(*b*) dependent on element values hence nothing can be predicated without a knowledge of element values
(*c*) a square wave varying between + V_{CC} and − V_{CC}.
(*d*) a simusodial wave of amplifier V_{CC}.

(*UPSC Engg. Services, 2002*)

15. The effect of a finite gain of an operational amplifier used in an integrator is that
(*a*) it would not integrate
(*b*) the slope of output will very with time
(*c*) the final vlaues of the output voltage will reduce
(*d*) there will be in stability in the circuit. (*UPSC Engg. Services 2001*)

16. In an PLL, lock occurs when the
(*a*) input frequency and the VCO frequency are the same
(*b*) phase error is 180°
(*c*) VCO frequency is double the input frequency.
(*d*) phase error is 90°. (*UPSC Engg Services, 2001*)

17. The Op-Amp circuit shown in Fig 36.74. is,

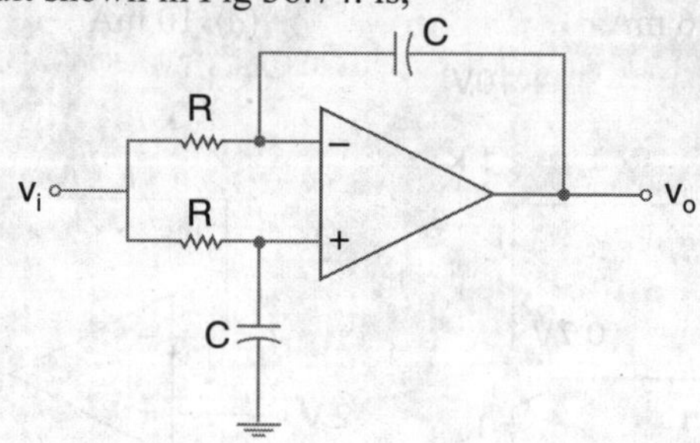

Fig. 36.74.

(*a*) a high-pass circuit (*b*) a low-pass circuit
(*c*) a band-pass circuit (*d*) an all-pass circuit
(*UPSC Engg. Services 2000*)

18. The essential blocks of a phase-locked loop (PLL) are phase detector, amplifier,
(*a*) high-pass filter and crystal control oscillator
(*b*) low-pass filter and crystal controlled oscillator
(*c*) high-pass filter and voltage controlled oscillator
(*d*) low-pass filter and voltage controlled oscillator (*UPSC Engg Services 2000*)

19. A second-order band-pass active filter can be obtained by cascading a low-pass second-order section having cut-off frequency, f_{OH} with a high-pass second-order section having cut-off frequency f_{OL}, provided
(*a*) $f_{OH} > f_{OL}$ (*b*) $f_{OH} < f_{OL}$ (*c*) $f_{OH} = f_{OL}$ (*d*) $f_{OH} \leq \frac{1}{2} f_{OL}$
(*UPSC Engg Services 1999*)

20. Higher order active filters are used for
(*a*) bandwidth (*b*) gain in the passband
(*c*) impedance (*d*) higher roll-off rate

21. In the circuit shown in Fig. 36.75, the output voltage will be
(*a*) 9 V (*b*) 10 V (*c*) 11 V (*d*) 12 V

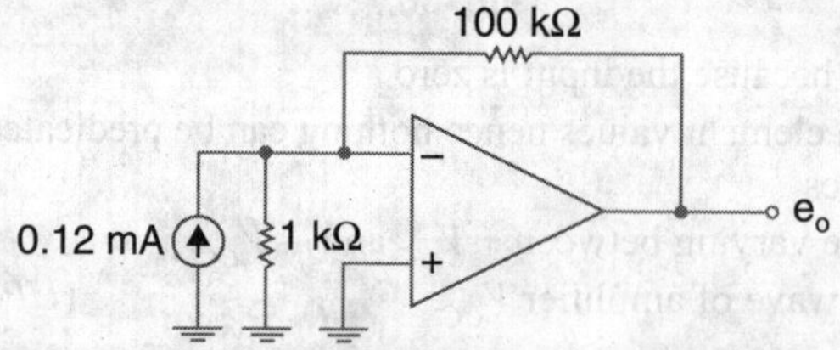

Fig. 36.75.

22. An Op-Amp is used in the circuit as shown in Fig. 36.76. The current I_O is,

(*a*) $V_{in} \times \dfrac{R_L}{R_S\,(R_L + R_S)}$ (*b*) $\dfrac{V_{in}}{R_S}$

(*c*) $\dfrac{V_{in}}{R_L}$ (*d*) $V_{in}\left(\dfrac{1}{R_S} + \dfrac{1}{R_L}\right)$

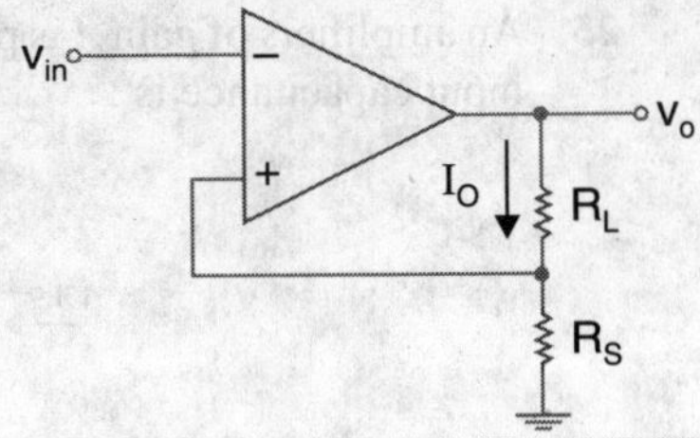

Fig. 36.76.

Hint. $V_o = V_{in} + I_o \,.\, R_L$

$\therefore$ $I_o = \dfrac{V_o - V_o}{R_L},$

But $A_V = 1 + \dfrac{R_L}{R_S} = \dfrac{V_o}{V_{in}}$

$\therefore$ $V_o = A_V \,.\, V_{in}$

and $I_o = \dfrac{(A_V - 1)\,V_{in}}{R_L} \approx \dfrac{A_v \cdot V_{in}}{R_L} = \dfrac{V_{in}}{R_S \cdot R_L}(R_S + R_L)$

23. In the Op-Amp circuit shown in Fig. 36.77. $V_i > O$ and $i\ I_o$. $e^{-\alpha v}$ the output, v_o will be proportional to,

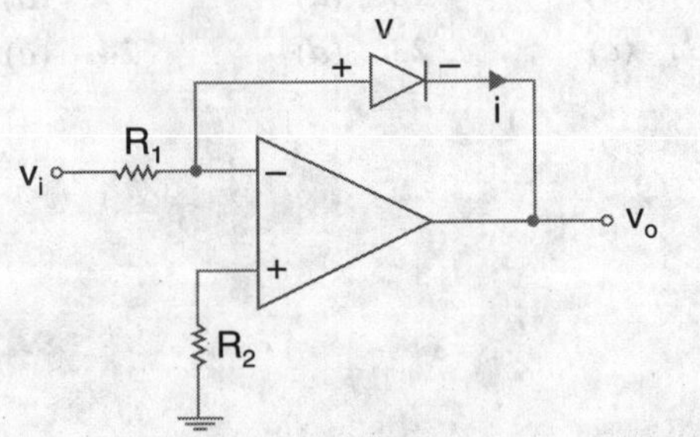

Fig. 36.77.

(*a*) $\sqrt{v_i}$ (*b*) v_i

(*c*) e^{kv_i} (*d*) $l_n\,(kv_i)$

24. In the circuit shown in Fig 36.78. the current through the resistance *R* is :

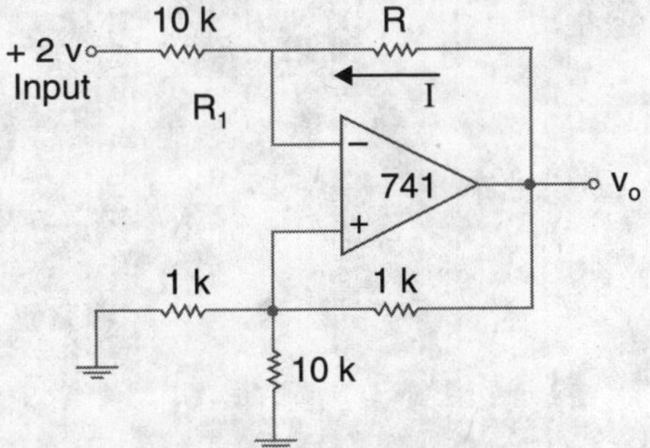

Fig. 36.78.

(*a*) 100 μA (*b*) – 100 μA (*c*) 1 μA (*d*) – 1 μA.

(*UPSC Engg. Services 2001*)

25. An amplifiers of gain *A* is pridged by a capacitance '*C*' as shown in Fig. 36.79. The effective input capacitance is :

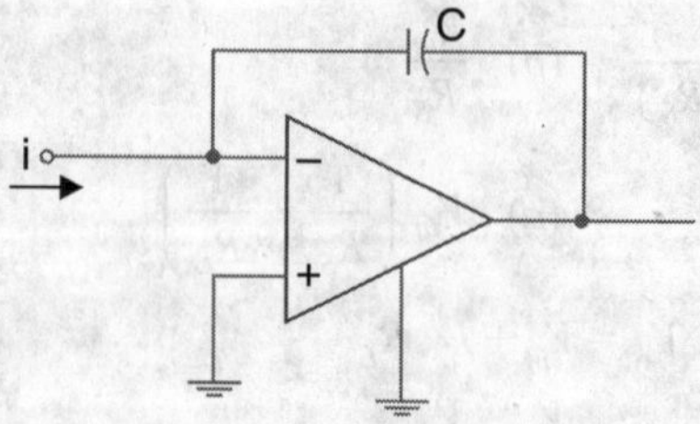

Fig. 36.79.

(*a*) *C*/*A* (*b*) *C* (1 – *A*) (*c*) *C* (1 + *A*) (*d*) *C A*

ANSWERS

1. (*c*)	**2.** (*d*)	**3.** (*a*)	**4.** (*b*)	**5.** (*b*)
6. (*a*)	**7.** (*b*)	**8.** (*c*)	**9.** (*b*)	**10.** (*d*)
11. (*d*)	**12.** (*a*)	**13.** (*b*)	**14.** (*c*)	**15.** (*a*)
16. (*a*)	**17.** (*d*)	**18.** (*d*)	**19.** (*a*)	**20.** (*d*)
21. (*d*)	**22.** (*c*)	**23.** (*d*)	**24.** (*c*)	**25.** (*a*)

APPENDIX

APPENDIX-A

Resistor and Capacitor Values

TYPICAL STANDARD RESISTOR VALUES

Ω	Ω	Ω	kΩ	kΩ	kΩ	MΩ	MΩ
—	10	100	1	10	100	1	10
—	12	120	1.2	12	120	1.2	—
—	15	150	1.5	15	150	1.5	15
—	18	180	1.8	18	180	1.8	—
—	22	220	2.2	22	220	2.2	22
2.7	27	270	2.7	27	270	2.7	—
3.3	33	330	3.3	33	330	3.3	—
3.9	39	390	3.9	39	390	3.9	—
4.7	47	470	4.7	47	470	4.7	—
5.6	56	560	5.6	56	560	5.6	—
6.8	68	680	6.8	68	680	6.8	—
—	82	820	8.2	82	820	—	—

TYPICAL STANDARD RESISTOR VALUES

pF	*pF*	*pF*	*pF*	μ*F*	μ*F*	μ*F*	μ*F*	μ*F*	μ*F*	μ*F*
5	50	500	5000		0.05	0.5	5	50	500	5000
—	51	510	5100		—	—	—	—	—	—
—	56	560	5600		0.056	0.56	5.6	56	—	5600
—	—	—	6000		0.06	—	6	—	—	6000
—	62	620	6200		—	—	—	—	—	—
—	68	680	6800		0.068	0.68	6.8	—	—	—
—	75	750	7500		—	—	—	75	—	—
—	—	—	8000		—	—	8	80	—	—
—	82	820	8200		0.082	0.82	8.2	82	—	—
—	91	910	9100		—	—	—	—	—	—
10	100	1000		0.01	0.1	1	10	100	1000	10000
—	110	1100		—	—	—	—	—	—	—
12	120	1200		0.012	0.12	1.2	—	—	—	
—	130	1300		—	—	—	—	—	—	—
15	150	1500		0.015	0.15	1.5	15	150	1500	
—	160	1600		—	—	—	—	—	—	
18	180	1800		0.018	0.18	1.8	18	180	—	
20	200	2000		0.02	0.2	2	20	200	2000	
22	220	2200		—	0.22	2.2	22	—	—	
24	240	2400		—	—	—	—	240	—	
—	250	2500		—	0.25	—	25	250	2500	
27	270	2700		0.027	0.27	2.7	27	270	—	
30	300	3000		0.03	0.3	3	20	300	3000	
33	330	3300		0.033	0.33	3.3	33	330	3300	
36	360	3600		—	—	—	—	—	—	
39	390	3900		0.039	0.39	3.9	39	—	—	
—	—	4000		—	—	4	—	400	—	
43	430	4300		—	—	—	—	—	—	
47	470	4700		0.047	0.47	4.7	47	—	—	

APPENDIX-B

Silicon Pin Diode

These devices are designed primarily for VHF band switching applications but are also suitable for use in general–purpose switching circuits. They are supplied in a cost–effective plastic package for economical, high–volume consumer and industrial requirements. They are also available in surface mount.

- Long Reverse Recovery Time
  t_{rr} = 300 ns (Typ)
- Rugged PIN Structure Coupled with Wirebond Construction for Optimum Reliability
- Low Series Resistance @ 100 MHz —
  R_S = 0.7 Ohms (Typ) @ I_F = 10 mAdc
- Reverse Breakdown Voltage = 200 V (Min)

Specifications Sheets of Selected Electronics Devices

MMBV3700LT1

SILICON PIN SWITCHING DIODE

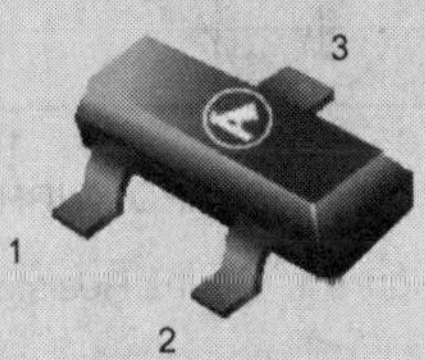

CASE 318–08, STYLES
SOT–23 (TO–236AB)

MAXIMUM RATINGS(EACH DIODE)

Rating	Symbol	MV21XX	MMBV21XXLT1	Unit
Reverse Voltage	V_R	200		Vdc
Device Dissipation @T_A = 25°C	P_D	280	200	mW
Derate above 25°C		2.8	2.0	mW/°C
Junction Temperature	T_J	+150		°C
Storage Temperature Range	T_{stg}	–55 to +150		°C

DEVICE MARKING

MMBV3700LT1=4R

ELECTRICAL CHARACTERISTICS (TA=25°C unless otherwise noted)

Characteristic	Symbol	Min	Typ	Max	Unit
Reverse Breakdown Voltage (I_R = 10μ Adc)	$V_{(BR)R}$	200	—	—	Vdc
Diode Capacitance (V_R = 20 Vdc,f=1.0 MHz)	C_T	—	—	1.0	pF
Series Resistance(figure5) (I_F = 10mAdc)	R_S	—	0.7	1.0	Ω
Reverse Leakage Current (V_R = 150Vdc)	I_R	—	—	0.1	μAdc
Reverse Recovery Time (I_F=I_R=10mAdc)	t_{rr}	—	300	—	ns

MMBV3700LT1 MPN3700

TYPICAL CHARACTERISTICS

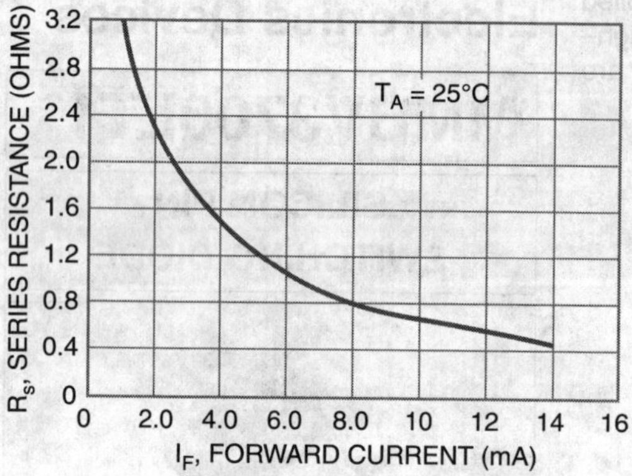

Figure 1. Series Resistance

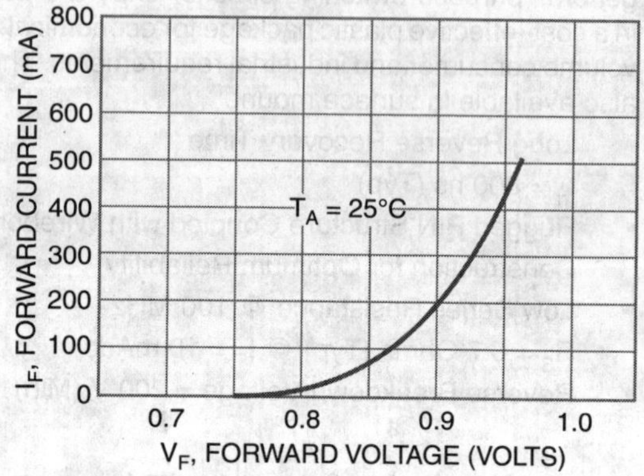

Figure 2. Forward Voltage

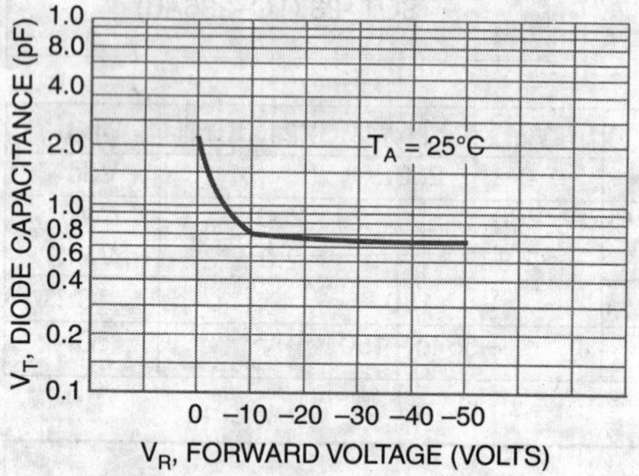

Figure 3. Diode Capacitance

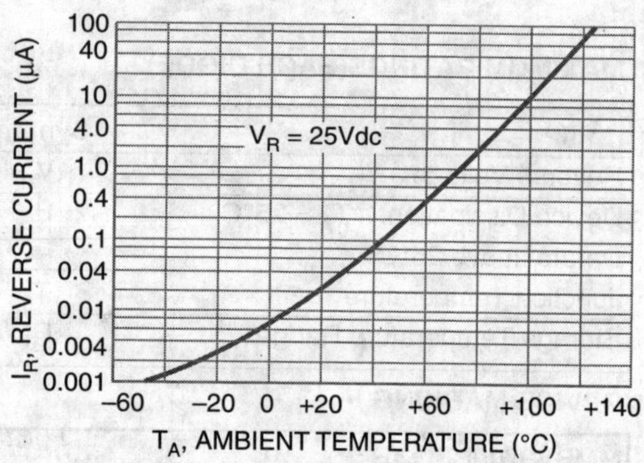

Figure 4. Leakage Current

VHF variable capacitance diode BB131

FEATURES

- Excellent linearity
- Very small plastic SMD package
- C28: 1 pF; ratio: 14.

APPLICATIONS

- Electronic tuning in satellite tuners
- Tunable coupling
- VCO.

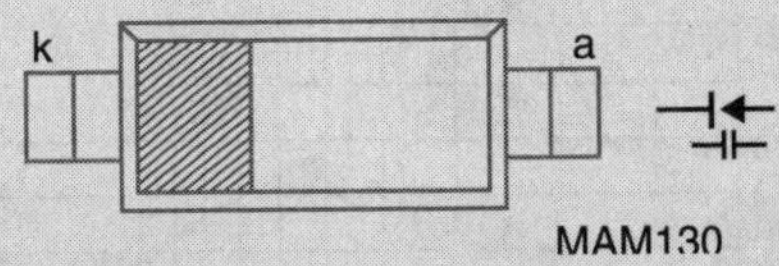

Marking code: P1.

Cathode side indicated by a bar.

Fig. 1. Simplified outine (SOD323) and symbol.

DESCRIPTION

The BB131 is a variable capacitance diode, fabricated in planar technology, and encapsulated in the SOD323 very small plastic SMD package.

LIMITING VALUES

In accordance with the Absolute Maximum Rating System (IEC 134)

SYMBOL	PARAMETER	MIN.	MAX.	UNIT
V_R	continuous reverse voltage	–	30	V
I_F	continuous forward current	–	20	mA
T_{stg}	storage temperature	–55	+150	°C
T_j	operating junction temperature	–55	+125	°C

ELECTRICAL CHARACTERISTICS

T_j = 25°C unless otherwise specified.

SYMBOL	PARAMETER	CONDITIONS	MIN.	MAX.	UNIT
I_R	reverse current	V_R = 30 V; see Fig.3	–	10	nA
		V_R = 30 V; T_j = 85 °C; see Fig.3	–	200	nA
r_s	diode series resistance	f = 470 MHz; note 1	–	3	Ω
C_d	diode capacitance	V_R = 0.5 V; f = 1 MHz; see Figs 2 and 4	8	17	pF
		V_R = 28 V; f = 1 MHz; see Figs 2 and 4	0.7	1.055	pF
$\frac{C_{d(0.5V)}}{C_{d(28V)}}$	capacitance ratio	f = 1 MHz	12	16	

Note

1. VR is the value at which Cd = 9 pF.

VHF variable capacitance diode BB131

GRAPHICAL DATA

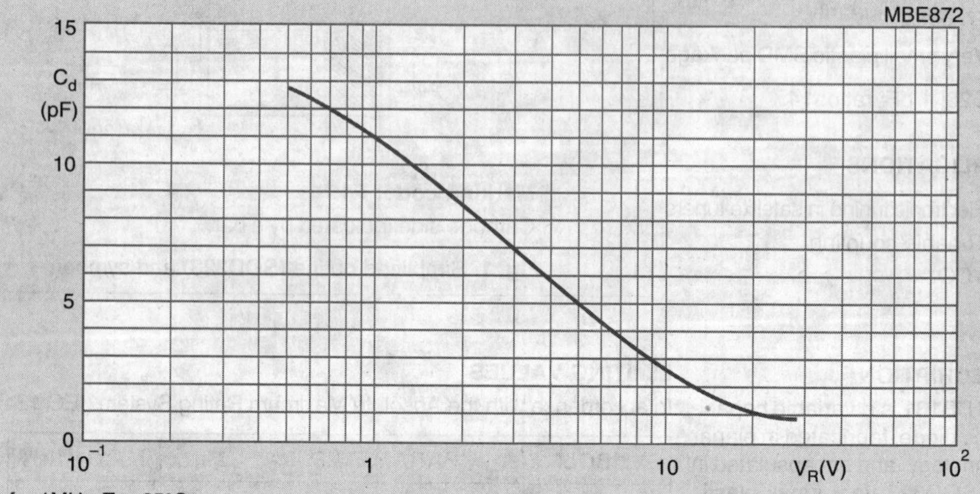

f = 1MHz; T_j = 25°C.

Fig. 2. Diode capacitance as a function of reverse voltage; typical values.

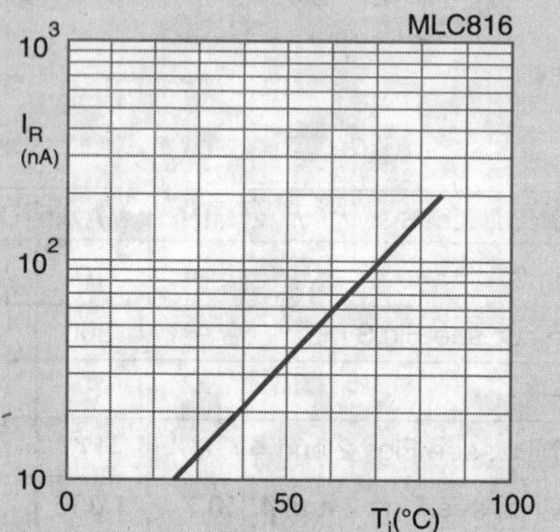

Fig. 3. Reverse current as a function of junction temperature; maximum values.

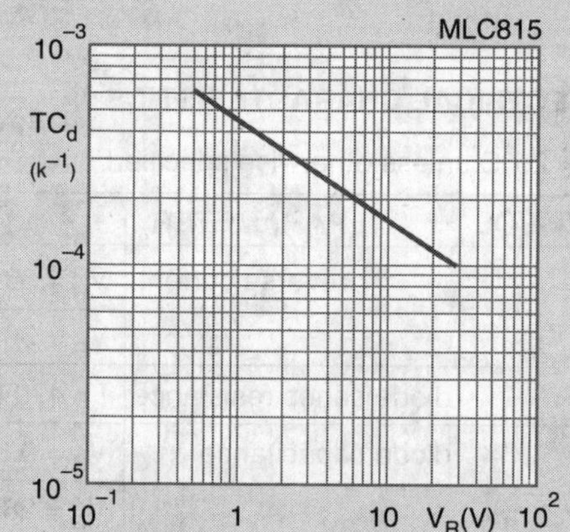

T_j = 0 to 85 °C.

Fig. 4. Temperature coefficient of diode capacitance as a function of reverse voltage; typical values.

FAIRCHILD
SEMICONDUCTOR ™

PN2222A MMBT2222A PZT2222A

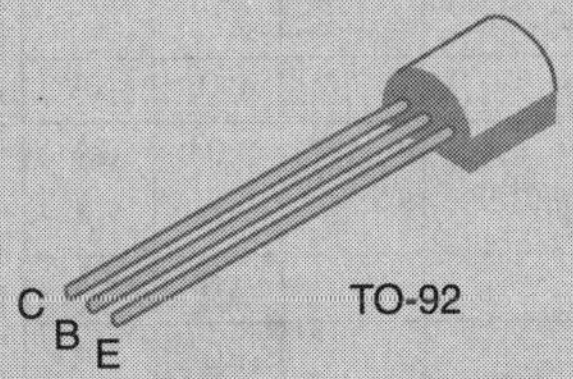

TO-92

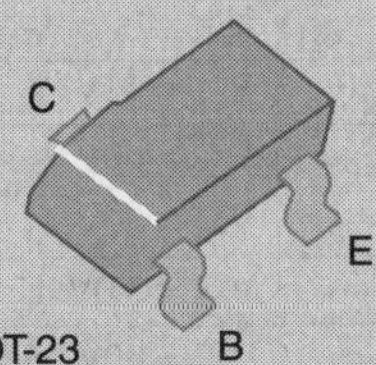

SOT-23
Mark:1P

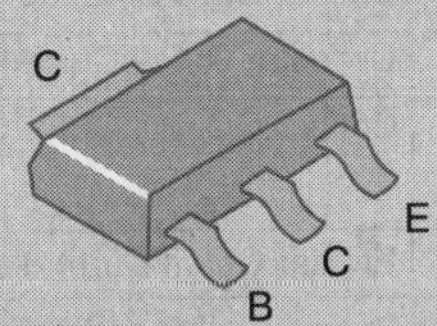

SOT-223

NPN General Purpose Amplifier

This device is for use as a medium power amplifier and switch requiring collector currents up to 500mA.
Sourced from process 19.

Absolute Maximum Ratings * T_A=25°C unless otherwise noted

Symbol	Parameter	Value	Units
V_{CEO}	Collector-Emitter Voltage	40	V
V_{CBO}	Collector-Base Voltage	75	V
V_{EBO}	Emitter-Base Voltage	6.0	V
I_C	Collector Current	1.0	A
T_J, T_{Stg}	Operating and Storage Junction Temperature Range	–55 to + 150	°C

* These ratings are limiting values above which the serviceability of any semiconductor device may be impaired

NOTES:

(1) These ratings are based on a maximum junction temperature of 150 degrees C.

(2) These are steady state limits. The factory should be consulted on applications involving pulsed or low duty cycle operations

Thermal Characteristics T_A=25°C unless otherwise noted

Symbol	Characteristic	Max			Units
		PN2222A	*MMBT2222A	**PZT2222A	
P_D	Total Device Dissipation 625 Derate above 25°C	350 5.0	1,000 2.8	mW 8.0	 mW/°C
$R_{\theta JC}$	Thermal Resistance, Junction to Case	83.3			°C/W
$R_{\theta JA}$	Thermal Resistance, Junction to Ambient	200	357	125	°C/W

* Device mounted on FR-4 PCB 1.6" × 1.6" × 0.06".

** Device mounted on FR-4 PCB 36mm × 18mm × 1.5mm; mounting pad for the collector lead min. 6cm^2.

PN2222A / MMBT2222A / PZT2222A

NPN General Purpose Amplifier

(continued)

Electrical Characteristics $T_A = 25°C$ unless otherwise noted

Symbol	Parameter	Test Conditions	Min	Max	Units
OFF CHARACTERISTICS					
$V_{(BR)CEO}$	Collector-Emitter Breakdown Voltage*	$I_C = 10$ mA, $I_B = 0$	40		V
$V_{(BR)CBO}$	Collector-Base Breakdown Voltage	$I_C = 10$ µA, $I_E = 0$	75		V
$V_{(BR)EBO}$	Emitter-Base Breakdown Voltage	$I_E = 10$ µA, $I_C = 0$ 6.0 V			
I_{CEX}	Collector Cutoff Current	$V_{CE} = 60V$, $V_{EB(OFF)} = 3.0V$		10	nA
I_{CBO}	Collector Cutoff Current	$V_{CB} = 60V$, $I_E = 0$		0.01	µA
		$V_{CB} = 60V$, $I_E = 0$, $T_A = 150°C$		10	µA
I_{EBO}	Emitter Cutoff Current	$V_{EB} = 3.0V$, $I_C = 0$		10	nA
I_{BL}	Base Cutoff Current	$V_{CE} = 60V$, $V_{EB(OFF)} = 3.0V$		20	nA
ON CHARACTERISTICS					
h_{FE}	DC Current Gain	$I_C = 0.1mA$, $V_{CE} = 10V$	35		
		$I_C = 1.0mA$, $V_{CE} = 10V$	50		
		$I_C = 10mA$, $V_{CE} = 10V$	75		
		$I_C = 10mA$, $V_{CE} = 10V$, $T_A = -55°C$ 35			
		$I_C = 150mA$, $V_{CE} = 10V$ *	100	300	
		$I_C = 150mA$, $V_{CE} = 10V$ *	50		
		$I_C = 500mA$, $V_{CE} = 10V$ *	40		
$V_{CE(sat)}$	Collector-Emitter Saturation Voltage*	$I_C = 150mA$, $I_B = 15V$		0.3	V
		$I_C = 500mA$, $I_B = 50V$		1.0	V
$V_{BE(sat)}$	Base-Emitter Saturation Voltage*	$I_C = 150mA$, $I_B = 10V$	0.6	1.2	V
		$I_C = 500mA$, $I_B = 10V$		2.0	V
SMALL SIGNAL CHARACTERISTICS					
f_T	Current Gain Bandwidth Product	$I_C = 20mA$, $V_{CE} = 20V$, f = 100MHz	300		MHz
C_{obo}	Output Capacitance	$V_{CB} = 10V$, $I_E = 0$, f = 1MHz		8.0	pF
C_{ibo}	Input Capacitance	$V_{EB} = 0.5V$, $I_C = 0$, f = 1MHz		25	pF
rb'Cc	Collector Base Time Constant	$I_C = 20$ mA, $V_{CB} = 20V$, f = 31.8MHz	150	pS	
NF	Noise Figure	$I_C = 100$ µA, $V_{CE} = 10V$, $R_S = 1.0$ KΩ, f = 1.0KHz		4.0	dB
Re(hie)	Real Part of Common-Emitter High Frequency Input Impedance	$I_C = 20mA$, $V_{CE} = 20V$, f = 300MHz		60	Ω
SWITCHING CHARACTERISTICS					
t_d	Delay Time	$V_{CC} = 30V$, $V_{EB(OFF)} = 0.5V$,		10	ns
t_r	Rise Time	$I_C = 150mA$, $I_{B1} = 15mA$		25	ns
t_s	Storage Time	$V_{CC} = 30V$, $I_C = 150mA$,		225	ns
t_f	Fall Time	$I_{B1} = I_{B2} = 15mA$		60	ns

* Pulse Test: Pulse Width ≤ 300 ms, Duity Cycle ≤ 2.0%

SPICE MODEL

NPN (Is = 14.34f Xti = 3 Eg = 1.11 Vaf = 74.03 Bf = 255.9 Ne = 1.307 Ise = 14.34 Ikf = .2847 Xtb = 1.5 Br = 6.092 Isc=0 Ikr = 0 Rc = 1 Cjc = 7.306p Mjc = .3416 Vjc = .75 Fc = .5 Cje = 22.01p Mje = .377 Vje = .75 Tr = 46.91n Tf = 411.1p Itf = .6 Vtf = 1.7 Xtf = 3 Rb = 10).

NPN General Purpose Amplifier
(continued)

Typical Characteristics

Typical Pulsed Current Gain vs Collector Current

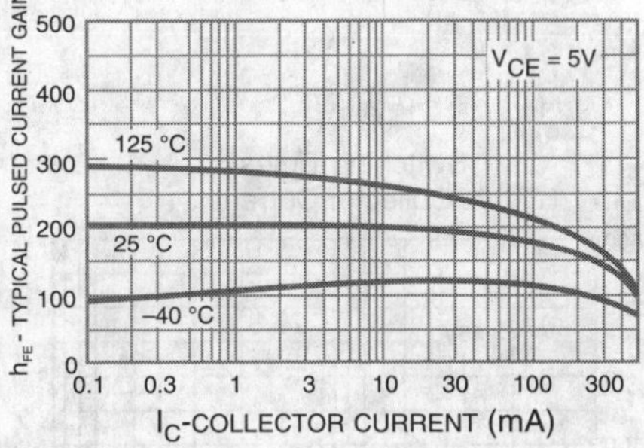

Collector-Emitter Saturation Voltage vs Collector Current

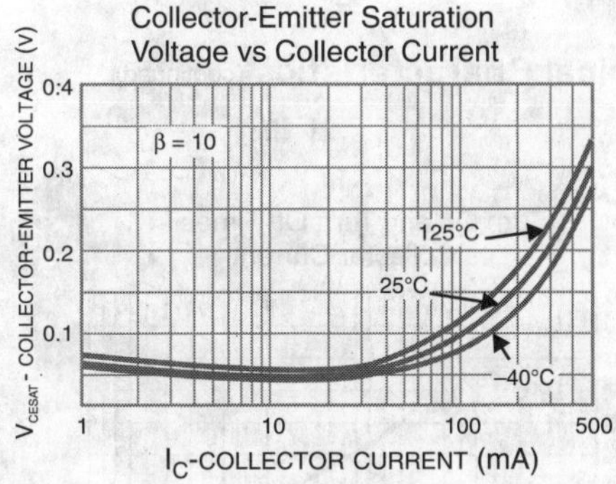

Base-Emitter Saturation Voltage vs Collector Current

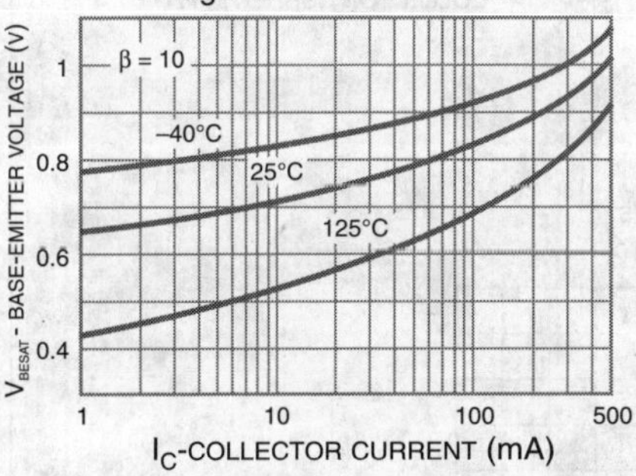

Base-Emitter ON Voltage vs Collector Current

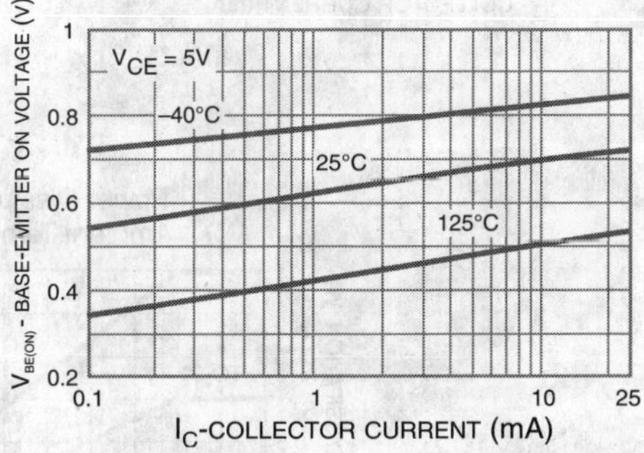

Collector-Cutoff Current vs Ambient Temperature

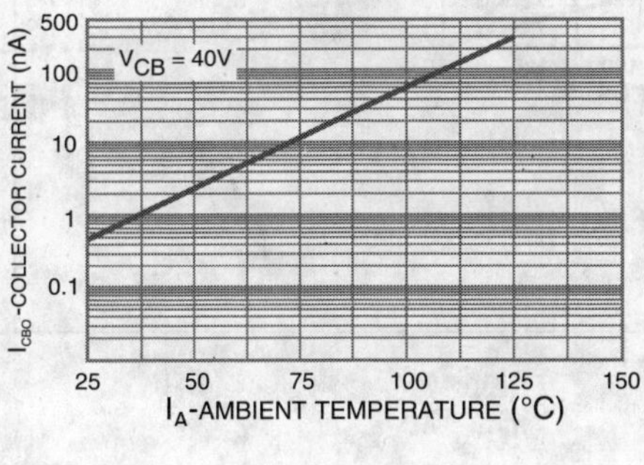

Emitter Transition and Output Capacitance vs Reverse Bias Voltage

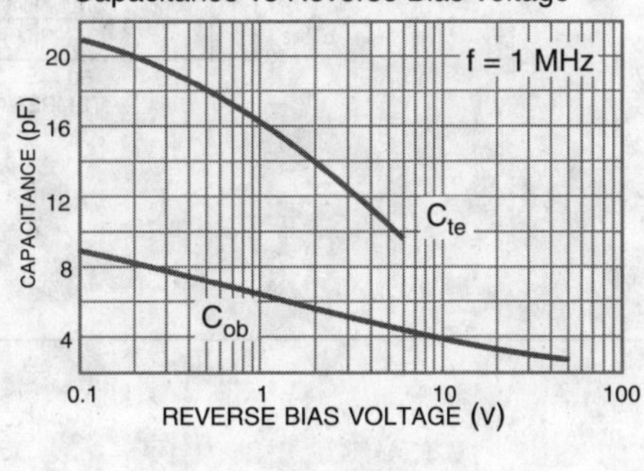

PN2222A / MMBT2222A / PZT2222A

NPN General Purpose Amplifier
(continued)

Typical Characteristics (continued)

Turn On and Turn Off Times vs Collector Current

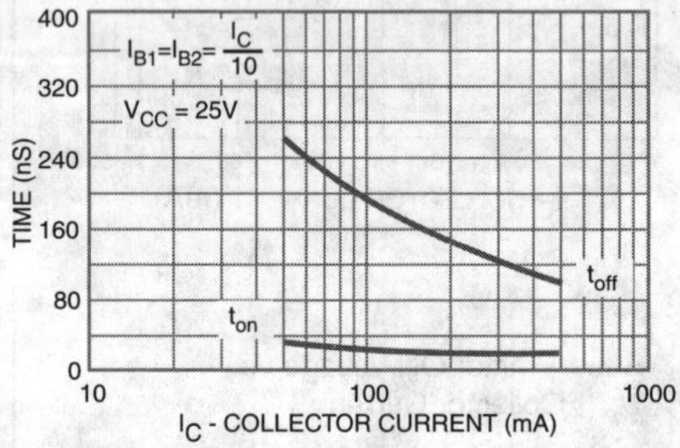

Switching Times vs Collector Current

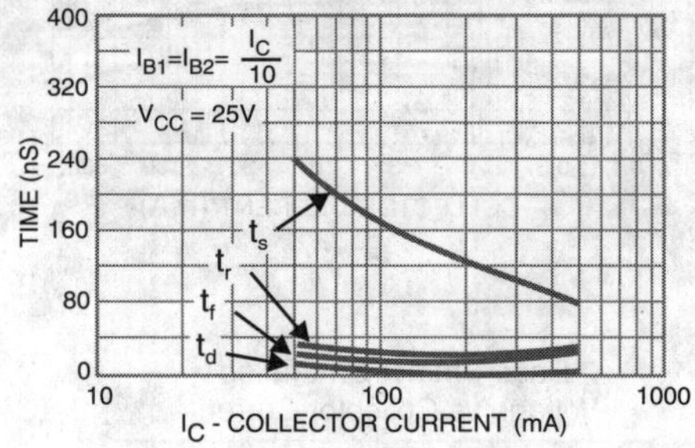

Power Dissipation vs Ambient Temperature

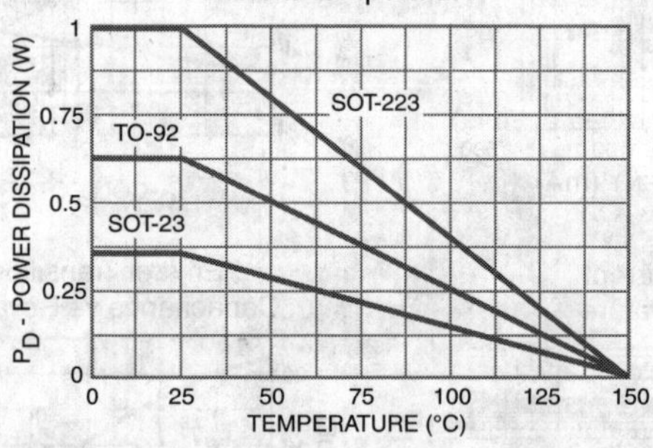

NPN General Purpose Amplifier
(continued)

Typical Common Emitter Characteristics (f = 1.0 kHz)

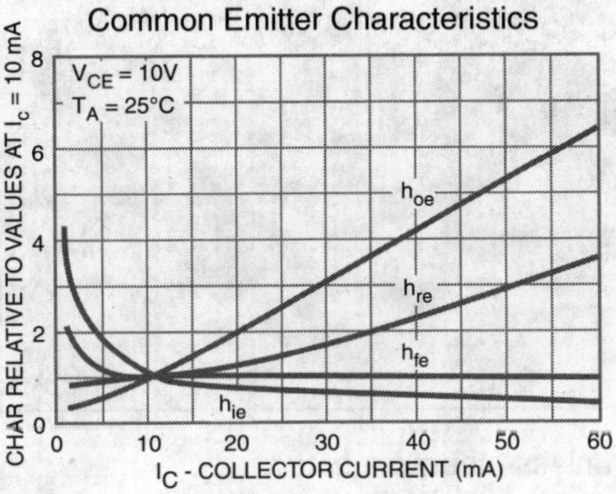

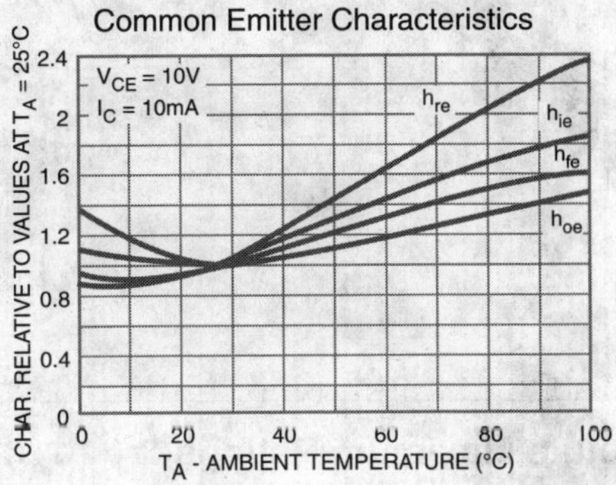

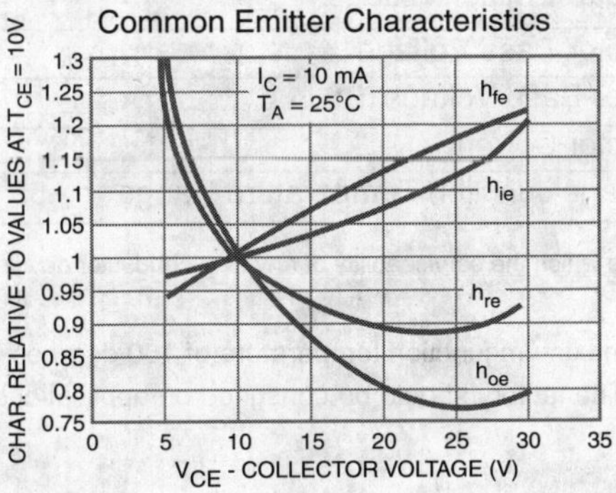

FAIRCHILD
SEMICONDUCTOR™

2N3904 / MMBT3904 / PZT3904

2N3904 MMBT3904 PZT3904

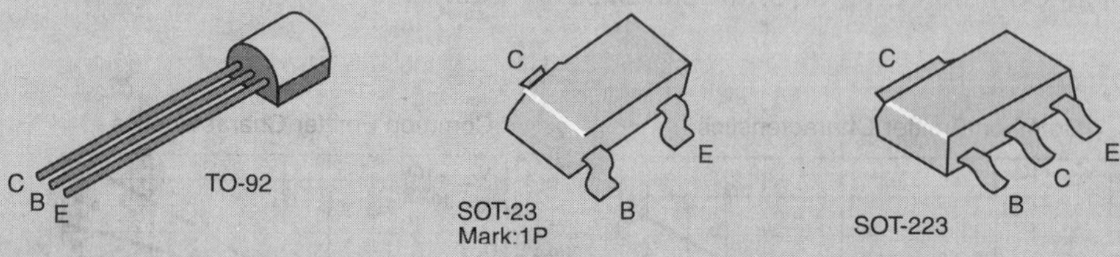

Absolute Maximum Ratings * T_A=25°C unless otherwise noted

Symbol	Parameter	Value	Units
V_{CEO}	Collector-Emitter Voltage	40	V
V_{CBO}	Collector-Base Voltage	75	V
V_{EBO}	Emitter-Base Voltage	6.0	V
I_C	Collector Current	1.0	A
T_J, T_{Stg}	Operating and Storage Junction Temperature Range	–55 to + 150	°C

* These ratings are limiting values above which the serviceability of any semiconductor device may be impaired

NOTES:

(1) These ratings are based on a maximum junction temperature of 150 degrees C.

(2) These are steady state limits. The factory should be consulted on applications involving pulsed or low duty cycle operations

Thermal Characteristics T_A=25°C unless otherwise noted

Symbol	Characteristic	Max			Units
		2N3904	*MMBT3904	**PZT3904	
P_D	Total Device Dissipation	625	350	1,000	mW
	Derate above 25°C	5.0	2.8	8.0	mW/°C
$R_{\theta JC}$	Thermal Resistance, Junction to Case	83.3			°C/W
$R_{\theta JA}$	Thermal Resistance, Junction to Ambient	200	357	125	°C/W

* Device mounted on FR-4 PCB 1.6" × 1.6" × 0.06".

** Device mounted on FR-4 PCB 36mm × 18mm × 1.5mm; mounting pad for the collector lead min. $6cm^2$.

PN2222A / MMBT2222A / PZT2222A

NPN General Purpose Amplifier

(continued)

Electrical Characteristics

$T_A = 25°C$ unless otherwise noted

Symbol	Parameter	Test Conditions	Min	Max	Units
OFF CHARACTERISTICS					
$V_{(BR)CEO}$	Collector-Emitter Breakdown Voltage*	$I_C = 10$ mA, $I_B = 0$	40		V
$V_{(BR)CBO}$	Collector-Base Breakdown Voltage	$I_C = 10\ \mu$A, $I_E = 0$	60		V
$V_{(BR)EBO}$	Emitter-Base Breakdown Voltage	$I_E = 10\ \mu$A, $I_C = 0$	6.0 V		
I_{BL}	Base Cutoff Current	$V_{CE} = 30V$, $V_{EB} = 3.0V$		50	nA
I_{CEX}	Collector Cutoff Current	$V_{CE} = 30V$, $V_{EB} = 3V$		50	nA
ON CHARACTERISTICS*					
h_{FE}	DC Current Gain	$I_C = 0.1$ mA, $V_{CE} = 1.0V$	40		
		$I_C = 1.0$ mA, $V_{CE} = 1.0V$	70		
		$I_C = 10$ mA, $V_{CE} = 1.0V$	100	300	
		$I_C = 50$ mA, $V_{CE} = 1.0V$	60		
		$I_C = 100$ mA, $V_{CE} = 1.0V$	30		
$V_{CE(sat)}$	Collector-Emitter Saturation Voltage	$I_C = 10$ mA, $I_B = 1.0mV$		0.2	V
		$I_C = 10$ mA, $I_B = 5.0V$		0.3	V
$V_{BE(sat)}$	Base-Emitter Saturation Voltage*	$I_C = 10$ mA, $I_B = 1.0V$	0.65	0.85	V
		$I_C = 500$ mA, $I_B = 10V$		0.95	V
SMALL SIGNAL CHARACTERISTICS					
f_T	Current Gain Bandwidth Product	$I_C = 10$ mA, $V_{CE} = 20V$ f = 100 MHz	300		MHz
C_{obo}	Output Capacitance	$V_{CB} = 5.0V$, $I_E = 0$, f = 1.0 MHz		4.0	pF
C_{ibo}	Input Capacitance	$V_{EB} = 0.5V$, $I_C = 0$, f = 1.0 MHz		8.0	pF
NF	Noise Figure	$I_C = 100\ \mu$A, $V_{CE} = 5.0V$,		5.0	dB
		$R_S = 1.0\ k\Omega$, f = 10 Hz to 15.7KHz			
SWITCHING CHARACTERISTICS					
t_d	Delay Time	$V_{CC} = 3.0V$, $V_{EB} = 0.5V$,		35	ns
t_r	Rise Time	$I_C = 10mA$, $I_{B1} = 1.0mA$		35	ns
t_s	Storage Time	$V_{CC} = 3.0V$, $I_C = 10mA$,		200	ns
t_f	Fall Time	$I_{B1} = I_{B2} = 1.0mA$		50	ns

* Pulse Test: Pulse Width ≤ 300 ms, Duity Cycle ≤ 2.0%

SPICE MODEL

NPN (Is = 6.734f Xti = 3 Eg = 1.11 Vaf = 74.03 Bf = 416.4 Ne = 1.259 Ise = 6.734 Ikf = 66.78 Xtb = 1.5 Br = .7371 Nc = 2 Isc = 0 Ikr = 0 Rc = 1 Cjc = 3.638p Mjc = .3085 Vjc = .75 Fc = .5 Cje = 4.493p Mje = .2593 Vje = .75 Tr = 239.5n Tf = 301.2p Itf = .4 Vtf = 4 Xtf = 2 Rb = 10).

NPN General Purpose Amplifier

(continued)

2N3904 / MMBT3904 / PZT3904

Typical Characteristics

Typical Pulsed Current Gain vs Collector Current

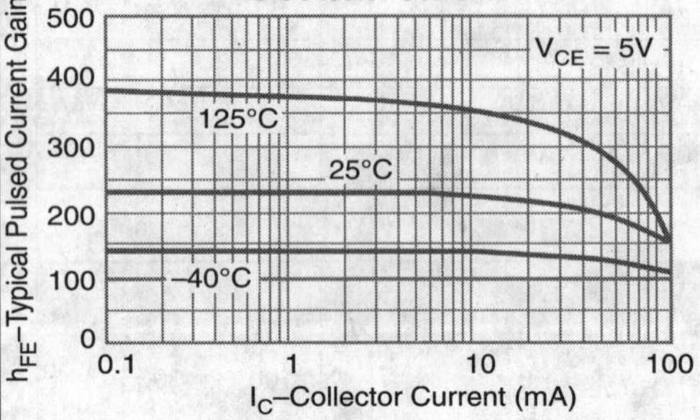

Base-Emitter Saturation Voltage vs Collector Current

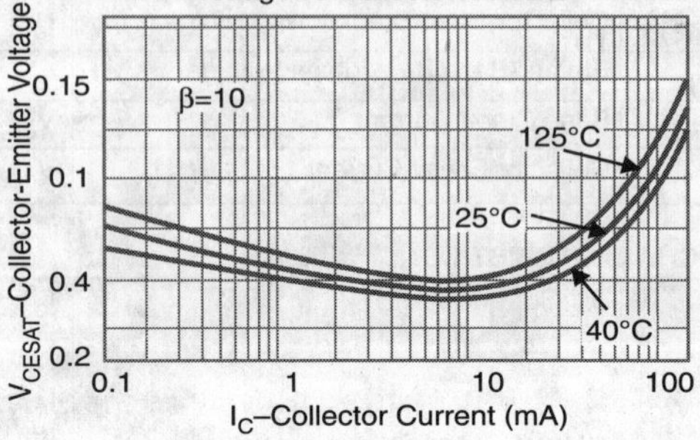

Base-Emitter Saturation Voltage vs Collector Current

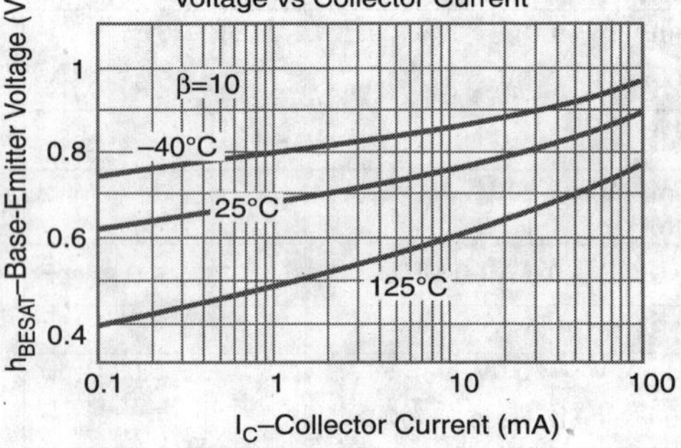

Base-Emitter ON Voltage Collector Current

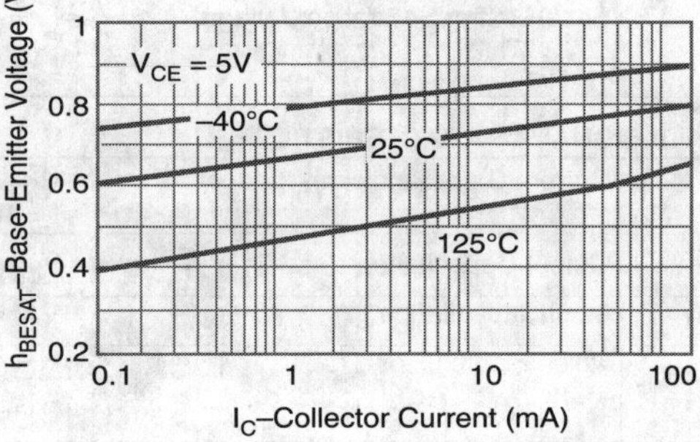

Collector-Cutoff Current Vs Ambient Temperature

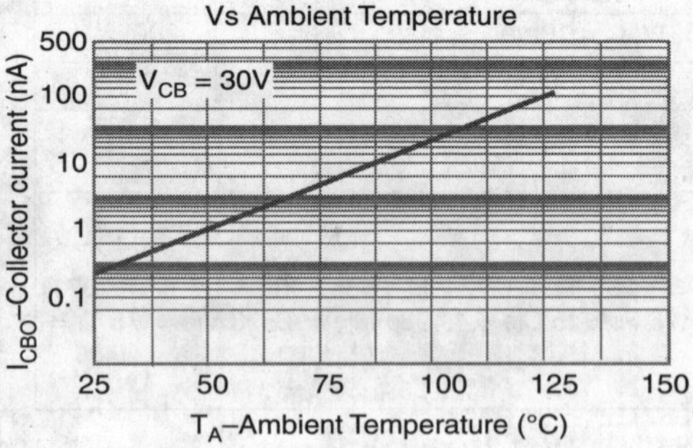

Capacitance vs Reverse Bias Voltage

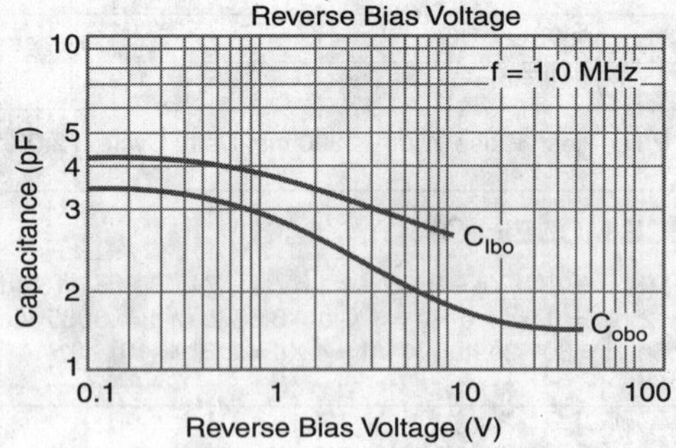

NPN General Purpose Amplifier

(continued)

Typical Characteristics (Continued)

Noise Figure vs Frequency

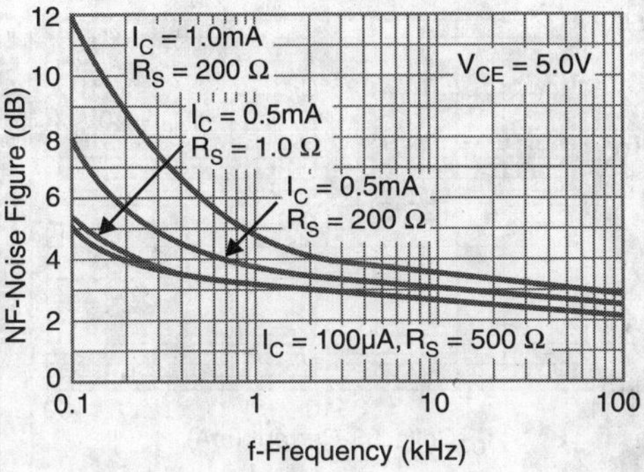

Noise Figure vs Source Resistance

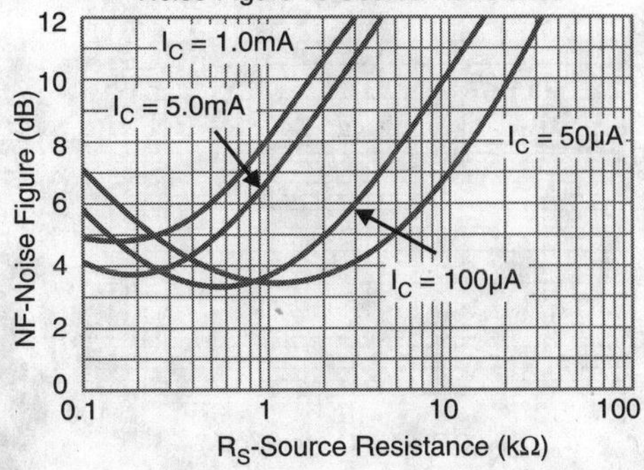

Current Gain and Phase Angle vs Frequency

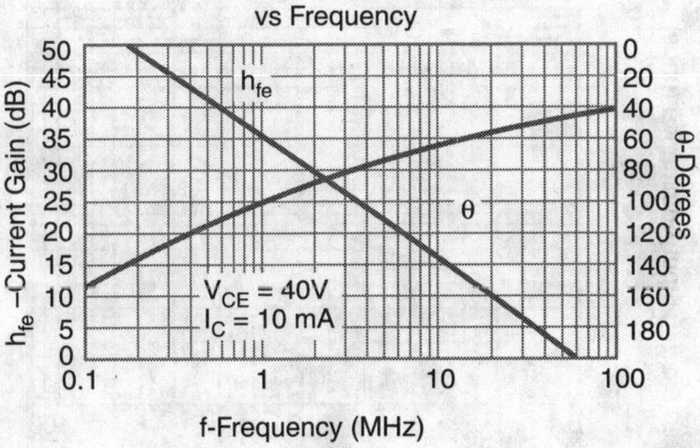

Power Dissipation vs Ambient Temperature

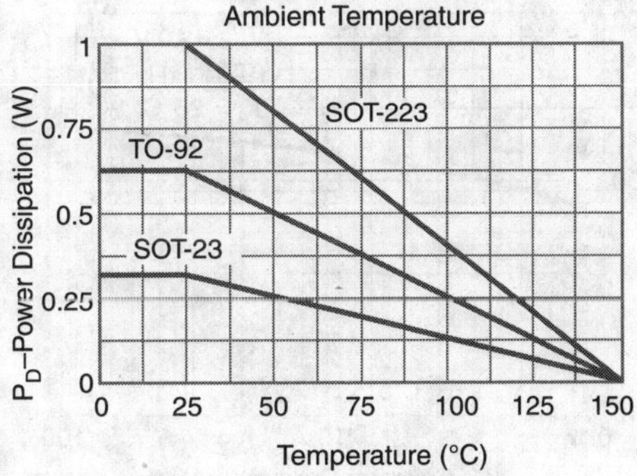

Turn-On Time vs Collector Current

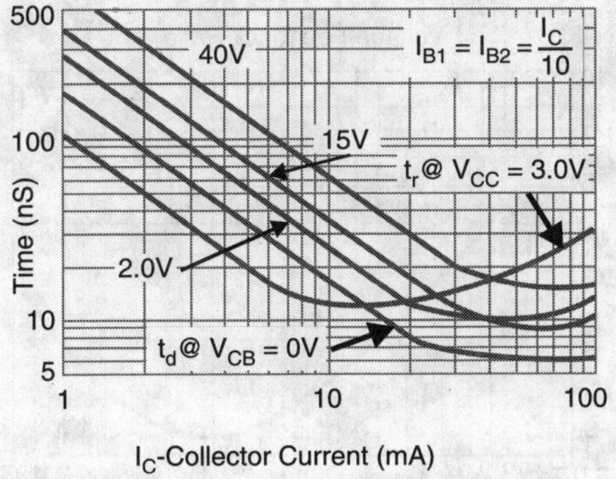

Rise Time vs Collector Current

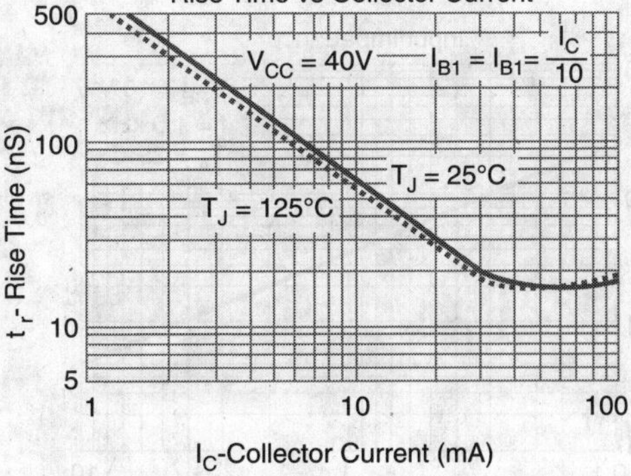

NPN General Purpose Amplifier

(continued)

Typical Characteristics (Continued)

Storage Time vs Collector Current

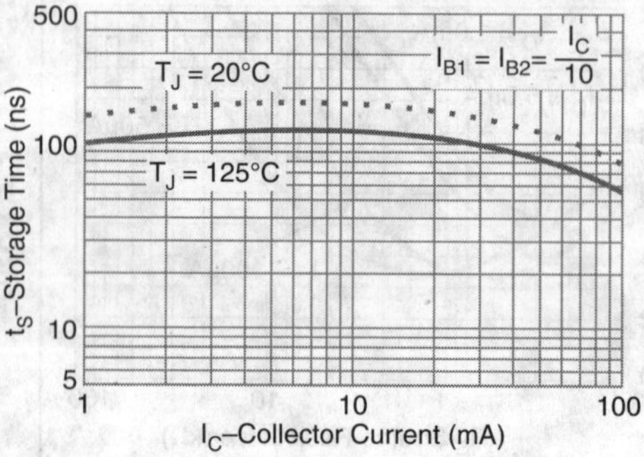

Fall Time vs Collector Current

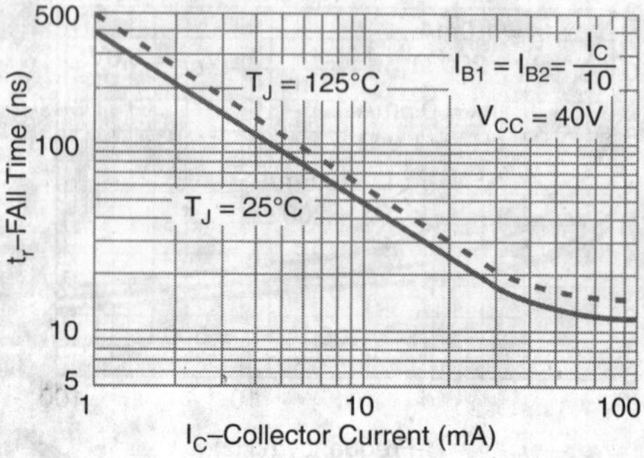

Current Gain

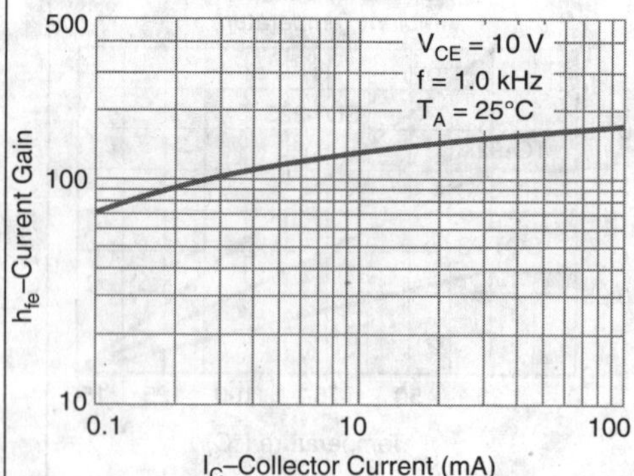

Output Admittance

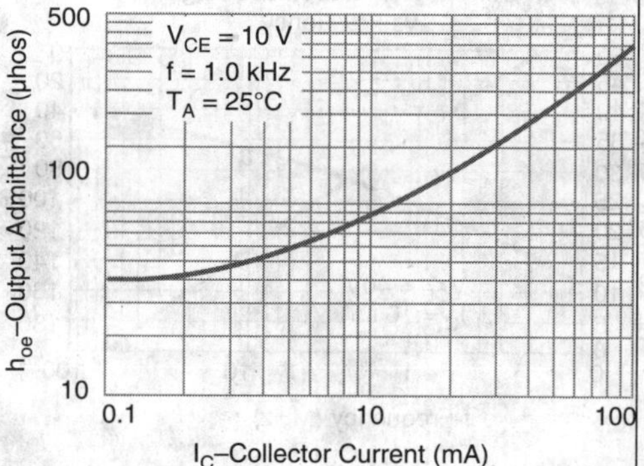

Input Impedance

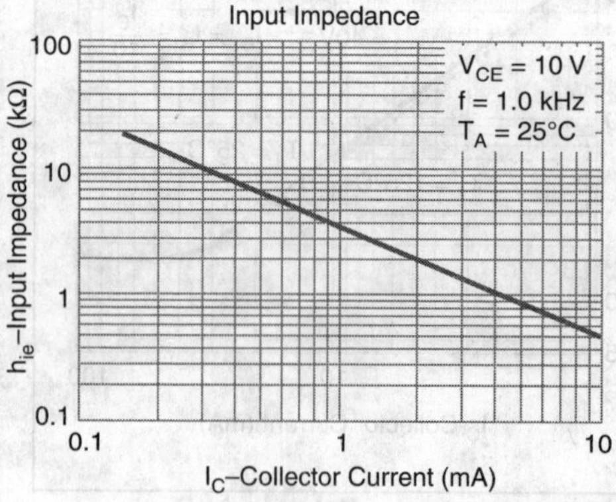

Voltage Feedback Ratio

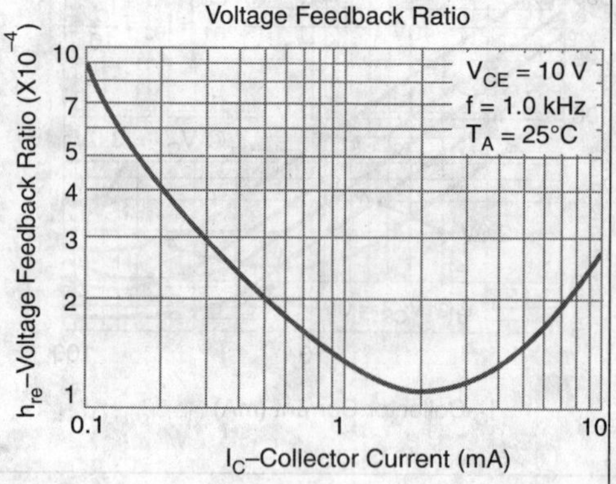

SGS-THOMSON MICROELECTRONICS

BFW43

HIGH VOLTAGE AMPLIFIER

DESCRIPTION

The BFW43 is a silicon planar epitaxial PNP transistors in Jedec TO-18 metal case. It is designed for use in amplifiers where high voltage and high gain are necessary. In particular, its feature a VCEO of 150V are specified over a wide range of curent.

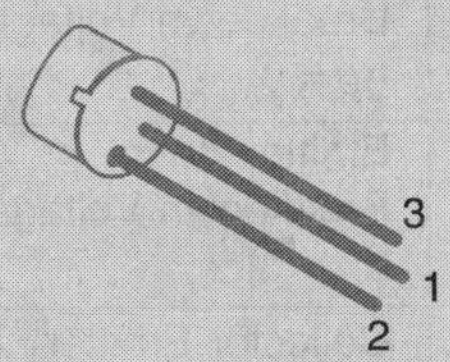

TO-18

INTERNAL SCHEMATIC DIAGRAM

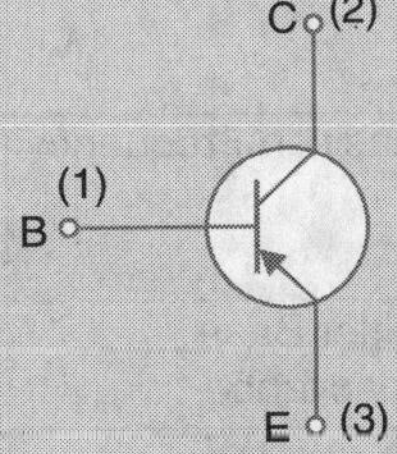

SCO8810

ABSOLUTE MAXIMUM RATINGS

Symbol	Parameter	Value	Unit
V_{CBO}	Collector-Base Voltage ($I_E = 0$)	– 150	V
V_{CEO}	Collector-Emitter Voltage ($I_B = 0$)	– 150	V
V_{EBO}	Emitter-Base Voltage ($I_C = 0$)	– 6	V
I_C	Collector Current	– 0.1	A
P_{tot}	Total Dissipation at Tamb ≤ 25 °C	0.4	W
	at Tcase ≤ 25 °C	1.4	W
T_{stg}	Storage Temperature	– 55 to 200	°C
T_j	Max. Operating Junction Temperature	200	°C

THERMAL DATA

$R_{thj\text{-}case}$	Thermal Resistance Junction-Case	Max	125	°C/W
$R_{thj\text{-}amb}$	Thermal Resistance Junction-Ambient	Max	438	°C/W

ELECTRICAL CHARACTERISTICS (Tcase = 25 °C unless otherwise specified)

Symbol	Parameter	Test Conditions	Min.	Typ.	Max.	Unit
I_{CBO}	Collector Cut-off Current (I_E = 0)	V_{CE} = –100 V V_{CE} = –100 V Tamb = 125 °C		–0.2 –0.03	–10 –10	nA μA
$V_{(BR)CBO*}$	Collector-Base Breakdown Voltage (I_E = 0)	I_C = –10 μA	–150			V
$V_{(BR)CEO*}$	Collector-Emitter Breakdown Voltage (I_B = 0)	I_C = –2 mA	–150			V
$V_{(BR)EBO*}$	Emitter-Base Breakdown Voltage (I_C = 0)	I_E = –10 μA	–6			V
$V_{CE(sat)*}$	Collector-Emitter Saturat ion Voltage	I_C = –10 mA IB = –1 mA		–0.1	–0.5	V
$V_{BE(sat)*}$	Base-Emitter Saturat ion Voltage	I_C = –10 mA IB = –1 mA		–0.74	–0.9	V
h_{FE*}	DC Current Gain	I_C = –1 mA V_{CE} = –10 V I_C = –10 mA V_{CE} = –10 V I_C = –10 mA V_{CE} = –10 V T_{amb} = –55 °C	40 40	85 100 30		
f_T	Transition Frequency	V_{CE} = –10 V f = 20 MHz I_C = –1 mA I_C = –10 mA	 60	 50		 MHz MHz
C_{EBO}	Emitter Base Capacitance	I_E = 0 V_{EB} = –0.5 V f = 1MHz		20	25	pF
C_{CBO}	Collector Base Capacitance	I_E = 0 V_{CB} = –5 V f = 1MHz		5	7	pF

* Pulsed: Pulse duration = 300 μs, duty cycle ≤ 1%

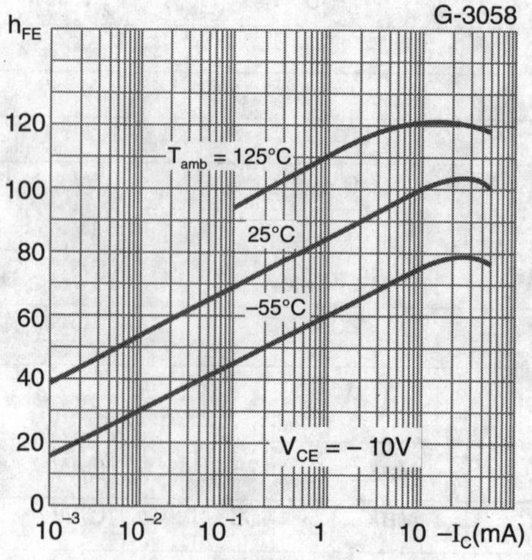

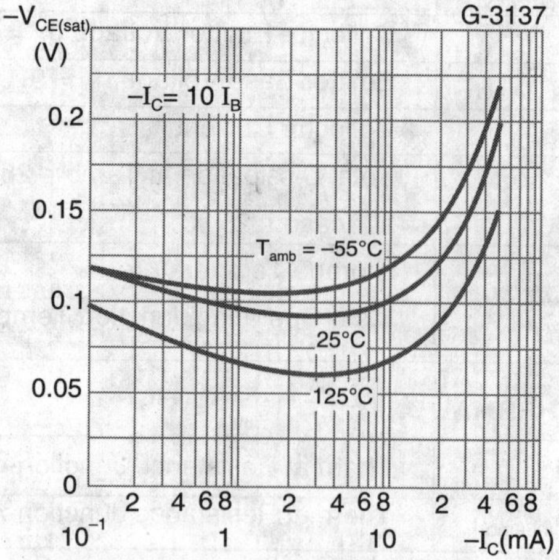

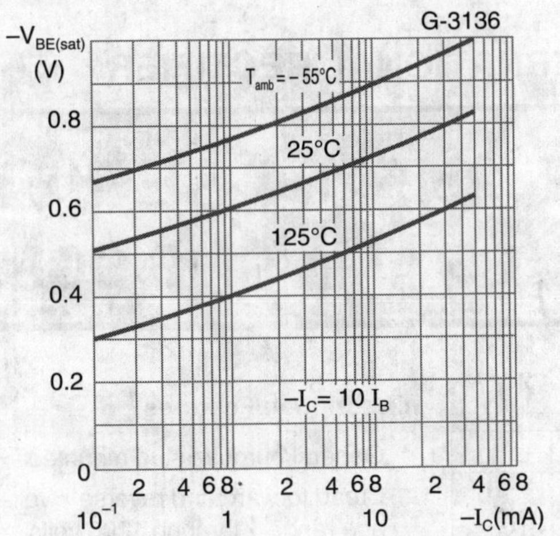
G-3136
$-V_{BE(sat)}$ (V)
$T_{amb} = -55°C$
25°C
125°C
$-I_C = 10\ I_B$
0.8
0.6
0.4
0.2
0
10^{-1}
1
10
$-I_C$(mA)

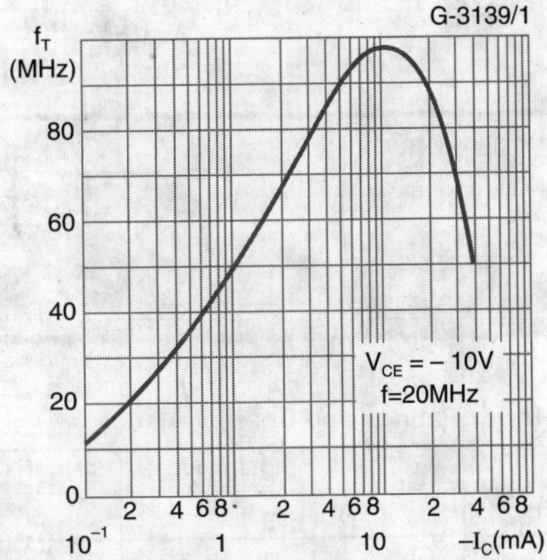
G-3139/1
f_T (MHz)
80
60
40
20
0
$V_{CE} = -10V$
f=20MHz
10^{-1}
1
10
$-I_C$(mA)

Data Sheet No. PD-3.081 A

INTERNATIONAL RECTIFIER

2N681 & 2N5204 SERIES
25 and 35 Amp RMS SCRs

Major Ratings and Characteristics

	2N681-92	2N5204-07	Units
$I_{T(RMS)}$	25	36	A
$I_{T(AV)}$	16*	22*	A
@T_C	– 65 to 65*	– 40 to 40	°C
I_{TSM}			
@ 50 Hz	145	285	A
@ 60 Hz	150*	300*	
$I2_t$			
@ 50 Hz	103	410	$A2_S$
@ 60 Hz	94	375	
IGT	40	40	mA
dv/dt	–	100*	V/μs
di/dt	75–100	100	A/μs
TJ	– 65 to 125*	–40 to 125*	°C
V_{RRM}, V_{DRM} range	25–800	600–1200	V

* JEDEC registered value

Description/Features

* General purpose stud mounted
* Broad forward and reverse voltage range – through 1200 volts
* Can be supplied to meet stringent military, aerospace and other high reliability requirements

CASE STYLE AND DIMENSIONS

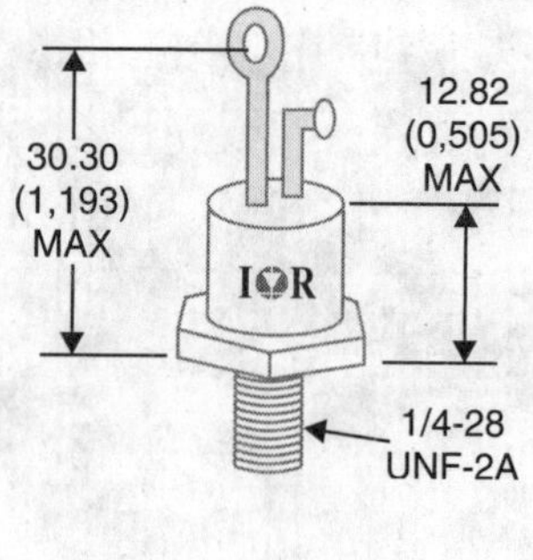

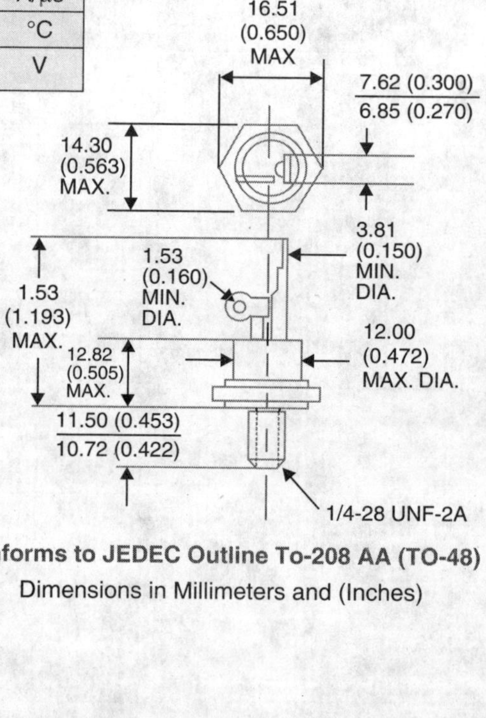

Conforms to JEDEC Outline To-208 AA (TO-48)

Dimensions in Millimeters and (Inches)

VOLTAGE RATINGS (Applied geta voltage zero or negative)

Part Numbers	V_{RRM}. V_{DRM} – Max. Repetitive Park Reverse and Off-State Voltage (V) $T_J = -65°C$ to 125°C	V_{RSM} Max. Non Repetitive Park Reverse Voltage $t_P \leqslant 5$ ms (V) $T_J = -65°C$ to 125°C
2N681	25*	35*
2N682	50*	75*
2N683	100*	150*
2N685	200*	300*
2N687	300*	400*
2N688	400*	500*
2N689	500*	600*
2N690	600*	720*
2N691	700*	840*
2N681	800*	960*
	$T_J = -40°C$ to 125°C	$T_J = -40°C$ to 125°C
2N5204	600	720
2N5205	800	960
2N5206	1000	1200
2N5207	1200	1440

ELECTRICAL SPECIFICATIONS

		2N681-92	2N5204-07	Units	Conditions	
	ON-STATE					
$I_{T(RMS)}$	Max. RMS on-stats curren:	25	35	A		
$I_{T(AV)}$	Max. avarage on-state	16*	22*	A	180° half sine wave conduction	
	currant @ T_C =	–65 to 65*	–40 to 40*	°C		
I_{TSM}	Max. peak one cycle. non-repetitive surge current	145	285	A	50 Hz half cycle sine wave or 6 ms rectangular pulss	Following any rated load condition, and with rated V_{RRM} applied following surgs.
		150*	300*		60 Hz half cycle sine wave or 5 ms rectangular pulse	
		170	340		50 Hz half cycle sine wave or 5 ms rectangular pulse	Same conditions as above except with VRRM applied following surge = 0
		180	35		60 Hz half cycle sine wave or 5 ms rectangular pulse	
I^2_t	Max. I^2_t capability, for fusing	103	410	A^2_s	t = 10 ms	Rated VRRM applied following surge, inltisl T_J = 125°C
		94	375		t = 8.3 ms	
I^2_t	Max. I^2_t capability, for individual device fusing	145	580	A^2_s	t = 10 ms	V_{RRM} = 0 following aurgs, initial T_J = 125°C
		135	530		t = 8.3 ms	
$I^2\sqrt{t}$	Max. $I^2\sqrt{t}$ capability, for individual devics fusing 1◯	1450	5800	$A^2\sqrt{s}$	t = 0.1 to 10 ma initisl $T_J \leqslant$ 125°C VRRM following surge = 0.	
V_{TM}	Max. park on-state voltage	2*	2.3*	V	T_J = 25°C, $I_{T(AV)}$ = 16A (50 A peak) 2N681 $I_{T(AV)}$ = 22A (70A peak) 2N5204	
I_H	Max. hlding current	20@25°C†	200*@–40°C	mA	Anode supply = 24V, Initial I_T = 1.0A.	
	BLOCKING					
dv/dt	Min. critical rate-or-rise of off-stste voltage	100†	100*	V/μs	T_J = 125°C Exponentisl to 100% rated V_{DRM}	Gate open circuited.
		250†	250		T_J = 125°C. Exponantial to 67% rated VDRM	

* JEDEC Registersd value.

I^2_t for time $t_X = I^2\sqrt{t} * \sqrt{t_x}$

† Typical

ELECTRICAL SPECIFICATIONS (Continued)

	BLOCKING (Continued)	2N681-92	2N5204-07	Units	Conditions
$I_{R(-1)}$ & $I_{D(-)}$	Max. reverse and off-state & current V_{RRM} & V_{DRM^*} SV	$I_{R(AV)}$ & $I_{D(AV)}$ (Average Values)	I_{RM} & I_{DM} (Peak Values)		
	25 to 150V	6.5*	–		
	200 & 250V	6.0*	–		
	300V	5.0*	–		
	400V	4.0*	–		
	500V	3.0*	–	mA	T_J = 125°C, gate open circuited.
	600	2.5*	3.3*		
	700V	2.25*	–		
	800V	2.0*	2.5*		
	1000V	–	2.0*		
	1200V	–	1.7*		
	SWITCHING				
t_d	Typical delay time	1	1	μs	T_C = 25°C, V_{DM} = rated V_{DRM}, I_{TM} = 10A dc resistive circuit. Gate pulse: 10V, 40Ω source. t_p = 6 μs, T_r = 0.1 μs.
dt/dt	Max. non-repetitive rate of rise of turned-on current V_{DM} = 25 to 600V	100	–		T_C = 125°C, V_{DM} = rated V_{DRM}, I_{TM} = 2 × di/dt. Gate pulse: 20V, 15Ω, tp = 6 ms, tyr = 0.1 ms max. Per JEDEC standerd RS-397, 5.2.2.6
	= 700 to 800V	75 –	– 100	A/ms	T_C = 125°C, V_{DM} = 600V, I_{TM} = 200A @ 400 Hz, max., Gate pulse: 20V, 15W tp = 6 ms, tr = 0.1 ms max. Per JEDEC standerd RS-397, 5.2.2.6
	TRIGGERING				
P_{GM}	Max.peak gate power	5*	60*	W	t_p ≤ 5 ms for 2N681 series; t_p ≤ 500 μs for 2N5204 series.
$P_{G(AV)}$	Max. sverage gate power	0.5*	0.5*	W	
$+I_{GM}$	Max. peak positive gate current	2*	2	A	
$+V_{GM}$	Max. peak positive gate voltage	10*	–	V	
$-V_{GM}$	Max. peak positive gate voltage	5*	5*	V	
I_{GT}	Max. required DC gate	80*	80*		T_C = min. rated value. Max. requined gate trigger cruuent is the lowest value which will trigger all units with +6V anode-to-cathods.
		40	40	mA	T_C = 25°C
		18.5	20		T_C = 125°C
	Typical DC gate current to trigger	30	30		TC = 25°C +6V anods-to-cathods
VGT	Max. required DC gatge	3*	3*		T_C = –65°C Max. required gate trigger voltage is the lowest value which will trigger all units with +6V anode-to-cathode.
		2	2	V	T_C = 25°C
	Typical DC gate voltage to trigger	1.5	1.5		T_C = 25° +6V anode-tocathode
V_{GD}	Max. DC gatge voltage not to trigger	0.25*	0.25*	V	T_C = 125°C. Max. gate voltage not to trigger is the maximum value which will not rigger any unit with rated V_{DRM} anode-to-cathode.

2N3870 thru 2N3873 2N6171 thru 2N6174

Silicon Controlled Rectifiers

Reverse Blocking Triode Thyristors

. . . designed for industrial and consumer applications such as power supplies: battery chargers: temperature, motor, light and welder controls.

- Econnomical for a Wide Range of Uses
- High Surge Current – I_{TSM} = 350 Amp
- Practical Level Triggering and Holding Characteristics – 4 and 5.2 mA (Typ) @ TC = 25°C
- Rugged Construction in Either Pressfit or Isolated Stud Package

SCRs
35 AMPERES RMS
100 thru 6C0 volts

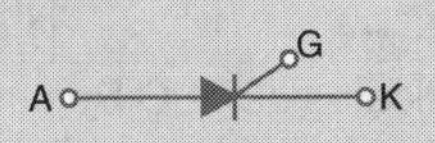

CASE 174-04
(TO-203AA)
STYLE 1
2N3870 thru 2N3873

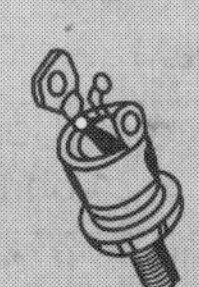

CASE 311-02
STYLE 1
(Stud Isolated)
2N6171 thru 2N6174

MAXIMUM RATINGS (T_J = 25°C uniess otherwise noted.)

Rating	Symbol	Value	Unit
*Peak Repetitive Forward or Reverse Blocking Voltage Note 1 (T_J = – 40 to + 100°C, 1/2 Sine Wave. 50 to 400 Hz, Gate Open)	V_{RRM} or V_{DRM}		Volts
2N3870, 2N6171		100	
2N3871, 2N6172		200	
2N3872, 2N6173		400	
2N3873, 2N6174		600	
*Peak Non-Repetitive Forward or Reverse Blocking Voltage (t ≤ 5 ms)	V_{RSM} or V_{DSM}		Volts
2N3870, 2N6171		150	
2N3871, 2N6172		330	
2N3872, 2N6173		660	
2N3873, 2N6174		770	
* Average On-State Current, Note 2	$I_{T(AV)}$		Amps
(T_C = – 40 to + 65°C)		22	
(T_C = + 85°C)		11	
* Peak Non-Repetitive Surge Current (One cycle, 60 Hz) (T_C = + 65°C)	I_{TSM}	350	Amps
Circuit Fusing (t = 8.3 ms)	I^2_t	510	A^2_s

*Indicates JEDEC Registered Data.

Note 1. V_{DRM} and V_{RRM} for all types can be applied on a continuous basis. R\atings apply for zero or negative gate voltage; however, positive gate votage shall not be applied concurrent with negative potential on the anode. Blocking voltaqges shall not be tested with a constant current source such that the voltage ratings of the devices are exceeded.

MAXIMUM RATINGS (T_C = 25°C unless otherwise noted.)

Rating	Symbol	Value	Unit
*Peak Gate Power	P_{GM}	20	Watts
*Average Gate Power	$P_{G(AV)}$	0.5	Watt
*Peak Forward Gate Current	I_{GM}	2	Amps
Peak Gate Voltage	V_{GM}	10	Volts
*Operating Junction Temperature Range	T_J	–40 to +100	°C
*Storage Temperature Range	T_{stg}	–40 to + 150	°C
Stud Torque	–	30	in. lb.

*Indicates JEDEC Registered Data.

***THERMAL CHARACTERISTICS**

Characteristic	Symbol	Max.	Unit
Thermal Resistance, Junction to Case	$R_{\theta JC}$		°C/W
2N3870 thru 2N3873		0.9	
2N6171 thru 2N6174		1	

*Indicates JEDEC Registcred Data.

ELECTRICAL CHARACTERISTICS (T_C = 25°C unless otherwise noted.)

Characteristic	Symbol	Min	Typ	Max	Unit
*Peak Repetitive Forward or Reverse Blocking Current	I_{DRM}, I_{RRM}				mA
(V_{AK} = Rated V_{DRM} or V_{RRM}, gate open, T_C = 100°C					
2N3870, 2N6171		–	1	2	
2N3871, 2N6172		–	1	2.5	
2N3872, 2N6173		–	1	3	
2N3873, 2N6174		–	1	4	
(Rated V_{DRM} or V_{RRM}, gate open, T_C = 25°C All Devices		–	–	10	μA
*Peak On-State Voltage (I_{TM} = 69 A Peak)	V_{TM}	–	1.5	1.85	Volts
*Gate Trigger Current (Continuous dc) (V_D = 12 V, R_L = 24 ohms) *T_C = –40°C	I_{GT}	–	9	80	mA
T_C = 25°C		–	4	40	
*Gate Trigger Voltage (Continuous dc) (V_D = 12 V, R_L = 24 ohms *T_C = –40°C	V_{GT}	–	0.9	3	Volts
T_C = 25°C		–	0.69	1.6	
*Holding Current (Gate Open) (V_D = 12 V, I_{TM} = 200 mA) *T_C = –40°	I_H	–	14	90	mA
T_C = 25		–	5.2	50	
*Gate Controlled Turn-On Time ($t_d + t_r$) (I_{TM} = 41 Adc, V_D = rated V_{DRM} I_{GT} = 40 mAdc, Rise Time ≤ 0.05 ms, Pulse Width = 10 μs)	t_{gt}	–	–	1.5	μs
Circuit Commutated Turn-Off Time	tq				μs
(I_{TM} = 10 A, I_R = 10 A)	–	25	–		
(I_{TM} = 10 A, I_R = 10 A, T_C = 100°C		–	35	–	
Forward Voltage Application Rate (T_C = 100°C, V_D = Rated V_{DRM}	dv/dt	–	50	–	V/μs

*Indicates JEDEC Registered Data.

MAC3010
MAC3020
MAC3030
MAC3040
Series

Triacs

Silicon Bidirectional Thyristors

. . . designed for full-wave ac power control app;ications, and specifically designed to be used in conjunction with MOC30XX opto coupiers in circuits similar to that shown on page 3-189.

- Blocking Voltages to 400 Volts
- Load Current Controlled Up to 40 A.
- Glass Passivated Junctions for Geeater Parameter Uniformity and Stability.
- Gate Triggering Guranteed in For Modes
- Designed for Use with MOC Series Optoisolators Having Triac Driver Outputs
- MAC3010/MAC3030 Are Recommended For Use With MOC3010/ MOC3030 Optoisolators
- MAC3010/MAC3030 Are Recommended For Use With MOC3020/ MOC3030 Optoisolators

TRIACs
4, 8, 15, 25 and 40
AMPERES RMS
250 thru 400 VOLTS

CASE 77-07
(TO-225AA)
STYLE 5
–4

CASE 221A-04
(TO-220AB)
STYLE 4
–8, –15, –25

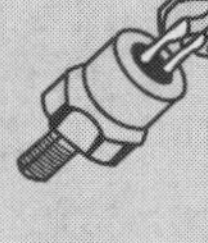

CASE 263-04
STYLE 2
–40

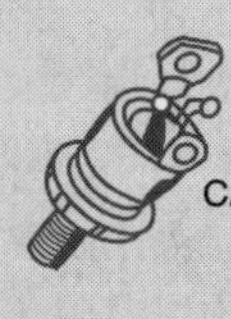

CASE 311-02
STYLE 2
–401

MAXIMUM RATINGS (T_J – 25°C unless Otherwise noted.)

Rating	Symbol	Current Ratoings					Unit
		–4	–8	–15	–25	–40 –401	
On-State RMS Current (see Fig. 1) (Fill Cycle Sine Wave 50 to 60 Hz)	$I_{T(RMS)}$	4	8	15	25	40	Amps
Peak Non-Repetitive Surge Current (One Full Cycle, 60 Hz, T_J = 110°C)	I_{TSM}	30	80	150	250	300	Amps
Circuit Fusing Considerations (T_J = –40 to + 110°C, t = 8.3 ms)	I^2_t	3.6	26	90	260	370	A^2_s
Peak Gate Voltage (t ≤ 2 μs)	V_{GM}	±5	±10	±10	±10	±10	Volts
Peak Gate Power (t ≤ 2 μs)	P_{GM}	10	20	20	20	20	Watts
Average Gate Power (T_C = 80°C, t ≤ 8.3 ms)	$P_{G(AV)}$	0.5	0.5	0.5	0.5	0.5	Watts
Peak Gate Current (t ≤ 2 μs)	I_{GM}	11	12	12	12	12	Amps
Operating Junction Temperature Range	T_J	*	–40 to + 125			*	°C
Storage Temperature Range	T_{stg}	–40 to + 150					°C
Mounting Torque	–	6	8	8	8	30	in.lb.
MAC3010/MAC3030 Note 1 (T_J = 25 to 125°C)	V_{DRM}	250	250	250	250	250	Volts
MAC3020/MAC3040		400	400	400	400	400	

Note 1. V_{DRM} for all types can be applied on a continuous basis. Blocking voltages shall not be tested with a constant current source such that the voltage ratings of the decvices are exceeded.

MAC3010, MAC3020, MAC3030, MAC3040 Series

– 8, – 15, – 25 Current Ratings

ELECTRICAL CHARACTERISTICS (T_C = 25°C, and Either Polarity of MT2 to MT1 Voltage uniess otherwise noted.)

Characteristic	Symbol	Min	Tup	Max.	Unit
Peak Forward or Reverse Blocking Current (Rated V_{DRM} or V_{RRM}) T_J = 25°C T_J = 125°C	I_{DRM}, I_{RRM}	 – –	 – –	 10 2	 μA mA
Peak On-State Voltage (I_{TM} = $\sqrt{2}$ $I_{T(RMS)}$ A Peak; Pulse Width ≤ 2 ms, Duty Cycle ≤ 2% MAC3030-8 MAC3030-15 MAC3030-25	V_{TM}	 – – –	 – – –	 1.6 1.6 1.85	Volts
Gate Trigger Current (Continuous dc) (V_D = 12 V, R_L = 100 Ohms) MT2 (+), G(+); MT2 (–), G(–) All Types	I_{GT}	–	–	40	mA
Gate Trigger Voltage (Continuous dc) (V_D = 12 V, R_L = 100 Ohms) MT2 (+), G(+); MTR2 (–), G(–) All Types (TJ = 125°C, RL = 10 k Ohms) MT2 (+), G (+); MT1 (–), G(–) All Types	V_{GT}	 – 0.2	 – –	 2 –	Volts
Holding Current (V_D = 12 V, I_{TM} = 200 mA, Gate Open)	I_H	–	–	40	mA
Gate Controlled Turn-On Time (I_{TM} = 2 $I_{T(RMS)}$ A Peak, I_G = 100 mA)	t_{gt}	–	1.5	–	μs
Critical Rate of Rise of Commutation Voltage (I_{TM} = $\sqrt{2}$ $I_{T(RMS)}$ A peak, Commutating di/dt = 0.36 I_{TM} A/ms, Gate Unenergized, T_C = 80°C	dv/dt(c)	–	5	–	V/μs
Critical Ratge of Rise of Off-State Voltage (Exponential Waveform, T_C = 125°C	dv/dt	–	40	–	V/μs

Triacs

Silicon Bidirectional Triode Thyristors

2N5571 thru 2N5574 2N6145 thru 2N6147 T4100M T4110M

. . . designed primarily for industrial and military applications for the control of ac loads in applications such as light dimmers, power supplies, heating controls, motor controls, welding equipment and power switching systems; or wherever full-wave, silicon gate controlled solid-state devices are needed.

- All Diffused and Glass Passivated Junctions for Greater Stability
- Pressfit, Stud and Isolated Stud Packages
- Gate Triggering Guaranteed In All 4 Qusdrants

TRIACs
15 AMPERES RMS
200 thru 600 VOLTS

CASE 174-04
(TO-203)
STYLE 3
2N5571
2N5572
T4100M

CASE 174-03
STYLE 3
2N5573
2N5574
T4110M

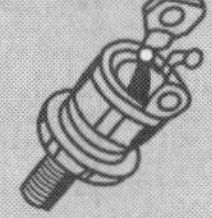

CASE 311-02
STYLE 2
2N76145
2N6146
2N6147

MAXIMUM RATINGS

Rating	Symbol	Value	Unit
*Peak Repetitive Off-State Voltage (T_J = –65 to +100°C 1/2 Sine Wave 50 to 60 Hz, Gate Open 2N5571, 2N5573, 2N6145 2N5572, 2N5574, 2N6146 T4100M, T4110M, 2N6147	V_{DRM}	 200 400 600	Volts
*Peak Gate Voltage	V_{GM}	20	Volts
*RMS On-State Current (T_C = – 65 to + 80°C) (T_C = + 85°C	$I_{T(RMS)}$	 15 10	Amps
*Peak Non-Repetitlve Surge Current (One Full cycle of surge current at 60 Hz, precededand followed by rated current, I_C = 85°C	I_{TSM}	100	Amps
Clrcuit Fusing (T_C = – 65 to + 80°C, t = 1 to 8.3 ms)	I^2_t	40	A^2_s
Peak Gate Power *(T_C = 80°C, Pulse Width = 1 µs) 2N5571 thru 2N5574 T4100M, T4110M *(T_C = 80°C, Pulse Width = 2 µs) 2N6145 thru 2N6147	P_{GM}	 16 16 20	Watts
*Average Gate Power (T_C = 80°C, Pulse Width = 8.3 ms)	$P_{G(AV)}$	0.5	Watt
*Peak Gate Current	I_{GM}	2	Amps
*Operating Junction Temperature Range	T_J	–65 to +100	°C
*Storage Temperature Range	T_{stg}	–65 to +150	°C
Stud Torque *2N5673, 2N5574, T4110M *2N6145, 2N6146, 2N6147	–	30	in. lb.

THERMAL CHARACTERISTICS

Characteristic	Symbol	Max.	Unit
*Thermal Resistance, Junction to Case	$R_{\theta JC}$	1	°C/W

*Indicates JEDEC Registered Data.

2N5571 thru 2N5574 • 2N6145 thru 2N6147 • T4100M • T4110M

Electrical Characteristics (T_C = 25°C, and Either Polarity of MT2 to MT1 Voltage uniese otherwise noted.)

Characteristic	Symbol	Min	Typ	Max.	Unit
*Peak Forward or Reverse Blocking Current	I_{DRM}, I_{RRM}				
(Rated V_{DRM} or V_{RRM}) T_C = 25°C		–	–	10	μA
T_C = 100°C		–	–	2	mA
*Peak On-State Voltage (I_{TM}=21 A Peak, Pulse Width=1 to 2 ms, Duty Cycle ≤ 2%)	V_{TM}	–	1.3	1.8	Volts
Gate Trigger Current (Continuous dc), Note 1 (V_D = 12 Vdc, R_L = 30 Ohms)	I_{GT}				mA
MT2(+), G(+); MT2(–), G(–)		–	–	50	
MT2(+), G(–); MT2(–), G(+)		–	–	80	
*MT2(+), G(+); MT2(–), G(–), T_C = – 65°C		–	–	150	
*MT2(+), G(–); MT2(–), G(+), T_C = – 65°C		–	–	200	
Gate Trigger Voltage (Continuous dc) (All Quadrants)	V_{GT}				Volts
(V_D = 12 Vdc, R_L = 30 Ohms) T_C = 25°C		–	–	2.5	
*T_C = – 65°C		–	–	4	
*(V_D = Rated V_{DRM}, R_L = 10 k ohms, T_C = + 100°C)		0.2	–	–	
Holding Current (V_D – 12 Vdc, Gate Open) (Initiating Current = 500 mA)	I_H				mA
T_C = 25°C		–	–	75	
*T_C = – 65°C		–	–	300	
Gate Controlled Turn-On Time (Rated V_{DRM}, I_{TM} = 21 A Peak, I_{GT} = 160 mA Rlse Time = 0.1 μs, Pulse Vidth = 2 μs)	t_{gt}	–	1	2	μs
*Critical Rate-of-Rise of Commutation Voltage (Rated VDRM, I_{TM} = 21 A Peak, Commutating di/dt = 8 A/ms, gate unenergized	dv/dt(c)				V/μs
T_C = 80°C 2N5571 thru 2N5574, T4100M, T4110M		2	10	–	
T_C = 75°C 2N6145 thru 2N6147		2	10	–	
Critical Rate-of-Rise of Off-State Voltage (Rated V_{DRM}, Exponential Voltage Rise, Gate Open, T_C=100°C	dv/dt				V/μs
*2N5571, 2N5573, 2N6145		30	150	–	
*2N5572, 2N5574, 2N6146		20	100	–	
T4100M, T4100M, 2N6147		10	75	–	

MOTOROLA

LM317

Three-Terminal Adjustable Output Positive Voltage Regulator

The LM317 is an adjustable 3–terminal Positive voltage regulator capable of supplying in excess of 1.5 A over an output voltage range of 1.2 V to 37 V. This voltage regulator is exceptionally easy to use and requires only two external resistors to set the output voltagve. Further, it employs internal current limiting, thermal shutdown and safe area compensation, making it essentially blow–out proof.

The LM317 serves a wide verety of applications including local, on card regulation. this device can also be used to make a programmable output regulator, or by connecting a fixed resistor between the adjustment and output, the LM317 can be used as a precision curerent regulator.

- Output Current in Excess of 1.5 A.
- Output Adjustable between 1.2 V and 37 V.
- Internal Thermal Overload Preotection.
- Internal Short Circuit Current Limiting Constant with Temperature.
- Output Transistor Safe–Area Compensation.
- Floating Operation for High Voltage Applications.
- Avaliable in Surface Mount D²PAK, and Standard 3–Lead Transistor Package
- Eliminates Stocking many Fixed Voltages.

THREE-TERMINAL ADJUSTABLE POSITIVE VOLTAGE REGULATOR

SEMICONDUCTOR TECHNICAL DATA

T SUFFIX
PLASTIC PACKAGE
CASE 221A

Healsink surface connected to Pin 2.

Pin **1.** Adjust
2. Vout
3. V_{in}

D2T SUFFIX
PLASTIC PACKAGE
CASE 936
(D²PAK)

Heatsink Surface (shown as terminal 4 in case outine drawing) is connected to Pin 2.

Standard Application

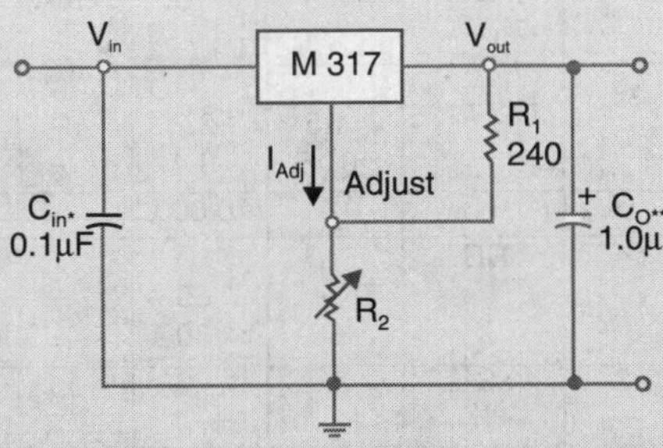

* C_{in} is required if regulator is located an appeciable distance from power supply filter.

** C_O is not needed for stability, however, it does improve transient respones.

$$V_{out} = 1.25V\left(1 + \frac{R_2}{R_1}\right) + I_{Adj}R_2$$

Since I_{Adj} is required to loess than 100 µA, the error associated with this term is negligible in most applications.

CRDERING INFORMATION

Device	Operating Temperature Range	Package
LM317BD2T	T_J = –40° to +125°C	Surface Mount
LM317BT		Insertion Mount
LM317D2T	T_J = 0° to +125°C	Surface Mount
LM317T		Insertion Mount

MAXIMUM RATINGS

Rating	Symbol	Value	Unit
Input–Output Voltage Differential $V_I–V_O$	40	Vdc	
Power Dissipation			
Case 221A			
T_A = 25°C	P_D	Internally Limited	W
Thermal Resistance, Junction–to–Ambient	θJA	65	°C/W
Thermal Resistance, Junction–to–Case	θJC	5.0	°C/W
Case 936 (D^2PAK)			
T_A = 25°C	P_D	Internally Limited	W
Thermal Rsistance, Junction–to–Ambient	θJA	70	°C/W
Thermal Resistance, Junction–to–Case	θJC	5.0	°C/W
Operating Junction Temperature Range	T_J	–40 to 125	°C
Storage Temperature Range	T_{stg}	–65 to +150	°C

ELECTRICAL CHARACTERSISTICS ($V_1–V_O$ = 5.0 V; I_O = 0.5 A for D2T and T packages; $T_J = T_{low}$ to T_{high} [Note 1]; I_{max} and P_{max} [Note 2]; unless otherwise noted.)

Characteristics	Figure	Symbol	Min	Typ	Max	Unit
Line Regulation (Note 3), T_A = +25°C, 3.0 V ≤ $V_I – V_O$ ≤ 40 V	1	Regline	–	0.01	0.04	%/V
Load Regulation (note 3), T_A = +25°C, 10 mA ≤ $I_O ≤ I_{max}$	2	Regload				
V_O ≤ 5.0 V			–	5.0	25	mV
V_O ≥ 5.0 V			–	0.1	0.5	& VO
Thermal Regulation, T_A = +25°C (Note 6), 20 ms Pulse		Regtherm	–	0.03	0.07	$\%V_O$/W
Adjustment Pin Current	3	I_{Adj}	–	50	100	µA
Adjustment Pin Current Change, 2.5 V ≤ $V_I – V_O$ ≤ V, 10 mA ≤ I_L ≤ Imax, $P_D ≤ P_{max.}$	1, 2	ΔI_{Adj}	–	0.2	5.0	µA
Reference Voltage, 3.0 V ≤ $V_I – V_O$ ≤ 40 V, 10 mA ≤ I_O ≤ Imax, P_D ≤ Pmax	3	V_{ref}	1.2	1.25	1.3	V
Line Regulation (Note 3), 10 mA ≤ $I_O ≤ I_{max}$	2	Regline	–	0.02	0.07	%V
Load Regulation (Note 3), 10 mA ≤ $I_O ≤ I_{max}$	2	Regload				
V_O ≤ 5.0 V			–	20	70	mA
V_O ≥ 5.0 V			–	0.3	1.5	% V_O
Temperature Stability ($T_{low} ≤ T_J ≤ T_{high}$)	3	T_S	–	0.7	–	$\%V_O$
Minimum Load Current to Maintain 3 Regulation ($V_I – V_O$ = 40 V)	I_{Lmin}	–	3.5	10	mA	
Maximum Output Current	3	I_{max}				A
$V_I – V_O$ ≤ 15 V, $P_D ≤ P_{max}$, T Package			1.5	2.2	–	
$V_I – V_O$ = 40 V $P_D ≤ P_{max}$, T_A = +25°C, T Package			0.15	0.4	–	
RMS Noise, % of V_O, T_A = +25°C, 10 Hz ≤ f ≤ 10 kHz		N	–	0.003	–	&V_O
Ripple Rejection, V_O = 10 V, f = 120 Hz (Note 4)	4	RR				dB
Without C_{Adj}			–	65	–	
C_{Adj} = 10 mF			66	80	–	
Long–Term Stability,m $T_J = T_{high}$ (Note 5), T_A = +25°C for Endpoint Measurements	3	S	–	-.3	1.0	%1.0 k Hrs.
Thermal Resistance Junction to Case, T Package		$R_{\theta JC}$	–	5.0	–	°C/W

Notes: **1.** T_{low} to T_{high} = 0° to +125°C, for LM317T, D2T. T_{low} to T_{high} = –40° to +125°C, for LM317BT, BD2T.

2. I_{max} = 1.5 A, P_{max} = 20 W.

3. Load and line regulation are specified at constant junction temperature. Changes in VO due to heating effects must be taken into account separately. Pulse testing with low duty cycle is used.

4. C_{Adj}, when use, is connected between the adjustment pin and ground.

5. Since Long-Term Stability cannot be measured on each device before shipment, this specification is an engineering estimate of average stability from lot to lot.

6. Power dissipation within an IC voltage regulator produces a ytemperature gradient on the die, affecting individual IC components on the die. These effects can be minimized by proper integrated circuit design and layout techniques. Thermal Regulation is the effect of these temperature gradients on the output voltage and is expressed in percentage of output change per watt of power change in a specified time.

APPLICATION INFORMATION

Basic Circuit Operation

The LM317 is a 3–terminal floating regulator. In operation, the LM317 develops and maintains a nominal 1.25 V refernce (V_{ref}) between its output and adjustment terminals. This reference voltage is converted to a programming current (I_{PROG}) by R_1 (see Figure 17), and this constant current flows through R_2 to ground.

The regulated output voltage is given by:

$$V_{out} = V_{ref}\left(1 + \frac{R_2}{R_1}\right) + I_{Adj} R_2$$

Since the current from the adjustment terminal (I_{Adj}) represents an error term in the equation, the LM317 was designed to control I_{Adj} to less than 100 μA and keep it constant. To do this, all quiescent operating current is returned to the output terminal. This mposes the requirement for a minimum load current. If the load current is less than this minimum, the output voltage will rise.

Since the LM317 is a floating regulator, it is only the voltage differential across the circuit which is important to performnce, and operation at high voltage with respect to ground is possible.

Fig. 17. Basic Circuit Configuration

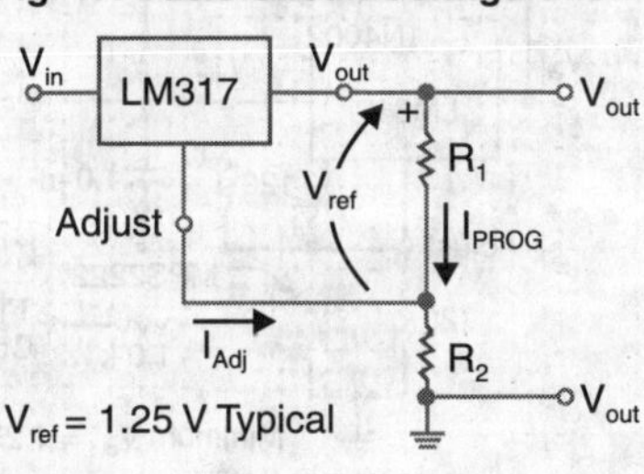

Load Regulation

The LM317 is capable of providing extremely good load regulation, but a few precautions are needed to obtain maximum performance. For best performace, the programming resistor (R_1) should be connedcted as close to the regulator as possible to minimize line drops which effectively appear in series with the reference, thereby degrading regulation. the ground end of R_2 can be returned near the load ground to provide remote ground sensing and improve load regulation.

External Capacitors

A 0.1 μF disc or 1.0 μF tantalum imput bypass capacitor (C_{in}) is recommended to reduce the sensitivity to input line impedance.

The adjustment terminal may be bypassed to ground to improve ripple rejection. This capacitor (C_{Adj}) prevents ripple from being amplified as the output voltage is increased. A 10 μF capacitor should improve ripple regjection about 15 dB at 120 Hz in a 10 V application.

Although the LM317 is stable with no output capacitance, like any feedback circuit, certain values of external capacitance can cause excessive rigning. An output capacitance (C_O) in the form of a 1.0 μF tantalum or 25 μF alumionum electrolytic capacitor on the output swamps this effect and insures stability.

Protection Diodes

When external capacitors are used with any IC regulator is is sometimes necessary to add protection diodes to prevent the capacitors from discharging through low current points into the regulator.

Figure 18 shows the LM317 with the recommended protection diodes for outpout voltages in excess of 25 V or high capacitance values (C_O 25 μF, C_{Adj} > 10 μF). Diode D_1 prevents C_O from discharging thru the IC during an input short circuit. Diode D_2 protects against capacitor C_{Adj} discharging through the IC during an output short circuit. The conmbination of diodes D_1 and D_2 prevents C_{Adj} from discharging through the IC during an input short circuit.

Fig. 18. Voltage Regulator with Protection Diodes

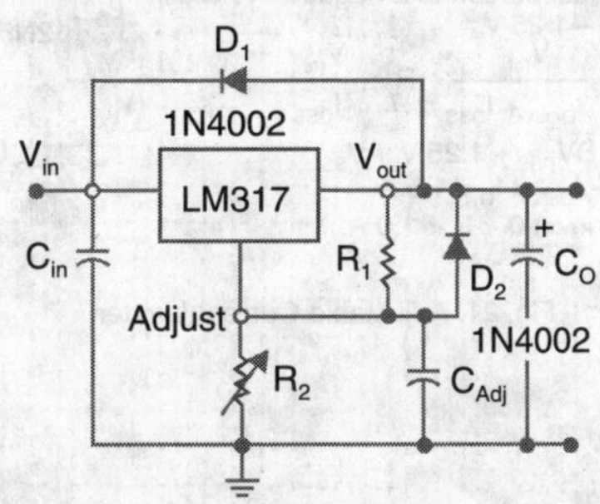

Fig. 19. D^2PAK Thermal Resistance and Maximum Power Dissipation versus P.C.B. Copper Length

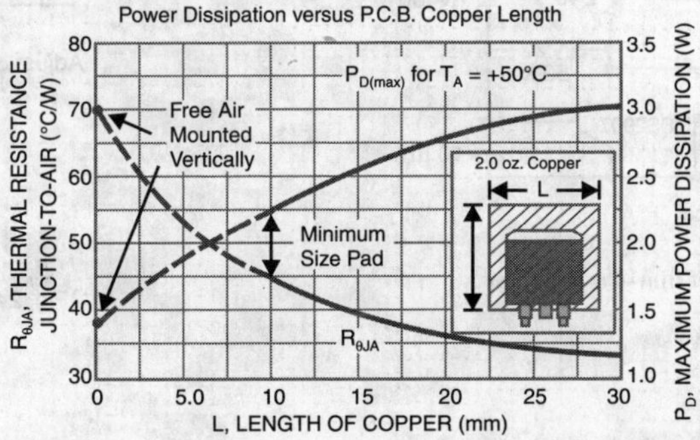

LM317

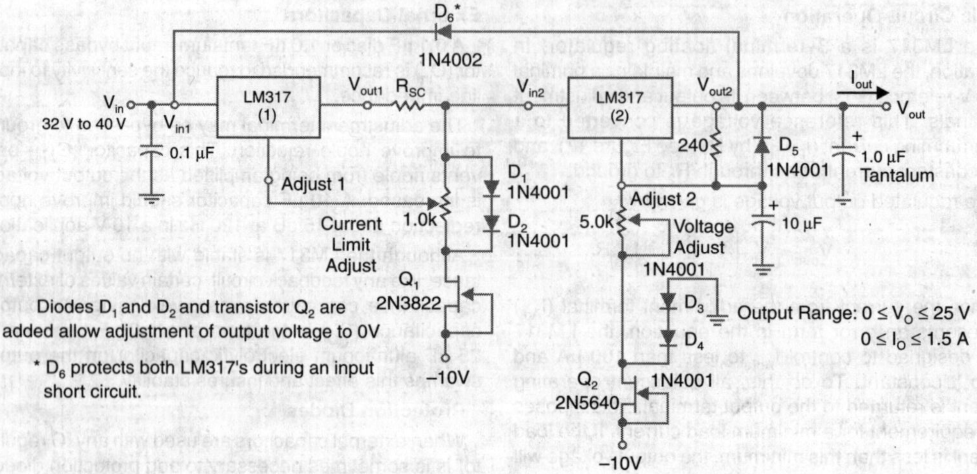

Fig. 20. "Laboratory" Power Supply with Adjustable Current Limit and Output Voltage

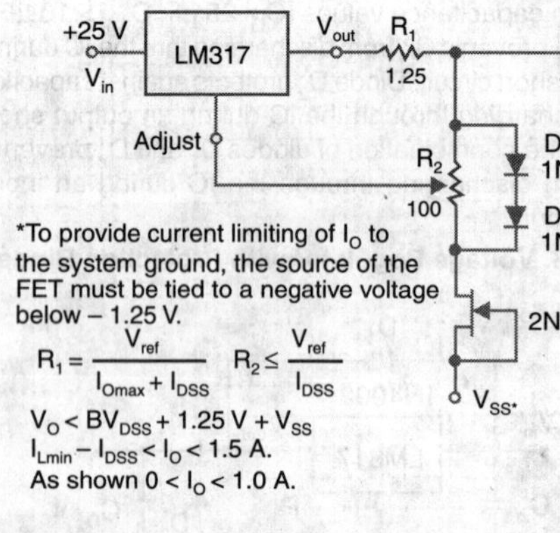

Fig. 21. Adjustable Current Limiter

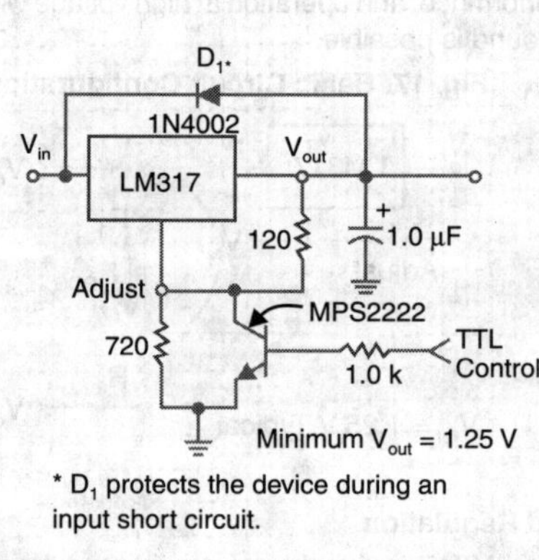

Fig. 22. 5.0 V electronic Shutdown Regualtor

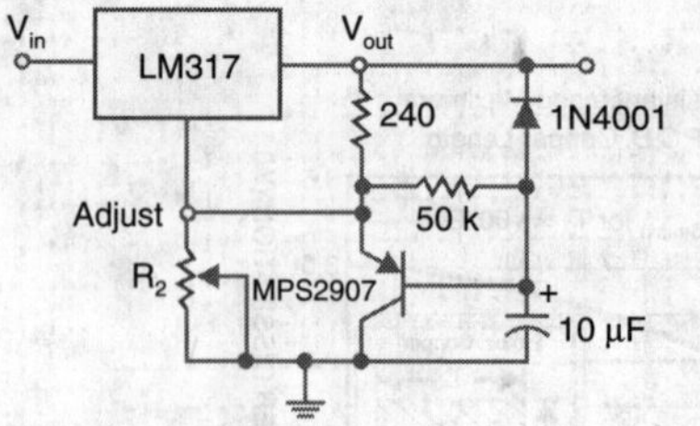

Fig. 23. Slow Trun–On regulator

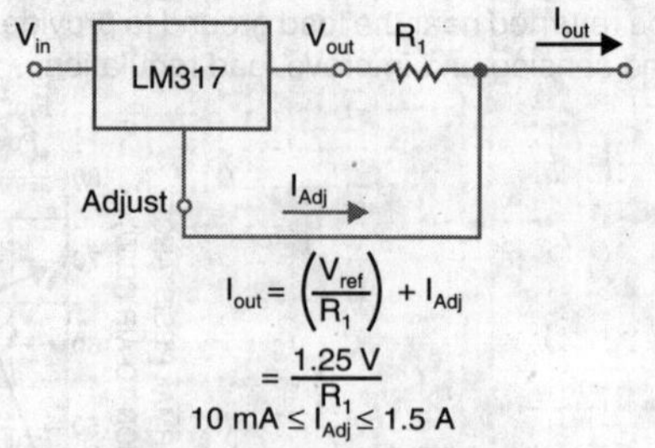

Fig. 24. Current Regulator

μA741, μA741Y
GENERAL-PURPOSE OPERATIONAL AMPLIFIERS

SLOS094A – NOVEMBER 1970 – REVISED JANUARY 1992

- Short-Circuit Protection
- Offset-Voltage Null Capability
- Large Common-Mode and Differential Voltage Ranges
- No Frequency Compensation Required
- Low Power Consumption
- No Latch-Up
- Designed to Be Interchangeable With Fairchild μA741

description

The μA741 is a general-purpose operational amplifier featuring offset-voltage null capability.

The high common-mode input voltage range and the absence of latch-up make the amplifier ideal for voltage-follower applications. The device is short-circuit protected and the internal frequency compensation ensures stability without external components. A low value potentiometer may be connected between the offset null inputs to null out the offset voltage as shown in Fig. 2.

The μA741C is characterized for operation from 0°C to 70°C. The μA741I is characterized for opera tion from –40°C to 85°C.The μA741M is charac terized for operation over the full military tempera ture range of –55°C to 125°C.

Symbol

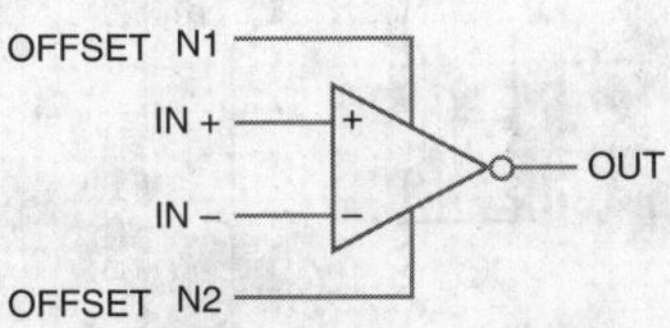

mA741M . . . J PACKAGE
(TOP VIEW)

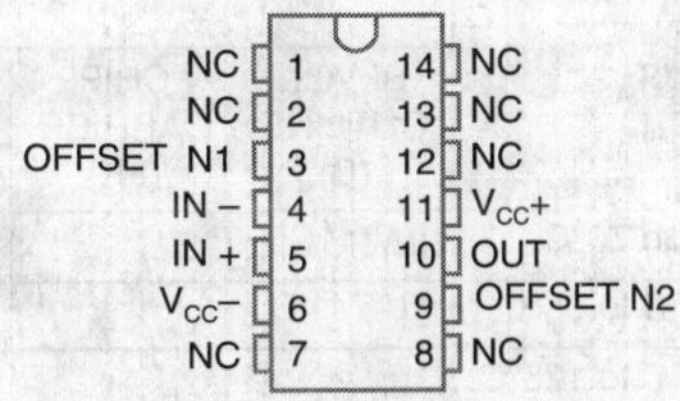

μA741M . . . JG PACKAGE
μA741C, μA741I . . . D,P, OR PW PACKAGE
(TOP VIEW)

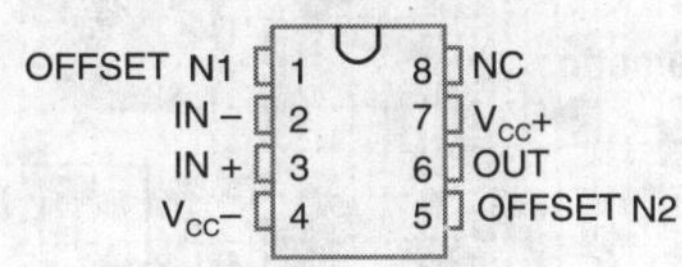

μA741M . . . U PACKAGE
(TOP VIEW)

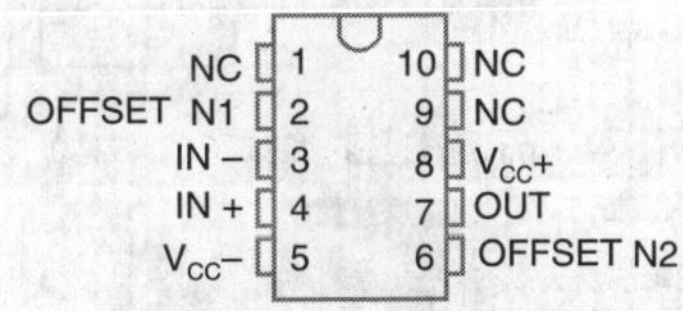

μA741M . . . FK PACKAGE
(TOP VIEW)

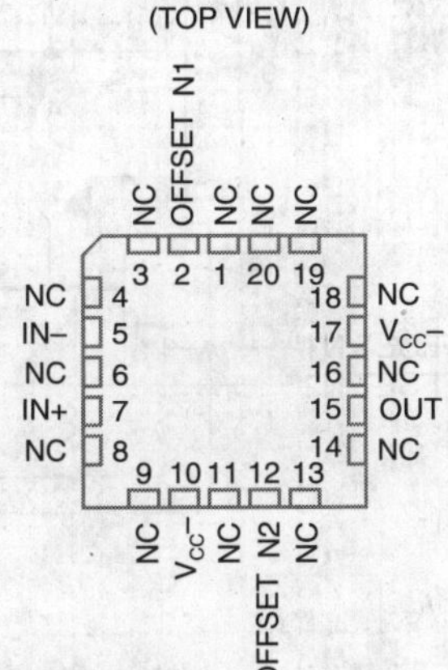

NC – No internal connection

Production Data information is current as of publication date. Products conform to specifications per the terms of Texas Instruments standard waranty. Production processing does not necesarily include testing of all parameters.

TEXAS INSTRUMENTS

POST OFFICE BOX 655303 • DALLAS, TEXAS 75265

μA741, μA741Y
GENERAL-PURPOSE OPERATIONAL AMPLIFIERS

SLOS094A – NOVEMBER 1970 – REVISED JANUARY 1992

AVAILABLE OPTIONS

T_A	PACKAGED DEVICES							
	SMALL OUTLINE (D)	CHIP CARRIER (FK)	CERAMIC DIP (J)	CERAMIC DIP (JG)	PLASTIC DIP (P)	TSSOP (PW)	FLAT PACK (U)	CHIP FORM (Y)
0°C to 70°C	μA741CD				μA741CP	μA741CPW		μA741Y
–40°C to 85°C	μA741ID				μA741IP			
–55°C to 125°C			μA741MFK	μA741MJ	μA741MJG		μA741MU	

The D package is available taped and reeled. Add the suffix R (e.g., μA741 CDR)

Schematic

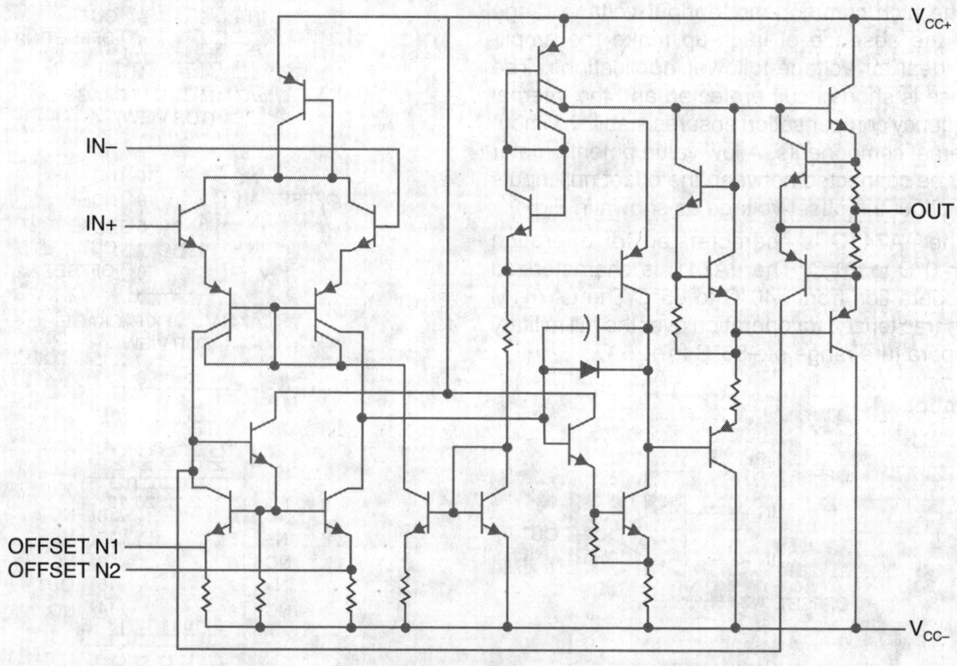

Component count	
Transistors	22
Resistors	11
Diode	1
Capacitor	1

μA741, μA741Y
GENERAL-PURPOSE OPERATIONAL AMPLIFIERS

SLOS094B – NOVEMBER 1970 – REVISED SEPTEMBER 2000

μA741Y chip information

This chip, when properly assembled, displays characteristics similar to the μA741C. Thermal compression or ultrasonic bonding may be used on the doped-aluminum bonding pads. Chips may be mounted with conductive epoxy or a gold-silicon preform.

DONDING PAD ASSIGNMENTS

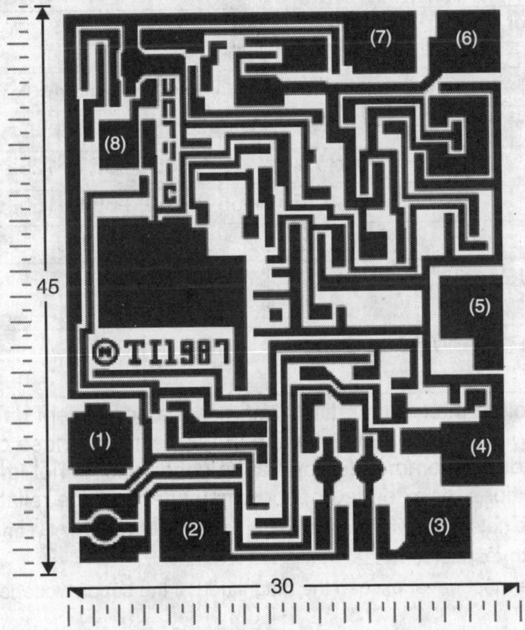

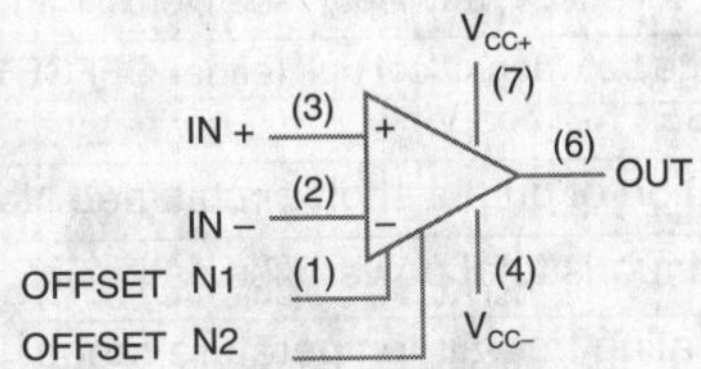

CHIP THICKNESS: 15 TYPICAL
BONDING PADS: 4 × 4 MINIMUM
$T_J$max = 150°C
TOLERANCES ARE ± 10%
ALL DIMENSIONS ARE IN MILS.

μA741, μA741Y
GENERAL-PURPOSE OPERATIONAL AMPLIFIERS

SLOS094B – NOVEMBER 1970 – REVISED SEPTEMBER 2000

absolute maximum ratings over operating free-air temperature range (unless otherwise noted)†

	μA741C	μA741I	μA741M	UNIT
Supply voltage, V_{CC+} (see Note 1)	18	22	22	V
Supply voltage, V_{CC-} (see Note 1)	−18	−22	−22	V
Differential input voltage, V_{ID} (see Note 2)	±15	±30	±30	V
Input voltage, V_I any input (see Notes 1 and 3)	±15	±15	±15	V
Voltage between offset null (either OFFSET N1 or OFFSET N2) and V_{CC-}	±15	±0.5	±0.5	V
Duration of output short circuit (see Note 4)	unlimited	unlimited	unlimited	
Continuous total power dissipation	See Dissipation Rating Table			
Operating free-air temperature range, T_A	0 to 70	−40 to 85	−55 to 125	°C
Storage temperature range	−65 to 150	−65 to 150	−65 to 150	°C
Case temperature for 60 seconds FK package			260	°C
Lead temperature 1,6 mm (1/16 inch) from case for 60 seconds J, JG, or U package			300	°C
Lead temperature 1,6 mm (1/16 inch) from case for 10 seconds D, P, or PW package	260	260		°C

† Stresses beyond those listed under "absolute maximum ratings" may cause permanent damage to the device. These are stress ratings only, and functional operation of the device at these or any other conditions beyond those indicated under "recommended operating conditions" is not implied. Exposure to absolute-maximum-rated conditions for extended periods may affect device reliability.

NOTES: **1.** All voltage values, unless otherwise noted, are with respect to the midpoint between V_{CC+} and V_{CC}.
2. Differential voltages are at IN + with respect to IN –.
3. The magnitude of the input voltage must never exceed the magnitude of the supply voltage or 15 V, whichever is less.
4. The output may be shorted to ground or either power supply. For the μA741M only, the unlimited duration of the short circuit applies at (or below) 125°C case temperature or 75°C free-air temperature.

DISSIPATION RATING TABLE

PACKAGE Power Rating	$T_A \le 25°C$ Factor	DERATING Above T_A	DERATE Power Rating	$T_A = 70°C$ Power Rating	$T_A = 85°C$ Power Rating	$T_A = 125°C$
D	500 mW	5.8 mW/°C	64°C	464 mW	377 mW	N/A
FK	500 mW	11.0 mW/°C	105°C	500 mW	500 mW	275 mW
J	500 mW	11.0 mW/°C	105°C	500 mW	500 mW	275 mW
JG	500 mW	8.4 mW/°C	90°C	500 mW	500 mW	210 mW
P	500 mW	N/A	N/A	500 mW	500 mW	N/A
PW	525 mW	4.2 mW/°C	25°C	336 mW	N/A	N/A
U	500 mW	5.4 mW/°C	57°C	432 mW	351 mW	135 mW

μA741, μA741Y
GENERAL-PURPOSE OPERATIONAL AMPLIFIERS

SLOS094B – NOVEMBER 1970 – REVISED SEPTEMBER 2000

electrical characteristics at specified free-air temperature, $V_{CC}\pm = \pm 15$ V (unless otherwise noted)

Parameter		TEST CONDITIONS	T_A† 25°C	μA741C MIN	μA741C TYP	μA741C MAX	μA741I, μA741M MIN	μA741I, μA741M TYP	μA741I, μA741M MAX	Unit
V_{IO}	Input offset voltage	$V_O = 0$	25°C		1	6		1	5	mV
		Full range			7.5				6	
$D_{VI(adj)}$	Offset voltage adjust range	$V_O = 0$	25°C		±15			±15		mV
I_{IO}	Input offset current	$V_O = 0$	25°C		20	200		20	200	nA
		Full range			300			500		
I_{IB}	Input bias current	$V_O = 0$	25°C		80	500		80	500	nA
		Full range			800			1500		
V_{ICR}	Common-mode input voltage range		25°C	±12	±13		±12	±13		V
			Full range	±12			±12			
		$R_L = 10$ kΩ	25°C	±12	±14		±12	±14		
V_{OM}	Maximum peak output voltage swing	R_L . 10 kΩ	Full range	±12			±12			V
		$R_L = 2$ kΩ	25°C	±10	±13		±10	±13		
		R_L . 2 kΩ	Full range	±10			±10			
A_{VD}	Large-signal differential voltage amplification	R_L . 2 kΩ	25°C	20	200	50	200			V/mV
		$V_O = \pm 10$V	Full range	15			25			
r_i	Input resistance		25°C	0.3	2		0.3	2		MΩ
r_o	Output resistance See Note 5	$V_O = 0$,	25°C		75			75		Ω
C_i	Input capacitance		25°C		1.4			1.4		pF
CMRR	Common-mode rejection ratio		25°C	70	90		70	90		dB
		$V_{IC} = V_{ICRmin}$	Full range	70			70			
k_{SVS}	Supply voltage sensitivity ($\Delta V_{IO}/\Delta V_{CC}$)	$V_{CC} = \pm 9$V	25°C		30	150		30	150	μV/V
			Full range			150			150	
I_{OS}	Short-circuit output current		25°C		±25	±40		±25	±40	mA
I_{CC}	Supply current	V_O=0, No load	25°C		1.7	2.8		1.7	2.8	mA
		Full nange			3.3			3.3		
P_D	Total power dissipation	V_O=0, No load	25°C		50	85		50	85	mW
		Full range			100			100		

† All characteristics are measured under open-loop conditions with zero common-mode input voltage unless otherwise specified. Full range for the μA741C is 0°C to 70°C, the μA741I is –40°C to 85°C, and the μA741M is –55°C to 125°C.

NOTE 5: This typical value applies only at frequencies above a few hundred hertz because of the effects of drift and thermal feedback.

Operating characteristics, $V_{CC\pm} = \pm 15$ V, $T_A = 25$°C

PARAMETER		TEST CONDITIONS		μA741C min	μA741C typ	μA741C max	μA741I, μA741M min	μA741I, μA741M typ	μA741I, μA741M max	UNIt
t_r	Rise time	$V_I = 20$ mV,	$R_L = 2$ kΩ		0.3			0.3		μs
	Overshoot factor	$C_L = 100$ pF,	See Figure 1		5%			5%		
SR	Slew rate at unity gain	$V_I = 10$ V,	$R_L = 2$ kΩ,		0.5			0.5		V/μs
		$C_L = 100$ pF,	See Figure 1							

μA741, μA741Y
GENERAL-PURPOSE OPERATIONAL AMPLIFIERS

SLOS094B – NOVEMBER 1970 – REVISED SEPTEMBER 2000

Electrical characteristics at specified free-air temperature, $V_{CC\pm} = \pm 15$ V, $T_A = 25°C$ (unless otherwise noted)

PARAMETER		TEST CONDITIONS	μA741Y, MIN	TYP	MAX	Unit
V_{IO}	Input offset voltage	$V_O = 0$		1	6	mV
$\Delta V_{IO(adj)}$	Offset voltage adjust range	$V_O = 0$		±15		mV
I_{IO}	Input offset current	$V_O = 0$		20	200	nA
I_{IB}	Input bias current	$V_O = 0$		80	500	nA
V_{ICR}	Common-mode input voltage range		±12	±13		V
V_{OM}	Maximum peak output voltage swing	$R_L = 10$ kΩ	±12	±14		V
		$R_L = 2$ kΩ	±10	±13		
A_{VD}	Large-signal differential voltage amplification	$R_L \geq 2$ kΩ	20	200		V/mV
r_i	Input resistance		0.3	2		MΩ
r_o	Output resistance	$V_O = 0$, See Note 5		75		Ω
C_i	Input capacitance			1.4		pF
CMRR	Common-mode rejection ratio	$V_{IC} = V_{ICR}$min	70	90		dB
k_{SVS}	Supply voltage sensitivity ($\Delta V_{IO}/\Delta V_{CC}$)	$V_{CC} = \pm 9$ V to ± 15 V		30	150	μV/V
I_{OS}	Short-circuit output current			±25	±40	mA
I_{CC}	Supply current	$V_O = 0$, No load		1.7	2.8	mA
P_D	Total power dissipation	$V_O = 0$, No load		50	85	mW

† All characteristics are measured under open-loop conditions with zero common-mode voltage unless otherwise specified.

NOTE 5: This typical value applies only at frequencies above a few hundred hertz because of the effects of drift and thermal feedback.

Operating characteristics, $V_{CC\pm} = \pm 15$ V, $T_A = 25°C$

PARAMETER		TEST CONDITIONS		μA741Y MIN	TYP	MAX	Unit
t_r	Rise time	$V_I = 20$ mV,	$R_L = 2$ kΩ		0.3		μs
	Overshoot factor	$C_L = 100$ pF,	See Figure 1		5%		
SR	Slew rate at unity gain	$V_I = 10$ V, $C_L = 100$ pF,	$R_L = 2$ kΩ, See Figure 1		0.5		V/μs

μA741, μA741Y
GENERAL-PURPOSE OPERATIONAL AMPLIFIERS

SLOS094B – NOVEMBER 1970 – REVISED SEPTEMBER 2000

PARAMETER MEASUREMENT INFORMATION

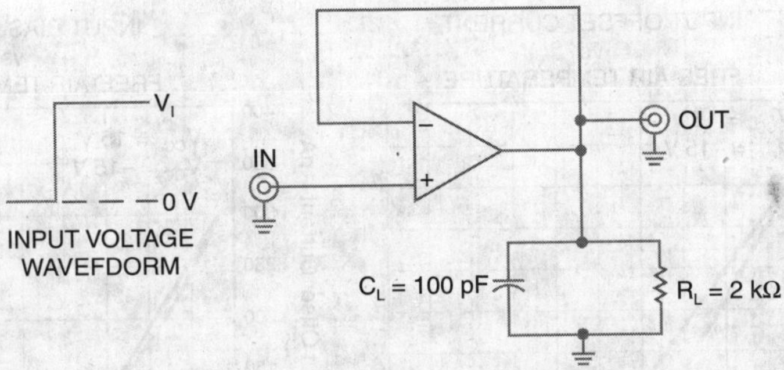

Fig. 1. Rise Time, Overshoot, and slew Rate

APPLICATION INFORMATION

Fig. 2. Shows a diagram for an input offset voltage null circuit.

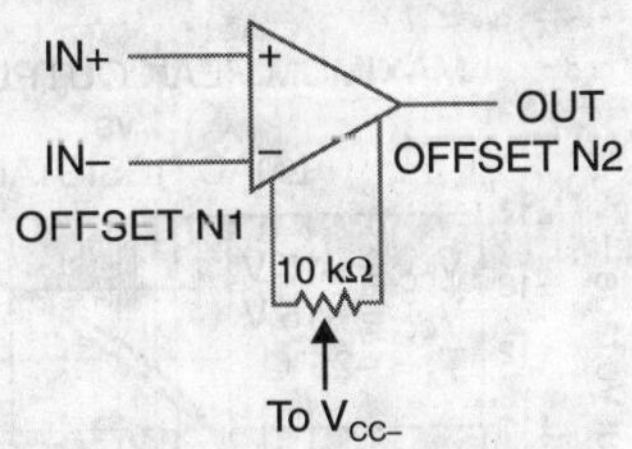

Fig. 2. Input Offset Voltage Null Circuit

μA741, μA741Y
GENERAL-PURPOSE OPERATIONAL AMPLIFIERS

SLOS094B – NOVEMBER 1970 – REVISED SEPTEMBER 2000

TYPICAL CHARACTERISTICS †

INPUT OFFSET CURRENT
vs
FREE-AIR TEMPERATURE

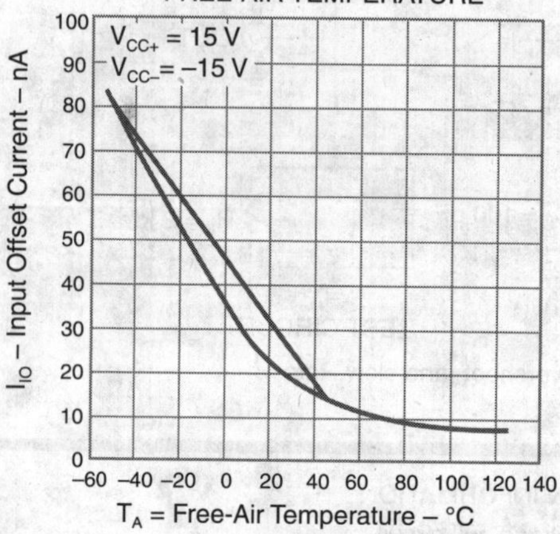

Fig. 3

INPUT BIAS CURRENT
vs
FREE-AIR TEMPERATURE

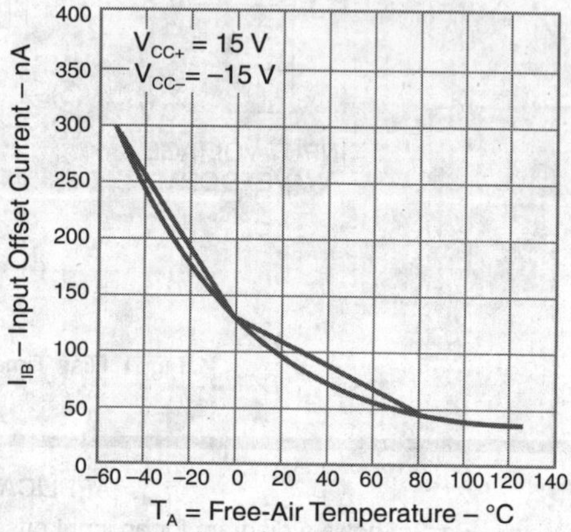

Fig. 4

MAXIMUM PEAK OUTPUT VOLTAGE
vs
LOAD RESISTANCE

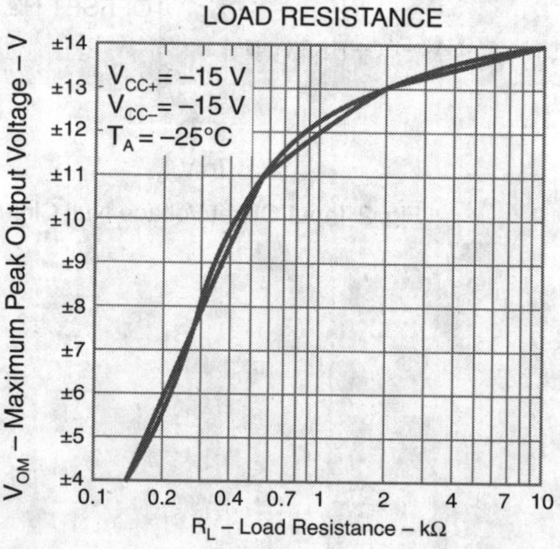

Fig. 5

† Data at high and low temperatures are applicable only withing the rated operating free-air temperature ranges of the various devices.

TEXAS INSTRUMENTS
POST OFFICE BOX 655303 • DALLAS, TEXAS 75265

μA741, μA741Y
GENERAL-PURPOSE OPERATIONAL AMPLIFIERS

SLOS094B – NOVEMBER 1970 – REVISED SEPTEMBER 2000

TYPICAL CHARACTERISTICS

MAXIMUM PEAK OUTPUT VOLTAGE
vs
FREQUENCY

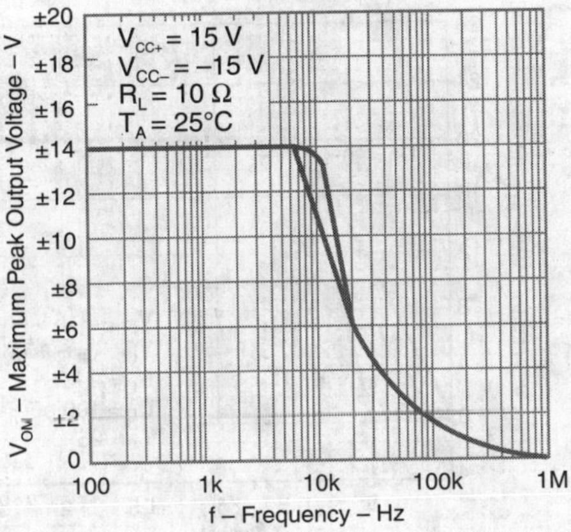

Fig. 6

OPEN-LOOP SIGNAL DIFFERENTIAL
VOLTAGE AMPLIFICATION
vs
SUPPLY VOLTAGE

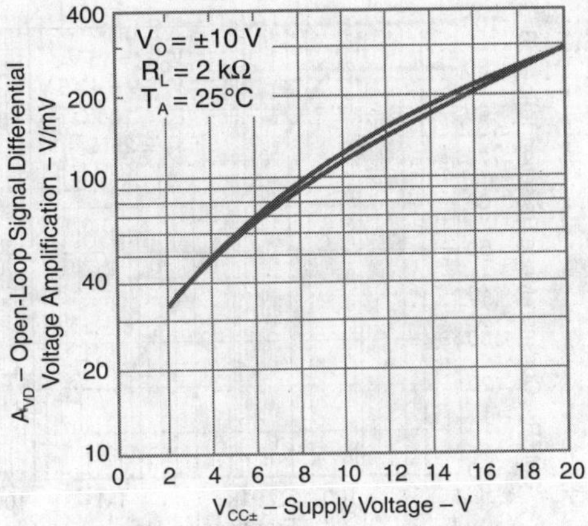

Fig. 7

OPEN-LOOP LARGE-SIGNAL DIFFERENTIAL
VOLTAGE AMPLIFICATION
vs
FREQUENCY

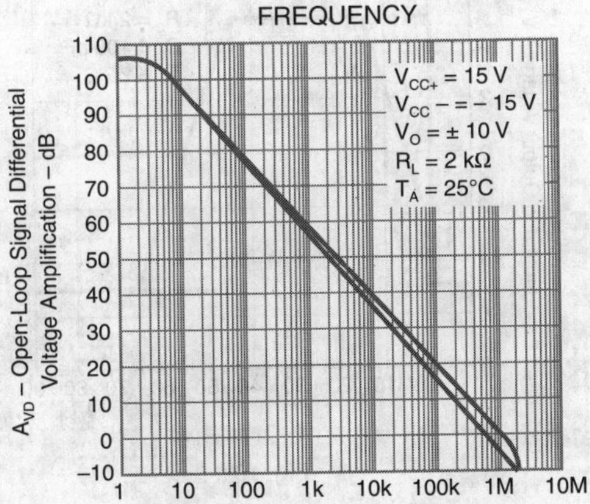

TYPICAL CHARACTERISTICS

COMMON-MODE REJECTION RATION
vs
FREQUENCY

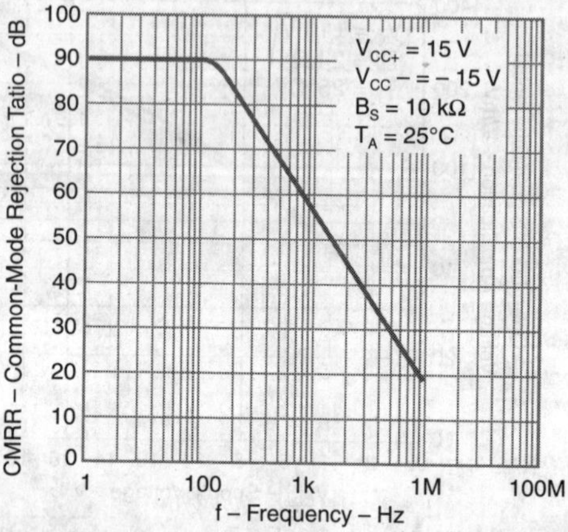

Fig. 8

OUTPUT VOLTAGE
vs
ELAPSED TIME

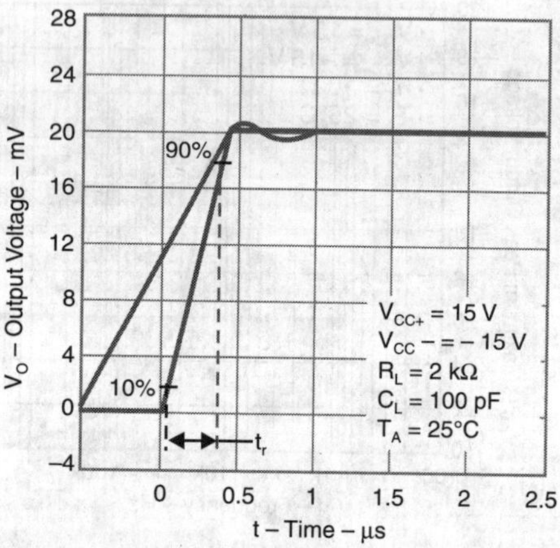

Fig. 9

VOLTAGE-FOLLOWER
LARGE-SIGNAL PULSE RESPONSE

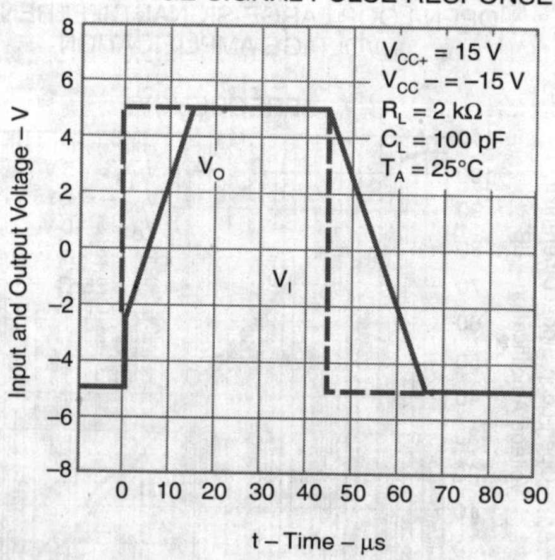

Fig. 10

APPENDIX-C

Transistor Part Numbers

All the electronic devices such as semiconductors, liquid crystal displays, sensor devices, electronic tubes, cathode ray tubes etc., are designated with certain part number.

The part number is assigned by Pro Electron (also called European Electronic Component Manufacturing Association (EECA). The objective of this company is to allow unambiguous identification of electronic parts even when made by several different manufactures.

Table shows the part numbers of few transistors. As seen from this table, the part number begins with two letters followed by a 3-digit sequence number. The first letter gives the type semiconductor material used. A for Germanium, B for Silicon and C for material like Gallium Arsenide (GaAs). The second letter is for intended use: A for diode, C for general purpose transistor etc.

Table : Part members

Prefix	Usage	Example
AC	Germanium small signal transistor	AC 126
AF	Germanium RF transistor	AF 117
BC	Silicon small signal transistor	BC 548B
BD	Silicon power transistor	BD 139
BF	Silicon RF (high frequency) transistor	BF 245
BS	Silicon switching transistor	BS 170
BL	Silicon high frequency, high power (for transmetters)	BLW 34
BU	Silicon high voltage (for CRT horizontal deflection circuits)	BU 508

JEDEC (Joint Electron Devices Engineering Council) is another independent standardization body from United States for developing part numbers. This company suggested device number for transistors that start with 2N, indicating a 3-terminal device followed by 2, 3 or 4 digit sequence member with no significance to device properties. For example 2N 2222 and 2N 3055.

APPENDIX-D

Quantify Symbols and SI Units

Quantity	Symbol	SI Unit	Symbol	Derivation
acceleration	a	ms^{-2}	–	velocity/time
acceleration due to gravity	g	ms^{-2}	–	velocity/time
amount of substance	n	mole	mol	mole fraction (n) used
amplification factor	μ	a ratio	–	–
angle	$\theta_1, \varphi_1, \alpha_1$	–	–	–
angle of incidence	i	degree or radian		° –
angle of refraction	r	degree or radian		° –
angle, Bragg	θ	number	–	–
angle, critical	c	degree or radian		° –
anode slope resistance	R_A	ohm	Ω	$\Delta V_a/\Delta I_a$
area	A	metres squared		m^2 $l \times b$
atomic number	Z	a number	–	nuimber of protons
Avogadro constant	L, N_A	Number	–	–
breadth	b	meter	m	fundamental unit
capacitance	C	farad	F	charge/p.d.
charge, electric	Q	coulomb	C	current × time
charge on electron	e	coulomb	C	1.6×10^{-19} C
conductance	G	ohm^{-1}	Ω^{-1}	reciprocal of resistance
current, electric	I	ampere	A	fundamental unit
decay constant	λ	a ratio	–	–
density	ρ	$kg\ m^{-3}$	–	m/V
distance along path	s	metre	m	fundamental unit
efficiency	η	a ratio	–	work output/work input
electrochemical equiv.	z	$g\ C^{-1}$	–	mass/charge
electromotive force	E	volt	V	energy/charge
electron	e	–	–	–
energy	E	joule	J	N m
energy, kinetic	E_k	joule	J	N m. $E_k = 1/2\ m\upsilon\ 2$
energy, potential	E_p	joule	J	N m. ep = mgh
Faraday constant	F	coulomb mol^{-1}	C mol^{-1}	96, 500 C mol^{-1}
field strength, electric	E	V m^{-1}	–	potential gradient: p.d./dist.
field-strength magnetic	H	ampere-trrns	–	corrent × no. of turns
flux. magnetic	Φ	weber	Wb	e.m.f/rate of change of flux
flux density	B	tesla	T	flux/area
focal length	f	metre	m	–

Quantity	Symbol	SI Unit	Symbol	Derivation
force	F	newton	N	kg m s^{-2}
free enery	ΔG	joule	J	–
frequency	f	hertz	Hz	ocillations/time
gas constant	r	joule	J	energy
half-life, radioactivity	$t_{1/2}$	second	s	fundamental unit
heat capacity	C	J K^{-1}	–	quantity of heat/temp. rise
heat of reaction	ΔH	joule	J	heat energy
heat capacity, specific	c	J K^{-1} kg^{-1}	–	heat capacity/mass
heat, quantity of	q	joule	J	energy
height	h	metre	m	fundamental unit
image distance	υ	metre	m	fundamental unit
inductance, mutual current	M	henry	H	induced e.m.f./rate of change of
inductance, self	L	henry	H	–
intensity of radiation	I	*a number*	–	–
latent heat	L	joule	J	quantity of heat
latent heat, specific	l	J kg^{-1}	–	quantity of heat
latent heat, molar	L_m	joule mol^{-1}	J	quantity of heat
length	l	metre	m	fundamental unit
magnetizing force	H	ampere-turns	–	–
magnetic moment	m	Wb m	–	torquen in unit magnetic field
magnification, linear	m	a ratio	–	–
mass	m	kilogramme	kg	fundamental unit
mass number	A	*a number*	–	number of neutrons + protons
molar volume	V_m	(dm^3)	–	volume of 1 mole
molar solution	M	a ratio	–	moles/dm^3
moment of force	–	N m	–	force × perp. distance
neutron number	N	*a number*	–	number of neutrons
number	n	–	–	–
number of molecules	N	–	–	–
number turns on coil	n	*a number*	–	–
number order spectrum	p	*a number*	–	–
object distance	u	metre	m	fundamental unit
peak current	I_O	ampere	A	see current
peak e.m.f.	E_O	volt	V	see e.m.f.
period	T	second	s	fundamental unit
permeability	μ	H m^{-1}	–	henry/metre
permeability, vacuum	μ_O	H m^{-1}	–	–
permeability, relative	mr	a ratio	–	$m = \frac{\mu}{\mu_o}$
permittivity	ε	F m^{-1}	–	farad/metre
permittivity, vacuum	ε_O	F m^{-1}	–	farad/metre
permittivity, relative	μ_r	*a ratio*	–	$\mu_r = \frac{\varepsilon}{\varepsilon_0}$

Quantity	Symbol	SI Unit	Symbol	Derivation
potential, electric	*V*	volt	V	energy/charge
potential differece	*V*	volt	V	energy/charge
power	*P*	watt	W	J s–1
pressure	*P*	pascal	Pa	N m^{-2}; force/area
radius	*r*	metre	m	fundamental unit
reactance	*X*	ohm	W	E_d/I_o
refractive index	*n*	*a ratio*	–	–
resistance	*R*	ohm	Ω	p.d./current
resistivity, electrical	ρ	ohm-metre	–	resistance × length
relative density	*d*	*a ratio*	–	ρ_{sub}/ρ_{water}
r.m.s. current	$I_{r.m.s}$	ampere	A	see current
r.m.s voltagve	$V_{r.m.s.}$	volt	V	see e.m.f.
slit separation	*s*	metre	m	fundamental unit
tension	*T*	newton	N	see force
temperature, celsius	θ	degree C	°C	from kelovin
temp. interval	θ	degree	° or K	–
temp. absolute	*T*	kelvin	K	fundamental unit
thickness	*d*	metre	m	fundamental unit
time	*t*	second	s	fundamental unit
torque	*T*	Nm	–	see moment
turns ratio	*T*	*a ratio*	–	n_{sec}/n_{prim}
(unit of electricity)	–	kW h	–	kilowatt × hour
velocity	*u*. υ	ms^{-1}	–	distance/time
velocity, angular	ω	$second^{-1}$	s^{-1}	angle/time
velocity, e.m. waves	*c*	m s^{-1}	–	–
velocity of sound	υ	m s^{-1}	–	–
volume	*V*	metre cubed	m^3	1 × b × h
wavelength	λ	metre	m	fundamental unit
work	*w*	joule	J	force × distance (Nm)
weight	*W*	newton	N	kg m s^{-2} or mg

Reproduced from *Electronic Engineers Master Catalogue*, courtesy of Hearst Business Communications. Inc.

INDEX

D

E

N

O

P

T